OXFORD DICTIONARY OF
Biochemistry and Molecular Biology

Editors

Teresa K. Attwood Professor of Bioinformatics, Faculty of Life Sciences & School of Computer Science, University of Manchester

Richard Cammack (Managing Editor) Professor of Biochemistry, King's College London

Peter N. Campbell (deceased) Emeritus Professor of Biochemistry and Honorary Research Fellow, University College London

J. Howard Parish Life Fellow, University of Leeds

Anthony D. Smith Emeritus Reader in Biochemistry, University College London

John L. Stirling Senior Lecture in Molecular Genetics, King's College London

Francis Vella Former Professor of Biochemistry, Faculty of Medicine, University of Saskatchewan, Saskatoon, Canada

OXFORD DICTIONARY OF
Biochemistry and Molecular Biology

REVISED EDITION

Managing Editor
Professor R. Cammack   King's College London
General Editors
Professor T.K. Attwood   University of Manchester
Professor P.N. Campbell   University College London
Dr J.H. Parish   University of Leeds
Dr A.D. Smith   University College London
Dr J.L. Stirling   King's College London
Professor F. Vella   University of Saskatchewan

OXFORD
UNIVERSITY PRESS

OXFORD
UNIVERSITY PRESS

Great Clarendon Street, Oxford OX2 6DP

Oxford University Press is a department of the University of Oxford.
It furthers the University's objective of excellence in research, scholarship,
and education by publishing worldwide in

Oxford New York

Auckland Cape Town Dar es Salaam Hong Kong Karachi
Kuala Lumpur Madrid Melbourne Mexico City Nairobi
New Delhi Shanghai Taipei Toronto

With offices in

Argentina Austria Brazil Chile Czech Republic France Greece
Guatemala Hungary Italy Japan Poland Portugal Singapore
South Korea Switzerland Thailand Turkey Ukraine Vietnam

Oxford is a registered trade mark of Oxford University Press
in the UK and in certain other countries

Published in the United States
by Oxford University Press Inc., New York

First edition published 1997
Revised edition 2000
Reprinted 2001, 2003
Second edition published 2006

British Library Cataloguing in Publication Data

Data available

Library of Congress Cataloging in Publication Data

Data available

Typeset by Market House Books Ltd.
Printed on acid-free paper by CPI Bath

ISBN 0-19-852917-1   978–0–19–852917-0

10 9 8 7 6 5 4 3 2 1

In memory of Peter Nelson Campbell (1921–2005)

Peter Campbell was the first person to synthesize a protein using components of the cell rather than complete cells. He was one of the group who first showed the importance of autoimmunity in human disease. Peter was Head of the Biochemistry Department in the University of Leeds from 1967 until 1975. He was a founder of the Federation of European Biochemical Societies (FEBS) and, among many international commitments, chaired the Education Committee of the International Union of Biochemistry and Molecular Biology (IUB). A great advocate of biochemistry teaching, he started the journal *Biochemical Education*. His books included *Biochemistry Illustrated*, with Tony Smith. He was also one of the main driving forces behind the creation of the first edition of the *Oxford Dictionary of Biochemistry and Molecular Biology*.

Preface

Preface to this edition

It is a decade since the first edition of the *Oxford Dictionary of Biochemistry and Molecular Biology*. It was a remarkable work of scholarship, arising from the work of journal editors and scientific writers. Since then the landscape of biochemistry has changed immeasurably. The genome sequences have laid out the blueprints of whole organisms, especially Man. They have revealed the diversity of gene expression, and the complex systems by which cellular molecules organize themselves. The molecular basis of many diseases has been revealed, and vital cellular components discovered.

The literature is more diverse than ten years ago. The identification of the genes has rushed ahead of the biochemical characterization of their functions. Many protein and nucleic acid factors have been discovered. While their functions are incompletely understood, they are referred to by laboratory shorthand abbreviations. These are well understood by the investigators who work on them, but the mass of them becomes very confusing to the student, or to those viewing biochemistry from the outside. New methods of bioinformatics have been developed to bridge the gap. Meanwhile the '-omics' projects have introduced new layers of complexity as we see the interactions between macromolecules leading to new emergent properties.

As predicted in the first edition, the influence of the Internet has expanded. Instead of searching for information in libraries, students now usually go first to a search engine. So, does such a dictionary have a role in the age of Google? In fact it has gained in sales and popularity. Evidently it fills a need for a source of reliable information that is not always so easy to find.

A revised version of the dictionary, with some additions and corrections was printed in 2000. At that time, the need for a complete revision was apparent. The work continued with a new team, recruited by the ever-enthusiastic Peter Campbell. We deeply regret that he did not live to see the completion of this task, having died on February 8th 2005 from complications after an accident.

In order to keep the dictionary as a handy reference volume, we have endeavoured to avoid it becoming much heavier. It is only by being selective that there are only about 20% more entries than the first edition. Most of the appendices have been removed, or their useful parts transferred. The listed Nobel prizewinners in biochemistry and molecular biology have been omitted except for eponymous entries, when they have lent their names to compounds or procedures. There has been a judicious removal of some older terms, though we found that surprisingly few have disappeared from the literature to such an extent that they are obsolete.

The literature abounds with laboratory shorthand names, database identifiers; TLA's (three-letter abbreviations) and other acronyms are extremely common, and a notorious source of ambiguity. We have cited these selectively, sometimes to indicate that a word or phrase has two meanings in different contexts. In the printed form we can show the full range of printed characters – boldface, italic, sub- and superscripts, Greek letters – that make up the syntax of many of the names, and that are difficult to find with search engines.

The dictionary is not intended to be a nomenclature document, and the terms that are in the entries are generally those that are in common use. We continue the practice of pointing the reader in the direction of recommended terminology and nomenclature. Nomenclature rules are applied less prescriptively these days; 'recommended' chemical nomenclature has become 'preferred'; 'recommended names' for enzymes have given way to 'common names'.

A great many of the new entries, on inherited diseases and much else besides are provided by Frank Vella, drawing on the eclectic collection of topical papers that he assembled for his columns in journals such as IUBMB Life. The entries on bioinformatics and genetics, which have assumed greater importance in BMB over the past decade, have been bolstered by the work of Terri Attwood and John Stirling. Finally it has been a pleasure to work with John Daintith and Robert Hine of Market House Books, whose expertise in chemistry and biology meant that their assembly of the book was an expert job.

The content of such a dictionary is necessarily selective. We have tried to ensure that the entries in the dictionary reflect current usage in biochemistry and molecular biology. As always, we are grateful to readers who point out errors in the present text.

Richard Cammack
March 2006

Preface to the revised first edition

It must be inevitable with any work of this nature that a number of imperfections and errors occur. So the opportunity provided by the need to reprint this dictionary has been taken to effect some improvements within the limitation imposed by retention of the original pagination. As well as the correction of a variety of minor misprints and other minor defects, over four hundred entries been either revised or completely rewritten, and fifty or so new entries have been provided, some to remedy deficiencies and others to provide additional terms that have become of topical interest. To help make way for the new ones, about half as many original entries have been deleted. In addition, Appendices B, C, and D have been updated, and Appendix B has been expanded and provided with all the relevant Internet addresses available at the time of writing.

Valuable comments on the original edition by a number of readers are gratefully acknowledged, and thanks are again due to Dr. H. B. F. Dixon for advice on aspects of nomenclature as well as to Oxford University Press and Market House Books for their much appreciated cooperation.

September 1999   A. D. S.

Preface to the first edition

Nearly twenty years ago one of us (S. P. D., soon joined by G. H. S.), began a distillation of the elements of biochemistry into an alphabetical arrangement. The task was formidable and eventually other editors were recruited, an editorial board was established, and now the work is offered as the *Oxford Dictionary of Biochemistry and Molecular Biology*. It is hoped that the dictionary will serve the needs of the research biochemist or molecular biologist, as well as teachers of the subject and their students. In addition, it should prove of value to practitioners of other fields of study or work seeking the meaning of a biochemical term.

An important function of a dictionary is to provide guidance on current usage in the field within its scope. The original 12-volume *Oxford English Dictionary* was compiled from about five million slips of paper bearing sentences or phrases extracted by some thousands of 'readers' from classical works of literature and those of the best contemporary authors. It was thus firmly based on good usage. In scientific subjects, specialist terminology is often codified in sets of recommendations regarding nomenclature, meaning, abbreviations, symbols, and so on. These have been agreed by international commissions (e.g. those of The International Union of Pure and Applied Chemistry and The International Union of Biochemistry and Molecular Biology) as a means of preserving order and facilitating communication between scientists. We have striven to conform as far as possible to the relevant international recommendations, but in some cases, where usage so frequently diverges from a recommendation that adherence to it would seriously detract from ease of use of the dictionary, we have kept to the principle that the dictionary should reflect usage (see the definitions of *lexicographer*). This does not extend, of course, to cases where usage, however widespread, contradicts sound scientific principles. The internationally agreed recommendation is always also listed. The various compilations of these recommendations that have been drawn upon are listed in Appendix B, together with a number of other sources of information on nomenclature.

Biochemistry is the discipline that embraces the study of the structure and function of life-forms at the molecular level. Molecular biology is a closely related discipline that originates in the study of DNA and its metabolism, and now embraces all those investigations that exploit the technology that has resulted from this work. Both disciplines aim to explain the behaviour of life-forms in molecular terms, and are so closely interrelated that separation is barely possible. It is inevitable that the content of this dictionary is to a degree arbitrary, but it is hoped that all important aspects of these subjects have received consideration. The compilers have attempted to offer a broad coverage of terms encountered in the literature of biochemistry and molecular biology by including an appreciable number from cognate sciences. Although the compilation is designed primarily to serve readers of contemporary material, the needs of those who turn to older literature have also been borne in mind. Some of the entries thus have a historical flavour, some obsolete terms are included (e.g. zymase), and in some cases a historical approach has been used as the best means of presenting an explanation of a term, as for example in the case of the entry for *gene*. The value of a scientific dictionary is enhanced by inclusion of contextual information as well as mere explanations of meaning or terminology. This dictionary will be found to have some of the attributes of an encyclopedia, although the extent to which it veers in this direction has varied with the whim of its compilers. It is our hope that in a single volume the reader has easy access to basic definitions as well as a generous helping of other information.

In the present-day world, we are assailed by floods of 'information'. It has been suggested that the average weekday edition of a newspaper of record (e.g. *The New York Times*) provides more information than Shakespeare and his contemporaries would have acquired in a lifetime. With the availability of much information through the Internet, it may be asked whether a dictionary in paper form is actually necessary. In answer, we note that the Internet can be slow, and is not readily accessible in some parts of the world; the databases may be inadequate, and although usually very up-to-date, the high cost of their maintenance restricts them to specialized knowledge in a limited number of fields. Moreover, books have a quality of their own, which is enabling them to maintain their popularity. It appears that the increasing use of the Internet is actually paralleled by the rate of publication of printed dictionaries; in an information-hungry age, there cannot be too many sources of good-quality information.

We are deeply indebted to the Leverhulme Trust for the award of an emeritus fellowship to one of us (A. D. S.), to University College London, which has provided us with friends and expert colleagues, and to Dr O. Theodor Benfey, Dr Mary Ellen Bowden, and Professor Arnold Thackray, The Beckman Center for History of Chemistry, Chemical Heritage Foundation, Philadelphia, and Dr John Edsall, Harvard University for assistance with biographical data.

Particular thanks are due to Dr H. B. F. Dixon for much advice on nomenclature and related matters. Help on questions of nomenclature from Dr G. P. Moss and Dr A. D. McNaught is also acknowledged. We are grateful to Dr D. H. Jenkinson for his help with the recommendations of the International Committee on Nomenclature in Pharmacology. We are also grateful for the valuable advice of Professor K. W. Taylor and Dr J. L. Crammer, on clinical topics, and Professor M. C. W. Evans, on plant biochemistry, and to Dr Margaret McKenzie, for reading the proofs.

During the earlier stages of the project, Mrs S. Gove gave much valuable assistance and Miss A. Straker was most helpful in suggesting terms for inclusion. We also wish to thank all those other friends and colleagues, in addition to those separately listed, who have unstintingly given us help and advice.

We are pleased to acknowledge the collaboration and material support given to us by Oxford University Press. We also acknowledge the very friendly cooperation of Market House Books, especially the patience and good humour of Dr John Daintith through all the complications of the production. The copy editors, Robert Hine and Jane Cavell, made a number of helpful suggestions.

The compilers offer no apology for their failure to include many deserving terms in the dictionary, but would be pleased to have their attention drawn to errors and to receive suggestions for additional entries in any future edition.

January 1997
A. D. Smith, S. P. Datta, G. H. Smith, P. N. Campbell, R. Bentley, H. A. McKenzie

This whole book is but a draught—nay, but the draught of a draught. Oh, Time, Strength, Cash, and Patience.

Herman Melville (1851) *Moby Dick, or The Whale* (ed. T. Tanner, 1988, p. 147, Oxford University Press).

Note on proprietary status

Guide to the Dictionary

1. Alphabetical order

1.1 Main order of headwords

Alphabetical order is determined on a letter-by-letter basis, not word by word; spaces are disregarded:

acid
acid anhydride
acid–base balance
acid–base catalysis
acid dissociation constant
acid dye
acidemia

1.2 Nonalphabetic characters

Numbers, hyphens, primes, and subscript/superscript text are ignored for the purpose of indexing; an example is the following sequence of entries:

FSH-RH
F1 sphere
F′ strain
F-type pentose phosphate pathway
ftz

1.3 Locants and modifiers

In chemical names, any locants and other hyphenated modifiers such as *cis*-, *trans*-, *p*-, and alphabetic Greek characters are not used to determine primary alphabetical order; hence the following entries all appear under the letter A:

***N*-acetylgalactosamine**
***p*-aminobenzoic acid**
***γ*-aminobutyrate shunt**
6-aminohexanoic acid

However, the unhyphenated letters 'c' in 'cDNA' and 'd' in 'dCTP', for example, are treated as integral parts of the word and *are* used to determine alphabetical order.

1.4 Secondary order involving locants

When such modifiers constitute the only difference between two headwords, they determine the alphabetical order of the entries:

benzodiazepine	**encephalitis**
***o*-benzoquinone**	**3′-end**
***p*-benzoquinone**	**5′-end**
benzoyl	**end+**

1.5 Format differences in headwords

The order for entries where the headword is identical except for format is

b, *b*, b-, *b*-, -b, -*b*, B, *B*, B-, *B*-, -B, -*B*

1.6 Subscripts and superscripts

Single letters with subscripts or superscripts are treated as single letters for the purposes of indexing, so entries for k_{cat} and K_m will both be found in the list of single-letter entries at the beginning of the letter K. The primary order of these single-letter entries is determined by their format (*see* section 1.5); where there is more than one entry with a given format (e.g. italic, lower case *k*), these are arranged by alphabetical order of their subscripts/superscripts.

1.7 Greek letters

- Where Greek letters form part of a chemical name, they are not used to determine alphabetical order (*see* section 1.3). Otherwise they are written out in full in the headword, e.g. **nu body, beta strand.**

- The names of the letters of the Greek alphabet, together with their English transliterations used in etymologies, are listed in Appendix A. The meanings assigned to Greek alphabetic characters used as symbols are also given in Appendix A.

- Greek characters are set in italic type when the character represents a variable or locant and in roman type when it represents a unit or subtype e.g. of a protein or particle.

2. Format of entries

2.1 Summary of typefaces

- The following distinguishing typefaces are employed

in addition to the text light serif typeface used for definitions:

large bold sans serif	headwords
text bold serif	alternative terms for and variant spellings of headwords; hidden entries; run-ons
text bold sans serif	cross-references
text italic serif	usage notes and field labels; parts of speech; foreign language terms (including scientific and medical Latin); symbols for physical quantities and fundamental physical constants; stereochemical prefixes and alphabetical locants

2.2 Headwords

- For each entry, the headword is in bold, sans serif type.

- Upper-case (capital) initial letters are used only for proper names (or terms derived from them) and for proprietary names. Abbreviations and symbols are printed in upper and/or lower case as appropriate.

- If a term would normally be set in bold type, this is indicated in the entry:

  > *B symbol for* **1** Napierian absorbance (*see* **absorbance**). **2** magnetic flux density (bold italic).

- Where the same basic term is used in different typefaces, such as roman/italic, or upper case/lower case, or as a prefix or suffix, each usage is given as a separate headword. For example, **h**, *h*, **H**, and *H* each have a separate entry.

- The order in which such entries are given is listed in section 1.5.

2.3 Alternative terms and variant spellings

2.3.1 Choice of headword

Where alternative terms for a headword, or variant spellings of it, exist (*see* section 1.3), the headword selected for the main entry is generally the recommended or preferred term, or the one judged to be the commonest. Exceptions to this generalization are those instances where the name of a Greek alphabetic character is written out for convenience of indexing:

> **beta globulin** *or* **β globulin**....

2.3.2 General

- Any alternative terms and alternative spellings are listed following the headword in bold, serif type:

  > **retrovirus** *or* **ribodeoxyvirus** *or* **RNA–DNA virus** any virus belonging to the family Retroviridae....

- Notes regarding the usage of these alternatives may be given in brackets and in italics; for example

> **DNA glycosylase** *or* (*sometimes*) **DNA glycosidase** any of a group of enzymes....
>
> **bacteremia** *or* (*esp. Brit.*) **bacteraemia** the presence of live bacteria in the blood.
>
> **bilirubin** *or* (*formerly*) **bilirubin IXα** *the recommended trivial name for* the linear tetrapyrrole....

- These alternative terms and spellings also appear as entries in the alphabetical sequence, with a cross-reference to the main entry where the term is defined, unless the variant would appear close to the main entry. Additional information is given where appropriate:

  > **demoxytocin** *an alternative name for* **deaminooxytocin**.
  >
  > **fructose-1,6-diphosphatase** *a former name for* **fructose-bisphosphatase**.
  >
  > **lipide** *a variant spelling of* **lipid**.
  >
  > **molecular exclusion chromatography** *a less common name for* **gel-permeation chromatography**.
  >
  > **oleomargarine** *an alternative name* (*esp. US*) *for* **margarine**.
  >
  > **penatin** *an obsolete name for* **glucose oxidase**.

2.3.3 Chemical names

- Synonyms may be given following the headword, in the order: other trivial names (if any); the semi-systematic or semi-trivial name(s); older systematic name in style, if still in widespread use; the systematic name in currently recommended style.

- The headword used to represent a chemical compound that can exist in ionized form(s) is in most cases the name of its physiologically predominant form. So, for example, an entry is headed 'succinate' rather that 'succinic acid'.

2.3.4 Enzyme names

Alternative names may be listed following the headword, which is normally the recommended name; otherwise alternative names include the recommended name (if the headword is the common name), the systematic name, and other names. The EC number is also given.

2.4 Multiple definitions

- Where a term has more than one meaning, the different senses are numbered with bold Arabic numerals.

  > **blockade 1** (*in pharmacology*) the saturation of a specific type of receptor with an antagonist to its normal agonist. **2** (*in immunology*) the overloading or saturation of the **reticuloendothelial system** with inert particles, such as carbon particles. **3** to impose any such blockade.

- The order of the numbered entries is generally determined by their biochemical significance.

- The different senses may be further subdivided into def. 1a, def. 1b, etc.

  > **di+** *comb. form* **1** (*in chemical nomenclature*) (*distinguish from* **bis+** (*def. 2*)) **a** indicating the presence in a molecule of two identical unsubstituted groups, e.g. diethylsulfide, 1,3-dihydroxyacetone. **b** indicating the presence in a molecule of two identical inorganic oxoacid residues in anhydride linkage, e.g. adenosine 5′-diphosphate. **2** *or* **bis+** (*def. 1*) denoting two, twofold, twice, doubled.

- Homographs are not distinguished.

2.5 Hidden entries

Hidden entries are terms that are not defined at their normal headword position. Instead, they are treated (implicitly or explicitly) at some other headword. They are set in bold serif type. In the following example, 'bentonite flocculation test' is the hidden entry:

> **bentonite** a colloidal, native hydrated aluminium silicate clay consisting principally of montmorillonite, a complex aluminosilicate, $Al_2O_3 \cdot 4SiO_2 \cdot H_2O$, which has marked adsorptive properties. It is used as an inhibitor of nucleases and also in the **bentonite flocculation test**, a passive agglutination test in which antigen-coated bentonite particles are used to detect specific antibody.

2.6 Other information

2.6.1 Plurals

The plural form (or forms) of a headword is (are) given in parenthesis following the headword if its formation is non-standard, e.g. for Latin headwords, or where there is more than one form of the plural.

> **medulla** (*pl.* **medullas** *or* **medullae**) the innermost part of an organ, tissue, or structure; marrow, pith. — **medullary** *adj.*

2.6.2 Affixes and combining forms

- In common with other dictionaries, this Dictionary lists and defines many word elements that are used to compose terms or to modify existing terms. These are either combining forms (which are derived from parent words) or affixes (infixes, prefixes, and suffixes, none of which have parents).
- The usual lexicographical convention is to add a hyphen to suffixes and combining forms when listing them as headwords, although generally the hyphen is omitted in formation of composite terms. However, chemical and biochemical terminology also includes a considerable number of specialized affixes that retain the hyphen in the formation of composite terms (e.g. '*meso-*' in '*meso*-cystine').

In order to make an explicit distinction between these alternatives, this Dictionary departs from the common convention by adding a hyphen to an affix in a headword only when a linking hyphen is retained in a combination:

> ***meso-*** *abbr.*: *ms-*; *prefix* (*in chemical nomenclature*) designating a substance whose individual molecules contain....

By contrast, combining forms (e.g., 'meso' in 'mesoderm') together with affixes producing unhyphenated composite terms, are listed with an added plus sign, placed after and/or before the word-element as appropriate:

> **meso+** *or* (*sometimes before a vowel*) **mes+** *comb. form* denoting middle, or intermediate.

> **+agogue** *or* (*US*) **+agog** *suffix* denoting an agent that elicits or enhances the secretion of....

2.6.3 Abbreviations and symbols

- Where a term may be abbreviated or indicated with a symbol, this is noted after the headword.

> **nuclear magnetic resonance** *abbr.*: NMR *or* nmr; the phenomenon that occurs when atomic nuclei....

> **electric potential** *or* potential *symbol*: *V or* ϕ; the work done in bringing unit electric charge....

- The distinction between an abbreviation and a symbol is a little blurred, since some abbreviations (e.g. lg) also may be used as symbols. In general, the term 'symbol' is used here for

  > units and their decimal prefixes (e.g. m, mol; μ, M)

  > physical quantities and constants (e.g. *a*, *H*; *k*, *R*)

  > mathematical functions (e.g. exp, ln)

  > chemical elements (e.g. K, Mg)

  > groups of letters that can be used in place of a chemical group or compound in an equation or formula (e.g. CoA, Me)

  > recommended abbreviations for nucleotides, bases, or amino acids.

- The symbols for SI base and derived units and their decimal prefixes are mandatory; all other symbols are recommendations of IUBMB or IUPAC. In conformity with these recommendations, symbols for physical quantities and fundamental physical constants are printed in a sloping (italic) typeface.

- No distinction is made between acronyms, contractions, abbreviations, etc. All are classed as abbreviations. Abbreviations formed from the initial letters of two or more words are printed without periods (full-stops), in line with contemporary practice, but abbreviations that are shortened forms of single words have a terminal period.

- In addition to recommended abbreviations, the Dictionary lists a selection of others commonly encountered in the scientific literature.

2.6.4 Derived terms

Derived terms not meriting separate definition are listed at the end of the entry for the parent term, preceded by a bold em dash and followed by an abbreviation indicating the part of speech.

> **bactericide** *or* **bacteriocide** any agent (biological, chemical, or physical) that destroys bacteria. —**bactericidal** *or* **bacteriocidal** *adj.*

2.6.5 Etymology

- Generally, the derivation of words is not explained in entries. The exceptions are for eponymous terms and other entries of particular etymological interest.

- The etymology is given within square brackets at the end of the entry.

> **ångström** *or* **Ångstrom** *symbol*: Å; a unit of length equal to 10^{-10} metres. ... [After Anders Jonas Ångström (1814–74), Swedish physicist noted for his work on spectroscopy.]

- Greek elements of etymologies are transliterated:

> **chirality** topological handedness; the property of non-identity of an object with its mirror image. ... [From Greek *kheir*, hand.]

2.6.6 Usage

- The field within which the term is used may be specified in italics and in parenthesis before the definition.

> **malonyl 1** *(in biochemistry)* the univalent acyl group, $HOOC–CH_2–CO–$, derived from malonic acid by loss of one hydroxyl group. **2** *(in chemistry)* the bivalent acyl group, $–CO–CH_2–CO–$, derived from malonic acid by loss of both hydroxyl groups.

- Notes may also be given regarding the use of alternative terms and variant spellings: *see* section 2.3.2.

2.7 Cross-references

2.7.1 Format

- Cross-references are set in bold sans serif type, e.g. **thiouridine**.
- Where a cross-reference refers to only one sense of a word with multiple definitions, this is indicated as in the following example:

> **siderophage** *an alternative name for* **siderophore** (def. 1).

2.7.2 Types of cross-reference

- There are cross-references from a variant spelling, or a less commonly used term, etc., to the entry where the term is defined. For examples, *see* section 2.3.2.
- Some cross-references are to related entries giving more information. These may be either embedded in the text:

> **octulose** any **ketose** having a chain of eight carbon atoms in the molecule.

or listed at the end of the entry:

> **vacuum evaporation** a technique for .... *See also* **shadow casting**.

- Cross-references may also be used to draw attention to contrasting terms:

> **heterochromatin** ... *Compare* **euchromatin**.

or to pairs of easily confused terms:

> **prolidase** *another name for* **X-Pro dipeptidase**. *Distinguish from* **prolinase**.
>
> **prolinase** *the recommended name for* **Pro-X dipeptidase**. *Distinguish from* **prolidase**.

3. Abbreviations

abbr.	abbreviation
adj.	adjective
adv.	adverb
Brit.	British
comb. form	combining form (*see* section 2.6.2)
3-D	three-dimensional
def.	definition
e.g.	[Latin, *exempli gratia*] for example
esp.	especially
etc.	etcetera
Fr.	French
i.e.	[Latin, *id est*] that is
max.	maximum
n.	noun

pl.	plural
sing.	singular
sp. or spp.	species (singular and plural respectively)
US	United States
vb.	verb

Other abbreviations are defined in the text itself.

4. Other conventions

4.1 Spelling and hyphenation

4.1.1 Spelling

- For chemical and biochemical terms, recommended international usage is followed; thus, for example, 'heme' is used rather than 'haem', 'estrogen' rather than 'oestrogen', 'sulfur' rather than 'sulphur', and 'oxytocin' rather than 'ocytocin'. All variants are listed as headwords, however, with cross-references to the corresponding main entries.
- For common terms, e.g. 'colour', British spelling is used.

4.1.2 Hyphenation

- Hyphens are used attributively:

> 'T cell' but 'T-cell receptor'
>
> 'amino acid' but 'amino-acid residues'

- This also applies to enzyme names; thus for example, there is no hyphen following the 'glucose' in 'glucose 6-phosphate', but where this substrate forms part of an enzyme name, it is hyphenated, e.g. in 'glucose-6-phosphatase' or 'glucose-6-phosphate isomerase'.

4.2 Nomenclature

In most cases, headwords conform with the recommendations of the various nomenclature bodies of IUB, IUBMB, and IUPAC. The phrase 'not recommended' has been used to indicate that certain forms are not the recommendation of one of these nomenclature bodies.

4.2.1 Drug names

The recommended international nonproprietary names are used (*International nonproprietary names (INN) for pharmaceutical substances.* World Health Organization, Geneva, 1992); hence, for example, main entries are found under epinephrine and norepinephrine rather than under adrenaline and noradrenaline.

4.2.2 Proprietary names

A few commonly used proprietary names are included; these may be listed at the end of an entry if considered to be of particular interest, especially to non-scientists:

> **acetaminophen** *or* **paracetamol** ... *Proprietary names*: Tylenol, Panadol. It inhibits ....

or may be the main headword:

> **Sephadex**.

4.2.3 Other substances

The main entry is under the name used most widely in the scientific literature. Where this is not the recommended name, a cross-reference is given from the recommended name to the main entry. For example, the name 'follicle-stimulating hormone (FSH)' is widely employed instead of the recommended name 'follitropin', hence the former name has been used as the main headword. In such cases there is a cross-reference from the recommended name back to the entry where the substance is defined:

> **follitropin** *the recommended name for* **follicle-stimulating hormone.**

4.3 Representation of chemical structures

4.3.1 Typeset formulae

In conformity with IUPAC nomenclature recommendations for typeset chemical formulae, parentheses (round brackets) indicate a side chain:

CH_3–$CH(NH_2)$–COOH,
HO–$C(CH_2$–$COO^-)_2$–COO^-

and square brackets indicate a condensed chain:

CH_3–$[CH_2]_8$–COOH

4.3.2 Carbohydrates

- The cyclic forms of monosaccharides are depicted by Haworth representations as are some other compounds; for clarity, the carbon atoms of the heterocyclic ring, and their attached hydrogen atoms, are not shown. *See* the **Haworth representation** entry for more detail.

- Where an abbreviated terminology is included for oligosaccharide chains, the extended or condensed forms described in the publication entitled *Nomenclature of carbohydrates (recommendations 1996)*) are variously used.

- Wherever possible, structure diagrams show absolute configurations.

4.4 Periodic table of the elements

The group numbers used in the text are those of the 18-column format of the table given in the 1990 edition of the IUPAC 'Red Book'. The correspondence between this and other versions of the table is described in the **periodic table** entry and shown below the table displayed on the endpapers.

4.5 Amino-acid sequences

- For peptide sequences of up to 15 amino-acid residues, the three-letter code is used; longer sequences are given in the one-letter code.

- Motifs are given in the one-letter code.

- The full sequences of many proteins can be found in protein sequence databases, and database codes are given to facilitate access to these. The database codes relate to a number of different databases. The style of the code gives an indication of the database from which the data originate, but if the user does not recognize the code, it is necessary to search for it in a composite database that integrates data from all the major databases.

4.6 Genes

- The accepted format of gene names (i.e., whether lower case or upper case or a mixture) varies between different organisms. Where an entry covers genes from various species, the convention for human genes is generally followed in the headword, i.e. all letters are given in upper case, e.g. '*JUN*'.

- However, when an entry refers only to a gene from a specified organism, the accepted convention for that organism is followed.

4.7 Names of organisms

- Where a binomial Latin name is repeated within an entry, the genus name is abbreviated after the first occurrence of the name; for example, the full form '*Escherichia coli*' is used when first mentioned in any entry, but subsequent references to this organism in the same entry are abbreviated to '*E. coli*'.

5. Appendices

Two appendices have been included after the main alphabetical text:

- Appendix A – The Greek alphabet and Greek characters used as symbols

- Appendix B – Sequence-rule priorities of some common ligands in molecular entities.

Aa

a 1 *abbr. for* adsorbed. **2** *symbol for* atto+ (SI prefix denoting 10^{-18}). **3** axial. **4** year.

a' *symbol for* pseudoaxial.

a *symbol for* **1** absorption coefficient. **2** acceleration (in vector equations it is printed in bold italic (***a***)). **3** activity (def. 3). **4** van der Waals coefficient. **5** *as subscript*, denotes affinity.

a_0 *symbol for* Bohr radius.

A *symbol for* **1** acid-catalysed (of a reaction mechanism). **2** a residue of the α-amino acid L-alanine (alternative to Ala). **3** a residue of the base adenine in a nucleic-acid sequence. **4** a residue of the ribonucleoside adenosine (alternative to Ado). **5** uronic acid. **6** ampere.

A *symbol for* **1** absorbance. **2** activity (def. 2). **3** affinity. **4** Helmholtz function. **5** mass number/nucleon number.

A_r *symbol for* relative atomic mass.

A_s *symbol for* area.

$[A]_{0.5}$ or **$[A]_{1/2}$** *symbol (in enzyme kinetics) for* the value of the concentration of a substrate, A, in mol dm^{-3}, at which the velocity of the reaction, v, is half the maximum velocity, V; i.e. when $v = 0.5V$.

$[A]_{50}$ *symbol for* the molar concentration of an agonist that produces 50% of the maximal possible effect of that agonist. Other percentage values ($[A]_{20}$, $[A]_{40}$, etc.) can be specified. The action of the agonist may be stimulatory or inhibitory. *Compare* **EC$_{50}$**.

2'-5'A *symbol for* any member of a series of oligonucleotides of the general formula p$_a$ A[2'p5'A]$_n$, where p and A are phosphoric and adenosine residues, respectively, and a and n are small integers (a = 1, 2, or 3 and n commonly = 2, 3, or 4). Potent inhibitors of protein biosynthesis *in vivo* and *in vitro*, they are believed to mediate the action of interferon on virus-induced cells.

A23187 *or* **calcimycin** a toxic and weakly antibiotic substance isolated from cultures of *Streptomyces chartreusensis*. It is a lipophilic 523 Da monocarboxylic acid of complex structure, two molecules of which form stable lipid-soluble complexes at pH 7.4 with one atom of certain divalent metal cations, especially Mn^{2+}, Ca^{2+}, and Mg^{2+}; monovalent cations are bound only weakly. It also forms lipid-soluble complexes with certain amino acids. It is used experimentally as a calcium **ionophore**.

Å *symbol for* ångström (unit of length equal to 10^{-10} m).

aa 1 *symbol for* an unknown or unspecified aminoacyl group when acting as a substituent on a base or internal sugar in a (poly)nucleotide. **2** *abbr. for* amino acid.

AA *(formerly) symbol for* an unknown or unspecified amino-acid residue. *See* **Xaa**.

AAA 1 a codon in mRNA for L-lysine. **2** *abbr. for* ATPase associated with varied activities. *See* **AAA protease**.

AAA protease *abbr. for* ATPase associated with varied activities; any member of a family of conserved ATP-dependent proteases that mediate degradation of nonintegrated membrane proteins in bacteria, mitochondria, and chloroplasts. They form large complexes composed of identical or homologous subunits. Each subunit contains two transmembrane segments, an ATP-binding domain, and a metal-dependent catalytic domain. Mitochondria contain a matrix-facing AAA protease (m-AAA protease) and an intermembrane space-facing AAA protease (i-AAA protease). The m-AAA protease is regulated by prohibitins. Paraplegin belongs to the AAA protease family.

AAC a codon in mRNA for L-asparagine.

Aad *symbol for* a residue of the α-amino acid L-α-aminoadipic acid, L-2-aminohexanedioic acid.

βAad *symbol for* a residue of the β-amino acid L-β-aminoadipic acid, L-3-aminohexanedioic acid.

AAG a codon in mRNA for L-lysine.

A antigen the antigen defining the A blood group. *See also* **blood-group substance**, **ABH antigens**.

aardvark a *Dictyostelium* orthologue of β-catenin with cytoskeletal and signal transduction roles. *See* **catenin**.

Aarskog–Scott syndrome *or* **Aarskog syndrome** *or* **faciogenital dysplasia** an extremely rare genetically heterogeneous developmental disorder in which individuals have widely spaced eyes, anteverted nostrils, a broad upper lip and a 'saddlebag' or 'shawl scrotum'. The X-linked form has been ascribed to mutations in the **FGD1** gene. [After Dagfinn Aarskog (1928–), Norwegian paediatrician, and Charles I. Scott Jr (1934–), US paediatrician.]

AAT *abbr. for* amino acid transporter.

AatII a type 2 **restriction endonuclease**; recognition sequence: GACGT↑C.

AAU a codon in mRNA for L-asparagine.

Ab *abbr. for* antibody.

abamectin *or* **avermectin B$_1$** a metabolite of *Streptomyces avermitilis* used as an acaricide, insecticide, and a veterinary anthelmintic.

A-band an anisotropic band in a **sarcomere**.

Abbe refractometer a **refractometer** in which the critical angle for total reflection at the interface of a film of liquid between two similar glass prisms is used in determining the refractive index of the liquid. [After Ernst Abbe (1840–1905), German physicist famous for his researches in optics.]

ABC *abbr. for* **1** antigen-binding capacity. **2** ATP-binding cassette (*see* **ABC transporter**).

ABC model a model for specification of floral organs especially in *Arabidopsis thaliana*. It views the floral primordium as comprising four whorls whose developmental fate is determined by the concentric and combinatorial activity of three classes of gene, denoted A, B, and C, which encode transcription factors. Class A determines the fate of whorls 1 and 2 (sepals and petals, respectively) and requires the *APETALA2* gene (*AP2*); class B determines whorls 2 and 3 (petals and stamens, respectively) and requires the *PISTILLATA* (*PI*) and *APETALA3* (*AP3*) genes; class C determines whorl 4 (carpels) and requires the *AGAMOUS* gene (*AG*). These genes are described as 'homeotic' even though they encode transcription factors that contain a **MADS box** instead of homeobox domains. Homologues of these genes occur in other plants.

ABCR *abbr. for* ATP-binding cassette transporter retina; *other name*: rim protein. A protein found in the disc membrane of the outer segment of photoreceptor cells of the retina. It consists of 2273 amino acids, and is presumed to function in the transport of retinoids. Mutations in the ABCR gene, at 1p21-p23, are associated with Stargardt and age-related macular dystrophies. *See* **Stargardt macular dystrophy**.

ABC transporter a membrane transport protein having the ABC molecular domain, named after ATP-binding cassette, characteristic of all members of a large superfamily of membrane transport proteins that hydrolyse ATP and transfer a diverse array of small molecules across membranes. *See also* **CFTR**, **MDR protein**, **sugar transporter**.

ABC transporter retina *see* **ABCR**; *see also* **Stargardt macular dystrophy**.

abductin an insoluble, rubber-like protein from the internal triangular hinge ligament of scallops.

Abe *symbol for* abequose.

abequose *symbol*: Abe; 3,6-dideoxy-D-*xylo*-hexose; 3,6-dideoxy-D-galactose; a deoxysugar that occurs, e.g., in O-specific chains of lipopolysaccharides in certain serotypes of *Salmonella*. For the L enantiomer *see* **colitose**.

abetalipoproteinemia *or* (*Brit.*) **abetalipoproteinaemia** an autosomal recessive disorder in which plasma lipoproteins lack **apolipoprotein B**. There is defective assembly and secretion both of chylomicrons in intestinal mucosa and of very-low-density lipoproteins in the liver. The cause is a deficiency of the 88 kDa subunit of **microsomal triglyceride transfer protein**.

ABH antigens one of the systems of **blood group** antigens having determinants associated with oligosaccharide structures. It is the basis of the **ABO system**, which was the first human blood group antigen system to be detected, by Austrian-born US pathologist Karl Landsteiner (1868–1943) in 1901, and it remains the most important in blood transfusion. Individuals having neither A nor B antigen express the H antigen, the product of an independent gene belonging to the Hh system. Antigens of the ABH system are oligosaccharide chains, in the erythrocyte carried on band 3 (the anion transporter) and band 4.5 (the glucose transporter), or on ceramide. A highly branched *N*-glycan, consisting of a trimannosyl-di-*N*-acetyl-chitobiosyl core with Gal(β1-4)GlcNAc(β1-3) repeats, forms the basis of ABH antigens. The H determinant is the precursor; antigen A is formed by addition of *N*-acetyl-D-galactosamine by fucosylgalactose α-*N*-acetylgalactosaminyltransferase (EC 2.4.1.40); antigen B is formed by addition of D-galactose by fucosylglycoprotein 3-α-galactosyltransferase (EC 2.4.1.37). The terminal sugar residues of importance are: H determinant, Fuc(α1-2)Galβ-R; A determinant, GalNAc(α1-3)(Fucα1-2)Galβ-R; B determinant, Gal(α1-3) (Fucα1-2)Galβ-R. The enzyme responsible for adding the terminal fucosyl residue of the H determinant is galactoside 2-α-L-fucosyltransferase (EC 2.4.1.69). *See also* **A-transferase, B-transferase**.

abietic acid a plant terpene acid present in the nonvolatile residue of pine oil.

ab initio *Latin* from first principles; literally it means 'from the beginning'.

ab initio gene prediction the prediction of genes in uncharacterized nucleotide sequences using only characteristics of the sequence (codon usage, compositional bias, etc.) – that is, without direct reference to other sequences.

ab initio protein structure prediction the prediction of the structure of proteins using only properties of the amino-acid sequence (solvation potentials, secondary structure propensities, etc.) – that is, without direct reference to the structure of known homologues.

abiogenesis *or* **spontaneous generation** the discredited doctrine that living organisms can arise from nonliving materials under current conditions. *Compare* **biogenesis** (def. 2).

abiotic characterized by the absence of life.

abl an oncogene from murine Abelson leukemia virus. The human equivalent is *ABL* (locus at 9q34), which encodes a tyrosine protein kinase. In humans, inappropriate activation of *ABL* occurs via a reciprocal translocation between chromosomes 9 and 22 in which *ABL* is joined at the **breakpoint cluster region** (bcr) of the *ph1* gene on chromosome 22(q11), resulting in an altered chromosome 22, referred to as the **Philadelphia chromosome** (*Ph^1*). The protein product of the spliced genes in the *Ph^1* chromosome is a molecule of 210 kDa, which has increased tyrosine kinase activity. The *Ph^1* chromosome occurs in most patients with chronic myelogenous leukemia. c-Abl can potentially regulate cell growth and may participate in growth regulation at multiple cellular locations, interacting with different cell components. It contains SH2 and SH3 domains (*see* **SH domains**) and also domains involved in binding to F-actin and DNA, and occurs in both cytoplasmic and nuclear locations. Its DNA-binding activity appears to be cell-cycle-regulated by Cdc2-mediated phosphorylation; it binds the **retinoblastoma protein** indicating involvement in transcriptional regulation.

ablation 1 (*in surgery*) the removal or destruction of tissue by a surgical procedure. **2** (*in genetics*) a technique for the removal of a tissue or a particular cell type during development. It depends on the tissue-specific expression of a toxin gene such as diphtheria A (*dipA*) in a transgenic organism.

ABM *abbr. for* 2-aminobenzyloxymethyl, a group used for derivatizing cellulose or paper. It is converted by diazotization into **DBM**.

abortive complex *or* **dead-end complex** *or* **nonproductive complex** any enzyme–substrate complex in which the substrate is bound to the enzyme in a manner that renders catalysis impossible so that products cannot be formed.

abortive infection infection of a bacterium by phage lacking phage DNA, e.g. in generalized **transduction**.

abortive transconjugate *see* transconjugate.

abortive transduction a type of **transduction** in which the donor DNA is not integrated with the recipient chromosome but persists as a nonreplicating fragment that can function physiologically and can be transmitted to one daughter cell at each cell division.

ABO system one of the systems of human **blood groups**, of great importance in blood transfusion because human plasma contains natural antibodies against A and B blood group antigens of the ABH system (*see* **ABH antigens**). The antigens on the red blood cells and the plasma antibodies corresponding to the various phenotypes are shown in the table. See also **A-transferase, B-transferase, O antigen** (def. 2).

Phenotype (blood group)	Antigen on red cells	Antibody in plasma
A	A	anti-B
B	B	anti-A
AB	A and B	neither
O	H	anti-A + anti-B

ABP1 *abbr. for* auxin-binding protein 1.

Abri a neurotoxic 34-residue polypeptide, derived from a mutant putative transmembrane precursor, that forms amyloid fibrils in the brain in familial British dementia.

abrin a highly toxic ~65 kDa glycoprotein obtained from the seeds of jequirity, or Indian liquorice (*Abrus precatorius* L.), a tropical Asian vine that also occurs in Florida. It consists of an ~30 kDa acidic A chain, and an ~35 kDa neutral B chain, held together by disulfide bonds. The A chain is a powerful inhibitor of protein synthesis, while the B chain functions as a carrier to bind abrin to the membrane, and perhaps to assist penetration of the A chain into the cell. One well-chewed seed can be fatal. The A and B chains are derived from a common 528 amino acid 59.24 kDa precursor. *Compare* **abrine, ricin**. *See also* **ribosome-inactivating protein**.

abrine *trivial name for* N^α-methyl-L-tryptophan, α-methylamino-amino-β-(3-indole)propionic acid; an imino acid obtained from seeds of jequirity (*Abrus precatorius*). Not to be confused with **abrin**.

abscisic acid *or* (*formerly*) **abscisin II** *or* **dormin** *abbr.*: ABA; 5-(1-hydroxy-2,6,6-trimethyl-4-oxocyclohex-2-en-1-yl)-3-methylpenta-2,4-dienoic acid; a chiral sesquiterpene. The naturally occurring form, the 2*Z*,4*E*,*S* isomer, also designated (*S*)-abscisic acid, is a phytohormone formed by the degradation of carotenoids. It controls abscission in flowers and fruit but probably not in leaves, and is also implicated in geotropism, stomatal closure, bud dormancy, dormancy of seeds requiring stratification (i.e. those that will only germinate after exposure to low temperatures), and possibly tuberization.

abscissa the horizontal or *x* coordinate in a plane rectangular (Cartesian) coordinate system. *Compare* **ordinate**.

abscission the natural shedding of leaves, fruits, and other parts by a plant.

absolute 1 pure, unmixed; e.g. absolute alcohol. **2** not relative; e.g. absolute configuration. **3** describing a measurement defined in fundamental units of mass, length, and time that does not depend on the characteristics of the measuring apparatus; e.g. absolute temperature.

absolute alcohol *the common name for* pure ethanol, i.e. ethanol that has been freed of water. It may contain small amounts of benzene that have been added to aid in removing water. Substances may be added to absolute alcohol to render it unfit for human consumption and hence free of excise duty: **industrial spirit** contains 5% v/v methanol, while **methylated spirit** also contains pyridine, petroleum oil, and methyl violet dye, and **surgical spirit** also contains castor oil, diethyl phthalate, and methyl salicylate.

absolute configuration the actual three-dimensional arrangement of the atoms in a chiral molecule.

absolute reaction rate theory a theory that sets out to predict the absolute reaction rate of a chemical reaction from the quantum mechanical description of the potential energy changes during the interaction between chemical species. It is most widely drawn upon in applying thermodynamic reasoning to equilibria between reactants in the ground state and chemical species in the activated state or transition state.

absolute temperature *see* **thermodynamic temperature**.

absolute zero zero thermodynamic temperature, i.e. 0 K or −273.15 °C.

absorb *see* **absorption**.

absorbance *symbol*: A; a measure of the ability of a substance or a solution to absorb electromagnetic radiation incident upon it. It equals the logarithm of the ratio of the radiant power of the incident radiation, Φ_0, to the radiant power of the transmitted radiation, Φ. For a solution, absorbance is expressed as the logarithm of the ratio of the radiant power of light transmitted through the reference sample to that of the light transmitted through the solution, the observations being made using identical cells. (Traditionally, radiant intensity was measured instead of radiant power, which is now the accepted form.) Two quantities are defined: **(decadic) absorbance** (*symbol*: A_{10} *or* A), and **napierian absorbance** (*symbol*: A_e *or* B).

$$A_{10} = \lg (\Phi_0/\Phi) = \lg T^{-1} = -\lg (1 - \alpha),\text{ and}$$

$$A_e = \ln (\Phi_0/\Phi) = \ln T^{-1} = -\ln (1 - \alpha),$$

where T is the (internal) **transmittance** and α is the **absorptance**. These definitions suppose that all the light incident upon the sample is either transmitted or absorbed, reflection and scattering being negligible. The more general term attenuation may be used when scattering is considerable, as when the quantity $\lg(\Phi_0/\Phi)$ is measured to estimate the cell density of a culture. The older terms **absorbancy**, **extinction**, and **optical density** should no longer be used. *Compare* **absorption coefficient**.

absorbancy *or* **absorbency** *(formerly)* an alternative term *(no longer recommended)* for **absorbance**.

absorbate a substance that is absorbed into another substance.

absorbed dose *(in radiation physics)* a measure of the energy deposition produced by **ionizing radiation** in any (specified) medium as a result of **ion-pair** formation. The CGS unit of absorbed dose is the **rad**; the SI derived unit is the **gray** (*symbol*: Gy); *compare* **exposure** (def. 3). *See also* **dose equivalent**, **dose rate**.

absorbence *a variant spelling of* **absorbance**.

absorbent 1 a substance that absorbs another substance. **2** having the capacity to absorb another substance.

absorptance *or* **absorption factor** *symbol*: α; the ratio of absorbed to incident radiant or luminous flux. A dimensionless physical quantity given by $\alpha = \Phi_{abs}/\Phi_0$, where Φ_0 and Φ_{abs} are the incident and absorbed radiant powers, respectively.

absorptiometer 1 an apparatus, frequently a photoelectric device, for measuring light **absorption** by solids, liquids, or gases. **2** an apparatus for measuring the amount of gas absorbed by a liquid.

absorption 1 the act or process whereby one substance, such as a gas or liquid, is taken up by or permeates another liquid or solid.

Compare **adsorption**. **2** the retention by a material of energy removed from electromagnetic radiation passing through the material. **3** the removal of any form of radiation, or the reduction of its energy, on passing through matter. **4** the process whereby a neutron or other particle is captured by an atomic nucleus. **5 a** *(in cellular physiology)* the uptake of fluids by living cells or tissues. **b** *(in animal physiology)* the totality of the processes involved in causing water, the products of digestion, and exogenous substances of low molecular mass such as drugs, salts, vitamins, etc. to pass from the lumen of the gastrointestinal tract into the blood and lymph. **c** *(in plant physiology)* the uptake of water and dissolved salts through the roots. **6** *(in immunology)* the process of removing a particular antibody (or antigen) from a mixture by adding the complementary antigen (or antibody) and discarding the antigen–antibody complex so formed. *Compare* **immunosorption**.

absorption band *or* **absorption line** a region of darkness or absorption of radiation in the spectrum of heterochromatic radiation that has passed through an absorbing material.

absorption coefficient four different coefficients are defined. The **(linear) decadic absorption coefficient** (*symbol*: a) is defined by $a = A_{10}/l$; units m^{-1}. The **(linear) napierian absorption coefficient** (*symbol*: α) is defined by $\alpha = A_e/l$; units m^{-1}. The **molar (decadic) absorption coefficient** (*symbol*: ε) is defined by $\varepsilon = a/c = A_{10}/cl$; units $m^2\ mol^{-1}$. The **molar napierian absorption coefficient** (*symbol*: κ) is defined by $\kappa = \alpha/c = A_e/l$. A_{10} and A_e are the decadic and napierian absorbances respectively (*see* **absorbance**), l = path length, and c = amount-of-substance concentration.

absorption cross-section the probability that a photon passing through a molecule will be absorbed by that molecule multiplied by the average cross-sectional area of the molecule. The **net absorption cross-section** (*symbol*: σ_{net}) is defined by $\sigma_{net} = \kappa/N_A$, where κ is the molar napierian **absorption coefficient** and N_A is the Avogadro constant.

absorption, distribution, metabolism, elimination, and toxicity *abbr.*: ADME/Tox; procedures for assessing how pharmaceutical entities are taken up by the body, where they go in the body, the chemical changes they undergo during these processes, how they are excreted, and the toxicological effects they might have. ADME/Tox is an essential component of drug-safety testing.

absorption factor *an alternative name for* **absorptance**.

absorption index *symbol*: k; it is given by $k = \alpha/4\pi\tilde{\nu}$, where α is the (linear) napierian **absorption coefficient** and $\tilde{\nu}$ the wavenumber in vacuum of the radiation.

absorption line *an alternative name for* **absorption band**.

absorption spectrometry the process of measuring an absorption spectrum with a **spectrometer**. **Absorption spectrophotometry** is a related process employing a spectrophotometer. *See also* **absorbance**, **absorptivity**.

absorption spectroscopy the **spectroscopy** of an absorption spectrum.

absorption spectrum a **spectrum** produced when electromagnetic radiation is absorbed by a sample. The frequencies of the radiation absorbed are those able to excite the atoms or molecules of the sample from their ground states to excited states. The frequency, ν, at which a particular absorption line occurs depends on the energy difference, ΔE, between that of a particular ground state and that of the corresponding excited state. It is given by $\nu = \Delta E/h$, where h is the Planck constant. *Compare* **emission spectrum**.

absorptivity a measure of the ability of a material to absorb electromagnetic radiation. It equals the absorptance of a sample of the material divided by the optical path-length. For very low attenuance, it approximates the absorption coefficient. Use of the term is not recommended.

Abu *symbol for* a residue of the α-amino acid L-2-aminobutanoic acid L-α-aminobutyric acid.

A$_2$bu *or* **Dab** *symbol for* a residue of the α,γ-diamino acid L-α,γ-diaminobutyric acid, L-2,4-diaminobutanoic acid.

abzyme *abbr. for* antibody enzyme (an antibody with enzyme activity; also known as catalytic monoclonal antibody).

a.c. *or* **AC** *or* **ac** *abbr. for* alternating current.

Ac *symbol for* **1** actinium. **2** the acetyl group, $CH_3CO–$.

ACA a codon in mRNA for L-threonine.

acanthosome an organelle of fibroblasts isolated from the dermis of hairless mice after chronic UV irradiation. It exists as a spinous membranous vesicle.

acarbose a pseudotetrasaccharide, *O*-4,6-dideoxy-4-[[[1*S*-(1*α*,4*α*,5*β*,6*α*)]-4,5,6-trihydroxy-3-(hydroxymethyl)-2-cyclohexen-1-yl]amino]-*α*-D-glucopyranosyl-(1→4)-*O*-*α*-D-glucopyranosyl-(1→4)-glucose, that inhibits *α*-glucosidase, thereby reducing gastrointestinal absorption of glucose. It is a putative antidiabetic agent.

acaricide a substance or mixture of substances intended to destroy or prevent infestation with mites and ticks.

acatalasemia or *(Brit.)* **acatalasaemia** or **Takahara disease** a rare, generally benign condition in which erythrocyte catalase activity is less than 1% of normal. It is sometimes associated with ulcerating lesions in the mouth. It is caused by a splice junction mutation in the catalase gene locus at 11p13.

ACC 1 a codon in mRNA for L-threonine. **2** *abbr. for* 1-aminocyclopropane-1-carboxylic acid.

***Acc*I** a type 2 **restriction endonuclease**; recognition sequence: GT↑[AC][GT]AC.

***Acc*II** a type 2 **restriction endonuclease**; recognition sequence: CG↑CG. *Fnu*DII is an **isoschizomer**.

***Acc*III** a type 2 **restriction endonuclease**; recognition sequence: T↑CCGGA. *Bsp*MII is an **isoschizomer**.

a.c. calorimetry a technique in which the thermal response of a sample to an oscillating heat signal is measured in the form of a temperature wave propagating through the sample. The technique allows the measurement of the heat capacity of the sample on both cooling and heating and the monitoring of its isothermal time-dependence. It is useful in the study of phase transitions in solids and in gel to liquid-crystal systems.

accelerator 1 *(in chemistry)* **catalyst**, especially one that increases the rate of a polymerization reaction. **2** *(in physics)* a device or machine used for imparting high kinetic energy to charged subatomic particles, e.g. electrons, protons, or alpha particles, by means of electric and/or magnetic fields.

accelerator globulin *an alternative name for* factor V. See **blood coagulation**.

accelerin *an alternative name for* factor Va. See **blood coagulation**.

acceptor 1 *(in chemistry)* a chemical entity that in a chemical reaction receives an electron, atom, or group of atoms. *Compare* **donor, donor atom**. See also **electron acceptor**. **2** *(in physiology)* a receptor that binds a hormone without a biological response being demonstrable. **3** *(in pharmacology)* a receptor that binds a drug but has no identified endogenous ligand.

accession number a systematic (computer-readable) number or code that uniquely identifies an entry in a particular database. Accession numbers are assigned when entries are first added to a database and should remain static between updates, providing a reliable means of locating them in subsequent releases. For example, P02699 identifies bovine rhodopsin in the **Swiss-Prot** database, and IPR000276 identifies the rhodopsin-like G protein-coupled receptor superfamily in InterPro.

accessory cell any one of various types of cell that assist in the immune response. The term includes **antigen-presenting cell**, **basophil**, **eosinophil**, **mast cell**, and **platelet**.

accessory chromosome *an alternative name for* **1** a **B chromosome**. **2** a **sex chromosome**.

accessory DNA surplus DNA present in certain cells or during certain stages of cell development owing, for example, to **gene amplification**.

accessory food factor *or* **accessory growth factor** a term originally used to describe any unknown substance – subsequently called **vitamin** – found in small amounts in some foods, such as milk, that was necessary for the normal growth of animals fed on diets of purified carbohydrates, fats, proteins, and salts.

accessory pigment any of the pigments, such as the yellow, red, or purple **carotenoids** and the red or blue **phycobiliproteins** in photosynthetic cells. The carotenoids are always present, whereas the phycobiliproteins occur only in algae belonging to the Rhodophyceae, the Cyanophyceae, and the Cryptophyceae. Strictly speaking, chlorophyll b is also an accessory pigment.

ACC oxidase an enzyme present in plant tissues that catalyses the Fe^{2+}- and ascorbate-dependent oxidation of **1-aminocyclopropane-1-carboxylic acid** (ACC) to ethylene, CO_2, HCN, and H_2O. It is a highly unstable monomer (35 kDa) that is inhibited by Co^{2+}. Its activity increases under conditions of stress and at certain developmental stages (e.g. during fruit ripening).

ACC synthase EC 4.4.1.14; *systematic name:* *S*-adenosyl-L-methionine methylthioadenosine lyase; an enzyme present in plant tissues that catalyses the pyridoxal phosphate-dependent conversion of *S*-adenosylmethionine to **1-aminocyclopropane-1-carboxylic acid** (ACC) and 5′-methylthioadenosine. It has been cloned from various fruits – zucchini (courgette), tomato, apple – and has 48–97% sequence identity in different plants. Tomato contains several genes for the enzyme that are differentially regulated and expressed in response to wounding, ripening, or various stresses.

accuracy a measure of the proximity of a measured value to a true value. Compare **precision**.

ACE *abbr. for* **1** amplification control element (a DNA sequence in vertebrates that functions as the origin for **amplification**). **2** **angiotensin converting enzyme**.

ACeDB or **acedb** *abbr. for* a *Caenorhabditis elegans* database; *see* **genome database**.

A cell or *(formerly)* **alpha cell** or ***α* cell** one of the three main histological cell types found in the islets of Langerhans of the pancreas, also found in the gastric oxyntic mucosa. A cells produce, store, and secrete the hormone **glucagon**.

A₁ cell *(formerly)* an alternative name for **D cell**.

aceruloplasminemia or *(Brit.)* **acaeruloplasminaemia** a rare autosomal recessive disorder in which plasma **ceruloplasmin** is severely deficient, characterized by neurological abnormalities and systemic hemosiderosis. Any of at least six mutations in a locus at 3q21-q24 can cause the disease.

Aces or **ACES** *abbr. for* *N*-(2-acetamido)-2-aminoethane sulfonic acid; 2-[(2-amino-2-oxoethyl)amino]ethane sulfonic acid; a **Good buffer substance**, pK_a (20°C) = 6.9.

acesulfame 6-methyl-1,2,3-oxathiazin-4(3*H*)-one 2,2′-dioxide; a sweet-tasting material that, as the potassium salt, has been used in foods and cosmetics.

acetal any member of a class of organic compounds having the general formula $R^1HC(OR^3)OR^4$ or $R^1R^2C(OR^3)OR^4$ – in a **thioacetal** the corresponding formulae are $R^1HC(SR^3)SR^4$ or $R^1R^2C(SR^3)SR^4$ – where R^3 and R^4 are alkyl groups (or R^4 is H in a **hemiacetal** or **hemithioacetal**). An acetal molecule is formed by the acid-catalysed combination of the carbonyl group of an aldehyde or ketone molecule with either one or two alcohol (or mercaptan) molecules (which may be the same or different), or with a diol (or dithiol), by a reaction of the following general type, where X is O (or S in a mercaptan or dithiol):

$$R^1R^2C=O + R^3XH = R^1R^2C(XR^3)OH.$$

The hemiacetal (or hemithioacetal) so formed may then undergo a further reaction:

$$R^1R^2C(XR^3)OH + R^3XH = R^1R^2C(XR^3)_2 + H_2O$$

or

$$R^1R^2C(XR^3)OH + R^4XH = R^1R^2C(XR^3)XR^4 + H_2O.$$

In carbohydrates such compounds are formed at the carbonyl group of the acyclic form of a saccharide or saccharide derivative. The terms 'ketal' (or 'thioketal') and 'hemiketal' (or 'thiohemiketal'), may be applied respectively to any acetal of general formula $R^1R^2C(XR^3)XR^4$ or $R^1R^2C(OH)XR^3$, i.e. to those derived from ketones. These terms, at one time abandoned, have recently been reintroduced as the respective names of subclasses of acetals and hemiacetals and as functional class names.

acetaminophen or **paracetamol** 4-acetamidophenol; *N*-acetyl-*p*-aminophenol; *N*-(4-hydroxyphenyl)acetamide; a drug widely used as an analgesic and antipyretic. It inhibits formation of prostaglandins within, but not outside the brain. It is metabolized within the liver mostly to glucuronide and sulfate conjugates. A small amount is oxidized to a highly reactive intermediate, *N*-acetylbenzoquinoneimine, that is normally detoxified by conjugation with glutathione. If it is produced in excess of the capacity of the liver to detoxify it, hepatic necrosis can result. It can be administered with methionine, which increases glutathione in the liver. *N*-Acetylcysteine is administered in cases of poisoning to act as a glutathione substitute. *Proprietary names include*: Panadol, Tylenol.

acetate 1 the traditional name for ethanoate; the anion, CH_3COO^-, derived from acetic acid (ethanoic acid). **2** any salt or ester of acetic acid.

acetate–CoA ligase EC 6.2.1.1; *systematic name*: acetate:CoA ligase (AMP-forming); *other names*: acetyl–CoA synthetase; acyl-activating enzyme; acetate thiokinase; acetyl-activating enzyme. An enzyme that catalyses a reaction between ATP, acetate, and CoA to form AMP, pyrophosphate, and acetyl-CoA. It is an important enzyme in organisms (e.g. *Escherichia coli*, many fungi, protozoans, algae) that utilize acetate as a carbon source. Distinguish from acetate–CoA ligase (ADP-forming), EC 6.2.1.13.

acetate thiokinase *see* **acetate–CoA ligase**.

acetazolamide an inhibitor of carbonic anhydrase that is useful as a diuretic. It acts by preventing bicarbonate reabsorption in the proximal tubules of the kidney.

(+)aceto+ *comb. form* denoting the acyl group derived from acetic acid.

acetoacetate–CoA ligase EC 6.2.1.16; *other name*: acetoacetyl-CoA synthetase; an enzyme that catalyses the formation of acetoacetyl-CoA from ATP, acetoacetate, and CoA with release of AMP and pyrophosphate. In bacteria that carboxylate acetone to acetoacetate, it activates the latter for further metabolism. It is also present in animals, but utilization of blood acetoacetate after its entry into tissues involves **3-oxoacid CoA-transferase**.

acetoacetyl acetyltransferase *see* **acetyl-CoA *C*-acetyltransferase**.

acetoacetyl-CoA synthetase *see* **acetoacetate–CoA ligase**.

acetoacetyl-CoA thiolase *see* **acetyl-CoA *C*-acetyltransferase**.

acetogenin any substance built up of two-carbon units that may formally be considered to derive from a polyacetyl chain intermediate; the carbon atoms derived from the carboxyl carbon atoms of acetic acid frequently remain oxidized. It is not a recommended term. *See* **polyketide**.

acetoin 3-hydroxy-2-butanone; a compound formed by action of acetolactate decarboxylase (EC 4.1.1.5) and, under some conditions, pyruvate decarboxylase (EC 4.1.1.1).

acetone body *see* **ketone body**.

acetone powder any preparation of ruptured cells obtained from a tissue or single-celled organisms that involves dehydration with acetone to form a powder. It is relatively stable, and is used in the preparation of some enzymes.

acetyl the acyl group ethanoyl, $CH_3CO–$, derived from acetic acid (= ethanoic acid).

1-*O*-acetyl-ADPribose *abbr*. OAADPr; a metabolite produced by SIRT2-like or other enzymes from acetylated histone and NAD with release of nicotinamide. Its function is not known. *See* **Sir**.

***N*-acetylaspartate** *abbr*.: NAA; a derivative of aspartic acid, synthesized by *N*-acetyl transferase and degraded by aspartoacylase, present at high concentration in brain grey matter. Its function is enigmatic but its distribution is similar to that of *N*-acetylaspartylglutamate, which is a putative neurotransmitter. Canavan disease results from mutations that decrease aspartoacylase activity and hence increase concentrations of NAA in cerebrospinal fluid and urine.

***N*-acetylaspartylglutamate** *see* ***N*-acetylaspartate**.

acetylation an **acylation** reaction in which an acetyl group, $CH_3CO–$, is introduced into an organic compound. —**acetylated** *adj*.

acetylation coenzyme *the original name for* **coenzyme A**.

acetylcholine *abbr*.: ACh; the acetyl ester of choline; it is a chemical transmitter in both the peripheral and central nervous system. *See* **neurotransmitters**.

acetylcholine binding protein *abbr*.: AChBP; a soluble protein that binds acetylcholine (ACh). It is homologous with, and has similar ligand-binding characteristics to, the extracellular domain of the alpha subunit of the acetylcholine receptor. It forms a homopentamer.

acetylcholine receptor *see* **cholinoceptor**.

acetylcholinesterase *abbr*.: AChE; EC 3.1.1.7; *systematic name*: acetylcholine acetylhydrolase; *other names*: true cholinesterase; cholinesterase I; an esterase enzyme that catalyses the hydrolysis of acetylcholine to choline and acetate; it also acts on a variety of acetic esters and catalyses transacetylations. It is found in or attached to cellular or basement membranes of presynaptic cholinergic neurons and postsynaptic cholinoceptive cells. A soluble form occurs in cerebrospinal fluid and within cholinergic neurons. It is inhibited by a number of drugs, e.g. physostigmine, and by several **organophosphates**. The 3-D structure is known for fragments obtained from the electric ray (fish).

acetylcholine transporter protein an integral membrane protein of synaptic vesicles of cholinergic neurons. It transports newly synthesized acetylcholine molecules into the synaptic vesicles in exchange for protons, thereby replenishing vesicular stores of the neurotransmitter.

acetyl-coA *abbr. for* **acetyl coenzyme A**.

acetyl-CoA *C*-acetyltransferase EC 2.3.1.9; *other names*: acetoacetyl acetyltransferase; acetoacetyl-CoA thiolase; an enzyme that catalyses the formation of two molecules of acetyl-CoA from CoA and acetoacetyl-CoA. During beta oxidation it catalyses the formation of acetyl-CoA from acetoacetyl-CoA, whereas it acts in the reverse direction to form acetoacetyl-CoA during ketogenesis.

It is important in regulating the metabolic pathways for the production of acids, i.e. acetate, butyrate, or solvents, i.e. acetone, butanol, ethanol, during the growth of *Clostridium acetobutylicum*. Reduced activity of the enzyme favours production of acetate and ethanol, while increased activity favours production of butyric acid, butanol, and acetone.

acetyl-CoA C-acyltransferase EC 2.3.1.16; *systematic name*: acyl-CoA:acetyl-CoA *C*-acyltransferase; *other names*: 3-ketoacyl-CoA thiolase; *β*-ketothiolase; an enzyme that catalyses the formation of acyl-CoA and acetyl-CoA from CoA and 3-oxoacyl-CoA. This is the concluding reaction of each cycle of the fatty acid oxidation pathway (**beta oxidation**). Different enzymes exist in the mitochondrion and peroxisome, both being included in the **thiolase** family.

acetyl-CoA carboxylase EC 6.4.1.2; *systematic name*: acetyl-CoA:carbon-dioxide ligase (ADP-forming); a multienzyme complex involved in the formation of **malonyl-CoA**, the first step in fatty-acid biosynthesis. It catalyses a reaction between ATP, acetyl-CoA, and HCO_3^- to form ADP, orthophosphate, and malonyl-CoA. Biotin is a cofactor. In bacteria it is a heterohexamer of biotin-carboxyl-carrier-protein, biotin carboxylase, and a 2:2 complex of the two subunits of carboxyl transferase. Biotin carboxylase (EC 6.3.4.14) catalyses the reaction between ATP, biotin-carboxyl-carrier-protein, and CO_2 to form ADP, orthophosphate, and carboxy-biotin-carboxyl-carrier-protein. The carbonyl group of the latter is then transferred to acetyl-CoA by carboxyl transferase, thus forming malonyl-CoA. In mammals the activity is part of a trifunctional polypeptide that contains carboxyl carrier protein, biotin carboxylase (EC 6.3.4.14), and acetyl-CoA carboxylase (EC 6.4.1.2) domains.

[acetyl-CoA carboxylase] kinase EC 2.7.1.128; an enzyme that catalyses the phosphorylation by ATP of [acetyl-CoA carboxylase] with release of ADP. This phosphorylation step is one of the regulatory mechanisms for acetyl-CoA carboxylase, causing that enzyme to dissociate from an active polymeric form to an inactive monomeric form.

[acetyl-CoA carboxylase] phosphatase EC 3.1.3.44; an enzyme that catalyses the hydrolysis of phosphate from [acetyl-CoA carboxylase] phosphate. It reverses the phosphorylation catalysed by [acetyl-CoA carboxylase] kinase.

acetyl coenzyme A *abbr.*: acetyl-CoA; a derivative of **coenzyme A** in which the sulfhydryl group is acetylated. Originally termed 'active acetate', it is an important metabolite, derived from pathways such as glycolysis, fatty-acid oxidation, and degradative metabolism of some amino acids. It is further metabolized by the **tricarboxylic-acid cycle** and represents a key intermediate in lipid and terpenoid biosynthesis and other anabolic reactions.

N-acetylcysteine *or* N-acetyl-L-cysteine a thiol-protecting agent used intravenously as an antidote in acetaminophen poisoning. It acts by enhancing glutathione synthesis, thereby increasing the capacity for detoxification and excretion of acetaminophen as a **mercapturic acid**. Methionine can be similarly used. It also has **mucolytic** properties, and is used in aiding the isolation of mycobacteria from sputum.

acetylene the nonsystematic name *for* **ethyne**.

N-acetylgalactosamine *symbol*: D-GalpNAc; *abbr.*: NAGA; the D isomer, 2-acetamido-2-deoxy-D-galactopyranose, is a common structural unit of oligosaccharides, such as the blood-group substances and O-linked glycoproteins, in which the sugar is in glycosidic linkage to a protein or serine residue, or, in the case of the blood-group substances, to a lipid hydroxyl group. The reactant in synthetic reactions is UDP-N-acetylgalactosamine, which is formed by epimerization of N-acetylglucosamine.

β-D-anomer

N-acetylgalactosamine-4-sulfatase EC 3.1.6.12; *other names*: arylsulfatase B; chondroitinsulfatase; chondroitinase; an enzyme that hydrolyses the 4-sulfate groups of the N-acetyl-D-galactosamine 4-sulfate units of chondroitin sulfate and dermatan sulfate. It is a lysosomal enzyme involved in the degradation of proteoglycans, which accumulate in Maroteaux–Lamy syndrome (**mucopolysaccharidosis VI**), a storage disease resulting from a deficiency of the enzyme.

N-acetylgalactosamine-6-sulfatase EC 3.1.6.4; *other names*: chondroitinsulfatase; chondroitinase; galactose-6-sulfate sulfatase; an enzyme that hydrolyses the 6-sulfate groups of the N-acetyl-D-galactosamine 6-sulfate units of chondroitin sulfate, and also the D-galactose 6-sulfate units of keratan sulfate. It is a lysosomal enzyme involved in the degradation of proteoglycans. Keratan sulfate and chondroitin 6-sulfate accumulate in Morquio A syndrome, a storage disease resulting from a deficiency of the enzyme.

N-acetylgalactosaminidase EC 3.2.1.53; either of two lysosomal enzymes that catalyse the hydrolysis of respectively *α*- and *β*-linked terminal nonreducing N-acetyl-D-galactosamine residues. Deficiency of *α*-N-acetylgalactosaminidase is associated with a storage disease (Schindler disease) in which sialyloligosaccharides are found in urine.

N-acetylglucosamine *symbol*: D-GlcpNAc; *abbr.*: NAG; the D isomer, 2-acetamido-2-deoxy-D-glucopyranose, is a common structural unit of plant glycoproteins and of many animal and bacterial glycoproteins. It is often the terminal sugar of an oligosaccharide moiety of a glycoprotein, linked glycosidically to the amide nitrogen of a protein asparagine residue. The acetyl group is introduced in a reaction between acetyl-CoA and glucosamine 6-phosphate to give N-acetylglucosamine 6-phosphate, which undergoes a mutase reaction to form N-acetylglucosamine 1-phosphate, from which UDP-N-acetylglucosamine is formed. The latter is the reactant in pathways of oligosaccharide synthesis and is also a precursor of N-acetylneuraminic acid (*see* **sialic acid**).

N-acetylglucosamine phosphotransferase EC 2.7.1.69; *systematic name*: protein-N^π-phosphohistidine: sugar N-*pros*-phosphotransferase; *other name*: enzyme II of the phosphotransferase system; an enzyme that catalyses the reaction:

protein N^π-phosphohistidine + sugar =
protein histidine + sugar phosphate.

It is a component of the phospho*enol*pyruvate-dependent sugar phosphotransferase system, a major carbohydrate active-transport system; the phosphoryl group from phospho*enol*pyruvate is transferred to phosphoryl carrier protein **HPR** by enzyme I, and from

β-D-anomer

phospho-HPR to the sugar by enzyme II. It is an integral membrane protein.

N-acetylglucosamine-6-sulfatase EC 3.1.6.14; an enzyme that catalyses the hydrolysis of sulfate groups of *N*-acetyl-D-glucosamine 6-sulfate units of heparan sulfate and keratan sulfate. It is a lysosomal glycoprotein. A deficiency is associated with the storage disease **mucopolysaccharidosis III**.

N-acetylglucosaminidase *abbr.*: NAG *(in clinical chemistry)* an alternative name for β-*N*-**acetylhexosaminidase**.

N⁴-(β-N-acetylglucosaminyl)-L-asparaginase EC 3.5.1.26; *other names*: aspartylglucosylamine deaspartylase; aspartylglucosylaminase; aspartylglucosaminidase; aspartylasparaginase; an enzyme that catalyses the hydrolysis of N^4-(β-*N*-acetyl-D-glucosaminyl)-L-asparagine, released from glycoproteins, to *N*-acetyl-β-glucosaminylamine and L-aspartate. A deficiency of the enzyme results in the lysosomal storage disease, **aspartylglucosaminuria**, in which there is an accumulation of the enzyme's substrate.

N-acetylglucosaminyl transferase any of various glycosyltransferase enzymes within the subclass EC 2.4.1 that transfer an *N*-acetylglucosaminyl residue from UDP-*N*-acetyl-glucosamine to an oligosaccharide, and which are important in oligosaccharide synthesis. An example is EC 2.4.1.144, β-1,4-mannosyl-glycoprotein β-1,4-*N*-acetylglucosaminyltransferase; *other name*: *N*-glycosyl-oligosaccharide-glycoprotein *N*-acetylglucosaminyltransferase III. It catalyses the addition of *N*-acetylglucosamine in β(1-4) linkage to the β-linked mannose of the trimannosyl core of *N*-linked sugar chains. It is a type II membrane protein of the Golgi stack. *See also* **N-acetyllactosamine synthase, lipopolysaccharide, N-acetylglucosaminyltransferase, UDP-N-acetylglucosamine-dolichyl phosphate-N-acetylglucosamine phosphotransferase.**

β⁶-N-acetylglucosaminyltransferase *see* β-**1,3-galactosyl-O-glycosyl-glycoprotein β-1,6-N-acetylglucosaminyltransferase.**

N-acetylglutamate synthase *(abbr. AGS) see* **amino-acid N-acetyltransferase.**

acetylglutamic acid the L isomer, *N*-acetyl-L-glutamic acid, is a key intermediate in ornithine formation in bacteria and plants. It is converted to *N*-acetyl-L-glutamic *c*-semialdehyde, from which *N*-acetyl-L-ornithine is formed in a transamination reaction. It activates **carbamoyl-phosphate synthase**, which catalyzes the synthesis of carbamoyl phosphate from ammonia and carbon dioxide, the first committed step in the urea cycle. It is formed from acetyl-CoA and glutamate by the action of amino acid *N*-acetyltransferase, EC 2.3.1.1.

β-N-acetylhexosaminidase EC 3.2.1.52; *other names*: β-hexosaminidase; hexosaminidase; an enzyme that catalyses the hydrolysis of terminal nonreducing *N*-acetyl-D-hexosamine residues in *N*-acetyl-β-D-hexosaminides.

N-acetyllactosamine synthase EC 2.4.1.90; *systematic name*: UDPgalactose:*N*-acetyl-D-glucosamine 4-β-D-galactosyltransferase; *other names*: *N*-acetylglucosamine β-(1→4)-galactosyltransferase; UDPgalactose-*N*-acetyl-glucosamine β-D-galactosyltransferase. An enzyme, located in the rough endoplasmic reticulum, that catalyses a reaction between UDPgalactose and *N*-acetyl-D-glucosamine to form UDP and *N*-acetyllactosamine. In humans the enzyme also has the activity of β-*N*-acetylglucosaminyl-glycopeptide β-1,4-galactosyltransferase (EC 2.4.1.38); *other names*: glycoprotein 4-β-galactosyltransferase;thyroidgalactosyltransfer-

ase; UDPgalactose– glycoprotein galactosyltransferase. It catalyses a reaction between UDPgalactose and *N*-acetyl-β-D-glucosaminyl-glycopeptide to form UDP and β-D-galactosyl-1,4-*N*-acetyl-β-D-glucosaminylglycopeptide. α-**Lactalbumin** is an allosteric regulator and converts this activity to **lactose synthase**.

N-acetylmuramoyl-L-alanine amidase *see* **autolysin**.

N-acetylneuraminic acid *see* **sialic acid**.

acetyloleoylglycerol *see* **oleoylacetylglycerol**.

acetylsalicylic acid *see* **aspirin**.

N-acetyl transferase EC 2.3.1.2; acetyl-CoA–L-aspartate *N*-acetyl hydrolase; an enzyme that catalyses the synthesis of *N*-acetylaspartate.

ACF *abbr. for* 1 ATP-utilizing chromatin assembly and remodeling factor of *Drosophila*. 2 APOBEC-1 complementation factor; a 65 kDa protein that contains three RNA-recognition motifs and is required for APOBEC-1 to edit apolipoprotein B pre-mRNA. *See* **apoB editing enzyme**.

ACG 1 a codon in mRNA for L-threonine. 2 *abbr. for* acycloguanosine *(see* **acyclovir***)*.

Ach *symbol for* the arachidoyl (i.e. eicosanoyl) group.

Δ₂Ach *symbol for* the (all-*Z*)-eicosa-8,11-dienoyl group; *see* **eicosadienoic acid**.

Δ₃Ach *symbol for* the (all-*Z*)-eicosa-5,8,11-trienoyl group; *see* **eicosatrienoic acid**.

Δ₄Ach *symbol for* the arachidonoyl (i.e.(all-*Z*)-eicosa-5,8,11,14-tetraenoyl) group; *see* **arachidonoyl**.

ACh *abbr. for* acetylcholine.

Achaete–Scute complex *see* **AS-C protein**.

achatin-1 an endogenous neuroexcitatory tetrapeptide, Gly-DPhe-Ala-Asp, isolated from the ganglia of the giant African snail, *Achatina fulica*.

AchBP *abbr. for* **acetylcholine binding protein**.

AChE *abbr. for* **acetylcholinesterase**.

achiral not chiral. —**achirality** *n*.

achlorhydria an inability to secrete gastric acid. It is a disorder, probably autoimmune, that is linked with pernicious anemia.

achondroplasia the most common form of dwarfism, inherited as an autosomal disorder. It is due mostly to one of two missense mutations in the fibroblast growth factor receptor-3 gene (*FGFR3*) locus at 4p16.3, which affect the transmembrane region of the receptor, causing activation of the receptor. Homozygosity is lethal in the neonatal period. Milder forms, called **hypochondroplasia**, are due to any of several missense mutations that affect the tyrosine kinase domain of this receptor and also activate it. **Pseudoachondroplasia** is caused by over 70 mutations at 19p13.1, within the gene for cartilage oligomeric matrix protein (COMP). These lead to accumulation of the mutant protein within chondrocytes.

achromic point the point in time during the action of amylase on starch at which the reaction mixture no longer gives a colour with iodine, i.e. the reaction has proceeded to the point when the starch has all been degraded at least as far as **achröodextrins**.

achröodextrin any **dextrin** that is small enough not to give a colour with iodine.

acid 1 in the Brønsted–Lowry concept, a molecular species having a tendency to lose a hydron forming a conjugate base, e.g.

$$A \rightleftharpoons H^+ + B;$$

$$HCl \rightleftharpoons H^+ + Cl^-;$$

$$RCOOH \rightleftharpoons H^+ + RCOO^-;$$

$$RNH_3^+ \rightleftharpoons H^+ + RNH_2.$$

2 in the Lewis concept, a substance capable of accepting from a base an unshared pair of electrons, which then form a covalent chemical bond, e.g.

$$F_3B + :NH_3 \leftrightarrow F_3B^{-}{-}^{+}NH_3.$$

acid anhydride any compound formed by the elimination of the elements of water from the acidic groups of two acids, e.g. acetic anhydride (two acetic acid molecules) or acetyl phosphate (one molecule each of acetic and phosphoric acids).

acid–base balance term descriptive of the hydrogen-ion status of the blood, the mechanisms that regulate it, and the causes of its deviation from normal.

acid–base catalysis catalysis of a chemical reaction in which either an acid or a base mediates the formation of a reactive intermediate.

acid–base titration a titration in which either acid or base is added to a solution and the progress of the titration is followed by pH measurements, either electrometrically or with the use of pH indicators.

acid box a peptide sequence that contains 4–8 acidic amino acid residues in a protein.

acid carboxypeptidase *see* **cysteine-type carboxypeptidase**.

acid ceramidase EC 3.5.1.23; *other name:* N-acylsphingosine deacylase; a lysosomal enzyme that catalyses the hydrolyis of ceramide to sphingosine and a fatty acid. Its activity requires sphingolipid activator proteins (i.e. saposins B and C) and negatively charged phospholipids. A genetic locus at 8p21.3-22 encodes a precursor that contains ≈395 amino acids and is proteolytically cleaved into an α subunit (≈13 kDa) and a β subunit (≈40 kDa) linked by a disulphide bridge. The β subunit is probably glycosylated. At least nine mutations in the gene are associated with various forms of a deficiency disease called Farber lipogranulomatosis. This is characterized by granuloma formation and lipid-laden macrophages in joints, subcutaneous tissue, larynx, and frequently also in liver, spleen, lungs, heart, and nervous system. *See* **saposin**.

acid dissociation constant *or* **acidity constant** *symbol:* K_a; the thermodynamic equilibrium constant for the dissociation of an **acid**. For a dilute solution of a weak acid, HA, dissociating in water according to the equilibrium:

$$HA + H_2O \rightleftharpoons H_3O^+ + A^-,$$

$$K_a = (a_{H_3O^+} \times a_{A^-})/a_{HA}$$

where a is the **activity** of the species designated by the subscripts. The activity of the water has been omitted from the equation since it may be taken as unity for a dilute aqueous solution. K_a is a measure of the strength of the acid, i.e. of its ability to donate hydrons to water. *Compare* **basic dissociation constant**. *See also* **pK**.

acid dye a **dye** containing an anionic acidic organic group that binds to and stains positively charged macromolecules.

acidemia *or (esp. Brit.)* **acidaemia** *(archaic)* a condition in which there is excessive acidity (i.e. increased hydrogen-ion concentration, lowered pH) of the blood. *Compare* **acidosis**, **alkalemia**.

acid-fast bacillus any bacterium able to resist decolorization by mineral acids after the application of specific basic aniline dyes; this property is possible due to the presence in these organisms of **mycolic acid**, together with a semipermeable membrane that allows the passage of the stain but not of the decolorizing acid.

acid β-glucosidase *see* **glucosylceramidase**.

α₁-acid glycoprotein *an alternative name for* **orosomucoid**.

acid-growth hypothesis the proposal that auxin-dependent acidification of plant cell walls promotes wall extensibility and cell growth. It is based on the demonstration that auxin causes acidification of the medium and that acid substitutes for auxin in causing the changes in th cell wall.

acid hydrolase any hydrolase enzyme that is active in mildly acidic conditions (pH 5–6); often found in **lysosomes**.

acidic 1 of, relating to, containing, or characteristic of an **acid**. *Compare* **basic** (def. 1). **2** having an acid reaction in water; of or relating to an aqueous solution having a pH < 7.0. *Compare* **basic** (def. 2).

acidic amino acid any amino acid containing more potentially anionic groups than potentially cationic groups. All such amino acids have a net negative charge at neutral pH: *e.g.*, aspartic acid and glutamic acid.

acidic-epididymal glycoprotein *abbr.:* AEG; *see* **CRISP**.

acid-labile sulfide a sulfido ligand, e.g. any of the **bridging ligands** in **iron–sulfur proteins**, that is released as H_2S at acid pH.

acid lipase a lysosomal acid triacylglycerol (triglyceride) and cholesterol esterase. It shows ≈60% sequence identity with human gastric and rat lingual lipase, and with them shares the Gly–X–Ser–X–Gly motif associated with esterase activity. Decreased activity is associated with Wolman disease and **cholesterol**

ester storage disease, and is due to mutations in a locus at 10q23.2-q23.3.

acid mucopolysaccharide any of a group of related heteropolysaccharides, found widely distributed in animal connective tissues, that contain N-acetylated hexosamine in its characteristic repeating disaccharide unit. They include chondroitin, **chondroitin sulfates**, **dermatan sulfates**, hyaluronic acid (*see* **hyaluronate**), and **keratan sulfates**.

acid number *or* **acid value** the mass, in milligrams, of potassium hydroxide required to neutralize the free fatty acid in one gram of fat; a measure of the mass of free acid in the sample.

acidophilic 1 staining readily with **acid dyes**. **2** (of organisms) preferring or thriving in a relatively acid environment.

acidosis a clinical condition in which excess acid or a base deficit tends to cause increased hydrogen-ion concentration (i.e. lowered pH) in the blood.

acidosome a non-lysosomal vesicle found in the ciliate protozoan *Paramecium*. The organelle is involved in acidification of digestive phagocytic particles through fusion.

acidotropic seeking an acid environment. The term is used e.g. in connection with Ser/Thr protein kinases that require Glu or Asp as part of the recognition site. The resulting Ser(P) or Thr(P) then acts as an acidic residue, extending the site with the result that further Ser or Thr residues are phosphorylated.

acid phosphatase *abbr. (in clinical biochemistry):* ACP; EC 3.1.3.2; *systematic name:* orthophosphoric monoester phosphohydrolase (acid optimum); *other names:* alkaline phosphomonoesterase; phosphomonoesterase; glycerophosphatase; a lysosomal enzyme (except in red cells). It catalyses the hydrolysis of orthophosphoric monoester to an alcohol and orthophosphate. Zinc and magnesium are cofactors. It is present in high concentrations in the prostate gland, and is also present in red cells, platelets, bone, liver, and spleen. Its measurement in blood may be of use clinically in monitoring progress in cases where prostatic cancer has metastasized, but not where cancer is confined to the prostate, being elevated in only about 30% of cases. Normal range in human plasma 4–11 IU L^{-1}.

acid proteinase *an older name for* enzymes of the sub-subclass **aspartic endopeptidase**, EC 3.4.23. It was suggested by their characteristic low optimum pH.

acid sphingomyelinase *see* **sphingomyelin phosphodiesterase**.

acinus (*pl.* acini) **1** a saclike structure that forms the terminal part of a gland. It comprises a cluster of cells surrounding a small duct. **2** one of the collection of small drupes making up an aggregate fruit such as a raspberry. —**acinar** *adj.*

Acinus a nuclear protein that is cleaved by caspase-3 and is required for chromatin condensation during **apoptosis**. It contains an N-terminal **SAP domain** and an RNA-recognition domain. Orthologues are present in vertebrates and in plants.

ackee *or* **akee** a tree, *Blighia sapida*, native to tropical Africa and widely cultivated in the West Indies, especially Jamaica, for its fruit, the fleshy aril of which is edible when cooked and forms an important item of local diet. Unripe fruits contain toxic amounts of **hypoglycin** and can cause **Jamaican vomiting sickness**.

Acm *abbr. for* acetamidomethyl-.

AcNeu (*formerly*) *symbol for* N-acetylneuraminic acid (*see* **sialic acid**).

acofriose 6-deoxy-3-O-methylmannose; the L enantiomer is a component of some **cardiac glycosides**.

α-L-anomer

aconitase EC 4.2.1.3; *other name*: aconitate hydratase; *systematic name*: citrate (isocitrate) hydro-lyase. A hydrolase enzyme that catalyses the reaction:

$$\text{citrate} = cis\text{-aconitate} + H_2O;$$

it also reversibly converts isocitrate into *cis*-aconitate + H_2O. An iron–sulfur protein, it removes H_R from the *pro-R*-CH$_2$–COOH group of citrate (*see* **citrate** for structure). Under kinetic conditions in which it forms isocitrate from citrate the product is (1R,2S)-1-hydroxypropane-1,2,3-tricarboxylate ((2R,3S)-isocitrate). The 3-D structure is known.

aconitate the *cis* isomer, (*Z*)-prop-1-ene-1,2,3-tricarboxylate, is an intermediate in the conversion of citrate to isocitrate in the tricarboxylic-acid cycle, by the action of **aconitase**.

cis-aconitate

aconitate hydratase *see* **aconitase**.

aconitate Δ-isomerase EC 5.3.3.7; *systematic name*: aconitate Δ^2–Δ^3-isomerase. An enzyme that catalyses the reaction:

$$trans\text{-aconitate} = cis\text{-aconitate}.$$

acoustic gene transfer a method of transforming (typically plant) cells by using ultrasound.

ACP *abbr. for* **1** acyl carrier protein. **2** (*in clinical biochemistry*) acid phosphatase.

εAcp (*formerly*) *symbol for* a residue of the ε-amino acid ε-caproic acid, now known as 6-aminohexanoic acid. εAhx is the preferred symbol.

acquired immune deficiency syndrome *abbr.*: AIDS; a collection of symptoms resulting from infection by a retrovirus (HIV-1 or HIV-2) that specifically attacks and destroys helper T lymphocytes, thereby impairing immunity and leading to a variety of other diseases (infections or neoplasms). Transmission is intrauterine, or by sexual contact, breast feeding, intimate contact with infected body fluids or tissues, and contaminated needles or syringes. The condition has reached epidemic proportions around the world. Treatment involves combinations of inhibitors of reverse transcriptase and HIV protease, and drugs to prevent microbial infections of brain, respiratory and alimentary tracts and vagina. *See* **HIV**.

acquired immunity *or* **adaptive immunity** immunity (active or passive, humoral or cellular) that is established during the life of an individual, as contrasted with innate or natural immunity. Such immunity is specific for the inducing agent and is marked by an enhanced response on repeated encounters with that agent. The key features are memory (*see* **memory cell**) and specificity.

acquired tolerance 1 (*in immunology*) tolerance to an antigen that is established during the life of an individual. Immunological tolerance can (rarely) be produced in an adult animal by prolonged injection of massive doses of antigen. The tolerance persists as long as antigen persists in the animal. *See also* **self tolerance**. **2** (*in pharmacology*) tolerance (generally to psychoactive compounds) that develops on prolonged or repeated drug administration. It can be either **pharmacokinetic** usually by increased drug metabolism, or **pharmacodynamic**. *See* **tachyphylaxis**, **tolerance**.

acrasin a chemotactic substance produced by the myxamoebae of the cellular slime mould *Dictyostelium discoideum*, now identified as cyclic AMP.

acridine the parent compound of a series of derivatives, e.g. 3,6-diaminoacridine (proflavin), that are potent mutagens. Some acridines are found in coal tar. They induce **frameshift mutations** during the replication of DNA by binding to DNA and distorting the double helix, causing additional bases to be incorporated or bases to be omitted. They are used as topical antiseptics and antimalarial agents.

acridine

Acridine Orange 3,6-bis(dimethylamino)acridine; a dye used as a probe of nucleic acids in microscopy and related techniques. When illuminated under UV light it yields a green (DNA) or reddish-orange (RNA) fluorescence.

Acridine Orange

acridinium esters esters of acridine-9-carboxylic acid. They have a quaternary nitrogen centre and are derivatized at the 9 position to give a labile phenyl ester moiety. Acridinium esters such as 2′,6′-dimethyl-4′-[*N*-succinimidyloxy carbonyl]phenyl-10-methylacridinium-9-carboxylate can be used as chemiluminescent labels for antibodies. Light emission is activated by adding hydrogen peroxide under alkaline conditions. In hybridization protection assays, single-stranded DNA probes labelled with acridinium esters are protected against rapid hydrolysis when hybridized with a target DNA molecule but not when they are in free solution.

acriflavin a mixture of the **acridine** derivatives 3,6-diamino-10-methylacridinium chloride (about 65%) and **proflavin**.

acrocentric describing a chromosome in which the **centromere** is very close to one end.

acrolein 3-propenal; CH_2=CH–CHO; an unstable flammable liquid with a pungent odour that irritates the eyes and mucosae. It polymerizes (especially under light) to form a plastic solid.

acrolein test a qualitative test for the presence of glycerol, either free or esterified, based upon its oxidative dehydration to acrolein when heated with solid potassium hydrogen sulfate.

acromegaly a chronic disease marked by the gradual enlargement of the bones of the hands, feet, head, and chest with thickening of the skin, lips, and vocal chords. It is caused by excessive secretion of, or increased responsiveness to, **somatotropin**, and is often due to a tumour of the somatotrope cells of the pituitary.

acrosin EC 3.4.21.10; a trypsin-like serine endopeptidase. A major proteinase of mammalian spermatozoa, synthesized in a zymogen form, proacrosin, and stored in the **acrosome**. It comprises a heavy chain (catalytic) and a light chain linked by two disulfide bonds; it is not inhibited by α$_1$-antitrypsin. It catalyses the hydrolysis of Arg-|-Xaa and Lys-|-Xaa bonds, with preferential cleavage in the order Arg-|-Xaa >> Lys-|-Lys >> Lys-|-Xaa.

acrosomal process a long thin actin-containing spike produced from the head of certain types of sperm when they make contact with the egg at fertilization. It is seen in sea urchins and other marine invertebrates having eggs surrounded by a thick gelatinous coat.

acrosome a modified lysosome in the head of a spermatozoon that contains acid hydrolases concerned in the breakdown of the outer membrane of the ovum during fertilization. It lies anterior to the nucleus just beneath the plasma membrane. *See also* **acrosin**.

ACRP30 *abbr. for* adipocyte complement-related protein of 30 kDa; *other names*: adipo-Q; adiponectin. An abundant serum protein that is synthesized and secreted by adipocytes (an adipokine) in response to insulin but is downregulated in obese mice and in human obesity. The mouse protein contains a globular domain similar to that of complement protein C1q. This domain, when injected into mice, induces weight loss via activation of fatty acid catabolism in muscle.

acrylamide *or* **acrylamide monomer** *the trivial name for* prope-namide, $CH_2=CH-CONH_2$; a water-soluble solid that is highly toxic and irritant, and readily polymerizes under the action of UV light or chemical catalysts into **polyacrylamide**.

ActA a bacterial actin-binding protein. The human pathogen *Listeria monocytogenes* uses actin filaments to transport itself within the host cell cytoplasm.

ACTH *abbr. for* adrenocorticotrop(h)ic hormone (**corticotropin**).

1–24 ACTH *abbr. for* tetracosactrin.

actidione *see* cycloheximide.

actin a major protein constituent of the thin filaments of muscle and of the **microfilaments** found in practically all eukaryotic cells – it comprises 5–10% of the protein of such cells (*see* **actin filament**). In solutions of low ionic strength, actin is a globular 42 kDa monomer, termed **G-actin**. At physiological ionic strengths, G-actin polymerizes into a fibrous form, termed **F-actin**, which resembles two strings of beads wound round each other. F-actin is a helix of actin monomers, with a helix diameter of about 7 nm, the structure repeating at intervals of 36 nm along the helix axis. When a solution of actin is mixed with a solution of the muscle protein myosin, a complex called **actomyosin** is formed, and the viscosity of the solution increases markedly. This increase in viscosity is reversed by the addition of ATP, which acts to dissociate the actomyosin complex. It is thought that the force of muscle contraction arises from an interaction of actin, myosin, and ATP.

actin-binding protein any of several proteins that associate with either actin monomers or actin filaments in cells and modify their properties. Many of these proteins are found in the cell cortex, an actin-rich layer just below the plasma membrane. Examples include **dystropin**, **profilin**, **spectrin** and **ankyrin**, **fimbrin** and α-actinin (*see* **actinin**), **filamin**, **gelsolin**, and the **myosins**. The term is sometimes applied specifically to filamin.

actin depolymerizing factor *see* destrin.

actin filament a two-stranded helical polymer of the protein **actin**. Actin filaments form the thin filaments of muscle and also the **microfilaments** of the cytoskeleton of eukaryotic cells. Hence they are a major component of the contractile apparatus of skeletal muscle, and one of the three types of protein filament that form the cytoskeleton, the others being **microtubules** and **intermediate filaments**. The filaments, comprising polymerized globular actin molecules, appear as flexible structures with a diameter of 5–9 nm. They are organized into a variety of linear bundles, two-dimensional networks, and three-dimensional gels. In the cytoskeleton they are most highly concentrated in the cortex of the cell just beneath the plasma membrane.

actinic describing electromagnetic radiation, especially higher frequencies of visible light and UV radiation, capable of initiating photochemical reactions.

actinidine a monoterpenoid alkaloid occurring in *Actinidia polygama*.

actinin a minor protein constituent of muscle, found to be concentrated in both the Z line and the I band. Two components of actinin have been identified: **α-actinin**, F-actin cross-linking protein, a dimer of ~200 kDa with an action similar to that of **actinogelin**.; and **β-actinin**, a dimer of ~70 kDa, similar in action to **gelsolin**.

actinogelin a protein factor that effects Ca^{2+}-sensitive gelation of actin filaments. It was first obtained from Ehrlich ascites tumour cells, and has a molecular mass of 100–120 kDa. α- Actinin (*see* **actinin**) has a similar action.

actinoid *or* **actinide** any member of the series of 15 metallic elements with proton numbers 89 (actinium) to 103 (lawrencium) inclusive that occur together in group 3 and period 7 of the IUPAC **periodic table**; sometimes the term is restricted to the 14 elements following actinium. Actinoid is now the preferred name. All actinoids are radioactive, and those of proton number 93 or greater are artificial. They are all electropositive, and their chemical properties are similar, due usually to the filling of an inner electron subshell (5f) progressively across the series. Like the **lanthanoids**, they thus represent a series of inner transition elements.

actinomycin any of a large group of antibiotics isolated from various species of *Streptomyces* bacteria and characterized by having a substituted phenoxazine ring linked to two cyclic **heterodetic** pep-

tides. The principal member of the group is **actinomycin D** (also called actinomycin C_1, actinomycin IV), which, at low concentrations, inhibits transcription without appreciably affecting DNA replication in both prokaryotic and eukaryotic cells. Actinomycin D binds tightly to duplex DNA thereby preventing it from being an effective template for RNA synthesis. Spectroscopic and hydrodynamic studies of complexes of actinomycin D and DNA suggest that the phenoxazine ring of actinomycin is intercalated between neighbouring base pairs in DNA. Other conformational studies indicate that actinomycin D recognizes the base sequence GpC in DNA.

actinomycin D

actinomyosin *(formerly)* the contractile system in muscle comprising **actin** and **myosin**. *Compare* **actomyosin**.

actin-related protein *see* Arp.

action potential the localized change of electrical potential across a nerve fibre or muscle membrane that marks the position of an impulse as it is propagated. It is caused by sodium ion fluxes across the membrane resulting from transitory opening of **sodium channels**.

action spectrum a graph or table showing the relative efficiencies of different frequencies of radiation in causing a chemical or biochemical reaction, e.g. in photosynthesis or gas exchange. Efficiency, or **quantum yield**, is given by the measurable effect divided by the intensity of the (monochromatic) incident radiation.

activate 1 to make or render active, reactive, or capable of action. **2** to make radioactive. —**activated** *adj.*; **activation** *n.*

activated alumina particles of alumina, aluminium oxide, that have been rendered more adsorbent by heating strongly.

activated amino acid an amino-acid residue in an aminoacyladenylate molecule. The amino acid reacts with ATP to form aminoacyl-AMP in a reaction catalysed by an aminoacyl-tRNA synthetase. It is the first metabolic step in the biosynthesis of a polypeptide. The same enzyme catalyses the transfer of the amino acid to an ester link with the 3′-terminal hydroxyl of tRNA.

activated complex term that may be used to denote the assembly of atoms at the transition state of a chemical reaction.

activation 1 the action or process of rendering an atom, molecule, or other substance reactive or more reactive, whether physicochemically, chemically, or biochemically. **2** the process of rendering material artificially radioactive: radioactivation. *See* **radioactivation analysis**. **3** the initial changes in an ovum during fertilization, covering the period from first contact with a sperm to dissolution of the nuclear membranes. **4** the initial changes in the conversion of a spore to a vegetative cell. This can be effected by various agents or processes and may involve alteration of one of the spore's outer layers.

activation analysis *an alternative term for* radioactivation analysis.

activation energy *see* Arrhenius equation.

activation hormone a polypeptide hormone synthesized in the neurosecretory brain cells of insects. It regulates the functioning of the whole endocrine system, stimulating the secretory activity of other glands.

activation-induced cytidine deaminase a putative RNA-editing enzyme, related to APOTEC-1, that is specific to antigen-activated B lymphocytes in germinal centres of lymph nodes. It is required for immunoglobulin (Ig) class switch recombination, somatic hypermutation, and gene conversion of immunoglobulin genes. Mutations are associated with an autosomal recessive form of hyper-IgM syndrome.

activator any substance that activates a chemical or enzymic reaction. The term was formerly used in biochemistry especially to describe metal-ion **cofactors** for enzymes.

activator constant *symbol*: K_A; the equilibrium **dissociation constant** of the reaction of an enzyme with an activator. For example, in the reaction:

$$E + A \rightleftharpoons EA,$$

$$K_A = [EA]/[E][A],$$

where [EA], [E], and [A] are the concentrations of enzyme– activator adduct, free enzyme, and free activator respectively.

active acetaldehyde *(sometimes)* the α-hydroxyethyl derivative, on C-2 of the thiazolidine ring, of thiamine diphosphate.

active acetate *(formerly) an alternative name for* **acetyl coenzyme A**.

active C$_1$ *abbr. for* active one-carbon.

active centre all the features of primary, tertiary, and quaternary structure of an enzyme – including the **active site** – that are required for substrate binding, specificity, and catalysis.

active-enzyme centrifugation a technique for determining the sedimentation and diffusion coefficients of an enzyme– substrate complex. A thin lamella of an enzyme solution is layered onto a substrate solution in an ultracentrifuge cell and, on rotation, as the enzyme molecules sediment in a band through the substrate solution they catalyse the enzymic reaction, the progress of which is observed optically (e.g. by spectrophotometry). The required coefficients are then calculated from either the rate of appearance of the product of the reaction or the rate of disappearance of the substrate. The centrifugation may be performed on impure enzyme preparations and at the very low concentrations of enzyme used in kinetic studies.

active fatty acid *(formerly) an alternative term for* an **acyl-CoA** derivative.

active formaldehyde *(formerly) an alternative term for* N^5,N^{10}-methylenetetrahydrofolate.

active immunity *or* **adaptive immunity** a type of immunity resulting from the stimulation of the immune response by an immunogen.

active immunization stimulation of the reactivity of the immune system of an individual towards an immunogen by administration of the immunogen, thereby producing a state of active immunity.

active methionine *or* **active methyl** *name sometimes used for* **S-adenosylmethionine**.

active methyl 1 *an alternative name for* **active methionine**. **2** one form of **active one-carbon** (e.g. 5-methyltetrahydrofolate).

active one-carbon any of the one-carbon units carried on tetrahydrofolate and concerned in a wide variety of biosynthetic reactions. They include the methyl-, methylene-, formimino-, formyl-, and methenyl- substituents.

active site 1 the general region of an enzyme molecule containing the **catalytic residues** identified with the binding and reaction of substrate(s). It includes those amino-acid residues that are, in the enzyme–substrate complex, either **contact amino acids**, i.e. those that at some point are only one bond distance removed from some point of the substrate molecule, or **auxiliary amino acids**, i.e. those that are not in such intimate physical contact with the substrate but nonetheless play a definite role in the action of the enzyme. *See also* **contributing amino acid**. **2** *(sometimes)* the portion of a peptide hormone responsible for its biological activity. Isolated fragments containing the active site may show some activity, but may function less efficiently than the intact hormone. *See also* **message sequence**.

active site-directed irreversible inhibitor a custom-made inhibitor of a given enzyme. Such inhibitors are typically trifunctional molecules containing: (1) a functional group able to bind to the enzyme's active site; (2) a nonpolar part that can interact with a nonpolar region on the enzyme, serving to align the inhibitor; and (3) an alkylating group capable of reacting with a susceptible group on the enzyme and irreversibly forming a covalent bond with it.

active sulfate *an alternative name for* **1** adenosine-5′-phosphosulfate. **2** 3′-phosphoadenosine-5′-phosphosulfate.

active transport any energy-dependent process by which molecules or ions are transported across membranes against a chemical potential gradient. *Compare* **facilitated diffusion**, **simple diffusion**.

activicin a structural analogue of glutamine that inhibits glutamine-PRPP amidotransferase (of purine biosynthesis) and carbomyl phosphate synthetase II (of pyrimidine biosynthesis) and is therefore a potential anticancer agent.

activin one of two gonadal glycoproteins related to **transforming growth factor**-β (the other is **inhibin**); present in two forms in human gonads, it exists as a dimer of inhibin β_A or β_B chains, linked by disulfide bonds. Activin A is a dimer of β_A chains; activin AB is a dimer of β_A and β_B chains. A potent selective stimulator of FSH secretion by the anterior pituitary gland, not via GnRH receptors, it modulates induction of hemoglobin accumulation and proliferation of erythroid progenitor cells in human bone marrow culture, and is important in embryonic axial development.

activin receptor-like kinase 1 *abbr.*: ALK1; a cell-surface receptor kinase of the transforming growth factor β (TGFβ) superfamily of ligands. It is expressed primarily in endothelial cells and highly vascularized tissues. It forms heterodimers with endoglin and signals through R-Smad (*see* **Smad**), which enters the nucleus to affect transcription. Mutations in the ALK1 gene at 12q11-q14 are associated with hereditary hemorrhagic telangiectasia.

activity 1 the natural or normal functioning of an enzyme, hormone, inhibitor, or other agent; or the intensity with which such an agent functions. *See also* **enzymic activity**. **2** the number of nuclear **transformations** (def. 5) that occur or that can be detected in a given quantity of radioactive material in unit time (*see also* **becquerel**, **curie**). This term is also often used in this sense for **radioactivity** and, loosely, for the radioactive material itself or its emitted radiation; see also **specific activity**. **3** *or* **relative activity** *symbol*: a; the apparent or effective concentration of a chemical substance as judged by the behaviour of the substance in a standard state. The activity of an entity B is defined by

$$a_B = \exp[(\mu_B - \mu_B^o)/RT]$$

where μ_B and μ_B^o are respectively the chemical potential and standard chemical potential of entity B, R is the gas constant, and T the thermodynamic temperature. **4** **absolute activity** *symbol*: λ; the absolute activity of an entity B is defined by

$$\lambda_B = \exp(\mu_B/RT),$$

where μ_B is the chemical potential of entity B, R is the gas constant, and T is the thermodynamic temperature. *See also* **activity coefficient**.

activity coefficient *symbol*: γ; the ratio of the **activity** (def. 3) of a component of a solution to its concentration. When expressed on an amount concentration basis it is denoted γ_c, when on a molality basis γ_m, and when on a mole fraction basis γ_x. *See also* **mean ionic activity coefficient**.

activity stain any reagent that develops a colour when acted on by a particular enzyme. Such a reagent is used to detect the presence of an enzyme in a gel or paper electrophoresis strip, in which it differentially stains the enzyme protein in question but not other proteins.

actobindin a monomeric protein capable of binding two molecules of monomeric (i.e., G-) actin.

actomyosin *see* actin.

ACU a codon in mRNA for L-threonine.

acumentin a Ca^{2+}-insensitive, actin-modulating protein, isolated from rabbit alveolar macrophages.

acute intermittent porphyria the commonest form of acute hepatic porphyria, which is caused by any of numerous mutations in **hydroxymethylbilane synthase** (porphobilinogen deaminase). It usually develops after puberty and is exacerbated by certain steroid hormones or drugs, or a diet that induces 5-aminolevulinic acid synthase in liver thus increasing formation of the porphobilinogen substrate. *See* porphyria.

acute myeloid leukemia *or (esp. Brit.)* **acute myeloid leukaemia** the most common form of acute **leukemia**, occurring within the first four weeks of life. Almost half of the cases have a translocation involving a chromosomal segment at 11q23 (the MLL gene – mixed lineage leukemia) and a segment of another chromosome, frequently 19p13.1 (for ELL), resulting in the production of a **fusion protein** that interferes with the normal control of cell division. The MLL gene encodes a protein of 3968 amino acids, which contains: in the N-terminal region, three A–T hooks for binding A–T base pairs in the minor groove of DNA; a segment homologous with the noncatalytic regions of DNA methyltransferase; in the central region, two zinc finger regions; and in the C-terminal region a 210-amino acid segment that, like the zinc finger regions, has significant homology to *Drosophila* trithorax. The resulting MLL-ELL fusion protein contains the N-terminal 1300 residues (up to but not including the zinc fingers) of MLL, and most of the sequence of ELL.

acute myeloid leukemia 1 *see* **AML1**.

acute-phase protein any of the non-antibody proteins that show raised plasma concentrations soon after the onset of infection, chemical or physical tissue injury, or malignant neoplasia. They include **complement** (def. 3) proteins, **C-reactive protein**, **fibrinogen** and other coagulation proteins, and **interferon**. Levels of some proteins (e.g. albumin) usually decrease. These responses in liver protein synthesis may have complex origins.

acute transfection the short-term infection of cells with DNA.

acyclic describing an organic compound devoid of a ring of atoms.

acycloguanosine *an alternative name for* **acyclovir**.

acyclovir or **acycloguanosine** 9-(2-hydroxyethoxymethyl)guanine; an antiviral agent widely used in the treatment of human herpes infections. It is selectively phosphorylated by herpesvirus-induced thymidine kinase and the phosphorylated compound is a potent inhibitor of herpesvirus-induced DNA polymerase. One proprietary name is Zovirax.

acyclovir

acyl *generic name for* any group formally derived by removal of a hydroxyl group from the acid function of an organic acid. Examples include any group of general structure R–CO– derived from a carboxylic acid; R–SO$_2$– from a sulfonic acid; and R–PO(OH)– from a phosphonic acid. However, unless the context indicates otherwise the term refers to such groups derived from carboxylic acids.

acyl-ACP dehydrogenase *see* **enoyl-[acyl-carrier protein] reductase (NADPH, A-specific)**.

acyl-activating enzyme *see* **long-chain-fatty-acid–CoA ligase**.

acylation the process of introducing an acyl group into a compound by substitution for a hydrogen atom. —**acylated** *adj.*

acyl carrier protein *abbr.*: ACP; any of the relatively small acidic proteins that are associated with the fatty acid synthase system of many organisms, from bacteria to plants. They contain one 4′-phosphopantetheine prosthetic group bound covalently by a phosphate ester bond to the hydroxyl group of a serine residue. The sulfhydryl group of the 4′-phosphopantetheine moiety serves as an anchor to which acyl intermediates are (thio)esterified during fatty-acid synthesis. In animals, the fatty acid synthase system is a 500 kDa polyfunctional enzyme containing two identical chains, each with one ACP unit. *Escherichia coli* ACP, a separate single protein, contains 77 amino-acid residues (8.85 kDa); the phosphopantetheine group is linked to serine 36.

acyl-CoA *generic name for* any derivative of **coenzyme A** in which the sulfhydryl group is in thioester linkage with a fatty-acyl group.

acyl-CoA dehydrogenase the first enzyme of fatty acid oxidation, responsible for catalysing a FAD-dependent alpha-beta dehydrogenation of a fatty acyl-CoA substrate. Four enzymes are recognized in humans. **Very-long-chain acyl-CoA dehydrogenase** (EC 1.3.99, *abbr.*: VLCAD) is a homodimer attached to the inner mitochondrial membrane in liver, and acts on fatty acids of chain length 14–20 carbon atoms. Its gene locus at 17p13 encodes a sequence of 655 amino acids. **Long-chain acyl-CoA dehydrogenase** (EC 1.3.99.13, *abbr.*: LCAD) is a homotetramer bound to the inner mitochondrial membrane that acts on fatty acids 12–18 carbon atoms long; it also has 2-enoyl-CoA hydratase activity and constitutes the α subunits of mitochondrial trifunctional protein (an α$_4$β$_4$ octamer). The β subunits of the trifunctional protein have 3-ketoacyl-CoA thiolase activity. Genetic loci for both subunits are at 2p23. **Medium-chain acyl-CoA dehydrogenase** (EC 1.3.99.3, *abbr.*: MCAD) is a homotetramer present in the mitochondrial matrix and acts on fatty acids of chain length 4–12 carbon atoms. Its gene locus is at 1p31. **Short-chain acyl-CoA dehydrogenase** (butgryl-CoA dehydrogenase, EC1.3.99.2, *abbr.*: SCAD), also a homotetramer, acts on fatty acids of 4–6 carbon atoms chain length. Its locus at 12q22-qter encodes a mature protein of 388 amino acids. Disease-causing mutations are known in all four enzymes.

acyl-CoA reductase EC 1.2.1.50; *recommended name*: long-chain-fatty-acyl-CoA reductase; *systematic name*: long-chain-aldehyde: NADP$^+$ oxidoreductase (acyl-CoA-forming). An enzyme that catalyses a reaction between a long-chain aldehyde, CoA, and NADP$^+$ to form long-chain acyl-CoA and NADPH. It is part of the bacterial **luciferase** system. Example: *see* **Lux proteins**.

acyl-CoA synthetase *see* **long-chain-fatty-acid–CoA ligase**.

acyl enzyme any intermediate formed in an enzymic acyl-transfer reaction in which a group at the active site of the enzyme is acylated.

acylglycerol 1 any mono-, di-, or triester of glycerol with (one or more) fatty acids (termed respectively monoacylglycerol, diacylglycerol, or triacylglycerol); formerly known as mono-, di-, or triglyceride. **2** acylglycerols (*plural*), also denotes any mixture of mono-, di-, and/or triacylglycerols whatever its degree of complexity, including any comprising a **neutral fat**.

2-acylglycerol *O*-acyltransferase EC 2.3.1.22; *other names*: acylglycerol palmitoyltransferase; monoglyceride acyltransferase; an enzyme that catalyses the acylation by acyl-CoA of 2-acylglycerol to form diacylglycerol and CoA. It is an enzyme involved in the **monoacylglycerol pathway** of triacylglycerol biosynthesis.

1-acylglycerol-3-phosphate *O*-acyltransferase EC 2.3.1.51; *systematic name*: acyl-CoA:1-acyl-*sn*-glycerol-3-phosphate 2-*O*-acyltransferase. An enzyme that catalyses the acylation by acyl-CoA of 1-acyl-*sn*-glycerol 3-phosphate (lysophosphatidic acid) to form 1,2-diacyl-*sn*-glycerol 3-phosphate (phosphatidic acid) with release of CoA. Acylated acyl carrier protein can also act as substrate. In animals, the enzyme specifically transfers unsaturated fatty acyl groups, a preference that contributes to the high proportion of unsaturated fatty acids on the 2-position of phosphoglycerides.

1-acylglycerophosphocholine *O*-acyltransferase EC 2.3.1.23; *other name*: lysolecithin acyltransferase; an enzyme that catalyses the acylation by acyl-CoA of 1-acyl-*sn*-glycero-3-phosphocholine to form 1,2-diacyl-*sn*-glycero-3-phosphocholine and CoA. It is involved in the resynthesis of phosphatidylcholine in situations where the latter has been degraded to lysophosphatidylcholine.

acylhomoserine *see* **O-succinylhomoserine**.

acyl migration any intramolecular rearrangement reaction in which, under certain conditions, an acyl group moves from one functional group to another, which may be of the same or of a different kind. N → O **acyl migration** (*also termed* N → O acyl shift, N → O acyl transfer, *or* N → O peptidyl shift) is a reaction that can occur in polypeptides and proteins subjected to hydrolysis by concentrated acid at room temperature or in the presence of an acid chloride (e.g. POCl$_3$), whereby the β-hydroxyl group of a serine or threonine residue displaces the amino group of the same residue from its amide (= peptide) linkage with the carboxyl group of the adjacent residue, thus forming an acid-sensitive β-ester linkage. The reaction may be used to advantage in partial acid hydrolysis of polypeptides and proteins. In the presence of alkali the ester reverts to the original amide by O → N **acyl migration**, a reaction that can also occur during the **Edman degradation** of a peptide. An acyl group, e.g. acetyl or maleyl, that happens to be present on the β-hydroxyl group of a serine or threonine residue, moves to the free amino group of such a residue when it becomes N-terminal, so preventing the next cycle of the degradation.

***N*-acylneuraminate glycohydrolase** *see* **sialidase**.

acylphosphatase EC 3.6.1.7; *systematic name*: acylphosphate phosphohydrolase. An enzyme that catalyses the hydrolysis of acyl phosphate to fatty-acid anion and orthophosphate.

acyl-protein synthase *see* **myelin-proteolipid *O*-palmitoyltransferase**.

acyl shift *see* **acyl migration**.

acylsphingosine deacylase *see* **ceramidase**.

***N*-acylsphingosine galactosyltransferase** EC 2.4.1.47; an enzyme that catalyses the formation of the cerebroside D-galactosylceramide from UDPgalactose and *N*-acylsphingosine with release of UDP.

acyltransferase any enzyme of sub-subclass EC 2.3.1. Such enzymes transfer acyl groups, forming either esters or amides, by catalysing reactions of the type:

$$\text{acyl-carrier} + \text{reactant} = \text{acyl-reactant} + \text{carrier}.$$

For example if acyl = acetyl, carrier = CoA, and reactant = choline, the reaction is that of **choline acetyltransferase**.

AD *(in clinical biochemistry)* *abbr. for* alcohol dehydrogenase (preferred to ADH because of possible confusion with antidiuretic hormone).

Ada or **ADA** *abbr. for* *N*-(2-acetamido)iminodiacetic acid; [(carbamoylmethyl)imino]diacetic acid; a **Good buffer substance**, pK_a (20 °C) = 6.6.

Adair equation an equation describing the oxygen saturation curve of hemoglobin:

$$R = \frac{K_1[p_{O_2}] + 2K_1K_2[p_{O_2}]^2 + 3K_1K_2K_3[p_{O_2}]^3 + 4K_1K_2K_3K_4[p_{O_2}]^4}{4(1 + K_1[p_{O_2}] + K_1K_2[p_{O_2}]^2 + K_1K_2K_3[p_{O_2}]^3 + K_1K_2K_3K_4[p_{O_2}]^4)}$$

where R is the fractional saturation of hemoglobin with oxygen, K_1, K_2, K_3, and K_4 are the stepwise intrinsic association constants for the four oxygen-binding sites, and p_{O_2} is the partial pressure of oxygen. For a generalized form of this equation *see* **Bjerrum formation function**. [Defined by Gilbert Smithson Adair (1896–1979), British biophysicist, in 1925.]

ADAM *abbr. for* a disintegrin and metalloproteinase; any of a large family of zinc-binding cell-surface transmembrane proteins that contain a **disintegrin** and a **metalloproteinase** domain. Such proteins are presumed to have roles in cell fusion, cell adhesion, and cell signaling.

ADAM-10 an α-secretase involved in amyloid peptide precursor proteolysis. It is an endopeptidase, EC 3.4.24.81. See **ADAM**.

ADAM-17 *an alternative term for* **TNF-α converting enzyme**. See also **ADAM**.

ADAM-33 a protein belonging to the **ADAM** family that is abundant in lung fibroblasts and bronchial smooth muscle cells. Mutations of its gene are associated with asthma.

adamalysin EC 3.4.24.46; *other name*: proteinase II; a metalloendopeptidase that catalyses the cleavage of Phe[1]-|-Val[2], His[5]-|-Leu[6], His[10]-|-Leu[11], Ala[14]-|-Leu[15], Leu[15]-|-Tyr[16], and Tyr[16]-|-Leu[17] in the insulin B-chain. Zn^{2+} and Ca^{2+} are cofactors. It is a snake venom proteinase whose natural substrate(s) is protease inhibitor(s), such as the **serpins**.

1-adamantanamine or **α-adamantanamine** *see* **amantadine**.

adamantane tricyclo[3.3.1.1^{3,7}]decane; a substance isolated from petroleum and said to have virostatic activity. *See also* **amantadine**.

ADAMTS *abbr. for* a disintegrin and metalloproteinase with a thrombospondin domain; any of a family of proteins that are related to the **ADAM** family of metalloproteinases but are distinguished by having a thrombospondin domain and by being secreted.

ADAMTS-2 a protein belonging to the **ADAMTS** family, deficiency of which is associated with a form of **Ehlers–Danlos syndrome**.

ADAMTS-13 a protein belonging to the **ADAMTS** family that cleaves **von Willebrand factor**. Deficiency leads to microvascular thrombus formation. Mutations in its gene are associated with a form of congenital purpura.

Adapt any of a family of proteins encoded by oxidant stress-inducible genes that provide partial protection against such stress. The genes are also inducible by calcium ionophores or by inceased intracellular calcium. **Adapt78** (*also called* DSCR1 *or* Down syndrome critical region 1) is a 25 kDa protein that specifically binds and inhibits calcineurin, which has a role in stress-induced apoptosis. Adapt78 gene is the most responsive to calcium of the various adapts; its gene is highly expressed in neurons of cerebral cortex and hippocampus, and is overexpressed in brains of Down syndrome sufferers.

adaptation *(in biology)* **1** any change in an organism's structure or function that allows it better to deal with its environmental conditions. **2** the mechanism whereby bacterial cells that approach cells of the opposite mating type, but fail to mate, can recover and continue dividing.

adapted gradient in **gradient elution**, a form of gradient in which the composition of the eluate is adjusted, on the basis of information obtained in a trial separation, to be optimal at each moment of the elution, so that the optimum resolution between the desired components of the sample is obtained.

adaptin a major coat protein of clathrin-coated vesicles. It functions as a component of the adaptor complexes that link clathrin to the receptors being taken into the cell in the coated vesicles, recognizing a motif of four amino acids (FRxY) in the cytoplasmic domain of the receptor.

adaptive describing changes in an organism, induced by environmental factors, that tend to increase its viability.

adaptive enzyme *see* **inducible enzyme**.

adaptive immunity *an alternative name for* **active immunity**.

adaptor **1** any of various devices useful for joining together two or more parts, otherwise incompatible, such as electrical connectors or pieces of glassware. **2** any synthetic single- or double-stranded oligodeoxynucleotide useful in recombinant DNA technology for joining two incompatible cohesive ends of restriction fragments. *Compare* **linker**. **3** or **adaptor molecule** the molecule that was postulated as carrying the amino acid to the (messenger) RNA in the **adaptor hypothesis** of protein synthesis, now known to be a tRNA molecule.

adaptor hypothesis the hypothesis, first postulated by the British molecular biologist Francis **Crick** (1916–), that during protein biosynthesis each amino-acid residue is carried to the RNA template by its appropriate small adaptor RNA molecule, and that the adaptor is the part that fits on to the messenger RNA. The hypothesis was subsequently confirmed by the discovery of **transfer RNA**.

adaptor protein *abbr.*: AP **1**. a protein that links together active signal components but is not itself catalytic. **2**. a component of protein complexes associated with clathrin in **clathrin-coated vesicles**.

ADAR *abbr. for* adenosine deaminase acting on RNA; any member of a family of enzymes that deaminate adenosine to inosine in RNA and share a common modular organization. This consists of a variable N-terminal region, one or several double-standed RNA-binding domains, and a zinc-containing catalytic domain. In vertebrates, ADAR 1 and ADAR 2 are expressed in most tissues and are involved in editing of pre-mRNA for mammalian ionotropic glutamate receptors and a subtype of serotonin receptors, whereas ADAR 3 is brain-specific but appears to lack catalytic activity. *See also* **adenosine deaminase**.

ADAT *abbr. for* adenosine deaminase acting on tRNA; *other name*: tRNA-specific adenosine deaminase; any member of a family of RNA-dependent adenosine deaminases that act on the anticodon in several tRNAs in higher eukaryotes, prokaryotes, and plant chloroplasts. The catalytic domain binds zinc and contains an essential glutamate residue.

ADCC *abbr. for* antibody-dependent cell-mediated cytotoxicity.

Addison's disease a disease due to deficiency of production of cortisol and other cortical steroids. It results from atrophy of the adrenal cortex, a condition having a variety of causes. [Described by Thomas Addison (1793–1860), British physician and endocrinologist, in 1849.]

addition 1 *(in chemistry)* the formation of an **adduct**. **2** an **addition reaction**.

addition compound *see* **adduct**.

addition reaction *or* **addition** any organic chemical reaction involving the combination of two or more substances to form a single product in which there are more groups attached to carbon atoms than there were in the original reactants. Such reactions thus involve a net reduction of bond multiplicity in one of the reactants, as in the example:

$$H_2C=CH_2 + Br_2 \rightarrow H_2BrC-CBrH_2.$$

address 1 that part of the information contained within the amino-acid sequence of a hormonal polypeptide, or its biosynthetic precursors, that determines the receptor-specific affinity of the hormone. *Compare* **message** (def. 4). **2** *(in computer technology)* a label, name, or number, identifying a device or a location in a memory.

addressin any of a group of glycoproteins, expressed specifically in a region of the vascular endothelium in peripheral and mucosal lymph nodes, that are involved in the homing of T lymphocytes to these sites.

address sequence *or* **address region** that segment of the amino-acid sequence of a polypeptide hormone, or prohormonal polypeptide, in which the address of the hormone resides, and through which the hormone is considered to bind to its specific receptor. *Compare* **message sequence**.

adducin a membrane skeleton protein (heterodimer) that interacts with a junctional complex that links **spectrin** assemblies. The complex in the red blood cell consists of **tropomyosin** and **actin** with band 4.1 (*see* **band**). Adducin probably binds to band 4.1.

adduct 1 any new chemical species, AB, formed by direct combination of two separate chemical species, A and B, in such a way that there is no change in **connectivity** of atoms within the moieties A and B. This term is preferred to **complex**, which is less explicit. **2** *(formerly)* the product of a reaction between molecules occurring in such a way that the original molecules or their residues have their long axes parallel to one another.

Ade *symbol for* a residue of the purine base adenine (alternative to A).

ADE2 a gene that encodes a multifunctional protein with activities for **phosphoribosylaminoimidazole carboxylase** and **phosphoribosylaminoimidazolesuccinocarboxamide synthase**. It has been studied as a model of position effect in gene expression. The gene is expressed in all cells if in its normal position near the middle of the chromosome, but if moved experimentally to the end of the chromosome it is silenced. The lack of the enzyme produces a red pigmentation in these colonies.

adenine *symbol*: A *or* Ade; 6-aminopurine; 1*H*-purin-6-amine; a purine base. It is one of the five main bases found in nucleic acids and a component of numerous important derivatives of its corresponding ribonucleoside, adenosine. It can exist in a tautomeric imino form.

adenine arabinoside *other names*: vidarabine, ara-A; 9-β-D-arabinofuranosyladenine; an antibiotic produced by *Streptomyces* bacteria that acts as an antiviral agent, causing mispairing of purines and pyrimidines in nucleic-acid synthesis.

adenine nucleotide translocase *see* **ADP,ATP carrier protein**.

adenine phosphoribosyltransferase *abbr.*: APRT; EC 2.4.2.7; *other names*: AMP pyrophosphorylase; transphosphoribosidase; an enzyme that catalyses the formation of AMP from 5-phospho-α-D-ribose 1-diphosphate and adenine with release of pyrophosphate. It is an enzyme of purine metabolism that salvages adenine released

9-β-D-form

by degradative enzymes, converting it back to a nucleotide. Several mutations in a locus at 16q24 produce **APRT deficiency type 1**, in which no activity is detectable in erythrocytes and the excess adenine forms the poorly soluble 2,8-dihydroxyadenine and leads to crystalluria. Other mutations produce **APRT deficiency type 2**, in which activity amounts to 10–25% of the wild type.

adeno+ *or (before a vowel)* **aden+** *comb. form* meaning of, pertaining to, or like a gland; glandular; found in glands.

adenohypophysis the glandular anterior lobe of the hypophysis (or **pituitary gland**). It produces corticotropin, gonadotropin, lipotropin, somatotropin, thyrotropin, and other hormones. *Compare* **neurohypophysis**. —**adenohypophyseal** *or* adenohypophysial *adj.*

adenoma any benign tumour formed by the multiplication of the epithelial cells that form the ducts and acini of glandular organs. The meaning has been broadened to include the benign tumours that arise from the solid masses of epithelium that form some of the endocrine glands. Adenomas often accurately reproduce the tissues from which they are derived and produce a secretion identical with or similar to that produced by the normal glandular tissue.

adenosine *symbol*: A *or* Ado; adenine riboside; 9-β-D-ribofuranosyladenine; a ribonucleoside found widely distributed in cells of every type as the free nucleoside and in combination in nucleic acids and various nucleoside coenzymes. It is a potent regulator of physiological transmission in both central and peripheral nervous systems, where it activates specific receptors (*see* **adenosine receptor**). It can inhibit or stimulate the release of a number of neurotransmitters, including acetylcholine, β-aminobutyrate, catecholamines, excitatory amino acids, and 5-hydroxytryptamine. The effect depends on whether the receptor involved inhibits adenylate cyclase (and thus inhibits release) or stimulates it.

adenosine aminohydrolase *see* **adenosine deaminase**.

adenosine 2′,3′-(cyclic)phosphate *see* **adenosine phosphate**.

adenosine 3′,5′-(cyclic)phosphate *see* **adenosine 3′,5′-phosphate**.

adenosine deaminase EC 3.5.4.4; *systematic name*: adenosine aminohydrolase; an enzyme that catalyses the hydrolysis of adenosine to inosine and NH_3. It is involved in the degradation of adenosine, the inosine thereby formed being then converted to hypoxanthine and thence to uric acid. The enzyme is also found on the outer surface of T cells, and a form that acts on double-stranded RNA (dsRNA) converts adenosines to inosines within dsRNA.

adenosine deaminase acting on RNA *see* ADAR.

adenosine deaminase acting on tRNA *see* ADAT.

adenosine deaminase deficiency an autosomal recessive disease in which deficiency of adenosine deaminase (ADA) causes a form of severe combined immunodeficiency (SCID). ADA activity is very low or undetectable in all tissues, and both cell-mediated (T-lymphocyte) and humoral (B-lymphocyte) forms of immunity are lacking. Successful therapy involves bone marrow transplantation. Numerous mutations of a locus at 20q13.11 lead to the deficiency. In partial ADA deficiency, ADA activity is absent in erythrocytes and partial in all other tissues, although immune function is essentially normal.

adenosine diphosphatase *see* apyrase.

adenosine 5′-diphosphate *symbol*: Ado5′*PP* or ppA; *the recommended name for* adenosine diphosphate (*abbr.*: ADP); 5′-diphosphoadenosine; 5′-adenylyl phosphate; adenosine 5′-trihydrogen diphosphate; a universally distributed nucleotide that occurs both in the free state and as a component of certain **nucleotide coenzymes**. It is a metabolic precursor of adenosine 5′-triphosphate, ATP, and is the product of many enzyme reactions in which ATP is hydrolysed by an **adenosine triphosphatase** or in which the terminal phosphoric residue of ATP is transferred to another organic compound by a **kinase** (including **adenylate kinase**).

adenosinediphosphoglucose or **adenosine (5′)diphospho(1)-α-D-glucose** *symbol*: Ado*PP*Glc or Ado-5′*PP*-Glc or A5′pp1Glc; the alternative recommended names for adenosine diphosphate glucose (*abbr.*: ADPG or ADP-Glc or ADPglucose); adenosine 5′-(α-D-glucopyranosyl diphosphate); α-D-glucopyranosyl 5′-adenylyl phosphate; a **nucleosidediphosphosugar** in which the distal phosphoric residue of adenosine 5′-diphosphate is in glycosidic linkage with glucose. ADPglucose is synthesized from glucose 1-phosphate and adenosine 5′-triphosphate, ATP, through the action of glucose 1-phosphate adenylyltransferase (EC 2.7.7.27). It is an intermediate in the incorporation of glucosyl groups into starch in plants and into storage glucans in bacteria.

adenosinediphosphoribose or **adenosine(5′)diphospho(5)-β-D-ribose** *symbol*: Ado*PP*Rib or A5′pp5Rib or (Rib5)ppA; *the alternative recommended names for* adenosine diphosphate ribose (*abbr.*: ADP-Rib or ADPribose); adenosine 5′-(β-D-ribofuranose 5-diphosphate); a **nucleoside diphosphosugar** in which the explicit sugar moiety is a second residue of D-ribose and is in ester linkage with the distal phosphoric residue of adenosine 5′-diphosphate atypically at C-5 instead of at C-1. ADPribose is synthesized from D-ribose 5-phosphate and ADP through the action of ribose-5-phosphate adenylyltransferase (EC 2.7.7.35), releasing orthophosphate. In addition, free ADPribose is released from nicotinamide-adenine dinucleotide, NAD+, on hydrolytic cleavage of the glycosylamine bond to nicotinamide through the action of NAD+ nucleosidase (EC 3.2.2.5). *See also* **cyclic adenosinediphosphoribose**.

adenosinediphosphoribosyl *abbr.*: ADPribosyl; the glycosyl group formally derivable from adenosinediphosphoribose by loss of the anomeric (i.e. C-1) hydroxyl group from its distal ribose moiety. The ADPribosyl group of nicotinamide-adenine dinucleotide may be utilized for **ADP-ribosylation**.

adenosine 5′-[β,γ-imido]triphosphate *symbol*: Ado*PP*[NH]*P* or p[NH]ppA; *the recommended name for* β,γ-imidoadenosine 5′-triphosphate (*abbr.*: ATP[β,γ-NH]); 5′-adenylyl imidodiphosphate; 5′-adenylyl iminodiphosphonate (*abbr.*: AMP-PNP); adenosine (5′→O³)-1,2-μ-imidotriphosphate; a synthetic analogue of adenosine triphosphate, in which the oxy group between the latter's intermediate and terminal phosphorus atoms is replaced by an imido group. It competitively inhibits ATP-dependent enzymes, including mitochondrial ATPase.

adenosine kinase EC 2.7.1.20; an enzyme that catalyses the reaction:

$$\text{ATP} + \text{adenosine} = \text{ADP} + \text{AMP}.$$

adenosine 5′-[α,β-methylene]diphosphate *symbol*: Ado*P*[CH₂]*P* or p[CH₂]pA; *the recommended name for* α,β-methyleneadenosine 5′-diphosphate (*abbr.*: ADP[α,β-CH₂]); 5′-adenylyl methylenephosphonate (*abbr.*: AMP-CP); adenosine (5′→O¹)-1,2-μ-methylenediphosphate; a synthetic analogue of adenosine 5′-diphosphate,

adenosine 5′-[β,γ-imido]triphosphate

ADP, in which the oxy group between the latter's two phosphorus atoms is replaced by a methylene group. It inhibits 5′-nucleotidase obtained from rat-heart membranes.

adenosine 5′-[α,β-methylene]triphosphate *symbol*: Ado*P*[CH₂]*PP* or pp[CH₂]pA; *the recommended name for* α,β-methyleneadenosine 5′-triphosphate (*abbr.*: ATP[α,β-CH₂]); 5′-adenylyl methylenediphosphate (*abbr.*: AMP-CPP); adenosine (5′→O¹)-1,2,-μ-methylenetriphosphate; a synthetic analogue of adenosine 5′-triphosphate, ATP, in which the oxy group between the latter's innermost and intermediate phosphorus atoms is replaced by a methylene group. It competitively inhibits rat-liver adenylate cyclase.

adenosine 5′-[β,γ-methylene]triphosphate *symbol*: Ado*PP*[CH₂]*P* or p[CH₂]ppA; *the recommended name for* β, γ-methyleneadenosine 5′-triphosphate (*abbr.*: ATP[β,γ-CH₂]); 5′-adenylyl methylenediphosphonate (*abbr.*: AMP-PCP); adenosine (5′→O¹)-1,2-μ-methylenetriphosphate; a synthetic analogue of adenosine 5′-triphosphate, ATP, in which the oxy group between the latter's intermediate and terminal phosphorus atoms is replaced by a methylene group. It inhibits the binding of ouabain to Na+/K+-transporting ATPase, the sodium pump.

adenosine monophosphate *abbr.*: AMP; *an alternative name for* any **adenosine phosphate**, but in particular for adenosine 5′-phosphate, especially when its distinction from adenosine (5′-) diphosphate and adenosine (5′-)triphosphate requires emphasis.

adenosine phosphate *symbol*: Ado*P*; adenosine monophosphate (*abbr.*: AMP); any phosphoric monoester or diester of adenosine. There are three monoesters – adenosine 2′-phosphate, adenosine 3′-phosphate, and **adenosine 5′-phosphate** – and two diesters – adenosine 2′,3′-phosphate and **adenosine 3′,5′-phosphate** – although adenosine 5′-phosphate is the ester commonly denoted (the locant being omitted if no ambiguity may arise). Adenosine 2′-phosphate (*symbol*: Ado2′*P*) is also named adenosine 2′-monophosphate (*abbr.*: 2′AMP) *or* 2′-adenylic acid *or* adenylic acid a, and adenosine 3′-phosphate (*symbol*: Ado3′*P*) is also named adenosine 3′-monophosphate (*abbr.*: 3′AMP) *or* 3′-adenylic acid *or* adenylic acid b *or* (*formerly*) yeast adenylic acid.

adenosine 2′-phosphate *see* adenosine phosphate.

adenosine 2′,3′-phosphate *see* adenosine phosphate.

adenosine 3′-phosphate *see* adenosine phosphate.

adenosine 3′,5′-phosphate *symbol*: Ado-3′,5′-*P*; *the recommended name for* cyclic adenosine 3′,5′-monophosphate (*abbr.*: cyclic AMP *or* cAMP); adenosine 3′,5′-cyclophosphate; adenosine 3′,5′-(cyclic)phosphate; a monophosphoric diester of adenosine. It is a universally distributed key metabolite, produced by the action of **adenylate cyclase** on adenosine 5′-triphosphate, ATP. The first compound to be named a **second messenger**, it mediates many effects in signal transduction pathways. It was first identified as a heat-stable activator of glycogen phosphorylase kinase, and is now known also to activate cyclic-AMP-dependent protein kinase and to regulate numerous other enzymic activities or physiological processes.

adenosine 5′-phosphate *or* **5′-adenylic acid** *or* **5′-phosphoadenosine** *or* **5′-O-phosphonoadenosine** *symbol*: Ado5′*P*; *alternative recommended names for* adenosine 5′-monophosphate (*abbr.*: 5′AMP); adenosine 5′-(dihydrogen phosphate); adenine (mono)ribonucleotide *or* (*formerly*) muscle adenylic acid. (The locant is commonly omitted if there is no ambiguity as to the position of phosphorylation.) It is a metabolic regulator, biosynthesized from inosine 5′-phosphate, 5′-inosinic acid; it is formed also by py-

rophosphatase-catalysed cleavage of (inorganic) diphosphate from ATP.

adenosine 3′-phosphate 5′-phosphosulfate *or* **3′-phospho-adenosine 5′-phosphosulfate** *symbol*: P*Ado*PS; *abbr.*: PAPS; *alternative recommended names for* active sulfate; 3′-phospho-5′-adenylyl sulfate; adenosine 3′-phosphate 5′-*P*-phosphatosulfate; 3′-phospho-5′-adenylic sulfuric monoanhydride; a naturally occurring mixed anhydride. It is synthesized from **adenosine 5′-phosphosulfate** by phosphorylation with ATP through the action of adenylylsulfate kinase (EC 2.7.1.25). It is an intermediate in the formation of a variety of sulfo compounds in biological systems. For example, in animals it is involved in sulfate transfer in the formation of sulfatides and in the synthesis of chondroitin sulfate and other sulfated polysaccharides, while in bacteria, by a process analogous to that involving adenosine 5′-phosphosulfate in plants, adenosine 3′-phosphate 5′-phosphosulfate interacts with reduced thioredoxin to yield adenosine 3′,5′-bis(phosphate) and sulfite; the latter can then undergo further reduction to sulfide, from which cysteine may be synthesized.

adenosine 5′-phosphosulfate *symbol*: Ado*PS*; *abbr.*: APS; *the recommended name for* 5′-adenylyl sulfate; adenosine 5′-*P*-phosphatosulfate; 5′-adenylic sulfuric monoanhydride; a naturally occurring mixed anhydride. It is synthesized from ATP and (inorganic) sulfate by the action of sulfurylase, sulfate adenylyltransferase (EC 2.7.7.4), and is an intermediate in the formation of **adenosine 3′-phosphate 5′-phosphosulfate**. In plants, by a process analogous to that involving adenosine 3′-phosphate 5′-phosphosulfate in bacteria, adenosine 5′-phosphosulfate can undergo reduction to yield sulfide, which may then be utilized for cysteine synthesis.

adenosine receptor *or* **P₁ purinoceptor** one of three types, A₁, A₂, and A₃, of membrane protein that bind adenosine or its analogues; they are **7TM proteins**. The selective agonist for A₁ is N^6-cyclopentyladenosine; that for A₂ is (2-*p*-carboxyethyl)phenylamino-5-*N*-ethyl-carboxamidoadenosine. Binding of agonist to A₁ causes inhibition of **adenylate cyclase**, opening of K⁺ channels, and inhibition of Ca²⁺ channels. Activation of A₂ brings about stimulation of adenylate cyclase. Activation of A₃ causes stimulation of adenylate cyclase. In some cases, particularly where A_1 receptors are involved, inositolphospholipid turnover may be stimulated. *See also* **purinoceptor**.

adenosine 5′-β-thiodiphosphate *symbol*: Ado*PP*[S] *or* [S]ppA; *abbr.*: ADP[S] *or* ADP-*β*-S; *the recommended name for* adenosine 5′-[β-thio]diphosphate; adenosine 5′-(2-thiodiphosphate); adenosine $(5′→O^2)$-1-thiodiphosphate; a synthetic analogue of adenosine 5′-diphosphate, ADP, in which an oxygen atom of its terminal phosphoric residue is replaced by a sulfur atom. It acts variously as a substrate or an inhibitor of ADP-dependent systems.

adenosine 5′-thiophosphate *symbol*: Ado*P*[S] *or* [S]pA; *the recommended name for* adenosine 5′-[α-thio]monophosphate (*abbr.*: AMP-S); a synthetic analogue of adenosine 5′-phosphate, AMP, in which an oxygen atom of its phosphoric residue is replaced by a sulfur atom. It acts variously as a substrate or an inhibitor of AMP-dependent systems.

adenosine 5′-γ-thiotriphosphate *symbol*: Ado*PPP*[S] *or* [S]pppA; *abbr.*: ATP[S] *or* ATP-*γ*-S; *the recommended name for* adenosine 5′-[γ-thio]triphosphate; adenosine $(5′→O^3)$-1-thiotriphosphate; a synthetic analogue of adenosine 5′-triphosphate, ATP, in which an oxygen atom of its terminal phosphoric residue is replaced by a sulfur atom. It can variously substitute for ATP or act as an inhibitor of ATP-dependent systems.

adenosinetriphosphatase *or* **ATP phosphohydrolase** *or* **ATPase**

any of several hundred enzymes catalyzing hydrolysis of ATP to ADP plus orthophosphate. Several types of ATPase are recognized.

1 ATPases functioning in the active transport of substances across membranes. These can be classified as: (a) P-type ATPases, which undergo phosphorylation at an aspartate residue during the transport cycle and transport mostly cations (H^+, Na^+, K^+, Ag^+, Ca^{2+}, Cd^{2+}, Cu^{2+}, Mg^{2+}), but also Cl^- and aminophospholipids (EC 3.6.3.1–3.6.3.13, 3.6.3.53 or TC 3.A.3.1.1–3.A.3.9.3); (b) ATPase exporting arsenite anions from bacteria (EC 3.6.3.16 or TC 3.A.4.1.1); (c) the multisubunit or two-sector ATPases transporting H^+ and Na^+ ions (EC 3.6.3.14–3.6.3.15 or TC 3.A.2.1.1–3.A.2.3.1); (d) the ABC-type ATPases (see also **ABC transporter**) transporting a variety of substances, from nutrient molecules in bacteria to xenobiotics in eukaryotic organisms (EC 3.6.3.17–3.6.3.49 or TC 3.A.1.1.1–3.A.1.210.5); (e) enzymes or enzyme complexes participating in the transport of macromolecules (EC 3.6.3.50–3.6.3.52 or TC families 3.A.5–3.A.11).

2 ATPases functioning in cellular and subcellular movement, such as myosin (EC 3.6.4.1), dynein (EC 3.6.4.2), and others (EC 3.6.4.3–3.6.4.11).

adenosine 5′-triphosphate *symbol*: Ado5′*PPP* or pppA; *the recommended name for* adenosine triphosphate (*abbr.*: ATP); 5′-triphosphoadenosine; 5′-adenylyl diphosphate; adenosine 5′-(tetrahydrogen triphosphate); a universally important coenzyme and enzyme regulator. It is formed from adenosine 5′-diphosphate by oxidative phosphorylation in coupled mitochondria, by photophosphorylation in plants, and by substrate-level phosphorylation. Reactions in which it participates are often driven in the direction leading to hydrolysis of ATP. The chemical energy so released may be utilized in active transport; it may be converted to mechanical energy (e.g. for muscular contraction, movement of cilia, etc.), to light energy (for bioluminescence), or to electrical energy (in electric fish); or it may be released as heat. ATP also participates in numerous synthetic reactions by the transfer to other metabolites of a phosphoric or a diphosphoric residue, of an adenosyl residue, or of an adenylyl residue.

adenosyl any chemical group formed by the loss of a 2′-, a 3′-, or a 5′-hydroxyl group from the ribose moiety of adenosine. *Compare* **adenylyl**.

S-adenosylhomocysteine the L enantiomer, *S*-(5′-adenosyl)-L-homocysteine (*symbol*: AdoHcy; *abbr.*: SAH), *S*-(5′-deoxyadenosine-5′-yl)-L-homocysteine, is formed from **S-adenosylmethionine**. It is a strong inhibitor of *S*-adenosylmethionine-mediated methylation reactions and is cleaved to adenosine and homocysteine.

S-adenosylmethionine the L enantiomer, active methionine, *S*-(5′-adenosyl)-L-methionine (*symbol*: AdoMet; *abbr.*: SAM), *S*-(5′-deoxyadenosine-5′-yl)-L-methionine, is an important intermediate in one-carbon metabolism, the methionine's methyl group (activated by the adenosyl moiety bonded through sulfonium) being donated to an acceptor molecule by transmethylation and *S*-adenosylhomocysteine being produced. It is important also as an intermediate in the production of ethylene from L-methionine in plants, being cleaved to 5′-methylthioadenosine and 1-aminocyclopropane-1-carboxylate. The latter compound is then fragmented under aerobic conditions to ethylene, formate, ammonium, and carbon dioxide.

S-adenosylhomocysteine

adenosylmethionine decarboxylase EC 4.1.1.50; *systematic name*: S-adenosyl-L-methionine carboxy-lyase. An enzyme of polyamine (and hence **trypanothione**) biosynthesis that catalyses the decarboxylation of S-adenosyl-L-methionine to (5′-deoxyadenosin-5′-yl)(3-aminopropyl)methylsulfonium salt; pyruvate acts as a cofactor. The product participates in reactions in which its 3-aminopropyl group is transferred to **putrescine** to form **spermidine**, and a second 3-aminopropyl group is then transferred to spermidine to form **spermine**. In most cases its subunits (α and β) are derived from a single proenzyme.

S-adenosylmethionine synthetase *see* **methionine adenosyltransferase**.

adenovirus any of a group of non-enveloped icosahedral viruses, the Adenoviridae, containing linear double-stranded DNA. They affect mammals or birds (usually being specific to one or a few closely related host species) and are often associated with disease of the respiratory tract.

adenyl *a misnomer for* **adenylyl** or **adenylate**.

adenylate 1 either the monoanion or the dianion of adenylic acid. 2 any mixture of the acid and its anions. 3 any salt or ester of adenylic acid.

adenylate cyclase or **adenylylcyclase** or *(incorrectly)* **adenyl cyclase** EC 4.6.1.1; *systematic name*: ATP pyrophosphate-lyase (cyclizing); a phosphorus–oxygen lyase enzyme that catalyses the elimination of diphosphate from adenosine 5′-triphosphate (ATP) to form **adenosine 3′,5′-phosphate** (cyclic AMP). It is an effector of signal transduction and other fundamental regulatory mechanisms, being regulated by some G_α subunits of G-proteins. In mammals it is a widely distributed membrane-bound glycoprotein with various isoforms (115–180 kDa). These are six-loop-six structures, the intracellular loop and the C-terminal domain being homologous to the guanylate cyclase catalytic domain. One of them bears an ATP-binding site. Type I, 1134 amino acids, is brain-specific, Ca^{2+}/calmodulin activated, and inhibited by G-protein βγ subunits. Types II (1088 amino acids) and IV (1064 amino acids) are **calmodulin** insensitive and activated by βγ subunits; type II is found in brain and olfactory tissue, type IV is widely distributed except in testis. Type III, from olfactory sensory neurons, is calmodulin sensitive and not regulated by βγ subunits. Types V and VI form a subgroup that is widely distributed and not stimulated by βγ subunits. *Saccharomyces cerevisiae* enzyme (2026 amino acids) has a weak structural relationship with the mammalian enzyme. It is positively regulated by the *RAS1* and *RAS2* gene products, but this property is not shared by a recombinant enzyme from *Schizosaccharomyces pombe*. A plant gene has been cloned, the deduced protein sequence having no similarity to prokaryotic counterparts, but showing striking similarity to the catalytic region of *S. cerevisiae* adenylate cyclase, and with the cytoplasmic domains of bovine adenylate cyclase and two mammalian guanylate cyclases. Soluble adenylate cyclases (*abbr.*: sAC) are not affected by G proteins but are activated by bicarbonate in kidney and choroid plexus. Some sACs are found in bacterial toxins.

adenylate deaminase EC 3.5.4.6; *other names*: AMP aminohydrolase; AMP deaminase; an enzyme that converts adenosine

monophosphate (AMP) to inosine monophosphate and ammonia. There are four isozymes in humans. The **M isozyme** (*also called* myoadenylate kinase) is specific to skeletal muscle where it is closely associated with the contractile proteins. It is encoded by the *AMPD1* gene at 1p13-p21. The **L isozyme** predominates in liver and brain. It is encoded by the *AMPD2* gene at 1p13.3. **E1** and **E2** are isoforms in erythrocytes, produced by differential splicing of the pre-mRNA encoded by the *AMPD3* gene at 11p13-pter. Myoadenylate deaminase deficiency can be inherited as an autosomal recessive trait and is associated with exercise-induced cramp or myalgia; alternatively it can be acquired, being associated with a variety of neuromuscular disorders.

adenylate energy charge (*or* **energy charge**) a measure of the phosphorylating power of an adenylate pool, equal to one half the average number of anhydride-bound phosphoric groups per adenosine moiety present in the pool. It may be defined in terms of the concentrations of AMP, ADP, and ATP in the pool and expressed by the quotient:

$$([\text{ATP}] + 0.5[\text{ADP}])/([\text{ATP}] + [\text{ADP}] + [\text{AMP}]).$$

Compare **phosphorylation state ratio**.

adenylate isopentenyltransferase EC 2.5.1.27; *other name*: cytokinin synthase; an enzyme that catalyses the reaction between Δ^2-isopentenyl diphosphate and AMP to form (N^6-Δ^2-isopentenyl)adenosine 5′-monophosphate, an intermediate in the formation of **zeatin** riboside, and pyrophosphate.

adenylate kinase *abbr. (in clinical biochemistry)*: AK; EC 2.7.4.3; *systematic name*: ATP:AMP phosphotransferase; *other name*: myokinase. An enzyme that catalyses the reaction:

$$\text{ATP} + \text{AMP} = \text{ADP} + \text{ADP}.$$

adenylate pool *or* **adenylate system** the total amount of AMP, ADP, and ATP in a cell, tissue, or organism.

adenyl cyclase *a misnomer for* **adenylate cyclase**.

adenylic acid the trivial name for any phosphoric monoester of adenosine. The position of the phosphoric residue on the ribose moiety of a given ester may be specified by a prefixed locant; *see* **adenosine phosphate, adenosine**. However, 5′-adenylic acid is the ester commonly denoted, its locant usually being omitted if no ambiguity may arise. 5′-Adenylic acid is also an alternative recommended name for **adenosine 5′-phosphate**.

adenylosuccinate lyase EC 4.3.2.2; *other name*: adenylosuccinate adenosine 5′-monophosphate lyase; an enzyme that catalyses the nonhydrolytic cleavage of fumarate from the succinyl moiety of 5-aminoimidazole-4-[n-succinyl carboxamide ribotide] (SAICAR) and of succinyl-adenosine 5′-monophosphate, in the synthesis of adenosine 5′-phosphate and of purines, respectively. Its gene locus, at 22q13.1-q13.2, encodes a protein of 484 amino acids, which forms a homotetramer that is widely distributed in tissues and organisms. Deficiency, associated with some 20 mutations, is inherited as an autosomal recessive disorder characterized by psychomotor retardation and often with epileptic seizures or autistic features.

adenylyl the adenosine[mono]phospho group; the acyl group derived from adenylic acid (*see* **adenosine 5′-monophosphate**). *Compare* **adenosyl**.

adenylylate to introduce adenylyl groups into a compound generally through the action of an adenylyltransferase. —**adenylylated** *adj.*; **adenylylation** *n*.

adenylyl cyclase *a misnomer for* **adenylate cyclase**.

adenylyl sulfate *see* **adenosine 5′-phosphosulfate**

adenylylsulfate kinase EC 2.7.1.25; *other name*: APS kinase; an enzyme that synthesizes 3′-phosphoadenosine-5′-phosphosulfate from ATP and adenylylsulfate with release of ADP. See **phosphoadenosine-5′-phosphosulfate**.

adenylyltransferase *generic name for* any of a number of enzymes within sub-class EC 2.7.7, nucleotidyltransferases, that are specific for the transfer of an adenylyl group from a donor (usually adenosine triphosphate, ATP) to an acceptor (such as a nucleotide, a polynucleotide, a protein, a sugar phosphate, or sulfate); e.g. FMN adenylyltransferase (EC 2.7.7.2), ribose-5-phosphate adenylyltransferase (EC 2.7.7.35).

ADH *abbr. for*: **1** antidiuretic hormone. **2** *(in clinical biochemistry)* alcohol dehydrogenase (preferred form: AD).

adhalin *or* **DAG2** a dystrophin-associated glycoprotein of skeletal muscle sarcolemma that is specifically deficient in severe childhood autosomal recessive muscular dystrophy. It has a 17-residue signal sequence and one transmembrane domain, and contains two sites for N-linked glycosylation.

adherens junction a **cell junction** in which the cytoplasmic face of the plasma membrane is attached to actin filaments.

adhesion *(in pathology)* the abnormal union of surfaces or parts, usually due to the formation of fibrous tissue following inflammation.

adhesion molecules molecules expressed on the surface of a cell that mediate the adhesion of the cell to other cells, or to the extracellular matrix. They bind to receptors, classed collectively as **integrins**. Adhesion molecules are grouped into classes: these include **selectins**; the immunoglobulin superfamily, containing **ICAM**, **MadCAM**, **NCAM**, **PECAM**, and **VCAM**; the **cadherins**; and **CD44**. Adhesion molecules play a part in morphogenesis (e.g., cadherins), and in the treatment of inflammation and wounds (e.g., selectins and the immunoglobulin superfamily).

adiabatic describing any thermodynamic process that occurs with neither gain nor loss of heat between the system and its surroundings. —**adiabatically** *adv.*

adiabatic calorimeter a calorimeter in which the temperature of the outer jacket is kept as close as possible to that of the inner compartment so that heat losses from the latter are minimized.

adiabatic system *(in thermodynamics)* any geometrically defined volume that does not exchange thermal energy with its surroundings. *Compare* **closed system, open system**.

adipocyte a cell of adipose tissue; a fat cell or lipocyte of an animal.

adipogenesis the formation of fat or adipose tissue.

adipokine any adipose tissue-derived protein hormone, such as **ACRP30** or leptin.

adipokinetic fat-mobilizing; lipotropic.

adipokinetic hormone *abbr.*: AKH; *an alternative name for* **lipotropin**.

adiponectin *an alternative name for* **ACRP30**.

adipo-Q *an alternative name for* **ACRP30**.

adipose tissue fat or fatty tissue. *See also* **brown adipose tissue, white adipose tissue**.

adipsin 1 *another name for* complement factor D. **2** a protein of the serine endopeptidase family secreted into serum by adipocytes. It is expressed abundantly in adipocytes and sciatic nerve. There are two forms, M_r 37 000 and 44 000.

adiuretin *(sometimes)* an alternative name for **antidiuretic hormone**.

adjuvant 1 *(in immunology)* any substance or mixture of substances that increases or diversifies the immune response to an antigen. *See also* **Freund's adjuvant**. **2** *(in pharmacology)* any remedy or drug that assists or modifies the action of other remedies or drugs.

adjuvant peptide *an alternative name for* **muramyl-dipeptide**.

ad lib without restraint. Short for the Latin *ad libitum*, meaning literally 'at pleasure'.

ADME/Tox *abbr. for* **absorption, distribution, metabolism, elimination, and toxicity**.

A-DNA *abbr. for* A-form of DNA.

Ado *symbol for* a residue of the ribonucleoside adenosine (alternative to A).

AdoHcy *symbol for* S-adenosylhomocysteine.

AdoMet *symbol for* S-adenosylmethionine.

adonitol *a former name for* **ribitol**.

adonose *a former name for* **ribulose**.

Ado*P* *symbol for* any adenosine phosphate.

Ado2′*P* *symbol for* adenosine 2′-phosphate.

Ado-2′,3′*P* *symbol for* adenosine 2′,3′-phosphate.

Ado3′*P* *symbol for* adenosine 3′-phosphate.

Ado-3′,5′-*P* *symbol for* adenosine 3′,5′-phosphate.

Ado5′*P* *symbol for* adenosine 5′-diphosphate *(alternative to pAdo)*.

Ado*P*[CH₂]*P* *symbol for* adenosine 5′-[α,β-methylene]diphosphate (alternative to p[CH₂]pA).

Ado*P*[CH₂]*PP* *symbol for* adenosine 5′-[α,β-methylene]triphosphate (alternative to pp[CH₂]pA).

Ado5′*PP* *symbol for* adenosine 5′-diphosphate (alternative to ppA).

Ado*PP*[CH₂]*P* *symbol for* adenosine 5′-[β,γ-methylene]triphosphate (alternative to p[CH₂]ppA).

Ado*PP*Glc *or* **Ado-5′***PP*-Glc *symbol for* adenosinediphosphoglucose (alternatives to A5′pp1Glc).

Ado*PP*[NH]*P* *symbol for* adenosine 5′-[β,γ-imido]triphosphate (alternative to p[N]ppA).

Ado5′*PPP* *symbol for* adenosine 5′-triphosphate (alternative to pppA).

Ado*PP*[S] *symbol for* adenosine 5′-β-thiodiphosphate (alternative to [S]ppA).

Ado*P*[S] *symbol for* adenosine 5′-thiophosphate (alternative to [S]pA).

Ado*PPP*[S] *symbol for* adenosine 5′-γ-thiotriphosphate (alternative to [S]pppA).

Ado*PP*Rib *symbol for* adenosine diphosphoribose (alternative to A5′pp5Rib *or* (Rib5)ppA).

Ado*PS* *symbol for* adenosine 3′-phosphate 5′-phosphosulfate.

adopted orphan receptor *see* nuclear receptor.

adoptive immunity immunity conferred by transfer of immunologically active cells from one individual to another.

ADP *abbr. for* adenosine 5′-diphosphate.

ADPase *abbr. for* adenosine diphosphatase (*see* apyrase).

ADP,ATP carrier protein *or* **adenine nucleotide translocase** an integral membrane protein of the inner mitochondrial membrane, responsible for the transport of ADP and ATP across the membrane. It exchanges one molecule of ADP^{3-} for one molecule of ATP^{4-}, and is thought to be driven by the proton gradient established across the inner membrane. It is a homodimer, each chain having three homologous domains. Different types occur in different mammalian tissues, but all are related.

ADP[α,β-CH₂] *abbr. for* α,β-methyleneadenosine 5′-diphosphate; i.e. adenosine 5′-[α,β-methylene]diphosphate.

ADPG *abbr. for* adenosinediphosphoglucose; ADP-Glc is preferred.

ADP-Glc *or* **ADPglucose** (*preferred*) *abbr. for* adenosinediphosphoglucose; *see also* **ADPG**.

ADPKD *abbr. for* autosomal dominant polycystic kidney disease.

ADP-Rib *or* **ADPribose** *abbr. for* adenosinediphosphoribose.

ADP-ribose (cyclic) *see* cyclic adenosine 5′-diphosphoribose.

ADPribosyl *abbr. for* the adenosinediphosphoribosyl group.

ADP-ribosylation the transfer of one or more **adenosinediphosphoribosyl** (*abbr.* ADPribosyl) groups from nicotinamide-adenine dinucleotide, NAD⁺, to a protein through the action of an **ADP-ribosyltransferase**. In eukaryotic cells the transfer occurs particularly to the α subunits of **G proteins**, and then is stimulated by treatment with certain bacterial toxins such as **cholera toxin** or **pertussis toxin**. Linear poly(ADPribosyl) attachments, which may consist of up to 50 ADPribose units, are found on nuclear proteins and also on some cytoplasmic proteins. In the stepwise generation of such attachments an incoming ADPribosyl group forms a (12′) glycosidic bond to the nucleosidic ribose moiety of the most recently attached ADPribose unit. ADP-ribosylation occurs also in cells of *Escherichia coli* infected with bacteriophage T4, where single ADPribosyl groups become attached to the host's RNA polymerase. —**ADP-ribosylated** *adj. See also* **ADP-ribosyltransferase, NAD(P)⁺–arginine ADP-ribosyltransferase**.

ADP-ribosylation factor a protein that acts as activator for **ADP-ribosyltransferase**.

ADP-ribosyl cyclase *abbr.*: ARC; an enzyme that catalyses the reactions:

(1) NAD⁺ → cADP-ribose + nicotinamide;
(2) NADP⁺ → 2′-phospho-cADP ribose + nicotinamide;
(3) NAD⁺ + nicotinate → NAAD + nicotinamide;
(4) NADP⁺ + nicotinate → NAADP + nicotinamide.

A physiological role has been assigned to cADP-ribose and NAADP (nicotinic acid adenine dinucleotide phosphate), but not to 2′-phospho-cADP ribose (also called cADPRP) or NAAD (nicotinic acid adenine dinucleotide). The enzyme from *Aplysia* has been cloned and is stimulated by the NO-CGMP pathway. In mammals the homologue is the lymphocyte antigen CD38 and the bone marrow stromal cell-surface molecule BST-1.

ADP-ribosyltransferase EC 2.4.2.30; *recommended name*: NAD$^+$ ADP-ribosyltransferase; *systematic name*: NAD$^+$: poly(adenine-diphosphate-D-ribosyl)-acceptor ADP-D-ribosyl-transferase; *other names*: poly(ADP) polymerase; poly(adenosine diphosphate ribose) polymerase; NAD$^+$ ADP-ribosyltransferase (polymerizing); poly(ADP-ribose) synthetase; *abbr.*: PARP *or* ADPRT. An enzyme that catalyses a reaction between NAD$^+$ and (ADP-D-ribosyl)$_n$-acceptor to form nicotinamide and (ADP-D-ribosyl)$_{n+1}$-acceptor. The enzyme acts on a number of nuclear proteins and thereby regulates events in differentiation and cell proliferation. The presence of DNA is necessary, and zinc is a cofactor.

ADP[S] *or* **ADP-β-S** *abbr. for* adenosine 5′-β-thiodiphosphate.

adren+ *or* **adreno+** *comb. form* denoting **1** the **adrenal gland(s)**. **2 epinephrine** or a related catecholamine.

adrenal 1 pertaining to or produced by the **adrenal glands**. **2** the adrenal gland itself.

adrenal androgen any of the C$_{19}$ steroid hormones produced in the cortex of the adrenal gland, including androstenedione and testosterone.

adrenal cortex *see* **adrenal gland**.

adrenal cortical hormone *or* **adrenocortical hormone** any (cortico)steroid hormone elaborated and secreted by the cortex of the **adrenal gland**.

adrenalectomize *or* **adrenalectomise** to carry out adrenalectomy. —**adrenalectomized** *or* **adrenalectomised** *adj.*

adrenalectomy surgical removal of one or (usually) both adrenal glands.

adrenal gland *or* (less commonly) **suprarenal gland** an endocrine organ in vertebrates. There is a single pair in mammals, one near each kidney; in other vertebrates there may be multiple adrenal glands. The gland has two components: an inner medulla, derived from the neural crest, that biosynthesizes and secretes epinephrine and norepinephrine; and an outer cortex, derived from the coelom, that is concerned in the biosynthesis and secretion of steroid hormones. The cortex consists in turn of three histologically defined zones: an outer **zona glomerulosa**, the cells of which are responsible for the biosynthesis of aldosterone and deoxycorticosterone; an intermediate **zona fasciculata**; and an inner **zona reticularis**. The cells of the zona fasciculata and zona reticularis are responsible for the biosynthesis of the glucocorticoid cortisol and the androgens dehydroepiandrosterone and androstenedione.

adrenaline *see* **epinephrine**.

adrenal medulla *see* **adrenal gland**.

adrenal medullary hormone any catecholamine hormone elaborated and secreted by the adrenal medulla. *See* **adrenal gland**.

adrenergic 1 describing a nerve or other cell, or cell receptor (*see* **adrenoceptor**) that is activated by epinephrine, norepinephrine, or an epinephrine-like substance. **2** any nerve that acts by releasing epinephrine, norepinephrine, or an epinephrine-like substance from its nerve ending. *See also* **cholinergic**, **GABAergic**, **noradrenergic**, **peptidergic**, **purinergic**, **serotonergic**. —**adrenergically** *adv.*

adrenergic receptor *an alternative term for* **adrenoceptor**.

β-adrenergic-receptor kinase *abbr.*: βARK, βAR kinase *or* BARK; EC 2.7.1.126; an enzyme that phosphorylates specifically only the agonist-occupied form of the β-adrenoceptor and closely related receptors at the C terminus. It appears to be important in mediating rapid agonist-specific (homologous) desensitization. Its cDNA codes for a precursor protein of 689 amino acids. The purified protein migrates as a single band of 80 kDa on electrophoresis. It has a protein kinase catalytic domain that has sequence similarity to protein kinase C and cyclic AMP-dependent protein kinase.

adrenic acid *see* **docosatetraenoic acid**

adreno+ *see* **adren+**.

adrenoceptor *or* **adrenergic receptor** *or* **adrenoreceptor** *or* **adrenotropic receptor** any receptor on an effector cell that is activated by epinephrine or related catecholamines. Structurally, they are all of the **7TM** type. Adrenoceptors may be classified phenomenologically into different types according to their sensitivities to agonists and antagonists.

α Adrenoceptors have a relative order of agonist potency: epi-

nephrine > norepinephrine > isoprenaline (isoproterenol), and a relative order of antagonist potency: phentolamine >> propranolol. They are associated with stimulatory effects, such as vasoconstriction and contraction of the iris, nictitating membrane, urinary bladder, seminal vesicles, and vas deferens, and with relaxation of propulsive smooth muscle in the gut. In some species, they mediate stimulation of gluconeogenesis and hepatic glycogenolysis. There are two groups, α$_1$ and α$_2$. α$_1$ receptors act through the phosphoinositide/Ca^{2+} second messenger system and are of four subtypes: α$_{1A}$, norepinephrine > epinephrine; α$_{1B}$, norepinephrine = epinephrine; α$_{1C}$, norepinephrine = epinephrine (different antagonist sensitivity from α$_{1B}$); α$_{1D}$. α$_{2A}$ receptors all inhibit the formation of cyclic AMP; they also open K$^+$ channels and inhibit Ca^{2+} channels. α$_{2B}$ receptors inhibit Ca^{2+} channels. α$_{2C}$ receptors have no effect on ion channels.

β Adrenoceptors have actions that can be ascribed to the activation of adenylate cyclase. They may be divided phenomenologically into three classes: (1) β$_1$ adrenoceptors, in which the relative order of agonist potency is isoprenaline > norepinephrine > epinephrine, and the relative order of antagonist potency is practolol > propranolol. They are associated with cardiac stimulation and glycogenolysis, lipolysis in white adipose tissue, and calorigenesis in brown adipose tissue; (2) β$_2$ adrenoceptors, in which the relative order of agonist potency is isoprenaline > epinephrine > norepinephrine, and the relative order of antagonist potency is propranolol > practolol; they are associated with skeletal muscle glycogenolysis, promotion of secretion of glucagon and insulin, vasodepression and bronchodilation; and (3) β$_3$ adrenoceptors, with agonist potency norepinephrine > epinephrine. *See also* **β-adrenergic receptor kinase**, **β-arrestin**.

adrenoceptor kinase *see* **β-adrenergic receptor kinase**.

adrenochrome an oxidation product of epinephrine that polymerizes at alkaline pH to form brown melanin-like pigments.

adrenocortical of, pertaining to, or derived from the cortex of the **adrenal gland**, e.g. adrenocortical hormone. *See* **adrenal androgen**, **corticosteroid**.

adrenocorticotropic hormone *or* adrenocorticotrophic hormone *abbr.*: ACTH; *an alternative name for* **corticotropin**.

adrenocorticotropic hormone-releasing factor an alternative name for **corticotropin releasing hormone**.

adrenocorticotropin *or* adrenocorticotrophin *an alternative name for* **corticotropin**.

adrenodoxin *or* (*sometimes*) **adrenoredoxin** a **ferredoxin** isolated from adrenal-cortex mitochondria that acts as an electron carrier in hydroxylase systems acting on steroids. It transfers electrons from adrenodoxin reductase to **cholesterol monooxygenase**. *See also* **NADPH:adrenodoxin oxidoreductase precursor**. *Compare* **putidaredoxin**.

adrenodoxin reductase *see* **NADPH:adrenodoxin oxidoreductase precursor**.

adrenoleukodystrophy protein *see* **ALDP**.

adrenomedullary of, pertaining to, or derived from the medulla of the **adrenal gland**.

adrenomedullin a hypotensive peptide that may function as a hormone in circulation control.

adrenoreceptor *a variant spelling of* **adrenoceptor**.

adrenosterone androst-4-ene-3,11,17-trione; a hormone with weak androgenic effect, originally called Reichstein's substance G.

adrenotropic receptor *an alternative name for* **adrenoceptor**.

adrex *or* **adx** *abbr. for* adrenalectomized.

adriamycin *former (generic) name for* **doxorubicin**.

adseverin *another name for* **scinderin**.

adsorb to undergo or elicit **adsorption**. —**adsorbable** *adj.*; **adsorbability** *n.*

adsorbate a substance that is adsorbed to the surface of another substance from either a gas or a liquid phase.

adsorbent 1 capable of **adsorption**. **2** a solid that adsorbs another substance from either a gas phase or a liquid phase.

adsorptiochromism the colour change that sometimes accompanies adsorption of organic compounds onto inorganic substances, e.g. onto alumina.

adsorption 1 any process in which a gas, liquid, or solute adheres to the exposed surfaces of a material, especially a solid, with which it is in contact. In **physical adsorption** the adhesion is through van der Waals forces of interaction, whereas in **chemisorption** (or chemical absorption) the adhesion is through formation of weak chemical bonds. *Compare* **absorption**. **2** (*in immunology*) the nonspecific attachment of an antigen (or antibody) onto the surfaces of red cells or inert particles so that the antibody (or antigen) to it may be detected by agglutination of the cells or particles. *Compare* **immunosorption**. **3** (*in microbiology*) the process of attachment of a phage or other virus to a cell.

adsorption chromatography any form of **chromatography** in which separation of the components of a mixture is based mainly on differences between the adsorption affinities of the components for the surface of an active solid.

adsorption coefficient in any adsorption equilibrium (of a substance from a solution), the mass of adsorbed substance per unit mass of adsorbent divided by the concentration of the substance in solution. It has the dimensions of reciprocal concentration. *Compare* **distribution coefficient**, **partition coefficient**.

adsorption isotherm any plot of the amount of solute adsorbed by an adsorbent (or of ligand bound by, e.g., a macromolecule, often expressed as the **saturation fraction**) versus the concentration of the free solute (or ligand), at constant temperature. *Compare* **Langmuir adsorption isotherm**.

advanced glycation end product *abbr.*: AGE; a product of the nonenzymatic glycation by glucose, fructose, or glyceraldehyde-3-phosphate, of extra- or intracellular proteins. High-affinity AGE receptors are present on monocytes, macrophages, liver, renal glomeruli and endothelial cells. AGEs contribute to age-dependent modification and crosslinking of tissue proteins, as in cataract formation, nephropathy, and vascular complications of diabetes mellitus.

adverse drug response *abbr.* ADR; any harmful or unintended effect of a medication, diagnostic test, or therapeutic intervention that occurs at a dose normally used for prophylaxis, disease diagnosis or therapy, or for the modification of physiological function. ADRs are one of the leading causes of hospitalization and death in the USA.

adx *see* **adrex**.

AEBSF *abbr. for* aminoethylbenzenesulfonyl fluoride.

AEG *abbr. for* acidic-epididymal glycoprotein. *See* **CRISP**.

+aemia *see* **+emia**.

aequorin a Ca^{2+}-dependent **photoprotein** responsible for luminescence by oxidation of the chromophore coelenterazine. It is obtainable from the hydrozoan jellyfish, *Aequorea forskaolea*. In the absence of any other cofactors or oxygen, it emits light on the addition of calcium ions, making it useful for the determination of free calcium ions.

aerate 1 to supply with or expose to air. **2** to pass air through a liquid. —**aeration** *n*.

aerenchyma plant tissue containing large, continuous extracellular air spaces. It is found in root cortical tissues of some plants (e.g. maize and rice) to transport oxygen from aerial structures to submerged roots. The spaces may form by the destruction of cells (lysigeny) or separation of cells (schizogeny), or a combination of both, and their formation is promoted by ethylene. –**aerenchymatous** *adj*.

aerobe any organism or class of organisms that can grow in the presence of dioxygen. **Facultative aerobes** are also capable of growing in the absence of dioxygen, whereas **obligate** (or strict) **aerobes** have an absolute requirement for dioxygen. *Compare* **anaerobe**.

aerobic 1 describing conditions in which gaseous or dissolved dioxygen is present. **2** describing an organism or process that requires or is able to use dioxygen. **3** of or produced by an aerobe. *Compare* **anaerobic**.

aerobiosis life in the presence of dioxygen. –**aerobiotic** *adj*.

aerolysin a channel-forming protein secreted by the human pathogen *Aeromonas hydrophila*. The cytolytic toxin is a dimer, but forms stable heptameric structures that insert into lipid bilayers to produce well-defined channels, leading to destruction of the membrane permeability barrier, and osmotic lysis.

aerosol a colloidal dispersion of solid particles or liquid droplets in air or another gas.

Aet *symbol for* the aminoethyl group, $-CH_2-CH_2-NH_2$.

aetiology *see* etiology.

AfaI a type 2 **restriction endonuclease**; recognition sequence: GT↑AC. *Rsa*I is an **isoschizomer**.

afamin a mammalian serum protein similar to albumin, α-fetoprotein, and vitamin D-binding protein.

afferent 1 conveying inwards to a part, organ, or centre, as of a blood vessel, nerve, or duct. **2** an afferent part, e.g. an afferent blood vessel or afferent nerve. *Compare* **efferent**.

affine the quality of having affinity.

affine gap penalty *see* **gap penalty**.

affinity 1 chemical attraction; the tendency of a chemical substance to combine with, bind to, or dissolve in other chemical substances. **2** any measure of such chemical attraction. **3** denoting biomolecular interaction that exhibits specificity.

affinity adsorbent any biospecific adsorbent used in affinity chromatography.

affinity chromatography a general chromatographic method that may, in principle, be used to isolate either of the components of a reversibly reacting chemical system provided that one component can be coupled to an insoluble matrix through a covalent linkage. The other component can then be bound to the immobilized component and the system eluted with a buffer that liberates the bound component. The technique has been applied to the separation of various substances, including enzymes, substrates, antigens, antibodies, nucleic acids, and even whole living cells. Pure antibodies can be prepared by this means: the antigen is covalently coupled to the dextran beads in the chromatography column, the antibody-containing solution is run into the column in neutral buffer, the specific antibody binds to the antigen, and the antibody is subsequently released with a buffer of high or low pH or with a denaturing reagent. The technique can also be used to isolate antigen. *See also* **affinity-elution chromatography**, **dye-ligand chromatography**. —**affinity-chromatographic** *adj*.

affinity constant *or* **binding constant** *an alternative name for* **association constant** (especially in relation to the binding of and/or to macromolecules, as in antigen–antibody, hormone– receptor, and enzyme–inhibitor reactions).

affinity cytochemistry a technique for detecting the distribution of specific cell-surface receptors. An easily visible (or electron-dense) material conjugated with a reagent specific for a particular cell-surface receptor is allowed to react with the cells in question and is then detected by light or electron microscopy.

affinity electrode a type of electrode useful for assaying specific proteins. It comprises a metal (e.g. titanium) wire whose surface has been oxidized and then covalently attached to a ligand capable of interacting biospecifically and reversibly with the protein in question. Binding of the complementary protein to the electrode results in a measurable change in electric potential relative to that given by a reference electrode. The latter is prepared in a similar way but lacks the specific ligand.

affinity electrophoresis *or* **affinoelectrophoresis** a type of **electrophoresis** in which the support medium contains an agent, immobilized by entrapment or by covalent linkage, that interacts specifically and selectively with certain of the components of the mixture to be analysed, thereby altering the electrophoretic mobility of those components.

affinity-elution chromatography a technique in which a compound that is nonspecifically bound to the matrix of a chromatographic column is specifically eluted by binding to a ligand in the eluting solvent. **Biospecific-elution chromatography** is a variant of this technique.

affinity gel any gel that serves as an **affinity matrix**.

affinity-isolated *an alternative term for* **affinity-purified**.

affinity label an active-site-directed irreversible inhibitor of an enzyme, antibody, or other protein. It is a chemically reactive compound that resembles a substrate or other specific ligand and bonds covalently to the active site or specific site on the protein. The affinity labelled groups can then be identified by **fingerprinting** and thus reveal the composition at the active site. It is sometimes termed **Trojan horse inhibitor**.

affinity matrix *or* **affinity support** any supporting material to which the biospecific reagent is attached in **affinity chromatography**.

affinity precipitation the precipitation of an enzyme by a homo- or hetero-bifunctional derivative of its coenzyme or/and a substrate or inhibitor. An example is the precipitation of **lactate dehydrogenase** by $N^{2'}$-adipodihydrazido-bis-(N^6-carbonylmethyl-NAD), a reactive derivative of its coenzyme, NAD.

affinity-purified *or* **affinity-isolated** describing a specified substance, usually a biological macromolecule such as an antigen or antibody, that has been purified (or isolated) by a technique such as **affinity chromatography** or **affinity electrophoresis**, thus implying high purity.

affinity support *an alternative term for* **affinity matrix**.

affinoelectrophoresis *an alternative term for* **affinity electrophoresis**.

afibrinogenemia *or (esp. Brit.)* **afibrinogenaemia** *see* **fibrinogen**.

aflatoxin any of a group of related and highly toxic secondary metabolites (**mycotoxins**) produced by strains of the moulds *Aspergillus flavus* or *A. parasiticus*, together with further metabolites of these mycotoxins. Their main structural feature is a fused coumarin–bis(dihydrofuran) ring system. The most important are aflatoxins B_1, B_2, G_1, and G_2 – so designated from the colour of their fluorescence (B, blue; G, green). The mycotoxins are produced naturally by the moulds growing on various seed crops, especially groundnuts (peanuts), and certain cereals (e.g. maize) during storage under moist conditions. They are acutely toxic and highly carcinogenic to many species of animals (including humans), and are responsible for turkey X disease. The main organ affected is the liver; aflatoxin B_1 is the most potent hepatocarcinogen known. **Aflatoxicosis** is a form of human hepatitis with jaundice, ascites and other signs of hepatic failure and has a high mortality. The term is derived as a contraction of *A. flavus* toxin.

aflatoxin B$_1$

AflII a type 2 **restriction endonuclease**; recognition sequence: C↑TTAAG.

AFLP *abbr. for* amplification fragment length polymorphism. The use of the **polymerase chain reaction** to amplify DNA in the study of **restriction fragment length polymorphism**.

A form 1 A form of DNA (*abbr.*: **A-DNA** *or* **DNA-A**); the molecular conformation adopted by fibres of the sodium salt of duplex DNA at a relative humidity of 75% or less. It consists of a right-handed double helix containing about 11 nucleotide residues per turn, with the planes of the base pairs inclined at about 70° to the axis of the helix. Unlike the **B form** of DNA it has a large hole (≈0.8 nm diameter) at the axis and a very deep major groove. *See also* **C form**, **Z form**. 2 A form of RNA (*abbr.*: **A-RNA**); the molecular conformation of double-stranded regions of RNA that is favoured at low-salt concentrations and moderate temperatures; it resembles the A form of DNA. (*see* def. 1).

AFP *abbr. for* α-fetoprotein.

African pygmyism a type of dwarfism, similar in appearance to the pituitary variety, due mainly to failure of the growth acceleration that is normal at puberty. It results from absence of the normal increase in levels of insulin-like growth factor-I (IGF-I) characteristic of this growth phase. Plasma somatotropin level is normal and there is a lack of responsiveness to exogenous somatotropin. *See* **insulin-like growth factor**.

Ag 1 *abbr. for* antigen. 2 *symbol for* silver.

AGA 1 a codon in non-mitochondrial mRNA for L-arginine. 2 a codon in human mitochondrial mRNA for chain termination. 3 *abbr. for* N-acetylglutamic acid.

agamic reproduction *see* **asexual reproduction**.

agammaglobulinemia *or (esp. Brit.)* **agammaglobulinaemia** a disease characterized by early onset of recurrent infections, profound hypogammaglobulinemia (of IgM, IgA, and IgG) and almost total absence of B lymphocytes in the peripheral circulation. An X-linked form is associated with over 400 mutations in a locus for Bruton's tyrosine protein kinase at Xq22. An autosomal recessive form is associated with at least four mutations in a locus at 14q32.3 encoding the mu heavy chain.

agamogeny the development of a new individual from a single cell. *Compare* **vegetative reproduction**.

agar *or* **agar-agar** a complex sulfated galactan extracted from certain seaweeds, especially *Gelidium* and related genera. The two main components are **agarose** and **agaropectin**. Agar forms an aqueous gel suitable for the solidification of microbiological culture media and for use as a support medium in zone electrophoresis or (immuno)diffusion techniques. It is not metabolized by most organisms. The gelling temperature varies from about 25 to 35 °C for different types of preparation. The gel is then stable to about 90 °C.

agar-diffusion method a method for determining the sensitivity of a microorganism to an antimicrobial drug. The zone of growth inhibition is measured around a ditch, hole, or a filter-paper disk containing the drug and located on an agar culture medium seeded with the microorganism in question.

agarobiose 4-O-β-D-galactopyranosyl-3,6-anhydro-L-galactose; a **disaccharide** that forms the basic unit of **agarose**.

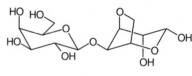

agarabiose

agaropectin a charged component of **agar** consisting of a mixture of polysaccharides containing D-galactose, 3,6-anhydro-L-galactose, and monoesterified sulfuric acid units. Samples from some algae may contain other components such as pyruvate, L-arabinose, and D-glucuronate units.

agarose an essentially uncharged component of **agar** comprising a ≈120 kDa alternating carbohydrate polymer. It consists of -3)D-Gal*p*-(β1-4)-3,6An-L-Gal*p*(α1 repeating units, containing small amounts of ionized sulfate and pyruvate groups. It is widely used as a matrix in zone and immunoelectrophoresis, immunodiffusion, and gel filtration and affinity chromatography. *See also* **agarobiose**.

AGC a codon in mRNA for L-serine.

AGE *abbr. for* advanced glycation product.

age-1 a gene encoding a phosphatidylinositol 3-kinase in *Caenorhabditis elegans* that functions downstream of the *daf-2* gene product (a receptor like that for insulin). Partial loss-of-function mutations in *age-1* and *daf-2* result in a long-lived phenotype with resistance to killing by several pathogenic bacteria.

agenized flour *see* **methionine sulfoximine**.

AGE receptors *see* **advanced glycation product**.

AGG 1 a codon in non-mitochondrial mRNA for L-arginine. 2 a codon in human mitochondrial mRNA for chain termination.

agglutinate 1 to adhere or cause to adhere. 2 to cause the clumping of cells, particles, etc., or to undergo such clumping.

agglutination 1 the act or process of adhering. 2 the process in which suspended cells or other **antigen** coated particles clump together when **antibody** is added due to an antigen–antibody reaction. 3 the result of such a process. *See also* **agglutinin**.

agglutinin an **antibody** that has the ability to agglutinate the corresponding antigen, e.g. when the antigen is present on the surface of a suspended cell or other particle.

agglutinogen an **antigen**, usually particulate, that stimulates the production of **agglutinin**.

aggrecan the major proteoglycan of cartilage. It has approximately one **glycosaminoglycan** chain for every 20 amino-acid residues, and a total mass of about 3 MDa. It assembles with glycosaminoglycans, especially hyaluronan, into aggregates several micrometres in diameter. *See also* **versican**.

aggregate 1 formed of separate units or particles collected into a whole or into larger units. **2** an assemblage or sum of many separate units or particles. **3** to form or be formed into a single body or larger body.

aggregation 1 the process of forming an **aggregate** (def. 2). **2** a cluster or group of particles held together into larger units.

aggregation number the number of monomers of an amphipathic substance that form a micelle under any particular conditions.

aggresome an inclusion body within cells that is formed in response to excessive misfolded protein.

aggressin a diffusible nontoxic substance produced by a microorganism that promotes the invasive power of the microorganism in the host.

aglucon or *(formerly)* **aglucone** an **aglycon** derived from a glucoside (i.e. where the sugar moiety was a glucose residue).

aglycon or **aglucon** or *(formerly)* **aglycone** or **aglucone** the part of any **glycoside** that remains after the sugar moiety has been chemically or enzymically removed.

agmatine 1-amino-4-guanidobutane; a putative endogenous ligand for **imidazoline receptors**, synthesized from arginine by the enzyme arginine decarboxylase (EC 4.1.1.19). In some invertebrates, e.g. the sponge *Geodia gigas,* and some cephalopods, the guanidine group can undergo phosphorylation to phosphoagmatine, which acts as a **phosphagen**.

agnotobiotic pertaining to the growth of organisms of a single species in the presence of one or more other species of which at least one is unknown. *Compare* **synxenic**.

+agogue or *(US)* **+agog** *suffix* denoting an agent that elicits or enhances the secretion of the indicated substance. [From the Greek *ag* + *gos*, leading, drawing forth.] —**+agogic** *adj.*

agonist any ligand, especially a drug or hormone, that binds to receptors and thereby alters the proportion of them that are in an active form, resulting in a biological response. A conventional agonist increases this proportion, whereas an **inverse agonist** reduces it. *See also* **full agonist, partial agonist**. *Compare* **antagonist**. —**agonistic** *adj.*

agouti signal protein *abbr.*: ASP; a protein (131 amino acids) encoded at 20q11.2 that stimulates pheomelanogenesis. It is an antagonist of melanocyte-stimulating hormone (MSH) at the latter's G-protein coupled receptor. *See* **pheomelanin**.

AGP *abbr. for* **arabinogalactan protein**.

Agranoff's turtle a zoomorphic mnemonic for avoiding the confusion that has tended to occur over the numbering of the carbon atoms of *myo*-inositol and its chiral derivatives, especially when their Haworth projections are converted into diagrams representing their normal chair conformations. The chair conformation of the parent compound is likened to a turtle, with its body corresponding to the C_6 ring, its coplanar limbs and tail representing the five equatorial hydroxyl groups, and its erect head the axial hydroxyl group. The turtle is viewed from above; then, for derivatives named as members of the d series, numbering of the turtle's appendages proceeds counterclockwise, commencing with its right front paw (i.e. its *dextro* 'hand'), which is designated 1d; its head becomes 2d, its left front paw is 3d, and so on around its body. Conversely, numbering of the l series of derivatives commences with the left front paw (the *laevo* 'hand'), designated 1l, and then proceeds clockwise, the head becoming 2l, and so on. [After its originator Bernard William Agranoff (1926–), who devised it in 1978.]

agrin a component of the synaptic basal lamina that causes aggregation of acetylcholine receptors and acetylcholinesterase on the surface of muscle fibres of the neuromuscular junction. It occurs in embryonic nervous tissue and muscle, especially during early development. At least five different forms arise by alternative splic-

ing; they differ in their acetylcholine receptor clustering activity. They contain **EGF**-like domains.

Agrobacterium a genus of Gram-negative aerobic rod-shaped soil bacteria. Most strains can initiate formation of galls in plants (*see* **crown-gall disease**). A feature of the infection of plants by *Agrobacterium* species is that the bacterium can subvert the host plant tissue to produce amino acids known as **opines**, which the bacterium can use as an energy, carbon, and nitrogen source. The type of opine produced is determined by the bacterial strain. Continued presence of the bacteria is not necessary for transformation of the cells of the host plant. Transformation by *Agrobacterium tumefaciens* has been shown to be due to large plasmids (140 – 235 kb). These are known as tumour-inducing, or Ti plasmids, and they have been exploited in the production of **transgenic** plants.

agrochemical a chemical that is used in agriculture or horticulture, especially as a biocide, fertilizer, etc.

agroclavine 8,9-didehydro-6,8-dimethylergoline; a nonpeptide **ergot** alkaloid.

agropine an **opine**; a rare amino-acid derivative that is produced by a certain type of crown-gall tumour. The genes responsible for its synthesis are part of the T-DNA from a **Ti plasmid**. *See also* **crown-gall disease**.

AGU a codon in mRNA for L-serine.

Ah *abbr. for* aromatic hydrocarbon; the name of a genetic locus in higher organisms that governs biological responses to some aromatic hydrocarbons. The proteins it encodes include cytochrome P450 1A1, a liver microsomal monooxygenase that oxidizes a variety of unrelated compounds, including xenobiotics such as the environmental carcinogens, benzo[a]pyrene and 3-methylcholanthrene. This enzyme is induced by these compounds and, more potently, by polychlorinated biphenyls such as 2,3,7,8-tetrachlorodibenzo-*p*-dioxin (TCDD), acting through the **Ah receptor**. The 5′-flanking region of the gene for this enzyme contains several short sequence motifs known as **xenobiotic response elements** (XREs). *See* **ARNT**.

ahnak a human gene that encodes a giant protein of ≈700 kDa with a large internal domain (≈4300 amino acids) of highly conserved repeated sequences, much of which are 128 amino-acid residues in length and contain a heptad repeat. The ahnak protein is found in human nuclei and is of unknown function. [From Hebrew *Ahnak*, giant.]

Ah receptor *abbr. for* aryl hydrocarbon receptor; a protein, encoded by the **Ah** gene, that binds a number of aryl hydrocarbons and mediates their biochemical and toxic effects. It is activated when it binds a ligand, and then translocates from the cytoplasm to the nucleus, where it is believed to enhance gene transcription by binding to the **xenobiotic-response element** sequence. It contains a basic helix-turn-helix motif; the DNA-binding form is a heterodimer of this protein and **ARNT**.

εAhx or *(formerly)* **εAcp** *symbol for* a residue of the ε-amino acid 6-aminohexanoic acid (formerly known as ε-aminocaproic acid).

aHyl *symbol for* a residue of the α-amino acid L-allohydroxylysine; *threo*-5-hydroxy-L_s-lysine.

AIB *abbr. for* 2-aminoisobutyric acid.

AICAR *abbr. for* 5-amino-4-imidazolecarboxamide ribonucleotide (an intermediate in **purine biosynthesis**).

AICAR transformylase *see* **phosphoribosylaminoimidazolecarboxamide formyltransferase**.

AIDS *abbr. for* acquired immunodeficiency syndrome.

AIF *abbr. for* **apoptosis inducing factor**.

alle (Alle) *symbol for* a residue of the α-amino acid L-alloisoleucine, (2*S*,3*R*)-2-amino-3-methylpentanoic acid.

AIR *abbr. for*. for 5-aminoimidazole ribonucleotide, an intermediate in **purine biosynthesis**.

AIR carboxylase *see* **phosphoribosylaminoimidazole carboxylase**.

AIRE *abbr. for* **autoimmune regulator**.

air-lift bioreactor a bioreactor in which the reaction medium is kept mixed and gassed by the introduction of air or another gas at the base of a column-like reactor.

***Air* RNA** a noncoding RNA required for silencing *Igf2r/Slc22a/Slc22a3* genes on the paternally imprinted mouse chromosome 17.

ajmaline ajmalan-17,20-diol; an alkaloid from the roots of *Rauwolfia serpentina*. It is used clinically as an antihypertensive and antiarrhythmic, having the effect of normalizing heart rhythm.

AK *(in clinical biochemistry) abbr. for* adenylate kinase.

AKAP *abbr. for* A-kinase anchor protein *or* cAMP-dependent protein kinase anchor protein; a protein that anchors the kinase to cytoskeletal and/or organelle-associated proteins, targeting the signal carried by cyclic AMP to specific intracellular effectors. The N-terminal region, which is highly basic, is required for interaction with **calmodulin**.

akaryotic describing a cell without a nucleus.

akee *a variant spelling of* **ackee**.

AKH *abbr. for* adipokinetic hormone.

A-kinase an enzyme that phosphorylates target proteins in response to a rise in intracellular **cyclic AMP**. *See* **cyclic AMP-dependent protein kinase**.

Akt *see* protein kinase B.

Akt the mouse or human homologue of v-*akt*, the oncogene of the transforming retrovirus AKT8. v-*akt* encodes a serine/threonine protein kinase that contains an SH2 domain.

AKT1 a protein that serves as an inward rectifier channel for K$^+$ ions in root cells of *Arabidopsis thaliana*. It is a member of the **shaker** superfamily of voltage-gated ion channels, and is a homotetramer in which each subunit contains six transmembrane segments.

al+ *prefix* denoting an acyclic monosaccharide or monosaccharide derivative. *See also* **aldehydo-**.

+al *suffix* denoting an unbranched acyclic mono- or dialdehyde.

Al *symbol for* aluminium.

Ala *symbol for* a residue of the α-amino acid L-alanine, 2-aminopropanoic acid (alternative to A).

βAla *symbol for* a residue of the β-amino acid β-alanine, 3-aminopropanoic acid.

ALA *or* **dALA** *abbr. for* δ-aminolevulinate.

Ala AT *or* **ALAT** *(in clinical biochemistry) (formerly) abbr. for* alanine aminotransferase. ALT is preferred.

ALA dehydratase *see* **porphobilinogen synthase**.

Alagille syndrome a rare autosomal recessive disorder characterized by intrahepatic cholestasis and developmental abnormalities in many structures including liver, skeleton, heart, eyes, and face. It is associated with over 35 microdeletions or other mutations in a locus at 20p12 for Jagged-1, a one-pass transmembrane protein that contains 16 EGF repeats in the extracellular region. Jagged-1 is a ligand for Notch receptor, which is crucial for cell fate development in *Drosophila* and mammals. *See* **Notch**. [After Daniel Alagille (1925–), French physician.]

alamethicin a linear ionophorous antibiotic polypeptide containing a high proportion of 2-methylalanine residues and blocked at both ends. *See also* **peptaibophols**.

Ac-2-MeAla-L-Pro-2-MeAla-L-Ala-2-MeAla-L-Ala-
L-Gln-2-MeAla-L-Val-2-MeAla-Gly-L-Leu-
2-MeAla-L-Pro-LVal-2-MeAla-2-MeAla-L-Glu-
L-Gln-L-phenylalaninol

alaninal the aldehyde obtained by reduction of the carboxyl group of alanine.

alaninate 1 alanine anion, the anion CH_3–$CH(NH_2)$–COO^-. **2** any salt containing alanine anion. **3** any ester of alanine.

alanine *the trivial name for* α-aminopropionic acid; 2-aminopropanoic acid; CH_3–$CH(NH_2)$–$COOH$; a chiral α-amino acid. L-alanine *(symbol*: A *or* Ala), (*S*)-2-aminopropanoic acid, is a coded amino acid found in peptide linkage in proteins; codon: GCA, GCC, GCG or GCU. In mammals, it is a non-essential dietary amino acid, and is glucogenic. Residues of D-alanine *(symbol*: D-Ala *or* DAla), (*R*)-2-aminopropanoic acid, are found in cell-wall peptidoglycans of various bacterial species, and in other materials, e.g. **cyclosporin**.

L-alanine

β-alanine *symbol*: βAla; *the trivial name for* β-aminopropionic acid; 3-aminopropanoic acid; H_2N–CH_2–CH_2–$COOH$; an achiral β-amino acid. It occurs free in brain, and in combination in pantothenate (and hence in coenzyme A), and in the isopeptides anserine and carnosine, but it is not a constituent of proteins. It has an inhibitory effect on the central nervous system.

alanine cycle *see* **glucose–alanine cycle**.

alanine–glyoxylate aminotransferase EC 2.6.1.44; a hepatic peroxisomal pyridoxal phosphate-dependent enzyme that catalyses the reaction:

alanine + glyoxylate = pyruvate + glycine.

Examples from six mammalian species show 78–89% sequence identity. Inactivating mutations in the gene for the human enzyme (392 amino acids) lead to hyperoxaluria type 1.

β-alanine–oxoglutarate aminotransferase *see* **4-aminobutyrate transaminase**.

alanine scanning mutagenesis a strategy for **site-directed mutagenesis** in which the amino acid alanine is substituted in each position that a change is desired.

alanine transaminase *abbr. (in clinical biochemistry)*: ALT; EC 2.6.1.2; *systematic name*: L-alanine:2-oxoglutarate aminotransferase; *other names*: glutamic–pyruvic transaminase; glutamic–alanine transaminase. An enzyme that catalyses the reversible reaction:

L-alanine + 2-oxoglutarate = pyruvate + L-glutamate.

Widely distributed in all tissues and organisms, it is a pyridoxal-phosphate enzyme. In clinical chemistry its assay alongside aspartate transaminase (AST), normally present in plasma in higher concentrations, may be useful as an indicator of liver damage, as levels of ALT are higher in liver than those of AST, in contrast to other tissues. In hepatitis, plasma levels of ALT may exceed those of AST.

alaninium alanine cation, CH_3–$CH(NH_3^+)$–$COOH$.

alanino the alkylamino group, $CH_3–CH(COOH)–NH–$, derived from alanine.

alaninol the alcohol obtained by the successive reduction of the carboxyl group of alanine.

alanyl the acyl group, $CH_3–CH(NH_2)–CO–$, derived from alanine.

alarmone any signal molecule that serves to reorient a cell's economy in response to stress. Such molecules include ppGpp, which is produced, e.g., in microorganisms, in response to growth-rate limitation caused by amino-acid stress and acts to correct this in various ways, and diadenosine tetraphosphate (ApppppA), which stimulates proliferation when DNA replication is halted at a replication fork.

ALA synthetase *see* **5-aminolevulinate synthase.**

ALAT *See* **Ala AT.**

albinism a heterogeneous group of disorders in which there is congenital reduction in the synthesis of **melanin** from tyrosine. It was first listed in 1908 by the British physician Archibald Garrod (1857–1936) in his lecture entitled 'Inborn Errors of Metabolism'. Melanism involving the melanocytes of the skin, hair and eyes is called **oculocutaneous albinism** (OCA), whereas melanism affecting primarily the retinal pigment epithelium is termed **ocular albinism** (OA). Three major types of OCA have been characterized. OCA1 (tyrosine-negative) results from numerous mutations in the locus for tyrosinase (**monophenol monooxygenase**) at 11q14-q21. The degree of pigmentation varies with the amount of residual enzyme activity. OCA2 (tyrosine-positive, P-related) results from many mutations in P protein (838 amino acids), a putative transmembrane protein of melanosomes, encoded at 15q11.12. OCA3, reported only in persons of African origin, is due to mutation in tyrosine-related protein 1 (537 amino acids, encoded at 9p23), which is important for stabilizing tyrosinase in melanocytes. OC1, the major form of ocular albinism, is associated with many mutations of a locus at Xp22 encoding a glycoprotein of the melanosomal membrane.

Albright syndrome *or* **hereditary osteoarthropathy** an uncommon autosomal dominant disorder. Type 1a is characterized by short stature, obesity, skeletal defects, and rounded facies. There is resistance to parathyroid hormone, thyrotropin, and gonadotropins, the activity of which is coupled to stimulation of adenylate cyclase in target cells. The alpha subunit of G-protein S (G_{Sa}) in affected persons is present at half the normal levels in tissues. This is due to any of some 30 mutations in a locus at 20q13.11, which encodes for the G_{Sa} subunit (comprising 394 residues). In type 1b, the physical appearance is normal and hormone resistance is limited to the effect on kidneys of parathyroid hormone. *See* **G-protein.** [After Fuller Albright (1900–1969), US physician.]

albumin any of a group of globular proteins that are soluble in distilled water and solutions of half-saturated ammonium sulfate, but insoluble in fully saturated ammonium sulfate solutions. **Serum albumin**, the major protein of serum, has a good binding capacity for water, Ca^{2+}, Na^+, K^+, fatty acids, hormones, bilirubin, and drugs. Its main function is to regulate the colloidal osmotic pressure of blood; it has structural similarity to α-**fetoprotein** and **vitamin D binding protein**. *See also* **endosperm albumin, lactalbumin, ovalbumin, serum albumin.** [The name originates from Latin *albus,* white, as it exists in nearly pure form in egg-white, from which it was easily obtained by the ancients.]

albuminoid 1 of, or relating to, an **albumin. 2** *an alternative name for* **scleroprotein.**

albuminuria the presence of excessive amounts of protein in the urine. It is usually a sign of renal disease.

albuterol *see* **salbutamol.**

Alcalase *the proprietary name for* a proteolytic enzyme preparation obtained by fermentation of a strain of *Bacillus subtilis*. It is characterized by stability and activity at high temperatures and alkaline pH values, and is used in enzymic laundering and various industrial processes.

alcapton 2,5-dihydroxyphenylacetic acid. *See* **alcaptonuria, homogentisate.**

alcaptonuria *or* **alkaptonuria** a rare autosomal recessive disorder, first studied by the British physician Archibald Garrod (1857–1936), caused by virtual absence of **homogentisate 1,2-dioxyge-**

nase and characterized by ochronosis and arthritis. There is inability to metabolize further the homogentisate produced by the normal metabolism of phenylalanine and tyrosine. Homogentisate therefore accumulates in the body and is excreted in the urine, which gradually turns dark as homogentisate is oxidized to a melanin-like product – a process speeded up at an alkaline pH. The deficiency is associated with more than 20 mutations of the gene locus at 3q21-q23.

alcohol 1 any of a class of alkyl compounds containing a hydroxyl group. An alcohol is called primary, secondary, or tertiary according to whether the carbon atom bearing the hydroxyl group is itself attached to one, two, or three other carbon atoms. **2** specifically, ethyl alcohol, ethanol. *See also* **absolute alcohol.** —**alcoholic** *adj., n.*

alcohol dehydrogenase any of the nicotinamide-nucleotide-linked dehydrogenase enzymes of groups EC 1.1.1.1 (NAD$^+$) or EC 1.1.1.2 (NADP$^+$) that catalyse the oxidation of alcohols to aldehydes. These are zinc proteins. Also EC 1.1.1.71, alcohol dehydrogenase [NAD(P)$^+$]; *other name*: retinal reductase; EC 1.1.99.8, alcohol dehydrogenase (acceptor), a quinoprotein.

alcoholic fermentation the conversion of glucose to ethanol by a group of reactions that are characteristic of yeast.

alcoholism a common disease caused by prolonged abuse of alcoholic (i.e. ethanol-containing) beverages. A genetic predisposition can be demonstrated in some individuals. Fatty liver develops rapidly after an episode of acute abuse, and chronic abuse leads to hepatic cirrhosis, cardiomegaly, premature atherosclerosis, and brain atrophy. Ethanol is metabolized by alcohol dehydrogenase (ADH) to acetaldehyde, which is then oxidized to acetate by aldehyde dehydrogenase (ALDH). An inactive form of an isozyme of ADH (ADH20) is transmitted as an autosomal dominant trait in a high percentage of Orientals and is caused by an inactivating polymorphism. The mitochondrial isozyme of ALDH (ALDH20) is also common in Orientals. Both defects increase the vulnerability to alcoholism of those who carry them.

alcohol oxidase *abbr.*: AOX *or* **methanol oxidase** *abbr.*: MOX; EC 1.1.3.13; a flavoprotein (FAD) enzyme that catalyses the oxidation by dioxygen of a primary alcohol to the corresponding aldehyde. It is important in methylotrophic yeasts, in which it is the first enzyme in the pathway for utilization of methanol, converting this to formaldehyde, thence to CO_2.

alcoholysis the solvolysis of a covalent derivative of an acid by a reaction in which one of the products combines with the H atom of an alcohol's hydroxy group and the other product with the alcohol's alkoxy group.

aldaric acid any dicarboxylic acid formed by oxidation of both terminal groups of an **aldose** to carboxyl groups. There are three tetaric acids (D-, L-, and *meso*-tartaric acids), four pentaric acids (D- and L-arabinaric acids, xylaric acid [*meso*], and ribaric acid [*meso*]), and ten hexaric acids (D- and L-glucaric acids, D- and L-idaric acids, D- and L-mannaric acids, D- and L-talaric acids, allaric acid [*meso*], and galactaric acid [*meso*]).

aldehyde 1 *an alternative name for* the **formyl** group, –CHO. **2** any organic compound with the formula R–CH=O. *Compare* **ketone.** —**aldehydic** *adj.*

aldehyde dehydrogenase any enzyme that catalyses the oxidation of an aldehyde group to a carboxylic acid. Many examples are known, acting specifically on a wide range of substrates. Those acting on aliphatic aldehydes are divided into four classes. Class 1, or Ald C, are tetrameric cytosolic enzymes, and include aldehyde dehydrogenase (NAD$^+$), EC 1.2.1.3; *systematic name*: aldehyde:NAD$^+$ oxidoreductase; an enzyme of wide specificity that contains molybdenum cofactor. This EC designation also includes Class 2 enzymes, or Ald M, which are tetrameric mitochondrial enzymes. Class 3, or Ald D, are dimeric cytosolic enzymes, and include aldehyde dehydrogenase [NAD(P)$^+$], EC 1.2.1.5; *systematic name*: aldehyde:NAD(P)$^+$ oxidoreductase.

aldehyde ferredoxin oxidoreductase an enzyme that catalyses the reversible oxidation of aldehydes to the corresponding carboxylic acid, using ferredoxin as an electron acceptor. Containing tungsten-pyranopterin and 4[Fe-45] clusters, the enzyme has remarkable thermostability.

aldehydo- *abbr.*: al-; *prefix* designating an uncyclized monosaccharide or monosaccharide derivative. It is used in (semi)systematic nomenclature to stress the acyclic nature of such compounds. *Compare* **keto-**.

aldimine any **imine** that is an analogue of an aldehyde; the general structure is RCH=NR where R may be any organyl group or H. *Compare* **ketimine**.

alditol any polyhydric alcohol derived from the acyclic form of a **monosaccharide** by reduction of its aldehyde or keto group to an alcoholic group. Older names for these compounds include glycitol and sugar alcohol.

aldo+ *prefix* indicating aldehydic; e.g. aldohexose.

aldoketose any monosaccharide derivative containing both a (potential) aldehydic carbonyl group and a (potential) ketonic carbonyl group.

aldol 1 any organic compound that is both an aldehyde and an alcohol, especially where the two functions are separated by two linked carbon atoms in accordance with the general structure: HO–C–C–CH=O. **2** *the trivial name for* acetaldol (3-hydroxybutanal), a compound formed by the self-condensation of two acetaldehyde molecules.

aldolase 1 *a generic name for* nearly every enzyme of the sub-subclass EC 4.1.2, the aldehyde-lyases, enzymes catalysing aldol condensations and their reversal. **2** *abbr. (in clinical biochemistry)*: ALS; *common name for* the enzyme fructose-bisphosphate aldolase; EC 4.1.2.13; *systematic name*: D-fructose-1,6-bisphosphate D-glyceraldehyde-3-phosphate-lyase; *other name*: fructose-1,6-bisphosphate triosephosphate-lyase. An enzyme that catalyses the reversible fission of D-fructose 1,6-bisphosphate to glycerone phosphate and D-glyceraldehyde 3-phosphate. It acts also on (3S,4R)-ketose 1-phosphates. In humans, aldolase A (genetic locus at 16q22-q24) occurs in erythrocytes, muscle, and fibroblasts; aldolase B (9q22.3) occurs in liver, kidney, and small intestine; and aldolase C (chromosome 17) occurs in brain. All three have homologous sequences, form homotetramers, and cleave D-fructose 1,6-bisphosphate and D-fructose 1-phosphate but at different rates. A and C are constitutive whereas B is under dietary control (low during fasting, increased by carbohydrate intake). Deficiency of aldolase A results in a nonspherocytic anemia or in a largely myopathic condition. Deficiency of aldolase B results in **hereditary fructose intolerance**.

aldol condensation the base-catalysed addition reaction of two aldehydes or an aldehyde and a ketone to form an **aldol** (def. 1).

aldonic acid a monocarboxylic acid having a chain of three or more carbon atoms and formally derived from an **aldose** by oxidation of the aldehydic group.

aldopentose any of the eight possible five-carbon-atom **aldoses**.

aldopyranose any **aldose** in the **pyranose** form.

aldose a monosaccharide in which the (potential) carbonyl group is terminal (i.e. aldehydic). The term is frequently modified to indicate the number of carbon atoms in the chain, as in aldotriose, aldotetrose, etc.

aldosterone 11β,21-dihydroxy-3,20-dioxopregn-4-en-18-al; the most powerful naturally occurring mineralocorticoid hormone, originally named **electrocortin** by its discoverers Sylvia Simpson and James Tait in 1954. The name was changed to aldosterone after its structure had been determined by **Reichstein** in 1953. It exists in solution as an equilibrium mixture of the aldehyde and the hemiacetal formed with the hydroxyl at position 11. Produced by the cells of the zona glomerulosa of the adrenal cortex, its main action is to increase reabsorption of sodium ions by the distal renal tubules and thus to regulate water and electrolyte metabolism. The biosynthesis of aldosterone is stimulated by angiotensin II (*see* **angiotensin, renin–angiotensin system**), **corticotropin**, and increased potassium-ion concentrations. 11-Deoxycorticosterone is converted into aldosterone by **aldosterone synthase**. Aldosterone exerts its effects through binding to cytosolic receptors. The resulting complex is transferred to the nucleus, where it acts to stimulate production of mRNA for sodium transport proteins. The action is thus on a timescale of many minutes.

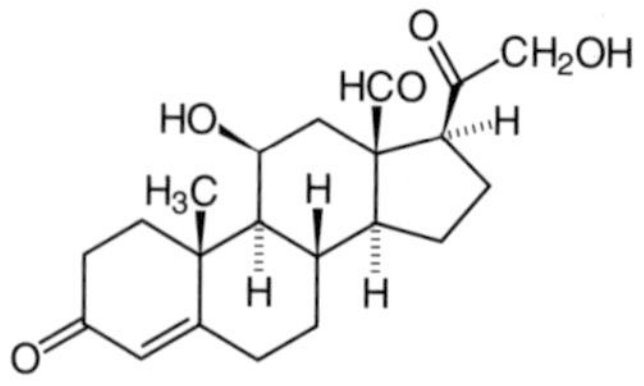

aldehyde form

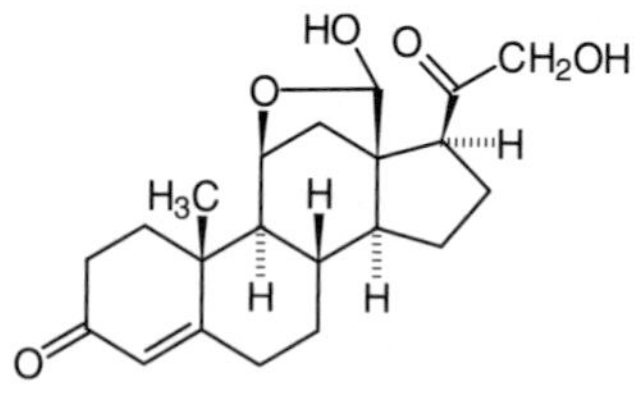

hemiacetal form

aldosterone synthase EC 1.14.15.4; *recommended name*: steroid 11β-monooxygenase; *other names*: cytochrome P450 11B2; steroid 18-hydroxylase; P450C18. A heme–thiolate enzyme that converts 11-deoxycorticosterone to aldosterone, by successive hydroxylations at the 11β and 18 positions. Reduced adrenodoxin is a reactant, being converted to oxidized adrenodoxin. The enzyme is normally expressed in the glomerulosa region of the adrenal cortex. Mutations in the CYP11B2 gene at 8q22 are associated with an autosomal recessive disorder that is potentially fatal in infants or may be asymptomatic in adults. ?

aldotetrose any of the four possible four-carbon-atom **aldoses**.

aldotriose either D- or L-glyceraldehyde; *see* aldose.

ALDP *abbr. for* adrenoleukodystrophy protein; a half-transporter of the ABC transporter superfamily that is encoded by a locus at Xp28 (gene product contains 745 amino acids) and is believed to function in transporting very-long-chain fatty acids across peroxisomal membranes. Over 300 mutations are associated with X-linked adrenoleukodystrophy, a condition in which tissues and body fluids have high levels of these fatty acids, particularly hexacosanoic acid (26:0).

aldrin 1,2,3,4,10,10-hexachloro-1,4,4a,5,8,8a-hexahydro-1,4:5,8-dimethanonaphthalene; an organochlorine compound formerly used as an insecticide, but now banned in certain countries on account of its toxicity.

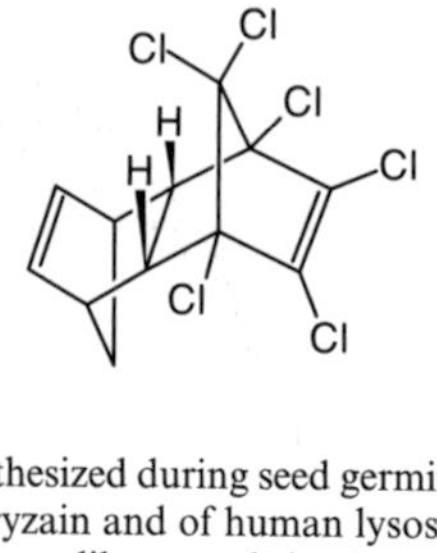

aleurain a protease synthesized during seed germination in barley. It is a homologue of γ-oryzain and of human lysosomal cathepsin H, and localizes in a lysosome-like vacuole in aleurone cells. The active mature protein (362 amino acids) is derived by proteolytic cleavge from a larger inactive precursor.

aleuron *or* **aleurone 1** granules of insoluble protein found in plant cells, especially in the seeds of cereals, where they appear to act as storage material. **2** the single layer of cells on the outside of the endosperm of cereal seeds containing such granules. Translocated **gibberellins** stimulate the cells to synthesize α-amylase, which is secreted into the endosperm to hydrolyse the stored starch.

aleuroplast any colourless **plastid** or **leukoplast** storing protein. Aleuroplasts are found in plant cells, particularly seeds.

Alexander disease a degenerative neurological disease of infantile, juvenile, or adult onset in which there is lack of myelin in the central nervous system and enlarged astrocytes rich in cytoplasmic granules that consist mainly of αB-**crystallin**. The gene at 11q13 corresponds to that for flavoprotein 1 of complex I (i.e. NADH dehydrogenase (ubiquinone)) of the electron transport chain.

alexin or **alexine** former term for **complement**; hence **alexinated** means treated with complement.

ALG abbr. for anti-lymphocyte globulin.

alga (pl. **algae**) any of a group of simple eukaryotic photosynthetic protists. They may be unicellular or multicellular, and are generally aquatic. The prokaryotes formerly called blue-green algae are now known as **cyanobacteria**. —**algal** adj.

algicide any chemical agent that selectively kills algae.

algin the sodium salt of **alginic acid**

alginate lyase EC 4.2.2.3; recommended name: poly(β-D-mannuronate) lyase; systematic name: poly(β-D-1,4-mannuronide) lyase. An enzyme that catalyses a reaction resulting in eliminative cleavage of polysaccharides containing β-D-mannuronate residues to give oligosaccharides with 4-deoxy-α-L-erythro-hex-4-enopyranuronosyl groups at their ends.

alginic acid a hydrophilic polysaccharide (≈240 kDa) occurring in brown algae (brown seaweeds), especially the Californian giant kelp (Macrocystis pyrifera) and horsetail kelp (Laminaria digitata). Sodium alginate is used commercially as an emulsifier and thickener in foodstuffs, pharmaceuticals, etc. It has an interrupted structure of stretches of α1-4-linked α-L-gulopyranosyluronic acid residues, stretches of β1-4-linked β-D-mannopyranosyluronic acid residues, and stretches where both uronic acids occur in alternating sequence. Similar glycans are secreted by some bacteria.

algorithm any systematic mathematical procedure that allows a problem to be solved in a finite number of steps. Compare **heuristic**, **stochastic**.

alicyclic describing an organic compound derived from a saturated cyclic hydrocarbon. Compare **aliphatic**, **aromatic**.

alignment 1 the process of comparing (aligning) linear sequences of amino acid residues or of nucleotide bases by sliding the sequences horizontally with respect to each other, often inserting gaps to bring equivalent regions into register. Sequences are usually aligned (manually or automatically) to assess their degree of similarity and their likely evolutionary relationships. **2** a horizontal stack of sequences, often including gaps, used to highlight regions of structural or functional similarity. Pairwise alignments compare two sequences, multiple alignments three or more. Alignments may be global (i.e. extend over the full length of the sequences) or local (i.e. extend only over short regions or domains); they underpin many sequence analysis methods.

alignment algorithm a method for automatically obtaining **alignments (def. 2)** of pairs of protein or nucleotide sequences. Alignment algorithms are generally either global or local – BLAST (basic local alignment search tool) and FastA are probably the best known implementations of local alignment algorithms, and are routinely used for database searching.

alignment editor software for manually creating and editing sequence **alignments** (e.g. to correct errors of multiple **alignment algorithms**). Such software typically includes options to add and remove gaps and/or sequences, to change the colour scheme, to select motifs, plot property profiles, view structures, etc. Examples include CINEMA (colour interactive editor for multiple alignments) and JalView.

aliphatic describing organic compounds in which the carbon atoms form open (noncyclic) chains. Compare **alicyclic**, **aromatic**.

aliquant one of a number of unequal parts of a whole.

aliquot one of number of equal parts of a whole; often used loosely, and erroneously, for a sample.

alizarin 1,2-dihydroxyanthraquinone; the **aglycon** of **ruberythric acid**.

ALK1 abbr. for **activin receptor-like kinase 1**.

alkalemia or (esp. Brit) **alkalaemia** a condition in which there is increased alkalinity (i.e. lowered hydrogen-ion concentration and hence raised pH) of the blood. Compare **acidemia**, **alkalosis**.

alkali any of a class of bases that neutralize acids and are themselves neutralized by acids, and form caustic and/or corrosive aqueous solutions. The term is applied in particular to hydroxides of the **alkali metals**, though the term is often extended to other substances with similar though weaker properties.

alkali metal any element of group 1 of the IUPAC **periodic table**; the group comprises lithium (Li), sodium (Na), potassium (K), rubidium (Rb), cesium (Cs), and francium (Fr).

alkaline 1 having the properties of an **alkali**; basic. **2** describing an aqueous solution having a pH > 7.

alkaline earth any metallic element belonging to group 2 of the IUPAC **periodic table**; the group comprises beryllium (Be), magnesium (Mg), calcium (Ca), strontium (Sr), barium (Ba), and radium (Ra).

alkaline lysis method a procedure for the recovery of plasmid DNA in which the alkaline conditions chosen for lysis of bacterial cells result in the denaturation of genomic DNA and proteins, which are precipitated when the lysate is neutralized with potassium acetate leaving a solution containing plasmid DNA.

alkaline phosphatase abbr.: AP; EC 3.1.3.1; systematic name: orthophosphoric monoester phosphohydrolase (alkaline optimum); other names: alkaline phosphomonoesterase; phosphomonoesterase; glycerophosphatase. A homodimeric or tetrameric zinc glycoprotein that is anchored to the plasma membrane by phosphatidylinositol glycan. Catalytic activity also requires magnesium. In human plasma the level is raised in the last trimester of pregnancy. Plasma levels may also be greatly elevated in cases of Paget's disease of bone, osteomalacia, and cirrhosis or biliary obstruction; levels may become moderately elevated in other types of bone disease. There are four genes for the enzyme in humans. That located at 1p36.1-p34 is for tissue-nonspecific AP, which occurs in liver, bone, and kidney. Three tissue-specific AP genes are located at 2q34-q37 and are expressed into placental AP (also called Regan isozyme), intestinal AP, and germ-cell AP. All four isozymes can be distinguished electrophoretically. **Hypophosphatasia**, a hereditary generalized deficiency of tissue-nonspecific AP, presents as rickets in infants and children and as osteomalacia in adults. Numerous mutations associated with this disorder have been reported.

alkaline tide the slight increase in plasma and urine pH that occurs after meals. It is believed to be due to withdrawal of hydrogen ions from the blood during the formation of gastric HCl.

alkali reserve or **alkaline reserve** a term formerly used to denote the bicarbonate-CO_2 extractable from blood plasma.

alkaloid any member of a broad group of nitrogen-containing basic organic compounds present in various dicotyledonous plants and some fungi. Although only about 5% of the world's plant species have so far been examined, they have yielded over 2000 different alkaloids. Heterocyclic alkaloids derived from amino acids are termed **true alkaloids**. Alkaloids with and without heterocyclic rings and not derived from amino acids are termed **pseudoalkaloids**; in these the carbon skeleton is usually isoprenoid derived. Alkaloids often have marked and specific pharmacological activity.

alkalosis a clinical condition in which total base excess or total acid deficit has the potential to cause decreased hydrogen-ion concentration (i.e. raised pH) in the blood, in the absence of compensating biochemical and physiological changes.

alkane any saturated aliphatic hydrocarbon compound.

alkanet the plant, Alkanna tinctoria, and its root, which contains the pigment **alkannin**.

alkannin (S)-5,8,-dihydroxy-2-(1-hydroxy-4-methyl-3-pentenyl)-1,4-naphthalenedione; a red pigment derived from the root of the alkanet plant, and used as a dye in cosmetics and food. The (+)(R) enantiomer is also a natural product, **shikonin**, and the racemate is known as shikalkin.

alkannin

alkaptonuria *a variant spelling of* **alcaptonuria**.

alkatriene any unsaturated aliphatic hydrocarbon containing two (conjugated or unconjugated) double bonds.

alkatrienyl any group derived from an alkatriene by removal of a hydrogen atom.

alkB a gene of *E.coli* that protects it against cell-killing by SN_2-alkylating agents. Its product is a nonheme-iron protein that catalyzes oxidative demethylation of N^1-methyladenine and N^3-methylcytosine in DNA. The reaction requires O_2 and 2-oxoglutarate and produces CO_2, formaldehyde, succinate, and demethylated substrate.

alkenyl any group derived from an alkene by removal of a hydrogen atom.

1-alkenyl-2-acylglycerol cholinephosphotransferase EC 2.7.8.22; an enzyme that catalyses the formation of the plasmalogen, plasmenylcholine (i.e. 1-alkenyl-2-acylglycerophosphocholine), from CDPcholine and 1-alkenyl-2-acylglycerol with release of CMP.

1-alkenylglycerophosphocholine *O*-acyltransferase EC 2.3.1.104; an enzyme that catalyses the formation of the plasmalogen, plasmenylcholine (i.e. 1-alkenyl-2-acylglycerophosphocholine) from acyl-CoA and 1-alkenylglycerophosphocholine with release of CoA.

alkenylglycerophosphocholine hydrolase EC 3.3.2.2; *other name*: lysoplasmalogenase; a phospholipase involved in the turnover of **plasmalogens**. It catalyses the hydrolysis of 1-(1-alkenyl)-*sn*-glycero-3-phosphocholine to an aldehyde and *sn*-glycero-3-phosphocholine.

1-alkenylglycerophosphoethanolamine *O*-acyltransferase EC 2.3.1.121; an enzyme that catalyses the formation of the **plasmalogen**, 1-alkenyl-2-acyl-glycerophosphoethanolamine, from acyl-CoA and 1-alkenylglycerophosphoethanolamine with release of CoA.

alkenylglycerophosphoethanolamine hydrolase EC 3.3.2.5; a phospholipase enzyme involved in the turnover of **plasmalogens**. It catalyses the hydrolysis of 1-(1-alkenyl)-*sn*-glycero-3-phosphoethanolamine to an aldehyde and *sn*-glycero-3-phosphoethanolamine.

alkyl any group derived from an alkane by the removal of one hydrogen atom. Alkyl groups are often designated by the symbol R.

alkylating agent any of a group of compounds that react with another compound so as to introduce an alkyl group into the second compound.

alkylation the process of replacing a hydrogen atom in a compound by an alkyl group.

1-alkylglycerophosphocholine *O*-acetyltransferase EC 2.3.1.67; an enzyme that catalyses the formation of 1-alkyl-2-acetyl-*sn*-glycero-3-phosphocholine (*see* **platelet-activating factor**) from acetyl-CoA and 1-alkyl-*sn*-glycero-3-phosphocholine with release of CoA. See also **1-alkylglycerophosphocholine *O*-acyltransferase**.

1-alkylglycerophosphocholine *O*-acyltransferase EC 2.3.1.63; an enzyme that catalyses the formation of the **ether lipid**, 1-alkyl-2-acyl-*sn*-glycero-3-phosphocholine, from acyl-CoA and 1-alkyl-*sn*-glycero-3-phosphocholine with release of CoA.

O^6-alkylguanine alkyltransferase see **O^6-methylguanine**.

alkyne any unsaturated aliphatic hydrocarbon compound containing one or more triple bonds.

alkynyl any group derived from an **alkyne** by the removal of one hydrogen atom.

All *symbol for* allose.

all+ *see* **allo+**.

all-alpha protein *or* **all-α protein** a member of a class of proteins whose predominant core secondary structure elements are alpha helical (*see* **alpha helix**), often packed into bundle-, folded-leaf- or hairpin-array-type folds. For example, DNA/RNA-binding proteins contain a core three-helical bundle with a right-handed twist; anti-sigma factor Asia contains a core of five helices in an orthogonal array; and members of the globin family contain a core of six alpha helices, with a partly opened folded-leaf topology. *Compare* **all-beta protein**.

allantoic acid *or* **diureidoacetate** the end product of **purine** metabolism in mammals and some fish, formed from **allantoin** by a hydrolytic reaction. Most fish metabolize allantoic acid further to urea and glyoxalate. It is widely distributed in plants as an important source of stored nitrogen.

allantoin (2,5-dioxo-4-imidazolidinyl)urea; 5-ureidohydantoin; an intermediate or end product of purine catabolism, formed from uric acid by urate oxidase. In certain animal groups it is converted to allantoic acid.

all-beta protein *or* **all-β protein** a member of a class of proteins whose predominant core secondary structure elements are beta strands or sheets, often packed into sandwich- or barrel-type folds (sandwiches comprise two aligned, twisted beta sheets; barrels comprise a single beta sheet folded back upon itself such that the first strand usually hydrogen-bonds to the last). For example, members of the immunoglobulin family have a sandwich-like fold, with seven strands in two sheets, whereas trypsin-like serine proteases have a duplicated eight-stranded closed barrel fold. *See* **beta configuration**. *Compare* **all-alpha protein**.

allele *or* **allelomorph** any of the forms of the same **gene** that occur at the same **locus** on a homologous chromosome but differ in base sequence. Two or more alleles are said to be **allelic** or **allelomorphic** to each other, and if more than two alleles exist in a population, the locus is said to show **multiple allelism**.

allele-specific oligonucleotide *abbr.*: ASO; an oligonucleotide that is constructed with a DNA sequence homologous to a specific allele. Two ASOs can be made so that they differ in sequence at only one nucleotide base, thereby distinguishing a mutant allele with a point mutation from its corresponding wild-type allele.

allelic exclusion the process by which a cell uses either the gene from its maternal chromosome or the one from the paternal chromosome, but not both. It seems to occur only in genes that encode antibodies and T-cell receptors. Individual B-lymphocytes display allelic exclusion of their heavy and light IgG genes.

allelomorph *another term for* **allele**.

allelopathic agent any plant excretory product that may be autotoxic or affect neighbouring plants, such as salicylate in *Quercus falcata*.

allelozyme any of two or more variants of a particular enzyme (with similar catalytic properties) whose amino-acid sequences are encoded in allelic structural genes (*see* **allele**); a class of **isoenzyme**. *Compare* **multilocus enzyme**.

allergen any **antigen** that stimulates an allergic reaction, inducing a type I hypersensitive reaction. *See* **allergy**, **hypersensitivity**.

allergic of, relating to, or involving **allergy**.

allergy *or* **allergic reaction 1** a state of altered (usually increased) re-activity of the body to foreign material. **2 hypersensitivity**.

allicin 2-propene-1-sulfinothioic acid *S*-2-propenyl ester; an antibac-terial compound derived from garlic (*Allium sativum*) and having an odour of garlic.

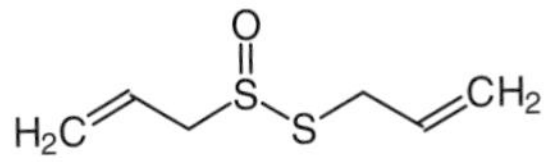

alliin 3-(2-propenylsulfinyl)-L-alanine; *S*-allyl-L-cysteine sulfoxide; a component of garlic (*Allium sativum*) and other plants that develops an odour of garlic following the action of alliinase (alliin lyase; EC 4.4.1.4).

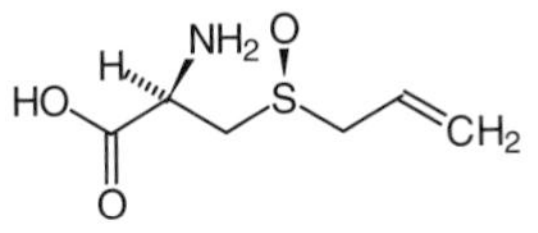

allitol a *meso*-hexitol; derived formally by reduction of the aldehyde group of either D- or L-allose.

allo+ *or (before a vowel)* **all+** *comb. form* **1** denoting other, dissimi-lar, different. *Compare* **hetero+** (def. 1), **homo+** (def. 1), **iso+** (def. 1), **xeno+** (def. 1). **2** referring to a dissimilar genome. *Compare* **homo+** (def. 2), **hetero+** (def. 2), **iso+** (def. 2), **xeno+** (def. 2). **3** referring to an isomeric form of an enantiomer of a compound that has more than one pair of enantiomers. *Compare* **iso+** (def. 3), **nor+** (def. 1).

allo+ *prefix* denoting the configuration of a set of four (usually) con-tiguous >CHOH groups, as in D- or L-allose. *See* **monosaccharide**.

alloalbumin an electrophoretic variant for serum albumin that gives a bisalbuminemia pattern. It is produced by one of various sense mutations in the gene.

alloantigen an antigen that is part of an animal's self-recognition system, e.g. **major histocompatability complex** molecules. When in-jected into another animal, they trigger an immune response aimed at eliminating them. *Compare* **alloreactivity**.

allocystathionine *see* **cystathionine**.

alloenzyme *an alternative name for* **allozyme**.

allogeneic *or* **allogenic** describing cells, tissues, organisms, etc. that are of different genetic constitution. *Compare* **heterogenic**, **homolo-gous**.

allograft any tissue graft between allogeneic individuals; i.e. from a donor of one genotype to a host of another genotype.

allohydroxylysine *see* **hydroxylysine**.

alloisoleucine *trivial name for* α-amino-β-methylvaleric acid; (2R*,3S*)-2-amino-3-methylpentanoic acid, CH_3–CH_2–CH–(CH_3)–$CH(NH_2)$–COOH; an α-amino acid with two chiral centres. Because molecules of alloisoleucine possess a second chiral centre, at C-3, in addition to the chiral centre at C-2 common to molecules of all α-amino acids other than glycine, the enantiomers L-al-loisoleucine (*symbol*: aIle), (2S,3R)-2- amino-3-methylpentanoic acid, and D-alloisoleucine (*symbol*: D-aIle), (2R,3S)-2-amino-3-methylpentanoic acid, are diastereoisomeric with those of **isoleucine**, (2R*, 3R*)-2-amino-3-methylpentanoic acid. L-Al-loisoleucine does not occur in peptide linkage in proteins. Residues of D-alloisoleucine are found in the peptidolipid of the actino-mycete, *Nocardia esteroides*, and various members of the actino-mycin group of antibiotics contain (usually) two residues per mol-ecule.

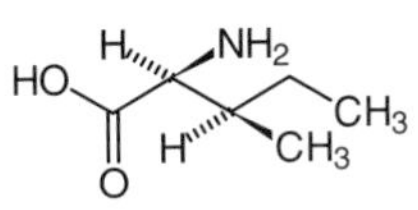

L-alloisoleucine pic

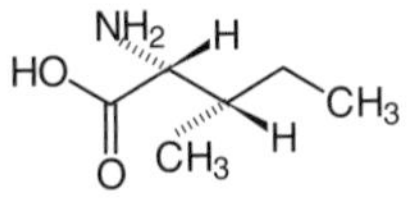

D-alloisoleucine

allolactose *trivial name for* β-D-galactopyranosyl-(1→6)-D-glucopy-ranose; an isomer of lactose and the natural intracellular inducer of the *lac* **operon** in *Escherichia coli*.

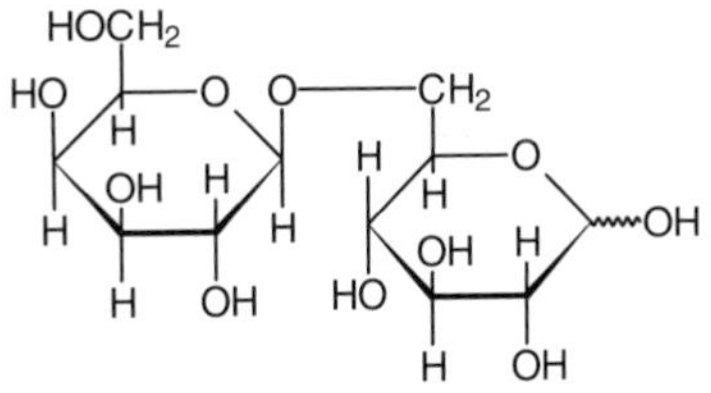

allomerism a state of similarity in the crystalline structures of sub-stances of different chemical composition. —**allomerous** *adj.*

allometric 1 differing in relative growth rates. **2** of or relating to al-lometry.

allometry 1 the study of the growth of parts of an organism relative to the growth of the whole organism. **2** change in the proportion of any part of an organism occurring during growth.

allomone any chemical substance produced or acquired by an or-ganism that, when it contacts an individual of another species in the natural context, evokes in the receiver a behavioural or develop-mental reaction adaptively favourable to the transmitter. *Compare* **kairomone**, **pheromone**.

allomorphism variability in the crystalline structure of certain sub-stances having the same chemical composition.

allophycocyanin a **phycobiliprotein** present in small amounts in red algae and cyanobacteria.

allopurinol 4-hydroxypyrazolo[3,4-*d*]pyrimidine; a constitutional isomer of hypoxanthine and a competitive inhibitor of xanthine ox-idase (EC 1.2.3.2). It is a synthetic drug used in the treatment of gout and other conditions characterized by raised blood urate con-tent. Patients with gout excrete less urate and more xanthine and hypoxanthine than normal.

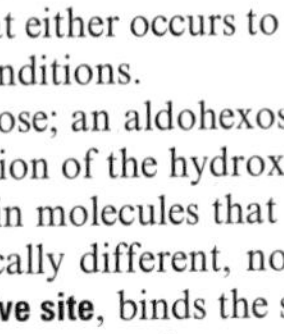

alloreactivity the T-cell response to non-self MHC molecules.

all-or-none describing a chemical reaction, or a pharmacological or physiological response, that either occurs to the fullest extent or not at all, depending on the conditions.

allose *symbol*: All; *allo*-hexose; an aldohexose that differs from glu-cose only in the configuration of the hydroxyl on C-3.

allosteric describing protein molecules that are assumed to possess two or more stereospecifically different, non-overlapping receptor sites. One of these, the **active site**, binds the substrate; the other, the **allosteric site**, is complementary to the structure of another, or the same, metabolite, the **allosteric effector**, which binds specifically and reversibly. The formation of the enzyme–allosteric effector complex does not activate or deactivate a reaction involving the effector it-self; it is assumed only to bring about a discrete reversible alteration of the molecular structure of the protein, called **allosteric transition**. This modifies the properties of the active site, changing one or several of the kinetic parameters that characterize the biological activity of the protein. *See also* **Monod–Wyman–Changeux model**.

allosteric activation any activation of an enzyme by a positive **allosteric effector**.

allosteric constant *symbol*: L; the equilibrium constant for the transition between two forms of an **allosteric** protein in the absence of ligand: $L = T_0/R_0$, where T_0 and R_0 are the concentrations of the less affine T-form and the more affine R-form of the protein, respectively. *See* **Monod–Wyman–Changeux model**.

allosteric effect an effect that arises when the reaction of ligands with one site of any polyvalent molecule affects the reaction of ligand(s) at one or more other sites as a result of conformational changes. The reactions may or may not both be reversible; when they are, and represent equilibria, they are **reciprocal effects** or **linked functions**.

allosteric effector a specific ligand that binds to the allosteric site of a protein with different affinities in the two states of the protein (R and T) in the **Monod–Wyman–Changeux model** of allosteric transition. An effector may be described as positive if it has an activating effect, or as negative if it is inhibitory. *See* **allosteric**. *Compare* **autosteric effector**.

allosteric enzyme designating any enzyme that: (1) contains a site topologically distinct from the active site; (2) has multiple subunits with active sites that interact cooperatively; (3) shows sigmoid kinetics; and (4) obeys the concerted model for allosteric proteins. Since all these properties may sometimes, but not always, go together, the term is somewhat imprecise.

allosteric inhibition any inhibition of an enzyme by a negative **allosteric effector**.

allosteric protein a protein that exhibits **allosterism**. *See also* **allosteric**.

allosteric regulation control of the rate of an enzyme or metabolic system by means of **allosteric activation** or **allosteric inhibition**.

allosteric site *see* **allosteric**.

allosteric transition *see* **allosteric, Monod–Wyman–Changeux model**.

allosterism *or* **allostery** a property of an enzyme or other macromolecule by which its principal biological reactivity is modified by the binding of an effector to a site other than the binding site of the principal reactant, thereby bringing about a conformational change in the macromolecule such as to alter its principal reactivity. *See* **allosteric**.

allothreonine *trivial name for* α-amino-β-hydroxybutyric acid; (2R*,3R*)-2-amino-3-hydroxybutanoic acid; CH₃–CH(OH)–CH(NH₂)–COOH; an α-amino acid with two chiral centres. Because molecules of allothreonine possess a second chiral centre, at C-3, in addition to the chiral centre at C-2 common to all α-amino acids other than glycine, the enantiomers L-allothreonine (*symbol*: aThr), (2S,3S)-2-amino-3-hydroxybutanoic acid, and D-allothreonine (*symbol*: D-aThr), (2R,3R)-2-amino-3-hydroxybutanoic acid, are diastereoisomeric with those of **threonine**, (2R*,3R*)-2-amino-3-hydroxybutanoic acid. [*Note*: the enantiomers of allothreonine may also be named semi-systematically as derivatives of erythrose: L$_s$-allothreonine in amino-acid nomenclature is synonymous with 2-amino-2,4-dideoxy-L$_g$-erythronic acid in carbohydrate nomenclature, and D$_s$-allothreonine with 2-amino-2,4-dideoxy-D$_g$-erythronic acid (the subscript letters d or g being added to the configurational prefixes where there might be uncertainty regarding the reference centre of chirality; *see* **D/L convention**).] L-Allothreonine does not occur in peptide linkage in proteins; however, one residue per molecule is present in the peptide antibiotic telomycin. Residues of D-allothreonine have been found in the glycolipid and peptidolipid of actinomycetes, and in mycobacteria.

allotopic 1 of, relating to, or having the property of **allotopy**. **2** (*in genetics*) describing a mutation that imparts to one tissue a particular metabolic characteristic normally found in another tissue.

allotopy the phenomenon of the formation of a membrane– enzyme complex resulting in alteration of the properties of both enzyme and membrane. For example, a change in enzymatic activity is sometimes observed when a soluble enzyme is adsorbed on an interface.

allotrope any of the different physical forms in which a chemical element can exist; e.g. diamond, graphite, and fullerenes are allotropes of carbon.

allotropy *or* **allotrophy** *or* **allotropism** the phenomenon of a substance, especially an element, existing in more than one physical form (**allotrope**), usually in the same phase; e.g. the different crystalline forms of carbon or phosphorus, or the different molecular forms of dioxygen and ozone. *Compare* **polymorphism** (def. 2). —**allotropic** *adj*.

allotype an antigenically distinct variant of a protein or other antigen arising from intraspecies genetic differences. Each individual has a particular variant at each immunoglobulin gene locus, which will often differ from those in other individuals of the same species.

allotypic specificity an antigenic specificity that is not the same in a given protein in all normal individuals of a species.

allotypy the property, possessed by certain proteins and other antigens, of existing in antigenically distinct forms or **allotypes**. Thus allotypes of the same protein, although not distinguished by usual physicochemical and chemical criteria, can elicit specific antibodies in some other members of the same species. They are those **idiotypes** that behave as **xenotypes** in other individuals of the same species.

alloxan mesoxalylurea; 2,4,5,6-(1H,3H)-pyrimidinetetrone; a compound used experimentally to induce diabetes through its action in selectively destroying pancreatic B-cells.

allozyme *or* **alloenzyme** any enzyme variant produced by a particular **allele**.

all-α protein *an alternative term for* **all-alpha protein**.

all-β protein *an alternative term for* **all-beta protein**.

allulose *trivial name (not recommended) for* **psicose**.

allurin a cysteine-rich secretory protein (CRISP) and sperm attractant from the egg jelly of *Xenopus* eggs.

allysine 6-oxonorleucine; 2-aminoadipaldehydic acid; an α-amino acid not normally found in proteins, but enzymically formed from lysine *in situ* as an intermediate in the creation of covalent cross-links between adjacent polypeptide strands during the maturation of certain connective-tissue proteins (e.g., collagen, elastin). The enzyme **lysyl oxidase** oxidizes the terminal amino-methylene groups of residues of either lysine or 5-hydroxylysine to aldehyde groups, forming residues of allysine or **hydroxyallysine** respectively. *See also* **desmosine, isodesmosine, lathyrism, noncoded amino acid, syndesine**.

L-allysine

L-allothreonine D-allothreonine

D-allysine

almond emulsin *see* **1,3-α-L-fucosidase**.

alopecia human hair loss, of which the most common is male pattern baldness (**androgenetic alopecia**). A rare complete loss of scalp and body hair (**alopecia universalis**) is associated with a recessive missense mutation in the hairless gene at 8p12. This encodes a putative single zinc finger transcription factor (of 1189 amino acids) that is expressed only in brain and skin, and shows 80% sequence identity with the equivalent mouse gene.

ALP *(in clinical biochemistry) abbr. for* alkaline phosphatase.

alpha *symbol*: α (lower case) *or* A (upper case); the first letter of the Greek alphabet. For uses *see* **Appendix A**.

$\alpha+\beta$ protein *or* **alpha+beta protein** member of a class of proteins that comprises both α- and β-secondary structures, in which the helical and strand components are segregated, the strand components mainly forming anti-parallel β-sheets: e.g., members of the lysozyme family have discrete α- and β-components, with a common $\alpha+\beta$ domain in the active site region; microbial ribonucleases have a single helix packed against an anti-parallel β-sheet.

α/β protein *or* **alpha/beta protein** member of a class of proteins that comprises both α- and β-secondary structures that alternate in closely-coupled α–β units, the strand components forming mainly parallel β-sheets or β-barrels: e.g., **TIM** barrels have a closed 8-stranded parallel β-sheet barrel topology; NAD(P)-binding **Rossmann folds** have a 3-layered α–β–α topology, with a core 6-stranded parallel β-sheet.

alpha carbon atom *or* **C-alpha** (*symbol*: Cα) the central tetrahedral carbon atom of an amino acid, to which is attached a hydrogen atom, an amino group, a carboxyl group and a side chain. *Compare* **beta carbon atom**.

alpha cell *an alternative term for* **A cell**.

alpha-1 cell *an alternative term for* **D cell**.

alpha chain *or* α **chain 1** the **heavy chain** of IgA **immunoglobulin** molecules. **2** one of the two types of polypeptide chain present in normal adult hemoglobin (Hb A) molecules.

alpha complementation a phenomenon in which the N-terminal sequence 1–56 of β-galactosidase from *Escherichia coli*, the α-peptide, is able to restore enzyme activity to a mutant form of the enzyme such as lacZΔM15 that has the sequence deleted.

alpha effect the enhancement of nucleophilicity that occurs when the atom adjacent to a nucleophilic site bears a lone pair of electrons. It has been invoked to explain the high toxicity of, e.g., hydroxylamine and the cyanide ion.

alpha-fetoprotein *see* α-**fetoprotein**.

alpha globulin *or* α-**globulin** any of a group of **plasma proteins** that migrate most anodically among the globulins in electrophoresis at pH values slightly above 7.

alpha-helical channels a subclass of transport proteins of the TC system, designated 1.A. It contains nearly 40 families, including voltage-gated ion channels, ATP-gated cation channels, **aquaporins**, **ligand-gated ion channels**, **connexins**, and **heat-shock proteins 70**. They are found in all domains of organisms.

alpha helix *or* α-**helix** a helical, or spiral, conformation of a polypeptide chain in which successive turns of the helix are held together by hydrogen bonds between the amide (peptide) links, the carbonyl group of any given residue being hydrogen- bonded to the imino group of the third residue behind it in the chain. This is the case for all of the carbonyl and amide groups of the peptide bonds of the main chain. The α-helix has 3.6 residues per turn, and the **translation** (def. 2), or pitch, along the helical axis is 1.5 Å per residue, 5.4 Å per turn. The helix may be left- or right-handed, the latter being much more common. It is one of the two basic elements of the **secondary structure** adopted by polypeptide chains (the other being **beta strand**). The α-helix was first described by Pauling and Corey in 1951, based on model building. *See also* **gamma helix**.

alpha-hemolysin a bacterial peptide creating pores in eukaryotic cell membranes; its structure is (2)$_2$ β-strands. It is classified in the TC system under number 1.C.3.

alpha hemolysis *or* α-**hemolysis** a type of hemolysis, characterizing certain strains of streptococci, in which a greenish tinge occurs around the bacterial colonies in a blood-agar plate. *Compare* **beta hemolysis**.

alpha MEM a modification of Minimal Essential Medium, used for the culture of a wide range of mammalian cells.

alphanumeric describing any coding system or data set that provides for letters, numbers, and other symbols such as punctuation. [From 'alphabetic' and 'numeric'.]

alpha oxidation *or* α-**oxidation 1** a series of reactions occurring in plant tissue by which a free fatty acid of chain length ranging from C_{18} to C_{13} is oxidatively degraded with the simultaneous release of a molecule of CO_2 from the carboxyl group and the formation of a free fatty acid containing one carbon atom fewer. NAD$^+$ is the only cofactor required. **2** In animals, the first reaction in the oxidation of phytanic acid in peroxisomes. It is catalysed by phytanoyl-CoA hydroxylase, deficiency of which produces Refsum disease (classical or adult).

alpha particle *or* α **particle** a fast-moving positively charged helium nucleus, ^{4}He^{2+}, emitted in the decay of certain **radionuclides**.

alpha-peptide the N-terminal peptide (1–58 or longer) of *E. coli* β-galactosidase. *See* **galactosidase**.

alpha radiation radiation consisting of fast-moving alpha particles.

alpha ray a directed stream of alpha particles.

alpha receptor *see* **adrenoceptor**.

Alport syndrome a progressive hereditary disease of renal glomerular membranes, sometimes also involving hearing loss and ocular lesions. The more common X-linked form is associated with over 300 mutations (mostly small deletions or insertions, splicing or missense mutations) in the gene at Xq22.3 for the type IV collagen α-5 chain. The less common autosomal form is associated with several mutations in the gene at 2q35 for the type IV collagen α-4 chain. Each of these chains is normally $\approx$1650 residues long and is highly glycosylated. [After Arthur Cecil Alport (1880–1959), South African physician.]

ALS *abbr. for* **1** antilymphocyte serum. **2** *(in clinical biochemistry)* (fructose-bisphosphate) aldolase, EC 4.1.2.13. **3** amyotrophic lateral sclerosis.

Alt *symbol for* altrose.

ALT *(in clinical biochemistry) abbr. for* alanine transaminase.

alternate pathway *see* **alternative pathway** (def. 2).

alternating axis of symmetry *see* **symmetry**.

alternating current *abbr.*: a.c. *or* AC; an electric current that varies with time in a sinusoidal manner. The abbreviation a.c. may also be applied to a voltage of varying polarity.

alternatively spliced form *an alternative name for* **splice variant**.

alternative oxidase *or* **cyanide-resistant oxidase** an enzyme of plants, many algae and fungi, and some protozoa that is associated with the inner mitochondrial membrane and transfers electrons from ubiquinol to oxygen, bypassing cytochrome *c* oxidase of the electron transport chain without generating a proton gradient or ATP. It is encoded in the nucleus, is insensitive to CO, cyanide and azide (inhibitors of cytochrome *c* oxidase), and to antimycin A and myxothiazol (inhibitors of complex III) but is inhibited by salicylhydroxamic acid and *n*-pyrogallate. Each subunit (32 kDa) of the dimer contains a central hydrophobic region that is membrane-intrusive and flanked by hydrophilic regions, the C-terminal region containing a di-iron centre. It is active in the thermogenesis associated with flowering in the voodoo lily and the skunk cabbage.

alternative pathway 1 a misleading term sometimes applied to any of the metabolic pathways by which glucose is broken down in animals, especially the **pentose (phosphate) pathway**. **2** (*or* **alternate pathway**) a pathway by which **complement** components C3–C9 are activated without a requirement for C1, C2, or C4.

alternative splicing the occurrence of alternative patterns of **splicing** of a primary RNA transcript of DNA to produce different mRNAs. After excision of **introns**, selection may determine which **exons** are spliced together to form the mRNA. An example is the production of mRNA for 19S IgM, which is secreted, and 7S IgM, which is inserted into the lymphocyte membrane, both products being coded for by the same gene, but the mRNA for 7S IgM having fewer of the original exons. Up to 50% of structural gene products in humans may be alternatively spliced.

altro+ *prefix* denoting a particular configuration of a set of four (usually) contiguous >CHOH groups, as in D- or L-altrose. *See* **monosaccharide**.

altrose *symbol*: Alt; *altro*-hexose; an aldohexose that differs from glucose in the configuration of the hydroxyl groups on C-2 and C-3.

***Alu*I** a type 2 **restriction endonuclease**; recognition sequence: AG↑CT. The M.*Alu*I modification site is C3.

alum-precipitated toxoid any preparation in which a toxoid is adsorbed onto an adjuvant containing an aluminium compound, e.g. aluminium hydroxide gel or an alum. The term is applied especially to diphtheria toxoid.

alum precipitation a technique in which a soluble immunogen is converted into particulate form by mixing with a solution of an alum (e.g. aluminium potassium sulfate) and adjusting the pH to near 6.5. The antigen is adsorbed on the aluminium hydroxide precipitate so formed, which acts as an adjuvant in immunization.

Alu sequence or ***Alu*I sequence** or **Alu repeat** or ***Alu*I repeat** any of various base sequences dispersed in human, rat, and mouse (and perhaps other) genomes of ≈0.3 kbp and spaced at approximately 5 kbp. Each Alu sequence comprises two similar 130 bp sequences (**Alu left** and **Alu right**) separated by a short A-rich region. Alu sequences are the major **SINE** in the human genome, being present ≈10^6 times. They are responsible for Alu-mediated deletions in several genes in which they occur (e.g. that for C1 inhibitor). These sequences may be sites for attachment of cohesin complexes that bind newly replicated chromosomes before they segregate into daughter cells. The name derives from the presence in such sequences of sites for the restriction enzyme *Alu*I. *See* **hereditary angioedema**.

alveolus (*pl.* **alveoli**) any small pit, cavity, or saclike dilatation, such as an air cell of the lungs, an **acinus** of a compound gland, or a tooth socket. —**alveolar** *adj.*

Alzheimer's disease or **Alzheimer's dementia** or **Alzheimer sclerosis** a genetically and phenotypically heterogeneous syndrome of progressive cognitive failure. It is the most common cause of late-life dementia (senile dementia) in developed nations, but early onset before the age of 65 (presenile dementia) is common in familial cases. All cases develop numerous extracellular deposits of the β-**amyloid peptide** in the brain, and almost all have intraneuronal bundles of abnormal filaments composed of highly phosphorylated forms of the microtubule-associated **tau protein**. Numerous mutations associated with early-onset forms occur at various loci, including: 21q21, which encodes the β-amyloid peptide precursor (a single-pass transmembrane glycoprotein of which an isoform containing 695 amino acids is expressed predominantly in brain); 14q21.3, which encodes presenilin 1 (a membrane protein with 5–10 putative transmembrane segments); and 1q42.1, which encodes presenilin 2 (a protein of 448 amino acids that has 60% overall sequence identity with presenilin 1). Inheritance of the E4 allele of the gene for **apolipoprotein E** (locus at 19q13) is associated with a dose-dependent increased risk for late-onset Alzheimer's disease. [After Alois Alzheimer (1864–1915), German neurologist who in 1907 described the condition of a patient referred to as Auguste D.]

Am *symbol for* americium.

Amadori rearrangement an acid- or base-catalysed chemical rearrangement reaction in which an *N*-substituted aldosylamine is converted into the corresponding *N*-substituted 1-amino-1-deoxy-2-ketose. It occurs, e.g., in the **Maillard reaction**, in the reaction of carbohydrates with phenylhydrazine, in the formation of hemoglobin A1$_c$, and in a step in tryptophan biosynthesis – isomerization of *N*-(5'-phosphoribosyl)-anthranilic acid into 1-(α-carboxyphenylamino)-1'-deoxyribulose-5'-phosphate.

***Amanita* toxin** *see* amatoxin.

amanitin any of various **amatoxins**, especially α-, β-, γ-, and ε-amanitin. These occur, together with other **amatoxins**, in the highly poisonous agaric commonly called the death-cap fungus or deadly agaric (*Amanita phalloides*), and in some related species.

α-amanitin

α-amanitin

amantadine *trivial name for* 1-adamantanamine; 1-aminoadamantane; $C_{10}H_{17}N$; a polycyclic antiviral agent that is especially effective as a prophylactic against influenza A virus. It appears not to affect viral attachment to cells but to block penetration or uncoating. *See also* **adamantane**.

amatoxin any member of a family of closely related bicyclic octapeptides present in the highly poisonous agaric commonly called the death-cap fungus or deadly agaric (*Amanita phalloides*), and in some related species. α-**Amanitin** is the best known member and the chief toxic constituent of *A. phalloides*. The majority of amatoxins are highly toxic, though much slower in action than the congeneric and less potent phallotoxins. For example, the ingestion by an adult man of 10 mg α-amanitin (the quantity in a single fruit body of *A. phalloides*) leads over the course of several days to death in the absence of treatment. The common elements of the molecular structure of amatoxins include residues of *trans*-4-hydroxy-L-proline, (usually mono- or dihydroxy-)L-isoleucine, and the bisfunctional amino acid (usually 6'-hydroxy-)L,L-tryptathionine *S*-oxide (which forms a sulfoxide bridge across the molecule; *see* structure at **amanitin**), together with two residues of glycine. α-Amanitin binds very tightly ($K_{diss} = 10^{-8}$ M) to eukaryotic RNA polymerase II, thereby blocking the formation of mRNA precursors and hnRNA. It also binds, but rather less tightly ($K_{diss} = 10^{-6}$ M), to eukaryotic RNA polymerase III, thereby blocking the formation of tRNA and 5S RNA. It has little action on RNA polymerase I. Hence the synthesis of RNA and protein is markedly inhibited in eukaryotic cells but not in prokaryotic cells.

amber 1 a fossil resin. **2** *see* **amber mutant**.

amber codon or **amber triplet** *symbol*: UAG; one of the three terminator codons or **nonsense codons** in an mRNA molecule.

Amberlite *proprietary name for* various cation- and anion-exchange resins.

amber mutant or **amber** a T4 bacteriophage carrying an **amber mutation**. *See also* **ochre mutant**, **opal mutant**.

amber mutation a temperature-sensitive conditionally lethal mutation, capable of occurring in almost any gene of a bacterial virus, that causes the synthesis of a particular polypeptide chain to terminate prematurely. It is due to the alteration of one codon in the corresponding messenger RNA to an **amber codon**, i.e. a termination codon.

amber suppressor any of a number of mutations in *Escherichia coli* that suppress an **amber codon** in mRNA thereby allowing insertion of one of several alternative amino acids into a polypeptide at that site. The best known are *supD* (Ser), *supE* (Gln), and *supF* (Tyr). The mutations result in changed anticodons of the tRNA molecules, consequent on single base substitutions in their DNA coding sequences.

ambi+ or **ambo+** *comb. form* denoting on both sides, both. *Compare* **amphi+**. [From Latin *ambo*, both.]

ambidentate describing ligands, especially metal ions, that can bond through either of two or more donor atoms; for example (NCS)⁻.

ambient of or relating to the immediate environment.

ambiquitous describing an enzyme that has reversibly convertible soluble and membrane-bound forms. —**ambiquity** *n*.

ambo+ *a variant form of* **ambi+**.

ambo- *prefix* designating chiral compounds in which both of the possible molecular configurations are present at an indicated chiral centre or residue but not necessarily in equimolar proportions. Such situations may arise especially with chiral compounds obtained by partial chemical synthesis from chiral precursors in asymmetric reactions. Examples: the amino-acid sequence Phe-*ambo*-Ala-Leu, representing a mixture of Phe-Ala-Leu and Phe-D-Ala-Leu; and the trivial name 4′-*ambo*-8′-*ambo*-α-tocopherol, representing a synthetic α-tocopherol in which C-2 has the natural, (*R*)-, configuration and C-4′ and C-8′ have both (*R*)- and (*S*)-configurations though not necessarily in equimolar proportions. *Compare* **DL-**, *RAC-*.

amboceptor *(formerly) an alternative term for* hemolytic antibody. [Named by German biochemist Paul Ehrlich (1854–1915), who pictured hemolysin as an antibody having specific receptors both for erythrocytes and for complement, thereby serving to link the two together.]

AMCA *abbr. for* 7-amino-4-methylcoumarin-3-acetic acid; a **fluorophor** (def. 2) that can be coupled to proteins or other molecules through the carboxylic acid group while retaining its fluorescence property of emitting light in the blue region of the spectrum.

ameba *an alternative US spelling of* **amoeba**.

amelogenin the predominant protein in dental enamel during early development.

amicyanin a bacterial **electron transfer protein** containing a **type 1 copper** site.

amidorphin an opioid peptide corresponding to residues 104–129 of bovine proenkephalin A, but which is amidated at its C terminus.

Ames test *see* **Salmonella mutagenesis test**.

amethopterin *another name for* **methotrexate**.

amidase 1. any enzyme that catalyses the hydrolysis of an ester. **2.** EC 3.5.1.4; acylamide amidohydrolase; an enzyme that catalyses the hydrolysis of a monocarboxylic acid amide to monocarboxylate and NH₃.

ω-amidase EC 3.5.1.3; ω-aminodicarboxylate amidohydrolase; an enzyme that catalyses the hydrolysis of an ω-amidodicarboxylic acid to dicarboxylate and NH₃.

amidate to convert any oxy acid into its corresponding amide. —**amidated** *adj.*; **amidation** *n*.

amide any compound containing one, two, or three acyl groups attached to a nitrogen atom. An amide may be derived formally or actually by condensation of an oxy acid with ammonia or a primary or secondary amine. Amides derived from carbon acids may be termed **carboxamides**, those from sulfonic acids **sulfonamides**, etc. Examples include: acetamide, methanamide, CH_3–CO–NH_2; sulfanilamide, 4-aminobenzenesulfonamide, H_2N–C_6H_4–SO_2–NH_2; and diethylformamide, formdiethylamide, H–CO–N(C_2H_5)₂. The term includes **imide** and **peptide**.

amidination the process or reaction of introducing an amidino group into a chemical entity.

amidine any compound of the type R–C(=NH)–NH_2.

amidino the monovalent group –C(=NH)–NH_2.

amidino+ *comb. form* denoting the presence of an **amidino** group.

N-amidinoglycine *see* **glycocyamine**.

amido+ *comb. form* denoting the presence of an amide group.

Amido Black *see* **Naphthalene Black**.

amidolysis any cleavage of an amide (including a peptide) to the parent oxy acid and ammonia or an amine.

amidophosphoribosyltransferase EC 2.4.2.14; *systematic name*: 5-phosphoribosylamine:pyrophosphate phospho-α-D-ribosyltransferase (glutamate-amidating); *other names*: glutamine phosphoribosylpyrophosphate amidotransferase; phosphoribosyldiphosphate 5-amidotransferase. An enzyme that catalyses the formation of 5-phospho-β-D-ribosylamine from L-glutamine, 5-phospho-α-D-ribose 1-diphosphate, and H_2O with release of L-glutamate and pyrophosphate. This is an early reaction in purine biosynthesis.

amidorphin an opioid peptide corresponding to residues 104–129 of bovine **proenkephalin** A, but which is amidated at its C terminus.

amidotransferase *see* **amidophosphoribosyltransferase**, **glutamine amidotransferase**.

amiloride N-amidino-3,5-diamino-6-chloropyrazinecarboxamide; a potent and specific inhibitor of Na⁺ entry into cells. It inhibits Na⁺,K⁺-ATPase at low concentrations. It is used as a potassium-sparing diuretic.

amiloride-binding protein *see* **amine oxidase (copper-containing)**.

amine any organic compound that is weakly basic in character and contains an **amino** or a substituted amino group. Amines are called primary, secondary, or tertiary according to whether one, two, or three carbon atoms are attached to the nitrogen atom. *See also* **quaternary ammonium compound**.

amine oxidase (copper-containing) EC 1.4.3.6; *other names*: diamine oxidase; diamino oxhydrase; histaminase; an enzyme that catalyses the oxidation by dioxygen of an amino group to an aldehyde, with release of ammonium and hydrogen peroxide. 6-Hydroxydopa and copper are cofactors. Example, human amiloride-sensitive amine oxidase (copper-containing) precursor (amiloride-binding protein; *abbr.*: ABP), an 85.27 kDa enzyme that binds amiloride.

amine oxidase (flavin-containing) *see* **monoamine oxidase**.

aminergic describing a receptor that is activated by a specific (biogenic) amine. The term is usually applied to nerves that act by releasing a monoamine (norepinephrine, dopamine, or 5-hydroxytryptamine) from their nerve endings. *Compare* **adrenergic**, **cholinergic**, **dopaminergic**, **peptidergic**, **purinergic**, **serotonergic**.

amino the chemical group –NH_2 in an organic molecule. It is basic in character and formally derived by the removal of a hydrogen atom from ammonia.

amino+ *comb. form* denoting the presence of an amino group.

amino acid any organic acid containing one or more amino substituents. The term is usually restricted to amino, especially α-amino, derivatives of aliphatic carboxylic acids, but it can also include β-amino derivatives.

amino-acid accepting RNA *see* **transfer RNA**.

amino-acid N-acetyltransferase EC 2.3.1.1; *other name*: N-acetylglutamate synthase (*abbr.*: AGS); an enzyme that catalyses the formation of N-acetyl-L-glutamate (*abbr.*: AGA) from acetyl-CoA and L-glutamate with the release of CoA. It is responsible for the formation of AGA, an obligate activator of **carbamoyl-phosphate synthase** (def. 1), the first step in the **ornithine–urea cycle**. The mammalian enzyme is present in mitochondrial matrix in liver, intestine, kidney, and spleen.

amino-acid analyser any instrument or apparatus for the automated analysis of mixtures of amino acids. The amino acids are separated individually and quantified spectrophotometrically.

amino-acid arm the base-paired segment of **transfer RNA** containing both the 3′ and the 5′ ends of the RNA molecule, on which the specific amino acid is attached to the 3′-hydroxyl at the 3′ end.

amino acid exchange matrix *an alternative name for* **substitution matrix**.

amino acid index a numerical scale that quantifies particular amino acid properties (e.g. size, polarity, weight, hydrophobicity). For example, the **aliphatic index** gives a measure of the relative volume of a protein occupied by aliphatic side chains. Amino acid indices are often used to predict the properties of the proteins in which they occur.

amino-acid oxidase any enzyme that catalyses the oxidation by dioxygen of the amino group of an α-amino acid to an oxo acid with release of ammonium and hydrogen peroxide. L-Amino-acid oxidase, EC 1.4.3.2, is specific for L enantiomers, and D-amino-acid oxidase, EC 1.4.3.3, is specific for D enantiomers. Both enzymes have FAD as coenzyme.

amino-acid residue that part of any amino-acid molecule that is present when the amino acid is combined in a (poly)peptide, i.e. the amino acid less the atoms (a hydrogen atom and/or a hydroxyl group) that have been removed during the formation of a peptide bond or bonds.

amino-acid sequence *or* **primary structure** the sequential order of amino-acid residues in a polypeptide or protein. It is usually given from the end carrying an α-amino group not in peptide linkage, i.e. from the N-terminal end.

α-amino-acid site *a less correct term for* **aminoacyl site**.

amino acid transporter *abbr.*: AAT; any of numerous membrane proteins involved in amino acid transport into cells or into membranous organelles such as mitochondria, chloroplasts, and vesicles. The human genome contains 46 genes for AATs, that of *Arabidopsis* has 53, and that of yeast 24. AATs fall into five superfamilies: amino acid–polyamine–choline (in bacteria, yeasts, plants, and animals); sodium-dicarboxylate symporter (only in animals); neurotransmitter (only in animals), amino acid transfer; and AATs within the major facilitator superfamily. These proteins are presumed to contain 10, 12, or 14 transmembrane segments. Mitochondrial AATs are presumed to contain six transmembrane segments.

amino acid–tRNA ligase *other name*: aminoacyl-tRNA synthetase. Any of the sub-subclass EC 6.1.1 of enzymes that catalyse the formation of aminoacyl-tRNA from ATP, amino acid, and tRNA with release of pyrophosphate and AMP. Each amino acid has its cognate amino acid–tRNA ligase and tRNA acceptor species; e.g. in the case of glutamate–tRNA ligase (EC 6.1.1.17), the tRNA is tRNA^{Glu} and the product is Glu-tRNA^{Glu}. Many contain zinc or putative zinc-binding motifs. Aminoacyl adenylate is normally an intermediate and the initial aminoacyl-tRNA product usually involves 2′-substitution with subsequent rearrangement. Specificity is not perfect but the enzymes have one or more mechanisms for correcting mischarged aminoacyl-tRNAs, which occur with structurally similar amino acids (e.g. Ile-tRNA^{Val}). Structurally the enzymes of *E. coli* belong to two classes. Class I are mostly monomeric, aminoacylate the tRNA on the 2′ –OH of the terminal ribose of the acceptor stem, require anticodon recognition for catalysis, and share two homologous sequences that are components of a nucleotide-binding fold. Class II are mostly dimeric or tetrameric, aminoacylate the 3′ –OH of the terminal ribose, do not require recognition of the anticodon for catalysis, and lack the homologous sequences of class I but share three sequences in the core of their catalytic domains.

amino aciduria a condition in which, on a normal dietary intake, amino acids appear in the urine. It may be restricted to one or a few amino acids, or be more generalized, and result from increased amounts in plasma (usually as a feature of an inherited disease of amino-acid metabolism) or from acquired damage or a hereditary reabsorption defect in the proximal renal tubules. *See* **Hartnup disease, maple syrup urine disease**.

aminoacyl the acyl group, R–CH(NH$_2$)–CO–, formed by removal of the hydroxyl group from the α-carboxyl group of an (unknown or unspecified) α-amino acid.

aminoacylase 1 EC 3.5.1.14; an *N*-acyl amino-acid aminohydrolase; an enzyme that catalyses any reaction of the general type:

N-acyl-L-amino acid (except *N*-acetylaspartate) + H$_2$O → fatty acid (anion) + amino acid.

2 EC 3.5.1.15; *recommended name*: aspartoacylase; an enzyme that hydrolyses *N*-acetylaspartate to acetate and aspartate. Mutations in a locus at 17p13-pter, which decrease the enzyme's catalytic activity, are associated with Canavan disease.

aminoacyl site *or* **A site** *or* (*correctly*) **aminoacyl-tRNA site** the site on a **ribosome** to which the incoming aminoacyl-tRNA is bound during protein synthesis. In bacteria this involves **ribosomal proteins** L1, L5, L7/L12, L20, L30, L33, both 16S and 23S rRNA (L7 and L12 differ in acetylation only).

aminoacyltransferase any enzyme of sub-subclass EC 2.3.2 that transfers aminoacyl groups, forming either an ester or an amide.

aminoacyl-tRNA the form in which amino acids are transported to the ribosomes during protein synthesis. The amino acid is esterified through its carboxyl group to the 3′-hydroxyl group of the 3′-terminal adenosine of a molecule of **transfer RNA**.

aminoacyl-tRNA synthetase *an alternative name for* **amino acid–tRNA ligase**.

α-aminoadipate 2-aminohexanedioate; an intermediate in the synthesis of lysine (*see* **aminoadipic pathway**). It is a competitive inhibitor of **glutamate receptors**.

aminoadipic pathway a metabolic pathway for the biosynthesis of lysine that occurs in some algae, *Euglena*, and fungi. It involves a two-carbon extension of 2-oxoglutarate to 2-oxoadipate (acetyl-CoA addition to form homocitrate, thence *cis*-homoaconitate, homoisocitrate, and oxidation to 2-oxoadipate), transamination to 2-aminoadipate, reaction of the latter with ATP to yield 5-adenylyl-2-aminoadipate, and further reductive reactions to 2-aminoadipic-5-semialdehyde, saccharopine, and finally L-lysine. Aminoadipate is also involved in lysine catabolism in mammals.

α-aminoadipic semialdehyde synthase a homotetrameric enzyme that takes part in the degradation of lysine. It catalyses the reactions:

(1) L-lysine + 2-ketoglutarate + NADPH = saccharopine + NADP$^+$;

(2) saccharopine = NAD$^+$ + H$_2$O = α-aminoadipic semialdehyde + glutamate + NADH.

In mammals it is bifunctional (human, 927 amino acids), but in bacteria and fungi it is represented by the separate enzymes lysine-ketoglutarate reductase (reaction 1) and saccharopine dehydrogenase (reaction 2). Deficiency of both activities leads to familial hyperlysinemia.

amino alcohol any aliphatic organic molecule or moiety containing both an amino and a hydroxyl substituent.

p-aminobenzoic acid *abbr.*: PABA; 4-aminobenzoic acid; a compound present in yeast as a factor in the B complex of vitamins, and a structural component of folic acid and related compounds. It is a substrate for the bacterial enzyme dihydropteroate synthase, EC 2.5.1.15, competitive inhibitors of which (e.g. **sulfonamides**) are effective antibacterial agents. It is used in sunscreen preparations. *De novo* synthesis is through the intermediate chorismate formed by the **shikimate pathway**. It is formed in a test of pancreatic function, the **PABA test**.

γ-aminobutyrate receptor *see* **GABA receptor**.

γ-aminobutyrate shunt *or* **GABA shunt** a variation of the **tricarboxylic-acid cycle** in which α-oxoglutarate is converted to L-glutamate either by transamination or by reductive amination. The glutamate is decarboxylated to form γ-aminobutyrate, which can re-enter the tricarboxylic acid cycle after being deaminated and oxidized to succinate. This pathway occurs in brain tissue, presumably to form and catabolize γ-aminobutyric acid, and is also prominent in green plants.

4-aminobutyrate transaminase *abbr.*: GABA-T; EC 2.6.1.19; *other names*: γ-amino-*N*-butyrate transaminase; GABA transaminase; β-alanine–oxoglutarate aminotransferase; an aminotransferase enzyme important in the catabolism of γ-amino-*n*-butyric acid (*abbr.*: GABA). It catalyses the reaction:

4-aminobutanoate (GABA) + 2-oxoglutarate = succinate semialdehyde + L-glutamate.

Pyridoxal phosphate is the coenzyme. The human enzyme is a homodimer present in the mitochondrial matrix. Deficiency of the enzyme results from a rare missense mutation and is accompanied by seizures and psychomotor retardation and elevated levels of GABA, β-alanine, and homocarnosine in plasma and cerebrospinal fluid.

α-amino-*n*-butyric acid 2-aminobutanoic acid; a component of **ophthalmic acid**.

γ-amino-*n*-butyric acid *abbr.*: GABA; 4-aminobutanoic acid, $H_2N–CH_2–CH_2–CH_2–COOH$; an amino acid not found in proteins, but occurring principally in the central nervous system, where it is an important inhibitory **neurotransmitter**. It is also found in some plants. *See also* **GABA receptor**.

ε-aminocaproic acid *symbol*: εAcp; *(formerly)* an alternative name *for* **6-aminohexanoic acid**.

7-aminocephalosporanic acid a compound obtained by the acid hydrolysis of **cephalosporin** C and used in the manufacture of semisynthetic cephalosporins.

4-amino-4′-chlorodiphenyl *abbr.*: pCPA; an irreversible inhibitor of **tryptophan 5-hydroxylase**, an enzyme in the pathway for synthesis of 5-hydroxytryptamine.

1-aminocyclopropane-1-carboxylic acid *abbr.*: ACC; a cyclic α-amino acid that is the immediate precursor of the plant hormone ethylene. ACC is produced by ACC synthase in the reaction:

S-adenosylmethionine = ACC + 5′-methylthioadenosine.

ACC produces ethylene in the reaction catalysed by ACC oxidase:

$ACC + \frac{1}{2}O_2$ = ethylene + CO_2 + HCN + H_2O.

ACC is also a potent and selective ligand for the glycine modulatory site of the NMDA receptor (*see* **glutamate receptor**).

aminodeoxysugar *or* **amino sugar** *or* **amino-saccharide** *or* **aminomonosaccharide** any monosaccharide or monosaccharide derivative in which an alcoholic hydroxyl group has been replaced by an amino group.

aminoethoxyvinylglycine *abbr.*: AVG; an inhibitor of ethylene formation in higher plants. It is produced by *Streptomyces* spp. It acts at the level of 1-aminocyclopropane-1-carboxylate synthase (*see* **1-aminocyclopropane-1-carboxylic acid**).

aminoethyl *symbol*: Aet; the $–CH_2–CH_2–NH_2$ group.

aminoethylbenzenesulfonyl fluoride *abbr.*: AEBSF; an inhibitor of serine endopeptidases. It is both soluble and stable in aqueous solutions, and is less toxic (LD_{50} 2.8 g kg^{-1}) than **phenylmethylsulfonyl fluoride**, but with the same action and similar potency. *Compare* **organophosphate**.

N-(2-aminoethyl)isoquinoline-5-sulfonamide *or* **H9** a selective

inhibitor of casein kinase I, protein kinase A, and, less potently, protein kinase C.

aminoglycoside antibiotic any member of a group of broad-spectrum antibiotics, of similar toxicity and pharmacology, that contain an aminodeoxysugar, an amino- or guanidino-substituted inositol ring, and one or more residues of other sugars. The group includes **streptomycin**, **neomycin**, framycetin, **kanamycin**, paromomycin, and **gentamicin**. An isolated form of deafness is induced by aminoglycosides in persons who carry one of two point mutations in the mtDNA for 12S rRNA.

aminoglycoside phosphotransferase a bacterial enzyme whose gene is carried on the transposon sequence Tn5. Expression of the enzyme in eukaryotic cells confers resistance to the **aminoglycoside antibiotics**. It is used as a **dominant selectable marker** in cells transfected with plasmids, where the presence of the gene is indicated by *neo* or *neo*ᴿ (denoting resistance to neomycin).

6-aminohexanoic acid *symbol*: εAhx; a compound that acts as an inhibitor of **streptokinase** and **urokinase** and prevents the synthesis of **plasmin**.

***para*-aminohippuric acid** *abbr.*: PAH; *N*-(4-aminobenzoyl)-glycine; it is used in renal clearance tests. Its clearance exceeds that of inulin, indicating that it is actively secreted by renal tubular cells. It shares this transport system with penicillin and competitively inhibits penicillin secretion. PAH clearance is taken as a measure of effective renal plasma flow and the inulin:PAH clearance ratio as the filtration fraction, normally about 18%.

α-amino 3-hydroxy-5-methylisoazolepropionic acid receptor *see* **AMPA repector**.

aminoimidazole carboxamide ribonucleotide *see* **AICAR**.

aminoimidazole ribonucleotide *see* **AIR**.

amino–imino tautomerism a form of **tautomerism** in which the amino and the imino forms of a compound are in equilibrium, the change corresponding to the migration of a hydrogen atom. *Compare* **keto–enol tautomerism**.

α-aminoisobutyric acid α-methylalanine, 2-amino-2-methyl-propanoic acid, $(CH_3)_2C(NH_2)COOH$; a non-metabolizable amino-acid analogue used in metabolic and other studies.

β-aminoisobutyric acid *or* **3-amino-2-methylpropanoic acid** a non-protein amino acid, the (*R*)-isomer of which is an intermediate in the degradation of thymine, and the (*S*)-isomer in the degradation of L-valine – both to the glycogenic succinyl-CoA. The (*R*)-isomer is formed in liver, and further utilization requires β-aminoisobutyrate–pyruvate aminotransferase (EC 2.6.1.40). When activity of this enzyme falls to below 10% of normal, the result is hyperbeta-aminoisobutyricaciduria, a benign condition that is inherited as an autosomal recessive trait. This polymorphism is found in about 10% of Caucasians and in 40–95% of Orientals.

δ-aminolevulinate *or* *(esp. Brit.)* **δ-aminolaevulinate** *abbr.*: ALA, δALA, *or* DALA; 5-amino-4-oxopentanoate; $H_2N–CH_2–CO–[CH_2]_2–COO^-$; a compound formed in animals and bacteria from glycine and succinyl-CoA (*see* **5-aminolevulinate synthase**); in plant chloroplasts it is formed from glutamate (*see* **glutamate-1-semialdehyde 2,1-aminomutase**). It is an intermediate in **tetrapyrrole** biosynthesis.

5-aminolevulinate synthase EC 2.3.1.37; *systematic name*: succinyl-CoA:glycine *C*-succinyltransferase (decarboxylating); *other names*: δ-aminolevulinate synthase; δALA synthetase. The first and regulatory enzyme in the porphyrin biosynthetic pathway. It catalyses a reaction in mitochondria between succinyl-CoA and glycine to form 5-aminolevulinate, CoA, and CO_2. There are two isozymes: housekeeping or ALAS1 (locus at 3p21.1, which encodes 640 amino acids, and is induced by certain drugs and repressed by hemin); and erythroid or ALAS2 (locus at Xp11.21, which encodes 587 amino acids, and is neither induced nor repressed). Both contain bound pyridoxal phosphate. Deficiency of ALAS2 is associated with over 20 inactivating mutations in congenital sideroblastic anemia.

δ-aminolevulinic acid dehydratase *an alternative name for* **porphobilinogen synthase**.

aminolysis 1 any chemical reaction in which an amine reacts analogously to ammonia in **ammonolysis**. **2** any hydrolytic deamination reaction in which an amino group is replaced by a hydroxyl group.

2-amino-4-methylhex-4-enoic acid a non-protein amino acid found in plants, e.g. *Aesculus californica* (California buckeye).

amino-monosaccharide *an alternative name for* **aminodeoxysugar**.

6-aminopenicillanic acid *abbr.*: 6-APA; a compound formally derived by the removal of the acyl side chain from **penicillin**. It is obtained industrially from fermentations in the absence of side-chain precursors but more usually by removal of the side chain using an amidase from *Escherichia coli* as an immobilized support. It is a very important intermediate for the manufacture of semisynthetic penicillins.

aminopeptidase any enzyme of sub-subclass EC 3.4.11, α-aminoacylpeptide hydrolases, that hydrolyse N-terminal amino-acid residues from oligopeptides or polypeptides.

4-aminophthalhydrazide *see* **luminol**.

aminopterin 4-aminofolic acid; a potent inhibitor of dihydrofolate reductase ($K_i < 10^{-9}$ M). It is used as an antineoplastic agent in the treatment of acute leukemia and choriocarcinoma. Aminopterin blocks nucleotide biosynthesis and thus also DNA biosynthesis, and has a similar action to **amethopterin**. It is a component of **HAT**, a selective medium used in tissue culture.

amino-saccharide *an alternative name for* **aminodeoxysugar**.

amino-terminal determination the analysis of a peptide or a protein to determine the nature of the amino-acid residue at the N terminus. The free terminal α-amino group is substituted with a reagent such as 2,4-dinitrofluorobenzene (Sanger's reagent), dansyl chloride, or phenylisothiocyanate (Edman's reagent). The substituted terminal residue is then removed by hydrolysis and identified by, e.g., chromatography.

amino terminus *or* **amino terminal** *an alternative name for* **N terminus**. *Compare* **carboxyl terminus**.

aminotransferase *or* **transaminase** any enzyme, of the large sub-subclass EC 2.6.1, that catalyses the reversible transfer of a nitroge-

nous group, usually an amino group, to an acceptor, generally a 2-oxo acid. Pyridoxal phosphate is a coenzyme, the intermediate pyridoxamine phosphate being formed in the reaction. *See also* **transamination**.

amitosis division of a cell nucleus by a process other than **mitosis**.

AML1 *symbol for* acute myeloid leukemia 1 gene; *also called* Runt-related transcription factor 1 (*RUNX1*); a gene, at 21q22.3, that encodes the α subunit of transcription factor PEBP2. The protein contains a DNA-binding domain that shows 70% homology with that of Runt and Lozenge (both of *Drosophila*). Chromosomal translocations t(8;21) that involve the *AML*1 gene result in oncogenic hybrid fusion proteins that are frequent in acute myeloid leukemia of children and young adults.

ammonification of nitrate *see* **nitrate reduction**.

ammonium 1 the NH_4^+ ion, formed by hydronation of a molecule of ammonia. It is formed from N_2 by root-nodule bacteria (*Rhizobium* spp.) in leguminous plants, and is an excretory product in **ammonotelic** animals. In higher animals it is toxic in elevated concentrations, and for excretion is converted to uric acid in **uricotelic** animals, and to urea in **ureotelic** animals. **2** the chemical group $–NH_3^+$.

ammonium sulfate ($(NH_4)_2SO_4$; a salt much used in the purification of proteins, which vary in their solubility in ammonium sulfate solutions. An early step in the purification of a protein is thus often an ammonium sulfate precipitation.

ammonolysis any chemical reaction in which a covalent bond is cleaved by the action of ammonia, one or more of the hydrogen atoms of the ammonia being replaced by other atoms or groups; e.g.

$$2NH_3 + CH_3CO–O–COCH_3 \rightarrow$$
$$CH_3CONH_2 + CH_3COO^-NH_4^+.$$

ammonotelic describing animal groups (e.g. crustaceans, marine mussels, octopods) in which ammonium is the principal end-product of the degradation of nitrogen-containing compounds.

amniocentesis the process of withdrawing a sample of amniotic fluid from the **amnion** by means of a hollow needle. It is particularly useful in the diagnosis of the status of a fetus with respect to proteins or DNA but, in the human, sampling can only be carried out during the second trimester of pregnancy.

amnion a membranous sac, filled with amniotic fluid, that surrounds and protects the embryo in higher animals. It is developed in reptiles, birds, and mammals but not in amphibians or fish.

amoeba *or (US)* **ameba** (*pl.* **amoebae** *or* **amebae**) any of a genus (*Amoeba*) of naked rhizopod protozoans in the order Amoebida, characterized by a thin pellicle and frequent alterations in shape due to the constant formation and retraction of pseudopodia. —**amoebic** *or (US)* **amebic** *adj.*

amoebapore any of a family of pore-forming peptidic toxins with about 100 amino acid residues, permitting passage of small solutes, including water. The amoebapores are produced by *Entamoeba histolytica*. They are classified in the TC system under number 1.A.35.

amoeboid *or (US)* **ameboid 1** of, pertaining to, or reminiscent of an amoeba or amoebae. **2** describing cells, etc. that move by forming pseudopodia.

amorph a mutant allele that is inactive and fails to produce a measurable effect. It may act as a genetic block to normal biosynthesis.

amorphous 1 lacking definite shape or form. **2** noncrystalline; devoid of real or apparent crystalline form.

amount concentration *see* **concentration**.

amount of substance *or* **chemical amount** *symbol*: *n*; a fundamental physical quantity – one of the seven SI base physical quantities – proportional to the number of specified elementary entities of a substance (i.e. atoms, molecules, ions, radicals, electrons, etc., or any specified group(s) of such particles). The SI base unit of the amount of substance is the **mole**. The proportionality factor is the same for all substances, and is the reciprocal of the **Avogadro constant**, N_A. It is given by: $n_B = N_B/N_A$, where n_B and N_B are the amount of substance B and the number of entities of substance B, respectively. The terms 'amount of substance' and 'chemical amount' may often usefully be abbreviated to 'amount'.

amount-of-substance concentration *see* **concentration**.

amp *abbr. for* ampere.

AMP *abbr. for* adenosine monophosphate; i.e. adenosine phosphate.

2'AMP *abbr. for* adenosine 2'-monophosphate; i.e. adenosine 2'-phosphate.

3'AMP *abbr. for* adenosine 3'-monophosphate; i.e. adenosine 3'-phosphate.

5'AMP *abbr. for* adenosine 5'-monophosphate; i.e. adenosine 5'-phosphate.

ampakine a drug that increases the efficiency of glutamate transmission at *α*-**amino-3-hydroxy-5-methylisoxazolepropionic acid receptors** (AMPA receptors) thereby allowing weaker stimuli to activate the receptors.

AMPA receptor *abbr. for* *α*-amino-3-hydroxy-5-methylisoxazolepropionic acid receptor; *former name*: quisqualate receptor. An excitatory amino acid receptor coupled directly to a transmembrane cation channel. It is expressed ubiquitously in neurons and glial cells and mediates the majority of fast excitatory synaptic transmission in the CNS. The receptor also has an excitatory glycine-binding site. It is a type of **glutamate receptor**.

AMP-CP *abbr. for* 5'-adenylyl methylenephosphonate; i.e. adenosine 5'-[*α*,*β*-methylene]diphosphate.

AMP-CPP *abbr. for* 5'-adenylyl methylenediphosphate; i.e. adenosine 5'-[*α*,*β*-methylene]triphosphate.

AMP deaminase *an alternative name for* **adenylate deaminase**.

ampere *symbol*: A; *abbr.*: amp; the SI base unit of electric current. It is defined as that constant current, which if maintained in two straight parallel conductors of infinite length, of negligible circular cross-section, and placed 1 metre apart in a vacuum, would produce between these conductors a force equal to 2×10^{-7} newton per metre of length. [After André Marie Ampère (1775–1836), French physicist.]

amperometric titration any titration in which the end point is detected by measuring the electric current flowing when a potential is applied between two electrodes in a solution.

AMPGD *abbr. for* 3-(4-methoxyspiro)-1,2-dioxetane-3,2'-tricyclo[3.3.1.1]decan-(4-yl)phenyl-*β*-D-galactopyranoside; a chemiluminescent substrate for *β*-galactosidase (*see* **chemiluminescence**). It is used to detect the activity of the *β*-galactosidase gene when the latter is employed as a reporter gene in transgenic plants, etc.

amphetamine (*R*,*S*)1-phenyl-2-aminopropane; (±)-*α*-methylphenethylamine; racemic desoxynorephedrine; a drug that has potent effects by stimulating the release of monoamines, inhibiting **monoamine oxidase**, and blocking neuronal reuptake of monoamines from the synapse. It is noted for its marked CNS stimulant actions. *See also* **dextroamphetamine**.

amphi+ *or (before a vowel)* **amph+** *prefix* **1** denoting on both sides of, of both kinds. **2** denoting around.

Amphibia a class of anamniote vertebrate animals in the superclass Tetrapoda. They are characterized by having gills during the larval stage; these are typically replaced by lungs in the adults. The skin is moist, glandular, and scaleless.

amphibian 1 a member of the class Amphibia. **2** adapted for life both on land and in water; amphibious.

amphibolic describing an enzyme or metabolic pathway that is concerned in both **anabolism** and **catabolism**. —**amphibolism** *n*.

amphile *an alternative name for* **amphiphile**.

amphion *or* **amphoion** any molecule that contains ionic groups of opposite charge, not necessarily in equal numbers; an ionized **ampholyte**. *Compare* **dipolar ion**. —**amphionic** *or* **amphoionic** *adj*.

amphipath *an alternative name for* **amphiphile**.

amphipathic *or* **amphiphilic** describing a molecule, such as a glycolipid or phospholipid, that contains both hydrophobic (i.e. lipophilic, nonpolar) and hydrophilic (i.e. aquaphilic, polar) parts; displaying **amphipathy**; being an **amphiphile**.

amphipathic helix an **alpha helix** with a specific topography consisting of opposite polar and non-polar forces, and a regular charge distribution. It is proposed as a basic structural element of the lipid-associating domains of apolipoproteins in very-low-density lipoproteins and high-density lipoproteins.

amphipathy *or* **amphiphilicity** the property (of a molecule, substance, etc.) of having affinity for both lipoidal and aqueous media; the extent or the degree to which a substance is amphipathic.

amphiphile *or* **amphile** *or* **amphipath** a substance whose molecules have an affinity for both aqueous and nonaqueous media; an **amphipathic** substance. [From Greek *amphi-*, of both kinds, + *philos*, loving.]

amphiphilic *an alternative term for* **amphipathic**.

amphiphysin an acidic protein that is localized in many synapses of avian and mammalian nervous systems, and also expressed in adrenal gland and anterior and posterior pituitary tissues. It is an autoantigen in paraneoplastic stiff-man syndrome, a generalized progressive muscular hypertonia.

amphipol a synthetic linear amphiphilic polymer that contains alternating hydrophilic and hydrophobic side chains. It solubilizes integral membrane proteins by wrapping around their hydrophobic segments.

amphiprotic describing a substance that can either gain or lose one or more protons.

amphiregulin a bifunctional transmembrane glycoprotein with a C-terminal EGF-like domain. It inhibits growth of several human carcinoma cells in culture, and stimulates proliferation of human fibroblasts and certain other tumour cells. It is induced by phorbol acetate.

amphitropism the property of proteins that allows them to interconvert between an inactive form in the cell cytosol and an active membrane-lipid-bound form, commonly in response to signals generated on the membrane. Their membrane affinity can be modulated in various ways including by phosphorylation, palmitoylation, prenylation, calcium, and nucleotides. –**amphitropic** *adj.*; **amphitropically** *adv*.

amphoion *a variant spelling of* **amphion**.

Ampholine *the proprietary name for* a series of mixtures of **carrier ampholytes**, which between them span the pH range 2.5–11. They consist of polyamino-polycarboxylic acids formed by reaction between acrylic acid and various polyethylene-polyamines. Each mixture contains a large number of ampholyte species of slightly differing isoelectric points. The mean and maximum apparent molecular masses are approximately 400 and 500 respectively. They are used in **isoelectric focusing**. They set up a pH gradient through the electrophoretic medium, and components being separated equilibrate to a point in the medium corrresponding to their isoelectric point.

ampholyte any amphoteric electrolyte; *see also* **carrier ampholyte**, **amphion**.

ampholyte-displacement chromatography a technique for the fractionation of protein mixtures by displacement chromatography on an ion exchanger using a mobile phase containing a mixture of amphoteric substances of closely spaced isoelectric points (i.e. carrier ampholytes). Separation of the components is thought to involve specific interaction between the proteins and the carrier ampholytes as well as competition between them for sites on the ion exchanger.

amphoteric describing a molecule that can react as acidic towards strong bases and as basic towards strong acids, e.g. an amino acid.

amphotericin B Fungizone; a polyene macrolide antifungal antibiotic produced by *Streptomyces* spp. It is useful against many mycotic infections, acting as a membrane-active lytic agent. At appropriate concentrations it permeabilizes cells, creating pores formed of five to ten molecules of antibiotic in association with cholesterol. These pores allow the passage of low M_r substances. Because it is active only against sterol-containing membranes, mitochondria are not affected. It interacts preferentially with ergosterol, giving it a high affinity for plant membranes; hence its value as a fungicide. *See also* **polyene antibiotic**.

ampicillin a semisynthetic antibiotic of the **penicillin** type, and having a $C_6H_5CH(NH_2)$– side chain.

ampicillin-resistant *symbol*: Ap^r; describing a cell, culture, organism, etc. that shows resistance to the lethal effects of the antibiotic **ampicillin**. This is a β-lactam antibiotic and resistance is often mediated by a class of enzymes called β-**lactamases**, which are secreted either into the periplasmic space of Gram-negative bacteria or into the medium of Gram-positive bacteria. The cloning vector pBR322 contains an ampicillin-resistance gene.

ampicillin-sensitive *symbol*: Ap^s; describing a cell, culture, organism, etc. that shows sensitivity to the antibiotic ampicillin.

AMPK *abbr. for* adenosine monophosphate-activated protein kinase; a protein serine/threonine kinase that consists of a catalytic α subunit, and regulatory β and γ subunits. Its substrates include glycogen synthase, acetyl-CoA carboxylase, and 3-hydroxy-3-methylglutaryl-CoA reductase. Mutation in a muscle isoform of the γ subunit (**PRKAG3**) in pig muscle results in a glycogen storage abnormality that is present in a high proportion of purebred Hampshire pigs.

amplicon a segment of chromosomal DNA that undergoes amplification or contains multiple copies of a gene.

amplification 1 the act, process, or result of amplifying. **2** the increase in the strength of an electromagnetic, chemical, or acoustic signal effected by an amplifier. **3** the production of additional copies of a stretch of genomic DNA, as a result of DNA either added to cells or originating in the chromosomes. These copies may result in tandem arrays in either an extrachromosomal or chromosomal location. This can result during selection of cells for resistance to certain agents; e.g. methotrexate blocks folate metabolism and may lead to amplification of dihydrofolate reductase genes. **4** the increase in the amount of a DNA with a specific sequence resulting from the **polymerase chain reaction**.

amplification refractory mutation system *abbr.*: ARMS; *a proprietary name for* a method for detection of mutations caused by single nucleotide changes. A pair of oligonucleotide primers are designed in which the 3′ end of one is complementary to the normal nucleotide at the position of the mutation and the other upstream or downstream from it to give a PCR product of a known size when normal DNA is amplified. Another pair of primers are designed in which one primer is complementary to the mutant nucleotide at its 3′ end and the other is placed upstream or downstream to give a PCR product that differs in size from the normal product when the mutant allele is amplified. The mismatch at the 3′ end of the normal primer with the mutant sequence makes it refractory to amplification, and vice versa. By electrophoretic separation of products obtained with both primer pairs it is possible to ascertain whether the individual is homozygous normal, heterozygous, or homozygous for the mutant allele.

amplifier 1 an electronic device used to increase the strength of an electromagnetic signal; especially such a device used to increase the strength of an acoustic signal. **2** a combination of a receptor and various enzymes used to increase the strength of a chemical signal, as in a **cascade sequence**.

amplify to enlarge or to increase in strength or effect, especially an electromagnetic, acoustic, or chemical signal.

amplimer the amplified product of a **polymerase chain reaction**.

amplitude 1 magnitude; greatness of extent. **2** *symbol*: *a*; the maximal displacement from zero or from a mean position of an oscillation or a curve.

AMP-PCP *abbr. for* 5′-adenylyl methylenediphosphonate; i.e. adenosine 5′-[β,γ-methylene]triphosphate.

AMPPD *abbr. for* adamantyl-1,2-dioxetane phosphate *or* 3-(4-methoxyspiro[1,2-dioxetane-3,2′-tricyclo-[3.3.1.^{3,7}]decan)-4yl)-phenylphosphate. A chemilumigenic substrate for alkaline phosphatase forming the basis of sensitive nonradioactive labelling systems used in association with Northern, Southern, and Western blotting.

AMP-PNP *abbr. for* 5′-adenylyl imidodiphosphonate; i.e. adenosine 5′-[β,γ-imido]triphosphate.

AMP-S *abbr. for* adenosine 5′-[α-thio]monophosphate; i.e. adenosine 5′-thiophosphate.

amrinone 5-amino-(3,4′-bipyridin)-6(1*H*)-one; an inhibitor of cardiac (type III) phosphodiesterase, resulting in increased cardiac contractility and vasodilatation. It is useful in the treatment of heart failure.

AMS *(in clinical biochemistry) abbr. for* α-amylase.

amu *or* **AMU** *symbol for* atomic mass unit. *See also* **unified atomic mass unit**.

amunine *another name for* **corticotropin-releasing factor**.

amygdalin (–)-D-mandelonitrile β-D-gentiobioside; a **cyanogenic** glycoside derived biogenically from phenylalanine and found principally in oil of bitter almonds, although it is present in other plants belonging to the Rosaceae.

amygdalin

amygdalose *former name for* **gentiobiose**.

amyl *former name for* the **pentyl** group.

amyl+ *see* **amylo+**.

amylase any enzyme able to hydrolyse *O*-glucosyl linkages in starch, glycogen, and related polysaccharides (*see* α-**amylase**, β-**amylase**).

α-amylase EC 3.2.1.1; *systematic name*: 1,4-α-D-glucan glucohydrolase. An enzyme that catalyses the endohydrolysis of 1,4-α-D-glucosidic linkages with the production of mainly α-maltose from amylose, but also limit dextrins from **glycogen** and **amylopectin** since it cannot catalyse hydrolysis of 1,6 linkages. It is typical of animal cells (especially pancreas), and saliva. It has five motifs. The plasma level of α-amylase is raised in acute pancreatitis, and assay of the pancreas-specific isoenzyme aids diagnosis. Normal levels in human plasma are <300 IU l^{-1} (this is very method-dependent). The α/β-amylase precursor from *Bacillus polymyxa* is cleaved proteolytically to form the α (551 amino acids) and β (419 amino acids) enzymes.

β-amylase EC 3.2.1.2; *systematic name*: 1,4-α-D-glucan maltohydrolase; *other names*: saccharogen amylase; glycogenase. An enzyme that removes successive maltose units from the nonreducing ends of carbohydrate chains; the anomeric group undergoes inversion so that the major product is β-maltose. In some organisms it shares the same biosynthetic precursor as α-amylase. β-Amylases are characterized by their formation as insoluble zymogens during the process of seed maturation, and are also found in bacteria.

amylin *or* **islet amyloid polypeptide** (*abbr.*: IAPP) a 37-residue peptide, sequence KCNTATCATQRLANFLVHSSNNFGAILSST-NVGSNTY–NH₂, isolated from pancreatic islet amyloid deposits of an insulinoma and from amyloid-rich pancreases of type 2 diabetic patients; it is found in pancreatic cells and secreted with insulin. Structurally, it closely resembles **calcitonin gene-related peptide** (CGRP), and has 43% and 46% amino-acid sequence similarity with human α- and β-CGRP respectively. Like CGRP, it is C-terminally amidated, and has an N-terminal disulfide bridge linking cysteines 2 and 7. It opposes the action of glycogen synthase in skeletal muscle, and may modulate insulin action. The peptide inhibits insulin-stimulated glucose metabolism in muscle, but not in adipocytes.

amylo+ *or (before a vowel)* **amyl+** *comb. form* denoting starch.

amylo-1,6-glucosidase EC 3.2.1.33; *other name*: dextrin 6-α-D-glucosidase; an enzyme that catalyses the endohydrolysis of 1,6-α-D-glucoside linkages at points of branching in chains of 1,4-linked α-D-glucose residues. The human enzyme also has glycogen debranching (4-α-glucanotransferase) activity.

amyloid 1 a waxy eosinophilic translucent material deposited in tissues in **amyloidosis**. Originally thought to be polysaccharide, it is now known to comprise proteinaceous fibrils with a twisted β-pleated sheet structure. It is insoluble, and exhibits a green fluorescence when stained with Congo Red and viewed through crossed Nicol prisms. Amyloid fibres are relatively inert, being resistant to solution in physiological solvents and to proteolytic enzymes. **2** pertaining to or characterized by the production of amyloid. **3** any non-nitrogenous substance resembling starch. **4** like or containing starch.

amyloid A protein *or* **serum amyloid A** (*abbr.*: SAA *or* AA) any of various plasma proteins synthesized by liver, and induced by **cytokine** stimulation. The extracellular accumulation in various tissues of amyloid proteins results in secondary amyloidosis in human and other species. These deposits are highly insoluble and resistant to proteolysis. They disrupt tissue structure and compromise function.

amyloidogenic describing a protein that contains or can form an antiparallel beta-sheet structure that aggregates to form nonbranching fibrils of **amyloid**. These fibrils stain with Congo Red and occur in extracellular spaces. Such proteins are associated with amyloidosis, and include amyloid peptide precursor, immunoglobulin light chain, transthyretin, and prion protein. –**amyloidogenically** *adv.*

amyloidosis any of various pathological states in which the deposition of **amyloid** (def. 1) occurs in tissues. It is often associated with chronic infections and immunocyte dyscrasias. In the absence of medical intervention amyloid accumulates implacably, causing pressure atrophy of the affected tissue(s). Death subsequently results from interference with the normal physiological processes of affected vital organs. The descriptive term β-fibrillosis has been suggested for amyloidosis, after the beta-pleated sheet structure of the amyloid fibrils. Amyloidosis may be systemic, with major organ systems involved, or localized and restricted to a single organ. Many cases have a hereditary basis. It is now classified in terms of the protein that accumulates. *See also* **amyloid A protein**.

β-amyloid peptide a glycoprotein associated with Alzheimer's disease, and derived from a precursor, A4, capable of several forms through alternative splicing.

β-amyloid peptide precursor *abbr.*: APP; a type 1 single-pass transmembrane protein that is largely extracellular and binds collagen, laminin, and heparan sulfate side chains of proteoglycans. Several isoforms result from alternative splicing of the pre-mRNA. APP695 predominates in brain, while APP751 and APP770 occur mainly in non-neuronal tissues. All contain the β-amyloid peptide sequence as part of the transmembrane segment. Cleavage by α-secretase (*see* **ADAM-10**) occurs within the β-amyloid peptide sequence, while cleavage by β- and γ-secretases releases this peptide. Several mutations in APP are associated with familial (autosomal dominant, early-onset) forms of Alzheimer's disease.

amyloid precursor-like protein *abbr.*: APLP; either of two proteins belonging to the amyloid peptide precursor family but lacking the amyloid peptide domain. Human APLP1 maps to chromosome 19 and APLP2 to chromosome 11.

amylolysis the process of converting starch into soluble oligo- and monosaccharides, especially by enzymic action. —**amylolytic** *adj.*

amylopectin a polydisperse highly branched **glucan** composed of chains of D-glucopyranose residues in α(1→4) glycosidic linkage. The chains are joined together by α(1→6) glycosidic linkages. A small number of α(1→3) glycosidic linkages and some 6-phosphate ester groups also may occur. The chains are of similar structure to those of glycogen but with fewer and longer branches. Those in amylopectin have been found variously to contain 24 to 30 glucose residues per nonreducing end-group. The size of the molecule varies greatly according to source and method of measurement, with average M_r values of 10^6 to 10^8 reported. Amylopectin occurs in starch, commonly as a mixture with about one-third its mass of **amylose**. However, some starches, notably the so-called waxy starch of maize (*Zea mays*), are almost pure amylopectin. Starch granules serve as the principal cellular store of glucose in most plants.

amylopectin 6-glucanohydrolase *see* **α-dextrin endo-1,6-α-glucosidase**.

amylopectinosis *see* **glycogen disease**.

amyloplast a colourless **plastid** or **leukoplast** in which starch is stored. It occurs in various plant tissues, e.g. cotyledons, endosperm, and storage organs such as potato tubers.

amylopsin *former term for* **pancreatic α-amylase**.

amylopullanase an amylolytic enzyme present in many microorganisms but especially in thermophilic bacteria and archaea. It cleaves α1,4- and α1,6-glycosidic bonds on the same substrate molecule.

amylose a form of starch comprising a long unbranched **glucan** with α(1→4) glycosidic linkages.

amylosucrase *an alternative name for* **sucrase** (def. 3).

amylo-(1,4–1,6)-transglycosylase *see* **branching enzyme**.

amyotrophic lateral sclerosis *abbr.*: ALS; a devastating progressive degeneration of lower motor neurons, usually occurring in adulthood, that leads to weakness and atrophy of the muscles of the face, limbs, and chest, and to death within ≈5 years. Some 10% of cases are familial, and about a fifth of these (ALS1) show autosomal dominant transmission and are due to numerous mutations in the gene for cytosolic superoxide dismutase at 21q22.1. A small subset (ALS2) are recessively transmitted; these are associated with mutations in a locus at 2q33-q35. However, most cases are sporadic. *See* **superoxide dismutase**.

Amytal *proprietary name for* the barbituric sedative 5-ethyl-5-isoamylbarbituric acid, 5-ethyl-5-isopentylbarbituric acid. It inhibits the **electron-transport chain** between flavoproteins and ubiquinone.

an+ *or (before a consonant)* **a+** *prefix* indicating not, without.

+an *suffix* used in forming the semisystematic name of a parent chemical structure from the trivial name of the key compound to which the parent structure is related if the parent structure contains a heteroatom and is saturated; e.g. glucosan, any internal anhydride of glucose.

An *symbol for* **anhydro+**.

anabiosis a return to life after apparent death; a state of suspended animation induced in an organism by, e.g., desiccation or freezing, and reversed by rehydration or warming.

anabolic pertaining to, involving, or exhibiting **anabolism**.

anabolic pathway any sequence of reactions involving **anabolism**; i.e. one in which energy is taken up. *Compare* **catabolic pathway**. *See also* **amphibolic**.

anabolic steroid any of various synthetic steroids that stimulate protein production *in vivo* by mimicking the action of **androgens**.

Drug design aims to minimize masculinizing effects when used in female patients.

anabolism the energy-requiring part of metabolism in which simpler substances are transformed into more complex ones, as in growth and other biosynthetic processes. *Compare* **catabolism**.

anabolite any product of an **anabolic pathway**.

anabolize *or* **anabolise** to effect **anabolism**.

anaemia *a variant spelling (esp. Brit.)* of anemia.

anaerobe any organism that can grow in an atmosphere devoid of dioxygen. A **facultative anaerobe** is also capable of growing in a dioxygen-containing atmosphere whereas an **obligate anaerobe** (or strict anaerobe) can only grow in the absence of dioxygen.

anaerobic 1 describing an organism that is able to grow in the absence of dioxygen. **2** of or produced by an **anaerobe**. **3** (of an atmosphere) devoid of dioxygen. **4** (of a process) characterized by the absence of dioxygen.

anaerobic fermentation any of several catabolic pathways used by living organisms to extract chemical energy from various organic substances in the absence of dioxygen. *Compare* **anaerobic respiration**.

anaerobic glycolysis the metabolic pathways by which glucose is degraded anaerobically to lactate or ethanol. *See* **glycolysis**.

anaerobic indicator any indicator of the attainment and maintenance of anaerobic conditions in an **anaerobic jar**. It may be based on a dye that is decolorized in anaerobic conditions or consist of a culture of a strict anaerobe in one of the dishes in the jar.

anaerobic jar *or* **McIntosh & Fildes jar** an apparatus used to hold culture dishes or tubes under anaerobic conditions. It usually consists of an upright jar with an airtight lid; the air is evacuated and replaced by dihydrogen, which combines with any remaining dioxygen, the reaction being catalysed by a palladium catalyst. [After James McIntosh (1882–1948) and (Sir) Paul (Gordon) Fildes (1882–1971), British bacteriologists who first described the apparatus in 1916.]

anaerobic respiration the metabolic processes by which organisms degrade organic compounds in the absence of dioxygen to yield energy. They make use of inorganic oxidizing agents or accumulate reduced coenzymes. *Compare* **anaerobic fermentation**.

anaerobiosis life in the absence of dioxygen or air.

anagenesis the progressive or 'upward' evolution of a species; any evolutionary process that results in biological improvement. *Compare* **cladogenesis**. —**anagenetic** *adj.*; **anagenetically** *adv.*

anaglyph a stereoscopic moving or still picture made up of two superimposed images of the same object, taken from slightly different angles and reproduced in complementary colours (usually red and bluish green) so as to produce a three-dimensional effect when viewed through spectacles having one lens of each colour.

AnalaR *proprietary name for* any analytical (chemical) reagent that conforms to the tests and purity standards in '*AnalaR*' *Standards for Laboratory Chemicals* (1977) AnalaR Standards Ltd.

analbuminemia *or (esp. Brit.)* **analbuminaemia** an autosomal recessive condition in which plasma **albumin** is missing, or present in very small amounts, and accompanied by compensatory increases in other plasma protein fractions and frequently hypercholesterolemia. It is associated with several splice junction or nonsense mutations in the gene for albumin at 4q11-q13.

analogous 1 having or showing similar or corresponding attributes. **2** *(in biology)* describing parts, organs, or tissues that have a similar function but differ in structure or evolutionary origin.

analogue *or (esp. US)* **analog 1** a chemical compound that is structurally similar to another. Such compounds can be used as probes to investigate specificity of, e.g. enzyme activity. **2** a molecule that has acquired similar characteristics to another (e.g. a similar structure or similar active site) from an unrelated ancestor via the process of convergent evolution. For example, the bakers' yeast triose phosphate isomerase and flavocytochrome b_2 C-terminal domain share a TIM-barrel architecture, but have no sequence similarities; the enzymes subtilisin and chymotrypsin are functional analogues that share a Ser–His–Asp catalytic triad with near identical spatial geometries, but have no other sequence or structural similarities.

analogue enrichment *or* **analog enrichment** enrichment of mi-

croorganisms capable of co-metabolic degradation of organic pollutants by application of biodegradable analogues of the pollutant to the microbial ecosystem. *See* **co-metabolism**.

analyser *or (US)* **analyzer 1** the **Nicol prism** of a polariscope or **polarimeter** that receives the light and indicates the degree and/or direction of rotation of its plane of polarization. *Compare* **polarizer**. **2** an instrument for routinely performing chemical analyses, often automatically and/or on a succession of samples. *See* **amino-acid analyser**, **autoanalyser**, **sequenator**.

analysis (*pl.* **analyses**) **1** the process of determining the nature of the chemical constituents of a mixture or of the chemical groups present in a compound (**qualitative analysis**), or the process of determining the amounts of the constituents of a mixture (**quantitative analysis**). **2** the results of such an analysis. **3** the branch of mathematics concerned with the concepts of algebra and calculus. —**analytic** *or* **analytical** *adj.*

analysis of covariance a statistical method for determining the extent of the variance of one variable that is due to variability in some other variable.

analysis of variance a statistical method for segregating the experimentally observed variance between two populations into portions traceable to specific sources.

analyte a component to be measured in an **analysis** (def. 1).

analytical chemistry the branch of chemistry concerned with analysing materials by chemical (or physical) methods.

analytical ultracentrifuge an **ultracentrifuge** equipped with optical systems designed to determine concentration distributions at any time during the measurement. It is used in determinations of **sedimentation coefficients** and **sedimentation equilibrium**.

anamnestic reaction any immunological response in which a second or subsequent exposure to an antigen causes a greater and more rapid reaction than that elicited by the initial exposure. It is a manifestation of immunological memory.

anandamide arachidonoyl ethanolamide; a specific endogenous agonist that serves as a neurotransmitter of the **cannabinoid receptor**. Ligand binding to the G-protein-associated receptor increases tyrosine kinase and decreases adenylate cyclase and cAMP-dependent protein kinase activities. Gap junction intercellular communication betwen striatal astrocytes is also inhibited. Anandamide is formed in neurons by phosphodiesterase cleavage of *N*-arachidonoyl-phosphatidylethanolamine, and is probably then released to the extracellular space. A series of analogues such as *N*-linoleoyl ethanolamide and *N*-palmitoyl ethanolamide are formed simultaneously from appropriate precursors, but their role remains unclear. It may also be formed by condensation of arachidonate with ethanolamine. It is inactivated by reuptake into neurons and hydrolysis by *N*-acyl-ethanolamine amidohydrolase to ethanolamine and arachidonate.

anaphase a stage in **mitosis** or **meiosis** during which the two sister chromatids of each chromosome separate and move towards opposite poles of the spindle.

anaphase promoting complex *abbr.*: APC; a protein complex that initiates the separation of sister chromatids and the onset of anaphase during cell division. It is a 20S particle that contains Cdc16, Cdc23, and Cdc 27 and acts as an E3 enzyme (ubiquitin–protein ligase) in the ubiquitination of cyclin B. *See* **cdc gene**, **cyclin**, **ubiquitin–protein ligase**.

anaphylactic shock a severe generalized form of **anaphylaxis** in which there is widespread release of histamine and other vasoactive substances causing edema, constriction of the bronchioles, heart failure, circulatory collapse, and sometimes death.

anaphylactoid reaction an acute syndrome resembling **anaphylac-**

tic shock though caused not by an immunological reaction but by injection of a variety of substances such as bee or snake venom.

anaphylatoxin any of a group of substances that mediate inflammation and are produced in serum during the fixation of **complement**, especially during the production of C3a and C5a. They cause mast cell degranulation and histamine release.

anaphylaxis extreme immunological sensitivity of the body or tissues to the reintroduction of an antigen. It is a form of **anamnestic reaction** and is accompanied by pathological changes in tissues or organs due to the release of pharmacologically active substances. — **anaphylactic** *adj.*

anaphylotoxin inactivator *see* **lysine(arginine) carboxypeptidase**.

anaplasia the reversion by a cell or tissue to a more primitive, embryonic, or undifferentiated form with loss of its characteristics and increased capacity for multiplication.

anaplerosis 1 the filling up of a deficiency; the repair or replacement of losses, especially of tissue. **2** *(in biochemistry)* the process in which an intermediate in the tricarboxylic-acid cycle is replenished through a distinct **anaplerotic metabolic pathway**. This prevents the cycle from ceasing to function because of withdrawal of the intermediate for some other metabolic purpose. [From Greek *ana*, again, back, up + *plerosin*, to fill.] —**anaplerotic** *adj.*

anaplerotic metabolic pathway any sequence of metabolic reactions that serves to maintain the operation of the central pathways of metabolism and which allows the operation and coordination of other metabolic pathways that enter and leave the central pathways.

ANC-1 *abbr. for* anchorage defective-1; a protein (8546 amino acids) of *Caenorhabditis elegans* that is involved in the positioning of nuclei and mitochondria within cells. It contains two actin-binding domains, six repeats (each of ≈903 residues) that form coiled coils, and a C-terminal nuclear envelope transmembrane domain. It is homologous with MSP-300 of *Drosophila* and the mammalian Syne proteins. Mutations in *anc-1* disrupt the positioning and shape of mitochondria.

ancestral sequence *(in molecular phylogeny)* a hypothetical biomolecular sequence reconstructed (using, say, **maximum parsimony** or **maximum likelihood** methods) from the relationships between modern sequences.

anchorage-dependent growth the form of growth shown by mammalian cells cultured from tissue explants that requires them to be attached to solid substrata in order to grow and divide. Cells vary in their requirements for attachment, with some thriving on plastic surfaces whereas others need components of the extracellular matrix such as collagens. Transformed cells such as those derived from tumours do not always exhibit anchorage-dependent growth and can be grown by suspending them in culture media.

anchorin CII *see* **annexin V**.

ancrod an enzyme, extractable from the venom of the Malayan pit viper, *Agkistrodon rhodostoma*, that produces unstable fibrin from fibrinogen by proteolysis. The resulting fibrinogen depletion results in an anticoagulant effect.

Andersen's disease *an alternative name for* type IV **glycogen disease**. [After Dorothy Hansine Andersen (1901–63), US physician.]

androgamone *see* **gamone**.

androgen any substance, natural or synthetic, that is able to stimulate the development of male sexual characteristics. Naturally occurring androgens are C_{19}-steroid hormones. They are produced especially by the testis (e.g. testosterone) and also by the adrenal cortex, ovary, and placenta. *Compare* **estrogen**. —**androgenic** *adj.*

androgen receptor any of the androgen-binding nuclear proteins that mediate the effects of androgens by regulating gene expression; they are **zinc-finger** proteins that bind discrete DNA sequences upstream of transcriptional start sites when the hormone–receptor complex is formed. The steroid-binding domain is in the C-terminal region. In humans, many variants are known that are associated with abnormalities, including testicular feminization syndrome, and complete or partial androgen insensitivity leading to external genitalia varying between female and nearly normal male.

androgen resistance *see* **testicular feminization**.

androstane the C_{19} parent compound from which androgens are structurally derived.

androstenediol 5-androstene-3β,17β-diol; a metabolite of dehydroepiandrosterone, from which it is formed by the action of 17β-hydroxysteroid dehydrogenase, EC 1.1.1.51.

androstenediol

androstenedione 4-androstene-3,17-dione; a precursor of **testosterone** and **estrone**, formed in the testis and ovary from 17α-hydroxyprogesterone by the action of 17α-hydroxyprogesterone aldolase (17,20-lyase), EC 4.1.2.30, or from dehydroepiandrosterone by the action of 3β-hydroxy-Δ^5-steroid dehydrogenase, EC 1.1.1.145. It is also formed in adrenal cortex, from 17-hydroxypregnenolone by an isoenzyme of the 17,20-lyase. Androstenedione is secreted by the adrenals into the circulation and serves as a precursor for sex-hormone formation in peripheral tissues.

androst-16-en-3α-ol 3α-hydroxy-5α-androst-16-ene; 'boar taint'; a major constituent of boar pheromone. It is also present in truffles, which may account for the ability of pigs to locate these underground delicacies.

androsterone 5α-androstan-3α-ol-17-one; a metabolite of **testosterone**, from which it is formed in the liver via androstenedione. The latter is converted to androstanedione by 3-oxo-5α-steroid 4-dehydrogenase, EC 1.3.99.5. Androstanedione in turn is acted on by 3α-hydroxysteroid dehydrogenase, EC 1.1.1.213, forming androsterone.

+ane *suffix (in systematic chemical nomenclature)* denoting **1** an alkane, e.g. pentane. **2** a saturated heterocyclic compound in which the heteroatom is other than nitrogen, e.g. thiolane. **3a** a hydride of boron, silicon, germanium, tin or lead, e.g. borane, silane; **b** a hydride of an unbranched chain of identical atoms of certain elements.

anemia or (esp. Brit.) **anaemia** any condition in which the blood has an abnormally low number of red cells (i.e. erythrocytes) or hemoglobin content, or is deficient in total volume. There are many causes, including hypersensitivity reactions to (e.g.) drugs (**hemolytic anemia**) and imbalance of Fe^{2+} metabolism (**iron deficiency anemia**). **Pernicious anemia** is characterized by defective production of erythrocytes and the presence of megaloblasts in the bone marrow; sometimes it is accompanied by neurological changes. It is caused by deficiency of vitamin B_{12} due either to a dietary deficiency of the vitamin or to a failure to produce **intrinsic factor** required for its absorption from the gut.

anergy (in immunology) any deficiency in the immune response to an agent that normally induces a response.

aneuploid describing a nucleus or organism in which the chromosome number is not an exact multiple of the haploid number (n). In humans (n = 23) examples include the presence of an extra copy of a single chromosome, e.g. trisomy 21 (Down's syndrome), or the absence of a single chromosome, e.g monosomy (as found in Turner's syndrome, 45,X). Compare **euploid**, **monosomic**, **nullisomic**, **trisomic**.

aneurin or **aneurine** a former name for **thiamine**.

ANGEL abbr. for activator nongenotropic estrogen-like; a synthetic estrogen-like compound (e.g. Estren, 4-estren-3α,17β-diol) that produces effects on certain parts of the body (e.g. osteoblasts) without affecting reproductive organs.

angiogenesis the formation of blood vessels, whether during embryogenesis, tissue repair, or invasive growth of tumours.

angiogenin a **cytokine**, produced by fibroblasts, lymphocytes, and colon epithelial cells, that induces neovascularization, and can serve as a substratum for endothelial and fibroblast cell adhesion. A monomeric protein, it is a member of the pancreatic ribonuclease family, and specifically hydrolyses tRNAs, thus inhibiting protein synthesis.

angioneurotic edema or (esp. Brit.) **angioneurotic oedema** an alternative name for **hereditary angioedema**.

angiopoietin either of two secreted glycoproteins that have considerable sequence homology, contain a coiled coil and a fibrinogen-like domain, and are major regulators of angiogenesis. Ang1 contains 498 amino acids, and Ang2 has 496 amino acids. Both bind to specific endothelial cell-specific receptor tyrosine kinases. Ang1 is widely expressed in embryonal and adult tissues, whereas Ang2 is widely expressed in the embryo but only in ovary, uterus, and placenta in adults.

angiosperm a flowering plant; any member of the class Angiospermae, division Spermatophyta (seed plants). Compare **gymnosperm**.

angiostatin an internal fragment of **plasminogen**, released by the action of elastase, that contains the first four kringle domains of plasminogen. It is a potent inhibitor of angiogenesis and of the growth of certain tumours.

angiotensin any of three related peptides, two of which regulate adrenal activity and also increase vasoconstriction and raise blood pressure. **Angiotensin I**, the decapeptide H–Asp-Arg-Val-Tyr-Ile-His-Pro-Phe-His-Leu–OH, is cleaved off **angiotensinogen** by the kidney enzyme **renin**; this peptide has no physiologically important action. **Angiotensin II**, the octapeptide formed from angiotensin I by removal of -His-Leu–OH by the action of **angiotensin converting enzyme**, is the most powerful pressor agent known, and the substance responsible for essential hypertension. It is also one of the most potent known stimulators of aldosterone production by the adrenal gland. It is rapidly destroyed by tissue peptidases, including **angiotensinase**. **Angiotensin III**, the heptapeptide formed by removal of the N-terminal Asp residue from angiotensin II (and the only one of its breakdown products to have activity), appears to be as active on the adrenal gland as angiotensin II, and may also have significant vasoconstrictor activity.

angiotensinase an enzyme with proteinase activity that inactivates angiotensin II. Angiotensinase A is an alternative name for **glutamyl aminopeptidase**, EC 3.4.11.7, and angiotensinase C is another name for **lysosomal Pro-X carboxypeptidase**, EC 3.4.16.2.

angiotensin converting enzyme abbr.: ACE; EC 3.4.15.1; recommended name: peptidyl dipeptidase A; other names: serum converting enzyme; peptidase P; peptidyl carboxypeptidase I; carboxycathepsin; kininase II. A Cl^--dependent zinc-containing dipeptidase present in lung, the brush borders of the proximal renal tubule, the endothelium of vascular beds, and plasma. It catalyses the release of a C-terminal dipeptide, -Xaa-|-Xbb-Xcc, when neither Xaa nor Xbb is proline, hence its conversion of angiotensin I to angiotensin II. It also acts on bradykinin. It is a transmembrane glycoprotein; two types (testis and somatic) arise by alternative splicing and different transcriptional start points. The presence (I) or absence (D) of a 287bp Alu repeat in intron 16 in a locus at 17q23 gives three genotypes: I/I, I/D, and D/D. The D allele is associated with increased plasma ACE levels and statistically is associated with hypertension in males, and hence has a bearing on the pathophysiology of cardiovascular disease.

angiotensin converting enzyme inhibitors compounds that prevent the synthesis of angiotensin II from its precursor, angiotensin I, e.g. **captopril**, **enalapril**. Such inhibitors are used in the treatment of cardiac failure and hypertension.

angiotensinogen a plasma α_2-globulin formed in liver, and the precursor of **angiotensin** I and III; it is similar to the **serpins**.

angiotensinogenase see **renin**.

angiotensin receptor any of several membrane proteins that bind angiotensin, and mediate its intracellular effects. Two types of receptor, AT_1 and AT_2, have been identified on the basis of their response to drugs. The AT_1 receptor (previously known as AII_1 or AII_α) has high affinity for **losartan**, and lower affinity for CGP42112A; the reverse is true for AT_2 receptors. The potency of endogenous ligands is angiotensin II > angiotensin III. In peripheral tissues, AT_1 receptors, but not AT_2 receptors, interact with G proteins, inhibit adenylate cyclase, and activate phospholipases C and A_2. AT_2 receptors may stimulate a phosphotyrosine phosphatase leading to inhibition of guanylate cyclase. Sulfhydryl reducing agents inhibit the binding of angiotensin II to AT_1 receptors, but enhance its binding to AT. Both types are G protein-coupled **7TM proteins**.

angle **1** the inclination between two lines or planes: **plane angle**. It is dimensionless and given in radians (symbol: rad), degrees (symbol: °; 1° = π/180 rad); minutes (symbol: ′) 1′ = π/10 800 rad; or seconds (symbol: ″); 1″ = π/64 800 rad. **2** the (three-dimensional) vertex of a cone: **solid angle**. It is dimensionless but may be expressed in **steradians** (symbol: sr).

angle head or **angle rotor** any rotor for a centrifuge in which the tubes containing the liquid are at a fixed angle to the axis of rotation. Compare **swinging bucket rotor**.

ångström or **Ångstrom** symbol: Å; a unit of length equal to 10^{-10} metres. It is used in atomic measurements and for wavelengths of electromagnetic radiation, and is temporarily approved for use with SI units. [After Anders Jonas Ångström (1814–74), Swedish physicist noted for his work on spectroscopy.]

angular velocity symbol: ω; the rate of rotation of an object expressed as its rate of motion through an angle about an axis. It is expressed in rad s^{-1}.

anhydrase see **carbonic anhydrase**.

anhydride a derivative of a substance that yields the substance when combined chemically with water.

anhydro+ symbol: An; prefix (in a chemical name) denoting the loss of the elements of water from within one molecule or residue. The positions from which removal has occurred may be indicated by a pair of locants, e.g. 2,5-anhydro-D-gluconic acid.

anhydrosialidase EC 3.2.1.138; other names: anhydroneuraminidase; sialidase L; an enzyme that catalyses the hydrolysis of α-sialosyl linkages in N-acetylneuraminic acid glycosides, releasing 2,7-anhydro-α-N-acetylneuraminic acid.

anhydrosugar any intramolecular ether formed by the elimination of water from two (indicated) alcoholic hydroxyl groups of a single molecule of a monosaccharide, whether aldose or ketose; i.e. an anhydromonosaccharide whose formation does not involve the reducing group.

anhydrous devoid of water, especially (of a substance or material) lacking water of crystallization or adsorbed water, or (of an environment) very dry. See also **dehydration**.

animal pole *see* **vegetal pole**.

animal protein factor a name formerly used for a growth factor for livestock, found to be present in meat and fish but not in feeds of plant origin. It was later shown to be identical with anti-pernicious anemia factor, i.e. vitamin B_{12}.

anion an ion with a net negative charge; during electrolysis of a solution anions migrate towards the **anode**. *Compare* **cation**. —**anionic** *adj.*

anion channel protein or **band 3 protein** or **anion exchanger** or **Cl⁻/HCO₃⁻ exchanger** either of two integral plasma membrane glycoproteins, both 911 residues long and showing ≈70% sequence homology. Their N-terminal 403 residues are cytoplasmic, the rest form 12–14 putative transmembrane segments. AE1 (locus at 17q21-q22), the erythroid form, is predominant in erythrocyte membranes, where it mediates exchange of Cl⁻ for HCO₃⁻ and is the major site for anchoring of the submembranous cytoskeleton. Mutations at this locus are associated with hereditary spherocytosis, elliptocytosis, ovalocytosis, or acanthocytosis, or with distal renal tubular acidosis. An isoform of AE1 that lacks the N-terminal 66 residues occurs in the basolateral membrane of cells of renal collecting ducts. AE2 (locus on chromosome 7) is the non-erythroid form. *See* **band protein**.

anion exchanger 1 or **anion-exchange resin** any **ion-exchange resin** that carries positively charged groups, e.g. quaternary ammonium groups, and binds anions. **2** an alternative name for **anion channel protein**.

anion gap *(in clinical chemistry)*, the difference, in meq l⁻¹, between the concentrations of serum sodium plus potassium and of the serum chloride plus bicarbonate; i.e.

$$\text{anion gap} = [Na^+ + K^+] - [Cl^- + HCO_3^-].$$

Normal values are 12 ± 2 meq L⁻¹. Variations in the anion gap provide a useful insight into the acid–base status of a subject.

anionic detergent a detergent in which the polar part of the molecule carries a negative charge. *Compare* **nonionic detergent**.

anisotropic describing a medium in which physical properties, such as the velocity of electromagnetic radiation transmission, electrical or heat conductivities, or compressibility, have different values when measured along axes in different directions. —**anisotropically** *adv.*; **anisotropy** or **anisotropism** *n.*

anisotropic motion rotation about a single axis, especially of a molecule or a **spin label**.

anisotropy the property of molecules and materials of exhibiting variations in physical properties along different axes of the substance.

ankyrin or **band 2.1 (protein)** or **nexin** or **syndein** a 200 kDa cytoskeletal protein that attaches other cytoskeletal proteins to integral membrane proteins. It is bound tightly to the cytoplasmic surface of the human erythrocyte membrane, to which it attaches the cytoskeletal protein **spectrin**. [After the Greek *ankyra*, anchor.]

ankyrin repeat a 33-amino acid sequence that occurs 24 times in the 'membrane' part of **ankyrin** and is also present in several hundred other proteins. It forms a beta hairpin-alpha helix-loop-alpha helix structure.

ANN *abbr. for* artificial neural network.

annealing 1 the renaturation of heat-denatured nucleic acids or proteins by slow cooling. **2** the formation of hybrid double-stranded nucleic acid molecules by the slow cooling of a mixture of denatured, single-stranded nucleic acids. It is used to detect complementary regions of different nucleic-acid molecules. When one of the nucleic acids is an oligonucleotide, perfect complementarity with a target sequence can be distinguished from mismatched duplexes by selecting an annealing temperature close to the **melting temperature** (T_m) for the oligonucleotide. **3** the second phase of a **polymerase chain reaction**. **4** the slow regulated cooling, especially of metals or glasses, to relieve the stresses caused by heating or other treatments. —**annealer** *n.*

annexin any of a ubiquitous family of Ca^{2+}-dependent phospholipid-binding proteins. They have been isolated from organisms as diverse as mammals, *Drosophila melanogaster*, and *Hydra vulgaris*. Characteristically they have a low intrinsic affinity for Ca^{2+} (K_d 25–1000 µm), although annexin VI has K_d ≈1 µm, but in the pres-

ence of phospholipid the affinity for Ca^{2+} increases up to 100-fold. The Ca^{2+}- and phospholipid-binding properties that typify the family derive from a region of common primary structure (identity in any other member of the family is 40–60%) in a conserved 34 kDa C-terminal region (the 'core'); each different type of annexin has a unique N-terminal domain known as the 'tail'. The C-terminal region in all cases except one consists of four repeats of a ≈70 amino-acid sequence, in which there is a high homology region known as the **endonexin fold**, with a characteristic motif GXGTDE; the exception is annexin VI, which has eight repeats. The true physiological roles of the annexins remain unknown, but their main characteristics, exhibited variously by different annexins and reflected in the alternative names they acquired in the course of discovery, include: the ability to aggregate membranes and vesicles (all annexins); inhibition of phospholipase A_2 (especially I, but also II to VI); formation of ion channels (especially I and V to VII); anticoagulant properties (especially I, and VIII); and acting as substrates for tyrosine kinases (I and II) and protein kinase C (I, II, IV, and XII). Alternative names are given under the individual entries.

annexin I *other names*: annexin A1; lipocortin I; p35 (EGF receptor substrate); calpactin II; chromobindin 9; GIF; lipomodulin; macrocortin; phospholipase A2 inhibitory protein; renocortin.

annexin II *other names*: annexin A2; calpactin I heavy chain; lipocortin II; p36 (pp60^src^ substrate); chromobindin 8; protein I; placental anticoagulant protein (PAP)-IV.

annexin III *other names*: annexin A3; lipocortin III; placental anticoagulant protein (PAP)-III; 35-α-calcimedin; calphobindin III; inositol 1,2-cyclic phosphohydrolase.

annexin IV *other names*: annexin A4; endonexin I; protein II; P32.5; lipocortin IV; chromobindin 4; placental anticoagulant protein (PAP)-II; PP4-X; 35-β-calcimedin; placental anticoagulant protein II; carbohydrate-binding protein P33/P41.

annexin V *other names*: annexin A5; lipocortin V; endonexin II; calphobindin I; CBP-I; placental anticoagulant protein (PAP)- I; PP4; thromboplastin inhibitor; vascular anticoagulant (VAC)-alpha; anchorin CII.

annexin VI *other names*: annexin A6; p68; p70; 73K; 67K calelectrin; lipocortin VI; protein III; chromobindin 20; 67 kDa calelectrin; calphobindin II; CPB-II.

annexin VII *other names:* annexin A7; **synexin**.

annexin VIII *other names*: annexin A8; vascular anticoagulant (VAC)-β.

annexin IX *other name*: annexin B9.

annexin X *other name*: *Drosophila melanogaster* annexin.

annexin XI *other names*: annexin A11; calcyclin-associated annexin 50; CAP-50; 56 kDa autoantigen.

annexin XII *other names*: annexin B12; a member of the annexin family obtained from *Hydra attenuata* (*Hydra vulgaris*). [

annexin XIII *other names*: annexin A13; intestine-specific annexin (ISA).

annexin XXI *other names*: annexin E1; a member of the annexin family obtained from *Giardia lamblia* (*Giardia intestinalis*).

annexin XXXI *other names*: annexin A9; annexin 31.

annexin 14 *other name*: annexin A10.

annihilation radiation electromagnetic radiation of wavelength 2.4 pm (energy 0.51 MeV) that is emitted as a pair of photons when a positron absorbed into matter reaches the end of its range and combines with an electron in mutual destruction.

annotation a set of descriptive notes (detailing, e.g., function, structure, disease associations, keywords, literature references) that are attached to database entries (such as protein or nucleic acid sequences, or protein families).

anode 1 the electrode of an electrolysis cell or electrophoresis apparatus towards which anions of the electrolyte migrate under the influence of an electric field applied between the electrodes. **2** the positive electrode, pole, or terminal of such a cell or apparatus, or of a discharge tube, solid-state rectifier, or thermionic valve; the electrode by which electrons flow out of the device. **3** the negative pole or terminal of an electric (primary or secondary) cell or battery (of cells). *Compare* **cathode**. —**anodal** or **anodic** *adj*; **anodally** or **anodically** *adv.*

anodic stripping voltammetry *abbr.*: ASV; an electrochemical procedure for estimating the concentration of metal ions in a solution. Exhaustive or partial (depending on the concentrations involved) deposition of the metal occurs on the cathode of an electrolytic cell at a constant applied electric potential (controlled potential deposition can be used for selectivity when mixtures are involved). The metal deposit is then anodically stripped by using a rapid sweep of voltage of opposite polarity and a peak-type current–voltage curve is obtained. The amount of metal deposited may be calculated coulometrically from the area under the peak, or by other means, and hence the concentration of metal ions in the initial solution may be determined.

anodic to during electrophoresis, migrating further towards the anode or less far towards the cathode than (the specified substance).

anolyte the electrolyte in immediate contact with the anode in isoelectric focusing.

anomalous dispersion 1 an X-ray crystallographic dispersion that occurs when the frequency of the X-rays used falls near an absorption frequency of an atom in the crystal. Anomalous dispersion techniques are frequently used in the determination of absolute configuration. **2** an optical rotatory dispersion that cannot be expressed by a simple one-term **Drude equation**.

anomer either of the two stereoisomers of the cyclic form of an **aldose** or a **ketose**, or an aldose or ketose derivative, dependent on the position of the free hydroxyl group belonging to the internal hemiacetal grouping. The anomer having the same configuration, in the **Fischer projection**, at the anomeric and the reference asymmetric carbon atom is designated α; the other anomer is designated β. In correct usage in regard to carbohydrates, the anomeric and configurational descriptors must always be linked together; e.g. α-D-glucopyranose, not α-glucose.

anomeric 1 of or pertaining to an **anomer**. **2** *or* **glycosidic** of or pertaining to the free hydroxyl group belonging to the internal hemiacetal grouping in the cyclic form of a monosaccharide or monosaccharide derivative.

anomeric effect the tendency for an electronegative substituent at C-1 of a pyranose to assume the axial orientation, in contrast to the equatorial orientation predicted on steric grounds.

anorexia absence of desire for food. In infants it is the failure to thrive. **Anorexia nervosa** is a feeding disorder, predominantly of females, that starts typically in adolescence and is associated with amenorrhea and regression of secondary sexual characteristics common in hypothalamic dysfunction.

anosmia loss of the sense of smell.

anoxia the absence or a deficiency of dioxygen. *Compare* **hypoxia**. — **anoxic** *adj.*

ANP *abbr. for* atrial natriuretic peptide. *See* **natriuretic peptide**.

ANPR *abbr. for* atrial natriuretic peptide receptor.

ANS *abbr. for* 1-anilino-8-naphthalene sulfonate; a common fluor used in extrinsic fluorescence studies of proteins.

ansamycin any antibiotic belonging to the **rifamycin** or the **streptovaricin** classes.

anserine *N*-β-alanyl-*N*$^{\pi}$-methyl-L-histidine; an isopeptide occurring in skeletal muscle and brain of some animals and humans (up to 20 mmol kg^{-1}). *Compare* **carnosine**.

ANSI *abbr. for* American National Standards Institute; an organization responsible for, among other standards, ANSI escape sequences for printing certain effects and available on some computers, and ANSI C (a standard for the C programming language).

Anson unit an old measure of the activity of a proteinase. It was based upon the extent to which a sample containing a particular proteinase can digest denatured hemoglobin, under standard conditions appropriate to the enzyme in question, with liberation of non-precipitable material giving the same colour value as one millimole of tyrosine on reaction with a diluted Folin–Ciocalteau reagent. [After Mortimer Lewis Anson (1901–68), US protein chemist, who described it in 1958.]

ant+ *a variant of* **anti+**.

Antabuse *proprietary name for* **disulfiram**.

antagonism 1 opposition; counteraction. **2** the mutual opposition of an agonist and an antagonist. **3** the interference with, or inhibition of, the growth of an organism by one of another kind, as by competing for nutrients or producing an antibiotic substance.

antagonist 1 anything that antagonizes. **2** any agent, especially a drug or hormone, that reduces the action of another agent, the **agonist**. Many act at the same receptor as the agonist. Such antagonists may be either surmountable or insurmountable, depending on prevailing conditions. Antagonism may also result from combination with the substance being antagonized (**chemical antagonism**), or the production of an opposite effect through a different receptor (**functional antagonism** or **physiological antagonism**), or as a consequence of competition for the binding site of an intermediate that links receptor activation to the effect observed (indirect antagonism). The term 'functional antagonism' is also used to describe a less well-defined category in which the antagonist interferes with other events that follow receptor activation. **3** a muscle that works in opposition to another (the agonist). *See also* **competitive antagonism**, **noncompetitive antagonism**. *Compare* **agonist**, **inverse agonist**, **synergist**. —**antagonistic** *adj.*; **antagonistically** *adv.*

antagonize *or* **antagonise** to act in opposition to; to counteract. — **antagonized** *or* **antagonised** *adj.*

antanamide a highly lipophilic homodetic cyclic decapeptide, cyclo(-Pro-Phe-Phe-Val-Pro-Pro-Ala-Phe-Phe-Pro-). It occurs in the death-cap fungus, *Amanita phalloides*, together with the **amatoxins** and **phallotoxins**. When administered to experimental animals (at 1 mg per kg of body weight), either before or at most a few minutes after a lethal dose of phallotoxin, it acts as an antidote to the toxin, apparently by competing for binding sites. *In vitro* it readily forms complexes with sodium and calcium ions, and can act as a sodium ionophore.

ANT-C *abbr. for* antennapedia complex; one of two major families of homeotic genes of *Drosophila* (the other being **BX-C**), mutation of which affects head and thoracic segments. Antennapedia protein is a penetratin.

ante+ *prefix* denoting before in time or position.

antegrade *a variant of* **anterograde**.

anteiso+ *prefix* denoting a branched-chain fatty acid having a methyl group at the *n*-3 position; e.g. 16-methyloctadecanoic acid, also named anteisononadecanoic acid.

antenna complex any of the light-harvesting pigment-protein complexes of photosynthetic organisms that funnel absorbed incident radiation to the photochemical reaction centres.

antennapedia complex *see* **ANT-C**.

anterograde *or* **antegrade 1** moving forwards; extending towards the front. **2** moving along the axon of a nerve cell in a direction away from the cell body (as in anterograde transport of vesicles). *Compare* **retrograde**.

antheridiol a steroid hormone produced by water moulds (e.g. *Achlya* spp.) to regulate their sexual reproduction.

anthocyanidin any member of a group of water-insoluble coloured polyhydroxylated 2-phenylbenzopyrylium compounds that occur as glycosides – anthocyanins – in flowers. They comprise one class of the widespread group of flavonoids. There are three coloured aglycones: the pelargonidins (of geraniums), which are orange, salmon, pink, or red; the cyanidins (of roses), which are magenta or crimson; and the delphinidins (of larkspur), which are purple, mauve, or blue. These aglycones differ by having one, two, or three hydroxyl groups, respectively, in the 2-phenyl nucleus.

anthocyanin any member of a group of intensely coloured soluble glycosides of anthocyanidins that occur in plants. They are responsible for most of the scarlet, purple, mauve, and blue colouring in higher plants, especially of flowers.

antho-K amide L-3-phenyllactyl-Phe-Lys-Ala–NH$_2$; a neuropeptide from the sea anemone, *Anthopleura elegantissima*.

anthracene a tricyclic aromatic compound, C$_{14}$H$_{10}$; it is a component of a number of **scintillation cocktails**.

anthralin 1,8-dihydroxyanthrone; a substance that inhibits **leukotriene** biosynthesis. It is used as an antipsoriatic drug.

anthranilate 2-aminobenzoate (anion, salt, or ester); an intermediate in the biosynthesis of tryptophan.

anthranilate synthase EC 4.1.3.27; *systematic name*: chorismate pyruvate-lyase (amino-accepting). An enzyme that catalyses the first step in tryptophan biosynthesis, in which chorismate and L-glutamine react to form anthranilate, pyruvate, and L-glutamate. It is a tetramer of two components, I and II: in the absence of II, I catalyses the formation of anthranilate using NH$_3$ rather than glutamine; II provides glutamine amidotransferase activity.

anthrax toxin a group of proteins secreted by *Bacillus anthracis* that causes anthrax, a highly contagious and potentially lethal disease.The toxin comprises protective antigen (PA), edema factor (EF), and lethal factor (LF). PA forms a hexamer that binds to cell-surface receptors on skin, lung, and gastrointestinal tract, and helps to translocate EF and LF into the cytoplasm. EF is a calmodulin-dependent adenylate cyclase, while LF is a zinc metalloproteinase uniquely specific for mitogen-activated protein kinase, which it inactivates by cleaving between the N-terminal region and the catalytic domain.

anthrone 9(10*H*)-anthracenone; a compound used in the colorimetric determination of carbohydrates, especially hexoses, with which it gives a green colour.

anti+ or (*sometimes before a vowel or h*) **ant+** *prefix* **1** acting against, counteracting, or inhibiting. **2** opposite to (in direction). **3** converse or reverse of; complementary to.

anti+ *prefix (obsolete)* denoting a *cis* isomer.

antiamoebin or **antiamebin** *see* **peptaibophol**.

antiarose the D isomer, 6-deoxy-D-gulose; occurs in glycosidic linkage in some **cardiac glycosides**.

β-D-pyranose form

antibiotic 1 any of numerous substances of relatively low M_r produced by living microorganisms (and also certain plants) that are able selectively and at low concentrations to destroy or inhibit the growth of other organisms, especially microorganisms. Also included are the many semi- or wholly synthetic organic compounds with similar antimicrobial properties. Many are useful chemotherapeutic agents. **2** of or relating to an antibiotic.

antibody any glycoprotein of the **immunoglobulin** family that is capable of combining noncovalently, reversibly, and in a specific manner with a corresponding antigen. Antibodies frequently, but not always, counteract the biological activity of antigens. They are produced in higher animals by cells of the lymphoid cell series, especially plasma cells, in direct response to the introduction of immunogens or autocoupling haptens. Antibodies were originally identified as a class of serum proteins, and most are present in the γ-globulin fraction of the serum. In the ultracentrifuge most antibodies have a sedimentation coefficient of 7S, except for the pentameric IgM at 19S. *See also* **cell-bound antibody**.

antibody-dependent cellular toxicity (**ADCT**) A cytotoxic reaction in which an antibody-coated target cell is directly killed by an Fc receptor-bearing **leucocyte**, such as a **NK cell**, a **macrophage**, or a **neutrophil**.

antibody-mediated hypersensitivity *or* **type II hypersensitivity** an immune response caused by the presence of antibody to cell-surface antigens and components of the extracellular matrix. These antibodies can sensitize the cells for antibody-dependent cytotoxic attack by **K cells** or **complement**-mediated lysis. Type II hypersensitivity is seen in the destruction of red cells in transfusion reactions and in hemolytic disease of the newborn.

anticholinesterase any agent, other than an antibody, that inhibits **cholinesterase** (def. 2). An example is **physostigmine**, used in the diagnosis and treatment of myasthenia gravis.

anticipation the phenomenon whereby the age of onset of a disease becomes earlier and the severity of disease symptoms increases with successive generations. It is common in disorders resulting from trinucleotide repeat mutations that increase in size through generations, such as **Huntington's disease**.

anticircular chromatography a technique of thin-layer chromatography in which the substances to be separated are applied just inside the perimeter of a circular thin-layer plate and are made to travel radially inwards by subsequent application of the developing solvent at the edge of the plate. It is very fast and is particularly useful for the resolution of substances whose R_f values approach unity. *Compare* **circular chromatography**.

anticlinal *see* **conformation** (def. 2).

anticoagulant 1 retarding or preventing coagulation. **2** any substance that retards or prevents coagulation, especially of blood or milk.

anticoagulate to treat (esp. blood) so as to prevent coagulation.

anticodon a triplet of nucleotides in any particular transfer RNA that is complementary to and therefore pairs with a given **codon** in messenger RNA. The pairing is antiparallel, 5′→3′ in the messenger, 3′→5′ in the anticodon. Thus the anticodon of 5′-ACG-3′ is 3′-UGC-5′. However, to maintain the convention of writing sequences in the 5′→3′ direction, the anticodon is thus written, but with a reverse arrow above: $\overleftarrow{\text{CGU}}$.

anticodon arm the part in the clover-leaf model of transfer RNA containing a loop bearing the anticodon.

anticooperativity negative **cooperativity** between binding sites on the same (macro)molecule by which binding of a ligand to one site makes the binding of a second ligand more difficult.

antidiuretic 1 describing any drug or other agent that decreases the excretion of urine. **2** an antidiuretic substance.

antidiuretic hormone *abbr.*: ADH; *an alternative name (esp. in clinical chemistry) for* **vasopressin**. It is so named from the hormone's most important biological action, viz. reducing the rate of urine output by stimulating the rate of water reabsorption by the kidney(s).

antidote a substance or other agent that limits or reverses the effects of a poison.

antifolate any of a class of drugs that interfere with the formation of tetrahydrofolate, and are used in the treatment of acute leukemia and choriocarcinoma. One of the best known is **methotrexate**.

antifreeze glycoprotein any of a group of glycoproteins found in the serum of certain cold-water fish, particularly notothanids and chaenichthyids, which inhabit polar seas; e.g., *Trematomus borchgrevinki* and *Dissostichus mawsoni*. These glycoproteins depress the freezing point of aqueous solutions about 200–500 times as effectively as sodium chloride on a molar basis. They differ only in mass (10.5 kDa, 17 kDa and 21.5 kDa), and are composed of repeating units of a diglycosyl tripeptide, Ala-Ala-Thr-*O*-disaccharide, in which the disaccharide consists of a β-galactosyl residue bound to an internal α-*N*-acetylgalactosaminyl residue.

antigen 1 (*or* **complete antigen**) any agent that, when introduced into an immunocompetent animal, stimulates the production of a specific antibody or antibodies that can combine with the antigen. It may be a pure substance, a mixture of substances, or particulate material (including cells or cell fragments). In this sense the term includes **agglutinogen** and **allergen** and is synonymous with **immunogen** (def. 1). **2** (*or* **incomplete antigen**) any substance that can combine with a specific antibody but is not an **immunogen** (def. 2), i.e. it is not by itself able to stimulate antibody production though it may do so if combined with a **carrier** (def. 8). In this sense the term includes **hapten**. —**antigenic** *adj*.

antigen–antibody complex *or* **immune complex** any specific macromolecular complex of antigen and antibody. It may be soluble, especially in **antigen excess**, or insoluble (as a precipitate) when antigen and antibody are present in optimal proportions. Components of **complement** may bind to antigen–antibody complexes. *See also* **immune complex-mediated hypersensitivity**.

antigen-binding capacity *abbr.*: ABC; the total amount of antibody in a preparation, e.g. serum, based on a determination of the amount of antigen bound by unit amount of the preparation.

antigen excess the condition when a mixture of antigen and antibody contains sufficient antigen to combine with all the combining sites of the specific antibody molecules, leaving some free uncombined antigenic determinants and resulting in the formation of soluble antigen–antibody complexes. This is the explanation for the type of 'precipitin' curve when increasing amounts of antigen are incubated with a fixed amount of antibody and the amount of precipitate is measured.

antigenic determinant any specific chemical structure within, and generally small in relation to, an antigen molecule that is recognizable by a combining site of an antibody, or T-cell or B-cell receptor, and at which combination takes place. It determines the specificity of the antibody–antigen reaction. Several different antigenic determinants may be carried by a single molecule of antigen. *See also* **epitope**.

antigenic drift any of the minor changes in the antigenicity of influenza virus that result from spontaneous mutation, with corresponding minor changes in the amino-acid sequence of the viral hemagglutinin. *Compare* **antigenic shift**.

antigenicity the capacity of an agent to stimulate the formation of specific antibodies to itself.

antigenic mimicry the sharing of antigenic determinants by microbial agents and body proteins as a result of similarity in amino acid sequences. This might play a role in autoimmune disease because the microbial sequences produce antibodies against the equivalent sequences in body proteins.

antigenic shift any of the major changes in the antigenicity of influenza virus that result from spontaneous mutation, with corresponding major changes in the amino-acid sequence of the viral hemagglutinin. *Compare* **antigenic drift**.

(antigen-independent) mitogenic factor a **lymphokine** that acts as a **mitogen** for lymphocytes.

antigen-presenting cell *abbr.*: APC; a cell, especially a macrophage or dendritic cell, that recognizes an antigen to be targeted for neutralization. It takes up the antigen and processes it, incorporating antigen fragments into its own membrane and presenting them in association with class II **major histocompatibility complex** molecules to T lymphocytes, which are then stimulated to mount a response.

antigibberellin any of the organic compounds that cause plants to grow with thick stems or with an appearance differing from that obtained with **gibberellin**. The effect can usually be reversed by gibberellin.

antiglobulin consumption test a serological method for demonstrating antibody (globulin) attachment to cells. It is based on showing that antiglobulin is removed from solution on addition of the cells in question.

antiglobulin test *or* **Coombs test** a test for the presence of incomplete **antibody** against red blood cells. It entails the addition of heteroantibody to immunoglobulin (i.e. antiglobulin) to cause the agglutination of red blood cells previously coated with nonagglutinating incomplete antibody. *Compare* **double-antibody method**. [After Robin Royston Amos Coombs (b. 1921), British immunologist and hematologist who (with others) described it in 1945.]

antihemophilic factor *or* antihemophilic factor **A** *or* antihemophilic globulin *an alternative name for* factor VIII; *see* **blood clotting factors**.

antihemophilic factor B *an alternative name for* factor IX; *see* **blood coagulation**.

antihemophilic factor C *an alternative name for* factor XI; *see* **blood coagulation**.

antihemophilic globulin *an alternative name for* **antihemophilic factor**.

antihistamine any drug or other agent that antagonizes an action of histamine on the body. Antihistamines are used in the treatment of immediate-type hypersensitivity.

antihormone any substance that acts to attenuate hormone-induced responses, regardless of the mechanism involved.

anti-idiotype an antibody that reacts with **epitopes** on the V region of the **T-cell receptor**.

anti-inducer any compound that inhibits **operon** induction by competing for inducer binding to free repressor and stabilizing the repressor–operator complex.

antilipolytic 1 describing any agent that inhibits **lipolysis**. **2** an antilipolytic agent.

antilipotropic 1 describing any agent that is active metabolically in deflecting methyl groups from the synthesis of choline. **2** an antilipotropic agent.

antilogarithm *abbr.*: antilog; *symbol*: (decadic) $\lg^{-1}$; (napierian) $\ln^{-1}$; the number that is represented by a **logarithm**.

antilymphocyte globulin the globulin fraction of an **antilymphocyte serum**.

antilymphocyte serum *abbr.*: ALS; any serum containing antibodies to lymphocytes. It is used as a selective immunosuppressive agent.

anti-messenger DNA *abbr.*: amDNA; *an alternative name for* **complementary DNA**.

antimetabolite any substance that inhibits the utilization of a metabolite, e.g. by acting antagonistically through competition in a transport system or at a key enzyme site.

antimicrobial 1 describing a drug, antibiotic agent, physical process, radiation, etc. that is inimical to microbes. **2** any substance having antimicrobial properties.

antimicrobial spectrum a list of the types of microbes against which an antimicrobial agent is effective.

antimitotic 1 describing any agent that decreases the rate of mitosis. **2** an antimitotic agent.

antimonial 1 of, relating to, or containing antimony. **2** a drug containing antimony.

antimorph 1 a mutant gene producing an effect opposite to the wild-type gene at the same locus. *Compare* **allele**. **2** *obsolete term for* an **enantiomorph**.

anti-Müllerian hormone *or* **Müllerian inhibiting factor** a glycoprotein, derived from a larger precursor and secreted by Sertoli cells of fetal testes, that causes regression of the Müllerian duct, which in normal female embryos develops into the uterus, Fallopian tubes, and vagina. Its receptor is a transmembrane serine/threonine kinase. Mutations in the gene for the hormone result in persistence of Müllerian duct derivatives (discovered during surgery) in normal males.

antimutagen any agent (often a purine nucleoside) that can decrease the rate of induced or spontaneous mutation.

antimutator gene any mutant gene that decreases the mutation rate.

antimycin any member of a group of antibiotics produced by *Streptomyces* spp. They associate strongly with mitochondria and block the passage of electrons from cytochrome *b* to cytochrome c_1; as little as 7 pmol antimycin A_1 per gram of mitochondrial protein is effective. They are used experimentally to distinguish between events in the earlier and later parts of the electron-transport chain.

antineoplastic 1 describing any drug or other treatment that prevents or limits the growth of a neoplasm. **2** an antineoplastic agent.

antinuclear factor *or* **antinuclear antibody** any antibody that reacts with a component of the cell nucleus.

antinutrient any natural or synthetic compound that interferes with the absorption of a nutrient. An example is phytic acid, which forms insoluble complexes with zinc, iron, and copper.

antioxidant any substance, often an organic compound, that opposes oxidation or inhibits reactions brought about by dioxygen or peroxides. Usually the antioxidant is effective because it can itself be more easily oxidized than the substance protected. The term is often applied to components that can trap **free radicals**, such as α-tocopherol (*see* **vitamin E**), thereby breaking the chain reaction that normally leads to extensive biological damage. The mechanism by which the free-radical scavenger is discharged of its free radical is not clear, but it is likely that ascorbate and/or glutathione are involved. Compounds such as **di-*tert*-butyl-*p*-cresol** act as antioxidants and are often added to protect labile compounds during storage or incubation. *See also* **quinhydrone**.

antiparallel 1 *(in biochemistry)* (describing of a pair of parallel linear structures, such as two polynucleotide or polypeptide chains, having directional polarity or asymmetry in opposite directions. **2** *(in quantum physics)* (of the spins of a pair of electrons occupying the same atomic or molecular orbital) described by different spin quantum numbers. *Compare* **parallel** (def. 2). —**antiparallelism** *n*.

antiperiplanar *see* **conformation**.

antipernicious anemia factor *former name for* **vitamin B_{12}**.

antiphospholipid antibodies autoantibodies against phospholipids that bind to soluble blood factors involved in activating the coagulation cascade. The resultant hypercoagulability can lead to deep vein thrombosis, pulmonary embolism, or complications in pregnancy. The syndrome can be primary (without evidence of other autoimmune disease); secondary (with evidence of such disease); or catastrophic (extensive and often fatal).

antiport *or* **countertransport** the process of coupled solute translocation in which two solutes equilibrate across an osmotic barrier, the translocation of one solute being coupled to the translocation of the other in the opposite direction. *Compare* **symport**.

antiporter any substance or structural feature, but usually a protein, that promotes the exchange diffusion of two specific substances across a membrane by **antiport**. The convention for denoting such a protein (e.g. a tetracycline-resistance protein) is tetracycline/H^+ antiporter.

antirachitic 1 describing any agent that opposes or prevents the development of **rickets**. **2** any agent with antirachitic properties.

antirachitic factor *or* **antirachitic vitamin** *former term for* any substance with **vitamin D** activity.

antirepressor any gene product that acts to decrease repression by a repressor of gene expression.

antirestriction the process whereby a host bacterium is prevented from inactivating foreign duplex DNA; i.e. the prevention of **restriction**. It is commonly achieved by **host-controlled modification** of the DNA, but in certain cases a specific bacteriophage-coded **antirestriction protein** is produced that directly inhibits the host's **restriction endonuclease**.

antirestriction protein *see* **antirestriction**.

anti-RNA single-stranded RNA complementary to mRNA. *See* **antisense RNA**.

antiscorbutic 1 describing an agent that opposes or prevents the development of **scurvy**. **2** any agent with antiscorbutic properties.

antiscorbutic factor *or* **antiscorbutic vitamin** *former term for* any substance with **vitamin C** activity.

antisense DNA *see* **noncoding strand**.

antisense oligonucleotide an oligoribonucleotide, or an analogue of one, having properties similar to those of **antisense RNA**.

antisense RNA *or* **messenger-RNA-interfering complementary RNA** *(abbr.: micRNA)* an artificial single-stranded RNA molecule complementary in sequence to all or part of a molecule of messenger RNA or of some other specific RNA transcript of a gene. It can therefore hybridize with the specific RNA and so interfere with the latter's actions or reactions. *Compare* **complementary RNA**.

antisense strand *see* **noncoding strand**.

antisepsis the elimination of pathogenic and other microorganisms using chemical or physical methods; the promotion of **asepsis**.

antiseptic 1 of, relating to, or bringing about **antisepsis**. **2** any chemical agent used in antisepsis, especially one that can be safely applied to the skin.

antiserum *(pl.* **antisera** *or (esp. US)* **antiserums**) serum that contains a high level of antibodies, usually antibodies against specific antigens.

antisigma factor a protein that interferes with the recognition of initiation sites by the **sigma factor** of DNA-dependent RNA polymerase.

antistasin *the generic name for* a family of potent inhibitors of factor Xa. Antistasins are isolated from leeches, the name having been first given to a 119-amino-acid protein present in salivary gland of the Mexican leech *Haementeria officinalis*; the protein has 20 cysteine residues, all in disulfide bridges (probably characteristic of the family). It appears to present a **bait**-like structure to factor Xa, which cleaves it at Arg^{34}. The 3-D structure has been partially solved, indicating a reactive site in domain I, sequence CRKTC. Antistasins have been considered as possible chemotherapeutic anticoagulants, but have strong metastatic activity. Example, **ghilanten** from the Amazonian leech *H. ghillianii*.

antistreptolysin-O *abbr.:* ASO; antibody formed against streptolysin-O. *See* **streptolysin**.

antisuppressor any mutation that diminishes the effects of a suppressor of gene expression.

antitemplate strand *see* **coding strand**.

antiterminator (*or* **antitermination factor**) a protein that prevents the termination of RNA synthesis. It is found as a regulatory device in, e.g., phage lambda, allowing a terminator to be masked from RNA polymerase so that distal genes can be expressed. Antiterminators are associated with switches from early to middle expression following induction (gene N product), and from early to late expression (gene Q product). The N system is referred to under **nus**. Antitermination is an important mechanism in the reproduction of some bacteriophages, allowing RNA polymerase to read through the terminator to the immediate early genes of the phage, and thus to initiate production of the mature phage, with consequent cell lysis.

antithrombin III *or* **heparin cofactor** *or* **factor Xa inhibitor** a 62 kDa protein inhibitor of serine proteases, normally present in plasma, that reacts with factor IIa (thrombin), and also factors IXa, Xa, XIa and XIIa, to form inert 1:1 stoichiometric complexes. The formation of complexes is slow in the absence of heparin, but very fast in its presence.

antitoxin any antibody to a **toxin** of a (micro)organism that neutralizes the toxin both *in vitro* and *in vivo*.

α_1-antitrypsin *or* **α_1-proteinase inhibitor** a major α_1-globulin, normally present in human plasma, that inhibits many serine proteases, including trypsin, but especially elastase, and is the archetypal **serpin**. Its concentration rises dramatically in response to infection or tissue injury (*compare* **acute-phase protein**). A 51 kDa glycoprotein, its prime physiological role is to inhibit elastase released by neutrophil leukocytes. In humans, its synthesis is controlled by an autosomal allelic system, and over 20 different phenotypes have been classified, one of which is associated with emphysema of early onset; in particular, the so-called Z mutant (Glu → Lys at position 342) is associated with a decreased level of α-antitrypsin in the blood. Smoking tends to cause oxidation of Met^{358}, by increasing hydrogen peroxide formation by lung neutrophils, and this leads to a less active α_1-antitrypsin, and hence reduced inhibition of elastase. Excessive activity of elastase may damage lung tissue, resulting in emphysema. Thus, smoking is associated with emphysema of early onset. Individuals who are homozygous for the Z mutant, and who also smoke, suffer decreased levels of a less active anti-elastase, and have a high probability of developing emphysema. The protein has two chains.

antitussive 1 relieving or suppressing coughing. 2 an agent with antitussive properties.

antivitamin any agent that antagonizes a specific vitamin. Sometimes it is a structural analogue of the vitamin and a competitive inhibitor of the vitamin's action.

antixerophthalmic factor *former term for* any substance with **vitamin A** activity.

antizyme a protein (26 kDa) that is induced by polyamines and binds specifically to form a tight 1:1 complex with **ornithine decarboxylase**. It inactivates the latter, and promotes its destruction by proteasomes.

antoxidant *a variant spelling of* **antioxidant**.

antrin a decapeptide, isolated from rat gastric antrum. It has the sequence Ala-Pro-Ser-Asp-Pro-Arg-Leu-Arg-Gln-Phe, and is identical to residues 25–34 of preprosomatostatin (*see* **somatostatin**).

anucleate *or* **anucleated** lacking a **nucleus**.

anucleolate lacking a **nucleolus**.

***Aor*13HI** a type 2 **restriction endonuclease**; recognition sequence: T↑CCGGA. *Bsp*MII and *Acc*III are **isoschizomers**.

***Aor*51HI** a type 2 **restriction endonuclease**; recognition sequence: AGC↑GCT. *Eco*47III is an **isoschizomer**.

AOX *abbr. for* alcohol oxidase.

AP *see* **AP endonuclease, AP protein**.

6-APA *abbr. for* 6-aminopenicillanic acid.

Apache freely distributed WWW server software that is very widely used on academic sites.

Apaf *abbr. for* apoptosis protease-activating factor.

Apaf1 *abbr. for* apoptosis protease activating factor 1.

***Apa*I** a type 2 **restriction endonuclease**; recognition sequence: GGGCC↑C. *Eco*47III is an **isoschizomer**.

***Apa*LI** a type 2 **restriction endonuclease**; recognition sequence: G↑TGCAC.

apamin a neurotoxin comprising about 2% of the dry mass of the venom of the honey bee (*Apis mellifera*). It is a highly basic octadecapeptide amide, structure

CNCKAPETALCAKKCEEH–NH$_2$,

containing two disulfide bridges (between Cys residues 1 and 11, and 3 and 15), and is the smallest neurotoxic peptide known. *Compare* **melittin**.

APC *abbr. for* 1 activated protein C. 2 antigen-presenting cell. 3 anaphase-promoting complex.

APC the human gene for adenomatous polyposis coli protein. The gene is a tumour suppressor, and mutations are associated with familial adenomatous polyposis, Gardner's syndrome, etc. and contribute to colorectal cancer. The protein product has serine phosphorylation, glycosylation, and myristoylation sites. It plays an important role in regulating the stability of free β-**catenin** by binding to it – in a complex with **actin** and **GSK-3** – and targeting it to ubiquitination and degradation by proteasomes during WNT signalling. It also regulates assembly of microtubules and binds directly to kinetochores. Deficient or dysfunctional APC protein (as occurs in many human tumours) permits β-catenin to act as a transcriptional enhancer for certain transcription factors. It is concentrated in the basolateral region of crypt epithelial cells. *See* **WNT family**.

Ape *symbol for* a residue of the α-amino acid L-α-aminovaleric acid, L-2-aminopentanoic acid.

AP endonuclease *abbr. for* apyrimidinic or apurinic nuclease; any of various DNA repair enzymes that make breaks at sites generated by DNA glycosylase. There are three types: (1) **deoxyribonuclease V**; (2) exodeoxynuclease III (endonuclease activity) *see* **deoxyribonuclease III**; and (3) DNA-(apurinic or apyrimidinic site) lyase, EC 4.2.99.18; *other names*: AP lyase; AP endonuclease class I; endodeoxyribonuclease (apurinic or apyrimidinic); deoxyribonuclease (apurinic or apyrimidinic); *Escherichia coli* endonuclease III; phage-T4 UV endonuclease; *Micrococcus luteus* UV endonuclease. They catalyse a reaction in which the C–O–P bond 3′ to the apurinic or apyrimidinic site in DNA is broken by a β-elimination

reaction, leaving a 3′-terminal unsaturated sugar and a product with a terminal 5′-phosphate. Endonuclease III from *E. coli* binds a Fe$_4$S$_4$ cluster, which is not important for the catalytic activity but is probably involved in the proper positioning of the enzyme along the DNA.

A peptide the octadecapeptide cleaved from the α chains of a fibrinogen molecule when it is converted into fibrin by the proteolytic action of thrombin. *Compare* **B peptide**.

aperture *(in physics)* 1 an opening in an optical or other similar instrument that controls the amount of radiation entering or leaving it. 2 the diameter of the refracting surface of a lens or reflecting surface of a mirror. *See also* **f-number**.

aphidicolin a potent antiviral and antimitotic agent, first isolated as an antibiotic from the fungus *Cephalosporium aphidicola*. It inhibits eukaryotic DNA polymerase.

Api *symbol for* D-apiose.

API *abbr. for* Application Programming Interface; a standard used for ensuring portability of computer software. Familiar APIs are the software development kits such as SDK (or JDK) used by Java programmers, but APIs can also include methods for validation of data transmission, etc.

apical surface the surface of an epithelial cell that faces the lumen.

apigenin 4′,5,7-trihydroxyflavone; the aglycon of **apiin**. It inhibits cyclooxygenase and estrogen synthase. *See* D-**apiose**.

apiin apigenin-7-apiosylglucoside, apioside; the name is derived from *Apium* (celery) from which apiin is isolated. *See also* **apigenin**, **apiose**.

apiose *symbol*: Api; the D isomer, 3-*C*-(hydroxymethyl)-D-*glycero*-aldotetrose is a branched carbon chain sugar that occurs widely in plants either in glycosides or polysaccharides. It is a component of **apiin**.

D-apiose

aplasia incomplete or deficient development of an organ or tissue.

aplastic 1 not showing normal growth or change in structure. 2 relating to or showing **aplasia**.

APLP *abbr. for* amyloid precursor-like protein.

Apm *symbol for* a residue of the α-amino acid α-aminopimelic acid, 2-aminoheptanedioic acid.

A$_2$pm *or* **Dpm** *symbol for* the di(α-amino acid) α, ε-diaminopimelic acid, 2,6-diaminoheptanedioic acid.

APMSF *p*-amidinophenylmethylsulfonylfluoride; a specific irreversible inhibitor of trypsin-like serine endopeptidases, with greater inhibitory activity than **phenylmethylsulfonylfluoride**.

apo+ *prefix* denoting detached, separate. It is used especially of enzymes and other proteins. For example, apocarboxylase is the protein part of the holoenzyme carboxylase, i.e. it lacks the covalently bound biotin residue(s). Apoferritin is a 450 kDa protein with 24 identical subunits, that combines with ferric hydroxide-phosphate to form **ferritin**. *See also* **apoenzyme**.

Apo-1 *an alternative term for* **Fas**.

Apo(a) or **lipoprotein(a)** a plasma glycoprotein serine endopeptidase (EC 3.4.21.–). It is related to **plasminogen**, having 37 repeats of plasminogen **kringle** domains and capable of cleaving **fibrinogen**. Its gene is closely linked to that of plasminogen at 6q21, from which it originated by gene duplication. It is mostly covalently linked (by disulfide bond) to apolipoprotein B-100, when it forms lipoprotein (a). Its plasma concentration is related to the incidence of atherosclerosis.

APOBEC-1 abbr. for apoB editing enzyme.

apoB editing enzyme abbr.: APOBEC-1; the catalytic subunit of the complex that converts the cytosine of codon 2153 (CAA for glutamine) in **apolipoprotein B** (apoB) pre-mRNA into uracil (UAA for termination) thus converting the mRNA for B-100 into one for B-48 in nuclei of human enterocytes. A locus at 12p13.1-p13.2 encodes 229 amino acids, which incorporate a deaminase domain that contains a zinc atom and a glutamate residue that are essential for activity. Complex formation requires APOBEC-1 complementation factor, which contains three RNA-recognition motifs. A cluster of APOBEC-1-related sequences has been reported on chromosome 22.

apocrine 1 describing a type of secretion that entails loss of part of the cytoplasm of the secretory cell. **2** describing a gland made up of such cells. Compare **eccrine**, **holocrine**, **merocrine**.

apoenzyme the protein part of an enzyme that forms a catalytically active enzyme (holoenzyme) when combined with an activator, a coenzyme, or a prosthetic group. It determines the specificity of the catalytic system.

APOE polymorphism polymorphism at the gene for apolipoprotein E (APOE). There are three major alleles at the APOE locus, APOE*ε2, APOE*ε3 and APOE*ε4, with a prevalence in the population of 6%, 78%, and 16% respectively. APOE*ε4 increases the risk of developing Alzheimer's disease. Individuals homozygous for APOE*ε4 are more likely to be affected and at an earlier age than heterozygotes, who in turn are more likely to develop the disorder than those who do not have this allele.

apoferritin see **ferritin**.

apolar 1 lacking poles, especially of nerve cells. **2** nonpolar.

apolar interaction any of the entropy-driven interactions between nonpolar parts of molecules when in polar solvents. It is important in maintaining the structure of protein molecules in polar solvents, in antigen–antibody interactions, in virus– protein and enzyme–subunit aggregation and disaggregation, and in determining enzyme specificity.

apolipophorin a protein component of **lipophorin**. This is an exchangeable apolipoprotein, found in many insect species, that functions in transport of diacylglycerol (DAG) from the fat body lipid storage depot to flight muscles in the adult life stage. An equilibrum is maintained between a soluble monomer and a bound lipoprotein form. Apolipophorin-III associates with lipophorin during lipid loading, until each particle contains 9 or 14 molecules of apolipophorin-III.

apolipoprotein the protein component of any lipoprotein, but especially of plasma lipoproteins. The latter are designated by the abbreviation 'Apo', e.g. ApoA (for apolipoprotein A), ApoB, etc., and are subdivided into subtypes, e.g. ApoA-I for apolipoprotein A, subtype I. See following entries for individual types of apolipoprotein.

apolipoprotein A-I abbr.: ApoA-I; the major protein of plasma **high-density lipoprotein** (HDL), also found in chylomicrons. It is synthesized in the liver and small intestine, and participates in the reverse transport of cholesterol from tissues to the liver by promoting cholesterol efflux from tissues; it acts as a cofactor for **lecithin–cholesterol acyltransferase**. Deficiency of ApoA-I is associated with **Tangier disease**.

apolipoprotein A-II abbr.: ApoA-II; an apolipoprotein associated with **high-density lipoprotein**, which it stabilizes. It is a homodimer (disulfide linked).

apolipoprotein A-IV abbr.: ApoA-IV; an apolipoprotein that is a major component of **high-density lipoprotein** and chylomicrons. It is involved in the catabolism of chylomicrons and very-low-density lipoprotein, being required for efficient activation of **lipoprotein li-**pase by apolipoprotein C-II. It is a potent activator of **lecithin–cholesterol acyltransferase**.

apolipoprotein B abbr.: ApoB; a major protein constituent of chylomicrons, very-low-density lipoprotein and low-density lipoprotein. It exists in two forms: B-100 (amino acids 28–4563 of the sequence) and B-48 (amino acids 28–2179); the latter is of intestinal origin, and derived from editing of mRNA to form an internal stop codon. Excessive plasma levels occur in familial **hypercholesterolemia**.

apolipoprotein C-I abbr.: ApoC-I; a component of chylomicrons and very-low-density lipoprotein (VLDL), synthesized mainly in liver. It modulates the interaction of apolipoprotein E with β-migrating VLDL, and inhibits binding of β-VLDL to low-density lipoprotein receptor-related protein.

apolipoprotein C-II abbr.: ApoC-II; a component of very-low-density lipoprotein (VLDL), and an activator of several triacylglycerol lipases. It has a reversible association with plasma chylomicrons, VLDL and high-density lipoprotein. Deficiency results in hypertriglyceridemia, xanthomas, and increased risk of pancreatitis and early atherosclerosis (hyperlipoproteinemia, type IB).

apolipoprotein C-III abbr.: ApoC-III; an apolipoprotein that constitutes 50% of the protein fraction of very-low-density lipoprotein. It binds to sugar and sialic-acid residues, and inhibits lipoprotein and hepatic lipases.

apolipoprotein D abbr.: ApoD; an apolipoprotein that occurs in a macromolecular complex with **lecithin–cholesterol acyltransferase** and in high-density lipoprotein. It is a **lipocalin**, probably involved in the transport and binding of bilin. The homodimer shows wide tissue distribution.

apolipoprotein E abbr.: ApoE; an apolipoprotein that constitutes 10–20% of very-low-density lipoprotein. It mediates binding, internalization and catabolism of lipoprotein particles, and is a ligand for the low-density lipoprotein (ApoB/E) receptor and for the ApoE receptor (chylomicron remnant) of hepatic tissues. It is present in brain, and may have a role in cholesterol transport. O-glycoside and sialic-acid residues occur, but it is de-sialylated in plasma. It has nine allelic variants, two of which are associated with hyperlipoproteinemia type III. The structure contains a heparin-binding motif.

apolipoprotein H abbr.: ApoH; other names: β$_2$-glycoprotein II precursor; APC inhibitor (see **activated protein C**). An apolipoprotein that binds to anions, such as heparin, and phospholipids. It may prevent activation of the intrinsic blood coagulation cascade by binding to phospholipids on the surface of damaged cells. See also **amyloid A protein**.

apolipoprotein J abbr: Apo J; other name: clusterin; an apolipoprotein associated with high-density lipoprotein (HDL) particles and produced in many tissues. A locus at 8p21 encodes a precursor, which is cleaved to form a two-chain disulfide-bonded glycoprotein in which ApoJα contains residues 1–205, and ApoJβ contains residues 206–427. Both contain amphipathic sequence repeats. ApoJ binds numerous proteins and seems to be cytoprotective.

apolipoprotein receptor see **LDL receptor**, **VLDL receptor**.

Apollon or **BRUCE** (abbr. for baculovirus repeat-containing ubiquitin-conjugating enzyme); an **IAP** (4845 amino acids) that is associated with membranes of the Golgi apparatus and vesicular system. It is expressed in most adult human tissues, and in cancer cells from various tissues including brain.

apomorphine 6aβ-aporphine-10,11-diol; 5,6,6a,7-tetrahydro-6-methyl-4H-dibenzo[de,g]quinoline-10,11-diol; a synthetic opiate. The (R)-enantiomer is a potent dopamine receptor agonist, noted for its powerful emetic effects.

apomucin the polypeptide part of a **mucin**.

apopain an apoptosis-specific cysteine protease of mammalian cells. It targets poly(ADP-ribose) polymerase, DNA-dependent protein kinase, sterol regulatory element-binding protein, and huntingtin.

apoplast the nonliving part of plant tissue, external to the plasmalemma and composed of the cell walls, the intercellular spaces, and the lumen of dead structures such as xylem cells.

apoprotein the protein part of a **conjugated protein**.

apoptosis or **programmed cell death** an intracellular controlled process characterized by cell shrinkage and detachment from the extracellular scaffolding, chromatin condensation and DNA fragmentation by activation of a calcium-dependent endonuclease, blebbing of plasma membrane, activation of **caspases**, and collapse of the cell into small membrane-enclosed fragments (called **apoptotic bodies**) that are removed by phagocytosis. There is no release of cell constituents and no inflammatory response. A cascade of caspases (which cleave at specific aspartic residues hence the name) is involved. In this cascade, caspases cleave procaspases to form additional caspases. Some of the caspases thus formed cleave other proteins, including deoxyribonucleases. Largely a phenomenon of multicellular organisms, apoptosis occurs under various physiological (e.g. during normal embryonal development, cell turnover) and pathological (nonlethal stress) conditions. It is distinguished from necrosis, in which the cell swells in response to lethal stress, bursts open and releases its contents into the surroundings, and elicits an inflammatory response. –**apoptotic** adj.; **apoptotically** adv. [From **apo+** + Greek ptosis, falling.]

apoptosis antigen 1 an alternative term for **Fas**.

apoptosis inducing factor abbr.: AIF; a flavoprotein of the mitochondrial intermembrane space that, when released by cellular stress, translocates to the nucleus and induces chromatinolysis and apoptosis.

apoptosis protease-activating factor 1 abbr.: Apaf-1; a mammalian cytosolic monomeric protein (≈130 kDa) involved in triggering **apoptosis**. It oligomerizes on binding cytochrome c (released from mitochondria) and binds procaspase 9, which becomes autocatalytically active and promotes apoptosis.

apoptosome a large complex of proteins that induces **apoptosis**. Of molecular mass ≈700–1400 kDa, it consists of several copies each of: cytosolic apoptosis protease-activating factor-1 (Apaf-1); cytochrome c (released from mitochondria); and cytosolic procaspase-9, which thereby becomes autocatalytically activated.

apoptotic bodies see **apoptosis**.

aporepressor the protein component of a complex that can act to inhibit gene expression. See also **repressor**.

APP abbr. for precursor protein.

apparent describing a physical quantity, especially an association constant, dissociation constant, equilibrium constant, etc., that has been determined under particular experimental conditions and hence is not the true (i.e. thermodynamic) constant. See **equilibrium constant**.

apparent exchangeable mass a term used in tracer kinetics in whole animals for a mass obtained by dividing the quantity of administered radioactivity that is retained by the specific activity of the tracer in the (blood) plasma. It equals the total exchangeable mass when the specific activities are equal throughout the system. Compare **compartment**.

apparent molecular weight former name for **apparent relative molecular mass**.

apparent relative molecular mass relative molecular mass that has been determined by a method that does not make allowance for any nonideality of the system.

appetite peptide or **orexigenic peptide** any of a group of peptides, such as neuropeptide Y, that enhance appetite for food.

apple domains sequences of ~90 amino-acid residues stabilized by three highly-conserved disulfide bonds, linking the first and sixth, second and fifth, and third and fourth cysteine residues (this cross-linking pattern can be drawn in the shape of an apple, hence the name). They may occur in multiple tandem repeats: e.g., four copies are present in human kallikrein and in human coagulation factor XI.

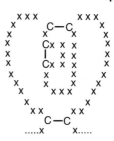

apple domains

applet a software application, written in Java, loaded from a server via HTML pages. Applets are executed on the local computer.

applicator any device or instrument, such as a comb, rod, spatula, or tube, for applying samples in a reproducible manner to a small defined area, e.g. in chromatography, or for applying something to a particular part of the body.

approach to (sedimentation) equilibrium method or **Archibald method** a method for determining the relative molecular masses of (macro)molecules by sedimentation in an ultracentrifuge, by measuring the depletion of the macromolecules from the meniscus and their accumulation at the cell bottom. Its advantage is that only very short periods of sedimentation are required. [After William James Archibald (1912–).]

AP protein abbr. for activator protein, one of two **transcription factors**; AP1, a product of c-jun (see **JUN**), interacts with Fos (product of c-fos) to form a dimer. It harbours two functional domains, one responsible for DNA binding and one for binding to Fos, and includes a leucine zipper. AP2 interacts with inducible viral and cellular enhancer elements to stimulate transcription of selected genes. It contains the consensus binding sequence CCCAGGC and a helix-span-helix motif.

A$_2$pr or **Dpr** symbol for a residue of the α,β-diamino acid L-α,β-diaminopropionic acid; L-2,3-diaminopropanoic acid.

aprotic describing a substance, especially a solvent, that is unable to act as a proton acceptor or donor.

aprotinin a proteinase inhibitor derived from animal tissues. It is a basic polypeptide of known sequence of 58 amino-acid residues (≈6.5 kDa) that exists as a homodimer. It is stable in acid or neutral media and at high temperature, and it inhibits kallikrein, trypsin, chymotrypsin, plasmin, and, to a lesser extent, papain. See also **Trasylol**.

APRT abbr. for adenine phosphoribosyltransferase.

APS abbr. for adenosine 5'-phosphosulfate.

APT abbr. for alum-precipitated toxoid.

aptamer a short strand of DNA that binds a target protein sequence. Decoy aptamers may activate or inactivate genes by inactivating DNA-binding proteins, such as transcription factors. –**aptameric** adj.

aptitude (in microbiology) the specific physiological state of a **lysogenic bacterium** that enables it, upon induction, to react so as to produce infectious bacteriophage particles.

apud or **APUD** abbr. for amine precursor uptake and decarboxylation; designating any of a series of cells that produce peptides and/or amines active as hormones or neurotransmitters, share a spectrum of cytochemical qualities, and appear in many cases to have a common embryological origin. They are thought to include the A, B, and D cells of the endocrine pancreas, the C cells of the thyroid, and the various hormone-producing cells of the pituitary and the gut.

apudoma any tumour of an **apud** cell, e.g. a gastrinoma, glucagonoma, insulinoma, or VIPoma.

apurinic acid a DNA derivative in which all the purine bases have been removed by selective hydrolysis, leaving the pyrimidine–deoxyribose bonds and the phosphodiester bonds of the backbone intact. It results from the gentle acid hydrolysis of DNA at pH 3.0.

apurinic endonuclease see **AP endonuclease**.

apyrase EC 3.6.1.5; *other names*: ATP-diphosphatase; adenosine diphosphatase (*abbr.*: ADPase); a Ca^{2+}-requiring enzyme that catalyses the hydrolysis of ATP to AMP and 2 orthophosphate. It also hydrolyses other nucleotides, especially ADP to AMP and orthophosphate. It was originally identified in plants, especially potatoes, but enzymes of this type have also been found in animals, e.g. as a surface enzyme on lymphoid cells identical to CD39, and as a secreted enzyme of the yellow-fever mosquito, *Aedes aegypti*. It is used in platelet incubations to destroy ADP.

apyrimidinic acid a DNA derivative from which the pyrimidine bases have been removed chemically, leaving the purine–deoxyribose bonds and the phosphodiester bonds of the backbone intact.

apyrimidinic endonuclease *see* **AP endonuclease**.

aq (*in chemistry*) *abbr. for* aqueous.

aqua+ *comb. form* **1** denoting acting on or reacting with water. **2** denoting presence of water as a neutral ligand in a coordination entity, e.g. aquacobalamin. The number of water molecules per entity is indicated by an approximate multiplying prefix, e.g. hexaquachromium trichloride.

aquacide a substance that when added to a mixture removes water molecules, usually by forming a compound with the elements of water. —**aquacidal** *adj.*

aquaflavin *or* **uroflavin** a fluorescent alloxazine derivative of riboflavin, present in the urine of most animals.

aqueous **1** of, having the characteristics of, or relating to water. **2** dissolved in or mixed with water. **3** *or* **aqueous humour** the fluid occupying the space between the crystalline lens and the cornea of the eye.

aquo+ *comb. form* **1** containing one or more water molecules in a coordination complex. **2** (*in chemical nomenclature*) derived from water.

aquo-ion a complex ion containing one or more water molecules held by coordination.

Ar *symbol for* **1** argon. **2** aryl.

Ara *symbol for* arabinose.

ara a plant gene related to *ras*.

araBAD an operon characterized in *Escherichia coli* and encoding L-ribulokinase, L-arabinose isomerase, and L-ribulose 5-phosphate 4-epimerase, three of the enzymes that metabolize arabinose. *See also* **araC**.

Arabidopsis thaliana a small Cruciferous plant (common wall cress) widely used in plant genetic research because of its small genome and very short generation time.

arabin+ *a variant (before a vowel) of* **arabino+**.

arabinan a branched homopolymer of L-arabinose found in plant pectins.

arabinaric acid *see* **aldaric acid**.

arabino+ *or (before a vowel)* **arabin+** *comb. form* indicating arabinose.

arabino- *prefix (in chemical nomenclature)* indicating a chemical compound containing a particular configuration of a set of three (usually) contiguous >CHOH groups, as in the acyclic form of D- or L-arabinose. *See also* **monosaccharide**.

arabinofuranosylcytosine cytosine arabinoside; *see* **cytarabine**.

arabinogalactan any member of a subgroup of plant hemicelluloses that contain arabinose and galactose and are particularly abundant in larch.

arabinogalactan proteins *abbr.*: AGP; one of the four major classes of **plant cell-wall proteins**. They contain ≈95% of their mass as carbohydrate (highly branched galactan chains decorated with arabinose moieties). They are present in the Golgi apparatus, plasma membrane, and cell wall. Enriched in proline/hydroxyproline, alanine, and serine/threonine, they are a diverse group related only by the glycan structure.

arabinonucleoside the 1-β-D-arabino-*N*-furanoside of a purine or pyrimidine base. It is a structural analogue of a **riboside**.

arabinose *arabino*-pentose; a monosaccharide. L-Arabinose occurs free, e.g. in the heartwood of many conifers, and in the combined state, in both furanose and pyranose forms, as a constituent of various plant hemicelluloses, bacterial polysaccharides, etc. D-Arabinose occurs to some extent in plant glycosides and is a constituent of the **arabinonucleosides**.

β-D-arabinose

D-arabinose isomerase EC 5.3.1.3; *systematic name*: D-arabinose ketol-isomerase. An enzyme that catalyses the reaction:

$$\text{D-arabinose} = \text{D-ribulose}.$$

See also **L-arabinose isomerase**.

L-arabinose isomerase EC 5.3.1.4; *systematic name*: L-arabinose ketol-isomerase. An enzyme that catalyses the reaction:

$$\text{L-arabinose} = \text{L-ribulose}.$$

It initiates the metabolism of L-arabinose. *See also* **araBAD**, D-**arabinose isomerase**.

arabinose-5-phosphate isomerase EC 5.3.1.13; *systematic name*: D-arabinose-5-phosphate ketol-isomerase. An enzyme that catalyses the reaction:

$$\text{D-arabinose 5-phosphate} = \text{D-ribulose 5-phosphate}.$$

arabinoside any glycoside derived from arabinose.

arabinoxylan any of various neutral polysaccharides that occur in association with acidic polysaccharides in cereal gums. They consist of a chain of (1→4)-linked β-D-xylan units with single α-L-arabinofuranosyl groups in (1→3) linkage.

arabitol *or* **arabinitol** *arabino*-pentitol; the pentitol formally derived by reduction of the aldehyde group of arabinose or lyxose. The D enantiomer is present in lichens and mushrooms.

ara-C *abbr. for* arabinosylfuranosylcytosine.

araC the gene for the araC protein, which acts as a regulator of expression of the *araBAD* operon and of *araC* itself. AraC acts as both a repressor (when intracellular arabinose levels are low), binding to *araO* operator sites, and an activator (in the presence of arabinose), when bound to the *araI* site, but only in the presence of cyclic AMP–CRP complex. It is a typical **helix-turn-helix** DNA-binding regulatory protein.

arachain an unclassified proteinase obtained from the peanut, *Arachis hypogaea*.

arachidic acid *the trivial name for* eicosanoic acid; CH_3–$[CH_2]_{18}$–COOH; a constituent of arachis oil (from the peanut, *Arachis hypogaea*) and other oils. *See also* **arachidoyl**.

arachidonate 5-lipoxygenase EC 1.13.11.34; *other names*: 5-lipoxygenase; leukotriene-A_4 synthase; a **lipoxygenase** (def. 2) enzyme that catalyses the reaction of arachidonate with dioxygen to form (6*E*,8*Z*,11*Z*,14*Z*)-(5*S*)-5-hydroperoxyeicosa-6,8,11,14-tetraenoate. Iron is a cofactor.

arachidonate 12-lipoxygenase EC 1.13.11.31; *other name*: 12-lipoxygenase; a **lipoxygenase** (def. 2) enzyme that catalyses the reaction of arachidonate with dioxygen to form (5*Z*,8*Z*,10*E*,14*Z*)-(12*S*)-12-hydroperoxyeicosa-5,8,10,14-tetraenoate. Iron is a cofactor.

arachidonate 15-lipoxygenase EC 1.13.11.33; *other names*: arachidonate ω-6 lipoxygenase; 6-lipoxygenase; a **lipoxygenase** (def. 2) enzyme that catalyses the reaction of arachidonate with dioxygen to form (5*Z*,8*Z*,11*Z*,13*E*)-(15*S*)-15-hydroperoxyeicosa-5,8,11,13-tetraenoate. Iron is a cofactor.

arachidonic acid *see* **eicosatetraenoic acid**.

arachidonoyl *symbol*: Δ_4Ach; *the trivial name for* (all-*Z*)-eicosa-5,8,11,14-tetraenoyl; CH_3–$[CH_2]_3$–$[CH_2–CH=CH]_4$– $[CH_2]_3$–CO– (all-*Z* isomer); the acyl group derived from arachidonic (i.e.(all-*Z*)-eicosa-5,8,11,14-tetraenoic) acid, a polyunsaturated, unbranched, acyclic, aliphatic acid. It occurs naturally in membrane phospholipids, and is a biochemical precursor of **prostaglandins** and **leukotrienes**. It is also present in mosses, algae, and ferns, but not in higher plants. Distinguish from **arachidonyl** and **arachidoyl**.

arachidonyl *the trivial name for* the alkyl group, CH_3–$[CH_2]_3$–$[CH_2$–$CH{=}CH]_4$–$[CH_2]_3$–CH_2–, derived from the alkenol analogue of arachidonic acid. Distinguish from **arachidonoyl**.

2-arachidonylglycerol an endocannabinoid, chemically similar to **anandamide**.

arachidoyl *or (formerly)* **arachidyl** *or* **arachyl** *symbol*: Ach; *the trivial name for* eicosanoyl; CH_3–$[CH_2]_{18}$–CO–; the acyl group derived from arachidic (*or* eicosanoic *or (formerly)* arachic) acid, a saturated, unbranched, acyclic, aliphatic acid. It occurs naturally as triarachin in certain seeds, oils, and butter fat. Arachidyl is now the name for the alkyl group, CH_3–$[CH_2]_{18}$–CH_2–. Distinguish from **arachidonyl**.

arachidyl *the trivial name for* the alkyl group, CH_3–$[CH_2]_{18}$–CH_2–, the alkanol analogue of arachidic acid. *See also* **arachidoyl**.

arachin the major protein of the groundnut, *Arachis hypogaea*, comprising a 345 kDa globulin with 12 subunits. Not all groundnuts contain the same arachin; there exist at least two forms, designated A and B, having similar amino-acid compositions and general properties, but differing in their subunit composition. *See also* **conarachin**.

arbovirus *abbr. for* arthropod-borne virus (a term formerly used for **togavirus**).

arc a continuous curved line, as part of a circle or ellipse; e.g. the line of antigen–antibody precipitate obtained in **immunodiffusion** or **immunoelectrophoresis**.

ARC *abbr. for* ADP-ribosyl cyclase.

Archaea one of three primary kingdoms of cells, together with **Eukarya** and **Bacteria** (formerly **Eubacteria**), defined when cells are classified on the basis of rRNA sequence homologies. Archaea include all methanogens, extreme halophiles and most extreme thermophiles. Representative genera are: *Halobacterium*, *Methanobacterium*, *Sulfolobus*, *Thermoplasma*, and *Purodictium* (growth optimum = 105°C); about 40 others are known. —**archaeal** *adj*.

Archaebacteria *former name for* **Archaea**.

Archibald method *see* **approach to (sedimentation) equilibrium method**.

Archimedes' principle the principle that when a body is partly or wholly immersed in a fluid there is an apparent loss of weight equal to the weight of the fluid displaced. [After Archimedes of Syracuse (?287–212 BC), Greek philosopher and applied mathematician.]

architectural gene a gene that determines the site of an enzyme within a cell.

A-region the presumptive initial oxidation site of a carcinogenic polycyclic aromatic hydrocarbon, on the terminal ring of the **bay region**, in the metabolic pathway leading to carcinogenesis (corresponding to what has sometimes been called the M-region). *Compare* **B-region**, **K-region**, **L-region**.

arenavirus any of a group of RNA viruses consisting of enveloped pleomorphic particles of 80–120 nm with helical nucleocapsids. The group includes Lassa virus and other viruses that spread to humans from rodents.

arene any monocyclic or polycyclic aromatic hydrocarbon.

arene oxide any epoxide formed by the addition of an atom of oxygen across any double bond of an **arene**.

ARF *abbr. for* ADP-ribosylation factor; a small monomeric highly conserved cytosolic **GTPase** that, when bound to GTP, becomes myristoylated and binds to the membranes of cells. It was originally discovered as a factor required for cholera toxin to ADP-ribosylate G_s, hence its name. It is a member of the **RAS** superfamily. Functions of ARF include regulation of membrane traffic in both endocytic and exocytic pathways, maintenance of organelle integrity, assembly of coat proteins (*see* **coatomer**), and activation of phospholipase D. *See* **phospholipase**.

Arg *symbol for* a residue of the α-amino acid L-arginine (alternative to R).

argentaffin cell a type of cell of the **apud** class, characterized by taking up silver stains with high affinity, without the addition of reducing agent. [Named from Latin *argentum*, silver, and affinity.]

argentation chromatography any method of chromatography in which an adsorbent impregnated with silver nitrate is used to separate lipid classes according to the degree and geometry of their saturation. The technique may be used with either thin-layer chromatography or column chromatography, or in conjunction with gas–liquid chromatography.

arginase *abbr. (in clinical biochemistry)*: ARS; *recommended name for* EC 3.5.3.1; *other names*: arginine amidinase; *systematic name*: L-arginine amidinohydrolase. An enzyme that catalyses the hydrolysis of L-arginine to L-ornithine and urea; it also acts on α-*N*-substituted L-arginines and canavanine. It is a key enzyme in the **ornithine–urea cycle**, occurring in high concentrations in the liver of ureotelic vertebrates and being absent or nearly so from the liver of uricotelic species. It has a very wide species distribution. In humans the liver (and erythrocyte) isozyme is a cytosolic homotrimer containing an Mn^{2+}–Mn^{2+} cluster in each subunit. Over a dozen mutations in the locus at 6q23 result in the rare autosomal recessive disorder argininemia, which is associated with hyperammonemia, progressive psychomotor and growth impairment, and spasticity. The renal isozyme shows some 57% sequence homology and occurs in mitochondrial matrix.

argininate **1** arginine anion; H_2N–C(=NH)–NH–$[CH_2]_3$–$CH(NH_2)$–COO^-. **2** any salt containing the arginine anion. **3** any ester of arginine.

arginine *trivial name for* N^5-amidino-ornithine; α-amino-δ-guanidinovaleric acid; 2-amino-5-guanidinopentanoic acid; H_2N–C(=NH)–NH–$[CH_2]_3$–$CH(NH_2)$–COOH; a chiral α-amino acid. L-Arginine (*symbol*: R *or* Arg), (*S*)-2-amino-5-guanidinopentanoic acid, is a coded amino acid found in peptide linkage in proteins; codon: AGA (not in mitochondria of fruit fly or mammals) or AGG (not in mitochondria of mammals); CGA, CGC, CGG, or CGU. It is an essential dietary amino acid in rats and probably in young humans, and is glucogenic in mammals. D-Arginine (*symbol*: D-Arg *or* DArg), (*R*)-2-amino-5-guanidinopentanoic acid, is not known to occur naturally.

L-arginine

arginine (base) the ionic form of an arginine residue when the guanidino group is unprotonated.

arginine carboxypeptidase *see* **lysine(arginine) carboxypeptidase**.

arginine kinase EC 2.7.3.3; an enzyme that catalyses a reaction between ATP and L-arginine to form ADP and N^ω-phospho-L-arginine. The latter compound may play a role in invertebrate muscle analogous to that of **creatine phosphate** in vertebrate muscle.

arginine phosphate *former name (not recommended) for* **phosphoarginine**.

arginine–urea cycle *see* **ornithine–urea cycle**.

[8-arginine]vasopressin *or (less correctly)* **arginine vasopressin** *abbr.*: AVP *or* [Arg⁸]vasopressin; the molecular form of **vasopressin** in which the variable eighth position in its amino-acid sequence is occupied by an arginine residue. This is the form present in humans and most other mammals.

arginine vasotocin *or* [Arg³]oxytocin a pituitary peptide found in nonmammalian vertebrates with activity resembling both oxytocin and vasopressin.

argininium **1** argininium(1+); the monocation of arginine. In theory, the term denotes any ion, or mixture of ions, formed from arginine, and having a net charge of plus one, although in practice the predominant species is generally H_2N–C(=NH_2^+)–NH–$[CH_2]_3$–$CH(NH_3^+)$–COO^-. **2** *the systematic name for* argininium(2+); the dication of arginine.

arginino **1** N^2-arginino; the alkylamino group, H_2N–C(=NH)–NH–$[CH_2]_3$–CH(COOH)–NH–, derived from arginine by loss of a hydrogen atom from its α-amino group. **2** N^ω-arginino; the alkylguanidino group, HOOC–$CH(NH_2)$–$[CH_2]_3$–NH–C(=NH)–NH–, derived from arginine by loss of a hydrogen atom from its ω-amino group.

argininosuccinase EC 4.3.2.1; *systematic name*: *N*-(L-argininosuccinate) arginine lyase; *recommended name*: argininosuccinate lyase.

An enzyme of the **ornithine–urea cycle** that catalyses the hydrolysis of *N*-(L-arginino)succinate to fumarate and L-arginine. The human enzyme is a homotetramer of hepatocyte cytosol with very strong similarity to bird and reptile δ-crystallin. Severe deficiency results from any of over a dozen mutations in a locus at 7cen-p21, and produces hyperammonemia and argininosuccinic aciduria.

argininosuccinate 2-(N^ω-arginino)succinate; L-argininosuccinate is an intermediate in the **ornithine–urea cycle**, where it is synthesized by the enzymic condensation of L-citrulline with L-aspartate. *See* **arginosuccinate synthase**.

argininosuccinate synthase EC 6.3.4.5; *systematic name*: L-citrulline:L-aspartate ligase (AMP-forming); *other names*: argininosuccinate synthetase; citrulline–aspartate ligase. An enzyme of the **ornithine–urea cycle** that catalyses a reaction between ATP, L-citrulline, and L-aspartate to form L-argininosuccinate with release of AMP and pyrophosphate. The human enzyme is a cytosolic homotetramer expressed mainly in hepatocytes. More than 20 mutations in a locus at 9q34 result in enzyme deficiency with consequent hypercitrullinemia and hyperammonemia.

arginyl the acyl group, H_2N–C(=NH)–NH–$[CH_2]_3$–CH(NH_2)–CO–, derived from arginine.

Argonaute any of a family of proteins that are important in diverse post-transcriptional RNA-mediated gene-silencing systems in plants, *Drosophila*, and fission yeast. They are components of **RISC**, are highly conserved in plants and animals, and have been implicated in methylation of histones and DNA.

argon detector a sensitive detector used in gas chromatography that depends on the unique ionization properties of argon whereby its long-lived metastable excited atoms can transfer their energy of excitation by collision to other gas molecules of lower excitation potential. Thus argon can be used as the chromatographic carrier gas, the detector being an ionization chamber containing a source of ionizing radiation. Detectable changes of the level of ionization occur when the argon issuing from the column has another gas mixed with it. The detector was invented by James E. Lovelock. Its great advantage is that it does not depend on the chemical nature of the substances being separated.

[Arg⁸]vasopressin *abbr. for* [8-arginine]vasopressin.

ARH protein a protein that is mutated in autosomal recessive hypercholesterolemia with normal low-density lipoprotein (LDL) receptors. The gene locus, at 1p35, encodes 308 amino acids. This product includes a 170-residue sequence homologous to the phosphotyrosine-binding domain of many adaptor proteins, and binds a cytoplasmic domain of several cell-surface tyrosine kinase receptors.

ARID protein *abbr. for* AT-rich interaction domain protein; any of various proteins containing a domain that binds AT-rich regions in DNA. ARID proteins are involved in embryonic development, cell-lineage gene regulation, and cell-cycle control. The ARID sequence is ≈80 residues long.

arithmetic mean *see* **mean**.

arm a base-paired segment of the clover-leaf model of transfer RNA. There are several such arms: **amino-acid arm**, **anticodon arm**, **dihydrouridine arm**, **extra arm**, and **TψC arm**.

armadillo any of a family of proteins that includes β-catenin, which forms a multimeric complex with presenilins. These proteins contain an imperfect repeat of ≈42 residues, which forms three alpha helices. *See* **catenin**.

ARNT *abbr. for* aryl-hydrocarbon receptor nuclear translocator; *other name*: dioxin receptor; a protein that mediates translocation of the **Ah receptor** from the cytoplasm to the nucleus after ligand has been bound.

A-RNA *abbr. for* A form of RNA.

aro1 a fungal gene for aromatic amino-acid biosynthesis. It encodes a multifunctional protein for, e.g., **3-dehydroquinate synthase**, 3-dehydroquinase, **shikimate 5-dehydrogenase**, **shikimate kinase**, **EPSP synthase**.

aroA–H, aroL *Escherichia coli* genes for **chorismate** synthesis.

arogenate *or* **pretyrosine** 3-(1'-carboxylato-4'-hydroxy-2',5'-cyclohexadienyl)alanine; an intermediate in phenylalanine and tyrosine biosynthesis in *Euglena gracilis* and some other microorganisms, and in plants. It is formed by transamination of **prephenate** and is subsequently decarboxylated and converted to tyrosine, or converted to phenylalanine by dehydration and decarboxylation.

aromatase *or* **P450 arom** *or* **CYP19** an enzyme complex responsible for converting testosterone to estradiol. It catalyses complete oxidation and removal of the angular methyl group at C19, dehydrogenation and isomerization of the A ring, and reduction of the 3α position. It occurs in estrogen-producing cells in ovarian granulosa tissue, testes, placenta, brain (hippocampus, hypothalamus, and amygdala), and adipocytes. It consists of cytochrome P450 aromatase (CYP19) and a ubiquitous flavoprotein NADPH-cytochrome P450 reductase, which contains FAD and FMN and binds the substrate. Mutations in the gene for CYP19 lead to aromatase deficiency, characterized by female pseudohermaphroditism.

aromatic describing any organic compound characterized by one or more planar rings, each of which contains (usually) three conjugated double bonds and $(4n + 2)$ delocalized pi-electrons, where n is a small integer. They undergo substitution reactions more readily than addition reactions. The simplest member of the class is benzene. The term was originally used to distinguish fragrant compounds from **aliphatic** compounds. —**aromaticity** *n*.

aromatic amino acid any amino acid containing an aromatic ring. In proteins, these are phenylalanine, tryptophan, and tyrosine.

aromatic-L-amino acid decarboxylase *abbr.*: AADC; EC 4.1.1.28; *systematic name*: aromatic-L-amino-acid carboxy-lyase; *other names*: dopa decarboxylase; tryptophan decarboxylase; hydroxytryptophan decarboxylase. An enzyme that catalyses the decarboxylation of L-tryptophan to tryptamine, and acts on 5-hydroxy-L-tryptophan to yield 5-hydroxytryptamine, and on dihydroxy-L-phenylalanine (dopa) to yield **dopamine**; pyridoxal 5'-phosphate is a coenzyme. AADC is present in brain, liver, and other tissues. Several mutations in a locus at 7p12.1-p12.3 lead to enzyme deficiency. The human enzyme is a homodimer, and structurally similar to certain other eukaryotic decarboxylases.

aromatic amino acid hydroxylase any of three monooxygenase enzymes: phenylalanine 4-monooxygenase (EC 1.14.16.1); tyrosine 3-monooxygenase (EC 1.14.16.2); and tryptophan 5-monooxygenase (EC 1.14.16.4). All require tetrahydrobiopterin as coenzyme (*see* **biopterin**), and are homologous in amino acid sequence and gene structure. In each the N-terminal region binds the substrate, while the C-terminal region binds the coenzyme and catalyses the monooxygenation.

Arp *abbr. for* actin-related protein; either of two proteins, Arp2 and Arp3, that with five other polypeptides form a complex (the **Arp2/3 complex**) that mediates initiation of a new actin filament at an actin branchpoint. The complex is necessary for the formation of actin-filament networks. It is activated by binding to **WASP** before binding to the side of an existing actin filament. It also binds **cortactin**.

ARP_a *abbr. for* assimilatory regulatory protein a, i.e. ferredoxin-thioredoxin reductase.

ARP_b *abbr. for* assimilatory regulatory protein b, i.e. **thioredoxin**.

arrestin *or* **S antigen** any of a protein family that includes the Ca^{2+}-binding protein of the retinal rod outer segments. It binds to photoactivated-phosphorylated rhodopsin, thereby apparently preventing the transducin-mediated activation of phosphodiesterase. It has been implicated in autoimmune uveitis. β-Arrestin is a member of the arrestin family, sharing similarity in the C-terminal part with α-transducin and with other purine nucleotide-binding proteins. β-Arrestins regulate β-adrenergic receptor function; they seem to bind phosphorylated β-adrenergic receptors, thereby causing a significant impairment of their capacity to activate G_S proteins.

(Arrhenius) activation energy *symbol*: E_a *or* E_A; *see* **Arrhenius equation**.

Arrhenius equation an equation representing the effect of temperature on the velocity of a chemical reaction. It may be expressed in the forms:

$$\mathrm{d}\ln k/\mathrm{d}T = E_a/RT^2 \ (1)$$

or

$$\mathrm{d}\ln k/\mathrm{d}(1/T) = -E_a/R \ (2)$$

or

$$k = Ae^{-Ea/RT} \ (3)$$

where k is the rate constant for the reaction, E_a is its **(Arrhenius) activation energy**, R is the gas constant, T is the thermodynamic temperature, and A is a constant termed the frequency factor or the pre-exponential factor. Version 2 may be integrated to:

$$\ln k = \ln A - E_a/RT \ (4).$$

Version 4 of the equation requires a plot of $\ln k$ against $1/T$, known as an **Arrhenius plot**, to be a straight line of slope $-E_a/R$, a relationship that has been shown to obtain for a considerable number of chemical reactions. *See also* **temperature coefficient of reaction**. [After Svante August Arrhenius (1859–1927), Swedish chemist and physicist.]

Arrhenius plot *see* **Arrhenius equation**.

ARS *abbr. for* **1** autonomously replicating sequence; any DNA sequence that can confer upon a plasmid the property of becoming a **replicon** in a particular host. The term is largely confined to fungal genetics. **2** *(in clinical biochemistry) abbr. for* arginase.

Arsenazo III 2,2′-(1,8-dihydroxy-3,6-disulfonaphthalene-2,7-bisazo)bisbenzenearsonic acid; an indicator substance used in the determination of low (micromolar) concentrations of free calcium ions. The free acid is purple; the calcium adduct is blue-violet.

arsenic *symbol*: As; a semi-metal element of group 15 of the IUPAC **periodic table**; atomic number 33; relative atomic mass 79.41. It exists in three forms: yellow, grey, and black. The yellow form, comparable to white phosphorus, is less stable than the other forms and is converted into the grey (metallic) form by heat. The black form is more stable than the yellow form but is metastable with respect to the grey form. Arsenic is required in trace amounts by some organisms, but generally it acts as an antimetabolite.

arsenical 1 of or containing arsenic. **2** a drug or other agent (e.g. an insecticide, rodenticide, or nerve gas) containing arsenic as an active principle.

artemesinin *see* **sesquiterpene lactone**.

Artemis *see* **genome annotation software**.

arteriosclerosis a pathological condition in which there is a thickening and loss of elasticity of the walls of the arteries, which may become calcified. This interferes with the blood supply to various organs and tissues, resulting in impaired function. It normally, though not invariably, results from **atherosclerosis**.

arthritis any abnormality of a joint in which heat, redness, swelling, tenderness, or loss of motion or deformity are present. The two main types are osteoarthritis (OA) and rheumatoid arthritis (OR). OA is defined pathologically as reduced joint space secondary to a loss of cartilage. RA is of unknown etiology but may be caused by an infective agent or by **autoimmunity** to **type II collagen** or to **IgG**.

Arthus reaction a **complement**-dependent hypersensitivity reaction caused when antigen reacts with precipitating antibody in the skin or other tissues, forming microprecipitates in and around the small blood vessels and, secondarily, damaging cells. In the **passive Arthus reaction** antibody is injected intravenously and the antigen applied locally; in the **reversed passive Arthus reaction** the opposite applies. An **active Arthus reaction** may occur when antigen is injected locally subsequent to previous injection of the same antigen. The inflammation that results is by type III mechanisms (*see* **hypersensitivity** (def. 2)). [Described by (Nicholas) Maurice Arthus (1862–1945), French physiologist and bacteriologist, in 1903.]

articulin any of a family of cytoskeletal proteins of the **epiplasm** of flagellate and ciliate organisms. They form filaments and larger sheets or tubes. They contain repeated valine- and proline-rich motifs with the consensus VPVP_ _V_ _ _V.

artificial intelligence *abbr.*: AI; the computational simulation of human intelligence. AI methods try to capture human expertise in explicit rules, or with **neural networks** trained to recognize complex patterns.

artificial neural network *abbr.*: ANN; *an alternative name for* **neural network**.

aryl any univalent organic radical derived from an **arene** by loss of one hydrogen atom.

arylene any bivalent organic group derived from an **arene** by loss of two hydrogen atoms.

arylsulfatase any of a group of enzymes that catalyse the hydrolysis of a phenol sulfate to a phenol and sulfate. All sulfatases contain a cysteinyl residue that has been modified to a formylglycinyl residue required for activity. Arylsulfatases A, B, and C are common names for **cerebroside-sulfatase**, **N-acetylgalactosamine-4-sulfatase**, and **steryl-sulfatase** respectively. A cluster of genes at Xp222.3 encodes arylsulfatases C, D, E, and F. All are associated with the endoplasmic reticulum and have a neutral-to-alkaline pH optimum. The gene for arylsulfatase E is involved in X-linked chondrodysplasia punctata.

As *symbol for* arsenic.

ASA dehydrogenase *see* **aspartate-semialdehyde dehydrogenase**.

ASAT *(in clinical biochemistry) abbr. for* aspartate aminotransferase. *See* **aspartate transaminase**.

AS-C *abbr. for* Achaete–Scute complex. *See also* **AS-C protein**.

ascarylose 3,6-dideoxy-L-mannose; 3,6-dideoxy-L-*arabino*-hexose; a component of glycolipids found in eggs of the nematode, *Ascaris*, and in the bacterium, *Yersinia pseudotuberculosis*, type V. For the D enantiomer *see* **tyvelose**.

ascending boundary the solute boundary that moves upwards in one arm of the cell during electrophoresis in a **Tiselius apparatus**.

ascending chromatography any of the techniques of **chromatography** in which the mobile phase moves upwards over the solid phase.

ASCII *abbr. for* American Standard Code for Information Interchange; a set of standard characters readable by all computers. Strictly ASCII uses characters in the range 0–127, but an 8-bit version (range 0–255) is used, the higher numbers representing characters not in the Latin or Carolingian alphabets (e.g. Å or é).

ascites the accumulation of serous fluid in the peritoneal cavity. The presence of ascites may be due to the growth of a tumour elsewhere in the body, the ascitic fluid then containing a suspension of single tumour cells. Administration of a **hybridoma** results in the formation of **monoclonal antibodies** which can be recovered from the ascitic fluid. —**ascitic** *adj*.

Ascoli test a precipitin test used in the serological diagnosis of anthrax. The antigens are extracted with boiling saline and detected by a ring test with immune serum.

ascorbic acid a hexose, the L enantiomer of which is found in fruit and vegetables. It has vitamin C activity in humans (most other primates can synthesize ascorbic acid) and guinea pigs. Humans are among the few higher animals that are unable to synthesize it, being deficient in L-**gulonolactone oxidase**, the enzyme catalysing the terminal step in L-ascorbic acid synthesis. Ascorbic acid contains an ene-diol group, one hydroxyl group of which is acidic (pK_a 4.04) and ionizes to give an anion ascorbate; the other ene-diol hydroxyl acts as a reducing group ($E^{o'}$ = +0.058 V, pH 7), and is oxidized on conversion to dehydroascorbate. Ascorbate is required together with ferric ion as a cofactor in the oxidation of prolyl residues in collagen to hydroxyprolyl residues and in other reactions. Glucose transporter 1 (GLUT1) mediates transport of dehydroascorbic acid across cell membranes.

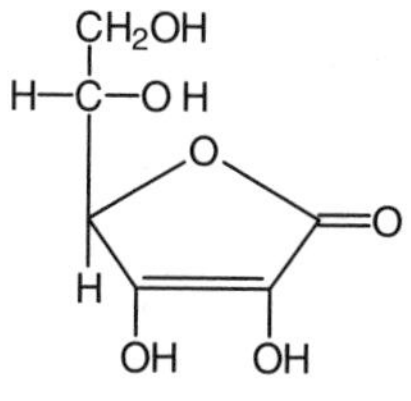

ascorbic acid

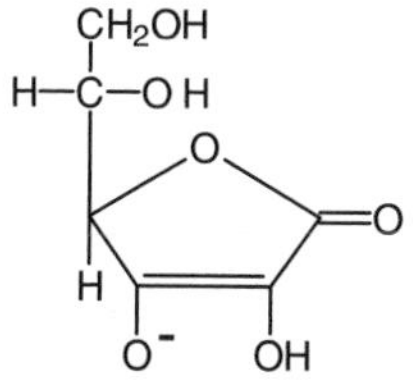

ascorbate ion

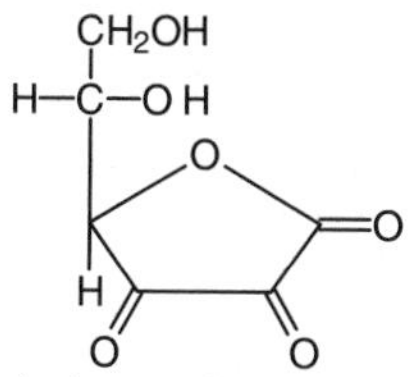

dehydroascorbic acid

ascospore a (usually sexually derived) haploid spore formed within an **ascus**.

AS-C protein *abbr. for* Achaete–Scute complex protein; any of a family of proteins encoded by the Achaete–Scute complex, a gene cluster first discovered in *Drosophila,* but with analogues in mammals and other organisms. They are involved in the determination of the neuronal precursors in the peripheral and central nervous systems. The T3, T4, and T5 proteins label strongly in the presumptive stomatogastric nervous system, while T8 is more prominent in the presumptive procephalic lobe. They resemble other members of the **MYC** family of helix-turn-helix transcription factors.

ascus (*pl.* **asci**) a sac-like fruiting body formed in ascomycete fungi, e.g. yeasts and *Neurospora*; it contains (typically eight) **ascospores** when mature.

+ase *suffix* denoting an enzyme; it is generally attached to a root indicating the substrate and/or the nature of the reaction catalysed. [from **diastase**.]

asemantic molecule any molecule that is not produced by an organism and therefore does not express any of the information that the organism contains. Absorbed molecules, although not derived from information contained within the organism, may nevertheless offer information about the organism relating to its absorption mechanisms. *See* **semantide**. *See also* **episemantic molecule**.

asepsis 1 the state of being **aseptic**. **2** the methods of making or keeping something aseptic. *Compare* **antisepsis**.

aseptic sterile; free from pathogenic microorganisms. *Compare* **antiseptic** (def. 2).

aseptic technique any technique incorporating measures that prevent the contamination of cultures, sterile media, etc. and/or the inappropriate infection of persons, animals or plants, by extraneous microorganisms.

asexual reproduction *or* **agamic reproduction** the development of a new individual from either a single cell or a group of cells without any sexual process. The term includes **agamogeny** and **vegetative reproduction**.

Ash protein *another name for* Grb2.

asialo+ *comb. form* denoting the absence from a compound of sialic-acid residues, e.g. after their deliberate enzymic removal from a naturally occurring **sialo-** (def. 2) compound.

A side the side of the nicotinamide ring of NAD(P)H from which projects the *pro-R* hydrogen atom (known as H_A) at the 4 position. *Compare* **B side**. *See also* **pro-R/pro-S** convention.

A site *abbr. for* aminoacyl-tRNA site (of a ribosome).

Asn *symbol for* a residue of the α-amino acid L-asparagine (alternative to N).

AsnC the asparagine biosynthetic operon regulatory protein, the product of *asnC*. In *Escherichia coli*, it is the activator of *asnA* expression, the autogenous regulator of *asnC* expression, and the repressor of *gidA* (a gene of *E. coli* associated with glucose-inhibiting division) at the post-transcriptional level. It is a **helix-turn-helix** (HTH) protein, and the prototype for the AsnC HTH family of regulatory proteins.

ASO *abbr. for* antistreptolysin-O; *see* **streptolysin**.

Asp *symbol for* a residue of the α-amino acid L-aspartic acid (alternative to D).

ASP *abbr. for* **agouti signal protein**.

asparaginase EC 3.5.1.1; *other names*: L-asparaginase; *systematic name*: L-asparagine amidohydrolase; an enzyme that catalyses the hydrolysis of L-asparagine to L-aspartate and NH_3. It is an effective antileukemic drug when injected into the bloodstream, its action being to deprive fast-growing tumour cells of the exogenous asparagine they require for rapid growth. Its clinical usefulness is limited because it causes damage to tissues with low asparagine synthetase activity. Several isoenzymes are found in several bacteria.

asparaginate 1 asparagine anion; the anion, $H_2N–CO–CH_2–CH(NH_2)–COO^-$, derived from asparagine. **2** any salt containing the asparagine anion. **3** any ester of asparagine.

asparagine *trivial name for* the β-amide of aspartic acid: aspartic 4-amide; α-aminosuccinamic acid; 2-amino-3-carbamoylpropanoic acid, $H_2N–CO–CH_2–CH(NH_2)–COOH$; a chiral α-amino acid. L-Asparagine (*symbol*: A *or* Asn; *(formerly)* also Asp(NH$_2$)), (*S*)-2-amino-3-carbamoylpropanoic acid, is a coded amino acid found in peptide linkage in proteins; codon: AAC or AAU. It is a non-essential dietary amino acid in mammals, and is glucogenic. One residue per molecule of D-asparagine (*symbol*: D-Asn *or* DAsn), (*R*)-2-amino-3-carbamoylpropanoic acid, may occur in the antibiotic bacitracin A as an alternative to one of D-aspartic acid. *Compare* **isoasparagine**.

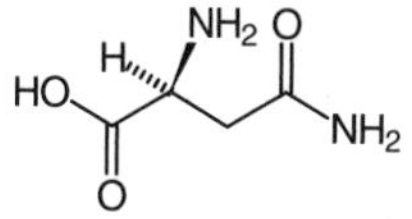

L-asparagine

asparagine repeat *or* **N repeat** a sequence in proteins that consists only of asparaginyl residues. For example, insulin receptor substrate 2 contains a sequence of eight such residues. Unlike glutamine repeats they are rare in mammalian proteins.

asparagine synthase (glutamine-hydrolysing) EC 6.3.5.4; *other name*: asparagine synthetase (glutamine-hydrolysing). An enzyme that catalyses a reaction between ATP, L-aspartate, and L-glutamine to form AMP, pyrophosphate, L-asparagine, and L-glutamate. *Compare* **aspartate–ammonia ligase**.

asparagino 1 N^2-asparagino; the alkylamino group, $H_2N–CO–CH_2–CH(COOH)–NH–$, derived from asparagine by loss of a hydrogen atom from its α-amino group. **2** N^4-asparagino; the acylamino group, $H_2N–CH(COOH)–CH_2–CO–NH–$, derived from asparagine by loss of a hydrogen atom from its amide group.

asparaginyl the acyl group, $H_2N–CO–CH_2–CH(NH_2)–CO–$, derived from asparagine. *Compare* **isoasparaginyl**.

aspartame *N*-L-α-aspartyl-L-phenylalanine methyl ester; a low-calorie artificial sweetener, about 160 times sweeter than sucrose in aqueous solution. Its sweetness is lost in cooking due to hydrolysis of the methyl ester. It is to be avoided in patients with phenylketonuria.

aspartase EC 4.3.1.1; *recommended name*: aspartate ammonia-lyase; *other name*: fumaric aminase. An enzyme that catalyses the hydrolysis of L-aspartate to fumarate and NH_3.

aspartate 1 aspartate(1–) *or* hydrogen aspartate; the monoanion derived from aspartic acid. In theory, the term denotes any ion or mixture of ions formed from aspartic acid and having a net charge of –1, although the species $^-OOC–CH_2–$ $CH(NH_3^+)–COO^-$ generally predominates in practice. **2** *the systematic name for* aspartate(2–); the dianion derived from aspartic acid. **3** any salt containing an anion of aspartic acid. **4** any ester of aspartic acid.

aspartate aminotransferase *see* **aspartate transaminase**.

aspartate–ammonia ligase EC 6.3.1.1; *other name*: asparagine synthetase; an enzyme that catalyses the reaction of ATP with L-aspartate and NH_3 to form AMP, pyrophosphate, and L-asparagine. *Compare* **asparagine synthase (glutamine-hydrolysing)**.

aspartate carbamoyltransferase *see* **aspartate transcarbamylase**.

aspartate kinase EC 2.7.2.4; *other name*: aspartokinase; an enzyme that catalyses the formation of 4-phospho-L-aspartate from ATP and L-aspartate with release of ADP. The reaction is the first step in the biosynthesis of lysine, methionine, and threonine in plants and bacteria. In *Escherichia coli*, the pathway for the synthesis of each amino acid is independently controlled by regulation of isoenzymes specific for each pathway. *See* **homoserine dehydrogenase** for another example.

aspartate-semialdehyde dehydrogenase EC 1.2.1.11; *systematic name*: L-aspartate-4-semialdehyde:NADP$^+$ oxidoreductase (phosphorylating); *other name*: ASA dehydrogenase; an enzyme that, although reversible, normally catalyses the formation of L-aspartate-4-semialdehyde, orthophosphate, and NADP$^+$ from L-4-aspartylphosphate and NADPH. Its product is a component of a pathway involved in the biosynthesis of lysine and methionine, and can be converted either into dihydropicolinate (a precursor of lysine) or into homoserine (a precursor of methionine). Example from *Escherichia coli*: DHAS_ECOLI, 367 amino acids (39.97 kDa).

aspartate transaminase *abbr.* (*in clinical biochemistry*): AST; EC 2.6.1.1; *systematic name*: L-aspartate:2-oxoglutarate aminotransferase; *other names*: transaminase A; glutamic– oxaloacetic transaminase; glutamic–aspartic transaminase. An enzyme that catalyses a reaction between L-aspartate and 2-oxoglutarate to form oxaloacetate and L-glutamate. It is a pyridoxal-phosphate enzyme, very widely distributed, performing a pivotal role in amino-acid metabolism. In mammals this includes the transfer of amino groups to aspartate, which then transfers them into the **urea cycle** through **argininosuccinate synthase**. Very high plasma concentrations of AST (more than 100 times normal) are found in cases of severe tissue damage including acute hepatitis, crush injuries, and tissue hypoxemia. In jaundice and myocardial infarction the levels reach 10 to 20 times normal; in myocardial infarction AST starts to rise about 12 h after the infarct, reaching a peak at 24 to 36 h, then declining over 2 to 3 days. The normal level in human plasma is in the range 10–50 IU L^{-1} (method-dependent; assumed presence of vitamin B$_6$ in the assay).

aspartate transcarbamylase *abbr.*: ATCase; EC 2.1.3.2; *recommended name*: aspartate carbamoyltransferase; *other names*: carbamylaspartotranskinase. An enzyme that catalyses a reaction between carbamoyl phosphate and L-aspartate to form *N*-carbamoyl-L-aspartate with release of orthophosphate. In many eukaryotes it is an activity of the **CAD protein**. In *Escherichia coli*, operon *pyrBI* codes for catalytic and regulatory chains; the catalytic chain consists of 311 amino acids (34.37 kDa). There are six motifs and the 3-D structure of the enzyme in several states is known. The protein has a very high α-helical content. Carbamoyl aspartate is an important component of the pathway for *de novo* synthesis of pyrimidines.

aspartic denoting a residue of aspartic acid, whether or not the β-carboxyl group is protonated.

aspartic acid *trivial name for* α-aminosuccinic acid; 2-aminobutanedioic acid, $HOOC–CH_2–CH(NH_2)–COOH$; a chiral α-amino acid. L-Aspartic acid (*symbol*: D *or* Asp), (*S*)-2-aminobutanedioic acid, is a coded amino acid found in peptide linkage in proteins; codon: GAC *or* GAU. In mammals it is a nonessential dietary amino acid and is glucogenic. Residues of D-aspartic acid (*symbol*: D-Asp *or* dAsp), (*R*)-2-aminobutanedioic acid, are found in the cell-wall material of bacteria of various species, and one residue per molecule may occur in the antibiotic bacitracin A as an alternative to one of D-asparagine. It is also formed by racemization of L-aspartic acid residues in long-lived proteins such as crystallins, dentine, and myelin.

L-aspartic acid

aspartic amide 1 aspartic 1-amide; *an alternative name for* **isoasparagine**. **2** aspartic 4-amide; *an alternative name for* **asparagine**.

aspartic andopeptidase *or* (*formerly*) **acid proteinase** *or* **aspartyl protease** *or* **carboxyl protease** any enzyme of the sub-subclass EC 3.4.23, comprising endopeptidases having a pH optimum below 5 by virtue of an aspartic residue being involved in the catalytic process. The group includes cathepsin D, chymosin, the pepsins, and renin.

aspartic proteinase any **endopeptidase** that contains an aspartyl residue that is essential for catalytic activity. Examples include pepsin, renin, rennin, and HIV protease.

asparto the alkylamino group, $HOOC–CH_2–$ $CH(COOH)–NH–$, derived from aspartic acid.

aspartoacylase an enzyme, EC 3.5.1.15, that catalyses the hydrolysis of *N*-acetyl-L-aspartate..

aspartokinase *see* **aspartate kinase**; *see also* **aspartokinase/ homoserine dehydrogenase**.

aspartokinase/homoserine dehydrogenase a multifunction enzyme, characterized in *Escherichia coli*, that catalyses two steps in the pathway from aspartate to diaminopimelate and lysine, to methionine, and to threonine and isoleucine. It is allosterically regulated by threonine. The *E. coli* protein is a homotetramer consisting of 820 amino acids (89.02 kDa); the first 149 amino acids form the aspartokinase domain, while the last 350 residues form the homoserine dehydrogenase domain.

aspartoyl the diacyl group, $–CO–CH_2–CH(NH_2)–CO–$, derived from aspartic acid.

α-aspartyl aspart-1-yl; the α-monoacyl group, $HOOC–CH_2–$ $CH(NH_2)–CO–$, derived from aspartic acid.

β-aspartyl *or* (*formerly*) **isoaspartyl** aspart-4-yl; the β-monoacyl group, $HOOC–CH(NH_2)–CH_2–CO–$, derived from aspartic acid.

β-aspartyl-*N*-acetylglucosaminidase EC 3.2.2.11; *systematic name*: 1-β-aspartyl-*N*-acetyl-D-glucosaminylamine L-asparaginohydrolase. An enzyme that catalyses the hydrolysis of 1-β-aspartyl-*N*-acetyl-D-glucosaminylamine to *N*-acetyl-D-glucosamine and L-asparagine. *See also* **aspartylglucosaminuria**.

aspartylglucosaminuria an autosomal recessive disorder of worldwide distribution associated with progressive mental retardation due to lysosomal accumulation of aspartylglucosamine (2-acetamido-1-β-L-aspartamido-1,2-dideoxy-β-D-glucose), which is excreted in abnormally high amounts in the urine. This compound is commonly found in several glycoproteins in normal individuals. The first sign of the condition is usually a delay in speech development (1–4 years). The enzymic effect is a deficiency of lysosomal aspartylglucosaminidase (N^4-(β-*N*-acetylglucosaminyl)-L-asparaginase, EC 3.5.1.26), which consists of two each of glycosylated A and B chains (130 and 205 residues, respectively), both derived from a precursor (346 residues) by proteolytic cleavage. A score of mutations in a locus at 4q34-q35 are associated with the disorder.

aspartylglucosylamine deaspartylase *see* N^4-(β-N-acetylglucosaminyl)-L-asparaginase.

β-aspartyl phosphate *or* 4-phospho-L-aspartate *or* **aspartyl-β-phosphate** an intermediate in the biosynthesis of lysine, threonine, and methionine in bacteria and plants. It is formed by **aspartate kinase**. It is converted to β-aspartyl semi-aldehyde, which is converted by the action of **homoserine dehydrogenase** to homoserine, a precursor of threonine or methionine. Alternatively addition of pyruvate to β-aspartyl phosphate leads to formation of dihydropicolinate, a precursor of lysine. A β-aspartyl phosphate residue is formed by phosphorylation of an aspartyl residue of Na^+,K^+-ATPase during the transport process.

aspartyl protease *an older name for* **aspartic proteinase**.

β-aspartyl semi-aldehyde $OHC–CH_2–CH(NH_2)–COOH$; *see* **β-aspartyl phosphate**.

aspartyl shift *or* α → β-**aspartyl shift** an intramolecular rearrangement reaction involving aspartic-acid residues that can occur during chemical synthesis or degradation of peptides. The β-carboxyl group of an aspartic residue displaces the α-carboxyl of the same residue from its amide (= peptide) linkage to the amino group of the adjacent amino-acid residue, thus forming an acid-sensitive β-amide (= isopeptide) linkage. Conditions must favour ionization of the β-carboxyl group. The reaction may be used to advantage in the partial hydrolysis of polypeptides and proteins by very dilute acid, but may occur to disadvantage during base-catalysed hydrolytic removal of protecting groups in the course of peptide synthesis.

Aspergillus a genus of filamentous fungi containing species of industrial and genetic importance. *A. niger* and *A. oryzae* are used in the production of citric acid, industrial enzymes, and fermented foods. The sexual species *A. nidulans* has been an important research tool in both biochemical and mitochondrial genetics. *A. flavus* is a source of **aflatoxins**.

asperlicin a naturally occurring antagonist of the peptide **cholecystokinin**, produced by various strains of *Aspergillus alliaceus*.

aspirin *or* **acetylsalicylic acid** 2-(acetyloxy)benzoic acid; an irreversible inhibitor of **prostaglandin-endoperoxide synthase**, through covalent acetylation of the α-amino group on the terminal serine active site. It is used as a drug for its antithrombotic activity and nonsteroidal anti-inflammatory and associated actions (antipyrexia, analgesia). *See also* **salicylic acid**.

Asp(NH₂) *(formerly) symbol for* L-asparagine.

Asp-NH₂ *symbol for* L-isoasparagine, an α-amino acid.

Asp(OMe) *symbol for* β-methyl aspartate; O^4-methyl hydrogen aspartate.

Asp-OMe *symbol for* α-methyl aspartate; O^1-methyl hydrogen aspartate.

assay *n.* **1** the determination of the activity, potency, strength, etc. of a substance, either on an absolute basis or in comparison with that of a standard preparation. **2** the determination of the relative amount(s) of one or more components of a mixture, or of the degree of purity of a substance. **3** to carry out such a determination. *See also* **bioassay, immunoassay, microbiological assay, radioassay**.

assembly *an alternative name for* **sequence assembly**.

assembly language *see* **programming language**.

assignment the identification of the part of a molecule that is responsible for a particular property, for example a cellular function or spectroscopic signal.

assimilation the absorption of simple foodstuffs (or of the products of their digestion) and their use in the biosynthesis of complex constituents of the organism; sometimes restricted to **photosynthesis**. —**assimilatory** *adj.*

assimilatory relating to autotrophic metabolism in which a substrate such as nitrate or sulfite is reduced and converted to organic compounds. *See* **nitrate reduction**. *Compare* **dissimilatory**.

assimilatory regulatory protein a *abbr.*: ARP_a; *obsolete name for* ferredoxin–thioredoxin reductase, obtained from chloroplasts.

assimilatory regulatory protein b *abbr.*: ARP_b; *obsolete name for* thioredoxin, obtained from chloroplasts.

association a reversible union between two chemical entities, whether alike or different, to form a more complex substance. *Compare* **dissociation** (def. 1).

association constant *symbol*: K_{ass} *or* K_a; the **equilibrium constant** for the reversible formation of a complex chemical compound from two or more simpler entities; the reciprocal of the **dissociation constant**, K_{diss}. Sometimes an **apparent** association constant, K'_{ass}, constrained with respect to stated values of certain variables (e.g. pH), is determined (*see* **equilibrium constant**). For an association of the type: A + B AB (the concentration) association constant is given by:

$$K_{ass} = [AB]/([A] \times [B]) = 1/K_{diss}$$

See also **affinity constant, off-rate, on-rate, stability constant**.

assortment *or* **reassortment** during meiosis, the distribution to opposing cell poles of the two members of each pair of chromosomes at anaphase I and of the two members of each pair of chromatids at anaphase II.

AST 1 *(in clinical biochemistry) abbr. for* aspartate transaminase. **2** *abbr. for* anion and sugar transporter, or sialin.

astacin EC 3.4.24.21; *other name*: *Astacus* proteinase; crayfish small-molecule proteinase; an enzyme that catalyses the hydrolysis of peptide bonds in substrates containing five or more amino acids, preferentially with Ala in the P1′ and Pro in the P2′ positions (*see* **peptidase P-sites**). Zinc is a cofactor.

astatin any of a subfamily of zinc metalloendopeptidases of unallocated number within sub-subclass EC 3.4.24, including **tolloid** (*Drosophila*) and bone morphogenetic protein 1 (i.e. C-proteinase of type I procollagen) (human), involved in differentiation.

astatine *symbol*: At; a halogen element of group 17 of the IUPAC **periodic table**; atomic number 85. It occurs as short-lived isotopes of mass numbers 215, 216, and 218 resulting from β-decay of other radioactive elements. Astatine-211, an α-emitter of half-life 7.5 h, results from α-bombardment of bismuth; it has potential in radiotherapy.

astaxanthin 3,3′-dihydroxy-β,β-carotene-4,4′-dione; a carotenoid pigment found mainly in animals (e.g. crustaceans, echinoderms) but also occurring in plants. It can occur free (as a red pigment), as an ester (e.g. the dipalmitate), or as a blue, brown, or green chromoprotein.

aster a system of cytoplasmic microtubules radiating from a **centriole**. It is present during cleavage of an ovum, during fusion of nuclei at fertilization, and in animal cells during cell division. *See* **importin, NuMA**.

asthma a common chronic and debilitating illness in which swollen and inflamed airways are produced by IgE-mediated allergic reactions to airborne allergens (most commonly from cockroaches, house-dust mites, or pets) or induced by exercise or cold . Two genes that seem to be involved are those for elevated IgE levels (locus at 5q31-q33) and for **ADAM-33** (locus at 20p). –**asthmatic** *adj.*, *n.*; –**asthmatically** *adv.*

astral microtubule *see* **prometaphase**.

astray *or* **Robo-2** a homologue (1513 residues) in zebrafish of **Robo**. Mutations in the gene cause disruption of retinal axon guidance.

Astrup technique a micro technique for the rapid determination of the acid–base status of a sample of blood. The pH of the blood as taken is measured and the remainder of the sample split into two and each part equilibrated with a different CO_2/O_2 mixture of known $p(CO_2)$ and their pH values measured. These last two pH values and the corresponding $p(CO_2)$ values are plotted on a Siggaard–Andersen nomogram and the points joined by a straight line. The $p(CO_2)$ of the original sample may be obtained from the point on this line corresponding to the pH of the blood as taken. Furthermore, the **standard bicarbonate value** of the blood may be obtained from the point where the straight line intersects with a scale on the $p(CO_2)$ axis at 5.33 kPa (40 mmHg).

ASV *abbr. for* anodic stripping voltammetry.

Asx *symbol for* a residue of either of the α-amino acids L-aspartic acid or L-asparagine when the state of amidation is uncertain (alternative to B).

asymmetric synthesis any synthesis of a compound that produces only one of two possible enantiomers of the product. This is usually the case for enzymic but not for nonenzymic reactions.

asymmetry absence of **symmetry**. A particular example of its meaning in biochemistry relates to the different phospholipid composition of the inner and outer leaflets of a lipid bilayer, said to confer asymmetry of composition. —**asymmetric** *adj.*

asymmetry potential the usually small potential across a membrane electrode, such as a glass electrode, when both sides are immersed in identical solutions. It is probably due to slight imperfections in the membrane.

asynapsis absence of **synapsis** (def. 1); i.e. the failure of homologous chromosomes to pair during meiosis.

asynteny the state when two genetic loci are situated on different, nonhomologous chromosomes. *Compare* **synteny**. –**asyntenic** *adj.* –**asyntenically** *adv.*

At *symbol for* astatine.

AT *abbr. for* **ataxia telangiectasia**.

atactic characterized by or showing irregularity in the spatial arrangement of the parts of a molecule.

atactic polymer any regular polymer in which the molecules display essential randomness with regard to the configurations at all main-chain sites of steric isomerism.

ataxia telangiectasia *abbr.*: AT; an autosomal recessive progressive cerebellar ataxia that presents in early childhood and is followed by variable degrees of telangiectasia and frequently immunodeficiency and cancer (usually lymphoid). The serum α-fetoprotein level is elevated in most patients, and X-ray sensitivity is also a feature of the disease. The gene locus at 11q22-q23 for AT mutated (ATM) encodes a protein containing 3056 residues, which includes regions of homology with yeast and *Drosophila* proteins involved in DNA repair and also with phosphatidylinositol kinases. Many mutations (most leading to truncated proteins) are associated with the disease.

ataxia with isolated vitamin E deficiency *abbr.*: AVED; an autosomal recessive neurodegenerative disease that is clinically similar to **Friedreich's ataxia** but characterized by very low plasma levels of tocopherol. Patients absorb dietary vitamin E normally by incorporating it into chylomicrons in intestinal mucosa but do not transfer it into very-low-density lipoproteins (VLDLs) in hepatocytes. Some 20 mutations in a gene at 8q13.1-q13.3 for α-tocopherol transfer protein (278 residues) are associated with the disease.

ataxin any of a group of proteins (of no known function) that contain glutamine repeats that expand in various forms of spinocerebellar ataxia (SCA). The gene for **ataxin-1** (at 6p23) encodes a polypeptide of 792–830 amino acids depending on the length (6–44 residues) of the repeat. Ataxin-1 is predominant in neuronal nuclei but also occurs in the cytoplasm in other tissues. The repeat length expands to 39–82 residues in SCA 1. **Ataxin-2** (≈140 kDa) is encoded by a gene at 12q24. It has a tissue distribution similar to that of ataxin-1. Its wild-type repeat of 15–32 residues expands to 36–63 in SCA 2. **Ataxin-3** (locus at 14q24.3-q31) has a wild-type repeat of 17–41 residues, which expands to 62–84 residues in SCA 3 (or Machado–Joseph disease). **Ataxin-7** (locus at 3p12-p13) is expressed ubiquitously but especially in the cerebellum, and is present in nuclei. The wild-type repeat of 4–35 residues expands to 37–306 in SCA 7. Ataxins-4, -5, -6, -8, -9, and -10 have different chromosomal loci and their repeat lengths also expand in the respective types of SCA.

+ate *suffix* **1** designating any anion (including mixtures of anion and free acid), salt, or ester of any **oxoacid** (def. 1) other than an +ous acid. **2** denoting the product of a process, e.g. distillate, hydrolysate.

ATF *abbr. for* activating (or activator) transcription factor; any of various transcription factors for eukaryotic RNA polymerase II promoters that bind to **CRE**. Example, ATF-A and ATF-A-δ (human): these form a dimer; they have different sequences (amino acids 114–134 missing in δ), and possess a leucine zipper.

(A + T)/(G + C) ratio the ratio of the sum of the adenine plus thymine bases to the sum of the guanine plus cytosine bases in a DNA molecule or preparation. The ratio is to some extent a characteristic of an individual type of DNA molecule. *Compare* **Chargaff's rules**.

athero-ELAMs endothelial leukocyte-adhesion molecules, i.e. **selectins**, expressed in atherosclerotic plaques.

atheroma a condition or process affecting blood vessels in which plaques form beneath the intima (i.e. inner lining). These start as in-

filtration by lipid-containing cells but later become invaded by fibrous tissue or become calcified. It is associated with hypercholesterolemia and defects in the **apolipoprotein B receptor** —**atheromatous** *adj.*

atherosclerosis an extremely common degenerative condition of the arteries that predisposes to myocardial infarction, cerebral thrombosis, ischemic gangrene of the extremities, and other serious complications. The arterial outer coat is thickened, the elastic lamina is fragmented and infiltrated by calcified deposits of cholesterol and other lipids, and the intima is hypertrophied and irregular. Hypertension, cigarette smoking, hypercholesterolemia, diabetes mellitus, and obesity are well-recognized risk factors.

aThr *symbol for* a residue of the α-amino acid L-allothreonine, (2*S*,3*S*)-2-amino-3-hydroxybutanoic acid.

ATM *abbr. for* ataxia telangiectasia mutated. *See* **ataxia telangiectasia.** .

atmosphere 1 the envelope of gas surrounding the Earth. **2** any local gaseous environment. **3** *symbol*: atm; a non-SI unit of pressure equal to 101.325 kPa; the pressure that will support a column of mercury 760 mm high at 273.15 K, sea-level, and latitude 45°.

atom a unit of matter consisting of a single **nucleus** surrounded by one or more orbital **electrons**. The number of electrons is normally sufficient to make the atom electrically neutral; adding or removing electrons converts the atom into a negative or positive ion, but this is regarded as a state of the same atom since the atom is characterized by its nucleus.

atomic absorption spectrophotometry a quantitative version of **atomic absorption spectroscopy** in which the absorption of electromagnetic radiation by the ground-state atoms in the flame is measured photoelectrically and the concentration of the element in question calculated by reference to the absorption observed with solutions of the elements of known concentration.

atomic absorption spectroscopy an analytical technique, based on the observation of the absorption of light energy by atoms, in which an atomic vapour of an element is formed by aspiration of a solution of the element into a flame, or otherwise, so that the large majority of the element's atoms remain in the non-emitting ground state. These ground-state atoms absorb electromagnetic radiation of their own specific resonance frequency, to an extent that is proportional to the density of the elemental atoms in the flame.

atomic mass constant *symbol*: m_u; a fundamental constant equal to the mass of a carbon-12 atom divided by 12; it is equal to the **unified atomic mass unit**.

atomic mass unit *abbr. (formerly)*: amu; *see* **unified atomic mass unit**.

atomic number *an older alternative name for* **proton number**.

atomic orbital the volume containing all the points within a free atom at which the Schrödinger wave-function of an electron is of appreciable magnitude.

atomic radius the distance between the centre of the nucleus and the outermost electron shell of an atom. *Compare* **van der Waals radius**.

atomic weight *abbr.*: at. wt.; *former name for* **relative atomic mass**.

atomic weight unit *abbr.*: awu; one-twelfth of the mean mass of the neutral atoms of naturally occurring carbon.

atomizer *or* **atomiser** an instrument for breaking up a solution into a spray of fine droplets.

atom percent the proportion of any nuclide in a mixture expressed as a numerical percentage of all the atoms of that element present irrespective of their nuclidic masses.

atom percent excess a measure of the abundance of a stable nuclide in a sample expressed in terms of the excess, in **atom percent**, over that naturally present. It is used to express the extent of enrichment or dilution of substances labelled with stable isotopes.

atopy *or* **atopic allergy** an IgE-mediated allergic response occurring in some individuals and produced by T lymphocytes, which secrete various cytokines when triggered by environmental antigens. The clinical result is asthma, eczema, or rhinitis. It is associated with a gene at 5q31-q33 for elevated IgE levels.

ATP *abbr. for* adenosine 5'-triphosphate.

ATPase *abbr. for* adenosine triphosphatase.

ATPase associated with varied activities *see* **AAA protease**.

ATP-dependent K⁺ channel *see* **KATP**.

ATP-diphosphatase *see* **apyrase**.

ATP[α,β-CH₂] *abbr. for* α,β-methyleneadenosine 5′-triphosphate; i.e. adenosine 5′-[α,β-methylene]triphosphate.

ATP[β,γ-CH₂] *abbr. for* β,γ-methyleneadenosine 5′-triphosphate; i.e. adenosine 5′-[β,γ-methylene]triphosphate.

ATP[β,γ-NH] *abbr. for* β,γ-imidoadenosine 5′-triphosphate; i.e. adenosine 5′-[β,γ-imido]triphosphate.

ATP(s) *abbr. for* adenosine 5′-γ-thiotriphosphate.

ATP synthetase *see* **H⁺-transporting ATP synthase**.

atractyloside an extremely toxic glycoside, obtained from the rhizomes of the Mediterranean thistle *Atractylis gummifera*, that produces hypoglycemia and convulsions in animals. It inhibits oxidative phosphorylation, in particular the translocation of mitochondrial adenine nucleotides. Carrier-bound adenine nucleotides are apparently replaced by atractyloside owing to the latter's higher affinity for a common translocation site. *Compare* **bongkrekic acid**.

A-transferase EC 2.4.1.40; *recommended name*: fucosylgalactose α-N-acetylgalactosaminyltransferase; *other names*: fucosylglycoprotein α-N-acetylgalactosaminyltransferase, [histo- blood group] A-transferase; an enzyme that can catalyse a reaction between UDP-N-acetyl-D-galactosamine and glycoprotein α-L-fucosyl-(1,2)-D-galactose to form UDP and glycoproteinN-acetyl-α-D-galactosaminyl-(1,3)-[L-fucosyl-(1,2)]-D-galactose. It thus catalyses the addition of N-acetylgalactosamine to the H antigen of the **ABH antigen** system, forming the A antigen. This and the **B-transferase** protein are products of alleles of the *ABO* gene, differing in only four residues.

atrial of or pertaining to an **atrium**.

atrial natriuretic peptide *see* **natriuretic peptide**.

atrial natriuretic peptide receptor *see* **natriuretic peptide receptor**.

atrioactivase a specific serine protease, present in bovine cardiac atria, that catalyses processing of the precursor of **natriuretic peptide** by selective cleavage at Arg⁸⁹.

atriopeptins biologically active peptides isolated from mammalian cardiac atria. Atriopeptin I (I) has 21 amino acids, with the structure

SSCFGGRIDRIGAQSGLGCNS

(3–19 disulfide link); atriopeptin II (II) is the same with additional Phe-Arg residues at the C terminus; atriopeptin III is the same as II but with an additional Tyr residue at the C terminus. I relaxes intestinal smooth muscle but not vascular smooth muscle, while II relaxes both. *Compare* **natriuretic peptide**.

atrium *or (formerly)* **auricle** (*pl.* **atria**) **1** either of the two chambers of the heart that lie above the ventricles. They receive blood from the afferent veins and pump it into the ventricles. **2** any of various anatomical chambers that receive the openings of other cavities.

atrolysin B EC 3.4.24.41; *other names*: *Crotalus atrox* metalloendopeptidase B; hemorrhagic toxin B (*abbr.*: HT-B); a snake-venom enzyme that catalyses the cleavage of His⁵-|-Leu⁶, His¹⁰-|-Leu¹¹, Ala¹⁴-|-Leu¹⁵, Tyr¹⁶-|-Leu¹⁷, and Gly²³-|-Phe²⁴ of the insulin B-chain. The reaction is identical to the cleavage of insulin B-chain by **atrolysin C**. The enzyme also cleaves Ser-|-Xaa bonds in glucagon. It is a zinc-requiring enzyme.

atrolysin C EC 3.4.24.42; *other names*: *Crotalus atrox* metalloendopeptidase C; hemorrhagic toxin C and D; a snake-venom enzyme from the Western diamondback rattlesnake that catalyses the cleav-

age of His⁵-|-Leu⁶, His¹⁰-|-Leu¹¹, Ala¹⁴-|-Leu¹⁵, Tyr¹⁶-|-Leu¹⁷, and Gly²³-|-Phe²⁴ of the insulin B-chain. With small-molecule substrates it prefers hydrophobic residues at the P2′ position and small residues such as Ala or Gly at the P1 position (*see* **peptidase P-sites**). It is a zinc-requiring enzyme.

atrolysin E EC 3.4.24.44; *other names*: *Crotalus atrox* metalloendopeptidase E; hemorrhagic toxin E; a snake-venom enzyme from the Western diamondback rattlesnake that catalyses the cleavage of Asn³-|-Gln⁴, Ser⁹-|-His¹⁰, and Ala¹⁴-|-Leu¹⁵ bonds in the insulin B-chain and Tyr⁴-|-Gln⁵ and Ala⁸-|-Ser⁹ in the A chain. It cleaves type IV collagen at Ala²⁵⁸-|-Gln²⁵⁹ in α₁-collagen and at Gly¹⁹¹-|-Leu¹⁹² in α₂-collagen. It is a zinc-requiring enzyme.

atrolysin F EC 3.4.24.45; *other names*: *Crotalus atrox* metalloendopeptidase F; hemorrhagic toxin F; a snake-venom enzyme that catalyses the cleavage of Val²-|-Asn³, Gln⁴-|-His⁵, Leu⁶-|-Cys⁷, His¹⁰-|-Leu¹¹, Ala¹⁴-|-Leu¹⁵, and Tyr¹⁶-|-Leu¹⁷ bonds in the insulin B-chain. It is a zinc-requiring enzyme.

atrophin a cytosolic protein that is widely expressed in tissues and contains glutamine, proline, and lysine repeats. It has no known function or homology to other proteins. The atrophin gene, at 12p13.31, encodes 1185 amino acids. The glutamine repeat (6–36 residues) expands to 49–84 residues (but the proline and lysine repeats are unchanged) in the form of spinocerebellar ataxia known as dentatorubropallidoluysian atrophy.

atrophy diminution in the size of a tissue or organ after full normal development has been attained. It is often a result of nutritional deficiency and/or decreased functional activity.

atropine (±)-hyoscyamine; tropyl dl-tropate; 1αH,5αH-tropan-3-ol (±)-tropate a tropane alkaloid obtained from deadly nightshade, *Atropa belladonna*; and other Solanaceae. The racemic mixture is a tertiary ammonium compound that acts as a **muscarinic receptor** antagonist. It is used as an anticholinergic, mydriatic, etc. *See also* **hyoscine**.

attachment proteins *general term for* proteins that are involved in binding other proteins to cell structures, e.g. **vinculin**, or α-actinin (*see* **actinin**), which bind actin to the plasma membrane.

attachment site *see* **active site** (def. 2).

A–T tailing **homopolymer tailing** of two types of duplex DNA molecule or fragment by the successive addition of adenine deoxynucleotide residues (A) to the 3′ ends of molecules of one type and the complementary addition of thymine deoxynucleotide residues (T) to the 3′ ends of molecules of the second type.

attB a bacterial attachment site that in the *E. coli* chromosome consists of a core of approx. 25 bp located between the *gal* and *bio* genes. In addition there are a number of secondary sites. Bacteriophage lambda DNA is integrated at this site through its attachment site ***attP*** (243 bp) in a reaction catalysed by the phage-encoded integrase (Int) and the integration host factor (IHF). On integration the prophage is bounded on its left side by the ***attL*** site, and on the right side by the ***attR*** site. Excision of the prophage takes place by recombination at the attL and attR sites in a reaction catalysed by Int, IHF, and excisionase (Xis). This recreates the ***attB*** site in the bacterial chromosome and the attP site in lambda DNA. attB and attP sites are used in plasmid vectors designed from cloning by site-specific recombination.

attenuance *symbol*: D; the logarithm to the base 10 of the ratio of the **radiant powers** of the incident radiation, P_0, and the transmitted radiation, P; thus:

$$D = \lg(P_0/P) = \lg T^{-1},$$

where T is the **transmittance**. Attenuance reduces to **absorbance** if the incident beam is only either transmitted or absorbed, but not reflected or scattered.

attenuated virus a virus that has become less pathogenic following passage in culture outside its natural host, or by the use of physical or chemical means.

attenuator a sequence in DNA that is located between the **operator** and the gene for the first protein in an **operon**. It can cause transcription termination. The operon for the synthesis of **tryptophan** in bacteria is an example of an operon that contains an attenuator.

atto+ *symbol*: a; *prefix* used with SI units denoting 10^{-18} times.

attractant *or* **chemoattractant** a substance that a motile cell or organism tends to move towards; i.e. one that elicits positive **chemotaxis**. For a responsive bacterium, attractants can be either nutritive or nonmetabolizable. An example is **cytotaxin**.

Au *symbol for* gold.

AUA 1 a codon in mRNA, excluding that of mammalian mitochondria, for L-isoleucine. **2** a codon in mammalian mitochondrial mRNA for L-methionine (and for *N*-formyl-L-methionine, chain initiation, in some species).

Aubergine a protein of *Drosophila* that is required for activation of mRNA for *oskar*-encoded protein. It is related to eIF2C.

AUC a codon in mRNA for L-isoleucine.

AUG a codon in mammalian mitochondrial mRNA, and in non-mitochondrial mRNA the only codon for L-methionine and usually for *N*-formyl-L-methionine, chain initiation.

Auger effect a process by which an orbital electron in an atom passes from an excited to a lower energy level. The X-ray so produced collides with, and ejects, another orbital electron, known as an **Auger electron**, with energy in the X-ray range. [After Pierre Victor Auger (1899–1993), French physicist.]

auracyanin a small blue copper-containing bacterial outer membrane glycoprotein (one Cu^{2+} ion per protein) that donates electrons to cytochrome c_{554}.

auranofin tetraacetylthioglucosato-*S*)(triethylphosphane)gold(I); a gold compound used in the treatment of rheumatoid arthritis.

aureomycin 7-chlortetracycline. *See* **tetracycline**.

auriculin either of two overlapping N-terminal fragments of atrial **natriuretic peptide**, termed auriculin A and auriculin B.

Auriculin *a proprietary name for* atrial natriuretic peptide.

aurosome an artificially induced organelle occurring in animal cells cultured in the presence of gold. It is an electron-dense lysosomal body containing gold particles.

Australia antigen *an alternative name for* hepatitis B antigen. *See* **hepatitis**.

aut+ *a variant form of* **auto+**.

autacoid any druglike principle that is produced in or can be extracted from the organs of internal secretion. Autacoids include both excitatory substances (**hormones**) and inhibitory substances (**chalones**).

auto+ *or (before a vowel)* **aut+** *comb. form* **1** self; one's own. **2** occurring within oneself or itself. **3** acting by itself; automatic; spontaneous.

autoagglutination the tendency of a suspension of bacteria or other cells to agglutinate spontaneously in a simple medium not containing an **agglutinin** or other similar reagent.

autoallergy *an alternative term for* **autoimmunity**.

autoanalyser *or (esp. US)* **autoanalyzer** any instrument for effecting analyses automatically.

autoantibody an **antibody** produced in an animal that reacts with a constituent of the animal's own tissues, which thus acts as a 'self' antigen or **autoantigen**.

autoantigen an antigen that is a normal constituent of an individual and has the capacity to produce an **immune response** in the same individual in certain circumstances. Examples include: thyroid peroxidase (in autoimmune thyroiditis); TSH receptor (in Graves' disease); skeletal muscle acetylcholine receptor (in myasthenia gravis); and myelin basic protein (in multiple sclerosis).

autocatalysis the catalysis of a chemical or biochemical reaction by one of the reaction products. —**autocatalytic** *adj.*

autochthonous derived from that part of the body where found; native. *Compare* **autologous**.

autoclave an instrument, used for sterilizing culture media, instruments, etc., that consists of an airtight chamber which can be filled with, or surrounded by, steam at high pressure.

autocoupling hapten a reactive compound of low molecular mass that, when injected into an animal, reacts with tissue antigens to form hapten–antigen compounds. These then lead to the formation of antibodies to the **hapten**.

autocrine describing an agent that acts on the cell in which it is produced.

autofluorogram *an alternative name for* **fluorogram** (when produced by **fluorography** (def. 2)).

autofluorography *an alternative term for* **fluorography** (def. 2).

autogenesis *or* **autogeny** *an alternative term for* **abiogenesis**. —**autogenetic** *adj.*

autogenous 1 self-produced. **2** originating within, especially within one's own body.

autogenous regulation the mechanism by which a protein directly controls the expression of its own structural gene.

autograft any graft or transplant of the subject's own tissue(s).

***Autographa californica* nuclear polyhedrosis virus** *abbr.*: AcNPV *or* AcMNPV; the baculovirus most commonly used as a gene expression vector. It is used for the biocontrol of the alfalfa looper (*Autographa californica*), a moth whose larvae cause crop damage.

autoimmune regulator *abbr.*: AIRE; a transcriptional activator protein that contains zinc finger domains and is expressed predominantly in the thymus but also in other tissues. It consists of 545 amino acids, and is encoded by a locus at 21q22.3. Several mutations in the gene result in autoimmune polyendocrinopathy and ectodermal dystrophy.

autoimmunity *or* **autoallergy** a state of immunological reactivity against constituents of the subject's own tissues. Autoimmune diseases include myasthenia gravis, multiple sclerosis, rheumatoid arthritis, lupus erythematous, Graves' disease, and thyroiditis.

autoimmunization *or* **autoimmunisation** the process of inducing **autoimmunity**.

autologous derived from the same organism. *Compare* **autochthonous**.

autolysate *or (esp. US)* **autolyzate** a product of **autolysis**, e.g. a suspension of broken cells.

autolysin EC 3.5.1.28; *other name*: *N*-acetylmuramoyl-L-alanine amidase; an enzyme that catalyses the hydrolysis of the link between *N*-acetylmuramoyl residues and L-amino-acid residues in certain bacterial cell-wall glycopeptides. The enzyme is required for cell separation, cell-wall turnover, etc. *See also* **lysostaphin**.

autolysis the process of self-digestion that occurs in plant and animal tissues after death of the organism or following separation of tissue from the rest of the organism. It is due to the action of the tissue's own enzymes. —**autolytic** *adj.*

autolysosome an autophagic vacuole.

automutagen any **mutagen** formed in an organism as a normal or abnormal metabolic product that is mutagenic in the same organism.

autonomic nervous system *abbr.*: ANS; those parts of the **peripheral nervous system** and **central nervous system** that govern homeostasis in vertebrates and are generally independent of voluntary control.

autonomous existing or able to exist independently; used, e.g., of a tumour cell that is free of host control or of an episome or plasmid that replicates independently of the chromosome.

autophagic vacuole an enlarged **lysosome** containing subcellular structures such as mitochondria; a secondary lysosome.

autophagosome a **phagosome** containing endogenous cellular debris. *Compare* **heterophagosome**.

autophagy a process, apparently ubiquitous among eukaryotes, by which cells digest parts of their own cytoplasm. *Compare* **heterophagy**. —**autophagic** *adj.*

autophagy factor *abbr.*: Apg; any of ≈16 ubiquitin-like modifier proteins of yeast that are required for autophagy induced by starvation. They show weak homology with ubiquitin.

autophosphorylation the phosphorylation by a protein of one of its own residues. It is important in signal transduction by growth hormone receptors (*see*, e.g., **epidermal growth factor receptor**), but occurs in the case of other proteins, e.g. **caldesmon**.

autoprothrombin I *an alternative name for* factor VII; *see* **blood coagulation**.

autoprothrombin II *an alternative name for* factor IX; *see* **blood coagulation**.

autoprothrombin IIA *an alternative name for* **protein C**.

autopsy *or* **necropsy** the examination of a dead body, commonly to determine the cause of death or the presence of disease processes; a post-mortem (examination). *Compare* **biopsy**.

autoradiogram an **autoradiograph**, especially of a chromatogram or electrophoretogram.

autoradiograph *or* **autoradiogram** the image produced by **autoradiography**.

autoradiography a method by which radioactive material can be localized in, e.g., a tissue, cell, a cell part, or molecule, or in a chromatogram or electrophoretogram. The sample containing a radioactive substance is put in direct contact with a thick layer of specially prepared photographic emulsion. The radiation emitted in the decay of the radionuclide(s) activates individual silver halide grains in the emulsion rendering them susceptible to conversion to metallic silver grains by a photographic developer. After fixation the silver grains may be located either visually or by microscopy.

autosomal of, relating to, or being an **autosome**.

autosomal disease disease that results from an abnormality affecting the **autosomes**.

autosomal dominant polycystic kidney disease *abbr.*: ADPKD; a disease that affects 1:1000 people and is characterized by development of numerous renal cysts and consequent renal failure. It is usually caused by mutation of the *PKD1* gene, locus 16p13.3-p13.12, encoding **polycystin** 1, but can also arise through mutation of the *PKD2* gene, locus 4q21-q23, encoding polycystin 2.

autosomal limb girdle dystrophy a genetically heterogeneous group of diseases occurring in both dominant and recessive forms. Limb girdle muscular dystrophies (LGMD) are characterized by muscle weakness of both upper arms and legs. Genes for five forms of autosomal dominant LGMD have been localized or identified, including the myotilin locus at 5q31 and the caveolin-3 locus at 3p25. Genes for nine forms of autosomal recessive LGMD have been localized or identified including five that involve the sarcoglycan complex. Dominant forms of the disease usually show adult onset whereas the recessive forms have childhood onset.

autosomal sex reversal *or* **campomelic dysplasia**. *See SOX*.

autosome any **chromosome** except the **sex chromosomes**. Hence the genes carried by the autosomes show **autosomal linkage** (or **autosomal inheritance**) according to the assortment of their respective autosomes to gamete cells during meiosis. Humans have 22 pairs of autosomes in addition to a pair of sex chromosomes.

autosteric effector any **effector** that binds to, and exerts its effect at, part of or immediately adjacent to the **active site** of an **allosteric** protein. It is distinguished from an **allosteric effector** operating at a distant site.

autotroph *or* **lithotroph** any organism that requires only simple inorganic substances to fulfil its nutritional requirements and for which gaseous or dissolved carbon dioxide is the sole source of carbon for the synthesis of cellular constituents. The term often includes any microorganism for which trace amounts of certain substances, e.g. vitamins, must also be supplied. Autotrophs may be subdivided into **chemoautotrophs** (*or* chemolithotrophs) and **photoautotrophs** (*or* photolithotrophs) according, respectively, to whether they are **chemotrophs** or **phototrophs**. *Compare* **heterotroph**. —**autotrophic** *adj.*; **autotrophy** *n*.

autoxidation *or* **autooxidation** an oxidation reaction that occurs spontaneously in the presence of an oxidant, usually oxygen.

AUU a codon in mRNA for L-isoleucine.

auxiliary amino acid any amino-acid residue in the **active site** (def. 1) of an enzyme that is more than one bond distance removed from the substrate molecule in the enzyme–substrate complex. *Compare* **contact amino acid**.

auxin any of a group of plant hormones, produced by regions of actively dividing and enlarging cells, that regulate aspects of plant growth. They include indole-3-acetic acid and indole-3-acetonitrile.

Auxins bind an auxin-response protein, which contains an N-terminal domain that binds to a 6 bp auxin-response element in auxin-inducible genes.

auxin-binding protein 1 *abbr.*: ABP1; a small family of proteins (≈23 kDa) that bind indole 3-acetic acid, naphthalene 1-acetic acid, and other molecules with auxin activity in plants. They contain a C-terminal HDEL putative ER-retention domain. Strong circumstantial evidence suggests that they may be auxin receptors. Plant cells contain a variety of enzymes and other proteins that also bind auxin.

auxochrome any saturated atomic grouping (e.g. -OH, -CH₃, -NH₂, -Cl, -SH) that, when attached directly to a **chromophore**, shifts the selective absorption of light in the direction of longer wavelengths and enhances the intensity of absorption.

auxostat in industrial fermentation, a **chemostat** in which the **dilution rate** can be varied, normally in response to some parameter such as growth rate or pH, so that growth rate is more readily maximized.

auxotroph any strain of microorganism (alga, bacterium, or fungus) that differs from the **wild-type** by requiring a supply of one or more growth factors. *Compare* **prototroph**. —**auxotrophic** *adj.*

auxotrophy the condition of being an **auxotroph**.

***Ava*I** a type 2 **restriction endonuclease** recognition sequence: C↑[TC]CG[AG]G.

***Ava*II** a type 2 **restriction endonuclease**; recognition sequence: G↑G[AT]CC.

avenacin A-1 a triterpenoid saponin present in roots of oat plants but not in those of wheat or barley. It is highly effective against a major pathogen of cereal roots, one that therefore affects wheat and barley but not oat plants. The terminal β-D-glucose (1→2 or 1→4) units are removed by the saponin-detoxifying avenacinase that is produced by a variety of this pathogen that affects oats.

avenacinase *see* avenacin A-1

avenic acid a phytosiderophore (or iron chelator) secreted by roots of oat plants (*Avena sativa*). It is a nonprotein amino acid derived from the α-aminobutyrate moieties of three S-adenosylmethionine molecules.

average *an alternative term for* (arithmetic) **mean**.

average life *see* **mean life**.

average molar mass *or (formerly)* **average molecular weight** any of a series of numeric characteristics of a macromolecular system that are useful for assessing the extent of its polydispersity. The three most important are the **number-average molar mass**, M_n, the **mass-average molar mass**, M_m, and the **Z-average molar mass**, M_z. These three characteristics are given by the expressions:

$$M_n = \Sigma n_i M_i / \Sigma n_i \text{ (kg mol}^{-1})$$

$$M_m = \Sigma n_i M_i^2 / \Sigma n_i M_i \text{ (kg mol}^{-1})$$

$$M_z = \Sigma n_i M_i^3 / \Sigma n_i M_i^2 \text{ (kg mol}^{-1})$$

where n_i is the amount of substance of species i and M_i is the molar mass of species i. Measurement of a **colligative property**, e.g. osmotic pressure, or end-group analysis, gives M_n; light scattering, dielectric dispersion, or fluorescence depolarization measurements give M_m; and data obtained by equilibrium ultracentrifugation will yield either M_m or M_z depending on how they are treated. For polydisperse or heterogeneous systems, $M_z > M_m > M_n$, whereas for monodisperse systems these three characteristics are equal (i.e. equal to the relative molecular mass, M_r, of the single species).

average molecular weight *former name for* **average molar mass**.

average radius of gyration *an alternative term for* **mean radius of gyration**.

average relative molecular mass *former name for* **average molar mass**.

avermectin B₁ *an alternative name for* **abamectin**.

avidin a 66 kDa glycoprotein comprising four essentially identical subunits, each of which consists of a single polypeptide chain of 128 amino-acid residues. Isolated from the white of the eggs of birds and amphibia, it binds extremely strongly to **biotin**, and produces biotin deficiency when fed to experimental animals.

avidity 1 the tendency of an antibody to form more-or-less stable complexes with a (macromolecular) antigen; a measure of this tendency. **2** *an alternative term for* **affinity**.

*Avi*II a type 2 **restriction endonuclease**; recognition sequence: TGC↑GCA. *Mst*I and *Fsp*I are **isoschizomers**.

avirulent lacking **virulence**.

avitaminosis a condition resulting from deficiency of one or more **vitamins**. The deficient vitamin may be specified, as in avitaminosis A, B, etc.

Avogadro constant *or (formerly)* **Avogadro's number** *symbol: L or* N_A; the number of molecular entities in one mole of any chemical species. It is a fundamental physical constant, of value 6.022 136 7(36) × 10²³ mol⁻¹. *Compare* **Loschmidt constant**.

Avogadro number *or* Avogadro's number the numerical value of the **Avogadro constant**.

Avogadro's hypothesis *or* **Avogadro's law** the principle that equal volumes of all ideal gases, at the same temperature and pressure, contain equal numbers of molecules. [After Lorenzo Romano Amadeo Carlo Avogadro, Count of Quaregna and Cerreto (1776–1836), Italian physicist and chemist who formulated it in 1811.]

avogram *see* **dalton**.

AVP *abbr. for* arginine vasopressin.

awk *see* **programming language**.

axenic pertaining to the growth of organisms of a single species in the absence of living organisms or living cells of any other species (also, strictly, in the absence of living cells of the organism itself except those in the intact living organism and its gametes). —**axenity** *n.*; **axenically** *adj.*

axenize *or* **axenise** to render axenic.

axenite an organism grown axenically.

axial 1 of, relating to, or being an **axis** (def. 1 or 2); located on or close to such an axis. **2** *(in stereochemistry) Symbol:* a. *See* **conformation**.

axial ratio the ratio of the length of the major axis to that of the minor axis of a prolate ellipsoid of rotation, a model often used for a protein or other macromolecule in solution.

axin any of a small family of proteins that bind glycogen synthase kinase 3 (a serine/threonine kinase), β-catenin, and APC protein (*see* **APC**) to form a complex that targets β-catenin for degradation (*see* **catenin**). It also interacts with the catalytic subunit of protein phosphatase A2 and is a negative regulator of Wnt signaling. Axin 2 is also called conductin or axil.

axis 1 a straight line about which a body or three-dimensional figure, e.g. a crystal, is considered to rotate. **2** one of the reference lines of a co-ordinate system. **3** the second vertebra of the neck in higher vertebrates.

axis of rotation *see* **rotation axis**.

axis of symmetry *see* **symmetry, axis of**.

axokinin a heat-stable phosphoprotein, M_r 56 000, present in flagella and cilia of vertebrate and invertebrate species. Its cyclic AMP-dependent phosphorylation is required for initiation and maintenance of flagellar motility.

axon the long process of a neuron that conducts nerve impulses, usually away from the cell body to the terminals which are the site of storage and release of neurotransmitter. *Compare* **dendrite**.

axonal transport the directed transport of organelles and molecules along the axon of a neuron. It can be anterograde (i.e. outwards from the cell body) or retrograde (i.e. towards the cell body).

axoneme the bundle of microtubules and associated proteins that forms the core of cilia and flagella in eukaryotic cells and is responsible for their movements.

axoplasm the cytoplasm of the axon of a neuron, especially after it has been extruded from the axon.

aza+ *prefix* **1** *(in systematic chemical nomenclature)* denoting the replacement of a designated carbon atom (together with one associated hydrogen atom) by a nitrogen atom. **2** *(in trivial chemical names)* denoting presence of one or more nitrogen atoms.

azaserine *trivial name for* serine diazoacetate; *O*-diazoacetyl-L-serine, N_2=CH–CO–O–CH$_2$–CH(NH$_2$)–COOH; an antibiotic substance produced by *Streptomyces* spp. It is an inhibitor of amidotransferase enzymes and interferes with glutamine utilization in purine synthesis; it also retards the growth of transplantable animal neoplasms and is a specific inhibitor of L-glutamate ferredoxin oxidoreductase (transaminating), EC 1.4.7.1, in ammonium assimilation in plants.

azathioprine 6(1-methyl-4-nitro-5-imidazolyl) mercaptopurine: 6-[(1-methyl-4-nitroimidazol-5-yl)thio]purine; a derivative of **mercaptopurine**, into which it is converted *in vivo*. It is useful as an immunosuppressive and antimitotic agent. *One proprietary name*: Imuran.

azelaic acid 1,7-heptanedicarboxylic acid, nonanedioic acid; a substance that acts as a substitute for biotin in the growth of a number of microorganisms and is proposed as an intermediate in biotin biosynthesis from oleic acid.

azeotrope a mixture of two or more liquids that distils without change in composition and has a constant boiling temperature.

AZF *abbr. for* azoospermia factor; a protein whose deficiency is associated with the absence of living sperm in semen (azoospermia). It consists of 366 amino acids, encoded by a gene at Yq11. The N-terminal region contains an 85-residue sequence homologous to one in many proteins that bind RNA or single-stranded DNA, and its C-terminal region includes seven tandem repeats of a 27-residue sequence. Deletion of the locus accounts for ≈13% of infertile men with azoospermia.

azide 1 any organic compound containing the monovalent azido group, –N=N⁺=N⁻. **2** the ion N$_3$⁻, derived from hydrazoic acid, HN$_3$. This ion blocks the electron transport chain by reacting with the ferric form of cytochrome aa_3 and thus acts as an inhibitor of respiration. **3** any salt or ester of hydrazoic acid.

azidothymidine *see* **AZT**.

azo the bivalent group –N=N–. It participates in noncyclic covalent linkage in organic molecules forming **azo compounds** or **diazo compounds**.

azocoll an insoluble bright-red collagen-rich proteinaceous material prepared by dyeing powdered cowhide or kangaroo-tail tendon with a dye prepared from tetrazolized benzidine, disodium 2-naphthol-3,6-disulfonate and sodium acetate. It is digested by most types of proteinase, the consequent solubilization of dye forming the basis of a simple nonspecific technique for the assay of proteolytic activity.

azo compound any organic compound with the general formula R–N=N–R'. Such compounds thus contain the bivalent azo group in covalent linkage, each nitrogen atom being linked to a carbon atom and replacing one hydrogen atom of each of the parent molecules RH and R'H. The latter may be the same or different, aliphatic or aromatic; e.g. azomethane, CH$_3$–N=N–CH$_3$; naphthalene-2-azobenzene, C$_{10}$H$_7$–N=N– C$_6$H$_5$. *Compare* **diazo compound**.

azoferredoxin *or* **protein II** *a former name for* **dinitrogenase reductase**.

azomethine strictly, any **imine** having the more limited general structure RR'C=NR where R and R' are any organyl groups, but often extended to include any compound where R' is H, the term then being considered to be synonymous with **Schiff base**.

azoospermia factor *see* **AZF**.

azoprotein any of the modified proteins formed by coupling an aromatic diazo compound to one or more of the tyrosine residues of a protein, e.g. azoalbumin, azocasein.

azotemia *former term for* the accumulation of abnormally large concentrations of nitrogenous compounds, especially urea, in the blood.

Azotobacter a genus of Gram-negative strictly aerobic soil bacteria, all members of which can effect **nitrogen fixation**.

azotoflavin a somewhat larger than normal **flavodoxin** that is believed to act as a one-electron carrier between ferredoxin and azoferredoxin during nitrogen fixation in some bacteria, e.g. *Azotobacter*.

AZT *abbr. for* azidothymidine; 3′-azido-3′-deoxythymidine; *other names*: zidovudine; Retrovir; an analogue of thymidine, the phosphorylated form of which is an inhibitor of **reverse transcriptase** in retroviruses; it also terminates DNA synthesis and depletes mitochondrial DNA. AZT undergoes phosphorylation in human T-cells to a nucleoside 5′-triphosphate, which competes with thymidine triphosphate and serves as a chain-terminating inhibitor of HIV reverse transcriptase. It is used clinically to treat patients with **HIV** infection and **AIDS**.

azurin any member of a group of brilliant-blue, copper-containing proteins of low M_r found in some bacteria, and thought to transfer electrons to cytochrome oxidase.

azurocidin *or* **CAP37** a **serpocidin** of 37 kDa that acts as a cationic antimicrobial protein in neutrophil azurophil granules.

azurophil(e) *or* **azurophilic** a blue-staining lysosome of neutrophil leukocytes that also contains **serpocidins**.

azurophil(e) granule *or* **primary granule** a blue-staining lysosome of neutrophil leukocytes.

Bb

b *symbol for* **1** molality. **2** one of two van der Waals coefficients. **3** breadth.

B *symbol for* **1** boron. **2** a residue of either of the α-amino acids L-aspartic acid or L-asparagine when the state of amidation is uncertain (alternative to Asx). **3** a residue of an incompletely specified base in a nucleic acid sequence that may be cytosine or guanine or either thymine (in DNA) or uracil (in RNA). **4** the ribonucleoside (5-)bromouridine (alternative to BrUrd). **5** the B blood group. *See* **ABO system**.

B *symbol for* **1** Napierian absorbance (*see* **absorbance**). **2** magnetic flux density (bold italic).

-B conformational descriptor designating the boat conformation of a six-membered ring form of a monosaccharide or monosaccharide derivative. Locants of ring atoms that lie on the side of the structure's reference plane from which the numbering appears clockwise are indicated by superscripts preceding the letter, and those that lie on the other side of the reference plane by subscripts following the letter; e.g. methyl 2,6-anhydro-α-D-altropyranoside-$^{2,5}B$. *See also* **conformation** (under *cyclic compounds*).

***B*$_{max}$** *symbol for* the amount of drug required to saturate a specific population of receptors in a membrane sample and hence an index of the density of receptor binding sites in the sample. It is usually derived from Scatchard analysis (*see* **Scatchard plot**) or from nonlinear regression analysis of data from a saturation binding assay.

Ba *symbol for* barium.

Baa helices *abbr. for* basic amphiphilic α helices; α helices that contain an array of basic amino-acid residues (His, Lys, or Arg) on one side and hydropathic residues on the other. They are found, e.g., in the calmodulin-binding regions of many proteins.

Babes–Ernst body *or* **Babes–Ernst granule** a **volutin granule** seen in *Corynebacterium diphtheriae*. [After Victor Babes (1854–1926), Romanian bacteriologist, and Paul Ernst (1859–1937), German pathologist.]

BAC *abbr. for* bacterial artificial chromosome, based on the F factor (**F plasmid**) of *Escherichia coli. Compare* **YAC**.

BACE2 *abbr. for* beta-site APP-cleaving enzyme 2. *See* **secretase**.

Bacillaceae a family of eubacteria that includes the genera *Bacillus* and *Clostridium*.

bacillary *or* **bacillar 1** of, relating to, caused by, or containing bacilli. **2** *an alternative term for* **bacilliform** (def. 1).

bacilliform 1 rod-shaped. **2** resembling a bacillus.

bacillus (*pl.* **bacilli**) any straight rod-shaped bacterial cell.

Bacillus a genus of large rod-shaped Gram-positive eubacteria belonging to the family Bacillaceae. Its members are aerobic or facultatively anaerobic, spore-bearing organisms. *B. subtilis* has become widely established as a vehicle for genetic engineering and many cloning **vectors** are available. *B. thuringiensis* synthesizes a toxin that is active against insects; different strains of the bacterium produce different forms of the toxin specific for different insect species.

bacitracin a cyclic antibiotic polypeptide complex produced by *Bacillus subtilis* and *B. licheniformis*. Commercial bacitracin is a mixture of at least nine bacitracins, mostly bacitracin A. Bacitracin A contains, in addition to several L amino acids, a number of D amino acids. Bacitracin inhibits bacterial cell wall synthesis by blocking the dephosphorylation of undecaprenyl diphosphate (undecaprenyl pyrophosphate) to the monophosphate form; it similarly inhibits the dephosphorylation of dolichyl diphosphate thereby blocking the formation of the core oligosaccharides of glycoproteins.

backbone 1 (*in biochemistry*) the sequence of bonded atoms of a polymer to which the side groups and/or the side chains are attached. **2** (*in zoology*) the vertebral column or spine.

backcross 1 any cross of a first-generation hybrid, F_1, with one parent or an individual genetically identical to one of the two parents. **2** the offspring of such a mating.

back-flush *an alternative term for* **back-wash**.

background any intrusive sound or electronic signal registered by a detector but not coming from the source being measured, e.g. the radioactivity registered by a counter in the absence of a radioactive sample.

back mutation *an alternative term for* **reverse mutation**.

back scattering *an alternative term for* **backward scattering**.

back titration an indirect titration procedure in which a measured excess of the reagent is added and the amount remaining after reaction with the analyte is titrated back to an end-point, the proportion consumed in the reaction being obtained by difference.

backward scattering *or* **back scattering** any scattering of radiation in a direction towards the source of the incident radiation.

back-wash *or* **back-flush** to pass a fluid through a chromatographic column in the direction opposite to that of the eluent flow. It is useful, e.g., to relieve bed-compression and to remove entrained gas.

baclofen γ-amino-β-(*p*-chlorophenyl)butyric acid; 3-(aminomethyl)-3-(4-chlorophenyl)propanoic acid; a 4-chlorophenyl derivative of γ-aminobutyric acid (GABA) that acts as a selective agonist for the **GABA$_B$ receptor** and inhibits the release of other neurotransmitters in the central nervous system. It is used for its antispastic (muscle-relaxing) effects.

bactenecin any of several highly cationic polypeptides, originally isolated from bovine neutrophil granules, but also found in sheep, that exert, *in vitro*, a potent antimicrobial activity, possibly owing to inhibition of respiratory chain function.

bacteremia *or* (*esp. Brit.*) **bacteraemia** the presence of live bacteria in the blood.

bacteria *plural of* **bacterium**.

Bacteria one of three superkingdoms (domains) of cellular organisms, the others being Archaea and Eukarya. Bacteria are unicellular and anucleate i.e. **prokaryotes**. They embrace a great diversity of forms, major divisions including the Cyanobacteria, Proteobacteria (which includes Gram-negative bacteria), and Gram-positive bacteria.

bacterial conjugation the process whereby genetic material is transferred from one bacterial cell (the donor) to another (the recipient). Material is usually transferred via a tubelike pilus, which extends from the donor cell to the recipient cell. The terms 'donor' and 'recipient' are preferred to 'male' and 'female' since the latter refer only to the presence or absence of **F plasmid**.

bacterial histone *see* **H-NS**.

bacterial/permeability-increasing protein *see* **BPI protein**.

bacterial photosynthetic reaction centre *see* **photosynthetic reaction centre**.

bacterial porin *see* **porin**.

bacterial recombinant *see* **transconjugant**.

bacterial virus *see* **bacteriophage**.

bactericide *or* **bacteriocide** any agent (biological, chemical, or physical) that destroys bacteria. —**bactericidal** *or* **bacteriocidal** *adj.*

bactericidin *or* **bacteriocidin** *or* **bacteriolysin** any antibody that, with **complement**, kills bacteria against which it is active.

bacteriochlorophyll any of the **chlorophylls** of photosynthetic bacteria. They differ structurally from the chlorophylls of higher plants. Bacteriochlorophylls *a* to *g* are known. Bacteriochlorophylls *a* and *b* are the best known, being the photosynthetic pigments of purple bacteria. Their purple colour results from the fact that they are reduced in both rings B and D, and thus may be regarded as tetrahydropyrroles. Tetrapyrrole carbon atoms and carbon rings are numbered according to the Fischer and IUPAC systems as indicated below. The structure of bacteriochlorophyll *a* is shown. Bacteriochlorophyll *b* has =CH–CH$_3$ in place of the ethyl group on ring B of bacteriochlorophyll *a*. In some bacteriochlorophylls *a* the phytyl group may be replaced by a geranylgeranyl group. Bacteriochlorophylls *c*, *d*, and *e* are antenna pigments of Chlorobiaceae, located in chlorobium vesicles (a core structure is shown). These are dihydropyrrole structures, being reduced in ring D only. The esterifying alcohol is farnesol. In these bacteriochlorophylls, the side chain R$_1$ (see core structure) on ring B varies, being ethyl, *n*-propyl, or isobutyl (in bacteriochlorophyll *d* neopentyl is also possible). On ring C the side chain R$_2$ is ethyl (in bacteriochlorophyll *d* methyl is also possible). R$_3$ is methyl in bacteriochlorophylls *c* and *e*, and H in bacteriochlorophyll *d*.

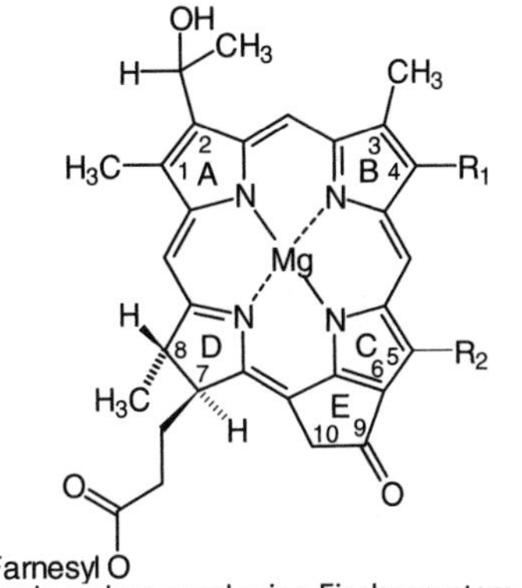

tetrapyrrole carbon numbering Fischer system

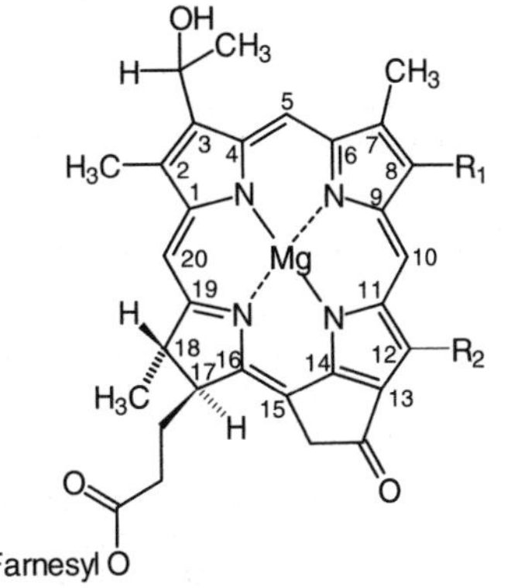

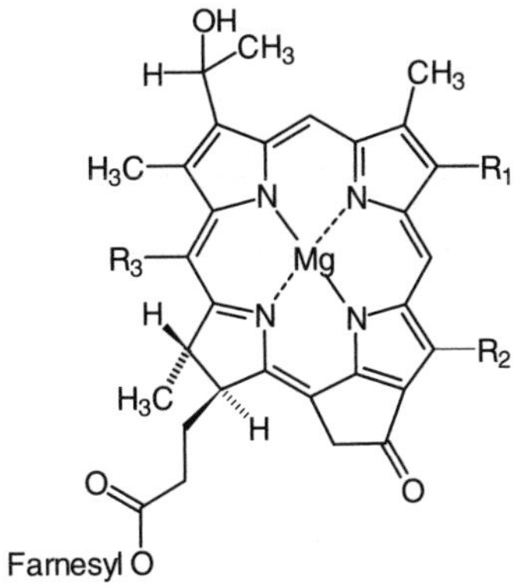

bacteriochlorophyll *c, d, e* core structure

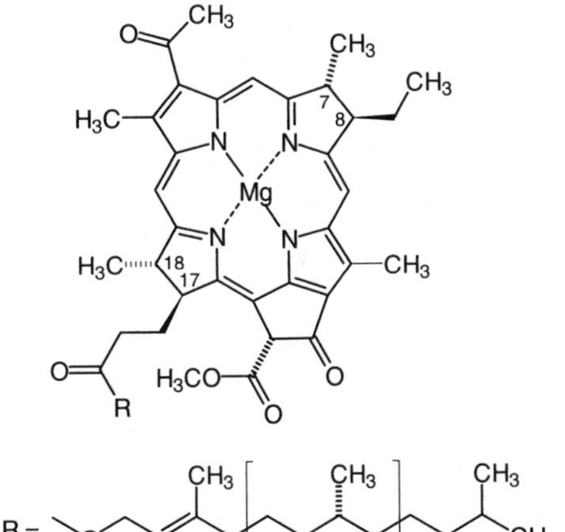

tetrapyrrole carbon numbering IUPAC system

bacteriochlorophyll *a*

plantarum; this is a channel-forming peptide dimer of α and β chains; the amino-acid sequence of the β chain is:

AYSLQMGATAIKQVKKLFKKWGW

the α chain lacks the N-terminal A. —**bacteriocinogenic** *adj*.

bacteriocinogenic factor any **plasmid** that carries genetic information for the production of a **bacteriocin**. Generally only a small proportion of cells containing the factor actually produce bacteriocin. *See also* **col plasmid**.

bacterioferritin *abbr*: BFR; *other names*: cytochrome b$_1$ or cytochrome b$_{557}$; an iron-storage protein comprising 24 identical subunits that pack to form a symmetrical, near-spherical shell, surrounding an ~8 nm central cavity. There is close structural similarity between BFR and the iron-storage **ferritin** found in eukaryotes and bacteria. BFRs store large quantities of iron within their hollow interior (13–20% w/w of iron), in the form of a hydrated ferric oxide mineral containing variable amounts of phosphate anion.

bacteriology the branch of science that deals with the study of **bacteria**.

bacteriolysin *an alternative term for* **bactericidin**.

bacteriolysis the lysis of bacteria. —**bacteriolytic** *adj*.

bacteriophaeophytin *a variant spelling of* **bacteriopheophytin**.

bacteriophage or **phage** or **bacterial virus** any virus that can infect and multiply in a bacterium. Phages parasitize almost every group of prokaryotes. Commonly they consist of a core of nucleic acid enclosed within a protein coat, the capsid; additional components, e.g. lipid, occur in some phages. Depending on the type of phage,

bacteriocide *a variant spelling of* **bactericide**.

bacteriocidin *a variant spelling of* **bactericidin**.

bacteriocin any of a heterogeneous group of polypeptide antibiotics that are secreted by certain bacterial strains, and are able to kill cells of other susceptible (frequently related) strains after adsorption at specific receptors on the cell surface. They include the colicins, and their mechanisms of action vary. Examples: colicin E1 from *Escherichia coli*; this is a channel-forming transmembrane protein that results in membrane depolarization; colicin E3 from *E. coli*; this is an rRNA ribonuclease; plantaricin A from *Lactobacillus*

the nucleic acid may be either DNA (single-stranded or double-stranded) or RNA (single-stranded), and may be either linear or circular. Phages may be filamentous, polyhedral, or polyhedral and tailed; the tubular tails, to which one or more tubular tail fibres are attached in certain circumstances, are involved in attachment of the phages to the bacterial surface and for the injection of the DNA into the host cells. The smallest phages, e.g. φX174, are ≈25 nm in diameter, whereas tailed polyhedral phages, such as the **T-even phages**, are 200–250 nm long, while filamentous phages measure roughly 800 nm × 5 nm. In general, a given bacteriophage can infect only one particular species, or strain, or group of closely related strains, though a given strain may be susceptible to infection by a number of different phages. Bacteriophages can bring about **transduction** (def. 1) between bacterial cells. *Compare* **cyanophage**. *See also* **(bacterio)phage conversion**, **lysogeny**, **phage induction**, **phage typing**, **prophage**, **temperate phage**, **virulent phage**.

(bacterio)phage conversion or **prophage-mediated conversion** or **lysogenic conversion** the introduction of new properties to a host bacterium by the genome of an infecting lysogenic prophage. Such properties may include resistance to lysis by related phages, changes in antigenic constitution, or the appearance of toxigenicity.

bacteriophage vector *see* **cloning vector**.

bacteriopheophytin or (*esp. Brit.*) **bacteriophaeophytin** a **bacteriochlorophyll** in which Mg^{2+} is replaced by 2 H^+.

bacteriorhodopsin a retinal-containing protein, resembling animal **rhodopsin**, formed by *Halobacterium halobium* and other halophilic archaebacteria, and inserted into patches of **purple membrane** in the cell surface. The purple membranes serve as light-operated proton pumps to translocate protons from the inside to the outside of the cells. The mature protein is a **7TM protein** of known 3-D structure; until the structure of bovine rhodopsin was determined, this provided a model for the architecture of **opsin** and related G protein-coupled receptors in eukaryotic cells.

bacteriostatic or **bacteristatic** describing a drug, antibiotic, or other agent that inhibits the growth and multiplication of bacteria. *Compare* **bactericide**. —**bacteriostasis** *n*.

bacterium (*pl.* **bacteria**) any of a vast and ubiquitous group of prokaryotic microorganisms that exist as single cells or in clusters or aggregates of single cells. Most authorities now place them in the exclusively prokaryotic kingdom Monera, along with the **cyanobacteria** (blue-green algae). The majority of bacteria possess a rigid cell wall; those lacking this feature are termed **Archaea**. Bacteria, Archaea, and Eukarya constitute the three primary Kingdoms (domains). —**bacterial** *adj*.

Bacteroides a genus of Gram-negative anaerobic rodlike bacteria. Mostly nonmotile, they are commonly present in the alimentary and urogenital tracts of mammals and may behave as opportunistic pathogens.

bactoprenol *an alternative name for* **undecaprenol**.

Bactrim *see* **sulfamethoxazole**.

baculovirus any of a group of DNA viruses that are known to multiply only in invertebrates and are now classified in the family Baculoviridae. Their genome consists of double-stranded circular DNA of 58–100 MDa. Because of their host range they have potential as pest-control agents. Baculovirus vectors are valuable as a means of expressing certain animal proteins (*see* **expression system**).

BAF *abbr. for* B-cell activating factor (*see* **interleukin 1**).

bafilomycin A1 an extremely hydrophilic macrolide antibiotic secreted by *Streptomyces*. It interacts with and specifically inhibits vacuolar H^+ ATPases.

baicalein 5,6,7-trihydroxyflavone; a pigment (greenish-brown in di-

lute alkali) found in the root of *Scutellaria baicalensis*. It is a selective 12-lipoxygenase inhibitor. Baicalein inhibits leukotriene biosynthesis and cellular Ca^{2+} uptake and mobilization.

bait a hybrid protein containing the GAL4 DNA-binding domain, used in the **yeast two-hybrid system**. *See also* **bait region**.

bait region a restricted region in the sequence of α_2-**macroglobulin** (α_2M) that is cleaved by proteases in the process leading to the conformational changes instrumental in the protease inhibitory function of α_2M. The bait region occurs in all α-macroglobulins, but shows great variation in length and amino-acid sequence between different α-macroglobulins. In human α_2M, it covers approximately residues 666–706. Outside the bait region, α_2Ms from different species exhibit very high sequence similarity. *See also* **trap hypothesis**.

bakers' yeast any of various strains of the yeast *Saccharomyces cerevisiae* that are used in baking.

baking soda *see* **soda**.

BAL *abbr. for* British anti-lewisite (*see* **dimercaprol**).

balance 1 (*in physiology*) the relation between the intake of a particular nutrient and its excretion (or the excretion of its metabolites). **2** any instrument for determining the equality in mass of two objects (or sets of objects). In cases where one (set) is of known mass the mass of the second (set) is thus obtained; e.g. an analytical balance. **3** any instrument designed to measure the weight of an object, e.g. spring balance, torsion balance.

balanced growth a type of growth such that over a time interval (within the exponential phase) every extensive property of the growing system increases by the same factor. Balanced growth for an individual cell requires that each cell after division is an exact replica of the one of the previous cycle.

balanced salt solution any solution designed to provide a normal environment, in terms of ionic composition, pH, and osmotic pressure, for cells in tissue culture.

balance study any study to determine **balance** (def. 1).

balata *see* **rubber**.

BALB/c a strain of inbred white mice that readily develop experimental myelomatosis.

Balbiani ring a large **puff** of giant chromosomes present during a greater portion of larval development in members of the Diptera. The structural modification of specific loci is characteristically large and ring-shaped. [After Eduard Gérard Balbiani (1823–99), French entomologist who reported it in 1881.]

*Bal*I a type 2 **restriction endonuclease**; recognition sequence: TGG↑CCA.

BAL 31 nuclease an exonuclease from *Alteromonas espejiana* used in making deletions by digestion from the ends of double-stranded DNA. It does not cleave DNA internally.

Baltimore classification a classification of animal viruses based on the nature of the genome and the mechanism for synthesis of mRNA, which is always regarded as the **plus strand** (def. 2). There are six classes:

Class I Viruses having a dsDNA genome that make mRNA by asymmetric transcription. In many, different mRNA species come from different DNA strands.

Class II Viruses having a ssDNA genome of the same polarity as the mRNA. Some have strands of both polarities in different particles.

Class III Viruses having a dsRNA genome that make mRNA by asymmetric transcription. Most have multiple pieces of dsRNA, each of which apparently contains the information for the synthesis of a single protein.

Class IV Viruses having an ssRNA genome that make mRNA of base sequence identical to the genomic RNA.

Class V Viruses having an ssRNA genome that make mRNA with a base sequence complementary to the genomic RNA.

Class VI **Retroviruses** having an ssRNA genome and a DNA intermediate in their growth. [After David Baltimore (1938–), US microbiologist and molecular biologist.]

*Bam*HI a type 2 **restriction endonuclease**; recognition sequence: G↑GATCC. The M.*Bam*HI modification site is C5.

band 1 an **absorption band** in an electromagnetic spectrum. **2** a zone of (macro)molecules such as that obtained in density-gradient centrifugation, zone electrophoresis, isoelectric focusing, chromatography, or other similar method. **3** any of the zones readily distinguishable by microscopy in a **sarcomere** of a muscle. **4** the specific association of a large number of **chromomeres** at the same level in somatically paired **polytenic** giant chromosomes of members of the Diptera. *See also* **cytochrome absorption bands**.

band centrifugation *see* **isopycnic centrifugation**.

band compression inadequate separation of the bands on a gel electrophoretogram, especially in DNA sequencing gels. It results in poor resolution and difficulties in interpretation.

band electrophoresis *an alternative name for* **zone electrophoresis**.

banding the formation of distinct bands or zones in density-gradient centrifugation, column chromatography, or other separation techniques.

banding density the density of the suspending medium at which a particular type of particle, e.g. a cell organelle, forms a band during centrifugation in a specified density gradient.

band intensity *(in NMR spectrometry)* the area under the band between the signal trace and the baseline. *See* **nuclear magnetic resonance**.

band-pass filter 1 an electric circuit or other device that transmits only frequencies within a selected band. **2** an **interference filter**.

band protein erythrocyte proteins are referred to as bands, numbered according to the position they occupy on electrophoresis gels. Some of the more important bands are: 1 and 2, α and β **spectrin**; 3, an anion channel protein; 4.1, always referred to as such, involved in spectrin junctions; 5, **actin**.

band 3 protein *an alternative name for* **anion channel protein**.

band width *(in NMR spectrometry)* the distance between the sides of the band at half the band height of a reasonably smooth and symmetrical band. *See* **nuclear magnetic resonance**.

Bangasome or **Bangosome** *(colloquially)* *an alternative name for* **liposome**. [After its inventor, Alec Douglas Bangham (1921–), British biophysicist and medical scientist.]

***Ban*II** a type 2 **restriction endonuclease**; recognition sequence: G[AG]GC[TC]↑C. *Hgi*JII is an **isoschizomer**.

Bantu siderosis *see* **iron overload**.

BAPTA 1,2-bis(*o*-aminophenoxy)ethane-*N*,*N*,*N*′,*N*′-tetraacetate; a Ca^{2+} chelator exhibiting a 10^5-fold greater affinity (110 nM) than for Mg^{2+}. Its UV absorption maximum is at 254 nm which, after Ca^{2+} is bound, shifts to 279 nm, a property that can be exploited in its use as an indicator of intracellular Ca^{2+} concentration. Its esterified form can enter cells where it is hydrolysed to BAPTA.

bar a unit of pressure temporarily approved for use with SI units. 1 bar = 10^5 Pa = 10^5 Nm^{-2}.

barbed end *see* **plus end**.

barbital *or* **barbitone** 5,5-diethylbarbituric acid; 5-diethylmalonylurea; 5,5-diethyl-2,4,6(1*H*,3*H*,5*H*)-pyrimidinetrione; a long-acting depressant of the central nervous system, widely used as a buffer substance; pK_a = 7.98 (25°C), 8.06 (20°C). *One proprietary name*: Veronal.

barbiturate 1 the tautomeric anion derived from barbituric acid (malonylurea; 2,4,6(1*H*,3*H*,5*H*)-pyrimidinetrione). **2** any mixture of free barbituric acid and its anion. **3** any salt of barbituric acid. **4** any of various pharmacologically active derivatives of barbituric acid, including barbitone, pentobarbitone, phenobarbitone, and thiopentone. Barbiturates are potent CNS depressants that potentiate the actions of GABA by binding to the **GABA_A** receptor. Pentobarbitone and thiopentone are used to induce anesthesia.

Barcroft apparatus *or* **Barcroft respirometer** a differential respirometer for studying gas exchange of cells, tissue slices, or tissue homogenates, now rarely used. It is a closed system comprising two flasks of equal volume connected by a U-shaped manometer (the Barcroft manometer). In operation, both flasks contain the same volumes of liquid and gas; one, the reaction flask, contains the cells or tissue; the other, the compensation flask, is free from cells or tissue and serves to compensate for changes in temperature and barometric pressure during the course of an experiment. The respirometers are arranged so that they may be shaken with the flasks in a constant-temperature bath. The reading of the apparatus is the difference between the levels of the manometer fluid in the two sides of the manometer. [After Joseph Barcroft (1872–1947), British physiologist, who described it in 1908.]

BARD1 *abbr. for* *BRCA1*-associated ring domain protein; a protein homologous with ***BRCA1*** in its N- and C-terminal regions. It interacts with the RING finger of *BRCA1* and is localized with *BRCA1* in the nucleus during the S phase of the cell cycle.

Barfoed's reagent a solution containing cupric acetate in dilute acetic acid that is reduced when heated with monosaccharides but not with disaccharides or oligosaccharides. [After Christen Thomsen Barfoed (1815–99), Swedish physician.]

barn *symbol*: b; a unit of area temporarily accepted for use with the SI. It is equal to 10^{-28} m^2. It is used as a unit of atomic nuclear cross section.

barnase *or* **ribonuclease BA** a small extracellular endoribonuclease from *Bacillus amyloliquefaciens*. Its action involves formation of a 2′:3′-cyclic phosphate (*see* **ribonuclease**). It has been studied as a model for protein folding. The sequence and 3-D structure are known.

baroceptor *a variant spelling of* **baroreceptor**.

barometer an instrument for measuring the pressure of the atmosphere.

barometric pressure the pressure exerted by the atmosphere at a given place and time. It is often expressed in terms of the supported height of a column of mercury, usually measured in millimetres. The SI unit is the **pascal**. Other non-SI units frequently used in addition to the **millimetre of mercury** are the **atmosphere**, the **bar**, and the **torr**.

baroreceptor *or* **baroceptor** any physiological receptor that is sensitive to changes in pressure, especially the sensory nerve endings in the carotid sinus that set off nerve impulses in response to alterations in blood pressure at that point.

Barr body *see* **sex chromatin**. [After Murray Barr (1908–1995), Canadian anatomist.]

barrel *see* **beta-barrel**.

barrier an extracellular protein, probably a protease, excreted by yeast cells of mating type **a**. Its role may be to act as an antagonist of the α-type mating pheromone, and to establish optimal pheromone concentration for conjugation.

Barth syndrome *or* **3-methylglutaconic aciduria type II** an X-linked mitochondrial myopathy with cardiomyopathy, growth retardation, and leukopenia. There may also be 3-methylglutaconic aciduria but with normal 3-methylglutaconyl-CoA hydratase activity. There is marked decrease in tetralinoleylcardiolipin in skeletal and cardiac muscle and in platelets. It is caused by mutations in the gene for **tafazzin**.

barwin a barley seed protein (possibly a lectin) involved in defence mechanisms. It is closely related to the C-terminal domain of proteins encoded by wound-induced plant genes. It gives its name to a family of such proteins.

basal 1 at, of, or being a **base** (def. 4). **2** at, of, or being the minimum level for maintaining the normal or essential functioning of an organism.

basal body 1 *(in prokaryotes)* a structure that apparently attaches the **flagellum** (def. 1) to the cell envelope. The proximal end of the flagellum, the hook, appears curved and thickened and leads into the basal body, which consists of parallel ring-shaped structures, arranged around a rod-shaped core. The rings of the basal body make contact with the layers of the cell envelope. **2** *(in eukaryotes)* a structure at the base of a cilium consisting of spherical granules or short rods arranged in rows below the cell surface. Electron microscopically they are cylindrical bodies, 300–500 nm long and 120–150 nm in diameter, open at one or both ends. They consist of nine sets of triplet microtubules, each triplet containing one complete microtubule fused to two incomplete microtubules.

basal granule *an alternative term for* **basal body** (esp. in ciliated epithelium).

basal lamina a thin sheet of proteoglycans and glycoproteins, especially **laminin**, secreted by cells as an extracellular matrix, forming a region between the cells and adjacent connective tissue. The basal lamina has important functions in the organization of tissues including an influence on cell polarity, cell differentiation, and cell migration. The basal lamina is subdivided into the *lamina lucida* (an electron-lucent region) immediately adjacent to the cell layer, and the *lamina densa* (an electron-dense region) external to it. Frequently there is a further outermost layer, the *lamina reticularis*, containing collagen fibrils. The three laminae are often collectively referred to as the **basement membrane**.

basal medium any culture medium that will support the growth of nutritionally undemanding chemoorganotrophs.

basal metabolic rate *abbr.*: BMR; the rate of energy metabolism of an animal at rest but not asleep. It is measured with the animal in a 'comfortable' ambient temperature, and in the postabsorptive state, at least 12 hours after taking food. It may be determined from the oxygen consumption and carbon dioxide (and urinary nitrogen) excretion, or from the heat production of the animal.

basal metabolism the minimum amount of energy required to maintain life. See **basal metabolic rate**.

base 1 *(in chemistry)* **a** a **Brønsted–Lowry base**. **b** a **Lewis base**. *See also* **acid** (def. 2). **2** *(in molecular biology)* any purine or pyrimidine that occurs as a component residue in polynucleotides or nucleic acids. Purines and pyrimidines are **nitrogen bases**, hence the terminology. **3** a unit of length of a polynucleotide or nucleic acid, equal to one nucleotide residue. Large molecules are usually measured in kilobases (10^3 bases; *symbol*: kb) or megabases (10^6 bases; *symbol*: Mb). *See also* **nitrogenous base**. **4** *(in mathematics)* **a** the number of units in a system of counting that is equivalent to one unit in the next higher counting place. **b** the number that when raised to a specified power, its **exponent**, has a logarithm equal to the specified power. **5** the bottom or supporting part of anything; the point of attachment of an organ or part to an organism.

base analogue *or (esp. US)* **base analog** any unnatural purine or pyrimidine base that can be incorporated *in vivo* into DNA but that, because of its different properties, causes altered base-pairing during incorporation or in subsequent DNA replication. Thus 5-bromouracil, an analogue of thymine, pairs with guanine thereby causing transitions of A–T → G–C. Similarly, 2-aminopurine, an analogue of adenine, pairs with cytosine thereby causing transitions of A–T → G–C. Some base analogues have been used as anticancer and antivirus agents.

base calling the automated reading of DNA-sequencing gels using a scanner and appropriate software.

base composition the relative amounts of the various purines and pyrimidines occurring in a specimen of polynucleotide or nucleic acid. It is often expressed as moles percent.

base excision repair see **excision repair**.

baseline an imaginary line or a standard or reference value, etc. by means of which measurements may be compared, especially the zero line in the tracing of a chromatogram, spectrum, ultracentrifuge pattern, etc.

basement membrane a thin layer of dense material found in various animal tissues interposed between the cells and adjacent connective tissue. It consists of the **basal lamina**, composed mainly of a tightly cross-linked meshwork of type IV **collagen** molecules interwoven with a laminin network, to which it is bound by **entactin** molecules. The collagen molecules are glycosylated, made up of largely triple-helical combinations of α chains, which may be of six types: α-1(IV) to α-6(IV). There is also an associated layer of reticulin fibres. Occurring in, e.g., capillary linings, kidney tubules, lung alveoli, and renal glomeruli, basement membrane has a supportive function in some tissues and may also act as a passive selective filter for substances diffusing in or out of the cells; e.g. in renal glomeruli

it retains protein molecules. It gives a strong periodic acid–Schiff reaction. *See* **Alport syndrome**.

base pair *symbol*: bp; any of the possible pairings between two bases in opposing strands of double-stranded DNA or RNA molecules. Adenine forms a base pair with thymine (in DNA) or uracil (in RNA) and guanine with cytosine, hence the number of adenine residues equals the number of thymine (and/or uracil) residues while the number of guanine residues equals that of the cytosine residues. *See also* **Chargaff's rules**. —**base-pair** *vb.*; **base-pairing** *n.*

base-pair ratio *see* **Chargaff's rules**.

base sequence the sequential order of nucleotide residues in a polynucleotide or nucleic acid molecule.

base stacking the arrangement of base pairs in parallel planes, at an angle to the helix axis, in the interior of a helical double-stranded polynucleotide or nucleic acid molecule.

base triplet any sequence of three bases in a nucleic acid that codes for an amino acid, or some other signal, in protein synthesis. *Compare* **anticodon**; **codon**.

basic 1 of, relating to, containing, or characteristic of a **base** (def. 1). *Compare* **acidic** (def. 1). **2** having an alkaline reaction; of or relating to an aqueous solution having a pH value >7.0. *Compare* **acidic** (def. 2).

basic amino acid any amino acid possessing a net positive charge at neutral pH.

basic azo-dye binding protein *abbr.*: B-ABP; a 45 kDa (Sephadex) protein, isolated from the soluble cell supernatant fraction from livers of rats given 4-dimethylaminoazobenzene. It was later named **ligandin**.

basic dissociation constant *or* **basicity constant** *symbol*: K_b; the thermodynamic dissociation constant for the dissociation of a **base** (def. 1, 2). For a dilute solution of a weak base, B, dissociating in water according to the equilibrium:

$$BOH \rightleftharpoons B^+ + OH^-, \quad K_b = a_{B^+} \times a_{OH^-} / a_{BOH}$$

where a is the **activity** of the species designated by the subscripts. The activity of the water has been omitted from the equation since it may be taken as unity in dilute aqueous solution. K_b is a measure of the strength of a base, i.e. of its ability to accept hydrons from water (*compare* **acid dissociation constant**; *see also* **pK**). The concept is no longer used since the conjugate acid, B^+, may be considered as an acid and an acid dissociation constant, K_a, for it defined. Also, in aqueous solutions

$$K_a \times K_b = [H^+][OH^-] = K_w,$$

where K_w is the ion product of water, whence K_b may be calculated from K_a or vice versa.

basic dye *or* **basic stain** any cationic dye that binds to, and hence stains, anionic macromolecules or other materials.

basic salt any salt containing hydroxyl or oxide groups in partial replacement of, or in addition to, ions or groups derived from an acid, e.g. basic lead acetate, $(CH_3COO)_2Pb \cdot Pb(OH)_2$.

basic stain *an alternative term for* **basic dye**.

basic zipper *see* **bZIP**.

basolateral surface the surface of an epithelial cell that adjoins underlying tissue. *Compare* **apical surface**.

basonuclin a **zinc finger** nuclear protein of epidermis, likely to be a transcription factor specific for squamous epithelium and its keratinocytes prior to terminal differentiation. It is present mainly in the nuclei of the basal cell layer.

basophil *or* **basophil(ic) leukocyte** a polymorphonuclear phagocytic leucocyte of the myeloid series that is distinguished by the presence of coarse cytoplasmic granules that stain with basic dyes. The granules are believed to contain histamine, heparin, and other vasoactive amines. Basophils are closely related to **mast cells** and constitute <0.5% of blood leukocytes.

basophilia 1 the property of being **basophilic**. **2** an increase in the number of **basophils** circulating in the blood.

basophilic staining readily with basic dyes.

basotropic seeking a basic environment; a term used especially in connection with protein kinases that have a requirement for basic residues near the target residue. *Compare* **acidotropic**.

bAT *abbr. for* basic amino acid transporter. *See* **cystinuria**.

batch a quantity of material produced or processed in one operation on a single occasion, as opposed to a continuous operation; hence batch adsorption, batch culture, etc.

batch elution a method of column chromatography in which conditions are chosen to ensure complete retention of the material under investigation, using the minimal amount of stationary phase, followed by elution with a solvent or buffer that ensures complete and rapid removal in the minimal possible volume.

batch process (in biotechnology) a process in which a bio-reactor is loaded with substrates inoculated with microorganisms or enzymes and allowed to run to completion, usually without removing biomass and products during the process.

bathochromic (in absorption spectroscopy) describing a shift of an absorption band in the direction of longer wavelengths. Compare **hypsochromic**.

batrachotoxin a steroidal alkaloid neurotoxin from the skin of Colombian arrow-poison frogs of genus *Phyllobates*. It is the most potent known venom. It binds specifically to the voltage-gated sodium channel and renders the membrane permeable to sodium ions. Compare **tetrodotoxin**, which reverses its depolarizing effect.

batroxobin see **venombin**.

battenin a protein of unknown function occurring in the membranes of lysosomes, mitochondria, nuclei, and synaptic vesicles. It consists of 438 amino acids, and is considered to contain 5–7 transmembrane segments. Some 20 mutations in the battenin locus at 16p12 are associated with Batten's disease (juvenile neuronal ceroid lipofuscinosis), an autosomal recessive neurodegenerative disease of childhood. This is characterized by the accumulation in lysosomes of storage bodies whose principal constituent is subunit c of mitochondrial ATPase. See **H⁺-transporting ATP synthase**.

Batten's disease see **H⁺-transporting ATP synthase**.

batyl alcohol 1-O-octadecyl-sn-glycerol; 3-octadecyloxy-1,2-propanediol; a hydrolysis product of **ether lipids**.

Bayesian statistics a statistical approach for estimating the probability of an observation given some prior knowledge or expectation. Bayesian statistics are often applied to the analysis of microarray data, phylogenetic tree data, evolutionary parameters, and so on, where they are used to estimate the likelihood (posterior probability) of a model being true given a particular data set, by estimating the probability of obtaining the data set if the model were true. [After Thomas Bayes (1702–61), English mathematician.]

bay region a concave exterior region, bordered by three phenyl rings, of a polycyclic aromatic hydrocarbon when an angularly fused terminal benzo-ring is present. For example, the sterically hindered area between carbon atoms 4 and 5 of phenanthrene, between carbon atoms 1 and 12 of benz[*a*]anthracene, or between carbon atoms 10 and 11 of benzo[*a*]pyrene. It has been suggested that dihydrodiol epoxides in the bay region of such compounds may be responsible for their mutagenicity and carcinogenicity. See also **A region**, **B region**, **K region**, **L region**.

BbeI a type 2 **restriction endonuclease**; recognition sequence: GGCGC↑C. *Nar*I is an **isoschizomer**.

BCAA abbr. for branched-chain amino acid.

B cell 1 or (formerly) **beta cell** the principal cell (60–80%) of the islets of Langerhans of the pancreas. It produces, stores (as B granules), and secretes the hormone **insulin**. 2 an alternative name for **B lymphocyte**.

B-cell activation factor abbr.: BAF; an alternative term for **interleukin 1**.

B-cell antigen receptor complex a complex consisting of membrane immunoglobulin associated non-covalently with two proteins known as mb-1 (other names: Igα, CD79a) and B29 (other names: Igβ, CD79b) that binds antigen and initiates B-cell transformation. mb-1 and B29 are often designated α and β respectively, and the complex consists of Ig(αβ)₂. Signalling occurs by activation of tyrosine kinases of the Src family (lyn, blk, fyn, or lck). α and β are themselves phosphorylated on activation. All five immunoglobulin isotypes (IgG, IgM, IgD, IgA, IgE) bind two α/β heterodimers and use a similar signalling pathway, but the glucosylation of the α chain differs according to the isotype with which it associates.

B-cell differentiation factor abbr.: BDF; an alternative term for **interleukin 1**.

BCG abbr. for bacille Calmette–Guérin; the bacillus from which is made a vaccine composed of live bovine-strain tubercle bacilli, *Mycobacterium bovis*, attenuated by long cultivation on a potato–glycerol–bile medium. [After Léon Charles Albert Calmette (1863–1933), French physician and bacteriologist, and Camille Guérin (1872–1961), French veterinary surgeon and bacteriologist, who described its preparation in 1908.]

B chromosome any of a heterogeneous group of accessory, extra, or supernumerary chromosomes found in many species of animals and plants. They differ from normal, or A chromosomes, by being usually smaller and frequently heterochromatic and telocentric, without great effect on the phenotype of their carriers. Both their numbers and their meiotic and mitotic behaviour are highly variable.

BCIP abbr. for 5-bromo-4-chloro-3-indolyl phosphate; a substrate for alkaline phosphatase. The soluble colourless 5-bromo-4-chloro-3-indolyl group is oxidized to give a purple-blue insoluble product. Redox coupling to the reduction of nitroblue tetrazolium (NBT) increases the sensitivity of detection.

BCKDH abbr. for branched-chain α-keto acid dehydrogenase.

Bcl a family of proteins that form pores in nuclear, endoplasmic reticulum, and outer mitochondrial membranes. Some members are antiapoptotic (e.g. Bcl-2, Bcl-XL) while others are proapoptotic. The structure of Bcl-XL is similar to that of the secreted bacterial pore-forming protein **colicin**. See **BCL**.

BCL any of several genes associated with B-cell leukemia and lymphoma. *BCL1* is identical to the gene for human **cyclin** D1; its product activates p34^cdc2 kinase and is phosphorylated in G₁ of the cell cycle. *BCL2* is a protooncogene encoding a protein (locus at 18q21) that is localized to nuclear, endoplasmic reticulum, and outer mitochondrial membranes, and prolongs survival of hematopoietic cells; its product is able to block **apoptosis**, and if overexpressed with Myc counteracts the oncogenic action of Myc. Translocation of the enhancer region of the gene for Ig heavy chain constant region from14q32.33 to the *BCL* locus is associated with B-cell leukemias and lymphomas. *BCL2* proteins contain four homology regions each ≈18 residues long. *BCL3* encodes a probable transcription factor of the I-κB family.

BcnI a type 2 **restriction endonuclease**; recognition sequence: CC↑[CG]GG. *Cau*II is an **isoschizomer**.

BCR abbr. for breakpoint cluster region, a human gene (locus at 22q11.21) for a 160 kDa member of the family of protein serine/threonine kinases, widely expressed in human hematopoietic and other cell lines. See also **abl**, **Philadelphia chromosome**.

BDF abbr. for B-cell differentiation factor interleukin (see **interleukin**).

B-DNA abbr. for B form of DNA.

BDNF abbr. for brain-derived neurotrophic factor.

Be symbol for beryllium.

Becker muscular dystrophy a mild form of Duchenne muscular dystrophy in which muscle **dystrophin** is reduced in amount or altered in structure by point mutations in the gene at Xp21.2. [After Peter Emil Becker (1908–2000), German geneticist.]

Beckmann thermometer a sensitive mercury-in-glass thermometer used for measuring small differences or changes of temperature. [After Ernst Otto Beckmann (1853–1923), German chemist.]

becquerel *symbol*: Bq; the SI derived unit of (radioactive) **activity** (def. 2) of a radionuclide; it corresponds to one **nuclear transformation** per second; i.e. 1 Bq = 1 disintegration s^{-1}. *See also* **curie**: 1 Ci = 3.7×10^{10} Bq or 1 Bq = 2.70×10^{-11} Ci. [After Antoine Henri Becquerel (1852–1908), French physicist.]

bed volume in column chromatography, the total volume of material, both solid and liquid, in the column; i.e. the volume of the support particles plus the **void volume**. It is synonymous with **column volume** for a packed column.

Beer–Lambert law *or* Beer–Lambert–Bouguer law *or* Lambert–Beer law a combination of **Beer's law** (def. 1) and **Lambert's law**; put into modern terms, it states that the **absorbance**, A, of a beam of collimated monochromatic radiation in a homogeneous isotropic medium is proportional to the absorption pathlength, l, and to the amount-of-substance concentration, c, (or, in the gas phase, to the pressure) of the absorbing species. The law may be expressed by the relations

$$A = \lg (\Phi_0/\Phi) = \varepsilon \, cl \text{ or } \Phi = \Phi_0 = 10^{-\varepsilon \, cl}$$

where Φ_0 and Φ are respectively the **spectral powers** of the incident and transmitted radiation beams, and the proportionality constant, ε, is the molar (decadic) absorption coefficient – *see* **absorption coefficient**.

Beer's law 1 *or (sometimes)* **Bouguer–Beer law**; a physical law extending **Bouger's law** to solutions: it states that when light passes through a given thickness of an absorbing solution of concentration c, the intensity of the transmitted light, I_{trs}, decreases exponentially as c increases linearly. It may be expressed as $I_{trs} = I_0 \, e^{-kc}$, where I_0 is the intensity of the incident light and k is a constant. The law applies in practice only to monochromatic light (or other electromagnetic radiation) and when factors such as reflection or scattering are negligible or can be corrected for. *See also* **Beer–Lambert law**. **2** name commonly but incorrectly applied to **Beer–Lambert law** [After August Beer (1825–63), German physicist who formulated it in 1857.]

beet sugar *an alternative name for* **sucrose** when this has been extracted from sugar beet.

Beh *symbol for* the behenoyl group.

behenoyl *symbol*: Beh; *trivial name for* docosanoyl, $CH_3-[CH_2]_{20}-CO-$, the acyl group derived from the aliphatic acid behenic acid (= docosanoic acid), which occurs naturally in animal tissues and most seed fats.

bek a gene related to that for the **fibroblast growth factor** (FGF) receptor. Named after bacterial expressed kinase, it encodes a plasma membrane protein tyrosine kinase of class IV with an insert in the kinase region that binds FGFA with high affinity.

Bence-Jones protein any of a group of **myeloma proteins**, occurring in the urine of patients with multiple myelomatosis and related conditions, that precipitate on heating the urine to 60 °C, but redissolve at higher temperatures. They consist of light chains of **immunoglobulins** synthesized by the myeloma cells. [After Henry Bence-Jones (1813–73), English physician who described them in 1847.]

Benedict's solution a reagent used to test for the presence of glucose and other reducing sugars. When heated with reducing sugars it is reduced and a red precipitate of cuprous oxide is formed. There are separate formulations for qualitative and quantitative analysis. [After Stanley Rossiter Benedict (1884–1936), US biochemist.]

benign describing a neoplasm that is harmless, localized, and non-metastasizing, although it may physically impede the function of a nerve or muscle.

bentonite a colloidal, native hydrated aluminium silicate clay consisting principally of montmorillonite, a complex aluminosilicate, $Al_2O_3{\cdot}4SiO_2{\cdot}H_2O$, which has marked adsorptive properties. It is used as an inhibitor of nucleases and also in the **bentonite flocculation test**, a passive agglutination test in which antigen-coated bentonite particles are used to detect specific antibody.

benzidine *trivial name for* 4,4′-diaminobiphenyl, a substance that is carcinogenic for humans. It was formerly used in a highly sensitive test for the presence of blood, giving a green or blue colour in acetic acid solution on addition of hydrogen peroxide. Another former application was in the chromogenic assay of peroxidases. In view of the carcinogenicity, it has been replaced by **2,7-diaminofluorene** and other compounds.

benzo+ *or (sometimes before a vowel)* **benz+** *comb. form* **1** denoting the fusion of a benzene ring to a cyclic organic molecule. **2** indicating derivation from benzene or benzoic acid; *compare* **phenyl**.

benzoate 1,2-dioxygenase EC 1.14.12.10; *systematic name*: benzoate,NADH:oxygen oxidoreductase (1,2-hydroxylating, decarboxylating); *other name*: benzoate hydroxylase. A bacterial dioxygenase that is involved in the utilization of aromatic compounds. It catalyses a reaction between benzoate, NADH, and dioxygen to form catechol, CO_2, and NAD^+. Iron is a cofactor. In *Acinetobacter calcoaceticus* the enzyme comprises three subunits, two of which, α and β, represent the hydroxylase, while the third, γ, is an electron-transport flavoprotein. The β subunit is probably involved in substrate specificity, and the γ subunit is similar to ferredoxin reductase with 2Fe–2S motifs.

benzodiazepine *abbr.*: BZ; **1** 2*H*-1,4-benzodiazepine; a bicyclic compound consisting of a benzene ring fused to a seven-membered partially unsaturated ring containing two nitrogen atoms. **2** any member of a large class of pharmacologically active substances containing the benzodiazepine ring system. They are CNS depressants of low toxicity, having sedative and anxiolytic effects. A number are useful therapeutically; one of the most widely used is **diazepam** (*one proprietary name*: Valium). Specific high-affinity **benzodiazepine binding sites** are present in the CNS. They have been classified as BZ$_1$, primarily associated with anxiolytic effects, and BZ$_2$. It has been suggested that the effects of benzodiazepines are mediated through enhancement of the inhibitory action of GABA on the brain, and they are known to modulate the effects of GABA at **GABA$_A$ receptors**. Cloning studies of the GABA$_A$ receptor subunits have indicated that the γ subunit confers benzodiazepine sensitivity, and that the combination of different subunit types affects the receptor specificity, the presence of α1 subunits in the complex conferring BZ$_1$ pharmacology.

1*H* form

benzodiazepine inverse agonist a drug or other compound that binds to benzodiazepine receptors but has effects that are the opposite of those produced by a benzodiazepine agonist, i.e. it causes anxiety and convulsions.

benzoic acid *or* benzene monocarboxylic acid a white crystalline solid found naturally in plants and also used as a food preservative. In plants it is derived metabolically from phenylalanine. It accumulates (with salicylic acid), free and as a glycoside, in the immediate vicinity of infection sites. In animals and humans it is excreted in urine after conjugation to benzoylglycine (or hippuric acid) in the liver.

***o*-benzoquinone** 3,5-cyclohexadiene-1,2-dione; 1,2-benzoquinone; *o*-quinone. The oxidized form of **pyrocatechol**.

***p*-benzoquinone** 2,5-cyclohexadiene-1,4-dione; 1,4-benzoquinone; (*p*-)quinone. The oxidized form of **hydroquinone**. *See also* **quinhydrone**.

benzoyl *symbol*: Bz *or (preferably)* PhCO; the C_6H_5-CO- group; the acyl group derived from benzoic (= benzenecarboxylic) acid.

benzyl *symbol*: Bzl *or (preferably)* PhCH$_2$; the $C_6H_5-CH_2-$ group; the alkyl group derived from benzyl alcohol (= benzenemethanol).

benzyloxycarbonyl *or (formerly)* **carbobenzoxy** *or* **carbobenzyloxy** *symbol*: Z *or* Cbz; the $C_6H_5-CH_2-O-CO-$ group, which is introduced into amino groups of peptides and proteins with reagents such as *N*-(benzyloxycarbonyloxy)succinimide.

beriberi a disease due to thiamine (vitamin B$_1$) deficiency and characterized by degenerative changes in the peripheral nerves, digestive system, and heart accompanied by edema.

Berkfeld filter a (candle) filter, made of diatomaceous earth, used to sterilize solutions. The pore size is relatively large and three porosities are graded V, N, and W (German *viel, normal*, and *wenig*) corresponding to coarse, normal, and fine. [After Wilhelm Berkefeld (1836–97), German chemist and manufacturer (the owner of the mine at Kieselguhr, where the earth was mined for production of the filter).]

Bernal chart a means of specifying the position of a spot on an X-ray diffraction photograph by two coordinates. [After John Desmond Bernal (1901–71), British crystallographer who described it in 1926.]

Bernard–Soulier syndrome *see* **von Willebrand factor**.

Bes *or* **BES** *abbr. for* 2-[bis(2-hydroxyethyl)amino]ethanesulfonic acid; a **Good buffer substance**, pK_a (20°C) = 7.15.

bestatin [(2S,3R)-3-amino-2-hydroxy-4-phenylbutanoyl]-L-leucine; an inhibitor of aminopeptidase B and leucine aminopeptidase with a variety of pharmacological functions. It is an inhibitor of enkephalin degradation in brain extracts, an anticarcinogenic agent, an immunomodulating agent, and an inhibitor of leukotriene-A$_4$ hydrolase.

bestrophin a protein of retinal pigment epithelium but with no known function. It comprises 585 amino acids, and contains four transmembrane segments. Numerous mutations of its gene locus at 11q13 are associated with Best macular dystrophy.

beta *symbol*: β (lower case) *or* B (upper case); the second letter of the Greek alphabet. For uses *see* **Appendix A**.

beta adrenergic *or* **β adrenergic** *see* **adrenergic**.

beta adrenoceptor *or* **β adrenoceptor** *see* **adrenoceptor**.

beta barrel *or* **β-barrel** a polypeptide super-secondary structure in which a single β-sheet folds back upon itself such that the first strand usually hydrogen-bonds to the last; such structures are found in both globular and membrane proteins.

beta-barrel porins members of a subclass of transport proteins of the TC system, designated 1.B. Their secondary structure is of the β-strand type, usually forming β-barrels. The subclass contains nearly 40 families, including general bacterial porins, alginate export porins, raffinose porins, two-partner secretion porins, secretins (not to be confused with the digestive hormone), and outer membrane factors. See also **porin**.

beta bend *or* **beta turn** *or* **β-bend** *or* **β-turn** a short stretch of polypeptide chain that allows the main direction of the chain to change. It consists of four amino-acid residues in which the CO group of residue *n* is hydrogen bonded to the NH group of residue *n* + 3. *See also* **reverse turn**.

beta blockade the process of inhibiting beta adrenoceptors with an antagonist.

beta blocker an antagonist of beta adrenoceptors.

beta carbon atom (*symbol*: Cβ) *or* **C-beta** the first carbon atom of an amino acid side chain, which is attached to the polypeptide chain via the main-chain **alpha carbon atom**.

beta cell *or* **β cell** (*formerly*) an alternative name for **B cell** (def. 1).

beta-cell tropin *or* **β-cell tropin** a peptide fraction obtained from the *pars intermedia* of mouse pituitary that stimulates insulin secretion by perfused rat pancreas or isolated pancreatic islets. It is immunochemically related to, but separable from, **corticotropin-like intermediary lobe peptide**.

betacellulin a potent mitogenic glycoprotein with EGF-like do-

mains, purified from the conditioned medium of mouse pancreatic beta tumour cells, retinal pigment epithelial cells, and vascular smooth muscle cells. It may play a role in the vascular complications associated with diabetes. The C-terminal region shares 50% similarity with rat **transforming growth factor** α. The precursor is a type I **membrane protein**, and the mature form is a secretory protein.

beta conformation *or* **β-conformation** a stable conformation of a polypeptide chain in which the chain is almost fully extended, but slightly puckered; the **torsion angle** of N-Cα-C-N in the backbone is about 120°, causing the side-chains of neighbouring residues to project in opposite directions from the backbone. Adjacent polypeptide chains may hydrogen bond, forming a **beta sheet**, the axial distance between adjacent amino-acid residues being ~0.35 nm; the chains may be either **parallel** (running in the same direction) or **antiparallel** (running in opposite directions).

betacyanin *see* **betalain**.

beta decay *or* **β decay** decay of a radionuclide by the emission of a **beta particle**.

beta emitter *or* **β emitter** any nuclide that emits either positive or negative beta particles.

beta globulin *or* **β globulin** any of a diverse group of **plasma proteins**, including lipoproteins and transferrin, that migrate slower than the **alpha globulins** during electrophoresis at slightly alkaline pH values.

betaglycan a cell-surface single-pass transmembrane protein that contains covalently bound chondroitin sulfate and heparan sulfate chains. It acts as a coreceptor for **transforming growth factor** β (TGFβ).

beta granin *or* **β granin** *see* **chromogranins**.

beta granule *an alternative name for* **B granule**.

beta helix *or* **β-helix** a polypeptide conformation in which a serial linkage of β-strands, connected via loop regions, coils into a wide helix. In a simple two-sheet β helix, each turn of helix contains two strands and two loops. In extracellular bacterial proteases, this simple unit is repeated three times, forming a right-handed, coiled structure comprising two adjacent three-stranded parallel β-sheets, enclosing a hydrophobic core. A more complex β-helical structure arises when the repeat unit is three short strands separated by loop regions, giving rise to a helix in which three parallel β-sheets are arranged such that two of the sheets are parallel to each other and perpendicular to the third; this structure was first seen in pectate lyase.

beta hemolysis *or* **β hemolysis** a type of hemolysis, characteristic of certain strains of streptococci, in which a clear zone is formed around the bacterial colonies when cultured on a blood agar plate. *Compare* **alpha hemolysis**.

betaine the *N*-trimethyl derivative of an amino acid; the term originally denoted **N-trimethylglycine**, now called **glycine betaine**. Glycine betaine acts as a methyl-group donor in some reactions; *see*, e.g., **betaine–homocysteine S-methyltransferase**.

betaine-aldehyde dehydrogenase EC 1.2.1.8; an enzyme involved in the formation of **betaine**. It catalyses the reaction:

betaine aldehyde + NAD$^+$ + H$_2$O = betaine + NADH.

betaine–homocysteine S-methyltransferase EC 2.1.1.5; an enzyme that catalyses the reaction:

trimethylammonioacetate (glycine **betaine**) +
L-homocysteine = dimethylglycine + L-methionine.

beta lactoglobulin *or* **β lactoglobulin** the principal protein of the whey of ruminants' milk, in which it constitutes 50–60% of the protein content. It has an M_r of ~40 000, and an isoelectric point of 5.2. Its function is unknown, but it binds retinol, and belongs to the family of transport proteins, known as **lipocalins**, that bind small hydrophobic molecules. *See also* **lactalbumin** (def. 1).

betalain any of a group of coloured alkaloids occurring widely in plants of the order Centrospermae. They are divided into two groups, **betacyanins** and **betaxanthins**, and both occur in plant vacuoles. The former impart a red-violet colour and the latter a yellow colour to the plant organs that contain them.

beta lipoprotein *or* **β lipoprotein** any of the plasma lipoproteins that behave electrophoretically as **beta globulins**, but normally designated **low-density lipoprotein**. *See also* **plasma protein**.

beta oxidation *or* **β oxidation** the metabolic oxidation of a long-chain fatty acid by successive cycles of reactions during each of which the fatty acid is shortened by a two-carbon-atom fragment removed as acetyl coenzyme A. The name derives from the fact that the β-carbon atom is the point of oxidation. *See also* **beta-oxidation system**, **Knoop's hypothesis**.

beta-oxidation system a mitochondrial complex that carries out the reactions of beta oxidation. It contains enzymes that fall into the following four categories:
(1) Acyl-CoA dehydrogenases. See **acyl-CoA dehydrogenase**.
(2) **enoyl-CoA hydratase** EC 4.2.1.17; this catalyses a reaction forming (3*S*)-3-hydroxyacyl-CoA from *trans*-2(or 3)-enoyl-CoA, hydrating the double bond introduced by the enzymes in category 1. It is a homohexamer and a mitochondrial matrix protein.
(3) **3-hydroxyacyl-CoA dehydrogenase** EC 1.1.1.35; this catalyses the oxidation of (3*S*)-3-hydroxyacyl-CoA by NAD$^+$ to 3-oxoacyl-CoA with formation of NADH. It is a homodimer and mitochondrial matrix protein.
(4) **3-ketoacyl-CoA thiolase** (*see* **thiolase**). This is a homotetramer and a mitochondrial protein.

beta particle *or* **β particle** either a (negatively charged) electron or a (positively charged) positron emitted in the **transformation** (def. 5) of a radionuclide.

beta-pleated sheet *or* **β-pleated sheet** *another name for* **beta sheet**.

beta propeller a toroidal arrangement, present in some proteins, that is made up of seven conserved repeats each of ≈60 residues that form seven beta sheets around a central axis so as to resemble a propeller. Each beta sheet contains four antiparallel beta strands. *See* **glycoprotein IIb/IIIa**.

beta radiation *or* **β radiation** an emission of **beta particles**.

beta receptor *or* **β receptor** *see* **adrenoceptor**.

beta sheet *or* **β-sheet** *or* **β-pleated sheet** an approximately planar array of two or more adjacent **beta strands** such that hydrogen bonds may be formed between C=O groups of one β-strand and NH groups of another. *See also* **beta conformation**.

beta strand *or* **β-strand** one of the two basic elements of the **secondary structure** adopted by a polypeptide chain of a protein (*see also* **alpha helix**). It consists of a single chain in a **beta conformation**. Two or more strands may interact to form a **beta sheet**.

beta structure *or* **β-structure** any part of the structure of a protein molecule consisting of one or more β-strands.

beta turn *or* **β turn** *an alternative name for* **beta bend**.

betaxanthin *see* **betalain**.

beticolin any of a family of eight nonpeptidic channel-forming toxins produced by the fungus *Cercospora beticola*. The structure of beticolin 2 is naphtho[2′,3′:1,2]cyclohepta[5,6-c]xanthene-14a-acetate. It is classified in the TC system under number 1.D.7.

B form of DNA *abbr.*: B-DNA *or* DNA-B; the molecular conformation adopted by fibres of the sodium salt of duplex DNA at high relative humidity and in aqueous solution. It consists of a right-handed double helix containing about ten nucleotide residues per turn with the planes of the base pairs normal to the axis of the helix. It is the basis of the **Watson–Crick model of DNA**, and is thought to resemble the conformation of most of the DNA *in vivo*. *See also* **A form** (def. 1), **C form**, **Z form**.

B/F ratio *abbr. for* bound-to-free ratio (in ligand binding). A plot of the B/F ratio versus the concentration of free ligand is known as a **Scatchard plot**.

BGG *abbr. for* bovine gamma globulin; the component of bovine serum represented by fraction II of a **Cohn fractionation**.

***Bgl*I** a type 2 **restriction endonuclease**; recognition sequence: GCC-NNNN↑NGG.

***Bgl*II** a type 2 **restriction endonuclease**; recognition sequence: CCCN$_1$↑NNN$_2$NGGC, in which N$_1$ and N$_2$ are complementary.

B granule *or* **beta granule** any of very numerous membrane-bound vesicles (granules) containing stored insulin that are found in the **B cells** of the pancreatic islets. The granules stain a deep bluish purple with Gomori's aldehyde fuchsin and Ponceau stain.

bhang *see* **cannabis**.

BHK cell *abbr. for* baby hamster kidney cell; a cultured cell derived from this organ.

BHT *abbr. for* butylated hydroxytoluene, di-*tert*-butyl-*p*-cresol, an **antioxidant**.

bi+ 1 *comb. form* meaning two, twice, or double. **2** *prefix* (in organic chemical nomenclature) indicating two identical groups or rings joined by a link; e.g. biacetyl, biphenyl. **3** *prefix* (in inorganic chemical nomenclature) signifying an acid salt of a dibasic acid; e.g. sodium bicarbonate (though now more correctly termed sodium hydrogencarbonate). **4** *(in enzymology)* denoting two kinetically important substrates and/or products of an enzymic reaction (*see* **reactancy of enzymes**). *See also* **bis+**, **di+**.

Bi *symbol for* bismuth.

BIAcore *proprietary name* of an optical biosensor instrument that measures real-time interactions of macromolecules using **surface plasmon resonance**. Ligands are immobilized on a dextran-coated gold surface. The resulting 'sensorgrams' can be used to derive kinetic parameters of the interactions.

bialaphos a linear tripeptide produced as a secondary metabolite by *Streptomyces hygroscopicus*. The structure is PT-Ala-Ala, where PT is the glutamate analog, phosphinothricin. After hydrolysis of bialaphos by intracellular nonspecific peptidases, PT acts as an inhibitor of **glutamate-ammonia ligase** (glutamine synthetase), and thereby derives herbicide and antibiotic activities.

Bial's test a colorimetric test for aldopentoses. *See* **orcinol test**. [After Manfred Bial (1870–1908), German physician.]

bibliographic database a database that houses bibliographic information, primarily in the form of citations, controlled vocabulary terms, index codes and abstracts. Examples include MEDLINE, which covers the worldwide biomedical literature; EMBASE Excerpta Medica, which covers the worldwide biomedical and pharmaceutical literature; and PASCAL, which covers the major international literature on science, technology and medicine.

bibrotoxin a vasoconstrictor peptide of the **endothelin/sarafotoxin** family, isolated from the venom of the burrowing wasp *Atractaspis bibroni*. The sequence,

CSCADMTDKECLYFCHQDVIW,

is identical to that of sarafotoxin B except that Ala^4 replaces Lys^4.

bicarbonate the HCO$_3^-$ ion; now more correctly termed hydrogencarbonate.

bichromatic analysis a term sometimes used to describe any analytical method involving observation at two wavelengths.

Bicine *or* **BICINE** *abbr. for* *N*,*N*-bis(2-hydroxyethyl)glycine; a **Good buffer substance**; pK_a (20 °C) = 8.35.

bicoid protein a segment-polarity protein, discovered in *Drosophila*, that provides positional cues for the development of head and thoracic segments. It regulates the expression of zygotic genes, possibly through its **homeodomain**, and inhibits the activity of other maternal gene products. It forms an anterior–posterior gradient in the embryo, and acts as a transcriptional activator. It binds to *caudal* mRNA. It shares similarity with other **paired**-type homeobox proteins.

bicuculline 6-(5,6,7,8-tetrahydro-6-methyl-1,3-dioxolo[4,5-g]isoquinolin-5-yl)furo[3,4-e]-1,3-benzodioxol-8(6*H*)-one; a naturally occurring alkaloid with several sources, including *Dicentra cucullaria* and *Adlumia fungosa*. It is a potent convulsant with effects probably mediated by its GABA antagonist activity at **GABA$_A$ receptors**.

bidentate 1 having two teeth or two toothlike processes. 2 describing a **ligand** molecule that chelates a metal ion by means of two donor atoms.

bifunctional having two functions.

bifunctional antibody an antibody having two combining sites for antigen.

bifunctional catalyst a catalyst that provides both acidic and basic catalytic functions.

bifunctional reagent a compound with two reactive functional groups that can interact with two groups in one molecule or with one group in each of two different molecules.

big-big gastrin *or* **big gastrin** *see* **gastrin**.

BigCHAP *N,N*-bis(3-D-gluconamidopropyl)cholamide; a nonionic detergent analogue of **CHAPS**, but with reduced electrostatic properties; aggregation number 10, CMC 3.4 mM.

BigDye terminator *a proprietary name for* reagents used in DNA sequencing in which the dideoxynucleoside triphosphates are each labelled with a different fluorescent dye to facilitate their identification when the products of sequencing reactions are separated by gel or by capillary electrophoresis.

big gastrin *see* **gastrin**.

big glucagon *or* **big plasma glucagon** material that is separable from normal human plasma and shows glucagon-like immunoreactivity. It has an estimated M_r of 160 000.

biglycan bone and cartilage proteoglycan I precursor; a connective tissue glycoprotein of the extracellular matrix related to **decorin** and fibromodulin.

bijou bottle a small glass screw-capped bottle of 5–7 ml capacity, used especially for liquid and solid cultures.

bikunin a plasma glycoprotein found both in the free state and complexed with the heavy chains of the inter-α-inhibitor family, forming the light chain. It is a serine protease inhibitor, having a tandem arrangement of Kunitz domains; it may participate in the control of events such as endothelial cell growth, and oocyte cumulus expansion and stabilization. It is synthesized from an α_1-microglobulin/bikunin precursor.

biladiene *the semisystematic name for* either of two linear tetrapyrroles in which the carbon bridges contain two more double bonds than **bilane** i.e. two of the three carbon bridges in the molecule are methine groups.

bilamellar having two **lamellae**; formed of two plates.

bilane *the systematic name for* a linear tetrapyrrole in which each of the three carbon bridges linking the pyrrole nuclei is saturated, i.e. is a methylene group, and in which, unless otherwise specified, each of the four nitrogen atoms also is saturated. It is the fundamental structure for the naming of linear tetrapyrroles, is defined without oxygen substituents, and is numbered (omitting C_{20}) to agree with the numbering of the unsubstituted porphyrin ring system. *Trivial name*: bilinogen.

bilatriene *an alternative (less-favoured) name for* **bilin**.

bilayer a layer that is two molecules thick, as in some membranes. *See* **lipid bilayer**, **unit membrane**.

bilayer lipid membrane *see* **black lipid membrane**.

bile a secretion produced by the vertebrate liver and conveyed through the bile duct to the duodenum. Its alkaline nature helps to neutralize acidic gastric secretions passing from the stomach, and the **bile salts** it contains are involved in the digestion of fat. Bile also serves as a medium for the excretion of various substances, notably cholesterol (either as such or in the form of bile salts to which it is converted), pigments (from hemoglobin breakdown), heavy metals, and other waste products (usually of >400 Da).

bile acid any member of a group of steroid carboxylic acids occurring in bile, where they are present as the sodium salts of their amides with glycine or taurine (*see* **bile salt**). The C_{24} acids are hydroxy or keto derivatives of cholanic or allocholanic acids; some C_{27} bile acids are known in eutherian mammals. In mammals, the C_{24} acids are found conjugated with glycine (*see* **glycocholate**) as well as with taurine (*see* **taurocholate**); in other animals only taurine conjugates have been found. Bile acids play an important role as emulsifiers in digestion and absorption of dietary lipids. In human adults 30 g are excreted daily, most of which are reabsorbed by an Na^+-dependent active-transport glycoprotein (locus on chromosome 14 encodes 349 residues) with several transmembrane segments. *See* **enterohepatic circulation**.

bile acid receptor *see* **farnesoid X receptor**.

bile alcohol any of a group of polyhydroxy derivatives of **cholestane**, present as their sulfuric acid esters in the bile of amphibians and some fishes.

bilene *the semisystematic name for* either of two tetrapyrroles in which the carbon bridges contain one more double bond than **bilane**; i.e. one of the three bridges in the molecule is a methine group.

bile pigment any bilirubinoid pigment present in bile; i.e. any linear tetrapyrrole derived from a **porphyrin** (which is a cyclic tetrapyrrole). For physiological reasons certain dipyrroles are also sometimes regarded as bile pigments.

bile salt the sodium salt of the conjugate of any **bile acid** with either glycine or taurine. Bile salts are potent surfactants; in the gut they emulsify ingested lipid, promote its hydrolysis by the activation of lipases, and facilitate its absorption, thereafter being excreted in the feces. *See also* **glycocholate**, **taurocholate**.

biliary of or relating to **bile**.

bilin 1 *the systematic name for* the linear tetrapyrrole in which the carbon bridges contain three more double bonds than **bilane**; i.e. all three bridges are methine groups. This name is preferred to bilatriene. 2 any coloured **bile pigment** formed by the oxidation of a colourless bile pigment, or **bilinogen** (def. 2). Thus the bilinogens urobilinogen and stercobilinogen form urobilin and stercobilin, respectively, on oxidation.

bilinogen 1 *trivial name for* bilane. 2 any colourless **bile pigment** that may be oxidized to a coloured bile pigment (*see* **bilin** (def. 2)).

biliprotein *an alternative name for* **phycobiliprotein**.

bilirubin *or (formerly)* **bilirubin IX**α *the recommended trivial name for* the linear tetrapyrrole 8,12-bis(2-carboxyethyl)-2,7,13,17-tetramethyl-3,18-divinylbiladiene-*ac*-1,19(21*H*,24*H*)-dione. It is produced in the reticuloendothelial system by the reduction of **biliverdin** and transported to the liver as a complex with serum albumin. The liver produces bilirubin from hepatic heme proteins and renders it water-soluble by conjugation of the carboxyethyl side-chains to form bilirubin bisglucuronide, most of which is excreted in the bile by an ABC-transporter called canalicular multiorganic anion transporter (or cMOAT). Bilirubin accumulates in the blood and tissues in jaundice, which may arise from acquired or hereditary causes. Unconjugated bilirubin is toxic to parts of the brain and is associated with kernicterus. *See* **Dubin–Johnson syndrome**.

bilirubin oxidase EC 1.3.3.5; an enzyme that catalyses the oxidation by dioxygen of bilirubin to biliverdin and H_2O.

biliverdin *or (formerly)* **biliverdin IX**α *the recommended trivial name for* the linear tetrapyrrole 8,12-bis(2-carboxyethyl)-2,7,13,17-tetramethyl-3,18-divinylbilin-1,19(21*H*,24*H*)-dione. It is formed in the reticuloendothelial system and in hepatocytes by the first step in heme degradation, catalysed by heme oxygenase (decyclizing), EC 1.14.99.3, a reaction in which the α-methene group of heme is oxidized to carbon monoxide. Biliverdin is then reduced by biliverdin reductase, EC 1.3.1.24, to **bilirubin**.

biliverdin reductase EC 1.3.1.24; an enzyme that catalyses the reaction of bilirubin with $NAD(P)^+$ to form biliverdin and NAD(P)H.

bimodal *(in statistics)* distribution with two **modes**.

bimolecular having a **molecularity** of two; relating to, consisting of, or involving two molecular entities.

bimolecular layer a layer that is two molecules thick.

bimolecular reaction any chemical reaction in which there is a transition state composed of the atoms of two separate molecular entities, which may be the same or different.

binal double; twofold.

binary 1 consisting of or forming two things or parts; dual. **2** describing a number system having the numeral 2 as its **base** (def. 4b). *See also* **bit**.

binary compound a compound consisting of only two chemical elements; e.g. ammonia, sodium chloride, water.

binary digit *see* **bit**.

binary fission the reproduction of a living cell by division into two equal, or near-equal, parts.

binary operator any arithmetic or logical operator (e.g. +, AND, OR) that operates on two quantities.

binary representation a notation by which **alphanumeric** symbols are represented by two digits, 0 and 1, in a known sequence. The displacement of one digit position to the left denotes multiplication by a power of 2.

BIND *abbr. for* biomolecular interaction network database; *see* **pathway database**.

binder protein I *see* **cortisol metabolite binding protein (large)**.

bindin a protein of 285 amino acids, M_r 35 000, extracted from the insoluble granular material of sea-urchin sperm, responsible for the attachment of the sperm to the vitelline layer of the egg. Bindins may mediate the species-specific recognition of eggs by sperm. *See also* **spermadhesin**, **zona pellucida sperm-binding protein**.

binding the act or process by which one molecule attaches to another by noncovalent forces; ligation *see* **ligate** (def. 3). *See* **ligand**. *See also* **binding site**.

binding assay *an alternative term for* **saturation analysis**.

binding capacity a measure of the quantity of ligand or the number of ligand molecules that can be bound by a given amount of a particular binding agent or system.

binding constant *an alternative term for* **affinity constant**.

binding curve any graphical representation of the extent of binding between two compounds at varying relative concentrations of the reactants.

binding energy the free energy change accompanying the binding of n_i ligand molecules to a macromolecule to produce a particular configuration of the liganded macromolecule.

binding fraction *an alternative term for* **saturation fraction**.

binding potential *symbol*: Π; a concept introduced to clarify the definition of linked functions, linkage groups, and linked sites in ligand binding. It is defined by: $\nu_i = \delta\Pi/\delta\mu_i$ where ν_i is the ratio of component i to that of a reference component and μ_i is the chemical potential of i.

binding protein *abbr.*: BiP; a **luminal protein** of the endoplasmic reticulum that is a member of the hsp70 family of **heat-shock proteins**, and is a typical **molecular chaperone**. It recognizes incorrectly-folded proteins, as well as protein subunits that have not yet assembled into their oligomeric complexes, and facilitates their oligomerization and folding. BiP activity requires the participation of ATP, which is hydrolysed to ADP and inorganic phosphate.

binding set the spatially organized collection of all the groups or residues of *one* protomer that are involved in its binding to *one* other protomer.

binding site a specific region in a macromolecule or other molecular entity, or in a membrane, that takes part directly in the combination with a **ligand**. Binding sites may be independent or interacting, depending on whether or not the binding of one ligand to one site affects the binding of other ligands to other sites on the same macromolecule, or membrane. *See also* **cooperative ligand binding**.

binomial 1 a mathematical expression made up of two terms, e.g. $2x + 5y$. **2** a two-part taxonomic name (*see* **binomial nomenclature**).

binomial coefficient the coefficient of any term arising from the expansion of the algebraic binomial $(a + b)^n$.

binomial distribution a statistical function in which the probabilities that an event will or will not occur are given by the successive coefficients of a binomial expansion of the form $(a + b)^n$, where a and b are the probabilities that an event will and will not occur, respectively.

binomial nomenclature a system of naming organisms, devised by the Swedish botanist Carolus Linnaeus (1707–78), that employs two Latin names, the first being the genus and the second the species to which the organism belongs.

binuclear *an alternative term for* **dinuclear**.

bio+ *comb. form* denoting life or living organisms, or systems derived from them.

bioaccumulation a progressive increase in the amount of a substance in an organism or part of an organism that occurs because the rate of intake exceeds the organism's ability to remove the substance.

bioactivation the activation of a drug or other substance in an organism or by means of a biological or biochemical system.

bioactive *an alternative term for* **biologically active.**

bioanalytical of or pertaining to biochemical analysis.

bioassay *or* **biological assay** any quantitative estimation of biologically active substances by measurement of the magnitude of their actions on living organisms or parts of organisms. It is performed under standard conditions and by reference to preparations of defined potency or activity.

bioautograph the record obtained in **bioautography.**

bioautography a method for detecting, in a complex mixture, small amounts of substances that are able to act as essential growth factors for a test organism(s). The components of the mixture are first separated by chromatography. The resulting chromatogram is then placed in contact with a culture of the test organism in a solid medium lacking a specific component. Hence, the organism will only grow where the component is present in the chromatogram, thereby establishing its presence in the original mixture.

bioavailability 1 the relative amount of the dose of a drug or other substance reaching the systemic circulation. **2** the rate and extent to which the therapeutic moiety of a drug is absorbed and becomes available to the site of drug action.

bioblast any very small mass of protoplasm having formative powers, formerly considered to be the ultimate elementary living particles; it is now thought the particles seen may have been **mitochondria.**

biocatalyst any agent, present in or obtained from living material, that can increase the rate at which a chemical reaction occurs without itself undergoing permanent change; a biological or biochemical **catalyst.** Biocatalysts are commonly **enzymes.** *See also* **ribozyme.** —**biocatalysis** *n.*

biochemical 1 of or pertaining to **biochemistry. 2** any chemical substance occurring in or isolated from an organism.

biochemical engineering *or* **bioengineering** the scientific discipline underlying the industrial processing of biological materials, e.g. fermentation-based pharmaceuticals, foods, and effluents.

biochemical evolution the evolution of the biochemical processes and components of living organisms, such as the structure and function of biologically important molecules, metabolic pathways, subcellular structures, and cells.

biochemical genetics the branch of genetics concerned with the chemical nature of hereditary determinants and how they function at the molecular level.

biochemical imprinting the adaptive increase in enzyme activities that can occur along specified metabolic pathways when there is a large and sustained traffic through them.

biochemical lesion any biochemical change, such as the absence or inactivation of an enzyme, that diminishes the fitness of an organism or leads to a pathological condition.

biochemical marker any specific character that may be detected by a biochemical test, e.g. the presence or absence of a particular enzyme.

biochemical oxygen demand *or* **biological oxygen demand** *abbr.*: BOD; a measure of pollution of water in lakes, rivers, etc. due to the presence of organic matter. Oxidation of the latter by microorganisms removes dissolved oxygen, hence reducing the ability of the water to sustain aquatic life. BOD is the amount of oxygen (mg) taken up by a 1-litre sample of water in 5 days (BOD_5) at 20°C in the dark; it may be necessary to add additional, uncontaminated, water to augment the amount of oxygen available. Sometimes incubation is for a longer time, e.g. 20 days, in which case it is indicated by BOD_{20}. *Compare* **chemical oxygen demand.**

biochemistry *or* **biological chemistry** the branch of science dealing with the chemical compounds, reactions, and other processes that occur in living organisms. Lehninger expressed the challenge to the biochemist as follows: 'Living things are composed of lifeless molecules. When these molecules are isolated and examined individually, they conform to all the physical and chemical laws that describe the behaviour of inanimate matter. Yet living organisms possess extraordinary attributes not shown by collections of inanimate molecules.' In this regard Horowitz has proposed a set of criteria for living systems: 'Life possesses the properties of replication, catalysis and mutability.' Biochemists are, therefore, concerned with the manner in which living organisms exhibit these properties. —**biochemist** *n.*

biochip 1 an electronic device in which electrical and logical functions are performed by protein molecules appropriately manipulated. *See* **nanotechnology. 2** *an alternative name for* **microchip.**

biocide any chemical substance that kills living cells or organisms. —**biocidal** *adj.*

biocompatible compatible with life; having no injurious effect on living organisms. —**biocompatibility** *n.*

bioconversion a technique in which the enzymic apparatus of an organism is used to convert or process a chemical substance, not normally present in the organism, into another, (often commercially) useful one.

biocytin ε-*N*-biotinyl-L-lysine; a naturally occurring derivative of **biotin,** found especially in autolysates of rapidly growing yeast. It is liberated from biotin holoenzymes by proteolysis, and is cleaved to biotin and lysine by biotinidase.

biodegradable capable of being broken down by the action of living organisms. —**biodegradability** *n.*

biodegradation 1 the processes by which exogenous, usually waste materials are broken down by living organisms. Sometimes this breakdown involves two or more kinds of organisms acting in cooperation. Such degradative abilities are employed to convert waste materials into more acceptable and manageable forms or to produce useful end products. **2** *an alternative term for* **catabolism.** *Compare* **biosynthesis.**

biodeterioration any undesirable change in the properties of a material brought about by the activities of organisms.

bioelectricity electric current or electric potential produced in living organisms. —**bioelectric** *adj.*

bioelectrochemistry the application of electrochemical models and techniques to biological problems.

bioelectronics the application of biomolecular principles to microelectronics, as in **biosensors** and **biochips.**

bioelement any element that is an essential component of living organisms. Bioelements include: as major elements, C, H, O, N, P, and S, found chiefly in combination in organic compounds, and the elements, Na, K, Mg, Ca, and Cl, found mainly as monatomic ions. Many other 'trace elements' occur in relatively minute amounts.

bioenergetics the study of energy transformations in living systems.

bioengineering 1 engineering relating to the operation on an industrial scale of biochemical processes, especially fermentation. This is usually now termed **biochemical engineering. 2** the application of the physical sciences and engineering to the study of the functioning of the human body and to the treatment and correction of medical conditions.

biofuel cell any fuel cell (a cell in which electricity is produced directly by oxidation of fuel) in which one or both electrode reactions are promoted or catalysed by a biological process.

Bio-Gel *proprietary name for* a group of materials used in gel filtration, consisting of either cross-linked agarose (Bio-Gel A) or porous polyacrylamide (Bio-Gel P).

biogenesis 1 the synthesis of a substance in a living organism; biosynthesis. **2** the principle that a living organism can originate only from a parent or parents that are similar to itself. *Compare* **abiogenesis.** —**biogenetic, biogenetical, biogenous** *adj.*

biogenic produced by living organisms.

biogenic amine any of various amines, especially those with neurological activity, isolated from plants or animals.

biohazard any organism or material derived from a biological source that is hazardous to health or life.

bioinformatics *see* **informatics.**

bioinformatics institute a research and service centre that creates and manages biological databases (e.g. of nucleic acid and protein sequences, protein families, macromolecular structures), and develops software (e.g. for analysing protein sequences and structures, genome sequences, microarray data) for free use by the community. The European Bioinformatics Institute (EBI) and the National Center for Biotechnology Information (NCBI) are the primary bioinformatics research and service centres for Europe and the USA respectively.

bioinorganic chemistry a branch of inorganic chemistry concerned with the interactions between inorganic substances (especially metallic ions or their complexes) and biological or biochemical systems, and with the properties of related model systems. *Compare* **inorganic biochemistry**.

bioinorganic motif *abbr.*: BIM; a structural feature shared by functionally related (not necessarily homologous) proteins, consisting of one or more metal atoms and first-coordination shell ligands.

bioleaching the extraction of metals from ores or soil by biological processes, mostly by lithotrophic bacteria.

biolistics descriptive of the administration of DNA in gold particles, using a device known as a 'gun' to deliver the particles to the cell interior. One use has been in the development of vaccines by the administration of suitable genes to foreign cells.

biolith a rock or other deposit (e.g. peat, humus) formed from organic material.

biological *adj.* **1** of or pertaining to **biology**. **2** of or pertaining to life or living organisms. **3** a biological product that is used prophylactically or therapeutically. —**biologically** *adv.*

biological activity any activity of a substance that is demonstrable in living organisms. Biologically active substances are often of biological origin themselves.

biological clock any biological mechanism that allows the expression of a certain biological structure (e.g. a gene) or a biological function (e.g. sleep) at periodic intervals. The term is also used colloquially to describe ageing, e.g. in respect to a woman's ability to bear a child.

biologically active *or* **bioactive** active in a living organism.

biological value *abbr.*: B.V.; a measure of the value of a protein in a foodstuff for the maintenance of growth and normal functioning of any given animal or species. It is based on the amino-acid composition of the protein, its digestibility, and the availability of its digestion products.

biology **1** the scientific study of living organisms, including their structure, functioning, development, distribution, interrelationships, and evolution. **2** the animal and plant life of a particular region considered as a unit. **3** the structure, functioning, etc. of a particular organism or group of organisms. —**biologist** *n.*

bioluminescence the production of light by certain enzyme-catalysed reactions in living organisms. —**bioluminescent** *adj.*

biolysis any decomposition of organic material by biological means. —**biolytic** *adj.*

biomarker an indicator signalling an event or condition in a biological system or sample and giving a measure of exposure, effect, or susceptibility. A biomarker may be a measurable chemical, biochemical, physiological, behavioural, or other alteration within an organism.

biomass **1** the total mass of living organisms, of a specified kind, forming a population inhabiting a given region. **2** the material produced by the growth of microorganisms, plants, or animals, especially as a product of or raw material for an industrial process, such as farming or fermentation.

biomedical science *or* **biomedicine** any branch of science that is relevant to medicine.

biomembrane any membrane surrounding a cell or a subcellular organelle, or derived from these structures.

biometry the statistical study of biology. —**biometric** *adj.*; **biometrically** *adv.*

biomimetic **1** describing a laboratory procedure designed to imitate a natural chemical process. **2** describing a compound that mimics a biological material in its structure or function.

biomimetic chemistry a branch of organic chemistry that aims to imitate natural reactions and enzymatic processes as a means to improve the power of organic chemistry.

biomineralization the synthesis of inorganic crystalline or amorphous mineral-like materials by living organisms. Among the minerals synthesized biologically in various forms of life are fluoroapatite ($Ca_5(PO_4)_3(F,OH)$), hydroxyapatite, magnetite (Fe_3O_4), and calcium carbonate ($CaCO_3$).

biomolecular engineering a term sometimes used to cover those branches of biotechnology dealing with enzyme engineering and genetic engineering.

biomolecule any molecule, especially a macromolecule, that occurs in or is formed by living organisms. —**biomolecular** *adj.*

bioorganic chemistry a branch of organic chemistry concerned with the study of organic molecules from living systems.

biophotolysis any type of **photolysis** reaction occurring in a biological system; e.g. the lysis of a water molecule in the reactions of **photosystem II** during photosynthesis.

biophysical chemistry the application of the techniques or methods of physical chemistry to biological systems. It is principally concerned with the conformations, shape, structure, conformational changes, dynamics, and interactions of macromolecules and complexes or arrays of macromolecules. *Compare* **biophysics**, **physical chemistry**.

biophysics the application of physical techniques and physical methods of analysis to biological problems. Traditionally, the discipline has focused on two main areas: first, the transmission of nerve signals and the maintenance of electrical potentials across membranes; and second, large-molecule crystallography and enzyme structure and mechanisms. For the latter X-ray crystallography has been supplemented by a number of physical techniques, including nuclear magnetic resonance spectrometry, mass spectrometry, fluorescence-depolarization measurements, and circular-dichroism studies. —**biophysical** *adj.*; **biophysicist** *n.*

biopolymer any **polymer**, such as a protein, nucleic acid, or polysaccharide, occurring in, or formed by, living systems.

biopotency the extent to which a material (e.g. drug or hormone) influences a biological process.

bioproduct any (usually) useful product produced by living organisms or by systems derived from living organisms.

biopsy examination of (usually) a sample of tissue, cells, or fluid from a living human or animal for diagnostic purposes. The term is also now commonly used for the process of removing such a sample. *Compare* **necropsy**.

biopterin 2-amino-4-hydroxy-6-(1,2-dihydroxypropyl)-pteridine; 1-(2-amino-4-hydroxy-6-pteridinyl)-1,2-propanediol; a growth factor for certain protozoans, e.g. *Crithidia fasciculata* and *Trypanosoma platydactyli*. It is widely distributed in tissues and functions in a reduced form, **tetrahydrobiopterin**, as a hydroxylation coenzyme, e.g. in the conversion of phenylalalnine to tyrosine, of tyrosine to 3-OH tyrosine (DOPA), and of tryptophan to 5-OH tryptophan. It is also a coenzyme for nitric oxide synthase. Biopterin is synthesized from GTP. It has been isolated from queen-bee jelly and is considered a growth factor for some insects. *See* **guanosine triphosphate cyclohydrolase**, **phenylketonuria**.

bioreactor any reactor containing immobilized enzymes or cells that is used to effect specific chemical reactions in the production of some economically desirable product.

bioregulator omnibus term sometimes used for any substance occurring within a cell or organism that has a specific and reversible controlling action on a particular biochemical event within that cell or organism; it may be, e.g., a hormone, an enzyme, an intermediate metabolite, an inorganic ion, or a cyclic nucleotide.

bioremediation the use of biological systems to remedy environmental damage.

bios *the former name for* various growth factors required by yeast. Originally believed to be a single factor, it was later fractionated into bios I (shown to be **inositol**), bios IIA (*β*-**alanine**), and bios IIB (**biotin**).

bioscience(s) collectively all the individual branches of science that are concerned with biology.

biose disaccharide; stem name used in the established trivial names for particular disaccharides whether or not they contain a single type of sugar residue and linkage, as in cellobiose and gentiobiose.

+biose *suffix* used in nomenclature of glycolipids to indicate a compound containing an oligosaccharide moiety composed of two sugar residues (which may be the same or different), as in galabiose.

biosemiotics the study of signs, of communication, and of information in living organisms.

biosensor a device that uses specific biochemical reactions mediated by isolated enzymes, immunosystems, tissues, organelles, or whole cells to detect chemical compounds, usually by electrical, thermal, or optical signals.

biospecific describing: **1** a reagent, e.g. an antibody or hormone, that can recognize and interact only with cells (e.g. within a mixed population of cells) displaying a particular surface characteristic, e.g. a corresponding antigen or receptor. **2** any reagent that, in a similar way, recognizes and interacts with a particular biospecifically occurring substance. **3** any reagent that interacts specifically with a particular biomolecule (e.g. an enzyme). **4** the reaction between two biospecific reagents.

biospecific-elution chromatography *an alternative name for* **affinity-elution chromatography** (when the ligand in the eluting solvent is a biospecific macromolecule).

biosphere the parts of the Earth's crust and atmosphere that contain living organisms.

biosynthesis the production of substances by processes occurring in living systems; **anabolism**; **biogenesis**. *Compare* **biodegradation**.

biosynthetic 1 of or relating to **biosynthesis**. **2** any substance produced by biosynthesis.

biosynthetic pathway any metabolic pathway effecting **biosynthesis**; an anabolic pathway.

biotechnology 'the integration of natural sciences and engineering sciences in order to achieve the application of organisms, cells, parts thereof and molecular analogues for products and services' (European Federation of Biotechnology General Assembly, 1989); a field of technological activity in which biochemical, genetic, microbiological, and engineering techniques are combined for the pursuit of technical and applied aspects of research into biological materials and, in particular, into biological processing. It includes traditional technologies such as fermentation processes, antibiotic production, and sewage treatment, as well as newer ones such as **biomolecular engineering**, and **single-cell protein** production. —**biotechnological** *adj.*

biotin *cis*-tetrahydro-2-oxothieno[3,4-*d*]imidazoline-4-valeric acid; hexahydro-2-oxo-1*H*-thieno[3,4-*d*]imidazole-4-pentanoic acid; vitamin H; *formerly known as* bios IIB and coenzyme RCO$_2$. A substance whose very (+) enantiomer is very widely distributed in cells and serves as a carrier in a number of enzymic β-carboxylation reactions. In biotin-containing enzymes it is covalently bound at the active site to the ε-amino group of a lysine residue. It is a growth factor for yeasts. *Compare* **biocytin**. *See also* **streptavidin**.

biotin carboxylase a carboxylase enzyme that contains biotin (as a **biocytin** residue) in its active site. In humans such enzymes include **pyruvate carboxylase**, **propionyl-CoA carboxylase**, and β-**methylcrotonyl-CoA carboxylase**, which are present in mitochondrial matrix, and **acetyl-CoA carboxylase**, which is cytosolic.

biotinidase EC 3.5.1.12; an enzyme that releases biotin from biocytin or biocytin peptides but not from biotin holocarboxylases. It is present in many human tissues. Serum biotinidase is a glycoprotein with about nine isoforms (67–76 kDa) and is derived from the liver. Many mutations in a locus at 3p25, which encodes a mature protein of 502 residues, result in biotinidase deficiency. This is an autosomal recessive disorder of highly variable clinical expression that commonly includes seizures, hypotonia, developmental delay, ketolactic acidosis, and organic aciduria. Serum biotinidase activity is below 10 to 30% of normal. *See* **holocarboxylase synthetase**.

biotinyl the acyl group derived from **biotin**.

biotinylation 1 the act or process of attaching a biotinyl residue by use of an appropriate reactive derivative, e.g. biotin 4-nitrophenyl ester. It is useful for labelling cell-surface functional groups or biologically active molecules. The presence of the label is subsequently detected, or the labelled material is separated, by means of the specific and high-affinity avidin–biotin reaction, using avidin or **streptavidin** conjugated with, e.g., a fluorochrome, an enzyme, an antibody, or an immobilizing agent. **2** the addition of biotin to the ε-amino group of a lysine residue, catalysed by holocarboxylase synthetase, to form a **biotin carboxylase**. —**biotinylate** *vb.*; **biotinylated** *adj.*

biotransformation the conversion or conjugation of a foreign compound introduced into an organism or a culture of cells, especially (*in biotechnology*) with formation of a substance of commercial value. *Compare* **detoxication**.

BiP *abbr. for* binding protein.

biphenylylphenyloxadiazole *abbr.*: PBD; 2-phenyl-5-(4-biphenylyl)-1,3,4-oxadiazole; a high-efficiency primary scintillator. The *tert*-butyl derivative, 2-(4-*tert*-butylphenyl)-5-(4-biphenylyl)-1,3,4-oxadiazole (*abbr.*: butyl-PBD) has greater solubility and only a slightly lower photon yield. *See also* **scintillation cocktail**.

biphosphate *an obsolete name for* the anion H$_2$PO$_4^-$ or PO$_2$(OH)$_2^-$, now correctly termed dihydrogenphosphate.

BIPM *abbr. for* Bureau International des Poids et Mesures.

BIR *abbr. for* baculovirus IAP repeat; a cysteine-rich protein sequence ($\approx$70 amino acids) that is present in several **IAPs** in humans. It is named after the baculovirus, in which it was first discovered.

biradical *or* **diradical** any molecular entity in an electronic state (described by a formula or by a combination of contributing structures) that contains two unpaired electrons in atomic orbitals on different atoms (*compare* **carbene**). Both singlet and triplet species are embraced.

bird flu a disease caused by the avian influenza virus with haemagglutin H5 and neuraminidase N1 and transmissible to the human species, leading to a dangerous respiratory infection..

birefringence *an alternative term for* **double refraction**. — **birefringent** *adj.*

bis+ *comb. form* **1** denoting two, twice, doubled; *see also* **bi+** (def. 1), **di+** (def. 1). *See* **multiplying prefix**. **2** (*in chemical nomenclature*) **a** indicating the presence in a molecule of two identical organic groups each substituted in the same way, e.g. **bis(2-chloroethyl)sulfide**. **b** indicating the presence in a molecule of two separate inorganic oxoacid residues, e.g. fructose 1,6-bisphosphate (formerly known as fructose 1,6-diphosphate). *Compare* **di+** (def. 1).

bisalbuminemia *or* (*esp. Brit.*) **bisalbuminaemia** a symptomless human condition in which the presence of a variant serum albumin leads to a double albumin band on electrophoresis of serum proteins.

bisbenzimide 4-[5-(4-methyl-1-piperazinyl)-2-benzimidazolyl]-2-benzimidazolyl phenol (Hoechst 33258) or its ethyl ether (Hoechst 33342); a substance that is useful as a DNA-binding fluorochrome for staining chromosomes within cell nuclei. It is believed to be specific for A–T base pairs.

bis(2-chloroethyl)sulfide *or* (*formerly*) **mustard gas** (ClCH$_2$–CH$_2$)$_2$S; a bifunctional alkylating agent that causes lethal cross-linking of DNA chains.

bisindolylmaleimide GF 109203X; a protein kinase inhibitor, structurally similar to **staurosporine**, that is highly selective for **protein kinase C** (PKC), though not between isotypes α, βI, βII, or γ. It acts as a competitive inhibitor at the ATP-binding site on PKC. It

bisbenzimide

exhibits a K_i of 14 nm for the purified enzyme, but its potency on whole cells is much lower (K_i 1–2 μm), possibly owing to high cellular ATP levels.

bismethylstyrylbenzene *abbr.*: bis-MSB; 1,4-bis(2-methylstyryl)-benzene; a secondary scintillator with good resistance to quenching and low reactivity with compounds in samples. *See also* **scintillation cocktail**.

bis(methylthio)gliotoxin a naturally occurring analogue of **gliotoxin**. It inhibits platelet activating factor (PAF)-induced platelet aggregation ($IC_{50} = 84$ μm) (but not platelet aggregation induced by arachidonic acid or ADP) and *in vivo* PAF-induced bronchoconstriction; it lacks the antiviral and immunomodulating effect of gliotoxin.

bis-MSB *abbr. for* bismethylstyrylbenzene.

bisphosphate any compound containing two independent phosphoric residues in ester linkages, at positions indicated by locants. *Compare* **diphosphate** (def. 3, 4).

bisphospho+ *prefix* to a chemical name indicating the presence of two independent phosphoric residues in ester linkages, at positions indicated by locants. *Compare* **diphospho+**.

2,3-bis(phospho)-D-glycerate *a recommended name for* (*R*)-[2,3-dihydroxypropanoate 2,3-bis(dihydrogenphosphate)] (formerly termed 2,3-diphosphoglycerate); *abbr.*: BPG; it decreases the dioxygen affinity of hemoglobin, its concentration in the erythrocyte being regulated by **bisphosphoglycerate mutase** (synthesis) and **bisphosphoglycerate phosphatase** (degradation). It does not pass through the erythrocyte membrane. *See also* **bisphosphoglycerate pocket**.

bisphosphoglycerate mutase *abbr.*: BPGM; EC 2.7.5.4; *systematic name*: 3-phospho-D-glycerate 1,2-phosphomutase; *other names*: 2,3-bisphosphoglycerate mutase; 2,3-bisphosphoglycerate synthase; diphosphoglycerate mutase; glycerate phosphomutase; bisphosphoglycerate synthase. In human reticulocytes and erythrocytes, an activity on a multifunctional homodimer (locus at 7q34-q22) that catalyses the reaction:

$$\text{3-phospho-D-glyceroyl phosphate} = \text{2,3-bisphospho-D-glycerate.}$$

The same protein also has **bisphosphoglycerate phosphatase** activity, and the two activities together regulate the dioxygen affinity of hemoglobin by controlling 2,3-bisphosphoglycerate concentration in erythrocytes. Several rare mutations lead to deficiency of these reactions and are associated with erythrocytosis. The protein also has a third activity, that of 3-phosphoglycerate mutase, which catalyses the reaction: 3-phosphoglycerate = 2-phosphoglycerate.

bisphosphoglycerate phosphatase EC 3.1.3.13; *systematic name*: 2,3-bisphospho-D-glycerate 2-phosphohydrolase; in human reticulocytes and erythrocytes, an activity on a multifunctional homodimer (locus at 7q34-q22) that catalyses the hydrolysis of 2,3-bisphosphoglycerate to 3-phosphoglycerate and orthophosphate; it thereby assists regulation of the concentration of 2,3-bisphosphoglycerate (in conjunction with the **bisphosphoglycerate mutase** activity of the same protein), and thus also the dioxygen affinity of hemoglobin.

bisphosphoglycerate pocket *abbr.*: BPG pocket; the cavity (pocket) between the two β chains of deoxyhemoglobin in which a single molecule of 2,3-bisphospho-D-glycerate (BPG) binds; the binding of this molecule reduces the affinity of hemoglobin for dioxygen. Residues Val[1] (terminal –NH$_3$[+]), His[2], Lys[82], and His[143] of each β chain participate in this binding. It plays an important part in establishing the difference in oxygen affinity that is found between maternal and fetal hemoglobin; in fetal hemoglobin, the γ chain has a serine in the position occupied by His[143] of the β chain of the adult hemoglobin, which weakens the affinity of fetal hemoglobin for BPG and thus results in a higher affinity for dioxygen than adult hemoglobin.

bis[2-succinimidooxycarbonyloxyethyl]sulfone a non cytotoxic, homobifunctional, lysine-specific cross-linking agent that reacts at 0 °C; the cross-links can be reversed at pH 11.6 at 37 °C.

2,5-bis(3,4,5-trimethoxyphenyl)-1,3-dioxolane a substance, the *trans*-isomer of which inhibits platelet activating factor-induced platelet aggregation. The *cis*-isomer is inactive.

Bistris or **bis-tris** *abbr. for* bis(2-hydroxyethyl)imino-tris(hydroxymethyl)methane; 2-[bis(2-hydroxyethyl)amino]-2-(hydroxymethyl)-propane-1,3-diol. A buffer substance useful for the calibration of glass electrodes and for the preparation of biochemical and biological buffer solutions; pK_a (25°C) = 6.5. It is inexpensive, stable, and readily crystallizable.

bis-tris propane *abbr. for* 1,3-bis[tris(hydroxymethyl)methylamino]propane; a water-soluble buffer substance used for the preparation of biochemical and biological buffer solutions; pK_a (20°C) = 6.8.

bit a single binary digit having the value of either 0 or 1; the smallest unit of information in a binary system. A group of (usually) eight bits comprises a **byte**, which can represent numbers between 0 and 11111111 in binary (equal to 0 and 255 in decimal). [*abbr. for* binary digit.]

bithorax complex *see* BX-C.

Bittner factor *or* **Bittner particle** *former name for* murine mammary tumour virus. [After John Joseph Bittner (1904–61), US biologist.]

biuret reaction the formation of a characteristic purple colour when biuret (carbamoylurea) or compounds containing two or more adjacent peptide bonds (e.g. proteins) react with copper sulfate in alkaline solution. It forms the basis of the **biuret test**, a colorimetric method for the qualitative or quantitative determination of proteins.

Bjerrum formation function a function used in studies of the binding of *n* ligand atoms, ions, or molecules to an acceptor atom, ion, or molecule. It is:

$$\nu = (K_1[L] + 2K_1K_2[L]^2 + ...nK_1K_2...K_n[L]^n)/(1 + K_1[L] + 2k_1k_2[L^2] + ...nK_1K_2...K_n[L]^n),$$

where ν is the average number of ligand molecules bound per molecule of acceptor; K_1, K_2, $...K_n$ are the stepwise formation (stability) constants of the *n* complexes; and $[L]$ is the free ligand concentration. This function expresses an unambiguous connection between the quantity ν and $[L]$. It is a generalized form of the **Adair equation**. [After Jannik Bjerrum (1909–92), Danish chemist who described it in 1941.]

Bjerrum plot a plot of ν, the average number of ligand molecules bound per molecule of acceptor, versus $-\log[L]$, where $[L]$ is the free ligand concentration. It is used for evaluating the association constants in a **Bjerrum formation function**.

Bk *symbol for* berkelium.

BK *abbr. for* bradykinin.

BK$_{1-8}$ *abbr. for* des-Arg⁹-bradykinin (*see* **bradykinin**).

BL21 a strain of *Escherichia coli* carrying a lambda phage DE(T7) lysogen. Inducible expression of T7 RNA polymerase gives high levels of transcription from plasmid vectors such as pET and high yields of protein product.

bla *symbol for* β-lactamase gene.

black lipid membrane *abbr.*: BLM; *other names*: black lipid film; bimolecular lipid membrane; bilayer lipid membrane; lipid bilayer. An experimental membrane formed by introducing a small amount of suitable lipid solution on to an opening in the wall of a hydrophobic support immersed in an aqueous solution. The thickness of the black membrane does not differ greatly from the length of two of the lipid molecules used and is often about 4.5 nm.

blank (determination) any analytical mixture from which analyte is deliberately omitted and which undergoes the same procedures as the complete mixture. It is used to ascertain the contribution of extraneous factors, e.g. reagent impurities or photometric absorption by the reagents, to the analytical result.

blast+ *a variant form of* **blasto+**.

+blast *suffix* denoting a formative cell, e.g. erythroblast, fibroblast, osteoblast.

BLAST *abbr. for* basic local alignment search tool; *see* **alignment algorithm**.

blast cell *or* **blastocyte** any undifferentiated embryonic cell. Such cells characteristically have a cytoplasm rich in RNA, and are actively synthesizing DNA.

blast crisis the appearance of heterogeneity in the cell surface characteristics of white blood cells when chronic myeloid leukemia evolves into acute leukemia. This process is virtually inevitable and often rapid.

blasticidin S a peptidylnucleoside antibiotic produced by *Streptomyces griseochromogenes* that inhibits protein synthesis through inhibition of peptide bond formation. Resistance to blasticidin is conferred by the expression of the blasticidin S resistance gene (*bsr*) from *Bacillus cereus* or the *BSD* gene from *Aspergillus terreus*. Both encode blasticidin S deaminase (BSD).

blasto+ *or (before a vowel)* **blast+** *prefix* denoting **1** an embryo in its early stages; e.g. blastocyst, blastula. **2** a germ cell; e.g. blastocyte.

blastokinin *see* uteroglobin.

blast transformation *see* **transformation** (def. 3).

bleb 1 a spherical vesicle, about 10 μm in diameter, formed when a chloroplast swells on suspension in very hypotonic medium. It is bounded by a single membrane, on the surface of which a part of the **thylakoid** system remains concentrated in a few patches. **2** animal membranes are said to form blebs under special conditions that cause small vesicles to pinch off.

blebbistatin a highly specific inhibitor of ATPase activity of myosin II, and of contraction of the cleavage furrow during cytokinesis. It rapidly and reversibly blocks formation of cell blebs. .

blender *or (sometimes, US)* **blendor** a device for disrupting tissue or any aggregate. It consists of blades rotating at the bottom of a glass or stainless-steel vessel (originally named a **Waring blendor**, from a trade name); in an alternative design, known as a top-drive blender, the blades rotate on a spindle from the top of the vessel. The method is undiscriminating and may damage organelle membranes through excessive force or generation of heat. For more precise work, a **Potter–Elvehjem homogenizer** or a **Dounce homogenizer** preserves organelle structure.

bleomycin any of a group of related glycopeptide antibiotics isolated from *Streptomyces verticillus*. Bleomycin acts to stop the cell cycle in the G_2 phase (*see* **cell-division cycle**). It is used to induce synchrony in cell cultures and as an antineoplastic agent, especially in lymphomas.

blepharoplast *an alternative term for* **basal body** of a flagellate.

Blk a gene encoding a protein tyrosine kinase of the Src family, specifically expressed in the B-cell lineage.

BLM *abbr. for* black lipid membrane.

BLM protein *or* **BLM helicase** *or* **RecQ-like protein 2** *or* **RECQL2** a protein of 1417 residues encoded by the locus for **Bloom syndrome** at 15q26.1. It contains a 350-residue domain that comprises seven sequences characteristic of DNA and RNA helicases and present in RecQ helicases. Numerous mutations are associated with Bloom syndrome. The *recQA* gene was first identified in an *E. coli* mutant resistant to thymineless death. RecQ protein is a DNA helicase. Apart from BLM protein, mammalian cells also contain the homologous RecQ-like protein 1 (RECQL1) of 652 residues, WRN (mutated in Werner syndrome), RECQL4, and RECQL5.

***Bln*I** a type 2 **restriction endonuclease**; recognition sequence: C↑CTAGG. *Avr*II is an **isoschizomer**.

block 1 (*in chemistry*) a portion of a polymer molecule comprising many **constitutional base units**; such a portion has a constitutional or configurational feature not present in the adjacent multiple-unit portions. **2** (*in physiology*) an interference in the normal metabolic or physiological functioning of an organ or tissue; e.g. metabolic block, heart block. **3** (*in bioinformatics*) an ungapped local **alignment**

(def. 2) derived from a conserved region in an aligned family of protein sequences, used to build a diagnostic familial signature. Within a block, segments are clustered to reduce multiple contributions from highly similar or identical sequences, each cluster being treated as a single segment, with a score giving a measure of its relatedness – the most distant segment is given a weight of 100. **4** to impede, retard, or prevent an action or activity, e.g. of an agonist, antigen, chemical reagent, enzyme, or receptor.

blockade 1 (*in pharmacology*) the saturation of a specific type of receptor with an antagonist to its normal agonist. **2** (*in immunology*) the overloading or saturation of the **reticuloendothelial system** with inert particles, such as carbon particles. **3** to impose any such blockade.

blocker something that blocks an action or activity; an antagonist or inhibitor.

blocking agent 1 a chemical reagent that introduces a **blocking group**. **2** a specific pharmacological antagonist.

blocking antibody 1 an incomplete antibody that does not agglutinate its antigen but attaches to it and so prevents its agglutination by normal, complete antibody. **2** antibody formed during specific **desensitization** in **atopy**.

blocking group 1 *or* **protecting group** a chemical group, often easily removable, that is substituted into a reactive part of a molecule to prevent its participation in subsequent chemical or enzymic reactions occurring at another part of the molecule. **2** a chemical group, not easily removable, that is substituted into the active site of an enzyme to prevent its action.

block polymer a polymer whose molecules contain two or more species of **block** (def. 1) attached linearly.

Blocks *see* **pattern database**.

Blocks-format-PRINTS *see* **derived database**.

blood a fluid tissue in animals that acts primarily as a transport medium. It is contained in vessels or spaces lined with endothelial cells. Metabolites, nutrients, and excretory substances are carried in solution or suspension from one part of the body to another. Respiratory pigments are often present to assist in the transport of dioxygen. It may also contain suspended cells.

blood–brain barrier the semipermeable membranous barrier that regulates the passage of dissolved materials from the blood into the cerebrospinal fluid that bathes the brain and spinal cord.

blood coagulation *or* **blood clotting** a mechanism that prevents blood loss at the site of an injury. It involves the formation of a semi-solid mass of material, the blood clot, which plugs the wound. The clot consists of aggregated platelets and a mesh of fibrin molecules. The fibrin components include 13 or 14 plasma proteins, at least one tissue protein, phospholipid membrane surfaces, Ca^{2+} ions, and platelets. The current concept of the mechanism has changed from one involving a linear sequence of activations of proenzymes to one involving protein–protein and protein–lipid–Ca^{2+} interactions between protease, protein substrates, and protein cofactors, to give discrete complexes and reaction steps. The following factors are involved in coagulation:

Factor I (*or* fibrinogen) concn. ~2–4 g L^{-1} human plasma; a soluble fibrous protein of 330 kDa, it has an axial ratio of ~5 and comprises six disulfide-linked polypeptide chains: two Aα chains (66 kDa); two Bβ chains (52 kDa); and two γ chains (47 kDa). After proteolysis by thrombin, it polymerizes to form an insoluble fibrin clot.

Factor II (*or* prothrombin) concn. 200 mg L^{-1} human plasma, 72.5 kDa; a vitamin K-dependent single-polypeptide-chain glycoprotein containing ten γ-carboxyglutamate residues. It is converted to **thrombin** (*or* fibrinogenase), EC 3.4.21.5, by the protease factor Xa.

Factor III (*or* thromboplastin *or* tissue factor) a factor that, when added to blood plasma, markedly shortens the time required for clot formation.

Factor IV (*or* calcium ion) calcium ions form bridges between vitamin K-dependent proteins and acid phospholipids of cell membranes. They also stabilize factor V and fibrinogen, and are involved in activation of factor XIII.

Factor V (*or* accelerator globulin *or* proaccelerin) a glycoprotein of ~330 kDa. After activation to **factor Va** (accelerin), it acts with the protease factor Xa in activating prothrombin.

Factor VI an obsolete term (originally named accelerin).

Factor VII (*or* proconvertin) concn. 2 mg L^{-1} human plasma; 50 kDa; a vitamin K-dependent single-polypeptide-chain glycoprotein containing γ-carboxyglutamate residues. It is the precursor of **factor VIIa**, convertin, a Ca^{2+}-dependent protease, EC 3.4.21.21, that activates factor X, and is the first component of the extrinsic pathway of coagulation.

Factor VIII (*or* antihemophilic factor); an accessory protein that participates with protease factor IXa in the activation of factor X in the intrinsic pathway of coagulation.

Factor IX (*or* Christmas factor) concn. 3–4 mg L^{-1} human plasma; 57 kDa (human), 55.4 kDa (bovine); a vitamin K-dependent single-polypeptide-chain glycoprotein containing 17% carbohydrate (human; 25% in bovine) and γ-carboxyglutamate residues. After proteolytic activation, it consists of two chains: **factor IXa**, a serine protease, EC 3.4.21.22, of 27 kDa that activates factor X in the intrinsic pathway of coagulation; and a light chain of 16 kDa that contains the γ-carboxyglutamate residues.

Factor X (*or* Stuart–Prower factor; *other names*: Stuart factor; thrombokinase; prothrombase; prothrombinase) concn. 6–8 mg L^{-1} human plasma; a vitamin K-dependent two-polypeptide-chain glycoprotein, containing 15% carbohydrate (human; 10% in bovine): the 40 kDa chain is a serine protease; the 15 kDa chain contains 12 γ-carboxyglutamate residues. It is the circulating precursor to the protease **factor Xa**, EC 3.4.21.6, that converts prothrombin into thrombin by selective cleavage of Arg-|-Thr then Arg-|-Ile bonds.

Factor XI (*or* plasma thromboplastin antecedent) 160 kDa (human), 124 kDa (bovine); a protein of two disulfide-linked chains of apparently equal molecular mass. It is the circulating precursor of the serine protease **factor XIa**, EC 3.4.21.27, to which it is converted by proteolysis: selective cleavage of Arg-|-Ala and Arg-|-Val bonds in factor IX, to form factor IXa, triggers the middle phase of the intrinsic pathway of blood coagulation by activating factor IX. Factor XIa converts factor IX into factor IXa.

Factor XII (*or* Hageman factor) 80 kDa (human), 74 kDa (bovine); a single-polypeptide-chain glycoprotein. It is the circulating precursor of the protease **factor XIIa**, which activates factor XI.

Factor XIII (*or* plasma transglutaminase) 320 kDa (plasma, tetrameric), 160 kDa (platelets, dimeric). An enzyme precursor whose subunits are proteolytically cleaved by thrombin to form the transglutaminase, **factor XIIIa**, EC 2.3.2.13, which covalently cross-links fibrin monomers into the fibrin polymer.

Prekallikrein a plasma protein, 88 kDa (bovine), that is the precursor of the protease **kallikrein**. This consists of a heavy chain (52 kDa) and a light chain (33–38 kDa), and cleaves factor XII to form factor XIIa.

High-molecular-weight kininogen a single-polypeptide-chain glycoprotein of ~150 kDa (human), 76 kDa (bovine). It is the source of **bradykinin**, which is released from it by **kallikrein**, and acts as an accessory protein in the activation of factor XI by factor XIIa.

Plasminogen 86 kDa (human); a single-polypeptide-chain protein that, on proteolysis, is converted into the protease **plasmin**, which degrades the fibrin clot. There are two pathways of coagulation: in the **intrinsic pathway**, which occurs during the clotting of platelet-poor plasma, activation of factor X to factor Xa is effected by the combination of factor IXa + factor VIIIa + Ca^{2+} + phospholipid; in the **extrinsic pathway**, the earlier stages of the cascade are bypassed, and the activation of factor X to factor Xa is effected by the combination of factor VIIa + thromboplastin; this second pathway occurs when tissue extracts are present in optimal amounts, and is much more rapid than the intrinsic pathway.

blood group any of various systems of classifying erythrocytes according to their surface isoantigens. In humans 14 different, genetically independent blood-group systems are known. They arise from naturally occurring antibodies and from the **blood-group substances**, antigenic substances present on the surface of the erythrocytes of only certain members of a species. These antigens are also found in a wide variety of other tissues, including endothelial and epithelial cells, and in certain body fluids. The specificities of the blood-group antigens arise in some cases from oligosaccharide chains linked to protein or lipid (ceramide); antigens that fall into this category are those associated with the ABH system, Lewis system, Ii system, and P1/P-related system. In other cases, antigenicity arises directly from the polypeptide chain composition of a surface protein; antigens that fall into this category are associated with the MN system, Ss system, Gerbich system, Rhesus system, Kell system, Duffy system, and Cromer-related system. The blood-group substances were discovered because of the presence in human blood of naturally occurring antibodies (i.e. not resulting from infection or other external cause) that react with these antigens on red blood cells causing them to agglutinate. Only the presence of a blood-group antigen during fetal life ensures that no antibody against it will be produced. Hence if blood from an individual (the donor) who lacks a particular antigen is transfused into an individual (the recipient) who possesses the same antigen, a transfusion reaction will result in the recipient owing to the presence in the donor's blood of antibody against the antigen in question. The blood-group substances have therefore been of interest largely in relation to blood transfusion, and in most cases the true physiological function of these substances is unknown.

blood-group substance any of numerous antigenic substances present on the surface of erythrocytes of certain members of a species but absent from those of other members of the same species. These same antigens are also found in a wide variety of tissues and in certain body fluids, e.g. saliva, milk, gastric juice, seminal fluid, urine, and ovarian-cyst fluid. The specificities of the A, B, and H antigens of the **ABO system** are determined by structures present at the nonreducing ends of oligosaccharides of both glycoproteins and glycolipids. The **Lewis antigens**, Le_a and Le_b, are associated with glycosphingolipids adsorbed from plasma lipoproteins on to adult cells and are not synthesized in the developing erythrocyte. Antigenic determinants of the **MN system** are associated with glycoproteins rich in sialic acid.

blood plasma 1 the fluid portion of blood in which the red blood cells and other cellular components (the 'formed elements') are suspended. Clotting may be prevented by the addition of disodium hydrogen citrate which removes ionized Ca^{2+} by the formation of an un-ionized complex. Fluoride or oxalate may also be used to precipitate the Ca as an insoluble salt. **Heparin** is often used. When whole blood containing such an anticoagulant is centrifuged, the formed elements sediment in a volume equal to about 45% of the total volume of blood. The supernatant is known as the plasma. In vertebrates it is clear and almost colourless and clots as easily as whole blood. *See also* **anticoagulant**. *Compare* **serum**.

blood sugar *old term for* the glucose concentration in the blood; when estimated by older methods it included small amounts of other reducing substances. *See also* **glycemia**.

Bloom syndrome an autosomal recessive condition characterized by profound disposition to neoplasias and associated with small body size, a sun-sensitive erythematous skin rash, characteristic facies and head shape, immunodeficiency, and chronic lung dissease. Heterozygotes have twice the normal risk of developing colorectal cancer. It is caused by many mutations in the gene for **BLM protein**. [After David Bloom (1892–), US dermatologist.]

BLOSUM *abbr. for* Blocks substitution matrix; *see* **substitution matrix**.

blotting *or* **blot transfer** a technique useful for identifying similar macromolecules, e.g. sets of DNA or RNA fragments, or mixtures of intact proteins, that are separable by gel electrophoresis. The separated components are then transferred from the gel to a sheet of an appropriate medium while preserving their two-dimensional relationships. This can be achieved either by capillary flow between an underlying wick loaded with buffer solution and an overlay of dry blotting paper, or by applying an electric potential to move separated components on to the medium, to which they bind firmly. Once on the transfer medium the substances may be detected with specific radioactive or fluorescent probes, e.g. complementary polynucleotides or antibodies. The original version of the technique, used for DNA fragments and often referred to as **Southern blotting**, has been extensively developed and adapted for other applications, e.g. **dot hybridization** (or dot blotting), **colony hybridization**, **electrophoretic transfer**, **Northern blotting**, and **Western blotting**.

BLOTTO *abbr. for* Bovine Lacto Transfer Technique Optimizer; a 5% solution of non-fat dried milk sometimes used in Southern hybridizations.

blue agarose any insoluble beaded form of agarose gel that is arti-

ficially cross-linked between the individual polysaccharide chains and conjugated with the dye Reactive Blue 2. Carboxymethyl (CM) or diethylaminoethyl (DEAE) groups may also have been introduced. It is useful in **triazine-dye affinity chromatography**.

blue copper protein an **electron transfer protein** containing a **type 1 copper** site. It is characterized by strong absorption in the visible region owing to coordination of the copper by cysteine sulfur.

blue dextran a soluble dextran, of average M_r 10^6, conjugated with the dye Reactive Blue 2. It is useful in gel filtration for checking column packing and determining the void volume.

blue-green algae *an alternative name for* **cyanobacteria**.

blue oxidase any of a small number of copper-containing oxidase enzymes in which the product of enzyme action in the reduction of dioxygen is water. Blue oxidases are the most complicated of copper proteins; the blue colour is due to a strong absorption band around 600 nm. Among the blue oxidases are laccase (EC 1.10.3.2), L-ascorbate oxidase (EC 1.10.3.3), and the weak oxidase **ceruloplasmin**.

blue shift any shift of the peaks of absorption or emission of a spectrum of electromagnetic radiation to shorter wavelengths, including from the visible to the ultraviolet region.

Blue Tetrazolium *or* **Tetrazolium Blue** 3,3'-dianisole bis[4,4'-(3,5-diphenyl)tetrazolium chloride]; a dye that yields a dark-blue diformazan pigment on reduction; it has a redox potential of approximately –0.08 V. It is useful as a stain for bacteria and moulds, in histochemical studies to demonstrate oxidoreductase activity, and in the colorimetric assay of some dehydrogenases.

blunt end (of DNA) *an alternative term for* **flush end** (of DNA).

B lymphocyte *or* **B cell** *or* *(archaic)* **bursacyte** any 'bursa-dependent' lymphocyte concerned in the synthesis of circulating antibody. B lymphocytes arise from primitive lymphoid cells of the bone marrow and develop into the **plasma cell** series. Their maturation is determined by processing within the **bursa of Fabricius** in birds, or equivalent hematopoietic tissue in other vertebrates. B lymphocytes may be differentiated from **T lymphocytes** by their surface markers which contain immunoglobulins. In humans the majority of B lymphocytes are derived from bone marrow stem cells, but a minor population, distinguished by the CD5 marker, appears to form a self-renewing set. Cells in this set respond to a number of common microbial antigens and sometimes generate **autoantibodies**.

BMAL1 a basic helix-loop-helix transcription factor that forms heterodimers with the transcription factors **clock** and **NPAS2**. Binding to a DNA motif (CACGTG) enhances expression of period proteins and cytochromes in the suprachiasmatic nucleus and forebrain nuclei. Transcription of genes activated by clock-BMAL1 dimers is repressed by CRY2 (*see* **cryptochrome**).

Bmi-1 a murine gene encoding a protein that is probably a transcriptional regulator. It is named from B-cell-specific Moloney murine leukemia virus integration site. The protein contains putative **zinc finger** and **helix-loop-helix** domains, as well as C-terminal PEST sequences. *See* **PEST hypothesis**.

BMI *abbr. for* body mass index; an anthropometric measurement equal to weight in kg/(height in m)^2. Values in the range 25.0–29.9 indicate that an individual is overweight, values >30 indicate obesity.

BMP *abbr. for* bone morphogenetic protein.

BMPR *abbr. for* bone morphogenetic protein receptor.

BMR *abbr. for* basal metabolic rate.

BmrR a protein of *Bacillus subtilis* that is a transcription regulator of the gene *bmr*, which encodes a multidrug transporter (**Bmr transporter**) and is expressed in response to a variety of structurally and chemically dissimilar hydrophobic cations, many of which are substrates of the Bmr transporter. It contains an N-terminal DNA-binding region specific for the promoter of *bmr*, and a C-terminal domain that contains a hydrophobic core and a buried glutamate residue and functions in inducer binding and in dimerization. The crystal structure of the C-terminal domain has been solved.

Bmr transporter *see* BmrR.

BNP *abbr. for* brain natriuretic peptide (*see* **natriuretic peptide**).

boat conformation any conformation of a nonplanar six-membered saturated ring compound when ring atoms in relative positions 1, 2, 4, and 5 lie in one plane, and those in relative positions 3 and 6 lie on the same side of that plane. Boat conformations of a monosaccharide or monosaccharide derivative may be designated by the conformational descriptor -*B*. See also cyclic compounds at **conformation**.

Boc *or* *t*-BOC *symbol for* the *tert*-butoxycarbonyl group, $(CH_3)_3C–O–CO–$, a **protecting group** used in polypeptide chemistry. It can be introduced using reagents such as di-*tert*-butyl dicarbonate (Boc anhydride).

BOD *abbr. for* biochemical oxygen demand.

body mass index *see* BMI.

body water the water content of the body of an organism. The **total body water** (*abbr.* TBW) of a man of average build (70 kg) has a volume of approximately 42 L and represents ≈ 60% of the body weight. It may be divided functionally into two parts: **intracellular fluid** and **extracellular fluid**.

Bohr coefficient (for hemoglobin) the ratio of the change in the logarithm of the partial pressure of oxygen when there is half-saturation of the hemoglobin to the change in pH; i.e. $\Delta \log P_{50}/\Delta pH$. See also **Bohr effect**. [After Christian Bohr (1855–1911), Danish physiologist.]

Bohr effect the variation of oxygen affinity of hemoglobin with pH (the oxygen affinity increases with increasing pH). It is one of the effects arising from oxygen-linked acid groups in hemoglobin and similar oxygen-carrying proteins, and also encompasses similar effects involving other acid-linked functions, e.g. oxidation loss of protons causing increased dioxygen binding. The effect was discovered by Christian Bohr *et al.* following observations that changes in the partial pressure of CO_2 influence the oxygen equilibrium of the blood.

Bohr magneton *symbol*: μ_B; the intrinsic magnetic moment of an electron given by $\mu_B = eh/4\pi m_e$, where e is the elementary charge, h is the Planck constant, and m_e is the rest mass of an electron. It is a fundamental physical constant, of value $9.2740154(31) \times 10^{-24}$ J T^{-1}. *Compare* **nuclear magneton**. [After Niels Bohr (1885–1962), Danish physicist and son of Christian Bohr.]

boivinose the D-enantiomer, 2,6-dideoxy-D-*xylo*-hexose, is a component of some **cardiac glycosides**.

Bollum's enzyme DNA nucleotidylexotransferase, EC 2.7.7.31. See **terminal transferase**. [After Frederick James Bollum (1927–), who described it in 1971.]

Bolton and Hunter reagent *or* **Bolton–Hunter reagent** 3-(4-hydroxyphenyl) propionic acid *N*-hydroxysuccinamide ester, iodinated with ^{125}I. This may be the di-iodo derivative, or the commercially available mono-iodo *N*-succinimidyl 3-(4-hydroxy-5-[^{125}I]iodophenyl)propionate; a reagent for **conjugation labelling** of polypeptides or proteins with iodine-125. It is specific for free amino groups, which under mild conditions become acylated to form (4-hydroxy-5-[^{125}I]iodophenyl)propionylamino groups. [After Anthony E. Bolton and William M. Hunter who described its use in 1972.]

Boltzmann constant *symbol*: k; the ratio of the gas constant, R, to the Avogadro constant, N_A. It is given by $k = R/N_A = 1.380658\ (12) \times 10^{-23}$ J K^{-1}. [After Ludwig Boltzmann (1844–1906), Austrian physicist.]

Boltzmann distribution law a law describing the population distribution of a system of particles in different energy states but at thermal equilibrium, given by: $n_i/n_0 = e^{-(E_i - E_0)/kT}$, where n_i is the number of particles in energy state i, n_0 is the number in the lowest energy state, E_i and E_0 are the energies of states of i and 0 respectively, T is the thermodynamic temperature, and k is the **Boltzmann constant**.

bolus **1** a soft, rounded mass, especially of chewed food. **2** a large pill, as used in veterinary or clinical medicine.

bomb calorimeter an apparatus with which the heat of combustion of a substance may be determined. The substance is placed in a metal bomb with dioxygen under pressure, and then ignited electrically. The heat evolved is measured by the rise in temperature of the water in a surrounding jacket.

bombesin a tetradecapeptide amide isolated from frog skin. It has an N-terminal 5-oxoprolyl residue, and shares striking sequence similarity with other peptides isolated from amphibian skin, all of which have a C-terminal methioninamide residue. The (toad) sequence is:

$$QQRLGNQWAVGHLM–NH_2,$$

In mammals, it is a potent releaser of, *inter alia*, gastrin and cholecystokinin, and bombesin-like immunoreactivity has been detected in mammalian gut and brain. It is mitogenic in a number of cell types, and stimulates smooth muscle contraction. The peptide is said to occur in gastrointestinal nerves of mammals, while in frogs and birds it occurs in endocrine cells.

bombesin receptor any of several membrane proteins that bind bombesin, and mediate its intracellular effects. Binding studies have indicated two receptor subtypes: on gastrin releasing peptide (GRP)-preferring receptors (e.g., rat pancreas), GRP is more potent than neuromedin B; on other receptors (e.g., rat esophagus), neuromedin B is more potent than GRP. All receptors are of the **7TM**, G protein-coupled type. The effector pathway involves the phosphoinositide/Ca^{2+} second messenger system (*see* **phosphatidylinositol cycle**).

bombolitin any of five structurally related peptides from the venom of the bumblebee, *Megabombus pennsylvanicus*, that are rich in hydrophobic amino-acid residues (*e.g.*, bombolitin I has sequence IKTITTMLAKLGKVLAHV). They all lyse erythrocytes and liposomes, release histamine from mast cells, and stimulate phospholipase A_2; their properties are similar to those of **melittin**.

bombyxin any of a family of brain secretory peptides of the silkmoth that activate prothoracic glands to produce ecdysone. They contain an A and B chain joined by disulfide bonds, and are homologous to vertebrate insulins.

bond 1 *or* **covalent bond** a region of high electron density between atoms that arises, at least partly, from sharing of electrons and gives rise to an attractive force and a characteristic internuclear distance between the atoms. See **coordination**, **dipolar bond**, **hydrogen bond**. **2** the representation of a covalent bond in an extended chemical formula. **3** to join or link together (esp. chemical entities); to **ligate** (def. 1).

bond angle the angle formed between two bonds at a given polyvalent atom.

bond-dissociation energy *see* dissociation energy.

bond energy 1 the average value of the gas-phase **bond dissociation energies** (usually at 298 K) of a given type of bond in the same chemical species: **mean bond energy**. **2** loosely, the free energy change occurring on hydrolysis of a hydrolysable chemical bond.

bonding the act or process of forming a covalent **bond** (def. 1); the state of being so linked. See also **domain of bonding**.

bond length the distance between the centres of the nuclei of two bonded atoms.

bone the hard, dense connective tissue that constitutes the skeleton of many animals. It consists of an organic matrix impregnated with bone mineral salts. The matrix is made up of 90–95% type I collagen, a small amount of proteoglycan, and a 49-residue protein containing three γ-carboxyglutamate residues, the formation of which is vitamin K-dependent. This protein binds strongly to the hydroxyapatite crystals of the bone mineral. The inorganic part of the bone consists largely of basic calcium phosphate, which is organized into small crystals of hydroxyapatite, 0.8–1.5 nm thick, 2–4 nm wide, and 20–40 nm long, of approximate composition $Ca_{10}(PO_4)_6(OH)_2$. Other anions present are carbonate, fluoride, hydroxide, and citrate. Most of the body's Mg^{2+}, about 25% of its Na^+, and a smaller proportion of its K^+ is found in bone.

bone GLA-protein (BGP) *see* **osteocalcin**.

bone marrow the soft tissue contained within the internal cavities of bones. **Red marrow** (*or* myeloid tissue), found in developing bone, ribs, vertebrae, and in parts of the long bones, is actively hemopoietic and contains all the cells of the circulating blood and their precursors together with megakaryocytes, reticulum cells, macrophages, and plasma cells. In adult animals the marrow of many bones, particularly limb bones, becomes filled with fatty tissue known as **yellow marrow**. In adult mammals, B lymphocytes develop and differentiate in the bone marrow.

bone-marrow-derived cell any cell derived from bone marrow tissue, including erythrocytes, lymphocytes, granulocytes, etc. The term is often used misleadingly to refer to the **B lymphocytes**.

bone morphogenetic protein *abbr*: BMP; any of several zinc metalloendopeptidases belonging to the sub-subclass EC 3.4.24, and having an **EGF** motif repeat; they are members of the **astacin** subfamily. BMP enzymes are disulfide-linked homodimers of the transforming growth factor-β family, and are related to **tolloid**. They are involved in induction of bone and cartilage formation.

bone morphogenetic protein receptor *abbr*.: BMPR; either of two receptors for **bone morphogenetic proteins**. They occur together on cell surfaces and are single-pass transmembrane proteins whose intracellular domains are serine/threonine protein kinases. They function via phosphorylation of Smad 5, which then dimerizes with Smad 4 before entering the nucleus to regulate DNA transcription. Many mutations in the BMPR II gene at 2q31-q32 are associated with primary pulmonary hypertension. *See* **Smad**.

bone sialoprotein 1 *see* **osteopontin**.

bongkrekic acid 3-carboxymethyl-17-methoxy-6,18,21-trimethyl-docosa-2,4,8,12,14,18,20-heptaenedioic acid; a toxic antibiotic formed by *Pseudomonas cocovenenans* grown on partially defatted coconut. It is named after 'bongkrek', a product of mouldy Indonesian coconut that becomes highly toxic when the *Pseudomonas* outgrows the mould. It is an inhibitor of oxidative phosphorylation, an effect ascribable to its inhibitory action on mitochondrial adenine nucleotide translocation, apparently by fixing the adenine nucleotides to the translocation sites thus rendering dissociation of the nucleotides from the translocator rate-limiting in the overall transport process. *Compare* **atractyloside**.

bookmark *(in computing)* a URL stored in a browser file, usually to allow quick access to frequently used WWW pages.

booster dose a second or subsequent dose of antigen, especially in the form of a vaccine, given after the priming dose with the object of stimulating rapid production of large amounts of antibody.

bootstrap 1 *(in computing)* to restart a computer, often after a crash. **2** *(in statistics)* a form of randomization test that involves sampling data with replacement from the original data set, used to estimate the reliability of a result. Bootstrapping is used, for example, in phylogeny to assess the reliability of the branching structure in phylogenetic trees.

Born equation an equation for the free energy change, ΔG, when a sphere of charge e and radius r is brought into a uniform medium of unvarying dielectric constant D:

$$-\Delta G = (1-1/D)(e^2/2r).$$

[After Max Born (1882–1970), German-born British physicist.]

borohydride any member of a group of compounds with the structure MBH_4, where M is an alkali metal. These compounds will reduce aldehydes, ketones, and Schiff bases in nonaqueous solvents, and also acids, esters, acid chlorides, and nitriles. ^3H-labelled borohydrides are useful for the facile reductive tritiation of the aforementioned compounds.

boronyl the group $-B(OH)_2$; it is used attached to a matrix to complex with *cis*-diols (as in some saccharides).

borosilicate (glass) a silicate glass containing at least 5% boric oxide. Such a material has a high melting point and a low coefficient of thermal expansion, and hence is useful for the manufacture of heat-resistant glassware.

boss symbol *for* bride-of-sevenless gene (*see* **Sevenless protein**).

bottom yeast the *popular name for* any of various strains of **brewers' yeast** that effect fermentation at a comparatively low temperature and tend to sediment to the bottom of the fermentation vessel. It is used for the manufacture of light beers, e.g. lager. Bottom yeast is generally distinguishable from **top yeast** by its ability to produce extracellular β-galactosidase, EC 3.2.1.23.

botulinum toxin a mixture of neurotoxins formed by *Clostridium botulinum*, the causal agent of botulism, that is a potent inhibitor of acetylcholine release from cholinergic neurons. Neurotoxins A and B are homologous zinc metalloproteases that cleave synaptobrevin and synaptosome-associated protein 25 (SNAP 25), respectively. Both consist of a catalytic light (or L) chain disulfide-linked to a heavy (or H) chain that mediates binding of the toxin to, and channel formation in, the host cell membrane. Local injection of the toxin (popularly termed 'Botox') is used to remove wrinkles and frown lines and to decrease muscle spasticity in wrist and fingers following cerebral thrombosis. The type A precursor consists of 1295 amino acids; residues 1–447 form the L chain and 448–1295 form the H chain.

Bouguer–Beer law *another name for* **Beer's law** (def. 1).

Bouguer's law a physical law stating the relation between the amount of light transmitted by an absorbing medium and its thickness, based upon the finding that successive equal thicknesses of sheets of glass absorbed equal fractions of the light incident upon them. It may be expressed in the form $I_{trs} = I_0 e^{-kd}$, where I_0 is the intensity of the incident light, I_{trs} is the intensity of the transmitted light, d is the thickness of the absorbing medium, and k is a constant. The law was later restated by Lambert — *see* **Lambert's law (of absorption)** — and extended to solutions by Beer — *see* **Beer's law** (def. 1). [After Pierre B. Bouguer (1698–1758), French astronomer and mathematician who discovered it in 1728.]

boundary any zone of transition, either between solvent and solution or between two solutions; i.e. a zone in which the composition of a solution changes.

boundary spreading the broadening of a boundary during ultracentrifugation of macromolecules due to the effects of diffusion.

bound-to-free ratio *abbr*.: B/F ratio; the ratio of the amount of a ligand bound by a specific macromolecule or receptor to the amount of ligand free in solution, in a specified system. Plotting [B]/[F] versus [F] yields a **Scatchard plot**.

boutique database *an alternative name for* **specialist database**.

bouvardin a cyclic hexapeptide isolated from the plant *Bouvarda ternifolia*, used as a drug against dysentery. It has anti-tumour properties and inhibits eukaryotic protein synthesis.

bovine of, belonging to, or characteristic of the ox tribe Bovini, which includes cattle, especially of the genus *Bos*.

bovine albumin *or* **bovine serum albumin** *see* **serum albumin**.

bovine pancreatic inhibitor *or* **bovine pancreatic trypsin inhibitor** *or* **basic protease inhibitor** a protein from cattle pancreas that inhibits trypsin, kallikrein, chymotrypsin, and plasmin. A popular molecule for studies on protein folding, etc., it contains three disulfide bonds. The sequence is:

RPDFCLEPPYTGPCKARIIRYFYNAKAGLCQT-FVYGGCRAKRNNFKSAEDCMRTCGGA.

(10.903 kDa). The 3-D structure is also known.

bovine spongiform encephalopathy *abbr.*: BSE; a transmissible dementia in cattle, sometimes also known as 'mad cow disease', characterized by a spongy degeneration of neuronal cells in the brain and also enlargement of the astrocytes therein. It appears to be caused by a **prion**, and original outbreaks were thought to be the result of feeding cattle on offal derived from sheep infected with **scrapie**.

Bowman–Birk inhibitor *see* **soybean trypsin inhibitor**.

Bowman's capsule *or* **glomerular capsule** the dilated, cup-shaped vesicle at the proximal end of a nephron that encloses a knot of blood capillaries constituting the glomerulus. It is the site of primary filtration of blood into the nephron. [After William Paget Bowman (1816–92), British histologist and ophthalmic surgeon.]

Boyden chamber an apparatus, consisting of two small chambers separated by a micropore filter, used in the study of chemotaxis.

Boyle's law *see* **gas laws**. [After Robert Boyle (1627–91), Irish chemist and physicist.]

bp *symbol for* base pair(s).

b.p. *abbr. for* boiling point.

53Bp1 *abbr. for* p53 binding protein 1; a protein that contains tandem *BRCA1* C-terminal sequence motifs in its C-terminal region, through which it binds the tumour suppressor protein p53. It colocalizes with DNA damage response proteins and is a critical regulator of ATM protein in cell cycle checkpoint signalling. Mice deficient in 53Bp1 show growth retardation, immune deficiency, and sensitivity to irradiation and are cancer prone.

BP *abbr. for* **1** blood pressure. **2** British Pharmacopoeia.

BP-80 a membrane protein from pea and *Arabidopsis thaliana* that binds the N-terminal propeptide of several vacuolar proteins. It may be a receptor for a subset of these proteins and is produced in these plants by a multigene family. Homologues occur in the Golgi apparatus and in clathrin-coated vesicles .

BPAG *abbr. for* bullous pemphigoid antigen.

BPC *abbr. for* British Pharmaceutical Codex.

B peptide *or* **fibrinopeptide B** the 20-residue peptide cleaved from the β chains of a fibrinogen molecule when it is converted into fibrin by the proteolytic action of thrombin.

BPG *abbr. for* 2,3-bis(phospho)-D-glycerate.

BPI protein *abbr. for* bacterial/permeability-increasing protein; a potent antimicrobial lysosomal protein of polymorphonuclear leukocytes that binds lipopolysaccharide of Gram-negative bacteria, thereby neutralizing its effects. Comprising 456 residues, it is homologous with the plasma proteins LPS-binding protein, CETP (cholesteryl ester transfer protein), and PLTP (phospholipid transfer protein).

BPOC *abbr. for* biphenylisopropyloxycarbonyl-; an acid-labelling group for amino acids, introduced using, e.g., 2-(4-biphenylyl)-prop-2-yl 4′-methoxycarbonylphenylcarbonate (reagent structure shown).

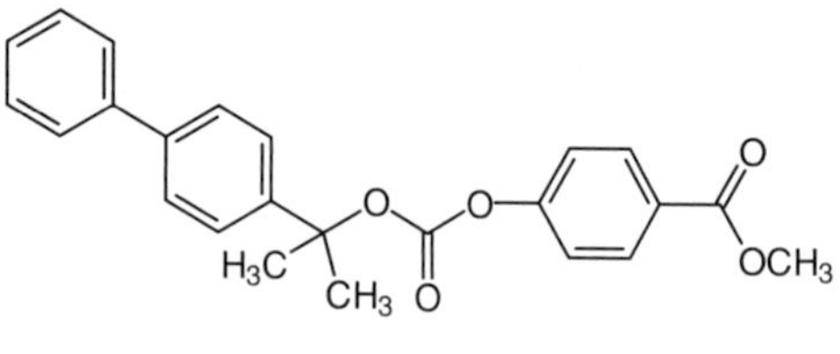

***Bpu*1102I** a type 2 **restriction endonuclease**; recognition sequence: GC↑TNAGC. *Esp*I is an **isoschizomer**.

Bq *symbol for* becquerel.

Br *symbol for* **1** bromine. **2** the butyryl group.

brachyurin EC 3.4.21.32; *other name*: collagenolytic proteinase; a serine proteinase that catalyses the hydrolysis of proteins with broad specificity for peptide bonds. It degrades native collagen at about 75% of the length of the molecule from the N terminus; it also shows low activity on small-molecule substrates of both trypsin and chymotrypsin.

Bradford assay a method for estimation of protein, based on binding of the dye **Coomassie brilliant blue** G-250. [After Marion M. Bradford, US biochemist.]

bradykinin *or* **kinin-9** *abbr.*: BK; a vasoactive nonapeptide, H-Arg-Pro-Pro-Gly-Phe-Ser-Pro-Phe-Arg-OH, formed by the action of **plasma kallikrein**, which hydrolyses the sequence out of the plasma globulin **kininogen**. Glandular kallikrein cleaves kininogen one residue earlier to give the decapeptide Lys-bradykinin (kallidin, *abbr.*: Lys-BK). Met-Lys-bradykinin is also formed, perhaps by the action of leukocyte kallikrein. Pharmacologically important analogues include des-Arg^9 or BK$_{1–8}$ and Ile-Ser-bradykinin (or T-kinin), [Hyp^3]bradykinin, and [Hyp^4]bradykinin. Bradykinin is formed in a variety of inflammatory conditions and in experimental anaphylactic shock. It is a powerful blood-vessel dilator, increasing vascular permeability and causing a fall in blood pressure; it is also a constrictor of smooth muscle. *See also* **lysine(arginine) carboxypeptidase**.

bradykinin receptor any membrane protein that binds bradykinin (BK) and mediates its intracellular effects. Two types of receptor are recognized: B$_1$ (previously BK$_1$), on which order of potency is des-Arg^9-bradykinin (*abbr.*: BK$_{1–8}$) = kallidin (Lys-BK) > BK; and B$_2$ (previously BK$_2$), with order of potency kallidin > BK >> BK$_{1–8}$. Hence, BK$_{1–8}$ is a powerful discriminator. B$_1$ receptors are considerably less common than B$_2$ receptors, which are present in most tissues. The rat B$_2$ receptor is a seven-transmembrane-domain protein which has been shown on activation to stimulate phosphoinositide turnover.

Bragg's law a law stating that when a crystal is pictured as a set of reflecting planes uniformly spaced at a distance d and a beam of X-rays of wavelength λ strikes the crystal at an angle θ, reinforcement of the reflected waves occurs when $\sin\theta = n\,\lambda/2d$, where n is an integer known as the **order of reflection**. [After William Lawrence Bragg (1890–1971), Australian-born British experimental physicist.]

brain-derived neurotrophic factor *abbr.*: BDNF; a protein that promotes the survival of neuronal populations located either in the central nervous system or directly connected to it. It is a key player in neuronal plasticity. Activity-dependent release of BDNF induces protein synthesis in dendrites via a receptor tyrosine kinase, and is important in induction of long-term potentiation.

brain natriuretic peptide *see* **natriuretic peptide**.

Brain tumor *see* **Pumilio**.

branch 1 a subdivision of the stem or the root of a plant; any smaller structure growing or arising from a larger one. **2** a side chain attached to the main chain of a molecular entity. **3** to form a branch; to bifurcate; to divide into offshoots. —**branched** *adj.*; **branching** *n.*

branch-and-bound *(in phylogeny)* a heuristic method for searching tree space for optimal trees. It is faster than an **exhaustive tree search**, but remains impractical for large numbers of taxa (e.g. more than 18).

branched-chain describing any molecular entity with one or more **branches** (def. 2).

branched-chain amino acid *abbr.*: BCAA; any of the neutral aliphatic essential amino acids L-leucine, L-isoleucine, and L-valine. They are incorporated into proteins or degraded oxidatively in mitochondria, especially in skeletal muscle, liver, and brain, and are ketogenic. BCAA infusions counteract the catabolic state of severe trauma and sepsis. Maple-syrup urine disease results from their defective oxidative decarboxylation.

branched-chain *α*-keto-acid dehydrogenase *abbr.*: BCKDH; EC 1.2.4.4; *recommended name*: 3-methyl-2-oxobutanoate dehydrogenase (lipoamide); *other name*: 2-oxoisovalerate dehydrogenase. An enzyme that catalyses the conversion of *α*-oxo acids to acyl-CoA and CO$_2$. Thiamine diphosphate is a coenzyme. The enzyme consists of a multienzyme complex of branched-chain *α*-keto-acid dehydrogenase (E1), lipoamide acyltransferase (E2), and lipoamide dehydrogenase (E3). The complex is associated with the matrix side on the inner mitochondrial membrane. In humans, E1 is a dimer of α (gene at 1p31) and β (gene at 7q31-q32) subunits; E2 is encoded by a gene at 1p31; and E3 is a homodimer (gene at 7q31-q32). Numerous mutations in each of these genes are associated with different forms of maple-syrup urine disease.

branched-chain *α*-keto-acid dehydrogenase kinase *see* [3-methyl-2-oxobutanoate dehydrogenase (lipoamide)] kinase.

branched-chain oxo-acid dehydrogenase phosphatase *see* [3-methyl-2-oxobutanoate dehydrogenase (lipoamide)]-phosphatase.

branched polymer any polymer in which there are side chains attached to a main chain.

brancher-deficiency amylopectinosis *an alternative name for* type IV **glycogen disease**.

branching *(in nuclear physics)* the simultaneous existence of more than one disintegration pathway for a particular radionuclide.

branching enzyme EC 2.4.1.18; *recommended name*: 1,4-α-glucan branching enzyme; *systematic name*: 1,4-α-D-glucan: 1,4-α-D-glucan 6-α-D-(1,4-α-D-glucano)-transferase; *other names*: amylo-(1,4–1,6)-transglycosylase; corn starch branching enzyme II; amylo-transglycosylase; glucanotransferase; starch branching enzyme II. Any 1,4-α-glucan branching enzyme that catalyses the transfer of a segment of 1,4-α-glucan chain to a primary C-6 hydroxyl group in a similar glucan chain. Branching enzymes fall into two types: the plant **Q-enzymes** and the mammalian enzymes, which are able to produce glycogen, a more highly branched molecule than the amylopectin produced in plants by Q-enzymes. The name should always be qualified, e.g. glycogen branching enzyme, amylopectin branching enzyme. Deficiency in humans results in glycogen disease type IV.

branch migration a proposed model to explain the occurrence of apparently branched DNA structures, as seen by electron microscopy, and their conversion into linear duplex DNA. The branched structures may form by association over parts of their lengths of three or more polynucleotide strands. The migration of the branch point is presumed to occur by displacement of a strand from its fellow in one branch by a strand in another. Such a process may occur in transcription or in recombination.

brassinolide *see* brassinosteroid.

brassinosteroid a class of steroids that contain 27, 28, or 29 carbon atoms, with different substituents on rings A and B and on the side chain. They occur in algae, ferns, gymnosperms, and angiosperms and induce a broad variety of biological responses. Brassinolide (a 28C steroid) is a plant growth regulator discovered in rape (*Brassica napus*) but widely distributed throughout the plant kingdom. It is the most active of the class; its immediate metabolic precursor being castasterone. The latter was discovered in insect galls of sweet chestnut (*Castanea crenata*) and has campesterol as a precursor.

BRCA either of two genes, *BRCA1* and *BRCA2*, implicated in susceptibility to breast cancer. Up to 80% of women who inherit mutated forms of either gene develop breast cancer, and *BRCA1* mutations imply a high risk of ovarian cancer. *BRCA1* (at 17q21) encodes 1863 amino acids. The protein includes a RING finger motif in the N-terminal region and two domains in the C-terminal region that are involved in DNA binding and in protein–protein interactions. It is implicated in transcriptional regulation of DNA damage-inducible genes that function in cell-cycle arrest. *BRCA2* (at 13q12-q13) encodes 3418 amino acids. The protein has histone acetyltransferase (HAT) activity and forms complexes with phosphorylated *BRCA1*, RAD51, and RAD52 for cell-cycle control and DNA repair through homologous recombination. Several hundred mutations in each gene have been reported.

brdU or **BrdU** *abbr. for* bromodeoxyuridine (the symbol BrdUrd is recommended).

BrdUrd *symbol for* bromodeoxyuridine. *Compare* BrUrd.

breakage and reunion model the classical and generally accepted model of chromosomal crossing-over; it involves breakage of the chromatids and reunion of the alternative fragments.

breakpoint cluster region term originally suggested for a limited region of 5–6 kb in the base sequence of chromosome 22, in which breakpoints involved in formation of the **Philadelphia chromosome** were found to cluster. Molecular cloning of specific areas of DNA (e.g. band 22q11) gave a probe specific for the breakpoint translocation domain. Subsequent work led to this region being identified as the *BCR* gene.

breathing *a colloquial term for* transient dynamic changes in the structure of macromolecules involving expansion and contraction of the 3-D structure.

brefeldin A a fungal metabolite, isolated from *Penicillium brefeldianum*, that inhibits a wide variety of membrane transformations

of importance in intracellular transport. It effects a rapid increase in the volume of Golgi cisternae and loss of recognizable Golgi apparatus in treated cells. There is a rapid reversible dissociation of a Golgi-associated peripheral protein identical with a coat protein subunit of non-clathrin-coated vesicles (β-COP). *See* coatomer.

B region the site of final epoxidation of carcinogenic polycyclic aromatic hydrocarbons. It is located on the terminal ring of the **bay region** on a bond adjacent to the bay region. *Compare* **A region**, **K region**, **L region**.

brei a finely and uniformly ground tissue suspension containing all the enzymes of the original material but with the normal spatial relationships between them disrupted. *Compare* homogenate.

bremsstrahlung the **X-rays** produced when an electrically charged particle, usually a β particle or an accelerated electron, is decelerated by the electric field of an atomic nucleus. The yield is directly dependent upon the magnitude of the atomic number of the target element. [German: braking radiation.]

brevetoxin any of the highly toxic lipid-soluble polyether compounds isolated from *Ptychodicus brevis* (*Gymnodinium breve*), the red-tide organism. They cause the death of fish and are responsible for shellfish poisoning in humans. They are activators of voltage-dependent Na^+ channels.

brevican a **proteoglycan** that contains linked chondroitin sulfate chains and is expressed primarily in cerebellar astrocytes. It contains an N-terminal hyaluronan-binding domain. The bovine protein contains 912 amino acids.

brevin *an alternative term for* gelsolin.

brevinin a secreted amphibian peptide with antibacterial and hemolytic activities. Example, brevinin-1 from *Rana brevipoda porsa* has sequence FLPVLAGIAAKVVPALFCKITKKC.

brewers' yeast *the common name for* strains of the yeast *Saccharomyces cerevisiae* that are used in brewing. In addition to their ability to produce, and remain viable under, appreciable levels of ethanol, such strains are able to ferment maltose, maltotriose, and other sugars in the wort, and to withstand its relatively high osmotic pressure. *See also* bottom yeast.

bridge *(in chemistry)* a valence bond, atom, or unbranched chain of atoms connecting two different parts of a molecule.

bridgehead *(in chemistry)* either of the two atoms, e.g. tertiary carbon atomks, that are connected by a **bridge**.

bridge-receptor any molecular entity with two or more specific binding sites able to interact with and join a corresponding number of complementary ligands, which may or may not be the same.

bridging ligand a ligand that binds to two or more atoms, usually metals, linking them together to produce a polynuclear coordination complex. Bridging is indicated by the Greek letter mu (μ) appearing before the ligand name and separated by a hyphen. An example is sulfide in an **iron–sulfur cluster**.

Briggs–Haldane theory an alternative to the Michaelis–Menten theory of the kinetics of enzyme reactions for cases where breakdown of the enzyme–substrate complex to products is not slow in comparison with its dissociation to enzyme and substrate. *Compare* **Michaelis kinetics**. [After George Edward Briggs (1893–1985), British plant physiologist, and John Burdon Sanderson Haldane (1892– 1964), British enzymologist, geneticist, philosopher, and popularizer of science, who put forward their theory in 1925.]

Brij *proprietary name for* any of a series of polyethylene ethers of higher fatty alcohols. They are surfactants useful for the solubilization of membrane fractions. Brij 35 is polyoxyethylene (23) lauryl ether, aggregation number 40, CMC 92 μM.

Brilliant blue *see* Coomassie blue.

British antilewisite *abbr.*: BAL; *another name for* **dimercaprol**.

British thermal unit *abbr.*: Btu; a non-SI unit given by the energy required to raise the temperature of 1 pound of water through 1 °F. Since this depends on the temperature of the water, the temperature range should be specified. The Btu(mean), defined as the 1/180 part of the energy required to raise the temperature of 1 pound of water from 32 °F to 212 °F, is equal to 1055.06 J.

Brix *abbr. for*: biogenesis of ribosomes in *Xenopus*; a protein involved in ribosome assembly, that colocalizes with fibrillarin in nucleoli, and associates with the RNA of large ribosomal subunits. Comprising 339 residues, it contains a central globular domain ($\approx$150–180 residues) found in about 50 other proteins. Knockout of the yeast orthologue is lethal and produces defects in synthesis of the large ribosomal subunit.

Brk *abbr. for* Bek-related kinase; *an alternative name for Cek2. See* **Cek**.

BRN a family of neurally expressed genes encoding brain-specific transcription factors. They contain a **POU domain**. Brn-3.1 and Brn-3.2 modulate the terminal differentiation of auditory and visual system development respectively.

broad host range relating to the ability of a plasmid to be subjugated as a vector in a broad range of hosts, especially Gram-negative bacteria. Such plasmids are also termed promiscuous.

broad-spectrum antibiotic any antibiotic that is active against a wide range of bacterial species.

Brockmann body a discrete organ, occurring in fish, that contains tissue corresponding to that found in the islets of Langerhans of mammals. One or more Brockmann bodies occur in various teleost species. They are free of pancreatic acinar tissue, which makes them particularly suitable for biochemical studies and a good source of fish insulin. The principal islet weighs in the region of 1–50 mg; smaller secondary islets may also occur.

Brodie's fluid a **manometer fluid** containing 23 g NaCl, 5 g sodium tauroglycerate, 100 mg Evan's Blue, and a few drops of an ethanolic thymol solution, in 1000 mL of water; $D = 1.033$. The standard pressure (760 mmHg) in mm of manometer fluid is 10 000.

brom+ a variant form of **bromo+**.

bromelain *or* **bromelin** either of two similar **cysteine proteinases**: **stem bromelain**, EC 3.4.22.32, isolated from the stem of the pineapple, *Ananas comosus*; and **fruit bromelain**, EC 3.4.22.33, from the same plant. They are broad-specificity proteinases, similar to other cysteine proteinases but with so far unique geometry and reactivity at the active site. The two types of enzyme are distinguished by the substrate Z-Arg-Arg-NHMec, which is hydrolysed by stem bromelain but not by fruit bromelain.

bromination the introduction of one or more bromine atoms (bromo residues) into an organic molecule, whether by addition or substitution. —**brominate** *vb*.

bromine *symbol*: Br; a halogen element of group 17 of the IUPAC **periodic table**; atomic number 35; relative atomic mass 79.909. Natural bromine consists of the two isotopes ^{79}Br and ^{81}Br at almost equal relative abundances.

bromine-82 the radioactive nuclide $^{82}_{35}$Br; half-life 35.3 h; it emits beta radiation (0.444 MeV) and gamma radiation of various energies.

bromo+ *or (sometimes before a vowel)* **brom+** *prefix* denoting a bromine atom in organic linkage.

bromocriptine *or* **bromoergocriptine** 2-bromoergocryptine; an ergotoxin **ergot** alkaloid derivative that acts as a dopamine (D$_2$) receptor agonist. It inhibits the secretion of prolactin and somatotropin by the pituitary gland, and has potential applications in the treatment of Parkinson's disease.

bromodeoxyuridine *symbol*: BrdUrd; 5-bromo-2'-deoxyuridine; 5-bromouracil 2'-deoxyriboside; a synthetic analogue of thymidine. It acts as an inhibitor of cell differentiation, and also has antiviral properties. *See* **bromouracil**.

bromodomain a domain present in some 75 proteins that forms a left-handed four-helix bundle whose pocket interacts with ε-*N*-acetyl-lysyl residues in the N-terminal segments of core histones. It is also present in many transcriptional regulators that have intrinsic histone acetyltransferase activity.

Bromophenol Blue a dye used as a pH indicator, changing from yellow to blue over the pH range 3.0–4.6

Bromophenol Red a dye used as a pH indicator, changing from yellow to red over the pH range 5.2–6.8

N-bromosuccinimide *abbr.*: NBS; an oxidizing and brominating agent, used for the oxidation of tryptophan residues in proteins. Tyrosine, histidine, and methionine may be oxidized to a lesser extent.

Bromothymol Blue a dye used as a pH indicator, changing from yellow to blue over the pH range 6.0–7.6

bromouracil *symbol*: BrUra; 5-bromouracil; 5-bromo-2,4(1*H*,3*H*)-pyrimidine-dione; a synthetic analogue of thymine with mutagenic activity. It is incorporated into DNA as **bromodeoxyuridine**, which replaces thymidine and induces transitions of G–C base-pairing to

A–T pairing. It can be used as a density marker for DNA. The free compound is an inhibitor of dihydrouracil dehydrogenase.

bromouridine *symbol*: B *or* BrUrd; 5-bromouridine; 5-bromouracil riboside; a synthetic analogue of ribothymidine.

Brønsted catalysis law an expression of the relation between the rate constant, k, of a catalysed reaction, and the pK of the catalyst: log$k = c + b$pK, where b and c are constants and the value of b provides a measure of the balance between nucleophilic and general base character of the catalysis. [After Johannes Nicolaus Brønsted (1879–1947), Danish physical chemist.]

Brønsted–Lowry acid any chemical species capable of donating a hydron to a base. Examples are H_2O, H_3O^+, CH_3COOH, $H_2PO_4^-$, NH_4^+. [After J. N. Brønsted (*see* **Brønsted catalysis law**) and Thomas Martin Lowry (1874–1936), British chemist.]

Brønsted–Lowry base any chemical species capable of accepting a hydron from an acid. Examples are OH^-, H_2O, CH_3COO^-, HPO_4^{2-}, NH_3.

Brookhaven database *see* **PDB**.

broth *(in bacteriology)* any liquid culture medium, especially nutrient broth.

brown adipose tissue a highly specialized tissue with a high content of lipid and cytochromes found in some animals, particularly hibernating animals and the newborn of some species. It is highly vascular and consists of small polygonal cells, each containing many separate lipid droplets and many mitochondria. Its function is thermogenesis during the arousal period after hibernation or, in the young, to provide heat before shivering has developed. It is active also in normal but not in obese humans. The colour is due to the high cytochrome content. Heat is generated by lipid oxidation through electron transport not coupled to oxidative phosphorylation. The uncoupling is mediated by **brown fat uncoupling protein**. *Compare* **white adipose tissue**.

brown fat uncoupling protein *abbr.*: UCP; *other names*: mitochondrial UCP; thermogenin; a dimeric integral mitochondrial inner membrane protein that forms a H^+-channel, is unique to brown adipose tissue mitochondria, and is important for thermogenesis of this tissue. By causing the membrane to leak protons, it abolishes the proton gradient that drives oxidative phosphorylation, so that electron transport results solely in heat production.

Brownian movement the peculiar random movements shown by microscopic particles in a disperse phase, i.e. when suspended in a liquid or gas. It is caused by the continuous irregular bombardment by the molecules of the surrounding medium. [After Robert Brown (1773–1858), British botanist who first observed it in 1827.]

browning reaction any of a group of complex reactions, both enzymic and nonenzymic, that occur in some foods during processsing and/or storage and cause a brownish discoloration. Members of this group include the oxidation of phenolic compounds (which is believed to be enzymic) together with caramelization, ascorbate decomposition, and the **Maillard reaction** (all of which are believed to be nonenzymic).

BRUCE *abbr. for* baculovirus repeat-containing ubiquitin-conjugating enzyme. *See* **Apollon**.

brucine 2,3-dimethoxystrychnidin-10-one; a highly toxic bitter-tasting alkaloid obtained from *Strychnos* seeds. It is used as a resolving agent for some racemates. *Compare* **strychnine**.

BRUNO a protein of *Drosophila* that is essential for repression of mRNA for the **oskar**-encoded protein as it is in transit from nurse cells to the posterior pole of the oocyte.

BrUra *symbol for* bromouracil.

BrUrd *symbol for* bromouridine (alternative to B). *Compare* **BrdUrd**.

brush border the dense covering of **microvilli** on the apical surface of epithelial cells in the intestine and kidney. The microvilli aid absorption by increasing the surface area of the cell.

Bruton's tyrosine kinase a cytoplasmic protein-tyrosine kinase expressed throughout B lymphocyte differentiation and in myeloid cells but not in the B lymphocyte lineage. The locus at Xq22 encodes a protein of 659 amino acids that contains: an N-terminal plekstrin homology domain that binds phosphoinositides, protein kinase C, and subunits of trimeric G-proteins; a region that is proline-rich and binds SH3 domains; and a C-terminal kinase domain. It is a mediator of radiation-induced apoptosis of lymphoma B lymphocytes. Over 400 mutations in the gene are associated with X-linked agammaglobulinemia. [After Ogden C. Bruton (1908–1994), US paediatrician.]

bryostatin any of a number of highly potent activators of protein kinase C found in bryozoans, especially *Bugula neritina*. They were isolated as the active anti-leukemic agent of bryozoan extracts. They block phorbol ester-induced differentiation of human promyelocytic leukemia cells and other actions of phorbol esters.

bryostatin 1

BSA *abbr. for* bovine serum albumin. *See* **serum albumin**.

BSE *abbr. for* bovine spongiform encephalopathy.

B side the side of the nicotinamide ring of NADH or NADPH at which projects the *pro-S* hydrogen atom (known as H_B) at the 4 position. *Compare* **A side**.

***Bsp*1286I** a type 2 **restriction endonuclease**; recognition sequence: G[AGT]GC[ACT]C. *Sdu*I is an **isoschizomer**.

***Bsp*1407I** a type 2 **restriction endonuclease**; recognition sequence: T↑GTACA.

BspT104I a type 2 **restriction endonuclease**; recognition sequence: TTT↑CGAA. *Asu*II and *Nsp*V are **isoschizomers**.

BspT107I a type 2 **restriction endonuclease**; recognition sequence: G↑G[TC][AG]CC. *Hgi*CI is an **isoschizomer**.

BssHII a type 2 **restriction endonuclease**; recognition sequence: G↑CGCGC. *Bse*PI is an **isoschizomer**.

Bst1107I a type 2 **restriction endonuclease**; recognition sequence: GTA↑TAC. *Sna*I is an **isoschizomer**.

BstPI a type 2 **restriction endonuclease**; recognition sequence: G↑GT-NACC. *Bst*EII and *Eco*065I are **isoschizomers**.

BstXI a type 2 **restriction endonuclease**; recognition sequence: CCANNNNN↑NTGG.

BTEE *abbr. for* N-benzoyl-L-tyrosine ethyl ester (a substrate for the assay of chymotrypsin).

B-transferase EC 2.4.1.37; *recommended name*: fucosylglycoprotein 3-α-galactosyltransferase; *other name*: [histo-blood group] B transferase; an enzyme that catalyses the reaction:

UDPgalactose + glycoprotein α-L-fucosyl-(1,2)-D-galactose = UDP + glycoprotein α-D-galactosyl-(1,3)-[α-L-fucosyl-(1,2)]-L-galactose,

thus adding galactose to the H antigen of the **ABH antigen** system, leading to formation of the B antigen. The protein is a product of one allele of the *ABO* gene. It has virtual identity of sequence with that catalysing **A-transferase** activity, differing only at residues 176 (A has Arg, B has Gly), 235 (A Gly, B Ser), 266 (A Leu, B Met), and 268 (A Gly, B Ala).

BT toxin *abbr. for Bacillus thuringiensis* toxin; any of a class of related protein toxins produced by different subspecies of the bacterium *Bacillus thuringiensis*. They are insecticidal and have been used, by spraying or by transgenic expression, to protect crops (e.g. cotton, maize, potato). The CryIA(a) toxin is active against lepidopterans, whereas CryIIIA toxin is active against coleopterans. The crystal structures of both have been elucidated.

Btu *abbr. for* British thermal unit.

Bu *symbol for* the butyl group.

BU (*sometimes*) *abbr. for* bromouracil (BrUra is recommended).

bubble column a column-shaped bioreactor in which the reaction medium is kept mixed and aerated by the introduction of air at the bottom.

Büchner funnel a cylindrical funnel for filtration, usually of porcelain or plastic, that includes a perforated plate on which a filter paper is placed. It is generally used with a vacuum. *See also* **Hartley filter funnel**.

bud 1 a small lateral or terminal protuberance on the stem of a plant that contains undeveloped foliage or floral leaves. **2** a budlike protuberance on the surface of a yeast cell or other simple organism. **3** to form a bud. **4** to reproduce asexually, as in yeasts, by the process of **budding**.

budding 1 the production of a bud or buds. **2** a form of asexual reproduction, occurring in certain bacteria and fungi (e.g. yeasts) and some primitive animals in which an individual arises from a daughter cell formed by pinching off a part of the parent cell. The budlike outgrowths so formed may sometimes remain attached to the parent cell.

BUdR or **BUDR** (*sometimes*) *abbr. for* bromodeoxyuridine (the symbol BrdUrd is preferred).

bufadienolide any of various naturally occurring doubly unsaturated lactones of certain steroids with important pharmacological effects on heart muscle (*see* **cardiac glycoside**). They are so named because they were originally found in the venomous secretion of the skin glands of some toads (*Bufo*), and are hence also known as **toad poisons**. They also occur in certain plants, e.g. *Digitalis*. *See also* **bufogenin B**.

buffer 1 any substance or mixture of substances that, when dissolved (usually in water), will maintain its solution at approximately constant pH despite small additions of acid or base. The commonest examples are moderately strong solutions containing both a weak acid and its conjugate base (or a weak base and its conjugate acid). A substance is useful as a buffer over a range of about one pH unit either side of its p*K*, but is most effective at or near the p*K*. Buffer substances used for biochemical or biological purposes

include: acetate, bicarbonate, bis-tris propane, borate, citrate, dimethylmalonate, glycinamide, glycylglycine, imidazole, phosphate, succinate, together with any **good buffer substances**. By extension, the term may be applied to agents controlling the activities of various other specified entities, e.g. redox buffer, carbon-dioxide buffer, metal-ion buffer. Also used attributively: e.g. buffer action; buffer salt; buffer solution. **2** a solution of a **buffer** (def. 1). **3** a short-term storage facility (e.g. as part of the memory of a computer), especially one whose patterns or rates of input and output can differ. **4** to treat with or to incorporate a buffer (def. 1); to act as a buffer. *See also* **buffering capacity**, **buffer value**. —**buffered** *adj.*; **buffering** *n.*

buffering capacity or **buffering power 1** the number of gram-equivalents of either hydrogen ions or hydroxide ions required to change the pH of 1 litre of 1 m buffer solution by one unit. Buffering capacity = $(1/m)$ (dn/dpH), where m is the number of moles of buffer, and dpH is the pH change produced by addition of dn equivalents of hydrogen ions or hydroxide ions. **2** *an alternative term for* **buffer value**.

buffer solution *see* **buffer** (def. 2).

buffer value or **Van Slyke buffer value** or **buffering capacity** symbol: β; the amount of acid or base, in gram-equivalents, needed to change the pH of 1 litre of a buffer solution by one unit at any pH; i.e. $\beta = db/dpH$, where b is the molar concentration of base in the solution. [After Donald Dexter Van Slyke (1883–1971), US biochemist, who described it in 1922.]

buffy coat the layer of white cells that forms between the layer of red cells and the plasma when unclotted blood is centrifuged or allowed to stand.

bufogenin B $3\beta,14,16\beta$-trihydroxy-5β-bufa-20,22-dienolide; a **bufadienolide** found in the Chinese drug Ch'an Su, prepared from Chinese toads (*Bufo asiaticus*).

bulge loop a structure in a polynucleotide duplex in which one strand contains a nonterminal extra sequence that is not able to base-pair with the second strand, thereby forming a bulge on one side of the duplex.

bullous pemphigoid antigen *abbr.*: BPAG; either of two proteins present in epithelia, to which autoantibodies are present in bullous pemphigoid (an acquired disease characterized by subepidermal blistering). BPAG1e (BPAG epithelial, 230 kDa) is an intracellular coiled-coil protein that attaches keratin intermediate filaments to hemidesmosomes. It is highly homologous with desmoplakin and plectin. BPAG1n is a neuronal isoform. The gene at 6p11-p12 encodes a protein of 2632 residues, and mutations in the mouse orthologue are associated with dystonia musculorum. BPAG2 (180 kDa) is a type XVII collagen and has a cytoplasmic N-terminal domain, one transmembrane segment, and an extracellular C-terminal domain that consists of 15 Gly–X–Y repeats interrupted by other short sequences. Mutations in the gene on 10q24.3 (for type XVII procollagen α1 chain) are associated with a benign form of epidermolysis bullosa.

BUN *abbr. for* blood urea nitrogen, an index of the blood-urea concentration.

bundle sheath a parenchymal sheath surrounding a vascular bundle in plants.

bungarotoxin any of various neurotoxins derived from the venom of the elapid snake, *Bungarus multicinctus*. The chief components are: (1) α bungarotoxin, a single polypeptide chain of 74 amino-acid residues; it is an irreversible antagonist of nicotinic cholinergic receptors and causes paralysis; (2) β bungarotoxin, a multicomponent protein composed of two polypeptide chains: a long chain (120 amino-acid residues) and a short chain (60 residues); it prevents the release of acetylcholine from cholinergic neurons.

Bunsen coefficient *symbol*: α; the absorption coefficient of a gas in a solution. It is defined as the volume of gas in litres, reduced to 273.15 K and 1 atm (101 325 Pa) pressure, that dissolves in 1 litre of liquid when the partial pressure of the gas in the gas phase is 1 atm. [After Robert Wilhelm Bunsen (1811–99), German chemist.]

bunyavirus any of a group of RNA animal viruses consisting of enveloped particles, 90–100 nm in diameter with helical nucleocapsids, and containing segmented RNA (minus strand) of 3–4 MDa. The group has been recognized as different from **togaviruses** and includes arthropod-borne viruses that can cause encephalitis.

buoyant density the density of a solute molecule as determined by density-gradient ultracentrifugation. It is the density of the solution, ρ, at the point in the gradient where: $\rho = 1/\gamma$, where γ is the **partial specific volume** of the solute in question.

buoyant force *symbol*: F_B; the force acting on an object suspended in a liquid due to the liquid displaced by the object. It is given by: $F_b = V_p\rho_m g$, where V_P is the volume of the object, ρ_m is the density of the displaced liquid m, and g is the acceleration due to gravity.

Bureau International des Poids et Mesures the agency in Sèvres, near Paris, that maintains the SI system of units.

burette *or (US)* **buret** a graduated (often glass) tube with a pinch-clamp, stopcock, or valve at one end used for measuring or dispensing known volumes of fluids, especially liquids. A graduated syringe used for the same purpose may also be termed a burette.

Burkitt's lymphoma a malignant tumour of the lymphatic system, most commonly affecting children in tropical Africa within 15° north and south of the equator. **Epstein–Barr virus** can be demonstrated in a proportion of cultured tumour cells though it is unlikely that the virus is the sole cause of the malignancy. [After Denis Parsons Burkitt (1911–93), British surgeon.]

Burnet's theory of immunity *see* **clonal selection theory**.

bursacyte *an alternative name (archaic) for* **B lymphocyte**.

bursa of Fabricius *or* **cloacal bursa** a saclike lymphoepithelial structure opening into the dorsal part of the cloaca in young birds. It usually degenerates as the birds reach maturity. It is associated with humoral immunity and the lymphocytes processed by it are termed **B lymphocytes**. In mammals it is likely that hemopoietic tissue itself fulfils the equivalent role, providing the appropriate microenvironment for the maturation of B lymphocytes from precursor stem cells.

bursectomy removal of the **bursa of Fabricius**, either by surgery *in ovo* or shortly after hatching, or by destruction *in ovo* by application of, e.g., testosterone. Bursectomy inhibits the formation of **B lymphocytes** and hence circulating antibody.

bursin the tripeptide Lys-His-Gly-NH₂; a selective B-cell differentiating hormone from the bursa of Fabricius of chickens.

burst 1 the initial pre-steady-state liberation of the first product, B, in an enzymic reaction of the type: $E + AB = EAB \rightarrow EA + B$ and $EA \rightarrow E + A$, where E is the enzyme and A is the second product. **2** the sudden release of phage particles accompanying lysis of a phage-infected bacterial cell. *See also* **burst size, respiratory burst**.

burst size the mean number of bacteriophage particles set free per infected bacterium upon lysis of phage-infected cells.

butanediol fermentation a type of fermentation effected by some members of the Enterobacteriaceae, e.g. *Enterobacter, Erwinia, Klebsiella,* and *Serratia*, in which glucose is fermented with the production of 2,3-butanediol and other substances.

butanoyl *another name for* **butyryl**.

butyl *symbol*: Bu; the alkyl group, $CH_3–[CH_2]_2–CH_2–$, derived from butane.

butylated hydroxytoluene *an alternative name for* **di-*tert*-butyl-*p*-cresol**.

butyl-PBD *see* **biphenylylphenyloxadiazole**.

butyrate 1 *trivial and preferred name for* **butanoate**; the anion $CH_3–[CH_2]_2–COO^-$, derived from butyric acid (butanoic acid), a saturated, unbranched, aliphatic acid. **2** any mixture of free butyric acid and its anion. **3** any salt or ester of butyric acid.

butyric acid fermentation a type of fermentation effected by some saccharolytic species of *Clostridium*, e.g. *C. butyricum*, in which glucose is fermented, with the production of acetic acid, butyric acid, carbon dioxide, and dihydrogen.

butyrophenone any of a group of antipsychotic drugs, e.g. **haloperidol**, that are more selective than the **phenothiazine** group. Butyrophenones interact rather more specifically with dopamine D_2 receptors and reduce the firing of dopaminergic neurons.

butyrophilin a type I **membrane protein** of mammary gland, probably involved in stimulating the secretion of droplets of fat in milk.

butyryl *symbol*: Br; *trivial and preferred name for* **butanoyl**; the acyl group, $CH_3–[CH_2]_2–CO–$, derived from butyric acid (butanoic acid).

BX-C *abbr. for* the bithorax complex, one of two (*see* **ANT-C**) major families of **homeotic genes** in *Drosophila*, mutations in which affect thoracic and abdominal segments.

byssus an elastomeric protein of bivalve molluscs, threads of which attach mussels to hard surfaces such as rocks. It consists of 882 amino acids; the N- and C-terminal regions are histidine-rich domains, while the central region is a collagen-like domain flanked by elastomeric domains.

byte a single unit of information handled by a computer; the binary unit required for storage of a character, usually comprising eight **bits**.

Bz *symbol for* the benzoyl group.

bZIP *abbr. for* basic zipper; a characteristic feature of a family of proteins that possess a basic region (b) and a **leucine zipper** domain (ZIP). The basic region harbours a DNA-contact surface, and the leucine zipper domain is essential for homodimer formation, or formation of heterodimers with other leucine zipper proteins. The family includes the **AP, ATF,** and **CREB protein** transcription factors.

Bzl *symbol for* the benzyl group.

BZLF1 one of a group of transcription factors related to the *jun* and *fos* gene products. It is involved in the switch from latency to virus-particle production following infection by **Epstein–Barr virus**.

Cc

c 1 *symbol for* **a** centi+ (SI prefix denoting 10^{-2}). **b cyclic** (def. 3) or **cyclo+** (denoting cyclic compounds, as in cAMP for cyclic AMP or Hx$_c$ for cyclohexyl. **c** *(obsolete)* **curie** (use Ci). **2** *abbr. for* **a** complementary (as in **cDNA**). **b** *cis* (not recommended). **c** *(obsolete)* cubic (i.e. to the third power).

c- *(in genetics) prefix* denoting cellular, as in c-*myc*, the cellular as opposed to viral (v-) version of a (proto)oncogene; *see*, e.g., **MYC**.

c *symbol for* **1** amount-of-substance concentration; that for a particular substance may be denoted by adding a subscript, e.g. c_B for amount concentration of a substance B. **2** speed (alternative to v, u, or w). **3** (bold italic) velocity (alternative to $\mathbf{v}$, $\mathbf{u}$, or $\mathbf{w}$).

c- *abbr. for* **cis-**.

c$_0$ *symbol for* speed of light *in vacuo*. the subscript may be omitted when there is no risk of ambiguity.

C 1 *symbol for* **a** carbon; *see* **carbon-11, carbon-12, carbon-13, carbon-14**. **b** designating a specific carbon atom, e.g. C-1, C-2. **c** a residue of the α-amino acid L-cysteine (or L-half-cysteine) (alternative to Cys). **d** a residue of the base cytosine in a nucleic acid sequence. **e** a residue of the ribonucleoside cytidine (alternative to Cyd). **f** coulomb. **g** *or (formerly)* C′ complement (the main components of complement are designated C1–C9). **2** *see* **programming language**.

°C *symbol for* degree Celsius *(formerly* degree centigrade). *See* **Celsius temperature**.

C$_H$ *symbol for* the constant region of an immunoglobulin **heavy chain**. The particular constant domain in question may be indicated as C$_H$1, C$_H$2, C$_H$3.

C$_L$ *symbol for* the constant region of an immunoglobulin **light chain**.

C *symbol for* **1** capacitance. **2** heat capacity (C_p at constant pressure, C_V at constant volume, C_m molar heat capacity). **3** number concentration (C_B for a substance B).

C- *(in chemical nomenclature)* a locant indicating substitution on carbon; a particular carbon atom may be specified by a right superscript.

-C a conformational descriptor designating the **chair conformation** of a six-membered ring form of a monosaccharide or monosaccharide derivative. Locants of ring atoms that lie on the side of the structure's reference plane from which the numbering appears clockwise are indicated by left superscripts and those that lie on the other side of the reference plane by right subscripts; e.g. $^1C_4$, $^4C_1$. *See also* **conformation** (def. 2).

$^1C_4$ *or (formerly)* **-1C** conformational descriptor for an aldopyranose in the chair conformation, with C2, C3, C5, and O5 in the structure's reference plane, and with C1 above and C4 below the plane; e.g. α-D-glucopyranose-$^1C_4$.

$^1C_4$

$^4C_1$ *or (formerly)* **-C1** a conformational descriptor for an aldopyranose in the chair conformation, with C2, C3, C5, and O5 in the structure's reference plane, and with C1 below and C4 above the plane; e.g. α-D-glucopyranose-$^4C_1$.

$^4C_1$

C++ *see* **programming language**.

Cα *symbol for* alpha carbon atom.

Cβ *symbol for* beta carbon atom.

Ca *symbol for* calcium.

CAA a codon in mRNA for L-glutamine.

CAAT box *or* **CAT box** a conserved sequence in DNA found within the **promoter** region of the protein-encoding genes of many eukaryotes. It has the consensus sequence GGCCAATCT and is believed to determine the efficiency of transcription from the **promoter**.

CAAX *abbr. for* Cys–aliphatic–aliphatic–X; a C-terminal tetrapeptide motif that is present in proteins that become prenylated. Following prenylation, the AAX is removed by a CAAX-prenyl endopeptidase (a metal-dependent integral membrane protein) and the new C-terminal cysteine is carboxymethylated.

Cab-O-Sil *proprietary name for* **fumed silica** supplied by Cabot Carbon Ltd.

CaBP3 *see* **calreticulin**.

CAC a codon in mRNA for L-histidine.

cachectic factor a 24 kDa proteoglycan isolated from the murine adenocarcinoma MAC16 that has been implicated in the production of **cachexia**. The peptide has the N-terminal sequence YD-PEAASAPGSGDPSHEA, and has *N*- and *O*-glycans. Intravenous injection of the proteoglycan induces rapid weight loss.

cachectin *see* **cachexia**, **tumour necrosis factor**-α.

cachexia a condition caused by chronic disease, such as cancer, and characterized by wasting, emaciation, feebleness, and inanition. It led to the name 'cachectin' for the protein now known as **tumour necrosis factor** but a 24 kDa proteoglycan has more probably been implicated in this. *See* **cachectic factor**.

CACNA1A *abbr. for* calcium channel alpha subunit 1A. *See* **hemiplegic migraine**.

cacodylate 1 dimethyl arsinate, the anion $(CH_3)_2$–$AsOO^-$ derived from cacodylic acid. It is useful as a buffer substance (pK_a = 6.2). **2** any salt of cacodylic acid.

CAD 1 *or* **CAD protein** a multifunctional protein found in many eukaryotes and containing domains for three enzymes of pyrimidine biosynthesis: **carbamoyl phosphate synthase** (glutamine-hydrolysing); **aspartate transcarbamylase**; and **dihydroorotase**. **2** *abbr. for* caspase-dependent DNase; a lysosomal enzyme that cleaves DNA in cells undergoing **apoptosis**.

cadaverine 1,5-pentanediamine; a substance formed by microorganisms in decaying meat and fish by the decarboxylation of lysine. It also occurs as an intermediate in the biosynthesis, via lysine, of some quinolizidine alkaloids (e.g., lupinine) in plants.

cadherin any member of a family of calcium-dependent cell adhesion proteins that preferentially interact in a homophilic manner in cell–cell interactions. Cadherins may thus contribute to the sorting of heterogeneous cell types; they are type I **membrane proteins**. The names of different classes indicate the tissues in which they were first found: E-cadherin is present on many types of epithelial cells; N-cadherin is present on nerve, muscle and lens cells; and P-cadherin is present on placental and epidermal cells. Typically, cadherins have five similar extracellular domains, the outermost three of which have Ca^{2+}-binding sites, and an intracellular C-terminal domain that interacts with the actin cytoskeleton.

cadmium mycophosphatin a cadmium-binding phosphoglycoprotein, M_r 12 000; it lacks sulfur-containing amino acids, but is rich in aspartic and glutamic acids, and phosphoserine.

CAD protein a multifunctional protein found in many eukaryotes, and containing domains for three enzymes of pyrimidine biosynthesis: **carbamoyl phosphate synthase** (glutamine-hydrolysing); **aspartate transcarbamylase**; and **dihydroorotase**.

cadystin the **phytochelatin** of *Saccharomyces pombe*.

Caenorhabditis elegans a nematode (roundworm), with a genome of about 97 million base pairs. Its cellular anatomy has been fully described, and the nucleotide sequence of the genome determined, revealing over 19 000 genes, 40% of the predicted protein products having significant matches in other organisms.

caerulein *the Brit. spelling of* **cerulein**.

caeruloplasmin *the Brit. spelling of* **ceruloplasmin**.

caesium *alternative spelling (esp. Brit.) for* **cesium**.

CAF1 *abbr. for* chromatin assembly factor 1; a complex of p150 and p60 subunits, present in *Drosophila* and human cells, that is responsible for chaperoning histones H3 and H4 to DNA during replication. The p60 gene maps to a locus on chromosome 21 that is strongly associated with the major features of Down syndrome. *See* **NAP1**.

caffeine 1,3,7-trimethylxanthine, 3,7-dihydro-1,3,7-trimethyl-1*H*-purine-2,6-dione; a substance found in certain plants, notably in tea and coffee and the beverages derived from them. It is an adenosine (A_1 and A_2) receptor antagonist and phosphodiesterase inhibitor, and a stimulant of the central nervous system, affecting the cardiovascular system and causing diuresis. Its actions are similar to those of **theophylline** but less potent.

caffetannic acid *an alternative name for* **chlorogenic acid**.

CAG a codon in mRNA for L-glutamine.

cage carrier *see* **ionophore**.

cage compound 1 *an alternative name for* **clathrate** (def. 1). **2** any compound possessing a nonplanar bicyclic or polycyclic molecular structure that encloses a cavity.

caged ATP a type of protected ATP analogue, e.g. adenosine-5′-triphospho-1-(2-nitrophenyl)ethanol (or the methanol-containing equivalent), that releases ATP in good yield when photolysed by a short pulse of light of 360 nm wavelength. Similarly, guanosine-5′-triphospho-1-(2-nitrophenyl)ethanol is used as **caged GTP**. These compounds can be introduced into cells prior to photolysis. Caged ATP has been used, e.g., to study muscle contraction on the millisecond timescale.

caged GTP *see* **caged ATP**.

Cahn–Ingold–Prelog rule *or* **Cahn–Ingold–Prelog system** *an alternative name for* **sequence rule**.

CAK *see* **Cdk-activating kinase**.

cal *symbol for* calorie.

calbindin *other names*: cholecalcin; calbindin D9K; CABP; vitamin D-dependent calcium-binding protein; an **EF-hand** Ca^{2+}-binding protein from intestine. It is present in a variety of cells, including absorptive cells of the duodenum (in which its synthesis is vitamin D_3-dependent), in hippocampal cells, and in kidney.

calcemic factor *or (esp. Brit.)* **calcaemic factor** the prohormone of **parathyrin**.

calcidiol *the recommended trivial name for* calcifediol, 25-hydroxy-cholecalciferol. It is formed in liver from cholecalciferol, and is the major store of the vitamin in the body, being present largely in plasma. It is the precursor of the hormonal form of the vitamin, calcitriol. *See also* **vitamin D**.

calciductin *a name proposed for a* 23 kDa sarcolemmal protein that, when phosphorylated by a cAMP-dependent protein kinase, causes an approximately two-fold stimulation of the slow Ca^{2+} inward current.

calcifediol 25-hydroxyvitamin D_3; *another name for* **calcidiol**. *See also* **vitamin D**.

calciferol *an alternative (older) name for* **ergocalciferol**, also known as vitamin D_2. *See also* **vitamin D**.

calcification the deposition of calcium salts in the tissues. It is part of the normal process of bone formation in which **hydroxyapatite** crystals are deposited, in the neighbourhood of osteoblast cells, in a pre-existing collagen-containing matrix.

calcified describing a tissue that has been solidified by deposition of calcium; having undergone **calcification**.

calcimedin *see* **annexin**.

calcimycin *see* **A23187**.

calcineurin *or* **protein phosphatase 2B** (*abbr.*: PP-2B) *or* **CaMBP₈₀** a major heat-labile calmodulin-binding protein isolated from bovine brain, but later found in all cells from yeast to mammals. It has protein serine/threonine phosphatase activity, and is classified as a protein phosphatase 2B (EC 3.1.3.16). It plays an important role in terminating synaptic transmission, and in T-cell activation. It is only weakly inhibited by **okadaic acid**, but potently inhibited by certain pyrethrin compounds. It consists of two subunits, A and B, the A subunit having the catalytic activity, the B subunit conferring calcium sensitivity.

calciol *an alternative trivial name for* **cholecalciferol**.

calciosome a discrete cytoplasmic organelle in non-muscle cells, and a candidate for the inositol 1,4,5-trisphosphate-sensitive Ca^{2+} store. It has a high content of a **calsequestrin**-like protein.

calciotropic describing any hormone or other agonist acting on calcium metabolism.

calciphorin a calcium ionophore polypeptide, M_r 3000, isolated from the inner membrane of calf mitochondria.

calcitetrol the recommended trivial name for 1α,24*R*,25-trihydroxy-cholecalciferol. It is the inactivation product of calcitriol, from which it is formed by calcitriol 24-hydroxylase. *See also* **vitamin D**.

calcitonin *or* **thyrocalcitonin** a 3.4 kDa polypeptide hormone that regulates the levels of calcium and phosphate in blood. In all species studied it consists of a single polypeptide chain of 32 amino-acid residues with a C-terminal prolinamide residue and an N-terminal seven-membered ring formed by a disulfide bridge between hemicystine residues at positions 1 and 7. The single letter code structure for the polypeptide is

CGNLSTCMLGTYTQDFNKFHTFPQTAIGVGAP–NH₂.

Calcitonin is secreted by the **C cells**, which in mammals occur primarily in the thyroid gland; in more primitive vertebrates they are found mainly in the ultimobranchial bodies and the lung. The hormone is concentrated in the hypothalamus of humans and some other vertebrates, and in the primitive brain of the chordate organism *Ciona intestinalis*. Calcitonin causes a rapid but short-lived drop in the level of calcium and phosphate in blood by promoting their incorporation in the bones. Overall its action is antihypercalcemic, opposing that of **parathyrin**. The response of plasma calcitonin to stimulation by administered **pentagastrin** or calcium may be used to screen for medullary cell carcinoma of the thyroid; the level rises to abnormally high values after stimulation, though normal in its absence. The precursor (locus at 11p15.2-p15.1) also contains **katacalcin**. Both the calcitonin precursor and **calcitonin gene-related peptide** precursor are formed by alternative splicing of the same gene.

calcitonin gene-related peptide *abbr.*: CGRP; a pleiotropic 3.8 kDa polypeptide of 37 amino-acid residues, found in parts of the nervous system and endocrine system, and in some other organs. It is a potent vasodilator and hypotensive agent, and is regarded as a neuromodulatory peptide. The mRNA is formed by **alternative splicing** of the transcript of the **calcitonin** gene, which has exons coding for a region common to both CGRP and calcitonin precursors together with specific calcitonin and CGRP exons. The polypeptide has the structure (rat):

SCNTATCVTHRLAGLLSRSGGVV-
KDNFVPTNVGSEAF-NH$_2$,

a form known as α-CGRP. β-CGRP is identical except for Lys instead of Glu at position 35, and is encoded by a separate but related gene. α-CGRP is present in trigeminal and spinal sensory ganglion cells, in olfactory and gustatory systems, and at neuromuscular junctions. CGRP is also found in the endocrine system – in a subset of adrenal medullary cells and in thyroid C cells – in bronchiolar cells, and in intestinal cells. Administration of synthetic rat CGRP causes a drop in blood pressure (CGRP is the most potent known hypotensive gene peptide), and increase in heart rate. *In vitro* it can relax arteries from many vascular beds taken from a variety of species, and is one of the most potent vasodilators known. It also inhibits gastrointestinal motility and causes catecholamine release and gastric hypoacidity. In Raynaud's disease there is deficiency of perivascular neurons that contain CGRP. CGRP is homologous to **amylin**.

calcitonin gene-related peptide receptor any of several membrane proteins that bind calcitonin gene-related peptide (CGRP) and mediate its effects. CGRP$_1$ and CGRP$_2$ subtypes have been proposed; the antagonist CGRP$_{8-37}$ is selective for CGRP$_1$ receptors, and [Cys(Acm)$^{2.7}$]CGRP for CGRP$_2$ receptor types. CGRP receptors are widely distributed in the nervous system, following the pattern of distribution of CGRP, except that receptor numbers are high in cerebellum. They are also widespread in the cardiovascular system, as well as in adrenal, pituitary, exocrine pancreas, kidney, and bone. CGRP receptor stimulation leads to cyclic AMP accumulation in a number of tissues (heart, liver, muscle, pancreatic acinar cells) indicating coupling to one or more G-proteins.

calcitonin receptor any of the membrane proteins that bind calcitonin and mediate its effects. They are seven-transmembrane-helix receptors, normally coupled to G-proteins. In many systems they can activate both adenylate cyclase and the phosphatidylinositol cycle.

calcitriol *the recommended trivial name for* 1α,25-dihydroxycholecalciferol. Produced from **cholecalciferol** by hydroxylation of C-25 in liver and C-1 in kidney, it increases calcium absorption by inducing synthesis of **calbindin** in duodenal mucosa, and promotes calcium deposition in bone by inducing **osteocalcin** in osteoblasts. *See* **vitamin D**.

calcium *symbol*: Ca; an alkaline-earth metal of group 2 of the IUPAC **periodic table**; relative atomic mass 40.08, atomic number 20; it occurs naturally only in an ionized (Ca^{2+}) or combined state, and is a mixture of stable nuclides of relative mass 40 (96.97 atom percent) and 44 (2.06 atom percent) with small proportions of nuclides of relative mass 43, 46, and 48. Calcium is the fifth most abundant element of the Earth's crust and is an essential component of all living material. It occurs in bone, shell, and teeth and low concentrations of ionic calcium play many important roles in the regulation of diverse cellular processes. The most abundant mineral in the human body, most of it is in the skeleton. It has especially important functions in bone, in the control of nervous, muscle, and other excitable tissue, and as a **second messenger** and regulator of enzyme activity. Calcium homeostasis depends on the action of: (1) **parathyroid hormone**, which increases tubular reabsorption of calcium, and releases calcium from bone, thereby having overall a hypercalcemic action; (2) **vitamin D** (calcitriol), which causes absorption of calcium from the gut and its release from bone, and also therefore has a hypercalcemic action; and (3) **calcitonin**, which reduces calcium resorption from bone. The range of plasma calcium in normal human adults is 2.2–2.6 mmol L^{-1}. *See also* **hypercalcemia, hypocalcemia**.

calcium-45 *symbol*: ^{45}Ca; an artificial radioactive nuclide of calcium, with a half-life of 165 days. It emits β$^-$-particles (i.e. electrons) of 0.252 MeV max. and no γ-radiation.

calcium-47 *symbol*: ^{47}Ca; an artificial radioactive nuclide of calcium, with a half-life of 4.54 days. It emits β$^-$-particles (i.e. electrons) of two energy ranges (1.98 and 0.67 MeV max.) and γ-radiation of three energies (1.31, 0.815, and 0.49 MeV).

calcium channel any of several proteins that permit the controlled (gated) passage of calcium ions through membranes. There are several types. **L-type channels** are involved in excitation–contraction coupling in muscle; they bind 1,4-dihydropyridine (DHP). The DHP receptor is a protein located in transverse (T-) tubules that specifically binds DHP compounds such as **nifedipine**. This modulates the slow L-type channel in skeletal muscle, neurons, and cardiac cells. **N-type channels**, identified only in neurons, participate in neurotransmitter release and are blocked by ω-conotoxin. **T-type channels** influence pacemaker activity in the heart and repetitive spike activity in neurons; they are blocked by octanol. **P-type channels**, identified in some CNS neurons and prominent in Purkinje cells, are blocked by ω-agatoxin, much less so by ω-conotoxin. The L-type channel from skeletal muscle consists of five subunits: α1, α2, β, γ, and δ. The largest subunit (α1) is responsible for many of the functional features, including DHP binding and the voltage-gated pore. It is structurally similar to the Na$^+$ channel α subunit, with four internal repeats of 200–300 amino acids exhibiting sequence homology. Each of these repeats contains six putative transmembrane segments, S1 to S6; segment S4 in all repeats has positively charged groups in -(K/R)XX- repeats (four to six K or R residues per segment), that have been implicated in sensing voltage changes; potential *N*-glycosylation sites and potential cAMP-dependent phosphorylation sites have been identified on the α$_1$ protein. Missense mutations in the gene at 19p13 for the α1A subunit are associated with familial hemiplegic migraine and with episodic ataxia-2, while expansion of the polyglutamine tract is associated with a form of spinocerebellar ataxia. *See also* **verapamil**.

calcium-dependent regulator protein or **calcium-dependent modulator protein** *abbr.*: CDR; *former name for* **calmodulin**.

calcium ion (Ca^{2+})-release channel *another name for* the **ryanodine receptor**.

calcium phosphate-mediated transfection a method for introducing exogenous DNA such as a plasmid into mammalian or other eukaryotic cells. Cells in culture are exposed to a calcium phosphate coprecipitate of DNA and a proportion of them take up the complex by endocytosis and express the transfected genes. Efficiency of transfection varies greatly with different cell lines.

calcium pump any system that transports calcium ions across a biological membrane from a lower to a higher concentration with the consumption of energy. The term is used especially of the pump in the sarcoplasmic reticulum of muscle of which a transmembrane **calcium-transporting ATPase** is an integral part.

calcium-regulated actin bundling protein a protein that is probably involved in the formation of filopodia.

calcium-sensing receptor a G-protein-coupled receptor that is expressed in parathyroid glands and renal tubules. It consists of 1078 amino acids, encoded by a gene at 3q13.3-q21. Loss-of-function mutations are associated with familial hypercalcemic hypocalciuria, and gain-of-function mutations with familial hypocalcemic hypercalciuria.

calcium-transporting ATPase EC 3.6.1.38; an integral membrane protein that forms an essential component of the **calcium pump** in sarcoplasmic reticulum. It catalyses the hydrolysis of ATP to ADP and orthophosphate, simultaneously transporting two calcium ions from the cytosol into the sarcoplasmic reticulum for each ATP hydrolysed. Probably comprising 10 transmembrane domains, it is related to other metal-ion-transporting ATPases. A number of types exist.

calcium-trigger protein any of a group of small calcium-binding **trigger** proteins, including **calmodulin**, troponin C (*see* **troponin**), and **parvalbumin**.

calculus (*pl.* **calculi**) a concretion of material that forms within the body. It often resembles a small pebble, hence is called a 'stone'. Calculi are most common in the gall bladder or kidney, and are composed variously of organic or inorganic salts, frequently of calcium; cholesterol calculi are gallstones of pure cholesterol.

calcyclin *other names*: prolactin receptor associated protein, growth factor-inducible protein 2A9, S100 calcium-binding protein A6; a small protein that copurifies with **prolactin receptor**. It is induced in fibroblasts by growth factors and is overexpressed in acute myeloid leukemia. It binds specifically **annexin XI**.

caldecrin a pancreatic protein that lowers serum calcium. It also has chymotrypsin-like activity, which is not needed for the calcium-lowering effect.

caldesmon an F-actin cross-linking protein in thin filaments in smooth muscle and in stress fibres in (non-muscle) fibroblasts. It also binds myosin and tropomyosin. It appears to be a homo- dimer comprising two polypeptide chains, each with a single actin-binding site and a single self-association site. It controls actin–myosin interactions, and inhibits the actin-activated ATPase of myosin. Its action is attenuated by calcium-calmodulin, and potentiated by tropomyosin. Caldesmon will interact with Ca^{2+}-calmodulin (but not with calmodulin alone), an interaction that causes caldesmon to dissociate from F-actin. *See also* **caldesmon kinase**, **caldesmon phosphatase**.

caldesmon kinase EC 2.7.1.120; the name given to a reaction now known to result from the autophosphorylation of **caldesmon** (by ATP) in the presence of calcium. This autophosphorylation abolishes the ability of caldesmon to bind actin.

caldesmon phosphatase EC 3.1.3.55; an enzyme that hydrolyses the phosphate from caldesmon phosphate.

caldolysin an extracellular protease isolated from *Thermus aquaticus* strain T-351, an extreme thermophile. It contains 13% carbohydrate by weight, and one atom of zinc; M_r 21 000.

calelectrin *see* **annexin IV**, **annexin V**, **annexin VI**.

calgizzarin an EF-hand Ca^{2+}-binding protein isolated from chicken gizzard smooth muscle. It is also abundant in rabbit lung and is found in other tissues. It is homologous to **annexin** II light chain, and is a homodimer (disulfide-linked). It is over-expressed in human colorectal tumours.

callipyge a mutation in sheep that produces enlarged hindquarters by enhanced expression of several genes of muscle

callose a linear (1,3)-β-D-glucan insoluble in water but soluble in dilute alkali. The glucan chains form stable double or triple helices. Callose is a ubiquitous higher plant polysaccharide occurring as a component of specialized, and often transient, cell walls, especially in reproductive tissues and is a component of the cell plate. Callose is synthesized at the outer plasma membrane surface of a few cell types. It is rapidly deposited as drops, plugs, or plates on wounding or physiological stress, or attempted penetration by fungal hyphae. Callose synthase is activated by Ca^{2+}, which increases in concentration on wounding. Callose deposits on cell walls stain specifically with Aniline Blue or its fluorochrome. Specific hydrolases, e.g. glucan endo-1,3-β-D-glucosidase (EC 3.2.1.39) for callose and membrane-bound callose synthetase (EC 2.4.1.34), are found in plants. Functionally callose may act as a temporary wall matrix, as a special permeability barrier or as a wall strengthening agent. Other (1,3)-β-D-glucans include yeast glucan, pachyman, lamarin, scleroglucan, curdlan, leucosin, mycolamarin, paramylon, chrysolamarin, and lentinan.

callus **1** a mass of relatively unspecialized tissue that develops at wound sites in plants, forming a protective covering. Callus cells are used in tissue culture as the starting material for the propagation of plant clones. **2** tissue formed during the healing of broken bone.

calmegin a spermatogenesis-specific chaperone protein that is homologous to **calnexin**, **calreticulin**, and CALNUC. Mouse knockouts produce sperm that cannot bind to the zona pellucida of ova.

calmitine a mitochondrial Ca^{2+}-binding protein of fast-twitch muscle fibres with a characteristic 28-residue signal. The mature protein is identical to **calsequestrin** of the sarcoplasmic reticulum.

calmodulin *abbr.*: CAM; a small, heat-stable, acidic, calcium-dependent modulator protein that binds four calcium ions per molecule and is then able to stimulate a variety of eukaryotic enzymes or enzyme systems. These include brain adenylate cyclase, cytosolic cyclic nucleotide phosphodiesterase, myosin light-chain kinase, erythrocyte Ca^{2+},Mg^{2+}-ATPase, plant NAD^+ kinase, and a number of calcium-dependent protein kinases including phosphorylase kinase (to which it contributes the δ subunits); these enzymes bind to calmodulin via basic amphiphilic alpha-helices they contain. Various other cellular processes such as membrane phosphorylation and microtubule disassembly also are stimulated. Calmodulin is a single-chain polypeptide, of which aspartic and glutamic account for about 30% of the total residues; it contains one residue of N^6-trimethyllysine per mole but no cysteine, hydroxyproline, or tryptophan residues. Calmodulin appears to be ubiquitous within the animal and plant kingdoms, to be without species or tissue specificity, and to be almost invariant in structure, being conserved nearly perfectly throughout evolution; it has **EF-hand** domains and is similar to troponin C from muscle. It has been known by a variety of names including modulator protein, phosphodiesterase activator protein, troponin C-like (modulator) protein, and calcium- (or Ca^{2+}-)dependent regulator (or modulator) protein (CDR protein).

calmodulin-dependent protein kinase II *recommended name*: Ca^{2+}/calmodulin-dependent protein kinase, EC 2.7.1.123; *other names*: Ca^{2+}-CAM kinase II; microtubule-associated protein 2 kinase; multifunctional Ca^{2+}-CAM kinase. An enzyme with broad specificity and activated by Ca^{2+} that catalyses the phosphorylation of Ser and Thr residues in proteins by ATP. It is involved in cell regulation. It is composed of α (M_r 60 000) and β (M_r 55 000) subunits in a dodecahedral structure. On activation by Ca^{2+}-CAM, autophosphorylation converts the enzyme to a Ca^{2+}-CAM-independent form. Substrates include synapsin I, tryptophan hydroxylase, skeletal muscle glycogen synthetase, and microtubule-associated proteins tau and MAP-2. A synthetic peptide substrate is PLSRTLSVSS.

calmodulin-like domain protein kinase *abbr.*: CDPK; EC 2.7.1.–; *other name*: Ca^{2+}-binding Ser/Thr protein kinase. An enzyme that is subject to autophosphorylation and resembles calmodulin-dependent protein kinases in its N-terminal region and contains four EF hands in the C-terminal region. It may be ubiquitous in plant cells, being attached to the cytoskeleton, by N-terminal myristoylation to the inner surface of the plasma membrane, or in the cytoplasm.

calnexin a calcium-binding protein of the endoplasmic reticulum. It appears to play a role in the processing of endoplasmic reticulum proteins, in monitoring assembly, and in retaining unassembled or incorrectly folded proteins. It has a single transmembrane helical region.

CALNUC a calcium-binding protein of Golgi apparatus, homologous with **calreticulin**.

calomel electrode the mercury–mercurous chloride (mercury–calomel) ($Hg;Hg_2Cl_2$, Cl^-) half-cell that is reversible to chloride ions. Because of its stability and simplicity, it is frequently used in conjunction with a saturated-KCl bridge as a reference electrode in pH meters, etc.

calorie *or (sometimes)* **gram calorie** *or* **small calorie** *symbol*: cal; any of several units of heat or energy in the CGS system. A calorie originally represented the quantity of heat required to raise one gram of water through one degree centigrade (i.e. Celsius). However, the energy represented by a calorie varies slightly according to the initial temperature of the water. Hence, the **International Table calorie** (*symbol*: cal_{IT}) is now taken to equal 4.1868 J exactly, while the **15°C calorie** (*symbol*: cal_{15}) is equal approximately to 4.1855 J. Thermochemical calculations have often used a slightly different value; one **thermochemical calorie** (*symbol*: cal_{th}) is now taken to equal 4.184 J exactly. In nutrition and physiology, the term 'calorie' is often used to mean the kilocalorie or Calorie (initial capital), equal to 10^3 cal_{IT}. Use of all these units is now deprecated, the joule now being preferred.

Calorie *or (sometimes)* **kilogram calorie** *or* **large calorie** the name used, especially in nutrition, for the kilocalorie, i.e. 10^3 cal_{IT}. *Compare* **calorie**.

calorific of, relating to, or generating heat.

calorific value the amount of heat produced by the complete combustion of a given mass of a substance, such as a fuel or foodstuff. It is usually expressed in $J kg^{-1}$.

calorigen any agent that increases heat production by increasing metabolism. *See* **salicylic acid**..

calorigenesis the production or increased production of heat in an organism. —**calorigenic** *adj*.

calorimeter an instrument for determining quantities of heat evolved, transferred, or absorbed.

calorimetry the measurement of the amount of heat evolved, transferred, or absorbed by a system. —**calorimetric** *adj*.

calpactin *see* **annexin**.

calpain *abbr.*: CAPN; EC 3.4.22.17; a Ca^{2+}-activated, neutral, thiol proteinase enzyme that preferentially cleaves the bonds: Tyr-|-Xaa, Met-|-Xaa, or Arg-|-Xaa with Leu or Val as the P2 residue. Calpains are cytosolic mammalian enzymes, of five main types: CAPN1 and 2 are ubiquitous; CAPN3 is muscle-specific; and CAPN4 and 5 are stomach-specific. These have an identical small regulatory **EF-hand** subunit and a large catalytic subunit, which contains an N-terminal domain that is cleaved by autocatalytic activation, a catalytic domain, and a C-terminal region that contains five EF hands. Mutations in CAPN3 (gene locus at 15q15.1-q21 encodes 996 amino acids) are associated with a form of limb girdle muscular dystrophy (LGMD2A).

calpastatin a protein, found in liver and erythrocytes of several mammals, that is a specific calpain inhibitor. Four characteristic domains are involved in the inhibitory action.

C-alpha *an alternative term for* **alpha carbon atom**.

calphostin C one of a group of compounds isolated from *Cladosporium cladosporioides* having a unique structural feature in possessing a perylene quinone. It was identified in a search for inhibition of protein kinases. It inhibits protein kinase C by action on the regulatory domain and has a relatively high degree of specificity, but acts equally on all isoforms having the common regulatory domain. It is

a highly potent and specific inhibitor (IC_{50} = 50 nM) of protein kinase C. At higher concentrations it inhibits other kinases, including myosin light chain kinase, and cyclic GMP- and cyclic AMP-dependent kinases.

calphotin a Ca^{2+}-binding protein found in the cytoplasm of photoreceptor cells of *Drosophila melanogaster*. The C-terminal region contains a leucine zipper uninterrupted by prolines. The protein contains >50% proline, alanine, and valine.

calponin a thin filament-associated protein implicated in regulation and modulation of smooth muscle contraction. It is found as an actin-, calmodulin-, and tropomyosin-binding protein present in many vertebrate smooth muscles, and is related to troponin T in immunological and biochemical characteristics.

calpromotin *an alternative name for* **torin**.

calregulin *see* **calreticulin**.

calreticulin *other names*: calregulin; Crp55; CaBP3; HACBP; erp60; a Ca^{2+}-binding protein of the endoplasmic reticulum lumen; its C terminus contains the **KDEL** sequence.

calsequestrin a 44 kDa calcium-binding protein of low isoelectric point, found on the interior surface of the membrane of the **sarcoplasmic reticulum** of muscle. It acts as an internal calcium store in muscle and releases Ca^{2+} at **calcium channels**. It has only a moderate affinity for calcium ions, but a high capacity (>40 Ca^{2+} ions per molecule); selective binding of Ca^{2+} in preference to other cations is promoted by ATP; it also exhibits protein kinase activity. Calsequestrin makes up some 19% of the protein of isolated sarcoplasmic vesicles, and is thought to assist in depleting the sarcoplasm of free calcium ions during relaxation. The human muscle isoform (390 amino acids, 44.53 kDa) is very acidic (47 Asp, 56 Glu), with a sequence of eight aspartic-acid residues at the C terminus.

calspectin *an alternative name for* **fodrin**.

calstorin a calcium-binding protein of the microsomal lumen of rat brain, similar to **calsequestrin**.

caltractin *or* **centrin** an **EF-hand** Ca^{2+}-binding protein of the centrosome of interphase and mitotic cells that plays a fundamental role in microtubule-organizing centre structure. It is present in the centrosome/basal body apparatus of the green alga *Chlamydomonas reinhardtii*. It contains four EF-hands, and belongs to the calmodulin/parvalbumin/troponin C superfamily.

caltrin *abbr. for* calcium transport inhibitor; *other name*: seminalplasmin; a small basic protein of male seminal vesicle fluids. It binds to calmodulin, and inhibits Ca^{2+} uptake by epididymal spermatozoa. It has antimicrobial properties.

Calvin cycle *see* **reductive pentose phosphate cycle**.

Calx a gene of *Drosophila* that encodes a Na^+/Cl^- exchanger channel protein.

Cam *symbol for* the **carbamoylmethyl** group.

CAM *abbr. for* **1** calmodulin. **2** crassulacean acid metabolism.

camalexin a sulfur-containing phytoalexin produced by plants of the *Brassica* genus and their relatives in response to attack by bacterial pathogens.

CaM-BP$_{80}$ *an alternative name for* **calcineurin**.

CAMDA *abbr. for* critical assessment of microarray data analysis; *see* **critical assessment competition**.

cAMP *abbr. for* cyclic AMP; *see* **adenosine 3′,5′-phosphate**.

campesterol (24*R*)-ergost-5-en-3*β*-ol; a sterol found in small amounts in rapeseed and other vegetable oils.

camphor 1,7,7-trimethylbicyclo[2.2.1]heptan-2-one; a compound obtained from the camphor tree, *Cinnamomum camphora*, which is indigenous to Taiwan. It is widely used in ointments and liniments. It is also used as a starting material in organic chemistry.

campomelic dysplasia *or* **autosomal sex reversal** a rare autosomal dominant chondrodysplasia caused by mutations in *SOX9* and fatal in early infancy. In most affected genotypic male patients it is associated with feminization.

cAMP receptor *see* **CAR**.

cAMP receptor protein *see* **catabolite (gene) activator protein**.

cAMP-regulatory protein *see* **catabolite (gene) activator protein**.

CaMV promoter *abbr. for* cauliflower mosaic virus promoter; a promoter used to obtain expression of exogenous genes in green plants.

canaline O^4-aminohomoserine; 2-amino-4-(aminooxy)-butanoic acid; $H_2N-O-[CH_2]_2-CH(NH_2)-COOH$; a basic α-amino acid found as the L-enantiomer in the jack bean, *Canavalia ensiformis*, and in some other legumes containing L-canavanine. It is formed from **canavanine** by deamidination or transamidination. It inhibits pyridoxal-dependent enzymes.

canavalin the major globular protein isolated from ripe seeds of the jack bean, *Canavalia ensiformis*. It is a seed storage protein, with sequence similarity to **phaseolin**, **vicilin**, and related proteins. *See also* **concanavalin A**.

canavanase *see* **arginase**.

Canavan disease an autosomal recessive degenerative disease due to a defect in the gene for aspartoacylase. Onset is in early infancy, consisting of blindness, psychomotor regression, enlarged head, optic atrophy, hypotonia, spasticity, and increased urinary excretion of *N*-acetylaspartate.

canavanine 2-amino-4-(guanidinooxy)butanoic acid; $H_2N-C(NH)-NH-O-[CH_2]_2-CH(NH_2)-COOH$; a basic amino acid, occurring as the L-enantiomer in certain legumes, esp. the jack bean, *Canavalia ensiformis*. In seeds of the jack bean, it constitutes up to 8% of their dry mass, and forms their main storage compound for nitrogen. On germination of the seeds, the nitrogen is released for synthetic purposes by hydrolysis of L-canavanine to L-canaline and urea, from each of which ammonium is then formed by further hydrolysis. It is a structural analogue of L-arginine, and a potent growth inhibitor of many organisms, in which it interferes with arginine metabolism. *See also* **canaline**.

cancellate *or* **cancellated** *or* **cancellous** having a porous or spongy structure; meshlike or lattice-like.

cancer any malignant **neoplasm**. Cancers are usually divided into **carcinomas**, derived from epithelial tissue, and **sarcomas**, derived from connective tissue. Some 300 genes are known to be involved in carcinogenesis. *See* **APC**, **BRCA**, **retinoblastoma**, **Wilms' tumour**.

cancer-associated retinopathy protein *see* **recoverin**.

candela *symbol*: cd; the SI base unit of **luminous intensity**, defined as the luminous intensity, in a given direction, of a source that emits monochromatic radiation of frequency 540×10^{12} hertz and that has a radiant intensity in that direction of 1/683 watt per steradian.

Candida a genus of (in many cases) dimorphic yeasts containing species of medical (*C. albicans*) and industrial importance: *C. cylindrica* is a source of **lipases** and is a rare example of a eukaryote with a non-standard genetic code for nuclear/cytoplasmic gene expression.

cane sugar *an old and common name for* **sucrose**.

cannabinoid any of about 30 derivatives of 2-(2-isopropyl-5-methylphenyl)-5-pentylresorcinol found in the Indian hemp, *Cannabis sativa*, among which are those responsible for the narcotic actions of the plant and its extracts. The most important cannabinoids are cannabidiol, cannabinol, *trans*-Δ^9-tetrahydrocannabinol, *trans*-Δ^8-tetrahydrocannabinol, and Δ^9-tetrahydrocannabinolic acid. *See also* **cannabis**.

cannabinoid receptor any of several membrane proteins that bind cannabinol and structurally similar compounds and mediate their intracellular action. Two types have been characterized, and both are G-protein-coupled receptors. CB1 receptors are found in brain and testis. CB2 receptors are found in spleen and immune tissues, but not in brain. For both types arachidonoylethanolamide (anandamide) is a putative endogenous ligand and both types are negatively coupled to adenylate cyclase decreasing intracellular cyclic AMP levels. CB3 receptors also occur in brain.

cannabinoid *see* **cannabinoid**.

cannabis the dried flowering or fruiting tops of Indian hemp, *Cannabis sativa*; numerous synonyms exist, including **marijuana**, **bhang**, and **maconha**. A resin, known variously as cannabin, hashish, or charas, is also obtained from the plant. Both preparations contain active principles, **cannabinoids**, and are used as recreational drugs, producing relaxation, euphoria, and enhanced awareness. Occasionally, anxiety or mental disturbance may result, and, rarely, loss of consciousness or even death. The major active ingredients are tetrahydrocannabinols. Cannabis is a controlled substance in the US Code of Federal Regulations.

Cannizzaro reaction the base-catalysed dismutation of aldehydes with no α hydrogen atoms into the corresponding acids and alcohols:

$$2R-CHO + OH^- \rightarrow R-COO^- + R-CH_2OH.$$

[After Stanislo Canizzaro (1826–1910).]

canonical form *or* **canonical structure** *an older name for* **contributing structure**.

canonical sequence a nucleotide or amino-acid sequence regarded as the archetype with which variants are compared.

cap 1 *or* **cap structure** a structural feature present at the 5′ end of most eukaryotic (cellular or viral) mRNA molecules and also of some virion mRNA molecules but not of bacterial mRNA molecules. It consists of a residue of 7-methylguanosine and a triphosphate bridge linking it 5′–5′ to the end of the polynucleotide chain; in addition, the first one or two nucleotide residues in the chain may possess a 2′-*O*-methyl group and the first, if adenylyl, a 6-*N*-methyl group. The cap structure is thought to protect the 5′ end of the mRNA from degradation by phosphatases or nucleases and to facilitate initiation of translation of mRNA by the eukaryotic (but not by the bacterial) ribosome. **2** a cluster or patch of aggregated proteins at one site on the surface of a cell, e.g. a lymphocyte. Surface components of the cell membrane are aggregated, e.g., by the action of polyvalent ligands, and may be swept to the posterior area of the cell when the cell moves, forming a cap at that site. **3** to form or place a cap on a molecule (or structure).

CAP *abbr. for* **1** catabolite (gene) activator protein. **2** chloramphenicol (use deprecated, full name preferred). **3** *Clostridium* (*histolyticum*) aminopeptidase, EC 3.4.11.13. **4** cyclic AMP receptor protein; *see* **CRP**.

capacitance *or* **electric capacitance** *symbol*: C; the ability of a system to store electric charge or a measure of this capacity for a particular system. $C = Q/U$ where Q is the quantity of charge stored to raise the system's potential by U volts. The SI derived unit of capacitance is the **farad**.

capacitance minimization method a simple method for the determination of asymmetric surface potentials in lipid bilayers, based on the dependence of bilayer capacitance on transmembrane voltage. The capacitance is measured by rectifying the 90° component of an applied a.c. current signal. A superimposed slow triangular wave results in a hysteresis-like time course of capacitance. The centre of the hysteresis figure is shifted along the voltage axis by an amount equal to the **capacitance minimization potential**, the difference between the dipole plus surface-charge potentials on the two sides of the membrane.

CAPB *see* **calbindin**.

cap-binding protein any protein (e.g. a 24 kDa protein obtainable from rabbit reticulocytes) that specifically binds to the **cap** (def. 1) of eukaryotic capped mRNA. Such proteins may facilitate the translation of capped mRNA molecules. In the nucleus a heterodimer is involved; in the cytosol a complex comprising the **initiation factors** eIF4A, eIF4E, eIF4GI, and eIF4GII is involved.

Capecitabine or **Xeloda** 5′-deoxy-5-fluoro-N-[(pentyloxy)carbonyl]-cytidine. The first tumour-activated drug. It acts by inducing the tumour to produce more thymidine phosphorylase, which causes the tumour to make more cytotoxic 5-fluorouracil, to which the drug is converted.

CAP factor see **catabolite (gene) activator protein**.

capillarity the phenomenon, resulting from surface tension, in which liquids rise up capillary tubes and which also causes them to form a concave or convex meniscus at their surface where it contacts a solid.

capillary 1 of, or relating to, a hair or hairlike structure. **2** of, or relating to, a tube with a very fine bore. **3** any of the very fine blood vessels that form a network between the arterioles and the venules throughout the body. **4** a capillary tube.

capillary column a chromatography column that has a very fine bore (0.1–0.25 mm), the inner surface of which may be coated with **stationary phase** or other support. Such columns are used in gas chromatography, supercritical fluid chromatography, or capillary electrophoresis. The column may be up to a few metres in length, coiled, and made of glass or fused silica. In high-performance liquid chromatography, columns of somewhat wider bore and shorter length are referred to as capillary columns.

capillary electrophoresis abbr.: CE; a very-high-resolution method of electrophoresis, also known as **capillary zone electrophoresis** (abbr.: CZE) or **high-performance electrophoresis**. It is based on the use of long (up to 100 cm), narrow (<100 μm) silica columns for separation of peptides, etc. by electro-osmotic flow. By using sensitive detectors (e.g. laser-induced fluoresence), sensitivity in the attomole range can be achieved.

capillary filtration a technique for **membrane filtration** in which the membrane is supported inside porous capillary tubes (capillary membranes). Bundles of such tubes make up a larger filtration unit. Compare **hollow-fibre technique**.

capillary viscometer an instrument for measuring the viscosity of a liquid by determining the rate of flow of the liquid through a capillary tube of known dimensions. An **Ostwald viscometer** is a simple capillary viscometer; an **Ubbelholde viscometer** is a viscometer with several different capillary tubes, so that measurements may be made at different shear values.

capped mRNA an mRNA molecule that has a **cap** (def. 1).

capping the act or process of forming or placing a **cap** on a molecule or structure.

capping enzyme any guanylyltransferase enzyme concerned in the formation of a **cap** (def. 1) on an RNA molecule.

caprate or **caprinate** a trivial name for **decanoate**, or for any salt or ester of decanoic acid.

capric acid or **caprinic acid** a trivial name for decanoic acid; see **decanoate**.

caprin a trivial name for any of the glyceryl esters of decanoic acid (see **decanoate**); these often occur in milk fat, especially the triester tricaprin, $(CH_3–[CH_2]_8–COO)_3C_3H_5$.

caprinate see **caprate**.

caprinyl (formerly) an alternative name for **caproyl** (def. 1).

caproate a trivial name for **hexanoate**, or any salt or ester of hexanoic acid.

caproic acid a trivial name for hexanoic acid (see **hexanoate**).

caproin a trivial name for any of the glyceryl esters of hexanoic acid (see **hexanoate**); these often occur in milk fat, especially the triester tricaproin, $(CH_3–[CH_2]_4–COO)_3C_3H_5$.

caproyl 1 or (formerly) **caprinyl**; symbol: Dco; trivial name for decanoyl, $CH_3–[CH_2]_8–CO–$, the univalent acyl radical derived from decanoic acid (formerly capric acid or caprinic acid). Use of the term caproyl is disfavoured because of confusion with **caproyl** (def. 2) and **caprylyl**. **2** an old trivial name for hexanoyl, the univalent group derived from hexanoic acid (formerly caproic acid), a saturated unbranched, acyclic, aliphatic acid (i.e. fatty acid). It occurs

naturally in acylglycerols in, e.g., coconut and palm nut oils, and milk fat. Use abandoned (see **caproyl** (def. 1)).

capryl 1 an old trivial name for **octyl**. **2** an old trivial name for sec-octyl. **3** symbol: Dec; a trivial name sometimes used for **decyl**.

caprylate former name for **octanoate** or for any salt or ester of octanoic acid.

caprylic acid a former name for octanoic acid; see **capryloyl**.

caprylic alcohol 1 or n-caprylic alcohol a former name for (n-)octyl alcohol (octan-1-ol). **2** or sec-caprylic acid a former name for sec-octyl alcohol (octan-2-ol). See also **caprylic acid**.

caprylin a former name for any of the glyceryl esters of **octanoic acid**; these occur in milk fat, especially the triester tricaprylin, $(CH_3–[CH_2]_6–COO)_3C_3H_5$.

capryloyl or **caprylyl** an old trivial name for **octanoyl**.

Caps or **CAPS** abbr. for 3-(cyclohexylamino)propanesulfonic acid; a compound with properties similar to a **Good buffer substance**, pK_a (20°C) = 10.4.

capsaicin (E)-N-(4-hydroxy-3-methyoxybenzyl)-8-methyl-non-6-enamide; the active principle in chilli pepper (Capsicum), paprika, and cayenne that causes a burning sensation by stimulation of nociceptive (pain sensory) neurons followed by desensitization of nociceptive responses and neuronal depletion of **substance P**. It binds to and activates the vanilloid receptor, an integral membrane protein with homology to a family of putative calcium channel proteins. The receptor is also activated by heat.

capsaicin

capsanthin (3R,3′S,5′R)-3,3′-dihydroxy-β,κ-caroten-6′-one; the main carotenoid pigment of red pepper (Capsicum annuum).

capsid the protein coat that surrounds the infective nucleic acid in some virus particles. It comprises numerous regularly arranged subunits, or **capsomeres**.

capsomere or **capsomer** any of the (protein) subunits that comprise the closed shell or coat (capsid) of certain viruses.

cap structure see **cap** (def. 1).

capsular antigen or **capsular polysaccharide** or **capsular substance** any of the antigens, usually polysaccharide in nature, that are carried on the surface of bacterial capsules. The term **K antigen** is used for those that mask somatic (O) antigens.

capsule 1 an outer layer, rich in polysaccharides, that is present in certain types of bacteria. Presence of a capsule is often linked with **virulence** and the capsular material is often immunogenic. **2** a membrane or sheath of connective tissue that encloses an organ or tissue.

capsulin or **epicardin** a member of the MyoD family of basic HLH transcription factors that control formation of skeletal muscles in vertebrates. Like the related **MyoR (musculin)**, capsulin is expressed transiently in specific facial muscle precursors derived from the first branchial arch. A null mutation in the mouse capsulin gene causes death very soon after birth from pulmonary hypoplasia but no obvious skeletal muscle abnormality. Mice homozygous for mutations in both MyoR and capsulin also die immediately after birth but lack the major muscles of mastication.

captopril D-3-mercaptomethylpropanoyl-L-proline, (S)-1-(3-mercapto-2-methyl-1-oxopropyl)-L-proline; an antihypertensive drug that acts as an inhibitor of **angiotensin converting enzyme**. It was designed to fit the known active site of carboxypeptidase A; the mercapto group binds to the enzyme zinc ion, the amide carbonyl to a hydrogen-bonding site, and the proline carboxylate to an electrophilic centre. A related compound, **enalapril**, is more potent when administered orally and has fewer side effects.

captopril

capture cross-section the probability of a particle that impinges on an atomic nucleus being captured by that nucleus multiplied by the cross-sectional area of the nucleus (measured in **barns**).

CapZ a heterodimer protein (subunits of 33 and 31 kDa) that interacts with α-actinin 2 and the barbed end of F-actin polymers at the Z disks of skeletal muscle.

CAR *abbr. for* **1.** coxsackie and adenovirus receptor; a single-pass transmembrane protein whose extracellular region contains two immunoglobulin-like domains and binds coxsackie virus and adenovirus 2 and 5. The gene locus on chromosome 21 encodes 351 amino acids, but the physiological ligand is unknown. **2.** cAMP receptor; a receptor of the slime mould *Dictyostelium* that senses extracellular cAMP gradients. **3.** constitutive androstane receptor; a receptor that activates the phenobarbital-inducible *CYP2B* gene.

carba+ *comb. form (in chemical nomenclature)* indicating the replacement of a heteroatom by a carbon atom.

carbamide *former name for* **urea.**

carbamido *an alternative term for* **ureido.**

carbamino the chemical group –NH–COO⁻, which is stable only in the salt form. Carbamino compounds are formed by spontaneous reversible reaction of CO_2 with unprotonated primary aliphatic amino groups; e.g. with α-amino acids to form carbamino acids, or with the N termini of proteins, as in the formation of carbamino hemoglobin from hemoglobin. *Compare* **carbamoyl.**

carbamoyl *or (formerly)* **carbamyl** *or* **carboxamido** *symbol*: Cbm; the acyl group, –CO–NH_2, corresponding to carbamic acid (an acid unknown in the free state). *See also* **carboxamide.**

carbamoylmethyl *symbol*: Cam; the chemical group 2-amino-2-oxoethyl, –CH_2–CO–NH_2.

carbamoyl-phosphate synthase 1 *or* **carbamoyl-phosphate synthase (glutamine-hydrolysing)** EC 6.3.5.5; *other names*: carbamoyl-phosphate synthetase (glutamine-hydrolysing); GD-CPSase. A glutamine-dependent enzyme that is the first step in the pathway for pyrimidine biosynthesis. It catalyses a reaction between 2 ATP, glutamine, CO_2, and H_2O to form 2 ADP, orthophosphate, glutamate, and carbamoyl phosphate; this enzyme is part of the **CAD protein** in many eukaryotes. **2** *or* **carbamoyl-phosphate synthase (ammonia)** EC 6.3.4.16; *other names*: carbamoyl-phosphate synthetase (ammonia); carbamoyl-phosphate synthetase I, carbon-dioxide–ammonia ligase. An enzyme that incorporates free ammonium ion into carbamoyl phosphate, which then enters the **ornithine–urea cycle**. It catalyses a reaction between 2 ATP, NH_3, CO_2, and H_2O to form 2 ADP, orthophosphate, and carbamoyl phosphate. It constitutes ≈20% of mitochondrial protein and ≈4% of total cell protein in liver. Rare mutations in the gene at 2q35 that lead to enzyme deficiency are associated with hyperammonemia, protein intolerance, and impaired mental and physical development from birth.

carbamyl *former name for* **carbamoyl.**

carbanion any anion containing an even number of electrons in which a significant part of the excess negative charge is located on one or more carbon atoms.

carbapenem antibiotic any of various broad-spectrum β-lactam antibiotics, such as thienamycin derived from *Streptomyces catt-*

leya, that inhibit peptidoglycan synthesis. Semisynthetic derivatives include imipenem.

carbene 1 any chemical species of the type R_2C: containing an electrically neutral, bivalent carbon atom with two nonbonding electrons. The spins of such electrons may be either antiparallel (in the singlet state) or parallel (in the triplet state). Carbenes are highly reactive short-lived groups formed as intermediates in the course of certain organic reactions. **2** *or (formerly)* **methylene** the chemical group H_2C:, the simplest of the carbenes.

carbenium ion any real or hypothetical **carbocation** with one important contributing structure containing a tervalent carbon with a vacant *p*-orbital. These structures have often been described as **carbonium ions**.

carbenoxolone glycyrrhetic acid hydrogen succinate, a 3-(3-carboxy-1-oxopropoxy)-11-oxoolean-12-en-29-oic acid, steroid agent with anti-ulcerative effects, the mechanism of which is unresolved.

carbobenzyloxy *or* **carbobenoxy** *former name for* **benzyloxycarbonyl.**

carbocation a cation with an even number of electrons with a significant portion of the excess positive charge located on one or more carbons. This general term includes **carbenium ions, carbonium ions**, etc.

carbocycle any compound or molecular structure that is **carbocyclic** (def. 1); a **carbocyclic** (def. 2).

carbocyclic 1 describing any cyclic molecular structure containing only carbon atoms in the ring or rings; describing any compound having such a stucture. *See also* **homocyclic** (def. 1). *Compare* **heterocyclic** (def. 1). **2** *or* **carbocyclic** any such compound.

carbodiimide any organic compound of general structure R–N=C=N–R, widely used for forming amide (peptide) or phosphodiester linkages in the laboratory. Water-soluble carbodiimides, e.g. 1-ethyl-3-(3-dimethylaminopropyl)carbodiimide, are used to couple carboxylic acid residues to ω-aminoalkyl agarose, while for reactions in nonaqueous media *N*,*N*'-dicyclohexylcarbodiimide is much used.

carbohydrase *former name for* **glycosidase.**

carbohydrate any of a group of organic compounds based on the general formula $C_x(H_2O)_y$. The group comprises the **monosaccharides, oligosaccharides**, and **polysaccharides**, and is usually extended to include their derivatives and, sometimes, the **cyclitols**. Some of the simplest members of the group may, notionally, be considered as hydrates of carbon. However, 2-deoxy-D-ribose ($C_5H_{10}O_4$) does not fit the empirical formulation. Carbohydrates are also components of nucleosides, nucleotides, RNA and DNA, glycoproteins, glycolipids, and glycosaminoglycans. Examples include **amylose, amylopectin, chitin**, and **glycogen.**

carbohydrate-deficient glycoprotein syndrome (*abbr.*: CDGS) *or* **glycoconjugate syndrome** *or* **congenital defect of glycosylation** any of several defects in N-glycan synthesis of glycoproteins. Such defects are usually identified by isoelectric focusing of serum transferrin, which shows a cathodal shift as a result of sialic acid deficiency. Type I defects are those where the defect is in the cytosol or in the endoplasmic reticulum. These include: **type Ia**, or **phosphomannomutase** deficiency; **type Ib**, or **mannose-6-phosphate isomerase** deficiency; **type Ic**, or α-1,3-glucosyl transferase deficiency; **type Id**, or α-1,3-mannosyl transferase deficiency; and **type Ie**, or dolichyl-phosphate-mannose synthase 1 deficiency. Type II disorders have a defect in the Golgi apparatus. These include: **type IIa**, or *N*-acetyl-glucosaminyl transferase deficiency; and **type IIb**, or **glucosidase** deficiency. In leukocyte adhesion deficiency II there is defective import of GDP-fucose into the Golgi apparatus.

carbon *symbol*: C; a nonmetallic element of group 14 of the IUPAC **periodic table**; atomic number 6; relative atomic mass 12.011. It exists in two main allotropic forms, diamond and graphite, and as a third allotrope, fullerene; there are also several amorphous forms, such as carbon black and charcoal. Carbon also forms a vast array of compounds, the organic compounds, based on chains and rings of covalently bonded carbon atoms. There are two stable nuclides, the predominant one being carbon-12. This has a relative abundance in natural state of 98.89 atom percent. Its mass is used as the reference standard for all relative atomic masses of nuclides. The other stable nuclide is **carbon-13**. There are also four radioactive nuclides: carbon-10, **carbon-11**, **carbon-14**, and carbon-15.

carbon-11 a radioactive nuclide of carbon, $^{11}_{6}C$. It emits a positron (β^{+}-particle) (0.97 Mev), no γ-ray, and has a half-life of 20.3 min.

carbon-13 or **heavy carbon** the stable nuclide $^{13}_{6}C$. Relative abundance in natural carbon 1.108 atom percent. It is used as a tracer in studies of reaction mechanisms and of carbon metabolism; now often replaced by **carbon-14** unless double labelling is required. Carbon-13 was/is used when a non-radioactive isotope is essential. See also **carbon-13 nuclear magnetic resonance spectroscopy**.

carbon-14 a radioactive nuclide of carbon, $^{14}_{6}C$. It emits an electron (β^{-}-particle) (0.156 Mev), no γ-radiation, and has a half-life of 5730 years, decaying to $^{14}_{7}N$. A minute proportion occurs in atmospheric CO_2 due to the action of cosmic rays, and it forms the basis of **carbon dating**. It is used very widely as a radioactive tracer, for which it is produced artificially.

carbonate dehydratase EC 4.2.1.1; systematic name: carbonate hydro-lyase; other names: carbonic dehydratase; carbonic anhydrase (abbr.: CA); an enzyme that catalyses the reversible hydration of carbon dioxide to carbonic acid (or to bicarbonate ion at certain pH values) in a wide range of species. Zinc is a cofactor. At least seven enzymatic forms are known, and are classified into: (1) cytosolic (CA-I in erythrocytes; CA-II in bone, kidneys, brain, and other tissues; CA-III in skeletal muscle; and CA-VII in salivary glands); (2) GPI-anchored to plasma membrane (CA-IV in microvasculature; CA-IX in gastrointestinal mucosa); (3) mitochondrial (CA-VA and CA-VB); and (4) salivary secretion (CA-VI). CA-I, CA-II, and CA-III are encoded by a gene cluster at 8q22, whereas the others have different loci. Deficiency of CA-I in erythrocytes has no pathological consequences. Deficiency due to inactivating mutations in the gene for CA-II is associated with a syndrome of osteopetrosis, renal tubular acidosis, and cerebral calcification. See also **dehydratase**.

carbon cycle the sum total of chemical and biochemical processes whereby carbon is cycled between carbon dioxide in the atmosphere (and oceans) and organic compounds in organisms and their remains. Atmospheric carbon dioxide is fixed during photosynthesis by green plants and other photoautotrophic organisms, using solar energy to split water molecules and liberate O_2. The carbon dioxide is thus converted into glucose and other organic compounds, which are used as metabolic fuel and building blocks by the majority of organisms. Aerobic organisms, in particular, are able fully to oxidize organic compounds during respiration to release carbon dioxide to the atmosphere. The carbon stored in the remains of dead organisms is liberated by the respiration of decomposers or, especially that in fossil fuels, by combustion. Hence, photosynthetic organisms and aerobic heterotrophic organisms exist in a state of syntrophy. However, green plants and other aerobic autotrophs also liberate carbon dioxide during respiration. Compare **oxygen cycle**.

carbon dating or **radiocarbon dating** a technique for determining the age of specimens of biological origin, e.g. wood. It is based on determining the amount of **carbon-14** remaining in the specimen, the known half-life of carbon-14, and the assumption that the abundance of this radionuclide in the atmosphere has remained constant since its incorporation into the material from atmospheric CO_2.

carbon-dioxide buffer any aqueous solution of a water-soluble salt, pK_a 9–10, that, when equilibrated with CO_2, is capable of maintaining a constant partial pressure of CO_2 (pCO_2) in the gas phase in a closed vessel. Carbon-dioxide buffers are used especially in maintaining a constant pCO_2 in manometric experiments on reactions involving the steady production of that gas.

carbon-dioxide electrode an electrode system for determining CO_2 tension in a gas or a liquid. It consists of a combination pH electrode bathed in a thin layer of a dilute bicarbonate solution, the whole being separated from the sample by a rubber or silicone-rubber membrane permeable to CO_2 but not to water or ionizable solutes. The pH of the layer of bicarbonate solution, and hence the electrode's response, varies logarithmically with the CO_2 tension in the sample.

carbon-dioxide fixation the process whereby the carbon atoms of carbon dioxide are incorporated into hexoses by the **reductive pentose phosphate cycle**.

carbon-dioxide snow solid carbon dioxide in powder form. See also **dry ice**.

carbonic anhydrase common name for **carbonate dehydratase**.

carbonium ion any real or hypothetical cation, containing an even number of electrons and at least one penta-coordinate carbon atom, in which a significant part of the excess positive charge is located on one or more carbon atoms. See also **carbenium ion**, **carbocation**.

carbon monoxide CO; a colourless, odourless, toxic gas that combines with hemoglobin to form **carboxyhemoglobin**; it also combines with and inhibits cytochrome oxidase. It is produced physiologically during conversion of heme to biliverdin by heme oxygenase. Like nitric oxide, it is a neurotransmitter in the brain, where it activates cytosolic guanylyl cyclase by binding to its heme prosthetic group.

carbon monoxide-dehydrogenase any enzyme that catalyses the oxidation of carbon monoxide to carbon dioxide. Enzymes of this type contain either a nickel–iron–sulfur cluster or a molybdenum–pyranopterin. The 'hydrogen' involved comes from water; the oxygen is added to the carbon monoxide. Examples are carbon-monoxide dehydrogenase (cytochrome b_{561}) (EC 1.2.2.4) and carbon-monoxide dehydrogenase (ferredoxin) (EC 1.2.7.4). The term is also applied to acetyl-CoA synthase.

carbon-13 nuclear magnetic resonance spectroscopy or **^{13}C-NMR spectroscopy** a variant of **nuclear magnetic resonance spectroscopy** in which resonances in **carbon-13** nuclei in natural carbon of organic molecules are examined instead of proton responses. Carbon-12 has no NMR spectrum, and with a natural abundance of carbon-13 of 1.11%, a carbon-13 NMR spectrum has a several thousandfold disadvantage in signal-to-noise ratio compared with a proton NMR spectrum. However, the carbon-13 spectrum is much simpler and therefore has a particular advantage in large molecules. Since ^{13}C–^{13}C coupling can be observed, a double labelling tracer technique is possible, using e.g. $^{13}CH_3$–$^{13}COOH$ as a precursor in studies of **polyketide** biosynthesis.

carbon skeleton the arrangement of the carbon atoms in an organic molecule.

carbonyl the bivalent group =C=O (often written =CO); it is a characteristic group in aldehydes, ketones, and carboxylic acids. It is also the name of the carbon monoxide ligand in metal complexes.

carboprost (15S)-15-methylprostaglandin $F_{2\alpha}$; an oxytocic agent used to terminate pregnancy. See also **prostaglandins**.

Carbowax proprietary name for a group of polyethylene glycols of general formula $H[OCH_2-CH_2]_n-OH$ and of average M_r 400 to 20000.

carboxamide an amide derived from a carboxylic acid. The carboxamide group is –CO–NH$_2$. See also **carbamoyl**.

carboxamido former name for **carbamoyl**.

carboxy+ prefix (in chemical nomenclature) denoting the presence of a carboxyl group.

carboxyethyl symbol: Cet; the –CH$_2$–CH$_2$–COOH group.

4-carboxyglutamic acid or **γ-carboxyglutamic acid** symbol: Gla; 1-aminopropane-1,3,3-tricarboxylic acid; a **noncoded amino acid**, one or more residues of which are present in a number of calcium-binding proteins, e.g. **osteocalcin** and **blood coagulation** factors II, VII, IX and X. It is formed subsequent to biosynthesis of the proteins by vitamin K-dependent carboxylation of glutamic acid residues.

carboxyhemoglobin a form of hemoglobin in which carbon monoxide has displaced oxygen from the heme group of hemoglobin, to which it binds with much greater affinity, and hence less reversibly than oxygen. The formation of carboxyhemoglobin thus reduces the oxygen-carrying capacity of the blood, and deprives tissues of oxygen. It is the basis of carbon monoxide poisoning.

carboxyl the chemical group –COOH in an organic molecule; it is acidic in character.

carboxylate 1 the anionic group –COO⁻, formed by dissociation of a **carboxyl** group. **2** any salt, anion, or ester formed from a **carboxylic acid**. **3** to perform or undergo **carboxylation**.

carboxylation the introduction of a molecule of CO_2 into an organic compound.

carboxylic acid any organic acid containing one or more carboxyl (–COOH) groups.

carboxyl proteinase or **acid proteinase** or **aspartic proteinase** any enzyme of the sub-subclass EC 3.4.23 (aspartic endopeptidases), a common feature of which is a low pH optimum and an aspartic acid residue in the active centre. It includes **pepsins**, **rennins**, and **cathepsin D**.

carboxyl terminus or **carboxyl terminal** an alternative name for the **C terminus** (also called, less correctly, carboxy terminus or carboxy terminal). —**carboxy-terminal** adj. Compare **amino terminus**.

carboxy-lyase any enzyme of sub-subclass EC 4.1.1 that catalyses a decarboxylation reaction.

carboxymethyl symbol: Cm; the –CH_2–COOH group.

carboxymethylation the process of introducing **carboxymethyl** groups into a substance; e.g. cellulose may be carboxymethylated to produce a cation-exchanger.

carboxypeptidase any enzyme of the sub-subclasses serine carboxypeptidases (EC 3.4.16) or metallo-carboxypeptidases (EC 3.4.17). Both types hydrolyse C-terminal amino-acid residues from oligopeptides or polypeptides. See **carboxypeptidase A**, **carboxypeptidase B**, **carboxypeptidase H**.

carboxypeptidase A EC 3.4.17.1; other name: carboxypolypeptidase. A **carboxypeptidase** enzyme that catalyses the hydrolysis of free C-terminal residues; zinc is a cofactor. The enzyme from bovine pancreas contains 307 amino acids, and its X-ray structure shows an eight-stranded mixed beta sheet.

carboxypeptidase B EC 3.4.17.2; other name: protaminase. A **carboxypeptidase** enzyme of bovine pancreas that catalyses the hydrolysis of C-terminal L-lysine (or L-arginine) to peptide and L-lysine (or L-arginine); zinc is a cofactor. It is also called thrombin-activated fibrinolysis inhibitor (TAFI) because it cleaves free C-terminal L-lysine from a fibrin clot.

carboxypeptidase C a **carboxypeptidase** enzyme, extracted from citrus leaves, that cleaves all free C-terminal residues.

carboxypeptidase D a **carboxypeptidase** enzyme, present particularly in trans Golgi network of glia cells and neurons, that plays a role in processing neuroendocrine peptides. It is specific for C-terminal L-lysine or L-arginine residues, and is a transmembrane protein with three carboxypeptidase domains, two of which are active.

carboxypeptidase E see **carboxypeptidase H**.

carboxypeptidase H EC 3.4.17.10; other names: carboxypeptidase E; enkephalin convertase; (wrongly) enkephalinase. A **carboxypeptidase** enzyme involved in prohormone processing. It catalyses the hydrolysis of free C-terminal L-lysine (or L-arginine) to peptide and L-lysine (or L-arginine); zinc is a cofactor. The enzyme (gene locus at 4q32) is expressed in brain, pituitary, and neurons, and is localized with prohormone convertases in peptide-containing secretory vesicles. A missense mutation (called fat mouse) in the mouse orthologue gene leads to maturity-onset obesity and infertility.

carboxypeptidase-tubulin see **tubulinyl-Tyr carboxypeptidase**.

carboxypeptidase Y a **carboxypeptidase** enzyme of yeast that cleaves all free C-terminal residues (glycine very slowly).

carboxy-terminal see **carboxyl terminus**.

carboxy terminus a less correct alternative term for the **C terminus** of a polypeptide chain. See also **carboxyl terminus**.

carcinoembryonic antigen abbr.: CEA; any member of a subfamily of immunoglobulin-like proteins on the outer cell surface and grouped in the cluster CD66 (see **CD markers**). **CD66e**, other name: **meconium antigen 100**, mass 180–200 kDa. It is a glycoprotein, attached to the membrane by a GPI-anchor. It is present in elevated concentrations in numerous nonmalignant conditions, and it can be used for monitoring therapy. **CD66d**, other names: **CGM1, CEA gene member 1**, is present on leukocytes and granulocytes, but not

lymphocytes, monocytes, or tumours. Several forms are produced by alternative splicing. **CD66b** (formerly CD67), other names: **CGM6, CEA gene member 6**, is expressed in leukocytes of chronic myeloid leukemia patients and bone marrow. It is attached to the membrane by a GPI-anchor.

carcinogen any agent that directly or indirectly induces the transformation of a normal cell into a neoplastic cell. Such agents include various substances of small M_r (often termed **chemical carcinogens**), **oncogenic** viruses, ionizing radiations, and ultraviolet light. See also **precarcinogen**, **proximate carcinogen**. —**carcinogenic** adj.

carcinogenesis the process(es) involved in the production of **cancers**, including the action of **carcinogens** on living cells. See also **oncogenesis**.

carcinogenicity the measure or extent of the potency of a chemical carcinogen.

carcinoid (tumour) or **argentaffin carcinoma** or **argentaffinoma** any tumour arising from **argentaffin cells**. Carcinoids arise from any site of such cells, notably in the ileocecal region, and produce large amounts of 5-hydroxytryptophan and 5-hydroxytryptamine, the former mainly by bronchial tumours and the latter by the decarboxylase-containing intestinal tumours. Histamine, kinins, and substance P may also be produced.

carcinoma any cancer that arises from epithelial tissue. Carcinomas form malignant tumours that tend to invade lymph spaces in surrounding connective tissue and to spread by lymphatic permeation and embolism; they also enter the bloodstream to form **metastases**. —**carcinomatous** adj.

CARD domain abbr. for caspase recruitment domain; a protein interaction module comprising a bundle of six α-helices. CARD-containing proteins are involved in the regulation of apoptosis and inflammation through their activation of CARD-containing caspases and **jNF-κB**, which typically involves assembly of multi-protein complexes that facilitate dimerization or trimerization, or that serve as scaffolds on which proteases and kinases are assembled and activated. The module is related in sequence and structure to **death domains** and death effector domains, and shows similar interaction properties. The domains may occur in isolation, or in combination with other domains (e.g. **leucine-rich repeats**, **SH domains**, **RING**, kinase, and death domains).

cardenolide any of a group of C_{23} cardiotonic steroids characterized by a 14β-hydroxyl group and an unsaturated γ-lactone ring on C-17. They occur free or as the aglycons of **cardiac glycosides** in plants and insects.

cardiac 1 of, relating to, or affecting the heart. **2** of, or relating to, the upper part of the stomach.

cardiac glycoside any of a group of toxic steroid glycosides, derived from plants, that in pharmacological doses increase the power of the contractions of the mammalian heart; some are inhibitors of Na^+,K^+-ATPase. The steroid aglycon carries a 17β unsaturated pentenyl lactone ring and a 3β-hydroxyl group to which the carbohydrate moiety is attached. Examples are **digitoxin** from Digitalis purpurea, **ouabain** from Strophanthus gratus, and **strophanthin** from Strophanthus kombe.

cardiac K channel see **HERG, KCNE1, KVLQT1, long QT syndrome**.

cardiac muscle the specialized muscle of which the walls of the vertebrate heart are made. It is composed of a network (syncytium) of branching elongated cells (fibres) with cross striations, and undergoes spontaneous rhythmical contractions.

cardiac Na channel see **long QT syndrome, SCN5A**.

cardiac puncture a convenient and quick method for obtaining reasonable volumes of blood from an animal by inserting a syringe needle directly into the heart.

Cardice proprietary name for solid carbon dioxide. See **dry ice**.

cardio+ or (before a vowel) **cardi+** comb. form denoting heart.

cardioid (in mathematics) a type of expression that yields a heart-shaped curve.

cardiolipin an immunogenic 1,3-bis(3-phosphatidyl)glycerol. It is synthesized within mitochondria from two molecules of phosphatidyl glycerol with release of one molecule of glycerol and one of cardiolipin. It is present exclusively in the inner mitochondrial membrane. Antibodies against cardiolipin are commonly present in patients with systemic lupus erythematosus. Prepared from beef heart, it is used as an active antigen in the Wassermann reaction and other serological tests for syphilis. It may carry antigenic determinants similar to *Treponema pallidum* or be similar to products of tissue necrosis formed when the spirochaete attacks human tissues.

cardiomyopathy a progressive disorder that impairs the structure or function of the muscles in the ventricles of the heart. It is frequently **secondary** to an underlying condition, for example viral myocarditis, hypertension, or ischemic heart disease, and is an important cause of morbidity and mortality worldwide. Much less frequently it is **primary** and associated with defects of fatty acid beta-oxidation or in oxidative phosphorylation (due to mutations in nuclear or in mitochondrial genes). **Dilated cardiomyopathy** is connected with mutations in more than a dozen genes, including those for lamin A/C, desmin, α-cardiac actin, and Duchenne and Becker muscular dystrophies. **Hypertrophic cardiomyopathy** is associated with mutations in at least ten genes, including those for cardiac forms of actin, troponin T and troponin I, α-tropomyosin, regulatory or essential myosin light chains, myosin heavy chain, and myosin-binding protein C.

cardiotonic having the effect of improving the tone of heart muscle. Cardiotonic steroids (**cardiac glycosides**) act directly on heart muscle to restore the power of the heart's contractions.

cardiovascular of, or relating to, the heart and the system of blood vessels.

cargo protein any protein that is carried within the vesicles of a cell's secretory system (i.e. from the endoplasmic reticulum to the Golgi apparatus to the plasma membrane). *Compare* **resident protein**.

carmustine 1,3-bis(2-chloroethyl)-1-nitrosourea; an antineoplastic derivative of chloroethylnitrosourea with immunosuppressive and antifungal activity.

carnitine γ-amino-β-hydroxybutyric acid trimethylbetaine, 3-carboxy-2-hydroxy-*N,N,N*-trimethyl-1-propanaminium hydroxide, inner salt; $(CH_3)_3N^+–CH_2–CH(OH)–CH_2–COO^-$; a compound that is found in striated muscle, liver, and other tissues and participates in the transfer of acyl groups across the inner mitochondrial membrane. It is synthesized in liver and kidney from *N*-trimethyllysine (released by degradation of certain proteins) by a pathway that involves two ascorbic acid-dependent hydroxylations. The rate of synthesis is low in newborn and premature infants. Deficiency results from hemodialysis and from excessive renal loss in various organic acidurias. It is considered to be vitamin-like for such individuals.

L-carnitine

carnitine/acylcarnitine translocase a 32 kDa transmembrane protein, found in the inner mitochondrial membrane, that has a high affinity for the exchange of $C_{12}–C_{16}$ acylcarnitine esters with **carnitine**. Several mutations of a gene at 2p21.3 lead to a deficient or nonfunctional translocase.

carnitine acyltransferase or (*sometimes*) **carnitine acylase** any of various enzymes that catalyse a reaction between acyl-CoA and carnitine to form CoA and *O*-acylcarnitine, which exchanges (via a carnitine/acylcarnitine translocase) across the inner mitochondrial membrane. This movement of carnitine is sometimes called the 'carnitine shuttle'. Carnitine *O*-acetyl transferase, EC 2.3.1.7, is found (in eukaryotes) in the mitochondrial matrix, peroxisomes, and the lumen of the endoplasmic reticulum. It is specific for short-chain fatty acyl-CoA ($C_2–C_{14}$) and provides a mechanism for the transport of acetyl groups, as acetylcarnitine, through the mitochondrial inner and peroxisomal membranes. Rare mutations in the gene at 9q34.1 result in a deficiency syndrome. Mitochondrial carnitine palmitoyltransferase, EC 2.3.1.21, exists as two isoforms. One isoform, CPT_1 or CPT-I, is a mitochondrial outer membrane protein involved in regulating fatty acid β-oxidation by controlling the rate of formation of carnitine acyl esters. Another isoform, CPT_2 or CPT-II, is located in the mitochondrial inner membrane and re-forms acyl-CoA esters in the mitochondrial matrix after the carnitine esters have been transported through the inner membrane by acylcarnitine:carnitine antiporter.

carnitine transporter an Na$^+$-dependent organic cation transporter (OCNT 2) present in the plasma membrane of cells of liver, muscle, and other tissues. The gene at 5q31.2-q32 encodes a protein of 557 amino acids, which shows ≈80% sequence homology with OCTN 1. Several mutations in the OCTN 2 gene result in systemic carnitine deficiency.

carnosine *N*-β-alanyl-L-histidine; a substance that occurs in the skeletal muscle of some animals, including humans (up to 30 mmol kg^{-1}). *Compare* **anserine**.

carnosine

carotene *see* **carotenoid**.

carotenoid any of the naturally occurring tetraterpenes that are widely distributed in plants and animals but only synthesized by higher plants, algae, fungi, and bacteria. They may be represented formally as consisting of eight isoprenoid (ip) residues and formed by the joining tail to tail of two units each comprising four ip residues joined head to tail: ip.ip.ip.ip.pi.pi.pi.pi. There are two groups: **carotenes**, which are hydrocarbons, and **xanthophylls**, which contain oxygen in various forms. They occur in all photosynthetic tissues, where they function as protectors against photosensitization and as antenna pigments in photosynthesis. α-Carotene is the precursor of vitamin A.

carrageenan or **carrageen** or **carrageenin** a mixture of sulfated galactans, also known as **Irish moss**, extracted from red seaweeds (class Rhodophyceae), particularly *Chondrus crispus* and *Gigartina stellata*. It shows blood anticoagulant activity and is used as a gelling and stabilizing agent in foods and pharmaceuticals. Three major types are known: ι-carrageenan is the most highly sulfated and contains alternate units of 3-linked β-D-galactopyranose 4-sulfate and 4-linked 3,6-anhydro-α-D-galactopyranose 2-sulfate; κ-carrageenan is precipitated by potassium chloride and contains only one sulfate per disaccharide unit; λ-carrageenan is largely devoid of the 3,6-anhydro component. ι-Carrageenan forms a kinked double-helical structure. [After Carragheen, near Waterford, Republic of Ireland.]

κ-carrageenase EC 3.2.1.83; an enzyme that catalyses the hydrolysis of 1,4-β-D-linkages between D-galactose 4-sulfate and 3,6-anhydro-D-galactose in various carrageenans.

carrier 1 any known or putative component of a biological membrane that effects the transfer of a specific substance or group of related substances from one side of the membrane to the other. The mechanism is supposedly a cyclic process, involving combination of the carrier with the substance(s) on one side of the membrane, diffusion of the combined form across the membrane, release of the substance(s) on the far side, and diffusion of the carrier back across the membrane in uncombined form. *See also* **facilitated diffusion**, **per-**

mease, translocation, transport. 2 any component of a biological fluid (e.g. a blood protein) that can combine with a specific substance (e.g. a hormone or metabolite) and transport it from one part of the organism to another. 3 *(in enzymology)* any substance undergoing reversible combination with a metabolite or reversible oxidation–reduction and thereby carrying the metabolite, electron(s), or hydrogen atom(s) between enzymes or cellular organelles, or into cells. Some **coenzymes** may act as carriers. *See* **electron carrier, hydrogen carrier. 4** a substance that is mixed with a reagent or other substance to facilitate the chemical or physical manipulation of the latter. The carrier is commonly of similar constitution and/or properties to the reagent, and present at relatively high concentration. The term is used especially for the nonradioactive chemical counterpart of a radionuclide or of a compound containing a radionuclide, whether deliberately added or present from the time of initial preparation of the radioactive substance. *Compare* **carrier-free. 5** *(in chromatography)* the **mobile phase** or a component thereof. *See* **carrier ampholyte, carrier displacement chromatography, carrier gas. 6** *or* **matrix** the inert support material to which a small molecule, biopolymer, cell fragment, or intact cell is artificially linked to form an immobilized or insolubilized derivative. *See also* **microcarrier. 7** any vesiculated structure within which a biologically active substance (e.g. a drug or enzyme) is artificially incorporated, commonly to protect the substance from degradation and/or to direct it to a target organ upon injection into the bloodstream. Such a carrier may be either artificial (e.g. a **liposome**) or of natural origin (e.g. an erythrocyte **ghost**). **8** *(in immunochemistry)* any substance of high M_r (e.g. a protein) to which a non- or poorly immunogenic substance (e.g. a **hapten**) is naturally or artificially linked to enhance its immunogenicity. **9** *(in genetics)* any individual who is heterozygous for a specified recessive allele and who thus does not normally display any characteristic (e.g. a disease) associated with that allele but who may, with appropriate mating, produce homozygous offspring that exhibit the recessive characteristic. **10** *(in pathology)* any individual who harbours a specified infectious microorganism without showing signs of disease, and is capable of transmitting the microorganism to others. *Compare* **vector** (def. 2). *See also* **vehicle.**

carrier ampholyte a mixture of amphoteric electrolytes used in **isoelectric focusing** to generate a linear and stable pH gradient, and in **isotachophoresis** and **ampholyte displacement chromatography**. The component electrolytes are commonly of similar constitution, low molecular mass, and closely spaced isoelectric points. *See also* **Ampholine.**

carrier displacement chromatography a variant of **displacement chromatography** in which substance(s), either related or unrelated, are added to the mixture being chromatographed to assist in the manipulation of the components separated.

carrier-free describing a preparation of a radionuclide, or of a radionuclide-labelled compound, to which no carrier has been added, and for which precautions have been taken to minimize contamination with other isotopic nuclides. It does not necessarily imply, however, 100% isotopic abundance. Material of high specific activity is often wrongly referred to as 'carrier-free'; more correctly it would be of high isotope abundance.

carrier gas the inert gas that forms the mobile phase in **gas–liquid chromatography** or gas–solid chromatography and that carries the eluted substances in gaseous form into the detector.

carrier lipid *see* **undecaprenol.**

carrier protein any protein that functions as a **carrier** (def. 8); e.g., **acyl carrier protein**. The term has been applied more specifically to describe **cytochrome b₅**, **ADP, ATP carrier protein**, and **GDC** (Graves' disease carrier protein).

Carr-Price reaction the reaction of an anhydrous solution of antimony trichloride in chloroform with retinol (vitamin A) to produce a transient blue colour, maximal at about 10 s. It can be used to estimate retinol.

CARS *abbr. for* **coherent anti-Stokes Raman spectroscopy.**

Cartesian of, or relating to, the French philosopher and mathematician René **Descartes** (1596–1650) or his works.

Cartesian coordinates a system of coordinates that defines the location of a point in a plane (two dimensions) or in a space (three dimensions) in terms of its perpendicular distance from each of a set (two or three) of mutually perpendicular axes.

Cartesian diver *or* **Cartesian devil** a device, devised as a toy, adapted as an ultramicrorespirometer, consisting of a small piece of capillary glass tubing open at one end and closed and slightly expanded at the other. A sample of tissue, in approximately 2 µL buffer, is placed in the vessel, which is then filled with gas and the neck sealed with a drop of oil. The diver is placed in a constant-temperature vessel filled with a solution in which the diver just floats. If the pressure in the vessel is reduced the diver will rise and if the pressure is increased the diver will fall. The buoyancy of the diver is determined by the gas volume within it, and changes in the latter may be measured by adjusting the pressure in the vessel so that the diver floats in the same position.

cartilage a tough, elastic type of **connective tissue**, found in most vertebrates. It is a major component of the embryonic skeleton, but in higher vertebrates it is mostly converted to bone during development. In adult humans, it is largely confined to the ears, nose, trachea, the anterior ends of the ribs, and the articular surfaces of bones. Cartilage consists of a firm resilient matrix, formed by chondroblasts (which become enclosed as chondrocytes), and composed of glycosaminoglycans including **chondroitin sulfate** A and C, hyaluronic acid, and keratosulfate II, together with varying amounts of **collagen**. The entire skeleton of some lower vertebrates, notably the cartilaginous fishes (class Chondrichthyes), consists of cartilage. —**cartilaginous** *adj.*

cartilage link protein a protein that stabilizes the aggregates of proteoglycan monomers with hyaluronic acid in the extracellular cartilage matrix.

cartilage oligomeric matrix protein *abbr.*: COMP; a pentameric disulfide-linked Ca^{2+}-binding **EF-hand** protein of the cartilage matrix.

cartilage-specific proteoglycan core protein a major proteoglycan of the extracellular matrix of cartilaginous tissues. The N-terminal region binds hyaluronic acid. Chondroitin sulfate is the main glycan component, but keratan sulfate and O-linked and N-linked oligosaccharides are also present. It also has an epidermal growth factor-like domain and a **Sushi domain**.

caryon *a variant spelling of* **karyon.**

caryopsis *see* **grain.**

CAS 1 *abbr. for* carbonic anhydrase. **2** *abbr. for* Chemical Abstracts Service.

cascade sequence *or* **waterfall sequence** a sequence of successive activation reactions involving either enzymes (**enzyme cascade**) or hormones (**hormone cascade**), or both. There are various examples: (1) the series of transformations of proteolytic enzymes in **blood coagulation** in which each enzyme activates the next until the final substrate, fibrinogen, is reached; (2) the series of reactions involved in the formation of, and mediation of the action of second messengers, e.g. in the activation of **glycogen phosphorylase**; (3) the process of activation of **steroidogenesis** in the adrenal gland by hypothalamic and pituitary hormones. The term has also been applied to other sequential phenomena, e.g. DNA → RNA → protein. Cascade sequences are characterized by a series of amplifications of a weak initial stimulus, and may be activated at more than one point along the sequence.

cascara the dried bark of the small tree *Rhamnus purshiana*. It is a potent purgative due to its constituent hydroxyanthracene derivatives (anthroglycosides), including cascarosides and related glycosides. These yield active anthracene derivatives after metabolism by colonic bacteria.

casein 1 *or (esp. Brit.)* **caseinogen** the principal protein fraction of milk; it is present as a colloidal suspension, and may be precipitated by acidification. It is a mixture of several related proteins (α-, β-, γ-, and λ-caseins), which differ in amino-acid composition and which may be separated electrophoretically. The α- and β-caseins are rich in phosphate, present mainly as *O*-phosphoserine residues, which bind calcium ions. Casein is easily prepared, and is valuable as a dietary supplement. **2** *or (esp. US)* **paracasein** the product formed by coagulation of milk (or **casein** (def. 1)) by treatment with chymosin (rennin). This enzyme removes a 64-residue carbohydrate-rich hydrophilic C-terminal moiety of λ-casein, thereby destroying the latter's protective action on the casein micelles, thus allowing them to be precipitated by the calcium ions present.

casein kinase I any of a family of ubiquitously expressed monomeric serine/threonine protein kinases found in a variety of subcellular locations in all eukaryotic organisms from protozoa to humans. There are seven isoforms in human cells, and all contain a kinase domain flanked by divergent N- and C-terminal extensions. They play roles in Wnt signalling, circadian rhythms, nuclear import, and progression of Alzheimer disease, and have a general preference for acidic substrates. The family is named after the first of two fractions to elute from an anion-exchange column with the ability to phosphorylate the acidic milk protein casein.

caseinogen an alternative name (esp. Brit.) for **casein** (def. 1).

CAS number abbr. for Chemical Abstracts Service Registry Number; a unique identifier for a compound.

casomorphin or **exomorphin** any of a number of opioid peptides isolated from an enzymatic digest of bovine casein; activity corresponds to residues 60–66 of bovine β-casein. See also **morphiceptin**.

CASP abbr. for **1** critical assessment of structure prediction; see **critical assessment competition**. **2** CTD-associated SR-like protein (carboxy terminal domain-associated splicing-recognition-like protein); any protein that contains a motif for sequence-specific RNA-binding and a domain for binding to the C-terminal domain of RNA polymerase II. See **SCAF**.

caspase any of a family of cysteine endopeptidases that cleave proteins at specific aspartic acid residues. The prototype is ICE (interleukin 1β-converting enzyme, now called caspase 1). Procaspases are activated in cascade fashion in response to a proapoptotic stimulus (e.g. DNA damage, heat stress, oxidative stress) by proteolytic cleavage at consensus sites. Activation is controlled by apoptosis inhibiting factors, e.g. Bcl-2 (see **BCL**), which is an integral protein of the outer mitochondrial membrane and regulates release of **apoptosis protease-activating factor 1** (Apaf-1) and cytochrome c from the intermembrane space. The cascade involves initiator (or upstream) components that contain a death effector domain. It starts with binding of procaspase 8 (also called FLICE) to death receptors, its activation, formation of a death-inducing signalling complex (DISC), and activation of several procaspases. A major effector (or executioner, or downstream) component is caspase 3, which is activated by caspase 9, Apaf-1, and cytochrome c. Procaspase 12 is associated with the endoplasmic reticulum. Caspases are essential for the proteolytic reactions of **apoptosis**..

cassette an informal name given to a functional segment of DNA, for example a dominant selectable marker such as neo^R under the control of an appropriate promoter. Cassettes may be used in making DNA constructs such as plasmids. Naturally occurring cassettes include those involved in mating type switching in yeasts.

castasterone 24-methylcholesterol the precursor of brassinolide in plants. See **brassinosteroid**.

castor oil an oil derived from seeds of the plant Ricinus communis. It contains triglycerides of ricinoleic, oleic, linoleic, palmitic, and stearic acids.

CAT abbr. for **1** computer of average transients. **2** chloramphenicol O-acetyltransferase. **3** computer-assisted tomography.

catabolic dead-end a point in a **catabolic pathway** where, in a particular organism, a metabolite is neither catabolized further nor excreted but accumulates intracellularly. Consequently the pathway is unable to function as an energy-yielding process. The enzyme that can form the metabolite thus lacks an obvious role in that organism.

catabolic pathway any degradative, usually exergonic, sequence of reactions. Compare **anabolic pathway**. See also **amphibolic**.

catabolism or **katabolism** any metabolic process involving the breakdown of complex substances into smaller products, including the breakdown of carbon compounds with the liberation of energy for use by the cell or organism. Compare **anabolism**. —**catabolize** or **catabolise** vb.; **catabolic** adj.

catabolite any product of a **catabolic pathway**.

catabolite (gene) activator protein abbr.: CAP; other names **cAMP receptor protein**, **cAMP-regulatory protein** (abbr.: CRP); a protein that binds to and is activated by cAMP. It then acts as a transcription-initiation factor, binding near the operator of bacter-

ial operons that are subject to **catabolite repression** and being necessary for their efficient transcription. For example, transcription of the lac operon is increased 50-fold in the presence of the CAP/cAMP complex. It belongs to the **CRP** family of transcription factors.

catabolite inhibition a general regulatory mechanism, found in Escherichia coli, that controls the activity of an early reaction in carbohydrate metabolism. For instance when E. coli is grown in synthetic medium with radioactive galactose or lactose as the carbon source, the addition of glucose rapidly inhibits utilization of the radioactive substrate. Subsequent removal of the glucose rapidly reverses the inhibition.

catabolite repression a mechanism of genetic regulation in bacteria in which the accumulation of catabolites in the cell represses the formation of enzymes that contribute to the formation of the catabolites. In E. coli the intracellular concentration of cAMP is low when cells are grown on media containing glucose and high when glucose is depleted. Binding of cAMP to **catabolite (gene) activator protein** (CAP) promotes transcription of lac and other operons such as ara.

catalase EC 1.11.1.6; a widely distributed homotetrameric enzyme that catalyses the formation of $O_2 + 2 H_2O$ from two molecules of H_2O_2; most commonly heme is a cofactor. It plays an important role in removing H_2O_2 formed in peroxisomes in animals and plants and in **glyoxysomes** in plants. In erythrocytes it is cytosolic. The molecule is an eight-stranded beta barrel and contains 13 alpha helices. A mutation in the gene locus at 11p13 is associated with acatalasemia. A microbial catalase contains a dinuclear manganese centre instead of heme.

catalase-peroxidase an enzyme with both **catalase** and **peroxidase** activity. The enzyme from Mycobacterium tuberculosis is required for sensitivity to isoniazid.

catalatic of, or pertaining to, **catalase**.

catalyse or (esp. US) **catalyze** to influence a reaction through **catalysis**.

catalysis an increase in the rate of a chemical reaction brought about by a **catalyst**. —**catalytic** adj.; **catalytically** adv.

catalysome **1** an enzyme-containing cap found on lipid droplets in the fat body of some insects. **2** a mitochondrion-like organelle specialized for lipid metabolism, found in adipose tissue.

catalyst any substance that increases the rate of a chemical reaction but is itself unchanged at the end of the reaction. Catalysts are usually present in very low concentrations relative to those of the substances whose reaction they are catalysing.

catalytic activity an index of the actual or potential activity of a catalyst. The catalytic activity of an enzyme or an enzyme-containing preparation is defined as the property measured by the increase in the rate of conversion of a specified† chemical reaction that the enzyme produces in a specified† assay system. [†This word, in one or other position, has been misprinted as 'specific' in certain published versions of this definition.] Catalytic activity is an extensive quantity (see **extensive property**) and is a property of the enzyme, not of the reaction mixture; it is thus conceptually different from rate of conversion although measured by and equidimensional with it. The unit for catalytic activity is the **katal**; it may also be expressed in mol s^{-1}. Dimensions: N T^{-1}. Former terms such as catalytic ability, catalytic amount, and enzymic activity are no longer recommended. Derived quantities are **molar catalytic activity**, **specific catalytic activity**, and **catalytic activity concentration**.

catalytic activity concentration or **catalytic concentration** the **catalytic activity** of an enzyme per unit volume, where volume refers to that of the original enzyme-containing preparation, not that of the assay system. It may be expressed in katals per litre.

catalytic amount the amount of a substance used in a chemical reaction as a catalyst, primer, or sparker. It is generally much smaller than the stoichiometric amounts of either reactants or products. For an enzyme the term is now obsolete and replaced by **catalytic activity**.

catalytic antibody a monoclonal antibody that emulates an enzyme activity. Typically it is raised against a compound that mimics the transition state of the reaction.

catalytic assay a technique for measuring low concentrations

($\approx 1\,\mu M$) of substrates in certain enzyme-catalysed reactions (e.g. NAD(P)$^+$ or coenzyme A), where simple kinetic methods are not feasible because too-rapid consumption makes reaction rates impossible to measure reliably. Conditions are arranged so that the concentration of the substrate in question is kept constant by a regeneration reaction, the substance being made to behave as an intermediate catalyst. The regeneration reaction may act also as the indicator reaction. For example, NAD$^+$ may be determined by coupling the reaction catalysed by alcohol dehydrogenase (EC 1.1.1.1) to the combined regeneration/indicator reaction catalysed by cytochrome c reductase (EC 1.6.99.3), according to the equations:

$$NAD^+ + C_2H_5OH \rightarrow NADH + CH_3CHO + H^+$$

and

$$NADH + H^+ + 2\,Cyt\ c^{3+} \rightarrow NAD^+ + 2\,Cyt\ c^{2+} + 2\,H^+.$$

The rate of increase in absorbance of Cyt c^{2+} at 550 nm is measured. If the coupled regeneration reaction does not act as an indicator reaction, the technique is known as **enzymic cycling**. *Compare* **coupled assay**.

catalytic concentration *an alternative term for* **catalytic activity concentration**.

catalytic constant *symbol*: k_{cat} *or* k_0; a measure of the catalytic potential of an enzyme, customary unit s^{-1}. In general it is the value of the velocity of the reaction, v, divided by the stoichiometric concentration of active sites, $[E]_0$, obtained by extrapolating all substrate concentrations to infinity; alternatively it can be expressed as $V_{max}/[E]_0$, where V_{max} is the **limiting rate**. k_{cat} is also known as the **turnover number**, because it represents the number of reactions catalysed per unit time by each active site.

catalytic domain *see* **domain** (def. 3).

catalytic exchange method a method for randomly labelling a compound with tritium. A solution of the compound in a tritiated hydroxylic solvent is allowed to stand in the presence of metal catalysts and either acid or base.

catalytic facilitation a process whereby the product formed by one enzyme in a complex is passed directly to the catalytic site of the next enzyme in a reaction sequence more efficiently than when the second substrate is supplied from the environment.

catalytic factor a term sometimes used of a specified enzyme to define the quantitative ratio between the rates of a chemical reaction in the presence and the absence of the enzyme but otherwise under the same conditions.

catalytic intermediate a chemical species produced during the course of a catalytic reaction, intermediate between reactants and products.

catalytic power a measure of the ability of an enzyme to accelerate a chemical reaction.

catalytic residue any of the amino-acid residues in an enzyme that are directly involved in making or breaking covalent bonds while the enzyme is acting on a substrate.

catalytic site the site on a catalyst that is concerned with catalysis, especially as part of the **active site** of an enzyme.

catalytic triad a group of three amino acid residues in an enzyme whose specific conformation and proximity in 3D space mediates the catalytic activity. For example, in serine proteases the catalytic triad Ser–His–Asp cleaves substrates at specific peptide bonds. Protein families with markedly different structures possess this triad, suggesting possible convergent evolution from independent origins.

catalyze *the US spelling of* **catalyse**.

cataplerosis the process by which an intermediate in the citric acid cycle is removed to prevent its accumulation in the mitochondrial matrix. –**cataplerotic** *adj*.

cataract opacity of the eye lens resulting from modification or degeneration of lens **crystallins**. It is the commonest form of blindness in many developing countries, and can result from biological aging, from other pathological processes (e.g. injury, radiation, diabetes mellitus), or as a consequence of primary ocular or dysmetabolic genetic disorders. It is described as congenital or infantile, juvenile, presenile, or senile, depending on age of onset.

CAT assay assay of **chloramphenicol *O*-acetyltransferase**, frequently employed to detect expression of its gene in cDNA technology.

catastrophe theory *see* **error catastrophe hypothesis**.
CAT box *see* **CAAT box**.
catechin the type compound for one of the 12 classes of flavanoids. It is characterized by oxidation state (OH at C-3) of the central pyran ring. Some 70 catechins are known; e.g. (+)-catechin is (2*R*,2*S*)-3,3′,4′,5,7-flavanpentol. Catechins are among the polyphenols widely distributed in vegetable foods, arising biosynthetically from the **shikimate** pathway. (+)-Catechin produces artificial cross-links in collagen *in vitro* and has anti-tumour activity. Catechin and **curcumin** have been found to be non-toxic, non-mutagenic, non-teratogenic and excellent anti-mutagens and anti-carcinogens.

(+)-catechin

catechol 1 *former name for* **pyrocatechol**. **2** any aromatic *o*-diol; any compound containing a pyrocatechol nucleus or substituent. *Compare* **hydroquinone** (def. 2), **quinone** (def. 2).

catecholamine any of a group of physiologically important **biogenic amines** that possess a **catechol** (def. 2) (i.e. 3,4-dihydroxyphenyl) nucleus and are derivatives of 3,4-dihydroxyphenylethylamine. They include **adrenaline** (epinephrine), **noradrenaline** (norepinephrine), and **dopamine**. Catecholamines have various physiological roles, particularly as neurotransmitters, and produce effects in the brain, the cardiovascular system, and other organs; they also help to regulate carbohydrate and fat metabolism.

catechol melanin any of various deep-brown or black plant pigments, so called because they yield catechol on alkali fusion. *See also* **melanin**.

catechol-*O*-methyltransferase EC 2.1.1.6; *systematic name*: *S*-adenosyl-L-methionine:catechol *O*-methyltransferase. An enzyme found primarily in non-neuronal tissues and catecholamine-releasing neurons, especially in the periphery; it effects *O*-methylation of catecholamines and their deaminated metabolites, but is less active against catechols. It catalyses the reaction between *S*-adenosyl-L-methionine and catechol to form *S*-adenosyl-L-homocysteine and guaiacol; in catecholamines it methylates the phenolic group at position 4. This inactivates catecholamine neurotransmitters and catechol hormones. In humans the membrane-bound protein consists of 271 amino acids; a cytosolic form (residues 52–271) lacks the transmembrane signal anchor.

catechol oxidase *see* **polyphenol oxidase**. *See also* **monophenol monooxygenase**.

catenane any type of compound containing two or more rings that are interlocked but not covalently joined; i.e. resembling the links of a chain. An example is an interlocked circular duplex DNA molecule found in mitochondria. *Compare* **concatemer**.

catenate 1 to interlock, without covalent bonding, two or more circular structures, e.g. cyclic molecules. **2** linked as the interlocked rings of a chain. *Compare* **concatenate**. —**catenated** *adj*.; **catenation** *n*.

catenin any of four intracellular attachment proteins of the armadillo family, designated α, β, γ, and δ. They bind actin microfilaments to the cytoplasmic domain of E-cadherin at adherens junctions of epithelial cells. **α-Catenin** (906 amino acids) is homologous with **vinculin**. **β-Catenin** (781 amino acids) is enriched in adherens junctions in cells lining the mammalian cerebral ventricles; it is homologous with **γ-catenin** (*or* **plakoglobin**). **δ-Catenin** is expressed in neurons, where it associates with presenilins. APC protein (*see* **APC**) plays an important role in regulating the stability of free β-catenin. A complex containing β-catenin, APC protein, **axin**, and **GSK-3** is targeted for ubiquitination and degradation by proteasomes. In the absence of functional APC protein (as occurs in many human cancers, including colorectal cancer) β-catenin is not degraded, enters the nucleus and enhances transcription by binding to members of the T cell-specific transcription factor/lymphoid enhancer-binding factor (TCF/LEF) family. Also, β-catenin acts as a transcriptional enhancer during Wnt protein signalling.

catestatin the 21-residue polypeptide that constitutes residues 352 to 372 of chromogranin A. It inhibits catecholamine release from the adrenal medulla.

cat-eye syndrome (*abbr.*: CES) or **Schmid–Fraccaro syndrome** a genetic disorder in which there is partial trisomy or tetrasomy of chromosome 22pter-22q11. It is named after a distinctive eye abnormality seen in some individuals in which there is partial absence of eye tissue (coloboma). Symptoms vary in type and severity but can include malformations of the skull and face, heart and kidneys. The anal canal may be absent and a fistula or abnormal passage formed between the rectum and the bladder or vagina of females, and the bladder or urethra of males.

CATH *abbr. for* class architecture topology homology; *see* **structure classification database**.

cathepsin any of a group of intracellular peptide hydrolases.

cathepsin A *an alternative name for* **protective protein/cathepsin A**.

cathepsin B EC 3.4.22.1; *other name*: lysosomal cysteine endopeptidase. An enzyme that catalyses the hydrolysis of proteins with broad specificity for peptide bonds; preferentially, it cleaves -Arg-Arg-|-Xaa- bonds in small substrates. It is a glycoprotein dimer, comprising a heavy chain and a light chain cross-linked by a disulfide bond.

cathepsin C EC 3.4.14.1; *recommended name*: dipeptidyl peptidase I. An enzyme that catalyses the hydrolysis of dipeptides from the N terminus of a polypeptide, except where the N-terminal residue is Arg or Lys, or where the adjacent amino acid is Pro. It also catalyses the transfer of dipeptide residues.

cathepsin D EC 3.4.23.5; a lysosomal aspartic endopeptidase that is active in intracellular protein breakdown. It has a specificity similar to, but narrower than, that of **pepsin A**. The molecule comprises a light chain and a heavy chain.

cathepsin E EC 3.4.23.34; an aspartic endopeptidase with similar activity to **cathepsin D**, but found in lymphoid-associated tissue. It may have a role in immune function.

cathepsin G EC 3.4.21.20; a glycoprotein serine endopeptidase with a specificity similar to that of **chymotrypsin**. It is found in the lysosomes of polymorphonuclear leukocytes and exhibits bactericidal activity against fungi and Gram-negative bacteria. It is therefore a **serpocidin**. Its gene is at 14q11.2.

cathepsin H EC 3.4.22.16; a lysosomal cysteine endopeptidase that also acts as an aminopeptidase on peptide substrates with a free N terminus. It is composed of a minichain and a large chain; the large chain may be split into heavy and light chains; all chains are held together by disulfide bonds.

cathepsin K a cysteine protease, with strong collagenase and gelatinase activity, that is selectively expressed in osteoclasts and plays a key role in bone growth and remodelling. The gene, at 1q21, encodes a protein of 329 amino acids, which is processed into a mature enzyme of 215 amino acids; this shows ≈60% sequence homology with cathepsins L and S. At least 10 mutations in the gene are associated with deficiency of cathepsin K and **pycnodysostosis**, a disorder of bone matrix remodelling.

cathepsin L EC 3.4.22.15; *other name*: major excreted protease (*abbr.*: MEP). A lysosomal cysteine endopeptidase that hydrolyses proteins but not acylamino acid esters, and has no peptidyl-dipeptidase activity – the activity is similar to that of **papain**. It is found in rat-liver and other mammalian lysosomes, and is inhibited by **leupeptin**. It consists of a heavy and a light chain linked by disulfide bonds. It shows ≈60% sequence homology with cathepsins K and S.

cathepsin S EC 3.4.22.27; a lysosomal cysteine endopeptidase with similar activity to **cathepsin L** but more activity for -Val-Val-Arg-|-Xaa. It shows ≈60% sequence homology with cathepsins K and L. Its gene is located near that for **cathepsin K** on 1q21, but unlike the latter is not associated with pycnodysostosis.

cathilicidin a pore-forming toxin permitting passage of small molecules across biomembranes. It is produced by certain mammals, such as horses, and contains 90–250 amino acid residues. Cathilicidin is classified in the TC system under number 1.C.33.

cathinone S(−)-2-amino-1-phenyl-1-propanone, the major central nervous system stimulant of the khat shrub, *Catha edulis*. Like amphetamine, it is a potent releaser of norepinephrine and dopamine from their intracellular stores.

cathode or (*formerly*) **kathode** 1 the electrode of an electrolysis cell or electrophoresis apparatus towards which cations of the electrolyte migrate under the influence of an electric field applied between the electrodes. 2 negative electrode: the electrode, pole, or terminal of a discharge tube, solid-state rectifier, thermionic valve, or other apparatus by which conventional electric current returns to an external source; i.e. the electrode through which electrons flow into such a device. 3 the positive pole or terminal of an electric (primary or secondary) cell or battery (of cells); i.e. the pole from which electrons emerge from such a device. *Compare* **anode**. — **cathodal, cathodic,** or **cathodical** *adj.*; **cathodally** or **cathodically** *adv.*

cathode-ray tube a vacuum tube in which a beam of electrons is emitted and focused by an electron gun onto a fluorescent screen. The beam can be deflected by horizontal and vertical electrostatic or magnetic fields to produce a visual display of electric signals, as in a television tube or cathode-ray oscilloscope.

catholyte in isoelectric focusing, the electrolyte that is in immediate contact with the cathode.

cation or (*formerly*) **kation** any **ion** with a net positive charge. In a solution being electrolysed, a cation migrates towards the **cathode**. *Compare* **anion**. —**cationic** *adj.*

cation exchange the process of **ion exchange** in which one cation is exchanged for another cation.

cation exchanger any **ion exchanger** that is able to exchange cations with the surrounding solution. *See also* **ion-exchange resin**.

cationic acid any acid that carries a positive charge, i.e. is a **cation**.

cationic detergent any **detergent** in which the polar part of the molecule carries one or more positive charges.

cationized ferritin a polycationic derivative of **ferritin** prepared by coupling horse-spleen ferritin with *N,N*-dimethyl-1,3-propanediamine, most of the carboxyl groups on the ferritin molecule thereby being converted into positively charged tertiary amino groups. It is useful for labelling negatively charged groups on cell surfaces, where the ferritin particles may easily be identified by electron microscopy because of their electron-dense iron cores.

cationmotive describing any enzyme, enzyme system, or other process that produces the vectorial movement of cations, especially across a biomembrane. The particular cation(s) may be specified, e.g. protonmotive, Na^+/K^+-motive.

catsper a protein of spermatozoa. Deficiency in the mouse leads to infertility and defective motility of spermatozoa.

CAU a codon in mRNA for L-histidine.

caudal a homeodomain transcription factor which forms a concentration gradient in the opposite direction to that of **bicoid protein**, possibly as a result of the translation control exerted on the caudal mRNA by bicoid.

cauliflower mosaic virus a double-stranded DNA virus that is a potential recombinant vector for plants.

caulimovirus any double-stranded DNA plant virus belonging to the *Caulimovirus* group, the type member of which is the **cauliflower mosaic virus**.

caustic soda *see* **soda**.

caveola (*pl.* **caveolae**) a flask-shaped invagination of the plasma membrane, rich in cholesterol, glycosphingolipids, caveolins, and glycosylphosphatidylinositol (GPI)-anchored proteins and proteins involved in transmembrane signalling (e.g. receptors, G-proteins, and protein kinases). Caveolae are clathrin-independent and do not undergo lysosomal degradation. They are sites for the concentration and uptake of small molecules that are ligands for GPI-anchored proteins. Folates, chemokines, and many pathogens or their products (e.g. HIV, cholera toxin) enter cells through caveolae.

Cholesterol efflux also is mediated by caveolae. They are prominent in fibroblasts and endothelial cells, but are found in many other cell types. Their formation is decreased in transformed cells. *See* **potocytosis**. [From Latin *cavea*, a small cavity.]

caveolin a protein that lines the cytoplasmic surface of a **caveola**.

C-beta *an alternative term for* **beta carbon atom**.

CBF *abbr. for* centromere binding factor (*see* **centromere binding proteins**).

CBG *abbr. for* cortisol binding globulin. *See* **transcortin**.

CBL a protooncogene related to v-*cbl*, the nuclear oncogene of Cas NS-1 retrovirus, named from Casitas B-lineage lymphoma. The function of the normal protein is unknown, but it may be a transcriptional regulator. v-*cbl* encodes a truncated form of Cbl, which has lost a C-terminal leucine zipper and a 208-residue proline-rich region.

Cbm *symbol for* carbamoyl.

CBN *abbr. for* (IUPAC-IUB) Commission on Biochemical Nomenclature; now succeeded by **JCBN**.

CBP *abbr. for* CREB protein-binding protein; a large nuclear multifunctional protein expressed in most vertebrate cell types that binds to a wide variety of transcription factors and has intrinsic histone acetyltransferase activity. The gene for CBP at 16p13.3 encodes a multidomain protein of 2442 amino acids, which contains domains for nuclear hormone receptor and **CREB protein** binding, four zinc fingers, a bromodomain, a glutamine-rich domain, and a histone acetyltransferase domain. It is thus a general transcriptional adaptor and integrator. Microdeletions and point mutations in the gene are responsible for Rubinstein–Taybi syndrome, which is characterized by abnormalities of the face, thumbs, and big toes, and physical and mental retardation. CBP is usually, but erroneously, lumped with p300 (adenovirus E1A-associated protein), from which it is almost indistinguishable functionally. However, mutations in the gene for p300 at 22p13 are not associated with the Rubinstein–Taybi syndrome.

Cbz *symbol for* the benzyloxycarbonyl group.

cc *symbol for* cubic centimetre (cm^3 is preferred).

CC *abbr. for* cholecalciferol (**vitamin D$_3$** is preferred).

CCA a codon in mRNA for L-proline.

CCA-adding enzyme a CTP/ATP:tRNA nucleotidyl transferase that builds or repairs the trinucleotide sequence CCA at the 3′ end of all tRNAs. The CCA is the site of attachment of amino acid residues for polypeptide synthesis. In the eubacterium *Aquifex aeolius* it consists of two subunits, one specific for CTP and one for ATP.

C$_{55}$ carrier lipid or **C$_{55}$ lipid carrier** *an alternative name for* **undecaprenol**.

CCC **1** a codon in mRNA for L-proline. **2** *abbr. for* β-chloroethyltrimethylammonium chloride; an inhibitor of gibberellin synthesis in plants.

cccDNA *see* **DNA**.

CCCP *abbr. for* carbonyl cyanide *m*-chlorophenylhydrazone, a powerful uncoupler of oxidative phosphorylation.

CCD *abbr. for* charge coupled device; an array of metal oxide semiconductor (MOS) capacitors that captures an optical image and converts it to an electronic digital form.

C cell the distinctive type of apud cell that secretes the polypeptide hormone **calcitonin**. In mammals C cells are found predominantly in the thyroid gland (when they are also known as **parafollicular cells**), and in some mammals and birds they also occur in the parathyroid gland and the thymus. In other vertebrates the cell type is present mainly in the discrete **ultimobranchial body** and the lung. [The name C cell was chosen to signify calcitonin-secreting cell.]

CCG a codon in mRNA for L-proline.

CCK *abbr. for* cholecystokinin.

CCK cell *see* **I cell**.

CCK-PZ *abbr. for* cholecystokinin–pancreozymin. *See* **cholecystokinin**.

cCMP *abbr. for* cyclic CMP.

C$_1$ compound or **one-carbon compound** or **single-carbon compound** any (real or hypothetical) organic compound that consists of a single carbon atom with attached hydrogen atom(s), and sometimes also an oxygen atom or imino group, and that functions as a unit in metabolism. Examples include the methyl group of *S*-adenosylmethionine or the iminomethyl-formimidoyl- (or formimino-), formyl-, hydroxymethyl-, methyl-, methylene-, and methenyl- groups of tetrahydrofolate coenzymes.

C3/C5 convertase *see* **complement**.

C5 convertase *see* **complement**.

CCP *abbr. for* Collaborative Computational Programs; an initiative, started in the UK, for the development and distribution of software for scientific applications. Programs include: CCP4 (crystallographic software), CCP5 (molecular dynamics), and CCP11 (sequence analysis).

CCU a codon in mRNA for L-proline.

C$_3$ cycle *an alternative name for* **reductive pentose phosphate cycle**.

C$_4$ cycle *an alternative name for* the **Hatch–Slack pathway**.

cd *symbol for* candela.

Cd *symbol for* cadmium.

CD *abbr. for* **1** circular dichroism. **2** cluster of differentiation *or* cluster determinant; *see* **CD marker**. **3** compact disc.

CD2 a T-cell-specific plasma membrane protein expressed early during T-cell development. It binds CD58, a PIGylated surface protein expressed on many cells, and thus facilitates adhesion of T cells to these other cells.

CD3 a **CD marker**, expressed exclusively in T lymphocytes, that consists of the γ (25 kDa), δ (20 kDa), and ε (20 kDa) polypeptides of the T-cell receptor complex. CD3 proteins are members of the immunoglobulin superfamily; each contains one transmembrane segment and an immunoreceptor tyrosine-based activation motif (ITAM). They chaperone the newly synthesized **T-cell receptor** to the cell membrane and help transduce the signal of antigen binding.

CD4 a **CD marker** that occurs on T-helper cells and is involved in MHC class II restricted interactions. It is a type I membrane glycoprotein of the immunoglobulin superfamily, acts as a receptor for HIV, and binds interleukin 16 (IL-16). Its intracellular domain binds the protein-tyrosine kinase Lck.

CD8 a **CD marker** that occurs on human cytotoxic T-cells and suppressor T-cells. It is associated with MHC class I restricted interactions. It is a heterodimer of an α and β chain and isoforms of the β chain result from alternative splicing. The α chain binds to the MHC α$_3$ domain. Both chains belong to the immunoglobulin protein superfamily and their intracellular domains (like those of CD4) bind Lck.

CD18 *see* **integrin**.

CD21 *see* **complement receptor**.

CD25 *see* **myelin protein**.

CD32 *or* **Fc receptor for IgG** *or* **FcγRII** a low-affinity receptor for the Fc region of antigen-bound IgG, present on B lymphocytes and other leukocytes. Its intracellular domain contains an immunoreceptor tyrosine-based inhibitory motif (ITIM), which binds a phosphoprotein phosphatase. *See* **Fc receptor**.

CD34 a **CD marker** displayed on the surface of stem cells.

CD35 *see* **complement receptor**.

CD36 *or* **platelet GPIV** *or* **CD36 activation antigen** *or* **Naka** a transmembrane multifunctional glycoprotein (88 kDa) present in platelets, endothelial cells, and many other cell types. It mediates adherence to collagen, thrombospondin, *Plasmodium falciparum*-infested erythrocytes, and oxidized low-density lipoproteins (LDL). About 3% of the apparently healthy Japanese population lacks this antigen (i.e. are Naka-negative) and have no problems with hemostasis.

CD38 *or* **T10** *or* **ADP-ribosyl cyclase/cyclic ADP-ribose hydrolase** a cell surface leukocyte antigen that catalyses multiple reactions, including NAD glycohydrolase, ADP-ribosyl cyclase, cyclic ADP ribose hydrolase, and base-exchange activities. Two of the reaction products, cyclic ADP-ribose and nicotinic acid adenine dinucleotide phosphate, are calcium messengers in a variety of cells. CD38 is a positive and negative regulator of cell activation and proliferation, depending on the cellular environment. It is involved in adhesion between human lymphocytes and endothelial cells.

CD40 a **CD marker** present on the plasma membrane of B lymphocytes and other cells including epithelial and endothelial cells. It is the receptor for **CD40L**, which is expressed on activated T lymphocytes. Binding of CD40L to CD40 induces growth, differentiation, activation, and immunoglobulin class switching of the CD40 cells.

CD44 *other names*: phagocyte glycoprotein I; lymph node homing receptor; HERMES antigen; extracellular matrix receptor III. A transmembrane glycoprotein (85–90 kDa) that is constitutively expressed on the surface of B and T cells, monocytes, neutrophils, epithelial cells, fibroblasts, glial cells, and myocytes. It plays a general role in cell adhesion, including lymphocyte homing; it binds to collagen and hyaluronate. Several splicing variants seem to be expressed preferentially in gastric tumour cells.

CD45 a **CD marker** present on T lymphocytes. It is the receptor for the membrane protein CD22, present in B lymphocytes. Binding of the ligand activates the protein tyrosine kinases Fyn and Lck.

CD55 *see* **complement decay-accelerating factor.**

CD58 *see* **CD2.**

CD59 a GPI-anchored glycoprotein that is expressed in erythrocytes, leukocytes, and endothelial cells, and is a potent inhibitor of complement membrane attack complex (MAC); it acts after C5b8 assembly (*see* **complement**). It is involved in signal transduction for T-cell activation, being complexed to a protein tyrosine kinase. It is a member of a protein family that includes the protooncogenes *Sis* and *Myc*. Mutation in the gene at 11p13 produces a CD59 deficiency that is manifest as paroxysmal nocturnal hemoglobinuria.

CD62 *see* **selectin.**

CD66 *see* **carcinoembryonic antigen.**

CD74 *an alternative name for* **invariant chain.**

CD79 *see* **B cell antigen receptor complex associated protein.**

CD95 *an alternative term for* **Fas.**

CD105 *see* **endoglin.**

CD2AP *abbr. for* CD2-associated protein; an SH3 domain protein present mainly in epithelial and lymphoid cells that is implicated in endocytosis and lysosomal sorting. Mutations in humans result in focal segmental glomerulosclerosis; mutations in mice lead to early death from massive proteinuria.

CD11b/18 *see* **complement receptor.**

CD11c/18 *see* **complement receptor.**

Cdc-activating kinase *abbr.*: CAK; a heterodimer of cyclin H and Cdk7. It is a protein serine/threonine kinase and a component of human transcription factor TFIIH. Among its substrates are Cdk subunits and the C-terminal domain of RNA polymerase II.

cdc **gene** *abbr. for* cell-division-cycle gene. Many genes encode proteins essential for cell division. These were initially identified in yeast, and named in a series *cdc1*, *cdc2*, etc. Homologues of these genes have since been found in many other species, but the *cdc* nomenclature is used specifically for the fission yeast, *Schizosaccharomyces pombe*. The convention Cdc followed by a numeral is used for the protein products of these genes. Many of these products are not directly involved in the control of cell division, being proteins such as DNA polymerases. Some of those involved in cell-division control are detailed below.

Cdc2, a major cell regulatory protein, with counterparts in all eukaryotic cells. *cdc2* is homologous to *CDC28* in budding yeast. Cdc2 is a Ser/Thr protein kinase, alternatively denoted p34^{cdc2}, and regulated by cyclin B. p34^{cdc2} and cyclin B comprise **maturation promoting factor** (MPF). Cdc2 is the patriarch of a family of cyclin-dependent kinases; Cdc2 is Cdk1. Cdc2 is important at the G_1/S and G_2/M phases of the cycle in yeast. The human homologue of Cdc2 phosphorylates retinoblastoma protein.

cdc25 encodes a protein phosphatase that positively regulates p34^{cdc2} by dephosphorylation of Tyr^{15}. Entry to mitosis is negatively regulated by *wee1*, a gene in *S. pombe* that encodes a protein tyrosine kinase. Its action on mitosis results from phosphorylation of Tyr^{15} of p34^{cdc2}. Note: *CDC25* from budding yeast encodes a guanine nucleotide exchange protein.

Cdc28, a Ser/Thr protein kinase required for completion of both the START transition of the cycle and also the G_2/M transition.

Cdc31, an **EF-hand** protein involved in microtubular organization.

Cdc46, a protein mobilized from the cytoplasm to the nucleus after mitosis, and a potential ATP-binding protein.

Cdc48, a putative ATP-binding protein involved in spindle formation.

Cdc42 a human homologue of a yeast protein, and a member of the Rho family of p21 GTPases. It is activated by several proteins including FGD1, which is a guanine nucleotide exchange protein. Cdc42 interacts with various protein kinases including WASP, **MRCK**, and PAK.

Cdc48 *or* **VCP/Cdc48** a hexameric AAA protein complex of yeast that is involved in endoplasmic reticulum (ER)-associated protein degradation and copurifies with 26S proteasomes. It interacts with proteasomes and with polyubiquitin chains of proteins targeted for degradation. The human homologue is VCP (valosin-containing protein).

Cdc53p *see* **cullin.**

CD66d *see* **carcinoembryonic antigen.**

CDD *abbr. for* conserved domain database; *see* **derived database.**

CD66e *see* **carcinoembryonic antigen.**

CDE *abbr. for* centromere DNA element; CDE-I is a group of DNA sequences, RTCACRTG, found in centromeres and some promoters. *See also* **centromere-binding proteins.**

CD59 glycoprotein precursor *see* **membrane attack complex inhibition factor.**

CDH *abbr. for* ceramide dihexoside (Cer(Hex)$_2$ is preferred).

c-di-GMP *abbr. for* cyclic diguanylic acid.

CDI *abbr. for* N,N'-carbonyldiimidazole; 1,1'-carbonylbis-1H-imidazole, a reagent used in the synthesis of nucleotide triphosphates and peptides, and for cross-linking proteins.

Cdk *abbr. for* cyclin-dependent protein kinase.

Cdk inhibitor *abbr.*: CKI; a protein that binds directly to, and inhibits, **Cdk** activity. In yeast some CKIs are inducible, whereas others are intrinsic to the mitotic cycle. In mammalian cells, one type (comprising p21, p27, and p57) inhibit most Cdks, whereas another type (called INK4 proteins) are specific for Cdk4 and Cdk6. Inactivating mutations in some of these CKIs are found in some human neoplasms.

CD40L *or* **CD154** *or* **gp39** *or* **TNFSF5** a type II transmembrane protein of the tumour necrosis factor (TNF) superfamily that is expressed in activated T lymphocytes. It is the ligand for **CD40** present on B lymphocytes and certain other cells. The gene for CD40L on Xq26 encodes a protein of 261 amino acids. Over a hundred mutations, of various types, cause X-linked hyper-IgM syndrome.

CD95 ligand *an alternative term for* **FasL.**

CD marker *(in immunology)* any of a series of antigenically distinct molecules occurring on the surface of leukocytes and other cell types, and used in the characterization of such cells. More than 80 individual CD marker molecules are recognized, as identified by **monoclonal antibodies**. The most frequently encountered CD molecules are those used to distinguish T cells. CD2 is present on all T cells and is involved in antigen-nonspecific activation. CD3 is also present on all T cells and is an invariant part of the T-cell antigen receptor involved in antigen-specific T-cell activation. **CD4** occurs on T-helper cells and is involved in MHC class II restricted interactions, **CD8** occurs on cytotoxic T cells and is involved in MHC class I restricted interactions. *See also* **CD44**, **CD59**. Other CD markers listed here under other names include CD16 (*see* **Fc receptor**), CD29 (*see* **VLA**), CD31 (*see* **PECAM**), CD32 (*see* **Fc receptor**), CD49a and CD51 (*see* **VLA**), CD50 and CD54 (*see* **ICAM**), CD62 (*see* **selectin**), CD64 (*see* **Fc receptor**), CD102 (*see* **ICAM**), and CD106 (*see* **VCAM**). [From 'cluster of differentiation'.]

cDNA *abbr. for* complementary DNA.

C-DNA *abbr. for* C form of DNA.

cDNA clone any bacterial cell transformed by a plasmid containing a **complementary DNA** copy of an RNA molecule.

cDNA library a collection of cloned recombinant cDNA, usually in bacteria or yeast. The clones represent cDNAs prepared from all the mRNAs in any particular species or organ.

CDP *abbr. for* **1** cytidine 5'-diphosphate *or* the group cytidine(5')-diphospho. **2** cardiodilatin related peptide; *see* **natriuretic peptide.**

CDPcholine *see* **cytidine(5')diphosphocholine.**

CDPdiacylglycerol *see* **cytidine(5')diphosphodiacylglycerol.**

CDPdiacylglycerol–inositol 3-phosphatidyltransferase EC 2.7.8.11; *other names*: phosphatidylinositol synthase; CDPdiacylglyceride–inositol phosphatidyltransferase; an enzyme that catalyses the formation of phosphatidyl-1-D-*myo*-inositol from CDPdiacylglycerol and *myo*-inositol with release of CMP. It is a component of the pathway for the biosynthesis of phosphatidylinositol and also for its resynthesis after activation of the **phosphatidylinositol cycle**.

CDPdiacylglycerol–serine *O*-phosphatidyltransferase EC 2.7.8.8; *systematic name*: CDPdiacylglycerol:L-serine 3-*O*-phosphatidyl-transferase; *other names*: phosphatidylserine synthase; CDPdiglyceride–serine *O*-phosphatidyltransferase. It is an enzyme of the pathway that effects *de novo* synthesis of phosphatidylserine. It catalyses the formation of *O*-sn-phosphatidyl-L-serine from CDPdiacylglycerol and L-serine with the release of CMP.

CDPethanolamine see **cytidine(5′)diphosphoethanolamine**.

CDP-glycerol pyrophosphorylase see **glycerol-3-phosphate cytidylyltransferase**.

CDPK *abbr. for* calmodulin-like domain protein kinase.

CDP-*Star* a proprietary chemilumigenic 1,2-dioxetane substrate for alkaline phosphatase giving a brighter signal than **AMPPD**.

CDR *abbr. for* calcium-dependent regulator protein (i.e. **calmodulin**).

CDS *abbr. for* coding sequence; *see* **coding region**.

CDTA *abbr. for trans*-cyclohexylene-1,2-diamine-*N,N,N′,N′*-tetraacetate, the anion of *trans*-1,2-diaminocyclohexane-*N,N,N′,N′*-tetraacetic acid. A **chelating agent** with a high affinity for Mg^{2+}; the abbreviation is also used for its partially or fully protonated forms.

Ce *symbol for* cerium.

C_xE_y a series of non-ionic detergents useful for the solubilization of membrane-bound proteins in their native state. They include $C_{10}E_6$, hexaethyleneglycol mono-*n*-decyl ether; $C_{10}E_8$, octaethyleneglycol mono-*n*-decyl ether; $C_{12}E_6$, hexaethyleneglycol mono-*n*-dodecyl ether; $C_{12}E_8$ octaethyleneglycol mono-*n*-dodecyl ether; and $C_{12}E_9$, nonaethyleneglycol mono-*n*-dodecyl ether; the latter two are constituents of the proprietary detergent Lubrol-PX, a detergent useful for the solubilization of membrane-bound proteins in their native state.

CEA *abbr. for* carcinoembryonic antigen.

CEA gene member 1 *see* **carcinoembryonic antigen**.

cecropin any of a group of small, basic polypeptides with potent antibacterial activity. Cecropins are active against *Escherichia coli* and several other Gram-negative species, and are isolated from the cecropia moth, *Hyalophora cecropia*, and several other insects.

CEF10 or **CEF-10** a secreted protein that is implicated in growth regulation. It is induced by v-*src*, and belongs to a family of growth regulators that includes **connective tissue growth factor**.

Cek a chick gene encoding receptor-type protein tyrosine kinases, and named from chicken embryo kinase. The protein product Cek-1 is the chicken **fibroblast growth factor** receptor; Cek-2 and Cek-3 are related to **Bek**; Cek-4 and Cek-5 are homologous to **Eph**.

Celebrex a drug that inhibits the enzyme **COX-2** (see **prostaglandin-endoperoxide synthase**) and hence is used in the treatment of arthritis. The active ingredient is celecoxib, 4-[-5-(4-methylpentyl)-3-trifluoromethyl-1*H*-pyrazol-1-yl]benzenesulfonamide.

celiac or (*esp. Brit.*) **coeliac** of, or belonging to, the abdominal cavity.

celiac disease or (*Brit.*) **coeliac disease** or **gluten enteropathy** or **nontropical sprue** a condition caused in genetically susceptible individuals by damage to the mucosa of the upper small intestine triggered by the α-gliadin component of **gluten** in wheat and rye. α-Gliadin is rich in glutamine residues, and the condition is marked by the presence of circulating antibodies to tissue transglutaminase. This enzyme, normally in inactive form in submucosal cells of the upper small intestine, is released by mechanical or inflammatory stress. The enzyme deamidates specific glutamine residues in peptides derived from α-gliadin thus producing peptides that activate the T cells of the submucosa. The disease is characterized by diarrhoea and malabsorption, but can be relieved by elimination of gluten from the diet.

Celite *proprietary name for* preparations of amorphous diatomaceous silica, used as a filter-aid and in partition chromatography.

cell 1 the basic structural unit of all living organisms; it typically comprises a small, usually microscopic, discrete mass of organelle-containing cytoplasm, bounded externally by a membrane. Each cell is capable of interacting with other cells and performing all the fundamental functions of life. In plants the cell usually also includes the cell wall. Originally the term referred to the bounding wall of a plant cell that had lost its living content. *See also* **+cyte**, **cyto+**. **2** a container for liquid during measurement or separation procedures, such as dialysis, electric conductivity measurement, electrolysis, microdiffusion, moving boundary electrophoresis, analytical ultracentrifugation, ultrafiltration, and (spectro)photometry; *see also* **cuvette**. **3** or **electric cell** a device for producing electric current by converting chemical energy into electric energy; it consists of two electrodes in contact with an electrolyte. **4** (*in crystallography*) *see* **unit cell**. —**cellular** *adj*.

cell-adhesion molecule *see* **adhesion molecules**.

cell body *an alternative term for* **perikaryon**.

cell-bound antibody or **cell-fixed antibody** any antibody molecule that is bound to the surface of a cell, e.g. an **incomplete antibody** (def. 1) or a **cytophilic antibody**.

cell coat *another name for* the **glycocalyx**.

cell cortex the part of the microtrabecular lattice lying just under the outer cell membrane of certain protozoans, e.g. paramecia.

cell counter any device for determining the number of cells in a known volume of a liquid suspension. *See* **Coulter counter**, **flow cytometer**, **hemocytometer**.

cell culture *see* **tissue culture**.

cell cycle the period between the formation of a cell by division of its mother cell and the time when the cell itself divides to form two daughter cells. *See also* **cell-division cycle**, **DNA cycle**.

cell-cycle engine all the biochemical components, and the reactions between them, that cause the periodic activation and inactivation of **M-phase-promoting factor** during cell maturation.

cell differentiation *see* **differentiation**.

cell-division cycle the pattern of DNA synthesis in dividing eukaryotic cells. It is divided into four periods: G_1 is the first 'gap' in DNA synthesis between **telophase** and the period of DNA synthesis, designated S. This S phase is followed by G_2, the second gap in DNA synthesis. This runs on into the period of **mitosis**, M, which lasts from early **prophase** to late **telophase**.

cell division protein kinase *abbr.*: CDK; any of various protein kinase enzymes involved in the regulation of **cyclin** activity, such as CDC20. *See also* **cdc**.

cell-fixed antibody *an alternative name for* **cell-bound antibody**.

cell fractionation the fractionation of disrupted cells into their subcellular components.

cell-free extract a fluid, made by breaking open cells, that contains most of the constituents of the cells in question but no intact cells.

cell-free system any experimental system composed of subcellular fractions and/or cell-free extracts. Cell-free systems for amino-acid incorporation (protein synthesis) generally contain ribosomes, natural or synthetic mRNA, tRNAs, enzymes, amino acids, an ATP generating system, GTP, buffer, certain inorganic salts, and some organic compounds.

cell fusion the formation of a single hybrid cell containing the nuclei and cytoplasms from different cells. It may be induced by treatment of a mixed cell population with certain **fusogens** (e.g. killed Sendai virus or polyethylene glycol). It is an important step in the technique of forming **hybridoma cells** for the production of **monoclonal antibodies**.

cell junction a specialized region of connection between two cells or between a cell and the extracellular matrix. *See also* **desmosome**, **gap junction**, **hemidesmosome**, **tight junction**.

cell line a population of cells cultured *in vitro* that are descended through one or more generations (and possibly subcultures) from a single primary culture. The cells of such a population share common characteristics.

cell locomotion *see* **cell migration**.

cell matrix *see* **extracellular matrix**.

cell-mediated immunity specific immunity that depends on the presence of T lymphocytes. It is responsible for, e.g., allograft rejection, delayed hypersensitivity, and tuberculin test reactions, and is important in the organism's defence against viral and some bacterial infections.

cell membrane any **membrane** found in a cell. The term is sometimes used loosely to indicate the **plasma membrane**. *See also* **cytoplasmic membrane**.

cell migration the active movement of a cell from one location to another, particularly the migration of a cell over a surface.

β-cellobiase *another name for* β-glucosidase; *see* **glucosidase**.

cellobiose 4-*O*-β-D-glucopyranosyl-D-glucose; a disaccharide that represents the basic repeating unit of **cellulose**; it is liberated by **cellulase**.

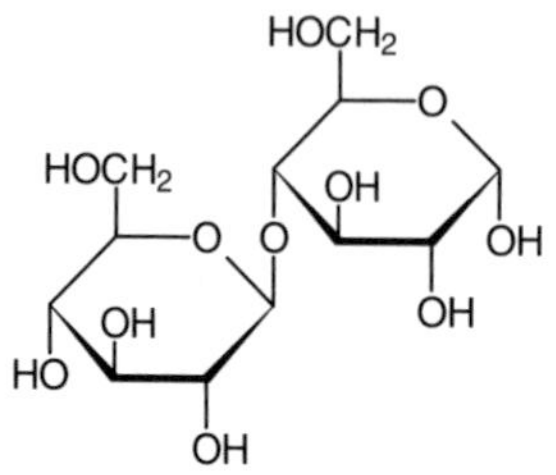

α-anomer

cellobiuronic acid 4-*O*-β-D-glucopyranuronosyl-D-glucose; *condensed form*: GlcA(β1-4)Glc. It is a component of, e.g., type III pneumococcal bacterial capsule polysaccharides.

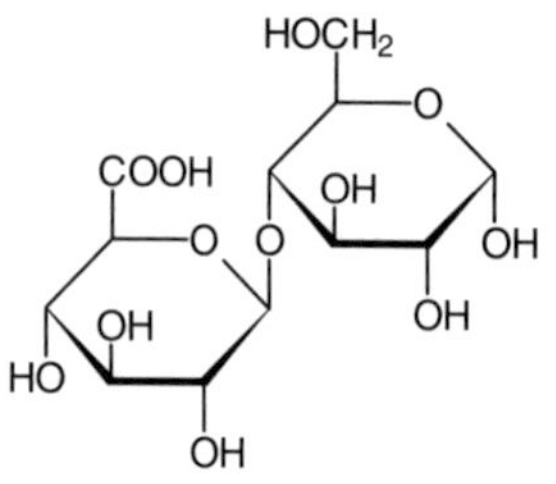

α-anomer

cellodextrin a glucan chain linked to **sitosterol**. It is formed by the transglucosylase components of cellulose synthase in plant cell plasma membranes, by transfer of the glucosyl moiety of UPD-glucose onto the sitosterol β-glucoside primer. The glucan is released by Korrigan cellulase to form the microfibrils of cellulose and to regenerate the sitosterol β-glucoside primer.

cellophane a form of transparent, flexible sheeting made from regenerated **cellulose**; originally a proprietary name.

Cellosolve *proprietary name for* a range of ethanediol monoalkyl ethers, especially 2-ethoxyethanol. They are useful as solvents, e.g. in liquid scintillation mixtures, to facilitate the dispersion of aqueous samples in organic solvents.

cellotetraose a tetrasaccharide composed of D-glucose units in β1-4 linkage.

cellotriose the trisaccharide *O*-β-D-glucopyranosyl-(1-4)-*O*-β-D-glucopyranosyl-(1-4)-D-glucose; *condensed form*: Glc(β1-4)Glc(β1-4)Glc.

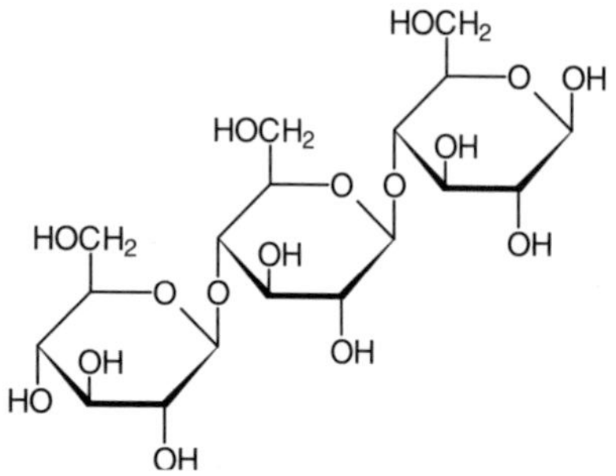

β-D-pyranose form

cell plate the nascent cell membrane and cell wall structure that forms between the two daughter nuclei near the centre of a dividing plant cell. It develops at the equatorial region of the **phragmoplast**. It grows outwards to join with the lateral walls and form two daughter cells.

cell sorter any **flow sorter** for sorting cells in liquid suspension, on the basis of some characteristic that is detectable by the sorter.

cell-surface protein *abbr.*: CSP; *an alternative name for* **fibronectin**.

cellubrevin a **synaptobrevin** homologue that is involved in membrane traffic as part of a constitutively recycling pathway. With VAMP proteins, it is a component of the machinery controlling docking and fusion of secretory vesicles with their target membrane. It is a ubiquitous substrate for **tetanus toxin** light chain.

cellugyrin *see* **gyrin**.

cellular affinity the tendency of cells to adhere specifically to cells of the same type, but not to cells of a different type. This property is lost in cancer cells.

cellular immunity 1 the increased ability of phagocytic cells to destroy parasitic organisms, i.e. macrophage immunity. **2** *an alternative name for* **cell-mediated immunity**.

cellulase 1 EC 3.2.1.4; *other names*: endoglucanase; endo-1,4-β-glucanase; carboxymethyl cellulase. An enzyme that catalyses the endohydrolysis of 1,4-β-D-glucosidic linkages in **cellulose**, thereby hydrolysing cellulose to cellobiose. It causes abscission in plants, and its synthesis is stimulated by ethylene. Microbial cellulases play an important part in the digestion of cellulose by ruminants. **2** any lytic enzyme whose substrate is cellulose. *See* **cellulose 1,4-β-cellobiosidase, glucosidase, Korrigan cellulase**.

cellule 1 (*archaic*) a minute cavity or very small **cell**. **2** *a former name for* a **liposome** bounded by a single lipid bilayer.

cellulose a linear β1-4 **glucan** of molecular mass 50–400 kDa with the pyranose units in the $-^4C_1$ conformation. It is predominantly a plant polysaccharide though it also occurs in some tunicates. Cellulose microfibrils are major components of plant cell walls. Cellulose is insoluble in water but will dissolve in ammoniacal solutions of cupric salts. It is often said to be the most abundant organic compound in the biosphere, but this distinction has also been accorded to the **hopanoids**.

cellulose 1,4-β-cellobiosidase EC 3.2.1.91; *other names*: exoglucanase; exocellobiohydrolase; 1,4-β-cellobiohydrolase. An enzyme that catalyses the hydrolysis of 1,4-β-D-glucosidic linkages in cellulose and cellotetraose, releasing cellobiose from the nonreducing ends of the chains.

cellulose synthase an enzyme complex present in plant cell plasma membranes that synthesizes cellodextrin by transfer of glucose from UDP-glucose to a sitosterol β-glucoside primer. The primer is first synthesized by a component of the complex by transfer of glucose from UDP-glucose to **sitosterol**. The elongating transglucosylation reactions are catalysed by several subunits that surround the primer synthase. Cleavage of the cellodextrin from the sitosterol β-glucoside primer is achieved by **Korrigan cellulase**, itself a component of the complex.

cellulosic 1 of, or pertaining to, cellulose; of the nature of cellulose; composed (primarily) of cellulose; derived from or made by chemical modification of cellulose. **2** a substance made from cellulose or a derivative of cellulose.

cellulolysis the process of hydrolysing cellulose. —**cellulolytic** *adj.*

cell wall the rigid or semi-rigid envelope lying outside the **cell membrane** of plant, fungal, and most prokaryotic cells maintaining their shape and protecting them from osmotic lysis. In prokaryotes it lies inside any capsule and slime layer, usually consists mainly of **peptidoglycan**, and can be removed by various techniques with retention of its three-dimensional form. In fungi the cell wall is composed largely of polysaccharides, while in plants it is made up of cellulose and, often, lignin.

Celsius temperature *symbol*: θ *or* t; a temperature defined as the excess of the thermodynamic temperature, T, over 273.15 K; i.e. $\theta = T - 273.15$, where θ is measured in °C (i.e. **degrees Celsius**, formerly called degrees centigrade) and T is measured in **kelvins**. It is identical with centigrade temperature. [After Anders Celsius (1701–44), Swedish astronomer.]

cenancestor the common ancestor of all living things.

CENP-A *abbr. for* centromere protein-A; a member of the histone H3 family of proteins. In mammalian cells it replaces histone H3 in the histone octamers of nucleosomes of **centromeres**.

centi+ *symbol:* c; SI prefix denoting 10^{-2}.

centigrade temperature *former name for* **Celsius temperature**.

centimetre–gram–second system *see* **cgs system**.

centimorgan *or* **centiMorgan** *abbr.*: cM; a unit of **map distance** used in genetic mapping, equal to 1% crossover found in recombination studies; 1 cM equals 1 map unit. Hence, if two genes are separated by 1 cM, there is a 1% probability of recombination in a single meiotic event.

central dogma a fundamental principle of molecular biology, first articulated by British molecular biologist Francis **Crick** (1916–) in 1958. He wrote that the central dogma states that once 'information' has passed into protein *it cannot get out again*. In more detail, the transfer of information from nucleic acid to nucleic acid, or from nucleic acid to protein may be possible, but transfer from protein to protein, or from protein to nucleic acid is impossible. Information means here the *precise* determination of sequence, either of bases in the nucleic acid or of amino-acid residues in the protein. The dogma has been taken to relate primarily to the following transfers of information that can occur in all cells: DNA → DNA, DNA → RNA, and RNA → protein. However, the transfer of information from RNA to RNA occurs in the replication of RNA viruses. The transfer of information from RNA to DNA occurs following infection of a cell with a retrovirus; the DNA is then incorporated into the host genome. When challenged that the reverse transfer of information from RNA to DNA by retroviruses contradicted the dogma, Crick refuted the challenge by drawing attention to the fact that as originally formulated the dogma permitted the transfer of information from (any) nucleic acid to (any other) nucleic acid (although the action of retroviruses was not known at that time) and that the essential point was that *sequence* information could not be passed from protein to nucleic acid, or from protein to protein.

central nervous system *abbr.*: CNS; the part of the nervous system of an animal that contains a high concentration of cell bodies and synapses and is the main site of integration of nervous activity. In higher animals it is isolated from the circulation by the **blood–brain barrier**. It consists essentially of a brain or cerebral ganglia, and a nerve cord, which may be dorsal or ventral, single or double. *Compare* **peripheral nervous system**.

centrifugal 1 moving, acting, or tending to act away from a centre or axis. **2** of, pertaining to, or operated by a **centrifugal force**. **3** *(in botany)* developing outwards from a centre, as the flowers of certain types of inflorescence. **4** *(in physiology)* sending nerve impulses from a central location to parts supplied by that nerve; **efferent**. *Compare* **centripetal**.

centrifugal analyser *or* **centrifugal fast analyser** an apparatus widely used in clinical chemical laboratories for the simultaneous and very rapid chemical, biochemical, or immunochemical determination of one particular constituent in each of a large number of samples. Basically it consists of an appropriately designed transfer disk into which reagent(s) and samples are discretely and automatically pipetted. When loaded, the transfer disk is placed in the centre of a cuvette rotor in the centrifuge, and the rotor is accelerated. Reagent(s) and each sample are mixed together as they are radially propelled along specially shaped channels into the cuvettes. Absorbance, fluorescence, or other measurements are then made automatically on the cuvettes as they are rotating. The whole process, including data handling, is controlled by a computer.

centrifugal elutriation a technique in which cells or other particles are separated by **elutriation** in a specially constructed centrifuge rotor; the increased gravitational field resulting from centrifugation speeds up separation of the particles.

centrifugal field any space in which a **centrifugal force** is operating, as in the spinning rotor of a centrifuge.

centrifugal force a force acting radially outwards on any body moving in a curved path, equal and opposite to the **centripetal force**. The centrifugal force on a body of mass m, moving in a circular path of radius r, with velocity v, is: mv^2/r.

centrifuge 1 an apparatus in which fluids may be rotated rapidly so that substances (solutes or dispersed particles) of different densities may be separated by centrifugal force. **2** to rotate rapidly in a centrifuge. —**centrifugation** *n*.

centrin *an alternative name for* **caltractin**.

centriole a cellular organelle, found close to the nucleus in many eukaryotic cells, consisting of a small cylinder with microtubular walls, 300–500 nm long and 120–150 nm in diameter. It contains nine short, parallel, peripheral microtubular fibrils, each fibril consisting of one complete **microtubule** fused to two incomplete microtubules. Cells usually have two centrioles, lying at right angles to each other. At division, each pair of centrioles generates another pair and the twin pairs form the poles of the mitotic spindle. *Compare* **basal body**.

centripetal 1 moving, acting, or tending to act towards a centre or axis. **2** of, pertaining to, or operated by a **centripetal force**. **3** *(in botany)* developing inwards from the outside, as the flowers of certain inflorescences. **4** *(in physiology)* sending nerve impulses towards a central location from peripheral parts supplied by that nerve; **afferent**. *Compare* **centrifugal**.

centripetal force a force acting radially inwards on any body moving in a curved path; it constrains the body to a curved path. It is equal and opposite to the **centrifugal force**.

centromere the region of a eukaryotic chromosome that is attached to the spindle during nuclear division. It is defined genetically as the region of the chromosome that always segregates at the first division of meiosis; i.e. the region of the chromosome in which no **crossing over** occurs. At the start of M phase, each chromosome consists of two sister chromatids with a constriction at a point that forms the centromere. During late prophase two **kinetochores** assemble on each centromere, one kinetochore on each sister chromatid. The DNA of centromeres is several megabase long and consists of tandem arrays of repetitive sequences (sometimes called alphoid DNA) with high affinity for specific nuclear proteins (e.g. **CENP-A**). In the human genome such repetitive sequences amount to ≈10% of total DNA.

centromere-binding proteins proteins from, e.g., the yeast **kinetochore** that are able to bind centromere DNA; examples from yeast: centromere-binding protein 1, a helix-turn-helix protein that binds to a conserved DNA sequence termed CDE-I, and is involved in chromosome segregation; centromere-binding protein 5, which is involved in mitotic chromosome segregation, and is essential for cell growth.

centrosome a structure occurring close to the nucleus during **interphase** in many eukaryotic cells. It comprises a pair of **centrioles**, satellite bodies, and a cytoplasmic zone from which the mitotic microtubule spindle is organized. The structure of the centrosome in animal cells changes continually during the **cell-division cycle**. The protein PCM-1 recruits centrosomal proteins such as centrin, pericentrin, ninein, and dynactin. On phosphorylation by Cdk2, the centrosomal nucleophosmin is released to initiate centrosomal duplication. In some cells, lack of p53 leads to multiple copies of the centrosome being generated during a single cell cycle.

centrosphere a more or less well-delineated part of the cytoplasm at the poles of the spindle (*see* **mitotic spindle**).

cephaeline an alkaloid derived from *Uragoga ipecacuanha*. It is a gastric irritant, inducing emesis. See also **emetine**, **ipecachuanha**.

cephalin *a former name for* **1 phosphatidylethanolamine**. **2 phosphatidylserine**.

cephalosporin any of a heterogeneous group of natural and semisynthetic, often β-lactam, antibiotics. Cephalosporins C, N, and P are obtained from cultures of moulds such as *Cephalosporium acremonium*. Cephalosporins are active against a range of Gram-positive and Gram-negative bacteria.

cephalosporinase *an alternative name for* β-**lactamase** II.

Cer *symbol for* a ceramide residue. *See also* **Cer(Hex)**.

ceramidase EC 3.5.1.23; *other name*: acylsphingosine deacylase; an enzyme involved in the degradation of sphingolipids. It catalyses the hydrolysis of *N*-acylsphingosine to a fatty acid and sphingosine.

ceramide *symbol*: Cer; any *N*-acylated **sphingoid**. Ceramide is released from sphingomyelin by sphingomyelinase (sphingomyelin phosphodiesterase). Activation of neutral sphingomyelinase (e.g. by tumour necrosis factor, FasL, cannabinoids, interleukin 1) releases ceramide, which activates several serine/threonine protein kinases, a serine/threonine phosphoprotein phosphatase, and other targets that influence apoptosis and the cell cycle. Membranes of the Golgi apparatus may also participate in ceramide signalling.

ceramide cholinephosphotransferase EC 2.7.8.3; an enzyme involved in the biosynthesis of sphingomyelin by the reaction of CDPcholine with *N*-acylsphingosine with the release of CMP.

ceramide glucosyltransferase EC 2.4.1.80; an enzyme that catalyses the formation of glucosylceramides. The reaction involves the formation of D-glucosyl-*N*-acylsphingosine from UDPglucose and *N*-acylsphingosine with the release of UDP.

cercidosome an atypical mitochondrion or oxidizing body, found, e.g., in trypanosomes, whose function may primarily be to regulate the NAD$^+$/NADH ratio of the cytosol. [From Greek *kerkis, kerkido*, weaver's shuttle, + **+some**.]

cerebellin *see* **precerebellin**.

cerebral 1 of, or relating to, the cerebrum. **2** *(loosely)* of, or relating to the brain.

cerebrodiene a lipid, resembling **sphingosine**, that is obtained from the brain of sleep-deprived cats.

cerebroglycan an integral membrane heparan sulfate proteoglycan unique to the developing nervous system, and expressed specifically during neuronal differentiation.

cerebronate 2-hydroxytetracosanoate; $CH_3(CH_2)_{21}CH(OH)COO^-$; the 2-hydroxy derivative of lignocerate. It is enriched in brain cerebrosides, the fraction containing this fatty acid being known as **phrenosin**.

cerebrose *see* **galactose**.

cerebroside any **glycosphingolipid** that contains a monosaccharide, normally glucose or galactose, in 1-*O*-β-glycosidic linkage with the primary alcohol of an *N*-acyl **sphingoid** (ceramide). In plants the monosaccharide is normally glucose; in animals it is normally galactose, though this may vary with the tissue, glucose being present in blood cerebroside. In plants, the sphingoid is usually phytosphingosine, in animals sphingosine or dihydrosphingosine (sphinganine). Cerebrosides frequently contain large amounts of longer-chain fatty acids such as behenic (22:0), lignoceric (24:0), and nervonic (24:1) acids, and 2-hydroxy fatty acids such as cerebronic (α-OH24:0) and 2-hydroxynervonic acids. Cerebroside synthesis may proceed either by acylation of psychosine or transfer of glucose or galactose to ceramide.

cerebroside-sulfatase EC 3.1.6.8; *other name*: arylsulfatase A; an enzyme involved in the catabolism of sulfatides. It catalyses the hydrolysis of a cerebroside 3-sulfate to a cerebroside and sulfate. It is a lysosomal enzyme of most tissues, active at acid pH, whose catalytic activity requires the noncatalytic activator protein saposin B. A formylglycinyl residue (formed by modification of a cysteinyl residue) is present in the active site. Over 60 mutations of the gene locus at 22q13.31-qter are associated with metachromatic leukodystrophy.

cerebrospinal fluid *abbr.*:CSF; a clear fluid, containing little protein and few cells, that fills the subarachnoid space and ventricles of the brain, and the central canal of the spinal cord. About 80% of the protein is derived from plasma, the rest is brain-specific and includes myelin basic protein, glial fibrillary acid protein, creatine kinase (BB isozyme), and neuronal enolase.

cerebrotendinous xanthomatosis a rare familial sterol storage disease caused by mutations that lead to deficiency of sterol 27-monooxygenase. There is accumulation of cholestanol and cholesterol in most tissues (but particularly in cerebellum, bile, and xanthomas), and of cholestanol, but not usually of cholesterol, in plasma. Clinical features include neurological dysfunction (cerebellar ataxia, dementia), xanthomas, early atherosclerosis, and cataracts.

Cerenkov counter *a variant spelling of* **Cherenkov counter**.

Cerenkov radiation *a variant spelling of* **Cherenkov radiation**.

Cer(Hex) *symbol for* ceramide monohexoside. The di- and trihexosides are symbolized by Cer(Hex)$_2$ and Cer(Hex)$_3$ respectively. *See also* **Cer**.

cerinic acid *see* **cerotoyl**.

ceroid an autofluorescent, insoluble lipopigment that accumulates in the livers of rats with nutritional cirrhosis. It is present in athero-

sclerotic plaques and builds up in lysosomes of neurons and other tissue cells in the ceroid lipofuscinoses.

ceroid lipofuscinosis (neuronal) *abbr.*: CLN; any of several progressive myoclonic epilepsies characterized by accumulation of **ceroid** in the lysosomes of neurons and other cell types. The diseases are distinguished largely by their age of onset and the mutated (loss-of-function) gene involved. **CLN1** is an infantile form caused by deficiency of palmitoylprotein thioesterase. **CLN2** is a late infantile form caused by deficiency of lysosomal tripeptidyl peptidase I. **CLN3** is a juvenile form caused by loss of the lysosomal transmembrane protein **battenin**. In **CLN5** and **CLN8** the function of the gene involved is unknown.

cerotoyl *symbol*: Crt; *the trivial name for* hexacosanoyl, $CH_3–[CH_2]_{24}CO–$, the univalent acyl group derived from hexacosanoic acid (*also called* **certoic acid** *or (formerly)* **cerinic acid**), a saturated, unbranched, acyclic, aliphatic acid, which occurs naturally in wool fat.

certoic acid *see* **cerotoyl**.

cerulein *or (Brit.)* **caerulein** a decapeptide, pGlu-Gln-Asp-Tyr(SO$_3$H)-Thr-Gly-Trp-Met-Asp-PheNH$_2$, isolated from the skin of various frogs, e.g. *Hyla caerulea*. It is often accompanied by the nonapeptide **phyllocerulein** (*or* **phyllocaerulein**). Both have the same C-terminal pentapeptide amide sequence as porcine **cholecystokinin**. Cerulein-like peptides have been isolated from the antrum and small intestine of amphibians, in which they stimulate strongly the secretion of gastric acid.

cerulenin (2*R*,3*S*)-epoxy-4-oxo-7*E*,10*E*-dodecadienamide; an antifungal antibiotic isolated from *Cephalosporium caerulens* and other sources. It inhibits fatty-acid biosynthesis by binding with 3-oxo-acyl-[acyl carrier protein] synthase (*see* **fatty acid synthase complex**) and also interferes with sterol biosynthesis by inhibiting **hydroxymethylglutaryl-CoA synthase** activity.

cerulin *a variant spelling of* **cerulein**.

ceruloplasmin *or (Brit.)* **caeruloplasmin** a bright-blue, copper-containing protein found in blood plasma. It has ferroxidase activity, EC 1.16.3.1, and catalyses the reaction:

$$4 \text{ iron(II)} + 4\,H^+ + O_2 = 4 \text{ iron(III)} + 2\,H_2O$$

It binds six or seven Cu^{2+} ions per molecule. It is important in iron metabolism, and is abnormally low in hepatolenticular degeneration (Wilson's disease). It is an acute phase protein, its level rising late in the acute phase.

ceryl *the common name for* hexacosyl, $CH_3–[CH_2]_{24}–CH_2–$, a saturated, unbranched, acyclic, alkyl group.

cesium chloride *or (esp. Brit.)* **caesium chloride** a salt that forms dense solutions in water and so is used in **isopycnic centrifugation** to separate DNA molecules of different densities.

Cet *symbol for* the carboxyethyl group, $–CH_2–CH_2–COOH$.

cetaceum *an alternative term for* **spermaceti**.

Cetavlon *a proprietary name for* **cetrimide**.

cetoleic acid *see* **docosenoic acid**.

CETP *abbr. for* cholesteryl ester transfer protein; a hydrophobic glycoprotein of plasma, secreted mainly from liver and associated largely with high-density lipoproteins (HDL). It effects the exchange of cholesteryl esters from HDL with triglyceride from chylomicrons and very-low density lipoproteins (VLDL), and is responsible for about half of the plasma phospholipid exchange. The gene locus at 16q21 encodes a protein (476 residues) with sequence homology to **BPI protein**, PLTP (phospholipid transfer protein), and LPS-binding protein. Deficiency of CETP is associated mostly with a splice junction or missense mutation, and is common in Japan. Heterozygotes have a modest increase, and homozygotes a massive increase, in HDL cholesterol content. Several missense mutations are also associated with increased HDL levels.

cetrimide *or* **cetrimidum** a detergent disinfectant consisting of a mixture of alkylammonium bromides, principally **cetyltrimethylammonium bromide**. *See* **quaternary ammonium compounds**.

cetyl *or* **palmityl** *the common name for* hexadecyl, $CH_3–[CH_2]_{14}–CH_2–$, a saturated, unbranched, acyclic, alkyl group (hexadecyl is preferred).

cetyltrimethylammonium bromide *abbr.*: CTAB; hexadecyltrimethylammonium bromide; *N,N,N*-trimethyl-1-hexadecanaminium bromide; a cationic detergent useful in the solubilization of membrane proteins; aggregation number 170, CMC 1 mM. It is the principal component of **cetrimide**.

Cf *symbol for* californium.

CFEM domain *abbr. for* common in fungal extracellular membrane proteins domain; a protein sequence (≈60 residues) that consists predominantly of hydrophobic residues and contains eight cysteine residues. It is present mainly in fungal and in some plant transmembrane or PIG-anchored proteins. The causal agent of rice blast disease contains at least eight proteins with this domain.

C form of DNA *abbr.*: C-DNA *or* DNA-C; the molecular conformation adopted by fibres of the lithium salt of duplex DNA at low relative humidity (below approximately 44%). It consists of a right-handed double helix containing about 9.3 nucleotide residues per turn; otherwise it is similar to the **B form**. *See also* **A form** (def. 1), **Z form**.

***Cfr*10I** a type 2 **restriction endonuclease**; recognition sequence: [AG]↑CCGG[TC].

***Cfr*13I** a type 2 **restriction endonuclease**; recognition sequence: G↑GNCC. *Asu*I is an **isoschizomer**.

CFTR *abbr. for* cystic fibrosis transmembrane conductance regulator.

CG *abbr. for* chorionic gonadotropin (*see* **choriogonadotropin**).

CgA1-40 the N-terminal 40 residues of chromogranin A, and of vasostatin. It increases secretion of calcitonin and calcitonin gene-related peptide (CGRP) from lung tumour cells but inhibits secretion of parathyroid hormone. *See* **chromogranin**.

cg clone any crown gall **clone** (def. 1); *see* **crown gall disease**.

CGM1 *see* **carcinoembryonic antigen**.

cGMP *abbr. for* cyclic GMP; *see* **guanosine 3′,5′-phosphate**.

CGRP *abbr. for* calcitonin gene-related peptide.

cgs system *or* **CGS system** *abbr. for* centimetre–gram–second system; a metric system of units based on the centimetre (length), gram (mass), and second (time). It has now been superseded by the **SI** system.

Chaikoff homogenizer a hydraulically operated tissue or cell homogenizer in which the sample is forced through an annulus by a piston. The piston diameter relative to that of the annulus is chosen to suit the size of the component that it is desired to isolate with minimal damage.

chain reaction a chemical reaction in which one or more reactive intermediates are continuously regenerated, often through a repetitive cycle of elementary 'propagation' steps.

chain-termination codon *an alternative name for* **termination codon**.

chain-termination method *or* **dideoxy method** *or* **Sanger method** a rapid technique for determining nucleotide sequences in DNA in which the 2′,3′-dideoxy analogues of the normal deoxynucleoside triphosphates are used as specific chain-terminating inhibitors of DNA polymerase. In the original method single-stranded DNA fragments are used as templates for the **Klenow enzyme** of DNA polymerase I in the presence of a primer and a mixture of a dideoxynucleoside triphosphate and the four normal deoxynucleoside triphosphates, one of which is labelled with ³²P. Separate incubations are carried out using analogous terminators for the other nucleotides. The effect is to yield radioactively labelled stretches of DNA corresponding in length to the number of bases between the 5′ end and the position of the base corresponding to the terminator, which will be incorporated once into each stretch of DNA in such a way that it yields an appropriate stretch of DNA for every position occupied by the base in question. The resulting products from each incubation are separated in parallel by gel electrophoresis and detected by autoradiography. From the pattern of radioactive bands obtained the sequence of the fragment can be read off. The method is reasonably accurate for sequences of 15 to about 200 nucleotides. Three modifications are made for automated DNA sequencing: (i) a different fluorescent label is covalently attached to each of the dideoxy chain terminators, or alternatively to the primer; (ii) **cycle sequencing** is used to generate dideoxy-terminated fragments; and (iii) a mixture of the four termination reactions is separated rapidly by capillary electrophoresis. Many hundreds of nucleotides can be read in a single run. [After Frederick Sanger (1918–), British biochemist.]

chair configuration (of a cyclobutadipyrimidine) *see* **pyrimidine dimer**.

chair conformation any **conformation** of a nonplanar, six-membered, saturated ring compound in which the ring atoms in relative positions 1, 2, 4, and 5 lie in one plane, and those in relative positions 3 and 6 lie on opposite sides of that plane. For a monosaccharide or monosaccharide derivative the conformational descriptor -*C* may be added to its name, and the reference plane is so chosen that the lowest-numbered carbon atom in the ring is exoplanar. *See also* ¹**C**₄, ⁴**C**₁. *Compare* **boat conformation**.

chalcone 1 phenyl styryl ketone;1,3-diphenyl-2-propen-1-one. **2** any of a group of hydroxylated derivatives of chalcone (def. 1); they are important intermediates in plants in the biosynthesis of flavanones, flavones, flavonols, and anthocyanidins. The chalcone derivative **isoliquivitigenin** has anti-platelet activity, inhibiting 12-lipoxygenase. It stimulates soluble guanylate cyclase. It has anti-tumour activity and potently prevents tumour promotion by phorbol esters.

challenge 1 *(in immunology)* the act or process of injecting antigenic material into an immunized animal to test for immunity or to provoke a further immune response. **2** *(in endocrinology)* the act or process of administering an agent to evaluate hormonal responses. It is used to test hormonal control and function.

chalone any inhibitor affecting the cell cycle before the onset of mitosis that is tissue-specific and present in the tissue of action; it need not be species-specific. [From Greek *khalon*, to make slack.]

channel 1 a passage, often tubular or trough-shaped, along or through which movement of substances, especially fluids, can occur. **2** the span between selected settings of two discriminators in a pulse counter, e.g. a scintillation counter, that defines the range of pulse intensities that will be recorded. **3** to guide into or cause to move through a particular channel or along a particular route; e.g. to channel a metabolite along a particular metabolic pathway. **4** descriptive of a membrane protein or a protein complex that transports solutes, mainly of ionic nature, across a cell membrane down their concentration or electrochemical potential gradient (in the TC system families 1.A.1–1.A.36 and 1.B.1–1.B.34). Channels are either closed or open to both membrane sides simultaneously and can be gated (regulated) (i) electrically (e.g. nerve Na⁺ and K⁺ channels), (ii) chemically (e.g. acetylocholine receptor-channel), (iii) mechanically (K⁺ channels in inner ear hair cells), or (iv) nongated (bacterial porins). *See also* **calcium channel**, **potassium channel**, **sodium channel**. —**channelled** *or* (US) **channeled** *adj*.

channel carrier *see* **ionophore**.

channel-forming integral membrane protein 28 *abbr.*: CHIP 28; *other name*: aquaporin 1; an integral membrane protein that forms a water-specific channel, conferring high permeability to water on the plasma membranes of red cells and kidney proximal tubules. It is a homotetramer.

channel gating the change that drives conformational changes in the structure of a membrane channel between closed and open states of the pore.

channelling *or* (US) **channeling** *(in column chromatography)* the uneven flow resulting from inhomogeneous packing of the column.

channel protein an **integral (cell membrane) protein** that permits the controlled (gated) movement of ions through a membrane. *See* **calcium channel**, **potassium channel**, **sodium channel**.

channel rhodopsin a light-gated protein channel in green algae. It consists of a seven-pass transmembrane protein (712 amino acids) whose hydrophobic core is homologous to that of **bacteriorhodopsin**.

channels-ratio method a method of quench correcting in liquid scintillation counting in which two channels are used to measure the average energies of beta particles both before and after quenching.

chaoptin an extracellular membrane glycoprotein required for *Drosophila melanogaster* photoreceptor cell morphogenesis, and containing a **leucine-rich repeat** domain. It is a cell-type specific adhesion molecule, exclusively localized to photoreceptor cells, linked to the outer surface of the membrane by a **GPI-anchor**.

Chaotropase *the proprietary name for* a mixture of endopeptidases isolated from *Streptomyces griseus*, so named for their stability and activity in the presence of the **chaotrope** 6 M guanidinium chloride.

chaotrope *or* **chaotropic agent** any substance that increases the transfer of apolar groups to water because of its ability to decrease the 'ordered' structure of water and to increase its lipophilicity. Chaotropes are usually ions, e.g. SCN^-, CNS^-, ClO_4^-, I^-, Br^-, CH_3-COO^-. They cause the dissolution of biological membranes, the solubilization of particulate proteins, changes in the secondary, tertiary, and quaternary structure of proteins, and denaturation of nucleic acids.

chaotropic dissociation assay a technique used to measure the heterogeneity of antibody affinities in, e.g., serum. Immune complexes are dissociated in buffers of varying strengths and pH or in chaotropic agents (*see* **chaotrope**). Low-affinity antibodies dissociate more readily.

chaperone any of a functional class of unrelated families of proteins that assist the correct non-covalent assembly of other polypeptide-containing structures *in vivo*, but are not components of these assembled structures when they are performing their normal biological functions. *See also* **chaperonin**, **heatshock protein**.

chaperonin any of a ubiquitous subclass of molecular **chaperones** implicated in the folding of other proteins. They include two kinds of protein: (1) chaperonin 60 kDa (*also called* chaperonin-60 *or* cpn 60; the counterpart in other structures of the product of the **groEL** gene of *Escherichia coli*); and (2) chaperonin 10 kDa (*also called* chaperonin-10 *or* cpn 10; the counterpart in, e.g., *Mycobacterium* of the product of the **groES** gene of *E. coli*). *See also* **heat-shock protein**.

CHAPS 3-[(3-cholamidopropyl)dimethylammonio]-1-propanesulfonate; a zwitterionic detergent combining the properties of sulfobetaine and bile-salt types of detergent. It is used for membrane solubilization. Aggregation number 4–14, CMC 6–10 mM.

$R = -(CH_2)_3-\overset{\overset{\displaystyle H_3C}{|}\ \overset{\displaystyle CH_3}{}}{\underset{+}{N}}-(CH_2)_3-SO_3^-$

CHAPSO 3-[(3-cholamidopropyl)dimethylammonio]-2-hydroxy-1-propanesulfonate; a zwitterionic detergent combining the properties of sulfobetaine and **bile-salt** types of detergent; it is more soluble than **CHAPS**. Aggregation number 11, CMC 8 mM.

$R = -(CH_2)_3-\overset{\overset{\displaystyle H_3C}{|}\ \overset{\displaystyle CH_3}{}}{\underset{+}{N}}-CH_2-CHOH-CH_2-SO_3^-$

characteristic group (*in chemistry*) any atom or group that is incorporated into a parent compound otherwise than by a direct carbon–carbon linkage, but including the groups –C=N and =C=X where X is O, S, Se, Te, NH, or substituted NH.

characteristic time a measure of the time required for a system to change from one state of motion to another, usually taken as the time required for the system to approach within one half of the new state of motion.

characteristic X-ray *see* X-rays.

Charcot–Marie–Tooth disease *abbr.*: CMT; a clinically and genetically heterogeneous polyneuropathy syndrome, affecting both motor and sensory peripheral nerves. It occurs in dominant, recessive, and X-linked forms and is one of the most common inherited neurological disorders, affecting 1:2500 people in the USA. Symptoms include weakness of the foot and lower leg muscles. The type I phenotype is due to demyelination of motor nerves and may be caused by overexpression (from a gene duplication) or underexpression (from a gene deletion) of peripheral myelin protein (PMP22), or by mutation in the genes for myelin protein zero, connexin 32, or EGR2. These are all proteins synthesized in myelinating Schwann cells. The type II phenotype is not well characterized at a molecular level. *See* **myelin protein**. [After Jean Charcot (1825–1893) and Pierre Marie (1853–1940), French neurologists; and Howard Tooth (1856–1925), British physician.]

Chargaff's rules a set of quantitative rules describing the base composition of duplex DNA: (1) [A] = [T] and (2) [G] = [C], where the square brackets denote the concentrations of the bases in moles per cent; minor bases, if present, are included with the appropriate major base. Three corollaries follow: (1) [A]/[T] = [G]/[C]; (2) total purines = total pyrimidines, i.e. [A] + [G] = [C] + [T]; and (3) total 6-aminobases = total 6-ketobases, [A] + [C] = [G] + [T]. A further consequence of these relationships is that the [A + T]/[G + C] ratio is a characteristic property of individual DNAs. [After Erwin Chargaff (1905–2002), US biochemist.]

charge 1 to load with the proper or appropriate quantity, e.g. to charge a tRNA molecule with an aminoacyl group, or to charge (an accumulator, capacitor, etc.) with electricity. **2** *see* **electric charge**. **3** the quantity of electricity held in, e.g., an accumulator or a capacitor. **4** a load. *Compare* **discharge**.

charge density *symbol* ρ; SI unit C m^{-3} (coulombs m^{-3}).

charged tRNA any **aminoacyl-tRNA** molecule.

charge-relay system a catalytic triad found in the serine proteases, comprising, in chymotrypsin, residues His^{57}, Ser^{195} and Asp^{102}, the histidine being hydrogen bonded to the carboxylate of the buried aspartate. Spatially, these residues occur in positions such that the electronegativity of the aspartate carboxyl group can be relayed through the histidine nitrogen atoms to the serine hydroxyl group, enhancing its reactivity.

chargerin II *other names*: A61; protein 8 of H$^+$-transporting ATP synthase (EC 3.6.1.34); one of the chains of the non-enzymic component (CF$_0$ subunit) of the mitochondrial ATPase complex (membrane bound). *See also* **H$^+$-transporting ATP synthase**.

charge-shift electrophoresis a method of **electrophoresis** in a detergent solution in which the net charges of the complexes between amphiphilic proteins and the detergent depend on the charge of the latter, while the net charges on hydrophilic proteins are unaffected. This effect is exploited in, e.g., SDS–polyacrylamide gel electrophoresis.

charge-transfer bond an absorption band in the UV-visible region observed in a charge-transfer complex. Such absorption bands are common in complexes of transition metals, and other redox-active molecules such as flavins.

charge-transfer chromatography a method of **liquid-gel chromatography** in which separation is due to electronic coupling between molecules being chromatographed and strong electron donor or acceptor groups attached to the supporting gel.

charge-transfer complex a complex that may be formed in an oxidation–reduction reaction, corresponding to electronic transition(s) to an excited state in which there is a partial transfer of electronic charge from the electron donor to the electron acceptor.

charging the covalent attachment of an aminoacyl group to a tRNA molecule to form an **aminoacyl-tRNA** molecule.

Charles's law *see* gas laws. [After Jacques Alexandre Cesar Charles (1746–1823), French physicist.]

charybdotoxin *abbr.*: ChTX; a 37-residue peptide isolated from the venom of the scorpion *Leiurus quinquestriatus hebraeus*. It is a potent inhibitor of some Ca^{2+}-activated K^+ channels and voltage-dependent K^+ channels, and probably acts at the outer face of the membrane, blocking the ion conductance channel. It has links between Cys residues 7–28, 13–33, and 17–35; amino-acid sequence:

XFTNVSCTTSKGCWSVCQRLHNTSRGKCMNKKCRCYS

(X = pyroglutamyl).

chase 1 the effective termination of incorporation of either a (radio)nuclide or a labelled compound into a substance, brought about by the addition of a large excess of an isotope of the (radio)nuclide or the unlabelled compound to the system, used especially after a pulse. **2** the quantity of isotope or unlabelled compound added to stop the incorporation of a (radio)nuclide or a labelled compound.

che the symbol for any of a class of genes that are implicated in chemotaxis in *Escherichia coli*. For example, *cheA* encodes a chemotaxis protein, protein phosphotransferase (EC 2.7.1.–), that is autophosphorylated and transfers its phosphate to CheB and CheC. Other two-component bacterial regulatory proteins are related. *cheB* encodes a chemotaxis protein, **protein-glutamate methylesterase** (EC 3.1.1.61), that catalyses the hydrolysis of protein L-glutamate O-methyl ester to protein L-glutamate and methanol. The protein is a **methyl-accepting chemotaxis protein** (MCP), the methylation state of which is crucial for sensory responses and adaptations. *cheR* encodes the chemotaxis protein methyltransferase (EC 2.1.1.80); it catalyses the reaction:

S-adenosyl-L-methionine + protein L-glutamate =
S-adenosyl-L-homocysteine + protein L-glutamate methyl ester.

The protein is an MCP, the methylation state of which is crucial for sensory responses and adaptations. *cheW* encodes a purine-binding chemotaxis protein that is involved in the transmission of sensory signals from chemoreceptors to the flagellar motors. *cheY* encodes a chemotaxis protein involved in the transmission of sensory signals from chemoreceptors to flagellar motors and the regulation of clockwise rotation. The structure is largely alpha helical, with some beta strands. *cheZ* encodes a chemotaxis protein thought to be involved in generating a regulating signal for bacterial flagellar rotation.

checkpoint a point in the eukaryotic **cell division cycle** where the cycle can be halted until conditions are suitable for the cell to proceed to the next stage – e.g. to allow for repair of damaged DNA.

checkpoint kinase a cell-cycle checkpoint protein kinase that is activated in response to DNA damage. Its action prevents the cell from entering mitosis.

CHEF *abbr. for* contour-clamped homogeneous electric field; one of several designs of apparatus for performing **pulsed field gel electrophoresis**.

chel *abbr. for* chelate effect.

chelate 1 any chemical species in which there is **chelation**. **2** of, or denoting, a chelate (def. 1). **3** to form a chelate compound; to effect **chelation**. A chelate compound is a chemical compound in which a metal ion is combined with a substance that contains two or more electron donor groups so that the resulting structure contains one or more rings. —**chelated** *adj.*

chelate effect *abbr.*: chel; the stability difference between a chelate complex and the corresponding complex with simple ligands. For the displacement reaction:

$$MA_n + mL \rightarrow ML_m + nA,$$

$$chel = \log \beta_m (ML_m) - \log \beta_n (MA_n),$$

where ML_m and MA_n are complexes containing an equal number of the same donor atoms. Ligand L is multidentate, containing several donor atoms of the same element as the unidentate ligand A, and β_m and β_n are the overall stability constants.

chelating agent *or* **chelator** a substance that is able to form a **chelate compound** with a metal ion.

chelation the formation or presence of bonds from two or more atoms within the same ligand to a single central (metal) atom.

chelation therapy the therapeutic use of chelating (metal-binding) agents for the removal of toxic amounts of metal ions from living organisms. The metal ions are sequestered by the chelating agents and are either rendered harmless or excreted. Chelating agents such as dimercaprol, EDTA, desferrioxamine, and penicillamine have been used effectively in chelation therapy for arsenic, lead, iron, and copper respectively.

Chelex 100 *the proprietary name for* a synthetic ion-exchange resin consisting of a styrene divinylbenzene copolymer containing paired iminodiacetate groups that act as chelating groups in binding metal ions, especially transition-metal or other divalent ions.

chemical adsorption *an alternative term for* **chemisorption**.

chemical antagonism antagonism resulting from the chemical combination of an **antagonist** with the substance being antagonized.

chemical carcinogen any carcinogenic substance of low relative molecular mass. *See* **carcinogen**.

chemical cleavage method 1 *or* **Maxam–Gilbert method** a rapid technique for determining nucleotide sequences in DNA (or in restriction fragments) that depends on labelling either end of a DNA strand with ^{32}P and the subsequent preferential, partial chemical cleavage of the labelled DNA specifically at positions occupied by one of the four possible bases. This results in a nested set of radioactive fragments extending from the ^{32}P label to each of the positions occupied by the particular base. The process is then repeated with cleavage, in turn, at each of the other three bases. The four sets of radioactive fragments are then separated, in adjacent lanes, by polyacrylamide gel electrophoresis, which arranges oligonucleotides in ascending order of the number of nucleotide residues contained. The base sequence of the original piece of DNA can be read directly from an autoradiograph of the resultant gel. The method can be used for sequences of 150 or more nucleotide residues. *Compare* **chain-termination method**. [After Allan M. Maxam and Walter Gilbert (1932–).] **2** a method for detecting the position of a mutation in DNA. In a heteroduplex formed between the normal and mutant DNA the deoxyribose phosphate backbone of DNA can be cleaved at the mismatched nucleotides by exposure to chemicals such as hydroxylamine and osmium tetroxide. The position of the mutation in polymerase chain reaction (PCR) products can be ascertained from the lengths of the resulting fragments.

chemical coupling hypothesis the hypothesis that the coupling of ATP synthesis to oxidation in oxidative phosphorylation is due to the formation of common 'high-energy' intermediates during electron transport that are subsequently used to phosphorylate ADP to ATP. It has been superseded by the **chemiosmotic coupling hypothesis**.

chemical equivalent the combining proportion of a substance, by mass, relative to a hydrogen standard. The chemical equivalent of an element is the number of grams of that element that will combine with or replace 1 g of hydrogen.

chemical exchange the changing magnetic environment of an atomic nucleus in a molecule that is rapidly undergoing reversible chemical or physical changes, which in turn produces changes in the **chemical shift** of that nucleus.

chemically induced dynamic nuclear polarization *abbr.*: CIDNP; a technique for the direct detection and identification of low concentrations of free radicals by **electron spin resonance**. It depends on strong polarization of certain nuclear spins by the unpaired electron during the molecule's existence as a free radical.

chemical messenger a term sometimes used loosely to mean variously a hormone, a neurocrine or paracrine transmitter, or a rheoseme. *See also* **second messenger**.

chemical oxygen demand *abbr.*: COD; a measure of the oxygen equivalent of that portion of the organic matter in a sample of water that is susceptible to oxidation by a strong chemical oxidant. Essentially the sample is boiled under a reflux condenser with potassium dichromate in strong sulfuric acid (with silver sulfate as catalyst); part of the dichromate is reduced by the contained organic matter and the remaining dichromate is titrated with ferrous sulfate. The result is expressed as milligrams oxygen absorbed from standard dichromate per litre of sample. *Compare* **biochemical oxygen demand**.

chemical pathology the branch of pathology dealing with the biochemistry of disease processes and the measurement of the amounts of substances present in body fluids and other samples as an aid to diagnosis.

chemical potential *symbol:* μ; the partial molar free energy of a chemical entity; the entity in question may be indicated by a subscript. For example, μ_B symbolizes the chemical potential of entity B in a mixture of entities B, C, ... etc. It is defined by:

$$\mu_B = (\partial G/\partial n_B)_{T,P,n_C},$$

where G is the Gibbs energy of the mixture, T is the thermodynamic temperature, P is the pressure, and n_B, n_C, ... etc. are the amounts of the entities B, C, ... etc. in the mixture.

chemical quenching *or* **impurity quenching** the **quenching** that occurs in liquid scintillation counting due to diffusion-controlled collisional interaction between the quencher molecules and the excited solvent and solute molecules. *Compare* **colour quenching**.

chemical reaction any single process or operation involving the interconversion of chemical species through changes in orbital electrons but not in the atomic nuclei.

chemical score a measure of the nutritional value of a protein. The limiting essential amino acid in the test protein is expressed as the percentage of the amount of the same amino acid present in egg albumin, the nutritionally perfect protein. Chemical score is numerically equal to **biological value**.

chemical shift the displacement of a resonance signal in NMR spectroscopy along the magnetic field coordinate resulting primarily from the magnetic effects of the chemical environment of the nuclei producing that signal. *See* **nuclear magnetic resonance**.

chemical shift anisotropy a technique based on nuclear magnetic resonance in which the chemical shift changes with the orientation of the sample in the magnetic field, used in molecular dynamics on proteins and in some other fields.

chemical species any set of chemically identical **molecular entities**, the members of which have the same composition and can explore the same set of molecular energy levels within the time scale of a particular experiment. Two conformational isomers may interconvert sufficiently slowly to be considered separate chemical species on an NMR time scale, but in a slow chemical reaction they may interconvert sufficiently quickly to be considered as a single chemical species.

chemical transmitter *(sometimes)* an alternative term for **neurotransmitter**.

chemiluminescence the production of visible light (luminescence) occurring as a result of a chemical reaction. It occurs, for example, during the **respiratory burst** in neutrophils and other phagocytic cells. The phenomenon is now exploited in sensitive enzyme assays and labelling methods in nucleic-acid hybridization, etc.

cheminformatics *an alternative name for* chemoinformatics; *see* **informatics**.

chemiosmotic coupling hypothesis a hypothesis proposed in 1961 by the British biochemist Peter Mitchell (1920–92) to explain ATP formation in mitochondria. Essentially it states that the energy-yielding reactions of the **respiratory chain** are coupled to the energy-requiring reactions of phosphorylation through the creation of a membrane electrochemical potential. An electrochemical gradient of H^+ ions, and a voltage gradient, is established across the mitochondrial inner membrane during electron flow through the respiratory chain. The electron carriers of the respiratory chain serve as an active-transport system or 'pump' to transport H^+ ions from the mitochondrial matrix across the inner membrane, which is an integral part of the coupling mechanism. The electrochemical gradient thus formed drives the synthesis of ATP from ADP and P_i. A similar mechanism is believed to drive oxidative phosphorylation in the chloroplast during the **light reactions of photosynthesis**.

chemism *(rare)* chemical action, operation, activity, or force.

chemisorption *or* **chemical adsorption** a type of **adsorption** in which the forces involved are valence forces of the same kind as those operating in the formation of chemical compounds. *Compare* **physisorption**.

chemoattractant *an alternative term for* **attractant**.

chemoautotroph *or* **chemolithotroph** any organism that is both an **autotroph** and a **chemotroph**. —**chemoautotrophic** *adj*.

chemoceptor *an alternative name for* **chemoreceptor** (esp. in immunology).

chemography an artefact in autoradiography due either to the creation of reduced crystals in the photographic emulsion by chemicals present in the specimen (**positive chemography**), or to the removal of latent images, also by chemicals in the specimen (**negative chemography**).

chemoheterotroph *or* **chemoorganotroph** any organism that is both a **chemotroph** and a **heterotroph**. —**chemoheterotrophic** *adj*.

chemoinformatics *or (sometimes)* **cheminformatics** *see* **informatics**.

chemokine *or* **intercrine** any of a subgroup of cytokines acting primarily as chemotactic agents that cause accumulation of leukocytes into tissues. They are secreted in response to inflammatory cytokines (e.g. interleukin 1, THFα), bacterial products, or viral infection. They are basic polypeptides (8–10 kDa), more than 40 of which are known and showing 20–70% homology. Subdivison into four families is on the basis of the number and relative position of their cysteine residues. More than a dozen G-protein-associated cytokine receptors activate phospholipase C. Cytokine receptor 5 is a coreceptor (with T-cell receptor) for HIV. A codon deletion in its gene (at 3p21) is present in $\approx$10% of white people and confers high resistance to HIV infection.

chemolithotroph *an alternative term for* **chemoautotroph**.

chemoorganotroph *an alternative term for* **chemoheterotroph**.

chemoreceptor *or* **chemoceptor** any **receptor** on a cell that receives chemical stimuli and reacts to them or binds them.

chemostat a device for the continuous culture of bacterial (and other) cells. Growth occurs in an aerated fermenter vessel and its rate is controlled by the rate of addition of fresh nutrient from a reservoir (i.e. the dilution rate); this in turn controls the rate of removal of cells (and culture medium). In a chemostat, as opposed to an **auxostat**, the dilution rate is constant. At equilibrium, the rate of production of new cells by multiplication is equal to the rate of removal of grown cells.

chemosynthesis the synthesis of organic chemical compounds by an organism using energy obtained from the oxidation of inorganic chemical compounds rather than from light. *Compare* **photosynthesis**. —**chemosynthetic** *adj*.

chemotactic factor for macrophages *see* **lymphokine**.

chemotaxin *(sometimes)* an alternative term for **cytotaxin**.

chemotaxis 1 a movement of a motile cell or organism *(see* **taxis***)* in response to a specific chemical concentration gradient. Movement may be towards a higher concentration (**positive chemotaxis**) or towards a lower concentration (**negative chemotaxis**). In *Escherichia coli* the *che* family of genes has been implicated in controlling certain aspects of chemotaxis, and **galactose/glucose binding protein** is also involved. *See also* **cytotaxis**. **2** *(in immunology)* the migration of polymorphonuclear leukocytes and macrophages towards higher concentrations of certain fragments of complement such as C3a and C5a. *See also* **lymphokine**. —**chemotactic** *adj*.

chemotaxonomy the classification of organisms on the basis of the nature, content, and/or the distribution of constituent chemical substances, e.g. DNA.

chemotherapy the treatment of disease, especially infections and neoplasms, with chemical agents (**chemotherapeutic agents**) that act specifically on infective organisms or tumours. The term also describes the treatment of psychiatric disease with chemical agents (as opposed to cognitive therapy). —**chemotherapeutic** *adj*.

chemotractant *see* **attractant**.

chemotroph any organism that derives its energy from exogenous chemical sources, independently of light. Chemotrophs may be subdivided into (1) chemoautotrophs (or chemolithotrophs), i.e. organisms that are both chemotrophs and **autotrophs**; and (2) chemoheterotrophs (or chemoorganotrophs), i.e. organisms that are both chemotrophs and heterotrophs. *Compare* **phototroph**. —**chemotrophic** *adj*.; **chemotrophy** *n*.

chemotropism the orientation of cells or organisms in response to chemical stimuli. The term is used especially of plants or plant organs. *See also* **tropism**. —**chemotropic** *adj*.

Cheng–Prussoff equation an equation relating the concentration, I_{50}, of a competitive ligand required to reduce the binding of a substrate to an enzyme by 50%, to the dissociation constant, K_i, of the enzyme–inhibitor complex; it is:

$$I_{50} = K_i (1 + S/K_m),$$

where S is the substrate concentration and K_m the Michaelis constant for the enzyme–substrate complex.

chenodeoxycholate *or* **chenodiol** the anion of $(3\alpha,5\beta,7\alpha)$-3,7-dihydroxycholan-24-oic acid; a major component of bile in some species (hens, geese) but a minor component in humans. It is used therapeutically to decrease the synthesis of cholesterol and to help dissolve cholesterol gallstones.

Cherenkov counter *or* **Cerenkov counter** a device for counting charged particles, or radionuclides emitting them, that depends on **Cherenkov radiation**.

Cherenkov radiation *or* **Cerenkov radiation** a radiation of bluish light, consisting of photons, emitted when charged particles, especially high-energy beta particles, pass through either a solid or liquid medium at velocities greater than that at which light passes through the same medium. The effect is analogous to the creation of a sonic boom that occurs when an object exceeds the speed of sound in a medium. [After Pavel Alekseyevich Cherenkov (1904–1990), Russian-born Soviet physicist.]

Ches *abbr. for* 2-(cyclohexylamino)ethanesulfonic acid; a compound with similar properties to a **Good buffer substance**; pK_a (25°C) = 9.3.

chewing gum *see* **chicle**.

chi *symbol:* χ (lower case) *or* X (upper case); the twenty-second letter of the Greek alphabet. For uses *see* **Appendix A**.

chiasma (*pl.* **chiasmata**) **1** (*in genetics*) a connection formed between **chromatids**, visible during **meiosis**, thought to be the point of the interchange involved in crossing-over. **2** *or* (*esp. US*) **chiasm** (*pl.* **chiasms**) (*in anatomy*) a decussation or intersection, as of two nerves; e.g. optic chiasma.

chicle a mixture of *cis*- and *trans*-1,4-polyprenoids obtained from the evergreen tree, *Achras sapota*; the original chewing gum base.

Chico an insulin receptor substrate-like protein in *Drosophilia* that links the insulin receptor to phosphatidylinositol 3-kinase and then protein kinase B and protein synthesis. A mutation of the corresponding gene results in fruit flies less than half the normal size.

chimera *or* (*esp. Brit.*) **chimaera** (*in genetics*) an organism comprising tissues of two or more **genotypes**. Chimeras can occur as a result of mutation, abnormal distribution of chromosomes, grafting, or genetic manipulation. For example, a chimera can be created by mixing the cells from embryos of two different animal species at the blastocyst stage of development and implanting the chimeric embryo in a surrogate mother to continue development. When the species are sheep and goats, the resulting chimeric progeny are called 'geep', or 'shoats'. Alternatively, in mice, genetically manipulated mouse embryonic stem (ES) cells are injected into the blastocoel of a blastocyst. Later stages of development are completed in the reproductive tract of a surrogate mother to produce chimeric mice. These are readily identified if the strain from which the ES

cells are derived has a different coat colour from that of the strain producing the blastocyst. Production of chimeras is a step in the generation of mice carrying gene knockouts produced by homologous recombination in ES cells. *See* **mosaic**, **transgenic**, **chimeric molecule**. —**chimeric** *or* **chimaeric** *adj.*

chimeraplast any RNA–DNA chimeric oligonucleotide used to induce base pair conversions at the genomic level and used in gene repair strategies for gene therapy. –**chimeraplastic** *adj.*

chimeric molecule **1** any of the hybrid DNA molecules formed when DNAs from different sources are digested with the same **restriction endonuclease** and the digests are mixed. The DNA fragments in the mixture become associated by hydrogen bonds between complementary sequences to form new arrangements, which may be converted into covalently linked molecules by the action of DNA ligase. This process can produce **chimeric plasmids** and other chimeric structures. **2** a **chimeric protein** (*or* **fusion protein**) is a protein obtained by the insertion or substitution of a partial sequence from one protein into another, typically using cDNA technology. The resultant protein thus has elements of both the original proteins. Such proteins frequently form naturally as a result of chromosomal translocations during neoplastic transformation. *See* **acute myeloid leukemia**. **3** a recombinant antibody that has, e.g., structural characteristics of both human and mouse antibodies.

chimerin any of several related proteins that function as GTPase-activating proteins, specific for the p21 Ras-related Rac GTPase. They include n-chimerin, and α- and β-chimerins. n-Chimerin is a cerebellar protein, and β-chimerin is expressed in testis. They are phorbol ester receptors, with an N-terminal domain similar to the zinc-finger region of protein kinase C, and a C-terminal domain similar to the product of **BCR**. Through their action on Rac, they lead to changes in cytoskeletal organization.

chimyl alcohol glycerol 1-hexadecyl ether, (+)-3-(hexadecyloxy)-1,2-propanediol; a hydrolysis product of **ether lipids**.

Chinese restaurant syndrome *see* **monosodium glutamate**.

ChIP *abbr. for* chromatin immunoprecipitation.

CHIP 28 *abbr. for* channel-forming integral membrane protein 28; *also called* AQP 1 (*see* **aquaporin**).

chiral describing a chemical compound that displays **chirality**.

chiral drug a single pure enantiomer of a specified drug rather than the racemate; the enantiomers may have very different pharmacological effects, e.g. in the case of propoxyphene and levopropoxyphene.

chirality topological handedness; the property of nonidentity of an object with its mirror image. A chiral chemical compound, i.e. one that possesses chirality, is one that cannot be superimposed on its mirror image, either as a result of simple reflection or after rotation and reflection. If superposition can be achieved then the molecule is said to be **achiral**. Chirality is most commonly due to the presence of one or more **chiral centres** (formerly referred to as asymmetric carbon atoms) or the presence of a **chiral axis** (as in allene structures, some of which are found in natural products). Rarely, chirality may also result from the presence of a chiral plane. *See also* **prochiral**. [From Greek *kheir*, hand.]

chiral methyl a carbon atom carrying the three hydrogen isotopes, i.e. $-C^1H^2H^3H$.

chiral recognition the differentiation of the enantiomers of a compound. This can be achieved by living organisms, chiral molecules, enzymes, drug receptors, etc.

chiroid (*rare*) any molecule that has **chirality**.

chiroptical describing any of the phenomena that depend on the ability of chiral and other intrinsically asymmetric molecules to rotate the plane of polarized light, known collectively as **optical activity**. Such phenomena include **optical rotation**, **optical rotatory dispersion**, and **circular dichroism**. The term is also applied to any of the techniques used for investigating these phenomena.

chi sequence *or* **chi site** a sequence of base pairs first discovered in a phage lambda mutant, called *chi*. These mutants were found to have single base pair changes creating sites that stimulate **recombination**. Characteristic of chi sites is a nonsymmetrical sequence of eight base pairs, consensus sequence:

5' GCTGGTGG 3'.

A recombinational hot spot in *Escherichia coli* and a binding site for **Rec A** is a chi site.

chi site *see* chi sequence.

chi-square test a statistical test to determine whether an observed series of values differs from a series of values expected on a hypothesis, to a greater degree than would be expected by chance. If m is the expected value and $(m + x)$ is the observed value then $\chi^2 = \Sigma(x^2/m)$. The goodness of fit may be found from available tables of χ^2 and the number of degrees of freedom, n, in which the observed series may differ from the hypothetical.

chi structure a structure formed when circular duplex DNA that is undergoing **recombination** is cut with a restriction enzyme. This yields two strands joined at a site intermediate between the ends, which by electron microscopy can yield a shape with the appearance of the Greek letter χ. The structure results from the fact that if two circular duplex DNAs undergo site-specific recombination, a single strand from each duplex will be opened and the two single strands ligated to each other to form a figure of eight. If each of the loops of this figure of eight is cut by the restriction enzyme the chi structure will result. *Compare* **Holliday junction**.

chitin a linear polysaccharide consisting of β-1,4-linked *N*-acetyl-D-glucosamine residues. It is found in annelid cuticle, arthropod exoskeleton, and in some plants and fungi. In fungi, chitin represents the microfibrillar component of the cell wall. *See also* **chitobiose**, **chitosan**.

chitinase EC 3.2.1.14; *other names*: chitodextrinase; 1,4-β-poly-*N*-acetylglucosaminidase; poly-β-glucosaminidase. An enzyme that catalyses the hydrolysis of the 1,4-β-linkages of *N*-acetyl-D-glucosamine polymers of chitin.

chitin synthase EC 2.4.1.16; *other name*: chitin–UDP *N*-acetylglucosaminyltransferase; an enzyme that catalyses the reaction between UDP-*N*-acetyl-D-glucosamine and [1,4-(*N*-acetyl-β-D-glucosaminyl)]$_n$ to form UDP and [1,4-(*N*-acetyl-β-D-glucosaminyl)]$_{n+1}$.

chitobiose 4-*O*-(2-amino-2-deoxy-β-D-glucopyranosyl)-2-amino-2-deoxy-D-glucose; *condensed form*: GlcpN(β1-4)GlcN; a disaccharide found in its acetylated form in chitin. This acetylated form, GlcpNAc(β1-4)GlcNAc, is the repeating unit of chitin, and should, logically, have the name chitobiose (by analogy with cellobiose, the repeating unit of cellulose). However, historically the nonacetylated form received the name chitobiose, a convention that has persisted.

α-anomer

chitosamine *see* glucosamine.

chitosan the cationic polymeric carbohydrate obtained by the deacetylation of **chitin**.

chitosome a body found in fungi and containing chitin synthase; it synthesizes chitin microfibrils.

chitotriose a triose formed of beta-1,4-linked *N*-acetyl-D-glucosamine.

chitotriosidase an enzyme able to hydrolyse the trisaccharide derived from chitin. Activities are greatly elevated in **Gaucher's disease**, in which it is used as a marker to assess whether enzyme replacement therapy with **glucosylceramidase** has been successful.

Chl *abbr. for* chlorophyll.

chlor+ *a variant of* **chloro+** (sometimes before a vowel).

chlorambucil 4-[bis(2-chloroethyl)amino]benzenebutanoic acid; an alkylating agent that reacts with DNA causing the formation of co-valent bonds that cross-link with the two strands of DNA, thereby preventing its replication.

chloramine T *N*-chloro-4-methylbenzenesulfonamide sodium salt; sodium 4-toluenesulfonchloramide hydrate; CH_3–C_6H_4– SO_2–NClNa•xH_2O. A white crystalline solid with a faint odour of chlorine. In aqueous solution it slowly decomposes to sodium hypochlorite and thus is a mild oxidant and antiseptic. It is widely used for the oxidation of radioiodide in the preparation at high specific radioactivity of radioiodine-labelled peptides and proteins for use in radioimmunoassay or other *in vitro* techniques. Microgram quantities of many proteins are labelled efficiently at or near pH 7; however, some proteins may suffer an unacceptable degree of oxidative damage.

chloramphenicol *abbr.*: CAP; a broad-spectrum, chlorine-containing antibiotic produced by *Streptomyces venezuelae*; it is now produced by chemical synthesis. The configurations at C-1 and C-2 are (*S*), as shown in the structure. It inhibits prokaryotic protein synthesis by attaching to the 50S ribosomal subunit and inhibiting peptidyltransferase (EC 2.3.2.12), thereby preventing the formation of peptide bonds. Chloramphenicol also inhibits protein synthesis in mitochondria, and this accounts for its toxic effect in causing aplastic anemia. It is much used for the treatment of typhoid fever.

chloramphenicol *O*-acetyltransferase *abbr.*: CAT; EC 2.3.1.28; a bacterial enzyme, not found in eukaryotes, that catalyses the reaction between acetyl-CoA and chloramphenicol to form CoA and chloramphenicol 3-acetate. This reaction forms the basis for acquired chloramphenicol resistance in certain bacterial strains. Also the gene for CAT is used as a **reporter** because its expression can be detected by a sensitive radiochemical assay. In cDNA technology, for example, the expression of a eukaryotic promoter can be characterized by assay of the enzyme if CAT cDNA is included in the genes to be expressed.

chloramphenicol amplification a technique whereby chloramphenicol is added to growing cultures of bacteria harbouring plasmids, to inhibit cell growth and division without interfering with plasmid replication. This results in an increased plasmid copy number per cell.

chlordiazepoxide 7-chloro-*N*-methyl-5-phenyl-3*H*-1,4-benzodiazepin-2-amine 4-oxide; a CNS depressant noted primarily for its effective anti-anxiety effects at nonsedative doses. It is the prototypical **benzodiazepine**, used therapeutically as the hydrochloride. It is a controlled substance in the US Code of Federal Regulations. One proprietary name of the hydrochloride is Librium.

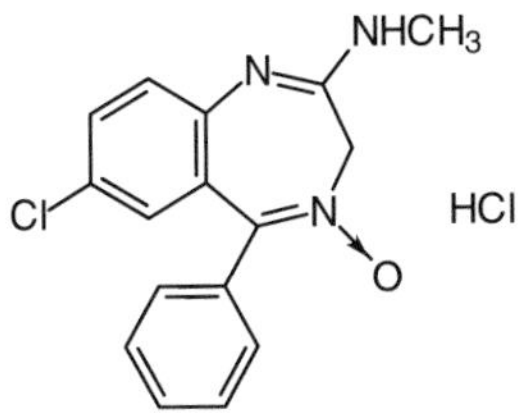

chlorogenic acid

chloride 1 *symbol*: Cl⁻; the anion of the element chlorine. **2** any salt of hydrochloric acid. **3** any compound containing a chlorine atom in organic linkage.

chloride channel any of a family of proteins that facilitate Cl⁻ transport through a membrane. These channels are of two types: (1) ligand-gated Cl⁻ channels associated with inhibitory neurotransmitter receptors, such as the **GABA receptor** or **glycine receptor**; and (2) voltage-gated Cl⁻ channels. These are structurally unrelated to other voltage-gated channels, and are believed to possess 10 or 12 transmembrane domains, with both N- and C-termini residing in the cytoplasm. They are found in a wide range of organisms, from bacteria and yeasts to plants and animals. Their functions in higher animals are thought to include regulation of cell volume, control of electrical excitability, and trans-epithelial transport. *See also* **cystic fibrosis transmembrane conductance regulator**.

chloride shift *or* **Hamburger shift** the phenomenon whereby, because of the **isohydric shift**, there is a higher concentration of hydrogencarbonate ions in the erythrocytes in blood leaving the tissues (venous blood) than in arterial blood, and hence the hydrogencarbonate/chloride ion ratio is higher in the erythrocytes than in plasma. This results in a movement of hydrogencarbonate ions out of the erythrocytes into the plasma and an equivalent movement of chloride ions from the plasma into the erythrocytes. In the lungs the reverse ion movements occur upon the oxygenation (arterialization) of the blood there.

chlorin a trivial name for 2,3-dihydroporphyrin.

chlorinate 1 to treat or react (a substance) with dichlorine. **2** to introduce one or more chloro groups into an organic compound, whether by addition or substitution. **3** to disinfect (especially water) with dichlorine. —**chlorination** *n*.

chlorine 1 *symbol*: Cl; a halogen element of group 17 of the IUPAC **periodic table**; atomic number 17; relative atomic mass 35.453. The 15th most abundant element of the Earth's crust, occurring in the combined state, mostly as inorganic chlorides. **2** *symbol*: Cl_2; the common name for the yellowish-green gaseous substance, correctly known as **dichlorine**.

chloro+ *or (sometimes before a vowel)* **chlor+** *comb. form* **1** denoting a chlorine atom in organic linkage. **2** denoting the colour green.

chlorocruorin any green, ~3 MDa, hemochrome respiratory pigment found in the blood of certain polychaete annelids; it contains 96 heme groups.

chlorogenic acid 3-*O*-(3,4-dihydroxycinnamoyl)-D-quinic acid; a tannin found in tea and coffee.

chlorolabe *or* **chloropsin** *or* **cyanopsin** *See* **cyanopsin**, **opsin**.

p-chloromercuribenzenesulfonate *abbr*.: PCMBS; 4-(chloromercuri)benzenesulfonate; $Cl–Hg–C_6H_4–SO_3^-$; a reagent that reacts with thiol groups; it is more soluble than *p*-chloromercuribenzoate.

p-chloromercuribenzoate *abbr*.: PCMB; 4-(chloromercuri)benzoate; $Cl–Hg–C_6H_4–COO^-$; a reagent that reacts with thiol groups. It is supplied either as *p*-chloromercuribenzoic acid or as *p*-hydroxymercuribenzoate sodium salt.

4-chloromercuriphenylsulfonate *see* **p-chloromercuribenzenesulfonate**.

chloroperoxidase EC 1.11.1.10; *other name*: chloride peroxidase.

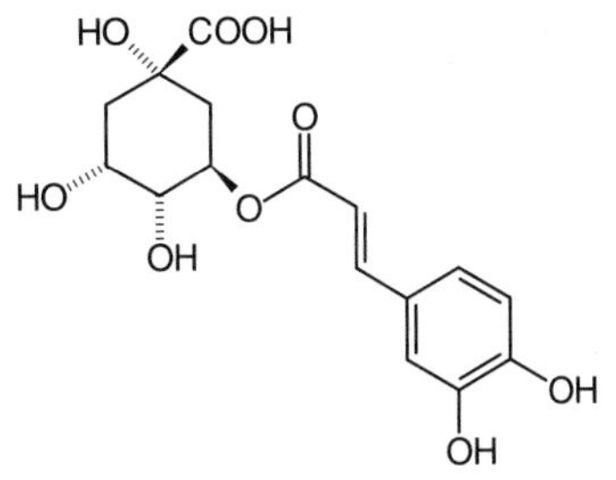

An enzyme that catalyses a reaction between H_2O_2, two molecules of Cl⁻ ion, and two molecules of an organic compound to form two molecules of an organic chloride and two molecules of H_2O. It is useful for the introduction of (radioisotopes of) chlorine, bromine, or iodine into tyrosine residues in proteins. It is involved in the synthesis of chlorine-containing compounds such as (1*S-trans*)-2,2-dichloro-1,3-cyclopentanedione (caldariomycin).

p'-chloro-p-phenylaniline *see* **4-amino-4'-chlorodiphenyl**.

chlorophyll any of the several green (or purple, *see* **bacteriochlorophyll**) pigments, found in plants and photosynthetic bacteria, that function in **photosynthesis** by absorbing light energy mainly in the red and violet-blue parts of the spectrum. Chlorophylls are magnesium complexes of various closely related porphyrins or chlorins. The main chlorophylls of land plants are chlorophylls *a* and *b*, and some algae contain chlorophylls *c*. Chlorophylls *a* and *b* are dihydroporphyrins, having no double bond at position 3 of ring B, while chlorophylls *c* are porphyrins. Photosynthetic bacteria contain various bacteriochlorophylls. The structure of chlorophyll *a* is shown. Chlorophyll *b* differs from chlorophyll *a* only in the presence of a formyl group in place of the methyl group at position 3 on ring B (Fischer system: *see* **bacteriochlorophyll**). Several chlorophylls *c* are known. They are accessory light-harvesting pigments found in eukaryotic algae that do not contain chlorophyll *b*. The structure of chlorophyll c_2 is shown, chlorophyll c_1 differing from this in having an ethyl group at position 4 of ring B, and chlorophyll c_3 differing from it in having a methylformyl group at position 3 of ring B. Most chlorophylls exist in **antenna complexes**, where their function is to absorb visible light and to transmit the energy so absorbed to a reaction centre in which other chlorophyll molecules are excited by

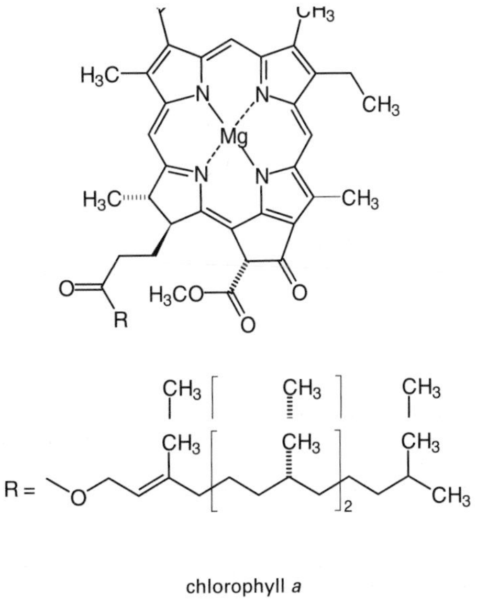

chlorophyll a

the accumulated energy to transfer an electron to an electron transfer system known as a **photosystem**. Differences in structure have significant effects on the absorption spectrum of the chlorophyll molecule, i.e. on the wavelength of the light absorbed, and thus on the energy available as a result of the absorption (*see* **Planck constant** for the relationship). To restore its electronic state, the oxidized chlorophyll molecule can accept an electron from a water molecule (or, in photosynthetic sulfur bacteria, from hydrogen sulfide), the transfer being catalysed by a manganese-rich protein complex that is part of the photosystem. This reaction is responsible for the dioxygen (or sulfur) that is formed during photosynthesis, one molecule of dioxygen, four protons, and four electrons being produced from two water molecules. *See also* **P680**, **P690**, **P700**. [From the Greek *khloros*, yellowish or pale green, and *phyllon*, leaf.]

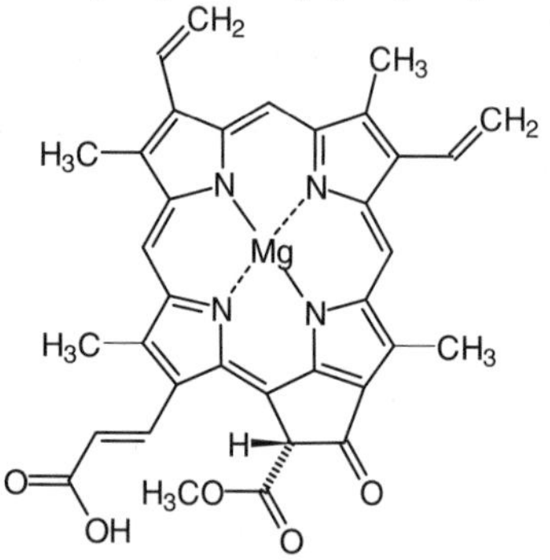

chlorophyll c_2

chlorophyllide a **chlorophyll** lacking the terpenoid side chain, usually phytol. The penultimate intermediate in chlorophyll biosynthesis. It is formed by the light-catalysed reduction of protochlorophyllide attached to a protein, protochlorophyllide holochrome.

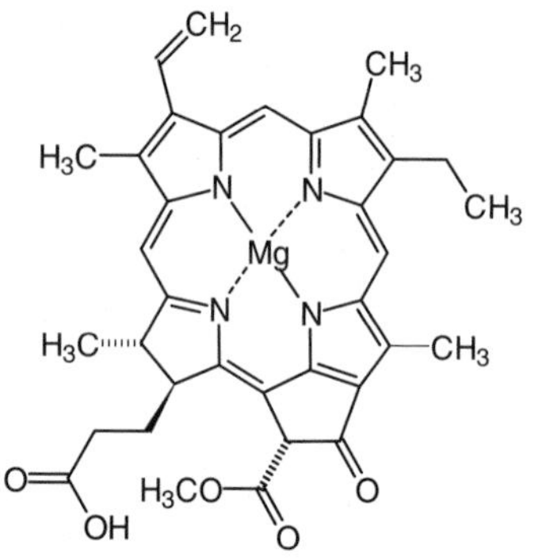

chlorophyllide *a*

chloroplast *or (sometimes)* **chloroplastid** a chlorophyll-containing **plastid**, found in cells of algae and higher plants, that is the site of **photosynthesis**. Chloroplasts are characterized by a system of membranes embedded in a hydrophobic proteinaceous matrix, or stroma. The basic unit of the membrane system is a flattened single vesicle called the **thylakoid**; thylakoids stack into grana (*sing.* **granum**). All the thylakoids of a granum are connected with each other, and the grana are connected by intergranal lamellae. Chloroplasts are rich in various proteases (e.g. serine, aspartic, ATP-dependent, light-activated, for oxidized or photodamaged proteins, for trafficking of protein precursors into the plastids). Some of these may have a role in senescence. *See* **Clp protease**.

chloroquine [N^4-(7-chloro-4-quinolinyl)-N^1,N^1-diethyl-1,4-pentane-diamine; a widely used synthetic quinoline antimalarial drug, with

action similar to **quinine**. It accumulates in the cells of the malaria parasites and prevents digestion of their host's hemoglobin; it may also possibly disrupt the parasite's ribonucleotide metabolism. *See also* **acridine**.

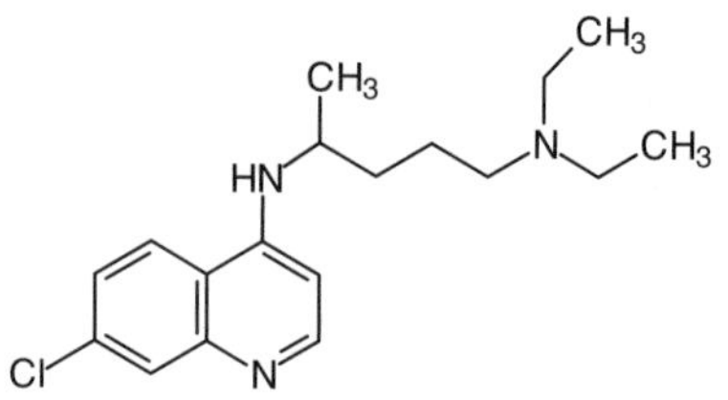

chlorosome a small compartment found in photosynthetic Chlorobiaceae bacteria. It contains **bacteriochlorophyll** *c* and is attached to the cytoplasmic membrane.

chlorosulfolipid a class of **sulfolipid** found in the alga, *Ochramonas danica*. They are based on *n*-docosane-1,14-diol disulfate; chlorine atoms have been located at C-2, C-13, C-15, and C-16.

chlorpromazine 2-chloro-10-(3-dimethylaminopropyl)phenothiazine; a **neuroleptic** drug with sedative and antiemetic actions. It acts as an antagonist at dopaminergic (D_2), α-adrenergic, and many other receptors (*see* **adrenoceptor**). It was formerly classed as a **major tranquillizer**, a term now deprecated. Largactil is one proprietary name of the hydrochloride.

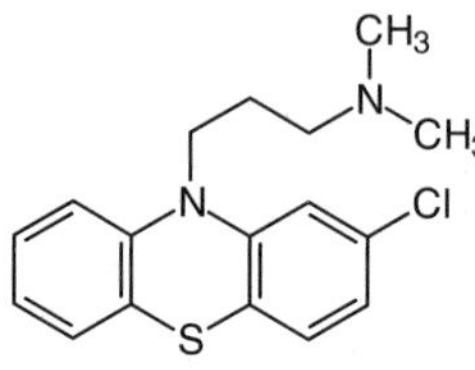

Cho *symbol for* a residue of choline.

CHO *abbr. for* Chinese hamster ovary (applied to cell cultures derived from this organ).

chole+ *or (before a vowel)* **chol+** *comb. form* denoting bile.

cholecalciferol *or* **calciol** *the recommended trivial name for* vitamin D_3; (5Z,7E)-(3S)-9,10-seco-5,7,10(19)-cholestatrien-3-ol. It is formed in skin by UV-irradiation of 7-dehydrocholesterol, and is metabolized to 1α,25-dihydroxycholecalciferol (calcitriol). The latter affects the absorption of calcium from the gastrointestinal tract and the metabolism of calcium in bone. *See* **vitamin D**.

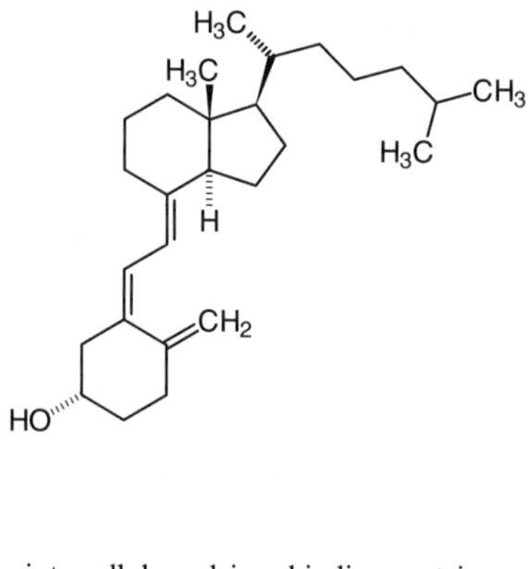

cholecalcin an intracellular calcium-binding protein.

cholecystokinin *abbr.*: CCK; *formerly called* cholecystokinin–pancreozymin (*abbr.*: CCK-PZ); a peptide hormone from **I cells** of the duodenum and proximal jejunum of mammals, also present in most parts of the brain and in the neurons of the gastrointestinal tract. It is identical with **pancreozymin**. The sequence of human cholecystokinin is

KAPSGRMSIVKNLQNLDPSHRISDRDY(-SO₃)MGWMDF-
NH₂.

Sulfation of the tyrosine is essential for activity. Highly heterogeneous, the various species are known as CCK-39, CCK-33 (e.g. human), CCK-12, CCK-8, CCK-4, etc. and are truncated from the N terminus. Their common C-terminal tetrapeptide amide sequence is identical to that in which the functional activity of **gastrin** largely resides. CCK is released by entry of food into the duodenum; it causes contraction of the gall bladder and secretion of enzyme precursors from pancreatic acinar tissue. It is also a neurotransmitter and decreases appetite.

cholecystokinin receptor either of two G-protein-associated receptors (called CCK-A and CCK-B) that bind the peptide hormone **cholecystokinin** (CCK) and stimulate phospholipasae C. CCK-A is present mostly in the alimentary tract and preferentially binds the sulfated form of CCK-8. CCK-B is widely distributed in the brain, binds both the sulfated and unsulfated forms of CCK-8, CCK-4, and gastrin.

choleragen (*sometimes*) *an alternative name for* **cholera toxin**.

choleragenoid a 57 kDa protein that occurs naturally in the cell-free culture fluid of *Vibrio cholerae* and is immunologically almost indistinguishable from **cholera toxin** but is nontoxic. It consists of the pentamer of the B subunit of the toxin without its A subunit.

cholera toxin *or* **choleragen** the antigenic and potently diarrhoeagenic **enterotoxin** produced by *Vibrio cholerae*. Pathogenic strains secrete the toxin, which is an 84 kDa multimeric protein of exceptional stability with two subunits: an A subunit that consists of an α (22 kDa) and a β (5 kDa) chain linked by a single disulfide bond; and a B subunit consisting of four to six noncovalently associated β chains (11.5 kDa each). Binding of the toxin to its receptor ganglioside (G$_{M1}$) on sensitive cells occurs via the B subunit; the A subunit then activates cellular adenylate cyclase; together with cellular **ADP-ribosylation factor** it constitutes an **ADP-ribosyltransferase**, which transfers ADP-ribose to the **G-protein** α$_s$ subunit, inhibiting its GTPase and converting it to an irreversible activator of adenylate cyclase.

cholestane the fully saturated carbon skeleton from which **cholesterol** is structurally derived; the configuration at C-5 can be α or β; 5β-cholestane is known as **coprostane**.

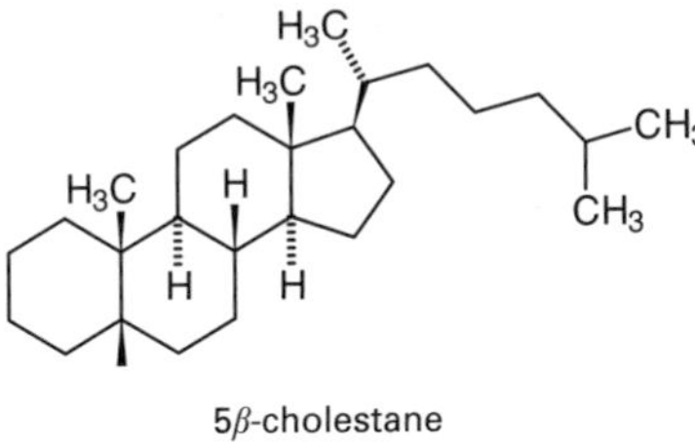

5β-cholestane

cholestanol 3β,5α-cholestan-3-ol; dihydrocholesterol; 3β-hydroxycholestane; a steroid that occurs in human feces, gallstones, and in eggs.

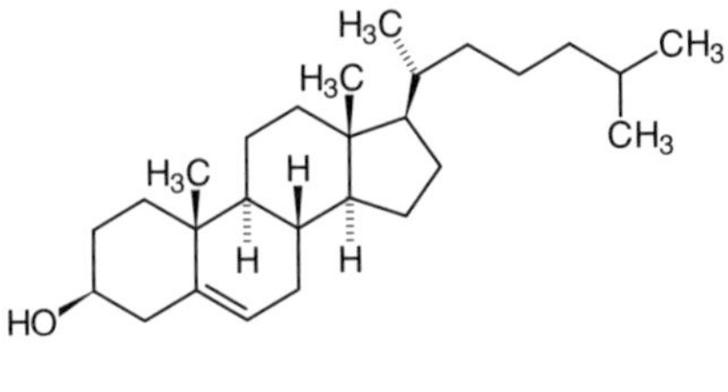

cholestanol

cholestasis the arrest (stasis) or impairment of bile flow. It may originate intrahepatically, due to hepatocellular defects (usually in the secretion of bile acids into the bile canaliculi), or extrahepatically, often due to obstruction of the bile duct by calculi. The former variety are often familial.

cholesteric of, relating to, or resembling cholesterol or a derivative of cholesterol. *See also* **liquid crystal**.

cholesterol cholest-5-en-3β-ol; the principal sterol of vertebrates and the precursor of many steroids, including bile acids and steroid hormones. It is found in all animal tissues, especially brain and spinal cord, in the bile, and in gallstones, being a component of the plasma membrane lipid bilayer. Both cholesterol and cholesterol esters are present in plasma lipoproteins and excessive levels in the plasma are implicated in atherogenesis. It is synthesized from acetyl-CoA via hydroxymethylglutaryl-CoA and squalene, by enzymes in the cytosol (acetyl-CoA → mevalonate), peroxisomes (mevalonate → farnesyl pyrophosphate), and endoplasmic reticulum (farnesyl pyrophosphate → cholesterol). Deficiency of cholesterol occurs in Smith–Lemli–Opitz syndrome. Cholesterol accumulates in hypercholesterolemia, cholesterol ester storage disease, Wolman disease, and some forms of Niemann–Pick disease.

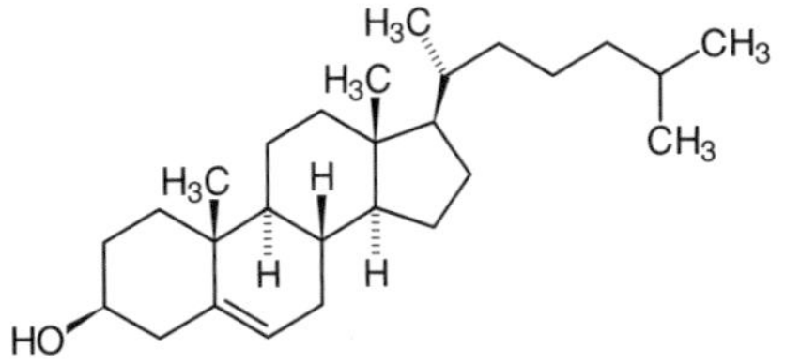

cholesterol acyltransferase *see* **sterol O-acyltransferase**.

cholesterol desmolase *see* **cholesterol monooxygenase**.

cholesterol ester storage disease an autosomal recessive disease, sometimes not detected until adulthood, in which there is widespread deposition of cholesterol esters and triacylglycerols in tissues. It is associated with a relatively common mutation in the gene for lysosomal acid lipase, which has ≈5% of the activity of the wild type. *See* **acid lipase**.

cholesterol ester transfer protein *see* **CETP**.

cholesterol 7α-hydroxylase *see* **cholesterol 7α-monooxygenase**.

cholesterol 7α-monooxygenase EC 1.14.13.17; *other names*: cholesterol 7α-hydroxylase; cytochrome P450 VII; a cytochrome P450 enzyme that catalyses the hydroxylation of cholesterol to form 7α-cholesterol. NADPH and dioxygen are reactants, NADP⁺ and H₂O being formed. This is the first step in the biosynthesis of bile acids from cholesterol and is rate-limiting.

cholesterol monooxygenase (side-chain cleaving) EC 1.14.15.6; *systematic name*: cholesterol,reduced-adrenal-ferredoxin:oxygen oxidoreductase (side-chain-cleaving); *other names*: CYPiiA cholesterol desmolase; cholesterol side-chain cleavage enzyme; cytochrome P450$_{scc}$; a cytochrome P450 enzyme that in association with adrenodoxin and adrenodoxin reductase catalyses the formation of pregnenolone from cholesterol, with release of 4-methylpentanal and H₂O. An integral membrane protein of the inner mitochondrial membrane, it is phosphorylated and activated by cAMP released in response to ACTH during the biosynthesis of steroid hormones from cholesterol in the adrenal cortex, gonads, and placenta. The reaction involves an oxidation by dioxygen, reduced adrenal ferredoxin being oxidized in the process. Enzyme activity is dependent on the entry of cholesterol into the mitochondrial matrix, a process that is facilitated by **StAR**.

cholesterol side-chain cleavage enzyme *see* **cholesterol monooxygenase (side-chain cleaving)**.

cholesteryl ester transfer protein *see* **CETP**.

cholestyramine resin *proprietary name*: Dowex 1-X2-Cl; a basic anion-exchange resin that consists predominantly of polystyrene trimethylbenzylammonium chloride. If administered orally, it exchanges anions of chloride with those of intestinal bile salts, which are then excreted as an insoluble complex. The resulting reduction in cholesterol absorption culminates in reduced circulating levels of low-density lipoproteins (LDLs), hence its therapeutic application in relieving hyperlipidemia and LDL-induced hypercholesterolemia.

cholic acid $3\alpha,7\alpha,12\alpha$-trihydroxy-5β-cholan-24-oic acid; a metabolic derivative of cholesterol, produced in the liver and secreted as a **bile acid**. It is used as a detergent, as the sodium salt; aggregation number 2, CMC 9–15 mM.

choline *symbol*: Cho; *abbr*.: Ch; N,N,N-trimethylethanolammonium (ion); 2-hydroxyethyltrimethylammonium (ion); 2-hydroxy-N,N,N-trimethylethanaminium; $(CH_3)_3N^+$–CH_2–CH_2–OH. An amino alcohol that functions as a methyl group donor in metabolism; it also occurs in the neurotransmitter acetylcholine and in some phospholipids. It is considered to be an essential dietary factor for humans.

choline acetyltransferase EC 2.3.1.6; *recommended name*: choline O-acetyltransferase; *systematic name*: acetyl-CoA:choline O-acetyltransferase. An enzyme that catalyses the reaction between acetyl-CoA and choline to form O-acetylcholine and CoA. It is instrumental in the synthesis of the neurotransmitter acetylcholine. *See also* **cholinoceptor**.

choline kinase EC 2.7.1.32; an enzyme that catalyses the phosphorylation by ATP of choline to form O-phosphocholine and ADP. Phosphocholine (choline phosphate) is an important intermediate in phospholipid biosynthesis, reacting with CTP to form CDP-choline. *See also* **choline-phosphate cytidylyltransferase**.

choline-phosphate cytidylyltransferase EC 2.7.7.15; *systematic name*: CTP:choline-phosphate cytidylyltransferase; *other name*: phosphorylcholine transferase. An enzyme that catalyses the formation of **cytidine(5′)diphosphocholine** (CDPcholine) from CTP and choline phosphate with release of pyrophosphate.

cholinergic describing nerve fibres that, when activated, release acetylcholine or an acetylcholine-like substance from their endings. *Compare* **adrenergic**, **peptidergic**, **purinergic**. *See also* **cholinoceptor**.

cholinesterase 1 *recommended name for* EC 3.1.1.8; *other names*: choline esterase II, pseudocholinesterase, butyrylcholinesterase. A glycoprotein homotetramer (gene locus at 3q26.1-q26.2, encodes 579 amino acids) that hydrolyses a variety of choline esters to choline and a carboxylic acid. It is found in most tissues including liver and brain. Its function is unknown. Serum activity is decreased in infancy, pregnancy, liver disease, thyroid deficiency, malnutrition, or as a result of gene mutation. In humans, individuals with abnormally low activity of cholinesterase fail to hydrolyse succinylcholine, used as a muscle relaxant in anesthesia with procaine and cocaine. They may suffer from procaine toxicity, prolonged succinylcholine apnea, and cocaine toxicity. The gene (E_1) encoding the enzyme can exist in at least five allelic forms, the normal (E_1^u), and four variants, E_1^a (atypical), E_1^f (fluoride resistant), E_1^s (silent), and E_1^K, which has 66% of normal activity and is the commonest variant. Homozygotes for the atypical gene have a dibucaine-resistant enzyme that is weakly active against most enzyme substrates, and homozygotes for the silent gene exhibit minimal enzyme activity. **2** *an alternative name for* **acetylcholinesterase**.

cholinoceptor any receptor for acetylcholine on the postjunctional membrane of synaptic clefts or on terminal axons that mediates its action. There are two types: (1) **muscarinic cholinoceptors** of smooth muscle, cardiac muscle, endocrine glands, and the central nervous system at which muscarine, but not nicotine, can readily mimic the action of acetylcholine; and (2) **nicotinic cholinoceptors** of skeletal

muscle, ganglia, and the central nervous system at which nicotine, but not muscarine, can readily mimic the action of acetylcholine.

Muscarinic acetylcholine receptors are G-protein-associated membrane proteins. M_1 and M_3 activate phosphoinositide-specific phospholipase C; M_2 and M_4 are negatively coupled to adenylate cyclase; M_2 also increases potassium channel conductance through the action of G-proteins.

Nicotinic acetylcholine receptors are glycosylated pentamers (i.e. α, β, γ, δ, and ε) with an intrinsic cation-permeable channel with little selectivity among cations. They respond by an extensive change in conformation that affects all subunits and leads to opening of an ion-conducting channel across the plasma membrane. *Torpedo* electric organ and embryonic muscle pentamers have $\alpha_2\beta\gamma\delta$, adult mammalian muscle $\alpha_2\beta\delta\varepsilon$, and mammalian neurons $\alpha_2\beta_3$ stoichiometry; each subunit contains four transmembrane segments. Bungarotoxin is a selective antagonist. The neuronal receptor has two subunits (α and non-α).

cholinomimetic describing a drug with an acetylcholine-like action.

cholyl the acyl group derived from **cholic acid**.

chondrioma *or* **chondriome 1** the entire mitochondrial content of a cell. **2** all the hereditary determinants that are localized in mitochondria.

chondriosome *a former name for* **mitochondrion**.

chondroblast a cell that secretes the matrix of **cartilage**.

chondrocalcin the C-terminal peptide formed from amino acids 1173–1418 of human procollagen α_1.

chondroclast a large, multinucleate cell that digests the cartilage matrix.

chondrocyte a cartilage cell, formed from a **chondroblast**.

chondroitinase *see* **N-acetylgalactosamine-4-sulfatase**, **N-acetylgalactosamine-6-sulfatase**.

chondroitinsulfatase *see* **N-acetylgalactosamine-4-sulfatase**, **N-acetylgalactosamine-6-sulfatase**, **iduronate-2-sulfatase**.

chondroitin sulfate any member of a group of 10–60 kDa glycosaminoglycans, widely distributed in cartilage and other mammalian connective tissues, the repeat units of which consist of $\beta(1-4)$-linked D-glucuronyl$\beta(1-3)N$-acetyl-D-galactosamine sulfate. They usually occur linked to a protein to form proteoglycans. Two subgroups exist, one in which the sulfate is on the 4-position (chondroitin sulfate A) and the second in which it is in the 6-position (chondroitin sulfate C). They often are polydisperse and often differ in the degree of sulfation from tissue to tissue. The chains of repeating disaccharide are covalently linked to the side chains of serine residues in the polypeptide backbone of a protein by a glycosidic attachment through the trisaccharide unit galactosyl–galactosyl–xylosyl. Chondroitin sulfate B is more usually known as **dermatan sulfate**.

chondromodulin a 28 kDA glycoprotein released by proteolytic cleavage of a larger precursor and localized in the interterritorial matrix of avascular zones of epiphyseal cartilage. It is abundant in fetal cartilage and inhibits angiogenesis. It is not produced in chondrosarcoma.

chondrosamine *former name for* the D-enantiomer of **galactosamine**, reflecting its presence in **chondrosine**.

chondrosine the disaccharide 2-amino-2-deoxy-3-O-β-D-glucopyranuronosyl-D-galactopyranose; *condensed form*: GlcpA($\beta1$-3)GalN. In sulfated form it is the core unit of **chondroitin sulfate**.

chopper 1 a device for chopping tissue e.g. the McIlwain chopper. **2** *(in optics)* a device for interrupting a light beam in a regular manner. For example, in a spectrophotometer it consists of a rotating wheel with alternate silvered and cut-out sectors that is placed in the light beam to allow the light to pass alternately through the sample and the reference cells, the photocell thereby generating an alternating current.

chordin a secreted protein of the notochord that contains cysteine-rich repeats. It binds and antagonizes **bone morphogenetic protein**.

choriocarcinoma a malignant tumour, typically occurring in the uterus, originating as a proliferation of chorionic villi. It may follow pregnancy or develop from a hydatidiform mole. **Choriogonadotropin** is an excellent marker for the presence of this tumour.

choriogonadotropin *or* **chorionic gonadotropin** *or* **choriogonadotrophin** *abbr.*: CG; *the recommended name for* a placental **glycoprotein hormone** that has the gonadotropic activities of both luteinizing hormone (LH) and follicle-stimulating hormone (FSH). It is a member of the family of glycoprotein hormones that includes LH and FSH, the β subunit being specific to choriogonadotropin. It is produced by the corpus luteum after fertilization of the ovum; the blood level rises to a detectable concentration 7–9 days after conception and CG may be detected in urine 1–2 days later, which yields a highly sensitive and specific test for pregnancy. Its administration to males causes increased secretion of testosterone; the lack of such a response indicates testicular failure. It is also a good marker for **choriocarcinoma**. It is a heterodimer of an α chain (identical with that of LH and FSH) and a β chain that confers specificity.

choriomammotropin *or* **chorionic somatomammotropin** *or* **choriomammotrophin** *other name*: placental lactogen (*abbr.*: PL); a placental hormone similar in action to **somatotropin** and **prolactin**. It shows ≈85% homology with the former and ≈13% with the latter. It is partly responsible for the rise in insulin-like growth factor I (IGF-I) level observed in pregnancy. Its absence does not affect fetal growth, pregnancy, or lactation. The gene at 17q22-q24 is near those for somatotropin and prolactin.

chorion 1 the highly vascular outermost embryonic membrane of higher vertebrates. In placental animals it has a villous part that enters the placenta. **2** the membrane of an animal egg, especially the hardened external membrane of an insect's egg.

chorionic gonadotropin *or* **chorionic gonadotrophin** *an alternative name for* **choriogonadotropin**.

chorionic somatomammotropin *or* **chorionic somatomammotrophin** *an alternative name for* **choriomammotropin**.

chorion villus sampling *abbr.*: CVS; a technique for a biopsy of the chorion frondosum during pregnancy to obtain fetal tissue for prenatal diagnosis.

chorismate the anion of (3*R-trans*)-3-[(1-carboxyethenyl)oxy]-4-hydroxy-1,5-cyclohexadiene-1-carboxylic acid; the unsymmetrical ether derived from phospho*enol*pyruvate and 5-phosphoshikimic acid (*see* **shikimate pathway**); it is formed as an intermediate in the biosynthesis of aromatic amino acids and many other compounds. The common pathway to aromatic amino acids bifurcates at chorismate: one branch, the prephenate pathway, leads to phenylalanine and tyrosine; the other branch, the anthranilate pathway, leads to tryptophan. There are many other branchpoints in the shikimate pathway. (Note: the configurations can be stated as 3*R*,4*R*.)

chorismate mutase EC 5.4.99.5; an enzyme that catalyses the conversion of chorismate to prephenate. It is the first enzyme in the branch of the **shikimate pathway** that leads to synthesis of phenylalanine. Isoenzymes are found in cytoplasm and in chloroplasts. The chloroplast isoenzyme is activated by tryptophan and inhibited by phenylalanine and tyrosine. For an example see **prephenate dehydratase**.

chorismate synthase EC 4.6.1.4; *other name*: 5-*enol*pyruvylshikimate-3-phosphate phospholyase; an enzyme that catalyses the formation of chorismate from 5-*O*-(1-carboxyvinyl)-3-phosphoshikimate with release of orthophosphate. *See also* **shikimate pathway**.

CHRAC *abbr. for* chromatin accessibility complex; an ATP-dependent chromatin remodelling complex that causes *cis*-displacement or sliding of a nucleosome along a stretch of DNA.

Christmas disease *or* **hemophilia B** a sex-linked inherited disorder of **blood coagulation** due to a deficiency of coagulation factor IX (Christmas factor) activity in the blood. It occurs in humans, dogs, cats, and other species, but is less common than classical **hemophilia** (hemophilia A). [After S. Christmas, the first described sufferer of the disease.]

Christmas factor *an alternative name for* **factor IX**. *See also* **blood coagulation**, **Christmas disease**.

chrom+ *see* **chromo+**.

chromacin the 22-residue polypeptide that constitutes residues 176 to 197 of chromogranin A. It is bacteriolytic and antifungal. *See* **chromogranin**.

chromaffin *or* **chromaffine** describing a tissue, cell, or subcellular granule that stains deeply with chromium salts.

chromaffin tissue a tissue made up of modified neural cells (chromaffin cells) that synthesize, store, and secrete catecholamines. The catecholamines are stored in special subcellular organelles, **chromaffin granules**, ready for release. Chromaffin tissue is found in all vertebrates, at various bodily locations but especially in the medulla of the adrenal gland. Tumours of this tissue, usually in the adrenal medulla, cause a syndrome of intermittent hypertension and excessive catecholamine secretion.

chromat+ *see* **chromato+**.

chromatic of, or related to, colour.

chromatid one of the two daughter strands of a duplicated chromosome that become apparent between early prophase and metaphase in mitosis and between diplotene and second metaphase in meiosis.

chromatin the complex of a double-stranded DNA fibre with histone and nonhistone proteins that makes up the **chromosomes** of the eukaryotic nucleus during interphase. It stains strongly with basic dyes. Some parts of the chromosomes are highly dispersed (called **euchromatin**) and other parts are highly condensed (called **heterochromatin**). Chromatin is made up of repeating units, each unit consisting of some 147 DNA base pairs and two each of **histones** H2A, H2B, H3, and H4 (thus forming a histone octamer); most of the DNA is wound around the outside of a core of histones to form **nucleosomes**. Adjacent nucleosomes are joined by a stretch of linker DNA (variable in length but usually ≈55 bp) to which a molecule of histone H1 binds, thereby contributing flexibility to the **chromatin fibre**.

chromatin accessibility complex *see* **CHRAC**.

chromatin assembly factor 1 *see* **CAF1**.

chromatin body a body, containing ribosomal DNA, that occurs during early oogenesis in some water beetles. It disappears at about the time of vitellogenesis and its contents become dispersed in the nuclear interior.

chromatin compaction the degree of packing of nucleosomes in **chromatin**. Increased compaction impedes accessibility to the nucleosomal DNA and represses its genomic activity, whereas decompaction increases the accessibility of DNA for transcription or replication. The conformational changes are facilitated by enzymes that act on the histones (kinases, methylases, acetylases) or on DNA (methylases), and by nucleosome-remodelling complexes and high-mobility group proteins (e.g. HMGN), which bind nucleosomes.

chromatin fingerprinting a two-dimensional electrophoretic technique for simultaneously mapping the total distributions of DNA sizes, protein compositions, and nuclease-mediated precursor-product relationships present in heterogeneous populations of mono- and polynucleosomes. Chromatin, partially digested with micrococcal nuclease, is electrophoretically separated in the first dimension in the presence of inactive DNase I. Subsequently the enzyme is activated by incubating the gel in a magnesium ion-containing buffer and the histones are then separated by electrophoresis in the second dimension.

chromatin immunoprecipitation *abbr.*: ChIP; a method used to demonstrate the association between transcription factors or other DNA-binding proteins and specific regions of genomic DNA. Chromatin is isolated from preparations of nuclei, chemically cross-linked, and sheared to provide short fragments. Antibodies are used to immunoprecipitate the specific DNA-binding protein and with it the fragment of DNA to which it is bound. Polymerase chain reaction (PCR) is then used to identify the presence of specific DNA fragments.

chromatin silencing the conversion of large regions of chromosomes into a form that does not permit transcription. Transcriptional silencers are required for initiation of chromatin silencing, a state that is correlated with decreased histone acetylation.

chromato+ *or (before a vowel)* **chromat+** *comb. form* **1** denoting colour or coloured. **2** denoting **chromatin**. **3** indicating **chromatogram**, **chromatograph**, or **chromatography**.

chromatofocusing a column-chromatographic technique of high capacity and high selectivity for separating proteins according to their isoelectric points. A pH gradient is formed in a column containing an anion exchanger, by use of a special eluting buffer containing a large number of differently charged species. Proteins emerge from the column as a series of sharply focused zones in descending order of their pI values, with peak widths in the range 0.04–0.05 pH unit.

chromatogram the result of a chromatographic separation. It may be directly visible, e.g. in the form of a column or strip of material, or, after processing of data, take the form of a graph.

chromatograph **1** to effect a separation by **chromatography**. **2** an apparatus with which to carry out a particular form of chromatography. —**chromatographic** *adj.*; **chromatographically** *adv.*

chromatography any technique, analytical or preparative, for separating the components of a mixture by differential adsorption of compounds to adsorbents, partition between stationary and mobile immiscible phases, ion exchange, or a combination of these. *See* individual entries for the many different types of chromatography; these include: adsorption, affinity, affinity-elution, ampholyte-displacement, argentation, ascending, biospecific-elution, charge-transfer, circular, countercurrent, covalent, descending, dye-ligand, electro, exclusion, frontal, gas, gas–liquid, gel-filtration, gel-permeation, high-performance liquid-affinity, high-performance liquid, high-pressure liquid, hydrophobic, ion-exchange, ion-exclusion, ionic-interaction, ion-moderated partition, ligand-mediated, liquid, liquid–liquid, metal-chelate affinity, molecular-exclusion, molecular-sieve, negative, paper, partition, permeation, positive, pseudo-affinity, reverse(d)-phase, salting-out, sievorptive, steric-exclusion, subunit-exchange, thermal-elution, thin-layer, triazine-dye affinity.

chromatophore **1** a pigment-containing cell, especially one in the integument of an animal, that is capable of changing the apparent skin colour by contracting or expanding. **2** any of the particles isolated from photosynthetic organisms that contain photosynthetic pigments, e.g. a **chromoplast**, or membrane fragments from photosynthetic bacteria.

chromatopile a stack of filter papers, of the same diameter, that is compressed in a chromatographic column in large-scale separations.

chromatoplate the plate that is covered with the support during **thin-layer chromatography**.

chromatosome a **nucleosome** lacking the linker DNA.

+chrome *comb. form* denoting colour, coloured, or pigment.

chromium *symbol*: Cr; the element of proton number 24. It is a paramagnetic, metallic transition element of group 6 of the IUPAC **periodic table**, with oxidation states of II, III, and IV. Natural chromium has a relative atomic mass of 51.966 and is a mixture of four stable nuclides with nucleon numbers of 50, 52, 53, and 54, the predominant one being chromium-52 (relative abundance 83.789 atom percent). A number of artificial radioactive nuclides are known, of which **chromium-51** is the longest lived. Chromium has an abundance of about 0.02% of the igneous rocks of the earth and it occurs in nature exclusively in the combined form, the mixed oxide

chromite $Fe_2Cr_2O_4$ being the chief ore. It is an essential dietary trace metal, but functions in the body primarily as a component of **glucose-tolerance factor**, the chief symptom of experimental chromium deficiency in animals being impaired glucose tolerance.

chromium-51 the artificial radioactive nuclide of chromium, $^{51}_{24}C$. It decays by electron capture, 9.8% of nuclei emitting γ-radiation (0.32 MeV); it also emits X-rays (0.005 MeV, 22%) but no β-particles; it has a half-life of 27.7 days. It is used as chromate for labelling blood cells, and as the 1:1 complex of Cr^{2+} with EDTA for measuring glomerular filtration rate.

chromo+ *or (before a vowel)* **chrom+** *comb. form* **1** denoting colour, coloured, or pigment. **2** indicating **chromium**.

chromobindin *see* annexin I, annexin II, annexin IV, annexin VI.

chromodomain a protein domain that recognizes the C-terminal tails of nucleosomal histones that contain specifically methylated lysine residues. It consists of an alpha helix and three beta strands. Chromodomains are present in the HP1 family of heterochromatin-associated proteins, in diverse chromatin remodelling factors (e.g. Mi2 of **NURD**), and in histone methyltransferase.

chromogen **1** the colourless precursor of a dye or pigment that is converted into the dye or pigment, e.g. by oxidation. **2** a compound containing a **chromophore** that is not itself coloured but becomes so when an auxochromic group (*see* **auxochrome**) is introduced into the molecule. **3** any microorganism that produces pigment.

chromogenic **1** producing colour. **2** of, or pertaining to, a **chromogen**.

chromogenic substrate a substrate that, when acted on by an enzyme, gives rise to a coloured product.

chromogranin *abbr.*: Cg; either of two soluble proteins, originally discovered in bovine chromaffin granules (*see* **chromaffin tissue**), but now known to be widely distributed in endocrine (including neuroendocrine) tissues. They are designated chromogranin A (CgA) and chromogranin B (CgB; *also called* **secretogranin I**). Both proteins are acidic (pI 4.5–5.0), but similarity is limited except in the N- and C-terminal regions, and there is a lack of immunological reactivity. Both have been detected immunologically in various mammals, amphibians, birds, and fishes; CgA also occurs in *Drosophila*. CgA is a 430 amino acid Ca^{2+}-binding glycoprotein, it is hydrophilic, with no hydrophobic stretches, and is co-secreted with many hormones from, e.g., pancreas and thyroid C cells. Both proteins have **dibasic cleavage sites** for endopeptidases, the action of which leads to the formation of several active peptides: CgA yields **parastatin** (porcine $CgA_{347–419}$), an inhibitor of low calcium-stimulated parathyroid hormone secretion; **pancreastatin** (porcine $CgA_{240–288}$), an inhibitor of glucose-induced insulin release from the pancreas; **chromostatin** (bovine $CgA_{124–143}$), an inhibitor of catecholamine release by chromaffin cells; **vasostatin** (bovine $CgA_{1–76}$); **beta-granin** (bovine $CgA_{1–114}$), an inhibitor of parahormone secretion; and a peptide WE-14. CgB is co-secreted from adrenal medulla with catecholamines, from sympathetic nerve with norepinephrine, and from some cell lines; proteolytic cleavage releases several peptides of unknown function.

chromomere any of the beadlike concentrations of **chromatin** that are arranged linearly along a **chromosome**.

chromonema (*pl.* **chromonemata**) *cytological term for* **1** all of the threads which make up the **nuclear reticulum**. **2** any of the smallest strands of DNA in a chromosome or chromatid. **3** a twisted chromatid thread within the chromosome.

chromophore any functional grouping within a molecular entity that gives rise to the colour (or characteristic absorption spectrum) of the substance.

chromoplast any yellow, orange, or red **plastid** containing carotenoid pigments but not chlorophyll, the colour depending on the relative amounts of carotenoids and xanthophylls present. Chromoplasts give the yellow to red colours of many flowers (e.g. daffodil), fruits (e.g. tomato), and some roots (e.g. carrot). They synthesize and store carotenoids in their membranes and when loaded with carotenoids are called **carotenoid bodies**. In ripening fruit tissues these are specifically associated with fibrillin-type proteins.

chromoprotein any of a class of proteins that contain a colour-producing group or **chromophore**. Both chlorophylls and carotenoids exist in chloroplasts as chromoproteins.

chromosomal inversion *see* **inversion** (def. 3).

chromosomal translocation the exchange of segments of non-homologous chromosomes. Translocations are apparent when the chromosome number is correct, but two chromosomes have abnormal sizes and banding patterns. Some are associated with malignancy.

chromosome a structure composed of a very long molecule of **DNA** (in humans 50–250 Mb long) and associated proteins (e.g. **histones**) that carries hereditary information. Chromosomes are especially evident in plant or animal cells undergoing mitosis or meiosis, where each chromosome becomes condensed into a compact, readily visible thread. In nondividing cells chromosomes typically assume a more dispersed form called **chromatin**. The number of chromosomes is characteristic for the species concerned. In a bacterium only one chromosome is evident as the cell is about to divide. This is a circular molecule of DNA, 4.6 Mb long in *E. coli*, with attached proteins. After DNA replication, the two new chromosomes attach to a specialized site on the bacterial plasma membrane for segregation to the two daughter cells. Mitochondria and chloroplasts contain several copies of a much smaller circular DNA molecule with some attached proteins. —**chromosomal** *adj.*

chromosome band any of numerous transverse bands along a chromosome that show differential intensity of staining. The pattern of such bands is highly characteristic for each chromosome of a given species, and is thus useful for large-scale physical mapping. The bands are an effect of staining, and were thought to be related to regional base composition. However, it is now believed that the intensity of staining depends on the packing density of the chromatin: more tightly packed regions contain a greater density of DNA to take up stain.

chromosome mutation a type of **mutation** (def. 1) involving alteration(s) in chromosome structure, including breakage and loss of chromosomal segments. Some deletions are associated with human genetic diseases (e.g. in chromosome 4, giving rise to Charcot–Marie–Tooth disease, and chromosome 22, leading to DiGeorge syndrome). *See also* **chromosomal translocation**, **inversion** (def. 3).

chromosome walking a method used to identify which clone in a gene bank contains a desired gene or sequence that cannot easily be selected for. The gene bank must contain the entire DNA sequence of the chromosome as a series of overlapping fragments. A series of **colony hybridizations** is carried out, starting with a cloned gene that has already been identified and that is known to be on the same chromosome as the desired gene. This identified gene is used as a probe to pick out clones containing adjacent sequences. These are then used in turn as probes to identify clones carrying sequences adjacent to them, and so on. At each round of hybridization one 'walks' further along the chromosome from the identified gene.

chromostatin a 20-amino-acid peptide that can be released *in vitro* by the proteolytic cleavage of **chromogranin** A. It potently inhibits catecholamine release by **chromaffin** cells induced by carbamoylcholine or by depolarizing K$^+$ concentration, and it inhibits secretagogue-induced calcium influx. Its physiological status is uncertain – it is not flanked by **dibasic cleavage sites** in the precursor.

chronic 1 continuing over a long time; repeatedly recurring. **2** *(of a disease)* developing slowly or lasting for a long time.

chronic external ophthalmoplegia *an alternative name for* **Kearns–Sayre syndrome**.

chronic granulomatous disease a rare condition characterized by suppurative bacterial and fungal infections starting in early childhood. Phagocytic cells from patients exhibit normal chemotaxis, degranulation, and phagocytosis but do not experience a respiratory burst that generates superoxide radical and other antimicrobial reactive oxygen species. This defect is caused by a hereditary dysfunction in any of four protein components of the NADPH oxidase complex.

chrono+ *or (before a vowel)* **chron+** *comb. form* denoting time.

chronobiochemistry the branch of biochemistry concerned with the biochemical aspects of **chronobiology**, the study of the inherent rhythmicity or periodicity of living organisms and their activities.

chrysolamarin *see* **callose**.

chrysotherapy the use of gold-containing substances in the treatment of disease.

CHS *(in clinical biochemistry)* abbr. *for* cholinesterase (def. 1).

ChTX *abbr. for* charybdotoxin.

CHX10 a human homeodomain transcription factor (361 amino acids) that is highly conserved in mammals and abundantly expressed in the adult retina. Homozygous null mutations affect only the eye and produce microphthalmia, cataracts, and blindness.

chyle the white lymph found in the lymph vessels of the intestine (lacteals) during the digestion of fats. It contains globules with a high fat content. *See also* **chylomicron**.

chylomicron a large lipoprotein particle, of molecular mass up to 10^7 kDa, measuring up to 100 nm in diameter, and of density < 1.006. Chylomicrons appear in plasma during absorption of a fat-containing meal and normally disappear by 12–15 hours. Their approximate composition (% by weight) is 2% cholesterol, unesterified; 5% cholesterol, esterified; 7% phospholipid; 84% triacylglycerol; 2% protein. Their apolipoprotein composition (% by weight total apolipoprotein) is 7.4% A-I; 4.2% A-II; 22.5% B-48; 66% C-I + C-II + C-III. Chylomicrons are formed in the mucosa of the small intestine, and contain mostly triacylglycerols re-esterified in the mucosal cells from dietary long-chain fatty acids associated with apolipoproteins. They then enter the lymphatic capillaries (lacteals) of the intestinal villi, and thence the bloodstream. Binding to lipoprotein lipase on the capillary endothelial surface deprives them of triglycerides, and converts them into chylomicron remnants. They are the vehicle by which fat and fat-soluble vitamins are absorbed. **Hyperchylomicronemia** results from deficiency of lipoprotein lipase. **Chylomicron retention disease** is a defect in secretion of chylomicrons characterized by fat malabsorption, absence of chylomicrons in plasma after ingestion of fat, and sometimes acanthocytosis and neurological abnormalities. There is accumulation in enterocytes of chylomicron-like particles that are rich in ApoB-48 (*see* **apolipoprotein B**).

chylomicron remnant a particle derived by the impoverishment of a **chylomicron** of triacylglycerols (triglycerides) by the action of lipoprotein lipase. Such remnants are enriched in cholesterol ester and contain less apolipoprotein A and apolipoprotein C than chylomicrons. In the liver, hepatic lipase degrades the remaining triacylglycerols, and remnants bind, through their apolipoprotein E, to low-density lipoprotein (LDL) receptors and to LRP in coated pits. They are then endocytosed and degraded in lysosomes. Chylomicron remnants accumulate – along with very-low-density lipoprotein (VLDL) remnants – in remnant (or type III) hyperlipidemia.

chymase EC 3.4.21.39; *other names*: mast cell protease I; skeletal muscle (SK) protease. An enzyme that catalyses preferential cleavage of the bonds: Phe-|-Xaa > Tyr-|-Xaa > Trp-|-Xaa > Leu-|-Xaa. It is a trypsin-like serine proteinase, and a homologue of chymotrypsin; it is found in mast cell granules (heart, lung, etc.).

chyme the semifluid mass of partially digested food that passes through the pylorus of the stomach into the small intestine.

chymopapain EC 3.4.22.6; *other name*: papaya proteinase II; the major endopeptidase enzyme of papaya (*Carica papaya*) latex; its specificity is similar to that of **papain**.

chymosin *or* **rennin** EC 3.4.23.4; an aspartic proteinase enzyme that cleaves a single bond in κ casein (Ser-Phe^{105}-|-Met-Ala) and is responsible for clotting of milk. It is produced from a precursor in the mucosa of the fourth stomach of calves. It is now produced by cDNA technology. (*Note*: use of the synonym rennin is now deprecated because of possible confusion with **renin**.)

chymotrypsin a serine endopeptidase, EC 3.4.21.1, formed from the proenzyme **chymotrypsinogen** by the action of trypsin. In vertebrates, chymotrypsins are contained in **pancreatic juice**, and hydrolyse protein in the small intestine. Chymotrypsinogen is synthesized in the acinar cells of the exocrine pancreas and stored in **zymogen** granules until released. **Chymotrypsin A** is formed from bovine and porcine chymotrypsinogen A. It hydrolyses peptides, amides, and esters, preferentially at the carbonyl end of Tyr, Trp, Phe, and Leu residues. **Chymotrypsin B**, formed from chymotrypsinogen B, is homologous with chymotrypsin A. **Chymotrypsin C**, EC 3.4.21.2, is formed from porcine chymotrypsinogen C and from bovine subunit II of procarboxypeptidase A. It preferentially cleaves at the carbonyl end of Leu, Tyr, Phe, Met, Trp, Gln, and Asn residues. Enzymes homologous to and with similar specificities to chymotrypsin have been isolated from many species. Chymotrypsinogens are synthesized as a single 245-residue polypeptide chain cross-linked by five disulfide bonds. This is fully activated when the bond between Arg^{15} and Ile^{16} is cleaved by trypsin to yield π-chymotrypsin; subsequent autocatalytic cleavage of more peptide bonds (to remove 14–15, 147–148) converts it into α-chymotrypsin. Amino acids 1–13 form the A-chain, 16–146 form the B-chain, and 149–245 form the C-chain of chymotrypsin, the three chains being covalently linked by disulfide bonds. The mechanism involves a **charge-relay system** of His^{57}, Asp^{102}, and Ser^{195}. Chymotrypsin is more active than trypsin against certain esters, e.g. N-acetyl-L-tyrosine ethyl ester. Chymotrypsin binds $α_1$-antitrypsin.

chymotrypsinogen see **chymotrypsin**.

chymotryptic of, or relating to, **chymotrypsin**; **chymotryptic digestion** is the partial hydrolysis of a protein by chymotrypsin.

cI857 a temperature-sensitive mutant of **lambda repressor** that has normal activity at 28°C but is denatured at 42°C. Bacteriophage lambda lysogens carrying this mutation can be induced by exposing them to a temperature of 42°C. See **induction**.

Ci 1 symbol for curie. **2** abbr. for cubitus interruptus.

cI, cII, cIII symbols for genes in bacteriophage lambda that give clear plaques when mutated. cI encodes the **lambda repressor**, which blocks the activities of lytic promoters. The cII gene product is necessary for transcription of repressor from the P_{RE} promoter and of the int gene. The cIII gene product inhibits cII proteolysis.

CI abbr. for Colour Index (especially as a prefix to the identifying number for a particular dye or stain).

CIDNP abbr. for chemically induced dynamic nuclear polarization.

CIE abbr. for counterimmunoelectrophoresis.

ciliary neurotrophic factor abbr.: CNTF; a protein that promotes survival of neurons of the parasympathetic ciliary ganglion and elsewhere. The richest sources of CNTF are Schwann cells of peripheral nerves and glial cells of the central nervous system. The **CNTF receptor** consists of a GPI-anchored α subunit, and two homologous transmembrane β subunits. Ligand binding activates JAK1 and JAK2, and STAT proteins (see **JAK**).

ciliatine see **phosphono+**.

cilium (pl. **cilia**) a specialized eukaryotic locomotor organelle that consists of a filiform extrusion of the cell surface. Each cilium is bounded by an extrusion of the cytoplasmic membrane, and contains a regular longitudinal array of **microtubules**, anchored basally in a **centriole**. Compare **flagellum**. See **immotile cilia syndrome**. —**ciliary** adj.

cimetidine N-cyano-N′-methyl-N″-[-2-[[(5-methyl-1H-imidazol-4-yl)methyl]thio]ethyl]guanidine; a competitive antagonist of histamine H_2 receptors used clinically as an anti-ulcer drug; it inhibits gastric acid secretion and reduces the output of pepsin. One proprietary name is Tagamet.

cIMP abbr. for cyclic IMP; see **inosine 3′,5′-phosphate**.

cin4 the gene for the enzyme **trans-cinnamate 4-monooxygenase** found in many plants.

cinchonine an alkaloid obtained from cinchona bark and possessing antimalarial properties. See also **quinine**.

CINEMA abbr. for colour interactive editor for multiple alignments; see **alignment editor**.

C1 inhibitor a plasma glycoprotein of molecular mass 105 kDa that is secreted primarily by the liver and acts as a serine proteinase inhibitor. It binds covalently to, and inhibits, C1r and C1s of the **complement** system. It also inhibits coagulation factors Xi and XII, and

kallikrein. It shares ≈20% homology with α1-antitrypsin. Deletions and null mutations in the gene at 11q11-q13.1 result in a severe form of hereditary angioedema, while point mutations result in a milder form.

cinnamate the anion of 3-phenyl-2-propenoic acid; trans-cinnamate (3-phenyl-2E-propenoate) is a compound formed from phenylalanine by phenylalanine ammonia-lyase, EC 4.3.1.5; it is the precursor of ferulic and sinapic acids in the pathway for **lignin** synthesis.

trans-cinnamate 4-monooxygenase abbr.: CA4H; EC 1.14.13.11; systematic name: trans-cinnamate,NADPH:oxygen oxidoreductase (4-hydroxylating); other names: cinnamic acid 4-hydroxylase; cinnamate 4-hydroxylase. An enzyme that catalyses a reaction between trans-cinnamate, NADPH, and dioxygen to form 4-hydroxycinnamate, $NADP^+$, and H_2O. It is an endoplasmic reticulum enzyme of high specificity – it does not act on amino acids, phenylacetate, or benzoate – and is the first enzyme in the pathway from **cinnamate** to **lignin**.

cinoxacin azolinic acid; 1-ethyl-1,4-dihydro-4-oxo[1,3]dioxolo[4,5-g]cinnoline-3-carboxylic acid; a **4-quinolone antibiotic** with activity against Gram-negative bacteria through the inhibition of DNA gyrase.

ciprofloxacin a fluorinated **4-quinolone antibiotic** active against both Gram-positive and Gram-negative bacteria, chlamydias, and mycobacteria. It inhibits prokaryotic type II DNA topoisomerase.

circadian describing biological activity (e.g. behavioural, physiological, metabolic) that exhibits an endogenous periodicity of approximately 24 hours independently of any daily variation in the environment. Compare **diurnal**, **infradian**, **ultradian**. [From Latin circa, about, + dies, day.]

Circe effect the phenomenon underlying any physicochemical description of the action of an enzyme upon a substrate. It refers to the strong attractive forces that bind the substrate into the active site of the enzyme, where it undergoes a radical transformation of form and structure. It is named after the sorceress, Circe, who according to Greek mythology transformed the men of Odysseus into beasts.

circular birefringence the **birefringence** produced by left and right circularly polarized light.

circular chromatography a technique of paper chromatography in which the substances to be separated are allowed to travel radially from the centre of a filter-paper disk. The material is applied to the centre of the disk from which a small sector is cut out and folded down to allow it to dip into the eluting solvent.

circular chromosome a circular DNA duplex, the form taken by the chromosomes of some prokaryotic organisms, e.g. Escherichia coli, Bacillus subtilis, and Pseudomonas aeruginosa.

circular dichroism abbr.: CD; the difference in absorption of left and right circularly polarized light. The shape and magnitude of the CD curve as a function of wavelength of solutions of proteins (and other macromolecules) are sensitive to changes in conformation of these solutes. See also **Cotton effect**, **ellipticity**, **molecular ellipticity**, **optical rotatory dispersion**.

circular DNA any single- or double-stranded DNA molecule with a circular (but not necessarily geometrically circular) structure. If single-stranded, the circle is closed covalently, but if double-stranded, one or both strands may be open. See also **covalent (closed) circle of DNA**.

circularly polarized light see **polarized light**.

circular peptide a naturally occurring polypeptide (of 14–70 residues) whose N- and C-termini are joined by a peptide bond. It is derived from a DNA-encoded protein precursor. For example, bacteriocin AS-48 of Enterococcus faecalis S-48 is a circular peptide that contains 70 residues, is hydrophobic, mostly alpha-helical, and forms pores in the plasma membrane of sensitive organisms. Microcin J25 of E. coli contains 21 residues of sequence unrelated to other microcins, consists mostly of beta-sheet structure, and interferes with cell division in sensitive organisms. Other examples include cyclotides of plants, and rhesus theta-defensin-1 (RTD-1) of leukocytes from Macaca mulatta, which contains 18 residues and three disulfide bonds, forms two beta strands, and is antibacterial.

circular permutation a linear DNA molecule whose sequence appears as if produced by cleaving circular molecules at different points. For example, circular permutations of ABCDEFG would be CDEFGAB, EFGABCD, etc.

cis *abbr.*: c; *conformational descriptor* denoting synperiplanar (use not recommended); *see* **conformation**.

cis- *abbr.*: c-; *prefix (in chemical nomenclature)* denoting that two specified substituents lie on the same side of a reference plane in the molecule. (*cis* means 'on this side'.) In systematic names it is denoted by the symbol *Z*. *See* **cis–trans isomerism**, **E/Z convention**. *Compare* **trans-**.

CIS *abbr. for* cytokine-inducible SH2 protein; any of a family of proteins that are induced by various **cytokines** (e.g. interleukins 2 and 3, erythropoietin) and contain an SH2 domain. They bind cytokine receptors and suppress cytokine signalling through **STAT proteins**.

cis-acting *(in genetics)* describing a regulatory genetic element whose effects are sensitive to its position relative to the gene being regulated, i.e. on the same molecule of DNA. Examples include bacterial **operators**.

cis configuration *(in genetics)* describing the configuration of two linked heterozygous genes in which both wild-type alleles occur on one of the paired homologous chromosomes while the mutant alleles occur on the other homologue; i.e. (++/*ab*). *Compare* **trans configuration**.

cis isomer *see* **cis–trans isomer**.

CISP *abbr. for* corticotropin-induced secreted protein; now called thrombospondin-2 (*see* **thrombospondin**).

cisplatin *cis*-diammineplatinum(II) dichloride; *cis*-diaminedichloroplatinum; a substance that interacts with, and forms cross-links between, DNA and proteins. It is used as a neoplasm inhibitor to treat solid tumours, primarily of the testis and ovary.

cisterna (*pl.* **cisternae**) **1** *(in cytology)* the flattened, saclike space enclosed between paired membranes of the **endoplasmic reticulum** and the **Golgi apparatus**. **2** *(in anatomy)* an enlarged space, such as the cisterna magna in the rear of the brain or the cisterna chyli between the crura of the diaphragm. —**cisternal** *adj.*

cis–trans complementation test a genetic test to determine whether two gene mutations (*a* and *b*) occur in the same functional gene (or **cistron**). It involves comparison of heterozygotes in which the relevant mutations are in the same chromosome, i.e. in the *cis* configuration (++/*ab*), with heterozygotes in which the mutations are in separate chromosomes, i.e. the *trans* configuration (*a*+/+*b*).

cis–trans isomer *or (sometimes)* **geometric isomer** *or* **geometrical isomer** either of a pair of **diastereoisomers** that differ only in the positions of atoms relative to a specified reference plane in the molecules, in cases where these atoms are, or are considered as if they were, parts of a rigid structure. The two isomers are designated the *cis* isomer and the *trans* isomer when the two particular atomic groupings lie respectively on the same or on opposite sides of the reference plane. *See also* **cis–trans isomerism**. —*cis–trans*-isomeric *adj.*

cis–trans isomerism *or (sometimes)* **geometrical isomerism** the phenomenon of the existence of *cis–trans* isomers in general, or the occurrence of paired *cis* and *trans* isomers in a particular molecule. Atoms or groups are termed *cis* or *trans* to one another when they lie respectively on the same or on opposite sides of a reference plane identifiable as common among stereoisomers. For compounds containing only doubly bonded atoms the reference plane is that containing the double-bonded atoms and is perpendicular to the plane containing these atoms and those directly attached to them. For cyclic compounds the reference plane is that in which the ring skeleton lies or to which it approximates. *See also* **E/Z convention**.

cistron a unit of DNA sequence that codes for a single polypeptide or protein. It may be a smaller unit than the **gene**, which contains the unit of DNA determining a single character. Whether or not a cistron is the same as a gene can be determined by the **cis–trans complementation test**, hence the name cistron (*cis* means 'on this side').

Cit *symbol for* a residue of the α-amino acid L-citrulline.

citraconate 1 *the trivial name for* methylmaleate, *cis*-1,2-propylenedicarboxylate, (*Z*)-2-methyl-2-butenedioate, the dianion of citraconic acid, (*Z*)-2-methyl-2-butenedioic acid. **2** any mixture of citraconic acid and its mono- and dianions. **3** any salt or ester of citraconic acid.

citraconic acid

citraconoyl the bivalent *cis*-acyl group, –CO–CH=C(CH₃)–CO–, derived from citraconic acid. *Compare* **citraconyl**.

citraconyl either of the isomeric univalent *cis*-acyl groups, HOOC–C(CH₃)=CH–CO– or HOOC–CH=C(CH₃)–CO–, derived from citraconic acid. *See also* **citraconylation**.

citraconylation introduction of one or more **citraconyl** groups into a substance by acylation. Similarly to **maleylation** (and succinylation), citraconylation of a protein with citraconic anhydride is used to acylate its free lysine (and other) amino groups in order to change their charge at neutral pH from positive to negative and to render the adjacent peptide bond resistant to hydrolysis by trypsin. Reaction occurs at pH 8 and the citraconyl groups can be removed at pH 3.5 (both operations at 20°C). Introduction of citraconyl groups is slower than is the case with maleyl groups, but their removal is faster and more complete. —**citraconylate** *vb.*; **citraconylated** *adj.*

citrase *see* **citrate lyase**.

citratase *see* **citrate lyase**.

citrate 1 *the trivial name for* β-hydroxytricarballylate, 2-hydroxy-1,2,3-propanetricarboxylate,

$$HO-C(CH_2-COO^-)_2-COO^-,$$

the trianion of citric acid, 2-hydroxy-1,2,3-propanetricarboxylic acid. **2** any mixture of free citric acid and its anions; the trianion is the predominant form at physiological pH values. It is widely distributed in nature and is an important intermediate in the **tricarboxylic-acid cycle** and the **glyoxylate cycle**. Citrate (alone or in mixtures) is useful as a buffer; pK_{a1} (25°C) = 3.13, pK_{a2} (25°C) = 4.76, pK_{a3} (25°C) = 6.40. It is useful also as a component of anticoagulant solutions and has many commercial applications in the chemical, food, pharmaceutical, and other industries. Industrial production of citrate is usually by fermentation of inexpensive sources of glucose or sucrose with selected strains of *Aspergillus niger*; world production in 1992 was 5×10^8 tons. **3** any (partial or full) salt or ester of citric acid.

citrate aldolase *see* **citrate lyase**.

citrate cleavage enzyme EC 4.1.3.8; *recommended name*: ATP–citrate (*pro-S*)-lyase. An enzyme that catalyses a reaction between ATP, citrate, and CoA to form acetyl-CoA, oxaloacetate, ADP, and orthophosphate. In mammals it participates in the release of acetyl-CoA in the cytosol, particularly as a precursor for fatty acid synthesis, there being no system for transporting excess acetyl-CoA out of the mitochondrion. The acetyl carbons are thus transported across the mitochondrial membrane as citrate and then released by this enzyme.

citrate cycle *another name for the* **tricarboxylic-acid cycle**.

citrate (*pro-3S*)-lyase EC 4.1.3.6; *other names*: citrase; citratase; citritase; citridesmolase; citrate aldolase; a bacterial enzyme that catalyses the formation of acetate and oxaloacetate from citrate. ATP–citrate lyase is another name for **citrate cleavage enzyme**.

citrate (*Re*)-synthase EC 4.1.3.28; an enzyme from some bacteria that forms citrate with the incoming –CH$_2$–COOH group in the *pro-R* position of citrate. The addition is to the *Re*-face of the carbonyl group of oxaloacetate.

citrate (S*i*)-synthase EC 4.1.3.7; *other names*: citrogenase; oxaloacetate transacetase; condensing enzyme; citrate condensing enzyme. An enzyme that catalyses a reaction between acetyl-CoA, H$_2$O, and oxaloacetate to form citrate and CoA. It synthesizes citrate from oxaloacetate and acetyl-CoA, placing the incoming –CH$_2$–COOH in the *pro-S* position in citrate (*see* **oxaloacetate**). The addition is to the *Si*-face of the carbonyl group of oxaloacetate. This is the usual mammalian citrate synthase; mitochondrial enzymes are closely related homodimers (six motifs).

citric acid *see* **citrate**.

citric-acid cycle *an alternative name for* **tricarboxylic-acid cycle**.

citridesmolase *see* **citrate lyase**.

citritase *see* **citrate lyase**.

citrovorum factor *(formerly)* an alternative name for **folinic acid**.

citrulline N^5-carbamoyl-L-ornithine; 2-amino-5-ureidovaleric acid; an α-amino acid, not found in proteins. L-Citrulline is an intermediate in the **urea cycle** in animals, and in a modified urea cycle in plants. It was first isolated from the juice of watermelon, *Citrullus vulgaris*.

citrullinemia *or* **citrullinuria** an autosomal recessive disease caused by deficiency of the enzyme **argininosuccinate synthase**. It is characterized by hyperammonemia, hypercitrullinemia, and citrullinuria. It has a neonatal onset and if untreated results in neurological damage and mental retardation.

CJD *abbr. for* Creutzfeldt–Jakob disease.

c-Jun N-terminal kinase *see* **JNK**.

CKI *abbr. for* **Cdk inhibitor**.

C-kinase *see* **protein kinase C**.

Cl *symbol for* chlorine.

clade 1 a monophyletic taxonomic group, i.e. a group that includes all descendants of a single ancestor. **2** a subtree of a rooted phylogenetic tree comprising a given node and all of its descendants.

cladinose 2,6-dideoxy-3-*C*-methyl-3-*O*-methyl-L-*ribo*-hexose; 3-*O*-methyl mycarose; a component of **erythromycins** A and B.

β-L anomer

cladistics phylogenetic studies aimed at establishing evolutionary relationships between different species or other groupings based on the identification of natural **clades**. —**cladistic** *adj.*; **cladistically** *adv.*

cladogenesis a process of adaptive evolution leading to the development of a greater variety of organisms; branching evolution. —**cladogenetic** *adj.*

cladogram a **phylogenetic tree** that illustrates ancestor-descendant relationships determined by cladistics.

***Cla*I** a type 2 **restriction endonuclease**; recognition sequence: AT↑CGAT. The M.*Cla*I modification site is A5.

clan *(in protein sequence analysis)* a group of families for which there is evidence of evolutionary relatedness (e.g. shared function) but between which there is no significant sequence or structural similarity. For example, the serine proteases have been grouped into six clans on the basis of structural and functional similarities, but the sequences and structures of the different clans appear to be unrelated.

clarify to make or become clear, especially to remove finely divided suspended material from a solution, e.g. by centrifugation or filtration. —**clarified** *adj.*; **clarification** *n.*

Clark and Lubs buffers one of the early series of buffer mixtures of predetermined pH, covering the range pH 2.2–10.0. [After William Mansfield Clark (1884–1964), US biochemist, and Herbert A. Lubs.]

Clark electrode a type of **oxygen electrode**. [After Leland Charles Clark (1918–).]

class a taxonomic category ranking below the **phylum** (or division) and containing several **orders** (or sometimes only a single order).

classical pathway of complement activation *see* **complement**.

class switch a change (switch) in the expression of B lymphocytes from one class of antibody to another.

clastic dividing into parts, especially of a chemical reaction in which one molecular species is divided into parts; e.g. phosphoroclastic, thioclastic.

clastogen an agent that is able to bring about breakage in chromosomes. —**clastogenic** *adj.*

clathrate 1 *or* **cage compound** *(in chemistry)* a solid complex or an addition compound in which the crystal lattice or structure of one component (the **host molecule**) completely encloses spaces in which a second component (the **guest molecule**) is located. There is no chemical bonding between the two components. *Compare* **inclusion complex**. **2** designating or relating to a clathrate. **3** *(in biology)* lattice-shaped.

clathrin a 180 kDa protein that is the main component of the coat of **coated vesicles** and **coated pits**, which are involved in the transfer of molecules between the membranous components of eukaryotic cells. Clathrin also occurs in **synaptic vesicles**. The clathrin coat is a polyhedral lattice of subunits; each subunit, or clathrin **triskelion**, comprises three heavy chains and three light chains.

clathrin uncoating protein *or* **uncoating ATPase** a protein from brain or erythrocyte cytosol that disassembles **clathrin** coats to **triskelions** in an ATP-dependent manner. It is thought that the uncoating ATPase is a chaperone from the hsp70 family, and that the dismantling process may be triggered by auxillin, a protein that can be found attached to clathrin-coated vesicles.

clavaminate synthase a nonheme iron-containing oxygenase, EC 1.14.11.21, from *Streptomyces* spp., that catalyses three steps in the biosynthesis of clavulanic acid: the first step (hydroxylation) is separated from the latter two (oxidative cyclization and desaturation) by the action of EC 3.5.3.22, proclavaminate amidinohydrolase.

clavulanic acid 3-(2-hydroxyethylidene)-7-oxo-4-oxa-1-azabicyclo[3.2.0]heptane-2-carboxylic acid; a weak antibacterial agent but potent β-lactamase inhibitor produced by *Streptomyces clavulingerus*. It protects lactamase-sensitive but otherwise potent antibiotics from deactivation, and is used in combination with β-lactam antibiotics to increase their efficacy. It has no antibacterial activity but by inactivating penicillinases it makes a combination active against penicillin-resistant bacteria. These include *Staphylococcus aureus*, 50% of many *Bacteroides*, and *Klebsiella* spp. A commonly used proprietary combination is Augmentin.

clearance *(in physiology)* symbol: *C*; a measure of the efficiency of the kidney in removing a substance from the blood. $C = UV/P$, where *U* and *P* are the concentrations of the substance in the urine and plasma, respectively, and *V* is the rate of urine flow in ml min^{-1}. Clearance is therefore a rate, with the dimensions of ml plasma per min.

clearing 1 *(in histology)* the process of preparing tissue samples for microscopical examination by removing the dehydrating agent and rendering the sample transparent. **2** *(in physiology)* the removal of a substance from the body; *see* **clearance**.

clearing factor *an alternative name for* **lipoprotein lipase**; an enzyme found bound to heparan sulfate on the capillary endothelium of various tissues, notably heart, lung, skeletal muscle, and adipose tissue, so called because it clears the opalescence from lipemic plasma. It is released into plasma by intravenous injection of heparin.

cleavage 1 the tendency of a crystal to split in certain preferred directions with the formation of smooth surfaces, or the act of so splitting a crystal; hence **cleavage plane**, **cleavage face**. **2** the rupture of a (usually specified) chemical bond in a molecule with the formation of smaller molecules. **3** *or* **segmentation** *(in embryology)* the series of mitotic divisions by which a fertilized animal ovum changes, without any overall change in size, into a ball of smaller cells constituting the primitive embryo.

cleavage division the mitotic division of a fertilized egg up to the point at which the regions of the egg shift relative to each other.

cleavage-polyadenylation specificity factor *see* **CPSF**.

cleave to split or cause to divide, especially along the line of a natural weakness; to split a chemical bond; to effect or undergo **cleavage**. —**cleavable** *adj*.

cleft lip/palate a dysmorphology of the head of mammals resulting from mutation of any one of numerous genes such as Nectin-1.

Cleland's reagent *see* **dithiothreitol**. [After William Wallace Cleland (1937–).]

Cl⁻/HCO₃⁻ exchanger *an alternative name for* **anion channel protein**.

climacteric 1 *(in medicine)* the female menopause, or the corresponding time of life in men. **2** *(in botany)* the marked increase in respiration in some fruits (e.g. tomato) just before and during ripening. This is usually associated with an increase in production of ethylene.

clindamycin (7S)-7-chloro-7-deoxylincomycin; a semisynthetic antibiotic that is useful against anaerobic bacteria; for structure and action *see* **lincomycin**, **lincosamide**.

clinical of, or pertaining to, the course of disease in an individual, or the examination and treatment of an individual directly, as opposed to the laboratory investigation of disease.

clinical chemistry a branch of applied biochemistry concerned with the nature and determination of chemical substances of interest in the investigation of diseases.

CLIP *abbr. for* **1** corticotropin-like intermediate peptide. **2** Class-II HLA-associated invariant-chain peptide (*see* Class II MHC antigens in **major histocompatibility complex**).

CLIP-170 *see* **restin**.

CLN *abbr. for* ceroid lipofuscinosis (neuronal).

clock 1 *abbr. for* circadian locomotor output cycles kaput; a basic helix-loop-helix (HLH) transcription factor expressed primarily in the suprachiasmatic nucleus of the hypothalamus . First detected in *Drosophila* as a key regulator of circadian rhythms, it occurs widely in the animal kingdom. It forms heterodimers with the basic HLH proteins **BMAL1** and **NPAS2** and binds to a specific DNA motif to regulate circadian activity, by stimulating expression of genes encoding period proteins and **cryptochromes**. Transcription of target genes for the clock-BMAL1 dimer is repressed by cryptochromes. **2** *see* **molecular clock**.

clofazimine *N*,5-bis(4-chlorophenyl)-3,5-dihydro-3-[(1-methylethyl)imino]-2-phenazinamine; an iminophenazine dye with anti-inflammatory, tuberculostatic, and leprostatic activity. It is used to treat multibacillary or sulfone-resistant leprosy.

clofibrate ethyl 2-(*p*-chlorophenoxy)-2-methylpropionate; 2-(4-chlorophenoxy)-2-methylpropanoic acid ethyl ester; a drug that reduces high plasma levels of triacylglycerols, very-low-density lipoproteins (VLDL), and (less so) cholesterol. Its actions are probably mediated by the stimulation of **lipoprotein lipase**, which reduces the triacylglycerol composition of VLDL and stimulates clearance of low-density lipoproteins by liver. It is used primarily in the treatment of type III (but also in type IV, type V, and type IIb) hyper-

lipoproteinemia. Clofibrate causes proliferation of peroxisomes in rat, but not in human, liver. *Compare* **gemfibrozil**.

clomifene *or* **clomiphene** 2-[4-(2-chloro-1,2-diphenylethenyl)phenoxy]-*N*,*N*-diethylethanamine; an antiestrogenic aminoether derivative of stilbene; it blocks estradiol receptors in the hypothalamus and may stimulate the secretion of gonadotropin-releasing hormone. It is active as an antitumour agent against mammary carcinomas, and can be used to induce ovulation in infertile women. Both its structure and actions are similar to those of **tamoxifen**.

clonal culture any culture of cells produced in such a way that an individual **clone** (def. 1) can be selected.

clonal selection theory *or* **Burnet's theory of immunity** a theory to account for the phenomenon of **acquired immunity** and the ability of the immune system to produce large quantities of highly specific antibody for any particular immunogen. According to the theory, proposed by the Australian physician (Sir) Frank Macfarlane Burnet (1899–1985), the ability to produce a given antibody is a pre-existing and genetically determined characteristic of a particular clone of B lymphocytes. An encounter with the antigen triggers proliferation of the corresponding clone, producing a greatly expanded population of cells that secrete specific antibody. Some members of the expanded clone assume the role of memory cells, which respond to any subsequent encounter with the same immunogen, which accounts for the secondary response being greater than the primary response.

clone 1 the descendants produced vegetatively or by apomixis from a single plant, or asexually or parthenogenetically from a single animal; a group of cells produced by division from a single cell. The members of a clone are of the same genetic constitution, unless **mutation** has occurred amongst them. **2** a homogeneous population of DNA molecules; *see* **molecular cloning**. **3** to propagate a selected organism or cell, or to replicate a particular DNA molecule so as to produce a clone. —**clonal** *adj.*; **cloning** *n*.

cloned library *an alternative name for* **gene library**.

clonidine 2,6-dichloro-*N*-2-imidazolidinylidenebenzenamine; an antagonist at α₂ (presynaptic) adrenergic receptors, P₁ purinergic receptors, and H₂ histamine receptors. It is used as a central antihypertensive drug; it also abolishes most symptoms of opiate withdrawal.

cloning vector *or* **cloning vehicle** the DNA of any transmissible agent (e.g. plasmid or virus) into which a segment of foreign DNA can be spliced in order to introduce the foreign DNA into cells of the agent's normal host and promote its replication and transcription therein. (1) **Plasmid vectors** contain (i) an origin of replication, so that the plasmid can be replicated; (ii) an antibiotic resistance gene to allow selection of transformed host cells; (iii) often, a 'polylinker' containing several different restriction enzyme recognition sites and a second marker gene (e.g. *lacZ*; *see* **X-gal**), which will be inactivated when foreign DNA is inserted therein, thus allowing identification of transformants bearing plasmids containing inserts, rather than 'empty' vectors. Examples include **pBR322** and the **pUC vectors**. *See also* **expression vectors**. (2) **Bacteriophage vectors**, usually based on **lambda phage**, can accommodate longer fragments of foreign DNA. In insertion vectors (e.g. λgt10) the foreign DNA is inserted into the phage DNA while in replacement vectors (e.g. λEMBL4) the foreign DNA is ligated between two phage fragments, or arms. (3) Hybrid vectors, **cosmids**, **phagemids** and **phasmids** (def. 1), with features of both plasmids and phage, have also been constructed. *See also* **molecular cloning**.

closed circle (of DNA) *see* **covalent (closed) circle (of DNA)**.

closed double-stranded DNA *see* **covalent (closed) circle (of DNA)**.

closed duplex DNA or **DNA I** a closed double-stranded DNA molecule in which the DNA helical structure is preserved.

closed-loop system a type of system, e.g. a metabolic pathway or an electrical circuit, in which control is exerted by **feedback**. *Compare* **open-loop system**.

closed system any (thermodynamic) system that cannot gain or lose matter, heat, or work. *Compare* **open system**.

close packing the structure of a solid in which nonbonded atoms (or molecules) are surrounded by other nonbonded atoms (or molecules) in such a way that the distance between them is close or equal to the sum of their van der Waals contact radii (*see* **van der Waals radius**).

clostridiopeptidase A the **collagenase** obtained from *Clostridium histolyticum*.

Clostridium a genus of spore-forming rod-shaped bacteria of the family Bacillaceae. Its members are chemoorganotrophic, obligately anaerobic, and typically Gram-positive. They are widespread in soil, mud, and in the intestinal tract of animals. Some species are pathogenic if they gain access to the tissues, and produce a range of toxins.

Clostridium histolyticum collagenase *see* **microbial collagenase**.

clostripain or **clostridiopeptidase B** EC 3.4.22.8; a cysteine proteinase enzyme from *Clostridium histolyticum*. It preferentially cleaves Arg-|-Xaa and Arg-|-Pro bonds and is inactivated by substituted arginine and lysine chloromethyl ketones. It is a heterodimer of α and β chains.

clot retraction the contraction of a blood clot after its formation, accompanied by the expression of serum.

clotrimazole 1-[(2-chlorophenyl)diphenylmethyl]-1*H*-imidazole; an **imidazole** used topically as an antifungal agent. It inhibits ergosterol synthesis and disrupts the transport of amino acids into the fungus.

clotting factor *an alternative term for* coagulation factor; *see* **blood coagulation**.

clotting time *an alternative term for* **coagulation time**.

clover-leaf structure one of the models for the secondary structure of **transfer RNA** (tRNA). The nucleotide sequences of different tRNAs can be arranged in similar two-dimensional representations, resembling a clover leaf in shape, in which there are four bihelical regions, containing the maximal number of base pairs, connected by three loops. Some tRNAs have a small extra loop. The clover leaf structure has been largely supported by the tertiary structure of those tRNAs for which structure has been determined by X-ray crystallography.

cloxacillin a semisynthetic penicillin that is resistant to penicillinase and is useful against certain staphylococci.

Clp protease or **ClpAP protease** a proteinase/ATPase from *Escherichia coli* with little protein specificity (EC 3.4.21.–). It is a **heat-shock protein** but supposedly a housekeeping proteinase; counterparts occur in many organisms (five motifs). The enzyme is a heterodimer with subunits named from the *clp* genes. The proteolytic subunit is ClpP, the ATP-binding subunit is one of ClpA or ClpX. A homologous protein in chloroplasts consists of two stacked heptameric rings of ClpP subunits (each a serine protease of 21 kDa), that are activated by a heptameric ring of ClpA subunits (each an ATPase of 81 kDa), which also unfolds the substrates for degradation. This enzyme is particularly important during leaf senescence.

clupein a **protamine** derived from herring sperm. Arginine constitutes 80–90% of its amino acid content.

Clustal *see* **multiple alignment program**.

cluster 1 (in chemistry) a number of metal centres grouped close together, which can have direct metal bonding interactions or interactions through a bridging ligand. **2** a group of iron and sulfur atoms, e.g. [2Fe-2S] or [4Fe-4S] , in an iron–sulfur protein. The term is sometimes extended to include other metal atoms, as in the Fe_7MoS_9 cluster in nitrogenase. **3** or **cluster of enzymes** or **enzyme cluster** the phenomenon of the emergence of new enzymes, or the upsurge of enzymes already present in low concentration, at certain (usually defined) periods along the age axis of the organism. **4** a set of sequences, genes, or gene products grouped according to given similarity criteria (e.g. distance-based, associative, correlative, probabilistic). Such clusters facilitate the recognition of patterns and reveal otherwise hidden relationships, especially in **high-throughput** data.

cluster analysis a method of statistical analysis used, e.g., in **phylogeny** deduced from phenetic characters.

cluster-crystal any of the globular aggregates of calcium oxalate crystals found in some plants.

cluster database a database whose entries are grouped by means of, and whose groupings depend on, specified clustering parameters, and hence may not map to precise biological roles. For example, CluSTr and ProDom are automatic similarity-based clusterings of sequences in the Swiss-Prot database and its supplement, TrEMBL; COG is a similarity-based protein clustering from an all-against-all comparison of sequenced genomes; UniGene is a dynamic mapping of overlapping sequences from GenBank into clusters representing single genes.

clusterin *an alternative name for* **apolipoprotein J**.

clustering the occurrence of a number of objects, like or unlike, in clusters; the term is used especially of components of cell membranes, such as receptors.

CluSTr *abbr. for* clusters of Swiss-Prot and TrEMBL proteins; *see* **cluster database**.

cm *symbol for* centimetre, i.e. 10^{-2} m.

Cm *symbol for* **1** the carboxymethyl group, $-CH_2-COOH$. **2** curium.

CM *abbr. for* carboxymethyl; e.g. CM-cellulose.

CMAR a gene, cloned from a colon cancer cell line, encoding an 82-amino-acid protein that enhances the attachment of cells to collagen and laminin; the gene is named from cell matrix adhesion regulator. The protein has an N-terminal myristoylation motif (underlined in the partial sequence) and a tyrosine phosphorylation site (bold in the partial sequence), this tyrosine being required for activity. Sequence:

MLRGSDMKGP...............TAL**RIVEI**LY.

CMC *abbr. for* **1** critical micellar concentration. **2** carboxymethylcellulose.

CMH *abbr. for* ceramide monohexoside (Cer(Hex) preferred).

CMP *abbr. for* cytidine monophosphate, i.e. cytidine phosphate.

CMR1 *abbr. for* cold and menthol receptor 1; a protein found in nerve endings that respond to cooling (to 15–25°C), and to menthol, in the mammalian skin and mouth, respectively. It belongs to the transient receptor potential (TRP) superfamily of calcium channels, and is a homotetramer of four 1104-residue subunits. Each subunit has long intracellular N- and C-terminal regions containing coiled-coil domains, and six transmembrane segments. On activation, the channel permits the entry of Ca^{2+} and Na^+ ions into the neuron. CMR1 is identical with TRPM8, which senses heat and capsaicin.

CMR (spectroscopy) *(sometimes) abbr. for* ^{13}C-nuclear magnetic resonance (spectroscopy).

CMU *abbr. for* 3-(*p*-chlorophenyl)-1,1′-dimethylurea; monuron; a herbicide that acts by blocking noncyclic photophosphorylation.

CMV abbr. for **cytomegalovirus**.

CMV promoter *abbr. for* the immediate early cytomegalovirus promoter; a strong promoter used to obtain expression of exogenous genes in mammalian and other higher eukaryotic cells.

Cn3d *see* **visualization software**.

^{13}C-NMR (spectroscopy) *abbr. for* carbon-13 nuclear magnetic resonance (spectroscopy).

CNP *abbr. for* C-type natriuretic peptide; it is similar to atrial **natriuretic peptide**.

CNS *abbr. for* central nervous system.

CNTF *abbr. for* ciliary neurotrophic factor.

co+ *prefix* denoting **1** together, joint or jointly; in partnership, an equality; to the same degree. **2** *(in mathematics)* of the complement of an angle.

Co 1 *symbol for* cobalt. **2** *abbr. for* coenzyme.

Co I *(obsolete) abbr. for* coenzyme I; *see* **nicotinamide adenine dinucleotide**.

Co II *(obsolete) abbr. for* coenzyme II; *see* **nicotinamide adenine dinucleotide phosphate**.

CoA *symbol for* (free or combined) coenzyme A.

coacervate any viscous phase obtained by **coacervation** of solutions of two or more polymers.

coacervate droplet any of the droplets that, it has been suggested, may have formed in the **primordial soup** by **coacervation** and, perhaps, entrapped primitive catalyst(s); they may have given rise to the first cells. *Compare* **protobiont**.

coacervation any process in which there is a spontaneous separation of a continuous one-phase aqueous solution of a highly hydrated polymer into two aqueous phases, one with a relatively high polymer concentration, the other with a relatively low polymer concentration.

coactone an **ecomone** active in the process of relationship between an active and directing organism and a passive and receiving organism.

coagulant any substance that produces, or aids the production of, coagulation.

coagulase any of a number of serologically distinguishable bacterial enzymes that catalyse the coagulation of citrated or oxalated blood plasma. The extracellular bacterial protein specifically forms a complex with human **prothrombin**; this complex can clot fibrinogen without any proteolytic cleavage of prothrombin.

coagulase/fibrinolysin precursor *see* **plasminogen activator**.

coagulate 1 to cause a fluid or part of a fluid to change into a solid or semisolid mass by the action of, e.g., heat or chemical substances; to clot. **2** the solid or semisolid mass produced by coagulation. —**coagulable** *adj.*; **coagulation** *n.*

coagulation factors *see* **blood coagulation**.

coagulation time *or* **clotting time 1** the time taken for a freshly drawn specimen of blood or blood plasma to coagulate (clot) when measured under standard, controlled conditions; it is an indication of the state of the coagulation processes in the specimen. **2** the time taken for a specimen of milk, or of a casein fraction, to coagulate in standard, controlled conditions.

coagulogen a ~16 kDa fibrinogen-like protein that is found in the hemocyte (amoebocyte) of horseshoe crab (*Limulus polyphemus*) hemolymph, and participates in hemostasis. Coagulogen consists of a single, basic polypeptide chain, and constitutes about 50% of the protein in horseshoe-crab hemocyte lysates.

coagulum (*pl.* **coagula**) any semisolid coagulated mass or clot, especially that formed by clotting of blood.

coarctation the state of being pressed together or constricted, especially of the aorta or other blood vessel.

coarse descriptive of **1** things that are rough in texture or structure. **2** particles that are large or powders composed of such particles. **3** threads, filaments, or fibres of relatively large diameter. **4** means of adjustment (as of instruments or processes) that provide only inaccurate or gross control, or the adjustments so made.

CoASAc (occasionally) *symbol for* acetyl-coenzyme A.

CoASH *symbol for* (free) coenzyme A.

coated pit an invagination of the cell membrane of many eukaryotic cells, concerned in receptor-mediated selective transport of many proteins and other macromolecules across the cell membrane. During endocytosis it is converted into a **coated vesicle**. The coat is of **clathrin**.

coated-tube method a solid-phase **radioimmunoassay** method in which the specific antibody is immobilized on the inner surface of a small test tube.

coated vesicle a vesicle, formed from a **coated pit**. Observed in the cytoplasm of many eukaryotic cells, coated vesicles measure 50–250 nm in diameter and are characterized by a 'bristle' coat on the outer surface of their lipid membrane. This coat is made up of a polyhedral lattice of **clathrin** subunits together with smaller amounts of other proteins, notably ones of about 100 kDa (**adaptins**), 50 kDa, and 20 kDa. Coated vesicles are concerned with the rapid and continuous transport of molecules between specific membranous organelles of the cell and to and from the cell membrane.

coatomer a large protein complex that forms part of the coat of specific Golgi intercisternal transport vesicles that are involved in constitutive vesicular transport between endoplasmic reticulum, Golgi apparatus, and plasma membrane. It has seven subunits known as COPs (from coat protein). Example: the yeast coatomer is a 700–800 kDa complex; one component is the transport protein product of *SEC21*.

coat protein 1 one of the outer structural proteins of a coated **virus**. **2** any of those subunits of the **coatomer** known as COPs.

cobalamin the *trivial name for* many members of the **vitamin B$_{12}$** group and their derivatives. A cobalamin is a **cobamide** in which 5,6-dimethylbenzimidazole is the aglycon that forms a link between the cobalt atom and the α1-position of the ribose residue. The sixth position of the cobalt atom may be (artefactually) filled by a variety of groups, e.g. a cyano group giving cyanocobalamin. The oxidation state of the cobalt may be indicated by using a Roman numeral, e.g. cob(ɪ)alamin for Co^{I}. *See also* **intrinsic factor**.

cob(I)alamin adenosyltransferase EC 2.5.1.17; *other name*: aquacob(ɪ)alamin adenosyltransferase; an enzyme that catalyses the formation of adenosylcobalamin from ATP, cob(ɪ)alamin, and H_2O, with release of orthophosphate and pyrophosphate. The enzyme requires manganese.

cob(II)alamin reductase EC 1.6.99.9; *other name*: vitamin B$_{12r}$ reductase; a flavoprotein enzyme that catalyses a reaction between NADH and two molecules of cob(ɪɪ)alamin to form two molecules of cob(ɪ)alamin and NAD$^+$.

cobamide any member of a group of compounds (including the **cobalamins**) that have molecules consisting of a highly substituted tetrapyrrole ring system (corrin) in which a cobalt atom is linked to the four pyrrole nitrogen atoms and with an α-D-ribofuranose 3-phosphate residue attached to one of the side chains.

cobamide coenzyme any of the biochemically active forms of **vitamin B$_{12}$**. They are **cobalamins** in which either a methyl or 5′-deoxy-5′-adenosyl residue is attached to the sixth position on the cobalt atom.

Cobol *see* **programming language**.

cocaine 2β-carbomethoxy-3β-benzoxytropane; an alkaloid obtained from dried leaves of the South American shrub *Erythroxylon coca* or by chemical synthesis. It is a cerebral stimulant and narcotic, and has been used as a local anaesthetic. Cocaine acts by blocking neuronal reuptake of, especially, **norepinephrine**, hence its potent vaso-constrictor effects, which can lead to tissue necrosis. Its use is restricted in many countries.

cocaine

cocarboxylase *a former name for* thiamine diphosphate (*see* **vitamin B$_1$**).

cocarcinogen any noncarcinogenic agent that enhances the effect of a **carcinogen**.

cochlin a protein produced in the cochlea and predicted to be secreted into the extracellular matrix. The gene at 14q12-q13 encodes 550 amino acids, which include two **von Willebrand factor**-like domains. Several missense mutations in the gene cause a form of hereditary deafness.

cochromatography a chromatographic technique in which an unknown substance is applied to a chromatographic support together with one or more known compounds, in the expectation that the relative behaviour of the unknown and known substances will assist in the identification of the unknown one.

Cockayne syndrome *abbr.*: CS; a rare syndrome of sensitivity to sunlight, dwarfism, skeletal abnormalities, and neurological dysfunction that is transmitted as an autosomal recessive trait. Cultured fibroblasts from patients are hypersensitive to UV irradiation but there is no increased predisposition to skin cancer. **CSA** is the more common phenotype, with late onset and slow progression. It is caused by deficiency or dysfunction of a 396-residue protein (encoded by a locus on chromosome 5), which contains five **WD40** repeats. **CSB** is an early-onset, rapidly progessive, more severe form. It is caused by deficiency or dysfunction of a DNA-dependent ATPase whose gene at 10q11-q21 encodes 1493 amino acids. This protein is normally a component of RNA polymerase II complex and functions as an RNA elongation factor. The CSA and CSB proteins are required for nucleotide **excision repair**. In rare cases, CS can also be associated with abnormality in XPB, XPD, and XPG components of nucleotide excision repair. [After Edward A. Cockayne (1880–1956), British physician.]

cocktail a mixture of substances prepared according to a recipe, especially reagent mixtures such as those for use in electrofocusing (**electrofocusing cocktails**), in liquid scintillation counting (**scintillation cocktails**), and in cell-free translation systems (**translation cocktails**). Many such cocktails are commercially available.

COD *abbr. for* chemical oxygen demand.

code 1 any system of symbols, together with the rules for their association, that can be used to represent or transfer information; e.g. the **genetic code**. **2** to contain, to be arranged in, or to be expressed through a code; used especially of DNA and mRNA. A gene is said to code for (or encode) its protein product. *Compare* **decoding**.

coded amino acid any of the 20 α-amino acids whose occurrence in proteins is under direct genetic control, i.e. for which a **codon** exists.

codehydrase I *or* **codehydrogenase I** *an obsolete name for* **nicotinamide adenine dinucleotide**.

codehydrase II *or* **codehydrogenase II** *an obsolete name for* **nicotinamide adenine dinucleotide phosphate**.

codeine methylmorphine; (5α,6α)-7,8-didehydro-4,5-epoxy-3-methoxy-17-methylmorphinan-6-ol; an opioid found in opium and manufactured by the methylation of **morphine**. It acts as an agonist at μ, δ, and κ-**opioid receptors** but is generally less potent than morphine, being metabolized in the liver to morphine and norcodeine. It is a narcotic analgesic with **antitussive** activity and is a controlled substance in the US Code of Federal Regulations.

coding ambiguity an adjustment to the prediction of British molecular biologist Francis **Crick** (1916–) that a single nucleotide triplet (codon) would not code for more than one particular amino acid, which was shown by subsequent work to require modification. Thus two codons, AUG and GUG, code not only for the amino acids, methionine and valine respectively, but also are important in chain initiation. Furthermore, there are differences between codon usage in the cytoplasm and mitochondria. In mitochondria UGA codes for Trp rather than chain termination, AUA for Met rather than Ile, and AGA for termination rather than Arg. There is also the problem of proteins containing selenocysteine, which is coded by a unique tRNA^{Sec}, the codon being UGA, which usually codes for termination.

'coding' problem the description given by British molecular biologist Francis **Crick** (1916–) and co-workers in 1957 to the crucial puzzle of the **genetic code**, namely how a sequence employing combinations of only four different nucleotides (i.e. genes) can determine sequences containing combinations of up to 20 different amino acids (i.e. proteins). Crick predicted that groups of three nucleotides would code for each amino acid providing the possibility of 64 codons, more than sufficient for the 20 amino acids. (A binary code of two nucleotides would only provide 16 codons.) The prediction was confirmed with the elucidation of the genetic code in the early 1960s.

coding region *or* **coding sequence 1** (of a DNA molecule) *an alternative name for* **exon**. **2** (of a messenger RNA molecule) the portion of its complete nucleotide sequence that is translatable into polypeptide. It starts with the **initiation codon** AUG and ends with one of the **termination codons** UAA, UAG, or UGA.

coding sequence *abbr.*: CDS; *an alternative name for* **coding region**.

coding single nucleotide polymorphism (*abbr.*: cSNP) *or* **coding SNP** a single nucleotide polymorphism (SNP) present in a coding sequence that may alter the encoded amino acid sequence.

coding SNP *see* **coding single nucleotide polymorphism**.

coding strand the strand in genomic duplex DNA that contains the same sequence of bases as messenger RNA transcribed from the DNA (excepting uracil in RNA in place of thymine in DNA). Its complement is the **noncoding strand**, on which transcription of RNA takes place. When DNA sequence is represented as a single strand it is usually the coding strand that is given. In some bacteria and viruses, alternate segments of both strands of duplex DNA may be coding, in some instances with overlap. *Other names*: antitemplate strand; codogenic strand; nontranscribing strand; plus strand; sense DNA; sense strand.

coding triplet *or* **coding unit** *an alternative name for* **1 codon**. **2 anticodon**.

cod-liver oil a fish oil obtained from the liver of cod, especially the Atlantic cod (*Gadus morhua*). It is a good source of long-chain *n*-3 **fatty acids**. The typical composition (% total) of major acids is: 16:0, 14%; 16:1 *n*-7, 12%; 18:1 *n*-9, 22%; 20:1 *n*-9, 12%; 22:1 *n*-11, 11%; 20:5 *n*-3, 7%; 22:6 *n*-3, 7%. *See also* **fish oil**.

codogenic strand *an alternative name for* **coding strand**.

codominance the state in which both alleles of a gene are expressed to an equal extent, for example the genes for the M and N, and A and B blood group antigens. —**codominant** *adj.*

codon any group of three consecutive nucleotide bases (base triplet) in a given messenger RNA molecule that, by its composition and sequence, specifies either a particular amino-acid residue in the polypeptide chain synthesized by translation of that messenger

RNA, or signals the beginning or the end of the message (i.e. **start codon**, **stop codon**). The term is also used loosely for base triplets in the genomic nucleic acid – in the case of duplex DNA, either on the **coding strand** or **noncoding strand**. For the full set of codons *see* **genetic code**.

codon matrix an amino acid **scoring matrix** based on the number of DNA bases that must be changed to cause a mutation from one amino acid to another.

codon optimization the systematic alteration of codons in recombinant DNA to be expressed in a heterologous system to match the pattern of **codon usage** in the organism used for expression. The intention is to enhance yields of the expressed protein.

codon usage the frequency with which the various alternative codons are used to specify an amino acid in a given organism. Most amino acids can be specified by more than one codon, and organisms differ in their pattern of codon usage for a particular amino acid.

codon usage bias the nonrandom use of synonymous codons in an organism. Bias varies between genomes, which have different synonym preferences for given amino acids, to some extent reflecting the background base composition of different genomes (and sometimes the relative abundance of alternative tRNA isoacceptors). Highly expressed genes tend to exhibit greater codon usage bias.

codon usage oligonucleotide design the selection of codons for incorporation into synthetic oligonucleotides that is based on the pattern of codon usage in the organism under investigation.

coefficient of danger (of a drug or other agonist) the percentage lethal responses to be expected at a dose level that is effective in 95% of trials.

coefficient of variation *or* **coefficient of variability** a parameter used in comparing dispersion in distributions. It is the ratio of the **standard deviation** to the **mean**, i.e. $sx/\bar{x}$. The coefficient of variation is an abstract number independent of the units of measurements; it may also be expressed as a percentage, i.e. $100sx/\bar{x}$.

coefficient of viscosity *see* **viscosity**.

coeliac a variant spelling (*esp. Brit.*) of **celiac**.

coenzyme any of various nonprotein organic **cofactors** that are required, in addition to an enzyme and a substrate, for an enzymic reaction to proceed. Compared with a **prosthetic group** a coenzyme is more easily removed from the **apoenzyme**. The term 'coenzyme' is frequently used imprecisely, although it is now possible to give more informative chemical names to such substances and to define their roles in specific enzymic reactions. *Compare* **cosubstrate**.

coenzyme I *an obsolete name for* **nicotinamide adenine dinucleotide**.

coenzyme II *an obsolete name for* **nicotinamide adenine dinucleotide phosphate**.

coenzyme A *symbol*: (combined form) CoA *or* (free form) CoASH; 3′-phosphoadenosine-(5′)diphospho(4′)pantetheine; a heat-stable compound that functions as an acyl carrier in a number of biochemical acylation and acyl-transfer reactions, in which the inter-

mediate is a thiol ester. It is synthesized in mammalian cells in a pathway that starts with the water-soluble vitamin pantothenate. It was originally known as the **acetylation coenzyme**, hence its present name.

coenzyme B *abbr.* CoB; *N*-(7-mercaptoheptanoyl)-L-threonine 3-*O*-phosphate; a compound involved in the methanogenic pathway in methanogenic archaea.

coenzyme C a name sometimes given to a polyglutamate derivative of **tetrahydrofolate**.

coenzyme F *abbr.*: CoF; a name sometimes given to **tetrahydrofolate** or some of its derivatives.

coenzyme F$_{420}$ the *N*-(*n*-L-lactyl-*γ*-glutamyl)-L-glutamic acid phosphodiester of 7,8-didemethyl-8-hydroxy-5-deazariboflavin that functions as a coenzyme for several oxidoreductases, e.g. methylenetetrahydromethanopterin dehydrogenase. It was first found in **methanogenic** Archaea, but later also in other Archaea, and some Bacteria and Eukarya. The oxidized form acts also as a photosensitizer in DNA photolyases. Reduction results in addition of hydrogen at C-1 and C-5, in all cases so far reported being stereospecific for the *Si*-face of the coenzyme.

oxidized form

reduced form (R = substituent at N$_{10}$)

coenzyme F$_{430}$ *see* factor F$_{430}$.

coenzyme factor *an alternative name for* **diaphorase**.

coenzyme M *abbr.*: CoM; 2-thioethanesulfonate, HSCH$_2$–CH$_2$SO$_3^-$; it is involved as a coenzyme in the pathway of methanogenesis in **methanogenic** bacteria. In the terminal step of the pathway, a methyl group is transferred from methyltetrahydromethanopterin (*see* **tetrahydromethanopterin**) to CoM to form CH$_3$–S–CoM, from which methane is formed in a reductive step involving **methyl CoM reductase**.

coenzyme Q *abbr.*: CoQ; *an obsolete name for* **ubiquinone** (the abbreviation Q is, however, frequently used in diagrams).

coenzyme R *(sometimes)* an alternative name for **biotin**.

CoF *abbr. for* coenzyme F; *see* **tetrahydrofolate**.

cofactor 1 any accessory, nonprotein substance, commonly of low molecular mass, that is necessary for the activity of an enzyme. Metal-ion cofactors have been called **activators**; organic cofactors that are easily removable from the enzyme (e.g. by dialysis) may be called **coenzymes**, but if they are tightly bound they are often termed **prosthetic groups**. *See also* **apoenzyme, holoenzyme. 2** any amino acid that must be present in the medium to allow certain phages to be adsorbed by host bacteria.

cofilin a pH-sensitive nuclear actin depolymerization factor, similar to **destrin**.

COG *abbr. for* clusters of orthologous groups; *see* **cluster database.**

cognitive neuroscience the area of neurobiology that aims to explain how the brain enables an animal to interact successfully with its environment. The term cognition has come to refer to that level of description that links molecular science to behaviour.

coherent *(in physics)* describing two or more waves having the same wavelength and the same phase, e.g. coherent light. The laser is a source of coherent light.

coherent anti-Stokes Raman spectroscopy *abbr.*: CARS; a nonlinear optical technique for obtaining **Raman spectra** with high efficiency. The signal levels, signal-to-noise ratios, resolution, and fluorescence rejection are far better than with normal Raman spectroscopy. CARS can be used for substances at about 1 mM concentration in aqueous solution.

cohesin a multimeric protein complex that binds to chromosomes from telophase to the onset of anaphase in the next cell cycle. It contains SMC1 and SMC3, and Scc1 and Scc3 (*see* **Scc**). The complex binds to more than half the Alu sequences on most of the human chromosomes. It is cleaved by **separase**, a cysteine protease that is inhibited by binding to **securin**.

cohesive end *or* **cohesive terminus** *or* **sticky end** the protruding (3′ or 5′) single-stranded terminus on a double-stranded DNA fragment that can associate with a complementary, similarly protruding sequence at the opposite end of the same DNA fragment, or on another fragment, to form a circular, or larger, DNA molecule.

Cohn fractionation a method for the fractionation of human blood plasma proteins to produce minimally altered proteins suitable for clinical use. It involves fractional precipitation with ethanol at temperatures near or below 0°C, with careful control of pH and ionic strength and additions of certain bivalent metal ions, especially Zn^{2+}. The numbered **Cohn fractions** contain the following main constituents: I, fibrinogen; II, γ-globulins; III, β-globulins; IV, α-globulins; V, albumin; VI, glycoproteins. The method has also been applied to the fractionation of plasma proteins of other animals. [After Edwin Joseph Cohn (1892–1953), US biochemist.]

coiled coil an arrangement of fibrous polypetide chains in which two alpha helices wind around each other as a consequence of a seven-residue (heptad) pseudorepeat, a-b-c-d-e-f-g, in which nonpolar residues predominate in positions a and d. The hydrophobic residues fall on roughly the same side of the helix, slightly out of phase, such that, when two or three heptads associate, the helices coil around each other in a left-handed supercoil. Coiled coils are often associated with specific functions, and examples include myosin heavy chains and intermediate filaments. *Compare* **leucine zipper.**

co-immunoprecipitation a technique for analysing protein–protein interactions in which antibodies to the protein of interest are used to precipitate it and any other proteins with which it forms stable associations.

coincidence 1 *(in physics)* the occurrence of radioactive nuclide disintegrations within a time span too short for them to be resolved by a radiation counter. **2** *(in genetics)* the ratio of the number of observed double crossovers to the number expected from the random combination of single crossovers among three or four linked genes.

coincidence correction a correction applied in counting radioactive events for **coincidence losses**, i.e. those events occurring at high count rates when two or more pulses arrive within the resolving time of the counter.

coincidence counter a counter for radioactive events that consists of two opposed detectors and appropriate electronic circuitry so that only those pulses observed by both detectors are registered. This arrangement, common in liquid scintillation spectrometers, considerably reduces the number of background counts attributable to random thermionic noise arising in the detectors.

co-ion any ion of low molecular mass present in a colloidal solution or substance that has a charge of the same sign as the (net) charge of a (specified) colloidal ion under particular conditions. *Compare* **counterion.**

colcemide *or* **demecolcine** a derivative of **colchicine** in which the amino group (attached to the non-tropolone seven-membered ring) is methylated instead of acetylated. It binds tubulin and inhibits formation of the mitotic spindle.

colchicine a major alkaloid from the autumn crocus, *Colchicum autumnale* L. It is used experimentally to inhibit the polymerization of **tubulin**, to which it binds specifically and tightly, and to encourage depolymerization of **microtubules**. This action inhibits spindle formation, arrests mitosis at metaphase, and promotes polyploidy in plants. Colchicine also disrupts neutrophil migration and phagocytosis, and inhibits leukotriene B$_4$ synthesis; it is useful in the treatment of gout. The molecule is noted for a tropolone ring, which is rather rare in natural products. *See also* **colcemide.**

cold 1 of low, or lower than normal, temperature. **2** *(in radiochemistry)* containing no, or an insignificant amount of, a radioactive nuclide; the nonradioactive counterpart of a radionuclide-labelled substance. *Compare* **hot.**

cold agglutinin *see* **cold autoantibody.**

cold autoantibody *or* **cold agglutinin** an antibody that attaches to red cells at low temperatures, reacting better at 4°C than at 37°C, and dissociates as the red cells are warmed. Cold autoantibodies are mainly directed against antigens of the **Ii system** of blood groups.

cold-insoluble globulin a **cryoglobulin** present in blood plasma or serum. It is normally present at only very low concentrations in human serum, but may be elevated in certain rare diseases such as systemic lupus erythematosis. *See also* **fibronectin.**

cold-labile enzyme *or* **cold-sensitive enzyme** any enzyme that, unlike most enzymes, is less stable at 0°C than at room temperature. In some, but not all, instances activity slowly returns on warming. The phenomenon can be due, e.g., to exothermic dissociation of a stabilizing cofactor from the enzyme or to dissociation into inactive subunits.

cold receptor *see* **CMR1.**

cold-sensitive mutation any mutation producing a gene that is functional at a high temperature (the permissive temperature) but inactive at a low temperature (the restrictive temperature). *See* **temperature-sensitive mutation.**

cold-shock protein any of various proteins, synthesized by microorganisms in response to a sudden drop in temperature. These are typically DNA-binding transcription activators, many of which have a characteristic **cold-shock domain.** *Compare* **heat-shock proteins.**

coleoptile the protective sheath covering the first leaves of seedling grasses.

ColE1 replicon an origin of DNA replication found in the naturally occurring colicin E1 plasmid of *Escherichia coli.* It has been used in the construction of certain synthetic plasmids. Synthesis of the leading strand of DNA is primed by RNAII.

colestipol a basic anion-exchange resin that when given orally as the Cl⁻ form, exchanges anions of chloride in the intestine with those of bile salts, which are then excreted and thus removed from the enterohepatic circulation as an insoluble complex. The resulting reduction in cholesterol absorption culminates in reduced circulating levels of low-density lipoproteins but can increase plasma triacylglycerols. It has therapeutic application in relieving hyperlipoproteinemia.

Col factor *(formerly) an alternative name for* **Col plasmid.**

colforsin *an alternative name for* **forskolin.**

colicin *or (formerly)* **colicine** any member of a range of 40–80 kDa proteins secreted by certain strains of the Enterobacteriaceae. They form pores on the outer surface of other susceptible strains within this family, or deliver DNAase or RNAase enzymes into the cytoplasm, to cause cell death. Colicins form a subcategory of **bacteriocin.** Different colicins are designated B, E₁, E₂, I, K, or V, and may be distinguished by their diffusibility and immunogenic and host specificities. They may be coded for by a **Col plasmid.**

colicin factor *or* **colicinogenic factor** *(formerly) an alternative name for* **Col plasmid.**

colicinogenic describing bacteria that are able to produce **colicins.**

colicinogenic factor *see* **Col plasmid.**

coliform 1 describing any bacterium belonging to the genera *Escherichia* or *Klebsiella.* **2** describing any Gram-negative, lactose-fermenting, enteric bacillus or, more loosely, any Gram-negative, enteric bacillus or related organism. **3** any coliform microorganism.

colipase a cofactor of **pancreatic lipase** that inhibits the surface denaturation of lipase and anchors the protein to the lipid–water interface of lipid micelles to prevent inhibition and displacement by bile salts (four motifs). The human precursor protein consists of 112 amino acids: amino acids 1–17 form the signal; 18–22 form enterostatin activation peptide; and 23–112 form the colipase.

coliphage any **bacteriophage** that infects a strain of *Escherichia coli.*

colistin *or* **polymyxin E** a **polymyxin** antibiotic derived from *Bacillus colistinus* and comprising colistins A, B, and C. It acts as a cationic surface-active detergent, disrupting microbial cell wall structure.

colitose 3,6-dideoxy-L-*xylo*-hexose; 3,6-dideoxy-L-galactose; a sugar present in lipopolysaccharide of some strains of *Escherichia coli.* For the D enantiomer, *see* **abequose.**

collagen a group of fibrous proteins of very high tensile strength that form the main component of connective tissue in animals. Collagen is highly enriched in glycine (up to one-third of its amino-acid residues) and somewhat less so in proline and 3- and 4-hydroxyproline (about 21%). The primary structure is characterized by a tripeptide repeating unit (G-x-P) that forms extended helical structures (prolines at the third position of the repeat unit are hydroxylated in some or all of the chains); three such chains twist together and are hydrogen bonded via the imino group of every third amino

acid and a carbonyl group in one of the other two chains. Collagen of bones and skin is metabolically stable, but that of organs has a high metabolic turnover. Collagens are products of a superfamily of closely related genes found in multicellular animals. The products have been classified into types I to XIII in the order in which they were purified and characterized, and grouped into classes 1 to 4 according to their properties.

Class 1 comprises types I to III, V and XI. All of these are fibreforming, of approximately the same length, having a a large C-terminal globular domain following a lengthy triple helical domain. Each is synthesized as a procollagen molecule. Several types of α chain are found variously in the different types of class 1 collagen. All of these aggregate into staggered arrays, first into microfibrils of five collagen molecules that then aggregate into fibres.

Class 2 comprises types IX and XII. Type IX was first isolated from pig hyaline cartilage. It comprises three different α chains, α-1, α-2 and α-3, and has three short triple-helical domains interrupted by non-collagenous domains. At the N-terminus, there is a large globular domain. A glycosaminoglycan molecule, chondroitin sulfate, is attached to the α-2 chain. Type XII was discovered by screening a DNA library and has structural similarities to type IX. The function of these molecules is not known.

Class 3 comprises types IV, VI, VII and X, which serve unique functions. Type IV is the major component of basement membranes. It has long triple-helical domains, with a complex globular domain at the C-terminal end. There are six α-x chain isoforms, numbered 1 to 6. Type VI is a short molecule, having a triple- helical domain, 105 nm in length, separating two globular domains. It aggregates into tetramers stabilized by disulfide bonds. Carbohydrate chains consisting of glucosamine, mannose, galactose, fucose, and sialic acid are associated with type VI, amounting to 10% of the total mass. Type VII is known as the **anchoring fibril network,** and is the largest collagen yet described, with a total molecular mass of the precursor of ~1050 kDa. It has an N-terminal globular domain, a triple-helical domain 424 nm long, and a highly complex C-terminal globular domain. Anchoring fibrils are found within the sub-basal lamina, and appear to anchor the lamina densa to the underlying matrix. Type X collagen has a restricted tissue distribution, being synthesized mainly by chondrocytes, and is 1438 nm in length.

Class 4 comprises types VIII and XIII. Type VIII was first observed as a cell culture product of endothelial cells. It is a 130–140 nm rod found in specialized tissues, including the sclera, periosta, perichondria, and cartilage growth plate. Type XIII was identified from a cDNA clone encoding short chains that form a triple-helical domain. Its function is unknown. *See also* **hydroxyproline.**

collagenase any of the proteinases that hydrolyse **collagen,** in particular *Clostridium histolyticum* collagenase (clostridiopeptidase A, EC 3.4.24.3), which preferentially cleaves at Xaa-|-Gly in the sequence

-Zaa-Pro-Xaa-Gly-Pro-Xaa.

It is widely used experimentally in the release of many types of cells from animal tissues, e.g. adipocytes and hepatocytes. *See also* **gelatinase, interstitial collagenase, microbial collagenase.**

collapsin *an alternative name for* **semaphorin.**

collectin one of a family of plasma **lectins** with a trimeric structure similar to complement C1q.

colligation the formation of a covalent bond by means of two combining groups donating one electron each; the reverse of unimolecular **homolysis.** *Compare* **coordination** (def. 2). —**colligate** *vb.*

colligative property the property of any substance or system (e.g. an ideal solution) that depends only on the number of particles (atoms, molecules, and ions) and not on their nature. For example, gaseous pressure and osmotic pressure are colligative properties.

collimator any device, generally embodying a system of slits and (sometimes) lenses (optical, electric, or magnetic), that is used to produce a nondivergent beam of electromagnetic radiation or particles.

collinear 1 lying in or on the same (straight) line. **2** having a (straight) line in common. **3** describing the state of congruence between the order of amino acids along a polypeptide chain and the order of the corresponding codons along the polynucleotide chain encoding the polypeptide. —**collinearity** n.

collision coupling hormone–receptor–enzyme model a model of enzyme activation summarized by the scheme:

$$H + R + E \underset{}{\overset{K_{mH}}{\rightleftharpoons}} HR + E \overset{k_1}{\rightarrow} HRE \overset{k_3}{\rightarrow} HR + E'$$

where H is the hormone, R is the receptor, E is the inactive enzyme, E′ is the activated enzyme, K_{mH} is the hormone–receptor dissociation constant, k_1 is the bimolecular rate constant governing HRE formation, and k_3 is the rate constant governing enzyme activation. In this model the enzymically inactive intermediate, HRE, never accumulates and constitutes only a small fraction of the total receptor and enzyme concentrations.

collision frequency or **collision number** the number of collisions per second when there is only one molecule of reactant per cubic centimetre of gas or liquid e.g. in kinetics.

collodion a solution of 4 g pyroxylin (mainly nitrocellulose) in 100 ml of a mixture of 25 ml ethanol and 75 ml diethylether. When exposed to the atmosphere in thin layers it evaporates leaving a tough, colourless film that may be used as a **dialysis membrane**.

colloid or **colloidal system 1** a state of subdivision of matter implying that the molecules or polymolecular particles (the dispersed phase) dispersed in a medium (the continuous phase) have, at least in one direction, a dimension roughly in the range 1 nm–1 μm or that in a system discontinuities are found at distances of this order. All three dimensions need not be in the colloidal range nor is it necessary for the units of a colloidal system to be discrete. Colloids may comprise many individual large molecules such as proteins or polysaccharides, or aggregations of smaller molecules such as colloidal gold. The colloids do not pass, or pass very slowly, through semipermeable membranes. **2** (in histology) the structureless substance seen in thyroid gland follicles. —**colloidal** adj.

colloidal dispersion any thermodynamically unstable colloidal system in which particles of colloidal size of any nature (e.g. solid, liquid, or gas) are dispersed in a continuous phase of different composition (or state), and which cannot easily be reconstituted after separation of the dispersed phase and the dispersion phase. Compare **colloidal solution**.

colloidal electrolyte any electrolyte that gives ions of which at least one is of colloidal size; it includes **polyelectrolytes**.

colloidal solution any thermodynamically stable solution consisting of colloidal macromolecules and solvent that can be easily reconstituted after separation of the macromolecules from the solvent. Compare **colloidal dispersion**.

colloid osmotic pressure that part of the osmotic pressure exerted by a solution that is due to dissolved colloids. See **osmotic pressure** (def. 1).

colon 1 (in vertebrates) the part of the large intestine that extends from the cecum to the rectum. **2** (in insects) the second part of the intestine.

colon bacillus a former term for **Escherichia coli**.

colony (in microbiology) a circumscribed group of cells, normally derived from a single cell or small cluster of cells, growing on a solid or semisolid medium.

colony hybridization or **colony hybridisation** a blot-transfer method (see **blotting**) whereby a very large number of colonies of a bacterium carrying different hybrid plasmids can be rapidly screened to determine which hybrid plasmids contain a specified DNA sequence or gene(s). The bacterial colonies are grown on an agar plate, then replica plated onto nitrocellulose filters. A reference set of the colonies is retained. Then the colonies on the filter are lysed and their DNA is denatured and fixed to the filter in situ. The resulting DNA-prints of the colonies are then hybridized to a radioactive RNA probe for the sequence or gene of interest. DNA that binds the probe is then assayed by autoradiography or another detection method, enabling the selection of corresponding colonies from the reference plate.

colony PCR or **colony polymerase chain reaction** a procedure performed on a suspension of bacteria from a single colony and used to check whether plasmids carried by the cells have a cloned insert

of interest. Oligonucleotide primers used for PCR are usually designed to anneal to vector sequences flanking the cloning site.

colony stimulating factor abbr.: CSF; any of the **cytokines** that control the differentiation of hemopoietic stem cells, both in the **bone marrow** and in the periphery. The group consists of granulocyte, macrophage, and granulocyte/macrophage CSFs. These promote the development of their specific subsets of leukocytes. **Interleukin-3** (IL-3), interleukin-5 (IL-5), and **erythropoietin** are also members of this group. IL-3 promotes the differentiation of all the leukocyte lineages, whereas IL-5 is necessary for eosinophil development. Erythropoietin promotes the expansion of erythroid colony forming units. **Granulocyte-colony stimulating factors** (G-CSFs) act in hemopoiesis by controlling the production, differentiation, and function of two related white cell populations: granulocytes and monocytes–macrophages. **Granulocyte/macrophage-colony stimulating factors** (GM-CSFs) are glycoproteins that act in hemopoiesis by controlling the production, differentiation, and function of the granulocytes and the monocytes–macrophages. **Macrophage-colony stimulating factor** (M-CSF) is produced by fibroblasts, endothelium, and epithelium; it stimulates growth of macrophage colonies. Membrane receptors for CSFs, like those for other cytokines, are type I transmembrane proteins that signal through JAKs and STATs. Several missense mutations in the gene for the G-CSF receptor (at 1p35, which encodes 813 amino acids) result in severe congenital neutropenia frequently associated with acute myeloid leukemia.

colorimeter a device for measuring colour intensity or differences in colour intensity, either visually or photoelectrically. It is used for quantitative determination of coloured compounds in solutions.

colorimetry any method of quantitative chemical analysis in which the concentration or amount of a compound is determined by comparing the colour produced by the reaction of a reagent with both standard and test amounts of the compound, often using a **colorimeter**.

colostrum the milk secreted by the mammary gland just prior to and for a short period after parturition. It has high levels of protein and immunoglobulin, and its ingestion confers passive maternal immunity on the offspring of some species.

Colour Index abbr.: CI; a catalogue of commercially available dyes, identifying each by chemical class, colour, and generic name, and giving their chemical composition where disclosed. Each dye is assigned a five-digit number, prefixed by 'CI'. The preferred trivial name of each dye, its commoner synonyms, and the name(s) of its known manufacturer(s) also are listed.

colour quenching the **quenching** that occurs in liquid scintillation counting due to the absorption of the scintillation emission by the specimen, resulting in a decrease in the light-collecting factor. It can often be reduced by prior bleaching or dilution of the specimen before introduction into the scintillator. Compare **chemical quenching**.

Col plasmid or (formerly) **Col factor** or **colicin factor** or **colicinogenic factor** any **plasmid** that carries genetic information for the production of a **colicin**; a subcategory of **bacteriocinogenic factor**.

Colton blood group antigen a human blood group antigen based on the presence or absence of a particular point mutation in an extracellular domain of aquaporin 1 of erythrocytes.

columbinic acid or **ranunculeic acid** (5E,9Z,12Z)-octadeca-5,9,12-trienoic acid; a fatty acid occurring in some seed oils. It can relieve many of the symptoms of essential fatty acid deficiency, although it does not act as a precursor for prostaglandin synthesis.

column an upright cylindrical shaft or tube, especially the tube used to contain the stationary phase in **column chromatography** or **column electrophoresis**, or to contain the packing material in fractional distillation. By extension, the term is used for the tube (whether vertical, horizontal, or spiral) used in **gas chromatography**.

column chromatography any method of **chromatography** in which the stationary phase is a porous solid contained in a column through which the liquid or gas phase percolates. The principal examples are **adsorption chromatography**, **gel-filtration chromatography**, **gas–liquid chromatography**, and **ion-exchange chromatography**.

column electrophoresis any method of **electrophoresis** in which the support material is contained in a column. Such a technique is usually used for preparative scale separations by **electrofocusing** or **isotachophoresis**.

column volume *symbol*: X; the volume (empty) of that part of a chromatographic column that contains the packing, calculated from the inner diameter and the height or length of the column occupied by the stationary phase under the conditions of use. *Compare* **bed volume**.

CoM *abbr. for* coenzyme M.

combination electrode an electrode comprising a glass electrode and a reference electrode in the same envelope providing a smaller, more compact instrument and ensuring the two electrodes are in the same environment.

combinatorial describing any process that is governed by a specific combination of factors, rather than by any single factor, with different combinations having different effects.

combinatorial chemistry *or* **virtual screening** *or* **in silico screening** mass screening and selection of chemical entities based on the desirability of their properties (usually as pharmaceuticals) according to a computational model.

combined test of pituitary function a test for suspected hypopituitarism. Insulin, luteinizing hormone-releasing hormone (LHRH), and thyrotropin releasing hormone (TRH) are administered and the response of pituitary hormones is compared with their levels before insulin administration. The lowering of the blood glucose level should stimulate a rise in the blood levels of cortisol and growth hormone; LHRH should increase follicle stimulating hormone (FSH) and luteinizing hormone (LH); and TRH should increase thyrotropin and prolactin.

combining site the site on an **antibody** molecule that combines with the corresponding antigen.

combrestatin a stilbene phytoalexin from *Arabidopsis thaliana* that has important antineoplastic activity.

combustion the act or process of burning, especially the quantitative conversion of organic compounds to carbon dioxide and water for elementary analysis, in the estimation of carbon-14 and tritium, or in the determination of a **heat of combustion**.

co-metabolism any oxidation of substances by a microorganism in which the energy yielded by the oxidation is not used to support microbial growth. The presence or absence of growth substrate therefore cannot be inferred from such an oxidation.

commensal 1 describing either of two different species of organism that live in close physical association with benefit to one partner but with little effect – either beneficial or harmful – on the other partner. **2** any organism in a commensal association. *Compare* **symbiosis**. —**commensalism** *n*.

commissureless a cell protein that regulates exposure of **Robo** receptors in growth cones of *Drosophila* axons that do not cross the midline. It alters intracellular trafficking of these receptors.

committed step *or* **committing reaction** a reaction in a multienzyme reaction sequence (e.g. a metabolic pathway) that once it occurs, causes all the ensuing reactions of the sequence to take place. The sole metabolic role of the committed step is to provide precursors of the final product of the pathway; it is usually a physiologically irreversible step.

common ancestor 1 the most recent ancestral form or species from which two different forms or species have evolved. **2** the node in a phylogenetic tree that is an ancestor of two given branches or leaves.

co-monomer the cross-linking agent added to the monomer in forming the gel used in **polyacrylamide gel electrophoresis**.

COMP *abbr. for* cartilage oligomeric matrix protein; a homopentameric glycoprotein (540 kDa) belonging to the thombospondin family, of unknown function and secreted by chondrocytes into their surrounding space. Each subunit contains an N-terminal region required for pentamer formation, four epidermal growth factor-like repeats, eight calcium-binding calmodulin-like repeats, and a globular C-terminal region. Over 70 mutations in the gene locus at 19p13.1 are associated with pseudoachondroplasia, in which the mutant protein forms large lamellar inclusions in the rough endoplasmic reticulum of chondrocytes and causes their death.

compactin an alternative name for **mevastatin**.

comparative genomics the systematic study of the similarities and differences in gene content and structure of whole genomes. Genome comparisons may shed light on the evolution of organisms, for example by locating 'missing' genes, pinpointing horizontal transfer events, and so on.

compartment 1 *(in tracer kinetics)* any anatomical, physiological, chemical, or physical subdivision of a system throughout which the concentration ratio of labelled to unlabelled substance is uniform at any given time. It is understood that the rate at which the tracer entering the compartment is mixed with tracer in the compartment is rapid compared with the rate at which the substance leaves the compartment. **2** any subdivision of a model used for the analysis of experimental observations.

compartmentation *or* **compartmentalization** the unequal distribution of a metabolite, enzyme, or metabolic pathway between different parts of a cell (e.g. in cell organelles) or of a whole organism. Consequently only a part of the total of any given chemical present is apparently available for a given reaction. It may refer to morphologically recognizable structures or to **metabolic pools**.

compatibility group *see* **incompatibility group**.

compatible solute a compound accumulated in plants and microorganisms to maintain osmotic balance, and to stabilize macromolecules against various forms of stress, including heat, freezing, and radical damage. Examples include trehalose and glycine betaine.

competence 1 the ability of a cell to take up a molecule of exogenously added DNA, and to be transformed by it (*see* **transformation**). **2** the ability of bacterial cells to incorporate phage DNA and to produce phage progeny (*see* **bacteriophage**). **3** the state of reactivity of a cell, tissue, or organism that allows it to respond to specific stimuli by forming new morphological structures or by synthesizing specific proteins (enzymes).

competence factor a secreted protein, or other substance added in culture, that is essential for **competence** in a bacterial or other cell. A 10 kDa basic protein of this type has been isolated from streptococci. It does not bind to DNA but interacts with the cell surface.

competition *(in chemistry)* rivalry between two or more different, but often similar, chemical species for a specific biochemical system, e.g. a receptor, enzyme, transport system, antibody molecule, or ion exchanger.

competition binding *see* **displacement binding**.

competitive antagonism the competition between an **agonist** and an antagonist for a receptor that occurs when the binding of agonist and antagonist is mutually exclusive. This may be because the agonist and antagonist compete for the same binding site, or combine with adjacent but overlapping sites. A third possibility is that different sites are involved but that they influence the receptor macromolecule in such a way that agonist and antagonist molecules cannot be bound at the same time. If the agonist and antagonist form only short-lived combinations with the receptor, so that equilibrium between agonist, antagonist, and receptor is reached during the presence of the agonist, the antagonism will be surmountable over a wide range of concentrations (*see* **reversible competitive antagonism**). In contrast, some antagonists, when in close enough proximity to their binding site, may form a stable covalent bond with it (*see* **irreversible competitive antagonism**); the antagonism becomes insurmountable when no spare receptors remain. More generally, the extent to which the action of a competitive antagonist can be overcome by increasing the concentration of agonist is determined by the relative concentrations of the two agents, by the association and dissociation rate constants for their binding, and by the duration of the exposure to each. The action of a competitive antagonist can therefore be surmountable under one set of experimental conditions and insurmountable under another.

competitive binding competition between two or more different chemical species for a limited number of binding sites in a biochemical system, such as a receptor, enzyme, transport system, or antibody molecule. *See also* **competition**.

competitive inhibition any inhibition of an enzyme activity by an inhibitor that binds to the same active site on the enzyme as does the substrate and when there is no interaction between the inhibitor and the enzyme–substrate complex, nor between the substrate(s) and the enzyme–inhibitor complex. Characteristically, analysis using a Lineweaver–Burk plot reveals that in the presence of a **competitive inhibitor** V_{max} is unchanged and K_m is increased.

competitive inhibitor any inhibitor that produces **competitive inhibition** of the activity of an enzyme. Inhibition produced by a given level of inhibitor can be overcome by an increase of substrate concentration.

competitive protein-binding assay *abbr.*: CPBA; a form of **saturation analysis** in which a naturally occurring proteinaceous binding substance is used.

complement or *(formerly)* **alexin(e)** a system of proteins and cell receptors, found in the plasma of vertebrates, that participates in immune defence against infection by microorganisms; the activation products of complement components cause lysis of antigenic cells, attract phagocytic cells to the site of activation, and assist the uptake and destruction of antigenic cells by phagocytes. Complement is also involved in immunological tissue injury. The plasma components comprise a recognition unit, three alternative pathways for activation of C3 (the classical pathway uses C2 and C4), and a terminal pathway that ends in the formation of the membrane attack complex (C356789). Synthesis occurs mainly in hepatocytes and that of several components is regulated by interleukins 1 and 6, tumour necrosis factor, γ interferon, and endotoxic polysaccharide. The plasma components amount to about 2000 mg L^{-1} (C1 ≈250, C3 ≈1200, C4 ≈400 mg L^{-1}). The components of complement, mostly β- or γ-globulins, are generally designated by the symbol C and an identifying suffixed numeral. There are three alternative pathways for activation of complement, and these probably work together. The classical pathway is followed below. The alternative pathway uses properdin and factors B and D to activate C3 and C5. The lectin pathway utilizes mannose-binding protein and its associated serine protease (MASP), and activates C3 as in the classical pathway but without a requirement for C1.

C1 is a Ca^{2+}-dependent complex of C1q, C1r, and C1s in the molar ratio of 1:2:2. C1q has a 'bunch-of-tulips' structure that consists of six collagen-like triple helices each ending in a globular C-terminal domain, which binds the Fc region of immune complexes. Each triple helix consists of homologous A, B, and C chains, which are encoded on neighbouring genes, and contain a collagen-like domain over ≈80% of their length, while the rest contributes to the C-terminal globular head. C1r and C1s are highly homologous serine protease zymogens (83 kDa per subunit). Two subunits of each bind C1q in a calcium-dependent but noncovalent manner. Binding of ligand to C1q causes a conformational change in, and autocatalytic activation of, C1r (EC 3.4.21.41), which then cleaves and activates C1s. The natural substrates for C1s are C2 and C4. Deficiency of A, B, or C chains of C1q results in the clinical phenotype of lupus erythematosus. Deficiency of C1r or C1s also leads to this phenotype or to glomerulonephritis.

C2 is a single-chain polypeptide (102 kDa) that is highly homologous with factor B. C1s releases a small peptide (S2b) and (active) C2a. C2a and C4b form a bimolecular enzyme called **C3 convertase**, a serine proteinase (EC 3.1.21.43). Deficiency of C2 results in rheumatic disorders and pyogenic infections.

C3 is a glycoprotein (185 kDa) that consists of an α chain linked by a disulfide bond to a β chain, both being derived from the same precursor, which is homologous with C4 and C5. The α chain contains an intrachain thiol ester that is important for the function of C3. C3 convertase (i.e. C4b2a) cleaves a small peptide, which is the N-terminal region of the α chain, to expose the thiol ester. C3a is an anaphylatoxin. The rest of the molecule (C3b) binds covalently (through the thiol ester) to cell surfaces or the immune complex, and in complex with C3 convertase forms C4b2a3b, which is called **C5 convertase**. Deficiency of C3 leads to rheumatic disorders and pyogenic infections.

C4 is a glycoprotein (200 kDa) that consists of disulfide bond-linked α, β, and γ chains, all derived from the same precursor, which is homologous to those of C3 and C5. The α chain contains an intrachain thiol ester, which is important for the function of C4. C1s releases C4a, the N-terminal portion of the α chain, and exposes the thiol ester. This binds covalently to cell surfaces or to immune complexes. The rest of the molecule is C4b, which with C2a forms C4b2a, or C3 convertase. In combination with C3b, C3 con-

vertase forms C4b2a3b, which is C5 convertase. Deficiency of C4 leads to a lupus erythematosus phenotype. The human form is polymorphic, the allotype being C4a. C4a alleles carry blood group Rogers and C4b alleles carry Chido. Total deficiency of C4a6 allotype is associated with hemolytic disease.

C5 is a glycoprotein (190 kDa) that consists of an α chain linked by a disulfide bond to a β chain, both derived from a precursor homologous to those of C3 and C4. C5 convertase (i.e. C4b2a3b) releases a small anaphylatoxic and leukocyte-attractant peptide (C5a), which is the N-terminal region of the α chain. Activation of C5 initiates the spontaneous assembly of the late complement components, C5–C9, into the membrane attack complex (*see* **MAC** (def. 1)). C5b recruits a molecule of each of C6, C7, and C8, which insert their hydrophobic regions into the target cell membrane. Deficiency of C5 leads to meningococcal infection.

C6, C7, C8, and **C9**, together with C5b, constitute the membrane attack complex; all are structurally related and similar to **perforin**. C6 is a single-chain polypeptide (128 kDa) that interacts with C5b, C7, and C8. C7 is a single-chain polypeptide that interacts with C5b, C6, and C8. C8 is a heterotrimer comprising α, β, and γ subunits. An α chain is linked by a disulfide bond to a γ chain, and a β chain then links to these two chains. C8 interacts with C5b, C6, and C7, the complex inserting into the target cell membrane. C9 is a single-chain polypeptide (70 kDa). From 1 to 18 molecules of C9 are recruited by the C5b678 complex to form a tubular transmembrane channel that disturbs the target cell membrane and increases its permeability. It is cleaved by thrombin to form C9a and C9b. Deficiency of C6, C7, C8, or C9 is associated with meningococcal infection.

Factor B, a zymogen of M_r 100 000, is activated by factor D, which selectively cleaves an Arg-|-Lys bond when factor B is complexed with C3b to form factor Bb. Following cleavage the C3b–Bb complex acts as **alternative-complement-pathway C3/C5 convertase** (EC 3.4.21.47), which can convert C3 to C3a and C3b by cleaving an Arg-|-Ser bond, and C5 to C5a and C5b by cleavage of an Arg-|-Xaa bond.

Factor D is a serine proteinase (EC 3.4.21.46) of M_r 24 000; its concentration in human plasma is approximately 1.5 mg L^{-1}.

Factor H is a cofactor for **factor I** (EC 3.4.21.45), a serine proteinase that cleaves the α chains of C4b and C3b in the presence of the cofactors for C4-binding.

Properdin, a glycoprotein of M_r 220 000, is composed of four, probably identical, noncovalently linked polypeptide chains. Two alternative reaction cascades for the activation of complement exist. In the so-called **classical pathway**, activation (which requires both Ca^{2+} and Mg^{2+}) is initiated by the binding of the C1q moiety of C1 to the C_H2 regions of antibody molecules in antibody–antigen aggregates or in antibody bound to cells, thus causing C1 to be converted to a proteinase, $\bar{C1}$. $\bar{C1}$ activates C4 and C2 leading to the formation of **C3 convertase**, a complex proteinase, C$\overline{42}$, consisting of the activated components C$\overline{4}$ and C$\overline{2}$. C$\overline{42}$ splits C3 to the chemotactic and anaphylatoxic fragment C3a and the fragment C$\overline{3}$, C3b, which associates with C$\overline{42}$ to give a C$\overline{42}$3 complex, **C5 convertase**. The latter is a complex proteinase that in turn splits C5 to C$\overline{5}$ and the chemotactic and anaphylatoxic fragment C5a. In the so-called **alternative pathway**, activation (which requires Mg^{2+} but not Ca^{2+}) is initiated by polysaccharides such as those present on bacterial and yeast cell walls. The proteinase $\bar{D}$ is analogous to $\bar{C1}$ but is normally present in serum, where it causes a continuous slow activation, which is greatly enhanced when polysaccharides are present. $\bar{D}$ acts on B and C3 to produce C$\overline{3B}$, an alternative C3 convertase. C$\overline{3B}$, which is stabilized by properdin, acts on C3 to produce (C3)$_n\bar{B}$, an alternative C5 convertase. In both pathways, C5 production is the last proteolytic step, and there follows a spontaneous association of C5 with C6, C7, C8, and C9 to form a lytic complex, C$\overline{56789}$. *See also* **lysine(arginine) carboxypeptidase**.

complementarity 1 the interrelationship between one or more units supplementing, being dependent on, or being in a polar position to another unit or units. **2** a reverse correspondence of part of one molecule to part of another molecule, especially of antigen to antibody, enzyme to substrate, or the partners in a **base pair**.

complementarity-determining region (*abbr.*: CDR) or **hyper-**

variable region the most variable (hypervariable) region of an antibody molecule (or other member of the immunoglobulin superfamily), comprising three loops (CDR1, CDR2, CDR3) from the heavy chain and three from the light chain. Together, the CDRs form a complementary surface to the antigen, and act as the antigen-binding site. Antibodies with different specificities have different CDRs; those with the same specificity have identical CDRs.

complementary base sequence a sequence in a polynucleotide chain in which all the bases are able to form **base pairs** with a sequence of bases in another polynucleotide chain.

complementary catalysis catalysis in the presence of a prototype molecule, leading to the formation of products with a structure that is in some way complementary to that of the prototype.

complementary DNA *abbr.*: cDNA; a DNA molecule that has a **complementary base sequence** to a molecule of a messenger RNA. The first strand, with the complementary nucleotide sequence, is synthesized by reverse transcriptase using the RNA as template. A second DNA strand is then synthesized by DNA polymerase using the first strand as template. The two strands together represent a double-stranded DNA copy of the mRNA. cDNA is used for **molecular cloning**, or to yield molecular **probes** (def. 3) in hybridization tests for a specific sequence in cellular DNA. It has the advantage that, unlike genomic DNA, it contains no introns. A population of cDNAs cloned in a suitable vector and representative of the mRNAs in a cell or tissue constitutes a **cDNA library**. cDNA must be distinguished from the complementary strand of duplex DNA (*see* **noncoding strand**).

complementary genes any genes that interact to produce a qualitative effect that is distinct from the effects of any one of them separately.

complementary RNA *abbr.*: cRNA; a single-stranded RNA molecule that has a **complementary base sequence** to a specific (chromosomal) DNA sequence (of either strand), from which it is produced by the action of DNA-directed **RNA polymerase**. Naturally occurring cRNA may represent **pre-messenger RNA**, **ribosomal RNA**, or **transfer RNA**. cRNA can also be produced by transcription *in vitro*, usually from a DNA template with a promoter for a bacteriophage RNA polymerase. *Compare* **antisense RNA**.

complementary strand (of duplex DNA) *see* **noncoding strand**.

complementation 1 any instance in which the interaction of two incomplete, nonfunctional systems results in a fully functional system. **2** the interaction between viral gene products or gene functions in cells that are multiply or mixedly infected with virus(es), leading to an increased yield of infective virus of one or more parental types. **3 genetic complementation**.

complementation test any experimental test for **genetic complementation**. *See* **cis–trans complementation test**.

complement-3 convertase *an alternative name for* C3 convertase; *see* **complement**.

complement-5 convertase *an alternative name for* C5 convertase; *see* **complement**.

complement decay-accelerating factor *abbr.*: DAF; a factor that interferes with catalysis of the conversion of component C2 and factor B of **complement**, and prevents formation of the amplification convertases of the complement cascade. DAF is expressed on the plasma membranes of all cell types that are in intimate contact with plasma complement proteins, and is a glycosylphosphatidylinositol-anchored protein. Human proteins DAF1 and DAF2 are formed by alternative splicing of a transcript of the same gene.

complement fixation activation of the **complement** system, characteristically *in vitro* by an antigen–antibody interaction, in which complement components C1 to C9 are activated sequentially.

complement-fixation inhibition test a serological test in which the presence of a substance of known structure either may or may not inhibit the interaction of antibody with complex antigen, and hence also the fixation of **complement**. *Compare* **complement-fixation test**.

complement-fixation test a serological test for the presence of antibody that can fix **complement** on interaction with antigen. The presence or absence of residual complement is detected by an indicator system, e.g. added sheep erythrocytes sensitized with an anti-body to sheep erythrocytes. If complement is still present, lysis of the erythrocytes occurs, but if no complement remains, no lysis takes place.

complement receptor any cell-surface receptor that binds activated **complement** components, e.g. the receptor for C3b found on macrophages, neutrophils, and B lymphocytes.

complete Freund's adjuvant *see* **Freund's adjuvant**.

complex 1 any distinct chemical species in which two or more identical or nonidentical chemical species (ionic or uncharged) are associated, usually in stoichiometric proportions, and bound together by weaker interactions than covalent bonds. The term is used for various entities, e.g. coordination complexes, antigen–antibody complexes, enzyme–substrate or enzyme–inhibitor complexes, multienzyme complexes, and receptor–agonist complexes; it is often employed where the precise nature of the interaction is uncertain. **2** to make a complex or incorporate into a complex, especially of a metal ion and a **chelating agent**; to chelate.

complex I, II, III, IV the cytochrome-containing protein complexes that, with associated ligands including ubiquinone, Fe–S clusters, FMN, FAD, and nicotinamide nucleotides, form the mitochondrial electron-transport chain or **respiratory chain**. Each complex can be isolated separately and with suitable substrates and electron acceptors can function independently in transporting electrons. Complexes I, III, and IV are also involved in proton translocation, a function that requires the complex to be embedded in an orientated environment such as that provided by the intact lipid bilayer of the mitochondrial inner membrane. Complex I (EC 1.6.5.3) – NADH dehydrogenase (ubiquinone) – contains ≈40 different polypeptides of which 7 are encoded in mtDNA. The complex includes one flavoprotein and seven polypeptide subunits including **NADH dehydrogenase (ubiquinone)**, together with FMN and a number of different iron–sulfur clusters containing nonheme iron in the forms Fe–S, Fe_2–S_2, and Fe_4–S_4, the iron of which undergoes oxidation–reduction between oxidation states II and III. The complex contains ≈31 hydrophobic polypeptides. The reaction catalysed is the oxidation of NADH by ubiquinone. Complex II (EC 1.3.5.1) contains four subunits of **succinate dehydrogenase (ubiquinone)**, FAD, and iron–sulfur. It catalyses the oxidation of succinate by ubiquinone. Complex III (EC 1.10.2.2) contains 11 polypeptides including the four redox centres: cytochrome b_{562}, cytochrome b_{566}, cytochrome c_1, and the Rieske protein, which contains an Fe_2–S_2 cluster. Cytochrome b is encoded in mtDNA. It catalyses the oxidation of ubiquinol by oxidized cytochrome c. The transfer of electrons to cytochrome c_1 involves the **protonmotive Q cycle**. Complex IV (EC 1.9.3.1) contains the 13 polypeptide subunits of cytochrome c oxidase, including cytochromes a and a_3, and $Cu^{2+.}$ Three polypeptides are encoded in mtDNA. It catalyses the oxidation of reduced cytochrome c by dioxygen. Numerous mutations in the genes in nuclear DNA and in mtDNA for protein components of these complexes produce a vast array of clinical manifestations in virtually all age groups and in most body tissues and systems.

complexation the act or process of forming a (chelate) **complex**, e.g. to render inorganic substances soluble in organic solvents.

complex closed DNA a DNA structure comprising catenated and concatenated circular polynucleotide oligomers. *See* **catenane**, **concatemer**.

complex formation *see* **complex** (def. 2), **complexation**.

complexity *(in molecular biology)* the number of different sequences in DNA as defined by hybridization kinetics.

complexometric titration any titration in which the titrant contains a complexing (chelating) agent that interacts with a metal ion or other chemical species in the test solution, the end point being determined either electrometrically or with an indicator.

complexometry any method for the determination of a metal ion (or other chemical species) involving the formation of a chemical **complex**.

complexone any complex-forming or chelating agent, especially any macrocyclic antibiotic that complexes metal ions.

Complexone *a proprietary name for* **EDTA**.

composite database a database that amalgamates different primary sources, often using criteria that determine the priority of inclusion of the sources and the level of redundancy retained. For example, nrdb (nonredundant database) combines sequences from GenBank (CDS translations), PDB, Swiss-Prot, PIR, and PRF; INSD (International Nucleotide Sequence Database) is an amalgam of nucleotide sequences in EMBL, GenBank, and DDBJ; UniProt (universal protein sequence database) is a collation of protein sequences from PIR-PSD, Swiss-Prot, and TrEMBL; wwPDB (worldwide PDB) is a composite of 3D structures in the RCSB (Research Collaboratory for Structural Bioinformatics), PDB, MSD, and PDBj. *See* **sequence database**.

compound any substance containing two or more identical or nonidentical chemical atoms in fixed numerical proportions and held together by one or more kinds of chemical bond.

compound A, B, C (etc.) *see* **Kendall's compounds**.

compound heterozygote an individual who has a recessive genetic disease resulting from the inheritance of two different mutant alleles of the same gene.

compressed file a computer file or collection of these in **archive format** that has been compressed by an appropriate algorithm such as that of Lempel-Zif. In UNIX and Linux such files have the extensions .Z or .gz. For PCs the extension is usually .zip and for Apple Macintosh 'stuffit'. The program that does the compression shows the file as an appropriate icon. Some compressed files may be programs that uncompress themselves. Others require decompression utilities such as WINZIP (Microsoft PCs) or uncompress or gunzip (UNIX and Linux). A good **software archive** will include notes (typically in a file called README or 00README) on decompression utilities that are required.

compression *see* **band compression**.

comproportionation any chemical reaction that is the reverse of **disproportionation**.

Compton effect *or* **Compton scattering** the scattering of photons by free particles that results in an increased photon wavelength (i.e. a decrease in photon energy) and a corresponding increase in the particles' velocity (i.e. an increase in the particles' kinetic energy). [After Arthur Holly Compton (1892–1962), US physicist.]

computer-assisted tomography *abbr.*: CAT; *see* **tomography**.

computer of average transients *abbr.*: CAT; an electronic device of many separate channels in which a set of signals (e.g. an NMR spectrum) is repeatedly scanned and the signals stored additively. The stored signals grow linearly with the number of scans but the noise, being random, grows only as the square root of the number of scans taken, thereby improving the signal-to-noise ratio.

COMT *abbr. for* catechol *O*-methyltransferase, EC 2.1.1.6.

comutagen any nonmutagenic agent that enhances the effect of a mutagen.

comutation the occurrence of a mutation near to, and at the same time as, another specified mutation.

con A *abbr. for* concanavalin A.

conalbumin *or* **ovotransferrin** a protein of ~77 kDa that comprises ~12% of the total solids of egg white. It is distinguishable from **ovalbumin** by its lower thermal coagulation point. It binds Fe^{3+} ions. The amino-acid sequence of conalbumin is identical to that of chicken serum **transferrin**, but their carbohydrate moieties are different.

conantokin G *also called* **sleeper peptide**; a toxic peptide from the venom of *Conus geographicus* (fish-hunting cone snail). It is an antagonist of brain **NMDA receptors**. It has the sequence

GEXXLQXNQXLIRXKSN,

where X = *γ*-carboxyglutamate (Gla). The five Gla residues bind Ca^{2+}.

conarachin a minor globulin occurring, together with **arachin** (def. 1), in seeds of the peanut, *Arachis hypogaea*.

conc. *abbr. for* concentrated.

concanavalin A *abbr.*: con A; an agglutinating and mitogenic protein (*see* **lectin**) that constitutes the major component of the subsidiary globulin fraction (originally named **concanavalin** because it accompanied **canavalin** in purification) obtained from the ripe seeds of the **jack bean**. It is separable from **canavalin** and from the minor component **concanavalin B** by differences in solubility. It is a crystallizable tetramer, with 27.5 kDa subunits, dissociating to a dimer below pH 5.8. Each subunit contains one binding site for Mn^{2+} (or

other transition metal ion), one for Ca^{2+}, and one for carbohydrate (specifically *α*-D-mannopyranoses or *α*-D-glucopyranoses with unmodified hydroxyl groups at C-3, C-4, and C-6); carbohydrate does not bind until the metal binding sites are filled. It agglutinates erythrocytes, and precipitates polysaccharides containing a sufficient number of appropriate terminal residues. It is mitogenic, transforming lymphocytes to blast cells, and can agglutinate some animal tumour cells and inhibit their growth. The precursor is processed in an unusual way: the mature chain consists of residues 164–281 followed by 30–148; to form a mature chain, the precursor undergoes further **PTM**, after removal of the signal sequence; cleavage after Asn at positions 148, 163, and 281 is followed by transposition and ligation (by formation of a new peptide bond) of residues 164–281 and 30–148.

concanavalin B a minor crystallizable protein obtained from ripe seeds of the **jack bean**, and separable from **canavalin** and **concanavalin A**.

concatemer *or* **concatenate** originally, an intermediate structure in the biosynthesis of certain DNA viruses, consisting of a number of genomes connected together end to end, and thought to be the immediate precursor to the mature viral genome. The term is now extended to include any linear or circular DNA structure composed of viral genomes joined end to end. The term is also used for circular DNAs which are physically inseparable, one being threaded through the other. *Compare* **catenane**. —**concatemeric** *adj*.

concatenate 1 to join or link together, end to end. **2** joined or linked together. **3** *an alternative term for* **concatemer**. *Compare* **catenate**. —**concatenated** *adj*.; **concatenation** *n*.

concatenated dimer *or* **circular dimer** any circular **concatemer** containing two DNA molecules.

concentrate 1 to reduce the volume of a solution or dispersion by removal of solvent or dispersing medium. **2** to make or become purer or more concentrated by removal of impurities from a substance or material. **3** a concentrated solution or material, especially of foodstuffs, e.g. a vitamin concentrate. —**concentrating** *n*.

concentrated 1 *abbr.*: conc.; *(of a solution, dispersion, or mixture)* containing a relatively high concentration (of a specified solute, kind of dispersed particle, or component); (of substance, particles, etc.) present at relatively high concentration (in a solution, dispersion, etc.). *Compare* **dilute** (def. 1). **2** *(of a solution, dispersion, or mixture)* increased in concentration and reduced in volume by removal of solvent, dispersing medium, or admixed material. *Compare* **diluted**.

concentration *abbr.*: concn.; **1** **(a)** **mass concentration** *or* **mass density** *symbol*: ρ *or* γ; an SI derived physical quantity; it equals the mass of a specified component per unit volume of the system in which it is contained. It is expressed in kg m^{-3}. **(b)** **amount-of-substance concentration** *or* **amount concentration** *or* *(in clinical chemistry)* **substance concentration** *symbol*: c; an SI derived physical quantity; it equals the number of moles of a specified component per unit volume of the system in which it is contained. Where there is no risk of confusion the term **concentration** may be used alone. It is expressed in mol m^{-3} or mol dm^{-3}. Amount concentration continues to be expressed often in terms of the litre (*symbol*: l *or* L), which is not an SI unit, although 1 L = 1 dm^3. The older, non-SI, term *molarity* is still commonly used for this quantity; a solution of, say, 1 mol dm^{-3} concentration may thus be referred to also as a 1 molar solution or denoted a 1 M solution, where M is treated as a symbol for mol dm^{-3} (or mol L^{-1}). *See* **molar** (def. 2). **(c)** **surface concentration** *symbol*: Γ; an SI derived physical quantity; it equals the number of moles of a specified substance per unit area of the surface at which it is present. It is expressed in mol m^{-2}. **(d)** **number concentration** *or* **number density** *symbol*: C *or* *(sometimes)* n; an SI derived physical quantity; it equals the number of specified particles or molecular entities per unit volume of the system in which it is contained. It is expressed in m^{-3}. *Note*: The concentration of a solute in solution may sometimes be expressed in more convenient units esp. in a percentage; e.g. (for solids) as the mass of solute in grams per 100 millilitres of solution (*abbr.*: % w/v), or (for liquids) as the number of volumes of solute per 100 volumes of solution (*abbr.*: % v/v). *See also* **activity** (def. 3), **molality**, **osmolality**, **osmolarity**. **2** the act or process of concentrating. *See* **concentrate** (def. 1, 2).

concentration equilibrium constant any **equilibrium constant** expressed in terms of concentrations rather than activities.

concentration gradient the change of concentration with distance, e.g. the change in concentration of a solute across a membrane or along a gradient.

concentration of enzymic activity (obsolete) see **enzymic activity**.

concentration polarization (in membrane ultrafiltration) the accumulation of retained macrosolutes on the membrane, forming a secondary membrane, that results in restriction of transport through the filter and alteration of selectivity, which may lead to rejection of normally permeating species.

concentration quenching the **quenching** of fluorescence that becomes significant when the fluor concentration is increased beyond a certain point. It is of importance, e.g., in liquid scintillation counting and in experiments involving the introduction of fluorescent groups (probes) into macromolecules.

concentration ratio or **dose ratio** symbol: r; the ratio of the concentration of an agonist that produces a specified response (often but not necessarily 50% of the maximal response to that agonist in an assay) in the presence of an antagonist, to the agonist concentration that produces the same response in the absence of antagonist.

concentric cylinder viscometer an alternative name for **Couette viscometer**.

conceptual translation the computational process of translating the sequence of nucleotides in mRNA, by means of the genetic code, to a sequence of amino acids.

concerted reaction any chemical reaction in which two chemical processes occur within the same reaction step; i.e. one in which a new chemical bond is formed at the same time as, and as a direct result of, the breaking of another chemical bond.

concerted symmetry model a model of ligand binding to oligomeric proteins in which the characteristic assumption is that molecular symmetry of the **oligomer** is conserved so that only two states are accessible to it: either with all the **protomers** in the first conformational state, or all the protomers in the second state. See **Monod–Wyman–Changeux model**.

concerted transition the simultaneous transition of all identical subunits (i.e., protomers) of an oligomeric molecule from the first conformational state to the second state, and vice versa, as is required in any **concerted symmetry model** for allosteric proteins.

concn. abbr. for concentration.

concordance the condition in which both members of a twin pair, or any two comparable individuals or species, possess the same gene, phenotype, or trait. —**concordant** adj.; **concordantly** adv.

condensate any product of a **condensation**, especially a liquid from a gas or vapour.

condensation 1 the act or process of condensing; the state of being condensed. **2** the product of condensing; a condensed mass. **3** any (usually stepwise) chemical reaction in which two or more identical or nonidentical molecular entities, or two parts of one entity, combine with the elimination of a molecule of water, ammonia, ethanol, hydrogen sulfide, or other simple substance. The combining entities or parts each (formally) contribute a moiety to the molecule eliminated.

condensation polymerization a polymerization by repeated condensation, i.e. with elimination of some simple substance, to form a **condensation polymer**.

condense 1 to make or to be made more compact or denser. **2** to change or cause to change a gas or vapour to a liquid or solid. **3** (of a compound) to undergo or cause to undergo a **condensation** (def. 3).

condensed 1 converted to a dense or denser form; e.g. **condensed DNA** in **nucleosomes** or in the early stages of maturation of T-even phages; or the **condensed form of mitochondria** in which the cristae are swollen, the matrix volume is much reduced, and the inner membrane space is increased. **2** reduced from the gaseous to the liquid state.

condenser 1 any device for converting a gas or vapour into a liquid (or solids) by cooling. **2** a lens or set of lenses that concentrates light into a small area. **3** a former name for capacitor; an electrical component that stores electric charge.

condensin a protein complex that binds to sister chromatids to cause their disentanglement. It consists of Smc2 and Smc4 (see **Smc**) and three other proteins.

condensing enzyme 1 citrate condensing enzyme, EC 4.1.3.7; see **citrate (si)-synthase**. **2** β-oxoacyl-ACP synthase, EC 2.3.1.41; 3-oxoacyl-[acyl-carrier-protein]synthase (see **fatty acid synthase complex**).

conditional association constant or **conditional stability constant** an **association constant** determined in conditions similar to those in which the reaction under consideration is taking place, e.g. in the internal milieu of a cell.

conditional lethal mutation any mutation that produces a mutant (**conditional lethal mutant**) whose viability depends on the conditions of growth. It grows normally in permissive conditions but in restrictive conditions it does not grow, thereby expressing its lethal mutation. See, e.g., **temperature-sensitive mutation**.

conduct to allow the passage of electricity, heat, or electromagnetic radiation.

conductance see **electric conductance**.

conductance water an alternative name for **conductivity water**.

conductimetric or **conductiometric** a variant spelling of conductometric; see **conductometry**.

conductimetry or **conductiometry** a variant spelling of conductometry.

conduction 1 the transfer of heat or electricity through a material. **2** the transmission of an impulse, either electrical or chemical, along a nerve fibre.

conductivity 1 electrical conductivity symbol: κ; a measure of the ability of a material to conduct an electric current or impulse. It is defined as the reciprocal of the resistivity of the material, and is expressed in siemens m^{-1}. See also **electrolytic conductivity**. **2** thermal conductivity.

conductivity water or **conductance water** very pure water, of electrolytic conductivity $<1 \times 10^{-4}$ siemens m^{-1}.

conductometric titration any titration in which the end point is indicated by an inflection in the curve of conductivity against volume of titrant added. For instance when an HCl solution is titrated with an NaOH solution, the conductivity decreases as the faster moving H_3O^+ ions are replaced by slower moving Na^+ ions; after the equivalence point the conductivity increases as the concentration of the fast-moving OH^- ions increases.

conductometry or **conductimetry** or **conductiometry** any method of chemical analysis based on measurements of electrical conductivity. —**conductometric** adj.

conductor any substance or system that allows the passage of electricity, heat, or electromagnetic radiation.

cone cell or **cone** one of the two types of light-sensitive cell of the vertebrate retina (compare **rod cell**). Cone cells are conical or flask-shaped, and contain the photopigment **iodopsin** or (in freshwater and migratory fishes and some amphibians) **cyanopsin**. They are responsible for photopic (daylight) vision, and are densest in the fovea and absent from the margin of the retina. Human cone cells are thought to be of three types, each containing an iodopsin of different spectral sensitivity; hence they are responsible also for colour vision.

cone/rod dystrophy any of the disorders of eyesight resulting mainly from mutation of the *CRX* (cone-rod homeobox containing) or *RPGR* (retinitis pigmentosum GTPase regulator) genes. See **CRX**.

cone-rod homeobox protein see **CRX**.

confidence limits (in statistics) the upper and lower limits of the range of values within which the true mean is expected to lie, on the basis of an estimate of the mean obtained from a sample of values. The confidence limits vary according to the (specified) probability that they will be correct.

configuration classically, the arrangement of the atoms of a molecule in space, without regard to arrangements that differ only as after rotation about one or more single bonds; the term is now sometimes limited so that no regard is paid to π-bonds or to bonds of a partial order between one and two, and sometimes further limited so that no regard is paid to rotation about bonds of any order, including double bonds. Interconversion of molecules of different configurations requires the formal making and breaking of covalent bonds, no account being taken of changes to hydrogen bonds.

Compare **conformation**. *See also* **absolute configuration, pyrimidine dimer**. —**configurational** *adj*.

configurational base unit (of a polymer) a **constitutional base unit** of the polymer whose configuration is defined at the minimum at one site of steric isomerism in the main chain of the polymer.

configurational block a **block** in a polymer containing only one species of **configurational base unit**.

configurational inversion *see* **racemize, Walden inversion**.

configurational isomer any of two or more isomers that differ only in stereochemical **configuration**.

confluence *(in cell culture)* the state reached when one or more cells or groups of cells, grown in culture on (so-called) solid medium, have multiplied to cover the surface of the medium completely but without overlying each other. —**confluent** *adj*.

confocal microscope a microscope that produces a series of images, often called optical sections, that have much better resolution than images of the same specimens obtained using conventional epifluoresence light microscopy. Such a microscope depends on laser light focused through a pinhole onto a single point in the specimen. From a succession of such images a 3-D image can be created. This is primarily used in biological imaging for mapping the distribution of fluorescently labelled macromolecules, such as antibodies, in various tissues.

conformation the various arrangements of the atoms of a molecule (of defined **configuration**) in space that differ only as after rotation about single bonds; the term is now sometimes extended to include rotation about π-bonds or bonds of partial order between one and two, and sometimes further extended to include also rotation about bonds of any order, including double bonds. Interconversion of molecules of different conformations does not involve the breaking and making of chemical bonds (except hydrogen bonds) or changes in **chirality**. The following special terms and symbols distinguish different possible conformations of various classes or portions of molecules.

(1) **Cyclic compounds**. Nonplanar six-membered rings may exist in **boat conformation, chair conformation, half-boat conformation, half-chair conformation, skew conformation** (def. 1), or **twist conformation** (def. 1); and nonplanar five-membered rings may exist in **envelope conformation** or **twist conformation** (def. 2). For a six-membered or five-membered ring form of a monosaccharide or monosaccharide derivative these conformations may be designated by the following conformational descriptors suffixed to their chemical names: *-B*, boat; *-C*, chair; *-E*, envelope; *-H*, half-chair; *-S*, skew; *-T*, twist; the locants of the ring atoms that lie above and/or below the reference plane, when the numbering appears clockwise, may be added to these symbols as left superscripts and/or right subscripts, respectively. A bond to a tetrahedral atom in a six-membered ring is termed **axial** (*abbr.*: a) if it or its projection makes a large angle with the plane containing a majority of the ring atoms, and **equatorial** (*abbr.*: e) if this angle is small; the same terms are applied to atoms or groups attached to such bonds. The corresponding terms for a bond from an atom directly attached to a doubly bonded atom in a monounsaturated six-membered ring are **pseudoaxial** (*abbr.*: a') and **pseudoequatorial** (*abbr.*: e'), respectively.

(2) **Nucleotides**. When atom C-2' or C-3' of the ribo- or deoxyribofuranose ring of a nucleotide lies out of the plane of the other four ring atoms, the ring conformation is designated by the descriptor *endo-* if either atom lies above the ring, i.e. towards the nitrogenous base, and by the descriptor *exo-* if it lies below the ring, i.e. away from the base. In addition the base and sugar rings may be in either of two relative positions: the conformation is said to be **anticlinal** (*abbr.*: anti-) if the groups at positions 2 and 3 of a pyrimidine residue or positions 1, 2, and 6 of a purine residue lie away from the sugar ring, and **synclinal** (*abbr.*: syn-) if these groups lie over the sugar ring.

(3) **Torsion angle**. If rotation about a given bond in a molecule is restricted for any reason, the conformation of atoms or groups about that bond may be described by the **torsion angle** (*symbol*: θ). When, in a **Newman projection**, two atoms or groups attached at opposite ends of a bond appear one more or less directly behind the other, i.e. $\theta \approx 0°$, these atoms or groups are termed **eclipsed**, and that portion of the molecule is described as being in the eclipsed conforma-

tion. If not eclipsed (in an ideal case, when all torsion angles equal 60°), the atoms or groups are termed **staggered**, and the molecule is described as being in the staggered conformation; such a conformation is termed **synperiplanar** (*abbr.*: *sp*) (or sometimes *cis* (*abbr.*: c)), **synclinal** (*abbr.*: *sc*) (or sometimes *gauche* (*abbr.*: g) or *skew*), **anticlinal** (*abbr.*: *ac*), or **antiperiplanar** (*abbr.*: *ap*) (or sometimes *trans* (*abbr.*: t)) according as the torsion angle is, respectively, within ±30°, ±60°, ±120°, or ±180° of 0°, the torsion angle being defined according to specified criteria. (*Note*: Because the terms *cis* and *trans* have other meanings, their use in this context is not recommended.) —**conformational** *adj*.

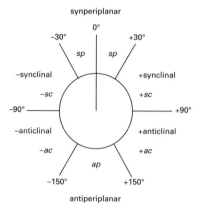

synperiplanar

antiperiplanar

conformational analysis the application of a set of specific rules to determine the conformation of a molecule in which the strain will be minimal, i.e. the most stable and preferred conformation of the molecule.

conformational change 1 any change in conformation. **2** any change in the three-dimensional structure of a macromolecule, membrane, or other entity.

conformational-coupling hypothesis a hypothesis (now superseded) that the energy yielded by the processes of electron transport, e.g. in mitochondria, is conserved in the form of a conformational change in an electron carrier protein or in the coupling factor (F_1 ATPase) molecule. The high-energy conformational state results from an energy-dependent shift in the number and location of weak bonds that maintain the three-dimensional conformation of the protein. The energy inherent in this 'energized' conformation is thought to be used to cause the formation of ATP from ADP and inorganic phosphate. This hypothesis was put forward after attempts to find a high-energy intermediate that coupled electron transport to oxidative phosphorylation, failed. It has been superseded by the **chemiosmotic coupling hypothesis**.

conformational formula a diagrammatic representation on a plane surface of the structural formula of a molecule that shows the relative dispositions in space of the constituent atoms or groups. Such representations are used especially to depict instances of conformational isomerism. For a monosaccharide or monosaccharide derivative, the conformational formula may be combined with a **Haworth representation**. *See also* **Newman projection**.

conformational isomer *or (sometimes)* **conformer** any of two or more isomers that differ only in their stereochemical **conformation**.

conformational map *an alternative name for* **Ramachandran plot**.

conformer 1 a molecule in a conformation into which its atoms return spontaneously after small displacements. **2** *an alternative term for* **conformational isomer**.

conformon a quantized energy package that, according to the **conformational-coupling hypothesis**, is associated with a localized conformational change in a protein in the mitochondrial membrane.

congener 1 any member of a (specified) group. **2** *(in taxonomy)* any member of a specified genus. **3** *(in chemistry)* **a** a chemical substance related to another, e.g. as a derivative or as an element of the same group of the periodic table. **b** a secondary product retained in an alcoholic beverage (or extract of a natural product) that contributes significantly to the final characteristics of the beverage (or the extract).

congenic 1 describing an inbred strain of animals in which a defined chromosomal segment has been transferred from a donor to a recipient strain. *Compare* **consomic**. **2** of, or pertaining to, a **congener**; **homogenic**.

congenital describing a character or trait (phenotype) that exists at or dates from birth. It neither includes all inherited conditions nor excludes characters resulting from environmental causes.

congenital adrenal hyperplasia a syndrome of abnormal adrenal steroid hormone biosynthesis, usually caused by deficiency of 21-hydroxylase (steroid 21-monooxygenase, EC 1.14.99.10). There is failure to synthesize cortisol, overproduction of ACTH (due to release from negative feedback by cortisol), hyperplasia of the adrenal cortex, and overproduction of 17-hydroxyprogesterone and androstene 3,17-dione by overstimulated alternative pathways. Deletion of the *CYP21* gene at 6p21.3, crossing over of this gene with the neighbouring pseudogene *CYP21P*, and various inactivating mutations in *CYP21* cause absence or marked deficiency of 21-hydroxylase activity. Deficiency (usually partial) of 11β-hydroxylase (steroid 11β-monooxygenase, EC 1.14.15.4) accounts for most other cases; in this condition, 11-deoxycorticosterone accumulates. Very rarely, deficiency of 17-hydroxylase (steroid 17α-monooxygenase, EC 1.14.99.9) or 18-hydroxylase (corticosterone 18-monooxygenase, EC 1.14.15.5) is the cause.

congenital lipid adrenal hyperplasia a rare condition in which there is a defect in production of all major steroid hormones in the adrenal cortex. The latter is grossly hyperplastic and its cells are engorged with cholesterol and cholesterol esters. A mutation in **StAR**, a protein essential for translocation of cholesterol into mitochondria in adrenal cortex and gonads, decreases or stops production of pregnenolone, which is the first step in corticosteroid synthesis.

conglutination the agglutination of bacterial or red blood cells, that have been sensitized with specific antibody, in the presence of (nonhemolytic) **complement** and **conglutinin**.

conglutinin a nonimmunoglobulin protein, present in the plasma of cattle and other bovids, that aggregates **complement**-coated immune complexes or cells.

conglycinin a 7S globulin storage protein of soybean (*Glycine max*) that is present as a glycosylated trimer in protein storage vacuoles in the embryo and aleurone layer. It is homologous with phaseolin and vicilin. All three globulins are synthesized as 45–50 kDa preproproteins.

Congo Red sodium diphenyldiazo-bis-α-naphthylaminesulfonate; CI 22120; Direct Red 28; an indicator dye that is blue-violet at pH 3.0 and red at pH 5.0. It is used for detecting and estimating free HCl in gastric contents, as an addition to culture media, and for staining **amyloid**.

Congo Red fibrin a preparation obtained by heating fibrin to 80°C in a slightly alkaline Congo Red solution and thoroughly washing. It was formerly used in the detection of proteolytic activity since the dye does not dissolve in aqueous solutions unless the fibrin is digested.

congruence *(in phylogeny)* agreement between characters or trees. –congruent *adj.*

coniferin 4-(3-hydroxy-1-propenyl)-2-methoxyphenyl β-D-glucopyranoside; the major **glycoside** present in conifers and some other plants.

coniferyl alcohol the aglycon of **coniferin**; it occurs as a building unit of plant **lignins**.

coniine 2-propylpiperidine; the toxic alkaloid of poison hemlock, *Conium maculatum*. The natural enantiomer has *S* configuration.

conjugase γ-Glu-X carboxypeptidase, EC 3.4.19.9 (formerly EC 3.4.22.12); γ-glutamyl hydrolase; an enzyme that hydrolyses γ-glutamyl bonds in *N*-pteroyl-γ-oligoglutamate, pteroyl-γ-diglutamate being a prominent product.

conjugate 1 to effect or undergo **conjugation**. **2** describing either of a pair of interconvertible substances, e.g. a **conjugate acid–base pair**. **3** the product formed as a result of conjugation.

conjugate acid–base pair the relationship between two chemical species, BH⁺ and B, that are related by the reaction: BH⁺ = B + H⁺. Then BH⁺ is the **conjugate acid** of the base B, B is the **conjugate base** of the acid BH⁺, and BH⁺ and B are a conjugate acid–base pair.

conjugate base *see* **conjugate acid–base pair**.

conjugated 1 describing a compound that is formed by the linking of two compounds. **2** displaying or having undergone **conjugation**.

conjugated double bond any of two or more double bonds in a molecule where each double bond is separated from the next one by one single bond. *Compare* **conjugation** (def. 2).

conjugated protein any protein that contains a non-protein component, often in stoichiometric proportion. The non-protein component may be a metal ion, a lipid, a carbohydrate, or a nucleic acid, and may be either loosely or tightly bound to the polypeptide chain(s).

conjugation 1 the act of joining together; the state of being joined together. **2** an alternating sequence of multiple and single chemical bonds containing at least two multiple bonds with delocalization of pi-electrons and resultant additional chemical stability. **3** the covalent or noncovalent joining together of one (larger) molecule, e.g. a protein or bile acid, with a second (smaller) molecule. **4** a process of sexual reproduction occurring in various types of unicellular organisms. In bacteria, e.g. *Escherichia coli*, it involves the transfer of DNA from a donor cell to a recipient cell via a sex **pilus**; in protozoa, e.g. *Paramecium aurelia*, a true exchange of DNA occurs between the participating cells, which belong to different mating types.

conjugation labelling a procedure for introducing a **label** into a large molecule of interest by covalently coupling it, in a specific chemical reaction, to a small molecule containing the label (*see* **conjugation** (def. 3)). It is particularly useful for labelling with radioiodine any protein or peptide that is sensitive to oxidative procedures or to noxious components of commercial radioiodine, that lacks tyrosine residues reactive to iodine, or that requires to be labelled at a residue other than tyrosine. *See also* **Bolton and Hunter reagent**.

conjugative plasmid any **plasmid** that can bring about the transfer of DNA by **conjugation**.

conjugon any genetic element that is required for bacterial **conjugation** (def. 4), e.g. **fertility factor**.

connective tissue any supporting tissue that lies between other tissues and consists of cells embedded in a relatively large amount of extracellular matrix.

connective tissue growth factor *abbr.*: CTGF; a growth factor of testis, spleen, kidney, lung, heart, and brain, that is similar to **CEF10**; one of a family of growth regulators that belong to a group of immediate-early genes expressed after induction by growth factors or certain oncogenes. The family includes **Cyr61**, βIG-M1 and βIG-M2.

connectivity *(in chemistry)* the information in any molecular formula or model regarding the order in which the constituent atoms of the molecule are linked, irrespective of the nature of the linkages.

connexin the main protein component of a **connexon**. Each connexin contains four putative membrane-spanning α-helices, and six connexins make up each connexon. A number of different subtypes of connexin exist within each species, contributing different functional behaviour to different connexons.

connexon the structural subunit of a **gap junction**, the structure that forms a bridge between adjacent cells in certain tissues of vertebrates. Seen by electron microscopy or low-angle X-ray diffraction, an individual gap junction consists of a number ($10-10^4$) of connexons, often in hexagonal array, embedded in and protruding from either side of each of the opposed cell membranes in register and linked to one another. Each connexon is a cylindrical structure, about 6–8.5 nm in diameter and 7.5 nm long, that consists of six rod-shaped, essentially rigid subunits (**connexins**). The diameter of the central opening can apparently be varied by radial displacement of the subunits at their cytoplasmic extremities. Each connexin spans one cell membrane, two with connecting ends being required to form a channel across both membranes. Such channels provide a regulable hydrophilic pathway permitting the passage of small molecules (up to about 800 Da) between adjacent cells; the specificity of each channel is determined by the type of connexin it contains.

conotoxin *abbr.*: ω-CT; any of several peptides of the family of ω-conotoxins isolated from the venoms of two marine snails, *Conus geographicus* and *C. magus*. All ω-conotoxins have a conserved pattern of cysteine residues linked by four disulfide bridges. They are neurotoxins that inhibit voltage-gated Ca^{2+} channels and neurotransmitter release. ω-Conotoxin GVIA has the structure:

[tricyclic ($1\rightarrow16$, $8\rightarrow19$, $15\rightarrow26$)(H-Cys^1-Lys-Ser-Hyp-Gly-Ser-Ser-Cys^8-Ser-Hyp-Thr-Ser-Tyr-Asn-Cys^{15}-Cys^{16}-Arg-Ser-Cys^{19}-Asn-Hyp-Tyr-Thr-Lys-Arg-Cys^{26}-Tyr-NH$_2$)].

consanguinity the relationship between individuals who are descended from the same ancestors and consequently are more likely to have mutant alleles in common. —**consanguineous** or **consanguine** *adj.*; **consanguineously** *adv.*

consensus sequence an idealized sequence of nucleotides, or their constituent bases, or amino acids, base, or amino acid that represents the nucleotide most likely to occur at each position in the sequence. Consensus sequences are used to identify RNA splicing sites, other sites, plasmids, and families of proteins.

consensus tree *(in phylogenetics)* a tree that summarizes relationships from multiple phylogenies, wherein groups of taxa that appear in all individual trees are shown, typically with multifurcations indicating disagreements between trees. Trees are often combined into a consensus phylogeny when tree-building tools generate more than one topology. *Compare* **majority-rule consensus tree**.

conservation the retention of structure by a macromolecule, or by a specified segment of one, with variation of circumstance (environmental, genetic, etc.). When used of primary structure, it can be synonymous with **sequence homology**. The degree of retention of structure is usually specified. *See also* **conserved**.

conservation of energy (law of the) *see* **thermodynamics**.

conservative base change *or* **conservative base substitution** any mutational change, including substitution of a base, in a particular base triplet in a DNA molecule such that either the amino acid encoded by that base triplet is not altered, or there is no major change

in the properties of the R group of the amino acids involved, e.g. Glu for Asp. *See* **genetic code**, **wobble hypothesis**.

conservative replacement *or* **conservative substitution** any replacement or substitution in a polypeptide chain of a particular amino-acid residue by another with similar properties: e.g., Arg for Lys, Phe for Tyr, Glu for Asp.

conserved describing a tendency to invariance in corresponding residues or sequences of residues of encoded macromolecules (e.g. proteins) obtained from specimens of genetically different sources. Macromolecules showing a high degree of invariance of their primary structure are said to be highly conserved; ones exhibiting considerable variation are described as poorly conserved. *See also* **conservation**, **homologous**.

consomic describing an inbred strain of animals in which an entire chromosome has been transferred from a donor to a recipient strain. *Compare* **congenic** (def. 1).

constant region the region of a chain of an **immunoglobulin** molecule that is characterized by invariability of the amino-acid sequence from molecule to molecule, apart from certain residues at allotypic marker sites. The constant region of the light chain (corresponding to the C-terminal part) is designated C_L, while the constant region of the heavy chain is designated C_H. Within a constant region, individual domains may be further designated C_H1, C_H2, etc. In a more precise designation of a heavy chain the 'H' subscript is replaced by the Greek letter corresponding to the type of heavy chain that defines the particular class of immunoglobulins. Hence, the constant region of an **immunoglobulin G**, in which the heavy chain is designated γ, can be designated $C_\gamma1$, $C_\gamma2$, etc.

constitutional base unit the smallest possible regularly repeated unit (i.e. identical group of monomers) of a regular **polymer**.

constitutional block a **block** of a polymer that contains only one species of **constitutional base unit**.

constitutional isomer *or* (*formerly*) **structural isomer** any of two or more **isomers** that differ in their molecular constitution, i.e. in the nature and sequence of bonding of their atoms rather than in the arrangement of their atoms in space. For example, ethanol, CH_3-CH_2-OH, and dimethyl ether, CH_3-O-CH_3, are constitutional isomers. *Compare* **stereoisomer**.

constitutive enzyme any enzyme that is formed at a constant rate and in constant amount in a given cell, regardless of the metabolic state of the cell or organism. *Compare* **induced enzyme**.

constitutive genes genes that are expressed following interaction between a promoter and RNA polymerase without additional regulation. *Also known as* **housekeeping genes**, they are expressed in virtually all cells since they are essential for the cell's activity.

constitutive morphogenetic protein 1 *see* **COP1**.

constitutive mutation any mutation that results in an increased constitutive synthesis (i.e. synthesis as if by a **constitutive gene**) by a prokaryote of several functionally related, inducible enzymes. In such mutations either an operator gene is modified so that the repressor cannot combine with it or the regulator gene is modified so that the repressor is not formed.

contact amino acid any of the amino-acid residues in the **active site** (def. 1) of an enzyme that are at some stage within one bond distance of some point on the substrate molecule in the enzyme–substrate complex. They may include both catalytic residues and specificity residues. *Compare* **auxiliary amino acid**.

contact hypersensitivity a form of immunological **hypersensitivity**, immediate or delayed, that is provoked in the skin or mucous surfaces by contact with specific agents acting as either antigens or haptens.

contact inhibition a phenomenon, seen in cultures of normal eukaryotic cells, in which cell division is inhibited by cell-to-cell contact. When one cell contacts another cell as a result of increased cell density, it first ceases to move, then metabolizes more slowly and ceases to divide. An outstanding feature of malignant cells is their loss of contact inhibition so that they multiply unchecked. *See also* **transformation** (def. 2).

contact residue *see* **contact amino acid**.

contact sensitivity the ability of an organism to respond to a biocide, e.g. an insecticide or a selective herbicide, applied to the integument.

containment the use of any structural or procedural safeguards to minimize the possible risks, to the experimenter or to the environment, in hazardous or potentially hazardous operations, especially genetic manipulations or the use of ionizing radiation.

contamination the presence of any contaminant; the act of introducing a contaminant; pollution; the presence of harmful or invasive microorganisms in a population of desirable ones. *Compare* **decontamination.**

context 1 any parts of a piece of writing or speech that precede or follow a specified word or passage and contribute to its full meaning. **2** (*in mycology*) the layers that develop between the hymenium and the true mycelium in certain fungi.

context effect any effect of nucleotide residues outside a codon in an mRNA molecule in influencing the efficiency of the reading of that codon by the appropriate tRNA molecule. *Compare* **context** (def. 1).

context mutation any mutation that brings about an alteration in the **context effect** of nucleotide residues outside a particular **codon** in an mRNA molecule.

contig one of a set of overlapping clones that represent a continuous region of DNA. Each contig is a genomic clone, usually in a **cosmid** or a **YAC.** They are used in **contig mapping.**

contig mapping a technique used in projects such as the Human Genome Project to enable the physical map of (part of) a chromosome to be determined. The method relies on the use of overlapping clones, referred to as **contigs.**

contiguous gene syndrome any of the complex disease phenotypes associated with a chromosomal microdeletion affecting multiple, unrelated genetic loci physically contiguous on a chromosome.

continuous culture any cell culture in which the cells are in the growth phase over a prolonged period of time. *See also* **chemostat.**

continuous distribution any nonquantized distribution; e.g. in a large population one measurement, x, may be considered to vary continuously, and the probability, y, of the occurrence of a particular value of x is given by $y = f(x)$.

continuous epitope an immunological determinant (i.e. **epitope**) on an antigen that is formed from a single linear stretch of amino acids. It is much less common than a **discontinuous epitope.**

continuous-flow centrifugation any preparative centrifugation technique, used for collecting materials from relatively large volumes of liquid, in which a liquid is continuously fed into a rotor, the sediment is collected, and the supernatant is withdrawn continuously as the rotor rotates. *See* **centrifugal elutriation, Sharples supercentrifuge, zonal centrifugation.**

continuous-flow electrophoresis a method of electrophoresis in which the protein (or other) solution is applied continuously to an inert support such as paper (then usually called **curtain electrophoresis**) or glass beads or is allowed to flow between two closely spaced parallel plates. The solution flows down the support by gravity and an electric potential applied at right angles causes migration of components away from the main stream. Fractions are collected at different positions across the bottom of the support.

continuous-flow technique a technique for following rapid chemical reactions in which two solutions, containing the reactants, are rapidly mixed and the mixture pumped at a predetermined speed through a tube of known dimensions that contains an observation chamber. By changing the rate of flow, the time between mixing and observation can be varied. Thus, given a method of analysis for reactants or products in the observation chamber, the extent of reaction as a function of time can be determined.

continuous process a method of cultivation in which nutrients are supplied and components of the culture medium are removed continuously at volumetrically equal rates, maintaining the cells in a condition of stable multiplication and constant growth rate.

continuous variation *or* **Job's method** a method for investigating metal–ligand (or other) equilibria in solution in which the mole ratio of, e.g., metal ion to ligand species is varied while the sum of their concentrations is kept constant and some property that varies in a fashion proportional to the concentration of one of the molecular entities present, e.g. absorbance or conductivity, is measured. The curve of the magnitude of the property observed versus the concentration ratio gives information about the stoichiometry of

the metal–ligand (or other) complex formed and allows the stability (association) constant to be calculated.

contour-clamped homogeneous electric field *see* **CHEF.**

contractile vacuole any of the small spherical vesicles, found in the cytoplasm of freshwater protozoans, that function to expel surplus water from the organism.

contrapsin a glycoprotein plasma serine protease inhibitor, similar to the **serpin**-type inhibitors. It specifically inactivates serine proteases of the trypsin class, but not of the chymotrypsin class. It shares 64% similarity with human α_1-antichymotrypsin.

contrast agent a substance used to improve the resolution in 'highlighting' an organ or tissue during X-ray tomography or magnetic resonance imaging. Analogous to the stains used in microscopy, it permeates the organ of interest or the surrounding medium. In X-ray examination, heavy nuclei such as iodine are used as contrast agents. For magnetic resonance imaging contrast agents are paramagnetic compounds such as lanthanide complexes.

contributing amino acid (in an enzyme) any amino-acid residue whose only function appears to be that of a foundation for the **contact amino acid** and **auxiliary amino acid** residues in the **active site.** They are required in the natural enzyme to maintain the framework to which the contact and auxiliary amino acids are attached.

contributing structure *or* (*formerly*) **canonical form** any one of two or more formal valence bond representations of a mesomeric molecular structure. *Compare* **mesomerism.**

control 1 any object or system in an experiment that is a standard of comparison. A control is prepared or carried out exactly as the other parts of an experiment but differs in respect of a single variable, allowing the significance of the latter to be assessed. *Compare* **experimental** (def. 3). **2** the regulation of a biochemical process such as a metabolic pathway or the expression of a gene. For metabolism, a distinction is sometimes made between control and **regulation.** Control is described as the power to change the state of metabolism in response to an external signal. This is a measure of the strength of response to external factors.

controllability coefficient *former name for* **elasticity coefficient.**

controlled pore glass a type of glass in the form of rigid particles honeycombed with a great number of pores of selected, precisely defined size. It is useful as a supporting matrix in a large variety of chromatographic techniques, including affinity, gel-permeation, and ion-exchange methods.

control point the point – corresponding to a particular enzyme – in a metabolic pathway or other biochemical sequence where control is exercised.

control region *or* **control sequence** the part of a DNA molecule containing the **promoter** and the **operator** region of an **operon.**

control strength *former name for* **flux control coefficient.**

convergence *see* **convergent evolution.**

convergent evolution *or* **convergence** evolution such as to produce an increasing similarity in some characteristic(s) between initially different groups of, e.g., organisms or **gene products.** *Compare* **divergent evolution.**

convertin *an alternative name for* factor VIIa (*see* **blood coagulation**).

convertin enzyme (factor) *see* **serum converting enzyme.**

convolution 1 a twisting together, a coil. **2** a folding, such as those that produce the irregular ridges on the surface of the cerebrum of the brain.

Conway microdiffusion method a microchemical analytical method applicable to substances from which a gas, e.g. ammonia or carbon dioxide, can be liberated quantitatively by a specific reagent or enzyme; the method can also be used to assay an enzyme. The technique employs a special microdiffusion dish, variously termed a **Conway cell, Conway dish, Conway unit,** or **Conway vessel.** This consists of a shallow circular glass vessel, 40–70 mm in diameter, containing a central well and with a closely fitting lid sealed with grease; the lid closes the outer annular space but does not touch the wall of the central well. The sample is placed in the outer annulus, and a gas-absorbing reagent is placed in the centre well. The specific reagent is added to the sample, the lid is closed, and diffusion of the evolved gas allowed to proceed. Subsequently the remaining gas-absorbing reagent is titrated. [After Edward Joseph Conway (1894–1968), Irish biochemist.]

Coomassie blue *or* **Brilliant blue** one of a series of naphthalene dyes, originally developed for wool, and commonly used for staining proteins on electrophoresis gels. It also forms the basis of the **Bradford assay** for proteins. [After the ancient Ashanti capital, Kumasi, now in Ghana.]

Coombs' test *an alternative name for* **antiglobulin test**. [After Robert Royston Amos Coombs (1921–), British immunologist.]

cooperative acting together or with others. *See also* **cooperativity**.

cooperative affinity the state of increased potency, receptor affinity, and resistance to destruction that is exhibited by agonist molecules covalently attached to a macromolecule in any situation where ligand and acceptor possess more than one mutual recognition or binding site, deployed in such a manner that simultaneous interaction is stereochemically feasible.

cooperative binding *see* **cooperative ligand binding**.

cooperative feedback inhibition the **feedback inhibition** of an enzyme by two or more end products of a reaction sequence such that the degree of inhibition caused by the mixture is greater than the sum of the effects of the individual end products.

cooperative hydrogen bonding the interaction between adjacent hydrogen bonds in a molecule such that the energy required to form them is less than the sum of the energies required to form the individual bonds, and the energy required to break these bonds is greater than the sum of the energies required to break the individual bonds. Such an interaction increases the stability of the molecular conformation in which it occurs.

cooperative ligand binding the binding of ligands to interacting sites on a macromolecule such that the binding of one ligand molecule to one site either increases (**positive cooperativity**) or decreases (**negative cooperativity**) the affinity of the other site(s) for ligand(s). Such effects may be either **homotropic**, i.e. one ligand species affects the binding of the same ligand species, or **heterotropic** (def. 1), i.e. one ligand species affects the binding of a different ligand species.

cooperativity *(in biochemistry)* any **cooperative** phenomenon, especially **cooperative ligand binding** or the situation in an enzyme that has more than one binding site for its substrate or allosteric effector where the binding of one molecule to an enzyme alters the affinity with which the others are bound.

cooperativity index *see* **saturation ratio**.

coordinate 1 to integrate or classify diverse elements or things. **2** *(in chemistry)* to be or become linked by **coordination** (def. 2). **3** *(in mathematics)* any of a set of values that defines the position of a point in space with reference to a system of axes.

coordinate bond *or* **coordinate covalence** *or* **coordinate link** *an alternative term for* **dipolar bond**.

coordinated enzymes enzymes whose rates of synthesis vary together; they are controlled by cistrons of the same **operon**.

coordinated enzyme synthesis the synthesis of enzymes that are controlled by the genes of one operon and are thus induced or repressed together.

coordinate induction/repression the simultaneous induction/repression of a number of enzymes that catalyse a sequence of either consecutive or related reactions and are controlled by scattered genes.

coordination 1 the balanced and effective interaction of processes. **2** the formation or existence of a covalent bond whose pair of electrons is regarded as originating from only one of the two parts of the molecular entity linked by it; linking by means of such a bond or bonds. *Compare* **colligation**. *See also* **dipolar bond, pi adduct**.

coordination number *or* **ligancy** the number of other atoms directly linked to a specified atom in a chemical species.

cooxidation *an alternative term for* **co-metabolism**.

COP *abbr. for* **1** coat protein; a protein subunit of a **coatomer**. **2** constitutive photomorphogenetic protein; a plant protein that represses photomorphogenesis.

COP1 *abbr. for* constitutive morphogenetic protein 1; a protein (675 amino acids) of *Arabidopsis* that contains a RING finger, a coiled-coil domain, and several WD40 repeats. It shuttles between nucleus and cytoplasm in a light-dependent manner. Accumulation in the nucleus requires the integrity of **COP9 signalosome**. In the nucleus it binds a number of proteins and is found in nuclear foci or bodies, where it might be responsible for recruiting specific transcription factors for ubiquitination and degradation. Its cell distribution, dynamics, and structure resemble those of **PML**. A mouse homologue of COP1 is known.

copeptin *see* **vasopressin–neurophysin 2–copeptin precursor**.

copigmentation the chemical association between a pigment molecule and another colourless molecule that results in altered absorption characteristics of the pigment and consequently a change in its shade or colour. Some flower petal colours are changed by hydrogen bonding between an anthocyanin pigment and a colourless flavonoid. For example, the anthocyanin in both purple and crimson roses is cyanidin 3,5-diglucoside, but in crimson flowers it is copigmented with large amounts of gallotannin. This results in a spectral shift of the absorption maximum by 5 nm to longer wavelengths.

copolymer the product of the polymerization of a mixture of more than one species of **monomer**.

copolymerization *or* **copolymerisation** the polymerization of a mixture of two or more species of **monomer**.

copper-chelate chromatography a type of **metal-chelate affinity chromatography** employing Cu^{2+} ions.

copro+ *or* *(before a vowel)* **copr+** *comb. form* indicating feces.

coprophagy the act of feeding on feces.

coproporphyrinogen III an intermediate in the biosynthesis of **metalloporphyrins** (hemes and chlorophylls).

coprostane 5β-cholestane; *see* **cholestane**.

coprostanol *or* **coprosterol** 5β-cholestan-3β-ol; a major **sterol** in feces. It is formed from cholesterol by bacterial action.

COP9 signalosome (*abbr.* CSN) *or* **signalosome** a protein complex first discovered in photomorphogenetic mutants of *Arabidopsis*, but widely distributed in plants and animals. Its eight subunits resemble in sequence those of the lid complex (19S) of the 26S **proteasome**. It may function in collaboration with **COP1**.

copy-choice hypothesis a hypothesis to explain interchromosomal genetic recombination, according to which copying to form a new DNA strand alternates from one parent strand to the other. If the parent strands contain different alleles, a recombinant chromosome may be produced possessing some alleles from one parental strand and some from the other.

copy error a mutation arising from an error in DNA replication.

copy number the number of copies of any gene or plasmid in a given cell. *See* **plasmid**.

CoQ *symbol for* coenzyme Q (obsolete; *see* **ubiquinone**).

CoQH$_2$ *symbol for* reduced coenzyme Q (obsolete; the reduced form of **ubiquinone**).

cord blood blood taken from the umbilical cord.

cord factor 6,6′-dimycolyl trehalose; a highly toxic glycolipid extractable with nonpolar solvents from the cell walls of certain, especially virulent, strains of *Mycobacterium* that characteristically grow in long cords in liquid media.

cordycepin 3′-deoxyadenosine; a substance found in culture fluids of the microfungus *Cordyceps militaris*. It inhibits poly(A) synthesis and interferes with the processing of much eukaryotic mRNA and hence with protein synthesis. The α-^{32}P-labelled 5′-triphosphate is used to label the 3′ ends of DNA fragments in DNA sequence analysis.

cordycepose 3-deoxy-D-ribose; 3-deoxy-D-*erythro*-pentose; the carbohydrate component of **cordycepin**.

core the central, innermost, or least easily destroyed or removed part of a structure; e.g. the core of a partially digested biopolymer, the protoplast core of an endospore, or the core of a virus.

core debranching enzyme *see* β-1,3-galactosyl-*O*-glycosyl-glycoprotein β-1,6-*N*-acetylglucosaminyltransferase.

core enzyme the part of bacterial (*Escherichia coli*) **RNA polymerase** lacking the σ subunit. It will catalyse chain elongation but not initiation.

core glycosylation the first of two stages in the biosynthesis of the carbohydrate moiety of some glycoproteins; it takes place in the endoplasmic reticulum.

core particle 1 the RNA-rich particle obtained from a **ribosome** by the gentle removal of some of the ribosomal proteins. The particle retains peptidyltransferase activity. **2** any of the eukaryotic **chromatin** particles containing 140 base pairs of DNA and an octamer of the four conservative **histones** (two each of histones H2A, H2B, H3, and H4) but not histone H1. These particles are obtained by the treatment of chromatin with DNase. In the intact chromatin they are linked by stretches of variable length of so-called linker DNA.

core polymerase *see* **RNA polymerase.**

corepressor any metabolite that by its specific combination with a **repressor** causes the activation of the latter.

core promoter or **basal promoter** a stretch of DNA upstream from the transcription start site of a eukaryote gene at which a complex of general transcription factors such as TFIID, TFIIB, and others is formed with RNA polymerase II prior to initiation of transcription.

core protein any of the many types of protein in glycosaminoglycans to which oligosaccharide side-chains are covalently attached, and which are themselves covalently attached to a long oligosaccharide backbone. Each glycosaminoglycan will typically contain many core protein molecules. An example is **heparan sulfate proteoglycan core protein.**

CoREST the corepressor of **REST.**

Cori cycle or **lactic acid cycle** an energy-requiring metabolic pathway in animals in which carbon atoms of glucose pass along the circular route: muscle glycogen → blood lactate → blood glucose → muscle glycogen. [After Carl Ferdinand Cori (1896–1984) and Gerty Theresa Cori (née Radnitz, 1896–1957), Czech-born US biochemists, who together formulated it in 1929.]

Cori ester *an alternative name for* α-D-glucopyranose 1-phosphate. [After C. F. and G. T. Cori who reported its isolation in 1936.]

corin a single-pass transmembrane protein that has a ≈970-residue extracellular region containing various domains, including one for serine protease. The human protein (1042 residues) is highly homologous to that of mouse and is present in heart muscle. *See* **TTSP.**

Cori's disease *an alternative name for* type III **glycogen disease.** [After G. T. Cori who first described it in 1953.]

CORN a mnemonic for remembering the absolute configuration of L amino acids. When viewed along the axis of the H–Cα, the carbonyl (CO), R, and amino (N) groups spell the word CORN in a clockwise direction..

cornea the transparent curved part of the front of the eyeball that refracts incident light onto the lens. It is composed of many layers, the thickest being the stroma, which is composed of uniformly fine collagen fibrils. These are kept in an orderly distribution by the expansive force between them of the contained glycosaminoglycan (mainly keratan sulfate I). The cornea has a high oxygen consumption and approximately 50% of the glucose utilized is oxidized via the phosphogluconate pathway. —**corneal** *adj.*

Cornelia de Lange syndrome a sporadic autosomal dominant disorder of unknown biochemical and genetic basis that is characterized by distinctive facial features, pre- and post-natal growth retardation, mental retardation, and upper limb abnormalities in many cases. [After Cornelia de Lange (1871–1950), Dutch paediatrician.]

cornifin or **small proline-rich protein I** a cross-linked envelope protein of keratinocytes that first appears in the cell cytosol, but ultimately becomes cross-linked to membrane proteins by transglutaminase. Its presence correlates with squamous differentiation. It is down-regulated by retinoids. Structurally, it is rich in proline, glutamine, and cysteine, and contains repeats of a consensus octapeptide, EPCQPKVP.

corn-steep liquor *or* **corn-steep water** a concentrated fluid obtained by steeping corn (maize) grains in water (containing 0.2% SO_2) for 36–40 hours at 46–50°C. It is used in the manufacture of **penicillin** to increase antibiotic yield.

corn sugar *a common name for* D-glucose.

corn syrup *or (sometimes)* **liquid glucose** a partial hydrolysate of corn (maize) starch containing **glucose** and glucose oligosaccharides.

coronand *an alternative name for* **crown compound.**

coronate *see* **crown compound.**

coronavirus any of a group of RNA animal viruses consisting of enveloped particles 80–120 nm long, with helical nucleocapsids. They contain the largest known viral RNA genomes (27–31 kb) and cause devastating epizootics (of respiratory or enteric disease) in livestock and poultry. Human coronaviruses cause upper respiratory tract infections and severe acute respiratory syndrome (SARS). *See* **main protease.**

corphin a tetrapyrrole ligand that combines the structural elements of porphyrins and corrins. **Factor F_{430}**, the prosthetic group for coenzyme-B sulfoethylthiotransferase, comprises a nickel ion coordinated by tetrahydrocorphin.

corpuscle any living metazoan cell, especially the red corpuscles (erythrocytes) and the white corpuscles (leukocytes) of the blood.

corpuscular radiation *a former term for* **particulate radiation.**

corpus luteum (*pl.* **corpora lutea**) a yellow, progesterone-secreting, glandular body formed in the mammalian ovary from a **Graafian follicle** after extrusion of an ovum.

correlated mutation a change in a (DNA, RNA or protein) sequence that is induced by a change in a different part of the sequence with which it interacts.

correlation *(in statistics)* the interdependence of two or more simultaneously variable quantities; or the nature or degree of this interdependence. If two variates both increase or both decrease together there is said to be **positive (direct) correlation** between them, whereas if they change in opposite directions to each other there is said to be **negative (inverse) correlation.**

correlation coefficient *(in statistics)* symbol: *r*; a measure of the correlation between two variates, *x* and *y*. If one variate has the values $x_1, x_2, ... x_n$ and the corresponding values of the other variate are $y_1, y_2, ... y_n$, with respective mean values of $\bar{x}$ and $\bar{y}$ then the correlation coefficient, *r*, is given by:

$$r = \sum_{i=1}^{n}(x_i - \bar{x})(y_i - \bar{y}) / [\sum_{i=1}^{n}(x_i - \bar{x})^2 \sum_{i=1}^{n}(y_i - \bar{y})^2]^{1/2}$$

If $r = 0$, *x* and *y* are completely independent. If $r = 1$ there is complete correlation and one variate may be calculated from the other variate.

correlation time *symbol:* τ_C ; the average time between molecular collisions for molecules in some state of motion. The reciprocal of the correlation time is given by the equation:

$$1/\tau_C = 1/\tau_R + 1/\tau_M + 1/\tau_S$$

where $1/\tau_R$ is the rate at which the molecule is tumbling (τ_R being the **rotational correlation time**), $1/\tau_M$ is the rate at which two dipoles approach each other (τ_M being the **residence time**), and $1/\tau_S$ is the electron spin lattice relaxation rate (τ_S being the time for the upper electron spin state to dissipate its excess energy).

correndonuclease any correctional **endonuclease** that specifically acts on damaged DNA initiating a correctional pathway *in vivo*. The name is misleading because nuclease activity itself does not lead to repair.

corrin $C_{19}H_{22}N_4$, the fundamental heterocyclic skeleton of the **corrinoids.** It consists of four reduced pyrrole rings joined into a macrocyclic ring by links between their α positions; three of these links are formed by a one-carbon unit (methylidyne group) and the other by a direct Cα–Cα bond. Corrin is so named because it is the core of the **vitamin B₁₂** molecule.

corrinoid any of a group of compounds containing the **corrin** skeleton. This group includes various B₁₂ vitamins, factors, and derivatives. Some are more unsaturated than corrin itself; many have a regular pattern of substituents on the methylene carbon atoms of the reduced pyrrole rings and a cobalt atom in the centre of the macrocyclic ring.

corrole *a trivial name for* octadehydrocorrin; *see* **corrin**.

cortactin a protein that stabilizes Arp2/3 complex-mediated branches in actin filament networks. Its N-terminal region binds Arp2/3, the central region contains several repeats of a 37-residue sequence, and the C-terminal region contains an SH3 domain and three tyrosyl residues that are targets for Src-related kinases. *See* **Arp**.

cortex (*pl.* **cortices**) **1** the outer layer of an animal organ, situated immediately beneath the capsule or outer membrane. It is usually morphologically distinct from the rest of the organ; e.g. adrenal cortex, cerebral cortex, renal cortex. **2** the unspecialized tissue in plant stems and roots lying between the vascular bundles and the epidermis; the outer layer of a stem. —**cortical** *adj.*

cortexillin an actin-binding protein discovered in the slime mould *Dictyostelium discoideum*. It has a C-terminal basic nonapeptide that binds **phosphatidylinositol 4,5-bisphosphate**.

corticoid *an alternative name for* **corticosteroid**.

corticoliberin *an alternative name for* corticotropin-releasing hormone (*see* **CRH**).

corticosteroid *or* **corticoid** any of a large group of C_{21} steroids synthesized from cholesterol in the **adrenal cortex**, especially the adrenocortical hormones. Usage is often extended to include synthetic analogues with hormonal activity. Corticosteroid hormones are divided into **glucocorticoids**, i.e. cortisol and related compounds, and **mineralocorticoids**, i.e. aldosterone and related compounds.

corticosteroid-binding globulin *or* **corticosteroid-binding protein** *an alternative name for* **transcortin**.

corticosterone 11β,21-dihydroxypregn-4-ene-3,20-dione; Kendall's compound B; Reichstein's substance H; a **glucocorticoid**, but on the biosynthetic route to aldosterone. Quantitatively it is the second most important corticosteroid of the normal human adrenal cortex. **Cortisol**, synthesized in the adrenal zona fasciculata, is the predominant corticosteroid in humans; corticosterone, synthesized in the adrenal zona fasciculata and glomerulosa, much less so, but it is the main corticosteroid in some rodents.

corticotrope a type of cell in the anterior pituitary that synthesizes **proopiomelanocortin**, a **polyprotein** from which **corticotropin** and a number of other peptide hormones are formed.

corticotrophin (*formerly*) *a variant spelling of* **corticotropin**.

corticotropin *or* **adrenocorticotropin** *or* **adrenocorticotropic hormone** (*abbr.*: ACTH) a 39-residue polypeptide hormone, secreted by the adenohypophysis of the pituitary, that stimulates growth of the adrenal cortex and the synthesis and secretion of various corticosteroids. It has the sequence (human):

[SYSMEHFRWGKPV]GKKRRPV-
KVYPNGAEDESAEAFPLEF.

The first 13 residues (in brackets) comprise, after proteolysis, α-**melanocyte-stimulating hormone**. Under normal conditions its own secretion is under feedback control from the plasma cortisol level, both directly and through hypothalamic regulatory factors. Biosynthesis occurs via the multihormone precursor peptide **proopiomelanocortin**. Hormonal activity is confined to the species-invariant N-terminal 24 amino acids (*see* **tetracosactide**). Corticotropin secretion is regulated by corticotropin-releasing hormone (**CRH**), is episodic (in bursts at intervals of a few hours), exhibits diurnal rhythm, and is increased by stress such as trauma, surgery, and hypoglycemia. The effects of fever may be mediated by cytokines 1, 2, and 6, which act indirectly by causing the release of CRH. *See also* **corticotropin-like intermediate lobe peptide, corticotropin receptor, melanotropin**.

corticotropin-induced secreted protein *abbr.*: CISP; *a former name for* thrombospondin-2 (*see* **thrombospondin**).

corticotropin-like intermediate lobe peptide *abbr.*: CLIP; a 22-residue peptide identical in sequence to the C-terminal moiety (residues 18–39) of a molecule of **corticotropin** from the same species. It is obtained from extracts of the pars intermedia of the pituitary of certain animals (rat, pig, dogfish), and has been found also in a carcinoid tumour from a human patient who was secreting ectopic corticotropin; it has not been found in extracts of human pituitary. It is derived from **proopiomelanocortin**. CLIP has B-cell tropin activity and acts to stimulate insulin release from B cells. *See also* **beta-cell tropin**.

corticotropin–lipotropin precursor *see* **proopiomelanocortin**.

corticotropin receptor one of a number of membrane proteins that bind corticotropin and couple to heterotrimeric G-protein-activated **adenylate cyclase**; they are seven-transmembrane-domain proteins and are distributed on cells of the adrenal cortex, on cells of the immune system, and in the brain. EC_{50} values for binding are near 10^{-8} M for corticotropin. *See also* **melanocortin receptor**.

corticotropin-releasing factor *an alternative name for* corticotropin-releasing hormone (*see* **CRH**).

corticotropin-releasing-factor binding protein a protein that binds and inactivates corticotropin-releasing hormone (**CRH**), to prevent inappropriate pituitary-adrenal stimulation in pregnancy.

corticotropin-releasing hormone *see* **CRH**.

cortisol *or* (*recommended name*) **hydrocortisone** 17-hydroxycorticosterone; 11β,17,21-trihydroxy-4-pregnene-3,20-dione; Kendall's compound F; Reichstein's substance M; the most powerful naturally occurring **glucocorticoid** hormone and the most abundant product of the adrenal cortex in humans. It is produced by the cells of the zona reticularis; its best known action is anti-inflammatory. The biosynthesis of cortisol is under the control of **corticotropin**.

cortisol-binding globulin *or* **cortisol-binding protein** *abbr.*: CBG; *an alternative name for* **transcortin**.

cortisone 11-dehydrocortisol; Kendall's compound E; Reichstein's substance F; a prodrug that is biologically inactive but is converted in the liver and, to a lesser extent, in other tissues, to cortisol, a substance with **glucocorticoid** activity. Cortisone was the first glucocorticoid to be isolated and made available for therapeutic use. It is not a primary secretory product of the adrenal cortex, being formed during tissue metabolism from cortisol by 11β-hydroxysteroid dehydrogenase. In fetal plasma, however, cortisone levels are about three times higher than those of cortisol.

cortoic acid any member of a group of acidic metabolites of 17-hydroxycorticosteroids in which the primary alcohol group at C-21 has been oxidized to a carboxylic-acid group. They are the principal acidic metabolites of **cortisol** in humans.

corynecin any of a number of antibiotic compounds produced by the cultivation of *Corynebacterium* KY 4339 in a medium containing C_{12} to C_{14} alkanes as the sole carbon source. Their antibacterial properties are similar to those of **chloramphenicol** but they are less potent.

coryneform 1 club-shaped. **2** describing any Gram-positive, asporogenous, club- or rod-shaped bacterium belonging to the genera *Corynebacterium* or *Brevibacterium*, possibly related to *Streptomyces*. Such bacteria are of biotechnological importance (amino-acid fermentations) and are amenable to genetic manipulation. **3** any coryneform microorganism.

cos *abbr for* **cosine.**

COS cell any member of a cell line, derived from monkey cells, that contains an integrated segment of **SV40** DNA coding for T antigen. Hence these cells will support replication of vector molecules containing the SV40 origin of replication, but no other DNA sequences from that virus.

cosec *abbr. for* cosecant.

cosecant *abbr.*: cosec or csc; the reciprocal of the **sine** of an angle.

cosine *abbr.*: cos; a trigonometric function defined, for a given angle of a right-angled triangle, as the ratio of the length of the side adjacent to the angle and the length of the hypotenuse. The cosine of an obtuse angle is numerically equivalent to that of its supplement but of opposite sign.

cosmid a hybrid **cloning vector** used for cloning DNA. Cosmids are derived from plasmids but also contain *cos* **sites** from phage lambda.

cos **site** the cohesive ends (sticky ends) of certain **phage** DNA molecules. The *cos* site usually referred to is from phage lambda.

costamere a vinculin-containing myofibril attachment site. Costameres form one of the major linkage sites of the skeletal muscle cell, encircling the cell and connecting the Z-discs to the sarcolemma.

co-stimulator *an alternative name for* **interleukin 2.**

cost-selectivity equation a relation between the accuracy gained by editing and the cost in terms of the wasteful hydrolysis of correct products in any enzyme system synthesizing informational macromolecules. It is useful in the rationalization of some observations and proposed editing mechanisms.

cosubstrate an imprecise term used for an organic cofactor that is necessary for the activity of some enzymes. Although not the substrate of principal interest, the cosubstrate acts as an acceptor or donor of atoms or of functional groups and participates stoichiometrically in the reaction. *Compare* **coenzyme.**

cosuppression *an alternative term for* **RNA silencing.**

COSY *abbr. for* correlated spectroscopy; a technique employed in the first two-dimensional **NMR** experiment, designed to elucidate which resonances of the NMR spectrum are connected by spin–spin J-coupling. J-couplings arise out of the perturbation of the nuclear energy levels of one nuclear spin by another nuclear spin that is propagated through the intervening electronic orbitals. In a one-dimensional NMR spectrum these J-couplings are manifested as a splitting in the Lorentzian lineshape. In 2D-COSY spectra the couplings are indicated by off-diagonal cross-peaks connecting the frequencies of the J-coupled spectra. Ordinarily the J-couplings are only large enough for pairs of ^{1}H nuclei separated by a maximum of two or three bonds to exhibit COSY cross-peaks.

Cot *see* $C_{o}t$.

cotransduction the **transduction** of two or more identifiable genes in a single event, probably because the transduced element contains more than one genetic locus.

cotransfection a technique in which two different plasmids are simultaneously introduced into mammalian, or other cells, by one of several methods of transfection. Relatively few cells take up DNA, but in those that do, both plasmids are taken up efficiently. One plasmid might carry a gene whose expression is of interest and the other a reporter gene used to monitor transfection efficiency.

cotranslation *(in recombinant DNA technology)* the synthesis of a hybrid molecule consisting of the desired protein product covalently linked to a normal protein secreted by the organism being used, through translation of a hybrid mRNA molecule generated on the recombinant DNA. —**cotranslate** *vb.*

cotranslational transport the transport of nascent protein chains through the membrane of the cisterna of the endoplasmic reticulum while translation is still proceeding at the ribosome. It is a common phenomenon in higher eukaryotes.

cotransmitter a substance, usually a peptide, that is released from neurons and modifies the actions of the substance regarded as the primary transmitter. For example, vasoactive intestinal peptide is a cotransmitter with acetylcholine, and neuropeptide Y performs a similar role with norepinephrine. Cotransmitters may also have direct effects on target cells.

cotransport 1 increased transport of a substance in excess to its concentration dependence, as found in positively cooperative systems. **2** *sometimes used as equivalent to* **symport**

co-trimoxazole an antibacterial drug that comprises a mixture of **sulfamethoxazole** and **trimethoprim** in a ratio of about 5:1. The combination may have synergistic effects.

Cotton effect an anomalous change in optical rotation with wavelength in the region of an absorption band. As the absorption band is approached there is a change in the rotation to a maximum, which may be either positive or negative, followed by a decrease to a minimum passing through zero at the wavelength of maximum absorption. Similarly there is a change in circular dichroism, either positive or negative, coinciding with the absorption band. The magnitudes and directions of these effects have been used to study macromolecular conformations. [After Aimé Cotton (1859–1951), French physicist.]

cotyledon the leaf-forming part of the embryo of seeds. —**cotyledonous** *adj.*

Couette viscometer an instrument for measuring viscosity, consisting of two concentric cylinders separated by a narrow annulus, which is filled with the sample. One cylinder is fixed and the other is rotated. From the angular velocity at a given torque and the dimensions of the cylinders, the viscosity of the sample can be calculated. The Couette viscometer is designed for use at relatively low shear gradients.

coulomb *symbol*: C; the SI derived unit of electric charge; the quantity of electricity transported in one second by a current of one ampere. Hence, 1 C = 1 A s. [After Charles Augustin de Coulomb (1736–1806), French physicist.]

coulomb force *or* **coulombic force** the electrostatic attraction or repulsion between two charged particles. See **Coulomb's law.**

Coulomb's law (of electrostatics) the law stating that the force, F, of attraction or repulsion between two point charges, Q_1 and Q_2, is proportional to the product of the charges and inversely proportional to the square of the distance, d, between them multiplied by the permittivity, ε, of the medium between them. Hence, $F = (Q_1Q_2)/\varepsilon d^2$. A similar law exists for magnetism.

coulometer *or* **coulombmeter** *or* **voltameter** an electrolytic cell for measuring a quantity of electricity passing through a circuit containing the cell by determining the amount of a metal (usually copper or silver) deposited on the cathode from a solution of a salt of the metal.

coulometry *or* **coulombmetry** *or* **voltametry** a method of chemical analysis based on the measurement of a quantity of electricity, in coulombs, in an electrode reaction. —**coulometric** *or* **coulombmetric** *adj.*; **coulometrically** *or* **coulombmetrically** *adv.*

Coulter counter a device for counting particles in a suspension, e.g. blood cells, microorganisms, or spores. Its functioning depends on changes in electric conductivity as the particles pass through a small aperture.

4-coumaric acid *trans*-4-hydroxycinnamic acid, an intermediate in the biosynthesis of flavonoids, and in the synthesis of lignin. 4-

coumarate is a chromophore of photoactive yellow protein from *Halorhodospira halophila*.

coumarin 1 1,2-benzopyrone; 2*H*-chromen-2-one; a fluorescent indicator. **2** any of a number of coumarin derivatives that are widespread in seed coats, leaves, and stems, and occur at high concentration in fruits and flowers. They have antimicrobial, antifeedant, UV-screening, and germination inhibitor properties. Coumarins from clover produce massive internal bleeding in mammals, which led to the synthesis of the rodenticide and *in vivo* anticoagulant warfarin. Some (e.g. 8-methylpsoralen) can cause photodermatitis and are used in treatment of skin conditions such as psoriasis and eczema. The coumarins are important in the perfume industry since they smell of new-mown hay and grass, though some are no longer used (acernocoumarol and phenindiene) because of adverse allergenic effects. Some are useful in forming fluorescent analytes. Others are antibiotics, e.g. **novobiocin**, or mycotoxins (*see* **aflatoxin**).

coumarin anticoagulants *see* **dicoumarol**.

counter any device for enumerating, especially one for detecting and determining particles and photons emitted from the disintegration of a radionuclide. *See*, e.g., **Cherenkov counter**, **Geiger counter**, **scintillation counter**.

countercurrent chromatography a form of **partition chromatography** that eliminates the need for a solid support, in which one of the components of a two-phase solvent system is allowed to flow through the second component in a coiled tube. The technique is claimed to have the advantages of both **countercurrent distribution** and **liquid–liquid chromatography** with the disadvantages of neither, and to be applicable to a broad spectrum of samples ranging from micromolecules to macromolecules, particles, and cells.

countercurrent distribution *or* **countercurrent separation** a method for the separation and purification of (usually organic) substances that depends on the repetitive distribution of solute between two immiscible liquid phases in a series of vessels in which the phases are in contact. The process is continued until homogeneous substances are distributed in various sets of vessels in accordance with their **distribution coefficients** between the two liquid phases. *See also* **Craig apparatus**.

countercurrent transport the movement of a solute in or out of a cell against a concentration gradient by an ATP-independent process. It is due to competition between the solute under consideration and a second solute, with an opposite concentration distribution, for the same membrane carrier protein.

counterelectrophoresis *see* **counterimmunoelectrophoresis**.

counterflow the movement of a substance across a membrane against a concentration gradient.

counterimmunoelectrophoresis *or* **counterelectrophoresis** *other names:* countermigration electrophoresis; immunoelectroosmophoresis; *abbr.*: CIE; a one-dimensional double **electroimmunodiffusion**; a technique wherein diffusion of the reactants towards each other, and hence the formation of precipitin lines, is accelerated by the combination of electrophoresis and electroendosmosis that occurs when an electric potential difference is applied across the gel between the antigen- and antibody-containing wells, with the latter positive to the former.

counterion any ion carrying a charge opposite to that of the charged chemical species with which it is associated; an ion of low M_r with a charge of opposite sign to that of the colloidal ion. *Compare* **co-ion**.

countermigration electrophoresis *an alternative name for* **counterimmunoelectrophoresis**.

counterstain 1 to apply a second stain to a preparation for microscopy. Counterstaining is used to stain in a contrasting way those parts of the preparation that have not been stained by the first stain. **2** any stain so used.

countertranscript *an alternative term for* **antisense RNA**.

countertransport 1 transient outward movement of a substance in equilibrium (radioactively labelled for experimental purposes) caused by addition of the same nonlabelled substance at higher concentration. It is most easily observed in systems of mediated diffusion. **2** *sometimes used, not quite correctly, for* **antiport**.

countertrypsin a glycoprotein trypsin inhibitor from mouse plasma and a member of the **fetuin** family.

counting efficiency the ratio of recorded radioactivity events to the number of actual radioactive disintegrations occurring in the same time and in the same sample. It is sometimes expressed as the percentage efficiency.

COUP *other names:* COUP-TF; v-erbA-related protein; EAR-3; a transcription factor named from one activity, chicken ovalbumin upstream promoter. In association with transcription factor IIB it stimulates initiation of transcription. It is similar to the steroid/thyroid/retinoic nuclear hormone receptors and **seven-up**.

couple 1 a pair of equal and opposite, parallel but not collinear, forces acting on a body, thus producing a turning effect. **2** to join or link (two things) together.

coupled assay a type of enzymic assay of a substrate (or of an enzyme) employed when the reaction in question does not give chromophoric or fluorescent products. A second enzyme reaction that acts on one of the products of the first reaction, and that gives an easily detectable product, is carried out simultaneously with the first reaction. Many coupled assays are based on the formation or disappearance of NADH, which can be easily detected by its absorbance at 340 nm.

coupled oxidation and phosphorylation an intimate relationship between oxidation of substrate and phosphorylation of ADP to ATP such that the two processes cannot proceed independently, as in mitochondria. *See also* **coupling** (def. 3).

coupled reaction either of two reactions, one endergonic and the other exergonic, that are linked energetically, occur simultaneously, and share a common intermediate, such that the overall free energy change for the two reactions is negative.

coupling 1 (*in chemistry*) the covalent linking of one chemical entity to another; *see* **covalent bond**. **2** (*in physics*) **a** an interaction between different parts of a system, e.g. between groups of atoms, electrons, or nuclei. **b** an interaction between two electronic circuits so that power is transferred from one to the other. **3** (*in biochemistry*) the linking of **electron transport** and **oxidative phosphorylation** in isolated mitochondria. Thus, if the **phosphorus:oxygen ratio** is high the mitochondria are said to be tightly coupled. *See also* **uncoupling**.

coupling constant the separation between any two bands of multiple peaks in nuclear magnetic resonance spectroscopy; it is proportional to the magnitude of **spin–spin coupling** constant.

coupling factor *abbr.*: CF; *archaic term for* **1** any of the proteins functioning in the coupling of ATP synthesis to electron transport in mitochondria. CF1–CF4 are obsolete names for the α–δ subunits of mitochondrial **H⁺-transporting ATP synthase**. **2** any of the proteins required for the coupling of ATP synthesis to the photoinduced electron transport in chloroplasts during photosynthesis. The concept has been replaced by the **chemiosmotic coupling hypothesis**.

coupling membrane any biological membrane in which energy-yielding and energy-requiring biochemical processes occur together.

covalence *or* **covalency** nonionic valence between two atoms in a chemical compound, characterized by their sharing of one or more electrons. —**covalent** *adj*.

covalent bond a chemical **bond** formed between two atoms in a molecule by the sharing of electrons, usually in pairs, by the bonded atoms.

covalent catalysis the catalysis of any reaction in which the substrate is modified by forming a covalent bond with the catalyst (enzyme).

covalent chromatography any technique of **column chromatography** in which the substance of interest binds covalently with groups, e.g. –SH groups, on the support medium and is subsequently displaced by an appropriate reagent.

covalent (closed) circle (of DNA) *or* **covalently closed circular DNA** *or* **closed double-stranded DNA** *abbr.*: cccDNA; any double-stranded DNA molecule that is circular (but not necessarily geometrically circular) and in which both polynucleotide strands are completely continuous. *Compare* **Hershey circle**, **open circle**.

covalent modification any of a diverse group of processes in which the initially synthesized structures of biopolymers, especially enzymes, proenzymes, or structural macromolecules, are enzymically modified by the breakage of covalent bonds or the addition of new covalently linked groups. It includes **post-translational modifica-**

tion. The term may sometimes be used in the more restricted sense of reversible interconversion of active and inactive forms of certain metabolic enzymes or other proteins by sets of control enzymes, often by phosphorylation–dephosphorylation.

covariance a measure of the association between two variables. For n pairs of values of two random variables, x and y, this is given by:

$$\text{Cov. } (x,y) = (x - \bar{x})(y - \bar{y})/(n-1)$$

where $\bar{x}$ and $\bar{y}$ are the means of the populations of x and y, respectively. *Compare* **correlation coefficient**, **regression coefficient**, **variance**.

covarion any of the codons in a given gene that may concomitantly vary, so resulting in favourable mutations or in mutations leading to amino-acid substitutions that have little or no effect. [From concomitantly variable codon.]

covirus any virus that exists as two or more separate particles all of which must be present together in the host organism for the complete replication cycle of the virus to occur.

covolume the amount by which the apparent volume of a molecule exceeds the sum of the volumes of the constituent atoms. It is usually 13–14 cm^3 mol^{-1} for protein molecules.

coxsackie and adenovirus receptor *see* **CAR**.

cozymase originally, the heat-stable, diffusible fraction of a crude aqueous extract of brewers' yeast that, if added to the heat-labile, nondiffusible fraction (zymase), would enable the alcoholic fermentation of glucose to occur in a cell-free system. The term was subsequently applied to one particular substance present in that fraction, now known as **nicotinamide adenine dinucleotide**.

C₃ pathway *an alternative name for* **reductive pentose phosphate cycle**.

C₄ pathway *an alternative name for* **Hatch–Slack pathway**.

CPBA *abbr. for* competitive protein-binding assay.

CPE *abbr. for* **1** cytopathic effect. **2** cytoplasmic polyadenylation element-binding protein; a protein that binds a specific sequence (called a CPE) at the 3′ untranslated region (UTR) of mRNA, and promotes elongation of the poly(A) tail of the mRNA. The elongation also requires binding of a cleavage-polyadenylation specificity factor (*see* **CPSF**) to the hexanucleotide sequence AAUAAA in the same region.

CPEB *abbr. for* CPE-binding protein. *See* **cytoplasmic polyadenylation element**. .

C peptide 1 a the inactive polypeptide excised from **proinsulin** during its conversion to insulin. It contains 31 amino-acid residues in the human but shows greater species variability in both length and amino-acid sequence than does insulin. The structure of porcine C-peptide is (including the flanking basic residues):

RREAQNPQAGAVELGGGLGGGLQALALEGPPQKR.

It is released into the bloodstream concomitantly with insulin, hence the blood level is useful in evaluation of B-cell function. **b** that segment of the amino-acid sequence of proinsulin lying between the two residues destined to become respectively the C-terminal residue of the B chain of insulin and the N-terminal residue of the A chain. It consists structurally of the above polypeptide plus two flanking pairs of basic residues. C peptide is named from connecting peptide, and perhaps also because, after A (chain) and B (chain), C (peptide) represents alphabetically the third major structural component of proinsulin. **2** *an alternative name for* C fragment (of β-lipotropin) (*see* **lipotropin**).

CpG island a region of 1–2 kb containing a high density of methylated cytosine residues and occurring immediately 5′ to G residues, i.e. in the sequence CpG. CpG islands are frequently found in animal genomes at the 5′ end of genes. In plants the methylated sequence is ...CpNpGp... where N can be any base. Methylation of DNA is a heritable phenomenon that reduces gene expression, probably by increasing the binding of a repressor. CpG islands are a site of high mutational frequency because spontaneous deamination of the methylated cytosine gives thymine, which is not recognized by DNA repair enzymes.

CPK Colours a colour scheme for molecules, conventionally used by chemists for molecular modelling software and based upon the colours of the popular plastic molecular models that were developed by Corey and Pauling, and later improved by Koltun. The as-

signment of elements to colours includes: carbon, black or grey; oxygen, red; hydrogen, white; nitrogen, blue; and sulfur, yellow.

C₃ plant any plant in which fixation of carbon dioxide occurs predominantly by the **reductive pentose phosphate cycle**; i.e. by incorporation initially into the three-carbon compound 3-phosphoglycerate.

C₄ plant any plant in which fixation of carbon dioxide occurs predominantly by the **Hatch–Slack pathway**; i.e. by incorporation initially into the four-carbon compound oxaloacetate. Such plants are found usually in semiarid conditions with high light levels. They include sugar cane, corn (maize), and many weeds.

cpm *or* **c.p.m.** *abbr. for* counts per minute.

CpoI a type 2 **restriction endonuclease**; recognition sequence: CG↑G[AT]CCG. *Rrs*II is an **isoschizomer**.

CPSF *abbr. for* cleavage-polyadenylation specificity factor; a protein that specifically binds the hexanucleotide sequence AAUAAA in the polyadenylation signal sequence during processing of the 3′ end of pre-mRNA. It also interacts with the general transcription factor TFIID and with the C-terminal domain of RNA polymerase II.

Cr *symbol for* chromium.

Crabtree effect the inhibition of respiration by glycolysis. The inhibition is small and occurs only in a few types of cells that possess a high glycolytic capacity, e.g. ascites tumour cells and other neoplastic tissues, renal medulla, leukocytes, and cartilage. *Compare* **Pasteur effect**. [Named after Herbert Grace Crabtree, English physician and biochemist, who reported it in 1929.]

Craig apparatus the most widely used type of apparatus for carrying out **countercurrent distribution** experiments. The earlier version (1944) was constructed of metal and the later version (1949) was of modular design and constructed of glass. [After Lyman Creighton Craig (1906–74).]

crambin a hydrophobic protein obtained from seeds of *Crambe abyssinica*. It comprises 46 amino acids, and has two isoforms. The protein belongs to the plant thionin family; its function is unknown.

cranin a form of **dystroglycan**.

Cre a member of the Int family of **recombinases** encoded in the genome of bacteriophage P1. The enzyme catalyses recombination of DNA at *loxP* sites. Linear DNA molecules with *loxP* sites in the same orientation are circularized in *Escherichia coli* strains expressing Cre. Cre can also be expressed in transgenic mammals under the control of a suitable promoter. *See* **Cre-LoxP system**.

CRE *abbr. for* cyclic AMP response element; a sequence,

GTGACGT[A/G][A/G],

that is present in many viral and cellular promoters. When it binds **CREB protein**, transcription of the genes regulated by such a promoter is turned on.

C-reactive protein *abbr.*: CRP; an acidic, crystallizable, heat-sensitive protein of ~118 kDa that is detectable in human or monkey blood serum early in the course of various infections, or when there is inflammation, tissue damage, or necrosis. One of the so-called **acute phase proteins**, it is normally undetectable. Not an immunoglobulin and of unknown function, it is named for its ability, in the presence of Ca^{2+} ions, to form a precipitate with a pneumococcal somatic C polysaccharide.

creatine *N*-(aminoiminomethyl)-*N*-methylglycine; an important metabolite in muscle, the precursor of **phosphocreatine**.

creatine kinase EC 2.7.3.2; *systematic name*: ATP:creatine *N*-phosphotransferase; an enzyme that catalyses the formation of **phosphocreatine** from ATP and creatine with release of ADP. The enzyme has M and B subunits, derived from different genes, and may exist as the homodimers MM and BB or the heterodimer MB. BB mainly occurs in brain and thyroid, while MM occurs in muscle; MB occurs to a much greater extent in cardiac muscle (30% of total) than skeletal muscle (1% of total) and is thus a useful diagnostic isoenzyme in myocardial infarction, an increased level of the MB isoenzyme occurring between 12 and 36 hours after the infarct. It is present in cytosol and in the intermembrane space of mitochondria, where it is coupled to the ATP/ADP translocator.

creatine phosphate *an alternative name (not recommended) for* **phosphocreatine**.

creatinine 2-amino-1,5-dihydro-1-methyl-4*H*-imidazol-4-one; an end product of creatine metabolism and a normal constituent of urine.

creatinuria the presence of excessive creatine in urine. It occurs during growth in children, in women especially during and after pregnancy, and in conditions of muscle breakdown.

CREB protein *abbr. for* cyclic AMP response element binding protein; a protein that, following phosphorylation by **cyclic AMP-dependent protein kinase** on a single serine residue, binds to the cyclic AMP response element (**CRE**). This then stimulates transcription of genes controlled by regulatory regions that contain the CRE. It regulates genes (responsive to nerve growth factor) for ion channels, for enzymes of neurotransmitter synthesis, and for cytoskeletal components, and is involved in learning, memory, and drug addiction. *See* **CBP**.

CREB protein-binding protein *see* **CBP**.

C region *abbr. for* constant region (of an immunoglobulin).

Cre-LoxP system a site-specific recombination system, derived from *Escherichia coli* bacteriophage P1, that allows the introduction of genetic modifications into specific genes. The method involves engineering short DNA sequences (*loxP* sites) that flank the target DNA, recombination between these sites via the P1-derived recombinase Cre, and excision of the intervening sequence when two *loxP* sites have the same orientation on the same DNA strand.

cretinism a condition of congenital **hypothyroidism** characterized by arrested mental and physical development.

Creutzfeldt–Jakob disease *abbr.*: CJD; a transmissible **spongiform encephalopathy** of humans. It has occurred in four forms: (1) familial (about 15% of cases), caused by mutations of the **prion** protein, in the form of an insertion of 2–9 octapeptide repeats (in addition to the 5 normally present), or a Glu → Lys substitution at codon 200; (2) iatrogenic or acquired, due to inadvertent exposure to CJD-contaminated material, notably human cadaveric growth hormone; (3) **kuru**, which is transmitted by ritualistic cannibalism; and (4) variant CJD (vCJD), an unusual form of the disease thought to be due to consumption of bovine products contaminated with the abnormal **prion** protein responsible for **bovine spongiform encephalopathy** (BSE). The latter is apparently able to cross species. They are all associated with the accumulation in affected brains of an abnormal isoform of a glycoprotein. The familial form shows an autosomal pattern of disease segregation. Forms (1) and (3) usually occur in the fifth or sixth decade of life and progress rapidly (within six months) to severe dementia and death typically in one year. [After Hans Gerhard Creutzfeldt (1885–1964) and Alfons Maria Jakob (1884–1931), German physicians.]

CRF *abbr. for* corticotropin-releasing factor; *see* **CRH**.

CRH *abbr. for* corticotropin-releasing hormone; *other names*: **corticotropin-releasing factor** *or* **corticoliberin** *or* **adrenocorticotropic hormone-releasing factor**. A 41-residue polypeptide released by the hypothalamus into the hypophyseal-portal circulation in response to neural and/or chemical stimuli. It controls the secretion by the anterior pituitary gland (i.e. adenohypophysis) of corticotropin, and possibly also of β-lipotropin and pro-γ-melanotropin – all three are derived from proopiomelanocortin. CRH binds to specific G-protein-associated receptors on the **corticotrope** and increases cyclic AMP levels in the cell, stimulating hormone release only in the presence of Ca^{2+}. It also enhances the rate of transcription of mRNA coding for corticotropin. CRH and its receptors occur in other parts of the brain, indicating that it might be a neuromodulator, and in other parts of the body, where it might have paracrine and immunoregulatory functions. CRH may be secreted by bronchial carcinoid tumours. *See also* **CRH test**.

CRH test a test that is useful in distinguishing **Cushing's disease** from ectopic corticotropin secretion. If **CRH** is given intravenously, there is an increase in plasma cortisol in normal individuals, while usually in Cushing's disease there is an exaggerated response; in ectopic corticotropin secretion there is usually no cortisol response.

Crick strand the lower, or minus, strand in a representation of double-stranded DNA. Open reading frames (ORFs) may occur on either strand of a genomic DNA duplex and it is convenient to distingish between the strands in order to identify ORFs. ORFs in the Crick strand of the yeast genome, for example, are given the suffix C as in the systematic name YLR452C for the *SST2* gene on yeast chromosome XII. *Compare* **Watson strand**.

Crigler–Najjar syndrome an autosomal recessive disease in which there is absence (type I, leading to kernicterus) or deficiency (type II, usually benign) of bilirubin glucuronidation, which results in predominantly unconjugated hyperbilirubinemia. A variety of mutations (deletion, insertion, missense, or nonsense) in a locus at 2q37 for UDP-glucuronosylytransferase I of endoplasmic reticulum of hepatocytes and other cell types is responsible. See **Gilbert syndrome**. [After John Fielding Crigler (1919–) and Victor Assad Najjar (1914–), US physicians.]

crinophagy a cellular process in which secretion granules fuse with lysosomes. It serves to dispose of excess secretory products in a cell when the stimulus for their discharge is lacking.

CRISP *abbr. for* cysteine-rich secretory protein; any of a small family of cysteine-rich secretory glycoproteins that escort or guide spermatozoa on their way from the testes to the egg. Examples include TPX1, which promotes adhesion of sperm to Sertoli cells in testes; AEG (acidic-epididymal glycoprotein), which binds sperm during maturation in the epididymis; and allurin, a sperm attractant from the egg jelly of *Xenopus* eggs.

crista (*pl.* **cristae**) **1** any of the inward folds of the inner mitochondrial membrane. Their number, extent, and shape differ in mitochondria from different tissues and organisms. They appear to be devices for increasing the surface area of the inner mitochondrial membrane, where the enzymes of electron transport and oxidative phosphorylation are found. Their shape can vary with the respiratory state of the mitochondria. **2** a sensory structure within the ampulla of a semicircular canal of the inner ear.

critical angle *symbol*: C; the least angle of incidence at which total internal reflection of a ray of light or other electromagnetic radiation occurs when passing through a medium and meeting the boundary with a medium of lesser density (e.g. of glass with air).

critical assessment competition a community-wide competition that aims to stimulate the production of better computational methods in a particular field. Examples include: CASP (Critical Assessment of Structure Prediction), which assesses the reliability of protein structure prediction tools; GASP (Genome Annotation Assessment Project), which assesses the reliability of gene prediction algorithms on eukaryotic genome sequences; and CAMDA (Critical Assessment of Microarray Data Analysis), which assesses the reliability of microarray data analysis tools.

critical electrolyte concentration *or* **critical salt concentration** the concentration of a particular small ion that will displace an ionic molecule (e.g. of a dye or detergent) of like charge from a polyelectrolyte macromolecule. For example, if a polymer P with z negative charges is stained with a dye R^+, the dyed polymer, $P^{z-}zR^+$, is formed according to the equation:

$$P^{z-}zM^+ + zR^+ \rightleftharpoons P^{z-}zR^+ + zM^+$$

where M^+ is a small cation. Hence the ratio of dyed to undyed polymer, D, is given by:

$$D = [P^{z-}zR^+]/[P^{z-}zM^+] = K[R^+]^z/[M^+]^z$$

where K is a constant. If z is large, there is a very sharp change in D at a critical value of $[R^+]/[M^+]$. Then if $[R^+]$ is held constant and

[M$^+$] varied, an 'all-or-none' change in D is observed at a well-defined [M$^+$]. This concept is used in the **critical electrolyte concentration method** for staining proteoglycans and polynucleotides, and for fractionation of nucleic acids and other polyanions.

critical micellar concentration *or* **critical micelle concentration** *abbr.*: CMC; *symbol*: c_M ; the concentration at which surfactant molecules in a lipid–water mixture begin to aggregate amongst themselves to form micelles in the mixture.

critical pressure the minimum pressure capable of liquefying a gas at its critical temperature.

critical temperature 1 the temperature at or above which a partially miscible two-component system exists as one phase and below which it separates into two phases. **2** the temperature above which liquefaction of a gas cannot occur however great the pressure.

Crk *symbol for* a protooncogene related to v-*crk*, the oncogene of avian sarcoma viruses CT10. The name derives from CT10 regulator of kinase, because the protein causes tyrosine phosphorylation but is not itself a tyrosine kinase. Human Crk contains SH2 and SH3 domains (*see* **SH domain**).

CRM *abbr. for* (immunologically) cross-reacting material.

CRM-positive describing a mutant strain of a microorganism that produces a substance that is immunologically cross-reactive with another (specified) substance, e.g. an enzyme or a toxin, produced by a nonmutant (wild-type) strain. Similarly, a **CRM-negative** strain does not produce such a cross-reactive substance.

cRNA *abbr. for* complementary RNA.

cro *symbol for* a regulatory gene in phage lambda, named from 'C-repressor and other things'. Its product, Cro protein, is a repressor that regulates the expression of the *cI* gene. Cro protein is a small globulin, with a **helix-loop-helix** motif; it occurs as a dimer when bound to the operator.

Crohn's disease *or* **regional enteritis** a chronic inflammatory disease that usually affects the ileum and colon, and frequently runs in families. The inflamed lesions are rich in the inflammatory cytokines, particularly interleukins 1, 6, and 12 and tumour necrosis factor α. Homozygosity for a mutation in a gene (called *NOD2*) on chromosome 16 increases the risk of the disease by a factor of 20–40. *NOD2* encodes an intracellular protein that contains 2 caspase-recruitment domains, a nucleotide-binding domain, and 10 leucine-rich sequence repeats in the C-terminal region. It may have a role in apoptosis or in the recognition of specific components of enteric flora. [After Burrill B. Crohn (1884–1983), US gastroenterologist.]

Cromer-related blood group a blood group antigen carried by **complement decay accelerating factor**.

cromolyn 1,3-bis(2-carboxychromon-5-yloxy)-2-hydroxypropane; a chromone complex that prevents mast cell degranulation; its mechanism of action is unclear. Commonly prepared as the disodium salt, **disodium chromoglycate**, it is used therapeutically to prevent hypersensitivity reactions, especially asthma.

cross-bridge any connection, seen in electron microcopy, between the contractile elements of muscle, specifically between the head of a myosin molecule and an actin filament.

crossed immunoelectrophoresis *or (sometimes)* **quantitative immunoelectrophoresis** two-dimensional single **electroimmunodiffusion**; a rapid and sensitive technique for separation with high resolution and for quantification of antigenic macromolecules. These are separated first by one-dimensional gel electrophoresis and second by gel electrophoresis orthogonally into a gel containing antibodies to one or more known substances, thereby forming a pattern of loop-shaped precipitation zones between antibodies and corresponding antigens. The height and shape of the loops give information about the quantity and quality of the antigens and of the precipitating antibodies. In **tandem-crossed immunoelectrophoresis** two antigen samples are run at the same time with the application holes placed in such a way that related precipitin peaks formed from the respective samples generate double peaks, permitting direct comparison between them. *Compare* **rocket immunoelectrophoresis**.

crossed immunoisoelectric focusing *or* **crossed immunoelectrofocusing** a two-dimensional separation technique for analysing complex mixtures of macromolecules, e.g. samples of body fluids, that combines the high resolution of **isoelectric focusing** with the selectivity of detection of **immunoelectrophoresis**. Isoelectric focusing in a thin layer of gel (polyacrylamide or, preferably, specially modified agarose) is followed by electrophoresis into an antibody-containing gel orthogonally to the first separation; a pattern of precipitation zones is formed by reaction of antibody with corresponding antigen as in **rocket immunoelectrophoresis**.

crossed-line immunoelectrophoresis a technique that combines the procedures of **crossed immunoelectrophoresis** and **line immunoelectrophoresis**. Antigens are separated in a one-dimensional run (as in crossed immunoelectrophoresis) and the resulting gel is applied to the edge of a spacer gel in which a strip of gel containing a uniform mixture of the antigens (as used in line immunoelectrophoresis) is embedded parallel to the edge; this combination of the spacer gel and the first gel is then applied to the edge of an antibody-containing gel into which the antigens are introduced by electrophoresis. The resulting sum pattern of crossed immunoelectrophoresis peaks, each standing on individual baselines, permits a direct comparison of crossed and line immunoelectrophoresis patterns and in addition provides a higher power of resolution.

cross-flow filtration a method of operating a filtration device in which retained fluid is circulated over the membrane surface thus preventing undue build-up of filtered material on the membrane.

cross-hybridization *or* **cross-hybridisation** the hybridization of a polynucleotide probe to another polynucleotide molecule.

crossing over the reciprocal exchange of segments at corresponding positions along pairs of homologous chromosomes by symmetrical breakage and crosswise rejoining of the chromatids. It results in the **recombination** of alleles. In eukaryotes, crossing over usually accompanies meiosis. Mitotic homologous recombination, while common in yeast, is rare in higher eukaryotes. *See also* **Holliday junction**.

cross-link a side bond between different chains (or parts of a single chain) of a polymer, increasing molecular rigidity. —**cross-linked** *adj.*; **cross-linking** *n.*

cross-matching a serological procedure used to select blood for transfusion. The donor's erythrocytes are mixed with the recipient's serum; agglutination of the erythrocytes indicates incompatibility, demonstrating that the recipient's serum contains antibodies to the donor's erythrocytes.

cross of isocline the dark cross formed in **flow birefringence** experiments between concentric cylinders when the polarizer and analyser are orthogonal. As the system has cylindrical symmetry, there are two regions where the average orientation is parallel to the polarizer direction and two regions, orthogonal to the first two, where the average orientation is parallel to the analyser. In these regions no light is transmitted and a dark cross appears. *See also* **extinction angle**.

crossover 1 *an alternative term for* **crossover chromatid**. **2** the individual resulting from a **crossing over**.

crossover chromatid a **chromatid** resulting from **crossing over** of corresponding segments of the chromatids of homologous chromosomes.

crossover effect *(in pharmacology)* the phenomenon in which a drug shows a positive effect at one concentration and a deleterious effect at another.

crossover method any method for investigating a metabolic (or electron transport) pathway by application of the **crossover theorem**.

crossover nuclease a **deoxyribonuclease** that is involved in recombination.

crossover point the point in a series of reactions in a metabolic (or electron transport) pathway, that, when one reaction is specifically inhibited, marks the boundary between reactants whose concentrations increase and reactants whose concentrations decrease. In particular, it is the point in the respiratory chain that, because of the specific action of an inhibitor, marks the boundary between respiratory catalysts that are more reduced and ones that are more oxidized (at steady-state levels). *See also* **crossover theorem**.

crossover theorem a theorem stating that when a series of reactions in a metabolic (or electron transport) pathway is inhibited at a specific reaction, the concentrations of reactants before the inhibited reaction increase above their steady-state values while the concentrations of the reactants after the inhibited reaction decrease below their steady-state values.

crossover unit *an alternative name for* **map unit**.

cross-react to react with a substance other than that for which the reagent is supposedly specific. *Compare* **cross-reaction**. —**cross-reactive** *adj*.

cross-reacting antibody any antibody that is able to react with an antigen that did not specifically stimulate its production. Such cross-reactions may be weaker than the reaction of the antibody with the antigen that caused its production.

cross-reacting antigen 1 any antigen that is able to react with an antibody produced in response to another antigen. This may be because the two antigens share the same determinants or carry determinants that are sufficiently alike stereochemically to enable the antibody to react with both of them. **2** any antigen that has an identical structure in two strains of microorganism, so that antibody raised against one strain will react with the second strain.

cross-reaction the immunological phenomenon in which an antigen reacts with an antibody raised against a different antigen. *See* **cross-reacting antibody**, **cross-reacting antigen**.

cross-reactivation the reappearance of activity in the progeny of a lethal mutant virus following mixed infection of a host cell with one or more active viruses. If the active viruses differ from each other in one or more of their genetic loci, the mutant virus is reactivated by genetic exchange leading to the replacement of its damaged nucleic acid.

cross-sensitivity a state of immunological hypersensitivity to one substance produced by priming an animal with another substance that bears **cross-reacting antigen**.

cross-tolerance 1 a state of immunological **tolerance** to one substance produced by priming an animal with another substance that bears **cross-reacting antigen**. **2** a state of physiological tolerance to one drug resulting from chronic administration of another.

croton oil an oil obtained from the seeds of *Croton tiglium*, a member of the family Euphorbiaceae. It is an irritant, and contains tumour-promoting **phorbol esters**.

crotoxin a neurotoxic protein of the venom of the South American rattlesnake *Crotalus durissus terrificus*. It has phospholipase A2 activity.

crown compound *or* **coronand** any synthetic macrocyclic polydentate nonplanar (generally uncharged) organic compound a molecule of which has three or more ring heteroatoms (usually nitrogen or oxygen) capable of ligation to a metal ion or other cationic entity. The resulting complex is termed a **coronate**. *See also* **crown ether**. *Compare* **cryptand**.

crown ether any of a large subclass of crown compounds that are synthetic macrocyclic polyethers, the molecules of which contain 9–60 atoms in the ring, including 3–20 oxygen atoms. Many crown ethers, especially those containing 5–10 ring oxygen atoms in the molecule, can act as ion-selective complexing agents and ionophores with metal cations; some will solubilize inorganic salts in nonpolar solvents.

crown-gall disease a tumorous disease of many dicotyledonous plants caused by the bacterium *Agrobacterium tumefaciens* and marked by enlargement of the stem near the root crown. The swelling, or **gall**, is produced by the transformation of cells at an infected wound through transfer of the **T-DNA**, part of a large (150–230 kb) plasmid (the **Ti plasmid**) found in oncogenic strains of the bacterium. Transformed tissue is distinguishable by its ability to grow in tissue culture in the absence of plant hormones and by the

synthesis of a characteristic **opine**, which is invariably a specific substrate for the particular inducing strain of the bacterium.

CRP *abbr. for* **1** C-reactive protein; **2** cyclic AMP receptor protein; a bacterial protein that complexes with cyclic AMP, and binds to specific DNA sites near the promoter to regulate the transcription of several catabolite-sensitive operons. The protein induces a severe bend in DNA, and acts as a negative regulator of its own synthesis. It binds to DNA as a dimer.

Crt *symbol for* the cerotoyl (i.e. hexacosanyl) group.

cruciform 1 cross-shaped. **2** a DNA conformation that can be generated from long **palindromes**, e.g.

crustacyanin a protein, found in the carapace of lobsters, that binds the carotenoid **astaxanthin**, which provides the blue colour. a-Crustacyanin is a dimer; ß-crustacyanin is a 16-mer. There are five types of subunit: A1, A2, A3, C1, and C2. Crustacyanin is a member of the **lipocalin** superfamily.

cruzaine *an alternative name for* **cruzipain**.

cruzipain *or* **cruzaine** a cysteine proteinase (EC 3.4.22.–) that hydrolyses chromogenic peptides at the carboxyl Arg or Lys. It requires at least one more amino acid, preferably Arg, Phe, Val, or Leu, between the terminal Arg or Lys and the amino-blocking group. The purified enzyme digests itself. The enzyme plays a role in development and differentiation of *Trypanosoma cruzi*.

CRX *abbr. for* cone-rod homeobox protein; a transcription factor protein (299 residues) that contains a homeodomain in its N-terminal region and is expressed specifically in developing and mature photoreceptor cells of the retina. It regulates expression of the genes for rhodopsin and other photoreceptor-specific proteins. Mutations in the gene at 19q13.3 result in a form of blindness called Leber's congenital amaurosis.

CRY *abbr. for* cryptochrome.

cryo+ *comb. form* indicating low temperature.

cryobiochemistry the study of biochemical substances and processes at subzero temperatures, i.e. below the freezing point of water. The field has expanded because of the development of **cryosolvents**. *See also* **cryoenzymology**.

cryobiology the science that is concerned with the study of the effects of very low temperatures on organisms.

cryoelectron microscopy *see* **electron microscopy**.

cryoenzymology the study of enzymes at low temperatures. By employing temperatures below 0°C and fluid **cryosolvents**, enzyme reaction rates can be reduced sufficiently for normally transient enzyme–substrate intermediates to be studied.

cryogen any substance used to produce a low temperature; e.g. a freezing mixture.

cryogenic of, or relating to, very low temperatures.

cryogenics the branch of science concerned with the production of very low temperatures and the study of phenomena occurring at these temperatures.

cryoglobulin any **globulin**, especially IgG or IgM, that forms a gel or flocculent precipitate, or spontaneously crystallizes, on cooling of a solution or a sample of serum containing it. *See also* **cold-insoluble globulin**.

cryomicrotome any **microtome** arranged in a **cryostat** and used to prepare thin sections of frozen tissue for microscopic examination.

cryoprecipitate the precipitate of a **cryoglobulin**.

cryopreservation the science of preserving cells, tissues, or organs over long periods with the aid of very low temperatures.

cryoscope any instrument or device for the determination of freezing points.

cryoscopy the measurement of freezing points, especially the determination of relative molecular mass or osmotic pressure by measurements of freezing point depressions. —**cryoscopic** *adj*.

cryosolvent any mixture of water and one of a select number of polar organic solvents that is used to maintain proteins in fluid solution at temperatures below 0°C without denaturation.

cryostat an apparatus used to produce and maintain a constant low temperature in an enclosure or liquid bath.

cryosublimation the process whereby water, or another solvent, is sublimed from a frozen sample and is collected in a cold trap. *Compare* **lyophilization**.

cryotolerant tolerant of low temperatures.

cryptand any synthetic macrooligocyclic organic compound, a molecule of which consists of two or more rings linked together by bridgehead (generally nitrogen) atoms to form a three-dimensional cagelike molecular structure having sufficient space within it to accommodate any one of various cations and thereby form a **cryptate**. *Compare* **crown compound**.

cryptate the adduct of a **cryptand** and a metal ion or other cationic entity. The cation is bound within the molecule by polydentate ligation to heteroatoms in the cryptand, an exceptionally stable compound being formed.

cryptic 1 hidden, secret. **2** not apparent, unrecognized, masked; e.g. of a medical condition. **3** (especially of the colouring of an animal) serving to conceal. **4** of a particle-bound enzyme, that it is not as accessible to substrate as when it has been solubilized.

crypticity *or* **latency** the property of being **cryptic**.

cryptic plasmid a **plasmid** to which no phenotypic traits have been ascribed.

cryptic splice site a randomly occurring site in the genome that contains the consecutive six-nucleotide consensus sequence for 5′ or 3′ intron splicing but is not normally used for that purpose. Such sites are expected to occur randomly every 4 kb or so on average. Mutations at splice sites may result in a nearby cryptic splice site being used instead, with the consequence that an abnormally spliced mRNA transcript is produced.

crypto+ *or* (before a vowel) **crypt+** *comb. form* meaning secret, hidden, concealed, unrecognized.

cryptochrome *abbr.*: CRY; an FAD-containing protein that is a receptor for blue light in plants and in *Drosophila* and other animals. In *Arabidopsis* it interacts with **COP1** to mediate photomorphogenetic development. In animals expression in the suprachiasmatic nucleus of the hypothalamus is enhanced by the clock-BMAL1 dimer, and in the forebrain by the NPAS2-BMAL1 dimer. CRY is regulated by a protein serine/threonine kinase. Humans possess genes for CRY1 and CRY2. *See* **BMAL1, clock, NPAS2**.

cryptotope a hidden immunological determinant. *Compare* **epitope**.

cryst. *abbr. for* crystalline, or crystallized.

crystal any three-dimensional solid aggregate in which the plane faces intersect at definite angles and in which there is a regular internal structure of the constituent chemical species.

crystal lattice the array of **unit cells** along parallel straight lines that makes up the structure of a crystal.

crystallin any of numerous water-soluble structural proteins found in the lens of the eye. Together, the crystallins account for about 90% of the total lenticular protein. Most vertebrate lenses contain three major classes of crystallin, designated α-, β-, and γ-; in birds and reptiles there is a fourth, designated δ-, which is homologous to the urea cycle enzyme argininosuccinate lyase. ε-Crystallin is a homologue of the enzyme **acetate dehydrogenase**. Although there are wide variations between species in the proportions of the various crystallins, and in the extent of their post-translational modification, or degree of aggregation, they display significant conservation of primary structure.

crystalline *abbr.*: cryst.; having the structure or characteristics of a **crystal**; made up of crystals; like a crystal.

crystallize *or* **crystallise** to form, or cause to form, crystals, e.g. by slowly removing solvent from a solution to make a solute form one or more crystals.

crystallized *or* **crystallised** *abbr.*: crys.; (of a substance) having been induced into a crystalline form.

crystallography the study of the geometric forms of crystals. *See also* **X-ray crystallography**. —**crystallographer** *n.*; **crystallographic** *adj.*

crystalloid 1 *an alternative term for* **crystalline**. **2** any solute that can pass through a semipermeable membrane that does not allow the passage of a **colloid**. **3** any of the protein crystals that occur in seeds and other plant storage organs.

crystalluria the presence of rather insoluble metabolic products (e.g. orotic acid, xanthine, uric acid, cysteine, oxalic acid) in crystalline form in urine, usually as a result of metabolic disorder. Calcium phosphate crystals are common especially in alkaline urine.

Cs *symbol for* cesium.

CS *abbr. for* choriomammotropin.

CSD *abbr. for* cold-shock domain (*see* **cold-shock protein**).

CSF *abbr. for* **1** cerebrospinal fluid. **2** colony stimulating factor.

cSNP *abbr. for* coding single nucleotide polymorphism.

CSP *abbr. for* cell surface protein; *see* **fibronectin**.

CSPCP *abbr. for* cartilage-specific proteoglycan core.

CstF *abbr. for* CTD-associated cleavage stimulatory factor; a protein that stimulates cleavage of eukaryotic pre-mRNA at a site situated 15–25 nucleotides 3′ of the AAUAAA sequence in the 3′ untranslated region. Like CPSF (*see* **polyadenylation element**), CstF binds the C-terminal domain of the largest component of RNA polymerase II and activates poly(A) polymerase to form the poly(A) tail of the mature mRNA.

$C_o t$ *or* Cot a measure of DNA kinetic **complexity** used in the renaturation analysis of DNA genomes. It is the product of initial DNA concentration, C_o, and time, t. Highly reiterated sequences will renature at low $C_o t$ values while unique sequences will renature at high $C_o t$ values. $C_o t_{1/2}$ is the value representing half complete renaturation.

CT *abbr. for* computerized tomography (*see* **tomography**).

CTAB *abbr. for* cetyltrimethylammonium bromide.

CTAC *abbr. for* cetyltrimethylammonium chloride; a detergent used for the solubilization of inclusion bodies.

CTAP-III *abbr. for* connective-tissue activating peptide III; *see* **LA-PF4**.

CTD *abbr. for* C-terminal domain; usually used specifically for the C-terminal domain of the largest subunit of RNA polymerase II, which contains 52 heptapeptide (YSPTSPS) repeats in which Y, S, and T are potential targets for protein kinases. In budding yeast, the domain contains 26 such repeats. The domain interacts with many proteins including those that contain a WW domain. It plays an intimate role in pre-mRNA capping, splicing, and polyadenylation. *See* **CstF**.

CTD kinase *see* **[RNA-polymerase]-subunit kinase**.

ctDNA *see* **DNA**.

C-terminal peptide amidation a type of post-translational modification, occurring in the Golgi apparatus, whereby a C-amidated peptide is formed and released from a precursor polypeptide. Many biologically active peptides and peptide hormones are only active in the C-amidated form. The amide group is derived from a glycine residue next to the peptide's C-terminus. Two enzymes are required for the reaction: peptidylglycine α-amidating monooxygenase and peptidyl α-hydroxyglycine α-amidating lyase. *See* **peptide amidation**.

C terminus *or* **C terminal** that end of any peptide chain at which the 1-carboxy function of a constituent *a*-amino acid is not attached in peptide linkage to another amino-acid residue. This function may bear an amide group, or other substituent, in nonpeptide linkage; more commonly, it is free (also unprotonated and negatively charged) – the chain end may then also be termed the **carboxy(l) terminus** or **carboxy(l) terminal**. *Compare* **N terminus**. —**C-terminal** *adj.*

C1-tetrahydrofolate synthase a trifunctional enzyme that consists of two major domains: an N-terminal domain, containing N^5,N^{10}-methylenetetrahydrofolate dehydrogenase and N^5,N^{10}-methenyltetrahydrofolate cyclohydrolase, and a C-terminal domain containing **formate–tetrahydrofolate ligase** domain. It is a homodimer.

CTF 1 *abbr. for* CAAT box-binding transcription factor *or* nuclear factor 1 (*abbr.*: NF-I); any of a class of **transcription factors** for the eukaryotic RNA polymerase II promoters, **CAAT box** sequences; CTF binds to DNA as a homodimer. **2** *abbr. for* competence transcription factor *or* competence protein K; a product of a bacterial regulatory gene required for expression of late competence genes.

CTGF *abbr. for* connective tissue growth factor.

CTH *abbr. for* ceramide trihexoside (Cer(Hex)$_3$ preferred).

CTP *abbr. for* cytidine 5′-triphosphate.

CTP synthase EC 6.3.4.2; *systematic name*: UTP:ammonia ligase (ADP-forming); *other names:* CTP synthetase; UTP– ammonia ligase. An enzyme responsible for *de novo* synthesis of CTP that in bacteria catalyses the reaction between ATP, UTP, and NH$_3$ to form CTP, ADP, and orthophosphate. In animals, the amino group is donated by glutamine. The enzyme bears the same EC number and the reaction is analogous to that of the bacterial enzyme, with glutamine replacing ammonium ion, and CTP and glutamate as products.

CTS (*in clinical biochemistry*) *abbr. for* catalase.

C-type lectin any of a group of carbohydrate-binding proteins of related structure that require Ca^{2+} for activity. An example is mannose-binding protein.

Cu *symbol for* copper.

CUA a codon in mRNA for L-leucine.

cubic symmetry the symmetry of one of the three **point groups** of a globular oligomeric protein; it may be **tetrahedral**, designated T_n, where n is 12; **octahedral**, designated O_n, where n is 24; or **icosahedral**, with 60 identical subunits.

cubitus interruptus *abbr.*: Ci; a transcription factor of *Drosophila* that is activated by the G-protein-coupled receptor Smoothened when this is released by binding of Hedgehog to Patched. Ci is a 155 kDa protein found in a multiprotein complex that is associated with microtubules. Its N-terminal region contains four zinc-finger domains, and its C-terminal region interacts with CREB protein, which appears to be a cofactor for Ci. In the inactive form, Ci is cytosolic, but on activation by protein kinase A it enters the nucleus. In vertebrates there are three Ci homologues called Gli1, Gli2, and Gli3. Several autosomal dominant syndromes that produce a number of malformations in humans are associated with truncating mutations in Gli3.

cubulin a peripheral membrane protein (≈400 kDa) that is the receptor for the intrinsic factor–cobalamin complex in the ileum, and facilitates its endocytosis. It copurifies with **LRP2** and colocalizes with it in intestinal and renal epithelium. The N-terminal region of cubulin contains 8 EGF repeats and 27 contiguous CUB domains (each of ≈110 residues) similar to those of certain development control proteins. Several mutations in the gene at 10p12.1 cause megaloblastic anemia from cobalamin deficiency.

CUC a codon in mRNA for L-leucine.

cucurbitacin any of a group of complex triterpenoids that are the bitter principles of the cucumber family, Cucurbitaceae.

CUG a codon in mRNA for L-leucine.

cullin *or* **Cdc53p** any of a small family of proteins homologous with Cdc53p of yeast and Cul1 of *Caenorhabditis elegans*. In yeast Cdc53p is present in complexes that target specific proteins for ubiquitin-mediated degradation. Cul1 regulates cell cycle exit in *C. elegans*. Several homologues are present in mammalian cells as components of complexes that contain E3 ubiquitin ligase.

culture 1 a a collection of cells, tissue fragments, or an organ that is growing or being kept alive in or on a nutrient medium (i.e. **culture medium**); *see also* **cell culture**, **tissue culture**. **b** any culture medium to which such living material has been added, whether or not it is still alive. **2** the practice or process of making, growing, or maintaining such a culture. **3** to grow, maintain, or produce a culture.

culture medium any nutrient medium that is designed to support the growth or maintenance of a **culture** (def. 1). Culture media are typically prepared artificially and designed for a specific type of cell, tissue, or organ. They usually consist of a soft gel (so-called solid or semisolid medium) or a liquid, but occasionally they are rigid solids.

cUMP *abbr. for* cyclic UMP; *see* **uridine 3′,5′-phosphate**.

cumulative inhibition the inhibition of an enzyme resulting from the action of a number of **effectors**, each of which reduces the activity of the enzyme by a fixed percentage irrespective of the degree of inhibition already extant as a result of the presence of other inhibitors.

cupin any of a superfamily of protein domains, with some 18 subclasses, found in archaea, eubacteria, and eukaryotes. They are present most often in enzymes with zinc-containing active sites based around a histidine cluster. The domain consists of a flattened beta barrel made of two sheets, each of five antiparallel beta strands, that form a cleft for the metal ion. It was first found in **germin**, where a manganese ion is coordinated to three histidines and a glutamate residue.

cuprein any of a group of copper proteins now known as **superoxide dismutase**.

Cuprolinic Blue *the proprietary name for* quinolinic phthalocyanine (*abbr.*: QPC; no CI number); a copper-containing, intensely blue dye used for the visualization of RNA and other polynucleotides. The size and shape of the molecule have been designed to provide an exactly complementary fit between any pair of adjacent isoindole rings in the dye and base pairs in nucleic acids.

cuproprotein any protein containing one or more copper atoms.

curare a poisonous extract from certain tropical South American vines of the genera *Strychnos* and *Chondodendron*, used by Amazon and Orinoco Indians as an arrow poison; it blocks neuromuscular transmission in voluntary muscle and is useful as a muscle relaxant, e.g. in surgery. *See* **(+)-tubocurarine** (the active principle) for the mechanism.

curarize *or* **curarise** to treat with (or as if with) curare; especially to cause muscular relaxation or paralysis. —**curarization** *or* **curarisation** *n.*

curculin a polypeptide present in the fruit of *Curculigo latifolia*, a stimulant herb that grows wild in western Malaysia. It is sweet tasting, and has taste-modifying activity. It exists as a homodimer.

curcumin 1,7-bis(4-hydroxy-3-methoxyphenyl)-1,6-heptadiene-3,5-dione; turmeric yellow; a condiment and dye obtained from dried and powdered rhizomes of *Curcuma domestica (C. longa)*. It provides the characteristic colour of curries, and is also an inhibitor of 5-lipoxygenase and **prostaglandin-endoperoxide synthase** cyclooxygenase activity. *See also* **catechin**.

curdlan *see* **callose**.

curie *abbr.*: Ci *(formerly* C); a non-SI unit of (radio)activity or of radioactive material. Originally (1910) it was defined as the quantity of radon in radioactive equilibrium with one gram of radium. Latterly (1968) it was defined as a unit of activity equal to 3.7×10^{10} disintegrations per second or, less correctly, as the quantity of any radioactive material having such activity. Hence, 1 Ci = 3.7×10^{10} becquerels. [After Pierre Curie (1859–1906), French physicist and chemist (not after Marie Curie as sometimes stated).]

curing *(in genetics)* the elimination of a **plasmid** from its host cell.

curli a class of proteins that form highly aggregated fibres on the outer surface of *E. coli* and *Salmonella* species. They are involved in colonization of inert surfaces and mediate binding to a variety of host proteins. They have amyloid-specific properties and are rich in beta-sheet structure.

curtain electrophoresis *see* **continuous-flow electrophoresis**.

cushingoid of, or relating to, **Cushing's disease** or **Cushing's syndrome**, or any of the signs and symptoms associated with these conditions.

Cushing's disease a pituitary-dependent, corticotropin-induced, bilateral adrenal hyperplasia that is one of several conditions embraced by **Cushing's syndrome**. [After Harvey Williams Cushing (1869–1939), US neurosurgeon.]

Cushing's syndrome the group of symptoms and signs resulting from prolonged exposure to inappropriately high plasma **cortisol** concentrations, caused either by primary disease (e.g. tumour) of the adrenal glands or by excessive production of **corticotropin** because of disease (e.g. basophil tumour) of the anterior pituitary. It may also be caused by excessive secretion of cortisol or corticotropin from ectopic sites, or by prolonged therapy with (large doses of) cortisol or other glucocorticoids. The syndrome is characterized by obesity of the trunk, polycythemia, osteoporosis, and glycosuria. The underlying disorder may be corticotropin-dependent or -independent. Of the corticotropin-dependent disorders, **Cushing's disease** is the most common, accounting for 68% of all types of Cushing's syndrome. The remainder (15% of all types) result from ectopic corticotropin-secreting tumours, that is, non-endocrine tumours such as lung, gut, ovarian, or carcinoid, which in some cases secrete corticotropin. Corticotropin-independent disorders include adrenocortical adenoma (5%) or carcinoma (3%) and nodular adrenal hyperplasia (9%).

cut 1 to sever, to make an incision, to divide (with a sharp instrument). **2** to separate into **fractions**. **3** a double-stranded scission in a polynucleotide duplex. *Compare* **nick**. **4** informal term for a **fraction**, such as one obtained in column chromatography or distillation.

cutaneous of, pertaining to, or in the skin.

cutin a polyester consisting of polyoxygenated C_{16} and C_{18} fatty acids esterified to each other to form a tough, inelastic, and hydrophobic meshwork that (like suberin) renders the walls of plant epidermal cells impermeable to water. It is usually associated with waxes on leaf and stem surfaces. The fatty acyl-CoAs are hydroxylated by oxidases in the endoplasmic reticulum and secreted into the cell wall space where esterification is catalysed by hydroxyacyl CoA:cutin transacylase.

cutinase an enzyme that catalyses the hydrolysis of the plant polyester, cutin, to allow penetration by phytopathogenic fungi.

CUU a codon in mRNA for L-leucine.

cuvette a small transparent vessel, often with parallel sides of specified separation, used to hold a liquid specimen, especially for optical or spectral measurements. *See also* **cell** (def. 2).

C-value the quantity of DNA present in a single haploid genome.

C-value paradox the lack of correlation between the amount of DNA in the haploid genome (i.e. **C-value**) and the complexity of the organism. The paradox was explained when it was discovered that a large proportion of some genomes is made up of repetitive DNA.

Cya *symbol for* a residue of the α-amino acid cysteic acid (an artefact in peptide sequences).

cyan+ *a variant form (before a vowel) of* **cyano+**.

cyanate 1 the anion, $N{\equiv}C{-}O^{-}$, derived from cyanic acid ($HOC{\equiv}N$).

2 any salt of cyanic acid. **3** any organic compound containing the monovalent group $-O{-}C{\equiv}N$. *Compare* **isocyanate**.

cyanelle a cyanobacterium-like structure, considered as a descendant of an ingested cyanobacterium, that exists as an endosymbiont in the biflagellate protist *Cyanophora paradoxa*. The cyanelle has photosynthetic properties resembling those of cyanobacteria but its genome size is about 10% of those reported for cyanobacteria.

cyanhydrin *a former name for* **cyanohydrin**.

cyanide 1 the anion, $N{\equiv}C^{-}$, derived from hydrocyanic acid (HCN). This ion blocks the electron transport chain by reacting with the ferric form of cytochrome aa_3 and thus acts as an inhibitor of respiration. **2** any salt of hydrocyanic acid. **3** any organic compound containing the monovalent cyano group, $-C{\equiv}N$; such a compound may also be known either as a nitrile or as a carbonitrile, depending respectively on whether its carbon atom is included in or excluded from the numbering of an attached chain.

cyanide-resistant oxidase *an alternative name for* **alternative oxidase.**.

cyanide-resistant respiration a respiratory pathway, occurring only in mitochondria of some plants, yeasts, and bacteria, that is unaffected by **cyanide**. Electron transport from NADH or succinate to O_2 involves a nonheme iron protein as the terminal oxidase; this is insensitive to cyanide. The cyanide-resistant pathway complements the more usual cyanide-sensitive pathway; this contains cytochrome oxidase, which is inhibited by cyanide. Cyanide-resistant respiration is not found in animals.

cyanidin *see* **anthocyanidin**.

cyano+ *or (before a vowel)* **cyan+** *comb. form* denoting **1** the cyanide ion or hydrocyanic acid. **2** the $-CN$ group in covalent linkage. **3** (dark) blue.

cyanobacteria *or (formerly)* **blue-green algae** a very heterogeneous group of prokaryotic photosynthetic organisms all of which contain chlorophyll *a*, carotenoids, and phycobiliproteins arranged in thylakoids. They lack chloroplasts, unlike true algae, and differ from photosynthetic bacteria, in, e.g., evolving oxygen during photosynthesis. Many can fix nitrogen.

cyanoborohydride any member of a group of compounds with the structure MBH_3CN, where M is an alkali metal (usually sodium). They are reducing agents similar in action to borohydride but more stable, more lipophilic, and more selective. They are especially useful for the facile reduction of aldehydes and ketones in acid conditions to the corresponding alcohols, for the reductive amination of aldehydes and ketones in the presence of an amine, and for the reductive methylation of amines (including amino acids and proteins) in the presence of formaldehyde.

cyanocobalamin the most usual form of **vitamin B_{12}**, in which the sixth position on the cobalt atom in **cobalamin** is filled by a cyanide group (the cyanide group being an artefact of the isolation procedure).

cyanogen bromide *or* **bromine cyanide** CBrN, or CNBr, or BrCN; a volatile solid (at ordinary temperatures) with two main uses in biochemistry: (1) in the sequencing of proteins, because (in 90% formic acid) it converts methionine residues in peptide linkage to homoserine residues, with simultaneous cleavage of peptide bonds whose carbonyls are contributed by methionine residues; also (in anhydrous heptafluorobutyric acid) it specifically cleaves tryptophanyl peptide bonds; and (2) for activating cross-linked agarose so that almost any molecule containing amino groups can be coupled to it to prepare a specific affinity adsorbent.

cyanogenesis the production or yielding of cyanide ions or hydrocyanic acid.

cyanogenic *or* **cyanogenetic 1** *or* **cyanophoric** describing any glucoside or other glycoside containing a cyano group that is released as hydrocyanic acid on acid hydrolysis; e.g. amygdalin. Such compounds occur in the kernels of various fruits. **2** describing any plant that synthesizes such a glycoside.

cyanoginosin a cyclic heptapeptide toxic agent responsible for numerous cases of poisoning among South African livestock that consume water contaminated with blooms of cyanobacteria. About seven variants are known.

cyanoglobin an unusual monomeric myoglobin-like hemoprotein found in multicellular cyanobacteria.

cyanohydrin or (formerly) **cyanhydrin** any member of a class of α-hydroxynitriles formed by the base-catalysed addition of a cyanide ion to an aldehyde; such compounds are formed also from some reactive ketones.

cyanolabe or **cyanopsin** or **chloropsin** or **chlorolabe** See **cyanopsin, opsin.**

cyanophage any virus whose host is a cyanobacterium.

cyanophoric an alternative term for **cyanogenic** (def. 1).

cyanophycin granule or **structured granule** a type of granule, without apparent limiting membrane, that is observed in most, if not all, cyanobacteria. These granules are variable in size and shape, with diameters up to about 500 nm, and contain polypeptides of 25–100 kDa with a 1:1 arginine:aspartic acid ratio. They are believed to act as a nitrogen reserve.

cyanopsin or **chloropsin** or **chlorolabe** a greenish-blue, light-sensitive, visual pigment occurring in the cone cells of the retina. It consists of a specific opsin combined (as a Schiff's base) with 3,4-didehydro-11-cis-retinal. See also **3,4-didehydroretinal, opsin.**

cyanosis the bluish appearance of the skin and mucous membranes due to insufficient oxygenation of the blood in the capillaries. —**cyanotic** adj.

cyanosome an alternative name for a **phycobilisome** in cyanophytes.

cybrid a portmanteau word used to denote a line of cultured mammalian cells that is a cytoplasmic hybrid of two other lines, as shown by inheritance of biochemical or other characteristics.

cycacin methylazoxymethanol β-D-glucoside; a toxic substance from the seeds of Cycas revoluta, the aglycon of which is carcinogenic.

cycl+ a variant form (before a vowel) of **cyclo+.**

cyclamate the anion of cyclamic acid, cyclohexylsulfamic acid. It is used in some countries as a sweetening agent, but is banned in the USA because of its possible carcinogenicity.

cyclase any enzyme that catalyses ring closure, e.g. **adenylate cyclase.**

cyclazocine 3-(cyclopropylmethyl)-1,2,3,4,5,6-hexahydro-6,11-dimethyl-2,6-methano-3-benzazocin-8-ol; a potent benzomorphan narcotic analgesic agent. It is an antagonist at μ-, and an agonist at δ- and especially κ-opioid receptors. Binding to σ-receptors may explain its psychotomimetic actions.

cycle 1 any recurring period of time in which certain events or operations occur, complete themselves, or repeat themselves in a regular sequence. **2** any sequence of changes occurring in a system in which the system is eventually restored to its initial state. **3** (in biochemistry) any closed sequence of metabolic reactions in which an end product acts as a reactant in the initiation of the cycle, e.g. the **tricarboxylic-acid cycle. 4** (in ecology) any closed sequence of large-scale processes that describes the nutritional interdependence of animals, plants, and microorganisms, e.g. the **nitrogen cycle. 5** to process through a cycle or cyclic system; to pass through cycles or to vary in a cyclical manner.

cycle sequencing a method for DNA sequencing based on dye-termination and employing repeated cycles of melting, annealing, and synthesis in a manner similar to that of **polymerase chain reaction.** See **dye-terminator sequencing.**

cyclic 1 of, relating to, moving in, or being a cycle. **2** (of an organic chemical compound) containing one (i.e. monocyclic) or more (i.e. polycyclic) rings of atoms. Compare **heterocyclic, homocyclic.** —**cyclically** adv.

cyclic adenosine 5′-diphosphoribose or **cyclic ADP-ribose** abbr.: cyclic ADPR or cADPR; a metabolite formed from NAD+ by ADP ribosyl cyclase (ARC) that may be responsible for mediating certain of the latter's regulatory effects. It consists of a ribose molecule linked by its C-5 hydroxyl to the β-phosphate of ADP and by its C-1 hydroxyl to the adenosine amino group. It may be involved in regulating intracellular Ca^{2+} concentrations and in activating the nonskeletal form of the ryanodine receptor Ca^{2+} channel.

cyclical a variant form of **cyclic** (def. 1).

cyclic AMP or **cAMP** abbr. for adenosine 3′,5′-phosphate. See also **adenylate cyclase.**

cyclic AMP-dependent protein kinase or **protein kinase A** (abbr.: PKA) a **protein kinase** that is regulated by cyclic AMP (cAMP). In **type I**, the inactive form of the enzyme is composed of two regulatory and two catalytic subunits; activation by cAMP produces active catalytic monomers and a regulatory dimer with four bound cAMP molecules. Expression of regulatory subunits in some tissues is constitutive and in others is inducible. In **type II**, regulatory subunits mediate membrane association by binding to anchoring proteins, including the MAP-2 kinase and **AKAP.** The inactive enzyme is composed of two regulatory and two catalytic subunits; activation by cAMP produces two active catalytic monomers and a regulatory dimer that binds four cAMP molecules.

cyclic AMP phosphodiesterase (number EC 3.1.4.17 shared with **cyclic GMP phosphodiesterase**); recommended name: 3′,5′-cyclic-nucleotide phosphodiesterase; an enzyme that catalyses the hydrolysis of a nucleoside 3′,5′-cyclic phosphate to a nucleoside 5′-phosphate. The intracellular concentration of cyclic AMP is regulated by low-affinity (type 1) and high-affinity (type 2) enzymes. The enzymes are monomers; type 2 are well conserved.

cyclic AMP receptor any of various receptor molecules that bind cyclic AMP, especially **CRP** (def. 2), a bacterial catabolite (gene) activator protein. Other examples are the signalling receptors of the slime mould, Dictyostelium discoideum, which are similar to other eukaryotic G-protein-linked receptors. Receptor 1 is involved in aggregation, and receptor 2 is expressed after aggregation in pre-stalk cells.

cyclic AMP response element see **CRE.**

cyclic analysis a procedure that is useful, e.g. in flame spectrophotometry, for minimizing the effects of any mutual interference between two or more components of a sample being analysed. First, the concentration of each such component is estimated. Next, standards for each of the components are prepared containing the other interfering components at their estimated concentrations and the estimations are repeated. Then fresh standards are prepared, using the revised values for the concentrations of the interfering components, and further estimations are made; and so on, until a constant and accurate value for each component is obtained. In practice, usually only two or three such cycles of analysis are necessary.

cyclic CMP see **cytidine phosphate.**

cyclic diguanylic acid bis(3′-5′)cyclic guanylic acid; a potent activator of cellulose synthase in the bacterium Acetobacter xylinum.

cyclic electron flow see **cyclic photophosphorylation.**

cyclic fatty acid any of a class of fatty acids that contain a carbocyclic unit. They include those with a cyclopropane unit, such as **lactobacillic acid** found in bacterial membranes, those with a cyclopropene unit such as sterculic acid (see structure), found in Sterculiaceae, and those with a cyclopentene unit, found in seed fats of Flacourtiaceae, these including manaoic acid (see structure). **Mycolic acids** are 2-alkyl-3-hydroxy acids with up to 80 carbons, and contain cyclopropane units. Fatty acids containing a cyclohexane unit are also known.

sterculic acid

manaoic acid

cyclic GMP *abbr. for* guanosine 3′,5′-phosphate.

cyclic GMP-dependent protein kinase *or* **PKG** a cytosolic protein serine/threonine kinase that contains a catalytic and a regulatory domain in a single polypeptide (80 kDa). Binding of cGMP causes activation by releasing part of the regulatory domain from the catalytic site.

cyclic GMP phosphodiesterase (EC 3.1.4.17, number shared with **cyclic AMP phosphodiesterase**); *recommended name:* 3′,5′-cyclic-nucleotide phosphodiesterase. An enzyme that catalyses the hydrolysis of a nucleoside 3′,5′-cyclic phosphate to a nucleoside 5′-phosphate. It is involved in the regulation of the concentration of cyclic GMP, and thereby involved in the transmission and amplification of the visual signal; *see* **guanosine 3′,5′-phosphate**. Example from mouse rod cells has three subunits: α and β are catalytic subunits (short form of β is β′ from alternative splicing), and γ is a regulatory subunit; α and β are typical cyclic nucleotide phosphodiesterases.

cyclic IMP *symbol:* cIMP *or* Ino-3′,5′-P; inosine 3′,5′-(cyclic)phosphate.

cyclic nucleotide any nucleotide in which the phosphate group is in diester linkage to two positions on the sugar residue, usually 3′,5′ but also sometimes 2′,3′.

cyclic peptide nanotube a stack of synthetic, cyclic, antibacterial peptides in which each monomer consists of 6–8 residues of alternating D- and L-amino acids. The peptides enter the disease-causing bacteria, insert as nanotubes into their membrane, and kill them.

cyclic photophosphorylation the processes that occurs in photosystem I in which the high-potential electron in bound ferredoxin can be transferred to cytochrome b_{563}, rather than to NADP$^+$, and then back to the oxidized form of P700 through cytochrome c_{552} and plastocyanin with the generation of one molecule of ATP. *See* **photosystem**.

cyclic symmetry the symmetry of a **point group**, designated C_n, having a geometric arrangement of the subunits such that a rotation of 360°/n transposes the structure into itself (i.e. it is indistinguishable from the original one). A dimeric molecule with chemically and spatially identical subunits must necessarily belong to the point group C_2. In general, a molecule with C_n symmetry will be a closed ring containing n subunits (e.g. of triangular, square, pentagonal, and hexagonal appearance for molecules containing three, four, five, and six subunits, respectively). If n is an odd number, C_n is the only symmetry possible.

cyclic UMP *symbol:* cUMP *or* Urd-3′,5′-P; uridine 3′,5′-(cyclic)phosphate.

cyclic voltammetry a form of **voltammetry** in which the working electrode potential is swept back and forth across the formal potential of the analyte. Repeated reduction and oxidation of the analyte causes alternating cathodic and anodic currents to flow at the electrode. Experimental results are usually plotted as a graph of current versus potential. The method is used to measure midpoint redox potentials of compounds, including proteins, and to follow intramolecular electron transfer. A three-electrode system is used, comprising a working electrode (such as platinum or glassy carbon), a counter-electrode, and a reference electrode.

cyclin any of a group of proteins that are involved in regulation of the eukaryotic **cell-division cycle**. Cyclins periodically rise and fall in concentration in step with the cycle. Individual cyclins activate particular **cyclin-dependent protein kinases** (Cdks) and thereby help to control progression from one stage of the cycle to the next (e.g. p34^{cdc2}/cyclin B regulates the G$_2$ to M transition in all eukaryotes). All the following examples are from *Saccharomyces cerevisiae*: (1) G$_1$/S-specific cyclin Cln1; this is essential for control of the cell cycle at the G$_1$/S (START) transition; it interacts with the Cdc28 protein

kinase to form S-phase promoting factor (**SPF**). (2) G$_1$/S-specific cyclin Cln2; this has an overlapping function with Cln1. Cln1 and Cln2 mRNAs fluctuate in the cell cycle. (3) G$_2$/mitotic-specific cyclin C1B1; this interacts with the Cdc2 protein kinase to form MPF. G$_2$/M cyclins accumulate steadily during G$_2$ and are abruptly destroyed at mitosis. G$_2$/mitotic-specific cyclins C1B2–4 are similar. (4) S-phase entry cyclins C1B5 and 6; these are maximally expressed just before the S-phase of the cell cycle. Animal cells have ≈10 cyclins, which dimerize with ≈8 Cdks. Cyclins C, D, and E function in G$_1$ phase of the cell cycle; cyclins A and B function in S and G$_2$/M phases; and cyclin H–Cdk7 is Cdk-activated kinase (CAK), also a component of the transcription factor TFIIH (*see* **TFII**). The cyclins contain a homologous nine-residue sequence near the N-terminus. This motif is called the **cyclin box**, or **destruction box**. It is recognized by a destruction box-recognizing protein, which targets cyclins to ubiquitination and proteosomal degradation. See also **cdc gene, cyclin regulatory protein**.

cyclin-dependent protein kinase Cdk; a protein kinase that is only active when complexed with a regulatory protein, a **cyclin**. Different Cdk–cyclin complexes are thought to trigger different steps in the cell-division cycle by phosphorylating specific target proteins. *See also* **cyclin regulatory protein**.

cyclin regulatory protein any protein involved in the regulation of **cyclin** gene expression. An example from *Saccharomyces cerevisiae* is Cdc68. This plays a role in general transcription, with both positive and negative effects on gene expression; it is required for the appropriate synthesis during heat shock, and for continued expression of cyclin genes. A further example, also from yeast, is the Ser/Thr protein kinase Cdc28 (*see* **cdc gene**).

cyclitol any cycloalkane containing one hydroxyl group on each of three or more ring atoms. Notable members are the **inositols** and their derivatives.

cyclo+ *or (before a vowel)* **cycl+** *comb. form* indicating a circle or ring, especially one in a cyclic organic chemical compound.

cycloartenol the first triterpene intermediate in the biosynthesis of sterols in photosynthetic tissues. It is formed from (*S*)-squalene-2,3-epoxide by a cyclization reaction catalysed by cycloartenol synthase, EC 5.4.99.8. The corresponding intermediate in animal tissues is **lanosterol**.

cyclobutadipyrimidine a type of **pyrimidine dimer** formed within a DNA strand between neighbouring pyrimidine residues.

cyclobuxine a steroidal alkaloid extracted from *Buxus microphylla*. The extracts are used as folk remedies for malaria and venereal diseases. It protects the heart from ischemia and reperfusion damage by inhibiting the release of ATP metabolites during reperfusion.

cyclobuxine D

cyclodeaminase *an informal name for* **formiminotetrahydrofolate** cyclodeaminase.

cyclodepsipeptide *see* **depsipeptide**.

cyclodextrin any of a group of cyclic nonreducing **dextrins** formed from starch or glycogen by the extracellular enzyme cyclomaltodextrin glucanotransferase (EC 2.4.1.19), produced by *Bacillus macerans*. They consist of cyclic polymers containing six, seven, or eight α-(1,4)-linked D-glucose residues. These were formerly known, respectively, as Schardinger α-, β-, or γ-dextrins, or α-, β-, or γ-cyclodextrins, but now are more correctly named cyclomaltohexaose, cyclomaltoheptaose, or cyclomaltooctaose. Cyclodextrins have a toroidal molecular structure enclosing a hydrophobic cavity, thereby permitting the formation of inclusion complexes with a wide variety of molecules.

cyclofenil bis-(*p*-acetoxyphenyl)cyclohexylidenemethane; an agent that prevents feedback inhibition of ovulation by inhibiting the binding of estrogen in the brain. The resulting increased secretion of gonadotropin-releasing hormone, gonadotropins, and estrogen promotes ovulation.

cycloheximide *or* **actidione** 3-[2(3,5-dimethyl-2-oxocyclohexyl)-2-hydroxyethyl]glutarimide; an antibiotic produced by some *Streptomyces* spp. It interferes with protein synthesis in eukaryotes by inhibiting peptidyltransferase activity of the 60S ribosomal subunit.

cyclohydrolase *an informal name for* the enzyme **methenyltetrahydrofolate cyclohydrolase**, EC 3.5.4.9.

cyclo-ligase any **ligase** enzyme of the sub-subclass EC 6.3.3 that catalyses the formation of a heterocyclic ring containing a C–N bond; e.g. dethiobiotin synthase, EC 6.3.3.3.

cyclomaltoheptaose *see* **cyclodextrin**.

cyclomaltohexaose *see* **cyclodextrin**.

cyclomaltooctaose *see* **cyclodextrin**.

cyclooxygenase *an alternative (informal) name for* **prostaglandin–endoperoxide synthase**.

cyclooxygenation the process that introduces one molecule of dioxygen at C-9 and a second dioxygen at C-15 of a molecule of arachidonic acid, accompanied by formation of a bond between C-8 and C-12 and an endoperoxide bridge across C-9 and C-11.

cyclophilin a protein of the immunophilin class that binds the immunosuppressant **cyclosporin A**. Cyclophilin possesses **peptidylprolyl isomerase** activity. Under physiological conditions it may have a role in the entry of proteins into, and protein folding within, the mitochondrial matrix. Cyclophilin binds to all six peroxiredoxins found in mammalian cells.

cyclophorase originally, a supposed integrated complex of enzymes that carries out the tricarboxylic-acid cycle, fatty-acid oxidation, oxidative phosphorylation, and terminal electron transport. This system was identified later with **mitochondria**. The cyclophorase complex and the mitochondria are thus the functional and structural sides of the same unit.

cyclophosphamide *N,N*-bis(2-chloroethyl)tetrahydro-2*H*-1,3,2-oxazaphosphorin-2-amine 2-oxide; an alkylating agent that is metabolized by the liver cytochrome P450 system to 4-hydroxycyclophosphamide, which is in equilibrium with aldophosphamide. These are transported to tissues, which convert them to the cytotoxic agent, phosphoramine mustard, and acrolein. Cyclophosphamide is useful as an immunosuppressant. *See also* **chlorambucil**.

cyclopropylamine *see* **monoamine oxidase inhibitors**.

D-cycloserine D-4-amino-3-isoxazolidinone; a tuberculostatic antibiotic found in cultures of *Streptomyces orchidaceus*. It is a potent inhibitor of alanine racemase (EC 5.1.1.1) and D-alanine–D-alanine ligase (EC 6.3.2.4; alanylalanine synthetase), two enzymes concerned in the biosynthesis of bacterial cell wall peptidoglycan.

cyclosporin A *or* **ciclosporin A** a cyclic undecapeptide isolated from the fungi *Cylindrocarpon lucidum* and *Tolypocladium inflatum* (formerly *Trichoderma polysporum*). It is a potent immunosuppressive drug that prevents the rejection of allografts, and is used extensively in human transplant operations. Its main target appears to be the T lymphocyte: it contains D-alanine and several nonprotein amino acids, and blocks a calcium-dependent signal from the **T-cell receptor** that normally leads to T-cell activation. When bound to **cyclophilin**, cyclosporin A binds and inactivates the key signalling intermediate **calcineurin**.

cyclosporin synthetase a multimodular enzyme protein (1.7 MDa, 15 281 amino acids) that catalyses the synthesis of cyclosporin. Each module is 1000–1200 residues long. This is an example of nonribosomal peptide synthesis in which the composition and sequence of the peptide product is determined by the sequence and specificity of the catalytic modules in the protein.

cyclotide any of a large group of circular polypeptides of plants. They contain ≈30 residues, of which 6 are conserved cysteines, and are derived from larger precursor proteins. They have diverse actions. Examples (from Rubiaceae family) include: circulins A and B, which are active against HIV; cyclopsychotride A, which inhibits neurotensin; and kalata B, which is uterotonic.

cyclotron a machine used to accelerate charged particles of atomic magnitudes to energies of the order of 25 MeV. The particles move in spiral paths under the influence of a uniform vertical magnetic field and are accelerated by a constant frequency electric field.

Cyd symbol for a residue of the ribonucleoside **cytidine** (alternative to C).

Cyd*P* symbol for cytidine phosphate (*see* **cytidine phosphate**).

Cyd2′*P* symbol for cytidine 2′-phosphate (*see* **cytidine phosphate**).

Cyd-2′,3′-*P* symbol for cytidine 2′,3′-phosphate (*see* **cytidine phosphate**).

Cyd3′*P* symbol for cytidine 3′-phosphate (*see* **cytidine phosphate**).

Cyd-3′,5′-*P* symbol for cytidine 3′,5′-phosphate (*see* **cytidine phosphate**).

Cyd5′*P* symbol for cytidine 5′-phosphate (*see* **cytidine phosphate**).

Cyd5′*PP* symbol for cytidine 5′-diphosphate (*see* **cytidine phosphate**).

Cyd5′*PPP* symbol for cytidine 5′-triphosphate (*see* **cytidine phosphate**).

Cy3 dye *a proprietary name for* an orange fluorescent cyanine dye having λ_{max} excitation 550 nm and λ_{max} emission 570 nm. It is used to label dideoxyribonucleotides in automated DNA sequencing and as a label for proteins.

Cy5 dye *a proprietary name for* a sulfoindocyanine dye having λ_{max} excitation 649 nm and λ_{max} emission 670 nm, and uses similar to those of **Cy3 dye**.

c-yes *see* **YES**.

D-cymarose 2,6-dideoxy-3-O-methyl-*ribo*-hexose; the D enantiomer is a constituent of some **cardiac glycosides**.

CYP19 *an alternative name for* **aromatase**.

cyproterone 6-chloro-6-dehydro-17α-hydroxy-1,2α-methylene-progesterone; a progesterone derivative with anti-androgen activity through partial agonist activity at androgen receptors. It inhibits the secretion of gonadotropins and spermatogenesis in humans.

Cyr61 a growth factor binding protein expressed from G_0/G_1 to mid-G_1 of the cell cycle. It is expressed in high amounts in lung. *Compare* **connective tissue growth factor**.

Cys *symbol for* a residue of the α-amino acid L-cysteine. When written

$$\underset{\text{Cys}}{|} \quad or \quad \overset{\text{Cys}}{|},$$

it represents L-half-cystine.

cyst 1 any saclike structure, generally of a pathological nature, comprising a central cavity lined with epithelium and containing fluid or semisolid material. **2** the capsule enclosing a dormant larva or spore. —**cystic** *adj.*

cystamine β,β′-diaminodiethyl disulfide; 2,2′-dithiobis(ethylamine); 2,2′-dithiobisethanamine; $H_2N–CH_2–CH_2–S–S–CH_2– CH_2–NH_2$; the decarboxylation product of cystine and an oxidation product of **cysteamine**.

cystate 1 the **cystine** anion; the dianion, $^-OOC–CH(NH_2)–CH_2–S–S–CH_2-CH(NH_2)–COO^-$, derived from cystine. **2** any salt containing the cystine anion. **3** any ester of cystine. *Distinguish from* **cysteate** and **cysteinate**.

cystathionine *trivial name for* S-(β-amino-β-carboxyethyl)homo-cysteine; (R*)-2-amino-4-[(S*)-2-amino-2-carboxyethylthio]butanoic acid; $HOOC–CH(NH_2)–CH_2–S–CH_2–CH_2–CH(NH_2)–COOH$; a di(α-amino acid) and a mixed thioether with two chiral centres. L-Cystathionine

$$(symbol: \overset{\text{Ala}}{\underset{\text{Hcy}}{|}}),$$

S-(L-alanin-3-yl)-L-homocysteine; (S)-2-amino-4-[(R)-2-amino-2-carboxyethylthio)butanoic acid – in which both moieties are in the L-configuration – does not occur in proteins but is an intermediate in the transfer of the sulfur atom from methionine, via homocysteine, to cysteine in mammals. (*Note:* The pair of (R*,R*)-enantiomers, in which one moiety is in the D-configuration and the other in the L-configuration, is sometimes known as **allocystathionine**.)

cystathionine γ-cysthathionase *see* **cystathionine** γ -**lyase**.

cystathionine γ-lyase EC 4.4.1.1; *systematic name*: L-cystathionine cysteine-lyase (deaminating); *other names*: homoserine deaminase; homoserine dehydratase; γ-cystathionase; cystine desulfhydrase; cysteine desulfhydrase; cystathionase; a multifunctional pyridoxal-phosphate enzyme that catalyses the formation of L-cysteine with release of 2-oxobutanoate and ammonium ion from L-cystathionine and water. It also acts on homoserine and cysteine to remove NH_3, forming oxobutanoate from homoserine or pyruvate and H_2S from cysteine. The enzyme carries out the final reaction in the pathway for the synthesis of cysteine from methionine.

cystathionine β-synthase EC 4.2.1.22; *systematic name*: L-serine hydro-lyase (adding homocysteine); *other names*: serine sulfhydrase; β-thionase; methylcysteine synthase. An enzyme that catalyses the reaction between L-serine and L-homocysteine to form cystathionine and water. This is the first step in the synthesis of cysteine from serine. Pyridoxal-phosphate is a cofactor. *See* **homocystinemia**.

cystathionine γ-synthase EC 4.2.99.9; *recommended name*: O-succinylhomoserine (thiol)-lyase. An enzyme in the pathway for the synthesis of methionine in plants and bacteria. It catalyses the for-mation of cystathionine from O-succinyl-L-homoserine and L-cysteine to form succinate. Pyridoxal-phosphate is a cofactor. The enzyme can utilize hydrogen sulfide or methanethiol as a thiol substrate.

cystathioninuria the excessive excretion of L-cystathionine in the urine in humans. It is usually benign, pyridoxine-responsive, and caused by an inherited deficiency of cystathionine γ-lyase.

cystatin any of a group of proteins, present in tissues and body fluids, that inhibit cysteine proteases. The cystatins constitute a single evolutionary superfamily, and are classified into three different families. Family 1 proteins contain ~100 amino-acid residues (M_r 11 000–12 000), and lack disulfide bonds. Family 2 proteins have ~120 residues (M_r 13 000–14 000), with two intrachain disulfide bonds. Family 3 proteins (*also called* **kininogens**) contain three cystatin-like domains, each of which has two disulfide bonds at positions homologous to those in family 2. Examples: cystatin A (*also called* stefin A); type 2 cystatin C (*also called* **neuroendocrine basic polypeptide** *or* **post-γ-globulin**); in humans, this is expressed in highest levels in the epididymis, vas deferens, brain, thymus, and ovary, and is found in sera from patients with Icelandic cerebrovascular type amyloidosis, and related diseases.

cysteamine thioethanolamine; β-mercaptoethylamine; 2-amino-ethanethiol; $HS–CH_2–CH_2–NH_2$; the decarboxylation product of cysteine. It readily oxidizes in air to **cystamine**. Pantetheine, a component of coenzyme A, is the amide formed between cysteamine and pantothenic acid.

cysteate 1 *the common name for* cysteate(1–); cysteic-acid monoanion; hydrogen cysteate. (*Note:* in theory, the term denotes any ion, or mixture of ions, formed from **cysteic acid**, and having a net charge of minus one, although the species $^-O_3S–CH_2–CH(NH_3{}^+)–COO^-$ generally predominates in practice.) **2** *the systematic name for* cysteate(2–); cysteate-acid dianion. **3** any salt containing an anion of cysteic acid. **4** any ester of cysteic acid. *Compare* **cystate**, **cysteinate**.

cysteic acid *the trivial name for* 3-sulfoalanine; a-amino-β-sulfopropionic acid; 2-amino-3-sulfopropanoic acid; $HO_3S–CH_2–CH(NH_2)–COOH$; a chiral a-amino acid, and an oxidation product of cysteine or cystine. Residues of L-cysteic acid (*symbol*: Cya), (R)-2-amino-3-sulfopropanoic acid, do not occur normally in proteins, although they are found in wool from the outer part of the fleece of sheep, following exposure to sunlight. They are commonly encountered during sequencing of peptides and proteins as the end product of oxidative cleavage of the disulfide bonds of cystine residues and concomitant oxidation of cysteine residues.

cysteinate 1 cysteinate(1–); cysteine monoanion; the anion, $HS–CH_2–CH(NH_2)–COO^-$, derived from **cysteine**. **2** cysteinate(2–); cysteine dianion; the anion, $^-S–CH_2–CH(NH_2)–COO^-$, derived from cysteine. **3** any salt containing an anion of cysteine. **4** any ester of cysteine. *Compare* **cystate**, **cysteate**.

cysteine *the trivial name for* β-mercaptoalanine; α-amino-β-thiol-propionic acid, 2-amino-3-mercaptopropanoic acid; $HS–CH_2–CH(NH_2)–COOH$; a chiral α-amino acid. In neutral or alkaline solution it is readily oxidized by air to **cystine**. L-Cysteine (*symbol*: C or Cys), (R)-2-amino-3-mercaptopropanoic acid, is a coded amino acid found in peptide linkage in proteins; codon: UGC or UGU. In mammals it is a nonessential dietary amino acid and is glucogenic. D-Cysteine (*symbol*: D-Cys *or* DCys), (S)-2-amino-3-mercaptopropanoic acid, occurs naturally in firefly **luciferin**. (*Note:* application of the **sequence rule** to the enantiomers of cysteine, results in their stereochemical designation as either R or S being the converse of that possessed by most chiral α-amino acids found in proteins.)

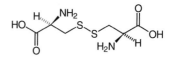

L-cysteine

L-cystine

cysteine (acid) the protonated form of a **cysteine** residue, i.e. with an undissociated thiol group.

cysteine desulfhydrase *see* **cystathionine γ-lyase**.

cysteine endopeptidase *or (formerly)* **thiol proteinase** any enzyme of the sub-subclass EC 3.4.22, which consists of proteinases characterized by having a cysteine residue at the active centre and by being irreversibly inhibited by sulfhydryl reagents such as iodoacetate. The group includes cathepsins B, H, and L. There are at least three superfamilies that are unrelated in structure. These are (1) papain-like proteases; (2) caspase-hemoglobinase proteases; and (3) adenovirus-type proteases. *See* **cathepsin B, cathepsin H, cathepsin L, clostripain, ficin, papain**. *See also* **apoptosis**.

cysteine-rich secretory protein *see* **CRISP**.

cysteine-type carboxypeptidase EC 3.4.18.1; *other names*: lysosomal carboxypeptidase B; cathepsin B₂; cathepsin IV; acid carboxypeptidase; an enzyme that catalyses the liberation of a C-terminal amino-acid residue from a protein. It is an enzyme with broad specificity, but lacks action on C-terminal proline.

cysteinyl the acyl group, HS–CH₂–CH(NH₂)–CO–, derived from **cysteine**. (*Note*: although this name is formed atypically, its use is recommended to avoid confusion with **cysteyl**.) *Compare* **cystyl, half-cystyl**.

cystein-S-yl the alkylthio group, HOOC–CH(NH₂)–CH₂–S–, derived from **cysteine**.

cysteyl 1 the acyl group, HO₃S–CH₂–CH(NH₂)–CO–, derived from **cysteic acid**. **2** *an alternative name for* **cysteinyl** (use now discouraged).

cystic fibrosis a heterogeneous disorder affecting all exocrine glands and probably other tissues. It is the most common lethal genetic disorder of Caucasians, but is also found in other races. There is an abnormality in the secretions of the exocrine glands, which have higher than normal concentrations of Na⁺, Cl⁻, nucleotides, Ca²⁺, etc.; a number of abnormal protein factors have been described in the exocrine glands of patients. It is due to mutation in the gene for **cystic fibrosis transmembrane conductance regulator**.

cystic fibrosis transmembrane conductance regulator *abbr.*: CFTR; a 168 kDa integral membrane protein encoded by the cystic fibrosis gene, and with structural similarity to membrane-associated ATP-dependent transporter protein (*see* **ABC transporters**), one of a family of proteins involved in the active transport of ions and small molecules across membranes. CFTR belongs to the subfamily. The CFTR protein appears to be involved in the functioning of a chloride ion channel. A mutation in this protein has been identified as being responsible for 70% of the cystic fibrosis chromosomes in Northern Europe.

cystine *the trivial name for* dicysteine; β,β′-dithiodialanine; *a,a′*-diamino-β,β′-dithiobis(propionic acid); (*R**,*R*'*)-3,3′-dithiobis(2-aminopropanoic acid);

$$HOOC–CH(NH_2)–CH_2–S–S–CH_2–CH(NH_2)–COOH;$$

a di(α-amino acid) with two chiral centres, formed by the oxidation of cystein. L-Cysteine,

$$[symbol: \begin{matrix} Cys \\ | \\ Cys \end{matrix}],$$

(*R*,*R*′)-3,3′-dithiobis(2-aminopropanoic acid), is found in peptide linkage in proteins, having been formed by post-translational oxidation of cysteine. In mammals it is a nonessential dietary amino acid and is glucogenic. As it is somewhat insoluble in neutral aqueous solutions, it sometimes forms urinary stones. *meso*-Cystine, (*R*,*S*′)-3,3′-dithiobis(2-aminopropanoic acid), is the internally compensated achiral diastereoisomer, in which D- and L-half-molecules are combined. *See also* note at **cysteine**.

cystine desulfhydrase *see* **cystathionine γ-lyase**.

cystine knot a structure, found in some proteins, that results from the formation of three disulfide bonds between six cysteine residues, positioned in such a way as to form a knot-like topology. It was originally found in transforming growth factor-ß and platelet-derived growth factor.

cystinosin a membrane protein of lysosomes that is encoded by the gene for cystinosis. It consists of 367 amino acids, 7 transmembrane segments, and 8 potential glycosylation sites.

cystinosis a genetic disease caused by mutation of the gene for cystinosin, a lysosomal cystine transport protein (gene locus 17p13). Cystine crystals are stored in lysosomes of many tissues including kidney, eyes, liver, muscles, pancreas, brain, and white blood cells. Without treatment the result is end-stage renal failure before the age of 10 years. —**cystinotic** *adj.*

cystinuria a disorder of cystine and diamino acid (arginine, lysine, and ornithine) transport in humans affecting the epithelial cells of the renal tubules and the gastrointestinal tract. It is characterized by the formation of cystine stones in the urinary tract and elevated levels of cystine and diamino acids in the urine. **Cystinuria type I** is a fully autosomal recessive disorder in which heterozygotes have normal urinary amino-acid excretion. It is caused by mutations in the gene (*SLC3A1*) for a four-transmembrane segment protein of 685 amino acids, called rBAT. **Cystinuria non-type I** (i.e. types II and III) is an incompletely autosomal recessive disorder in which heterozygotes have abnormal urinary excretion of cystine and basic amino acids. It is caused by mutation in the gene (*SLC7A9*) for a 12-transmembrane segment protein called bAT. Both genes are expressed in kidney and small intestine.

cystyl the diacyl group, –OC–CH(NH₂)–CH₂–S–S–CH₂–CH(NH₂)–CO–, derived from **cystine**. *Compare* **cysteinyl, half-cystyl**.

Cyt *symbol for* a residue of the pyrimidine base cytosine (alternative to C).

cytarabine β-cytosine arabinoside; 1-β-D-arabinofuranosylcytosine; *abbr.*: ara-C; a cytotoxic drug used as an antineoplastic and antiviral agent. It is effective in the treatment of acute myeloblastic leukemia. It is converted metabolically to 1-β-D-arabinofuranosylcytidine 5′-triphosphate (*abbr.*: ara-CTP), a potent inhibitor of DNA synthesis.

+cyte *comb. form* indicating a **cell** (def. 1).

cytidine *symbol*: C *or* Cyd; cytosine riboside; 1-β-D-ribofuranosyl cytosine; a widely distributed **nucleoside**.

cytidine 2′,3′-(cyclic)phosphate *see* **cytidine phosphate**.

cytidine 3′,5′-(cyclic)phosphate *or* **cytidine 3′,5′-cyclophosphate** *see* **cytidine phosphate**.

cytidine deaminase EC 3.5.4.5; *systematic name*: cytidine amino-hydrolase. An enzyme that catalyses the reaction:

$$\text{cytidine} + H_2O = \text{uridine} + NH_3.$$

It is a homodimer, with bound Zn^{2+}. Specificity is poor, and deoxycytidine is also a substrate.

cytidine 5′-diphosphate *symbol*: Cyd5′PP *or* ppC; *the recommended name for* cytidine diphosphate (*abbr.*: CDP), 5′-diphosphocytidine, 5′-cytidylyl phosphate, cytidine 5′-(trihydrogen diphosphate), a universally distributed nucleotide that occurs both in the free state and as a component of certain **nucleotide coenzymes**. It is a metabolic precursor of cytidine 5′-triphosphate, CTP. It is a product of nucleoside triphosphatase, EC 3.6.1.15, and is also formed in a few cases when CTP is involved in phosphotransferase reactions (e.g. dolichol kinase, EC 2.7.1.108); it is converted to deoxycytidine 5′-diphosphate by **ribonucleoside-diphosphate reductase** (EC 1.17.4.1).

cytidine(5′)diphosphocholine *or* **cytidine diphosphate choline** an intermediate in phospholipid synthesis, especially of phosphatidylcholines and sphingomyelins.

cytidine(5′)diphosphoethanolamine *or* **cytidine diphosphate ethanolamine** an intermediate in the synthesis of phosphatidylethanolamines.

cytidine monophosphate *abbr.*: CMP; an alternative name for

any cytidine phosphate, but in particular for cytidine 5′-phosphate, especially when its distinction from cytidine (5′-)diphosphate and cytidine (5′-)triphosphate requires emphasis.

cytidine phosphate *symbol*: CydP; cytidine monophosphate (*abbr.*: CMP); any phosphoric monoester or diester of cytidine. There are three monoesters: cytidine 2′-phosphate, cytidine 3′-phosphate, and **cytidine 5′-phosphate**, and two possible diesters: cytidine 2′,3′-phosphate and cytidine 3′,5′-phosphate, although cytidine 5′-phosphate is the ester commonly denoted (the locant being omitted if no ambiguity may arise). Cytidine 2′,3′-phosphate (*symbol*: Cyd-2′,3′-P), also named cytidine 2′,3′-(cyclic)phosphate or cyclic cytidine 2′,3′-monophosphate (*abbr.*: 2′,3′-cyclic CMP), is formed as an intermediate during the alkaline hydrolysis of ribonucleic acid. This diester then readily hydrolyses to a mixture of the two monoesters cytidine 2′-phosphate (*symbol*: Cyd2′P), *also named* cytidine 2′-monophosphate (*abbr.*: 2′CMP) *or* 2′-cytidylic acid *or* cytidylic acid a, and cytidine 3′-phosphate (*symbol*: Cyd3′P), also named cytidine 3′-monophosphate (*abbr.*: 3′CMP) *or* 3′-cytidylic acid *or* cytidylic acid b. Cytidine 3′,5′-phosphate (*symbol*: Cyd-3′,5′-P), also named cytidine 3′,5′-(cyclic)phosphate *or* cyclic cytidine 3′,5′-monophosphate (*abbr.*: 3′,5′-cyclic CMP), is a product of cytidylate cyclase, EC 4.6.1.6.

cytidine 2′-phosphate *see* **cytidine phosphate**.

cytidine 2′3′-phosphate *see* **cytidine phosphate**.

cytidine 3′-phosphate *see* **cytidine phosphate**.

cytidine 5′-phosphate *or* **5′-cytidylic acid** *or* **5′-phosphocytidine** *or* **5′-O-phosphonocytidine** *symbol*: Cyd5′P; *alternative recommended names for* cytidine 5′-monophosphate (*abbr.*: 5′CMP), cytidine 5′-(dihydrogen phosphate), cytidine (mono)-ribonucleotide. (The locant is commonly omitted if there is no ambiguity as to the position of phosphorylation.) It is formed by pyrophosphatase-catalysed cleavage of (inorganic) diphosphate from CTP, and also as a product of a variety of reactions, e.g. that in which phosphatidylcholine is formed from CDPcholine and diacylglycerol (diacylglycerol cholinephosphotransferase, EC 2.7.8.2).

cytidine 5′-triphosphate *symbol*: Cyd5′PPP *or* pppC; *recommended name for* cytidine triphosphate (*abbr.*: CTP), 5′-triphosphocytidine, 5′-cytidylyl diphosphate, cytidine 5′-(tetrahydrogen triphosphate), an important coenzyme. It is formed by **CTP synthase**. It participates as coenzyme in reactions in which a cytidylyl group is transferred, e.g. in lipid metabolism, to choline phosphate (EC 2.7.7.15, to form CDPcholine) and similar reactions, or in carbohydrate metabolism to form CDPglucose (EC 2.7.7.33), or in the phosphotransferase dolichol kinase (EC 2.7.1.108).

cytidyl any chemical group formed by the loss of a 2′-, a 3′-, or a 5′-hydroxyl group from the ribose moiety of cytidine. *Compare* **cytidylyl**.

cytidylate **1** either the mono- or the di-(phosphoric) anion of **cytidylic acid**. **2** any mixture of cytidylic acid and its anions. **3** any salt or ester of cytidylic acid.

cytidylic acid *the trivial name for* any phosphoric monoester of cytidine. The position of the phosphoric residue on the ribose moiety of a given ester may be specified by a prefixed locant. *See* **cytidine phosphate**. However, 5′-cytidylic acid is the ester commonly denoted, its locant usually being omitted if no ambiguity may arise. 5′-

cytidylic acid is also an alternative recommended name for **cytidine 5'-phosphate**.

cytidylyl the cytidine[mono]phospho group, the acyl group derived from **cytidylic acid**. *Compare* **cytidyl**.

cytidylylation the process of introducing a cytidylyl group into a substance.

cyto+ *comb. form* indicating a **cell** (def. 1).

cytochalasin any of a group of structurally related fungal metabolites (designated A, B, C, D, E, F, etc.) that share a number of unusual and characteristic effects on cultured mammalian cells. The most notable are inhibition of cytoplasmic cleavage following nuclear division, induction of nuclear extrusion, and inhibition of cell motility. They bind to the growing plus ends of actin filaments, preventing further addition of actin molecules, and thus affect functions that involve assembly and disassembly of actin filaments. The basic structure is a benzyl-substituted hydroisoindolone nucleus to which is fused a small macrolide-like ring. [From Greek *cytos*, cell; *chalasis*, relaxation.]

cytochemical bioassay a technique for the assay of various polypeptide hormones, based on function rather than structure. It involves measuring changes in the biochemical activity of a specific target cell, appropriate to the physiological effect of the hormone in question, by quantitative cytochemical methods. It is claimed to be more sensitive and more specific than equivalent radioimmunoassays.

cytochemistry the chemistry of living cells, especially the cellular localization of biochemical substances and processes; microscopical biochemistry. —**cytochemical** *adj.*

cytochrome any member of a class of hemoproteins whose characteristic mode of action involves the transfer of electrons in association with a reversible change in oxidation state (between Fe(II) and Fe(III), and sometimes Fe(IV)) of the heme iron. The name embraces all intracellular hemoproteins except hemoglobin, myoglobin, peroxidases, catalase, tryptophan 2,3-dioxygenase, nitrite and sulfite reductases, and heme-thiolate proteins. There are four major established subclasses; well-established members of each subclass are generally distinguished by consecutive numerical subscripts; remaining members are designated by a suffixed number based upon the wavelength (in nanometres) of the α-band (e.g. cytochrome c_1, cytochrome c_{555}); *see* **cytochrome absorption bands**. (1) In **cytochromes *a*** the heme prosthetic group contains a formyl side-chain, i.e. heme *a*. (2) **Cytochromes *b*** contain protoheme or a related heme (without a formyl group) as prosthetic group, not covalently bound to the protein. The group includes **cytochrome *o*, cytochrome P-450**, and **helicorubin**, as well as **ubiquinol-cytochrome *c* reductase, cytochrome *b*, cytochrome b_2, cytochrome b_5, cytochrome b_{245}, cytochrome b_{599}, cytochrome b_6**, and **cytochrome b_6–*f* complex**. (3) In **cytochromes *c*** there are covalent linkages, not necessarily thioether linkages, between the heme side-chains and the protein (*see also* **cytochrome *c*, cytochrome c_1**). (4) **Cytochromes *d*** contain a tetrapyrrollic chelate of iron as prosthetic group, in which the degree of conjugation of double bonds is less than in porphyrin. *See also* **cytochrome *f***.

cytochrome absorption bands three principal absorption bands in the absorption spectrum of many ferrohemoproteins, being the α-band, β-band, and γ-band (*or* Soret band), resulting from the absorption of radiation by the heme groups. They are useful in the characterization and study of these proteins. Characteristic absorption maxima for cytochromes *a*, *b*, *c*, and c_1 are for the α-band: 600, 563, 550, and 554 nm respectively; for the β-band: (none for *a*) 532 (*b*), 531 (*c*), and 524 (c_1) nm; and for the γ-band 439, 429, 415, and 418 nm.

cytochrome *b* a component of the eukaryotic mitochondrial **respiratory chain** complex III. It is an integral membrane protein that contains eight transmembrane segments and two heme groups, b_{562} and b_{566}. It is encoded in mitochondrial DNA. Several mutations in it result in mitochondrial myopathy or cardiomyopathy.

cytochrome b_2 EC 1.1.2.3; an enzyme (*recommended name*: L-lactate dehydrogenase (cytochrome)) that catalyses the oxidation of L-lactate by 2 ferricytochrome *c* to form pyruvate and 2 ferrocytochrome *c*. Flavin mononucleotide (FMN) and protoheme IX are cofactors. Cytochrome b_2, is a mitochondrial intermembrane-space protein induced by lactate during respiratory adaptation (in yeast).

cytochrome b_5 an electron carrier for several oxygenases, present in animal microsomes and in erythrocytes. It is reduced by **cytochrome b_5 reductase**. Two forms are synthesized by alternative splicing.

cytochrome b_5 reductase EC 1.6.2.2; a flavoprotein (FAD) enzyme that catalyses the reaction:

$$\text{NADH} + 2 \text{ ferricytochrome } b_5 = \text{NAD}^+ + 2 \text{ ferrocytochrome } b_5.$$

It is cytosolic in erythrocytes and a myristoylated peripheral membrane protein of 275 amino acids in hepatocytes. Point mutations in the gene cause hereditary enzymopenic methemoglobinemia.

cytochrome b_6 *or* **cytochrome b_{563}** a *b*-type cytochrome that contains two heme moieties and forms part of cytochrome $b_6 f$ complex functioning between photosystems II and I of photosynthesis.

cytochrome $b_6 f$ complex a complex, present in thylakoid membranes of chloroplasts, that consists of cytochrome b_6, cytochrome *f*, and the **Rieske protein**. It transmits electrons between photosystems II and I of photosynthesis.

cytochrome b_{245} *or* **cytochrome b_{558}** an integral membrane component of the phagocyte NADPH oxidase complex. The protein is a dimer (human) of α (the glycoprotein gp91-phox, which contains two heme and one FAD moieties) and β (the protein p22-phox) subunits. Deficiency or structural variation in the α subunit causes X-linked chronic granulomatous disease, while similar defects in the β subunit cause an autosomal recessive form of the same disease.

cytochrome b_{599} a cytochrome of photosystem II of photosynthesis. It exists as a dimer of α and β chains. *See* **photosystem**.

cytochrome bd a terminal quinol oxidase in bacteria such as *E. coli*, containing heme and copper.

cytochrome bo a terminal quinol oxidase in bacteria such as *E. coli*, containing heme and copper.

cytochrome *c* a soluble protein of the intermembrane space of mitochondria that associates with the inner mitochondrial membrane and functions to transfer electrons between **respiratory chain** complexes III and IV. The sequence from many (>100) species has been determined. About 26 residues are completely invariant, especially those with side chains packing against the heme. The apoprotein is nucleus-encoded, synthesized on free ribosomes, and transported into the mitochondrion, for which the action of heme lyase (holocytochrome-*c* synthase) is essential. It is released into the cytosol after induction of apoptosis, binds to Apaf1, and activates procaspase 9.

cytochrome c_1 an inner mitochondrial membrane protein component of **respiratory chain** complex III. Comprising 241 amino acids (bovine heart), its C-terminal globular heme-containing domain extends into the intermembrane space, while its C-terminal region forms a transmembrane segment that anchors it into the inner mitochondrial membrane. It transfers electrons directly to cytochrome *c*.

cytochrome *c* oxidase *or* **cytochrome oxidase** *or* **cytochrome aa_3** EC 1.9.3.1; an oxidoreductase, containing heme and copper, that catalyses the oxidation of cytochrome *c* by dioxygen with the production of water. The reaction is:

$$4 \text{ ferrocytochrome } c + O_2 = 4 \text{ ferricytochrome } c + 2 \text{ H}_2\text{O}.$$

It is an integral mitochondrial membrane protein comprising a complex of 13 subunits; subunits I, II, and III are encoded in the mitochondrion and the remaining 10 are nucleus encoded. Electrons originating in cytochrome *c* are transferred via heme *a* and Cu_A to the binuclear centre formed by heme a_3 and Cu_B. Subunit I binds the two hemes and Cu_B. Subunit II binds Cu_A and cytochrome *c*.

cytochrome *c* reductase see **NADH dehydrogenase**.

cytochrome *f* *or* **cytochrome b_{552}** a cytochrome, designated after the Latin *frons* (leaf), that occurs in high concentration in leaves (up to 25% of the concentration of chlorophyll). It is a typical *c*-type cytochrome, and a component of the **cytochrome $b_6 f$ complex** of thylakoid membrane, transferring electrons from photosystem II to I. *See also* **cytochrome b_6**.

cytochrome *h* a *b*-type cytochrome found in the hepatopancreas of the snail *Helix pomatia*.

cytochrome *m* *an alternative name for* **cytochrome P450**.

cytochrome *o* any *b*-type prokaryotic cytochrome containing protoheme as the prosthetic group that, unlike a typical cytochrome *b*, serves as a terminal electron acceptor (cytochrome oxidase) and is autoxidizable by dioxygen. Use of the term is discouraged.

cytochrome P450 *or (sometimes)* **cytochrome *m*** *abbr.:* (of protein) CYP, (of gene) *CYP*; any *b*-type cytochrome that has a cysteine sulfur atom ligated to the heme iron and that, when reduced, forms carbon monoxide complexes – the latter are pigments with a major absorption peak (Soret band) near 450 nm. These cytochromes are mixed-function oxidase enzymes, including unspecific monooxygenase (EC 1.14.14.1), camphor 5-monooxygenase (EC 1.15.15.1), alkane 1-monooxygenase (EC 1.14.15.3), steroid 11β-monooxygenase (EC 1.14.15.4), cholesterol monooxygenase (side-chain-cleaving) (EC 1.14.15.6), prostacyclin synthase (EC 5.3.99.4), and thromboxane synthase (EC 5.3.99.5). They catalyse the reaction:

$$RH + NADPH + H^+ + O_2 = ROH + NADP + H_2O.$$

This reaction requires a specific cytochrome P450 and cytochrome P450 reductase (which contains FAD and FMN). They occur in most animal cells, mainly in smooth endoplasmic reticulum, and in plants and microorganisms, and their chief function is the monooxygenation of lipophilic substances (RH can be steroid, fatty acid, or xenobiotics), as in the case of cytochrome P450$_{SCC}$, cholesterol monooxygenase (side-chain-cleaving). In mammalian liver many cytochromes P450 are inducible by drug and other treatment, e.g. the major phenobarbital-inducible cytochrome in rabbit liver microsomes, known as P450$_{LM2}$.

Over 1000 genes for cytochrome P450 (450–530 amino acids) are known in animals, bacteria, fungi, and plants. They are classed into 27 families, of which 10 are found in all mammals. Each family consists of a group of proteins that exhibit 40% or greater sequence identity, and the family is described by an Arabic number. Within these families, subgroups of proteins having greater than 55% identity are classified into subfamilies, designated by a capital letter, and individual proteins within subfamilies are arbitrarily assigned numbers. Under this nomenclature, P450$_{LM2}$ is known as CYP2B4, or P4502B4, and the gene as *CYP2B4*. Mutations in CYP21A2 cause congenital adrenal hyperplasia, and in CYP1B1 cause congenital glaucoma.

cytochrome P450 reductase a flavoprotein enzyme (76.5 kDa) that contains one molecule each of FAD and FMN and is homologous to **nitric-oxide synthase**. Its N-terminal region is hydrophobic and keeps the enzyme attached to the endoplasmic reticulum, while the rest of the protein projects into the cytosol. It catalyses transfer of electrons from NADPH, via FAD and FMN, to the heme of **cytochrome P450**, and then on to O$_2$ to form water during hydroxylation reactions involving lipophilic substrates.

cytocuprein *an alternative name for* Cu/Zn **superoxide dismutase**.

cytoflav(e) *or* **cytoflavin** *a historical name for* the yellow pigment(s) discovered in heart muscle and other tissues. In some cases these pigments were subsequently shown to be riboflavin or a flavin-containing enzyme, especially **yellow enzyme**.

cytofluorometry a technique for the detection, localization, and estimation of substances in or attached to cells, using naturally fluorescent substances or those to which a fluorescent label has been attached. The technique has many applications, including specific labelling of cells using fluorescent antibodies, using fluorescent **probes** to report intracellular events or membrane dynamics, or following the fate of a protein with a fluorescent label after injection into a cell.

cytogenetics the branch of genetics concerned with correlating inherited traits with cytological features, especially the appearance, structure, and behaviour of chromosomes. —**cytogenetic** *adj.*

Cytohelicase *the proprietary name for* a commercial enzyme preparation consisting predominantly of a 1,3-β-D-glucanase. It is useful for the isolation of protoplasts because it lyses cell walls.

cytohemolysin any of a family of pore-forming toxins of molar mass of 50–82 kDa, produced, e.g., by *Vibrio cholerae*. Cytohemolysins are classified in the TC system under number 1.C.14.

cytokeratin *see* **keratin**.

cytokine any protein that is released by mammalian cells and has

autocrine or paracrine activity. Cytokines elicit a variety of responses depending on the cytokine and target cell. Their actions include control of cell proliferation and differentiation, regulation of immune responses, hemopoiesis, and inflammatory responses. Cytokines produced by lymphocytes are known also as **lymphokines**; those produced by monocytes are also called **monokines**. Other types of cytokine include **chemokines**, growth factors, **colony stimulating factors**, **transforming growth factors**, **interferons**, and **tumour necrosis factors**. *See also* **angiogenin**, **epidermal growth factor**, **erythropoietin**, **fibroblast growth factor**, **heparin-binding EGF-like growth factor**, **hepatocyte growth factor**, **insulin-like growth factor**, **interleukin**, **leukemia inhibitory factor**, **macrophage inflammatory protein**, **nerve growth factor**, **oncostatin M**, **platelet-derived endothelial growth factor**, **platelet-derived growth factor**, **RANTES**, **stem cell factor**, **vascular endothelial growth factor**.

cytokine-inducible SH2 protein *see* **CIS**.

cytokine receptor any of a group of membrane receptors for a variety of interleukins and some colony stimulating factors. They consist of specific α or β (or both) type I transmembrane proteins that signal through JAK1 and STATs, and a common γ type I transmembrane protein that signals through JAK3 and STAT5.

cytokinesis 1 the changes that occur in the cytoplasm of a cell during nuclear division. **2** the division of the cytoplasm of a parent cell into daughter cells after nuclear division. *Compare* **karyokinesis**.

cytokinin *or* **phytokinin** any of a group of plant hormones that have a stimulatory effect on the division of plant cells and a retarding effect on leaf senescence. They are purine derivatives. The receptor is a histidine kinase that targets ARR1 (*Arabidopsis* response regulator 1), a transcription factor that binds DNA.

cytolemma *an alternative name for* **cell membrane**.

cytology the branch of science dealing with the origin, structure, function, biochemistry, and life history of cells. It includes the (microscopic) examination of cells, particularly in the detection and diagnosis of disease.

cytolysin any protein, especially an antibody, that can induce **cytolysis**.

cytolysis the breakdown of cells, especially by destruction of their outer membranes; the lysis of cells. —**cytolytic** *adj.*

cytomegalovirus (CMV) a common virus of the Herpesviridae family that rarely causes illness unless the immune system is compromised. The double-stranded DNA genome is packed into an icosahedral capsid that is surrounded by an amorphous layer called the 'tegument', which includes 15 distinct proteins. The tegument is enclosed in a cell-derived lipid membrane envelope containing virus encoded **glycoproteins** that are important for the attachment of the virus particle to the host cell. CMV causes the retrotranslocation of **Class I MHC** proteins from the **endoplasmic reticulum** to the **cytosol** where they are degraded by **proteasomes**.

cytometer 1 a **cell counter**. **2** a device for measuring cells.

cytometry the counting and/or measuring of cells in a fluid suspension.

cyton *or* **cytone** the body of a cell, especially a nerve cell, as distinct from its processes.

cytopathic damaging to cells.

cytopathic effect the morphological changes that cells may show in response to toxic agents or viruses.

cytopempsis the intracellular phase of **transcytosis**, whereby droplets, molecules, or particles are transported across a polarized cell from one surface to another.

cytophilic having an affinity for cells.

cytophilic antibody any antibody that binds to the surface of cells by means other than by combination of the antibody-combining site with antigenic determinants on the cell surface. Antibody thus fixed to a cell is still able to bind antigens in the vicinity; i.e. the cells become able specifically to absorb antigen.

cytophotometry a technique that employs a combination of microscopy and spectrophotometry for the detection, localization, and determination of substances in cells. It may be done on stained or unstained preparations.

cytoplasm all the **protoplasm** of a living cell excluding the nucleus and the **plasma membrane** but including the other intracellular organelles and structures. Some authors currently also exclude mitochondria, chloroplasts, or other structures containing independently replicating DNA. The term is not synonymous with **cytoplasmic fraction**. *See also* **cytosol**. —**cytoplasmic** *adj.*

cytoplasmic droplet a large (1–8 μm) piece of membrane-bound cytoplasm, containing the usual cytoplasmic organelles (but no nucleus) and one or more large lipid droplets, found in skimmed freshly secreted goat's milk. The droplets are capable of triglyceride biosynthesis.

cytoplasmic fraction the material obtained after the removal of unbroken cells, cell debris, and nuclei from a **homogenate** of eukaryotic cells or tissue. It may be further resolved into a **mitochondrial fraction**, a **microsomal fraction**, and a soluble fraction. The term is not synonymous with **cytoplasm**. *See also* **cytosol**.

cytoplasmic gene any extranuclear gene located in certain eukaryotic cellular organelles, e.g. mitochondria and chloroplasts.

cytoplasmic inheritance the inheritance of characters or traits whose determinants are not located on a chromosome. The determinants may be, e.g., viral, mitochondrial, or on plasmids.

cytoplasmic matrix *an alternative name for* **groundplasm**.

cytoplasmic membrane 1 any of the membranous structures within the **cytoplasm** of eukaryotes, such as the rough and smooth **endoplasmic reticulum** and the **Golgi apparatus**. *See also* **microsomal fraction, microsome. 2** *an alternative term for* **cell membrane** (especially in relation to bacteria).

cytoplasmic polyadenylation element *abbr.*: CPE; a uridine-rich sequence in the 3′ untranslated region of eukaryotic mRNA that is required for cytoplasmic polyadenylation. It binds **CPE-binding protein** (CPEB), which recruits other proteins (e.g. maskin, XPum).

cytoplasmic polyadenylation element-binding protein *see* **cytoplasmic polyadenylation element**.

cytoplast 1 an anuclear eukaryotic cell resulting, e.g., from treatment with cytochalasin B (*see* **cytochalasin**). **2** the apparent single structural and functional unit of the eukaryotic cell, consisting of the **microtrabecular lattice** linking the subcellular organelles and structural fibres.

cytosegresome *an alternative name for* **autophagic vacuole**.

cytosine *symbol*: Cyt; 4-amino-2-hydroxypyrimidine; 4-amino-2(1*H*)-pyrimidinone; a pyrimidine derivative that is one of the five main bases found in nucleic acids; it occurs widely in **cytidine** derivatives.

β-cytosine arabinoside *see* **cytarabine**.

+cytosis *suffix* indicating an increase in the number of cells (in an indicated tissue or organ).

cytoskeleton any of the various filamentous elements within the cytoplasm of eukaryotic cells that remain after treatment of the cells with mild detergent to remove membrane constituents and soluble components of the cytoplasm. The term embraces **intermediate filaments, microfilaments, microtubules,** and the **microtrabecular lattice**. The various elements of the cytoskeleton not only serve in the maintenance of cellular shape but also have roles in other cellular functions, including cellular movement, cell division, endocytosis, and movement of organelles. (*Note:* The term was coined to describe hypothetical structural cytoplasmic mosaics envisaged as providing physical support for, and thus spatial coordination of, enzymes effecting sequences of reactions, and as imposing restriction of mobility of the cell. Subsequently the cytoskeleton was regarded also as a possible subcellular site of action of various hormones.) —**cytoskeletal** *adj.*

cytosol the part of the **cytoplasm** that does not appear to contain membranous or particulate subcellular components. The term is not synonymous with **soluble fraction**. —**cytosolic** *adj.*

cytosol aminopeptidase *another name for* **leucyl aminopeptidase**.

cytosome any cytoplasmic structure bounded by a single **unit membrane** and of dubious identity.

cytostatic any agent that suppresses cell growth and multiplication.

cytotactin *another name for* **tenascin**.

cytotaxigen any substance that acts indirectly to induce **cytotaxis** by inducing **cytotaxin** formation. For instance, the fixing of complement by antigen–antibody complexes causes the liberation of cytotaxins, and hence cytotaxis. —**cytotaxigenic** *adj.*

cytotaxin *or (sometimes)* **chemotaxin** any chemotactic substance acting directly to induce **cytotaxis**. *See also* **cytotaxigen**.

cytotaxis 1 positive or negative **chemotaxis** of motile cells in response to diffusible substances (cytotaxins) emitted by other cells. **2** the rearrangement of cells leading to the ordering and arranging of new structures. —**cytotactic** *adj.*

cytotoxic causing cell death.

cytotoxicity the attribute of being cytotoxic.

cytotoxicity test any test of cell viability after exposure of cells to (potentially) toxic agents, e.g. antibody and complement. Loss of viability may be demonstrated by a **dye exclusion test**.

cytotoxic T-cell any of a subset of T lymphocytes, bearing the CD8 (human) or Ly2 (mouse) surface marker, that have the ability to kill infected cells. This requires cooperation from T-helper cells. Each cytotoxic T-cell specifically recognizes one antigen, which must be associated with a syngeneic class I MHC molecule.

cytotrophoblast the inner cellular layer of the **trophoblast** of a mammalian conceptus, from which chorionic villi arise. *Compare* **syncytiotrophoblast**.

cytotropism the mutual attraction of two or more cells. —**cytotropic** *adj.*

cytovillin *another name for* **ezrin**.

Dd

dyskerin d *symbol for* **1** deci+ (SI prefix denoting 10^{-1} times). **2** a deuteron. **3** day.

d+ 1 *abbr. for* deoxy+ (def. 1); used as a prefix especially to form the symbol dRib for deoxyribose. **2** deoxy+ (def. 2); used as a prefix to form symbols for 2′-deoxyribonucleosides (e.g. dAdo *or* dA for deoxyadenosine) and abbreviations for 2′-deoxyribonucleotides (e.g. dGMP for deoxyguanosine monophosphate).

d *symbol for* **1** diameter, distance, or thickness. **2** relative density.

d- *abbr. for* dextro- used *(formerly) as symbol* denoting dextrorotatory; (+)- should now be used. *See* **optical isomerism.**

D *symbol for* **1** a residue of the α-amino acid L-aspartic acid. **2** a residue of an incompletely specified base in a nucleic-acid sequence that may be adenine, guanine, or either thymine (in DNA) or uracil (in RNA). **3** a residue of the ribonucleoside (5,6-)dihydrouridine. **4** debye. **5** deuterium (use deprecated).

D600 gallopamil, 5-methoxyverapamil, α-(3- [2-(3,4-dimethoxyphenyl)ethyl] methylamino propyl)-3,4,5-trimethoxy-α-(1-methylethyl) benzeneacetonitrile; a drug used in the laboratory to block transport of calcium ions in biological membranes; it is more potent than **verapamil.**

2,4-D *abbr. for* 2,4-dichlorophenoxyacetic acid, a synthetic **auxin** used as a herbicide and to increase latex output from old rubber trees. When sprayed onto foliage at appropriate concentrations it kills broad-leaved plants but not grasses.

D *symbol for* **1** diffusion coefficient. **2** (decadic) attenuance. **3** absorbed dose. **4** dilution rate.

D_i *symbol for* **1** electric polarization. **2** internal transmission density (*formerly called* optical density).

D- *prefix* denoting one of two possible **absolute configurations** around a chiral element in certain classes of compounds. *See* D/L **convention.**

2D two-dimensional, e.g. 2D gel electrophoresis, 2D NMR spectroscopy.

da *symbol for* deca+ (the SI prefix denoting 10 times).

dA a residue of the deoxynucleoside deoxyadenosine (alternative to dAdo).

Da *symbol for* dalton.

DA *abbr. (sometimes) for* dopamine.

Dab *see* **A₂bu.**

DAB *abbr. for* diaminobutyric acid.

DABA *abbr. for* diaminobenzoic acid.

DABITC *abbr. for* dimethylaminobenzene isothiocyanate, 4-N,N-dimethylaminoazobenzene 4′-isothiocyanate; a sensitive and versatile chromophoric reagent used in sequencing polypeptides by the **Edman degradation.**

dabsyl *abbr. for* dimethylaminobenzenesulfonyl, 4-dimethylaminobenzene-4′-sulfonyl; a group used as a chromophoric label for free amino groups in amino acids, peptides, and proteins. It is intro-

duced by reaction with dabsyl chloride, 4-dimethylaminobenzene-4′-sulfonyl chloride.

dAdo *symbol for* a residue of the deoxynucleoside deoxyadenosine; 2′-deoxyribosyladenine; adenine 2′-deoxyriboside (alternative to dA).

dAdo5′P *symbol. for* deoxyadenosine 5′-phosphate.

daf-2 a *Caenorhabditis elegans* gene encoding a receptor like that for insulin. Partial loss-of-function mutations result in a long-lived phenotype with resistance to killing by several bacterial pathogens. *See* **age-1.**

DAF *abbr. for* complement decay-accelerating factor.

DAG *abbr. for* diacylglycerol.

DAHP *abbr. for* 3-deoxy-D-*arabino*-(2-)heptulosonate 7-phosphate.

daidzein an isoflavone that, like genestien, is synthesized by soybean and induces *nod* genes in some *Rhizobium* species.

dal *(sometimes) symbol for* dalton (Da is usual).

Dale's principle the principle that a given neuron is likely to release the same active substance from all its active terminals. Sometimes aggrandized into **Dale's law** – one neuron, one neurotransmitter – its only subsequent modification has been the recognition that a neuron may release more than one active substance. *See* **cotransmitter.** [After (Sir) Henry Hallett Dale (1875–1968), British physiologist.]

dalton *symbol*: Da; a non-SI unit of atomic and molecular mass, equal to one-twelfth of the mass of the nuclide carbon-12. It is identical with the **unified atomic mass unit.** It is widely used in biochemistry. Sometimes it is used incorrectly as a supposed unit of the dimensionless terms **relative atomic mass** and **relative molecular mass.** [After John Dalton (1766–1844), British chemist and physicist.]

Dalton's law of partial pressures a statement that the total pressure of a mixture of gases is equal to the sum of the partial pressure of the constituent gases.

Dalziel coefficient any of the coefficients of an equation expressing the initial rate, v, of a sequential two-substrate enzymic reaction of the form:

$$v = EAB / [AB \phi_0 + B \phi_1 + A \phi_2 + \phi_{12}]$$

where E is the enzyme concentration, A and B are the substrate concentrations, and ϕ_0, ϕ_1, ϕ_2, and ϕ_{12} are the Dalziel coefficients. The method can be extended to the evaluation of the eight Dalziel coefficients needed to describe the rate of a three-substrate reaction. [After Keith Dalziel (1921–94), British biochemist.] *Compare* **Michaelis kinetics.**

dam a gene of *Escherichia coli* encoding DNA-adenine methylase. *See* **DNA methylase.**

damascenine 3-methoxy-2(methylamino) benzoic acid methyl ester; a **protoalkaloid** from the seeds of *Nigella damascena* (love-in-a-mist) and the odoriferous principle of the oil of nigella.

dam methyltransferase an enzyme that introduces a methyl group at the C5 position of the inner cytosine of the sequence 5′CCAGG3′ or 5′CCTGG3′. This prevents cleavage by the *Eco*RII restriction endonuclease. *Compare* **dam.**

dAMP *abbr. for* deoxyadenosine monophosphate, the common name for 2′-deoxyadenosine 5′-phosphate; 2′-deoxy-5′-adenylic acid; adenine 2′-deoxyriboside 5′-phosphate.

danazol 17α-pregna-2,4-dien-20-ynol[2,3-*d*]isoxazol-17β-ol; a derivative of ethisterone that is an inhibitor of anterior pituitary function. It inhibits secretion of gonadotropins and synthesis of sex steroids and acts as a ligand for their receptors.

Dane particle *see* **hepatitis**.

Danielli–Davson model *see* **Davson–Danielli model**.

dansyl *symbol*: Dns; *abbr. for* 5-dimethylaminonaphthalene-1-sulfonyl, a group used in **extrinsic fluorescence** studies of proteins. *See also* **dansylate**.

dimethylaminonaphthalene sulfonyl chloride

dansylate to derivatize with a dansyl group by reacting with dansyl chloride, 5(dimethylamino)naphth-1-ylsulfonyl chloride. The method is used to acylate free amino groups in protein end-group analysis. The dansyl amino acids, isolated after hydrolysis of the protein, are highly fluorescent and amounts as small as 1nmol may be detected. —**dansylated** *adj.*; **dansylation** *n.*

Danysz's effect *or* **Danysz's phenomenon** the variation in the toxicity of toxin–antitoxin mixtures dependent on whether toxin is added to equivalence in one lot (nontoxic mixture produced) or in smaller lots over a time period (toxic mixture produced). When added in small lots the toxin reacts with more than one equivalent of antitoxin (antibody) so that insufficient antibody is left to neutralize all the subsequent lots of toxin. [After its discoverer, Jean Danysz (b. 1860), Polish pathologist working in Paris.]

DAP *abbr. for* diaminopimelate.

DAP pathway *see* **diaminopimelic-acid pathway**.

dapsone 4,4′-diaminodiphenyl sulfone; bis(4-aminophenyl)sulfone; an antibacterial and antiprotozoal substance used especially against *Mycobacterium leprae*, the causative agent of leprosy, and as an adjunct in the treatment of malaria. It acts as a competitive antagonist of *p*-aminobenzoic acid and an inhibitor of folate synthesis.

DARC *abbr. for* Duffy antigen receptor for chemokines. *See* **Duffy system**.

dark adaptation the process by which the rods of the retina gradually become fully responsive to dim light when no longer exposed to bright light. —**dark-adapted** *adj.*

dark band *see* **light band**.

dark current the electric current observed in any photoelectric device in the absence of incident radiation. It is caused by thermal effects.

dark-field microscope *or* **dark-ground microscope** a form of light microscope in which the specimen is illuminated by the apex of a hollow cone of light. On diverging, the light does not enter the aperture of the objective lens, so that unstained or transparent specimens appear as bright objects on a black background.

dark reactions of photosynthesis traditional term for the reactions of the **reductive pentose phosphate cycle** that utilize the ATP and NADPH produced by the light-driven reactions of photosynthesis. The term misleadingly indicates that these reactions occur in the dark. This is true only in the sense that, over very short time intervals, they can occur in the absence of light provided ATP and NADPH are present. However, they are subject to regulatory mechanisms that are stimulated by light, and inhibited by the dark. In growing plants, these reactions occur during daylight hours. The reactions of photosynthesis are used to convert CO_2 into hexoses and other organic compounds by the reactions of the reductive pentose phosphate cycle.

dark reactivation *or* **dark repair** any light-independent enzymic mechanism for the repair of damaged DNA molecules. *Compare* **photoreactivation**.

DARPP-32 *abbr. for* dopamine and cAMP-regulated phosphoprotein 32 kDa; a cellular glycoprotein produced by phosphorylation of a threonine residue by protein kinase A when dopamine binds the dopamine receptor. It is dephosphorylated mainly by calcineurin, and is involved in the psychomotor effects of caffeine.

database a collection of information organized within a (usually relational or object-oriented) management system (the term is also often inaccurately applied to data collected within a flat-file). Hundreds of biodatabases are now available, containing biomolecular sequences or structures, complete genomes, genetic disease information, biochemical pathway data, protein–protein interactions, biomedical literature, etc. Many are listed in the online catalogue **DBCAT**.

database division a data subset created to ease management and to facilitate rapid, specific searches of large databases. For example, EMBL, GenBank, and DDBJ are split into divisions that are largely taxonomic (e.g. prokaryotes, fungi, plants, mammals), although some divisions are based on the types of data being held – e.g. sequence and mapping data from expressed sequence tags (ESTs); sequence-tagged sites (STSs); genome survey sequences (GSSs); high-throughput genomic sequences (HTGs); and unfinished high-throughput cDNA sequences.

database ID *an alternative name for* **database identification code**.

database identification code *or* **database ID** a (usually human readable) code that identifies a particular entry in a particular database. Database IDs are assigned when entries are first added to a database but may change between updates, so are not a reliable means of locating them in subsequent releases. Some databases use systematic IDs, some use fixed-length mnemonics, and others use the same code as the accession number. For example, OPSD_BOVIN identifies bovine rhodopsin in Swiss-Prot; GPCR_Rhodpsn identifies the rhodopsin-like G-protein-coupled receptor superfamily in InterPro; and O96585 identifies the long-wavelength rhodopsin from the bumblebee (*Bombus terrestris*) in TrEMBL.

database-search program 1 any software used to search a database with a user query. 2 *(in sequence analysis)* software for protein or nucleotide sequence searches, such as FastA, BLAST and its variants, e.g. PSI-BLAST (position-specific iterated-BLAST) searches databases such as nrdb or Swiss-Prot iteratively by creating a profile from an alignment of the best-scoring hits and refining the profile with the results of each new search. PHI-BLAST (pattern hit initiated BLAST) searches iteratively for sequences that contain a user-specified pattern, each time refining the results via PSI-BLAST. 3 *(in structure analysis)* software for protein structure searches. For example, VAST (vector alignment search tool) searches for structurally similar proteins in MMDB (molecular modelling database). *See* **structure database**.

data cleaning the removal from, or correction of, erroneous data in a data set.

data mining 1 the process by which implicit information is extracted from data. 2 the process of searching for statistically significant relationships and global patterns in databases.

DATD *abbr. for* diallyltartardiamide, a cross-linking agent.

dative bond *an alternative term for* **dipolar bond**.

dATP *abbr. for* deoxyadenosine triphosphate, the common name for 2'-deoxyadenosine 5'-triphosphate, 2'-deoxy-5'-adenylyl diphosphate, adenine 2'-deoxyriboside 5'-triphosphate.

Daudi a human cell line established from a patient with Burkitt's lymphoma.

daughter *(as modifier)* 1 denoting a cell (or cell nucleus) produced by the fission of the original (mother) cell or nucleus. 2 denoting either of the two strands of DNA that originate from the parental duplex DNA molecule at replication. 3 denoting a nuclide produced by the nuclear disintegration (either spontaneous or induced) of another nuclide.

daunomycin *or* **daunorubicin** an anthracycline antibiotic and antineoplastic isolated from fermentation broths of *Streptomyces peucetius* widely used in the treatment of cancer. It is a glycoside formed from the tetracyclic aglycon daunomycinone ($C_{21}H_{18}O_8$) and the aminohexose **daunosamine**. It inhibits DNA replication by intercalation into duplex DNA and also inhibits RNA transcription. *Compare* **doxorubicin**.

daunosamine 3-amino-2,3,6-trideoxy-L-*lyxo*-hexose; the aminodeoxy sugar component of **adriamycin**, **daunomycin**, **doxorubicin**, and other antibiotics

α-L anomer

Davson–Danielli model *or (sometimes)* **Danielli–Davson model** a structural model of the **plasma membrane** consisting, as originally proposed, of a lipid bilayer covered on both surfaces by a film of adsorbed globular proteins. Essentially it was a lamellar model: two sheets of protein separated by a sheet of lipid. The lipid layer had a hydrophobic core and more hydrophilic surfaces. To account for the permeability properties of the plasma membrane the model was subsequently modified by the proposal that pores allow polar solutes to penetrate the lipid layer, and that these pores are lined by protein molecules, providing hydrophilic tubes through the membrane. The model has been totally superseded by more recent concepts. *See* **cell membrane**. [After Hugh Davson (1909–96), British biochemist and physiologist and J. F. Danielli (1911–84), British biochemist.]

day *symbol*: d; a non-SI unit of time equal to 24 hours or 86 400 s.

dbcAMP *abbr. for* dibutyryl cyclic AMP; $N^6,O^{2'}$-dibutyryladenosine 3',5'-(cyclic)phosphate; a fat-soluble analogue of cyclic AMP, **adenosine 3',5'-phosphate**.

DBCAT *abbr. for* the Catalogue of Databases; an online catalogue of biomolecular, genetic, genomic, and literature databases.

D-bifunctional enzyme *see* **peroxisomal bifunctional enzyme**.

DBL a gene, isolated and named from a human diffuse B-cell lymphoma, that encodes a protein that contains a Dbl domain of ≈200 amino acids, which is active as a Rho guanine nucleotide exchange factor, and a pleckstrin domain for binding phosphoinositides. It is a protooncogene that when mutated results in hemopoietic tumour.

Dbl domain a protein domain comprising a sequence of ≈200 amino acids, for which the prototype is the Rho guanine nucleotide exchange factor (*see* **RhoGEF**) of DBL. It shows 29% sequence identity with yeast Cdc24 (which is an activator of yeast Cdc42), and 25–30% homology with domains in other RhoGEFs. It seems to be usually accompanied by a pleckstrin homology domain. These proteins help transduce signals from receptors for growth factors, cytokines, and adhesion factors, and from G-protein-coupled receptors.

DBM *abbr. for* 2-diazobenzyloxymethyl, $-CH_2-O-CH_2-C_6H_4-N_2^+$; a group that introduces reactive diazonium groups into cellulose or paper forming **DBM-cellulose** or **DBT-paper** in **blot transfer** of proteins or nucleic acids. DBM is formed by diazotization of 2-aminobenzyloxymethyl (ABM) groups previously coupled to the support.

DBMIB *abbr. for* 2,5-dibromo-3-methyl-6-isopropyl-*p*-benzoquinone; a competitor of **plastoquinone** in photosynthetic electron flow.

DBP *see* **vitamin D binding protein**.

dbSNP a repository of genetic variations – mostly single nucleotide polymorphisms (SNPs) and a few small insertions and deletions – that includes, for example, sequence information around the polymorphism, descriptions of the population containing the variation, and frequency information by population or individual genotype.

d.c. *or* **DC** *abbr. for* 1 direct current. 2 a voltage of constant polarity.

dC *symbol for* a residue of the deoxynucleoside deoxycytidine (alternative to dCyd).

DCC *or* **DCCD** *or* **DCCI** *abbr. for* N,N'-dicyclohexylcarbodiimide, a coupling agent useful in peptide synthesis and other coupling reactions. *See also* **carbodiimide**.

DCC *abbr. for* deleted in colonic carcinoma; a human gene that encodes a transmembrane protein that is produced in intestinal mucosa and brain, and in small amounts in other tissues. Its N-terminal extracellular segment contains four immunoglobulin-like and seven fibronectin repeats, and shows about 30% homology with netrin receptor. The gene, at 18q21.2 and comprising ≈1.4Mb, is deleted in over 70% of cancers of the colon and rectum. Hence, *DCC* is classed as a tumour suppressor gene

D cell *or (formerly)* A_1 **cell** *or* $α_1$ **cell** a type of endocrine cell of the **apud** series, found in gut mucosa and the pancreatic islets. It was formerly thought to produce gastrin, but is now believed to produce **somatostatin**.

dCMP *or* **dCyd5'P** *abbr. for* deoxycytidine monophosphate, the common name for 2'-deoxycytidine 5'-phosphate; 2'-deoxy-5'-cytidylic acid; cytosine 2'-deoxyriboside 5'-phosphate.

dCMP hydroxymethylase *systematic name*: 5,10-methylenetetrahydrofolate:deoxycytidylate 5-hydroxymethyl transferase. EC 2.1.2.8; *other names*: deoxycytidylate hydroxymethylase; deoxycytidylate hydroxymethyltransferase; an enzyme that catalyses the reaction:

5,10-methylenetetrahydrofolate + H_2O + deoxycytidylate = tetrahydrofolate + 5-hydroxymethyldeoxycytidylate.

It is a gene product of T-even bacteriophage; the DNA of these phages contains glycosylated 5-hydroxymethylcytosine residues in place of cytosine residues. There is similarity to *Escherichia coli* **thymidylate synthase**.

DCMU *abbr. for* 3-(3,4-dichlorophenyl)-1,1-dimethylurea; a herbicide that blocks electron transport on the reducing side of photosystem II. One proprietary name is Diuron.

Dco *symbol for* the decanoyl group, $CH_3-[CH_2]_8-CO-$.

DCoH *abbr. for* dimerization cofactor for HNF-1α; an activity of a bifunctional protein (PCD/DCoH; 103 amino acids) that is required for dimerization of **hepatocyte nuclear factor-1α**. *See* **pterin-4α-carbinolamine dehydratase**.

DC-SIGN *abbr. for* dendritic cell-specific ICAM-3-grabbing nonintegrin; either of two type II transmembrane proteins that are specific to dendritic cells (i.e. antigen- presenting macrophages, mainly of spleen and lymph nodes) and bind T lymphocytes by ICAM-3 on their surfaces. DC-SIGN1 is also found on endothelial cells and shows 77% homology with DC-SIGN2, which is also found on liver sinusoidal cells. Their extracellular C-terminal regions contain seven 23-residue repeats in tandem, and a Ca^{2+}-dependent lectin domain that binds mannose-containing ligands. Both proteins bind gp120 of the HIV-1 envelope.

dCTP *abbr. for* deoxycytidine triphosphate, the common name for 2'-deoxycytidine 5'-triphosphate, 2'-deoxy-5'-cytidylyl diphosphate, cytosine 2'-deoxyriboside 5'-triphosphate.

dCyd *symbol for* a residue of the deoxynucleoside deoxycytidine; 2'-deoxyribosylcytosine; cytosine 2'-deoxyriboside (alternative to dC).

dCyd5′P *symbol for* deoxycytidine 5'-phosphate.

dd+ *abbr. for* dideoxy+ (def. 2); **1** used as a *prefix* to the appropriate three-letter or one-letter symbols for ribonucleosides in order to form symbols for 2',3'-dideoxyribonucleosides (e.g. ddNuc *or* ddN, *symbol for* any (unknown or unspecified) 2',3'-dideoxyribonucleoside; ddAdo *or* ddA, *symbol for* dideoxyadenosine). **2** used similarly to form abbreviations for 2',3'-dideoxyribonucleoside 5'-triphosphates (e.g. ddNTP, *abbr. for* any (unknown or unspecified) 2',3'-dideoxyribonucleoside 5'-triphosphate; ddCTP, *abbr. for* dideoxycytidine triphosphate).

DDBJ *abbr. for* DNA databank of Japan; *see* **sequence database**.

DDT *abbr. and common name for* dichlorodiphenyltrichloroethane; 1,1,1-trichloro-2,2-bis(*p*-chlorophenyl)ethane; a contact insecticide that has proved particularly valuable in controlling insect disease vectors, especially mosquitoes (*Anopheles* spp.). However, it is toxic to higher animals as well. It has a very long persistence of activity from residual deposits which has caused it to be banned in many countries.

de+ *prefix (in organic chemical nomenclature)* denoting replacement of a specified group or atom in a specified chemical compound by a hydrogen atom. An exception is **deoxy+** when applied to hydroxy compounds, which denotes replacement of a hydroxyl group by a hydrogen atom. *See also* **dehydro+, des+**.

deacylase *informal name for* any enzyme that catalyses a deacylation reaction; it may be a hydrolase or a transferase.

deacylation the act or process of removing an acyl group from a chemical compound.

DEAD box a highly conserved motif in a family of putative RNA **helicases**, named after the single-letter code for the amino-acid sequence involved, i.e. Asp-Glu-Ala-Asp. The first DEAD-box protein shown to have RNA-helicase activity was the eukaryotic translational initiation factor eIF4-A. Other proteins have since been included in the family, based on sequence similarities; some have been shown to have ATP-dependent RNA-helicase activity, while in others this remains to be experimentally confirmed. *See also* **DEAH family**,

dead-end complex *an alternative name for* **abortive complex**.

dead ringer an **ARID protein** of *Drosophila* whose DNA-binding sequence is the prototype for the dead ringer domain that is present in human and mouse jumonji proteins. Mutations in dead ringer are lethal at the embryonic stage in *Drosophila* and mouse.

DEAE *abbr. for* diethylaminoethyl, $(C_2H_5)_2N–CH_2–CH_2–$, a weakly basic group widely used as a substituent in hydroxylated chromatographic support matrices, e.g. cellulose, Sephadex, to form anion exchangers. *Compare* **QAE**.

DEAE-dextran *abbr. for* diethylaminoethyl-dextran; a polycationic derivative of dextran used in the transfection of mammalian cells with exogenous DNA.

de-aerate to remove air from a liquid or a slurry, commonly by heating and/or reduction of pressure.

DEAH family yeast proteins that are structurally similar to **DEAD-box** proteins, but larger. They differ fom DEAD-box proteins in parts of the conserved region, especially in having a conserved DEAH sequence rather than DEAD. Some of them are known to be involved in RNA splicing, e.g. in *trans*-esterification steps and spliceosome release.

deamidase *name sometimes used for* an enzyme that catalyses the hydrolytic cleavage of the C–N bond in a carboxamide, but that is correctly termed an **amidase**.

deamidation the process of removing an amino group from an amide, either by hydrolysis or by a transfer reaction.

deaminase any of the enzymes that catalyse the nonoxidative removal of amino groups, with production of ammonia. They are usually aminohydrolase enzymes (sub-subclass EC 3.5.4), which hydrolytically deaminate amino-substituted cyclic amidines including various (free or combined) pterins, purines, or pyrimidines. Certain hydro-lyase (sub-subclass EC 4.2.1) and hydrolytic or nonhydrolytic ammonia-lyase (sub-subclass EC 4.3.1) enzymes also are commonly called deaminases.

deamination the process of removing an amino group from an organic compound. *Compare* **transamination**.

deaminooxytocin *or* **demoxytocin** a synthetic analogue of **oxytocin**, in which the N-terminal half-cystine residue has been replaced by a residue of 3-mercaptopropanoic acid. It is more potent than the parent compound.

de-ashing a procedure in which interfering inorganic (i.e., ash-forming) components are removed from samples of carbohydrates being analysed by liquid chromatography by passing them first through anion- and cation-exchange resins.

death domain a region of limited similarity, comprising ~80 residues close to the intracellular C terminus of certain cell-membrane receptors (e.g., the **TNF receptor**), that is essential for the receptors to generate a signal leading to apoptosis.

death-inducing signalling complex *see* **DISC**.

death receptor a transmembrane receptor protein involved in inducing apoptosis. It oligomerizes on binding of its ligand and contains an intracellular **death domain** that recruits a cytosolic death domain protein (e.g. **FADD**) to initiate apoptosis. Examples include **Fas** and TNF receptor-1.

7-deaza-dGTP an analogue of dGTP used when sequencing DNA with a high content of GC base pairs.

deblock to remove a **blocking group**. —**deblocking** *n*.

debrancher deficiency *an alternative name for* type III **glycogen** (storage) **disease**.

debranching enzyme any enzyme that catalyses the cleavage of branch points in particular branched biopolymers, e.g. **isoamylase**, **oligo-1,6-glucosidase** and **pullulanase**.

debye *symbol*: D; a non-SI unit of molecular dipole moment: $1D \approx 3.335\ 64 \times 10^{-30}$ Cm.

Debye–Hückel theory a theory used to calculate the electric potential at a point in a solution in terms of the concentrations and charges of the ions and the properties of the solvent. It yields the logarithm of the mean rational activity coefficient:

$$\lg f^{+/-} = (A|z_1 z_2|\ I^{1/2})/(1 + BaI^{1/2}),$$

where the constants A and B involve the temperature and the dielectric constant of the solvent, a is an adjustable 'ion-size' parameter, I is the **ionic strength**, and $|z_1 z_2|$ is the absolute value of the valence product of the electrolyte under consideration. [After Petrus ('Peter') Debye (1884–1966), Dutch-born US physicist and physical chemist, and Eric Hückel (1896–1980), German physicist.]

dec+ *see* **deca+** (def. 2).

Dec *symbol for* the decyl group, $CH_3–[CH_2]_8–CH_2–$.

deca+ **1** *symbol*: da; the SI prefix denoting 10 times. **2** *or (before a vowel)* **dec+** *comb. form* denoting ten, tenfold, or ten times; formerly sometimes written as deka+ *or* dek+.

decamethonium N,N,N,N^1,N^1,N^1-hexamethyl-1,10-decanediaminium a nicotinic receptor ligand that causes sustained activation of the receptors, with consequent blockade of synaptic transmission (termed 'depolarizing block'). *See also* **suxamethonium**.

decanoate 1 the anion, $CH_3-[CH_2]_8-COO^-$, derived from decanoic acid (formerly capric acid). 2 any mixture of decanoic acid and the anion. 3 any salt or ester of decanoic acid.

decanoyl *symbol*: Dco; the univalent acyl group, $CH_3-[CH_2]_8-CO-$, derived from decanoic acid. *See also* **caproyl** (def. 1).

decant to pour off a supernatant liquid from a precipitate or other sediment. —**decantation** *n.*

decap to (enzymically) remove a **cap** (def. 1) from the 5′ end of a capped mRNA molecule. —**decapped** *adj.*

decapentaplegic *abbr.*: dpp; a secreted protein of the transforming growth factor β superfamily that triggers various morphogenetic processes in *Drosophila*. It participates in an intercellular signalling pathway that activates a specific transcription factor required for normal embryonal development. A homologue of dpp occurs in mammals.

decapsidate to remove the capsid from a virus. —**decapsidation** *n.*

decarboxylase any of the enzymes of the sub-subclass EC 4.1.1, carboxy-lyases; they catalyse the removal of carboxyl groups from carboxylic acids.

decarboxylation the act or process of removing the carboxyl group from a carboxylic acid as carbon dioxide. The reaction may be enzyme-catalysed by a **decarboxylase** or, in some instances, particularly with 2-oxo acids, it may be spontaneous. —**decarboxylate** *vb.*; **decarboxylated** *adj.*

decay 1 a decrease in activity, force, quantity, or other physical attribute, especially with a characteristic time constant. 2 (a) **nuclear decay**: the spontaneous change of one nuclide into another, its daughter, with a different proton number or nucleon number. (b) **radioactive decay**: nuclear decay in which particulate or electromagnetic radiation (or both) is emitted, or in which spontaneous fission of the nucleus occurs. *See also* **disintegration**. 3 the decline of an atom or molecule from an excited state. 4 the biochemical decomposition of plant and animal remains, due chiefly to the activities of microorganisms.

decay accelerating factor *see* **complement decay-accelerating factor**.

decay constant *an alternative term for* **disintegration constant**.

deci+ 1 *symbol*: d; the SI prefix denoting 10^{-1} times. 2 *comb. form* denoting one tenth, one in ten.

decimolar *(informal)* one-tenth molar; i.e. denoting a solution of concentration 0.1 mol dm^{-3}.

decinormal *(informal; not now recommended)* one-tenth normal; i.e. denoting a solution of concentration 0.1 equivalents per litre (*see* **normal** (def. 2)).

Declomycin *proprietary name for* **demeclocycline**. *See also* **tetracycline**.

decoding *(informal) an alternative term for* **translation** (def. 1).

decontaminant an agent for removing or neutralizing a contaminant.

decontamination the act or process of removing or neutralizing any toxic or potentially toxic materials, or of reducing their concentrations to nonhazardous levels. The term is used especially in regard to microbes, toxic chemicals, carcinogens, and radioactive substances. *Compare* **contamination**.

decorin *or* **fibromodulin** *or* **bone proteoglycan II precursor** a collagen-binding protein of the extracellular matrix, so called because it 'decorates' collagen fibrils. The core protein has a single **glycosaminoglycan** chain, together with other oligosaccharide substituents.

decose any **aldose** having a chain of ten carbon atoms in the molecule.

dCTP deaminase EC 3.5.4.13; *systematic name*: dCTP aminohydrolase; *other name*: deoxycytidine triphosphate deaminase. An enzyme that hydrolyses the amino group of dCTP to form dUTP and NH_4^+.

deculose any **ketose** having a chain of ten carbon atoms in the molecule.

decyl *symbol*: Dec; the alkyl group, $CH_3-[CH_2]_8-CH_2-$, derived from decane.

n-decyl-β-D-glucopyranoside a nonionic detergent, useful for solubilizing membrane proteins; CMC 2–3 mm.

n-decyl-β-D-maltopyranoside a nonionic detergent, useful for solubilizing membrane proteins; CMC 1.6 mm.

dedifferentiation the loss of the characteristics of a specialized cell and regression to a simpler state. —**dedifferentiate** *vb.*

defective virus a virus that is unable to reproduce itself whilst infecting a cell without the presence of a **helper virus**.

defensin any of a family of small cationic peptides, containing six cysteines in disulfide linkage, that have broad-spectrum antibiotic action, and contribute to host defence against microorganisms. They are abundant in phagocytes, in the small intestinal mucosa of humans and other mammals, and in the hemolymph of insects. Defensins adopt multimeric pore-forming complexes in membranes, thereby rendering the membrane permeable. They are distinguished by having a predominantly **beta sheet** structure.

deficiency disease any disease resulting from the deficiency of one or more essential nutrients, e.g. a vitamin, an essential amino acid, or an essential mineral.

deficient lacking some essential component; inadequate in amount. The term can be used, e.g., for a deficient (culture) medium, deficient diet, or deficient strain (of an organism).

defined medium any culture medium in which the concentrations of all the constituents, including trace substances, are quantitatively known.

deformylase *see* **formylmethionine deformylase**.

degeneracy 1 *(in a code)* the existence of two or more synonyms specifying a particular item or function; *see also* **degenerate code**. 2 *(in a quantized system)* the existence of two or more linearly independent **eigenfunctions** with the same **eigenvalue**, especially the condition of an atom or molecule. 3 *(of an atom or molecule)* the condition in which the energy levels of different quantum states have the same energy level. —**degenerate** *adj.*

degenerate code a code in which more than one symbol represents the same entity. One such is the **genetic code**, wherein a specific amino acid may be encoded by two or more nucleotide triplets, or **codons** present in a nucleic acid or polynucleotide.

degenerate oligonucleotides a mixture of oligonucleotides corresponding to a known protein sequence and accommodating all the possible coding sequences that arise from the degeneracy of the genetic code.

degeneration the process whereby a cell, tissue, or organ changes to a less specialized or functionally less active form.

degradation *(in chemistry)* 1 the gradual stepwise and deliberate conversion of a molecule into smaller chemical entities, commonly to elucidate its chemical structure. 2 any undesired breakdown of a molecule or material with impairment or loss of its characteristic properties. 3 depolymerization; decomposition. 4 *(in biochemistry)* catabolism.

degradosome a multienzyme complex involved in the degradation of mRNA in *E. coli*. It contains RNase E (the microbe's main endoribonuclease), polynucleotide phosphorylase, the glycolytic enzyme enolase, and a DEAD-box RNA helicase. Under certain conditions it may also contain poly(A) polymerase, polyphosphate kinase, and another DEAD-box RNA helicase. *Compare* **exosome**.

degranulation 1 any process by which granular cells lose their granules. 2 the stripping of polysomes from rough endoplasmic reticulum *in vitro*.

degree *see* **angle, temperature**.

degree Celsius *symbol*: °C; the SI derived unit of **Celsius temperature**. It denotes the same interval of temperature as the **kelvin** on a scale with different reference points.

degree of polymerization the number of **monomeric units** in the molecule of a polymer.

degrees of freedom 1 *(in statistics)* the number of unrestricted independent variables constituting a statistic; thus in a sample of *n* values of x there are n deviations $(x_i - \bar{x})$, where $\bar{x}$ is the estimate of the mean, used in calculating the sample **variance**, s^2, but they are not all independent, since there is a linear relation connecting them, namely:

$$\sum_{i=1}^{n} (x - \bar{x}) = 0$$

hence only $(n - 1)$ of the quantities are independent, and there are $(n - 1)$ degrees of freedom in the calculation of s^2. **2** *(in physics)* the minimum number of parameters necessary to the description of the state of a system. **3** the number of independent equations by which the energy of a system may be expressed. **4** *(in chemistry)* the number of independently variable intensive properties that must be specified in order to describe completely the state of the system. *See* **phase rule**.

dehydrase *obsolete term for* **dehydratase** *or* **dehydrogenase**.

dehydratase *or (formerly)* **dehydrase** any hydro-lyase enzyme, of sub-subclass EC 4.2.1, that catalyses the (reversible) breakage of a carbon–oxygen bond leading to the formation of an unsaturated product and the elimination of water; e.g. citrate dehydratase, EC 4.2.1.4. When named after the alternative, i.e. unsaturated, substrate, the enzyme is termed a **hydratase**; e.g. aconitate hydratase, EC 4.2.1.3. *See also* **carbonate dehydratase**.

dehydration 1 the removal of combined, or adsorbed, water from a chemical compound; *see also* **anhydrous**. **2** the process or result of losing water, especially the abnormal loss of water from an organism. **3** the replacement of water by solvents.

dehydro+ *prefix (in chemical nomenclature)* denoting the loss of two hydrogen atoms. In common usage it is often used in place of **didehydro+**; e.g. 7,8-didehydrocholesterol is often termed dehydrocholesterol. *Distinguish from* **anhydro+**.

7-dehydrocholesterol reductase EC 1.3.1.21; *systematic name*: 3β-hydroxysteroid-Δ^7-reductase; an enzyme of the endoplasmic reticulum that reduces 7-dehydrocholesterol to cholesterol, using NADPH as coenzyme. The gene at 11q12-q13 encodes a multipass transmembrane protein of 475 amino acids. Inactivating mutations are responsible for Smith–Lemli–Opitz syndrome.

2-dehydro-3-deoxyphosphooctonate aldolase *another name for* **3-deoxy-D-manno-octulosonic acid 8-phosphate synthase**.

dehydroepiandrosterone *abbr.*: DHEA; androstenolone, *trans*-dehydroandrosterone (*abbr.*: DHAS), 3β-hydroxyandrost-5-en-17-one; an intermediate in androgen and estrogen biosynthesis, formed from 17α-hydroxypregnenolone in testis, ovary, and adrenal cortex by the action of 17α-hydroxyprogesterone aldolase, EC 4.1.2.30. It is the immediate precursor of androstenedione, which is formed from it by the action of 3β-hydroxy-Δ^5-steroid dehydrogenase, EC 1.1.1.145 and it can be converted to androst-5-en-17-one 3β-sulfate (DHEA-S), in which form it is measured in plasma, there being about 1000 times more DHEA-S than DHEA in plasma.

dehydrogenase any donor:acceptor **oxidoreductase** enzyme, of the class EC 1, when named after the alternative, i.e. more reduced, substrate and when dioxygen is not the acceptor.

dehydrogenate to oxidize an organic compound by the removal of one or more hydrogen atoms. —**dehydrogenation** *n*.

3-dehydroquinate dehydratase EC 4.2.1.10; *systematic name*: 3-dehydroquinate hydro-lyase; *other name*: 3-dehydroquinase; an enzyme that catalyses the reaction:

3-dehydroquinate = 3-dehydroshikimate + H_2O.

The reaction forms part of the **shikimate pathway**. For example, *see* **aro1**. *See also* **pentafunctional arom polypeptide**.

3-dehydroquinate synthase EC 4.6.1.3; an enzyme that catalyses the hydrolysis of 7-phospho-3-deoxy-*arabino*-heptulosonate to 3-dehydroquinate and orthophosphate. *See* **pentafunctional arom polypeptide**; **shikimate pathway**.

(3-)dehydroretinal *an alternative name for* **3,4-didehydroretinal** (not recommended, except for nutritional usage).

(3-)dehydroretinoic acid *an alternative name for* **3,4-didehydroretinoic acid** (not recommended, except for nutritional usage).

(3-)dehydroretinol *an alternative name for* **3,4-didehydroretinol** (not recommended except for nutritional usage).

deionize *or* **deionise** to remove all ions from a solution. *See also* **desalt**. —**deionization** *or* **deionisation** *n*.

deionizer *or* **deioniser** a device used to remove ions from solutions, usually by the use of **ion exchangers**.

Dejerine–Sottas syndrome *see* periaxin.

deka+ *or (before a vowel)* **dek+** *former spelling of* **deca+** (def. 2).

delayed early gene *see* early gene.

delayed hypersensitivity *or* **type IV hypersensitivity** *(in immunology)* a **hypersensitivity** reaction that arises more than a day after encounter with the antigen and is mediated by antigen-sensitized CD4$^+$ T cells that release **cytokines**, attracting macrophages to the site and activating them. This type of hypersensitivity is seen in skin contact reactions and in response to some chronic pathogens such as those that cause leprosy, tuberculosis, and schistosomiasis. *Compare* **immediate hypersensitivity**.

deletion the loss of a segment of genetic material from a chromosome. A deletion can vary in size from a single nucleotide residue to a segment containing a number of genes. A **terminal deletion** occurs at one end of the chromosome, whereas an **intercalary deletion** occurs elsewhere along the chromosome. Some deletions are associated with genetic diseases (e.g. in human chromosome 4, giving rise to Charcot–Marie–Tooth disease, and in human chromosome 22, leading to DiGeorge syndrome).

DELFIA *the registered abbr. for* Dissociation Enhanced Lanthanide Fluoroimmunoassay; a collection of proprietary diagnostic methods based on heterogeneous immunoassays and time-resolved fluorescence.

delipidation the removal of lipid from a sample.

deliquescent *(of certain salts)* tending to absorb water from the surrounding atmosphere at ordinary temperature and pressure and then to dissolve in the water so taken up. *Compare* **hygroscopic**. —**deliquesce** *vb.*; **deliquescence** *n*.

DELLA protein any of a family of highly conserved proteins that repress plant growth. They are present in plant cell nuclei and, in response to gibberellin, become phosphorylated, polyubiquitinated, and targeted for degradation by proteasomes.

delocalization *or* **delocalisation** a quantum-mechanical concept implying that the unpaired electron density in a molecule is spread over several atoms. Analogously, delocalization of the spin of an unpaired electron can occur.

delphinidin *see* anthocyanidin.

delta *symbol*: δ (lower case) *or* Δ (upper case); the fourth letter of the Greek alphabet. For uses *see* **Appendix A**.

Delta *see* Notch.

deltahedron (*pl.* **deltahedra**) any polyhedron whose faces are all equilateral triangles. Some viruses disaggregate to form particles that are deltahedra. —**deltahedral** *adj*.

deltakephalin (D-Thr^2)-Leu-enkephalin-Thr; a highly potent and specific synthetic hexapeptide δ-opiate receptor agonist; structure Tyr-D-Thr-Gly-Phe-Leu-Thr. *See also* **enkephalin**.

delta-sleep-inducing peptide *or* **sleep peptide** a nonapeptide, Trp-Ala-Gly-Gly-Asp-Ala-Ser-Gly-Glu, found in the plasma of rats and rabbits, that shows a several-fold increase in concentration during sleep. It rapidly induces slow-wave (i.e. delta) sleep on injection into awake animals.

deltex a conserved positive regulatory protein for Notch signalling in humans. It contains an N-terminal globular region that includes two divergent sequence repeats of ≈80 residues each, called **WWE motifs** for their most conserved residues. These motifs are predicted to mediate protein interactions in ubiquitin and ADP-ribose conjugation systems. The protein's C-terminus has a RING finger domain.

dematin a protein involved in the formation of actin bundles, abundantly expressed in brain, heart, skeletal muscle, kidney, and lung, whose activity is abolished upon phosphorylation by cAMP-dependent protein kinase. It is a member of the **villin** family.

demeclocycline 6-demethyl-7-chlorotetracycline; *see* **tetracycline**.

demethylation any reaction involving the removal of one or more methyl groups from a molecule, whether by oxidation or transfer to an acceptor molecule.

demineralization *or* **demineralisation 1** the process of removing inorganic salts, e.g. from water or other liquid; deionization. **2** excessive loss of mineral salts from the tissues, especially bone, as occurs in certain diseases.

demoxytocin *an alternative name for* **deaminooxytocin**.

denaturation 1 *(of a protein)* a process in which the three-dimensional shape of a molecule is changed from its **native state** without rupture of peptide bonds. It is sometimes taken to include disulfide bond rupture or chemical modification of certain groups in the protein if these processes are also accompanied by changes in its overall three-dimensional structure. Denaturation is frequently irreversible and accompanied by loss of solubility (especially at the isoelectric point) and/or of biological activity. **2** *(of a nucleic acid)* a process whereby a molecule is converted from a firm, two-stranded, helical structure to a flexible, single-stranded structure. *Compare* **renaturation**.

denaturation loop a non-base-paired interstrand loop formed in the more thermally unstable regions (i.e., of high A+T content) of a duplex DNA molecule during **melting** (def. 3).

denaturation mapping the identification of regions of low thermal (or alkali) stability (i.e., of high A+T content) in a duplex DNA molecule, by trapping the partly melted structure and blocking renaturation, e.g. with formaldehyde (which preferentially couples with the amino groups of the single-stranded regions), and subsequently examining the specimen by electron microscopy.

denaturing gradient gel electrophoresis *abbr.*: DGGE; a method used to detect sequence differences in relatively short double-stranded (ds) DNA molecules (<500 bp). Electrophoresis of dsDNA through increasing concentrations of urea and formamide causes dissociation of the duplex in a manner that depends on sequence domains. The resulting single-stranded DNA has a much lower electrophoretic mobility than dsDNA. The method is used to detect heteroduplexes formed after polymerase chain reaction (PCR) by annealing of wild-type and mutant strands, and in separation of mixtures of PCR products generated by amplification of 16S RNA genes in naturally occurring populations of bacteria.

denaturing high-performance liquid chromatography *abbr.*: DHPLC; a technique used to distinguish between short double-stranded (ds) DNA molecules whose strands are complementary and those that contain one or more mismatched deoxynucleotides. The latter are heteroduplexes and have a lower melting temperature and behave differently on chromatography. The technique is used in screening for individuals who are heterozygotes at a specific genetic locus.

dendrite a freely branching protoplasmic process of a nerve cell that transmits nerve impulses to the body of the cell and that communicates functionally with dendrites or bodies of other nerve cells.

dendritic cell a type of cell derived from bone marrow that is characterized by branching projections and present in lymphoid tissue, skin, and squamous epithelium. It is the major antigen-presenting cell in the human body. This it does by engulfing protein antigens, cleaving them, and displaying the fragments on the cell surface by means of the **major histocompatibility complex** (MHC) molecules.

dendrogram any branching treelike diagram that illustrates the relationships between organisms or objects.

Denhardt's solution a solution containing Ficoll, polyvinylpyrrolidone, and bovine serum albumin; 1% (w/v) of each component gives 50× concentrated Denhardt's reagent. It is a constituent of solutions used as blocking reagents for nitrocellulose membranes in Northern and Southern hybridizations. [After David Tilton Denhardt (1939–), US biochemist.]

denitrification the reduction, especially by denitrifying bacteria, of simple inorganic nitrogen compounds, such as nitrates and nitrites, to gaseous products such as nitric oxide, nitrous oxide, and/or dini-

trogen. *See also* **nitrate reduction**. *Compare* **nitrification**, **nitrogen fixation** (def. 1).

denitrifying bacteria soil bacteria that, in the absence of dioxygen, reduce nitrates and nitrites to gaseous products. *Compare* **nitrifying bacteria**.

de novo *Latin* anew, from the beginning; e.g. *de novo* synthesis.

densimeter *an alternative name for* **densitometer** (def. 2).

densitometer 1 *or* **photodensitometer** an apparatus for measuring the extent to which a material absorbs light (*see* **absorptiometer** (def. 1)) or reflects light (*see* **reflectometer**). It is useful for scanning chromatograms or electrophoretograms, for determining the opacity of solutions or suspensions, or for measuring the blackening of photographic films. **2** *or* **densimeter** an instrument for measuring the mass density or relative density of liquids or gases. —**densitometry** *n.*; **densitometric** *adj.*

density 1 a *or* **mass density** *symbol*: ρ; the mass of a substance per unit volume. **b** *or* **linear density** *symbol*: ρ_l; the mass of a substance per unit length. **c** *or* **surface density** *symbol*: ρ_A *or* ρ_S; the mass of a substance per unit area. *Compare* **relative density**. **2** the distribution of a scalar physical quantity per unit area of space (e.g. radiant energy density) or per unit volume (e.g. charge density). **3** the distribution of a vector physical quantity per unit area of space (e.g. electric current density, magnetic flux density). **4** optical density. *See* **absorbance**.

density functional theory a method of calculating the chemical properties of complex materials such as proteins. The complexity of using quantum wave-function calculations is simplified by the local density approximation, which substitutes the exchange-correlation energy density of an inhomogeneous system by that of an electron gas evaluated at the local density.

density-gradient centrifugation *see* **isopycnic centrifugation**.

density-gradient sedimentation equilibrium *see* **isopycnic centrifugation**.

density-marker bead any small coloured bead of accurately known mass density that is used for calibrating density gradients. Generally a set of several such beads, covering a range of densities, is used.

dentate having teeth or toothlike projections. *See also* **multidentate**.

dentatorubralpallidoluysian atrophy *abbr.*: DRPLA; a neurodegenerative genetic disease that shows genetic **anticipation** due to expansion of a trinucleotide repeat (CAG) in the atrophin 1 gene at locus 12p13.31.

denticity the number of distinct but linked atoms of a specified ligand binding to the central atom in a coordination entity. *Compare* **ligancy**.

dentine *or* **dentin** the hard, dense, calcified tissue of a tooth lying between the enamel and the pulp. It is 75% mineral, mainly hydroxyapatite.

dentinogenesis imperfecta *abbr.*: DGI; a group of genetically and clinically heterogeneous conditions characterized by abnormal dentine formation. DI type I is associated with **osteogenesis imperfecta**; types II and III are autosomal dominant conditions linked to the chromosomal locus 4q12-21 and are caused by the same mutation of the dentine sialophosphoprotein (*DSPP*) gene. It is likely that DGI type I and DGI type II are not separate diseases but rather the phenotypic variation of a single disease.

D-enzyme *or* **plant glucosyltransferase** an enzyme of plants that condenses short α1-4-glucans in the reaction α-glucan$_m$ + α-glucan$_n$ = α-glucan$_{(m+n-1)}$ + glucose. The product is long enough to be a substrate for starch phosphorylase.

deoxy+ *or (formerly)* **desoxy+** *prefix (in organic chemical nomenclature)* **1** denoting the replacement of a hydroxyl group by a hydrogen atom; *see also* **de+**. **2** (*abbr.*: d+) indicating the presence of one or more residues of 2-deoxy-D-ribose. *Compare* **ribo+** (def. 2). **3** used especially with hemoglobin, to signify absence or removal of dioxygen. *Compare* **oxy+**.

5′-deoxy-5′-adenosylcobalamin *or* **adenosylcobalamin** *see* **cobamide coenzyme**.

3-deoxy-D-*arabino*-(2-)heptulosonate 7-phosphate *or (less formally)* **3-deoxy-D-*arabino*-heptulosonate 7-phosphate** *abbr.*: DAHP; an early intermediate in aromatic amino acid biosynthesis by the **shikimate pathway** in autotrophic organisms.

α-pyranose form

deoxy-BigCHAP *N,N-bis*-(3-D-gluconamidopropyl)-(7-deoxy)cho-
lamide; a nonionic detergent analogue of **CHAPS**, **CHAPSO**, and
BigCHAP; aggregation number 8–16; CMC 1.1–1.4 mm.

deoxycholic acid 3*α*,12*α*-dihydroxy-5*β*-cholan-24-oic acid; one of
the **bile acids** occurring as its conjugate with glycine or taurine in the
bile of most mammals, including dog, goat, human, ox, sheep, and
rabbit. The sodium salt, sodium deoxycholate, is used as a detergent. Aggregation number 3–12; CMC 2–6 mm.

deoxycorticosterone *abbr.*: DOC; Kendall's desoxy compound B,
Reichstein's substance Q, 21-hydroxyprogesterone, 21-hydroxy-
pregn-4-ene-3,20-dione; a **mineralocorticoid** hormone with little glucocorticoid activity, synthesized by the adrenal gland. For clinical
purposes it may be administered as its ester, deoxycorticosterone
acetate (*abbr.*: doca *or* DOCA).

deoxycytidine deaminase *see* **cytidine deaminase**.
deoxycytidylate deaminase *an alternative name for* dCMP deaminase.

3-deoxy-D-*manno*-octulosonic acid 8-phosphate synthase
EC 4.1.2.16; *recommended name*: 2-dehydro-3-deoxyphosphoocto-
nate aldolase; *other name*: phospho-2-keto-3-deoxyoctonate (*abbr.*:
phosphoKDO) aldolase; an enzyme that catalyses the formation of
3-deoxy-D-*manno*-octulosonic acid 8-phosphate from D-arabinose
5-phosphate and phospho*enol*pyruvate with release of orthophosphate. It is the first step in the biosynthesis of the bacterial
lipopolysaccharide, lipid A, and is required for cell growth.
deoxynojirimycin an antibiotic produced by *Bacillus* species, the
reduced form of **nojirimycin**; it inhibits formation of N-linked complex oligosaccharides in culture.
deoxynucleoside *an alternative name for* **deoxyribonucleoside**.
deoxynucleotide *an alternative name for* **deoxyribonucleotide**.
deoxynucleotidyltransferase EC 2.7.7.31; *recommended name*:
DNA nucleotidylexotransferase; *other names*: terminal addition enzyme; terminal transferase; terminal deoxyribonucleotidyltransferase. It is a Mg^{2+}-dependent polymerase enzyme that extends the
3′ end of a DNA strand one nucleotide at a time, independent of a
template. It catalyses the reaction (if *n* nucleotides are added in all):

$$n \text{ (deoxynucleoside triphosphate)} + \text{(deoxynucleotide)}_m =$$
$$n \text{ (pyrophosphate)} + \text{(deoxynucleotide)}_{m+n}.$$

deoxypentose any monosaccharide in which one alcoholic hydroxyl group of a pentose has been replaced by a hydrogen atom.
deoxypentose nucleic acid *former term for* **deoxyribonucleic acid**.
deoxypolymeric tailing extension of the 3′ end of a strand of
DNA, or of each strand of (a fragment of) duplex DNA, by a tail
comprising a homooligomer or homopolymer of a deoxyribonucleotide. It is effected by the enzyme **deoxynucleotidyl transferase**.
The process is useful for forming cohesive termini to DNA fragments in the preparation of **recombinant DNA**. *See also* **cohesive end**.
deoxyribonuclease *abbr.*: DNase (*or sometimes* DNAse *or*
DNAase); any enzyme within subclass EC 3.1, esterases, that catalyses the hydrolytic cleavage of phosphodiester linkages in DNA. If
such an enzyme catalyses endonucleolysis (i.e. hydrolysis of nonterminal phosphodiester linkages) it is termed an **endodeoxyribonuclease**; if it catalyses exonucleolysis (i.e. hydrolysis of terminal linkages) it is termed an **exodeoxyribonuclease**. *See also* **AP
endonuclease, crossover point, deoxyribonuclease I, deoxyribonuclease
II, deoxyribonuclease III, deoxyribonuclease IV (phage T4-induced), de-
oxyribonuclease V, deoxyribonuclease X, deoxyribonuclease KI, deoxyri-
bonuclease (pyrimidine dimer), endonuclease, exodeoxyribonuclease I,
exodeoxyribonuclease V, exodeoxyribonuclease VII, exodeoxyribonuclease (phage SP3-induced), myt5, restriction enzyme**.
deoxyribonuclease I EC 3.1.21.1; *other names*: pancreatic DNase;
DNase; thymonuclease. An enzyme that catalyses the endonucleolytic cleavage of DNA to 5′-phosphodinucleotide and 5′-phospho-
oligonucleotide end products. It shows a preference for double-
stranded DNA. *See also* **streptodornase**.
deoxyribonuclease II EC 3.1.22.1; *other names*: DNase II; pancreatic DNase II. An enzyme that catalyses endonucleolytic cleavage
of DNA to 3′-phosphomononucleotide and 3′-phosphooligonu-
cleotide end products.
deoxyribonuclease III EC 3.1.11.2; *other name*: exonuclease III.
An enzyme that catalyses exonucleolytic degradation of double-
stranded DNA progressively in the 3′- to 5′-direction, releasing 5′-
phosphomononucleotides.
deoxyribonuclease IV (phage T4-induced) EC 3.1.21.2; *other
names*: endodeoxyribonuclease IV (phage T4-induced); endonuclease IV. An enzyme that catalyses endonucleolytic cleavage of DNA
to 5′-phosphooligonucleotide end products.
deoxyribonuclease V EC 3.1.22.3; *other name*: endodeoxyribonuclease V. An enzyme that catalyses endonucleolytic cleavage at
apurinic or apyrimidinic sites within DNA to yield products with a
3′-phosphate. It is involved in DNA repair. The enzyme from *Escherichia coli* consists of three subunits, RECB, RECC, and RECD,
also known as the β chain, γ chain, and α chain, respectively.
deoxyribonuclease X EC 3.1.22.5; an enzyme that catalyses endonucleolytic cleavage of supercoiled plasmid DNA to produce linear DNA duplexes.
deoxyribonuclease K1 *or* *Aspergillus* **deoxyribonuclease K1** EC
3.1.22.2; *other name*: *Aspergillus* DNase K1. An enzyme that cata-

lyses endonucleolytic cleavage of DNA to 3′-phosphomononucleotide and 3′-phosphooligonucleotide end products.

deoxyribonuclease (pyrimidine dimer) EC 3.1.25.1; *other name*: endodeoxyribonuclease (pyrimidine dimer). An enzyme involved in DNA repair; it can induce single-strand breaks in DNA on the 5′-side of pyrimidine dimers in the strand containing the lesion, after formation of the dimers as a result of, e.g., ultraviolet radiation.

deoxyribonucleate 1 *abbr.*: DNA; any anionic form of deoxyribonucleic acid. **2** any salt of deoxyribonucleic acid.

deoxyribonucleic acid *former names*: desoxyribonucleic acid *or* thymonucleic acid; *also* deoxypentose nucleic acid *or* deoxyribose nucleic acid *or* desoxyribose nucleic acid *or* thymus nucleic acid. *See* **DNA**.

deoxyribonucleoprotein *abbr.*: DNP; any conjugated protein that contains DNA as the nonprotein moiety.

deoxyribonucleoside *or* **deoxynucleoside** any **nucleoside** (def. 1) in which the glycose moiety is 2-deoxy-β-D-ribofuranose.

deoxyribonucleoside monophosphate kinase any of several enzymes within sub-subclass EC 2.7.4 that catalyse the transfer of a phosphoric residue from a nucleoside triphosphate to a deoxyribonucleoside monophosphate, e.g. from ATP to a deoxyribonucleoside phosphate to form ADP and a deoxyribonucleoside diphosphate.

deoxyribonucleotide *or* **deoxynucleotide** a deoxyribonucleoside in ester linkage to phosphate, commonly at the 5′ position of its deoxyribose moiety. *See* **nucleotide**.

deoxyribose 1 *symbol*: dRib; *the trivial name for* 2-deoxyribose; 2-deoxy-*erythro*-pentose. The D enantiomer, as 2-deoxy-β-D-ribofuranose, forms the glycose moiety of all deoxyribonucleosides, deoxyribonucleotides, and deoxyribonucleic acids. **2** 3-deoxyribose (*see* **cordycepose**).

2-deoxy-α-D-ribofuranose

deoxyribose nucleic acid *former name for* **deoxyribonucleic acid**.

deoxyriboside any glycoside or glycosylamine in which deoxyribose forms the glycose moiety, especially a deoxyribonucleoside.

deoxyribosome *see* **nu body**.

deoxyribosylthymine *an alternative name for* thymidine.

deoxyribotide any deoxyriboside esterified with phosphate at the 5′ position on its ribose moiety, especially a deoxynucleotide.

deoxyribovirus *an alternative name for* **DNA virus**.

deoxyribozyme *see* **DNAzyme**.

deoxysaccharide *or* **deoxysugar** any saccharide in which one alcoholic hydroxyl group has been replaced by a hydrogen atom.

depactin a protein isolated from starfish eggs, able to depolymerize F-actin, and to inhibit the extent of actin polymerization. It binds to actin monomers from filaments and in solution.

DEPC *abbr. for* diethylpyrocarbonate.

dependent form of glycogen synthetase the phosphorylated form of the enzyme glycogen synthetase that is 'dependent' for its activity on the presence of glucose 6-phosphate, a positive effector.

dependent variable any variable in a mathematical statement or equation whose value depends on the values taken by the related independent variables.

dephospho-[reductase kinase] kinase EC 2.7.1.110; *other name*: reductase kinase kinase; an enzyme that catalyses the phosphorylation of dephospho-[[3-hydroxy-3-methylglutaryl-CoA reductase (NADPH)] kinase] by ATP with release of ADP. It is involved in the regulation of **hydroxymethylglutaryl-CoA reductase** (HMGCoA reductase). By phosphorylating reductase kinase, it activates that enzyme to phosphorylate HMGCoA reductase, thereby inactivating the latter.

dephosphorylation any reaction in which phosphate groups are re-

moved chemically or enzymatically. In gene cloning the removal of 5′ phosphate from cloning vectors by treatment with calf intestinal phosphatase (CIP) or bacterial alkaline phosphatase (BAP) prevents religation of the vector ends and increases the probability of cloning a DNA insert.

depolarization *or* **depolarisation 1** a reduction in the electric potential measured across a cell membrane, especially that brought about in a nerve or muscle cell by an electrical stimulus or a drug. **2** the decrease of the angle of polarization of polarized light so that it appears altered when viewed through an analyser. *See also* **fluorescence depolarization**.

depolarization ratio *or* **depolarisation ratio** *(in light scattering)* the ratio of the horizontally to the vertically polarized components of the light scattered by scattering particles when illuminated by unpolarized light. If the scattering particles are isotropic the depolarization ratio will be zero; if they are anisotropic it will not be zero.

deproteinization *or* **deproteinisation** the process of removing protein from a biological sample, either by precipitation of protein from solution or by hydrolysis of protein with proteolytic enzymes.

depside any of a family of natural or artificial compounds derived from two or more molecules of various trihydroxybenzoic acids by esterification of a carboxyl group on one molecule with a hydroxyl group on another molecule (which may or may not be of the same structure as the first). By analogy with peptides, depsides derived from two, three, etc., or several trihydroxybenzoic acid units may be termed **didepside**, **tridepside**, etc., or **polydepside**, respectively. Depsides occur widely in lichens and also as components of some tannins.

depsidone any compound derived from two phenolcarboxylic acid molecules (which may or may not be the same molecules) that are joined together by both an ester and an ether link. Such substances are found in lichens.

depsipeptide any linear or macrocyclic compound containing (alternating) amino-acid and hydroxy-acid residues, linked by amide, *N*-methylamide, and ester bonds. Among the naturally occurring **cyclodepsipeptides** (sometimes alternatively termed peptolides) are the antibiotics **valinomycin**, the **enniatins**, and the monamycins.

depth of coverage the number of times a genome is sequenced to ensure data accuracy and to help assembly of the DNA fragments into the correct order. Half-shotgun coverage implies 4–5-fold sequencing, to provide a robust framework for fragment assembly; full-shotgun coverage implies 8–9-fold sequencing, to reduce ambiguities and errors, and to close gaps. *See also* **draft genome sequence**, **finished genome sequence**, **shotgun sequencing**.

depurinate to remove purine residues from DNA or another polynucleotide by exposure to weak acid. The reaction may be used in preparation for **Southern blotting** when partial depurination of large DNA molecules (>15 kb) followed by treatment with a strong base causes hydrolysis of the phosphodiester backbone at sites of depurination. The resulting DNA fragments are more readily transferred from the agarose gel to the membrane by capillary transfer. —**depurinated** *adj.*; **depurination** *n.*

depyrimidinate to remove pyrimidine residues from a polynucleotide. —**depyrimidinated** *adj.*; **depyrimidination** *n.*

derepression the phenomenon in which a repressor is prevented from interacting with the operator component of an **operon** and an increase in the synthesis of a gene product thereby occurs. With an inducible enzyme, the inducer derepresses the operon. Derepression may also result from mutations, either in the **regulator(y) gene**, thus blocking repressor synthesis, or in the operator gene, thus rendering it insensitive to the repressor.

deRib *(sometimes) symbol for* 2-deoxy-D-ribose (dRib is the correct symbol).

derivative 1 any compound that may, at least theoretically, be formed from another compound to which it is structurally related. **2** *see* **derivative of a function**.

derivative of a function *(in mathematics)* the limit of the ratio of an increment of a dependent variable to the corresponding increment of an associated independent variable as the latter increment tends to zero.

derivatize *or* **derivatise** to convert a chemical compound into a derivative. —**derivatization** *or* **derivatisation** *n.*

derived database a database derived from other resources but including relationships or data not found in those resources. There are a number of examples as follows. Swiss-Prot was originally derived from sequences in the PIR-PSD, but included additional annotation and an EMBL-like format. Blocks-format-PRINTS is derived from motifs in PRINTS, but includes Blocks scoring methods and a Swiss-Prot-compatible format. eMOTIF is derived from motifs in Blocks and PRINTS using the emotif-maker algorithm. CDD (conserved domain database) is derived from protein domains in Pfam, SMART (simple modular architecture research tool), and COG, with links to the Cn3D structure viewer and domain searches via BLAST. InterPro is a composite of protein families from sources such as PROSITE, PRINTS, Pfam, and ProDom, providing sibling and partonomic links between families, with functional links to GO (gene ontology) and structural links to PDB (protein database). *See also* **composite database, partonomy**.

dermaseptin any of a series of antimicrobial peptides derived from amphibian skin. Dermaseptins s1–5 constitute a family of cationic lysine-rich amphipathic antifungal peptides of 28–34 residues. Presumed to protect the naked frog skin from infections, they have wider application, being the first vertebrate peptides to show lethal effects against the filamentous fungi that can cause infections in immunodeficiency syndrome, or during the use of immunosuppressive agents. Dermaseptin is classified in the TC system under number 1.C.52. *Compare* **magainin**.

dermatan sulfate *or* **chondroitin sulfate B** any of a group of glycosaminoglycans with repeats consisting of $\beta1\rightarrow4$-linked L-iduronyl-$(\beta1\rightarrow3)$-*N*-acetyl-D-galactosamine 4-sulfate units. They are important components of ground substance or intercellular cement of skin and some connective tissues (*see* **matrix** (def. 3)). They usually occur attached to proteins as proteoglycans, the mode of linkage being the same as for **chondroitin sulfate**. A hereditary deficiency in the ability to degrade dermatan sulfate characterizes certain types of **mucopolysaccharidosis**.

dermatosparaxis a disorder of cattle, sheep, cats, and dogs characterized by generalized skin fragility, joint laxity, and early death from sepsis. In cattle it is caused by inactivating mutations in procollagen N-endopeptidase.

dermis *or* **corium** the thick layer of tissue forming part of the skin and lying beneath the epidermis. It consists of loose connective tissue in which are blood capillaries, smooth-muscle fibres, sweat glands and sebaceous glands with their ducts, hair follicles, and sensory nerve endings. —**dermal, dermic** *adj.*

dermorphin any of a group of heptapeptide amides of the general structure H-Tyr-DAla-Phe-Gly-Tyr-Xaa-Ser-NH$_2$, isolated from the skin of frogs of the genus *Phyllomedusa,* and from some other amphibians; they show high- and long-lasting opiate-like activity.

DES *abbr. for* diethylstilbestrol.

des+ *prefix* **1** (in the trivial name of a polycyclic compound, e.g. a steroid) indicating the removal of a terminal ring, with addition of a hydrogen atom at each junction arm with the adjacent ring; it should be followed by the italic capital letter designating the terminal ring in question, e.g. des-*A*-androstane. **2** (in the trivial name of a polypeptide) indicating the removal of an amino-acid residue, terminal or otherwise; it should be followed by the name or symbol for the amino acid in question and an arabic number indicating its position in the normal polypeptide, e.g. des-7-proline-oxytocin, or des-Pro^7-oxytocin. **3** indicating the removal of a specified atom or group from a molecule of an indicated substance; *see* **desamido+, desthio+**; *see also* **descarboxy-clotting factor**. **4** *(formerly) a variant of* **de+** (still preferred in some languages, e.g. French, German).

desalinate to remove salt, especially from seawater to render it suitable for drinking or irrigation. *Compare* **desalt**. —**desalination** *n.*

desalt to remove small, usually inorganic ions from a sample. Methods that may be used include electrodialysis, electrophoresis, gel filtration, and ion-exchange chromatography. *See also* **deionize**. *Compare* **desalinate**.

desamidation the hydrolysis of one or more carboxamide groups in a molecule to carboxyl groups.

desamido+ *prefix* indicating **desamidation** of a molecule, especially of a peptide.

desaturase *general name for* any enzyme catalysing a desaturation reaction. The name specifically often refers to any of several fatty-

acyl-CoA desaturases, enzymes coming within sub-subclass EC 1.14.99; in animals these are associated with the endoplasmic reticulum, require dioxygen, and insert double bonds into saturated and *Z*(i.e. *cis*)-unsaturated fatty acids. They also occur in higher plants, protozoa, and fungi. These enzymes are specific for position in the fatty-acyl chain, and are termed Δ^5-desaturase, Δ^9-desaturase, and so on, according to the distance from the carboxyl carbon atom to the first carbon atom of the double bond. In mammals, there is a lack of any desaturase that inserts a double bond nearer to the terminal methyl group of the fatty acid than nine carbon atoms, hence the need for *n*-3 and *n*-6 precursor fatty acids (the **essential fatty acids**). *See also* **linoleic family, linolenic family**.

desaturation **1** any process or reaction in which an organic compound becomes **unsaturated** (def. 2), e.g. by removal of two hydrogen atoms, or a hydrogen atom and a hydroxyl group, from adjacent carbon atoms. **2** any process in which ligands are removed from a macromolecule so that all the binding sites for that ligand are no longer occupied. —**desaturate** *vb.*; **desaturated** *adj.*

descarboxy-clotting factor any of the abnormal blood clotting factors containing glutamic- instead of γ-carboxyglutamic-acid residues. They are formed in animals deficient in vitamin K, or during administration of vitamin K antagonists. *Compare* **PIVKA**.

descending chromatography any chromatographic technique in which a liquid mobile phase runs downwards through the supporting matrix.

descriptive name name of an organic substance that is more descriptive of action or function and often more convenient than its **trivial name, semisystematic name, systematic name**, or other type of name permitted by the rules of chemical nomenclature.

desensitization *or* **desensitisation 1** a spontaneous decline in response resulting from continuous application of agonist, or to repeated applications or doses; such attenuation may result within minutes of prior stimulation, or may take hours to develop. It is recommended that this term be used where **fade** or **tachyphylaxis** are considered to involve the receptor itself, or to be a direct consequence of receptor activation. **Homologous desensitization** results from prior stimulation with the receptor's own agonist, whereas **heterologous desensitization** results from the activity of other agonists acting at other receptors. On the longer time scales, desensitization may result from receptor internalization in **coated pits**, and over shorter time scales from modification of the receptor in a way that uncouples it from signal transduction systems, e.g. by phosphorylation (*see* **adrenergic receptor kinase**). **2** the modification of an enzyme, either by mutation or chemically, so that it no longer responds to **effectors** but retains its catalytic activity. **3** the suppression of an established (immunological) hypersensitivity by injection of antigen, usually in very small doses. In delayed hypersensitivity the antigen reacts with sensitized cells, so neutralizing them. In immediate hypersensitivity the antigen either causes blocking antibody production (in atopy) or forms nontoxic complexes with existing antibodies (in anaphylaxis). —**desensitize** *vb.*; **desensitized** *adj.*

deserpidine   desmethoxyreserpine,   11-desmethoxy-*O*-(3,4,5-

trimethoxybenzoyl)reserpic acid methyl ester; an alkaloid derived from roots of *Rauwolfia canescens* with pharmacological actions similar to the parent methoxy compound, **reserpine**.

desferrioxamine a siderophore of the hydroxamate type derived from *Streptomyces pilosus* that forms a chelate complex specifically with Fe(III) ions, yielding **ferrioxamine**. It is used parenterally to treat acute iron poisoning and to reverse iron and aluminium overload. Its natural function is to transport iron into the microbial cell and/or to make iron available for the synthesis of heme.

desiccant a substance such as $CaCl_2$ or P_2O_5 that has a high affinity for water and is used to remove moisture from a gas or another substance; a drying agent.

desiccate to remove (most of) the water from a substance or material containing moisture, especially with the aid of a **desiccant**; to dry. —**desiccation** *n*.

desiccator an apparatus in which to effect desiccation.

desmin or (*formerly*) **skeletin** a 50–55 kDa protein found in the Z disk of skeletal and cardiac muscle cells, and in **intermediate filaments** in smooth muscle and non-muscle cells.

desmo+ or (*before a vowel*) **desm+** *comb. form* denoting ligament or bond.

desmofibrin a stabilized insoluble fibrin polymer formed from soluble fibrin polymer (*see* **fibrin**) in the final stage of **blood coagulation** by the action of factor XIII (fibrin-stabilizing factor) (*see* **blood coagulation**).

desmoglein a type I **membrane protein** of the mature desmosomal junction, involved in the interaction of plaque proteins and intermediate filaments, mediating cell–cell adhesion. It contains **cadherin**-like repeats.

desmolase *former term for* any enzyme that catalyses the formation or rupture of a carbon–carbon bond.

desmoplakin any of several intracellular desmosome proteins that are probably involved in attaching intermediate filaments (e.g., keratin filaments in epithelial cells, desmin filaments in heart cells) to the desmosome. *See also* **plakoglobin** (for desmoplakin III).

desmopressin 1-(3-mercaptopropanoic acid)-8-D-arginine vasopressin, desamino-[8-D-arginine]vasopressin; a synthetic analogue of **vasopressin** with greater antidiuretic activity and less pressor activity than the parent compound.

desmosine 4-(4-amino-4-carboxybutyl)-1-(5-amino-5-carboxypentyl)-3,5-bis(3-amino-3-carboxypropyl)pyridinium; a tetra(α-amino acid) isolated from hydrolysates of the fibrous protein

elastin. Desmosine and **isodesmosine** covalently cross- link two or more polypeptide chains in this protein, the postulated mode of formation of both compounds being condensation of two residues of **allysine** in one polypeptide chain with one allysine and one lysine residue in a second chain.

desmosome *or* **macula adherens** a patchlike intercellular junction found in vertebrate tissue. It consists of parallel zones of two cell membranes, separated by an interspace of 25–35 nm and having dense fibrillar plaques in the subjacent cytoplasm. The interspace is continuous with the intercellular space and contains a dense, central plaque of material rich in protein. These proteins are known as linker proteins and consist of cell–cell adhesion molecules of the **cadherin** type. Desmosomes are important in cell-to-cell adhesion and are particularly numerous in stratified squamous epithelium that is subject to mechanical stress. A similar type of intercellular junction, the **zona adherens**, sometimes also referred to as a desmosome, encircles the cell like a belt. —**desmosomal** *adj*.

desmotubule *see* **plasmodesma**.

desorb to remove adsorbed substances from a surface. —**desorption** *n*.; **desorbed** *adj*.

desosamine 3,4,6-trideoxy-3-(dimethylamino)-D-*xylo*-hexose; an aminodeoxy sugar component of several **macrolide** antibiotics, including **erythromycins** A, B, and C.

β-D anomer

desoxy+ *an obsolete variant of* **deoxy+**.

destain to remove dye in excess of that required to stain specifically the material in question, e.g. in staining proteins in an electrophoretogram. —**destaining** *n*.

destainer any apparatus used for removing excess dye electrophoretically from stained gels.

desthio+ *prefix* indicating the replacement of a sulfur atom by two hydrogen atoms.

destrin an **actin binding protein** similar to **cofilin** with wide tissue distribution. It severs actin filaments and binds to actin monomers.

Desulfovibrio a genus of Gram-negative, obligately anaerobic, asporogenous, chemotrophic bacteria that obtain energy by anaerobic respiration using reducible inorganic sulfur compounds.

desumoylation *see* **SUMO**.

detergent 1 a cleansing agent, especially a soap or (usually synthetic) soaplike substance that, when used in aqueous solution, cleanses by reducing surface tension. **2** having the properties of a detergent. *See also* **BigCHAP, Brij, $C_x E_y$, cetyltrimethylammonium bromide, CHAPS, CHAPSO, cholic acid, n-decyl-β-D-glucopyranoside, n-decyl-β-D-maltopyranoside, deoxyBigCHAP, deoxycholic acid, digitonin, dodecylglucopyranoside, dodecylmaltoside, n-dodecyl-N, N-dimethylglycine, Genapol, glycocholate, glycodeoxycholate, heptyl glucoside, heptyl thioglucoside, hexyl glucoside, lauryldimethylamine oxide, sodium dodecylsulfate, Lubrol, Mega-8, nonyl glucoside, NP-40, octyl glucoside, octyl thioglucoside, taurocholate, taurodeoxycholate, Triton, Tween, Zwittergent**.

detergent gel electrophoresis *an alternative name for* **SDS-polyacrylamide gel electrophoresis**.

deteriosome a microvesicle that is found in eukaryotic cells and is rich in phospholipase A_2 (*see* **phospholipase**). It is believed to represent a stage in membrane degradation.

deterministic a property or relationship that holds to some defined rule or formula. For example, the equation $v = f\lambda$ describes the relationship between the velocity (v), frequency (f) and wavelength (λ) of light. –**deterministically** *adv*.

detoxication *or (sometimes)* **detoxification**; the act or process of counteracting or rendering innocuous a poisonous substance. In an organism this involves various enzymic reactions by which harmful compounds, whether produced in the organism or introduced into it, are converted into (normally) less harmful substances and into more readily excretable products. Enzymic detoxification is brought about by chemical modification (frequently by oxidation) and/or conjugation with a normal metabolite. The enzymic process does not always result in a less toxic product and the term is sometimes restricted to those instances in which a less toxic product actually occurs, as distinct from **intoxication**. Enzymic detoxification may be divided into two phases. Phase 1 concerns metabolic transformation by a wide variety of oxidations, reductions, hydrolyses, etc., usually resulting in the introduction of functional groups that increase the polarity of the molecule and act as centres for phase 2 reactions. Phase 2 involves conjugation, in which the foreign compound itself or the compound resulting from phase 1 reactions is combined with endogenous substances, such as acetate, glucuronate, sulfate, or an amino acid, or is alkylated by introduction of methyl, or other groups, to form the more water-soluble ultimate excretory metabolite. *See also* **lethal synthesis**. —**detoxicate** *vb.*

detoxification 1 the removal of poison. **2** *(sometimes)* **detoxication**.

detoxify 1 to remove a poison or the effect of a poison. **2** to bring about **detoxication**.

deuterate to combine or label with **deuterium**. —**deuteration** *n.*; **deuterated** *adj.*

deuterium *symbol*: ^{2}H *or (deprecated)* D; a stable isotope of hydrogen, relative atomic mass 2.014, sometimes known as heavy hydrogen. Its relative abundance in natural hydrogen is 0.015 atom percent. It was much used as a tracer in studies of the chemistry and biochemistry of compounds containing nonlabile hydrogen before the ready availability of tritium and carbon-14; its use continues in tracer experiments with detection by NMR spectroscopy, and where the use of radioactive compounds is contra-indicated. *See also* **heavy water, protium, tritium**.

deuterium (discharge) lamp a **hydrogen (discharge) lamp** filled with dideuterium in place of dihydrogen; this substitution increases the intensity 3-to 5-fold for the same power consumption and extends the upper limit of the continuum further into the visible range.

deutero+ *or (before a vowel)* **deuter+** *comb. form* **1** denoting that an indicated chemical compound or class of compound contains deuterium. **2** denoting second or secondary.

deuteron 1 *symbol*: ^{2}H$^+$; the cation derived from an atom of deuterium. *Compare* **hydron**. **2** *symbol*: d; a particle of nucleon number two, having a charge equal and opposite to that of an **electron** and having a mass of 2.0136 dalton.

develop 1 to process an exposed photographic film with a **developer** in order to render the image visible. **2** to render visible the colourless substances on a chromatogram or electrophoretogram by treatment with a stain or chromogenic reagent. **3** to cause a solvent to flow through a chromatographic support in order to separate the components of an applied mixture. **4** *(of a cell, organ, or organism)* to undergo or cause to undergo a series of orderly changes whereby a mature state is attained. —**development** *n.*

developer 1 a solution of a reducing agent that renders visible the latent image on an exposed photographic film or paper by reducing to metallic silver those areas of silver halide that had been exposed to light. **2** a solution of a stain or chromogenic reagent used to render visible the colourless substances on a chromatogram or electrophoretogram.

Dewar flask *or* **Dewar** a double-walled vessel or flask, usually of silvered glass but sometimes of metal, in which the space between the walls is evacuated to prevent conduction and convection of heat between the inner and outer walls. It is used to maintain the contents at temperatures either higher or lower than that of the environment. [After its inventor Sir James Dewar (1842–1923), British chemist and physicist.]

dexamethasone 9α-fluoro-16α-methylprednisolone; a synthetic **glucocorticoid** with negligible mineralocorticoid activity that has actions similar to adrenal corticosteroids.

DEXD/H box an element of protein sequence that is a variant of the **DEAD box** in which the third residue is variable and the fourth residue can be D or H. It is present in several putative NTP-dependent RNA helicases that frequently form part of large ribonucleoprotein assemblies (e.g. spliceosomes), and are involved in many aspects of RNA turnover and in replication of many viruses. *See* **Dicer**.

dextr+ *see* **dextro+**.

dextran any glucan of very high relative molecular mass and consisting of linear chains of $\alpha(1{\rightarrow}6)$-linked D-glucose residues, often with $\alpha(1{\rightarrow}2)$- or $\alpha(1{\rightarrow}3)$-branches. Native dextran, produced by a number of species of bacteria of the family Lactobacilliaceae, is a polydisperse mixture of components of M_r <10^3 to >10^7. Specific fractions of (partially hydrolysed) native dextran that have a narrow and well-defined size distribution may be designated by a numeric suffix equal to 10^{-3} × mass-average molar mass; e.g. Dextran T 500. Dextrans used in medicine are acid-degraded native dextrans.

dextranosome a **lysosome** loaded with dextran. It has higher than normal density.

dextransucrase EC 2.4.1.5; *systematic name*: sucrose:1,6-α-D-glucan 6-α- D-glucosyltransferase; *other name*: sucrose 6-glucosyltransferase. An extracellular bacterial enzyme that catalyses the reaction:

$$\text{sucrose} + (1,6\text{-}\alpha\text{-D-glucosyl})_n = \text{D-fructose} + (1,6\text{-}\alpha\text{-D-glucosyl})_{n+1}.$$

Streptococcus mutans dextransucrases are of several types: dextransucrase S produces water-soluble glucans ($\alpha(1{\rightarrow}6)$-linked glucose); dextransucrase I produces water-insoluble glucans ($\alpha(1{\rightarrow}3)$-linked glucose and some $\alpha(1{\rightarrow}6)$-linkages); and dextransucrase SI produces both types; the sequences are related. *See also* **sucrase** (def. 3).

dextrin any of the poly-D-glucosides of intermediate chain length that are formed during the hydrolytic degradation of starch or glycogen by enzymes, acid, or heat. *See also* **limit dextrin, Schardinger dextrin**.

α-dextrin endo-1,6-α-glucosidase EC 3.2.1.41; *other names*: pullulanase; pullulan 6-glucanohydrolase; limit dextrinase; debranching enzyme; amylopectin 6-glucanohydrolase; the starch-debranching enzyme, which hydrolyses $(1{\rightarrow}6)$-α-D-glucosidic linkages in pullulan and starch to form maltotriose.

dextro- *symbol*: d-; a prefix, no longer used, to denote a compound that is **dextrorotatory**; the dextrorotatory enantiomer of compound A is now designated as (+)-A.

dextro+ *or (before a vowel)* **dextr+** *comb. form* denoting on or towards the right. *Compare* **levo+**.

dextroamphetamine (*S*)-methylbenzeneethanamine; the potent (–)-stereoisomer of **amphetamine**. One proprietary name is Dexamphetamine.

dextrorotatory *or (US)* **dextrorotary 1** *symbol*: (+)- *or (formerly)* *d*-; describing a chemical compound that rotates the plane of polarization of a transmitted beam of plane-polarized light to the right, i.e. in a clockwise direction as viewed by an observer looking toward the light source. **2** describing any rightwards or clockwise rotation. *Compare* **levorotatory**. —**dextrorotation** *n*.

dextrorphan *see* **levorphanol**.

dextrose *traditional name (still used in pharmacy)* for D-glucose, D-(+)-glucopyranose, the dextrorotatory component of **invert sugar**. *Compare* **levulose**.

DFP *abbr. for* diisopropyl fluorophosphate.

dG *symbol for* a residue of the deoxynucleoside deoxyguanosine (alternative to dGuo).

DGGE *abbr. for* denaturing gradient gel electrophoresis.

dGMP *abbr. for* deoxyguanosine monophosphate, the common name for 2′-deoxyguanosine 5′-phosphate; 2′-deoxy-5′-guanylic acid; guanine 2′-deoxyriboside 5′-phosphate.

dGuo *symbol for* a residue of the deoxynucleoside deoxyguanosine; 2′-deoxyribosylguanine; guanine 2′-deoxyriboside (alternative to dG).

dGuo5′P *symbol for* deoxyguanosine 5′-phosphate.

dGTP *abbr. for* deoxyguanosine triphosphate, the common name for 2′-deoxyguanosine 5′-triphosphate, 2′-deoxy-5′-guanylyl diphosphate, guanine 2′-deoxyriboside 5′-triphosphate.

DH+ *abbr. for* dehydro+ or dihydro+ or dihydroxy+, sometimes prefixed to chemical names or alternatives. Its use is deprecated.

DH5 alpha a strain of *Escherichia coli* used in gene cloning using pUC and similar plasmid vectors and permitting **alpha complementation**.

DHAS *abbr. for trans*-dehydroandrosterone, i.e. **dehydroepiandrosterone**.

DHEA *abbr. for* dehydroepiandrosterone.

DHFR *abbr. for* dihydrofolate reductase.

DHPLC *abbr. for* denaturing high-performance liquid chromatography.

DHPR *abbr. for* dihydropteridine reductase.

DHU arm *abbr. for* dihydrouridine arm of transfer RNA.

dhurrin L-*p*-hydroxymandelonitrile-*β*-D-glucopyranoside; a glucoside from the fruit of young *Sorghum vulgare*. *See also* **amygdalin**.

di+ *comb. form* **1** *(in chemical nomenclature) (distinguish from* **bis+** (def. 2)) **a** indicating the presence in a molecule of two identical unsubstituted groups, e.g. diethylsulfide, 1,3-dihydroxyacetone. **b** indicating the presence in a molecule of two identical inorganic oxoacid residues in anhydride linkage, e.g. adenosine 5′-diphosphate. **2** *or* **bis+** (def. 1) denoting two, twofold, twice, doubled.

diabetes any clinical condition characterized by thirst and the passing of large volumes of dilute urine, especially **diabetes mellitus**. *See also* **diabetes insipidus**. —**diabetic** *adj., n*.

diabetes-associated peptide *abbr.*: DAP; *another name for* **amylin**.

diabetes insipidus a disease characterized by thirst and the excretion of large volumes of hypotonic (dilute) urine, caused by any lesion, nervous or endocrine, that interferes with the normal secretion of **vasopressin** by the posterior pituitary gland. *See also* **nephrogenic diabetes insipidus**.

diabetes mellitus *or* (commonly) **diabetes** a disorder in which the level of blood glucose is persistently above the normal range. Two main forms of the disease are generally recognized. **Type 1 diabetes mellitus**, the so-called juvenile type, often manifests itself in children and young adults. There is a marked failure to release **insulin** from the B cells of the **islets of Langerhans** in the pancreas, and frequently an almost complete absence of insulin in the B cells. The only effective treatment is administration of insulin, hence the frequent designation of this form as **insulin-dependent diabetes mellitus** (*abbr*.: IDDM). Type 1 diabetes mellitus is thought to arise in many patients from autoimmune attack on the B cells. Without insulin a severe **ketoacidosis** rapidly develops, with tissue wasting resulting from breakdown of muscle protein and hydrolysis of triacylglycerols in fat depots. **Type 2 diabetes mellitus** or **maturity onset diabetes** tends to manifest itself in adults, especially if obese, though the distinction between types 1 and 2 is blurred. In type 2 there may often be secretion of significant amounts of insulin, but for reasons not well understood these are not fully effective. There is by definition a failure to lower blood glucose adequately after a **glucose tolerance test**; patients of this type seldom develop severe ketoacidosis. **Tolbutamide** and related sulfonylureas can be used to stimulate insulin release in type 2. In both types of diabetes mellitus there may develop serious complications, including neuropathy, retinopathy, atherosclerosis, peripheral vascular disease, and nephropathy. Type 2 diabetes mellitus is also referred to as **non-insulin-dependent diabetes**.

diabetes mutation a mutation first identified in mice that spontaneously develop diabetic symptoms. It appears to be a mutation in the **leptin receptor** gene.

diabetic a sufferer from diabetes.

diabetogenic causing or being caused by **diabetes**.

diabetogenic hormone *former term for* **somatotropin**.

Diablo *abbr. for* direct IAP-binding protein with low pI; *other name*: second mitochondria-derived activator of caspases (*abbr*.: Smac); a mammalian proapoptotic protein that is released, like apoptosis inducing factor (AIF) and cytochrome *c*, from the intermembrane space of mitochondria on induction of **apoptosis**. It binds **IAP** and abrogates the inhibition of caspases 3 and 9.

diabolic a term suggested to categorize enzymes that appear to serve a complex or multiple metabolic function. *See also* **amphibolic**.

diacetylchitobiose 4-*O*-(2-acetamido-2-deoxy-*β*-D-glucopyranosyl)-2-acetamido-2-deoxy-D-glucose; the repeating unit of **chitin**. *See also* **chitobiose**.

diacetylmorphine *other names*: diamorphine, heroin; a semisynthetic narcotic analgesic derived from **morphine** by acetylation and used either as the free base or its hydrochloride. It has more potent analgesic activity than the parent compound but otherwise its actions are similar. Noted for its dependence-inducing actions and potential for abuse, it is a controlled substance under the US Code of Federal Regulations; import and manufacture of diacetylmorphine and its salts are forbidden in the USA and other countries. *See also* **codeine**.

diacylglycerol *or (formerly)* **diglyceride** any (1,2- or 1,3-) diester of glycerol and two fatty acids; the latter may be identical or different.

diacylglycerol O-acyltransferase EC 2.3.1.20; *other name*: diglyceride acyltransferase; an enzyme that catalyses the synthesis of triacylglycerol from acyl-CoA and 1,2-diacylglycerol with release of CoA.

diacylglycerol cholinephosphotransferase EC 2.7.8.2; *other names*: phosphorylcholine–glyceride transferase; alkyl-acylglycerol cholinephosphotransferase; 1-alkyl-2-acetylglycerol cholinephosphotransferase; an enzyme of the pathway for the *de novo* synthesis of phosphatidylcholines. It catalyses the formation of a phosphatidylcholine from CDPcholine and 1,2-diacylglycerol with release of CMP.

diacylglycerol kinase *abbr.*: DGK; EC 2.7.1.107; *systematic name*: ATP:1,2-diacylglycerol 3-phosphotransferase; *other name*: diglyceride kinase. An enzyme that catalyses the reaction between ATP and 1,2-diacylglycerol to form ADP and 1,2-diacyl-*sn*-glycerol 3-phosphate (phosphatidic acid). It plays an important part in the response of cells to agonists that stimulate the **phosphatidylinositol cycle**, being involved in the resynthesis of inositolphospholipids. The enzyme is stimulated by Ca^{2+} and phosphorylated by protein kinase C; it is an **EF-hand** protein.

diacytosis the process of transporting material across a cell, especially an epithelial cell, by **pinocytosis** at one surface, movement of the **pinocytic vesicle** across the cell, and **exocytosis** of its contents at another part of the cell surface.

diafiltration the process of separating microsolutes, e.g. salts, from a solution of larger molecules (or of exchanging them for different microsolutes) by **ultrafiltration** with continuous addition of solvent (or of a solution of the new microsolutes). This rapidly removes the original microsolutes from the solution, whose volume remains constant.

diagnostic performance or **diagnostic power** a measure of the ability of a discriminator to distinguish true from false matches at a given scoring threshold.

diagnostic power an alternative term for **diagnostic performance**.

diagonal chromatography a two-dimensional chromatographic technique used to determine the sensitivity of a constituent of a mixture to some (photo)chemical process, e.g. oxidation. The sample is chromatographed in one direction, the process carried out in situ, and the specimen is then rechromatographed at right angles. Compounds unmodified by the treatment all lie on a diagonal line across the chromatogram.

diagonal electrophoresis a two-dimensional electrophoretic technique for determining the position of disulfide bonds in a polypeptide. The polypeptide is partially hydrolysed and the resulting smaller peptides separated by paper electrophoresis in one direction. After treatment on the paper with performic acid vapour, they are subjected to a second electrophoresis step at right angles. Those peptides not falling on a diagonal line on the paper contain cysteic acid residues formed by the oxidation of disulfide groups.

diakinesis the final phase of prophase I in **meiosis**.

dialdose any monosaccharide derivative containing two (potential) aldehydic carbonyl groups. Compare **aldose**, **diketose**.

diallyltartardiamide abbr.: DATD; N,N′-diallyl-L-tartramide, $[CH(OH)–CO–NH–CH_2–CH=CH_2]_2$; a compound sometimes used as a cross-linking agent in the preparation of polyacrylamide gels for electrophoresis or related separative procedures. Its structure contains a 1,2-diol grouping, which is susceptible to **periodate oxidation**. Such gels may thereby be solubilized after use to facilitate recovery of separated substances.

dialysable or (esp. US)) **dialyzable** capable of being purified or separated by **dialysis**. The term has been used also, variously, to describe material that either can or cannot traverse a dialysis membrane. Some authorities prefer the term **diffusible** for material that can traverse a dialysis membrane. See also **dialysate**.

dialysate or (esp. US) **dialyzate** the solution resulting from **dialysis** that is free from colloidal material, i.e. free from material unable to pass through a dialysis membrane; the opposite of **retentate**. In the past the term was used confusingly to describe both the material retained by the membrane and the material that has passed through a membrane. Some authorities prefer the term **diffusate** for the resulting colloid-free material and the term **analysis residue** for the retentate. See also **dialysable**.

dialyser or (esp. US) **dialyzer** an apparatus consisting of two or more compartments separated by **dialysis membrane(s)**, in which to carry out **dialysis**.

dialysis a process in which solute molecules are exchanged between two liquids through a membrane in response to differences in chemical potentials between the liquids. It is used especially for separating macromolecules (colloids) in solution from smaller dissolved substances (crystalloids) by selective **diffusion** through a **semipermeable membrane**, which allows the passage of all components of the system except the macromolecules. See also **diafiltration**, **diasolysis**, **electrodecantation**, **electrodialysis**, **equilibrium dialysis**, **reverse dialysis**, **thin-film dialysis**, **ultrafiltration**. [From Greek dialusis, a separation or dissolution.] —**dialyse** or (esp. US) **dialyze** vb.; **dialytic** adj.; **dialytically** adv.

dialysis bag the significant component of a simple **dialyser**, commonly made from a convenient length of **dialysis tubing** by knotting or tying off one or both ends. When filled with the colloidal solution to be dialysed it is immersed or suspended in an appropriate colloid-free aqueous medium.

dialysis cell any small device for use as a **dialyser**.

dialysis membrane any semipermeable membrane suitable for **dialysis**, often made of cellulose acetate, cellulose nitrate, or regenerated cellulose. It may be fashioned into flat sheets or into continuous tubing. Most dialysis membranes are permeable to molecules of up to 12–14 kDa.

dialysis residue an alternative name for **retentate**.

dialysis tubing seamless tubular **dialysis membrane**.

diamagnetic pair a diamagnetic dimer produced by spin coupling of two paramagnetic (molecular or atomic) species.

diamagnetism the property displayed by substances that have a very small negative magnetic susceptibility, due to the changes induced in the orbital motions of atomic electrons by an applied external magnetic field; such changes generate a magnetic field in opposition to the applied field. Whereas such a phenomenon occurs in all substances it is often masked by the opposing and stronger effects of **ferromagnetism** or **paramagnetism**. —**diamagnetic** adj.

diamine any compound containing two amino groups or substituted amino groups.

diamine oxidase see **amine oxidase (copper-containing)**.

diamino+ prefix (in chemical nomenclature) indicating the presence of two (unsubstituted or substituted) amino groups.

diamino acid any organic acid, usually an alkanoic acid, that contains two (unsubstituted or substituted) amino groups.

di(α-amino acid) any compound whose molecules may be derived from two (identical or differing) molecules of α-amino- alkanoic acid, however they are linked. The term comprehends (but is not limited to) any dipeptide.

diaminobenzoic acid abbr.: DABA; 3,5-diaminobenzoic acid; a compound that gives fluorescent products when heated in mineral acid solution with aldehydes. It is used for the microfluorimetric determination of DNA (as it does not react with RNA) and in the analysis of sialic acid.

diaminobutyric acid symbol: A_2bu; abbr.: Dab; α,γ-diaminobutyric acid; 2,4-diaminobutanoic acid; a diamino acid not occurring in proteins. The L enantiomer is sometimes found replacing meso-diaminopimelic acid in peptidoglycan of bacterial cell walls, and several residues per molecule are present in the peptide antibiotics of the polymyxin group (together with one residue of the D enantiomer in polymyxin B_1). It also occurs in higher plants as a non-protein amino acid.

diaminocaproic acid see **lysine**.

2,7-diaminofluorene a dye used in the chromogenic assay of peroxidases. It replaces benzidine, which was formerly used but is now known to be carcinogenic.

diaminopimelic acid symbol: A_2pm or Dpm; abbr.: DAP; α,ε-diaminopimelic acid, 2,6-diaminoheptanedioic acid; a α,ε-diaminodicarboxylic acid with two chiral centres. meso-Diaminopimelic acid is the stereoisomer occurring in peptidoglycan in the cell walls of Escherichia coli and all other Gram-negative bacteria. See also **diaminopimelic-acid pathway**.

diaminopimelic-acid pathway or **DAP pathway** a metabolic pathway for the synthesis of L-lysine from L-aspartic acid via α,ε-diaminopimelate. Found in bacteria, some lower fungi, and green plants, it involves an initial condensation between pyruvate and aspart-4-al, yielding, following dehydration, 2,3-dihydropicolinate. This substance is then reduced; ring opening occurs followed by succinylation to form α-(succinylamino)-ε-oxo-L-aminopimelic acid. Transamination, deacylation, and epimerization yield the corresponding meso-α,ε-diaminopimelic acid; then decarboxylation completes the overall synthesis of L-lysine. Compare **aminoadipic pathway**.

diamorphine see **diacetylmorphine**.

diaphanous a gene that is homologous in Drosophila, mouse, and human. It encodes a protein (≈1250 amino acids) that binds profilin and regulates polymerization of actin in the cytoskeleton of hair cells of the inner ear. A protein-truncating mutation in the human gene diaphanous 1 (at 5q31) leads to an autosomal dominant form of deafness. The locus for diaphanous 2 (at Xq22) is the same as that for an X-linked form of deafness.

diaphorase name formerly applied indiscriminately to any enzyme capable of catalysing the oxidation of either NADH or NADPH in the presence of an electron acceptor such as a dye (e.g. Methylene Blue, or 2,6-dichlorophenolindophenol), ferricyanide, a quinone, or a cytochrome but not dioxygen. Also formerly known as **coenzyme factor**.

diaphragm 1 *(in chemistry)* a semipermeable membrane, or porous plate, used to separate two solutions in **dialysis** or **osmosis. 2** *(in anatomy)* any separating membrane, especially the thin, domed fibromuscular partition separating the thorax from the abdomen. Flattening of the diaphragm, accompanied by expansion of the rib cage, causes air to be sucked into the lungs. **3** *(in optics)* any device, sometimes adjustable (e.g. iris diaphragm), that can limit or vary the aperture of a lens or optical system. **4** any other thin dividing membrane.

diarylpropane peroxidase *see* **ligninase.**

diasolysis a membrane permeation process for the separation of organophilic from hydrophilic solutes, akin to **dialysis** but differing in that separation depends primarily on the relative solubilities of the solutes in the membrane rather than on their molecular dimensions. Such a process may play a part in the selective passage of solutes through membranes of living cells.

diastase 1 *obsolete name for* α-amylase (EC 3.2.1.1). **2** the crude mixture of amylases obtained commercially as a yellowish white amorphous powder from malt. [From Greek *diastasis*, a separation.] —**diastatic** *adj.*

diastatic activity the enzymic activity of α-amylase.

diastereoisomer *or* **diastereomer** any **stereoisomer** that is not **enantiomeric**; it may be either chiral or achiral. The term includes *cis–trans* **isomer, configurational isomer, conformational isomer,** and **epimer.** —**diastereoisomeric** *or* **diastereomeric** *adj.*; **diastereoisomerism** *or* **diastereomerism** *n.*

diastereotopic 1 chemically-like ligands in constitutionally equivalent locations that are not symmetry related (cannot be interchanged by rotation about an axis (C_n) or alternating axis (S_n) of symmetry) are diastereotopic; the two ligands are in a stereochemically different, nonmirror-image environment. Separate replacement of each ligand by a different achiral ligand yields two products that are in a diastereoisomeric relationship. Example: replacement of the H_R hydrogen of the methylene group in (R)–CH_3–CH_2–$CHBr$–CH_3 with chlorine yields $(2R,3R)$-2-bromo-3-chlorobutane; the similar replacement of the H_S hydrogen yields the diastereoisomeric $(2R,3S)$-2-bromo-3-chlorobutane. A biochemically important example is that both of the methylene groups of citric acid have diastereotopic hydrogens. **2** the two faces of a double bond or of a planar cyclic ring system that are not related by any symmetry operation (axis, plane, centre, or alternating axis) are diastereotopic; the two faces show stereochemically different, nonmirror-image environments. Separate addition of an achiral reagent to each face yields two diastereoisomeric products. Example: hydrogenation of the C=O bond in $(3S)$–CH_3–CH_2–$CH(CH_3)$–CO–$COOH$ (in absence of any chiral influence such as a chiral catalyst) yields the two diastereoisomers, $(2R,3S)$-2-hydroxy-3-methylpentanoic acid and $(2S,3S)$-2-hydroxy-3-methylpentanoic acid. A biochemical example is the two diastereotopic faces, *A* and *B*, of the (dihydro)pyridine ring of nicotinamide-adenine coenzymes (*see* **nicotinamide-adenine dinucleotide (reduced)**). *Compare* **enantiotopic.**

diastole 1 the passive dilatation of the chambers of the heart that occurs between the rhythmical contractions. **2** the rhythmical expansion of a pulsating vacuole. *Compare* **systole.** — **diastolic** *adj.*

diastrophic dysplasia sulfate transporter *abbr.*: DTDST; a 739-residue integral membrane protein that contains 12 transmembrane segments, is produced in many mammalian tissues, and functions as a sulfate–chloride antiporter. Many mutations in the gene at 5q32-q33.1 result in insufficient sulfation of cartilage glycoproteins and are associated with various autosomal recessive skeletal dysplasias, some of which are lethal.

diatomaceous earth *or* **kieselguhr** a friable, whitish material consisting mostly of silica, derived from diatomite, a rock consisting largely of the remains of diatoms. It is used as a filter aid and as an absorbent in column chromatography. *See also* **Berkefeld filter.**

diatomic *(of a molecular entity)* composed of only two atoms. —**diatomicity** *n.*

diatrizoate the anion 3,5-diacetamido-2,4,6-triiodobenzoate. Its sodium salt forms solutions of low viscosity and high relative density. A mixture with Ficoll provides a solution having the optimal density and osmolarity for the rapid, one-step isolation of lympho-

cytes from small volumes of blood by centrifugation. It is also used as a radiopaque material.

diauxy the adaptation of microorganisms to culture media containing two different carbohydates. Growth occurs in two phases separated by a period of less rapid or zero growth. In the first phase, the organism utilizes the carbohydrate for which it possesses constitutive enzymes. During the interlude the organism synthesizes the required induced enzymes for the metabolism of the other carbohydrate, which it then proceeds to exploit in the second growth phase. —**diauxic** *adj.*

diazepam 7-chloro-1,3-dihydro-1-methyl-5-phenyl-2*H*-1,4-benzodiazepin-2-one; one of the most widely used **benzodiazepine** drugs, noted for its CNS depressant effects. Used as an anti-anxiety–hypnotic agent at subsedative doses, its actions are similar to those of chlordiazepoxide, but it has greater potency. One proprietary name is Valium.

diazepam binding inhibitor a 10 kDa, 86-residue polypeptide, that acts as an endogenous ligand for a mitochondrial receptor (formerly regarded as a peripheral benzodiazepine binding site) in steroidogenic cells (i.e., adrenocortical, glial, and Leydig cells), and regulates stimulation of steroidogenesis by tropic hormones. It also binds to the benzodiazepine binding site on the $GABA_A$ receptor, and modulates glucose-dependent insulin secretion and synthesis of acyl-CoA esters.

diazo compound any organic compound, other than an **azo compound**, that contains the bivalent **azo** group –N=N– (or =N_2) in covalent linkage. It may be (1) any aliphatic compound of general formula R–CH=N_2, in which both nitrogen atoms are linked to the same carbon atom by replacing two hydrogen atoms of the parent molecule, e.g. diazomethane, CH_2=N_2; or (2) any aromatic compound of general formula Ar–N=N–X in which one nitrogen atom is linked to a carbon atom of the aromatic ring by replacing one hydrogen atom of the parent molecule, and the other nitrogen atom is linked covalently to a group X through an atom other than carbon (except when X = CN), e.g. benzenediazohydroxide, C_6H_5–N=N–OH.

diazonium compound any aromatic organic compound of general formula Ar–N_2^+, containing the univalent cationic **diazonium group**, –N^+≡N, linked to a carbon atom of an aromatic ring system by replacing one hydrogen atom of the parent molecule. The positive charge is balanced by an anion to form a **diazonium salt**, e.g. benzenediazonium chloride, C_6H_5–$N_2^+Cl^-$.

diazotize *or* **diazotise** to cause a primary aromatic amine to form a

diazonium compound by reaction with nitrous acid in the presence of a strong acid, according to the equation:

$$Ar\text{-}NH_2 + HONO + H^+ = Ar\text{-}N_2^+ + 2H_2O.$$

—**diazotized** *adj.*; **diazotization** *n.*

dibasic cleavage site a site within a proprotein structure at which endopeptidases act specifically to cleave the protein to one or more active polypeptides. It comprises a connected pair of basic amino-acid residues (Arg–Arg, Arg–Lys, Lys–Arg or Lys–Lys). The cleavage is effected by a family of cellular endoproteases; these have catalytic domains related to **subtilisin,** and go under the general name mammalian **prohormone convertases** (PCs). Examples are **kex2**, first described in yeast, PC2 in human insulinoma, PC3, and **furin**. Dibasic cleavage sites occur, e.g., in **proinsulin**, **proopiomelanocortin**, and **chromogranins**.

dibenzazepine any of a group of compounds with a tricyclic structure related to phenothiazines that prevent the neuronal reuptake of norepinephrine or 5-hydroxytryptamine from the synaptic cleft, and are useful as drugs in the treatment of depression, e.g. imipramine, clomipramine.

dibenzocycloheptene any of a group of agents that prevent neuronal reuptake of norepinephrine and 5-hydroxytryptamine from the synaptic cleft and are useful as drugs in the treatment of depression, e.g. amitriptyline.

dicarboxylate transport protein *former name for* **oxoglutarate/malate carrier protein.**

dicarboxylic acid any organic compound containing two carboxyl groups.

dicaryon *a variant spelling of* **dikaryon.**

DICE *abbr. for* differentiation control element; a nucleotide sequence in the 3′ untranslated region of mRNA for 15-lipoxygenase. In erythroid precursor cells DICE binds and is repressed by the hnRN proteins K and E1. The repression is removed in the terminal stages of erythrocyte differentiation and the encoded lipoxygenase mediates mitochondrial breakdown.

Dicer a multidomain endonuclease of ribonuclease III-type that is involved in the generation of small interfering RNAs (siRNA) and other forms of microRNA (miRNA) species. It is important for normal development in *Caenorhabditis elegans*. The human and *Drosophila* enzymes are homologous and contain a DEXH-box ATP-dependent RNA helicase domain, tandem RNase III motifs, and a C-terminal dsRNA-binding domain. In animals, Dicer and its cofactors are absent in differentiated cells.

2′,7′-dichlorofluorescein a dye useful as a detection reagent for lipids, yielding fluorescent spots on chromatograms under UV light.

5,6-dichloro-1β-D-furanosyl benzimidazole *see* **DRB.**

dichroic ratio the ratio of the absorbance of light polarized perpendicularly to the axis of a nonisotropic sample (e.g. the fibre axis of a protein or nucleic-acid molecule or one axis of a crystal) to the absorbance of light polarized parallel to the sample axis. If the dichroic ratio is greater than unity, the vibrational transition observed is said to exhibit **perpendicular dichroism**; if less than unity, it is said to show **parallel dichroism.**

dichroism 1 the directional effect observed in the absorption of electromagnetic radiation resulting from the relative orientations of the absorbing chromophore and the direction of the electric vector in the polarized electromagnetic radiation. Thus the absorption of such radiation may be restricted only to atoms or groups that vibrate in a specific direction or to a component of the radiation that is polarized in a specific direction. *See* **circular dichroism, dichroic ratio, electric dichroism.** 2 concentration- and path-length-dependent changes in the frequencies of electromagnetic radiation absorbed by solutions of certain substances. —**dichroic** *adj.*

dicoumarol *or* **dicumerol** *or (esp. US)* **dicumarol** (*formerly* bishydroxycoumarin);3,3′-methylenebis(4-hydroxycoumarin), 3,3′-methylenebis(4-hydroxy-2H-1-benzopyran-2-one); a substance originally isolated from spoiled sweet clover but now synthetic, that prevents the normal formation of prothrombin and other blood coagulation factors. It interferes with the action of vitamin K in γ-carboxylation of certain glutamic residues in the precursors of the coagulation proteins. It is used therapeutically as an anticoagulant, and is also used as a rodenticide. **Warfarin** is one of the **coumarins**.

dictyosome 1 *or* **golgiosome** *an alternative name for* **Golgi apparatus,** especially in cells of invertebrates and plants; there are usually several scattered dictyosomes per cell. 2 a collar-like structure that may form an anchor for the base of the flagellum in uniflagellate fungi.

Dictyostelium discoideum a cellular **slime mould**. It exists in the vegetative form as unicellular, haploid **myxamoebae**, 8–12 μm maximum dimension, using bacteria as food. In the absence of bacteria, chemotactic aggregation of myxamoebae occurs, mediated by myxamoebal secretion of acrasin (i.e., cyclic AMP), to form an elongated, finger-like, multicellular **pseudoplasmodium**, 1–4 mm maximum dimension. Aggregation is assisted by the synthesis of a multivalent carbohydrate-binding lectin, **discoidin**, found in the outer membrane of starved cells. The pseudoplasmodium differentiates to form a single, unbranched stalk, the **sorocarp**, which carries a spheroidal or ovoid mass of yellow spores, each commonly 5–10 μm × 3–6 μm with a thick cellulose-containing wall. Germination of spores only occurs in the presence of an adequate supply of amino acids.

dicyclohexylcarbodiimide *see* **DCC**. *See also* **carbodiimide**.

didehydro+ *prefix (in chemical nomenclature)* denoting the loss of two hydrogen atoms. *See also* **dehydro+**.

3,4-didehydroretinal *recommended trivial name for* (3-)dehydroretinal; (3-)dehydroretinaldehyde; retinal$_2$; vitamin A$_2$ aldehyde; retinene$_2$; (2E,4E,6E,8E)-3,7-dimethyl-9-(2,6,6-trimethylcyclohex-1,3-dien-1-yl)nona-2,4,6,8-tetraen-1-al; the didehydro derivative of **retinal** (def. 2). This all-*E*-isomer, 3,4-didehydro-all-*trans*-retinal, often termed all-*trans*-retinal$_2$, plays a role in the visual process in freshwater fish and some amphibians analogous to that of all-*trans*-retinal in other vertebrates; *see* **retinal** (def. 2). Its isomerization product, the (2E,4Z,6E,8E)-stereoisomer, generally termed 11-*cis*-retinal$_2$, combines (as a **Schiff base**) with an opsin to form (in cone cells) **cyanopsin** or (in rod cells) **porphyropsin**. *Note*: The numbering system of the systematic name is different from that of the trivial name, the latter following the rules used for carotenoids.

(all-*E*) isomer

(2E,4Z,6E,8E) isomer

3,4-didehydroretinoic acid *recommended trivial name for* (3-)dehydroretinoic acid; vitamin A_2 acid; (2E,4E,6E,8E)-3,7-dimethyl-9-(2,6,6-trimethylcyclohex-1,3-dien-1-yl)nona-2,4,6,8-tetraenoic acid; the didehydro derivative of **retinoic acid**.

3,4-didehydroretinol *recommended trivial name for* (3-)dehydroretinol; retinol$_2$; vitamin A_2; vitamin A_2 alcohol; (2E,4E,6E,8E)-3,7-dimethyl-9-(2,6,6-trimethylcyclohex-1,3-dien-1-yl)nona-2,4,6,8-tetraen-1-ol; the didehydro derivative of **retinol**, being characterized by a second double bond in the terminal ring of the molecule. It is the predominant form of vitamin A in freshwater fishes. *See also* **vitamin A**.

dideoxy+ *prefix (in organic chemical nomenclature)* **1** denoting the replacement of two hydroxyl groups by two hydrogen atoms. **2** (*abbr.*: dd+) indicating the presence in a nucleoside of a residue of 2,3-dideoxy-D-ribose.

dideoxyadenosine *symbol*: ddAdo *or* ddA; *abbr.*: DDA; 2',3'-dideoxyadenosine; an adenosine analogue with antiviral activity.

dideoxycytidine *symbol*: ddCyd *or* ddC; *abbr.*: DDC; 2',3'-dideoxycytidine; a cytidine analogue with antiviral activity.

9,11-dideoxy-9α,11α-epoxymethanoprostaglandin F$_{2α}$ U46619; a stable **thromboxane** A_2 mimetic.

dideoxyinosine *symbol*: ddIno *or* ddI; *abbr.*: DDI; 2',3'-dideoxyinosine; an **inosine** analogue with antiviral activity.

dideoxy method *or* **dideoxy(nucleotide) sequencing analysis** *an alternative name for* **chain-termination method** (for sequencing DNA molecules).

dideoxynucleoside triphosphate *abbr.*: ddNTP; any artificial nucleoside triphosphate in which both of the hydroxyl groups on C-2' and C-3' of the pentose moiety have been replaced by hydrogen atoms. They are used in sequencing DNA molecules by the **chain-termination method**.

DIDS *abbr. for* 4,4'-diisothiocyano-2,2'-disulfonic acid stilbene; a powerful anion-transport inhibitor

didymium glass a glass with a bluish-pink coloration due to the inclusion of a mixture of rare-earth elements, chiefly neodymium and praseodymium. It has intense, narrow, light-absorption bands, and is used for calibrating spectrometers. An example is a glass known as Corning 5120, which has absorption bands at 441, 528.7, 585, 684.8, 743.5, 754, 808, 883, and 1067 nm.

dielectric 1 a substance or medium that can sustain an electric field but does not conduct electricity; an insulator. **2** of, relating to, or being a dielectric.

dielectric absorption the strong absorption of microwave electric fields shown by polar liquids, particularly water. It is especially troublesome in **electron spin resonance spectroscopy**.

dielectric constant *an obsolete term for* **relative permittivity**.

dielectric dispersion variation of the **relative permittivity** with the frequency of the applied electric field.

dielectric increment the change in the **relative permittivity** of a solution with the change in the concentration of the solute. For low concentrations of solute in polar solvents, the dielectric increment, $\delta\ D/\delta\ c$, is equal to $(D - D_0)/c$, where D and D_0 are the relative permittivities of the solution and solvent, respectively, and c is the solute concentration.

dielectrophoresis the translational motion of neutral suspended particles relative to that of the suspending medium when the suspension is subjected to a nonuniform electric field. The movement is the result of polarization induced in the particles, the degree of which is dependent on the ratio of the relative permittivities (*formerly* dielectric constants) of particles and medium. In **positive dielectrophoresis** (that most commonly observed), movement of the particles is towards the region of highest field intensity, irrespective of its sign (*compare* **electrophoresis**). **Negative dielectrophoresis** may be observed in certain frequencies when an alternating field is used, the particles then being attracted less than the medium to the region of most intense field; at such frequencies the polarizability of the particles is not greater than that of the medium. Dielectrophoresis is used, e.g., in the concentration of certain biological products, the characterization of single-celled organisms or cellular organelles by their characteristic yield spectrum (i.e. yield vs. field frequency), and the determination of the diffusion coefficients of proteins in solution.

diene any **alkadiene** or substituted alkadiene.

+diene(+) *or (before a vowel)* **+dien+** *(in chemical nomenclature) infix or suffix* in systematic names denoting the presence in a molecular structure of an unsaturated aliphatic hydrocarbon chain containing two (conjugated or unconjugated) double bonds. The position of each double bond may be indicated by a locant for its lowest-numbered carbon atom; e.g. cyclohexa-1,3-diene, eicosa-8,11-dienoic acid.

dienoic indicating a compound having an aliphatic chain with two double bonds and a carboxyl group.

dienoyl *general name for* any acyl group derived from an unsaturated fatty acid containing two (conjugated or unconjugated) double bonds.

2,4-dienoyl-CoA reductase (NADPH) EC 1.3.1.34; *systematic name*: *trans*-2,3-didehydroacyl-CoA:NADP$^+$ 4-oxidoreductase; *other name*: 4-enoyl-CoA reductase (NADPH). An enzyme of the pathway for oxidation of unsaturated fatty acids that catalyses the reduction by NADPH of *trans,trans*- 2,3,4,5-tetrahydroacyl-CoA to *trans*-2,3-didehydroacyl- CoA, a normal substrate of β oxidation. For the bacterial enzyme, this is a direct product of the enzyme, but for the mammalian enzyme the product is *trans*-3,4-didehydroacyl- CoA, which must undergo isomerization to *trans*-2,3-didehydroacyl- CoA. *See also* **unsaturated fatty acid oxidation**.

diesterase *see* **phosphodiesterase**.

dietary deficiency a nutritional deficiency due to inadequate intake of one or more specific and necessary constituents of the diet, even though the diet may in other respects be adequate quantitatively.

dietary fibre 1 all ingested food that reaches the large intestine in an essentially unaltered state. **2** the structural parts of plant foods that are poorly digested in the gastrointestinal tract. These consist mostly of lignin and the polysaccharides celluloses, hemicelluloses, glucans, xylans, gums, and pectins. The whole-grain cereals wheat, maize, and rice are rich sources of the insoluble celluloses and lignin. Oats, barley, and rye are rich in soluble gums and pectins. Dietary fibre binds bile acids and cholesterol, delays the postprandial rise in blood glucose, and therefore decreases insulin secretion; it also retains water, which promotes the formation of larger and softer feces. Much dietary fibre is digested in the large intestine by anaerobic microorganisms into short-chain fatty acids – which are absorbed into the portal vein and used as an energy source – and into the gases carbon dioxide, hydrogen, and methane. Low fibre

intake is associated with increased incidence of constipation, diverticular disease, cancer of the large bowel, diabetes mellitus type II, and coronary artery disease.

diethyldithiocarbamate a chelating agent used for estimation of copper. The iron complex is used as a spin-trap for nitric oxide.

diethyl ether *see* **ether**.

diethylpyrocarbonate *abbr.*: DEPC; a powerful protein denaturant used to treat water to be used for dissolving RNA. DEPC treatment inactivates ribonucleases whose activity would be detrimental to RNA integrity.

diethylstilbestrol *or (esp. UK)* **diethylstiboestrol** 4,4′-(1,2-diethyl-1,2-ethenediyl)bisphenol; a nonsteroidal compound that possesses the activity of an **estrogen**. It is used therapeutically to replace natural hormones, and was formerly used as a growth promoter in livestock (this use is now prohibited in many countries, including the UK and USA). It was originally used for the treatment of prostatic cancer.

difference spectrophotometry a spectrophotometric method for investigating the effects of potential perturbants on a chemical substance or system in solution. Two samples are prepared containing identical solutions except that one contains a potential perturbant, e.g. a protein denaturant. The absorbances of the two samples at each frequency are subtracted one from another generating a **difference spectrum** and thereby highlighting any small differences between the spectra of the normal and perturbed systems.

differential centrifugation centrifugation of a suspension in a succession of increasing gravitational fields so that particles of differing sizes and/or relative densities may be separated and collected.

differential counting counting of the pulses produced in a **scintillation counter** that fall within two or more differing energy ranges selected for with a pulse-height analyser. The technique is used for counting the disintegrations due to two or more radionuclides in the same sample.

differential display PCR a type of real-time polymerase c hain reaction (RT-PCR) for comparing differences in mRNA expression between cells or tissues. It is based on the generation of a representative set of cDNA fragments corresponding to the 3′ ends of mRNAs from each source, which are run side by side on electrophoresis gels. Increases or decreases in the abundance of fragments separated by gel electrophoresis indicate changes in the abundance of mRNAs. Excision of bands of interest from the gel, cloning, and sequencing permits identification of genes whose expression has changed.

differential labelling a method of determining the part of a molecule that reacts in a particular way by blocking one part with a specific reagent and investigating whether a particular reaction will still occur. It is used, e.g., for investigating the combining sites of antibodies.

differential medium any medium used for growing microorganisms that reveals specific properties of the organism(s) grown and hence aids in identification of the organism(s). *See also* **selective medium**.

differential scanning calorimetry *abbr.*: DSC; a form of **differential thermal analysis** in which the temperatures of both the sample and the reference substance are maintained at either the same or a constant difference; the amount of heat that must flow into the sample to maintain the temperature differential during a transition is then measured.

differential thermal analysis *abbr.*: DTA; a method for analysing temperature-induced transitions in a sample. The sample and an inert reference material are heated or cooled at the same rate and the difference in temperature between them is monitored. This difference is zero or constant until a thermally induced transition occurs in the sample, when the temperature difference changes. The direction of the change shows whether the transition is endo- or exothermic.

differential titration the titration of a group, e.g. an acid group, before and after another group in the same molecule has reacted with some other reagent. It is used to determine the extent, if any, of **linkage** between the functions of the groups in question. *See* **titrate**.

differentiating-stimulating factor *see* **leukemia inhibitory factor**.

differentiation the process whereby relatively unspecialized cells, e.g. embryonic or regenerative cells, acquire specialized structural and/or functional features that characterize the cells, tissues, or organs of the mature organism or some other relatively stable phase of the organism's life history.

differentiation antigen any cell-surface antigen whose expression varies during successive developmental stages of a particular cell type.

diffraction the bending of light, or other waves, by an obstacle or aperture, into the region of the geometrical shadow of the object; any redistribution in space of the intensity of waves resulting from the presence of an object in their path causing variations of either their amplitude or their phase. The effect is particularly marked when the size of the object is of the same order as that of the wavelength of the waves. Diffraction accounts for the phenomenon observed when light passes through an aperture and falls on a screen; this takes the form of patterns of light and dark bands (with monochromatic light) or coloured bands (with heterochromatic light) near the edges of the beam and extending into the shadow. Diffraction is also responsible for the propagation of radio waves over the surface of the globe and of sound waves around solid objects. *See* **diffraction grating, electron diffraction, X-ray diffraction.**

diffraction grating a glass or quartz plate on the surface of which there is a large number of equidistant parallel black lines (500 or more per millimetre). When light or other electromagnetic radiation passes through such a grating it is diffracted in a manner dependent on the wavelength of the radiation. Other plane surfaces, e.g. metals, may be used.

diffusate a product of a process of diffusion, especially the material that enters into or traverses a barrier of restricted or selective permeability. In **dialysis** the term is sometimes preferred to **dialysate.** *Compare* **nondiffusible material.**

diffusible 1 capable of diffusing. **2** (*sometimes*) a preferred alternative term for **dialysable** (i.e. capable of passing through a semipermeable membrane used in dialysis).

diffusion 1 the net flow of molecules from a region of high concentration, or high chemical potential, to one of low concentration, or low chemical potential, due only to random thermal motion. *See also* **facilitated diffusion. 2** the irregular reflection or transmission of light or other electromagnetic radiation; scattering. **3** the act or process of dispersing or of being dispersed widely. —**diffuse** *vb., adj.*

diffusion cell *or* **diffusion chamber 1** a cell or chamber with porous walls that allow the passage of dissolved substances but not of biological cells. It is used in studying materials released by cells. **2** a device for measuring the diffusion of macromolecules and consisting of two chambers separated by a porous membrane of nearly parallel channels. The channels are large compared with the size of the macromolecules but small enough that the bulk flow of liquid through them is negligible.

diffusion coefficient *or* **diffusion constant** *symbol:* D; a coefficient of proportionality between flux and concentration gradient for a solute diffusing in a system of constant temperature and pressure in accordance with **Fick's first diffusion law** and **Fick's second diffusion law.** For an ideal solute it is a constant, but in practice usually varies with concentration. The cgs unit for D is $cm^2 s^{-1}$ (*see also* **Fick unit**) and the SI unit is $m^2 s^{-1}$.

diffusion-controlled *or* **diffusion-limited** describing a chemical reaction (or other process) whose velocity is dependent on the rate at which one or all of the reactants (or participants) can diffuse to the site at which the reaction (or process) occurs, as to the active site of an enzyme catalysing the reaction.

diffusion gradient any gradient e.g. in the concentration of a dissolved solute that forms if the solute is capable of diffusing from a region of higher concentration to a region of lower concentration until its concentration in the solution is uniform.

diffusion-limited *see* **diffusion-controlled.**

diffusion potential an electric potential difference across a membrane due to the diffusion through the membrane of more cationic charges than anionic charges, or vice versa.

diffusive characterized by diffusion. —**diffusively** *adv.*

DIFP *see* **DFP.**

DiGeorge syndrome a variety of phenotypes mostly arising from deletion of a portion of chromosome 22q11.2. The CATCH phenotype includes **c**ardiac **a**bnormality, abnormal facies, **T**-cell deficit due to hypoplasia of the thymus, **c**left palate, and **h**ypocalcemia. [After Angelo DiGeorge (1921–), US paediatrician.]

digest 1 to subject material to enzymic or chemical breakdown. **2** to subject food to **digestion** (def. 1). **3** to break down nitrogenous compounds by hot sulfuric acid to convert them to ammonium salts. **4** to treat biological materials with powerful oxidants to facilitate determination of trace elements present. **5** the mixture of compounds obtained from the enzymic or chemical degradation of a macromolecule or other substance. —**digester** *n.* an apparatus that facilitates digestion procedures.

digestion 1 the whole of the physical, chemical, and biochemical processes carried out by living organisms to break down ingested nutrients into components that may be easily absorbed and directed into metabolism. **2** the process of breaking down complex substances or materials into smaller parts either enzymically or chemically, especially in the course of their analysis.

digestive enzyme any hydrolytic enzyme that brings about or assists in **digestion** (def. 1).

digestive juice any of the secretions entering the digestive tract that contain digestive enzymes. They include saliva, **gastric juice, intestinal juice,** and **pancreatic juice.**

digestive system the digestive tract together with related organs, such as the pancreas and liver.

digestive tract *or* **alimentary tract** the tubelike passage in higher animals, extending from the mouth to the anus, in which the food is digested, from which absorption of nutrients occurs, and which serves to eliminate waste products.

diginose the D enantiomer, 2,6-dideoxy-3-*O*-methyl-D-*lyxo*-hexose, is a component of some **cardiac glycosides**

digital imaging the process of recording an image directly in digital form using a digital camera or similar device.

digitalis 1 any plant of the genus *Digitalis.* **2** a preparation of powdered dried leaves of the purple foxglove (*Digitalis purpurea*). It contains the cardiac glycosides **digitoxin, gitoxin** (a glycoside of 3,14,16-trihydroxycard-20(22)-enolide), and **gitaloxin** (16-formylgitoxin), and is used therapeutically to increase the strength of the heart's contraction and to slow auricular–ventricular conduction.

digitalose the D enantiomer, 3-methyl-D-fucose, 6-deoxy-3-*O*-methyl-D-galactose, is a component of some **cardiac glycosides,** e.g. digitoxin.

digitogenin (25*R*)-5α-spirostan-2α,3β,15β-triol; the aglycon of the glycosides (saponins) present in **digitonin.**

digitonin a mixture of at least four steroid saponins obtained from the seeds of the purple foxglove, *Digitalis purpurea.* It is useful in the fractional precipitation and analysis of cholesterol and other steroids and as a nonionic detergent in the solubilization of membrane proteins. Aggregation number 60. Its aglycon is **digitogenin.**

digitoxigenin 3β,14-dihydroxy-5β-card-20(22)-enolide; the aglycon of **digitoxin.**

digitoxin a secondary glycoside, or mixture of glycosides comprised mainly of digitoxin, that is extracted from *Digitalis purpurea*. It has actions similar to, but longer-lasting than, **digoxin**. Its aglycon is digitoxigenin. *See also* **cardiac glycoside**.

digitoxose the D enantiomer, 2,6-dideoxy-D-*ribo*-hexose, is a component of some **cardiac glycosides**.

diglyceride *former name for* **diacylglycerol** (it is still quite commonly used, as in 'diglyceride fraction', but its use is discouraged for such purposes as diglyceride is properly a generic term for any compound having two glyceryl residues, e.g. **phosphatidylglycerol**).

digoxigenin *or* **digoxygenin** $3\beta,12\beta,14$-trihydroxy-5β-card-20(22)-enolide; the aglycon of **digoxin**. In molecular biology digoxigenin is used as a hapten for labelling RNA and DNA probes. Uridine triphosphate and deoxyuridine triphosphate are modified by attaching digoxigenin at position 5 of the pyrimidine ring by a spacer arm. These analogues can then be incorporated into probes by RNA or DNA polymerases respectively. Detection of the labelled probes depends on anti-digoxigenin antibodies coupled to an enzyme such as alkaline phosphatase.

digoxin a secondary glycoside extracted from the Austrian (or woolly) foxglove, *Digitalis lanata*. It acts on the heart to potentiate contractility and reduce conductivity mainly through inhibition of Na^+,K^+-ATPase, hence its use in congestive heart failure and atrial dysrhythmia. Its aglycon is digoxigenin. *See also* **cardiac glycoside**.

dihedral having or contained by two planes or plane faces.

dihedral angle the inclination of two planes that meet at an edge.

dihedral symmetry the symmetry of a **point group**, designated D_n, containing an n-fold axis with n twofold axes perpendicular to it.

dihomo-(6,9,12)-linolenate dihomogammalinolenate, dihomo-γ-linolenate, (*all-Z*)-eicosa-8,11,14-trienoate; an intermediate in the synthesis of arachidonate from linoleate, formed by Δ^6 desaturation and chain elongation. It acts as a precursor for the PG_1 series of **prostaglandins**. *See also* **eicosatrienoic acid**.

dihydric describing a chemical compound containing two hydroxyl groups per molecule. It is used especially of **alcohols**. *Compare* **dihydroxy+**.

dihydro+ *prefix (in chemical nomenclature)* indicating the presence of two additional hydrogen atoms.

dihydrobiopterin reductase EC 1.6.99.7; an NADH-requiring enzyme of 244 amino acids, present in liver, brain, and adrenal medulla, that converts dihydrobiopterin into tetrahydrobiopterin.

It is a homodimer and essential for supplying tetrahydrobiopterin to the aromatic amino acid hydroxylases that require it. Deficiency arises from many mutations in the gene at 4p15.3 and causes a phenylketonuria phenotype that is characterized by convulsions and progressive mental deterioration. This can be treated by maintaining the patient on a low-phenylalanine diet supplemented with DOPA, 5-hydroxytryptophan, and folinic acid.

dihydrofolate *abbr.*: H_2folate (DHF not recommended); **1** 7,8-dihydrofolate, 7,8-dihydropteroylglutamate; *the trivial name for* the 7,8-dihydro derivative of **folate** (def. 1). **2** any **folate** (def. 2) that is a dihydro compound, especially any of the respective 7,8-dihydro intermediates in the conversion of nonreduced folates to their corresponding folate coenzymes (i.e. tetrahydrofolates) by the enzyme dihydrofolate reductase. *See also* **antifolate**, **tetrahydrofolate**.

dihydrofolate reductase *abbr.*: DHFR; EC 1.5.1.3; *systematic name*: 5,6,7,8-tetrahydrofolate:NADP$^+$ oxidoreductase; *other name*: tetrahydrofolate dehydrogenase. An enzyme that catalyses the interconversion of 5,6,7,8-tetrahydrofolate and 7,8-dihydrofolate, and also slowly interconverts 7,8-dihydrofolate and folate; both reactions are linked to the reduction/oxidation of NADP$^+$/NADPH. It is inhibited by **aminopterin**, amethopterin (*also termed* **methotrexate**), and **trimethoprim**.

dihydrofolate synthase EC 6.3.2.12; *other name*: dihydrofolate synthetase; an enzyme of the pathway for the *de novo* synthesis of dihydrofolate. It catalyses the formation of dihydrofolate from ATP, dihydropterate, and L-glutamate with release of ADP and orthophosphate.

dihydrogen *systematic name for* the diatomic compound H_2, commonly termed (molecular) hydrogen.

dihydrolipoamide *S*-acetyltransferase EC 2.3.1.12; *other names*: lipoate acetyltransferase; thioltransacetylase A; an enzyme that catalyses the formation of acetyl-CoA and dihydrolipoamide from *S*-acetyldihydrolipoamide and CoA. A lipoyl group is required as a coenzyme, the enzyme having three covalently linked lipoyl groups. This is the E_2 component of the **pyruvate dehydrogenase complex**. Example: dihydrolipoamide acetyltransferase component (E_2) of a human pyruvate dehydrogenase complex precursor (PDC-E_2): (incomplete sequence) ODP2_HUMAN, 615 amino acids (65.32 kDa).

dihydrolipoamide dehydrogenase EC 1.8.1.4; *other names*: lipoamide reductase (NADH); lipoyl dehydrogenase; dihydrolipoyl dehydrogenase; diaphorase; the E_3 component of α-ketoacid dehydrogenase complexes, is an FAD-flavoprotein enzyme that catalyses the oxidation by NAD$^+$ of dihydrolipoamide to lipoamide and NADH. Example (precursor) from human: DLDH_HUMAN, 509 amino acids (54.09 kDa). *See also* **oxoglutarate dehydrogenase complex**; **pyruvate dehydrogenase complex**.

dihydrolipoamide *S*-succinyltransferase EC 2.3.1.61; an enzyme that catalyses the formation of succinyl-CoA and dihydrolipoamide from *S*-succinyldihydrolipoamide and CoA. It is the E_2 component of the **oxoglutarate dehydrogenase complex**. The human precursor protein consists of 453 amino acids; the enzyme is formed from residues 68–453, 1–67 forming the transit peptide.

dihydroorotase EC 3.5.2.3; *systematic name*: (*S*)-dihydroorotate amidohydrolase; *other names*: carbamoylaspartic dehydrase; DHOase. An enzyme that catalyses the formation of *N*-carbamoyl-

L-aspartate from (S)-dihydroorotate. In many eukaryotes this is an activity of **CAD protein**.

dihydropyridine receptor a voltage-gated Ca^{2+} channel of skeletal muscle transverse tubules that specifically binds 1,4-dihydropyridine compounds. It is an integral membrane protein containing four homologous repeats that form six transmembrane helices each. It is in close proximity to the foot region of the ligand-gated **ryanodine receptor** of the sarcoplasmic reticulum.

5α-dihydrotestosterone 4,5-dihydrotestosterone; 5α-androstan-17β-ol-3-one; a potent metabolite of testosterone, formed by either of the steroid 5α-reductases. It mediates many of the differentiating, growth-promoting, and functional actions of testosterone by binding to the **androgen receptor**. Many mutations that inactivate steroid 5α-reductase 2 cause an autosomal recessive form of **testicular feminization**.

dihydrouracil 5,6-dihydrouracil, a minor pyrimidine base. It occurs in the **dihydrouridine arm** of **transfer RNA**.

dihydrouridine *symbol*: D *or* hU; 5,6-dihydrouracil riboside; 1-β-D-ribofuranosyl-5,6-dihydrouracil; a rare nucleoside occurring in the **dihydrouridine arm** of **transfer RNA**.

dihydrouridine arm *abbr*.: DHU arm; the base-paired segment of the **clover leaf structure** of **transfer RNA** that has attached to it the loop containing 5,6-dihydrouracil (*see* **dihydrouracil**).

dihydroxy+ *prefix (in chemical nomenclature)* indicating the presence of two substituent hydroxyl groups.

dihydroxyacetone *older name for* **glycerone**.

dihydroxyacetone phosphate *older name for* **glycerone phosphate**.

dihydroxyacetone synthase *see* **formaldehyde transketolase**.

dihydroxyacetone transferase *see* **transaldolase**.

2,8-dihydroxyadenine an abnormal metabolite of adenine that is formed in excess by xanthine oxidase when there is a deficiency of **adenine phosphoribosyltransferase**. It is protein-bound in the circulation and actively secreted into the urine, where it is very insoluble and results in crystalluria and possibly calculus formation.

dihydroxyeicosatetraenoate *abbr*.: diHETE; any of several products resulting from dioxygenation of arachidonate at two sites,

normally resulting from **lipoxygenase** action on **hydroxyeicosatetraenoates**. (6E,8Z,10E,14Z)-(5S,12S)-5,12-dihydroxyeicosa-6,8,10,14-tetraen-1-oate (*abbr*.: 5(S),12(S)-diHETE) is synthesized in human leukocytes by two successive lipoxygenase reactions and acts as a leukotriene B₄ antagonist. *See* **leukotriene B**. (6E,8Z,11Z,13E)-(5S,15S)-5,15-dihydroxyeicosa-6,8,11,13-tetraen-1-oate (*abbr*.: 5(S),15(S)-diHETE) is synthesized in human neutrophils and enhances neutrophil degranulation. (5Z,9E,11Z,13E)-(8S,15S)-8,15-dihydroxyeicosa-5,9,11,13-tetraen-1-oate (*abbr*.: 8(S),15(S)-diHETE) is synthesized in human leukocytes, eosinophils, and platelets and is chemotactic for human polymorphonuclear leukocytes.

3,4-dihydroxymandelate an intermediate in the formation of 4-hydroxy-3-methoxymandelate from **norepinephrine**.

3,4-dihydroxyphenylalanine *see* **dopa**.

3,4-dihydroxyphenylethylamine *see* **dopamine**.

3,4-dihydroxyphenylglycolaldehyde the product of the action of monoamine oxidase on **norepinephrine**. It is converted by an NAD⁺-dependent aldehyde dehydrogenase to 3,4-dihydroxymandelate.

1,25-dihydroxyvitamin D₃ *another name for* **calcitriol**.

diiodotyrosine *abbr*.: DIT; 3,5-diiodotyrosine; iodogorgoic acid, 3,5-diiodo-4-hydroxyphenylalanine, 2-amino-3-(3,5-diiodo-4-hydroxyphenyl)propanoic acid. Residues of the L enantiomer, together with residues of the monoiodo compound, 3-iodo-L-tyrosine, are formed by iodination of L-tyrosine residues in thyroglobulin as precursors of thyroid hormones. *See* **thyroperoxidase**.

diisopropylfluorophosphonate diisopropylfluorophosphate; dyflos; *abbr*.: DIPF *or* DFP; an inhibitor of **cholinesterase** and other **serine proteases**. It phosphorylates the serine in the enzyme's active site, causing long-lasting inhibition. Recovery of enzymic activity requires *de novo* enzyme synthesis.

diisopropylphospho *symbol*: Dip; *abbr*.: DIP; the doubly esterified phosphoric acyl group, $[(CH_3)_2CH–O–]_2P(O)–$, derived from diisopropyl fluorophosphate by loss of a fluorine atom.

dikaryon *or* **dicaryon** a living cell containing two nuclei, or a structure (e.g. a mycelium) composed of such cells. It may contain nuclei of the same genetic constitution (i.e. **homokaryons**) or of differing genetic constitutions (i.e. **heterokaryons**). —**dikaryotic** *or* **dicaryotic** *adj.*

diketose any monosaccharide derivative containing two (potential) ketonic carbonyl groups. *Compare* **ketose**.

dikinase either of the two enzymes belonging to the sub-subclass EC 2.7.9, comprising phosphotransferases with paired acceptors. For example, pyruvate,orthophosphate dikinase, EC 2.7.9.1, catalyses the reversible reaction:

$$\text{ATP} + \text{pyruvate} + \text{orthophosphate} = \text{AMP} + \text{phospho}enol\text{pyruvate} + \text{pyrophosphate}.$$

dil. *abbr. for* dilute (def. 2).

dilate to expand or cause to expand; to make larger or wider; e.g. of the pupil of the eye or a chamber of the heart. —**dilation** *or* **dilatation** *n.*

dilatometer any instrument for measuring small changes in the volume of liquids. One design consists of a (glass) bulb fitted with a graduated hollow stem in which changes in volume may be observed. In another design, the (glass) bulb is fitted with a stopper through which runs a capillary tube so that any increased volume of the contained liquid may escape; the change in mass of the bulb and its contained liquid may then be determined. —**dilatometry** *n.*

diluent a diluting agent; *see* **dilute** (def. 1).

dilute 1 a to lower the concentration of a solute in a solution (or of particles in a suspension). **b** to mix one fluid with another thereby lowering the concentration of the first. **2** *abbr.*: dil.; describing a solution (dispersion or mixture) having a relatively low concentration (of a specified solute, type of dispersed particle, or component). *Compare* **concentrated** (def. 1).

diluted (of a solution or suspension) lowered in concentration and increased in volume by addition of solvent or other diluent. *Compare* **concentrated** (def. 2).

dilution 1 the act or process of making more **dilute** (def. 2). **2** a diluted solution or suspension. **3** the (specified) extent to which a solution or suspension has been diluted.

dilution end-point analysis the determination of the concentration of antibody in a solution by allowing a given amount of antigen to react with increasing dilutions of the antibody. The most dilute solution that produces a detectable effect, e.g. precipitation, is taken as the end point.

dilution law *see* **Ostwald dilution law**.

dilution quenching the lowering of **counting efficiency** of a liquid scintillation counter by the addition of solvent to the contents of the scintillation vial. *See also* **quenching**.

dilution rate *symbol*: D; the rate, F, at which existing medium is replaced with fresh medium in a **continuous culture**, divided by the volume, V, of the culture: $D = F/V$. Under steady-state conditions the dilution rate numerically equals the specific growth rate of the culture. *See also* **auxostat**, **chemostat**.

dilution value of a buffer *symbol*: $(\Delta pH)_{1/2}$; the change in pH that occurs when a buffer solution is diluted with an equal volume of water.

dimer 1 any molecular structure of any size in which two initially identical chemical entities have become covalently combined. **2** any macromolecular structure in which two (either identical or non-identical) subunits are noncovalently associated. *See also* **heterodimer**, **homodimer**. —**dimeric** *adj.*

dimercaprol *or* **British anti-lewisite** (*abbr.*: BAL); 2,3-dimercaptopropanol; an antidote to the arsenical war gas, lewisite, that is also useful as a chelating agent and detoxicant for heavy-metal ions and as a protective agent for sulfhydryl groups in enzymes, coenzymes, or the like.

dimerize *or* **dimerise** cause a **dimer** to be formed. —**dimerization** *or* **dimerisation** *n.*

dimerizer hypothesis *or* **dimeriser hypothesis** an unsubstantiated model of facilitated diffusion proposed to explain the rates of entry of sugars into human erythrocytes from various mixtures of sugar species. According to this hypothesis, permeation occurs as a result of pairs of sugar molecules interacting with a membrane compo-

nent, followed by the association of the sugars into dimers, a form in which a reduced number of free hydrogen-bond-forming groups would be available.

dimethoxytrityl *abbr.*: DMT; 4,4'-dimethoxytriphenylmethyl; a group used for protecting hydroxyl groups (especially the 5'-hydroxyl groups of nucleosides) during chemical synthesis or other procedures. It is usually introduced by reaction with dimethoxytrityl chloride, 4,4-dimethoxytriphenylmethyl chloride. *Compare* **trityl**.

4,4-dimethoxytriphenylmethyl chloride

dimethylallyl *trans***transferase** EC 2.5.1.1; *systematic name*: dimethylallyl-diphosphate:isopentenyl-diphosphate dimethylallyl-*trans*transferase; *other names*: prenyltransferase; geranyl-diphosphate synthase. An enzyme that catalyses the reaction of dimethylallyl diphosphate with isopentenyl diphosphate to form pyrophosphate and geranyl diphosphate. It participates in the pathway for the formation of **polyprenyl** groups, sesqui- and higher terpenes, and **cholesterol**.

dimethylaminoazobenzene isothiocyanate *see* **DABITC**.

dimethylaminoazobenzene sulfonyl chloride *see* **dabsyl**.

dimethylaminobenzaldehyde *see* **Ehrlich's reagent**.

dimethylaminonaphthalene sulfonyl chloride *see* **dansyl**, **dansylate**.

dimethylformamide *see* **DMF**.

dimethyl sulfoxide an unpleasant-smelling colourless liquid that is water- and alcohol-soluble. It is used as a cryoprotectant when cells are frozen for storage and for dissolved substances that are not readily water-soluble but are to be applied to cultured cells.

dimethyltryptamine *N,N*-dimethyltryptamine; 3-[2-(dimethyl-lamino)ethyl]indole; a hallucinogenic agent extracted from several plant species including *Piptadenia peregrina*. It is also a metabolite of 5-hydroxytryptamine. It is a controlled substance under the US Code of Federal Regulations.

dimorphism 1 the existence of a chemical compound in two distinct physical forms, e.g. two different crystalline forms. **2** the existence of individuals belonging to two distinct forms within the same species of organism; e.g. in **sexual dimorphism** the males and females of the same species are markedly different. **3** the occurrence in certain plants of two distinct forms of leaves or other parts in the same individual. —**dimorphic** *or* **dimorphous** *adj.*

dinactin *see* **nonactin**.

dinitrofluorobenzene *abbr.*: DNFB; *a (less correct) name for* fluorodinitrobenzene. *See* **Sanger's reagent**.

dinitrogen *systematic name for* the diatomic compound N_2, commonly termed (molecular) nitrogen.

dinitrogenase reductase or **nitrogenase**, **Fe-protein** or **component II** of nitrogenase. A very oxygen-sensitive iron–sulfur protein that is a component of the **nitrogenase** enzyme system. It is a dimer of identical 30 kDa peptide chains. It is the electron carrier responsible for the reduction of molybdenum in **molybdoferredoxin**.

dinitrophenate the strongly yellow-coloured anion, $(NO_2)_2–C_6H_3–O^-$, formed from dinitrophenol.

dinitrophenol *abbr.*: DNP; 2,4-dinitrophenol; a pale-yellow crystalline solid. It is a powerful uncoupler of **oxidative phosphorylation**, and hence toxic.

dinitrophenyl *symbol*: Dnp; the group formed by loss of the phenolic hydroxyl group from 2,4-dinitrophenol. *See* **Sanger's method**.

dinitrophenylate to introduce a dinitrophenyl group into a compound. —**dinitrophenylation** *n*.

dinoprost a generic name for prostaglandin $F_{2\alpha}$. *See* **prostaglandin**.

dinor+ *prefix (in chemical nomenclature)* indicating shortening of a carbon chain or contraction of a ring by two methylene groups. *See also* **nor+**.

DInR *abbr. for Drosophila* insulin receptor.

Dintzis procedure a procedure for investigating the direction and rate of biosynthesis of a polypeptide. It requires the addition to a protein-synthesizing system *in vitro* of a **pulse** of a labelled form of an amino acid. After various periods of elapsed time completed molecules of the polypeptide are isolated. Over all the samples, a gradient of radioactivity (increasing from the N terminus to the C terminus of the chain) is found, indicating that the pulse tends to enter the more C-terminal positions and that the more N-terminal region has already been synthesized. By these means it was clearly shown for the first time that polypeptide chains are assembled from amino acids on a template and that synthesis takes place from its N terminus. [After Howard Marvin Dintzis (1927–), US biochemist, who conceived the procedure and described its use in 1961.]

dinuclear or **binuclear** describing a complex containing two metal atoms.

dinucleotide 1 any **oligonucleotide** consisting of two mononucleotides in 3′–5′ phosphodiester linkage. **2** any **nucleotide coenzyme** in which either of the mononucleotides adenosine 5′-phosphate or adenosine 2′,5′-bisphosphate is in 5′–5′ diphosphate (i.e. phosphoric anhydride, pyrophosphate) linkage with nicotinamide mononucleotide or riboflavin 5′-phosphate.

dinucleotide fold an alternative name for **nucleotide binding fold**.

diode 1 a semiconductor device, containing one p–n junction, used in circuits to rectify **alternating current** to a **direct current**: semiconductor diode. **2** a thermionic valve containing an anode and a cathode and functioning in a similar manner: **thermionic diode**.

dioestrus *see* **estrous cycle**.

diol any organic compound (or portion of such a compound) containing two hydroxyl groups.

diol lipid any of various neutral lipids that are fatty-acid esters of dihydroxy alcohols (i.e. are not glycerides). They occur in small amounts in seed oils; e.g. 1-palmitoleoyl-2-*trans*-vaccenoyl-*erythro*-butane-2,3-diol, which is found in seeds of *Coix lachrima*.

dioxin any of a family of over 200 chlorinated heterocyclic compounds that are persistent in the environment. The term is most commonly applied to 2,3,7,8-tetrachlorodibenzo-*p*-dioxin (TCDD), which is an impurity in the defoliant Agent Orange and in the pesticide 2,4,5-T. Dioxins are also produced when chlorinated materials such as plastics are burned. They are known to cause cancer, birth defects, miscarriages, and skin diseases (chloracne).

TCDD

dioxin receptor *see* **ARNT**.

dioxygen *systematic name for* the diatomic compound O_2, commonly termed (molecular) oxygen. *Compare* **trioxygen**.

dioxygenase any oxidoreductase enzyme of the sub-subclass EC 1.13.11. Such enzymes catalyse reactions involving the incorporation of both oxygen atoms of dioxygen into a second substrate; *see* e.g. **ethylene-forming enzyme**, **benzoate 1,2-dioxygenase**.

DIP 1 *abbr. for* the diisopropylphospho group. **2** *abbr. for* database of interacting proteins; *see* **specialist database**.

dipentene a racemic mixture of the (+) and (–) enantiomers of **limonene**.

dipeptide any **oligopeptide** consisting of two (α-)amino-acid residues.

dipeptide mimic any synthetic dipeptide that inhibits **angiotensin-converting enzyme**. An example is **captopril**.

dipeptidyl-peptidase *abbr.*: DPP; any member of a group of enzymes belonging to the sub-subclass EC 3.4.14, dipeptidylpeptide and tripeptidylpeptide hydrolases, that release an N-terminal dipeptide from an oligo- or polypeptide. **Dipeptidyl-peptidase I** (*or* cathepsin C), EC 3.4.14.1, is a lysosomal cysteine-type peptidase that acts on polypeptides, but not if the terminal amino-acid residue is Arg or Lys, and not if either of the adjacent residues is Pro. **Dipeptidyl-peptidase II**, EC 3.4.14.2, is a lysosomal serine-type peptidase that acts preferentially on tripeptides. **Dipeptidyl-peptidase III**, EC 3.4.14.4 (*other names*: red cell angiotensinase; enkephalinase B), is a cytosolic serine-type peptidase that acts on tetra- or longer peptides. **Dipeptidyl-peptidase IV**, EC 3.4.14.5 (*other names*: dipeptidyl aminopeptidase IV; Xaa-Pro-dipeptidyl-aminopeptidase; Gly-Pro naphthylamidase; post-proline dipeptidyl aminopeptidase IV), catalyses the release of an N-terminal dipeptide, Xaa-Xbb-|-Xcc, from a polypeptide, preferentially when Xbb is Pro, provided Xcc is neither Pro nor hydroxyproline. The protein is a serine proteinase, type II transmembrane glycoprotein, and a homodimer. **Dipeptidyl-dipeptidase**, EC 3.4.14.6, is a thiol-activated peptidase from *Brassica oleracea* that acts preferentially on tetrapeptides.

dip-F *symbol for* diisopropyl fluorophosphate.

DIPF *abbr. for* diisopropylphosphofluoridate. *See* **diisopropyl fluorophosphonate**.

diphenylamine reaction the reaction of an acidic solution of diphenylamine with deoxypentoses to give a blue colour. It is used in the quantitative determination of DNA, and alone or in combination with other reagents as a detection reagent.

2,5-diphenyloxazole *abbr.*: PPO; a compound used as a primary scintillant in scintillation counting. *See* **scintillation counter**. *See also* **scintillation cocktail**.

2,5-diphenyloxazole

diphosphate 1 the tetravalent anion, $P_2O_7^{4-}$, derived from diphosphoric acid (earlier known as pyrophosphoric acid), heptaoxodiphosphoric(V) acid, $(HO)_2PO–O–PO(OH)_2$. **2** any mixture of free diphosphoric acid and its anions. *see also* **inorganic diphosphate**, **pyrophosphate**. **3** any (partial or full) salt or ester of diphosphoric acid. **4** any organic compound containing two phosphoric residues linked by an oxygen atom. **5** name of the univalent group $–O–PO(OH)–O–PO(OH)_2$ (irrespective of its state of ionization). **6** *formerly*, any organic compound containing two independent phosphoric residues attached to different parts of the molecule; now incorrect, use **bisphosphate**.

(+)diphospho+ 1 *prefix to (or infix in) a chemical name indicating the presence in ester (or diester) linkage of a diphosphoric residue. **2** *formerly*, prefix to a chemical name indicating the presence of two independent phosphoric residues in ester linkages; now incorrect, use **bisphospho+**.

diphosphoglycerate *former name for* **1** 2,3-bisphospho-D-glycerate. **2** 3-phospho-D-glyceroyl phosphate.

diphosphoglycerate pocket *abbr.*: DPG pocket; *see* **bisphosphoglycerate pocket**.

diphosphoinosotide *former name for* **phosphatidylinositol 4-phosphate**.

diphosphopyridine nucleotide (oxidized) *abbr.*: DPN; *obsolete name for* the oxidized form of nicotinamide adenine dinucleotide; *see* **NAD+**.

diphosphopyridine nucleotide (reduced) *abbr.*: DPNH; *obsolete name for* the reduced form of nicotinamide adenine dinucleotide; *see* **NADH**.

diphosphoric *symbol*: *PP* or (in the one-letter convention for nucleotides) pp; *general name for* a residue of diphosphoric acid, $(HO)_2PO-O-PO(OH)_2$, whether singly or multiply attached through oxygen to other groups and whether or not any residual hydroxyl groups are dissociated.

diphthamide a modified histidine residue on EF-2. *See* **elongation factor**.

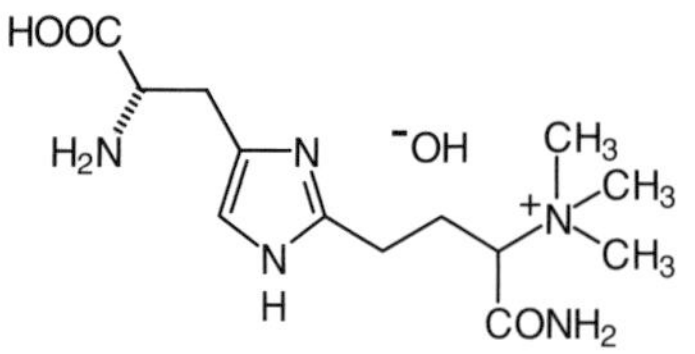

diphtheria toxin an enzyme toxin produced by strains of *Corynebacterium diphtheriae*, which are lysogenic for corynephage, whose genome contains the *tox* structural gene (carried by phage β*tox+*). The toxin has the activity of EC 2.4.2.36; *recommended name*: NAD+-diphthamide ADP-ribosyltransferase; *systematic name*: NAD+:peptide-diphthamide *N*-(ADP-D-ribosyl)transferase. It catalyses the reaction between NAD+ and peptide diphthamide to form nicotinamide and peptide *N*-(ADP-D-ribosyl)diphthamide. The toxin molecule is a proenzyme comprising a single polypeptide chain of ~62 kDa; the toxin is lethal for susceptible animals at around 100 ng per kg of body mass. On cleavage of one peptide bond and one disulfide bond, the toxin splits into an N-terminal fragment, A (21.15 kDa), the enzyme that specifically inhibits protein synthesis by catalysing transfer of the ADP-ribosyl moiety of NAD+ to eukaryotic EF-2 (*see* **elongation factor**) in the cytoplasm, and a C-terminal fragment, B (~40 kDa), that is required for recognition of specific surface receptors on sensitive cell membranes. Fragment B is necessary for the toxicity of fragment A, as it allows the latter to cross the plasma membrane.

dipicolinate 2,6-pyridinedicarboxylate; a substance that is abundant (10%) in bacterial spores. Its production is a striking feature of sporulation and it may play a part in making spores resistant to heat.

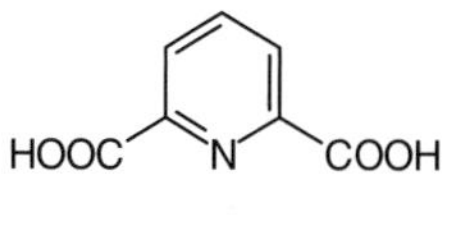

dipicolinic acid

diplochromosome a 'double chromosome', comprising the arms of two daughter chromosomes attached to a single centromere. It arises due to failure of the centromere to divide following duplication of a chromosome.

diploid describing a cell or nucleus having two sets of homologous **chromosomes**; i.e. containing twice the **haploid** number. —**diploidy** *n*.

diploid state the chromosome state of a cell in which each type of chromosome, except for the sex chromosomes, is always represented twice.

diplotene the fourth phase of prophase I in **meiosis**.

dipolar bond *or* **coordinate bond** *or* **coordinate covalence** *or* **coordinate link** *or* **dative bond** *or* **semipolar double bond** a covalent bond formed (actually or conceptually) between two initially uncharged moieties by **coordination** (def. 2). The consequent generation of opposing charges on the linked structures creates a **dipole**; e.g.

$$R_3N: +\ddot{O}: \rightarrow R_3N^+-\ddot{O}:^-.$$

The term is preferred to the alternatives, which are now either obsolescent or obsolete. *Compare* **polar bond**.

dipolar correlation time time taken for an atomic spin system to rotate through one radian from its previous orientation, the rotation being caused by interactions with other, neighbouring spins.

dipolar interaction the interaction between two **magnetic moments** due to the effect of the magnetic field of one on the other.

dipolar ion *or* **zwitterion** a molecule containing ionic groups of opposite charge in equal numbers. *Compare* **amphion**.

dipole a pair of equal and opposite electric charges or magnetic poles (especially in a molecular entity) separated by a, usually, small distance. —**dipolar** *adj.*

dipole moment *or* **electric dipole moment** *symbol*: *p* or *μ*; a vector quantity equal to the product of the absolute magnitude of either charge in a dipole (the two are equal in an uncharged molecule) multiplied by the distance separating them. If a molecule possessing a dipole moment is placed in a uniform electric field, it will be subject to torque tending to line it up in the direction of the field. $p = \Sigma Q_i r_i$ coulomb metres, where Q_i is the charge and r_i is the position vector.

dipyridamole an inhibitor of cyclic AMP phosphodiesterase and 5-lipoxygenase, used as a coronary vasodilator.

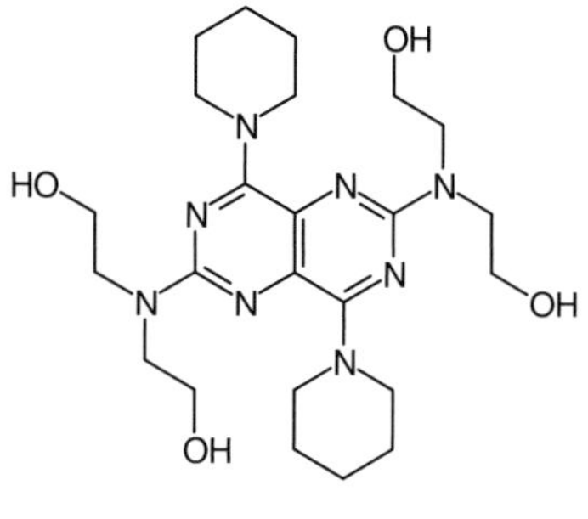

4 dipyridamole

diradical *an alternative term for* **biradical**.

direct current *abbr.*: d.c. *or* DC; an electric current that flows in only one direction.

directional cloning a procedure that ensures the insertion of a DNA insert into a plasmid or other cloning vector in a known orientation with respect to the flanking vector sequences. It is accomplished by providing each end of the insert with different restriction endonuclease sites that are also present in the multiple cloning site of the vector.

directional selection any genetic **selection** that shifts the population mean in the direction desired by the breeder.

dis+ *prefix* indicating **1** reversal. **2** deprivation, exclusion, or lack. **3** negation. **4** removal or release.

dis2 a gene in fission yeast that is the equivalent of *DIS2 S1* in budding yeast.

disabled strain any strain of organism, especially a microorganism, that because of some character, e.g. a nutritional requirement, is unable to reproduce in the normal environment, including the normal host, and can only survive in laboratory conditions.

disaccharidase any enzyme within sub-subclass EC 3.2.1 that hydrolyses a disaccharide to its constituent monosaccharides.

disaccharidase deficiency *see* **lactase deficiency**.

disaccharide any **oligosaccharide** consisting of two monosaccharide residues. It may be a **nonreducing disaccharide** or a **reducing disaccharide**, depending on the mode of linkage between the residues.

disalicylidenepropanediamine *abbr.*: DSPD; *N,N'*-disalicylidene-1,3-diaminopropane; an inhibitor of photosynthetic electron flow at the ferredoxin level.

disassociation *a variant form of* **dissociation** (def. 1).

disc *a variant spelling (esp. Brit.) of* **disk**.

DISC *abbr. for* death-inducing signalling complex; the complex formed by **FasL**, **Fas**, **FADD** and procaspase 8, that initiates the caspase cascade and apoptosis.

DISC1, DISC2 human genes at locus 1q42.1 that are disrupted in schizophrenia with a balanced translocation t(1;11)(q42.1;q14.3).

discharge **1** *(in physics)* the release of an electric charge, e.g. from a condenser or a battery, or the equalization of an electric potential difference. **2** *(in physics)* the intermittent or continuous flow of electricity through a gas by the formation of electrons and ions. **3** *(in medicine)* an abnormal secretion from a body cavity, or the secretion from a wound.

discoidin any of a group of endogenous carbohydrate-binding proteins (lectins) produced by cells of the slime mould *Dictyostelium discoideum* during differentiation. Similar to **pallidin**, they may be involved in cell adhesion.

discoidin domain *or* **discoidin-like domain** a protein domain with a jelly-roll fold first described in the lectin discoidin from the cellular slime mould *Dictyostelium discoideum*. Proteins with discoidin domains are implicated in interactions with cell-surface carbohydrates.

discoidin domain receptor any of the family of membrane receptors with a discoidin domain. They are tyrosine kinases that are activated by binding to collagen in the extracellular matrix and differ in this respect from other receptor tyrosine kinases.

discontinuous epitope an immunological determinant (i.e. **epitope**) on an antigen that comprises different sections of a protein molecule brought together by folding of the polypeptide chain. *Compare* **continuous epitope**.

discordance *(in genetics)* the state in which a character or gene is present in only one of two comparable individuals or species; for example, when only one of a pair of twins has a particular character. *Compare* **concordance**. –**discordant** *adj.* –**discordantly** *adv.*

discriminator **1** any (electronic) device that gives an output signal only when the input signal is above a certain preset level or has certain preset characteristics. **2** any criterion used in sorting objects, experimental results, signals, etc.

disease any anatomical abnormality or impairment of the normal functioning of an organism or of any of its parts other than one arising directly from physical injury. It may be caused by environmental factors (e.g. malnutrition, toxic agents, etc.), infective agents (bacteria, viruses, etc.), inherent defects in the organism (e.g. genetic disease), or any combination of these factors.

disequilibrium assay any radioimmunoassay in which the reagents are not added at the same time and are not allowed to reach thermodynamic equilibrium. *Compare* **equilibrium assay**.

Disheveled a transduction component of Wnt/Frizzled signalling, that protects β-catenin against phosphorylation and degradation via ubiquitination and proteasomes. *See* **catenin**.

disinfectant a substance that destroys or inhibits pathogenic microorganisms, but not necessarily bacterial spores, in the inanimate environment. *Compare* **sanitizer**, **sterilant**.

disinfection the destruction or removal of all organisms capable of giving rise to infection. *Compare* **sterilize** (def. 1).

disintegration *or* **nuclear disintegration** any process of nuclear **decay** involving a splitting into more nuclei or the emission of particles (either spontaneously or as a result of a collision).

disintegration constant *or* **decay constant** *symbol*: λ; for a ra-

dionuclide, the probability of the nuclear decay of one of its nuclei in unit time. It is given by:

$$\lambda = -(1/t)\ln(A/A_0),$$

where A_0 is the radioactivity at zero time and A is the activity at time t. *See also* **average life**.

disintegrin any of a group of compounds that act as inhibitors of platelet–fibrinogen interaction and other aggregations. They are found in certain snake venoms.

disjunction the separation of chromosomes at **anaphase** during nuclear division.

disk *or (esp. Brit.)* **disc 1** a circular, flattened structure, or any object having or resembling such a shape. *See also* **retinal rod**. **2**. *(in computing)* a disk-shaped device on which files are stored electromagnetically.

disk (gel) electrophoresis a technique of separation by electrophoresis in which a tube is loaded first with the sample, applied in a gel support (usually polyacrylamide), followed by a spacer gel and finally by the running gel, in which separation occurs. The gels are so formed as to give a gradient of progressively smaller pore size, and a discontinuous buffer system is used. This arrangement causes the separating components of the applied mixture to stack as a series of very thin bands and permits very high resolution.

dismembrator an ultrasonic device for rapidly disrupting cells in bacteriological and cytological studies. It consists of a probe oscillating at ≈20 kHz and an acoustical power of up to 150 W that is dipped into a cell suspension, contained in either a stainless-steel or glass beaker.

dismutase *see* **superoxide dismutase**.

dismutation a chemical reaction in which a single compound serves both as an oxidizing agent and as a reducing agent; e.g. the reaction of two molecules of pyruvate and one of water giving rise to one molecule each of lactate, acetate, and carbon dioxide. *Compare* **disproportionation**.

disodium chromoglycate *see* **cromolyn**.

disodium etidronate *see* **etidronate**.

disome a **polysome** in which two ribosomes are attached to a strand of messenger RNA. *Compare* **monosome**, **nucleodisome**.

Dispase *proprietary name for* a highly stable neutral metalloproteinase obtained from *Bacillus polymyxa*. It is so named from its use, alone or with collagenase, for gentle enzymic disaggregation and dispersal of animal tissues into separated cells for primary **cell culture** and for redispersal of cells on subculturing. *See also* **Chaotropase**, **Pronase**.

disperse phase *or* **dispersed phase** the phase in a two-phase system that consists of the particles in a colloid, emulsion, or suspension.

dispersion **1** the selective separation of an inhomogeneous emission according to some criterion (especially frequency, particle mass, speed, or energy); e.g. the separation of heterochromatic light into its components. **2** any system in which particles of any nature (e.g. solid, liquid, gas, or composite) are dispersed in a solid, liquid, or gas of different composition. **3** *(in statistics)* the extent to which values of a statistical frequency distribution are scattered around a mean or median value.

dispersion force *or* **London force** a long-range attractive interaction occurring between atoms or molecules whether or not they have permanent charges or dipole moments. Since the electrons in an atom or molecule are moving, an instantaneous dipole moment is induced in the atom or molecule; this averages to zero over a short time. This instantaneous dipole moment will induce an oppositely oriented dipole moment in a neighbouring atom or molecule in a manner that does not time-average to zero. This interaction energy, $V(R)$, is given approximately by:

$$V(R) = -\frac{3}{2}\frac{h\nu_A\nu_B}{\nu_A + \nu_B}\frac{\alpha_A\alpha_B}{(4\pi\varepsilon_0)}\frac{1}{R^6}$$

where ν_A and ν_B are the approximate frequencies of the first electron transitions of entities A and B, α_A and α_B are their polarizabilities, ε_0 is the permittivity of vacuum, h is the Planck constant, and R the distance between the entities. [After Fritz London (1900–54), German-born US physicist.]

dispersive replication of DNA *see* **replication of DNA**.

displacement binding a technique in which the displacement of radioligand bound to a population of receptors is measured as the concentration of an unlabelled ligand, which binds to the same receptors, is increased. It is used to distinguish different affinity states of the receptor population or receptor selectivity of the radioligand.

displacement chromatography an analytical chromatographic technique in which a mixture is applied to a chromatographic column and the components of the mixture are then successively displaced by elution with a solution containing another substance of higher affinity for the column material than that of the most firmly held component. The relative proportions of the components of the mixture are determined by **frontal analysis** of the eluate.

displacement electrophoresis *an alternative term for* **isotachophoresis**.

displacement loop *an alternative name for* **D-loop**.

displacement reaction a chemical reaction in which an atom or group is displaced from its covalent attachment to a central (frequently a carbon) atom following an attack by either an **electrophile** or a **nucleophile**.

disproportionation any chemical reaction of the type:

$$A + A \rightarrow A' + A'',$$

where A, A', and A'' are different chemical species. *Compare* **dismutation**.

disRNA *abbr. for* **Dicer**-substrate RNA **shRNA**s with 29 bp stem and 2 nt 3′ overhangs. They are thought to be more potent inducers of **RNAi** than shorter hairpins.

DIS2 S1 a gene of budding yeast encoding protein phosphatase I (its equivalent in fission yeast is *dis2*). In the absence of gene function **mitosis** is arrested. The product in *Saccharomyces cerevisiae* consists of 312 amino acids (35.93 kDa).

dissimilation *a less common alternative term for* **catabolism**. *Compare* **assimilation**. —**dissimilatory** *adj*.

dissimilatory nitrate reduction *see* **nitrate reduction**.

dissociation 1 *or* **disassociation** a reversible splitting of a chemical substance into simpler entities; *compare* **association**. In this sense, the term may be applied to: **a** the **heterolysis** of a covalent bond, yielding, in the case of an initially uncharged molecule, positively and negatively charged ions: **ionic dissociation**; **b** the disaggregation of a macromolecular homopolymer or heteropolymer into its constituent units (i.e. **protomers**), usually by a stepwise process through components of intermediate size (i.e. **oligomers**) and without rupture of covalent bonds; **c** the breakdown of an enzyme–substrate complex, which may or may not involve cleavage of a covalent bond; or **d** the separation of a complex of two or more molecules into the constituent molecules. **2** a splitting of a chemical substance by the **homolysis** of a covalent bond, yielding **radicals**. **3** translocation between a large chromosome and a small supernumerary one, in effect splitting the large chromosome into two chromosomes. **4** separation (induced or spontaneous) of the nuclear components of a heterokaryotic **dikaryon**. —**dissociated, dissociable, dissociative** *adj*.; **dissociate** *vb*.

dissociation constant *symbol*: K_{diss} *or* K_d; the **equilibrium constant** for the reversible breakdown of a complex or a chemical compound into two or more simpler entities; the reciprocal of the **association constant**, K_{ass}. In some cases, an **apparent dissociation constant** (K'_d) is determined (for the distinction *see* **equilibrium constant**). It is also known as the **ionization constant** in cases where a substance dissociates into ions. For a dissociation (or ionization) of the general type $AB \rightleftharpoons A + B$, the dissociation (or ionization) constant is given by:

$$K_{diss} = [A][B]/[AB] = 1/K_{ass}.$$

dissociation energy *symbol*: D; the enthalpy (per mole) required to break a given chemical bond in a specified chemical entity by homolysis.

distal situated furthest from an origin, central line, or point of attachment.

distal protein a protein that binds late in the assembly of complex structures, such as ribosomes. *Compare* **proximal protein**.

distance matrix a set of pairwise distances between taxa, based on some observation such as the number of observed nucleotide differences, the reconstructed distances between sequences (based on some model of sequence evolution), or observed morphological, immunological, ecological, or behavioural differences. Distance matrices are often used to construct dendrograms.

distil *or (US)* **distill** to undergo, to subject to, to purify, or to separate by **distillation**.

distillate the liquid (or solid) product of **distillation**.

distillation the act or process of evaporating a liquid by boiling so that its components, which are vaporized at different temperatures or at different rates, can be separated and condensed back to a liquid (or solid). It is used to purify one liquid from a mixture of liquids or from (dissolved) solids. *See* **fractional distillation**.

distribution 1 *(in chemistry) an alternative term for* **partition**. **2** *(in biochemistry)* the pattern of occurrence of a substance within or between cells or tissues, or within a group of organisms, taxa, etc. **3** *(in biology)* the geographical range or pattern of occurrence of a particular taxon or genetic variant. **4** *(in statistics)* the frequency of occurrence of a variable at each of a number of discrete values. *See also* **normal distribution**.

distribution coefficient 1 *an alternative term for* **partition coefficient**. **2** in **gel chromatography**, the fraction of the stationary phase that is available for diffusion of a given solute species.

distribution law *an alternative name for* **partition law**.

distributive 1 of, relating to, or characterized by distribution. **2** *(in molecular biology)* or **nonprocessive** describing (the action of) any enzyme or catalytic complex that progressively synthesizes or degrades a biopolymer by effecting several or many cycles of the same type of reaction or reaction sequence and dissociates from the template or intermediate product after each catalytic event. *Compare* **processive**. —**distributively** *adv*.

disulfide *or* **disulphide** *(in organic chemistry)* any compound containing two bivalent sulfur atoms linked by a covalent bond and each attached to a carbon atom.

disulfide bond *or* **disulfide bridge** *or* **disulfide link** a covalent bond between two sulfur atoms. Such bonds in biomolecules are usually formed by the oxidation of **sulfhydryl groups** in two neighbouring cysteine molecules or residues. Disulfide bonds may link two half-cystine residues in the same polypeptide chain, as in ribonuclease, or in different peptide chains, as in insulin or oxidized glutathione, and they often are important in maintaining the three-dimensional structure of polypeptides and proteins. They may also be reversibly formed in oxidation–reduction reactions as in the interconversion of lipoamide and dihydrolipoamide. Also: **disulphide bond** *or* **disulphide bridge** *or* **disulphide link**.

disulfide interchange *or* **disulphide interchange** any chemical reaction in which there is an interchange in the groups attached to two or more disulfide bonds; e.g.:

$$R–S–S–X + Y–S–S–Z \rightleftharpoons R–S–S–Y + X–S–S–Z,$$

where r, X, Y, and Z are different polypeptide chains or other SH-bearing molecules. *See also* **protein disulfide isomerase**.

disulfide knot *or* **disulphide knot** a region at the centre of a **fibrinogen** molecule where the six polypeptide chains are linked together by eight disulfide bonds. *Compare* **cystine knot**.

disulfiram *trivial name for* tetraethylthiuram disulfide; bis(diethylthiocarbamoyl) disulfide; a drug used to deter alcohol abuse in the treatment of alcoholism. Alcohol ingestion after disulfiram causes vasomotor disturbances, nausea, vomiting, and even unconsciousness and death. It acts by inhibiting the enzyme acetaldehyde dehydrogenase and hence slows the removal of acetaldehyde. It occurs naturally in the otherwise edible fruit body of the agaric mushroom *Coprinus atramentarius*. One proprietary name is Antabuse.

DIT *abbr. for* diiodotyrosine.

di-*tert*-butyl-*p*-cresol *or* **butylated hydroxytoluene** 2,6-di-*tert*-butyl-4-methylphenol; a common antioxidant used in foods containing fats.

dithioerythritol *abbr.*: DTE; *erythro*-1,4-dimercapto-2,3-butanediol; both this and **dithiothreitol** are known as **Cleland's reagent**. It has similar properties to dithiothreitol but is less often used.

dithiothreitol *abbr.*: DTT; *threo*-1,4-dimercapto-2,3-butanediol; both this and **dithioerythritol** are known as **Cleland's reagent**. A water-soluble solid compound which, because of its low redox potential (−0.33 V, pH 7.0), is capable of maintaining monothiols completely in the reduced state and of reducing disulfides quantitatively, so becoming oxidized to a cyclic disulfide; it has little tendency to be oxidized directly by air.

diuresis increased or excessive flow of urine.

diuretic 1 producing an increase in the volume of urine. **2** any agent that increases the volume of urine.

diurnal 1 occurring during the day. **2** repeating every day; having a periodicity of approximately 24 hours. *Compare* **circadian**.

Diuron *see* DCMU.

divalent metal transporter 1 *abbr.*: DMT1; *an alternative name for* NRAMP2. *See* **Nramp1**.

divergent evolution *or* **divergence** biological evolution that tends to produce an increasing difference in some characteristic(s) between initially similar groups of, e.g., organisms or gene products. *Compare* **convergent evolution**.

divergent transcription the transcription of adjacent genes from opposite strands of duplex DNA and in opposite directions.

dixenic *see* **synxenic**.

Dixon plot a graphical method for the presentation of enzyme-kinetic data by which a Michaelis constant, K_m (*see* **Michaelis kinetics**), or an **inhibitor constant**, K_i, may be determined. The method is primarily used as a means of readily obtaining K_i. the reciprocal of the initial velocity (i.e. $1/V$) is plotted against a series of inhibitor concentrations [I], at constant substrate concentration, [S]. When this is done for a number of values of [S], the resulting lines intersect at a point corresponding to K_i. The value of [I] at which this occurs is to the left of the ordinate (for competitive inhibition) and corresponds to −K_i. For non-competitive inhibition the lines meet on the abscissa. [After Malcolm Dixon (1899–1985), British enzymologist, who described it in 1972.]

dizygotic describing twin offspring formed from two separate ova fertilized by two separate sperms. *Compare* **monozygotic**.

DJ-1 a human gene (locus at 1p36) encoding a 189-residue protein that is widely distributed in tissues, including brain, is highly conserved, and may be involved in the response to oxidative stress. Mutations in it are associated with an autosomal recessive early-onset form of Parkinson's disease.

DL *prefix (in chemical nomenclature)* denoting an equimolar mixture of D and L enantiomers. *See* **D/L convention**.

dl- *prefix (in chemical nomenclature)* formerly used to denote **racemic**; now obsolete, use (±)-, *racemo*-, or *rac*-.

D/L convention a convention of symbols used to designate the **absolute configuration** around a chiral element in α-amino acids, cyclitols, monosaccharides, and derivatives of these classes of compounds. Within these classes it can be applied to any molecule of the type R_1CHXR_2 that can be oriented in a **Fischer projection** with the most highly oxidized atom at the top; then if X is on the right the configuration is designated by the prefix D-, whereas if X is on the left the configuration is designated by the prefix L-, relative to D-

(+)- or L-(−)-glyceraldehyde. Racemic (i.e. equimolar) mixtures of D and L enantiomers are designated by the prefix DL-. α-Amino acids are designated D- or L- according to the configuration at the α-carbon atom. Cyclitols are designated D- when the formula is drawn in such a way that the substituent on the lowest-numbered asymmetric carbon atom is above the plane of the ring and the numbering is anticlockwise, and L- when it is clockwise; the locant of the defining centre should precede the prefix D- or L-, e.g. 1D-1-*O*-methyl-*myo*-inositol. For monosaccharides the highest-numbered chiral centre (i.e. carbon atom) is the reference atom determining whether the monosaccharide is D or L. When a compound contains more than one class of residue the configuration designators are D_s or L_s if they apply to an amino-acid residue, D_c or L_c to a cyclitol residue, and D_g or L_g to a monosaccharide residue (the subscript s denoting the reference α-amino acid serine, the subscript c denoting cyclitol, and the subscript g denoting the reference monosaccharide glyceraldehyde. The absolute configurations of chiral compounds of classes other than those listed are designated by application of the **sequence rule**. *Compare* **ambo-, d-, l-, dl-, isomerism**.

D-loop *or* **displacement loop** a DNA structure generated when an additional strand of DNA is taken up by a DNA duplex, so that one strand of the original duplex is displaced, forming a D-shaped loop. Formation of a D-loop is favoured in negatively supercoiled DNA and can occur spontaneously. Such D-loops are formed in covalently closed mammalian mitochondrial DNA, and are prominent as intermediates in genetic recombination, in which their formation is catalysed by strand-exchange enzymes, the best characterized of which is the RecA protein of *Escherichia coli*.

DMEM *abbr. for* Dulbecco's Modification of Eagle's Medium; one of many formulations of culture media for the growth of mammalian cells.

DMF *or* **DMFA** *abbr. for* *N,N*-dimethylformamide; a neutral liquid of low melting point (−61°C) and a reasonably high boiling point (153°C), miscible with water and with most common organic solvents, hence sometimes termed the **universal solvent**.

DMR *abbr. for* differentially methylated region. In the epigenetic regulation of parentally imprinted genetic loci, one allele is methylated at **Cp G islands** within the DMR and is not transcribed, while the other allele is unmethylated and is transcribed. DMRs can map to the promoter regions of imprinted genes but are also found in non-promoter sequences. *IGF2* is an example of a maternally imprinted, paternally expressed gene that contains a DMR.

DMSO *abbr. for* dimethylsulfoxide (use discouraged, Me$_2$SO preferred).

DMT *abbr. for* the dimethoxytrityl group.

dna *symbol for* any of various bacterial genes that are involved in DNA replication. The following applies to *Escherichia coli*: *dnaA* encodes a 52 kDa protein (DnaA) of which up to 30 subunits bind to the *oriC* region and unwind it at the start of replication; *dnaB* encodes a helicase that, as a 300 kDa homohexamer of DnaB proteins, causes unwinding beyond the *oriC* region complexed to DnaA proteins; *dnaC* encodes DnaC (29 kDa), required for binding of DnaB; *dnaG* encodes DnaG, which with DnaB constitutes the ≈600 kDa assembly called a **primosome**.

DNA *abbr. and common name for* **deoxyribonucleate** (def. 1), or **deoxyribonucleic acid**; one of the two main types of **nucleic acid**, consisting of a long, unbranched macromolecule formed from one, or more commonly, two, strands of linked deoxyribonucleotides, the 3′-phosphate group of each constituent deoxyribonucleotide being joined in 3′,5′-phosphodiester linkage to the 5′-hydroxyl group of the deoxyribose moiety of the next one. *Compare* **RNA**. The absence of a 2′-hydroxyl group on each deoxyribose moiety renders these phosphodiester linkages resistant to hydrolytic attack by alkali, in contrast to those of RNA. A DNA strand, unless circular, has polarity, with one 5′ end and one 3′end, the two strands in duplex DNA having opposed polarities. Two purines, adenine and guanine, and two pyrimidines, cytosine and thymine, are the major bases present. The linear sequence of the purine and pyrimidine bases carries genetic information, whereas the deoxysugar and phosphoric residues play a structural role; the base sequence is written in the 5′→3′ direction. *See also* **double helix**. Specific forms, functions, molecules, or preparations of DNA may be designated by prefixes or suffixes, thus:

A-DNA *or* DNA-A, A form of DNA; amDNA, anti-messenger DNA; B-DNA *or* DNA-B, B form of DNA; cDNA, complementary DNA; cccDNA, covalently closed circular DNA; C-DNA *or* DNA-C, C form of DNA; ctDNA, chloroplast DNA; dsDNA, double-strand(ed) DNA (*see* duplex DNA); iDNA, intercalary DNA; msDNA, multicopy single-strand(ed) DNA; mtDNA, mitochondrial DNA; nDNA, nuclear DNA; pDNA, plasmid DNA; rDNA, ribosomal DNA; recDNA, recombinant DNA; rfDNA, replicative-form DNA; rtDNA, recombinant DNA; scDNA, single-copy DNA; ssDNA, single-strand(ed) DNA; tDNA, transfer DNA; T-DNA, transfer DNA of the **Ti plasmid**. *See* full entry for **T-DNA**. Z-DNA *or* DNA-Z, Z form of DNA.

DnaA protein *see* **dna**.

DNAase *or* **DNASe** *abbr.* sometimes used for **deoxyribonuclease**; DNase is more usual.

DNA-binding protein F/G *see* **matrin**.

DNA blot transfer *see* **blot-transfer technique**.

DnaB protein *see* **dna**.

DNA chip *an alternative name for* DNA **microarray**.

DNA cloning *see* **molecular cloning**.

DnaC protein *see* **dna**.

DNA cycle *see* **cell-division cycle**.

DNA-dependent DNA polymerase *see* **DNA polymerase**.

DNA-dependent protein kinase *see* **DNA-PK**.

DNA-dependent RNA polymerase *see* **RNA polymerase**.

DNA-directed DNA polymerase *recommended name for* **DNA polymerase**, EC 2.7.7.7.

DNA-directed RNA polymerase *recommended name for* **RNA polymerase**, EC 2.7.7.6.

DNAfingerprinting a term sometimes used to refer to **genetic profiling**.

Dna G protein *see* **DNA**.

(DNA) glycosylase *or* **DNA glycosidase** any of a group of glycohydrolases that initiate repair of DNA by hydrolysing N-glycosyl bonds linking deoxyribose residues with modified or incorrect bases to generate apurinic or apyrimidinic sites in the DNA and release the bases. They include DNA-3-methyladenine glycosidase I (EC 3.2.2.20), DNA-3-methyladenine glycosidase II (EC 3.2.2.21), and formamidopyrimidine-DNA glycosidase (EC 3.2.2.23). The enzymes form part of the pathway of base-excision repair, which acts on oxidative or nonoxidative damage, as determined by the specificity of individual enzymes. Sequences tend to be related; examples from *Escherichia coli* are: DNA-3-methyladenine glycosidase I, which is constitutive and specific to the named base; and DNA-3-methyladenine glycosidase II, which is inducible and releases any of 3-methyladenine, 3-methylguanine, 7-methylguanine, *O*-2-methylthymine, and *O*-2-methylcytosine.

DNA gyrase *see* type II **DNA topoisomerase**.

DNA helicase an enzyme that catalyses unwinding of the DNA double helix during DNA replication and repair. It requires ATP. An example is the *rep* gene product from *Escherichia coli*.

(DNA) joinase an alternative name for **DNA ligase**.

DnaK *or* **heat-shock protein 70** *abbr.*: hsp70; a **heat-shock protein** that plays an essential role in the initiation of DNA replication in **lambda phage**, and, in combination with DnaJ protein, acts to release proteins O and P.

DNA ligase *or* **(DNA) joinase** *or* **DNA repair enzyme** either of two enzymes that restore broken phosphodiester bonds in deoxyribonucleic acids. (1) **DNA ligase (ATP)** EC 6.5.1.1; *systematic name*: poly(deoxyribonucleotide):poly(deoxyribonucleotide) ligase (AMP-forming); *other names*: polynucleotide ligase; DNA repair enzyme; (DNA) joinase; *former name*: sealase. An enzyme that catalyses the reaction:

$$ATP + (deoxyribonucleotide)_n + (deoxyribonucleotide)_m = $$
$$AMP + pyrophosphate + (deoxyribonucleotide)_{n+m}.$$

(2) **DNA ligase (NAD⁺)** EC 6.5.1.2; *systematic name*: poly(deoxyribonucleotide):poly(deoxyribonucleotide) ligase (AMP-forming, NMN-forming); *other names*: polynucleotide ligase (NAD⁺); DNA repair enzyme; (DNA) joinase. An enzyme that catalyses the reaction:

$$NAD^+ + (deoxyribonucleotide)_n + (deoxyribonucleotide)_m = $$
$$AMP + nicotinamide\ mononucleotide + (deoxyribonucleotide)_{n+m}.$$

DNA-like RNA *abbr.*: D-RNA *or* dRNA; a minor fraction of eukaryotic cellular RNA that has a base composition much closer to that of DNA than the bulk of RNA. It is mainly of high molecular mass, short half-life, and confined to the nucleus, and is probably the same as **giant messenger-like RNA** and **pre-messenger RNA**.

DNA-3-methyladenine glycosidase *see* **(DNA) glycosylase**.

DNA methylase either of two enzymes involved in **modification** and other roles. (1) **Site-specific DNA-methyltransferase (adenine-specific)** EC 2.1.1.72; *other names*: N-6 adenine-specific DNA methylase; modification methylase; restriction–modification system; it catalyses the reaction:

$$S\text{-adenosyl-L-methionine} + DNA\text{-adenine} = $$
$$S\text{-adenosyl-L-homocysteine} + DNA\text{-6-methylaminopurine.}$$

In *E. coli* the enzyme recognizes the base sequence GATC and is part of the system for postreplicative mismatch repair. (2) **Site-specific DNA-methyltransferase (cytosine-specific)** EC 2.1.1.73; *other names*: C-5 cytosine-specific DNA methylase; modification methylase; restriction–modification system; it catalyses the reaction between *S*-adenosyl-L-methionine and DNA-cytosine to form *S*-adenosyl-L-homocysteine and DNA-5-methylcytosine. In *E. coli* it recognizes the base sequence CCWGG and methylates both strands.

DNA methyltransferase one of a group of eukaryotic enzymes such as DNMT1 – DNA (cytosine-5-) methyltransferase – that functions in establishing and regulating tissue-specific patterns of methylated cytosine residues. CpG methylation is an epigenetic modification that is important for embryonic development, imprinting, and, X-chromosome inactivation. Mutation of the DNMT3B DNA methyltransferase gene causes the autosomal recessive disorder ICF (immunodeficiency, centromeric instability, and facial anomalies).

DNA microarray *or* **DNA chip** *see* **microarray**.

(DNA) nicking–closing enzyme *or* **nicking-and-closing enzyme** *alternative names for* type I **DNA topoisomerase**.

DNA nucleotidylexotransferase *recommended name for* **terminal deoxynucleotidyltransferase** .

DNA photolyase EC 4.1.99.3; *recommended name*: deoxyribodipyrimidine photo-lyase; *other name*: photoreactivating enzyme. An enzyme that splits cyclobutadipyrimidine (in DNA) into two pyrimidine residues (in DNA), thereby repairing the DNA after light-induced **pyrimidine dimer** formation. It is light dependent, requiring the cofactors reduced flavin and either (in different enzyme types) a folate coenzyme or a modified flavin.

DNA-PK *abbr. for* DNA-dependent protein kinase; a mammalian nuclear protein serine/threonine kinase involved in DNA replication, transcription, repair of dsDNA breaks, and in V(D)J recombination. It consists of a catalytic subunit and a DNA-binding heterodimer (subunits of 70 and 80 kDa) called Ku that binds the dsDNA breaks.

DNA polymerase *common name for* any enzyme that catalyses the synthesis of DNA from deoxyribonucleoside triphosphates in the presence of a nucleic-acid primer. There are two families based on structural similarities. In family A are *Escherichia coli* and various other bacterial polymerases I, *Thermus aquaticus* taq polymerase, some bacteriophage polymerases, and yeast mitochondrial polymerase γ. In family B are many polymerases of higher eukaryotes, yeast polymerases I to III, and polymerase REV3, *E. coli* polymerase II, archaebacterial polymerases, and many viral polymerases.

(1) **DNA-directed DNA polymerase** EC 2.7.7.7; *other name*: DNA nucleotidyltransferase (DNA-directed); it is required for DNA replication and repair and has 3′-exonuclease activity (*see* **proofreading**). Eukaryotic DNA-directed DNA polymerases are of five types: (a) DNA polymerase-α (also called DNA polymerase I *or* Kornberg enzyme) was first detected in *E. coli*; in association with **DNA primase** it is replicative. The human enzyme belongs to the polymerase B family, with six conserved regions, four of which are involved in substrate binding. Several DNA viruses have type B polymerases; examples include adenoviruses and herpesviruses. (b) DNA polymerase-β (*or* DNA polymerase II) is a low-molecu-

lar-mass (<50 kDa) enzyme recovered almost entirely in nuclear extracts.

(c) DNA polymerase-γ (or DNA polymerase III) is an enzyme that copies synthetic polyribonucleotides with high efficiency, but does not copy DNA well.

(d) DNA polymerase-δ is a zinc finger protein with both polymerase and 3′-exonuclease activities; it is a heterodimer of 125 kDa and 50 kDa subunits.

(e) DNA polymerase-mt is the mitochondrial DNA polymerase.

E. coli has three DNA polymerases: (a) pol I, which has 5′- and 3′-exonuclease activities; the 5′-exonuclease activity is involved in DNA replication; the polymerase activity is employed during DNA repair. (b) pol II, which is regulated by **lexA** and belongs to family B (above). (c) pol III, the major replication polymerase, has an α subunit with polymerase activity, 1160 amino acids (129.76 kDa), and a total subunit composition of $(\alpha,\varepsilon,\theta)_2$-$\tau_2$-$(\gamma,\delta,\delta',\psi,\chi)_2$-$\beta_4$. The complex $(\alpha,\varepsilon,\theta)$, the 'core', is assembled first.

(2) **RNA-directed DNA polymerase** is the EC recommended name for **reverse transcriptase**.

DNA primase a DNA-directed **RNA polymerase** (EC 2.7.7.–) of the bacterial **primosome**. It associates transiently with the primosome and synthesizes short RNA primers for the **Okazaki fragments** on both template strands at replication forks.

DNA print the pattern formed by denaturing the DNA in colonies of bacteria grown on a support and fixing the DNA to the support.

DNA provirus see **provirus**.

DNA repair enzyme any of the various enzymes involved in the different stages of DNA repair. Three steps are involved, each catalysed by a different set of enzymes: (1) the altered portion of DNA is removed by repair nucleases; (2) **DNA polymerase** fills the gap with a complementary copy; (3) the break is sealed by **DNA ligase**. There are two other related repair pathways, called (base-)**excision repair** and nucleotide-excision repair. See also **repair enzyme**.

DNA replication see **replication of DNA**.

DNA–RNA hybrid a double helix consisting of one DNA chain and one complementary RNA chain, the two chains being held together by hydrogen bonds between the complementary **base pairs**.

DNA–RNA hybridization or **DNA–RNA hybridisation** the formation of a **duplex** (def. 2) nucleic-acid structure that occurs when DNA and RNA molecules that have complementary base sequences are brought together. This phenomenon is exploited in a technique used to localize genes in a chromosome that depends on allowing radioactively labelled RNA to hybridize with denatured, single-stranded DNA from which the RNA was transcribed and then determining the position on the DNA where hybridization has taken place by autoradiography. It also forms the basis of **Northern blotting**.

DNase usual abbr. for deoxyribonuclease. Compare **DNAase**.

(DNA) sealase an alternative name for **polynucleotide ligase** (def. 1).

DNase I hypersensitivity the phenomenon whereby treatment of chromatin with deoxyribonuclease I (DNase I) results in cleavage of DNA at sites that are poorly protected by histones. These 'hypersensitive sites' correspond with sites of active transcription in the genome.

DNA sequencing determination of the order in which deoxynucleotides occur in DNA. Fragments of DNA, often cloned in a plasmid vector, are subjected to either the **chemical cleavage method** or, much more commonly, the **chain-termination method**.

DNA superhelix see **supercoil**.

(DNA) swivelase a former name for type I **DNA topoisomerase**.

DNA topoisomerase any enzyme that alters superhelix density in supercoiled DNA. Topoisomerases are found in all cell types, from microorganisms to those of humans, and in some viruses e.g. bacteriophage T4 and vaccinia. All known topoisomerases relax negatively supercoiled DNA, converting it to a less supercoiled form. Type I makes a transient single-strand break, whereas type II makes a transient double-strand break. **Type I DNA topoisomerase**, EC 5.99.1.2; other names: DNA topoisomerase, relaxing enzyme, untwisting enzyme, swivelase, nicking–closing enzyme, ω-protein; it catalyses the ATP-independent breakage of a single strand of DNA, which may be followed by passage (i.e., in the case of duplex DNA, transfer of the unbroken strand through the broken strand)

and rejoining. This has the effect of relaxing one negative supercoil. This enzyme can relax only negatively supercoiled DNA, whereas some enzymes, e.g. that from calf thymus, can relax both negatively and positively supercoiled DNA. **Type II DNA topoisomerase**, EC 5.99.1.3; other name: DNA topoisomerase (ATP-hydrolysing); it catalyses ATP-dependent breakage, passage (in this case, of a double-strand through a double-strand break) and rejoining of the double-stranded DNA. This enzyme can thus decatenate circular duplex DNA. All type II enzymes can relax both positively and negatively supercoiled DNA. Bacterial **DNA gyrase** is a special example of this class of enzyme because it catalyses the ATP-dependent introduction of negative superhelical turns into a closed circular DNA. It is an A_2B_2 tetramer. The energy-free topoisomerase activity of the A subunit, which breaks the DNA, is inhibited by 4-quinolone antibiotics such as nalidixic acid and ciprofloxacin. The energy-transducing activity of the B subunit is inhibited by novobiocin and other coumarin antibiotics. This enzyme can introduce negative superhelical turns into a closed circular DNA molecule with the coupled hydrolysis of ATP. See also **linkage number**, **supercoil**, **winding number**, **writhing number**.

(DNA) untwisting enzyme an alternative name for **(DNA) swivelase**.

DNA vaccines plasmids containing foreign genes under the control of a strong promoter induce transient expression of the foreign genes in tissue cells when introduced into animals. Immune responses are induced when the gene is taken up and expressed in antigen-presenting cells. DNA may be given by parenteral injection, by shooting DNA-coated particles into the skin using compressed gas (a 'gene gun'), or by the nasal or oral routes. Cellular and humoral immune responses are induced but with different patterns of **cytokine** production and **immunoglobulin** classes depending on the method of administration. An advantage of the method is that antigens of an organism can be screened for immunogenicity by immunising with expression libraries without the labour of purifying each antigen. DNA immunisation alone is not yet proven in man but may be more effective when combined with other methods ('prime-boost regimes'). DNA has the advantage that it is stable and simple to produce and plasmids can be engineered to contain **adjuvant** activity. Side effects of DNA immunisation are few. Although insertional mutagenesis remains a possibility, attempts to demonstrate this experimentally have shown it to be extremely rare.

DNA virus or **deoxyribovirus** any virus in which the genome consists of DNA, either double-stranded (in class I viruses) or single-stranded (in class II viruses). See **Baltimore classification of viruses**.

DNAzyme or **deoxyribozyme** a synthetic DNA molecule with catalytic properties.

DNFB abbr. for (2,4-)dinitrofluorobenzene. See **Sanger's reagent**.

Dnm1 a dynamin-like protein of yeast that is involved, with Fis1 and Mdv1, in forming the contractile ring around the outer membrane of mitochondria that brings about their fission. Homologues occur in animals (e.g. Drp1) and in plants.

Dnp symbol for **dinitrophenyl** group.

DNP abbr. for 1 (2,4-)dinitrophenol. 2 deoxyribonucleoprotein (use deprecated).

Dns symbol for 5-(dimethylamino)naphth-1-ylsulfonyl; see also **dansyl**.

dNTP abbr. for any deoxynucleoside (5′-)triphosphate (the deoxynucleoside being either unspecified or unknown).

DOC abbr. for (use discouraged) 1 deoxycholate (see **deoxycholic acid**). 2 deoxycorticosterone.

doca or **DOCA** abbr. for deoxycorticosterone acetate (see **deoxycorticosterone**).

Docetaxel a proprietary name for a semisynthetic analogue of **taxol**.

Dock an adaptor protein for the insulin receptor in *Drosophila*. It contains SH3 domains through which it binds the C-terminal region of the receptor.

docking the binding of any macromolecule or part of such a molecule to its specific harbouring site on another molecular structure. See also **signal recognition particle receptor**.

docking protein another name for **signal recognition particle receptor**.

docosa+ or (before a vowel) **docos+** prefix indicating twenty-two or twenty-two times.

docosadienoic acid any straight-chain fatty acid having 22 carbon atoms and two double bonds. The (5Z,13Z)-form has been reported to occur in *Limnanthes douglasii*.

docosahexaenoic acid any straight-chain fatty acid having 22 carbon atoms and six double bonds. Only the (all-Z)-4,7,10,13,16,19-isomer, of the n-3 family, occurs naturally, being present in substantial amounts (10–15% of total fatty acids) in fish oils, and in variable amounts (a few percent of total) in animal glycerophospholipids. It can be formed metabolically from α-linolenic acid, is present in highest concentration in the retina, and is a major component of brain lipids.

docosanoic acid *or* **behenic acid** a straight-chain saturated fatty acid having 22 carbon atoms that occurs as a constituent of cerebrosides and some seed oils (especially of *Lophira* spp.); in some of the latter it accounts for 20–30% of total fatty acids.

docosapentaenoic acid any straight-chain fatty acid having 22 carbon atoms and five double bonds. The (all-Z)-4,7,10,13,16-isomer is a member of the n-6 family and can be formed metabolically from linoleic acid; it is a constituent of animal glycerophospholipids. The (all-Z)-7,10,13,16,19-isomer, of the n-3 family, can be formed metabolically from α-linolenic acid; it is a precursor of 4,7,10,13,16,19-docosahexaenoic acid, and is present in small amounts (a few percent) in animal glycerophospholipids and in larger amounts (up to 10%) in fish oils.

docosatetraenoic acid any straight-chain fatty acid having 22 carbon atoms and four double bonds. Adrenic acid is found in animal glycerophospholipids, especially in brain and heart of animals fed sunflower or corn oil, and is presumed to be the (all-Z)-7,10,13,16-isomer.

docosatrienoic acid any straight-chain fatty acid having 22 carbon atoms and three double bonds. The (all-Z)-7,10,13-isomer occurs in animal lipids, especially during essential fatty acid deficiency.

docosenoic acid any straight-chain fatty acid having 22 carbon atoms and one double bond. The (11Z)-isomer, cetoleic acid, occurs in marine oils and rapeseed oil. The (13Z)-isomer, erucic acid, is an important constituent of rapeseed and other oils of plants belonging to the family Cruciferae. The possibility that it may be harmful has prompted efforts to breed crop varieties that yield oils containing low levels or none of this fatty acid.

Dod *symbol for* the dodecyl group.

dodeca+ *or (before a vowel)* **dodec+** *prefix* indicating twelve.

dodecandrin a type I **ribosome-inactivating protein**.

dodecanoic acid *systematic name for* lauric acid, a straight-chain saturated fatty acid having 12 carbon atoms. *See also* **laurate**.

dodecanoyl *symbol* Lau; *systematic name for* lauroyl.

dodecenoic acid any straight-chain fatty acid having 12 carbon atoms and one double bond.

dodecenoyl the acyl group derived from any dodecenoic acid.

dodecenoyl-CoA Δ-isomerase EC 5.3.3.8; *systematic name*: dodecenoyl-CoA Δ³-*cis*–Δ²-*trans*-isomerase; *other names*: Δ³-*cis*-Δ²-*trans*-enoyl-CoA isomerase; acetylene–allene isomerase; 3,2 *trans*-enoyl-CoA isomerase. An enzyme that is involved in **unsaturated fatty acid oxidation**; it catalyses the isomerization of 3-*cis*-dodecenoyl-CoA to 2-*trans*-dodecenoyl-CoA, thereby producing a substrate for **enoyl-Coa hydratase**. *See also* **peroxisomal bifunctional enzyme**.

dodecyl *symbol*: Dod; *preferred name for* the alkyl group, CH₃–[CH₂]₁₀–CH₂– (*also known as* **lauryl**), derived from dodecane.

n-dodecyl-N,N-dimethylglycine lauryldimethylbetaine; a zwitterionic betaine detergent, CMC 1.6–2.1 mm. *Proprietary name*: Empigen BB.

n-dodecyl-β-D-glucopyranoside a water-soluble nonionic detergent, CMC 0.13 mm.

n-dodecyl-β-D-maltoside a water-soluble nonionic detergent, aggregation number 98, CMC 0.1–0.6 mm.

dodecylsulfate the CH₃–[CH₂]₁₀–CH₂–O–SO₃⁻ anion. *See* **sodium dodecyl sulfate**.

doghouse configuration (of a cyclobutadipyrimidine) *see* **pyrimidine dimer**.

DOGS *abbr. for* dioctadecylamidoglycylspermine; a reagent used in lipofection.

dolichol *the recommended term for* any 2,3-dihydropolyprenol derived from four or more linked isoprene units, i.e. any prenol deriv-

ative of general structure H–[–CH₂–C(CH₃)=CH–CH₂–]ₙ –CH₂–CH(CH₃)–CH₂–CH₂–OH where n is greater than 3; in naturally occurring dolichols n is commonly in the range 13–23. (*Note*: Because the isoprene unit carrying the hydroxyl group is saturated in dolichols, they are not true **prenols** or **polyprenols**, though often classed as such. It is therefore recommended that, unless qualified, these collective terms are not used to include dolichols.) Dolichol is synthesized from mevalonate in peroxisomes. The phosphoric esters of dolichol (dolichol phosphate *or* phosphodolichol *or* dolichyl phosphate) and the diphosphoric esters (dolichol diphosphate *or* diphosphodolichol *or* dolichyl diphosphate) function in eukaryotes as carriers of mono- and oligosaccharide residues in the glycosylation of lipids and proteins in the endoplasmic reticulum membrane. First, glycosyl groups are transferred from soluble nucleoside-diphosphosugars such as GDPmannose, UDP-N-acetylglucosamine, UDPglucose, or UDPxylose to form dolichyl phospho- or diphosphosugars. Then dolichyl diphosphooligosaccharides are built up from these by stepwise transfer of additional sugar residues. Finally the completed oligosaccharide moiety is transferred to a lipid or to a growing polypeptide chain with release of dolichyl diphosphate.

dolichol kinase EC 2.7.1.108; an enzyme that catalyses the formation of dolichyl phosphate from CTP and dolichol with release of CDP. Dolichyl phosphate is a precursor of dolichyl diphosphooligosaccharides, dolichyl phosphoglucose, and dolichyl phosphomannose, intermediates in N-linked glycoprotein oligosaccharide biosynthesis.

dolichol O-acyltransferase EC 2.3.1.123; an enzyme that catalyses the formation of dolichyl palmitate from palmitoyl-CoA and dolichol with release of CoA. Acylated dolichols are found in some cell membranes, though their function is obscure.

dolichyl the alkyl group derived from a **dolichol** by loss of its hydroxyl group.

dolichyl diphosphooligosaccharide a precursor for the transfer of N-linked oligosaccharide chains to protein in the synthesis of glycoproteins. It consists of dolichol esterified to diphosphate with an oligosaccharide chain linked to the diphosphate by the C-1 of the sugar at the reducing end of the chain.

dolichyl-diphosphooligosaccharide–protein glycosyltransferase EC 2.4.1.119; an enzyme that catalyses the reaction of dolichyl diphosphooligosaccharide with protein L-asparagine to form a glycoprotein with the oligosaccharide chain attached by glycosylamine linkage to protein L-asparagine, with release of dolichyl phosphate.

dolichyl-phosphate β-D-glucosyltransferase EC 2.4.1.117; an enzyme that catalyses the reaction of UDPglucose with dolichyl phosphate to form dolichyl β-D-glucosyl phosphate (an intermediate in the biosynthesis of N-linked glycoprotein oligosaccharides) and UDP. It is a type II membrane protein of the endoplasmic reticulum involved in the glycosylation pathway.

dolichyl-phosphate-mannose–protein mannosyltransferase EC 2.4.1.109; an enzyme that catalyses the reaction of dolichyl phosphate D-mannose with protein to form O-D-mannosylprotein and dolichyl phosphate. *See also* **dolichyl-phosphate β-D-mannosyl**transferase.

dolichyl-phosphate β-D-mannosyltransferase EC 2.4.1.83; an enzyme that catalyses the formation of dolichyl D-mannosyl phosphate, an intermediate in the biosynthesis of N-linked glycoprotein oligosaccharides, from GDPmannose and dolichol phosphate with release of GDP. *See also* **dolichyl-phosphate-mannose–protein manno**syltransferase.

domain **1** any topological region having specific characteristics, contained within certain limits, and/or under individual control. **2** *or* **structural domain** a compact, globular region in the structure of a single protein molecule, which may consist of several such globular regions held together by more flexible parts of the polypeptide chain. It has been suggested that the word be reserved for large subassemblies that would be stable if the polypeptide chain connecting them to the rest of the protein molecule were to be cleaved, and that the term **folding unit** be used to define small assemblies of secondary structure segments in a protein molecule. **3** a region of a protein molecule delimited on the basis of sequence or function, without knowledge of its molecular substructure, e.g., that binds to a receptor or substrate, that possesses a catalytic function (forming a **catalytic domain**), or that traverses a membrane (forming a **transmembrane domain**). Such a region may contain more than one structural domain (*see* def. 2). **4** a poorly characterized length of chromosomal DNA, ~50–200 kbp, that comprises all the (coding or non-coding) sequences required for the formation of mRNA for any one (specified) protein. **5** a zone within a cellular membrane consisting of one class of component, e.g. lipid. **6** a diverse set of cellular events, processes and metabolic reactions controlled or affected by a specific agent, e.g. a hormone or another messenger. *See* **metabolic code**. **7** a mathematical aggregate to which a variable is confined. **8** (*in physics*) any region in a ferromagnetic solid in which all the atoms have the spins of their unpaired electrons aligned in the same direction: **magnetic domain**.

domain family a structurally and functionally diverse group of proteins that share a local sequence region that mediates a particular biological role (e.g. a kringle or SH2 domain). Different processes underpin the evolution of domain families and gene families, with different functional consequences.

domain of bonding the two linked **binding sets** through which a pair of associated **protomers** are associated.

domain of unknown function *abbr.*: DUF; part of a (usually) protein sequence that is conserved in a family of sequences but whose function is unknown.

dominant (*in genetics*) describing a gene that produces the same character or phenotype whether present in the homozygous state or in the heterozygous state (i.e. together with a different specified allele). The ineffective heterozygous allele is described as **recessive** to the dominant one. Dominant inherited diseases fall into one of three categories: (1) '**change of function**' in which the abnormal protein acquires a novel activity that is deleterious to the cell, (2) '**dominant negative**' in which the abnormal protein forms heterooligomeric complexes with the protein from the normal allele thereby knocking out the function of the entire protein complex, (3) '**haplo-insufficiency**' in which a single copy of the gene has lost function and the 50% of the normal amount of protein that is produced from the non-mutated allele is not adequate to preclude clinical symptoms. Examples: type (1), sodium channel disorders; type (2), mutations in collagen genes which lead to osteogenesis imperfecta; type (3), loss of function in one copy of the low-density-lipoprotein receptor gene resulting in familial hypercholesterolemia.

dominant selectable marker (*in molecular genetics*) a marker, usually an enzyme, that confers resistance to the toxic effects of a drug or other pharmacologically active compound used for selection. Such markers are used to confer an advantage on mammalian and other cells that have been transfected with exogenous DNA constructs. For example, the gene for aminoglycoside phosphotransferase is commonly incorporated in plasmids to confer resistance (denoted as *neo*^{R}) to antibiotics such as G418. Dominance refers to the fact that a single copy of the marker is sufficient to confer the resistant phenotype.

Donnan effect *or* **Donnan (membrane) equilibrium** the effect of the presence of charged macromolecules upon one side of a semipermeable membrane on the distribution of small, permeable ions across the membrane. At equilibrium, the small counterions will be more concentrated on the macromolecules' side of the membrane and the small ions of the same charge as the macromolecules will be less concentrated there, whereas on the side of the membrane with no macromolecules the concentrations of the small ions of opposite charge will be equal. It can be shown that, if the small ions are A$^+$ and B$^-$,

$$[A^+]_{(1)}/[A^+]_{(2)} = [B^-]_{(2)}/[B^-]_{(1)} = r_D,$$

where r_D is the **Donnan ratio** and the subscripts 1 and 2 indicate the two sides of the membrane. [After Frederick George Donnan (1870–1956), British physical chemist.]

donor (*in chemistry*) any chemical species that is able to donate one or more electrons, atoms, or groups to another species, the acceptor.

donor atom 1 an atom that supplies two electrons in forming a **dipolar bond**. **2** an atom, added as an impurity to a semiconductor, that

provides additional electrons in the conduction band. *See also* **electron donor**.

donor cell any living cell that contributes genetic information to another, **recipient** cell. *See* **conjugation** (def. 3), **transduction**, **transformation**.

dopa *or* **DOPA** *abbr. for* 3-hydroxytyrosine; 3,4-dihydroxyphenylalanine; 2-amino-3-(3,4-dihydroxyphenyl)propanoic acid; the naturally occurring enantiomer is 3-hydroxy-L-tyrosine (*abbr.*: L-dopa, also known as levodopa because it is levorotatory; $[\alpha]^2{}_5{}^0{}_{46} \approx 13.0$). L-Dopa is formed from L-tyrosine by tyrosine 3-monooxygenase, EC 1.14.16.2, and is an immediate metabolic precursor of the neurotransmitter **dopamine** and other **catecholamines**. L-Dopa is used in the treatment of **Parkinson's disease**.

dopac *or* **DOPAC** *abbr. for* dihydroxyphenylacetate.

dopachrome Δ-isomerase EC 5.3.3.12; *systematic name*: dopachrome Δ7–Δ2-isomerase; an enzyme that catalyses the conversion of dopachrome (2-carboxy-2,3-dihydroindole-5,6-quinone) to 5,6-dihydroxyindole-2-carboxylate. Dopachrome is formed nonenzymatically from dopaquinone, the product of **monophenol monooxygenase**. These are reactions in the pathway of melanin synthesis. A deficiency in the enzyme results from a mutation in mice known as *slaty*. The enzyme contains zinc.

dopa decarboxylase *see* aromatic-L-amino-acid decarboxylase.

dopamine *an alternative name for* 3,4-dihydroxyphenylethylamine; 3-hydroxytyramine; 3,4-dihydroxyphenylethylamine; 4-(2-aminoethyl)-1,2-benzenediol; a **catecholamine** neurotransmitter formed by aromatic-L-amino-acid decarboxylase, EC 4.1.1.28, from 3,4-dihydroxy-L-phenylalanine. A metabolic precursor of **norepinephrine** and **epinephrine**, it is found in **dopaminergic** nerves in the brain and in the adrenal medulla. *See also* **Parkinson's disease**.

dopamine β-hydroxylase EC 1.14.17.1; *recommended name*: dopamine β-monooxygenase; *systematic name*: 3,4-dihydroxyphenethylamine,ascorbate:oxygen oxidoreductase (β-hydroxylating). An enzyme that catalyses hydroxylation of dopamine to norepinephrine in aminergic neurons releasing norepinephrine and epinephrine; ascorbate is simultaneously oxidized to dehydroascorbate. The enzyme is a copper protein, stimulated by fumarate. It is absent from dopaminergic neurons.

dopamine β-monooxygenase *see* dopamine β-hydroxylase.

dopamine receptor any protein that binds dopamine and mediates its intracellular effects. Types D$_1$ to D$_5$ have been recognized; D$_1$ and D$_5$ (also called D$_{1A}$ and D$_{1B}$) comprise one subfamily; D$_2$, D$_3$, and D$_4$ (also called D$_{2A}$, D$_{2B}$, and D$_{2C}$) comprise a second subfamily. D$_3$ receptors seem to be insensitive to guanine nucleotides and regulate conductance of Ca^{2+}. All are G-protein-associated receptors. D$_1$ and D$_5$ elevate cyclic AMP, but vary in binding characteristics; D$_2$ has a short and a long form originating from alternative splicing; in humans, the short form (D$_{2S}$) has 414 amino acids, and the long form (D$_{2L}$) has 443 amino acids. D$_2$ lowers cyclic AMP levels and regulates ion channels, opening K$^+$ and inhibiting Ca^{2+} channels. D$_3$ and D$_4$ have binding characteristics similar to, but distinct from D$_2$; their action is not mediated by G-protein interaction with adenylate cyclase. *See* **DARPP-32**.

dopaminergic describing a nerve that is activated by dopamine.

The term is applied to nerves that act by releasing dopamine at their nerve endings. *Compare* **adrenergic, aminergic, cholinergic, peptidergic, purinergic, serotonergic.**

dopamine transporter an integral membrane protein of dopaminergic nerve terminals that permits re-entry of dopamine in a Na^+- and Cl^--dependent manner. Cocaine binds strongly to, and inhibits, the transporter. A variety of mood-modifying drugs act similarly.

DOPA-responsive dystonia *or* **hereditary progressive dystonia**; an autosomal dominant progressive dystonia that starts in childhood and results from deficiency of tetrahydrobiopterin, the cofactor for tyrosine hydroxylase. Caused by mutations in guanosine triphosphate cyclohydrolase, it responds dramatically to treatment with DOPA or dopamine.

DOPE *abbr. for* dioleoylphosphatidylethanolamine; a phospholipid used in conjunction with cationic lipids for **lipofection.**

Doppel a mouse protein that resembles a truncated version of cellular **prion protein.** It lacks the N-terminal region up to and including the transmembrane segment. It is not normally abundant in brain but is abundant in testes. When overexpressed it causes cerebellar cell death and ataxia.

Doppler effect the apparent change in the frequency of acoustic or electromagnetic waves due to relative motion between the source and the observer. The frequency appears to increase when the source and the observer move towards each other, and to decrease when they move away from each other. [After Christian Johann Doppler (1803–53), Austrian mathematician and physicist.]

dorsal protein embryonic developmental morphogenetic nuclear protein; a protein whose concentration in the nucleus of a cell during the blastoderm stage of development determines the cell's lateral or ventral identity. It is similar to the product of the vertebrate proto-oncogene c-*rel*.

dosage compensation the epigenetic inactivation of one of the two X chromosomes of female cells that results in them having only one functional copy, thereby making them comparable with X-chromosome dosage in male cells, which have only one.

dose 1 a measured quantity of a drug, microorganism, radiation, etc. administered to an organism, often at specified intervals. In experimental pharmacology, the units and routes of administration should be specified and the quantity of the drug expressed per unit of animal mass, e.g. $mol\ kg^{-1}$, $mg\ kg^{-1}$, $mg\ min^{-1}\ kg^{-1}$. **2** to prescribe or administer a dose or doses. —**dosage** *n.*

dose equivalent *symbol*: H; a measure of the biological effects of **ionizing radiation** upon humans (or other mammals) used in radiation dosimetry and health physics. It is usually equal to the **absorbed dose** multiplied by the **quality factor.** In the case of radiation arising from radioactive material within a person's body, the absorbed dose must be multiplied also by a distribution factor (normally unity, but in certain instances given a value of 5) in order to take account of any nonuniform distribution of radioactivity. The CGS unit of dose equivalent is the **rem**; the corresponding SI derived unit is the **sievert** (def. 1).

dose rate in radiation dosimetry, the amount of **ionizing radiation** (expressed in grays, rads, rems, röntgens, or sieverts) received or receivable by an irradiated object or organism in unit time. For personnel, it is usually expressed in sieverts per hour. (There is no separately named unit of dose rate.)

dose ratio *an older name for* **concentration ratio.**

dose–response curve any graphical representation of the relationship between the size of an administered dose and the extent of the response of the organism, or other test system, to the dose. The response may be some biochemical or physiological change in one or more organisms, tissue samples etc., or the proportion of organisms in a group that show a response.

dosimeter *or* **dose meter** an instrument for measuring the quantity of ionizing radiation administered to an organism or an object, or the intensity of a source of such radiation.

dosimetry the process of measuring administered quantities of ionizing radiation, drugs, microorganisms, etc.

DOSPA *abbr. for* 2,3-dioleoyloxy-N-[2(sperminecarboxamido)-ethyl]-N,N-dimethyl-1-propaniminium trifluoroacetate; a polycationic lipid used for **lipofection.**

DOSPER *abbr. for* 1,3-dioleoyloxy-2-(6-carboxyspermyl)-propylamide; a dicationic lipid used for **lipofection.**

DOTAP *abbr. for* N-[1-(2,3-dioleoyloxy)-propyl]-N,N,N-trimethylammonium methyl sulfate; a monocationic lipid used for **lipofection.**

dot hybridization *or* **dot hybridisation** *or* **dot blotting** a development of **blotting** or blot transfer, in which as little as 1 pg of a specific nucleic-acid sequence, bound to nitrocellulose as a small spot, can be detected using a ^{32}P-labelled probe as a 'dot blot'.

DOTMA *abbr. for* N-[1-(2,3-dioleoyloxy)-propyl]-N,N,N-trimethylammonium chloride; a monocationic lipid used for **lipofection.**

dotplot a graphical depiction of the similarity between pairs of sequences, sometimes used as the basis for sequence alignment. Identical regions form straight lines along, or parallel to, the plot diagonal, their offset from the diagonal indicating the number of gaps needed to bring these regions into register.

double-antibody method any technique in which a specific antibody, used to detect or measure its corresponding antigen, as in **immunoassay,** is isolated from the reaction mixture by combination with heterologous antibody to the specific antibody. *Compare* **antiglobulin test.**

double-beam spectrophotometer a **spectrophotometer** in which the incident light beam is split into two halves, one half passing through the sample and the second half passing through the reference so that direct measurement of the difference in absorption can be made. In one type the two emergent beams fall on different detectors and the resulting signal is a direct current; in another type the incident beam passes alternately through the sample and the reference, and the two emergent beams fall on the same detector resulting in an alternating current signal.

double-blind testing a method of evaluating the efficacy of a substance (e.g. a drug) in which neither the subjects nor the experimenter knows the identity of the samples at the time of the evaluation.

double bond a **covalent bond** that consists of two electron pairs shared by two atoms.

doublecortin a tubulin-binding protein (40 kDa) expressed in the developing nervous system. It is homologous to a mouse protein that contains a Ca^{2+}-dependent protein kinase domain. Several mutations in the gene at Xq22.3-q23 produce X-linked lissencephaly in males and double cortex syndrome in females.

double-displacement (enzyme) mechanism *or* **ping-pong mechanism** *or* **substituted-enzyme mechanism** any enzymic mechanism in which an enzyme reacts with one substrate to give a covalently modified enzyme together with the release of one product, and the modified enzyme then reacts with the second substrate with the subsequent release of the second product and the return of the enzyme to its initial state.

double helix *or* **Watson–Crick model of DNA** the main features of this model are that DNA consists of two antiparallel helical polynucleotide chains coiled around the same axis to form a double helix. Deoxyribose–phosphate backbones are on the outside of the helix and purine and pyrimidine bases lie approximately at right angles to the axis on the inside of the helix. The diameter of the helix is 2.0 nm and there is a residue on each chain every 0.34 nm. The angle between each residue on the same chain is 36°, so that the structure repeats after 10 residues (3.4 nm) on each chain. The two chains are held together by hydrogen bonds between pairs of bases, each member of the pair belonging to a different polynucleotide chain. Adenine is always paired with thymine and guanine with cytosine. The two chains are therefore complementary.

double heterozygote an individual who is heterozygous at two separate genetic loci.

double immunodiffusion *or* **double diffusion** a technique of **immunodiffusion** in which both the antibody and the antigen diffuse towards each other through a gel to form lines of precipitation. *See* **Ouchterlony technique.**

double labelling an experimental strategy in which either a single chemical compound is labelled with two different nuclides or two compounds are each labelled with a different nuclide. It is used to follow the simultaneous transformations of two substances or two parts of one substance, or to distinguish identical molecules synthesized at different times. —**doubly labelled** *adj.*

double layer *see* **electric double layer**.

double membrane *see* **unit membrane**.

double-reciprocal plot a plot of $1/x$ versus $1/y$, where x and y are two variables. If the plot of y versus x has the shape of a **rectangular hyperbola** the corresponding double-reciprocal plot will be a straight line. Note: the x and y axes are not the asymptotes of the rectangular hyperbola. The **Lineweaver–Burk plot** is a double-reciprocal plot. *See also* **hyperbolic kinetics curve**.

double refraction *see* **birefringence**.

double refraction of flow *see* **flow birefringence**.

double-sector cell a **cell** (def. 2), used in analytical ultracentrifugation experiments, with two radial sector-shaped cavities, allowing the concentration distribution of a solute in one sector to be related to that of the solvent alone in the other sector: i.e. the second sector provides a reference for the first.

doublesex a gene of *Drosophila* involved in sex determination. It encodes two zinc-binding transcription factors derived by alternative splicing of the pre-mRNA transcript. The proteins, called DsxF and DsxM, determine whether the embryo is female or male, respectively.

double-sieve hypothesis a hypothesis invoked to account for the unexpectedly low level of error (misreading) observed in protein biosynthesis. According to this hypothesis, the discrete synthetic and hydrolytic sites that exist on certain aminoacyl-tRNA synthetases provide a two-stage stereochemical sorting mechanism – likened in action to a pair of sieves – for the elimination of inappropriate amino acids. The synthetic site first rejects amino acids larger than that genetically specified; the hydrolytic site then preferentially destroys reaction products of amino acids smaller than, or isosteric with, the one specified. The second process has also been termed **editing** or proofreading. Such a mechanism would have particular importance in discriminating between a pair of very similar amino acids such as isoleucine and valine. *Compare* **kinetic proofreading**.

double-stop terminator two consecutive **termination codons** in a molecule of mRNA.

double-stranded *or* **double-strand** describing a **duplex** form of a nucleic acid. *See* **double-stranded RNA**, **duplex DNA**.

double-stranded RNA *abbr.*: dsRNA; a structure that occurs when complementary base sequences in single-stranded RNA form a **duplex** (def. 2). The Rna may fold back on itself to generate an antiparallel duplex structure, known as the stem, with a loop of unpaired bases at the end. The loop and stem together are called a **hairpin**. The two complementary sequences can result from an inverted repeat. *See also* **palindrome**.

double-stranded template double-stranded DNA that can be used as a template for sequencing. DNA cloned in plasmids can be sequenced without preparing single-stranded DNA as a template. Denaturation of double-stranded templates exposes sufficiently long stretches of the template strand to enable sequencing reactions to take place.

doublet 1 any sequence of two nucleotides in a polynucleotide strand. **2** a closely spaced pair of spectral lines or peaks. *Compare* **twin**.

doubling dilution a serial dilution effected so that the concentration in each tube of the particular cell, substance, etc. of interest is half that in the preceding tube of the series.

doubling time the average time taken for the number of cells in a population to double. It will equal the **generation time** only if every cell in the population is able to form two daughter cells and there is no cell lysis.

Dounce homogenizer *or* **Dounce homogeniser** a hand-operated tissue homogenizer, usually of glass, consisting of a ball that is operated as a piston in a tube. The clearance is adjusted to give a fine-particle size reduction without damage to cell nuclei or mitochondria. [After Alexander Latham Dounce, US biochemist (1909–).]

Dowex *proprietary name for* a group of synthetic ion-exchange resins made by Dow Chemical Company. *See also* **cholestyramine resin**.

down regulation 1 a decrease in response of a cell to a hormone brought about either by a decrease in receptor number or by the uncoupling of an effector component. **2** a decrease in the activity of an enzyme or the amount of a protein resulting from the action of an effector molecule on gene expression or protein synthesis. —**down-regulated** *adj*.

Down's syndrome *or* **Down syndrome** *or (formerly)* **mongolism** a type of mental abnormality frequently associated with **trisomy** of chromosome 21. The condition is characterized by a small stature, a rounded head with obliquely slanted ('Mongol-like') eyes and high cheekbones, a fissured tongue with enlarged papillae, and a characteristic palmprint. The degree of mental defect varies considerably. [First described in 1866 by John Langdon Haydon Down (1828–96), British physician.]

downstream 1 *(in molecular biology)* that part of a strand of DNA lying towards the 3′ end from the recognition sequence for a particular restriction enzyme. **2** towards the end of a multistage process, or at a stage nearer to the endpoint of the process than a key stage or one under consideration. *Compare* **upstream**.

doxorubicin a cytotoxic anthracycline antibiotic from *Streptomyces peucetius* var. *caesius*; its carbohydrate is **daunosamine**. It intercalates with DNA and prevents the actions of type II **DNA topoisomerase**. It also inhibits reverse transcriptase and acts on cell membranes. It has anti-tumour activity and has been used to treat some forms of cancer.

doxycycline 5-hydroxy-6-deoxytetracycline; an antibiotic having a wide range of activity against both Gram-positive and Gram-negative organisms. It is closely similar to **tetracycline** in structure and mechanism of action. One proprietary name is Vibramycin.

DPG *or* **2,3-DPG** *abbr. for* (2,3-)diphosphoglycerate (former name for **2,3-bis(phospho)**-D-glycerate).

DPG pocket *abbr. for* diphosphoglycerate pocket. *See* **bisphosphoglycerate pocket**.

dpm *or* **d.p.m.** *abbr. for* disintegrations per minute.

Dpm *see* **A₂pm**.

DPN *abbr. for* diphosphopyridine nucleotide when its oxidation state is unknown or unspecified (obsolete; NAD now used).

DPN⁺ *abbr. for* diphosphopyridine nucleotide (oxidized) *(obsolete name for* **NAD⁺** *)*.

DPNH *abbr. for* reduced diphosphopyridine nucleotide (obsolete; NADH now used).

DPP *abbr. for* **1** dipeptidyl peptidase. **2** deafness-dystonia protein; a protein (97 amino acids) of the intermembrane space of mitochondria. It is homologous to a yeast Tim protein and seems to be involved in importing proteins into mitochondria. Deficiency of DPP causes an X-linked syndrome of deafness, dystonia, dementia, and blindness. *See* **TIM**.

Dpr *see* **A₂pr**.

dps *abbr. for* disintegrations per second.

DPT *abbr. for* 2-diazophenylthioether, *the short name of* the group –O–CH₂–CH(OH)–CH₂–O–[CH₂]₄–O–CH₂–CH(OH)– CH₂–S–C₆H₄–N₂⁺, used to introduce reactive diazonium groups into cellulose or paper, forming **DPT-cellulose** or **DPT-paper**. These are used in **blot transfer** of proteins or nucleic acids. DPT is formed by diazotization of 2-aminophenylthioether (*abbr.*: APT) groups previously coupled to the support.

draft genome sequence a genome that has been sequenced 4–5 times (4–5-fold coverage), providing a framework for DNA fragment assembly and potentially allowing up to 95% of genes to be identified. Draft sequence is mostly in the form of 10000-bp fragments whose approximate chromosomal locations are known; the fragments may contain gaps and their true order and orientation may not be known. *Compare* **finished genome sequence**.

DRAGON *abbr. for* Database Referencing of Array Genes Online; an internet site providing free access to a number of bioinformatics tools including analysis of microarray data.

***Dra*I** a type 2 **restriction endonuclease**; recognition sequence: TTT↑AAA. *Aha*III is an **isoschizomer**.

DRB *abbr. for* 5,6-dichloro-1β-D-furanosyl benzimidazole; a protein kinase inhibitor that affects synthesis of most mRNAs by inhibiting phosphorylation of the C-terminal domain of the largest subunit of RNA polymerase II. See **DRB-sensitivity inducing factor**.

DRB-sensitivity inducing factor *abbr.* DSIF; a protein kinase that phosphorylates the C-terminal domain of the largest subunit of RNA polymerase II and functions as a positive transcriptional elongation factor. It is sensitive to **DRB**.

DREAM *abbr. for* downstream regulatory element antagonistic modulator; a protein that acts as a Ca²⁺-regulated transcription factor by binding the 3′ regulatory element of the gene for preprodynorphin and inhibiting its transcription.

D region the (third) hypervariable region occurring between the V region and the J region in the heavy chains (but not in the light chains) of **immunoglobulin** molecules. It is coded for by the D (for 'diversity') gene segment.

dRib *symbol for* the monosaccharide deoxyribose, 2-deoxy-D-ribose.

Drk *abbr. for* **1** a small protein that bears SH2 domains (*see* **SH domain**), and acts as a link between **sevenless** protein and **son of sevenless** protein in the *Drosophila* photoreceptor system (downstream of receptor kinases). **2 Drk1** a mammalian integral membrane protein forming a voltage-dependent potassium channel of the delayed rectifier class. The voltage sensor, probably segment S4, is characterized by a series of positively charged amino-acid residues at every third position.

D-RNA *or* **dRNA** *abbr. for* DNA-like RNA.

dropping-mercury electrode an electrode used in **polarography** in which mercury drops from a reservoir through a capillary tube into the solution under study. It has the advantage of constantly presenting a clean, unpolarized mercury surface to the solution.

Drosha a nuclear RNase III that is specific for dsRNA. It contains a dsRBM and 2 RNase III domains and processes pri-miRNA to pre-miRNA. It interacts with **Pasha** in the nuclear Microprocessor complex.

drosomycin a peptide from *Drosophila* that is highly homologous with a defensin (RS-AFP1) from radish plants. Both exhibit antifungal activity and have a similar tertiary structure.

Drosophila a genus of small dipterous insects that includes the fruit fly, *D. melanogaster*, also known as vinegar fly or banana fly. This is much used in genetic research because of the giant chromosomes, which occur in the cells of its salivary glands, and because of its short life cycle.

Drp1 an animal homologue of **Dnm1**, a yeast protein involved in forming the contractile ring on the outer mitochondrial membrane that brings about fission of a mitochondrion. Knockout of the Drp1 gene in mice leads to highly interconnected tubular mitochondria.

DRPLA *abbr. for* dentatorubralpallidoluysian atrophy.

Drude equation an equation describing the variation of optical rotation with wavelength of light. It is useful in the interpretation of measurements of **optical rotatory dispersion**, when it may be expressed in the form:

$$[m']_\lambda = \frac{96\,\pi\,N_A}{h\,c} \times \frac{R\lambda_0^2}{\lambda^2 - \lambda_0^2}$$

where N_A is the Avogadro constant, h is the Planck constant, c is the speed of light, R is the rotational strength, λ_0 is the wavelength of the centre of an absorption band, λ is the wavelength of the observation, and $[m']_\lambda$ is the reduced mean residue rotation at that wavelength, as defined at **molar optical rotation**. *See also* **Moffitt–Yang equation**. [After Paul Karl Ludwig Drude (1863–1906), German physicist, noted for, *inter alia*, his work on refraction of light.]

drug any naturally occurring or synthetic substance, other than a nutrient, that, when administered or applied to an organism, affects the structure or functioning of the organism; in particular, any such substance used in the diagnosis, prevention, or treatment of disease.

drug discovery the process of identifying compounds (often via high-throughput methods, such as **combinatorial chemistry** and **high-throughput** screening) that interact with, and modulate the activity of, specific proteins or genes, and the subsequent production of candidate drugs for testing.

druggable target *see* **drug target**.

drug-metabolizing enzyme *or* **xenometabolic enzyme** any enzyme that acts on a drug or other foreign compound but is not known to have any action on a normal metabolite. Many enzymes previously included in this category have subsequently been found to act also on normally occurring substrates.

drug resistance the relatively enhanced resistance of an organism to the action of a drug. It may be caused by induction of an enzyme acting on the drug, by mutation, or by the acquisition of a plasmid coding for drug resistance.

drug target a biological entity (usually a protein or gene) that interacts with, and whose activity is modulated by, a particular compound. If its biochemical profile makes it a suitable candidate for drug testing it is deemed a **druggable target**.

dry ice solid carbon dioxide. It is a convenient cooling agent because, at ordinary atmospheric pressures, heat converts it directly to the gaseous form (sublimation point 194.65K at 101325 Pa; –78.5°C at 760 Torr).

ds *abbr. for* double-stranded.

DsbA a 21 kDa protein that is the bacterial functional equivalent in *Escherichia coli* of the eukaryotic **protein disulfide-isomerase**; however, there is little sequence similarity between the two proteins. The structure of DsbA has been determined, and shown to be similar, in part, to **thioredoxin**.

DSC *abbr. for* differential scanning colorimetry *or* differential scanning calorimeter.

dsDNA *abbr. for* double-strand(ed) DNA (*see* **duplex DNA**).

DSIF *abbr. for* DRB-sensitivity inducing factor.

DSPD *abbr. for* disalicylidenepropanediamine.

dsRBM *abbr. for* dsRNA-binding motif; a protein sequence of ≈70 amino acids that interacts with dsRNA, partially duplexed RNA, and in some cases RNA-DNA hybrids, generally without obvious specificity for the RNA nucleotide sequence. It is predicted to consist of three beta strands flanked by two alpha helices, and may occur up to five times in a protein (e.g. in **Staufen**). Numerous proteins from viruses, bacteria, and eukaryotes that contain the motif fall into nine major groups.

dsRNA *abbr. for* double-strand(ed) RNA.

dsRNA-binding motif *see* **dsRBM**.

DSS *abbr. for* 2,2-dimethylsilapentane 5-sulfonic acid; a compound used as a water-soluble reference substance in **nuclear magnetic resonance spectroscopy**.

dsx symbol for *doublesex*; a gene in *Drosophila melanogaster* that encodes both female-specific and male-specific polypeptides whose synthesis is regulated by alternative sex-specific splicing of the *dsx* transcript. The gene products are transcription factors that repress expression of terminal differentiation genes specific to the opposite sex.

dT *symbol for* a residue of the deoxynucleoside thymidine (alternative to dThd).

dTDP *abbr. for* deoxyribosylthymine diphosphate *or* thymidine diphosphate, the common names for 2′-deoxyribosylthymine-5′-diphosphate; thymidine 5′-diphosphate; 5′-thymidylyl phosphate; thymine 2′-deoxyriboside 5′-diphosphate.

DTDST *abbr. for* diastrophic dysplasia sulfate transporter.

DTE *abbr. for* 1,4-dithioerythritol.

dThd *symbol for* a residue of the deoxynucleoside thymidine; 2′-deoxyribosylthymine; thymine 2′-deoxyriboside (alternative to dT).

dThd5′P *symbol for* thymidine 5′-phosphate.

dThd5′PP *symbol for* thymidine 5′-diphosphate (alternative to ppdT).

dThd5′PPP *symbol for* thymidine 5′-triphosphate (alternative to pppdT).

dTMP *abbr. for* deoxyribosylthymine monophosphate *or* thymidine monophosphate, the common names for 2′-deoxyribosylthymine 5′-phosphate, thymidine 5′-phosphate, 5′-thymidylic acid, thymine 2′-deoxyriboside 5′-phosphate.

DTNB *abbr. for* 5,5′-dithiobis(2-nitrobenzoic acid); *see* **Ellman's reagent**.

DTT *abbr. for* 1,4-dithiothreitol.

dTTP *abbr. for* deoxyribosylthymine triphosphate *or* thymidine triphosphate, the common names for 2′-deoxyribosylthymine 5′-triphosphate; thymidine 5′-triphosphate; 5′-thymidylyl diphosphate; thymine 2′-deoxyriboside 5′-triphosphate.

dU *symbol for* a residue of the deoxynucleoside 2′-deoxyuridine (alternative to dUrd).

dual luciferase reporter assay system a plasmid carrying the genes for firefly and *Renilla reniformis* luciferases with provision for cloning a mammalian promoter upstream from the firefly luciferase reporter gene. The *Renilla* luciferase is under the control of a constitutive promoter and is used to monitor transfection efficiency. Both luciferases can be assayed in cell extracts by the bioluminescence generated using their respective substrates.

dual specificity mitogen-activated protein kinase kinase *see* **MAP kinase kinase**.

dual specificity protein phosphatase an enzyme capable of hydrolysing phosphate monoesters in peptides that contain phosphotyrosine, or phosphoserine/threonine. At least 20 such phosphatases have been detected in mammalian cells. They include Cdc25 and a vaccinia H1-related (VHR) phosphatase (184 amino acids) of humans. The crystal structure of the latter consists of seven alpha helices and five beta strands, and shows a cysteine that is essential for activity.

Dubin–Johnson syndrome a rare inherited defect of liver function characterized by elevated levels of (mainly conjugated) serum **bilirubin**, with accumulation of a melanin-like pigment in hepatocyte lysosomes. It is transmitted as an autosomal recessive trait, and caused by mutations in a gene at 10q23-q24 for canalicular multiorganic anion transporter (cMOAT), an ABC-transporter of 1545 amino acids. [After Nathan Dubin (1913–1980), US pathlogist, and Frank Johnson (1919–), US pathologist.]

Dubnoff shaker a controlled-temperature water bath with a uniformly oscillating load platform and facilities for maintaining a controlled atmosphere. It is used for studies of the metabolism of tissues and microorganisms and for enzyme and blood-coagulation studies. [After Jacob William Dubnoff (1909–), US biochemist.]

Duboscq colorimeter an early, accurate visual colorimeter for determination of small quantities of substances in solution, in which the intensities of colour of two solutions, standard and unknown,

were compared by varying the path-lengths of light passing through them from a single source until their apparent colours matched. This was achieved by placing the solutions in flat-bottomed vessels (termed cups) and adjustment of their positions in relation to two identical, fixed, transparent plungers through which the light passed, via prisms, to a divided-field eyepiece. The path-length in each solution was then read from a vernier scale. [After Jules Duboscq (1817–86), French optician.]

Duchenne muscular dystrophy *see* **dystrophin**.

ductless gland *an alternative name for* **endocrine gland**.

dUDP *abbr. for* deoxyuridine diphosphate, the common name for 2′-deoxy-5′-uridylyl phosphate; 2′-deoxyuridine 5′-diphosphate; uracil 2′-deoxyriboside 5′-diphosphate.

DUF *abbr. for* domain of unknown function.

Duffy system *symbol*: Fy; a blood-group system for which the antigenic determinant depends on the structure of an erythrocyte membrane glycoprotein, of molecular mass 35–45 kDa, named GPD. This functions as a chemokine receptor (Duffy antigen receptor for chemokines; *abbr.*: DARC) and as a receptor for the malaria parasite *Plasmodium vivax*. The parasite is unable to enter Duffy-negative erythrocytes. GPD is also a receptor for several proinflammatory cytokines and is expressed in other cell types besides erythrocytes. The system is encoded by a locus at 1q21-q22, and there are three major alleles: FY*A, FY*B, and FY*O.

dulcitol *another (older) name for* **galactitol**.

dUMP *abbr. for* deoxyuridine monophosphate, the common name for 2′-deoxyuridine 5′-phosphate; 2′-deoxy-5′-uridylic acid; uracil 2′-deoxyriboside 5′-phosphate.

duodenase a type II transmembrane serine protease produced by duodenal mucosa. It has been proposed as the activator of enteropeptidase, which has very low autocatalytic activity.

duodenum that part of the small intestine extending from the pylorus of the stomach to the jejunum. In mammals it lies around the head of the pancreas and receives the duct from the exocrine pancreas and the gall bladder. **—duodenal** *adj*.

Duolite *proprietary name for* a group of ion-exchange resins. It includes those formerly known by the names Biodeminrolit, De-Acidite, Zeo-Karb, and Zerolit.

duplex 1 having two parts or elements, especially of similar form or function. **2** describing a molecular structure (or part of one) in which two polynucleotide strands lie side by side in a head-to-tail orientation and are linked together along their length; *see* **duplex DNA**. **3** describing a telegraphic system, computer channel, etc. that permits two messages to be sent simultaneously in opposite directions.

duplex DNA double-stranded **DNA** in which the two polynucleotide chains are linked together by hydrogen bonds between complementary **base pairs** along their lengths, with the 3′→5′ phosphodiester bonds of the two chains running in opposite directions. Such a molecule is usually coiled into a double helix and it may or may not be covalently closed into a circular molecule and/or formed into superhelical DNA (*see* **supercoil**) *See also* **double helix**.

duplicate *(in biology)* to reproduce by division into two equal parts; *see* **binary fission**.

duplication *see* **gene duplication**.

dUrd *symbol for* a residue of the deoxynucleoside deoxyuridine; 2′-deoxyribosyluracil; uracil 2′-deoxyriboside (alternative to dU).

dUrd5′P *symbol for* deoxyuridine 5′-phosphate.

dUrd5′PP *symbol for* deoxyuridine 5′-diphosphate (alternative to ppdU).

dUrd5′PPP *symbol for* deoxyuridine 5′-triphosphate (alternative to pppdU).

DUST *see* **sequence filter**.

dUTP *abbr. for* deoxyuridine triphosphate, the common name for 2′-deoxyuridine 5′-triphosphate; 2′-deoxy-5′-uridylyl diphosphate, uracil 2′-deoxyriboside 5′-triphosphate.

dUTPase EC 3.6.1.23; *recommended name*: dUTP pyrophosphatase; *systematic name*: dUTP nucleotidohydrolase; *other name*: deoxyuridine-triphosphatase. An enzyme of nucleotide metabolism in eukaryotes that prevents the incorporation into DNA of deoxyuridylyl residues from dUTP by catalysing the hydrolysis of the latter to dUMP and pyrophosphate. Genes for dUTPase are found in several viruses, e.g. herpes simplex type 1.

Duve *see* **de Duve**.

dwarfism atypically short stature reaching adult heights of generally less than 1.25 m. There are various causes. A common one is an acquired or inherited defect related to the synthesis, secretion, or function of somatotropin (e.g. pituitary dwarfism, Laron dwarfism, and pigmyism). Other causes include congenital hypothyroidism, resistance to parathyroid hormone or fibroblast growth factor, and a wide variety of achondroplasias and metabolic disorders.

Dy *symbol for* dysprosium.

dyad 1 two units treated as one. **2** a meiotic chromosome in which the two homologous members of the tetrad have separated. **3** a pair of cells resulting from the first meiotic division. **4** a bivalent atom, molecule, or group. *See also* **dyad symmetry**.

dyad axis a twofold **rotation axis**.

dyad-related describing a phenomenon that occurs at a specific position with respect to a **dyad axis**.

dyad symmetry 1 the type of symmetry occurring in molecular structures in which there is repetition after every 180° rotation around a dyad axis. **2** *(in molecular biology)* the formal description of a palindromic sequence in which the axis of symmetry is at the centre of the palindromic sequence. *See* **palindrome**.

dye a natural or synthetic substance that strongly absorbs certain wavelengths of visible or near ultraviolet light and is used to colour material by becoming firmly attached to it. *See also* **stain**.

dye-exclusion test a test for cell viability based on the phenomenon that living cells exclude some dyes, e.g. Eosin and Trypan Blue, whereas dead cells take them up and become stained.

dye-laser a device for obtaining high-intensity light of chosen wavelength by passing the monochromatic emission beam of a **laser** through a stream of liquid containing a fluorescent dye, and then passing the emitted broad-band fluorescent light through a monochromator. By variation of the dye employed a wide range of wavelengths of high-intensity light can be obtained from the single-wavelength laser emission. The light obtained by this device is necessarily of longer wavelength than that of the laser emission.

dye-ligand chromatography *or* **pseudo-affinity chromatography** *or* **pseudo-ligand affinity chromatography** a technique similar to **affinity chromatography** in which the immobilized ligand is any one of certain reactive dyes (coupled usually to cross-linked agarose), the dye commonly being a triazinyl compound, e.g. Reactive Blue 2 or Reactive Red 120. The technique is thus sometimes referred to (inappropriately) as **triazine-dye affinity chromatography**. The fixed dye molecules selectively bind particular enzyme or other protein molecules; the specificity of the interaction varies widely with both protein and dye, although frequently the dye appears to mimic a natural ligand of the protein. Examples of use are the isolation of nucleotide-requiring enzymes (*see also* **nucleotide-binding fold**) and the removal or purification of albumin from serum. The underlying mechanism is not fully understood but appears to involve both hydrophobic (i.e. apolar) and strongly ionic interactions. Many uses may in reality be examples of **affinity-elution chromatography**.

dye-primer sequencing DNA sequencing reactions in which the primer is labelled with a fluorescent dye. Four reactions are carried out, each with a different dideoxynucleoside triphosphate and a primer labelled with one of four distinctive fluorescent dyes.

dye-terminator sequencing DNA sequencing reactions in which the four dideoxynucleoside triphosphates are each labelled with a different fluorescent dye.

dynactin the largest protein component of the **dynactin complex**, a 20S multiprotein assembly of total mass ~1.2 MDa, that activates dynein-based activity *in vivo*. A large structural component of the complex is an actin-like 40 nm filament composed of actin-related protein, to which other components attach. Vertebrate dynactin is a ubiquitously expressed cytoplasmic protein; several isoforms, of mass ranging from 117 kDa to 160 kDa, are produced by alternative splicing. It shares a high degree of similarity with the product of the *Drosophila* gene *glued*, essential during embryogenesis, some mutations of which kill homozygous embryos.

dynamic of, or concerned with, motion or force, or with dynamics.

dynamic programming a programming method that finds the optimal alignment between pairs of protein or nucleotide sequences by exploring all possible alignments and choosing the best, based on a given scoring matrix. Dynamic programming solves complex problems by building up from smaller subproblems using a recursion relation. For example, the **Needleman–Wunsch alignment algorithm** is a dynamic global alignment algorithm, whereas the **Smith–Waterman alignment algorithm** is a dynamic local alignment algorithm.

dynamics 1 the branch of a science, especially mechanics, concerned with the motions of objects and the forces that produce such motions. It includes **kinetics**. **2** those forces that produce change or movement in a system or molecule; the motions (e.g. of residues within a macromolecule) so resulting.

dynamic state of body constituents the phenomenon, discovered in the early 1940s, that body components, especially proteins, are in a state of continuous degradation and resynthesis (i.e. of dynamic equilibrium). It was subsequently found to be applicable to other body constituents, although genomic DNA is a notable exception.

dynamin a microtubule-associated protein able to bind and hydrolyse GTP. The dynamin family includes two types, neuron-specific dynamin 1, and the widely expressed dynamin 2, which exhibit a high degree of sequence similarity in the 300 amino acid N-terminal region containing the GTP-binding site, but show divergence in a 100 amino acid proline-rich C- terminal region that in dynamin 1, but not dynamin 2, is a site for regulation of GTPase activity by phosphorylation–dephosphorylation. Dynamins are activated by protein kinase C and have a **pleckstrin homology domain**, which binds acidic (especially inositol-) phospholipids, with activation of the GTPase activity. Dynamins are involved in formation of microtubule bundles, and in vesicular trafficking; purified dynamin 2 in solution forms helical structures that *in vivo* may facilitate clipping of vesicles budding from the plasma membrane, and accelerate endocytic recycling. The *Drosophila* protein (a dynamin 2) is a product of *shibire*, a gene known to be involved in endocytosis.

dyne *symbol*: dyn; the CGS unit of force, defined as the force required to give one gram an acceleration of one centimetre per second per second. A gram weight thus represents a force of about 980 dyn. 1 dyn = 10^{-5} N.

dynein a large, multisubunit protein complex with ATPase activity, that is associated with microtubules, and functions as a molecular motor to drive the movement of a diverse range of intracellular structures, such as endosomes, lysosomes, and mitochondria, along microtubules towards the centrosome. It is also attached to the peripheral fibrous components of eukaryotic flagella and cilia, and is responsible for their movement; it may also play a role in the separation of chromosomes at cell division. The complex consists of two heavy chains (~500 kDa), three or four intermediate chains (~70 kDa), and four light chains (~50 kDa).

dynorphin an opioid tridecapeptide of high potency found in pig brain, duodenum and pituitary, having the structure

Tyr-Gly-Gly-Phe-Leu-Arg-Arg-Gln-Phe-Lys-Val-Val-Thr,

the N-terminal pentapeptide sequence of which is identical to **[5-leucine]enkephalin**. It is synthesized as **preprodynorphin** (*also called* preproenkephalin B), from which its precursor protein **prodynorphin** (*also called* β-neoendorphin *or* proenkephalin B), a 236-residue polypeptide, is formed. *See also* **endorphin**.

dys+ *prefix* indicating abnormal, diseased, impaired, painful, or difficult.

dysferlin a protein (2080 amino acids) of heart and skeletal muscle. It has no known function and is homologous with the spermatogenesis factor fer-1 of *Caenorhabditis elegans*. Nonsense mutations in the gene at 2p13 produce dysferlin deficiency and limb-girdle muscular dystrophy type 2B.

dysfunction any abnormality or disturbance in the function of an organ, tissue, cell, or part of a cell. —**dysfunctional** *adj*.

dysgenic genetically deleterious.

dyskeratosis congenita a rare syndrome of skin defects (leukoplakia, atrophy and pigmentation, nail dystrophy), heightened susceptibility to epithelial cancers, and bone marrow failure. It is caused by mutations in the gene for **dyskerin**.

dyskerin a putative pseudouridine synthase that catalyses site-specific isomerization of uridine residues in RNA to pseudouridine. It is associated with snoRNA and binds to the RNA component of telomerase. Mutations in the gene at Xq28 result in **dyskeratosis congenita**.

dysmorphology 1 the abnormal development of form or shape. **2** an anatomical feature that is abnormal. Some are caused by mutations, such as cleft lip and palate. **3** the study of abnormal development. **–dysmorphological** *adj.* **–dysmorphologically** *adv.*

dysplasia the abnormal development of tissue. **–dysplastic** *adj.*

dystrobrevin a phosphoprotein of human and murine muscle that is associated with the C-terminal region of **dystrophin**. It is a homologue of an 87 kDa protein of the postsynaptic membrane of the electric organ of electric rays (*Torpedo* spp.).

dystroglycan a protein precursor yielding a 43 kDa trans-membrane glycoprotein and an extracellular 156 kDa dystrophin-associated glycoprotein in skeletal muscle; the 156 kDa component binds **laminin**, acting as a laminin receptor that links the extracellular matrix and **sarcolemma** in skeletal muscle.

dystrophin a protein present in small amounts in normal muscle, but absent or abnormal in patients with **Duchenne muscular dystrophy**. It interacts with syntrophins, and may play a role in anchoring the cytoskeleton to the plasma membrane. The DMD gene is the largest in humans, with 2.4 million bp; it contains 79 exons, and takes more than 16 hours to be transcribed and co-transcriptionally spliced. *See also* **dystroglycan, utrophin**.

dystrophin-related protein (*abbr* DRP) *see* **utrophin**.

dystrophy any disorder, sometimes genetic, of the structure and functions of an organ or a tissue, as of muscle or bone. **—dystrophic** *adj.*

Ee

e *symbol for* **1** electron; it may be written as e⁻ to indicate its charge; e⁺ denotes a positron. **2** *(in mathematics)* the transcendental number equal to the limiting value of the series $[1 + (1/n!)]$ for $n = 1$ to infinity; its value is 2.71828182846. It is used as the base of natural (Napierian) logarithms. **3** equatorial; *see* **conformation**.

e′ *symbol for* pseudo-equatorial; *see* **conformation**.

e *symbol for* **1** elementary charge (or proton charge) **2** efficacy (of a drug).

E *symbol for* **1** a residue of the α-amino acid L-glutamic acid (alternative to Glu). **2** elimination reaction. **3** electromeric effect (in electron displacement). **4** exa+ (SI multiplicative prefix denoting 10^{18}).

[E]₀ *or* **[E]ₜ** *or* **[E]ₛₜₒᵢ𝒸ₕ** *(in enzyme kinetics) symbol for* the total or stoichiometric concentration of active (catalytic) centres.

E₂

E₁ **1** *abbr. for* estrone. **2** the pyruvate dehydrogenase (lipoamide) component of **pyruvate dehydrogenase complex**. **3** *see* **ubiquitin-activating enzyme E1**.

E₂ **1** *abbr. for* estradiol. **2** the dihydrolipoamide *S*-transacetylase component of **pyruvate dehydrogenase complex**. **3** *see* **ubiquitin-conjugating enzyme**.

E₃ **1** the dihydrolipoamide dehydrogenase component of **pyruvate dehydrogenase complex**. **2** *see* **ubiquitin–protein ligase**.

E74 a protein, expressed during development of *Drosophila*, that correlates with an ecdysone-induced activity of puff 74EF. It is encoded by *Drosophila EIP74Ef*, an ecdysone-inducible gene that encodes two *ETS*-related protein products, E74A and E74B, that differ mainly in their N-terminal regions, and have substantial poly-Ala, poly-Asn, poly-Gln, poly-Gly and poly-Ser domains.

E *symbol for* **1** *(in chemical nomenclature)* a specific configuration around a double bond; *see* **E/Z convention**. **2** electromotive force. **3** electric potential difference (of a galvanic cell). **4** energy; kinetic energy and potential energy may be denoted by the symbols E_k and E_p respectively. **5** extinction coefficient (no longer recommended). **6** *(bold italic)* electric field strength *(see* **electric field***)*.

(E)- prefix *(in chemical nomenclature)* denoting a geometric isomer in which the highest priority ligands according to the Cahn–Ingold–Prelog rules (*see* **sequence rule**) are located on opposite sides of a double bond. *See also* **E/Z convention**.

-E conformational descriptor denoting the envelope conformation of a monosaccharide or monosaccharide derivative.

E° *or* **E⁻θ** *symbol for* standard electrode potential.

E′ *or* **E⁻θ′** *symbol for* standard electrode potential at a specified pH.

Eₐ *or* **E_A** *symbol for* (Arrhenius) activation energy.

Eₑₐ *symbol for* electron affinity (def. 2).

Eadie–Hofstee plot a graphical representation of enzyme kinetic data in which the reaction velocity divided by the substrate concentration, $v/[S]$, is plotted as ordinate against the reaction velocity, v, as abscissa. If a straight line is obtained, the intercept on the abscissa gives the maximum velocity, V, and that on the ordinate V/K_m. The slope is $-1/K_m$. [After George Sharpe Eadie (1895–1976) and Barend Hendrik Jan Hofstee (1912–80). US biochemists, who independently described it in 1942 and 1952, respectively.]

EaeI a type 2 **restriction endonuclease**; recognition sequence: [TC]↑GGCC[AG]. *Cfr*I is an **isoschizomer**.

Eagle's medium any of various growth or maintenance media used in tissue culture and essentially comprising a balanced salt solution, usually **Earle's balanced salt solution** or **Hanks's balanced salt solution**, supplemented with amino acids, vitamins, serum, and antibiotics.

Eam1105I a type 2 **restriction endonuclease**; recognition sequence: GACNNN↑NNGTC.

EAR-3 *see* **COUP**.

Earle's balanced salt solution a solution used in tissue culture to provide a normal ionic, pH, and osmotic environment for cell growth. It contains, per 100 ml distilled water, 0.68 g NaCl, 0.04 g KCl, 0.02 g $CaCl_2$, 0.02 g $MgSO_4{\cdot}7H_2O$, 0.0125 g $NaH_2PO_4{\cdot}H_2O$,

0.22 g $NaHCO_3$, 0.1 g glucose, and 0.002 g phenol red; it has a pH of 7.6–7.8.

early enzyme an enzyme that is transcribed from an **early gene** (def. 2) of a virus. *Compare* **late enzyme**.

early gene 1 *or* **immediate gene** any of a number of genes involved in the earliest responses of cells to factors that initiate the transition from quiescence to proliferation. Early genes include those that encode transcription factors, such as the c-*jun*, c-*fos*, and c-*egr* genes, and those coding for structural proteins such as actin. Early genes are further subdivided into immediate-early and delayed-early genes. **2** any viral gene that is transcribed early after the virus infects a host cell. Such genes probably code for proteins that are necessary for viral nucleic acid replication. *Compare* **late gene**.

early growth response protein any of a family of proteins that are expressed early in response to growth factors and function as transcriptional regulators, binding to the DNA sequence 5′-CGCC-CCCGC-3′.

early protein a protein that is transcribed from an **early gene** (def. 2) of a virus. *Compare* **late protein**.

early quitter any incomplete polypeptide formed in an *in vitro* translation system.

Easson–Stedman model a model to explain the differences in biological activity of two enantiomeric molecules; it assumes that three of the groups attached to an asymmetric carbon atom in an agonist (or substrate) are involved in its attachment to a specific receptor (or enzyme), one enantiomer will therefore be a better agonist (or substrate) than the other. The concept is similar to, but considerably pre-dates, that developed by Ogston (*see* **Ogston concept**).

EBI *abbr. for* European Bioinformatics Institute; *see* **bioinformatics institute**.

4E-BP *abbr. for* eIF4E-binding protein; any of a group of small proteins that bind eIF4E (the mRNA 5′ cap-binding protein) thus controlling its availability for initiating translation. The state of phosphorylation of 4E-BP controls its binding to eIF4E. Under conditions that induce **apoptosis**, interaction of 4E-BP with eIF4E increases, so that initiation of translation decreases.

EBV *or* **EB virus** *abbr. for* Epstein–Barr virus.

EC *abbr. for* **1** Enzyme Commission; it is used as a prefix in the numerical designation of an enzyme (*see* **enzyme classification**). **2** electron capture. **3** enterochromaffin (as in **EC cell**).

EC₅₀ *abbr. for* half effective concentration; the molarity of an agonist that produces 50% of the maximal possible effect of that agonist. Other percentage values (EC₂₀, EC₄₀, etc.) can be specified. The action of the agonist may be stimulatory or inhibitory.

EC cell *abbr. for* enterochromaffin cell.

eccrine 1 describing a secretory cell that discharges its product without loss of cytoplasm. **2** describing a gland made up of such cells. *Compare* **apocrine**, **holocrine**, **merocrine**.

ECD *abbr. for* **1** electron capture detector. **2** electrochemical detection. *See* **polarography**.

ecdysis the periodic shedding of the exoskeleton and construction of a new cuticle, as in insects and crustaceans, or the shedding of the outer layer of the skin, as in reptiles. —**ecdysial** *adj.*

ecdysone 1 α-ecdysone (22R)-2β,3β,14,22,25-pentahydroxycholest-7-en-6-one; the **ecdysteroid** that is synthesized in and secreted by the prothoracic glands of immature insects and the ovaries of the adult females. First isolated from the silkworm moth (*Bombyx mori*), it is the inactive prohormone of the moulting hormone **ecdysterone**; it may also have intrinsic hormonal activity at other stages of insect development. **2** β-ecdysone *an alternative name for* **ecdysterone**. **3** *a former name for* **ecdysteroid**.

ecdysone-inducible mammalian expression system a system for the expression of exogenous transgenes in mammalian cells in which the timing of expression is dependent on the administration of the insect hormone ecdysone.

ecdysone receptor a receptor for ecdysone, located in the cell nucleus and related to the steroid/thyroid/retinoic acid family of nu-

ecdysone

clear hormone receptors. It consists of a heterodimer of a subunit named EcR (ecdysone receptor) and another named USP (ultra-spiracle), the insect homologue of vertebrate retinoid X receptor. The heterodimer binds ecdysone with high affinity, and the ternary complex binds to DNA.

ecdysteroid or (formerly) **ecdysone** any of a group of polyhydroxy-lated ketosteroids of which α-ecdysone is the parent substance for the purpose of nomenclature (see **ecdysone**). They are almost invariably C_{27} steroids, differing only in the number and/or steric arrangement of their hydroxyl groups. Ecdysteroids are ubiquitous in insects and other arthropods, in which they initiate postembryonic development, including the metamorphosis of immature forms and the development of the reproductive system and the maturation of oocytes in adult females. They have been found also in annelids, molluscs, and nematodes. Cognate substances, often known as **phytoecdysones**, have been identified in plants. The principal ecdysteroids in insects are α-ecdysone (see **ecdysone** (def. 1)) and **ecdysterone**.

ecdysterone or β**-ecdysone** or **crustecdysone** or (insect) **moulting hormone** 20-hydroxyecdysone; (22R)-2β,3β,14,20,22,25-hexahydroxy-cholest-7-en-6-one; the **ecdysteroid** isolated both from insects and from crustaceans. In insects, ecdysterone is formed in peripheral tissues from α- ecdysone through the action of ecdysone 20-monooxygenase (EC 1.14.99.22). The hormone causes puffing of polytene chromosomes, and initiates moulting of larval and pupal forms by stimulating the epidermal cells to enlarge, divide, and secrete degradative enzymes that dissolve the protein and chitin components of the cuticle. Moreover, in the adult females of certain insects, e.g. mosquitoes (Aëdes spp.) and Drosophila melanogaster, ecdysterone formed from α-ecdysone in the ovaries stimulates vitellogenin synthesis in the fat body.

e-cell a computer model of a cell generated using genomic, proteomic, transcriptomic, and metabolomic data.

ECF abbr. for extracellular fluid.

ECG or (US) **EKG** abbr. for **1** electrocardiograph. **2** electrocardiogram.

echinocandin any of a group of cyclic hexapeptides that contain an N-acyl aliphatic or aryl side chain. They inhibit synthesis of 1,3β-glucans of fungal cell walls and are effective against Candida spp., Aspergillus spp., and Pneumocystis carinii. Examples include caspofungin, micafungin, and anidulafungin.

echinochrome A 2-ethyl-3,5,6,7,8-pentahydroxy-1,4-naphthalene-dione; a red pigment of sea urchin eggs.

echinoderm any of the phylum Echinodermata of marine invertebrates that contains sea urchins, sea cucumbers, starfish, etc. They are more or less radially symmetrical, with calcareous skeletal plates in their skin and a well-developed coelom. Locomotion and gaseous exchange are accomplished by means of retractable tube feet. They are related to the same ancestral stock that gave rise to the vertebrates.

ECL abbr. for enhanced chemiluminescence.

eclipsed conformation see torsion angle.

eclosion hormone any peptide hormone that programmes the death of certain muscles and neurons during metamorphosis of insects.

Eco52I a type 2 **restriction endonuclease**; recognition sequence: C↑GGCCG. XmaIII is an **isoschizomer**.

Eco81I a type 2 **restriction endonuclease**; recognition sequence: CC↑TNAGG. SauI is an **isoschizomer**.

EcoCyc see pathway database.

E. coli abbr. for Escherichia coli.

ecology the branch of biology dealing with the relations of living organisms to their surroundings. —**ecological** adj.; **ecologically** adv.; **ecologist** n.

ecomone any nontrophic molecule that ensures a flux of information between organisms in an **ecosystem**.

EcoO65I a type 2 **restriction endonuclease**; recognition sequence: G↑GTNACC. BstEII and BstPI are **isoschizomers**.

EcoO109I a type 2 **restriction endonuclease**; recognition sequence: [AG]G↑GNCC[TC]. DraII is an **isoschizomer**.

EcoRI a type 2 **restriction endonuclease**; recognition sequence: G↑AATTC. The M.EcoRI modification site is A3.

EcoRV a type 2 **restriction endonuclease**; recognition sequence: GAT↑ATC.

ecosystem a unit of the environment together with the organisms it contains. There is a constant interchange between living organisms and their chemical and physical environment.

EcoT14I a type 2 **restriction endonuclease**; recognition sequence: C↑C[AT][AT]GG. StyI is an **isoschizomer**.

EcoT22I a type 2 **restriction endonuclease**; recognition sequence: ATGCA↑T. AvaIII is an **isoschizomer**.

ecotin a monomeric periplasmic protein of Escherichia coli that inhibits the pancreatic serine proteases chymotrypsin, trypsin, and elastase, allowing the organism to survive in the presence of these enzymes. It also inhibits blood coagulation factors Xa and XIIa, and kallikrein, thereby having anticoagulant effects.

ECP abbr. for eosinophil cationic protein.

ecstasy see 3,4-methylenedioxymethamphetamine.

ECTEOLA cellulose abbr. for a weakly basic (pK ≈7.5) anion-exchange material of uncertain structure, prepared by the condensation of epichlorhydrin, triethanolamine (hence e,c + t,e,ola), and cellulose. It is useful for separations of proteins, nucleoproteins, and nucleic acids.

ecto+ comb. form indicating external, outer.

ectocrine describing or relating to a metabolite that, when released from the organism in which it was made, differentially affects other organisms of the same or a different species. Such a metabolite might be harmful to some members of a community and beneficial to others. The term includes **allomone**, **ectohormone**, **kairomone**, and **pheromone**. Compare **endocrine** (def. 1), **exocrine** (def. 1).

ectodysplasin A abbr.: EDA-A; a type II transmembrane protein of the tumour necrosis factor (TNF) family that contains a furin recognition sequence. It is present in keratinocytes, hair follicles, and sweat glands, and regulates development and differentiation of epidermal structures. There are two isoforms: EDA-A1 (391 amino acids) and EDA-A2 (389 amino acids). Their receptors activate NF-κB. Mutations in the furin recognition sequence cause X-linked anhidrotic ectodermal dysplasia.

ectoenzyme any enzyme that is attached to the external surface of the plasma membrane of a cell.

ectohormone any **ectocrine** substance whose production and release benefits either the organism producing it or other members of the same species. *Compare* **endohormone**.

ectoparasite any **parasite** that lives on the exterior of its host organism.

ectopic occurring in an unusual place or in an unusual form or manner. For example, an **ectopic protein** is a protein produced by a neoplasm derived from a tissue that does not normally produce that protein. An example is the production of **vasopressin** and **corticotropin** by small cell lung cancer. Ectopic pregnancy occurs when the fertilized ovum implants outside the uterus e.g. in a fallopian tube. *Compare* **entopic**.

ectoplasm the outer, relatively rigid part of the **cytoplasm**. *Compare* **endoplasm**. —**ectoplasmic** *adj*.

ectoplast the part of the plasma membrane of a plant cell that is in contact with the cell wall.

ectoprotein any individual protein found on the exterior of cells. Such proteins may function, e.g., as mediators of cell–cell or cell–surface interactions, or as receptors for substances with regulatory actions on cells.

ectosymbiont a partner in a symbiotic relationship that remains outside the tissues and cells of the other partner, often occupying a body cavity; e.g. one of the cellulose-metabolizing microorganisms occurring in the digestive tract of ruminants. *Compare* **endosymbiont**. *See also* **symbiosis**.

ED$_{50}$ *abbr. for* median effective dose; i.e. the dose of a drug or other agent that produces, on average, a specified all-or-none response in 50% of a test population, or, if the response is graded, the dose that produces 50% of the maximal response to that drug or agent.

edeine any of several pentapeptide-amide antibiotics elaborated by a strain of *Bacillus brevis* and effective against Gram-positive and Gram-negative bacteria, some fungi and other eukaryotic cells, and some neoplastic cells. Edeine A$_1$, N^a[(N^2-{N^2-[N-(b-tyrosyl)-isoseryl]-2,3-diaminopropionyl}-2,6-diamino-7-hydroxyazela-9-yl)-glycyl]-spermidine, and its N^x-amidinospermidine analogue, edeine B$_1$, are highly active; their respective constitutional isomers, edeines A$_2$ and B$_2$, have low intrinsic activity. In intact bacteria, low concentrations of edeine reversibly inhibit DNA synthesis and enhance RNA synthesis but do not affect protein synthesis. In cell-free systems, edeine A$_1$ interacts with both ribosomal subunits (preferentially the smaller one) and strongly inhibits translation.

edeine A$_1$

EDEM a stress-inducible endoplasmic reticulum (ER) membrane protein that is homologous with α-mannosidase but is catalytically inactive. Through its transmembrane segment it interacts with calnexin and recognizes the folding status of glycoproteins in the ER.

edema *or (esp. Brit.)* **oedema** swelling of an organ or tissue due to the accumulation of fluid. —**edematous** *or (esp. Brit.)* **oedematous** *adj*.

edema factor *or (esp. Brit.)* **oedema factor** a protein component of the *Bacillus anthracis* toxin (responsible for anthrax) that causes tissue swelling.

edestin a 300 kDa globulin obtained from hemp seed, castor-oil beans, and certain other seeds. It readily forms polymorphic crystals and will support the growth of animals in the absence of other dietary proteins. [From Greek *edestos*, edible.]

edetate *see* EDTA.

EDF *abbr. for* erythroid differentiation protein; *see* **activin**.

Edg *abbr. for* endothelial differentiation gene; a gene that encodes a G-protein-coupled receptor present on endothelial cells. A major ligand for the receptor is sphingosine 1-phosphate, which on binding activates nitric oxide synthase.

editing 1 the process or act of altering or adapting erroneous molecular structures, especially nucleotide sequences in polynucleotides, so as to preserve fidelity in the transfer and expression of genetic information. For example, the 3′→5′ exonuclease activity of DNA polymerase I has an editing function in DNA polymerization by removing mismatched residues at the primer terminus of the growing chain. Another editing function is shown by aminoacyl-tRNA synthetases. In the presence of ATP and the 'wrong' amino acid, these enzymes do not catalyse the formation of the incorrect aminoacyl-tRNA but act as ATP pyrophosphatases, hydrolysing ATP to AMP and pyrophosphate. The term is applied also to the processes of conversion of RNA precursors, e.g. hnRNA, to the mature form, mRNA. **2** the insertion, deletion, or substitution of nucleotides within nascent RNA transcripts to produce RNA molecules with sequences that differ from those coded genomically. Most examples have been found in organelle-encoded RNAs, e.g. mitochondrial RNAs, and include ApoB, several **glutamate receptor** mRNAs, and **Wilms tumour** susceptible gene mRNA. In the case of ApoB, a cytidine residue is deaminated to a uridine residue so that the codon for glutamine (CAA) becomes a stop codon (UAA); hence, translation results in a shortened form of the protein.

editosome a macromolecular complex involved in the editing of RNA transcripts. Specific ribonucleotide complexes include guide RNA (*see* **gRNA**), which specifies the edited sequence.

Edman degradation a procedure used in sequencing (poly)peptides in which amino-acid residues are removed stepwise from the N terminus by reaction with phenylisothiocyanate, C_6H_5–N=C=S, to form the phenylthiocarbamyl-peptide (*abbr.*: PTC-peptide). This is then cleaved in anhydrous acid (which favours attack by the sulfur atom on the carbonyl carbon of the terminal peptide) to give a thiazolinone intermediate and the release of the remainder of the peptide chain. The thiazolinone intermediate is hydrolysed to the phenylthiocarbamyl derivative of the N-terminal amino acid, which then cyclizes to form the phenylthiohydantoin derivative (*abbr.*: PTH-amino acid). The PTH-amino acid is extracted into organic solvents and identified by chromatography. These derivatives have a strong UV-absorption at about 259 nm, which permits them to be visualized against a fluorescent screen using an ultraviolet light source. As the remainder of the peptide chain is liberated intact, the procedure may be repeated – up to about ten amino-acid residues may be sequenced in this way, or more in favourable cases. The procedure may be automated, e.g. in a **sequenator**. [After Pehr Victor Edman (1916–77), Swedish protein chemist, who described the method in 1956.]

Edman reagent phenylisothiocyanate; C_6H_5–N=C=S. *See* **Edman degradation**.

EDR *abbr. for* equi-effective dose ratio.

EDRF *abbr. for* endothelium-dependent releasing factor (*see* **nitric oxide**).

EDTA ethylenediaminetetraacetate; edetate; any anion or salt, esp. the (hydrated) disodium salt, of edetic acid, ethylenediamine-N,N,N',N'-tetraacetic acid, (ethylenedinitrilo)tetraacetic acid. Commonly used as the disodium salt, it is a powerful chelating agent for divalent metal ions.

HOOC—CH2

HOOC—CH2

N—CH2—CH2—N

—CH2—COOH

—CH2—COOH

ethylenediaminetetraacetic acid

EEDQ *abbr. for* N-ethoxycarbonyl-2-ethoxy-1,2-dihydroquinoline; an agent for peptide condensation that causes little or no racemization; it is useful in coupling ligands to insoluble polymers. It also causes progressive and irreversible inhibition of mitochondrial F_1-ATPase.

eEF-2 a eukaryotic **elongation factor** that is equivalent to the prokaryotic **EF-G**.

eEFsec *an alternative name for* **SelB**.

EET *(formerly) abbr. for* epoxyeicosatetraenoate.

EF *abbr. for* elongation factor.

EF-1 *see* **EF-T**, **elongation factor**.

EFE *abbr. for* ethylene-forming enzyme.

EFEMP *abbr. for* EGF-containing fibrillin-like extracellular matrix protein; either of two glycoproteins that contain multiple EGF repeats and belong to the fibrillin family of extracellular matrix proteins. A missense mutation in the EFEMP1 gene is common in an autosomal dominant form of retinal dystrophy. EFEMP2 is encoded at a different locus.

effective dose *see* **ED₅₀**.

effective theoretical plate number *symbol*: N; a number indicative of chromatographic column performance when resolution is taken into account: $N = 16R_s^2/(1-\alpha)^2$, where R_s is the peak resolution and α the separation factor (i.e. the ratio of the distribution ratios or coefficients for the two substances being resolved when measured under identical conditions); by definition α is greater than unity. *Compare* **height equivalent to an effective theoretical plate**. *See also* **theoretical plate**.

effectomer the part of a two-component agonist that brings about the biological effect but only when bound to a **haptomer**, which enables it to interact with the cell membrane and exert its toxic action on intact cells. For example, the A fragment of diphtheria toxin can only exert its toxic effects on a target cell when bound to the B fragment.

effector 1 *or* **modifier** *or* **modulator** *(in molecular biology)* any small molecule or ligand that interacts with an enzyme thereby changing its catalytic behaviour but that is not itself changed during the enzyme action. A **positive effector** enhances catalytic activity while a **negative effector** reduces it. *See also* **allosteric effector**. **2** *(in physiology)* a cell or organ that produces a physiological response when stimulated by the nervous system.

effector site *or* **regulatory site** any site on an enzyme molecule that binds an **effector**.

efferent 1 conveying outwards from a part, organ, or centre; e.g. of a blood vessel, nerve, or duct. **2** an efferent part, e.g. a blood vessel or nerve. *Compare* **afferent**.

efficacy *symbol*: e; a quantitative index of drug action related to the magnitude of a tissue response generated at a given level of receptor occupancy with (full) agonists (of high efficacy) generating the maximum possible response. Drugs with lower efficacy that cannot generate a maximal response even when all receptors are occupied are known as **partial agonists**; **antagonists** have an efficacy of zero.

The overall response to a drug will be a function of its efficacy and receptor binding affinity, which determines the proportion of receptors occupied at a given concentration of drug. It may be formulated as the combination of an agonist with its receptors that results in a signal or a **stimulus**, S, equal to the product of the efficacy of the agonist, e_A, and the proportion of receptors occupied, P_{AR}; i.e. $S_A = e_A P_{AR}$. *See also* **intrinsic efficacy, maximal agonist effect**.

efficiency the ratio of the useful output of a machine, device, system, etc. to the input, whether in terms of thermal, mechanical, or radiation energy, or of biological or chemical conversions.

efficiency of counting *see* **counting efficiency**.

efficiency of plating *abbr.*: EOP; a quantification of the relative efficiencies with which different cells can be infected by viruses, and support viral replication. It is the ratio of the plaque count to the number of virions in the inoculum.

efflux *see* **flux**.

EF-G *or* **translocase** *abbr. for* an **elongation factor** of *Escherichia coli* that promotes the GTP-dependent translocation of the nascent protein chain from the A-site to the P-site of the ribosome. An equivalent factor in eukaryotes is named eEF-2.

EF-hand a **helix-turn-helix** motif that binds Ca^{2+}; it is found in many Ca^{2+}-binding membrane proteins and also in certain others, e.g. **myosin** light chain, and is so called because it involves helices E and F of the Ca^{2+}-binding protein. To envisage the motif, use the right hand with thumb and forefinger extended at ~90°, with the remaining three fingers clenched. The thumb will then point towards the C terminus of helix F, and the forefinger will point along helix E in the N-terminal direction, the clenched fingers tracing the course of the E–F loop about the bound calcium ion. Two apposed right-hands, representing the E and F and the C and D helices of the molecule, are related by a 2-fold axis of symmetry.

efrapeptin *or* **efrastatin** *or* **A23871** a hydrophobic peptide fungal antibiotic that inhibits oxidative phosphorylation by binding to the soluble component, F_1, of mitochondrial **H⁺-transporting ATP synthase**.

EF-T *or* **transfer factor** a cytoplasmic protein of *Escherichia coli* and other prokaryotes that functions in protein synthesis to promote the GTP-dependent binding of aminoacyl-tRNAs to the A-site of the ribosome during protein chain elongation. EF-T can be dissociated into two polypeptides, EF-Tu and EF-Ts, so termed because, on initial isolation, one was heat-labile or unstable (Tu), and the other heat-stable (Ts); the complex is the main form in bacterial cells. Only Tu functions directly in the GTP-dependent binding of the aminoacyl-tRNA to the A-site of the ribosome, but Ts is required for regenerating Tu·GTP after the binding reaction, which results in the hydrolysis of GTP to GDP. The factors were originally isolated from both *E. coli* and *Pseudomonas fluorescens*. The 3-D structure of Tu shows high alpha-helix content, with identified binding sites for GTP and aminoacyl-tRNA. The Tu from *E. coli* is also a subunit of RNA polymerase of phage Qβ (Q beta). *See also* **elongation factor**.

EF-Ts *now designated* **EF-1B**; *see* **EF-T**.

EF-Tu *now designated* **EF-1A**; *see* **EF-T**.

egasyn an amphipathic glycoprotein, found in microsomal (i.e., endoplasmic reticulum) membranes of mouse liver, that forms a complex with microsomal (but not lysosomal) β-D-glucuronidase (EC 3.2.1.31), and thereby anchors it to the microsomal membranes. It may similarly anchor other polar proteins. It hydrolyses carboxylic esters to an alcohol and carboxylic anion. It has a C-terminal consensus sequence, HXEL, that retains the proteins in the endoplasmic reticulum. [From e (endoplasmic) + g (glucuronidase) + Greek *syn*, with.]

EG cell *or* **L cell** an **enteroglucagon**-producing endocrine cell. Such cells are found in the basal part of the mucosal glands in the lower intestine; the highest concentrations occur in the terminal ileum and colon. They contain large, dense, secretory granules and are easily distinguishable from other endocrine cells in the same area.

EGF *abbr. for* epidermal growth factor.

EGFP *abbr. for* enhanced green fluorescent protein.

egg the reproductive structure of certain animals, e.g. reptiles, birds, and insects, consisting of an **ovum** together with nutritive and protective tissues, and from which, when fertilized, a young offspring emerges.

egg albumin *the common name for* **ovalbumin.**

egg cell *an alternative name for* **ovum.**

egg white the white part of an egg (especially a bird's egg) surrounding the yolk. Its principal organic constituent is egg albumin (*see* **ovalbumin**).

EGR2 *abbr. for* early growth response 2; a C_2H_2-type of zinc-finger protein that is a transcription factor involved in myelination in the peripheral nervous system. Mutations in the gene on 10q21-q22 cause a Charcot–Marie–Tooth type 1 disease. *See* **myelin protein.**

EGTA *abbr. for* ethylene glycol-*O*,-*O*′-bis(2-amino-ethyl)-*N*,*N*,*N*′,*N*′-tetraacetic acid, [ethylenebis(oxyethylenetrinitrilo)]tetraacetic acid; also for any anion or salt thereof. It is a chelating agent with a high affinity for Ca^{2+} ions.

Ehlers–Danlos syndrome a heterogeneous group of inherited human disorders sharing many phenotypic features. Classically the clinical picture includes hyperelastic, fragile skin, poor wound healing, hyperextensible joints, and easy bruising. Ten distinct genetic forms have been described. Type I, which is severe, is inherited as an autosomal dominant; and is caused by mutations in collagen type V. Types II and III are inherited as autosomal dominants; they are due to mutations in collagen type V and type I, respectively. Type IV is inherited as an autosomal dominant and is caused by mutations in collagen type III. Type V shows X-linked inheritance but the cause is unknown. Type VI is inherited as an autosomal recessive, and is caused by deficiency of lysyl hydroxylase (EC 1.14.11.4). Types VIIA and VIIB are autosomal dominant and caused by mutations in collagen type I. Type VIIC is autosomal recessive and is caused by a deficiency of procollagen peptidase. Type VIII is autosomal dominant and of unknown cause. Type IX shows X-linked inheritance and is a variant of Menkes disease. Type X is autosomal recessive and associated with a defect in fibronectin. [After Edvard Ehlers (1863–1937), German dermatologist, and Henri Alexandre Danlos (1844–1932), French dermatologist, whose respective descriptions of the disorders were published in 1901 and 1908.]

Ehrlich's reagent a solution of *p*-dimethylaminobenzaldehyde in conc. HCl. It forms coloured complexes with a number of compounds, such as indoles, aromatic amines, ureides, and hydroxyproline, and is useful in the detection and estimation of these compounds. [After Paul Ehrlich (1854–1915), German medical scientist.]

EIA *abbr. for* enzyme immunoassay.

eicosa+ *or (before a vowel)* **eicos₊** *a variant spelling (esp. US) of* **icosa+** (*comb. form* denoting twenty or twenty times). This spelling is universally used in naming **eicosanoids**, although not officially recommended.

eicosadienoic acid any straight-chain fatty acid having twenty carbon atoms and two double bonds per molecule. The *all-Z*-(8,11)-isomer is a normally minor metabolite of oleic acid in mammals, but is important as an intermediate in the pathway synthesizing *all-Z*-eicosa-5,8,11-trienoic acid (**Mead acid**) during essential fatty acid deficiency. The (5,11)-isomer occurs in gymnosperms, and several others have been synthesized. The (11*Z*,14*Z*)-isomer occurs as an intermediate in the synthesis of arachiodonate from linoleate via dihomo-*γ*-linolenate, but this is not the major pathway for this biosynthesis.

eicosanoic acid *the systematic name for* **arachidic acid.**

eicosanoid any C_{20} polyunsaturated fatty acid, commonly arachidonic acid (*see* **eicosatetraenoic acid**), or an **eicosapentaenoic acid**, or their derivatives. They include the **prostanoids** (i.e. prostaglandins,

prostacyclins, and thromboxanes), **leukotrienes,** and **lipoxins. They** are noted for their widespread biological activity, such as contraction or relaxation of smooth muscle, platelet aggregation, and the inflammatory response. Produced mostly from arachidonate, they are released from phospholipids of the inner plasma-membrane leaflet by the action of phospholipase A_2, and modified by enzymes of the smooth endoplasmic reticulum. They are secreted and act as paracrine hormones that bind G-protein-associated receptors on neighbouring cells. Their catabolism occurs primarily in peroxisomes.

eicosapentaenoic acid any straight-chain fatty acid having twenty carbon atoms and five double bonds per molecule. The most noteworthy is the *all-Z*-(5,8,11,14,17)-isomer, which occurs in animal phospholipids, especially those of many marine species; fish oils are a good source, with eicosapentaenoic acids often representing up to 10% of total fatty acids. Eicosapentaenoic acids act as precursors of the PG_3 series of **prostaglandins**. The suggestion has been made that a higher than normal content of this acid in the diet can protect against heart disease.

eicosatetraenoic acid any straight-chain fatty acid having twenty carbon atoms and four double bonds per molecule. The *all-Z*-(5,8,11,14)-isomer is **arachidonic acid**. The (8,11,14,17)-isomer is an important intermediate in the formation of long-chain *n*-3 fatty acids from (9,12,15)-linolenic acid, being formed from that acid by desaturation at C-6 and chain elongation. The (5,11,14,17)-isomer is another naturally occurring form.

eicosatetraynoic acid any straight-chain fatty acid having twenty carbon atoms and four triple bonds per molecule. eicosa-5,8,11,14-tetraynoic acid (*abbr*.: ETYA) is a relatively nonspecific inhibitor of reactions involving arachidonate, including phospholipase A_2, **prostaglandin-endoperoxide synthase**, and **lipoxygenase.**

eicosatrienoic acid any straight-chain fatty acid having twenty carbon atoms and three double bonds per molecule. The *all-Z*-(5,8,11)-isomer (**Mead acid**) is synthesized in mammals in essential fatty-acid deficiency. The *all-Z*-(8,11,14)-iso mer, known also as dihomo-(6,9,12)-linolenic acid (*or* dihomo-*γ*-linolenic acid), is a precursor of arachidonic acid and of the PG_1 series of **prostaglandins**. The (5,11,14)-isomer has been found in higher plants and algae, cats, marine organisms, and insects.

eicosatriynoic acid any straight-chain fatty acid having twenty carbon atoms and three triple bonds per molecule. 5,8,11-eicosatriynoic acid (*abbr*.: ETI) is an agent that selectively inhibits 5- and 12-**lipoxygenase** without inhibiting **prostaglandin-endoperoxide synthase.**

eicosenoic acid any straight-chain fatty acid having twenty carbon atoms and one double bond per molecule. The 9*E*-isomer is gadelaidic acid (*see* **gadelaidate**); the 9*Z*-isomer is gadoleic acid (*see* **gadoleate**), a constituent of fish oils; and the 11*Z*-isomer is gondoic acid, present in seeds and fish oils.

EID *abbr. for* electroimmunodiffusion.

eIF *abbr. for* eukaryotic initiation factor (*see* **initiation factor**).

eIF4E-binding protein *see* **4E-BP.**

eIF2 kinase *an alternative name for* **PERK.**

eigenfunction a solution of a differential equation that has solutions only for particular values of some parameters.

eigenvalue any one of the possible values for a parameter of an equation for which the solution will be compatible with the boundary conditions.

einstein a non-SI unit used to express the energy associated with one mole of photons during a photochemical reaction. 1 einstein = $N_A h \nu$, where N_A is the Avogadro constant, *h* is the Planck constant, and ν is the particular frequency of the electromagnetic radiation effecting the reaction. [After Albert Einstein (1879–1955), German-born Swiss then US mathematical physicist]

Einstein–Sutherland relation a relation between the diffusion coefficient at infinite dilution, D_0, and certain molecular parameters. $D_0 = kT/f$, where *T* is the thermodynamic temperature, *k* is the Boltzmann constant, and *f* is the frictional coefficient of the diffusing particle.

EKG *abbr. (US) for* **1** electrocardiograph. **2** electrocardiogram.

elafin *another name for* **SKALP.** [From elastase f inhibitor. The 'f' was inserted to avoid confusion with elastin and gives an association to the Latin *finire*, to terminate.]

elaidate 1 *numerical symbol*: 18:1(9); *the trivial name for* (*E*)-octadec-9-enoate; CH_3–$[CH_2]_7$–CH=CH–$[CH_2]_7$–COO⁻ (*trans* isomer); the anion derived from elaidic acid, (*E*)-octadec-9-enoic acid, a monounsaturated straight-chain higher fatty acid. **2** any mixture of free elaidic acid and its anion. **3** any salt or ester of elaidic acid. *Compare* **oleate**.

elaidinization *or* **elaidinisation** the *cis* to *trans* isomerization of mono- and polyunsaturated fatty acids, as in the isomerization of oleic acid to elaidic acid brought about by treatment with nitrous acid or by the action of rumen bacteria.

elaidoyl *the trivial name for* (*E*)-octadec-9-enoyl; CH_3–$[CH_2]_7$–CH=CH–$[CH_2]_7$–CO– (*trans* isomer); the acyl group derived from elaidic acid, (*E*)-octadec-9-enoic acid. It occurs in acylglycerols to an appreciable extent in the body fats of ruminants, possibly having arisen from oleic acid by elaidinization. *Compare* **oleoyl**.

elaioplast any oil-storing **leukoplast** that occurs in plant cells.

ELAM *abbr. for* endothelial leukocyte adhesion molecule; i.e. E-selectin (*see* **selectin**).

elasmobranch 1 any member of the subclass Elasmobranchii, comprising fishes with cartilaginous skeletons such as sharks, skates, and rays. **2** of, pertaining to, or relating to the Elasmobranchii.

elastase historically, any proteinase that hydrolyses elastin. However, the name is now applied to a group of enzymes that vary in specificity. **pancreatic elastase**, EC 3.4.21.36 (*other names*: pancreatopeptidase E; pancreatic elastase I), catalyses the hydrolysis of proteins, including elastin, with preferential cleavage at Ala-|-Xaa. **Pancreatic elastase II**, EC 3.4.21.71, preferentially cleaves the carbonyl side of Leu, Met, and Phe. **leukocyte elastase** EC 3.4.21.37 (*other names*: lysosomal elastase; neutrophil elastase; bone marrow serine protease; medullasin), catalyses the hydrolysis of proteins, including elastin, with preferential cleavage at Val-|-Xaa > Ala-|-Xaa. Pancreatic elastase and leukocyte elastase, in common with most elastases, are serine proteinases, but others, e.g. elastase (EC 3.4.24.26) from *Pseudomonas aeruginosa*, are zinc-containing metalloproteinases. **Elastase III** is pancreatic endopeptidase E (EC 3.4.21.70; *other name*: cholesterol-binding proteinase); it shows preferential cleavage at Ala-|-Xaa, but does not hydrolyse elastin.

elastatinal a peptide produced by actinomycetes that inhibits serine proteineases, particularly elastase.

elastic (*of a body or a substance*) capable of returning to its original shape after deformation. —**elasticity** *n*.

elasticity coefficient *symbol*: ε_s^v; for a metabolic system in a steady state, the fractional change in the velocity, *v*, of an enzyme step caused by a fractional change in substrate, effector, or product concentration, *S*. It is given by:

$$\varepsilon_S^v = \frac{S}{v}\left(\frac{\partial v}{\partial S}\right)_{y,z....} = \frac{\partial \ln v}{\partial \ln S}$$

Since the elasticity coefficient is defined as the partial derivative of a velocity, *v*, with respect to *S* (multiplied by the scaling factor, *S*/*v*) all other variables that are able to influence the rate through the enzyme are held constant. There are, for any reaction, as many elasticity coefficients as there are metabolites and effectors that interact with the enzyme in question. *See also* **controllability coefficient, response coefficient, sensitivity coefficient**.

elastin a major structural protein of mammalian connective tissues, especially elastic fibres, and found at sites such as the aorta, nuchal

ligament, and lung. Glycine makes up one third of the amino-acid residues in elastin; glycine, proline, alanine, and valine together account for over 80% of its residues. The polymeric chains are cross-linked together into an extensible 3-D network, lysinonorleucine and desmosine cross-links being formed between allysine and lysine residues. A soluble precursor of elastin, **proelastin**, has been isolated from copper-deficient pigs in which the formation of allysine is blocked.

elastonectin a protein, M_r 120 000, from human skin fibroblasts, associated with adhesion of mesenchymal cells to elastic fibres.

elective theory of immunity *an alternative name for* **selective theory of immunity**

electrical coupling the phenomenon of ion transport across a membrane, in the absence of a cation–anion cotransport system, when the movement of anions and cations, following the law of electroneutrality, is governed only by the passive permeabilities of the membrane to the ions in question.

electric birefringence a form of **birefringence**, also known as the **Kerr effect**, caused by the alignment of (macro)molecules in an electric field.

electric capacitance *see* **capacitance**.

electric cell an apparatus consisting of two reversible electrodes dipping into the same solution, or into separate solutions electrically connected through an electrolyte bridge. It is used to generate an electromotive force, for electrolysis, or to measure the chemical activity of a chemical species in one of the solutions. *See also* **half-cell**.

electric charge *or* **quantity of electricity** *symbol*: *Q* or *q*; the attribute of certain elementary particles that is responsible for electric phenomena. Electric charge may be of two kinds, positive or negative, which interact electromagnetically such that like charges repel one another while unlike charges are attracted one to another. The SI derived unit of electric charge is the coulomb. *See also* **charge** (def. 3).

electric conductance *or* **conductance** *symbol*: *G*; a measure of the readiness with which an electric current can flow through a given component of a circuit. For a direct-current circuit it is equal to the reciprocal of the resistance; for an alternating-current circuit it is equal to the resistance divided by the square of the impedance. The SI derived unit of conductance is the siemens. *See also* **mho**.

electric current *or* **current** *symbol*: *I* or *i*; one of the seven SI base physical quantities – flow of electric charge through a conductor. The magnitude of a current is given by the amount of charge flowing per unit time. The SI base unit of electric current is the ampere.

electric dichroism the form of **dichroism** (def. 1) caused by orientation of (macro)molecules in an electric field.

electric dipole moment *see* **dipole moment**.

electric double layer *an alternative name for* **electrochemical double layer**.

electric field a field in which a stationary electric charge, *Q*, experiences a force, *f*, proportional to the magnitude of the charge. The electric field strength, *e*, is given by: *e* = *f*/*Q*; it is expressed in V m⁻¹. If the field is produced by other stationary charges it is called an **electrostatic field**. *Compare* **Coulomb's law**.

electricity 1 any of the forms of energy associated with moving or stationary electrons, protons, or other charge-bearing entities (whether molecular or otherwise). **Static electricity** is associated with stationary charges, **current electricity** with moving charges. **2** the science dealing with electricity.

electric potential *or* **potential** *symbol*: *V* or *ϕ*; the work done in bringing unit electric charge from infinity to a specified point in an electric field. The SI derived unit of electric potential is the volt.

electric potential difference *or* **potential difference** *symbol*: *E* or *U* or ΔV or Δu; *abbr*.: pd; the work done when a unit electric charge is moved between two points having different electric potentials. If the points are joined by a conductor an electric current will flow between them. The SI derived unit of electric potential difference is the volt. *Compare* **electromotive force**.

electric resistance *or* **resistance** *symbol*: *R*; the impediment to the flow of an electric current through a component of a circuit, a medium, or a substance. The SI derived unit of resistance is the ohm.

electro+ *or (sometimes before a vowel)* **electr+** *comb. form* indicating electric, electrically, electrolytic.

electroblot *see* **electrophoretic transfer**.

electroblotting *an alternative name for* **electrophoretic transfer**.

electrochemical analysis *or* **electrochemical assay** a chemical analysis or assay performed by means of electrolysis or electrodeposition.

electrochemical double layer any double layer of charges formed in a solution by the adsorption of a layer of ions of opposite charge at the surface of another charge-carrying phase, e.g. the walls of the vessel or the surface of a macromolecule.

electrochemical equivalent the mass of an ion deposited or liberated by the passage of 1 coulomb of electricity. In grams it is 1/96487 of the **chemical equivalent**.

electrochemical potential *symbol:* $\bar{\mu}$; the change in free energy occurring when a charged solute is transported either up or down a concentration gradient. It is the sum of the free energy changes due both to the concentration gradient and to the electric potential. It is expressed in J mol^{-1}.

electrochemical series *an alternative name for* **electromotive series**.

electrochemistry the branch of science concerned with the study of electrolysis, electrolytic cells, and the applications and properties of ions in solution. —**electrochemical** *adj.*

electrochromatography the form of **chromatography** in which separation of the constituents of a mixture is due to the effect of a constant electric field. *See also* **zone electrophoresis**.

electrochromism any reversible change in optical absorption or emission spectra of a molecule in the condensed phase due to an external electric field. *Compare* **photochromism**, **thermochromism**. —**electrochromic** *adj.*

electrocompetent describing *Escherichia coli* or other bacterial cells that have been suspended in a medium of low conductivity such as water or 10% glycerol to prepare them for transformation by electroporation.

electrocortin *the former name for* **aldosterone** (before establishment of its structure).

electrocrystallization *or* **electrocrystallisation** **electrodeposition** that results in crystal formation.

electrode any conductor, such as metal, metal covered with one of its salts, carbon, or a thin glass or other membrane, used to establish electric contact with a solution or other nonmetallic part of an electric circuit. The potential assumed by the electrode may depend on the (thermodynamic) activity of one of the solutes in the solution with which it is in contact. *See also* **anode**, **calomel electrode**, **cathode**, **enzyme electrode**, **hydrogen electrode**, **ion-selective electrode**, **oxygen electrode**, **quinhydrone electrode**, **reference electrode**.

electrodecantation *or* **electrophoresis convection** a technique for separating and fractionating proteins and other sols that exploits local differences in concentration and density that may occur during **electrodialysis**. Under the influence of gravity such density differences lead to large-scale separation of sols of high and of low concentrations. This results in a gradual increase in the concentration of a sol in the lower layers of the solution and a decrease in the upper layers.

electrodeposition the deposition of material from a dissolved or suspended state onto an electrode by the application of an electric field; it includes **electrocrystallization**.

electrode potential *symbol:* E; the electric potential between a reversible electrode and the solution in which it is immersed. It may be due to the loss or gain of electrons, if the solution contains a redox couple, or to the formation of ions from unchanged atoms or of atoms from ions, if the solution contains ions of a chemical element present as unchanged atoms in the electrode.

electrodialysis a form of **dialysis** conducted in the presence of an electric field across the membrane(s). It is useful for quickly removing ionic microsolutes from a solution containing nondiffusible material.

electroelution a technique in which application of an electric potential causes biomolecules to be transferred from an electrophoresis gel to a membrane during **blotting** procedures. *See also* **transfer membrane**.

electroendosmosis *a former term for* **electroosmosis**. —**electroendosmotic** *adj.*

electrofocusing *a newer name for* **isoelectric focusing**.

electrofuge *(in chemistry)* a leaving group in a reaction that does not carry away the bonding electron pair.

electrogenesis the generation of an electrochemical potential gradient, esp. across a membrane.

electrogenic pump any biological system that generates an electrochemical potential across a membrane by pumping one ion across the membrane without the concomitant movement of another ion of the same charge in the opposite direction.

electroimmunoassay *the original name for* **rocket immunoelectrophoresis**.

electroimmunodiffusion *abbr.:* EID; any technique for the analysis of antigenic macromolecules in which a combination of electrophoresis and immunodiffusion is employed. Four types may be distinguished: (1) one-dimensional single, i.e. **rocket immunoelectrophoresis**; (2) one-dimensional double, i.e. **counter immunoelectrophoresis**; (3) two-dimensional single, i.e. **crossed immunoelectrophoresis**; and (4) two-dimensional double, not so far individually named or much used.

electroimmunodiffusogram the immunoprecipitation pattern produced by any type of **electroimmunodiffusion**.

electrokinetic potential *or* **zeta potential** *symbol:* ζ; the potential drop across the mobile part of an **electrochemical double layer** that is responsible for electrokinetic phenomena. The electrokinetic potential is positive if the potential increases from the bulk of the liquid phase towards the interface.

electrolectin a lectin of the **galaptin** family from *Electrophorus electricus*, and the name of a group of β-D-galactoside-binding lectins found in teleosts, amphibians, and mammals.

electrolysis the chemical change produced by passing a direct electric current through a solution of an electrolyte, or through a fused electrolyte. Positively charged ions (cations) move towards the cathode and negatively charged ions (anions) move towards the anode, thus carrying the electric current. At the electrodes the positively charged ions gain electrons and negatively charged ions give up electrons, forming uncharged atoms, molecules, or radicals, which may either be deposited on the electrode or react with the electrode, the solvent, or each other. Alternatively, atoms of the electrode material may ionize and go into solution.

electrolyte any substance that, when in solution or molten, can undergo (partial or complete) dissociation into ions and hence is able to conduct an electric current by the movement of the ions.

electrolyte balance *(in physiology)* the sum total of the reactions and processes concerned with the maintenance of a constant internal environment in the body with respect to the distribution of ions between the various fluid compartments and excretions.

electrolytic **1** of, pertaining to, or produced by **electrolysis**. **2** of or pertaining to an **electrolyte**.

electrolytic conductivity *or (formerly)* **specific conductance** *symbol:* κ; the ability of an electrolyte to permit the flow of electric current, defined by the equation: $\kappa = j/e$, where j is the electric current density and e is the electric field strength. In SI units, κ is in S m^{-1}, j in A m^{-2}, and E in V m^{-1}. Electrolytic conductivity is usually measured in a rigid conductivity cell whose cell constant has been determined by means of a reference potassium chloride solution.

electromagnetic **1** of, relating to, or involving both electricity and magnetism; having both electrical and magnetic properties. **2** of or relating to **electromagnetic radiation**. **3 a** of, relating to, containing, or operated by an electromagnet. **b** of or relating to electromagnetism.

electromagnetic radiation *abbr.:* emr *or* EMR; radiation consisting of transverse waves of energy with associated electric and magnetic fields that can be represented by vectors at right angles to each other and to the direction of its propagation. The radiation results from the acceleration of electric charges and can be propagated through free space (i.e. it requires no supporting medium). The characteristics of the radiation vary with the frequency (or the wavelength) of the wave motion, the range of possible frequencies constituting the **electromagnetic spectrum**. From the low-frequency end of the spectrum to the high end are ranged: **radiofrequency radiation**, **infrared radiation**, visible radiation or **light**, ultraviolet radiation, X-rays, and **gamma-radiation**. Radiation of all frequencies travels at the same velocity through space. The speed of light in vacuum, c_0, is defined as 299792458 ms^{-1}.

electromagnetic unit *abbr.*: emu *or* EMU; any of the electrical units based on the magnetic properties of electric currents. The force, F, between two magnetic poles of strength m_1 and m_2 placed a distance d apart in a medium of permeability μ is given by $F = m_1m_2/\mu \ d^2$. If F, μ, and d each be unity and if $m_1 = m_2 = m$, then m_1 and m_2 are poles of unit strength. Units based on this definition of m are electromagnetic units and are denoted by the prefix 'ab+'.

electromeric effect an obsolescent term for a molecular polarizability effect occurring by intramolecular electron displacement characterized by the substitution of one electron pair for another in the same atomic octet of electrons.

electrometer any instrument of very high input impedance that is used for detecting or measuring (changes in) the magnitude of an electric potential difference or electric charge esp. if of small value. Hence it is useful also for the (indirect) measurement of a very low electric current or a very high electric resistance, and for the detection or measurement of ionizing radiation. —**electrometric** *adj.*

electrometric titration 1 *an alternative name for* **potentiometric titration**. 2 *a name sometimes used generically for* **conductometric titration** and **potentiometric titration**.

electromotive force *abbr.*: emf *or* EMF; *symbol*: E; the energy supplied by a source of electric current in driving unit electric charge through an electric circuit. The SI derived unit of electromotive force is the volt. *Compare* **electric potential difference**.

electromotive series *or* **electrochemical series** a series in which the metals are arranged in decreasing order of their tendency to form cations by a reaction of the type: $M \rightleftharpoons M^+ + e^-$.

electromyogram *abbr.*: EMG; the tracing produced by an **electromyograph**.

electromyograph an instrument for detecting and recording the electrical activity of a muscle or group of muscles.

electron 1 *symbol*: e *or* e$^-$ *or* β$^-$; a negatively charged elementary particle of rest mass, m_e, ≈9.109 39 × 10^{-31} kg and electric charge 1.602 177 33(49) × 10^{-19} C. It is sometimes termed a **negatron**, especially to distinguish it from its antiparticle, the positron. Electrons are present orbiting in all atomic nuclei and are involved in the formation of bonds between atoms. Specific designations of electrons in atoms and molecules include: σ-electrons, which participate in single bonds; π-electrons, which participate in double or triple bonds; and p-electrons, which participate in unshared electron pairs. 2 *(sometimes)* either a negatron (i.e. **electron** (def. 1)) or a **positron**. *See also* **beta-particle**.

electron acceptor 1 any substance to which an electron may be transferred. 2 *an alternative name for* **Lewis acid** (use discouraged). *Compare* **electron donor**.

electron affinity 1 the degree to which any molecular entity attracts one or more additional electrons. 2 *symbol*: E_{ea}; the minimum energy, expressed in joules, needed to remove an electron from a negatively charged molecular entity, or the reciprocal of the energy required to introduce an electron into a molecular entity.

electron-affinity spectroscopy a sensitive ionization method for the qualitative analysis of small quantities of organic compounds separated by gas chromatography, based on the differences in electron-capture ability that exist between the principal classes of organic compounds. The substances to be detected are passed through an ionized gas; the detector is sensitive to the resulting reduction in current.

electron capture 1 *or* K-capture *abbr.*: EC; a mode of decay of radionuclides in which an electron from an inner orbit of a decaying atom is absorbed into its nucleus, wherein a proton is converted to a neutron. Rearrangement of the orbital electrons then leads to the emission of characteristic X-radiation (*see* **X-radiation**). Generally the captured electron is derived from the K-shell, hence the alternative name. 2 *see* **electron-affinity spectroscopy**.

electron-capture detector *abbr.*: ECD; the type of detector used in **electron-affinity spectroscopy**.

electron carrier any molecular entity that serves as an electron acceptor and electron donor in an electron transport system.

electron-dense describing any structure, material, or compound

that scatters an impinging electron beam, thus becoming apparent in **electron microscopy**, etc.

electron-dense label any atom group or substance (e.g. ferritin) that is **electron-dense** and that has been or may be attached in a specific manner to compounds or structures so as to show them up by electron microscopy, etc.

electron density the electron probability distribution in any molecular entity. If $P(x,y,z)$ is the electron density at the point with coordinates x,y,z, then the probability of finding an electron in the volume element dxdydz is $P(x,y,z)$ dxdydz. In, e.g., X-ray scattering experiments the scattering at this position is proportional to this probability. The term is often wrongly applied to negative **charge density**.

electron-density map a three-dimensional representation of the structure of a molecule or substance derived from X-ray diffraction data and made by the superposition of the electron-density contours of a series of imaginary parallel planes through the specimen being examined.

electron diffraction an effect observed when a narrow beam of electrons passes through a layer of material, e.g. a crystal, and is deflected by this material. The effect is due to the wavelike nature of electrons.

electron donor 1 any substance that is able to transfer an electron to another chemical species. 2 *an alternative name for* **Lewis base** (use discouraged). *Compare* **electron acceptor**. *See also* **donor atom**.

electron donor–acceptor chromatography *see* **charge-transfer chromatography**.

electron donor–acceptor complex *(sometimes)* an alternative name for 1 **charge-transfer complex** (use not recommended). 2 **Lewis adduct** (use not recommended).

electronegative 1 describing an atom or a group of atoms that tends to attract electrons, especially in the formation of a covalent bond. 2 describing any chemical or other entity that carries a negative charge and hence tends to move to the anode in electrophoresis.

electronegativity a measure of the power of an atom or group of atoms to attract electrons to itself from other parts of the same molecular entity.

electron equivalent the **equivalent** of any oxidizing (or reducing) agent that is equivalent to one entity of electrons.

electroneutrality condition the condition that in any solution the sum of the negative electric charges on ions must be equal to the sum of the positive charges on ions.

electron-exchange resin any synthetic resin containing groups able to undergo reversible oxidation–reduction reactions; such resins are useful as insoluble oxidizing or reducing agents.

electron gun a device used in electron microscopes, mass spectrometers, and other instruments to generate a beam of high-velocity electrons. It consists of a white-hot tungsten filament, as the electron source, and an anode kept at a potential of several tens of kilovolts above that of the filament. The electrons are accelerated towards the anode, which contains a small orifice through which some of the fastest electrons pass to form the electron beam.

electron hole the hole in the valency structure of an atom created by the absence of an electron when the energy level of an electron is raised from the valency band to the conduction band. The hole has a great tendency to recapture an electron. *See also* **hole**.

electronic 1 of or relating to an electron or electrons. 2 of, relating to, or operated by **electronics** (def. 2).

electronics 1 *(functioning as sing.)* the branch of science and technology concerned with electronic devices and circuits and with their development and applications. 2 *(functioning as pl.)* the circuits, devices, and equipment that function by the passage of electrons through semiconductors, transistors, and thermionic valves.

electronic transition the passage of an electron in a chemical entity from one energy level to another.

electron lens any arrangement of coils and/or electrodes producing a magnetic and/or an electric field used to focus an electron beam, e.g. in an electron microscope.

electron micrograph any photograph of an object taken with an **electron microscope**.

electron microscope *abbr.*: EM *or* e.m.; an instrument in which a magnified image of a sample is produced using a beam of high-energy electrons. The electrons are accelerated by means of an **electron gun**, using voltages in the range 40–100 kV, and the apparatus is evacuated to minimize electron scatter by air. In the **transmission electron microscope** (*abbr.*: TEM) the electron beam is focused onto a thin sample by a combination of magnetic and electric lenses and, after passing through the sample, the electrons are focused onto a fluorescent screen or photographic plate. The resolving power can be as great as 0.2–0.5 nm. In **cryoelectron microscopy**, lightly fixed frozen sections are employed. If these are used for the detection of antigens with antibodies coupled to colloidal gold, the technique is called **cryoimmunoelectron microscopy**. In the **scanning electron microscope** (*abbr.*: SEM) a narrow beam of electrons is made to scan to and fro across the sample by a varying field, and emitted **secondary electrons** are focused onto a screen to produce an image of the specimen. The specimen may be thicker than in the transmission electron microscope and, though resolution is lower with the scanning instrument, the depth of field is greater and the image gives a three-dimensional effect.

electron multiplier an electronic device for detecting, and counting, electrons, using **secondary emission**. It consists of an evacuated tube containing a series of electrodes, each held at a higher positive electric potential than the preceding one. When an electron hits the first electrode two or more **secondary electrons** are emitted; these are accelerated to the second electrode, where more electrons are produced, and so on.

electron nuclear double resonance *abbr.*: ENDOR; a phenomenon in nuclear magnetic resonance spectroscopy in which saturation of an electron resonance by the simultaneous application of a microwave field causes, under appropriate conditions, an improvement of the resolution of nuclear hyperfine structures.

electron optics the branch of science concerned with the effects of electric and magnetic fields on the movement of electrons, especially with the focusing and deflection of electron beams in **cathode-ray tubes**, **electron microscopes**, and similar devices.

electron paramagnetic resonance spectroscopy *abbr.*: EPR (spectroscopy); a technique, also called **electron spin resonance spectroscopy** (*abbr.*: ESR (spectroscopy)) and analogous to **nuclear magnetic resonance spectroscopy**, that is used to investigate paramagnetic centres in the system under study; only electrons whose spin is not paired with the oppositely directed spin of another electron give an ESR signal. Useful information is obtained by this technique about certain transition-metal ions (notably Cu^{2+}), (free) radicals, and free electron centres such as may be produced by X-irradiation of (macro) molecules. A **probe** (def. 2) giving an ESR signal may be incorporated into membrane lipids or attached to proteins to enable otherwise inaccessible systems to be studied by this technique. A distinction is sometimes made between EPR as applied to transition-metal complexes, in which the paramagnetism often comprises a strong orbital contribution, and ESR as to radicals, in which the g-factors are close to the free electron value.

electron probe X-ray microanalysis a highly sensitive technique for elemental microanalysis based on the frequencies and/or intensities of the characteristic X-rays emitted by each element present in a sample when it is excited by a beam of accelerated electrons.

electron sink *or* **electron trap** an **electrophilic** atom or group of atoms that can capture an electron from another part of a molecular system.

electron spectroscopy for chemical analysis *abbr.*: ESCA; *see* **photoelectron spectroscopy**.

electron spin *see* **spin** (def. 3).

electron spin-echo envelope modulation *abbr.* ESEEM; a pulsed technique in EPR spectroscopy, used to measure anisotropic nuclear hyperfine coupling. The intensity of the electron spin-echo resulting from the application of two or more microwave pulses is measured as a function of the temporal spacing between the pulses. The echo intensity is modulated as a result of interactions with the nuclear spins. The frequency-domain spectrum corresponds to hyperfine transition frequencies. The method gives analogous results to **ENDOR**, and is particularly effective for nuclei with a **quadrupole moment**.

electron spin-echo spectroscopy a technique giving precise values for electron spin relaxation times, in which intense microwave

pulses at a frequency satisfying the resonance condition are applied to a paramagnetic sample in a magnetic field. Phase coherence in the spins is lost through relaxation when the pulse is switched off but can be restored by a second pulse 180° out of phase. The restoration of phase coherence gives rise to a nuclear induction signal (the echo).

electron spin-lattice relaxation time (*in* electron spin resonance spectroscopy) a measure of the time taken for the spin population to return to its equilibrium value by interaction with the fluctuating internal electric fields surrounding it.

electron spin resonance spectroscopy *abbr.*: ESR (spectroscopy); *see* **electron paramagnetic resonance spectroscopy**.

electron spin–spin relaxation time (*in* electron spin resonance spectroscopy) a measure of the time taken for phase coherence to be lost and equilibrium to be re-established through interaction with neighbouring spins. It is inversely proportional to the line width of the signal.

electron-transfer flavoprotein *or* electron-transferring flavoprotein *abbr.*: ETF; a protein containing flavin adenine dinucleotide (FAD) that in mammalian mitochondria and some bacteria, forms an intermediary electron carrier between several primary flavoprotein dehydrogenases (e.g. acyl- CoA dehydrogenase, EC 1.3.99.3) and the terminal respiratory chain. In mitochondria it consists of an α subunit (30 kDa, containing one FAD) and a β subunit (28 kDa, containing one adenosine 5′-monophosphate). *See* **glutaric acidemia**.

electron-transferring flavoprotein dehydrogenase *abbr.*: ETF dehydrogenase EC 1.5.5.1; an enzyme (pig heart, 64 kDa) that catalyses the oxidation of reduced ETF by ubiquinone to form ETF and ubiquinol. The ubiquinol is then oxidized by **ubiquinol–cytochrome c reductase**. It is an iron–sulfur flavoprotein involved in coupling the **beta-oxidation system** to the **respiratory chain**. *See* **glutaric acidemia**. *See also* **electron-transferring flavoprotein**.

electron-transferring fluoroprotein *see* **azurin**.

electron transport 1 respiratory electron transport; the process by which pairs of electrons derived from intermediates of the **tricarboxylic-acid cycle** and other substrates flow down the **respiratory chain** to dioxygen, the ultimate electron acceptor in respiration. **2** photosynthetic electron transport; the process by which pairs of electrons are transported from H_2O to $NADP^+$ in the light phase of photosynthesis during noncyclic **photophosphorylation**.

electron-transport chain an alternative name for **respiratory chain**.

electron trap an alternative name for **electron sink**.

electronvolt *or* electron volt *symbol*: eV; a non-SI unit of energy equal to the kinetic energy acquired by an electron when accelerated through an electric potential difference of 1 volt. 1 eV = $e \times V$ ≈ 1.602 18 × 10^{-19} J, where e is the **elementary charge** and V is the volt.

electroosmosis *or* (*formerly*) **electroendosmosis** the motion of a liquid through a membrane (or plug or capillary) consequent upon the application of an electric field across the membrane. A similar phenomenon may occur in electrophoresis, where many of the supporting media used, e.g. paper or agar, acquire negative charges during electrophoresis at alkaline pHs and, since the medium cannot move, H_3O^+ ions move towards the cathode, giving the effect of an osmotic movement of solvent towards the cathode and making electrically neutral molecules appear to be cationic. —**electroosmotic** *adj.*

electropherogram *a variant spelling of* **electrophoretogram**.

electrophile *or* electrophilic reagent any chemical species that is preferentially attracted to a region of high electron density in another species during a chemical reaction. Such reagents normally are positively charged or contain electron-deficient chemical groups. They tend to react with electron-rich or negatively charged chemical species. *Compare* **nucleophile**.

electrophilic 1 of, pertaining to, or being an **electrophile**; having or involving an affinity for regions of high electron density in a chemical reactant. **2** describing a chemical reaction in which an electrophile participates.

electrophilic catalysis catalysis by a Lewis acid, i.e. any chemical species that abstracts an electron pair from the reactant.

electrophilic displacement *an alternative term for* **electrophilic substitution reaction.**

electrophilicity the relative reactivity of an **electrophile**, measured by the relative rate constants of different electrophiles towards a common reactant.

electrophilic reagent *an alternative name for* **electrophile.**

electrophilic substitution reaction *or* electrophilic displacement a chemical reaction in which an **electrophile** effects heterolytic substitution in another reactant, both bonding electrons being supplied by that other reactant.

electrophoresis 1 the phenomenon of the movement of ions (including macromolecular ions) or charged particles or ions through a fluid under the influence of an electric field applied to the fluid. A number of different media have been used as the fluid support, including paper, cellulose acetate, starch gel, and polycrylamide gel. Ions or particles bearing a net positive charge tend to move towards the negative pole of the electric field and vice versa, the rate of movement of a particular variety of ion or particle depending, *inter alia*, on its charge-to-mass ratio. The phenomenon has been widely applied in separating proteins, nucleic acids, and other charged molecular species for analytical or preparative purposes, and also in the analytical or preparative fractionation of heterogeneous populations of dispersed cells or other types of macroscopic particles. **2** the act or process of causing ions or charged particles so to migrate; any technique based upon such a phenomenon, e.g. **continuous flow electrophoresis, immunoelectrophoresis, moving boundary electrophoresis, paper electrophoresis, polycrylamide gel electrophoresis, zone electrophoresis.** *See also* **electrodecantation.** —**electrophoretic** *adj.*

electrophoresis convection *an alternative term for* **electrodecantation.**

electrophoretic effect the phenomenon of decreased **electrophoretic mobility** of a charged macromolecule caused by the movement of counter ions and/or solvent molecules in the opposite direction to that of the macromolecule.

electrophoretic mobility *symbol*: u; the **electrophoretic velocity**, v, of a charged particle expressed per unit field strength; hence, $u = v/e$, where e is the field strength. The value of u is positive if the particle moves towards the pole of lower potential and negative in the opposite case. The electrophoretic mobility depends only on molecular parameters.

electrophoretic mobility shift assay (*abbr.*: EMSA) *or* **gel retardation assay** a method to evaluate the binding of transcription factors to response elements in which the electrophoretic mobility of short, radiolabelled double-stranded DNA is reduced by the bound protein.

electrophoretic molecular sieving (*sometimes*) an alternative term for **polyacrylamide (gel) electrophoresis.**

electrophoretic titration curve the pH-mobility curve of an ampholyte, e.g. a protein, generated by subjecting a zone of it to electrophoresis in a gel slab at right angles to a preformed, stationary pH gradient. *Compare* **isoelectric focusing.**

electrophoretic transfer *or* electroblotting a development of the technique of blot transfer, in which proteins or nucleic acids are transferred from a separation gel to nitrocellulose or diethylaminoethyl- (DEAE-) cellulose membranes or to diazobenzyloxymethyl- (DBM)- or diazophenylthioether- (DPT-) paper by electrophoresis, rather than by capillary flow, with a consequent decrease in the time required for the transfer. The membrane or paper bearing a resultant pattern of separated substances has been termed an **electroblot**. *see* **blotting.**

electrophoretic velocity *symbol*: v; the velocity of a charged particle during electrophoresis. It is normally proportional to the electric field strength. *Compare* **electrophoretic mobility.**

electrophoretogram *or* electropherogram the result of a zone-electrophoretic separation, either directly visible or after staining or processing to produce a graph.

electrophysiology the part of physiology concerned with the electrical phenomena associated with bodily processes, such as nervous and muscular activity.

electroporate to create momentary pores in the membranes of living cells, without loss of their viability, by exposing them to a sequence of brief electrical pulses of high field strength. The reversible breakdown of the cell membranes thus caused enables treated cells to take up exogenous material (e.g. drugs or foreign DNA). —**electroporated** *adj.*; **electroporation** *n.*

electroporator an apparatus or device for effecting electroporation.

electropositive 1 describing an atom or group of atoms that tends to give up electrons, especially in the formation of a covalent bond. **2** describing any chemical or other entity that carries a positive charge and hence tends to move to the cathode in **electrophoresis.**

electropositivity a measure of the power of an atom or group of atoms to give up electrons to other parts of the same molecular entity.

electrospray a technique used in **mass spectrometry** in which a dilute acidic solution of the macromolecule is sprayed from a metal syringe needle maintained at +5000 v, forming fine highly charged droplets from which the solvent rapidly evaporates.

electrostatic of or pertaining to static electricity or electrostatics.

electrostatic bond any valency linkage between atoms arising from the transfer of one or more outer-shell electrons of one atom to the outer shell of another atom, leading to more complete outer shells in both atoms. The dissociation of an electrostatic bond leads to the production of ions.

electrostatic field any **electric field** produced by stationary charges.

electrostatic interaction any of the attractive or repulsive forces between atoms and/or groups of atoms and/or molecules that are due to the presence of ionized chemical entities and to the electronegative and electropositive properties of these atoms, groups, or molecules. *Compare* **electric field.**

electrostatic precipitation the removal of small particles suspended in a gas by electrostatic charging followed by precipitation onto a highly charged collector.

electrostatics the branch of physics concerned with static electricity.

electrostatic units *abbr*: esu or ESU; a system of electrical units, used in the cgs system, based upon the electrostatic unit of electric charge, i.e. the quantity of electricity that will repel an equal quantity of electricity, 1 cm distant from it in a vacuum, with the force of 1 dyne.

electrostriction the reversible change in dimensions of a dielectric when an electric field is applied to it. For example, the electrostatic field associated with a dissolved electrolyte causes a shrinkage in the volume occupied by the solvent.

electrotaxis the directional movement of an organism in an electric field. —**electrotactic** *adj.*

electrotropism any orientation response of a sessile organism to an electrical stimulus.

electrovalency *or* (*esp. US*) **electrovalence 1** the type of **valency** characterized by the transfer of one or more electrons from one atom to another atom, with the formation of ions. **2** the number of negative or positive charges acquired by an atom by the respective gain or loss of electrons. —**electrovalent** *adj.*; **electrovalently** *adv.*

electrovalent bond *an alternative term for* **ionic bond.**

electroviscosity the, usually minor, change in viscosity that occurs when an electric field is applied to some polar liquids.

eledoisin a **tachykinin** with the structure Glp-Pro-Ser-Lys-Asp-Ala-Phe-Ile-Gly-Leu-Met-NH_2; it is obtained from the posterior salivary glands of small octopus (*Eledone* spp.). It is more potent than **physalaemin** and **substance P** in some assay systems. It is useful as a lachrymal and salivary secretagogue and in the study of receptors for substance P.

element any basic and distinct component of matter that is not resolvable into simpler components with differing chemical properties. Each consists exclusively of atoms with the same unique **proton number** (although such atoms may not necessarily have the same **nucleon number** or **relative atomic mass**). Ninety-three chemical elements are known to occur naturally, either in the free state or in combination with others; to date, a further 16 have been produced artificially. —**elemental** *or* **elementary** *adj.*

elementary analysis any quantitative chemical analysis of the amounts of different chemical elements present in a sample.

elementary charge *or* **proton charge** *symbol*: e; a fundamental physical constant representing the electric charge on one electron (or proton). Its recommended value is $1.602\ 177\ 33\ (49) \times 10^{-19}$ C.

elementary particle 1 *or* **fundamental particle** any of the particles that form the base units of which all matter is composed. The stable elementary particles are protons, electrons, and neutrinos; they combine with neutrons to form stable atoms. Other, short-lived elementary particles are known. **2** *or* **Fernández–Morán particle** *or* **oxysome** *or* **stalked particle** any of the knoblike structures, 8–9 nm in diameter, seen in the electron microscope on the inner surface of the inner membrane of unfixed mitochondria negatively stained with phosphotungstate.

elementary rest mass *symbol*: m_e; the mass of an electron, it is a fundamental atomic constant, $9.109\ 3897(54) \times 10^{-31}$ kg.

eleostearate 1 *numerical symbol*: 18:3(9,11,13); *the trivial name for* octadeca-9,11,13-trienoate; $CH_3-[CH_2]_3-[CH=CH]_3-[CH_2]_7-COO^-$; the anion derived from eleostearic acid, octadeca-9,11,13-trienoic acid, a triunsaturated straight-chain higher fatty acid. It is a constitutional isomer of (6,9,12)-linolenate with conjugated double bonds. **2** any mixture of free eleostearic acid and its anion. **3** any salt or ester of eleostearic acid. *See also* **eleostearoyl**.

eleostearoyl *symbol*: eSte; *the trivial name for* octadeca-9,11,13-trienoyl; $CH_3-[CH_2]_3-[CH=CH]_3-[CH_2]_7-CO-$; the acyl group derived from eleostearic acid. The $9Z,11E,13E$-isomer occurs naturally as acylglycerols in some seed oils, especially tung oil.

ELF *abbr. for* embryonic liver fodrin.

ELF1 a gene encoding transcription factors, related to the protooncogene *ETS* and named from E74-like factor (*see* **E74**), a *Drosophilia* protein that binds a regulatory element of the dopa decarboxylase gene.

elicitor a substance that induces the production of **phytoalexins** in higher plants. Elicitors may be exogenous (e.g. produced by potentially pathogenic bacteria or fungi) or endogenous, such as cell-wall degradation products.

elimination reaction a chemical reaction in which two groups are removed from a molecule of an organic compound without being replaced by other groups. In most such reactions the groups are lost from adjacent carbon atoms: one of the groups eliminated is commonly a hydron and the other a nucleophile, and the elimination results in the formation of a multiple bond. The reverse process is an **addition reaction**.

ELISA *abbr. for* enzyme-linked immunosorbent assay.

eLiXiR an agonist or antagonist of the oxysterol receptor.

Elk a family of cell-surface proteins similar to those encoded by the protooncogene **ETS**, and with presumed similar function (named from eph-like kinase); they are protein tyrosine kinases, restricted to brain and testis, having a single transmembrane domain and two fibronectin-like domains.

Elk-1 a gene-regulating protein, found in lung and testis, that binds to DNA at purine-rich sites. It is a substrate for a **MAP kinase** in the phosphorylation cascade, originating with activated protein kinase C or Ras; on phosphorylation, it forms a ternary complex with the transcription factor **SRF**, which binds to the **serum response element** of DNA, turning on *FOS* transcription.

ellagic acid 4,4′,5,5′,6,6′-hexahydroxydiphenic acid 2,6,2′,6′-dilactone; a commonly occurring plant polyphenol that inhibits glutathione S-transferase. It is present in many berries, is an antioxidant, and may prevent certain cancers.

ellipse any closed plane figure whose perimeter is the locus of a point that moves so that the sum of its distances from two fixed

points, the foci, is constant. An ellipse has two axes of symmetry, the major axis at the furthest distance between opposing points on its circumference and the minor axis at the nearest distance.

ellipsoid any surface or solid figure whose plane sections are all either ellipses or circles. —**ellipsoidal** *adj.*

ellipsoid of rotation *or* **ellipsoid of revolution** a solid figure generated by the rotation of an ellipse about one of its axes. If the rotation is about the major axis the ellipsoid of revolution is said to be **prolate**, whereas if the rotation is about the minor axis it is said to be **oblate**. *See also* **equivalent ellipsoid of rotation**, **spheroid** (def. 2).

ellipsosome a compartment in the retinal cones of fish that contains cytochrome-like pigment.

elliptical of, relating to, or having the shape of an **ellipse** or **ellipsoid**. —**elliptically** *adv.*

elliptically polarized *or* **elliptically polarised** describing light (or other electromagnetic radiation), whose electric vector appears to follow an elliptical path when viewed along the light beam. *See also* **polarized light**.

ellipticity 1 the ratio of the lengths of the major to the minor axis of an **ellipse** or **ellipsoid**. **2** the arc tangent of the ratio of the lengths of the minor to the major axis of an ellipse. *See also* **circular dichroism**, **Cotton effect**, **molecular ellipticity**.

Ellman's reagent bis(4-nitro-5-carboxyl phenyl)disulfide, (formerly 5,5′-dithiobis(2-nitrobenzoic acid)) (*abbr.*: DTNB); a reagent used for labelling thiol (SH) groups in protein side-chains. The reagent reacts quantitatively with thiol groups, forming mixed disulfides and releasing the anion 5-sulfido-2-nitrobenzoic acid; this absorbs light at 412 nm, thus permitting the determination of the number of thiol groups in the protein sample. [After George Leon Ellman (1923–).]

elongase any of the long-chain fatty-acyl-coenzyme A elongating enzymes, found in mammalian microsomes, that act on saturated and unsaturated fatty-acyl derivatives having 16 or more carbon atoms per molecule of fatty acid.

elongation factor *abbr.*: EF; any of a group of soluble proteins required for chain elongation during polypeptide synthesis at the ribosome. In *Escherichia coli* there are three elongation factors: EF-Tu, EF-Ts (*see* **EF-T**), and **EF-G**. EF-Tu helps to bind aminoacyl-tRNA, via a tertiary complex, aminoacyl-tRNA•EF-Tu•GTP, to the ribosomal A-site. After EF-Tu-dependent GTP hydrolysis, the EF-Tu•GDP complex dissociates from the ribosome and is recycled to EF-Tu•GTP, probably via an EF-Tu•EF-Ts intermediate. *E. coli* EF-Tu and GTP form ternary complexes with all aminoacyl-tRNAs except the initiator, $tRNA^{Met}_f$. EF-G catalyses the GTP-dependent movement of the ribosome and the codon–anticodon-linked mRNA•peptidyl-tRNA complex relative to each other, resulting in the removal of a deacylated tRNA and the location of peptidyl-tRNA from the ribosomal A-site to the P-site. In eukaryotes three proteins, EF-1α, EF-1β, and EF-2, are required for the elongation events. EF-1α, which occurs in different tissues as aggregates of varying size of a polypeptide chain of M_r 53000, functions during chain elongation, like EF-Tu, to bring the aminoacyl-tRNA to the ribosomal A-site in a reaction in which GTP is hydrolysed. EF-1β, of M_r 30000, functions as EF-Ts. EF-2, like EF-G, catalyses the translocation of the peptidyl-tRNA from the ribosomal A-site to the P-site in a GTP-dependent reaction. Rat liver EF-2 has an M_r of 110000, while wheatgerm EF-2 has an M_r of 70000. *Compare* **initiation factor**, **release factor**.

elongin a heterotrimeric protein complex that is highly conserved in eukaryotes. It consists of elongin A, the catalytic subunit (773 residues); and the regulatory subunits elongin B (118 residues), which contains a ubiquitin-homology domain, and elongin C (112 residues). It enhances transcriptional elongation by suppressing pausing by eukaryotic RNA polymerase II. Elongins B and C also form part of a ubiquitin ligase complex and of the **VHL** protein complex.

Elson–Morgan reaction a colorimetric reaction for the estimation of combined and free hexosamines. The specimen is heated in alkaline solution with acetylacetone to form a pyrrole derivative, which then gives a red colour with *p*-dimethylaminobenzaldehyde on acidification. [After Leslie Alderman Elson and Walter Thomas James Morgan (1900–), British biochemist, who together described the reaction in 1933.]

eluant *a variant spelling of* **eluent**.

eluate 1 *(in chromatography)* the solution flowing from a chromatographic column; the desired material in such a solution. **2** *(in immunology)* a solution of material derived from an insoluble or particulate antigen–antibody complex, e.g. by heating or by treatment with an appropriate buffer solution.

eluent *or* **eluant** a liquid used in **elution**, particularly in **chromatography**.

eluent strength function *(in chromatography)* the order of an **eluotropic series** defined as the adsorption energy per unit area of standard activity of the solvent:

$$\log K = \log V_a + \alpha(S^o - A_s \varepsilon^o),$$

where K is the distribution coefficient in mass/volume of adsorbent and is independent of the solute concentration, V_a is the adsorbent surface volume, α is an adsorbent function proportional to the average surface energy of the adsorbent, S^o is a solute parameter equivalent to the adsorption energy of the solute from solution, A_s the area of absorbant occupied by solvent, and ε^o is a solvent strength parameter; the larger ε^o, the stronger the solvent.

eluotropic series a series of solvents arranged in the order of their relative abilities to effect **elution**.

elute to wash out or extract retained material, particularly from a chromatography column.

eluting 1 of or relating to **elution**, e.g. an eluting solvent. **2** the process of carrying out an **elution**.

elution the process of washing out or extracting adsorbed material from a chromatographic column.

elution volume *(in chromatography)* the volume of mobile phase that must pass through a chromatographic column, after sample application, to produce a peak in the concentration of a particular solute in the effluent or eluate.

elutriation the process of separating and purifying particles in a powder or suspension according to their rates of sedimentation. It may be effected by repeatedly washing, decanting, and settling, or by allowing a liquid or gas containing the particles to flow upwards (against gravity), when particles whose sedimentation rates are less than the rate of upward flow are washed away from the larger, heavier particles. *Compare* **centrifugal elutriation**, **electrodecantation**.

elutriator a device for grading finely divided material according to the sizes and weights of the particles by means of a stream of liquid or gas.

elymoclavine an **ergot alkaloid** produced by *Claviceps purpurea* and other fungi.

em+ *prefix; a variant of* **en+** *(sometimes before b, m, and p)*.

EM *or* **e.m.** *abbr. for* electron microscope *(or electron microscopy)*.

e-mail *abbr. for* electronic mail.

e-mail address an address for e-mail of the form <name>@<address> where the <address> is an **IP address** in its named format, for example, president@whitehouse.gov.

EMBASE *see* **bibliographic database**.

Embden–Meyerhof pathway *or* **Embden–Meyerhof–Parnas pathway** *alternative names for* **glycolytic pathway**. [After Gustav Embden (1874–1933) and Otto **Meyerhof** (1884–1951), German biochemists, and Jacob Karol Parnas (1884–1949), Polish biochemist.]

EMBL 1 *abbr. for* European Molecular Biology Laboratory; an international research facility based at Heidelberg in Germany. **2** a major database of nucleic acid and protein sequences maintained by EMBL. *See* **sequence database**.

EMBL4 a bacteriophage lambda replacement vector.

EMBO *abbr. for* European Molecular Biology Organization.

embolus *(pl.* **emboli***)* a thrombus fragment that is transported in the circulation to distant tissues, where it can cause ischemic tissue damage.

EMBOSS *abbr. for* European Molecular Biology Open Software Suite; *see* **sequence analysis package**.

embryo the structure that develops from a **zygote**, up to the time of birth or hatching in eutherian animals, or of germination in plants. In mammals the term is restricted to the structure present in the early part of gestation that develops into a **fetus**. —**embryonic, embryonal** *adj.*

embryogenesis *or* **embryogeny** the formation of an embryo from an ovum and the processes of its development. —**embryogenic** *adj.*

embryology the branch of biology concerned with the study of embryos and their development. —**embryological** *or* **embryologic** *adj.*; **embryologically** *adv.*; **embryologist** *n.*

embryonated differentiated into, or having an embryo, e.g. an embryonated hen's egg.

embryonic stem cell *abbr.*: ES cell; a **totipotent** cell that can be cultured from an early embryo.

embryonin *another name for* bovine α_2-**macroglobulin**.

emerimicin *see* **peptaibophols**.

emerin a nuclear membrane protein (254 amino acids) that is present in most tissues and is linked to the lamins of the nuclear lamina. It is serine-rich, homologous with lamina-associated protein 2 (LAP2) and contains a C-terminal transmembrane segment. During mitosis it becomes dispersed throughout the cell and participates in formation of the daughter nuclei. In the heart, it is also localized with the intercalated discs. Mutations in the gene at Xq24.3 cause the X-linked form of Emery–Dreifuss muscular dystrophy.

Emerson enhancement the phenomenon in photosynthesis in which the **quantum yield** of light in the far-red region of the spectrum is enhanced by weak background illumination with shorter wavelength light. *Compare* **enhancement spectrum**. [After Robert Emerson (1903–59).]

emetic 1 having the power to cause vomiting. **2** any compound with emetic properties.

emetine 6′,7′,10,11-tetramethoxyemetan; an alkaloid from *Cephaelis ipecacuanha* that is a gastrointestinal irritant and potent emetic and used as an expectorant and antiamoebic agent. It inhibits protein synthesis by preventing translocation of peptidyl-tRNA from the A- to P-site on the ribosome.

emf *or* **EMF** *abbr. for* electromotive force.

emg *abbr. for* electromyogram.

+emia *or (esp. Brit.)* **+aemia** *suffix* denoting blood, especially some abnormal condition of the blood.

emilin *abbr. for* elastin microfibril interface located protein; an extracellular matrix glycoprotein, gp115. It is a component of elastin fibres, preferentially located at the elastin–microfibril interface.

emiocytosis a process whereby a secretory substance is released from a cell; a form of **exocytosis**. A secretory granule moves to the cell surface, where the membrane sac enclosing the granule fuses with the plasma membrane, ruptures, and liberates the granule into the extracellular space.

emission *(in physics)* the release of energy in the form of radiation or particles.

emission spectrophotometry *or* **emission photometry** the processes of measuring an **emission spectrum**, either photometrically or photographically.

emission spectroscopy the **spectroscopy** of an **emission spectrum**. This differs from emission spectrophotometry in being a generic term. Thus measurement may be photometric or photographic, for example.

emission spectrum any **spectrum** of emitted electromagnetic radiation produced by a sample after its atoms or molecules have been put into excited states by the absorption of energy, each line in the emission spectrum being produced by the decay of a particular species of excited atom or molecule to some lower energy level. The exciting energy may be supplied as heat, bombardment by particles, or irradiation with other electromagnetic radiation. *Compare* **absorption spectrum, excitation spectrum**.

EMIT *abbr. and trademark for* enzyme-multiplied immunoassay technique; *see* **homogeneous enzyme immunoassay**.

emodin 1,3,8-trihydroxy-6-methylanthraquinone; a compound found (as its rhamnoside) in rhubarb root and other plants.

eMOTIF *see* **derived database**.

EMPIGEN BB *proprietary name for* dodecyl-**N**,**N**-dimethylglycine.

EMR *abbr. for* **1** equi-effective molarity concentration ratio. **2** *or* **emr** electromagnetic radiation.

EMSA *abbr. for* electrophoretic mobility shift assay.

emu *or* **EMU** *abbr. for* electromagnetic unit.

emulsifier 1 *an alternative name for* **emulsifying agent**. **2** a machine for making emulsions.

emulsify to form or to convert into an **emulsion**. —**emulsifiable** *or* **emulsible** *adj.*; **emulsification** *n*.

emulsifying agent *or* **emulsifier** any amphiphile or surfactant, such as a phospholipid or a soap, small quantities of which promote the formation of, and stabilize, an **emulsion**.

emulsin a ferment (i.e. an enzyme preparation) obtained from sweet or bitter almonds and originally studied in 1837 by the German chemists Justus von Liebig (1803–73) and Friedrich Wöhler (1800–82) for its ability to degrade **amygdalin**. Together with invertin, it played an important role in Fischer's work on enzyme specificity leading to his **lock and key model**. In modern terms, it was an enzyme mixture containing mainly β-glucosidase activity (EC 3.2.1.21).

emulsion 1 a temporary or permanent dispersion of an oil or other hydrophobic material in an aqueous solution, or vice versa, forming an **oil-in-water emulsion** or a **water-in-oil emulsion** respectively. **2** *short for* **photographic emulsion**.

EMX2 one of several homeodomain proteins that are expressed in forebrain neurons and are important in brain regionalization. Mutations in the gene at 10q26 lead to the inherited epilepsy syndrome designated **schizencephaly**.

en- *or (sometimes before b, m, and p)* **em+** *prefix* **1** *(forming verbs)* put in or into, e.g. encapsulate, encode; surround or cover with, e.g. enmesh; to make into a certain condition, e.g. enrich. **2** *(forming nouns and adjectives)* in, into, e.g. endemic.

enabled a protein of *Drosophila* that suppresses the gene for D-Abl (i.e. Abelson protein tyrosine kinase).

ENaC *abbr. for* epithelial Na channel; a trimer of similar subunits, each of which has two transmembrane segments and intracellular N- and C- termini. The C-terminal tails of the β and γ subunits bind Nedd-4 with great affinity. Mutations that eliminate the C-terminal tail of β or γ subunits result in Liddle syndrome.

enalapril (*S*)-1-[*N*-[1-(ethoxycarbonyl)-3-phenylpropyl]-L-alanyl]-L-proline; an inhibitor of **angiotensin converting enzyme** (ACE) and a valuable antihypertensive drug. It has the advantage over **captopril** that it is more easily absorbed and has fewer side effects, and also that enalaprilate, to which it is converted after absorption by re-

moval of the ethyl ester group, is a much more potent ACE inhibitor than captopril. ACE inhibitors were developed from compounds such as succinylproline, an early inhibitor. Conversion of the free succinyl carboxyl group to a sulfhydryl, and addition of a methyl group, yielded captopril with greatly enhanced binding properties. In enalaprilate there was addition of further binding affinity by a phenylethyl group, restoration of carboxyl in place of sulfhydryl and introduction of a secondary amino group. These drugs lower angiotensin II levels which lowers blood pressure, both because of a reduction in direct action of angiotensin as a vasoconstrictor and also by reducing its actions in stimulating aldosterone and norepinephrine release, both of these latter substances having hypertensive effects.

enamel the hard, white, calcified material that forms the outer covering of the crown of a tooth.

enamilin *see* **tuftelin**.

enamine *or* **alkenylamine** any organic molecule having the general structure $R^1R^2NCR^3{=}CR^4R^5$ (where the R groups may be any hydrocarbyl group or H, and the same or different).

enanthate *see* **heptanoate**.

enantiomer either of a pair of **stereoisomers** whose molecules as a whole display **chirality**, i.e. are mirror images of each other and thus not superposable. They are sometimes referred to as optical isomers or optical antipodes; these terms are not recommended. *See also* **enantiomorph** (def. 2), **racemate** (def. 1). *Compare* **diastereoisomer**.

enantiomeric of, or pertaining to, an **enantiomer**; pertaining to the phenomenon of, or displaying **enantiomerism**. Chiral groups that are mirror images of one another are termed **enantiomeric groups**. *See also* **enantiomorphic**, **racemic**.

enantiomeric purity a measure of the proportion of one enantiomer in an enantiomeric mixture expressed as either a fraction or a percentage. *Distinguish from* **optical purity**.

enantiomerism the phenomenon of the existence of **enantiomers** in general, or the existence of enantiomeric molecules in a particular instance; it includes **optical isomerism**.

enantiomorph 1 either of two objects, especially crystals, that are mirror images of each other and thus are not superposable. **2** either of the two crystalline forms exhibited by a pair of enantiomers. Use of the term to mean **enantiomer** is deprecated.

enantiomorphic *or* **enantiomorphous** of, or pertaining to, an **enantiomorph**; pertaining to the phenomenon of, or displaying **enantiomorphism**. The term is often used synonymously with **enantiomeric** (enantiomeric molecules frequently form enantiomorphic crystals).

enantiomorphism the phenomenon of being related as between an object and its nonsuperposable mirror image. The term is used especially in relation to enantiomorphic crystals.

enantiomorphous *see* **enantiomorphic**.

enantiotopic 1 when chemically-like ligands in constitutionally equivalent locations (generally the two *a* ligands in C*aabc*) are related by a centre or plane of symmetry, or by an alternating axis of symmetry (but not by a simple axis of symmetry), they are enantiotopic. The two ligands are in a stereochemically different, mirror-image environment. If each *a* ligand of C*aabc* is replaced separately by a different, achiral ligand, *d*, the products are the two enantiomers of C*abcd*. Example: the methylene hydrogens of ethanol are enantiotopic; if ethanol is written as a Fischer projection structure with OH at the top, H–C–H in the middle, and CH₃ at the bottom, the left-hand hydrogen of the central methylene is H*S* , while that at the right is H*R* (*see* **pro-R/pro-S convention**). Replacement of ¹H*R* by ²H yields (+)-(*R*)-[1-²H₁]ethanol and the same replacement of ¹H*S* yields the enantiomer, (–)-(*S*)-[1-²H₁]ethanol. In another important compound, citric acid, the two CH₂COOH groups are also enantiotopic. **2** the two faces of a double bond or of a planar cyclic ring system that are related by a symmetry plane but not by a C_2 axis (i.e., a two-fold axis of symmetry) are enantiotopic; the two faces show stereochemically different, mirror-image related environments. Separate addition of the same achiral reagent to the two faces (*see* **Re/Si convention**) gives enantiomeric products. Example: the simple addition of HCN to CH_3–CHO yields a racemic mixture of the (*R*) and (*S*) cyanohydrins, CH_3–CH(OH)–CN, with both faces of C=O being involved. The reduction of the C=O bond of CH_3–CHO to form ethanol by alcohol dehydrogenase requires addition of a hydride ion from NADH at the C atom and a hydron at the O atom. Thus, reduction of CH_3–CHO with NAD^2H at its *A* face (*see* **diastereotopic** (def. 2)) yields (*R*)-[1-$^2H_1$]ethanol and reduction of CH_3–C^2HO with *A*-NADH yields (*S*)-[1-$^2H_1$]ethanol. The enzymatic reduction is stereospecific and only one of the enantiotopic faces of C=O is attacked; it is the same one (the *Re* face) in both of these situations. *Compare* **diastereotopic**.

encapsidate to surround (a particle of viral nucleic acid) with a **capsid**. —**encapsidation** *n*.

encapsis the association of myofibrils into bundles and the further association of these bundles into larger bundles, etc.

encephalin *a variant spelling of* **enkephalin**.

encephalitis inflammation of the brain.

encephalomyocarditis virus a cardiovirus in the family Picornaviridae that is pathogenic in pigs.

3′ end the end of a linear polynucleotide strand at which the 3′-hydroxyl group of the terminal nucleoside residue is normally not phosphorylated.

5′ end the end of a linear polynucleotide strand at which the 5′-hydroxyl group of the terminal nucleoside residue is normally phosphorylated.

end+ *a variant form of* **endo+** *(sometimes before a vowel)*.

endA an *Escherichia coli* gene encoding endonuclease A, an enzyme that digests plasmid DNA. Strains of *E. coli* lacking this enzyme consequently give higher yields and better quality plasmids.

end capping *(in chromatography)* the blocking of residual silanol groups on the surface of silica where these remain exposed after the bonding of C_{18} or other alkyl chains to the silica in the formation of reversed-phase stationary phases for column chromatography. For this purpose hydrocarbyl silanes (*see* **silane** (def. 3)) having small alkyl (usually methyl) groups are used so that they can penetrate between the main bonded-phase groups.

endemic present in or peculiar to a more or less localized area, e.g. an endemic disease. *Compare* **enzootic**.

endergonic describing a process or reaction on which work must be done, i.e. one requiring an energy input, for it to take place. At constant pressure and temperature the free energy content of such a system increases. *Compare* **exergonic**. [From **endo+** plus Greek *ergon*, work.]

end group any residue at an extremity of a branched or linear macromolecule.

end-group analysis determination of both the nature and the number of terminal groups in a macromolecule, e.g. in proteins, the N- and C-terminal amino-acid residues; in polynucleotides, the 3′- and 5′-terminal nucleotide residues.

end labelling the introduction of a radiolabelled molecule such as ^{32}P at the end of a DNA chain by an enzyme such as **polynucleotide kinase**. Alternatively a deoxynucleotide labelled with a fluorescent dye can be introduced using **terminal deoxynucleotidyltransferase**.

endo+ *or (sometimes before a vowel)* **end+** *comb. form* meaning within, inner, absorbing, containing. *Compare* **exo+**. *See also* **intra+**.

endo- *prefix (in chemical nomenclature)* denoting insertion (of the additional constituent(s) specified) into the structure of (a named compound); e.g. endo-4a-glycine-[5-leucine]enkephalin; endo-Gly^{4a}-[Leu^5]enkephalin; Tyr-Gly-Gly-Phe-Gly-Leu; a synthetic polypeptide in which a glycine residue has been inserted between residues 4 and 5 of [5-leucine]enkephalin.

endo- *prefix (in stereochemistry)*. *See* **conformation**.

endoamylase any **amylase** that hydrolyses nonterminal glycosidic linkages; it is a subcategory of **endoglycosidase**.

endocannabinoid any endogenous compound that binds a **cannabinoid receptor**, such as anandamide and 2-arachidonylglycerol. In the brain these function as retrograde synaptic messengers and may af-

fect memory, cognition, and pain perception by suppressing neurotransmitter release. They are degraded by intracellular enzymes.

endocrine 1 describing or relating to any gland or other group of cells that synthesizes **hormones** and secretes them directly into the blood, lymph, or other intercellular fluid. **2** describing or relating to a secretion of endocrine tissue. **3** a secretory product of endocrine tissue; a hormone. Originally known as **internal secretion**. *Compare* **exocrine**.

endocrine gland *or* **ductless gland** any of the ductless glandular structures that secrete (one or more) hormones directly into the bloodstream.

endocrinology the science concerned with the endocrine organs, their products, and the effects of these products. — **endocrinological** *adj.*

endocytic 1 situated within a living cell but not belonging to the cell itself; intracellular. **2** *an alternative term for* endocytotic (*see* **endocytosis**).

endocytosis the uptake of external materials by cells through the mechanism of **phagocytosis** or **pinocytosis**. The term is often used interchangeably with pinocytosis. *Compare* **exocytosis**, **transcytosis**. *See also* **internalize**, **viropexis**. —**endocytic** *or* endocytotic *adj.*; **endocytose** *vb.*

endocytotic vesicle *see* **pinocytotic vesicle**.

endodeoxyribonuclease *see* **deoxyribonuclease**.

endoenzyme 1 any intracellular enzyme. *Compare* **ectoenzyme**, **exoenzyme** (def. 1). **2** any enzyme that catalyses **endohydrolysis**. It may be an **endoglycosidase**, an **endonuclease**, or an **endopeptidase**. *Compare* **exoenzyme** (def. 2).

endogenous arising or developing within an organism, tissue, or cell, and excluding any consequences of externally added agents or materials. —**endogenously** *adv.*

endoglin a major glycoprotein of vascular endothelium that may be important in the binding of endothelial cells to integrins. It forms a heteromeric complex with the signalling receptors for transforming growth factor β (TGF-β). It has an **RGD** integrin-recognition motif, and is a homodimer of disulfide-linked subunits.

endoglycosidase any enzyme within subclass EC 3.2, glycosidases, that hydrolyses nonterminal glycosidic linkages in oligo- or polysaccharides. Many activities of this type are known, e.g. from *Flavobacterium meningosepticum*.

endohormone any hormone acting within the individual organism that produces it. *Compare* **ectohormone**.

endohydrolysis the hydrolysis, esp. by an **endoenzyme**, of any linkage between residues in a biopolymer. For example, **endopeptidases** attack neither the C-terminal nor the N-terminal peptide linkages of an oligo- or polypeptide, and **endoglycosidases** attack the terminal glycosidic linkages at either the reducing or nonreducing end of an oligo- or polysaccharide.

endolyn-78 a glycoprotein, M_r 78 000, present in substantial amounts in membranes of endosomes and lysosomes, but occurring only at low levels in plasma membrane and peripheral tubular endosomal compartment.

endolysins phage-encoded peptidoglycan-degrading enzymes that affect some bacterial cell walls. Access to the peptidoglycan layer is made possible by one or more of the **holins**.

endomembrane system a hypothetical integrated membrane system of eukaryotic cells, proposed by Morré and Mollenhauer, that represents a developmental and functional continuum. It comprises the endoplasmic reticulum, nucleus, Golgi apparatus and vesicles, plasmalemma, tonoplast, and outer membranes, but not the inner membranes of the mitochondria and chloroplasts.

endometrium the mucous membrane that lines the uterus. It becomes progressively thicker and more glandular in the later stages of the estrous (menstrual) cycle, which prepares it for embryo implantation. If pregnancy occurs the endometrium becomes the decidua, which is shed after birth. If pregnancy does not occur the endometrium returns to its original state; in primates, including humans, much of the endometrium breaks down and is lost in menstruation. —**endometrial** *adj.*

endomitosis the replication of chromosomes without cellular or nuclear division. It is a form of polyploidization that is fairly common in differentiated or differentiating tissues. It is characterized by an increase in nuclear DNA content.

endonuclease any enzyme of a large group of phosphoric diester hydrolases, forming sub-subclasses EC 3.1.21–31, that catalyses the hydrolysis of nonterminal diester linkages in polynucleotides to yield oligonucleotides. Examples include certain of the **deoxyribonucleases** and **ribonucleases**. *Compare* **exonuclease**. *See also* **restriction endonuclease**.

endopeptidase *or* endoproteinase *or (formerly)* **proteinase** any enzyme within subclass EC 3.4, peptide hydrolases, that hydrolyses nonterminal peptide linkages in oligopeptides or polypeptides, and comprising any enzyme of sub-subclasses EC 3.4.21–99. They are classified according to the presence of essential catalytic residues or ions at their active sites. Four distinct sub-subclasses are: (1) serine proteinases (EC 3.4.21); (2) cysteine proteinases (EC 3.4.22); (3) aspartic proteinases (EC 3.4.23), and (4) metalloproteinases (EC 3.4.24). There are two major families of serine proteinases, the **chymotrypsins** and the **subtilisins**. Aspartic proteinases contain two aspartic residues at their active site. The metalloproteinases contain metal ions, usually zinc. The proteinases can be distinguished by inhibitors, which are usually specific for a particular class or type.

endopeptidase La the enzyme EC 3.4.21.53; *other names*: ATP-dependent serine proteinase; ATP-dependent protease La; a serine proteinase that catalyses the hydrolysis of large proteins such as globin, casein, and denatured serum albumin in the presence of ATP. It is seemingly the sole member of its own superfamily.

endoperoxide a cyclic peroxide formed from a long-chain polyunsaturated fatty acid, especially arachidonic acid, through the action of **prostaglandin-endoperoxide synthase**.

endoplasm the inner, relatively fluid, part of the **cytoplasm**. *Compare* **ectoplasm**. —**endoplasmic** *adj*.

endoplasmic reticulum *abbr*.: ER; the irregular network of unit membranes, visible only by electron microscopy, that occurs in the cytoplasm of many eukaryotic cells. The membranes form a complex meshwork of tubular channels, which are often expanded into slitlike cavities called **cisternae**. The ER may be rough (or granular), with ribosomes adhering to the outer surface, or smooth (with no ribosomes attached). When cells are disrupted by homogenization, the cisternae break up into small closed vesicles called rough **microsomes** or smooth microsomes. The ribosomes attached to the rough ER are the site of translation of the mRNA for those proteins which are to be retained within the cisternae (**ER-resident proteins**), targeted to the **lysosomes**, or destined for export from the cell. Glycoproteins undergo their initial glycosylation within the cisternae. The smooth ER is the recipient of the proteins synthesized in the rough ER. Those proteins to be exported are passed to the Golgi complex (*see* **Golgi apparatus**), the resident proteins are retained in the rough ER, and the lysosomal proteins after phosphorylation of their **mannose** residues are passed to the lysosomes. Glycosylation of the glycoproteins also continues. The smooth ER is the site of synthesis of lipids, including the phospholipids. The major membrane phospholipids are phosphatidylcholines, phosphatidylethanolamines, phosphatidylserines, and sphingomyelins. The ER also produces cholesterol, triacylglycerols, eicosanoids, and ceramides, and contains enzymes that catalyse reactions to detoxify lipid-soluble drugs and harmful products of metabolism. Large quantities of certain compounds such as phenobarbital cause an increase in the amount of the smooth ER.

endoproteinase *or (formerly)* **proteinase** an alternative name for **endopeptidase**.

ENDOR *abbr. for* electron nuclear double resonance.

endorphin any endogenous peptide with morphine-like activity. The term includes the pentapeptide **enkephalins**, and the peptides of the pituitary gland, including α-endorphin, which has the structure Tyr-Gly-Gly-Phe-Met-Thr-Ser-Glu-Lys-Ser-Gln-Thr-Pro-Leu-Val-Thr; γ-endorphin consists of α-endorphin with an additional Leu at the C terminus; β-endorphin consists of α-endorphin with -Leu-Phe-Lys-Asn-Ala-Ile-Ile-Lys-Asn-Ala-His-Lys-Lys-Gly-Gln-OH at the C terminus. Endorphins have the same N-terminal tetrapeptide sequence as the enkephalins. Preproopiomelanocortin is the precursor of β-endorphin, both in brain and other tissues, and [Met^5]enkephalin; preprodynorphin (preproenkephalin B) is the

precursor of dynorphin; [Leu^5]enkephalin is formed from preproopiomelanocortin and from preproenkephalin A. Tissue endorphins are larger polypeptides (about 7 kDa) found in pancreas, placenta, and adrenal medulla. *Compare* **exorphin**. [From endo(genous) + morphine.]

endo-α-sialidase EC 3.2.1.129; *other names*: endo-*N*-acetylneuraminidase; endoneuraminidase; poly(α-2,8-sialosyl) endo-*N*-acetylneuraminidase; an enzyme that catalyses the endohydrolysis of α-2,8-sialosyl linkages in oligo- or poly(sialic) acids.

endoskeleton any skeleton lying within an organism, such as the axial skeleton of vertebrates. *Compare* **exoskeleton**.

endosmosis the osmotic flow of water or of an aqueous solution into a cell, organism, or vessel from a surrounding aqueous medium. *Compare* **exosmosis**. —**endosmotic** *adj*.

endosome a membrane-bound organelle in animal cells that carries materials newly ingested by **endocytosis**. It passes many of the materials to lysosomes for degradation.

endosperm the nutritive tissue, found in the seeds of most angiosperms, that surrounds and nourishes the embryo.

endosperm albumin the proteins occurring in the endosperm of, e.g., barley; *see* **protein Z**.

endospore 1 an asexual spore formed within a cell, especially one produced by some bacteria and algae. **2** *or* **endosporium** the innermost coat of the wall of a spore or pollen grain.

endostatin a monomeric (≈20 kDA) cleavage product that consists of the C-terminal domain of either of the two endostatin-forming **collagens** (types XV and XVII), which are expressed only in endothelial cells. They are the most potent inhibitors of angiogenesis and of endothelial cell migration. On binding to a specific receptor endostatin activates a protein tyrosine kinase and causes expression of specific genes.

endosymbiont a partner in a symbiotic relationship that penetrates the tissues or cells of the other partner; e.g. any of the nitrogen-fixing species of bacteria that occur in the root nodules of legumes. *See* **symbiosis**. *Compare* **ectosymbiont**.

endosymbiotic infection an infection of cells by viruses that replicate without cytopathic effect.

endothelial leukocyte adhesion molecule *abbr*.: ELAM; *another name for* **selectin**.

endothelin *abbr*.: ET; any one of three 21-residue peptides, called endothelins 1, 2, and 3 (*abbr*.: ET-1, ET-2, ET-3). Endothelins 2 and 3 are homologous with ET-1, which is known to be made by the endothelial cell. ET-1 is the most potent vasopressor known, being ten times more potent than angiotensin II. They have disulfide bonds, between cysteine residues 1 and 15, and 3 and 11; these hold the structure in a conical spiral shape. The amino-acid sequence of ET-1 is bicyclic(1→15, 3→11) [Cys^1-Ser-Cys^3-Ser-Ser-Leu-Met-Asp-Lys-Glu-Cys^{11}-Val-Tyr-Phe-Cys^{15}-His-Leu-Asp-Ile-Ile-Trp]; ET-2 differs in having -Trp-Leu- at positions 6,7; ET-3 has the substitutions (position in parentheses) Thr(2), Phe(4), Thr(5), Tyr(6), Lys(7), and Tyr(14). The endothelins are products of different genes, and are synthesized as part of a large precursor molecule, **preproendothelin** (prepro-ET), which is then proteolytically processed. The 2.2 kb mRNA for human prepro-ET-1 is encoded in five exons, the 5′-flanking region containing response elements for phorbol ester (Fos/Jun-inducible) and nuclear factor 1 (TGF-β-induced expression; *see* **transforming growth factor**), and its expression is stimulated by vasopressor hormones such as epinephrine, angiotensin II, and arginine vasopressin, as well as TGF-β (from aggregating platelets), thrombin, and interleukin-1. Prepro-ET-1 is a 203-residue peptide, which is cleaved to the 92-residue pro-ET (**big endothelin**), from which ET-1 is formed by **endothelin converting enzyme**. Related peptides are **vasoactive intestinal contractor** and **sarafotoxin**, a peptide produced by *Atractaspsis engaddensis*. Only ET-1 is detected in vascular endothelial cells; it is also expressed in nonvascular cells in other tissues, including brain, kidney, and lung. ET-2 and ET-3 are expressed in tissues such as brain, kidney, adrenal, and intestine. Mutations in the ET-3 gene occur in some patients with Hirschsprung disease.

endothelin receptor any of several membrane proteins that bind **endothelins** and mediate their intracellular effects. They are G-protein-coupled receptors. ET$_A$ (human, 427 amino acids) has affinity

ET-1 > ET-2 >> ET-3, while ET_B (human, 442 amino acids) is equipotent for all three. Loss-of-function mutations in the endothelin receptor B gene occur in Waardenburg syndrome with Hirschsprung disease.

endothelium the single layer of thin, flattened cells of mesoblastic origin that lines the blood vessels and some body cavities, e.g. those of the heart. *Compare* **epithelium, mesothelium.** —**endothelial** *adj.*

endothelium-derived relaxing factor *abbr.*: EDRF; *see* **nitric oxide.**

endothermic describing a process or reaction that absorbs heat, i.e. a process or reaction for which the change in enthalpy, ΔH, is positive at constant pressure and temperature. *Compare* **exothermic.**

endothiapepsin EC 3.4.23.22; *other name*: *Endothia* aspartyl protease. An enzyme, similar to other aspartic proteases, that catalyses the hydrolysis of proteins with broad specificity similar to that of **pepsin** A; it prefers hydrophobic residues at P1 and P1′, but does not cleave the Ala–Leu linkage in the B chain of insulin, or the protected dipeptide Cbz-Glu-Tyr; it clots milk.

endotoxin any microbial toxin that cannot easily be separated from the structure of the cell. *Compare* **exotoxin.**

end point *or* **endpoint** the point in a titration that should (but may not always) correspond to the theoretical **equivalence point.**

end product the final chemical substance formed in a sequence of metabolic (or chemical) reactions.

end-product inhibition the inhibition of a sequence of metabolic reactions by the end product of the sequence, usually by action on a reaction which is at the beginning of the sequence.

end-window counter a **Geiger counter** or **proportional counter** so constructed that its base is relatively transparent to the radiation to be detected.

+ene *suffix (in chemical nomenclature)* indicating the presence of one or more carbon–carbon double bonds in an organic compound.

enediol any acyclic organic compound in which there is a hydroxyl group attached to each of two carbon atoms that are linked by a double bond. *Compare* **enol.**

energy *symbol*: E; the capacity of a system for doing work. There are various forms of energy – potential, kinetic, electrical, chemical, nuclear, and radiant – which can be interconverted by suitable means. The SI unit of energy is the joule.

energy balance the algebraic balance of the various energy inputs versus energy outputs of a system; it is positive if energy is released and negative if it is absorbed. In physiology it is the relation of the amount of energy taken into the organism to the amount used for internal work, external work, and for the growth and repair of tissues.

energy barrier *see* **potential energy barrier.**

energy charge *see* **adenylate energy charge.**

energy coupling the coupling of ATP synthesis to electron transport in the **respiratory chain.**

energy diagram a diagram representing the energy contents of various states of the reactants, activated complexes, and products in a chemical reaction; or of the nuclear energy levels of an atom; or of the electronic energy levels of an atom or molecule.

energy flow *(in ecology)* the transfer of energy between organisms in an ecosystem.

energy level any of the stable energy states that a molecular entity can take up. In **quantum mechanics** only certain discrete energy levels are possible and continuous variation of the energy level of a molecular entity is excluded.

energy metabolism the metabolic reactions of a cell or organism concerned with energy transformations.

energy of activation *see* **activation energy.**

energy-poor *term sometimes used to describe* **1** a compound whose hydrolysis under standard conditions manifests a small negative free energy change. *See also* **low-energy compound. 2** a chemical group with a low **group potential. 3** the linkage of such a chemical group to some other group.

energy requirement the amount of energy needed to maintain a cell or organism.

energy sink a molecule or a group in a molecule that is able easily to accept energy transferred to it from another component of the system.

energy transduction *see* **transduction** (def. 2).

energy transfer the transfer of excitation energy from one chromophore or one molecular entity to another by a process not involving radiation; the energy may then be dissipated in a variety of ways, e.g. by fluorescence. Such energy transfer is very dependent on the distances involved and is useful in studies of structural relationships between groups on a macromolecule.

energy trapping the processes by which energy released in a catabolic reaction is coupled to the synthesis of another compound, often a nucleoside triphosphate such as ATP.

engineered 1 (of a cell or organism, or a strain of cells or organisms) *colloquial term for* **genetically engineered. 2** (of a nucleotide sequence) *jargon term, usually followed by 'into'* artificially inserted (into a longer sequence, such as a vector). **3** (of a protein) *jargon term* describing an altered protein synthesized by a cell or organism (or a strain of cells or organisms) whose genetic material has been manipulated to that end.

engrailed any of various protein products of the **homeotic gene** *engrailed*; in *Drosophila*, this is a segment-polarity gene.

enhanced chemiluminescence *abbr.*: ECL; an increased intensity of light output for a greatly extended period in chemiluminescence reactions involving the peroxidase-catalysed oxidation of luminol. The effect is achieved by the addition of molecules such as 6-hydroxybenzothiazole.

enhanced green fluorescent protein *abbr.*: EGFP; a modified form of green fluorescent protein (GFP) in which the λ_{max} excitation is shifted to 488 nm making it more suitable for analysis by flow cytometry. The λ_{max} emission is unchanged at 509 nm.

enhanced yellow fluorescent protein *abbr.*: EYFP; a proprietary green-yellow variant of green fluorescent protein (GFP) optimized for brighter fluorescence, and higher expression, in mammalian cells. It has λ_{max} excitation 513 nm and λ_{max} emission 527 nm.

enhancement 1 the improvement of the effects of ionizing radiation on tissue by dioxygen or other chemical agents. **2** the increased yield of one virus from cells infected by another virus; the term is often used when the mechanism of the process is not known. *Compare* **complementation** (def. 2), **interference. 3** *see* **Emerson enhancement. 4** *see* **fluorescence enhancement.**

enhancement of fluorescence *an alternative term for* **fluorescence enhancement.**

enhancement spectrum a spectrum showing the enhancement of photosynthetic **quantum yield** from that produced at a fixed-wavelength background illumination to that produced by illumination with a variable-wavelength beam. Such spectra show peaks that can sometimes be reasonably well identified with specific pigments in the system. *Compare* **Emerson enhancement.**

enhanceosome a protein complex that forms on an enhancer site of DNA and influences the rate of transcription at a promoter, which may be thousands of base pairs away. For example, that for the interferon β gene in mammals contains high-mobility group protein 1 (HMG1) and recruits two histone acetyl transferases, which acetylate core histones on neighbouring nucleosomes and also HMG1 of the complex. Another example is the disclike structure formed by a 180 bp loop of DNA and dimers of the *Xenopus* transcription factor xUBF. This is involved in rRNA transcription.

enhancer *(in molecular biology)* a eukaryotic control element that can increase the expression of a gene when transcription factors are bound to it. Enhancers may be located thousands of base pairs from the gene and may be either upstream, downstream, or in the gene itself.

enhancer trap vector a vector designed to identify genes expressed during development. It consists of a linear DNA molecule in which the *lacZ* reporter gene is placed under the control of a weak eukaryotic promoter and alongside a **dominant selectable marker** such as *neo*[R] under the control of its own strong promoter. In mice, for example, random integration of the vector into the genome of ES cells allows clones of recombinants expressing *neo*[R] to be isolated. These in turn can be used to create mosaic animals in which reporter activity is expressed if integration is, fortuitously, close to an enhancer. DNA from the region of integration can be cloned thereby allowing identification of the enhancer that is active during development. *Compare* **gene trap vector**.

enkephalin any pentapeptide **endorphin** with the sequence Tyr-Gly-Gly-Phe-Xaa. There are two similar naturally occurring enkephalins, present in brain, spinal cord, and gut; their recommended trivial names are: [5-leucine]enkephalin (*abbr.*: [Leu^5]enkephalin *or* [Leu]enkephalin), Tyr-Gly-Gly-Phe-Leu (I); and [5-methionine]enkephalin (*abbr.*: [Met^5]enkephalin *or* [Met]enkephalin), Tyr-Gly-Gly-Phe-Met (II). (II has the same sequence as residues 61–65 of β-**lipotropin**.) The common precursor of I and II, proenkephalin – itself formed from preproenkephalin by removal of the signal peptide – contains four copies of II and one each of I, the heptapeptide [Met^5]enkephalinyl-Arg-Phe, and the octapeptide [Met^5]enkephalinyl-Arg-Phe-Leu. (*Note*: Designations such as Leu-, leucyl-, or leucine-enkephalin have commonly been given to I, with corresponding terms for II, but all of these incorrectly imply N-terminal extension with a residue of leucine or methionine, respectively.)

enkephalinase *see* **carboxypeptidase H**.

enkephalin convertase a specific carboxypeptidase that converts enkephalin precursors into enkephalin in adrenal chromaffin granules. It is now known as **carboxypeptidase H**.

enniatin any of various antibiotics from *Fusarium* spp. that function as **ionophores**. They are cyclodepsipeptides (*see* **depsipeptide**).

enol any acyclic organic compound with a hydroxyl group attached to either of two carbon atoms that are linked by a double bond. *Compare* **enediol**. *See also* **keto–enol tautomerism**.

enolase EC 4.2.1.11; *recommended name*: phosphopyruvate hydratase; *other name*: 2-phosphoglycerate dehydratase. A cytosolic enzyme that catalyses the conversion of 2-phospho-D-glycerate to phospho*enol*pyruvate and H$_2$O; magnesium is a cofactor. There are three isoenzymes in human tissues: α2 is ubiquitous; β2 is in skeletal muscle; and γ3 is in the nervous system. A rare deficiency in which enolase activity is ≈50% of normal in erythrocytes may be accompanied by hemolytic anemia.

enology *or (esp. Brit.)* **oenology** the study of wines.

5-enolpyruvylshikimate-3-phosphate pholyase *see* **chorismate synthase**.

eNOS *see* **nitric-oxide synthase**.

enoyl an acyl group derived from any alkenoic acid.

enoyl-[acyl-carrier protein] reductase (NADPH, A-specific) EC 1.3.1.39; *other name*: acyl-ACP dehydrogenase; an enzyme of the **fatty acid synthase complex** in liver. It catalyses the reduction by NADPH of *trans*-2,3-dehydroacyl-[acyl-carrier protein] to form acyl-[acyl-carrier protein] and NADP$^+$.

enoyl-CoA hydratase 1 EC 4.2.1.17; *systematic name*: (3*S*)-3-hydroxyacyl-CoA hydro-lyase; *other names*: enoyl hydrase, unsaturated acyl-CoA hydratase; an enzyme of the **beta-oxidation system**, present in mitochondria and peroxisomes. It catalyses the hydration of (3*S*)-3-hydroxyacyl-CoA to *trans*-2(or 3)-enoyl-CoA. It is a mitochondrial matrix protein, a homohexamer; a characteristic transit peptide sequence is found in the precursor. **2** EC 4.2.1.74; *recommended name*: long-chain-enoyl-CoA hydratase; an enzyme that catalyses a similar reaction to the above; it does not act on crotonoyl-CoA. *See also* **beta-oxidation system**.

enoyl-CoA isomerase *see* dodecenoyl-CoA *Δ*-isomerase.

3,2-*trans*-enoyl-CoA isomerase *see* dodecenoyl-CoA *Δ*-isomerase.

enrich to increase in content or abundance; e.g. to increase the abundance of one (usually stable) nuclide of an element above its naturally occurring level; to increase the content in a foodstuff of one or more specific nutrients (such as vitamins or minerals) above the natural level. —**enrichment** *n.*

enrichment medium a selective culture medium that favours the growth of a desired microorganism.

Ensembl *see* **genome annotation software**.

entactic describing an enzyme poised for catalytic action in the absence of substrate.

entactin *or* **nidogen** a sulfated Ca^{2+}-binding protein occurring in basement membrane, and involved in cell adhesion; it binds **laminin** and type IV collagen. The protein has two globular domains, an **EF-hand** domain, EGF-like domains, and a thyroglobulin type I domain.

enteric *or* **enteral** of or pertaining to the intestine or gut.

enteric-coated describing a drug or medicament that is prepared in a form enabling it to pass through the stomach unaltered and be released in the intestine.

entero+ *or (before a vowel)* **enter+** *comb. form* denoting the intestine.

enterobactin *or* **enterochelin** *N′,N′,N″*-tris(2,3-dihydroxybenzoyl)-2,6,10-trioxo-1,5,9-trioxacyclododecane-3,7,11-triamine; a **siderochrome** of the catechol-derivative variety produced by certain members of the Enterobacteriaceae, e.g. *Escherichia coli*, *Salmonella* spp. It is the cyclic self-triester (trilactide) of *N*-(2,3-dihydroxybenzoyl)-L-serine, and a product of the **shikimate pathway**.

enterochelin *an alternative name for* **enterobactin**.

enterochromaffin cell *abbr.*: EC cell; any gut endocrine cell containing biogenic monoamines that gives a positive **chromaffin** reaction and displays a characteristic yellow formaldehyde-induced fluorescence. Such cells may be divided immunocytochemically and ultrastructurally into functionally distinct types.

enterocrinin a putative peptide hormone, crystallizable from extracts of intestine of certain mammals, that is held to stimulate duodenal and jejunal secretion of digestive enzymes.

enterocyte any of the columnar epithelial cells on the luminal surface of the villi of the small intestine. Enterocytes are immature in the crypts, mature in the middle of the villi, and senile at the exfoliation area at the villous tips. They have numerous microvilli on their luminal surface (the brush border) and are responsible for the synthesis of digestive enzymes and for the absorption of materials from the gut.

enterogastrone the putative hormone from the duodenum that inhibits gastric activity. Some or all of the actions attributed to enterogastrone are probably due to **glucose-dependent insulinotropic peptide**.

enterogastrone effect *see* **gastric inhibitory peptide**.

enteroglucagon 1 a collective term for a small family of polypeptides derived from **proglucagon** by post-translational processing in the L cells (or **EG cells**) of the distal small intestine and colon. One component is GLP-1$^{7–37}$ (*see* **GLIP**). **2** *or* **glucagon-37** *or* **oxyntomodulin** a polypeptide isolated from porcine jejuno-ileum that binds glucagon receptors and activates adenylate cyclase in hepatocytes. It specifically inhibits gastric acid secretion. Its N-terminal 29 residues (of a total of 37) are identical to those of glucagon.

enterohepatic recirculation the cycle of secretion from the gall bladder into the upper intestinal tract with reabsorption in the lower tract and transport to the liver followed by resecretion. *See* **bile acids**.

entero-insular axis the interrelationship of endocrine function between the gastrointestinal tract and the pancreatic islets in which signals arising in the gut after ingestion of nutrients effect endocrine responses by the islets.

enterokinase *a former name for* **enteropeptidase**.

enteropeptidase *or (formerly)* **enterokinase** the enzyme EC 3.4.21.9; a serine proteinase enzyme of the brush border of the upper small intestine that activates **trypsinogen** by the selective cleavage of the Lys^6-|-Ile^7 bond in the latter. The bovine protein consists of 1035 amino acids; residues 1–800 form the non-catalytic heavy chain, and residues 801–1035 the catalytic light chain. *See* **duodenase**.

enterostatin a pentapeptide generated by tryptic hydrolysis of procolipase in the intestinal lumen. It has the sequence APGPR in human, and VPDPR in pig, horse, and rat. It decreases fat intake and secretion of insulin and pancreatic enzymes in rodents. It has been found in rodent serum and brain.

enterotoxin any bacterial **exotoxin**, whether ingested or produced within the intestine, that has an action upon the intestinal mucosa, usually giving rise to diarrhoea and other unpleasant symptoms. **Cholera toxin** is an example. Other enterotoxins are produced by *Escherichia coli* and some Staphylococci. Of the *E. coli* enterotoxins, LT (from labile, i.e. heat-labile, toxin) is similar in structure to cholera toxin and has a similar mode of action. ST (from heat-stable toxin) has an effect similar to atrial **natriuretic peptide**, activating guanylate cyclase. It has 13 amino acids. Other enterotoxins produced by *E. coli* include **hemolysin** and vero toxin, a Shiga-like toxin similar to that of *Shigella dysenteriae* that has *N*-glycosidase activity, and cleaves an adenine residue from the 28S rRNA of the 60S ribosome subunit.

enthalpimetry the measurement of the **enthalpy** of a body or system.

enthalpy *or* **heat content** *symbol*: *H*; a physical quantity and a thermodynamic property defined as the sum of the internal energy of a system, *U* and the product of the volume, *V*, multiplied by the pressure, *p*, i.e. $H = U + pV$. Enthalpy is related to the **Gibbs energy**, *G*, by the equation $G = H - TS$, where *T* denotes the thermodynamic temperature and *S* denotes **entropy**.

Entner–Doudoroff enzyme the 'catabolic' D-glucose-6-phosphate dehydrogenase (EC 1.1.1.49) from *Pseudomonas fluorescens*. It is a 220 kDa protein composed of four, apparently identical, polypeptide chains. *See also* **Entner–Doudoroff pathway**.

Entner–Doudoroff pathway a metabolic pathway occurring in *Pseudomonas* spp. in which D-glucose 6-phosphate is first oxidized to 6-phospho-D-gluconate, which is then acted on by 6-phosphogluconate dehydratase (EC 4.2.1.12) to form 2-dehydro-3-deoxy-6-phospho-D-gluconate. This is then split by 2-dehydro-3-deoxyphosphogluconate aldolase (EC 4.1.2.14) to give pyruvate and D-glyceraldehyde 3-phosphate. [After Nathan Entner (1920–) and Michael Doudoroff (1911–).]

entomology the science of the study of insects.

entopic occurring in the usual place. *Compare* **ectopic**.

Entrez *see* **information retrieval software**.

entropic union a polymerization, or other reaction, resulting from an energetically unfavourable interaction of a surface of the molecule in question with water molecules; the polymerization causes the removal of water from the molecules' surfaces and a hence a decrease in the free energy content of the system.

entropy *symbol*: *S*; a physical quantity and a thermodynamic property indicating the amount of disorder in a system; i.e. the amount of energy in a system that is unavailable for doing work. In any irreversible process the total entropy of all systems concerned is increased. In a reversible process the total increase in entropy in all systems concerned is zero, while the increase in entropy, d*S*, of any individual system, or part of a system, is equal to the heat that it absorbs, dq_{rev}, divided by the thermodynamic temperature, *T*; i.e. d*S* = dq_{rev}/*T*. Entropy is related to **Gibbs energy**, *G*, and **enthalpy**, *H*, by the equation $G = H - TS$. The SI unit for entropy is JK^{-1} —**entropic** *adj*.

entry exclusion the phenomenon in which a resident plasmid interferes with the entry of genetic material to a cell by (bacterial) **conjugation** (def. 4).

enucleate 1 to remove the nucleus from (a cell). **2** (of a cell) deprived of its nucleus. —**enucleation** *n*.

ENU mutagenesis *abbr. for* *N*-ethyl-*N*-nitrosourea mutagenesis; ENU is an alkylating agent and powerful mutagen used to induce mutations in living organisms. AT base pairs are predominantly modified and a high proportion of the mutations are missense.

enuresis the involuntary passing of urine. —**enuretic** *adj*.

env a viral gene for **envelope protein**.

envelope 1 any covering or enclosing structure, e.g. a membrane, capsule, shell, or skin. **2** the cell membrane of a bacterium, together with all the structures external to it including the cell wall and, sometimes, the capsule. **3** *or* **peplos** an outer lipoprotein coat of a virion that occurs in an **enveloped virus**. **4** *see* **nuclear envelope**.

envelope conformation any **conformation** of a nonplanar five-membered saturated ring compound when four of its ring atoms lie in one plane and the remaining atom lies outside that plane. For a monosaccharide or monosaccharide derivative the conformational descriptor -*E* may be added to its name.

enveloped virus any virus in which a nucleoprotein core is surrounded by a lipoprotein envelope consisting of a closed bilayer of lipid derived from that of the host cell's membrane(s), with glycoprotein on the outside and matrix protein or nucleocapsid protein on the inside. Enveloped viruses include herpesviruses, negative-strand RNA viruses, retroviruses, and togaviruses.

envelope protein any protein (usually a glycoprotein) of the envelope of a virus. Example: envelope polyprotein gp160 precursor of human immunodeficiency virus (HIV) type 1: residues 1–29 are the signal peptide; 30–509 form exterior membrane glycoprotein gp120;

510–855 form transmembrane glycoprotein gp41. The variable (very rapidly evolving) antigenic regions of HIV are in the gp120 sequence.

envelysin EC 3.4.24.12; *other names*: sea-urchin-hatching proteinase; hatching enzyme; an enzyme so named because it dissolves (lyses) the fertilization envelope of the sea-urchin embryo. It is a metalloendopeptidase that catalyses the preferential cleavage of proteins on the amino side of bulky hydrophobic residues, Leu, Ile, Phe, and Tyr. The enzyme requires zinc and calcium.

envoplakin a precursor (2033 amino acids) of the crosslinked protein (called cornified envelope) present under the plasma membrane of keratinocytes. It is closely related to plectin, desmoplakin, and bullous pemphigoid antigen.

Enzacryl *the proprietary name for* a group of synthetic carriers for the immobilization of enzymes and other organic biomolecules that are copolymers of acrylamide and various derivatives of *N,N′*-methylene diacrylamide.

enzootic present in or peculiar to animals in a more or less localized area; e.g. an enzootic disease. *Compare* **endemic**.

enzymatic *an alternative term for* **enzymic**.

enzyme any naturally occurring or synthetic macromolecular substance composed of protein, that catalyses, more or less specifically, one or more (bio)chemical reactions at relatively low temperatures. RNA that has catalytic activity (*see* **ribozyme**) is often regarded as enzymic. The substances upon which enzymes act are known as **substrates**, for which the enzyme possesses a specific binding or **active site**. *See also* **enzyme classification**, **isoenzyme**. [From German *enzym*, possibly from modern Greek *enzumos*, leavened, or more probably from Greek *en*, in, + *zum ē*, leaven, yeast.] —**enzymic** *or* **enzymatic** *adj*.

enzyme I EC 2.7.3.9; enzyme I of the **phosphotransferase** system; phospho*enol*pyruvate–protein phosphotransferase; *systematic name*: phospho*enol*pyruvate:protein-L-histidine *N*$^{\pi}$-phosphotransferase. A soluble bacterial enzyme, part of a system for the transport of hexoses across the cell membrane, that catalyses the phosphorylation by phospho*enol*pyruvate of a low M_r, heat-stable protein, **HPr**. *Compare* **enzyme II** (def. 1).

enzyme II 1 EC 2.7.1.69; enzyme II of the **phosphotransferase** system; protein-*N*$^{\pi}$-phosphohistidine–sugar phosphotransferase; *systematic name*: protein-*N*$^{\pi}$-phosphohistidine:sugar *N*$^{\pi}$-phosphotransferase. Any of a group of related membrane-bound bacterial enzymes, part of the system for the transport of hexoses across the cell membrane, that catalyse the phosphorylation of hexoses by phospho-HPr (*see* **enzyme I**). Enzyme II is responsible for the specificity of the transport process with respect to the sugar. **2** *a former name for* **acyl-carrier protein**.

enzyme-activated irreversible inhibitor *an alternative name for* **suicide inhibitor**.

enzyme activation the generation of a catalytically active enzyme from an inactive or poorly active form or from a biosynthetic precursor (proenzyme). Activation may be by enzymic or chemical covalent modification or by addition of a specific activator.

enzyme adaptation *see* **inducible enzyme**.

enzyme cascade *see* **cascade sequence**.

enzyme classification the systematic arrangement and naming of enzymes by the **Enzyme Commission**. The basis of the classification is the reaction classified; enzymes that catalyze the same reaction but have different genetic origins, protein structure, and cofactors, are generally all given the same EC number. Each enzyme is denoted by the abbreviation EC followed by a set of four numbers separated by stops. The first number denotes one of the six main divisions: EC 1, oxidoreductases; EC 2, transferases; EC 3, hydrolases; EC 4, lyases; EC 5, isomerases; and EC 6, ligases. The second number denotes the subclass, the third number denotes the sub-subclass; the fourth number is the serial number of the particular enzyme. The most recent edition of the classification at

http://www.cehm.qmul.ac.uk/iubmb.enzyme

lists approximately 4000 enzymes.

enzyme cluster *or* (*sometimes*) **multienzyme cluster 1** any physiologically significant system of two or more enzymes in physical (i.e. noncovalent) association. The term embraces any **enzyme complex**,

multienzyme complex, or membrane-bound enzyme array. Clustered enzymes usually display different kinetic and/or regulatory features from their unassociated counterparts; they may be encoded in a cluster-gene. *See also* **multienzyme system**. **2** *an alternative term for* **cluster** (def. 2).

Enzyme Commission *or the* **International Commission on Enzymes** a body established in 1956 by the International Union of Biochemistry (IUB) to consider the classification and nomenclature of enzymes and coenzymes, their units of activity and standard methods of assay, together with the symbols used in the description of enzyme kinetics. The Commission, which worked closely with the Biological Chemistry Nomenclature Commission of the International Union of Pure and Applied Chemistry (IUPAC), was dissolved in 1961 and its work has been carried on in turn by the Standing Committee on Enzymes of IUB, by the IUPAC/IUB Joint Commission on Biochemical Nomenclature (JCBN), by an Expert Committee on Enzymes, and most recently by the Nomenclature Committees of IUB (NC-IUB) or the International Union of Biochemistry and Molecular Biology (NC-IUBMB). *See also* **enzyme classification**.

enzyme complex an operational term for any structural and functional entity composed of a number of dissociable enzymes that catalyse a sequence of closely related chemical reactions. *Compare* **multienzyme**.

enzyme detergent any detergent preparation incorporating an enzyme to assist its cleansing action. The enzymes used in such detergents are usually proteinases of high thermal and alkaline stability, e.g. **Alcalase**.

enzyme differentiation the process whereby, during the development of an organism, each tissue acquires its own characteristic quantitative pattern of enzymes, which underlies the physiological functions and morphological features of the tissue. From a fairly uniform enzymic make-up in the cells of the early embryo, the enzyme patterns of different tissues become progressively more differentiated as development into the mature organism proceeds.

enzyme electrode any electrode, incorporating an enzyme into its structure, that responds to the concentration of one of the substrates or products of the reaction catalysed by the enzyme. The enzyme is trapped within a gel matrix surrounding the electrode or is kept in contact with the electrode by a semipermeable membrane.

enzyme engineering *or* **enzyme technology** the branch of biomolecular engineering concerned with processes designed to produce, isolate, purify, and immobilize enzymes and to use them for the catalysis of specific chemical reactions.

enzyme immunoassay *or* **enzymoimmunoassay** *abbr.*: EIA; any **immunoassay** in which an enzyme-catalysed reaction is used as the indicator. *See also* **heterogeneous immunoassay**, **homogeneous immunoassay**.

enzyme induction the synthesis of an enzyme in a cell or organism at a markedly increased rate in response to the presence of an **inducer**. The inducer is thought to combine with a **repressor** thereby preventing the latter from blocking an **operator**, which controls the translation of the structural gene for the enzyme.

enzyme-inhibition immunoassay a variation of **enzyme immunoassay** in which the inhibition of an enzyme-catalysed reaction is used as the indicator.

enzyme labelling a method used to detect or locate (and sometimes estimate) an antigen in, e.g., a tissue section. The section is exposed to a complementary antibody that has been covalently linked to an enzyme; the antibody binds to the antigen, and its location (and its amount) is determined by an assay dependent on the catalytic activity of the linked enzyme. *See also* **enzyme immunoassay**.

enzyme-linked immunosorbent assay *abbr.*: ELISA; a form of quantitative **immunoassay** based on the use of antibodies (or antigens) that are linked to an insoluble carrier surface, which is then used to 'capture' the relevant antigen (or antibody) in the test solution. The antigen–antibody complex is then detected by measuring the activity of an appropriate enzyme that had previously been covalently attached to the antigen (or antibody).

enzyme membrane any (semipermeable) membrane to which an enzyme has been covalently bound. Such membranes are useful in constructing **enzyme electrodes**.

enzyme-multiplied immunoassay technique *abbr. and proprietary name*: EMIT; *an alternative name for* **homogeneous enzyme immunoassay**.

enzyme-paper graft an enzyme immobilized on (filter) paper, which is frequently impregnated with indicators. It is useful for making analytical devices.

enzyme reactor a device for using immobilized enzymes or enzyme systems for synthetic or other processing reactions, especially on an industrial scale.

enzyme recruitment the exploitation of substrate-ambiguous enzymes or transport proteins in the evolution of new biochemical pathways.

enzyme repression inhibition of the formation of an enzyme by a compound formed in or taken in by a cell or organism.

enzyme specificity *see* **specificity**.

enzyme–substrate complex *abbr.*: ES; the stoichiometric complex of an enzyme molecule and a substrate molecule bound at the enzyme's active site.

enzyme technology *an alternative term for* **enzyme engineering**.

enzyme unit *symbol*: U; *abbr.*: EU; a unit of activity of enzymes, defined as the amount of enzyme that will catalyse the transformation of one micromole of the substrate per minute under standard conditions. It is not recommended, having been superseded by the **katal**, but is still widely used: $1 \text{ kat} = 6 \times 10^7 \text{u}$; $1\text{u} = 16.7 \text{ nkat}$.

enzymic *or* **enzymatic** by, of, involving, or relating to an enzyme or enzymes; catalysed by an enzyme or by enzymes. —**enzymically** *or* **enzymatically** *adv*.

enzymic activity the rate of reaction of substrate that may be attributed to catalysis by an enzyme. The concept is now obsolete, having been superseded by **catalytic activity**. The unit of enzymic activity was initially the **enzyme unit** and subsequently the **katal** (as originally defined). The derived quantity concentration of enzymic activity was defined as activity divided by volume and was expressed as katals per litre. *See also* **catalytic activity concentration**, **molar (enzymic) activity**, **specific (enzymic) activity**.

enzymic cycling a form of **catalytic assay** in which there is no coupled indicator reaction; instead, the regenerating system is allowed to cycle for a set time, after which the reactions are stopped by destruction of all enzymes and the most easily determined product is measured in a separate assay. The coupled system thus acts as a chemical amplifier in the determination of the intermediate catalyst.

enzymo+ *comb. form* denoting an **enzyme** or enzymic activity.

enzymoimmunoassay *an alternative name for* **enzyme immunoassay**.

enzymology the science of the study of enzymes and enzyme-catalysed reactions. —**enzymological** *adj.*; **enzymologist** *n*.

enzymolysis the enzyme-catalysed splitting of a chemical compound into smaller ones. The term commonly refers to enzymic hydrolysis. —**enzymolytic** *adj*.

eobiogenesis the first occurrence of the formation of living matter from nonliving material.

eobiont any system showing some characteristics of living systems but not enough to be generally accepted as living.

eosin any of a number of similar red acidic dyes, derivatives of fluorescein, especially Eosin Y (yellowish), 2′,4′,5′,7′-tetrabromofluorescein disodium salt, Acid Red 87 (CI 45380), or Eosin B (bluish), 4′,5′-dibromo-2′,7′-dinitrofluorescein disodium salt, Acid Red 91 (CI 45400). They are widely used as stains in histology and hematology.

2′,4′,5′,7′-tetrabromofluorescein

eosinophil(e) 1 *or* **eosinophil leukocyte** *or* **eosinophilic leukocyte** a polymorphonuclear leukocyte, present in blood and other connective tissues, that has numerous large cytoplasmic granules that stain readily with eosin. The granules contain numerous hydrolytic enzymes including acid phosphatase, arylsulfatase, cathepsins, glucuronidase, peroxidase, and ribonuclease. The relative number of eosinophils in the blood, normally low, rises markedly in certain allergic conditions and parasitic infections but their function is poorly understood. 2 any cell whose cytoplasm stains readily with eosin, especially an eosinophil cell of the anterior pituitary that secretes prolactin and somatotropin.

eosinophil cationic protein *abbr.*: ECP; a highly basic zinc-containing **ribonuclease** that binds avidly to negatively charged surfaces and is particularly effective at damaging the tegument of schistosomes.

eosinophilic having an affinity for **eosin**.

eosinotactic having an attractive or repulsive action on eosinophils. [A contraction of eosinophil + chemotactic.]

eosome the suggested primary particulate biogenetic precursor, via the **neosome**, of the **ribosome**. Eosomes of *Escherichia coli* have an average sedimentation coefficient of 14S.

eotaxin any chemokine that attracts eosinophils. An eotaxin receptor is present only in these leukocytes.

EPA *abbr. for* 1 eicosapentaenoic acid. Normally indicating the (all-*Z*)-5,8,11,14,17 isomer. 2 erythroid potentiating activity (*see* **TIMP**).

EpETrE *abbr. for* epoxyeicosatrienoate.

EPH a gene encoding a subfamily of protein tyrosine kinases. It is named from erythropoietin-producing hepatocellular carcinoma cell line, from which the probe used to isolate the gene was obtained. Related proteins are Elk (Eph-like kinase), Eek (Eph and Elk-related kinase), Eck (epithelial cell kinase), and Erk (Elk-related kinase). Overexpressed *Eph* has tumorigenic potential; the ligand of Eph is unknown.

ephedrine (1*R*,2*S*)-1-phenyl-1-hydroxy-2-methylaminopropane; an alkaloid obtained from several species of gymnosperm shrubs belonging to the genus *Ephedra*. A structural analogue of epinephrine, it mimics central and peripheral effects of noradrenergic and adrenergic neurons through direct and indirect actions, hence its CNS-stimulant effects and its use as a decongestant in upper respiratory tract infections.

ephexin a guanine nucleotide exchange factor (GEF) regulated by the **ephrin** class A receptors.

ephrin any axon guidance protein that is a membrane-bound ligand (analogous to the semaphorins) for the Eph family of receptor tyrosine kinases (*see* **EPH**). Ephrins regulate cell migration, axon guidance, regionalization in the nervous system, and morphogenesis. Mammals produce 8 ephrins and 14 Eph receptors. Class A ephrins are GPI-anchored and bind Eph A receptors. Class B ephrins have one transmembrane segment, bind Eph B receptors in the nervous system, are important in development, and influence synapse assembly.

epi+ *or (before a vowel)* **ep+** *prefix* denoting 1 on, upon, above, over, beside, near, close to, in addition to. 2 a chemical compound or group related in some way to a specified chemical compound or group, e.g. an **epimer**. 3 a chemical compound or group distinguished from a specified chemical compound or group by having a bridge connection in the molecule, e.g. an **epoxide**.

epiallele any of two or more genetically identical alleles that are epigenetically distinct owing to methylation. –**epiallelic** *adj*.

epiarginase *see* **epiprotein**.

epichlorohydrin 3-chloropropylene oxide; chloromethyloxirane; 1-chloro-2,3-epoxypropane; a colourless liquid widely used in industry as a solvent and an intermediate in the manufacture of glycerol, glycerol ethers and epoxy resins. In the presence of potassium hydroxide it reacts with compounds containing alcoholic hydroxyl groups to form diethers of glycerol and hence finds application in the manufacture of cross-linked derivatives of polysaccharides such as the proprietary materials sephacel, sephadex, and sepharose.

epidemic 1 describing a disease that affects many persons simultaneously in a more or less restricted area. 2 the widespread occurrence of a disease in a human population. *Compare* **endemic**.

epidemiology the study of the nature and spread of a disease in a human population.

epidermal growth factor *abbr.*: EGF; any member of a group of heat-stable, hormonal, ≈6 kDa proteins that consist of a single polypeptide chain of 49–53 residues with three intrachain disulfide bonds; they are classified as **cytokines**. One form (mEGF), the tooth-lid factor, can be isolated from submaxillary glands of the male mouse (*see* **nerve growth factor**). A similar protein (hEGF), extracted from human pregnancy urine and named **urogastrone**, is highly homologous to EGF (37 of 53 residues identical, disulfide bonds preserved). EGF *in vivo* stimulates the growth of epidermal and epithelial cells, and inhibits the secretion of gastric acid. It also has a potent mitogenic action on many types of cultured mammalian cells and displays a number of insulin-like properties towards them (*see* **insulin-related growth factors**). It is synthesized as a large preprotein, EGF membrane glycoprotein precursor, which contains, as well as the mature EGF protein, nine EGF-like repeats and urogastrone. Many proteins have one or more EGF-like sequence repeats. The action of EGF is mediated by the **epidermal growth factor receptor**.

epidermal growth factor receptor *or* **EGF receptor** the mediator of the biological signal of **epidermal growth factor** (EGF) and also of **transforming growth factor**-α (TGF-α). It is a membrane protein with a single transmembrane domain, and a **tyrosine kinase** activity in the cytoplasmic domain; the extracellular domain, which binds EGF, is highly glycosylated. The v-*erbB* oncogene codes for a protein highly homologous to a truncated EGF receptor, and the gene for the receptor is the protooncogene c-*erbB* (*see* **erbB**). Binding of EGF leads to the induction of tyrosine kinase activity, formation of dimers, and autophosphorylation; stimulation of cell DNA synthesis and cell proliferation follows. Internalization of the EGF–receptor complex leads to its degradation in lysosomes.

epidermis the outermost layer of cells of an animal or plant. —**epidermal** *adj*.

epienzymatic control a form of control of arginine metabolism, occurring in some *Saccharomyces* spp., in which arginases bind stoichiometrically to ornithine carbamoyltransferase (EC 2.1.3.3), inhibiting the latter's activity but without substantially modifying the arginase activity. *See also* **epiprotein**.

epigenesis the concept that an organism develops from an originally undifferentiated mass of living material through the appearance of structures and functions not originally present.

epigenetic describing any of the mechanisms regulating the expression and interaction of genes, particularly during the development process. These include changes that influence the phenotype but have arisen as a result of mechanisms such as inherited patterns of DNA methylation rather than differences in gene sequence: **imprinting** is an example of this.

epigenetics the study of factors that influence gene expression but do not alter genotype, such as chromatin methylation and acetylation involved in tissue-specific patterns of gene expression, or the parental imprinting of genes.

epigenomics a genomic approach to studying environmental or developmental epigenetic effects, primarily DNA methylation, on gene function. Thus, epigenomics focuses on those genes whose function is determined by external factors.

epiligrin an extracellular matrix protein, secreted by cultured epidermal keratinocytes. It is found in epithelial basement membranes, but not membranes of endothelia, nor in muscle or nerve cells. It is a ligand for cell adhesion via **integrins**, with three chains linked by disulfide bonds, and is also an adhesive ligand for T lymphocytes. *See also* **laminin**.

Epilim *a proprietary name for* sodium valproate (*see* **valproic acid**).

epimer *(in stereochemistry)* either of two **diastereoisomers** that differ in configuration at only one chiral centre. The term includes **anomer**.

epimerase any enzyme within subclass EC 5.1, that catalyses **epimerization** (note that subclass EC 5.1 also includes **racemases**). Some epimerases, e.g. UDParabinose 4-epimerase (EC 5.1.3.5), contain tightly bound NAD, which undergoes oxidation and reduction during the reaction.

epimeric of or relating to an epimer or the chiral centre that specifies an epimer.

epimerization or **epimerisation** the process of converting an epimer into its diastereoisomer by altering the configuration at the epimeric chiral centre. —**epimerize** or **epimerise** vb.

epimino+ an alternative term for **imino+** (def. 2).

epinephrine or **adrenaline** (–)-1-(3,4-dihydroxyphenyl)-2-(methylamino)ethanol; a hormone secreted by the adrenal medulla and a neurotransmitter secreted by neurons in the brainstem. It is synthesized by the methylation of **norepinephrine**. It is an agonist for **adrenoceptors**, through which it has powerful glycogenolytic and lipolytic effects (both of which are mediated by cyclic AMP) and also affects the activity of smooth muscle (notably of the cardiovascular system and bronchi) and glandular tissue. Its mode of action critically depends on whether it acts through α or β adrenoceptors. Chromaffinoma of the adrenal medulla or other chromaffin tissue is a rare cause of intermittent hypertension and other signs of catecholamine excess. The hormone was named adrenalin by Takamine in 1901.

epinephrine

epiphase the upper, less dense layer of a two-phase system. Compare **hypophase**. —**epiphasic** adj.

epiplasm a membrane skeleton of certain protists, that contributes to cell shape and patterning of the species-specific cortical architecture. This cytoskeleton contains proteins known as **articulins**.

epiprotein a specific protein, produced in certain circumstances in some *Saccharomyces cerevisiae* strains, that binds stoichiometrically to ornithine carbamoyltransferase (EC 2.1.3.3), thereby inhibiting its activity. It has subsequently been shown to have arginase activity and so is sometimes known as **epiarginase**. See also **epienzymatic control**.

episemantic molecule any molecule that is synthesized under the control of a tertiary **semantide**. All molecules built by enzymes in the absence of a template are included because, although they do not express extensively the information contained in the semantide, they are a product of this information. See also **asemantic molecule**.

episome 1 a genetic element that is sometimes found in cells, especially those of bacteria, and that can replicate either when integrated into or independently of the host chromosome. **2** any fragment of DNA that exists in a cell as an extrachromosomal element. This DNA may or may not be replicated and passed to daughter cells. The **plasmid** is an example of an episome. DNA elements that use transposases involved in phage lambda integration are also examples of episomes.

epistasis the interaction between two or more genes at different loci such that phenotypic expression of one depends on expression of another. For example, a mutation preventing an early step in a biochemical pathway is epistatic to a mutation that prevents a step later in the pathway, because mutation of the first gene would mask expression of the second gene. –**epistatic** adj. –**epistatically** adv.

epitempin a protein of unknown function that is produced mainly in brain and other nervous tissue. It contains three leucine-rich repeats in the N-terminal region and seven specific repeats in the C-terminal region. Mutations in the gene at 10q24 produce an autosomal dominant form of familial lateral temporal epilepsy.

epithelium 1 (in animal anatomy) a sheet of closely packed cells, arranged in one or more layers, that covers the outer surfaces of the body or lines any internal cavity or tube (except the blood vessels, heart, and serous cavities). Compare **endothelium**, **mesothelium**. **2** (in plant anatomy) any of certain layers of parenchymal cells that line an internal cavity or tube. — **epithelial** adj.

epithin a type II transmembrane serine protease of mouse that is the orthologue of human **MT-SP1**.

epitope any immunological determinant group of an antigen. —**epitopic** adj.

epitope tagging the practice of adding an epitope to the N- or C-terminus of proteins expressed in prokaryotic or eukaryotic cells. Antibodies to epitope tags such as 6xHis or c-myc can then be used to detect the intracellular location of the expressed protein, or to purify it and any other proteins that might be bound to it in immunoprecipitation experiments.

epitype any class or group of related epitopes.

EPO abbr. for **erythropoietin**

epoprostenol international nonproprietary name for prostaglandin I_2; see **prostacyclin**.

epothilone see **taxol**.

epoxidation the formation of an **epoxide** by addition of an oxygen atom as a bridge across the double bond of an alkene.

epoxide any compound where molecules contain an epoxy group attached to adjacent carbon atoms to form a saturated three-membered ring system.

epoxy+ prefix (in chemical nomenclature) indicating the presence in a molecule of an oxygen atom directly attached (by single covalent bonds) to two (adjacent or nonadjacent) carbon atoms of a carbon chain or ring system. Epoxy compounds are cyclic ethers.

epoxyeicosatrienoate abbr.: EpETrE or (formerly) EET; any of various metabolites of arachidonate that contain an epoxy group. They are formed by the action of cytochrome P450 – this occurs in numerous tissues and has a high degree of stereospecificity. The products are predominantly present esterified to membrane-bound phosphatidylcholine. They have a wide spectrum of biological activities, including stimulation of peptide hormone release, inhibition of Na^+,K^+-ATPase, mobilization of microsomal Ca^{2+}, and inhibition of prostaglandin-endoperoxide synthase. Examples are (all-Z)-5S,6R-epoxyeicosa-8,11,14-trien-1-oate and (all-Z)-11S,12R-epoxyeicosa-5,8,14-trien-1-oate. EpETrEs can be further metabolized to *vic*-diols, diepoxides, and other products.

(all-Z)-5S,6R-epoxyeicosa-8,11,14-trien-1-oate

(all-Z)-11S,12R-epoxyeicosa-5,8,14-trien-1-oate

epoxy resin or **epoxide resin** any of a class of synthetic resins, containing ether linkages and epoxy groups, formed by the copolymerization of an epoxide (e.g. 1-chloro-2,3-epoxypropane) and a polyphenol (e.g. 2,2-bis(4-hydroxyphenyl)propane). When mixed with certain accelerators, e.g. an amine or an anhydride, such resins produce hard, clear, resistant, thermosetting polymers that are used as adhesives and coatings and for embedding materials for microscopy, especially electron microscopy.

Eppendorf pipette any of a range of push-button, plunger-operated, precalibrated, dispensing micropipettes with disposable polypropylene tips. [After the name of the manufacturer.]

Eppendorf tube the first polypropylene microcentrifuge tube, introduced by the company that became Eppendorf AG.

EPPS an alternative name for **Hepps**.

EPR or **epr** abbr. for electron paramagnetic resonance spectroscopy.

epsilon symbol: ε (lower case) or E (upper case); the fifth letter of the Greek alphabet. For uses see **Appendix A**.

epsilon chain or **ε chain** the **heavy chain** of IgE **immunoglobulin** molecules.

epsin a protein involved in vesicle budding in the plasma membrane. It binds to clathrin, synaptojanin, and adaptor protein-2.

EPSP abbr. for 5-enolpyruvylshikimate phosphate; an intermediate in aromatic amino-acid biosynthesis by the shikimate pathway (see **shikimate**).

EPSP synthase an alternative name for **3-phoshoshikimate 1-carboxyvinyltransferase**. See also **shikimate**.

Epstein–Barr virus human (gamma) herpesvirus 4; a DNA virus, first isolated from specimens of tumour tissue obtained from African children affected by Burkitt's lymphoma, that causes infectious mononucleosis (glandular fever) in young adult humans. [After (Sir) Michael Anthony Epstein (1921–), British pathologist and immunologist, and Yvonne M. Barr.]

eq. or **Eq.** abbr. for equivalent (def. 4).

equation of state any equation connecting the pressure, volume, and temperature of a substance or a mixture.

equatorial 1 of, relating to, or being an equator; located at or near an equator. Compare **axial** (def. 1). **2** (in stereochemistry) symbol: e; see **conformation**.

equi+ comb. form denoting equal or equality.

equi-effective dose ratio abbr.: EDR; the ratio of the doses of test and reference substances that produce the same biological effect (whether activation or inhibition).

equi-effective molarity ratio abbr.: EMR; the ratio of the molarity of test and reference substances that produce the same biological effect (whether activation or inhibition).

equilenin the trivial name for 3-hydroxyestra-1,3,5,7,9-pentaen-17-one; a weakly estrogenic steroid hormone isolated from the urine of pregnant mares. Compare **equilin**.

equilibrate to bring to equilibrium, or cause (something) to be in equilibrium (with some other thing), especially with the environment or with a (specified) system. —**equilibration** n.

equilibrium 1 a state of balance between or among opposing forces or processes that results in the absence of net change. **2** (in chemistry) a state of dynamic balance in a reversible chemical reaction when the reaction velocities in both directions are equal.

equilibrium assay a type of **radioimmunoassay** in which the reagents used are allowed to reach thermodynamic equilibrium. Compare **disequilibrium assay**.

equilibrium constant symbol: K; an expression of the position of the **equilibrium** (def. 2) of a reversible chemical reaction under specified physical conditions (e.g. temperature, pressure, nature of solvent, ionic strength, etc.). For a reaction of the generalized form:

$$aA + bB + ... \rightleftharpoons mM + nN + ...$$

when the thermodynamic **activity** of each of the components A, B, M, N, etc. is written as (A), (B), (M), (N), etc., and a, b, m, n, etc.

are the respective numbers of molecular entities participating in the reaction, the equilibrium constant is given by:

$$K = [(M)^m (N)^n ...]/[(A)^a (B)^b ...]$$

Frequently in biochemical systems it is not possible to evaluate the activities of all the components; the **concentration equilibrium constant** (symbol: K_a) rather than the true or **thermodynamic equilibrium constant**, K being expressed in terms of the molarities of each of the components in place of their activities. When the specified concentration of a component includes more than one chemical species (e.g. ionized plus un-ionized forms, or chelated plus unchelated metal ion) in unknown proportions, an **apparent (concentration) equilibrium constant** (symbol: K′) constrained with respect to certain variables (e.g. pH, or total concentration of metal ion), is written. See also **affinity constant**, **association constant**, **dissociation constant**, **pK**.

equilibrium density-gradient centrifugation a method for the separation of cells, cell organelles, macromolecules, or other particles of different densities by centrifugation in a solution that increases in solute concentration, and hence in density, from the top to the bottom of the centrifuge tube. At equilibrium, particles of the same density are found as a band or zone in the density gradient. Compare **differential centrifugation**.

equilibrium dialysis a technique used to measure the binding of a **microsolute** or ions to a **macrosolute**. A solution of the macrosolute is placed inside a dialysis bag (through which the macromolecules will not pass); this represents phase α. The bag is suspended in a solution containing the microsolute (phase β). At equilibrium, any excess concentration of the microsolute in phase α is taken as evidence for binding. From a measurement of the excess concentration of the microsolute and knowledge of the concentration of the macrosolute the extent of the binding may be determined.

equilin the trivial name for 3-hydroxyestra-1,3,5(10),7-tetraen-17-one; a weakly estrogenic steroid hormone isolated from the urine of pregnant mares. Compare **equilenin**.

equimolar of equal **molarity**.

equimolecular containing equal numbers of molecular entities.

equinatoxin a pore-forming toxin, permitting passage of small molecules across membranes. It is produced by the sea anemone *Actinia tenebrosa*; it contains 150–250 amino acid residues. Equinatoxin is classified in the TC system under number 1.C.38.

equine of, relating to, or being a member of the family Equidae (horses, zebras, and asses); of, belonging to, characteristic of, obtained from, or relating to, a horse or horses; resembling a horse.

equiv. abbr. for equivalent (def. 3).

equivalence 1 the state of being equal, equivalent, or interchangeable. **2 equivalency. 3** the point or zone in a precipitin test at which antigen and antibody are present in optimal proportions for combination and precipitation.

equivalence factor symbol (for a chemical species X): $f_{eq}(X)$; a number pertaining to a given reacting component of a specified titrimetric reaction derived from consideration of the overall stoichiometry of the reaction. For a reaction of the following type: νA + νB → products, where νA and νB are the respective numbers of reacting entities, then the equivalence factor of species B is given by: $f_{eq}(B) = νB/νA$. When νA > νB and $f_{eq}(A)$ is taken as unity, then $f_{eq}(B)$ is unity or less than unity. In the case of a reaction that can be clearly identified as acid–base or oxidation–reduction, the equivalence factor must be related to one entity of titratable hydrons or of transferable electrons, respectively.

equivalence point or **stoichiometric point** or **theoretical end point** the point in a titration at which the amount of titrant added is chemically equivalent to the amount of substance titrated.

equivalency or **equivalence 1** the state of having equal valencies. **2** the state of having equal **equivalence factors**.

equivalent 1 equal or interchangeable in amount, importance, meaning, or value. **2** (in chemistry) having equal valencies. **3** something that is equivalent (to something else). **4** (in chemistry) **a** abbr.: equiv. or eq. or Eq.; a unit of amount of substance, defined as the entity of a chemical species that in a specified reaction would combine with, displace, or in any other appropriate way be equivalent to, one entity of titratable hydrons (in an acid–base reaction) or one entity of electrons (in an oxidation–reduction reaction). It is the arithmetic product of the **molar mass** of the chemical species and its **equivalence factor. b** *(formerly)* **equivalent weight.**

equivalent activity a measure of the radiation output of a real radiation source containing a particular radionuclide, defined as being equal to the **activity** (def. 1b) of a hypothetical point source of the same radionuclide that would give the same exposure rate at the same distance from the centre of the source. It is customarily used to express the strengths of most sources emitting high-energy gamma radiation.

equivalent ellipsoid of rotation the **ellipsoid of rotation** having the same volume as an actual hydrodynamic unit of a macromolecule in solution (which consists of the macromolecule and tightly bound solvent).

equivalent weight *abbr.*: equiv. wt.; an obsolete term defined as the weight in grams of an element, compound, or group that, in a specified reaction, will combine with or displace 8 grams of oxygen, or the equivalent weight of another chemical species. *See also* **equivalent** (def. 4).

ER *abbr. for* endoplasmic reticulum.

ERAD *abbr. for* ER-associated degradation; the process by which many structurally flawed proteins and orphan subunits are restricted from traversing the entire secretory pathway through the endoplasmic reticulum. Such proteins are degraded by proteasomes in the cytosol.

ER-associated degradation *see* ERAD.

erbA an oncogene (v-*erbA*) originally found in avian erythroblastosis virus; the corresponding protooncogene (c-*erbA*) encodes thyroid hormone receptor. The product of v-*erbA* requires the activity of other oncogenic agents for tumorigenesis. *See also* **erbB**.

erbB an oncogene (v-*erbB*) from avian erythroblastosis virus; the corresponding protooncogene (c-*erbB*) encodes the **epidermal growth factor receptor**. In tumours the extracellular ligand-binding domain is deleted, contributing to a system with a constitutively active receptor protein. The transforming protein from the virus is a polyprotein product from *gag*, *erbA*, and *erbB*. erbB is a family of highly homologous genes that become oncogenic due to gene amplification or overexpression of the mRNA. Of the products, erbB1 is the epidermal growth factor receptor; erbB2 is also called HER2 (*abbr. for* heregulin receptor 2)/neu; and erbB3 and erbB4 are also called Her3 and HER4, respectively. erbB1 and erbB2 are overexpressed (probably from gene amplification) in 20–40% of breast cancers, and erbB2 in ≈10% of stomach cancers. erbB4, on glial cell surfaces, interacts with neuregulin and participates in glial-mediated neuronal migration. In mice, loss of any of these four genes results in embryonal death and defects in brain, heart, lung, and gastrointestinal tract, depending on which gene is affected.

ercalcidiol *or (formerly)* **25-hydroxyergocalciferol** *the recommended trivial name for* (5Z,7E,22E)-(3S)-9,10(19),22-ergostatetraene-3,25-diol; a secosteroid with **vitamin D** activity. It is formed in the liver from **ercalciol** and converted to **ercalcitriol** in the kidney.

ercalciol *or* **ergocalciferol** *the recommended trivial name for* calciferol, vitamin D$_2$, (5Z,7E,22E)-(3S)-9,10-seco-5,7,10(19),22-ergostatetraen-3-ol; a secosteroid with **vitamin D** activity, obtained synthetically by UV-irradiation of **ergosterol**. It is the form of vitamin

D most commonly used as a dietary supplement and for treatment of vitamin D deficiency. *International nonproprietary name*: ergocalciferol.

ercalcitriol *or (formerly)* **1α,25-dihydroxyergocalciferol** *the recommended trivial name for* (5Z,7E,22E)-(1S,3R)-9,10-seco-5,7,10(19),22-ergostatetraene-1,3,25-triol; a secosteroid with **vitamin D** activity. It is the biologically active metabolite of **ercalcidiol** and **ercalciol**.

ERCC *abbr. for* excision repair cross complementing; any of 10 or so genes or their products that are involved in nucleotide excision repair. Mutation in any of them leads to xeroderma pigmentosum.

ERE *abbr. for* estrogen response element (*see* **response element**).

erepsin a once-supposed single enzyme secreted by the mucosa of the small intestine and thought to be responsible for completing the breakdown in the gut of partially digested proteins into amino acids. It is now known to be a mixture of **peptidases**.

erg the cgs unit of work, defined as the work done by a force of 1 dyne acting over a distance of 1 cm; equivalent to 10^{-7} J.

ERG a gene encoding a family of transcription factors and related to the protooncogene *ETS*; it is named from 'ets-related gene'. It is expressed in some tumour-derived cell lines.

ergastoplasm *a term formerly applied to* the basophilic, fibrillar structures seen especially in pancreatic secretory cells; now termed rough **endoplasmic reticulum**.

+ergic *suffix (in pharmacology)* indicating agonist activity on a receptor in a neuron. See, e.g., **adrenergic**.

ergo+ *(comb. form)* **1** *or (before a vowel)* **erg+** denoting work. **2** *or (before a vowel)* **ergot+** denoting obtained from **ergot** or derived from a substance occurring naturally in ergot.

ergocalciferol *the alternative recommended trivial name and the international nonproprietary name for* **ercalciol**.

ergoflavin the major pigment from **ergot**.

ergometrine *see* **ergonovine**.

ergonovine D-lysergic acid L-2-propanolamide; ergometrine; an **ergot alkaloid** noted for inducing sustained uterine contractions and used in the treatment of postpartum hemorrhage.

ergosome *(sometimes)* a **polyribosome**.

ergosterol *trivial name for* (22E)-ergosta-5,7,22-trien-3β-ol; (22E)-(24R)-24-methylcholesta-5,7,22-trien-3β-ol; a sterol found in ergot, yeast, and moulds. It is usually obtained industrially from yeast, which synthesizes it from simple sugars. It is the most important of the provitamins D. On irradiation with ultraviolet light it is converted to ercalciol (vitamin D_2).

ergot *the common name for* the sclerotial phase of the fungus *Claviceps purpurea*, which parasitizes the ovaries of cereals such as rye, and other grasses, leading to ergot-contaminated seeds. Ergot yields various medically useful alkaloids, including ergotamine and ergometrine, and is a source of lysergic acid. However, ingestion of contaminated seeds by livestock or humans can result in poisoning (**ergotism**).

ergot alkaloid any of a group of about 30 indole alkaloids obtainable from **ergot**. All are derivatives of **lysergamide** and are biosynthesized from L-tryptophan.

ergotamine 12′-hydroxy-2′-methyl-5′α-(phenylmethyl)ergotaman-3′,6′,18-trione; an **ergot alkaloid** with vasoconstrictor activity, used in the treatment of migraine.

ERIC *abbr. for* enterobacterial repetitive intergenic consensus sequence; any of a number of highly conserved partly palindromic sequences. They are reminiscent of **REPs** but are longer (120 bp) and are in extragenic sites.

ERK *abbr. for* from extracellular signal-regulated kinase; any of a sub-subclass (EC 2.7.–), protein kinases, also (more frequently) known as **MAP kinases**. ERK1 (*or* insulin-stimulated MAP2 kinase) phosphorylates MAP2 and myelin basic protein. ERK2 (*or* mitogen-activated protein kinase 2 *or* MAP kinase 2) is similar, as is ERK3 (*or* MAP kinase isoform p63).

ER α-mannosidase I EC 3.2.1.113; *other names*: ER Class 1 α1,2-mannosidase *or* Man$_9$GlcNAc$_2$-specific processing α-mannosidase; an enzyme that acts in the first step in trimming a single mannose from Man^{9}GlcNAc$_2$ oligosaccharides in glycoproteins that are destined for transport through the secretory pathway.

erp60 *see* calreticulin.

ER-resident protein any protein that is retained by the endoplasmic reticulum. Their retention is regulated by the heterotetramer of a phosphoprotein (**calnexin**), a phosphoglycoprotein (*see* **Ssr** α), and two glycoproteins (*see* **Ssr** β). *See also* **KDEL**.

error catastrophe the threshold at which the rate of error accumulation becomes deleterious. For example, it may be the point at which mutation becomes harmful, or at which the error rate in databases becomes damaging.

'error catastrophe' hypothesis a hypothesis stating that since the protein synthetic machinery of the cell is itself made of protein, any imprecision in protein synthesis will lead to the formation of protein synthetic machinery of lowered accuracy, which will lead in turn to the production of even less precise machinery. If this positive feedback of error into the protein synthetic machinery is suffi-

ciently high it will cause instability and an accelerating deterioration of the cell, resulting in cell death or loss of function.

error-prone repair the processes of DNA repair, following damage, that give rise to 'fixed' mutations. Whether or not a particular damaged site will eventually lead to a mutation is dependent both on the genetic constitution of the cell and on environmental factors. It has been suggested that such mutations are due to errors in filling gaps in daughter DNA strands.

erucic acid *the trivial name for* (*Z*)-docos-13-enoic acid (22:1 *n*-9). It is present in rapeseed oil. In rats, the incidence of myocardial lesions increases with the amount of the acid in the diet.

ERYF1 *see* GATA-1.

erythritol *i*-erythritol; *meso*-erythritol; *ms*-1,2,3,4-tetrahydroxybutane; *erythro*-1,2,3,4-butanetetrol; a polyol configurationally isomeric with (active) **threitol** but optically inactive through internal compensation. It occurs in certain fungi (where it appears to act as a storage carbohydrate), and is found also (as an ester of orsellinic acid) in some green algae and lichens. It is about twice as sweet as sucrose.

erythro- the configurational prefix to the systematic name of a polyhydric alcohol, especially of a monosaccharide, used to indicate the particular stereochemical configuration of a set of two contiguous CHOH groups that occurs in D- or L-erythrose; e.g. D-*erythro*-2-pentulose for D-ribulose. *Compare* **threo-**.

D-erythrose   D-erythro

erythro+ *or (before a vowel)* **erythr+** *or (sometimes)* **eryth+** *comb. form* **1** indicating red-coloured. **2** pertaining to red blood cells.

erythroblast any of the nucleated red blood cells that are intermediates in the formation of a nonnucleated erythrocyte from a hemocytoblast. —**erythroblastic** *adj.*

erythrocruorin any of a group of respiratory pigments of invertebrates in the range of 0.4–6.7 MDa and containing 30–400 heme groups per molecule.

erythrocuprein *an alternative name for* Cu/Zn **superoxide dismutase**.

erythrocyte a mature red blood cell; in mammals it is non-nucleated and lacks mitochondria. Erythrocytes contain, but are no longer capable of synthesizing, **hemoglobin** and they function in the transport of oxygen. Mammalian erythrocytes obtain energy from anaerobic glycolysis and also metabolize glucose via the phosphogluconate pathway. —**erythrocytic** *adj.*

erythrocyte sedimentation rate *abbr.*: ESR; the rate at which erythrocytes sediment as measured under defined conditions. An increased ESR is almost always indicative of organic disease and is associated with an increased concentration of fibrinogen, γ-globulin, α_2-macroglobulin, or γ_1-macroglobulin in the plasma.

erythrodextrin any of the larger **dextrins** that give a red colour with diiodine.

erythrogenesis *an alternative term for* **erythropoiesis**.

erythro-β-hydroxyaspartic acid an amino acid formed by PTM of an aspartyl residue at position 41 of the light chain of **protein C**.

erythroid **1** red or reddish in colour. **2** of or relating to **erythrocytes** or their precursors.

erythroid differentiation factor *see* inhibin.

erythrolabe *or* **erythropsin** *or* **porphyropsin** *See* opsin.

erythromycin any of several wide-spectrum macrolide antibiotics obtained from *Streptomyces erythreus*. Three components, A, B, and C, are produced during fermentation, the major component being erythromycin A. They inhibit protein synthesis in prokaryotes by binding to the 50S ribosomal subunit and inhibiting translocation. They are synthesized by multimodular nonribosomal peptide synthetases.

erythromycin resistance factor any of several proteins that confer resistance to the antibiotic **erythromycin**. There are three mechanisms of resistance: (1) erythromycin esterase (EC 3.1.1.), which is itself of two types (I and II); (2) a transmembrane export system (e.g., *erm* genes of *Arthrobacter* and certain Gram-positive bacteria); (3) rRNA (adenine-N^6-)-methyltransferase (EC 2.1.1.48), an enzyme that catalyses the reaction between *S*-adenosyl-L-methionine and rRNA to form *S*-adenosyl-L-homocysteine and rRNA containing N^6-methyladenine. For example, the protein from *Staphylococcus aureus* produces a dimethylation of the adenine residue at position 2058 in 23S rRNA, resulting in reduced affinity between ribosomes and all macrolide, lincosamide, and streptogramin B-type antibiotics.

erythron *a collective term for* the circulating **erythrocytes** and their precursors.

erythrophore a type of chromatophore, coloured red and found especially in some fishes and crustaceans. The pigment, a carotenoid, is contained in bright red granules, which normally are dispersed throughout the cytoplasm, but which aggregate rapidly in response to epinephrine leading to near discoloration of the cell.

erythropoiesis *or* **erythrogenesis** the formation of red blood cells, in bone marrow or elsewhere. —**erythropoietic** *or* **erythrogenetic** *adj.*

erythropoietic porphyria 1 *or* **congenital erythropoietic porphyria** an autosomal recessive abnormality of heme synthesis caused by deficiency of **uroporphyrinogen-III synthase** and characterized by excessive amounts of uroporphyrin I and coproporphyrin I in urine, and cutaneous photosensitivity. The onset and clinical severity are highly variable. Two cytosolic isoforms of the enzyme (each of 265 amino acids, called the 'housekeeping' and the 'erythroid-specific' isoforms) are derived from the gene (locus at 10q25.3) by alternative splicing. Many mutations lead to the deficiency. **2** an autosomal dominant disease caused by deficiency of the inner mitochondrial membrane enzyme **ferrochelatase**. It is characterized by excessive levels of protoporphyrin in erythrocytes, bone marrow, and plasma, cutaneous photosensitivity, and hepatobiliary complications. Many mutations in the gene (at 18q21.3) lead to the deficiency. *See* **porphyria**.

erythropoietin a 46 kDa mammalian glycoprotein **cytokine**, formed in the kidneys and liver of mammals, that stimulates cellular differentiation of bone-marrow stem cells at an early stage of erythropoiesis, accelerates the proliferation and maturation of terminally differentiating cells into erythrocytes, and maintains a physiological level of circulating erythrocyte mass. The protein is now prepared from cDNA expressed in CHO cells and other cells. It is now established as the accepted treatment of chronic renal failure. Its receptor is a type I membrane protein.

erythrose *the trivial name for* the aldotetrose *erythro*-tetrose; it has D and L enantiomers, which are respectively diastereoisomeric with those of threose. *See* **erythro-** for structure.

erythrosome an artificial membrane preparation in which detergent-extracted, glutaraldehyde-fixed erythrocyte ghosts are coated with phospholipid.

erythrulose *the trivial name for* the ketotetrose *glycero*-tetrulose; it has D and L enantiomers.

ES *abbr. for* enzyme–substrate complex.

esc a regulatory gene in *Drosophila* that controls expression of *BX-C* genes in embryos. It is named from extra sex combs.

ESCA *abbr. for* electron spectroscopy for chemical analysis. *See* **photoelectron spectroscopy**.

ES cell *abbr. for* embryonic stem cell.

Escherichia a genus of Gram-negative, rod-shaped, usually motile, chemoorganotrophic bacteria belonging to the family Enterobacteriaceae. It contains a single species, the colon bacillus, ***Escherichia coli***. [After Theodor Escherich (1857–1911), German physician, who (in 1885) isolated the colon bacillus and gave it its original name of *Bacterium coli commune*.]

Escherichia coli *or* **colon bacillus** *abbr.*: *E. coli*; the sole member of the bacterial genus *Escherichia* and arguably the most widely used experimental cell system in biochemistry and molecular biology. The cells are straight, round-ended rods, commonly 0.5–1 μm × 1–4 μm, and usually occurring singly or in pairs. The organism is present in the intestinal tract of humans and other animals and is common in soil and water. The numerous strains, some of which are en-

teropathogenic, are commonly distinguished serologically by their O, K, and H antigens. Strain O157 is a life-threatening enteropathogen with genes seemingly derived (at least in part) from *Shigella* spp. It produces both hemolysin and enterotoxin. *E. coli* takes part in bacterial conjugation and other forms of genetic transfer, and can be infected with some bacteriophages and plasmids. *See also* **toxin**.

Eschweiler–Clarke reaction *see* **reductive amination**.

esculetin 6,7-dihydroxycoumarin; an agent that selectively inhibits 5- and 12-lipoxygenase, without inhibiting prostaglandin-endoperoxide synthase.

ESEEM *see* **electron spin-echo envelope modulation**.

E-selectin *see* **selectin**.

eserine *an alternative name for* **physostigmine**.

E-site the deacylated tRNA binding site in ribosomes; in bacteria this involves **ribosomal proteins** of the large subunit and 23S rRNA.

eskimo 1 an *Arabidopsis thaliana* mutant that is constitutively freeze-tolerant and accumulates sucrose and other sugars (as is usual in the development of tolerance to low temperatures) and also proline.

ESR *or* **esr** *abbr. for* **1** electron spin resonance. **2** erythrocyte sedimentation rate.

essential 1 of or being the essence of something; inherent. **2** absolutely necessary, basic, fundamental, indispensable, vitally important. **3** necessary for the normal growth of an organism but either not synthesized by that organism or synthesized at an inadequate rate. **4** *(in pathology)* describing a clinical condition or disorder that is not attributable to any discernible cause; **idiopathic** (def. 1). **5** relating to or derived from an extract of a plant, especially as an essence.

essential amino acid any amino acid that cannot be synthesized by a cell or an organism in an amount corresponding to need. The extent to which a particular amino acid is essential in any one species of organism depends on the stage of development of the subject and its physiological state. The following L-amino acids are nutritionally essential for the maintenance of nitrogen equilibrium in an adult human: isoleucine, leucine, lysine, methionine, phenylalanine, threonine, tryptophan, and valine. In addition, arginine and histidine are essential in growing children. The initial letters of the L-amino acids needed for optimal growth of infants or rats may be remembered using the mnemonic 'Many Very Hairy Little Pigs Live In The Torrid Argentine'. *Compare* **nonessential amino acid**.

essential fatty acid any polyunsaturated fatty acid that cannot be synthesized by a cell or an organism in an amount corresponding to need or from any dietary precursor. For mammals, linoleic acid (*see* **linoleate**) and (9,12,15)-linolenic acid (*formerly collectively known as* **vitamin F**) are nutritionally essential. *Compare* **nonessential fatty acid**.

essential fatty acid deficiency a condition that arises from the absence of **essential fatty acids** in the diet. A deficiency of fatty acids of the *n*-6 family (e.g. linoleic acid) in the diet of young animals results during growth in decreased body weight and, after some months, in increased skin permeability and capillary fragility, and dermatitis. If prolonged into adulthood, poor litter sizes (e.g. in rats) and infertility can result. Deficiency of fatty acids of the *n*-3 family, in the presence of *n*-6 fatty acids, does not result in such obvious manifestations, partly because linoleic acid's fatty acids are tenaciously retained in tissues, and partly because *n*-6 fatty acids (e.g. 22:5 *n*-6) not normally present in significant amounts are synthesized and appear to compensate for the deficiency of *n*-3 fatty acids of similar length and degree of unsaturation. That the effects of deficiency are not solely due to lack of substrates for prostanoid synthesis is indicated by the ability of **columbinic acid** to relieve some of the effects.

essential fructosuria a benign asymptomatic disorder in which hyperfructosemia and fructosuria follow the ingestion of nutrients that contain fructose or sucrose. It is caused by hepatic fructokinase deficiency, the result of mutation in the gene at 2p23.2. *See* **fructokinase** (def. 2).

essential oil any of a group of volatile, odorous, natural oils obtained from plants. Such oils are usually benzene derivatives or **terpenes**, and are useful as flavouring agents or perfumes.

essential pentosuria *see* **pentosuria**.

EST *abbr. for* expressed sequence tag.

EST1 a gene involved in telomere maintenance, possibly coding for **telomerase**. It is named from 'even shorter telomeres'. It encodes a DNA-binding protein with RNA-directed DNA polymerase activity.

established cell line cultured cells of a single origin that have the potential to be subcultured indefinitely whilst maintaining stable characteristics.

eSte *symbol for* the eleostearoyl group, $CH_3-[CH_2]_3-[CH=CH]_3-[CH_2]_7-CO-$.

ester any organic compound formed, either actually or formally, by the elimination of the elements of water between a hydroxyl group of an **oxoacid** (def. 1) and a hydroxyl group of either an alcohol or a phenol. Esters formed by condensation of one, two, or three molecular proportions of alcohol or phenol with one of a tribasic oxoacid may be termed monoesters, diesters, or triesters, respectively.

esterase any enzyme that catalyses the hydrolysis of an ester. Esterases form subclass EC 3.1 of the class EC 3, hydrolases, and are divided according to the nature of their substrates into the following sub-subclasses: EC 3.1.1, carboxylic ester hydrolases; EC 3.1.2, thiolester hydrolases; EC 3.1.3, phosphatases, phosphoric monoester hydrolases; EC 3.1.4, phosphodiesterases, phosphodiester hydrolases, phosphoric diester hydrolases (other than nucleases; see EC 3.1.11–3.1.31 below); EC 3.1.5; triphosphoric monoester hydrolases; EC 3.1.6, sulfatases, sulfuric ester hydrolases; EC 3.1.7, pyrophosphatases, diphosphoric monoester hydrolases; EC 3.1.8, phosphoric triester hydrolases; EC 3.1.11, exodeoxyribonucleases producing 5′-phosphomonoesters; EC 3.1.13, exoribonucleases producing 5′-phosphomonoesters; EC 3.1.14, exoribonucleases producing other than 5′-phosphomonoesters; EC 3.1.15, exonucleases active with either ribo- or deoxyribonucleic acids and producing 5′-phosphomonoesters; EC 3.1.16, exonucleases active with either ribo- or deoxyribonucleic acids and producing other than 5′-phosphomonoesters; EC 3.1.21, endodeoxyribonucleases producing 5′-phosphomonoesters; EC 3.1.22, endodeoxyribonucleases producing other than 5′-phosphomonoesters; EC 3.1.25, endodeoxyribonucleases specific for altered bases; EC 3.1.26, endoribonucleases producing 5′-phosphomonoesters; EC 3.1.27, endoribonucleases producing other than 5′-phosphomonoesters; EC 3.1.30, endonucleases active with either ribo- or deoxyribonucleic acids and producing 5′-phosphomonoesters; EC 3.1.31, endonucleases active with either ribo- or deoxyribonucleic acids and producing other than 5′-phosphomonoesters.

esteratic of, by means of, or having the activity of an esterase.

esterification the act or process of forming an ester from an oxoacid by condensing it with an alcohol or a phenol, or vice versa.

esterify to effect or to undergo esterification.

ester value an alternative term for saponification number.

17β-estradiol *or (esp. Brit.)* **17β-oestradiol** 1,3,5-estratriene-3,17β-diol; the most potent of the major **estrogens**. It is synthesized by the ovary, from which it is secreted directly into the circulation. Its formation depends on the presence of both **luteinizing hormone** and FSHRH; it is synthesized from testosterone by the aromatase complex, which is localized in ovarian granulosa cells, and is stimulated by follitropin; lutropin, on the other hand, stimulates production of testosterone, the aromatase substrate, by ovarian thecal cells. Apart from being important during development in inducing female characteristics, 17β-estradiol is important during the first half of the menstrual cycle (i.e. before ovulation) in stimulating proliferation of the epithelial and stromal layers of the endometrium in preparation for ovulation; it also stimulates production by the liver of proteins that bind steroids and thyroxine. During pregnancy, synthesis of estrogens is greatly increased, most of this increase being due to

synthesis by the fetal adrenal gland of the precursors dehydroepiandrosterone and 16-hydroxydehydroepiandrosterone.

estramustine *or (esp. Brit.)* **oestramustine** estradiol 3-bis(2-chloroethyl)carbamate; an ester formed between estradiol and the carbonic-acid analogue of nitrogen mustard (mustine); it has antineoplastic activity.

estriol *or (esp. Brit.)* **oestriol** 16α-estradiol; 1,3,5-estratriene-3,16α,17β-triol; a relatively weak estrogen produced mainly during pregnancy. It is synthesized by the placenta from 16α-hydroxydehydroepiandrosterone sulfate, which is formed in fetal liver from dehydroepiandrosterone sulfate secreted by the fetal adrenal gland. Urinary estriol is measured to monitor fetal well-being in high-risk pregnancies. Low levels are also associated with the inherited X-linked disorder, placental sulfatase deficiency.

estrogen *or (esp. Brit.)* **oestrogen** any substance, natural or synthetic, that produces changes in the female sexual organs similar to those produced by 17β-estradiol, the principal hormone of the vertebrate ovary. Naturally occurring estrogens are C_{18}-steroid hormones; they are produced also by the placenta, adrenal cortex, and testis, as well as by many species of plants. *Compare* **androgen**. —**estrogenic** *or* **oestrogenic** *adj.*

estrone *or (esp. Brit.)* **oestrone** 1,3,5-estratrien-3-ol-17-one; a major estrogen, also called folliculin, synthesized and secreted by the ovary, in which it is formed from androstenedione by the aromatase complex. However, estrone is mainly synthesized in peripheral tissues, from androstenedione taken up from the circulation. Its actions are similar to those of **17β-estradiol**.

estrone sulfotransferase EC 2.8.2.4; an enzyme involved in the inactivation of estrogens. It catalyses the reaction of 3′-phosphoadenylylsulfate with estrone to form the inactive estrone 3-sulfate with release of adenosine 3′,5′-bisphosphate. Example from *Bos taurus*: SUOE_BOVIN, 295 amino acids (34.60 kDa).

estrophile *(sometimes)* estrogen receptor.

estrous cycle *or (esp. Brit.)* **oestrous cycle** the hormonally controlled cycle of activity of the reproductive organs of nonpregnant, sexually mature females of many species of mammal. It may be continuous or seasonal, depending on species. There are four principal phases. (1) **Follicular phase** *or* **estrus** – ovarian Graafian follicles mature, estrogen secretion by the ovary is at its highest, and the endometrium proliferates; the phase culminates in ovulation; in seasonal animals it is the time of sexual receptivity. (2) **Luteal phase** *or* **metestrus** – the corpus luteum develops in the ovary, estrogen secretion diminishes, and progesterone secretion reaches its maximum. (3) **Diestrus** – the corpus luteum and endometrium regress, new follicular growth begins, and progesterone secretion diminishes. (4) **Proestrus** – the corpora lutea involute, Graafian follicles re-emerge, and gonadal hormone secretion is at its lowest. The cycle then recommences.

estrus *or (esp. Brit.)* **oestrus** the first phase of the **estrous cycle**. —**estrous** *or (esp. Brit.)* **oestrous** *adj.*

esu *or* **ESU** *abbr. for* electrostatic unit.

Et *symbol for* the ethyl group, $CH_3–CH_2–$.

ET *abbr. for* endothelin.

eta 1 *symbol*: η (lower case) *or* H (upper case); the seventh letter of the Greek alphabet. For uses see **Appendix A**. **2** *(in chemical nomenclature)* *a less common alternative name for* **hapto** .

ET cloning a method for engineering DNA in *E. coli* using homologous recombination catalysed by RecE and RecT recombinases.

ETF *abbr. for* electron-transfer flavoprotein *or* electron-transferring flavoprotein.

ethambutol (+)-*N*,*N*′-bis(1-hydroxymethylpropyl)ethylenediamine; an antitubercular drug.

ethanol *or (formerly)* **ethyl alcohol** $CH_3–CH_2–OH$; a colourless, water-miscible, flammable liquid. It is produced by **alcoholic fermentation** and is thereby probably the single most important product of bioindustry in economic terms. However, most ethanol (not destined for human consumption) is now manufactured from ethylene as a by-product in the petroleum industry. Commonly known as alcohol.

ethanolamine-phosphate cytidylyltransferase EC 2.7.7.14; *other name*: phosphorylethanolamine transferase; an enzyme that catalyses the reaction of CTP and ethanolamine phosphate to form **CDPethanolamine** with release of pyrophosphate.

ethanolaminephosphotransferase EC 2.7.8.1; an enzyme of the pathway for *de novo* synthesis of phosphatidylethanolamines. It catalyses the reaction of CDPethanolamine with 1,2-diacylglycerol to form phosphatidylethanolamine with release of CMP. *Compare* **phosphatidylserine decarboxylase**.

ethanol fermentation *an alternative term for* **alcohol fermentation**.

ethanolic of, containing, or derived from ethanol; dissolved in pure or aqueous ethanol.

ethanol precipitation a common method for concentrating DNA from dilute solutions. Addition of two volumes of ethanol to a DNA solution containing 0.3 mol L^{-1} sodium acetate causes all the DNA to be precipitated.

ethanolysed cellulose a powdered and highly purified form of cellulose, formerly in use as a supporting medium in preparative zone electrophoresis of proteins, e.g. serum proteins. It is prepared by heating fibrous cotton (cotton wool) under reflux for approximately 24 h with acid-ethanol (absolute ethanol made 1 molar with respect to HCl either by introduction of dry hydrogen chloride gas or by reaction with acetyl chloride). By this means, partial alcoholysis of the cellulose occurs and acidic groups are neutralized by esterification, the cotton fibres being partially degraded into nonclumping particles; at the same time, light-absorbing and other interfering impurities are extracted from the cellulose.

ether 1 any anhydride, of general formula $R^1–O–R^2$, formed between two (identical or nonidentical) organic hydroxy compounds.

2 *the common name for* **diethyl ether** (*or* ethyl ether *or* ethoxyethane), $C_2H_5–O–C_2H_5$.

ethereal *(in chemistry)* of, containing, or dissolved in an ether, especially diethyl ether.

ethereal sulfate *an obsolete term for* any biogenic half-ester of sulfuric acid with an aromatic hydroxy compound. Ethereal sulfates are found in the urine as end products of sulfur metabolism and metabolism of some aromatic compounds.

ether lipid any lipid that contains (normally) one fatty alcohol in ether linkage to one of the carbon atoms (normally C-1) of glycerol. They are found in low concentrations in glycerophospholipid fractions, the constituent at C-2 of the glycerol moiety then being an acyl group. **Platelet-activating factor** is an important physiologically active ether glycerophospholipid. Ether linkages in neutral acylglycerols are known, and contain one fatty alcohol linked to glycerol by an ether linkage (usually at C-1), with fatty acids in ester linkage at the other two positions. The ether linkage resists hydrolysis, and typical hydrolysis products are the glyceryl ethers such as **batyl alcohol**, **chimyl alcohol**, and **selachyl alcohol**.

ethernet low-level protocols and associated hardware such as cabling for computer communication.

ethernet address *an alternative name for* **machine address**.

ethidium the 3,8-diamino-5-ethyl-6-phenylphenanthridinium cation, usually used as the bromide salt. It binds by **intercalation** to double-stranded regions of DNA and RNA molecules with marked enhancement of fluorescence. It is used as a trypanocide and to reveal double-stranded DNA and RNA in gel electrophoresis.

ethinylestradiol *the international nonproprietary name for* ethynylestradiol; 17α-ethynyl-1,3,5(10)-estratriene-3,17β-diol]; (17α)-19-norpregna-1,3,5(10)-trien-20-yne-3,17-diol; a synthetic steroid with potent estrogenic activity, commonly used in oral contraceptive preparations.

ethionine *S*-ethyl-L-homocysteine; a synthetic, carcinogenic analogue of methionine. Its toxicity is due to the fact that it competes with methionine for the enzyme **methionine adenosyltransferase**, with the result that *S*-adenosylmethionine is formed. This leads to a deficiency of ATP.

ethoxy *symbol*: OEt *or* Eto; the alkoxy group, $CH_3–CH_2–O–$, derived from ethanol by loss of a hydrogen atom.

ethyl *symbol*: Et; the $CH_3–CH_2–$ alkyl group, derived from ethane.

ethyl alcohol *the former name for* **ethanol**.

ethylene $CH_2=CH_2$; a colourless gas and the first member of the olefin (or alkene) series of hydrocarbons. It occurs naturally as a phytohormone, with a number of physiological effects, the most important being stimulation of ripening of fleshy fruit and stimulation of abscission of leaves. It is synthesized by the action of **ACC synthase**, which converts the α-aminobutyrate moiety of *S*-adenosylmethionine to 1-aminocyclopropane-1-carboxylic acid (ACC), followed by ACC oxidase (or **ethylene-forming enzyme**), which converts ACC to ethylene.

ethylenediaminetetraacetate *see* **EDTA**.

ethylene-forming enzyme *abbr.*: EFE; *see* **ACC oxidase**.

ethylene receptor 1 *abbr.*: ETR1; a single-pass transmembrane protein (75 kDa) of *Arabidopsis thaliana* that, as a dimer, binds ethylene and elicits its effects. Its intracellular C-terminal region is homologous to the bacterial two-component system hybrid kinase. Other receptors include: ethylene response sensor (ERS); never ripe

(Nr), which is developmentally regulated in tomato; and LeTae1 (a homologue of ETR1 of tomato), which is expressed during flower and fruit senescence.

ethylenic containing one or more aliphatic carbon–carbon double bonds, as in **ethylene**.

N-ethylmaleimide *abbr.*: NEM; an agent that reacts irreversibly with thiol groups in proteins; it is useful as an inhibitor in transport studies and of thiol enzymes.

N-ethylmaleimide-sensitive fusion protein *abbr.*: NSF protein; a homotetrameric cytoplasmic protein required for vesicle-mediated transport, and transport from the endoplasmic reticulum to the Golgi stack. *See also* **SNAP** (def. 3).

ethyne, C_2H_2, $CH_2{\equiv}CH_2$, former name acetylene. A substrate used in the assay of **nitrogenase**.

ethynylestradiol *see* **ethinylestradiol**.

ETI *see* **eicosatriynoic acid**.

etidronate any anion of etidronic acid, (1-hydroxyethylidene)bisphosphonic acid, an analogue of diphosphoric acid in which the latter's bridging oxy group is replaced by a 1-hydroxyethylidene group. Disodium etidronate is useful clinically for reducing the turnover rate of bone, e.g. in osteoporosis or in Paget's disease of bone.

etidronic acid

etiology *or (esp. Brit.)* **aetiology** the cause or causes of disease and their study.

Etn *symbol for* a residue of ethanolamine, $H_2N{-}CH_2{-}CH_2{-}O{-}$.

EtO *or* **OEt** *symbol for* the ethoxy group, $CH_3{-}CH_2{-}O{-}$.

ES cell *abbr. for* embryo-derived stem cell.

Etoposide a semisynthetic lignan derivative prepared from **podophyllotoxin**. It is useful as an antitumour agent in multidrug therapies, exerting its effect probably through interaction with topoisomerase II, which it inhibits; it also inhibits nucleoside transport in mammalian cells.

ETS *abbr. for* expression tagged site; a jargon phrase from genome mapping projects, etc. meaning a DNA database code (usually short) corresponding to an actively transcribed gene (i.e. cDNA from mRNA) of known sequence but, in general, of unknown func-

tion. Large ETS libraries are used in the human genome project and similar projects on other organisms.

ETS a protooncogene related to *v-ets*, one of the oncogenes of the acutely transforming avian erythroblastosis virus E26; is named from 'E twenty-six specific'. It encodes a family of transcription factors: Ets1 activates stromelysin and collagenase genes and binds to the enhancer sequence of the TCRα gene; it is expressed in tumours of the peripheral nervous system and in Ewing's sarcoma; Ets1 and Ets2 are kinase substrates.

ETYA 5,8,11,14-eicosatetraynoic acid (*see* **eicosatetraynoic acid**).

eu+ *prefix* **1** indicating well, pleasant, good. **2** indicating normal, true, typical. In medicine, sometimes used interchangeably with **normo+**.

e.u. *symbol for* entropy unit.

Eu *symbol for* europium.

EU *abbr. for* enzyme unit (obsolete).

Eubacteria *a name suggested for* one of three major lineages of cellular organisms, the others then being **Archaebacteria** and **eukaryotes**, defined when organisms are classified on the basis of sequence homologies of their ribosomal RNA. It comprised all typical bacteria, the major subdivisions being the cyanobacteria, the Gram-positive bacteria, and the Gram-negative bacteria. This classification has now been superseded: *see* **Archaea**, **Bacteria**, and **Eukarya**.

eucaryon *a variant spelling of* **eukaryon**.

eucell any cell of a **eukaryote**.

euchromatin the dispersed less dense form of **chromatin** in the interphase **nucleus**. About 10% is in the form of transcriptionally active chromatin, which is the least condensed, while the rest is inactive euchromatin which is more condensed than active chromatin but less condensed than **heterochromatin** which represents about 10% of the chromatin. *See* **nucleosome code**.

eucollagen **1** a highly modified form of **collagen** that can be easily and completely transformed into gelatin. **2** a hypothetical limiting structure for collagen in which Gly occurs at intervals of three residues and Hyp at intervals of ten residues; it is considered to be responsible for the 28.6 repeat distance in X-ray diffraction patterns.

eugenics *(functioning as sing.)* the study of methods of improving the hereditary characters of the human race, especially by selective breeding. —**eugenic** *adj.*

Euglena a genus of fusiform, photosynthetic protozoa (sometimes classified as algae) of the division Euglenophyta. They are useful in biochemical research as unicellular photosynthetic organisms.

euglobulin any 'true' globulin of blood plasma that is soluble in isotonic salt solutions, insoluble at low ionic strengths (or in pure water), and precipitated by addition of saturated ammonium sulfate solution to a final concentration of 33%. *Compare* **pseudoglobulin**.

euglycemia *or* **normoglycemia** the condition or state in which the blood glucose level is within the normal range. *See also* **glycemia**. —**euglycemic** *adj.*

eukarya one of the three major kingdoms of cellular organisms, the others being archaea and bacteria. The eukarya are characterized by the presence of a **eukaryon**.

eukaryon *or (sometimes)* **eucaryon** (*pl.* **eukarya** *or* **eucarya**) the type of cell nucleus bounded by a nuclear membrane and containing true chromosomes. It is characteristic of all multicellular and unicellular organisms except bacteria, actinomycetes, and cyanobacteria. *Compare* **prokaryon**. —**eukaryous** *or* **eucaryous** *adj.*

eukaryosis *or (sometimes)* **eucaryosis** the condition of possessing a **eukaryon**. —**eukaryotic** *or* **eucaryotic** *adj.*

eukaryote *or (sometimes)* **eucaryote** any organism whose cells contain a **eukaryon** (or eukarya) and undergo meiosis. *Compare* **prokaryote**.

eukaryotic porin *see* **porin**.

eumelanins *see* **melanin**.

eupeptide *a name suggested for* any physiologically active gastrointestinal polypeptide.

eupeptide bond *a name suggested for* a peptide bond formed specifically between C-1 of one amino-acid residue and N-2 of another. *Compare* **isopeptide bond**.

euploid having a chromosome number that is an exact multiple of the monoploid number; i.e. each chromosome of the set is present in the same number. *Compare* **aneuploid**. —**euploidy** *n.*

euthyroid having a normally functioning thyroid gland and normal levels of thyroid hormones.

eutopic binding *(sometimes) an alternative term for* **productive binding.**

eutrophic describing a habitat, esp. a lake or other mass of water, that contains abundant nutrients, both organic and inorganic, and hence favours excessive growth of aerobic plants and microorganisms. Such growth results in depletion of dissolved dioxygen and, ultimately, extinction of aerobic life. *Compare* **dystrophic** (def. 2), **oligotrophic.**

eutrophication the process of becoming or rendering eutrophic.

eV *symbol for* electronvolt.

e-value *abbr. for* expectation value.

E(var) *abbr. for* enhancing variegation; any of a group of genes in *Drosophila* and yeast that enhance euchromatin/heterochromatin variegation in the higher structure of chromatin. They include genes for components of ATP-dependent nucleosome remodelling complexes, such as SW1/SNF, and act in opposition to **Su(var)** gene products.

EVD *abbr. for* extreme value distribution.

event marker a feature of some chart recorders by means of which a mark is made on the chart to record the occurrence of a specifiable event or series of events, e.g. the stepwise operation of a fraction-collecting device.

EVH1 domain *abbr. for* enabled, VASP, Homer1 domain; a protein homology sequence present in the N-terminal region of *Drosophila* proteins enabled, VASP, Homer 1, and WASP. It is similar in structure, but not in sequence, to the pleckstrin homology domain, and interacts with a proline-rich motif frequently present in proteins involved in remodelling of the actin cytoskeleton.

evolution 1 *(in biology)* the process of cumulative change occurring in the form and mode of existence of a population of organisms in the course of successive generations related by descent. **2** *(in chemistry)* the liberation of a gas, or of heat, in the course of a chemical reaction. —**evolutionary** *or* **evolutional** *adj.*

evolve 1 to develop gradually; to cause to develop. **2** *(in biology)* to undergo **evolution** (def. 1). **3** *(in chemistry)* to give off a gas or heat.

ex+ 1 *prefix* indicating out of, away from, outside of; without, lacking; former. **2** *comb. form* a variant of **exo+** *(sometimes before a vowel)*.

exa *symbol*: E; the SI prefix denoting 10^{18} times.

EXAFS *abbr. for* extended X-ray absorption fine structure (spectroscopy).

excelsin a crystallizable globulin of ~290 kDa obtained from the seed of the brazil-nut tree, *Bertholettia excelsa*. Structurally, it resembles arechin, edestin, and glycinin.

exchange 1 to cause two different objects to change places with each other. **2** the act or process of exchanging. **3** *(in genetics)* the reciprocal exchange of chromatid segments between chromosomes at meiosis or mitosis that results in genetic recombination.

exchange diffusion the exchange of ions or molecules across a (biological) membrane; it can be by simple diffusion or be carrier-mediated.

exchange labelling the catalysed exchange of one nuclide for another radioactive or stable nuclide of the same element in a chemical compound in order to produce an isotopically labelled form. The procedure is useful esp. in the preparation of tritium-labelled compounds.

exchanger a substance upon which, or a device by which, an **exchange** (def. 2) can be effected; e.g. an ion exchanger, or a heat exchanger.

exchange reaction a chemical reaction that results in the production of a substance that is chemically identical with the starting material. It is not detectable by traditional chemical analysis but (often) is demonstrable with the help of (radio)isotopes.

excimer an adduct formed between a molecular entity that has been excited by a photon and an identical unexcited molecular entity; the adduct exists until it fluoresces by emission of a photon. An excimer is recognized by the production of a new fluorescent band at a longer wavelength than that of the usual emission spectrum. Its for-

mation is distinguishable from resonance energy transfer in that the excitation spectrum is identical with that of the monomer.

excitation: $A + photon \rightarrow A^*$

excimer formation: $A^* + A \rightarrow D^*$

excimer fluorescence: $D^* \rightarrow A + A + photon$

excinuclease a DNA-repair endonuclease. An example is the ABC excision nuclease of *Escherichia coli*; this comprises three subunits, named from the corresponding *uvr* genes. UvrA is an ATPase and DNA-binding protein that preferentially binds single-stranded or UV-irradiated double-stranded DNA; it belongs to the **ABC transporter** family. UvrB stimulates the activities of UvrA. UvrC is a nuclease component.

excisase an enzyme involved in lambda phage induction. *See* **integrase.**

excise to cut out or remove, as of a tumour, organ, or part of a linear polymer.

excision 1 the act or process of excising, especially the enzymic removal of an oligonucleotide segment from a nucleic-acid molecule. **2** *(in genetics)* the separation of one or more replicons from a cointegrate.

excisionase *see* **Xis protein.**

excision repair one of the intracellular mechanisms for the repair of DNA lesions (single-strand breaks, damaged bases, etc.). It occurs in the following stages: (1) recognition of the damaged region; (2) removal of the damaged oligonucleotide by two enzymic nucleolytic reactions (excision); (3) synthesis by DNA polymerase of the excised oligonucleotide using the second (intact) DNA strand as template; and (4) covalent joining by DNA ligase of the newly synthesized segment to the existing ends of the originally damaged DNA strand. The process is light-independent. *Compare* **photoreactivation**, **post-replication repair**. *See also* **short-patch repair.**

excitability proteins *a collective name for* intrinsic membrane proteins that are ion channels, receptors, and ion pumps.

excitable capable of responding to a stimulus; used especially of a biological membrane, neuron, or other living matter. —**excitability** *n.*

excitable membrane any biological membrane capable of responding to a specific chemical or physical stimulus. Such membranes are commonly characterized by: (1) the presence of a highly specific, integral protein receptor; (2) the occurrence of a conformational change in the receptor in response to the specific stimulus with a consequential change in membrane permeability or in the activity of a membrane-bound enzyme; and (3) reversibility of the alterations in conformation of the receptor and functional activity of the membrane.

excitation–contraction coupling a general term often used to refer to the coupling of an excitatory stimulus to the contraction of muscle; the process by which the fibres of a muscle are caused to contract by the stimulation of a neuron.

excitation spectrum any **spectrum** of electromagnetic radiation that, when applied to a sample, causes the atoms and molecules of the sample to become excited by the absorption of energy and, when reverting to their normal energy levels, to emit radiation of different frequencies to that of the absorbed radiation. *Compare* **absorption spectrum, emission spectrum.**

excitatory able to excite; stimulatory.

excitatory amino-acid receptor a receptor for any excitatory amino-acid in the central nervous system. These comprise two main groups: ionotropic, regulating an intrinsic ion channel, or metabotropic, coupled to either the cAMP or inositol phospholipid second messenger systems. The three main types of ionotropic receptors all have four transmembrane domains and are characterized pharmacologically by selective agonists: *N*-methyl-D-aspartate (NMDA), α-amino-3-hydroxy-5-methyl-isoxazole (AMPA), and kainate. NMDA receptors have a further binding site for glycine, which modulates receptor activation. All such ionotropic receptors regulate Na^+ and K^+ conductance; NMDA receptors additionally regulate Ca^{2+} conductance. seven subtypes of **metabotropic** receptor have been cloned, all with seven transmembrane domains (mGlu1 to mGlu7): mGlu1 and mGlu5 modulate inositol phospholipid turnover while mGlu2, mGlu3, mGlu4, mGlu6, and mGlu7 are all negatively coupled to adenylate cyclase.

excite 1 to raise a molecular entity from its ground state to an excited state by the input of energy. **2** to cause an increase in activity or other response in an organism or part of an organism. —**excitation** *n*.

excited state any state of a molecular entity that has a higher energy level than its ground state.

exciter any device that excites, or raises a substance to an excited state.

exciton any nonconducting nonlocalized neutral entity that occurs in semicrystalline semiconductors, consisting of an excited electron bound to a positive hole, the combination forming a concentration of energy with certain properties characteristic of a particle.

exciton splitting the splitting of electronic absorption bands of a population of identical chromophores that are arranged in space such that sets of two or more are in close proximity. The splitting arises from electronic interactions between excited and nonexcited chromophores within the sets.

exclusion chromatography *an alternative name for* **permeation chromatography**.

exclusion limit an attribute of a specific gel used in **gel-permeation chromatography**. It is the relative molecular mass of the largest molecule of a particular shape that can be effectively fractionated by its use.

exclusion principle *see* **Pauli exclusion principle**.

exclusion reaction the reaction of a phage-infected bacterium that prevents the entry of additional phages. It is brought about by the strengthening of the bacterial cell envelope.

exclusion volume the volume of solvent retained by the column in **gel-permeation chromatography** that is outside the gel particles. *Compare* **void volume**.

excrete to eliminate waste materials (from a body or organism). —**excretion** *n*.; **excretive** *or* **excretory** *adj*.

exendin any of several peptides belonging to the glucagon family that have been isolated from *Heloderma* spp.: exendin-1 (*or* **helospectin I**), exendin-2 (*or* **helodermin**), and exendin-4 from *Heloderma suspectus*; and exendin-3 from *Heloderma horridus horridus*. They have biological activity similar to vasoactive intestinal peptide and secretin. Exendin-4 shares 53% similarity in its primary structure with glucagon-like peptide 1 (GLP-1) (7–36) amide; the latter binds to exendin receptors on pancreatic acinar cells. Exendin (9–39) (from exendin-3 or -4) is a GLP-1 receptor antagonist that blocks the inhibitory effect of GLP-1 on food intake of fasted rats. Exendin-4 is

HGEGTFTSDLSKQMEEEAVRLFIEWLKNGGPSSGAPPPS

Exendin-3 is similar except for residues 1–3, which are HSD.

exergonic describing any process or reaction that can produce work; i.e. one that, at constant pressure and temperature, results in a negative change in free energy content. *Compare* **endergonic**.

exhaustive methylation the maximal methylation of all groups in a substance that are capable of being methylated; it is useful in structural studies of carbohydrates, alkaloids, etc.

exhaustive tree search *(in phylogeny)* a method that systematically examines all possible relationships between taxa or sequences. Its time-consuming nature renders it infeasible for data sets much larger than ~10.

exo+ *or (sometimes before a vowel)* **ex+** *comb. form* indicating outside, external, outer. *Compare* **endo+**. *See also* **exo-**.

exo- *prefix (in stereochemistry) see* **conformation**.

exocellular describing cellular components that are external to the cell membrane but attached to its outer surface. *Compare* **extracellular**.

exocrine 1 describing or relating to a gland or other group of cells that discharges its secretion through a duct. **2** relating to a secretion of exocrine tissue. **3** a product of exocrine tissue. *Compare* **endocrine**.

exocytic 1 situated outside a living cell and not belonging to the cell itself; extracellular. **2** *an alternative term for* exocytotic (*see* **exocytosis**).

exocytose to effect exocytosis. —**exocytosed** *adj*.

exocytosis the discharge by a cell of intracellular materials to the exterior by **emiocytosis** or reverse **pinocytosis**. *Compare* **endocytosis**, **transcytosis**. *See also* **externalize**. —**exocytic** *or* **exocytotic** *adj*.

exodeoxyribonuclease *see* **deoxyribonuclease**.

exodeoxyribonuclease I EC 3.1.11.1; *other names*: exonuclease I; *Escherichia coli* exonuclease I. An enzyme that catalyses exonucleolytic cleavage of DNA in the 3′- to 5′-direction, releasing 5′-phosphomononucleotides. It shows a preference for single-stranded DNA.

exodeoxyribonuclease V EC 3.1.11.5; *other names*: exonuclease V; *Escherichia coli* exonuclease V; RecBC. An enzyme that catalyses exonucleolytic cleavage of DNA (in the presence of ATP) in either the 5′- to 3′- or the 3′- to 5′-direction, to yield 5′-phosphooligonucleotides. The subunits (products of genes *recA*, *recB*, and *recC*) are RecA, RecB, and RecC.

exodeoxyribonuclease VII EC 3.1.11.6; *other names*: exonuclease VII; *Escherichia coli* exonuclease VII. An enzyme that catalyses exonucleolytic cleavage of DNA, in either the 5′- to 3′- or the 3′- to 5′-direction, to yield 5′-phosphomononuc-leotides.

exodeoxyribonuclease (phage SP3-induced) EC 3.1.11.4; *other names*: phage SP3 DNase; DNA 5′-dinucleotidohydrolase. An enzyme that catalyses exonucleolytic cleavage of DNA in the 5′- to 3′-direction to yield 5′-phosphodinucleotides.

exoenzyme 1 any enzyme that occurs attached to the outer surface of a cell (i.e. an **ectoenzyme**) or in the periplasmic space, or one that is secreted by a cell into the medium. *Compare* **endoenzyme** (def. 1). **2** any enzyme that cleaves a linkage to the terminal residue of a biopolymer. *Compare* **endoenzyme** (def. 2).

exogenous originating outside an organism, tissue, or cell, e.g. nutrients; of or pertaining to external factors that affect an organism, e.g. light. —**exogenously** *adv*.

exomorphin *see* **casomorphin**.

exon any coding or messenger sequence of deoxynucleotides; i.e. any intragenic region of DNA in eukaryotes that will be expressed in (mature) mRNA or rRNA residues. An arrangement of exons alternating with **introns** constitutes a **transcriptional unit**. —**exonic** *adj*.

exon shuffling the recombination of different coding regions of eukaryotic structural genes (i.e. **exons**) through crossing-over, a mechanism proposed to explain the mosaic structure of eukaryotic genes. Such a mechanism is also responsible for the generation of, e.g., antibody diversity.

exonuclease any enzyme of sub-subclasses EC 3.1.11–16, that catalyses the hydrolysis of terminal diester linkages in polynucleotides to yield mononucleotides, the degradation of the chain occurring sequentially from one end. **Exonuclease III** (*abbr*.: ExoIII) catalyses the stepwise removal of mononucleotides from the 3′OH end of double-stranded DNA that is blunt-ended (the end recessed), or from nicks in one DNA strand. Overhangs of more than four nucleotides at the 3′ end are protected from ExoIII activity. *Compare* **endonuclease**. *See also* **deoxyribonuclease**, **nuclease**, **ribonuclease**.

exopeptidase any enzyme of subclass EC 3.4 that catalyses the hydrolysis of peptide bonds adjacent to the terminal amino or carboxyl group of an oligo- or polypeptide. Exopeptidases include: EC 3.4.11, aminopeptidases; EC 3.4.13, dipeptidases; EC 3.4.14, di- and tripeptidyl peptidases; EC 3.4.15, peptidyldipeptidases; EC 3.4.16, serine-type carboxypeptidases; EC 3.4.17, metallocarboxypeptidases; EC 3.4.18, cysteine-type carboxypeptidases; and EC 3.4.19, omega peptidases.

exophthalmic goitre *or* **Graves' disease** a form of hyperthyroidism characterized by enlargement of the thyroid gland and protrusion of the eyeballs.

exopolysaccharide any of a range of extracellular homo- or heteropolysaccharides produced by microorganisms, either as insoluble capsules or as soluble slimes. They serve for protection and possibly also in recognition by host organisms.

exoribonuclease *see* **ribonuclease**.

exorphin any peptide with morphine-like activity (*see* **opioid peptide**) that originates from outside the body. Such substances occur, e.g., in proteolytic digests of certain dietary proteins. *Compare* **endorphin**.

exo-α-sialidase EC 3.2.1.18; *other names*: sialidase; neuraminidase; *N*-acylneuraminate glycohydrolase; α-neuraminidase; an enzyme that catalyses the hydrolysis of α-2,3-, α-2,6-, and α-2,8-glycosidic linkages joining terminal nonreducing N- or O-acylneuraminyl residues to galactose, oligosaccharides, glycoproteins, or glycolipids.

exoskeleton any skeleton covering the outside of an organism or lying in the skin, such as the hard chitinous cuticle of an arthropod or the shell of a mollusc. *Compare* **endoskeleton**.

exosmosis the osmotic flow of water or of an aqueous solution from a cell, vessel, or organism into the surrounding aqueous medium. *Compare* **endosmosis**. —**exosmotic** *adj*.

exosome 1 a fragment of exogenous DNA that, when taken up by a cell, is not readily integrated into the chromosome but can replicate, and be expressed. **2** a multienzyme complex involved in turnover of RNA in yeast and probably in other eukaryotes. The exosome core consists of six phosphorolytic and two hydrolytic exoribonucleases (*see* **Rrp**) and another protein of undetermined function. In the cytoplasm, the core is also associated with four Ski proteins and degrades mRNA. In the nucleus it contains another hydrolytic ribonuclease (Rrp6p) and a DEAD-box RNA helicase, and is involved in processing rRNA precursors and a variety of snRNAs and snoRNAs. *Compare* **degradosome**.

exothermic describing a process or reaction that is accompanied by the evolution of heat; i.e. a process or reaction for which the change in enthalpy, ΔH, is negative at constant pressure and temperature. *Compare* **endothermic**.

exotic describing an introduced (i.e. foreign), nonendemic, nonacclimatized organism.

exotoxin any toxin formed by a microorganism and secreted into the surrounding medium. *Compare* **endotoxin**.

exp *(in mathematics)* symbol for exponential; *see* **exponential function**.

expansin any of a group of proteins, located within the cell walls of plants (both dicotyledons and grasses), that play an essential role in loosening cell walls during cell growth. They are hydrophobic, nonglycosylated proteins, M_r ~30 kDa.

expectation value *or* **expect value** *abbr*.: e-value; a statistical measure of the significance of an alignment obtained by searching a database with a query sequence. The e-value denotes the number of alignments with scores better than or equal to the obtained alignment score that are expected to occur by chance in a database of the same size and composition, using the same scoring system. The lower the e-value, the more significant the score, its value decreasing exponentially as the alignment score increases.

experiment *abbr*.: expt.; **1** any procedure or series of procedures, carried out in defined conditions and designed to obtain new information about substances, materials, organisms, or natural phenomena, or to test or confirm a hypothesis. **2** *an alternative term for* **experimentation**. **3** to carry out an experiment or experiments. —**experimenter** *n*.

experimental *abbr*.: exptl.; **1** of, relating to, or based on the results of, an experiment or experimentation. **2** describing a disease produced deliberately in laboratory plants or animals for the purposes of study or as a model for the natural disease condition. **3** any particular object or system in an experiment that includes, or is subjected to, one of the variables. *Compare* **control** (def. 1). —**experimentally** *adv*.

experimentation the act, process, or practice of carrying out an experiment or experiments.

explant 1 to transfer a piece of living tissue from its normal situation to a culture medium. **2** any piece of tissue treated in this way. —**explanted** *adj*.

exponent *or* **index** *(in mathematics)* any number (or expression, or symbol for a number) indicating the power to which another number (or expression or symbol) is raised; e.g., in the expression a^n, n is the exponent, indicating that a is multiplied by itself n times.

exponential *(in mathematics)* **1** of, containing, or involving one or more exponents, e.g. an exponential equation. **2** describing something approximately expressible by an exponential equation.

exponential function *(in mathematics)* any function of the type e^x or Ae^{ax}, where A and a are constants, e is an irrational number that is the base of natural logarithms, and x is a variable; e^x is sometimes written exp$x$.

exponential growth *an alternative term for* **logarithmic growth**.

exponential phase *an alternative term for* **logarithmic phase**.

export *informal term* describing a substance synthesized in, and then secreted from, a cell or type of cell in a multicellular organism; e.g. export protein.

exportins proteins that facilitate the export of a variety of different proteins, ribonucleoproteins and RNAs through the nuclear pore complex to the cytoplasm. Exportins bind their cargo cooperatively with RanGTP in the nucleus and release it following hydrolysis of RAN-bound GTP in the cytoplasm. CRM1/Exportin 1 has the broadest substrate range and often recognises a short leucine-rich nuclear export signal in the protein.

exposure 1 *(in photography or radiography)* the act or process of exposing sensitized photographic material, e.g. a film or plate, to light or to ionizing radiation. **2** *(in photography)* **a** a measure of the amount of light admitted into a photographic device expressed in terms of the lens aperture (f number) and time. **b** *or* **light exposure** *symbol*: H; a physical quantity equal to the product of the illuminance and the time, Δt, for which the area is exposed. The SI unit is the lux second. **3** *(in radiation dosimetry)* a measure of the amount of X- or gamma radiation to which a subject or object is exposed, expressed in terms of the quantity of electric charge of the ions of one sign produced when all the electrons of both signs liberated in a volume of air of unit mass are completely stopped. The SI unit is C kg^{-1}, which has replaced the **röntgen**. *Compare* **absorbed dose**.

express 1 *(in genetics)* **a** to allow or cause the information in a gene to become manifest. **b** to activate the cellular functions involved in **gene expression**. **c** *(in cDNA technology)* to effect the synthesis of protein from the corresponding cloned gene using an **expression vector** in an **expression system**. **2** to indicate by means of a symbol, formula, or equation. **3** to force or squeeze out a liquid from an object.

expressed sequence tag *abbr*.: EST; a partial sequence of a clone picked at random from a cDNA library and used in the identification of genes being expressed in a particular tissue. The technique exploits advances in automated DNA sequencing and sequence data handling, and a remarkably high number of ESTs turn out to represent previously unknown genes. These are identified by the predicted primary structure of the proteins that would be expressed and their relation to proteins of known structure. ESTs have proved extremely valuable in mapping the human genome, and there has been controversy surrounding attempts to patent them.

expression 1 the act or process of allowing information to become manifest. *See* **gene expression**, **expression vector**, **expression system**. **2** *(in mathematics)* a variable function or some combination of constants, variables, and/or functions. **3** the act or process of forcing or squeezing fluid from or out of something.

expression profile 1 the range of genes expressed at particular stages of cell development. **2** the level and duration of expression of one or more genes in a particular cell or tissue (detected, e.g., via sample sequencing, serial analysis, or microarray-based methods).

expression system a combination of an **expression vector** and a host cell host selected to enable the production of protein from a cloned gene. Ideally, it also contains the enzymes needed for post-translational modification of the synthesized protein. Expression in *Escherichia coli* or other bacterial cells may be inadequate for eukaryotic protein expression, but expression in mammalian cell lines may be inconvenient. Use of a baculovirus expression vector in insect cells is one technique that has been used successfully for eukaryotic protein expression.

expression tagged site *see* **ETS**.

expression vector any vector designed to enable the expression of a cloned gene. Vectors for expression in *Escherichia coli* may use the *lac* promoter, for example. Vectors for expression in mammalian cells are more complex; they may have the CMV (cytomegalovirus) promoter, which is active in a wide range of cell types, and other features such as termination sequences and sometimes an intron. *See also* **expression system**.

expressor a chemically undefined positive regulator of eukaryotic gene expression.

expt. *abbr. for* experiment (def. 1).

exptl. *abbr. for* experimental (def. 1, 2).

extein a peptide counterpart of an exon. *See* **intein**, **splicing**.

extended X-ray absorption fine structure (spectroscopy) *abbr*.: EXAFS (spectroscopy); a method for determining short-range order and local structure around specific atoms in noncrystalline materials. In this technique the X-ray absorption spectrum of the material is measured over an extended energy range near the absorption edge of one particular element in the material and structural details may be obtained from the observed modulation of the X-ray absorption on the high-energy side of the absorption edge. It is useful for elucidating the environment of metal ions in biological macromolecules.

extensin a glycoprotein found in plant cell walls in association with pectin. It resembles collagen of animal cells, being rich in *trans*-4-hydroxy-L-proline residues; it also contains L-arabinofuranose and D-galactopyranose residues attached in oligosaccharide units to the hydroxyl groups of the protein.

extension sequence a terminal sequence of amino-acid residues that is present in a nascent polypeptide as generated by translation of mRNA but is absent from any mature polypeptide or protein formed from it. An N-terminal extension sequence may be a signal sequence.

extensive property any of the properties of a substance or object that depend on the amount of the substance or the size of the object being considered, e.g. internal energy, mass, volume. *Compare* **intensive property**.

externalize *or* **externalise** to render external, especially to secrete a substance from a cell or organism. —**externalization** *or* **externalisation** *n*.

external standard any standard that is added to a sample at some point in an assay and serves as a reference by which the unknown being investigated can be quantified or identified.

external-standard method of quench correction *or* **external standardization** a method for correcting for quenching in **liquid scintillation counting** in which a source of gamma radiation external to the sample is brought into proximity with it after the sample has been counted and the sample is then counted again. The observed increase in count rate is correlated with that obtained using a set of differently quenched standards and the appropriate correction to the sample-count rate obtained. The method is simple and can be automated but on its own is less accurate than other methods. *Compare* **internal-standard method of quench correction**.

extinction *an older term (no longer recommended) for* **absorbance**.

extinction angle in flow birefringence between concentric cylinders, with polarizer and analyser at right angles, the angle between the **cross of isocline** and the cross formed by the polarizer and the analyser. It ranges from 45° (no orientation) to zero (complete orientation).

extinction coefficient *symbol E; an alternative term (no longer recommended) for* molar (decadic) absorption coefficient. *See* **absorption coefficient**.

extra+ *prefix signifying outside or beyond.*

extra arm *or* **variable arm** the base-paired segment of variable length in the cloverleaf model of some tRNA molecules. *See* **transfer RNA**.

extracellular present outside a cell, expelled from a cell, or happening outside a cell. *Compare* **exocellular, intracellular**.

extracellular fluid *abbr.*: ECF; the portion of the total liquid content that occurs outside the cells, esp. as distinct from **intracellular fluid**. In higher animals it comprises the **interstitial fluid**, the **lymph**, and the **blood plasma**. In a 70 kg man the volume of ECF is approximately 15.5 L and it represents approximately 22% of body mass. *Compare* **intracellular fluid**.

extracellular matrix a layer consisting mainly of proteins (especially collagen) and glycosaminoglycans (mostly as proteoglycans) that forms a sheet underlying cells such as endothelial and epithelial cells. The constituent substances are secreted by cells in the vicinity, especially fibroblasts.

extracellular space that part of a multicellular organism outside the cells proper – usually taken to be outside the plasma membranes – and occupied by fluid.

extrachromosomal describing structures, including DNA molecules, that are not part of a chromosome, or processes that proceed outside the chromosomes.

extrachromosomal inheritance inheritance controlled by extrachromosomal genes, such as those in mitochondrial or chloroplast DNA or in plasmids.

extract 1 to remove from or separate; to obtain a substance from a material, mixture, organism, or part of an organism by some chemical and/or physical process. **2** something extracted; a solution containing an active principle that has been extracted from plant or animal material.

extraction 1 the act or process of extracting. **2** something that has been extracted. *Compare* **extract** (def. 2).

extra piece *an alternative name for* **signal peptide**.

extrapolate *(in mathematics)* to estimate a value or the value of a function beyond the data values already obtained by either extending a graph or by calculation. *Compare* **interpolate**. —**extrapolation** *n*.

extreme value distribution *abbr.*: EVD; an asymmetric unimodal distribution whose righthand tail decays more gradually at highthan at low-score values. For example, in sequence analysis, local similarity scores of ungapped alignments follow the EVD, as a best local alignment has a score that is the maximum of a large number of random alignment scores.

extremophile an organism that requires extreme physicochemical conditions for its optimum growth and proliferation. Such organisms include **thermophiles** or **psychrophiles**, **halophiles**, **alkalophiles** or **acidophiles**, **osmophiles**, and **barophiles**, based on their growth at extremes of temperature, salt concentration, pH, osmolarity, or pressure, respectively.

extrinsic originating or acting from the outside. —**extrinsically** *adv*.

extrinsic factor *a former name for* **vitamin B$_{12}$**.

extrinsic fluorescence fluorescence caused by a molecule of a fluor of small molar mass attached by either chemical coupling or simple binding to a (macro)molecule under study.

extrinsic pathway (of blood coagulation) *see* **blood coagulation**.

extrinsic protein *an alternative name for* **peripheral (cell membrane) protein**.

extrude to squeeze or force out; used especially of a solid or semisolid material. —**extrusion** *n*.

extrusome an organelle excreted as an anterior vacuole by a parasitic protist. It contains substances used for penetration of host cells.

ex vivo describing any biological process, reaction, or experiment in which the nature or magnitude of a change occurring in living tissues in intact organisms is subsequently measured *in vitro* following the excision of affected tissues. *Compare* **in vivo** [Latin, meaning from life.]

eye *(in molecular biology)* a region of DNA undergoing replication, in which the separated strands give the appearance of an eye.

EYFP *abbr. for* enhanced yellow fluorescent protein.

***E/Z* convention** a convention of stereochemical descriptors for an unsaturated compound showing *cis–trans* isomerism that indicate unequivocally the steric relations around a (given) double bond. The **sequence rule** is applied in turn to the pair of atoms or groups attached to one of the doubly bonded atoms and to the pair attached to the other doubly bonded atom, and the order of preference in each case is established. If the two preferred atoms or groups are on opposite sides of the reference plane of the molecule the arrangement is called *E* [from German *entgegen*, opposite], and if they are on the same side of the plane it is called *Z* [from German *zusammen*, together]. For use as a prefix (in chemical nomenclature) the descriptor is placed in parentheses and connected with a hyphen, e.g. (*E*)-butenedioate and its stereoisomer (*Z*)-butenedioate (the respective systematic names for *trans*-1,2-ethylenedicarboxylate (= fumarate) and for *cis*-1,2-ethylenedicarboxylate (= maleate)). When a molecule of an unsaturated compound contains more than one double bond the prefix is composed of multiple symbols, each preceded if appropriate by the lower-numbered locant of the relevant double bond; e.g. (2*E*,4*Z*)-2,4-hexadienoate. The convention is now generally preferable to that using the structural prefixes *cis*- and *trans*-, especially for olefinic compounds having more than two substituents at a double bond (where it avoids possible ambiguity as to which pair of substituents is being designated). *See also* **pro-*E*/pro-*Z* convention**.

ezrin *or* **cytovillin** *or* **villin-2** *or* **P81** a protein involved in connections of major cytoskeletal structures to the plasma membrane, having domains in common with **talin**. It undergoes tyrosine phosphorylation after activation of specific growth factor receptors, and functions as a membrane-to-cytoskeleton linker that is involved in assembly of apical microvilli in parietal cells. It participates in the regulation of acid secretion.

f *symbol for* femto+ (SI prefix denoting 10^{-15} times).

f *symbol for* **1** frequency (alternative to ν). **2** activity coefficient referenced to Raoult's law. **3** fugacity (alternative to ⁻p). **4** frictional coefficient. **5** furanose form (added after the symbol for a monosaccharide, e.g. Ara*f* for arabinofuranose). **6** *(in mathematics)* function.

F *symbol for* **1** a residue of the α-amino acid L-phenylalanine (alternative to Phe). **2** fluorine. **3** the fluoro group in an organic compound. **4** *(in genetics)* filial generation. **5** farad. **6** Fick unit. **7** *abbr. for* fertility; *see* **F plasmid**. **8** *abbr. sometimes used* (not recommended) to designate **folate** or folic acid or their reduced derivatives.

F1, F2a1, F2a2, F2b, F2c, F3 *older nomenclature for* **histones**. F signifies fraction; in early work, the histones were extracted from the nuclei with sulfuric acid solutions, and separated according to their solubility.

F$_{420}$ *see* **coenzyme F$_{420}$**.

F$_{430}$ *see* **factor F$_{430}$**.

°F *symbol for* degree Fahrenheit. *see* **Fahrenheit temperature scale**.

F$^+$ *symbol for* a donor bacterial cell containing an **F plasmid**; F$^-$ denotes a recipient bacterial cell lacking an F plasmid.

F *symbol for* **1** Faraday constant. **2** force (bold italic in vector equations). **3** fluence (alternative to H).

F9 a continuous line of mouse teratocarcinoma cells that retain the capacity for differentiation. They can give rise to chimeric mice when injected into the blastocoel of mouse blastocysts that are then allowed to complete the developmental process *in utero*.

FA *abbr. for* **1** fatty acid. **2** folic acid or folate (not recommended). **3** filtrable agent.

Fab fragment a $\approx$45 kDa protein fragment obtained (together with **Fc fragment** and **Fc ' fragment**) by papain hydrolysis of an **immunoglobulin** molecule. It consists of one intact light chain linked by a disulfide bond to the N-terminal part of the contiguous heavy chain (the **Fd fragment**). Two Fab fragments are obtained from each IgG antibody molecule; each fragment contains one antigen-binding site. The term Fab is derived from ['fragment antigen-binding']. *Compare* **F(ab)$_2$ fragment**. *See also* **Fv fragment**.

F(ab')$_2$ fragment a $\approx$90 kDa protein fragment obtained (together with **pFc ' fragment**) by pepsin hydrolysis of an **immunoglobulin** molecule. It consists of that part of the immunoglobulin molecule N-terminal to the site of pepsin attack and contains both **Fab fragments** held together by disulfide bonds in a short section of the **Fc fragment** (the hinge region). One F(ab')$_2$ fragment is obtained from each IgG antibody molecule; it contains two antigen-binding sites but not the site for complement fixation. *Compare* **Fab fragment**.

FABP *abbr. for* fatty acid binding protein; **albumin** is the major FABP of plasma, while a 40 kDa plasma membrane FABP of liver appears to be involved in uptake of oleate in an ATP- and Na$^+$-dependent manner. Small cytosolic FABPs ($\approx$130 amino acids) have been characterized from liver, heart, skeletal muscle, intestine, and other tissues. The crystal structure of FABP from rat intestine shows 10 antiparallel beta strands in two sheets that form a hydrophobic binding pocket; four motifs are characteristic of this function.

Fabry's disease *or* **Fabry disease** an X-linked inborn error of human metabolism due to defective lysosomal α-D-galactosidase A activity. It is a **sphingolipidosis** or sphingolipid lysosomal storage disease. The enzyme defect leads to the progressive deposition in most visceral tissues of neutral glycosphingolipids with terminal α-D-galactosyl residues, principally globotriaosylceramide, Gal(α1-4)Gal(β1-4)Glc(β1-1)cer. [After Johannes Fabry (1860–1930), German physician.]

facilitated diffusion diffusion (def. 2) across a biological membrane through the participation of specific transporting agents (transporters) or carriers. The equilibrium distribution reached is the same as that achieved by **simple diffusion** but at any one site facilitated diffusion is mediated by a transport protein that exhibits specificity for the transported species. The existence of the transporter mediates passage across the membrane of molecules to which it would otherwise be impermeable. *Sometimes termed* **mediated transport** to distinguish it from **active transport**.

faciogenital dysplasia protein *see* **FGD1**.

FACIT *abbr. for* fibril-associated collagens with interrupted triplex helices; a group of collagens (types IX, XII and XIV) composed of triple-helical regions interrupted by non-triple-helical regions of 8–42 amino acids.

FACS *abbr. for* fluorescence-activated cell sorter.

F-actin the form of **actin** found in filaments.

factor 1 any component or cause that contributes to an effect or result; a term often used to denote an uncharacterized (or incompletely characterized) component of a biological system. **2** *(in mathematics)* any integer or polynomial that can be divided exactly into another integer or polynomial.

factor I to factor XIII, factor Va, etc. names of the **blood coagulation** factors.

factor B a component of **complement**.

factor D̄ a component of **complement**.

factor F an *alternative name for* **initiation factor**.

factor F$_{430}$ a nickel-containing tetrapyrrole compound, the prosthetic group of EC 2.8.4.1 coenzyme B sulfoethylthiotransferase (methyl **coenzyme M** reductase), which catalyses the terminal reductive step of methanogenesis in **methanogenic** bacteria, in which methane is formed from methyl-coenzyme M.

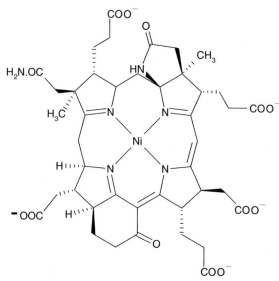

factor G an *alternative name for* EF-G; *see also* **elongation factor**.

factor H a component of **complement**.

factor I a component of **complement**.

factorial *symbol:* !; the product of all positive integers from 1 up to and including the integer in question, n; hence

$$n! = n(n-1)(n-2)(n-3)\ldots 3 \times 2 \times 1.$$

factor IF an *alternative name for* **initiation factor**.

factor R an *alternative name for* **release factor**.

factor S a tetrapeptide, of unknown structure, isolated from cerebrospinal fluid of goats, that slowly induces delta (slow wave) sleep in rats on infusion into the cerebral ventricles.

factor T an *alternative name for* EF-T; *see also* **elongation factor**.

facultative able to live under more than one set of environmental conditions.

facultative aerobe *see* **aerobe**.

facultative anaerobe *see* **anaerobe**.

FAD *abbr. for* flavin-adenine dinucleotide.

FADD *abbr. for* Fas-associated protein with death domain; *other name*: MORT1; a cytosolic adaptor protein that binds **Fas** (via the **death domain** on each) and procaspase 8 (via the death effector domain on each). The resulting complex forms part of the death-inducing signalling complex, or **DISC**

fade the waning of a response in the continued presence of an agonist. *Compare* **desensitization**.

FADH₂ *abbr. for* flavin-adenine dinucleotide (reduced).

F agent *see* sex factor.

Fahrenheit temperature scale a temperature scale in which 32°F is set equal to 273.15 K and 212 °F is set equal to 373.15 K. Thus 1 Fahrenheit degree = 5/9 × K, and Fahrenheit temperature (°F) = 9/5 × (Celsius temperature) + 32. [After G. D. Fahrenheit (1686–1736), German scientist.]

falcipain a cysteine protease of *Plasmodium falciparum*. It may be involved in host cell invasion by the parasite.

fall curve the curve describing the decrease in the colour intensity of a sample with time, obtained in a colorimetric analysis using an autoanalyser.

falling-drop method a method for the accurate determination of relative densities of liquids in which a drop of the liquid is allowed to fall to its equilibrium position in a column containing a defined density gradient of an immiscible liquid. It was formerly used for determination of the deuterium oxide content of samples of water.

falling-mercury electrode *an alternative term for* **dropping-mercury electrode**.

falling-sphere viscometer an instrument for measuring the viscosity of a liquid from the time taken for a solid sphere to fall through a given distance in a column of the liquid compared with the time taken for it to fall through the same distance in a liquid of known viscosity.

Fallopian tube an **oviduct** of a mammal.

false negative a true member of a data set that is not recognized by a discriminator for that data set.

false positive a match that is recognized by a discriminator for a given data set but is not a true member of that data set.

familial 1 of or pertaining to a given family. **2** describing something transmitted between members of a family, especially a familial character, condition, or disease.

familial betadysproteinemia *an alternative name for* **remnant hyperlipidemia**.

familial British dementia *see* **Abri**.

familial hypercholesterolemia a common autosomal dominant genetic condition resulting in elevated levels of low-density lipoprotein in blood. Fatty deposits accumulate in the arteries and skin, and the condition predisposes to coronary heart disease. It results from mutation of the **low-density lipoprotein receptor** gene resulting in haploinsufficiency in the heterozygous state.

familial hyperlysinemia an apparently benign genetic human disorder in which hyperlysinemia and hyperlysinuria result from decreased activity of the bifunctional enzyme α-**aminoadipic semialdehyde synthase**.

familial Mediterranean fever an autosomal recessive disorder characterized by periodic fever and systemic amyloidosis with renal involvement. Prominent in Sephardic Jews and Armenians, it is caused by many mutations in the gene (locus at 16p13) for a protein (781 amino acids) of unknown function called pyrin or marenostrin.

family 1 *(in taxonomy)* a taxonomic category of related organisms ranking below the **order** and above the **genus**. **2** *(in chemistry and biochemistry)* an imprecise term for a group of substances closely related structurally. *See also* **superfamily**. **3** *(in molecular biology) see* **protein family**.

FAN *abbr. for* factor associated with neutral sphingomyelinase activation; a protein of the **WD40** family that couples the p55 TNF receptor and the cannabinoid receptor to neutral sphingomyelinase to release ceramide as a signalling molecule from **sphingomyelin** on the inner leaflet of plasma membranes.

Fanconi anemia or *(esp. Brit.)* **Fanconi anaemia** a rare autosomal recessive, highly variable disorder, characterized by bone marrow failure, developmental abnormalities, and increased susceptibility to cancer. It usually proves fatal by the second decade of life. Cells from patients show increased susceptibility to crosslinking (alkylating) agents. Eight complementation groups are recognized: A, B, C, D1, D2, E, F, and G. The gene products for A, C, E, F, and G form a constitutive nuclear protein complex that is required for monoubiquitination and activation of D2 following DNA damage or during the S phase of the cell cycle. The activated D2 assembles with nuclear foci that contain BRCA1 and BRCA2. Mutations in any of these gene products leads to the disease. Cell lines from groups B and D1 contain truncated BRCA2 proteins. [After Guido Fanconi (1892–1979), Swiss paediatrician.]

Fanconi syndrome or **renal Fanconi syndrome** a human disorder characterized by generalized dysfunction of proximal renal tubules, manifested as glycosuria, amino aciduria, phosphaturia, acidosis, and vitamin D-resistant bone disease manifested as rickets in children or osteomalacia in adults. It may be acquired as a results of various toxic agents (including heavy metals or certain drugs) or be associated with a variety of inherited metabolic disorders. *See also* **renal rickets**.

farad *symbol*: F; the SI derived unit of electric capacitance. A capacitor has a capacitance of one farad when a charge of one coulomb raises the potential between its plates to one volt: i.e. $1F = 1CV^{-1}$. [After Michael Faraday (1791–1867), British physicist.]

faradaic current an electric current flowing through an electrode that corresponds to the electrolytic oxidation or reduction of one or more chemical species.

Faraday constant *symbol*: F; a fundamental constant, defined as the quantity of electricity required to deposit one mole of a univalent ion from a solution of an electrolyte. It is the arithmetic product of the Avogadro constant, L (or N_A), and the elementary charge, e; hence

$$F = Le \text{ (or } F = N_A e) = 9.648\ 530\ 9(29) \times 10^4 \text{ C mol}^{-1}.$$

Faraday effect the rotation of the plane of vibration of polarized light, or of other polarized electromagnetic radiation, when passing through an isotropic, transparent medium in a magnetic field that has a component in the direction of the radiation.

Farber's disease or **Farber's lipogranulomatosis** *an alternative name for* **ceramidase** deficiency.

farnesoid X receptor *(abbr.*: FXR) or **bile acid receptor** an orphan receptor restricted to the enterohepatic system, kidney, and adrenal cortex. It forms an obligate heterodimer with retinoid X receptor (which binds 9-*cis*-retinoic acid), and the dimer binds to specific hormone response elements on DNA. Bile acids are the endogenous ligands for FXR.

farnesol the trivial name for any of four possible stereoisomers of 3,7,11-trimethyl-2,6,10-dodecatrien-1-ol; a sesquiterpene alcohol. The 2*E*,6*E*-isomer is found in many essential oils. The 2*Z*,6*E*-isomer is a minor constituent of some essential oils.

2*E*,6*E*-isomer

farnesyl the (2*E*,6*E*)-3,7,11-trimethyl-2,6,10-dodecatrien-1-yl group; the alkenyl group derived from the sesquiterpene alcohol **farnesol**. 2*E*,6*E*-Farnesyl diphosphate (= *trans,trans*-farnesyl diphosphate) is an intermediate in carotenoid, sesquiterpene, squalene, and sterol biosynthesis and is also a substrate in the addition of the farnesyl group to proteins (*see* **prenylation**).

farnesyltransferase EC 2.5.1.21; *recommended name*: farnesyl-diphosphate farnesyltransferase; *systematic name*: farnesyl-diphosphate:farnesyl-diphosphate farnesyltransferase; *other name*: presqualene-diphosphate synthase. An enzyme that catalyses the formation of presqualene diphosphate from two molecules of farnesyl diphosphate with release of pyrophosphate. It occurs in the pathway for the synthesis of **squalene** and derivatives (cholesterol, sesquiterpenes). In eukaryotes it is a monomeric microsomal enzyme.

farnesyltranstransferase EC 2.5.1.29; *systematic name*: *trans,trans*-farnesyl-diphosphate:isopentenyl-diphosphate farnesyltranstransferase; *other name*: geranylgeranyl-diphosphate synthase. An enzyme that catalyses the formation of geranylgeranyl diphosphate from *trans,trans*-farnesyl diphosphate and isopentenyl diphosphate with the release of pyrophosphate. It occurs in the pathway for sesquiterpene and cholesterol synthesis and forms the **prenyl** derivative geranylgeranyl diphosphate. In some fungi and plants this activity is part of a multifunctional protein that also has the activities of geranylgeranyl-diphosphate synthase (*see* **dimethylallyltranstransferase**) and **geranyltranstransferase**.

farnoquinone *an alternative name for* menaquinone-6; *see* **menaquinone**.

Farr test a **radioimmunoassay** technique for measuring the absolute antigen-binding capacity of an antiserum. The antibody is allowed to react with radiolabelled antigen and the antigen–antibody complex is then precipitated by addition of saturated ammonium sulfate solution to a final concentration of 40 w/v. The antigen that has reacted can be determined in the precipitate.

far ultraviolet *see* **ultraviolet**.

far Western analysis a modification of **Western blotting** in which one of a complex mixture of proteins, separated by either denaturing or non-denaturing gel electrophoresis, is detected by its interaction with a bait protein. The bait protein may be provided with an **epitope tag** so that it can be readily detected with anti-tag antibody labelled with an enzyme.

Fas *or* **Apo-1** *or* **CD95** *or* **APT 1** *or* **apoptosis antigen 1** *or* **TNFRSF6** a transmembrane receptor protein of the TNF receptor superfamily that forms a homotrimer and is the receptor for the ligand **FasL**. The gene at 10q24.1 encodes a protein of 325 amino acids, expressed in many cell types including T lymphocytes. Fas has an N-terminal extracellular segment containing cysteine-rich repeats, and its C-terminal intracellular segment contains a **death domain**. On binding of FasL, the death domain of Fas binds to a homotypic domain present in **FADD** to promote apoptosis. Doxorubicin enhances expression of Fas in many tumours. Antibodies to a soluble N-terminal segment are present in some systemic lupus erythematosus patients. A deletion and other mutations in the Fas gene cause autoimmune lymphoproliferative syndrome (also called Canale–Smith syndrome).

Fas-associated protein with death domain *see* **FADD**.

fascicle *or* **fibre tract** a tight parallel bundle of nerve fibres that are either axons or dendrites. It is formed during the growth of nerves, and is mediated in part by cell **adhesion molecules**.

fasciclin any of the proteins of a related group that are involved in nerve **fascicles** (bundles). They include fasciclin I, a neuronal cell **adhesion molecule**, and fasciclin II, which is a neuronal recognition molecule related to NCAM (*see* **adhesion molecules**).

fascin a protein involved in the formation of actin bundles and actin polymerization.

FasL *or* **CD95 ligand** *or* **CD95L** *or* **TNFSF6** a transmembrane protein of the TNF receptor superfamily that is the ligand for **Fas**. Encoded by a gene at 1q23, it consists of an N-terminal extracellular region that contains cysteine-rich repeats, a transmembrane segment, and an intracellular C-terminal region. It forms a homotrimer and is expressed in epithelial cells, neurons, T lymphocytes, and various tumour cell lines. Binding to Fas activates membrane sphingomyelinase with release of ceramide, which activates a serine/threonine kinase that targets several proteins. Doxorubicin enhances expression of FasL in many tumours. Processing by membrane metalloproteases releases soluble FasL, raised levels of which are found in alcoholic hepatitis and malignant melanoma.

Fas ligand a homotrimeric protein on the cytotoxic T cell surface which binds to transmembrane receptor proteins on the target cell called Fas. The binding alters the Fas proteins so that their clustered cytosolic tails recruit procaspase-8, which leads to **apoptosis** .

Fas ligand a homotrimeric protein on the cytotoxic T cell surface which binds to transmembrane receptor proteins on the target cell called Fas. The binding alters the the Fas proteins so that their clustered cytosolic tails recruit procaspase-8, which leads to **apoptosis** .

Fas receptor interacting protein *see* **RIP**.

fast 1 (*of a component of a mixture or specified substance*) one that migrates further or furthest from the origin in chromatography or electrophoresis. **2** uninfluenced by a specified agent, e.g. by light, acid, etc. *See also* **acid-fast bacilli**. **3** to abstain from, or to be prevented from, taking nourishment for at least 24 hours (in the case of laboratory animals). In clinical chemistry, the requirement is often for an overnight fast, normally at least 10 hours. —**fasted** *adj.*; **fasting** *n*.

FastA *see* **alignment algorithm**.

fast component an unusual component of a mixture, e.g. a hemoglobin variant, that moves in a specified buffer system more rapidly in chromatography or electrophoresis than does the normal component.

fast reaction any reaction, or step in a reaction sequence, that has a large rate constant and proceeds rapidly. In a reaction sequence, a fast reaction is not the rate-limiting step.

fat 1 any triacylglycerol or mixture of triacylglycerols that is solid below 20°C; those that are liquid at such temperatures are usually referred to as oils. **2** *an alternative name for* **lipid**. **3** *an alternative name for* adipose tissue; *see* **brown adipose tissue**, **white adipose tissue**.

fatal familial insomnia a rare inherited neurological disease, first described in 1986 among inhabitants of mountainous regions of Italy. It is characterized by intractable insomnia and other neurological abnormalities. It is one of three familial forms of human **prion** disease. Neuropathologically, there is neuronal loss with associated gliosis. It is associated with a pathogenic mutation of the prion gene at codon Asn^{178} (shared with Creutzfeldt–Jakob disease, but for unknown reasons of different phenotype) linked to codon Met^{129}, a site that exhibits Met/Val polymorphism.

fat body any of the fat-containing cellular structures that serve as energy reserves in amphibians, insects, and reptiles.

fat cell *or* **fat-cell** *or* **lipocyte** *or* (*in animals*) **adipocyte** any living cell containing noticeable amounts of lipid, primarily as fat or oil.

fat index the mass of diethylether-extractable fat per unit mass of nonfat material (on a dry-weight basis) in, e.g., a foodstuff or tissue.

F$_1$-ATPase the globular catalytic domain of the **H$^+$-transporting ATP synthase**.

F$_o$ATPase the intrinsic domain of the **H$^+$-transporting ATP synthase**. The 'o' originally referred to 'oligomycin-sensitive'.

fat-soluble vitamin any of a diverse group of **vitamins** that are soluble in organic solvents and relatively insoluble in water. The group includes vitamins A, D, E, and K.

fat solvent any **lipid solvent** e.g. chloroform or diethylether, that will extract triacylglycerols from biological or other materials.

fatty derived from or containing fat.

fatty acid any of the aliphatic monocarboxylic acids that can be liberated by hydrolysis from naturally occurring fats and oils. Fatty acids are predominantly straight-chain acids of 4 to 24 carbon atoms, which may be saturated or unsaturated; branched fatty acids and hydroxy fatty acids also occur, and very long chain acids of over 30 carbons are found in waxes. *See also* **fatty-acid nomenclature**, **polyunsaturated fatty acid**.

fatty-acid activation the conversion of a fatty acid molecule to its fatty acyl-coenzyme A thioester, the first step in the reactions of **beta oxidation**. Fatty acyl-coenzyme A can be formed in a reaction catalysed by the acid-thiol ligases EC 6.2.1.1, 6.2.1.2, and 6.2.1.3, or by a CoA-transferase of sub-subclass EC 2.8.3.

fatty-acid nomenclature a system of symbols for describing fatty acids. The basic symbolism comprises the number of carbon atoms in the molecule, followed by the number of double bonds; the two numbers are separated by a colon. Thus, 16:0 represents palmitic acid, 16:1 palmitoleic acid, etc. To avoid ambiguity, double bond positions in unsaturated fatty acids should be indicated; 18:3 (9,12,15) represents α-linolenic acid, 18:3 (9,11,13) represents eleostearic acid. These symbols may also include the *E/Z* configurational descriptors (see *E/Z* **convention**). Although not now recommended, a still used system numbers the first double bond from the carboxyl group using the Greek letter *Δ*. Thus, oleic acid is 18:1 *Δ*9 or 18:1 Δ^9. This system persists in some enzyme names; e.g., the systematic name for linoleate isomerase (EC 5.2.1.5) is linoleate Δ^{12}-*cis*-Δ^{11}-*trans*-isomerase. *See also* **fatty acid oxidation complex**. Double bonds may also be designated from the end of the chain remote from the carboxyl group. If *n* is the total number of carbons, *n*-3 represents CH_3–CH_2–CH=CH–, etc. Since in many unsaturated fatty acids the double bonds are successively separated by methylene groups, only the position of the first double bond may be given. Thus linoleic acid [18:2 (9,12)] is also represented as 18:2 *n*-6. The Greek letter ω has also been used in the same way; e.g. for linoleic acid, 18:2 ω-6. However, it is not recommended that a double bond position, e.g. that of oleic acid, should be designated as ω9. *See also* **fish oil, linoleic family, linolenic family, oleic family, palmitoleic family**.

fatty-acid oxidation complex a bacterial multifunction protein complex that catalyses the **beta oxidation** of fatty acids. The example from *Escherichia coli* comprises α and β subunits. The α subunit contains **decenoyl-CoA Δ-isomerase, enoyl-CoA hydratase, 3-hydroxyacyl-CoA dehydrogenase**, and 3-hydroxybutyryl-CoA epimerase. The β subunit contains 3-ketoacyl-CoA thiolase. For a mammalian system *see* **beta-oxidation system**.

fatty acid synthase complex a large multienzyme complex that catalyses a spiral set of reactions whereby fatty acids are synthesized from one molecule of acetyl coenzyme A and successive molecules of malonyl coenzyme A. In most bacteria (including *Escherichia coli*) and in chloroplasts the complex consists of six α subunits (each catalysing three activities) and six β subunits (each catalysing three activities and containing an ACP domain). In animals, the enzyme complex (EC 2.3.1.85) produces a long-chain fatty acid (typically palmitate), whereas in yeast the product is a long-chain-fatty-acyl-CoA. The synthase complex of yeast has a mass of 2300 kDa and appears in the electron microscope as an ellipsoid, 25 μm long and 21 μm in cross section. The yeast enzyme comprises α and β subunits. It is termed fatty-acyl-CoA synthase (EC 2.3.1.86), and catalyses the overall reaction:

acetyl-CoA + *n* malonyl-CoA + 2*n* NADH + 2*n* NADPH = a long-chain-acyl-CoA + *n* CoA + *n* CO_2 + 2*n* NAD^+ + 2*n* $NADP^+$;

To synthesize palmitoyl-S-[ACP domain] requires seven repetitions of the following set of reactions, where X = -S-[ACP domain]:

(1) R–CH_2–CO–X + ⁻OOC–CH_2–CO–X →
R–CH_2–CO–CH_2–CO–X + CO_2 + H–X

(2) R–CH_2–CO–CH_2–CO–X + NADPH →
R–CH_2–CH(OH)–CH_2–CO–X + $NADP^+$

(3) R–CH_2–CH(OH)–CH_2–CO–X →
R–CH_2–CH=CH–CO–X + H_2O

(4) R–CH_2–CH=CH–CO–X + NADH →
R–CH_2–CH_2–CH_2–CO–X + NAD^+

The hydroxacyl product of reaction (2) has the $3R$ configuration and the enoylacyl product of reaction (3) has E configuration. Before the assembly begins, acetyl and malonyl units are transferred from their CoA derivatives to acyl carrier protein (ACP) by the activities of two transferases, EC 2.3.1.8 and EC 2.3.1.9. Representative enzyme activities for the four reactions are named as follows: reaction (1), EC 2.3.1.41, 3-oxoacyl-[ACP] synthase; reaction (2), EC 1.1.1.100, 3-oxoacyl-[ACP] reductase; reaction (3), EC 4.2.1.61, 3-hydroxypalmitoyl-[ACP] dehydratase; reaction (4), EC 1.3.1.9, enoyl-[ACP] reductase (NADH).
Similar enzymes with different chain length specificities may also be involved. For the first round, R = H, the two synthase substrates thus being acetyl-[ACP] and malonyl-[ACP]; the product is 3-oxobutanoyl-[ACP]. After seven repetitions of the four reactions, R = $C_{12}H_{25}$, and the product is palmitoyl-[ACP].
The vertebrate complex is a cytosolic dimer of identical subunits (260 kDa). Each human subunit contains 2504 amino acids. The sequence shows ≈80 identity with that of rat, and ≈63 identity with that of chicken. Since the product is palmitic acid itself a further enzyme is present: EC 3.1.2.14 (oleoyl-[ACP] hydrolase). Each polypeptide chain is folded into three domains: domain 1 contains

the acyltransferase activities and catalyses reaction (1); domain 2 catalyses reactions (2), (3), and (4); domain 3 catalyses the hydrolysis of palmitoyl-[ACP] to palmitate. In liver, reaction (4) is catalysed by EC 1.3.1.39, **enoyl-[acyl-carrier-protein] reductase (NADPH,A-specific)**.

fatty-acid thiokinase *an alternative name for* 1 butyryl-CoA ligase (EC 6.2.1.2). 2 long-chain-fatty-acid-CoA ligase; *see* **fatty-acyl-CoA ligase**.

fatty-acyl-CoA dehydrogenase any of the acyl-CoA dehydrogenase components of the **beta-oxidation system**. Mammalian liver contains a family of these enzymes that are distinguished by their chain-length specificity into: very-long-chain (C_{20}–C_{14}), long-chain (C_{18}–C_{12}), medium-chain (C_{12}–C_4), and short-chain (C_6–C_4) acyl-CoA dehydrogenases. These share 30–40 sequence identity within a species, and 87–90 sequence identity for each enzyme between species. Several crystal structures have been solved.

fatty-acyl-CoA desaturase *see* **stearoyl-CoA desaturase**.

fatty-acyl-CoA ligase either of two enzymes involved in fatty-acid activation.
(1) Long-chain-fatty-acid-CoA ligase (EC 6.2.1.3); *other names*: acyl-activating enzyme; acyl-CoA synthetase; fatty acid thiokinase (long-chain); lignoceroyl-CoA synthase. It catalyses a reaction between ATP, a long-chain carboxylic acid, and CoA to form an acyl-CoA and pyrophosphate. In mammals there are tissue isoforms.
(2) Medium-chain-fatty-acid-CoA ligase (EC 6.2.1.–); *other name*: medium-chain-fatty-acid-CoA synthase. It catalyses the same reaction as above but with medium-chain fatty acids.

fatty-acyl-CoA synthase 1 the multifunctional protein responsible for the synthesis of long-chain fatty acyl moieties in some eukaryotes, e.g. yeast. *See* **fatty acid synthase complex**. 2 an enzyme of the endoplasmic reticulum, EC 6.2.1.3, long-chain-fatty-acid-CoA ligase, that carries out a reaction between ATP, CoA, and a long-chain fatty acid to form the corresponding fatty-acyl CoA, AMP, and pyrophosphate.

fatty-acyl group any acyl group derived from a fatty acid.

fatty alcohol *an alternative term for* **long-chain alcohol**.

fatty aldehyde:NAD^+ oxidoreductase an NAD^+-dependent long-chain-aldehyde dehydrogenase, EC 1.2.1.48, that catalyses conversion of fatty aldehydes (6C to 24C long) to the corresponding fatty acids. It is anchored to the membrane of the endoplasmic reticulum through its C-terminal region and accompanies fatty alcohol:NAD^+ oxidoreductase. Several isozymes occur, the longest containing 488 amino acids. Many mutations lead to enzyme deficiency and the Sjøgren–Larsson syndrome.

fatty degeneration *or* **fatty infiltration** the deterioration of a tissue due to the deposition of abnormally large amounts of fat, usually as globules, in its cells.

fatty liver a pathological condition of liver tissue that has undergone **fatty degeneration**. It results from administration of various poisons, especially chlorinated hydrocarbons, or as a result of a dietary deficiency of choline or threonine.

FATZ *abbr. for* filamin-, actinin-, and telethonin-binding protein of Z disc; any of three isoforms (FATZ1, 2, or 3) of a protein of the Z disc of skeletal muscle. It binds γ-filamin, α-actinin, telethonin, and calcineurin.

favism a disease of humans in which hemolysis is caused by eating fava beans (*Vicia faba*) or by certain drugs, e.g. sulfonamides, primaquine, and nitrofuran antibiotics. The basic defect is an inherited deficiency of glucose-6-phosphate dehydrogenase in the red blood cells; about 130 variants are known. *See also* **glucose-6-phosphate dehydrogenase deficiency**.

f1 bacteriophage a filamentous single-stranded DNA virus that infects only male strains of *Escherichia coli*.

FbaI a type 2 **restriction endonuclease**; recognition sequence: T↑GATCA. *Bcl*I is an **isoschizomer**.

FBP *abbr. for* fructose 1,6-bisphosphate.

FBS *abbr. for* fetal bovine serum.

FCCP *abbr. for* carbonylcyanide *p*-trifluoromethoxyphenylhydrazone; an uncoupling agent.

F cell *a former name for* PP cell.

F_1-CF_0 the ε subunit of the **H^+-transporting ATP synthase** complex in the chloroplasts of *Chlamydomonas reinhardtii*, in which the subunit arrangement is different from that in higher plants.

Fc fragment a crystallizable protein fragment obtained (together with **Fab fragment**) by papain hydrolysis of an **immunoglobulin** molecule. That obtained from human IgG is a 50 kDa protein consisting of the C-terminal halves of two heavy chains linked by two disulfide bonds. The Fc fragment has no antigen-binding activity but carries sites for complement fixation. [Fc is from 'fragment-crystallizable'.] *Compare* **Fc′ fragment, Fd fragment**.

Fc′ fragment a protein fragment obtained in small amounts by papain hydrolysis of an **immunoglobulin** molecule, in addition to **Fab fragments** and **Fc fragment**. That obtained from human IgG is a 24 kDa protein; it forms noncovalently bonded dimers consisting of the C-terminal moieties of two Fc fragments minus their C-terminal tridecapeptide segments. It normally occurs in small quantities in human urine. *See also* **pFc′ fragment**.

Fc receptor any of a number of proteins that bind the Fc region of immunoglobulins A, E or G, and facilitate their recognition and uptake, especially by phagocytic cells (*see* **phagocyte**). Several types have been described; all of the proteins involved have a single transmembrane helical region. Fcγ receptor I (*abbr*.: FcγRI) *or* CD64, is a high affinity receptor for IgG, capable of binding monomeric antibody. It is a characteristic marker of mononuclear phagocytes, but may also be expressed on activated neutrophils. FcγRII *or* CD32 is a low affinity receptor for IgG present on monocytes, neutrophils, eosinophils, B cells and platelets. Crosslinking of the surface antibody and the receptor leads to downregulation of the B cell. Binding of immune complexes to this receptor leads to degranulation of platelets and release of inflammatory mediators. Another low affinity IgG receptor, FcγRIII *or* CD16 is found on macrophages, polymorphs, eosinophils and NK cells. On large granular lymphocytes it is a transmembrane glycoprotein that can link the cells to antibody-sensitized target cells, while on macrophages and neutrophils it has a **GPI anchor**. Two types of IgE receptor are known. (1) A high-affinity receptor, FcεRI, found on basophils and mast cells, which on activation leads to the release of histamine and the manifestation of allergy. It is a tetramer of one α chain, one β chain and two disulfide-linked γ chains. (2) A low-affinity receptor, FcεRII, found on lymphocytes and monocytes and involved in IgE production and differentiation of B cells. An IgA receptor is found on myeloid cells.

FCS *abbr. for* fetal calf serum.

Fd *symbol for* ferredoxin.

Fd₅₄₀ *symbol for* a membrane-bound **ferredoxin** with a redox potential of –0.54V.

Fd₅₉₀ *symbol for* a membrane-bound **ferredoxin** with a redox potential of –0.59V.

FDA *abbr. for* Food and Drug Administration (of the USA).

Fd fragment a protein fragment obtained by papain hydrolysis of an **immunoglobulin** molecule followed by reduction of the disulfide bonds. It consists of the part of the **heavy chain** N-terminal to the papain hydrolysis site. The N-terminal moiety is variable and forms part of the original antigen-combining site; the C-terminal moiety is constant. *Compare* **Fc fragment**.

FDNB *abbr. for* fluorodinitrobenzene; **Sanger's reagent**.

FDP *abbr. for* fructose 1,6-diphosphate (now named **fructose 1,6-bisphosphate**).

FdU *abbr. for* fluorodeoxyuridine (the symbol FdUrd is recommended).

FdUrd *symbol for* (a residue of) the deoxyribonucleotide fluorodeoxyuridine (*see* **flurouracil**).

Fe *symbol for* iron.

Fe(II) *symbol for* ferrous (divalent) iron.

Fe(III) *symbol for* ferric (trivalent) iron.

FEBS *abbr. for* Federation of European Biochemical Societies.

fecal fat fat in the feces; the output of fecal fat is normally below about 18 mmol 24 h⁻¹ in the adult human. Tests that measure the fat content of feces may assist in diagnosis of generalized malabsorption syndrome; an alternative method is the [^{14}C]**triolein breath-test**.

fed batch process a process during which one or more nutrients are supplied to a bioreactor, usually without removing biomass and products. *See* **batch process**.

Federation of European Biochemical Societies *abbr*: FEBS; an international scientific organization founded in 1964 and now representing some 32 countries, including all those of mainland Europe plus Israel. FEBS is governed by a Council composed of one delegate from each of the adhering national biochemical societies plus the members of the Executive Committee. FEBS organizes an annual meeting, advanced courses, and supports fellowships and three annual prizes. It publishes two journals, *FEBS Letters* and *European Journal of Biochemistry*.

feedback 1 any feature of a system, e.g. of a metabolic pathway or of an electronic circuit, whereby information about the output of the system is used to influence the input to that system. Such feedback may increase the input to the system – **positive feedback** – or reduce the input to the system – **negative feedback**. A cyclic system of components participating in feedback forms a **feedback loop**, and the system is known as a **closed-loop system**. *Compare* **open-loop system**. **2** that part of the output of an electronic, mechanical, or other system that is returned to the input, so modifying its characteristics. *Compare* **feedforward**. —**feed back** *vb*.

feedback control control of the activity of any system by a mechanism involving **feedback**.

feedback inhibition *or* **retroinhibition** inhibition of the activity of an enzyme by the accumulation of a substance produced further along a metabolic pathway of which the enzyme is a constituent. *Compare* **enzyme repression, feedback repression**.

feedback repression an inhibitory feedback system in which the accumulation of a substance produced in a metabolic pathway represses the synthesis of an enzyme that is required at an earlier stage of the pathway. *Compare* **enzyme repression, feedback inhibition**.

feeder cell an irradiated cell that is capable of metabolizing but not of dividing. Such cells are sometimes added to cultures of unirradiated cells to help in the provision of nutrients.

feeder layer a layer of **feeder cells** in a cell culture.

feeder pathway any (minor) metabolic pathway that supplies metabolites to another (major) metabolic pathway.

feedforward 1 any feature of a system, e.g. a metabolic pathway, whereby information about the input to the system is used to influence the output of the system. *Compare* **feedback** (def. 1). **2** that part of the input of a system that influences the output of the system.

Fehling's solution a test reagent for glucose and other reducing substances. It is based on the reduction of copper(II) sulfate in alkaline solution. When boiled with a reducing substance a yellow or red precipitate of copper(I) oxide is formed. *Compare* **Benedict's solution**. [After Hermann Christian von Fehling (1812–85), German organic chemist.]

feline of, pertaining to, or resembling cats.

felix a protein designed for theoretical modelling studies, and expressed from a synthetic gene in *Escherichia coli*. [From four-helix bundle.]

female *symbol*: C; **1** of, pertaining to, or designating the sex that only produces gametes that can be fertilized by male gametes. **2** describing a flower that lacks stamens or has nonfunctional stamens.

female sex hormones any of the **estrogens** (estradiol-17β, estrone, and estriol) that are concerned with the normal growth and development of the mammalian female reproductive tract and, together with **progesterone**, control the estrous and reproductive cycles.

FeMo-co *abbr. for* iron–molybdenum cofactor (of component I of nitrogenase).

FeMo protein any iron–molybdenum-containing protein.

FEMS *abbr. for* Federation of European Microbiological Societies.

femto+ *symbol*: f; SI prefix denoting 10⁻¹⁵ times.

femtomolar *(informal)* describing a solution containing 10^{-15} mol dm⁻³ (of a specified solute).

femtomole *symbol*: fmol; one 10^{-15}th part of a mole (of a specified substance).

Fenton reaction the oxidation of α-hydroxy acids with hydrogen peroxide and ferrous salts (Fenton's reagent) to 2-oxo acids. The term is now used in descriptions of oxidative stress: iron-dependent decomposition of dihydrogen peroxide generates the highly reactive hydroxyl radical, probably via an oxoiron(IV) intermediate. Addition of a reducing agent such as ascorbate sets up a cycle that leads to damage to biomolecules. *Compare* **Haber–Weiss reaction**. [after H.J.H. Fenton]

F episome *an alternative name for* **F plasmid**.

Fe protein *any iron-containing protein.*

fer-1 *see* otoferlin.

feral wild; having escaped from cultivation or domestication and reverted to the wild state.

FERM domain *abbr. for* band F ezrin-radixin-moesin homology domain; a common membrane-binding module involved in localization of proteins to the plasma membrane. FERM domain-containing proteins include cytoskeletal proteins (*e.g.*, talin, the ezrin-radixin-moesin (ERM) protein family), several protein tyrosine kinases and phosphatases, and the neurofibromatosis 2 tumour suppressor protein, merlin. The ERM family proteins crosslink the actin filaments of cytoskeletal structures to the plasma membrane.

ferment 1 any agent or substance that causes fermentation. **2** *a former name for* **enzyme**. **3** *an alternative name for* **fermentation**. **4** to undergo or to effect **fermentation**. —**fermentable** *adj.*; **fermentability** *n*.

fermentation 1 (or ferment) the **anaerobic** decomposition of chemical substances, brought about by **ferments** (def. 1), resulting in the production of simpler substances and, often, of energy; an instance of this, especially the breakdown of glucose to lactate or ethanol. **2** (*in biotechnology*) the use of microorganisms or cultured cells to produce useful materials, such as antibiotics, beverages, enzymes, and some commodity chemicals.

fermenter 1 *or* **fermentor** *or* **biofermenter** an apparatus or a vessel in which to conduct **fermentation** under more or less controlled conditions. **2** a person or other agent that causes or is able to cause fermentation. **3** a genus, species, or strain of microorganism that ferments (or does not ferment) glucose or a specified sugar; e.g. **lactose fermenter**, **non-lactose fermenter**.

fermentor *a variant spelling of* **fermenter** (def. 1).

Fernández–Morán particle another name for **elementary particle** (def. 2), attached to the cristae of the mitochondrion.

ferralterin a soluble non-heme, iron–sulfur-containing chloroplast protein, mediating the light activation of chloroplast fructose 1,6-bisphosphatase.

ferredoxin *symbol*: Fd; any simple, non-enzymic **iron–sulfur protein** that is characterized by having equal numbers of atoms of iron and labile sulfur (releasable as hydrogen sulfide by acidification). The iron and sulfur atoms are present in one or two clusters of two or four atoms of each. Ferredoxins are of M_r 6000–24 000; many have a particularly low redox potential ($E_o = -0.2$ V to -0.6 V). They act as electron carriers in a variety of oxidation–reduction systems, and have been found in a wide range of microorganisms, in chloroplasts, and in mitochondria from adrenal cortex (*see* **adrenodoxin**) and heart muscle. *Compare* **rubredoxin**. *See also* **high-potential iron–sulfur protein**, **nitrogenase**.

ferredoxin-NADP⁺ reductase *another name for* **NADPH:adrenodoxin oxidoreductase**.

ferredoxin–nitrite reductase *see* nitrite reductase.

ferri+ *comb. form* denoting containing ferric iron.

ferric of or containing trivalent iron, Fe^{3+} or Fe(III).

ferrichrome any of a group of growth-promoting Fe(III) chelates formed by various genera of microfungi, e.g. species of *Aspergillus*, *Neurospora*, and *Ustilago*. They are homodetic cyclic hexapeptides made up of a tripeptide of glycine (or other small neutral amino acids) and a tripeptide of an N^4-acyl-N^4-hydroxy-L-ornithine (e.g. the N^4-acetyl compound). They are thus **siderochromes** of the hydroxamic-acid-derivative category.

ferricytochrome any **cytochrome** in which the iron atom is in the ferric form.

ferriheme *or (esp. UK)* **ferrihaem** any Fe(III)–porphyrin coordination complex.

ferriheme chloride *another name for* **hemin**.

ferriheme hydroxide *another name for* **hematin**.

ferrihemochrome any **hemochrome** containing Fe(III).

ferrihemoglobin *or* **methemoglobin** (*abbr.*: MetHb) an oxidation product of hemoglobin in which all its four iron atoms are in the ferric, Fe(III), state, the sixth coordination valencies of which are occupied, in the acid form, by water molecules or, in the alkaline

form, by hydroxide ions; these two forms have characteristic absorption spectra. Unlike hemoglobin, it cannot bind and transport dioxygen. The 3-D structure is known (Brookhaven file 1HGB.PDB). *Compare* **ferrimyoglobin**.

ferrimyoglobin *or* **metmyoglobin** (*abbr.*: MetMb) an oxidation product of myoglobin in which its iron atom is in the ferric, Fe(III), state, the sixth coordination valency of which is occupied, in the acid form, by a water molecule, or, in the alkaline form, by a hydroxide ion. Unlike myoglobin, it cannot bind and store dioxygen. The 3D structure is known (Brookhaven file 1MBC.PDB). *Compare* **ferrihemoglobin**.

ferrioxamine *see* desferrioxamine.

ferritin any of a group of electron-dense, major iron-storage proteins that are widely distributed in animal and plant tissues; the iron is stored in a soluble, readily available form. Ferritins consist of a hollow protein shell; the functional molecule, called **apoferritin**, comprises 24 chains and is roughly spherical, with an external diameter of 12.4–13 nm and an internal diameter of 7–8 nm. The central cavity is able to accommodate a core of up to 4500 Fe(III) atoms, mainly as ferric hydroxide-phosphate, $(FeO(OH))_8$-$(FeO \cdot PO_4H_2)$. There are two types of ferritin subunit: L (light) and H (heavy); the major chain can be L or H, depending on species and tissue type. In clinical chemistry, low plasma ferritin is an indication of a decrease in the body's iron stores; a concentration of less than $12~\mu g~L^{-1}$ indicates a complete absence of stored iron.

ferritin-labelling the use of **ferritin** as an electron-dense label by covalently linking ferritin molecules with the aim of making their position in microscopic specimens apparent by electron microscopy.

ferro+ *comb. form* denoting containing **ferrous** iron.

ferrochelatase EC 4.99.1.1; *systematic name*: protoheme ferrolyase; *other names*: protoheme ferro-lyase; heme synthetase; an enzyme of the pathway for heme biosynthesis that is present on the matrix side of the inner mitochondrial membrane. It catalyses the formation of protoheme from protoporphyrin and Fe^{2+} with release of 2 H^+. The human enzyme (369 amino acids) shows 88 sequence identity with the murine enzyme, and 46 identity with the yeast enzyme. The mammalian enzymes contain a 2Fe-2S binding site at the C-terminus. Numerous mutations in the gene at 18q21.3 result in ferrochelatase deficiency and erythropoietic porphyria. In bacteria the enzyme is part of **siroheme synthase**.

ferrocytochrome any **cytochrome** in which the iron atom is in the Fe(II) state.

ferroheme *or (esp. Brit.)* **ferrohaem 1** heme. **2** any Fe(II)–porphyrin coordination complex.

ferrohemochrome any **hemochrome** containing Fe(II).

ferromagnetism the property displayed by certain substances, typically metallic iron, that have a high magnetic susceptibility and that show increasing magnetization with increasing applied magnetic field strength. Some ferromagnetic materials retain their magnetization in the absence of an applied magnetic field, i.e. they become permanent magnets. Ferromagnetism is due to alignment, temporary or permanent, of the spins of the unpaired electrons in domains in the material. *Compare* **diamagnetism**, **paramagnetism**. —**ferromagnetic** *adj*.

ferroportin *or* **iron-regulated transporter** (*abbr.*: IREG); an integral membrane protein of the basolateral surface of mucosa of the upper small intestine. In association with **hephaestin** it transfers Fe^{2+} from cytosol to transferrin in portal blood. Rare mutations in its gene cause an autosomal dominant form of hemochromatosis.

ferroprotoporphyrin *see* heme.

ferrous of or containing divalent iron, Fe^{2+} or Fe(II).

ferroxamine *see* desferrioxamine.

ferroxidase *see* ceruloplasmin.

ferryl of or containing trivalent iron, Fe(IV)=O. This state is stabilized in iron-containing proteins such as peroxidases and some oxygenases.

fertile 1 capable of producing offspring; capable of producing gametes. **2** capable of undergoing growth and development, e.g. fertile seeds or fertile eggs.

fertilin a mouse sperm **ADAM**. Gene knockout results in sperm that are defective in binding the zona pellucida of oocytes.

fertility the state of being **fertile**.

fertility factor *an alternative name for* **F plasmid**.

fertility vitamin *a former name for* **vitamin E**.

fertilization *or* **fertilisation 1** the fusion of two gametes of opposite sex to form a zygote. **2** the act or process of fertilizing.

fertilize *or* **fertilise 1** to provide a female gamete with a male gamete to effect fertilization. **2** to supply soil or water with mineral and/or organic nutrients to assist the growth of plants.

fertilizer *or* **fertiliser 1** any substance or mixture of substances that is added to soil or water to assist the growth of plants. **2** any object or organism, e.g. an insect, that fertilizes an animal or plant.

fertilizin a mucopolysaccharide from the jelly-coat of the eggs of some species that plays a part in sperm recognition and attracts sperms of the same species.

ferulic acid 4-hydroxy-3-methoxycinnamic acid; 3-(4-hydroxy-3-methoxyphenyl)-2-propenoic acid; an aromatic acid widely distributed in plants. It is a component of **suberin**.

[2Fe-2S] an **iron-sulfur cluster** of two iron and two sulfide atoms, in an **iron-sulfur protein**. The oxidized and reduced states are written thus: $[2Fe-2S]^{2+}$ / $[2Fe-2S]^{+}$.

[4Fe-4S] an **iron-sulfur cluster** of four iron and four sulfide atoms, in an **iron-sulfur protein**.

FES/FPS a family of genes encoding nonreceptor tyrosine kinases; v-*fps* is an oncogene of Fujinami-PRCII sarcoma, while v-*fes* is a cognate gene of feline sarcoma virus. FES/FPS proteins contain an SH2 domain (*see* **SH domain**) adjacent to the kinase domain. Overexpression of c-*fes* transforms fibroblasts.

Fe–S protein *abbr. for* iron–sulfur protein.

fet+ *a variant form of* **feto+** (before a vowel).

fetal alcohol syndrome a combination of severe birth defects (low body weight, short length, facial defects, and microcephaly) affecting infants born to mothers who drink alcohol excessively, especially during the early part of their pregnancy. The offspring may also have defects in joints, heart, and muscle. Children that survive have low intelligence or are mentally retarded.

fetal bovine serum (*abbr.*: FBS) *or* (*sometimes*) **fetal calf serum** a serum prepared from the blood of a fetal calf, widely used in tissue culture media intended for the growth of mammalian cells.

fetal calf serum *abbr.*: FCS; *an alternative term for* **fetal bovine serum**.

fetal hemoglobin *an alternative name for* hemoglobin F (*see* **hemoglobin**).

fetal-lung maturity test a test for maturity of the fetal lung based on its content of dipalmitoylphosphatidylcholine (dipalmitoyllecithin). Dipalmitoyllecithin appears in lung surfactant in embryonic life only when the lung reaches maturity. Its concentration in amniotic fluid reflects its production in the lung and can be measured to determine lung maturity as a prenatal test to aid the decision when to induce labour. Formerly, the lecithin/sphingomyelin ratio was measured.

feto+ *or* (*before a vowel*) **fet+** *comb. form* denoting fetus.

fetomodulin *see* β-**thromboglobulin**.

α-fetoprotein *abbr.*: afp *or* AFP; a protein formed in fetal liver, and found in amniotic fluid. In humans, increased levels in the amniotic fluid between the 14th and 17th weeks of gestation reliably predict most cases of anencephaly and spina bifida cystica, and AFP determination is performed as an antenatal screening test for such defects. Increased serum levels may be found in adults with hepatoma and teratoma; AFP is a valuable marker for liver tumours, blood levels being elevated in about 70% of cases. The normal level in adult human plasma is <10 kU L^{-1}, but rises in normal pregnancy. It is similar to **albumin** (serum) and **vitamin D binding protein**.

fetuin *see* α$_2$**HS-glycoprotein**.

fetus *or* **foetus** the embryo of a mammal in the later stages of development, from the time when the main characteristics of the mature animal can be recognized. In humans the products of conception are termed the fetus from the end of the eighth week of pregnancy until birth. —**fetal** *or* **foetal** *adj*.

Feulgen reaction a specific cytochemical reaction to indicate the presence of DNA. It is based on converting DNA, but not RNA, to apurinic acid by acid hydrolysis with release of the aldehyde functions; the aldehyde functions then give a magenta colour on treatment of the specimen with **Schiff reagent**. Tissue giving such a reac-

tion is termed **Feulgen positive**, whereas the absence of such a reaction characterizes **Feulgen-negative** tissues. The reaction is not suitable for living cells. [After Robert Joachim Feulgen (1884–1955), German chemist.]

FFA *abbr. for* free fatty acid(s). *Compare* **NEFA**.

F factor *an alternative name for* **F plasmid**.

F′ factor *an alternative name for* **F′ plasmid**.

F$_0$F$_1$ complex *see* **H$^+$-transporting ATP synthase**.

(F$_1$ + F$_2$) fragment *see* **heavy meromyosin**.

F$_1$ fragment the separated globular head portion of a **myosin** molecule.

F$_2$ fragment the separated fibrous part of **heavy meromyosin**.

F$_3$ fragment *see* **light meromyosin**.

FGAM synthase *see* **phosphoribosylformylglycinamidine synthase**.

FGAR *abbr. for* formylglycinamide ribonucleotide; an intermediate in **purine biosynthesis**.

FGD1 *or* **faciogenital dysplasia protein** a polypeptide of 761 amino acids, encoded by a gene at Xp11.21, that acts as a guanine nucleotide exchange factor (GEF) for the cell-division cycle protein **Cdc42**. It contains an SH3 domain, a RhoGEF domain, a pleckstrin domain, and a **FYVE** domain, and is activated by binding to Cdc42. Inactivating mutations are the cause of faciogenital dysplasia, or **Aarskog–Scott syndrome**.

FGF *abbr. for* fibroblast growth factor.

FGR a gene encoding a nonreceptor tyrosine kinase and belonging to the **src** family; v-*fgr* is the oncogene of the feline sarcoma virus Gardner-Rasheed. *FGR* transforms most mammalian fibroblasts, but not epithelial cells.

FH$_2$ *abbr.* (*not recommended*) *for* dihydrofolate or dihydrofolic acid.

FH$_4$ *abbr.* (*not recommended*) *for* tetrahydrofolate or tetrahydrofolic acid.

FHA domain *abbr. for* forkhead-associated domain; a phosphopeptide-recognition domain found in many regulatory proteins. It is largely specific for phosphothreonine-containing epitopes, but also recognizes phosphotyrosine. It contains ~80-100 amino acid residues, folded into an 11-stranded β-sandwich, sometimes with small helical insertions between the connecting loops. The domain occurs in a range of eubacterial and eukaryotic proteins (*e.g.*, kinases, phosphatases, kinesins, transcription factors, RNA-binding proteins and metabolic enzymes) that mediate many different cellular processes, including DNA repair, signal transduction, vesicular transport, and protein degradation.

FHA domain *abbr. for* forkhead-associated domain; a phosphopeptide-recognition domain found in many regulatory proteins. It is largely specific for phosphothreonine-containing epitopes, but also recognises phosphotyrosine. It contains ~80-100 amino acid residues, folded into an 11-stranded β-sandwich, sometimes with small helical insertions between the connecting loops. The domain occurs in a range of eubacterial and eukaryotic proteins (*e.g.*, kinases, phosphatases, kinesins, transcription factors, RNA-binding proteins and metabolic enzymes) that mediate many different cellular processes, including DNA repair, signal transduction, vesicular transport, and protein degradation.

FIA *abbr. for* fluorescence immunoassay.

FIAU *abbr. for* 1-(2-deoxy-2-fluoro-β-D-arabinofuranosyl)-5-iodouracil; a drug used for the treatment of herpes virus infections.

fiber *the US spelling of* **fibre**.

fibre *or* (*US*) **fiber** any natural filamentous structure, typically pliable and often strong; by extension the term is applied to filamentous glass or filamentous synthetic polymeric materials. —**fibrous** *adj*.

fibre diagram the X-ray diffraction pattern obtained from a stretched, hence orientated, sample of an amorphous, solid, naturally occurring or synthetic fibrous linear polymer when analysed by the rotating crystal method.

fibre optics 1 a flexible glass or plastic fibre, or a bundle of such fibres, used as a light guide for the transmission of light, especially as images or digitized information. **2** the study of the optical properties of transparent fibres.

fibriform having the form of a fibre or fibres.

fibril any small thread or fibre, bundles of which may constitute a fibre. —**fibrillar, fibrilliform** *adj*.

fibrillar collagen see collagen.

fibrillarin a component of a nucleolar small nuclear ribonucleoprotein particle (snRNP) thought to participate in the first step in processing pre-rRNA, and associated with U3, U8, and U13 snRNP RNAs. It is the nucleolar scleroderma antigen, rich in N^G,N^G-dimethylarginine, and common to the major family of nucleolar snRNPs.

fibrillin a large glycoprotein isolated from fibroblast cell cultures. It is a structural Ca^{2+}-binding component of connective tissue microfibrils, and contains 34 six-cysteine (EGF-like) repeats and five eight-cysteine (TGF-β1 binding protein-like) repeats. Defects are associated with the autosomal dominant disorder, Marfan's syndrome.

β-fibrillosis see amyloidosis.

fibrin the product(s) formed from **fibrinogen** when blood plasma clots. The initial stage of this process is the formation of a **fibrin monomer** (~340 kDa) through the action of **thrombin**, which cleaves negatively charged **fibrinopeptides** A and B from fibrinogen. In intact fibrinogen, fibrinopeptides A and B prevent self-association of fibrinogen. Their removal allows the resultant molecule to self-associate readily in a staggered side-by-side arrangement, forming the fibrous **soluble fibrin polymer**. This in turn is converted into **insoluble fibrin polymer** (see desmofibrin) through the action of plasma transglutaminase (factor XIIIa, EC 2.3.2.13), which, in the presence of calcium ions, creates covalent cross-links between the monomers by transamidating glutamine residues with the amino groups of lysine side chains in other monomers. See also **blood coagulation**.

fibrinase an alternative name for **plasmin**.

fibrin monomer see fibrin.

fibrinogen or factor I a highly soluble, 340 kDa, elongated protein of blood plasma involved in clot formation. It is converted into **fibrin** monomer by the action of **thrombin**. Fibrinogen is a hexamer, 46 nm long and 9 nm maximal diameter, containing two sets of three non-identical chains (α, β, and γ), linked to each other by disulfide bonds. The N termini of all chains are contained in a central nodule, diverging from which are two three-chain coiled coils; these connect the central nodule to the distal nodules containing the distal domains; there are long C termini of the α chains. Thrombin cleaves four Arg–Gly bonds, releasing two each of **fibrinopeptides** A and B from the α and β chains, and forming fibrin monomer (the intact chains in fibrinogen being referred to as the Aα and Bβ chains). This exposes the N-terminal polymerization sites responsible for the formation of the soft clot, which is converted into the hard clot by **factor XIIIa**. This catalyses the ε-(γ-glutamyl) lysine cross-linking between γ chains (stronger) and between α chains (weaker) of different monomers. Fibrinogen also acts as a cofactor in platelet aggregation (a binding site on the γ chain participates). See also **blood coagulation**.

fibrinogenase an alternative name for **thrombin**.

fibrinolysin an alternative name for **plasmin**.

fibrinolysis the process of the dissolution of fibrin in blood clots resulting from the proteolytic action of **plasmin** or other enzymes, e.g. **streptokinase** or the proteases in snake venom. —**fibrinolytic** adj.

fibrinopeptide either of the types of negatively charged peptide cleaved off **fibrinogen** when it is converted to **fibrin** by the action of **thrombin**. Human **fibrinopeptide A** is a 16-residue peptide cleaved from each of the fibrinogen Aα chains; structure (human): ADSGEGDFLAEGGGVR; **fibrinopeptide B** is a 14-residue peptide cleaved from each of the Bβ chains; structure QGVNDNEEGFFSAR.

fibrin polymer see fibrin.

fibrin stabilizing factor or fibrin stabilising factor abbr.: FSF; an alternative name for factor XIIIa; see blood coagulation.

fibroblast or fibrocyte a flattened, irregular, branched, motile cell found distributed throughout vertebrate connective tissue. Such cells form, secrete, and maintain the extracellular collagen and mucopolysaccharide of this tissue. Similar cells occur in many invertebrates.

fibroblast growth factor abbr.: FGF; any of several growth factors that have high affinity for heparin and are mitogenic for mesodermal and neuroectodermal cells but not for endodermally derived cells. Two major subtypes are recognized: an acidic group (aFGFs) and a basic group (bFGFs). FGFs have angiogenic properties, may function as inducing agents during embryonic development, and are useful as a partial or total substitute for serum in tissue culture. Those of the acidic group, mainly from brain or retina, include acidic brain fibroblast growth factor (brain aFGF), astroglial growth factor-1 (AGF-1), retina-derived growth factor (RDGF), eye-derived growth factor-2 (EDGF-2), endothelial cell growth factor (ECGF), and heparin-binding growth factor (HGFα and HGFβ). Those of the basic group, isolated from brain, retina, and cartilage, include basic brain fibroblast growth factor (brain bFGF), eye-derived growth factor-1 (EDGF-1), astroglial growth factor-2 (AGF-2), and cartilage-derived growth factor (CDGF). FGF-3, so named because aFGF and bFGF had already been identified, is a monomer of 239 amino acids expressed during embryonic development and by mammary tumour epithelial cells; it is the product of the int-2 oncogene. FGF-4 is the product of the hst/KS3 gene, comprises 205 amino acids, is expressed during development and by some tumours, is mitogenic for fibroblasts and endothelial cells, and is angiogenic in vivo. FGF-5 (identified by DNA probes in human bladder carcinoma, hepatoma, and endometrial carcinoma) is a monomer of 267 amino acids; it is expressed by fibroblasts, epithelium, some tumour cells, and during embryonic development, and is mitogenic for fibroblasts and endothelial cells. FGF-6 is the product of hst-2 and is a monomer of 208 amino acids; it is expressed during embryonic development and in adult testis, heart, and skeletal muscle, and is mitogenic for fibroblasts but only weakly for endothelial cells. FGF-7 (keratinocyte growth factor) is a monomer of 194 amino acids; it is expressed by epithelial tissue and is mitogenic for epithelial cells but not for fibroblasts or endothelial cells. FGF-23 is present in normal plasma and is secreted in large amounts by human tumours that cause osteomalacia with hyperphosphaturia. It is also increased in patients with X-linked hypophosphatemia caused by mutations in the gene encoding PHEX (a putative endopeptidase). This suggests that PHEX may be responsible for degradative inactivation of FGF-23. An Na^+-phosphate cotransporter of renal tubules may mediate the action of FGF-23. Activating mutations in the gene for FGF-23 on 12p13 cause autosomal dominant hypophosphatemic rickets. See also **KS/hst**.

fibroblast growth factor receptor abbr.: FGFR; any of the plasma membrane **tyrosine kinase** proteins that bind **fibroblast growth factors** and mediate their intracellular effects. All have a single transmembrane helical domain of 21 amino acids, an extracellular domain of about 350 amino acids with three immunoglobulin domains (but variants with fewer are known), and a cytoplasmic domain of about 410–425 amino acids within which is a tyrosine kinase domain which is activated by ligand binding. Six to nine potential N-linked glycosylation sites are present in the extracellular domain. FGFRs are classified into five types. FGFR1 was isolated as a partial cDNA clone encoding an fms-like gene (flg) and is also known as flg. Common variants have only two immunoglobulin (Ig) domains. FGFR2 was isolated as a tyrosine kinase gene encoding a bacterial expressed kinase (**bek**). Common variants of this also have deletion of the first Ig domain. FGFR3 is the product of cek2 (see cek). FGFR4 was isolated as a clone from human erythroleukemia cells. A fifth type of FGFR is known as flg-2. FGFR1 and 2 are expressed by a variety of species, tissues, and cell lines, but FGFR3, FGFR4, and flg-2 have been isolated from a limited number of sources. Heparan sulfate proteoglycans promote binding of FGFs to FGFRs. Binding of ligand to the extracellular domain of FGFRs induces oligomerization of receptors, autophosphorylation, and phosphorylation of intracellular proteins that associate with the activated receptor through their SH2 domains (see **SH domain**). Prominent among these is phospholipase Cγ1 (see **phospholipase**).

fibroblast surface antigen a variant form of surface fibroblast antigen; see fibronectin.

fibrocyte an alternative name for fibroblast.

fibroglycan see syndecan-2.

fibroin or silk fibroin the protein of the silk fibres that are secreted by the posterior gland of the larva of the silk moth, Bombyx mori. It comprises two chains, heavy and light, linked by disulfide bond(s).

Both heavy and light chains, assembled as a mature protein, are needed for efficient intracellular transport and secretion of fibroin; the secondary structure is largely beta sheet.

fibronectin any of a group of related adhesive glycoproteins of high molecular mass found on the surface of animal cells, in connective tissue matrices, and in extracellular fluids. They bind to cell surfaces, collagen, fibrin, heparin, DNA, actin, etc. and are involved in cell adhesion, cell motility, opsonization, wound healing, and maintenance of cell shape. Fibronectin is a disulfide-bonded heterodimer. There are two immunologically cross-reacting forms: **plasma fibronectin** (*also called* cold-insoluble globulin *or* cell attachment protein *or* cell adhesion factor *or* cell spreading factor *or* anti-gelatin factor *or* opsonic α_2-SB-glycoprotein); and **cellular fibronectins** (*also called* cell surface protein *or* large external transformation sensitive (LETS) protein *or* surface fibroblast antigen *or* galactoprotein A *or* zeta protein). Plasma fibronectins are soluble dimers secreted by hepatocytes. Cellular fibronectin is secreted by fibroblasts, epithelial and other cells, and is deposited as multimeric crosslinked fibrils. It is greatly reduced or absent when cells are transformed by oncogenic viruses, chemical carcinogens, or temperature-sensitive viruses, and several changes associated with transformation can be partially reversed by addition of purified cellular fibronectin. A single human gene produces 20 isoforms by alternative splicing. These contain three types of fibronectin domain: I is $\approx$40 amino acids long and contains two disulfide bonds; II is $\approx$60 amino acids long and contains two disulfide bonds; and III is $\approx$90 amino acids long, lacks disulfide bonds, and forms a seven-stranded beta sheet. The longest isoform contains 12, 2, and 15, respectively, of these domains. The rare human disorder Ehlers–Danlos type X is associated with a defect in fibronectin. Fibronectin binds to **integrins** $\alpha_4\beta_1$, $\alpha_v\beta_1$, $\alpha_v\beta_3$, $\alpha_v\beta_5$, $\alpha_v\beta_6$, and $\alpha_4\beta_7$. [From Latin *fibra*, fibre and *necto*, to attach.]

fibrosome an artificial macromolecular complex of fibronectin layered on an agar-coated substrate.

fibrous protein any protein able to aggregate into fibres or filaments, and serving mainly as a structural protein. The polypeptide chains in such proteins are either extended or coiled in one dimension, and their structure is maintained largely by interchain hydrogen bonds.

fibulin a Ca^{2+}-binding glycoprotein of the extracellular matrix with nine EGF-like repeats. Four forms (A–D) differ only in their C-terminal region, and arise by alternative splicing.

ficain *or* **ficin** a cysteine endopeptidase, EC 3.4.22.3, from *Ficus* spp. with preferential cleavage at Lys-, Ala-, Tyr-, Gly-, Asn-, Leu-, and Val-. The name is also used for the crude dried latex from *Ficus* spp., but extracts of only a small proportion of these (in one study only 13 out of 43 species) contain proteolytic activity. Its action is similar to that of **papain**.

ficin *another name for* ficain.

Fick's first diffusion law a mathematical description of the physical process of solute diffusion; the flux (mass transferred) of particles, *J*, diffusing across a unit plane in unit time is given by:

$$J = -D(\partial c/\partial x),$$

where the mass concentration gradient at *x* is $\partial c/\partial x$ and *D* is the **diffusion coefficient** for the solute. [After Adolph Eugen Fick (1829–1901), German physiologist.]

Fick's second diffusion law a mathematical description of the rate of change in mass concentration, *c*, at point *x* of the diffusing particles with time, $(\partial c/\partial t)$, during solute diffusion, for unsteady conditions $(\partial c/\partial x \neq 0)$; it is given by:

$$(\partial c/\partial t) = D(\partial^2 c/\partial x^2),$$

where *D* is the **diffusion coefficient**, assumed to be independent of the concentration.

Fick unit *symbol*: F; a non-SI unit sometimes used to express a **diffusion coefficient**; it is equal to 10^{-7} cm^2 s^{-1}.

ficolin a multimeric protein, named after its fibrinogen- and collagen-like domains, originally isolated as a TGF-β1-binding protein. Recombinant ficolin-α and ficolin-β expressed from cDNA do not bind TGF-β1. Ficolin-α is expressed preferentially in uterus, lung, and placenta, whereas ficolin-β RNA is most abundant in skeletal muscle. The function is at present unknown.

Ficoll *or* **Ficoll 400** a synthetic copolymer of sucrose and epichlorhydrin. It has a branched structure, a high content of hydroxyl groups, which renders it very water-soluble (up to 50 w/v), and a mass average M_r of 400 000 $\pm$ 100 000. Solutions of Ficoll have low osmotic pressures and are useful in preparing density gradients; Ficoll is widely used in preparing continuous or discontinuous gradients for separating cell types. Ficoll 70, with an M_r of 70000, is useful in perfusion studies.

Ficoll-Paque *the proprietary name for* an aqueous solution of **Ficoll** and diatrizoate sodium with a density of 1.077 $\pm$ 0.001 g cm^{-3}. It is useful for the one-step separation of lymphocytes from a sample of anticoagulant-treated blood and for the preparation of other cell types by density-gradient centrifugation.

fidelity the degree to which the output of any system accurately describes or reflects the input to the system. The term is used especially of the processes of replication, translation, and transcription, and of electronic amplifiers.

field 1 a region of space in which a force is exerted on an object because of its charge (**electric field**), magnetic dipole (**magnetic field**), mass (**gravitational field**), or other attribute. **2** a region of space through which (ionizing) radiation is passing. **3** *or* **field of view** the area within which an object is observable with a microscope or other optical instrument.

Fieser's solution a solution for removing dioxygen from any other nonacidic, nonoxidizing gas. It contains potassium hydroxide, anthraquinone β-sulfonate, and sodium thiosulfate. Red when fresh, it becomes colourless when exhausted. [After Louis Frederick Fieser (1899–1977), US chemist.]

FIGLU *abbr. for* formiminoglutamate.

figure of merit the square of the efficiency divided by the background count for a given liquid scintillation system (*see* **liquid-scintillation counting**). It is useful for optimizing the scintillator composition and the discriminator settings for the counting channel(s).

filaggrin a protein that aggregates keratin intermediate filaments, and promotes disulfide-bond formation amongst them during terminal differentiation of mammalian epidermis. It is initially synthesized as a large, insoluble, highly phosphorylated precursor containing multiple tandem repeats of a 324-amino acid domain. The precursor is deposited as keratohyalin granules; during terminal differentiation, it is dephosphorylated and proteolytically cleaved.

filament any very thin, threadlike structure. *See also* **actin**, **microfilament**, **thick filament**, **thin filament**. —**filamentary** *or* **filamentous** *adj*.

filamentous phage any bacteriophage, such as f1, fd, and M13, that has a single-stranded DNA genome and is capable of infecting only male (F^+) strains of *Escherichia coli*. M13 particles $\approx$900 nm long and have a diameter of 6.5 nm. The replicative form (RF) of the genome is a circular double-stranded DNA molecule that generates single-stranded copies by the rolling-circle mode of replication. These are packaged in capsid proteins and released from the cell. Infected cells are not lysed.

filamin a protein, extractable from smooth muscle and other tissues, that induces calcium-ion-insensitive gelation of actin. It promotes orthogonal branching of actin filaments, and links actin filaments to membrane glycoproteins. It exists as a homodimer, and contains an actin-binding domain.

filensin a membrane- and cytoskeleton-associated protein. It is an intermediate filament type III, unique to lens fibre cells. It has primary and secondary structural similarity to other intermediate filament proteins, and serves as an anchorage site for **vimentin** intermediate filaments.

filial generation *symbol*: F; any of the generations of living organisms following a particular parental generation. The generation resulting from a mating of the parents is the **first filial generation** *symbol*: (F$_1$), while that resulting from crossing of F$_1$ individuals is the **second filial generation** *symbol*: (F$_2$), and so on.

filipin a neutral polyene antibiotic from *Streptomyces filipinensis*. It is similar in action to **amphotericin B**.

film badge a masked photographic film worn by workers exposed to ionizing radiation to indicate the extent of their exposure to radiation.

filter **1** any piece or layer of material used or useful for (completely or partially) removing selected components, especially in suspension, from liquid or gaseous mixtures, e.g. a sheet of paper, a stratum of sand (or similar material), or a piece of porous ceramic or plastic. *See also* **membrane filter**, **ultrafilter**. **2** a device incorporating such a piece or layer of material. **3** a block or sheet of material for reducing the intensity of particulate radiation. **4** a screen for selectively absorbing or attenuating electromagnetic radiation or sound waves of some particular or all frequencies. **5** a passive electronic circuit or device that selectively attenuates or allows the passage of certain frequencies of an electrical signal. **6** to remove or separate (suspended particles, macromolecular solutes, selected frequencies of electrical signal or electromagnetic radiation, etc.) from (a heterogeneous or composite fluid, beam, signal, etc.) by means of a filter; to pass (something) through a filter. **7** (*as modifier*) containing or using a filter or filters, or used for filtering, as in a filter fluorometer or filter photometer.

filterable *or* **filtrable** capable of being filtered; able to pass through a (specified) filter.

filterable agent *a former name for* bacteriophage, virus, or other submicroscopic noncellular microorganism.

filterable virus *a former name for* a virus the particles of which are capable of passing through a filter that will retain bacteria and other cellular microorganisms.

filter membrane *an alternative term for* **membrane filter**.

filter paper porous, unsized paper used for filtration; a disk, piece, or sheet of such paper. Such paper may be chemically treated for special applications, e.g. for use in quantitative analysis; it is also used in **paper chromatography**.

filter press a device for effecting filtration, consisting of a series of (metal, plastic, or wooden) frames the two sides of which are covered with filter cloth. The frames are clamped together and the liquid to be filtered is pumped into them in such a way that the solid residue forms a cake between the cloths and the liquid filtrate is conducted away separately.

filter pump a type of vacuum pump used to assist filtration, in which a jet of water forced through a narrow orifice entraps air from the system being evacuated.

filtrable *a variant spelling of* **filterable**.

filtrate the liquid or gas that has passed through a **filter** (def. 1).

filtration the act or process of separating or (completely or partially) removing selected components (of a mixture) by means of a **filter** (def. 1), to produce a **filtrate**. The term includes **gel filtration**, **membrane filtration**, and **ultrafiltration**.

filtration enrichment a method for isolating fungal auxotrophs. Mutagenized spores are placed in a minimal medium in which only normal spores germinate and produce a mycelial network. This mycelial network is then removed by filtration and the remaining auxotrophic spores are provided with an enriched medium to allow their germination, growth, and propagation.

filtration fraction (*in physiology*) the fraction of the plasma that is filtered through the renal glomeruli. It is equal to the **glomerular filtration rate** divided by the renal plasma flow and is frequently taken to be equal to the ratio of the inulin clearance to p-aminohippurate clearance.

fimbria (*pl.* **fimbriae**) **1** (*in botany and zoology*) a fringe or fringelike edge or margin, e.g. especially at the opening of a fallopian tube. **2** (*in bacteriology*) *an alternative term for* **pilus** (especially a common pilus). —**fimbrial** *adj.*

fimbriate *or* **fimbriated** characterized by or possessing fimbriae; fringed at the edge or margin.

fimbrin an **actin-binding protein** involved in the development and maintenance of cell polarity.

fine structure *or* **microstructure** any structural detail observable in genetic material, microscopic specimens, spectra, etc. when examined at high magnification or resolution. *See also* **hyperfine structure**.

fingerprinting any pattern obtained by a variety of means that is characteristic of a biological entity, including **1** the electrophoretic or chromatographic pattern obtained after a process in which a protein is digested, usually enzymically, in defined conditions to produce a mixture of peptides, the nature of the mixture being dependent on the particular protein and the conditions of digestion.

The distribution pattern of spots is rarely, if ever, duplicated if a different protein is used. **2** a set of one or more amino-acid **motifs** (def. 2) obtained by aligning conserved regions of proteins found in protein sequence databases to provide a sequence or set of sequences characteristic of a particular function e.g. the rat metabotropic glutamate receptor 2 precursor fingerprint has seven motifs, the first of which is the 22-residue

VGPVTIACLGALATLFVTVIFI.

The technique involves alignment followed by maximization of sequence information through iterative scanning, with multiple motifs, of a large composite database. The results are held in the **PRINTS** database. **3** the result of **genetic profiling**.

finished genome sequence a genome that has been sequenced 8–9 times (8–9-fold coverage) to reduce ambiguities in the draft sequence, to close gaps, and to allow for only a single error every 10 000 bases. *Compare* **draft genome sequence**.

firefly lantern *or* **lantern** the luminescent abdominal organ, or photophore, of the American firefly, *Photinus pyralis*, a common source of the enzyme **luciferase** and its luminogenic substrate **luciferin**.

firefly luciferase *see* **luciferase**.

firefly luciferin *see* **luciferin**.

first law of photochemistry *an alternative name for* **Grotthus and Draper law**.

first law of thermodynamics *see* **thermodynamics**.

first messenger *an alternative term* sometimes used for **hormone** to distinguish it from **second messenger**. *See also* **chemical messenger**, **pheromone**.

first-order kinetics the kinetics of a first-order reaction; *see* **order of reaction**.

first-order reaction *see* **order of reaction**.

first-pass clearance *or* **first-pass elimination** *or* **first-pass metabolism** the effect of passage of a drug administered by mouth through the liver before it reaches the systemic circulation. Some drugs, in spite of their rapid uptake through the gut mucosal membrane, appear in low concentration in the systemic circulation and hence more has to be administered by mouth than by intravenous injection. As little as 10–20% of the drug may reach the systemic circulation unchanged as a result of their metabolism, mainly by the liver. Some authors extend the term to include metabolism also by the cells of the gut mucosal membrane.

Fis1 a mitochondrial fission protein of yeast that forms an anchor on which **Dnm1** and **Mdv1** are recruited to form the contractile ring on the outer membrane.

Fischer conventions conventions used in the construction of plane projection formulas and first formulated by German organic chemist Emil Hermann Fischer (1852–1919). The first convention may be stated as 'All of the tetrahedral apices that connect asymmetric carbon atoms lie in a straight line in the plane of the paper. This line shall contain the lower edges of all these tetrahedra, their other apices being above the plane and on opposite sides of the line' (*see* **Fischer projection**). The second convention is the assumption of one specified configuration for (+)saccharic acid and hence for glucose. At the time of Fischer's work, absolute configurations were not known. Fortunately, he guessed the correct configuration for (+)glucose (now designated D-glucose).

Fischer ester *former name for* glyceraldehyde 3-phosphate (*see* **d-glyceraldehyde 3-phosphate**). [After Hermann Otto Laurenz Fischer (1888–1960), German organic chemist and son of Emil H. Fischer.]

Fischer projection a diagrammatic representation in two dimensions (as on paper) of the three-dimensional structure of an open-chain chiral molecule of the type R^1CHXR^2, useful particularly for monosaccharides (X = OH) and their derivatives and for amino acids (X = NH_2) and their derivatives. According to the first **Fischer convention**, the carbon chain is written vertically with the most highly oxidized carbon atom (number 1) at the top, the bonds between the carbon atoms being either in or below the plane of the paper; the hydrogen atoms and the X groups, written to the left and right of the carbon atoms to which they are attached, are considered as being in front of the plane of the paper. If X is on the right, the configuration at that carbon atom is designated by the prefix D- (from Latin *dexter*, turned to the right), whereas if X is on the left, the configuration is designated L- (from Latin *laevus*, turned to the left). *See also* **d/L convention**.

FISH *abbr. for* fluorescence *in situ* hybridization.

fish-odour syndrome a condition of humans characterized by an offensive odour of rotting fish, due to abnormal excretion of trimethylamine in the breath, urine, sweat, saliva, and vaginal secretions, especially after eating foods containing trimethylamine (e.g. sea fish) or choline. It results from a reduced ability to oxidize trimethylamine *N*-oxide, normally effected by liver microsomal monooxygenases.

fish oil oil obtained from the liver or flesh of fish; it is comparatively rich in fatty acids of the *n*-3 family (*see* **fatty-acid nomenclature**). Generally, the flesh of so-called oily fish, such as herring, mackerel, salmon, and trout, has a higher content of C_{20} and C_{22} *n*-3 acids than that of white fish such as cod or plaice. Recently, increasing the dietary content of long-chain *n*-3 fatty acids has been promoted as a prophylactic measure in the prevention of heart disease, though the mechanisms suggested are conjectural. *See also* **cod-liver oil**.

Fiske and SubbaRow method a colorimetric method for determination of inorganic phosphate, based on the production of a blue colour when the sample is treated with ammonium molybdate and 1-amino-2-naphthol-4-sulfonic acid. [After Cyrus Hartwell Fiske (1890–1978) and Yellapragada SubbaRow (1896–1948).]

FISP-12 *another name for* mouse **connective tissue growth factor**.

fissile tending to undergo **fission**.

fission 1 *or* **nuclear fission** (*in physics*) a nuclear reaction in which a heavy atomic nucleus splits into two (or, rarely, three or four) approximately equally sized nuclides, with release of a considerable amount of energy. **2** (*in biology*) cleavage of cells or division of unicellular organisms into two (or more) parts.

fission product any of the stable or unstable nuclides resulting from **fission** (def. 1).

FITC *abbr. for* **1** fluorescein isothiocyanate. **2** the fluorescein isothiocarbamoyl group.

fitness (*in biology*) the survival value and the reproductive capability of a given genotype as compared with the average of the population or of other genotypes in the population.

fix 1 to render, or to become, stable, firm, or insoluble; to attach (to something). **2** (*in chemistry*) to convert a free element into one of its compounds; especially to convert atmospheric dinitrogen into nitrogen-containing substances (*see* **nitrogen fixation**); to convert a volatile or fluid substance into a nonvolatile or solid form; to combine an inorganic substance with an organic substance. **3** (*in biology*) to preserve (and harden) cells, tissues, or organisms for subsequent (microscopic) study. **4** (*in photography*) to treat a film or plate with fixer so as to render permanent the developed image. —**fixation** *n*.

fixative a protein-denaturing fluid, e.g. formaldehyde, that is used to **fix** (def. 3) biological material for (microscopic) study.

fixed antibody *see* **cell-bound antibody**. *Compare* **humoral antibody**.

fixed nitrogen any product of **nitrogen fixation**.

fixer (*in photography*) a chemical solution used to remove unexposed, hence unreduced, silver halide granules from a photographic emulsion after the image has been developed.

FiXeR an agonist or antagonist of the **farnesoid X receptor** (FXR).

FIZZ a family of cysteine-rich secreted proteins. FIZZ 1 is present in inflamed lung tissue, and FIZZ 3 is the same as **resistin**.

FK506 an immunosuppressant, isolated from *Streptomyces tsukubaensis*. It is a macrolide antibiotic of M_r 803, and binds to **FKBP**-binding proteins. It is a potent inhibitor of T-cell activation and prevents allograft rejection. The *in vivo* function of this **immunophilin** is not clear but in a complex with FKBP it blocks the action of **calcineurin** in the liver. It is also a powerful neuroprotective agent in a model of focal cerebral ischemia and may have clinical potential for the treatment of stroke. FK506 also reduces **nitric oxide** production by inhibiting the calcineurin-mediated dephosphorylation of **nitric-oxide synthase** (EC 1.14.13.39). One proprietary name is Tacrolimus.

FKBP *abbr. for* FK506-binding protein; any of an extensive family of small proteins to which the immunosuppressant **FK506** binds. They are found in many species, from bacteria to humans. Like **cyclophilin**, they possess **peptidyl–prolyl isomerase** activity (EC 5.2.1.8), but have no similarity in structure. Both types belong to the class of proteins termed **immunophilins**.

FLAG a proprietary synthetic epitope tag consisting of eight amino acids (Asp–Tyr–Lys–Asp–Asp–Asp–Asp–Lys) and readily detected using anti-FLAG antibodies.

flagella *the plural of* **flagellum**.

flagellate 1 describing unicellular organisms that have a **flagellum** or flagella. **2** resembling a flagellum. **3** an organism that propels itself by means of a flagellum or flagella, especially any member of the protozoan superclass Mastigophora.

flagellin any member of a group of similar monomeric soluble globular proteins that constitute the subunits of bacterial flagella. They also comprise the major antigens of several bacteria, including *Salmonella* spp., which alternate between the production of two antigenic forms of flagella, termed phase 1 and phase 2.

flagellum (*pl.* **flagella** *or* **flagellums**) **1** the specialized locomotory appendage of bacteria of certain taxonomic groups, consisting of a noncontractile, filiform extension through the cell surface, borne singly, in groups, or as a covering layer. It is commonly 3–20 μm long and 12–25 nm in diameter, is built up of several (often three) longitudinally arranged chains of **flagellin** subunits (often in a spiral), and is anchored to the cell envelope by a **basal body** (def. 1), which imparts rotatory motion to it. Bacterial flagella carry the H (or flagellar) antigens, which are useful in some instances, e.g. in *Salmonella* spp., for classification. **2** the specialized locomotory appendage of the motile cells of unicellular eukaryotes of certain taxonomic groups (e.g. some algae and protozoa) and also of the male gametes (spermatozoa) of most groups of animals. Usually borne singly or in small groups, it is a whiplike structure, similar to, but generally longer than a **cilium**, commonly about 40 μm long and 200–300 nm in diameter. It is built up of an outer membranous sheath, which is a continuation of the cell membrane, and an inner **axoneme**, which terminates within the cell at a **basal body** (def. 2). Eukaryotic flagella often exhibit undulatory motion, which in multiflagellate organisms can occur as coordinated beating. —**flagellar** *adj.*

flame emission spectrophotometry a type of **atomic emission spectrophotometry** in which excitation is produced by a flame.

flame ionization detector *or* **flame ionisation detector** a sensitive detector used in **gas chromatography** that depends on changes in the electrical conductivity of a flame (burning a mixture of dihydrogen and dinitrogen in air or dihydrogen and dioxygen), brought about by ionization of the vapours of organic substances emerging from the column.

flame photometer an instrument for **flame photometry**.

flame photometry an earlier and simpler, often direct-reading, version of atomic **emission spectrophotometry** in which isolation of the required emission line(s) is achieved by the use of an appropriate filter.

flame spectrophotometry any form of **atomic absorption spectrophotometry** or atomic **emission spectrophotometry** in which excitation is produced by a flame.

flanking marker (DNA) a DNA marker on either side of a locus; a DNA sequence on either side of a gene.

flash evaporation a technique for concentrating a solution at a relatively low temperature thus preventing denaturation of non-volatile solutes. The solution is placed in a round-bottomed flask or other vessel, which is then rapidly rotated and evacuated; the vapour removed from the solution is reliquefied in a condenser and retained in a cold trap. The rotation of the vessel causes the liquid to form a comparatively thin film, which increases its surface-to-volume ratio, thereby assisting evaporation.

flash evaporator an apparatus for **flash evaporation**.

flash photolysis a technique for investigating short-lived primary or subsequent intermediates, e.g. radicals, in a **photochemical reaction**. The sample is subjected to a brief but powerful flash of light of appropriate wavelength and the **photolysis** products are identified by their absorption or fluorescence spectra, using an independent source of radiation. The kinetics of very fast reactions can be investigated by this technique.

flash point the lowest temperature at which the vapour above a volatile combustible substance will ignite to produce a momentary flash when a small flame is applied under specified conditions.

flat-bed indicating that a technique is carried out in a horizontal plane, e.g. flat-bed chromatography.

flat-file a human-readable data file that has no structured interrelationships. It is a convenient format for data interchange or analysis; flat-files may be created as output from relational databases in a format suitable for loading into other databases or for access by analysis or query software.

flavan 2-phenylchroman; 2,3-dihydro-2-phenylbenzopyran; the parent ring structure on which the structures of flavanols, flavanones, flavones, flavonols, and other classes of aglycons of flavonoids are based.

flavanol any hydroxylated derivative of **flavan**. Glycosides of flavan-3-ol and of flavan-3,4-diol respectively form two of the classes of **flavonoid** (def. 1).

flavanone 1 2,3-dihydroflavone; 2-phenylchroman-4-one; 2,3-dihydro-2-phenylbenzopyran-4-one; it is the 4-oxo derivative of **flavan** and the 2,3-dihydro derivative of **flavone**. **2** any of a group of hydroxylated derivatives of flavanone. Flavanone glycosides are colourless; they are found in flowering plants and form one of the classes of **flavonoid** (def. 1).

flavin or (formerly) **flavine** the stem name and trivial name for any member of a group of yellow-coloured, 10-substituted derivatives of 7,8-dimethylisoalloxazine, as in riboflavin, **flavin-adenine dinucleotide**, etc.

flavin-adenine dinucleotide abbr.: FAD; adenosine diphosphate riboflavin; adenosine(5′)diphospho(5′)riboflavin; riboflavin 5′-(adenosine 5′-diphosphate); 7,8-dimethyl-10-[5-(adenosine(5′)-diphospho)-D-ribityl]isoalloxazine; an atypical dinucleotide, not only in that it contains, as does **nicotinamide-adenine dinucleotide**, a phosphoric anhydride linkage between the component mononucleotide units, which in this instance are **adenosine 5′-phosphate** and **flavin mononucleotide**, but also because the latter is not strictly a nucleotide in any case, although considered as one (see **nucleotide** (def. 3)). It forms the coenzyme or the prosthetic group of various flavoprotein oxidoreductase enzymes, in which it functions as an electron acceptor by being reversibly converted to **flavin-adenine dinucleotide (reduced)**. See also **riboflavin**.

flavin-adenine dinucleotide (reduced) abbr.: FADH$_2$; adenosine diphosphate dihydroriboflavin; 1,5-dihydro-7,8-dimethyl-10-[5-(adenosine(5′)diphospho)-D-ribityl]isoalloxazine. The reduction occurs at the two unprotonated nitrogen atoms at positions 1 and 5 of FAD. see **flavin-adenine dinucleotide**.

flavin coenzyme a general name sometimes used for a **flavin nucleotide** when not covalently linked to an apoenzyme.

flavin enzyme see **flavoenzyme**.

flavin mononucleotide abbr.: FMN; riboflavin 5′-phosphate, 7,8-dimethyl-10-(5-phospho-D-ribityl)isoalloxazine. Although not strictly a mononucleotide, since it contains an alditol moiety rather than a glycose moiety, it nevertheless is considered as one (see **nucleotide** (def. 3)), and as such is a component of **flavin-adenine dinucleotide**. It forms the coenzyme (or occasionally the prosthetic group) of various flavoprotein oxidoreductase enzymes, where it functions as an electron acceptor by being reversibly converted to **flavin mononucleotide (reduced)**. See also **riboflavin**.

flavin mononucleotide (reduced) *abbr.* FMNH₂; dihydroriboflavin 5′-phosphate, 1,5-dihydro-7,8-dimethyl-10-(5-phospho-D-ribityl)isoalloxazine. The reduction occurs at the two unprotonated nitrogen atoms at positions 1 and 5 of FMN. *See* **flavin mononucleotide**.

flavin nucleotide a general name sometimes used for flavin-adenine dinucleotide and flavin mononucleotide, and their reduced forms.

flavin reductase *(in luminescent bacteria)* the product of the *luxG* gene in several marine luminescent bacteria. It is presumed from sequence similarity to be an FAD/NAD(P) flavoprotein oxidoreductase.

flavo+ *comb. form* **1** yellow. **2** containing a **flavin nucleotide**.

flavocytochrome any protein that is both a **flavoprotein** and a **cytochrome**. Examples include: (1) yeast L-lactate dehydrogenase (cytochrome), (EC 1.1.2.3), also known as **cytochrome b_2**; and (2) *Bacillus megatherium* flavocytochrome P450 BM3, a fatty acid hydroxylase (119 kDa).

flavodoxin any of a group of small protein electron carriers, widely distributed in cyanobacteria and anaerobic- and photosynthetic bacteria, that contain covalently-linked flavin mononucleotide. They readily form stable blue semiquinones, and can substitute functionally for **ferredoxin**.

flavoenzyme *or* **flavin enzyme** any of a large group of yellow **flavoprotein** enzymes that contain FAD or FMN as cofactor. They frequently have an α/β barrel topology.

flavolans *see* **tannin**.

flavone 1 2-phenylchromen-4-one; 2-phenyl-4*H*-1-benzopyran-4-one; a potent inhibitor of **prostaglandin-endoperoxide synthase**. **2** any of a group of hydroxylated derivatives of flavone. Flavone glycosides occur as pigments, usually yellow, in flowering plants and form one of the classes of **flavonoid** (def. 1). *Distinguish from* **flavanone**.

flavonoid 1 any of a very large and widespread group of water-soluble phenolic derivatives, often brightly coloured and usually found in the vacuoles of plant cells. They are glycosides with the structures of their aglycon moieties based on the **flavan** skeleton, and are classified according to the oxidation state of its pyran ring. Some 10–12 classes are recognized. **2** any flavone, isoflavone, or neoflavone, or any of their derivatives, including bioflavonoids. *See also* **apigenin**, **myricetin**, **quercetin**. **3** of, relating to, or resembling flavone, isoflavone, or neoflavone.

flavonol 1 3-hydroxyflavone. **2** any of a group of hydroxylated derivatives of flavonol. Flavonol glycosides occur as pigments, usually yellow, in flowering plants, and form one of the classes of **flavonoid** (def. 1). *Distinguish from* **flavanol**.

flavopiridol a chlorinated flavone, which is a potential anti-cancer therapeutic agent. It inhibits Cdk1,Cdk2 and Cdk4. The drug docks in the ATP binding site. The crystal structure of the flavone at-

tached to Cdk2 has been determined. The flavone is derived by chemical synthesis from a parent structure obtained from *Dysoxylum binectariferum* a plant native to India.

flavopiridol a chlorinated flavone, which is a potential anti-cancer therapeutic agent. It inhibits Cdk1,Cdk2 and Cdk4. The drug docks in the ATP binding site. The crystal structure of the flavone attached to Cdk2 has been determined. The flavone is derived by chemical synthesis from a parent structure obtained from *Dysoxylum binectariferum* a plant native to India.

flavoprotein *abbr.*: FP; any enzyme or other protein of which a flavin nucleotide is a coenzyme or prosthetic group. They occur in virtually all cells, where they act as oxidoreductases or as electron carriers in the terminal electron transport chains.

Fletcher factor *see* **plasma kallikrein**.

flexible linker a sequence of amino acids that is structurally flexible, allowing the domains either side of it to move relative to each other. Linker regions are often disordered in protein crystal structures.

flg *see* **fibroblast growth factor receptor**.

flickering cluster model a model for the structure of liquid water in which small, tetrahedrally coordinated, icelike regions, held together by hydrogen bonds, are thought to exist for short periods, continually breaking up and re-forming.

flightin the enzyme **glycerol 3-phosphate dehydrogenase (NAD⁺)** obtained from certain insects. It is so called because this and mitochondrial **glycerol 3-phosphate oxidase** play a major role in the transfer of energy needed for flight in insects.

FLIM *abbr. for* fluorescence lifetime imaging microscopy; a technique that depends on the fact that each fluorescent dye has its own lifetime in the excited state, so that different dyes with the same fluorescent colour can be distinguished. Fluorescence lifetime is influenced by concentration and by molecular interactions such as occur in **FRET**.

FLIM-FRET a technique that makes it possible to measure the interaction between two proteins labelled with a pair of fluorescent dyes. **FRET** reduces the fluorescence lifetime of the donor fluorophore and therefore can be detected by donor fluorescence alone using **fluorescence lifetime imaging microscopy**. Experiments can be carried out with saturating amounts of acceptor molecules since acceptor fluorescence is not detected.

flip-flop 1 *(in electronics)* a circuit or device having two stable states and two inputs corresponding to these states. The device remains in either state until caused to change to the other state by application of an appropriate signal to the inputs. **2** *(in enzymology)* a mechanism in which a functional interrelationship between distinct active sites on identical monomers, which are parts of an oligomeric enzyme showing Michaelian kinetics, is mediated by a structural modification, the **flip-flop transition**. *Compare* **half-of-the-sites reactivity**. **3** *(in genetics)* the informal name for a reversible, reciprocal exchange betweeen recombining segments of DNA in site-specific recombination systems in prokaryotes or eukaryotes. *See also* **transposable element**. **4** *(in lipid biochemistry)* the passage of phospholipid molecules from one leaflet of a membrane to the other.

flippase EC 3.6.3.1 a protein that binds a phospholipid molecule on one leaflet of a membrane and transports it to the other leaflet, with hydrolysis of ATP.

flipping 1 *a jargon term used in* **nuclear magnetic resonance spectroscopy** *for* the change in orientation of an atomic nucleus from an angle $\theta°$ to $(180 - \theta)°$, with respect to the direction of the magnetic field, caused by absorption of radiation energy of a specific radiofrequency. **2** The rotation of a protein with associated phospholipid from one leaflet of a membrane to the other.

flocculate to aggregate or cause to aggregate into 'woolly' clusters or masses.

flocculation 1 the act or process of flocculating (*see* **flocculate**). **2** the product of such an act or process.

flocculation reaction *or* **flocculation test** a **precipitin** test in which the precipitate appears as floccules.

floccule a small aggregated mass of flocculent material.

flocculent 1 having the appearance of wool or down. **2** *(in chemistry)* consisting of or occurring as small woolly or downy aggregates. **3** *(in biology)* covered with waxy flakes or tufts resembling wool.

flora 1 *collective term for* all the plants, or plant life, characteristic of a given place or period. **2** microbial flora; the microbial population of part of the body, e.g. of the gut, skin, etc. **3** a comprehensive treatise or list of all the plants or plant life of a particular region or period.

Florisil *the proprietary name for* a type of activated magnesium silicate. It is an adsorbent and is useful for column chromatographic separation of lipids.

flotation coefficient a name sometimes given to a negative **sedimentation coefficient**.

flotillin an integral membrane protein associated with caveolae and present primarily in the developing nervous system.

flow birefringence *or* **double refraction of flow** the **birefringence** induced in a fluid by shear, which causes orientation of any asymmetric molecules in the fluid.

flow cabinet another name for **safety cabinet**.

flow cell any **cell** (def. 2) so constructed that appropriate measurements (as of absorbance) may be made on a flowing fluid.

flow cytofluorimeter a **flow cytometer** in which fluorescence of the objects being counted is monitored.

flow cytometer an apparatus for **flow cytometry** in which cells, or subcellular components (e.g. isolated chromosomes), in aqueous suspension are made to flow at high speed, in single file, through a sensing region where optical or electric signals indicative of some important physical or chemical properties of the cells, or components, are generated. These signals are analysed and accumulated in a computer for quantitative evaluation. Typical flow rates that can be achieved are in the region of 1000 cells per second. *See also* **flow sorter, fluorescence-activated cell sorter**.

flow cytometry the technique for counting or measuring some property of cells, or subcellular components, using a **flow cytometer**, often with labelling with a fluorescent marker.

flow dichroism **dichroism** (def. 1) induced in a fluid by shear, which causes orientation of any asymmetric molecules in the fluid.

flower the reproductive structure of angiosperm plants. It consists generally of sepals, petals, and stamens and/or carpels. It is basically a highly modified leafy shoot.

flow-signal *an earlier term for* **rheoseme**.

flow sorter (for cells) a **flow cytometer** to which has been added a device for physically separating cells, or subcellular components, of desired characteristics in order to provide subpopulations of high homogeneity. Typically the cell suspension is forced out of an orifice, 50–100 μm in diameter, forming a high-speed liquid jet in air. Optical sensing of the jet takes place near the orifice and the jet is then broken up into uniform droplets by applied ultrasonic vibration. The droplets then traverse a region of high-intensity constant electric field where the decision-making and charging circuits, activated by the sensors, cause only droplets containing the wanted cells to be charged and deflected by the electric field into a separate collector.

FLP recombinase a yeast enzyme that catalyses site-specific recombination of double-stranded DNA at a 34-nt sequence called the FLP recombinase target (FRT) site. If two FRT sites are facing in the same direction the DNA between them is excised and the flanking sequences joined.

fluctuation test a method of statistical analysis used to determine whether selected variants, e.g. phage- or drug-resistant bacterial mutants, arose spontaneously prior to exposure to the selective agent or were caused by the agent.

fludrocortisone 9α-fluorohydrocortisone (11β)-9-fluoro-11,17, 21-trihydroxypregn-4-ene-3,20-dione; a **corticosteroid** with about 10 times the glucocorticoid action of cortisol, but with 300–600 times greater mineralocorticoid action. It is used topically in ointments.

fludrocortisone

fluence *symbol*: *F* or *H*; *(in photochemistry)* the energy delivered in a given time interval, d*t*; $F = \int I dt$ where *I* is the radiant intensity and *t* is time. It is expressed in J m^{-2}.

fluence rate **fluence** per unit time; it is commonly referred to (strictly incorrectly) as **flux**.

fluid 1 any substance that can flow; any portion of matter in which the components (molecules or particles) continuously change their relative positions. In physics, the term comprehends any liquid or gas; in biology (and in common usage) it generally denotes only a liquid, especially one present in, or secreted by, a living organism. **2** able to flow or to change shape easily; not solid or rigid; liquid. **3** of, concerned with, or using fluid, a fluid, or fluids.

fluidity *(in physics)* the reciprocal of the **viscosity** of a fluid.

fluid mosaic model a model of cell-membrane structure in which the proteins that are integral to the membrane are considered as a heterologous set of globular molecules, each arranged in an amphipathic structure, i.e. with their ionic and highly polar groups protruding from the membrane into the aqueous phase and their nonpolar groups largely buried in the hydrophobic interior of the membrane. These globular molecules are partially embedded in a matrix of phospholipid, the bulk of which is organized as a discontinuous, fluid bilayer, although a small fraction of the lipid may interact specifically with the membrane proteins. Thus the fluid mosaic structure is formally analogous to a two-dimensional oriented solution of integral proteins (or lipoproteins) in a viscous phospholipid bilayer solvent.

Fluon *a proprietary name for* **polytetrafluorethylene**.

fluor *or* **lumiphor** *or* **scintillator** any substance that displays **fluorescence** when excited by electromagnetic or particulate radiation. Fluors are useful especially in **scintillation counters**.

fluor+ *a variant form of* **fluoro+** (sometimes before a vowel).

fluoracetate *see* **fluoroacetate**.

fluorenylmethoxycarbonyl *symbol*: Fmoc; *abbr.*: FMOC; fluoren-9-ylmethoxycarbonyl; a group used for linkage to amino groups for the purpose either of forming fluorescent amino-acid derivatives that can readily be detected after column chromatography, or to protect the amino groups of amino acids or nucleotides while other functional groups are undergoing reaction. Reagents useful for introducing the group are 9-fluorenylmethyl chloroformate and 9-fluorenylmethyl succinimidyl carbonate.

fluorescamine 4-phenylspiro[furan-2(3*H*),1'-(3'*H*)-isobenzofuran]-3,3'-dione; a nonfluorescent compound forming fluorescent structures on reaction with primary amines. *Proprietary name*: Fluram.

fluorescein　9-(2-carboxyphenyl)-6-hydroxy-3*H*-xanthen-3-one; Acid Yellow 73; CI 45350; an orange-red dye, the sodium salt of which is freely soluble in water to give a solution with yellowish-green fluorescence (to a dilution of 0.02 ppm) under UV light. Certain derivatives, e.g. fluorescein isothiocyanate, are useful for **fluorescein labelling** and others, e.g. the nonfluorescent fluorescein diacetate, are useful as substrates for the detection and assay of esterases.

fluorescein labelling the introduction of fluoresceinyl groups (*see* **fluorescein**) into biopolymers for fluorescence polarization studies and fluorescence immunoassay techniques.

fluorescence **1** a type of luminescence that consists of the emission by a substance of electromagnetic radiation, especially visible light, as a result of and immediately (10–100 ns) after the absorption of energy derived from exciting radiation of another, usually shorter, wavelength or from incident subatomic particles (especially electrons or alpha particles); the property of emitting such radiation. **2** the radiation emitted during **fluorescence** (def. 1). *Compare* **phosphorescence**. —**fluorescent, fluorescing** *adj.*; **fluoresce** *vb.*

fluorescence-activated cell sorter *or* **fluorescence-assisted cell sorter** *abbr.*: FACS; a type of **flow sorter** in which cells are characterized and sorted by the intensity of the fluorescence they emit when passing through an exciting laser beam.

fluorescence depolarization *or* **depolarization of fluorescence** the phenomenon occurring when fluorescence is excited by plane polarized radiation; the emitted radiation is always partially depolarized (*see* **fluorescence polarization**). The extent of such depolarization is due, *inter alia*, to the motion of the absorber (i.e. Brownian motion) and to energy transfer between like chromophores. Fluorescence depolarization can be used to follow the changes in the motion (rotation) of an entire macromolecule, or of a part of the molecule, to which fluorescent chromophores are attached; it can also be used to indicate the environment of a **fluorescent probe** in, e.g., a membrane – an environment that results from less tightly packed structures permits greater movement of the probe molecule, thus leading to greater depolarization. *See also* **photon counter, time-resolved fluorescence polarization**.

fluorescence energy transfer a process of energy transfer between two **fluorophors**, which can occur when the emission spectrum of the first fluorophor overlaps the absorption spectrum of the second fluorophor. **Quenching** of the emission from the first compound occurs but the excitation energy is absorbed by the second compound, which then emits its own characteristic fluorescence. If the two fluorophors are attached at different positions to a macromolecule, observations of fluorescence energy transfer between them may be used to determine the distance between the two attachment positions.

fluorescence enhancement *or* **enhancement of fluorescence** a phenomenon occurring with a number of **fluorophors** whose fluorescence is markedly dependent on the polarity of the medium, in which fluorescence is greatly enhanced in a nonpolar or rigid environment, but strongly quenched in an aqueous environment. Examples are 8-anilinonaphthalene sulfonate (**ANS**) when bound to an antibody or trapped in the hydrophobic environment within a protein; and ethidium when intercalated into double-helical regions of a nucleic acid.

fluorescence immunoassay *or* **fluorescent immunoassay** *or* **fluoroimmunoassay** *abbr.*: FIA; a method of assaying very small amounts of material that is similar in principle to **radioimmunoassay** but uses a fluorescent rather than a radioactive label for the antigen. Various techniques can be used for determination of the ratio of bound to free labelled antigen that do not necessitate a separation of the two; these may depend instead on **fluoresence energy transfer, fluorescence enhancement, fluorescence depolarization, fluorescence quenching**, or other phenomena.

fluorescence *in situ* **hybridization** *abbr.*: FISH; a procedure that involves the use of DNA probes to locate in a tissue section specific regions of DNA in the chromosomes. The probes are tagged with a selection of fluorescent molecules that are both excited by and emit light of different wavelengths and hence different colours. The structure and behaviour of individual chromosomes at all stages of the cell cycle can thereby be studied.

fluorescence lifetime imaging microscopy *see* **FLIM**.

fluorescence microscopy a type of light microscopy by means of which fluorescent substances in a specimen may be detected and examined. Exciting light, commonly UV and of appropriate wavelengths, is isolated by means of a suitable filter placed between the light source and the condenser, and emitted radiation is selected by a different filter placed between the objective lens and the eye-piece. The technique is useful for visualizing certain components that are difficult to see by direct light, for localizing fluorescent substances bound to specific components of the specimen, and for determining the orientation of a specimen using the **fluorescence depolarization** technique.

fluorescence photobleaching recovery *an alternative name for* **fluorescence redistribution after photobleaching**.

fluorescence polarization *or* **polarization of fluorescence** the phenomenon that the light emitted from a fluorescent molecules is always partially polarized. This is true whether or not the exciting radiation is plane polarized. In consequence, when plane polarized radiation is used to excite fluorescence, **fluorescence depolarization** occurs.

fluoresence probe *or* **fluorescent probe** any small molecule that undergoes changes in one or more of its fluorescence properties as a result of noncovalent interaction with a protein or other macromolecular structure. Such probes are useful for studying the properties and behaviour of macromolecules. *See* **probe** (def. 1).

fluorescence quantum yield in any fluorescent system, the ratio of the number of photons emitted to the number of photons absorbed; the fraction of excited singlets that become de-excited by fluorescence. *See also* **quantum yield** (def. 1).

fluorescence quenching the reduction in the **fluorescence quantum yield** of a system due to de-excitation of the fluorescent molecules by environmental factors such as collision or complex formation with quenchers in the system, or by **fluorescence energy transfer** to them. The phenomenon is useful in studies of the binding of small molecules to proteins and other macromolecules.

fluorescence redistribution after photobleaching *abbr.*: FRAP; a technique, also termed **fluorescence photobleaching recovery**, for measuring two-dimensional lateral mobility of fluorescent particles in small ($\approx$10 μm^2) regions of a cell surface. A small spot on the surface is photobleached by a brief exposure to an intense, focused laser beam, and the subsequent recovery of fluorescence, due to replenishment of intact fluorophor in the bleached spot by lateral transport from the surrounding surface, is followed.

fluorescence resonance energy transfer *abbr.*: FRET; *see* **Forster transfer**.

fluorescence spectrophotometry the measurement, by means of a **spectrofluorometer**, of the wavelengths and the intensities of the light emitted from a fluorescent sample that is excited by more or less monochromatic exciting radiation.

fluorescence spectrum the **emission spectrum** of a fluorescent system that is excited by radiation of a given wavelength. *Compare* **excitation spectrum.**

fluorescent antibody technique an **immunofluorescence** technique in which antibody conjugated with a fluorophor is used to locate corresponding antigen in tissue sections or smears.

fluorescent immunoassay *a variant spelling of* **fluorescence immunoassay.**

fluorescent probe *a variant spelling of* **fluorescence probe.**

fluorescent screen a screen coated with a substance, e.g. zinc sulfide or calcium tungstate, that fluoresces when irradiated with electromagnetic radiation of certain wavelengths or with particulate radiation.

fluori+ *a variant form of (sometimes before a vowel)* **fluoro+.**

fluoridation the addition of up to approximately 1 ppm of fluoride ion to a drinking water supply with the object of reducing tooth decay in consumers of the water. —**fluoridate** *vb.*

fluorimeter *a variant spelling of* **fluorometer.**

fluorinate 1 to treat or react (a substance) with difluorine. **2** to introduce (one or more) fluorine atoms into (a molecule of an organic compound), whether by addition or by substitution. —**fluorinated** *adj.*; **fluorination** *n.*

fluoro+ *or (sometimes before a vowel)* **fluor+** *comb. form* **1** denoting a fluorine in organic linkage. **2** *or* **fluori+** indicating fluorescence.

fluoroacetate the anion of fluoroacetic acid (= fluoroethanoic acid); a toxic substance found in the leaves of *Dichapetalum cymosum* and other plants that in animals is converted enzymically into **fluorocitrate.** Fluoroacetate and fluoroacetamide have been used as rodenticides.

fluorochrome any substance exhibiting fluorescence, especially one used to impart fluorescence to another molecule or to a specific structure in a biological specimen.

fluorocitrate the trianion of 1-fluoro-2-hydroxy-1,2,3-propanetricarboxylic acid; an extremely toxic substance formed in animal tissues from fluoroacetate. Fluorocitrate is a potent inhibitor of aconitase; it is believed that the fluorine atom interacts strongly with the Fe^{2+} ion at the active centre of the enzyme thereby inhibiting its activity. *See also* **lethal synthesis.**

fluorodeoxyuridine *see* **fluorouracil.**

fluorodeoxyuridylate *see* **fluorouracil.**

fluorodinitrobenzene *see* **Sanger's reagent.**

fluorogen the nonfluorescent precursor of a **fluorophor.**

fluorogenic 1 producing fluorescence. **2** of or pertaining to a **fluorogen**; e.g. a fluorogenic substrate is a nonfluorescent material that can be acted on by an enzyme to produce a fluorescent substance.

fluorogram 1 *or (sometimes)* **autofluorogram** an autographic image derived by **fluorography** (def. 1). **2** *or* **fluorograph** a photographic image produced by **fluorography** (def. 2).

fluorography 1 *or* **autofluorography** *or* **scintillation autography** *or* **scintillation autoradiography** an elaboration of autoradiography that shows a considerable increase in the efficiency of detection of low-energy-beta-particle-emitting nuclides, e.g. carbon-14, sulfur-35, and (especially) tritium, in preparations such as thin-layer chromatograms, gel electrophoretograms, tissue or whole-body sections, etc. The otherwise poor response of autoradiography in these situations results from internal absorption of most of the emitted radiation by the specimen in question. In fluorography this effect is diminished by impregnating the specimen (which preferably should be translucent and colourless) with fluor(s) prior to placing it in contact with an appropriately sensitive photographic emulsion or film, which then responds both to direct emission of beta particles from the specimen and to emission of photons of UV light (scintillations) generated by beta particles, exciting the fluor within the specimen. **2** photography of a fluorescent body, especially of an image on a fluorescent screen. —**fluorograph** *n.*, *vb.*; **fluorographic** *adj.*

fluoroimmunoassay *abbr.*: FIA; *a variant spelling of* **fluorescence immunoassay.**

fluorometer *or* **fluorimeter 1** a device for measuring the intensity and duration of fluorescence. **2** a device that may be attached to a **fluoroscope** to enable the position of a sought object to be deter-

mined with precision. —**fluorometric** *or* **fluorimetric** *adj.*; **fluorometry** *or* **fluorimetry** *n.*

5-fluoroorotic acid *abbr.*: 5-FOA; a substrate for orotidine-5′-phosphate decarboxylase, which converts it to 5-fluorouracil, an inhibitor of DNA synthesis. In yeast genetics it is used to select for strains lacking a functional **ura3** gene.

fluorophor *or* **fluorophore 1** a group of atoms in a molecule that renders the latter fluorescent. **2** a fluorescent substance.

fluoroscope 1 a screen coated with a substance that fluoresces when excited by X-rays; it is used to view the X-ray image of an object. **2** a device used to compare the fluorescence of a solution with that of a standard.

fluorouracil *symbol*: FUra (*not* FU); 5-fluorouracil; 5-fluoro-2,4(1*H*,3*H*)-pyrimidinedione; a synthetic analogue of uracil that is enzymically converted via its β-D-ribofuranoside, fluorouridine (*symbol*: FUrd), to fluorodeoxyuridine (*symbol*: FdUrd) and then to fluorodeoxyuridylate, which is a potent inhibitor of **thymidylate synthase** and, hence, of DNA synthesis. Thus fluorouracil is a **suicide inactivator** of thymidylate synthase. Fluorouracil and fluorodeoxyuridine are anticancer drugs, commonly used in combination therapy against fast-growing tumours.

fluoxetine (±)-*N*-methyl-γ-[4-(trifluoromethyl)phenoxy]benzenepropanamine; a serotonin uptake inhibitor, used as an antidepressant, particularly as the hydrochloride (*proprietary name*: Prozac).

* = racemic

flush end *or* **blunt end** an end of a duplex DNA fragment in which the two strands are coextensive, i. e. terminate at the same point. Such ends may be formed by scission of duplex DNA with some types of endonucleases. *Compare* **cohesive end.** —**flush-ended** *adj.*

flux 1 the rate of passage of energy or matter (usually in crossing a given area or passing through a given volume) under steady-state conditions. *See* **diffusion coefficient, luminous flux, radiant power.** **2** the number of particles passing through unit cross-sectional area of beam per unit time. **3** the rate of (chemical or physical) transformation of a substance within a living cell or tissue; the rate of translocation of a substance within a living cell or tissue, especially across a cell membrane. Such processes are usually measured in moles per second. *See also* **metabolic flux.** **4** the strength of a field of force orthogonal to unit area. *See* **magnetic flux.** **5** *an alternative term for* **flux density** (def. 2).

flux control coefficient *symbol*: C_E^J; in a metabolic sequence at steady state, the response, δ*J*, of the original system flux, *J*, to a small change, δE, in either the concentration or one of the kinetic parameters of a specified enzyme, E; it is given by:

$$C_E^J = (\frac{\partial J/J}{\partial E/E}) \, p, q, r, ...$$

and similarly for every other enzyme in the system. This is the flux control coefficient of the enzyme E on the flux *J*, and is also called **sensitivity coefficient** or **control strength.** Similarly one can calculate a flux control coefficient for the substrate concentration, *S*, on the flux, *J* (this is also called **sensitivity** or **net sensitivity**) and a flux control coefficient of an effector concentration, *I*, on the flux, *J* (this is also called **control strength**).

flux density 1 (of energy or field strength) the **flux** per unit cross-sectional area; *see* **luminous flux density, magnetic flux density.** **2** *or (sometimes)* **flux** (of particulate radiation, including photons) the arithmetic product of the particle density and the mean velocity of the particles; **neutron flux (density).**

fluxional describing a chemical species that undergoes rapid degenerate rearrangements, generally detectable by methods such as nuclear magnetic resonance or X-ray crystallography.

flux-ratio method a method for investigating the mechanisms of enzymic reactions in which simultaneous measurements are made of the rates of enzyme-catalysed isotopic transfers from one product to two substrates while the reaction system is in a steady state. It provides relatively unambiguous information about the order of binding of substrates.

FlyBase *see* **genome database**.

Fm *symbol for* fermium.

fMet *symbol for* a residue of the α-acylamino acid *N*-formyl-L-methionine.

FMN *abbr. for* flavin mononucleotide.

FMN adenylyltransferase EC 2.7.7.2; *other name*: FAD pyrophosphorylase; an enzyme of the pathway for biosynthesis of FAD. It catalyses the reaction of ATP with FMN to form FAD with release of pyrophosphate.

FMNH$_2$ *symbol for* flavin mononucleotide (reduced).

Fmoc *symbol for* the fluorenylmethyloxycarbonyl group.

FMOC *abbr. for* the fluorenylmethyloxycarbonyl group.

fmol *symbol for* femtomole (10^{-15} mole).

FMR1 *symbol for* fragile site mental retardation 1 gene; a genetic locus on the X chromosome (Xq27.3) involved in fragile X syndrome, the most common form of inherited mental retardation affecting males. Mutations take the form of expanded triplet repeats $(CGG)_n$, or $(CCG)_n$ on the complementary strand. *See also* **FMRP**.

FMRP *abbr. for* fragile X mental retardation protein; any isoform of a protein (the longest contains 655 amino acids) generated by alternative splicing of the product encoded by the gene at Xq27.3, which is associated with fragile X syndrome. Expressed mostly in brain and testis, FMRP contains RNA-binding sequences (called KH domains) and clusters of the tripeptide sequence RGG. It is localized mostly to ribosomes active in translation and appears to shuttle between cytoplasm and nucleus. In the syndrome the number of the CGG trinucleotide repeat in the 5′ untranslated region of the mRNA expands from <60 to >200.

FMS *symbol for* either the retroviral **oncogene**, v-*fms*, which is carried by the McDonough strain of feline sarcoma virus, or its related protooncogene, c-*fms*, which encodes the M-CSF-1 receptor precursor (*see* **colony stimulating factor**). The latter is a type I transmembrane protein of the immunoglobulin superfamily. It possesses intrinsic protein tyrosine kinase activity within its cytoplasmic domain, which is activated by the binding of CSF-1 (*see* **colony stimulating factor**) to the extracellular domain. Its transforming counterpart, v-*fms*, encodes a constitutively active receptor kinase that can transform both fibroblasts and hematopoietic cells in a CSF-1-independent manner. Mutations in the c-*fms* gene as well as a critical alteration of the 3′ coding sequences appear to be responsible for fully activating its latent transforming potential. In the v-*fms* protein product, 40 C-terminal amino acids of the normal CSF-1 receptor are replaced by 11 unrelated residues, suggesting that residues near the C terminus normally negatively regulate kinase activity in the absence of CSF-1.

f-number the numerical value of the relative aperture of a lens; it is the ratio of the focal length of a lens to its diameter.

foam 1 a dispersion in which a large proportion of gas by volume, in the form of gas bubbles usually >1 μm in diameter, is dispersed in a liquid, solid, or gel. **2** to produce or cause to produce such a foam.

foam fractionation a method of separation in which a component of the bulk liquid is preferentially adsorbed at the liquid/vapour interface and is removed by foaming.

foaming agent a **surfactant** that when present in small amounts facilitates the formation of a foam and/or stabilizes it.

focal adhesion kinase *abbr.*: FAK; a highly conserved, cytosolic, protein-tyrosine kinase involved in cell–cell and cell–matrix interaction and responsible for formation of the focal adhesion complex. It is widely expressed throughout development. FAK contains an N-terminal integrin-binding domain, a central kinase domain, two proline-rich sequences, and others for targeting to focal adhesions.

focal contact *or* **focal adhesion** any of the areas on the surface of fibroblasts or other cells that are anchored to extracellular matrix where actin-containing microfilaments terminate. The contact is mediated by transmembrane proteins such as integrins, which are then linked to actin microfilaments in the cytoplasm. The filaments are involved both in the attachment of the cells to the substrate and in cell movement.

fodrin *or* **brain spectrin** *or* **calspectin** a protein from bovine brain, immunologically related to **spectrin**, that is involved in secretion, interacts with calmodulin in a calcium-dependent manner, and thus is a candidate for the calcium-dependent movement of the cytoskeleton at the membrane. It contains 20 repeats of a 106-amino-acid domain.

foetal *a variant spelling of* fetal (*see* **fetus**).

foeto+ *a variant spelling of* **feto+**.

foetus *an etymologically misconceived variant spelling of* **fetus**.

FOG1 *abbr. for* friend of GATA-1; a transcription regulator protein that binds erythroid gene-specific transcription factor GATA-1. It contains four C_2H_2 and five CCHC zinc fingers.

FokI a type 2 **restriction endonuclease**; recognition sequence: GGAT-GNNNNNNNNN↑.

folacin *a common name for* **folic acid**.

folate 1 *the recommended trivial name for* pteroylglutamate; *N*-[4-[(2-amino-3,4-dihydro-4-oxopteridin-6-yl)methyl]amino benzoyl]-L-glutamate; the anion of **folic acid** (def. 1). **2** any member of the family of pteroylglutamates, irrespective of the level of reduction of the pteridine ring, of the nature of any one-carbon substituent, and of the number of glutamate residues. *See also* **folic acid** (def. 2).

folate-binding protein *or* **folate receptor** (*abbr.*: FR); either of two GPI-anchored membrane glycoproteins that bind folate with high affinity. FRα and FRβ are encoded by a gene at 11q13.3-q13.5 and have similar coding regions. They provide entry into cells for folate by receptor-mediated endocytosis.

folate coenzyme any of various interconvertible tetrahydrofolates that bear a one-carbon substituent and function as coenzymes in metabolic reactions involving the transfer of one-carbon units. The substituent may be 5-formimino, 5-formyl, 10-formyl, 5,10-methenyl, 5-methyl, or 5,10-methylene.

folate receptor *an alternative term for* **folate-binding protein**.

Folch–Lees protein the most abundant type of protein that is extractable from central-nervous-system myelin by chloroform: methanol (2:1 v/v); it is of at least 150 kDa, and contains a high proportion of aliphatic amino acids. *Compare* **Wolfgram proteolipid protein**.

Folch method a method for the fractionation and isolation of lipids from either tissues or fluids. It involves extraction with various chloroform–methanol–water mixtures.

fold a qualitative decription of a protein tertiary structure; folds are classified using different criteria in various **structure classification databases**.

foldase an enzyme, such as peptidyl proline isomerase or disulfide isomerase, that catalyses isomerization of a polypeptide conformation or exchange of disulfide-bond partners, to facilitate protein folding.

foldback DNA *an alternative term for* hairpin DNA. *See* **hairpin**.

folding problem the problem of determining how a protein folds into its final 3D form given only the information encoded in its primary structure.

folding unit a locally ordered region within the core of a protein molecule, formed by 3-D interaction between two or three segments of secondary structure (α-helix and/or β-strand) that are near one another along the polypeptide chain. Examples are: (αα), (αβ), (βββ), (βαβ). *See also* **domain**.

fold recognition a method for predicting protein tertiary structure that detects 3D folds that are compatible with a particular sequence by comparing the probe sequence with known fold templates housed in a fold library. The most likely match is detected using statistical scoring functions (e.g. solvation potentials or pair potentials), predicted secondary structure, sequence profile information, and so on.

fold template a representative known protein fold stored in a library for use in **fold recognition**.

folic acid *abbr.*: (not recommended) FA *or* F; **1** *the recommended trivial name for* pteroylglutamic acid, the corresponding acid of **folate** (def. 1). **2** *or* **folacin** the name under which any **folate** (def. 2) is generally known in medicine, nutrition, etc. Folic acids are widely distributed members of the vitamin B complex, particularly plentiful in green leafy vegetables, liver, and yeast. Folates are essential for the synthesis of purines and pyrimidines, functioning as carriers of one-carbon units. A deficiency in humans results in megaloblastic anemia and in birth defects in children born to deficient mothers. Folic acids are also growth factors for certain microorganisms. *See also* **folinic acid**.

folic acid conjugate the corresponding acid of any **folate** (def. 2) containing two or more glutamyl residues.

folic acid reductase *an alternative name for* **dihydrofolate reductase**.

Folin–Ciocalteu reagent a phosphomolybdotungstic acid reagent that gives a blue colour with phenols. The reagent has been used in the determination of tryptophan and tyrosine in proteins and peptides. *See also* **Lowry method**.

folinic acid 1 *or (formerly)* **citrovorum factor** (*abbr.*: CF) *or (formerly)* **leucovorin** *or* **5-formyltetrahydrofolic acid** 5-formyl-5,6,7,8-tetrahydropteroyl-L-glutamic acid; the common name for the corresponding acid of **folinate** (5-formyltetrahydrofolate), a folate coenzyme discovered to be a growth factor for the bacterium *Leuconostoc citrovorum* (hence its former synonyms). Folinate is an intermediate in the metabolic conversion of formiminoglutamate pppnpn to glutamate, ammonium, and **active one-carbon**. The calcium salt, calcium folinate (*or formerly* calcium leucovorin), is useful therapeutically, e.g. as an antidote to doses of methotrexate or other folate antagonists. **2** *name sometimes used for* certain other folates, e.g. tetrahydrofolate, especially as growth factors for microorganisms.

Folin–Lowry method a sensitive method of protein assay in which protein reacts both with an alkaline copper tartrate solution and with the **Folin–Ciocalteu reagent** to give a blue colour, the absorbance of which is measured at 750 nm. *Compare* **Lowry method**.

Folin reaction a colorimetric reaction used for the determination of ammonia and amino-acid nitrogen in which a red colour is produced when an alkaline solution of amino acids is treated with 1,2-naphthoquinone-4-sulfonate. [After Otto Knute Olof Folin (1867–1934).]

Folin–Wu reagent a phosphomolybdic acid solution, containing a little tungstate, used in the determination of glucose. It gives a blue colour with the copper (I) oxide produced when the sample containing glucose is heated with an alkaline copper (II) solution. [After Otto Knute Olof Folin (1867–1934) and Hsien Wu (1893–1959).]

folliberin *the recommended name for* follicle-stimulating hormone releasing hormone (*abbr.*: FSH-RH or FRH). Luteinizing hormone releasing hormone (*abbr.*: LHRH) stimulates not only release of luteinizing hormone from the anterior pituitary but also release of follicle-stimulating hormone. This activity is commonly named **gonadotropin-releasing hormone**. There is, however, evidence that a specific FSH-RH may exist in addition.

follicle 1 any small cavity or recess. **2** a cavity that has a secretory or excretory function, e.g. the **Graafian follicle**.

follicle-stimulating hormone *abbr.*: FSH; *recommended name*: follitropin; one of two gonadotropic **glycoprotein hormones** secreted by the anterior pituitary; of 34 kDa (human), 33 kDa (ovine), or 29 kDa (porcine), it consists of α and β subunits, the α subunit being almost identical within a given species to that of other glycoprotein hormones and the β subunit conferring hormonal specificity; the carbohydrate content is 16 (human), 8–9 (ovine), or 7–8 (porcine). In the female it stimulates the growth of Graafian follicles in the ovaries, whereas in the male it stimulates the epithelium of the seminiferous tubules to increased spermatogenesis. Secretion of FSH is stimulated by activin and inhibited by inhibin. Mutation in the FSH β subunit causes delayed puberty in females. FSH acts via a G-protein-associated receptor and activation of adenylate cyclase. Mutation in the receptor causes primary gonadal failure in females.

follicle-stimulating hormone releasing factor (*abbr.*: FRF *or* FSH-RF) *or* **follicle-stimulating hormone releasing hormone** (*abbr.*: FRH *or* FSH-RH) *see* **gonadotropin-releasing hormone**.

follicular 1 like, pertaining to, consisting of, or provided with a follicle or follicles. **2** affecting or involving follicles.

follicular phase *see* **estrous cycle**.

folliculate(d) having or consisting of follicles.

follipsin a serine proteinase from pig ovarian fluid. It is a heterodimer, and cleaves Arg-|-Xaa bonds.

follistatin an **activin**-binding glycoprotein that regulates biosynthesis and secretion of **follicle-stimulating hormone**. Alternative isoforms are formed by alternative splicing.

follitropin *the recommended name for* **follicle-stimulating hormone**.

folyl *the trivial name for* pteroylglutamyl, the acyl group derived from **folic acid** (def. 1). Use of the name is no longer recommended.

fomecin any of several antibiotic substances produced by certain strains of the basidiomycetous fungus *Fomes juniperinus*, especially fomecin A, 2,3,4,-trihydroxy-6-(hydroxymethyl) benzaldehyde.

food any material that can be ingested by an organism and metabolized to provide energy and precursors for biosynthetic reactions.

foot and mouth disease virus the seven serotypes of foot-and-mouth disease (FMD) virus, together with equine rhinovirus 1 form the aphthovirus genus of the family *Picornaviridae*. The **virion** is non-enveloped, 25-30 nm in diameter, icosohedral shaped (5:3:2 symmetry), and contains an infectious single strand of positive sense RNA, approximately 8.4 kb in length. The viral capsid consists of 60 copies of the four structural proteins, 1A (VP4), 1B (VP2), 1C (VP3) and 1D (VP1). The genome also contains the genes of 8 non-structural proteins. Cattle, pigs, sheep, goats and all other cloven hoofed animals, and some other species such as rats, hedgehogs, and coypu are susceptible to FMD. The disease is characterized by fever and vesicles in the mouth and on the feet and mammary glands of adolescent and adult animals and sudden death in young animals of susceptible species. Transmission of the virus is usually by close contact between an infected and susceptible animal, but can be by consumption (usually by pigs) of animal products contaminated by live virus or mechanical transfer of virus by people or vehicles. Pigs infected with some strains of FMD virus can produce large amounts of aerosol virus which, depending on weather conditions and topography, can form discrete plumes over long distances. In 1981, cattle on the Isle of Wight in the UK were infected by aerosol virus produced by infected pigs in Brittany, France, a distance of approximately 250 km.

footprinting a technique for identifying sites where there is protein bound to DNA. Protein–DNA complexes are resistant to the action of nucleases, so that undegraded DNA (the 'footprint') remains after treatment with a nuclease.

For *symbol for* the formyl group, H–CO– (alternative to HCO).

Forbes' disease *an alternative name for* type III **glycogen disease**. [After Gilbert Burnett Forbes (1915–), US physician.]

forbidden clone a hypothetical clone of immunologically competent cells with specificity for some autoantigen that, in accord with the clonal selection theory, has been suppressed in the fetus but may be reactivated in adult life and cause an autoimmune disease.

forbidden transition any transition between energy states in an atom or molecule that is forbidden by the selection rules of quantum mechanics (but can occur nevertheless!).

force *symbol*: *F* (*in physics*) a vector quantity, the external influence that changes or tends to change the state of rest or uniform motion of a body, or the shape of a fixed body. The SI derived unit of force is the **newton**.

forkhead *or* **winged helix/forkhead-like protein** any of a family of transcription factors related to the *Drosophila* developmental gene *forkhead*. These proteins are implicated in developmental control of many tissues, often in response to signalling by members of the transforming growth factor/bone morphogenetic protein (TGF/BMP) family. They bind to DNA through a domain (≈108 amino acids) that resembles a winged helix (a helix-turn-helix with two additional loops, or wings). The gene for human FKHL7 (forkhead homologue-like protein 7) at 6p25 encodes a protein of 553 residues. Several mutations in this gene result in Axenfeld–Rieger anomaly of the eye.

forkhead box protein *see* **FOXP**.

formaldehyde dismutase EC 1.2.99.4; *systematic name*: formaldehyde:formaldehyde oxidoreductase. An enzyme that catalyses a dismutation of two molecules of formaldehyde to formate and methanol. The enzyme functions to detoxify formaldehyde thus rendering the organism resistant to formaldehyde. Its synthesis is induced by addition of formaldehyde to the medium.

formaldehyde gel electrophoresis a modification of agarose gel electrophoresis in which formaldehyde is incorporated in the gel as a denaturing agent for RNA. When denatured in this way RNA molecules migrate as a function of molecular mass.

formaldehyde transketolase EC 2.2.1.3; *systematic name*: D-xylulose-5-phosphate:formaldehyde glycolaldehydetransferase; *other names*: dihydroxyacetone synthase; DHAS; glycerone synthase. An enzyme that catalyses a reaction between D-xylulose 5-phosphate and formaldehyde to form glyceraldehyde 3-phosphate and glycerone; thiamine diphosphate is a coenzyme. The enzyme is important in organisms (some fungi and many prokaryotes) that can utilize methanol as a carbon source in converting formaldehyde, produced by methanol oxidation, into carbohydrate, xylulose 5-phosphate acting as C_2 donor.

formalin *or* **formol** a 40 w/v aqueous solution of formaldehyde stabilized with approximately 12 v/v methanol. It is useful as a general disinfectant and for fixation and preservation of biological specimens.

formate 1 *trivial and preferred name for* **methanoate**, the anion, H–COO⁻, derived from formic acid (methanoic acid). **2** any mixture of formic acid and its anion. It is useful as a buffer over the pH range 2.6–4.8; pK_a (25 °C) = 3.75. **3** any salt or ester of formic acid.

formate–dihydrofolate ligase EC 6.3.4.17; *systematic name*: formate dihydrofolate ligase (ADP-forming); an enzyme that catalyses the formation of 10-formyldihydrofolate from ATP, formate, and dihydrofolate with release of ADP and orthophosphate.

formate–tetrahydrofolate ligase EC 6.3.4.3; *systematic name*: formate:tetrahydrofolate ligase (ADP forming); *other name*: formyltetrahydrofolate synthetase; an enzyme involved in one of the transformations between **folate coenzymes**. It catalyses the formation of 10-formyltetrahydrofolate from ATP, formate, and tetrahydrofolate with release of ADP and orthophosphate. It is one of three enzymes (*see also* **methylenetetrahydrofolate dehydrogenase (NADP⁺)** and **methenyltetrahydrofolate cyclohydrolase**) necessary for the biosynthesis of purines, thymidylate, methionine, histidine, pantothenate, and formyl tRNA-met. It is one activity of the trifunctional enzyme **C1-tetrahydrofolate synthase**. *See* **methylenetetrahydrofolate dehydrogenase (NADP⁺)** for an example.

formation constant *an alternative name for* **stability constant**.

formation function *see* **Bjerrum formation function**.

formazan 1 the compound NH_2–N=CH–N=NH. **2** any substitution derivative of this compound. Many are brightly coloured and may be reversibly oxidized into the corresponding colourless tetrazolium compounds (*see also* **Blue Tetrazolium**). Formazans prepared by coupling the phenylhydrazones of acyclic aldoses with diazonium compounds are useful in characterizing such sugars.

formed elements of the blood a general term for the **erythrocytes**, **leukocytes**, and **platelets**.

formimidoyl the acyl group, CH(=NH)–, derived from (the hypothetical) formimidic acid; the latter may also be regarded as the

imino tautomer, CH(=NH)–OH, of formamide; hence also known as **formimino**.

formimino *see* **formimidoyl**.

formiminoglutamate *abbr.*: FIGLU; *N*-formimidoyl-L-glutamate; formamidino-L-glutarate; an intermediate in the catabolism of histidine. Its formimino (= formimidoyl) group is transferred to tetrahydrofolate to produce **5-formiminotetrahydrofolate**. In folate (or vitamin B_{12}) deficiency FIGLU accumulates in the tissues and its excretion in the urine increases. Estimation of the level of FIGLU in the urine is useful in assessing the extent of folate (or vitamin B_{12}) deficiency.

5-formiminotetrahydrofolate a derivative of tetrahydrofolate that is an intermediate in the metabolism of one-carbon units. It is formed from tetrahydrofolate and **formiminoglutamate**, and with **cyclodeaminase** yields 5,10-methenyltetrahydrofolate.

formin one of several developmental nuclear phosphoproteins found in several tissue types. They are formed by alternative splicing of the mouse *ld* (limb deformity) gene. Mutations in this locus result in defects of growth and patterning of the limb and kidney during embryonic development.

formol *(formerly)* a proprietary name for **formalin**.

formolize *or* **formolise** to react the amino groups of amino acids, peptides, or proteins with formaldehyde.

formol titration the pH titration, with alkali, of amino acids, peptides, or proteins in the presence of formaldehyde. The formaldehyde reacts with the amino groups and lowers their apparent pK values, allowing identification of the region in the titration curve where they are being titrated.

formula (*pl.* **formulas** *or* **formulae**) **1** *(in chemistry)* a collection of symbols of atoms and numbers giving information about the composition and structure of a molecular chemical entity. **2** *(in mathematics)* a general relationship, principle, or rule expressed in symbols, frequently as an equation. **3** a recipe, or prescription, for the preparation of a medicine, food, etc. giving the constituents, their amounts, and the method to be used.

formula weight *a former name for* **relative formula mass**.

formycin either of two related **nucleoside antibiotics** synthesized by the actinomycete *Nocardia interforma* that display antitumour activity. **Formycin A**; 8-aza-9-deazaadenosine; 7-amino-3-(*β*-D-ribofuranosyl)-1*H*-pyrazolo[4,3-*d*]pyrimidine; an analogue of adenosine that inhibits *de novo* purine biosynthesis in tumour cells and is incorporated into RNA. It is deaminated in erythrocytes to **formycin B**, a competitive inhibitor of NAD⁺ in the action of NAD⁺ ADP-ribosyltransferase and an inhibitor of purine nucleoside phosphorylase.

formycin A

formyl *symbol*: For or HCO; the acyl group, H–CO–, derived from formic acid.

formylate to introduce (one or more) formyl groups into an organic compound. —**formylation** *n*.

N-formylmethionine *symbol*: fMet; *codon*: AUG (when used as start codon for initiation) or GUG (when present as the first codon; within genes GUG encodes valine); a modified amino-acid residue which is found at the N terminus of nascent polypeptide chains in prokaryotes and bacteriophages. It is also present in the mitochondria and chloroplasts of eukaryotes, supporting the hypothesis that they are derived from prokaryotes.

formylmethionine deformylase EC 3.5.1.31; *systematic name*: N-formyl-L-methionine amidohydrolase; *other name*: **polypeptide deformylase**; an enzyme, found in prokaryotes, that removes the formyl group from the N-terminal N-formylmethionine of many polypeptides produced by translation of mRNA; example from *Escherichia coli*: DEF_ECOLI, 169 amino acids (19.33 kDa). *See also* **post-translational modification**.

5-formyltetrahydrofolate *see* **folinic acid**.

5-formyltetrahydrofolate cyclo-ligase EC 6.3.3.2; *other name*: 5,10-methenyltetrahydrofolate synthetase; an enzyme that catalyses the reaction between ATP and 5-formyltetrahydrofolate to form 5,10-methylenetetrahydrofolate with release of ADP and orthophosphate.

formyltetrahydrofolate synthetase *see* **formate–tetrahydrofolate ligase**.

5-formyltetrahydrofolic acid *an alternative name for* **folinic acid**.

forskolin colforsin; 7β-acetoxy-8,13-epoxy-1α,6β,9α-trihydroxy-labd-14-en-11-one; a diterpene isolated from the roots of the Indian labiate plant *Coleus forskohlii*. It stimulates adenylate cyclase, has positive inotropic activity, and relaxes smooth muscle; at higher doses it is hypotensive and vasodilatory. The water-solubility of the active 7-O-hemisuccinyl-7-deacetyl- or (less potent) 7-deacetyl-6-(N-acetylglycyl)-derivatives is useful; the 1,9-dideoxy-derivative is inactive and can be used as a control.

Forssman antibody an antibody against **Forssman antigen**; it causes agglutination of sheep erythrocytes.

Forssman antigen *or* **Forssman hapten** *abbr*.: F antigen; a heterophil antigen, present on the tissue cells of many species, including horse, sheep, and mouse, but absent in human and rabbit. The glycolipid hapten (from horse) is a ceramide pentasaccharide, GalNAc(α1-3)GalNAc(β1-3)Gal(α1-4)Gal(β1-4)Glc(β1-1)Cer; when incorporated in liposomes the nonpolar ceramide end anchors the hapten in the bilayer leaving the polar pentasaccharide region accessible to Forssman antibodies.

Förster theory of fluorescence energy transfer a theory giving the rate of dipole–dipole energy transfer in fluorescence as:

$$k_T = 8.71 \times 10^{23}(r^{-6}K^2 Jn^{-4}k_F)$$

and the efficiency of transfer as:

$$E = r^{-6}/(r^{-6} + R_0^{-6}),$$

where R_0, the distance at which transfer efficiency is 50· is given by:

$$R_0 = 9.71 \times 10^3 (JK^2 Q_0 n^{-4})^{1/6} \text{Å},$$

in which r is the distance between the centres of donor and acceptor chromophore, K^2 is the orientation factor for a dipole–dipole interaction, J is the spectral overlap integral, n is the refractive index of the medium between the donor and acceptor, k_F is the rate constant for fluorescence emission by the energy donor, and Q_0 is the quantum yield of fluorescence of the energy donor in the absence of the acceptor. For energy transfer to be efficient the fluorescence emission spectrum of the donor must overlap the absorption spectrum of the acceptor, i.e. the donor and acceptor must be in resonance, as measured by the spectral overlap integral J, which, in cm^{3}M^{-1}, is:

$$J = (\int F(\lambda)\varepsilon(\lambda)\lambda^4 d\lambda)/(\int F(\lambda)d\lambda),$$

where $F(\lambda)$ is the fluorescence intensity in arbitrary units of the energy donor at wavelength λ in cm, and $\varepsilon(\lambda)$ is the extinction coefficient of the energy acceptor in cm^{-1}M^{-1}. The orientation factor, K^2, is given by:

$$K^2 = (\cos\alpha - 3\cos\beta\cos\gamma)^2,$$

where α is the angle between the donor and acceptor transition moments and β and γ are the angles between the donor moment and the acceptor moment and the line joining the centres of the donor and acceptor, respectively. [After T. Förster.]

Förster transfer the nonradiative, resonant energy transfer that occurs when electronic excitation of one chromophore in a molecule, or other structure, elicits fluorescence from a different chromophore located up to 5–7 nm distant. For example, excitation of a tyrosine residue in a protein molecule can lead to fluorescence from a tryptophan residue in the same protein, and excitation of a tryptophan residue can cause fluorescence of dyes attached to the protein surface or to an embedded coenzyme. Resonant energy transfer is important in photosynthesis; many of the chlorophyll molecules in a chloroplast transfer the energy they have absorbed to other chlorophyll molecules at reaction centres.

fortimycin any of several aminoglycoside antibiotics from the actinomycete *Micromonospora olivoasterospora*.

Fortran *see* **programming language**. [Contraction of formula translator.]

forward mutation any gene mutation from wild-type to the mutant phenotype, as opposed to a **back mutation**.

forward scattering scattering of radiation in the direction of the beam of radiation, i.e. away from the radiation source.

FOS a gene family encoding nuclear transcription factors; v-*fos* is the oncogene of the FBJ and FBR murine osteosarcoma viruses and Fujinami (avian) sarcoma virus. *FOS* is rapidly (within 5 minutes) but transiently expressed following stimulation of cells by mitogens and also during the onset of differentiation and by wounding. The gene product, FOS, and JUN (from the related gene family *JUN*) form homo- and heterodimers that bind to DNA; the heterodimers bind to the AP-1 consensus site (*see* **JUN**); FOS homodimers bind to the TPA response element (TRE) if Glu[168] is substituted by Lys. The proteins are members of the **leucine zipper** and **helix-turn helix** superfamilies.

fosfomycin *another name for* **phosphonomycin**.

Fouchet's test a colour test for the presence of bilirubin in the urine. A green colour is produced when the sample is treated with iron (III) chloride solution and trichloroacetic acid.

founder effect *(in genetics)* a high frequency of a particular allele in a population resulting from the presence of the allele in one or more of a small number of individuals from whom the population is descended.

Fourier analysis the fitting of the terms of a **Fourier series** to a set of periodic data.

Fourier series the expression of any ordinary, single-valued (univalent) periodic function:

$$y = f(x) \quad (1)$$

as the sum of an infinite number of sine waves of successively higher orders. Substituting $n\theta$ for x, to be appropriate to circular functions,

$$y = a_0 + a_1\cos\theta + a_2\cos2\theta + a_3\cos3\theta + ... + b_0 + b_1\sin\theta + b_2\sin2\theta + b_3\sin3\theta + ... \quad (2)$$

The coefficients a_n and b_n can be uniquely determined by equating equations (1) and (2) point by point for successive values of x and solving the simultaneous equations so found. [After Jean Baptiste Joseph Fourier (1768–1830), French mathematician.]

Fourier synthesis the process of computing the form of a function from the values of the coefficients of its **Fourier series**.

FOXP *abbr. for* forkhead box protein; any of a family of homologous proteins implicated in embryonal development. Each contains a forkhead-winged-helix domain of ≈ 108 amino acids. Mutations in the FOXP2 (715 amino acids) gene at 7q31 lead to speech and language disabilities. FOXP3 (or scurfin) is expressed preferentially in regulatory T lymphocytes and regulates immunity. Mutations in FOXP3 cause an X-linked multiorgan autoimmune and inflammatory syndrome in humans, and a similar condition called scurfy in mice.

FP *abbr. for* flavoprotein.

F6P *abbr. for* fructose 6-phosphate.

F-pilus a type of sex pilus, one or two of which may be present in each cell of 'male' strains of *Escherichia coli*. Each pilus is approximately 8 nm in diameter and up to 20 µm long. F-pili are consistently associated with the presence in the bacterium of the **F ′ plasmid**, which directs their synthesis; they are believed to serve as conductors of nucleic acids in various circumstances, e.g. during bacterial conjugation.

F plasmid *or* **F factor** the prototype 'fertility factor' responsible for conjugation in the K-12 strains of *Escherichia coli* used in the early studies of bacterial mating. The letter F refers specifically to this plasmid and thus should not be used to refer to other naturally occurring **conjugative plasmids**.

F′ plasmid *or* **F′ factor** an **F plasmid** derivative incorporating a segment of the bacterial chromosome. An organism in which an F′ plasmid has arisen and that carries a chromosomal deletion corresponding to the segment incorporated by the plasmid is termed a **primary F′-containing strain**, and an organism into which an F′ plasmid has been transferred is termed a **secondary F′-containing strain**.

FPS *see* **FES/FPS**.

Fr *symbol for* francium.

FR *abbr. for* folate receptor. *See* **folate-binding protein**.

fraction **1** *(in chemistry)* any one of several portions of a mixture that can be separated by a fractional process, e.g. by fractional distillation or chromatography, and consisting either of a mixture or of a pure compound. **2** to separate or divide into portions; to **fractionate**.

fraction I–VI (of blood plasma) *see* **Cohn fractionation**.

fractional **1** of, relating to, or being a fraction. **2** of or denoting any process by which parts of a mixture are separated by exploiting differences in their physical properties, e.g. their boiling points, solubility, or chromatographic characteristics.

fractional centrifugation *an alternative name for* **differential centrifugation**.

fractional crystallization the separation of substances with nearly the same solubilities. It involves a series of recrystallizations at each stage of which the crystals will be enriched in one component and the mother liquor will be enriched in the other component.

fractional distillation the process of separating the components of a mixture by heating it and separately condensing portions of

vapour at different temperatures, corresponding to the boiling points of the components of the mixture.

fractional precipitation the stepwise separation of substances, especially macromolecules, from a solution by exploiting differences in their solubilities in various conditions of pH, ionic strength, dielectric constant, etc. Changes in these conditions are effected by the addition of acid, alkali, salt (e.g. ammonium sulfate), ethanol, or other substances.

fractional sterilization *an alternative name for* **tyndallization**.

fractionate **1** to separate (a mixture) or cause (a mixture) to separate into portions or fractions, e.g. by distillation or crystallization. **2** to obtain (a part of a mixture) by a **fractional** process. —**fractionation** *n.*

fractionating column a vertical tube or hollow column frequently forming part of an apparatus for **fractional distillation** and usually filled with packing or intersected with perforated plates. An internal reflux takes place during distillation resulting in a gradual separation between the high-boiling-point fractions and the low-boiling-point fractions, the lowest-boiling-point fraction distilling over first.

fraction collector *or (sometimes)* **fraction cutter** an apparatus for the automatic collection of consecutive, commonly equal-sized, fractions of a flowing liquid, e.g. of the eluate from a chromatographic column. The size of the fractions may be predetermined by volume, mass, time, or number of drops.

fraction cutter **1** the device in a **fraction collector** that separates the flow of eluate into fractions of predetermined size, which then may be individually collected. **2** *(sometimes) an alternative name for* **fraction collector**.

fraction I protein *a former name for* RuBP carboxylase; *see* **ribulose bisphosphate carboxylase**.

fragile X mental retardation protein *see* **FMRP**.

fragile X syndrome an X-linked inherited disorder, with mental retardation and connective tissue abnormalities, arising from expansion of a **trinucleotide repeat** of CGG in the 5′ untranslated region of the mRNA for fragile X mental retardation protein (**FMRP**), which leads to nonexpression of that protein. In normal individuals the repeat number is 5–60, in the premutation 60–200, and in the full mutation >200. This is the most common form of inherited mental retardation. It is named for the folate-sensitive fragile site at Xq27.3. *See* **FMR1**.

fragment-antigen binding *see* **Fab fragment**.

fragmentation **1** *(in cell biology)* **a** the separation of a small piece (i.e. a fragment) of a chromosome from the rest. **b** the division of a cell nucleus by simple splitting. **2** *(in chemistry)* the breakdown of a radical into a diamagnetic molecule and a smaller radical; the breakdown of a radical ion in a mass spectrometer to form an ion of lower molecular mass and a radical.

fragment-crystallizable *see* **Fc fragment**.

fragment ion *(in mass spectrometry)* any ion produced by the loss of one or more pieces from a parent **molecular ion** (or larger fragment ion).

fragmentography *see* **molecular fragmentography**.

fragmin a Ca^{2+}-sensitive protein, M_r 43 000, isolated from *Physarum polycephalum*, that regulates the length of actin filaments *in vitro*.

frame *short name for* **reading frame**.

frameshift an alteration in the reading sense of DNA resulting from an inserted or deleted base, such that the **reading frame** for subsequent codons is shifted according to the number of changes made. Frameshifts may arise, for example, via random mutations or errors in reading sequencing output.

frameshifting *(in ribosomal function) see* **programmed ribosomal frameshifting**.

frameshift mutagen any mutagen that is capable of causing **frameshift mutations**.

frameshift mutation *or* **phaseshift mutation** any mutation involving the addition or deletion of nucleotide residues in a DNA molecule that results in out-of-phase translation, at positions beyond the site of mutation, of the messenger RNA formed from it. Since the genetic code is read as unpunctuated base triplets, additions or deletions in the message that are not multiples of three bases result in a shift in the reading frame.

frameshift suppressor any of a class of external suppressor mutations that specifically suppress a frameshift mutation by changing (correcting) the phase of translation. Frameshift suppressors are genetic alterations outside the mutated cistron. Some give rise to altered tRNA molecules, **frameshift suppressor tRNAs**, that recognize a base quadruplet instead of a base triplet.

framework regions in **immunoglobulin G** the heavy and light chains contain in all six hypervariable regions of amino acid residues that form the antigen-binding site of the antibody molecule. The other domains exhibit far less variation and are named the **framework regions**. These form the basic beta-pleated sheet structure, which forms a scaffold for the hypervariable regions. The 3-D structure of the framework regions is very similar in all antibodies.

framework regions in **immunoglobulin G** the heavy and light chains contain in all six hypervariable regions of amino acid residues that form the antigen-binding site of the antibody molecule. The other domains exhibit far less variation and are named the **framework regions**. These form the basic beta-pleated sheet structure, which forms a scaffold for the hypervariable regions. The 3-D structure of the framework regions is very similar in all antibodies.

Franck–Condon principle the principle that the time taken for electronic transitions in molecules is very short in comparison with the period of vibration of the atomic nuclei, with the consequence that the nuclei do not appreciably alter their positions during the emission or absorption of a photon. [After James Franck (1882–1964) and Edward Uhler Condon (1902–74), US physicists.]

FRAP *see* **mTor**.

FRAT *abbr. and proprietary name for* free radical assay technique (*see* **spin immunoassay**).

frataxin a 210-residue protein identified from a mutation found in **Friedreich's ataxia** (*abbr.*: FRDA). Some FRDA patients have point mutations in the gene encoding frataxin, but the majority are homozygous for an unstable GAA trinucleotide expansion in the first intron. The protein is involved in iron storage and iron–sulfur cluster assembly in mitochondria.

fraudulent DNA a DNA molecule containing **fraudulent nucleotides**.

fraudulent nucleotide any (artificial) nucleotide that contains a non-naturally occurring base that is an analogue of a purine or pyrimidine found in natural DNA.

frayed end (of DNA) an end region of a double-stranded polynucleotide in which there is imperfect complementarity of the base pairs.

fredericamycin A an antibiotic produced by *Streptomyces griseus* that contains a *spiro* ring system and is active against Gram-positive bacteria and fungi. It is cytotoxic against some transplantable tumours.

fredericamycin A

free *(in chemistry)* not chemically combined (with another substance); not physically bound (to another substance).

free diffusion diffusion in a gas phase or across a boundary between two solutions (or between a solution and pure solvent), as distinct from diffusion across a membrane or some other porous barrier.

free-draining coil a model for polymer molecules in solution that are not well approximated by a solid particle. In this model the polymer is considered as a string of beads through which the solvent streams during viscous flow.

free electrophoresis *an alternative name for* **moving boundary electrophoresis** (def. 1).

free energy the thermodynamic potential of a system. For a change (such as a chemical reaction or series of reactions) to be feasible the change in free energy must be negative. Free energy contains contributions from energy and entropy terms. *See* **Gibbs energy**, **Helmholz free energy**.

free energy change *an alternative name for* **Gibbs free energy change**.

free fatty acid *abbr.*: FFA; a nonesterified fatty acid; the term is often used of such acids in the blood.

free-flow electrophoresis *an alternative name for* **moving boundary electrophoresis** (def. 1).

free radical *see* **radical** (def. 1).

free-radical scavenger *the former term for* **radical scavenger**.

free rotation sterically unhindered rotation of atoms or groups about a single covalent bond.

freeze-clamp to halt abruptly the metabolic processes in a piece of tissue, or other specimen, by clamping it into a thin, frozen wafer between two aluminium (or other metal) plates, previously cooled in liquid nitrogen. The frozen specimen then may be pulverized and treated with a protein precipitant, e.g. acetone at –70°C. The technique is useful in the determination of the concentrations of various metabolites that otherwise are rapidly consumed by enzymic action continuing during an investigation.

freeze-cleaving *an alternative term for* **freeze-fracturing**; *see also* **freeze-etching**.

freeze-dry to dry in a frozen state at very low pressure (i.e. in a high vacuum) so that ice, or other frozen solvent, sublimes rapidly, leaving a porous solid; to lyophilize. Used, e.g. in histochemistry, as frozen biological material can be preserved by removal of water under vacuum without damage to cell structures.

freeze-dryer any apparatus used for freeze-drying.

freeze-etching *or* **frozen replica method** a technique of preparing, for electron-microscopic examination, a surface that has been exposed by fracturing or cutting a deep-frozen specimen. The sample is rapidly frozen to about –150°C and placed in a vacuum, where it is cleaved or sliced with a cold knife – the process known as **freeze-fracturing**. Etching is then achieved by evaporation of some of the solvent from the surface in a vacuum at about –100°C to produce a textured field, in which nonvolatile macromolecules protrude from a frozen aqueous layer. A replica of the surface is then prepared by shadowing with carbon and/or a heavy metal while the sample is still in the vacuum. This replica is then lifted off, rinsed, and placed on an electron microscope grid for examination. *See also* **replica formation**.

freeze-fracturing *(in electron microscopy)* a method of visualizing the interior of cell membranes in which cells are frozen at the temperature of liquid nitrogen in the presence of antifreeze and then the frozen block is cracked with a knife blade. The fracture plane often passes through the hydrophobic middle of lipid bilayers, thereby exposing the interior of cell membranes. The resulting fracture faces are shadowed with platinum, the organic material is dissolved away and the replicas floated off for electron microscopy.

freeze-stop to stop a reaction rapidly by freezing the reaction mixture; for **freeze-clamp**.

freeze–thaw to disrupt cells in a suspension, or to disperse emulsions, by repeated freezing and thawing.

freezing 1 the change of a liquid into a solid by a lowering of temperature. **2** the stopping of a reaction or other process at a particular point, often by a lowering of temperature. **3** *or* **cryopreservation** the use of temperatures below 0°C in order to preserve biochemicals and foodstuffs, or to maintain viable populations of (micro)organisms.

freezing-microtome a microtome in which the tissue sample is frozen and maintained in a frozen state with liquid carbon dioxide. It is useful for (rapidly) obtaining sections of unfixed soft tissue. *Compare* **cryomicrotome**.

freezing point the temperature at which **freezing** (def. 1) occurs.

French chalk *see* **talc**.

French pressure cell a device to disrupt chloroplasts and bacteria, yeasts, or other cells. A cell or chloroplast suspension is placed in a cylinder fitted with a piston and a narrow outlet controlled by a needle valve. After removal of air, the piston is forced into the cylinder with a hydraulic press to the desired operating pressure (<20000 pound-force per square inch, 140 MPa). The valve is then opened. Extrusion through the narrow orifice together with the sudden release of pressure creates shear forces that break the suspended particles. *See also* **Hughes press**. [After Charles Stacy French (1907–), US plant physiologist.]

Freon *the proprietary name for* any of a group of fluorocarbon or chlorofluorocarbon compounds that are clear, colourless liquids, chemically inert, and of low toxicity. They are useful as refrigerants, aerosol propellants, cleaning fluids, and solvents. These substances have been shown to be important in the chemical removal of ozone by chlorine-catalysed cycles in the polar region during winter and spring. Nitric oxide and ice-clouds are formed in the polar stratosphere during winter. Heterogeneous chemical reactions take place on the surface of these particles, dramatically shifting the balance between inactive and active forms of chlorine and increasing the efficiency of chlorine in catalytic ozone removal. The use of Freon is being phased out in many countries; it is being replaced by substances such as tetrafluoroethane (ICI Trademark KLEA) which are claimed not to cause depletion of ozone.

frequency *symbol*: ν or f; the number of complete cycles of any periodic process occurring in unit time. The SI derived unit of frequency is the **hertz**. The frequency of a wave is equal to its velocity divided by its wavelength.

frequency distribution a table, graph, or histogram constructed from a series of measurements to show the classes into which the measurements are divided and the numbers in each of these classes.

fresnel a non-SI unit of frequency of electromagnetic radiation, equal to 10^{12} Hz.

Fresnel diffraction the diffraction occurring when electromagnetic radiation encounters abrupt discontinuities in refractive index, as in a **Fresnel lens** or in the examination of some specimens in (electron) microscopy. [After Augustin Jean Fresnel (1788–1827), French physicist.]

Fresnel lens a lens whose surface consists of a concentric series of lens sections (or one convex surface and one stepped surface). This arrangement allows a thin lens of short focal length and large diameter to be easily made, and finds application in searchlights, viewing devices, and overhead projectors.

fret a broad membrane-bound channel interconnecting the grana of **chloroplasts**.

FRET *abbr. for* fluorescence resonance energy transfer.

Freund's adjuvant a widely used immunological **adjuvant** (def. 1), used to produce water-in-oil emulsions with aqueous antigens or antigen preparations. It consists of a mixture of mineral oil and emulsifier (**incomplete Freund's adjuvant**) in which killed mycobacteria have been dispersed (**complete Freund's adjuvant**). Omission of mycobacteria may modify the duration, intensity, and nature of the responses. [After Jules Freund (1891–1960).]

Frey–Wyssling particle a double-membrane particle (4–6 μm in diameter) representing a minor component of rubber latex. [After Albert Frey-Wyssling (1900–).]

FRF *abbr. for* follicle-stimulating-hormone-releasing factor; *see* **gonadotropin-releasing hormone**.

FRH *abbr. for* follicle-stimulating hormone releasing hormone; *see* **gonadotropin-releasing hormone**.

friable easily crumbled or pulverized.

friction the resistance to the relative motion of one body sliding, rolling, or flowing over another body with which it is in contact. — **frictional** *adj.*

frictional coefficient *or* **frictional constant** *symbol*: f; the frictional force opposing the motion of a body in a fluid medium divided by the velocity of movement. It may be expressed in different ways. The **translational frictional coefficient**, f, for a rigid sphere of radius r moving at low velocity through a medium of viscosity η, is given by: $f = 6\pi\eta r$ (**Stokes' law of viscosity**). For particles of other shapes additional terms must be included. The **rotational frictional coefficient**, f_{rot}, is a parameter relating the velocity of rotation of a rigid

sphere of volume V in a fluid medium of viscosity η. It is given by: $f_{rot} = 6\eta V$. *See also* **frictional ratio**.

frictional force the force exerted on a particle moving through a fluid medium as a result of friction. It is the product of the velocity, relative to the medium, of the particle and its **frictional coefficient**.

frictional ratio the ratio of the experimentally determined translational **frictional coefficient**, f, of a hydrated, anisotropic macromolecule to the calculated frictional coefficient, f_0, of a nonhydrated spherical macromolecule of the same molecular mass. The ratio f/f_0 gives some indication of the asymmetry of the macromolecule.

Friedreich's ataxia an autosomal recessive inherited disease that involves the central and peripheral nervous systems and the heart. It is the most common hereditary ataxia, and is associated with defects in the gene encoding **frataxin**. [After Nikolaus Friedreich (1825–1882), German neurologist.]

Friend cell a cell of a cultured cell line of leukemic mouse erythroblasts obtained by transformation with Friend leukemia virus. In such cells development is arrested before any globin synthesis occurs, but treatment with certain chemicals, e.g. dimethylsulfoxide, causes a resumption of development and hemoglobin synthesis.

Friend leukemia virus a strain of mouse type C oncovirus that specifically affects immature erythroid cells in mice to cause an erythroleukemia. [After Charlotte Friend (1921–).]

friend of GATA-1 *see* **FOG1**.

frit filter *or* **fritted filter** *or* **fritted glass filter** *an alternative name for* **sintered glass** filter.

Frizzled any of a family of G-protein-coupled plasma membrane receptors present throughout the animal kingdom. The N-terminal extracellular segment contains a cysteine-rich domain (called the FRZ domain) of ≈120 amino acids, which binds a Wnt protein. The domain also occurs in secreted Wnt antagonists such as WIF-1.

Fromm's method a kinetic approach for studying the mechanism of action of enzyme systems that utilize three substrates. Such systems may be divided into two groups on the basis of their initial rate expressions: the first group is made up of sequential pathways of enzyme and substrate interaction alone, the second is the ping-pong type. The mechanisms in each category may further be separated either by initial rate measurements or from studies with competitive inhibitors. [After Herbert Jerome Fromm (1929–), US biochemist.]

frontal analysis any analytical technique by which the progress of the separation of components of a mixture, as by frontal chromatography, moving boundary electrophoresis, or analytical ultracentrifugation, is followed, and the relative proportions of the components determined, by examination of the individual boundaries, or fronts, formed by each component with the solvent.

frontal chromatography a technique for chromatographic separation in which the sample (liquid or gas) is fed continuously into the chromatographic bed and only the component that emerges first is obtained in pure form.

fronting *(in chromatography)* the asymmetry of a peak such that, relative to the baseline, the front is less steep than the rear. In paper chromatography and thin-layer chromatography it refers to the distortion of a zone showing a diffuse region preceding the zone in the direction of flow.

frontside attack the mechanism of a chemical displacement reaction in which the configuration is retained.

frozen replica method *an alternative name for* **freeze-etching**.

FRS *abbr. for* Fellow of the Royal Society (of London).

FRSE *abbr. for* Fellow of the Royal Society of Edinburgh.

Fru *symbol for* a residue (or sometimes a molecule) of the ketohexose **fructose**. In the condensed system of symbolism of sugar residues, Fru signifies specifically a fructose residue with the common configuration and ring size, i.e. D-fructofuranose.

fructan any **glycan** composed only of D-fructose residues. Fructans occur commonly as food reserves in the roots, stems, leaves, and seeds of many plants, e.g. in tubers of dahlias and Jerusalem artichokes. There are two main types: (1) inulin-type, in which the fructofuranose residues are linked by $\beta2\rightarrow1$ glycosidic linkages; and (2) levan-type, in which the linkages are $\beta2\rightarrow6$, which were formerly referred to as **fructosans**.

β-fructofuranosidase EC 3.2.1.26; *other names*: invertase; saccharase; an enzyme that catalyses the hydrolysis of terminal nonreducing β-D-fructofuranoside residues in β-D-fructofuranosides, e.g. sucrose.

fructokinase 1 *the recommended name for* the enzyme ATP:D-fructose 6-phosphotransferase; EC 2.7.1.4; this catalyses the phosphorylation of D-fructose by ATP to form D-fructose 6-phosphate and ADP. The enzyme occurs in plants and in schistosomes, and is highly specific for D-fructose. In bacteria it is important in fructose utilization. Structurally it is similar to other carbohydrate kinases. **2** *an alternative name for* **ketohexokinase** (EC 2.7.1.3). This usage is misleading since ketohexokinase yields D-fructose 1-phosphate.

fructosaccharase *a trivial name for* any **saccharase** that attacks the fructose-containing end of appropriate oligosaccharides, liberating fructose.

fructosan 1 *(formerly) a trivial name for* 2,6-anhydrofructofuranose. **2** *an (older) incorrect name for* **fructan**.

fructose *symbol*: Fru; *the trivial name for* the ketohexose *arabino*-2-hexulose; there are two enantiomers. D-(−)-fructose (*symbol*: D-Fru), commonly known as fructose, and formerly known as **fruit sugar**, is levorotatory, hence formerly known also as **levulose** or (*esp. UK*) laevulose; it is the sweetest of the sugars. Aqueous solutions of D-fructose at room temperature contain an equilibrium mixture of β-D-fructopyranose, β-D-fructofuranose, α-D-fructofuranose, and the open-chain form, whereas in crystals the sugar is present entirely as β-D-fructopyranose. Combined D-fructose is almost invariably in the furanose form, however. D-fructose is found free in a large number of fruits, in honey, and (as the sole sugar) in bull and human semen. In combined form it occurs at one residue per molecule of the disaccharide **sucrose** and the trisaccharides **gentianose**, **melezitose**, and **raffinose**, and as the sole component of the polysaccharide **fructans**.

β-D-fructofuranose

fructose-bisphosphatase EC 3.1.3.11; *systematic name*: D-fructose-1,6-bisphosphate 1-phosphohydrolase; *other names*: hexosediphosphatase; *(formerly)* fructose-1,6-diphosphatase. A phosphoric monoester hydrolase that hydrolyses D-fructose 1,6-bisphosphate to D-fructose 6-phosphate and orthophosphate. It is important in **gluconeogenesis** and in regulation of the **reductive pentose phosphate cycle** in chloroplasts.

fructose-2,6-bisphosphatase an enzyme that hydrolyses fructose 2,6-bisphosphate to fructose 6-phosphate; it is the same protein as phosphofructokinase-2. *See* **phosphofructokinase** (def. 2).

fructose-1,6-bisphosphatase deficiency *see* **hereditary fructose-1,6-bisphosphatase deficiency**.

fructose 1,6-bisphosphate or *(formerly)* fructose 1,6-diphosphate *abbr.*: FBP; the D enantiomer is a metabolic intermediate in **glycolysis** and **gluconeogenesis**.

fructose 2,6-bisphosphate the D enantiomer is an important regulator of the **glycolytic** and **gluconeogenic pathways**. It inhibits fructose 1,6-bisphosphatase and activates phosphofructokinase, and its concentration is controlled by the **phosphofructokinase** (def. 2) and fructose 2,6-bisphosphatase activities of a 49 kDa protein.

fructose-bisphosphate aldolase *see* **aldolase** (def. 2).

fructose-2,6-bisphosphate 2-phosphatase EC 3.1.3.46; *other name*: fructose-2,6-bisphosphatase; an enzyme that catalyses the hydrolysis of D-fructose 2,6-bisphosphate to D-fructose 6-phosphate and orthophosphate, being part of the system for regulating the concentration of fructose 2,6-bisphosphate. This enzyme is the phosphorylated form of the protein. If dephosphorylated, it functions as a 2-kinase for fructose 6-phosphate. *See also* **phosphofructokinase** (def. 2).

fructose-1,6-diphosphatase *a former name for* **fructose-bisphosphatase**.

fructose 1,6-diphosphate *a former name for* **fructose 1,6-bisphosphate**.

fructose disease *see* **essential fructosuria**, **hereditary fructose intolerance**, **hereditary fructose-bisphosphatase deficiency**.

fructose intolerance *an alternative name for* **essential fructosuria**.

fructose 6-phosphate *abbr.*: F6P; the D enantiomer is an important intermediate in glycolysis, gluconeogenesis, and fructose metabolism.

fructosuria the presence of abnormal amounts of fructose in the urine. *See also* **essential fructosuria**.

Fru*f* *symbol for* a fructofuranose residue.

fruit the structure that develops from the fertilized ovary of a flowering plant and contains the ripening seed or seeds. It may be fleshy or dry; the term is often extended to include other associated parts such as a fleshy receptacle.

fruit fly a common name for species of flies belonging to the genus *Drosophila*.

fruit-ripening hormone *see* **ethylene**.

fruit sugar *a former name for* the D enantiomer of **fructose**.

FseI a type 2 **restriction endonuclease**; recognition sequence: GGCCGG↑CC.

FSF *abbr. for* fibrin-stabilizing factor (i.e. factor XIIIa); *see* **blood coagulation**.

FSH *abbr. for* follicle-stimulating hormone.

FSH-RF *abbr. for* follicle-stimulating hormone releasing factor; *see* **gonadotropin-releasing hormone**.

FSH-RH *abbr. for* follicle-stimulating hormone releasing hormone; *see* **gonadotropin-releasing hormone**.

F₁ sphere a submitochondrial structure containing the catalytic site for ATP synthesis.

F′-strain a bacterial strain containing an F′ plasmid.

FtsZ a bacterial GTPase that is crucial for assembly of the **Z ring** in the mid-region of the plasma membrane inner surface. It is the first protein to localize to the future site of cell division. Highly conserved in eubacteria and archaea, it also occurs in most groups of land plants and in many algae, and resembles eukaryotic tubulin. Its correct localization requires the Min system.

F-type pentose phosphate pathway *see* **pentose phosphate pathway**.

ftz the gene for Fushi Tarazu protein, a nuclear DNA-binding homeobox segmentation protein from *Drosophila melanogaster*.

FU *abbr. for* fluorouracil (use proscribed; the symbol FUra is recommended).

Fuc *symbol for* a residue (or sometimes a molecule) of the deoxyaldohexose **fucose**. In the condensed system of symbolism of sugar residues, Fuc signifies specifically a fucose residue with the common configuration and ring size, i.e. L-fucopyranose.

fucan a **glycan** composed only of fucose residues.

fucoidan or **fucoidin** any of a group of complex branched, sulfated heteropolysaccharides that are composed of residues of L-fucose (usually sulfated) together with small numbers of residues of various other monosaccharides. Fucoidans occur in the cell walls, intercellular matrix, and exuded mucilage of many (especially intertidal) species of marine algae, in which they appear to assist in water retention during periodic exposure.

fucosamine 2-amino-2,6-dideoxygalactose; an aminodeoxysugar, both enantiomers of which occur in lipopolysaccharides of *Salmonella* and other bacterial genera.

fucose *symbol*: Fuc; *the trivial name for* 6-deoxygalactose; there are two enantiomers. D-fucose (*symbol*: D-Fuc), 6-deoxy-D-galactose,

β-D-fucose

occurs in glycosides in various plants belonging to the Convolvulaceae. L-fucose (*symbol*: L-Fuc), 6-deoxy-L-galactose, occurs in polysaccharides (fucans) in seaweeds, especially *Fucus* spp., and in the cell wall matrix of higher plants. In animals fucosyl residues occur in a number of glycoproteins and glycolipids; the presence of α-1,4-linked L-fucosyl residues at specific positions in the highly branched oligosaccharides of human erythrocyte surface antigens is controlled by the *H* and *Le* blood-group genes.

fucose-1-phosphate guanylyltransferase EC 2.7.7.30; *other name*: GDPfucosepyrophosphorylase; an enzyme that catalyses the formation of GDP-L-fucose from GTP and L-fucose 1-phosphate with release of pyrophosphate.

α-**fucosidase** any of several enzymes that hydrolyse α-fucosides. **1** α-fucosidase, EC 3.2.1.51; *systematic name*: α-L-fucoside fucohydrolase; it catalyses the hydrolysis of an α-L-fucoside to an alcohol and L-fucose. **2** 1,3-α-L-fucosidase, EC 3.2.1.111; *other names*: almond emulsin; fucosidase I; an enzyme that catalyses the hydrolysis of 1,3-linkages between α-L-fucose and *N*-acetylglucosamine residues in glycoproteins. **3** 1,6-α-L-fucosidase, EC 3.2.1.127; an enzyme that catalyses the hydrolysis of 1,6-linkages between α-L-fucose and *N*-acetyl-D-glucosamine in glycopeptides such as immunoglobulin G glycopeptide and fucosyl-asialo-agalacto-fetuin.

fucosidosis a rare autosomal recessive disorder characterized by progressive psychomotor and growth retardation; urinary excretion and lysosomal accumulation of glycolipids, glycopeptides, and oligosaccharides that contain terminal fucose moieties; and α-fucosidase deficiency. Mutations in the gene at 1p24 are responsible.

FUdR or **FUDR** *abbr. for* fluorodeoxyuridine (use deprecated; the symbol FdUrd is recommended).

fugu a teleost fish, *Takifugu rubripes*, used as a model experimental organism. Sequencing of the fugu genome has enabled valuable comparisons with other organisms. Fugu has a similar number of genes to mammals, but the fugu genome is much smaller at 390 Mbp, one-eighth the size of the human genome.

fugu toxin *see* tetrodotoxin.

fukuatin a protein (461 amino acids, gene at 9q31-q33) implicated in Fukuyama type congenital muscular dystrophy, which is associated with dysplasia of cerebral cortex, epilepsy, and severe mental retardation. Its localization and function are still unclear.

full agonist an **agonist** that is able to elicit the maximal response of a tissue even when the tissue has spare receptors. In such a situation, several different agonists may be able to elicit the same maximal response, albeit at different receptor occupancies. Moreover, a full agonist in one tissue may be a **partial agonist** in another.

fuller's earth a naturally occurring, porous, absorptive aluminium–magnesium silicate clay. It is useful for decolorizing oils and other liquids and as a filtering medium. It is so-called because it was originally used to full (i.e. cleanse and thicken) wool and woollen cloth.

full-shotgun coverage *see* depth of coverage.

fully overlapping code *see* overlapping code.

fumarase EC 4.2.1.2; *recommended name*: fumarate hydratase. An enzyme of the **tricarboxylic-acid cycle** that catalyses the reaction:

$$(S)\text{-malate} = \text{fumarate} + H_2O.$$

There are two different enzymes in facultative bacteria (e.g. *Escherichia coli*). In humans a cytosolic and a mitochondrial isoform are derived from the gene at 1q42.1. A missense mutation produces deficiency of both isoforms and results in neurological impairment and neuropathy, fumaric aciduria, and increased urinary excretion of succinate, 2-oxoglutarate, and citrate. Mutations are also associated with hereditary fibroids of skin and uterus, suggesting that the protein is a tumour suppressor.

fumarate 1 *the trivial name for trans*-1,2-ethylenedicarboxylate, (*E*)-butenedioate, the dianion of fumaric acid; the diastereoisomer of **maleate** and a key intermediate in metabolism. In the tricarboxylic-acid cycle it is formed from succinate by **succinate dehydrogenase**, and is converted to malate by **fumarase**. **2** any mixture of fumaric acid and its mono- and dianions. **3** any salt or ester of fumaric acid.

fumarate hydratase *see* fumarase.

fumarate reductase an enzyme in the respiratory chains of anaerobic bacteria that catalyses the reduction of fumarate by menaquinol to form succinate and menaquinone. *E. coli* fumarate reductase is an iron–sulfur flavoprotein (FAD), containing at least four different subunits: a flavoprotein, an iron–sulfur protein, and two hydrophobic anchor proteins.

fumaroyl the bivalent *trans*-acyl group, –CO–CH=CH–CO–, derived from fumaric acid.

fumaryl the univalent *trans*-acyl group, HOOC–CH=CH–CO–, derived from fumaric acid.

fumed silica a preparation of pure silicon dioxide as an extremely fine, white powder with an external surface area of up to 400 m^2g^{-1}. It is useful for thickening and forming thixotropic gels. Fumed silica is used as a suspending and gelling agent in liquid-scintillation counting; a 3–5 concentration in dioxan, toluene, or xylene-based liquid scintillators will form an almost transparent, rigid gel, capable of holding an insoluble solid sample in suspension. *Proprietary name*: Cab-O-Sil.

fume hood a work space contained under a hood from which air is extracted to remove harmful vapours.

function 1 *(in chemistry)* a **functional group**. **2** *(in mathematics)* **a** either of two variables so related to each other that for any value of one variable there is a corresponding value of the other. **b** a variable or expression that can take a set of values each of which is associated with the magnitude of an independent variable or variables. **3** *(in physiology)* the normal vital activity of a cell, organ, or tissue within an organism. —**functional** *adj*.

functional antagonism 1 antagonism arising by the antagonist's producing an opposite effect to the substance being antagonized through the antagonist's action at a different receptor. **2** antagonism arising when the antagonist interferes with events that follow receptor activation by the substance being antagonized.

functional cloning a cloning strategy in which knowledge of a gene's product (function) is used to clone the gene. An example is the gene for the β chain of hemoglobin responsible for sickle-cell anemia. This is in contrast to **structural cloning** where the function of the gene is not known. A common procedure would be to construct a DNA oligonucleotide based on the amino-acid sequence of the peptide and use this as a probe to isolate the gene from a DNA library. In cases where one protein is dominant in a cell, such as hemoglobin in the reticulocyte, the preparation of a cDNA from the mRNA also provides a useful probe for gene isolation.

functional genomics the systematic analysis of genes, their encoded proteins, and the relationships between gene activity and cell function at the level of whole genomes.

functional group or **function** *(in chemistry)* any **group** (def. 1) in a molecule of an organic compound that confers characteristic chemical and physical properties on the molecule and that may be considered to have replaced a hydrogen atom in a molecule of the corresponding hydrocarbon, e.g. the hydroxyl group in an alcohol.

functional polymer 1 a polymer that bears specified chemical groups. **2** a polymer that has specified physical, chemical, biological, pharmacological, or other uses. Examples of functions include: catalytic activity; selective binding of particular species; capture and transport of electric charge carriers or energy; and transport of drugs to a particular organ in which the drug is released.

fungicide any biological, chemical, or physical agent able to destroy fungi. —**fungicidal** *adj*.

fungistat any drug, antibiotic, or other agent that inhibits the growth and multiplication of fungi. —**fungistatic** *adj*.

fungus (*pl.* **fungi**) any member of the taxonomic kingdom Fungi, a group of diverse and widespread unicellular, multicellular, or cenocytic eukaryotic organisms that do not contain chlorophyll. Fungi may be parasitic, saprophytic, or symbiotic. They reproduce through spores. *See also* **microfungus**. —**fungal** *adj*.

FUR *abbr. for* fluorouridine (use proscribed; the symbol FUrd is recommended).

FUra *symbol for* (a residue of) the synthetic pyrimidine base **fluorouracil**.

Fura 2 1-[2-(5-carboxyoxazol-2-yl)-6-aminobenzofuran-5-oxyl]-2-(2′-amino-5′-methylphenoxy)-ethane-*N*,*N*,*N*′,*N*′-tetraacetic acid; a highly sensitive calcium ion chelator that exhibits a spectral shift on binding Ca^{2+}. Calcium ion concentration can be measured directly, independent of dye concentration, by measuring the ratio of fluorescence at 510 nm after excitation at 340 nm and then at 380 nm. It is frequently used as the acetoxymethyl ester (Fura 2/AM, Fura 2-pentakis(acetoxymethyl)ester), which is a membrane-permeable derivative that can enter cells, after which it is rapidly hydrolysed by cytoplasmic esterases to Fura 2, which remains trapped within the cell and can thus report the intracellular Ca^{2+} concentration. *See also* **Quin-2**.

furan *the trivial name for* oxa-2,4-cyclopentadiene, C_4H_4O. *See also* **furanose**.

furanoid furan-like, i.e. describing a molecular structure that consists of or contains a ring of four carbon atoms and one oxygen atom (as in furan).

furanose *symbol*: *f*; any monosaccharide or monosaccharide derivative whose molecule contains a furanoid ring. *Compare* **pyranose**, **septanose**.

furanoside any glycoside in which the sugar moiety is in the furanose form.

FUrd *symbol for* (a residue of) the ribonucleoside fluorouridine (*see* **fluorouracil**).

furin the mammalian counterpart of **kexin**.

furyl either 2-furyl or 3-furyl, the two oxa-2,4-cyclopentadienyl groups derived from a molecule of **furan** by loss of a hydrogen atom from C-2 and C-3 respectively.

fusaric acid 5-butylpyridine-2-carboxylic acid; an antibiotic originally isolated from the fungus *Fusarium heterosporium* and obtained also from *Gibberella fujikuroi*. It inhibits **dopamine β-hydroxylase** and has hypotensive activity.

fused ring *(in chemistry)* any ring of atoms in a molecule that is joined to another ring of atoms in such a way that two atoms are common to both rings.

fused rocket immunoelectrophoresis a modification of **rocket immunoelectrophoresis** designed to follow the distribution of proteins among the fractions obtained in a separation experiment. Small volumes of fractions obtained in the separation experiment are placed in a row of sample wells in an antibody-free agarose gel and the samples are allowed to diffuse into the gel for 30–60 min. The proteins are then forced at right angles into an antibody-containing gel by electrophoresis; the resulting precipitation lines (rockets) show the elution profiles for each individual protein in the separation experiment and the distance between the precipitation line and the corresponding well is proportional to the amount of antigen precipitated.

fusel oil a colourless, oily liquid of variable composition obtained as a by-product in the distillation of alcoholic fermentation liquors. It contains chiefly isopentanol, 2-methyl-1-butanol, isobutanol, and propanol. It is considerably more toxic than ethanol. [From German *fuseln*, to bungle.]

fusicoccin a toxin produced by the fungus *Fusicoccus amygdali* that infects peach and almond trees. Its receptor is a 14.3.3 protein that binds proteins containing a conserved sequence that includes a phosphorylated serine residue. It activates the plasma membrane H^+ ATPase in a variety of plant cell types, induces stomatal opening and consequent wilting of leaves, enhances growth rates, and breaks seed dormancy.

fusidic acid 3α,11α,16β-trihydroxy-4α,8,14-trimethyl-18-nor-5α,8α,9β,13α,14β-cholesta-17(20),24-dien-21-oic acid 16-acetate; a steroid antibiotic, structurally related to cephalosporin P (*see* **cephalosporin**) and isolated from the fermentation broth of *Fusidium coccineum*. Fusidic acid is active particularly against Gram-positive organisms, especially *Staphylococcus* spp. It prevents translocation during protein synthesis and inhibits the accumulation of ppGpp.

fusiform spindle-shaped; elongated and tapering at both ends.

fusin *or* **HIV entry cofactor**; a putative G-protein-coupled membrane receptor (352 amino acids) that cooperates with CD4 on immune cell surfaces and permits them to fuse with the HIV surface. It shows 37% sequence identity with the interleukin 8 (IL-8) receptor.

fusion 1 the act or process of melting. 2 the act or process of melting together or uniting. 3 the process whereby two membranes are joined together, as in **cell fusion**. 4 the process whereby nuclei of light elements are united to form nuclei of heavier elements with the release of energy: **nuclear fusion**. 5 the state of being fused. 6 something produced by fusion. 7 *(in molecular biology)* the act, process, or result of artificially linking genes that code for two different proteins, with the aim of generating a **fusion protein**.

fusion protein an expression product resulting from the fusion of two genes. Such a protein may be produced, e.g., in recombinant DNA expression studies or, naturally, in certain viral **oncogenes** in which the oncogene is fused to **gag**. Their production sometimes results from the need to place a cloned eukaryotic gene under the control of a bacterial promoter for expression in a bacterial system; sequences of the bacterial system are then frequently expressed linked to the eukaryotic protein. Fusion proteins are used for the analysis of structure, purification, function, and expression of heterologous gene products.

fusogen any agent, or set of conditions, that gives rise to fusion of membranes, including cell membranes (and hence of cells). Killed Sendai virus was formerly much used for this purpose but now polyethylene glycol is commonly used as a fusogen particularly in the preparation of hybridomas for the production of monoclonal antibodies. —**fusogenic** *adj.*

futile cycle any metabolic cycle that, if not controlled, acts as an ATPase and hence converts chemical energy (stored as ATP) to heat. For example, a coupling of phosphofructokinase, which converts fructose 6-phosphate to fructose 1,6-bisphosphate at the expense of ATP, with fructose-bisphosphatase, which hydrolyses fructose 1,6-bisphosphate to fructose 6-phosphate and (inorganic) phosphate, can form a cyclic process whose net effect is merely the hydrolysis of ATP to ADP and phosphate. A futile cycle may be an important control mechanism in some physiological conditions.

fuzzy coat *or* **fuzzy layer** the indistinct layer seen on electron microscopy outside the cell coat of various eukaryotic cells, e.g. cells of mammalian intestinal epithelium and a number of single-celled amoebae. It is composed principally of glycoproteins.

fuzzy onions *abbr.*: fzo; a mutant of *Saccharomyces cerevisiae* in which mitochondria undergo abnormal fission with fragmentation.

Fv fragment the N-terminal part of the **Fab fragment** of an immunoglobulin molecule, consisting of the variable portions of one light chain and one heavy chain.

FXR *abbr. for* farnesoid X receptor.

FYN a gene family encoding nonreceptor tyrosine kinases of the Src family (*see* **src**) that are implicated in the control of cell growth. The gene product, p59^{fyn}, associates with the p85 subunit of **1-phosphatidylinositol 3-kinase**. The proteins undergo myristoylation and phosphorylation on serine and tyrosine residues; overexpression transforms fibroblasts. The alternative gene symbols are *SYN* (from *src/yes*-related novel gene) and *SLK*.

FYVE a protein sequence of ≈70 amino acids that contains two zinc atoms coordinated by eight cysteine residues, the third cysteine being surrounded by a highly conserved basic sequence. It forms two pairs of antiparallel beta strands stabilized by coordination with the zinc atoms, termed the **FYVE finger**. The name is based on the first letters of four proteins containing this motif, although this is not found in all proteins of this type. The basic residues form a shallow groove that can accommodate the phosphate group of phosphatidylinositol 3-phosphate. It occurs in many proteins of numerous species (from yeast to mammals) and is involved in membrane trafficking. *See* **FGD1**.

Gg

g *symbol for* **1** gram. **2** gaseous state. **3** gluon.

g **1** *symbol for* acceleration of free fall (formerly known as acceleration due to gravity). **2** *abbr. for* gauche. **3** *symbol for* the electron g-value.

g_n *symbol for* standard acceleration of free fall, a fundamental physical constant equal by definition to 9.806 65 m s^{-2}.

g_N nuclear g-value; a constant for any nucleus that defines the energy of interaction with a magnetic field, e.g. in **NMR spectroscopy**.

G *symbol for* **1** giga+ (SI prefix denoting 10^9 times). **2** gauss (alternative to Gs). **3** a residue of the α-amino acid glycine (alternative to Gly). **4** a residue of the base guanine in a nucleic-acid sequence. **5** a residue of the ribonucleoside guanosine (alternative to Guo). **6** a residue of the monosaccharide glucose (alternative to Glc). **7** ganglioside.

G *symbol for* **1** Gibbs energy. **2** gravitational constant. **3** weight (alternative to *W* or *P*). **4** electric conductance.

G418 *a proprietary name for* the aminoglycoside antibiotic geneticin. Resistance to the antibiotic is conferred by the bacterial gene *neo*^{R}, which encodes aminoglycoside-3′-phosphotransferase,

Ga *symbol for* gallium.

Ga$_2$ *symbol for* galabiose (alternative to GaOse$_2$).

Ga$_3$ *symbol for* galatriaose (alternative to GaOse$_3$).

G$_M$ *abbr. for* monosialoganglioside.

GA *abbr. (sometimes) for* **1** glyceric acid. **2** glutamic acid. **3** glucuronic acid.

GAA a codon in mRNA for L-glutamic acid.

GABA *abbr. for* γ-amino-*n*-butyric acid or γ-aminobutyrate.

GABAergic **1** describing a nerve or other cell that is acted upon by γ-aminobutyrate. **2** describing nerves that act by releasing γ-aminobutyrate.

GABA-gated channel *see* **GABA receptor**.

GABA-modulin a brain protein that down-regulates the high-affinity binding site for γ-amino-*n*-butyric acid (GABA) on synaptic membranes.

GABA receptor any of several membrane proteins that bind γ-aminobutyrate (GABA) and mediate its effects as an inhibitory neurotransmitter. GABA$_A$ receptors, known also as GABA-gated channels, function as a Cl$^-$ channel; GABA$_B$ receptors are G-protein-coupled receptors. **Baclofen** is a potent GABA mimetic that is selective for the GABA$_B$ receptor; **muscimol** is its counterpart for the GABA$_A$ receptor. The GABA$_A$ receptor is multimeric, with α (types α1–α6), β (types β1–β4), γ (types γ1–γ3), δ, and ρ subunits. Subunits α and β have four putative transmembrane segments that show some homology with those seen in the nicotinic acetylcholine receptor (nAChR) (*see* **cholinoceptor**). Increased Cl$^-$ ion flux brings about hyperpolarization of an excitatory neuron, which inhibits it from firing on receipt of a normal impulse. **Benzodiazepine** drugs interact with GABA$_A$ receptors at the γ subunit. GABA$_B$ receptors are coupled through G-proteins to stimulation of outward K$^+$ current, inhibition of inward Ca^{2+} channels, inhibition of adenylate cyclase, and stimulation of the phosphatidylinositol–Ca^{2+} system. *See also* **chloride channel**.

gabase *an informal name for* aminobutyrate aminotransferase (EC 2.6.1.19).

GABA shunt *abbr. for* γ-aminobutyrate shunt.

GABA transaminase *see* **4-aminobutyrate transaminase**.

GAC a codon in mRNA for L-aspartic acid.

G-actin the monomeric form of **actin**.

G$_{M2}$-activator protein a glycosylated protein (162 amino acids), encoded as a precursor by a gene at 5q32-q33, that is essential for hydrolysis by lysosomal β-hexosaminidase A (*see* **hexosaminidase**). Deficiency can result from any of four mutations, and is the cause of the AB variant of **Tay–Sachs disease**.

GADD *abbr. for* growth arrest and DNA damage; either of two proteins (GADD45 and GADD153) that are produced in response to oxidative stress, induce temporary arrest of the cell cycle, and may

be involved in DNA repair. Expression of the GADD45 gene is modulated by p53.

Gaddum equation a relationship, suggested to replace the **Hill–Langmuir equation**, that describes the situation when two ligands, A and B, are in equilibrium with a common binding site. It is:

$$P_{AR} = [A]/K_A(1 + [B]/K_B) + [A],$$

where P_{AR} is the proportion of the binding sites occupied by ligand A; the square brackets indicate concentrations. *Compare* **Schild equation**. [After (Sir) John Henry Gaddum (1900– 67), British pharmacologist.]

gadelaidate *the trivial name for* (*E*)-eicos-9-enoate, CH$_3$–[CH$_2$]$_9$–CH=CH–[CH$_2$]$_7$–COO (*trans* isomer), the anion derived from gadelaidic acid, (*E*)-eicos-9-enoic acid. *Compare* **gadoleate**.

gadoleate *the trivial name for* (*Z*)-eicos-9-enoate, CH$_3$–[CH$_2$]$_9$–CH=CH–[CH$_2$]$_7$–COO$^-$ (*cis* isomer), the anion derived from gadoleic acid, (*Z*)-eicos-9-enoic acid. *Compare* **gadelaidate**.

gag a gene that encodes the glucosaminoglycan core protein for a **retrovirus**. It is named from group-specific antigen gene, or glucosamino glycan gene. *See also* **erbA**, **erbB**.

GAG **1** a codon in mRNA for L-glutamic acid. **2** *abbr. for* glucosaminoglycan.

GAGA factor a DNA-sequence-specific transcription factor in *Drosophila* that binds to several (GA)$_n$ repeats in the promoter for the Hsp70 gene. It acts in concert with **NURF** to disrupt nucleosome spacing in the promoter region.

Gaia a theory of the role of biota in the maintenance of a climatic homeostasis, proposed by British biochemist James Lovelock (1919–) in 1979. The basic concept is that all of the living organisms that inhabit the earth can be regarded as a single vast organism capable of manipulating the atmosphere, geosphere, and hydrosphere to suit its needs. Lovelock named this organism Gaia, after the Greek goddess of the earth.

Gal *symbol for* a residue (or sometimes a molecule) of the aldohexose galactose. In the condensed system of symbolism of sugar residues, it signifies specifically a galactose residue with the common configuration and ring size, i.e. D-galactopyranose.

GAL4 a gene involved in the regulation of galactose utilization in yeast. The GAL4 protein has a zinc-finger motif for binding to its upstream activator sequence (UAS), which it then activates, and a site for binding the complex of the GAL80 protein – encoded by the *GAL80* gene – and galactose. When this is bound, activation by GAL4 protein is prevented.

GalA *symbol for* a residue (or sometimes a molecule) of the uronic acid galacturonic acid.

galabiose *symbol*: Ga$_2$ or GaOse$_2$; *the trivial name for* the disaccharide Gal(α1-4)Gal.

galactan any glycan composed solely of galactose residues. The galactans thus form a subgroup of plant hemicelluloses.

galactitol or *meso*-galactitol or (*formerly*) **dulcitol** the trivial name for the hexitol formally derived by reduction of the aldehyde group of either D- or L-galactose. It occurs widely especially in seaweeds and higher plants. It is excreted in human galactokinase deficiency and causes cataracts.

galacto- prefix (in chemical nomenclature) indicating a particular configuration of a set of (usually) four contiguous CHOH groups, as in the acyclic form of D- or L-galactose. *See* **monosaccharide**.

galactoceramide *see* **galactosylceramide**.

galactocerebroside *see* **galactosylceramide**.

galactokinase *abbr.*: GLK; EC 2.7.1.6; *systematic name*: ATP:D-galactose 1-phosphotransferase. An enzyme that catalyses the phosphorylation by ATP of D-galactose to D-galactose 1-phosphate, with release of ADP; this initiates the metabolism of galactose. There are two human galactokinases. GLK1, a monomer of 392 amino acids encoded at 17q24, is the major enzyme in various tissues. Mutations in its gene lead to galactokinase deficiency with galactosemia, urinary excretion of galactose and galactitol, and cataracts. GLK2 is encoded on chromosome 15 and shows 35% sequence homology with GLK1. Its metabolic role is unclear.

galactolipid any **glycolipid** containing one or more residues of galactose and/or N-acetylgalactosamine. The term includes **galactosylceramide** and **galactosyldiacylglycerol**.

galactomannan any heteroglycan containing D-galactose and D-mannose residues; the structures have not been fully elucidated. They occur in seeds of various plants and trees, probably as storage polysaccharide. See also **guaran**.

galactoprotein A abbr.: Gap A; an alternative name for **fibronectin**.

galactosaemia a variant spelling (esp. Brit) of **galactosemia**.

galactosamine symbol: GalpN; the trivial name for the aminodeoxysugar 2-amino-2-deoxygalactopyranose; there are two enantiomers, D- and L-galactosamine. D-Galactosamine (symbol: D-GalpN), formerly known as **chondrosamine**, is a constituent (commonly as its N-acetyl derivative) of some glycolipids and of chondroitin sulfate, dermatan sulfate, and many bacterial glycosaminoglycans.

galactose symbol: Gal; the trivial name for the aldohexose galactohexose; there are two enantiomers, D- and L-galactose. D-Galactose (symbol: D-Gal), commonly known as galactose, and formerly known as **brain sugar** or **cerebrose**, rarely occurs free but is widely distributed in combined form in plants, animals, and microorganisms as a constituent of many oligo- or polysaccharides; it occurs also in **galactolipids** and as its glucoside in **lactose** and **melibiose**. Residues of both D-galactose and 3,6-anhydro-L-galactose (symbol: 3,6An-L-Gal) occur in the disaccharide **agarobiose** and in polysaccharides such as **agaropectin** and **agarose**.

β-D-galactose

galactose/glucose binding protein a signal transducing protein for bacterial chemotaxis.

galactosemia or (esp. Brit.) **galactosaemia** comparatively rare condition in humans in which there is a supranormal concentration of D-galactose in the blood and a specific inability to metabolize galactose. It is caused either by (1) a hereditary autosomal recessive trait in which there is a virtual absence of **UDPglucose–hexose-1-phosphate uridylyltransferase** activity; or less frequently (2) an inherited autosomal recessive trait in which there is a deficiency in **galactokinase** activity.

galactose oxidase EC 1.1.3.9; a copper-containing enzyme that catalyses the oxidation by dioxygen of D-galactose to D-galactohexodialdose and H_2O_2.

galactose-1-phosphate uridylyltransferase see **UDPglucose–hexose-1-phosphate uridylyltransferase**.

galactose-6-sulfate sulfatase see **N-acetylgalactosamine-6-sulfatase**.

galactosialidosis a lysosomal storage disorder, transmitted as an autosomal recessive trait, that results from deficiency of three lysosomal enzymes: protective protein/cathepsin A (PPCA), β-galactosidase, and neuraminidase. This causes accumulations of sialooligosaccharides in lysosomes and their excretion in urine. Three clinical phenotypes are recognized on the basis of age of onset and clinical severity: early infantile; late infantile; and juvenile or adult. At least 12 point mutations in the gene at 20q13.1 for PPCA are responsible for the condition.

galactosidase either of two types of enzyme: α-galactosidase EC 3.2.1.22; systematic name: α-D-galactoside galactohydrolase; other name: melibiase. An enzyme that hydrolytically removes terminal nonreducing α-D-galactose residues in α-galactosides, including

galactose oligosaccharides, galactomannans, and galactolipids. β-galactosidase EC 3.2.1.23; systematic name: β-D-galactoside galactohydrolase; other name: lactase. An enzyme that removes terminal β-D-galactose residues in β-galactosides by hydrolysis.

galactoside any **glycoside** in which the sugar moiety is galactose.

galactoside acetyltransferase EC 2.3.1.18; recommended name: galactoside O-acetyltransferase; systematic name: acetyl-CoA:β-D-galactoside 6-acetyltransferase; other name: thiogalactoside transacetylase. An enzyme that catalyses the acetylation by acetyl-CoA of β-D-galactoside to 6-acetyl-β-D-galactoside with release of GDP. Such an enzyme is the lacA gene product in Escherichia coli (see lac operon). Its function remains unclear.

galactoside 3(4)-L-fucosyltransferase EC 2.4.1.65; systematic name: GDP-L-fucose:β-D-galactosyl-N-acetyl-D-glucosaminyl-R 4-L-fucosyltransferase; other name: Lewis transferase. An enzyme that catalyses the transfer by GDP-L-fucose of fucose to 1,3-β-D-galactosyl-N-acetyl-D-glucosaminyl-R to form 1,3-β-D-galactosyl-(α-1,4-L-fucosyl)-N-acetyl-D-glucosaminyl-R with release of GDP; this forms the blood group determinant Le^a of **Lewis antigen**. The enzyme also acts on the corresponding 1,4-galactosyl derivative, forming 1,3-L-fucosyl links. It is a type II membrane glycoprotein, and a product of the Lewis blood group locus.

β-galactoside permease any of several transport proteins in bacteria that facilitate the uptake of β-galactosides. An example is melibiose carrier protein from Klebsiella pneumoniae. This is responsible for melibiose transport; it is capable of using hydrogen and lithium cations as coupling cations for cotransport, depending on the particular sugar transported (e.g. H$^+$-melibiose, Li$^+$-lactose), but cannot recognize Na$^+$ ions although it is an integral membrane protein of the sodium:galactoside symporter family. See also **lactose permease**.

galactosylceramidase EC 3.2.1.46; an enzyme that catalyses the hydrolysis of D-galactosyl-N-acylsphingosine to D-galactose and N-acylsphingosine.

galactosylceramide the recommended term for any D-galactosyl-N-acylsphingosine, i.e. any **cerebroside** in which the sugar moiety is D-galactose (less correctly termed **galactoceramide** or **galactocerebroside**).

galactosylceramide lipidosis or Krabbe's disease or globoid cell leukodystrophy one of a number of **sphingolipidoses** or sphingolipid lysosomal storage diseases. A rapidly progressive, invariably fatal, hereditary (autosomal recessive) disease of human infants caused by deficient activity of galactocerebroside β-galactosidase (galactosylceramidase; EC 3.2.1.46). Clinical symptoms usually start between the ages of 3 and 6 months and the condition is characterized by severe mental and motor deterioration and the presence of numerous globoid cells containing galactosylceramide in the white matter of the brain.

galactosylceramide sulfotransferase EC 2.8.2.11; an enzyme involved in **sulfatide** biosynthesis that catalyses the reaction between a galactosylceramide and 3'-phosphoadenylylsulfate to form galactosylceramide sulfate and adenosine 3',5'-bisphosphate.

galactosyldiacylglycerol any of a class of lipids in which one or two galactose residues are in glycosidic linkage with a 1,2-diacylglycerol (loosely termed mono- and digalactosyldiacylglycerols), more specifically as, e.g., 1,2-diacyl-sn-glycerol 3-β-D-galactoside or (preferred, as glycosphingolipids are named this way) 1,2-diacyl-3-β-D-galactosyl-sn-glycerol. They are widely distributed in plants and are the main constituent lipids of chloroplasts, where they occur in the thylakoid membrane. They are also present in mitochondria.

β-1,3-galactosyl-O-glycosyl-glycoprotein β-1,6-N-acetylglucosaminyltransferase EC 2.4.1.102; other names: O-glycosyl-oligosaccharide-glycoprotein; N-acetylglucosaminyltransferase I; β^6-N-acetylglucosaminyltransferase; core debranching enzyme; an enzyme that catalyses the reaction between UDP-N-acetyl-D-glucosamine and β-D-galactosyl-1,3-N-acetyl-D-galactosaminyl-R to form β-D-galactosyl-1,3-(N-acetyl-β-D-glucosaminyl-1,6)-N-acetyl-D-galactosaminyl-R with release of UDP. It introduces branches into O-linked glycoprotein oligosaccharides.

galactosyltransferase any enzyme belonging to the sub-subclass hexosyltransferases, EC 2.4.1, for which the glycosyl group is galactosyl; the acceptor is typically another carbohydrate molecule or lipid. *See also* **B-transferase**, **lactose synthase**, **Lewis transferase**.

galactowaldenase *a former name for* **UDPglucose 4-epimerase**.

galacturonan *or (formerly)* **polygalacturonide** any glycan composed solely of galacturonic-acid residues; a specific kind of glycuronan.

galacturonate **1** the anion of galacturonic acid. **2** any salt or ester of galacturonic acid.

galacturonic acid *symbol*: GalA *or (formerly)* GalU *or* GalUA; the uronic acid formally derived from galactose by oxidation of the hydroxymethylene group at C-6 to a carboxyl group. There are two enantiomers, D- and L-galacturonic acid; D-galacturonic acid is a component of plant gums and bacterial cell walls.

β-D-galacturonic acid

galanin a biologically active neuropeptide, with wide distribution in the central neurons of several mammalian species. Human galanin has the 30-residue sequence

GWTLNSAGYLLGPHAVGNHRSFSDKNGLTS.

The peptide effects contraction of smooth muscle of the gastrointestinal and genitourinary tracts, and regulates somatotropin release; it may be involved in the control of adrenal secretion, and is also produced in the islets of Langerhans from stimulated intra-islet nerve terminals. Galanin inhibits insulin release *in vivo* and *in vitro*, and inhibits formation of cyclic AMP in cells stimulated by **glucagon-related peptide**. A fragment, GWTLNSAGYLLGPHAI, is an agonist at the hippocampal galanin receptor.

galaptin any of various proteinaceous, *β*-galactoside-specific, animal lectins of relatively low molecular mass, that are isolated from a number of developing and adult tissues, and are believed to mediate intercellular adhesion in some developmental systems.

galectin a secreted carbohydrate-binding protein that binds CD45 and inhibits its receptor tyrosine phosphatase activity.

gall **1** an abnormal swelling or growth produced by a plant in response to the activities of various fungi, mites, and insects, especially members of the Cynipidae and Cecidomyidae. Insect eggs and larvae are often found in galls. **2** *a former name for* **bile**.

gallate **1** the anion of gallic acid, 3,4,5-trihydroxybenzoic acid, a phenolic acid. **2** any salt or ester of gallic acid. The esters and polyesters are very widely distributed in angiosperms, usually as **gallotannins**.

gall bladder a hollow, muscular organ, found in many vertebrates, that receives dilute **bile** from the liver, concentrates and stores it, and discharges it into the duodenum. *See* **gall stone**.

gallotannic acid *see* **tannic acid**.

gallotannin *or* **hydrolysable tannin** any **tannin** consisting of a molecule of a polyhydric alcohol (usually glucose) esterified with several gallic-acid molecules, each of which may itself be esterified with a further gallic acid molecule, and so on.

gallstone *or* **cholelith** a concretion that forms in the gall bladder. The most common type are light-coloured stones, usually multiple and multifaceted, consisting mostly of cholesterol with some protein and calcium bilirubinate. These usually form as a result of bile stasis, supersaturation with cholesterol, and/or presence of nucleation factors such as glycoprotein in mucus, bacteria, or epithelial cells. They occur in about 20% of women and about 5% of men between 50 and 65 years of age, are usually asymptomatic but may produce epigastric pain and obstructive jaundice. They are usually associated with obesity. Less common are brown or black gritty pigment stones consisting mostly of calcium bilirubinate. These are likely to be associated with chronic hemolytic anemia, and form at an earlier age.

Gal*p* *symbol for* a galactopyranose residue.

Gal*p*A *symbol for* a residue (or sometimes a molecule) of galactopyranuronic acid.

Gal*p*N *symbol for* a residue (or sometimes a molecule) of the aminodeoxysugar **galactosamine**.

Gal*p*NAc *symbol for* a residue (or sometimes a molecule) of the acetamidodeoxysugar *N*-acetylgalactosamine.

GalU *or* **GalUA** *former symbol for* **galacturonic acid** (GalA now recommended).

galvanic cell *an alternative name for* **cell** (def. 3).

gamete a mature reproductive cell whose nucleus (and often cytoplasm) fuses with that of another gamete of similar origin but of opposite sex to form a **zygote**, which develops into a new individual. Gametes are haploid and are differentiated into male (+) and female (−).

gamma *symbol*: γ (lower case) *or* Γ (upper case); the third letter of the Greek alphabet. For uses *see* **Appendix A**.

gamma chain *or* **γ chain** **1** the **heavy chain** of an IgG immunoglobulin molecule. **2** one of the two types of subunit in a human fetal hemoglobin (Hb F) molecule.

gamma globulin *or* **γ-globulin** any of a band of serum proteins that migrate most slowly on electrophoresis. In humans, this normally constitutes quantitatively the largest fraction of the total globulins. As the γ-globulins are predominantly antibodies, and are not synthesized in the liver as are virtually all the other serum proteins, the albumin/globulin ratio has traditionally been a useful indicator of liver function. The γ-globulins may be separated from the other serum globulins in pooled adult serum, and may then be used for administration by intramuscular injection to provide passive immunity in premature infants or for those in contact with cases of infectious hepatitis. Such γ-globulin preparations tend to form aggregates spontaneously; these can lead to severe anaphylactic reactions when administered intravenously, owing to their ability to aggregate platelets, and to activate complement and generate C3a and C5a anaphylatoxins. In certain diseases, such as Bruton's congenital agammaglobulinemia, the production of immunoglobulin in affected males is grossly depressed. In patients with myeloma, the γ-globulin band on electrophoresis of serum proteins is markedly more prominent, indicating the presence of a raised antibody against an unknown antigen. *See also* **globulin**.

galaptin any of various proteinaceous, *β*-galactoside-specific, animal lectins of relatively low molecular mass, that are isolated from a number of developing and adult tissues, and are believed to mediate intercellular adhesion in some developmental systems.

gamma helix *or* **γ-helix** a polypeptide chain conformation in which each amide group forms hydrogen bonds with the group's five residues away in each direction along the chain. It has 5.1 amino-acid residues per turn; the translation along the helical axis is 0.99 Å per residue, 5.03 Å per turn.

gamma-hydroxybutyrate *see* γ-hydroxybutyrate.

gamma-peptidase *see* X-Pro dipeptidase.

gamma radiation *or* **γ-radiation** electromagnetic radiation with wavelengths in the range 0.1–100 pm (frequencies 3–3000 EHz) emitted by some atomic nuclei during **transformation** (def. 3). Gamma rays have shorter wavelengths than X-rays and higher photon energies.

gammopathy any pathological increase in production of an immunoglobulin (usually IgM, IgA, or IgG) or Bence-Jones protein. Gammopathies may be monoclonal or polyclonal and are associated with increased immunoglobulin synthesis.

gamone any of a group of biological agents secreted by gametes that act on gametes of the opposite sex to initiate the processes leading to fertilization. They may be **gynogamones**, liberated from female gametes, or **androgamones**, liberated from male gametes.

ganciclovir *or* **gancyclovir** 9-[(1,3-dihydroxy-2-propoxy)methyl]-guanine; a nucleoside analogue of **acyclovir** that is more potent against cytomegalovirus. When phosphorylated by virally encoded thymidine kinase it inhibits DNA synthesis through inhibition of DNA polymerase.

ganglion (*pl.* **ganglia**) a cluster of nerve cells and associated glial cells located outside the central nervous system.

ganglioside any ceramide oligosaccharide carrying, in addition to other sugar residues, one or more sialic residues. Gangliosides were first isolated from ganglion cells (neurons) of the grey matter of the brain but are in fact widely distributed in vertebrate tissues. They are *N*-acetyl- (or *N*-glycolyl-) neuraminosyl-(X)osylceramides, where (X) stands for the root name of the neutral oligosaccharide to which the sialosyl residue is attached; the acyl groups in the ceramide moiety are commonly stearoyl or lignoceroyl. Gangliosides are designated by G for ganglioside, plus subscript M, D, or T for mono-, di-, or trisialo, respectively, followed by a subscript arabic numeral to indicate the sequence of migration in thin-layer chromatograms. The principal gangliosides of human brain are:

(1) monosialogangliosides: G_{M1}, Gal(β1-3)GalNAc(β1-4)(NeuAc(α2-3))Gal(β1-4)Glcβ1Cer (G_{A1} denotes the asialo derivative of G_{M1}); G_{M2}, GalNAc(β1-4)(NeuAc(α2-3))Gal(β1-4)Glcβ1Cer (G_{A2} denotes the asialo derivative of G_{M2}); G_{M3}, NeuAc(α2-3)Gal(β1-4)Glcβ1Cer;

(2) disialogangliosides: G_{D1a}, NeuAc(α2-3)Gal(β1-3) GalNAc(β1-4)(NeuAc(α2-3))Gal(β1-4)Glcβ1Cer; G_{D1b}, Gal(β1-3)GalNAc(β1-4)(NeuAc(α2-8)NeuAc(α2-3))Gal(β1-4)Glcβ1Cer; G_{D3}, NeuAc(α2-8)NeuAc(α2-3)Gal(β1-4)Glcβ1Cer;

(3) trisialogangliosides: G_{T1a}, NeuAc(α2-8)NeuAc(α2-3)Gal(β1-3)GalNAc(β1-4)(NeuAc(α2-3))Gal(β1-4)Glcβ1Cer; G_{T1b}, NeuAc (α2-3)Gal(β1-3)GalNAc(β1-4)(NeuAc(α2-8)NeuAc(α2-3))Gal (β1-4)Glcβ1Cer.

ganglioside galactosyltransferase EC 2.4.1.62; *other name*: UDPgalactose–ceramide galactosyltransferase; an enzyme that catalyses the reaction between UDPgalactose and *N*-acetyl-D-galactosaminyl-(*N*-acetylneuraminyl)-D-galactosyl-D-glucosyl-*N*-acylsphingosine to form the **ganglioside** G_{M1}, with release of UDP.

gangliosidosis any of a group of inherited human diseases involving defects in **ganglioside** breakdown. They are characterized by: progressive mental and motor deterioration with fatal outcome; autosomal recessive inheritance; neuronal lipidosis secondary to storage of gangliosides and structurally related glycolipids, polysaccharides, or glycoproteins; and absence or severe deficiency of specific lysosomal glycohydrolases. The principal gangliosidoses are G_{M1} and G_{M2} gangliosidoses (the names indicating which ganglioside accumulates). G_{M1} gangliosidoses are due to a striking deficiency of β-galactosidase activity when tested using G_{M1} or galactose-containing glycoproteins as substrates. Type 1 or infantile G_{M1} gangliosidosis is a generalized gangliosidosis with bony involvement and onset of signs from birth or shortly after. Type 2 or juvenile G_{M1} gangliosidosis is of slower onset than type 1 and with milder bony abnormalities. There are three types of G_{M2} gangliosidosis. Type 1 G_{M2} gangliosidosis (*or* **Tay–Sachs disease**) is the commonest gangliosidosis, occurring especially in Ashkenazi Jews. It is caused by a severe deficiency of hexosaminidase A. Type 2 G_{M2} gangliosidosis (*or* **Sandhoff's disease**) occurs much less frequently than Tay–Sachs disease and is caused by severe deficiency of both hexosaminidases A and B; globoside accumulates in addition to G_{M2} in this disease. Type 3 or juvenile G_{M2} gangliosidosis is caused by either severe or partial deficiency of hexosaminidase A.

gangliotetraose *symbol*: Gg$_4$ or GgOse$_4$; *a trivial name for* the tetrasaccharide Gal(β1-3)GalNAc(β1-4)Gal(β1-4)Glc.

gangliotriaose *symbol*: Gg$_3$ or GgOse$_3$; *a trivial name for* the trisaccharide GalNAc(β1-4)Gal(β1-4)Glc.

GaOse$_2$ *symbol for* galabiose (alternative to Ga$_2$).

GaOse$_3$ *symbol for* galatriaose (alternative to Ga$_3$).

gap 1 a space introduced into the written form of a sequence of residues in a biopolymer to bring part of that sequence into register with a similar region of sequence in another polymer; typically used to effect protein or nucleotide sequence alignment. **2** a discontinuity in one of the two strands of duplex DNA resulting from enzymic elimination of one or more nucleotide residues. *Compare* **nick**.

GAP *abbr. for* **1** GnRH-associated peptide (*see* **gonadotropin-releasing hormone**). **2** *abbr. for* GTPase activating protein; a protein that activates the hydrolysis of GTP bound to certain monomeric proteins,

especially p21^{ras} (Ras, *see* **RAS**), but also for other Ras-related proteins, e.g. Rho (*see* **Rho protein**) and Ran (*see* **RAN**). Since GTP-bound Ras is the active form, binding to GAP inactivates it. The protein has SH2 and SH3 domains (*see* **SH domains**) and **pleckstrin homology domains**. RGS proteins are GAPs for the α subunit of certain trimeric G-proteins.

Gap a *abbr. for* galactoprotein A (i.e. plasma **fibronectin**).

GAPDH *abbr. for* glyceraldehyde-3-phosphate dehydrogenase. It is used as an internal standard in Northern blotting and PT-PCR because its mRNA is believed to be expressed constitutively in mammalian tissues.

gap gene any of at least six genes of *Drosophila* that act early in embryogenesis to define the coarsest subdivisions of the embryo. Mutations in these genes result in gaps in the segmentation pattern; the **Krüppel** and **hunchback** proteins are products of gap genes.

gap junction *or* **nexus** any specialized area of the plasma membranes of apposed vertebrate cells where the membranes are 2–4 nm apart and penetrated by a **connexon** that bridges the extracellular space and provides open means of communication between the cytoplasm of one cell and that of the other cell. *Compare* **tight junction**.

Gap1p *abbr. for* general amino acid permease protein 1; a membrane protein of *Saccharomyces cerevisiae* that permits uptake of amino acids. It is induced by nitrogen starvation. If the yeasts are given a nitrogen-rich source, Gap1p is targeted for degradation by polyubiquitination and its synthesis is repressed.

gap penalty a numerical cost for inserting a gap in a sequence alignment. Depending on the alignment algorithm used, there may be a single gap penalty, or an **affine gap penalty** containing separate costs for initiating and extending the gap. With affine gap penalties the extension penalty is usually much lower than the insertion penalty. This penalizes gap insertions while recognizing that, where insertions exist, they often span more than one residue (hence it costs more to open a gap than it does to extend it once the gap has been opened).

gap phase either of the two phases, designated G_1 and G_2, in the cell-division cycle during **interphase** in growing eukaryotic cells during which there is no DNA synthesis.

gap repair the process of repair of a **gap** (def. 2) in one strand of a duplex DNA molecule.

garosamine 3-deoxy-4-*C*-methyl-3-(methylamino)-L-arabinose; a component of **gentamicins**.

gas a substance whose physical state is such that it always occupies the whole of the space in which it is contained, irrespective of the amount present. *See also* **gas laws**; **ideal gas**.

gas chromatography *abbr.*: GC; any form of chromatography in which the mobile phase is a gas. The term commonly refers to **gas–liquid (partition) chromatography** in which the stationary phase is a liquid, normally deposited on a solid support in a coiled column of glass or stainless steel, or the wall of a capillary column, but it also includes **gas–solid chromatography**, in which the stationary phase is a solid. The column is normally heated in an oven.

gas chromatography/mass spectrometry *abbr.*: GC/MS; an analytical technique that combines the separation process of **gas chromatography** with the highly selective detection technique of **mass spectrometry**.

gas constant *symbol*: R; a fundamental physicochemical constant; $R = 8.314\,471 \pm 0.000\,014 \; JK^{-1}mol^{-1}$.

gas exchange (*in physiology*) the exchange of gases between an organism and its environment. It includes oxygen uptake and carbon dioxide elimination in the respiration of all aerobic organisms, including plants, and carbon dioxide uptake and oxygen release in photosynthetic processes.

gas flow counter any gas-filled radiation detector in which the gas filling is renewed continuously. It may be operated in either the Geiger or the proportional regions of the voltage–current curve. Such devices are especially useful for detection and counting of alpha and soft-beta emitters, as no window is present to absorb the radiation.

gas laws either of two laws describing the effects of changes in pressure, p, and thermodynamic temperature, T, on the volume, V, of an ideal gas. **Boyle's law** (*or* **Mariotte's law**) states that at constant temperature the volume of a given mass of gas is inversely proportional to the pressure; i.e. pV = a constant. **Charles's law** states that at constant pressure the volume of a given mass of gas is directly proportional to its thermodynamic temperature; i.e. V/T = a constant. These two laws may be combined in the expression: $pV \propto T$, or, for one mole of gas: $pV = RT$, where R is the **gas constant**.

gas–liquid (partition) chromatography an analytical method based on the partition chromatographic separation of the components of a mixture through the use of a stationary liquid phase held in place on a solid support and a mobile gas phase flowing over the liquid phase in a controlled manner. Typically, the column is heated to temperatures in the range 50–300°C and the technique is applicable to compounds that partition appreciably into the vapour phase in that range. Detection of the eluents is commonly by a **flame ionization detector**.

GASP abbr. for genome annotation assessment project; see **critical assessment competition**.

gas–solid chromatography see **gas chromatography**.

gastr+ a variant form of **gastro+** (before a vowel).

gastric of, pertaining to, or derived from the stomach.

gastric acid the hydrochloric acid that is secreted from parietal (or oxyntic) cells of the wall of the stomach and forms part of the **gastric juice**.

gastric inhibitory peptide abbr.: GIP; or **glucose-dependent insulinotropic peptide** or **glucose-dependent insulin-releasing peptide** a 42-residue polypeptide hormone produced by the K cells of the proximal small intestine in response to glucose in the intestine. It is structurally related to glucagon, secretin, and vasoactive intestinal polypeptide having the sequence (human) YAEGTFISDYSIAMD-KIHQQDFVNWLLAQKGKKNDWKHNITQ, porcine being similar but with Arg^{18} and Ser^{34}. It inhibits gastric secretion and motility (**enterogastrone effect**), and stimulates insulin release when administered during hyperglycemia (**insulinotropic effect**); in the latter action GIP may be synonymous with **incretin**.

gastric inhibitory peptide receptor abbr.: GIP receptor; any of the G-protein-associated membrane receptors that bind **gastric inhibitory peptide** (GIP) and activate adenylate cyclase. The rat protein consists of 455 amino acids (52.26 kDa), of which residues 1–18 are the signal peptide, and 19–455 are the receptor.

gastric juice the thin, clear, acidic fluid secreted by glands of the stomach mucosa in response to the ingestion of food. It contains mucus, intrinsic factor, hydrochloric acid (about 155 mm in humans) from the parietal cells, and **pepsinogen**, the precursor of pepsin proteolytic enzymes.

gastricsin EC 3.4.23.3; other names: pepsin C, pig parapepsin II; an enzyme of the pepsin family that appears to be present in the gastric juice of most vertebrates. It effects preferential cleavage at Tyr-|-Xaa bonds. It is a single-chain phosphoprotein formed by limited proteolysis of pepsinogen C with a more restricted specificity than pepsin A but high activity towards hemoglobin.

gastrin any of various members of a group of interrelated peptide hormones from the **G cells** of the pyloric antral mucosa of the stomach, and of the proximal duodenum, that strongly stimulate acid secretion by the stomach mucosa. There are three principal, biologically active forms: **big gastrin** (or G-34); **little gastrin**, (or G-17 or gastrin I); and **mini gastrin** (or G-14); these have 34, 17, and 14 amino-acid residues, respectively. The sequence of human big gastrin (gastrin I in parentheses, but with pE for the terminal Q) is

pELGPQGPPHLVADPSKK(QGPWLEEEEEA-YGWMDF-NH₂),

where pE indicates pyroglutamyl. Each has a counterpart, e.g. G-17S (or gastrin II), in which the tyrosyl residue is sulfated. There are other, larger immunoreactive forms, known as **big-big gastrins**, that elute from a gel column before G-34. The entire physiological activity is possessed by the C-terminal tetrapeptide amide sequence that ends in a phenylalaninamide residue. The C-terminal pentapeptide sequence of gastrin is the same as that of **cholecystokinin** and **cerulein**. See also **gastrinoma**, **pentagastrin**, **Zollinger–Ellison syndrome**.

gastrinoma a tumour of the gastrin-producing **G cells** of the pancreas, stomach, or duodenum. Gastrinomas are associated with high gastrin levels in the circulating blood, gastric hypersecretion,

and fulminant gastroduodenojejunal ulcerations (**Zollinger–Ellison syndrome**).

gastrin receptor see **cholecystokinin receptor**.

gastrin-releasing protein a growth factor secreted by some small-cell lung cancers.

gastro+ or (often before a vowel) **gastr+** comb. form meaning stomach. [From Greek gast ē r, belly.]

gastroenteric an alternative term for **gastrointestinal**.

gastrointestinal or **gastroenteric** of, or pertaining to, the stomach and intestines.

gastrointestinal hormone any member of a diverse group of peptides with hormone or hormone-like actions that occur in specific cells of the gastrointestinal tract, the pancreatic islets, and other locations. The best characterized include **cholecystokinin**, **gastric inhibitory polypeptide**, **gastrin**, **glucagon**, **insulin**, **motilin**, **neurotensin**, **pancreatic polypeptide**, **secretin**, **substance P**, and **vasoactive intestinal polypeptide**. See also **gut hormone**.

gastrula an early stage of development of an animal embryo, following the blastula, in which the cells are invaginating to form the rudiment of a gut cavity.

gas vacuole any membrane-bound, gas-containing, intracellular organelle. Gas vacuoles provide aquatic prokaryotes with buoyancy. See also **gas vesicle**.

gas vesicle the submicroscopic component of the gas vacuole in certain aquatic prokaryotes, especially cyanobacteria. It is constructed of molecules of a single protein (≈15 kDa), which form a monolayer enclosing a hollow space into which gases diffuse. Typically it is cylindrical with conical ends, 3–300 nm in diameter and 300–1000 nm in length.

GATA-1 or **ERYF1** or **GF-1** or **NF-E1** a transcriptional activator protein that serves as a general switch factor for erythroid development. It binds to DNA sites with the consensus sequence [A/T]GATA[A/G] within regulatory regions of globin genes and other genes expressed in erythroid cells. Other transcription factors with similar sequences are termed **GATA-type**. GATA-1 binds the zinc finger-rich transcription regulator FOG-1.

GATase see **glutamine amidotransferase**.

gate (in electronics) **1** a logic circuit that gives an output signal only in response to a combination of two or more input signals. **2** a signal that allows or controls the passage of another signal.

gated channel a channel in the membrane of a cell (especially an excitable cell) that allows entry of ions into the cell only under specific conditions. See also **ligand-gated**, **voltage-gated ion channel**.

gating the opening and closing of a gated channel.

gating current an electric current arising from the movement of charged groups in the protein of the sodium-ion channels of excitable nerve membranes, during the passage of an action potential, prior to the opening of the channel.

g-atom abbr. for gram-atom or gram-atomic weight.

GAU a codon in mRNA for L-aspartic acid.

gauche abbr.: g; (in chemical nomenclature) a conformational descriptor indicating a synclinal conformation. See **conformation**.

Gaucher activator or **saposin C**; a glycosylated polypeptide derived by appropriate proteolytic cleavage of prosaposin. It is an activator of lysosomal **glucosylceramidase**. Deficiency of prosaposin or saposin C, due to mutation in the prosaposin gene, produces a variant of Gaucher's disease type 3. See **saposin**.

Gaucher's disease an autosomal recessive lysosomal storage disease characterized by **glucosylceramide** accumulation, especially in macrophages in spleen, bone marrow, and liver, with splenomegaly and bone lesions. Type 1 is the commonest, does not affect the nervous system (i.e. is non-neuronopathic), and is frequently diagnosed in adulthood. Type 2 is acute, neuronopathic, and usually fatal in the first two years of life. Type 3 is subacute, neuronopathic, and associated with mental retardation. Numerous mutations in the gene at 1q31 result in various degrees of deficiency of **glucosylceramidase** (EC 3.2.1.45). Prosaposin or saposin C deficiency produces variants of the disease. See also **saposin**. [After Philippe Charles Ernest Gaucher (1854–1918), French physician.]

gauss symbol: G or Gs; the cgs unit of magnetic flux; it is equal to a magnetic flux density of one maxwell per cm², one line of magnetic induction per cm^2, or 10^{-4} tesla. It is not accepted for use with the SI.

GAUSSIAN a computer program, based on quantum mechanics, which calculates the energies, structures, and vibrational frequencies of molecules.

Gaussian distribution *an alternative name for* **normal distribution**.

gavage the administration of food or other material in liquid form through a tube directly into the stomach; forced feeding in this manner.

Gb$_3$ *symbol for* globotriaose (alternative to GbOse$_3$).

Gb$_4$ *symbol for* globotetraose (alternative to GbOse$_4$).

G-banding *or* **Giemsa banding** a commonly used procedure in which **Giemsa's stain** is used to identify chromosomal bands in a karyotype.

GbOse$_3$ *symbol for* globotriaose (alternative to Gb$_3$).

GbOse$_4$ *symbol for* globotetraose (alternative to Gb$_4$).

GBP *abbr. for* glucose bisphosphate (*formerly called* glucose diphosphate).

G-1,6-BP *abbr. for* glucose 1,6-bisphosphate (*formerly called* glucose 1,6-diphosphate).

Gc *symbol for* the glycoloyl group.

GC *abbr. for* gas chromatography.

GCA a codon in mRNA for L-alanine.

GCAP *abbr. for* guanylate cyclase activator protein; either of two Ca^{2+}-binding proteins (GCAP-1 and GCAP-2) of the calmodulin family that bind and activate retinal guanylyl cyclase protein (*see* **RETGC**) in the membrane of retinal cone cells.

GC box a nucleotide sequence element common in eukaryotic promoters: consensus GGGCGC.

GCC a codon in mRNA for L-alanine.

G cell a gastrin-producing cell found in the antrum of the stomach and to a lesser extent in the mucosa of the duodenum, though the distribution varies in different species. In humans, G cells are small and pyriform while in dogs they are larger and triangular in shape. They can be identified by immunofluorescence using a gastrin-specific antibody.

GCG 1 a codon in mRNA for L-alanine. **2** *abbr. for* Genetics Computer Group; *see* **sequence analysis package**.

GC/MS *abbr. for* gas chromatography/mass spectrometry.

GCN4 a gene for a yeast transcription factor that binds, as a dimer, to the sequence TGA[C/G]TCA. The protein has a **leucine zipper**, and resembles **FOS** and **JUN**.

GCRDB *abbr. for* G protein-coupled receptor database; *see* **specialist database**.

G-CSF *abbr. for* granulocyte colony stimulating factor.

GCU a codon in mRNA for L-alanine.

Gd *symbol for* gadolinium.

GDC *abbr. for* Graves' disease carrier protein; it is similar in structure and some other properties to **ADP,ATP carrier protein**.

G4 DNA *abbr. for* DNA containing a **G-quartet**.

GDNF *abbr. for* glial cell line-derived neurotropic factor; a glycosylated, disulfide-bonded homodimer that promotes survival and morphological differentiation of dopaminergic neurons. Human and rat GDNFs share 93% similarity. *Compare* **brain-derived neurotropic factor**.

GDP *abbr. for* **1** guanosine diphosphate. **2** glucose diphosphate (now called glucose bisphosphate).

G-1,6-DP *abbr. for* glucose 1,6-diphosphate (now called glucose 1,6-bisphosphate).

GDP-fucosepyrophosphorylase *see* **fucose-1-phosphate guanylyltransferase**.

GDPM *or* **GDPmannose** *abbr. for* guanosinediphosphomannose.

GDPmannose 6-dehydrogenase EC 1.1.1.132; an enzyme that catalyses the reaction between GDP-D-mannose, 2 NAD$^+$ and H$_2$O to form GDP-D-mannuronate and 2 NADH. Example from *Pseudomonas aeruginosa*: ALGD_PSEAE, 436 amino acids (47.58 kDa).

GDPmannose 3,5-epimerase EC 5.1.3.18; an enzyme that catalyses the interconversion of GDP-D-mannose and GDP-L-galactose.

GDS (*in clinical chemistry*) *abbr. for* guanine deaminase.

Ge *symbol for* germanium.

Geiger counter *or* **Geiger–Müller counter** a device for detecting and measuring ionizing radiation (alpha or beta particles or gamma-ray photons). It consists of a glass envelope containing low-pressure gas and an anode and a cathode, between which a large electric-potential difference is maintained. The ions produced in the gas by the incoming ionizing radiation are accelerated towards their appropriate electrodes, causing a momentary drop in the potential between the latter. The voltage pulses so produced may be counted with a suitable electronic circuit. [After Hans Wilhelm Geiger (1882–1945), German physicist, and Erwin Wilhelm Müller (1911–77), German-born US physicist.]

Geiger region *or* **Geiger–Müller region** the part of the curve of applied voltage against number of ions collected for a given incoming radiation by a **Geiger counter**, where all possible ion-pairs, both primary and secondary, are collected; i.e. the signal is relatively independent of the applied voltage. *Compare* **proportional region**.

Gek *abbr. for* Genghis Khan.

gel 1 a colloidal system, with the semblance of a solid, in which a solid is dispersed in a liquid. A gel has a finite, usually rather small, yield stress. **2** to become, or cause to become, a gel.

gelatin a heterogeneous mixture of water-soluble proteins of high average M_r, derived from collagen by partial hydrolysis in either acid (type A) or alkali (type B). Collagen obtained by boiling skins, ligaments, tendons, etc. with water is also termed gelatin, as also is collagen denatured by heating for 15 minutes at 45 °C, used as a substrate for **gelatinase**. It is colourless, or slightly yellow, and transparent, and occurs as brittle sheets, flakes, or coarse powder; these swell in water to form gels below 35–40 °C, containing 5–10 times as much water as gelatin by weight. Nutritionally, gelatin is an incomplete protein, containing only small amounts of certain essential amino acids, especially tryptophan. It is widely used in foods, and in the manufacture of adhesives, photographic films and plates, and light filters.

gelatinase EC 3.4.24.24; *recommended name*: gelatinase A; *other names*: 72 kDa gelatinase; matrix metalloproteinase 2; type IV collagenase. An enzyme that catalyses the cleavage of **gelatin** and collagen types IV, V, VII, and X; it cleaves the collagen-like substrate sequence Pro-Gln-Gly-|-Ile-Ala-Gly-Gln; zinc and calcium are cofactors. It is a metallo (zinc) endopeptidase with **fibronectin** domains, and is produced by fibroblasts. The human precursor (660 amino acids) has a large pre-pro region; the enzyme starts at amino acid 110.

gelation 1 the (usually reversible) act or process of forming a **gel** (def. 1) from a **sol**. *Compare* **solation**. **2** the solidification of a liquid by cold. —**gelate** *vb*.

gel chromatography *see* **liquid–gel chromatography**.

gel diffusion (*in immunology*) a technique in which antibodies and antigens are allowed to diffuse independently through a gel, often of agar, and form a precipitate within the gel where homologous antigens and antibodies meet in optimal proportions.

gel electrophoresis a type of **zone electrophoresis** in which the supporting medium is a gel of uniform concentration, commonly of polyacrylamide, agarose, or starch. *See* **polyacrylamide (gel) electrophoresis**. *Compare* **gradient (acrylamide) gel electrophoresis**.

gel filtration *or* **gel-filtration chromatography** *see* **gel-permeation chromatography**.

gel-filtration chromatography *see* **gel-permeation chromatography**.

gel-filtration high-performance liquid chromatography a technique of **gel-permeation chromatography** that uses a support material of high mechanical strength and chemical inertness thereby enabling separations to be effected under the conditions, and with the advantages of **high-pressure liquid chromatography**.

gel fluorography the application of **fluorography** (def. 2) to the detection of substances separated in gels, as by gel electrophoresis.

gellify to cause to become a **gel** (def. 1); to become gel-like; to **gel** (def. 2).

gelonin *or* **ribosome-inactivating protein** *or* **rRNA N-glycosidase** EC 3.2.2.22; *recommended name*: rRNA N-glycosidase; an extremely stable, potent toxin extracted from seeds of *Gelonium multiflorum* (a plant of the spurge family). It is a glycoprotein of 28–30 kDa containing terminal mannose residues. It acts as a powerful inhibitor of protein synthesis in cell-free systems but not in intact cells; it inactivates the 60S ribosomal subunit but not the 40S subunit. It hydrolyses a specific N-glycosidic bond of an adenosine on the 28S rRNA. in cell-free protein-synthesizing systems gelonin acts like the A chains of **abrin**, **ricin**, and **modeccin**, suggesting that it lacks the ability to bind to cell surfaces and enter cells.

gel-permeation chromatography a method for the separation of substances in solution, *commonly known as* **gel filtration** *or* **gel-filtration chromatography**. The separation is based mainly on exclusion effects, such as differences in molecular size and/or shape or in charge, when the stationary phase is a swollen gel. The gel used is commonly a water-swollen insoluble carbohydrate polymer from which macromolecules are excluded and migrate without retention in the interstitial fluid. Substances of low or intermediate relative molecular mass penetrate into the gel particles to an extent that is, in most instances, determined by their molecular dimensions and the degree of cross-linking of the gel. The larger molecules therefore pass through the bed more rapidly than smaller molecules. The technique can be used to determine the relative molecular mass (M_r) of unknown materials, since, for a particular column, elution occurs at a volume determined by the equation

$$V_R/V_0 = 1 - \rho \log(M_r/M_{r0})$$

where V_R is the elution volume of the unknown, V_0 is the volume of the mobile phase in the column, M_r is the M_r of the unknown, M_{r0} is the M_r of the largest totally permeating molecule, and ρ is a factor that is a property of the column. Thus each column must be calibrated using standard proteins, and a plot made of the elution volumes against $\log M_r$ of the standards, from which the unknown M_r can be determined.

gel retardation assay *an alternative name for* **electrophoretic mobility shift assay**.

gelsolin a heat-labile, monomeric protein, of apparent M_r 90 000, isolated from macrophages, platelets, and plasma. It appears to promote the gel–sol transformation of **actin**, and may thereby be important in the control of locomotion, secretion, and endocytosis in these and other cells; β-actin has a similar action. There are two forms of gelsolin, plasma and cytoplasmic, generated by alternative splicing. The cytoplasmic, calcium- regulated, actin-modulating form binds to the barbed ends of actin monomers or filaments, preventing monomer exchange; it also promotes nucleation, and binds with high affinity to fibronectin.

gem- *prefix (in chemical nomenclature)* denoting the presence of **geminal** substituents.

gemeprost 16,16-dimethyl-*trans*-Δ^2-prostaglandin E_1 methyl ester; 11,15-dihydroxy-16,16-dimethyl-9-oxoprosta-2,13-dien-1-oic acid methyl ester; an analogue of **prostaglandin** E_1 with uterine stimulant activity.

gemfibrozil 2,2-dimethyl-5-(2,5-xylyloxy)valeric acid; an antihyperlipoproteinemic agent with actions similar to **clofibrate**.

gemin 3 a mammalian DEAD-box helicase associated with the SMN (survival of motor neuron) complex. Gemin 3, gemin 4, eIF2C2, and miRNA are present in this 15S ribonucleoprotein complex. This may be a mammalian RNA-induced silencing complex (RISC).

geminal *(in chemistry)* describing two like groups or atoms attached to the same atom in a molecule, i.e. the geminal groups or atoms are separated by two bonds attached to the same atom. The presence of geminal substituents is denoted by the prefix *gem*- attached to the name of the compound. *Compare* **vicinal** (def. 2).

geminate occurring in pairs; doubled, twin.

geminin a protein (35 kDa) of metazoan animals that is present in the cell nucleus from the S phase of the cell cycle to mitosis, being degraded on completion of mitosis. It binds and inhibits two proteins (65 kDa and 130 kDa) that are required for initiation of DNA replication.

geminivirus a plant virus in which the virion morphology is characterized by the possession of a pair of isometric particles, each containing circular or linear single-stranded DNA. Transmission occurs largely via leafhoppers or whitefly, but sometimes transmission is effected mechanically. Some geminiviruses cause economically important diseases in cultivated plants. [From Latin *gemini*, twins.]

+gen *or (after a consonant)* **+ogen** *comb. form* signifying producing or capable of producing (either directly or indirectly). In biological sciences it is used especially (1) with word-stems relating to physiological or pathological processes, states, conditions, etc., to denote a causative agent (e.g. antigen, lactogen, mutagen, estrogen); (2) with names of certain proteinases or blood components, to denote their inactive precursors or substances from which they can be derived by enzymic action (e.g. angiotensinogen, caseinogen, fibrinogen, trypsinogen) (*see also* **pro+** (def. 3), **zymogen**); (3) with names or name-roots of certain metabolites, to denote either a storage substance (e.g. glycogen, phosphagen) or a biosynthetic precursor (e.g. porphobilinogen); (4) with word-stems relating to physical properties (e.g. chromogen, luminogen). —**+genic** *or* **+genous** *adj. suffix.*

Genaminox KC *the proprietary name for* detergents similar to lauryldimethylamine oxide, but with different alkyl chain lengths (9–13).

Genapol *the proprietary name for* a series of nonionic polyoxyethylene detergents of the general formula $CH_3(CH_2)_x – O(CH_2CH_2O)_y$ –H; Genapol X-080 has $x = 12$, $y = 8$, CMC 0.06–0.15 mm; Genapol X-100 has $x = 12$, $y = 10$, aggregation number 88, CMC 0.15 mm. *Compare* **$C_x E_y$, Lubrol, Triton, Tween**.

GenBank *see* **sequence database**.

gene in classical genetics, a statistical entity that correlates with a particular phenotypic characteristic; the functional unit of heredity. Before their biochemical nature was understood, genes were defined in terms of units of mutation and/or recombination. Discovery of the role of DNA in genetic processes, followed by elaboration of the **central dogma** of protein synthesis, led to enunciation of the 'one gene–one enzyme' hypothesis, i.e. that a gene consisted of DNA that coded for a protein that performed the functions associated with the phenotypic expression of the gene. In current molecular genetics, the concept requires modification in a number of ways. First, the genomic DNA that codes for a polypeptide is associated with regulatory sequences such as promoters. Second, the polypeptide resulting from translation of mRNA may be subsequently split to give several polypeptides with different functions (*see* **polyprotein**). Third, the RNA transcribed may give rise to several different proteins as a result of **alternative splicing**. Three types of genes are now distinguished: those that are both transcribed into mRNA and translated into polypeptides (**structural genes**); those that are only transcribed into RNA (e.g. rRNA, tRNA); and those that function as regulators of the expression of the other two types (**regulator genes**). In diploid organisms a gene may occur in alternative forms (**alleles**). The term gene is sometimes used interchangeably with **cistron**.

gene amplification the selective, repeated replication of a certain gene or genes without a proportional increase in other genes in the genome. It occurs, e.g., in the DNA puffs of *Rhynchosciara* and some other flies and in the genes encoding ribosomal RNA (so-called ribosomal DNA) in amphibian oocytes. The mammalian dihydrofolate reductase gene is amplified in cells selected for resistance to increasing concentrations of methotrexate.

gene bank *an alternative name for* **gene library**.

GeneChip array a standardized, reproducible, proprietary technology for simultaneous measurement of the expression levels of large numbers of genes. The method involves *in situ* synthesis of DNA oligonucleotide probes on glass slides by means of photolithography, which allows several hundred thousand spots (with copies of specific DNA sequences) to be placed on a single slide. Single mRNA samples are then fluorescently labelled and hybridized to the chip, scanned with a fluorescent scanner, and the data collected and automatically analysed to quantify the relative amount of fluorescence and hence the relative levels of expression of each gene. *See also* **microarray**.

gene cloning *see* **molecular cloning**.

gene cluster *or* **gene complex** any group of two or more functionally related genes that are closely linked on a chromosome. The genes of a gene cluster are often structural genes coding for the enzymes that catalyse the various steps of a metabolic pathway.

gene complex *an alternative name for* **gene cluster**.

gene conversion a genetic event involving transfer of DNA between genes, usually as a result of unequal crossing over during meiosis. If the transferred material contains mutations, or if it disrupts the coding sequence of the gene, it may be a mechanism for mutation.

gene desert a region of a chromosome that contains few if any genes.

gene dosage the number of times a particular gene is present in the genome.

gene duplication the phenomenon, occurring in some higher organisms, in which there is duplication of the DNA sequences representing particular genes. This process is thought to have played a decisive part in the evolution of these organisms, through the occurrence of different mutations in the two duplicated genes.

gene expression the process by which the information carried by a gene or genes becomes manifest as the phenotype. It involves **transcription** of the gene into complementary RNA sequences and, for structural genes, subsequent **translation** of mRNA into polypeptide chains and their assembly into the ultimate protein products. Gene expression is tightly regulated by **promoters**, **enhancers**, and **transcription factors**.

gene expression array *see* **microarray**.

gene expression profile a characteristic pattern or inventory of gene expression from particular cell types, organs, or tissues, under particular physiological or disease conditions. Methods for analysing expression profiles depend on the technology used to measure expression intensities, but are usually classed as sequence tag counting – e.g. expressed sequence tags (ESTs), SAGE – or hybridization methods (e.g. microarrays).

gene family a group of closely related genes that encode similar protein products. Such genes tend to arise by duplication of an ancestral gene, and subsequently evolve independently via random mutational events. This results in independent genes, either clustered together on a chromosome or dispersed throughout the genome, that encode proteins with slightly different functions.

gene frequency a measure of the proportion of an allele in a given population, equal to the number of loci at which a given allele occurs, divided by the total number of loci at which it could occur.

gene fusion *or* **translational fusion** *or* **protein fusion** the joining of DNA from two genes as a result of translocation or inversion. It may give rise either to a chimeric protein product or to the misregulation of transcription of one gene by the *cis*-regulatory elements of the other.

gene identification *an alternative term for* **gene prediction**.

gene jungle a region of a chromosome that has a high gene density.

gene library *or* **cloned library** *or* **gene bank** *or* **shotgun collection** a random collection of DNA fragments cloned in a **vector** (def. 3), and which may include all the genetic information of a particular species. It may be prepared from a variety of sources, including an extract of mRNA, in contrast to a **genomic library** which is prepared from genomic DNA.

gene mutation any **mutation** (def. 1) occurring within a single gene.

gene ontology *abbr.*: GO; a controlled vocabulary (based on model eukaryotic organisms, including *Drosophila melanogaster*, *Saccharomyces cerevisiae*, *Mus musculus*, etc.) that has been developed to relate gene and protein names, functional groupings of genes and gene products, and location of action on a community-wide basis

of these entities, into a coherent, standardized whole. The three organizing principles of GO are molecular function, biological process, and cellular component.

gene pool the sum of the genetic information in the reproductive members of a population of sexually reproducing organisms.

gene prediction *or* **gene identification** the prediction of genes in an uncharacterized nucleotide sequence based on its similarity to known gene sequences. *Compare* ***ab initio* gene prediction**.

gene product any of the types of RNA (transcription products) or any of the proteins or protein subunits (translation products) synthesized biochemically on the basis of the information encoded in a genome.

general acid-catalysis homogeneous catalysis in which the catalysts are various hydron donors (acids). *Compare* **specific acid-catalysis**.

general base-catalysis homogeneous catalysis in which the catalysts are various hydron acceptors (bases). *Compare* **specific base-catalysis**.

generally labelled describing the labelling of a molecule in such a way that a radionuclide may be present at any or all (but not necessarily all) possible positions. *Compare* **uniformly labelled**.

general transcription factor any of the proteins whose assembly around the **TATA box** is required for the initiation of transcription of most eukaryotic genes. Their activity is enhanced by transcription activators that bind at distal enhancer sites. *See* **promoter**.

generate *(in mathematics)* to conceive a point, line, or surface to be moving in a specified way so as to form a line, surface, or solid, respectively.

generation 1 the act or process of producing or reproducing, naturally or artificially. **2** the phase in a life cycle that extends from one to the immediately successive reproduction. **3** any group comprising all those members of a population who are equally removed from a common ancestor or from coeval ancestors.

generation of diversity *(in immunology)* the process by which a large number of variable regions are generated in the immunoglobulins. The **stem cell** genome contains multiple variants of L-chain V (variable) and J genes and of the H-chain V, J, and D (diversity) genes. As the stem cells differentiate, the maturing lymphocyte constructs particular L and H genes of virtually unique structure by a recombination process that randomly selects one out of each set of gene segments and assembles them, together with a C (for the constant region) gene into a mature H or L gene. This, combined with recombinational inaccuracies, somatic point mutations, and the varied combinations of l and h chains found in immunoglobulins results in a repertoire of millions of lymphocytes each with H and L genes encoding unique molecules. *See also* **immunoglobulin**.

generation time the time between division of a cell and that of its daughter cells, averaged over a whole cell population.

gene redundancy the presence in a cell of many copies of a single gene. The multiple copies may be inherited or result from selective **gene duplication** during development.

gene-regulatory protein any protein that binds to a specific DNA sequence to alter the expression of a gene.

+genesis *comb. form* denoting beginning or origin; development; generation. *See also* **+gen**. —**+genetic** *or* **+genic** *adj. comb. form.*

gene splicing biochemical and/or chemical manipulations with the object of attaching one DNA molecule to another. The neatest method is to cleave the DNA to be inserted (foreign DNA) with a **restriction endonuclease** that yields single-stranded ends that are complementary to each other. The **cloning vector** into which the foreign DNA is to be inserted is treated with the same endonuclease so that the complementary ends of the two DNAs specifically associate under annealing conditions and are subsequently covalently joined (spliced) through the action of **DNA ligase**. If the endonuclease produces fragments with 'blunt' or **flush ends** then a similar procedure is adopted. If the foreign DNA and the cloning vector have no common restriction sites then terminal **deoxynucleotidyl transferase** may be used which adds nucleotides to the 3′-terminal OH group of a DNA chain. For this purpose tails of poly(dT) and poly(dA) may be used. The transformed cells are then grown and those containing the spliced DNA are selected (cloning). The preferred method of splicing is one that allows the foreign DNA to be removed easily by means of a restriction endonuclease, preferably the same one that was initially used in splicing. By this means large amounts of the foreign DNA can be synthesized and used for many different purposes. Alternatively the cloned cells containing the foreign DNA can be used to permit the expression of a foreign protein which is then recovered.

gene therapy the treatment of disease by incorporating into a person's genome DNA that encodes a specific therapeutic protein or that corrects deficiency of a normal protein due to a gene mutation.

genetic *or* **genetical** of or pertaining to **genetics**, **genes**, or the origin of something. —**genetically** *adv.*

genetic ablation a method applied in transgenic organisms in which a tissue-specific promoter is used to enable the expression of a gene for a toxin, such as diphtheria toxin A (*dipA*), to cause destruction of the particular tissue during development.

genetic algorithm an algorithm that exploits evolutionary principles. Given an initial population of solutions, a genetic algorithm evaluates which have the best fit to some desired property as measured by a fitness function. More optimal solutions are then generated by a heuristic process that mimics genetic selection, employing, for example, crossover, mutation, and replication over a number of generations.

genetically engineered describing a cell, strain, or organism whose phenotype has been altered by manipulation of its genetic material. *See* **genetic engineering**.

genetically modified organism *abbr.*: GMO; an organism that has been modified by the addition (from the same or different species) or deletion of genetic material, or by some other genetic alteration, to enhance a natural characteristic or to provide one that it does not have naturally.

genetic block a reduction in the activity of a particular enzyme in a metabolic pathway as a result of gene mutation. A genetic block is termed **complete** when the particular enzyme activity is absent, or **incomplete** (*or* leaky) when the enzyme formed is defective and of limited activity.

genetic carrier *see* **carrier** (def. 9).

genetic code all the regularities in nucleotide sequence according to which genetic information for the sequences of all the polypeptides synthesized by transcription and translation is encoded in DNA (or RNA in some viruses). A sequence of three nucleotide residues (a codon) is required to code for one amino-acid residue, and since there are four different kinds of base in nucleic acids (apart from minor constituents), 4^3 i.e. 64 different nucleotide triplets can exist. But only 20 different amino-acid residues in polypeptides can be coded for; therefore the code is degenerate in that most amino acids can be specified by more than one codon. In terms of the sequence of nucleotides in mRNA, i.e. the nucleotides complementary to those in the genetic material, the genetic code, reading from the 5′ end to the 3′ end of each codon, is given in the table. (The letter A symbolizes a nucleotide residue containing adenine; G, guanine; C, cytosine; and U, uracil; each amino-acid residue is represented by its usual three-letter symbol.)

UUU: Phe	UCU: Ser	UAU: Tyr	UGU: Cys
UUC: Phe	UCC: Ser	UAC: Tyr	UGC: Cys
UUA: Leu	UCA: Ser	UAA: Ter†	UGA: Ter†
UUG: Leu	UCG: Ser	UAG: Ter†	UGG: Trp
CUU: Leu	CCU: Pro	CAU: His	CGU: Arg
CUC: Leu	CCC: Pro	CAC: His	CGC: Arg
CUA: Leu	CCA: Pro	CAA: Gln	CGA: Arg
CUG: Leu	CCG: Pro	CAG: Gln	CGG: Arg
AUU: Ile	ACU: Thr	AAU: Asn	AGU: Ser
AUC: Ile	ACC: Thr	AAC: Asn	AGC: Ser
AUA: Ile	ACA: Thr	AAA: Lys	AGA: Arg
AUG: Met‡	ACG: Thr	AAG: Lys	AGG: Arg
GUU: Val	GCU: Ala	GAU: Asp	GGU: Gly
GUC: Val	GCC: Ala	GAC: Asp	GGC: Gly
GUA: Val	GCA: Ala	GAA: Glu	GGA: Gly
GUG: Val	GCG: Ala	GAG: Glu	GGG: Gly

† termination codon. UAA was once called **ochre codon**; UAG, **amber codon**; UGA, **opal codon**.

‡ initiation codon. The methionine codon AUG is the most common starting point of translation of a genetic message but GUG sometimes serves.

The genetic code is almost universal but differences have been found in the DNA of mitochondria from a number of organisms. For example, in human mitochondria UGA codes for Trp and not for termination; AUA codes for Met and not for Ile; AGA and AGG are termination codons and do not code for Arg; AUA and possibly AUU act as initiation codons as well as AUG.

genetic colonization the introduction by a parasite of genetic information into a host such as to cause the host to synthesize products that only the parasite can use. This occurs in some plants, e.g., in cases of parasitism by *Agrobacterium* spp.

genetic complementation the phenomenon whereby two different mutations produce a wild-type phenotype in a double heterozygote. It thus provides evidence that the mutations do not affect the same gene.

genetic cross 1 any mating of two organisms that results in a **genetic recombinant**. **2** the offspring of two (or more) parent organisms produced by mating or other means.

genetic defect any alteration in the genetic information carried in an individual organism, the effect of which is to produce a phenotype that is at a disadvantage in competing with a normal organism of the same species or in meeting the challenges of its environment. *See also* **inborn error of metabolism**, **molecular disease**.

genetic disease any disease ultimately caused by a genetically determined biochemical defect. In the strict sense a genetic disease is only slightly influenced by environmental factors.

genetic drift 1 any change with time in gene frequency in a population. It may be either directed, **steady drift**, or undirected, **random drift**. **2** irregular, random fluctuations in gene frequency in a (relatively) small population caused by statistical effects.

genetic engineering the genetic manipulations used to produce individuals having a new combination of inherited properties. Such manipulation may be of two kinds: (1) cellular manipulation, involving the culturing of (haploid) cells and the hybridization of somatic cells (*see* **cell fusion**); and (2) molecular manipulation, involving the construction of artificial recombinant DNA molecules, their insertion into a **vector** (def. 3), and their establishment in a host cell or organism. The latter approach has been called **recombinant DNA technology**.

genetic erosion any natural or human-engendered process that, with time, results in loss of genetic diversity in populations of the same species, or reduction of the genetic base of a species, or loss of an entire species.

Geneticin *a proprietary name for* **G418**.

genetic information the information carried in a sequence of nucleotides in a molecule of DNA or RNA.

genetic locus *see* **locus**.

genetic map any diagram showing the linear order of, and the relative distances between, mutable sites on a chromosome as deduced from genetic recombination experiments.

genetic mapping any method that may be used to determine the positions of, and the relative distances between, genes of a linkage group or of sites within a gene (fine-scale mapping). In classical genetics, mapping relied on a study of recombination frequency and was measured in **morgans**. In molecular biology, mapping relies on the sequencing of DNA in a chromosome. Current work is aimed at mapping entire genomes. *See also* **genetic map**, **map unit**.

genetic marker 1 any genetically controlled phenotypic difference that can be used in analysing the genetic make-up of an organism. **2** any gene difference in either one or both alleles used in the detection of genetic recombination events.

genetic material the molecular carrier of primary genetic information that serves as a template for its own replication and provides the structural and regulatory information for the processes of protein synthesis and cell development. It consists of double-stranded DNA in higher organisms, bacteria, and most bacteriophages, single-stranded DNA in some bacteriophages, and RNA in the RNA viruses.

genetic pollution contamination of the genetic make-up of wild populations or species with foreign genes (usually as a result of genetic engineering).

genetic polymorphism the regular and simultaneous existence in the same population of two or more discontinuous variants or genotypes, at a frequency existing in more than 1% of the population, that cannot be due to recurrent mutations. The different human blood groups are an example.

genetic profiling *or* **DNA fingerprinting** a technique for providing profiles of DNA fragments (resulting from digestion with restriction enzymes) that characterize a genome. The human genome is scattered with minisatellites, regions of DNA consisting of tandem repeats of short base sequence which can show extensive variation in the number of repeats, leading to multi-allelic variation and high degrees of heterozygosity. This extreme level of variation renders minisatellite or VNTR (**variable number tandem repeats**) loci invaluable as genetic markers. The DNA is digested with a suitable restriction endonuclease and the fragments are electrophoresed through agarose gel and transferred by blotting to a nylon filter. Hybridization is with a minisatellite probe labelled with ^{32}P. The use of the polymerase chain reaction (**PCR**) to amplify hypervariable loci, including very short tandem-repetitive microsatellites, has greatly increased the sensitivity of DNA typing systems and the ability to type degraded human DNA. The method can be used to identify an individual's DNA and also to trace the parentage of an individual. The method is not confined to humans but has been applied to a wide range of species.

genetic recombinant any organism produced by **genetic recombination**.

genetic recombination the processes by which a new genotype is formed by reassortment of genes resulting in gene combinations different from those that were present in the parents. In eukaryotes genetic recombination can occur by chromosome assortment, intrachromosomal recombination, or nonreciprocal interchromosomal recombination. Intrachromosomal recombination occurs by **crossing over**. In bacteria it may occur by genetic transformation, conjugation, transduction, or F-duction. *See also* **Holliday junction**.

genetics 1 the branch of biology concerned with heredity and variation in organisms. It embraces the phenomenology and physiology of heredity; the nature of genetic information; the storage of genetic material; the replication, mutation, transmission, and recombination of genetic material; and the way it is translated into systems that control development and metabolism, and determine the reappearance of parental characters among progeny. **2** the genetic features and constitution of any single organism, species, or group of organisms. *See also* **reverse genetics**.

genetic screening the testing of individuals or groups of individuals for some specific gene product or resulting metabolite for the purpose of identifying genetic defects or diseases.

genetic signature *or* **genetic fingerprint 1** the nucleotide sequences that are specific to an individual. **2** a gene expression pattern characteristic of a particular cell type. **3** a genetic trait characteristic of an ethnic group.

genetic transformation *see* **transformation** (def. 1).

genetic variance that part of the phenotypic variance caused by the varying genotypes of the individuals within a population.

gene trap vector a DNA construct incorporating a selectable marker such as *neo*^{R}, a promoterless reporter such as *lacZ*, and an upstream splice acceptor site. The vector is used to transfect mouse ES cells or other cells by electroporation. If the vector is integrated into a functioning gene, expression of the reporter gene will occur. Flanking DNA can be recovered and sequenced to reveal the identity of the gene. *Compare* **enhancer trap vector**.

Genghis Khan *abbr.*: Gek; a protein serine/threonine kinase of *Drosophila* that is remarkably homologous to the human myotonic dystrophy protein kinase.

genic of, or relating to, a gene or genes.

genin 1 *the class name for* the noncarbohydrate residue of steroid glycosides. It is used as a suffix in forming trivial names of the naturally occurring glycoside precursors; e.g. digitogenin (from digi-

tonin), sapogenin (from saponin). **2** *a former name for* spirostan, 16,22:22,26-diepoxycholestane.

genistein 4,5,7-trihydroxyisoflavone; 5,7-dihydroxy-3-(4-hydroxyphenyl)-4*H*-1-benzopyran-4-one; a yellow pigment of the broom plant *Genista tinctoria*. It inhibits protein-tyrosine kinases and other protein kinases, being a competitive inhibitor at the ATP binding site. *See* **isoflavanoid**.

genome the whole of the genetic information of an organism. It is contained as DNA in eukaryotes and prokaryotes, and as either DNA or RNA in viruses. A given organism has only one genome regardless of whether the organism is haploid, diploid, or polyploid. The term was originally used to denote one haploid set of chromosomes in a eukaryote organism. —**genomic** *adj*.

genome annotation software software that facilitates the creation and maintenance of genome sequence annotation. Examples include Artemis, which allows visualization and annotation especially of the compact genomes of bacteria, archaea, and lower eukaryotes; and Ensembl, which produces and maintains (within a database system) automatic annotation principally for the human genome, but also for the genomes of other vertebrates, with which it provides comparative views.

genome database a database that houses information specific to the genome of a particular organism. For example, ACeDB, FlyBase, and SGD, house information on the *Caenorhabditis elegans, Drosophila melanogaster,* and *Saccharomyces cerevisiae* genomes respectively.

genome mutation any **mutation** (def. 1) occurring by alteration of the number of chromosomes in the genome.

genome project an initiative (often via international collaboration) to map and sequence the entire genome of a particular organism. The first complete prokaryotic and eukaryotic genomes to be sequenced were those of *Haemophilus influenzae* in 1995 and *Saccharomyces cerevisiae* in 1997. The Human Genome Project was completed in 2003, but numerous other genome projects are currently underway. *See* **GOLD**.

genome-regulatory protein any site-specific protein involved in the regulation of transcription, such as a polymerase, repressor, or activator.

genomes online database *abbr.*: GOLD; a Web resource that monitors the status of worldwide genome sequencing efforts, giving access to information about both completed and ongoing projects.

genome survey sequence *abbr.*: GSS; a division of the GenBank database that is genomic in origin (as opposed to cDNA), and includes: single-pass genomic data, exon-trapped sequences, cosmid/BAC/YAC end sequences, and Alu PCR sequences.

genome walking a process in which one end of a genomic DNA clone is used to identify an overlapping clone from a genomic DNA library. Repetition of this procedure several times over results in the assembly of a series of overlapping cloned fragments, or contig.

genomic blotting *an alternative name for* **Southern blotting**.

genomic DNA all of the DNA comprising one genome.

genomic imprinting *see* **imprinting**.

genomic library a set of clones containing DNA fragments that *in toto* embrace an entire genome. It is prepared by firstly randomly cutting the genomic DNA to give fragments of varying length. The average fragment length is controllable by discrimination in the choice of restriction enzymes used for cutting, and is selected to optimize the length such that fragments will contain entire open reading frames without excessive lengths of additional DNA at either end. Secondly the DNA fragments thus obtained are ligated into a suitable **vector** (def. 3), which is cloned in a host of choice.

genomics the systematic sequencing and characterization of complete genomes. *See also* **comparative genomics, functional genomics, +omics, structural genomics**.

genophore the structural equivalent of a genetic **linkage group** in any organism. The term includes the linear chromosome of the (enveloped) eukaryon of all animal and plant cells, the circular chromosome of the (nonenveloped) prokaryon of cells of bacteria, the duplex DNA of mitochondria, chloroplasts, kinetoplasts, plasmids, and episomes, and the nucleic acid, whether DNA or RNA, single-stranded or double-stranded, of viruses.

genotoxin an agent that reacts with and interferes with the action of genetic DNA.

genotype 1 the total genetic constitution of an organism. It comprises the genetic information carried both in the chromosomes and extrachromosomally. *Compare* **phenotype**. —**genotypic** *adj*.

gentamicin *or* **gentamycin** any of a group of aminoglycoside antibiotics produced by fermentation of some *Micromonospora* spp. They are inhibitors of protein synthesis, and are particularly effective in the treatment of *Pseudomonas aeruginosa* infections.

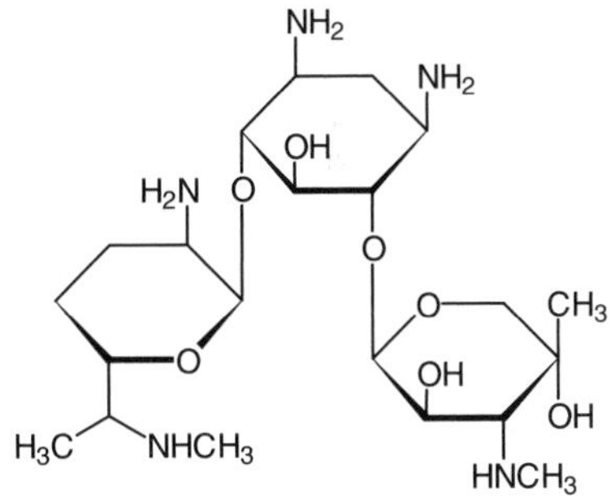

gentamicin C_1

gentianose the trisaccharide, O-β-D-glucopyranosyl-$(1\rightarrow6)$-O-α-D-glucopyranosyl-$(1\rightarrow2)$-β-D-fructofuranoside. It is found in the rhizomes of many species of *Gentiana*.

gentiobiose *or (formerly)* **amygdalose** the disaccharide, O-β-D-glucopyranosyl-$(1\rightarrow6)$-D-glucose. It is a component of some glycosides, e.g. **amygdalin**.

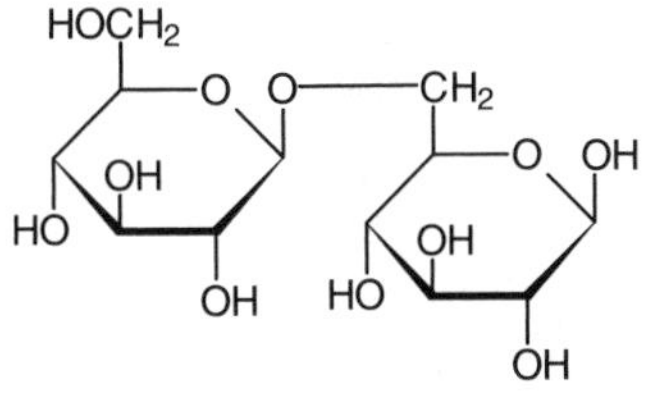

β-gentiobiose

gentosamine 3-deoxy-3-methylamino-D-xylose; a component of gentamicin A, one of the structural variants of **gentamicin**.

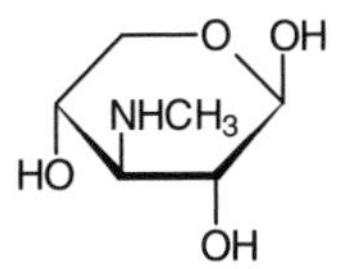

genus (*pl.* **genera**) (*in taxonomy*) a category used in biological classification consisting of species with a common phylogenetic origin that appear to be clearly differentiated from other species.

+geny *comb. form* denoting origin or mode of development. —**+genic** *adj*.

geometric isomer *or* **geometrical isomer** an alternative term for *cis–trans* isomer.

geometric isomerism *or* **geometrical isomerism** an alternative term for *cis–trans* isomerism.

geometric mean *or* **geometric average** for *n* positive numbers, the *n*th root of the product of the numbers. Thus for a set of positive numbers $x_1, x_2, x_3, \dots x_n$, the geometric mean is $(x_1 \cdot x_2 \cdot x_3 \dots x_n)^{1/n}$. *Compare* **mean**.

geotaxis the movement of an organism or a cell under the influence of a gravitational field. In **positive geotaxis** the organism moves in the same direction as the gravitational field, while in **negative geotaxis** the movement is in a direction opposite to the gravitational field. —**geotactic** *adj*.

gephryin a cytoplasmic protein that associates with microtubules, and is required for clustering of glycine receptors in spinal cord. It has a role in molybdoenzyme function in non-neuronal tissue.

g-equiv *abbr. for* gram-equivalent or gram-equivalent weight.

GERAD *abbr. for* glycoprotein ER-associated degradation; a disposal system for glycoproteins in the endoplasmic reticulum that is signalled by: (1) removal of the outermost mannose residue from the middle branch of an asparagine-linked oligosaccharide core; and (2) misfolding of the associated protein.

geraniol (2*E*)-3,7-dimethyl-2,6-octadien-1-ol, a monoterpene alcohol with a sweet rose odour, present in oil of rose and other essential oils.

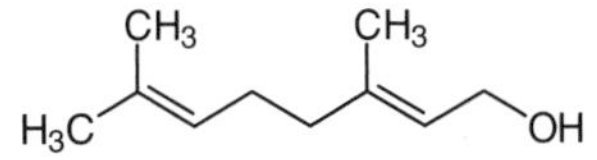

geranyl the (2*E*)-3,7-dimethyl-2,6-octadien-1-yl group. The 1-diphosphate derivative is an intermediate in the synthesis of cholesterol and in the formation of **geranylgeranyl** diphosphate.

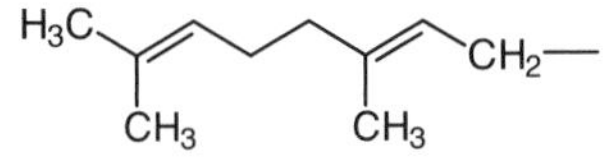

geranyl-diphosphate synthase *see* dimethylallyl*trans*transferase.

geranylgeranyl the (2*E*,6*E*,10*E*)-3,7,11,15-tetramethyl-2,6,10,14-hexadecatetraen-1-yl group. The 1-diphosphate derivative is a substrate for enzymes introducing the geranylgeranyl group into proteins. *See also* **prenylation**.

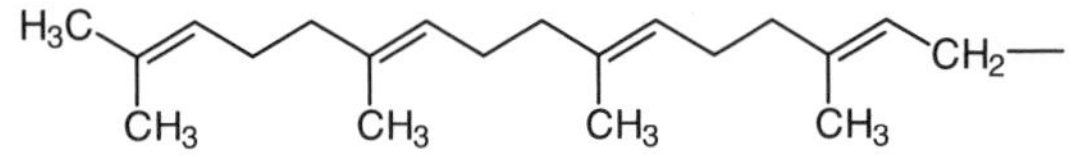

geranylgeranyl pyrophosphate synthetase *see* farnesyl-*trans*transferase.

geranyltranstransferase EC 2.5.1.10; *systematic name*: geranyl-diphosphate:isopentenyl-diphosphate geranyl *trans*transferase; *other names*: farnesyl-diphosphate synthase; farnesyl pyrophosphate synthetase (*abbr.*: FPP synthetase). An enzyme of the pathway synthesizing **prenyl** groups, **sesquiterpenes**, and **cholesterol**. It catalyses the reaction between geranyl diphosphate and isopentenyl diphosphate to form *trans,trans*-farnesyl diphosphate with release of pyrophosphate.

Gerbich system *abbr*.: Ge; a blood-group system whose antigenic determinants are located on the extracellular, glycosylated domain of **glycophorins** C and D (GPC and GPD). The polypeptide sequences of these give rise to several Ge phenotypes: on GPD the antigenic site for Ge2 resides between residues 1 and 17, and that for Ge3 resides between residues 19 and 34; on GPC, the antigenic site for Ge4 is at the N terminus, and that for Ge3 between residues 40 and 55.

GERL *abbr. for* golgi, endoplasmic reticulum, lysosomes; a term formerly used to describe a specialized region of smooth **endoplasmic reticulum** with high levels of acid phosphatase and other acid hydrolase activities. It is thought to be a region where acid hydrolases are concentrated and packaged into lysosomes. It may also be concerned in packaging other materials, including epinephrine granules in the adrenal medulla, tyrosinase and other materials in premelanosomes in mouse melanomas, peroxidase, etc. in secretory granules in cells of parotid and submaxillary glands, and un-iodinated thyroglobulin in thyroid epithelial cells.

germ cell any cell of the **germ line**, including **gametes**.

germ-free describing any group of animals or plants that is bred and maintained in such conditions that they contain no microorganisms, either normal symbionts or abnormal, infective ones. *Compare* **gnotobiotic**.

germin a pepsin-resistant water-soluble glycoprotein of the walls of germinated plant embryos that may alter the properties of cell walls during germinative growth. Its synthesis coincides with the onset of growth in germination. The protein has oxalate oxidase activity.

germ line any line of cells that gives rise to gametes and is continuous through the generations.

Gerstmann–Straussler syndrome a transmissible **spongiform encephalopathy** of humans, usually familial, characterized by widespread degeneration of the nervous system. *See also* **prion**. [After Joseph Gerstmann (1887–1969), Austrian neurologist.]

GF-1 *see* **GATA-1**.

g-factor or *g*-value, a parameter in EPR spectroscopy, which defines the field B_0 and frequency v position of resonance, according to the relation $hv = g\mu_B B_0$, where h is the Planck constant and μ_B the Bohr magneton.

GFP *abbr. for* **green fluorescent protein**.

GFR *abbr. for* (renal) glomerular filtration rate.

Gg$_3$ *symbol for* gangliotriaose (alternative to GgOse$_3$).

Gg$_4$ *symbol for* gangliotetraose (alternative to GgOse$_4$).

GGA a codon in mRNA for a glycine residue.

GGC a codon in mRNA for a glycine residue.

GGG a codon in mRNA for a glycine residue.

GgOse$_3$ *symbol for* gangliotriaose (alternative to Gg$_3$).

GgOse$_4$ *symbol for* gangliotetraose (alternative to Gg$_4$).

GGU a codon in mRNA for a glycine residue.

GH *abbr. for* 1 growth hormone (i.e. **somatotropin**). 2 glycohydrolase.

ghilanten *see* **antistasin**.

ghost an empty red-blood-cell membrane, usually obtained by lysis of red cells by the controlled reduction in the osmotic pressure of the suspending medium followed by restoration of normal osmotic conditions causing the membranes to reseal. Ghosts are extensively used in studies of the properties of membranes.

ghrelin a 28-residue acylated polypeptide hormone secreted by endocrine cells of the stomach and duodenum in the human and rat. It strongly stimulates appetite and secretion of somatotropin. Plasma levels rise shortly before and shortly after a meal. It is derived from a precursor protein, proghrelin, from which another appetite-suppressing hormone, **obestatin**, is derived. The two hormones activate distinct receptors. In humans the sequence is

GSSFLSPEHQRVQQRKESKKPPAKLQPR;

n-octanoylation at serine 3 is essential for activity.

GHRF *abbr. for* growth hormone releasing factor (*see* **GHRH**).

GHRH *abbr. for* growth hormone releasing hormone; *other names*: growth hormone releasing factor (*abbr.*: GHRF) *or* somatoliberin; any of a family of peptide hormones that act on the anterior pituitary to stimulate the secretion of growth hormone and exert a trophic effect on the gland. The bovine hormone, for example, consists of 44 amino acids (5.109 kDa), and is derived from a precursor protein. *See also* **GHRH receptor**.

GHRH receptor *abbr. for* growth hormone releasing hormone receptor; *other name*: growth hormone releasing factor receptor (*abbr.*: GRFR); any of a family of G-protein-associated membrane receptors that bind GHRH and stimulate adenylate cyclase. A de-

fect in this receptor (Asp^{60} → Gly^{60}) is associated with hypoplasia of the anterior pituitary in mice.

GH-RIH *abbr. for* growth hormone release inhibiting factor (i.e. **somatostatin**).

GHRp *abbr. for* **growth hormone-releasing peptide**.

giantin an integral membrane protein of the Golgi apparatus that contains a 376 kDa cytoplasmic domain consisting almost exclusively of heptad repeats for coiled-coil formation, and a C- terminal transmembrane segment.

gibberellin any of a widely distributed, multimembered group of highly modified C_{20} (and C_{19}) terpene plant hormones, the first of which was isolated from plants infected with the fungus *Gibberella fujikuroi*. The most important is gibberellin A$_3$, also known as **gibberellic acid**. Gibberellins accelerate shoot growth, and can actively stimulate RNA synthesis in dwarf mutant plants, making them grow like normal plants. Other physiological effects, e.g. overcoming the need for vernalization in long-day plants, have also been recorded.

gibberellic acid

Gibbs–Duhem equation an equation relating the chemical potential and concentration of a chemical species in a mixture at constant temperature and pressure. If n_i moles of species i, having chemical potential μ_i, are present in a mixture, then

$$\sum_i n_i \mathrm{d}\mu_i = \sum_i x_i \mathrm{d}\mu_i = 0$$

where $x_i = n_i/\sum_i n_i$ is the mole fraction and $\mathrm{d}\mu_i$ is the change in chemical potential for $\mathrm{d}n_i$, a small change in the amount of species i present, in moles. [After Josiah Willard Gibbs (1839–1903), US physical chemist, and Pierre Duhem (1861–1916), French physical chemist.]

Gibbs energy *or* **Gibbs free energy** *or* **Gibbs function** *symbol*: G; a thermodynamic property of a system; it is defined as $H - TS$, where H is the enthalpy, S the entropy, and T the temperature. For a molecule in solution, Gibbs energy is concentration-dependent. The Gibbs free energy (G_i) of the ith component is equal to the standard free energy of the component (G°_i, the free energy at unit activity) plus a term to adjust for concentration. Thus

$$G_i = G^\circ_i + RT\ln a_i,$$

where R is the gas constant, and a_i the activity of the ith component. The molar free energy is referred to as the **chemical potential**. *See also* **Gibbs free energy change**.

Gibbs energy diagram *or* **Gibbs energy profile** for a chemical reaction, a diagram showing the relative standard Gibbs energies of reactants, transition states, reaction intermediates, and products in the same sequence as they occur in the reaction.

Gibbs energy of activation *or* **Gibbs free energy of activation** *symbol*: $\Delta^\ddagger G^\circ$; the standard Gibbs energy difference between the transition state of a chemical reaction (either elementary or stepwise) and reactants.

Gibbs free energy change the difference in Gibbs free energy (G) at any instant in time between the reactants and products of any reaction. Thus for a reaction A + B $\rightleftharpoons$ C + D, at any particular concentration of reactants and products, the free energy change, ΔG, for that state is calculated as

$$G_C + G_D - G_A - G_B.$$

Since, for each substance,

$$G_X = G_X^\circ + RT \ln a_X$$

(*see* **Gibbs energy**), this leads to

$$\Delta G = \Delta G^{\circ} + RT \ln a\{_C a_D / a_A a_B\};$$

$$\Delta G^{\circ} = \Delta G^{\circ}_C + \Delta G^{\circ}_D - \Delta G^{\circ}_A - \Delta G^{\circ}_B,$$

and is the **standard free energy change**. All chemical reactions proceed in the direction of negative free energy change until equilibrium is reached, at which point ΔG is zero. Since at equilibrium $a_C a_D / a_A a_B$ is the equilibrium constant K, it follows that the standard free energy change $\Delta G^{\circ} = -RT \ln K^{\circ}$ where K° is the standard equilibrium constant.

Gibbs–Helmholtz equation an equation that may be expressed as

$$\Delta G - \Delta H = T[\delta\{\delta G / \delta T]_P$$

where ΔG is the change in Gibbs energy during a reaction, ΔH the increase in enthalpy, P the pressure, and T the temperature.

Giemsa's stain a stain used for staining chromosomes, in which it reveals bands known as **G-bands** (from the name). These are rich in A–T nucleotide pairs. Other compounds, e.g. the antibiotic olivomycin, stain G–C-rich bands known as **R-bands** (because they give the 'reverse' of the G-bands). [After Gustav Giemsa (1867–1948), German parasitologist.]

GIF abbr. for growth hormone-inhibitory factor (i.e. **somatostatin**).

giga+ symbol: G; **1** the SI prefix denoting 10^9 times. **2** (in computing) prefix used to indicate a multiple of 2^{30}, e.g. gigabyte.

GIGANTEA a gene of Arabidopsis thaliana that is controlled by the circadian clock mechanism and regulates flowering. Its product is predicted to contain several transmembrane segments. It is a key link between clock genes and flowering genes.

gigantism a condition arising from the excessive release of **somatotropin** leading to an individual of excessive size. Compare **acromegaly**.

Gilbert syndrome a very mild form of jaundice in which there is increased bilirubin in plasma, predominantly in unconjugated form. Usually diagnosed in young adults, it is transmitted as an autosomal recessive trait. It is caused by an abnormality in the TATA box within the promoter region at 2q37 for UDPglucuronsyltransferase I, which results in reduced expression of the gene, and decreased conjugation of bilirubin as a glucuronoside. See **UDPglucuronsyltransferase**. [After Nicholas A. Gilbert (1858–1927), French physician.]

Gilson pipette a proprietary name for one of a range of piston-operated pipettes designed to function with disposable polypropylene tips. Individual pipettes are identified by the maximum volume in microlitres they are designed to deliver, for example P2, P20, P200, P1000. Precise volumes within each range can be delivered. [Named after the manufacturer, Gilson Inc.]

Gim complex an alternative name for **prefoldin**.

GIP abbr. for **1** gastric inhibitory peptide. **2** general insertion pore. See **TOM**.

GIPL abbr. for glycoinositolphospholipid (see **GPI anchor**).

GIRK abbr. for G-protein gated inward rectifier K^+.

GKRP abbr. for glucokinase regulatory protein.

Gla symbol for a residue of the α-amino acid L-4-carboxyglutamic acid.

Glanzmann thrombasthenia see **von Willebrand factor**.

glass electrode an electrode made of glass whose potential, when immersed in a solution, is a function of the hydrogen-ion activity, i.e. pH, of the solution. A common form of glass electrode consists of a thin glass bulb inside which is mounted a reference electrode, often a silver/silver chloride electrode, immersed in a solution of constant pH and containing the ion to which the inner reference electrode is reversible. A number of special glasses have been made in an effort to extend the range of pH over which the electrode responds linearly, to reduce errors caused by other ions in solution, to extend its useful temperature range, etc.

glassmilk an aqueous suspension of a proprietary silica matrix used in the purification of DNA.

glaucoma a group of heterogeneous diseases in which vision is impaired through damage to the optic nerve, and usually associated with increased intraocular pressure. It is the leading cause of irreversible blindness in the world. Depending on the mechanism impeding outflow of aqueous humor from the anterior chamber of the eye, glaucoma is classified into open-angle, closed-angle, or congen-

ital forms. An autosomal dominant primary open-angle glaucoma results from mutations in the gene for **myocilin**. Primary congenital glaucoma is autosomal recessive and commonly caused by mutations in the gene for a cytochrome P450 designated CYP1B1.

Glc symbol for a residue (or sometimes a molecule) of the aldohexose **glucose**. In the condensed system of symbolism of sugar residues, Glc signifies specifically a glucose residue with the common configuration and ring size, i.e. D-glucopyranose. (Note: Glu is the symbol for a glutamic residue.)

GLC or **glc** abbr. for gas–liquid chromatography.

GlcA 1 symbol for a residue (or sometimes a molecule) of the uronic acid glucuronic acid. **2** former symbol for a residue (or sometimes a molecule) of the aldonic acid **gluconic acid** (no symbol now recommended).

Glcf symbol for a glucofuranose residue.

GlcN symbol for a residue (or sometimes a molecule) of the aminodeoxysugar **glucosamine**.

GlcNAc symbol for a residue (or sometimes a molecule) of the N-acetylated aminodeoxysugar **N-acetylglucosamine**.

Glcp symbol for a glucopyranose residue.

GlcU or **GlcUA** former symbol for glucuronic acid (GlcA now recommended).

Gleevec a drug approved for treatment of chronic myeloid leukemia. The drug inhibits a tyrosine kinase enzyme p210 produced by the BCR-ABL oncogene as a result of a translocation from chromosome 22 to 9. A single mutation in the p210 BCR-ABL gene is a source of resistance to Gleevec in blast crisis cells, which leads to reactivation of the enzyme.

Gli any of three vertebrate homologues of the Drosophila zinc-finger transcription factor **cubitus interruptus**. Mutations in the Gli3 gene (at 7p13) are associated with the multiple-malformation Pallister–Hall syndrome.

GLI abbr. for glucagon-like immunoreactant (or glucagon-like immunoreactivity).

glia or **neuroglia** the special connective tissue of the central nervous system constituting some 40% of the total volume of the brain and spinal cord. It has supportive and nutritive functions and is made up of **glial cells** – oligodendrocytes, astrocytes, ependymal cells, and microglial cells.

gliadin or (formerly) **gliadine** any of a group of proline-rich proteins (**prolamins**) found in the seeds of wheat and rye. All gliadins are mixtures of simple proteins of similar composition and properties. They are differentiated from **glutenin**, with which they are associated in gluten, by their solubility in 70% aqueous ethanol. With water, gliadins form a sticky mass, which binds flour into a dough. They constitute the major seed storage proteins in wheat. Gliadins fall into two groups: α/β gliadins (also prolamin or gliadin AI–V) are divided into five classes, by similarity; there are more than 100 copies of the gene per haploid genome; γ-gliadins are divided into three classes (BI–III). For both groups, sequence divergence between the respective classes is the result of single base substitutions, and of duplications or deletions within or near direct repeats. [French gliadine; from Greek glia, glue.]

glial cell-derived neurotrophic factor abbr.: GDNF; a homodimeric protein of the transforming growth factor β (TGFβ) family that is secreted in the nervous system and other tissues and promotes neuronal survival during development. Both subunits contain 134 residues and three disulfide bonds, and are derived from a 211-residue precursor. GDNF is structurally related to neurturin and persephin, and like these binds a specific GPI-anchored membrane protein that recruits and signals through **RET** and the adaptor protein GRB-10. Numerous inactivating mutations in RET and a few mutations in GDNF are associated with **Hirschsprung disease**.

glial cells see **glia**.

glial derived nexin a glycoprotein serine protease inhibitor, similar to **serpin**, that promotes neurite extension. See also **nexin**.

glial fibrillary acidic protein abbr.: GFAP; a class III **intermediate filament** and a cell-specific marker that, during the development of the central nervous system, distinguishes astrocytes from other glial cells. It is similar to other intermediate filament proteins.

glibenclamide see **glyburide**.

glicentin a peptide originally obtained as a component of porcine enteroglucagon, and known at first as **porcine gut GLII**. It reacts fully with antisera to the N-terminal moiety of glucagon from pancreas. The known amino-acid sequence includes the entire sequences of **glicentin-related pancreatic peptide** (at residues 1–30) and pancreatic **glucagon** (at residues 33–61), and is identical to the N-terminal moiety of the sequence of proglucagon. It was initially thought to comprise 100 amino-acid residues, hence the name.

glicentin-related pancreatic peptide or **glicentin-related pancreatic polypeptide** *abbr.*: GRPP; originally, any polypeptide material present in extracts of porcine pancreas and in the secretory granules of the pancreatic A cell that reacted with certain antisera to glicentin, but not with any antisera to glucagon. The term is now usually applied to a specific component of such material that is a 30-residue peptide identical in sequence to the N-terminal 30-residue sequence of **proglucagon**, is secreted by the pancreas concomitantly with glucagon, and thus is probably a cleavage product of proglucagon.

GLI-I *see* **glicentin**.

GLIMA38 *abbr. for* glycosylated islet-cell-membrane antigen 38 kDa; a cell surface antigen of B cells of pancreatic islets that, with **glutamate decarboxylase** (GAD65) and insulinoma antigen-2 (IA-2), elicits autoantibodies in the early phase of the autoimmune response in the development of diabetes mellitus type 1.

gliostatin *see* **platelet-derived endothelial cell growth factor**.

GLIP *abbr. for* glucagon-like insulinotropic peptide; a gastrointestinal peptide that potentiates release of insulin. It is released during a meal and after oral glucose administration. It is GLP-1$^{7-37}$. *See* **glucagon**.

Gln *symbol for* a residue of the α-amino acid L-**glutamine** (alternative to Q or Z).

global alignment *see* **alignment**.

global similarity similarity that spans the full extent of a pair or set of aligned sequences (i.e. within a global **alignment**).

globin the colourless and basic protein moiety (and biosynthetic precursor) of **hemoglobin** and **myoglobins**. In normal human adult hemoglobin, the globin component comprises two non-identical pairs of polypeptide chains, whereas in myoglobin there is only one polypeptide chain. Pharmaceutical globin is prepared by adding ox hemoglobin to acetone–HCl; this causes the heme groups to separate and remain in solution, while the globin precipitates. It is used in the preparation of globin–insulin, which is absorbed more slowly than soluble insulin, and has a prolonged action.

globo+ *prefix* for an oligosaccharide, signifying the core structure GalNAc(β1-3)Gal(α1-4)Gal(β1-4)Glc (I), present in **globoside** from which the term originated. Also, by derivation it signifies Gal(α1-4)Gal(β1-4)Glc, residues 2–4 of I. *Compare* **isoglobo+**.

globoside or **cytolipin K** *symbol*: GbOse$_4$Cer or Gb$_4$Cer; globotetraosylceramide; ceramide globotetraoside; the major neutral glycosphingolipid in normal kidney and erythrocytes. The name originated from the discovery of this lipid on red cells, and led to the use of **globo+** as a root signifying this core carbohydrate structure. *Compare* **globotriaosylceramide**. *See also* **globotetrasylceramide**, **P system**.

globotetraose *symbol*: GbOse$_4$ or Gb$_4$; the tetrasaccharide GalNAc(β1-3)Gal(α1-4)Gal(β1-4)Glc. *See also* **P system**, **globo+**.

globotetraosylceramide *symbol*: GbOse$_4$Cer or Gb$_4$Cer; a glycolipid having the structure

GalNAc(β1-3)Gal(α1-4)Gal(β1-4)GlcCer.

It is the P determinant in the **P system** of blood groups. *See also* **globo+**.

globotriaose *symbol*: GbOse$_3$ or Gb$_3$; the trisaccharide Gal(α1-4)Gal(β1-4)Glc. *See also* **globo+**.

globotriaosylceramide *symbol*: GbOse$_3$Cer or Gb$_3$Cer; a glycolipid **globoside** having the structure Gal(α1-4)Gal(β1-4)GlcCer. It is the major neutral glycosphingolipid in leukocytes, platelets, liver, spleen, and most other non-neural tissues. It is the P^k determinant in the **P system** of blood groups. *See also* **globo+**.

globular protein any protein whose polypeptide chain(s) are folded so as to give the whole molecule a rounded shape; the term is often used loosely to mean soluble as distinct from **membrane proteins**.

globulin any simple, globular protein that is insoluble or sparingly

soluble in water, but soluble in dilute salt solutions, and can be precipitated from solution by half-saturating the solution with neutral ammonium sulfate. Globulins may be distinguished by, e.g., electrophoresis into α-, β-, and γ-globulins, or by ultracentrifugation into 7S globulins, 19S globulins, etc. *See also* **gamma globulin**, **immunoglobulin**.

Globus a freely available toolkit (but not limited to APIs) for grid computing.

glomerular filtrate the filtrate, free of cells and major plasma proteins, that passes from the blood through a renal glomerulus into the lumen of a nephron in the kidney.

glomerular filtration rate *abbr.*: GFR; the rate at which the **glomerular filtrate** is produced in the kidney(s). It is a measure of renal efficiency and is usually expressed in terms of the renal clearance of some substance, e.g. inulin, that is freely filterable by the glomerulus and is neither reabsorbed nor excreted by the renal tubules.

glomerulus (*pl.* **glomeruli**) **1** a knot of blood capillaries projecting into the capsular end of a nephron and from which the **glomerular filtrate** is formed. *See also* **Malpighian body**. **2** any compact tuft or tangled mass of processes, nerve fibres, or blood vessels. —**glomerular** *adj.*

glp a **regulon** of *Escherichia coli* encoding proteins required for the utilization of glycerol 3-phosphate and its precursors. Operons for enzymes of these pathways are controlled negatively by *glp* repressor and positively by the cyclic AMP–cyclic AMP receptor protein complex.

Glp *symbol for* a residue of the α-imino acid L-5-oxoproline (often known as pyroglutamic acid or pyrrolidinecarboxylic acid).

GLP *abbr. for* glucagon-like peptide (*see* **glucagon-related polypeptide**).

Glu *symbol for* a residue of the α-amino acid L-**glutamic acid** (alternative to E). (*Note*: Glc is the symbol for a glucose residue.)

gluc+ *a variant* (*sometimes, before a vowel*) *of* **gluco+**.

glucaemia *a variant spelling* (*esp. UK*) *of* glucemia; *see* **glycemia**.

glucagon a 29-residue polypeptide hormone synthesized in the A cells of pancreatic islets and mammalian (except human) gastric mucosa. Formerly known as hyperglycemic-glycogenolytic factor (*abbr.*: HGF), or hyperglycemic factor. The amino-acid sequence is identical in human and all other mammalian glucagons (except guinea-pig):

HSQGTFTSDYSKYLDSRRAQDFVQWLMNT;

It is secreted in response to falling blood glucose levels and raises the glucose concentration in the blood by either (1) stimulation of adenylate cyclase or (2) activation of inositolphospholipid-specific phospholipase C with increase in intracellular Ca^{2+} concentration. It stimulates glycogenolysis and gluconeogenesis in liver, increases lipolysis in adipose tissue, and increases insulin secretion. Three peptides (glucagon itself and glucagon-like peptides 1 and 2 (GLP-1 and -2)) are roughly 50% homologous in sequence. While GLP-1 is the major form in the pancreas, GLP-1$^{7-36}$-amide predominates in the intestine. This stimulates insulin secretion and inhibits gastric secretion. Exendin$^{9-39}$ is a GLP-1 receptor antagonist. GLP-1 injected into cerebral ventricles inhibits feeding in the starving rat. This effect is blocked by exendin$^{9-39}$. *See also* **enteroglucagon**, **glicentin**, **glucagon receptor**, **proglucagon**.

glucagon-37 *see* **enteroglucagon**.

glucagon-like immunoreactant *abbr.*: GLI; any member of a group of peptides found in extracts of mammalian gastrointestinal tract that react with N-terminal-specific anti-glucagon antibodies. A particular component of such material may be denoted by means of a suffixed roman numeral, e.g. GLI-I, and the relative molecular mass of the component in question may be indicated by a superscript, e.g. ^{8000}GLI. The principal molecular species recorded include one of $\approx$12 kDa (^{12000}GLI), a second of $\approx$8 kDa (^{8000}GLI), and a third of $\approx$4.5 kDa (^{4500}GLI), the first two being the predominant forms. All have glucagon-like biological activities; e.g. they bind to glucagon receptors of liver plasma membrane, activate hepatic adenylate cyclase, and cause the release of insulin. *See also* **glicentin**, **proglucagon**.

glucagon-like insulinotropic peptide *see* **GLIP**.

glucagon-like peptide receptor any of the membrane proteins that bind glucagon-like peptides (*abbr.*: GLPs) and mediate their intracellular effects. A receptor for GLP-1 has been characterized. It is a seven-transmembrane-segment protein that is coupled to a G-protein activating adenylate cyclase.

glucagonoma a **neoplasm** (benign or malignant), originating usually from the A cells of pancreatic islets, that autonomously secretes **glucagon** or **glucagon-like immunoreactant**, frequently with associated diabetes mellitus.

glucagon receptor any G-protein-associated membrane receptor that binds glucagon and activates adenylate cyclase and phospholipase C. A missense mutation in the gene at 17q25 has been found in a small fraction of diabetes mellitus type 2 patients.

glucagon-related polypeptide or **glucagon-like peptide** any peptide found, e.g. in gastrointestinal tissue and blood, that reacts with anti-glucagon antibodies and contains part or all of the amino-acid sequence of **glucagon**. Such peptides include: **enteroglucagon**, **glicentin**, **glucagon-like immunoreactant**, and **proglucagon**.

glucan any glycan composed solely of glucose residues. *See also* **dextran**.

glucan 1,4-α-glucosidase EC 3.2.1.3; *other names*: glucoamylase; amyloglucosidase; α-amylase; lysosomal α-glucosidase; acid maltase; exo-1,4-α-glucosidase. An enzyme that hydrolyses terminal 1,4-α-D-glucosidic bonds (at the nonreducing end) of polysaccharides, producing β-D-glucose with inversion of the configuration. It is less active towards oligosaccharides. Some examples have a raw-starch binding domain. That from *Aspergillus niger* is used commercially in production of glucose. The example from *Neurospora crassa* contains a **kex2** site. Mutations in the human lysosomal enzyme, a glycoprotein of 952 amino acids, cause type II glycogen storage disorder. *See also* **glucosidase**.

α-glucan phosphorylase a **phosphorylase** with affinity for α-glucans.

glucarate 1 the dianion of glucaric acid, the **aldaric acid** derived from either glucose or gulose. There are two enantiomers L- and D-glucarate; D-glucarate, *formerly known as* **saccharate**, is derived from either D-glucose or L-gulose (and vice versa for L-glucarate). **2** any salt or ester of glucaric acid.

$$\begin{array}{c}
\text{COO}^- \\
| \\
\text{H}-\text{C}-\text{OH} \\
| \\
\text{HO}-\text{C}-\text{H} \\
| \\
\text{H}-\text{C}-\text{OH} \\
| \\
\text{H}-\text{C}-\text{OH} \\
| \\
\text{COO}^-
\end{array}$$

D-glucarate

glucemia or (*esp. UK*) **glucaemia** an alternative term for **glycemia**.

glucitol *see* **sorbitol**.

gluco+ or (*sometimes before a vowel*) **gluc+** comb. form indicating glucose. *See also* **glyco+** (def. 2).

gluco- prefix (*in chemical nomenclature*) indicating a particular configuration of a set of (usually) four contiguous CHOH groups as in the acyclic forms of D- or L-glucose. *See* **monosaccharide**.

glucoamylase another name for **glucan 1,4-α-glucosidase**. *See also* **glucoamylase complex**.

glucoamylase complex EC 3.2.1.20; an enzyme that splits 1,4-α-D-glucosidic bonds from the nonreducing end of amylose, amylopectin, glycogen, and oligosaccharides including maltose. The human complex is a homodimer (1857 amino acids per subunit), which may also associate with homodimers of sucrase-isomaltase. It shares 59% sequence identity with the latter, and is a type II membrane protein anchored into the small intestinal brush border membrane by an N-terminal transmembrane segment.

β-glucocerebrosidase *see* **glucosylceramidase**.

glucoconjugate a former term for **glycoconjugate**.

glucocorticoid or **glucocorticosteroid** any natural or synthetic hormonal **corticosteroid** that acts primarily on carbohydrate and protein metabolism, e.g. by promoting hepatic glycogen deposition, and that has an anti-inflammatory effect. Glucocorticoids are produced in the middle layer (zona fasciculata) of the adrenal cortex, and include **cortisol**, **corticosterone**, and **cortisone**. In the rat, corticosterone is the major product; in humans and most other mammalian species, cortisol is the major product – in humans it is secreted at a rate of ≈8–25 mg per day, compared with corticosterone at 1–4 mg per day. Secretion of glucocorticoids is enhanced during stress, including hypoglycemia, hypotension, trauma (including surgery), and illness. Their main actions include stimulation of liver glucose output (by glycogenolysis and gluconeogenesis), inhibition of glucose utilization by tissues, mobilization of amino acids by breakdown of muscle protein, and mobilization of fatty acids from adipose tissue. Anti-inflammatory actions include stabilization of lysosomal membranes, preventing the release of degradative enzymes, depression of phagocytosis and suppression of **T-cell** activity, suppressing the synthesis of **interleukin-1**, and stimulating the synthesis of **lipocortin**. Glucocorticoid actions are mediated in large part through **glucocorticoid receptor** molecules, which regulate gene transcription. *Compare* **mineralocorticoid**.

glucocorticoid receptor a mammalian **transcription factor** involved in the regulation of eukaryotic gene expression and affecting cellular proliferation and differentiation in target tissues. It is a cytosolic, soluble glucocorticoid-binding protein (90 kDa) in a complex with heat-shock protein hsp90 (total mass ≈300 kDa). Binding of steroid induces DNA-binding function and translocation to the nucleus where the receptor–steroid complex interacts with DNA at a **glucocorticoid response element** to regulate gene transcription. The receptor contains a modulating N-terminal domain, two DNA-binding zinc fingers in the central region, and a glucocorticoid-binding domain in the C-terminal region. Two isoforms, α (777 residues) and β (742 residues), which differ only in the C-terminal region, are probably generated by alternative splicing of a single gene. The glucocorticoid receptor is similar to other steroid/thyroid/retinoic nuclear hormone receptors.

glucocorticoid response element *abbr.*: GRE; a specific regulatory DNA sequence that mediates the effects of glucocorticoids on gene transcription through interaction with the activated glucocorticoid receptor. It is located upstream of a particular gene. A consensus sequence for one such GRE is the almost palindromic structure GGTACAnnnTGTTCT.

glucocortin a protein, M_r 17 000, induced by glucocorticoids in adipocytes, liver, fibroblasts, and thymus, and common to all glucocorticoid target cells examined.

glucogenesis the production of glucose in an organism. This may be from glucose storage forms such as glycogen or starch, from interconversion of other hexoses, or from synthesis from noncarbohydrate precursors.

glucogenic describing a substance, some or all the carbon atoms of which can be used to produce glucose in an organism. A glucogenic amino acid is any amino acid that gives rise to increased urinary glucose excretion when fed to a diabetic animal.

glucokinase 1 the recommended name for the enzyme ATP: D-glucose 6-phosphotransferase (EC 2.7.1.2), which catalyses the phosphorylation by ATP of D-glucose to D-glucose 6-phosphate with release of ADP. The enzyme occurs in invertebrates and microorganisms. Unlike **hexokinase**, which catalyses the same reaction, it is highly specific for D-glucose. **2** a name frequently applied to an isoenzyme of hexokinase (EC 2.7.1.1; hexokinase IV) that is characteristic of mammalian liver and pancreatic B cells. Numerous loss-of-function mutations in the gene for this isozyme result in the commonest form of maturity-onset diabetes of the young (MODY2). A rare activating mutation results in congenital hyperinsulinism.

glucokinase regulatory protein *abbr.*: GKRP; a protein that sequesters glucokinase in an inactive state in the cell nucleus.

glucon or (*formerly*) **glucone** any **glycon** in which the sugar involved is glucose.

gluconate 1 the anion of **gluconic acid**. **2** any salt or ester of gluconic acid.

gluconeogenesis the metabolic formation of carbohydrates from non-carbohydrate organic precursors in liver. An important source of precursor molecules is amino acid released from muscle protein during fasting. This is under hormonal control, especially by glucocorticoids. Alanine is an important precursor, and there is extensive conversion in muscle of other amino acids to alanine, which is then transported to the liver, where it undergoes transamination to form pyruvate. *See also* **gluconeogenic pathway**.

gluconeogenic pathway the pathway in liver by which non-carbohydrate precursors are converted to glucose 6-phosphate and thence to glucose or other carbohydrates. Precursor molecules must first be converted to pyruvate, thence successively to oxaloacetate, phosphoenolpyruvate, 2-phosphoglycerate, 3-phosphoglycerate, 1,3-bisphosphoglycerate, glyceraldehyde 3-phosphate plus glycerone phosphate, fructose 1,6-bisphosphate, fructose 6-phosphate, glucose 6-phosphate, and glucose (or other sugars). The enzymes involved are in many cases the same as those involved in the corresponding reverse step in the **glycolytic pathway**. The enzymes specific to the gluconeogenic pathway are pyruvate carboxylase, phospho*enol*pyruvate carboxykinase, fructose 1,6-bisphosphatase and glucose 6-phosphatase (see individual entries). The pathway is hormonally regulated, especially by glucagon, which increases cyclic AMP levels, thereby stimulating protein kinase activity. Control is effected by phosphorylation of liver pyruvate kinase, which inactivates the enzyme and thus restricts glycolysis, and phosphorylation of phosphofructokinase-2, which converts its action to that of a phosphatase which hydrolyses the 2-phosphate of fructose 2,6-bisphosphate, thereby inactivating phosphofructokinase.

gluconic acid the aldonic acid derived from glucose.

gluconolactone *the trivial name for* the γ-lactone (D-glucono-1,4-lactone) or δ-lactone (D-glucono-1,5-lactone) of D-gluconic acid.

D-glucono-1,5-lactone

glucophore the common structural feature postulated to be present in molecules of all substances that possess a sweet taste, and thus to confer sweetness on them. The essential components are considered to be a polarized bond A–H (e.g. O–H in a sugar or N–H in a sweet amino acid), an electronegative atom B (typically =O), and a polarizable hydrophobic group X (such as H or Cl). The approximate distances between them are: A–B, 260 pm; A–X, 350 pm; B–X, 550 pm. The groups A–H and B are considered to bind respectively to similar B and A–H groups in the receptor, and the group X to bind to the receptor by charge-transfer or dispersion bonding.

glucoprotein *a former name for* **glycoprotein**.

glucosaccharase *the trivial name for* any **saccharase** that attacks the glucose end of appropriate oligosaccharides, liberating glucose.

glucosamine *symbol*: GlcpN; *the trivial name for* the aminodeoxysugar 2-amino-2-deoxyglucopyranose; there are two enantiomers. D-Glucosamine (*symbol*: D-GlcpN), *formerly known as* **chitosamine**, occurs in combined form in chitin, in mucoproteins, and in mucopolysaccharides, and is one of the most abundant natural monosaccharides.

glucosamine *N*-acetyltransferase EC 2.3.1.3; *other name*: glucosamine acetylase; an enzyme forming *N*-acetyl-D-glucosamine from acetyl-CoA and D-glucosamine while releasing CoA.

glucosamine-6-phosphate isomerase EC 5.3.1.10; *systematic name*: D-glucosamine-6-phosphate ketol-isomerase (deaminating). An enzyme that catalyses the hydrolysis of D-glucosamine 6-phosphate to D-fructose 6-phosphate and NH_3. The amino-acid sequences of bacterial and fungal types are similar.

glucosaminic acid *the common biochemical name for* 2-amino-2-deoxygluconic acid.

glucosan 1 *an obsolete name for* **glucan**. **2** *an old trivial name* (estab-

lished by usage but not recommended) *for* 1,6-anhydroglucopyranose.

glucose *symbol*: Glc; *the trivial name for* the aldohexose *gluco*-hexose; there are two enantiomers, D- and L-glucose. D-(+)-Glucose (symbol: D-Glu), commonly known as glucose, and formerly known as **grape sugar** or **corn sugar**, is dextrorotatory, hence also formerly (and sometimes still) known as **dextrose**; it is 0.74 times as sweet as sucrose. Aged aqueous solutions of D-glucose contain an equilibrium mixture of α-D-glucopyranose, β-D-glucopyranose, and the open-chain form (*see also* **mutarotation**), whereas in crystals α-D-glucopyranose monohydrate is the stable form below 50°C. Combined D-glucose may have either the α or the β configuration, but is invariably in the pyranose form. D-Glucose is an important source of energy for living organisms. It is found free in fruits and other parts of plants, in honey, and in animals, especially in the blood (≈5 mm in human blood). In combined form it occurs in many homo- and hetero-oligosaccharides and polysaccharides, especially in the animal storage polysaccharide glycogen and in the plant storage polysaccharides cellulose and starch.

β-D-glucose

glucose–alanine cycle *or* **alanine cycle** a metabolic cycle in which alanine is formed peripherally, e.g. in muscle, by transamination of glucose-derived pyruvate, and is transported to the liver where its carbon skeleton is reconverted to glucose.

glucose-assimilation coefficient *symbol*: *K*; *see* **glucose-tolerance test**.

glucose-dependent insulinotropic peptide *or* **glucose-dependent insulin-releasing peptide** *see* **gastric inhibitory peptide**.

glucose effect the ability of glucose in the growth medium to inhibit the synthesis of certain enzymes in bacteria growing on the medium. *See also* **catabolite repression**.

glucose fatty-acid cycle a concept to explain how changes in plasma levels of glucose and fatty acids influence one another to provide a homeostatic control mechanism. Its essential features are: (1) the release, for oxidation, of fatty acids from muscle or adipose tissue acylglycerols restricts glucose metabolism in muscle; and (2) glucose uptake by muscle or adipose tissue restricts the release of fatty acids from acylglycerols in these tissues. The cycle thus provides a hormone-independent mechanism tending to maintain a constant plasma glucose level in animals that feed intermittently. Insulin modifies this mechanism by enhancing glucose uptake by muscle and adipose tissue, inhibiting fatty-acid release by adipose tissue, and increasing fatty-acid esterification in both muscle and adipose tissue.

glucose-galactose malabsorption a rare autosomal recessive disorder caused by mutations in the gene for Na^+-glucose transporter 1. *See* **glucose transporter**.

glucose isomerase a bacterial enzyme preparation that catalyses the reaction:

$$\text{D-glucose} = \text{D-fructose}.$$

It is used on a large scale in industry to convert glucose-containing materials to the much sweeter mixture of glucose plus fructose. The enzymes are usually immobilized on a variety of supports. The product is known as **high-fructose corn syrup** and is widely used as a sweetening agent. No enzyme for which this is the main function has been isolated. EC 5.3.1.18 (glucose isomerase) is now a deleted entry.

glucose oxidase EC 1.1.3.4; *other names*: glucose oxyhydrase; notatin; *systematic name*: β-D-glucose:oxygen 1-oxidoreductase; a flavoprotein (FAD) enzyme that catalyses the oxidation by dioxygen of β-D-glucose to D-glucono-1,5-lactone and H_2O_2. It has application in the experimental determination of glucose concentration (e.g. in medicine) and has some industrial uses.

glucose-1-phosphatase EC 3.1.3.10; *systematic name*: D-glucose-1-phosphate phosphohydrolase. An enzyme that catalyses the hydrolysis of D-glucose 1-phosphate to D-glucose and orthophosphate.

glucose-6-phosphatase EC 3.1.3.9; *systematic name*: D-glucose-6-phosphate phosphohydrolase. An enzyme that catalyses the hydrolysis of D-glucose 6-phosphate to D-glucose and orthophosphate. It has rather wide specificity. It is the final enzyme in the **gluconeogenic pathway**. The human enzyme is an integral membrane protein (357 amino acids, nine transmembrane segments) of the endoplasmic reticulum (ER), particularly in liver, kidney, and pancreatic B cells. The active site is in the ER lumen. Its activity is increased by starvation and diabetes mellitus, and decreased by insulin. It forms part of a system that includes ER transporters for: glucose-6-phosphate (429 amino acids, 10 transmembrane segments), for glucose, and for phosphate. Deficiency of any of these four components leads to a specific form of type I glycogen storage disease.

glucose-6-phosphate dehydrogenase *abbr.*: G6PDH *or (in clinical chemistry)* GPD; EC 1.1.1.49; *recommended name*: glucose-6-phosphate 1-dehydrogenase; *systematic name*: D-glucose-6-phosphate:NADP$^+$ 1-oxidoreductase; *other name (obsolete)*: Zwischenferment. An enzyme that catalyses the conversion of D-glucose 6-phosphate and NADP$^+$ to D-glucono-1,5-lactone 6-phosphate and NADPH. This is an early step in the **pentose phosphate pathway** and the **Entner–Doudoroff pathway**, and thus in **heterolactic fermentation**. The enzyme has the greatest known variability of all enzymes in humans. Over 130 molecular variants that result in deficient activity are associated with a drug-induced hemolytic anemia, **favism**. This arises because the enzyme is involved in maintaining reduced glutathione in red cells. *See also* **glucose-6-phosphate dehydrogenase deficiency**.

glucose-6-phosphate dehydrogenase deficiency an X-linked deficiency of **glucose-6-phosphate dehydrogenase** that affects mostly males. It is associated with a hemolytic anemia, but also confers some protection against malaria, since the parasite is unable successfully to pass through the schizont stage in red cells lacking this enzyme activity; hence there is a particularly high incidence in tropical Africa and the Middle and Far East. Lack of the enzyme in red cells results in failure to reduce NADP$^+$ to NADPH in adequate amounts (the **pentose phosphate pathway** playing a major role in this in the red cell), as a result of which oxidized glutathione cannot be reduced by **glutathione reductase**. Consequently, there can be hemolysis, as hemoglobin is not maintained in the ferrous state and the peroxides produced as a result of drug action promote red-cell destruction. The antimalarial drug **primaquine** and related compounds generate oxidative stress. Over 130 mutations result in some degree of deficiency. *See also* **favism**.

glucose-6-phosphate isomerase *abbr. (in clinical chemistry)*: GPI; EC 5.3.1.9; *systematic name*: D-glucose-6-phosphate ketol-isomerase; *other names*: phosphoglucose isomerase; phosphohexose isomerase; phosphohexomutase; oxoisomerase; hexosephosphate isomerase; phosphosaccharomutase; phosphoglucoisomerase; phosphohexoisomerase. A homodimeric cytosolic enzyme of the glycolytic pathway that catalyses the interconversion of glucose 6-phosphate and fructose 6-phosphate. Loss-of-function mutations in the enzyme (558 amino acids) result in an autosomal recessive chronic hemolytic anemia that may be associated with mental retardation and myopathy. Monomeric GPI is identical with neuroleukin.

glucose-tolerance curve the curve relating blood glucose concentration to time that is obtained in a **glucose-tolerance test**, either oral or intravenous.

glucose-tolerance factor *abbr.*: GTF; *also called*: chromodulin. A water-soluble, relatively stable organic chromium complex of M_r ≈500 that is believed to be required in the diet for normal glucose tolerance in animals, including humans. It is a coordination complex between chromium (III) ions, nicotinate, glycine, glutamate, and cysteine or glutathione.

glucose-tolerance test any procedure designed to assess the response of an individual to a loading dose of glucose, widely used in the diagnosis of diabetes mellitus. The standard **oral glucose tolerance test** is performed in the morning after 12 hours of fasting following at least 3 days' unrestricted diet, physical activity, and freedom from medication. After collection of the fasting blood sample, 75 g (or 1.75 g per kg of body weight for children) glucose is ingested in 5 minutes in 200–500 mL water. Blood samples are then collected at 30-minute intervals for 2 hours, and urine samples are collected at 0, 1, and 2 hours. All samples are assayed for glucose. For capillary blood samples, a maximum glucose level below 10 mmol L^{-1} (180 mg/100 mL) and a 2-hour value less than 7.5 mmol L^{-1} (135 mg/100 mL) are the criteria for normality. If venous blood is used the corresponding values are 8.9 mmol L^{-1} (160 mg/100 mL) and 6.1 mmol L^{-1} (110 mg/100 mL). The urine samples should be glucose free. The **intravenous glucose tolerance test** may be used when abnormal carbohydrate absorption from the gut is suspected. A common procedure is to take a fasting blood sample and further samples at 10-minute intervals for 1 hour following intravenous injection of 50 mL of 50% w/v glucose solution over a period of 2 minutes. Glucose is assayed in the blood samples and the logarithms of the values are plotted against time. The slope of the resulting straight line, multiplied by 100, is the **glucose-assimilation coefficient**, K, which represents the percentage fall in blood glucose per minute. Values of K in excess of 1.1 per cent per minute are taken as normal.

glucose transporter any sugar-transport protein for which glucose is a substrate. Apart from certain ABC types (e.g., in *Escherichia coli*), many glucose transporters belong to the 12-helix superfamily of integral transmembrane proteins. They resemble one another and also the typically sugar–H$^+$ symport proteins of bacteria, even though the mammalian examples are usually facilitated-diffusion proteins. The human type 4 (insulin responsive, muscle and brown fat) glucose transporter is a candidate for the factor responsible for certain post-receptor defects in non-insulin-dependent diabetes mellitus. *See also* **sugar transporter**.

glucosidase any **glycosidase** that hydrolyses O-glucosyl-, N-glucosyl-, or S-glucosyl-compounds.

α**-glucosidase** EC 3.2.1.20; *systematic name*: α-D-glucoside glucohydrolase; *other names*: maltase; glucoinvertase; glucosidosucrase; maltase–glucoamylase. An enzyme that hydrolyses maltose and terminal (nonreducing end) 1,4-α-D-glucosidic bonds of oligosaccharides to yield α-D-glucose with retention of configuration. Maltases from some sources (especially intestine) act also on polysaccharides and more slowly hydrolyse (1→6)-α-D-glucosidic linkages as well. Most active with oligosaccharides, the enzyme catalyses the final step in the degradation of starch and is widely distributed in plants. *See also* **glucan 1,4-α-glucosidase**.

β**-glucosidase** EC 3.2.1.21; *systematic name*: β-D-glucoside glucohydrolase; *other names*: cellobiase; gentiobiase. An enzyme that hydrolyses terminal (nonreducing end) 1,4-β-D-glucosidic bonds to yield β-D-glucose with retention of configuration.

glucoside *or (formerly)* **glycoside** any **glycoside** (def. 1) in which the sugar moiety is a glucose residue.

glucosinolate a class of sulfur-containing glycosides produced by brassica plants and relatives as part of their defence against pathogens. Glucosinolates are grouped according to the nature of their side chains, which are derived from aliphatic, indolyl, or aralkyl α-amino acids. Tissue damage causes decompartmentation with release of the thioglycosidase myrosinase. This releases D-glucose and an unstable aglycone from the glucosinolates. The aglycone then forms toxic isothiocyanate, thiocyanate, or nitrile derivatives.

glucosuria the presence of glucose in the urine. It is sometimes loosely referred to by the older term **glycosuria**, which now strictly implies the presence of any type of sugar in urine. Glucosuria occurs in **diabetes mellitus** and is diagnostically important. It may also occur because of renal defects, as in **Fanconi syndrome**. *See also* **fructosuria**.

glucosyl any glycosyl group formed by detaching the anomeric hydroxyl group, on C-1, from a cyclic form (pyranose or furanose) of glucose or a glucose derivative.

glucosylated hemoglobin *a misnomer for* **glycated hemoglobin**.

glucosylation glycosylation with one or more glucosyl groups. — **glucosylated** adj.

glucosylceramidase EC 3.2.1.45; other names: β-glucocerebrosidase; acid β-glucosidase; systematic name: D-glucosyl-N-acylsphingosine glucohydrolase; a lysosomal glycoprotein enzyme that catalyses the hydrolysis of D-glucosyl-N-acylsphingosine to D-glucose and N-acylsphingosine. The enzyme requires saposin C for activity. A deficiency of this enzyme is the cause of **Gaucher's disease**, in which there is an accumulation of glucocerebroside. See **saposin**.

glucosylceramide the recommended term for any D-glucosyl-N-acylsphingosine, i.e. any **cerebroside** in which the sugar moiety is D-glucose.

glucosyloxy any of the glycosyloxy groups formed by detaching the hydrogen atom from the anomeric hydroxyl group, on C-1, from a cyclic form (pyranose or furanose) of glucose.

glucosyltransferase any hexosyltransferase enzyme (EC 2.4.1) for which the glycosyl group transferred is glucosyl. The acceptor may be orthophosphate (as for e.g. **phosphorylase**), but more often is another carbohydrate molecule (as for e.g. **glycogen synthase**) and sometimes a lipid (as in the synthesis of 3-D-glucosyl-1,2-diacylglycerol). See **D enzyme**. See also **B-transferase**, **lactose synthase**, **Lewis transferase**.

glucuronate 1 the anion of glucuronic acid. 2 any salt or ester of glucuronic acid.

glucuronate pathway or **glucuronate oxidation pathway** the metabolic pathway by which D-glucuronate is converted, via L-gulonate, L-xylulose, xylitol, and D-xylulose, to D-xylulose 5-phosphate, an intermediate in the pentose phosphate pathway. In **essential pentosuria** there is defective conversion of L-xylulose to xylitol. In a side branch of the pathway L-ascorbate is formed, via L-gulonolactone, from L-gluconate. Primates and the guinea pig lack the ability to convert L-gulonolactone to L-ascorbate (see **ascorbic acid**).

glucuronate reductase EC 1.1.1.19; systematic name: L-gulonate:NADP⁺ 6-oxidoreductase. An enzyme that reduces the aldehyde group at C-1 of D-glucuronate thus forming L-gulonate. See **glucuronate pathway**.

glucuronic acid abbr. (sometimes): GA; symbol: GlcA (or formerly) GlcU or GlcUA); the uronic acid formally derived from glucose by oxidation of the hydroxymethylene group at C-6 to a carboxyl group. There are two enantiomers; D-glucuronic acid is widely distributed in plants and animals, where it usually occurs as glucuronides.

β-D-glucuronic acid

β-glucuronidase abbr. (in clinical chemistry): GRS; EC 3.2.1.31, an enzyme that catalyses the hydrolysis of a β-D-glucuronoside to D-glucuronate and the corresponding hydroxy compound, which may be an alcohol, a phenol, or a saccharide. Deficiency of β-glucuronidase occurs in human type VII mucopolysaccharidosis. It is a homotetramer, and a lysosomal enzyme.

glucuronide former name for **glucuronoside**.

glucuronoside or (formerly) **glucuronide** any compound formed by combination in (α or β) glycosidic linkage of a hydroxy compound with the anomeric carbon atom of glucuronate. The hydroxy compound may be a saccharide, an alcohol, or a phenol. β-D-Glucuronosides are formed in so-called **detoxication** reactions.

glucuronosyl the glycosyl group formed by removal of the anomeric hydroxyl group from glucuronic acid or glucuronate.

glucuronosyltransferase another name for **UDPglucuronosyltransferase**.

glucuronoyl the univalent acyl group derived from glucuronic acid.

glued see **dynactin**.

Glu(NH₂) (formerly) symbol for a residue of the α-amino acid L-glutamine (Gln or Q now recommended).

Glu-NH₂ symbol for a residue of the α-amino acid L-isoglutamine.

Glu(OEt) symbol for a residue of γ-ethyl glutamate; O^5-ethyl hydrogen glutamate.

Glu-OEt symbol for a residue of α-ethyl glutamate; O^1-ethyl hydrogen glutamate.

GluR receptor see **glutamate receptor**.

GLUT an abbr. commonly used for **glucose transporter**.

glutaconyl-CoA decarboxylase EC 4.1.1.70; an enzyme that catalyses the decarboxylation of glutaconyl-CoA (pent-2-enoyl-CoA) to but-2-enoyl-CoA, releasing CO_2. Biotin is a coenzyme. The enzyme is dependent on Na⁺, and functions as a sodium pump in the bacterial membrane.

glutamate 1 (in biochemistry) glutamate(1–); glutamic-acid monoanion; hydrogen glutamate. In theory, the term denotes any ion, or mixture of ions, formed from **glutamic acid**, and having a net charge of –1, although the species _OOC–[CH₂]₂–CH(NH₃⁺)–COO_ generally predominates in practice. 2 (in chemistry) the systematic name for glutamate(2–); glutamic-acid dianion. 3 any salt containing an anion of glutamic acid. 4 any ester of glutamic acid.

glutamate N-acetyltransferase EC 2.3.1.35; other names: ornithine acetyltransferase; ornithine transacetylase; an enzyme of the pathway in bacteria for the formation of ornithine from glutamate. It catalyses a reaction between N^2-acetyl-L-ornithine and L-glutamate to form L-ornithine and N-acetyl-L-glutamate (abbr.: AGA). AGA is then converted to N-acetylglutamate 5-semialdehyde, which undergoes transamination to N^2-acetylornithine. In Bacillus stearothermophilus it is a bifunctional enzyme (other name: AGS/OCTase) containing activities for this and also **amino-acid N-acetyl transferase**.

glutamate–ammonia ligase EC 6.3.1.2; systematic name: L-glutamate:ammonia ligase (ADP-forming); other name: glutamine synthetase. An enzyme that catalyses a reaction between ATP, L-glutamate, and NH_3 to form ADP, orthophosphate, and L-glutamine. This is the first step in the fixation of ammonia in plants, the ATP-dependent incorporation of NH_4^+ into the amide position of glutamine. The second step involves L-glutamate synthase (ferredoxin) and the combined effect of the two enzymes is to produce L-glutamate from 2-oxoglutarate and ammonium (NH_4^+), a major NH_4^+ assimilatory route. Glutamate–ammonia ligase has a molecular mass of 33 kDa and its eight monomers are arranged in two stacked tetramers. The protein from Escherichia coli consists of 12 subunits in two hexagons. The activity is controlled by adenylylation of a tyrosine residue in the presence of ATP by [**glutamate-ammonia-ligase**] **adenylyl transferase**. The fully adenylylated dodecamer is inactive; transcription is regulated by **nitrogen regulatory protein**.

[glutamate–ammonia-ligase] adenylyltransferase EC 2.7.7.42; other name: glutamine-synthetase adenylyltransferase; an enzyme that catalyses the adenylylation of [L-glutamate:ammonia ligase (ADP-forming)] by ATP to form adenylyl-[L-glutamate:ammonia ligase (ADP-forming)] with release of pyrophosphate. The adenylylation regulates enzyme activity (see **glutamate–ammonia ligase**).

glutamate/aspartate transporter any membrane protein that transports glutamate and aspartate. Two types are known. (1) A H⁺-symport protein of the bacterial (inner) membrane; it is probably a 12-helix integral membrane protein. Although Na⁺ is not involved, there is a sequence resemblance to the second type (below). (2) The sodium-dependent transporter of brain; it is similar to sodium-dicarboxylate symporters. Putatively, it is a glycosylated integral membrane protein of brain (high levels in Purkinje cell layer).

glutamate dehydrogenase (NADP⁺) EC 1.4.1.4; systematic name: L-glutamate:NADP⁺ oxidoreductase (deaminating). An enzyme that catalyses the oxidation by NADP⁺ of L-glutamate to form 2-oxoglutarate, NH_3, and NADPH. Two forms exist: one is a homotetramer; the other form, found in bacteria and plants, mainly proceeds in the direction of glutamate synthesis, as an NH_3-assimilatory enzyme; it is the product of the am gene, and is a homohexamer. Isoforms of this enzyme provided the first demonstration of **interallelic complementation**.

glutamate–cysteine ligase EC 6.3.2.2; *other name*: γ-glutamylcysteine synthetase; the first cytosolic enzyme for the formation of glutathione. It catalyses the formation of γ-L-glutamyl-L-cysteine from ATP, L-glutamate, and L-cysteine with release of ADP and orthophosphate. The mammalian enzyme is a heterodimer. Deficiency in humans produces glutathione deficiency associated primarily with chronic hemolysis.

glutamate decarboxylase *abbr*.: GAD; EC 4.1.1.15; a decarboxylase enzyme necessary for the production of γ-amino-*n*-butyric acid (GABA) from glutamate, with pyridoxal phosphate as coenzyme. GAD65 (585 amino acids; gene at 10p11) is present in small vesicles in brain and in pancreatic islets. Autoantibodies to it occur early in the development of diabetes mellitus type 1, being present in the plasma in ≈90% of such patients. They are also present in ≈60% of patients with stiff-man syndrome. GAD65 shows sequence homology over 22 residues with a sequence in Coxsackie virus P2-C protein. The gene for GAD67 at 2q31 encodes 594 amino acids.

glutamate dehydrogenase EC 1.4.1.2; *systematic name*: L-glutamate:NAD$^+$ oxidoreductase (deaminating). An enzyme that catalyses the reaction:

$$\text{L-glutamate} + H_2O + NAD^+ =$$
$$\text{2-oxoglutarate} + NH_3 + NADH.$$

This oxidative removal of the amino group from glutamate avoids the need for an oxo-acid acceptor, and in **ureotelic** animals is a major route for the removal of nitrogen as NH_4^+, which is then converted to urea in the **ornithine–urea cycle**. *See also* **glutamate dehydrogenase (NADP$^+$)**.

glutamate dehydrogenase (NADP$^+$) EC 1.4.1.4; *systematic name*: L-glutamate:NADP$^+$ oxidoreductase (deaminating). An enzyme that catalyses the oxidation by NADP$^+$ of L-glutamate to form 2-oxoglutarate, NH$_3$, and NADPH. Two forms exist. One is a homotetramer. The other form, found in bacteria and plants, mainly proceeds in the direction of glutamate synthesis, as an NH$_3$-assimilatory enzyme. It is the product of the *am* gene, and is a homohexamer. Isoforms of this enzyme provided the first demonstration of **interallelic complementation**.

glutamate–oxaloacetate transaminase *see* **aminotransferase**.

glutamate–pyruvate transaminase *see* **aminotransferase**.

glutamate receptor *abbr*.: GluR; any membrane protein that binds glutamate and mediates its effects in neurotransmission. Over 20 genes encode human GluRs, which may be ionotropic and act as ligand-gated ion channels or metabotropic G-protein-associated membrane proteins. Some GluR gene products are subject to RNA editing. **Ionotropic GluR receptors** are subdivided into three classes, named after their agonists: (1) *N*-methyl-D-aspartate (NMDA) receptors, which play a key role in memory acquisition, learning, and neurological disorders and are the target for **conantokin-G**; (2) α-amino-3-hydroxy-5-methyl-4-isoxalone (**AMPA**) receptors, formerly designated quisqualate receptors; and (3) kainate receptors. Most of the fast excitatory synaptic transmission in the vertebrate central nervous system is mediated by GluR channels. Generally they have a broad ionic selectivity to monovalent ions; NMDA channels are also five to ten times more permeable to Ca^{2+} ions than to Na$^+$ or K$^+$ ions. Subunits (designated α, β, γ, δ, ε, and ζ) of these channels contain four putative transmembrane segments. The members of the α subfamily constitute AMPA-selective GluR channels; the β and γ subfamilies include the subunits of the kainate-selective GluR channel; NMDA receptor channels are composed of εζ heteromers. **Metabotropic GluR receptors** (mGluR) include mGluR$_1$, of which two alternatively spliced versions exist (1199 and 906 amino acids), and mGluR$_2$ (872 amino acids); both activate phospholipase C.

glutamate 1-semialdehyde glutamic α-semialdehyde; the L enantiomer, L-glutamate 1-semialdehyde, (*S*)-4-amino-5-oxopentanoate, is the precursor of 5-aminolevulinate (*see* **glutamate-1-semialdehyde 2,1-aminomutase**). It is synthesized in plants from glutamate bound to a specific tRNA; after reduction of the 1-carboxyl group, the tRNA is removed to yield glutamate 1-semialdehyde. *Compare* **glutamate 5-semialdehyde**.

glutamate 5-semialdehyde glutamic γ-semialdehyde; the L enantiomer, L-glutamate 5-semialdehyde, (*S*)-2-amino-5-oxopentanoate, is an intermediate in the synthesis of proline from or-

nithine. In plants and bacteria it is formed in a pathway in which L-glutamate is first converted to *N*-acetyl-L-glutamate, which is reduced to *N*-acetyl-L-glutamate 1-semialdehyde, from which the acetyl group is then removed. In animals it is formed from ornithine by transamination of the δ-amino group. *See* **proline**. *Compare* **glutamate 1-semialdehyde**.

glutamate-1-semialdehyde 2,1-aminomutase EC 5.4.3.8; *other name*: glutamate-1-semialdehyde aminotransferase. The enzyme synthesizing 5-aminolevulinate in plants; it catalyses the conversion of (*S*)-4-amino-5-oxopentanoate (i.e. **glutamate 1-semialdehyde**) to 5-aminolevulinate.

glutamate synthase (ferredoxin) EC 1.4.7.1; *systematic name*: L-glutamate:ferredoxin oxidoreductase (transaminating). An enzyme that catalyses the second step in the fixation of ammonia in plants, i.e. the transfer of the NH$_2$ group from the amide group of glutamine to 2-oxoglutarate to form L-glutamate. The reaction is between L-glutamine, 2-oxoglutarate, and two molecules of reduced ferredoxin; the products are two molecules of L-glutamate and two of oxidized ferredoxin. In plants the combination of the action of **glutamate–ammonia ligase** and this enzyme results in the production of one molecule of glutamate from one of 2-oxoglutarate and one of NH$_4^+$. *See also* **glutamate synthase (NADH)**, **glutamate synthase (NADPH)**.

glutamate synthase (NADH) EC 1.4.1.14; *systematic name*: L-glutamate:NAD$^+$ oxidoreductase (transaminating). An enzyme that catalyses a reaction between L-glutamine, 2-oxoglutarate, and NADH to form two molecules of L-glutamate and NAD$^+$. FMN is a cofactor.

glutamate synthase (NADPH) EC 1.4.1.13; *systematic name*: L-glutamate:NADP$^+$ oxidoreductase (transaminating). An enzyme that catalyses the formation of two molecules of L-glutamate and one of NADP$^+$ from L-glutamine, 2-oxoglutarate, and NADPH. It is an iron–sulfur flavoprotein. It brings together nitrogen and carbohydrate metabolism in bacteria, since in association with **glutamate–ammonia ligase** it functions as the second step in a major NH$_3$-assimilatory pathway. It exists as an aggregate of four heterodimers of large and small subunits; glutamine binds to the large subunit and transfers –NH$_2$ from its amido group to 2-oxoglutarate, which apparently binds to the small subunit.

glutamic *the general name for* a residue of glutamic acid, whether or not the γ-carboxyl group is protonated.

glutamic acid *abbr*. (*sometimes*): GA; *the trivial name for* α-aminoglutaric acid; 2-aminopentanedioic acid; HOOC–[CH$_2$]$_2$–CH(NH$_2$)–COOH; a chiral α-amino acid. L-Glutamic acid (*symbol*: E *or* Glu), (*S*)-2-aminopentanedioic acid, is a coded α-amino acid found in peptide linkage in proteins; codon: GAA or GAG (anticodon: CUU). In mammals, it is a non-essential dietary amino acid, and is glucogenic. Residues of D-glutamic acid (*symbol*: D-Glu *or* DGlu), (*R*)-2-aminopentanedioic acid, are found in the cell-wall material of various bacterial species, and (as poly-D-glutamic acid) in the capsular substances of *Bacillus anthracis*, *B. subtilis*, and other bacteria. One residue of D-glutamic acid per molecule occurs in the antibiotic bacitracin A (*see* **bacitracin**) and also in **glutathione**.

L-glutamic acid

glutamic amide **1** glutamic 1-amide (i.e. **isoglutamine**). **2** glutamic 5-amide (i.e. **glutamine**).

glutaminase EC 3.5.1.2; *other and systematic name*: L-glutamine amidohydrolase; an enzyme that catalyses the hydrolysis of L-glutamine to L-glutamate and NH$_3$.

glutaminate **1** the glutamine anion; the anion, H$_2$N–CO–[CH$_2$]$_2$–CH(NH$_2$)–COO$^-$, derived from glutamine. **2** any salt containing glutamine anion. **3** any ester of glutamine.

glutamine *the trivial name for* the γ-amide of glutamic acid; glutamic 5-amide; α-aminoglutaramic acid; 2-amino-4-carbamoylbutanoic acid; $H_2N–CO–[CH_2]_2–CH(NH_2)–COOH$; a chiral α-amino acid. L-Glutamine (*symbol*: Q *or* Gln *or (formerly)* Glu(NH_2)), (*S*)-2-amino-4-carbamoylbutanoic acid, is a coded α-amino acid found in peptide linkage in proteins; codon: CAA or CAG (anticodon: GUU). In mammals, it is a non-essential *a*-amino acid, and is glucogenic. D-Glutamine (*symbol*: D-Gln *or* DGln), (*R*)-2-amino-4-carbamoylbutanoic acid, is not known to occur naturally. *Compare* **isoglutamine**.

L-glutamine

glutamine amidotransferase *abbr.*: GATase; *a nonsystematic name for* carbon–nitrogen ligases that employ glutamine as the amido-N-donor; i.e. members of the sub-subclass EC 6.3.5. These enzymes transfer the –NH_2 group from glutamine to another reactant, e.g. **GMP synthase**. The enzyme activity is also an activity associated with **CAD protein** in mammals. *See also* **glutamate synthase (ferredoxin), glutamate synthase (NADH), glutamate synthase (NADPH)**.

glutamine–pyruvate aminotransferase EC 2.6.1.15; *other names*: glutaminase II; glutamine–oxo-acid transaminase; glutamine transaminase L; an enzyme that catalyses the reaction:

L-glutamine + pyruvate = 2-oxoglutaramate + L-alanine.

Pyridoxal-phosphate is a cofactor. *See also* **transamination**.

glutamine repeat a protein sequence consisting exclusively of glutaminyl residues. Expansion of the length of the repeat in several proteins, for example ataxins, androgen receptor, and huntingtin, is associated with various forms of spinocerebellar ataxia, spinobulbar muscular atrophy, and Huntington's disease, respectively. The length expansion leads to protein misfolding and aggregation, formation of cellular inclusion bodies, and neurotoxicity.

glutamine synthetase *see* **glutamate–ammonia ligase**.

glutamine-synthetase adenylyltransferase *see* **[glutamate–ammonia-ligase] adenylyltransferase**.

glutamine transaminase L *see* **glutamine–pyruvate aminotransferase**.

glutamino 1 N^2-glutamino the alkylamino group, $H_2N–CO–[CH_2]_2–CH(COOH)–NH–$, derived from glutamine by loss of a hydrogen atom from its α-amino group. 2 N^5-glutamino the acylamino group, $H_2N–CH(COOH)–[CH_2]_2–CO–NH–$, derived from glutamine by loss of a hydrogen atom from its amide group.

glutaminyl the acyl group, $H_2N–CO–[CH_2]_2–CH(NH_2)–CO–$, derived from glutamine. *Compare* **isoglutaminyl**.

glutamo the alkylamino group, $HOOC–[CH_2]_2–CH(COOH)–NH–$, derived from glutamic acid.

glutamoyl the diacyl group, $–CO–[CH_2]_2–CH(NH_2)–CO–$, derived from glutamic acid.

α-glutamyl glutam-1-yl; the α-monoacyl group, $HOOC–[CH_2]_2–CH(NH_2)–CO–$, derived from glutamic acid.

γ-glutamyl *or (formerly)* isoglutamyl, glutam-5-yl; the γ-monoacyl group, $–CO–[CH_2]_2–CH(NH_2)–COOH$, derived from glutamic acid.

glutamyl aminopeptidase EC 3.4.11.7; *other names*: aminopeptidase A; angiotensinase A; differentiation antigen gp160; an enzyme that removes an N-terminal glutamyl residue from a peptide. It can also hydrolyse N-terminal aspartase, and can thus inactivate **angiotensin** II, being expressed in kidney, as well as on cells of the capillaries and intestine, and on lymphocytes.

γ-glutamyl cycle a cycle of chemical events that uses the γ-carboxyl group of glutamate to transport amino acids actively across (mammalian) cell membranes. Glutathione reacts by transpeptidation with the amino acid to be transported. The resulting γ-glutamylamino acid enters the cell, where it is split into the free amino acid and 5-oxoproline. The latter is then converted to glutamate in an ATP-dependent reaction. The cysteinylglycine is hydrolysed by a

peptidase. Glutathione is then regenerated in two ATP-dependent steps. *See also* **glutamyltransferase**.

glutamyl endopeptidase γ-glutamylcysteine synthetase *see* **glutamate–cysteine ligase**.

glutamyl endopeptidase EC 3.4.21.19; *other names*: staphylococcal serine proteinase; V8 proteinase; protease V8; endoproteinase Glu-C. An enzyme that catalyses the preferential cleavage: Asp-|-Xaa, Glu-|-Xaa.

glutamyltransferase *abbr. (in clinical chemistry)*: GMT; EC 2.3.2.2; *recommended name*: γ-glutamyltransferase; *systematic name*: (5-L-glutamyl)-peptide:amino acid 5-glutamyltransferase; *other name*: γ-glutamyl transpeptidase. An ectoenzyme of plasma membrane that reversibly transfers a γ-L-glutamyl residue (linked to the substrate by an isopeptide bond) from the N terminus of a peptide, normally glutathione, to the amino group of an amino acid. It is present in high concentrations in liver, kidney, prostate, and pancreas, and is useful diagnostically in liver or biliary disease, levels being five times normal or more in cholestasis, hepatitis, and cirrhosis, and also in pancreatitis; it is also elevated in the plasma of alcoholics. The level in normal human (male) plasma is <60 IU L^{-1}. *See also* γ-**glutamyl cycle**.

glutaraldehyde pentanedial; a compound that is useful as a tissue fixative in electron microscopy, as a crosslinking agent for compounds containing amino groups, and as a sporicide.

glutaramic acid *the trivial name for* 4-carbamoylbutanoic acid; $H_2N–CO–[CH_2]_3–COOH$; the monoamide of glutaric acid (*see* **glutarate**).

glutarate 1 *trivial name for* propane-1,3-dicarboxylate; pentanedioate; $^-OOC–[CH_2]_3–COO^-$; the dianion of glutaric acid (pentanedioic acid). 2 any mixture of free glutaric acid and its mono- and dianions. 3 any (partial or full) salt or ester of glutaric acid.

glutaredoxin a monomeric polypeptide with a similar function to **thioredoxin**. It can replace thioredoxin, or may be the main reactant, in the pathway for nucleotide reduction.

glutaric acidemia *or (esp. Brit.)* **glutaric acidaemia** the abnormal elevation of levels of glutaric and hydroxyglutaric acids in plasma and their excessive excretion in urine. **Type I** glutaric acidemia is associated with severe infantile dyskinesis and dystonia, and results from numerous mutations that produce deficiency of **glutaryl-CoA dehydrogenase**. **Type II** is associated with metabolic acidosis and hypoglycemia, and results from mutations in **electron-transfer flavoprotein** or **electron-transferring flavoprotein dehydrogenase**.

glutaryl-CoA dehydrogenase EC 1.3.99.7; a flavoprotein decarboxylase that catalyses the reaction between glutaryl-CoA and electron transfer flavoprotein (ETF) to form crotonoyl-CoA, CO_2, and reduced ETF. *See* **glutaric acidemia**.

glutathione *abbr.*: GSH; *symbol*: Glu-(Cys-Gly); γ-glutamylcysteinylglycine, a tripeptide that is widely distributed in most if not all cells. It acts as a coenzyme for some enzymes and as an antioxidant in the protection of sulfhydryl groups in enzymes and other proteins; it has a specific role in the reduction of hydrogen peroxide and oxidized ascorbate, and it participates in the γ-**glutamyl cycle**. Oxidation links two molecules by a disulfide bond (represented as GSSG). For clarity, glutathione is sometimes termed **reduced glutathione** (*or* the reduced form of glutathione). It is involved in the synthesis of certain leukotrienes.

glutathione dehydrogenase (ascorbate) EC 1.8.5.1; an enzyme that catalyses the reduction by two molecules of glutathione of dehydroascorbate to form oxidized glutathione and ascorbate.

glutathione peroxidase EC 1.11.1.9; *systematic name*: glutathione:hydrogen-peroxide oxidoreductase. An enzyme that catalyses the reaction:

$$2 \text{ glutathione} + H_2O_2 = \text{oxidized glutathione} + 2H_2O.$$

It has a UGA-encoded selenocysteine residue at its active site. Its main role is to protect hemoglobin from oxidative breakdown: the oxidized glutathione is reduced by **glutathione reductase**, the system thus contributing to a reduction of peroxide levels in the cell.

glutathione reductase abbr. (in clinical chemistry): GTD; EC 1.6.4.2; recommended name: glutathione reductase (NADPH); systematic name: NADPH:oxidized-glutathione oxidoreductase. A homodimeric enzyme that catalyses the reaction:

$$\text{NADPH} + \text{oxidized glutathione} = \text{NADP}^+ + 2 \text{ glutathione}.$$

It has FAD as cofactor, and is important in maintaining glutathione in the reduced state.

glutathione synthase EC 6.3.2.3; other name: glutathione synthetase; the second cytosolic enzyme of the formation of glutathione from ATP, γ-L-glutamyl-L-cysteine, and glycine with release of ADP and orthophosphate. The mammalian enzyme is a homodimer. Deficiency in humans produces glutathione deficiency, (5)-oxoprolinuria, hemolysis, and central nervous system symptoms.

glutathione transferase EC 2.5.1.18; systematic name: RX:glutathione R-transferase; other names: glutathione S-alkyltransferase; glutathione S-aryltransferase; S-(hydroxyalkyl)glutathione lyase; glutathione S-aralkyltransferase; glutathione S-transferase. Glutathione transferases catalyse various glutathione-dependent reactions:
(1) nucleophilic substitution, e.g.

$$RX + GSH \rightarrow RSG + HX$$

where RX is an aryl- or alkyl-halide (EC 2.5.1.18); (2) addition of GSH across activated unsaturated carbon–carbon or carbon–nitrogen bonds (often α,β-saturated ketones/ thiones) or oxirane, e.g.

$$R-CH-CH_2 + GSH = R-CHOH-CH_2SG;$$
$$\underset{O}{\diagdown\diagup}$$

(3) organic hydroperoxide reduction (selenium-independent GSH peroxidase):

$$ROOH + 2GSH \rightarrow ROH + GSSG + H_2O;$$

(4) organic nitrite reduction

$$RONO + GSH \rightarrow GSNO + ROH;$$

(5) isomerizations in which GSH acts as cofactor, e.g. conversion of prostaglandin H to prostaglandin D. The proteins also bind a variety of hydrophobic anions, e.g. hematin, fatty acid and steroid metabolites, and various xenobiotics/metabolites including the GSH conjugates which may be their catalytic products. Such binding is responsible for their retention by S-linked GSH-Sepharose, hence the widely used simple purification of **GST fusion proteins**. GSH transferases were originally identified as binding proteins (see **ligandin**) and are widely distributed, abundant dimeric soluble enzymes or trimeric membrane-bound forms. They are generally considered an important part of phase II metabolism of xenobiotics. Five classes of soluble GSH transferase subunits have been distinguished (α, μ, π, θ, and σ) and heterodimers occur at least within classes yielding, for example, about 20 different isoenzymes in the rat. Distinct soluble mitochondrial and bacterial forms also occur.

glutelin any member of a group of simple proteins found especially in the seeds of cereal plants. They are insoluble in water, dilute salt solutions, and ethanol, but are soluble in dilute acids or alkalis. The group includes the **glutenins** of wheat, the hordenins of barley, and the oryzemins of rice. They serve as storage proteins, and have Gln-rich domains. Compare **prolamin**.

gluten a proteinaceous fraction obtained from certain cereal seeds, especially wheat, consisting of a considerable number of proteins that are all either **glutelins** or **prolamins**. It forms a sticky mass when wet, and gives dough its characteristic consistency. See also **celiac disease**.

glutenin any **glutelin** of the endosperm of wheat seeds. Glutenins form aggregates of high molecular mass in which they are linked together by disulfide bridges; they are thought to be responsible for the viscoelastic properties of wheat dough. The sequences contain a central low-complexity region, rich in glutamine, proline, glycine, tyrosine, threonine, and serine residues; these cluster within several different types of repeating unit, but all sequences include QQPGQG and QQGYYPTS motifs. Compare **gliadin**.

Glx general symbol for a residue of any α-amino (or imino) acid that yields L-glutamic acid on acid hydrolysis. It was formerly used only for L-glutamic acid (specific symbol: Glu) or L-glutamine (specific symbol: Gln), but is now used to represent any such residues, including those of L-4-carboxyglutamic acid (specific symbol: Gla) and 5-oxoproline (= pyroglutamic acid) (specific symbol: Glp); (alternative to Z).

Gly symbol for a residue of the achiral (α-)amino acid glycine (alternative to G).

glyburide 1-[[p-[2-(chloro-o-anisamido)ethyl]phenyl]sulfonyl]-3-cyclohexylurea; glibenclamide; a sulfonylurea with oral hypoglycemic actions similar to **tolbutamide**.

glyc+ a variant form of **glyco+** (before a vowel).

glycaemia a variant spelling (esp. UK) of **glycemia**.

glycamine any ω-amino-ω-deoxyalditol, i.e. any compound formally derived from an **alditol** by replacement of a terminal hydroxyl group by an amino group.

glycan or **polysaccharide** any polymer consisting of more than about 10 monosaccharide (i.e. glycose) residues joined by glycosidic linkages. A **homoglycan** (or **homopolysaccharide**) contains only one type of monosaccharide residue, whereas a **heteroglycan** (or **heteropolysaccharide**) contains two or more different kinds of monosaccharide residue. A homoglycan is named, generally, after the constituent monosaccharide, with the +ose ending replaced by +an; hence, cellulose, dextran, dextrin, and starch are all glucans, since they consist of glucose. Compare **glycosan**.

glycaric acid an obsolescent name for **aldaric acid**.

glycated hemoglobin or **glycohemoglobin** any derivative of hemoglobin that is formed nonenzymically by reaction at the N terminus with glucose. In the normal adult human such derivatives constitute a few percent of the total erythrocyte hemoglobin, the most abundant being **hemoglobin A$_{1c}$**, which increases severalfold in concentration in diabetes mellitus, and is assayed to monitor control of diabetes. Glycation of other proteins such as lens crystallins, collagen and low-density lipoproteins has been implicated in the pathogenesis of diabetic complications. It is sometimes improperly termed **glucosylated hemoglobin** or **glycosylated hemoglobin**. See also **glycation**.

glycated proteins glycation is the post-translational modification of a protein by the covalent attachment of a sugar residue, resulting from the spontaneous amino–carbonyl reaction (the **Maillard reaction**) between these two moieties. The reaction occurs principally between the free amino groups of proteins and the carbonyl group of reducing sugars, as first described by Louis-Camille Maillard in 1912. The initial condensation to form a Schiff's base is a spontaneous, but reversible, reaction that is rapidly followed by a complex series of irreversible reactions, ultimately leading to generation of Amadori products of brown pigments, and cross-linking of the polypeptide chains involved. The importance of the Maillard reaction has long been recognized in the food industry, where it causes spoilage, and produces colour and off-flavour in cooked foods. In 1977, Cerami identified an Amadori product in variants of human hemoglobin. Glycation of various proteins has now been implicated in the etiology of various diseases, such as the development of Alzheimer-type pathology. *See* **Alzheimer's disease**. *See also* **glycation**.

glycation any chemical reaction, whether enzyme-catalysed or not, that links a glycose (i.e. a sugar) to a peptide or protein and thereby forms a glycopeptide or glycoprotein. The term thus includes (enzymic or nonenzymic) **glycosylation** (and hence **glucosylation**) of a peptide or protein. However, it is usefully applied in addition to any (nonenzymic) reaction of glucose (or another sugar) with a peptide or protein where the product is not a glycoside but results from condensation of the sugar with free amino groups to form a **Schiff base** which then undergoes an **Amadori rearrangement**. —**glycated** *adj*.

glycemia *or* **glucemia** *or (esp. UK)* **glycaemia** *or* **glucaemia** **1** the presence of glucose (strictly sugar) in the blood. **2** the concentration of (strictly) sugar but, in fact, glucose in the blood. When this concentration is within normal limits the condition of the blood (or of the corresponding individual) is termed **normoglycemia**. An abnormally elevated concentration, **hyperglycemia**, may result in thirst, whereas an abnormally depressed concentration, **hypoglycemia**, may produce sweating, trembling, hunger, irritability, and in extreme cases confusion or coma. *See also* **blood sugar**, **glucose-tolerance test**, **insulin-tolerance test**. —**glycemic** *or* **glycaemic** *or* **glucemic** *or* **glucaemic** *adj*.

glycemic index *or (esp. UK)* **glycaemic index** the area under the blood glucose response curve for a food expressed as a percentage of the area produced by ingesting the same amount of carbohydrate as glucose. It is usually determined as the mean for 5–10 individuals. Examples include: carrots 92 ± 20; bread (white) 69 ± 5; whole milk 34 ± 6; peas (blackeye) 33 ± 4; fructose 20 ± 5.

glyceral *an alternative name for* **glyceraldehyde**.

glyceraldehyde *symbol*: Gra; glyceral; 2,3-dihydroxypropanal; *formerly sometimes known as* **glyceric aldehyde**; *sometimes (not recommended) called* **glycerose**. It is the simplest chiral aldose and a constitutional isomer of **glycerone**. The naturally occurring form is D-glyceraldehyde, the absolute configuration of which is known and is taken as the configurational reference standard for the D-, and hence also for the L-, forms of monosaccharides.

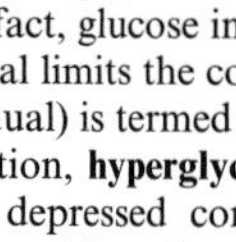

$$\begin{array}{c} \text{CHO} \\ | \\ \text{H–C–OH} \\ | \\ \text{CH}_2\text{OH} \end{array}$$

D-glyceraldehyde

D-glyceraldehyde 3-phosphate *abbr*.: D-glyceraldehyde-3-*P*; an important intermediate in glycolysis. It is isomeric with **glycerone phosphate**. (phosphorylating) (phosphorylating)

glyceraldehyde-3-phosphate dehydrogenase (NADP⁺) (phosphorylating) EC 1.2.1.13; *other name*: triosephosphate dehydrogenase (NADP⁺); an enzyme that catalyses the conversion of D-glyceraldehyde 3-phosphate, orthophosphate, and NADP⁺ to 3-phospho-D-glyceroyl phosphate and NADPH.

glyceraldehyde-3-phosphate dehydrogenase (phosphorylating) *abbr*.: GAPDH; EC 1.2.1.12; *systematic name*: D-glyceraldehyde-3-phosphate:NAD⁺ oxidoreductase (phosphorylating); *other name*: triosephosphate dehydrogenase. An homotetrameric enzyme that catalyses a reaction between D-glyceraldehyde 3-phosphate, orthophosphate, and NAD⁺ to form 3-phospho-D-glyceroyl phosphate and NADH. It is important in the pathways of glycolysis and gluconeogenesis. GAPDH is used as an internal standard in Northern blotting and real-time polymerase chain reaction (RT-PCR) because its mRNA is expressed constitutively in mammalian tissues.

glycerate *symbol*: Gri; **1** 2,3-dihydroxypropanoate; $HOCH_2$–$CH(OH)$–COO^-; the anion derived from glyceric acid. The naturally occurring form is D-glycerate. **2** any salt or ester of glyceric acid.

glyceric aldehyde *former name for* **glyceraldehyde**.

glyceride *the former name for* **acylglycerol**; i.e. any ester of glycerol,

especially the mono-, di-, or triesters of glycerol with fatty acids, formerly termed mono-, di-, or triglycerides respectively. These are now known as mono-, di-, or tri-*O*-acylglycerols.

glycerin(e) *see* **glycerol**.

glycero+ *prefix or infix (in chemical nomenclature)* **1** denoting a residue of glycerol. **2** denoting the acyl group derived from glyceric acid.

glycero- *prefix (in chemical nomenclature)* indicating the presence of a single CHOH group, in the D or L configuration as specified.

glycerokinase *see* **glycerol kinase**.

glycerol *symbol*: Gro; 1,2,3-propanetriol; a sweet, hygroscopic, viscous liquid, widely distributed in nature as a constituent of many lipids (*see* **acylglycerol**). The older name **glycerin(e)** is now obsolete except in non-technical usage and sometimes in pharmacy.

glycerol channel *or* **glycerol facilitator** (*abbr*.: GlpF) an integral membrane protein of bacteria and fungi that mediates facilitated diffusion of glycerol. The crystal structure shows eight transmembrane segments that form a channel lined with amphipathic and hydrophobic residues on one side and polar residues on the other, and that permits glycerol molecules to pass through it in single file.

glycerol kinase EC 2.7.1.30; *systematic name*: ATP:glycerol 3-phosphotransferase; *other name*: glycerokinase. An enzyme that catalyses the phosphoryation by ATP of glycerol to form *sn*-glycerol 3-phosphate with release of ADP. The amino-acid sequence is remarkably conserved from prokaryotes to mammals. In the latter, the enzyme is cytosolic and also bound to porin of the outer mitochondrial membrane. It is induced, e.g. in *Escherichia coli*, by *sn*-glycerol 3-phosphate; it is a homotetramer.

glycerol phosphate *or* **glycerophosphate** any of three isomeric phosphoric monoesters of glycerol; of these, two are chiral and form an enantiomeric pair, whereas the third is achiral. The earlier nomenclature of the chiral esters has been confusing; ambiguities are avoided by the adoption of **stereospecific numbering** (*symbol*: *sn*-). (1) *sn*-glycerol 3-phosphate *or* *sn*-glycero-3-phosphate *or* 3-phospho-*sn*-glycerol *or* 3-*O*-phosphono-*sn*-glycerol (*abbr*. *sn*-glycerol-3-*P*) are alternative recommended names for the chiral ester (*R*)-[glycerol 1-(dihydrogen phosphate)]; (*R*)-[2,3-dihydroxypropyl dihydrogen phosphate]. It was previously known as L-glycerol 3-phosphate or D-glycerol 1-phosphate, and originally named D-glycerol 1-phosphate *or* L-*a*-glycerophosphate *or* L-*a*-glycerophosphoric acid. It is a component of many phospholipids. The absolute configuration is illustrated. (2) *sn*-glycerol 1-phosphate *or* *sn*-glycero-1-phosphate *or* 1-phospho-*sn*-glycerol *or* 1-*O*-phosphono-*sn*-glycerol (*abbr*. *sn*-glycerol-1-*P*) are alternative recommended names for the chiral ester (*S*)-[glycerol 1-(dihydrogen phosphate)]; (*S*)-[2,3-dihydroxypropyl dihydrogen phosphate]. It was previously known as L-glycerol 1-phosphate *or* D-glycerol 3-phosphate, and originally named L-glycerol 1-phosphate *or* D-*a*-glycerophosphate *or* D-*a*-glycerophosphoric acid. This enantiomer is rarely encountered. (3) glycerol 2-phosphate *or* glycero-2-phosphate *or* 2-phosphoglycerol *or* 2-*O*-phosphonoglycerol are alternative names for the achiral ester glycerol 2-(dihydrogen phosphate); di(hydroxymethyl)methyl dihydrogen phosphate. It was originally named β-glycerophosphate *or* β-glycerophosphoric acid. It is an artefact occurring in hydrolysates of glycerophospholipids that arises by migration of the phosphate group from C-1 to C-2 of the glycerol moiety. *Distinguish from* **phosphoglycerate**.

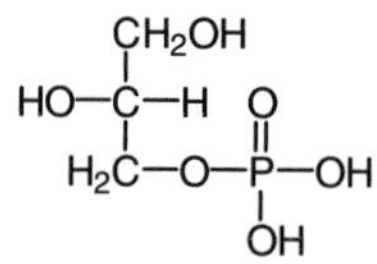

$$\begin{array}{c} \text{CH}_2\text{OH} \\ | \\ \text{HO–C–H} \quad \text{O} \\ | \quad\quad || \\ \text{H}_2\text{C–O–P–OH} \\ | \\ \text{OH} \end{array}$$

sn-glycerol 3-phosphate

glycerol-3-phosphate cycle *or* **glycerol-3-phosphate shuttle** *or* **α-glycerophosphate cycle** an intracellular metabolic cycle that can transfer reducing equivalents from the cytosol to the mitochondria.

In the cytosol NAD-linked **glycerol-3-phosphate dehydrogenase (NAD+)** catalyses the reduction of glycerone phosphate (formed together with glyceraldehyde 3-phosphate in glycolysis) to *sn*-glycerol 3-phosphate. The latter enters the mitochondria where it is oxidized to glycerone phosphate by the FAD-linked mitochondrial glycerol-3-phosphate dehydrogenase. The glycerone phosphate then returns to the cytosol to complete the cycle. The overall result is the oxidation of cytosolic NADH + H+ to NAD+ and the reduction of mitochondrial enzyme–FAD to enzyme–FADH₂.

glycerol-3-phosphate cytidylyltransferase EC 2.7.7.39; *other name*: CDPglycerol pyrophosphorylase; an enzyme that catalyses the formation of CDPglycerol from CTP and *sn*-glycerol 3-phosphate with release of pyrophosphate.

glycerol-3-phosphate dehydrogenase *abbr.*: G3PDH; EC 1.1.99.5; a flavoprotein enzyme that catalyses the oxidation by an electron acceptor of *sn*-glycerol 3-phosphate to glycerone phosphate. It is part of the **glycerol 3-phosphate cycle**. In *Escherichia coli* there are two unrelated enzymes: anaerobic G3PDH, which is an iron–sulfur flavoprotein comprising two subunits; and aerobic G3PDH.

glycerol-3-phosphate dehydrogenase (NAD+) EC 1.1.1.8; *systematic name*: *sn*-glycerol-3-phosphate:NAD+ 2-oxidoreductase. An enzyme that catalyses the reaction:

sn-glycerol 3-phosphate + NAD+ =
glycerone phosphate + NADH.

Its role is especially important in the provision of glycerol 3-phosphate for lipid synthesis, using glycerone phosphate derived from glucose by the **glycolytic** pathway, and as a component of the **glycerol-3-phosphate cycle**. *See* **flightin**.

glycerol-3-phosphate *O*-acyltransferase EC 2.3.1.15; *systematic name*: acyl-CoA:*sn*-glycerol-3-phosphate 1-*O*-acyltransferase; it catalyses the acylation by acyl-CoA of glycerol 3-phosphate to form 1-acyl-*sn*-glycerol 3-phosphate (lysophosphatidic acid) with release of CoA. This is the first step in phospholipid biosynthesis and may be involved in regulation of membrane biogenesis. Plants exhibit oleate-selective acyltransferases; the enzyme from chilling-resistant plants discriminates against palmitic acid (a fatty acid with high transition temperature, and selects oleic acid (leading to more fluid membranes), whereas the acyltransferase from sensitive plants accepts both fatty acids.

glycerol-3-phosphate oxidase EC 1.1.3.21; *systematic name*: *sn*-glycerol-3-phosphate:oxygen 2-oxidoreductase. An enzyme that catalyses the oxidation by dioxygen of *sn*-glycerol 3-phosphate to glycerone phosphate and hydrogen peroxide; FAD is a cofactor. It is a mitochondrial protein. In insects it plays an important role in the transfer of energy for flight, along with **glycerol-3-phosphate dehydrogenase (NAD+)** (flightin).

glycerol trinitrate *see* **trinitroglycerine**.

glycerone *symbol*: Grn; *the currently recommended trivial name for* dihydroxyacetone; 1,3-dihydroxy-2-propanone. It is the simplest ketose, is achiral, and is a constitutional isomer of glyceraldehyde.

glycerone phosphate *the currently recommended trivial name for* dihydroxyacetone phosphate; 1,3-dihydroxy-2-propanone 1-phosphate; the monoester formed by phosphorylation of dihydroxyacetone, glycerone. Glycerone phosphate and D-glyceraldehyde 3-phosphate are together described as triose phosphates and are interconvertible by triose-phosphate isomerase, EC 5.3.1.1. They are important components of the glycolytic pathway. Glycerone phosphate is a precursor of the glycerol moiety of glycerolipids.

glycerone transferase *see* **transaldolase**.

glycerophosphate *an alternative name for* **glycerol phosphate**.

α-glycerophosphate cycle *an alternative name for* **glycerol-3-phosphate cycle**.

glycerophosphate phosphatidyltransferase EC 2.7.8.5; *recommended name*: CDPdiacylglycerol–glycerol-3-phosphate 3-phosphatidyltransferase. An enzyme that catalyses a reaction between CDPdiacylglycerol and *sn*-glycerol-3-phosphate to form 3-(3-*sn*-phosphatidyl)-*sn*-glycerol 1-phosphate, with release of CMP. It is a membrane protein of endoplasmic reticulum and component of the pathway for the biosynthesis of acidic phospholipids; its product

loses phosphate to form phosphatidylglycerol from which **cardiolipin** is formed.

glycerophospholipid *general name for* any derivative of glycerophosphate that contains at least one *O*-acyl, *O*-alkyl, or *O*-alkenyl group attached to the glycerol residue. Formerly known as phosphoglyceride.

glycerophosphoric acid *see* **glycerol phosphate**.

glycerose *alternative name* (not recommended) *for* **glyceraldehyde**.

glyceroyl 2,3-dihydroxypropanoyl; the acyl group, HOCH₂–CH(OH)–CO–, derived from glyceric acid.

glyceryl any group derived from glycerol by replacing or removing one or more of its hydroxyl groups.

glycidol *the trivial name for* 3-hydroxypropylene oxide; 2,3-epoxy-1-propanol.

S(–)-glycidol

glycidoyl 2,3-epoxypropanoyl; the acyl group derived from glycidic acid.

(*R*)-glycidoyl group

glycidyl 2,3-epoxypropyl; the alkyl group: derived from glycidol by removal of its hydroxyl group.

glycinamide *the trivial name for* the amide of glycine, aminoacetamide; aminoethanamide; H₂N–CH₂–CO–NH₂. It is predominantly cationic at neutrality and is useful as a buffer substance; pK_a (20°C) = 8.20.

glycinamide ribonucleotide synthetase (*abbr.* GARS) *see* phosphoribosylamine–glycine ligase.

glycinamide ribotide *N*-glycyl-5-phosphoribosylamine; an early intermediate in the biosynthesis of purines.

glycinate 1 glycine anion; the anion, NH₂–CH₂–COO⁻, derived from **glycine**. **2** any salt containing glycine anion. **3** any ester of glycine.

glycine *symbol*: G or Gly; *the trivial name for* aminoacetic acid; aminoethanoic acid; H₂N–CH₂–COOH; an achiral (*α*-)amino acid, and the simplest naturally occurring amino acid. It is a coded amino acid found in peptide linkage in proteins; codon: GGA, GGC, GGG, or GGU (anticodon: CCU). In mammals, it is a non-essential dietary amino acid, and is **glucogenic**. It is sweet-tasting, hence its name. [From **glyc(o)+** (def. 1) + **+ine** (def. 1)]

glycine amidinotransferase EC 2.1.4.1; an enzyme that catalyses the formation of guanidinoacetate, a precursor of **creatine**, from L-arginine and glycine, also forming L-ornithine.

glycine betaine *N*-trimethylglycine. *See* **betaine**.

glycine cleavage enzyme *or* **glycine oxidase** (when working in the reverse direction); a multienzyme complex that resembles the pyruvate dehydrogenase complex but contains the following elements: **P protein**, a pyridoxal phosphate-dependent glycine decarboxylase; **H protein**, a lipoamide-containing aminomethyltransferase; **T protein**, an N^5,N^{10}-methylene tetrahydrofolate synthesizing enzyme; and **L protein**, an NAD-dependent, FAD-requiring lipoamide dehydrogenase. It provides the major mechanism for glycine degradation in mammalian tissues.

glycine hydroxymethyltransferase EC 2.1.2.1; *systematic name*: 5,10-methylenetetrahydrofolate:glycine hydroxymethyltransferase;

other names: serine hydroxymethylase; threonine aldolase; serine aldolase. An enzyme that catalyses the reaction:

$$5,10\text{-methylenetetrahydrofolate} + \text{glycine} + H_2O =$$
$$\text{tetrahydrofolate} + \text{L-serine.}$$

It is a pyridoxal phosphate-requiring enzyme that interconverts glycine and serine; it transfers a hydroxymethyl group from N^5,N^{10}-methylenetetrahydrofolate to a glycine-pyridoxal phosphate Schiff base or from a serine-pyridoxal phosphate Schiff base to tetrahydrofolate.

glycine receptor a complex of integral membrane proteins that binds glycine and mediates its effects as an inhibitory neurotransmitter in lower brain areas and spinal cord. It is a pentamer of α (four isoforms, α1–α4) and β subunits, forms an intrinsic Cl^- channel, blocked by strychnine (distinguishing it from the $GABA_A$ Cl^- channel). All subunits have four transmembrane segments, M1–M4, M3 and M4 being separated by a cytoplasmic hydrophilic loop; this is comparable to nicotinic acetylcholine receptors (*see* **cholinoceptor**) and $GABA_A$ receptors (*see* **GABA receptor**), with which some sequence similarities also exist. Glycine also binds to a site on the NMDA (glutamate) receptor, which must be occupied by glycine to enable activation of the receptor by NMDA, and is also blocked by strychnine.

glycine-rich proteins *abbr*.: GRP; one of the four major groups of **plant cell-wall proteins**. Glycine-rich proteins can contain over 70% glycine residues, are predicted to form beta sheets, and are thought to form a platelike structure at the interface between the plasma membrane and cell wall, perhaps becoming crosslinked to the latter. They are not heavily glycosylated.

glycinin a 350 kDa **globulin** that is the chief protein constituent of the soya bean, *Glycine max*. Structurally, it resembles **arachin**, **edestin**, and **excelsin**.

glycinium glycine cation; the cation, H_3N^+–CH_2–COOH, derived from **glycine**.

glycino the alkylamino group, HOOC–CH_2–NH–, derived from **glycine**.

glycitol *an obsolescent name for* **alditol**.

glyco+ *or (before a vowel)* **glyc+** *comb. form* **1** indicating sweet-tasting. **2 a** indicating sugar. **b** (*in medicine*) indicating glucose. *See also* **gluco+**. **3** (*in chemical nomenclature*) indicating covalently linked to carbohydrate. **4** indicating combination with glycine; e.g. **glycocholate**, **glycocyamine**.

glycoamino acid any compound in which a saccharide is combined with an amino acid through any kind of covalent linkage. The term includes **glycosylamino acids**.

glycobiology the branch of biochemistry dealing with **glycoconjugates**, including their structure, analysis, function, and metabolism.

glycocalicin a carbohydrate-rich hydrophilic fragment, M_r 130 000, cleaved from the external portion of the α chain of **glycoprotein Ib** by protease action.

glycocalyx *or* **cell coat** a zone surrounding many plant and animal cells that is rich in glycoproteins and proteoglycans.

glycocholate 1 the anion of glycocholic acid; *N*-cholylglycine; a bile acid. The anions of similar glycine conjugates of other bile acids are known, e.g. glycochenodeoxycholate, **glycodeoxycholate**, glycolithocholate. **2** any salt of glycocholic acid. Sodium glycocholate is the chief **bile salt** in humans and herbivorous animals. It is useful as a detergent; aggregation number 2.1, CMC 7.1 mm.

glycocholic acid

glycocoll *an obsolete name for* **glycine**.

glycoconjugate *or (formerly)* **glucoconjugate** any large and complex biological molecule that contains a carbohydrate moiety. The term comprehends glycolipid, glycopeptide, glycoprotein, peptidoglycan, proteoglycan, and heteroglycan. *See also* **glyco+** (def. 3), **conjugate** (def. 5).

glycoconjugate syndrome *an alternative name for* **carbohydrate-deficient glycoprotein syndrome**.

glycocyamine *an older name for* *N*-amidinoglycine; guanidinoacetic acid. In polychaete worms, the guanidine group can undergo phosphorylation to phosphoglycocyamine, which acts as a **phosphagen**.

glycodeoxycholate 1 glycodeoxycholic acid; *N*-deoxycholylglycine; a bile acid. **2** any salt of glycodeoxycholic acid; sodium glycodeoxycholate is useful as a detergent; aggregation number 2, CMC 2.1 mm.

glycoform any of several differently glycosylated variants of a specified glycoprotein. Glycoforms may vary in the type, number and/or position of the sugar residues, differ in their physiological properties (e.g., rate of action), and occur together in a given tissue.

glycogen a polydisperse, highly branched glucan or polyglucose, composed of chains of D-glucose residues in $\alpha(1\rightarrow4)$ glycosidic linkage, joined together by $\alpha(1\rightarrow6)$ glycosidic linkages; a small number of $\alpha(1\rightarrow3)$ glycosidic linkages and some cumulative $\alpha(1\rightarrow6)$ links also may occur. Its structure is similar to that of **amylopectin** but with more, though rather shorter, branches; those in glycogen variously contain 8 to 12 glucose residues. The size of the molecule varies greatly according to source and method of measurement, with average M_r values of 2.5×10^5 to upwards of 1.5×10^7 reported. Glycogen serves as a cellular store of glucose; it occurs, frequently as granules, in all animal tissues, especially liver and skeletal muscle, and also in some bacteria and yeasts. The glycogen molecule is linked to the protein **glycogenin** which is required for the *ab initio* synthesis of glycogen by **glycogen synthase**.

glycogenase *an alternative name for* either α- or β-amylase.

glycogen branching enzyme *see* **branching enzyme**.

glycogen disease *or* glycogen storage disease *or* glycogenosis any human disease in which there is an inherited abnormality in glycogen metabolism. Such diseases have been classified into eight types. Type I glycogen disease (*or* von Gierke's disease) involves a deficiency of hepatic **glucose-6-phosphatase** or of the endoplasmic reticulum (ER) transporter for glucose-6-phosphate, for glucose, or for phosphate. It is an autosomal recessive disease of infants, characterized by massive hepatomegaly, failure to thrive, severe hypoglycemia, and increased hepatic content of normal glycogen. Type II glycogen disease (*or* **Pompe's disease**) involves a generalized deficiency of lysosomal α-**glucosidase**. It is an autosomal recessive disease of infants, characterized by massive cardiomegaly, hypotonia, and an increased content of normal glycogen in virtually all tissues. Type III glycogen disease (*or* **limit dextrinosis** *or* **Cori's disease**) involves a generalized deficiency of amylo-1,6-glucosidase (debrancher enzyme) (EC 3.2.1.33). It is an autosomal recessive disease of infants that is similar to, but milder than, type I glycogen disease. It is characterized by an elevated content of glycogen in muscle, liver, and erythrocytes, the glycogen having abnormally short outer branches. A number of subtypes have been recognized. Type IV glycogen disease (*or* amylopectinosis *or* **Andersen's disease**) is a rare condition of infants in which there is a generalized deficiency of 1,4-α-glucan branching enzyme (EC 2.4.1.18). It is characterized by hepatic cirrhosis, hepatosplenomegaly, ascites, and glycogen with abnormally long outer branches; the glycogen may not accumulate excessively. Type V glycogen disease (*or* **McArdle's disease**) involves a deficiency of glycogen phosphorylase (EC 2.4.1.1) in voluntary muscle. The disease is probably autosomal recessive. It is characterized by myoglobinuria and painful cramps occurring during strenuous exercise, symptoms that do not usually appear until 20 years of age. There is no hypoglycemia but muscular exercise fails to elevate the blood lactate concentration. Type VI glycogen disease (*or* **Hers's disease**) involves an increased hepatic glycogen content and a lowered hepatic glycogen phosphorylase (EC 2.4.1.1) activity but a normal phosphorylase-activating system. The clinical course is mild. Type VII glycogen disease involves a deficiency of muscle 6-**phosphofructokinase**. The disease is probably autosomal recessive and is clinically identical to Type V glycogen disease. There are raised contents in muscles of normal glycogen, glucose 6-phosphate, and fructose 6-phosphate, together with a lower than normal concentration of fructose 1,6-bisphosphate. Type VIII glycogen disease involves a reduced **phosphorylase kinase** activity in the liver and in leukocytes. Thought to be inherited as an X-linked character, the disease is characterized by mild hepatomegaly, increased liver glycogen content, and mild hypoglycemia. Type IX glycogen disease results from phosphorylase kinase deficiency. There are several forms depending on which subunit (or isoform) is affected. At least five other types exist, but are very rare.

glycogenesis the metabolic formation of glycogen from sugar (i.e. glucose).

glycogenetic of or relating to the production of glycogen. (*Note*: The term is often used confusingly in the senses of **glycogenic** (def. 1, 2).)

glycogenic 1 of, relating to, or involving glycogen. **2** of or relating to the formation of sugar (i.e. glucose), especially in animals; glucogenic. (*Note*: the term is often used confusingly in the sense of **glycogenetic**.)

glycogenin EC 2.4.1.186; *recommended name*: glycogenin glucosyltransferase; a protein that is required for the initiation of glycogen synthesis by glycogen synthase. Glycogenin catalyses formation of a glycosidic link between a glucose and one of its own tyrosines (Tyr[194] in human), and then catalyses addition of a further five glucose molecules. UDPglucose is the donor of the glucose units. This forms a primer, which is then extended with further glucose molecules by **glycogen synthase**. Glycogen synthase is only catalytically efficient when it is bound to glycogenin. This has two consequences: first, the number of glycogen granules is determined by the number of molecules of glycogenin, and second, elongation stops when the synthase is sterically prevented from contact with glycogenin, which forms the core of the particle. Hence, the glycogenin–synthase interaction limits the size of glycogen granules. Glycogenin is a monomeric glycoprotein (via its reactivity), tightly bound to the 86 kDa catalytic subunit of glycogen synthase.

glycogenolysis the breakdown of glycogen into glucose, either by hydrolysis or via glucose 1-phosphate and glucose 6-phosphate. The term is sometimes restricted to the phosphorolysis of glycogen to glucose 1-phosphate. —**glycogenolytic** *adj*.

glycogenosis an alternative name for **glycogen disease**.

glycogen phosphorylase *see* **phosphorylase**.

glycogen phosphorylase kinase EC 2.7.11.1; *recommended name*: ATP:phosphorylase-*b* phosphotransferase, other names glycogen phosphorylase kinase; PHK; phosphorylase b kinase. An enzyme that phosphorylates a specific serine residue in each of the subunits of the dimeric phosphorylase *b*. Requires Ca^{2+} and calmodulin for activity.

glycogen storage disease an alternative term for **glycogen disease**.

glycogen synthase EC 2.4.1.11; *recommended name* (example): glycogen (starch) synthase (the nature of the synthetic product should be included in the name); *systematic name*: UDPglucose:glycogen 4-α-D-glucosyltransferase; *other name*: UDPglucose–glycogen glucosyltransferase. An enzyme that catalyses the reaction between UDPglucose and $(1,4-\alpha$-D-glucosyl$)_n$ to form $(1,4-\alpha$-D-glucosyl$)_{n+1}$ with release of UDP. The mammalian enzyme requires glucosylated **glycogenin** as a primer (product of glycogenin glucosyltransferase, EC 2.4. 1.186), and is susceptible to allosteric activation by glucose 6-phosphate, and phosphorylation by a cyclic AMP-dependent kinase. *See also* **dependent form of glycogen synthase, glycogenin, starch synthase**.

[**glycogen-synthase-D**] **phosphatase** EC 3.1.3.42; an enzyme that hydrolyses phosphate from glycogen synthase D, thereby activating it.

glycogen synthase kinase-3 *abbr*. **GSK-3**; a cytosolic serine/threonine protein kinase that (like several other protein kinases) phosphorylates one or more of nine serine residues in each subunit of **glycogen synthase**. It may be regulated by inactivation by protein kinase B in response to insulin. It is an important component of **Wnt** protein signalling during embryonal patterning and development and also thereafter. Wnt signalling inactivates GSK-3 by targeting

it (in a complex with APC protein, β-catenin, and axin) for ubiquitination and proteosomal degradation. Human GSK-3 (420 amino acids) is highly homologous to SHAGGY (514 amino acids) of *Drosophila*. *Arabidopsis thaliana* contains ten GSK-3/SHAGGY-like kinase genes.

glycogen synthase α kinase *see* **non-specific serine/threonine protein kinase**.

glycoglycerolipid any **glycolipid** containing one or more glycerol residues.

glycohemoglobin or (*esp. UK*) **glycohaemoglobin** *an alternative term for* **glycated hemoglobin**.

glycoinositolphospholipid *see* **GPI anchor**.

glycol 1 ethylene glycol; ethane-1,2-diol; $HOCH_2-CH_2OH$. **2** or **glychol** any dihydric alcohol, i.e. any compound in which two alcoholic hydroxyl groups are on different carbon atoms (which are usually but not necessarily adjacent).

glycolaldehyde hydroxyacetaldehyde; hydroxyethanal; $HOCH_2-CHO$.

glycolate or (*formerly*) **glycollate 1** *the trivial name for* hydroxyacetate; hydroxyethanoate; the anion, $HOCH_2-COO^-$, derived from **glycolic acid**. It is an important metabolite in C_3 plants. **2** any mixture of glycolic acid and its anion. **3** any salt or ester of glycolic acid. It is formed from phosphoglycolate, a product of **ribulose bisphosphate carboxylase** when this is acting as an oxidase. It is converted to glyoxylate by glycolate oxidase (*see* **2-hydroxy-acid oxidase**).

glycolate oxidase an FMN-containing enzyme that oxidizes glycolate to glyoxylate in plant peroxisomes. It is now included with **2-hydroxy-acid oxidase**.

glycolic acid or (*formerly*) **glycollic acid** hydroxyacetic acid; hydroxyethanoic acid; $HOCH_2-COOH$; a constituent of sugar-cane juice.

glycolipid any compound containing (usually) one to four linked glycose residues joined by a glycosyl linkage to a lipid moiety (e.g. to a mono- or diacylglycerol, to a long-chain base such as sphingosine, or to a ceramide) or (in a bacterial glycolipid) present in ester linkage to one or more fatty-acid molecules. The term embraces **galactolipid**. *See also* **cerebroside, ganglioside, lipopolysaccharide**.

glycollate an obsolete variant spelling of **glycolate**.

glycoloyl *symbol*: Gc; the acyl group, $HOCH_2-CO-$, derived from **glycolic acid**.

glycolytic pathway or **Embden–Meyerhof(–Parnas) pathway** the widely distributed, cytosolic, metabolic pathway by which sugars, sugar phosphates, or their precursors are broken down with an overall negative free energy change favourable to formation of ATP; the pathway yields usually pyruvate aerobically or lactate anaerobically in animal tissues. By extension, the term is also used of similar processes in other organisms yielding, sometimes, other products, e.g. ethanol, glycerol, etc. Overall there is a net yield of 2 ATP for each glucose metabolized, since phosphorylation of the hexose substrates requires 2 ATP, but conversion to pyruvate of the two 3-carbon substrates formed yields 2 ATP for each. The pathway from glucose involves its conversion to glucose 6-phosphate, thence successively to fructose 6-phosphate, fructose 1,6-bisphosphate, glyceraldehyde 3-phosphate plus glycerone phosphate (which is converted to glyceraldehyde 3-phosphate in situations in which it is further degraded), 1,3-bisphosphoglycerate, 3-phosphoglycerate, 2-phosphoglycerate, phospho*enol*pyruvate, and pyruvate. The enzymes involved are hexokinase, glucose-6-phosphate isomerase, phosphofructokinase, fructose-bisphosphate aldolase, triose-phosphate isomerase, glyceraldehyde-3-phosphate dehydrogenase, phosphoglycerate kinase, phosphoglycerate mutase, enolase, and pyruvate kinase (*see* individual entries). Glycogen breakdown (**glycogenolysis**) by phosphorylase yields glucose 1-phosphate, which is converted by phosphoglucomutase to glucose 6-phosphate, which in muscle is then degraded by the pathway outlined above. The pathway is subject to hormonal regulation, especially by insulin, glucagon, and epinephrine. Flux through the pathway is under powerful regulation by phosphofructokinase, the main modulator of which is **fructose 2,6-bisphosphate**.

glycolytic oscillator any metabolic activity that produces periodic regular oscillations in the concentrations of intermediates of the **glycolytic pathway, as** observed in glycolysing yeast cells and extracts of skeletal muscle and heart. These oscillations reflect periodic changes in the activity of **phosphofructokinase** in response to changes in the levels of its activators and inhibitors.

glycome the full array of polysaccharide chains in a cell, tissue, organ, or organism. *Compare* **proteome**.

glycomics the study of all the carbohydrate molecules that are produced by an organism.

glycon or *(formerly)* **glycone** the carbohydrate moiety of any **glycoside** that also contains a noncarbohydrate component.

glyconeogenesis 1 *an alternative name for* **gluconeogenesis**. **2** *(sometimes)* the formation of glycogen from noncarbohydrate precursors.

glyconic acid *an obsolescent name for* **aldonic acid**.

glycopeptide any compound in which carbohydrate is covalently attached to an oligopeptide composed of residues of L- and/or D-amino acids. The term usually denotes a product of proteolytic degradation of a glycoprotein, but includes glycated peptide. It should be distinguished from **peptidoglycan**, for which it is sometimes considered to be a synonym.

glycophorin any of several type I membrane proteins of the erythrocyte membrane. They are carbohydrate-rich sialoglycoproteins (SGPs – hence the nomenclature for glycophorins: αSGP, etc.), and stain with periodic acid–Schiff (PAS) reagent (hence the nomenclature PAS 1, etc. for bands on electrophoresis). The glycosylated moiety serves as an antigenic determinant and as a receptor for viruses and plant agglutinins. A glycophorin was the first membrane protein to have its full sequence elucidated. There is a single membrane-spanning domain. **Glycophorin A** (*or* αSGP) is the major sialoglycoprotein of the human erythrocyte membrane. On electrophoresis, it occurs in PAS 1 (dimer). It has a molecular mass of ~31 kDa, contains 60% carbohydrate by weight, and consists of a 131- residue polypeptide chain that spans the cell membrane; all the carbohydrate is attached to the N-terminal half, and is exposed to the exterior of the cell. Glycophorin A displays either M or N blood group activity (*see* **MNSs system**), depending on which of two alternative pairs of amino-acid residues occurs at positions 1 and 5 in the sequence. It binds influenza virus. **Glycophorin B** (*or* δSGP), also from the human erythrocyte membrane, is a minor sialoglycoprotein, similar to glycophorins A and E. It occurs in PAS 3, and has an amino-acid sequence identical in its N-terminal segment to that of glycophorin A$_N$. It carries the immunodeterminants for the N and Ss antigens (*see* **MNSs system**). **Glycophorin C** (*or* βSGP) occurs in PAS 2, glycoconnectin; it is a minor sialoglycoprotein that probably contains blood group Gerbich antigens, and receptors for *Plasmodium falciparum* (malaria) merozoites. **Glycophorin E** is similar to glycophorins A and B.

glycoprotein a conjugated protein in which the non-protein group is a carbohydrate; the sugar moiety occurs most commonly as oligosaccharide or fairly small polysaccharide, but occasionally as monosaccharide. Glycoproteins are widespread in animals and in the primary cell walls of plants. The term includes glycated protein, and is now considered also to include **proteoglycan**. *Compare* **mucoprotein**.

glycoprotein hormone any of **follicle-stimulating hormone**, **luteinizing hormone**, and **thyroid-stimulating hormone** produced by the pituitary, and **chorionic gonadotropin**, produced by the placenta, all of which are heterodimers of glycosylated α and β subunits. In any given species, the amino-acid sequences of the α subunits are identical, or nearly so, whereas the β subunits have greater differences in amino-acid sequence and confer hormonal specificity on the molecules.

glycoprotein Ib *abbr.*: GP-Ib; a surface membrane protein of platelets that participates in the formation of platelet plugs by binding to von Willebrand factor. It is a disulfide-linked heterodimer of α and β subunits.

glycoprotein IIb/IIIa *abbr.*: GpIIb/IIIa; a major glycoprotein complex of the platelet surface. GpIIb consists of a heavy chain (containing seven homologous repeats that form a beta propeller and including four Ca^{2+}-binding sites) linked by a disulfide bond to a membrane-anchored light chain. Both chains are derived by cleavage from a precursor (1039 amino acids). GpIIb (762 amino acids) contains five cysteine-rich regions and a transmembrane segment.

On activation, e.g. by thrombin, the complex binds fibrinogen, fibronectin, and **von Willebrand factor**. Numerous mutations in GpIIb and in GpIIIa cause Glanzmann thrombasthenia.

glycoprotein N-palmitoyltransferase EC 2.3.1.96; an enzyme that catalyses a reaction between palmitoyl-CoA and a glycoprotein to give *N*-palmitoylglycoprotein with release of CoA.

glycosaemia *a variant spelling (esp. Brit) of* glycosemia (*see* glycemia).

glycosamine *an older name for* **aminodeoxysugar**, now restricted to some trivial names. *Distinguish from* **glycosylamine**.

glycosaminoglycan any glycan (i.e. polysaccharide) containing a substantial proportion of aminomonosaccharide residues.

glycosaminopeptide *an alternative name for* **peptidoglycan**.

glycosan *an old term (not recommended) for* any anhydrosugar whose formation involves its reducing group. *Distinguish from* **glycan**.

glycose 1 *a less frequently used term for* any monosaccharide. **2** *an obsolete name for* **glucose** (*see also* **glyco+** (def. 2)). **3** *(sometimes)* an alternative name for* **corn syrup**.

glycosemia or *(esp. Brit)* **glycosaemia** *an alternative name for* **glycemia**.

glycosidase any hydrolase of subclass EC 3.2, enzymes that hydrolyse glycosidic linkages. Glycosidases are subdivided into those that hydrolyse *O*-glycosyl compounds (EC 3.2.1), *N*-glycosyl compounds (EC 3.2.2), or *S*-glycosyl compounds (EC 3.2.3). Some glycosidases can also transfer glycosyl residues to oligosaccharides, polysaccharides, or other acceptor alcohols. *See also* **(DNA) glycosylase**, **glucosidase**.

glycoside or *oside* **1** any compound in which a **glycosyl** group is substituted into a hydroxyl group, a thiol group, or a selenol group in another compound; such a substance may be named, respectively, *O*-glycoside, *S*-glycoside (or thioglycoside), or *Se*-glycoside (or selenoglycoside). A glycoside is either a mixed acetal (if derived from a cyclic form of an aldose) or a mixed ketal (if derived from a cyclic form of a ketose). By extension, any compound formed by attachment of a glycosyl group to a hydrocarbyl group has been termed a *C*-glycoside and that by attachment of a glycosyl group to an amino or substituted amino group has been termed also *N*-glycoside (or nitrogen glycoside), but these terms are misnomers and should not be used; the preferred terms are **C-glycosyl compound** and **glycosylamine**, respectively. *See also* **cardiac glycoside**, **nucleoside**. **2** *an obsolete term for* **glucoside**.

glycosidic 1 of, pertaining to, or being a **glycoside**. **2** *an alternative term for* **anomeric** (def. 2).

glycosidic bond the covalent bond between the anomeric carbon atom of a saccharide and some other group or molecule with which it forms a glycoside.

glycosiduronate 1 the anion of a **glycosiduronic acid**. **2** any salt or ester of a glycuronic acid.

glycosiduronic acid or *(formerly)* **glycuronide** a glycoside of any **uronic acid**.

glycosome a microbody-like organelle present in *Trypanosoma brucei* and containing nine enzymes involved in glycolysis, and glycerol kinase.

glycosphingolipid any compound containing residues of a sphingoid and at least one monosaccharide. Glycosphingolipids are subdivided into: (1) neutral glycosphingolipids, comprising monoglycosyl- and oligoglycosylsphingoids and monoglycosyl- and oligoglycosylceramides; and (2) acidic glycosphingolipids, comprising sialosylglycosylsphingolipids (**gangliosides**) and sulfoglycosylsphingolipids (formerly known as sulfatides).

glycostatic maintaining a constant level of sugar, e.g. of glucose or glycogen. *Compare* **glycotropic**.

glycosuria the presence of abnormally high concentrations of glucose and/or other sugars in the urine. *See also* **glucosuria**, **fructosuria**.

glycosyl any chemical group formed by detaching the anomeric, i.e. glycosidic, hydroxyl group from the cyclic form of a monosaccharide or monosaccharide derivative, or, (by extension) of a lower oligosaccharide or oligosaccharide derivative.

glycosylamine any compound formed by the replacement of the anomeric, i.e. glycosidic, hydroxyl group of a cyclic form of a monosaccharide or monosaccharide derivative by an amino group or a substituted amino group. (The term **nitrogen glycoside** (*or N*-glycoside) is a misnomer for glycosylamine and should not be used.) *Distinguish from* **glycosamine**.

glycosylamino acid any **glycoamino acid** in which the saccharide moiety is in glycosyl (*O*-, *N*- or *S*-) linkage to the amino-acid moiety. These are obtained by enzymic or chemical cleavage of glycoproteins, or by chemical or enzymic synthesis.

glycosylase *see* **(DNA) glycosylase**.

glycosylasparaginase *see* ***N*^{4}-(*β-N*-acetylglucosaminyl)-L-asparaginase**.

glycosylated hemoglobin *a misnomer for* **glycated hemoglobin**.

glycosylation substitution of one or more **glycosyl** groups into a chemical compound or group. *See also* **glycation**. —**glycosylated** *adj*.

***C*-glycosyl compound** any compound formed by the replacement of the anomeric, i.e. glycosidic, hydroxyl group of a cyclic form of a monosaccharide or monosaccharide derivative by an organyl group. The term *C*-glycoside is a misnomer for *C*-glycosyl compound and should not be used.

glycosyloxy any chemical group formed by detaching the hydrogen atom from the anomeric, i.e. glycosidic, hydroxyl group of the cyclic form of a monosaccharide or monosaccharide derivative.

glycosylphosphatidylinositol *see* **GPI anchor**.

glycosyltransferase *abbr*.: GT; any enzyme transferring a glycosyl group. Glycosyltransferases belong to subclass EC 2.4, which includes hexosyltransferases (EC 2.4.1.1 to 2.4.1.206); pentosyltransferases (EC 2.4.2.1 to 2.4.2.37); and enzymes transferring other glycosyl groups (EC 2.4.99.1 to 2.4.99.11).

glycotropic increasing the concentration of glucose in the blood. *Compare* **glycostatic**.

glycuronan or *(formerly)* **polyuronide** any glycan composed entirely of uronic-acid residues.

glycuronic acid an obsolete term for **uronic acid**.

glycuronide *the former term for* **glycosiduronic acid**.

glycyl the acyl group, NH_2–CH_2–CO–, derived from **glycine**.

glycyl chain *a name sometimes applied to the* **A chain** of any of those species of insulin whose amino-acid sequences were the first to be determined. It is so-called by virtue of the N-terminal glycine residue occurring in those instances.

glycylglycine a dipeptide useful for the preparation of buffers in the pH range around its pK_a (20 °C) of 8.40.

glycyrrhiza *also called* **liquorice** or *(esp. US)* **licorice**; a preparation obtained from the dried roots of the leguminous shrub, *Glycyrrhiza glabra*, with flavouring and medical uses.

glycyrrhizic acid 20*β*-carboxy-11-oxo-30-norolean-12-en-3*β*-yl-2-*O*-*β*-D-glucopyranuronosyl-*α*-D-glucopyranosiduronic acid; a glycosiduronic acid extracted from *glycyrrhiza (see* **glycyrrhiza** *)*; it has an intensely sweet taste.

glyoxal ethanedial; OHC–CHO; *also called* diformyl; the dialdehyde analogue of glycol and oxalic acid. It is sometimes used to replace formaldehyde in **denaturing agarose gel electrophoresis** of RNA. The basis of denaturation is the formation of an adduct with the imino groups of guanosine.

glyoxalase I EC 4.4.1.5; *recommended name*: lactoylglutathione lyase, *other name*: methylglyoxalase. An enzyme catalysing the reaction:

$$(R)\text{-}S\text{-D-lactoylglutathione} = \text{glutathione} + \text{methylglyoxal.}$$

glyoxalase II EC 3.1.2.6; *recommended name*: hydroxyacylglutathione hydrolase. An enzyme catalysing the reaction:

$$S\text{-(2-hydroxyacyl)glutathione} + H_2O =$$
$$\text{glutathione} + \text{a 2-hydroxy-acid anion.}$$

glyoxalate *a former name for* **glyoxylate**.

glyoxaline *an alternative name for* **imidazole**.

glyoxylate or *(formerly)* **glyoxalate 1** the anion, OHC–COO$^-$, derived from glyoxylic acid. **2** any salt or ester of glyoxylic acid.

glyoxylate cycle or **glyoxylate pathway** a modification of the **tricarboxylic-acid cycle** occurring in some plants and microorganisms, in which isocitrate, instead of being dehydrogenated, is cleaved as described under **isocitrate lyase** to glyoxylate and succinate. This serves to generate glyoxylate from the acetyl moiety of acetyl-CoA. The glyoxylate can then react with another molecule of acetyl-CoA to form malate, a reaction catalysed by **malate synthase**. These two reactions taken together allow a four-carbon molecule, malate, and hence carbohydrate, to be formed from acetyl-CoA, which, in turn, may be formed from fatty acids. In some plants the enzymes of the glyoxylate pathway are present in **glyoxysomes**.

glyoxysome a cellular organelle, of density 1.25 gcm^{-3}, found in the organs of fatty seedlings where fat is being converted to carbohydrate. It contains the enzymes of the **glyoxylate cycle** and most of the particulate beta-oxidation activity.

glyphosate *N*-(phosphonomethyl)glycine; a potent, systemically acting, nonselective herbicide, present as its isopropylamine salt in some commercial weedkillers. It functions by inhibiting 5-*enol*pyruvylshikimate-3-phosphate (EPSP) synthase. *See also* **shikimate**. *Proprietary names* include Roundup and Tumbleweed.

glypiation the attachment or process of attachment of glycosylphosphatidylinositol to proteins. *See* **GPI anchor**.

glypican any of a family of heparan sulfate proteoglycans that are linked to the cell surface via a GPI anchor. Their core proteins (60–70 kDa) have attachment sites for heparan sulfate chains within the 50 C-terminal residues and, N-terminal to these sites, a conserved globular cysteine-rich domain. They are involved in development and morphogenesis. At least six members occur in mammals and two in *Drosophila*. Mutation in glypican-3 causes the X chromosome-linked Simpson–Golabi–Dahmel syndrome type 1, characterized by overgrowth, dysmorphism, and anomalies of the skeleton, heart, gastrointestinal tract, and central nervous system.

GM or **G-M** *abbr. for* Geiger–Müller (e.g. in **GM counter** for Geiger–Müller counter).

GM-CSF *abbr. for* granulocyte/macrophage colony stimulating factor. *See* **colony stimulating factor**.

GMD *(in clinical chemistry)* abbr. for **glutamate dehydrogenase (NAD(P)$^+$)**.

GMP *abbr. for* guanosine monophosphate i.e. guanosine phosphate.

GMP synthase EC 6.3.5.2; *recommended name*: GMP synthase (glutamine-hydrolysing); *systematic name*: xanthosine-5-phosphate:L-glutamine amido-ligase (AMP-forming); *other names*: GMP synthetase (glutamine-hydrolysing); glutamine amidotransferase. An enzyme that catalyses a reaction between ATP, xanthosine 5′-phosphate, L-glutamine, and H_2O to form AMP, pyrophosphate, guanosine monophosphate, and L-glutamate. It is inhibited by azaserine and 6-diazo-5-oxonorleucine, in common with other glutamine amidotransferases. It has four motifs and a region conserved in amidotransferases (one motif).

GMT *(in clinical chemistry)* abbr. for **glutamyltransferase**.

GNAT *abbr. for* GCN5-related N-acetyltransferase. The GNAT superfamily is one of five families of histone acetyltransferase, a class of enzyme that catalyses the transfer of an acetyl group from acetyl-CoA to the lysine ε-amino groups on the N-terminal tails of histones. GNATs share various functional domains (e.g., a variable-length N-terminal region, an acetyltransferase domain, a coactivator interaction region, and a C-terminal bromodomain) and are involved in transcription and DNA repair, and hence play important roles in the regulation of cell growth and development.

gnotobiotic describing an organism, especially a higher animal, or its environment, that has been deliberately rendered free of microorganisms and other organisms, including parasites, commensals, and symbionts, that it would normally harbour, or that harbours only a few, known, such organisms.

GnRH *abbr. for* gonadotropin-releasing hormone.

GNTR a transcriptional repressor for the gluconate (*gnt*) operon in *Bacillus subtilis*. The *gnt* operon is induced by gluconic acid *δ*-lactone; this was the first of a family of *gnt* repressors with **helix-turn-helix** motifs.

GNU *abbr. for* Gnu is Not Unix; a humorous recursive title for a very large open source software project that is an important source of academic software and the GNU Public Software License (known as 'copyleft', in contrast to copyright). This is widely used for free software distribution on terms that are well tested particularly under US jurisdiction.

GO *abbr. for* gene ontology.

goblin a protein of avian erythrocyte plasma membranes, M_r 260

000. It undergoes phosphorylation, and is highly similar to **ankyrin**, of which it is the avian equivalent.

GOD 1 *(sometimes) abbr. for* glucose oxidase. **2** *abbr. for* generation of diversity.

godnose a name proposed for ascorbic acid by A. Szent-Györgyi, as an alternative to **ignose**, before he knew its structure.

GOGAT *abbr. for* glutamine (amide):2-oxoglutarate aminotransferase (oxidoreductase NADP+). It comprises two separate enzymes, named **glutamate synthase (ferredoxin)** and **glutamate–ammonia ligase**.

goitre *or (US)* **goiter** any swelling of the thyroid gland. Although common in areas where dietary iodine is insufficient, the condition is not related to thyroid state and patients with goitre may be euthyroid, hyperthyroid, or hypothyroid.

goitrogen any agent or factor causing goitre.

GOLD *abbr. for* genomes online database.

Gold book *informal name for* the IUPAC *Compendium of Chemical Terminology*, in recognition of the contribution of Victor Gold (1922–1985), who initiated work on the first edition.

Golgi apparatus *or* **Golgi body** *or* **Golgi complex** *or* **Golgi** a compound membranous cytoplasmic organelle of eukaryotic cells, consisting of flattened, ribosome-free vesicles arranged in a more or less regular stack. It differs from the endoplasmic reticulum in often having slightly thicker membranes, appearing in sections as a characteristic shallow semicircle so that the convex side (*cis* or entry face) abuts the endoplasmic reticulum, secretory granules emerging from the concave side (*trans* or exit face). In vertebrate cells there is usually one such organelle, while in invertebrates and plants, where they are known usually as **dictyosomes**, there may be several scattered in the cytoplasm. The Golgi apparatus processes proteins produced on the ribosomes of the rough endoplasmic reticulum; such processing includes modification of the core oligosaccharides of glycoproteins, and the sorting and packaging of proteins for transport to a variety of cellular locations, e.g. to storage granules, lysosomes, or secretory vesicles. The Golgi is a major site of synthesis of polysaccharides, including the **pectin** and hemicellulose of the plant cell wall and most of the glycosaminoglycans of the extracellular matrix in animals. Three different regions are recognized in terms of structure and function: *cis*, in the vicinity of the *cis* face, *trans*, in the vicinity of the *trans* face, and *medial*, lying between the *cis* and *trans* regions. While the **luminal proteins** of the endoplasmic reticulum leave it together with the proteins to be secreted, they are retrieved from the *cis* network and returned to the endoplasmic reticulum. Thus transport occurs in both directions. **Brefeldin A** blocks protein secretion through the Golgi but does not appear to block the return pathway. **Galactosyltransferase** is a marker enzyme for the *trans* membranes while *N*-**acetylglucosaminyltransferase** III serves this role for the medial membranes. [After Camillo Golgi (1843–1926), Italian anatomist, cytologist, neurologist, and pathologist, who, in 1898, was the first to describe it.]

Golgi mannosidase EC 3.2.1.114; *recommended name*: mannosyl-oligosaccharide 1,3-1,6-α-mannosidase; *systematic name*: 1,3-(1,6-)mannosyl-oligosaccharide α-D-mannohydrolase; *other name*: mannosidase II. An enzyme that catalyses the hydrolysis of the terminal 1,3- and 1,6-linked α-D-mannose residues in the mannosyloligosaccharide Man₅GlcNAc₃. It is a type II membrane protein, and a glycosyl hydrolase enzyme in all tissues, mostly in adrenal and thymus. It belongs to a family of glycosyl hydrolases.

golgin any protein from the Golgi apparatus that is identified as an antigen in autoimmune conditions. The normal function of golgins may be to maintain the structure of the apparatus or to participate in membrane trafficking. *See* **giantin**.

golgiosome *an alternative name for* **dictyosome**.

GoLoco motif a 20-residue protein sequence in some proteins that regulate G-protein signalling via G$_{α0}$ and G$_{αi}$. It is present in two proteins required for asymmetric spindle positioning in *Caenorhabditis elegans* embryos.

gonad any animal organ that produces gametes. In some species the gonads also produce hormones. —**gonadal** *adj.*

gonadoliberin *recommended name for* **gonadotropin-releasing hormone**.

gonadotropic hormone *or* **gonadotrophic hormone** *a former name for* **gonadotropin**.

gonadotropin *or (formerly)* **gonadotrophin** *or* **gonadotropic hormone** *or* **gonadotrophic hormone 1** any mammalian glycoprotein hormone that stimulates gonadal function. Strictly, gonadotropins comprise the adenohypophyseal hormones **follicle-stimulating hormone** (FSH) and **luteinizing** hormone (LH) together with the placental hormone **choriogonadotropin**, but the term is often extended to include also the adenohypophyseal hormone **prolactin** (which is not a glycoprotein but has gonadotropic activity). **2** any single substance having the combined activities of FSH and LH, such as choriogonadotropin or the gonadotropic hormones of cold-blooded vertebrates. **3** any impure hormonal preparation displaying the separate activities of FSH and LH, such as (human) menopausal gonadotropin (i.e. urogonadotropin) and **pregnant mare's serum gonadotropin**. —**gonadotropic** *adj.*

gonadotropin-releasing factor *abbr.*: GnRF; *another name for* **gonadotropin-releasing hormone**.

gonadotropin-releasing hormone (*abbr.*: GnRH) *or* **gonadoliberin** (recommended name) *or* **luteinizing hormone/follicle-stimulating hormone releasing factor** (*abbr.*: LH/FSH-RF) a family of decapeptide amides containing at least seven members and present in many vertebrate species from jawless fish (lamprey) to higher mammals. The structure is highly conserved; the sequence Glp-His-Trp-Ser-Tyr-Gly-Leu-Arg-Pro-Gly-NH₂ is common to all mammals; among other species, position 8 is most variable, positions 1, 2, 4, 9, and 10 being invariant. In mammals it is released by the hypothalamus, where it localizes with galanin, into the hypophyseal portal circulation in response to neural and/or chemical stimuli, and causes the release of FSH and LH by the anterior pituitary. It is also expressed in the placenta, and may act as a neurotransmitter, neuromodulator, or local hormone. It is derived from a larger precursor. For example, the rat precursor protein consists of 93 amino acids; residues 24–33 form GnRH and residues 37–92 form a peptide (GnRH-associated peptide, *abbr.*: GAP) that is poorly conserved and of unclear function. *See* **KAL**.

gonadotropin-releasing hormone receptor *abbr.*: GnRHR; any of a family of G-protein-associated membrane receptors that bind gonadotropin-releasing hormone (GnRH) and stimulate the phosphatidylinositol Ca²⁺ second messenger system. They are present in pituitary, ovary, testis, breast, and prostate.

gonane the parent 17-carbon tetracyclic hydrocarbon of steroids and bile acids.

gondoic acid *see* **eicosenoic acid**.

gonosome any motile animal germ cell, e.g. a spermatozoon.

Good buffer substance *or* **Good's buffer (substance)** any member of a collection of **zwitterionic buffer** substances selected or devised for suitability in experimental biological systems according to a number of predetermined criteria. These include: ease of preparation and purification; high solubility in water; low temperature coefficient of pK_a; and high transparency to light or ultraviolet light of wavelengths greater than 230 nm. The original lists included substances having the following trivial names (pK_a values at 20°C are in parentheses): **Mes** (6.15), **Ada** (6.6), **Pipes** (6.8), **Aces** (6.9), **Bes** (7.15), **Mops** (7.2), **Tes** (7.5), **Hepes** (7.55), **Hepps** (8.0), **Tricine** (8.15), **Bicine** (8.35), and **Taps** (8.4); **Ches** (9.3 at 25°C) and **Caps** (10.4) have similar properties. [After Norman Everett Good (1917–), US plant pathologist.]

gossypol a pigment found in cottonseed and containing two naphthalene rings with multiple –OH and other groups. It has antispermatogenic activity, and has been used (in China) as a male contraceptive. It is a potent inhibitor of contractions of strips of guinea-pig lung induced by platelet-activating factor or leukotriene, and it inhibits 5- and 12-lipoxygenase.

GOT *(in clinical chemistry) abbr. for* glutamate-oxaloacetate transaminase (AST preferred). *See* **transaminase**.

Goto-Kakizaki rat a strain of laboratory rats that have an inherited predisposition to develop diabetes. [Named after Y Goto and M Kakizaki, Tohoku University, Japan.]

gout a condition characterized by inflammation of the joints and cartilage due to hyperuricemia and consequent deposition of monosodium urate crystals in the joints, often within polymorphonuclear leukocytes. The hyperuricemia results from overproduction of uric acid, which may be due to elevated activity of ATP phosphoribosyltransferase (PRPP synthetase; EC 2.4.2.17), an abnormal **amidophosphoribosyltransferase**, or deficiency of hypoxanthine phosphoribosyltransferase (EC 2.4.2.8). **Allopurinol** is useful in treatment.

gp *abbr. for* glycoprotein used in conjunction with the relative molecular mass (e.g. gp115).

gp32 an envelope glycoprotein from HIV-related viruses, including **SIV** types. *See* **HIV protein**.

gp96 *abbr. for* glycoprotein of 96 kDa; *other name*: glucose-regulated protein of 94 kDa (*abbr.*: Grp94); an abundant molecular chaperone glycoprotein endoplasmic reticulum, and an orthologue of heat shock protein 90 (Hsp90).

G1P *or* **G-1-P** *abbr. for* glucose 1-phosphate.

G6P *or* **G-6-P** *abbr. for* glucose 6-phosphate.

G-patch a conserved protein sequence of ≈48 residues that contains six highly conserved glycine residues. It is present in some eukaryotic RNA-processing proteins and certain retroviral polyproteins. It usually occurs singly, and is often associated with other RNA-binding sequences.

GPCR *abbr. for* G-protein-coupled receptor.

GPD (*in clinical chemistry*) *abbr. for* glucose-6-phosphate dehydrogenase.

G$_0$ phase G-'zero' phase; a state of withdrawal from the eukaryotic **cell-division cycle**. It occurs if no stimulus for cell division occurs during G$_1$ phase, leaving the cell in a quiescent G$_1$ phase. It is often seen in differentiated cells.

G$_1$ phase Gap 1 phase; the phase of the eukaryotic **cell-division cycle** between the end of **cytokinesis** and the start of DNA synthesis, i.e. the gap between M phase and S phase.

G$_2$ phase Gap 2 phase; the phase of the eukaryotic **cell-division cycle**, between the end of DNA synthesis and the beginning of mitosis, i.e. the gap between S phase and M phase.

GPI (*in clinical chemistry*) *abbr. for* glucose-6-phosphate isomerase.

GPI anchor *abbr. for* glycosylphosphatidylinositol anchor; a molecular mechanism for attaching a membrane protein to the lipid bilayer of a cell membrane. Structurally it consists of a molecule of phosphatidylinositol (*abbr.*: PI) to which is linked, via the C-6 hydroxyl of the inositol, a carbohydrate chain. This chain is in turn linked to the protein through an ethanolamine phosphate moiety, the amino group of which is in amide linkage with the C-terminal carboxyl of the protein chain, the phosphate group being esterified to the C-6 hydroxyl of the terminal mannose of the core carbohydrate chain. The carbohydrate chain varies in composition between species, but invariably contains a core tetrasaccharide, the first component of which is a glucosamine residue in glycosidic linkage with the inositol C-6 hydroxyl. The tetrasaccharide has the structure Man(α1-2)Man(α1-6)Man(α1-4)GlcNH$_2$. Variation frequently occurs in the presence of additional α-mannose residues on the anchors of *Trypanosoma cruzi*, yeast, slime mould, and mammalian proteins. Anchors of *Trypanosoma brucei* have more elaborate carbohydrate side-chains containing galactose, glucosamine, and sialic-acid residues. In the case of metazoan eukaryotes, additional ethanolamine phosphate substituents may occur on mannose residues, and the inositol is often palmitoylated. Biosynthesis begins with attachment of the glucosamine moiety to phosphatidylinositol, followed by sequential addition of other sugars; the GPI moiety is attached to protein during processing in the endoplasmic reticulum, the signal being a hydrophobic sequence at the C terminus, which is removed during the attachment process. GPI-anchored proteins are released from membranes by PI-specific phospholipases. While the general function of the GPI anchor is to anchor the protein to the membrane, wider functions may include: location of the protein at specific sites, particularly the apical membrane of polarized epithelial cells; promotion of sequestration into membrane microdomains, these proteins being excluded from clathrin-associated regions but found in **caveolae**; anchoring of proteins of the glycocalyx, and cell signalling and cell recognition proteins. In addition to protein-linked GPIs, *Leishmania* spp. also synthesize two classes of free GPIs, the polydisperse **lipophosphoglycans** (LPGs) and the low molecular mass **glycoinosi-**

tolphospholipids (GIPLs). LPGs are restricted in occurrence (e.g. promastigote stage) but GIPLs are relatively abundant on the cell surface of promastigote and amastigote stages. GIPLs are also found on the surface of many other trypanosomatid parasites, e.g. *T. cruzi*, *Leptomonas*, *Herpetomonas*, and *Phytomonas*.

gp22-phox *see* phagocyte NADPH oxidase.

gp91-phox *see* phagocyte NADPH oxidase.

gp 43 processing protease *see* oviductin.

G protein guanine nucleotide binding protein; any heterotrimeric GTP-binding and hydrolysing protein belonging to a superfamily of GTPases that includes the monomeric proteins EF-Tu (*see* **EF-T**) and p21^{ras} (*see* **RAS**). G-proteins participate in activation or inhibition of **adenylate cyclase** and other **second messenger** systems, e.g. **phosphatidylinositol cycle**. They are associated with membrane receptors that contain seven transmembrane helices and bind hormones, growth factors, and cytokines. The system is unique to eukaryotes. G-proteins consist of three subunits; the α subunit (39–46 kDa) contains the guanine nucleotide binding site and possesses GTPase activity; the β (37 kDa) and γ (8 kDa) subunits are tightly associated and function as a βgamma; heterodimer. There are 23 types (including some splicing isoforms) of α subunits, 6 of β, and 11 of γ. The classes of G-protein and subunits are subscripted: thus the α subunit of G-protein S (which activates **adenylate cyclase**) is G$_{s\alpha}$; other G-proteins include G$_i$, which differs from G$_S$ structurally (different type of α subunit) and inhibits **adenylate cyclase**; G$_o$, a G-protein detected by cDNA probes, of wide tissue distribution in mammals but of unclear role; G$_p$, a G-protein that activates phospholipase C (*see* **phospholipase**), release of Ca^{2+} and activation via **calmodulin** and **protein kinase C** (*see also* **phosphatidylinositol cycle**). Binding of G$_{\alpha\beta\gamma}$ to liganded receptor occurs when GDP bound to the G-protein ($\alpha\beta\gamma$-GDP) exchanges for GTP, which causes the heterotrimer to dissociate into α-GTP and βgamma; units, either of which may interact with an effector. The GTPase activity of α yields α-GDP, which inactivates it. α-GDP has a high affinity for βgamma; thus reforming $\alpha\beta\gamma$-GDP, which becomes available for association with an activated receptor. G$_T$, *see* **transducin**. In some tissues heterotrimeric G-proteins may be present in tight complexes that include: membrane receptor, adenylate cyclase, cAMP-dependent protein kinase, calcium channel, and a phosphoprotein phosphatase.

G protein-coupled receptor (*abbr.*: GPCR) *or* **G protein-associated receptor** *or* **G protein-linked receptor** member of a diverse family of cell-surface proteins that has adapted a common heptahelical framework to mediate a range of physiological processes (e.g., vision, olfaction, chemotaxis). Their functional diversity is achieved via interactions with a variety of ligands (e.g., peptides, glycoproteins, small molecule messenger molecules, vitamin derivatives), the effects of which stimulate a range of intracellular pathways through coupling to different G proteins. GPCRs are medically very important – they are involved in several major pathophysiological conditions (e.g., cardiovascular diesase, cancer, obesity), and are the targets of the majority of prescription drugs. *See* **G protein**.

gpt *symbol for* a prokaryotic gene for xanthine–guanine phosphoribosyltransferase. It is used as a dominant selectable marker in transfection of mammalian cells with exogenous DNA. Its expression enables mammalian cells to grow in the presence of **mycophenolic acid** in a medium supplemented with xanthine.

GPT (*in clinical chemistry*) *abbr. for* glutamate–pyruvate transaminase (EC 2.6.1.2) (ALT preferred).

G-quartet *or* **G-tetrad** a cyclic H-bonded array of guanines that can occur in both RNA and DNA to form a quadruple helix structure. Sequences that form such structures are found in telomeric sites, the switch regions of immunoglobulin heavy-chain genes, and elsewhere. *See* **KEM1**.

Gra *symbol for* a residue of **glyceraldehyde**.

Graafian follicle a spherical vesicle occurring in the mature mammalian ovary and in which the ovum develops. When the ovum matures it is discharged after rupture of the follicle, which is then transformed into a **corpus luteum**. [After Reijnier (Regnier) de Graaf (1641–73), Dutch physician and anatomist.]

grade 1 a degree or position in a scale, as of quality, size, etc., e.g.

analytical grade, commercial grade. **2** a unit of angle equal to one-hundredth of a right angle, i.e. 0.9°.

gradient the change in the value of a specified (chemical or physical) property or value per unit distance in a specified direction.

gradient elution a chromatographic elution procedure in which the concentration of a component of the eluent is increased continuously. The gradient may be linear, concave – i.e. increasing slowly at first and more rapidly as elution proceeds – or convex – i.e. increasing rapidly at first and more slowly as elution proceeds. More complex or discontinuous (stepped) elution gradients are sometimes used.

gradient former or **gradient mixer** any device for producing solutions of varying, but defined, composition containing one or more components.

gradient gel electrophoresis or **gradient acrylamide gel electrophoresis** or **polyacrylamide gradient gel electrophoresis** (*abbr.*: PAGE) a type of **zone electrophoresis** in which the supporting medium is a pore-gradient gel, i.e. one where an increasing concentration of polyacrylamide in the direction of electrophoretic mobility decreases the sizes of the pores in the gel in that direction to the point where molecules of more than a certain size are prevented from moving further in the electric field. The concentration gradient may be either discontinuous, i.e. stepped, or continuous. Advantages of the technique over electrophoresis in uniform gels (*see* **gel electrophoresis**) are sharper zones and an ability to resolve, on a single gel, mixtures of macromolecules of a wider range of sizes.

gradient layer or **gradient packing** a chromatographic layer, or column packing, incorporating a continuous change of some property that affects separations, e.g. a pH gradient or a pore gradient.

graft 1 any part of an organism, animal or plant, that is inserted into and unites with a larger part of another or the same organism. **2** to transplant (tissue) to another organism or to another site on the same organism.

graft versus host disease a condition where immunocompetent donor cells recognize and react against the recipient's tissues, because the recipient is immunosuppressed or cannot recognize the different genetic constitution of the donor cells.

Graham's law of diffusion a law stating that at equal temperatures and pressures the rates of diffusion of gases are inversely proportional to the square roots of their densities and hence to the square roots of their relative molecular masses (M_rs). Thus for two gases A and B,

$$J_A/J_B = \bar{v}_A/\bar{v}_B = M_{rB}^{1/2}/M_{rA}^{1/2},$$

where J_A and J_B, $\bar{v}_A$ and $\bar{v}_B$, and M_{rA} and M_{rB} are the fluxes, mean velocities, and relative molecular masses of A and B, respectively. [After Thomas Graham (1805–69), Scottish chemist.]

grain 1 any small, hard particle, e.g. a grain of sand. **2** any simple single-seeded indehiscent fruit in which the seed wall fuses with the carpel wall during development. Technically called a **caryopsis**, it is typical of the cereals and other grasses. **3** *(in photography)* any of the many particles of silver halide in a photographic emulsion. The resolution possible in the image is inversely proportional to the grain size in the emulsion. **4** any of the dried insect bodies present in dyes such as cochineal.

gram *symbol*: g; a unit of mass equal to a thousandth part of the **kilogram** in both the Système International (SI) and the metric system (cgs or mks) of units.

+gram *comb. form* indicating a drawing, something written or recorded; the record produced by an instrument, e.g. spectrogram; the visible result of a procedure, e.g. autoradiogram, chromatogram.

gram-atom or **gram-atomic weight** *abbr.*: g-atom; the atomic weight of an element expressed in grams; it has now been replaced by amount of substance, measured in **moles**.

gram-equivalent or *(formerly)* **gram-equivalent weight** *abbr.*: g-equiv; an **equivalent** (def. 3) expressed in grams.

gramicidin any of a group of linear polypeptide antibiotics that are active against Gram-positive microorganisms and are isolated from cultures of *Bacillus brevis*. The commercial material is a mixture of gramicidins A, B, C, and D, all of which consist of chains of 15 amino-acid residues in a pattern of alternating D- and L-forms; all

have a formyl group at the N terminus and an ethanolamine residue at the C terminus. Gramicidins increase the permeability of biological membranes to protons and alkali-metal cations by forming transient dimeric ionophoric channels through the membrane. Gramicidins are classified in the TC system under number 1.D.1. *See also* **gramicidin S**.

gramicidin S a homodetic cyclic decapeptide antibiotic, *cyclo*(-Val-Orn-Leu-DPhe-Pro-Val-Orn-Leu-DPhe-Pro-), produced by a strain of *Bacillus brevis* and active against Gram-positive bacteria. Gramicidin S is more closely related to **tyrocidins** in its biological and chemical properties than to true **gramicidins**. Its biosynthesis occurs by a mechanism that is not dependent on the presence of ribosomes or mRNA. It was the first peptide antibiotic to be chemically synthesized, and is so named because it was first isolated in Soviet Russia, where it was widely used.

gram-molecule or **gram-molecular weight** *symbol*: g-mol; the molecular weight of a compound (or element) expressed in grams; it has now been replaced by the **mole**.

Gram-negative describing a bacterial cell that, during **Gram staining**, does not retain the basic dye in its cell wall. The walls of Gram-negative organisms are thin, about 10 nm across, but complex, having a lipopolysaccharide layer outside the peptidoglycan layer; they are also resistant to digestion by lysozyme. *Compare* **Gram-positive**.

Gram-positive describing a bacterial cell that, during **Gram staining**, retains the basic dye in its cell wall. The walls of Gram-positive organisms are relatively thick, 15–80 nm across, and consist of a network of peptidoglycan; they are readily digested by lysozyme. *Compare* **Gram-negative**.

Gram staining or **Gram stain** a staining procedure used for bacteria in which the cells are first heat-fixed and stained with a basic dye, e.g. Crystal Violet, then treated with I_2–KI solution to fix the stain, washed with ethanol or acetone, and finally counterstained with another dye, e.g. carbolfuchsin or safranin. Cells that retain the basic dye after this procedure are called **Gram-positive**; those that do not are called **Gram-negative**. [After Hans Christian Joachim Gram (1853–1938), Danish physician and bacteriologist, who described the method in 1884.]

grancalcin a Ca^{2+}-binding protein containing four **EF hands**, and abundant in neutrophils and monocytes. It shares strong similarity with **sorcin**. It is a homodimer.

β granin *see* **chromogranin**.

granular or **granulose** or **granulate** being composed of, or having the character of small particles or **grains** (def. 1).

granular endoplasmic reticulum or **rough endoplasmic reticulum** *see* **endoplasmic reticulum**.

granulate 1 to make (something) into **grains** (def. 1). **2** (of a wound, ulcer, etc.) to form **granulation tissue**. **3** *another word for* **granular**.

granulation 1 the act or process of subdividing a solid into small, usually rounded **grains** (def. 1). **2** the formation of one or more small, rounded lumps (grains (def. 1)) in a tissue. **3** the development of **granulation tissue**.

granulation tissue the mass of young connective tissue and blood vessels formed on the surface of a wound or ulcer in the early stages of healing.

granule any small **grain** (def. 1), especially any small intracellular particle containing stored material that can be seen by (electron) microscopy, e.g. insulin-containing β granules in the B cells of the islets of Langerhans, starch granules in potato tubers.

granulocyte a category of white blood cells distinguished by the presence of conspicuous cytoplasmic granules. Granulocytes include **neutrophils**, **basophils**, and **eosinophils**.

granulocyte-colony stimulating factor *see* **colony stimulating factor**.

granulose 1 *another word for* **granular**. **2** *an obsolete name for* **starch** or amylose.

granum (*pl.* **grana**) a stack of disclike **thylakoids** resembling a pile of coins. Grana occur in **chloroplasts**, especially in higher plants.

granzyme any of three related serine proteases (EC 3.4.21) that occur (with perforin) in cytoplasmic vesicles of killer T-lymphocytes and induce death in target cells.

granzyme A enzyme of sub-subclass EC 3.4.21; *other names*: cytotoxic T-lymphocyte protease 1; Hanukkah factor (HF); granzyme

1; CTL tryptase; fragmentin 1; trypsin-like serine protease. An enzyme necessary for target-cell lysis in cell-mediated immune responses; it cleaves on the carbonyl side of Lys or Arg. It occurs in cytoplasmic granules, and may be involved in **apoptosis**. The molecule is a disulfide-linked homodimer.

grape sugar *a former name for* D-glucose. *See* **glucose**.

graph 1 a diagram showing the relation between two variable quantities, or the relation between a mathematical function and different values of one of the variables in the function, usually plotted between axes at right angles to each other. **2** to plot or draw a graph.

+graph *comb. form* **1** denoting (something) drawn, recorded, or written (in the way specified), e.g. autoradiograph, photograph. **2** denoting an instrument or device that draws or records (observations), e.g. spectrograph. **3** denoting the act or process of producing such a drawing or record. *Compare* **+gram**. —**+graphic** *or* **+graphical** *adj. comb. form.*

graphics *(functioning as sing.)* **1** the use of diagrams in calculation. **2** the production of diagrams, representations of molecular structures, etc., often by computer.

+graphy *comb. form* denoting a method, process, or technique that produces a drawing or record, e.g. chromatography. *See* **+graph**.

grating *see* **diffraction grating**.

gratuitous inducer any compound that causes the induction of gene expression and is not itself metabolized by the enzymes so induced. A well-known example is the induction of the *lac* operon by isopropylthiogalactoside (IPTG).

Graves' disease *or* **Basedow's disease** a disease characterized by diffuse goitre, thyrotoxicosis, and exophthalmia. It is an autoimmune condition associated with serum antibodies against the thyrotropin receptor. [After Robert James Graves (1796–1853), Irish physician.]

Graves' disease carrier protein see **GDC**.

gravimetric analysis quantitative analysis of the chemical composition of substances or materials based on the separation of each component of interest and weighing either the purified compound or a derivative of it.

gravimetry measurement by weighing. —**gravimetric** *or* **gravimetrical** *adj.*

gravitation the attractive force that bodies exert on one another as a result of their mass; any process or result caused by such an attractive force. —**gravitational** *adj.*

gravitational constant *symbol*: G; the proportionality constant relating the attractive force, F, between two bodies of mass m_1 and m_2, and the distance between them, s, in Newton's law of gravitation:

$$G = Fs/m_1m_2 = 6.67259(85) \times 10^{-11} \text{ m}^3\text{kg}^{-1}\text{s}^{-2}.$$

gravitational field any **field** (def. 1) associated with a mass by which a force is exerted on a second mass placed in this field.

gravitational interaction any interaction between bodies or particles that results from their masses.

gravitational mass *see* **mass**.

gravity 1 the attractive force that moves or tends to move bodies towards the centre of any celestial body such as the earth or the moon. **2** the attribute of having weight. **3 specific gravity**.

gray *symbol*: Gy; the SI derived unit of **absorbed dose** of radiation, equal to one joule per kilogram of living tissue;

$$1 \text{ Gy} = 1 \text{ Jkg}^{-1} = 1 \text{ m}^2\text{s}^{-2}.$$

Compare **sievert**. *See also* **rad**. [After Louis Harold Gray (1905–65), British physicist and radiobiologist.]

Grb2 growth factor receptor-bound protein 2; *also called* **Ash protein**; a cytosolic adaptor protein, related to sem-5 (*see* **sem-5**), that participates in downstream signalling after activation of epidermal growth factor and platelet-derived growth factor receptors; it can also be recruited by G-protein-coupled receptors through tyrosine phosphorylation of Shc (*see* **SHC**). Grb2 binds to activated Shc and the complex then recruits Sos (*see* **sos**). The Shc–Grb2–Sos complex binds to the plasma membrane where Sos catalyses **Ras** guanine nucleotide exchange, leading to MAP kinase activation. Grb-10 is activated by RET.

GRE *abbr. for* glucocorticoid response element.

green fluorescent protein *abbr.*: GFP; a protein isolated from the jellyfish, *Aequorea victoria,* that fluoresces when exposed to excitation light (there are several spectral variants, including yellow, cyan, blue, and red). GFP has important applications in research: the gene can be expressed in other living organisms, making it possible to create gene fusions and hence to visualize the localization of specific gene products inside living cells. GFP comprises 238 amino acids arranged in a helical basket motif, within which three amino acids form an annular structure that mediates its fluorescent properties. *See also* **enhanced green fluorescent protein**.

GRF *abbr. (sometimes) for* growth hormone releasing factor (i.e. **GHRH**).

Gri *symbol for* glyceric acid *or* glycerate.

grid 1 a framework made up of fine wires arranged in two sets at right angles to each other, especially a device for supporting the specimen in an electron microscope. **2** an electrode controlling the current flowing through a thermionic valve.

Grid the basis for the rapidly developing field of grid computing. It enables scalable virtual organizations (**VO**s), has some parallels with the World Wide Web, but differs in that the members of the VO can not only access data but also run applications on computers in the grid. The Grid has its own **protocol**s and **API**s and has enormous implications for bioinformatics.

Grim a protein of *Drosophila* that functions, like mammalian **Diablo**, in promoting apoptosis. It binds **IAP**, releasing any associated caspase, and thereby is essential for apoptosis.

griseofulvin [2S]-*trans*-7-chloro-2′,4,6-trimethoxy-6′-methylspiro[benzofuran-2(3H),1′-(2)cyclohexene]-3,4′-dione; a chlorine-containing **polyketide** metabolite of *Penicillium griseofulvum* and *P. janczewskii*. It has antifungal properties and is used therapeutically, e.g. for treatment of ringworm. It inhibits spindle formation in mitosis, microtubule function, and possibly nucleic-acid synthesis.

GRK *abbr. for* G-protein coupled receptor kinase; a widely expressed protein kinase that can phosphorylate many G-protein-coupled receptors. There are three isoforms: GRK2 and GRK3 contain an N-terminal RGS domain (*see* **RGS protein**), a central protein kinase domain, and a C-terminal pleckstrin homology (PH) domain. GRK1 lacks the PH domain.

Grn *symbol for* glycerone.

gRNA *abbr. for* guide RNA; small RNA molecules, 60–80 nucleotides long, that may be involved in editing of RNA by supplying information concerning nucleotide deletion or addition. *See also* **editosome**.

gro the symbol for any of a group of related genes found in bacteria (especially *Escherichia coli*), mutations in which affect the growth of phages.

Gro *symbol for* glycerol.

groEL symbol for any of a family of bacterial genes encoding **chaperonin**s. Eukaryotic homologs are the hsp60 and Cpn60 proteins respectively of mitochondria and chloroplasts. Example of a *groEL* protein from *Escherichia coli*: oligomer of 14 subunits composed of two stacked rings of seven subunits; it has ATPase activity.

groES symbol for any of a family of bacterial genes encoding cochaperonins that functions with *groEL*. Eukaryotic homologs are the hsp 10 and Cpn 10 proteins respectively of mitochondria and chloroplasts, that cycle on and off the hsp60 or Cpn60 cylinder in response to its ATPase activity. Example of a *groES* protein product from *Escherichia coli*: seven subunits arranged in a ring. *See* **chaperonin**.

Grotthus and Draper law a law stating that only those radiations absorbed by a system are effective in producing chemical change in the system.

groundplasm *or* **cytoplasmic matrix** the sub-electron-microscopic material of the **cytoplasm**, excluding the cell organelles, cell particles, and other microscopically resolvable structures. It contains the cytoplasmic enzymes and metabolites together with components of the **cytoskeleton**.

ground state the state of lowest energy of any system.

ground substance *an alternative name for* the **matrix** (def. 2) of connective tissue.

ground tissue *a collective term for* plant tissue other than the epidermis or peridermis and vascular tissue.

group *(in chemistry)* or *(formerly)* **radical 1** any defined continuously linked collection of atoms within a molecular entity. In organic chemical nomenclature the meaning within this general sense is restricted to any covalently bonded collection of atoms that behaves as a unit in chemical reactions; **functional group**, **substituent group**. **2** any of the columns of chemical elements in the **periodic table**. **3** a **blood group**.

group p*K* the p*K* of a specific group in a molecule with two or more acid groups. In general the term applies only to a particular ionic form of the molecule, because the p*K* of a group changes when any other group in the molecule changes its ionization state, and thus changes as the substance is titrated. *Compare* **molecular p*K*, titration p*K*.**

group potential the tendency of a **group** (def. 1) to escape from a molecular entity.

group-transfer potential the Gibbs energy change (free energy change, ΔG) occurring when a **group** (def. 1) is transferred from a donor compound to the nucleophile water, i.e. when the group is removed by hydrolysis of the donor. It reflects the thermodynamic tendency for the group to be transferred.

group-transfer reaction any chemical reaction in which a **group** (def. 1) is transferred from a donor compound to an acceptor compound. Such reactions may be catalysed by a **transferase** enzyme of EC class 2.

group translocation any process in which a substance to be transported across a biological membrane undergoes covalent modification concomitant with crossing the membrane. The best-known example is the group of bacterial phosphotransferases, which catalyse phosphorylation of some mono- and disaccharides, and of sugar alcohols during their import into cells.

growth curve the change, as a function of time, in the number of cells of a growing culture or in some measure of the size of an organism.

growth factor 1 any specific substance that must be present in a culture medium for multiplication of the cultured cells to take place. **2** a specific substance that is a necessary constituent of the diet for the normal growth of a particular animal, or is required in the soil for the normal growth of a particular plant.

growth factor receptor-bound protein 2 *see* Grb2.

growth hormone *another name for* **somatotropin**.

growth hormone inhibitory factor *abbr.*: GHIF *or* GIF; *a former name for* **somatostatin**.

growth hormone release-inhibiting hormone *abbr.*: GH-RIH; *a former name for* **somatostatin**.

growth hormone releasing factor *abbr.*: GHRF *or* GRF; *another name for* **GHRH**.

growth hormone releasing hormone *see* GHRH.

growth hormone-releasing peptide *abbr*.: GHRp; a synthetic hexapeptide that stimulates release of somatotropin by binding to a highly conserved G-protein-coupled receptor in the pituitary for which the physiological ligand is unidentified.

Grp94 *an alternative term for* **gp96**.

GRP *see* **Grb2**.

GRP 75 *see* **mortalin**.

GRS *(in clinical chemistry) abbr. for* β-glucuronidase.

Grunstein–Hogness procedure *an alternative name for* **colony hybridization**.

GSH *abbr. for* glutathione.

GSK-3 *abbr. for* **glycogen synthase kinase-3**.

GSS *abbr. for* genome survey sequence.

GSSG *abbr. for* oxidized glutathione (*see* **glutathione**).

GST *abbr. for* glutathione S-transferase (*see* **glutathione transferase**).

GST fusion protein any protein that has been genetically engineered to have glutathione-*S*-transferase (GST) fused at its N-ter-

minal end. A range of **pGEX vectors** are designed to facilitate such expression in *E. coli*. The fusion protein can be purified by affinity chromatography on columns of immobilized glutathione.

GT *abbr. for* glycosyltransferase.

GT–AG rule a rule stating that in RNA there is a requirement for a GT at the 5′ side of a splice junction and an AG at the 3′ side.

GTBP *abbr. for* G/T binding protein; a human homologue (1292 amino acids) of *E. coli* **MutS** protein that is essential for DNA G/T mismatch recognition and binding. It forms a heterodimer with MutS homologue 2 (hMSH2). Mutations in GTBP are present in many hereditary, and some sporadic, colorectal cancers.

GTD *(in clinical chemistry) abbr. for* glutathione reductase.

G-tetrad *an alternative name for* **G-quartet**.

GTF *abbr. for* glucose-tolerance factor.

GTP *abbr. for* guanosine triphosphate.

GTPase any enzymic activity that hydrolyses GTP to GDP and orthophosphate. It is associated with **G-proteins**, **EF-Ts**, and **HIV protein**.

GTP-binding protein any of various proteins that bind GTP as a regulatory molecule; this binding is frequently associated with a **GTPase** activity. GTP-binding proteins include elongation and initiation factors (e.g., **EF-Tu**), G proteins, G25K GTP- binding protein, **Ras proteins**, and yeast CDC42 (*see* **CDC markers**). Binding of GTP normally activates the protein, an effect that is reversed by hydrolysis of the GTP to GDP.

GTP[S] *or* **GTPγS** *abbr. for* guanosine 5′-γ-thiotriphosphate.

GTR1 a representative 12-transmembrane domain transport protein that effects the facilitated diffusion of glucose in certain mammalian tissues.

GTT *abbr. for* glucose tolerance test.

Gu *symbol for* guanidine.

Gua *symbol for* a residue of the purine base guanine (alternative to G).

GUA a codon in mRNA for L-valine.

guanidine *symbol*: Gu; the strongly basic substance, iminourea; aminomethanamidine; $(H_2N)_2C=NH$.

guanidinium the cation, $[C(NH_2)_3]^+$, derived from guanidine by uptake of a hydron.

guanidinium isothiocyanate a powerful **chaotrope** used as a protein denaturant in the extraction of RNA from biological materials. Both the guanidinium cation and the isothiocyanate anion are strong chaotropic agents.

guanidino the amidinoimino group, –HN–C(=NH)–NH₂, derived from guanidine.

guanidinoacetate *N*-methyltransferase EC 2.1.1.2; an enzyme of the pathway for *de novo* synthesis of **creatine**, by reaction of *S*-adenosyl-L-methionine with guanidinoacetate releasing *S*-adeno-syl-L-homocysteine.

guanine *symbol*: G *or* Gua; 2-amino-6-hydroxypurine; 2-amino-1,7-dihydro-6*H*-purin-6-one; a purine that is one of the five main bases found in nucleic acids and a component of a number of phosphorylated guanosine derivatives whose metabolic or regulatory functions are important.

guanine nucleotide binding protein *see* **GTP-binding protein**.

guanine nucleotide releasing protein *abbr.*: GNRP; *or* **guanine nucleotide releasing factor**, *abbr.*: GRF; a protein that binds to a Ras protein (*see* **Ras family of proteins**) or Ras-related **GTP-binding protein**, and activates it by stimulating it to release its bound GDP and to bind GTP in its place. The C-terminal 310 amino acids share similarity with Cdc25 (*see* **cdc gene**) and Sos, the product of **sos**.

guanosine *symbol*: G *or* Guo; guanine riboside; 9-β-D-ribofura-nosylguanine; a **nucleoside** with a wide species distribution.

guanosine

guanosine 2′,3′-(cyclic)phosphate see **guanosine phosphate**.
guanosine 3′,5′-(cyclic)phosphate or guanosine 3′,5′-cyclophosphate see **guanosine 3′,5′-phosphate**.

guanosine 5′-diphosphate symbol: Guo5′PP or ppG; the recommended name for guanosine diphosphate (abbr.: GDP), 5′-diphosphoguanosine, 5′-guanylyl phosphate, guanosine 5′-(trihydrogen diphosphate), a universally distributed nucleotide that occurs in both the free state and as a component of **guanosinediphosphosugars**. It is a metabolic precursor of guanosine 5′-triphosphate, GTP, and is a substrate, a product, or a regulatory agent in many metabolic systems.
guanosine 3′-diphosphate 5′-diphosphate symbol.: ppGpp; see **relA**, **spoT**.
guanosinediphosphomannose or **guanosine(5′)diphospho(1)-α-**D-mannose symbol: GuoPPMan; the alternative recommended names for guanosine diphosphate mannose (abbr.: GDPM or GDP-mannose); guanosine 5′-(α-D-mannopyranosyl diphosphate); α-D-mannopyranosyl 5′-guanylyl phosphate; the most important **guanosinediphosphosugar**. It is synthesized from α-D-mannose 1-phosphate and guanosine 5′-triphosphate by the action of **mannose-1-phosphate guanylyltransferase** (EC 2.7.7.13). It is an intermediate in the synthesis of mannose-containing homo- and heteroglycans in plants. In animal cells GDPmannose is an intermediate in the incorporation of residues of both mannose and fucose into N-linked oligosaccharides of glycoproteins, its conversion to GDP-L-fucose having first taken place in the case of fucose.
guanosinediphosphosugar or **guanosine diphosphate sugar** abbr.: GDPsugar; any of several **nucleosidediphosphosugars** in which the distal phosphoric residue of guanosine 5′-diphosphate is in glycosidic linkage with a monosaccharide (or a derivative of a monosaccharide). Included are GDP-L-fucose, GDP-D-glucose, GDP-L-guluronic acid, GDP-D-mannose (see **guanosinediphosphomannose**), and GDP-D-mannuronic acid.
guanosine 5′-[β,γ-imido]triphosphate symbol: Guo-PP[NH]P or p[NH]ppG; the recommended name for β,γ-imidoguanosine 5′-triphosphate (abbr.: GTP[β, γ-NH]), 5′-guanylyl imidodiphosphate (abbr.: GMP-PNP), 5′-guanylyl iminodiphosphonate, guanosine

($5′→O^3$)-1,2-μ-imidotriphosphate, a synthetic nonhydrolysable analogue of guanosine triphosphate, GTP, in which the oxy group between the latter's intermediate and terminal phosphorus atoms is replaced by an imido group. It stimulates G-proteins and systems activated thereby.

guanosine 5′-[β,γ-methylene]triphosphate symbol: GuoPP-[CH$_2$]P or p[CH$_2$]ppG; the recommended name for β,γ-methyleneguanosine 5′-triphosphate, (abbr.: GTP[β,γ-CH$_2$]), 5′-guanylyl methylenediphosphonate (abbr.: GMP-PCP), guanosine ($5′→O^3$)-1,2-μ-methylenetriphosphate, a synthetic nonhydrolysable analogue of guanosine 5′-triphosphate, GTP, in which the oxy group between the latter's intermediate and terminal phosphorus atoms is replaced by a methylene group. It is an effective activator of ADP-ribosylation factor but not of G$_S$ (see **G-protein**).

guanosine monophosphate abbr.: GMP; an alternative name for any guanosine phosphate, but in particular for **guanosine 5′-phosphate**, especially when its distinction from guanosine (5′-)diphosphate and guanosine (5′-)triphosphate requires emphasis.
guanosine phosphate symbol: Guo$P$; guanosine monophosphate (abbr.: GMP); any phosphoric monoester or diester of guanosine. There are three monoesters – guanosine 2′-phosphate, guanosine 3′-phosphate, and **guanosine 5′-phosphate** – and two diesters – guanosine 2′,3′-phosphate and **guanosine 3′,5′-phosphate** – although guanosine 5′-phosphate, guanylic acid is the ester commonly denoted (the locant being omitted if no ambiguity may arise). Guanosine 2′,3′-phosphate (symbol: Guo-2′,3′-P), also named guanosine 2′,3′-(cyclic)phosphate or cyclic guanosine 2′,3′-monophosphate (abbr.: 2′,3′-cyclic GMP), is formed as an intermediate during the alkaline hydrolysis of ribonucleic acid. This diester then readily hydrolyses to a mixture of the two monoesters guanosine 2′-phosphate (symbol: Guo2′P), also named guanosine 2′-monophosphate (abbr.: 2′GMP) or 2′-guanylic acid or guanylic acid a, and guanosine 3′-phosphate (symbol: Guo3′P), also named guanosine 3′-monophosphate (abbr.: 3′GMP) or 3′-guanylic acid or guanylic acid b.
guanosine 2′-phosphate see **guanosine phosphate**.
guanosine 2′,3′-phosphate see **guanosine phosphate**.
guanosine 3′-phosphate see **guanosine phosphate**.
guanosine 3′,5′-phosphate symbol: Guo-3′,5′-P; the recommended name for cyclic guanosine 3′,5′-monophosphate (abbr.: 3′,5′-cyclic GMP or cyclic GMP or cGMP), guanosine 3′,5′-cyclophosphate, guanosine 3′,5′-(cyclic) phosphate, a monophosphoric diester of

guanosine found in practically all mammalian tissues. It is a cyclic nucleotide similar in structure and the regulation of its metabolism to **adenosine 3′,5′-phosphate**, cyclic AMP, but the levels in tissues are commonly less than 5 percent of those of cyclic AMP. Also like cyclic AMP, it acts as a **second messenger** in many systems, and it activates a number of distinct cyclic-GMP-dependent protein kinases that share some of their properties with their cyclic AMP counterparts; in some instances it antagonizes the action of cyclic AMP. Cyclic GMP is produced by the action of **guanylate cyclase** on guanosine 5′-triphosphate, GTP, and it mediates signal transduction for systems that activate this enzyme.

guanosine 5′-phosphate *or* **5′-guanylic acid** *or* **5′-phosphoguanosine** *or* **5′-O-phosphonoguanosine** *symbol*: Guo5′P; *alternative recommended names for* guanosine 5′-monophosphate (*abbr.*: 5′GMP), guanosine 5′-(dihydrogen phosphate), guanine (mono)ribonucleotide. (The locant is commonly omitted if there is no ambiguity as to the position of phosphorylation.) It is biosynthesized from inosine 5′-phosphate, 5′-inosinic acid, via xanthosine 5′-phosphate and also from free guanine by the **salvage pathway**.

guanosine 5′-β-thiodiphosphate *symbol*: Guo5′PP[S] *or* [S]ppG; *the recommended name for* guanosine 5′-[β-thio]diphosphate (*abbr.*: GDPβS), guanosine 5′-(2-thiodiphosphate) (*abbr.*: GDP[S]), guanosine (5′→O²)-1-thiodiphosphate, a synthetic nonhydrolysable analogue of guanosine 5′-diphosphate, GDP, in which an oxygen atom of its terminal phosphoric residue is replaced by a sulfur atom. It is a competitive inhibitor of the activation of G-proteins by guanosine 5′-triphosphate and of the stimulation of adenylate cyclase by guanosine 5′-triphosphate and fluoride.

guanosine 5′-γ-thiotriphosphate *symbol*: Guo*PPP*[S] *or* [S]pppG; *the recommended name for* guanosine 5′-[γ-thio] triphosphate (*abbr.*: GTP-γ-S), guanosine 5′-(3-thiophosphate) (*abbr.*: GTP[S]), guanosine (5′→O³)-1-thiotriphosphate, a synthetic nonhydrolysable analogue of guanosine 5′-triphosphate, GTP, in which an oxygen atom of its terminal phosphoric residue is replaced by a sulfur atom. It is used in membrane studies to stimulate G-proteins and in studies of DNA synthesis.

guanosinetriphosphatase a GTP-hydrolyzing enzyme belonging to group EC 3.6.1 and involved in transduction of information (EC 3.6.1.46, EC 3.6.1.49), protein synthesis (EC 3.6.1.48), and movement of intracellular particles (EC 3.6.1.47, EC 3.6.1.50, EC 3.6.1.51).

guanosine 5′-triphosphate *symbol*: Guo5′*PPP or* pppG; *the recommended name for* guanosine triphosphate (*abbr.*: GTP), 5′-triphosphoguanosine, 5′-guanylyl diphosphate, guanosine 5′-(tetrahydrogen triphosphate), an important coenzyme and enzyme regulator present in all cells. It is formed from guanosine 5′-diphosphate by substrate-level phosphorylation or by transfer of the terminal phosphoric residue from **adenosine 5′-triphosphate** through the action of nucleoside-diphosphate kinase (EC 2.7.4.6). Reactions in which GTP participates are often driven in the direction leading to hydrolysis, the free energy change then being utilized for energy-requiring purposes, e.g. microtubule assembly or protein biosynthesis. GTP also participates in various synthetic reactions by the transfer of a phosphoric residue to other metabolites, e.g. to oxaloacetate to form phospho*enol*pyruvate in gluconeogenesis.

guanosine triphosphate cyclohydrolase EC 3.5.4.16; the first enzyme in the biosynthesis of tetrahydrobiopterin. It catalyses the reaction:

GTP + 2H₂O = 7,8-dihydroneopterin trisphosphate + HCOOH.

The enzyme, whose subunits each contain 250 amino acids and form a stack of two pentamers, occurs mainly in monoaminergic neurons, adrenal medulla, and liver. Deficiency is rare and results from numerous mutations associated with dopa-sensitive dystonia. *See* **biopterin**.

guanosyl any chemical group formed by the loss of a 2′-, a 3′-, or a 5′-hydroxyl group from the ribose moiety of guanosine. *Compare* **guanylyl**.

guanyl *misnomer for* 1 **guanylyl**. 2 **guanylate**.

guanylate 1 the mono- or the di-anion of **guanylic acid**. 2 any mixture of guanylic acid and its anions. 3 any salt or ester of guanylic acid.

guanylate cyclase EC 4.6.1.2; *systematic name*: GTP pyrophosphate-lyase (cyclizing); *other names*: guanylyl cyclase (not recommended but often used); guanyl cyclase (incorrect). An enzyme that catalyses the elimination of pyrophosphate from guanosine 5′-triphosphate, GTP, by formation of guanosine 3′,5′-phosphate, cyclic GMP. There are two types: (1) a heterodimer of α and β chains; and (2) a membrane-bound form, typical of some receptors. A retina-specific receptor guanylate cyclase (retGC-1) is present in the outer segment of cone cells and is activated by retina-specific activator proteins. Many mutations in the gene for retGC-1 at 17p13.1 result in Leber's congenital amaurosis (a form of blindness). *See* **GCAP**.

guanyl cyclase *incorrect name for* **guanylate cyclase**.

guanylic acid *the trivial name for* any phosphoric monoester of guanosine. The position of the phosphoric residue on the ribose moiety of a given ester may be specified by a prefixed locant. *see* **guanosine phosphate**. However, 5′-guanylic acid is the ester commonly denoted, its locant usually being omitted if no ambiguity may arise. 5′-Guanylic acid is also an alternative recommended name for **guanosine 5′-phosphate**.

guanylin a pentadecapeptide paracrine hormone that contains four disulfide bonds and represents the C-terminal segment of proguanylin (115 amino acids). It is secreted by enterochromaffin cells along the intestinal tract, secretory cells in bronchioles, and probably the kidney. Its receptor is a membrane receptor guanylate cyclase that also binds the heat-stable enterotoxin of several Gram-negative bacteria. It acts to increase secretion of NaCl.

guanylyl the guanosine[mono]phospho group; the acyl group derived from **guanylic acid**.

guanylylation the process of introducing guanylyl groups into a substance.

guanylyl cyclase *see* **guanylate cyclase**.

guaran a mucilaginous **galactomannan** of ≈220 kDa that serves as a storage polysaccharide in the seeds of guar and is the prime constituent of **guar gum**. It consists of long chains of (1→4)-linked β-D-mannopyranose residues, to every alternate one of which a single α-D-galactopyranose residue is attached in (1→6)-linkage.

guard cell either of a pair of crescent-shaped cells that surround a **stoma** in plants.

guard column a small column fitted in front of the main column as a device to prolong the life of columns in high-performance liquid chromatography by removing from the sample any particulate matter, aggressive agents, or components that bind irreversibly to the column packing.

guar gum a greyish-white powder consisting predominantly of the polysaccharide **guaran** and obtained by grinding the endosperm of the seeds of guar, an Indian leguminous plant, *Cyamopsis tetragonoloba* (*psoralioides*), also called cluster bean. It hydrates and disperses readily in cold water to give colloidal solutions of very high viscosity that are stable over a very wide pH range and have very good suspending and lubricating properties. It is used for sizing paper and textiles, and as a stabilizing, suspending, and thickening agent in the food and pharmaceutical industries, etc. It has also been suggested as a nonabsorbable dietary additive to improve glycemic control in diabetes mellitus, possibly by slowing the uptake of ingested absorbable carbohydrate.

GUC a codon in mRNA for L-valine.

guessmer an oligonucleotide probe assembled on the basis of partial protein sequence information coupled with intelligent guesswork guided by knowledge of codon usage. Such a probe is used to screen a cDNA library for the corresponding gene.

GUG a codon in mRNA for L-valine.

Guggenheim plot a procedure that makes it possible to obtain rate constants for first-order reactions when neither the initial nor the final reactant concentrations are known. Measurements of the concentration of the reactant (or of some physical quantity that measures it), c_1, c_2, c_3,...c_n, are made at times t_1, t_2, t_3,...t_n and also at times $(t_1 + \delta t)$, $(t_2 + \delta t)$, $(t_3 + \delta t)$,...$(t_n + \delta t)$, when the concentrations of the reactant are c_1', c_2', c_3',...c_n', where δt is a constant time difference, then

$$\ln(c_i - c_i') = \ln(c_0 - c_\infty)(1 - e^{-k\delta t}) - kt,$$

or

$$\ln(c_i - c_i') = a \text{ constant} - kt_i,$$

hence a plot of $\ln(c_i - c_i')$ versus t_i has a slope of $-k$. [After Edward Armand Guggenheim (1901–70), British physical chemist.]

guide RNA *see* **gRNA**.

guinea pig any of several species of wild or domesticated hystricomorph rodents, or cavies, in the genus *Cavia*, especially *Cavia porcellus*, which is widely used in biomedical research and testing.

Guinier plot a method of plotting the results of small-angle X-ray scattering to obtain the radius of gyration of a molecule, r_G, in the absence of information about the relative molecular mass of the component in question. It can be shown that

$$\ln(I_\theta /I_0) = -(4\pi \, r_G \sin\theta / \lambda)^2/3,$$

where I_θ and I_0 are the intensities of the scattered radiation at angles θ and 0 respectively, and λ is the wavelength of the radiation; hence a plot of $\ln(I_\theta /I_0)$ versus $-(4\pi\sin\theta / \lambda)^2/3$ will yield r_G.

Gul *symbol for* a residue (or sometimes a molecule) of the aldohexose gulose.

GulA *symbol for* a residue (or sometimes a molecule) of the uronic acid guluronic acid.

gulo- *prefix (in chemical nomenclature)* indicating a particular configuration of a set of (usually) four contiguous >CHOH groups, as in the acyclic form of D- or L-gulose. *See* **monosaccharide**.

L-gulonolactone oxidase EC 1.1.3.8; *other name*: L-gulono-γ-lactone oxidase; a flavoprotein enzyme that catalyses the oxidation by dioxygen of L-gulono-1,4-lactone to L-*xylo*-hexulonolactone (I) and H_2O_2. I isomerizes spontaneously to L-ascorbate. The flavin is FAD. The enzyme is essential for the formation of ascorbate, and a deficiency of the enzyme, as in humans, results in dependency on a dietary supply of ascorbate (*see* **ascorbic acid**). Example from rat: GGLO_RAT, 439 amino acids (50.42 kDa).

gulose *symbol*: Gul; *the trivial name for* the aldohexose *gulo*-hexose; a stereoisomer of glucose in which the configurations at both C-3 and C-4 are inverted. There are two enantiomers. (Illustrated opposite.)

gulose (*β*-D-pyranose form)

guluronic acid *symbol*: GulA; the **uronic acid** resulting from the oxidation of the hydroxymethylene group at C-6 of gulose to a carboxyl group. L-Guluronic acid is a component of **alginic acid**.

gum any of numerous water-soluble or water-dispersible heteroglycans, frequently containing uronic acids, that are produced as exudates from certain plants, especially trees, following physical injury. Gums are tacky, slimy substances that harden when dry. *See also* **gum arabic**, **gum ghatti**.

gum arabic a **gum** exuded from trees of the genus *Acacia*. Structurally it is a heteroglycan containing D-galactose, L-arabinose, L-rhamnose, and D-glucuronic acid. It is used in industry, e.g. in the manufacture of thickening agents, emulsifiers, inks, and pills. It is also used in glues and pastes.

gum ghatti a **gum** from the tree *Anogeissus latifolia*. It has a complex structure and contains as major components D-mannose, D-galactose, and L-arabinose, the latter in both furanose and pyranose arrangement. Small amounts of D-xylose and D-glucuronic acid are also present.

GUN4 a nuclear-encoded protein (266 amino acids) of *Arabidopsis thaliana* chloroplasts that controls synthesis of Mg-protoporpyhrin IX through its action on Mg-chelatase. GUN4 binds the product and the substrate of Mg-chelatase and represses many nuclear genes that control plastid differentiation. The name derives from the chlorophyll-deficient mutant *gun4*. A similar protein is present in other plants.

Gunn rat any member of a mutant strain of Wistar rats with hereditary hyperbilirubinemia. The defect affects the rate of glucuronidation of a number of substrates, but not equally. [After C. H. Gunn.]

Güntelberg–Müller charging process a method of calculating the electrical energy of an ion. According to the process, only the ion under consideration is supposed to receive a charge; the surrounding ions are assumed to be already charged, and the only change that they undergo during the process is a gradual redistribution as the central ion receives its charge.

Guo *symbol for* a residue of the ribonucleoside guanosine (alternative to G).

Guo*P* *symbol for* guanosine phosphate.

Guo2′*P* *symbol for* guanosine 2′-phosphate; *see* **guanosine phosphate**.

guo-2′,3′-*P* *symbol for* guanosine 2′,3′-phosphate; *see* **guanosine phosphate**.

Guo3′*P* *symbol for* guanosine 3′-phosphate; *see* **guanosine phosphate**.

Guo-3′,5′-*P* *symbol for* guanosine 3′,5′-phosphate.

Guo5′*P* *symbol for* guanosine 5′-phosphate.

Guo5′*PP* *symbol for* guanosine diphosphate (alternative to ppG).

Guo*PP***[CH₂]***P* *symbol for* guanosine 5′-[β,γ-methylene] triphosphate (alternative to p[CH₂]ppG).

Guo*PP***Man** *symbol for* guanosinediphosphomannose.

Guo*PP***[NH]***P* *symbol for* guanosine 5′-[β,γ-imido]triphosphate (alternative to p[NH]ppG).

Guo5′*PPP* *symbol for* guanosine 5′-triphosphate (alternative to pppG).

Guo5′*PPP***[S]** *symbol for* guanosine 5′-γ-thiotriphosphate.

Guo5′*PP***[S]** *symbol for* guanosine 5′-β-thiodiphosphate.

Gurken a transforming growth factor α-related protein produced in the dorsal region of *Drosophila* embryos, where it is involved in dorsoventral specification. It binds a membrane tyrosine kinase receptor that activates Ras.

gustducin a taste-cell-specific G protein α subunit closely related to the transducins, and comprising 354 amino acids.

gut a term variously used to describe the whole of the alimentary canal or only the intestine, i.e. the part from the stomach to the anus.

gut glucagon *(sometimes) an alternative term for* **enteroglucagon**.

gut hormone *a general, inaccurate, but widely used term for* all the various regulatory peptides of the gut. They are neither confined to the gut, nor are they exclusively hormones.

Guthrie test a microbiological test in which a strain of *Bacillus subtilis* that requires phenylalanine for growth is used to test for excessive levels of plasma phenylalanine in conditions such as **phenylketonuria**. [After R. Guthrie (1916–).]

gutta-percha a *trans*-1,4-polyisoprene derived by coagulation of the latex of various Malayan trees of the genus *Palaquium*. It has a lower molecular mass than rubber and is thermoplastic.

GUU a codon in mRNA for L-valine.

g-value *or* **g-factor** a factor that specifies the position at which the absorption of a paramagnetic species occurs in an **EPR** spectrum. For a free electron in a vacuum, g has the value 2.002 319 304 4.... In a paramagnetic molecule, the energy difference between the two states is afected by interaction of the unpaired electron spin with electron orbitals. For the purposes of EPR spectroscopy, these interactions are taken into account by allowing the g-value to be a spectroscopic variable. For a field B_O and frequency υ, resonance occurs when $h\upsilon = g\mu_B B_O$, where h is the Planck constant and μ_B is the Bohr magneton.

G value (in nuclear chemistry) the number of specified chemical events in an irradiated substance produced per 100 eV of energy absorbed from ionizing radiation.

Gy *symbol for* gray.

gynaminic acid *formerly, a name for* a constituent of human milk, now known to be identical with **sialic acid**.

gynogamone *see* **gamone**.

gyrase *see* type II **DNA topoisomerase**.

gyrate atrophy of choroid and retina *see* **OAT**.

gyrin any of several membrane proteins, including synaptogyrins 1 to 4, that occur as components of transport vesicles and neurotransmitter vesicles. They contain four transmembrane segments, have intracellular N- and C-termini, and resemble the physins in both their architecture and function. **Synaptogyrin 2** (*also called* cellugyrin) is also present in vesicles that contain glucose transporter 4 (GLUT4). Homologues occur in *Drosophila* and *Caenorhabditis*.

gyromagnetic ratio *or* **magnetogyric ratio** *symbol*: γ; the ratio of the magnetic moment, μ, of a given atomic particle to its angular momentum, L; i.e. $\gamma = \mu/L$. *See also* **proton magnetogyric ratio**.

Hh

h *symbol for* **1** hecto+ (SI prefix denoting 10^2 times). **2** hour (supersedes hr.).

h *symbol for* **1** Planck constant. **2** Hill coefficient (alternative to n_H). **3** height.

h+ *or (not recommended)* **H+** *prefix* to abbreviations for peptide hormone (names) denoting human, i.e. of (or as if of) human origin; e.g. hGH (for human growth hormone).

H *symbol for* **1** hydrogen; *see also* **protium, deuterium, tritium**. **2** a residue of the α-amino acid **histidine** (alternative to His). **3** a residue of an incompletely specified base in a nucleic-acid sequence that may be adenine, cytosine, or either thymine (in DNA) or uracil (in RNA).
4 henry. **5** histone. **6** histamine (*see* **histamine receptor**).

H+ *see* **h+**.

H1 a DNA-binding protein from *Escherichia coli* belonging to the **H-NS** family of proteins.

H7 1-(5-isoquinolinylsulfonyl)-2-methylpiperazine; an inhibitor of protein kinase C ($K_i = 6.0\ \mu M$), protein kinase A ($K_i = 3.0\ \mu M$), and cyclic GMP-dependent protein kinase ($K_i = 5.8\ \mu M$).

H₂ *symbol for* **1** dihydrogen. **2** the prefix **dihydro+** in a trivial chemical name, e.g. H₂folate for dihydrofolate.

H₄ *symbol for* the prefix **tetrahydro+** in a trivial chemical name, e.g. H₄folate for tetrahydrofolate.

H *symbol for* **1** fluence (alternative to *F*). **2** enthalpy. **3** magnetic field strength (light italic in nonvector equations; bold italic in vector equations).

-H *conformational descriptor* designating the half-chair conformation of a six-membered ring form of a monosaccharide or monosaccharide derivative. Locants of ring atoms that lie on the side of the structure's reference plane from which the numbering appears clockwise are indicated by left superscripts and those that lie on the other side of the reference plane by right subscripts; e.g. methyl 2,3-anhydro-5-thio-β-L-lyxopyranoside-^5H_S. *See also* **conformation**.

H- *or (sometimes)* **H** *(in chemical nomenclature) symbol for* indicated hydrogen (at the position specified by a prefixed locant).

Habc domain *abbr. for* helix a, b, and c domain; a protein sequence in the N-terminal region of syntaxins that forms a triple alpha-helix bundle that folds back on the SNARE domain (which consists of heptads) and prevents it from forming complexes with other SNAREs.

Haber–Weiss reaction the iron-catalysed reaction of a superoxide anion with hydrogen peroxide to form oxygen, a hydroxyl radical, and a hydroxyl ion. It is thought to proceed as follows:

$$O_2^- + Fe^{3+} \rightarrow O_2 + Fe^{2+}$$

$$H_2O_2 + Fe^{2+} \rightarrow OH^{\bullet} + OH^- + Fe^{3+}$$

[After Fritz Haber (1868–1934), German chemist, and Joseph Weiss.]

HAc *abbr. for* acetic acid.

HACBP *see* **calreticulin**.

HaeII a type 2 **restriction endonuclease**; recognition sequence: [AG]GCGC↑[TC].

HaeIII a type 2 **restriction endonuclease**; recognition sequence: GG↑CC. The M.*Hae*III modification site is C3.

haem *a variant spelling (esp. Brit.) of* **heme**.

haema+ *or (before a vowel)* **haem+** *variant spellings (esp. Brit.) of* **hema+**.

haemadin a 57-amino-acid anticoagulant peptide from the Indian leech, *Haemadipsa sylvestris*. It is a slow, tight-binding inhibitor of thrombin, but does not inhibit trypsin, chymotrypsin, factor Xa, or plasmin. .

haemato+ *or (before a vowel)* **haemat+** *variant spellings (esp. Brit.) of* **hemato+**.

haemo+ *a variant spelling (esp. Brit.) of* **hemo+**.

haemoglobin *a variant spelling (esp. Brit.) of* **hemoglobin**.

Hageman factor *an alternative name for* factor XII; *see* **blood coagulation**.

hairpin any part of a linear molecular structure, e.g. of a polynucleotide strand or a prostaglandin, in which two adjacent segments of the molecule are folded back one on the other and held in that conformation by secondary molecular forces such as hydrogen bonds or van der Waals interactions. **Hairpin DNA** (*or* foldback DNA *or* loopback DNA) contains **inverted repeats**; when denatured it renatures extremely rapidly by intrachain base-pairing between complementary sequences of the inverted repeat. Similar structures occur in large RNA molecules. The hairpin conformation of a prostaglandin molecule is thought to be necessary for expression of its hormonal activity; in this conformation the α and κ chains of each molecule are extended, parallel to one another, at van der Waals contact distance for their full lengths. *See also* **ribozyme, beta bend**.

hairy *a pair-rule gene of Drosophila*.

Hakai a protein belonging to the E3 ubiquitin ligase family that specifically binds to phosphotyrosyl residues on E-cadherin, targeting it for ubiquitination and endocytosis. [Japanese for 'destruction'.]

Haldane coefficient the difference, per bound dioxygen molecule, in the number of hydrons bound by oxy- and deoxyhemoglobin, at constant pH. [After John Scott Haldane (1860–1936), British physiologist.]

Haldane effect the observation that oxygenated blood absorbs less carbon dioxide than deoxygenated blood. It is the reciprocal of the **Bohr effect**.

Haldane gas-analysis apparatus an apparatus once used for the estimation of the carbon dioxide and dioxygen contents of a gas sample.

Haldane relation a relation for reversible enzyme reactions. In the single-substrate, essentially reversible reaction:

$$E + S \underset{k_{-1}}{\overset{k_1}{\rightleftharpoons}} ES \underset{k_{-2}}{\overset{k_2}{\rightleftharpoons}} EP \underset{k_{-3}}{\overset{k_3}{\rightleftharpoons}} E + P$$

the overall equilibrium constant,

$$K = C_P/C_S = k_1 k_2 k_3/k_{-1} k_{-2} k_{-3},$$

where C_S and C_P are the concentrations of substrate, S, and product, P, respectively. It can further be shown that

$$K = V_S K_{mP}/V_P K_{mS},$$

where V_S and V_P are the maximum velocities of the reactions $S \rightarrow P$ in the absence of P and of $P \rightarrow S$ in the absence of S, and K_{mS} and K_{mP} are the respective Michaelis constants for these reactions. [After John Burdon Sanderson Haldane (1892–1964), British biochemist, physiologist, and geneticist.]

half-boat conformation either of the **conformations** of a nonplanar monounsaturated six-membered ring compound when the two ring atoms not directly bound to the doubly bonded atoms lie on the same side of the plane containing the other four (adjacent) atoms.

half-cell one half of an **electric cell**, consisting of one reversible electrode inserted into a solution. Electrical connection between this solution and that in another half-cell is required to complete the electric circuit.

half-chair conformation 1 either of the **conformations** of a nonplanar monounsaturated six-membered ring compound when the two ring atoms not directly bound to the doubly bonded atoms lie on opposite sides of the plane containing the other four (adjacent) atoms. **2** any conformation of the six-membered ring form of a monosaccharide or monosaccharide derivative when four adjacent ring atoms are coplanar and the other two lie on opposite sides of the plane. The conformational descriptor *-H* may be added to the name of such compounds.

half-cystine *or* **hemicystine** *symbol:* $_{Cys}^{|}$ *or* $^{Cys}_{|}$; the group formally derived by removal of a hydrogen atom from the thiol group on the side chain of cysteine (or of a cysteine residue), or by homolysis of the dithio linkage of cystine (or of a cystine residue). It is used to describe or depict the content or linear sequence of amino-acid residues in a polypeptide when it contains more than two apparent cysteine residues, and the positions of any dithio linkages between them are not known.

half-cystyl *or* **hemicystl** the acyl group, ·S–CH$_2$–CH(NH$_2$)–CO–, derived from the **half-cystine** group. *Compare* **cysteinyl**.

half-ester any monoester of a dibasic acid.

half-life 1 *symbol:* $t_{1/2}$ *or* $T_{1/2}$; the time for one half of the atoms of an amount of radionuclide to undergo radioactive decay. **2** a similar measure of the stability (i.e. rate of decay) of an excited atom or molecule, a radical, an unstable elementary particle, etc. **3** the time for one half of the amount of an administered substance to be metabolized or excreted. If the substance is radioactive, the time required for one half of the dose to be eliminated biochemically or physiologically is termed the **biological half-life** and that required for one half to disappear by radioactive decay as well as by the elimination is the **effective half-life**. **4** *or* (*sometimes*) **half-time** *symbol:* $t_{1/2}$; the time required for the concentration of a reactant in a chemical reaction to reach a value that is the arithmetic mean of its initial and final (equilibrium) values. **5** the time for one half of the number of cells in a tissue or organ to be replaced by new cells.

half-of-the-sites reactivity *or* **half-site reactivity** a phenomenon shown by many enzymes that exhibit **negative cooperativity**, in which the maximal stoichiometric yield of either an enzyme–substrate intermediate or a product in a single turnover amounts to only half the number of apparently equivalent active sites. This may be caused by the binding of substrate to one site of a dimeric enzyme, e.g., inducing a conformational change in the enzyme so as to abolish completely the binding affinity of the other site.

half-reaction either of the two coupled chemical changes that together constitute an oxidation–reduction reaction. In one half-reaction there is a gain of electrons and in the other half-reaction a corresponding loss of electrons.

half-shotgun coverage *see* **depth of coverage**.

half-site reactivity *an alternative name for* **half-of-the-sites reactivity**.

half-thickness the thickness of a piece of a specified material that will reduce the intensity of a beam of transmitted electromagnetic radiation to one half that of the incident beam.

half-time *an alternative name for* **half-life** (def. 4) for a first-order process.

half-time of exchange the time taken for one half of the exchangeable atoms (or molecules) to be exchanged in a reaction involving such an exchange.

half-transporter a member of the **ABC transporter** superfamily usually found in peroxisomal or in microbial membranes. It contains a membrane-spanning region (with six transmembrane segments) and a nucleotide-binding domain.

half-wave potential the electrical potential at the mid-point of the current–voltage curve in **polarography**.

halide *or* **halogenide 1** an anion of any halogen. **2** a salt of any halogen acid (halogen hydride). **3** any compound containing halogen atoms in organic linkage.

Hallervorden–Spatz syndrome an early-onset neurodegenerative syndrome of parkinsonism, dystonia, and iron accumulation in the brain. It is caused by deficiency of **pantothenate kinase**. [After Julius Hallervorden (1882–1965) and Hugo Spatz (1888–1969), German neurologists.]

hallucinogen any drug or other substance that causes hallucinations. —**hallucinogenic** *adj.*

halo+ *or* (*sometimes before a vowel*) **hal+** *comb. form* **1** relating to or containing a halogen. **2** relating to the sea or salt.

halochromism the phenomenon or property of the formation of strongly coloured compounds on the addition of strong acids or certain metallic salts to colourless or faintly coloured substances.

halofantrine   1,3-dichloro-α-[2-(dibutylamino)-ethyl]-6-(trifluoromethyl)-9-phenanthrenemethanol; a derivative of 9-phenanthrenemethanol that is effective against chloroquine-resistant malarial infections.

halogen any of the monovalent chemical elements of group 17 of the IUPAC **periodic table**: fluorine, chlorine, bromine, iodine, and astatine.

halogenate to treat or react (a substance) with a halogen; to introduce one or more halo groups into an organic compound. —**halogenation** *n.*

halogenide *an alternative name for* **halide**.

haloperidol   4-[4-(4-chlorophenyl)-4-hydroxy-1-piperidinyl]-1-(4-fluorophenyl)-1-butanone; one of the butyrophenone group of **neuroleptic** drugs.

haloperoxidase one of a group of enzymes that halogenate organic substrates using hydrogen peroxide and a halide ion (Cl$^-$, Br$^-$, or I$^-$). Haloperoxidases contain either heme or vanadium.

halophile 1 an organism that grows in or tolerates saline conditions, especially concentrations of sodium chloride equal to or greater than that in sea water. **2** describing such an organism. —**halophilic** *adj.*

halorhodopsin a pigment identified in membranes of *Halobacterium halobium*. It is similar to **bacteriorhodopsin** and, upon absorption of a photon, catalyses inward transport of chloride ions.

halothane *the trivial name for* 2-bromo-2-chloro-1,1,1-trifluoroethane, a widely used, volatile, lipid-soluble general inhalational anesthetic agent.

hamamelose 2-*C*-(hydroxymethyl)-D-ribose; a substance found principally in **tannin** of witch hazel (*Hamamelis virginiana*).

hamartin a protein (1164 amino acids) that is widely expressed, contains a potential transmembrane segment and a predicted coiled-coil domain, and has no homology to any other protein. It may be a tumour suppressor, and forms a mainly cytosolic complex with **tuberin**. Many mutations in the gene at 9q34 produce the autosomal dominant tuberous sclerosis type 1 (TSC1).

Hamburger shift *an alternative name for* **chloride shift**. [Named after Hartog Jakob Hamburger (1859–1924), Dutch physiologist.]

Hamiltonian operator *symbol*: H; a function used to express the total energy of a system in joules: $H = T + V$, where T is the kinetic energy operator and V the potential energy operator. For a particle of mass m and momentum p,

$$H = p^2/2m + V.$$

[After William Rowan Hamilton (1805–65), Irish astronomer and mathematician.]

hamlet a protein that acts as a binary genetic switch between single- and multi-dendrite neuron morphology in *Drosophila* embryos. It is predicted to contain 990 amino acids, with six zinc fingers in the N-terminal half and three more towards the C-terminus. It may be a nuclear transcription factor.

hammerhead *see* **ribozyme**.

Ham's F10 medium, Ham's F12 medium formulations of culture media for the propagation of mammalian cells based on saline F. [Named after Richard G. Ham.]

handedness a form of terminology (based on the human left and right hands) for describing objects and compounds that are mirror images of each other and hence nonsuperimposable. A pair of **enantiomers** display molecular handedness. *See* **chirality**.

handle *a colloquial term for* any chemical group attached to a specified substance with the purpose of making it more easily identified, more reactive, or more easily operated upon, or one that may serve in targeting or specific binding. The attachment of a handle may be brought about chemically or metabolically.

handle technique a technique used to separate excess alkylating agents from products in peptide synthesis. For instance, the C terminus of a growing peptide chain can be protected with a basic function, e.g. by forming a 4-picolyl ester of the carboxyl group, thereby allowing separation of the products on cation-exchange resin columns or by extraction into aqueous acidic solutions.

Hanes plot *or* **Hanes–Woolf plot** a graphical method for treating enzyme kinetic data using the following form of the Michaelis equation (*see* **Michaelis kinetics**):

$$s/v = K_m/V + s/V$$

where s is the substrate concentration, v is the velocity of the reaction, and V is the limiting rate. If s/v is plotted against s, the slope of the resulting straight line is $1/V$, the intercept on the s/v axis is K_m/V, and the intercept on the s axis is $-K_m$. [After Charles Samuel Hanes (1903–90) and Barnet Woolf (1902–1983), British biochemists.]

hanging drop method a technique used in protein crystallization in which a droplet of protein solution, buffer, and precipitant is allowed to equilibrate with a reservoir of buffer and precipitant at higher concentration in a closed system. The name comes from the fact that the droplet hangs from a glass coverslip suspended over the reservoir.

Hanks' balanced salt solution *abbr.*: HBSS; a balanced salt solution used in tissue culture to provide a suitable ionic and osmotic environment for cell growth and development.

Hansch equation an equation relating the differences in the partitions of variously substituted compounds between an organic and an aqueous phase to the **hydrophobicity constant**, π, of the substituent. If P_0 is the ratio of the solubility of the parent compound in an organic phase to that in the aqueous phase, and P is the corresponding ratio for the substituted compound, then: $\pi = \log(P/P_0)$.

Hansenula *see* **methylotrophic yeasts**.

H antigen *see* **blood-group substance**.

***Hap*II** a type 2 **restriction endonuclease**; recognition sequence: C↑CGG. *Hpa*II and *Msp*I are **isoschizomers**. The M.*Hap*II modification site is C2; M.*Hpa*II is an isoschizomer.

haploid describing a cell, an organism or a nucleus of a cell having a single genome or a single set of homologous chromosomes; i.e. containing half the **diploid** number; having a **ploidy** of one. —**haploidy** *n*.

haplotype a set of genes located on a single chromosome; the term is used also to denote the characteristics dependent on those genes. In outbred populations the maternal and paternal chromosomes usually differ, so an individual has two haplotypes, one derived from each parent.

'happiness hormone' *a jocular name for* any **endorphin** or **enkephalin**, so-named because of the supposed ability of such substances to promote a sense of well-being.

hapt+ *a variant form of* **hapto+** (before a vowel).

hapten any, usually small, substance that can combine specifically with antibody but that is not immunogenic unless bound to a **carrier** (def. 8). Most haptens carry only one or two antigenic determinants. —**haptenic** *adj*.

hapten-inhibition test a serological test for characterizing an antigenic determinant. It involves the use of haptens of known structure to block the combining site of a particular antibody.

hapten-sandwich labelling a variant of the **sandwich technique** for amplification of an antigen–antibody reaction. In this modification, the primary antibody, specific for the antigen of interest, is combined with a number of residues of a hapten, e.g. *p*-azobenzene arsonate. The secondary (labelled or marked) antibody is a specific anti-hapten antibody; on reaction of secondary antibody molecules with a molecule of primary antibody, the hapten residues become sandwiched between the large molecules of antibody. By the use of different haptens and two or more non-cross-reactive anti-hapten antibodies bearing distinguishable markers, two or more antigens on the same or different cells may simultaneously be labelled.

hapticity *symbol*: η; the attribute of a ligand entity or group that combines with a central (usually metal) atom to form a coordination entity in which the central atom is bonded to two or more contiguous atoms of the ligand rather than to a specific atom. Such a coordination entity is frequently (but not exclusively) a **pi adduct**. *See also* **hapto**.

hapto or (*less commonly*) **eta** (*in chemical nomenclature*) the name by which to read the symbol η- (Greek eta), when prefixed to the name of the ligand group in a coordination entity (with added locants if appropriate) to indicate hapticity.

hapto+ or (*before a vowel*) **hapt+** *comb. form* indicating ability to bind or combine.

haptocorrin a blood cobalamin-binding glycoprotein that may prevent uptake of bacterial cobalamin analogues.

haptoglobin *abbr.*: Hp; any α_2 globulin of blood plasma that can combine with free oxyhemoglobin to form a stable, ~300 kDa, complex. Such combinations occur when hemoglobin is liberated in the blood, and the complexes so formed, which have weak peroxidase activity and cannot be filtered by the kidney, are rapidly removed and degraded in the tissues. Three human haptoglobin phenotypes are known, designated Hp 1-1, Hp 2-2, and Hp 2-1, each of which contains two pairs of non-identical polypeptide chains, α and β, joined by disulfide bonds. There are two kinds of α chain, α_1 of 84 amino-acid residues (9 kDa) and α_2 of 143 residues (17 kDa), and a single kind of β chain, of 244 residues (in human). The α_2 chain, found solely in human haptoglobin, is identical in sequence to that of residues 1–71 or 1–72 of an α_1 chain linked to residues 11 or 12 to 84. The basic subunit structure of Hp 1-1 is $(\alpha_1\beta)_2$, that of Hp 2-2, $(\alpha_2\beta)_2$, and that of Hp 2-1, $(\alpha_2\beta)(\alpha_1\beta)$. However, the α_2 chain has an odd number of half-cystine residues; consequently, larger oligomers of Hp 2-2 and Hp 2-1 may arise by formation of disulfide bonds with α_2 chains of other molecules or of $(\alpha_2\beta)$ half-molecules.

haptomer any substance (such as a lectin or the B fragment of diphtheria toxin) that can interact with the cell membrane and that binds or can be bound to an **effectomer** (such as gelonin or the A fragment of diphtheria toxin), thereby allowing the effectomer to exert its toxic action on intact cells. —**haptomeric** *adj*.

haptophore a term introduced by Ehrlich to denote the specific chemical group in a toxin (i.e. antigen) molecule that attaches it to an antitoxin (i.e. antibody) molecule. Unlike the **toxophore**, it is not destroyed on conversion of the toxin to a toxoid.

H$^+$(aq) *symbol for* the hydrated hydrogen ion, the normal state of the hydrogen ion in aqueous solutions, consisting of a hydron surrounded by a shell of water molecules held by ion–dipole interaction. *Compare* **oxonium** (def. 1).

hard *(in radiation physics)* describing corpuscular or electromagnetic ionizing radiation (especially beta particles or X- or gamma-radiation) having relatively high energy, hence high ionizing power and long-range or high penetrating power. Hard X- or gamma-radiation has short wavelengths. *Compare* **soft** (def. 1).

Harden–Young ester *a former name for* D-fructose 1,6-bisphosphate (before its chemical structure had been elucidated). [After (Sir) Arthur Harden (1865–1940) and William John Young (1878–1942), British biochemists.]

Hardy–Weinberg law a mathematical statement about the relation between gene frequencies and genotype frequencies within a population. If the genotypes are AA, Aa, and aa, and p is the frequency of allele A in the population, and q the frequency of allele a, so that $p + q = 1$, then the relative proportions of the three genotypes in the population, AA:Aa:aa, is given by $p^2:2pq:q^2$. [After Godfrey Harold Hardy (1877–1947), British mathematician, and Wilhelm Weinberg (1862–1937), German physician.]

Hardy–Weinberg population a population of randomly mating individuals in which there are no changes in allele frequencies from generation to generation. *See* **Hardy–Weinberg law**.

harmaline 4,9-dihydro-7-methoxy-1-methyl-3*H*-pyrido[3,4-*b*]indole; dihydroharmine; a cardioactive alkaloid from Syrian rue, *Peganum harmala* that also has hallucinogenic properties. *See also* **harmine**.

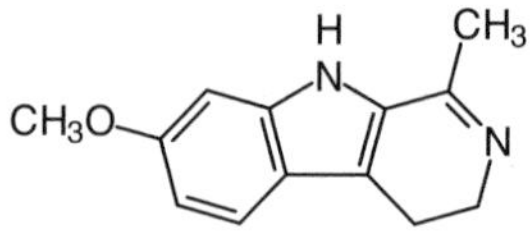

harmine 7-methoxy-1-methyl-9*H*-pyrido[3,4-*b*]indole; a β-carboline alkaloid found in, e.g., *Peganum harmala* (Syrian rue). It inhibits monoamine oxidase and is a central nervous system stimulant. *See also* **harmaline**.

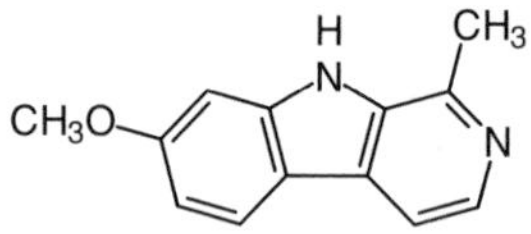

harmonin a cytosolic protein, encoded by a gene at 11p15.1, mutations of which cause the autosomal recessive **Usher syndrome** type IC. It contains a common, highly conserved sequence of ≈90 residues (called a **PDZ domain**) that forms a hydrophobic pocket capable of binding the intracellular hydrophobic C-terminal residue of several transmembrane proteins and ion channels.

HART *abbr. for* hybrid-arrested translation.

Hartley filter funnel a development of the **Büchner funnel** comprising three pieces, the reservoir, a perforated plate to support the filter disk, and a conical piece to collect the filtrate. [After (Sir) Percival Hartley (1881–1957), British bacteriologist.]

Hartnup disorder a rare, autosomal recessive condition of humans in which there is a specific hyperaminoaciduria and indoluria, due to diminished renal reabsorption of monoamino-monocarboxylic acids, especially tryptophan. There may also be an impaired intestinal absorption of the same amino acids. It is often associated with a pellagra-like skin rash, which may be resolved by oral nicotinic acid. [After E. Hartnup, the first patient with the condition to be described.]

hashish a resinous extract obtained from the dried flower tops of the hemp plant (*Cannabis sativa*). The term may sometimes be used for various other extracts. *See* **cannabis**.

HAT *abbr. for* **1** hypoxanthine-aminopterin-thymidine (medium); a medium devised to select against cells deficient in hypoxanthine/guanine phosphoribosyltransferase (HGPRT⁻ cells). Aminopterin blocks *de novo* synthesis of purines and HGPRT⁻ cells are unable to utilize the salvage pathway, and hence are killed. It is of particular use in the production of **hybridomas**. **2** human airway trypsin-like protease; a serine protease type II transmembrane protein (418 amino acids) of the trachea. **3** histone acetyltransferase.

hatching enzyme *see* **envelysin**.

Hatch–Slack pathway *or* **C₄ cycle** *or* **C₄ pathway** a metabolic pathway for carbon-dioxide translocation occurring in **C₄ plants**. In the mesophyll cells **phospho*enol*pyruvate carboxylase** catalyses the condensation of CO_2 with phospho*enol*pyruvate to form oxaloacetate; this is then reduced to malate by light-generated NADPH. The malate passes into the bundle-sheath cells where it is oxidatively decarboxylated by the **malic enzyme** to pyruvate and CO_2; the pyruvate returns to the mesophyll cells and is reconverted to phospho*enol*pyruvate by ATP and pyruvate dikinase; the CO_2 remains in the bundle-sheath cells where it enters the **reductive pentose phosphate cycle**. The overall effect of the pathway is to carry CO_2 from the mesophyll cells, which are in contact with air, to the bundle-sheath cells, which are the major sites of photosynthesis.

Haworth representation a diagrammatic way of unambiguously representing the structural formulae of the cyclic forms of monosaccharides to show on a plane surface the relative configurational arrangements of the atoms and groups in space. Pyranoses are represented by a hexagon (five carbons and one oxygen) and furanoses by a pentagon (four carbons and one oxygen). A Haworth representation is derived from a **Fischer projection** as follows. The monosaccharide in question is depicted with the carbon-chain horizontal and in the plane of the paper, the potential carbonyl group being to the right. The oxygen bridge is then depicted as being formed behind the plane of the paper. The heterocyclic ring is therefore located in a plane approximately perpendicular to the plane of the paper and the groups attached to the carbon atoms of that ring are above and below the ring. The edge of the ring nearer the viewer is indicated by thickened lines. For clarity, the carbon atoms of the ring and their attached hydrogen atoms are not shown. Groups that, in the Fischer projection, appear to the right of the vertical chain then appear in the Haworth representation below the plane of the ring. However, at the asymmetric carbon atom involved via oxygen in ring formation (e.g. C-5 in glucopyranose, C-4 in glucofuranose), a rotation about the bond to the preceding carbon atom must be envisaged in order to bring that oxygen atom into the correct orientation to link to the carbon atom of the carbonyl group. Thus, if the hydroxyl group that is engaged in ring formation lies to the right in the Fischer projection, the group that lies below it must point upwards in the Haworth representation, whereas if it lies to the left, it must point down in the Haworth representation. In representing the α and β anomers in the Haworth convention, it should be noted that for the α anomer, the anomeric hydroxyl points in the opposite direction to the R group (see structure); for D-sugars, the R group points upwards, so the α-hydroxyl must point down; conversely, if the R group points down (as in α-L-glucose), the α anomeric hydroxyl must point upwards. The above rules only obtain if the ring is numbered clockwise from the anomeric carbon, as is frequently the case, and it is important to appreciate that Haworth representations, unlike Fischer projections, cannot be rotated in the plane of the paper but must be reorientated in space. [After (Sir) Walter Norman Haworth (1883–1950), British organic chemist.]

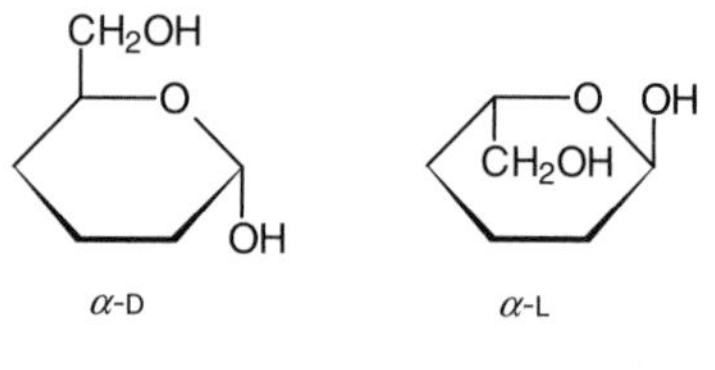

α-D   α-L

Haymann nephritis antigen *an alternative name for* **LRP2**.

Hb *abbr. and symbol for* hemoglobin; hence, e.g., HbO₂ symbolizes oxyhemoglobin, and HbCO symbolizes carbonmonoxyhemoglobin.

Hb A *abbr. for* adult hemoglobin.

H band *an alternative name for* H zone (*see* **sarcomere**).

HBD (*in clinical chemistry*) *abbr. for* α-hydroxybutyrate dehydrogenase, the LD_1 isoenzyme of **lactate dehydrogenase**, or 'heart-specific' lactate dehydrogenase, which is assayed using 2-hydroxybutanoate as substrate, a substance that does not occur naturally.

Hb F *abbr. for* fetal hemoglobin.

Hb S *abbr. for* sickle-cell hemoglobin; *see* **sickle cell**.

HBSS *abbr. for* Hanks' balanced salt solution.

hCG *or* **HCG** *abbr. for* human choriogonadotropin (*see* **choriogonadotropin**).

H chain *see* **heavy chain**.

HCO *symbol for* the formyl group (alternative to For).

HCR *abbr. for* host-cell reactivation.

Hcy *symbol for* a residue of the α-amino acid L-homocysteine.

HDAg *abbr. for* hepatitis delta antigen. *See* **hepatitis**.

HDEL a four-residue motif, His-Asp-Glu-Leu, of similar function to the **KDEL** motif.

HDH *abbr. for* **histidinol dehydrogenase**.

HDL *abbr. for* high-density lipoprotein.

HDP *abbr. for* helix-destabilizing protein.

He *symbol for* helium.

head 1 the foremost or uppermost part of the body of an animal; in vertebrates it comprises the skull containing the mouth, brain, and the organs of hearing, sight, smell, and taste. **2** the part of a molecular structure that bears a specific functional group or that is polar, reactive, bulky, or in some other characteristic way differs from the remainder of the molecule. **3** *a common name for* a centrifuge rotor.

headgroup of a polar lipid, that part of the molecule that expresses its polar character. For a phospholipid, this comprises the phosphate group together with any polar entity attached to it. For a glycolipid, it comprises the carbohydrate moiety.

headspace analysis a technique used in the identification of microorganisms involving gas-chromatographic analysis of the vapour in the headspace above the specimen or culture.

heat *symbol*: q *or* Q; the energy possessed by a system in the form of kinetic energy of atomic or molecular translation, rotation, or vibration. It is measured in joules or calories. Heat energy can be transferred from points of higher temperature to points of lower temperature by conduction, convection, or radiation. When the temperature of a system changes, its enthalpy (heat content) changes by an amount equal to the product of its mass, specific heat capacity, and its change in temperature.

heat capacity *symbol*: C; the amount of heat required to raise the temperature of a body or a system by 1 kelvin; it is usually measured in joules per kelvin. The **heat capacity at constant pressure**, C_p, is given by: $C_p = (\partial H/\partial T)_p$ J K^{-1} and the **heat capacity at constant volume**, C_v, is given by: $C_v = (\partial U/\partial T)_v$ J K^{-1} where H is the enthalpy, U the internal energy, and T the thermodynamic temperature.

heat content *another name for* enthalpy.

heat-labile describing molecules that decompose at elevated temperatures or that lose biological activity when heated to a moderate temperature, e.g. 50°C.

heat of activation the difference in heat content between a transition state and the state of its reactants.

heat of combustion the amount of heat evolved when a given amount, usually one mole, of a substance is completely burnt in dioxygen.

heat of evaporation *an alternative name for* heat of vaporization.

heat of formation the amount of heat absorbed or evolved when one mole of a substance is formed from its elements.

heat of fusion *or* **heat of melting** the amount of heat absorbed or evolved when a specified quantity of substance, usually one mole, is converted from the solid to the liquid state at a specified pressure and a specified temperature. If the pressure is 1 atm the temperature is the melting point of the substance.

heat of neutralization the amount of heat absorbed or evolved when one mole of an acid or base is completely neutralized.

heat of reaction the amount of heat absorbed or evolved in the course of a (bio)chemical reaction, usually expressed per mole of reacting substances.

heat of solution the amount of heat absorbed or evolved when one mole of a substance is dissolved in a large volume of specified solvent.

heat of vaporization *or* **heat of evaporation** the amount of heat absorbed or evolved when a specified amount of substance, usually 1 mole, is converted from the liquid to the vapour state at a specified pressure and a specified temperature. If the pressure is 1 atm the temperature is the boiling point.

heat-shock protein *abbr.*: hsp; any of a group of specific proteins that are synthesized by both prokaryotic and eukaryotic cells on exposure to temperatures that are higher than normal, or to other stresses, such as free-radical damage. Many members of the hsp family are not induced, but are normally present in all cells. Human cells in culture also produce heat-shock proteins, but induction of heat-shock proteins has not been clearly demonstrated in the body. Heat-shock proteins are classified, according to their size, into three types: Hsp60, Hsp70, and Hsp90. They are characterized by their role as molecular **chaperones**. A major chaperone in the lumen of the endoplasmic reticulum is **binding protein** (BiP), a member of the Hsp70 class. These proteins are highly conserved, those from *Escherichia coli* and humans being 50% identical in primary structure. The Hsp70 proteins possess ATPase and peptide-binding domains. The other members of the Hsp70 class are Grp75 in mitochondria, and **DnaK** in bacterial cytosol. Hsp60 is also found in the mitochondrial matrix; the prokaryotic orthologue of this class is GroEL (*see* **groEl**).

heat-shock regulatory element *or* **heat-shock response element** *abbr.*: HSE; a base sequence in prokaryotic and eukaryotic genomes responsible for regulating gene expression in response to heat-shock and related stress situations.

heat-shock response element *an alternative name for* **heat-shock regulatory element**.

heat-stable describing something that retains its (bio)chemical activity when heated to a moderate temperature, e.g. 50°C.

heavy (*in physics*) of more than the usual mass. *Compare* **light** (def. 2).

heavy atom *or* **heavy isotope** *or* **heavy nuclide** an isotopic form of an atom that contains more than the common number of neutrons and is thus of greater relative atomic mass than the most abundant or most commonly observed isotope.

heavy carbon *see* **carbon-13**.

heavy chain *or* **H chain** the heavier of the two types of polypeptide chain found in **immunoglobulin** molecules. Each heavy chain is linked, usually by disulfide bonds, to a **light chain** and to another, identical heavy chain. Heavy chains differ in relative molecular mass according to the type of immunoglobulin: in humans the M_r is ≈50 000 in IgG and ≈70 000 in IgM. Each heavy chain consists of an **Fc fragment** and an **Fd fragment**. Heavy chains carry the antigenic determinants that differentiate the various immunoglobulin classes.

heavy-chain disease a rare disease of humans in which there are tumours of lymphoid tissue associated with the presence of free immunoglobulin fragments in plasma and urine.

heavy-chain switch *or* **H-chain switch** the change in the expression of the immunoglobulin **heavy chain** (H chain) gene system as manifested by the predominance of IgM (μ chain class) in the primary immune response and by the predominance of IgG (γ chain class) in the **secondary immune response**.

heavy hydrogen *an alternative name for* **deuterium**.

heavy meromyosin *abbr.*: H-meromyosin *or* HMM; a 350 kDa fragment produced when **myosin** is subjected to tryptic digestion. It contains part of the helical, rod-shaped tail and both globular heads of the original myosin molecule, and hence retains both the ATPase and the actin-binding activity of myosin.

heavy metal a term variously applied to the salts of certain metallic elements of atomic number greater than 11, usually referring to their toxicity and preference for covalent bonding (hence calcium and magnesium are omitted). There is no agreed and consistent definition of a heavy metal.

heavy nitrogen the stable nuclide $^{15}_{7}N$ (nitrogen-15); relative abundance in natural nitrogen 0.37 atom percent. It is used as a tracer in studies of nitrogen metabolism.

heavy oxygen the stable nuclide $^{18}_{8}O$ (oxygen-18), usually accompanied by the stable nuclide $^{17}_{8}O$ (oxygen-17); relative abundances in natural oxygen: $^{18}_{8}O$, 0.20 atom percent; $^{17}_{8}O$, 0.037 atom percent. $^{18}_{8}O$ is used as a tracer in studies of reaction mechanisms and of oxygen metabolism.

heavy strand *or* **H strand 1** any polynucleotide chain labelled with a heavy isotope, e.g. nitrogen-15. **2** any naturally occurring chain in a polynucleotide duplex that is heavier or has a greater density than the complementary strand.

heavy water deuterium oxide; *symbol*: ^2H_2O *or* D_2O; water in which the hydrogen atoms in all of the molecules have been replaced by deuterium; or water in which the hydrogen is appreciably enriched in deuterium. The M_r of pure ^2H_2O is 20.028. Heavy water is prepared from natural water by exchange techniques or by fractional distillation or electrolysis. It is used as a tracer in metabolic studies, in investigation of reaction mechanisms, and as a solvent in resonance spectroscopy.

Hechtian stand a threadlike structure that connects protoplasts of plasmolysed plant cells to their cell wall. Hechtian stands are tubes of cytoplasm delineated by plasma membrane that retain tight contacts with the cell wall, possibly via integrin-type receptors or **WAK1**.

hect+ *see* **hecto+** (def. 2).

hecto+ 1 *symbol*: h; an SI prefix denoting 10^2 times. **2** *or (before a vowel)* **hect+** denoting 100 or 100 times.

hedgehog protein a transmembrane protein involved in segment polarity and cell-to-cell signalling during embryogenesis and metamorphosis in *Drosophila melanogaster*. *See also* **sonic hedgehog proteins**.

HEDTA *abbr. for* N-(2-hydroxyethyl)ethylenediamine triacetate, or its trisodium salt, or for its partially or fully protonated forms. It is a chelating agent, especially for ferric ions in the pH range 7.0–10.0.

height equivalent to an effective theoretical plate *abbr.*: HEETP; *symbol*: *H*; *(in chromatography)* the column length divided by the **effective theoretical plate number**.

height equivalent to a theoretical plate *or* **plate height** *abbr.*: HETP; *symbol*: *h*; *(in chromatography)* the column length divided by the **theoretical plate** number.

Heisenberg uncertainty principle the principle that the simultaneous precise determination of the velocity (or any related property, e.g. momentum or energy) of a material particle and its position is impossible; the smaller the particle the greater the degree of uncertainty. [After Werner Karl Heisenberg (1901–76), German physicist; Nobel Laureate in Physics (1932).]

HeLa cell an established tissue culture strain of human epidermoid carcinoma cells containing 70–80 chromosomes per cell (compared with 46 in normal cells). It is much used for biochemical work. HeLa cells have been in continuous culture since 8 February 1951, and were derived from tissue removed from a patient named Henrietta Lacks.

helical cross *a name sometimes given to* the X-shaped pattern of X-ray scattering from a continuous helical molecule.

helical wheel *(in protein sequence analysis)* a circular graph depicting five turns of helix (≈20 residues), around which residues in a protein sequence are plotted 100° apart (3.6 residues per 360° turn) for an alpha helix. Different parameters are used to depict helices with different pitches (e.g. 120° for a 3_{10} helix, 160° for beta strands). Helical wheels are particularly useful for depicting amphipathic character (recognized by the clustering of hydrophilic and hydrophobic residues in distinct polar and nonpolar arcs); however, they are not predictive.

helicase *a name sometimes given to* the protein (**repA**) that promotes the ATP-driven unwinding of the parental DNA duplex during DNA replication. *See also* **DNA helicase**, **RNA helicase**.

Helicobacter pylori a Gram-negative bacterium that colonizes the human gastric mucosa and is acquired orally in infancy. There are several types and subtypes, with distinct geographical distributions. It is a common cause of gastric ulcer and gastric cancer. The major virulence factor in ulcer formation is the toxin Vac A, which, *in vitro*, disrupts gastric epithelial cell integrity in many ways. In the mouse, Vac A gastric injury is mediated by protein tyrosine phosphatase receptor z (Ptprz) of epithelial cells.

helicorubin a *b*-type cytochrome that occurs in the hepatopancreas of the snail *Helix pomatia* and related species.

helix (*pl.* **helices**) **1** a coiled or spiral structure, e.g. the thread of a bolt or a coil (tubular) spring. *See also* **alpha helix**, **beta helix**, **gamma helix**. **2** the curve traced on the surface of a cylinder or cone by a point crossing its right sections at a constant angle. **3** a space curve with turns of constant angle to the base and constant distance from the axis. —**helical** *adj.*; **helicity** *n.*

helix breaker an amino-acid residue that has a propensity to interrupt α-helical structures when it occurs in a polypeptide chain; e.g., proline (in globular proteins – in membrane proteins, proline introduces a distinctive kink, but does not break the helix), hydroxyproline. *Compare* **helix former**.

helix-destabilizing protein *abbr.*: HDP; any of a group of proteins that bind tightly and preferentially to single-stranded DNA. Theoretically, they: (1) lower the melting temperature, T_m, for thermally induced **helix-to-coil transition**; (2) lower the optimum temperature for duplex renaturation; (3) sterically hinder base pairing and degradation of single strands by nucleases; (4) induce a conformation for the single-stranded DNA that is optimal for its function in some processes, e.g. as a template in replication; and (5) interact with other proteins (enzymes) to potentiate processes involving single-stranded DNA and these proteins.

helix former an amino-acid residue that has a propensity to promote α-helical structures when it occurs in a polypeptide chain. *Compare* **helix breaker**.

helix-loop-helix *abbr.*: HLH; *an alternative term for* **helix-turn-helix**.

helix-to-coil transition the transition of a macromolecule (polypeptide or polynucleotide) from an ordered to a disordered structure; it is equivalent to **denaturation**. Helix-to-coil transitions are usually detected by monitoring a change in some physical property of the macromolecule such as intrinsic viscosity, optical absorbance, or sedimentation coefficient. *See also* **melting temperature**.

helix-turn-helix (*abbr.*: HTH) *or* **helix-loop-helix** (*abbr.*: HLH) a structural feature of several DNA-binding regulatory proteins consisting of two helices separated by a **beta turn**.

helix winding number *see* **supercoil**.

Helmholz free energy *or* **Helmholz function** *abbr.*: *F* or *A*; free energy at constant volume.

helodermin *see* **exendin**.

helospectin I *see* **exendin**.

helper cell *abbr.*: T_H cell; a **T** lymphocyte, carrying the surface markers CD4 and CD5 in the human (L3T4 and Ly1 in the mouse), that acts as an inducer of the effector cells for both **humoral** and **cell-mediated immunity**. Helper cells recognize and bind to antigen in combination with class II **major histocompatibility complex** molecules on the surface of antigen-presenting cells, and generate lymphokines that stimulate effector cytotoxic cells and B cells. Other cells of the helper phenotype, e.g. $CD4^+$ cells, are known as T-**suppressor** cells and are responsible for suppression, exerting negative feedback control on the helper cells.

helper peak-1 *abbr.*: HP-1; *a former name for* **interleukin 1**.

helper phage a bacteriophage that is capable of infecting *E. coli* cells and directing the synthesis of proteins necessary for phage replication but is itself unable to replicate owing to mutations it carries. Cloning vectors, such as lambda ZAP, which have an M13 origin of replication, can be induced to replicate in single-stranded form if the *E. coli* cells that harbour them are superinfected with helper phage.

helper virus any virus that supplies one or more of the functions that a **defective virus** is unable to perform. When a cell is infected with both a defective virus and a helper virus, the former is able to replicate, a property that can be exploited in various molecular biological techniques.

hem+ *or (esp. Brit.)* **haem+** *a variant form of* **hemo+** (before a vowel).

hema+ *or (esp. Brit.)* **haema+** *a variant form of* **hemo+**.

hemadsorption the adsorption of an agent or substance to or on the surface of a red blood cell.

hemagglutination *or (esp. Brit.)* **haemagglutination** the agglutination of red blood cells.

hemagglutinin *or (esp. Brit.)* **haemagglutinin 1** any **agglutinin** of red blood cells. **2** any nonantibody substance, e.g. a lectin or a surface component of some virus particles, that can agglutinate red blood cells. **3** a viral envelope glycoprotein responsible for the attachment of virus to cell receptors and the initiation of infection.

hemat+ *or (esp. Brit.)* **haemat+** *a variant form of* **hemato+** (before a vowel).

hematin *or (esp. Brit.)* **haematin 1** *or* **ferriheme hydroxide** *the trivial name for* ferriprotoporphyrin hydroxide (*formerly called* ferriprotoporphyrin IX hydroxide); hydroxo(protoporphyrinato) iron(III); hydroxoiron(III) protoporphyrinate. *Compare* **heme** (def. 1), **hemin** (def. 1). **2** *the generic name for* any hydroxo(porphyrinato)iron(III) coordination complex.

hemato+ *or (before a vowel)* **hemat+** *comb. form* denoting blood. *Also (esp. Brit.)*: **haemato+, haemat+.** *See also* **hemo+**.

hematocrit *or (esp. Brit.)* **haematocrit 1** *or* **packed red cell volume** the proportion of the volume of a sample of blood that is represented by the red blood cells. **2** an apparatus, essentially a graduated centrifuge tube, used to determine the proportion of the volume of a sample of blood that is represented by the red blood cells.

hematocrit technique a method for the rapid isolation of mitochondria or other particles, in which they are separated from the medium by centrifugation through a layer of silicone fluid. The latter is immiscible with water, and of a density intermediate between that of the particles to be separated and that of the medium. A high-density aqueous 'fixative' beneath the silicone layer can be used to stabilize labile compounds in the mitochondria or other particles. *Compare* **hematocrit** (def. 2).

hematology *or (esp. Brit.)* **haematology** the branch of medicine concerned with diseases of the blood or of the blood-forming tissues.

hematolysis *or (esp. Brit.)* **haematolysis** *an alternative name for* **hemolysis**.

hematopoiesis *or (esp. Brit.)* **haematopoiesis** *an alternative name for* **hemopoiesis**.

hematoporphyrin *or (esp. Brit.)* **haematoporphyrin** *or (formerly)* **hematoporphyrin IX** *or (esp. Brit.)* **haematoporphyrin IX** *the trivial name for* 2,7,12,18-tetramethyl-3,8-bis(1-hydroxyethyl) porphyrin-13,17-dipropionic acid; the porphyrin formed from **heme** (def. 1) or a heme-containing protein on treatment with hydrogen bromide in glacial acetic acid.

hematoside *or (esp. Brit.)* **haematoside** a **ganglioside**, NeuAc(α2-3)Gal(β1-4)Glc(β1-1)Cer, found in red blood cells.

hematoxylin *or (esp. Brit.)* **haematoxylin** 7,11b-dihydrobenz[*b*]indeno[1,2-*d*]pyran-3,4,6*a*,9,10(6*H*)-pentol; a dye isolated from the heartwood of logwood, *Haematoxylon campechianum* L. Colourless or yellowish, it turns red on exposure to light, and is used as a stain in microscopy.

hematuria *or (esp. Brit.)* **haematuria** the presence of red blood cells in the urine. It can be due to lesions in the urinary tract, but also to glomerular disease of the kidney.

heme *or (esp. Brit.)* **haem 1** *or* **ferroheme** *or* **ferrohaem** *or* **protoheme** *the trivial name for* **ferroprotoporphyrin** (*formerly called* ferroprotoporphyrin IX); 2,7,12,18-tetramethyl-3,8-divinylporphyrin- 13,17-dipropionic acid iron(II) coordination complex. It occurs free and as the prosthetic group of a number of hemoproteins, e.g. hemoglobins, erythrocruorins, myoglobins, some peroxidases, catalases, and cytochromes *b*. *Compare* **hematin** (def. 1), **hemin** (def. 1). **2** *the generic name for* any iron–porphyrin coordination complex irrespective of the valence state of the iron atom (which may be specified). The name of an individual heme is derived from that of the corresponding porphyrin; e.g., cytoheme, from cytoporphyrin, and ferroprotoheme, from protoporphyrin. For biosynthesis, *see* **tetrapyrrole**.

heme (def. 1)

heme–heme interaction *or (esp. Brit.)* **haem–haem interaction** the cooperative interaction between dioxygen binding sites that occurs when dioxygen binds to hemoglobin; as oxygenation proceeds, the combination of additional dioxygen molecules is made easier.

heme oxygenase *abbr.*: HO; EC 1.14.99.3; a microsomal enzyme, mostly of liver and spleen, that converts heme (in the presence of O_2) to α-hydroxyhemin, which converts noncatalytically to biliverdin and carbon monoxide. It requires NADPH and NADPH–hemoprotein reductase (EC 1.6.2.4). Three isozymes are known of which HO 1 is heme-inducible. Deficiency of HO 1 (encoded at 22q22.1) is very rare and is associated with severe anemia, high levels of circulating heme, enhanced endothelial cell damage, and early death.

heme pocket *or (esp. Brit.)* **haem pocket** a hydrophobic crevice in the tertiary structure of myoglobin, or hemoglobin subunit, into which the heme group fits. The non-polar vinyl side chains of the heme are buried in the interior of the pocket, while the hydrophilic propionate side-chains project out of the pocket towards the surface of the molecule.

heme protein *or (esp. Brit.)* **haem protein** *an alternative name for* **hemoprotein**.

hemerythrin *or (esp. Brit.)* **haemerythrin** an oxygen carrier found in a few groups of invertebrates, e.g. sipunculid worms, certain molluscs and crustaceans. It is a non-heme, iron-containing protein, the subunits of which each contain about 113 amino-acid residues. Each monomer has an active site containing two atoms of Fe(II), 0.34 nm apart, between which a dioxygen molecule is thought to fit. The molecule is a homooctamer.

heme synthetase *see* **ferrochelatase**.

heme–thiolate protein *or (esp. Brit.)* **haem–thiolate protein** *see* **P450**.

hemi+ *prefix* denoting half, or affecting one half. *Compare* **semi+**.

hemiacetal *see* **acetal**.

hemicellulose any or all of the cell-wall polysaccharides that are extractable from wood or other plant material with aqueous alkali (in which cellulose is insoluble). Hemicelluloses are chiefly xylans, but other homoglycans or heteroglycans containing hexose and/or pentose residues, uronic acid-containing xylans, and other **pectic substances** may also be present.

hemicystine *an alternative name for* **half-cystine**.

hemicystyl *an alternative name for* **half-cystyl**.

hemidesmosome a specialized cell junction between an epithelial cell and its underlying basal lamina. These resemble **desmosomes** morphologically but are chemically and functionally distinct. Between the cell membrane and the underlying connective tissue is a specialized mat of extracellular matrix sometimes known as the linker. The transmembrane linker proteins belong to the **integrin** family of extracellular matrix receptors rather than the **cadherin** family of cell–cell adhesion proteins found in desmosomes.

hemidiaphragm a muscle preparation widely used for metabolic studies *in vitro*, consisting of either of the two excised and separated left and right halves of the diaphragm, usually from a rat.

hemiketal *see* **acetal**.

Hemimetabola *or* **Exopterygota** one of the two divisions of the subclass Pterygota, comprising those insects, e.g. cockroaches, locusts, and grasshoppers, in which the juvenile forms resemble the adults, except for their lack of fully developed wings. Such insects undergo incomplete metamorphosis. *Compare* **Holometabola**.

hemin *or (esp. Brit.)* **haemin 1** *or* **ferriheme chloride** *or* **protohemin** *the trivial name for* ferriprotoporphyrin chloride (*formerly called* ferriprotoporphyrin IX chloride); chloro(protoporphyrinato)iron(III); chloroiron(III) protoporphyrinate. *Compare* **heme** (def. 1), **hematin** (def. 1). **2** *the generic name for* any chloro(porphyrinato)-iron(III) coordination complex.

hemiplegic migraine a rare familial autosomal dominant form of migraine that is marked by transient hemiplegia during the aura. It is caused by several point mutations in a gene at 19p13 for the α1A subunit (CACNA1A, calcium channel alpha subunit 1A) of a voltage-gated calcium channel present in neurons. The protein contains 2261 amino acids, with four repeated domains each containing four transmembrane segments, and a polyglutamine tract in the C-terminal region. Expansion of this tract leads to episodic ataxia and to spinocerebellar ataxia type 6.

hemiport a model of translocation of solutes across membranes, based on the possibility that there is interaction through the membrane between superficially disposed protein subunits. These subunits are envisaged as being embedded within the bimolecular lipid layer, anchored to the aqueous phase, and perhaps mobile in the plane of each face of the membrane. The subunits bind the solute in question and, when associated with the symmetrical protein at the opposite face of the membrane, transfer the solute across the membrane.

hemithioacetal *see* **acetal**.

hemithioketal *see* **acetal**.

hemizygous describing an organism in which a gene is present only once in a genotype, as a gene in a haploid cell or organism, a sex-linked gene in the heterogametic sex, or a gene in a segment of chromosome in a diploid cell or organism where its partner segment has been deleted. —**hemizygosity** *n*.

hemo+ *or* **hema+** *or (before a vowel)* **hem+** *comb. form* denoting blood. *Also (esp. Brit.)*: **haemo+, haema+, haem+**. *See also* **hemato+**.

hemochromatosis *or (esp. Brit.)* **haemochromatosis** an iron-overload disease of humans characterized by widespread, massive accumulation of tissue iron and pathological changes of tissue structure and function. **Primary hemochromatosis** results from increased absorption of iron from a normal diet. This is usually caused by autosomal recessive mutations in the gene for HFE, an HLA-type protein (321 amino acids) that binds β_2-microglobulin and associates with the transferrin receptor in the gastrointestinal tract. About 10% of persons of European descent are heterozygotes for such mutations. More rarely, mutations in the transferrin receptor, or autosomal dominant mutations in the genes for ferritin H subunit or ferroportin, are the cause. **Secondary hemochromatosis** results from an increased intake and accumulation of iron secondary to other causes. *Compare* **hemosiderosis**.

hemochrome *or (esp. Brit.)* **haemochrome** any iron–porphyrin coordination complex with one or more strong-field axial ligands (e.g., pyridine).

hemochromogen *or (esp. Brit.)* **haemochromogen** any compound of heme with a nitrogenous base, such as a protein (whether native or denatured) or pyridine, typically with two absorption bands in the green part of the visible spectrum. *See also* **chromogen**.

hemocuprein *or (esp. Brit.)* **haemocuprein** *an alternative name for* **superoxide dismutase**.

hemocyanin *or (esp. Brit.)* **haemocyanin** a blue, copper-containing oxygen carrier present in many molluscs and arthropods. It is a non-heme protein that binds one dioxygen molecule for two Cu(I) atoms; the dioxygen molecule is thought to form a bridge between the two copper atoms. The oxygenated compound is bright blue, with an intensity of light absorption 5–10 times that of other known copper complexes. For example, hemocyanin from lobster, *Panulirus interruptus,* is a hexamer of a number of different chains, of which A, B (88% identity between these two), and C have been identified.

hemocyte *or (esp. Brit.)* **haemocyte** any blood cell, especially of an invertebrate animal.

hemocytometer *or (esp. Brit.)* **haemocytometer** an apparatus for counting the number of cells in a known volume of blood. It consists of a slide with a chamber of known depth; the base of the chamber is ruled with a graticule over which the cells can be enumerated under a microscope.

hemodialysis *or (esp. Brit.)* **haemodialysis** the process of separating low-molecular-mass solutes from blood by **dialysis** through a semipermeable membrane, as in an artificial kidney machine.

hemoglobin *or (esp. Brit.)* **haemoglobin** *symbol*: Hb; any of a group of red, iron-containing, oxygen-carrying pigments of the blood of vertebrates and some invertebrates; hemoglobin also occurs in the root nodules of leguminous plants. All vertebrate hemoglobins consist of two pairs of associated **globin** polypeptide chains, each chain carrying a heme prosthetic group bound non-covalently, the iron atom of which is in the ferrous state and forms a coordination complex with the pyrrole nitrogens. All normal human hemoglobins contain one pair of 15.7 kDa α chains, of 141 amino-acid residues, and a pair of varying 16.5 kDa polypeptide chains, of 146 amino-acid residues and of similar amino-acid sequence. In adult human blood, 98% of the total hemoglobin is Hb A (or Hb A$_1$), containing a pair of β chains (i.e., it has composition $\alpha_2\beta_2$) and 2% is Hb A$_2$, containing a pair of δ chains (i.e., $\alpha_2\delta_2$). In fetal blood, from about 10 to 30 weeks of gestation, >90% of the hemoglobin is Hb F, containing a pair of γ chains (i.e., $\alpha_2\gamma_2$). In embryonic blood, the second pair of chains is ε chains (i.e., the structure is $\alpha_2\epsilon_2$). Each of the four heme groups in a hemoglobin molecule is able to combine reversibly with one dioxygen molecule; the oxygen-saturation curve of hemoglobin is sigmoidal, showing that there is positive cooperativity between the subunits. The hemoglobin molecule has a two-fold axis of symmetry, each half containing one α chain and one non-α chain; the overall shape of the molecule is globular, with the heme groups buried in pockets in the polypeptide chains. There are eight helical regions, designated A to G. The histidine at position F8 (i.e., the eighth residue in helix F) is the fifth ligand of the heme Fe(II). Upon oxygenation, dioxygen becomes the sixth ligand of Fe(II), and is also hydrogen bonded to HisE7, with an accompanying conformational change. In the absence of the correct bonding with the globin side chains, Fe(II) autooxidizes to Fe(III) and can no longer bind dioxygen. Among human α chain variants are Luxembourg, Ann Arbor, and Hirosaki. Among the **hemoglobinopathies** associated with β-chain variants is **sickle-cell anemia**. In those invertebrates that have hemoglobin (not **hemocyanin**), the protein is polymeric: *e.g.*, in earthworm hemoglobin, 12 subunits are arranged in a hexagonal bilayer structure of 3.8 MDa, each subunit of which is composed primarily of disulfide-linked trimers (chains A, B, and C) and monomers (chain D). *See also* **hemoglobin A$_{1c}$**.

hemoglobin A₁c *or (esp. Brit.)* **haemoglobin A$_{1c}$** *abbr.*: Hb A$_{1c}$; the most abundant of the **glycated hemoglobins**, which are present in normal adult hemoglobin as minor components (together, they represent 3–5% of the total hemoglobin of the erythrocyte). Hb A$_{1c}$ is formed from Hb A$_1$ subsequent to biosynthesis by glycosylation of one of the two β chains, by a non-enzymic mechanism involving formation of an N-substituted aldimine (Schiff's base) between the amino group of the peptide chain and the aldehyde group of glucose, followed by an **Amadori rearrangement** to the corresponding N-substituted ketoimine (aminodeoxyfructose derivative) (*see also* **Maillard reaction**). In inadequately controlled **diabetes mellitus**, the steady-state concentration of Hb A$_{1c}$ in serum is 2–3 times higher than normal, possibly reflecting an abnormally high mean blood glucose concentration during the preceding few months. Other glycosylated hemoglobins change very little in concentration in diabetes; thus, measurement of serum levels of Hb A$_{1c}$ is considered to be useful in assessing the degree of control of the condition.

hemoglobinopathy *or (esp. Brit.)* **haemoglobinopathy** any genetic disorder involving globin, the protein moiety of **hemoglobin**. These are classified into **thalassemia** syndromes, caused by defective synthesis of the component polypeptides of globin, and conditions that arise from mutations leading to amino-acid substitutions in the polypeptide chains. Of the latter there are over 850 known variants, including the classically important **sickle cell** hemoglobin. Many changes lead to increased oxygen affinity, e.g. Hb Rainier (β-145, Tyr → Cys), while in others decreased oxygen affinity occurs, e.g. Hb Kansas (β-102, Asn → Thr). Many substitutions lead to an unstable hemoglobin molecule, as in Hb Hammersmith (β-42, Phe → Ser) where the important heme contact of the phenylalanine is lost, leading to the formation of **ferrihemoglobin**, or in Hb Bristol (β-67, Val → Asp) where replacement of the nonpolar valine by aspartic acid causes distortion of the E helix. Some substitutions lead to congenital methemoglobinemia, as in Hb MBoston (α-58, His → Tyr) and Hb MSaskatoon (β-63, His → Tyr). Some variant globins are elongated whereas others are shortened.

hemolin *or (esp. Brit.)* **haemolin** an insect hemolymph protein of the immunoglobulin superfamily, containing four internal immunoglobulin-type repeats. It is induced by bacteria.

hemolymph *or (esp. Brit.)* **haemolymph** a fluid occurring in the secondary body cavity (coelom) of some invertebrates, considered as functionally equivalent to the blood and lymph of higher animals.

hemolysate *or (esp. Brit.)* **haemolysate** a preparation obtained by lysis of erythrocytes.

hemolysin *or (esp. Brit.)* **haemolysin 1** any substance, e.g. an antibody, that causes hemolysis. **2** any bacterial exotoxin that ruptures blood-cell membranes. In hemolytic *Escherichia coli* strains that cause predominantly nongastrointestinal (e.g. urinary tract) infections, four genes are involved: *hlyA* encodes hemolysin (Ca^{2+}-binding toxin); *hlyB* encodes an **ABC transporter** for HlyA export; *hlyC* encodes a protein involved in hemolysin export; and *hlyD* encodes a transmembrane protein component of the HlyA export system. *Staphylococcus aureus* secretes **α-hemolysin**, a protein (293 amino acids) that binds erythrocytes and some other cells. It forms a mushroom-shaped homoheptameric channel to which each subunit contributes two antiparallel beta strands. It is responsible for the hemolysis seen around colonies grown on blood agar plates.

hemolysis *or (esp. Brit.)* **haemolysis** the lysis of red blood cells, either *in vivo* or *in vitro*. —**hemolyse** *or* **haemolyse** *vb*.

hemolytic *or (esp. Brit.)* **haemolytic 1** of, pertaining to **hemolysis**. **2** an agent or condition that causes hemolysis.

hemopexin *or (esp. Brit.)* **haemopexin** a 57 kDa β$_1$-glycoprotein of human blood serum with one high-affinity binding site for **heme** (def. 1). The heme complex can be reduced with dithionite to give a three-band absorption spectrum characteristic of ferrohemochrome. Hemopexin may be necessary for heme to be taken up and degraded in the liver.

hemophilia *or (esp. Brit.)* **haemophilia** any of various hereditary disorders in which there is a deficiency or defect in certain of the **blood coagulation** factors resulting in prolonged bleeding following injury. Classical hemophilia, or **hemophilia A**, is a condition in which the blood has a prolonged clotting time due to an abnormally slow conversion of prothrombin to thrombin. It is caused by a functional deficiency of antihemophilic factor (factor VIII), inherited as an X-linked recessive trait. Other deficiencies in the blood coagulation system, such as **Christmas disease**, or **hemophilia B**, due to a deficiency of factor IX, produce similar clinical pictures.

hemopoiesis *or* **hematopoiesis** the process of blood formation, especially the formation of blood cells, occurring mainly in bone marrow. *Also (esp. Brit.)*: **haemopoiesis, haematopoiesis**. —**hemopoietic, hematopoietic** *or* **haemopoietic, haematopoietic** *adj*.

hemoprotein *or* **heme protein** any protein to which an iron–porphyrin compound is linked in a stoichiometric manner. Hemoproteins include hemoglobins, myoglobins, cytochromes, catalase, and some peroxidases. *Also (esp. Brit.)*: **haemoprotein, haem protein**.

hemorphin *or (esp. Brit.)* **haemorphin** any of a group of hemoglobin-derived peptides with affinity for opioid receptors.

hemorrhage *or (esp. Brit.)* **haemorrhage** bleeding; the escape of blood from a ruptured blood vessel, either externally or internally. —**hemorrhagic** *or (esp. Brit.)* **haemorrhagic** *adj*.

hemorrhagic toxin C and D *see* atrolysin C.

hemosiderin *or (esp. Brit.)* **haemosiderin** an insoluble, granular, ill-defined complex formed by the denaturation of **ferritin**, with associated loss of apoferritin and micellar aggregation; about one third of its mass is iron. Hemosiderin occurs in liver, spleen, and red bone marrow, its amount increasing with increased iron content of the body. It is responsible for the histochemical staining of tissue by the Prussian Blue reaction, a deep-blue ferric ferrocyanide (Fe$_{III}$[Fe$_{II}$(CN)$_6$]). The distribution of stored iron is shifted from ferritin to hemosiderin in direct correlation with ascorbic acid deficiency. *See* **iron overload**.

hemosiderosis *or (esp. Brit.)* **haemosiderosis** an abnormal increase in tissue iron content due to some defect in the control of iron turnover in the body or to increased iron ingestion. In a rare pulmonary form, recurrent bleeding into the lungs loads the macrophages with hemosiderin and produses diffuse fibrosis. Hepatic hemosiderosis is common in alcoholic cirrhosis. *Compare* **hemochromatosis**.

hemostasis *or (esp. Brit.)* **haemostasis 1** the stopping of bleeding or the arrest of the circulation to an organ or part. **2** the stagnation of blood in an organ or part.

hemostat *or (esp. Brit.)* **haemostat** a device or a chemical agent that stops or retards bleeding.

hemostatic *or (esp. Brit.)* **haemostatic 1** of, relating to, or functioning to cause **hemostasis** (def. 1). **2** a drug or other agent that retards or stops bleeding.

Henderson–Hasselbalch equation an equation relating pH to the composition of buffer solutions:

$$pH = pK_A + \log([\text{conjugate base}]/[\text{acid}]),$$

where pK_A is the common logarithm of the acid dissociation constant of the buffer acid. The following approximation is often used:

$$pH = pK_A + \log([\text{salt}]/[\text{acid}]).$$

[After Lawrence Joseph Henderson (1878–1942), US physician and biochemist, and Karl A. Hasselbalch (1874–), Danish biochemist, who together formulated the equation in 1910.]

henna a preparation obtained from the dried and powdered leaves of *Lawsonia alba*, *L. inermis*, or *L. spinosa*, widely used for dyeing hair and fingernails. The active ingredient is **lawsone**.

henry (*pl.* **henrys** *or* **henries**) *symbol*: H; the SI derived unit of inductance, equal to the inductance of a closed circuit with a **magnetic flux** of 1 weber per ampere of current,

$$H = Wb \, A^{-1} \, s = m^2 \, kg \, s^{-2} \, A^{-2}$$

[After Joseph Henry (1797–1878), US physicist.]

Henry's function a function used in the theoretical treatment of the electrophoretic mobility of macromolecules that is dependent on the radius of the macromolecule, r, and its reciprocal ion-atmosphere radius, κ (derived from **Debye–Hückel theory**). It varies between 1.0 and 1.5 as κr goes from zero to infinity.

Henry's law the principle that the concentration of a gas in solution, at constant temperature, is directly proportional to the partial pressure of the gas in the vapour phase in equilibrium with that solution. It holds with high precision for gases of low solubility at pressures below approximately 10^5 Pa. It may be expressed by: $c_G = kp_G$, where c_G is the molar concentration of the gas in solution, p_G is the partial pressure of the gas in the gas phase, and k is a proportionality constant. [After William Henry (1774–1836), British chemist.]

HEPA filter a high-efficiency particulate air filter used, for example, in removing bacteria and other particulates from air in safety cabinets.

heparan any polysaccharide derived by desulfation of **heparan sulfate** or **heparin**.

heparan sulfate any member of a group of glycosaminoglycans that have repeat units consisting of alternating $\alpha1\rightarrow4$-linked hexuronic acid and glucosamine residues, the former being a mixture of sulfated and nonsulfated D-glucuronic and L-iduronic acids, and the latter being either sulfated or acetylated on its amino group as well as sulfated on one of its hydroxyl groups. They usually occur attached to protein to form proteoglycans, the mode of attachment being through a xylose residue as in **chondroitin sulfates**. Heparan sulfate accumulates in a number of the mucopolysaccharidoses, e.g. Hunter syndrome, Hurler syndrome, and Sanfilippo syndrome. *See* **mucopolysaccharidosis**. *See also* **heparin**.

heparan sulfate proteoglycan core protein *abbr.*: HSPG core protein; any of a group of modular heparan sulfate proteoglycan proteins with three heparan sulfate chains. Also known as **perlecans**, they contain domains that share similarity with the low-density lipoprotein receptor, epidermal growth factor, laminin and neural cell adhesion molecule, and have 15 immunoglobulin-like domains.

heparin any glycosaminoglycan found mainly in mast cells. Heparins are similar to **heparan sulfates** but somewhat larger (6–20 kDa) and with fewer *N*-acetyl groups and more *N*-sulfate and *O*-sulfate groups; they are proteoglycans. They consist predominantly of alternating $\alpha1\rightarrow4$-linked D-galactose and *N*-acetyl-*D*-glucosamine-6-sulfate residues. Heparins are secreted into the bloodstream by the mast cells of liver, lung, and other tissues; they act as inhibitors of blood clotting by activating **antithrombin III**. Commercial heparins are protein-free.

heparin cofactor II a plasma glycoprotein that is homologous with **antithrombin III** but, of the proteases involved in coagulation, only inhibits thrombin. Its physiological role is obscure, but its activity is stimulated by dermatan sulfate and by heparin. The gene at 22q11 encodes 480 amino acids.

heparinize *or* **heparinise** to treat with a preparation of **heparin**, usually as its sodium or lithium salt, in order to prevent clotting, either *in vivo* or *in vitro*, of blood or blood plasma. —**heparinization** *or* **heparinisation** *n*.

hepat+ *a variant form of* **hepato+** (before a vowel).

hepatectomy surgical removal of the liver. In **partial hepatectomy**, only some of the lobes of the liver are removed; this causes the remaining lobes to regenerate so that the liver *in toto* regains its former weight. —**hepatectomize** *or* **hepatectomise** *vb*.

hepatic of, or pertaining to, the liver.

hepatic lipase EC 3.1.1.3; an enzyme that hydrolyses triacylglycerols and phosphoproteins present in chylomicron and very-low-density lipoprotein remnants. It is a glycoprotein synthesized and secreted by hepatocytes, and binds to proteoglycans on hepatocyte surfaces and on sinusoidal endothelial cells. It also binds LDP (LDL receptor-like protein). The gene at 15q21 encodes 499 amino acids. Rare inactivating mutations in the gene result in an autosomal recessive syndrome of hepatic lipase deficiency.

hepatitis any inflammatory disease of the liver. In humans the most common cause is infection with a virus or an amoeba. Hepatitis A is caused by a single-stranded RNA **picornavirus**, for which there is a killed virus vaccine, virus having been grown in tissue culture. Hepatitis B is caused by a double-stranded DNA virus of the Hepadnaviridae family forming a spherical virion often referred to as the **Dane particle**. It is especially virulent and is associated with the development of primary liver tumours – the incidence of these is particularly high in China. The antigens may be in the form of spherical particles designated HBsAg; principal components include gp27, which is a glycosylated form of p24, and a 42 kDa protein. HBsAg is produced in recombinant yeast, the particles being similar to those found in the plasma of infected individuals and providing a very effective vaccine. Hepatitis C, formerly called 'non-A non-B', is caused by a single-stranded RNA virus, for which there is no vaccine. Hepatitis D virus is a satellite virus of hepatitis B virus and contains a circular single-stranded RNA genome (≈1700 nt). Its replication requires the antigen HDAg – the only protein encoded by the virus – and RNA polymerase II. HDAg-S (195 amino acids) activates replication of the virus. HDAg-L is the former with a 19-residue C-terminal extension; it inhibits replication and directs assembly of the virion. HDAg-S shares 21% sequence identity with NELF-A. Other forms are hepatitis E, F, and G.

hepato+ *or (before a vowel)* **hepat+** *comb. form* denoting of or pertaining to the liver.

hepatocarcinogen any **carcinogen** that acts specifically or primarily on liver.

hepatocellular of, or pertaining to **hepatocytes**.

hepatocyte *or (formerly)* **parenchymal liver cell** the major (but not the only) cell type of the liver. They are arranged in folded sheets facing blood-filled spaces called sinusoids. Hepatocytes are responsible for the synthesis, degradation, and storage of a wide range of substances. They are the site of synthesis of all the plasma proteins, except for antibody, and are the site of storage of **glycogen**.

hepatocyte growth factor *abbr.*: HGF; *other name*: scatter factor; a **cytokine** produced by platelets, fibroblasts, macrophages, endothelial cells, and smooth muscle cells that is a potent **mitogen** for hepatocytes, epithelial cells, keratinocytes, melanocytes, and hematopoietic precursor cells. It is a glycoprotein, with four **kringle** domains and no known protease activity. The molecule is a disulfide-linked dimer of α (463 amino acids) and β (234 amino acids) subunits, which are derived from a common precursor.

hepatocyte growth factor receptor a type I membrane protein that binds **hepatocyte growth factor** and mediates its intracellular action by a protein tyrosine kinase (EC 2.7.1.112) activity on its intracellular domain. It is encoded by the *MET* protooncogene, and comprises a dimer of two subunits, α and β. *See also* **tpr-met**.

hepatocyte nuclear factor *abbr.*: HNF; any of several transcription factors involved in activation of numerous liver-specific genes. Some are also present in pancreatic islet cells. **HNF-1α** is a homeobox transcription factor that requires dimerization cofactor (DCoH; *see* **pterin-4α-carbinolamine dehydratase**) for dimerization and stabilization. Numerous mutations in HNF-1α are associated with maturity-onset diabetes of the young (MODY) type 3. **HNF-**

1β is homologous with HNF-1α, and also requires DCoH for dimerization and stabilization. Several mutations are associated with MODY type 5. **HNF-4α** belongs to the nuclear receptor superfamily, and is a major regulator of hepatic gene expression and of HNF-1α. Mutations are associated with MODY type 1.

hepatoflavin *or* **hepatoflavine** *a former name for* **riboflavin**.

hepatolenticular degeneration *an alternative name for* **Wilson's disease**.

hepatoma *any* **carcinoma** derived from liver cells.

hepatopancreas a digestive gland, found in many invertebrates, that performs functions similar to those of the liver and the pancreas in vertebrates.

hepatosis any noninflammatory disease of the liver.

hepatotoxin any toxin that acts specifically or primarily on the liver.

hepcidin a peptide hormone with antimicrobial activity that regulates uptake and storage of iron. Elevated levels of hepcidin are associated with iron deficiency.

Hepes *or* **HEPES** *abbr. for (and trivial name of)* 4-(2-hydroxyethyl)-1-piperazine-ethanesulfonic acid; a **Good buffer substance**, pK_a (20°C) = 7.55.

HepG2 a cell line derived from a patient with primary liver cancer.

hephaestin a membrane-bound homologue of ceruloplasm that is required, with **ferroportin**, for the transport of iron across the basolateral plasma membrane of mucosal cells of the upper small intestine. It appears to act as a ferroxidase ($Fe^{2+} \rightarrow Fe^{3+}$) in the process. A hereditary sex-linked anemia occurs in mice in which the gene is mutated.

HEPN domain *abbr. for* higher eukaryotes and prokaryotes nucleotide-binding domain; a protein sequence (≈110 amino acids) whose prototype is in the C-terminal region of **sacsin**. It is present in proteins from several vertebrates and many eubacteria and archaea. Because it is similar to a domain present in many kanamycin nucleotidyltransferases it is thought to mediate nucleotide binding.

hepoxilin *abbr.*: Hx; any of various epoxyhydroxy metabolites of 12-hydroperoxyeicosatetraenoate, e.g. 13-hydroxy-14,15-epoxy-eicosatetraenoate. A number of sites of formation have been reported, with activities including modulation of neurotransmission in brain and release of insulin in pancreatic islets.

hepoxilin-epoxide hydrolase EC 3.3.2.7; an enzyme that converts (5Z,9E,14Z)-(8ζ,11R,12S)-11,12-epoxy-8-hydroxyeicosa-5,9,14-trienoate (**hepoxilin** A₃) to the trioxilin, (5Z,9E,14Z)-(8ζ,11ζ,12S)-8,11,12-trihydroxyeicosa-5,9,14-trienoate (trioxilin A₃).

Hepps *or* **HEPPS** *or* **EPPS** *abbr. for (and trivial name of)* 4-(2-hydroxyethyl)-1-piperazinepropanesulfonic acid; a **Good buffer substance**, pK_a (20°C) = 8.0.

Heppso *or* **HEPPSO** *abbr. for (and trivial name of)* 4-(2-hydroxyethyl)-1-piperazine-(2-hydroxypropanesulfonic acid); a **Good buffer substance**, pK_a (25°C) = 7.8.

hepsin EC 3.4.21.–; a serine protease, similar to other trypsin-like

enzymes, that is a type II membrane protein. It is required for growth and maintenance of cell morphology in many tissues.

hepta+ *or (before a vowel)* **hept+** *comb. form* denoting seven.

heptad a seven-residue pseudorepeat in an alpha-helical segment of a polypeptide. In coiled coils, positions a and d of the heptad contain nonpolar residues. In leucine zippers, position g contains leucine.

heptanoate 1 the anion, CH_3–$[CH_2]_5$–COO^-, derived from heptanoic acid (formerly enanthic acid). **2** any salt or ester of heptanoic acid.

heptanoyl *symbol*: Hpo; the univalent acyl group, CH_3–$[CH_2]_5$–CO–, derived from heptanoic acid.

heptitol any **alditol** having a chain of seven carbon atoms in the molecule.

heptose any **aldose** having a chain of seven carbon atoms in the molecule.

heptulose any **ketose** having a chain of seven carbon atoms in the molecule.

heptyl *symbol*: Hp; the alkyl group, CH_3–$[CH_2]_5$–CH_2–, derived from heptane.

heptyl glucoside n-heptyl β-D-glucopyranoside; a nonionic detergent, CMC 79 mM.

heptyl thioglucoside n-heptyl 1-thio-β-D-glucopyranoside; a nonionic detergent; the CMC, 30 mM, remains unchanged between 1 and 20°C.

HER2 *see* **neu**.

herbicide any chemical substance that destroys plants or inhibits their growth, especially one used to control weeds.

hereditary of, pertaining to, or caused by **heredity**; inherited.

hereditary angioedema *or* **angioneurotic edema** *or* **C1 inhibitor deficiency** an autosomal dominant condition characterized by episodes of swelling, primarily of the face, extremities, larynx, or abdomen. The attacks may be triggered by emotional stress and can prove fatal. It is caused by deficiency of **C1 inhibitor**, which may be absent (in type I disease) or some 30% of normal (in type II). The C1 inhibitor gene at 11q11 contains 17 Alu sequences, which mediate the gene deletions responsible for most type I cases. Mutations that lead to a dysfunctional protein are common in type II disease.

hereditary disease any pathological condition that is caused by the presence in the organism of an abnormal gene.

hereditary fructose-bisphosphatase deficiency an inherited autosomal recessive condition in humans characterized by less than 20% of normal activity of **fructose-bisphosphatase** (EC 3.1.3.11). It is characterized in infants by episodic spells of hyperventilation, apnoea, hypoglycemia, ketosis, and lactic acidosis. Surviving infants appear to become normal, but gluconeogenesis in their livers is severely impaired, and when glycogen stores are depleted, amino acids, lactate, and ketones accumulate. Subjects are not sensitive to fructose intake. *Compare* **hereditary fructose intolerance**.

hereditary fructose intolerance a condition in humans, inherited as an autosomal recessive character, in which there is a deficiency of the aldolase-B isoenzyme of **fructose-bisphosphate aldolase** (EC 4.1.2.13) (there is less than 15% normal activity with fructose 1-phosphate as substrate but activity is less depressed with fructose 1,6-bisphosphate as substrate) while the levels of aldolases A and C are normal (*see* **aldolase** (def. 2)). The condition is characterized by severe hypoglycemia and vomiting shortly after fructose intake. In this condition prolonged fructose ingestion in infants leads to poor feeding, vomiting, hepatomegaly, jaundice, hemorrhage, proximal-renal-tubule syndrome, and death. The hypoglycemia ensuing after fructose ingestion is caused by fructose 1-phosphate inhibition of glycogenolysis at the phosphorylase level and of gluconeogenesis at the mutant aldolase level. *Compare* **hereditary fructose-bisphosphatase deficiency**.

hereditary hemorrhagic telangiectasia *see* **activin receptor-like kinase 1**.

hereditary multiple exostoses *abbr.*: HME; *other names*: multiple osteochondromata; osteochondromatosis; diaphyseal aclasis; familial bony spurs. An inherited disease in which numerous exostoses (bony spurs) develop (usually around the knee, but also at the shoulder, wrist, and on the ribs) as a result of a disorder of the growth plate.

hereditary persistence fetal hemoglobin a benign condition in which there is continued production of hemoglobin F in adults. Some cases are caused by deletion of the β- and δ-globin loci. Nondeletion phenotypes are characterized by point mutations in the promoter region of one or the other γ-globin genes.

heredity 1 the transmission from one generation to another of genetic factors that determine the characteristics of an organism. **2** all of the inherited factors in an organism or all the characteristics determined by them.

heregulin *abbr.*: HRG; one of several glycoproteins that bind to the transmembrane tyrosine kinase ErbB2 p185, and stimulate its activity. They are secreted by human breast cancer cells in culture.

hermone *an alternative term for* **hormone** (suggested by E. Schäfer in 1913 as better expressing the concept of chemical messengers).

heroin *see* **diacetylmorphine, morphine.**

herpes virus any of a family (Herpesviridae) of enveloped, icosahedral, double-stranded DNA viruses, many of which infect animals. The family includes human herpesviruses types 1 and 2, Epstein–Barr virus, zoster (varicella) virus, pseudorabies virus, cytomegalovirus, Lucke virus, and Marek's disease virus.

herring any marine fish of the family Clupeidae, the flesh of which is comparatively rich in oils with a high content of long-chain *n*-3 **fatty acids**; typical composition (range of major fatty acids, expressed as % of total fatty acids): 16:0, 12–18%; 16:1(*n*-7), 6–8%; 18:1(*n*-9), 11–25%; 20:1(*n*-9), 7–19%; 20:5(*n*-3), 11–15%; 22:6(*n*-3), 5–8%.

Hershey–Chase experiment a seminal experiment, reported in 1952, in which it was demonstrated that when an *Escherichia coli* cell is infected with T2 bacteriophage, the phage DNA is injected into the cell whereas the phage coat protein remains on the outside of the infected bacterial cell. This was shown by demonstrating that most of the ^{35}S-labelled phage protein could be removed from the surface of the bacterial cells by application of a shearing force without removing the ^{32}P-labelled nucleic acid and without interfering with the course of productive infection. The phage nucleic acid could be shown to have moved into the bacterial cell. It was therefore concluded that only the phage DNA was required for infection, indicating that DNA must be the hereditary material. This confirmed earlier experiments in 1944 by Avery and colleagues (*see* **transforming principle**), but was somewhat more readily accepted when Watson and Crick deduced the structure of DNA. [After Alfred Day Hershey (1908–97), US geneticist, and Martha Chase (1927–).]

Hershey circle a circular duplex DNA structure in which each strand contains one interruption and in which the interruptions are not opposite each other. It is formed by some linear, double-stranded DNA molecules (e.g. from *Escherichia coli* phage λ) containing short, complementary single strands (cohesive ends) at each end when these cohesive ends base-pair with each other.

Hers's disease *an alternative name for* type VI **glycogen disease.**

hertz *symbol*: Hz; the SI derived unit of **frequency** of a periodic phenomenon, equal to 1 cycle per second; i.e. $1\ \text{Hz} = 1\ \text{s}^{-1}$. [After Heinrich Hertz (1857–94), German physicist.]

Hess's law of constant heat summation the statement that the enthalpy change in a chemical reaction is the same whether the reaction takes place in one stage or in more than one stage. [After Germain Henri Hess (1802–50), Genevese-born Russian chemist; known in Russia as German Ivanovich Gess.]

HETE *abbr. for* hydroxyeicosatetraenoate; *see also* **dihydroxyeicosatetraenoate.**

hetero+ *comb. form* **1** other, different, unusual. *Compare* **allo+** (def. 1), **homo+** (def. 1), **iso+** (def. 1), **xeno+** (def. 1). **2** for, from, or directed towards a different species. *Compare* **allo+** (def. 2), **homo+** (def. 2), **iso+** (def. 2), **xeno+** (def. 2). **3** containing atoms, groups, linkages, residues, or subunits of different kinds. **4** being or pertaining to an atom other than the predominant or significant one, especially of a nitrogen, oxygen, or sulfur atom in a ring containing mainly carbon atoms. *Compare* **homo+** (def. 4).

heteroantibody any antibody (including any autoantibody) that is able to react with an antigen derived from another species. *Compare* **isoantibody.**

heteroantigen any antigen derived from one species that is able to stimulate an immune response in another species. *Compare* **isoantigen.**

heteroatom any atom in the chain or ring of an organic compound that is not a carbon atom. —**heteroatomic** *adj.*

heterobifunctional describing any reagent that carries two differing reactive groups. *Compare* **homobifunctional.**

heterocaryon *a variant spelling of* **heterokaryon.**

heterochromatic 1 relating to or possessing more than one colour; relating to light or other electromagnetic radiation of more than one frequency. *Compare* **monochromatic. 2** of, relating to, or possessing **heterochromatin.**

heterochromatin a condensed form of **chromatin**, occurring in the nucleus at interphase, that stains strongly with basophilic dyes. The DNA of heterochromatin is replicated at a later stage in the cell-division cycle than **euchromatin.** 'Satellite' DNA is associated with heterochromatin, and the sex chromosomes of animals, especially the Y chromosome, contain large regions of heterochromatin. *Compare* **euchromatin.**

heteroconjugation the association between a base and the conjugate acid of a different base through a hydrogen bond. *See* **conjugate acid–base pair.** *Compare* **homoconjugation.**

heterocycle any compound or molecular structure that is **heterocyclic** (def. 1); a **heterocyclic** (def. 2).

heterocyclic 1 describing any cyclic molecular structure containing atoms of at least two different elements in the ring or rings; describing any compound having such a structure. *Compare* **carbocyclic** (def. 1), **homocyclic** (def. 1). **2** *or* **heterocycle** any such compound.

heterocyst any of the clear cells that occur at intervals on the filaments of certain cyanobacteria. They are usually the sites of nitrogen fixation in species that carry out the reaction. They have thickened cell walls and lack photosystem II and oxygenic photosynthesis.

heterodetic describing a cyclic peptide in which the ring consists of amino-acid residues only, but the linkages forming the ring are not solely eupeptide bonds; one (or, if non-proximate, more than one) is an isopeptide, disulfide, ester, or other bond. *Compare* **homodetic.**

heterodimer any molecular structure in which two nonidentical subunits are associated. *Compare* **dimer, homodimer.**

heteroduplex 1 any double-stranded DNA in which the two strands do not have completely complementary base sequences. In a **mutational heteroduplex** a mutational event has occurred in one of the polynucleotide chains; such a molecule does not produce uniform progeny on replication. A **recombinational heteroduplex** is a double-stranded DNA molecule that gives rise to one parental and one recombinant molecule of DNA in the reproductive cycle that immediately follows its production. Heteroduplex DNA may also arise from annealing DNA single strands *in vitro*. **2** a DNA–RNA hybrid. *Compare* **homoduplex.**

heteroduplex analysis a method of detecting mutations by the formation of heteroduplex DNA from complementary but nonidentical DNA strands. When a single-stranded DNA molecule anneals to one in which there is a mismatch at one or more base pairs, the resultant heteroduplex has an electrophoretic mobility that is lower than those of homoduplexes formed from the wild-type or the mutant sequences.

heteroduplex mapping the physical mapping of the base sequence homology of two DNA molecules. It is accomplished after the *in vitro* formation of a hybrid double-stranded molecule, in which one polynucleotide strand is from one of the DNA molecules and the complementary polynucleotide strand is from the other. Unpaired single-stranded regions are located and measured by electron microscopy.

heterogametic describing the sex whose cell nuclei contain a pair of dissimilar **sex chromosomes**. The heterogametic sex is usually male in mammals but is female in the Lepidoptera, birds, reptiles, some amphibians and fishes, and a few plants. *Compare* **homogametic.**

heterogeneous 1 differing in kind, as in a heterogeneous population. **2** describing a group of molecules that is not uniform with respect to size, charge, structure, or other property. **3** describing a system consisting of two or more phases. *Compare* **heterogenous.** —**heterogeneity** *n.*

heterogeneous catalysis any catalytic system in which the catalyst constitutes a different phase from that of the reactants. *Compare* **homogeneous catalysis**.

heterogeneous immunoassay any form of **immunoassay** that involves physical separation, at some stage of the procedure, of antibody-bound antigen from remaining free antigen. It includes **radioimmunoassay**. *Compare* **homogeneous immunoassay**.

heterogeneous nuclear ribonucleoprotein *abbr.*: hnRNP; particulate complexes of **heterogeneous nuclear RNA** (hnRNA) with proteins, which are cell-specific and themselves heterogeneous. The protein component may play a role in the processing of the hnRNA to mRNA.

heterogeneous nuclear RNA *abbr.*: hnRNA or HnRNA or H-RNA; a heterogeneous mixture of RNA molecules of high M_r with a rapid turnover rate that occurs in cell nuclei during protein synthesis. It is the form of RNA synthesized in eukaryotes by RNA polymerase II, i.e. that which is translated into protein. It represents the transcription products of both the exons and the introns of a gene or genes and is processed to form mRNA by the removal of the nontranslated portions (corresponding to the introns in the DNA). So-named because heterogeneity in size was one of the first characteristics used to distinguish it from other forms of nuclear RNA.

heterogenesis the production of offspring having different characteristics in successive generations; e.g. the alternation between generations that produce gametes and those that reproduce through agametes or by parthenogenesis.

heterogenetic describing chromosome pairing during meiosis when the pairing partners are derived from different original ancestors. *Compare* **homogenetic**.

heterogenic 1 describing a population or a gamete containing more than one allele of a particular gene. **2** describing two genetic elements that are not known to have a common ancestry. **3** *an alternative word for* **heterogenous**. *Compare* **homogenic**.

heterogenize *or* **xenogenize** to introduce a foreign, especially a more potent, antigen into a cell surface in order to enhance the response by T lymphocytes to other cell-surface antigens. —**heterogenization** *n*.

heterogenous *or* **heterogenic** not originating within the organism in question; of foreign origin. *Compare* **heterogeneous**.

heteroglycan *or* **heteropolysaccharide** any polysaccharide (i.e. glycan) that contains residues of two or more kinds of monosaccharide (i.e. glycose) molecule. *Compare* **homoglycan**.

heterograft *or* **heterologous graft** *an alternative name for* **xenograft**. *Compare* **homograft**.

heteroimmune describing a pair of bacteriophages in which each is sensitive to its own repressor but not to that of the other.

heterokaryon *or* **heterocaryon** any cell with more than one nucleus, in which the nuclei are not all of the same genetic constitution, or a tissue composed of such cells. *Compare* **homokaryon**. —**heterokaryotic** *or* **heterocaryotic** *adj*.

heterokaryosis *or* **heterocaryosis** the association of genetically different nuclei in a common cytoplasm.

heterolactic fermentation a type of fermentation of glucose in which the products include lactate, acetate and/or ethanol, and carbon dioxide. A key enzyme in the pathway is **phosphoketolase**, which cleaves xylulose 5-phosphate (produced by the pentose phosphate pathway) to D-glyceraldehyde 3-phosphate and acetyl phosphate, the former being then converted to lactate while the latter can form acetate and/or ethanol. Organisms that ferment in this manner include *Leuconostoc* spp. and some species of *Lactobacillus*; some produce D(–)-lactate while others produce DL-lactate. *Compare* **homolactic fermentation**.

heterolipid *a former name for* any complex lipid.

heterologous 1 consisting of different elements or of like elements in different proportions. **2** *or* **heterospecific** derived from another species, e.g. heterologous serum or tissue. *Compare* **homologous**, **isologous**.

heterologous association describing any association between two **protomers** in which the **domain of bonding** is made up of two different **binding sets**. *Compare* **isologous**.

heterologous desensitization *see* **desensitization**.

heterologous graft *or* **heterograft** *an alternative name for* **xenograft**.

heterolysin any antibody or other hemolysin formed in response to the introduction of erythrocytes from another species into the bloodstream of an animal. *Compare* **autolysin**, **homolysin**.

heterolysis 1 *(in chemistry)* the cleavage of a bond in such a manner that both bonding electrons remain within one of the two fragments between which the bond is broken; e.g. A–B $\rightarrow$ A$^+$ + B$^-$. *Compare* **homolysis**. **2** *(in biochemistry)* the lysis of cells or tissues by the action of exogenous agents, e.g. enzymes, detergents, or lysins. *Compare* **autolysis**. —**heterolytic** *adj*.

heterolysosome any early **secondary lysosome** concerned with the digestion of exogenous material (i.e. **heterophagy**). *See* **heterophagosome**.

heteromorphic *or* **heteromorphous 1** differing from the normal form, e.g. in size, shape, or function. **2** describing the members of any pair of homologous chromosomes that differ from each other in size or shape. **3** *(especially of insects)* having differing forms at different stages of the life cycle. **4** *an alternative term for* **polymorphic**. *Compare* **isomorphic**. —**heteromorphism** *or* **heteromorphy** *n*.

heteronuclear NMR a form of **nuclear magnetic resonance spectroscopy** that exploits the spin–spin interactions between different nuclear spins in a molecule. For determination of structures of proteins, the isotopes ^{13}C, ^{15}N are introduced into the sample, and the interactions between the spins of these nuclei and ^{1}H are probed by sequences of pulses at the appropriate frequencies.

heterooligomer any **oligomer** made up of two or more kinds of constitutional repeating unit. —**heterooligomeric** *adj*.

heterophagosome any **phagosome** containing material exogenous to the cell. Fusion with a lysosome leads to the formation of a heterolysosome. *Compare* **autophagosome**.

heterophagy the process by which cells digest exogenous material that they have taken up. *Compare* **autophagy**. —**heterophagic** *adj*.

heterophilic binding *(of adhesion molecules)* binding of an adhesion molecule in one cell to a non-identical adhesion molecule in an adjacent cell. *Compare* **homophilic binding**.

heteroplasmy the presence of more than one type of mitochondrial DNA in a cell. A mutant mitochondrial DNA may be present in varying amounts so that phenotypic variation between cells is possible.

heteroploid describing a cell that contains any number of chromosomes except the **haploid** or **diploid** number. The term includes all **aneuploid** or **polyploid** numbers of chromosomes.

heteropolymer any polymer made up of two or more kinds of constitutional repeating unit. Proteins are typical heteropolymers. —**heteropolymeric** *adj*.

heteropolysaccharide *an alternative term for* **heteroglycan**.

heteroside *a former name for* any glycoside containing a noncarbohydrate moiety (aglycon).

heterospecific having specificity for or derived from a different species; **heterologous** (def. 2).

heterotetramer any molecular structure that can be dissociated into four (or two pairs of) nonidentical monomers. —**heterotetrameric** *adj*.

heterotope any of two or more chemical elements having atoms of different atomic number and occupying different places in the periodic table. The term includes **isobar** (def. 1). *Compare* **isotope**.

heterotopia *or* **heterotopy** the displacement, or unusual position, of a cell, organ, tissue graft, etc.

heterotopic 1 *(in chemistry)* of, relating to, or being a **heterotope** or heterotopes. *Compare* **isotopic**. **2** *(in biology and medicine)* *or* **heterotopous** displaced or in an unusual anatomical position.

heterotopic transplant any tissue graft that is transplanted from its normal position to another in the same organism. *Compare* **orthotopic transplant**.

heterotopous *a variant form of* **heterotopic** (def. 2).

heterotopy *a variant form of* **heterotopia**.

heterotrimer any molecular structure that is dissociable into three nonidentical monomers. —**heterotrimeric** *adj*.

heterotroph *or* **heterotrophe** *or* **organotroph** any organism whose nutritional requirements are not satisfied solely by simple inorganic substances and for which a supply of organic carbon is required for the synthesis of cellular constituents. Heterotrophs may be subdivided into **chemoheterotrophs** (or chemoorganotrophs) and **photo-heterotrophs** (or photoorganotrophs) according, respectively, to whether they are **chemotrophs** or **phototrophs**. *Compare* **autotroph**. —**heterotrophic** *adj.*; **heterotrophy** *n.*

heterotropic 1 describing an **allosteric** effect in which interaction occurs between nonidentical ligands; the effect may be either cooperative or antagonistic. The term is applied also to such an interaction, to an allosteric enzyme for which different substances act as the substrate and the effector, to the regulation of such an allosteric system, etc. *Compare* **homotropic**. **2** describing a chromosome that does not pair with another at meiosis, especially a sex chromosome in the heterogametic sex.

heterotypic pertaining to, or being the first, reductive division in **meiosis**. *Compare* **homeotypic**.

heterozygosis 1 the union of genetically different gametes to form a **heterozygote**. **2** *or* **heterozygosity** the state or condition of being a heterozygote. *Compare* **homozygosis**.

heterozygosity *an alternative term for* **heterozygosis** (def. 2).

heterozygote any cell or organism having two different genes (alleles) at one or more corresponding loci on homologous chromosomes. With respect to such loci a heterozygote will produce two different types of gametes. *Compare* **homozygote**.

heterozygous *or* **heterozygotic** of, pertaining to, or being a **heterozygote**; hybrid.

HETP *abbr. for* height equivalent to a theoretical plate.

heuristic describing any method for solving mathematical or other problems for which no **algorithm** (or closed method) exists. It involves narrowing down the field of search for a solution by inductive reasoning from past experience of similar problems. *Compare* **stochastic**.

HEV *abbr. for* high-walled endothelium of the post-capillary venules.

hevein an *N*-acetyl-D-glucosamine- and *N*-acetyl-D-neuraminic acid-binding lectin; it is present in the latex of the rubber tree, *Hevea brasiliensis*, and shares 56% similarity with wheat germ agglutinin.

hexa+ *or (before a vowel)* **hex+** *comb. form* denoting six, sixfold, six times.

hexabrachion a six-armed structure, characteristic of the **tenascin** molecule. Regions of the protein radiate outwards like spokes of a wheel. [From L. *brachium* arm.]

hexadecadienoic acid any straight-chain fatty acid having sixteen carbon atoms and two double bonds per molecule. (*all-Z*)-Hexadeca-6,9-dienoic acid is found in small amounts in some animal lipids.

hexadecanoic acid *the systematic name for* palmitic acid; *see* **palmitate**.

hexadecenoic acid any straight-chain fatty acid having sixteen carbon atoms and one double bond per molecule. The most common are the 9*Z*-isomer (palmitoleic acid; *see* **palmitoleate**) and the 11*Z*-isomer, but the 3*E*-, 6*E*-, 6*Z*-, 9*E*-, and 10*Z*-isomers are also variously found in animals or plants; the 9*E*-isomer (i.e. palmitelaidic acid) is a pheromone.

hexaethyleneglycol mono-*n*-decyl ether a nonionic detergent; *see* $C_x E_y$.

hexaethyleneglycol mono-*n*-dodecyl ether a nonionic detergent; *see* $C_x E_y$.

hexagonal phase (of phospholipids) a phase in which the polar headgroups are orientated towards the interior of a phospholipid aggregate, with the fatty acyl carbon chains pointing outwards; a form of inverted micelle. It can occur in lipid bilayers, especially at higher temperatures. Certain phospholipids, such as phosphatidylethanolamine, have a greater tendency to adopt this phase than, e.g., phosphatidylcholine. A hexagonal phase with the polar headgroups oriented towards the exterior of the micelle can also exist. *See also* **liquid-crystalline phase**, **phase transition**.

hexakisphosphate *see* **multiplying prefix**.

hexamer 1 any oligomer consisting of six monomers. **2** the group of six capsomeres in an icosahedral virus capsid.

hexanoate *or* **caproate 1** the anion, $CH_3-[CH_2]_4-COO^-$, derived from hexanoic acid. **2** any salt or ester of hexanoic acid.

hexanolamino-PAF 1-*O*-hexadecyl-2-*O*-acetyl-*sn*-glycero-3-phospho-(*N*,*N*,*N*-trimethyl)-hexanolamine; a substance that inhibits the effects induced by **platelet activating factor**, including: platelet aggregation, secretion, and inositol phospholipid turnover; superoxide ion and H_2O_2 release from macrophages; hepatic glycogenolysis; and vasoconstriction.

hexanoyl *or (formerly)* **caproyl** *symbol*: Hxo; the univalent acyl group, $CH_3-[CH_2]_4-CO-$, derived from hexanoic acid. *See* **caproyl** (def. 2).

hexaric acid an **aldaric** acid formally derived from a hexose by oxidation at both C-1 and C-6.

hexitol any **alditol** having a chain of six carbon atoms in the molecule.

hexokinase EC 2.7.1.1; *systematic name*: ATP:D-hexose 6-phosphotransferase; *other name* (used for hexokinase IV): glucokinase; the phosphotransferase enzyme that catalyses the phosphorylation by ATP of D-hexose to form D-hexose 6-phosphate and ADP. In vertebrates there are four major isoenzymes, types I–IV, or A–D. All are relatively nonspecific with respect to sugar, and phosphorylate a number of hexoses such as glucose, fructose, and mannose. Thus the name glucokinase for type IV may mislead. Types I–III all have $M_r \approx 100$ kDa and are sensitive to glucose 6-phosphate (*abbr.*: Glc-6-*P*). They differ from yeast hexokinases (≈ 50 kDa), which are not inhibited by Glc-6-*P*. Type III is inhibited by excess glucose. Type IV is a monomer associated with cytoskeletal actin filaments and shows cooperativity with respect to glucose. In hepatocytes, at less than 5 mM glucose it is inactive, being bound to a regulatory protein; a glucose concentration of 15 mM causes translocation and activation of the enzyme in the nucleus, where it regulates expression of several genes. For example, hexokinase type IV from human mitochondrial outer membrane has a sequence with similar N-terminal and C-terminal halves. The regulatory function is associated with the N-terminal half. Type II is the major isoenzyme of skeletal muscle and adipose tissue. There are strong sequence similarities between the different types. Mutations in glucokinase produce maturity-onset diabetes of the young (MODY).

hexon a **capsomere** that is surrounded by six other capsomeres in an icosahedral virus capsid.

hexonic acid any monocarboxylic acid (i.e. aldonic acid) formally derived from a hexose by oxidation at C-1.

hexosamine any aminodeoxysugar with six carbon atoms.

hexosaminidase EC 3.2.1.52; *recommended name*: β-*N*-acetylhexosaminidase; *systematic name*: β-*N*-acetyl-D-hexosaminide *N*-acetylhexosaminohydrolase; *other name*: β-hexosaminidase. An enzyme that catalyses the hydrolysis of terminal nonreducing *N*-acetyl-D-hexosamine residues in *N*-acetyl-β-D-hexosaminides. It is a lysosomal enzyme in mammals. In humans there are two precursors, α and β. The α chain, a glycoprotein, comprises amino acids 109–529 of its precursor. The β precursor gives rise to three glycoproteins (all similar to α): β (amino acids 122–556), β-B (amino acids 122–311), and β-A (amino acids 315–556). There are three forms of the enzyme: form A, responsible for degradation of GM_2 gangliosides, is a trimer composed of one α chain, one β-A chain, and one β-B chain; form B is a tetramer of two β-A and two β-B chains; form S is a homodimer of two α chains. **Tay–Sachs disease** is due to defects of the α chain and Sandoff disease due to defects of the β chain.

hexosan *a former name for* any of a class of polysaccharides that yield only hexoses on hydrolysis.

hexose any **aldose** having a chain of six carbon atoms in the molecule.

hexose diphosphate pathway *an alternative name for* **Embden–Meyerhof pathway**.

hexose monophosphate pathway *or* **hexose monophosphate shunt** *an alternative name for* **pentose phosphate pathway**.

hexose phosphate transport protein an integral membrane transport protein responsible for sugar phosphate uptake in bacteria.

hexose-1-phosphate uridylyltransferase *see* **UDP-glucose–hexose-1-phosphate uridylyltransferase**.

hexulose any **ketose** having a chain of six carbon atoms in the molecule.

hexuronic acid 1 any monocarboxylic acid (i.e. uronic acid) formally derived from a hexose by oxidation at C-6. **2** *a historical name originally used for* ascorbic acid.

hexyl *symbol*: Hx; the alkyl group, $CH_3–[CH_2]_4–CH_2–$, derived from hexane.

hexyl glucoside hexyl β-D-glucopyranoside; a mild nonionic detergent, CMC 250 mM.

Hf *symbol for* hafnium.

HFBA *abbr. for* heptafluorobutyric acid; a reagent used in automatic peptide sequence analysis. It renders the action of cyanogen bromide specific to the cleavage of tryptophanyl peptide bonds.

HFCS *abbr. for* high fructose corn syrup.

Hfr *abbr. for* high-frequency recombinant.

HFT *abbr. for* high-frequency transduction.

Hg *symbol for* mercury.

HGF *abbr. for* **1** hepatocyte growth factor. **2** hyperglycemic-glycogenolytic factor; *see* **glucagon**.

HGF receptor *see* **hepatocyte growth factor receptor**.

HGG *abbr. for* human gamma-globulin.

hGH *or* **HGH** *abbr. for* human growth hormone (i.e. human **somatotropin**).

HGI *abbr. for* human gene index.

HGPRT *abbr. for* hypoxanthine/guanine **phosphoribosyltransferase**. *See* **hypoxanthine phosphoribosyltransferase**.

HhaI a type 2 **restriction endonuclease**; recognition sequence: GCG↑C.

HHH syndrome *see* **ornithine transporter 1**.

HHT *abbr. for* hydroxyheptadecatrienoic acid.

5-HIAA *abbr. for* 5-hydroxyindoleacetic acid.

Hid a proapoptotic protein of *Drosophila* and a functional homologue of mammalian **Diablo**. It binds and inactivates **IAP** thereby releasing any associated caspase. *See* **apoptosis**.

hidden (immunological) determinant *or* **cryptodeterminant** any antigenic determinant so positioned that it is cryptic, i.e. normally not accessible for recognition by lymphocytes or antibody, but that becomes accessible after some stereochemical change occurs, e.g. by breakage, decomposition, or denaturation of the antigen. *See also* **cryptotope**.

hidden Markov model *abbr.*: HMM; (*in sequence alignment*) a probabilistic model that encodes the residue conservation within an aligned domain, used to provide a diagnostic signature for that domain. HMMs are linear chains of "states" in which a match state allows a sequence to match a conserved column in an alignment; an insert state allows insertions relative to match states; and delete states allow match positions to be skipped (i.e. denote deletions in a sequence relative to the alignment). HMMs form the basis of **Pfam** (*see* **pattern database**). *Compare* **profile**.

hide the skin of an animal, especially the tough, thick skin of larger mammals. Powdered, dried hide is used in assays of some proteinases.

HIDS *abbr. for* hyper-IgD syndrome.

HIF *abbr. for* hypoxia-inducible factor; a transcription factor that is stable under conditions of oxygen deficiency but is targeted for degradation in the presence of oxygen. **HIFα** contains an oxygen-dependent degradation domain in which a specific prolyl residue becomes modified by a HIF prolyl-4-hydroxylase. It forms a heterodimer with **HIFβ** (i.e. ARNT) and binds a 5 bp response element in hypoxia-inducible genes responsible for increasing oxygen delivery or facilitating adjustment to hypoxic stress. *See* **HIF-PH**.

HIF-PH *abbr. for* hypoxia-inducible factor prolyl hydroxylase; a prolyl-4-hydroxylase enzyme that hydroxylates a specific proline residue in the HIFα subunit (*see* **HIF**) in the presence of O_2. The reaction promotes binding of HIF to the tumour suppressor VHL protein and thereby polyubiquitination and degradation of HIF by proteasomes.

high-density lipoprotein *abbr.*: HDL; one of the classes of lipoprotein found in blood plasma in many animals (data normally relate to humans). They are also known as α lipoproteins, from having the highest electrophoretic mobility of the lipoproteins. HDL particles are the smallest of the blood lipoproteins: diameter 7.5–10 nm; solvent density for isolation 1.063–1.21 g mL^{-1}; hydrated density 1.21 g mL^{-1}. Their approximate composition (% by weight) is 6% unesterified cholesterol, 13% esterified cholesterol, 28% phospholipid, 3% triacylglycerol, 50% protein. Their apolipoprotein composition (% by weight total apolipoprotein) is 67% A-I, 22% A-

II, 5–11% C-I + C-II + C-III, 1–2% E-II + E-III + E-IV, trace of D. They are synthesized in liver as precursor molecules, which undergo modification in plasma to the mature molecules, especially as a result of the action of **lecithin-cholesterol acyltransferase**. They appear to function in reverse transport of cholesterol from tissues to liver. *See also* **apolipoprotein** for individual apolipoproteins.

high-energy bond *symbol*: ~; *a term formerly sometimes used by biochemists for* a chemical bond whose hydrolysis is accompanied by an unusually large (>20 kJ mol^{-1}) negative standard free energy change. Although long used, the term has been recognized as incorrect and misleading, since it suggests that such a bond actually contains an unusually large amount of energy and fails to take into account the various factors that contribute to the large negative free energy change.

high-energy compound *a term formerly sometimes applied by biochemists to* any compound containing one or more so-called **high-energy bonds**, e.g. ATP, phospho*enol*pyruvate, acetyl-coenzyme A, etc.

high-frequency recombinant any bacterial strain that has a high frequency of recombination because the **F plasmid** has become incorporated into the bacterial chromosome. *See also* **Hfr**.

high-frequency transduction *abbr.*: HFT; any **transduction** in which the transducing phage(s) constitute a large proportion of the total phage population.

high-fructose corn syrup *abbr.*: HFCS; a mixture of fructose and glucose used as a sweetener in soft drinks, etc.; it is sometimes called **high fructose syrup** or, especially in Europe, **isoglucose**. The breakdown of corn starch to glucose (plus oligosaccharides) is followed by the action of **glucose isomerase** yielding a mixture consisting mainly of fructose and glucose. The product thus manifests the greater sweetness of fructose compared with glucose. The content is typically 42% fructose, 51% glucose, 5% maltose, and 2% other oligosaccharides.

highly conserved *see* **conserved**.

highly repetitive DNA a form of DNA that consists of short tandemly repeated sequences. It is found outside coding regions, and its role is unknown. Such DNA forms a significant proportion of most mammalian genomes.

high mobility group proteins *abbr.*: HMG proteins; small nuclear proteins that bind to nucleosomal DNA, and probably maintain genes in a transcribable conformation. The name derives from their behaviour on electrophoresis.

high-performance capillary electrophoresis electrophoresis through a long coiled capillary column that gives very high resolution of e.g. peptides. *See also* **capillary electrophoresis**.

high-performance liquid affinity chromatography *abbr.*: HPLAC; a technique that combines the use of bioaffinity supports and **high-pressure liquid chromatography**.

high-performance liquid chromatography *abbr.*: HPLC; *an alternative name for* **high-pressure liquid chromatography**.

high-potential iron–sulfur protein *abbr.*: Hipip *or* HiPIP; a type of **ferredoxin**, originally found in the photosynthetic bacterium *Chromatium vinosum*, that has a very much higher redox potential (E_o ~0.3 V) and different magnetic properties compared with other ferredoxins. It contains a [4Fe-4S] cluster that can be oxidized to the 3+ oxidation state. Other proteins, to which the term has been mistakenly applied, contain [3Fe-4S] clusters; like the *Chromatium* protein, they are able to achieve a paramagnetic higher oxidation level, but do not have high redox potentials.

high-pressure liquid chromatography *or* **high-performance liquid chromatography** *abbr.*: HPLC; a technique of **column chromatography** that is rapid and provides high resolution. It can be used with the various modes of liquid chromatography, the liquid being forced at high pressure, normally up to 420 kg cm^{-2} (6000 psi), through the column (which is usually of stainless steel) and thence through a detector into a fraction collector. Because of the ability to use high pressures, the particles of the support materials used for the stationary phase can be of very small diameter (3–10 μm), which greatly improves resolution.

high-scoring segment pair *abbr.*: HSP; in BLAST searches, an ungapped local pairwise alignment between arbitrary but equal-length fragments of a query and a database sequence that scores above a user-defined threshold. HSPs are the fundamental unit of BLAST output. *See* **alignment algorithm**.

high-speed supernatant *a jargon term for* the supernatant resulting from high-speed centrifugation ($\approx$100 000 g).

high-spin complex any of the complex transition-metal ions having a maximal number of unpaired electrons in the d- or f-shell. This state occurs when pairing energy is higher than the crystal field splitting energy. *Compare* **low-spin complex**.

high-throughput describing a process that is scaled up, usually via increased levels of automation using robots. For instance, **high-throughput screening** refers to the rapid *in vitro* screening of large numbers of compound libraries (of tens to hundreds of thousands of compounds), using robotic screening assays. **High-throughput sequencing** involves the application of rapid sequencing technology at the scale of whole genomes.

high-throughput genomic sequence *abbr.*: HTG; a division of the GenBank database that provides access to unfinished genomic sequence. Its records might typically contain all of the first-pass sequence data generated from a single cosmid, BAC, YAC, or P1 clone.

high-throughput screening *see* **high throughput**.

high-throughput sequencing *see* **high throughput**.

high-voltage electron microscope a type of electron microscope that employs accelerating voltages of 1 MeV, ten times that of previous instruments, enabling images of thick sections and even intact cells to be produced without significant loss of resolution.

high-voltage electrophoresis *abbr.*: HVE; a technique of paper or thin-layer electrophoresis that results in better and more rapid separations (10–60 min) than the standard technique, which suffers from minimal diffusion of low-M_r components. The high-voltage technique employs currents of up to 500 mA and potential gradients of up to 200 V cm^{-1}. The high field strength requires cooling of the support and adequate protection devices to be incorporated into the apparatus.

Hildebrand solubility parameter *symbol*: δ; a measure of the polarity of a solvent; nonpolar solvents have low δ values while polar solvents have large values. In liquid–liquid chromatography the Hildebrand parameter indicates the relative position of a solvent in an eluotropic series. The parameter is the sum of a number of factors:

$$\delta = \delta_d + \delta_o + \delta_a + \delta_h,$$

where δ_d quantifies interactions due to **dispersion forces**, δ_o quantifies interactions due to dipole interactions, δ_a quantifies the ability of the solvent to interact as a hydrogen acceptor, and δ_h quantifies its interactions as a hydrogen donor. [After Joel Henry Hildebrand (1881–1983), US chemist.]

Hill coefficient *or* Hill constant *symbol*: h or n_H; *see* **Hill equation** and **Hill plot**.

Hill equation 1 an equation,

$$y = K[x]^h/(1 + K[x]^h),$$

used to express the binding of a ligand to a (macro)molecule. It may be rearranged to give:

$$\log[y/(1 - y)] = -\log K + h\log[x],$$

where y, the **fractional saturation**, is the fraction of the total number of binding sites occupied by ligand, $[x]$ is the free (unbound) ligand concentration, K is a constant, and h is the Hill coefficient. **2** *(in pharmacology)* an equation derived from the above relating the effect of a drug to its concentration. For an effect, E, produced when an agent, A, is applied at a concentration $[A]$, then the relationship between E and $[A]$ may often be described empirically by the equation

$$E/E_{max} = [A]^h/([A]_{50}{}^h + [A]^h),$$

where E_{max} is the maximal value of E and $[A]_{50}$ is the concentration that produces an effect that is 50% of E_{max}. [After (Sir) Archibald Vivian Hill (1886–1977), British physiologist and biophysicist.]

Hill–Langmuir equation *(in pharmacology)* a variation of the **Langmuir absorption isotherm**; for a ligand, L, binding to a receptor, R, according to the equilibrium

$$L + R \underset{}{\overset{K_L}{\rightleftharpoons}} LR,$$

$$p_{LR} = [L]/(K_L + [L]),$$

where p_{LR} is the fraction (proportion) of binding sites occupied by a ligand at equilibrium and K_L is the dissociation equilibrium constant.

Hill plot a plot of the **Hill equation** in the form $\log[y/(1 - y)]$ against $\log[x]$, where y is the fractional saturation with ligand and $[x]$ is the free (unbound) ligand concentration. The slope at any one point, h, is the Hill coefficient at that value of y. The values of h along the plot, which usually do not exceed the number of binding sites on each molecule and which tend to unity as y tends to 0 or 1, give an indication of the nature and degree of cooperativity (interaction) between the various binding sites on the molecule. [After (Sir) Archibald Vivian Hill (1886–1977), British physiologist and biophysicist.]

Hill reaction the phenomenon in which isolated chloroplasts evolve dioxygen when illuminated in the presence of a suitable electron acceptor, e.g. ferricyanide or ferrioxalate, with a concomitant reduction of the electron acceptor. [After Robert ('Robin') Hill (1899–1991), British plant biochemist.]

Hill reagent any electron acceptor that functions in a **Hill reaction**, e.g. ferrioxalate or ferricyanide.

*Hinc*II a type 2 **restriction endonuclease**; recognition sequence: GT[TC]$\uparrow$[AG]AC. *Hinc*II is an **isoschizomer**.

*Hind*III a type 2 **restriction endonuclease**; recognition sequence: A$\uparrow$AGCTT. The M.*Hind*III modification site is C1.

*Hinf*I a type 2 **restriction endonuclease**; recognition sequence: G$\uparrow$ANTC.

hinge region 1 a flexible, proline-rich region of an **immunoglobulin** molecule, adjacent to the two disulfide bonds linking the two heavy chains together. It probably acts as a hinge on which the **Fab fragments** can rotate allowing the immunoglobulin molecule to take up a Y-shape when it reacts with antigen. The hinge region is near the sites of action of papain and pepsin. **2** the region in the helical, rod-shaped part of a **myosin** molecule that is susceptible to cleavage by trypsin to give light meromyosin and heavy meromyosin. **3** a region in any protein about which rigid body motions occur.

*Hin*II a type 2 **restriction endonuclease**; recognition sequence: G[AG]$\uparrow$CG[TC]C. *Acy*I and *Bbi*II are **isoschizomers**.

Hipip *or* HiPIP *abbr. for* high-potential iron–sulfur protein.

hippocalcin a neuron-specific Ca^{2+}-binding protein with **EF-hands**, similar to recoverin. It is expressed exclusively in the pyramidal layer of the hippocampus.

Hirschsprung disease a congenital megacolon that may occur alone or be part of a syndrome. *See* **glial cell-derived neurotrophic factor**. [After Harald Hirschsprung (1830–1916), Danish physician.]

hirudin a dried extract obtained from leeches, and possessing anticoagulant activity. The name is also applied to the purified active component, a 65-amino-acid polypeptide with three disulfide bridges; this has highly specific antithrombin properties, inactivating thrombin by blocking substrate-binding groups.

hirudisin a recombinant **hirudin** variant engineered to inhibit α-thrombin and exhibit **disintegrin** activity. Hirudisins have antiplatelet aggregation activity. Replacement of the hirudin sequence Ser-Asp-Gly-Glu (residues 32–35) by Arg-Gly-Asp-Ser (RGDS) yields hirudisin, or by Lys-Gly-Asp-Ser yields hirudisin-1. The integrin-binding RGD motif of hirudisin is implicated in its antiplatelet action.

hirulog any synthetic peptide based on **hirudin** and designed as an inhibitor of the thrombin catalytic site, and exhibiting specificity for the anion-binding site of thrombin.

hirustasin an **antistasin**-like serine proteinase inhibitor isolated from the medicinal leech, *Hirudo medicinalis*.

His *symbol for* a residue of the α-amino acid L-**histidine** (alternative to H).

hist+ *a variant form of* **histo+** (before a vowel).

His-tag *abbr. for* histidine tag; a short sequence of six histidine residues placed at either the N- or C-terminus of recombinant proteins expressed in bacteria or mammalian cells. Proteins tagged in this way can be purified by Ni^{2+} chelate chromatography.

His-tag antibody a preparation of monoclonal or polyclonal antibodies raised to 6 × His sequences used in the immunodetection and immunoprecipitation of His-tagged proteins. *See* **His-tag**.

histaminase *see* **amine oxidase (copper-containing)**.

histamine 2-(4-imidazolyl)ethylamine, formed by the decarboxylation of L-histidine and present in many mammalian tissues, including the brain and spinal cord, with especially high concentrations in lung, skin, and intestine; it is stored in the granules of tissue mast cells and circulating basophil cells. Histamine is a potent vasodilator, increases capillary permeability, causes contraction of smooth muscle, regulates gastric secretion, and mediates allergic and anaphylactic conditions.

histamine receptor any membrane receptor in mammalian tissues that binds specifically with **histamine**. **H1 receptors** mediate the stimulating effects of histamine on smooth muscle in the gut and bronchi and most of the depressor effects on blood pressure. They are G-protein-associated receptors that activate phospholipase C, and are antagonized by such classical **antihistamines** as pyrilamine (mepyramine) and promethazine. **H2 receptors** mediate the stimulation of gastric acid secretion, the increase in the contraction frequency of cardiac muscle, the increase in cyclic AMP stores in basophil and mast cells, the inhibition of release of histamine by mast cells, and some depressor effects on blood pressure. They are G-protein-associated receptors that stimulate adenylate cyclase, and are antagonized by, e.g., **cimetidine**. **H3 receptors** are G-protein-associated receptors linked to Ca^{2+} channels. They occur predominantly in presynaptic nerve terminals, attenuate the release of neurotransmitters (e.g. histamine, acetylcholine, and norepinephrine), and are antagonized, e.g., by thioperamide and clobenpropit.

histamine-releasing factor *abbr.*: HRF; a protein (172 amino acids) produced by the lymphocytes of atopic children and present in the blood of allergic individuals. It causes the release of histamine only from basophilic cells that have IgE bound to them, and may have an important role in chronic allergic diseases.

histatin any member of a family of histidine-rich peptides found in human saliva and having antibacterial and antifungal activity.

histidase *an alternative name for* **histidine ammonia-lyase**.

histidinal α-amino-β-4-imidazolepropionaldehyde; 2-amino-3-(1*H*-imidazol-4-yl)propanal; $(C_3H_3N_2)CH_2$–CH(NH$_2$)–CHO; the α-amino-aldehyde analogue of histidine. It is the immediate biosynthetic precursor of histidine.

histidinase *an alternative name for* **histidine ammonia-lyase**.

histidinate 1 histidine anion; the anion, $(C_3H_3N_2)CH_2$–CH(NH$_2$)–COO$^-$, derived from histidine. **2** any salt of histidine anion. **3** any ester of histidine anion.

histidine *the trivial name for* β-4-imidazolylalanine; α-amino-β-imidazole- 4-propionic acid; 2-amino-3-(1*H*-imidazol-4-yl)propanoic acid; $(C_3H_3N_2)CH_2$–CH(NH$_2$)–COOH; a chiral α-amino acid. L-Histidine (*symbol*: H *or* His), (*S*)-2-amino-3-(1*H*-imidazol-4-yl)-propanoic acid, is a coded amino acid found in peptide linkage in proteins; codon: CAC or CAU (anticodon: GUG). It is also a component of the isopeptide **carnosine**. In mammals, it is an essential dietary amino acid, and is glucogenic. D-Histidine (*symbol*: D-His *or* DHis), (*R*)-2- amino-3-(1*H*-imidazol-4-yl)propanoic acid, is not known to occur naturally. N-1 and N-3 of the imidazole ring are designated τ and π respectively.

L-histidine

histidine ammonia-lyase EC 4.3.1.3; *systematic name*: L-histidine ammonia-lyase; *other names*: histidase; histidine α-deaminase; histidinase. An ammonia-lyase (deaminase) enzyme of liver and (in some mammals) skin that catalyses the deamination of L-histidine to urocanate. It is unusual in having dehydroalanine, probably derived from a serine residue, in its active site. It is widely distributed in animals, plants, and microorganisms; in plants it is present in **glyoxysomes**. The rat, mouse, and human enzymes each contain 657 amino acids. The activity of this enzyme is defective in **histidinemia**.

histidine (base) a histidine residue in which the imidazole group is unprotonated.

histidine decarboxylase *abbr.*: HDC; EC 4.1.1.22; *systematic name*: L-histidine carboxy-lyase. A pyridoxal phosphate homodimeric enzyme that catalyses the decarboxylation of L-histidine to histamine. It is present in mast cells, enterochromaffin-like cells, and histaminergic neurons that project from the posterior hypothalamus to most parts of the brain and spinal cord. It is similar to dopa and glutamate decarboxylases.

histidine kinase *abbr.*: HK; any member of a superfamily of protein kinases specific for histidine residues. In prokaryotes and lower eukaryotes they are involved in two-component signal transduction. Most are integral membrane homodimeric proteins that contain a periplasmic N-terminal sensing domain of variable sequence, and a highly conserved cytoplasmic C-terminal kinase domain. They usually autophosphorylate (using ATP as the phosphoryl group donor) one of their own histidine residues and then transfer the group to an aspartate residue of a response regulator (RR) protein. **Hybrid histidine kinases** contain an RR – or histidine-containing phosphotransfer (HPt) – domain as part of their own structure. HKs belong to a structural class distinct from that of serine/threonine and tyrosine kinases.

histidinemia *or (esp. Brit.)* **histidinaemia** a rare, autosomal recessive, generally benign metabolic disorder in which abnormally high levels of histidine are present in the blood and urine. It is due to deficient activity of **histidine ammonia-lyase**.

histidine tail a sequence of about six histidine residues that is deliberately introduced at the C terminus of certain recombinant proteins to allow purification of the recombinant product by a type of affinity chromatography using a nickel chelating resin. Only the histidine-tailed proteins are retained by the column and these may then be eluted using a buffer containing imidazole.

histidinium 1 *a term normally used to mean* histidinium(1+); the (imidazolinium) monocation of histidine. In theory, the term denotes any ion, or mixture of ions, formed from histidine, and having a net charge of +1, although the species $(C_3H_4N_2^+)CH_2$–CH(NH$_3^+$)–COO$^-$ generally predominates in practice. **2** *the systematic name for* histidinium(2+); the dication of histidine.

histidino 1 N^2-histidino; the alkylamino group, $(C_3H_3N_2)CH_2$–CH(COOH)–NH–, derived from histidine by loss of a hydrogen atom from its α-amino group. **2** either of the two possible groups, N^π-histidino or N^τ-histidino, HOOC–CH(NH$_2$)–CH$_2$–(C$_3$H$_2$N$_2$)–, derived from histidine by loss of a hydrogen atom from its position on one or other of the nitrogen atoms of the imidazole ring.

histidinol 4-(2-amino-3-hydroxypropyl)imidazole; 2-amino-3-(1*H*-imidazol-4-yl)propanol; $(C_3H_3N_2)CH_2$–CH(NH$_2$)–CH$_2$OH; the α-amino-alcohol analogue of histidine. It is an intermediate, together with its phosphate ester, in the biosynthesis of histidine, and a powerful reversible inhibitor of protein synthesis.

histidinol dehydrogenase *abbr.*: HDH; EC 1.1.1.23; L-histidinol:NAD$^+$ oxidoreductase. A homodimeric enzyme that catalyses the reaction:

$$\text{L-histidinol} + 2NAD^+ = \text{L-histidine} + 2NADH$$

this being the last step in the biosynthesis of histidine. The *his* operon of *Salmonella typhimurium* encodes a bifunctional protein that also has an aldehyde dehydrogenase site, histidinol being an intermediate. *See also* **Salmonella mutagenicity test**.

histidyl the acyl group, $(C_3H_3N_2)$–CH$_2$–CH(NH$_2$)–CO–, derived from histidine.

histo+ *or (before a vowel)* **hist+** *comb. form* denoting tissue.

histochemistry the chemistry of the tissues, as studied with a combination of the methods of chemistry and histology.

histocompatibility the extent to which an organism will tolerate a graft of a foreign tissue. —**histocompatible** *adj.*

histocompatibility antigen any of the genetically determined **isoantigens** present on the surface of many animal cells that determine the immune reactions to tissue grafts. If tissue is grafted onto a recipient of the same species that does not carry the same histocompatibility antigens as the graft material, an immune response may be provoked in the recipient leading to graft rejection. *See also* **HLA histocompatibility system, major histocompatibility complex**.

histocompatibility gene any gene determining the formation of a **histocompatibility antigen**. *See* **HLA histocompatibility system**, **major histocompatibility complex**.

histoelectrofocusing a technique in which an unfixed frozen section of tissue, 25–40 μm thick, is applied to a support gel and then subjected to **electrofocusing**. The proteins and enzymes present in the tissue move into the gel and may be revealed by staining or zymographic techniques.

histogram a diagram representing a frequency distribution. It consists of a number of contiguous rectangles whose widths are proportional to the class interval under consideration and whose heights are proportional to the frequency associated with each class.

histohematin or *(esp. Brit.)* **histohaematin** a term first used in 1884 by Charles Alexander MacMunn (1852–1911) for the pigments now known as **cytochromes**. It included **myohematin**.

histone any of a group of evolutionarily highly conserved basic proteins, molecular mass 11–21 kDa, that constitute about half the mass of the chromosomes of all eukaryotic cells, except spermatozoa. They comprise single polypeptide chains with a lysine-plus-arginine content of about 25%, and are concerned in the packing of DNA in **chromatin**. The principal histones may be classified into five types, as shown in the table.

Type	Other names	Lys/Arg ratio	M_r	Location
H1	F1,Ia	20.0	21 000	linker
H2A	F2a2,IIb1	1.25	14 500	core
H2B	F2b,IIb2	2.25	13 800	core
H3	F3,III	0.72	15 300	core
H4	F2a1,IV	0.79	11 300	core

The octomer (H2A,H2B,H3,H4)$_2$ forms the **nucleosome** core. H1 is associated with nucleosomes, especially the linking DNA. *See also* **H-NS**.

histone acetyltransferase EC 2.3.1.48; an enzyme that transfers an acetyl group from acetyl-CoA to a lysine residue in the N-terminal region of a core histone in a nucleosome. Type A enzymes are cytosolic and acetylate freshly synthesized histones before they enter the nucleus. Type B are nuclear and reside in large multiprotein complexes, being recruited by, and acting as cofactors for, many DNA sequence-specific transcription factors. Examples include p300/CBP (which targets histones H2A, H2B, H3, and H4) and **PCAF** (which targets histones H3 and H4).

histone code the hypothesis that the pattern of acetylation, methylation, phosphorylation, and ubiquitination of the N-terminal regions of core histones within nucleosomes is recognized by proteins (or complexes) that control chromatin modelling and transcriptional regulation. This is considered to extend the potential of the genetic code.

histone deacetylase *abbr.* HDAc; an enzyme that catalyses the hydrolysis of acetyl groups from histones, and thereby regulates DNA-protein interactions. Histone deacetylases are part of transcriptional corepressor complexes. Inhibitors of histone deacetylases induce hyperacetylation of histones that modulate chromatin structure and gene expression, and induce growth arrest, cell differentiation, and apoptosis of tumor cells.

histone fold the interaction motif involved in homodimerization of the core histones (H2A and H2B, H3 and H4) and their assembly into an octomer. It comprises three alpha helices linked by two loops. The surface residues mediate octomer formation and interaction with the DNA that wraps around the octomer. The fold also

occurs in many transcription factors and in the chromatin accessibility complex (**CHRAC**).

histone modification the post-translational enzymatic modification of the N-terminal region of the core histones H2A, H2B, H3, and H4 by acetylation, methylation, phosphorylation, or ubiquitination. The extent and pattern of modification regulates chromosome condensation and is the basis of the **histone code** and the **nucleosome code**.

hit-and-run mutagenesis a form of mutagenesis by homologous recombination in mammalian cells in which a vector carrying the mutated homologous DNA sequence is inserted into the gene of interest thereby duplicating part of it. In a second step, intrachromosomal recombination between the native and homologous mutated sequences results in introduction of the mutation at the desired location.

hitchhiking 1 *a jargon term for* the sequence of events in which a genetic element (or **transposon**) inserts into an active secondary site on a replicon, which is then transferred to a recipient, the transposon then coming immediately to occupy its normal primary site in the DNA of the recipient organism. **2** the spread of a neutral allele through a population due to its linkage with a beneficial allele that is subject to selection.

HIV *abbr. for* human immunodeficiency virus; a **retrovirus**, the causative agent of **AIDS**. HIV-1 is found worldwide; HIV-2 is found mainly in West Africa. The gene structures of the **viral envelope sequences** (*see* **HIV protein**) of HIV-1 and HIV-2 are not closely related. HIV-2 is closely related to a **simian immunodeficiency virus** (SIV) common in sooty mangabeys and probably originated there. HIV-1, which probably originated from a non-human primate in Africa, is closely related to SIV from chimpanzees. However, it is not clear that chimpanzees represent the natural host of these viruses. The reverse transcriptase enzyme of HIV is highly error-prone, leading to a high mutation rate. Thus these viruses evolve very quickly even within single patients, and drug-resistant mutations appear rapidly. For HIV-1 to enter its target cells, it must bind to the cell-surface protein **CD4** and to one or more chemokine receptors. One of these, named CCR5, is effective in the early stages of infection. Patients with defective CCR5 cannot be infected with HIV-1. During the later stages of infection, another chemokine receptor, CXCR4 (*former name*: fusin) dominates.

HIV entry cofactor *an alternative term for* **fusin**.

HIV-1 protease an aspartyl protease that is a homodimer of two subunits each containing 99 amino acids, with cysteines at positions 67 and 95. Its proteolytic function is essential for the life cycle of human immunodeficiency virus (HIV) and is, therefore, a target for the design of inhibitors as therapeutic drugs, several examples of which are in use.

HIV protein any of the proteins encoded by the human immunodeficiency virus (HIV). The HIV-1 genome (9.5 kb) is transcribed to give 12 different mRNAs, generated by alternative splicing. The proteins are conventionally named for the gene and/or their size (kDa), with the convention that p = protein and gp = glycoprotein. The primary translation products are: Gag precursor (p53); Pol precursor (gp160); Nef (p27; negative regulator); Vif (p23); Tat (p14; regulatory protein); Rev (p19; regulatory protein); Vpr (p18); and Vpu (p15). The Gag precursor is cleaved to myristoylated Gag protein (p17), structural protein (p24), and p15; p15 is the precursor for RNA-binding protein (p7) and proline-rich protein (p9). The Pol precursor is cleaved to protease (p10), reverse transcriptase (p51), and endonuclease (p32). The Env precursor is cleaved to extracellular protein (gp120) and transmembrane protein (gp41). Protease, reverse transcriptase, and endonuclease are synthesized as part of a polyprotein. *See also* **retrovirus**.

HK *abbr. for* **histidine kinase**.

HKD catalytic motif *see* **PhD enzyme**.

HL-60 a continuous line of promyelocytic cells that can be made to differentiate into macrophage-like cells by treatment with the phorbol ester TPA or into granulocytes by treatment with dimethylsulfoxide (DMSO) or all-*trans* retinoic acid.

HLA *abbr. for* human leukocyte-associated antigen. *See also* **histocompatibility antigen, major histocompatibility complex**.

HLA histocompatibility system a family of histocompatibility antigens encoded by the complex of genetic loci known as the **major histocompatibility complex** (MHC). It is the major histocompatibility system in humans, equivalent to the H-2 histocompatibility system in mice. The MHC genes control the presence of HLA **isoantigens** on cell surfaces. These antigens are of primary importance in tissue grafting since HLA-incompatible grafts are more likely to be rejected than incompatible grafts of other isoantigen groups.

HLH abbr. for helix-loop-helix (see **helix-turn-helix**).

hly **genes** see **hemolysin**.

HMC abbr. for 5-hydroxymethylcytosine.

H-meromyosin abbr. for heavy meromyosin.

hMG or **HMG** abbr. for human menopausal gonadotropin (i.e. human urogonadotropin).

HMG 1 abbr. for 2-hydroxy-2-methylglutaryl. **2** abbr. for high mobility group of nuclear proteins. **3** see **hMG**.

HMG-CoA abbr. for hydroxymethylglutaryl-CoA.

HMG-CoA lyase abbr. for hydroxymethylglutaryl-CoA lyase.

HMG-CoA reductase abbr. for hydroxymethylglutaryl-CoA reductase.

HMG-CoA synthase abbr. for hydroxymethylglutaryl-CoA synthase.

HMM abbr. for **1** heavy meromyosin. **2** hidden Markov model.

HMMA abbr. for hydroxymethoxymandelate.

HMP abbr. for hexose monophosphate.

HMP pathway or **HMP shunt** abbr. for hexose monophosphate pathway (shunt) (i.e. **pentose phosphate pathway**).

HNF abbr. for hepatocyte nuclear factor.

HNPCC abbr. for hereditary nonpolyposis colon cancer; an autosomal dominant disease associated with mutations in several genes encoding proteins (e.g. MutL, MutS, GTBP) that may normally take part in mismatch repair of DNA.

hnRNA or **HnRNA** abbr. for heterogeneous nuclear RNA (alternatives to H-RNA).

hnRNP abbr. for heterogeneous nuclear ribonucleoprotein.

hnRNP particles abbr. for heterogeneous nuclear ribonucleoprotein particles; particles in the nucleus containing **hnRNA** bound to protein.

H-NS basic DNA-binding proteins of several bacterial families, characterized by an N-terminal tetrapeptide, consensus Ser-Glu-Xaa-Leu . Although their precise role remains unclear, they may be the prokaryotic counterparts of **histones** in eukaryotic chromatin. They have the ability to bind to curved DNA, and may have a role in transcriptional regulation. The name derives from an isolation procedure extracting them from the 30S ribosomal native subunit.

Ho symbol for holmium.

HO abbr. for **heme oxygenase**.

Hoechst 33342 see **bisbenzimide**.

Hoechst 33258 a bis-benzimidazole fluorescent dye used for measuring DNA concentration. See **bisbenzimide**.

Hoff see **van't Hoff**.

Hofmeister series or **lyotropic series** an arrangement of either the simple anions or the simple cations (or of the salts of such ions) in order of their ability to remove lyophilic substances from colloidal solutions. Each series is an index of, *inter alia*, the relative abilities of the ions to salt-out proteins from solution. The order for anions is: citrate^{3-} > tartrate^{2-} > acetate$^-$ > NO_3^- > ClO_4^- > CNS$^-$, and that for cations is: Mg^{2+} > Ca^{2+} > Sr^{2+} > Ba^{2+} > Li^+ > Na^+ > K^+ > Rb^+ > Cs^+. [After Franz Hofmeister (1850–1922).]

Hofstee plot see **Eadie–Hofstee plot**.

Hogness box an alternative name for **TATA box**.

hol+ a variant form of (before a vowel) **holo+**.

holandric inherited solely in the male line. Compare **hologynic**.

holdback agent or **holdback carrier** any nonradioactive substance that is added to a sample to prevent either coprecipitation or adsorption of some soluble radioactive substance.

hold-up volume symbol: V_M; (in chromatography) the volume of eluent required to elute a component whose concentration in the stationary phase is negligible relative to that in the mobile phase. In gas chromatography, the volume of carrier gas (eluent) is specified at the same temperature and pressure as the total retention volume.

hole (in physics) the absence of an electron in a normally filled va-

lency structure of a crystalline **semiconductor** that behaves as a carrier of charge and is mathematically equivalent to a positron.

holin a protein encoded by dsDNA bacteriophages that permeabilizes the plasma membrane of host cells to permit exit of the enzyme **endolysin** that degrades the murein matrix to enable the release of progeny phage. Holins constitute a subclass of transport proteins of the TC system, designated 1.E. They contain between 45 (TC 1.E.6) and 250 (TC 1.E.8) amino acid residues.

Holliday junction or **Holliday structure** one of the junctions between four strands of DNA that are important intermediates in **genetic recombination**. They are part of the mechanism for **crossing over** during the long prophase of meiotic cell division I, when four strands of DNA are aligned. The process involves breaking the DNA double helix in a maternal chromatid and in a homologous paternal chromatid and then exchanging fragments between the two non-sister chromatids in a reciprocal fashion. At this point, in each four strands, a paternal strand links across to a maternal strand, forming the junction known by the name of its proposer. [After Robin Holliday (1932–), British biologist.]

hollow-cathode discharge lamp a radiation source used in **atomic absorption spectrophotometry** to produce a spectrum of radiation specific to the element being assayed. A sample of the element in question is held in a metal cup-shaped cathode, which, together with a tungsten anode, is enclosed in a glass or quartz envelope filled with argon at low pressure. A high voltage is used to produce an arc spectrum of the element.

hollow-fibre technique a technique in which bundles of semipermeable, hollow fibres with pores of controlled dimensions that function as molecular sieves are used for concentrating, desalting, dialysing, and fractionating macromolecules in solution. Compare **capillary filtration**.

holo+ or (before a vowel) **hol+** comb. form whole, entire, complete; wholly, entirely, completely.

holocarboxylase synthetase an enzyme that biotinylates the ε-amino group of lysine residues to form **biocytin** at the active site of biotin carboxylases (see **acetyl-CoA carboxylase**). In humans, numerous mutations of the gene at 21q22.1 (which encodes 726 residues) are associated with neonatal multiple carboxylase deficiency, which is clinically similar to biotinidase deficiency.

holocellulose the complex mixture of polysaccharides remaining after the removal of lignin from tree-wood by treatment with sodium chlorite solution. It consists of a mixture of **cellulose** and **hemicellulose**.

holocrine 1 describing a type of secretion that involves disintegration of the entire secretory cell to release the secretory products. **2** describing a gland made up of such cells. Compare **apocrine**, **eccrine**, **merocrine**.

holoenzyme a catalytically active complex comprising the protein part of an enzyme (**apoenzyme**) combined with the appropriate cofactor or cofactors.

hologram 1 the pattern produced when light, or other electromagnetic radiation, that has been reflected, diffracted, or transmitted by an object placed in a coherent beam, interferes with a reference beam related in phase to the first beam. **2** a photograph of such a pattern.

holograph 1 to make a **hologram** (def. 1, 2). **2** any book or other document handwritten by its author.

holography the process of making and using **holograms** (def. 1, 2). —**holographic** adj.

hologynic inherited solely in the female line. Compare **holandric**.

Holometabola or **Endopterygota** one of the two divisions of the subclass Pterygota, comprising those insects, e.g. butterflies, moths, flies, and beetles, in which the juvenile forms (larvae) are very different from the adults. Such insects undergo complete metamorphosis during a resting stage (pupa). Compare **Hemimetabola**.

holoprosencephaly see **hedgehog protein**.

holoprotein the functional form of a protein containing a protein part (**apoprotein**) together with any appropriate ligand or ligands.

holorepressor a functional repressor protein consisting of an **aporepressor** plus **corepressor** complex.

homeo+ or **homoeo+** or **homoio+** comb. form denoting like or similar.

homeobox a short (180 bp) conserved DNA sequence that encodes a DNA-binding motif famous for its presence in genes that are involved in orchestrating the development of a wide range of organisms. *See also* **homeotic gene**.

homeodomain the DNA-binding motif of 60 amino acids encoded by a **homeobox**.

homeomorphism *or* **homoeomorphism** *(in chemistry)* the property shown by certain substances of having the same crystal form but different compositions. —**homeomorphic, homeomorphous,** *or* **homoeomorphic, homoeomorphous** *adj*.

homeostasis *or* **homoeostasis** the maintenance of a relatively constant internal environment in the bodies of higher animals by means of a series of interacting physiological and biochemical processes. —**homeostatic** *or* **homoeostatic** *adj*.

homeotic gene *or* **homeodomain-encoding gene** *or* **homeobox-encoding gene** *or* **Hox gene** any of the genes that are allelic for mutations resulting in the conversion of one body part into another. Homeotic genes were first discovered in *Drosophila*; a mutation in such a gene could, e.g., result in a leg replacing an antenna. While *Drosophila* has only two *Hox* clusters, in nematodes there is one, and in vertebrates four clusters of 9 to 11 genes each located on a separate chromosome and spanning more than 100 kb. The vertebrate *Hox* genes are expressed in specific patterns and at particular stages in embryogenesis. They are activated by **retinoic acid** and **sonic hedgehog protein**. In the limb these are active in concert with **fibroblast growth factor**. The homeobox portion of the *Hox* genes in different species is remarkably conserved. The structure of some homeodomains has been determined, as has their interaction with DNA.

homeothermic *the usual US spelling of* **homoiothermic**.

homeotic selector any of a family of genes in *Drosophila* that select the expression of **homeotic genes**. They encode homologous regulatory proteins that contain a **homeobox** sequence.

homeotypic pertaining to, or being the second, reductive division in **meiosis**. *Compare* **heterotypic**.

Homer any protein belonging to a family whose members have an N-terminal EVH1 homology sequence that interacts with a proline-rich sequence motif (called Homer ligand), which is present in some metabotropic glutamate receptors (mGluRs) and on the cytosolic region of the inositol trisphosphate (IP₃) receptor. It also contains a C-terminal coiled-coil region for homodimer formation. Homer helps localize mGluRs to synapses and mediates their effects on IP₃ receptors.

Homer ligand *see* **Homer**.

homo+ *comb. form* **1** the same, similar, alike. *Compare* **allo+** (def. 1); **hetero+** (def. 1); **iso+** (def. 1); **xeno+** (def. 1). **2** from, or directed towards, the same species. *Compare* **allo+** (def. 2); **hetero+** (def. 2); **iso+** (def. 2); **xeno+** (def. 2). **3** containing atoms, groups, linkages, residues, or subunits of the same kind. *Compare* **hetero+** (def. 3). **4** denoting the next higher homologue of a specified chemical compound, especially one whose molecules contain, in a hydrocarbon chain or ring, one more methylene, –CH₂–, than those of of the parent compound, e.g. homocysteine, homoserine, homo-5α-pregnane. *Compare* **nor+** (def. 2).

Homo a genus of primates that includes modern man (*Homo sapiens*) and a number of more primitive, fossil species.

homoarginine *(sometimes)* a trivial name for N^6-amidino lysine.

homobifunctional describing a chemical reagent that carries two identical reactive groups. *Compare* **heterobifunctional**.

homocarnosine *(sometimes)* a trivial name for N^α-(4-aminobutyryl)histidine. It is present in brain but its function is unclear.

homochromatic *a less common term for* **monochromatic** (def.1).

homochromatography a chromatographic technique for the separation of labelled oligonucleotides, in which a mixture of unlabelled oligonucleotides is used to displace the labelled components according to differences in size.

homocitrate 2-hydroxybutane-1,2,4-tricarboxylate; the (*R*) enantiomer is a component of the **aminoadipic pathway** and is formed by **homocitrate synthase**.

homocitrate synthase EC 4.1.3.21; *systematic name*: 2-hydroxybutane-1,2,4-tricarboxylate 2-oxoglutarate-lyase (CoA-acetylating). An enzyme that catalyses the formation of **homocitrate** from 2-oxoglutarate and acetyl-CoA and H_2O. *See also* **aminoadipic pathway**.

homoconjugation the association between a base and its conjugate acid through a hydrogen bond. *See* **conjugate acid–base pair**. *Compare* **heteroconjugation**.

homocyclic 1 describing any cyclic molecular structure containing atoms of a single element in the ring or rings; describing any compound having such a structure. The term is preferred to its synonym **isocyclic**; it includes **carbocyclic** (def. 1). *Compare* **heterocyclic** (def. 1). **2** any such compound.

homocysteine *symbol*: Hcy; α-amino-γ-mercaptobutyric acid; 2-amino-4-mercaptobutanoic acid; an important intermediate in the metabolic reactions of its *S*-methyl derivative, methionine. Its concentration in human serum is raised after stroke.

L-homocysteine

homocysteine methyltransferase one of several enzymes, e.g. the two alternative forms of enzyme synthesizing methionine: (1) 5-methyltetrahydrofolate–homocysteine *S*-methyltransferase. (EC 2.1.1.13); *other name*: tetrahydropteroylglutamate methyltransferase; it catalyses a reaction between 5-methyltetrahydrofolate and L-homocysteine to form tetrahydrofolate and L-methionine; cobalamin is a cofactor. (2) 5-methyltetrahydropteroyltriglutamate–homocysteine *S*-methyltransferase (EC 2.1.1.14); *other name*: tetrahydropteroyltriglutamate methyltransferase; it catalyses a reaction between 5-methyltetrahydropteroyltri-L-glutamate and L-homocysteine to form tetrahydropteroyltri-L-glutamate and L-methionine.

homocystine 2,2′-diamino-4,4′-dithiobis(butanoic acid); the disulfide oxidation product of **homocysteine**.

homocystinemia *or (esp. Brit.)* **homocystinaemia** a usually inherited biochemical abnormality of humans, characterized by an abnormally high concentration of homocystine in the blood and in the urine (**homocystinuria**). The condition may be due to: (1) a deficiency of cystathionine β-synthase (EC 4.2.1.22) activity; (2) a deficiency of methylene tetrahydrofolate reductase; or (3) a deficiency of cobalamin. Homocystinemia is an independent risk factor for thromboembolic events.

homodetic describing a cyclic peptide in which the ring consists solely of amino-acid residues in eupeptide linkage. *Compare* **heterodetic**.

homodimer any macromolecular structure in which two identical subunits (monomers or protomers) are noncovalently associated. *See also* **dimer**. *Compare* **heterodimer**.

homoduplex hybrid DNA involving two strands that are complementary. *Compare* **heteroduplex**.

homoeo+ *a variant form of* **homeo+**.

homogametic describing the sex whose cell nuclei contain a pair of similar **sex chromosomes**. The homogametic sex is usually female, but is male in the Lepidoptera, birds, some amphibians and fishes, and a few plants. *Compare* **heterogametic**.

homogenate any disrupted tissue preparation in which cell disruption is maximal while the destruction of intracellular organelles is kept to the minimum.

homogeneic *an alternative term for* **allogeneic**. *Compare* **heterogeneic**.

homogeneous 1 being uniform in kind, as in a homogeneous population. **2** (*describing a group of molecules*) uniform in structure or composition. **3** (*describing a system*) consisting of only a single chemical phase. *Compare* **heterogeneous, homogenous**. —**homogeneity** *n*.

homogeneous catalysis any catalytic action or process in which the catalyst is uniformly distributed in the same chemical phase as the reactants. *Compare* **heterogeneous catalysis**.

homogeneous enzyme immunoassay *or* **enzyme-multiplied immunoassay technique** (*abbr.*: EMIT) a technique of **homogeneous immunoassay** in which an enzyme is covalently coupled to the substance to be assayed. The enzymic activity of the complex is inhibited by an antiserum specific for the substance and this inhibition is relieved by the presence of free molecules of the substance to be assayed. The enzyme acts as a molecular amplifier.

homogeneous immunoassay any **immunoassay** that does not involve physical separation of antibody-bound antigen from antigen molecules that remain free. Labels that may be attached to the antigen include: enzymes, cofactors, prosthetic groups or substrates of enzymes, chemiluminescent or fluorescent molecules, stable free radicals. *Compare* **heterogeneous immunoassay**.

homogeneously staining region a region in a chromosome that stains homogeneously, rather than with the banding pattern normally seen. The region contains massively amplified numbers of copies of a small segment of the genome. The amplified DNA may often consist of multidrug resistance gene DNA (*see* **MDR**).

homogenesis the production of offspring having the same characteristics in successive generations. *Compare* **heterogenesis**.

homogenetic describing chromosome pairing during meiosis when the pairing partners are derived from one of the original ancestors. *Compare* **heterogenetic**.

homogenic *or* **congenic 1** describing a gamete that contains only one allele of a particular gene. **2** describing two genetic elements that are descended from a common ancestor by a known sequence of steps. **3** *an alternative word for* **homogeneous**. *Compare* **heterogenic**.

homogenize *or* **homogenise** produce a **homogenate**. —**homogenization** *or* **homogenisation** *n*.

homogenizer *or* **homogeniser** any apparatus for producing a **homogenate**.

homogenous 1 of, pertaining to, or exhibiting **homogeny**. **2** *an alternative word for* **homogeneous**.

homogentisate *or* (*formerly*) **alcapton** 2,5-dihydroxyphenylacetate; an intermediate in the catabolism of tyrosine and in the biosynthesis of plastoquinone and tocopherol. It is excreted in abnormally large quantities in the urine in **alcaptonuria**. *See also* **homogentisate 1,2-dioxygenase**.

homogentisate 1,2-dioxygenase EC 1.13.11.5; *systematic name*: homogentisate:oxygen 1,2-oxidoreductase (decyclizing); *other names*: homogentisicase; homogentisate oxygenase; homogentisic acid oxidase. An enzyme with highest activity in liver that catalyses the oxidation by dioxygen of homogentisate to 4-maleylacetoacetate; iron(II) is a cofactor. In the catabolism of L-tyrosine, **homogentisate** is formed from 4-hydroxyphenylpyruvate; maleylacetoacetate is converted to fumarylacetoacetate, which is split to fumarate and acetoacetate. Mutations in the gene result in enzyme deficiency and the rare metabolic disorder alkaptonuria.

homogeny (*in biology*) similarity in the structure of organisms or parts of organisms because of common ancestry.

homoglycan *or* **homopolysaccharide** any polysaccharide (i.e. glycan) that contains residues of only one kind of monosaccharide (i.e. glycose) molecule. *Compare* **heteroglycan**.

homograft *an old term for* **allograft**. *Compare* **heterograft**.

homoio+ *a variant form of* **homeo+**.

homoiosmotic describing organisms with a constant internal osmotic pressure. *Compare* **poikilosmotic**.

homoiothermic *or* **homoiothermal** *or* (*esp. US*) **homeothermic** *or* **homeothermal** describing an organism (**homoiotherm**) that sustains a relatively constant body temperature, usually higher than that of its

surroundings. *Compare* **poikilothermic**. —**homoiothermy** *or* (*esp. US*) **homeothermy** *n*.

homokaryon *or* **homocaryon** any cell with more than one nucleus, and in which the nuclei are all of the same genetic constitution; a tissue composed of such cells. *Compare* **heterokaryon**. —**homokaryotic** *or* **homocaryotic** *adj*.

homokaryosis *or* **homocaryosis** the condition of having a **homokaryon** or homokaryons. *Compare* **heterokaryosis**.

homolactic fermentation a type of fermentation of glucose to lactate via the **Embden–Meyerhof pathway**. *Compare* **heterolactic fermentation**.

homolog (*sometimes*) the US spelling of **homologue**.

homologous 1 having a related or similar position, structure, etc.; corresponding; exhibiting **homology**. **2** (*in chemistry*) describing compounds that form a series with successive constant differences in composition. **3** (*in biology*) of common ancestry; especially of organs and tissues that have a similar anatomical position and structure in different species by virtue of their common evolutionary origin, even though their functions may have come to differ; e.g. the wing of a bird and the forelimb of a reptile. *Compare* **analogous**. **4** (*in genetics*) describing chromosomes that pair during meiosis, each member of a homologous pair being a duplicate of one of the chromosomes contributed at syngamy by the mother or the father. **5** (*in biochemistry*) **a** (of sequences of residues in encoded macromolecules) having the same or similar residues at corresponding positions (*see also* **conserved**). With respect to proteins the term is used to imply a common evolutionary origin. Specifically this requires evidence based on gene structure and not merely a similarity of protein structure. **b** (of proteins from different species) having identical or similar functions. *Compare* **heterologous**. —**homology** *n*.

homologous antibody the antibody elicited by a specified antigen.

homologous antigen the antigen that has elicited a specified antibody.

homologous desensitization *see* **desensitization**.

homologous recombination genetic recombination between two DNA molecules of identical or nearly identical sequence when contained in the same cell. It occurs naturally during meiosis, and is exploited when performing gene knockouts in yeast, embryonic stem cells, and other cells.

homologue *or* (*sometimes*) US **homolog 1** (something) exhibiting homology. **2** (*in chemistry*) any member of a **homologous** (def. 2) series of compounds. **3** (*in biology*) any one of a series of **homologous** (def. 3) organs or structures. **4** (*in genetics*) **a** either members of a pair of **homologous** (def. 4) chromosomes. **b** either one of two genes in corresponding loci in homologous chromosomes. **5** (*in sequence analysis*) any of two or more nucleotide or protein sequences that are evolutionarily related to each other by descent from a common ancestor. Homologous sequences may be orthologues or paralogues.

homology domain any of the regions in an immunoglobulin involved in forming the **immunoglobulin fold**.

homology modelling a computational method for predicting protein structure based on similarity to a sequence of known structure. The accuracy of homology-derived models depends on the level of similarity between the target and model sequences, and hence on the quality of their alignment: the greater the similarity, the better the alignment and the better the model is likely to be. The resolution of structures modelled in this way is usually low.

homolysis (*in chemistry*) the cleavage of a covalent bond in such a manner that each of the fragments between which the bond is broken retains one of the bonding electrons; e.g. A–B → A• + B•. —**homolytic** *adj*.

homooligomer any **oligomer** made up of only one kind of constitutional repeating unit. —**homooligomeric** *adj*.

homophilic binding (*of adhesion molecules*) binding of an adhesion molecule in one cell to an identical molecule in an adjacent cell. *Compare* **heterophilic binding**.

homoplasy (*in phylogeny*) the occurrence of similar characters, independently of lineage, via parallel or convergent evolution (sometimes these processes are distinguished, sometimes the difference is considered irrelevant) or reversion. Such characters cannot therefore be used to construct phylogenetic trees.

homopolymer any polymer made up of only one kind of constitutional repeating unit. For example, cellulose contains only glucose as the monomeric unit. —**homopolymeric** *adj.*

homopolymer tailing *a jargon term for* a procedure useful in joining two types of duplex DNA molecules to form mixed dimeric circular DNA. In this procedure, homopolymer sequences of one type, e.g. poly(dA), are added to the 3′-ends of one of the populations of DNA molecules, while complementary homopolymer sequences, e.g. poly(dT), are added to the 3′-ends of the other population of DNA molecules. The two types of molecules are then annealed to form mixed dimeric circles.

homopolysaccharide *an alternative name for* **homoglycan**.

homoserine *symbol*: Hse; α-amino-γ-hydroxybutyric acid; an intermediate in the biosynthesis of cystathionine, threonine, and methionine. It also occurs in bacterial peptidoglycans and, as its *O*-guanidino derivative, in canavanine. The term normally implies the L enantiomer.

homoserine deaminase *see* **cystathionine γ-lyase**.

homoserine dehydratase *see* **cystathionine γ-lyase**.

homoserine dehydrogenase EC 1.1.1.3; *systematic name*: L-homoserine:NAD(P)$^+$ oxidoreductase; an enzyme of methionine and threonine biosynthesis. It catalyses the formation of L-homoserine and NAD(P)$^+$ from L-aspartate 4-semialdehyde and NAD(P)H. *Escherichia coli* has a homotetrameric bifunctional enzyme consisting of 820 amino acids. Residues 1–249 form the aspartokinase I domain, 250–470 form a large 'interface', and 471–820 form homoserine dehydrogenase; regulation is by L-Thr. *See also* **aspartokinase/homoserine dehydrogenase**.

homoserine lactone *symbol*: Hsl; α-amino-γ-butyrolactone; a substance formed by the cleavage of methionine-containing peptides by cyanogen bromide.

homosterism *or* **homostery** the phenomenon in which a second molecule of normal substrate, or a structurally similar compound, combines at the catalytic site of an enzyme leading to a modification in the reaction of the bound intermediate. *Compare* **allosterism**. —**homosteric** *adj.*

homostery *an alternative term for* **homosterism**.

homothallic describing species, e.g. of certain fungi and algae, in which a sexual spore can result from the fusion of nuclei that are genetically distinct (i.e. not necessarily homozygous), but are derived from the same thallus; thus the species is self-fertile.

homothallic switching endonuclease a sequence-specific endonuclease, with a site on yeast chromosome III, that is involved in mating-type switching. It is a **zinc finger** protein.

homotherm *an alternative word for* **homoiotherm**.

homotope any residue that can take the place of another residue in a particular position in a polymer; e.g., the serine and glycine residues at position 9 of the A chain of bovine and ovine insulin, respectively, are homotopes.

homotransplant 1 *an alternative term for* **homograft**. **2** to effect such a transplant.

homotropic describing an **allosteric** effect in which interaction occurs between identical ligands; the effect appears to be always cooperative. The term is applied also to such an interaction in the case of an allosteric enzyme for which the same substance acts as both the substrate and the allosteric effector, etc. *Compare* **heterotropic** (def. 1).

homovanillic acid *abbr.*: HVA; 4-hydroxy-3-methoxyphenylacetic acid; a metabolite of catecholamine metabolism, found in human urine. It is the principal urinary metabolite of dopa and dopamine. It is used as a reagent in the fluorometric determination of glucose oxidase and other oxidases.

homozygosis 1 the union of gametes identical for one or more pairs of genes to form a **homozygote**. **2** *or* **homozygosity** the state or condition of being a homozygote. *Compare* **heterozygosis**.

homozygosity *an alternative term for* **homozygosis** (def. 2).

homozygote any cell or organism having identical genes (alleles) at one or more corresponding loci on homologous chromosomes. With respect to such loci, a homozygote will produce two identical gametes. *Compare* **heterozygote**. — **homozygotic** *or* **homozygous** *adj.*

Hoogsteen base pair a nucleic-acid base pair that differs from the Watson–Crick base pair. In the Hoogsteen adenine– thymine base pair the 6-NH$_2$ and N-7 of the adenine are hydrogen bonded respectively to the 4-O and H-1 of the thymine. The Hoogsteen guanine–cytosine pair requires that N-1 of the cytosine is protonated. In this base pair the 6-O and N-7 of the guanine are hydrogen bonded respectively to the 4-NH$_2$ and the protonated N-1 of cytosine. In Watson–Crick base pairs the glycosidic bond of both nucleotides has the *anti* conformation, but in the Hoogsteen pairs the glycosidic bond of the purine nucleotide has the *syn* conformation. Hoogsteen base pairs are particularly important in the structure of DNA triple helices, where the third strand forms Hoogsteen pairs to bases in the DNA duplex. [After Karst Hoogsteen (1923–), Dutch–American biochemist, who first described it.]

hopane the pentacyclic triterpene hydrocarbon parent of hopanone and derivatives found in plant resins. *See also* **hopanoids**.

hopanoids pentacyclic sterol-like molecules based on the **hopane** nucleus. They are found in bacteria, some plants and lichens, and in sediments and crude oils, and are said to be the most abundant organic molecules in the biosphere (estimated total biomass 10^{11} to 10^{12} tonnes), thus outranking **cellulose**.

Hopkins–Cole reaction a colour reaction for free or combined tryptophan, in which a violet colour is produced when a solution containing tryptophan is treated with glyoxylic acid (present in impure glacial acetic acid) and subsequently layered onto a solution of pure, concentrated sulfuric acid. [After (Sir) Frederick Gowland Hopkins (1861–1947), British biochemist, and Sydney William Cole (1877–1951).]

HOQNO *or* **HQNO** *abbr. for* 2-*n*-heptyl-4-hydroxyquinoline *N*-oxide; an inhibitor of the mitochondrial respiratory chain at cytochrome bc_1 and of photosynthetic electron flow immediately before cytochrome b_{559}LP.

hordein a type of **glutelin** found in barley.

horizontal gene transfer the transfer of genes from one organism to another (as distinct from vertical gene transfer, i.e. from parent to offspring).

hormone any substance formed in very small amounts in one specialized organ or group of cells and carried (sometimes in the bloodstream) to another organ or group of cells, in the same organism, upon which it has a specific regulatory action; an **endocrine** (def. 2). The term was originally applied to agents with a stimulatory physiological action in vertebrate animals (as opposed to a **chalone**, which has a depressant action). Usage is now extended to regulatory compounds in lower animals and plants, and to synthetic substances having comparable effects. *Compare* **autacoid, hermone**. *See also* **secretin**. —**hormonal** *adj*.

hormone cascade *see* **cascade sequence**.

hormone receptor any protein of the plasma membrane, or present intracellularly, that binds a hormone and mediates its physiological effects. Hydrophilic hormones (i.e. those that are proteins, peptides, or other amino acid derivatives) are ligands of plasma membrane receptors, which modulate the activity of intracellular intermediaries. Hydrophobic hormones (e.g. steroids, thyroid hormones, vitamin D derivatives) are ligands of cytosolic or nuclear receptors, which modulate the activity of responsive genes by binding to their **hormone response elements**. Examples include the **glucocorticoid receptor** and **insulin receptor**.

hormone resistance the state of decreased or absent sensitivity to a hormone. It may be primary and due to a defect in a specific receptor (e.g. for glucocorticoids, androgens, estrogens, vitamin D derivatives, thyroid hormones, thyroid stimulating hormone, parathyroid hormone, antidiuretic hormone, insulin) or due to post-receptor defects (e.g. in diabetes mellitus type 2). Acquired resistance may result from damage to a target tissue, in the course of therapy with the hormone, or by immunological mechanisms.

hormone response element *abbr*.: HRE; a specific nucleotide sequence in the promoter region of a structural gene that binds to the DNA-binding domain of a nuclear hormone receptor. The HRE usually consists of two hexanucleotide sequences that are contiguous or separated by three nucleotides, and may be in tandem or in palindromic arrangement.

hormone-sensitive lipase any enzyme of the sub-subclass EC 3.1.1 that, when activated in adipose tissue, catalyses the mobilization of fatty acids into the circulation. Such enzymes are activated by cyclic AMP-dependent phosphorylation under the influence of catecholamines, and are inactivated by insulin-dependent dephosphorylation.

hormonotoxin any conjugate of a peptide hormone and a peptide toxin, e.g. ovine luteinizing hormone–gelonin.

horse radish peroxidase *abbr*.: HRP; a hemoprotein enzyme that is widely used to label antibodies and other proteins in enzyme immunoassays and related procedures such as immunodetection following Western blotting.

host 1 any organism in which another organism, especially a **parasite** or **symbiont**, spends part or all of its life cycle and from which it obtains nourishment and/or protection. **2** any organism that harbours a pathogenic or nonpathogenic infectious agent. **3** the recipient of a transplanted tissue or organ graft. **4** a cell or organism that contains recombinant DNA.

host cell any cell whose metabolism is used for the growth and reproduction of a virus.

host-cell reactivation *abbr*.: HCR; the restoration of activity in a UV-damaged DNA bacteriophage by the excision-repair mechanisms of the host cell in which it multiplies.

host-controlled modification *or* **host-controlled variation** *or* **host-induced modification** *or* **host-induced variation** any nonheritable change in the properties of a DNA bacteriophage resulting from the phage DNA assuming the DNA **modification** pattern characteristic of the host bacterium. It is usually recognized by **restriction** in the efficiency of plating of the newly modified phage on its former host strain. Similar phenomena have been described for viruses infecting eukaryotic cells.

host-induced modification *or* **host-induced variation** *see* **host-controlled modification**.

host–guest chemistry a branch of chemistry in which a larger and stucturally concave organic molecule (the host) selectively binds to a smaller molecule or ion (the guest). This may mimic a biological effect, or prevent a biological interaction.

host range the spectrum of host organisms that can be infected by a specified infectious agent or parasite.

host–vector system a compatible combination of host and **vector** (def. 3) that allows propagation of DNA.

hot *(in physics) an informal term* describing a substance or material

that contains a radionuclide, especially if **carrier-free** or at a dangerously high activity or specific activity. *Compare* **cold**.

hot laboratory *or* **hot lab** *an informal term for* a laboratory reserved and equipped for the manipulation of radioactive substances or materials of (dangerously) high activity.

hot room 1 a room (or laboratory) maintained at a constant, higher-than-ambient temperature. **2** *an informal term for* a room (or laboratory) that is specially equipped to deal with high levels of radioactivity.

hot spot any site at which mutations are observed with an unusually high frequency. Such sites have a highly enhanced, inherent susceptibility to mutation that cannot be ascribed to the type of base pair involved, the way in which it is altered, or the resulting change in any protein encoded by the DNA molecule.

hot start PCR a type of polymerase chain reaction (PCR) designed to minimize nonspecific amplification. Reagents are heated to a temperature that does not permit hybridization between oligonucleotides and the target sequence before the addition of a thermostable DNA polymerase to start the reaction.

hour *symbol*: h *or* hr; a non-SI unit of time equal to 3600 s.

housekeeping genes *see* **constitutive genes**.

Houssay animal a pancreatectomized animal which has also undergone **hypophysectomy**. The latter ameliorates the diabetic state that results from pancreatectomy by abolishing the secretion of **somatotropin**. [After Bernardo Alberto Houssay (1887–1971), Argentinian physiologist and endocrinologist.]

***Hox* gene** *see* **homeotic gene**.

Hp 1 *symbol for* the heptyl group, $CH_3-[CH_2]_5-CH_2-$. **2** *abbr. for* haptoglobin.

HP-1 *abbr. for* helper peak-1 (now called **interleukin 1**).

***Hpa*I** a type 2 **restriction endonuclease**; recognition sequence: $GTT\uparrow AAC$.

***Hpa*II** a restriction endonuclease that cuts at the sequence CCGG when the first C is unmethylated. ***Msp*I** is unaffected by methylation at this position. Comparison of restriction digests with both enzymes is informative about the methylation of DNA.

HPCE *abbr. for* high performance capillary electrophoresis.

HpETE *abbr. for* hydroperoxyeicosatetraenoate.

hPL *or* **HPL** *abbr. for* human placental lactogen.

HPLAC *abbr. for* high-performance liquid affinity chromatography.

HPLC *abbr. for* high-pressure liquid chromatography *or* high-performance liquid chromatography.

Hpo *symbol for* the heptanoyl group, $CH_3-[CH_2]_5-CO-$.

HpODE *abbr. for* hydroperoxyoctadecadienoate.

HPR *abbr. for* H protein; a low-M_r, heat-stable, cytoplasmic phosphocarrier protein found in some bacterial species, and concerned in the phospho*enol*pyruvate (PEP)-hexose transferase system for the transport of hexoses across the cell membrane. The phosphate of PEP is transferred transiently to HPR. *See* **enzyme I, enzyme II**.

HPRD *abbr. for* human protein reference database; *see* **pathway database**.

H protein *see* **glycine cleavage enzyme**.

HPRT *abbr. for* hypoxanthine phosphoribosyltransferase.

H⁺ pump *see* **hydrogen ion pump**.

HQNO *see* **HOQNO**.

hr *abbr. for* hour (now superseded by h).

Hrb a Rab protein (*see* ***RAB***) that is essential for formation of the acrosome of mouse spermatozoa. It is associated with the cytosolic surface of proacrosomic vesicles that fuse to create a single large acrosome, and is required for docking and/or fusion of these vesicles.

HRE *abbr. for* hormone response element.

H1, H2, H3 receptors *see* **histamine receptor**.

H-RNA *abbr. for* heterogeneous nuclear RNA (alternative to hnRNA or HnRNA).

HRP *abbr. for* horse radish peroxidase.

HRT *abbr. for* hybrid-released translation.

HSA *abbr. for* human serum albumin.

Hse *symbol for* a residue of the α-amino acid L-homoserine.

HSE *abbr. for* heat-shock regulatory element or heat-shock response element.

α₂-HS-glycoprotein *or* **fetuin** an α-globulin found in fetal calf serum, where it accounts for up to 45% of the total protein. It has a low isoelectric point and contains up to 35% carbohydrate, having three oligosaccharide chains, each of M_r 3400; it comprises various glycoproteins that differ in M_r. It is a growth factor for a variety of cell types; it also promotes endocytosis, possesses opsonic properties, and influences the mineral phase of bone. The amino-acid sequence has **cystatin** features; two peptides are cleaved from a precursor to form chains held together by a single disulfide bond.

Hsl *symbol for* a residue of the α-amino lactone L-homoserine lactone.

HSL *abbr. for* hormone-sensitive lipase.

hsp *or* **Hsp** *abbr. for* heat-shock protein.

HSP *abbr. for* **1** heat-shock protein. **2** high-scoring segment pair.

Hsp90 a highly conserved molecular chaperone. Other family members are Grp94/gp96 in the endoplasmic reticulum and HtpG in the prokaryotic cytosol. Unlike Hsp70, which promiscuously recognizes a large number of proteins, Hsp90 is characterized by the specific nature of its protein-protein associations. The majority of Hsp90 substrates (usually referred to as 'client' proteins) are involved in regulatory processes, the classical examples being signalling kinases and steroid hormone receptors. Maintenance of the biochemical activity of the clients is entirely dependent on continued interaction with Hsp90..

HSPG core protein *abbr. for* heparan sulfate proteoglycan core protein. *See also* **syndecan**.

HSSP *abbr. for* homology-derived secondary structure prediction; *see* **structure database**.

hst/KS *see* **KS/hst**.

H strand *abbr. for* heavy strand.

HSV *abbr. for* herpes simplex virus.

5-HT *abbr. for* 5-hydroxytryptamine (i.e. **serotonin**).

HTG *abbr. for* high-throughput genomic sequence.

HTH *abbr. for* helix-turn-helix.

HTLV *abbr. for* human T-cell lymphotropic virus; a **retrovirus** that causes a rare leukemia.

HTML *abbr. for* hypertext markup language; *see* **markup language**.

Htr either of two transmembrane proteins (HtrI or HtrII) that extend far into the cytoplasm of halobacteria and mediate phototaxis signalling from a sensory rhodopsin (SRI or SRII, respectively) to the flagellar motor.

H⁺-transporting ATP synthase EC 3.6.1.34; *systematic name*: ATP phosphohydrolase (H⁺-transporting); *other names*: H⁺-transporting ATPase; mitochondrial ATPase; chloroplast ATPase; coupling factors $F_0 \cdot F_1$, $C_0 \cdot F_1$. The enzyme that is responsible for the phosphorylation of ADP to ATP in mitochondria, during **oxidative phosphorylation**. *In vitro* it can function as an ATPase. The 3-D structure is known for the major components of bovine heart mitochondrial enzyme. The enzyme comprises an intrinsic membrane domain, F_0, linked by a stalk to a globular catalytic domain, F_1, and is hence sometimes termed the F_0F_1 complex. Energy released by hydron flux through F_0 leads to ATP synthesis by F_1; the stoichiometry is approximately 3H⁺ per ADP phosphorylated. F_1 (371 kDa) has the subunit composition: $3\alpha,3\beta,\gamma,\delta,\epsilon$. The ϵ and γ subunits form a shaft inserted at one end into F_0 and at the other into F_1. The α and β (catalytic) subunits are arranged round it alternately; the three β subunits have different conformations during the synthesis cycle. The N- and C-terminal regions of γ form an antiparallel coiled coil; the α and β proteins are essentially beta barrels. There are several proteins in the F_0 component, the main subunits of which in eukaryotes are named A to G. Twelve C subunits form a channel within the membrane. The C chain is the protein that accumulates in ceroid lipofuscinosis (**Batten's disease**). It is a proteolipid encoded by two genes which code for two precursor proteins, P_1 and P_2, having different signal peptides for locating the proteins in mitochondria, but the same structure in the mature protein. Two B subunits are anchored, with an A subunit, into the membrane, and through a δ subunit bind F_1. The complex formed by $AB_2\delta\alpha_3\beta_3$ is called the stator; that formed by $C_{12}\epsilon\gamma$ is called the rotor, because it rotates during the flow of hydrons through the F_0 channel.

H⁺-transporting pyrophosphatase an integral membrane protein (80 kDa) that contains 16 transmembrane segments and is

ubiquitous in vacuolar membranes in plants. It is also present in some species of archaea. In the presence of K⁺ it hydrolyses dimagnesium pyrophosphate of the cytosol and transports H⁺ into the vacuole. Ca^{2+} is a potent inhibitor, as is aminomethylenediphosphonate. In immature tissues it is more active than the V-type H⁺ ATPase, the opposite being true in mature tissues.

huangosome *a colloquial term for* a bilayered **liposome**. *Compare* **bangosome**.

Hudson's convention a convention, no longer used, for designating anomeric configurations in monosaccharides. The anomer in the D-series with the highest positive rotation (or lowest negative rotation) was designated the α form and the anomer with the lowest positive rotation was called the β form. The reverse applied in the L-series, the anomer with the lowest positive rotation (or the highest negative rotation) being designated the α form. The α and β descriptors now refer to the absolute structural relationships. *See* **anomer, Haworth representation**.

Hughes press a device to disrupt microorganisms. It consists of a steel cylinder in which a paste of the microorganisms in question, mixed with abrasives, is subjected to sudden blows by a steel piston driven by a fly-press; the piston forces the crushed material out of the cylinder into a reservoir. At temperatures below –20°C no abrasives are needed, ice crystals taking their place. *See also* **French pressure cell**. [After D. E. Hughes (1922–), who published the design in 1951.]

HUGO *abbr. for* Human Genome Organization; an international organization of scientists involved in the **Human Genome Project**, the global initiative directed at mapping and sequencing the human genome. It was established in 1989, and carries out a coordinating role, supporting data collection for constructing genetic and physical maps of the human genome, organizing the Single Chromosome Workshops and other workshops and promoting consideration of ethical, legal, social, and intellectual property issues. It fosters exchange of data and biomaterials. The Human Genome Project is, however, not under the central control or direction of any individual, group, or organizing committee, the work being done in many countries and coordinated by the scientists carrying out the work.

human *abbr. (in bionomenclature)*: h *or (formerly)* H; of, relating to, characteristic of, obtained from, or being a member of the species *Homo sapiens*. For names of substances or products beginning with the word human (e.g. human choriogonadotropin, human serum albumin) see at the corresponding general entry.

human gene index *abbr.*: HGI; a database of The Institute for Genome Research (TIGR) that integrates data from international human gene research projects, including gene expression patterns, cellular roles, functions, and evolutionary relationships.

Human Genome Project *see* **HUGO**

human immunodeficiency virus *see* **HIV**.

humanize to produce chimeric monoclonal antibodies that, by use of recombinant DNA technology, contain sequences translated from components of human immunoglobulin genes.

humanized antibody a modified antibody; since human monoclonal antibodies cannot be generated in the same way as in mice, and the body reacts to a mouse antibody the 6 complementary determining regions of a high affinity rodent monoclonal antibody are grafted on to a completely human IgG framework without loss of specific activity.

human T-cell lymphotropic virus *see* **HTLV**.

humectant **1** producing moisture. **2** any substance added to another substance or material to keep it moist.

humic acid any of the mixture of complex macromolecules having polymeric phenolic structures and extractable from soils and peat. It is produced by the oxidative degradation of lignin and has the ability to chelate metals, especially iron.

humidity cabinet *or* **humidity chamber** an enclosure in which the humidity is maintained at a constant, often high level. It is useful for keeping experimental samples or materials moist or at a predetermined moisture content.

humin the very dark-brown or black insoluble material that is produced when carbohydrate-containing proteins are subjected to acid hydrolysis.

humor *the US spelling of* **humour**.

humoral of, contained in, or involving the blood or other body fluid; consisting of, relating to, or involving some chemical agent or substance contained in a body fluid, especially the blood.

humoral immunity any specific immunity mediated by antibodies (i.e. **humoral antibodies**) present in blood plasma, lymph, and tissue fluids, and that may also become attached to cells.

humour or *(US)* **humor** *a former name for* any of the body fluids, e.g. blood, lymph.

humulin 1 human insulin made using cDNA technology. **2** the bitter aromatic principle of hops (lupulin).

humus the dark-brown or black complex mixture of colloidal organic material that results from the decomposition of plant and animal material in the soil. It is of great importance for plant growth.

hunchback a segmentation protein in insects. Similar to **ikaros**, it is a zinc-finger protein and is encoded by a **gap gene**. *Compare* **Krüppel**.

Hunter syndrome *an alternative name for* mucopolysaccharidosis II. *See* **mucopolysaccharidosis**. [After Charles Hunter (1872–1955), British physician.]

huntingtin *see* **Huntington's disease**.

Huntington's disease or **Huntington's chorea** an autosomal dominant genetic disease characterized by involuntary (choreic) movements and dementia over a 10–20 year course. It is inevitably fatal. Linking studies using a variety of RFLPs have localized the defective gene to a site on the short arm of chromosome 4. There is a selective loss of a group of neurons; the biochemical basis for their loss is unknown. The gene, which encodes a 3145-residue product, **huntingtin**, of unknown function, contains a polymorphic trinucleotide repeat of the codon for glutamine, CAG, within the coding sequence. Normals have 35–37 of these repeats. In no case has any individual with more than 41 repeats been found to be free of the disease. [After George Huntington (1850–1916), US physician.]

HU protein an abundant bacterial histone-like protein that has the ability to wrap and bend DNA *in vitro*, and is thought to play a role in chromosome DNA topology, probably by facilitating the action of gyrase. *See* **DNA topoisomerase** type II.

Hurler–Scheie compound syndrome *an alternative name for* **mucopolysaccharidosis I H/S**.

Hurler syndrome *an alternative name for* mucopolysaccharidosis I H; *see* **mucopolysaccharidosis**. [After Gertrude Hurler (1889–1965), Austrian physician.]

HUT *(in clinical chemistry)* *abbr. for* UDPglucose-hexose-1-phosphate uridylyl transferase (EC 2.7.7.12).

Hutchinson–Gilford syndrome a rare form of premature ageing characteristic of children and associated with truncation of **lamin** A. [After Jonathan Hutchinson (1828–1913), British surgeon, and Hastings Gilford (1861–1941), British physician.]

HVA *abbr. for* homovanillic acid.

HVE *abbr. for* high-voltage electrophoresis.

Hvidt–Nielsen mechanism a proposed mechanism to account for the kinetics of hydrogen exchange between protein molecules in solution and solvent water molecules. It is considered that protein molecules exist in solution in more than one conformational state and that hydrogen atoms in the interior of the protein molecule in the most stable conformational state are inaccessible to exchange; these hydrogen atoms may become transiently externalized, and hence accessible to exchange, in one or more of the other conformational states. [After Aasa Hvidt and S. O. Nielsen.]

Hx *abbr. for* hepoxilin.

Hx^c *symbol for* the cyclohexyl group (alternative to cHx).

hyal+ *a variant form of* **hyalo+** (before a vowel).

hyaladherin *a name sometimes used for* proteins in the **extracellular matrix** that bind to **hyaluronate**.

hyaline clear, transparent; with no inclusions.

hyalo+ or *(before a vowel)* **hyal+** *comb. form* denoting glassy, transparent.

hyaluronan *an alternative name for* **hyaluronate**.

hyaluronate or **hyaluronan** the naturally occurring anionic form of hyaluronic acid; any member of a group of glycosaminoglycans, the repeat units of which consist of $\beta1\rightarrow4$-linked D-glucuronyl-($\beta1\rightarrow3$)-N-acetyl-D-glucosamine. Most preparations contain 1–2% by mass of protein, and have M_r values in the range 5×10^4 to 8×10^6 depending on source and methods of preparation and determination.

Hyaluronates are widely distributed in many tissues, especially in the vitreous humour of the eye, synovial fluid, and umbilical cord. They have an important role in the formation of proteoglycan aggregates. Hyaluronates form very viscous solutions and gels in water, a property that may be important in protecting animal organisms against invasion by bacteria.

hyaluronidase or **mucinase** or **spreading factor** *a trivial name for* either of the glycosidases hyaluronoglucosaminidase (EC 3.2.1.35) and hyaluronoglucuronidase (EC 3.2.1.36), and for the carbon–oxygen lyase hyaluronate lyase (EC 4.2.2.1). These enzymes have in common the ability to depolymerize hyaluronidate by cleavage of glycosidic bonds; e.g. hyaluronoglucosaminidase catalyses the random hydrolysis of 1,4-linkages between N-acetyl-β-D-glucosamine and D-glucuronate residues in hyaluronate. Hyaluronidases occur in certain animal tissues, e.g. sperm acrosome, and in some bacteria, e.g. *Clostridium* spp., and enhance the penetration of various substances and pathogens through the extracellular matrix (hence the name 'spreading factor').

hyaluronidase deficiency *an alternative name for* **mucopolysaccharidosis IX**.

hyamine hydroxide methylbenzethonium hydroxide; a water-soluble organic base, useful biochemically because it and some of its salts are soluble in organic solvents such as chloroform and toluene. It is therefore used to absorb radioactive CO_2 released during metabolic incubations; the resultant hyamine carbonate can then be dissolved in a scintillation cocktail for counting. It is also used as a topical anti-infective agent.

hybrid 1 something derived from heterogeneous sources or composed of different or incongruous elements. **2** *(in genetics)* the offspring of two animals or plants of different species, or (less strictly) different varieties. **3** *(in chemistry)* a bond or valence orbital obtained by the linear combination of two or more different atomic orbitals.

hybrid antibody any artificially produced immunoglobulin molecule in which the combining sites are of different specificities.

hybrid-arrested translation *abbr.*: HART; a method for analysing the relationship between a DNA sequence, its corresponding messenger RNA (mRNA), and the protein for which it codes, based on the observation that mRNA when hybridized with its complementary DNA is not translated in eukaryotic cell-free systems whereas heat-induced dissociation of the hybrid completely restores translational activity. The method has been used to map precisely specific protein-coding regions in **restriction fragments** of DNA. *See also* **hybrid-released translation**.

hybrid cell any cell containing the nuclei and cytoplasms from differing cells, formed artificially by **cell fusion**.

hybrid DNA 1 artificial duplex DNA produced by collinear (i.e. covalent) union of two (or more) fragments of duplex DNA (i.e. **recombinant DNA**; *see also* **plasmid chimera**). **2** artificial duplex DNA produced by collateral (hydrogen bonded) union of two incompletely complementary, single strands of DNA. *Compare* **DNA–RNA hybrid**.

hybridization or **hybridisation 1** the act or process of forming a genetic **hybrid** (def. 2) by cross-breeding, or other unnatural recombination. **2** the act or process of forming a macromolecular **hybrid** (def. 1) by the artificial recombination of subunits, e.g. of polynucleotide strands or polypeptide chains. **3** the mathematical combination of atomic orbitals to form an orbital **hybrid** (def. 3).

hybridization analysis a method of testing for complementarity in the base sequences of two polynucleotides from different sources, based on the ability of complementary single-stranded DNA molecules to form duplex DNA molecules on being annealed together, and the ability of single strands of DNA and RNA similarly to form **DNA–RNA hybrids**.

hybridization probe *see* **probe** (def. 3).

hybrid myeloma *see* **hybridoma**.

hybridoma any permanent **cell line** derived from an artificial **somatic cell hybrid** formed between a cultured neoplastic lymphocyte and a normal primed B or T lymphocyte, all the cells of which express the specific immune potential of the normal parental cell. The term is derived from a contraction of 'hybrid' and 'myeloma'. Hybridomas are produced by the technique of **cell fusion**, using, e.g., a suspension of mouse myeloma or thymoma cells from a selected established cell line to which is added a preparation of spleen cells (predominantly normal lymphocytes) from mice or other animals that have been immunized against a specified antigen. After cell fusion, selection is aided by the use of **HAT** medium, which kills HGPRT⁻ myeloma cells, and lymphocytes that have not fused do not survive in culture. This treatment thus selects for the immortalized hybridoma cells that retain the lymphocyte HGPRT. Such hybrid cells are both permanently adapted to growth in culture and capable of forming specific antibody-producing tumours *in vivo*. A B-cell hybridoma continuously secretes pure **monoclonal antibody** of a predetermined specificity and may thus be used for the facile large-scale production of such antibody.

hybrid-released translation *abbr.*: HRT; a method in which cloned DNA is bound to a nitrocellulose filter and hybridized with unfractionated mRNA, or even total cellular RNA. After washing the filter, any mRNA that has hybridized with the DNA is eluted and translated in a cell-free translation system, thus revealing the product that is encoded by the DNA. *See also* **hybrid-arrested translation**.

hydantoin *generic name for* 2,4-imidazolidinedione or its derivatives. A hydantoin is formed by the reaction of an isocyanate, commonly phenylisocyanate, with the free amino group of an amino acid, and the cyclization of the hydantoic acid derivative initially formed. When an isocyanate reacts with the N-terminal amino group of a peptide, the corresponding hydantoin is formed, with simultaneous cleavage of the substituted N-terminal amino-acid residue from the peptide.

hydr+ *a variant form of* **hydro+** (sometimes before a vowel).

hydrase *an obsolete term for* **hydratase**.

hydratase *see* **dehydratase**.

hydrate 1 any adduct of water with another compound or element, usually in definite molecular proportions. **2** to form a **hydrate** (def. 1); to undergo **hydration**. **3** to cause to take up or to combine with water. **4** to maintain or restore the normal proportion of water in the body (of someone or something). —**hydrated** *adj.*

hydration 1 the act or process by which water molecules surround and bind to molecular entities. **2** the addition of H and OH to a chemical species. **3** the formation of a **hydrate** (def. 1). **4** the quality or state of being hydrated. **5** the introduction of additional water into the body (of something or someone). *Compare* **anhydrous, dehydration, solvation**.

hydration shell the layer of water molecules surrounding an ion or molecule in aqueous solution.

hydraulic 1 operated by pressure transmitted through a liquid, e.g. through water or oil. **2** of or concerning hydraulics (i.e. fluid mechanics).

hydrazine $H_2N–NH_2$; an oily liquid, fuming in air, that is very toxic and possibly a carcinogen. It is available as a hydrate and as various salts. *See also* **hydrazinolysis**.

hydrazinolysis the cleavage of amide and peptide bonds by heating at 100°C with **hydrazine** in anhydrous conditions to form the corresponding acyl hydrazides. This reaction is useful in the determination of C-terminal amino-acid residues in peptides and proteins as all the other amino-acid residues will be converted to their hy-

drazides and only the C-terminal amino acid will be set free in an underivatized form.

hydrazone any condensation product of the type R′R″C=N–NHR, formed between carbonyl compounds (R′R″=CO) and hydrazine or a substituted hydrazine. *See also* **osazone, phenylhydrazone**.

hydride 1 the ion H⁻, hydrogen anion. It has an independent existence only in the salt-like hydrides formed by strongly electropositive elements (*see* def. 2) but is also sometimes regarded as participating in the formation of complex hydrides (*see* def. 4) and in certain oxidation–reduction reactions (*see* **hydride-ion transfer**). **2** any stoichiometric binary compound of hydrogen with one of many (but not all) of the elements in groups 1, 2, and 3 of the IUPAC **periodic table**, e.g. NaH or CaH_2. Such hydrides are salt-like ionic compounds, are solids at ambient temperatures, and on fusion can be electrolysed (dihydrogen then being liberated at the anode). They are readily decomposed by water and are potent reducing agents. **3** Any stoichiometric compound of hydrogen with one (or sometimes more than one) of many of the elements in other groups of the periodic table, and also with boron. Such hydrides are volatile covalent compounds containing one atom of the element per molecule, e.g. PbH_4, or sometimes more than one, e.g. Si_3H_8; in the latter type certain of the hydrogen atoms may be present as bridges, as in B_2H_6. **4** any salt-like compound containing a complex anion that may be considered as formed by coordination of a hydride ion to an uncharged hydride, e.g. KBH_4 (*see* **borohydride**) or $LiAlH_4$; such compounds are widely used in organic chemistry as reducing agents or for introduction of tritium, ³H. **5** any (frequently nonstoichiometric) product formed by uptake of dihydrogen by certain transition metals, especially palladium.

hydride-ion transfer *term conventionally often applied to* a chemical or enzyme-catalysed oxidation–reduction reaction involving the transfer from one molecule to another of a hydrogen atom together with its associated pair of bonding electrons. Such a reaction may notionally be considered to involve the transfer of a hydride ion without such an ion ever existing as a free entity. Examples include the **Cannizzaro reaction** and reactions of the general type

$$NADH + M + H^+ \rightleftharpoons NAD^+ + MH_2,$$

where M is a substrate molecule undergoing enzymic reduction, to which a hydride ion is transferred from NADH.

hydrindantin the reduced form of **ninhydrin**. It is sometimes used in conjunction with ninhydrin in the chromogenic determination or visualization of amino or imino groups.

hydrion *a former name for* **hydrogen ion**.

hydro+ *or (sometimes before a vowel)* **hydr+** *comb. form* **1** indicating or denoting water, liquid, or fluid. **2** indicating the presence of hydrogen in a compound.

hydrocarbon 1 any organic compound composed exclusively of hydrogen and carbon. **2** of, pertaining to, or being a **hydrocarbon** (def. 1) or a chemical group or side chain composed of hydrogen and carbon only.

hydrocarbyl any univalent group formed from a hydrocarbon by removal of a hydrogen atom; e.g. methyl (from methane).

hydrocolloid any colloidal substance for which the dispersing liquid is water or an aqueous solution.

hydrocortisone *an alternative name for* **cortisol**.

hydrodynamic 1 of or concerned with the mechanical properties of fluids. **2** of, pertaining to, or concerned with **hydrodynamics**.

hydrodynamic particle the entity formed by any (macro)molecule in solution together with its solvation shell.

hydrodynamic radius the radius of a (macro)molecule or other particle in solution as determined by a hydrodynamic method, e.g. sedimentation equilibrium.

hydrodynamics the branch of physics concerned with the forces acting on or exerted by fluids.

hydrodynamic shear the microscopic shear force caused by the motion of different layers in a fluid, where the layers move at different velocities. This acts to break any molecule sufficiently long to bridge the layers, e.g. DNA.

hydrodynamic volume *(of a molecule in solution)* the sum of the time-average of the molecular volume and the volume of the solvent molecules associated with it.

hydrogel any **gel** (def. 1) in which water is the liquid component. *Compare* **hydrosol**.

hydrogen *symbol*: H; the lightest of the elements and the most abundant in the universe. It exists as an odourless, colourless, flammable diatomic gas, dihydrogen, H_2, and forms compounds with most of the elements, being present in water and all organic compounds. There are two naturally occurring isotopes, hydrogen-1 (**protium**) and hydrogen-2 (**deuterium**), and one artificial isotope, hydrogen-3 (**tritium**), which is radioactive.

hydrogenase an enzyme that catalyses the production or consumption of dihydrogen. Hydrogen is produced by hydrogenases during the photosynthetic cleavage of water, and by fermentation processes in many anaerobic organisms. Hydrogen is also produced by **nitrogenase** as a by-product of the nitrogen-reducing reaction, but this enzyme not defined as a hydrogenase. Hydrogen is consumed by hydrogen-uptake hydrogenases in many bacteria, with the concomitant reduction of compounds such as oxygen, which is reduced to water, sulfate to sulfide, or carbon monoxide to methane. One of the most common hydrogen-consuming hydrogenases is hydrogen:quinone oxidoreductase, EC 1.12.7.2; in nitrogen-fixing bacteria, this enzyme recycles H_2 produced by nitrogenase to increase the production of ATP and to protect nitrogenase against inhibition or damage by O_2. In order to react with other electron carriers the hydrogenases employ various cosubstrates. Hydrogen dehydrogenase, EC 1.12.1.2, catalyses the reaction: H_2 + NAD^+ = H+ + NADH and contains FMN. This enzyme is structurally related to mitochondrial Complex I. EC 1.12.1.3 is similar, but reduces $NADP^+$. Cytochrome-C_3 hydrogenase, EC 1.12.2.1 is used by sulfate-reducing bacteria. Almost all hydrogenases are iron–sulfur proteins. The defining characteristic is an active site containing iron that has carbonyl and usually cyanide ligands. Two main nonhomologous classes are recognized: (1) hydrogenases containing an H cluster comprising a dimer of iron atoms linked to a [4Fe-4S] cluster,known as [Fe]- or [FeFe]-hydrogenases (*see* **iron hydrogenase**), and those containing nickel and iron (*see* **nickel–iron-hydrogenase**). In addition there is a third type, containing iron and no iron–sulfur clusters, of which the only example known so far is 5,10-methenyltetrahydromethanopterin hydrogenase (EC 1.12.98.2).

hydrogenate to undergo or cause to undergo reduction by addition of the hydrogen atoms of dihydrogen, usually with the aid of a catalyst such as finely divided forms of nickel, palladium, or platinum. The process is of industrial importance in petroleum refining, the petrochemical industry and the hardening of fats. —**hydrogenated** *adj.*; **hydrogenation** *n.*

hydrogen bond an association between an electronegative atom, e.g. fluorine, oxygen, nitrogen, or sulfur, and a hydrogen atom attached to another such electronegative atom. Although **hydrogen bonding** is due to interaction between dipoles, the force of attraction is large enough to permit formation of aggregates of small molecules or to stabilize the conformation of many macromolecules. The spatial relation of the donor and acceptor atoms is such that the hydrogen atom lies very close to the straight line between them.

hydrogen carrier *a name sometimes used for* any electron carrier that undergoes oxidation–reduction reactions by the loss or gain of hydrogen atoms, e.g. NADH, FADH.

hydrogen dehydrogenase *the recommended name for* the enzyme, EC 1.12.1.2, *also known as* **hydrogenase**, that catalyses the reaction:

$$H_2 + NAD^+ = H^+ + NADH.$$

It is a flavoprotein; iron and nickel are cofactors; the molecule is heterotetrameric. It may form part of a system involving **hydrogenase** (def. 1). For example, in *Alcaligenes eutrophus*, the α and γ subunits form one dimer (FMN, 4Fe–4S cluster, NADH oxidoreductase); and the β and δ subunits form the other dimer, which exhibits **hydrogenase** activity with artificial electron acceptors.

hydrogen–deuterium exchange *see* **hydrogen exchange**.

hydrogen electrode an electrode that is reversible to hydrogen ions and dihydrogen molecules. It is formed by bubbling pure dihydrogen gas over a metal wire or a piece of foil whose surface is able to catalyse the reaction:

$$0.5H_2(g) = H^+ + e \tag{1}$$

where e represents an electron. The most useful catalyst is platinum black on a platinum substrate. Hence the catalyst establishes an equilibrium between dihydrogen molecules (or hydrogen atoms) and the hydrogen ions (i.e. hydrons) in the solution in which the metal electrode is immersed. If the partial pressure (in atm) of dihydrogen gas at the electrode is p_{H_2}, the half-cell potential, E_H, of reaction (1) is given by:

$$E_H = E^o{}_H - (RT/F)\ln(a_{H^+}/p_{H_2}) \tag{2}$$

where R is the gas constant, F the faraday constant, T the thermodynamic temperature, and a_{H^+} the hydrogen-ion activity. $E^o{}_H$, the standard potential of the hydrogen electrode, is, by convention, zero at all temperatures, hence at $p_{H_2} = 1$, the usual standard state, equation (2) becomes:

$$E_H = -(RT/F)\ln a_{H^+}.$$

Thus the hydrogen electrode is the ultimate reference electrode for the determination of pH values. It should be noted that the potential of any single electrode cannot be determined by itself; another electrode must be connected to the solution.

hydrogen exchange *or* **hydrogen–deuterium exchange** a technique useful for demonstrating conformational differences between samples of the same protein and for providing insight into the dynamics of protein conformations in solution. It can be observed using nuclear magnetic resonance spectroscopy. The method is based on observation of the kinetics of the noncatalysed exchange of hydrogen atoms in peptides and proteins with protons or other hydron isotopes in aqueous solutions and the effects of pH (or p²H), temperature, etc. on the process.

hydrogen ion *or (formerly)* **hydrion** the cation, H^+, derived from the hydrogen atom. In aqueous solution it is mostly in the hydrated form, oxonium (H_3O^+). In nonaqueous solution it is likely, by analogy, that it is attached particularly to one solvent molecule, e.g. ROH_2^+ in an alcohol, R_2COH^+ in a ketone, RNH_3^+ in an amine, etc. *See also* **hydron**, **proton** (defs 1, 2).

hydrogen-ion concentration *see* **pH**.

hydrogen-ion pump any of a number of membrane proteins that transport hydrogen ions through membranes. The H^+- transporting ATPase of fungi and plants (EC 3.6.1.35) has an action similar to that of the **H^+-transporting ATP synthase** of mitochondria, but without coupling factors; it is an integral membrane glycoprotein. The proton gradient it generates drives the active transport of nutrients by H^+-symport. The **vacuolar ATPase** is another type.

hydrogen lamp *or* **hydrogen discharge lamp** a high-voltage discharge tube with a quartz window and containing dihydrogen at low pressure, in which the glow produced by excitation of the dihydrogen molecules is concentrated in a long capillary tube joining the two electrodes; this arrangement gives increased intensity. It produces a continuous emission spectrum over the range 180–350 nm. *See also* **deuterium (discharge) lamp**.

hydrogenlyase *see* **hydrogenase** (def. 1).

hydrogenosome an organelle with a double membrane occurring in the cytoplasm of some anaerobic flagellate protozoans. They are chromatic granules of high density containing enzymes, including hydrogenase, that catalyse the conversion of pyruvate to acetate, CO_2, and H_2. These organelles may substitute for mitochondria (which are absent from such organisms).

hydrogen peroxide H_2O_2; the product of reactions catalysed by NADPH oxidase of phagocytes, cytochrome oxidase of mitochondria, acyl-CoA dehydrogenase of peroxisomes, glycolate oxidase of glyoxysomes, and xanthine oxidase of liver and plasma. It is rapidly eliminated by catalase, glutathione peroxidase, and thioredoxin-linked reactions. Although H_2O_2 is poorly reactive, it is rapidly converted into the highly reactive hydroxyl radical (–OH) by transition metal ions, especially iron, and by ultraviolet light. In some plants it is a second messenger in the response to hormones, pathogenic infection, and abiotic stress. .

hydrolase any enzyme of class EC 3 that catalyses the hydrolysis of various bonds, e.g. C–O, C–N, C–C, phosphoric anhydride bonds, etc. Often the name of a hydrolase is formed from the name of the substrate and the suffix '-ase'.

hydrolysate *or (sometimes US)* **hydrolyzate** any substance or mixture of substances produced by hydrolysis.

hydrolyse *or (US)* **hydrolyze** to subject to, to undergo, or to effect **hydrolysis**. —**hydrolysable** *or (US)* **hydrolyzable** *adj.*; **hydrolysation** *or (US)* **hydrolyzation** *n.*

hydrolysis (*pl.* **hydrolyses**) the rupture of one or more chemical bonds by reaction with, and involving the addition of, the elements of water; **solvolysis** by water.

hydrolytic of, pertaining to, producing, or produced by **hydrolysis**.

hydrolyzate *an alternative US spelling of* **hydrolysate**.

hydrolyze *an alternative US spelling of* **hydrolyse**.

hydrometer an instrument used to measure the **density**, or **relative density**, of a liquid. It usually consists of a glass (or plastic) bulb, weighted at the bottom so that it floats upright, and fitted with a stem carrying a scale. The point on the scale that is level with the surface of the liquid indicates the (relative) density of the liquid.

hydron *symbol*: H^+; *the recommended name for* the cation derived from an atom of hydrogen in its natural isotopic abundance. *See also* **hydrogen ion, proton** (defs 1, 2).

hydronate to combine with a **hydron** or hydrons. —**hydronation** *n.*

hydronium ion *an alternative name (now disfavoured) for* **oxonium** (def. 1).

hydropathy a measure of hydrophobicity. *See* **hydrophobic**. –**hydropathic** *adj.*

hydropathy profile a graphical representation of the hydropathic nature of a protein sequence. Within a profile, hydropathy values are calculated in a sliding window and plotted for each residue in the sequence. Resulting graphs show characteristic peaks and troughs, corresponding to the most hydrophobic and hydrophilic regions of the sequence respectively. They are most frequently used to predict the presence of **transmembrane domains**.

hydroperoxide any organic compound in which one hydrogen atom of a hydrocarbon is replaced by the –O–OH group.

hydroperoxyeicosatetraenoate *abbr.*: HpETE; any of various related products of the action of **lipoxygenase** on arachidonate, (all-*Z*)-eicosa-5,8,11,14-tetraenoate. The most notable are 5(*S*)-HpETE, (6*E*,8*Z*,11*Z*,14*Z*)-(5*S*)-5-hydroperoxyeicosa-6,8,11,14-tetraenoate; 12(*S*)-HpETE, (5*Z*,8*Z*,10*E*,14*Z*)-(12*S*)-12-hydroperoxyeicosa-5,8,10,14-tetraenoate; and 15(*S*)-HpETE, (5*Z*,8*Z*,11*Z*,13*E*)-(*S*)-15-hydroperoxy-5,8,11,13-eicosatetraenoate. Their species distribution follows that of the lipoxygenases. HpETEs function as precursors of hydroxyeicosatetraenoates, leukotrienes, dihydroxyeicosatetraenoates, hepoxilins, and lipoxins, but they also have biological activity in their own right, e.g. regulation of electrolyte flux, release of histamine, regulation of eicosanoid and corticosterone synthesis, and regulation of oocyte maturation. 12-HpETE inhibits platelet thromboxane synthase and prostaglandin-endoperoxide synthase, and acts as a second messenger in *Aplysia* sensory neurons.

5(*S*)-HpETE

12(*S*)-HpETE

hydroperoxyoctadecadienoate *abbr.*: HpODE; any of various related products of the action of **lipoxygenase** on linoleate, (all-*Z*)-octadeca-9,12-dienoate. The most notable are 9(*S*)-HpODE, (10*E*,12*Z*)-(9*S*)-octadeca-10,12-dienoate; and 13(*S*)-HpODE, (9*Z*,11*E*)-(13*S*)-octadeca-9,11-dienoate.

hydrophilic 1 tending to dissolve in water; having a strong affinity for water; readily mixing with or wetted by water. *See also* **lyophilic**. **2** polar. *Compare* **hydrophobic**. —**hydrophilicity** *n.*

hydrophobic 1 tending not to dissolve in water; having a low affinity for water; not readily mixing with or wetted by water. **2** nonpolar. *Compare* **hydrophilic**. *See also* **lyophobic**. —**hydrophobicity** *n.*

hydrophobic bond *a term sometimes used incorrectly for* a **hydrophobic interaction**.

hydrophobic chromatography *or* **hydrophobic affinity chromatography** *or* **hydrophobic interaction chromatography** a technique for the separation and purification of proteins and other substances by **adsorption chromatography** on polysaccharide or other gels containing lipophilic substituents (which may or may not bear terminal ionizable groups). These hydrophobic centres on the gel interact with nonpolar amino-acid residues exposed on the surface of the protein molecules being separated.

hydrophobic interaction *or (discouraged)* **apolar bond** the tendency of hydrocarbons (or of lipophilic hydrocarbon-like groups in solutes) to form intermolecular or intramolecular aggregates in polar solvents. *See also* **apolar interaction**.

hydrophobic pocket *(in a protein)* a binding site (crevice, pocket) that contains mostly hydrophobic amino acids. For example, the binding site for the aromatic side chains of the specific substrates in **chymotrypsin** is a hydrophobic pocket, consisting of a groove in the enzyme, 100–120 nm deep, and 35–40 nm by 55–65 nm in cross section, lined with non-polar amino-acid side chains. *See also* **heme pocket**.

hydrophobic residue any building block of a polymer, the residue of which in that polymer is hydrophobic. The term is applied especially to amino-acid residues exhibiting such properties in polypeptides (e.g., valine, leucine, phenylalanine).

hydrophobin any of a family of membrane proteins present in some fungi that are involved in cell adhesion and morphogenesis. They have a characteristic domain that contains eight cysteine residues.

hydroquinone 1 1,4-benzenediol; *p*-dihydroxybenzene; quinol. *Compare* **p-benzoquinone**, **pyrocatechol**. *See also* **quinhydrone**. **2** any member of a class of aromatic *p*-diols derivable from *p*-quinones (*see* **quinone** (def. 2)) by reduction of two CO groups to COH groups with any necessary rearrangement of double bonds; any compound containing a quinol nucleus. *See also* **quinol** (def. 2), **semiquinone**. *Compare* **catechol** (def. 2).

hydrosol any **sol** in which water is the liquid component. *Compare* **hydrogel**.

hydrous *(of a chemical compound)* containing combined water (usually in stoichiometric proportions).

hydroxide 1 *or (formerly)* **hydroxyl** the ion OH^-. **2** any inorganic or organic compound containing or dissociating to give this ion. Hydroxides that are soluble in water, and can dissociate, give aqueous solutions with an alkaline reaction (pH > 7).

hydroxo+ *comb. form* indicating the presence of a OH^- ligand in a coordination compound.

hydroxocobalamin *or* **vitamin B_{12b}** a form of vitamin B_{12} in which a hydroxide ion is in coordinate linkage to the sixth (i.e. *β*) position on the cobalt atom in a molecule of **cobalamin**; the conjugate base of aquacobalamin.

hydroxonium ion *an alternative name (now disfavoured) for* **oxonium** (def. 1).

hydroxy denoting the presence in a compound of one or more **hydroxyl** groups.

hydroxy+ *comb. form (in chemical nomenclature)* denoting the presence of one or more **hydroxyl** groups; the number of hydroxyl groups may be specified, e.g. dihydroxy+ (two), trihydroxy+ (three), etc.

hydroxy acid any organic acid containing one or more **hydroxyl** groups.

2-hydroxy-acid oxidase EC 1.1.3.15; *systematic name*: (*S*)-2-hydroxy-acid:oxygen 2-oxidoreductase; *other name*: glycolate oxidase; an enzyme that catalyses the oxidation by dioxygen of (*S*)-2-hydroxy-acid to 2-oxo-acid and H_2O_2. It is a peroxisomal enzyme in plants that is involved in the conversion of glycolate to glycine in **photorespiration**. The coenzyme is FMN.

3-hydroxyacyl-CoA dehydrogenase EC 1.1.1.35; *systematic name*: (*S*)-3-hydroxyacyl-CoA:NAD$^+$ oxidoreductase; *other names*: β-hydroxyacyl dehydrogenase; β-keto-reductase; an enzyme of the **beta-oxidation system**. It catalyses the oxidation by NAD$^+$ of *s*-3-hydroxyacyl-CoA to 3-oxoacyl-CoA with formation of NADH. *See also* **fatty-acid oxidation complex, peroxisomal bifunctional enzyme**.

hydroxyallysine 5-hydroxy-6-oxonorleucine; 2-amino-5-hydroxyadipaldehyde. An α-amino acid not normally found in proteins, but enzymically formed from 5-hydroxylysine *in situ* as an intermediate in the creation of covalent cross-links between adjacent polypeptide strands during the maturation of certain connective tissue proteins, e.g. collagen, elastin. *See also* **allysine**.

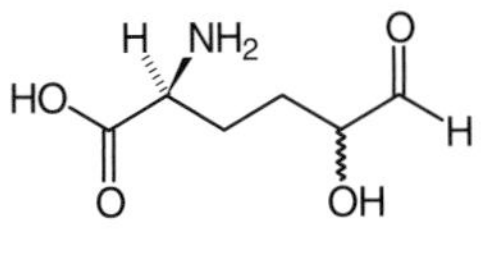

11β-hydroxy-4-androstene-3,17-dione 4-androsten-11β-ol-3,17-dione; a 17-oxo steroid formed from androstenedione by steroid 11β-monooxygenase (EC 1.14.15.4).

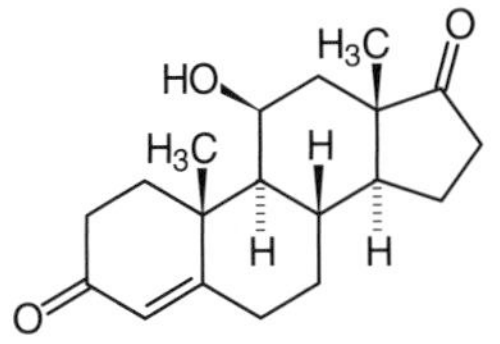

hydroxyapatite or **hydroxylapatite** a crystallized form of calcium orthophosphate hydroxide, $Ca_{10}(PO_4)_6(OH)_2$, virtually insoluble in water and in many aqueous solvents; it occurs as a mineral in phosphate rock. The main mineral component of bone, cartilage, and tooth approximates in composition to hydroxyapatite, small proportions of Ca^{2+} and OH^- being replaced respectively by Mg^{2+} and F^-; trace amounts of other ions may be present as well. Hydroxyapatite may also occur in granules in mitochondria. Chemically prepared hydroxyapatite is useful in the separation of proteins or nucleic acids by **adsorption chromatography**. It is used in orthopedic and dental prostheses and in the prevention of osteoporosis.

N^ω-hydroxy-L-arginine *abbr.*: L-NHA; a derivative of L-arginine in which one of the guanidine nitrogens is oxidized to an *N*-hydroxy group. It is an intermediate in the formation of nitric oxide from L-arginine; it may function as an intercellular NO carrier, and as a vasodilator in its own right.

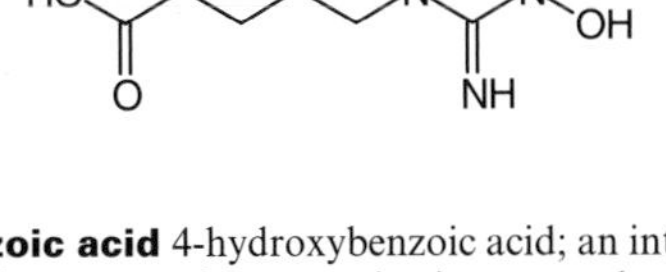

p-hydroxybenzoic acid 4-hydroxybenzoic acid; an intermediate in the synthesis of **ubiquinone** from tyrosine in mammals.

γ-hydroxybutyrate or **gamma-hydroxybutyrate** a rapid-acting central nervous depressant. It is abused because it produces euphoric and hallucinogenic states and is alleged to produce muscle growth. It is one of the 'date rape' drugs.

3-hydroxybutyrate cycle or **3-hydroxybutyrate shuttle** an intracellular metabolic cycle that can transfer reducing equivalents from the cytosol to the mitochondria in liver. Extramitochondrial NADH is oxidized to NAD$^+$ by acetoacetate, which is reduced to D(–)-3-hydroxybutyrate; the latter then passes into the mitochon-

dria where it is reoxidized to acetoacetate by intramitochondrial NAD$^+$. The acetoacetate passes out into the cytosol thereby completing the cycle.

3-hydroxybutyrate dehydrogenase EC 1.1.1.30; *other name*: D-β-hydroxybutyrate dehydrogenase; an enzyme that catalyses the reaction:

(*R*)-3-hydroxybutanoate + NAD$^+$ = acetoacetate + NADH.

α-hydroxybutyrate (2-hydroxybutanoate) dehydrogenase *an alternative name for* the LD$_1$ isoenzyme of **lactate dehydrogenase** (EC 1.1.1.27).

4-hydroxybutyric aciduria an autosomal recessive disorder caused by mutation in the gene for succinic-semialdehyde dehydrogenase. The excess succinic semialdehyde is reduced by 4-hydroxybutyrate dehydrogenase (EC 1.1.1.61) and the product is excreted in urine.

hydroxyeicosatetraenoate *abbr.*: HETE or mono-HETE; any of various related products arising, either spontaneously or by the action of glutathione peroxidases, from reduction of **hydroperoxyeicosatetraenoates**. The main forms are the 5-, 12-, and 15-HETEs, resulting from the action of 5-, 12-, and 15-**lipoxygenase** enzymes, but 8-, 9-, 11-, 16-, 17-, 18-, 19-, and 20-HETEs are known. The hydroxy group in most HETEs is of the *S* configuration, but those of *R* configuration are known, e.g. 12(*R*)-HETE from psoriatic lesions; 12(*R*)-HETE and 11(*R*)-HETE are found in sea-urchin eggs. The three main HETEs are: (6*E*,8*Z*,11*Z*,14*Z*)-(5*S*)-5-hydroxyeicosa-6,8,11,14-tetraen-1-oate, 5(*S*)-HETE; (5*Z*,8*Z*,10*E*,14*Z*)-(12*S*)-12-hydroxyeicosa-5,8,10,14-tetraen-1-oate, 12(*S*)-HETE; and (5*Z*,8*Z*,11*Z*,13*E*)-(15*S*)-15-hydroxyeicosa-5,8,11,13-tetraen-1-oate, 15(*S*)-HETE. Their tissue distribution follows that of **lipoxygenases**. 15(*S*)-HETE can act as a precursor for 5(*S*),15(*S*)-dihydroxyeicosatetraenoate and **lipoxins**. 5-HETE and 12-HETE are chemoattractants, 12(*R*) being more potent than 12(*S*), but 5(*S*) being more potent than 5(*R*); 8(*R*)-HETE, synthesized by starfish oocytes, is orders of magnitude more potent than 8(*S*)-HETE in inducing oocyte maturation.

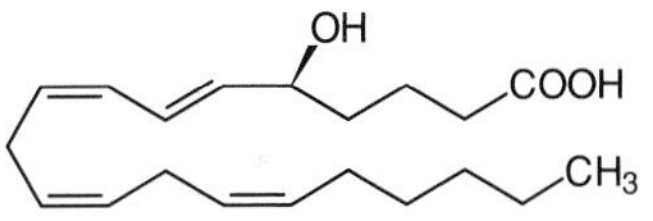

5(*S*)-HETE

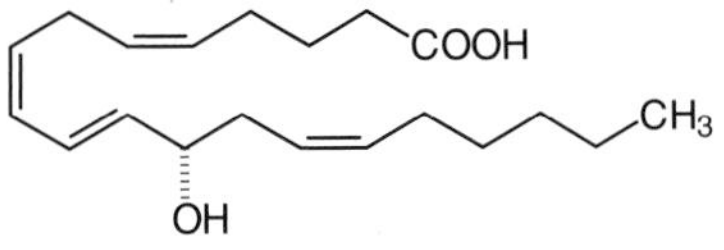

12(*S*)-HETE

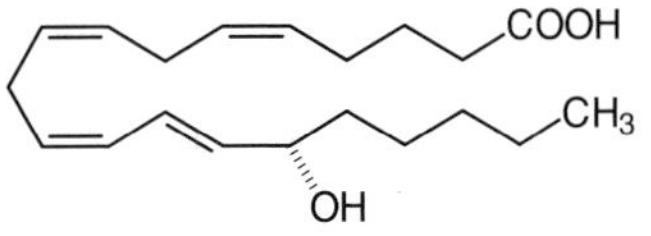

15(*S*)-HETE

hydroxy fatty acid any member of a class of fatty acids having a hydroxyl group, usually at the 2- or 3-position. 2-D(–)-hydroxy fatty acids are widespread minor constituents of plant and some animal (wool, skin) lipids; 3-L(+)-hydroxy fatty acids are found in low amounts in plant and bacterial lipids (e.g. in **mycolic acids**). Mid-chain hydroxy acids and polyhydroxy acids also occur in plant lipids.

5-hydroxyindoleacetaldehyde *see* **5-hydroxyindole metabolism.**

5-hydroxyindoleacetic acid *abbr.*: 5-HIAA; a metabolite of **5-hydroxytryptamine**, produced by oxidative deamination catalysed by the enzyme **monoamine oxidase**, followed by oxidation to 5-HIAA. *See also* **5-hydroxyindole metabolism.**

5-hydroxyindole metabolism the metabolism of 5-hydroxytryptophan and its derivatives; it is of particular interest because it is concerned with the synthesis and degradation of the neurotransmitter 5-hydroxytryptamine (i.e. **serotonin**). Serotonin is converted to 5-hydroxy-3-indoleacetaldehyde by **monoamine oxidase**, and thence to either 5-hydroxytryptophol (3%) or 5-hydroxy-3-indoleacetic acid (5-HIAA) (97%), and these are excreted in the urine. Urinary 5-HIAA is measured as a diagnostic aid in cases of carcinoid syndrome, as the tumour converts large amounts of dietary tryptophan to serotonin.

hydroxyl 1 the (covalently linked) –OH group in any chemical compound. **2** *a former name for* the hydroxide ion; *see* **hydroxide** (def. 1).

hydroxylamine 1 NH_2–OH; a chemical mutagen that causes the deamination of cytosine residues in DNA so that, during subsequent replication of the DNA, the altered cytosine pairs with adenine rather than with guanine, causing G–C to A–T transitions. Hydroxylamine is an intermediate in the oxidation of ammonium in nitrifying bacteria, and in the reverse reaction in denitrifying bacteria. It may be used to cleave specifically Asp-|-Gly bonds in peptides. **2** any organic compound of the type RNH–OH.

hydroxylapatite *an alternative name for* **hydroxyapatite.**

hydroxylase the common name for many monooxygenase and a few **dioxygenase** enzymes that catalyse hydroxylation reactions.

hydroxylate to introduce one or more **hydroxyl** (def. 1) groups into a compound, often by the replacement of hydrogen atom(s). —**hydroxylation** *n.*

hydroxylation-induced migration *an alternative name for* **NIH shift.**

hydroxylysine the trivial name for δ-hydroxylysine; δ-hydroxy-α, ε-diaminocaproic acid; 5-hydroxy-2,6-diaminohexanoic acid; H_2N–CH_2–$CHOH$–$[CH_2]_2$–$CH(NH_2)$–$COOH$; a chiral α-amino acid. Because the molecule of hydroxylysine possesses two chiral centres, four isomeric forms are possible, comprising an enantiomeric (D_s and L_s) pair of each of two (*erythro* and *threo*) diastereoisomers (the pair of *threo* enantiomers are sometimes known as **allohydroxylysine**). L-Hydroxylysine (*symbol*: 5Hyl or Hyl or Lys(5-OH) or (*formerly*) Hylys), *erythro*-5-hydroxy-L_s-lysine, is a **noncoded amino acid** found in peptide linkage in proteins (mainly collagen). In collagen, 15–20% of the lysine residues are enzymically converted to hydroxylysine residues while the growing peptide chains are still attached to ribosomes; some of the hydroxyl groups of the hydroxylysine side chains are subsequently glycosylated.

L-hydroxylysine

L-allohydroxylysine

hydroxy-methoxymandelate *abbr.*: HMMA; 4-hydroxy-3-methoxymandelate; *often (incorrectly) known as* vanillylmandelate

---

(*abbr.*: VMA); the L enantiomer is the end product of **catecholamine** catabolism, and is excreted in the urine. In clinical chemistry the measurement of HMMA in urine is used to detect excessive biosynthesis of catecholamines, as, e.g., in **pheochromocytoma**.

hydroxymethylbilane synthase EC 4.3.1.8; *systematic name*: porphobilinogen ammonia-lyase (polymerizing); *other names*: porphobilinogen deaminase; pre-uroporphyrinogen synthase. An enzyme that catalyses the conversion of four molecules of porphobilinogen to hydroxymethylbilane and four molecules of ammonia; dipyrromethane is a cofactor. Together with **uroporphyrinogen-III synthase** it is responsible for the step in heme biosynthesis in which four porphobilinogen molecules are condensed to form uroporphyrinogen III.

5-hydroxymethylcytosine *abbr.*: HMC; a minor base occurring in some types of DNA, e.g. in T4 bacteriophage.

hydroxymethylglutaric aciduria an autosomal recessive disease characterized by hypoglycemia, metabolic acidosis, hyperammonemia, and excessive urinary excretion of hydroxymethylglutarate. It can lead to coma and death in infancy. It is caused by inactivating mutations in the gene (at 1p35.1) for hydroxymethylglutaryl-CoA lyase.

hydroxymethylglutaryl-CoA *abbr.*: HMG-CoA; β-hydroxy-β-methylglutaryl-CoA; (S)-3-hydroxy-3-methylglutaryl-CoA; an important intermediate in the synthesis of cholesterol, farnesyl and geranyl derivatives, and ketone bodies, and in the catabolism of leucine. *See also* **hydroxymethylglutaryl-CoA reductase, hydroxymethylglutaryl-CoA lyase.**

hydroxymethylglutaryl-CoA lyase *abbr.*: HMG-CoA lyase; EC 4.1.3.4; *systematic name*: (S)-3-hydroxy-3-methylglutaryl-CoA acetoacetate lyase. An enzyme that catalyses the formation of acetyl-CoA and acetoacetate from (S)-3-hydroxy-3-methylglutaryl-CoA. It requires Mg^{2+} or Mn^{2+} for activity. In mammals it is a homodimer (each subunit in humans contains 298 amino acids) mostly of mitochondrial matrix but also present in peroxisomes that converts hydroxymethylglutaryl-CoA to acetoacetate in the synthesis of **ketone bodies**. Deficiency leads to hydroxymethylglutaric aciduria.

hydroxymethylglutaryl-CoA reductase (NADPH) *abbr.*: HMG-CoA reductase; EC 1.1.1.34; *systematic name*: (R)-mevalonate:NADP+ oxidoreductase (CoA-acylating). An enzyme that catalyses a reaction between (S)-3-hydroxy-3-methylglutaryl-CoA and two molecules of NADPH to form (R)-mevalonate, CoA, and two molecules of NADP+. Its function is to reduce HMG-CoA to (R)-mevalonate. In mammals it is an integral membrane glycoprotein of endoplasmic reticulum and peroxisomes. It catalyses a rate-limiting step and is an important regulatory enzyme in pathways for the synthesis of cholesterol and other sterols. It is the target for therapeutic strategies for reducing plasma cholesterol. *See also* **dephospho-[reductase kinase] kinase.**

[hydroxymethylglutaryl-CoA reductase (NADPH)] kinase EC 2.7.1.109; *other name*: reductase kinase; an enzyme that catalyses the phosphorylation by ATP of [3-hydroxy-3-methylglutaryl-CoA reductase (NADPH)] with release of ADP. It thereby inactivates the latter enzyme.

[hydroxymethylglutaryl-CoA reductase (NADPH)]-phos-

phatase EC 3.1.3.47; *other name*: reductase phosphatase; an enzyme that hydrolyses phosphate from [hydroxymethylglutaryl-CoA reductase (NADPH)] phosphate. It thereby activates the latter enzyme.

hydroxymethylglutaryl-CoA synthase *abbr.*: HMG-CoA synthase; EC 4.1.3.5; *systematic name*: (S)-3-hydroxy-3-methylglutaryl-CoA acetoacetyl-CoA-lyase (CoA-acetylating). A mostly hepatic enzyme that catalyses the reaction:

acetoacetyl-CoA + acetyl-CoA + H_2O → hydroxymethylglutaryl-
CoA + CoA.

5-hydroxynorleucine *an older name (no longer recommended)* for 2-amino-5-hydroxyhexanoic acid; a non-protein (non-coded) amino acid found in plants, e.g. *Crotolaria juncea*.

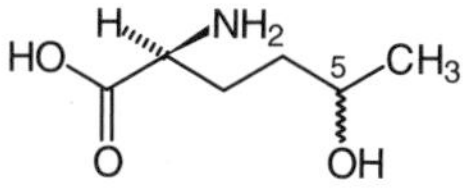

(R)-4-hydroxyphenyllactate dehydrogenase EC 1.1.1.222; *other name*: (R)-aromatic lactate dehydrogenase; an enzyme that catalyses the reaction:

(R)-3-(4-hydroxyphenyl)lactate + NAD(P)$^+$ =
3-(4-hydroxyphenyl)pyruvate + NAD(P)H.

hydroxyphenylpyruvate reductase EC 1.1.1.237; an enzyme that catalyses the reaction:

3-(4-hydroxyphenyl)lactate + NAD$^+$ =
3-(4-hydroxyphenyl)pyruvate + NADH.

17α-hydroxypregnenolone 3β,17-dihydroxy-5-en-20-one; an intermediate in the synthesis of androgens, estrogens, and glucocorticoid hormones from cholesterol in the adrenal cortex. It is formed from **pregnenolone** by steroid 17α-monooxygenase (EC 1.14.99.9).

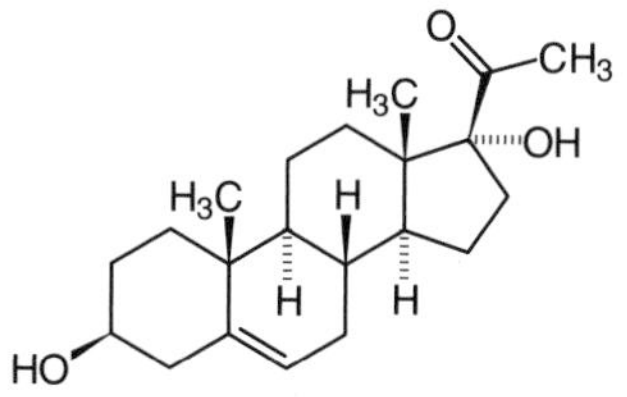

17α-hydroxyprogesterone 17-hydroxypregn-4-one; an intermediate in the synthesis of androgens, estrogens, and glucocorticoid hormones from cholesterol in the adrenal cortex. It is formed from **progesterone** by steroid 17α-monooxygenase (EC 1.14.99.9).

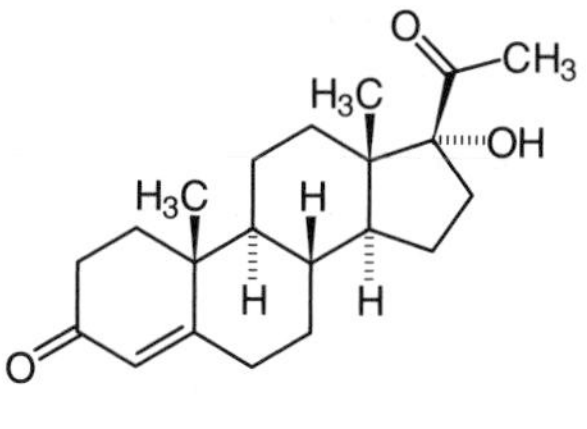

hydroxyproline *symbol*: Hyp or *(formerly)* Hypro; either of the two constitutional isomers 3-hydroxyproline and 4-hydroxyproline, HO–(C$_4$H$_6$NH)–COOH, both of which are chiral cyclic N-alkylated α-amino acids. Each constitutional isomer possesses two chiral centres; each can therefore exist as an enantiomeric (D and L) pair of each of two (*cis* and *trans*) diastereoisomers, making eight possible isomeric structures for hydroxyproline in all. Both *trans*-3-hydroxy-L-proline (*symbol*: 3Hyp or Pro(3-OH)), (2S,3S)-3-hydrox-

ypyrrolidine-2-carboxylic acid, and *trans*-4-hydroxy-L-proline (*symbol*: 4Hyp or Pro(4-OH)), (2S,4R)-4-hydroxypyrrolidine-2-carboxylic acid, are **noncoded amino acids** found in peptide linkage in proteins. In collagen, nearly 50% of the proline residues are hydroxylated to 4Hyp, and a small proportion to 3Hyp; hydroxyproline residues are also found in elastin, enamel of the teeth, in the C1a moiety of the C1 component of **complement**, and in **extensin**. In collagen formation, enzymic hydroxylation of proline residues occurs while the growing peptide chains are still attached to ribosomes. Residues of both the *cis*-(2S,3R)- and the *trans*-(2S,3S)-isomers of 3-hydroxy-L-proline, one of each per molecule, occur in the peptide antibiotic telomycin. One residue per molecule of *cis*-4-hydroxy-L-proline, the (2S,4S)-isomer, occurs in every member of the **amatoxin** group of toxic fungal octapeptides. None of the D series of isomers is known to occur naturally.

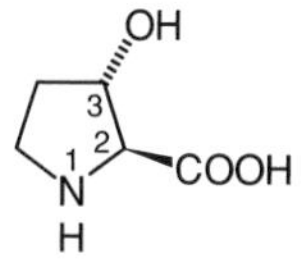

trans-(2S,3S)-3-hydroxy-L-proline

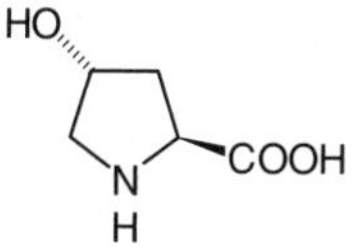

trans-(2S,4R)-4-hydroxy-L-proline

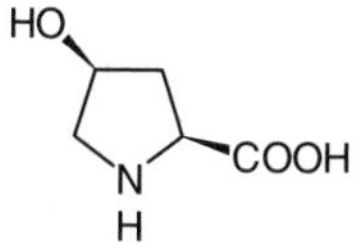

cis-(2S,4S)-4-hydroxy-L-proline

hydroxyprolinemia *or* **hyperhydroxyprolinemia** a metabolic disorder of humans characterized by high levels of free 4-hydroxyproline in the blood plasma and urine. It is of no clinical consequence and is probably due to a deficiency in the activity of so-called hydroxyproline oxidase, i.e. the enzyme 4-oxoproline reductase (EC 1.1.1.104).

hydroxyproline-rich glycoproteins *abbr.*: HPRG; one of the four major types of **plant cell-wall proteins**. The best studied example is **extensin**, which is extensively glycosylated and forms a polyproline type of rodlike structure.

15-hydroxyprostaglandin dehydrogenase (NAD⁺) EC 1.1.1.141; an enzyme involved in the inactivation of prostaglandins E$_2$, F$_{2α}$, and B$_1$, by oxidizing the 15-hydroxyl group to a keto group. It catalyses a reaction between (5Z,13E)-(15S)-11α,15-dihydroxy-9-oxoprost-5,13-dienoate and NAD$^+$ to form (5Z,13E)-11α-hydroxy-9,15-dioxoprost-5,13-dienoate (15-keto PGE$_2$) and NADH.

3β-hydroxysteroid dehydrogenase an enzyme that converts the 3β-OH group of the substrate to a 3-keto group and isomerizes the C5,6 double bond to the C4,5 position in the following reactions on C$_{21}$ and C$_{19}$ steroids:

    (1) pregnenolone + NAD$^+$ = progesterone + NADH;
    (2) 17-hydroxypregnenolone + NAD$^+$ = 17-hydroxyprogesterone + NADH;
    (3) dehydroepiandrosterone + NAD$^+$ = androstenedione + NADH.

It is the only steroidogenic enzyme of the endoplasmic reticulum that is not a cytochrome P450. In humans, two genes at 1p13 result in two isozymes: type 1 is expressed in nonsteroidogenic tissues, e.g. liver, breast, prostate, and skin; type 2 is expressed in steroidogenic tissues, i.e. adrenal cortex and gonads. Mutations in the gene for the type 2 isozyme are a rare cause of a salt-losing congenital adrenal hyperplasia in which there is deficiency of aldosterone, cortisol, and adrenal androgens (the three final products of the above reactions, respectively).

17β-hydroxysteroid oxidoreductase an enzyme of the endoplasmic reticulum that catalyses the reaction:

androstenedione + NADP$^+$ = testosterone + NADPH.

There are three homologous human isozymes of which type 3 (gene locus at 9q22) is expressed predominantly in the testes. Inactivating mutations in this gene result in enzyme deficiency, which leads to testosterone deficiency and male pseudohermaphroditism.

3β-hydroxysteroid-Δ⁵-oxidoreductase/isomerase an enzyme of liver and fibroblasts that catalyses the reaction:

7α-hydroxycholesterol + NAD$^+$ = testosterone + NADH.

It differs from the 3β-**hydroxysteroid dehydrogenase**, for which the substrates are C_{21} and C_{19} steroids. A very rare deficiency produces a syndrome of jaundice, hepatosplenomegaly, pale stools, and dark urine (from abnormal bile acid excretion). If untreated the condition progresses to hepatic cirrhosis.

N-hydroxysuccinimide ester or **NHS ester** any compound in which *N*-hydroxysuccinimide (NHS) has reacted with a carboxy group to form an ester. NHS esters react under mild conditions (>pH 7) with primary amino groups such as those at the N-terminus of proteins or ε-amino groups of lysine to form amide bonds with the release of NHS. NHS esters are used for labelling proteins with fluorescent tags or for crosslinking two molecules (often proteins) by covalent bonds.

5-hydroxytryptamine *abbr.*: 5-HT; *an alternative name for* **serotonin**.

5-hydroxytryptamine receptor any membrane protein that binds 5-hydroxytryptamine (i.e. **serotonin**) and mediates its intracellular effects. The receptors belong to seven families. Members of the 5-HT$_1$ family are G-protein-associated receptors and inhibit adenylate cyclase, but the 5-HT$_{1A}$ receptor also increases K$^+$ conductance. Those of the 5-HT$_2$ family are G-protein-associated receptors and stimulate phospholipase C. 5-HT$_3$ opens a Cl$^-$ channel; 5-HT$_4$ has four transmembrane segments and activates adenylate cyclase; 5-HT$_5$ is G-protein-associated but its effector is unknown. 5-HT$_6$ and 5-HT$_7$ are G-protein-associated receptors that activate adenylate cyclase.

5-hydroxytryptophan a non-protein amino acid found in plants; it is formed in legumes by hydroxylation of tryptophan. It is also formed in animals. *See also* **serotonin**.

hydroxyurea NH$_2$CONHOH; an analogue of **urea** with antineoplastic activity, possibly through inhibition of DNA synthesis.

hygro+ *comb. form* denoting moisture, especially that due to water.

hygrometer any instrument for measuring the relative humidity of a gas.

hygromycin B an aminoglycoside antibiotic from *Streptomyces hygroscopicus*. It kills bacterial, fungal, and higher eukaryotic cells by inhibiting protein synthesis. The inclusion of genes conferring resistance to hygromycin B in certain fungal recombinant DNA vectors allows selection of recombinants when these are plated out on medium containing this antibiotic.

hygromycin B phosphotransferase an enzyme from *Streptomyces hygroscopicus* that confers resistance to the toxic effects of hygromycin B. It is used as a dominant selectable marker in eukaryotic cells transfected with DNA constructs that incorporate the gene.

hygroscopic *(of a substance)* tending to absorb water from the surrounding atmosphere. *Compare* **deliquescent**.

Hyl or *(formerly)* **Hylys** *symbol for* a residue of any of the hydroxy derivatives of the α-amino acid L-lysine, especially 5-hydroxy-L-lysine. *See* **hydroxylysine**.

5Hyl *symbol for* a residue of the α-amino acid 5-hydroxy-L-lysine *(preferred alternative to* Lys(5-OH) *or* Hyl). *See* **hydroxylysine**.

Hylys *former symbol for* a residue of (5-)hydroxy-L-lysine (5Hyl or Hyl now recommended).

hyoscine or **scopolamine** an alkaloid found, e.g., in *Datura stramonium* (thorn apple) and related to **atropine** in its structure and effects; it is a tertiary ammonium compound noted for its actions as a **muscarinic receptor** antagonist. It has found use as a sedative and preanesthetic and for prevention of motion sickness.

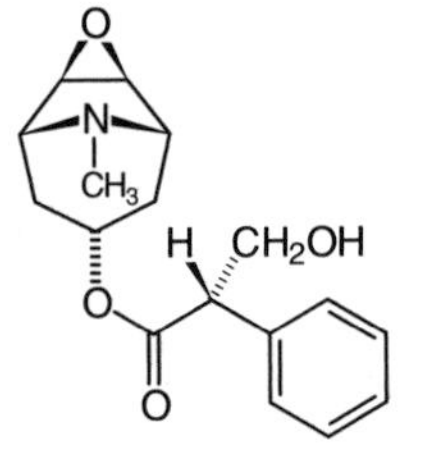

hyoscyamine 3α-tropanyl S-(–)-tropate; an anticholinergic material obtained from *Atropa belladonna* and other plants. It undergoes partial racemization on isolation; this process is completed by alkali treatment, and the racemic mixture is known as atropine.

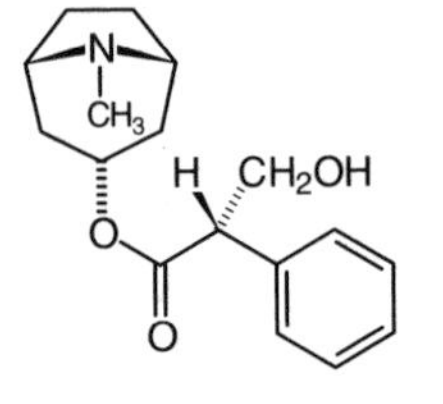

(–)-hyoscyamine

hyp+ *a variant form of* **hypo+** *(before a vowel)*.

Hyp 1 or *(formerly)* **Hypro** *symbol for* a residue of either the 3-hydroxy or the 4-hydroxy derivative of the α-amino (strictly, α-imino) acid L-proline, usually 4-hydroxy-L-proline. *See* **hydroxyproline**. *See also* **3Hyp**, **4Hyp**. **2** *symbol for* a residue of the purine base **hypoxanthine**.

3Hyp *symbol for* a residue of the α-amino acid 3-hydroxy-L-proline *(preferred alternative to* Pro(3-OH) *or* Hyp). *See* **hydroxyproline**.

4Hyp *symbol for* a residue of the α-amino acid 4-hydroxy-L-proline *(preferred alternative to* Pro(4-OH) *or* Hyp). *See* **hydroxyproline**.

hypelcin A a 20-residue peptide of microbial origin containing 10 α-aminoisobutyric acid residues. The N terminus is acylated and the C terminus amidated with **leucinol**. It modifies the permeability of phospholipid bilayers.

hyper+ *prefix* indicating over, beyond, excessive, or in excess; above the normal range. *Compare* **hypo+**, **normo+**.

hyperaldosteronism an elevated level of **aldosterone** in the blood. In primary aldosteronism, there is overproduction of aldosterone due to an adrenal tumour, as in Conn's syndrome or hyperplasia of the zona glomerulosa. In secondary aldosteronism the overproduction is caused by excessive levels of stimulatory hormones, especially increased renin secretion.

hyperammonemia or *(esp. Brit.)* **hyperammonaemia** the presence of excess ammonia in the blood (the normal range is 10–47 μmol L^{-1}). It occurs in inherited disorders of the urea cycle, organic acidemias, liver disease, and severe systemic illness. It is accompanied by cerebral edema with clouding of consciousness that may lead to coma and death. In these conditions it may be precipitated or aggravated by high protein intake. *See also* **ornithine–urea cycle**.

hyperbaric of, relating to, or operating at higher than normal pressures.

hyperbilirubinemia or *(esp. Brit.)* **hyperbilirubinaemia** the presence of higher than normal concentrations of **bilirubin** in the blood;

it may involve unconjugated or conjugated bilirubin. It occurs in liver disease or biliary obstruction, and is characteristic of neonatal hemolytic **jaundice**, where the bilirubin is mainly unconjugated. — **hyperbilirubinemic** or **hyperbilirubinaemic** *adj.*

hyperbola one of the conic sections; a plane curve in which the focal distance of any point bears to its distance from the directrix a constant ratio greater than unity. It has two foci and two asymptotes. *Compare* **rectangular hyperbola**. —**hyperbolic** *adj.*

hyperbolic function any one of a group of functions of an angle, analogous to the trigonometric functions, expressed as a relationship between the distances of a point on a hyperbola to the origin and to the coordinate axes. The group includes sinh (hyperbolic sine), defined as $\sinh x = (e^x - e^{-x})/2$; cosh (hyperbolic cosine), defined as $\cosh x = (e^x + e^{-x})/2$; and tanh (hyperbolic tangent), defined as $\tanh x = (\sinh x)/(\cosh x)$. The corresponding reciprocal functions are cosech (hyperbolic cosecant), sech (hyperbolic secant), and coth (hyperbolic cotangent), respectively.

hyperbolic kinetics curve *a description sometimes used of* the $v/[S]$ plot obtained for enzymes exhibiting **Michaelis–Menten kinetics**. The axes of the plot are not those of the rectangular hyperbola that results, but are rotated 45° (and displaced) from the true axes of the hyperbola.

hypercalcemia or *(esp. Brit.)* **hypercalcaemia** the presence of higher than normal concentrations of **calcium** in the blood. Clinical features include tiredness, muscle weakness, anorexia, weight loss, cardiac arrhythmia, and hypertension. It arises from a multiplicity of causes including malignancy, primary hyperparathyroidism, hyperthyroidism, acute adrenal insufficiency, familial hypocalciuric hypercalcemia, vitamin D or A overdose, or sarcoidosis. —**hypercalcemic** or **hypercalcaemic** *adj.*

hypercapnia or **hypercarbia** the presence of greater than normal amounts of carbon dioxide in a vertebrate or in its blood. *Compare* **hypocapnia**. —**hypercapnic** or **hypercarbic** *adj.*

hyperchlorhydria the presence of greater than normal amounts of hydrochloric acid in the gastric juice. —**hyperchlorhydric** *adj.*

hypercholesterolemia or *(esp. Brit.)* **hypercholesterolaemia** the presence of higher than normal concentrations of cholesterol in the blood. *See also* **hyperlipidemia**. —**hypercholesterolemic** or **hypercholesterolaemic** *adj.*

hyperchromatic 1 of, relating to, or exhibiting **hyperchromatism**. **2** *an alternative word for* **hyperchromic** (def. 1).

hyperchromatism the condition or property (of a cell or cell organelle) of staining more intensely than normal; the presence of a greater than normal amount of chromatin.

hyperchromic 1 or **hyperchromatic** more highly coloured than normal; intensely coloured. **2** of, pertaining to, exhibiting, or resulting from **hyperchromism**. **3** *(of a substance)* displaying a **hyperchromic effect**. **4** describing erythrocytes that contain, or appear to contain, more hemoglobin than normal.

hyperchromic effect the large increase in absorption of ultraviolet light, usually measured at 260 nm, shown by a solution of any natural or synthetic polynucleotide with a hydrogen-bonded structure when it is denatured or degraded. The effect is due to alterations in the electronic interactions between the initially stacked and hydrogen-bonded bases.

hyperchromicity a measure of the increase in absorption of electromagnetic radiation at a specific wavelength, usually that of the absorption maximum, due to **hyperchromism**, or to the **hyperchromic effect**. It is equal to the reciprocal of the **hypochromicity** minus unity.

hyperchromism the increased absorption of electromagnetic radiation exhibited by an ordered structure above that predicted on the basis of its constitution.

hyperfine splitting the splitting of a spectral line into multiplets of closely spaced lines. In electron spin resonance spectra hyperfine splitting is due to the interaction of unpaired electrons with neighbouring nuclei and can be used to determine the structure of a free radical or to identify the ligands of a paramagnetic ion and to measure the degree of covalent binding that exists between them. In nuclear magnetic resonance spectra, hyperfine splitting, sometimes called **spin–spin splitting**, is due to the interaction of the nuclear magnetic moment with those of neighbouring nuclei and can be

used to assign particular nuclear resonances and to determine molecular conformations.

hyperfine structure the presence in a spectrum of multiplets of closely spaced lines resulting from **hyperfine splitting**.

hyperforin the major active ingredient of the herb St John's wort, which is used as a herbal in a variety of ailments. It binds to a transcription factor (pregnane X receptor) and enhances expression of cytochrome P4503A4, which degrades certain drugs (e.g. cyclosporin and the HIV protease inhibitor indinavir).

hyperglycemia or *(esp. Brit.)* **hyperglycaemia** an abnormally high blood glucose concentration, especially in relation to the fasting value. *See also* **glycemia**. —**hyperglycemic** or **hyperglycaemic** *adj.*

hyperglycemic factor or **hyperglycemic-glycogenolytic factor** *a former name for* **glucagon**.

hyperhydroxyprolinemia *see* **hydroxyprolinemia**.

hyper-IgD syndrome *abbr.*: HIDS; a rare autosomal recessive disease in which recurrent bouts of fever are associated with increased plasma levels of IgD and mevalonic aciduria. Activity of the enzyme mevalonate kinase is 5–15% of normal.

hyper-IgM syndrome a condition in which recurrent respiratory infections in early childhood are associated with very low concentrations of IgG, IgA, and IgE but normal or elevated concentrations of IgM in serum. It may be due to X-linked and autosomal recessive traits. *See* **activation-induced cytidine deaminase**.

hyperimmune describing the state of an animal in which very high concentrations of specific antibody are present in the serum, brought about by **hyperimmunization**.

hyperimmunization or **hyperimmunisation** any method of immunization designed to stimulate the production in the animal immunized of very large quantities of specific antibody. It typically involves the repeated administration of immunogen, often in increasing amounts.

hyperinsulinemia syndrome or **(metabolic) syndrome X** a syndrome of impaired glucose tolerance, fasting hyperinsulinemia, hypertriglyceridemia, hypertension, and increased predisposition to atherosclerosis. If untreated it often leads to diabetes mellitus type 2. *See* **insulin resistance**.

hyperkalemia or *(esp. Brit.)* **hyperkalaemia** the presence in the blood of an abnormally high concentration of **potassium**. There are many causes including excessive K^+ intake (infusion, transfusion), increased transcellular movement (tissue damage, systemic acidosis), and decreased K^+ loss (renal failure, mineralocorticoid deficiency). —**hyperkalemic** or **hyperkalaemic** *adj.*

hyperlactatemia or *(esp. Brit.)* **hyperlactataemia** a persistent raised blood lactate concentration, usually below 5 mmol L^{-1}, and not accompanied by a lowered blood pH. The normal range is between 0.5 and 2.2 mmol L^{-1}. *Compare* **lactic acidosis**. —**hyperlactatemic** or **hyperlactataemic** *adj.*

hyperlipidemia or **hyperlipemia** or *(esp. Brit.)* **hyperlipidaemia** or **hyperlipaemia** the presence in the blood of an abnormally high concentration of fats (or lipids). Hyperlipidemias have been classified into various types according to which lipoprotein class is found to be elevated in the blood. The Fredrickson classification scheme designates: type I, elevated chylomicrons; type II, elevated low-density lipoprotein; type III, elevated intermediate-density lipoprotein; type IV, elevated very-low-density lipoprotein; type V, elevated chylomicrons and very-low-density lipoprotein. A simpler distinction, of use in treatment, is between elevated cholesterol levels (type II, **hypercholesterolemia**) and elevated triacylglycerol levels (types I, III–V, **hypertriglyceridemia**, an older term still often used). Hyperlipidemia may be secondary to other conditions; several drugs cause or exacerbate hyperlipidemia, including thiazides, **beta blockers** lacking intrinsic sympathomimetic activity, and **corticosteroids**. Estrogens may lower hypercholesterolemia, but may cause or exacerbate hypertriglyceridemia; hypothyroidism is commonly associated with hypercholesterolemia; high alcohol intake causes hypertriglyceridemia, but modest alcohol intake may have a beneficial effect on cholesterol status, as it tends to increase **high-density lipoprotein** concentration. —**hyperlipidemic, hyperlipemic** or **hyperlipidaemic, hyperlipaemic** *adj.*

hypermodified base any extensively modified nucleic-acid base occurring adjacent to the 3′ end of the anticodon in many tRNA molecules. Examples are N^6-(δ_2-isopentenyl)adenosine (*symbol*: iA) and N^6-(threonylcarbamoyl)adenosine (*symbol*: tA). Hypermodified bases are thought to be important for proper binding of tRNA molecules to ribosomes.

hypernatremia *or (esp. Brit.)* **hypernatraemia** the presence in the blood of an abnormally high concentration of sodium. It may be associated with conditions in which body water is depleted. —**hypernatremic** *or* **hypernatraemic** *adj.*

hyperornithinemia *see* **OAT**.

hyperoxaluria the presence of oxalate in the urine in abnormally high amounts. It is associated with the formation of renal calculi. Type 1, the commonest, results from overproduction of oxalate from glyoxylate because of deficiency of alanine–glyoxylate aminotransferase (EC 2.4.1.44) activity. Type 2 results from deficiency of D-glycerate dehydrogenase/glyoxylate reductase (EC 1.1.1.29). Type 3 results from a poorly characterized increased absorption of oxalate from the intestine.

hyperoxia the presence of greater than normal amounts of dioxygen in a vertebrate or in its blood. *Compare* **hypoxia**. —**hyperoxic** *adj.*

hyperparathyroidism the increased production of **parathyroid hormone**. It may be primary, as a result of a tumour of the parathyroid gland, or secondary, as a result of a disturbance in calcium metabolism.

hyperphosphatemia *or (esp. Brit.)* **hyperphosphataemia** an excessive level of phosphate in the blood. It occurs most commonly in renal insufficiency, but may also accompany hypoparathyroidism, vitamin D intoxication, or excessive administration of phosphate. —**hyperphosphatemic** *or* **hyperphosphataemic** *adj.*

hyperplasia the increase in size of a tissue or organ resulting from an increase in the total number of cells present. The part thus affected retains its normal form. *Compare* **hypertrophy**. —**hyperplastic** *adj.*

hyperprolactinemia *or (esp. Brit.)* **hyperprolactinaemia** the presence in the blood of abnormally high levels of **prolactin**, due to excessive secretion of the hormone. The most common example of pituitary hyperfunction, it results from pituitary adenoma, and also from treatment with dopamine receptor blockers. It can be treated with dopamine agonists, e.g. **bromocriptine**. —**hyperprolactinemic** *or* **hyperprolactinaemic** *adj.*

hypersecretion excessive secretion by any endocrine or exocrine gland.

hypersensitive having an abnormally great sensitivity, especially to an allergen, drug, or other agonist.

hypersensitive sites regions of chromatin that are unprotected by histones and other DNA-binding proteins and are sites of active transcription in the genome. The presence of such sites is revealed by treating intact nuclei with DNase I and analysing the resulting fragments of DNA.

hypersensitivity 1 the state or condition of being **hypersensitive**. **2** *(in immunology)* an immune response that occurs in an exaggerated or inappropriate form. Such responses may result from agents including pollen or drugs, or from genuine pathogens. Also of importance are the different kinds of tissue damage seen in autoimmune diseases. In general terms the reactions may be classified into four types, although they do not occur in isolation from each other: type I, **immediate hypersensitivity**; type II, **antibody-mediated hypersensitivity**; type III, **immune complex-mediated hypersensitivity**; and type IV, **delayed hypersensitivity**.

hypersensitivity mapping a method for identifying sites in chromatin at which active transcription is taking place. Cloning and sequencing of DNA flanking DNase I **hypersensitive sites** permits the identification of genes being transcribed.

hypersharpening the formation of a 'sharper' boundary in sedimentation and moving boundary electrophoresis experiments than would be expected as a result of diffusion-broadening, because molecules in the trailing portion of the boundary tend to move at a greater velocity per unit field than molecules on the leading edge or in the plateau region.

hypertension the state or condition of having a higher than normal arterial blood pressure. —**hypertensive** *adj.*

hyperthermophile an organism growing at a temperature above 80°C. Almost all thermophiles belong to the kingdom **Archaea**. *Compare* **thermophile**.

hyperthyroidism *or* **thyrotoxicosis** a clinical state resulting from excessive secretion of thyroid hormones. The most common cause is Graves' disease; other causes include goitre and solitary toxic adenoma. The associated nervousness, irritability, fatigue, heat intolerance, and (not in all) exophthalmia derive from exaggeration of normal responses to thyroid hormones. *See also* **thyroxine**. —**hyperthyroid** *adj.*

hypertonic 1 *(of a solution)* having a higher osmotic pressure than that of some given solution, especially that of the fluid in or surrounding a given type of cell or a body fluid. **2** *(of a muscle)* showing or characterized by excessive tone or tension. *Compare* **hypotonic**. —**hypertonicity** *n.*

hypertriglyceridemia *or (esp. Brit.)* **hypertriglyceridaemia** *see* **hyperlipidaemia**.

hypertrophy the increase in the size of a tissue or organ resulting from an increase in the size of the cells present. *Compare* **hyperplasia**. —**hypertrophic** *adj.*

hyperuricemia *or (esp. Brit.)* **hyperuricaemia** the presence of elevated amounts of uric acid in the blood. This may be due to increased purine synthesis arising from a metabolic disorder, inherited or otherwise, or to reduced excretion – e.g. renal disease, the effect of drugs (e.g. salicylates), or the presence of organic acids (e.g. lactate) that competitively inhibit urate excretion. Hyperuricemia is associated with **gout**.

hypervariable region *an alternative name for* complementarity-determining region.

hypo *a common name for* sodium thiosulfate, especially when used in photographic processing.

hypo+ *or (before a vowel)* **hyp+** *prefix* denoting under, beneath, below, less than, or in deficit; below the normal range. *Compare* **hyper+**, **normo+**.

hypoalphalipoproteinemia *see* hypolipoproteinemia.

hypobaric of, relating to, or operating at pressures lower than normal.

hypobetalipoproteinemia *see* hypolipoproteinemia.

hypocalcemia *or (esp. Brit.)* **hypocalcaemia** the presence of a lower than normal concentration of **calcium** in the blood. It may be associated with disorders of vitamin D metabolism, **hypoparathyroidism**, magnesium deficiency, or pancreatitis. Clinical features include behavioural disturbances, stupor, numbness, paresthesia, muscle cramps or spasms, and convulsions. —**hypocalcemic** *or* **hypocalcaemic** *adj.*

hypocapnia *or* **hypocarbia** the presence of less than the normal amount of carbon dioxide in a vertebrate or in its blood. *Compare* **hypercapnia**. —**hypocapnic** *or* **hypocarbic** *adj.*

hypochlorhydria the presence of less than the normal amount of hydrochloric acid in the gastric juice. —**hypochlorhydric** *adj.*

hypochlorous acid HOCl; a strong oxidizing and halogenating agent and bactericide, used as the active ingredient of bleach. It is generated by the enzyme **myeloperoxidase** in azurophilic granules of neutrophils and lysosomes of monocytes, in response to the entry of bacteria.

hypocholesterolemia *or (esp. Brit.)* **hypocholesterolaemia** the presence of a lower than normal concentration of cholesterol in the blood. —**hypocholesterolemic** *or* **hypocholesterolaemic** *adj.*

hypochondroplasia *see* achondroplasia.

hypochromatic 1 of, relating to, or exhibiting **hypochromatism**. **2** *an alternative word for* **hypochromic** (def. 1).

hypochromatism the condition or property (of a cell or cell organelle) of staining less intensely than normal; the presence of an abnormally low amount of chromatin.

hypochromic 1 *or* **hypochromatic** less highly coloured than normal; weakly coloured. **2** of, pertaining to, exhibiting, or resulting from **hypochromism**. **3** *(of a substance)* displaying a **hypochromic effect**. **4** describing erythrocytes that contain, or appear to contain, less hemoglobin than normal.

hypochromic effect the observed decrease in absorption of certain frequencies of electromagnetic radiation by solutions of some macromolecules when the structure of the molecules becomes more ordered.

hypochromicity a measure of the decrease in absorption of electromagnetic radiation at a specific wavelength, usually that of the absorption maximum, due to **hypochromism**.

hypochromism the decreased absorption of electromagnetic radiation exhibited by an ordered structure below that predicted on the basis of its constitution.

hypocretin or **orexin** either of two polypeptides (of 39 and 29 amino acids) that occur in the lateral hypothalamus and increase food intake. They arise from the same precursor by differential cleavage and act on G-protein-coupled membrane receptors that regulate feeding behaviour.

hypoglycemia or (esp. Brit.) **hypoglycaemia** the presence of an abnormally low blood glucose concentration, especially in relation to the fasting value. See also **glycemia**. —**hypoglycemic** or **hypoglycaemic** adj.

hypoglycin or **hypoglycine A** β-(methylenecyclopropyl)alanine; 2-amino-3-(methylenecyclopropane)propanoic acid. A nonprotein α-amino acid identified as the hypoglycemic and toxic principle of the unripe fruit of the **ackee** (*Blighia sapida*), where it is present both in the fleshy aril and in the seeds. In the latter it occurs also as the *N*-L-glutamyl derivative, **hypoglycin B**. Certain other fruits also contain hypoglycin. It is degraded in animals to methylenecyclopropylacetate, either free or conjugated to glycine or coenzyme A. Methylenecyclopropylacetyl-CoA is a potent and specific inhibitor of the oxidation of short-chain acyl-CoAs, particularly of butyryl-CoA and glutaryl-CoA. In hypoglycin poisoning butyrate and glutarate accumulate in the blood and lead to metabolic acidosis and excessive urinary excretion of these compounds. See also **Jamaican vomiting sickness**.

Hyp(3-OH) *symbol for* a residue of the α-amino acid 3-hydroxy-L-proline (alternative to 3Hyp or Hyp). See **hydroxyproline**.

Hyp(4-OH) *symbol for* a residue of the α-amino acid 4-hydroxy-L-proline (alternative to 4Hyp or Hyp). See **hydroxyproline**.

hypokalemia or (esp. Brit.) **hypokalaemia** the presence in the blood of an abnormally low concentration of potassium. It may result from alkalosis, insulin administration, primary or secondary aldosteronism, or administration of drugs that are aldosterone antagonists. —**hypokalemic** or **hypokalaemic** adj.

hypolipoproteinemia the partial or total absence of any lipoprotein, or the presence of a defective lipoprotein, in the plasma. Several types of this condition are recognized. **Hypobetalipoproteinemia** is arbitrarily defined as a low-density lipoprotein (LDL) cholesterol level below the fifth percentile for a given sex- and age-matched population. Such patients also have low triacylglycerol levels. In this condition, LDL cholesterol is always detectable, in contrast to homozygous **abetalipoproteinemia**, in which it is entirely absent. Hypobetalipoproteinemia appears to be due to a low rate of synthesis of LDL from very-low-density lipoprotein. Abetalipoproteinemia is a rare autosomal recessive disorder. It probably results from an abnormality in the synthesis or secretion of apo B, and it is accompanied by fat malabsorption, as chylomicron formation requires apo B. In the homozygote, levels of triacylglycerols, cholesterol, and phospholipids are extremely low. Heterozygotes may present with hypobetalipoproteinemia. **Hypoalphalipoproteinemia** is in some cases familial, and may result from defective apo A-I. In another disorder, known as fish-eye disease (due to severe corneal opacity), the structure of apo A-I is normal, but high-density lipoprotein (HDL) levels are only 10% of normal. In analphalipoproteinemia, or **Tangier disease**, HDL is virtually absent.

hyponatremia or (esp. Brit.) **hyponatraemia** the presence in the blood of an abnormally low concentration of sodium. It can occur as a result of incorrect administration of fluid intravenously or in parenteral feeding, after administration of diuretics, in congestive heart failure, or in protein imbalance. —**hyponatremic** or **hyponatraemic** adj.

hypoparathyroidism underactivity of the **parathyroid glands**.

hypophase the lower layer of a two-phase system. Compare **epiphase**. —**hypophasic** adj.

hypophosphatasia see **alkaline phosphatase**.

hypophosphatemia or (esp. Brit.) **hypophosphataemia** the presence of an abnormally low level of phosphate in the blood. It occurs in vitamin D deficiency, primary **hyperparathyroidism**, incorrect par-

enteral nutrition, during recovery from diabetic ketoacidosis, and in hereditary defects of uptake of renal phosphate. See **fibroblast growth factor-23**, **PHEX**.

hypophysectomy surgical removal of the pituitary gland (i.e. hypophysis). —**hypophysectomize** or **hypophysectomise** vb.

hypophysiotropic factor or **hypophysiotropic hormone** an alternative name for **hypothalamic (regulatory) factor**.

hypophysis or **hypophysis cerebri** an alternative name for **pituitary gland**. See also **adenohypophysis**, **neurohypophysis**. —**hypophyseal** or **hypophysial** adj.

hypopituitarism total or (more often) partial loss of pituitary function, usually due to destructive lesions of the pituitary.

hypoplasia underdevelopment or defective formation of an organ or tissue. —**hypoplastic** adj.

hypoprothrombinemia or (esp. Brit.) **hypoprothrombinaemia** an abnormally low (inadequate) level of plasma prothrombin (factor II), commonly caused by vitamin K deficiency.

hyposecretion lesser than normal secretion by any endocrine or exocrine gland.

hypotension the state or condition of having a lower than normal arterial blood pressure. —**hypotensive** adj.

hypothalamic (regulatory) factor or **hypothalamic hormone** or **hypophysiotropic factor** or **hypophysiotropic hormone** any peptidic substance that is synthesized in the mammalian hypothalamus and released into the hypophyseal–portal circulation in response to neural and/or chemical stimuli, and that regulates the secretion (and perhaps also the synthesis) of a specific polypeptide hormone by the anterior lobe of the pituitary gland (i.e. adenohypophysis). These specific factors may be either releasing or release-inhibiting factors (or hormones); their recommended names are now formed by replacing the word-ending 'tropin' (or '-in' in prolactin) in the name of the corresponding pituitary hormone by the word-ending '-liberin' (for a releasing factor) or '-statin' (for a release-inhibiting factor).

hypothalamus (pl. **hypothalami**) the region of the vertebrate brain lying below the thalamus and around the floor of the third ventricle, just posterior to the attachment of the cerebral hemispheres. It lies just above the **pituitary gland**, to which it supplies various **hypothalamic (regulatory) factors**. —**hypothalamic** adj.

hypothetical protein a translation product that cannot be related (e.g. by means of database similarity searches) to any previously characterized protein.

hypothyroidism deficient secretion by the thyroid gland. In infants it leads to cretinism; in adults the consequences include low metabolic rate, lethargy, menstrual disorders, and in extreme cases **myxedema**. —**hypothyroid** adj.

hypotonic 1 (of a solution) having a lower osmotic pressure than that of some given solution, particularly the fluid in or surrounding a given type of cell or a body fluid. **2** (of a muscle) showing or characterized by diminished tone or tension. Compare **hypertonic**. —**hypotonicity** n.

hypovolemia or (esp. Brit.) **hypovolaemia** a reduction in the volume of circulating blood. It may be due to hemorrhage or to redistribution of fluid from the plasma to the extravascular tissues and spaces.

hypox abbr. for hypophysectomized. See **hypophysectomy**.

hypoxanthine symbol: Hyp; purin-6(1*H*)-one; 6-hydroxy purine; an intermediate in the degradation of adenylate. Its ribonucleoside is known as **inosine** and its ribonucleotide as **inosinate**. Compare **xanthine**. See also **hypoxanthine phosphoribosyltransferase**.

hypoxanthine/guanine phosphoribosyltransferase abbr.: HGPRT; see **hypoxanthine phosphoribosyltransferase**.

hypoxanthine phosphoribosyltransferase EC 2.4.2.8; systematic name: IMP:pyrophosphate phospho-D-ribosyltransferase; other names: hypoxanthine/guanine phosphoribosyltransferase (abbr.: HGPRT); IMP pyrophosphorylase. An enzyme that catalyses a reaction between 5-phospho-α-D-ribose 1-diphosphate and either hypoxanthine or guanine to form IMP or GMP respectively, with release of pyrophosphate. It is important for the metabolic activation of purines, e.g. 6-mercaptopurine, and forms part of the purine salvage pathway. Partial deficiency is associated with **Lesch–Nyhan syndrome** and **gout**.

hypoxia the presence of less than normal amounts of dioxygen in a vertebrate or in its blood. *See* **ARNT**, **HIF**. *Compare* **hyperoxia**. —**hypoxic** *adj.*

hypoxia-inducible factor *See* **HIF**.

Hypro *(formerly) symbol for* a residue of (4-)hydroxy-L-proline (4Hyp or Hyp now recommended).

hypsochromic describing the shift of an absorption band in the direction of shorter wavelengths. *Compare* **bathochromic**.

hypusine N^{ε}-[4-amino-2-hydroxybutyl]lysine; an amino acid uniquely formed by post-translational modification of one of the lysyl residues of eukaryotic translation initiation factor 5A (eIF-5A). It is formed by deoxyhypusine synthase, an enzyme that transfers the butylamine portion of spermidine to the ε-NH$_2$ group of a specific lysine of the eIF-5A precursor, followed by addition of the hydroxyl group by deoxyhypusine hydroxylase.

hysteresis the delay between the effect on a system and the cause producing the effect. —**hysteretic** *adj.*

hysteretic enzyme any enzyme that responds slowly (in terms of some kinetic characteristic) to a rapid change in ligand (either substrate or modifier) concentration. Such slow changes, defined in terms of their rates relative to the overall catalytic reaction, result in a lag in the response of the enzyme to changes in the ligand level.

Hz *symbol for* hertz.

H zone *see* **sarcomere**.

Ii

i *symbol for* **1** iso (def. 3); (as a prefix or superscript in symbols and abbreviations for iso compounds or groups; e.g. isopropyl can be symbolized as iPr or Pr^i). **2** the chemical group 2-isopentenyl; 3,3-dimethylallyl (as a prefix to a single-letter symbol for a nucleoside residue to designate substitution of the base, as in **iA**). **3** the square root of -1.

i *symbol for* **1** electric current (alternative to *I*). **2** van't Hoff factor.

I *symbol for* **1** iodine. **2** an iodo group in an organic compound. **3** a residue of the α-amino acid L-isoleucine (alternative to Ile). **4** a residue of the ribonucleoside inosine (alternative to Ino).

I *symbol for* **1** electric current (preferred alternative to *i*). **2** intensity; the subscripts e (for energetic), p (for photon), or v (for visible) may be added to distinguish between radiant intensity, photon intensity, and luminous intensity, respectively. **3** ionic strength; the subscripts c or m may be added to indicate whether it is expressed based on concentration or molality respectively. **4** the inductive effect of a particular chemical group in an organic compound.

iA *symbol for* N^6-(2-isopentenyl)adenosine, a **hypermodified base**.

IAA *abbr. for* indole-3-acetic acid *or* **indoleacetate**.

IAN *abbr. for* indoleacetonitrile.

IAP *abbr. for* **1** islet-activating protein; *see* **pertussis toxin**. **2** inhibitor of apoptosis protein; *other name*: apoptosis inhibitory protein (*abbr.*: AIP); any of a group of proteins that inhibit apoptosis. They bind tumour necrosis factor (TNF) receptor-associated proteins 1 and 2. Mutations in a neuronal IAP (*abbr.*: NIAP), a 294-residue protein encoded at 5q12.2-q13.3, cause various forms of spinal muscular atrophy. *See* **TNF receptor-associated protein 2**.

iatrogenic describing a condition or disease induced unintentionally by a physician through his or her diagnosis, manner, or therapy. —**iatrogenicity** *n*.

IκB *abbr. for* inhibitor of NF-κB; a cytosolic protein that binds and inhibits **NF-κB** in most cell types. It is released from binding following phosphorylation by the **IκB kinase** complex in response to binding of tumour necrosis factor α and interleukin-1 to their receptors. The phosphorylation targets IκB for ubiquitination and degradation by proteasomes.

I-band *abbr. for* isotropic band. The I-bands of striated muscle contain the **thin filaments** and correspond to the light bands. The name derives from the fact that they are isotropic in polarized light. *See also* **sarcomere**.

IκB kinase a heterotrimeric complex ($\approx$700 kDa) that acts on IκB to release NF-κB and target IκB for ubiquitination and degradation by proteasomes. It consists of two catalytic subunits (α and β), which share $\approx$50% sequence identity, and a regulatory subunit (γ). It is activated by phosphorylation by multiple signalling pathways, including tumour necrosis factor α and interleukin-1.

IBMX *abbr. for* isobutylmethylxanthine.

ibuprofen 2-(4-isobutylphenyl)propionic acid; a nonsteroidal anti-inflammatory agent of the substituted propionic acid type; others of this type are flurbiprofen, **ketoprofen**, and naproxen. They inhibit the cyclooxygenase activity of **prostaglandin-endoperoxide synthase**, reversibly over short time intervals, followed by time-dependent irreversible inactivation, due probably to conformational rather than covalent changes. The racemic mixture is present in many over-the-counter drugs, but the (*S*)-enantiomer is the active form. *Proprietary names include*: Advil; Brufen; Motrin.

(*S*)-ibuprofen

ic *or* **i.c.** *abbr. for* intracutaneous or intracutaneously.

IC *abbr. for* internal conversion.

IC$_{50}$ **1** the median inhibitory concentration (in mol L^{-1}) of an antagonist, i.e. the concentration that reduces a specified response to 50% of its former value; *compare* **EC$_{50}$**. **2** the median inhibitory concentration (in mol L^{-1}) of an agent (agonist or antagonist), i.e. the concentration that causes a 50% reduction in the specific binding of a radioligand.

ICAM *abbr. for* intercellular adhesion molecule; any of several type I membrane glycoproteins of the immunoglobulin superfamily. They are ligands for leukocyte adhesion to target cells, in conjunction with **LFA-1**; in fact LFA-1–ICAM links mediate adhesion between many cell types. **ICAM-1** (*or* CD54), of 90–115 kDa, are expressed on B and T lymphocytes, endothelial, epithelial, and dendritic cells, fibroblasts, keratinocytes, and chondrocytes. They are inducible in 12–24 hours by cytokines such as gamma interferon, interleukin-1β, and tumour necrosis factor-α. **ICAM-2** (*or* CD102), of 55–65 kDa, are constitutively expressed on endothelial cells, some lymphocytes, monocytes, and dendritic cells. **ICAM-3** (*or* CD50), of 116–140 kDa, are constitutively expressed on monocytes, granulocytes, and lymphocytes; and on stimulation are rapidly and transiently phosphorylated on serine residues.

ICAT *abbr. for* **isotopically coded affinity tag**.

ICD (*in clinical chemistry*) *abbr. for* isocitrate dehydrogenase (EC 1.1.1.41).

ice the solid form of water that can exist at temperatures below the triple point of water (273.16 K at 101 325 Pa).

ICE *abbr. for* **1** imprint control element; a CpG rich region of genomic DNA that attracts *de novo* methylation in one parental gamete and avoids methylation in the other parental gamete. It is necessary for imprinted expression: when unmethylated it silences mRNA and activates non-coding RNA (ncRNA); it is inactive when methylated. An example is the 3.7 kb element within the *Igf2r* gene responsible for paternal-specific gene silencing of *Igf2r*, *Slc22a2*, and *Slc22a3* genes on mouse chromosome 17. **2** interleukin 1β convertase.

iceberg a metaphor used to describe an interpretation of the anomalous entropies of solution of noble gases and other nonpolar substances in water, suggesting that water tends to organize itself into quasi-solid supramolecular structures around the molecules of such substances. In the case of alkyl compounds, this tendency increases markedly with the length of the alkyl chain.

Iceberg a human protein of the CARD family that binds caspase-1 tightly. It inhibits the generation of interleukin 1β.

ICE-like protease any of a family of endopeptidases that structurally resemble **interleukin-1 β convertase** (ICE). They are involved in **apoptosis**, being implicated in the proteolysis that causes cell death.

I cell **1** *or* **CCK cell** any of a group of cells, widely distributed in the duodenal and jejunal mucosa, that produce **cholecystokinin**. So named because their histological features are intermediate between those of **S cells** and **L cells**. **2** *abbr. for* inclusion cell. *See also* **I-cell disease**.

I-cell disease *or* **inclusion-cell disease** *or* **mucolipidosis II** an autosomal recessive disease in which most of the lysosomes in the connective tissue (fibroblasts) contain large inclusions of glycoaminoglycans and glycolipids as a result of the absence of several lysosomal hydrolases. These enzymes, which are synthesized in the endoplasmic reticulum, are secreted into the extracellular medium rather than being directed to the lysosomes. This is due to the absence of a mannose 6-phosphate marker on the carbohydrate moieties of these hydrolases because of a deficiency in an enzyme required for mannose phosphorylation. The failure of the phosphorylation in the *cis* Golgi network means that the enzymes are not segregated by the **mannose 6-phosphate receptors** into the appropriate transport vesicles in the *trans* Golgi network. The patients have an abnormally high level of lysosomal enzymes in their sera and body fluids. A milder form of I-cell disease is pseudo-Hurler polydystrophy.

ice point the temperature at which ice melts. It is taken as the temperature (273.15 K) at which ice and water are in equilibrium at standard pressure (101 325 Pa). It was used as a reference temperature on the Celsius scale, but the kelvin is based instead on the temperature at the **triple point** of water (273.16 K).

ICF *abbr. for* intracellular fluid.

ichthyosis a human condition in which the skin is thickened and scaly, as a result of a disorder of keratinization in the outer layers of the epidermis. An X chromosome-linked form is caused by mutations that produce deficiency of **steryl-sulfatase**. An autosomal dominant form results from mutations in the genes for keratin 1 or keratin 10. An autosomal recessive form is caused by deficiency of transglutaminase K. –**ichthyotic** *adj.*

ichthyotocin *an alternative name for* **isotocin**.

I-clip any intramembrane-cleaving protease, which may be a metalloprotease, serine protease, or aspartate protease.

icosa+ *or (before a vowel)* **icos+** *comb. form* recommended for denoting twenty or twenty times. *Also (formerly)*: **eicosa+**, **eicos+**. (*Note*: the eicos(a)+ variant is still always used for the C_{20} fatty acids and skeletally related compounds, e.g. the eicosanoids.)

icosadeltahedron (*pl.* **icosadeltahedra**) any solid geometrical figure obtained by dividing the triangular faces of an **icosahedron** into smaller triangles. If divided into more than four smaller triangles, the original faces will not be flat, in which case the 12 pentameric vertices will be supplemented by hexameric vertices. —**icosadeltahedral** *adj.*

icosahedral symmetry a form of **cubic symmetry**, being the crystallographic symmetry displayed by an **icosahedron** and by the capsid of some types of virus.

icosahedron (*pl.* **icosahedra**) any solid geometrical figure having 20 triangular faces, 30 edges, and 12 pentameric vertices. A regular icosahedron has faces that are congruent plane equilateral triangles; there is twofold rotational symmetry about an axis through the midpoint of each edge, threefold rotational symmetry about an axis through the centre of each face, and fivefold rotational symmetry about an axis through each vertex. —**icosahedral** *adj.*

icosanoate *or (formerly)* **eicosanoate** *the systematic name for* **arachidate**.

icosanoid *the recommended name for* **eicosanoid**, but not generally used.

ICRO *abbr. for* International Cell Research Organization.

ICSH *abbr. for* interstitial cell-stimulating hormone (i.e. **luteinizing hormone**).

ICSU *abbr. for* International Council of Scientific Unions.

icterus (*in pathology*) *an alternative name for* **jaundice**.

icterogenic causing **jaundice**.

id *or* **i.d.** *abbr. for* **1** intradermal, or intradermally (alternative to ID). **2** inside diameter.

Id *abbr. for* inhibitor of differentiation; any of four mammalian helix-loop-helix (HLH) proteins (ID1 to ID4) that are negative regulators of cell differentiation. They form heterodimers, which do not bind to DNA, with certain ubiquitously expressed basic HLH proteins, and regulate tissue-specific gene expression.

ID *abbr. for* **1** (*in clinical chemistry*) iditol dehydrogenase (EC 1.1.1.14). **2** intradermal, or intradermally (alternative to id).

ID$_{50}$ *or* **ID50** *abbr. for* median infectious dose.

idaric acid the **aldaric acid** derived from idose.

IDDM *abbr. for* insulin-dependent diabetes mellitus (*see* **diabetes mellitus**).

+ide *suffix (in chemical nomenclature)* denoting a **lactide**. It replaces the terminal '+ic acid' of the trivial name of the parent hydroxy acid, and a multiplying prefix indicates the number of molecules of hydroxy acid involved; e.g. dilactide (formed from two molecules of lactic acid).

ideal describing any object, process, or system that is perfect, or that behaves exactly in accordance with some simple law or theory. *See also* **ideal gas**, **ideal solution**. —**ideality** *n.*

ideal gas any gas whose behaviour is accurately described by the **gas laws**.

ideal solution any solution in which, for each component,

$$d\mu_i = RT\, d\ln x_i,$$

where μ_i is the chemical potential of the ith component and x_i is its mole fraction, R is the universal gas constant, and T is the thermodynamic temperature. Integration of the above equation gives:

$$\mu_i = \mu_i^{\circ} + RT \ln x_i,$$

where μ_i° is the standard chemical potential of the ith component.

identity (*in sequence analysis*) the extent to which compared sequences have identical bases or residues at equivalent positions, usually expressed as a percentage.

identity matrix *or* **unitary matrix** *see* **scoring matrix**.

idio+ *comb. form* denoting **1** peculiar, distinct, one's own, separate. **2** self-produced, arising within.

idiogram *or* **karyogram** a diagrammatic representation of the **karyotype** of a cell, based on measurement of the chromosomes in several or in many cells.

idiom (*in immunology*) the group of **idiotypes** to which the **epitopes** carried by components of one individual animal belong.

idiopathic 1 (*in pathology*) describing a disease arising by itself, not consequent upon, or symptomatic of, another disease; of the nature of a primary morbid state; essential. **2** of the nature of a particular affection or susceptibility. —**idiopathy** *n.*

idiopathic autonomic neuropathy a severe subacute disorder caused by autoantibodies to ganglionic acetylcholine receptors (i.e. not those of neuromuscular junctions as in **myasthenia gravis**).

idiopathic pentosuria *see* **pentosuria** (def. 1).

idiotope 1 the group of **epitopes** carried by components of one individual animal. **2** as originally defined by Danish immunologist Niels Jerne (1911–94), an individual determinant (i.e. site) on an antibody molecule; together, the idiotopes constitute the idiotype.

idiotype 1 an antigenic specificity, particularly of antibodies directed against a single antigen. **2** any class or group of related antigenic characteristics or **idiotopes**; an **epitype** that is characteristic of a particular animal. —**idiotypic** *adj.*

iditol the hexitol derived formally by reduction of the aldehyde group of idose. L-Iditol is a rare natural product accompanying sorbitol in the fermented juice of berries of mountain ash (*Sorbus aucuparia*).

IDL *abbr. for* intermediate-density lipoprotein.

iDNA *abbr. for* intercalary DNA.

IDNPAP *abbr. for* isethionyl-3-(*N*-2,4-dinitrophenyl)amino propioimidate; a reagent useful in the isolation of plasma membranes. It reacts with, but does not penetrate, the membranes and leaves the net charge on the proteins unaltered.

ido- *prefix (in chemical nomenclature)* indicating a particular configuration of a set of four (usually) contiguous >CHOH groups, as in the acyclic form of D- or L-idose. *See* **monosaccharide**.

Ido *symbol for* a residue (or sometimes a molecule) of the aldohexose idose.

IdoA *symbol for* a residue of iduronic acid.

idose *symbol*: Ido; *ido*-hexose; an aldohexose whose D- and L-enantiomers are epimeric at C-2, C-3, and C-4 with D- and L-glucose, respectively.

β-D-idose

idoxuridine 5-iodo-2′-deoxyuridine; a pyrimidine nucleoside analogue, metabolized to the triphosphate, that inhibits DNA replication by substituting for thymidine in viral and mammalian DNA incorporation. It is used as an antiviral agent.

IDP *abbr. for* **1** the ribonucleotide inosine (5′-)diphosphate. **2** isopentenyl diphosphate (*formerly called* Δ³-isopentenyl pyrophosphate).

IDU *abbr. for* 5-iodo-2′-deoxyuridine (*see* **idoxuridine**).

IdUrd *symbol for* 5-iodo-2′-deoxyuridine (*see* **idoxuridine**) (preferred to IDU, IUdR, or IUDR).

iduronate 1 the anion of **iduronic acid**. **2** any salt or ester of iduronic acid.

iduronate-2-sulfatase EC 3.1.6.13; *other name*: chondroitin sulfatase; an enzyme that hydrolyses the 2-sulfate groups of the L-iduronate 2-sulfate units of dermatan sulfate, heparan sulfate, and heparin.

iduronic acid *symbol*: IdoA; the **uronic acid** formally derived from idose by oxidation to a carboxyl group of the hydroxymethylene group at position 6. In nature α-linked L-iduronic-acid residues, sulfated at position 2, occur in terminal positions in dermatan sulfate and heparan sulfate; these iduronic-acid residues are formed by epimerization of the configuration at C-5 of D-glucuronic-acid residues in these heteropolysaccharides.

L-iduronidase EC 3.2.1.76; an enzyme that hydrolyses α-L-iduronosidic linkages in desulfated dermatan. Deficiency leads to mucopolysaccharidosis I H.

iduronosyl the glycosyl group formed by removal of the anomeric hydroxyl group from **iduronic acid** or iduronate.

IE *abbr. for* **1** immunoelectrophoresis (alternative to IEP). **2** information extraction.

IEC *abbr. for* ion-exchange cellulose.

IED *abbr. for* individual effective dose. IED$_{50}$ denotes the median of the IEDs within a group of subjects.

IEF *abbr. for* isoelectric focusing.

IEP *abbr. for* **1** immunoelectrophoresis (alternative to IE). **2** *or* **i.e.p.** isoelectric point (pI preferred).

IF *abbr. for* initiation factor. Suffixed numerals are added to designate individual factors, e.g. **IF-1**, **IF-2**, IF-2a, **IF-3**. Eukaryotic initiation factors are generally denoted by the prefix 'e', e.g. eIF-2, eIF-3.

IF-1 an *Escherichia coli* **initiation factor**.

IF-2 an *Escherichia coli* **initiation factor**.

IF-3 an *Escherichia coli* **initiation factor**.

IFCC *abbr. for* International Federation of Clinical Chemistry.

IFN *abbr. for* interferon. Suffixed Greek letters are added to differentiate the main classes, e.g. IFN-α, IFN-β, IFN-γ.

Ig *symbol for* immunoglobulin.

IgA *symbol for* immunoglobulin A.

IgA nephropathy a form of primary glomerulonephritis that is common throughout the world. There is reduced hepatic clearance of IgA, which deposits in the renal glomeruli.

IgA protease *another name for* **IgA-specific serine endopeptidase**.

IgA-specific serine endopeptidase EC 3.4.21.72; *other name*: IgA protease; a secreted bacterial serine protease that catalyses the cleavage of immunoglobulin A molecules at certain Pro-|-Xaa bonds in the hinge region. A signal peptide guides the enzyme precursor to the periplasmic space, and the C-terminal helper domain forms a pore in the outer membrane for exit of the protease domain. The helper domain is then released by autoproteolysis.

IG cell *abbr. for* intestinal gastrin cell; a small, granulated, gastrin-producing cell of the human upper small intestine. IG cells are smaller than, and ultrastructurally different from, CCK-producing **I cells**, gastrin-producing **G cells** of the gastric pylorus, and **TG cells**.

IgD *symbol for* immunoglobulin D.

IgE *symbol for* immunoglobulin E.

igensin a neutral cytosolic serine proteinease, activated by sodium dodecyl sulfate and found in a number of mammalian tissues.

IGF *abbr. for* insulin-like growth factor.

IGFBP *abbr. for* insulin-like growth factor binding protein.

IgG *symbol for* immunoglobulin G. Suffixed numerals are added to designate individual subclasses, e.g. IgG1, IgG2, IgG3, IgG4.

Ig-Hepta a rat glycoprotein (1389 amino acids) that is a G-protein-coupled membrane receptor with a long N-terminal extracellular region containing two C2-type Ig-homology units. It occurs on lung cell surfaces, and probably functions in cell adhesion and intracellular signalling. It defines a subgroup of G-protein-coupled receptors designated **LNB-TM7**.

Ig-like domain *see* **immunoglobulin fold**.

IgM *symbol for* immunoglobulin M.

β IG-M1 *see* **connective tissue growth factor**.

β IG-M2 *see* **connective tissue growth factor**.

ignose the name first proposed for ascorbic acid by A. Szent-Györgyi, before he knew its structure. *See also* **godnose**, .

IGT *abbr. for* impaired glucose tolerance.

Ii *see* **invariant chain**.

Ii system a blood group system in which the antigenic determinant is an oligosaccharide related to the ABH system (*see* **ABH antigens**). Anti-I and anti-i are the major **cold autoantibodies**; the former occur in the transient anemia accompanying viral or *Mycoplasma pneumoniae* infections; the latter are characteristic of the chronic cold agglutinin disease, accompanying chronic lymphoproliferative diseases of the non-Hodgkin type. The structures of the determinants are: i determinant, GlcNAc(β1-3)Gal(β1-4)GlcNAcβ1-R; I determinant, Gal(β1-4)[Gal(β1-4)GlcNAc(β1-6)]GlcNAc(β1-3)-Gal(β1-4)GlcNAcβ1-R.

ikaros a protein that binds and activates the **enhancer** of a gene *CD3*-δ involved in T-lymphocyte specification and maturation. Similar to **hunchback**, it is a zinc-finger protein.

IL *abbr. for* interleukin. Various interleukins are distinguished by suffixed numerals, e.g. IL-1, IL-2, IL-3, etc.

ILA *abbr. for* insulin-like activity.

Ile *symbol for* a residue of the α-amino acid L-isoleucine (alternative to I).

ileum the lower part of the small intestine, extending from the jejunum to the large intestine. —**ileal** *adj.*

illegitimate transcription transcription at a low level of a normally tissue-specific transcribing gene in nonspecific cells.

Im *abbr. for* immunity protein.

IM *or* **i.m.** *abbr. for* intramuscular, or intramuscularly.

image analysis analysis of a digital image, characterizing its content in terms of, e.g., physical measurements, anomalous features, tagged pixels, or 2D or 3D representations suitable for computer graphics rendering.

imaging the production of an image of all or part of the body using techniques such as X-radiography, ultrasound scanning, nuclear magnetic resonance scanning, radionuclide scanning, or thermography and particularly used for gene expression and proteomic data. *Compare* **tomography**.

imidate 1 any salt or ester of an **imidic acid**. **2** to convert an oxo diacid into its corresponding **imide**. —**imidation** *n.*

imidazole *or (formerly)* **glyoxaline** *or* **iminazole** 1,3-diazole; 1,3-diaza-2,3-cyclopentadiene; a weak base, pK_a ~6.95, used as a **buffer** (def. 1). The numbering and tautomerism of the imidazole ring are described at **histidine**.

imidazole 4-acetate a product of histamine metabolism; it results from the action of monoamine oxidase on histamine to form imidazole 4-acetaldehyde, which is then oxidized by an NAD⁺-dependent aldehyde dehydrogenase to imidazole 4-acetate. *See also* **histidine**.

imidazoline receptor any mitochondrial or plasma membrane binding site for compounds from the generic groups imidazoles, imidazolines, imidazolidines, guanidines, and oxazolines, and having a putative endogenous ligand, **agmatine**. Imidazoline receptors occur throughout the brain, being possibly associated with glial cells, and in peripheral tissues. I_1 **receptors** have high affinity for *p*-aminoclonidine and are localized in plasma membranes. I_2 **receptors** have high affinity for idazoxan and are localized in mitochondrial membranes. The latter are known alternatively as nonadrenergic imidazoline binding sites (NAIBSs). Photoaffinity labelling to at least three proteins occurs: two, of 55 kDa and 61 kDa, respectively, are found in rat liver and differ in their affinity for the guanidium, amiloride; a third, of 61 kDa, is found in PC-12 cells and has high affinity for amiloride but low affinity for idazoxan. These proteins may represent different I_2 receptors. Subtypes of I_2 receptors are also suggested pharmacologically. The function of imidazoline receptors is not known, but many effects of imidazoline compounds, formerly attributed to their binding to α_2 receptors (e.g. hypotension), are now attributed to NAIBSs.

imidazolium 1 the cation formed by addition of a proton (hydron) to a molecule of **imidazole**. **2** the cationic chemical group formed by addition of a proton (hydron) to an **imidazolyl** group.

imidazolyl the chemical group derived formally by removal of one hydrogen atom from **imidazole**. The position from which the hydrogen atom has been removed may be indicated, e.g. 2-imidazolyl. The imidazolyl group is present in **histidine** and its derivatives.

imide any secondary organic amide (i.e., any diacyl derivative of ammonia or a primary amine), especially any that is a cyclic compound in which both of the acyl groups (which may be the same or different) are derived formally or actually from a diacid. (*Note*: non-cyclic symmetrical secondary organic amides are generically termed **diacylamines**.)

+imide *suffix* designating any organic compound that is an **imide**. The nitrogen atom may be substituted with a named alkyl or other group. An example is succinimide:

succinimide

imidic acid any derivative of a carboxylic acid in which the carbonyl oxygen atom of the carboxyl group has been replaced by an **imino** (def. 1) or alkylimino group; i.e., any containing the –C(=NR)–OH group. *Distinguish from* **imino acid**.

imido+ *comb. form* designating any inorganic oxo acid (including the anions, salts, and esters) in which =O has been replaced by =NH or =NR (where R is a named substituent), or any derived di-, tri-, etc. acid in which a bridging oxygen atom, –O–, has similarly been replaced by –NH– or –NR–; e.g., **adenosine 5′-[β,γ-imido]triphosphate**.

imidodipeptiduria *or* **prolidase deficiency** the presence in urine of peptides of structure X-Pro or X-Hyp in which X is frequently glycine. It results from deficiency of X-Pro dipeptidase and is accompanied by severe ulcers, mostly of the feet and hands, and frequently by mental retardation.

Imigran *see* **sumatriptan**.

iminazole *a former name for* **imidazole**.

imine 1 any organic compound containing one or more **imino** groups or substituted imino groups. **2** *an obsolete term for* azacycloalkene.

iminium the bivalent quaternary nitrogen-containing cationic group =⁺NR₂ attached to carbon in a chemical compound, where the R groups may each be hydrogen or any hydrocarbyl group.

imino the bivalent group =NH attached to carbon in a chemical compound.

imino+ *comb. form* denoting the presence in a chemical compound

of one or more **imino** groups; the number of imino groups in the molecule may be specified, e.g. diimino+.

imino acid any carboxylic acid in which an imino substituent replaces two hydrogen atoms. In biochemistry, the term is commonly applied also to certain cyclic alkylamino (especially *α*-alkylamino) derivatives of aliphatic carboxylic acids, e.g. **proline**, although such compounds are now preferably classed as azacycloalkane carboxylic acids. *Compare* **imidic acid**.

iminodipeptidase *see* **Pro-X dipeptidase**.

iminoglycinuria the secretion in urine of excessive amounts of glycine, proline, and hydroxyproline, due to a defect in their shared transport in renal tubules. The condition is clinically benign.

imipenem a carbapenem β-lactam antibiotic that is a semisynthetic derivative of thienamycin produced by *Streptomyces cattleya*. Its broad-spectrum antibiotic actions resemble those of penicillins, i.e. inhibition of bacterial peptidoglycan synthesis.

immediate gene *or* **immediate-early gene** *see* **early gene**.

immediate hypersensitivity *or* **type I hypersensitivity** a type of hypersensitivity responsible for asthma, hayfever, and some types of eczema. It occurs within minutes of exposure to antigen and depends on activation of **mast cells** and the release of mediators of acute inflammation. The mast cells bind **immunoglobulin E** (IgE) at their surface **Fc receptors** and when antigen cross-links the IgE, the mast cells degranulate releasing vasoactive amines that mediate the symptoms.

immobilize *or* **immobilise** to render any agent, whether a micro- or macrosolute, a particle, or an intact cell, nondispersible in an aqueous medium with retention of its specific ligating, antigenic, catalytic, or other properties. Immobilization may be achieved by, e.g., encapsulation, entrapment in a small-pore gel, adsorption on, or covalent linkage to, an insoluble supporting material (matrix), or through formation of aggregates by cross-linkage.

immobilized enzyme *see* **solid-phase technique**.

immortalization *or* **immortalisation** the **transformation** of a cell population with a finite life span to one possessing an infinite life span. It is a characteristic of cancer cells, and is of practical importance in, e.g., the creation of **monoclonal antibodies**. Immortalization of eukaryotic cells may be effected experimentally by infection with certain viruses, or by fusion of the cell with a neoplastic cell to form a **hybridoma**. —**immortalize** *or* **immortalise** *vb*.

immotile cilia syndrome a genetically heterogeneous human disease characterized by impaired mobility or complete immobility of the cilia of airways and elsewhere and of sperm tails. Electron microscopy of such ciliated cells usually shows a lack of dynein arms between the longitudinal microtubular doublets, but other abnormalities may also be present.

immune 1 relating to or possessing **immunity**. **2** regarding or describing an animal that has been immunized. **3** *an alternative word for* **immunological**.

immune adherence the **complement**-dependent adherence of antibody–antigen complexes, or of antibody-coated bacteria (or other particles), to primate erythrocytes, causing the agglutination of the latter.

immune clearance *an alternative term for* **immune elimination**.

immune complex *an alternative term for* **antigen–antibody complex**.

immune complex-mediated hypersensitivity *or* **type III hypersensitivity** a type of **hypersensitivity** caused by the deposition of antigen–antibody complexes in tissue and blood vessels. The complexes activate **complement** and attract polymorphs and macrophages to the site.

immune cytolysis the **complement**-dependent lysis of cells by antibody molecules. When the cells concerned are erythrocytes the process is termed **immune hemolysis**.

immune elimination *or* **immune clearance** the accelerated, exponential removal of antigen from an immune animal as a result of complexing with antibody molecules.

immune globulin *(sometimes) an alternative name for* **immunoglobulin**.

immune hemolysis *see* **immune cytolysis**.

immune response the total immunological reaction of an animal to an immunogenic stimulus. It includes antibody formation, the development of hypersensitivity, and immunological tolerance.

immune serum *an alternative term for* **antiserum**.

immunity freedom and protection from 'infection'. Two forms of immunity can be recognized: (1) innate, meaning non-specific antimicrobial systems (e.g. phagocytosis) that are innate in that they are not intrinsically affected by prior contact with the infectious agent; and (2) the state of an animal that has an enhanced ability, above the nonimmune state, to respond to some (specific) antigen in which the antigen is bound and rendered inactive or eliminated from the body. *See also* **active immunity**, **passive immunity**.

immunity protein *abbr.:* Im; a protein, cosynthesized in *E. coli* with **colicin**, that binds to the nuclease domain of the colicin and inhibits its activity. Ims specific for RNase domains contain 84 amino acids, which form mostly beta strands. Those specific for DNase domains contain 86 amino acids and form four alpha helices.

immunization *or* **immunisation** any procedure carried out on an animal, e.g. the administration of antigen or antibody, that leads to an increased reactivity of the animal's immune system towards an antigen or antigens. —**immunize** or **immunise** *vb*.

immuno+ *comb. form* indicating immune or immunity.

immunoadsorbent any insoluble preparation of an antigen (or antibody) suitable for use in **immunoadsorption**. *See also* **immunosorbent**.

immunoadsorption the use of an insoluble antigen (or antibody) to remove specifically unwanted antibodies (or antigens) from a mixture so as to make the antibody (or antigen) more specific. *Compare* **adsorption** (def. 2). *See also* **immunosorption**.

immunoaffinity chromatography a type of **affinity chromatography** in which one of the two components of an antigen–antibody system is coupled to an insoluble matrix and used for the separation and purification of the other component.

immunoaffinity column a chromatographic column used in **immunoaffinity chromatography**.

immunoassay 1 any technique for the measurement of specific biochemical substances, commonly at low concentrations and in complex mixtures such as biological fluids, that depends upon the specificity and high affinity shown by suitably prepared and selected antibodies for their complementary antigens. A substance to be measured must be antigenic – either an immunogenic macromolecule or a haptenic small molecule. To each sample a known, limited amount of specific antibody is added and the fraction of the antigen combining with it, often expressed as the bound:free ratio, is estimated, using as indicator a form of the antigen labelled with a radioisotope (**radioimmunoassay**), fluorescent molecule (**fluoroimmunoassay**), stable free radical (**spin immunoassay**), enzyme (**enzyme immunoassay**), or other readily distinguishable label. The amount of the antigen in the sample is found by comparison with standards containing known amounts. *See also* **heterogeneous immunoassay**, **homogeneous immunoassay**, **saturation analysis**. **2** *an alternative name for* **radioimmunoassay**. **3** to carry out or to estimate (something) by an **immunoassay** (def. 1, 2).

immunobiology the branch of biology dealing with the activities of the cells of the immune system and their relationship to each other and to their environment. —**immunobiological** *adj*.

immunoblotting a technique for the detection of specific polypeptides separated by **polyacrylamide gel electrophoresis** (PAGE). The bands are transferred from the gel to a nylon or nitrocellulose membrane by Western blotting, followed by immunological detection of the immobilized antigen.

immunochemical 1 of or relating to **immunochemistry**. **2** any specific immunological reagent that consists of or incorporates an antigen or an antibody, especially one of commerce.

immunochemistry the branch of biochemistry dealing with the chemical nature of antigens, antibodies, and their interactions, and with the chemical methods and concepts as applied to immunology.

immunochromatogram the pattern of precipitation bands formed in **immunochromatography**.

immunochromatography any of various techniques for separating and identifying soluble antigens. In one method, analogous to **immunoelectrophoresis**, the antigens are first fractionated through a thin-layer chromatogram, e.g. of Sephadex; the fractions are then diffused against a trough containing a solution of antibodies (antiserum). In another method a small drop of antigen–antiserum mixture is applied to filter paper, or resin-impregnated paper, and the chromatogram developed with buffer; any antigen–antibody complex remains at the point of application and can be detected with a protein stain, while the other proteins are washed away from the origin.

immunocompetent cell an immunologically competent cell.

immunocyte any **immunologically competent cell**.

immunocytochemistry cytochemistry using appropriately labelled antibody preparations to detect specific cellular components. —**immunocytochemical** *adj*.

immunodeficiency any condition in which there is a deficiency in the production of humoral and/or cell-mediated immunity. —**immunodeficient** *adj*.

immunodeterminant *an alternative name for* **antigenic determinant**.

immunodiagnosis the diagnosis of disease by immunological methods.

immunodiagnostic 1 of or relating to **immunodiagnosis**. **2** any specific immunological reagent used in diagnosis.

immunodiffusion any of various analytical methods by which components of soluble antigen or antibody mixtures may be distinguished. Essentially, the antigens and antibodies are allowed to diffuse towards each other in a translucent gel, wherein they react to gives lines or bands of precipitation in characteristic positions. Such methods include: **double immunodiffusion**, **electroimmunodiffusion**, **immunochromatography**, **immunoelectrophoresis**, immunorheophoresis, and **single radial immunodiffusion**. —**immunodiffuse** *vb*.

immunodominance the attribute of a part of an **epitope** that contributes a disproportionately great portion of the binding energy; e.g. the immunodominance shown by a monosaccharide residue in determining the antigenic specificity of a polysaccharide. —**immunodominant** *adj*.

immunoelectrofocusing *or* **immunoisoelectric focusing** a technique that is analogous to **immunoelectrophoresis** but with the preliminary fractionation of one of the reactants being effected by electrofocusing instead of electrophoresis.

immunoelectron microscopy a form of electron microscopy in which structures are stained with specific antibodies labelled with an electron-dense material.

immunoelectropherogram *(sometimes)* the record of an immunoelectrophoresis experiment, either the electrophoretic support itself or a tracing derived therefrom.

immunoelectrophoresis *abbr.:* IE *or* IEP; a technique for separating and identifying soluble antigens. It consists of **zone electrophoresis**, in a reasonably transparent gel, in one direction followed by **immunodiffusion** against a solution of antibodies (antiserum) placed in a trough parallel to the direction of electrophoresis. In suitable conditions each antigen shows up as an arc of precipitation in a characteristic position. —**immunoelectrophoretic**, **immunoelectrophoretical** *adj*.

immunoenzymatic of or pertaining to the particular techniques used in **immunoenzymology**.

immunoenzymology the subspeciality of immunochemistry in which the activity of enzymes coupled to antigens or antibodies is utilized as a molecular amplifier of antigen–antibody reactions.

immunofixation *or* **immunofixation electrophoresis** a variant of **immunoelectrophoresis**, *or* **immunoelectrofocusing**, in which proteins of a single immunological species are anchored to the support by treatment with monospecific antibodies, allowing the other proteins to be washed out of the support. It is useful in the identification of proteins in complex mixtures or of minor proteins obscured in electrophoretic patterns.

immunofluorescence *or* **immunofluorescence microscopy** a technique in which an antigen or antibody is made fluorescent by conjugation to a fluorescent dye and then allowed to react with the complementary antibody or antigen in a tissue section or smear. The location of the antigen or antibody can then be determined by observing the fluorescence by microscopy under ultraviolet light. —**immunofluorescent** *adj.*

immunogelfiltration a variant of **radial single immunodiffusion** in which antigen is first fractionated according to its molecular size by permeation, in a buffer solution, into a thin layer of an appropriate cross-linked dextran gel. The dextran layer is then covered with a layer of agarose gel containing antibody. Rings or spots of antigen–antibody precipitate develop in the agarose gel over the complementary antigens.

immunogen 1 any substance that, when introduced into the body, elicits humoral or cell-mediated immunity, but not immunological tolerance. **2** any substance that is able to stimulate an immune response, as distinct from substances (antigens) that can combine with antibody. **3** any substance that is able to stimulate protective immunity against a pathogen, as distinct from an immune response that is of no intrinsic value to the animal immunized. —**immunogenic** *adj.*

immunogenetics *(functioning as sing.)* the branch of biology that combines immunology and genetics. It includes the use of immunological methods and knowledge in the study of genetics, and the use of genetics in the study of immunological phenomena and substances. —**immunogenetic** *adj.*

immunogenicity *an alternative term for* **antigenicity**.

immunoglobulin *symbol*: Ig; any protein occurring in higher animals as a major component of the immune system. Immunoglobulins are produced by lymphocytes, and virtually all possess specific antibody activity. Each immunoglobulin comprises four polypeptide chains, two identical **heavy chains** and two identical **light chains**, linked together by disulfide bonds; all contain in addition differing amounts of attached oligosaccharide. There are five classes, IgA, IgD, IgE, IgG, and IgM. The primary structures of the heavy chains differ among the various classes, being designated α, δ, ε, γ, and μ, respectively, and there are two types of light chain, κ and λ. IgA and IgM molecules contain multiples of the four-chain unit. The homology units form a characteristic structure, the **immunoglobulin fold**. *See also* **immunoglobulin A, immunoglobulin D, immunoglobulin E, immunoglobulin G, immunoglobulin M.**

immunoglobulin A *symbol*: IgA; an **immunoglobulin** found in human plasma, mostly as a 170 kDa monomer of the basic four-chain immunoglobulin structure or, as a 400 kDa dimer, together with an extra 15 kDa polypeptide, the **J chain**, linked to **secretory piece**. It is the major immunoglobulin of seromucous secretions, in which it protects external body surfaces against attack by microorganisms. It has the structure $(\alpha_2 l_2)_n$ J, where α is the heavy chain, l is the light chain (κ or λ), J is the J chain, and n is 1 or 2. *See also* **IgA-specific serine endopeptidase**.

immunoglobulin amyloidosis the commonest form of systemic **amyloidosis** in humans. The amyloid fibrils consist of λ and κ Ig light chains that are overproduced by a clone of B lymphocytes.

immunoglobulin D *symbol*: IgD; a 185 kDa **immunoglobulin** that is very susceptible to proteolytic degradation and has a very short half-life in plasma. It is found in an integral membrane protein form, together with immunoglobulin M, on the surface of B lymphocytes, where both function as antigen receptors and control lymphocyte activation and suppression. It has the structure $\delta_2 l_2$ where δ is the heavy chain and l is a light chain (κ or λ). *See* **hyper-IgD syndrome**.

immunoglobulin E *symbol*: IgE; a 200 kDa **immunoglobulin** of high carbohydrate content, normally found in very low concentrations in plasma, although its concentration there is elevated in allergic conditions. IgE is synthesized mainly in lymphoid tissue of gut and respiratory tract. It binds strongly to mast cells and contact with antigen leads to degranulation of the mast cells and release of inflammatory mediators. It has the structure $\varepsilon_2 l_2$ where ε is a heavy chain and l is a light chain (κ or λ).

immunoglobulin fold a series of β strands, arranged as two antiparallel β sheets packed tightly against each other, that is highly characteristic of immunoglobulin homology units. In the constant region, the two β sheets consist of four and three β strands respectively. In the variable region there are nine β strands (five and four). This type of structure is found in many other proteins, especially cell surface receptors, and is often referred to as an **Ig-like domain**.

immunoglobulin G *symbol*: IgG; the principal **immunoglobulin** of human plasma and other internal body fluids. It is a 150 kDa protein and is the major immunoglobulin synthesized during the secondary response to an antigen. In the tissue fluids it neutralizes bacterial toxins and binds to microorganisms thereby enhancing their phagocytosis. Complexes of IgG with microorganisms activate **complement** by the classical pathway. IgG binds to macrophages and polymorphonuclear leukocytes and can cross the placenta. It has the structure $\gamma_2 l_2$ where γ is a heavy chain and l is a light chain (κ or λ). There are subclasses of IgG depending on different types of γ chain.

immunoglobulin M *symbol*: IgM; a 900 kDa **immunoglobulin** consisting of a star-shaped polymer of five basic four-chain structures, each heavy chain bearing an extra constant domain. Polymerization is dependent on the presence of **J chain**, which may act to stabilize the sulfhydryl groups in the constant portion of the heavy chains during IgM synthesis. IgM is produced early in the immune response and is mainly found intravascularly. Because of its high antigen-combining valency it is very effective in binding microorganisms and mediating **complement**-dependent cytolysis. It has the structure $(\mu_2 l_2)_5$J where μ is a heavy chain, l is a light chain (κ or λ), and J is a J chain. Virgin B lymphocytes have a transmembrane sequence at the C terminus of μ and the switch is effected by alternative splicing. In **heavy chain disease** the protein has no C and C_H1 regions due to deletion. *See* **hyper-IgM syndrome**.

immunoglobulin superfamily a superfamily of proteins that have similarities of sequence and structure to immunoglobulins. There are three sets, based on immunoglobulin domains: one with variable-like domains, the V set, and two with variants of the constant-like domains, the C1 and C2 sets.

immunogold labelling a method of labelling specific molecules in electron-microscopic thin sections using electron-dense gold particles attached to specific antibody molecules.

immunogram *(sometimes)* the pattern of bands or lines of precipitation formed in a gel by **immunodiffusion**.

immunohistochemistry a form of **histochemistry** in which appropriately labelled antibody preparations are used to detect specific structures in tissues. —**immunohistochemical** *adj.*

immunoliposome a **liposome** bearing a chemically coupled monoclonal antibody.

immunolocalization *or* **immunolocalisation** the use of appropriately labelled antibody (or antigen) preparations to detect the position of specific structures or components that contain or are the complementary antigens (or antibodies).

immunological *or* *(esp. US)* **immunologic** of, concerned with, or pertaining to immunity or immunology. —**immunologically** *adv.*

immunologically competent (of cells, tissues, etc.) qualified for and capable of giving an **immune response**.

immunological paralysis *an alternative name for* **acquired tolerance**.

immunological rejection the destruction, by a specific immune reaction, of foreign cells or tissues transplanted or inoculated into a recipient animal from a donor animal.

immunological response any specific response to an antigen. It includes **cell-mediated immunity, humoral immunity,** and **immunological tolerance**.

immunological surveillance the postulated monitoring of the cells of large, long-lived animals by their immunological systems so that aberrant cells arising from somatic mutations, and so containing new antigens, are destroyed.

immunological tolerance a type of immunological response in which there develops a specific nonreactivity of the lymphoid tissues towards a given antigen, one that in other circumstances is able to induce cell-mediated or humoral immunity. Immunological tolerance may follow contact with the antigen in fetal or early postnatal life or the administration of the antigen to adults (**acquired tolerance**). The reaction towards other, unrelated antigens is unaffected.

|

immunology the study of **immunity** and related phenomena. The science of immunology grew out of the study of resistance to infectious disease, and now encompasses the study of antigens, antibodies, and their interactions both *in vivo* and *in vitro*, and the cellular phenomena of recognition of, and responsiveness to, foreign substances. Related disciplines include **immunobiology** and **immunochemistry**.

immunometry the measurement of amounts of substances by the use of specific antigen–antibody reactions. —**immunometric** *adj.*; **immunometrically** *adv.*

immunomodulation the alteration of an immune response by an agent other than the antigen.

immunomodulator any agent that alters the extent of the immune response to an antigen, by altering the antigenicity of the antigen or by altering in a nonspecific manner the specific reactivity or the nonspecific effector mechanisms of the host. Immunomodulators include adjuvants, immunostimulants, and immunosuppressants.

immunoosmophoresis *an alternative term for* **counter(immuno)electrophoresis**.

immunophenotyping the **typing** of cells with immunological markers such as monoclonal antibodies.

immunophilin one of a conserved group of proteins that bind the immunosuppressant drugs cyclosporin A, FK506, and rapamycin. All have peptidylprolyl isomerase activity and play major roles in protein folding and trafficking within cells. Cyclosporin A binds to a cyclophilin, while FK506 and rapamycin bind to FK506-binding proteins (FKBPs). Immunophilins inhibit S6 kinase and several cyclin-dependent kinases. They also bind to the Hsp90 component of unliganded steroid hormone receptors. *See* **mTor**.

immunoprecipitate **1** the precipitate formed in an antigen–antibody reaction. **2** to precipitate (something) by reaction with a specific antibody or antigen. —**immunoprecipitation** *n.*

immunoradiometric assay *abbr.*: IRMA; an alternative method to **radioimmunoassay** for the measurement of very small amounts of nonradioactive material. An excess amount of a specific antibody labelled with a radioactive isotope is added to the sample containing the substance to be assayed. After equilibration, unreacted antibody is removed and the amount of radioactive material remaining is measured.

immunoreactive capable of reacting with a specific antibody. The term is used especially of substances that react with an antibody directed against a particular peptide hormone (or part of such a hormone) but that do not necessarily exhibit the expected physiological or pharmacological activity. —**immunoreactivity** *n.*

immunosorbent any sorbent used in, or useful for, **immunosorption**. *See also* **immunoadsorbent**.

immunosorption the use of antigens (or antibodies) immobilized on a solid matrix to remove specifically the complementary antibodies (or antigens) from a mixture and their subsequent elution from the solid phase. It is used in, and is useful for, purification of antibodies (or antigens). *See also* **immunoadsorption**.

immunostimulant any agent that nonspecifically enhances the immunologically specific reactivity of an animal to an antigen and also the animal's nonspecific effector mechanisms. —**immunostimulatory** *adj.*

immunosuppressant *or (sometimes)* **immunosuppressive** any agent that causes **immunosuppression**.

immunosuppressed describing an organism that is incapable of showing an immune response to an antigen.

immunosuppression the suppression of immune responses to antigens. This can be achieved by various means, including physical (e.g. X-irradiation), chemical (e.g. antimetabolic drugs), or biological (e.g. antilymphocyte serum). After transplant surgery **azathioprine** is commonly used. This drug has a preferential effect on T-cell-mediated reactions. It is degraded in the body first to mercatopurine and is then converted to the active agent, the ribotide, which inhibits nucleic-acid synthesis. **Methotrexate**, cyclophosphamide, **cyclosporin A**, and **FK506** are also highly effective immunosuppressants, the latter two being used extensively. Infection by HIV-1 eventually leads to a reduction in the population of helper T-cells and macrophages. This immunosuppression of the system is associated with occurrence of **Kaposi's sarcoma** and allows infection by agents such as *Pneumocystis carini*, cytomegalovirus, and tuberculosis. *Compare* **immunodeficiency**. *See also* **suppressor cell**.

immunosuppressive **1** able to cause **immunosuppression**. **2** *an alternative term for* **immunosuppressant**.

immunosympathectomy the destruction of some or all of the neurons of the sympathetic nervous system by injections into a newborn animal of an antiserum to the appropriate **nerve growth factor**.

immunotoxin any of a class of therapeutic drugs consisting of a monoclonal antibody linked to a toxic protein (e.g. ricin, diphtheria toxin).

IMP *abbr. for* **1** inosine monophosphate; i.e. inosine phosphate. **2** ion-moderated partition.

impaired glucose tolerance *abbr.*: IGT; the term recommended to denote a clinical state or condition of humans that is intermediate between the normal one and the overtly diabetic. Characteristically, in a standard oral **glucose-tolerance test** in non-pregnant adults, the glucose concentration in capillary whole blood or venous plasma taken at 2 hours is in the range 8–11 mmol L^{-1} (1.4–2.0 gL^{-1}); the corresponding values for venous whole blood are 7–10 mmol L^{-1} (1.2–1.8 gL^{-1}). In addition, the fasting plasma glucose must be below 8 mmol L^{-1}, and plasma glucose during the test must be greater than 11 mmol L^{-1} at 30, 60, and 90 min.

IMP cyclohydrolase EC 3.5.4.10; *other names*: IMP synthetase; inosinicase; an enzyme of the pathway for *de novo* purine biosynthesis that cyclizes 5-formamido-1-(5′-phosphoribosyl)imidazole-4-carboxamide to IMP with elimination of H_2O. For an example, *see* **phosphoribosylaminoimidazolecarboxamide formyltransferase**.

IMP dehydrogenase EC 1.1.1.205; *other name*: inosine-5′-monophosphate dehydrogenase; an enzyme that catalyses the conversion of inosine 5′-phosphate, NAD^+, and H_2O to xanthosine 5′-phosphate and NADH. The reaction involves the introduction of an oxo group at the purine C-2. Xanthosine 5′-phosphate is then converted to guanosine 5′-phosphate by **GMP synthase** (glutamine hydrolysing). In humans there are two isoenzymes.

impedance *symbol*: Z; the quality determining the amplitude of the current flowing in a circuit for a given applied alternating electric potential. It depends upon the electric resistance, self-inductance, and capacitance of the circuit.

impeller **1** the vaned, rotating disk of a centrifugal pump. **2** any device incorporating an **impeller** (def. 1).

impermeable (of a substance or structure) not permitting the passage through it of a fluid or of a particular solute in a fluid; not permeable. —**impermeability** *n.*

impermeant (of a fluid or a solute in a fluid) incapable of passing through a particular substance or structure.

importin *or* **karyopherin** a heterodimeric protein essential for the import of proteins from the cytoplasm to the nucleus. The α subunit binds the nuclear localization signal of the cargo protein, and the β subunit binds to nucleoporins of the nuclear pore. Entry into the nucleus is mediated by the GTPase Ran, and the α and β subunits subsequently exit the nucleus separately. Importin is also involved in aster formation during cell division.

imprinting *(in genetics)* the phenomenon whereby the expression of genes is determined by the parent who contributed them. For example, the onset of Huntington's disease is earlier if the defective allele is inherited from the father rather than the mother. A maternal allele can be expressed exclusively because the paternal allele is silenced by methylation or other modification, or vice versa. Some 70 imprinted genes have so far been discovered in both mouse and human, and of these roughly half are maternally imprinted and half paternally.

impulse **1** *(in physics)* a vector quantity equal to the integral of the force, F, acting on a body with respect to the time, t, over which the force acts: $\int F dt$. If the force is constant the expression reduces to the product Ft. It also equals the total change of momentum of the body induced by the applied force. **2** *(in physics)* an electrical surge of unidirectional polarity. **3** *(in physiology)* or **nerve impulse** an all-or-none signal propagated along the axons of neurons by which information is transmitted rapidly and precisely through the nervous system. It consists of a sequence of changes in ionic permeability of the neuronal membrane that gives rise to ionic currents and associated changes in transmembrane electrical potential.

impure not pure, lacking purity; (of a substance or material) containing contaminants.

impurity 1 the state or quality of being impure. **2** any or all of the admixed components in an impure substance or material.

impurity quenching chemical quenching due to an **impurity** (def. 2).

In *symbol for* indium.

inappropriate secretion a clinical syndrome in which the secretion of a specified hormone, commonly **antidiuretic hormone**, is abnormally high relative to the state of the normal regulatory mechanisms. It is so termed because it results from an unknown or unmistakably abnormal cause and serves no recognizable homeostatic function.

inborn error of metabolism any genetically determined biochemical variation occurring in humans. Such variations are highly specific and represent many diverse metabolic phenomena, resulting in very varied effects on the viability of the individual. The first such errors were publicized in 1909 by the British physician Archibald Garrod (1857–1936) in his book *Inborn Errors of Metabolism*.

inbred strain *or* **inbred line** a group of organisms obtained by repeated inbreeding over several successive generations, e.g. by self-pollination in plants or by repeated brother × sister matings in animals. Eventually, all individuals will possess identical sets of **autosomes**, and all the chromosome pairs will become homozygous.

INCHI *abbr. for* International chemical identifier.

inclusion *(in cytology)* **1** any discrete body or particle within a cell, especially a passive product of cell activity such as a starch or volutin granule. **2** *an alternative (less common) name for* **inclusion body**.

inclusion body *or* **inclusion 1** any discrete assembly of virions and/or viral particles that may be visualized in virus-infected cells, either in the nucleus or in the cytoplasm. **2** *(in bacteriology)* **a** any one of a number of bodies found within bacterial cells, such as storage granules (e.g. of glycogen, poly-β-hydroxybutyrate, polyphosphate, or sulphur), gas vacuoles, carboxysomes, and magnetosomes. **b** an insoluble complex of protein stored within the cytoplasm of bacterial cells and resulting from the overexpression of a foreign protein.

inclusion cell *abbr.*: I cell; a type of cell that is seen in tissues from patients with mucolipidosis II (**I-cell disease**) and that contains inclusions consisting of very large lysosomes whose contents are heterogeneous, comprising mucopolysaccharide, whorls of membrane, and other material.

inclusion-cell disease *or* **mucolipidosis II** *see* **I-cell disease**.

inclusion complex *or* **inclusion compound** any chemical complex in which one component (the host molecule) forms a crystal lattice containing tunnel- or channel-shaped spaces in which molecular entities of a second species (the guest molecule) are located. There is no bonding between the host and the guest molecules.

incompatibility 1 a criterion for classifying bacterial **plasmids**; two plasmids of the same incompatibility group cannot coexist in one host cell. The molecular basis is that the incompatible plasmids share sites during plasmid **segregation** (def. 3). **2** *(in phylogeny)* an *alternative name for* **incongruence**. **3** the state of being **incompatible**.

incompatibility group a group to which plasmids are said to belong if they cannot coexist in the same bacterial host over several generations without selection.

incompatible 1 *(in immunology)* having antigenic nonidentity between a donor and a recipient, e.g. in blood transfusion or tissue transplantation. **2** *(in therapeutics)* (of two drugs) incapable of being used together or combined; antagonistic.

incomplete antibody 1 any antibody whose activity in an agglutination test can only be demonstrated indirectly, e.g. by an antiglobulin test. **2** *(sometimes)* a univalent antibody fragment produced enzymically, e.g. a **Fab fragment**.

incomplete Freund's adjuvant *see* **Freund's adjuvant**.

incongruence *or* **incompatibility** *(in phylogeny)* difference of characters or trees suggesting different groups of relationships, owing to the presence of **homoplasy**.

incorporation 1 the act or process of including (something) as a part of a whole. **2** *(in biochemistry)* the inclusion of (isotopically labelled) metabolites, by covalent linkage, into biopolymers by living cells or cell-free preparations; often taken as an index of biosynthesis. —**incorporate** *vb*.

incretin a putative insulinogenic principle secreted from the duodenum and jejunum in response to the presence of glucose. Its effects are now generally considered to be attributable to **glucose-dependent insulinotropic peptide**.

incubate 1 to maintain at an appropriate temperature so as to favour growth, development, or continued survival (e.g. of cells); by extension, to maintain under specified conditions in a controlled or artificial environment (e.g. in studies of enzymic or antigen–antibody reactions), especially at a particular ambient temperature (which may be above or below that of the environment). **2** a material or preparation that has been incubated. —**incubation** *n*.

incubation mixture any reaction mixture that is maintained at a controlled temperature and/or in a controlled, artificial environment.

incubation period 1 *(in pathology)* the time interval between the invasion of an organism by a pathogenic virus or microorganism and the overt manifestation of disease. **2** *(in microbiology)* the period of development of a culture of bacteria or other microorganisms.

incubator any closable, heat-insulated cabinet that is maintained at a constant internal temperature (and, sometimes, humidity and/or atmosphere) and that is used for the growth or maintenance of cells, cultures, tissues, or organisms, or for the hatching of eggs.

indel an inserted or deleted position within a DNA or protein sequence **alignment**. The term reflects the difficulty in knowing whether the base or residue represents an insertion in one sequence or a deletion in the other relative to an ancestral sequence.

Inderal *see* propranolol.

index (*pl.* **indices** *or* **indexes**) *(in informatics)* a table of relations (or addresses) between or within data sets that enables rapid file searching. Searches are directed only to specific data subsets, obviating the need to search the data set serially. For example, SRS (Sequence Retrieval System) operates by searching indexed flat-file databases. *See* **information retrieval software**.

index of refraction *see* **refractive index**.

Indian tobacco *see* lobelia.

indican 1 metabolic indican indol-3-yl hydrogensulfate; 3-indoxyl-sulfuric acid; a so-called ethereal sulfate, formed in the gut by bacterial action on tryptophan and excreted in the urine, where its level may be taken as an index of intestinal stagnation. **2** plant indican indol-3-yl-β-D-glucopyranoside; indoxyl-β-D-glucoside; a substance isolated from *Indigofera* spp., members of the Leguminosae, and *Polygonium tinctorium*. The glucoside is hydrolysed during extraction by water or dilute acid, and the indoxyl is liberated spontaneously, oxidizing to the blue dye indigo.

indicated hydrogen *symbol*: H- *(or sometimes H)* (with a prefixed locant); *(in chemical nomenclature)* an atom of hydrogen whose position in the molecular structure of an unsaturated cyclic organic compound or group requires to be explicitly stated in order to make the name of the compound or group specific to a particular structural isomer (or tautomer) if otherwise the name would apply equally to two or more isomers (or tautomers). The compound or group in question commonly consists of or is derived from a condensed ring system with the maximum number of non-cumulative double bonds, although the notation is applied also to certain monocyclic systems. The symbol (with its locant) ordinarily precedes the name or the relevant part of the name, e.g. 1*H*-purin-6-amine or 6-amino-1*H*-purine (alternative semisystematic names for one of the tautomers of adenine); however, where the insertion of a substituent into a cyclic structure has necessitated the addition of a hydrogen atom at another position the symbol is placed in parentheses after the locant for the substituent, e.g. purin-6(1*H*)-one (*a semisystematic name for* one of the tautomers of hypoxanthine).

indicator 1 any substance used in a chemical operation to indicate by a colour change the completion of a reaction or the attainment of a desired state. **2** any substance that by a characteristic colour change indicates the presence of another particular substance. **3** *(in saturation analysis)* the labelled substance whose distribution between the reactants of the system is used to determine the amount of analyte present. **4** an isotope, often a radioactive one, that is used as a tracer.

indicator enzyme *an alternative name for* **marker enzyme**.

indicator organism *or* **indicator species** any species of organism whose presence or population characteristics in a particular habitat are used to provide information about the nature of that habitat, especially regarding its degree of pollution.

indicator yellow *the original name for N-*retinylidene opsin. It was so called because its colour changes between yellow and orange according to the pH of its aqueous solution. It spontaneously hydrolyses into opsin and (all-*trans-*)retinal. *See* **rhodopsin**.

indigo a blue dye prepared from the same plants as plant **indican**. Its major blue component is **indigotin**.

indigotin [$\Delta^{2,2'}$-biindoline]-3,3′-dione; the major blue component of **indigo**.

indirect antagonism a form of **antagonism** (def. 2) arising as a consequence of competition by the antagonist for the binding site of an intermediate that links receptor activation to the observed effect.

INDO-1 1-[2-amino-5-(6-carboxyindol-2-yl)phenoxy]-2-(2-amino-5-methylphenoxy)ethane-*N,N,N′,N′*-tetraacetic acid; a fluorescent calcium chelator, named from the indolyl moiety, that can be used as an indicator of Ca^{2+} concentration; it has a large shift in fluorescence emission, from 480 to 400 nm, on binding Ca^{2+}.

indole 2,3-benzopyrrole; in plants and microorganisms, a precursor in the biosynthesis of tryptophan. In animals it is formed by the degradation of tryptophan by bacteria in the intestinal tract.

indoleacetate *abbr.*: IAA; the anion of indole-3-acetic acid. Formed from L-tryptophan, it is a plant hormone, and the main **auxin** of higher plants.

indoleacetonitrile *abbr.*: IAN; an **auxin**, second in abundance to indoleacetate.

indole-3-glycerol-phosphate synthase *see* **trp genes**.

indole test a test used for the identification of certain microorganisms, e.g. members of the Enterobacteriaceae. It determines the ability of the organism to produce indole from tryptophan.

indolicidin a bactericidal and fungicidal 13-residue peptide amide from the cytoplasmic granules of bovine neutrophils. It contains five tryptophan residues, hence its name. Some therapeutic success has been achieved in animals against the fungus, *Aspergillus fumigatus*, using liposomally encapsulated indolicidin, indicating that neutrophil-derived antimicrobial peptides may have therapeutic value.

indoluria the presence in (human) urine of compounds containing an indole ring.

indomethacin 1-(*p*-chlorobenzoyl)-5-methoxy-2-methylindole-3-acetic acid; an inhibitor of **prostaglandin-endoperoxide synthase**. It is useful as an anti-inflammatory drug.

INDOR *abbr. for* internuclear double resonance spectrometry.

induced enzyme *or* **inducible enzyme** any enzyme that is synthesized in a cell only in very small amounts except when induced by the presence of its substrate or a closely related compound. *Compare* **constitutive enzyme**.

induced-fit theory an explanation, first proposed by D. E. Koshland, of enzyme specificity. It holds that reaction between an enzyme and its substrate occurs only following a change in the enzyme's structure, induced by the substrate. It proposes that: (1) a precise orientation of catalytic groups is required for enzyme action; (2) the substrate may cause a change in the three-dimensional relationship of the amino-acid residues at the active site; and (3) the changes in enzyme structure will bring the catalytic groups into the proper orientation for reaction, whereas a nonsubstrate will not. *Compare* **lock-and-key model**.

inducer (*in genetics*) a molecule, generally small, that acts by binding either to an activator or to a repressor protein to stimulate gene expression.

inducible capable of being induced.

inducible enzyme *an alternative name for* **induced enzyme**.

induction 1 *see* **Jacob–Monod model**. **2** (*in physics*) any physical effect brought about (i.e. induced) in an object by the action of a field without any direct physical contact with the inducer. It may be electrostatic, magnetic, or electromagnetic. *See also* **induced enzyme**. **3** (*in genetics*) the mechanism whereby a bacteriophage in the lysogenic state is prompted to enter the lytic mode of replication. Induction occurs under adverse nutritional conditions and classically by exposure of the lysogen to ultraviolet irradiation.

induction coil a device for producing a rapid succession of large pulses of induced emf in a secondary coil from an intermittent low emf in a primary coil.

induction motor an electric motor in which an alternating current is supplied to the winding of the stator, arranged in such a manner as to produce in effect a rotating magnetic field. This in turn induces electric currents in the winding of the rotor, and interaction between these currents and the magnetic flux exerts a torque on the rotor. Induction motors are useful in situations where sparks caused by moving electric contacts are undesirable.

inductive effect (*in chemistry*) *symbol*: *I*; an experimentally observable effect (on reaction rates, etc.) caused by the transmission of charge through a chain of atoms by electrostatic induction. It is symbolized by +*I* or –*I*, according to whether the group in question is left with a small positive or a small negative charge.

+ine *suffix* designating **1** an organic compound that is either an amine (including any monoamino monocarboxylic acid other than tryptophan), or a nitrogen-containing heterocycle other than one containing four or five atoms in the ring. **2** a halogen. **3** any of a group of certain inorganic binary hydrogen compounds, namely arsine, bismuthine, phosphine, and stibine; also hydrazine.

inert 1 having little or no ability to react chemically or physiologically. **2** having no inherent ability to move or resist being moved. —**inertness** *n.*

inertia 1 (*in physics*) the tendency of an object to resist changes to its state of rest or uniform motion; by analogy extended to other physical qualities that resist change. **2** the state of being inert or inactive. —**inertial** *adj.*

inertial mass *see* **mass**.

infantile Refsum disease *see* **Refsum disease**.

infantile sialic acid storage disease *abbr.*: ISSD; the fulminant form of **Salla disease** and, like it, caused by mutations in sialin.

infect 1 to cause infection in (an organism, wound, etc.), especially with pathogenic microorganisms. **2** to affect or become affected with a communicable disease.

infection 1 the invasion of a cell or organism by other (especially pathogenic) organisms. **2** the act or process of infecting; the state of being infected. **3** any agent or process that infects.

infectious 1 (*of a disease*) caused by pathogenic (micro)organisms. **2** causing or transmitting infection. **3** (*of a disease*) capable of being transmitted to another organism. —**infectiously** *adv.*; **infectiousness** *n.*

infectious mononucleosis *see* **Epstein–Barr virus**.

infectious viral nucleic acid a purified viral nucleic acid that, when it infects a host cell, causes the production of progeny viral particles.

infective 1 capable of causing infection. **2** *an alternative (less common) word for* **infectious.** —**infectively** *adv.*; **infectiveness** *or* **infectivity** *n.*

infinite having no limits in extent, magnitude, space, or time; immeasurably numerous or great. —**infiniteness** *n.*; **infinitely** *adv.*

infinite dilution a hypothetical state (of a solution) in which the solute concentration is considered to be zero and the activity coefficient unity.

infinitely thick *(of a radioactive sample)* having a thickness no less than **infinite thickness** for the particular radionuclide to be measured.

infinitely thin *(of a radioactive sample)* having a thickness no greater than **infinite thinness** for the particular radionuclide to be measured.

infinite thickness the thickness of a sample containing a particular radionuclide, especially a soft beta-emitter, beyond which, because of self-absorption, any increase results in no further increase in recordable disintegrations. Under these conditions the (radio)activity is proportional to the specific (radio)activity of the substance.

infinite thinness the thickness of a sample containing a particular radionuclide at or below which self-absorption is negligible. Under these conditions the (radio)activity is proportional to the amount of the radionuclide present.

inflammable very liable to catch fire and produce flame; flammable.

inflammation the immediate defensive reaction of vertebrate tissue to infection or to injury by chemical or physical agents. The part affected is characterized by pain, heat, redness, swelling, and loss of function; there is local vasodilatation, extravasation of plasma into the intercellular spaces, and accumulation of white blood cells and other macrophages in the injured part. Plasma enzyme systems are important sources of inflammatory mediators. These include the **complement**, **blood coagulation**, fibrinolytic, and **kinin** systems. Also active are the mediators released by mast cells, basophils, and platelets, as well as the eicosanoids generated by many cells at inflammatory sites. —**inflame** *vb.*

influenza virus any one of several orthomyxoviruses that cause respiratory disease in warm-blooded vertebrates. Influenza viruses are negative strand RNA viruses with segmented genomes comprising eight RNA molecules per virion. Influenza viruses are characterized by their antigenic properties. The matrix serotypes are referred to as A, B, and C. Influenza B and C types are only found in the human species whereas type A viruses infect serveral species and some can cross a species barrier. They are conventionally sub-typed according to their haemagglutinin (H) and neuraminidase (N) antigens. A complete designation is, for example, A.Hong Kong/1/68 (H3N2) meaning it is a type A, isolated in Hong Kong, was the first such virus isolated there in 1968, and is H3N2.

informatics 1 the science concerned with the structure and properties of scientific information, its theory, history, methodology, and organization. **2** the application of computational and statistical techniques to the management and analysis of information. For example, bioinformatics relates to the management and analysis of biological information; medical informatics to medical information; neuroinformatics to neurobiological data; and chemoinformatics to chemical entities. —**informatical** *adj.*; **informatician** *n.*

informatin *see* informofer.

information 1 communicated or communicable factual knowledge; data; news. **2 genetic information. 3** *(in computing)* the results obtained by processing raw data according to programmed instructions.

informational 1 of, pertaining to, giving, or carrying information. **2** *(in biochemistry)* describing any biological macromolecule that contains or transmits **genetic information** in the form of specific sequences of nucleotides or amino acids.

information extraction *abbr.*: IE; *(in text mining)* a method for identifying specific query terms (words or phrases), and the relationships between them, within a passage of text, a complete document, or within sets of retrieved documents. By recognizing explicit concepts and relationships, and relating specific parts of a docu-

ment to the subject of interest, IE sheds light on the semantic structure of the text. *Compare* **information retrieval.**

information retrieval *abbr.*: IR; *(in text mining)* a method for identifying from a large set of documents those that relate to user-specified query terms (the method does not identify the query terms within those texts). *Compare* **information extraction.**

information retrieval software software that retrieves information relevant to a user query from unstructured text, such as occurs in the literature or in databases. For example, PubMed is a Web-based information retrieval service for NCBI's MEDLINE bibliographic database; Entrez retrieves information from an integrated set of NCBI databases; SRS (Sequence Retrieval System) is a commercial product that can be customized to search any user-defined **flat-file** databases.

information theory the quantitative theory of the coding and transmission of signals and information.

informofer a specific, 30S particle consisting of the protein **informatin.** Informofers are present in the cell nucleus, tightly packed along molecules of **heterogeneous nuclear RNA.**

informosome a cytoplasmic mRNA-carrying ribonucleoprotein particle of relatively low buoyant density. Informosomes are believed to function in the protection of mRNA from nuclease attack and in the transport of mRNA from the nucleus to the cytoplasm.

infra+ *prefix* denoting below, beneath, coming after, within. *Compare* **sub+**, **supra+**, **ultra+** (def. 1).

infradian describing any biological activity that occurs, or varies cyclically, more often than once every 24 hours. *Compare* **circadian**, **ultradian.**

infranatant 1 lying below. **2** a solid or liquid lying below a **supernatant** liquid. *Compare* **subnatant.**

infrared of, pertaining to, or being **infrared radiation.**

infrared dichroism the **dichroism** of polarized infrared radiation. It is used in studies of polypeptide structure.

infrared radiation *abbr.*: IR *or* i.r.; electromagnetic radiation of wavelengths in the range 730 nm to 1 mm (frequencies 300 GHz to 410 THz). This range lies between those of visible light and microwaves.

infrared spectrophotometer any instrument for **infrared spectrophotometry.**

infrared spectrophotometry the measurement of the absorption or emission of infrared radiation at particular wavelengths or frequencies. *See* **infrared spectrum.**

infrared spectrum any absorption spectrum or emission spectrum of infrared radiation. It includes photons that are absorbed or emitted during vibrational and rotational transitions in a molecule. Infrared spectra are widely used in organic chemistry to identify particular groups of atoms by their characteristic vibrational frequencies.

infuse 1 to introduce gradually, as of a solution into a vein or body cavity. **2** to make an **infusion** (def. 2).

infusion 1 the process of infusing. **2** an extract, e.g. of meat or leaves, obtained by soaking in water.

ingest to take material, e.g. food or liquid, into a cell or organism. —**ingestible** *adj.*; **ingestion** *n.*; **ingestive** *adj.*

inhibin either of the glycoproteins inhibin or **activin** of the transforming growth factor β family and secreted by the gonads. They participate in differentiation and growth of diverse cell types. Inhibin inhibits secretion of **follicle-stimulating hormone** by the pituitary, whereas activin stimulates its secretion. Inhibin is present in seminal plasma and follicular fluid. It consists of two subunits crosslinked by disulfide bridges. Inhibin A is a dimer of α and $β_A$ subunits, and inhibin B is a dimer of α and $β_B$ subunits. The $β_A$ and $β_B$ subunits are also constituents of activin. Their precursors are isoforms derived from the same gene. Inhibin is normally secreted throughout the menstrual cycle and during pregnancy, but not after the menopause. Plasma levels increase in certain ovarian cancers.

inhibit 1 to restrain or hinder. **2** to stop, prevent, or reduce the rate of (some process, e.g. the growth or functioning of an organism or part of an organism, or a chemical or enzymic reaction).

inhibition constant *symbol*: K_i; the equilibrium (dissociation) constant for the reaction: EI $\rightleftharpoons$ E + I. It is given by $K_i = [E][I]/[EI]$, where E is the enzyme and I the inhibitor, the square brackets indicating concentrations.

inhibition index an inverse measure of the potency of an antimetabolite. It is the ratio of the concentration of antimetabolite that is required to inhibit the biological effect of unit concentration of corresponding metabolite.

inhibitor 1 any agent that inhibits. **2** *(in biochemistry)* any substance that inhibits an enzymic reaction. *See also* **competitive inhibition, noncompetitive inhibition, uncompetitive inhibition.**

inhibitor of apoptosis protein *see* IAP.

inhibitor of differentiation *see* Id.

initial velocity or **initial steady-state rate** *symbol*: v_0 *or* v; the reaction velocity in the earliest stage of an enzyme-catalysed reaction. It is given by the tangent at the origin to the curve of reaction velocity as a function of time. In practice it is often measured over the period when the initial substrate concentration has diminished by <5%.

initiation 1 the formation of **initiation complex(es)** at the start of ribosome-mediated translation of mRNA into polypeptide. **2** the first stage in carcinogenesis, in which a normal cell is converted to a precancerous one. **3** the first step in a chain reaction. —**initiate** *vb.*; **initiatory** *adj.*

initiation codon or **initiator codon** the codon AUG (or, sometimes in prokaryotes, GUG) that binds *N*-formylmethionyl-tRNA$_f^{Met}$ in prokaryotes or initiator methionyl-tRNA$_f^{Met}$ in eukaryotes, and serves as a signal for the start of polypeptide synthesis at the ribosome.

initiation complex any of the complexes formed at the start of ribosome-mediated translation of mRNA into polypeptide. They contain mRNA, initiation factors, initiator fMet-tRNA$_f$ or Met-tRNA$_f^{Met}$, one or two ribosomal subunits, and sometimes GTP. *See* **initiation factor.**

initiation factor *abbr.*: IF (in prokaryotes) *or* eIF (in eukaryotes); any soluble protein that functions in the initiation of ribosome-mediated translation of mRNA into polypeptide. *Escherichia coli* has three principal initiation factors, each consisting of a single polypeptide chain. **IF-1** is a basic protein (9 kDa) that is a stimulator of IF-2 and IF-3. There are two forms of **IF-2**: IF-2a, which is acidic, unstable, and of 115 kDa; and IF-2b, formed by limited proteolysis of IF-2a. It is a GTP-binding protein, protects formylmethionyl-tRNA$_f^{Met}$ from spontaneous hydrolysis, and promotes its binding to the 30S ribosomal subunits; it is also involved, with hydrolysis of GTP, in the formation of the 70S ribosomal complex. By using alternative initiation codons in the same reading frame, the eIF-2b gene is translated into three isozymes, α, β, and β', having different N termini. **IF-3** is a basic, stable protein of 22 kDa. It binds to the 30S ribosomal subunit and shifts the equilibrium between 70S ribosomes and their subunits in favour of the free subunits, thus enhancing the availability of 30S subunits. It is also a translational repressor of its own gene. IF-1, IF-2, and IF-3 bind cooperatively to a 30S ribosomal subunit to form a 30S•IF-1•IF-2•IF-3 particle in which the three IFs are bound contiguously near the 3' end of 16S RNA in a 30S ribosomal subunit. This particle subsequently binds mRNA at a site that includes the initiation codon AUG (or GUG), initiator fMet-tRNA$_f^{Met}$, and GTP with concomitant release of IF-3 to form a 30S initiation complex, 30S•IF-1•IF-2•mRNA•GTP•fMet-tRNA$_f^{Met}$. Joining of a 50S ribosomal subunit with concomitant release of IF-1, IF-2, GDP, and P$_i$ forms a 70S initiation complex, mRNA•70S• fMet-tRNA$_f^{Met}$.

Eukaryotes have more initiation factors. Many have been characterized in lysates of rabbit reticulocytes. They include: eIF-1, a monomer of 15 kDa; eIF-2, a heterotrimer of 150 kDa; eIF-3, a nonamer of 700 kDa; eIF-4A, a monomer of 49 kDa; eIF-4B, a monomer of 80 kDa; eIF-4C, a monomer of 17.5 kDa; eIF-4D, a monomer of 16.5 kDa; eIF-5, a monomer of 190 kDa; Co.eIF-2, a monomer of 22 kDa; and cap-binding protein, a monomer of 24 kDa. eIF-2, initiator Met-tRNA$_f^{Met}$, and GTP form a ternary complex; eIF-3 promotes ribosomal dissociation and stabilizes binding of ternary complex to a 40S ribosomal subunit to form the complex 40S•eIF-2•eIF-3•Met-tRNA$_f^{Met}$•GTP. Then eIF-1, eIF-4A, eIF-4B, eIF-4C, and cap-binding protein promote binding of mRNA and ATP-dependent formation of a **40S initiation complex,** 40S•eIF-2•eIF-3•Met-tRNA$_f^{Met}$•mRNA. Factor eIF-5 is required for formation of an **80S initiation complex,** 80S•Met-tRNA$_f^{Met}$• mRNA, from a 40S complex, together with hydrolysis of the bound GTP and release of initiation factors. Co.eIF-2 stimulates ternary complex formation.

initiation signal 1 any of the purine-rich regions of mRNA on the 5' side of the initiation codon that serve to distinguish the latter from other similar codons. **2** any of the promoter regions of DNA on the 5' side of the start of transcription to form mRNA.

initiation switch *see* ISWI.

initiator 1 a gene product that interacts specifically with the replicator site on a replicon resulting in the initiation of a new round of DNA replication. **2** a starting point for either nucleic-acid transcription or translation. **3** a carcinogen that initiates the first stage of carcinogenesis. **4** any compound that, when added to monomer, initiates a polymerization reaction.

initiator codon *see* **initiation codon.**

initiator transfer RNA *abbr.*: tRNA$_f^{Met}$; any of the tRNA molecules that, when aminoacylated, specifically binds to an initiation codon (usually AUG) in mRNA at the start of ribosome-mediated translation of mRNA into polypeptide. In prokaryotes it is *N*-formylmethionyl-tRNA$_f^{Met}$, and in eukaryotes it is methionyl-tRNA$_f^{Met}$; neither is able to promote the insertion of methionine residues at internal sites.

INK4 *or* **CDI with 4 ankyrin repeats** any of four structurally related proteins (i.e. p15, p16, p18, and p19), each of which contains four **ankyrin** repeats. Each is a specific inhibitor of Cdk4 and Cdk6. Somatic inactivating mutations in the gene at 9p21 for CDIp16 occur in sporadic melanomas, whereas germline mutations increase the risk of melanoma and possibly other neoplasms.

innate being an essential characteristic of an organism or thing; inborn.

innate immunity an immune response in both vertebrates and invertebrates to a **pathogen** that involves the pre-existing defences of the body – the **innate immune system** – such as barriers formed by skin and mucosa, antimicrobial molecules and **phagocytes.** Such a response is not specific for the pathogen. The agents are either small proteins or peptides many of which have been detected in insects, which lack the immune system of higher animals. Examples are **cecropin** and **attracin** of the moth *Cecropia*, which kill *E.coli*. **Drosocin** kills fungi present in the fruit fly. Mammals make similar peptides called **defensins.**

innervation the nerve supply to an organ or part.

Ino *symbol for* a residue of the ribonucleoside inosine (alternative to I).

inoculate 1 to introduce cells, microorganisms, or viruses into or onto a culture medium. **2** to introduce causative agents of disease (alive, dead, or attenuated) into an animal or person for the purpose of immunization. —**inoculation, inoculator** *n.*; **inoculative** *adj.*

inoculum (*pl.* **inocula**) a quantity or suspension of cells, microorganisms, or viruses used to start a new culture or to infect another culture; something used to inoculate.

Ino-3',5'-P *symbol for* inosine 3',5'-phosphate.

Ino5'P *symbol for* inosine 5'-phosphate.

Ino5'PP *symbol for* inosine 5'-diphosphate (alternative to ppI).

Ino5'PPP *symbol for* inosine 5'-triphosphate (alternative to pppI).

inorganic not having the characteristics of living organisms; not organic.

inorganic biochemistry the branch of biochemistry concerned with the role of inorganic substances in biochemical systems. *Compare* **bioinorganic chemistry.**

inorganic chemistry the branch of chemistry dealing with the elements and all their compounds other than ones containing carbon, except for carbon itself and some simple compounds of carbon, e.g. oxides, sulfides, and cyanides.

inorganic diphosphate *symbol*: PP$_i$; any of the anions formed from diphosphoric acid, $(HO)_2PO-O-PO(OH)_2$, or any mixture of the acid and its anions. The older term **inorganic pyrophosphate** is less preferable.

inorganic phosphate or **inorganic orthophosphate** *symbol*: P$_i$; any of the anions formed from [ortho]phosphoric acid, PO(OH)$_3$, or any mixture of the acid and its anions.

inorganic pyrophosphatase an enzyme, EC 3.6.1.1, often less precisely termed pyrophosphatase; *systematic name*: pyrophosphate phosphohydrolase. It catalyses the hydrolysis of pyrophosphate to two molecules of orthophosphate.

inorganic pyrophosphate *a less preferred term for* **inorganic diphosphate**.

iNOS *see* **nitric-oxide synthase**.

inosinate 1 either the mono- and/or the di-anion of **inosinic acid**. **2** any mixture of the acid and its anions. **3** any salt or ester of inosinic acid.

inosine *symbol*: I *or* Ino; hypoxanthosine; hypoxanthine riboside; 9-β-D-ribofuranosylhypoxanthine; a nucleoside found free, especially in meat, sugar beet, and yeast, and in the anticodons of some transfer RNAs. It may be formed by the action of adenosine deaminase (EC 3.5.4.4) in the first step of the breakdown of adenosine to uric acid, and it occurs as an intermediate in the recycling of adenosine and inosine 5'-phosphate by **salvage pathways**.

inosine 3′,5′-(cyclic)phosphate *or* **inosine 3′,5′-cyclophosphate** *see* **inosine 3′,5′-phosphate**.

inosine 5′-diphosphate *symbol*: Ino5′PP *or* ppI; *recommended name for* inosine diphosphate (*abbr.*: IDP), 5′-diphosphoinosine, 5′-inosinyl phosphate, inosine 5′-(trihydrogen diphosphate). It is an intermediate in the formation of inosine 5′-triphosphate from inosine 5′-phosphate.

inosine monophosphate *abbr.*: IMP; *an alternative name for* any **inosine phosphate**, but particularly for **inosine 5′-phosphate**, especially when its distinction from inosine (5′-)diphosphate and inosine (5′-)triphosphate requires emphasis.

inosine-5′-monophosphate dehydrogenase *see* **IMP dehydrogenase**.

inosine phosphate inosine monophosphate (*abbr.*: IMP); any phosphoric monoester or diester of inosine. Of the three possible monoesters – the 2′-phosphate, the 3′-phosphate, and the 5′-phosphate – only **inosine 5′-phosphate** is of any importance (the locant being omitted if no ambiguity may arise). The only diester of significance is **inosine 3′,5′-phosphate**.

inosine 3′,5′-phosphate *symbol*: Ino-3′,5′-P; *the recommended name for* cyclic inosine 3′,5′-monophosphate (*abbr.*: 3′,5′-cyclic IMP *or* cyclic IMP *or* cIMP), inosine 3′,5′-cyclophosphate, inosine 3′,5′-(cyclic)phosphate; a monophosphoric diester of inosine. It is a cyclic nucleotide similar in structure and metabolism to **adenosine 3′,5′-phosphate** (i.e. cyclic AMP) but only rarely encountered.

inosine 5′-phosphate *or* 5′-inosinic acid *or* 5′-phosphoinosine *or* 5′-*O*-phosphonoinosine *symbol*: Ino5′P; *alternative recommended names for* inosine monophosphate (*abbr.*: IMP), inosine 5′-(dihydrogen phosphate), hypoxanthine ribonucleotide. (The locant is commonly omitted if there is no ambiguity as to the position of phosphorylation.) It is the initial biosynthetic product containing an intact purine nucleus, its immediate precursor being 4-carbamoyl-5-formamidoimidazole riboside 5′-phosphate. Hence IMP is a key intermediate in the biosynthesis of all other purine nucleotides and in the formation of purine nucleosides and free purines. In addition, resynthesis of IMP from hypoxanthine can be effected via a **salvage pathway** by the action of hypoxanthine phosphoribosyltransferase (EC 2.4.2.8) (*see* **phosphoribosyltransferase**).

inosine 5′-triphosphate *symbol*: Ino5′PPP *or* pppI; *the recommended name for* inosine triphosphate (*abbr.*: ITP), 5′-triphosphoinosine, 5′-inosinyl diphosphate, inosine 5′-(tetrahydrogen triphosphate). ITP can substitute for adenosine 5′-triphosphate or guanosine 5′-triphosphate in certain enzymic reactions. It is formed from inosine 5′-diphosphate by transfer of the terminal phosphoric residue from adenosine 5′-triphosphate through the action of nucleoside-diphosphate kinase (EC 2.7.4.6).

inosinic acid *the trivial name for* any phosphoric monoester of inosine. The position of the phosphoric residue on the ribose moiety of a given ester may be specified by a prefixed locant (*see* **inosine phosphate**). However, 5′-inosinic acid is the ester commonly denoted, its locant usually being omitted if no ambiguity arise. 5′-Inosinic acid is also an alternative recommended name for **inosine 5′-phosphate**.

inosinyl the inosine [mono]phospho group; the acyl group derived from **inosinic acid**.

inositol 1 *symbol*: Ins; *the common name for* the cyclitol *myo*-inositol (*formerly called* meso-inositol *or* i-inositol *or* bios I *or* muscle sugar); 1,2,3,5/4,6-cyclohexanehexol; a growth factor for some animals and microorganisms, widely distributed in nature. Inositol occurs free, as various phosphates, or as a constituent of certain phospholipids (*see* **phosphatidylinositol**); in plants it also occurs as its hexaphosphate phytate. Note: whereas *myo*-inositol itself is achiral, its phosphorylated derivatives are chiral and the numbering of their carbon atoms can be confusing. For a proposed solution of the problem, *see* **Agranoff's turtle**. **2** *generic name for* any member of a subclass of cyclitols comprising the nine possible 1,2,3,4,5,6-cycloxanehexols, of which inositol (def. 1) is the most prevalent and most important.

inositol-1,3-bisphosphate 3-phosphatase EC 3.1.3.65; an enzyme that hydrolyses D-*myo*-inositol 1,3-bisphosphate to D-*myo*-inositol 1-monophosphate and orthophosphate.

inositol-1,4-bisphosphate 1-phosphatase EC 3.1.3.57; *systematic name*: D-*myo*-inositol-1,4-bisphosphate 1-phosphohydrolase; *other name*: inositol polyphosphate 1-phosphatase; an enzyme that hydrolyses D-*myo*-inositol 1,4-bisphosphate to D-*myo*-inositol 4-phosphate and orthophosphate.

inositol glycan any compound that consists of inositol linked to a glycan. The best known is the one that forms part of the membrane anchor of PIG-tailed proteins. It is rapidly released by insulin-sensitive cells *in vitro* in response to insulin and by the action of phosphoinositol-specific phospholipase C. Its physiological role is unclear.

inositol phosphate any phosphoric ester of an inositol, but normally referring to the esters of *myo*-inositol (*abbr.*: Ins). Of particular interest are those formed by the hydrolysis of membrane phosphoinositides following activation of phosphoinositide-specific phospholipase C (*see* **phospholipase**). The generalized abbreviation for phosphorylated *myo*-inositols follows the convention Ins(numbers of the carbons bearing phosphorylated hydroxyls)P_x, where P_x indicates the number of phosphates. They include inositol 1,4,5-trisphosphate (*abbr.*: Ins(1,4,5)P_3), formed by phospholipase C action on phosphatidylinositol 4,5-bisphosphate. Ins(1,4,5)P_3 is important in the **phosphatidylinositol cycle**. It can be phosphorylated to Ins(1,3,4,5)P_4, which is also physiologically active in calcium metabolism. Other inositol phosphates are intermediates in the degradation of Ins(1,4,5)P_3 and Ins(1,3,4,5)P_4. Ins(1,4,5)P_3 is hydrolysed to Ins(1,4)P_2, which is hydrolysed to Ins(4)P and thence to Ins. Ins(1,3,4,5)P_4 is hydrolysed to Ins(1,3,4)P_3, which is hydrolysed to either Ins(1,3)P_2 or Ins(3,4)P_2, these being hydrolysed to Ins(1)P or Ins(3)P respectively and thence to Ins.

inositol-1,3,4,5-tetrakisphosphate 3-phosphatase EC 3.1.3.62; an enzyme that hydrolyses D-*myo*-inositol 1,3,4,5-tetrakisphosphate to D-*myo*-inositol 1,4,5-trisphosphate and orthophosphate.

inositol-1,4,5-trisphosphate 1-phosphatase EC 3.1.3.61; an enzyme that hydrolyses D-*myo*-inositol 1,4,5-trisphosphate to D-*myo*-inositol 4,5-bisphosphate and orthophosphate.

inositol-1,4,5-trisphosphate 5-phosphatase EC 3.1.3.56; *systematic name*: D-*myo*-inositol-1,4,5-trisphosphate 5-phosphohydrolase; *other names*: inositol trisphosphate phosphomonoesterase; inositol polyphosphate 5-phosphatase; 5PTase; an enzyme that hydrolyses D-*myo*-inositol 1,4,5-trisphosphate to D-*myo*-inositol 1,4-bisphosphate and orthophosphate. The enzyme has two types: type I (but not type II) hydrolyses inositol 1,3,4,5-tetrakisphosphate to inositol 1,3,4-trisphosphate; neither enzyme acts on the latter or on inositol 1,4-bisphosphate.

inositol trisphosphate receptor *see* **IP₃ receptor**.

inosyl any chemical group formed by the loss of a 2'-, 3'-, or 5'-hydroxyl group from the ribose moiety of **inosine**.

inotropic influencing contractility of muscles, especially cardiac muscle.

Ins *symbol for* a residue of the cyclitol 1-D-*myo*-inositol (*see* **inositol** (def. 1)).

Inscuteable a protein of *Drosophila* that is involved in differentiation of a neuroblast into a ganglion mother cell. It is present in the apical cortical crescent and is essential for correct segregation of the basal cortical crescent from the former into the latter cell.

INSD *abbr. for* International Nucleotide Sequence Database; *see* **composite database**.

insect any small air-breathing arthropod of the class Insecta, typically having a segmented body with an external, chitinous covering, three pairs of legs, and in most groups two pairs of wings.

insecticide any substance or agent that kills insects. —**insecticidal** *adj*.

(insect) moulting hormone (*or US*) **(insect) molting hormone** a common name for **ecdysterone**.

insectorubin any red or red-brown eye pigment of insects derived from oxidation products of tryptophan. *See also* **ommin**.

insectoverdin any green pigment of insects formed of mixtures of carotenoids and biliverdin.

insert *(in genetics)* a fragment of DNA that has been linked into a cloning vector. The term can also be used as a modifier as, e.g., in **insert DNA**.

insertional inactivation the inactivation of a gene in a plasmid or other cloning vector by insertion of foreign DNA at an endonuclease site. If the gene is *lacZ* its inactivation can be detected by the inability of bacterial colonies containing the recombinant plasmid to hydrolyse **X-gal**. Such colonies remain white whereas plasmids religated without an insert give blue colonies.

insertional mutagenesis the induction of a mutation by the insertion of DNA into a gene using recombinant DNA techniques. One approach, known as restriction enzyme-mediated integration (REMI), utilizes a plasmid that is cut with a restriction enzyme. Then the cut plasmid, together with the restriction enzyme, is introduced into cells (e.g. *Dictyostelium*) by electroporation. The restriction enzyme cuts the genomic DNA at sites into which the plasmid can integrate through its **sticky ends**. If the plasmid contains a suitable marker, transformants can be isolated. The gene affected can be identified because it is effectively tagged with the plasmid. If the plasmid also contains an additional marker conferring drug resistance, the gene from the transformant can be readily cloned.

insertion mutation any mutation caused by insertion of an extranumerary nucleotide residue between two successive nucleotide residues in a DNA molecule.

insertion sequence *abbr.*: IS; *see* **signal sequence, transposon**.

insertion vector *(in recombinant DNA technology)* any **vector** (def. 3) that has been cut with a restriction enzyme having a single target site in the centre of the vector, so that two 'arms' are generated between which foreign DNA may be inserted.

insertosome an 800–1400 bp stretch of DNA that can insert itself randomly into the chromosome of *Escherichia coli*, thereby causing polar mutations analogous to those caused by phage Mu-1.

inside-out vesicle any vesicle formed from disrupted cell or other membranes that has sealed up so that the face of the membrane that was originally on the outside is now on the inside.

in silico *Latin (of any biological process, reaction, or experiment)* made to occur by means of a computer (e.g. a simulation in the e-cell). Literally, it means 'in silicon'. *Compare* **in vitro**, **in vivo**.

in silico screening *an alternative name for* **combinatorial chemistry**.

in situ *Latin* in the normal, natural, original, or appropriate position.

in situ hybridization *or* **hybridization in situ** a technique that uses labelled nucleic acid probes to reveal the location of genes on chromosomes, or of specific mRNA molecules within tissue sections, small organisms, or embryos. *See* **fluorescence in situ hybridization**, **hybridization analysis**.

insolubilize *or* **insolubilise** to render **insoluble** (def. 1), especially by coupling covalently to an insoluble matrix. —**insolubilization** *or* **insolubilisation** *n*.

insoluble 1 incapable of being dissolved (in water unless otherwise specified); of extremely low solubility. **2** incapable of being solved; having no solution.

Ins(3,4)P_2 *symbol for* 1D-*myo*-inositol 3,4-bisphosphate.

Ins(1,3,4)P_3 *symbol for* 1D-*myo*-inositol 1,3,4-trisphosphate.

Ins(1,4,5)P_3 *symbol for* 1D-*myo*-inositol 1,4,5-trisphosphate.

Ins(1,3,4,5)P_4 *symbol for* 1D-*myo*-inositol 1,3,4,5-tetrakisphosphate.

instructive theory of immunity a theory ascribing to an antigen the function in immunity of instructing or directing uncommitted cells to form new antibodies that reflect the nature of the antigen itself. *Compare* **selective theory of immunity**.

insulate to prevent, or reduce, the passage of electricity, heat, or sound to or from (something) by interposition of a (relevant) nonconducting material.

insulaxin a synthetic porcine insulin that instead of the A chain Ile[10], and B chain Ser[9], His[10], and Glu[13] contains, respectively, Gly, Arg, Glu, and Arg, as in porcine **relaxin**. Unlike the native insulin, it binds to the porcine relaxin receptor.

insulin *or (formerly)* **insuline** the hypoglycemic, antidiabetic, and anabolic polypeptide hormone of the endocrine portion of the mammalian pancreas or of homologous organs of lower vertebrates. In mammals, it is elaborated in the B cells of the islets of Langerhans and stored within them in the B granules prior to release by exocytosis. Insulin is a slightly acidic protein that readily aggregates to form dimers, hexamers, and higher multimers. It is very poorly soluble in aqueous media around its isoelectric point (pH ≈ 5.4), especially in the presence of certain divalent metal ions, typically Zn^{2+}, with which it can form crystals, usually containing two or four metal ions per insulin hexamer, depending on conditions. It does, however, dissolve without denaturation in organic solvents in an acid medium, a property that can be exploited in its isolation. The monomer of insulin is one of the smallest known proteins, with M_r values such as 5733 (bovine), 5777 (porcine), or 5807 (human). Bovine insulin was the first protein to have its primary sequence determined, by British biochemist Frederick Sanger (1918–), in 1955. The sequences of a large number of insulins are now known, all of which consist of two polypeptide chains, commonly termed the **A chain** and the **B chain**, with three disulfide bridges formed between half-cystine residues; two of these bridges are interchain, forming A7–B7 and A19–B20 links, and the third is intrachain, linking positions A6 and A11 to form a heterodetic ring in the A chain. The numbers of amino-acid residues vary slightly with the species of origin, commonly being 21 and 30 in the A and B chains, respectively; six amide groups are present per molecule of the monomer if insulin is carefully isolated. The three-dimensional structure is known: chains A and B have two motifs each. Insulin is structurally related to **relaxin** and **insulin-like growth factor**. Insulin is biosynthesized by post-translational modification of its single-chain precursor **preproinsulin**. In preproinsulin, residues 1–24 are the signal sequence. When this is removed, proinsulin is formed. In the Golgi apparatus, prohormone convertases PC2 and PC3 cleave proinsulin on the carboxyl side of a pair of basic amino-acid residues (Arg^{55}-Arg^{56} and Lys^{88}-Arg^{89}). This excises the C peptide (residues 57–87), leaving the A and B chains (two Arg residues (55 and 56) must first be cleaved from the B chain by carboxypeptidase H for the mature B chain to be formed). Residues 55–87 are known as the connecting peptide. Human insulin consists of 110 amino acids; the A chain is derived from residues 90–110, and the B chain from residues 25–54.

Insulin stimulates glucose uptake by muscle and adipose tissue, and promotes glycogenesis, lipogenesis, and synthesis of protein and nucleic acid. Deficiencies in the secretion or action of insulin in humans result in the overproduction and increased urinary excretion of glucose (*see* **glucose-tolerance test, diabetes mellitus**); continuing therapeutic administration of insulin is necessary in type I (insulin-dependent) diabetes mellitus. Commercial insulin for therapeutic use is extracted mostly from bovine or porcine pancreas; human insulin is now also being made by recombinant DNA technology. The international unit for insulin is defined as the activity contained in 0.041 67 mg of the Fourth International Standard for Insulin, Bovine and Porcine, for Bioassay (which is defined as crystalline zinc insulin derived from a mixture of 52% bovine insulin and 48% porcine insulin).

insulinase EC 3.4.99.45; *other names*: insulin-degrading enzyme; insulin protease; insulysin. An enzyme that progressively cleaves a number of peptide bonds in insulin, but only Arg^{17}-|-Arg^{18} in glucagon and some other unrelated hormones and cytokines. It is a zinc thioendopeptidase of the inverzincin family and is widely expressed in mammalian tissues (including some that are not insulin-sensitive). It shows little sequence homology with classical metallo or cysteine proteases.

insulin-dependent diabetes mellitus *abbr.*: IDDM; *see* **diabetes mellitus**.

insulin-like growth factor *abbr.*: IGF; any member of a group of polypeptides that are structurally homologous to **insulin** and share many of its biological activities but are immunologically distinct from it; they are mitogens of the insulin/relaxin family. They include substances now named insulin-like growth factor-I (IGF-I), which is a monomer of 70 amino acids, and insulin-like growth factor-II (IGF-II), which is a monomer of 67 amino acids; together these were formerly known as **nonsuppressible insulin-like activity** (NSILA). The insulin-like growth factors also include **somatomedin** C, which is IGF-I, and possibly also somatomedins A_1 and A_2, and **multiplication-stimulating activity**. They appear to be more potent than insulin as growth-promoting factors but less potent in stimulating glucose utilization. Brain IGF is a truncated form of IGF-I, in which the N-terminal tripeptide sequence is lacking. *See also* **insulin-related growth factor, somatomedin**.

insulin-like growth factor binding protein *abbr.*: IGFBP; any of several proteins that bind insulin-like growth factors (IGFs), and are found in serum, amniotic, and other body fluids, and in conditioned media from a variety of cell types. They are believed to prolong the half-life, and regulate the endocrine effects, of IGFs. Six IGFBPs (IGFBP-1 to -6) have been identified in humans, their complementary DNA having been cloned and sequenced. IGFBP-1 was originally identified as a placental protein, placental protein-12 (PP-12) and as α-1 pregnancy-associated globulin; it is a non-glycosylated protein, M_r 25 000; IGFBP-2 has been identified in human and rat central nervous systems; it is a non-glycosylated protein, M_r 31 000. IGFBP-1 and IGFBP-2 both have the RGD motif, a recognition site for **integrins**. IGFBP-3 is the principal IGFBP in human serum; it is a glycoprotein, M_r 28 500, and circulates as a complex with IGF peptide (γ subunit). For IGFBP-5, *see* **skeletal growth factor**.

insulin-like growth factor receptor an integral membrane protein that binds and mediates the effects of **insulin-like growth factor** (IGF). **IGF-I receptor** is a glycoprotein of composition $\alpha_2\beta_2$, the subunits being linked by disulfide bonds. It has an extracellular ligand-binding domain and a cytoplasmic protein tyrosine kinase domain. It is widely expressed in embryonic but not in adult tissues. **IGF-II receptor** is the cation-independent **mannose-6-phosphate receptor**. It mediates endocytosis and degradation of excess IGFII. The insulin receptor is the main receptor for IGFII.

insulinogenesis the formation of **insulin**.

insulinogenic resulting from the administration of **insulin**.

insulinoma any neoplasm originating in the B cells of the pancreatic islets (or sometimes from nonislet tissue), that autonomously secretes insulin and/or biosynthetic precursors of insulin. Such lesions are frequently recognized by their metabolic consequences.

insulinotropic stimulating **insulin** secretion.

insulinotropic effect *see* **gastric inhibitory peptide**.

insulinotropin a peptide containing residues 7–37 of glucagon-like peptide I, present with glucagon in preproglucagon. It is a potent stimulator of insulin release from the perfused rat pancreas. *See* **glucagon-like insulinotropic peptide**.

insulin receptor a type I membrane protein with tyrosine kinase (EC 2.7.1.112) activity. The protein is derived from a precursor; after being transported from the endoplasmic reticulum to the Golgi apparatus, the single glycosylated precursor is further glycosylated and then cleaved to α and β chains. In the plasma membrane the receptor is a disulfide-bonded heterotetramer, $\alpha_2\beta_2$; the α chains are entirely extracellular, while each β chain has one transmembrane domain. The ligand binds to the α-subunit extracellular domain and the kinase is associated with the β-subunit intracellular domain; it is partly regulated by autophosphorylation. Associated diseases include various forms of insulin resistance, including type 2 **diabetes mellitus** and **leprechaunism**. The human precursor protein consists of 1382 amino acids; amino acids 28–762 form the α subunit, while 763–1382 form the β subunit.

insulin receptor substrate *abbr.*: IRS; any of four proteins (IRS-1, -2, -3, and -4) that bind the activated tyrosine kinase domain of the insulin receptor and become phosphorylated by it. They share certain structural features: in the N-terminal region, a **plekstrin** homology region (≈100 residues that bind phosphatidylinositides) and a phosphotyrosine binding region (also ≈100 residues); and in the C-terminal region multiple tyrosine phosphorylation sites. All contain binding sites for SH2 domains in other proteins. Two polymorphisms in IRS-1 (1242 residues) are associated with a subset of diabetes mellitus type 2 patients. Mouse knockouts of IRS-1 and IRS-2 develop insulin resistance.

insulin-related growth factor (*sometimes*) any vertebrate and invertebrate protein or polypeptide hormone or growth factor (including **relaxin**, **nerve growth factor**, **insulin-like growth factors**, silkworm prothoracicotropic hormone, and possibly others) that shares with insulin certain features of primary and three-dimensional structure and may also display some similarities to insulin as a regulator of growth and/or metabolism.

insulin resistance a decreased ability of insulin to produce its biological effects across a wide range of concentrations. It is usually associated with hyperinsulinemia and is promoted by obesity, a sedentary lifestyle, hyperlipidemia, hypertension, and increasing age. In the majority of cases its molecular basis is largely unknown. It is usually reversed by exercise, weight loss, and diet. *See* **hyperinsulinemia syndrome**.

insulin-tolerance test a sensitive and accurate test of the integrity of the hypothalamic–pituitary–adrenal axis and its ability to respond to stress, and also of somatotropin responsiveness. The test is performed after an overnight fast with measurement of plasma cortisol, glucose, and somatotropin levels before and periodically after intravenous administration of insulin. The hypoglycemia induced by insulin elicits a potent stress response via the central nervous system, stimulating the release of **corticotropin releasing hormone** and consequently, of **corticotropin**.

Int the product of phage gene *int*, involved in the integration of phage DNA into host DNA during transition from the lytic to the lysogenic states. *Compare* **Int-2**. *See also* **retroregulation**.

Int-2 an oncogene first identified following the integration of mouse mammary tumour virus into the mouse DNA. Its product is a **fibroblast growth factor** (FGF3). *Compare* **Int**.

integer any rational whole number.

integral 1 of, or being an essential part of a whole; entire, complete. **2** (*in mathematics*) of, or denoted by, an **integer**; of or involving an **integral** (def. 3). **3** (*in mathematics*) the sum of a large number of infinitesimally small quantities, denoted ∫, summed either between specified limits (**definite integral**) or within indeterminate limits (**indefinite integral**).

integral (cell membrane) protein *or* **intrinsic protein** any protein that can only be dissociated from a cell membrane by drastic treatment with a reagent such as detergent, bile salt, protein denaturant, or organic solvent. Such proteins have one or more domains that traverse the cell membrane. *Compare* **peripheral (cell membrane) protein**.

integrase a protein of lambdoid phages that catalyses integration of phage DNA during establishment, probably by forming a transient DNA–protein link. The same enzyme is one subunit of the **excisase** activity that reverses this; the other subunit is the Xis protein, known as **excisionase**.

integrate 1 to combine or be combined into a whole; to complete (something imperfect) by the addition of parts. **2** *(in mathematics)* to find the **integral** (def. 3) of (quantities, expression, etc.). —**integrable, integrative** *adj.*; **integrability** *n.*

integrated rate equation a general expression for the interpretation of a single enzyme-catalysed reaction. The expression for the velocity of a reaction at time *t* is:

$$v = d[P]/dt = V([S]_0 - [P]_t)/\{K_m + ([S]_0 - [P]_t)\}.$$

This can be integrated with respect to time to give:

$$Vt = [P]_t + K_m \ln\{[S]_0/([S]_0 - [P]_t)\},$$

or

$$\ln\{[S]_0/([S]_0 - [P]_t)\}/t = V/K_m - [P]_t/K_m t,$$

where [S] and [P] are the concentrations of substrate and product, *v* is the velocity of the reaction, *V* is the limiting rate, and K_m is the Michaelis constant. The subscripts indicate the concentrations at zero time and at time *t*. A plot of $\ln\{[S]_0/([S]_0 - [P]_t)\}/t$ against $[P]_t/t$ will have a slope of $-1/K_m$ and an intercept of V/K_m.

integration 1 the act or process of uniting or adding parts to make a whole. **2** the process of determining and/or indicating the mean value or total sum of (some physical quantity, area, etc.). **3** *(in mathematics)* the process of determining the **integral** (def. 3) of a function or a variable. *Compare* **differentiation**. **4** *(in genetics)* the act or process of forming a **cointegrate**.

integrative genomics the study of complex interactions between genes, organisms, and the environment.

integrator any device (mechanical or electronic) or any person that determines the value of an **integral** (def. 3).

integrin any member of the large family of transmembrane proteins that act as receptors for **cell-adhesion molecules**. Integrins are heterodimeric molecules in which the α and β subunits are noncovalently bonded. The subunits are named from α_1 to α_8, α_L (for **LFA-1**), α_M (for **Mac-1**), α_v (for **vitronectin** receptor), and from β_1 to β_7. Both subunits penetrate the cell-membrane lipid bilayer; the α subunit has four Ca^{2+}-binding domains on its extracellular chain, and the β subunit bears a number of cysteine-rich domains extracellularly. The integrins are grouped into families: the VLA family, having the β_1 subunit; the LEUCAM family, which includes LFA-1 and Mac-1, having the β_2 subunit; and a group of other integrins having subunits β_3–β_8. The type of integrin expressed on the cell surface determines which adhesion molecules (and thus which other cells) the cell will bind, and can be varied in different circumstances. For example, transforming growth factor-β increases expression of $\alpha_1\beta_1$, $\alpha_2\beta_1$, $\alpha_3\beta_1$, and $\alpha_5\beta_1$ integrins on fibroblasts and the $\alpha_v\beta_3$ integrin on fibroblasts and osteosarcoma cells; interleukin-1β enhances β_1 expression on osteosarcoma cells; and in response to wounding, the keratinocyte, which normally expresses integrin $\alpha_6\beta_4$, will express $\alpha_5\beta_1$ (VLA-1, the fibronectin receptor) so that the keratinocyte will then migrate over fibronectin in the cell matrix, thus covering the wound. *See also* **adhesion molecules** (for details of the integrins to which they bind).

intein an internal peptide sequence of a protein precursor that is spliced out by transpeptidation during processing to form the mature protein. The peptide sequences that are spliced together are termed **exteins**. The terms are derived, respectively, by analogy from **introns** and **exons**.

intensify *(in photography)* to increase the opacity of (a film or plate). —**intensification, intensifier** *n.*

intensifying screens thin sheets of calcium tungstate used in autoradiographic detection of strong beta-emitting nuclides such as

^{32}P. Strong beta-particles pass through X-ray film and cause the screen to emit blue light, which is captured by silver halide grains in the film. Sensitivity of detection is increased fivefold when the film is exposed at $-70°C$ because the decay of the latent image is decreased.

intensity *(in physics)* the magnitude of an energy flux, field strength, or force. *See* **luminous intensity, radiant intensity**.

intensity millicurie *an obsolete name for* **sievert** (def. 2).

intensive property any of the properties of a substance or object that describe a specific characteristic of the substance or object in a given state. Intensive properties are independent of the amount of the substance or the size of the object being considered, e.g. density, specific volume, temperature, etc. *Compare* **extensive property**. *See also* **intensive**.

inter+ *comb. form* signifying between, within, or among. *Compare* **intra+**

interaction 1 any reciprocal or mutual action or effect. **2** *(in physics)* the transfer of energy between elementary particles, between particles and a field, or between two or more fields. —**interact** *vb.*

interaction broadening the concentration-dependent broadening of lines in an electron spin resonance spectrum caused by magnetic interactions between paramagnetic molecular or atomic species.

interactive *(of two or more forces, fields, processes, etc.)* acting reciprocally or in close and/or continuing relation with each other; interacting.

interallelic complementation interaction between products of two mutant alleles to form a functional protein. It can occur in the case of multimeric proteins, for which subunits are encoded by two alleles. If products from only one mutant allele compose the protein, an inactive protein results, but in certain cases, if the products from the two alleles (each of which codes for an inactive protein) are combined in the protein, an active protein results. This was first shown in *in vitro* experiments by taking two inactive proteins of alleles encoding glutamate dehydrogenase (NADP⁺), and recombining them by freezing and thawing in the presence of NaCl to form an active enzyme. There is some evidence that it can also occur *in vivo*.

interband any DNA-poor region between two (DNA-rich) bands in a polytene chromosome.

intercalary deletion *see* **deletion**.

intercalary DNA *abbr.*: iDNA; moderately repetitive DNA sequences occurring between structural genes in chromosomal DNA.

intercalation *(in molecular biology)* the insertion of additional material between the parts of an existing series, especially the insertion of molecules of antibiotic substances (e.g. actinomycin D) or acridine dyes (e.g. proflavine) between adjacent base pairs in a DNA duplex, thereby interfering with replication, etc. —**intercalate** *vb*; **intercalation** *n.*; **intercalary** or **intercalative** *adj.*

intercalator *or* **intercalating agent** *(in molecular biology)* any substance capable of being intercalated.

intercellular between cells. *Compare* **intracellular**.

intercept *(in mathematics)* a point at which two figures intersect; the point at which a line, curve, or surface cuts a coordinate axis.

interchain *(of any interaction or linkage)* between two chains of a biopolymer.

interconvertible enzyme any enzyme that exists in at least two well-defined reversibly convertible forms, produced by covalent modifications of amino-acid side chains under biological conditions (covalent modifications occurring as intermediates in the catalytic process are excluded). *Compare* **isoenzyme**.

intercrine *an alternative name for* **chemokine**.

interface 1 any surface forming a common boundary between two bodies, molecules, liquids, or phases. **2** any device, instrument, or method connecting two systems or processes; used especially of connections between computer systems. **3** to connect with or by an **interface** (def. 2). —**interfacial** *adj.*

interfacial tension the surface tension occurring at the interface between two mutually saturated liquids that tends to contract the area of the interface.

interference 1 *(in physics)* the process in which two or more wave trains of the same frequency and phase are superimposed to form a wave train in which there are alternate bands (called **interference fringes**) of no energy (due to **destructive interference**) or higher energy than that of the individual wave trains (due to **constructive interference**). **2** *(in virology)* the partial or complete prevention of the replication of one virus in a cell or tissue by the concomitant interaction between the given cell or tissue and another virus.

interference (light) filter a type of (light) filter with which a transmitted band-width of 10–20 nm can be obtained with a peak transmission of about 40%. It is possible to make interference filters in which the position of the centre of the transmittted band can be set within 1–2 nm.

interference fringe *see* **interference** (def. 1).

interference optics a system of optics, based on the Rayleigh interferometer, that is sometimes fitted to analytical ultracentrifuges. A double-sector ultracentrifuge cell is used, one sector containing only solvent and the other the solution being examined. The pattern of interference fringes produced by light passing through both sectors is a function of the refractive index at each point in the cell, each fringe tracing a curve of the refractive index against distance. Interference optics allows smaller changes in concentration to be detected than is possible by the **Schlieren method**.

interfering RNA *see* **RNA interference**.

interferogram a photographic record of a pattern of interference fringes.

interferometer an instrument that divides a beam of electromagnetic radiation into a number of beams and reunites them to produce **interference** fringes. It is useful for accurate measurement of the wavelength of the radiation, examination of the hyperfine structure of spectral lines, the testing of the flatness of surfaces, etc. —**interferometric** *adj.*; **interferometrically** *adv.*; **interferometry** *n.*

interferon *abbr.*: IFN; any member of a group of proteins that form a closely related group of non-viral proteins, M_r in the range 15 000 to 30 000, that are produced and liberated by animal cells following exposure to a variety of inducing agents; they are not normally present in uninduced cells. They exert non-specific, antiviral activity, at least in homologous cells, through cellular metabolic processes involving the synthesis of both RNA and protein. Viruses are the most potent inducing agents, but interferons can also be induced by exposure of cells to a wide variety of microorganisms, bacterial endotoxins, phytohemagglutinins, and other substances. Interferons do not prevent viral penetration into cells, but induce within the cells a complex of inhibitors of protein synthesis that block translation of viral mRNA, but not the cell's own mRNA. Among the interferon inhibitors are a protein kinase, which phosphorylates an initiation factor in protein synthesis, and an oligonucleotide synthetase, which catalyses the synthesis of pppA2′p5′A2′p5′A (*abbr.*: 2,5-A) – this trinucleotide activates a latent cellular endonuclease that degrades the viral mRNA. Interferons are classified into three types. **IFN-α** *(formerly* leukocyte interferon) comprises many examples; they are produced by macrophages, have antiviral activities, and stimulate production of two enzymes: a protein kinase and an oligoadenylate synthetase; they exist as monomers. **IFN-β** *(formerly* fibroblast interferon) has antiviral, antibacterial, and anticancer activities. It is related to the α family. **IFN-γ** *(formerly* immune interferon) is produced by lymphocytes activated by specific antigens or mitogens. In addition to antiviral activity, it has important immunoregulatory functions: it is a potent activator of macrophages, it has antiproliferative effects on transformed cells, and can potentiate the antiviral and antitumour effects of the type I interferons (α and β). It is a glycoprotein and exists as a homodimer. Interferons are produced commercially either by *in vitro* culture of human leukocytes or by recombinant DNA or RNA technology, γ interferon requiring a system that effects glycosylation.

interferon-γ receptor *abbr.*: Iγ-R; a widely expressed membrane receptor that consists of two single-pass transmembrane subunits that dimerize on binding of interferon γ (IFN-γ). One (Iγ-R1) is a glycoprotein of 472 amino acids, binds IFN-γ, becomes phosphorylated intracellularly, and activates JAK1 and STAT 1. The other (Iγ-R2) activates JAK2. Ligand binding also stimulates membrane sphingomyelinase to release ceramide. Several mutations in Iγ-R1 result in a rare autosomal recessive susceptibility to weakly pathogenic mycobacteria.

intergenic DNA regions of chromosomes between contiguous genes, whether or not they are encoded on the same or complementary strands of the duplex.

intergenic suppression *see* **suppression**.

interjacent RNA a 26S single-stranded RNA found in cells infected with certain togaviruses, e.g. Semliki forest virus. It codes for all the virus coat proteins.

interleukin *abbr.*: IL; any member of a heterogeneous group of **cytokines** that have the ability to act as signalling molecules between different populations of leukocytes. Several classes have been defined; each is designated by a suffixed arabic numeral: e.g., **interleukin 1** (IL-1), **interleukin 2** (IL-2), etc.

interleukin 1 *abbr.*: IL-1; *former names*: lymphocyte-activating factor (LAF); mitogenic protein (MP); helper-peak-1 (HP-1); T-cell-replacing factors III and m (TRF-III; TRF-m); B-cell-activating/differentiation factor (BAF; BDF). An interleukin of M_r 12 000–18 000, produced mainly by activated macrophages. It is not able to promote and maintain *in vitro* long-term cultures of T cells, but stimulates thymocyte proliferation by inducing **interleukin 2** release. It is involved in the inflammatory response, and is identified as an endogenous pyrogen.

interleukin 2 *abbr.*: IL-2; *other names*: thymocyte-stimulating factor (TSF); thymocyte mitogenic factor (TMF); T-cell growth factor (TCGF); co-stimulator; killer-cell helper factor (KHF); secondary cytotoxic T-cell-inducing factor (SCIF). An interleukin of M_r 30 000–35 000, produced by T cells in response to antigenic or mitogenic stimulation. Translocation involving the IL-2 gene causes one form of acute lymphoblastic leukemia. IL-2 promotes and maintains *in vitro* long-term cultures of T cells.

interleukin 3 *abbr.*: IL-3; *other names*: multipotential colony-stimulating factor; hematopoietic growth factor; P-cell-stimulating factor; mast-cell growth factor. A class of interleukins that are granulocyte/macrophage colony-stimulating cytokines; they act in hematopoiesis by controlling the production, differentiation and function of granulocytes and monocytes/macrophages. They are glycoprotein monomers.

interleukin 4 *abbr.*: IL-4; *other names*: B-cell stimulatory factor 1; lymphocyte stimulatory factor 1. An interleukin that participates in several B-cell activation processes, as well as those of other cell types. It is a co-stimulator of DNA synthesis, and is a glycoprotein.

interleukin 5 *abbr.*: IL-5; *other names*: T-cell-replacing factor; eosinophil differentiation factor. An interleukin that induces terminal differentiation of late-developing B cells to immunoglobulin-secreting cells. It is a homodimer, with subunits linked by a disulfide bond.

interleukin 6 *abbr.*: IL-6; *other names*: B-cell stimulatory factor 2; interferon β-2; hybridoma growth factor. An interleukin that induces myeloma and plasmacytoma growth, and nerve-cell differentiation. In hepatocytes, it induces acute-phase reactants. It is a glycoprotein.

interleukin 7 *abbr.*: IL-7; *other name*: hematopoietic growth factor. An interleukin that stimulates proliferation of lymphoid progenitor cells. It is produced by monocytes, T cells, and NK cells.

interleukin 8 *abbr.*: IL-8; *other names*: monocyte-derived neutrophil chemotactic factor (MDNCF); T-cell chemotactic factor; neutrophil-activating protein 1 (NAP-1); lymphocyte-derived neutrophil-activating factor; protein 3-10C. An interleukin that acts as a chemotactic factor, attracting neutrophils, basophils, and T cells, but not monocytes. It is released from several cell types in response to inflammatory stimuli.

interleukin 9 *abbr.*: IL-9; *other names*: T-cell growth factor; p40 precursor. An interleukin that supports IL-2/IL-4-independent growth of helper T cells.

interleukin 10 *abbr.*: IL-10; *other names*: cytokine synthesis inhibitory factor (CSIF); cytokine glycoprotein. An interleukin that appears to be a suppressor factor for Th1 immune responses.

interleukin 11 *abbr.*: IL-11; *other name*: adipogenesis inhibitory factor (AGIF). An interleukin that acts as both cytokine and growth factor. It stimulates plasmacytoma proliferation, and the T-cell-dependent development of immunoglobulin-producing B cells.

interleukin 12 *abbr.*: IL-12; *other name*: cytotoxic lymphocyte maturation factor. An interleukin that can act as a growth factor for activated T and natural killer (NK) cells, enhance the lytic activity of NK/lymphokine-activated killer cells, and stimulate the production of interferon-γ by resting PMBC. It is a disulfide-linked heterodimer of 35 kDa (α) and 40 kDa (β) subunits. NK stimulatory factor (NKSF) is essentially a complex of cytokine and soluble receptor. Both subunits are glycoproteins; α has some similarity to **interleukin 6**; β has an immunoglobulin-like domain.

interleukin 13 *abbr.*: IL-13; an interleukin that inhibits inflammatory cytokine production. It synergizes with **interleukin 2** in regulating interferon-γ synthesis.

interleukin 15 *abbr.*: IL-15; an interleukin produced in various tissues that has biological functions similar to those of **interleukin 2**. Its receptor consists of an α subunit that activates JAK1 (*see* **JAK**), and a γ subunit of interleukin 2 receptor (IL-2R) that activates JAK3.

interleukin 16 *abbr.*: IL-16; an interleukin that binds T lymphocytes through **CD4**.

interleukin 1β convertase EC 3.4.22.36; *other name*: interleukin-1β converting enzyme (*abbr.*: ICE); a proteolytic enzyme that cleaves the interleukin-1β precursor specifically at Asp^{116}-|-Ala^{117} and Asp^{27}-|-Gly^{28} in the precursor, producing the mature cytokine. It is a cysteine endopeptidase.

intermediary metabolism the chemical pathways by which molecules and structures found in living cells are synthesized from molecules that are taken into the cells, and subsequently broken down to molecules that are eliminated from the cells.

intermediate 1 occurring, or situated between two things in time, place, or order. **2** *(in chemistry or biochemistry)* any compound that takes part in a reaction, or sequence of reactions, occurring between the starting materials and the products. In metabolism, intermediates occur between the nutrients taken in and cellular material synthesized, or between cellular material and excretory products.

intermediate body *a former name for* **amboceptor**.

intermediate-density lipoprotein *abbr.*: IDL; one of the classes of lipoprotein found in blood plasma in many animals (data normally relate to humans). These lipoprotein particles have diameter 19.6–22.7 nm, solvent density for isolation (g mL^{-1}) 1.006–1.019, hydrated density (g mL^{-1}) 1.003. Their approximate composition (% by weight) is 8% unesterified cholesterol, 22% esterified cholesterol, 25% phospholipid, 30% triacylglycerol, 15% protein. Their apolipoprotein composition (% by weight total apolipoprotein) is 50–70% B-100, 5–10% C-I + C-II + C-III, 10–20% E-II + E-III + E-IV. They are formed in the plasma from very-low-density lipoproteins as a result of the action of **lipoprotein lipase**, and represent an intermediate stage in the conversion of very-low-density lipoproteins to low-density lipoproteins. *See also* **apolipoprotein** for individual apolipoproteins.

intermediate filament a distinct elongated structure, characteristically 10 nm in diameter, that occurs in the cytoplasm of higher eukaryotic cells. Intermediate filaments form a fibrous system, composed of chemically heterogeneous subunits and involved in mechanically integrating the various components of the cytoplasmic space. Intermediate filaments may be divided into five chemically distinct classes: **keratin** filaments of mammalian epithelial cells; **desmin** filaments found in cells of muscles of all types; **vimentin** filaments found in mesenchymal cells; neurofilaments of neurons; and glial filaments found in all types of glial cells. *Compare* **microfilament, microtuble.**

intermediate lobe or **intermediate part** or **pars intermedia** (hypophyseos) the part of the **pituitary gland** lying between the anterior and the posterior lobes. It is absent in adult humans but is present in fetal humans and in certain other species.

intermedin *a former name for* β-melanocyte-stimulating hormone (so-called because of its occurrence in the intermediate lobe of the pituitary gland). *See* **melanocyte-stimulating hormone.**

internal compensation the phenomenon of the existence of **internally compensated** molecular entities in general, or the property of such a molecular entity in a particular instance.

internal conversion *abbr.*: IC; a mode of radioactive decay of an excited atomic nucleus (formed by a nuclear transformation) in which, in a proportion of instances, excess energy released from the nucleus is not radiated from the atom as a quantum of low-energy gamma radiation but is imparted to one of the inner orbital electrons, which is thereupon ejected as a **conversion electron**; the electron emission is homoenergetic. The filling of the vacant orbital in the resultant excited ion causes emission also of either an Auger electron or a quantum of the characteristic X-radiation of the daughter atom.

internal energy *symbol*: U or E; a function of state that describes the sum of the kinetic and potential energy of a system. For closed systems, a change in internal energy, ΔU, usually results in a transfer of work or heat between the system and the surroundings. Thus $\Delta U = q + w$, where q is the heat gained by the system and w is the work done on the system.

internalin a surface protein of invasive forms of *Listeria monocytogenes*. It belongs to the invasin family, contains leucine-rich repeats, and binds to the epithelial transmembrane protein E-cadherin.

internal indicator any indicator substance used to follow a reaction, or other process, that is placed in the reaction mixture, or system.

internalize *or* **internalise** to render internal; especially to take a substance into a cell or organism. —**internalization** *or* **internalisation** *n*.

internally compensated describing any molecular entity that possesses equal numbers of structurally identical groups of opposite chirality. If no other chiral groups are present the overall entity is achiral and is designated by the prefix *meso+*.

internal ribosomal entry *abbr.*: IRE; the initiation of protein synthesis by certain viral and cellular mRNAs (e.g. those for cMyc, Apaf1, and Xiap) by a mechanism that is independent of recognition of the 5′ cap of the mRNA by the initiation factor eIF4E.

internal secretion *a former name for* **endocrine** (def. 2).

internal standard any standard used as a reference in an assay that is added to the samples and subjected to the whole of the procedures applied to them.

internal-standard method (of quench correction) *or* **internal standardization** a method of correcting for quenching in **liquid scintillation counting**. After the sample has been counted a measured small amount of a (nonquenching) standard preparation, of known activity, of the radionuclide in question is added to it and the sample is recounted. From the increment in the count rate the counting efficiency within the sample, and hence the degree of quenching, may be calculated. The method is simple and reliable and corrects adequately for all types of quenching but is more time-consuming than the **external-standard method** and has other disadvantages.

International Biological Standard any of the numerous internationally agreed standards of activity or potency of antisera, hormones, vitamins, antibiotics, and other biologicals. Each has been defined in terms of a mass of an appropriate standard preparation, prepared and stored under specified conditions.

International chemical identifier *abbr.*: INCHI; a unique, computer-readable identifier for a chemical compound. It consists of an alphanumeric string that describes the chemical structure.

International Practical Temperature Scale *abbr.*: IPTS; *see* **temperature.**

International System of Units the English translation of Système International d'Unités; *see* **SI.**

International Unit *abbr.*: IU (preferred) *or* i.u. *or* u; any unit of activity or potency of an antiserum, hormone, vitamin, antibiotic, or other biological defined in terms of the activity of an appropriate **International Biological Standard** or International Biological Reference Preparation. International Units are useful for quantifying a specific biological activity of a material before the chemical nature of the substance responsible for the activity has been fully characterized, or before pure preparations with reproducible activity are available.

Internet a protocol for allowing communication between computers, including those at distant sites. More loosely 'the internet' is used to refer to the worldwide network of such computers, including the providers of database services, discussion forums, and other communication media.

α-internexin *see* **peripherin.**

internuclear double resonance spectrometry *abbr.*: INDOR; a technique used in **nuclear magnetic resonance spectrometry** as an aid to assignment of resonances. It involves monitoring a single frequency, f_1 (usually one corresponding to a single transition), while sweeping a perturbing (decoupling) frequency, f_2, through a selected range of frequencies. Signals, negative or positive, will only appear in the INDOR spectrum when f_2 passes through a transition that has an energy level in common with the transition at f_2, making it possible to determine otherwise inaccessible chemical shifts and coupling constants.

interoperability *(in computing)* the ability of different computers, networks, operating systems, and applications to communicate with one another seamlessly (i.e. without technological barriers).

interphase the state of a eukaryotic cell when not undergoing mitosis or meiosis. It covers the G_1, S, and G_2 phases of the **cell-division cycle**.

interpolate *(in mathematics)* to estimate a value of a function between data values already obtained or known either graphically or by calculation. *Compare* **extrapolate**. —**interpolation** *n*.

InterPro *see* **derived database**.

intersectin a protein partner of dynamin in the endocytosis machinery.

intersegment transfer a proposed mechanism for the transfer of a repressor of gene expression in a nonspecifically bound state to its target site. It involves the rapid and direct transfer of repressor from one segment of a DNA molecule to another as a consequence of the relative diffusion of these segments within the domain of the molecule.

interstice any minute space or crevice that intervenes between structures, e.g. between cells, organs, or tissues, or between atoms in a crystal lattice. —**interstitial** *adj*.

interstitial cell a connective tissue cell lying in the interstices between other cells in a tissue and serving to bind the tissue cells. Examples include the neuroglial cells, and the testosterone-producing Leydig cells lying between the seminiferous tubules in the testis of vertebrates.

interstitial cell-stimulating hormone *abbr*.: ICSH; *a former name for* **luteinizing hormone**.

interstitial collagenase EC 3.4.24.7; *other names*: vertebrate collagenase; matrix metalloproteinase 1; an enzyme that cleaves preferentially one bond in native collagen and catalyses the cleavage of the triple helix of collagen at about three-quarters of the length of the molecule from the N terminus, at Gly^{775}-|-Ile^{776} in the α-1(I) chain. It cleaves synthetic substrates and α-macroglobulins at bonds where P1′ is a hydrophobic residue. Zinc is a cofactor.

interstitial deletion the loss of DNA or part of a chromosome that occupies an interstitial (i.e. nonterminal) position.

interstitial fluid the portion of the **extracellular fluid**, consisting mainly of water, that occurs outside the blood vessels and the lymphatics. In a 70 kg man its volume is approximately 12 L and it represents approximately 17% of body mass. It has a higher content of sodium, chloride, and bicarbonate and a lower content of potassium, magnesium, sulfate, and phosphate than **intracellular fluid**.

intertrabecular space the space within cells lying between the elements of the **microtrabecular lattice**.

intervening sequence an alternative name for **intron**.

intimin an actin-polymerization-inducing protein produced by enteropathogenic and enterohemorrhagic strains of enteric bacteria. It is an outer membrane protein of *Escherichia coli*, necessary for attachment of the bacterium to epithelial cells.

intoxication 1 any metabolic process that converts a nontoxic **xenobiotic** to a toxic product. **2** poisoning, especially with ethanol, manifested in behavioural change of the subject.

intra+ *prefix* signifying within, on the inside, or during. *See also* **endo+**. *Compare* **inter+**.

intracellular within a cell or cells.

intracellular fluid *abbr*.: ICF; the portion of the total fluid content of a multicellular organism that occurs within the cells, especially as distinct from **extracellular fluid**. In a 70 kg man the volume of ICF is approximately 26.5 L, and it represents approximately 38% of body mass. It has a higher content of potassium, magnesium, sulfate, and phosphate and a lower content of sodium, chloride, and bicarbonate than **interstitial fluid**.

intrachain *(of any interaction or linkage)* between different parts of a single chain of a biopolymer.

intradermal *abbr. (in medicine)*: ID *or* i.d.; within the layers of the skin.

intragenic suppression *see* **suppression**.

intramolecular within the same molecule.

intramolecular chaperone a domain within a protein that mediates conformational change of that protein (i.e., exercises a **chaperone** function). **Trypsinogen**, e.g., contains such a domain.

intramuscular *abbr. (in medicine)*: IM *or* i.m.; within a muscle or muscles; used especially of an injection.

intranet a computer network isolated from the Internet, often by means of a firewall, that offers typical Internet facilities to the local community (e.g. Web servers, email, bulletin boards).

intraperitoneal *abbr*.: IP *or* i.p.; within the peritoneum; used especially of an injection, infusion, or aspiration.

intrapleural within the pleura or pleurae; used especially of an injection or aspiration.

intrathecal within the intradural or subarachnoid space; used especially of an injection or aspiration.

intravenous *abbr. (in medicine)*: IV *or* i.v.; within a vein; used especially of an injection, infusion, transfusion, or aspiration.

intrinsic 1 of or pertaining to the essential nature of something; inherent. **2** *(in morphology)* peculiar to, or situated in a part. —**intrinsically** *adv*.

intrinsic activity *an alternative name for* **maximal agonist effect**.

intrinsically unstructured describing a protein or protein domain that lacks a folded structure but instead displays a highly flexible, random-coil-like conformation under physiological conditions. Examples include **calpastatin**, **casein**, and **tau**.

intrinsic association constant *or* intrinsic binding constant *see* **intrinsic constant**.

intrinsic birefringence the optical anisotropy of an individual molecule.

intrinsic constant any association, binding, or dissociation constant that describes the association, binding, or dissociation of a ligand to (or from) a particular site on a polyvalent molecule, when all the sites of this type are identical, present in the same molecule, and noninteracting.

intrinsic dissociation constant *see* **intrinsic constant**.

intrinsic efficacy *symbol*: ε; an expression introduced by Furchgott that denotes the notional efficacy of a drug associated with a single receptor. It is given by $\varepsilon = e/[R_t]$, where e is the **efficacy** and $[R_t]$ indicates the total concentration of active receptors of a certain type.

intrinsic factor *abbr*.: IF; a glycoprotein, secreted by the parietal cells of the gastric mucosa, that is necessary for the adequate absorption of dietary **cobalamins**. Intrinsic factor binds cobalamins in the intestinal lumen to form a complex that is subsequently bound by cubulin on the mucosa of the ileum. Cubulin is associated with LRP2 in the ileal mucosa and together they cause endocytosis of the IF–cobalamin complex. Cobalamin is bound to transcobalamin II in the portal blood. Congenital pernicious anemia is caused by absence of IF, a defect in cubulin, or deficiency in transcobalamin II of intrinsic factor.

intrinsic pathway (of blood coagulation) *see* **blood coagulation**.

intrinsic protein *less common term for* **integral (cell membrane) protein**.

intrinsic viscosity *symbol*: η_{int}; the limiting value of the **specific viscosity**, η_{sp}, of a solution as the concentration of the solute tends to zero. The units are $cm^3\,g^{-1}$. Its value depends on the specific volume of the (solvated) solute, V, and on a shape-dependent factor, ν, as follows: $\eta_{int} = \nu V$. The value of ν is 2.5 for a sphere and its value increases for solutes of other shapes. The name **limiting viscosity number** has been recommended for this attribute.

intron *or* intervening sequence any intragenic region of DNA in eukaryotes that is not expressed in a mature RNA molecule. Such regions occur between coding sequences (**exons**); this arrangement of introns alternating with exons constitutes a **transcription unit**. By customary usage, the term is extended to the corresponding regions of the **primary transcript** of RNA. In RNA, four general types of introns are distinguishable: (1) yeast nuclear tRNA; (2) group I introns; (3)

group II introns; and (4) nuclear mRNA precursor. Group I and II introns were first described for fungal mitochondrial genes; group I introns are characterized by a lack of conserved sequences at splicing junctions but have short conserved sequences internally, while group II introns have conserved sequences at the splicing junctions. The introns of different RNA types are spliced by different mechanisms (*see* **splicing**). It has been customary to describe introns as 'junk DNA', i.e. DNA for which no function can be found, but this concept must be modified. Many 'self-splicing introns' in the pre-rRNA of *Tetrahymena* double as mRNAs from which proteins are expressed that assist self-splicing and/or mediate intron translocation to new DNA sites. In the expression of a gene coding for small, stable RNAs that accumulate in the nucleolus, certainly of humans and mouse, it is the introns that give rise to RNA rather than the exons. Hence the terms intron and exon should simply signify RNA sequences that become physically separated during the process of RNA splicing. —**intronic** *adj.*

inulin a slightly soluble polysaccharide, of ≈6 kDa, found in members of the Compositae and Campanulaceae, in which it partially or completely replaces starch as a reserve carbohydrate. It is a linear glycan composed of about 35 (β2→1)-linked D-fructofuranose residues terminating in an (α1→2)-D-glucopyranosyl residue. Inulin is useful in the measurement of renal clearance, as inulin is neither reabsorbed nor secreted by the renal tubular cells. It is also useful in the measurement of extracellular fluid volume (**inulin space**), since it can pass through the blood capillary walls but is not taken up by cells.

inulobiose the disaccharide 2-*O*-β-D-fructofuranosyl-D-fructose.

inulosucrase EC 2.4.1.9; an enzyme that participates in inulin biosynthesis. *See also* **sucrase** (def. 3).

in utero *Latin* within the uterus.

in vacuo *Latin* in a vacuum.

invaginate 1 to infold, especially to form a sheathlike or saclike enclosure. **2** describing an organ or part that is folded back upon itself. —**invagination** *n.*

invariant chain *or* **CD74** *or* **Ii** a protein (31 kDa) of the endoplasmic reticulum (ER) lumen that binds newly formed class III MHC heterodimeric proteins so as to prevent binding of endogenous peptides to their groove. It acts as a chaperone for these molecules until they leave the Golgi apparatus and enter the endosome pathway. It is degraded proteolytically but leaves a fragment (called CLIP) bound within the groove. In acid vesicles, after leaving the Golgi apparatus, CLIP exchanges with peptides derived from exogenous antigens.

invariant residue any amino-acid residue that always occurs in equivalent positions in the polypeptide chains of homologous proteins from different species, or in genetic variants of the protein from a given species.

invasin 1 *an alternative name for* **hyaluronidase**. **2** any of a family of surface proteins of various intracellular bacterial pathogens by which the bacteria invade nonphagocytotic host cells (e.g. those of the gut epithelia). All contain leucine-rich repeats. *See* **internalin**.

invasive *(in pathology)* tending to penetrate into healthy tissue and grow within the host away from the site of entry or origin. It is used especially of a pathogenic organism, cell, or group of cells, e.g. cancer cells. —**invasiveness** *n.*

inverse agonist any ligand that by binding to receptors reduces the fraction of them in an active conformation, thereby having biological effects opposite to those produced by an **agonist** (as in **benzodiazepine inverse agonist**). This can occur if some of the receptors are in the active form even in the absence of a conventional agonist. *Compare* **partial inverse agonist**.

inverse PCR a variant form of **polymerase chain reaction** (PCR) in which DNA of unknown sequence but flanking a known sequence can be amplified. Digestion of the DNA with a restriction enzyme followed by ligation at low DNA concentrations favours the formation of closed circular molecules. PCR primers based on the known sequence are designed so that their 5' ends are adjacent on the two strands of the duplex. The PCR product is a linear DNA molecule containing known sequence at each end with unknown sequence between.

inverse-square law a law according to which the magnitude of a physical quantity, e.g. the intensity of electromagnetic or other radiation, is proportional to the reciprocal of the square of the distance from the source of the property.

inversion 1 configurational inversion the process of changing the **configuration** about a chiral centre into that with the opposite configuration. *See also* **Walden inversion**. **2** the hydrolysis of sucrose to an equimolar mixture of glucose and fructose. It is so-called because of the accompanying reversal of direction of optical rotation by the solution. **3 chromosomal inversion** the excision of a chromosomal segment followed by its reinsertion at the same position after having been turned through 180°. **Paracentric inversions** are confined to one or other arms of the chromosome, whereas **pericentric inversions** span the centromere. Both types create changes in gene order, and sometimes cause disordered meiosis, genetic abnormalities, and reduced fertility.

inversion isomer *an alternative name for* **invertomer**.

invert 1 to turn or cause to turn inward, or upside down. **2** to reverse in direction, sequence, effect, etc. *See also* **inversion**.

invertase *or* **saccharase** EC 3.2.1.26; *recommended name*: β-fructofuranosidase; *systematic name*: β-D-fructofuranoside fructohydrolase. An enzyme from yeast and other sources that catalyses the **inversion** (def. 2) of sucrose. *See also* **saccharase**.

invertebrate 1 any animal lacking a vertebral column (backbone) and internal skeleton. **2** of, pertaining to, or designating any such animal.

inverted repeat *or* **indirect repeat** *abbr.*: IR; either of a pair of sequences in duplex DNA that exhibit two-fold rotational symmetry around a centre, the two sequences being complementary to each other. The sequence of the two repeats together is called a **palindrome**, e.g.

GGTACC
CCATGG.

The repeats need not be contiguous, e.g.

GGTNNACC
CCANNTGG.

Inverted repeats promote the formation of **cruciform** structures. Their function is unknown. However, they are found flanking transposable elements, which may indicate an associated role. *See also* **cruciform, dyad symmetry, hairpin**.

invertin an enzyme preparation from yeast. The major component is an α-glucosidase activity. *See also* **emulsin**.

invertomer *or* **inversion isomer** either of two **conformers** that interconvert by inversion at an atom possessing a nonbonding electron pair.

invert sugar an equimolar mixture of D-glucose and D-fructose formed by the hydrolysis, i.e. **inversion** (def. 2), of sucrose. It is sweeter than its parent disaccharide. *See also* **high-fructose corn syrup**.

inverzincin a family of conserved metalloproteases that contain an inverted zinc-binding site (i.e. His–Xaa–Xaa–Glu–His). Examples include insulinase and *E. coli* protease III.

in vitro *Latin (of any biological process, reaction, or experiment)* occurring or made to occur outside an organism, e.g. in extracts or cultures; literally it means 'in glass'. *Compare* ***in vivo***.

***in vitro* packaging** the assembly of head and tail proteins of bacteriophages with recombinant bacteriophage DNA to generate infectious phage particles.

***in vitro* transcription** the generation of a sense or antisense RNA copy of a cDNA molecule usually carried in a plasmid or other vector. This is accomplished by using an RNA polymerase from T3, T7, or SP6 phage that initiates transcription from a cognate promoter situated upstream from the cloned DNA insert.

***in vitro* translation** the generation of a polypeptide whose sequence is specified by mRNA added to extracts of reticulocytes or wheatgerm. Such extracts have all the components necessary for translation, including ribosomes, tRNAs, and amino acids. One or more of the amino acids can be radioactively labelled. This method is sometimes combined with *in vitro* transcription in the same reaction mixture.

in vivo *Latin (of any biological process, reaction, or experiment)* occurring or made to occur within a living organism; literally it means 'in life'. *Compare* **in vitro**.

involucrin a protein of keratinocytes and other stratified squamous epithelia that first appears in the cell cytosol, but ultimately becomes cross-linked to membrane proteins by the action of transglutaminase.

involute 1 involved, intricate. **2** spiral, having whorls wound closely around the axis. **3** having edges that roll under or inwards. **4** to be subjected to a process involving the inward rolling of the surrounding tissue.

iod+ *a variant form of* **iodo+** (*sometimes before a vowel*).

iodate 1 the anion, IO_3^-, derived from iodic(v) acid. **2** any salt of iodic(v) acid.

iodide 1 the anion, I^-, of the element iodine. **2** any salt of hydriodic acid. **3** any compound containing an iodine atom in organic linkage.

iodide-chloride transporter *or* **Pendrin** a human integral membrane protein (780 amino acids) predicted to contain 11 transmembrane segments. It may be the iodide transporter on the colloid surface of thyroid follicle cells. More than 30 mutations in the gene lead to the Pendred syndrome of thyroid goitre with congenital deafness, or simple, isolated deafness.

iodide pump *or* **Na^+/I^- symporter** (*abbr.*: NIS) an integral membrane protein that cotransports Na^+ and I^- into the thyroid follicles. It is dependent on sodium/potassium ATPase activity, and is stimulated by thyroliberin. It contains 643 amino acids and is predicted to contain 12, 13, or 14 transmembrane segments. It is also present in salivary glands, gastric mucosa, choroid plexus, and mammary glands, and shares 30% sequence homology with the Na^+/glucose transporter 1. Several mutations lead to congenital hypothyroidism as a result of a defect in entry of iodide into cells of thyroid follicle.

iodide trapping defect *see* **iodide pump**.

iodimetry a volumetric method of analysis applicable to substances with a redox potential appreciably lower than that of the iodine–iodide system (E^o = +0.535 V at 25°C). Such substances are oxidized by iodine and may therefore be titrated in acid solution with a standard solution of iodine. *Compare* **iodometry**. —**iodimetric** *adj.*; **iodimetrically** *adv.*

iodinatable capable of being converted into an iodo derivative. The term may be used of an organic compound or a particular part of one, such as a residue in a peptide.

iodinate 1 to treat or react (a substance) with iodine. **2** to introduce one or more iodo groups into (an organic compound), whether by addition or substitution. —**iodination** *n.*

iodine 1 *symbol*: I; a nonmetallic element of group 17 of the IUPAC **periodic table** and one of the halogens; proton number 53; it exists naturally as a single stable nuclide, **iodine-127**, of relative atomic mass 126.904. It is widely distributed in nature, although never found as the free element, and is an essential dietary element. Numerous (about 29 unstable and seven metastable) artificial radioactive nuclides are known. Of these, the most widely used in biochemistry and molecular biology are **iodine-125** and **iodine-131**. **2** *symbol*: I_2; *the common name for* the substance correctly known as diiodine, a bluish-black solid that sublimes at room temperature into a violet-coloured irritating vapour. An ethanolic solution is useful as a mild topical antiseptic.

iodine-125 the artificial radioactive nuclide $^{125}_{53}I$. It decays to the stable nuclide tellurium-125 by capture to the nucleus of an orbital electron from the K (i.e. innermost) shell, emitting low-energy gamma radiation (35.5 keV, 7% emitted and 93% internally converted) and characteristic tellurium K X-radiation (27–32 keV, 140%), but no beta radiation. It has a half-life of 59.6 days.

iodine-127 the stable nuclide $^{127}_{53}I$. Its relative abundance in natural iodine is 100 atom percent.

iodine-131 the artificial radioactive nuclide $^{131}_{53}I$. It decays to the stable nuclide xenon-131, emitting a moderate-energy electron or beta particle of various energies (0.25–0.81 MeV max., mainly 0.61 MeV max.) plus gamma radiation of various energies (0.08–0.72 MeV, mainly 0.36 MeV), 1.3% of nuclides decaying via the metastable nuclide xenon-131m. It has a half-life of 8.04 days.

iodine-labelling a widely used procedure for labelling a compound

or material by introducing into it, directly or indirectly, one or more atoms per molecule of iodine, especially radioactive isotopes (i.e. radioiodine). *See also* **radioiodinate**.

iodine number *or* **iodine value** a measure of the degree of unsaturation of a fat or fatty acid, defined as the number of grams of iodine capable of reacting with 100 g of the sample. It is usually determined indirectly by measuring the amount of an iodine-containing reagent, e.g. iodine monobromide, that remains after excess has been added and reaction is complete. The iodine number of saturated fatty acids is zero, that of oleic acid is 90, of linoleic acid 181, and of linolenic acid 274.

iodo of, relating to, or describing a compound that contains an iodine atom in organic linkage.

iodo+ *or* (*sometimes before a vowel*) **iod+** *comb. form* denoting **1** an iodine atom in organic linkage. **2** violet-coloured. *See also* **iodo**.

iodoacetamide $I–CH_2–CONH_2$; a compound that acts as an inhibitor of an enzyme containing –SH groups, Enz–SH, by reacting irreversibly to give $Enz–S–CH_2–CONH_2$, with elimination of HI.

Iodo-Gen *a proprietary name for* 1,3,4,6-tetrachloro-3α,6α-diphenylglycouril; a solid-phase reagent that oxidizes low concentrations of inorganic iodide to iodine and hence is useful for the rapid and efficient labelling of proteins with radioactive iodine. Being very stable and virtually insoluble in water, it can be used in the form of a prior coating on the walls of a reaction vessel containing aqueous solutions of iodide ion and protein.

iodogorgoic acid *a former name for* 3,5-**diiodotyrosine**.

iodometry a volumetric method of analysis applicable to substances with a redox potential appreciably higher than that of the iodine–iodide system ($E^{o'}$ = +0.535 V at 25°C). Such substances are reduced by iodide, which is added in excess to acidified solutions and the liberated iodine titrated with a standard solution of thiosulfate. *Compare* **iodimetry**. —**iodimetric** *adj.*; **iodimetrically** *adv.*

iodopsin any of three light-sensitive visual pigments in the cone cells of the retinas of vertebrates, other than freshwater and migratory fishes, and some amphibians. They consist of different cone-type opsins combined (as Schiff's bases) with 11-*cis*-**retinal**. Each individual cone cell contains one of the three pigments, which have absorption maxima at ~430 nm, ~540 nm, and ~575 nm. The iodopsins were once thought to be a single pigment, termed visual purple or visual violet. *See also* **cyanopsin**, **retinal**.

iodotyrosine deiodinase an NADPH-dependent flavoprotein enzyme that deiodinates monoiodotyrosine and diiodotyrosine derived from proteolytic degradation of thyroglobulin within lysosomes of thyroid follicle cells. The released iodide (I^-) is recycled by the thyroperoxidase of those cells. Enzyme deficiency leads to excessive loss of the iodotyrosines and to congenital goitre and hypothyroidism.

ion any atom or other molecular entity that has acquired an electric charge by loss or gain of one or more electrons. An anion is negatively charged, and a cation is positively charged.

ion atmosphere the region surrounding a charged molecular entity in which there is a statistical preference for ions of opposite charge.

ion chamber *another name for* **ionization chamber**.

ion channel any channel in cell membranes that allows the passage of specific inorganic ions through the membrane. Ion channels may be **ligand-gated**, **voltage-gated**, or **mechanically gated channels**. For ligand-gated channels, the ligand may be extracellular such as a neurotransmitter (**transmitter-gated channel**), or intracellular such as an ion (**ion-gated channel**), or a nucleotide. Ion channels are common direct or indirect targets for the action of drugs, which may either potentiate or inhibit channel opening.

ion exchange the process in which ions of like charge, which may be the same or different chemically, are exchanged between two phases, such as two solutions separated by a semipermeable membrane or a solution and an insoluble material, e.g. a resin.

ion-exchange cellulose *abbr.*: IEC; any derivatized cellulose containing cationic or anionic substituents used in **ion exchange**, especially for the separation of charged biopolymers.

ion-exchange chromatography any form of chromatography that depends on the process of **ion exchange** to effect the separation.

ion-exchange polymer an insoluble polymer that has ionic groups and is able to exchange counterions with the ionic components of a solution. Cation exchangers carry anionic groups such as sulfonate or carboxyl, whereas anion exchangers carry cationic groups such as primary or quaternary ammonium groups. An ion-exchange polymer in ionized form may also be referred to as a **polyanion** or a **polycation**.

ion exchanger any **anion exchanger** or **cation exchanger**.

ion-exchange resin *a former term (now discouraged) for* **ion-exchange polymer**.

ion-exclusion chromatography a chromatographic technique by which strong electrolytes can be separated from nonelectrolytes and weak electrolytes by passage of the mixture through an ion-exchange resin in a buffer of ionic strength and pH that can be varied in the course of the separation. At low salt concentrations particularly, ionic repulsive forces tend to exclude from the resin those ions that have the same charge as the functional groups on the resin so that these ions, together with an equivalent number of counterions, pass rapidly through the column while nonelectrolytes and weak electrolytes enter the resin and thus pass more slowly through the column at a rate determined in part by the pK of their ionizable groups.

ion-gated channel *see* **ion channel**.

ionic 1 of, pertaining to, existing as, or characterized by **ions**. **2** using or operated by ions.

ionic atmosphere the cloud of ions of one charge type that tend to accumulate by interionic attraction around an ion of the other charge type in a solution of an electrolyte. The cloud is symmetrical about the charge on the central ion unless the solution is subjected to a shearing force conducive to flow or is exposed to an applied electric field. In the latter event the central ion and its surrounding counterions migrate in opposite directions, the resultant asymmetric disposition of the ionic atmosphere effecting retardation of the central ion in a concentration-dependent manner.

ionic bond any bond formed by attraction between a positively and a negatively charged group on different parts of the same molecule or on different molecules.

ionic conduction the passage of electricity due to the movement of ions.

ionic detergent any **anionic detergent** or **cationic detergent**. *Compare* **nonionic detergent**.

ionic double layer the region surrounding a charged molecular entity together with the layer of **counterions** that surrounds it.

ionic mobility a measure of the velocity at which an ion moves through a liquid under the influence of an applied electric potential.

ionic product *see* **ion product**.

ionic strength *symbol*: I or I_m (molality basis) *or* I_c (concentration basis); a physical quantity equal to half the summation over all ions in a solution of the concentration of each ion multiplied by the square of its charge. Hence,

$$I_m = 0.5(\Sigma m_i z_i^2)$$

and

$$I_c = 0.5(\Sigma c_i z_i^2),$$

where m_i and c_i are the molal and molar concentrations of the *i*th ion, respectively, and z_i is its charge.

ionizable group *or* **ionisable group** any uncharged group in a molecular entity that is capable of dissociating by yielding an ion (usually an H^+ ion) or an electron and itself becoming oppositely charged, or of associating with an ion (usually an H^+ ion) and itself assuming the ion's charge. The extent to which an ionizable group becomes an ionizing group depends on the constitution of the molecular entity and upon the environment.

ionization *or* **ionisation** any of the processes by which one or more ions are generated. It may occur, e.g., by loss or gain of a hydron by a solute in response to changes in pH, or by the loss of an electron from a neutral chemical species by unimolecular heterolysis of such a species into two or more ions. Further, a gas may become ionized by the passage through it of fast charged particles or electromagnetic radiation.

ionization chamber *or* **ionisation chamber** *or* **ion chamber** a chamber containing a gas and two electrodes between which an electric potential difference is maintained. It is used to detect ionizing particles or electromagnetic radiation by virtue of the current that flows following ionization of the gas.

ionization constant *or* **ionisation constant** the **dissociation constant** for a reversible reaction in which the products are ionic.

ionization energy *or* **ionisation energy** *symbol*: E_I; the minimum energy (enthalpy) required to remove an electron from an isolated molecular entity (in its vibrational ground state) in the gaseous phase. It was formerly, and misleadingly, termed ionization potential.

ionization potential *or* **ionisation potential** *a former term for* **ionization energy**.

ionize *or* **ionise** to convert, or to become converted, wholly or in part, into ions; to cause **ionization** of (something). —**ionizable** *adj.*

ionizing particle *or* **ionising particle** any charged particle of sufficient energy to enable it to dislodge an orbital electron from an atom or other molecular entity. *See* **ionizing radiation**.

ionizing power *or* **ionising power** *or (sometimes)* **polarity** a qualitative term used to denote the tendency of a particular solvent to promote ionization of an uncharged solute.

ionizing radiation *or* **ionising radiation** any radiation, corpuscular or electromagnetic, that is capable of causing **ionization** (directly or indirectly) of matter through which it passes. For ionization to occur, the kinetic or photon energy, respectively, must be greater than the **ionization energy** of the irradiated substance. Ultraviolet radiation, X-radiation, alpha particles, and high-energy electrons (including beta particles) are more effective than gamma radiation or neutrons. It is commonly known simply as radiation.

ionogen 1 any compound that can dissociate into ions in a suitable environment; an electrolyte. **2** any atom or group that is capable of becoming ionized chemically. —**ionogenic** *adj.*

ionography *an alternative name for* **zone electrophoresis**.

ionomycin a nonfluorescent antibiotic ionophore derived from *Streptomyces conglobatus*. It is highly specific for divalent cations and exhibits greater selectivity for Ca^{2+} over Mg^{2+} than **A23187** (calcimycin). It complexes with Ca^{2+} between pH 7 and 9.5 with intense UV absorption, and thus may be used as a carrier of Ca^{2+} and for measurement of cytoplasmic Ca^{2+} concentration.

ionone either of two terpene compounds, α-ionone and β-ionone, found in a number of plants. β-Ionone is an intermediate in the synthesis of vitamin A; its odour is reminiscent of cedarwood or, in dilute alcohol solution, of violets.

α-ionone

β-ionone

ionophore *or* **ionophorous agent** any member of a varied group of both natural and synthetic substances, cyclic or linear and of greatly differing relative molecular masses, that are capable of binding metal ions in solution and transporting them across lipid barriers in natural or artificial membranes. They may be classed as **cage carriers**, when the ionophore surrounds the metal ion during transport, or as **channel carriers**, when the ionophore forms an ion-conducting pore in the membrane. —**ionophorous** *or* **ionophoric** *adj.*

ionophoresis electrophoresis, especially of relatively small ions. *See also* **iontophoresis**.

ionotropic describing a type of receptor that mediates its effects by regulating ion channels, in distinction from receptors that act by regulating enzymes such as adenylate cyclase or inositide-specific phospholipase C. *Compare* **metabotropic**. *See also* **excitatory amino-acid receptor**.

ion pair a pair of oppositely charged ions held together by coulombic attraction. *See also* **ionic bond**.

ion product *or (sometimes)* **ionic product** *symbol*: K_i; a measure of the tendency of an amphiprotic solvent, XH, to take up or to lose a hydron in the reaction:

$$XH + XH = XH_2^+ + X^-.$$

In such a reaction, $K_i = a_{XH_2^+} \times a_{X^-}$ where $a_{XH_2^+}$ and a_{X^-} are the activities of XH_2^+ and X^-, respectively. The ion product of water, K_W, is given by

$$K_W = a_{H_3O^+} \times a_{OH^-}; \ K_W \approx 1 \times 10^{-14} \ mol^2 \ dm^{-6} \ at \ 25°C.$$

It has a large positive temperature coefficient.

ion pump a **transporter** that actively transports ions across a membrane against an **electrochemical potential**. *See* **active transport**.

ion-selective electrode *or* **ion-selective microelectrode** any electrode (or microelectrode) whose potential responds in a reproducible manner to changes in the concentration (strictly, in the activity) of a specific kind of ion in the solution in which the electrode is immersed and that is insensitive, or essentially so, to changes in the concentrations (strictly, in the activities) of all other kinds of ion in the solution. The electrode has at its surface a thin membrane of electrically conducting material, separating the internal and external solutions, across which an electric potential difference develops. The membrane consists of or contains an electroactive material responsive to the ions to be determined.

iontophoresis 1 the introduction of drugs and other substances through plasma membranes of cultured cells by the transfer of ions brought about by the application of a small electrical potential through electrodes placed on the membrane. 2 *a less common word for* **ionophoresis**.

iota *symbol*: ι (lower case) *or* I (upper case); the ninth letter of the Greek alphabet. For uses *see* **Appendix A**.

iotatoxin any of a family of peptidic pore-forming toxins of 150–800 amino acid residues, permitting the passage of small molecules across membranes. Iotatoxins are produced by *Clostridium perfringens*. They belong to the TC 1.C.42 family.

IP 1 *or* **i.p.** *abbr. for* intraperitoneal *or* intraperitoneally. 2 *abbr. for* Internet protocal. *See* **protocol**.

IP₂ *abbr. for* 1D-*myo*-inositol 3,4-bisphosphate.

IP₃ *abbr. for* 1D-*myo*-inositol 1,3,4- (or 1,4,5-) trisphosphate.

IP₄ *abbr. for* 1D-*myo*-inositol 1,3,4,5-tetrakisphosphate.

IP address a unique software address for an Internet computer. Although actually a binary number it is usually represented in dotted decimal notation, e.g. '130.88.203.71'. Associated with this is a 'named' version (e.g. www.oup.co.uk corresponding to the dotted decimal address above at the time of writing). The conventions for named addresses include a two-letter abbreviation for a country (e.g. uk for the United Kingdom, ca for Canada, in for India, de for Germany) except for addresses in the USA: because IP is American, the default country is the USA. The 'co' in the example above stands for 'commercial'; other possibilities include ac (academic), gov (government), or org (organization). In the USA the corresponding abbreviations are all of three letters (com for commercial, edu for educational, mil for military; org and gov are also used in the USA). Some countries (e.g. Germany) do not usually have this field.

ipecacuanha *or* **ipecac** the dried root of the South American shrub, *Cephaelis ipecacuanha*; it contains the alkaloids **cephaeline** and **emetine**. Ipecacuanha preparations are used as emetics and expectorants.

IPP *abbr. for* Δ³-isopentenyl pyrophosphate. This compound is now preferably called isopentenyl diphosphate (*abbr.*: IDP).

iPr *symbol for* the isopropyl group, $(CH_3)_2CH-$ (alternative to Pr^i).

IP₃ receptor a receptor/Ca^{2+} channel of the endoplasmic reticulum (ER). On binding of inositol 1,4,5-trisphosphate (IP₃), released by the action of phospholipase C from phosphatidyl inositol 3,4-bisphosphate on the inner leaflet of the plasma membrane, the channel opens to release Ca^{2+} ions from the ER lumen into the cytosol. The receptor's subunits each contain six transmembrane segments. Homer1 mediates the effect of ligand binding to metabotropic glutamate receptors on the IP₃ receptor.

iproniazid 1-isonicotinoyl-2-isopropylhydrazine; a derivative of **isoniazid** that is a **monoamine oxidase** inhibitor.

Ips *symbol for* the pipsyl group.

IPTG *abbr. for* isopropyl β-D-thiogalactoside; a gratuitous inducer of the *lac* operon in *E. coli*.

IPTS *abbr. for* International Practical Temperature Scale.

Ir *symbol for* iridium.

IR 1 *or* **i.r.** *abbr. for* infrared (radiation). 2 *abbr. for* inverted repeat. 3 *abbr. for* information retrieval.

IRAK *abbr. for* IL-1 receptor-associated kinase; a human cytosolic serine/threonine protein kinase of 752 amino acids that binds to liganded interleukin 1 (IL-1) receptor. Its function is unknown.

IRE *abbr. for* iron response element.

I₁, I₂ receptors *see* **imidazoline receptor**.

IREG *abbr. for* iron-regulated transporter. *See* **ferroportin**.

IRES *abbr. for* internal ribosome entry site; a sequence found immediately upstream from a cistron in polycistronic mRNA – such as produced by transcription of encephalomyocarditis and other viral genomes – that allows ribosomes to initiate translation from a secondary site within the transcript rather than assembling at the 5' cap, as is usual with the translation of eukaryotic mRNAs.

IRG *abbr. for* immunoreactive glucagon.

Ir genes the genes, including those within the **MHC**, that together determine the overall level of immune response to a given **antigen**.

IRMA *abbr. for* immunoradiometric assay.

iron *symbol*: Fe; a metallic element of group 8 of the IUPAC **periodic table**; proton number 26; natural iron has a relative atomic mass of 55.847 and is a mixture of stable nuclides of nucleon numbers 54 (5.8 atom per cent), 56 (91.7 atom per cent), 57 (2.2 atom per cent), and 58 (0.3 atom per cent). About 12 artificial radioactive nuclides are known; of these, **iron-59** is the one most widely used in biochemistry and molecular biology. Iron is a **transition element** and hence forms ionic salts and numerous complexes in the oxidation states +2, i.e. iron(II) (ferrous iron), or +3, i.e. iron(III) (ferric iron). Iron is one of the most important elements in biology; it is essential in substantial quantities to all cellular organisms from bacteria to mammals, particularly for the synthesis of hemoproteins such as **cytochromes**, **hemoglobins** and related proteins, **myoglobin**, **catalase**, and some **peroxidases**. Important iron-transport proteins include **ferritin** and **transferrin**; iron is taken up into mammalian cells, bound to transferrin, by the **transferrin receptor**. There is no physiological excretory mechanism for iron, but it is lost from females during menstruation.

iron-59 the artificial radioactive nuclide $^{59}_{26}Fe$. It decays to the stable nuclide cobalt-59 with emission of a beta particle (i.e. an electron) in two moderate-energy ranges (0.27 MeV max., 48%, and 0.48

MeV max., 51%) plus gamma radiation of various energies (mainly 1.10 MeV, 56%, and 1.29 MeV, 43%). It has a half-life of 44.5 days.

iron hydrogenase *abbr.*: [Fe]- or [FeFe]-hydrogenase; any of a family of hydrogenases (EC 1.12.) present in some bacteria, anaerobic fungi, ciliates, flagellates, and green algae, within the cytosol, the cell membrane, or in hydrogenosomes. Two nonhomologous classes are recognized: (1) hydrogenases containing an H cluster comprising a dimer of iron atoms linked to a [4Fe-4S] cluster; and (2) hydrogenases containing a single iron atom.

iron–molybdenum cofactor *abbr.*: FeMo-co; *see* **nif**, **nitrogenase**.

iron overload a condition of humans in which iron is taken up in excess over a prolonged period. As there is no excretory mechanism for iron, it accumulates in Kupffer cells in the liver as **hemosiderin**, a form of **ferritin** complexed with other proteins and iron. The condition is known as **hemosiderosis**. It also occurs as a result of repeated blood transfusions, e.g. in treatment for the hemoglobinopathies. A form of the condition is referred to as **Bantu siderosis**, being very prevalent in the Bantu people of southern Africa; it was formerly thought to be always caused by use of cast-iron vessels for cooking, but in some families a hereditary susceptibility has been found.

iron-regulated transporter *abbr.*: IREG; *an alternative name for* **ferroportin**.

iron regulatory protein *abbr.*: IRP; a cytoplasmic protein that binds an iron-response element present in mRNAs for proteins involved in iron metabolism.

iron response element *or* **iron regulatory element** *abbr.*: IRE; a base sequence in eukaryotic 5' or 3' untranslated regions of mRNAs, that either blocks translation or enhances stability of the mRNA for proteins involved in iron metabolism (e.g. ferritin, transferrin receptor). The IRE is modulated by an iron regulatory protein that binds to it.

iron–sulfur cluster a unit comprising two or more iron atoms and bridging sulfide ligands in an **iron–sulfur protein**. The recommended designation of a cluster consists of the iron and sulfide content, in square brackets, for example [2Fe–2S], [3Fe–4S]. The possible oxidation levels are indicated by the net charge excluding the ligands, for example: a [4Fe–4S]$^{2+}$; [4Fe–4S]$^{1+}$ (or [4Fe–4S]$^{2+;1+}$) cluster.

iron–sulfur protein *abbr.*: Fe–S protein; any member of a group of proteins to which iron is bound via sulfur-containing ligands (excluding hemoproteins containing iron linked only through axial sulfur ligands). Iron–sulfur proteins are characterized by the possession of non-heme bound iron, as well as one or more of the following: (1) acid-labile sulfur; (2) absorption bands between 300 and 600 nm; (3) the capability of transferring electrons at low redox potential; and (4) paramagnetism in one or more oxidation states. They may be classified as **simple** if they contain one or more Fe–S clusters only (*see* **ferredoxin**, **rubredoxin**), and **complex** where additional active groups, such as flavin, heme, or another metal, are present.

IRP *abbr. for* iron regulatory protein.

irradiate to subject to or treat with electromagnetic radiation or beams of particles. —**irradiation** *n.*

irreversible competitive antagonist *(in pharmacology)* an antagonist for which the dissociation rate is so low that antagonist molecules cannot be replaced by increasing the concentration of an agonist for the same receptor.

irreversible process any process occurring in an isothermal system without change of internal energy but with an increase in entropy.

IRS *abbr. for* insulin receptor substrate.

IS *abbr. for* insertion sequence.

ischemia *or (esp. Brit.)* **ischaemia** a disorder of function caused by insufficient blood flow. Resumption of the blood supply may lead to **ischemia-reperfusion injury** due to the formation of reactive oxygen species.

islet any small group of cells that is structurally distinct from the surrounding tissue, especially the **islets of Langerhans**.

islet-1 *abbr.*: ISL-1; *other name*: insulin-related protein; a transcription factor that is expressed during vertebrate embryogenesis in several neuronal and other cell types (including pancreatic B cells). It contains a homology region of 50–60 amino acids, called LIM, that is also present in Lin-II ISL-1, and Mec-3 of *Caenorhabditis elegans*.

It forms a double zinc finger and may be involved in protein, rather than DNA, binding.

islet-activating protein *abbr.*: IAP; *a former name for* **pertussis toxin** (so-called because it stimulates release of insulin from the islets of Langerhans).

islets of Langerhans *or (sometimes)* **islands of Langerhans** *or* **pancreatic islets** the numerous small, spherical or oval, highly vascular clusters of endocrine cells that lie scattered throughout the acinar (i.e. exocrine) tissue of the pancreas of gnathostome (i.e. jaw-bearing) vertebrates and are readily distinguishable microscopically from it. The normal adult human pancreas contains about 2 million islets, which together comprise under 2% of its total volume. Islets commonly consist of three main types of cells, all granule-containing: the glucagon-secreting **A cells**; the insulin-secreting **B cells**; and the somatostatin-secreting **D cells**. Some other types of hormone-secreting cells may be present also. [After Paul Langerhans (1847–88), German physician and anatomist, who first described such clusters of cells (1869).]

iso+ *or (sometimes before a vowel)* **is+** *comb. form* **1** equal, the same, uniform, homogeneous. *Compare* **allo+** (def. 1), **homo+** (def. 1), **hetero+** (def. 1), **xeno+** (def. 1). **2** *(in immunology and genetics)* **a** in or of some individual or ones that are genetically identical (e.g. **isogenic**). **b** in or of another, genetically differing individual of the same species (e.g. **isoantigen**); **allo+** (def. 2) is now usually preferred. **c** in or of all normal individuals of the same species (e.g. **isotypic**). *Compare* **allo+** (def. 2), **hetero+** (def. 2), **homo+** (def. 2), **xeno+** (def. 2). **3** *(in chemical nomenclature)* **a** signifying isomer of (formerly 'iso' was italicized and often hyphenated). **b** *symbol*: i (as a prefix or superscript); signifying the isomer of a straight-chain alkane or alkane derivative that differs only in having a terminal (CH$_3$)$_2$CH– group (leucine and isoleucine are exceptions). *Compare* **allo+** (def. 3), **nor+** (def. 1). **4** *(in names of certain classes of macromolecules)* signifying irrespective of location or origin, one of two or more structural variants with the same specificity of activity or action (e.g. **isoacceptor tRNA**, **isohormone**; *see also* **isoenzyme**).

ISO *abbr. for* International Standards Organization.

isoacceptor tRNA *or* **isoaccepting tRNA** any transfer RNA species obtainable from a given organism that can be acylated by the same amino acid; they may differ in their anticodons.

isoagglutinin *other names*: isohemagglutinin; isoantibody; a naturally occurring IgM antibody that is specific for one of the ABO blood group antigens on human erythrocytes. Type O blood has antibodies to A and B antigens; type A has antibodies to B antigen; type B has antibodies to A antigens; and type AB has neither anti-A nor anti-B antibodies. The presence of isoagglutinins in blood explains the disastrous effect of incompatible blood transfusions.

isoalloxazine benzo[g]pteridine-2,4(1H,3H)dione; the hypothetical tricyclic parent compound of any N^{10}-substituted alloxazine in riboflavin, flavin mononucleotide, or flavin-adenine dinucleotide.

isoamylase EC 3.2.1.68; *other name*: debranching enzyme; an enzyme that catalyses the hydrolysis of 1,6-α-D-glucosidic branch linkages in glycogen, amylopectin, and their β-limit dextrins. It does not attack pullulan, and is limited in action on α-limit dextrins.

isoantibody any antibody formed in one individual in response to an antigen derived from another individual of the same species. *See also* **isoagglutinin**.

isoantigen any antigen occurring in an individual that is capable of eliciting an immune response in genetically different individuals of the same species but not in the individual itself.

isoasparagine *the trivial name for* the α-amide of aspartic acid, aspartic 1-amide; β-aminosuccinamic acid; 3-amino-3-carbamoylpropanoic acid; H₂N–CO–CH(NH₂)–CH₂–COOH; a chiral α-aminoacylamide. L-Isoasparagine (*symbol*: Asp-NH₂) can occur as an N-terminal amidated residue of proteins. *Compare* **asparagine**.

isoasparaginyl *the acyl group*, H₂N–CO–CH(NH₂)– CH₂–CO–, derived from **isoasparagine**. *Compare* **asparaginyl**.

isoaspartyl *the former name for* **β-aspartyl**.

isobar 1 any of two or more nuclides having the same mass number (nucleon number) but different atomic numbers (proton numbers) and hence of differing chemical properties. *See also* **heterotope**. **2** any curve or equation relating quantities measured at the same pressure; any line joining points on a graph or map that have equal pressures. —**isobaric** *adj.*

isobutylmethylxanthine *abbr.*: IBMX; 3-isobutyl-1-methylxanthine; an inhibitor of **cyclic AMP phosphodiesterase**; it raises cyclic AMP levels in cells and tissues.

isocaloric *(of foodstuffs)* having the same **calorific value**.

isochore a stretch of DNA that is over 300 kb long and has a relatively homogeneous GC composition.

isochorismate 5-[(1-carboxylatoethenyl)oxy]-6-hydroxy-1,3-cyclohexadiene-1-carboxylate; the dianion of isochorismic acid; an isomer of **chorismate** and an important precursor to some compounds in the **shikimate pathway**.

isochorismic acid

isochromosome any chromosome having homologous, genetically identical arms. It results from abnormal splitting of a chromosome at the centrosome during mitosis so that one arm is lost and the other becomes duplicated.

isocitrate 1 any of the anionic forms of isocitric acid, 1-hydroxy-1,2,3-propanetricarboxylic acid. (1R,2S)-1-hydroxypropane-1,2,3-tricarboxylate (*also termed* (2R,3S)-isocitrate or *threo*-D_s-isocitrate) is an important intermediate in the **tricarboxylic-acid cycle** and the **glyoxylate cycle**. **2** any mixture of free isocitric acid and its ionized forms. **3** any (partial or full) salt or ester of isocitric acid.

(1R,2S)-1-hydroxypropane-1,2,3-tricarboxylate

isocitrate dehydrogenase 1 isocitrate dehydrogenase (NAD⁺) EC 1.1.1.41; *other names*: isocitric dehydrogenase; β-ketoglutaric-isocitric carboxylase; an enzyme that catalyses the oxidation of isoci-

trate to 2-oxoglutarate and CO₂, with reduction of NAD⁺ to NADH. The isomer of isocitrate involved is (1R,2S)-1-hydroxypropane-1,2,3-tricarboxylate. In mammalian tissues, this enzyme is localized exclusively in the mitochondria. **2 Isocitrate dehydrogenase (NADP⁺)** (EC 1.1.1.42); an enzyme that is localized in both mitochondria and cytosol.

[isocitrate dehydrogenase (NADP⁺)] kinase EC 2.7.1.116; *other names*: isocitrate dehydrogenase kinase/phosphatase (*abbr.*: IDH kinase/phosphatase); an enzyme that catalyses the phosphorylation by ATP of isocitrate dehydrogenase (NADP⁺) with formation of ADP. The phosphorylated protein is inactive. In *Escherichia coli* a bifunctional enzyme contains this and isocitrate dehydrogenase kinase/phosphatase (EC 3.1.3.–) activities.

isocitrate lyase EC 4.1.3.1; *systematic name*: isocitrate glyoxalate-lyase; an enzyme, present in the glyoxysomes of seeds, that converts (1R,2S)-1-hydroxypropane-1,2,3-tricarboxylate (the isomer of isocitrate involved) to succinate and glyoxylate. *See* **glyoxylate cycle**.

isocratic *(in chemistry)* describing any chromatographic procedure or separation in which the composition of the eluent is maintained constant. *Compare* **gradient elution**.

isocyanate any organic compound containing the univalent group –N=C=O. *Compare* **cyanate**, **isothiocyanate**.

isocyclic see **homocyclic** (def. 1).

isodalt a type of very-high-resolution two-dimensional electrophoresis system for the analysis of small (1–2 mg) samples of complex mixtures of proteins. It employs isoelectric focusing in cylindrical gels in the first dimension, followed by slab-gel electrophoresis in the presence of sodium dodecyl sulfate in the second dimension. Such systems are considered to be capable of resolving mixtures containing 10⁴ or more individual proteins within the range 10–200 kDa.

isodesmosine the 2-(4-amino-4-carboxybutyl)-1-(5-amino- 5-carboxypentyl)-3,5-bis(3-amino-3-carboxypropyl)pyridinium ion; a tetra(α-amino acid) isolated from hydrolysates of the fibrous protein elastin. *See* **desmosine**.

isodiametric 1 *(of cells)* having similar diameters in all planes. **2** *(of a crystal)* having three equal axes.

isodynamic *(in physics)* having equal force or strength.

isoelectric 1 having the same electrical potential; containing or indicating no potential difference. **2** *(of particles or molecular entities)* having no net electric charge.

isoelectric focusing *or* **electrofocusing** *abbr.*: IEF; a technique in which solutes of different isoelectric points are caused to form stationary bands in an electric field, which is superimposed on a (stable) pH gradient, the pH increasing from the anode to the cathode. In practice the pH gradient is most conveniently formed by electrolysing a solution containing a mixture of **carrier ampholytes** of low molecular mass and slightly differing isoelectric points, each of which will move to its isoelectric region in the electric field and remain there.

isoelectric-focusing electrophoresis a separation technique in which electrophoresis is effected in a gel slab at right angles to a preformed, stationary pH gradient generated by isoelectrically focused carrier ampholytes (*see* **isoelectric focusing**).

isoelectric fractionation any technique for fractionating proteins or other ampholytes using **isoelectric focusing**.

isoelectric pH *an alternative term for* **isoelectric point**.

isoelectric point *or* **isoelectric pH** *abbr.*: IEP *or* i.e.p.; *symbol*: pI; the pH of the solution in which a protein or other ampholyte has zero mobility in an electric field; hence the pH at which the protein or other ampholyte has zero net charge, i.e. no charges or an equal number of positive and negative charges including those due to any extraneous ions bound to the ampholyte molecule. The pH value of the isoelectric point may depend on other ions, except hydrogen and hydroxide ions, present in the solution. *Compare* **isoionic point**.

isoelectric precipitation the precipitation of a protein at its **isoelectric point**, at which proteins are generally least soluble. It is useful in the fractionation of mixtures of proteins.

isoelectronic describing two or more chemical entities having the same number of electrons, especially valence electrons. *Compare* **isosteric**.

isoenergetic taking place at constant energy.

isoenzyme *or* **isozyme** any one of the multiple forms of an enzyme arising from a genetically determined difference in primary structure. Isoenzyme is an operational term to denote any one of a group of enzyme proteins with similar catalytic properties but separable by suitable methods, e.g. by electrophoresis, where knowledge of the nature of the multiplicity may be lacking. If the different enzyme proteins are derived from a common gene, then the term 'multiple forms' rather than 'isoenzymes' should be used. The term embraces **allelozyme** and **multilocus enzyme**. *Compare* **interconvertible enzyme**.

isoflavone 7-*O*-glucosyltransferase EC 2.4.1.170; an enzyme that catalyses the reaction between UDPglucose and isoflavone to form isoflavone 7-*O*-β-D-glucoside and UDP.

isoflavonoid any of various yellow- or orange-coloured hydroxyl or alkoxyl derivatives of isoflavanone. They are plant pigments, present as glycosides or esters with tannic acid, have no recognized nutritional function, and are rapidly excreted in urine after ingestion. An example is genistein.

isoflurane 2-chloro-2-(difluoromethoxy)-1,1,1-trifluoroethane; a widely used nonflammable volatile anesthetic agent.

isoform any of several multiple forms of the same protein that differ in their primary structure, but retain the same function. They can be produced by different genes or by **alternative splicing** of RNA transcripts from the same gene.

isogeneic *or* **isogenic** describing two or more individuals that possess exactly the same genotype. *See also* **syngeneic**.

isoglucose *see* **high-fructose corn syrup**.

isoglutamine *a trivial name for* the α-amide of glutamic acid; glutamic 1-amide; γ-aminoglutaramic acid; 4-amino-4-carbamoylbutanoic acid; $H_2N–CO–CH(NH_2)–[CH_2]_2–COOH$; a chiral α-amino acylamide. L-Isoglutamine (*symbol*: Glu-NH$_2$) can occur as an N-terminal amidated residue of proteins. Residues of D-isoglutamine (*symbol*: D-Glu–NH$_2$) are found in *Staphylococcus aureus* peptidoglycan. *Compare* **glutamine**.

isoglutaminyl the acyl group, $H_2N–CO–CH(NH_2)–[CH_2]_2–CO–$, derived from **isoglutamine**. *Compare* **glutaminyl**.

isoglutamyl *a former name for* **γ-glutamyl**.

isograft *an alternative term for* **syngenic graft**.

isohormone any of two or more structural variants of a polypeptide hormone that have the same hormonal action and are produced within the same organism, or at least within the same species of organism. —**isohormonal** *adj.*

isohydric describing solutions that have the same hydrogen-ion concentration, hydrogen-ion activity, or pH.

isohydric shift the opposing and counterbalancing changes in hydrogen-ion concentration occurring in peripheral tissues when dioxygen leaves the blood and carbon dioxide enters it. The hydration of CO_2 to H_2CO_3, catalysed by erythrocyte carbonic anhydrase, has a tendency to lower the blood pH. This tendency is counteracted in some degree by the simultaneous deoxygenation of hemoglobin; deoxyhemoglobin binds hydrogen ions more strongly than oxyhemoglobin, thus tending to raise the blood pH.

isoionic pH *an alternative name for* **isoionic point**.

isoionic point *or* **isoionic pH** the pH of a solution of a protein or other ampholyte from which all other ions have been removed, except hydrogen and hydroxide ions formed by dissociation of water or of the protein or ampholyte itself. *Compare* **isoelectric point**.

isoionic solution any solution of a protein or other ampholyte that contains no other ions except those arising from dissociation of the solute and the solvent.

isokinetic gradient any concentration and viscosity gradient used

in centrifugation that allows particles (molecules) of the same density to sediment at a constant velocity at all distances from the centre of rotation.

isolate 1 to separate (a pure substance, a cell type, or a sample of a subcellular component) from a mixture or from naturally occurring material and then (usually) to characterize it; to separate and put into pure culture (a particular species or strain of a microorganism) from a mixture, sample, or biological specimen. **2** to remove a viable tissue or organ for the purpose of study or experiment. **3** (*in physics*) to disconnect (part of a circuit), or to prevent interaction between (circuits or components). **4** a group of similar microorganisms, especially a pure culture, obtained by isolation for study, experiment, or as an aid to diagnosis; something isolated. —**isolable** *or* **isolatable** *adj.*; **isolability, isolator** *n.*

isolectin any of two or more molecular forms of lectin, of the same origin, apparently with the same biological properties, and of very similar molecular masses, but of differing electrophoretic mobilities. They may be the products of closely related genes, or formed prior to or during isolation. In glycoprotein lectins, their differences may lie in the carbohydrate side chains.

isoleucinate 1 isoleucine anion; the anion, $CH_3–CH_2–CH(CH_3)–CH(NH_2)–COO^-$, derived from **isoleucine**. **2** any salt containing isoleucine anion. **3** any ester of isoleucine.

isoleucine *the trivial name for* α-amino-β-methylvaleric acid; (2*R*,3*R*)-2-amino-3-methylpentanoic acid; $CH_3–CH_2–CH(CH_3)–CH(NH_2)–COOH$; an α-amino acid with two chiral centres. Because molecules of isoleucine possess a second chiral centre, at C-3, in addition to the chiral centre at C-2 common to all α-amino acids other than glycine, the enantiomers L-isoleucine, (2*S*,3*S*)-2-amino-3-methylpentanoic acid (*symbol*: I *or* Ile) and D-isoleucine, (2*R*,3*R*)-2-amino-3-methylpentanoic acid (*symbol*: D-Ile *or* DIle), are diastereoisomeric with those of **alloisoleucine**, (2*R**,3*S**)-2-amino-3-methylpentanoic acid. L-isoleucine is a coded amino acid found in peptide linkage in proteins; codon: AUA (not in mitochondria of fruit fly, mammals, or yeast) (anticodon: UAU); AUC or AUU (anticodon: UAG). In mammals, it is an essential dietary amino acid, and is both glucogenic and ketogenic. D-Isoleucine is not known to occur naturally.

L-isoleucine

isoleucine biosynthesis a pathway that commences with formation of α-aceto-α-hydroxybutyrate (I) from pyruvate and α-ketobutyrate (2-oxobutanoate, formed by removal of ammonia from threonine by serine dehydratase). I is converted via α,β-dihydroxy-β-methylvalerate to α-keto-β-methylvalerate (II). Isoleucine is then formed from II as a result of transamination with glutamate.

isoleucinium isoleucine cation; the cation, $CH_3–CH_2–CH(CH_3)–CH(NH_3^+)–COOH$, derived from **isoleucine**.

isoleucino the alkylamino group, $CH_3–CH_2–CH(CH_3)–CH(COOH)–NH–$, derived from **isoleucine**.

isoleucyl the acyl group, $CH_3–CH_2–CH(CH_3)–CH(NH_2)–CO–$, derived from **isoleucine**.

isoliquivitigenin *see* **chalcone**.

isolog *the US spelling of* **isologue**.

isologous 1 of, relating to, or being two or more closely related chemical compounds in a series whose successive members differ in composition in a regular manner, other than by a difference of one methylene group. The term is now applied especially to compounds that have different atoms of the same valency at one or more positions in the molecule but are otherwise structurally identical, e.g. methionine and selenomethionine. **2** *an alternative term for* **isogeneic**.

isologous association *or* **isologous binding** any association between two identical protein subunits (i.e., **protomers**) in which the **domain of bonding** comprises two identical **binding sets**.

isologue or (*US*) **isolog** any of two or more **isologous** (def. 1) compounds.

isology the state or condition of being **isologous** (def. 1); chemical kinship among molecules or compounds.

isomaltase EC 3.2.1.10; *recommended name*: oligo-1,6-glucosidase; *systematic name*: dextrin 6-α-D-glucanohydrolase. An enzyme that hydrolyses 1,6-α-D-glucosidic bonds in dextrins and isomaltose produced after α-amylase degradation of starch and glycogen. The enzyme from intestinal mucosa also catalyses the reaction of EC 3.2.1.48 (*see* **sucrase**).

isomaltose the disaccharide 6-*O*-α-D-glucopyranosyl-D-glucose; it exists as the branching unit in **amylopectin**, **glycogen**, and some bacterial **dextrans**. *Compare* **maltose**.

isomaltulose the disaccharide 6-*O*-α-D-glucopyranosyl-D-fructofuranose.

isomer 1 (*in chemistry*) **molecular isomer** any of two or more compounds that have identical molecular formulas but differ in the nature of bonding of their atoms (**constitutional isomers**) or in the arrangement of their atoms in space (**stereoisomers**). 2 (*in physics*) **nuclear isomer** any of two or more nuclides that have the same proton number and the same mass (nucleon) number but different energy states.

isomerase any enzyme of EC class 5 that catalyses geometric or structural changes within one molecule. According to the type of isomeric change they catalyse, they may be called racemases or epimerases (EC subclass 5.1); *cis–trans*-isomerases (EC subclass 5.2); intramolecular oxidoreductases (EC subclass 5.3); intramolecular transferases (mutases) (EC subclass 5.4); or intramolecular lyases (EC subclass 5.5); other isomerases are placed in EC subclass 5.99.

isomeric of or pertaining to an **isomer**; pertaining to the phenomenon of, or displaying, **isomerism**.

isomeric state any of the states of an atomic nucleus having a different energy and observable half-life from those of other states of the same nucleus; the condition of being an **isomer** (def. 2).

isomeric transition *abbr.*: IT; the change of one nuclear isomer (*see* **isomer** (def. 2)) into another without change in either proton number (atomic number) or nucleon number (mass number), energy being lost in the form of gamma radiation; e.g. the decay of the metastable nuclide technetium-99m (half-life 6 h) to the unstable nuclide technetium-99 (half-life 2.1×10^5 years), which in turn decays to ruthenium-99 with emission of an electron.

isomerism the phenomenon of the existence of **isomers** in general, or the occurrence of molecular or nuclear isomers in a particular instance.

isomerization or **isomerisation** the act of changing or the process of change of one molecular isomer into another one. —**isomerize** or **isomerise** *vb.*

isometric 1 having identical dimensions. 2 (*in physiology*) of or pertaining to a contraction of a muscle that does not cause shortening of the muscle. 3 (*in crystallography*) having three mutually perpendicular axes of equal length; cubic.

isomorph a substance or organism that is **isomorphic** (def. 1) to another substance or organism.

isomorphic 1 or **isomorphous** of identical or similar form, shape, or

structure. 2 describing a type of alternation of generations in which morphologically identical generations occur in both diploid and haploid phases. It occurs mainly in algae. *Compare* **heteromorphic**.

isomorphism 1 (*in biology*) similarity of form in organisms of different ancestry. 2 (*in chemistry*) similarity of crystalline form and structure between substances of similar composition or substances that can form a more or less continuous series of solid solutions.

isomorphous 1 an alternative term for **isomorphic** (def. 1). 2 displaying **isomorphism** (esp. def. 2).

isomorphous replacement method a method used to solve the phase problem in **X-ray diffraction** analysis of macromolecular structures. It involves combining one or more **heavy atoms** (def. 2) (which are strong scatterers) in a specific manner with the macromolecule in question and investigating their effects on the diffraction pattern. The heavy atom must not modify either the conformation or the crystal form of the macromolecule, i.e. the replacement must be isomorphous.

isoniazid 4-pyridinecarboxylic acid hydrazide; an antitubercular drug that may act by inhibiting the synthesis of the mycolic acids that are vital for the construction of the cell wall of the pathogen, *Mycobacterium tuberculosis*. Like related hydrazides it is highly hepatotoxic and no longer used. Its euphoric side effects led to the discovery of monoamine oxidase inhibitors.

isoosmotic a variant spelling of **isosmotic**.

isopartitive (*in chromatography*) describing two or more solvents that bring about similar partitions of a given substance between the solvent and the stationary phase and hence result in R_f values that are the same or nearly so.

isopentenyl the trivial name for 3-methylbut-3-enyl, $CH_2=C(CH_3)-CH_2-CH_2-$, the alkenyl group formally derived from isopentenyl alcohol (i.e. 3-methylbut-3-en-1-ol) by loss of its hydroxyl group. *Compare* **prenyl**.

isopentenyl diphosphate *abbr.*: IDP; *formerly called* Δ^3-isopentenyl pyrophosphate (*abbr.*: IPP); an isomer of dimethylallyl diphosphate and the key precursor of all **isoprenoids** (def. 2). *See also* **dimethylallyl*trans*transferase**, **isopentenyl-diphosphate Δ-isomerase**, **isoprene rule**.

isopentenyl-diphosphate Δ-isomerase EC 5.3.3.2; *systematic name*: isopentenyl diphosphate Δ^3–Δ^2-isomerase; *other name*: IPP isomerase. An enzyme in the pathway for the synthesis of cholesterol, terpenes, and prenyl groups; it catalyses the reversible reaction:

isopentenyl diphosphate = dimethylallyl diphosphate.

isopeptidase any peptidase catalysing the cleavage of an **isopeptide bond**.

isopeptide bond any peptide bond other than a **eupeptide bond**, i.e. any amide bond formed between a carboxyl group of one amino-acid molecule or residue and an amino group of another, where either group occupies a position other than α (or both do so). Examples include peptide bonds formed from the β-carboxyl group of aspartic acid, the γ-carboxyl group of glutamate from the ε-amino group of lysine, and from either the carboxyl group or the amino group of β-alanine.

isopiestic of, pertaining to, or characterized by equal pressures.

isopiestic vapour pressure method a method for the determination of the activity of water in a solution of a nonvolatile solute. The solution in question is equilibrated with a reference solution containing a nonvolatile solute for which the activity of water at different solute concentrations has been determined with high precision. The two solutions are placed in an enclosure at constant temperature until their vapour pressures are equal, i.e. the activity of water in both solutions is equal. They are then analysed for their solute contents, from which the activity of water in both of them can be calculated.

isoprenaline *an alternative name for* **isoproterenol**.

isoprene 2-methylbuta-1,3-diene; $CH_2=C(CH_3)-CH=CH_2$; a liquid hydrocarbon hemiterpene isolated from products of the pyrolysis of natural rubber, turpentine, etc., and widespread in plants although generally found only in extremely small amounts.

isoprene residue the branched-chain bivalent group, 3-methyl-2-butenylene; 3-methylbut-2-en-1,4-diyl; $-CH_2-C(CH_3)=CH-CH_2-$. It is formally derived by bond rearrangement from **isoprene** and constitutes the repeating unit of the molecular structures of prenols and many other isoprenoids.

isoprene rule a rule, formulated by Wallach in 1887, that stated that many natural products are built up from isoprene carbon skeletons. It was modified to the **biogenetic isoprene rule** by Croatian-born Swiss chemist Leopold Ružička (1887–1976), who proposed that all terpenoids, steroids, etc. are synthesized from a precursor called 'active isoprene'. This precursor is now known to be **isopentenyl diphosphate**.

isoprene unit *(in an isoprenoid)* an **isoprene residue** or any branched five-carbon-atom moiety formally derived from an isoprene residue by chemical modification, e.g. by hydrogenation, hydroxylation, etc. It represents isopentenyl alcohol (i.e. 3-methylbut-3-en-1-ol), the diphosphate of which is the universal biogenetic precursor of isoprenoids.

isoprenoid 1 of, pertaining to, or being **isoprene** and its chemical relatives; having a molecular structure composed of, containing, or derived from linked **isoprene residues**. **2** any isoprenoid compound. Isoprenoids comprise a large and varied family of natural products, including carotenoids, phytol, prenols, rubber, steroids, terpenoids, and tocopherols. Many substances contain both isoprenoid and nonisoprenoid components, e.g. the vitamins K.

isoprenyltransferase any of the enzymes within subclass EC 2.5.1, existing as α–β heterodimers, that add C_{15} or C_{20} carbon chains by **prenylation** near the carboxyl terminus of proteins. Examples include the nuclear lamins and Ras.

isopropyl *symbol*: iPr or Pr^i; the alkyl group $(CH_3)_2CH-$, derived from propane. *Compare* **propyl**.

isopropylmalate either of the isomeric substances 2-isopropylmalate (i.e. 3-carboxy-3-hydroxy-4-methylpentanoate) or 3-isopropylmalate (i.e. 3-carboxy-2-hydroxy-4-methylpentanoate); they are enzymatically interconvertible intermediates in the biosynthesis of leucine in plants. *See also* **3-isopropylmalate dehydratase**.

(2*R*,3*S*)-3-isopropylmalic acid

3-isopropylmalate dehydratase EC 4.2.1.33; *systematic name*: 3-isopropylmalate hydro-lyase; an iron–sulfur protein, containing a [4Fe-4S] cluster, that catalyses the isomerization of 3-isopropylmalate to 2-isopropylmalate, by dehydration to 2-isopropylmaleate and rehydration to generate the isomer, in the pathway of L-leucine biosynthesis in plants. *See* **leucine and valine biosynthesis**.

3-isopropylmalate dehydrogenase EC 1.1.1.85; *systematic*

name: 3-carboxy-2-hydroxy-4-methylpentanoate:NAD$^+$ oxidoreductase; an enzyme that catalyses the oxidation of 3-**isopropylmalate** by NAD$^+$ (followed by loss of CO_2) to 4-methyl-2-oxopentanoate, in the pathway of L-leucine biosynthesis in plants. *See* **leucine and valine biosynthesis**.

2-isopropylmalate synthase EC 4.1.3.12; *systematic name*: 3-carboxy-3-hydroxy-4-methylpentanoate 3-methyl-2-oxobutanoate-lyase (CoA-acetylating); an enzyme that condenses acetyl-CoA with 3-methyl-2-oxobutanoate to form 2-isopropylmalate, in the pathway of L-leucine biosynthesis in plants. *See* **leucine and valine biosynthesis**.

isopropyl β-thiogalactoside *abbr*.: IPTG; isopropyl β-D-1-thiogalactopyranoside; an inducer of β-galactosidase in bacteria; it also induces the *lac* operon by binding and inactivating the *lac* repressor. It is used in cloning techniques to enhance expression of DNA cloned into vectors containing the *lac* operon.

isoproterenol *or* **isoprenaline** 3,4-dihydroxy-α-[(isopropylamino)methyl]benzyl alcohol; the synthetic isopropyl analogue of epinephrine. It is a selective agonist of β adrenoceptors, but does not discriminate between $β_1$ and $β_2$ subtypes.

isopycnic of, pertaining to, or characterized by equal densities.

isopycnic banding the formation during **isopycnic centrifugation** of one or more bands composed of molecules that have the same density as that of the solvent at that position in the gradient; the process of separating or purifying substances by forming such bands.

isopycnic centrifugation *or* band centrifugation *or* density-gradient centrifugation *or* density-gradient sedimentation equilibrium a separation technique in which macromolecules are subjected to a centrifugal field in a solvent containing a density gradient. The macromolecules will sediment or float and arrange themselves in bands (isopycnic bands) in the column of solvent where the density of the solvent is the same as that of the particular macromolecule. *Compare* **isokinetic gradient**.

isoquinoline 2-benzazine; benzo[*c*]pyridine; a bicyclic, aromatic, nitrogen-containing compound. Many alkaloids contain an isoquinoline ring system.

isoquinolinesulfonylmethylpiperazine *see* **H7**.

isosbestic point any wavelength at which the molar **absorption coefficients** of two or more particular chemical species are equal, i.e. their absorption spectra intersect. All mixtures (or solutions) of two or more such species, having the same total concentration, will then exhibit absorption spectra that pass through the isosbestic point, provided the species do not interact.

isoschizomer any of two or more restriction endonucleases from different sources that recognize the same sequence in a DNA molecule and display identical specificity of cleavage within that sequence. The term is sometimes extended to include those enzymes that cleave at different sites within the sequence recognized, or that respond to methyl groups on different bases in the sequence. Examples are *Msp*I and *Hpa*II, both of which cleave the unmethylated sequence CCGG. —**isoschizomeric** *adj*.

isosemantic substitution the incorporation of an amino-acid residue into its normal position in a polypeptide chain in response to a mutated but synonymous codon.

isosmotic or **isoosmotic** *(of two or more solutions)* having equal osmotic pressure; having the same osmotic pressure as another solution. *See also* **isotonic** (def. 2).

isostere or **isoster** any of two or more **isosteric** chemical species.

isosteric *(of two or more chemical species)* having the same, or similar, shape and size, e.g. L-threonine and L-valine; having the same number of valence electrons arranged in a similar manner. *Compare* **isoelectronic**. —**isosterism** *n*.

isotachophoresis or **isotachoelectrophoresis** or **displacement electrophoresis** a method for separating ion species in which all ions under observation are given the same migration velocity; moreover, they all have the same sign and a common counterion. The sample ions are introduced as a zone between a leading electrolyte, whose ions have a higher mobility than any of the sample ions, and a terminating electrolyte, whose ions have a lower mobility than any of the sample ions.

isotactic of, relating to, or having stereochemical regularity in the repeating units of a polymer. *Compare* **syndyotactic**.

isotherm or **isothermal** any line on a graph, or diagram, that connects points of equal temperature.

isothermal 1 *(of a process or reaction)* occurring at constant temperature. 2 of or pertaining to an isotherm. 3 *an alternative name for* isotherm.

isothiocyanate any organic compound containing the monovalent group –N=C=S. *Compare* **thiocyanate**.

isotocin or *(formerly)* **isotocine** [Ser^4,Ile^8]oxytocin; the somewhat less potent structural variant of **oxytocin**. It is present in many bony fish in place of oxytocin, hence it is sometimes known as **ichthyotocin**. *See also* **mesotocin**.

isotone any of two or more nuclides having the same number of neutrons but, usually, different numbers of protons and therefore differing in atomic number.

isotonic 1 of, or having the same tension; used especially of muscular contraction occurring in the absence of significant resistance and accompanied by marked shortening of the muscle fibres. *Compare* **isometric** (def. 2). 2 having the same osmotic pressure as some particular solution, especially one in a cell or body fluid; **isosmotic**. *Compare* **hypertonic**, **hypotonic**. 3 of, relating to, or being an **isotone** or isotones.

isotonically under **isotonic** (def. 1) conditions.

isotonicity the state or quality of being **isotonic** (def. 2); the degree of osmotic pressure.

isotope any of two or more **nuclides** that have the same **proton number**, i.e. that are nuclides of the same chemical element and thus display qualitatively the same chemical properties (for quantitative differences, *see* **isotope effect**). Isotopes with a given proton number usually have different **nucleon numbers** and thus characteristically usually have different relative atomic masses. In some instances, however, two isotopes of an element are isomeric, i.e. they have the same nucleon number (and hence the same relative atomic mass) but differ instead in their nuclear energy levels, the isotope with the higher level being **metastable** (def. 3). For the majority of elements, isotopes occur naturally and at more or less constant relative abundances; additional isotopes can be produced artificially. In general, any given element possesses both stable and unstable isotopes, although there are no stable ones for elements of high proton number; conversely, the naturally occurring isotopes of elements of lower proton number are, with a few exceptions, all stable. Metastable and unstable isotopes undergo radiative decay and are collectively termed radioactive isotopes, or radioisotopes. (*Note*: In common parlance the term isotope is often used loosely as a syn-

onym of nuclide irrespective of its proton number.) *Compare* **heterotope**, **isobar**, **isotone**. —**isotopic**, *adj*.

isotope abundance the number of atoms of a particular isotope in a mixture of the isotopes of an element in a sample, expressed as a percentage of all the atoms of that element in the sample. *Compare* **isotope ratio**.

isotope dilution analysis an accurate method for measuring the concentration of an element, or a compound of that element, in a system by adding a known amount of a different isotope, or a labelled compound, and then measuring its concentration in a sample withdrawn from the system after mixing.

isotope effect any variation in some chemical, physicochemical, or biochemical property of a molecular entity that is detectable on isotopic substitution of an atom or atoms of the entity, e.g. variation in reaction rate, in chromatographic mobility, or in binding to macromolecules. For a given pair of isotopes, 'light' and 'heavy', the effect may be expressed quantitatively either (in the **kinetic isotope effect**) as the ratio of the rate constants, k_{light}/k_{heavy}, or (in the **equilibrium isotope effect**) as the ratio of the equilibrium constants, K_{light}/K_{heavy}, for two chemical reactions that differ only in the isotopic composition of one or more of their otherwise chemically identical components. The magnitude of the effect depends on the relative difference in atomic mass of the isotopes in question; thus it is the greatest by far for isotopes of hydrogen.

isotope exchange any chemical reaction in which the reactants and products are chemically identical but of different isotopic composition. In such a reaction the isotope distribution tends towards equilibrium.

isotope farming *an informal term sometimes used to describe* biological techniques for the large-scale preparation, usually for sale, of radioisotope-labelled biochemicals, especially when a number of such compounds are obtainable in the same operation, e.g. the simultaneous preparation of a number of ^{14}C-labelled compounds from ^{14}CO$_2$ by photosynthesis in whole leaves or whole plants.

isotope ratio or **isotopic ratio** the ratio between the numbers of atoms of any two isotopes of a given element in a particular preparation, sample, or substance. *Compare* **isotope abundance**.

isotope-ratio mass spectrometer a mass spectrometer designed specifically for the determination of **isotope ratios**.

isotopic 1 of, relating to, or being an **isotope** or isotopes. 2 being or containing a less common or artificial isotope, especially as a label or tracer. 3 *(especially of techniques)* using or depending on isotopes. —**isotopically** *adv*.

isotopically coded affinity tag abbr.: ICAT; a reagent used for quantitation of differential protein expression. The proteins in two samples to be compared are labelled with a reagent containing a thiol-specific reactive group that binds to the side chain of reduced cysteines, with a linker group that is either deuterated or non-deuterated, and a group such as biotin that allows affinity isolation of the ICAT-labelled peptides from complex mixtures using affinity chromatography. The isolated and tagged peptides are analysed by mass spectrometry.

isotopy the phenomenon of the existence of isotopes in general; or the fact or condition of being isotopic in a particular instance.

isotropic or **isotropous** 1 *(in physics)*: exhibiting uniform properties when measured along axes in all directions. *Compare* **anisotropic**. 2 *(in biology)* not having predetermined axes of growth. —**isotropically** *adv*.; **isotropy** or **isotropism** *n*.

isotropic band or **I-band** *see* **sarcomere**.

isotropic motion the random tumbling of a chemical species about its three axes.

isotropous *an alternative word for* **isotropic**.

isotypic specificity any antigenic specificity that is the same in a given protein in all normal individuals of the same animal species. For example anti-rabbit gamma globulins prepared in another animal species do not distinguish between gamma globulins from different rabbits.

isotypic variation structural variability of any one antigen that is common to all members of the same species.

isovaleric acidemia a hereditary metabolic disease of humans, due to various mutations in **isovaleryl-CoA dehydrogenase**. Of variable severity, it is characterized by metabolic acidosis and excessive urinary excretion of isovaleric acid or isovaleryl conjugates.

isovaleryl-CoA dehydrogenase *abbr.*: IVD; EC 1.3.99.10; a mitochondrial matrix flavoprotein enzyme of the pathway for degradation of leucine. It catalyses the oxidation of 3-methylbutanoyl-CoA to 3-methylbut-2-enoyl-CoA. The flavin is FAD.

isozyme *an alternative word for* **isoenzyme**.

isozymogen any precursor of an isozyme.

ISSD *abbr. for* infantile sialic acid storage disease.

ISWI *abbr. for* initiation switch; an ATPase protein component of **NURF** that with GAGA factor initiates expression of the Hsp70 gene in *Drosophila*.

IT *abbr. for* isometric transition.

itaconate methylenesuccinate; methylenebutanedioate; the dianion $^-OOC-CH_2-C(=CH_2)-COO^-$; a constitutional isomer of **citraconate**. It is a metabolite of strains of *Aspergillus terreus*, which can yield it in large amounts by decarboxylation of *cis*-aconitate. It is a comparatively reactive compound, with properties useful in the manufacture of synthetic resins and plastics.

ITAM *abbr. for* immunoreceptor tyrosine-based activation motif; a consensus sequence for src-family tyrosine kinases. ITAMs are found in the cytoplasmic domains of several signalling molecules including the signal transduction units of lymphocyte antigen receptors and of Fc receptors.

iterative process a process in which a set of steps or instructions is repeatedly executed until a predefined or natural end point is reached. For example, in sequence analysis PSI-BLAST (position-specific iterated-BLAST) runs the BLAST program multiple times until no further matches are identified above a significance threshold.

iteron any sequence of nucleotides in the duplex DNA of certain lambdoid phages that is reiterated several times with only minor changes. These repeated sequences, of 13–19 base pairs, are usually separated by short spacer sequences, each having internal dyad symmetry.

ITF *abbr. for* intestinal trefoil factor.

ITIM *abbr. for* immunoreceptor tyrosine-based inhibitory motif; any of various sequences that are present in the cytoplasmic domains of certain cell surface molecules e.g. Fcγ, inhibiting **NK cell** receptors that mediate inhibitory signals.

+itis *suffix forming nouns* denoting inflammation of the (specified) organ, tissue, etc.

ITP *abbr. for* **1** inosine (5′-)triphosphate. **2** isotachophoresis.

itraconazole a complex derivative of triazole, useful as an antimycotic. It inhibits cytochrome P450-dependent enzymes, thereby impairing the synthesis of ergosterol.

+ity *suffix forming nouns* meaning a property of a substance; e.g. absorptivity, density, solubility.

IU *abbr. for* International Unit *(preferred alternative to* **i.u.** *or* **u** *)*.

IUB *abbr. for* International Union of Biochemistry; renamed (1992) International Union of Biochemistry and Molecular Biology (IUBMB).

IUBMB *abbr. for* International Union of Biochemistry and Molecular Biology.

IUdR *or* **IUDR** *abbr. for* 5-iododeoxyuridine (IdUrd preferred).

IUPAB *abbr. for* International Union of Pure and Applied Biophysics.

IUPAC *abbr. for* International Union of Pure and Applied Chemistry.

iupaciubin alanyl-lysyl-glutamyl-tyrosyl-leucine; a pentapeptide containing one residue each of the five main types of amino acid found in peptides and proteins, i.e. small neutral, basic, acidic, aromatic, and lipophilic. It exists only on paper, and was designed as a model to illustrate nomenclature recommendations relating to the modification of named peptides. Its name symbolizes the cooperation between IUPAC and IUB.

IUPAP *abbr. for* International Union of Pure and Applied Physics.

IUPHAR *abbr. for* International Union of Basic and Clinical Pharmacology.

IV *or* **i.v.** *abbr. for* intravenous or intravenously.

IVD *abbr. for* isovaleryl-CoA dehydrogenase.

ivermectin 22,23-dihydroabamectin; 22,23-dihydroavermectin B_1; a semisynthetic material derived from **abamectin** and used as an anthelmintic in both clinical and veterinary practice, especially against onchocerciasis. It is used extensively as a once-a-month pill against heartworm in dogs.

component B_{1a}, R = C_2H_5
component B_{1b}, R = CH_3

ivermectin

iVI *symbol for* the isovaleryl group, $(CH_3)_2CH-CH_2-CO-$ (alternative to VI^i).

Jj

j *symbol for* electric current density (alternative to *J*).

J *symbol for* **1** joule. **2** J chain (of an immunoglobulin).

J *symbol for* **1** coupling constant. **2** flux (def. 2). **3** electric current density (preferred alternative to *j*).

jack bean **1** an annual tropical American leguminous plant, *Canavalia ensiformis*, grown in the USA for forage. **2** the seed of this plant. It is the source of the globulin **canavalin**, the toxic amino acid L-canavanine, the lectin **concanavalin A**, and the enzyme **urease**.

jack-knife *(in phylogeny)* a statistical method for assessing confidence levels of inferred phylogenetic relationships and the variance of estimated evolutionary parameters, in which values are iteratively and randomly removed, each new data set being used to recalculate the relationships. A **consensus tree** is built after multiple repetitions, and the degree of variance of the sample is determined by comparing each branch of the original tree with the number of times it was found by the jack-knife process.

Jacob–Monod model a model of genetic regulation of protein synthesis in prokaryotes whereby the structural genes that determine the organization of the proteins are controlled by other regions of DNA upstream from the structural genes. The latter consist of a regulator gene and a **control site**, which comprises the **promotor** (*p*) and **operator** (*o*). The regulator gene produces a **repressor** (*r*), which interacts with the operator. The operator and its associated genes are called an **operon**. The structural genes are transcribed into a single mRNA coding for the several proteins. An mRNA coding for more than one protein is known as a **polygenic** or **polycistronic** transcript. The first example was the *lac* **operon** but the model has been used to elucidate many other such phenomena, e.g. the **tryptophan** and **arabinose** operons. The repressor can be either inactivated (**induction**) or activated (**repression**) by specific metabolites. While the model has been helpful in explaining the mode of expression of genes in eukaryotes, operons have not been described in them. [After François Jacob (1920–) and Jacques Monod (1910–76), French molecular biologists.]

Jagged-1 *see* **Alagille syndrome**.

JAK *abbr. for* Janus kinase; any of three cytosolic protein tyrosine kinases that bind the intracellular domains of certain membrane receptors. **JAK1** is activated by binding on the non-γ polypeptide of receptors for the interleukins IL-2, IL-4, IL-7, IL-9, and IL-15. It phosphorylates **STAT** proteins other than STAT 5a and 5b. **JAK2** is activated on binding receptors for somatotropin, several cytokines, and leptin. It then phosphorylates STAT proteins, which dimerize and enter the nucleus to activate specific genes. **JAK3** is activated on binding the γ polypeptide of receptors for IL-2, IL-4, IL-7, IL-9, and IL-15. It phosphorylates STAT 5a and 5b. Several mutations in the gene at 19p13 for JAK2 account for ≈7% of patients with severe combined immunodeficiency (SCID).

JalView *see* **alignment editor**.

Jamaican vomiting sickness a syndrome of sickness, vomiting, and poisoning of the nervous system, associated with profound hypoglycemia, metabolic acidosis, and excessive urinary excretion of butyrate and glutarate, induced by consumption of the unripe fruit of the **ackee**, the toxic principle of which is **hypoglycin**.

Janus kinase *see* **JAK**.

jasmonic acid a plant hormone that consists of a $C_{12:1}$ fatty acid that is unsaturated between C-9 and C-10 and contains a 6-oxocyclopentane ring formed by C-3 to C-7. Derived metabolically from α-linolenic acid, it inhibits growth processes in many tissues, modulates genes involved in development, and plays a role in plant resistance to insects and disease. It is metabolized by hydroxylation and conjugation to the branched chain *α*-amino acids. As methyl jasmonate and the structurally related *cis*-jasmone, it is the fragrant component of oil of jasmine, derived from jasmine plants (*Jasminum* spp.).

jaundice *or* **icterus** a yellowing of the skin and/or of the whites of the eyes indicating an excess of bilirubin in the blood. Jaundice may be classified into three types depending on cause: **obstructive jaundice**, in which the passage of bile from the liver to the intestine is obstructed; **hepatocellular jaundice**, in which there is disease and/or inflammation of liver cells thus diminishing their ability to secrete bile pigments into the bile passages; and **hemolytic jaundice**, in which there is an excessive breakdown of erythrocytes in the body and hence an excess of bile pigments.

Java *see* **programming language**.

javascript *see* **programming language**.

JCBN *abbr. for* (IUPAC–IUB *or* IUPAC–IUBMB) Joint Commission on Biochemical Nomenclature. *See also* **CBN**.

J chain *or* **J piece** *abbr. for* joining chain *or* joining piece; a 15 kDa, cysteine-rich polypeptide occurring in **immunoglobulin A** (IgA) and **immunoglobulin M** (IgM) molecules. It cross-links the monomeric immunoglobulin moieties to form a dimer or pentamer in IgA and IgM, respectively. It is not to be confused with the J segment of an immunoglobulin gene (*see* **joining gene**).

jejunum the part of the small intestine of mammals between the duodenum and the ileum. —**jejunal** *adj.*

jekyll a zebrafish mutant phenotype that shows defective initiation of heart-valve formation. The mutation involves a homologue of the *Drosophila sugarless* gene that encodes a UDP-glucose dehydrogenase required for production of heparan sulfate, chondroitin sulfate, and hyaluronic acid.

jelly roll a protein fold consisting of two antiparallel β-sheets placed one above the other to form a sandwich structure. The polypeptide chain follows a route in which successive β-strands alternate between the two sheets. If the β-strands are numbered β $_1$, β $_2$, etc., and the β-sheets numbered in parentheses, as (1), (2), etc., the structure of the motif can be represented as loop-β $_1$(1)-loop-β $_2$(2)-loop-β $_3$(1)-loop-β $_4$(2)- loop-β $_5$(1), etc. The jelly-roll fold was first observed in viral coat proteins, but has since been found in other proteins.

Jerne's theory of immunity an *alternative name for* **natural-selection theory of immunity**.

jervine a steroidal alkaloid, found in plants of the genus *Veratrum*, that inhibits cholesterol biosynthesis. Pregnant ewes that ingest these plants give birth to lambs with congenital malformations.

J gene *abbr. for* joining gene.

JIP *abbr. for* JNK-interacting protein; *other name*: kinesin-binding protein; any of three proteins (JIP1 to JIP3) that form part of the **JNK** signalling pathway and are present in cells and organisms that contain **kinesin** light chains (KLC). JIP1 and JIP2 are highly homologous and interact through their C-terminal regions with tetratricosapeptide sequences of KLC. JIP3 is not homologous with the others and interacts through internal sequences with the same sequences in KLC as the others. They are adaptor molecules between kinesins and their cargo. They also bind MAP kinase, MAP kinase kinase, and MAP kinase kinase kinase.

JIPID *abbr. for* Japan International Protein Information Database; *see* **sequence database**.

JIT *abbr. for* Just In Time (compiler); a program that converts **Java** to local machine code at runtime.

JM101 one of a series of *E. coli* strains that support the growth of bacteriophage M13 vectors. The F′ episome carries *lacZM15* making them suitable for alpha-complementation with plasmids encoding the N-terminal peptide of *β*-galactosidase.

JM109 an *E. coli* strain from the same series as **JM101** that carries *recA1* and *endA1* mutations.

JmjC the C-terminal domain of jumonji protein. *See* **jumonji**.

JmjN the N-terminal domain of jumonji protein. *See* **jumonji**.

JNK *abbr. for* c-Jun N-terminal kinase; *other name*: stress-associated protein kinase (*abbr.*: SAPK); a nuclear protein serine/threonine kinase, belonging to the **MAP kinase** family, that phosphorylates residues on the N-terminal region of the transcription factor c-Jun (also known as AP1). It is activated by binding to **Cdc42** and regulates cell growth and differentiation, and apoptosis. It is also activated by UV light, thermal stress, and tumour necrosis factor α.

JNK-interacting protein *see* JIP.

Job's method *an alternative name for* **continuous variation**.

Johnston–Ogston effect a phenomenon occurring in sedimentation velocity experiments when two or more solutes are present in a mixture and their sedimentation coefficients are mutually concentration-dependent. It follows that the sedimentation coefficient of the more slowly sedimenting component is less in the plateau region containing both species than it is in the plateau region between the boundaries, resulting in a higher concentration of the more slowly moving component in the plateau region between the boundaries. As a consequence, the apparent concentration of the faster moving species, based on the concentration difference between the two plateaus, is too low, while the apparent concentration of the more slowly moving species is too high. [After Alexander George Ogston (1911–96) and Joseph Percy Johnston (1923–76), British biochemists.]

joinase *or* **joining enzyme** *a former name for* DNA ligase (ATP) and DNA ligase (NAD⁺). *See* **DNA ligase**.

joining gene *or* **J segment** *abbr.*: J gene; any of approximately five germ-line genes concerned in coding for **immunoglobulin** synthesis. The J genes code for the fourth framework region of both light and heavy immunoglobulin chains and for part of the third hypervariable region of the light chain. *Compare* **J chain**.

joining peptide an acidic, 18-residue peptide situated between γ_3-MSH and ACTH in **proopiomelanocortin**.

Jones–Mote sensitivity a weak, delayed-type hypersensitivity of the skin seen on challenge a few days after priming with soluble protein in aqueous solution or in incomplete Freund's adjuvant. The cells infiltrating the lesion characteristically include a high proportion of basophils.

joule *symbol*: J; the SI derived unit of energy, work, or heat, equal to a force of one newton acting over a distance of one metre; i.e. 1 J = 1 N m = 1 W s. The joule is equivalent to 10^7 ergs and is the energy dissipated by one watt in one second. [After James Prescott Joule (1818–89), British physicist.]

J piece *an alternative name for* **J chain**.

J segment *an alternative name for* **joining gene**.

jumonji a gene of mice and humans that is required for neural-tube formation. The protein contains an N-terminal JmjN domain, a dead ringer domain, and a C-terminal JmjC domain. The two Jmj domains form a functional unit. The JmjC domain also occurs alone in hairless, retinoblastoma protein-binding protein 2, and several putative chromatin-binding proteins. In the mouse, mutations in *jumonji* cause embryonic lethality.

jumping gene *an informal term for* **transposon**.

JUN a gene family encoding nuclear transcription factors; v-*jun* is the oncogene from avian sarcoma virus ASV17 (named from *junana*, Japanese for seventeen). *JUN* encodes transcription factor AP1; it belongs to the *fos/jun* family, and is rapidly and transiently expressed on stimulation of cells by mitogens. The product, Jun, forms heterodimers with Fos, and these, together with Jun homodimers, bind to the AP1 consensus site. Jun is a leucine zipper protein, which also contains basic sequences essential for DNA binding. In resting cells Jun is phosphorylated on four serines and one threonine; removal of one or more of these phosphates leads to increased binding to the AP1 promoter site. Jun has five motifs.

junction potential *see* **liquid-junction potential**.

'junk' DNA *see* **selfish DNA hypothesis**.

juvenile hormone *or* **neotenin** any insect hormone, produced in the corpora allata glands, that functions to keep the insect in the larval (juvenile) stage. At least three are known; all are derivatives of farnesoic acid. They have potential as insect control agents.

juvenile-hormone esterase *see* **methylesterase**.

juvenile macular degeneration *an alternative name for* **Stargardt macular dystrophy**.

juxtacrine stimulation a mode of intercellular communication established by the binding of a membrane-anchored growth factor on one cell to its receptor on an adjacent cell, e.g. in the case of pro-transforming growth factor-α (*see* **transforming growth factor**). Unlike endocrine, paracrine, or autocrine modes there is no diffusible factor.

Kk

k *symbol for* kilo+ (SI prefix denoting 10^3).

k *symbol for* **1** rate constant; the various associated terms are distinguished by subscripts and/or superscripts (see individual entries below). **2 Boltzmann constant**. *See also* $k(T)$.

$\bar{k}$ *symbol for* pH-independent value of the rate constant k.

k_0 *symbol for* catalytic constant (of an enzyme); alternative to k_{cat}.

k_A, k_B *symbols for* the specificity constants of substances A and B.

k_{app} *symbol for* the apparent value of the rate constant k.

k_{cat} *symbol for* catalytic constant (of an enzyme); alternative to k_0.

k_i, k_{-i} *symbols for* the forward and reverse rate constants, respectively, for the ith step in an enzyme reaction.

k_{ij} *symbol for* the rate constant for the step from E_i to E_j in an enzyme reaction.

K *symbol for* **1** potassium. **2** kelvin. **3** a residue of the α-amino acid L-lysine (alternative to Lys). **4** a residue of an incompletely specified base in a nucleic-acid sequence that may be guanine, or either thymine (in DNA) or uracil (in RNA). **5** phylloquinone. **6** (*in computing*) indicating multiplied by 1024 (= 2^{10}).

K562 an erythromegakaryoblastic leukemia cell line. Cells in culture can be induced into erythroid development by treating them with sodium butyrate.

K *symbol for* **1** (thermodynamic) equilibrium constant (of a chemical reaction); the various associated terms are distinguished by subscripts and/or superscripts (see individual entries below). **2** kinetic energy. **3** kerma (kinetic energy released in matter). **4** luminous efficacy.

K' *symbol for* apparent (concentration) equilibrium constant.

K^O or **$K^{\ominus}$** *symbol for* standard equilibrium constant.

$K_1, K_2, \ldots K_n$ *symbols for* the first, second, ... nth equilibrium (dissociation or association) constants, respectively, of the same substance.

K_a, K'_a *symbols for* the thermodynamic and apparent **acid dissociation constants**, respectively.

K_A *symbol for* activator constant.

K_b, K'_b *symbols for* the thermodynamic and apparent **basic dissociation constants**, respectively.

K_c *symbol for* equilibrium constant with respect to concentration (concentration equilibrium constant).

K_i *symbol for* inhibition constant (inhibition type unspecified); the reactant to which it relates may be indicated by an appropriate subscript, e.g. K_{iA}.

K_{ic} *symbol for* competitive inhibition constant.

K_{iu} *symbol for* uncompetitive inhibition constant.

K_m *symbol for* Michaelis constant (*also called* **Michaelis concentration** *or, more appropriately* **Michaelis–Menten constant**); the reactant to which it relates may be indicated by an appropriate subscript, e.g. K_{mA}.

K_s *symbol for* substrate dissociation constant; the reactant to which it relates may be indicated by an appropriate subscript, e.g. K_{sA}.

K_s' *symbol for* salting out constant.

K_w *symbol for* the ion product of water.

kainate receptor a type of neuroexcitatory membrane receptor characterized by its specific agonist ligand, kainate (*see* **glutamate receptor**), and for which glutamate may be the endogenous ligand. The receptor is coupled to Na^+/K^+ channels in the cell membrane, which are opened when activated by agonist.

kainic acid (2S,3S,4S)-2-carboxy-4-(1-methylethenyl)-3-pyrrolidineacetic acid; digenic acid; a cyclic analogue of L-glutamic acid

obtained from *Digenea simplex* (a red alga). It is an amino-acid agonist for the **kainate receptor**, is a neurotoxin, and has anthelmintic properties.

kairomone any of a heterogeneous group of chemical messengers that are emitted by organisms of one species but benefit members of another species; they are commonly nonadaptive or maladaptive to the emitter. They include attractants, phagostimulants, and other substances that mediate the positive responses of, e.g., predators to their prey, herbivores to their food plants, and parasites to their hosts. *Compare* **allomone**, **pheromone**.

KAL a human X-chromosome gene that when mutated produces Kallmann syndrome of hypogonadotropic hypogonadism and anosmia. It encodes a glycoprotein (680 amino acids) of the extracellular matrix that is responsible for migration of gonadotropin-releasing hormone (GnRH)-producing neurons and of olfactory neurons. It contains an N-terminal four-disulfide-bond core domain, similar to that of some protease inhibitors, and in the C-terminal region two fibronectin domains (type III) similar to those present in a variety of N-CAMs.

kalata B a uterotonic circular peptide of 29 amino acids extracted from the African plant *Oldenlandia affinis* and several members of the Rubiaceae and Violaceae families. It contains three disulfide bonds and forms only beta strands.

kalinin 1 *another name for* laminin 5 (*see* **laminin**). **2** a filament-associated protein of the **laminin** family, e.g. (precursor) human kalinin B1, also known as nicein B1 or (most commonly) laminin B1k.

kalirin a multidomain cytoskeletal protein that functions as a guanine nucleotide exchange factor. It contains 2959 amino acids, which form 87 spectrin repeats, two Dbl and two pleckstrin homology regions, two Ig-like regions, two SH3 regions, and a kinase domain.

kallidin *an alternative name for* lysylbradykinin; *see* **bradykinin**.

kallikrein any of two groups of serine endopeptidases that are widely distributed in mammalian tissues and body fluids, including blood. **Plasma kallikrein** (EC 3.4.21.34; *also called* kininogenin or Fletcher factor or serum kallikrein) cleaves Arg-|-Xaa and Lys-|-Xaa bonds, including Lys-|-Arg and Arg-|-Ser bonds in kininogen to produce bradykinin, and activates blood coagulation factors XII and VII; it is formed from prekallikrein by factor XIIa. **Tissue kallikrein** (EC 3.4.21.35; *other names*: glandular kallikrein; kininogenin) catalyses the preferential cleavage of Arg-|-Xaa bonds in small molecule substrates. Its highly selective action to release kallidin (lysyl-bradykinin) from **kininogen** involves hydrolysis of Met-|-Xaa or Leu-|-Xaa. The rat enzyme is unusual in liberating bradykinin directly from autologous kininogens by cleavage at two Arg-|-Xaa bonds.

kallistatin *or* protease inhibitor 4 an acidic protein (in plasma) of the **serpin** family, expressed widely in human tissues. It inhibits tissue **kallikrein**, forming a 1:1 complex with it.

Kallmann syndrome *see* **KAL**.

kanamycin an aminoglycoside antibiotic complex produced by *Streptomyces kanamyceticus*. It is made up of three components: kanamycin A (in which $R_1 = NH_2$, $R_2 = OH$; see structure), the major component, often designated as kanamycin; and two minor congeners, kanamycin B (R_1 and $R_2 = NH_2$) and kanamycin C ($R_1 = OH$, $R_2 = NH_2$). Kanamycin causes misreading during protein synthesis in bacteria, but its target site on the ribosome is apparently different from that of streptomycin. *See also* **kanosamine**.

kanosamine 3-amino-3-deoxy-D-glucose; an amino sugar present in **kanamycin** and other antibiotics, and as free carbohydrate in fermentation broths of *Bacillus aminoglucosidicus*.

K antigen *see* **capsular antigen**.

kaolin *or* **china clay** a finely powdered hydrated aluminium silicate; it is useful as an adsorbent.

Kaposi's sarcoma a skin cancer endemic to equatorial Africa and often occurring in AIDS patients. [After Moritz Kaposi (1837–1902), Austrian dermatologist.]

kanamycin

kappa *symbol*: κ (lower case) *or* K (upper case); the tenth letter of the Greek alphabet. For uses *see* **Appendix A**.

kappa B *or* κB an element common in eukaryotic promoters. The consensus sequence is GGRNNYYCC, e.g. GGGGACTTTCC in mouse and human.

kappa chain *or* κ chain *symbol*: κ; one of the two types of **light chain** found in human immunoglobulins, the other type being a lambda chain. *See also* **Bence-Jones protein**.

kappa convention a way of denoting single ligating atom attachments of a polyatomic ligand to a coordination centre such as a metal ion. Ligands are indicated by the italic element symbol preceded by a Greek kappa, κ. A right superscript numerical index indicates the number of such attachments. *See* **donor atom symbol**.

kappa factor a protein that causes termination of transcription of various bacteriophage DNAs, including those of T4, T5, and T7, by *Escherichia coli* DNA-dependent RNA polymerase in a DNA site-specific manner. It is a dimer of two apparently identical 17 kDa peptide chains.

kappa particle a cell of any of several species of *Caedibacter* present in the cytoplasm of killer strains of *Paramecium aurelia*. (*Caedibacter* is a genus of Gram-negative bacterial obligate endosymbionts of this protozoan.) When kappa particles (and/or toxic constituents or products of them) are released by killer paramecia they may be lethal to sensitive paramecia (which lack both the endosymbiont and a protective genetic factor).

Karplus curve a curve describing the relation of the nuclear magnetic resonance spin–spin coupling constant and the dihedral angle. [After Martin Karplus (1930–).]

Karplus relation an equation describing the relation between the magnitude of the nuclear magnetic resonance spin–spin coupling constant, J, across three bonds and the dihedral angle, Θ, about the central bond. It is:

$$J = A + B\cos\Theta + C\cos^2\Theta,$$

where A, B, and C are coefficients that depend on the electronegativity of the substituents.

karyo+ *or* **caryo+** *comb. form* denoting the nucleus of a cell.

karyogram *an alternative name for* **idiogram**.

karyoid *an alternative name for* **nucleoid** (def. 1).

karyokinesis 1 *an alternative name for* **mitosis**. **2** division of the cell nucleus during mitosis. *Compare* **cytokinesis**.

karyon *or* **caryon** the **nucleus** of a cell.

karyopherin *an alternative name for* **importin**.

karyophilic proteins soluble proteins that accumulate in the nucleus on microinjection into the cytoplasm of amphibian oocytes. They are not bound to chromatin or other nuclear structures. They include **nucleoplasmin** and the acidic nuclear proteins N1 to N4 of the *Xenopus* oocyte nucleus.

karyosome a compartment, located on the nucleolus of some plant species, that may support the rRNA genes as they extend from the nucleolar organizer region.

karyotype the visual appearance of the set of chromosomes of a typical somatic eukaryotic cell of a given species, individual, or cell strain. It is expressed in terms of chromosomal sizes, shapes, and number. —**karyotypic** *adj.*

kasugamine 2,4-diamino-2,3,4,6-tetradeoxy-D-*arabino*-hexose; a component of **kasugamycin** and other antibiotics.

kasugamycin an aminoglycoside antibiotic produced by a strain of *Streptomyces kasugaensis* and containing **kasugamine**. It interacts stoichiometrically with the 30S ribosomal subunit, preventing initiation of protein biosynthesis in prokaryotes; it also blocks translation in fungal systems.

kat *symbol for* katal.

katabolism *a variant spelling of* **catabolism**. —**katabolic** *adj.*

katacalcin a 21-residue peptide derived from the protein precursor of calcitonin, in which it flanks calcitonin on the C-terminal side. It is potent in lowering plasma calcium levels. It is present in normal plasma in concentrations equimolar with calcitonin, and is released rapidly by calcium infusion.

katal *symbol*: kat; a derived coherent unit for expression of enzyme activities, defined as that **catalytic activity** of an enzyme that will raise the rate of conversion of a specified chemical reaction by one mole per second in a specified assay system. There are several important points regarding usage. (1) This definition supersedes a version expressed in terms of the intensive quantity 'rate of reaction' rather than the extensive quantity 'rate of conversion', as well as earlier versions in which catalytic activity was identified with the overall rate of reaction (including the spontaneous rate) rather than with the increase in rate. (2) The katal was adopted as a derived SI unit in 1999. (3) The katal is not used to express rates of conversion themselves (for which the unit is mol s⁻¹). (4) The katal is independent of assay conditions (although the measured catalytic activity of a given system depends on assay conditions). (5) *Compare* **enzyme unit**; 1 kat ≈ 6 × 10⁷ U; 1 U ≈ 16.7 nkat).

katanin a microtubule-stimulated ATPase that severs and disassembles microtubules to tubulin dimers.

Kat.f. an obsolete unit for expressing the specific activity of catalase preparations, originally defined as the rate constant for the catalysed reaction under certain conditions divided by the number of grams of enzyme in the total test volume. The rate constant is calculated from the standard equation for a unimolecular reaction using decadic logarithms and is expressed per minute. Later (apparently unintended) variations of this definition sometimes were used. [From the German *Katalasefähigkeit*, catalase capability.]

katharometer an apparatus used to determine the composition of a gas mixture by measuring its thermal conductivity.

kathode *a variant spelling of* **cathode**. —**kathodic** *adj.*

kation *a variant spelling of* **cation**.

KATP *abbr. for* ATP-dependent K⁺ channel; *other name*: Kir 6.2; a K⁺ inward rectifier channel of pancreatic B cells that is regulated by the ATP-binding sulfonylurea receptor.

Kauffmann–White scheme a classification scheme for the many serotypes of the bacterial genus *Salmonella*, defined in terms of their somatic (surface) antigens and, where appropriate, their capsular and flagellar antigens.

Kautsky effect the characteristic changes with time in the fluorescence emission of chlorophyll *a* that occur when dark-adapted cells or isolated chloroplasts are illuminated. The changes may be distinguished into a fast change lasting only a few seconds, and a slower change lasting a few minutes. [After Hans Kautsky (1891–1966).]

kayser a unit of **wave number** expressed as the number of wavelengths of electromagnetic radiation in 1 cm.

Kazal inhibitor an electrophoretically heterogeneous proteinaceous trypsin inhibitor and anticoagulant found in the pancreas and in pancreatic juice of a number of mammalian species. *Compare* **pancreatic trypsin inhibitor**. [After Louis Anthony Kazal (1912–), US physiologist.]

kb *symbol for* kilobase.

K capture *see* **electron capture**.

K cell 1 an immunocytochemically distinguishable, granule-containing cell, found in the duodenum and jejunum, that produces **glucose-dependent insulinotropic peptide**. **2** *or* killer cell a type of large granular mononuclear lymphocyte that can kill target cells sensitized with antibody. The K cells engage the antibody via their **Fc receptors**. They are probably identical with natural killer (NK) cells.

K chromophore any **chromophore** that gives rise to high-intensity absorption bands due to π–π conjugation. *Compare* **R chromophore**.

KCNE1 the gene at 21q22.1-q22.2 for a cardiac K channel β subunit. It encodes minK, a protein of 129 amino acids, including one transmembrane segment. Numerous loss-of-function mutations are associated with the **long QT syndrome**. *See also* **KVLQT1**.

KCNQ *abbr. for* voltage-gated K$^+$ channel.

kDa *symbol for* kilodalton.

KDEL a tetrapeptide sequence (in the single-letter amino-acid code) at the C terminus of proteins that are to be retained in the endoplasmic reticulum (ER). An equivalent sequence is HDEL, especially in plants. KDEL binds the **KDEL receptor**, and retention is thought to be mediated by continual retrieval from a post-ER compartment. *See also* **luminal proteins**.

KDEL receptor a transmembrane protein of the endoplasmic reticulum that binds the **KDEL sequence**, which constitutes the C terminus of proteins destined to be retained in the endoplasmic reticulum. Such receptors are seven-transmembrane-helix proteins. An equivalent receptor in yeast for the HDEL sequence is a product of the gene *ERD2*.

KDO *abbr. for* ketodeoxyoctanoate.

ke- *abbr. for* keto-.

Kearns–Sayre syndrome (*abbr.*: KSS) *or* **chronic external ophthalmoplegia** (*abbr.*: CPEO) a mitochondrial disease of humans in which ptosis and paralysis of the external eye muscles may be accompanied by myopathy. It is frequently due to a large deletion (5 kbp) in mitochondrial DNA (mtDNA), and less commonly to point mutations in mtDNA for isoleucyl-, asparaginyl-, or leucyl-tRNA. [After Thomas Kearns (1922–) and George Pomeroy Sayre (1911–), US ophthalmologists.]

KEGG *abbr. for* Kyoto Encyclopedia of Genes and Genomes; *see* **pathway database**.

Keilin–Hartree heart-muscle preparation a preparation of submitochondrial particles from bovine heart muscle that is capable of catalysing the reoxidation by dioxygen of NADH and succinate. [After D. Keilin (1887–1963) and E. F. Hartree (1910–94).]

Kell system a blood-group system whose antigenic determinant is part of an erythrocyte membrane glycoprotein of 93 kDa that has an extracellular C-terminal domain with 15 cysteine residues, an N-terminal cytoplasmic domain, and a zinc binding site of the type found in neutral endopeptidases.

kelvin *symbol*: K; the SI base unit of thermodynamic temperature, defined as the fraction 1/273.16 of the thermodynamic temperature of the triple point of water. The symbol K is used both for thermodynamic temperature and for thermodynamic temperature interval, and supersedes °K (degree Kelvin). [After William Thomson, Lord Kelvin (1824–1907), British mathematician, physicist, and inventor, who first defined the absolute scale of temperature.]

KEM1 the symbol of a gene of yeast that encodes a nuclease specific for **G-quartets**.

Kendall's compounds a series of naturally occurring steroids of the adrenal cortex and their metabolites; they were provisionally designated by US physiological chemist Edward Calvin Kendall (1886–1972), using letters of the alphabet, before elucidation of their structures. They are: Kendall's compound A, 11-dehydrocorticosterone; Kendall's compound B (Reichstein's substance H), corticosterone; Kendall's compound C (Reichstein's substance C), 5α-pregnane-3α,11β,17α,21-tetrol-20-one; Kendall's compound D (Reichstein's substance A), 5α-pregnane-3β,11β,17α,20β,21-pentol; Kendall's compound E (Reichstein's substance Fa), cortisone; Kendall's compound F (Reichstein's substance M), cortisol; Kendall's compound G (Reichstein's substance D), 5α-pregnane-3β,17α,21-triol-11,20-dione; Kendall's compound H (Reichstein's substance N), 5α-pregnane-3β,21-diol-11,20-dione; Kendall's desoxy compound B (Reichstein's substance Q), cortexone, 11-deoxycorticosterone.

Kennedy's disease *an alternative name for* **spinal and bulbar muscular dystrophy**.

kephalin *a variant spelling of* **cephalin**.

kerasin *or* **cerasin** a **cerebroside** in which the fatty acyl moiety is lignoceroyl.

keratan sulfate *or (formerly)* **keratosulfate** any member of a group of glycosaminoglycans with repeat units consisting of (β1→3)-linked D-galactopyranosyl-(β1→4)-N-acetyl-D-glucosamine 6-sulfate and containing variable amounts of fucose, sialic acid, and mannose units. In keratan sulfate I, isolated from cornea, the carbohydrate is attached to the polypeptide backbone through N-acetylglucosaminyl–asparaginyl linkages, while in keratan sulfate II, isolated from cartilage and bone, the carbohydrate appears to be attached to the polypeptide backbone through N-acetylgalactosamine in O-glycosidic linkage to either serine or threonine residues. The keratan sulfates are often attached to the same polypeptide backbone as the chondroitin sulfates.

keratein the soluble product formed by reductive scission of disulfide cross-links in wool keratin on treatment with sulfhydryl compounds, such as thioglycolate.

keratin any member of a class of structural fibrous scleroproteins that occur in vertebrate skin and in various specialized epidermal structures, e.g. feathers, nails, hair, hooves, horns, and quills, as well as in the cytoskeleton. They are classified into **hard keratins**, specific to hair, nails, and feathers, and **cytokeratins**, which form part of the cell cytoskeleton. The hard keratins are typically sulfur-rich proteins cross-linked by disulfide bonds, having M_r in the region of 6000–20 000; the cytokeratins are larger (40 000–56 000) low-sulfur proteins. The hard keratins of mammals are largely α-helical ('α keratins'), whereas those of birds and reptiles are largely β-sheet ('β keratins'). However, α keratin can be converted to a β-sheet conformation by reduction and heating (as in hairdressing). Cytokeratins are intermediate- filament proteins (similar to other intermediate-filament proteins), and are of two types: I (acidic) and II (neutral to basic) (40–55 and 56–70 kDa respectively); these form microfibrillar filaments in epithelial cells. Cytokeratin is composed of complexes formed by the aggregation of at least one of each type.

keratinization *or* **keratinisation 1** the process by which epidermal cells synthesize keratin. **2** the elaboration of structures containing keratin, especially in skin. —**keratinize** *or* **keratinise** *vb.*

keratinocyte *or* **keratocyte** any epidermal cell that produces keratin.

keratohyalin a material seen in microscopic sections in granular cells of the skin. It is thought to consist mainly of **filaggrin**.

keratosulfate *a former name for* **keratan sulfate**.

kernicterus an encephalopathy associated with degeneration and yellow pigmentation of basal ganglia and other nerve cells in the spinal cord and brain, caused by the severe unconjugated bilirubinemia occurring in newborn infants and occasionally in older children with hereditary unconjugated hyperbilirubinemia.

kerosene *or* **kerosine** *an alternative name (esp. US) for* **paraffin** (def. 2).

Kerr effect *an alternative name for* **electric birefringence**. [After John Kerr (1824–1907), Scottish physician.]

kestose 1 *or (sometimes)* **6-kestose** *O*-β-D-fructofuranosyl-(2→6)-*O*-β-D-fructofuranosyl-(2→1)-α-D-glucopyranose; a trisaccharide widely distributed in plants, especially in the bulbs and tubers of monocotyledons, and formed artificially when yeast invertase is incubated with a 50% solution of sucrose. The name is derived from the fact that it was discovered at the Tate and Lyle research laboratories, then at Keston, Kent, UK. **2** *or* **1-kestose** isokestose; inulobiosyl glucose; *O*-β-D-fructofuranosyl-(2→1)-*O*-β-D-fructofuranosyl-(2→1)-α-D-glucopyranose; an isomer and congener of 6-kestose. *See also* **neokestose**.

ket+ *a variant form of* **keto+** (sometimes before a vowel).

ketal *see* **acetal**.

ketamine (±)-2-(2-chlorophenyl)-2-(methylamino)cyclohexanone; a dissociative anesthetic with cataleptic and analgesic effects, possibly through acting as an *N*-methyl-D-aspartate (excitatory amino acid) receptor antagonist.

ketanserin 3-[2-[4-(4-fluorobenzoyl)-1-piperidinyl]ethyl]-2,4[1H,3H]-quinazolinedione; an antagonist to α_1 **adrenoceptors**, H_1 **histamine receptors**, and 5-HT$_2$ **5-hydroxytryptamine receptors**. A clinically effective hypotensive agent, it differentiates between 5-HT$_1$ and 5-HT$_2$ receptors with high specificity; within adrenoceptors and histamine receptors it is less specific.

ketene 1 ethenone; CH_2=C=O. 2 any substituted **ketene** (def. 1).

ketimine any **imine** that is an analogue of a ketone; the general structure is R_2C=NR′ where R is any organyl group and R′ may be any organyl group or H. *Compare* **aldimine**.

keto 1 carbonyl group, =CO, in any ketone. 2 describing a class of chemical compounds of the general formula R–CO–R′ (where R, R′ are not H). The term is often used to distinguish such compounds from isomeric compounds of another class, e.g. keto sugar. 3 a nonpreferred term, often still used, to denote the presence of a substituent in an organic compound of a double-bonded oxygen atom that is part of a ketonic structure. The recommended term is **oxo**.

keto+ 1 or (sometimes before a vowel) **ket+** comb. form indicating ketonic or derived from a ketone; e.g. ketal, ketoaldose, ketol. 2 or (sometimes before a vowel) **keton+** comb. form denoting of or pertaining to a **ketone body** or bodies; e.g. ketoacidosis, ketogenesis, ketonuria. 3 prefix (in chemical nomenclature) denoting a ketonic derivative of an indicated organic compound; **oxo** now recommended.

keto- abbr.: ke-; an optional prefix to the (semi)systematic name of a monosaccharide or monosaccharide derivative containing an uncyclized ketonic carbonyl group, used to stress its acyclic nature. *Compare* **aldehydo-**, **keto**, **keto+**.

keto acid any organic (carboxylic) acid containing a **keto** (def. 3) group; the recommended term is an oxo acid.

ketoacidosis a type of **acidosis** caused by enhanced production of **ketone bodies** by the ketogenic pathway. This excessive production lowers blood pH and promotes excretion of K^+ and Na^+ by the kidney, leading to cation depletion. It results most often from hormonal imbalance, especially insulin deficiency, when fatty acids are mobilized from adipose tissue, transported to the liver, and converted to ketone bodies. In humans, a plasma level of up to 8 mmol L^{-1} is considered safe and may be beneficial, but in disease states levels may reach up to 20 mmol L^{-1}, contributing to life-threatening situations. *See* **ketogenesis**. —**ketoacidotic** adj.

2-ketoadipic acid dehydrogenase EC 1.2.4.1; an enzyme of mitochondrial matrix that catalyses the reaction:

$$2\text{-ketoadipic acid} + \text{CoA} + \text{NAD}^+ = \text{glutaryl-CoA} + \text{NADH} + \text{H}^+ + CO_2.$$

It is believed to be similar to the pyruvate dehydrogenase complex. Deficient activity gives rise to 2-**ketoadipic acidemia**.

2-ketoadipic acidemia or (esp. Brit.) **2-ketoadipic acidaemia** a rare, largely benign condition in which 2-keto, 2-amino, and 2-hydroxy derivatives of adipic acid accumulate and are excreted in excessive amounts in urine. These metabolites are derived from catabolism of lysine, hydroxylysine, and tryptophan. The defect is considered to involve some component of the 2-**ketoadipic acid dehydrogenase** complex.

ketoaldonic acid any monosaccharide formally derived from an aldonic acid by oxidation of a secondary alcoholic hydroxyl group to an oxo group.

ketoaldose any monosaccharide containing both an aldehydic and a ketonic carbonyl group.

ketoconazole a complex derivative of imidazole, clinically useful as an antimycotic. It impairs ergosterol synthesis in fungal cell membranes, thereby increasing their permeability.

ketodeoxyoctanoate abbr.: KDO; 2-keto-3-deoxy-D-octanoate; 3-deoxy-D-*manno*-2-octulosonate; an acid present in lipopolysaccharides of the outer membranes of certain Gram-negative bacteria.

keto–enol tautomerism a type of **tautomerism** in which the keto and enol forms of a compound are in equilibrium, the change corresponding to the migration of a hydron and a shift of bonding electrons. *Compare* **amino–imino tautomerism**.

ketofuranose any **ketose** in the **furanose** form.

ketogenesis the metabolic production of **ketone bodies**. The pathway starts with the formation of acetoacetyl-CoA (I) from two molecules of acetyl-CoA by **acetyl-CoA C-acetyltransferase** (EC 2.3.1.9); I then reacts with a further molecule of acetyl-CoA, in a reaction catalysed by **hydroxymethylglutaryl-CoA synthase**, to form (S)-3-hydroxy-3-methylglutaryl-CoA (II); II is then cleaved to acetoacetate and acetyl-CoA by **hydroxymethylglutaryl-CoA lyase**. These reactions occur in higher animals and are localized entirely in mitochondria; the liver is the main site of synthesis, but kidney, and possibly also small intestine, and adipose tissue may play a part in some situations.

ketogenic forming, or having the quality of being convertible into ketone bodies by metabolic processes.

α-ketoglutarate a nonpreferred (though often used) name for 2-oxoglutarate; 2-oxo-1,5-pentanedioate; a compound that has important roles in carbohydrate and amino-acid metabolism, especially in transamination reactions and as a component of the **tricarboxylic-acid cycle**.

α-ketoglutaric dehydrogenase see **oxoglutarate dehydrogenase (lipoamide)**.

2-keto-L-gulonate or (preferably) **2-oxo-L-gulonate** L-*xylo*-2-hexulosonate; a compound obtained from sorbitol and an important intermediate in the manufacture of **ascorbic acid**.

ketoheptose any **ketose** having a chain of seven carbon atoms in the molecule. Eight enantiomeric pairs of such compounds are possible.

ketohexokinase EC 2.7.1.3; systematic name: ATP:D-fructose 1-phosphotransferase; other name: hepatic fructokinase. An enzyme that catalyses the phosphorylation by ATP of D-fructose to D-fructose-1-phosphate with release of ADP. The enzyme occurs in cytosol of liver cells, in muscle, intestine, and renal cortex, and in bacteria. It also phosphorylates some other ketoses (e.g. 5-dehydro-D-fructose, D-sorbose, D-tagatose). 'Fructokinase' is the recommended name for the enzyme EC 2.7.1.4, which forms D-fructose 6-phosphate (*see* **fructokinase**). In mammals, some fructose can be phosphorylated to fructose 6-phosphate by **hexokinase**. The rat liver enzyme consists of 299 amino acids. Two human isoforms are derived by alternative splicing from one gene. Inactivating mutations lead to hepatic fructokinase deficiency (also called essential fructosuria).

ketohexose any **ketose** having a chain of six atoms in the molecule. Four enantiomeric pairs of such compounds are possible, these being the D and L isomers of fructose, psicose, sorbose, and tagatose.

α-ketoisocaproate or (preferably) **α-oxoisocaproate** 2-methyl-2-oxopentanoate; an intermediate in the biosynthesis of L-leucine. *See* **leucine and valine biosynthesis**.

α-ketoisovalerate or (preferably) **α-oxoisovalerate** 3-methyl-2-oxobutanoate; an intermediate in the biosynthesis of L-leucine, L-valine, and pantothenic acid. *See* **leucine and valine biosynthesis**.

ketol 1 any organic compound that is both a ketone and an alcohol, especially an α-ketol (i.e. acyloin), one in which the two functions are adjacent. 2 the strongly reducing group, HO–CH_2–CO–, that is characteristic of an α-ketol. It is present, e.g., in ketoses and certain corticosteroids.

ketol-acid reductoisomerase EC 1.1.1.86; systematic name: (R)-2,3-dihydroxy-3-methylbutanoate:NADP$^+$ oxidoreductase (isomerizing); other names: dihydroxyisovalerate dehydrogenase (isomerizing); acetohydroxy-acid isomeroreductase. An enzyme of chloroplasts that catalyses the reaction:

(*R*)-2,3-dihydroxy-3-methylbutanoate + NADP⁺ =
(*S*)-2-hydroxy-2-methyl-3-oxobutanoate + NADPH.

It is a component of the pathways for isoleucine and valine biosynthesis. The protein from *Spinacia oleracea* (spinach), for example, is a tetramer of similar but nonidentical chains.

ketol condensation the enzyme-catalysed formation of certain **ketols** (def. 1) by addition of an acyl moiety of one *α*-oxo acid to another *α*-oxo acid to form an *α*-hydroxy-*β*-oxo acid. It occurs, e.g., in the condensations catalysed by acetolactate synthase (EC 4.1.3.18); these involve either two molecules of pyruvate condensing to form *α*-acetolactate and CO₂, or one molecule of pyruvate and one of *α*-oxobutyrate condensing to form *α*-aceto-*α*-hydroxybutyrate and CO₂. The main products are biosynthetic precursors of valine and leucine, and of isoleucine, respectively.

ketolysis *a term sometimes used to* describe the oxidation, for energy production, of ketone bodies by extrahepatic tissues.

***α*-keto-*β*-methylvalerate** *or (preferably)* ***α*-oxo-*β*-methylvalerate** 3-methyl-2-oxopentanoate; the biosynthetic precursor of, and catabolic product of, L-isoleucine.

keton+ *a variant form of* **keto+** (def. 2) (*sometimes before a vowel*).

ketone 1 any compound containing a carbonyl group, >C=O, joined to two carbon atoms. **2** *the functional class name for* any **ketone** (def. 1) that does not have a trivial name, e.g. ethyl methyl ketone.

ketone body any of the three substances: (1) acetoacetate, formed enzymically from acetyl-CoA; (2) D-3-hydroxybutyrate (*β*-hydroxybutyrate), formed by liver mitochondrial 3-hydroxybutyrate dehydrogenase (EC 1.1.1.30) from acetoacetate; or (3) acetone, formed by spontaneous decarboxylation of acetoacetate. Ketone bodies may accumulate in the body in starvation, diabetes mellitus, or in other defects of carbohydrate metabolism. They are formed in the post-absorptive state and during more prolonged fasting, and can be used as an alternative energy source to glucose, sparing glucose and thus indirectly muscle protein, since the carbon source for glucose synthesis under such conditions derives from muscle amino acids. Biosynthesis involves the formation of **hydroxymethylglutaryl-CoA**, which is cleaved to acetoacetate and acetyl-CoA by **hydroxymethylglutaryl-CoA lyase**; it occurs largely in the liver, but to a small extent also in kidney. Utilization of ketone bodies in peripheral tissues involves conversion by **3-oxo-acid CoA-transferase** of acetoacetate to acetoacetyl-CoA, which is then converted to two molecules of acetyl-CoA by **acetyl-CoA *C*-acetyltransferase**. 3-Hydroxybutyrate is classed as a ketone body because it exists in an equilibrium with acetoacetate that depends on the redox state in the liver mitochondria.

ketonemia *or (esp. Brit.)* **ketonaemia** an abnormally high concentration of **ketone bodies** in the blood.

ketonic of, relating to, or having the properties of a **ketone** or a **keto** (def. 1) group.

ketonize *or* **ketonise** to convert into or become converted into a ketone. —**ketonization** *or* **ketonisation** *n.*

ketonuria a condition in which there is an abnormally large excretion of ketone bodies in the urine.

ketopentose any **ketose** having a chain of five carbon atoms in the molecule. Two enantiomeric pairs of such compounds are possible, these being the D and L isomers of ribulose and xylulose.

ketoprofen 2-(3-benzoylphenyl)propionic acid; a nonsteroidal anti-inflammatory agent; it inhibits prostaglandin-endoperoxide synthase (EC 1.14.99.1). *See also* **ibuprofen**.

(*S*)-ketoprofen

ketopyranose any **ketose** in the **pyranose** form.
ketose any monosaccharide in the acyclic form of which the actual or potential carbonyl group is nonterminal, i.e. ketonic. The sys-

tematic names and many of the trivial names of ketoses have the ending '+ulose'. The position of the (potential) carbonyl group is indicated by a prefixed locant (e.g. 3-ketose, 5-nonulose), which may be omitted from the name of a 2-ketose when no ambiguity can arise. The term is frequently modified by an infix indicating the number of carbon atoms in a chain in the molecule; thus **ketopentose**, **ketotetrose**, etc. *Compare* **aldose**.

ketosis any clinical condition in which there are abnormally high concentrations of **ketone bodies** in the tissues. The condition is usually accompanied by acidosis. *See also* **ketogenesis**. —**ketotic** *adj.*

ketosteroid *a former name for* **oxosteroid**.

keto sugar *a common name for* any **ketose**.

ketotetrose either of two **ketoses** having a chain of four carbon atoms in the molecule, these being the enantiomeric pair D- and L-*glycero*-tetrulose.

***β*-ketothiolase** *see* **acetyl-CoA *C*-acyltransferase**.

ketotriose the single achiral **ketose** having a chain of three carbon atoms in the molecule and known as dihydroxyacetone, or glycerone.

kettin a 500–700 kDa protein of *Drosophila* muscle. Its N-terminal region lies in the Z disk and binds α-actinin. It contains Ig-like regions through which it binds to the thin filaments.

kex2 a structural gene for pre-pro-alpha mating type polypeptide in *Saccharomyces cerevisiae*. It gives rise to the designation Kex2 site, a cleavage site (Arg-|-Arg or Lys-|-Arg) for **kexin** in the processing of the mating polypeptide and several extracellular fungal proteins.

kexin EC 3.4.21.61; *other names*: yeast Kex2 protease; proteinase YSCF; prohormone-processing endoprotease; paired-basic endopeptidase. An enzyme that catalyses the cleavage of Lys-Arg-|-Xaa and Arg-Arg-|-Xaa bonds to process yeast α-factor pheromone and killer toxin precursors. The human enzyme with the same specificity (and EC number) is neuroendocrine convertase 1. *See also* **furin**, ***kex2***, **proopiomelanocortin**.

keyhole limpet hemocyanin *or (esp. Brit.)* **keyhole limpet haemocyanin** *abbr.*: KLH; a respiratory protein from *Megathura crenulata* that elicits a strong T-cell immune response. It is frequently used as a carrier protein to which peptides or other haptens are coupled in order to induce an immune response when used as an immunogen.

kg *symbol for* kilogram.

KH domain *abbr. for* ribonucleoprotein K homology domain; a peptide sequence (40–60 amino acids) present in proteins believed to bind RNA. It consists of several invariant hydrophobic regions (containing Ile, Leu, Met, and Gly) and is highly conserved in the three proteins encoded by *FMR1* gene family members, each of which contains two KH domains. The protein **vigilin**, found in domestic fowl, contains 14 KH domains.

KHF *abbr. for* killer cell helper factor (i.e. **interleukin 2**).

kicksorter a multichannel pulse-height analyser, especially one used to distinguish between radioisotopes by sorting them according to the energies (i.e. 'kicks') of their emitted radiations.

kidney either of a pair of organs of vertebrates concerned with osmoregulation and the elimination of nitrogenous waste products. Kidneys of different species vary in structure according to the animal's mode of life, but their basic design and function is the same. Each kidney is enclosed in a capsule and is made up of an outer cortex and an inner medulla, and contains numerous **nephrons** and their associated blood supply. The nephrons filter constituents from the blood to form urine, which is discharged from the kidney via the ureter.

kieselguhr *an alternative name for* **diatomaceous earth**.

kifnensine an alkaloid produced by the actinomycete *Kitasatosporia kifunense* that inhibits **ER α-mannosidase I**.

killed vaccine any suspension of killed (or otherwise inactivated) microorganisms or virus used as an antigen to produce immunity.

killer cell any cell that is cytotoxic to *in vitro* cell cultures, e.g. killer T-lymphocytes and macrophages. Antibody-dependent cell-mediated cytotoxicity may be displayed by a number of cell types, e.g. by human monocytes or interferon-activated neutrophils, which can kill antibody-coated tumour cells, and by **NK cells**, which kill hybridoma cells through the NK cell FcγRII receptor. *See also* **K cell**.

killer-cell helper factor *abbr.*: KHF; *an alternative name for* **interleukin 2**.

killer factor an RNA molecule responsible for the lethal property of **killer strains** of yeast.

killer strain any strain of cells that is able to kill sensitive cells of the same species, e.g. some strains of yeast and *Paramecium*. *See also* **kappa particle, killer factor.**

killer T-lymphocyte any T lymphocyte that is able to kill target cells. Killer T-lymphocytes are produced in response to viral infection or an allogeneic tissue graft, and are specifically cytotoxic to host cells infected with virus or to grafted cells carrying the major histocompatibility antigens of the donor, respectively.

kilo+ *symbol*: k; an SI prefix denoting 10^3 times.

kilobase *symbol*: kb; a unit of length of a polynucleotide equal either to 1000 base residues in a single-stranded polynucleotide or to 1000 base pairs in a double-stranded polynucleotide.

kilocalorie *or* **kilogram calorie** *or* (*esp. in nutrition*) **Calorie** a non-SI unit of heat or energy equal to 10^3 cal$_{IT}$. *See* **calorie.**

kilodalton *symbol*: kDa; a unit of molecular mass equal to 10^3 **daltons.**

kilogram *symbol*: kg; the SI base unit of mass, and also the fundamental unit of mass in the metric system; it is equal to the mass of the International Prototype of the Kilogram, a cylinder of 90% platinum, 10% iridium, which is kept at the BIPH. It is the only base SI unit that is still defined in terms of an artifact.

kilogram calorie *an alternative name for* **kilocalorie.**

kinase any phosphotransferase enzyme that transfers a phosphate group, usually from ATP. Kinases are divided into EC sub-sub-classes according to the nature of the phosphate-accepting group: EC 2.7.1 – phosphotransferases with an alcohol group as acceptor; EC 2.7.2 – phosphotransferases with a carboxyl group as acceptor; EC 2.7.3 – phosphotransferases with a nitrogenous group as acceptor; EC 2.7.4 – phosphotransferases with a phosphate group as acceptor. The name is also used for enzymes that transfer a phosphate group from other nucleoside triphosphates. *Compare* **dikinase, pyrophosphokinase, thiokinase.**

+kinase *suffix (in the trivial name of an enzyme)* **1** designating a kinase or, by extension in certain instances, one that transfers some other inorganic oxo-acid group, e.g. **sulfokinase.** **2** designating an enzyme that activates a zymogen by proteolysis, e.g. **enterokinase, streptokinase, thrombokinase.**

kinasing *a jargon term for* the act or process of labelling the 5′ end of a nucleic-acid molecule or fragment with phosphorus-32 by reaction with adenosine 5′-[α-^{32}P]triphosphate and the enzyme T4 polynucleotide kinase. This is a step in the **chemical cleavage method** for sequencing DNA and RNA.

kinectin a kinesin-binding protein (1356 amino acids) of endoplasmic reticulum and various other organelles. It contains an N-terminal transmembrane segment and an extensive region for coiled-coil formation in the C-terminal region. Its central region binds RhoGTPase.

kinematic viscosity *symbol*: ν; the dynamic viscosity, η, of a fluid divided by its density, ρ, i.e. $\nu = \eta/\rho$.

kinesin microtubular motor protein, composed of two heavy chains and several light chains, that uses the energy of ATP hydrolysis (ATPase activity in the head of the heavy chain) to move along microtubules, mostly to the plus end, but in some cases to the minus end (e.g., in *Drosophila*). Kinesins are involved in organelle transport, and in mitosis and meiosis. *See also* **dynein.**

kinesin-binding protein *an alternative name for* JNK-interacting protein. *See* **JIP.**

kinetensin *or* **neurotensin-related peptide** (*abbr.*: NRP) the nonapeptide Ile-Ala-Arg-Arg-His-Pro-Tyr-Phe-Leu; it has **neurotensin**-like immunoreactivity, and is isolated from human, bovine, and canine plasma.

kinetic 1 of or relating to the motions of material bodies and the forces and energies associated therewith; of or relating to motion. **2** of or relating to the rate(s) of change of physical or chemical systems. **3** of or relating to **kinetics.**

kinetic energy *symbol*: E_k *or* T *or* K; the energy possessed by a material body by virtue of its motion, determined by the work necessary to bring it to rest.

kinetic proofreading a suggested mechanism of twice using the same Michaelis kinetic ability of an enzyme system to distinguish between two substrates C and D. The essential features of the mechanism are contained in the following reaction scheme, shown for substrate C, but identical in form if substrate C is replaced by D.

$$C + E \underset{k_{-1}}{\overset{k_1}{\rightleftharpoons}} EC \xrightarrow{\quad} EC^* \overset{k_3}{\underset{k_4 \Vert k_{-4}}{\rightarrow}} E + product$$
$$E + C^* \overset{}{\underset{k_5}{\rightarrow}} C$$

The second step is enzymically coupled to an energy source, typically the hydrolysis of a nucleoside triphosphate, and is strongly enough driven to be essentially irreversible. EC*, a high-energy intermediate, may *either* react to give free enzyme plus product *or* diffuse into the solution where it decomposes to free enzyme plus the nonproduct C*, which can further decompose to C. According to kinetic proofreading, the incorrect intermediate, ED*, diffuses into the solution and decomposes faster than it reacts to give the wrong product, i.e. for ED* $k_3 \ll k_{-4}$, while for EC* $k_3 > k_{-4}$. In this way, the better binding of C is used twice, once in the formation of EC and again in the binding of the intermediate EC*. If the discrimination in the formation of the initial complex is f, a second factor of f may theoretically be attained, giving a total discrimination of f^2.

kinetics 1 the branch of chemistry concerned with studying rates of chemical reactions. **2** the branch of mechanics concerned with the effects of forces on the motions of material bodies.

kinetin 6-furfurylaminopurine; N^6-furfuryladenine; an artefactual compound formed by autoclaving DNA. It exhibits **cytokinin** activity in plants and was originally thought to be a naturally occurring substance. *Compare* **Kinetin.**

Kinetin *a proprietary name for* hyaluronidase. *Compare* **kinetin.**

kinetochore a specialized assembly of proteins that bind to the **centromere** and also attach to **microtubules** growing out from **centrosomes** during mitosis. They are involved in the separation of chromosomes into the daughter cells. They are composed of **centromere-binding proteins.**

kinetochore microtubule *see* **mitotic spindle, prometaphase.**

kinetoplast a DNA-rich organelle located within an expanded region of a mitochondrion lying close to the **basal body** in members of the order **Kinetoplastida.** The kinetoplast is self-replicating and divides before the nucleus in cell division.

Kinetoplastida an order of protozoans whose members are characterized by the presence of a **kinetoplast** in the mitochondrion. It includes the genera *Trypanosoma*, *Leishmania*, and *Crithidia*.

kinetosome *an alternative name for* **basal body.**

King–Altman method a general method for obtaining steady-state expressions for enzymic mechanisms with n enzyme-containing intermediates $EX_1 \ldots EX_n$. First, a simple geometrical pattern describing the mechanism is drawn. Second, all possible patterns, consisting of $(n-1)$ lines, that interconnect the possible enzyme forms are drawn; closed loops are forbidden. In general there will be $m!/(n-1)!(m-n+1)!$ patterns, where m is the number of interconversion steps. Third, n expressions for $(d[EX_i]/dt/[E_{tot}]) = 0$ in terms of the rate constants are set out and solved by determinant methods, using the fact that

$$\sum_{i=1}^{n} ([EX_i]/[E_{tot}]) = 1$$

where $[E_{tot}]$ is the total enzyme concentration. [After E. L. King and C. Altman, who described it in 1956.]

kingdom the highest level of taxonomic classification. All organisms fall into one of three currently accepted kingdoms, each of which has unique and unifying characteristics: eukaryotes, prokaryotes, and archaea.

kinin 1 *or* **kinin-hormone** any member of a varied group of linear polypeptides that constitute a class of so-called tissue or local hormones. Each is locally released from diffuse stores in the body, not having specialized glands of secretion, and is rapidly inactivated at the site of release. Among the most important are **angiotensins, bradykinin, substance P**, colostrokinin, urinary kinins, and **secretin. 2 kinin-9** *an alternative name (sometimes) for* bradykinin. **3 kinin-10** *an alternative name (sometimes) for* kallidin (lysylbradykinin). *See* **bradykinin. 4** *an alternative name for* **cytokinin.**

kininase any enzyme concerned in the natural breakdown of a **kinin** (def. 1). Included are kininase I (*or* lysine carboxypeptidase, EC 3.4.17.3), a carboxypeptidase circulating in plasma that inactivates bradykinin and possibly also kallidin, and kininase II (peptidyl-dipeptidase A, EC 3.4.15.1), *also known as* **angiotensin-converting enzyme**.

kininogen any inactive precursor of a **kinin** (def. 1).

kininogenin *an alternative name for* **kallikrein**.

kinked-helix model of duplex DNA a speculative model of the structure of double-stranded DNA in chromatin, proposed in 1975 by British molecular biologists Francis Crick (1916–2004) and Aaron Klug (1926–) to account for the high degree of folding of the DNA (to about one seventh of its length). In this model, based on the B form of duplex DNA, all the base pairs of the double helix remain intact but a sharp turn, or kink, of 95–100° occurs every 20 base pairs along the duplex. At this point one base pair has become unstacked from an adjacent pair through change of the conformation of the deoxyfuranose ring of a nucleotide residue from one of its possible staggered conformations to another. It was suggested that a kink might render several base pairs more easily available for specific interaction with a protein, and that a similar type of folding might occur in double-stranded DNA in other situations, especially when closely associated with protein.

kinomere *an alternative term for* **centromere**.

Kirkwood–Riseman theory of hydrodynamic properties of complex structures a theory providing an approximate method for computing the friction of a structure made up of identical subunits. A linear or coiled polymer is approximated as a string of beads, each with identical hydrodynamic properties, while an oligomeric protein is modelled as a cluster of identical subunits each with identical hydrodynamic properties.

Kirkwood theory of dielectric polarization of polar liquids an extension of the Onsager theory in which the dielectric constant of the central molecule and its first shell of neighbours is considered to differ from its macroscopic value. [After John Gamble Kirkwood (1907–59).]

Kirkwood theory of ion–dipole interactions a model in which the electrical contribution to the chemical potential of an ion having an arbitrary charge distribution is calculated with the aid of the Debye–Hückel theory and in which the Born relation between the free energy of solvation of a spherical ion and the dielectric constant of the solvent is generalized to include ions of arbitrary charge distribution.

Kirkwood–Westheimer theory of electrostatic effects on acid dissociation constants an extension and amplification of the Bjerrum theory of the influence of substituents on acid dissociation constants, the molecules and ions entering into ionization equilibria being treated as cavities of low dielectric constant. [After John Gamble Kirkwood (1907–59) and Frank Henry Westheimer (1912–), US biochemist.]

KIRs killer cell immunoglobulin-like receptors; receptors that are found on **NK cells**, and some γδ and αβ **T-cells**. They recognize **MHC class I** molecules, and, like the C-type **lectin** receptors also found in these cells, can either inhibit or activate the killer cells. If **ITIM** sequences are present in their cytoplasmic domain they are inhibitory. KIRs lacking ITIMs can associate with ITAM-containing adaptor molecules in which case they can activate the killer cell.

kirromycin *or* **mocimycin** an antibacterial antibiotic isolated from *Streptomyces collinus* and others. It specifically interacts with **elongation factor** EF-Tu (*see* **EF-T**) and prevents the release of EF-Tu from the bacterial ribosome after GTP hydrolysis. In the absence of ribosomes, the antibiotic induces an EF-Tu-dependent hydrolysis of GTP.

kis- *see* **multiplying prefix**.

kistrin a protein isolated from the venom of the Malayan pit viper, *Agkistrodon rhodostoma*. It contains the **RGD** sequence, and binds to a platelet adhesion site. It is a potent platelet aggregation inhibitor and antagonist of a family of glycoproteins that mediate aggregation. *For example, see* **disintegrin**.

kit a commercially available collection of biochemical reagents and apparatus to carry out a procedure, such as an assay.

KIT a family of genes encoding plasma membrane class III receptor tyrosine kinases. The viral gene, v-*kit*, is the oncogene of the Hardy–Zuckerman 4 strain of acutely transforming feline sarcoma virus. The product of the protooncogene c-*kit* has five immunoglobulin-like loops in the extracellular domain; the cytoplasmic region contains a tyrosine kinase domain with an insert region, and is myristoylated. Interaction with its ligand, steel factor (*see* **stem cell factor**), brings about dimerization, leading to enhanced autophosphorylation and binding of phosphatidylinositol 3-kinase and phospholipase Cγ. Kit and its ligand are highly expressed in small-cell lung cancer and a significant proportion of testicular germ cell tumours.

Kjeldahl flask a small pear-shaped glass flask with a long neck, as used in the **Kjeldahl method**.

Kjeldahl method a method for the determination of the total nitrogen in an organic compound or mixture of compounds. The sample, in a **Kjeldahl flask**, is digested in hot, concentrated sulfuric acid with a catalyst, usually one containing mercury, copper, or selenium; the nitrogen in the sample is thereby converted to ammonium hydrogen sulfate. The ammonium formed may be estimated by neutralization and steam distillation of the resultant ammonia into a known amount of acid, the residual acid being determined by titration, or it may be estimated by **nesslerization** or with an ammonium electrode. [After Johan Kjeldahl (1849–1900), Danish chemist.]

Kleinschmidt technique a technique for preparing monolayers of DNA for electron microscopy in which the DNA molecules are bound to a film of denatured basic protein, e.g. cytochrome *c*, on the surface of an aqueous solution. The DNA molecules are consequently brought from a three-dimensional conformation to a two-dimensional one by adsorption to the protein layer. The DNA–protein complex is then transferred to a solid surface, e.g. an electron microscope grid or collodion film, dried, and contrasted with heavy metals by shadowing or staining. The tendency of **ssDNA** to fold back on itself with the formation of intramolecular base pairs can be overcome by using formamide (which denatures short base-paired regions) or by replacing cytochrome *c* with DNA-unwinding proteins, which have high affinity for ssDNA. The method may also be applied to RNA. [After Albrecht K. Kleinschmidt (1916–), German microbiologist.]

Klenow enzyme *or* **Klenow fragment** the larger (75 kDa) of the two fragments formed by the treatment of *Escherichia coli* DNA polymerase I with subtilisin in the presence of DNA. It retains the polymerase activity and the single-strand-specific 3′→5′ exonuclease activities but lacks the 5′→3′ exonuclease activity of DNA polymerase I. It is useful in sequencing single-stranded DNA. [After Hans Klenow (1923–), Danish biochemist.]

KLF *abbr. for* Krüppel-like factor.

KLH *abbr. for* keyhole limpet hemocyanin.

Klinefelter syndrome a genetic condition in which males have an extra X chromosome (XXY). It occurs as frequently as 1–2 per 1000 male births, and the chief sign is infertility caused by androgen deficiency and absence of sperm. Stature is also often affected, with long limbs and above-average height, and secondary sexual characteristics can be underdeveloped. The term **XXY male** is preferred because many are otherwise normal individuals who do not have the syndrome as originally described. [After Harry F. Klinefelter (1912–), US physician.]

Klotz plot *a name sometimes given to* a plot of the reciprocal form of an equation describing ligand binding at multiple independent sites on, e.g., a macromolecule. It has the form:

$$1/\bar{v} = (1/nk)(1/[A]) + 1/n,$$

where $\bar{v}$ is the average number of ligand molecules bound per molecule of the macromolecule, n is the number of binding sites per macromolecule, $[A]$ is the molar concentration of free ligand molecules, and k is the intrinsic association constant. If the equation correctly describes the equilibria involved, a plot of $1/\bar{v}$ versus $1/[A]$ gives a straight line of slope $(1/nk)$ and with an intercept on the $1/\bar{v}$ axis of $1/n$ and on the $1/[A]$ axis of $-1/k$. [After Irving Myron Klotz (1916–), US biochemist.]

k-means clustering an algorithm for clustering data that allows generation of user-specified numbers of clusters from a given data set based on distance metrics for similarity. The method is dependent both on the specified value of k, which determines the number of clusters that are created, and on the distance measure employed. It is widely used in the analysis of microarray data.

KMEF group *abbr. for* keratin-myosin-epidermin-fibrinogen group; a group of proteins that have in common an α-type X-ray diffraction pattern. Some of these may show a β-type pattern in some conditions. Silk fibroin, which normally occurs with a β structure, is often included in the KMEF group.

knallgas bacteria *or* **hydrogen bacteria** any of a number of bacterial species that can obtain energy from the **knallgas reaction**, i.e. the oxidation of dihydrogen to water by dioxygen. They include some *Alcaligenes* and *Pseudomonas* spp. and *Paracoccus denitrificans*. [From the German, *Knallgas*, an explosive mixture of hydrogen and oxygen.]

knallgas reaction the energy-yielding direct oxidation of dihydrogen by dioxygen occurring in certain bacteria (*see* **knallgas bacteria**). The reaction is:

$$2 H_2 + O_2 \rightarrow 2 H_2O + \text{energy}.$$

knockdown in gene expression the experimental reduction in the amount of a specific gene product. It is usually achieved by the use of siRNA to decrease the abundance of its mRNA or inhibit its translation. *Compare* **knockout**.

knockin denoting an animal in which a gene function has been introduced by genetic engineering techniques. [Opposite to 'knock-out'.]

knockout indicating an animal in which function of a gene has been deleted by genetic engineering techniques.

Knoop's hypothesis the hypothesis that fatty acids are metabolically oxidized by the successive removal of two-carbon fragments as acetic acid. *See also* **beta oxidation**. [After Franz Knoop (1875–1946), German physiologist and biochemist.]

Knop solution an aqueous solution containing the major inorganic constituents required for the growth of cells of higher plants, namely calcium nitrate, potassium nitrate, magnesium sulfate, potassium dihydrogen phosphate, and potassium chloride. [After J. A. L. W. Knop (1817–1901), German chemist.]

Knox plot (*in high-performance liquid chromatography*) a plot of relative separation efficiency versus flow rate, useful in determining the optimal conditions for a desired separation. [After John H. Knox.]

Knudson's hypothesis (**two-hit hypothesis**) the hypothesis, formulated in the early 1970s, that hereditary retinoblastoma results from the inheritance of one mutant copy of the *RB1* gene and mutation of the second normal allele during the life of the patient. Such a situation, in which somatic mutation occurs alongside one inherited mutant allele, is known as loss of heterozygosity (LOH). [After Alfred Knudson (1922–), US geneticist.]

Koch phenomenon a cell-mediated hypersensitivity reaction in which inflammation of guinea-pig skin, when injected with living or dead *Mycobacterium tuberculosis*, is very much greater in animals actively infected with the organism than in uninfected animals injected with dead organisms. [After (Heinrich Hermann) Robert Koch (1843–1910), German bacteriologist, who isolated the bacillus in 1882.]

kojibiose the disaccharide 2-*O*-α-D-glucopyranosyl-D-glucopyranose.

kojic acid 5-hydroxy-2-(hydroxymethyl)-4*H*-pyran-4-one; a metabolite from *Aspergillus oryzae* and *A. terreus*. It can be obtained in large amounts and has some industrial application.

Kok effect the sharp difference in yield of photosynthesis above and below the compensation point. Below compensation, where respira-

tion can supply intermediates at a greater rate than the rate of photosynthesis, the yield is about twice as great as above compensation, where CO_2 has presumably to be drawn into the process. [After Bessel Kok (1918–81) Dutch-born US plant physiologist.]

Kornberg enzyme *an alternative name for* DNA polymerase I. *See* **DNA polymerase**. [After Arthur Kornberg (1918–), US biochemist.]

Korrigan cellulase a subunit of the **cellulose synthase** complex found in plasma membranes of plant cells. It cleaves off the glucan chain from cellodextrin with release of the sitosterol β-glucoside primer and the glucan chain, which then forms the microfibrils of cellulose.

Korsakoff's psychosis *see* **Wernicke–Korsakoff syndrome**.

Koshland model *see* **induced-fit theory**.

Kozak sequence a consensus sequence (optimally CCA/GC-CAUGG) located around the **translation start site** and involved in its recognition by the ribosome.

*Kpn*I a type 2 **restriction endonuclease**; recognition sequence: GGTAC↑C.

Kr *symbol for* krypton.

Krabbe's disease *an alternative name for* **galactosylceramide lipidosis**. [After Knudd Haraldsen Krabbe (1885–1961), Danish physician.]

KRB *abbr. for* Krebs Ringer bicarbonate; *see* **Krebs mammalian Ringer**.

Krebs–Cohen dismutation The anaerobic reaction summarized as:

$$\text{2-oxoglutarate} + NH_4^+ \rightarrow$$
$$\text{succinate} + \text{glutamate} + CO_2 + H^+.$$

It has been shown to occur in pigeon heart and breast muscle, rat and guinea-pig renal cortex, and rat-liver mitochondria. It is the sum of two reactions:

$$\text{2-oxoglutarate} + H_2O + NAD^+ \rightarrow$$
$$\text{succinate} + CO_2 + NADH + H^+,$$

and

$$\text{2-oxoglutarate} + NADH + NH_4^+ \rightarrow$$
$$\text{glutamate} + NAD^+ + H_2O.$$

[After (Sir) Hans Adolf Krebs (1900–81), German-born British biochemist, and Philip Pacy Cohen (1908–93), US biochemist.]

Krebs cycle *an alternative name (sometimes) for* **1** tricarboxylic-acid cycle. **2** ornithine–urea cycle.

Krebs–Henseleit cycle *an alternative name for* the **ornithine–urea cycle**. [After (Sir) Hans Adolf Krebs (1900–81), German-born British biochemist, and Kurt Henseleit (1908–73).]

Krebs–Henseleit Ringer bicarbonate *see* **Krebs mammalian Ringer**.

Krebs–Kornberg cycle *an alternative name for* **glyoxylate cycle**. [After (Sir) Hans Adolf Krebs (1900–81) and Hans Leo Kornberg (1928–), German-born British biochemists.]

Krebs mammalian Ringer any of several media devised for the suspension of mammalian tissues. In the first version (1932), **Krebs–Henseleit Ringer bicarbonate**, the buffer was bicarbonate-CO_2, while in the second (1933), **Krebs original Ringer phosphate**, it was phosphate. In improved versions (1950), often referred to respectively as **Krebs Ringer bicarbonate** (*abbr.*: KRB) and (calcium-free) **Krebs Ringer phosphate** (*abbr.*: KRP), a mixture of organic acids and glucose was substituted for part of the NaCl in the first two versions, to resemble as closely as possible the state of the tissues in the intact animal. Also, an additional medium, low in phosphate, bicarbonate, and CO_2, was devised. *See also* **Ringer's solution**.

Krebs' manometer fluid a solution containing 44 g NaBr, 0.3 g Triton X-100, and 0.3 g Evans' Blue in 1000 mL of water; $D = 1.033$. The standard pressure (760 mmHg) in mm of manometer fluid is 10 000. Other detergents may be substituted for the Triton X-100 in equivalent proportions.

K region a region in a carcinogenic polycyclic aromatic hydrocarbon molecule in which there is a carbon–carbon bond having ordinary double-bond rather than aromatic character and that takes part in addition reactions; e.g. the bond between atoms C-9 and C-10 in phenanthrene. It has been suggested that carcinogenicity in such molecules is determined by the existence in them of an active K region. *Compare* **L region**.

kringle a protein module of about 85 amino-acid residues that is held in a characteristic 3-D conformation by three internal disulfide bridges. It is so-named because its 2-D representation resembles that of the kringle, a Scandinavian pastry of characteristic shape. It was first described in prothrombin, but has since been found in plasminogen and (both tissue-type and urokinase-type) plasminogen activators.

Kronig–Kramers transform any of a set of mathematical relations describing the response of a system to a perturbation such as light. These relations allow the calculation of optical rotary dispersion from circular dichroism (ellipticity) and vice versa, and describe the relationship between dispersion and absorption. [After R. de L. Kronig and Hendrik Anthony Kramers (1894–1952), Dutch physicist, who independently devised them in 1926 and 1927, respectively.]

KRP *abbr. for* Krebs Ringer phosphate; *see* **Krebs mammalian Ringer**.

Krüppel a protein involved in the segmentation of the embryo and in the differentiation of the malpighian tubules in insects (e.g., *Drosophila*). It is a nuclear, chromatin-associated protein, encoded by a **gap gene**, and is a member of a subfamily of C_2H_2-type zinc-finger proteins. Mutant *krüppel* larvae lack eight body segments. [From German *Krüppel*, cripple.]

Krüppel-like factor *abbr.*: KLF; any DNA-binding transcriptional regulator that is homologous to **Krüppel** protein of *Drosophila*. At least 15 human KLFs are recognized.

KS/hst a human oncogene, with no known viral counterpart, whose product is related to **fibroblast growth factor** and is expressed in **Kaposi's sarcoma**.

k(T) *symbol for* rate constant (alternative to *k*, used when distinction of a rate constant from the Boltzmann constant is necessary).

K-tup *an alternative term for* **K-tuple**.

K-tuple *or* **K-tup** the number of sequence characters that are considered as an indivisible unit (or word) by database search algorithms such as FastA.

Ku the dsDNA-binding component of **DNA-PK**.

Kunitz and Northrop inhibitor *see* pancreatic trypsin inhibitor.

Kunitz inhibitor the best known of the three **soybean trypsin inhibitors**. The term refers to any type of protease inhibitor that contains consensus sequences known as **Kunitz domains**. These include some serine protease inhibitors. Such domains have also been identified in some α-amyloid precursor protein isoforms.

Kunitz (tryptic) unit a measure of tryptic activity based on the extent to which trypsin-containing samples can digest casein under standard conditions, often used to determine the relative amounts of Kunitz and Bowman–Birk type trypsin inhibitors present in samples, comparison being made between tryptic activity and chymotryptic activity. Crystalline trypsin has an activity of 6 Kunitz units per mg.

Kupffer cell a type of nonmotile, mononuclear phagocytic cell. Such cells line the sinusoids of the liver. [After Karl Wilhelm von Kupffer (1829–1903), German anatomist.]

kuru a transmissible **spongiform encephalopathy** of humans, characterized by degeneration of the nervous system and found among the Fore people of Papua New Guinea. It is a **prion** disease, transmitted through funerary practices thought to have involved endocannibalism. The brains of victims of this and **Gerstmann–Sträussler syndrome** contain so-called kuru plaques, which have been shown to stain positively with antibodies against prion protein in patients with Gerstmann–Sträussler syndrome. It has become extremely rare since cannibalism ceased in the late 1950s.

KVLQT1 the gene at 11q15.5 for a cardiac K voltage-gated channel α subunit. It encodes 676 amino acids, including six transmembrane segments and extensive intracellular N- and C-terminal domains. Numerous loss-of-function mutations are associated with the **long QT syndrome**. *See also* **KCNE1**.

kwashiorkor a form of severe malnutrition of children occurring in tropical or near-tropical regions. It is generally believed to be caused by a dietary deficiency of protein, but not of other energy-producing foodstuffs. The condition commonly develops when, after prolonged breast feeding, the child is weaned onto a diet inadequate in protein. *Compare* **marasmus**.

kynurenate 4-hydroxyquinaldate; 4-hydroxy-2-quinoline carboxylate; the anion of kynurenic acid and an intermediate in the conversion of **kynurenine** to quinaldate.

kynureninase EC 3.7.1.3; *systematic name*: L-kynurenine hydrolase; a pyridoxal-phosphate-containing enzyme that hydrolyses L-kynurenine to anthranilate and L-alanine. It also acts on 3′-hydroxykynurenine and some other (3-arylcarbonyl)-alanines.

kynurenine 3-anthraniloylalanine; L-kynurenine is a metabolite on the degradative pathway of tryptophan, is found in the urine especially in pyridoxal (vitamin B_6) deficiency, and is an intermediate in the synthesis of nicotinic acid from tryptophan.

kyotorphin the analgesic dipeptide, L-Tyr-L-Arg, isolated from bovine brain; D-kyotorphin is L-Tyr-D-Arg.

LI

l *symbol for* **1** litre (alternative to L). **2** liquid.

l *symbol for* length (alternative to *L*).

l- *prefix (in chemical nomenclature)* denoting levorotatory; now obsolete, use (−)-; *see* **optical isomerism**.

L *symbol for* **1** litre (alternative to l). **2** a residue of the α-amino acid L-leucine (alternative to Leu). **3** an atom of hydrogen that may be one of *either* protium *or* deuterium. **4** linking number. **5** light chain (def. 1) (of an immunoglobulin).

L *symbol for* **1** Avogadro constant (alternative to N_A). **2** latent heat. **3** radiance.

L- a configurational prefix used in designating the absolute configuration of chiral compounds of certain classes; *see* **D/L convention**.

La *symbol for* lanthanum.

label 1 *or (sometimes informally)* **tag** any (radioactive or stable) isotope substituted in a proportion of the molecules of a particular compound so that the translocation and/or the chemical or biochemical transformation of the compound may be followed. *See also* **tracer**. **2** *or (sometimes)* **marker** an easily recognizable chemical group (which can be radioactive) attached to a molecule or incorporated within a structure. **3** to attach or apply such a label. The following particular uses of the terms label or labelled have individual entries (*see also* the list at **labelling**): **affinity label, electron-dense label, immunoassay, mixed labelled, multiply labelled, nonselectively labelled, photoaffinity label, radioactive label, selectively labelled, singly labelled, specifically labelled, spin label, uniformly labelled.**

labelled compound *or (US, usually)* **labeled compound** a compound bearing a **label** (def. 1, 2). An isotopically labelled compound is a compound in which one or more atoms of a proportion of the molecules have been replaced by a (radioactive or stable) isotope; i.e. it is a mixture of an isotopically unmodified compound with one or more analogous isotopically substituted compound(s). In some cases, the amount of isotopically unmodified compound may be small. This is particularly the case for ^2H compounds, when near 100% replacement is possible. *See also* **carrier-free**.

labelling *or (US, usually)* **labeling 1** the attribute of being labelled. **2** the act or process of chemically introducing a **label** (def. 1, 2) into a compound either by substitution or by synthesis, or of incorporating such a label into a material or structure. The following particular uses of the term labelling have individual entries (*see also* the list at **label**): **conjugation labelling, double labelling, enzyme labelling, ferritin labelling, fluorescein labelling, generally labelled, hapten-sandwich labelling, iodine labelling.**

labile 1 describing a chemical compound that is prone to spontaneous change or decomposition; relatively unstable. **2** describing an atomic or molecular grouping in a chemical compound that usually readily dissociates from the compound; loosely attached. **3** describing a chemical bond that very easily ruptures or is ruptured. **—lability** *n.*

labile factor *an alternative name for* factor V; *see* **blood coagulation**.

labile phosphate the amount or proportion of any phosphate that is readily liberated from a phosphate-containing compound by hydrolysis in 1 M HCl at 100°C within 7–10 min (*hence also sometimes known as* **seven-minute phosphate** *or* **ten-minute phosphate**). These conditions were selected originally for estimating the two terminal phosphate groups of ATP. The following compounds are included in the group (percentage hydrolysis in parentheses): phosphocreatine (100), ATP (67), ADP (50), aldose 1-phosphate (100), pyrophosphate (100).

labilize *or* **labilise** to render **labile** (*esp.* def. 3). **—labilization** *or* **labilisation** *or* **labilizer** *or* **labiliser** *n.*

lac any of various resinous substances secreted by certain plants or insects, used in the manufacture of lacquers.

lac *symbol for* the genetic system of *Escherichia coli* that controls the organism's ability (*lac⁺*) or inability (*lac⁻*) to metabolize **lactose**. *See also* **lac operon, lac repressor**.

laccase any of a group of copper–protein enzymes (EC 1.10.3.2) of low specificity, that catalyse the oxidation of both *o*- and *p*-quinols

by dioxygen to form the corresponding semiquinones and two molecules of water. They are so-called because such an enzyme was first found in the lac (or lacquer) tree (*Rhus verniciflua*); however, they are now known to be widely distributed in plants and fungi.

lacI the structural gene for **lac repressor**, the repressor of the *lac* operon.

lacI^q a mutation in the promoter of the *lacI* gene of *E. coli* that results in increased transcription and higher levels of **lac repressor** within cells.

lac operon *or* **lactose operon** a genetic system, or **operon**, of *Escherichia coli* that controls the synthesis of the inducible enzyme β-D-galactosidase (*see* **galactosidase**). It is probably the most intensively studied group of *E. coli* genes. It consists of a linear sequence (5′ end to 3′ end of the coding strand) of two control sites, designated *p* and *o*, followed by three structural genes, designated *z*, *y*, and *a*; *p* is the promoter (concerned with RNA-polymerase binding), *o* is the operator (binding site for repressor), *z* is the gene for β-galactosidase, *y* is the gene for the **permease**, and *a* is the gene for thiogalactoside transacetylase (an enzyme of uncertain physiological function). Immediately preceding the *p* site is a regulatory gene designated *i* (concerned with synthesis of **lac repressor**).

lac repressor a protein encoded by the *lacI* gene, that controls the expression of structural genes coding for enzymes involved in lactose metabolism in the *Escherichia coli* genome; in the absence of inducer it binds with high affinity to the operator site of the **lac operon**. It is a homotetramer and specifically recognizes a 23 base-pair DNA sequence in the *lac* operator by means of its four N-terminal domains (headpieces), each of 51 amino-acid residues; each subunit also has one allosteric binding site for an inducer. The repressor is a typical **helix-turn-helix** protein and gives its name to a subfamily.

lact+ *a variant form of* **lacto+** *(before a vowel)*.

lactacin any of a family of pore-forming peptidic toxins with approximately 75 + 65 amino acid residues, permitting passage of small solutes, including water. They are produced by *Lactobacillus johnsonii* and are classified in the TC system under number 1.A.26.

lactadherin a protein produced by smooth muscle cells of the aorta and the precursor of the polypeptide **medin**.

lactalbumin 1 the fraction of (normally cow's) milk proteins remaining after removal of casein. It consists principally of **beta-lactoglobulin** and α-lactalbumin. **2 α-lactalbumin** a Ca^{2+}- binding protein that occurs as a minor constituent of milk of many mammals; its 123 amino-acid sequence is highly similar to that of lysozyme. It is a non-catalytic regulatory protein whose presence is required for **lactose synthase** to transfer a galactose residue from UDPgalactose to D-glucose; hence it is also known as B protein of lactose synthase. It causes the substrate for *N*-acetyllactosamine synthase to change from *N*-acetylglucosamine to glucose, to give lactose and UDP. Protein B had, therefore, been called a **'specifier' protein**. The genes for lysozyme and α lactalbumin arose evolutionarily from a common ancestral gene.

lactam any intramolecular cyclic amide produced by the formal removal of a molecule of water between the amino and the carboxyl groups of an amino acid, other than an α-amino acid. A prefix (as in β-lactam) may be used to designate the position of the amino group in the parent compound. A lactam commonly exists in tautomeric equilibrium with its corresponding **lactim**.

β-lactam antibiotic any member of a class of natural or semisynthetic antibiotics whose characteristic feature is a strained, four-membered β-lactam ring (*see* **lactam**). They include the **penicillins** and many of the **cephalosporins**.

β-lactamase any member of a group of constitutive or inducible enzymes (EC 3.5.2.6); *systematic name*: β-lactamhydrolase. They are produced by certain strains of various bacterial species and can catalyse the hydrolysis of the **lactam** linkage in one or more β-lactam antibiotics, leaving a substituted β-amino acid; they thereby inactivate the antibiotic. The enzymes are of varying specificity, some acting more rapidly on penicillins (**penicillinases**), some more rapidly on cephalosporins (**cephalosporinases**); the latter were formerly classified separately as EC 3.5.2.8. The β-lactamases can be inhibited by various substances, such as **clavulanic acid**. There are several classes: class A (or type I) enzymes have seven motifs; class B (or type II) enzymes have a zinc cofactor. For example, the class A Men-1 from *Escherichia coli* has a broad spectrum of activity, hydrolysing penicillins and many cephalosporins.

lactam–lactim tautomerism a form of **tautomerism** occurring in lactam–lactim interconversion as a result of migration of a hydron between the nitrogen atom and the oxygen atom. It represents a special case of amide–imidol tautomerism.

lactase *a trivial name for* any β-D-**galactosidase**. Some lactases hydrolyse other glycosides, including α-L-arabinosides, β-D-fucosides, and β-D-glucosides.

lactase deficiency a condition of humans in which there is a congenital or acquired deficiency of lactase (i.e. β-**galactosidase**) in the mucosa of the small intestine leading to failure to digest lactose and to **lactose intolerance**. The condition is more common in healthy adolescents and adults, and its frequency varies considerably: it is more prevalent in native populations of Australia and Oceania, east and south-east Asia, tropical Africa, and the Americas (45–95%) than in those of central and northern European ancestry (<25%).

lactate 1 α-hydroxypropionate; 2-hydroxypropanoate; the anion, CH_3–CH(OH)–COO⁻, derived from **lactic acid**. It occurs naturally as D(–)-lactate (i.e. (*R*)-2-hydroxypropanoate) and L(+)-lactate (i.e. (*S*)-2-hydroxypropanoate). L(+)-lactate is formed by anaerobic glycolysis in animal tissues, and DL-lactate is found in sour milk, molasses, and certain fruit juices. Bacterial **lactic-acid fermentations** produce either the D(–) or L(+) enantiomer or the racemate (DL) depending on the individual organism. 2 any salt or carboxylic ester of lactic acid. 3 to secrete milk; *see* **lactation**.

lactate dehydrogenase any enzyme that interconverts **lactate** and pyruvate. L-**lactate dehydrogenase** (EC 1.1.1.27; *abbr.*: LD *or* LDH) acts on L-lactate and NAD⁺. It occurs as homotetramers or heterotetramers of the heart muscle (H) and skeletal muscle (M) forms (the five isoenzymes are: LD_1, H_4; LD_2, H_3M; LD_3, H_2M_2; LD_4, HM_3; and LD_5, M_4). In clinical chemistry the assay of LD_1 (*known also as* α-hydroxybutyrate dehydrogenase *or* HBD) can be of use in diagnosing myocardial infarction, in which it peaks at 2–3 days after the infarct, and in hemolytic crisis of sickle-cell anemia, since this isoenzyme is the predominant form in cardiac muscle and red cells. D-**lactate dehydrogenase** (EC 1.1.1.28) acts on D-lactate and NAD⁺. L-**lactate dehydrogenase (cytochrome)** (EC 1.1.2.3) acts on L-lactate and ferricytochrome *c* and is a flavohemoprotein (FMN is coenzyme). D-**lactate dehydrogenase (cytochrome)** (EC 1.1.2.4) acts on D-lactate and ferricytochrome *c* and is a flavoprotein (FAD is coenzyme). D-**lactate dehydrogenase (cytochrome *c*-553)** (EC 1.1.2.5) acts on D-lactate and ferricytochrome *c*-553 and is present in *Desulfovibrio vulgaris*. These enzymes are very widely distributed.

lactation 1 the secretion of milk by the mammary gland. 2 a complete period of milk secretion from about the time of parturition to that of weaning. *See also* **lactate** (def. 3).

lacteal 1 of, relating to, or resembling milk; milky. 2 (of lymph vessels) containing or conveying **chyle**. 3 any blind-ended lymph vessel that carries chyle from a villus of the small intestine to the thoracic duct.

lactic pertaining to, or derived from, milk.

lactic acid α-hydroxypropionic acid; 2-hydroxypropanoic acid; a reasonably strong acid ($pK_a = 3.86$ at 25°C). In biological systems it normally exists largely as its anion, **lactate** (def. 1). It is also useful as a preservative of foodstuffs.

lactic-acid bacterium any bacterium in which lactate is the main metabolic product. Such bacteria belong chiefly to the genera *Lactobacillus* and *Streptococcus*, and some are of commercial importance, e.g. in the manufacture of fermented foods.

lactic-acid cycle *an alternative name for* **Cori cycle**.

lactic-acid fermentation *a general name for* **heterolactic fermentation** and/or **homolactic fermentation**.

lactic acidosis a clinical condition of persistent raised blood lactate concentration, usually above 5 mmol L⁻¹, with a lowered blood pH. It may be caused by severe exercise, cardiovascular insufficiency, hypoxia, shock, drug intoxication, and some inborn errors of metabolism. *Compare* **hyperlactatemia**.

lacticin any of a family of pore-forming peptidic toxins with about 50 amino acid residues, permitting passage of small solutes, including water. They are produced by *Lactobacillus lactis* and are classified in the TC system under number 1.A.27.

lactide any intermolecular cyclic ester formed by self-esterification from two (or more) molecules of a hydroxy carboxylic acid. *Compare* **lactone**.

lactifer a special cell, found in the bark of the rubber tree (*Hevea*), in which rubber latex accumulates.

lactim any intramolecular cyclic carboximidic acid produced by the formal removal of a molecule of water between the amino and the carboxyl groups of an amino carboxylic acid other than an α-amino acid. A prefix (as in β-lactim) may be used to designate the position of the amino group in the parent compound. A lactim commonly exists in tautomeric equilibrium with its corresponding **lactam**.

lactitol 4-*O*-β-D-galactopyranosyl-D-glucitol; a disaccharide used as a sweetener.

lacto+ *or (before a vowel)* **lact+** *comb. form* indicating milk.

lactobacillic acid (1*R*-*cis*)-2-hexylcyclopropanedecanoic acid; 11,12-methyleneoctadecanoic acid; a lipid constituent of various microorganisms, including *Lactobacillus arabinosus*.

Lactobacillus a genus of Gram-positive, anaerobic or facultatively aerobic, straight or curved bacilli or coccobacilli. Their metabolism is fermentative and all species form lactate as the major end product of glucose fermentation. Various species of lactobacilli are important in the preparation of certain fermented foods, e.g. cheese, yoghurt, sauerkraut.

***Lactobacillus bulgaricus* factor** *an alternative name for* **pantetheine**.

***Lactobacillus casei* factor** *an alternative name for* **folic acid**.

lactobionic acid the **aldonic acid** obtained by the oxidation of lactose.

lactochrome *a former name for* **riboflavin**.

lactocin any of a family of pore-forming peptidic toxins with about 25 amino acid residues, permitting passage of small solutes, including water. They are produced by *Lactobacillus sake* and are classified in the TC system under number 1.A.23.

lactococcin designation of three different families of peptidic pore-forming toxins, produced by Gram-positive bacteria; they contain 65–75 amino acid residues (lactococcin A family, TC 1.C.22), or 40+35 amino acid residues (lactococcin G family, TC 1.C.25) or 50–100 amino acid residues (lactococcin 972 family, TC 1.C.37).

lactoferrin *or* lactotransferrin a **transferrin** present in milk. Its level in human plasma may be elevated in pancreatitis.

lactoflavin *or* lactoflavine *a former name for* **riboflavin**.

lactogen 1 any agent that stimulates the secretion of milk. 2 placental lactogen *an alternative name for* **choriomammotropin**. —**lactogenic** *adj.*

lactogenic hormone *an alternative name for* **prolactin**.

β-lactoglobulin *see* **beta lactoglobulin**.

lactonase 1 EC 3.1.1.17; *recommended name*: gluconolactonase; *sys-*

tematic name: D-glucono-1,5-lactone lactonohydrolase; *other name*: aldonolactonase. An enzyme that catalyses the reaction:

D-glucono-1,5-lactone + H_2O = D-gluconate;

it also acts on a wide range of hexono-1,4-lactones. **2 1,4-lactonase** *or* **γ-lactonase** EC 3.1.1.25; an enzyme that specifically hydrolyses 1,4-lactones having 4–8 carbon atoms to the corresponding 4-hydroxyacids.

lactone any intramolecular cyclic ester produced by the formal removal of a molecule of water between the hydroxyl and the carboxyl groups of a molecule of a hydroxy carboxylic acid other than an *α*-hydroxy acid. A prefix (as in *β*-lactone) may be used to designate the position of the hydroxyl group in the parent compound. *Compare* **lactide**.

lactoperoxidase *a special name for* **peroxidase** (def. 2) from milk. It also catalyses the oxidation of iodide to iodine, hence it is useful in radioiodine labelling of proteins and other biological materials under mild conditions.

lactose *or (formerly)* **milk sugar** *the trivial name for* the **disaccharide** 4-*O*-*β*-D-galactopyranosyl-D-glucose. It constitutes roughly 5% of the milk of almost all mammals (human milk 6.7%; cow's milk 4.5%). It occurs rarely in plants, e.g. in the anthers of *Forsythia* spp. *See also* **allolactose**.

lactose fermenter *jargon term for* any genus, species, or strain of microorganism that ferments lactose.

lactose intolerance a syndrome occurring in humans with congenital or acquired intestinal **lactase deficiency** in response to the ingestion of lactose (in milk). The failure to hydrolyse lactose leads to its accumulation in the intestine and consequent osmotically induced net fluid secretion into the gut, rapid intestinal transit, and acid stools due to bacterial fermentation of lactose. The condition is particularly prevalent in the Orient.

lactose operon *see* **lac**.

lactose permease an H^+–lactose symporter in *Escherichia coli*. It has 12 transmembrane helices but is not very closely related to most other members of the superfamily.

lactose synthase EC 2.4.1.22; *systematic name* UDPgalactose:D-glucose 4-*β*-D-galactotransferase; *other name*: UDPgalactose–glucose galactosyltransferase. An enzyme that catalyses the reaction:

UDPgalactose + D-glucose = UDP + lactose.

In mammary gland the protein comprises *N*-acetyllactosamine synthase, the A protein, and α-lactalbumin (*see* **lactalbumin**), the B protein, both of which are required for lactose synthesis.

lactose-tolerance test a procedure used in the investigation of suspected **lactase deficiency**. Blood-glucose concentrations are estimated before an oral dose of lactose (1 g per kg body mass) and thereafter at 30-min intervals for 2 h. In normal individuals an increase of at least 30 mg per 100 ml (0.167 mM) in the blood-glucose concentration above the fasting value is observed.

lactosuria the presence of lactose in the urine. It is not uncommon during late pregnancy and lactation, or if the flow of milk is prevented.

lactosyl either of the glycosyl groups formed from lactose by detaching the anomeric (*α* or *β*) hydroxyl group on C-1.

lactotransferrin *an alternative name for* **lactoferrin**.

lactotroph any of a class of irregular, granulated, acidophilic cells of the anterior pituitary gland that secrete prolactin. Their number increases greatly during pregnancy.

lactotrophic describing any agent that induces normal growth and development of the mammary gland and sustains it.

lactotropic describing any agent that acts on the mammary gland to induce changes in its metabolism.

lactotropic hormone *an alternative name for* **prolactin**.

lactotropin *an alternative name for* **prolactin**.

lactoyl the acyl group, CH_3–CH(OH)–CO–, derived from lactic acid.

lactulose *the trivial name for* 4-*O*-*β*-D-galactopyranosyl-D-fructofuranose; a semisynthetic disaccharide, prepared from lactose, that is sweeter than lactose but less sweet than sucrose. Concentrated solutions of lactulose are useful clinically as an osmotic laxative.

lactyl *a former term for* **lactoyl**.

*lac***Y** a gene in *Escherichia coli* for **lactose permease**.

*lac***Z** a gene in the *lac* operon of *E. coli* encoding β-galactosidase, it is widely used as a reporter gene.

*lac***Zdelta M15** a mutant form of the *E. coli* β-galactosidase gene in which the sequence encoding the N-terminal region corresponding with the α-peptide is deleted. *E. coli* strains carrying this deletion are used to test for a-complementation.

LAD *abbr. for* **leukocyte adhesion deficiency**.

ladder *informal*, one of many types of DNA molecular size markers, an example is the 1kb ladder.

L1 adhesion molecule a type I membrane glycoprotein involved in neuron–neuron adhesion, neurite fasciculation, outgrowth of neurites, etc. It is a member of the immunoglobulin superfamily, and has fibronectin domains.

Laetrile *see* **amygdalin**.

laevo- *a variant spelling (esp. Brit.) of* **levo-**.

laevo+ *a variant spelling (esp. Brit.) of* **levo+**.

laevodopa *a variant spelling (esp. Brit.) of* **levodopa**.

laevorotation *a variant spelling (esp. Brit.) of* **levorotation**.

laevulose *a variant spelling (esp. Brit.) of* **levulose**.

LAF *abbr. for* lymphocyte-activating factor (i.e. **interleukin 1**).

laforin a protein-tyrosine phosphatase present in many tissues. Mutations lead to Lafora disease, a progressive myoclonus epilepsy with intellectual decline, accumulation of glycogen in brain, skin, and liver, and early death.

lagging strand the DNA strand that is synthesized by the joining together of short **Okazaki fragments**, giving the effect of continuous 3′ to 5′ synthesis in the direction of fork movement during the discontinuous replication of duplex DNA. *Compare* **leading strand**.

lag phase 1 the initial phase of the growth of a bacterial culture during which the rate of increase of cell numbers remains static, before rising to a value determined by environmental conditions. The extent of the lag phase is determined by the cultural history of the cells of the inoculum and the chemical and physical conditions of the medium into which they are placed. **2** a period of comparatively low activity after the start of an enzymic reaction that involves enzymes that act at a surface, e.g. phospholipases. It is due to the slow formation of the complex between the enzyme and the substrate interface. Once this has been formed the reaction accelerates.

lake to lyse erythrocytes, especially by suspending them in a hypotonic medium.

Laki–Lorand factor *abbr.*: LLF or L-L factor; *an alternative name for* factor XIII; *see* **blood coagulation**.

lambda *symbol*: λ (lower case) *or* Λ (upper case); the eleventh letter of the Greek alphabet. For uses *see* **Appendix A**.

lambda chain *or* λ **chain** *symbol*: λ; one of the two types of **light chain** of human immunoglobulins, the other type being a kappa chain.

lambda DASH a proprietary bacteriophage λ replacement vector used for cloning DNA fragments of up to 20 kb.

lambda FIX a proprietary bacteriophage λ replacement vector used for cloning DNA fragments of up to 20 kb.

lambda gt10 a bacteriophage λ insertion vector used for cloning DNA fragments of up to 8 kb.

lambda gt11 a bacteriophage λ insertion vector used for cloning DNA fragments of up to 8 kb and designed to allow expression of proteins encoded by DNA inserts.

lambda insertion vector one of a number of bacteriophage λ cloning vectors in which the cloned sequence is inserted into a single restriction endonuclease site. The size constraints on packaging bacteriophage λ recombinant DNA limits the size of insert to <8 kb, consequently making them most suitable for cloning cDNA.

lambda phage *or* λ **phage** a **temperate phage** that infects *Escherichia coli* and contains linear duplex DNA, which circularizes during infection; the genome is 48 514 bp. Important gene clusters include those for the head and tail genes, recombination (*att, int, xis, α, β, γ*), lysis, and regulatory genes. Of the regulatory genes (*cIII, N, cI, cro, cII, O, P, Q*), two are immediate-early genes, *cro* and *N*; *cro* prevents synthesis of the repressor (*see* **prophage**) and turns off expression of immediate-early genes, including itself, after its product, cro, binds to *O*. Lambda phage is an example of the antitermination mechanism of phage control (*see* **antiterminator**); *N* codes for an antitermination factor that allows transcription to proceed through the terminator to the delayed-early genes, *cII* and *cIII*, and *Q* codes for an antitermination factor that allows host RNA polymerase to proceed to the late (lytic) genes. The lytic cycle is repressed by expression of *cI*, the product of which acts at the operator *O* to prevent use of the promoter *P*. Lambda phage has been widely used as a vector in cDNA technology; foreign DNA can be inserted into the chromosome, and some foreign genes can be expressed using the lambda promoters. The linear DNA has cohesive ends that, when the DNA circularizes, associate to form the **cos site** (cohesive end site), which is important in packaging. The *cos* site has been exploited in the formation of **cosmids**.

lambda pipette *colloquial term for* a pipette used for the transfer of volumes in the microlitre range.

lambda replacement vector one of a number of bacteriophage λ cloning vectors designed for the cloning of DNA inserts of up to 20kb. The size constraints on packaging bacteriophage λ recombinant DNA necessitates the removal of a 'stuffer fragment' located between a pair of identical restriction endonuclease sites before the insertion of the DNA to be cloned. They are used for cloning fragments of genomic DNA.

lambda repressor *or* λ **repressor** the product of gene *cI*; it is responsible for maintenance of phage λ **prophage**. Its product binds to the operator associated with the RNA polymerase promoter, thus preventing RNA polymerase from initiating transcription. The phage genome then cannot enter the lytic cycle. Mutants in this gene cannot maintain **lysogeny**.

lambda ZAP a proprietary bacteriophage λ insertion vector in which DNA fragments of up to 8 kb are cloned into the multiple cloning site of the pBluescript plasmid incorporated between the left and right arms of the λ vector. The vector permits the excision of the plasmid with its insert in the form of bacteriophage M13 when *E. coli* strains infected with λZAP are superinfected with M13 helper phage.

Lambert–Beer law *another name for* **Beer–Lambert law**.

Lambert's law (of absorption) a physical law in which **Bouger's law** was re-expressed; put into modern terms, it states that the fraction of monochromatic light (or other electromagnetic radiation) absorbed by a system is independent of the incident **radiant power**, Φ_0. The law holds only if Φ_0 is small and other factors such as reflection, scattering, and photochemical reactions are negligible. *See also* **Beer–Lambert law**. [After Johann Heinrich Lambert (1728–77), German mathematician and physicist who stated it in 1760.]

lamella (*pl.* **lamellae**) any thin layer, membrane, or platelike structure, e.g. one of the series of double membranes that carry the photosynthetic pigments in chloroplasts, one of the thin layers of calcified bone matrix, or one of the spore-bearing gills of the fruiting bodies of mushrooms or related fungi. —**lamellar** *or* **lamellate** *adj*.

lamellar phase one type of **liquid-crystalline phase**.

lamin any of a group of intermediate-filament proteins that form the fibrous matrix (**nuclear lamina**) on the inner surface of the nuclear envelope. They are classified as lamins A, B, and C. *See also* **gliadin**.

lamina (*pl.* **laminae**) **1** any thin plate or layer, e.g. of bone or mineral. **2** (*in botany*) **a** the blade of a leaf or petal. **b** the flattened part of the thallus of certain algae. *See also* **basal lamina, nuclear lamina**. —**laminar** *adj*.

laminar flow a form of nonturbulent flow of a fluid in which parallel layers (laminae) of the fluid are moving in the same direction but with different velocities.

laminar-flow cabinet *another name for* **safety cabinet**.

laminaribiose the disaccharide, 3-*O*-β-D-glucopyranosyl-D-glucose; a unit of the seaweed polysaccharide **laminarin**.

pyranose form

laminarin a β-1,3-glucan reserve polysaccharide in some brown algae (*Laminaria* spp.). It contains **laminaribiose**.

laminin a glycoprotein that is the major constituent of basement membranes. It consists of three chains, bound to each other by disulfide bonds into a cross-shaped molecule having the appearance of one long and three short arms, with globules at each end (the three chains are wound together in a helical conformation for two thirds of their length, but diverge at the N-terminal end into the three arms that form the apex and side arms of a cross). The molecule interacts with procollagen type IV. Laminin mediates the attachment, migration and organization of cells into tissues during embryonic development by interacting with other extracellular-matrix components. Laminins of varying composition and sequence have been categorized as laminins 1 to 5. Laminin 5 is also known as **kalinin**.

laminin receptor any protein, especially an **integrin** (*see also* **VLA**), that binds laminin, interacting with it to promote cell adhesion and migration.

Lamm equation a general hydrodynamic equation describing the mass transport of a two-component system during ultracentrifugation. It is:

$$dc_2/dt = (1/x) \, d\{x[D(dc_2/dx) - c_2 s \, \omega^2 x]\}/dx,$$

where c_2 is the concentration of solute molecules at distance x from the axis of rotation, D is the translational diffusion constant, ω is the angular velocity, and s is the sedimentation coefficient.

LAMP *abbr. for* **1** lysosomal-associated membrane protein. One of three membrane glycoproteins that shuttle between lysosomes, endosomes and the plasma membrane. They are heavily glycosylated and are thought to protect lysosomal membranes from attack by lysosomal enzymes. LAMP-1 deficient mice have a mild phenotype but LAMP-2 deficient mice have severe symptoms arising from the accumulation of autophagic vesicles in numerous tissues. LAMP-2 deficiency is the primary defect in Danon disease, a lysosomal glycogen storage disease with normal acid maltase activity. LAMP-3 is CD63 a member of the **tetraspanin** superfamily. **2** Linux Apache MySQL PHP; a popular software raft for a WWW server, relational database, and interrogatory software.

lampbrush chromosome any of the large diplotene chromosomes found in the oocytes of many animals. They are up to 1mm long and are covered with loops projecting in pairs from most chromomeres; each loop has a core of DNA and is the site of active gene expression.

lamprin a self-aggregating extracellular cartilage protein of lampreys, aggregates of which form fibres rich in beta structure. It is a major structural protein isolated from the annular cartilage of the lamprey, similar in amino-acid composition to **elastin**, but it does not contain desmosine or isodesmosine.

Lan *symbol for* lanthionine.

lan+ *comb. form* indicating wool. [From Latin *lana*, wool.]

Landsteiner–Weiner antigen *abbr.*: LW antigen; *see* **Rhesus factor**.

lane any of a set of parallel tracks resulting from the simultaneous separation of similar samples by one-dimensional zone electrophoresis or chromatography in thin-layer media.

Langendorff perfused heart a nonworking rodent heart preparation maintained *in vitro* by perfusion into the aorta with an oxygenated fluid so that the fluid circulates primarily into the coronary arteries. It is useful for studying the metabolism of heart muscle. A number of improvements of the original preparation have been made since its introduction. [After O. Langendorff (1853–1909), German physiologist.]

Lang–Levy pipette a type of constriction micropipette (5–500 μL) that is self-adjusting and very reproducible.

Langmuir adsorption isotherm an equation describing the fraction of the surface of an adsorbent covered with a monomolecular layer of an adsorbed gas, arrived at from a consideration of the number of gas molecules striking and leaving the surface in a given time. It is:

$$V/V_m = k_1 p/(1 + k_2 p),$$

where V is the volume of gas adsorbed, V_m is the volume of gas required to cover the whole surface, p is the gas pressure, and k_1 and k_2 are constants for the system. *Compare* **Hill–Langmuir equation**. [After Irving Langmuir (1881–1957), US physical chemist and inventor.]

Langmuir–Blodgett multilayer a thin film of oriented amphiphilic molecules, consisting of superposed monolayers built up by successive dipping and withdrawing of a solid substrate, e.g. a glass slide, through an insoluble monolayer of an appropriate amphiphilic substance (or mixture of such substances) spread on the surface of an aqueous solution and kept at constant pressure. [After Irving Langmuir and Katharine B. Blodgett (1898–1979).]

Langmuir trough an apparatus for measuring the dependence of the surface force exerted by a film of a substance (of low vapour pressure), floating on a liquid with which it is immiscible, on the area of the film. The apparatus consists of a rectangular tank containing the subnatant liquid, e.g. water, on which a known amount of the immiscible substance, e.g. an oil or a higher fatty acid, is spread between two parallel barriers to form a film. One of the barriers is fixed and the other may be moved so as to compress the film to a predetermined area. The force exerted on the film to achieve any given area is measured by a system of levers and either a torsion wire or balancing masses.

lanolin *or* **wool fat** the fatlike secretion of the sebaceous glands of sheep that is deposited onto the wool fibres. It is a wax consisting of esters and polyesters of over 30 aliphatic, steroid, and triterpenoid alcohols and a similar number of fatty acids. Lanolin normally contains 25–30% water and is widely used as an ointment base. *See also* **lan+**.

lanosterol kryptosterol; lanosta-8,24-dien-3β-ol; a steroid alcohol (triterpene) originally obtained from **lanolin**. It occurs in the biosynthetic pathway from squalene 2,3-epoxide to cholesterol in animals and fungi, but not in photosynthetic tissues, where it is replaced by **cycloartenol**. *See also* **lanosterol synthase**.

lanosterol synthase EC 5.4.99.7; *other names*: 2,3-epoxysqualene–lanosterol cyclase; oxidosqualene–lanosterol cyclase (*abbr.*: OSC); an enzyme of the pathway for the synthesis of sterols. It catalyses the cyclization of (*S*)-2,3-epoxysqualene to lanosterol.

lantern *see* firefly lantern.

lanthanoid *or* **lanthanide** *or* (*formerly*) **lanthanon** any member of the series of 15 metallic elements with atomic numbers 57 (lanthanum) to 71 (lutecium) inclusive that occur together in group 3 and period 6 of the IUPAC **periodic table**; sometimes the term is restricted to the 14 elements following lanthanum. Lanthanoid is now the preferred name. Together with the elements scandium and yttrium, the lanthanoids are known also as **rare-earth elements** *or* (*more loosely*) **rare earths** (the latter term referring strictly to their oxides). They mostly occur widely in nature, albeit in small amounts. Apart from the element of atomic number 61 (promethium), all lanthanoids have at least one stable nuclide and occur naturally, the principal source being the mixed phosphate ore monazite. All nuclides of promethium, however, are radioactive, and (except for traces possibly occurring as a fission product of natural uranium) all are artificial. The lanthanoids are strongly electropositive, and their chemical properties are similar, due usually to the filling of an inner electron subshell (4f) progressively across the series. Like the **actinoids**, they thus represent a series of inner transition elements within the series of **transition elements** embraced by groups 3 to 11 inclusive. Some paramagnetic lanthanoid ions induce strong shifts in the positions of many NMR spectral peaks and have been used as relaxation and shift probes in structural studies employing NMR spectroscopy.

lanthionine (2*S*,6*R*)-lanthionine, *meso*-lanthionine, *S*-(alanin-3-yl)-L-cysteine, 2,2′-diamino-3,3′-thiobis(propionic acid); *symbol* (recommended):

$$\begin{array}{c}\text{Ala}\\|\\\text{Cys}\end{array}$$

or Lan; a di(α-amino acid) found in **lantibiotic** polypeptides. It is formed as a post-translational modification, the first step involving dehydration of a Ser to the α,β-unsaturated amino-acid residue, 2,3-didehydroalanine. A bond is then formed between the =CH₂ group thus formed and the sulfur atom of a cysteine, inversion of configuration of the α carbon of the Ser-derived moiety occurring in the process. Chick embryo protein contains the (2*R*,6*R*) diastereoisomer. Lanthionine is also formed artefactually by the mild alkaline treatment of wool, hence its name (*see* **lan+**).

(2*S*,6*R*)-lanthionine

lantibiotic any of a number of polypeptides, of 1.89–4.63 kDa, that contain **lanthionine** or methyllanthionine (*threo*-β-methyllanthionine, (2*S*,3*S*,6*R*)-3-methyllanthionine), and have antibiotic action. They are produced by, and primarily act on, Gram-positive bacteria. In addition to lanthionines, they contain the amino acids 2,3-didehydroalanine (*abbr.*: Dha) and 2,3-didehydrobutyrine (2-aminocrotonic acid, (*Z*)-2-aminobut- 2-enoic acid, *abbr.*: Dhb), resulting from dehydration of serine and threonine residues respectively. These amino acids are formed post-translationally from normally-coded amino acids. Lantibiotics are broadly divided into type A, elongated amphiphilic peptides, and type B, compact globular peptides. Type-A peptides kill bacterial cells by forming pores in the cytoplasmic membrane, type-B peptides possibly acting by enzyme inhibition.

LAP *abbr. for* **1** lamin-associated protein; any of the integral proteins of the inner nuclear membrane that bind to **lamins**. LAPs include several isoforms of LAP1 and LAP2, lamin B receptor, and emerin. **2** (*in clinical chemistry*) leucine aminopeptidase (LAS preferred).

LA-PF4 *abbr. for* low-affinity platelet factor 4; *other name*: connective-tissue activating peptide III (*abbr.*: CTAP-III). A protein (83 amino acids) derived from **platelet basic protein precursor** (94 amino acids). It stimulates DNA synthesis, mitosis, glycolysis, intracellular cyclic AMP accumulation, prostaglandin E₂ secretion, and synthesis of hyaluronic acid and sulfated glycosaminoglycan.

Laplace transform method a method that enables solutions of linear differential equations with constant coefficients to be found by reference to tables of Laplace transform pairs, with a minimum of mathematical work. The Laplace transform of a function $y = F(t)$ is defined by:

$$F(s) = \int_0^\infty e^{-st} F(t)\, dt$$

in which *s* is usually a complex quantity and must have a positive real term, so that e^{-st} will be zero when $t = \infty$. [After Pierre Simon, Marquis de Laplace (1749–1827), French astronomer and mathematician.]

Largactil *see* **chlorpromazine**.

large calorie *an alternative name for* **kilocalorie** *or* (*in nutrition*) **Calorie**.

large-patch repair (of DNA) *a name sometimes given to* the process of excision repair of duplex DNA in some mutant organisms, e.g. *polA⁻ Escherichia coli*, in which the newly synthesized polynucleotide 'patch' is considerably larger than in the wild type.

large T-antigen *see* **T antigen** (def. 1).

lariat a lasso-shaped structure formed from the fragment of RNA to be excised during the process of splicing out the introns in the primary transcript of eukaryotic genes.

lariat debranching enzyme an enzyme that cleaves the 2′,5′-phosphodiester bond formed during pre-mRNA splicing between the 5′-terminal guanylyl residue of the intron and an adenylyl residue near the 3′ splice site, at the branch point of excised intron lariats. It is involved in the turnover of excised introns.

Larmor frequency the angular frequency of **Larmor precession** of an atomic electron.

Larmor precession the **precession** of a charged particle moving under the influence of a central force, e.g. of an orbiting electron in an atom, when acted on by an external magnetic field. [After Joseph Larmor (1847–1942), Irish-born physicist.]

LAS (*in clinical chemistry*) *abbr. for* leucine aminopeptidase (preferred alternative to LAP).

lasalocid *or* **X-537A** any of five closely related lipophilic antibiotics, lasalocids A–E, isolated from cultures of *Streptomyces lasaliensis*. Lasalocid A ($M_r = 591$) is a carboxylic ionophore for virtually every known cation, but transports monovalent cations about ten times more effectively than divalent cations.

lased light *an imprecise term for* laser light; *see* **laser** (def. 2).

laser 1 any device for producing an intense, coherent, monochromatic, parallel beam of light (or other electromagnetic radiation) by stimulated emission. A laser consists of an optically transparent cylinder containing an active medium bounded by a full mirror at one end and by a partial mirror at the other end. Some chemical species of the medium are excited above their ground state by application of electromagnetic radiation from an external source; the excited species revert to their ground state with the emission of a photon, which in turn can collide with other excited species to produce stimulated emission of other photons. **2** having the characteristic quality of, being, or using, light emitted by a **laser** (def. 1), as in laser densitometry, laser ionization (micro) mass spectrometry, laser light, laser microprobe mass analysis, laser nephelometry, laser photolysis, laser (light) scattering, laser spectrophotometry, and laser velocimetry. [Acronym from light amplification by stimulated emission of radiation.]

Latapie mincer a device for mincing tissues for metabolic studies in which both the rate of rotation of the cutting disks and the rate at which the tissue is forced against these disks may be controlled by varying the rates at which two cranks are turned, thus enabling minces of different degrees of fineness to be obtained. Sterile conditions may be maintained during the mincing if desired.

late enzyme *see* **late gene**.

late gene any gene of a bacteriophage or other viral genome that is expressed significantly, as a **late enzyme** or other **late protein**, only after replication of the genome has started. *Compare* **early gene**.

latency 1 *or* **crypticity** the state or quality of being **latent**. **2** *an alternative term for* **latent period**.

latent 1 existing in a potential, dormant, or suppressed form but usually capable of being expressed or evoked. **2** (*in pathology*) (of an infection or disease) not yet manifest or apparent.

latent enzyme any enzyme whose activity only becomes manifest when the conditions are changed. The term is used especially of a particulate enzyme that becomes active when the particles are disrupted.

latent heat *symbol*: L; the change in the heat content of a (given amount of) substance when it changes phase without change in its temperature.

latent image (*in photography*) the invisible image produced by the action of light, etc. on the photosensitive emulsion that is made visible by development.

latent period *or* **latency 1** the time between the arrival of a stimulus at an irritable tissue and the beginning of the appropriate response. **2** the time between the adsorption of virulent bacteriophage onto a bacterium and lysis with liberation of newly formed virions.

late protein *see* **late gene**.

lateral gene transfer *or* **horizontal gene transfer** the transfer of genes between species, e.g. for antibiotic resistance in bacteria.

lateral phase separation the phenomenon of coexistence of solid and liquid regions in a lipid bilayer at temperatures between the **pre-transition temperature** and the **transition temperature**.

lathyrism any abnormal state that results from ingesting any of several toxins found in plants of the genus *Lathyrus*. In humans one kind is characterized by a spastic paraplegia of the lower extremities associated with degenerative changes in the lateral pyramidal tracts of the spinal cord. It is caused by ingestion of neurotoxins, e.g. β-cyano-L-alanine and L-α,γ-diaminobutyric acid. These are also present in the common vetch (*Vicia sativa*). In experimental animals a defect of cross-linking of collagen is caused by feeding sweet pea (*Lathyrus odoratus*) seeds; the toxic substance was identified as β-(γ-L-glutamyl)aminopropionitrile but β-aminopropionitrile alone is fully active. Turkeys fed *L. odoratus* died of hemorrhage from rupture of the aorta.

lathyrogen any substance that on ingestion induces the disease **lathyrism**.

Latin square (*in statistics*) a square array of *n* columns and *n* rows containing *n* different elements and in which no element occurs more than once in any row or column. It is used as a basis of experimental procedures when it is desired to allow for two sources of variability while investigating a third.

LATS *abbr.* for long-acting thyroid stimulator.

lattice 1 any regular array of objects or points in two or three dimensions. **2** (*in crystallography*) any regular, three-dimensional array of fixed points about which atoms, ions, or molecules vibrate in a crystal. **3** (*in immunology*) the three-dimensional aggregate suggested to be built up in an antigen–antibody reaction (*see* **lattice hypothesis**).

lattice energy (of a crystal) the energy that would be required to separate the ions of a crystal to an infinite distance from each other.

lattice hypothesis a hypothesis describing antigen–antibody reactions in terms of the building up of large antigen–antibody aggregates or three-dimensional lattices. The hypothesis requires both antigen and antibody to have valencies greater than one.

Lau *symbol for* the dodecanoyl (i.e. **lauroyl**) group.

Laue equation any of a set of three equations describing the occurrence, positions, and intensities of X-ray beams after diffraction by a crystal. For three crystal axes, *x*, *y*, and *z*, with separations of chemical entities along them of *a*, *b*, and *c*, respectively, the equations are:

$$a(\cos\alpha - \cos\alpha_0) = h\lambda,$$

$$b(\cos\beta - \cos\beta_0) = k\lambda,$$

and

$$c(\cos\gamma - \cos\gamma_0) = l\lambda,$$

where α_0, β_0, and γ_0 are the angles that the incident beam makes with the *x*, *y*, and *z*, axes, respectively, and α, β, and γ are the corresponding angles for the diffracted beam; λ is the wavelength of the radiation and *h*, *k*, and *l* are integers corresponding to the **Miller indices** of the planes considered to be diffracting the radiation. *See also* **Bragg's law**. [After Max Theodor Felix von Laue (1879–1960), German physicist.]

Laue photograph *or* **Laue pattern** the spot diagram produced when a heterogeneous beam of X-rays is diffracted by passage through a crystal.

laurate 1 *the trivial name for* dodecanoate, $CH_3–[CH_2]_{10}–COO^-$, the anion derived from lauric acid (i.e. dodecanoic acid), a saturated straight-chain higher fatty acid. **2** any mixture of free lauric acid and its anion. **3** any salt or ester of lauric acid. Dodecyl laurate is one of the main components of laurel oil, the oil obtained from berries of the laurel plant, *Laurus nobilis*. Laurate is present in appreciable amounts in hydrolysates of phosphoglycerides of plants and animals.

Laurell technique *an alternative name for* **rocket immunoelectrophoresis.** [After C.-B. Laurell, Swedish biochemist.]

lauroyl *or (formerly)* **lauryl** *symbol*: Lau; *the trivial name for* dodecanoyl, $CH_3-[CH_2]_{10}-CO-$, the acyl group derived from lauric acid (i.e. dodecanoic acid). It occurs in natural lipids, mainly as acylglycerols.

lauryl 1 *symbol*: Dod; *a common name for* dodecyl, $CH_3-[CH_2]_{10}-CH_2-$, the alkyl group derived from dodecane; use deprecated. **2** *a former trivial name for* dodecanoyl (lauroyl now used).

lauryldimethylamine oxide *abbr.*: LDAO; *N,N*-dimethyl-*N*-dodecylamine oxide; a water-soluble, zwitterionic (at pH 7 or above) detergent, aggregation number 76, CMC 1–2 mM.

$$CH_3(CH_2)_{11}-\underset{\underset{CH_3}{|}}{\overset{\overset{O}{\parallel}}{N}}-CH_3$$

lauryl sulfate, sodium salt *an alternative name for* **sodium dodecyl sulfate.**

lavendustin any substituted amino salicylic acid that is used as a protein tyrosine kinase inhibitor. Lavendustin A is 5-amino-[(*N*-2,5-dihydroxybenzyl)-*N'*-2-hydroxybenzyl]aminosalicylic acid and lavendustin C is 5-(*N*-2,5-dihydroxybenzyl)aminosalicylic acid. Lavendustin A inhibits protein-tyrosine kinase activity with little effect on cyclic AMP-dependent kinase or protein kinase C. Lavendustin C is similar but also inhibits Ca^{2+}/calmodulin kinase II.

lavendustin A

lawn a layer of microorganisms uniformly distributed in or over the surface of a solid culture medium.

law of conservation of energy *see* **thermodynamics, laws of.**

law of independent assortment *see* **Mendel's laws of heredity.**

law of mass action *see* **mass action.**

laws of thermodynamics *see* **thermodynamics.**

lawsone 2-hydroxy-1,4-naphthoquinone; the active dyeing compound of **henna.**

layer line one of the parallel lines of spots obtained on the photographic plate in the rotating crystal method of X-ray diffraction analysis.

lb *symbol for* pound.

LB *abbr. for* Luria-Bertani medium a synthetic culture medium commonly used for the growth of bacteria.

LBF *abbr. for Lactobacillus bulgaricus* factor (*see* **pantetheine**).

L-bifunctional enzyme *see* **peroxisomal bifunctional enzyme.**

LC *abbr. for* liquid chromatography.

LCAD *abbr. for* long-chain acyl-CoA dehydrogenase; *see* **beta-oxidation system.**

LCAT *abbr. for* lecithin–cholesterol acyl transferase.

LCD *abbr. for* liquid-crystal display.

L cell *an alternative name for* **EG cell.**

L chain *see* **light chain** (def. 1).

LCK a gene family encoding nonreceptor tyrosine kinases (e.g. p56^{lck} or lymphocyte-specific protein-tyrosine kinase) of the Src family (*see* **src**), first detected in the LSTRA murine lymphoma cell line. The product is lymphocyte-specific and is activated by binding to the intracellular domains of CD3 and CD4 in T lymphocytes. The mitogen-activated protein (MAP) kinase is a substrate for p56^{lck}, which is a myristoylated protein with SH2 and SH3 domains. A rare mutation in the gene at 1p35-p34.3 leads to Lck deficiency and a form of severe combined immunodeficiency.

LCR *abbr. for* **1** locus control region. **2** ligase chain reaction.

LD *abbr. for* **1** lethal dose. **2** *(in clinical chemistry)* L-lactate dehydrogenase (alternative to LDH); *see also* **HBD**. **3** low density.

LD$_{50}$ *abbr. for* median lethal dose.

LDH *abbr. for* L-lactate dehydrogenase (alternative to LD).

LDL *abbr. for* low-density lipoprotein.

LDL receptor a cell-surface receptor that binds apolipoprotein B and so internalizes **low-density lipoprotein** (LDL) from the plasma, leading to processing that regulates cholesterol and LDL synthesis, and thence plasma cholesterol concentration. It is a type I membrane protein. The number of LDL receptors on liver cells is regulated by cholesterol synthesis: when cholesterol levels rise, the number of receptors falls. Defects in this receptor are associated with certain forms of type II **hyperlipidemia**. The amino end of the extracellular domain contains seven or eight 40-residue repeats; each repeat has about six Cys residues, all of which are involved in disulfide bonds; following these repeats is a region of about 350 residues that is homologous with part of the epidermal growth factor precursor.

LDL receptor-related protein *see* **LRP.**

Le *abbr. for* Lewis antigen.

LE392 a strain of *E. coli* favoured for the propagation of bacteriophage lambda.

leaching (microbial) a process for the solubilization of metals, mostly from poor ores, by lithotrophic bacteria. *See also* **lithotroph.**

lead *symbol*: Pb; a very soft, bluish-white metallic element of group 14 of the IUPAC **periodic table**. It has a low melting point and high relative density; proton number 82; relative atomic mass 207.2. It is commonly used for shielding from ionizing radiation.

lead compound 1 *(in pharmacology)* a molecule that serves as the starting point for development of a drug, typically by variations in structure for optimal efficacy. **2** *(in chemistry)* a compound of lead (Pb).

leader peptidase EC 3.4.99.36; *other name*: signal peptidase I, phage-procoat-leader peptidase; an enzyme that catalyses the cleavage of N-terminal leader sequences from secreted proteins.

leader peptide *an alternative name for* **signal peptide.**

leader sequence *or* **lead sequence** *an alternative name for* signal sequence (*see* signal peptide).

leading strand the DNA strand that is synthesized continuously, in the 5' to 3' direction, towards the replication fork during the discontinuous replication of duplex DNA. *Compare* **lagging strand.**

lead screw a threaded rod the rotation of which is used to drive the tool carriage of a lathe, the plunger of a syringe, etc.

leaf (*pl.* **leaves**) **1** the principal organ of photosynthesis and transpiration in higher plants. It typically consists of a flattened bladelike lamina, often relatively broad, attached to the plant stem directly or by a stalk. **2** any thin, flattened object that resembles a leaf. **3** *or* **terminal tip** *(in phylogeny)* the end of a terminal branch of a phylogenetic tree. Collectively the leaves represent the contemporary taxa in a data set. *See also* **leaflet.**

leaflet 1 a any small or young **leaf** (def. 1). **b** an individual segment of a compound leaf. **2** any small, thin, flattened object or structure that resembles the leaf of a plant, especially a lipid bilayer or a portion of a biological membrane.

leak a cellular process whereby ions or other compounds escape though a membrane, dissipating a chemical potential.

leakage see **leak**.

leaky *(in genetics)* describing any mutation that fails to shut off completely the activity of a gene so that some residual expression of the gene remains.

Le antigen *symbol for* Lewis antigen.

LEA protein *abbr. for* late embryogenesis abundant protein; any of the proteins expressed in seeds during maturation and desiccation. Five families have been described. These proteins are hydrophilic, cytoplasmic, and rich in glycine and alanine but lacking in cysteine and tryptophan. Their concentration increases in vegetative parts of plants exposed to water deficit. In rice, overexpression increases resistance to such stress.

least squares a statistical method for determining the 'best' value of an unknown quantity relating one or more sets of measurements or observations. It is based on minimizing the sum of the squares of the deviations of the experimentally determined values from the calculated values, and is used especially to find a curve that best fits a given set of data.

leaving group any atom or group, charged or uncharged, that departs during a substitution or displacement reaction from what is regarded as the residual or main part of the substrate of the reaction.

Lebedev-Saft *or* Lebedev juice an aqueous extract of air-dried yeast. It will ferment glycerone (dihydroxyacetone) and, in the presence of phosphate, produces a hexose phosphate. [After Alexander Nikolaevich Lebedev (1881–1938), Russian botanist who described the preparation in 1911.]

Leber's hereditary optic neuropathy *abbr.*: LHON; a mitochondrial disease of humans manifested as painless, subacute, bilateral visual loss with abnormal colour vision. It is associated with at least eight genes in mitochondrial DNA for components of certain complexes of the electron transport chain. [After Theodor Leber (1840–1917), German ophthalmologist.]

Le Chatelier principle *or* **Le Chatelier–Braun principle** a principle stating that when a system in equilibrium is subject to stress, i.e. to a change in its conditions, the system will adjust itself to annul, as far as possible, the effect of the stress. [First formulated (in 1885) by Henry (Louis) Le Chatelier (1850–1936), French chemist, and independently in 1886 by Karl Ferdinand Braun (1850–1918), German physicist.]

lecithin *an old trivial name, still in frequent use, for* any 3-*sn*-**phosphatidylcholine**.

lecithinase any of a number of distinct enzymes acting on lecithins that are now usually considered among the **phospholipases**.

lecithin–cholesterol acyl transferase *abbr.*: LCAT; EC 2.3.1.43; *recommended name*: phosphatidylcholine–sterol *O*-acyltransferase; *other name*: phospholipid–cholesterol acyltransferase. An enzyme glycoprotein, secreted into plasma by the liver, that catalyses a reaction between phosphatidylcholine and cholesterol to form a cholesteryl ester and 1-acylglycerophosphocholine. This enzyme is central to the extracellular metabolism of plasma lipoproteins; it acts on free cholesterol in the high-density lipoprotein (HDL) precursor, transferring a fatty acyl group to cholesterol, thus contributing to the altered class composition of the mature HDL; some of the resulting cholesterol esters are partly taken up by low-density lipoprotein. It is activated by apolipoprotein A1, and is deficient in Norum and fish-eye diseases.

lectin any of a group of specific agglutinins and other antibody-like (glyco)proteins of non-immune origin, defined (by IUBMB) as 'sugar-binding protein or glycoprotein of non-immune origin that agglutinates cells and/or precipitates glycoconjugates'. Lectins are widely distributed in nature, being found mainly in seeds, but also in other parts of certain plants, and in many other organisms, from bacteria to mammals. Lectins bear at least two sugar-binding sites; they bind specific sugars, and thereby precipitate certain polysaccharides, glycoproteins, and glycolipids, and/or agglutinate animal and plant cells. Some can distinguish between normal and malignant cells, and many, e.g. phytohemagglutinin, are mitogenic. They are widely used experimentally, especially **concanavalin A**, as tools in carbohydrate biochemistry, for studying cell surfaces and for inducing **transformation** (def. 3) of lymphocytes. Plant lectins are also known as **phytoagglutinins**.

LED *abbr. for* light-emitting diode.

leeching a form of blood letting in which blood is removed by leeches. *Compare* **leaching**.

leghemoglobin *or (esp. Brit.)* **leghaemoglobin** a hemoglobin-like red pigment found in the root nodules of leguminous plants. An autoxidizable hemoprotein, it sequesters dioxygen to prevent inhibition of **nitrogenase** in the bacteroids. It is thus essential for symbiotic nitrogen fixation.

legumain EC 3.4.22.34; *other names*: bean endopeptidase; vicilin peptidohydrolase; phaseolin; an endopeptidase from legumes that hydrolyses Asn-|-Xaa bonds in small molecule substrates such as Boc-Asn-|-OPhNO$_2$.

legumin a major storage protein of the seeds of both leguminous and non-leguminous plants. It consists of two chains, derived from a common precursor: an acidic α chain and a basic β chain – these are linked by a single disulfide bond.

Leigh syndrome *or* **necrotizing encephalopathy** a devastating heterogeneous neurodegenerative disease with severe lactic acidosis. It is commonly associated with nuclear mutations in protein components of **complex I, II, III, IV**, and with a mitochondrial mutation in subunit 6 of the F_0F_1 complex of the inner mitochondrial membrane (*see* **H$^+$-transporting ATP synthase**). Mutations in other proteins of the mitochondrion (e.g. pyruvate dehydrogenase complex, pyruvate carboxylase) produce a very similar clinical state. [After Denis Leigh (1915–), British psychiatrist.]

LEM *abbr. for* light-effect mediator.

length *symbol*: *l* or *L*; a scalar physical quantity indicating linear extent; one of the seven SI base physical quantities. The SI base unit of length is the **metre**.

lens 1 a piece of glass or other transparent and refracting material, or two or more such pieces joined together, for converging or diverging a beam of light in order to form optical images. **2** any electrostatic or electromagnetic device for focusing or otherwise altering the direction of movement of a beam of electrons or other elementary charged particles; any similar device acting on sound waves. **3** *(in zoology)* any of various transparent structures responsible for focusing light onto photoreceptors. The **crystalline lens** of vertebrates is a near-spherical structure in the eye, lying in the aqueous humour behind the pupil. It focuses or assists in focusing light rays onto the retina. Enclosed within a collagenous capsule, it consists largely of cells containing proteins called **crystallins**.

lenticular 1 having the shape of a biconvex **lens** (def. 1). **2** of or pertaining to a **lens**.

lentinan a polysaccharide isolated from the edible mushroom *Lentinus edodes*. It is a β-1,3-glucan, having for every five β-1,3 linear linkages, two glucopyranoside units in β-1,6 linkage.

lentinarin see **callose**.

lentropin a protein, present in the vitreous humour of the eye, that stimulates lens-fibre differentiation. It is related to **insulin-like growth factors**.

Lepidoptera a large order of insects that includes butterflies, skippers, and moths. The adults are characterized by a prominent coiled proboscis and two pairs of large, scaly, membranous wings. The larvae are caterpillars. —**lepidopteran** *adj*.

leporine of, pertaining to, or resembling a rabbit or hare.

leprachaunism an autosomal recessive inherited disease characterized by small stature and elfin-like appearance, with motor and mental retardation and severe failure to thrive. Most patients die in early childhood, though a few, who are compound heterozygotes for two different alleles, survive childhood. The condition can be caused by any of a number of mutations in the gene encoding the insulin receptor.

leptin a protein encoded by a gene, *ob*, first identified as the *obese* mutation (*see* **ob/ob**) in mice. Leptin may act as a signal in the regulation of adipose mass, possibly by regulating appetite and energy expenditure. It appears to be expressed exclusively in white adipose tissue. The sequence is largely hydrophilic, and has structural features indicative of a secreted protein. *See also* **leptin receptor**.

leptin receptor any member of a widely expressed single-pass transmembrane receptor family that binds **leptin** and exert its effects through the JAK/STAT signalling pathway. It is expressed at high levels in the hypothalamus relative to other tissues. Mutations in the receptor are responsible for *db* mice, corpulent mice, and Zucker rats, all of which are grossly obese. *See also* **diabetes mutation**, *ob/ob*.

leptocene the first phase of prophase I in **meiosis**.

leptotene *see* meiosis.

Lesch–Nyhan syndrome an X-chromosome-linked, inherited disorder of male humans in which there is an almost complete deficiency of **hypoxanthine phosphoribosyltransferase** (EC 2.4.2.8), a key enzyme of the **salvage pathway** of purine metabolism. The condition is characterized by hyperuricemia, excessive uric-acid biosynthesis, and certain neurological features including self-mutilation, spasticity, and mental retardation. Over 200 mutations in the gene for **HGPRT** cause the syndrome. Death usually occurs in the second or third decade of life. [After M. Lesch (1939–), US paediatrician, and William Leo Nyhan (1926–), US physician and biochemist.]

lesion 1 *(in pathology)* a zone of tissue with impaired function resulting from damage by disease or wounding. **2** *(in biochemistry)* any deleterious disturbance of a metabolic pathway resulting from chemical interference or genetic abnormality.

+less *suffix forming adjectives and adverbs* without, lacking; unable to do or to be done. In biochemistry it is used especially of auxotrophic mutants to indicate the substance they cannot synthesize, e.g. thymineless mutant.

let23 a gene in *Caenorhabditis elegans* for a receptor **tyrosine kinase** of the epidermal growth factor family. It is probably activated by the *lin3* product. The product of *let23* is a type I membrane glycoprotein.

lethal dose *abbr.*: LD; the amount of a drug or other agent that when administered to an organism causes the organism's death. *See also* median lethal dose.

lethal factor any **lethal mutation**.

lethal gene any gene carrying a **lethal mutation**. Such mutations may only be fatal in the homozygous state.

lethal mutant any organism carrying a **lethal mutation**.

lethal mutation any **mutation** that may bring about the death of the organism before its maturity. The consequences of such a mutation may depend on whether the organism is homozygous or heterozygous for the gene in question, and possibly also on the conditions (*see* **conditional lethal mutation**).

lethal synthesis any process by which a highly toxic compound is synthesized in an organism from a nontoxic precursor. The classic example is the formation of **fluorocitrate** when e.g. kidney homogenates are incubated with **fluoroacetate**. The fluorocitrate thus formed inhibits **aconitase**.

LETS protein *abbr.* for large external transformation sensitive (glyco)protein (i.e. cellular **fibronectin**).

Leu *symbol for* a residue of the α-amino acid L-leucine (alternative to L).

leuc+ *a variant spelling of* **leuk+**. *See* **leuko+**.

leucinate 1 leucine anion; the anion, $(CH_3)_2CH–CH_2–CH(NH_2)–COO^-$, derived from **leucine**. **2** any salt containing leucine anion. **3** any ester of leucine.

leucine the trivial name for α-aminoisocaproic acid; α-amino-γ-methylvaleric acid; 2-amino-4-methylpentanoic acid; $(CH_3)_2CH–CH_2–CH(NH_2)–COOH$; a chiral α-amino acid. L-leucine (*symbol*: L or Leu), (S)-2-amino-4-methylpentanoic acid, is a coded amino acid found in peptide linkage in proteins; codon: CUA, CUC, CUG, or CUU; UUA or UUG. In mammals, it is an essential amino acid, and is ketogenic. Residues of D-leucine (*symbol*: D-Leu or DLeu), (R)-2-amino- 4-methylpentanoic acid, occur in a number of peptide antibiotics, e.g. gramicidins A, B, C, and D, and polymixins D_1, D_2, E_1, and E_2.

L-leucine

leucine aminopeptidase *abbr.* *(in clinical chemistry)*: LAP (preferred) *or* LAS. *See* **leucyl aminopeptidase**.

leucine and valine biosynthesis the biosynthesis of leucine and

valine initially follows a common pathway that commences with the formation of acetolactate from two molecules of pyruvate, a reaction catalysed by acetolactate synthase. Acetolactate is converted via α,β-dihydroxyisovalerate (2,3-dihydroxy-3-methylbutanoate) to α-ketoisovalerate (2-oxo-3-methylbutanoate). At this point the pathway branches. α-Ketoisovalerate can be converted by transamination with glutamate to valine. Alternatively, it can be converted successively to α-isopropylmalate (2-isopropylmalate), β-isopropylmalate (3-isopropylmalate), and α-ketoisocaproate (4-methyl-2-oxopentanoate), and thence by transamination with glutamate to leucine. *See also* **3-isopropylmalate dehydratase**, **3-isopropylmalate dehydrogenase**.

[5-leucine]enkephalin *abbr.*: [Leu⁵]enkephalin *or* [Leu]enkephalin; the recommended trivial name for the pentapeptide Tyr-Gly-Gly-Phe-Leu. *See* **enkephalin**.

leucine-rich repeat *abbr.*: LRR; short sequence motif in proteins, of average length 24 residues, containing a high proportion of leucines, typically five to seven at positions 2, 5, 7, 12, 16, 21, and 24 within the repeat. Such arrays form an amphipathic sequence, with leucine as the predominant hydrophobic residue. They are believed to participate in protein–protein or protein–lipid interactions. LRRs form elongated non-globular structures, often flanked by cysteine-rich domains. They were first described in α₂-glycoprotein, but they have since been found in other proteins. The number of repeats can vary from one, in platelet glycoprotein Ibβ, to 30, in, e.g., **chaoptin**. The 3-D structure of ribonuclease inhibitor has been shown to contain 15 repeats, and has revealed LRRs to represent a new class of alpha/beta fold.

leucine zipper a domain in a protein structure in which two alpha helices containing leucines (normally four or five) repeating every seventh amino-acid residue, interdigitate in zipper fashion to stabilize the structure. Such an arrangement was originally found in DNA-binding proteins, but it is now known to occur also in other proteins, especially adjacent to proposed transmembrane regions, mediating dimerization. The abbreviation ZIP is used, e.g., in designating the **bZIP** family of proteins. *See also* **potassium channels**.

leucinium leucine cation; the cation, $(CH_3)_2CH–CH_2–CH(NH_3^+)–COOH$, derived from **leucine**.

leucino the alkylamino group, $(CH_3)_2CH–CH_2–CH(COOH)–NH–$, derived from **leucine**.

leuco+ *a variant spelling of* **leuko+**.

leucocyte *a variant spelling of* **leukocyte**.

leucyl the acyl group, $(CH_3)_2CH–CH_2–CH(NH_2)–CO–$, derived from **leucine**.

leucyl aminopeptidase EC 3.4.11.1; *other names*: leucine aminopeptidase; peptidase S; cytosol aminopeptidase; an enzyme that catalyses the hydrolysis of an aminoacyl-peptide to an amino acid and the residual peptide. Its action is most rapid when the N-terminal amino acid is leucine; it liberates other residues at varying rates, and acts poorly at an Xaa-Pro bond. It requires zinc, and is widely distributed in animal tissues, its level in the serum being useful clinically in monitoring hepatobiliary disease processes and placental well-being.

[Leu⁵]enkephalin *or* **[Leu]enkephalin** *abbr. for* [5-leucine]-enkephalin.

leuk+ *a variant form of* **leuko+** (before a vowel).

leukemia *or (esp. Brit.)* **leukaemia** *or (sometimes)* **leukosis** any of a group of neoplastic diseases of the leukocytes, usually involving increase in the number of leukocytes in the blood. Leukemias are classified according to the type of leukocyte affected and whether the disease is acute or chronic.

leukemia inhibitory factor *abbr.*: LIF; *also called* **differentiating-stimulating factor** *or* **differentiation-inducing factor**; a **cytokine** produced by fibroblasts, T cells, and macrophages. It is a monomer of 179 amino acids that maintains the pluripotent phenotype of embryonic stem cells, and potentiates interleukin-3-dependent proliferation of hemopoietic progenitors.

leuko *or* **leuco** describing the colourless derivative formed reversibly from a coloured substance, often by reduction; e.g. leuko compound, leuko dye.

leuko+ *or* **leuco+** *or (before a vowel)* **leuk+** *or* **leuc+ 1** white or colourless. **2** denoting of, pertaining to, or acting upon a leukocyte or leukocytes.

leukocidal *or* **leucocidal** describing an agent that kills leukocytes.

leukocidin *or* **leucocidin** any extracellular bacterial product that can kill leukocytes of certain animal species and that may also be toxic or lytic towards other cell types. Leukocidins are produced by pathogenic strains of *Staphylococcus* and *Streptococcus*.

leukocyte any white blood cell, i.e. one not containing an oxygen-transporting pigment such as hemoglobin. Vertebrate leukocytes include the **polymorphonuclear leukocytes**, **lymphocytes**, and **monocytes**; they are distinguished by size, shape, and staining characteristics. —**leukocytic** *or* **leucocytic** *adj*.

leukocyte adhesion deficiency (*abbr.*: LAD) *or* **leucocyte adhesion deficiency** a condition of severe leukocytosis and recurrent infections in which leukocytes show defective migration and phagocytosis. **LAD I** is associated with mutations in a locus at 21q22.3 that cause lack of or dysfunction of **integrins** containing the β subunit (also called CD18). **LAD II** is accompanied by unusual facial appearance, short stature, and encephalopathy, and is a form of carbohydrate-deficient glycoprotein syndrome that results from defective import of GDP-fucose into the Golgi body.

leukocytosis *or* **leucocytosis** a gross increase in the number of leukocytes in the blood. It often occurs as a response to an infection. *See also* **leukemia**.

leukokinin *or* **leucokinin** a specific cytophilic immunoglobulin G (IgG) that constitutes about 2% of the total IgG. It binds specifically and reversibly to autologous polymorphonuclear leukocytes, causing 2- to 2.5-fold stimulation of their phagocytic activity. A proteinase, **leukokininase**, cleaves from leukokinin a tetrapeptide, **tuftsin**, which is thought to bear all the activity.

leukopenia *or* **leucopenia** an abnormal decrease or deficiency of leukocytes circulating in the blood. *Also (esp. Brit.)* **leukopaenia** *or* **leucopaenia**.

leukoplast *or* **leucoplast** *or* **leukoplastid** *or* **leucoplastid** any of the colourless **plastids** found in plant cells. They include **aleuroplasts**, **amyloplasts**, and **elaioplasts**. Leukoplasts may develop into **chloroplasts**.

leukopoiesis *or* **leucopoiesis** the processes by which leukocytes are formed in the body. —**leukopoietic** *or* **leucopoietic** *adj*.

leukosis *or* **leucosis** **1** a condition in which there is an abnormal proliferation of one or other of the leukocyte-forming tissues. **2** *an alternative name for* **leukemia**.

leukosulfakinin a sulfated myotropic 11-residue peptide isolated from head extracts of the cockroach. It is a homologue of human gastrin II and cholecystokinin.

leukotaxin *or* **leucotaxin** a nitrogen-containing material, found in injured tissue, that when injected into an animal causes increased capillary permeability and migration of leukocytes to the site of injection.

leukotoxin *or* **leucotoxin** **1** any compound toxic to leukocytes. **2** cytochrome P450-derived linoleic acid peroxide; any of a group of substances observed in high concentrations in burned skin samples that may have a bacteriocidal function. **Leukotoxin A** is (12Z)-9,10-epoxyoctadec-12-enoic acid, and **leukotoxin B** is (9Z)-12,13-epoxyoctadec-9-enoic acid (*abbr.*: EpODE) or, its methyl ester (*abbr.*: EpOME).

leukotriene *abbr.*: LT; any member of a family of biologically active substances derived from polyunsaturated fatty acids, notably arachidonic acid, some of which contain a peptide moiety based on cysteine. Leukotrienes are synthesized in response to specific stimuli, are formally derived from eicosanoic acid, and contain a set of three conjugated double bonds (hence 'triene'); one or two additional double bonds, not conjugated to the others, may also be present – a subscript numeral after the abbreviation indicates the total number of double bonds, depending on the polyunsaturated fatty acid from which each is derived, e.g. LTA₄ from arachidonic acid, LTB₅ from eicosapentaenoic acid. They are products of the 5-lipoxygenase pathway. LTC₄ consists of glutathione linked to an eicosanoid; loss of glutamic acid from this leads to LTD₄, and loss of glycine then leads to LTE₄. These three, sometimes collectively known as **peptidoleukotrienes**, are potent spasmogenic agents that make up the active components of **slow-reacting substance of anaphylaxis** (SRS-A).

Leukotrienes exert their effects by binding to G-protein-associated membrane receptors. *See also* individual entries below.

leukotriene A *abbr.*: LTA; any of a group of **leukotrienes** characterized as unstable 5,6-epoxides. Leukotriene A₄ (*abbr.*: LTA₄) is (7E,9E,11Z,14Z)-(5S,6S)-5,6-epoxyeicosa-7,9,11,14-tetraen-1-oate. Leukotriene A₃ (*abbr.*: LTA₃) is similar, but lacks the double bond at the C-14, while leukotriene A₅ (*abbr.*: LTA₅) has an additional, 17Z, double bond. LTA₄ is derived from arachidonate via 5-hydroperoxyeicosatetraenoate, and is the precursor either of leukotriene B₄ (*see* **leukotriene B**) by the action of highly specific **leukotriene-A₄ hydrolase** (EC 3.3.2.6), or of **leukotriene C₄** by the action of leukotriene-C₄ synthase (EC 2.5.1.37).

leukotriene A₄

leukotriene-A₄ hydrolase *abbr.*: LTA₄ hydrolase; EC 3.3.2.6; an enzyme that converts (7E,9E,11Z,14Z)-(5S,6S)-5,6-epoxyeicosa-7,9,11,14-tetraenoate (LTA₄) to (6Z,8E,10E,14Z)-(5S,12R)-5,12-dihydroxyeicosa-6,8,10,14-tetraenoate (LTB₄).

leukotriene B *abbr.*: LTB; any of a group of **leukotrienes** characterized as (5S,12R)-5,12-dihydroxy polyunsaturated acids. Leukotriene B₄ (LTB₄) is (6Z,8E,10E,14Z)-(5S,12R)-5,12-dihydroxyeicosa-6,8,10,14-tetraen-1-oate. Leukotriene B₅ (LTB₅) has an additional, 17Z, double bond. LTB₄ is a product of leukotriene-A₄ hydrolase (*see* **leukotriene A**); it is a major product of the 5-lipoxygenase pathway, and is a potent aggregatory, chemokinetic, and chemotactic agent for polymorphonuclear leukocytes (PMNLs) *in vitro* and induces increased vascular permeability and PMNL infiltration *in vivo*.

leukotriene B₄

leukotriene C₄ *abbr.*: LTC₄; a **leukotriene** comprising a glutathione moiety linked through its thio group to C-6 of an eicosanoid. LTC₄ is formed from LTA₄ and glutathione by leukotriene-C₄ synthase (EC 2.5.1.37). It is a potent mediator of smooth muscle contractility, and of vascular tone and permeability, being a potent vasoconstrictor in a variety of vascular beds including the coronary and cerebral circulations. It is the predominant leukotriene in the central nervous system.

leukotriene D₄ *abbr.*: LTD₄; a **leukotriene** comprising a cysteinylglycine moiety linked through its thio group to an eicosanoid moiety. It is a metabolite of **leukotriene C₄** (LTC₄), resulting from the action of membrane-bound γ-glutamyl transferase, which removes glutamic acid. Its actions are similar to those of LTC₄.

leukotriene E₄ *abbr.*: LTE₄; a **leukotriene** comprising a cysteine moiety linked to an eicosanoid moiety. It is derived metabolically from **leukotriene D₄** (LTD₄) by the action of specific membrane-bound dipeptidases, which remove glycine. It has a somewhat longer metabolic half-life than LD_4, but is metabolized to more polar products, notably those in which the terminal methyl group of the eicosanoid moiety is oxidized to a carboxyl group. It activates receptors in human airways and has been implicated in airway hyper-responsiveness in humans.

leukovirus or **leucovirus** any tumorigenic **retrovirus**.

leukovorin or **leucovorin** see folinic acid.

leumorphin an endogenous 29-amino-acid opioid peptide that corresponds to residues 228–256 of preproenkephalin B; the N-terminal 13 residues correspond to rimorphin (dynorphin B), and the N-terminal five to Leu-enkephalin (see enkephalin).

leupeptin any modified tripeptide protease inhibitor produced by various actinomycetes. Commonly *N*-acetyl-Leu-Leu-argininal is used in the laboratory; other variants have propionyl in place of acetyl, and valine or isoleucine in place of one or both leucines.

Leu-prolide an analogue of **gonadotropin-releasing hormone** in which residues 5 and 10 (glycine and glycinamide) are replaced by D-Leu and ethylamine respectively.

levan a common name for any (2→6)-β-D-fructofuranan; such compounds are produced mostly by bacteria. As in **inulin** each chain terminates in a (1→2)-α-D-glucopyranosyl residue, but it may also contain (2→1)-β branch points.

levansucrase the recommended name for sucrose 1-fructosyl transferase, EC 2.4.1.10; an enzyme that catalyses the reaction:

sucrose + (2,6-β-D-fructosyl)$_n$ = glucose + (2,6-β-D-fructosyl)$_{n+1}$.

See also **sucrase** (def. 3).

levarterenol see norepinephrine.

levigate (in chemistry) **1** to grind (something) into a smooth powder or paste. **2** to separate fine from coarse particles by grinding and suspending them in a fluid. —**levigation** n.

levo- or (esp. Brit.) **laevo-** symbol: *l*-; a prefix, no longer used, to denote a compound that is **levorotatory**; the levorotatory enantiomer of compound A is now designated as (–)-A.

levo+ or (before a vowel) **lev+** comb. form denoting on or towards the left. Also (esp. Brit.) **laevo+, laev+**. Compare **dextro+**.

levodopa or (Brit.) **laevodopa** (–)-3-hydroxy-L-tyrosine; a naturally occurring amino acid that has been used in the treatment of **Parkinson's disease** easing symptoms such as rigidity, postural problems, and slowness of movement. See **3,4-dihydroxyphenylalanine**.

levorotatory or (esp. Brit.) **laevorotatory 1** symbol: (–)- or (formerly) *l*-; describing a chemical compound that rotates the plane of polarization of a transmitted beam of plane-polarized light or other electromagnetic radiation to the left, i.e. in an anticlockwise direction as viewed by the observer looking towards the light source. **2** describing any leftwards or anticlockwise rotation. Compare **dextrorotatory**. —**levorotation** or **laevorotation** n.

levorphanol 17-methylmorphinan-3-ol; a synthetic analogue of morphine. It is an orally active analgesic that binds stereospecifically with high affinity to opioid receptors (K_d 10⁻⁹ M). Five to six times more active than morphine, it is a controlled substance under the US Code of Federal Regulations. **Dextrorphan**, the dextrorotatory form, does not bind to opioid receptors, is not analgesic, but is a very effective antitussive.

levulose or (esp. Brit.) **laevulose** an old name, still sometimes used in pharmacy, for D-(–)-fructose, the levorotatory component of invert sugar. Compare **dextrose**.

Lewis acid any chemical species that is able to accept an electron

pair and is thus able to react with a **Lewis base** to form an adduct (i.e. Lewis adduct). For example:

$$(CH_3)_3B + :NH_3 \rightarrow (CH_3)_3B^-N^+H_3$$

or

$$(CH_3)_3C^+ + Cl^- \rightarrow (CH_3)_3C–Cl;$$

i.e.

Lewis acid + Lewis base → Lewis adduct.

[After Gilbert Newton Lewis (1875–1946), US chemist.]

Lewis adduct the adduct formed between a **Lewis acid** and a **Lewis base**.

Lewis antigen abbr.: Le antigen; symbol: Le; one of the human **blood-group substances**. Lewis specificities are carried on glycosphingolipids and glycoproteins. Primarily a system of soluble antigens in secretions (e.g. saliva) and plasma, the Lewis antigens on red cells are adsorbed passively from plasma, which must be constantly present for them to be maintained. Three Lewis phenotypes are common amongst Caucasians, namely Le(a⁺b⁻), Le(a⁻b⁺), and Le(a⁻b⁻); the very rare Le(a⁺b⁺) is detected in Polynesians and Orientals. The carbohydrate structures providing Le^a and Le^b specificity are shown below; these partial structures are linked through other carbohydrate moieties to a **ceramide**, thus forming a glycosphingolipid, or to a protein, thus forming a glycoprotein.

Le^a Gal(β1-3)[Fuc(α1-4)]GlcNAc(β1-3)Gal...

Le^b Fuc(α1-2)Gal(β1-3)[Fuc(α1-4)]GlcNAc(β1-3)Gal...

The structures of the Le^c, Le^d, Le^x, and Le^y determinants are also known. The Lewis antigens are named after a Mrs H. D. G. Lewis, in whom they were first identified. See also **galactoside 3(4)-L-fucosyltransferase**.

Lewis base any chemical species that is able to supply an electron pair and is thus able to react with a **Lewis acid** to form an adduct (i.e. Lewis adduct).

Lewis formula or **Lewis structure** a depiction of molecular structure in which the valency electrons are shown as dots so placed that each pair of dots represents two electrons. When placed between two atoms each pair of dots also represents a single covalent bond, whereas dots placed adjacent to only one atom represent the non-bonded outer-shell electrons considered to be associated with that atom alone. (e.g. H:C̈l:). A double bond is represented by two pairs of dots (e.g. Ö: :Ö), and so on. Formal charges are attached to atoms as appropriate.

Lewis transferase see **galactoside 3(4)-L-fucosyltransferase**.

Lewy body an aggregate of fibrillar ubiquitinated α-**synuclein** occurring in the substantia nigra in Parkinson's disease. Fibril formation is inhibited in vitro by dopamine. [After Frederic H. Lewy (1885–1950), German-born US neurologist.]

lexA a gene in *Escherichia coli* encoding a repressor that regulates the genes induced by **SOS repair**. In the latter, following DNA damage, the *lexA* product, LexA, is hydrolysed by RecA protein (see **recA**), thereby allowing the SOS genes to be activated. *lexA* is itself regulated autogenously. LexA binds to operators as a dimer.

Leydig cell a type of steroid-secreting, interstitial cell that occurs in large numbers between the seminiferous tubules of the testis. The cells secrete androgens, especially testosterone, and this activity is stimulated by luteinizing hormone, for which the cells have specific receptors. [After Franz von Leydig (1821–1908), German anatomist and zoologist.]

LFA 1 a leukocyte cell-surface glycoprotein involved in the adhesion of leukocytes to target cells and thus in a variety of immune phenomena, including leukocyte–endothelial cell interaction, cytotoxic T-cell-mediated killing, and antibody-dependent killing by granulocytes and monocytes. It binds to **ICAM**. LFA 1 is a type I membrane glycoprotein belonging to the **LEUCAM** family of **integrins**; it is a dimer of the α$_L$ and β$_2$ subunits.

L form (of a bacterium) a defective bacterial cell, of indefinite or spherical shape, in which the cell wall is defective or absent. L forms may develop spontaneously or in response to a variety of stimuli, e.g. thermal or osmotic shock, or the presence of antibiotics that inhibit cell-wall synthesis. They may revert to their original condition on removal of the stimulus or they may reproduce as L forms. On solid media, L forms have a characteristic colonial appearance resembling that of **mycoplasmas** (with which they were originally confused).

LFS *abbr. for* Li–Fraumeni syndrome.

lg *symbol for* decadic, or common or Briggsian, **logarithms** (alternative to $\log_{10}$).

LH *abbr. for* luteinizing hormone.

LHC *abbr. for* light-harvesting complex.

LHCP *abbr. for* light-harvesting chlorophyll protein.

LH/FSH-RF *abbr. for* luteinizing hormone/follicle-stimulating hormone releasing factor (i.e. **gonadotropin releasing hormone**).

LHRF *or* **LH-RF** *abbr. for* luteinizing hormone releasing factor (*see* **gonadotropin-releasing hormone**).

LHRH *or* **LH-RH** *abbr. for* luteinizing hormone releasing hormone (*see* **gonadotropin releasing hormone**).

Li *symbol for* lithium.

+liberin *word ending* denoting a hypothalamic factor (or hormone) that promotes the release (and perhaps also the biosynthesis) of the particular pituitary hormone indicated; e.g. luliberin, thyroliberin. *Compare* **+statin**.

library of genes *see* gene library.

librate to oscillate about an equilibrium position. —**libration** *n*.

Librium *see* chlordiazepoxide.

licensing factor *see* **MCM protein**.

licorice *see* glycyrrhiza.

Liddle syndrome a rare autosomal dominant disease characterized by early and severe hypertension, hypokalemic metabolic alkalosis, and decreased secretion of aldosterone. It results from mutations that remove the C-terminal tails of β or γ subunits of **ENaC**. [Named after Grant Liddle, US physician who described it in 1963.]

Liebermann–Burchard reaction the reaction of acetic anhydride in chloroform with unsaturated steroids (e.g. cholesterol) or triterpenes in the presence of concentrated sulfuric acid to produce a blue-green colour. It is used especially in the estimation of cholesterol.

LIF *abbr. for* leukemia inhibitory factor.

life cycle the totality of functional and morphological stages through which any organism passes between any one developmental stage and the identical stage in the next generation. Less commonly it refers to the totality of the various stages through which an individual organism passes between birth and death.

Li–Fraumeni syndrome *abbr.*: LFS; a rare familial predisposition to cancer characterized by presence of a bone or soft-tissue sarcoma before 45 years of age in the proband, and of a cancer or sarcoma in at least one first- or second-degree relative also before the age of 45 years. Germline mutations in the gene for **p53** at 17p13.1 are inherited in an autosomal dominant manner in the majority of LFS families. [After Frederick P. Li and Joseph F. Fraumeni (1933–), US physicians.]

Lig *symbol for* the lignoceroyl group, CH_3–$[CH_2]_{22}$–CO–.

ligancy *a less common alternative name for* **coordination number**.

ligand 1 (in a simple inorganic compound) any individual atom, group, or molecule that is attached covalently to the characteristic or central atom or moiety. **2** (in an organic compound) any individual atom or chemical group that is attached covalently to a specified carbon atom; *see also* **point ligand**. **3** (in a coordination entity) **a** any individual atom, ion, or molecule that is attached coordinately to one central metal atom (**unidentate ligand**). **b** an ion, molecule, or molecular grouping that has more than one coordinating group (**bi-**, **tri-**, or **multidentate ligand**). In both cases the ligand is an electron donor in the formation of one or more coordinate bonds. **4** an ion of either sign that can associate reversibly with one or more (charged or uncharged) atoms or characteristic molecular groupings in another (usually larger) molecule (as in the hydronation of a conjugate base, the coordination of a metal cation by a protein, etc.). **5** a molecule (or a class of molecules) that has become linked covalently to another molecule to form a conjugate, whether in a single-step reaction or by stepwise addition of its components (e.g. an oligosaccharide moiety in a glycoprotein). **6** a molecule (or part of one) that is bound or is able to bind selectively and stoichiometrically, whether covalently or not, to one or more specific sites on another molecule (as in the combination of antigen with antibody, of hormone with receptor, of substrate with enzyme, etc.).

liganded bound by or as a ligand or ligands; ligated; *see* **ligate** (def. 2).

ligand exchange (chromatography) a column-chromatographic technique in which a cation-exchange resin is loaded with a complexable metal ion and the resulting loaded resin, which retains the capability of the metal ion to coordinate to an appropriate complexing agent or ligand, is used for the sorption of that ligand from a solution or a gas phase. Elution is then effected by displacement with, i.e. exchange for, another ligand. The resin is commonly a chelating resin (e.g. a styrene–divinylbenzene copolymer to which iminodiacetate groups are attached) and the ion is usually of a transition metal (e.g. copper or nickel), such ions tending to be bound strongly to these resins. Separation of components of mixtures occurs by virtue of the differences in the stabilities of the various metal–ligand complexes formed.

ligand-field theory *(in inorganic chemistry)* a theory, closely related to the valence-bond theory, that deals with the effects of ligands on the energy levels of the central atom or ion in complexes and crystals. It is especially useful in interpretation of the spectra of inorganic complexes.

ligand-gated describing an ion channel in a cell membrane that is caused to open by the action of an agonist binding to the receptor that governs the channel. Channel opening can be direct, when the receptor acts as an ionophore, or indirect, involving a second messenger regulated by a G-protein.

ligandin *a name formerly given to* a basic dimeric protein of ~25 kDa per dimeric subunit (*see* **gluthathione transferase**). It is abundant in rat and human liver, kidney, and small intestine. It binds with high affinity to heme, bilirubin and other organic anions, polycyclic aromatic carcinogens, and various metabolites. Its presence in the urine, **ligandinuria**, has been used to follow the evolution of various forms of kidney injury, while its presence in serum provides a sensitive indicator in liver injury. Also variously called Y protein, organic anion binding protein, basic azo-dye binding protein, cortisol metabolite binding protein (binder protein I), glutathionetransferase B.

liganding binding by or as a ligand or ligands; ligating; *see* **ligate** (def. 2).

ligandinuria *see* ligandin.

ligand-mediated chromatography *a variant of* **affinity chromatography** that employs an additional component, the affinity ligand, which is a ligand both for the reactive component attached to the insoluble matrix and for the substance to be purified.

ligand–receptor assay a form of **saturation analysis** for a hormone or other agonist wherein the specific binding agent used is an isolated preparation of the naturally occurring cellular receptor substance for the agonist. An advantage of the assay is that the values obtained are considered to be a direct reflection of the biological properties of the substance assayed.

ligase 1 *or (formerly)* **synthetase** *the systematic class name for* enzymes of class EC 6, all of which catalyse the ligation of molecules of two substances with concomitant breaking of a diphosphate linkage in a nucleoside triphosphate. The recommended names of particular ligases are formed in some instances, especially those of enzymes of subclass EC 6.4, by using the generic name carboxylase, e.g. pyruvate carboxylase, EC 6.4.1.1. In the remaining instances the generic names ligase (def. 2) or synthase are used, according to whether the name of an individual enzyme is to be based on the names of the reactants or on the name of the product, respectively. **2** *a generic name for* many enzymes of class EC 6 (*see* **ligase** (def. 1)); when added to the names of the two reactants whose ligation is catalysed by a particular enzyme it forms the recommended name of that enzyme; e.g. tyrosine–tRNA ligase (EC 6.1.1.1), acetate–CoA ligase (EC 6.2.1.1). *See also* **synthetase** (def. 2). **3** *see* **DNA ligase**.

ligase chain reaction *abbr.* LCR. a method of detecting single base mutations based on thermal cycling and involving a thermostable DNA ligase. Two contiguous oligonucleotides annealed to a single stranded DNA molecule are ligated if there is perfect complementarity between them and the template. When performed with two pairs of complementary oligonucleotides, one pair for each strand of the DNA duplex, the ligated oligonucleotides act as templates in the next and subsequent cycles of melting, annealing, and ligation leading to amplification in a manner similar to PCR. The small size of the products makes them suitable for rapid analysis by capillary electrophoresis.

ligate 1 to join together, especially with a ligature. 2 *(in chemistry)* to join (molecules or molecular fragments) together with a bond; to coordinate. 3 to bind as or to a **ligand** or ligands. 4 *(in surgery)* to tie off (a blood vessel or duct) with a ligature so as to occlude. —**ligation** *n.*; **ligatable** *adj.*; **ligative** *adj.*

ligatin a filamentous plasma membrane protein for the attachment of peripheral glycoproteins to the external cell surface. It acts as a trafficking receptor for phosphoglycoproteins, localizing them, after internalization, within endosomes.

light 1 electromagnetic radiation capable of producing a visual sensation in the eye. It comprises wavelengths in the range 380–780 nm (frequencies 385–789 THz), although the term is sometimes loosely extended to include some ultraviolet and infrared radiation in adjacent parts of the spectrum. 2 *(in chemistry)* describing any (especially metallic) element whose density is relatively small (usually <5000 kg m^{-3}). 3 *(in physics)* of less than the usual mass. *Compare* **heavy**.

light band a region of a muscle **sarcomere** that stains less densely in light microscopy and low-magnification electron microscopy relative to the **dark bands**; together, these differentially staining bands give the striated appearance of skeletal muscle. The light bands are relatively isotropic in polarized light; they represent the regions occupied by **thin filaments**. *See also* **thick filament**, **Z line**.

light chain 1 *or* **L chain** *or (formerly)* **B chain** *symbol*: L; the shorter of the two main types of polypeptide chain of an **immunoglobulin** of any class. Of $\approx$22 kDa (human), each light chain is linked at its C-terminal cysteine residue by a disulfide bond to the constant region of a **heavy chain**. In any given animal species, light chains may be distinguished serologically into two types, κ and λ, which occur in immunoglobulin molecules of every class in proportions that vary only with the species (about 65% κ in the human), and which have different but homologous sequences. The C-terminal half of each type of light chain consists of a **constant region** and the N-terminal half is a **variable region** possibly involved in the antibody-combining site. *See also* **Bence-Jones protein**. 2 any of the four 15–27 kDa polypeptide chains that, with two heavy chains of $\approx$200 kDa, constitute a molecule of **myosin**.

LightCycler proprietary name for an instrument used for real-time PCR. Monitoring DNA abundance in each cycle is made possible by measuring the intensity of fluorescence yielded by an intercalating dye such as Sybr Green.

light-effect mediator *abbr.*: LEM; any of a number of substances that, following their light-induced reduction, activate some of the enzymes of the Calvin cycle (*see* **reductive pentose phosphate cycle**) in photosynthesis. Not all of these mediators have been identified, but it is thought that they are cyst(e)ine-containing proteins that become oxidized in the dark and reduced again in the light. An important member of the group is **thioredoxin**.

light-harvesting chlorophyll protein *or* **light-harvesting chlorophyll-binding protein** *abbr.*: LHCP; a chlorophyll-binding protein that is abundant in green plants, particularly in those grown at low light intensities. It consists of a 25 kDa polypeptide, encoded by a nuclear gene and synthesized in the cytoplasm, to which is bound most of the chlorophyll *b* and xanthophylls, together with some of the chlorophyll *a*. It appears to be capable of transferring the energy of absorbed light photons to both photosystem I and photosystem II. *See also* **light-harvesting complex**.

light-harvesting complex *abbr.*: LHC; a **thylakoid** membrane complex of chlorophylls *a* and *b* together with chlorophyll *a-b* binding proteins. LHCs contain a number of other proteins, the function of which is speculative, and accessory pigments. The LHCs capture and transfer energy to photosystems I and II, and are regulated by phosphorylation on threonine residues. *See also* **light-harvesting chlorophyll protein**.

light meromyosin *abbr.*: L meromyosin *or* LMM; the smaller, 120 kDa, fragment, also called F3 fragment, produced when **myosin**

is subjected to brief treatment with trypsin or certain other peptidases. It contains much of the helical, rod-shaped tail (>90% α helical) and forms large, ordered aggregates at low ionic strength. Light meromyosin has no ATPase activity and cannot bind actin.

light path 1 the path followed by a beam of light when passing through an optical device or a particular part of one. 2 *or* **path length** the length of such a path.

light pen an input/output device for use with a computer that, when pointed at a spot on a visual display unit, can sense whether or not the spot is illuminated; the output current from the light pen can then be used to control the operations of the computer.

light reactions of photosynthesis *or* **light phase of photosynthesis** those reactions of **photosynthesis** in higher plants and algae in which light energy is used to generate NADPH and ATP.

light repair *an alternative name for* **photoreactivation**.

light-scattering a technique used for the determination of the relative molecular mass, M_r, in solution of a macromolecule whose size is large relative to the wavelength of the light used. The value of M_r of the macromolecule can be calculated from the wavelength of the incident light, the angle and intensity of the scattered light, the refractive indices of the solvent and solute, and the concentration of the solute. *See also* **Rayleigh scattering**.

light strand *or* **L strand** 1 a polynucleotide strand not labelled with heavy isotope. 2 the strand in a naturally occurring polynucleotide duplex that is of a lower density than its complementary strand. *Compare* **heavy strand**.

light trapping 1 the process of capturing the energy of incident light in a photosynthetic pigment system. 2 the act or process of using focused light in a liquid medium so arranged as to trap phototactic organisms in a restricted part of that medium.

lignan any member of a class of phenylpropanoid dimers in which the phenylpropane units are linked tail-to-tail and thus have a 2,3-dibenzylbutane skeleton (*compare* **neolignan**). Lignans occur in higher plants, and in humans and other mammals as products of the action of gut flora on dietary lignans of plant origin. Guaiaretic acid is structurally one of the simplest; **podophyllotoxin** and more particularly its semisynthetic derivatives **Etoposide** and **Teniposide** have defined actions in clinical medicine, with antitumour and possibly antiviral actions. *See also* **lignin**.

lignify to make or become woody; to accumulate or deposit lignin in the cell walls (of plant tissues). —**lignification** *n.*

lignin any random phenylpropanoid polymer formed in higher plants by the dehydrogenative radical polymerization of various 4-hydroxycinnamyl alcohols, in which the residues are joined in differing proportions and by several different linkages, some of which are nonhydrolysable. Lignin is insoluble in water, and occurs in the cell walls of vascular plants (especially of the supporting and conducting tissues), on which it confers strength, rigidity, and resistance to degradation. It is one of the most abundant biopolymers. *See also* **lignan**.

ligninase an enzyme that catalyses breakdown of lignin using hydrogen peroxide formed from **oxygenases** to cleave carbon–carbon bonds – i.e. **lignin peroxidases**. An example is diarylpropane peroxidase, EC 1.11.1.14; *systematic name*: diarylpropane:oxygen,hydrogen peroxide oxidoreductase (C–C-bond-cleaving); *other name*: ligninase I. It is a hemoprotein that catalyses a reaction between 1,2-bis(3,4-dimethoxyphenyl)propane-1,3-diol and H_2O_2 to form 3,4-dimethoxybenzaldehyde, 1-(3,4-dimethylphenyl)ethane-1,2-diol, and $4H_2O$.

lignin peroxidase *see* **ligninase**.

lignocellulose a covalent adduct of **lignin** and **cellulose**, found in the walls of xylem cells in woody tissues of plants. —**lignocellulosic** *adj.*, *n.*

lignocerate 1 *the trivial name for* tetracosanoate, $CH_3-[CH_2]_{22}-COO^-$, the anion derived from lignoceric acid (i.e. tetracosanoic acid), a saturated straight-chain higher fatty acid. 2 any mixture of lignoceric acid and its anion. 3 any salt or ester of lignoceric acid.

lignoceroyl *symbol*: Lig; *the trivial name for* tetracosanoyl, $CH_3-[CH_2]_{22}-CO-$, the acyl group derived from lignoceric acid (i.e. tetracosanoic acid). It occurs naturally as acylglycerols in peanut and rapeseed oils, and in certain sphingomyelins.

lignoceroyl-CoA synthase *see* **long-chain-fatty-acid–CoA ligase.**

LIM domain *see* **islet-1.**

limit *(in mathematics)* a value to which a function or the sum of a series approaches as an independent variable approaches a specified value or infinity.

limit dextrin the highly branched core that remains after exhaustive treatment of amylopectin or glycogen with α- and/or β-amylases. It is formed because these enzymes cannot hydrolyse the $(1{\rightarrow}6)$ glycosidic linkages present.

limit dextrinase *see* α-**dextrin endo-1,6-**α-**glucosidase, oligo-1,6-glucosidase.**

limit dextrinosis *an alternative name for* type III **glycogen disease.**

limit digest a digest in which no further degradation can occur under the particular conditions adopted.

limiting current *(in polarography)* the maximal value of the current reached in circumstances when the steady-state current/potential characteristic has a positive slope.

limiting rate *symbol*: V; *the recommended name for* the extrapolated rate of an enzyme-catalysed reaction when the enzyme is saturated with substrate(s). It is equal to the product of the stoichiometric concentration of active centres, $[E]_0$, and the catalytic constant, k_0. It can be determined by a number of methods, including graphical methods such as the **Lineweaver–Burk plot**. *Other names*: maximum rate; maximal rate; maximum velocity; maximal velocity.

limiting viscosity number *the recommended name for* **intrinsic viscosity.**

LIM kinase *abbr.*: LIMK; a family of protein serine/threonine kinases that contain two N-terminal LIM domains (*see* **islet-1**). They are strongly expressed in neuronal tissues.

limonene p-mentha-1,8-diene; a compound that occurs in many essential oils (e.g. oils of lemon, orange, and dill); both enantiomers and the racemic mixture (i.e. dipentene) occur naturally.

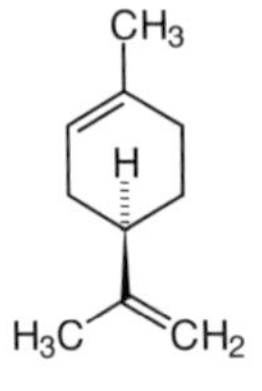

(R)-(+) form

LIMS *abbr. for* laboratory information management systems; normally, systems involving automatic electronic data capture and computerized manipulation of data.

Lin *symbol for* the linoleoyl group, $CH_3-[CH_2]_3-[CH_2CH{=}CH]_2-[CH_2]_7-CO-$ (*all-Z* isomer).

lin3 a gene in *Caenorhabditis elegans* that encodes an **epidermal growth factor**-like type I membrane protein. The protein has two isoforms, produced by alternative splicing, and is essential for vulval development.

lin14 a gene involved in embryonic development in *Caenorhabditis elegans* that codes for a nuclear protein. Regulatory sequences in the 3′ untranslated region generate a temporal switch during development. There are three forms of the product: A (537 amino acids); B1 (539 amino acids); and B2 (515 amino acids).

linalool 3,7-dimethyl-1,6-octadien-3-ol; a compound that occurs in linaloe oil and other essential oils; both enantiomers occur naturally. It is used in perfumery.

lincomycin an antibiotic produced by *Streptomyces lincolnensis*. It consists of an amide formed between a substituted pyrrolidine-2-carboxylic acid and methyl 1-thio-α-lincosamine. Lincomycin is active mainly against Gram-positive bacteria; it inhibits protein synthesis by interacting with the 50S ribosomal subunit, inhibiting the early steps of peptide bond formation prior to the formation of polysomes.

lincosamides *generic name for* a group of antibiotics structurally related to **lincomycin.**

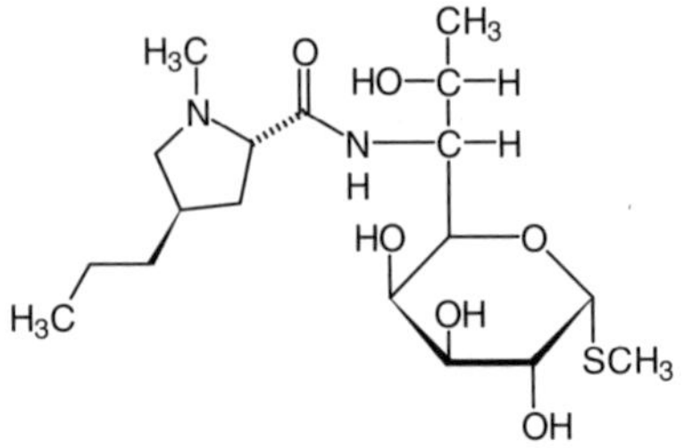

lincomycin

lincosamine 6-amino-6,8-dideoxy-D-*erythro*-D-*galacto*-octose; the carbohydrate component of **lincomycin.**

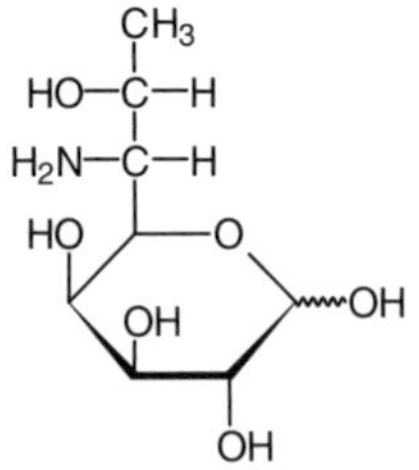

line 1 *(in spectroscopy)* a very narrow band of frequencies of electromagnetic radiation that is noticeably more intense or less intense than adjacent parts of the spectrum; the image of such a line in a spectrogram. **2** *(in biology)* several generations of a family. *See also* **cell line, lineage.** —**lineal** *or* **linear** *adj*.

LINE *abbr. for* long interspersed nucleotide element (or sequence or repeat); part of the moderately repetitive DNA of mammalian genomes. Any highly repeated DNA sequence interspersed in a mammalian genome and characteristically several kilobases in length. LINEs are **retroposons** resulting from the action of RNA polymerase II (*see* **RNA polymerase**). Mammalian genomes may contain 10^4 to 10^5 copies of particular LINEs. They are regarded as processed **pseudogenes**. The function of LINEs is unknown. *Compare* **SINE**. *See also* **Alu sequence.**

lineage lineal descent, especially from a common ancestor.

linear accelerator a device for producing high-energy particles, in which charged particles are accelerated in a straight line along a long evacuated tube by potential differences applied to a series of circular electrodes along the tube.

linear chain a chain of atoms in a molecule, or a chain of residues in a macromolecule, that is neither branched nor covalently closed.

linear correlation any relationship between two variables such that a graphical plot of one variable against the other is a straight line.

linear density gradient a gradient of linearly and continuously increasing density.

linear dichroism the **dichroism** (def. 1) occurring when linearly polarized light is absorbed by a sample containing partially or completely oriented anisotropic molecules.

linear equation any equation containing any number of variables that is of the first degree, i.e. one in which the variables are all raised to the power of one.

linear growth the growth of a culture such that the number of cells (or the cell mass) increases in direct proportion to the duration of growth.

linearity the condition of being in a straight line.

linearize *or* **linearise 1** to make linear, i.e. into a straight line. **2** to convert a branched or cyclic form of a (macro)molecule to a linear form, i.e. one with only two ends. —**linearization** *or* **linearisation** *n*.

linearly polarized *or* **plane-polarized** describing light (or other electromagnetic radiation) in which the electric vectors of the vibrations all lie in a given plane containing the light beam. *See* **polarized light**.

line formula a two-dimensional representation of a molecular entity in which atomic symbols are joined by lines representing single bonds, without any indication or implication of stereochemistry.

line immunoelectrophoresis a one-dimensional single **electroimmunodiffusion** technique, useful for the qualitative and quantitative comparison of the antibody spectra of different polyvalent antisera.

line spectrum any spectrum consisting of discrete spectral lines.

Lineweaver–Burk plot a method of plotting enzyme kinetic data. The Michaelis equation is rearranged to give:

$$1/v = K_m/V[S] + 1/V,$$

where K_m is the Michaelis constant, $[S]$ the concentration of the substrate S, v the (initial) rate of reaction at a given value of $[S]$, and V the limiting rate of reaction at infinite $[S]$. If the Michaelis equation holds for the system, a plot of $1/v$ against $1/[S]$ gives a straight line of slope K_m/V, with intercepts on the $1/v$ axis at $1/V$ and on the $1/[S]$ axis at $-1/K_m$. *See* **Michaelis kinetics**. [After Hans Lineweaver and Dean Burk.]

linewidth the width of a spectral line, usually taken as the distance, in terms of wavelength, frequency, or magnetic flux difference, between the two points of half-peak intensity (half-maximum height).

link *(in chemistry) an informal term for* **bond** (def. 1). *See also* **linkage** (def. 2).

linkage 1 *(in genetics)* an association of two or more nonallelic genes so that they do not show independent **assortment**; i.e. they tend to be transmitted together usually due to physical association on the same chromosome. **2** *(in chemistry) an informal term for* **bond** (def. 1). *See also* **link**.

linkage disequilibrium the preferential association of linked genes/DNA markers in a population; i.e. the tendency for some alleles at a locus to be associated with certain alleles at another locus on the same chromosome with frequencies greater than would be expected by chance alone.

linkage group any group of genes whose members show a marked tendency to segregate as a unit. *See also* **genophore**.

linkage map a map of a chromosome showing in linear order the relative positions of the known genes on that chromosome.

linked genes genes that exhibit **linkage** (def. 1).

linker 1 any small synthetic single-chain oligodeoxynucleotide that is six or more residues long and has a self-complementary base sequence. When self-annealed it forms a stable duplex with blunt ends and containing a **restriction site**. Such linkers are useful for gene splicing. **2** *see* **cross-link**.

linker DNA *see* **chromatin, nucleosome**.

linking number *symbol*: L; a topological property of closed circular duplex DNA, defined as the number of turns one strand makes about the other. It must be an integer, and is positive if the turns are right-handed and negative if they are left-handed. It can change only when one or both strands are nicked. It comprises the sum of the **writhing number** (W) and the **twist number** (T), so that a change (ΔL) in the linking number (L) is equal to ΔW + ΔT.

linoleate 1 *the trivial name for* (*all-Z*)-octadeca-9,12-dienoate, CH_3-$[CH_2]_3$-$[CH_2$-CH=$CH]_2$-$[CH_2]_7$-COO^-, (*all-Z* isomer) the anion derived from linoleic acid, (*all-Z*)-octadeca-9,12-dienoic acid, a diunsaturated straight-chain higher fatty acid essential in the diet of mammals. It occurs naturally as acylglycerols in many vegetable oils and in the lipids of many animal tissues. **2** any mixture of free linoleic acid and its anion. **3** any salt or ester of linoleic acid. *See also* **linoleic family**.

linoleate synthase *see* **phosphatidylcholine desaturase**.

linoleic family (of polyunsaturated fatty acids) a series of polyenoic acids occurring in mammals, in which the hydrocarbon chain terminates with the alkenyl grouping, CH_3-$[CH_2]_4$-CH=CH- (as in linoleic acid; *see* **linoleate**) and that are synthesized from linoleic acid by chain elongation and/or further desaturation (but which in mammals cannot be synthesized from the corresponding saturated acids or from (9,12,15)-linolenic, oleic, or palmitoleic acids – the latter represent three other series of polyenoic acids). The linoleic family is also known as the $\omega6$, or n-6, family because of the position of the double bond nearest the terminal methyl group. The se-

ries includes (6,9,12)-linolenic acid, eicosa-11,14-dienoic acid, eicosa-5,11,14-trienoic acid, eicosa-8,11,14-trienoic acid, and arachidonic acid.

linolenate 1 *or* **(9,12,15)-linolenate** *or (formerly)* **α-linolenate** *the alternative trivial names for* (*all-Z*)-octadeca-9,12,15-trienoate; CH_3-$[CH_2$-CH=$CH]_3$-$[CH_2]_7$-COO^- (*all-Z* isomer); the anion derived from (9,12,15)-linolenic acid, (*all-Z*)-octadeca-9,12,15-trienoic acid, a triunsaturated straight-chain higher fatty acid synthesized by plants (notably found in linseed oil) but essential in the diet of mammals (*see* **linolenic family**). The name equally refers to any mixture of (*all-Z*)-octadeca-9,12,15-trienoic acid and its anion, or any salt or ester of this acid. **2** **(6,9,12)-linolenate** *or (formerly)* **γ-linolenate** *the trivial name for* (*all-Z*)-octadeca-6,9,12-trienoate; CH_3-$[CH_2]_3$-$[CH_2$-CH=$CH]_3$-$[CH_2]_4$-COO^- (*all-Z* isomer), the anion derived from (6,9,12)-linolenic acid (*all-Z*)-octadeca-6,9,12-trienoic acid, a triunsaturated straight-chain higher fatty acid belonging to the **linoleic family**. It occurs naturally as acylglycerols in certain plant-seed oils, especially that of the evening primrose (*Oenothera* spp.), and also in small amounts in the lipids of many animal tissues, especially human breast milk. It is the biosynthetic precursor of arachidonate and the prostaglandins. The name refers equally to any mixture of (*all-Z*)-octadeca-6,9,12-trienoic acid and its anion or any salt or ester of the acid.

linolenic family (of polyunsaturated fatty acids) a series of polyenoic acids occurring in mammals, in which the hydrocarbon chain terminates with the alkenyl grouping CH_3-CH_2-CH=CH- (as in (9,12,15)-linolenic acid; *see* **linolenate**). The linolenic family is also known as the $\omega3$ or n-3 family because of the position of the terminal double bond three carbon atoms from the methyl end. The longer-chain members can be synthesized in mammals only from dietary (9,12,15)-linolenic acid by chain elongation and/or further desaturation (but not from the corresponding saturated acids or from linoleic, oleic, or palmitoleic acids – the latter represent the other three series of polyenoic acids). If animals are deprived of these fatty acids, certain tissues, such as the eye, retain existing stores of these fatty acids tenaciously, even in offspring of such animals when the restricted diet is continued for more than one generation. Under these conditions, long-chain acids of the **linoleic family**, e.g. (4,7,10,13,16,)-docosapentaenoic (C22:5 n-6) acid, are synthesized in appreciable amounts and presumably partially support the function supplied by long-chain $\omega3$ acids, though the nature of this is unknown. The series includes eicosapentaenoic (C20:5 $\omega3$) and docosahexaenoic (C22:6 $\omega3$) acids, which are found in high concentration in fish oils; these $\omega3$ acids may have value in protecting against heart disease, but in amounts that greatly exceed those needed to fulfil their role as essential fatty acids.

linolenoyl 1 *or* **(9,12,15)-linolenoyl** *or (formerly)* **α-linolenoyl** *symbol*: αLnn *or (sometimes)* Lnn; *the alternative trivial names for* (*all-Z*)-octadeca-9,12,15-trienoyl; CH_3-$[CH_2$-CH=$CH]_3$-$[CH_2]_7$-CO- (*all-Z* isomer); the acyl group derived from (*all-Z*)-octadeca-9,12,15-trienoic acid. **2** **(6,9,12)-linolenoyl** *or (formerly)* **γ-linolenoyl** *symbol*: γLnn; *the trivial names for* (*all-Z*)-octadeca-6,9,12-trienoyl; CH_3-$[CH_2]_3$-$[CH_2$-CH=$CH]_3$-$[CH_2]_4$-CO- (*all-Z* isomer); the acyl group derived from (*all-Z*)-octadeca-6,9,12-trienoic acid. *See* **linolenate**.

linoleoyl *symbol*: Lin; *the trivial name for* (*all-Z*)-octadeca-9,12-dienoyl; CH_3-$[CH_2]_3$-$[CH_2$-CH=$CH]_2$-$[CH_2]_7$-CO- (*all-Z* isomer); the acyl group derived from linoleic acid. *See* **linoleate**.

Linux a widely used open-source free operating system similar to UNIX and incorporating UNIX commands and much software from the **GNU** project. The word Linux is constructed in part from the name of the original author, Linus Torvalds, and there is disagreement about the pronunciation of Linux. The original author's family comes from the Swedish-speaking minority in Finland, and Linus is pronounced 'Leenus', of which Linus with a short i (as in 'bit') is the nearest English equivalent. This is favoured by Linus Torvalds himself, but there are those who point out that the name Linus in English has the first syllable sounding like 'line' (e.g. Linus Pauling) and insist on the pronunciation 'line-ux'.

liothyronine a pharmaceutical name for L-3,5,3'-**triiodothyronine**.

lip+ a variant form of **lipo+** (before a vowel).

lipaemia a variant spelling (esp. Brit.) of **lipemia**.

lipase 1 *a general term for* any acylglycerol carboxylic ester hydrolase, of EC sub-subclass 3.1.1. The term may be incorporated in the enzyme name as the suffix '+lipase'. **2** *abbr. (in clinical chemistry)*: LPS; *an alternative name for* **triacylglycerol lipase**. *See also* **hormone-sensitive lipase, lipoprotein lipase, pancreatic lipase.**

lipemia *or* **lipidemia** the condition when there is a greater than normal content of lipid in the blood, e.g. after a fatty meal or in certain pathological states. *Also (esp. Brit.)* **lipaemia, lipidaemia.**

lipid *or (formerly)* **lipide** *or* **lipoid** *or* **lipin(e) 1** any member of a large and diverse group of oils, fats, and fatlike substances that occur in living organisms and that characteristically are soluble in **lipid solvents** but only sparingly soluble in aqueous solvents. Lipids constitute one of the four major classes of compounds found in living tissues (the others being carbohydrates, proteins, and nucleic acids) and they include: (1) fatty acids; (2) neutral fats (i.e. triacylglycerols), other fatty-acid esters, and soaps; (3) long-chain (or fatty) alcohols and waxes; (4) sphingoids and other long-chain bases; (5) glycolipids, phospholipids, and sphingolipids; and (6) carotenes, polyprenols, sterols (and related compounds), terpenes, and other isoprenoids. See individual entries for details. **2** any oily, fatty, or fatlike material that occurs in living organisms. **3** of, concerned with, for, or relating to **lipid** (def. 2); consisting of or containing **lipid** (def. 1, 2); **lipoid** (def. 1, 2) or lipoidal. —**lipidic** *adj.*

lipid A the glycolipid moiety of bacterial **lipopolysaccharide.**

lipidaemia *a variant spelling (esp. Brit.)* of lipidemia; *see* **lipemia.**

lipid bilayer any layer, two molecules thick, of amphipathic lipid molecules. If it is surrounded by a polar environment, the polar parts of the lipid molecules are oriented outwards towards the environment and the nonpolar parts inwards into the interior of the bilayer. The two layers of lipid are referred to as **leaflets**, e.g. the inner and outer leaflets of the bilayer. *See also* **black lipid membrane.**

lipide *a variant spelling of* **lipid.**

lipidemia *an alternative term for* **lipemia.**

lipidosis *or* **lipoidosis** (*pl.* **lipidoses** *or* **lipoidoses**) any clinical disorder, *also called* **lipid storage disease**, that results in the excessive deposition within specific tissues of particular types of lipids. Many lipidoses are inherited disorders of galactosphingolipid metabolism.

lipid raft a type of lipid microdomain in the plasma membrane of cells that sequesters signalling proteins of certain receptors such as those expressed by **T** and **B lymphocytes** that move into the outer leaflet of the **plasma membrane** lipid bilayer. **Caveolae** are a subset of lipid rafts.

lipid-soluble *a vague term applied to* various classes of nonpolar substances, e.g. sterols, certain vitamins, carotenoids, and polyprenoids, that are commonly present in small amounts in association with the major lipid.

lipid solvent any relatively nonpolar solvent that can be used to extract lipids from tissues or other materials.

lipid storage disease *an alternative term for* **lipidosis.**

lipid transport protein any of the small proteins that transport lipids and facilitate their exchange between membranes. In plants they are located in the cell wall space and transfer wax and cutin monomers from the plasma membrane to the cell surface.

lipin *or* **lipine** *a former term for* **lipid** (def. 1), especially one containing nitrogen and/or phosphorus.

lipo+ *or (before a vowel)* **lip+** *comb. form* denoting lipid, lipid-containing, or lipid-like.

lipoamide *or (sometimes)* **lipoyllysine** a **lipoyl** group when in amide linkage to the ε-amino group of a specific lysine residue in dihydrolipoamide acetyltransferase (EC 2.3.1.12) and dihydrolipoamide succinyl transferase (EC 2.3.1.61), which form, respectively, parts of the **pyruvate dehydrogenase complex** and the **2-oxoglutarate dehydrogenase complex.**

lipoamide reductase (NADH) EC 1.8.1.4; *recommended name*: dihydrolipoamide dehydrogenase; *other names*: E3 component of α-ketoacid dehydrogenase complexes; lipoyl dehydrogenase; dihydrolipoyl dehydrogenase; diaphorase. An enzyme that catalyses the oxidation by NAD^+ of dihydrolipoamide to form lipoamide and NADH. FAD is a coenzyme. It is typically a homodimer, occurring in the mitochondrial matrix (in eukaryotes), with a redox-active S–S bond similar to glutathione reductase.

lipoate 1 1,2-dithiolane-3-pentanoate, the anion derived from **lipoic acid** (i.e. thioctic acid). **2** any salt or ester of lipoic acid.

lipocaic a substance extractable from the pancreas that was said to be lipotropic in pancreatectomized animals.

lipocalin any of a large group of chiefly extracellular ligand-binding proteins, particularly one that binds hydrophobic ligands, e.g. biliverdin. The family includes retinol-binding protein, α-microglobulin, insect bilin-binding proteins, and **beta lactoglobulin.** The ligand is enclosed by the protein as a flower is enclosed by the calyx (i.e. the sepals), hence the name. Lipocalins have an eight-stranded beta-barrel structure; different ligand-binding properties are determined by loops on the periphery.

lipochrome *a former name for* any of various lipid-soluble biological pigments.

lipocortin *see* annexin.

lipofection a lipid-mediated DNA-transfection technique that makes use of unilamellar liposomes that contain a cationic lipid and spontaneously react with DNA.

lipofuscin any of various brownish lipid-soluble pigments that are deposited in animal cells, especially in senility.

lipogenesis the formation of fat. The term is applied especially to the processes by which higher fatty acids are synthesized from nonlipid sources. —**lipogenic** *adj.*

lipogenic amino acid any amino acid that can provide carbon atoms for the biosynthesis of higher fatty acids.

lipoic acid *or* **α-lipoic acid** *symbol*: Lip(S₂); 1,2-dithiolane-3-pentanoic acid; thioctic acid; the naturally occurring form is the (*R*)-(+)-enantiomer. Lipoic acid provides the reversibly reducible moiety, **lipoyl**, of **lipoamide** and is a growth factor for certain microorganisms. The reduced form is dihydrolipoic acid, $HS–CH_2–CH_2–CH(SH)–[CH_2]_4–COOH$ (*symbol*: Lip(SH)₂); the redox potential of the system is –0.29 V at pH7. Lipoic acid is an essential component of the α-ketoacid dehydrogenase complexes. *See* **pyruvate dehydrogenase complex, 2-oxoglutarate dehydrogenase complex.**

lipoid 1 *or* **lipoidal** being or resembling an oil or a fat; oily, fatty. **2** *a former term for* **lipid** (def. 1, 2).

lipolysis the breakdown of lipids, especially the (enzymic) hydrolysis of acylglycerols. —**lipolytic** *adj.*

lipolytic hormone *an alternative name for* **lipotropin.**

lipomodulin *or* **annexin I** a 40 kDa regulatory protein isolated from rabbit neutrophils that, in its nonphosphorylated form, is an inhibitor of phospholipase A_2 and that on phosphorylation by a cyclic-AMP-dependent protein kinase loses this inhibitory property. It has been suggested that lipomodulin thereby regulates the release of arachidonate from leukoytes, or other cells, that may occur in response to chemoattractants and other external stimuli.

lipophilic tending to dissolve in, having a strong affinity for, or readily mixing with lipids or substances of low polarity. *See also* **hydrophilic, lyophilic.** —**lipophilicity** *n.*

lipophilin *another name for* myelin proteolipid. *See* **myelin protein.**

lipophorin any member of the major class of lipid-transporting proteins found in the hemolymph of insects. In locusts, the apoprotein is synthesized in the fat body as an 85 kDa polypeptide. *See also* **apolipophorin.**

lipophosphoglycans *see* **GPI anchor.**

lipopolysaccharide *abbr.*: LPS; any of a group of related, 10 kDa, structurally complex components of the outer leaflet of the outer membrane of Gram-negative bacteria. Lipopolysaccharide molecules consist of three covalently linked regions: lipid A, core oligosaccharide, and an O side chain. The innermost layer, lipid A, which is responsible for the toxicity of the lipopolysaccharide, consists of six fatty acyl chains (sometimes hydroxylated) linked in various ways to two glucosamine residues. The branched core oligosaccharide contains ten saccharide residues, several of them unusual, and has a structure that appears to be similar in closely related bacterial strains. The outermost O side chain, which is highly variable and determines the antigenic specificity of the organism, is made up of many (≈50) repeating units of a branched tetrasaccharide containing further unusual sugar residues.

lipopolysaccharide *N*-acetylglucosaminyltransferase EC 2.4.1.56; an enzyme of the lipopolysaccharide core biosynthetic pathway in enteric bacteria; it catalyses a reaction between UDP-*N*-acetyl-D-glucosamine and lipopolysaccharide to form UDP and *N*-acetyl-D-glucosaminyllipopolysaccharide. *See also* **glycolipid**.

lipopolysaccharide-binding protein a plasma protein that binds bacterial lipopolysaccharide and is homologous with phospholipid transfer protein and cholesterol ester transfer protein. *See* **plasma lipid transfer protein**.

lipoprotein any conjugated, water-soluble protein in which the nonprotein moiety consists of a lipid or lipids. The lipid may be triacylglycerol, cholesterol, or phospholipid, or a combination of these. *Compare* **proteolipid**.

lipoprotein(a) one or two molecules of apolipoprotein(a) linked by a disulfide bridge to **apolipoprotein B**-100; it may be a physiological inhibitor of plasminogen activation. Levels in human plasma (which range from undetectable to 100 mg dL^{-3}) show correlation with the incidence of atherosclerosis.

lipoprotein-associated phospholipase A2 *or* **platelet-activating factor acetylhydrolase** (*abbr.*: PAF-AH); EC 3.1.1.47; an enzyme that is mainly associated with plasma low-density lipoprotein (LDL) particles and is released by a varietyof leukocytes. It catalyses hydrolytic removal of the 2-acetyl moiety of platelet-activating factor (PAF). Deficiency of plasma PAF-AH activity is associated with atherosclerotic occlusive disease in Japan.

lipoprotein lipase EC 3.1.1.34; *abbr.*: LPL; *other names*: clearing factor lipase; diglyceride lipase; diacylglycerol lipase. An enzyme that catalyses the hydrolysis of triacylglycerol to diacylglycerol and fatty-acid anion. Its function, after uptake of dietary lipid, is to hydrolyse triacylglycerols in chylomicrons and very-low-density lipoproteins, to diacylglycerols and fatty-acid anions; the fatty acids are then taken into peripheral tissues for futher metabolism; it will also hydrolyse diacylglycerols. It acts in the presence of apolipoprotein C-II on the luminal surface of vascular endothelium. It is a **GPI-anchor** dimer, and is released by heparin, etc.

lipoprotein signal peptidase *see* **premurein-leader peptidase**.

liposome 1 a natural lipid globule suspended in the cytoplasm of some cells. **2** any small, roughly spherical, artificial vesicle consisting of a continuous bilayer or multibilayer of complex lipids enclosing some of the suspending medium. Liposomes are formed by allowing complex lipids to 'swell' in aqueous solution sometimes with the aid of sonication. They are used experimentally as models of biological membranes and therapeutically for entrapment of drugs, enzymes, or other agents with a view to their more effective delivery to target cells. Originally, such a structure was termed a cellule (when bilayered), smectic mesophase, spherule, or spherulite. *See also* **liquid crystal**.

lipotaurine the 7,13-dihydroxy-2-*trans*-octadecenoyl amide of taurine. It is found in the lipid fraction of *Tetrahymena thermophila*.

lipoteichoic acid any of a group of **teichoic acids** that contain lipid, are present in the membranes of all species of Gram-positive bacteria so far examined, and, unlike wall teichoic acids, are extractable with hot water or phenol.

lipotropic *or* **lipotrophic** tending to prevent the accumulation of, or to remove, abnormal amounts of lipid (in a tissue or organ, e.g. adipose tissue, liver).

lipotropic agent *or* **lipotropic substance** any substance capable of preventing or correcting fatty infiltration of the liver caused by choline deficiency.

lipotropic hormone *or* **lipotrophic hormone** *a former name for* **lipotropin**.

lipotropin *or* **lipotropic hormone** (*abbr.*: LPH) either of two polypeptides, β- and γ-lipotropin, of the anterior and intermediary lobes of mammalian pituitary glands, that elicit lipolysis in adipose tissue. β-Lipotropin corresponds in sequence to the 91 C-terminal amino-acid residues of **proopiomelanocortin**, and γ-lipotropin corresponds

to the 58 N-terminal residues of β-lipotropin. In fact, they are of no greater lipotropic potency than corticotropin and α-melanotropin, and whether they have any intrinsic physiological function is still uncertain. Both β- and γ-lipotropins are found in normal human blood. *Other names*: lipolytic hormone; adipokinetic hormone; *former names*: lipotrophin; lipotrophic hormone.

lipotropism the state or quality of being lipotropic.

lipovitellin a ~135 kDa phospholipoprotein that, together with phosvitin, constitutes the bulk of the yolk proteins of the eggs of oviparous vertebrates. These two proteins are synthesized and secreted by the liver as a single large molecule, **vitellogenin**, which is taken up by the ovary and split into the two components.

lipoxin *abbr.*: LX; any of a group of eicosanoids containing a conjugated tetraene structure and having three hydroxyl groups. They are generated by the action of **lipoxygenase** enzymes on polyunsaturated fatty acids (arachidonate and eicosapentaenoate). The major lipoxins, A_4 and B_4, are derived from arachidonate; lipoxin A_4 (LXA_4) is (7*E*,9*E*,11*Z*,13*E*)-(5*S*,6*R*,15*S*)-5,6,15-trihydroxy-eicosa-7,9,11,13-tetraen-1-oate; lipoxin B_4 (LXB_4) is (6*E*,8*Z*,10*E*,12*E*)-(5*S*,14*R*,15*S*)-5,14,15-trihydroxy-eicosa-6,8,10,12-tetraen-1-oate. LXA_4 causes contraction of guinea-pig lung strips, but not ileum; LXA_4 and LXB_4 cause dilation of arterioles.

lipoxin A_4

lipoxin B_4

lipoxygenase 1 EC 1.13.11.12; *systematic name*: linoleate:oxygen 13-oxidoreductase; *other names*: lipoxidase; carotene oxidase. An enzyme that catalyses the oxidation by dioxygen of linoleate to (9*Z*,11*E*)-(13*S*)-13-hydroperoxyoctadeca-9,11-dienoate (*abbr.*: 13(*S*)-HpODE); iron is a cofactor. **2** any dioxygenase enzyme that catalyses the oxidation of polyunsaturated fatty acids to a particular corresponding hydroperoxide. Such enzymes are widely distributed in plants and animals, and occur in leukocytes, mast cells, platelets, and lung, their positional specificities varying from tissue to tissue; notably a 12-lipoxygenase predominates in platelets, and a 5-lipoxygenase in neutrophils. They are important in the conversion of arachidonate to 5- and 12-hydroperoxyeicosatetraenoates (the 5-hydroperoxy compound being the biosynthetic precursor of the leukotrienes). 5-lipoxygenase (EC 1.13.11.34) is important in yielding precursors for **leukotriene** biosynthesis; 15-lipoxygenase (EC 1.13.11.33) yields precursors for the **lipoxins**. All three, however, are involved in the synthesis of hydroperoxy- and dihydroxy-eicosatetraenoates. Compounds that inhibit both lipoxygenase and prostaglandin-endoperoxide synthase include **curcumin**, **eicosatetraynoic acid**, and phenidone (1-phenyl-3-pyrazolidinone). Compounds that selectively inhibit lipoxygenases without inhibiting prostaglandin-endoperoxide synthase include **eicosatriynoic acid** and **esculetin**. *See also* **dihydroxyeicosatetraenoate**, **hydroperoxyeicosatetraenoate**, **hydroperoxyoctadecadienoate**, **hydroxyeicosatetraenoate**.

lipoyl the acyl group, 1,2-dithiolane-3-pentanoyl, derived from **lipoic acid**. It provides the reversibly reducible moiety of **lipoamide**.

lipoyllysine an alternative name for **lipoamide**.

Lip(S₂) *symbol for* a residue of (the oxidized form of) **lipoic acid**.

Lip(SH)₂ *symbol for* a residue of dihydrolipoic acid, the reduced form of **lipoic acid**.

lipstatin an irreversible inhibitor of several lipases, e.g. pancreatic, gastric, carboxyl ester, or bile-stimulated lipase of milk, isolated from *Streptomyces toxytricini*. It blocks fat absorption selectively.

lipuria the presence of lipid or oily droplets in the urine.

liquefy *or* **liquify** to become, or cause to become, liquid. —**liquefaction** *or* **liquification** *n.*

liquid 1 a state of matter, intermediate between a solid and a gas, in which the molecules are relatively free to move with respect to each other but are restricted by intermolecular forces strong enough to prevent spontaneous expansion or significant compression. **2** any substance that is liquid at normal room temperature and pressure. **3** of, pertaining to, being, or behaving as a liquid. *Compare* **fluid**.

liquid chromatography *abbr.*: LC; any form of **chromatography** in which a liquid is used as the mobile phase.

liquid crystal *or* **mesomorphic state** *or* **mesophase** any new phase, resembling both liquid and solid phases, that appears when certain pure liquids are cooled, or pure solids are heated. These new phases have a translucent or cloudy appearance. In liquids of asymmetric molecules the molecular axes are arranged at random. In liquid crystals, however, there is some alignment, which occurs in three main distinguishable modes (certain such mesomorphic substances displaying more than one). In **nematic liquid crystals** the long axes of the molecules are aligned parallel to each other but the molecules are not arranged in layers. These scatter light strongly and appear translucent. In **cholesteric liquid crystals** the molecular axes are aligned and the molecules are arranged in layers but the orientation of the axes changes in a regular way on going from one layer to the next. Because the distance between any two layers with the same molecular orientation is of the order of the wavelength of visible light such crystals display vivid iridescence. **Smectic liquid crystals** are formed by certain molecules that have chemically dissimilar parts, e.g. polar and nonpolar; the chemically similar parts attract one another and the molecules form layers and align themselves in one direction. *See also* **liquid-crystalline phase**.

liquid-crystal display *abbr.*: LCD; any display of characters or numbers in a calculator, clock, or other device, based on **liquid crystal** cells whose reflectivity is changed by application of an electric field

liquid-crystalline phase any of several types of phase adopted by polar lipids at moderately elevated temperatures. Two major types exist, the lamellar and hexagonal phases. In the lamellar phase, also known as the L_α phase, the lipids form sheets of bilayers with the polar groups oriented outwards and the fatty-acyl moieties oriented into the hydrophobic core between the two bilayers. Such a phase is likely to include many such bilayers stacked on top of each other, with their polar groups apposed. In the hexagonal phase, the lipids form a spherical aggregate, in which the polar headgroups may be oriented either into the centre of the sphere, when this is known as the H_1 phase, or towards the surface of the sphere, the H_2 phase. In liquid-crystalline phase, the lipid adopts a looser packing, with greater mobility of the fatty-acyl chains than in the crystalline state. At low temperatures, in the absence of water, lipids exist in a crystalline state. If the lipid is heated to a point that attains the **transition temperature**, a temperature characteristic of each pure compound, the lipid undergoes a transition to one of the liquid-crystalline phases. On further heating, it melts completely. Addition of water at low temperature to the crystalline state causes formation of a gel phase, with water layers between the apposed polar headgroups. This gel phase will, on heating to the transition temperature, undergo transition to a liquid-crystalline phase similar to those described above.

liquid-gel chromatography a type of liquid chromatography in which the stationary phase is a gel. It includes **gel-permeation chromatography** and **ion-exchange chromatography**.

liquid glucose an alternative name for **corn syrup**.

liquid-junction potential the electric potential, not experimentally measurable, that arises across the boundary of two solutions of different compositions, e.g. between two half-cells. It is a complex function of the activities and transference numbers of the several ions in the transition layer.

liquid–liquid chromatography *abbr.*: LLC; a type of liquid chromatography in which the stationary phase is a liquid supported on a solid.

liquid medium any liquid **culture medium**.

liquid nitrogen nitrogen in the liquid state. It has a b.p. of –195.79°C (77.36 K), and is useful as a cooling and freezing agent.

liquid oxygen oxygen in the liquid state. It has a b.p. of –182.96°C (90.19 K).

liquid paraffin a colourless, odourless fraction of higher liquid hydrocarbons obtained from petroleum. Light liquid paraffin has a density of 0.83–0.86 g mL⁻¹ and heavy liquid paraffin has a density of 0.86–0.89 g mL⁻¹. It is used for (hot) oil baths, and the purified material is used as a suspending medium in infrared spectroscopy.

liquid-scintillation counting *abbr.*: LSC; a technique for counting particulate radiation. The (radioactive) sample is dissolved or suspended in, or placed in close approximation to, a solvent containing one or more substances that are fluorescent and emit a pulse of light (scintillation) when excited by an incident ionizing particle or photon. The light is detected by photomultipliers, in a liquid-scintillation counter.

liquid-scintillation spectrometry liquid-scintillation counting of particulate radiation of one or more selected ranges of energies.

liquid-solid chromatography *abbr.*: LSC; a type of liquid chromatography in which the mobile phase is a liquid and the stationary phase a solid.

liquor 1 (in pharmacy) a (usually aqueous) solution of a pure substance. **2** a solution in a chemical manufacturing process. **3** the water used in brewing.

liquorice *or* (*esp. US*) **licorice** *see* **glycyrrhiza**.

Lisp *see* **programming language**. [Acronym from 'list processing'.]

liter the US spelling of **litre**.

lithiasis (in pathology) the presence of calculi (stones), e.g. in the biliary or urinary tracts.

lithium *symbol*: Li; an element of group 1 of the IUPAC **periodic table**; atomic number 3; relative atomic mass 6.939. It is a soft, silvery white, alkali metal of such high reactivity that it must be stored in an inert atmosphere or under oil. It is not found free in nature, occurring in the minerals spodumene and lepidolite. Lithium carbonate, Li_2CO_3 (commonly referred to as 'lithium') is used to control the symptoms of manic-depressive illness. Li⁺ inhibits inositol phosphate phosphatase, thereby inhibiting the recycling of inositol, and reducing its availability for phosphatidylinositol synthesis.

lithogenesis (in pathology) the processes leading to formation of calculi (stones). —**lithogenic** adj.

lithotroph any member of the chemoautotrophs (see **chemotroph**) or **photolithotrophs** that uses CO_2 as a carbon source.

litre *or* (*US*) **liter** *symbol*: L *or* l; a non-SI unit of volume, now identical to the cubic decimetre; i.e. $1\ L = 1\ dm^3 = 10^{-3}\ m^3$.

little gastrin *see* **gastrin**.

liver one of the largest organs of the vertebrate body, situated in the abdominal cavity and formed as a diverticulum of the gut. It is highly vascular; venous blood carrying the products of digestion enters via the hepatic portal vein, while arterial blood enters via the hepatic artery. Blood leaves via the hepatic vein. The liver functions as a gland by secreting **bile** and is an important site of metabolism of carbohydrates, proteins, and fats; it stores carbohydrate as **glycogen**, is active in **gluconeogenesis**, and is important in the regulation of the blood glucose level. It is also an important site for the synthesis of **urea** and **ketone bodies**. The liver synthesizes **fibrinogen**, **prothrombin**, **serum albumin**, and other blood proteins and plays an important part in **detoxication**. The major cells in the liver are the **hepatocytes**. The liver also contains **Kupffer cells** of the reticuloendothelial system, cells derived from the bile duct cannaliculae, and fibroblasts, which are particularly important in the pathogenesis of liver cirrhosis. *See also* **regeneration** (def. 2).

liver X receptor *abbr.*: LXR; an alternative name for **oxysterol receptor**.

livin *or* **melanoma inhibitor of apoptosis protein** (*abbr.*: ML-IAP); a protein expressed in melanoma cells that inhibits apoptosis. Livin α is not expressed in fetal tissues, whereas livin β is. They bind and inactivate caspases 3, 7, and 9.

LLC *abbr. for* liquid–liquid chromatography.

LLD factor *abbr. for* Lactobacillus lactis Dorner factor; a growth factor for this organism, identical to **vitamin B₁₂**.

LLF *or* **L-L factor** *abbr. for* Laki–Lorand factor, i.e. factor XIII (*see* **blood coagulation**).

lm *symbol for* **lumen** (def. 1).

L meromyosin *abbr. for* light meromyosin.

LMM *abbr. for* light meromyosin.

ln *symbol for* natural, or Napierian, **logarithm** (alternative to logₑ).

L-NAME *abbr. for* N-ω-nitro-L-arginine methylester; an analogue of L-arginine that is a competitive inhibitor of this amino acid on NO synthase.

LNB-TM7 *abbr. for* long N-terminal of group B transmembrane 7; a subgroup of the G-protein-coupled membrane receptor family. Its members share a stalk region (consisting of ≈20% serine and threonine residues) and a characteristic sequence, containing four conserved cysteine residues, that is located immediately N-terminal to the first transmembrane segment. Several members also contain other homology regions in the extracellular region. *See* **lg-Hepta**.

αLnn *or* (*sometimes*) **Lnn** *symbol for* the linolenoyl or (9,12,15)-linolenoyl or (*formerly*) α-linolenoyl group; *see* **linolenoyl** (def. 1).

γLnn *symbol for* the (6,9,12)-linolenoyl or (*formerly*) γ-linolenoyl group; *see* **linolenoyl** (def. 2).

lnRNA *abbr. for* low-molecular-weight nuclear RNA (*see* **small nuclear RNAs**).

load (*in metabolic studies*) the amount (of a specified substance) given in a metabolic test.

loading test *or* **load test** a metabolic test in which the subject is given a dose of a specific substance and the concentration in blood and/or urine of the substance, or of some related metabolite, is determined as a function of time. Such a test is useful for determining the capacity of a particular metabolic pathway in the subject.

lobelia 1 any plant of the genus *Lobelia*. **2** *or* **Indian tobacco** the dried leaves, shoot tips, and seeds of *Lobelia inflata*, which contain the alkaloid **lobeline**.

lobeline 2-[6-(β-hydroxyphenethyl)-1-methyl-2-piperidyl]acetophenone; an alkaloid derived from **lobelia** (def. 2) with actions on the central and peripheral nervous systems that resemble those of **nicotine**. It is responsible for some of the properties of herbal cigarettes.

(–)-lobeline

Lobry de Bruyn–van Ekenstein transformation a nonenzymic transformation occurring in carbohydrates, usually in alkaline conditions, that includes epimerization of both aldoses and ketoses and aldose–ketose isomerization. [After Cornelius Adrian Lobry de Bruyn and W. Alberda van Ekenstein (1857–1937), who first described it in 1895.]

local alignment *see* **alignment**.

localized melting the (reversible) destabilization of tertiary structure in a limited region of an ordered biopolymer through interaction with a specific agent, e.g. an enzyme.

local similarity similarity that spans only short regions of a pair or set of aligned sequences (i.e. within a local alignment). *See* **alignment**.

locant a prefix or infix used in organic chemical nomenclature to identify the position of a substituent in a parent structure or of an isotopic modification to a compound. Locants are commonly Ara-

bic numerals, but Greek letters and italicized symbols of heteroatoms are also used. In isotopically modified compounds the italicized name of the function containing the modification may be used if it has no number, e.g. in [*methyl*-¹⁴C]methionine.

loci *the plural of* **locus**.

lock-and-key model a model for the mechanism of an enzyme–substrate combination, of a hormone–receptor interaction, or of an antibody–antigen reaction. In this model, which is based on the analogy of the complementarity of a lock and its key, one reactant is envisaged binding to the other on a site to which it is structurally complementary.

Locke's solution *or* **Locke solution** *or* **Ringer–Locke solution** any one of several balanced salt solutions modified from **Ringer's solution**, initially for the perfusion of mammalian heart and other tissues in physiological experiments. Such solutions have usually contained (in grams per litre, approx.) NaCl (9.0–9.5), KCl (0.20–0.42), CaCl₂ (anhydrous, 0.20–0.24), NaHCO₃ (0.1–0.3), and glucose (1.0–2.5). [After Frank Spiller Locke (1871–1949), British physiologist, who devised them in 1894 and 1900.]

lock-washer structure a putative intermediate structure formed during the assembly of subunits of the coat protein of tobacco mosaic virus into the helically organized structure of the fully formed virion. It is believed to arise after the primary sub-assembly, a two-layered disk with 17 subunits per layer, has self-aggregated into a cylinder; a dislocation in each disk then occurs so that it becomes a two-turn flattened protohelix, of similar shape to a lock washer of the sprung type.

Loco a protein of *Drosophila* that is homologous to the mammalian RGS12, which regulates G-protein signalling via G$_{\alpha 0}$ and G$_{\alpha i}$.

loco disease *see* **locoweed**.

locoweed any of various plants belonging to the genera *Astragalus* or *Oxytropis*, found mainly in western USA. They contain L-2-amino-3-(methylseleno)propionic acid (methylselenocysteine), a nonprotein amino acid, and may cause poisoning (**loco disease**) when ingested by livestock. [From Spanish *loco*, mad.]

locus (*pl.* **loci**) **1** (*in genetics*) the specific position, in all homologous chromosomes, of a particular gene or one of its alleles. **2** (*in mathematics*) the curve or line traced by a point moving so as to satisfy a particular set of conditions.

locus control region *abbr.*: LCR; a group of five regions upstream of the structural genes for the β chain of hemoglobin in human and mouse. These regions are essential for the expression of all the genes of the β-like gene cluster.

lod score *abbr. for* log of the odds score; a statistical test to determine whether a set of linkage data indicates linkage or not, the log₁₀ of the odds favouring linkage. For genetic disorders that are not X-linked, a lod score of +3 (1000:1 odds of linkage) indicates linkage, whilst a score of –2 is odds of 100:1 against linkage.

log 1 *symbol for* logarithm. **2** *abbr. for* logarithmic.

logarithm *abbr. and symbol*: log; the power to which a fixed number (the base) must be raised in order to produce a specified number; i.e. in the expression $a = b^n$, a is the specified number, b is the base, and n is the logarithm. In **decadic (common)**, or **Briggsian**, **logarithms** the base is 10; the common logarithm of a is written as log₁₀a, or lg a. In **natural**, or **Napierian**, **logarithms** the base is e (= 2.718 28...); the natural logarithm of a is written as logₑa or ln a. In **binary logarithms** the base is 2, and the binary logarithm of a is written as log₂a or lb a. In converting between them, logₑa = 2.303 × log₁₀a, and log₂a = 3.322 × log₁₀a.

logarithmic *or* **logarithmical** *abbr.*: log; **1** of, pertaining to, using, or containing **logarithms** of numbers or variables. **2** describing a relationship in which one variable is proportional to the logarithm of another.

logarithmic growth a type of growth pattern shown by bacteria or other cells in which the number of cells, or the cell mass, increases logarithmically (exponentially) with time. The rate of increase at any time is proportional to the number of cells, or the cell mass, present.

logarithmic phase *abbr.*: log phase; the phase of growth of bacterial or other cells during which they undergo **logarithmic growth**.

logarithmic plot *abbr.*: log plot; a graph in which one axis has a **logarithmic** (def. 1) scale and the other a linear scale.

logical operation an operation on **binary** numbers or statements used in computing and also in applications such as databases and library searches. There are three common binary operators (AND, OR, XOR) and one **unary operator** (NOT).

log–log plot double logarithmic plot; a graph on which both variables are plotted logarithmically.

log-odds score a score calculated as the logarithm of the likelihood of an event relative to its likelihood under a null model. Positive log-odds scores indicate that the event is more likely than it would be under the null model.

LOH *abbr. for* **loss of heterozygosity**.

Lohmann reaction *a name sometimes given to* the reversible reaction, catalysed by **creatine kinase**, in which ATP and creatine are formed from ADP and phosphocreatine. [After Karl Lohmann (1898–1978), German biochemist.]

Lohmann's enzyme *an alternative name for* **creatine kinase**.

lombricine the 2-guanidoethanol phosphoester of phosphoserine. In earthworms, the guanidine group can undergo phosphorylation to phospholombricine, which acts as a **phosphagen**.

lomustine 1-(2-chloroethyl)-3-cyclohexyl-1-nitrosourea; a chloroethylnitrosourea derivative with actions resembling **carmustine**.

London forces *see* **dispersion forces**.

lone pair *or* **nonbonding electrons** a pair of electrons of opposite spins occupying the same orbital in the valence shell of an atom.

long-acting thyroid stimulator *abbr.*: LATS; an IgG immunoglobulin found in the serum of a large proportion of thyrotoxic patients. It binds competitively with thyrotropin to thyrotropin receptors in thyroid plasma membrane, stimulates membrane adenylate cyclase, and causes longer-lasting stimulation of thyroid-hormone secretion than thyrotropin itself.

long-chain describing an aliphatic compound of any type with a chain length greater than ten carbon atoms (i.e. C_{10}).

long-chain alcohol *or* **fatty alcohol** any aliphatic compound with a chain length greater than C_{10} that possesses a terminal hydroxymethyl group. They are components of **ether lipids**, and esterified with long-chain fatty acids they form wax esters, often major components of plant cuticular waxes.

long-chain-alcohol *O*-fatty-acyltransferase EC 2.3.1.75; *other name*: wax synthase; an enzyme that catalyses the formation of a long-chain fatty acyl ester, i.e. a wax, from acyl-CoA and a long-chain alcohol with release of CoA.

long-chain base any aliphatic compound with a chain length greater than C_{10} that possesses a terminal basic group.

long-chain fatty acid any aliphatic compound with a chain length greater than C_{10} that possesses a terminal carboxyl group.

long-chain-fatty-acid–CoA ligase EC 6.2.1.3; *other names*: acyl-activating enzyme; acyl-CoA synthetase; fatty acid thiokinase (long-chain); lignoceroyl-CoA synthase; an enzyme that catalyses the reaction between ATP and a long-chain carboxylic acid and CoA to form an acyl-CoA with release of AMP and pyrophosphate. It activates free fatty acids formed, e.g., by phospholipase action for further metabolism.

long-chain-3-hydroxyacyl-CoA dehydrogenase EC 1.1.1.211; *other name*: β-hydroxyacyl-CoA dehydrogenase; an enzyme of the **beta-oxidation system** that catalyses the oxidation by NAD^+ of (*S*)-3-hydroxyacyl-CoA to 3-oxoacyl-CoA with the formation of NADH. It forms part of a trifunctional protein. *See* **3-hydroxyacyl-CoA dehydrogenase**.

longitudinal relaxation or **spin-lattice relaxation**, a process in magnetic resonance spectroscopy, such as EPR and NMR, that describes the return of the z component of longitudinal (z-) magnetization to its equilibrium value. The characteristic spin-lattice relaxation time is given the symbol T_1. Its reciprocal is the spin-lattice relaxation rate.

longitudinal relaxation time *an alternative name for* **spin-lattice relaxation time**.

long-patch repair *see* **short-patch repair**.

long PCR a form of PCR designed for the amplification of DNA molecules of >10kb or longer and using a mixture of proofreading and non-proofreading thermostable DNA polymerases such as Pfu and Taq.

long QT syndrome a syndrome of prolongation of the QT interval on surface electrocardiogram, recurrent fainting, episodic tachyarrhythmia, and sudden (or aborted sudden) death. It is associated with numerous loss-of-function mutations in several genes, including: *SCN5A* (encoding the α subunit of cardiac Na channel); *HERG* (encoding the α subunit of cardiac K channel I_{Kr}); *KCNE1* (encoding the β subunit of cardiac K channel I_{Ks}); and *KVLQT1* (encoding the α subunit of cardiac K channel I_{Ks}). *SCN5A* (at 3p21-p24) encodes a protein of 2016 amino acids containing four regions, each of which has six transmembrane segments. Gain-of-function mutations are associated with idiopathic ventricular tachycardia. *HERG* (for human ether-a-go-go-related gene) is at 7q35-q36 and encodes 1159 amino acids, including six transmembrane segments and extensive intracellular N- and C-terminal domains. *See also* ***KCNE1***, ***KVLQT1***.

long-range interaction describing an interaction between residues that are situated relatively far apart in a polypeptide or polynucleotide sequence.

long terminal repeat *abbr.*: LTR; a double-stranded DNA sequence, generally several hundred base pairs long, that is repeated at the two ends of the DNA of retroviruses and retrotransposons.

loop 1 the round or oval figure produced by a line, wire, thread, etc. that curves round to cross (or nearly cross) itself. **2** *(in electronics)* any closed electric or magnetic circuit through which a signal can circulate. **3** *(in computing)* any set of instructions in a program that is carried out repeatedly until some specified condition is satisfied.

loopback DNA *an alternative name for* hairpin DNA; *see* **hairpin**.

loop injector a device for introducing samples into a high-pressure liquid chromatographic column without interruption of the solvent flow to the column. It consists of a loop of metal tubing of small volume, which is filled with the sample; this can then, by an appropriate valve, be flushed by eluent directly onto the column.

loop of Henle *or* **Henle's loop** the U-shaped part of a nephron lying in the renal medulla. It comprises a thin descending tubule and an ascending tubule formed of both a thin and a thick segment. It plays a role in the selective reabsorption of fluid and solutes. [After Friedrich Gustav Jakob Henle (1809–85), German anatomist and pathologist.]

lophotoxin a cyclic diterpenoid, isolated from Pacific gorgonians (sea fans and whips) of the genus *Lophogorgia*, that irreversibly inactivates the nicotinic acetylcholine receptor on intact BC3H-1 cells.

Lorentz correction factor a correction factor for the effect of solvent polarizability on optical rotation by solutions of proteins or other macromolecules. It is given by: $3/(n^2 + 2)$, where n is the refractive index of the solvent. [After Hendrik Antoon Lorentz (1853–1928).]

Lorentzian lineshape function an equation for the lineshape of a nuclear magnetic resonance absorption signal:

$$g(\omega) = T_2/[1 + (\omega - \omega_0)^2 T_2^2],$$

where $g(\omega)$ is a measure of the intensity of the resonance at a frequency ω, ω_0 is the actual Larmor frequency of the resonating nucleus, and T_2 is the spin–spin relaxation time.

loricrin the major protein of the cornified cell envelope of terminally differentiated epidermal keratinocytes. It is a substrate for transglutaminase, and exists as monomers cross-linked by disulfide and *N*-(γ-glutamyl)lysine isodipeptide bonds.

losartan 2-butyl-4-chloro-1-(*p*-(*O*-1*H*-tetrazol-5-ylphenyl)benzyl)-imidazol-5-methanol; an antihyperactive drug that blocks the binding of angiotensin II (*see* **angiotensin**)to the AT_1 receptor. Affinity for this receptor is much greater than for the AT_2 receptor. It is oxidized by a cytochrome P450 to a carboxylic-acid metabolite that is largely responsible for most of its angiotensin II antagonistic effects.

Loschmidt constant *symbol*: N_L; the number of molecular entities in unit volume of an ideal gas at 1 atm and 273.15 K. It is equal to the **Avogadro constant** divided by the molar volume, and has the value $2.686\ 763 \times 10^{25}$ m^{-3}. [After Josef Loschmidt (1821–95), Austrian physicist.]

Loschmidt number *or* **Loschmidt's number** the numerical value of the **Loschmidt constant**.

loss of heterozygosity *abbr.*: LOH; in genetics, the situation where one of the mutant alleles of an autosomal gene is inherited, and the mutation of the other allele arises by somatic mutation. Cells with the somatic mutation at that locus have lost heterozygosity as they now have two mutant alleles. LOH in tumor supressor genes such as *RB1* and *p53* is of importance in cancer genetics.

lovastatin *or* **mevinolin** 6α-methylcompactin; a metabolite from some fungi (e.g. *Monascus ruber*, *Aspergillus terreus*) that is a potent inhibitor of **hydroxymethylglutaryl-CoA reductase**. It is useful clinically as an antihypercholesterolemic; it reduces plasma levels of total low-density-lipoprotein (LDL), and very-low-density-lipoprotein cholesterol, but it can increase high-density-lipoprotein cholesterol. Its effects result in the induction of up-regulation of LDL receptors on hepatocytes. Two proprietary names are: Mevacor; Mevinacor.

low-complexity region a region in the amino-acid sequence of a protein in which a particular amino acid, or a small number of different amino acids, are enriched. *See*, for example, **leucine zipper**.

low-density lipoprotein *abbr.*: LDL; one of the classes of lipoprotein found in blood plasma in many animals (data normally relate to humans). LDLs are also known as β lipoproteins, owing to their electrophoretic mobility intermediate between high-density or α lipoprotein, and the origin. LDL particles have diameter 19.6–22.7 nm, solvent density for isolation (g mL^{-1}) 1.019–1.063, hydrated density (g mL^{-1}) 1.034. Their approximate composition (% by weight) is 13% unesterified cholesterol, 39% esterified cholesterol, 17% phospholipid, 11% triacylglycerol, 20% protein. Their apolipoprotein composition (% by weight total apolipoprotein) is 98% B-100, with traces of different C and E apolipoproteins. They are synthesized in plasma from very-low-density lipoproteins and intermediate density lipoproteins through the action of lipoprotein lipase; they transport cholesterol to peripheral tissues. They have the highest content of cholesterol of any plasma lipoprotein, and have been linked to the incidence of coronary heart disease in epidemiological studies. *See also* **apolipoprotein** (for individual apolipoproteins) and **LDL receptor**.

low-density-lipoprotein receptor *see* **LDL receptor**.

low-energy compound a name sometimes applied improperly to certain compounds that undergo reaction leading to release or transfer of, e.g., phosphate, where the standard free energy change for the hydrolysis of the bond is <20 kJ mol^{-1}. It is used especially in making a distinction from a **high-energy compound**. The use of the term does not meet the highest standards of thermodynamic correctness.

low-lipid lipoprotein *(sometimes)* an alternative name for **high-density lipoprotein**.

Lowry method a colorimetric method for the estimation of small amounts (down to 10 μg) of protein. It is a combination of the **biuret reaction** for peptide bonds and the colour formed by reaction of **Folin–Ciocalteau reagent** with tyrosine and tryptophan residues in the protein. [After Oliver Howe Lowry (1910–1996), US pharmacologist and biochemist.]

low-spin complex any of the ligand complexes of transition-metal ions having a maximal number of paired *d*- or *f*-shell electrons in the lower energy levels. This state occurs when the pairing energy is less than the crystal-field splitting energy. *Compare* **high-spin complex**.

Lozenge a protein of *Drosophila* that is involved in the determination of photoreceptor identity during eye development. Its DNA-binding domain is homologous with that of Runt (of *Drosophila*) and AML1 (of human).

LPG *abbr. for* lipophosphoglycan (*see* **GPI anchor**).

LPH *abbr. for* lipotropic hormone (i.e. **lipotropin**).

LPL *abbr. for* lipoprotein lipase.

L protein *see* **glycine cleavage enzyme**.

LPS *abbr. for* **1** lipopolysaccharide. **2** lipase (def. 2).

LPS-binding protein *see* **BPI protein**.

Lr *symbol for* lawrencium.

L region any region in a (carcinogenic) polycyclic aromatic hydrocarbon that consists of two *para*-carbon atoms, e.g. the *meso*-9,10-carbon atoms of anthracene, which display the highest free-valence indices. The lack of such a region has been suggested to correlate with a lack of carcinogenicity of some hydrocarbons. *Compare* **K region**.

LRF *abbr. for* luteinizing hormone releasing factor (*see* **gonadotropin-releasing hormone**).

LRH *abbr. for* luteinizing hormone releasing hormone (*see* **gonadotropin-releasing hormone**).

LRP *abbr. for* LDL receptor-related protein; *other names*: apoE/α₂-macroglobulin receptor; E/α₂-M receptor. A multifunctional plasma membrane receptor primarily of hepatocytes. It is a transmembrane protein of 4525 residues that is homologous to **LDL receptor**. Its extracellular domain contains 31 cysteine-rich repeats and 22 EGF repeats (versus 7 and 3, respectively, in LDL receptor). It binds chylomicron remnants, hepatic lipase, α₂-macroglobulin, plasminogen activator-inhibitor complexes, lactoferrin, and *Pseudomonas* exotoxin A. *See also* **LRP2**.

LRP2 *abbr. for* LDL receptor-related protein 2; *other names*: megalin; gp330; Haymann nephritis antigen. A large transmembrane glycoprotein of 4635 residues that is homologous with LRP and binds most of the latter's ligands. It is abundant in renal glomeruli and proximal tubules but also occurs in other tissues. It appears to be a drug receptor that mediates uptake of aminoglycosides, polymixin B, and gentamicin. *See also* **LRP**.

LRr *abbr. for* leucine-rich repeat.

LSC *abbr. for* **1** liquid-scintillation counting. **2** liquid–solid chromatography.

LSD *abbr. for* lysergic acid diethylamide; *see* **lysergic acid**.

L7/L12 site a component of EF-Tu translocase (*see* **EF-T**) activity involving in bacteria the **ribosomal proteins** L7 and L12.

L strand *abbr. for* light strand.

LT *abbr. for* **1** leukotriene; the subtype may be indicated by a third letter, as in LTA (leukotriene A). **2** lymphotoxin.

LTQ *abbr. for* lysine tyrosylquinone; a quinone cofactor derived by post-translational linking of a lysine residue with a topaquinone in some quinoproteins. It is the cofactor of lysyl oxidase.

LTR *abbr. for* long terminal repeat.

LTRPC *abbr. for* long transient receptor potential channel. *See* **TRP channel**.

LTRPC2 *see* **TRP channel**.

LTRPC7 *see* **TRP channel**.

L-type pentose pathway *see* **pentose phosphate pathway**.

Lu *symbol for* lutetium.

Lubrol *a proprietary name for* a polyoxyethylene nonionic detergent; the structure of Lubrol PX is the same as $C_{12}E_9$ (see **C_xE_y**). It is useful for solubilizing membrane-bound enzymes such as adenylate cyclase.

LUCA *abbr. for* last universal common ancestor.

LUCA 15 a multidomain RNA-binding protein that is generally conserved in eukaryotes and is deleted in certain human lung cancers. It contains two N-terminal RNA-recognition sequences, a C2H2 zinc finger, and a C-terminal glycine-rich sequence. The rat homologue is associated with hnRNPs.

luciferase any member of a group of monooxygenase enzymes that catalyse bioluminescent reactions. **Firefly luciferase** (EC 1.13.12.7; *recommended name*: *Photinus* luciferin 4-monooxygenase (ATP hydrolysing)) occurs in the lantern of the American firefly, *Photinus pyralis*, and may be used with firefly **luciferin** for estimation of low concentrations (10^{-6}–10^{-10} mol L^{-1}) of ATP. It catalyses a reaction between *Photinus* luciferin, O_2, and ATP to form oxidized *Photinus* luciferin, CO_2, H_2O, AMP, pyrophosphate, and $h\nu$; magnesium is a cofactor. It covalently binds the adenylyl group (AMP) after reaction with ATP and release of pyrophosphate. It is a peroxisomal enzyme. **Renilla luciferase** (EC 1.13.12.5; *recommended name: Renilla*-luciferin:oxygen 2-oxidoreductase (decarboxylating)) occurs in the sea pansy *Renilla reniformis* and catalyses the reaction between *Renilla* luciferin (2,8-dibenzyl-6-(4-hydroxyphenyl)imidazo[1,2-*a*]pyrazin-3(7*H*)-one) and O_2 to form oxidized *Renilla* luciferin, CO_2, and a luciferin-binding protein (BP-LH2). Bioluminescence is triggered by calcium ions, and the excited state transfers energy to an accessory protein, *Renilla* green fluorescent protein, which emits green light. **Bacterial luciferase** (EC 1.14.14.3; *recommended name*: alkanal monooxygenase (FMN-linked)) occurs in *Photobacterium fischeri* and catalyses a reaction between RCHO, reduced FMN, and O_2 to form RCOOH, FMN,H_2O, and $h\nu$. It may be used together with NAD(P)H dehydrogenase (FMN), EC 1.6.8.1, from luminescent bacteria for estimation of low concentrations (10^{-7}–10^{-9} mol L^{-1}) of NAD(P)H. *See also* **lux proteins**.

luciferin any bioluminescent substance that is a substrate for a **luciferase** enzyme. Luciferins from different sources often bear no structural resemblance to one another. Firefly luciferin, obtained from the lantern of the American firefly, *Photinus pyralis*, is D-(–)-4,5-dihydro-2-(6-hydroxy-2-benzothiazoloyl)-4-thiazolecarboxylic acid.

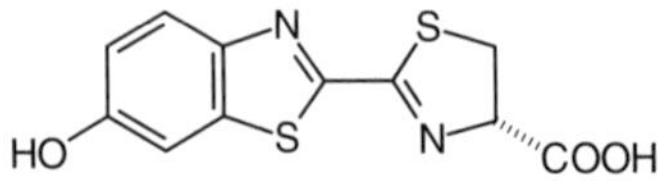

Lucifer Yellow one of several dyes based on the structure 6-amino-2,3-dihydro-1,3-dioxo-benz[*d,e*]isoquinoline-5,8-disulfonic acid. Lucifer Yellow CH has been used as a marker for movement between cells, and for ultrastructure tracing; Lucifer Yellow VS covalently labels proteins rapidly under mild conditions.

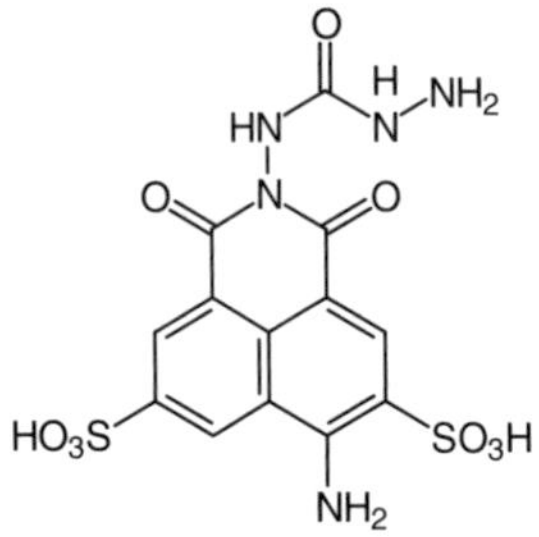

Lucifer Yellow CH

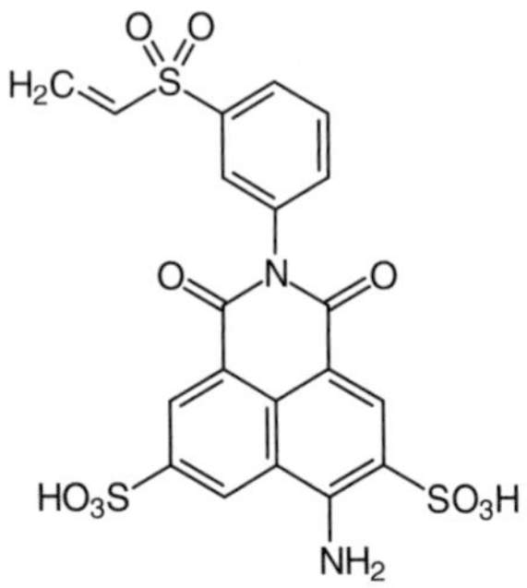

Lucifer Yellow VS

Lucite *a proprietary name for* **poly(methyl methacrylate)**.

lufenuron *N*-[2,5-dichloro-4-(1,1,2,3,3,3-hexafluoropropoxy)-phenylaminocarbonyl]-2,6-difluorobenzamide; an inhibitor of insect development, used as a once-per-month pill to control flea populations on domestic dogs, by inhibiting egg development.

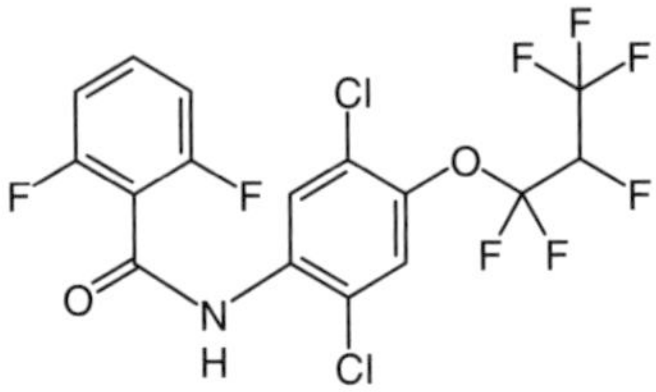

luffin a type I **ribosome-inactivating protein**.

luliberin *the recommended name for* a hypothalamic decapeptide amide now known to be identical with **gonadotropin-releasing hormone**.

lumen 1 (*pl.* **lumens**) *symbol*: lm; the SI derived unit of **luminous flux**, equal to the quantity of visible electromagnetic radiation emitted per second in a cone of unit solid angle (one steradian, sr) by a uniform point source of unit luminous intensity (one candela, cd); i.e. 1 lm = 1 cd sr. **2** (*pl.* **lumina**) the cavity or passageway of a tubular organ, such as a blood vessel, duct, or length of gut. **3** the volume enclosed within the membranes of a mitochondrion or of the endoplasmic reticulum, or of microsomes prepared from cells. **4** the space enclosed by the walls of a plant cell that no longer contains a protoplast. **5** the bore of a catheter, hollow needle, or other tubular device. —**luminal** *adj.*

lumichrome 7,8-dimethylalloxazine; a compound showing a blue fluorescence, formed by the photolysis of riboflavin in acid or neutral solution.

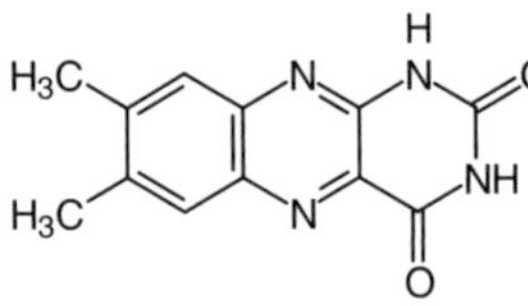

lumiflavin 7,8,10-trimethylisoalloxazine; a compound showing a yellow-green fluorescence, formed by the photolysis of riboflavin in alkaline solution.

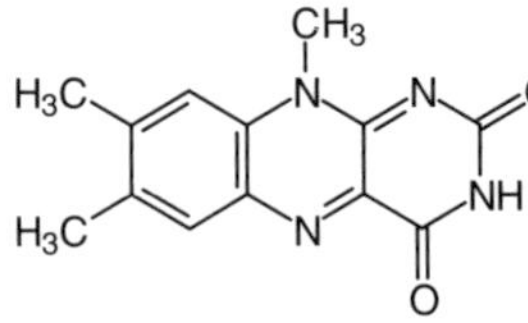

lumigenic substrate any substrate capable of emitting light when acted on by an appropriate enzyme.

lumina *the plural of* **lumen** (def. 2).

luminal protein *or (sometimes)* **ER-resident protein** any protein that is retained within the cisternae of the **rough endoplasmic reticulum**, e.g. protein disulfide isomerase. Such proteins possess at their C terminus the tetrapeptide sequence Lys-Asp-Glu-Leu (**KDEL**) in animal cells and His-Asp-Glu-Leu (**HDEL**) in yeast. The proteins move to the Golgi and are returned to the endoplasmic reticulum.

luminescence the production of visible light by any mechanism that does not depend upon the system having a temperature appreciably above the ambient. It comprises **fluorescence** and **phosphorescence**. —**luminescent** *adj.*

luminogenic 1 producing **luminescence**. **2** of or pertaining to a non-luminescent material that can be acted on by an enzyme to produce a luminescent substance.

luminol 5-amino-2,3-dihydro-1,4-phthalazinedione; a substance useful in the detection or determination of trace amounts of oxidizing agents by virtue of the intense bluish luminescence it emits on oxidation in alkaline solution.

luminometer an instrument for measuring **luminescence**. Given appropriate conditions, it may be used also to measure the concentrations of enzymes, substrates, etc. that participate in, or can be coupled to, light-producing reactions.

luminometry the measurement of **luminescence** with a luminometer. —**luminometric** *adj*.

luminophore 1 any luminescent substance, whether a fluor or a phosphor. 2 a molecular grouping responsible, or thought to be responsible, for the **luminescence** of a larger molecule; a luminescent chromophore.

luminous 1 emitting visible light; bright, shining. 2 *(in photometry)* (of a physical quantity relating to light) evaluated by the visual sensation of brightness to the eye of an observer rather than by an absolute measurement of radiant energy.

luminous flux *symbol*: Φ_v or Φ or P_v or P; a measure of the rate of flow of radiant energy within the visible region of the electromagnetic spectrum. The SI derived physical quantity is the arithmetic product of the **luminous intensity** of a source of visible light and the solid angle, expressed in steradians, into which the light is emitted. The SI derived unit of luminous flux is the **lumen** (def. 1).

luminous flux density the **luminous flux** per unit cross-sectional area of a beam of light.

luminous intensity *symbol*: I_v or I; one of the seven SI base physical quantities, defined as the amount of visible light emitted per second per steradian by a point source in a given direction. The SI base unit of luminous intensity is the **candela**.

lumirhodopsin an intermediate in the light-induced modification of **rhodopsin** prior to its breakdown into opsin and all-*trans*-retinal. It has an absorption maximum at 497 nm.

lumisome a membrane-bound particle, ≈200 nm in diameter and present in homogenates of various species of luminescent marine coelenterates, that contains all the proteins necessary for the production of the typical green bioluminescence observable *in vivo*. It is probably the cytoplasmic organelle responsible for emission of light by such organisms.

lumisterol 9β,10α-ergosta-5,7,22-trien-3β-ol; a compound produced by the UV irradiation of **ergosterol**, from which it differs in the spatial arrangement of the methyl group at C-10.

Lundh test a test for pancreatic function in which the pancreas is stimulated by a test meal (corn oil, milk powder, and glucose), after which trypsin activity is measured in duodenal juice.

lupus erythematosus an autoimmune disease characterized by recurrent red patches on the skin, usually accompanied by episodes of inflammation of joints, tendons, and other connective tissues and organs. Antinuclear and various other antibodies are usually present in serum.

luteal of, relating to, or characterized by the development of a corpus luteum or corpora lutea.

luteal phase *see* **estrous cycle**.

lutein 3,3′-dihydroxy-α-carotene, one of the most widespread naturally occurring xanthophylls (i.e. carotene alcohols; *see* **carotenoid**), found in many higher plants, some algae, egg yolk, and corpora lutea. It is an antioxidant.

luteinize *or* **luteinise** to cause the formation of corpora lutea in; to undergo transformation into a corpus luteum. —**luteinization** *or* **luteinisation** *n*.

luteinizing hormone *or* **luteinising hormone** *abbr*.: LH; *recommended name*: lutropin; *former name*: interstitial cell-stimulating hormone (*abbr*.: ICSH). One of the two gonadotropic **glycoprotein hormones** secreted by the anterior pituitary. The human hormone (28.5 kDa) consists of an α subunit that is almost identical to that of other glycoprotein hormones, and a β subunit that confers hormonal specificity; the carbohydrate content is 15.5%. In the female it promotes ovulation and luteinization of the ovarian follicle, where it stimulates the conversion of progesterone to pregnenolone, thereby increasing the formation of further progesterone. In the male it stimulates testicular interstitial (Leydig) cell function by promoting the production of androgens, especially testosterone. LH binds to a G-protein-associated membrane receptor. Mutations in the receptor lead to hypogonadism or to precocious puberty in males. Mutations in the β subunit of LH also lead to hypogonadism in males.

luteinizing hormone/follicle-stimulating hormone releasing factor *abbr*.: LH/FSH-RF *see* **gonadotropin-releasing hormone**.

luteinizing hormone releasing factor *abbr*.: LHRF *see* **gonadotropin-releasing hormone**.

luteinizing hormone releasing hormone *abbr*.: LHRH *or* LH-RH *or* LRH; *see* **gonadotropin-releasing hormone**.

luteo+ *comb. form* indicating 1 yellow. 2 of the **corpus luteum**.

luteolin a flavone that is present in alfalfa and induces *nod* gene expression in some *Rhizobium* species.

luteolysin *an alternative name for* **prostaglandin** F$_{2\alpha}$.

luteolysis degeneration or regression of the **corpus luteum**. —**luteolytic** *adj*.

luteotropic hormone *or* **luteotrophic hormone** *abbr*.: LTH; *a former name for* **prolactin**.

luteotropin *or* **luteotrophin** *a former name for* **prolactin**.

lutropin *the recommended name for* **luteinizing hormone**.

lutropin receptor a G-protein-coupled membrane receptor that mediates the effects of lutropin via activation of adenylate cyclase. It has a long extracellular N-terminal region (≈340 amino acids) that is glycosylated.

lux *symbol*: lx; the SI derived unit of illuminance, equal to 1 lumen per square metre; i.e. $1 \, lx = 1 \, l \, mm^{-2}$.

Lux proteins the counterpart in marine bacteria of the **luciferase** system. They mediate bioluminescence, and are products of genes regulated by the *lux* operon. In *Vibrio* spp., LuxA (catalytic subunit) and LuxB constitute bacterial luciferase (EC 1.14.14.3); it catalyses a reaction between reduced FMN, RCHO, and O_2 to form FMN, RCOOH, H_2O, and *hν*. LuxC is an **acyl-CoA reductase** required to form the 'RCHO' above. LuxD is an acyl transferase component of the fatty acid reductase system required for aldehyde biosynthesis (EC 2.3.1.–). LuxE is an acyl-protein synthetase (EC 6.2.1.19) that activates tetradecanoic acid. LuxG is the flavin reductase. LuxI and LuxJ are part of the regulatory system; they are required for the synthesis of OHHL (*N*-(3-oxohexanoyl)-L-homoserine lactone), an autoinducer molecule that binds to LuxR. LuxR is the transcriptional activator.

LVP *abbr. for* lysine vasopressin.

LW antigen *abbr. for* Landsteiner–Weiner antigen. *See* **Rhesus factor**.

lx *symbol for* lux.

LX *abbr. for* lipoxin.

LXR *abbr. for* liver X receptor. *See* **oxysterol receptor**.

lyase 1 *the systematic name for* any enzyme of EC class 4 that catalyses the cleavage of C–C, C–O, C–N, and other bonds, leaving double bonds, or conversely adding a group to a double bond. **2 17,20-lyase** *the common name for* 17α-hydroxyprogesterone aldolase (EC 4.1.2.30).

lyate ion the anion produced by the autoprotolysis of a protogenic solvent; e.g. the hydroxide ion is the lyate ion of water.

lycopene the characteristic red pigment of tomatoes and some other ripe fruit. It is an acyclic carotenoid, and the biosynthetic precursor of the cyclic carotenes in plants.

lycopodium 1 any clubmoss of the genus *Lycopodium*. **2** the spores of various clubmosses, e.g. *L. clavatum*. They are used as a covering for pills and suppositories, in pyrotechnics, and in foundry work.

lymph 1 the pale-yellow, clear, or cloudy fluid that is contained within the vessels of the **lymphatic system**. It contains a little protein, and derives from the interstitial fluid, which is drained by the lymphatic vessels and ultimately discharged into the bloodstream, e.g. via the thoracic duct. The composition of lymph varies in different parts of the body, that from the liver and intestine containing more protein, and the latter more fat in the form of chylomicrons. Lymph also contains varying numbers of cells, mostly small lymphocytes, and coagulation factors. **2** the exudation from a sore.

lymphatic 1 of, pertaining to, or containing lymph; of or pertaining to the lymphatic system. **2** a lymphatic vessel.

lymphatic system the extensive network of capillary-like lymphatic vessels in vertebrates that convey the lymph from the peripheral tissues to the venous system. **Lymph nodes** occur at various points along the vessels.

lymph node *or* **lymph gland** any of numerous small swellings distributed throughout the lymphatic system of vertebrates. Each node consists largely of lymphocytes, together with macrophages and dendritic cells held together in a loose reticulum. Lymph nodes receive lymph from the peripheral tissues via afferent lymphatic vessels and pass it on via efferent vessels to more centrally placed lymph nodes, and ultimately to the thoracic duct. The lymph nodes act as filters through which the lymph must pass and where foreign material comes into intimate contact with macrophages and lymphocytes. They are active centres for phagocytosis and the site of initiation of humoral and cell-mediated immune reactions.

lympho+ *or (before a vowel)* **lymph+** *comb. form* indicating lymph or the lymphatic system.

lymphoblast any **blast cell** of the **lymphoid cell series**. —**lymphoblastic** *adj.*

lymphocyte a type of **leukocyte** found in vertebrates and having various functions in specific immunity. It is a spherical cell, 7–12 μm in diameter, with a large, round nucleus, often indented when seen by electron microscopy, and scanty cytoplasm containing scattered ribosomes but little endoplasmic reticulum and few other organelles. Lymphocytes occur in lymph nodes, other tissues, lymph, and blood, and they circulate between these tissues. Mature, differentiated lymphocytes comprise two main classes, **B lymphocytes** and **T lymphocytes**. Lymphocytes without the characteristics of either of these classes are termed **null lymphocytes**.

lymphocyte-activating factor *abbr.:* LAF; *an alternative name for* **interleukin 1**.

lymphocyte-specific protein-tyrosine kinase *see* **Lck**.

lymphocyte transformation *see* **transformation** (def. 3).

lymphocytosis an abnormal increase in the number of lymphocytes, especially in blood.

lymphoid of or resembling lymph, or lymphatic tissue.

lymphoid cell series a cell series comprising all cells that morphologically resemble **lymphocytes**, their precursors (e.g. lymphoblasts), and cells that may be derived from them (e.g. plasma cells).

lymphoid tissue any vertebrate tissue that is made up predominantly of **lymphocytes**, e.g. lymph, lymph nodes, spleen, thymus, Peyer's patches, adenoids, pharyngeal tonsils, and, in birds, bursa of Fabricius and cecal tonsils.

lymphokine any soluble nonimmunoglobulin released by primed lymphocytes on a second interaction with the specific activator (which may be a soluble antigen, a cell-associated antigen, or a mitogen). They modulate **cell-mediated immunity** by modulating the behaviour of other types of cell within the same organism (e.g. macrophages, polymorphonuclear leukocytes), and are recognized as a subgroup of the **cytokines**. Specific lymphokines include: chemotactic factor for macrophages, lymphotactin, lymphotoxin (i.e. **tumour necrosis factor** β), macrophage-activating factor (MAF), macrophage migration enhancement factor, **macrophage migration inhibitory factor** (MIF), (antigen-independent) mitogenic factor, skin-reactive factor, and **transfer factor**. Lymphokines that can act between different populations of leukocytes are known also as **interleukins**. *See also* **colony-stimulating factor**.

lymphoma any malignant tumour, or condition allied to a tumour, arising from some or all of the cells of lymphoid tissue. The term includes lymphatic leukemias, Hodgkin's disease, and reticuloses.

lymphoreticular tissue *an old general term for* tissues consisting largely of lymphocytes and macrophages.

lymphotoxin *abbr.:* LT; *an alternative name for* **tumour necrosis factor** β; *see also* **lymphokine**.

lyn a protooncogene encoding a **tyrosine kinase** belonging to the *src* family.

Lynx a very powerful free open-source text-only **browser** originally designed for use with character-cell, nongraphics terminals but of great value in rapid access to HTML pages and for checking such pages for compliance with guidelines for visually impaired users.

lyo+ *comb. form* indicating looseness, dispersion, or dissolution.

lyochrome *a former name for* **flavin**.

lyocytosis cell lysis by extracellular digestion, e.g. during insect metamorphosis.

lyoenzyme *an old term for* any enzyme existing in the cell in soluble form.

lyoglycogen *an old name for* the fraction of a sample of glycogen that is extractable by water on a steam bath.

lyolysis *an alternative term for* **solvolysis**.

Lyon hypothesis *(in genetics)* the hypothesis that gene dosage compensation in mammals is accomplished by random inactivation of one of the two X chromosomes in all somatic cells of females. The first inactivation occurs early in embryogenesis. [After Mary Francis Lyon (1925–), British geneticist.]

lyonium ion any cation produced by the protonation of a solvent molecule, e.g. H_3O^+, $CH_3OH_2^+$.

lyophile 1 *(in chemistry)* a **lyophilic** compound. **2** having lyophilic properties.

lyophilic *or* **lyophile** (of a substance or surface) solvent preferring; easily wetted or solvated by the solvent present; the term comprehends **hydrophilic** when the solvent is water. *Compare* **lyophobic**.

lyophilic sol any sol in which the disperse phase has an attraction for the continuous phase.

lyophilization *or* **lyophilisation** strictly, the process of rendering lyophile or lyophilic. The term is now usually specifically applied to freeze-drying in which, by use of a vacuum, water is removed from ice by sublimation. This use bears little relationship to the true meaning of lyophile, except in so far as it maintains proteins subjected to it in a lyophilic state. —**lyophilize** *or* **lyophilise** *vb.*

lyophilizer *or* **lyophiliser** an apparatus for effecting lyophilization; a freeze-drier.

lyophobic *or* **lyophobe** (of a substance or surface) solvent rejecting; resistant to being wetted or solvated by the solvent present; the term comprehends **hydrophobic** when the solvent is water. *Compare* **lyophilic**.

lyophobic sol any sol in which the disperse phase has no attraction for the continuous phase. Lyophobic sols cannot be formed by spontaneous dispersion in the medium.

lyotropic *or* **lyotrope** of or associated with the difference between the internal pressure of a solution compared with that of the solvent caused by the solute.

lyotropic mesophase any mesophase (i.e. **liquid crystal**) whose phase transitions are readily effected by changes in concentration.

lyotropic series *or* **Hofmeister series** a series of ions arranged in order of their **lyotropic** effects, especially their ability to cause precipitation of a lyophilic sol (to salt-out).

lypressin *see* **[8-lysine]vasopressin**.

Lys *symbol for* a residue of the α-amino acid L-lysine (alternative to K).

lys+ *a variant form of* **lyso+** (before a vowel).

lysate any solution or other preparation containing the products of **lysis** of cells. The term is applied especially to lysed bacterial or blood cells.

+lysate *comb. form* denoting a mixture resulting from decomposition, breakdown, or lysis (of or by something indicated); e.g. autolysate, cytolysate, hemolysate, hydrolysate.

Lys-bradykinin *abbr. for* lysylbradykinin (i.e. kallidin); *see* **bradykinin**.

lyse to effect **lysis** (of a cell, etc.); to undergo lysis.

+lyse *comb. form* denoting decompose, breakdown, or lyse (of or by something indicated); e.g. autolyse, cytolyse, electrolyse, hemolyse, hydrolyse.

lysenin any of a family of pore-forming peptidic toxins with about 300 amino acid residues, permitting passage of small solutes, including water. They are produced by *Eisenia foetida* and are classified in the TC system under number 1.A.43.

lysergamide lysergic acid amide. All known **ergot alkaloids** are derivatives of D-lysergamide, the simplest being **ergonovine**.

lysergic acid 9,10-didehydro-6-methylargoline-8β-carboxylic acid; an alkaloid obtained from **ergot**; it is a tetracyclic indole derivative. Lysergic acid diethylamide (*abbr.*: LSD) is a powerful hallucinogenic substance.

lysigeny *(in botany)* the disintegration of cells that occurs when new structures (e.g. aerenchyma or oil glands) are differentiating.

lysin any agent, such as an antibody, enzyme, toxin, etc., that is able to lyse cells.

+lysin *comb. form* denoting an (indicated) agent that effects, or can effect, lysis, specifically the lysis of cells (cytolysis).

lysinate 1 lysine anion; the anion, $H_2N–[CH_2]_4– CH(NH_2)–COO^-$, derived from **lysine**. **2** any salt containing lysine anion. **3** any ester of lysine.

lysine the trivial name for α,ε-diaminocaproic acid; 2,6-diaminohexanoic acid; $H_2N–[CH_2]_4–CH(NH_2)–COOH$; a chiral α-amino acid. L-Lysine (*symbol*: K *or* Lys), (*S*)-2,6-diaminohexanoic acid, is a coded amino acid found in peptide linkage in proteins; codon: AAA or AAG. In mammals, it is an essential dietary amino acid, and is ketogenic. D-Lysine (*symbol*: DLys *or* D-Lys), (*R*)-2,6-diaminohexanoic acid, is a substrate for **lysine racemase** and D-lysine 5,6-aminomutase. The conversion of D-lysine to L-pipecolic acid, and of the latter to L-lysine, occurs in *Nicotiana glauca* and in *Neurospora crassa*.

L-lysine

D-lysine 5,6-aminomutase EC 5.4.3.4; *other name*: D-α-lysine mutase; an enzyme that catalyses the reaction:

D-lysine = 2,5-diaminohexanoate;

cobalamin is a cofactor.

lysine(arginine) carboxypeptidase EC 3.4.17.3; *other names*: carboxypeptidase N; arginine carboxypeptidase; kininase I; anaphylotoxin inactivator; an enzyme that hydrolyses peptidyl-L-lysine (or L-arginine) to peptide and L-lysine (or L-arginine). It inactivates bradykinin complement components, C3a and C5a, by removal of the C-terminal Arg.

lysine (base) a lysine residue when the ε-amino group is unprotonated.

lysine hydroxylase *see* **procollagen-lysine 5-dioxygenase**.

lysine 6-monooxygenase *see* **monooxygenase**.

lysine-2-oxoglutarate reductase *an alternative name for* **saccharopine dehydrogenase (NAD⁺, L-lysine forming)**

lysine racemase EC 5.1.1.5; an enzyme catalysing the reaction:

L-lysine = D-lysine.

lysine tyrosylquinone *see* **LTQ**.

lysine vasopressin *see* **[8-lysine]vasopressin**.

[8-lysine]vasopressin *abbr.*: [Lys⁸]vasopressin; *nonproprietary name*: lypressin; the molecular form of **vasopressin** in which the variable eighth position in its amino-acid sequence is occupied by a lysine residue. This is the form of the hormone present in pigs and a few other related mammals. It is sometimes incorrectly designated as lysine vasopressin.

lysinium 1 *the usual term for* lysinium(1+); the monocation of **lysine**. In theory, the term denotes any ion, or mixture of ions, formed from lysine, and having a net charge of +1, although the species $H_3N^+–[CH_2]_4–CH(NH_3^+)–COO^-$ generally predominates in practice. **2** *the systematic name for* lysinium(2+); the dication of lysine.

lysino 1 N^2-**lysino** the alkylamino group, $H_2N–[CH_2]_4–CH(COOH)–NH–$, derived from lysine by loss of a hydrogen atom from its α-amino group. **2** N^6-**lysino** the alkylamino group, $HOOC–CH(NH_2)–[CH_2]_4–NH–$, derived from lysine by loss of a hydrogen atom from its ε-amino group.

lysinuric protein intolerance a genetically determined metabolic defect of humans in which the ingestion of relatively large amounts of lysine (in protein) causes hyperlysinemia, hyperammonemia, and coma. The defect is probably in a cationic amino-acid transporter.

lysis 1 *(in chemistry)* the breaking down of a substance by the splitting of its molecules into two parts; the rupture of a covalent bond. **2** *(in biology)* the disintegration of cells or cell organelles by rupture of their outer membranes. **3** *(in medicine)* the gradual diminution in the severity of the symptoms of a disease; remission.

+lysis *comb. form* denoting breaking down, decomposition, disintegration, or dissolution. The first element of the word may indicate: (1) the agent, e.g. electrolysis, hydrogenolysis, biolysis; (2) the substance or object affected, e.g. glycolysis, lipolysis; or (3) some other characteristic, e.g. autolysis, catalysis, heterolysis, homolysis. —**+lytic** *adj*.

lyso+ *or (before a vowel)* **lys+ 1** *comb. form* indicating a dissolving or loosening; denoting lysis or the products of lysis. **2** *prefix* (to the name of a phospholipid) denoting removal of one of its two acyl groups; a locant may be used to designate the site of (hydro)lysis, e.g. 2-lyso designates deacylation at position 2. The term originated from the fact that such compounds are hemolytic, but it is now used to indicate a limited deacylation.

lysochrome any substance that will dissolve in lipids and colour them.

lysogen 1 any **lysogenic bacterium**. **2** any antigen that stimulates **lysin** production.

lysogenesis 1 *an alternative term for* **lysogeny. 2** the production of a lysin or lysis.

lysogenic 1 pertaining to, or capable of producing or undergoing **lysis** (def. 1). **2** of, or relating to **lysogeny.** —**lysogenicity** *n.*

lysogenic bacterium any bacterium carrying a temperate bacteriophage as a **prophage** integrated into the bacterial genome. It has the potential to produce and release infective bacteriophages as a stable, heritable trait, which may become activated spontaneously or in response to certain stimuli. Moreover, it is immune to lytic infection by the same or closely related phages.

lysogenic conversion *see* **bacteriophage conversion.**

lysogenic virus any virus capable of becoming a **prophage.**

lysogenize *or* **lysogenise** to convert a nonlysogenic bacterial strain into a lysogenic strain by infection with a temperate bacteriophage, with resultant incorporation of the phage into the bacterial genome. —**lysogenization** *or* **lysogenisation** *n.*

lysogeny *or* **lysogenesis** the state created in a bacterial cell when the prophage of a temperate bacteriophage is incorporated into the bacterial genome as a stable, heritable character. Such a cell can be activated to produce and release mature phage, with accompanying cell lysis. *See also* **lysogenic bacterium.**

Lys(5-OH) *symbol for* a residue of the α-amino acid 5-hydroxy-L-lysine (less-preferred alternative to Hyl or 5Hyl).

Lysol *the proprietary name for* a disinfectant consisting of a mixture of *o*-, *m*-, and *p*-cresols solubilized with an excess of a potassium soap.

lysolecithin *a former name for* **lysophosphatidylcholine.**

lysolecithin acyltransferase *see* **1-acylglycerophosphocholine O-acyltransferase.**

lysophosphatidic acid any **phosphatidic acid** deacylated at positions 1 or 2.

lysophosphatidyl any **phosphatidyl** group deacylated at positions 1 or 2.

lysophosphatidylcholine *or (formerly)* **lysolecithin** any **phosphatidylcholine** (i.e. lecithin) deacylated at positions 1 or 2.

lysophospholipid any **phospholipid** deacylated at positions 1 or 2.

lysoplasmenic acid any derivative of *sn*-glycero-3-phosphate that has an *O*-(1-alkenyl) residue on carbon-1, the hydroxyl group in position 2 being unsubstituted.

lysoplasmenyl any **plasmenyl** group in which the hydroxyl group in position 2 is unsubstituted.

lysosomal carboxypeptidase B *see* **cysteine-type carboxypeptidase.**

lysosomal carboxypeptidase C *see* **lysine pro-X carboxypeptidase.**

lysosomal Pro-X carboxypeptidase EC 3.4.16.2; *other names*: proline carboxypeptidase; angiotensinase C; lysosomal carboxypeptidase C; an enzyme that catalyses the cleavage of a Pro-|-Xaa bond to release a C-terminal amino acid.

lysosomal storage disease any of a group of hereditary diseases, with widely varying manifestations, in which there is deficiency of a single lysosomal enzyme associated with abnormal deposits (of substrate) within membrane-bound vesicles. The group is subdivided into the following classes: type II **glycogen disease, mucopolysaccharidoses**, sphingolipidoses, **mucolipidoses**, and a few others.

lysosome any of a group of related cytoplasmic, membrane-bound organelles that are found in most animal cells and that contain a variety of hydrolases, most of which have their maximal activities in the pH range 5–6. The contained enzymes display latency if properly isolated. About 40 different lysosomal hydrolases are known and lysosomes have a great variety of morphologies and functions. —**lysosomal** *adj.*

lysostaphin EC 3.5.1.–; an enzyme that hydrolyses polyglycine links in peptidoglycans. Zinc is a cofactor. Its physiological role is likely to be in the remodelling of the bacterial sacculus during growth. The C terminus has some similarity to **autolysin.**

lysozyme EC 3.2.1.17; *systematic name*: peptidoglycan N-acetylmuramoylhydrolase; *other name*: muramidase. An enzyme that hydrolyses β-1,4-linkages between N-acetylmuramic acid and 2-acetamido-2-deoxy-D-glucose residues in peptidoglycan heteropolymers (mucopolysaccharide or muropeptide) of prokaryote cell walls. It occurs in tears, other exocrine secretions, and in very large amounts in the white of eggs of birds. The antibacterial action of lysozyme depends on the cleavage of the $\beta(1->4)$glycosidic linkage between alternating units of N-acetylmuramic acid and N-acetylglucosamine that form long-chain mucopolysaccharides in the cell walls of bacteria such as *Micrococcus luteus*. There are a number of types of lysozyme, designated A–G. Lysozyme was the first enzyme whose 3-D structure was determined.

[Lys⁸]vasopressin *abbr. for* [8-lysine]vasopressin.

lysyl the acyl group, $H_2N–[CH_2]_4–CH(NH_2)–CO–$, derived from **lysine.**

lysylbradykinin *or* **Lys-bradykinin** *an alternative name for* kallidin; *see* **bradykinin.**

lysyl endopeptidase EC 3.4.21.50; *other names*: Achromobacter proteinase I; lysyl bond specific proteinase. An extracellular bacterial serine proteinase that preferentially cleaves at Lys-|-Xaa, including Lys-|-Pro.

lysyl oxidase EC 1.4.3.13; *recommended name*: protein-lysine 6-oxidase; *systematic name*: protein-L-lysine:oxygen 6-oxidoreductase (deaminating). A quinoprotein enzyme that catalyses a reaction between peptidyl-L-lysyl-peptide, H_2O and dioxygen to form peptidyl-allysyl-peptide, NH_3, and H_2O_2. It contains lysine tyrosylquinone as a cofactor. *See also* **allysine.**

+lyte *comb. form* indicating a substance that can be decomposed, broken down, or lysed.

lytic 1 of, pertaining to, or resulting from **lysis. 2** of, or pertaining to a **lysin.**

+lytic *see* **+lysis.**

lyticase *a jargon term for* a combination of enzymes, with minimal amounts of nucleases, that displays two major activities: β-1,3-glucanase and a highly specific alkaline protease. It is prepared from *Oerskovia xanthineolytica* and is useful in removing cell walls from yeasts and fungi, causing extremely rapid cell lysis.

lytic cycle the sequence of events occurring when a sensitive bacterium is infected by a virulent bacteriophage. The phage diverts the metabolism of the cell to the exclusive production of progeny phage, resulting in the lysis of the host cell with liberation of phage.

α-lytic endopeptidase EC 3.4.21.12; a bacterial serine proteinase, a homologue of chymotrypsin, that hydrolyses proteins, especially at bonds adjacent to L-Ala and L-Val residues in bacterial cell walls, elastin, and other proteins.

lytic virus any virus including bacteriophages whose multiplication in a host cell leads to lysis of that cell.

Lyx *symbol for* a residue (or sometimes a molecule) of the aldopentose **lyxose.**

lyxo+ *prefix (in chemical nomenclature)* indicating a particular configuration of a set of three (usually) contiguous >CHOH groups, as in the acyclic forms of D- or L-lyxose. *See* **monosaccharide.**

lyxose *symbol*: Lyx; *a trivial name for* the aldopentose, *lyxo*-pentose; the C-2 epimer of **xylose**, rare in nature.

Mm

m *symbol for* **1** metre. **2** milli+ (SI prefix denoting 10^{-3} times).

m+ *prefix (in peptide hormone nomenclature)* denoting mouse, murine; derived, or as if derived, from a mouse.

m$_r$ *symbol for* relative electrophoretic mobility.

m *symbol for* **1** mass; the various associated terms are distinguished by subscripts: m_e, *symbol for* electron rest mass; m_n, *symbol for* neutron rest mass; m_p, *symbol for* proton rest mass; m_u, *symbol for* atomic mass constant. **2** molality (alternative to *b*). **3** (bold italic) magnetic moment.

m- *prefix (in chemical nomenclature)* denoting **meta-**. *Compare* **o-**, **p-**.

M *symbol for* **1** mega+ (SI prefix denoting 10^6 times). **2** a residue of the α-amino acid L-methionine (alternative to Met). **3** a residue of an incompletely specified base in a nucleic-acid sequence that may be either adenine or cytosine. **4** mesomeric effect. **5** a metal (in chemical formulae). **6** molar (concentration), i.e. moles per litre (alternative to M or *M*; mol dm^{-3} is recommended). **6** 6-mercaptopurine ribonucleoside, 6-thioinosine.

M *symbol for* molar (concentration), i.e. moles per litre (alternative to M or *M*; mol dm^{-3} is the recommended form).

α$_2$-M *abbr. for* α$_2$-macroglobulin.

M *symbol for* **1** mass, especially molar mass; related terms are designated by subscripts: M_m, *symbol for* mass-average molar mass (*see* **average molar mass**); M_n, *symbol for* number-average molar mass (*see* **average molar mass**); M_o, *symbol for* mean relative residue mass; M_r, *symbol for* relative molecular mass; M_w, symbol for weight-average molar mass; M_z, *symbol for* Z-average molar mass (*see* **average molar mass**). **2** molar (concentration), i.e. moles per litre (alternative to M or M; no longer recommended; use mol dm^{-3}). **3** (or M_v) luminous exitance. **4** (or M_e) radiant exitance. **5** (bold italic) magnetization. **6** (bold italic) moment of a force.

M. *abbr. for* the **modification enzyme** associated with a type 2 **restriction endonuclease**. For example, M.*Hin*dIII is the enzyme associated with *Hin*dIII.

M9 a solution of inorganic salts used as the basis of minimal growth media for bacteria.

MAA *abbr. for* macroaggregated (human serum) albumin.

Mab *abbr. for* monoclonal antibody.

Mabthera *a proprietary name for* **Rituximab**.

MAC *abbr. for* **1** membrane attack complex. **MAC 1** is a cell-surface glycoprotein of monocytes, macrophages, and granulocytes that is implicated in adhesive interactions of these cells and in mediating the uptake of complement-coated particles. It probably recognizes the RGD peptide in **complement** component C3b. It is a heterodimer of α_M and β_2 chains (*see* **integrin**). **2** mammalian artificial chromosome.

machine address a binary number unique to a computer (more exactly its network hardware) that is used by a **DNS** (domain name server). It should be transparent to the average end-user. The number is usually represented as a colon-delimited hexadecimal, for example 00:01:02:1a:45:66.

machine language *see* **programming language**.

MACIF *abbr. for* membrane attack complex inhibition factor.

Mac-IP *abbr. for* Mac-inhibitory protein (i.e. **CD59**).

mackerel any marine fish of the subfamily Scombroidei. These fish are of importance as food fish, and their flesh is rich in oils with a relatively high content of *n*-3 **fatty acids**; typical composition (major fatty acids, expressed as percentage of total fatty acids): 14:0, 8%; 16:0, 16%; 18:0, 2%; 16:1 (*n*-7) 8%; 18:1 (*n*-9), 8%; 18:1 (*n*-7), 4%; 18:4 (*n*-3), 2%; 20:1 (*n*-9), 11%; 20:5 (*n*-3), 8%; 22:1 (*n*-11), 13%; 22:6 (*n*-3), 8%.

maconha *see* **cannabis**.

macro+ *or (before a vowel)* **macr+** *comb. form* **1** denoting large, long, or great in size or duration; large-scale. **2** denoting larger than normal, overdeveloped. *Compare* **micro+**.

macroaggregate any macromolecular aggregate large enough to be visible to the unassisted eye.

macrocycle any **macrocyclic** compound.

macrocyclic 1 *(in chemistry)* describing any compound containing a ring structure consisting of a large number (usually 15 or more) atoms. **2** *(in biology)* having a long life cycle.

macrocytic anemia anemia in which the circulating erythrocytes are larger than normal. It is often due to deficiency of vitamin B$_{12}$ (cobalamin) or folic acid.

macroglobulin any plasma globulin of >400 kDa. Macroglobulins include **immunoglobulin M**, α$_2$-**macroglobulin** and many lipoproteins.

α$_2$-macroglobulin *abbr.*: α$_2$-M; a glycoprotein found in vertebrate plasma that strongly inhibits proteases of all classes. In humans, it constitutes about one third of the α$_2$-globulins; its synthesis is enhanced in hypoalbuminemia, when it constitutes a major proportion of the increased α$_2$-globulin band on electrophoresis strips. Human α$_2$-M is a 726 kDa molecule comprising two pairs of disulfide-linked, identical 185 kDa subunits, which also contain heterodetic peptide rings containing reactive thiol ester linkages, each formed between the γ-carboxyl group of a glutamic residue and the thiol group of a cysteine residue in the same peptide chain. It can inhibit all four classes of **proteinase** by a mechanism involving a **bait region**. This contains specific sites, cleavage of which induces a conformational change that results in trapping of the protease: following cleavage in the bait region, a thiol-ester bond is hydrolysed, and mediates the covalent binding of the protein to the protease; subsequently, ε-amino groups of the protease react with the thiol-ester linkages in the inhibitor to form stable amide links. The entrapped protease is then able to act on substrates of low molar mass only.

macroglobulinemia *or (esp. Brit.)* **macroglobulinaemia** a condition in humans in which there is an increase in the concentration of macroglobulins, especially immunoglobulin M, in the blood. It is often associated with lymphocyte tumours.

macro histone 2A *abbr.*: macro H2Aa; the counterpart of histone H2A that replaces this in certain nucleosomes, with an implied role in nucleosome positioning.

macroion an ion of a macromolecule, especially one that does not pass through a dialysis membrane.

macrolide any antibiotic produced by *Streptomyces* spp. that contains a large lactone ring (*see* **+olide**) with few or no double bonds and no nitrogen atoms, linked to one or more sugar moieties. The macrolides include the carbomycins, the **erythromycins**, oleandomycin, **oligomycins**, and the spiramycins. They are active mainly against Gram-positive bacteria and they inhibit the early stages of protein synthesis.

macroligand any macromolecular ligand.

macrometabolite any metabolite present in a system in reasonably high concentration, especially as opposed to a trace metabolite, i.e. one present in extremely small concentrations.

macromolecule any molecule composed of a very large number of atoms, operationally defined as any molecule of mass greater than about 10 kDa (and ranging up to 10^2 MDa or more) that is incapable of passing through the pores of dialysis tubing as generally used. The term includes nucleic acids and most polysaccharides and proteins. *Compare* **micromolecule**. —**macromolecular** *adj.*

macronutrient any element that is required by living organisms in relatively large quantities for normal growth. *Compare* **micronutrient**.

macrophage any cell of the **mononuclear phagocyte system** that is characterized by its ability to phagocytose foreign particulate and colloidal material. Macrophages occur in connective tissue, liver (Kupffer cells), lung, spleen, lymph nodes, and other tissues. They contain prominent lysosomes, stain with vital dyes, and play an important part in nonspecific immune reactions.

macrophage-activating factor *abbr.*: MAF; *see* **lymphokine**.

macrophage-colony stimulating factor *see* **colony stimulating factor**.

macrophage inflammatory protein *abbr.*: MIP; a cytokine produced by B and T cells, monocytes, mast cells, and fibroblasts. MIP 1α is a chemoattractant for monocytes, T cells, and eosinophils; it also inhibits hematopoietic stem cell production. MIP 1β is a chemoattractant for monocytes and T cells and is an adhesion molecule, binding to $β_1$ (**VLA** family) **integrins**.

macrophage migration enhancement factor *see* **lymphokine**.

macrophage migration inhibitory factor *abbr.*: MIF; *other name*: delayed early response protein 6; a **lymphokine** that is released by primed lymphocytes on incubation with the priming antigen. It inhibits the migration of macrophages out of macrophage-rich tissue such as spleen, and is probably a mediator of macrophage participation in inflammation.

macrophage scavenger receptor 1 a glycoprotein membrane receptor of macrophages and sinusoidal endothelial cells that binds and mediates uptake of oxidized low-density lipoproteins (LDL). It is implicated in deposition of cholesterol into arterial walls during atherogenesis. **2** a membrane receptor of hepatocytes and steroid hormone-producing cells that binds high-density lipoproteins (HDL), selectively removes cholesterol, and mediates its uptake.

macrophage system *see* **mononuclear phagocyte system**.

macropipette (*sometimes*) a pipette of similar construction to a standard plunger-type micropipette but of capacity in the range 0.25–5 mL.

macropore any pore (in a gel or porous solid) whose width is greater than about 50 nm. *Compare* **mesopore**, **micropore**. —**macroporous** *adj*.

macroreticular describing a three-dimensional polymer with a very open structure, the cross links being widely spaced.

macroscopic 1 of a size that is visible to the unaided eye. *Compare* **microscopic** (def. 1). **2** concerned with large units. *Compare* **microscopic** (def. 2). **3** relating to the behaviour of atoms or molecules in bulk. *Compare* **microscopic** (def. 3). —**macroscopically** *adv*.

macroscopic equilibrium constant any constant describing the overall equilibrium between two chemical entities that may be interconverted through two or more alternative intermediates. *Compare* **microscopic equilibrium constant**.

macrosolute *an operational term describing* any solute of a size larger than will pass through a membrane of a specified pore size or permeability limit.

macrotetrolide or **macrotetralide** any antibiotic that contains a 32-membered ring built up of four hydroxy-acid residues and containing four ether and four ester bonds; the group includes **nonactin** and related antibiotics. Some similar compounds have an open-chain structure, e.g. **monensin** and **nigericin**.

macula adherens *an alternative term for* **desmosome**.

MAD 1 a basic helix-loop-helix transcription factor capable of forming dimers with MYC and MAX. **2** *abbr. for* **multi-wavelength anomalous dispersion**.

MAD2 a protein in yeast that is an essential component of the spindle checkpoint during mitosis and determines the fidelity of chromosome distribution. Mutation leads to a large increase in the rate of chromosome misaggregation during meiosis.

MadCAM *abbr. for* mucosal addressin cell adhesion molecule; MadCAM 1, M_r 58–66 kDa, is a member of the immunoglobulin superfamily of **adhesion molecules**. It is constitutively expressed on endothelial cells of post-capillary venules. It mediates lymphocyte homing to mucosal lymphoid organs and lamina propria venules, and binds L-**selectin** and **integrin** $α_4β_7$.

mad cow disease *an informal name for* **bovine spongiform encephalopathy**.

madder the ground root of *Rubia tinctoria*, used as a plant dyestuff; the principal component is **ruberythric acid**.

MADS box a highly conserved amino acid sequence (≈56 amino acids) present in the floral 'homeotic' transcription factors APETALA1, APETALA3, PISTILLATA, and AGAMOUS of *Arabidopsis thaliana* and in many other proteins expressed in flowers of other plant species. *See* **ABC model**.

MAF *abbr. for* macrophage-activating factor; *see* **lymphokine**.

magainin any antibiotic peptide from *Xenopus* skin whose pore-forming activity (*see* **PFP**) results in the permeabilization of bacterial membranes. Magainins have a wide spectrum of action against bacteria, protozoa, and fungi. Magainin I has the sequence

GIGKFLHSAGKFGKAFVGEIMKS;

magainin II is (Lys^{10},Asn^{22})-magainin I. An analogue with greater antibiotic potency is $(Ala^{8,13,18})$-magainin II amide. Magainins are classified in the TC system under number 1.C.16. *Compare* **dermaseptin**.

MAGE *see* **visualization software**.

magic spot I *see* **magic spot (compound)**, **ppGpp**.

magic spot II *see* **magic spot (compound)**, **pppGpp**.

magic spot (compound) *abbr.*: MS; either of the guanosine polyphosphates produced when stringent strains of *Escherichia coli* are starved of an amino acid required for protein synthesis. MS I is guanosine 5′-diphosphate 2′-(or 3′)-diphosphate (*symbol*: ppGpp); MS II is guanosine 5′-triphosphate 2′-(or 3′)-diphosphate (*symbol*: pppGpp).

magnesium *symbol*: Mg; a metallic element – an alkaline earth metal – of group 2 of the IUPAC **periodic table**; atomic number 12; relative atomic mass 24.305. It occurs naturally only in the combined state and is a mixture of stable nuclides of mass 24, 25, and 26. Magnesium is one of the most abundant elements of the Earth's crust (2.1%) and an essential component of all living material. It forms a divalent cation, Mg^{2+}, which is a component of chlorophyll, occurs in bone, and is an essential cofactor for many enzymes, including the majority of enzymes utilizing ATP. It is a major biological cation, the fourth most abundant cation in the human, much of it being in the skeleton; the range in normal human plasma is 0.8–1.2 mmol L^{-1}. An excretion mechanism in the kidney is partly regulated by parathyroid hormone. *See also* **magnesium-28**.

magnesium-28 the artificial radioactive nuclide of magnesium, $^{28}_{12}Mg$, of half-life 21.2 h. It emits a moderate-energy beta particle (i.e. electron) (0.46 MeV max.) and gamma radiation of various energies (1.35, 0.95, 0.40, and 0031 MeV).

magnesium porphyrin *a general term for* **chlorophyll**.

magnet any body that can produce an appreciable **magnetic field** external to itself. Magnets may be either permanent or temporary. —**magnetic** *adj*.

magnetic beads microscopic synthetic polymer beads incorporating a magnetite core and chemically modified to enable the covalent attachment of proteins such as antibodies. Beads coated with streptavidin will bind proteins or DNA labelled with biotin. Such beads can be concentrated from suspension by exposing them to a magnetic field thereby providing an alternative to centrifugation.

magnetic circular dichroism spectroscopy measurement of circular dichroism of a coloured material which is induced by a magnetic field applied parallel to the direction of the measuring light beam. Even achiral materials exhibit MCD (the Faraday effect). Variable-temperature MCD is used for identifying and assigning electronic transitions originating from paramagnetic *chromophores*.

magnetic constant *symbol*: $μ_0$; *an alternative name for* **permeability of vacuum**.

magnetic dipole moment *symbol*: m or $μ$ (italic bold); a vector quantity indicating the strength of a magnet: the product of a magnet's dipole moment and the ambient **magnetic flux density** gives the torque on the magnet. The SI unit is the A m^2 (ampere square metre = joule per tesla (J T^{-1})).

magnetic field 1 a field of force existing in the neighbourhood of either a permanent magnet or a circuit carrying an electric current. Another magnet will experience a couple when placed in a magnetic field. **2** *an alternative name for* **magnetic field strength**.

magnetic field strength or **magnetic field** or (*formerly*) **magnetic intensity** *symbol*: H (italic bold); a vector quantity, the magnetizing force at any point in a medium. It is measured by a couple that would act on a small magnet of unit moment placed at that point with its axis at right angles to the direction of the magnetic field. It is the **magnetic flux density**, B, divided by the **permeability**, $μ$; i.e. $H = B/μ$. The SI derived unit $H =$ A m^{-1}.

magnetic flux *symbol*: $Φ$; a measure of the total strength of a magnetic field, defined as the scalar product of the magnetic flux density and the area. The SI derived unit is the **weber**.

magnetic flux density or **magnetic induction** *symbol*: B (italic bold); a pseudovector quantity, the **magnetic flux** passing through unit area of a magnetic field orthogonal to the magnetic force. The SI derived unit is the **tesla**.

magnetic force the force exerted by a magnetic field on a magnetic pole or a moving electric charge. It may be either attractive or repulsive.

magnetic immunoassay *abbr.*: MAIA; a variant of **radioimmunoassay** in which the antibody is bound to a particulate, inert, magnetic support material – e.g. cellulose containing embedded grains of iron oxide, Fe_3O_4 – and separation of antigen–antibody complexes from free antigen is effected with the aid of a powerful magnet.

magnetic induction *an alternative term for* **magnetic flux density**.

magnetic intensity *a former name for* **magnetic field strength**.

magnetic lens *an alternative name for* a magnetic **electron lens**.

magnetic moment 1 *symbol*: μ; a property of a charged particle arising from its spin; $\mu = eL/2mc$, where c is the velocity of light, L is the angular momentum of the particle, e is its charge, and m is its mass. The SI unit is joule per tesla. The **electron magnetic moment** (*symbol*: μ_e) is equal to $9.284\ 7701 \times 10^{-24}$ J T^{-1}. **The proton magnetic moment** (*symbol*: μ_p) is equal to $1.410\ 607\ 61 \times 10^{-26}$ J T^{-1}. **2** *an alternative name for* **magnetic dipole moment**.

magnetic permeability *see* **permeability, permeability of vacuum**.

magnetic pole either of the two regions of a magnet into which the lines of force of the magnetic field converge or from which they diverge. They are designated north (or north-seeking) and south (or south-seeking) according to whether they are attracted to the Earth's north or south pole.

magnetic resonance *see* **electron-spin resonance spectroscopy, nuclear magnetic resonance, electron paramagnetic resonance spectroscopy**.

magnetic resonance imaging *abbr.*: MRI; an imaging technique that uses **nuclear magnetic resonance** (NMR) to investigate the state of tissues in the intact body; it is so-called to avoid the use of the term 'nuclear' in the treatment of patients. It has the advantage that it is non-invasive. A magnet large enough to surround a part of the body is used in conjunction with computer analysis to provide an image resulting from the NMR signals. The technique can distinguish extracellular water from that inside cells on the basis of protein spin relaxation rates; this distinction is used to reveal certain types of tissue damage.

magnetic shielding 1 protection from the effects of an external magnetic field. **2** (in **nuclear magnetic resonance spectroscopy**) the reduction of the applied magnetic field at an atomic nucleus (as compared with that applied to the whole sample) brought about by the electron cloud of the molecule.

magnetic stirrer any apparatus for stirring a liquid in which a follower – usually a small bar magnet coated in glass or plastic – may be caused to oscillate or rotate within the liquid under the influence of a more powerful magnet outside the container. Motion of the external magnet is controlled by a variable-speed electric motor. The term is sometimes applied specifically to the follower.

magnetic susceptibility *symbol*: X or (*sometimes*) X_m or K; the ratio of the magnetization, M, produced in a substance to the magnetic field strength, H, applied to it, i.e.

$$X = M/H = \mu_r - 1,$$

where μ_r is the relative permeability of a vacuum. Paramagnetic materials have positive magnetic susceptibilities, diamagnetic materials have negative magnetic susceptibilities. (Note: the symbol μ_m is reserved for molar magnetic susceptibility.)

magnetism 1 the property of producing, or being affected by, magnetic flux. **2** the branch of physics concerned with magnets and magnetic phenomena.

magnetite *or* **magnetic iron oxide** *or* **black iron oxide** ferrosoferric oxide; iron(II)–diiron(III) oxide; Fe_3O_4; a strongly magnetic black mineral form of iron oxide.

magnetization *or* **magnetisation 1** the act or process of magnetizing. **2** *symbol*: M (italic bold); the magnetic moment per unit volume of a magnetized body; i.e. the difference between the ratio of the **magnetic flux** to the **permeability of vacuum** and the **magnetic field strength**. The SI unit is the ampere per metre.

magnetization-transfer nuclear magnetic resonance spectroscopy a procedure in **nuclear magnetic resonance spectroscopy** in which one of the resonances is saturated by a strong radiofrequency

field, and the changes in the other resonances are observed. It is useful in measuring the rates of chemical exchange processes.

magnetize *or* **magnetise** to induce a magnetic flux density in (something); to cause (something) to exhibit **magnetism** (def. 1).

magnetogyric ratio *see* **gyromagnetic ratio**.

magneton a unit of magnetic moment. The **Bohr magneton** (*symbol*: μ_B) refers to an electron of mass m_e; the **nuclear magneton** (*symbol*: μ_N) refers to a proton of mass m_p. They are related by the equation: $\mu_N = (m_e/m_p)\mu_B$.

magnetophoresis the movement of magnetizable particles through a fluid under the action of a magnetic field. It is useful for separating of dispersed cells to which ferromagnetic microspheres have been conjugated. —**magnetophoretic** *adj*.

magnetotactic referring to bacterial cells that align along a magnetic field. They contain crystals of magnetite.

magnification *(in microscopy)* the apparent linear enlargement of an object when viewed through a lens, system of lenses (as in a microscope, telescope, etc.), or other instrument. It is given by the ratio of the apparent diameter of the object, as seen using the lens, to its real diameter, as seen unaided.

Magnogel *a proprietary name for* a bead-form support material useful in immunosorption and related procedures. It is similar to **Ultrogel** but contains embedded particles of iron oxide, Fe_3O_4, that enable the beads to be drawn to and held on the walls of the container by magnetism.

MAIA *abbr. for* magnetic immunoassay.

Maillard reaction a nonenzymic reaction in which a reducing sugar combines at its anomeric carbon atom with an amino group of an amino acid, peptide, or protein to form first a Schiff base and then, by an **Amadori rearrangement**, the corresponding ketoamine. It is the first of a series of reactions that may occur during the processing and/or storage of food, resulting in a loss of nutritive value and the formation of a brown coloration; hence it is also known as the **browning reaction**. It may also be the reaction leading to the formation of hemoglobin A_{1c} in the erythrocyte. [After Louis Camille Maillard (1878–1936).]

main protease the protease that cleaves the polyprotein encoded in the RNA genome of coronaviruses. The human virus's main protease (305 amino acids) contains two six-stranded beta barrels and a cluster of five alpha helices. It cleaves on the carboxyl side of glutamine at about a dozen sites that contain Leu–Gln–Ala/Ser/Gly sequences.

major facilitator superfamily *abbr.* MFS, one of the two largest families of membrane transporters (the other being the ABC transporters). Transporters of this family are present ubiquitously in bacteria, archaea, and eukarya. They transport a wide range of metabolites and drugs, including sugars, amino acids, nucleosides, and a large variety of anions and cations. All permeases of the MFS possess 12–14 putative transmembrane α-helical regions, and exhibit a well-conserved motif present between transmembrane regions 2 and 3.

major groove the larger and deeper of the two spiral grooves on the surface of a double-helical b-DNA molecule. *Compare* **minor groove**.

major histocompatibility complex *abbr.*: MHC; a complex of genetic loci occurring in higher vertebrates that encodes a family of cellular antigens, known in mice as **histocompatibility-2 antigens** (*abbr.*: H-2 antigens; *see* **histocompatibility antigen**) and in humans as human leukocyte-associated antigens (*abbr.*: HLA antigens; *see* **HLA histocompatibility system**). The MHC antigens are cell-surface glycoproteins and may be divided into two classes, designated I and II. In transplantation reactions, cytotoxic T lymphocytes respond mainly against foreign MHC glycoproteins of class I together with antigen, while the response against class II and antigen is mainly by helper T lymphocytes.

Class I MHC antigens (*or* transplantation antigens) are on the surface of most nucleated somatic cells and enable the immune system to distinguish self from foreign cells. They are encoded by three separate loci (H-2K, H-2D, and H-2L in mice; HLA-A, HLA-B, and HLA-C in humans). The antigenic activity resides in a transmembrane polypeptide chain, the α subunit, of ≈45 kDa (roughly 345 amino acids), which has a short hydrophilic C-terminal segment inside the cell and a large glycosylated N-terminal segment. The latter is folded into three separate domains (two of which contain an intrachain disulfide bond) and exposed to the cell exterior, where it is noncovalently linked with the β subunit (of 11.5 kDa and also called β_2 microglobulin), which is homologous with a single Ig-like domain (*see* **immunoglobulin fold**). In humans, these proteins are also referred to as **HLA class I**.

Class II MHC antigens are found on the surface of most B lymphocytes, some T lymphocytes, and some macrophages and macrophage-like cells. They are encoded by murine I-region (*abbr.*: Ir) genes and human HLA-D genes, and contain a ≈33 kDa α chain and a 28 kDa β chain; the C-terminal regions of both subunits have transmembrane and intracellular domains. Celiac disease is due to an allelic variant in one class-II gene.

majority-rule consensus tree (*in phylogenetics*) a tree in which groups of taxa that appear in 50% or more of the component trees are shown. *Compare* **consensus tree**.

major tranquillizer *a term, now deprecated, for* any drug used to treat major psychosis, e.g. schizophrenia or mania; the term **neuroleptic** is now recommended.

MAK *abbr. for* methylated albumin-kieselguhr; a substance used as a medium for chromatography.

Makefile a file read by the UNIX utility make. It includes instructions for the computer and is used largely, but not exclusively, for compiling programs from source code. The advantage of make is that instructions on files that have not been altered since the last use of make are not repeated.

Mal 1 (when linked to one other group) *symbol for* the univalent maleyl group. **2** (when linked to two other groups) *symbol for* the bivalent maleoyl group.

malate 1 *a trivial name for* hydroxysuccinate; hydroxybutanedioate; $^-$OOC–CH$_2$–CH(OH)–COO$^-$; the dianion of malic acid, a chiral hydroxydicarboxylic acid. The (+)-enantiomer, (*S*)-malate (also designated L-malate), is an important intermediate in metabolism as a component of both the tricarboxylic-acid cycle and the glyoxylate cycle. **2** any mixture of free malic acid and its mono- and dianions. **3** any salt or ester of malic acid.

(*S*)-malate

malate–aspartate cycle *or* **malate cycle** an intracellular metabolic cycle, also known as the malate–aspartate shuttle or malate shuttle, that can transfer reducing equivalents from the cytosol to the mitochondria in liver and other tissues. Oxaloacetate formed inside the mitochondria is there converted by a transaminase into aspartate, which passes through the mitochondrial membrane on a specific carrier; a similar transaminase in the cytosol converts this aspartate back into oxaloacetate, which then reacts with cytosolic NADH, in a reaction catalysed by cytosolic **malate dehydrogenase**, to form (*S*)-malate and NAD$^+$. The malate thereupon passes back into the mitochondria, where it is dehydrogenated again to oxaloacetate, with concomitant reduction of mitochondrial NAD$^+$ to NADH and completion of the cycle.

malate dehydrogenase 1 *abbr.*: MDH *or* (*in clinical chemistry*) MD; EC 1.1.1.37; *recommended name*: malate dehydrogenase; *systematic name*: (*S*)-malate:NAD$^+$ oxidoreductase; *other name*: malic dehydrogenase. A widespread enzyme that interconverts malate and oxaloacetate according to the reaction:

$$(S)\text{-malate} + NAD^+ = \text{oxaloacetate} + NADH.$$

In yeast, e.g., there are three isoenzymes: cytosolic, glyoxysomal, and of the mitochondrial matrix. **2** EC 1.1.1.82; *recommended name*: malate dehydrogenase (NADP$^+$); *systematic name*: (*S*)-malate:NADP$^+$ oxidoreductase. A chloroplast enzyme required for both C$_3$ and C$_4$ pathways (*see* **reductive pentose phosphate cycle**, **Hatch–Slack pathway**) to allow plants to circumvent the problem of photorespiration. It catalyses the reaction:

$$(S)\text{-malate} + NADP^+ = \text{oxaloacetate} + NADPH.$$

See also **malate dehydrogenase (decarboxylating)**.

malate dehydrogenase (decarboxylating) any enzyme that brings about the oxidation of malate with simultaneous decarboxylation of the product to pyruvate. Such enzymes include: (1) Malate dehydrogenase (oxaloacetate decarboxylating), EC 1.1.1.38; *systematic name*: (*S*)-malate:NAD$^+$ oxidoreductase (oxaloacetate decarboxylating); *other names*: malic enzyme; pyruvic–malic carboxylase; malate oxidoreductase; it catalyses the reversible reaction:

$$(S)\text{-malate} + NAD^+ = \text{pyruvate} + CO_2 + NADH.$$

(2) Malate dehydrogenase (decarboxylating), EC 1.1.1.39; *systematic name*: (*S*)-malate:NAD$^+$ oxidoreductase (decarboxylating); *other names*: malic enzyme; pyruvic–malic carboxylase; it catalyses a reaction between (*S*)-malate and NAD$^+$ to form pyruvate, CO$_2$, and NADH. (3) Malate dehydrogenase (oxaloacetate decarboxylating) (NADP$^+$), EC 1.1.1.40; *systematic name*: (*S*)-malate:NADP$^+$ oxidoreductase (oxaloacetate decarboxylating); *other names*: malic enzyme; pyruvic–malic carboxylase; it catalyses a reaction between (*S*)-malate and NADP$^+$ to form pyruvate, CO$_2$, and NADPH. It is important in generating NADPH in the cytosol of mammals. Thus types 1 and 2 utilize NAD$^+$, while only types 1 and 3 decarboxylate added oxaloacetate, which leads to three activities being distinguished. (4) D-Malate dehydrogenase (decarboxylating), EC 1.1.1.83; *systematic name*: (*R*)-malate:NAD$^+$ oxidoreductase (decarboxylating); it catalyses a reaction between (*R*)-malate and NAD$^+$ to form pyruvate, CO$_2$, and NADH.

malate synthase EC 4.1.3.2; *systematic name*: L-malate glyoxylate-lyase (CoA-acetylating); *other names*: malate synthetase; malate condensing enzyme; glyoxylate transacetylase. An enzyme that catalyses the formation of L-malate and CoA from acetyl-CoA, H$_2$O, and glyoxylate. This is the second step in the glyoxylate bypass, an alternative to the complete tricarboxylic-acid cycle (in bacteria and plants).

MALDI *abbr. for* matrix-assisted laser desorption/ionization; *see* **tandem mass spectrometry**.

maleate 1 *the trivial name for cis*-1,2-ethylenedicarboxylate; (*Z*)-butenedioate; the dianion of maleic acid, the diastereoisomer of fumarate. **2** any mixture of maleic acid and its mono- and dianions. **3** any salt or ester of maleic acid.

maleoyl *symbol*: Mal; the bivalent *cis*-acyl group, –CO–CH=CH–CO–, derived from maleic acid.

maleyl *symbol*: Mal; the univalent *cis*-acyl group, HOOC–CH=CH–CO–, derived from maleic acid.

maleylation the introduction of one or more **maleyl** groups into a substance by acylation, e.g. with maleic anhydride. Maleylation of a protein with maleic anhydride is used to acylate its free lysine (and other) amino groups in order to change their charge at neutral pH from positive to negative and to render the adjacent peptide bonds resistant to hydrolysis by trypsin; it can also lead to disaggregation of a multimeric protein. The reaction occurs at pH 8.5 (usually at 2°C) and the maleyl groups can be removed at pH 3.5 (usually at 60°C). Maleylation of thiol groups may also occur. *Compare* **citraconylation, succinylation**. —**maleylate** *vb.*

malformin A a cyclic pentapeptide from *Aspergillus niger* that causes curvature and malformations of various plants. The structure contains D-enantiomers of cysteine and leucine.

malic enzyme *or* **pyruvic–malic carboxylase** *a trivial name applied to* any of three enzymes, found in most organisms, that catalyse the oxidative decarboxylation of (*S*)-malate with concomitant production of pyruvate. *See* **malate dehydrogenase (decarboxylating)**.

malignant *(in pathology)* **1** (of a tumour) invasive, spreading, resistant to therapy. **2** (of a disorder) becoming progressively more serious if not treated.

malignant hyperthermia a genetic disorder of humans and pigs that manifests as a fulminant episode of muscle rigidity, high fever, and acidosis, which is frequently fatal. In humans there is an autosomal dominant predisposition, with general anesthesia (frequently involving halothane inhalation) as the trigger. Most cases are associated with hyperexcitability of the ryanodine receptor. In pigs, the predisposition is autosomal recessive with stress as the trigger, and is associated with a point mutation in the same receptor.

malnutrition any imbalance, usually a deficiency, between the nutrients taken in by an organism and the amounts necessary to maintain that organism's normal development and health.

malonate 1 *the trivial name for* propanedioate; methanedicarboxylate; $^-OOC-CH_2-COO^-$, the dianion of malonic acid. It is a potent competitive inhibitor of **succinate dehydrogenase**. **2** any mixture of free malonic acid and its mono- and dianions. **3** any salt or ester of malonic acid.

malonate semialdehyde *the recommended trivial name for* malonate semialdehyde; formylethanoate; $OHC-CH_2-COO^-$. It is an intermediate in the oxidation of propionyl-CoA to acetyl-CoA in the terminal stages of the beta oxidation of fatty acids with an odd number of carbon atoms in plants.

malondialdehyde a compound produced by oxidation of unsaturated fatty acids. *See* **TBARS**.

malonyl 1 *(in biochemistry)* the univalent acyl group, $HOOC-CH_2-CO-$, derived from malonic acid by loss of one hydroxyl group. **2** *(in chemistry)* the bivalent acyl group, $-CO-CH_2-CO-$, derived from malonic acid by loss of both hydroxyl groups.

malonyl-CoA a key intermediate in fatty-acid biosynthesis, formed by reaction of carbon dioxide with acetyl-CoA, with hydrolysis of ATP to ADP and inorganic phosphate. This is catalysed by **acetyl-CoA carboxylase**; biotin is an essential cofactor. The plant enzyme also carboxylates propanoyl-CoA and butanoyl-CoA.

malonyl-CoA–acyl carrier protein transacylase EC 2.3.1.39; *recommended name*: [acyl-carrier-protein] *S*-malonyltransferase; *systematic name*: malonyl-CoA:[acyl-carrier-protein] *S*-malonyltransferase. An enzyme that catalyses a reaction between malonyl-CoA and ACP to form CoA and malonyl-ACP. It is a component of fatty-acid synthase in bacteria (*see* **fatty-acid synthase complex**).

maloyl the bivalent acyl group, $-OC-CH(OH)-CH_2-CO-$, derived from malic acid by loss of the hydroxyl groups from both of its carboxyl groups.

Malpighian body *or* **Malpighian corpuscle** *or* **renal corpuscle** the part of a nephron comprising a cluster of blood capillaries (i.e. the renal glomerulus) together with the enclosing **Bowman's capsule**. [After Marcello Malpighi (1628–1694), Italian anatomist.]

malt 1 barley or other cereal grain that has been soaked in water, allowed to germinate, and then dried. In this process the starch of the cereal is largely degraded to maltose and other oligosaccharides. **2** to make into or become malt; to make or treat (something) with malt or an extract of malt.

maltase a trivial name for α-glucosidase; *see* **glucosidase** (def. 2).

malto+ *comb. form (in chemical nomenclature)* designating an oligosaccharide or polysaccharide (e.g. maltose, maltotriose, maltoheptaose) consisting of D-glucopyranose residues in (α1→4)-linkage (as in maltose).

maltol 3-hydroxy-2-methyl-4-pyrone; a flavour enhancer and modifier with a characteristic smell of candyfloss or burnt bread. Ferric(III) trimaltol is effective in the treatment of iron deficiency anemia.

maltoporin a bacterial outer membrane transmembrane protein, with possibly 16 helices, having two roles in *Escherichia coli*: it transports maltodextrins (including maltose) into the cell, and acts as a receptor for lambda and some other phages.

maltose *or (formerly)* **malt sugar** *the trivial name for* the disaccharide 4-*O*-α-D-glucopyranosyl-D-glucopyranose. It is an intermediate in the enzymic breakdown of glycogen and starch.

maltotriose the trisaccharide Glc(α1-4)Glc(α1-4)Glc; it is formed by the action of α-amylase on starch.

malt sugar *a former name for* **maltose**.

malyl either of the isomeric univalent acyl groups, HOOC–CH(OH)–CH_2–CO– or HOOC–CH_2–CH(OH)–CO–, derived from malic acid by loss of a hydroxyl group from one or the other of the carboxyl groups.

MAML *abbr. for* microarray markup language; *see* **markup language**.

mamma *(pl. mammae)* a mammary gland. —**mammary** *adj*.

mammal any animal of the Mammalia, a large class of warm-blooded tetrapod vertebrates characterized by the possession of sweat glands in the skin and, generally, an insulating layer of hair. Female mammals characteristically suckle their young at modified sweat glands – mammary glands – that secrete milk. Like birds, mammals possess a four-chambered heart, but their thoracic diaphragm is a uniquely mammalian feature. —**mammalian** *adj*.

mammalian artificial chromosome *abbr.*: MAC; a still hypothetical mammalian counterpart of a **yeast artificial chromosome** (YAC).

mammalian target of rapamycin *see* **mTor**.

mammary gland *or* **mamma** the milk-secreting organ of a mature female mammal. The mammary glands occur in one or more pairs on the ventral surface.

mammotropic *or* **mammotrophic** *an alternative term for* **lactotropic**.

mammotropic hormone *or* **mammotrophic hormone** *an older name for* **prolactin**.

mammotropin *or* **mammotrophin** *an older name for* **prolactin**.

Man *symbol for* a residue (or a molecule) of the aldohexose mannose. In the condensed system of symbolism of sugar residues it signifies specifically a mannose residue with the common configuration and ring size, i.e. D-mannopyranose.

manaoic acid *see* **cyclic fatty acid**.

Mancini technique *see* **single radial immunodiffusion**.

mandelate α-hydroxybenzeneacetate; the anion, C_6H_5- CH(OH)–COO⁻, derived from mandelic acid, the α-carbon atom of which is chiral. **2** any salt or carboxylic ester of mandelic acid.

L-(*S*)-(+)-form

mandelonitrile α-hydroxybenzeneacetonitrile; benzaldehyde cyanohydrin; a compound that occurs free and as β-glucosides of (+)-, (–)-, and (±)-mandelonitrile.

manganese *symbol*: Mn; a ferromagnetic, metallic transition element of group 7 of the IUPAC **periodic table**; atomic number 25; relative atomic mass 54.938. There is one stable nuclide of mass number 55. Manganese is widely distributed and abundant in the Earth's crust and is an essential component of all living material. It forms manganese(II) (manganous) and less stable manganese(III) (manganic) compounds, including ionic compounds, and complexes; it can also exist in higher oxidation states. Mn^{2+} ions act as activators of various enzymes. A number of artificial radioactive isotopes are known.

manifold a vessel or tube with a number of inlets or outlets used to collect or distribute gases or liquids in an apparatus.

manna a sweet-tasting, white to yellow dried exudate of the manna ash tree (*Fraxinus ornus*) and other plants. *See also* **manninotriose**.

mannan the main **hemicellulose** of soft (coniferous) wood, made up of D-mannose, D-glucose, and D-galactose in the ratio of 3:1:1. Glucomannans occur in ferns and in certain higher plants and algae where they may function as the main structural component of the cell wall. Highly branched mannans can be extracted from the cell walls of some yeasts.

mannanase *or* **mannase** *see* **mannosidase** (def. 2).

Mannich reaction any chemical reaction in which an imminium ion of the type $CR'_2=N^+R_2$ reacts with a nucleophilic carbon atom, such as that of an enolate anion derived from a ketone or that of an activated double bond as in a phenol, aryl ether, pyrrole, or indole. Mannich reactions are believed to occur widely in biosynthetic pathways of polycyclic compounds, especially alkaloids and porphyrins. [After Carl Mannich (1877–1947).]

manninotriose the trisaccharide Gal(α1-6)Gal(α1-6)Glc; a component of **stachyose**, it also occurs free in **manna** from ash trees.

mannite *a former name for* **mannitol** (use deprecated).

D-mannitol the alditol derived from D-mannose by reduction of the –CHO group. It is widespread in higher plants, plant exudates, and algae, and is obtained from the manna ash (*Fraxinus ornus*) and seaweeds. Mannitol is metabolized by a number of bacteria including many strains of *Escherichia coli* and *Salmonella* spp. Clinically it is administered intravenously as an osmotic diuretic.

manno+ *comb. form* of mannose.

manno- *prefix (in chemical nomenclature)* indicating a particular configuration of a set of (usually) four contiguous >CHOH groups, as in the acyclic forms of D- or L-mannose. *See* **monosaccharide**.

mannobiose the disaccharide Man(β1-4)Man; a constituent of D-gluco-D-mannans isolated from hardwoods, and of polysaccharides from iris, orchid, and lily families.

mannoheptulose *manno*-2-heptulose; the D enantiomer (formerly known as D-*manno*-ketoheptose) occurs in the avocado.

mannokinase EC 2.7.1.7; an enzyme that catalyses the phosphorylation by ATP of D-mannose to D-mannose 6-phosphate with release of ADP. It initiates the further metabolism of mannose, e.g. after uptake by bacteria. Hexokinase has a wide specificity for sugars and catalyses this reaction, e.g., in mammalian liver.

mannosamine *symbol*: ManN; *a trivial name for* 2-amino-2-deoxy-mannose.

mannose *symbol*: Man; *a trivial name for* the aldohexose *manno*-hexose, the C-2 epimer of **glucose**. The D-(+)-form is widely distributed in mannans and hemicelluloses and is of major importance in the core oligosaccharide of N-linked oligosaccharides of glycoproteins, especially at branch points. A commonly occurring structure is Man(α1-6)[Man(α1-3)]Man(β1-4)GlcNAc(β1-4)GlcNAc-Asn.

mannose-binding protein (*abbr.*: MBP) *or* **mannose-binding lectin**

a **collectin** secreted by hepatocytes as an acute-phase response protein into human plasma. The clusters of **lectin** domains bind mannose residues on microbial and fungal surfaces in a Ca^{2+}-dependent manner. It activates (in the third pathway) the C3 component of **complement** after binding to C1q receptors, and its structure resembles that of C1q. It acts through the serine protease **MASP** associated with it. Macrophages have a membrane-bound MBP. Hereditary deficiency or dysfunction results in increased frequency of bacterial infections in infants and children, and is a frequent finding in lupus erythematosus and rheumatoid arthritis patients.

mannose 6-phosphate an intermediate in **mannose** metabolism, formed from mannose by **mannokinase** or from fructose 6-phosphate by **mannose-6-phosphate isomerase**. Apart from acting as a precursor in GDPmannose formation, it is a component of lysosomal enzymes, playing an important role in sorting the enzymes into lysosomes. *See also* **I-cell disease**, **mannose 6-phosphate receptor**.

mannose-1-phosphate guanylyltransferase EC 2.7.7.13; *other name*: GTP–mannose-1-phosphate guanylyltransferase; an enzyme that catalyses the reaction between GTP and α-D-mannose 1-phosphate to form GDPmannose with release of pyrophosphate.

mannose-1-phosphate guanylyltransferase (GDP) EC 2.7.7.22; *systematic name*: GDP:D-mannose-1-phosphate guanylyltransferase. An enzyme involved in M-antigen biosynthesis in salmonellae; it catalyses a reaction between GDP and D-mannose 1-phosphate to form orthophosphate and GDPmannose. This protein belongs to the **mannose-6-phosphate isomerase** family 2.

mannose-6-phosphate isomerase EC 5.3.1.8; *systematic name*: D-mannose-6-phosphate ketol-isomerase; *other names*: phosphomannose isomerase; phosphohexoisomerase; phosphohexomutase. An enzyme that catalyses the reaction:

$$\text{D-mannose 6-phosphate} = \text{D-fructose 6-phosphate;}$$

zinc is a cofactor. The sequences belong to two families. Most enzymes (including human) belong to family 1. Family 2 is represented by an enzyme involved in alginate metabolism. *See also* **mannose-1-phosphate guanylyltransferase**.

mannose 6-phosphate receptor a protein on the surface of membranes in the Golgi apparatus and late endosomes that recognizes mannose 6-phosphate residues on lysosomal enzymes and is responsible for sorting them into late endosomes, which then become lysosomes. Lysosomal enzymes bearing phosphomannosyl residues bind specifically to mannose-6-phosphate receptors in the Golgi apparatus, and the resulting receptor–ligand complex is transported to an acidic prelysosomal compartment where the low pH mediates dissociation. There are two types of receptor protein. (1) The cation-dependent type, or M6P receptor, is a type I membrane glycoprotein homodimer involved in transport of phosphorylated lysosomal enzymes from the Golgi apparatus and the cell surface to lysosomes. It binds one molecule of M6P, does not bind insulin-like growth factor II (IGFII), and has an N-terminal region homologous to the repeats of the second type. (2) The cation-independent type, or M6P/IGFII receptor, is a type I glycoprotein membrane receptor that binds two molecules of M6P, and at a different site, IGFII. Its N-terminal region contains 15 homologous repeats (each of ≈145 amino acids) homologous to that in the N terminus of the first type. *See also* **I-cell disease**.

mannosidase 1 α-mannosidase; EC 3.2.1.24; *systematic name*: α-D-mannoside mannohydrolase; an enzyme that catalyses hydrolytic removal of terminal, nonreducing residues in α-D-mannosides. **2** β-mannosidase EC 3.2.1.25; *systematic name*: β-D-mannoside mannohydrolase; *other names*: mannanase; mannase; an enzyme that catalyses hydrolytic removal of terminal, nonreducing residues in β-D-mannosides.

Both enzymes function in lysosomes at an acid pH, in the degradation of glycoproteins. Deficiency of either activity leads to a form of mannosidosis. *See also* **ER mannosidase**, **Golgi mannosidase**.

mannoside any **glycoside** in which the sugar moiety is a **mannose** residue.

α-mannosidosis a genetic disorder of humans characterized by abnormal storage and urinary excretion of mannose-rich oligosaccharides due to deficiency of lysosomal α-**mannosidase** activity. Type 1 is a severe or infantile form. Type 2 is milder and appears in juveniles and adults.

β-mannosidosis a genetic disorder of humans, less frequent than α-mannosidosis, characterized by abnormal accumulation and urinary excretion of the disaccharide Manβ-1,4-*N*-acetylglucosamine, due to deficiency of lysosomal β-mannosidase.

mannosyl any of the glycosyl ligands formed by detaching the anomeric (α- or β-) hydroxyl group, on C-1, from a cyclic form (pyranose or furanose) of mannose.

α-1,3-mannosyl-glycoprotein β-1,2-*N*-acetylglucosaminyl-transferase EC 2.4.1.101; *other name*: *N*-glycosyl-oligosaccharide-glycoprotein *N*-acetylglucosaminyltransferase I (*abbr*.: Glc-NAc-TI *or* GNT-I); an enzyme involved in the synthesis of oligosaccharides, particularly those that are N-linked in glycoproteins. It catalyses the reaction between UDP-*N*-acetyl-D-glucosamine and α-D-mannosyl-1,3-(R)-β-D-mannosyl-R′ to form *N*-acetyl-β-D-glucosaminyl-1,2-α-D-mannosyl-1,3-(R)-β-D-mannosyl-R′ with release of UDP.

mannuronate 1 the anion of mannuronic acid, the compound formed by oxidation of the C-6 hydroxymethylene group of mannose to a carboxyl group. **2** any salt or ester of mannuronic acid. D-Mannuronic acid is a component of **alginic acid**.

β-D-pyranose form

manoalide a naturally occurring marine sesterterpenoid. It is an irreversible inhibitor of phospholipases A$_2$ and C, and a potent inhibitor of Ca^{2+} mobilization.

manometer an instrument for measuring pressure differences. The simplest type consists of a U-tube filled with a suitable liquid (often mercury), and compares a pressure delivered to one side with atmospheric pressure on the other, open, side of the tube. The difference in the level of liquid in the two arms of the tube gives a measure of the unknown pressure. In biochemistry the term is often used to mean a manometric **respirometer**.

manometer constant a predetermined factor, calculated for each individual constant-volume manometric **respirometer**, that when multiplied by the manometer reading immediately gives the volume of gas evolved in cubic millimetres at **standard** temperature and pressure. The factor is itself dependent upon temperature, liquid volume, density of the manometer fluid, and nature of the gas phase.

manometer fluid any fluid used to fill a **manometer**. In biochemistry this usually refers to the fluid used to fill the manometer of a respirometer. Such fluids are generally easily visible, wet the glass walls of the manometer, have a density that will give an easily calculated manometer constant, and have a low vapour pressure. Examples include **Brodie's fluid**, and **Krebs' manometer fluid**.

manometric 1 of, pertaining to, or measured with a **manometer**. **2** (*in biochemistry*) of, or pertaining to, a manometric respirometer. —**manometrically** *adv*.

manometry 1 the measurement of gas pressure by means of a manometer. **2** (*in biochemistry*) the use of manometric respirometers in studies of chemical and biochemical reactions.

Man*p* *symbol for* a mannopyranose residue.

mantissa the fractional part of any **logarithm**.

MAO *abbr. for* monoamine oxidase.

MAO inhibitor *abbr. for* monoamine oxidase inhibitor.

map 1 any representation of the surface features of an object, especially the spatial relationships of physical or chemical landmarks, as in **electron-density map**, **genetic map**, and **restriction map**. By extension the term is applied to certain two-dimensional patterns obtained in analysis, e.g. **peptide map**. **2** to make a **map** (def. 1). **3** to occur, or be placed, at a specific position on a **map** (def. 1).

MAP *abbr. for* microtubule-associated protein.

map distance (*in genetics*) the distance, in terms of percentages of crossing-over (i.e. 100 × no. of recombinants/total no. of progeny), between particular linked genes. *See also* **map unit**.

MAP kinase *abbr. for* mitogen-activated protein kinase; a family of protein kinases that perform a crucial step in relaying signals from the plasma membrane to the nucleus. They are turned on by a wide range of proliferation- or differentiation-inducing signals. For example, the MAP kinase FUS3 of *Saccharomyces cerevisiae* is required for the transition from mitosis into conjugation. The name does not indicate that this is a kinase for MAPs – rather that it is a protein kinase that is mitogen-activated. Nevertheless, some MAPs can be substrates for this enzyme. *See also* **ERK**.

MAP kinase kinase *abbr*.: MAPKK; EC 2.7.1.–; a dual-specificity mitogen-activated protein kinase kinase; *other names*: erk activator kinase; extracellular signal-regulated kinase; an enzyme that catalyses the concomitant phosphorylation of threonine (T) and tyrosine (Y) residues in a TEY sequence in **MAP kinases**. It requires activation by the serine/threonine kinase, MAP kinase kinase kinase (*see* **RAF**).

maple-syrup urine disease *or* **branched-chain ketonuria** an autosomal recessive disease of humans characterized by neurological signs and the onset of feeding problems in the first week of life, and associated with a maple-syrup odour of the urine. The concentrations of branched-chain amino acids and branched-chain keto acids are elevated in the blood of patients, due to deficiency of the enzyme responsible for the oxidative decarboxylation of branched-chain keto acids.

MAPTAM 1,2-bis-(2-amino-5-methylphenoxy)ethane-*N*,*N*,*N*′,*N*′-tetraacetic acid tetraacetoxymethyl ester; a substance that readily penetrates cell membranes and undergoes intracellular hydrolysis with the formation intracellularly of the 5-methyl form of **BAPTA**. It is useful as an indicator of intracellular calcium concentrations.

map unit *or* **Morgan's unit** a unit of genetic length along a chromosome and defined in terms of recombination frequency, expressed as a percentage; a recombination frequency of 1 per cent (that is one recombinant individual in a hundred progeny) equals one map unit. 1 map unit = 1 centimorgan. *Compare* **morgan**.

maquette a protein designed *de novo*, to investigate properties such as electron transfer.

MAR 1 *abbr. for* matrix attachment region. **2** *abbr. for* matrix-associated region; a region of DNA that binds to the nuclear protein network and segregates chromosomal subdomains.

marasmus a wasting away of the body, especially of infants, caused by chronic deficiency of intake of energy-producing foodstuffs. *Compare* **kwashiorkor**.

marcks *abbr. for* myristoylated alanine-rich C kinase substrate; a filamentous (F) actin cross-linking protein and the most prominent substrate for protein kinase C. It also binds calmodulin, actin, and synapsin and is displaced from the membrane by phosphorylation by protein kinase C.

Marcus theory a set of equations that describe the rates of electron transfer between molecules, also applicable to transfer between electron carriers within a protein. The rate depends on the energy barrier to electron transfer, which involves reorganization of the structures so that the potentials of donor and acceptor are equal. The theory leads to the counterintuitive conclusion that above a certain driving force (the inverted region) a greater driving force leads to a decreased rate of electron transfer. [After Rudolph A. Marcus, winner of the 1992 Nobel Prize in Chemistry for his contributions to the theory of electron transfer reactions in chemical systems.]

marenostrin *an alternative name for* pyrin. *See* **familial Mediterranean fever**.

margarine a processed fat used as a substitute for butter. Modern margarines are prepared from hydrogenated natural fats, the trend in recent years having been to preserve a high content of nutritionally desirable **polyunsaturated fatty acids**. Apart from the oil or fat content, margarine contains a water phase that is 16–20% of content and includes water, cream, whole milk, skim milk (powder), whey, and buttermilk in various ratios; flavouring, which may include diacetyl, fatty acids, and ketones; colouring agents, mainly based on carotenes; additional vitamins A and D; and **antioxidants** such as di-*tert*-butyl *p*-cresol. The process was invented in 1869 by H. Mège-Mouriès on the order of Napoleon III to produce a cheap stable butter substitute. Mège-Mouriès churned a beef tallow fraction, known as **oleomargarine**, with cream.

marijuana *see* **cannabis**.

Mariotte bottle *or* **Mariotte flask** a device for providing a flow of liquid maintained at a constant pressure head; it is used, e.g., in column chromatography. It consists of a sealed flask, closed to the atmosphere, with an outlet at or near the bottom and an air inlet tube open to the atmosphere. The height of the bottom of the inlet tube is adjustable. When the flask is filled with liquid to a level above the bottom of the inlet tube the pressure of the liquid flowing from the flask is determined by the height of the bottom of the inlet tube above the outlet point. [After Abbé Edmé Mariotte (1620–84), French physicist.]

Mariotte's law *an alternative name (esp. Fr.) for* Boyle's law (*see* **gas laws**).

marker 1 an appropriate reference substance or material included in a sample being subjected to a separation procedure such as chromatography, electrophoresis, or density-gradient centrifugation. **2** *see* **biochemical marker**. **3** *see* **genetic marker**. **4** *an alternative term for* **label** (def. 3).

marker enzyme any enzyme of known intracellular location that can be assayed as an aid in following the isolation and purification of subcellular fractions.

marker gene any gene of known function and known location on a chromosome. *Compare* **genetic marker**.

marker rescue a process whereby one or more genetic markers from a UV-irradiated virus are integrated into an unirradiated helper virus in infected bacteria; such genes are thereby 'rescued' by recombination.

Markham still a simple, all-glass, steam-jacketed apparatus for the rapid and efficient steam distillation of a succession of small samples of volatile substances. It was designed especially for the distillation of ammonia as part of the micro-Kjeldahl procedure for the determination of nitrogen, but has been found to be prone to carry-over from one sample to the next. [After Roy Markham (1916–79), British biochemist.]

Markovnikov('s) rule a generalization stating that when a molecule of water, or a hydrogen halide, or some other polar substance, is added heterolytically across an asymmetrically substituted carbon–carbon multiple bond, the hydrogen atom or other more electropositive part of the molecule becomes attached usually to that carbon atom carrying the larger number of hydrogen atoms while the hydroxyl, halo, or other more electronegative part of the molecule becomes attached to that carbon atom carrying the smaller number of hydrogen atoms. [After Vladimir Vasilyevich Markovnikov (1838–1904), Russian chemist, who formulated the rule on empirical grounds (1870).]

markup language a language used to structure and label documents in a logical fashion. The labels may provide format information on how the data should be presented visually (e.g. by a Web browser) or semantic information concerning the kind of data being represented. Markup languages facilitate text-based data exchange and can be understood by humans and machines. Examples include HTML (hypertext markup language), XML (extensible markup language), UML (universal markup language), and MAML (microarray markup language).

Markush structure a chemical structure in which multiple 'functionally equivalent' chemical entities are allowed in one or more parts of the compound. Markush structures are used for example in patents to protect an invention of related compounds, without requiring the inventor to prepare and test every possible compound. [After Eugene A. Markush, the first inventor to include them successfully in a U.S. patent, in 1924.]

Maroteaux–Lamy syndrome *an alternative name for* **mucopolysaccharidosis VI**.

marsh gas *another name for* **methane**.

MAS a human transforming gene that encodes a neuronal-type angiotensin III receptor. The latter is a G-protein-associated membrane receptor that activates phospholipase C.

maser a device for producing intense beams of microwave radiation of defined frequency. *Compare* **laser**. [From microwave amplification by stimulated emission of radiation.]

Mash *abbr. for* mammalian Achaete-Scute homologue; either of two mammalian genes that encode basic helix-loop-helix transcription factors and are related to the *Drosophila Achaete–Scute* genes. Mash-1 is expressed in precursor cells for the central and autonomic nervous systems and olfactory epithelium. Mash-2 is expressed in oocytes, preimplantation embryos, and placenta. *See* **AS-C protein**.

mask (*in sequence analysis*) to remove repeated or low-complexity regions from a sequence in order to improve the sensitivity of database-search programs. *See* **sequence filter**.

masked residue any residue in a biopolymer that is not accessible to, and cannot react with, a reagent specific for that type of residue, either because of the position of the residue or because of the nature of its immediate environment.

maskin a protein of *Xenopus laevis* that is involved in regulating the translation of mRNA during embryonal development. It interacts with cytoplasmic polyadenylation-element binding protein bound to the polyadenylation element at the 3′ untranslated region of mRNA.

MASP *abbr. for* MBP-associated serine protease; a serine protease (100 kDa) that is associated with **mannose-binding protein** (MBP). It shares ≈39% homology with C1r and C1s of **complement**. Binding of MBP to microbial polysaccharides enhances activation of MASP.

maspin a serpin-like tumour suppressor of epithelial cells.

mass 1 *symbol*: m or M; one of the seven SI base physical quantities; the SI base unit of mass is the **kilogram**. **2** *a general term for* the quantity of matter in a body, a characteristic that is both a measure of the resistance of the body to changes in its velocity, i.e. **inertial mass**, and a measure of the force experienced by the body in a gravitational field, i.e. **gravitational mass**. Mass is commonly confused with weight. According to the theory of relativity, inertial and gravitational masses are equal, mass increases with velocity, and mass and energy are interconvertible.

mass action, law of a law stating that, for a homogeneous system, the rate of reaction of a chemical reaction is proportional to the active masses of the reacting substances; i.e. the rate is proportional to the product of the activities of the reactants, each raised to a power equal to the number of reactant molecules of that type that participate in the reaction.

mass-average molar mass *see* **average molar mass**.

mass concentration *see* **concentration** (def. 1a).

mass density *see* **concentration** (def. 1a).

mass fraction *symbol*: w; the mass of a specified component of a system divided by the mass of the system; it is dimensionless.

mass fragmentography *a common but less rigorous name for* **selected-ion monitoring**.

mass number *an older name for* **nucleon number**.

mass spec. *abbr. (colloquial) for* mass spectrometer, mass-spectrometric, or mass spectrometry.

mass spectrograph a **mass spectrometer** in which the ions fall on a photographic plate.

mass spectrometer *abbr.*: MS *or (colloquial)* mass spec.; any instrument for **mass spectrometry**.

mass spectrometry *abbr.*: MS *or (colloquial)* mass spec.; an analytical method for determining the relative masses and relative abundances of components of a beam of ionized molecules or molecular fragments produced from a sample in a high vacuum. It is applicable to any sample that is gaseous or volatile at low pressure, or that can be so rendered by derivatization. In the commonest version, the ions, generated usually by electron bombardment of the sample, are formed into a beam and then accelerated to a uniform velocity by means of an arrangement of electrically charged slits. The beam is then dispersed into a spectrum of particles of different mass-to-charge ratios, m/e, by application of a suitably disposed magnetic field. The ions of any particular m/e value may be focused on a collecting electrode by appropriate adjustment of the accelerating potential or magnetic field and the resulting ion currents measured electrometrically. With a suitably designed mass spectrometer the method can be used to determine the relative abundances of isotopes of an element or the relative molecular masses of two or more compounds, or used to aid the identification of a complex compound by the mass spectrum of the fragments generated from it on electron bombardment. *See also* **gas chromatography/mass spectrometry, selected-ion monitoring. —mass-spectrometric** *adj.*

mass spectrum a tracing, photographic record, or diagram of the relative numbers of different ions (representing molecules or molecular fragments) obtained (from a specimen) by **mass spectrometry**, ranked in order of their mass-to-charge ratios.

mass-to-charge ratio *(in mass spectrometry)* the ratio of the mass, m, of an ion to the charge, e, on the ion. *See also* **mass spectrometry**.

mass transfer the spontaneous and irreversible process of transfer of mass across nonhomogeneous fields, e.g. between liquid phases, or from one side of a membrane to another.

mass unit *see* **atomic mass unit, dalton**.

mast cell a type of leukocyte, of the subclass granulocyte; it is characterized by having numerous cytoplasmic basophilic granules rich in histamine together with 5-hydroxytryptamine and heparin. Degranulation of mast cells, with release of the vasoactive amines, occurs by reaction, through a specialized region of the Fc piece, of antigen with homocytotropic antibody (reagin) bound to the cell. This leads to an immediate hypersensitivity reaction.

mast cell degranulating peptide *abbr.*: MCDP; a major neurotoxic 22-residue peptide of bee venom: IKCNCKRHVIKPHICR-KICGKN–NH$_2$. It stimulates degranulation of mast cells, and selectively blocks the neuronal Ca^{2+}-dependent K^+ channel.

mast cell growth factor *see* **stem cell factor**.

mastoparan a tetradecapeptide neurotoxin from wasp venom: Ile-Asn-Leu-Lys-Ala-Leu-Ala-Ala-Leu-Ala-Lys-Lys-Ile-Leu-NH$_2$. It functions like **melittin** and **mast cell degranulating peptide**, in generating ion channels in membranes. It is classified in the TC system under number 1.C.16.

MATα2 yeast mating-type protein α-2 repressor; a regulatory DNA-binding protein that acts as a master switch in yeast differentiation by controlling gene expression in a cell-type-specific fashion. In α haploids and **a/α** diploids (*see* **mating type**) it represses the transcription of α-cell-specific genes. In **a/α** diploids it also forms a complex with MATα1 protein that represses haploid-specific gene expression.

matchmaker proteins *see* **molecular matchmakers**.

maternal-effect gene any gene expressed in females whose mRNA transcript is found in the egg. Notable examples from *Drosophila* are *oskar* and *bicoid* (*see* **bicoid protein**), which determine egg polarity.

maternal imprinting the epigenetic silencing of an allele inherited from the mother. *See* **imprinting**.

Math 1 a basic helix-loop-helix transcription factor that determines epithelial cell commitment in the mouse intestine. It is a downstream target in the notch signalling pathway, and is also involved in cell fate determination in the nervous system.

mathematics 1 *functioning as sing.* the group of related sciences concerned with number, space, shape, and quantity, and their inter-

relationships. **2** *functioning as sing. or pl.* the use of mathematical processes or operations in calculations or the solution of problems.

mating type one of the two types of *Saccharomyces cerevisiae*, α and **a**, that are necessary for mating, a process in which two haploid cells fuse. The resulting diploid cell, **a/α**, cannot mate, but can produce spores, giving rise to haploid cells by meiosis. The mating type of the haploid cell is determined by a single locus, the mating-type (*MAT*) locus. In **a** cells, this encodes a regulatory protein **a**1, and in an α cell it encodes two such proteins, α1 and α2. In the **a/α** cell, the regulatory proteins **a**1 and α2 combine to generate a completely new pattern of gene expression. *See also* **MATα2**.

matricin *an alternative name for* **matrix metalloprotease**.

matrilysin EC 3.4.24.23; *other names*: matrin; uterine metalloendopeptidase; matrix metalloproteinase 7 (*abbr.*: MMP7); putative (or punctuated) metalloproteinase-1 (*abbr.*: PUMP-1); an enzyme that catalyses the cleavage of Ala^{14}-|-Leu^{15} and Tyr^{16}-|-Leu^{17} in the B chain of insulin. It has no action on collagen types I, II, IV, or V. It cleaves casein and fibronectin and shows preference for gelatin chain α-2(I) over α-1(I). It is secreted as the precursor, promatrilysin, and is activated by autocatalytic cleavage at Glu^{77}-|-Tyr^{78}. Zinc and calcium are cofactors.

matrin *or* **DNA-binding protein F/G** a **zinc-finger** DNA-binding protein of the fibrogranular areas of the nucleus. Matrins comprise a family of eight major polypeptides of nuclear matrices of rat liver and tissue culture cells, highly conserved between humans and mouse. The name has sometimes been used as a synonym for **matrilysin** and matrin F/G is therefore used to avoid confusion.

matripsin *or* **matrix-type serine protease 1** (*abbr.*: MT-SP1) EC 3.4.21; a type 2 transmembrane serine protease of gastrointestinal tract, spleen, and prostate. The mouse orthologue shares 81% sequence identity with the human enzyme (855 amino acids).

matriptase *an alternative name for* membrane type-serine protease 1. *See* **MT-SP1**.

matrix (*pl.* **matrices** *or* **matrixes**) **1** a medium or place wherein something is formed, develops, or is embedded or enclosed. **2** the medium that forms, or supports, the stationary phase in certain separative procedures, as in chromatography or zone electrophoresis; the material that forms, or is used to form, the substratum of an immobilized reagent in affinity chromatography. **3** *(in animal histology) see* **extracellular matrix**. **4** *(in plant histology)* the meshwork of soluble substances, largely polysaccharide in nature, occurring in cell walls and in which insoluble microfibrils of cellulose are embedded. **5** *see* **mitochondrial matrix**. **6** cytoplasmic matrix; *see* **groundplasm**. **7** the substratum, living or nonliving, in or on which a fungus or lichen grows. **8** *(in mathematics)* any rectangular array of numbers or symbols manipulable algebraically as a single entity and used according to specific rules to facilitate solving of certain kinds of problems, e.g. a set of simultaneous linear equations.

matrix attachment region *abbr.*: MAR; an AT-rich, ≈200-bp sequence of DNA that binds to the nuclear scaffold (the site of RNA synthesis) in the interphase nucleus. MARs anchor genes near the nuclear membrane; they are believed to help regulate transcription by controlling the chromatin state of DNA, mediating chromatin unfolding prior to expression.

matrix Gla protein a protein that mediates association of the organic matrix of bone and cartilage. It contains γ-carboxyglutamate (Gla) residues, formation of which is vitamin-K-dependent.

matrix metalloproteinase *abbr.*: MMP; any proteolytic enzyme associated with the **extracellular matrix**. The group includes: MMP1, interstitial collagenase; MMP2, gelatinase A; MMP3, stromelysin-1; MMP8, neutrophil collagenase; MMP10, stromelysin-2; MMP12, metalloelastase. They have been implicated in facilitating cell migration, and in the ability of malignant cells to invade tissues and metastasize. They convert tumour necrosis factor (TNF) precursor into active TNF and may be implicated in other diseases such as rheumatoid arthritis or osteoarthritis. For these reasons, there is current interest in matrix metalloproteinase inhibitors (*abbr.*: MMPIs) as potential drugs. *See also* **TIMP**.

matrix metalloproteinase inhibitor *see* **MMPI**.

matrix processing peptidase *abbr.*: MPP; *an alternative term for* mitochondrial processing peptidase.

matrix protein 1 the major protein component of the outer membrane of *Escherichia coli*. It is a transmembrane protein that forms hydrophilic pores permitting the passage of microsolutes; in association with **lipopolysaccharide** (def. 4) it forms a receptor for phages. It is a member of the **porin** family, and is a product of the gene *DMPC*. **2** any of a class of proteins that coat the inside of the lipid bilayer in the envelope of some groups of enveloped viruses.

matrix space *or* **(mitochondrial) lumen** the innermost compartment of a mitochondrion, containing the **mitochondrial matrix**.

matrix vesicle any of the minute, membrane-bound, extracellular particles (≈100 nm in diameter) occurring in the interstitial matrix of calcifying cartilage. The vesicles contain a number of enzymes involved with local calcification mechanisms.

maturation 1 the process of coming to full development, ripening, or becoming mature. **2** *or* **maturation division** the processes of formation of gametes (sperm or ova) from primary gametocytes involving meiosis of the nucleus with reduction of chromosome number from diploid to haploid. **3** *(in virology)* the overall process leading to the incorporation of a viral genome into a capsid and the development of a complete virion.

maturation-promoting factor *an alternative name for* **M-phase-promoting factor**.

maturation protein *or* **A protein** one of the three proteins encoded by certain single-stranded RNA bacteriophages, one molecule of which is necessary for proper encapsulation of the RNA and for binding of the phage to a pilus on the host cell.

mature mRNA the functional form of **messenger RNA** (mRNA) that results from capping, splicing, and polyadenylation of the primary transcript.

maturity-onset diabetes *a former name for* type 2 **diabetes mellitus**.

maturity-onset diabetes of the young *abbr.*: MODY; an autosomal dominant form of type 2 (non-insulin-dependent) diabetes mellitus that has early onset. Mutations of six genes have been implicated: MODY 1, HNF-4α (locus 12q24.31); MODY 2, glucokinase (locus 7p13); MODY 3, HNF-1α (locus 20q13.12); MODY 4, IPF-1 (locus 13q12.2); MODY 5, TCF2 (locus 17cen-q21.3); MODY 6, NEUROD1 (locus 2q32). MODY 3 and MODY 2, the most prevalent forms, account for more than 80% of Caucasian patients.

max *abbr. for* maximal or maximum.

MAX a gene family coding for DNA-binding proteins that function in cell growth control. The product, Max, forms heterodimers with Myc (*see **MYC***) that bind specifically to CACGTG; it also forms homodimers of similar specificity. Max contains **helix-turn-helix** and heptad repeats.

Maxam–Gilbert method (for DNA sequencing) *see* **chemical cleavage method**.

maxicell *or* **maxi cell** a bacterial cell that has been heavily irradiated with UV light and in which the chromosomal DNA is extensively degraded. Any plasmid(s) present usually escape damage and the genes they encode can be expressed.

maxicircle DNA any large, closed circular DNA molecule, as in the **kinetoplast** of *Leishmania tarantolae*.

maxima a plural of **maximum** (def. 1).

maximal *abbr.*: max; of, pertaining to, or reaching a maximum; being the greatest possible in size, duration, etc.; **maximum** (def. 2). —**maximally** *adv.*

maximal agonist effect *or* **intrinsic activity** *symbol*: α; the maximal effect that an agonist, whether conventional or inverse, can elicit in a given tissue under particular experimental conditions, expressed as a fraction of that produced by a full agonist acting through the same receptors under the same conditions.

maximal rate *or* **maximum rate** *another name for* **limiting rate**. Symbol V (recommended) or V_{max}. *See* **Michaelis kinetics**.

maximal velocity *or* **maximum velocity** *an older name for* **limiting rate**.

maximum *abbr.*: max; **1** (*pl.* **maxima** *or* **maximums**) the highest or greatest possible amount, degree, etc.; the highest value of a variable quantity. *Compare* **minimum**. **2** of, being, or exhibiting a maximum or maxima; **maximal**.

maximum likelihood a statistical method for parameter estimation using a prespecified model and a set of empirical data. The parameter values are chosen to maximize the probability of observing the data, given the hypothesis. In phylogeny, the method is used to find the tree topology with maximum likelihood, given the data and the model of sequence evolution, and to estimate branch lengths and other evolutionary parameters.

maximum parsimony the principle that a simpler solution is preferred to a more complex one. For example, in phylogenetics, a tree that requires the fewest evolutionary events to explain the data is preferable to a tree that requires more events.

maximum-scoring segment pair *abbr.*: MSP; the highest scoring of high-scoring segment pairs (HSPs) that result from the alignment-extension step in BLAST. *See* **alignment algorithm**.

Maxwell relation a relation between the dielectric constant, ε, and the refractive index, n, of a nonabsorbing, nonmagnetic material such that $\varepsilon = n^2$. [After James Clerk Maxwell (1831–79), British physicist.]

Mb 1 *abbr. for* myoglobin; hence MbO_2, oxymyoglobin; MbCO, carbonmonoxymyoglobin; etc. **2** *symbol for* megabase, i.e. = 10^6 bases. **3** *(in computing) abbr. for* megabyte (i.e. 2^{20} bytes; alternative to MB).

MBOAT *abbr. for* membrane-bound *O*-acyltransferase.

***Mbo*I** a type 2 **restriction endonuclease**; recognition sequence: ↑GATC. *Sau*3AI is an **isoschizomer**.

***Mbo*II** a type 2 **restriction endonuclease**; recognition sequence: GAAGANNNNNNNN↑.

MBP *abbr. for* mannose-binding protein.

MBP-associated serine protease *see* MASP.

MBS *abbr. for meta*-maleimidobenzoyl-*N*-hydroxysuccinimide ester; a heterobifunctional reagent useful, *inter alia*, for coupling an enzyme to an antibody, without appreciable loss of either activity.

MBSA *abbr. for* methylated bovine serum albumin.

mc *(formerly) abbr. for* millicurie (mCi now used).

Mc₃ *symbol for* mucotriaose (McOse₃ preferred if space permits).

Mc₄ *symbol for* mucotetraose (McOse₄ preferred if space permits).

MCAD *abbr. for* medium chain acyl-CoA dehydrogenase; *see* **beta-oxidation system**.

McArdle's disease *an alternative name for* type V **glycogen disease**.

MCC the symbol for a gene (standing for 'mutated in colorectal cancer') that is closely linked to the *APC* gene at 5q21 in humans. It encodes a protein (829 amino acids) that may be a tumour suppressor. Somatic mutations of *MCC* occur in 5–10% of colorectal cancers.

MCD *abbr. for* **1** magnetic circular dichroism. **2** mean corpuscular (erythrocyte) diameter.

MCDP *abbr. for* mast cell degranulating peptide.

M cell *or* **microfold cell** a type of epithelial cell of intestinal mucosa that overlies patches of lymphoid tissue (Peyer's patches). M cells have an invaginated base in which lymphocytes and macrophages can aggregate. The cells present bacterial and viral protein antigens to the underlying lymphocytes, inducing proliferation of the latter and the secretory immune response.

McFarlane's method a method for the radioiodination of proteins with high efficiency in which the protein to be labelled is reacted with iodine monochloride that has been previously equilibrated with radioiodide. The extent of substitution can be precisely controlled by use of an appropriate quantity of iodine monochloride. [After Arthur Sproul McFarlane (1905–78), British biophysicist.]

MCH *abbr. for* melanin concentrating hormone.

mCi *symbol for* millicurie.

McIlwain chopper *or* **McIlwain–Buddle chopper** a hand- or motor-driven machine for rapidly cutting a piece of animal tissue into slices of uniform, preselected thickness (≥0.07 mm) and then, if desired, into prisms by making a second series of cuts. Such prisms can yield easily dispensed suspensions that retain those metabolic characteristics that tend to be lost on homogenization of tissue. [After Henry McIlwain (1912–92), British biochemist, and H. L. Buddle.]

McIntosh and Fildes jar *see* **anaerobic jar**.

McLeod gauge a mercury-in-glass vacuum pressure gauge that is capable of measuring down to 10^{-5} mmHg (1.33×10^{-3} Pa). A known large volume of gas is compressed into a small volume, in which the now much larger pressure is measured.

MCM protein any of six related ATPase proteins in budding yeast that are essential for initiation of replication. They assemble into a large ring complex around the DNA at the origin of replication complex (**ORC**), being stabilized by Cdc45. They appear in the nucleus during G_1 phase of the cell cycle, and are exported into the cytoplasm from S to G_2 phases. They are considered to function as a **licensing factor**, permitting one round of replication to occur per cell cycle. Mutants of the MCM homologue PROLIFERA in *Arabidopsis* are lethal.

McOse$_3$ *symbol for* mucotriaose (preferred alternative to Mc$_3$).

McOse$_4$ *symbol for* mucotetraose (preferred alternative to Mc$_4$).

MCP *abbr. for* methyl-accepting chemotaxis protein.

MCS *abbr. for* multiple cloning site.

Md *symbol for* mendelevium.

MD *(in clinical chemistry) abbr. for* **malate dehydrogenase** (def. 1).

MDa *abbr. for* megadalton.

MDCK cell *abbr. for* Madin–Darby canine kidney cell; a cell of a heteroploid cell line.

MDH *abbr. for* **malate dehydrogenase** (def. 1).

Mdm2 *or* **MDM2** *abbr. for* murine double minute chromosome protein 2; a protein, present mainly in mammalian cell nuclei, that is an E3 ubiquitin-protein ligase. It binds to **p53** and targets it to ubiquitin-mediated degradation in proteasomes.

MDMA *abbr. for* 3,4-methylenedioxymethamphetamine.

MDP *abbr. for* muramyl-dipeptide.

MDR *abbr. for* multiple drug resistance *or* multidrug resistance; the acquired simultaneous resistance of tumour cells or pathogenic organisms to a wide spectrum of drugs arising from the administration, typically over long periods, of one or a limited number of drugs. Such resistance may extend to drugs with little structural or even functional similarity to the original drug(s), and results in reduced efficacy of all the drugs concerned. In long-term chemotherapy of cancer patients it is due to the induced expression of plasma membrane glycoproteins known as P-glycoproteins, or **MDR proteins**. In bacteria it is due to induction of multidrug transporters. *See* **BmrR**.

MDR protein any transmembrane glycoprotein that is involved in (multiple) drug resistance (*see* **MDR**). MDR proteins belong to the ATP-binding cassette (ABC) class of the multidrug transporter superfamily, and are ATP-dependent. MDR1 (or P-glycoprotein 1) occurs in polarized epithelia of intestine, kidney, liver, and blood–brain barrier. MDR2 (or P-glycoprotein 2) is the phosphatidylcholine translocator of the bile canaliculi. MDR3 (or P-glycoprotein 3) is an efflux pump that is not capable of conferring drug resistance by itself. However, its absence leads to severe liver disease. MDR-associated protein participates in the active transport of drugs into subcellular organelles. MDR protein (or P-glycoprotein) of *Leishmania tarentolae* is a protein responsible for acquired methotrexate resistance, and is encoded by a gene located in the 'H-circle', a 68 kb duplex DNA circle containing a 30 kb inverted repeat.

Mdv1 a dynamin-like protein of yeast that contains WD repeats and is required, with Fis1 and Dnm1, in forming the contractile ring around the outer mitochondrial membrane that brings about fission of the organelle.

Me *symbol for* the methyl group, $CH_3–$.

Mead acid a name sometimes given to (*all-Z*)-eicosa-5,8,11-trienoic acid. It can be synthesized *de novo* by mammals via palmitate and oleate, and is thought to substitute partially for the long-chain polyunsaturated fatty acids lacking in essential fatty acid deficiency. [After J. F. Mead, who first identified it as an acid found in elevated concentrations in mammals during essential fatty acid deficiency.]

mean *or* **arithmetic mean** *or* **average** the quotient of the arithmetic sum of any set of two or more numbers, measurements, values, etc. and the number of components of the set. *See also* **geometric mean**. *Compare* **median** (def. 1).

mean deviation the arithmetic average of the absolute values of the differences between the individual values of a set of numbers, values, etc. from their mean.

mean free path the average distance travelled by a molecular entity between collisions. In a gas, the mean free path between molecular entities is inversely proportional to the pressure of the gas.

mean ionic activity coefficient *symbol*: $\gamma_\pm$; the average rational activity coefficient of the ions of an electrolyte that dissociates in solution into cations of charge z^+ and anions of charge z^-, given, according to the limiting law of the Debye–Hückel theory, by:

$$-\log\gamma_\pm = z^+z^-AI^{0.5},$$

where A is a constant whose magnitude depends on temperature, dielectric constant, etc., having a value near 0.5, and I is the ionic strength of the solution.

mean life *or* **average life** *abbr*.: τ; the mean of the lengths of time required for the disintegration of all atoms of a given radionuclide. It is related to the **half-life**, $t_{1/2}$, and to the **disintegration constant**, λ, of the radionuclide by the expressions:

$$\tau = t_{1/2}/\ln 2 = 1.443t_{1/2},$$

and

$$\tau = 1/\lambda.$$

mean relative residue mass *or (formerly)* **mean residue weight** *symbol*: M_o; the quotient of the relative molecular mass of a polymer and the number of residues it contains.

mean residue rotation the average molar optical rotation for a residue in a polymer. It is obtained by substituting the mean relative residue mass, M_o, for the relative molecular mass, M_r, of the polymer in the equation defining **molar optical rotation**.

mean residue weight *a former name for* **mean relative residue mass**.

Mec *abbr. for* the 4-methylcoumarin-7-yl group.

mechanical concerned with, executed or operated by, or as by, a machine; relating to mechanics. —**mechanically** *adv*.

mechanically gated channel *see* **ion channel**.

mechanics 1 the science concerned with the equilibrium or motion of bodies in a given frame of reference. **2** the science of machines and/or machinery.

mechanism 1 the chain of events in a particular process; the action or process by which an effect is produced or an action performed. **2** *(in chemistry)* the detailed description of the pathway leading from reactants to products of a reaction, including where possible a characterization of the composition, structure, and other properties of any reaction intermediates and transition states. **3** any system or structure of moving parts that performs some function; a machine.

mechanistic of, pertaining to, or explained by mechanics; of, pertaining to, or resembling a mechanism. *Compare* **stochastic**.

mechanochemical coupling the coupling of a chemical process with a mechanical process in the same structure. *See also* **motor protein**.

mechanotransduction the activation of an intracellular effector pathway by a mechanical force, as in stretch receptors in the walls of the heart and blood vessels.

meconium antigen *see* **carcinoembryonic antigen**.

MECP2 *abbr. for* methylCpG-binding protein 2.

media 1 a plural of **medium**. **2** *(in morphology) or* **tunica media** the middle layer of the wall of a blood or lymph vessel.

medial 1 of, situated in, or towards the middle. **2** of, or pertaining to, a **media** (def. 2).

median 1 *(in statistics)* the centre value of any series of numbers, measurements, values, etc. when these are arranged in order from the lowest value to the highest value (with an even-numbered series, the mean of the two central values is taken); i.e. half the members of the series have values greater than that of the median and half have values smaller than that of the median. *Compare* **mean**. **2** of, pertaining to, situated in, or pointing towards the middle. **3** *(in biology)* of or pertaining to the plane that divides an organism, organ, etc. into two symmetrical parts.

median effective dose *abbr.*: ED_{50} *or* ED50; the dose of a drug, agonist, or other agent that is effective in 50% of the test objects.

median infectious dose *abbr.*: ID_{50} *or* ID50; the dose of bacteria, viruses, or other infective agents that produces infection in 50% of the test objects.

median lethal dose *abbr.*: LD_{50} *or* LD50; the dose of bacteria, viruses, toxic substances, or other agents that causes death of 50% of the test objects.

mediated diffusion *or, sometimes,* **facilitated diffusion** a specific membrane transport process on a transporter (carrier) that does not require coupling with energy and, in the case of nonelectrolytes, that leads to equal concentrations of the transported substance on both sides of the membrane.

mediated transport the transport of a substance across a (biological) membrane that requires the participation of one or more transporting agents (or **carriers** (def. 3)). *See also* **facilitated transport**.

mediator *or* **MED** a multiprotein complex (20–30 polypeptides, ≈2 MDa) that acts as a transcriptional coactivator of yeast. It bridges general initiation factors that bind to promoters of protein genes, and the C-terminal domain of the largest subunit of RNA polymerase II. The complex is distinct from that composed of general transcriptional activators, or of upstream stimulatory activity (USA) factors. Several mammalian complexes that have mediator-like transcriptional activity have been characterized. The human equivalents are **TRAP** and **SMCC**.

medical of, or pertaining to, the science or practice of medicine.

medical informatics *see* **informatics**.

medical subject headings *abbr.*: MeSH; the National Library of Medicine's controlled vocabulary of medical and scientific terms used to index articles in the bibliographic database MEDLINE. MeSH terms provide a consistent method of information retrieval when different terminology has been applied to the same concepts.

medicinal pertaining to or possessing therapeutic properties.

medicine 1 the science or practice of preventing, diagnosing, alleviating, and curing disease. **2** *a common term for* a pharmaceutical substance or mixture.

medin a 50-residue polypeptide, derived from the protein lactadherin, that is produced by aortic smooth muscle cells. It is the commonest form of human amyloid protein but unlike other amyloidogenic peptides it is not degraded by insulinase.

medium *(pl. media or mediums)* **1** the surrounding substance in which organisms or cells are studied or preserved. **2** the continuum in which particular chemical entities and their reactions are studied, commonly the solvent together with any nonreacting solutes. **3** any of the materials in which a chromatographic or an electrophoretic separation is effected. **4** the continuum through which electromagnetic radiation is considered to be transmitted. **5** *see* **culture medium**.

medium-chain acyl-CoA dehydrogenase deficiency the most common defect of mitochondrial beta oxidation in humans, due to deficiency of acyl-CoA dehydrogenase (*see* **beta-oxidation system**). It is characterized by hypoglycemic coma provoked by fasting, hepatic dysfunction, and sudden death. There is abnormal accumulation and urinary excretion of C_6–C_{12} fatty acids and their dicarboxylic acids. Many mutations at the locus 1p31 cause the deficiency.

MEDLINE *see* **bibliographic database**.

medulla *(pl. medullas or medullae)* the innermost part of an organ, tissue, or structure; marrow, pith. —**medullary** *adj.*

MEF *abbr. for* myocyte enhancement factor.

mega+ 1 *symbol*: M; an SI prefix denoting 10^6 times. **2** a prefix used in computing to indicate a multiple of 2^{20} (i.e. 1 048 576), as in megabyte. **3** *a general prefix* denoting large size.

Mega-8 *or* **Omega** the octanoyl amide of *N*-methylglucamine (i.e. 1-methylamino-1-deoxy-D-glucitol); a water-soluble, nondenaturing detergent (CMC = 58 mM) useful for membrane research. It is one of a series, which also comprises Mega-9, the nonanoyl derivative (CMC = 19–25 mM), and Mega-10, the decanoyl derivative (CMC = 6–7 mM).

Mega-8 $n = 6$
Mega-9 $n = 7$
Mega-10 $n = 8$

megabase *symbol*: Mb; a unit of length of a polynucleotide equal to 10^6 base residues in a single-stranded polynucleotide or 10^6 base pairs in a double-stranded polynucleotide.

megadalton *symbol*: MDa; a unit of mass equal to 10^6 **daltons**.

megakaryocyte a large (diameter 40–150 μm), polyploid cell that gives rise to the blood platelets by budding of its cytoplasm. It contains, on average, four times the diploid number of chromosomes in a large, usually lobulated nucleus. Megakaryocytes are found in bone marrow, lung, and to a lesser extent in other tissues.

megalin an alternative name for **LRP2**.

megaloblastic anemia any of a group of disorders characterized by an abnormality of red cell development in the bone marrow. There is an accumulation of primitive cells, with failure of nuclear development, while cytoplasmic development, including hemoglobinization, continues; oval macrocytes appear in the blood. Vitamin B_{12} or folate deficiency, or abnormality of vitamin B_{12} or folate metabolism, is a frequent cause, but drug treatment (with antifolates, e.g. methotrexate, or drugs that interfere with DNA synthesis, e.g. cytarabine (β-cytosine arabinoside), hydroxyurea, AZT) or certain leukemias may also be causes.

megatomic acid *see* **tetradecadienoic acids**.

meiosis *or* **reduction division** a process by which the nucleus of a diploid cell divides twice with the formation of four daughter haploid cells, each containing half the number of chromosomes of the parent nucleus. Meiosis occurs during the formation of gametes from diploid cells and at the beginning of haplophase in those organisms that exhibit alternation of generations. Meiosis involves two rounds of cell division. Prophase in the first division proceeds as follows. Chromosomes first become visible as elongated threads during **leptotene**; by this time each chromosome has already replicated to form two identical sister chromatids. During **zygotene** each chromosome pairs with its homologue; the two become aligned and the **synaptonemal complex** develops between them. When synapsis is complete **pachytene** is said to have begun; it is during this stage that **crossing over** occurs between a chromatid in one partner and another chromatid in the homologous chromosome. The points of crossing over, or **chiasmata**, first become visible as the two chromosomes begin to separate in **diplotene**, when the synaptonemal complex dissolves. The final phase of prophase I, **diakinesis**, is the transition into metaphase I, which is analogous to mitotic metaphase and is followed by anaphase I and telophase I, producing two daughter cells each with one copy of each chromosome (each still made up of two sister chromatids). A second division quickly follows, comprising an interphase (without chromosome replication), and mitosis-like prophase II, metaphase II, anaphase II, and telophase II to separate the two chromatids in each chromosome. *Compare* **mitosis**. —**meiotic** *adj.*

MEK *abbr. for* **1** methyl ethyl ketone. **2** mitogen-activated ERK-activating kinase; a protein kinase with dual specificity for threonine and tyrosine. During intracellular signalling by insulin it phosphorylates these residues in MAP kinases (*see* **ERK**).

Mel-18 *or* **Zfp144** a gene, first identified in a melanoma, that encodes a nuclear RING finger protein. It is related to *Polycomb* of *Drosophila*, which controls expression of homeotic genes. The gene is widely expressed during mouse embryogenesis.

melanin any of the black or brown pigments of high molecular mass formed by the action of oxidase enzymes on phenols. They are widely distributed in plants and animals, are derived from tyrosine, and are usually bound to protein. Plant melanins are called **catechol melanins** because they yield catechol on alkali fusion; animal melanins, termed **eumelanins**, arise from the formal polymerization of indole-5,6-quinone and thus, in contrast to plant melanins, contain nitrogen. **Pheomelanin** is red and/or yellow and is derived from tyrosine and cysteine, as an alternative to eumelanin production in melanosomes. —**melanic** *adj*.

melanin concentrating hormone *abbr*.: MCH; a neuropeptide (19 amino acids, one disulfide bond) of the intermediate lobe of the pituitary gland of teleost fishes and of the lateral hypothalamus of mammals. Infused into mice it stimulates feeding behaviour. The precursor of mammalian MCH also contains a 13-residue peptide homologous to CRH, and a 19-residue peptide homologous to GHRH. The physiological role of MCH is unclear. It acts on a G-protein-coupled membrane receptor that activates phospholipase C.

melanocortin *a generic term for* either of the peptide hormones **melanotropin** or **corticotropin**, derived from their common precursor **proopiomelanocortin**. Although they have distinct functions (melanotropin in control of melanocyte growth and pigment formation, corticotropin in stimulating the adrenal cortex to produce glucocorticoids and aldosterone), they also share functions in, e.g., immunomodulation and antipyretic activity; because corticotropin can regulate melanocyte function, it is possible also to regard it as a melanotropin.

melanocortin receptor any of a family of G-protein-associated membrane receptors that on binding of ligand activate adenylate cyclase. Several types of receptor have been identified. **Melanocortin-1 receptor** (*other names*: melanocyte-stimulating hormone receptor; melanotropin receptor) and **melanocortin-2 receptor** (*other name*: adrenocorticotropic hormone receptor) are both located on melanocytes and adrenocortical tissue. **Melanocortin-3 receptor** and **melanocortin-4 receptor** are both located on brain, placental, and gut tissue, whereas **melanocortin-5 receptor** is expressed in brain but not melanocytes. Human ACTH receptor is 39% identical with the human MSH receptor. Receptors 1–3 and 5 respond to α-, β-, and γ-MSH and ACTH, melanocortin-4 receptor being specific to a common heptapeptide core sequence MGHFRWG. However, despite similarities in structure and action on melanocortin receptors, physiologically MSH has no ACTH-like activity. These are more closely related to the Δ^9-tetrahydro-cannabinol receptor (*see* **cannabinoid receptor**) than to other peptide and amine G-protein-associated receptors. Mutations that lead to deficiency of melanocortin-4 receptor lead to severe obesity in childhood.

melanocyte a type of pigmented cell that can synthesize and store **melanin** in melanosomes. Melanocytes occur in the skin and in the pigmented layer of the eye. *See also* **melanophore**.

melanocyte regulatory factor *an alternative name for* **melanotropin regulatory factor**.

melanocyte-stimulating hormone *abbr*.: MSH; any of three peptide hormones, α-MSH, β-MSH, and γ-MSH, produced primarily by the intermediate lobe of the pituitary gland, that cause dispersal of melanosomes in **melanophores** of poikilothermic vertebrates, but whose physiological function in mammals is not understood. Their sequences are all contained within that of **proopiomelanocortin** (the common precursor of corticotropin and β lipotropin). α-MSH is a tridecapeptide amide (usually acetylated on the N-terminal serine residue) that corresponds to residues 1–13 of corticotropin and has the same sequence in almost all animals studied. β-MSH (*formerly called* **intermedin**), not believed to be present in humans, corresponds to residues 41–58 of β lipotropin and shows greater structural variability. γ-MSH, a putative hormone, is a peptide of variable length that corresponds to part of the so-called cryptic region of proopiomelanocortin. *See also* **melanotropin**.

melanocyte-stimulating hormone receptor *see* **melanotropin receptor**.

melanocyte-stimulating-hormone regulatory factor *or* **melanocyte-stimulating-hormone regulatory hormone** *an alternative name for* **melanotropin-regulatory factor**.

melanocyte-stimulating-hormone release-inhibiting factor (*abbr*.: MSHIF *or* MSHRIF) *or* **melanocyte-stimulating-hormone release-inhibiting hormone** (*abbr*.: MSHIH *or* MSHRIH) *an older name for* **melanostatin**.

melanocyte-stimulating-hormone releasing factor (*abbr*.: MSHRF) *or* **melanocyte-stimulating-hormone releasing hormone** (*abbr*.: MSHRH). *See* **MSHRH**.

melanoliberin *the recommended name for* melanocyte-stimulating-hormone releasing hormone (*see* **MSHRH**).

melanophilin a protein found on the inner surface of the plasma membrane of melanocytes that, with the small GTPase Rab27a, forms a receptor for myosin-Va and mediates movement of melanosomes.

melanophore a type of **melanocyte** that occurs especially in the skin of poikilothermic vertebrates. The cell's contractile properties allow rapid internal translocation of melanosomes, thereby bringing about rapid adaptive changes in the skin coloration of such creatures.

melanopsin an opsin-based photopigment present in retinal ganglion cells of various animals (frog, mouse, rat). It colocalizes with pituitary adenylyl cyclase-activating peptide (PAPEP), and is a component of the photoreceptive system for circadian rhythms.

melanosome a type of cell organelle that occurs in **melanocytes** and contains tyrosinase and other enzymes required for synthesis of melanin. Melanosomes are specialized lysosomes, sharing with them many proteins unrelated to melanin synthesis.

melanostatin a peptide that inhibits release of melanotropin, and is identical with **neuropeptide Y**. *Other names*: melanotropin release-inhibiting factor (*abbr*.: MIF *or* MRIF); melanotropin release-inhibiting hormone (*abbr*.: MIH *or* MRIH); melanocyte-stimulating-hormone release-inhibiting factor (*abbr*.: MSHIF *or* MSHRIF); melanocyte-stimulating-hormone release-inhibiting hormone (*abbr*.: MSHIH *or* MSHRIH).

melanotransferrin *see* **transferrin**.

melanotropic *or* **melanotrophic** describing any agent that causes dispersal of melanosomes within **melanophores**.

melanotropin *or* (*formerly*) **melanotrophin** *a generic term for* any melanotropic hormone of the anterior pituitary that is derived from **proopiomelanocortin**. The term embraces α, β, and γ **melanocyte-stimulating hormones**. *See also* **melanocortin**.

melanotropin receptor any of a family of receptors that bind **melanotropins**. *See also* **melanocortin receptor**.

melanotropin-regulatory factor *or* **melanotropin-regulatory hormone** *an older generic name for* either **MSHRH** or **melanostatin**. *Other names*: melanocyte-regulatory factor; melanocyte-stimulating-hormone regulatory factor; melanocyte-stimulating-hormone regulatory hormone.

melanotropin release-inhibiting factor (*abbr*.: MIF *or* MRIF) *or* **melanotropin release-inhibiting hormone** (*abbr*.: MIH *or* MRIH) *an older name for* **melanostatin**.

melanotropin-releasing factor (*abbr*.: MRF) *or* **melanotropin-releasing hormone** (*abbr*.: MRH) *see* **MSHRH**.

MELAS *abbr. for* mitochondrial encephalomyopathy, lactic acidosis, and strokelike episodes; a mitochondrial disease of humans commonly associated with point mutations in mitochondrial DNA encoding a leucyl-tRNA.

melatonin *N*-acetyl-5-methoxytryptamine; a neuroendocrine tryptophan derivative synthesized in the pineal gland from 5-hydroxytryptamine; its synthesis exhibits a circadian photoperiodism. In amphibia, it has an action opposite to that of **melanocyte-stimulating hormone**. It stimulates the aggregation of melanosomes in melanophores, thus lightening the skin. In mammals it seems to regulate circadian rhythms, sleep, and mood, and may have effects on the reproductive system. It acts through G-protein-associated membrane receptors of which type 1 inhibit adenylate cyclase, and type 2 activate phospholipase C. It is inactivated by 6-hydroxylation and is excreted in conjugated form. When used as a drug it is reputed to control jet lag.

melezitose the trisaccharide, *O-α-*D-glucopyranosyl-(1→3)-*O-β-*D-fructofuranosyl-(2→1)-*α-*D-glucopyranoside; a component of the sweet exudations from many plants and of the leaves of some trees after insect attack.

melibiase *see* **glucosidase**.

melibiose the disaccharide, 6-*O-α-*D-galactopyranosyl-D-glucose; it occurs as a constituent of the trisaccharide, **raffinose**, and in the exudates and nectaries of a number of plants.

melissyl *an alternative name for* **myricyl**.

melitriose *see* **raffinose**.

melittin a strongly basic, sulfur-free, hemolytic, 26-residue peptide amide, often formylated on its N-terminal glycine residue. It comprises 40–50% of the dry mass of the venom of the honey bee, *Apis mellifera*. The hemolytic activity of melittin appears to be related to the ordering of its amino-acid residues, which renders the polypeptide markedly surface active.

melting 1 the (reversible) act or process of liquefaction of a solid through the action of heat (or change in pressure). **2** the increase (reversible and usually fairly abrupt) of molecular mobility of lipids in a biological membrane or artificial lipid bilayer with increase of temperature above the transition temperature. **3** *or* **thermal denaturation** the transition (usually fairly abrupt but not necessarily reversible) of a biopolymer from a highly ordered to a less highly ordered state with increase in temperature. Examples are the heat-induced change from a helical to a random coil structure in a protein or a duplex nuleic acid. *See also* **localized melting**, **melting temperature**.

melting curve *an alternative name for* **thermal denaturation profile**.

melting point *abbr.*: m.p.; the temperature at which the solid and liquid phases of a substance are in equilibrium at a given pressure. Melting points are usually given for standard atmospheric pressure, i.e. 101 325 Pa.

melting temperature *symbol*: T_m or t_m or Θ_m; the temperature at which a conformational transition brought about by change in temperature (e.g. of a macromolecule in solution) is 50% complete. For example, it is the temperature (°C) at which 50% of a target DNA sequence is hybridized to oligonucleotides present in molar excess. It follows therefore that $T_m = \Delta H_{ot}/\Delta S_{ot}$, where ΔH_{ot} and ΔS_{ot} are, respectively, the standard enthalpy difference and the standard entropy difference for the transition. In the denaturation of DNA the melting temperature is taken as the midpoint of the **helix-to-coil transition**.

MEM *abbr. for* **1** minimum essential medium. **2** minimum Eagle's medium.

memapsin2 *see* **secretase**.

membrane any thin sheet or layer, often separating two compartments. Membranes may be natural or artificial, and are sometimes soft and pliable. *See also* **cell membrane**, **semipermeable membrane**, **unit membrane**. —**membranous** *adj.*

membrane attack complex *see* **MAC** (def. 1).

membrane attack complex inhibition factor (*abbr.*: MACIF) *or* **CD59 glycoprotein precursor**; *other names*: MAC-inhibitory protein, protectin, membrane inhibitor of reactive lysis (*abbr.*: MIRL); a surface antigen of T cells that potently inhibits the membrane attack complex (*see* **MAC**). It is complexed with a GPI anchor.

membrane-bound describing any component that is not readily dissociable from preparations of cellular membranes *in vitro* and/or that appears to be an integral part of a membrane in an intact cell.

membrane-bound O-acyltransferase *abbr.*: MBOAT; a widely distributed superfamily of integral membrane proteins that have *O*-acyltransferase activity. They typically have 8–10 transmembrane segments and transfer organic acids (especially fatty acids) to the hydroxyl group of a membrane-embedded acceptor. MBOATs share a region of conserved polar residues within and adjacent to the membrane, and an invariant histidine within a hydrophobic sequence. Examples include **sterol O-acyltransferase** (yeast and human), **diacylglycerol O-acyltransferase** (mouse and *Arabidopsis*), and porcupine protein (*Drosophila*).

membrane channel a protein complex that specifically facilitates the passive diffusion of small molecules through a membrane; particularly applied to **ion channels**.

membrane electrode any electrode that incorporates a membrane in its structure, e.g. a glass electrode, an ion-exchange electrode, or an oxygen electrode.

membrane filter *or* **filter membrane** any membrane consisting of a thin polymeric structure with pores of a defined and very uniform size that is used for filtration; a disk or piece of such membrane.

membrane filtration the act or process of separating suspended particles from a liquid, or of separating macromolecules from their solvent, with a **membrane filter**.

membrane-intrusive describing a peptide loop, situated between two neighbouring transmembrane segments of an integral membrane protein, that is forced back into the membrane but without penetrating it. An example is the **P domain** of shaker-type subunits of ion channels.

membrane-organizing spike protein *see* **moesin**.

membrane osmometer *see* **osmometer**.

membrane potential the electric potential existing across any membrane arising from charges in the membrane itself and from the charges present in the media on either side of the membrane. Plant and animal plasma membranes maintain a potential, the cell interior being more negative than the exterior. In plants this arises mainly from the activity of electrogenic pumps, whereas in animals passive ion movements predominate in creating this potential. The resting potential varies from –20 mV to –200 mV according to cell type; for a neuron it is about –70 mV. In plants, **H⁺-transporting ATPases** pump protons out of the cell, thus maintaining a pH gradient across the plasma membrane. This is involved in symport of carbohydrates and amino acids into the cell. In animals, the Na⁺,K⁺-ATPase exchanges intracellular Na⁺ for extracellular K⁺, maintaining the high intracellular K⁺ concentration and low intracellular Na⁺ concentration. K⁺-leak channels in the plasma membrane allow K⁺ to escape until counteracted by the electrical potential and excess of intracellular negative charge that results. An electrochemical gradient across the mitochondrial inner membrane is created by proton-pumping cytochromes. *See also* **chemiosmotic coupling hypothesis**, **Nernst equation**.

membrane protein any protein that is normally closely associated with a cell membrane, owing to the properties of the protein and the method by which it is synthesized. **Integral proteins** contain hydrophobic regions that traverse the lipid bilayer; in some cases, the protein has only one such region, with hydrophilic domains on either side of the membrane; in other cases, where several such regions are present, a polypeptide may traverse the membrane several times (*see* **seven-transmembrane-domain proteins**). Other proteins are bound to the membrane by a **gpi anchor**, and others by the hydrophobic moiety of a fatty-acyl group (*see* **myristoylation**) or a prenyl group (*see* **prenylation**). Other proteins may be more loosely attached, mainly by ionic forces (peripheral proteins). **Type I membrane proteins** have a single (monotopic) transmembrane domain, with the C terminus on the cytoplasmic side. **Type II membrane proteins** have a single transmembrane domain with an N terminus that is cytoplasmic. **Type III membrane proteins** traverse the membrane several times (polytopic). **Type IV membrane proteins** are multimers of subunits that when assembled make a transmembrane channel. Proteins attached to the membrane only by covalently bound lipids are classified as **type V**, and those that are anchored by one or more transmembrane domains and also by a GPI anchor are designated **type VI**. The structures of only a few membrane proteins have been solved at high resolution (7 Å or better). Except for these it is not known whether the transmembrane domain is helical.

membrane receptor any protein, situated in or on a cell membrane, that binds extracellular signalling molecules (e.g. hormones, growth factors) and transduces the signal to produce intracellular effects.

membrane skeleton the protein network of an erythrocyte membrane, defined operationally as the insoluble residue that remains after extraction of either intact red cells or their isolated membranes with the nonionic detergent Triton X-100.

membrane teichoic acid any **teichoic acid** found in the protoplast of a bacterial cell as opposed to any found exclusively in the cell wall or in capsular material. Such substances commonly are glycerol teichoic acids, and they occur, covalently attached to lipid, in the cell membrane and probably also in mesosomes.

membrane transport any movement of substances across a (biological) membrane.

membrane type-serine protease 1 *see* **MT-SP1**.

memory cell any of a population of long-lived **T** or **B** lymphocytes that have previously been stimulated by antigen, and make an accelerated response to that antigen if they encounter it again. Memory B cells carry surface immunoglobulin G as their antigen receptor; this is of higher affinity than on virgin lymphocytes. Memory T cells may be distinguishable by high expression of CD44.

MEN *abbr. for* multiple endocrine neoplasia; any of two distinct genetic syndromes in which multiple endocrine neoplasias occur. MEN1 is due to numerous loss-of-function mutations in the nuclear protein **menin**, with usually benign tumours of anterior pituitary, parathyroid, and enteropancreatic neuroendocrine cells. MEN2 is due to gain-of-function mutations in the *ret* protooncogene, which is expressed as a cell surface glycoprotein that is a member of the receptor protein kinase family. It usually presents as medullary thyroid carcinoma, often associated with pheochromocytoma or parathyroid adenoma.

menadione 2-methyl-1,4-naphthalenedione; 2-methyl-1,4-naphthoquinone; vitamin K_3; a synthetic naphthoquinone derivative possessing the physiological properties of **vitamin K**.

menaquinone *abbr.*: MK; *the class name for* a series of 2-methyl-3-all-*trans*-polyprenyl-1,4-naphthoquinones, also known as vitamins K_2 (*see* **vitamin K** for function). They are antihemorrhagic vitamins, present in bacteria, and possess 1–13 prenyl residues, the number of prenyl residues being indicated by a suffixed Arabic number, e.g. menaquinone-6 (MK-6; *also known as* **farnoquinone**). [From methyl naphthaquinone.]

menaquinone

Mendelian 1 of or relating to Gregor Johann Mendel (1822–84), Austrian monk and botanist, or following his laws of heredity. **2** one who adheres to or supports **Mendel's laws of heredity**.

Mendelian population a community of sexually interbreeding individuals.

Mendel's laws of heredity the principles of heredity proposed by Gregor Johann Mendel (1822–84), Austrian monk and botanist, on the basis of his researches on hybridization of peas to explain the behaviour of factors (now called genes) that govern the transmission of inherited characters from generation to generation. On the basis of modern knowledge, Mendel's first law, the **law of segregation**, states that every somatic cell carries a pair of allelic genes for each character and that the two genes in each pair separate during meiosis so that each gamete carries only one gene from each pair. Mendel's second law, the **law of independent assortment**, states that the separation of members of any pair of allelic genes occurs independently of every other pair provided they are not linked.

menin a ubiquitously expressed nuclear protein (human, 610 amino acids) that contains two nuclear localization signals at the C-terminus. It binds and inhibits the transcription factor JunD and may be a tumour suppressor. Numerous mutations are associated with the autosomal dominant multiple endocrine neoplasia type 1 (MEN1). *See* **MEN**.

meniscus (*pl.* **menisci**) the curved upper surface of a liquid column that is concave when the containing walls are wetted by the liquid (contact angle <90°) or convex when not (contact angle >90°), sometimes extended to include a flat surface (contact angle = 90°). —**meniscoid** *adj.*

meniscus depletion (sedimentation equilibrium) method *or* **Yphantis method** a variant of the sedimentation equilibrium method for determining the relative molecular mass of a macromolecule in which centrifugation is effected at a speed high enough to render the concentration of the solute at the meniscus essentially zero; the concentrations of solute along the length of the cell are proportional to the difference in refractive index at any point from that at the meniscus. The method is especially useful for monodisperse systems.

Menkes ATPase *see* **Menkes disease**.

Menkes disease *or* **kinky-hair syndrome** *or* **steely-hair syndrome** a rare X-linked disorder characterized by cerebral and cerebellar degeneration, abnormalities of connective tissue, and kinky hair. It results from mutations in the gene for the Menkes ATPase, a copper transporter involved in the intestinal absorption of copper. [After John H. Menkes (1928–), US neurologist.]

menopausal gonadotropin *or* **human menopausal gonadotropin** (*abbr.*: hMG *or* HMG) *another name for* **menotropin**.

menotropin *or* **menotrophin** *other name*: human menopausal gonadotropin; *abbr.*: hMG *or* HMG; a gonadotropin with **follicle-stimulating hormone** and **luteinizing hormone** activity, isolated from the urine of postmenopausal women and used to stimulate ovulation in infertility. One proprietary name is Pergonal.

MEP *abbr. for* major excreted protease; *see* **cathepsin L**.

meprin A EC 3.4.24.18; *other names*: endopeptidase-2; endopeptidase MEP-1; a type 2 transmembrane serine protease of kidney and intestinal brush borders. Zinc is a cofactor. It catalyses the hydrolysis of proteins and peptides on the C-terminal side of hydrophobic residues, and is a homotetramer of α or β subunits.

meq *or* **mEq** *symbol for* milliequivalent (use deprecated, mmol recommended).

mer *or* **constitutional unit** the repeating unit of a polymeric molecule.

+mer *suffix forming nouns (in chemical nomenclature)* denoting specific kinds or classes of polymer (e.g. hexamer, oligomer) or isomer (e.g. enantiomer, epimer, tautomer). —**+meric** *adj.*; **+merism** *n.*

mer the symbol for genes of the *mer* operon whose products confer (or regulate) mercury resistance. For example, the bacterium *Serratia marcescens* has several *mer* products. (1) MerA is mercury(II) reductase (EC 1.16.1.1); *systematic name*: Hg:NADP$^+$ oxidoreductase; one of the very rare enzymes with metal substrates, it catalyses the reduction by NADPH of Hg^{2+} to form Hg and NADP$^+$. It is a flavoprotein (FAD) with redox-active disulfide bonds, and is responsible for volatilizing mercury as Hg(0). (2) MerB is alkylmercury lyase (EC 4.99.1.2); *systematic name*: alkylmercury mercuriclyase; it catalyses a reaction between RHg$^+$ and H$^+$ to form RH and Hg^{2+}. This enzyme cleaves the carbon–mercury bond of organomercurials and the Hg^{2+} product is detoxified by MerA. (3) MerR is a **helix-turn-helix** regulatory protein. In the absence of mercury, MerR represses transcription by binding tightly to the *mer* operator region; when mercury is present the dimeric complex binds a single ion and becomes a potent transcriptional activator, while remaining bound to the *mer* site. *mer* operons are encoded by a number of plasmids that have been cloned into *Escherichia coli*, e.g. Tn21 of *Shigella flexneri*, Tn501 from *Pseudomonas aeruginosa* (these two lack the organomercury lyase gene and so confer resistance only against inorganic mercury salts), and pDU1358 from *S. marcescens* (which includes *merB* and thus encodes a broad-spectrum *mer* operon).

mer+ *prefix (in coordination compound nomenclature)* denoting a **meridional** (def. 2) arrangement of ligands.

MER *abbr. for* medium reiteration frequency repeat; *other usage*: medium reiteration frequency interspersed repetitive elements (*abbr.*: MER elements). They consist of repeated base sequences in eukaryotic genomes. More than fifty human repetitive families are known, of which most represent MERs. The latter have copy numbers per genome ranging from hundreds to thousands, and length ranging from tens to a few hundred base pairs. They are of unknown function, but some may play a role in regulation of gene expression and others may affect genome stability and gene rearrangements.

mercapt+ *a variant form of* **mercapto+** *(sometimes before a vowel)*.

mercaptan *an old name for* **thiol**.

mercapto+ *or (sometimes before a vowel)* **mercapt+** *comb. form (in chemical nomenclature)* indicating the presence of an unsubstituted thio group, –SH.

2-mercaptoethanol HO–CH$_2$–CH$_2$–SH; 2-hydroxyethylmercaptan; a sulfhydryl reducing agent often used in buffers to maintain protein sulfhydryl groups in the reduced state.

mercaptopurine *or* **6-mercaptopurine** *abbr.*: 6-MP; *symbol*: Shy; thiohypoxanthine; 6-purinethiol; an immunosuppressive and antineoplastic drug useful in organ transplantation and in the treatment of some leukemias. It is a substrate for guanine-hypoxanthine phosphoribosyltransferase, which converts it to **6-mercaptopurine ribonucleoside** (or 6-thioinosine; *symbol*: Sno *or* M *or* sI).

mercapturic acid any *S*-alkyl- or *S*-aryl-*N*-acetylcysteine; such compounds are formed as detoxification products in mammals. The initial reaction involves conjugation with the thiol group of glutathione catalysed by a **glutathione transferase** to form the *S*-radyl-glutathione (*see* **radyl**). Glutamic acid and glycine are then removed in the kidney, and the resulting mercapturic acid is excreted in the urine.

mercurate to treat or mix with mercury; to introduce mercury into an organic compound. —**mercuration** *n.*

mercurial 1 of, containing, like, or pertaining to mercury. 2 any compound containing mercury, especially one used in medicine and/or agriculture; *see also* **organomercurial**.

mercuric perchlorate Hg(ClO$_4$)$_2$; a substance used in solution to trap ethylene.

mercury II reductase EC 1.16.1.1; product of the *Serratia marcesens merA* gene (*see* **mer**).

mercury-vapour lamp *or* **mercury-discharge lamp** *or (informally)*

mercury lamp a lamp that emits bluish-green light rich in ultraviolet radiation, produced by passing an electric discharge through mercury vapour, often at low pressure. It is useful for provoking or investigating fluorescence, as a light source for a **polarimeter**, and as a source of sterilizing radiation.

meridian 1 any line on the surface of a sphere or spheroid that is orthogonal to its equator, or any corresponding line in a projection of part of the surface of a sphere or spheroid onto a planar or cylindrical surface. 2 in a fibre diagram produced by X-ray diffraction, the vertical axis on the film that corresponds to the *z* axis in reciprocal space.

meridional 1 of, along, pertaining to, or resembling a meridian. 2 describing the geometrical isomer of a coordination compound of octahedral configuration in which three ligands in the coordination sphere are in such a relationship that one is *cis* to the two others, which are themselves *trans*. Such an isomer is designated by prefixing *mer-* to the chemical name.

meristem a region of tissue in a plant that is composed of one or more undifferentiated cells that are capable of undergoing mitosis and differentiation, and thereby of effecting growth and development of a plant by giving rise either to more meristem or to specialized tissue. In vascular plants, meristems may be apical (located at the tips of growing roots and shoots) or lateral (arranged circumferentially and constituting the cambium). —**meristematic** *adj.*

merlin *or* **neurofibromatosis type 2 (tumour suppressor)** (*abbr.*: NF2) *or* **Schwannomin** a membrane-stabilizing protein from fetal brain and other tissues, such as kidney, lung, and breast. It is similar to **ezrin** and **talin**. It is a tumour suppressor of the **TERM** family. In humans, it is encoded by a gene on chromosome 22q12.

mero+ *comb. form* denoting part, partly, or partial.

merocrine 1 describing a type of secretion that does not entail disintegration of the secretory cell. 2 describing a gland made up of such cells. *Compare* **apocrine**, **eccrine**, **holocrine**.

merodiploid a haploid organism that is diploid for a small region of the chromosome, i.e. a partial diploid.

meromyosin either of the two fragments, **heavy meromyosin** and **light meromyosin**, produced by brief treatment of myosin with trypsin or certain other proteases.

MEROPS *see* **specialist database**.

mero-receptor the smallest subunit of a receptor for a (steroid) hormone that contains an intact hormone-binding site. The biological significance of mero-receptors is unclear; they may be artefacts of experimental cellular disruption or products formed by proteolysis associated with the termination of (steroid) hormone action.

merosin a tissue-specific-basement-membrane protein, and the M chain of **laminin**. It is covalently associated with laminin B chains and, like laminin, is cruciform. It shares similarity with the C-terminal part of the laminin A chain.

merozygote any bacterial cell that is partly diploid and partly haploid, formed by conjugation, transduction, or transformation. It contains one complete genome, that of the recipient organism, and a fragment of a second genome, derived from the donor.

MERRF *abbr. for* myoclonus epilepsy with ragged-red fibres; a mitochondrial disease of humans characterized by abnormally shaped mitochondria (which sometimes contain paracrystalline inclusions) in skeletal muscle fibres. It is commonly associated with point mutations in the mitochondrial DNA for lysyl-tRNA.

Merrifield method *an alternative name for* **solid-phase peptide synthesis**. [After Robert Bruce Merrifield (1921–), US peptide chemist and biochemist.]

Merrifield resin a beaded reactive synthetic resin designed for use in the Merrifield method (i.e. **solid-phase peptide synthesis**).

mersalyl *pharmaceutical name of* sodium *o*- [3-(hydroxymercuri)-2-methoxypropyl]carbamoyl phenoxyacetate; a mercury-containing derivative of salicylamide. It is a diuretic drug useful as an inhibitor of the exchange of phosphate and hydroxide ions across the inner mitochondrial membrane.

mes+ *a variant form of* **meso+** *(sometimes before a vowel).*

Mes *or* **MES** *abbr. for and trivial name of* 2-(*N*-morpholino)ethane-sulfonic acid; a **Good buffer substance**, pK_a (20°C) = 6.15.

mesaconic acid (*E*)-2-methyl-2-butenedioic acid; methylfumaric acid; an intermediate in the fermentation of glutamate to acetyl-CoA by *Clostridium tetanomorphum*.

mescaline 3,4,5-trimethoxyphenylethylamine; a hallucinogenic and sympathomimetic drug isolated from the flowering heads of the mescal cactus (*Lophophora williamsii*), also known as peyotle or peyote.

Meselson–Stahl experiment a landmark experiment, reported by Meselson and Stahl in 1958, that provided convincing evidence for semiconservative replication of DNA in bacteria. Cultures of *Escherichia coli* were grown in a medium containing ^{15}NH$_4$Cl (96.5% isotopic purity) as sole nitrogen source for many generations so as to label the DNA with heavy nitrogen. The cells were then transferred to a medium containing ^{14}NH$_4$Cl and the culture sampled for several subsequent generations. Cell lysates were subjected to density-gradient centrifugation and the positions of the DNA bands in the gradients located by UV absorption. Initially the DNA appeared as a single band corresponding to DNA containing heavy nitrogen; subsequently a second band of DNA, containing equal amounts of heavy and light nitrogen, appeared, and one generation after transfer to natural ammonium chloride all the DNA was this hybrid form. After two generations only hybrid and natural DNA was present. Furthermore, the separated strands of this hybrid DNA proved to contain all heavy nitrogen or all normal nitrogen only. [After Matthews Meselson (b. 1930), US molecular biologist, and Franklin William Stahl (b. 1929), US geneticist.]

mesenchyme immature, unspecialized connective tissue found in the early embryo of animals. It consists of cells embedded in a tenuous extracellular matrix. —**mesenchymal** *adj.*

mesentery 1 the double-layer fold of peritoneum serving to hold the abdominal viscera in position. **2** any of the muscular partitions extending inwards from the body wall in coelenterates. —**mesenteric** *adj.*

MeSH *abbr. for* medical subject headings.

Me₃Si *symbol for* the trimethylsilyl group.

meso describing a bridging atom or group between two rings in a polycyclic molecule.

meso+ *or (sometimes before a vowel)* **mes+** *comb. form* denoting middle, or intermediate.

meso- *abbr.*: ms-; *prefix* (*in chemical nomenclature*) designating a substance whose individual molecules contain equal numbers of enantiomeric groups, identically linked, but no other chiral group; i.e. a substance whose molecules are optically inactive because of internal compensation; e.g. *meso*-compound, *meso*-cystine. *Compare* **racemo-**.

meso-carbon atom *a term suggested for* any prochiral carbon atom in an organic molecule whose valence bonds are linked to two identical groups, *a*, and to two other dissimilar groups, *b* and *d*. The resulting molecule, C*aabd*, has a plane of symmetry and thus the carbon atom is not chiral; however, the two *a* groups are not geometrically equivalent but bear a mirror-image relationship to each other, and they may react at different rates with chiral reagents or in enzymic reactions. The term is now obsolete. See **prochiral**.

mesoderm the middle of the three primary germ layers of a triploblastic animal embryo, lying between the ectoderm and endoderm. It is the source of bone, muscle, connective tissue, the inner layer of the skin, and other structures. —**mesodermal** *adj.*

meso-inositol *a former name for* **myo-inositol**.

mesomeric effect (*in chemistry*) an experimentally observable effect (on reaction rates, etc.) of a substituent caused by the overlap of its p- or π-orbitals with the p- or π-orbitals of the rest of the molecular entity, **resonance** (def. 2) (or mesomerism) being thereby introduced into, or extended within, the molecular entity in question. *Compare* **inductive effect**.

mesomerism (*in chemistry*) a term used to imply that the correct representation of the structure of a particular molecular entity is intermediate between two or more formal valence-bond representations (known as **contributing structures**); **resonance** (def. 2). —**mesomeric** *adj.*

mesomorphic 1 (*in chemistry*) existing in, designating, or pertaining to an intermediate state between a true liquid and a true solid that exists in a **liquid crystal**. **2** (*in biology*) having a form, size, or structure that is average, normal, or intermediate between extremes.

mesophase any phase of matter intermediate between a true liquid and a true solid that exists in a **liquid crystal**.

mesophil *or* **mesophile** any organism that thrives at moderate temperatures, between 20 and 40°C. —**mesophil, mesophile,** *or* **mesophilic** *adj.*

mesophyll the internal parenchyma of a foliage leaf between the two epidermal layers; it is usually photosynthetic. —**mesophyllic** *adj.*

mesophyte any plant requiring moderate amounts of water for optimum growth. —**mesophytic** *adj.*

mesopore any pore (in a gel or porous solid) intermediate in size between a micropore and a macropore, i.e. of width in the range roughly 2.0–50 nm. —**mesoporous** *adj.*

mesosecrin a glycoprotein, M_r 46 000, secreted by human mesoderm-derived and endothelial cells in culture.

mesosome an intracellular, often complex, membranous structure found in many bacteria and probably formed by invagination of the plasma membrane. It is more complex and larger in Gram-positive organisms than in Gram-negative ones. Its function is unclear. —**mesosomal** *adj.*

mesothelium (*pl.* **mesothelia**) a layer of flat cells derived from mesoderm and found lining various body cavities. —**mesothelial** *adj.*

mesotocin [Ile⁸]oxytocin; a neurohypophyseal hormone present in some nonmammalian vertebrates, so-named because it is intermediate in structure between **isotocin** and **oxytocin**. It may represent an ancestral molecule, which has been preserved in these groups of vertebrates but has undergone a single residue replacement in mammals.

mesotrophic describing an environment providing a moderate amount of nutrients.

message 1 a specific piece of information, or something containing information, that is transmitted from a donor to a recipient by some means or agent. **2** a jargon term used, sometimes without an article, as the name of a substance, to denote the particular **messenger RNA** (mRNA) of interest in a specified context, or any preparation containing that mRNA. **3** the segment of mRNA (or any particular segment of a polycistronic mRNA) that consists of a sequence of codons carrying the information determining the sequence of amino-acid residues of a specific polypeptide to be synthesized at a ribosome. **4** the part of the information contained within the amino-acid sequence of a hormonal polypeptide, or of its biosynthetic precursors, that elicits a specific physiological response to the hormone. *Compare* **address** (def. 2).

message-dependent (of a partially purified, cell-free, protein-synthesizing system) capable of a significant response to added mRNA.

message sequence *or* **message region** the segment (or segments) of the complete sequence of amino-acid residues of a molecule of a hormonal or prohormonal polypeptide in which the **message** (def. 4) resides. *Compare* **address sequence**.

messenger 1 something, or someone, that bears or carries a **message** (def. 1), or that is able to do so. **2** a molecule or substance that carries genetic information; *see* **messenger RNA**. **3** a molecule or substance that carries physiological instructions; a **rheoseme**. *See also* **chemical messenger, first messenger, second messenger**.

messenger ribonucleoprotein *abbr.*: mRNP; any complex of messenger RNA (mRNA) and protein. Included are: nuclear ribonucleoprotein that contains mRNA sequences; cytoplasmic ribonucleoprotein that is associated with polyribosomes; and free cytoplasmic ribonucleoprotein particles, also known as **informosomes**.

messenger RNA *or* **template RNA** *abbr.*: mRNA; a class of naturally occurring RNA molecules that carry the information embodied in the genes of DNA to the ribosomes, where they direct protein synthesis. Molecules of mRNA are single-stranded, relatively short-lived, and formed by the process of **transcription** from the template strand of genomic DNA. They carry genetic information in the form of a succession of **codons**. Mature eukaryotic mRNAs contain a 5′ cap and a 3′ poly(A) tail. At the ribosomes, by the process of **translation**, they orchestrate the synthesis of the corresponding polypeptide(s). The plus strand of certain RNA viruses can act as messenger RNA without transcription, as can a synthetic polyribonucleotide. The primary transcript of DNA is often referred to as **hnRNA**, which contains the **introns** that have to be removed by excising and splicing together the **exons** to form the mature mRNA. mRNA may also undergo 'editing' whereby some bases are changed. *See also* **mRNA binding site, splicing.**

messenger-RNA-interfering complementary RNA *abbr.*: micRNA; *an alternative name for* **antisense RNA.**

Messing vector any one of certain cloning vectors derived from bacteriophage M13 by insertion of a synthetic multipurpose cloning site and which are especially suitable for the cloning and rapid sequence determination of small (≈500 base pair) fragments of DNA. These have developed into the **pUC vectors**. [After Joachim Messing (1946–).]

met+ *a variant form of* **meta+** (usually before a vowel or letter h).

Met *symbol for* a residue of the α-amino acid L-methionine (alternative to M).

MET a gene, first identified in an *N*-methyl-*N*′-nitrosoguanidine-treated cell line, that encodes a receptor tyrosine kinase with disulfide-linked α and β subunits. The receptor binds **hepatocyte growth factor** and undergoes autophosphorylation of the cytoplasmic domain. Cells co-transfected with *MET* and *HGF* are highly tumorigenic. Germ-line missense mutations lead to a familial form of renal cancer. *See also* **hepatocyte growth factor receptor.**

meta+ *or (before a vowel or letter h)* **met+** *prefix* **1** indicating alteration, alternation, or change of condition or position; after, distal to, behind, beyond; more highly specialized form of. **2** *(in inorganic chemistry)* denoting the least hydrated acid of a series of oxoacids formed from any one of certain anhydrides, or the anion or a salt formed from such an acid; *Compare* **ortho+** (def. 2), **pyro+** (def. 2). It is now approved only for the polymeric acids: metaboric acid, $(HBO_2)_n$, metaphosphoric acid, $(HPO_3)_n$, and metasilicic acid, $(HSiO_3)_n$; and for anions and salts of these acids (*but see* **metabisulfite**). **3** *(in organic chemistry)* distinguishing a particular polymeric form of the substance named, e.g. metaldehyde, metaformaldehyde.

meta- *symbol*: m-; *prefix (in organic chemical nomenclature)* denoting the isomer of a disubstituted monocyclic aromatic compound in which the substituents are separated by one intervening unsubstituted ring-carbon atom. *Compare* **ortho-, para-.**

meta-analysis a quantitative systematic review in which the results of many studies dealing with the same topic are combined. It is used in the study of multifactorial diseases.

metabisulfite *an old name, still frequently used, esp. in commerce, for* inorganic salts of the general formula $M_2S_2O_5$, where M is a monovalent metallic element.

metabolic of, pertaining to, occurring in, produced by, or liberated during **metabolism**. —**metabolically** *adv.*

metabolic acidosis any clinical condition in which hydrogen ions accumulate in the blood plasma through an excessive production of acidic metabolites (as in ketoacidosis or lactic acidosis), the excessive ingestion of acids, or a defect in the excretion of hydrogen ions by the kidneys.

metabolic activation bioactivation, usually with particular reference to the conversion of procarcinogens to carcinogens in the body.

metabolic alkalosis any clinical condition in which there is a deficiency of hydrogen ions in the blood plasma caused by the excessive

ingestion of alkalis, extrarenal loss of acid (as with vomiting), or therapy with diuretic agents.

metabolic balance *see* **balance** (def. 1).

metabolic block the complete or partial interruption of a specific reaction in a metabolic pathway caused by the absence or greatly reduced activity of the enzyme catalysing that reaction (for genetic reasons or due to the action of a specific agonist). The consequences of a metabolic block include: (1) the complete or partial absence of the final product of the pathway; (2) the accumulation of a metabolic intermediate of the pathway before the block, for which no significant alternative pathway exists; (3) the accumulation of products of an alternative, normally minor, metabolic pathway. *See also* **inborn error of metabolism.**

metabolic code a model for the evolution of biological regulation and the origin of hormone-mediated intercellular communication. It is characterized by a set of symbols and their corresponding domains. A **symbol** is an intracellular molecule, e.g. a hormone or other messenger, that indicates to the cell the existence of a particular state of the environment, e.g. lack of glucose or of amino acids. The symbol regulates a diverse set of biochemical events and reactions in the cell, the nature and number of which are the **domain** of that symbol.

metabolic compartmentation *or* **metabolic compartmentalization** the concept that certain metabolic reactions occur in separate compartments of the cell (or organism) and are therefore not necessarily in equilibrium with each other.

metabolic control *an alternative term for* **metabolic regulation.**

metabolic cycle any closed sequence of metabolic reactions in which an end product is a reactant in the initiation of the cycle, e.g. the **tricarboxylic-acid cycle.**

metabolic disease any disease caused by a recognizable abnormality of metabolism.

metabolic flux the rate of passage of material, or of a specified substance or molecular fragment, through a given metabolic pathway.

metabolic map a diagram representing the sequence and relationships of enzyme-catalysed reactions in metabolism. It typically consists of the names and/or structures of metabolites, connected by arrows representing enzyme-catalysed reactions.

metabolic pathway any series of connected (enzymic or other) reactions occurring in a cell or organism, whether anabolic, catabolic, or amphibolic.

metabolic pool the sum total of all the different amounts of a given substance, group of similar substances, or specified molecular fragment, irrespective of their distribution in a cell, organ, tissue, or organism, that are in complete equilibrium with each other in regard to participation in a specified set of metabolic reactions.

metabolic quotient *symbol*: Q_x ; a measure of the rate at which a given substance (x) is taken up by, or discharged from, a cell, organ, tissue, or organism.

metabolic rate the rate of metabolism of an animal measured under given conditions. It may be determined from the dioxygen consumption and carbon dioxide (and urinary nitrogen) excretion, or from the total energy output of the animal. *Compare* **basal metabolic rate.**

metabolic reconstruction the process of modelling the metabolism of an organism (e.g. using biochemical, sequence, and phenotypic data) in order to identify which pathways are present and which genes are involved in which functional roles.

metabolic regulation the coordinated action of the many systems that control the activity of metabolic pathways. The systems embrace gene regulation and control of protein biosynthesis, as well as the control of the rate of enzyme reactions by phosphorylation of the enzyme protein or by the action of allosteric effectors. The action of second messengers is important in the transduction of signals from outside a cell to the interior.

metabolic state (of mitochondria) *an alternative term for* **respiratory state.**

metabolic turnover the continual exchange of cellular components by degradation and synthesis. Even for cell components for which the content and composition remain relatively constant, a continuous breakdown and resynthesis of individual molecules is occuring.

metabolism 1 the totality of the chemical reactions and physical changes that occur in living organisms, comprising **anabolism** and **catabolism**. **2** the totality of the chemical reactions and physical processes undergone by a particular substance, or class of substances, in a living organism.

metabolite any substance, either endogenous or exogenous, that is formed or changed by metabolism. *See also* **primary metabolite, secondary metabolite.**

metabolize *or* **metabolise 1** to affect by metabolism. **2** to effect metabolism. —**metabolizable** *or* **metabolisable** *adj.*

metabolome the universe of low-molecular-weight metabolites present in cells in a particular physiological or developmental state.

metabolomics the quantitative analysis of all the small-molecular-weight metabolites present inside a cell (or other sample).

metabolon a supramolecular complex of consecutively acting metabolic enzymes and cellular structural elements.

metabotropic describing a type of receptor (e.g. certain **glutamate receptors**) that mediates its effects by activation of enzymes, especially **adenylate cyclase** or the **phosphatidylinositol cycle**, in distinction from receptors that regulate ion channels. *Compare* **ionotropic**. *See also* **excitatory amino-acid receptor**.

metachromasia *or* **metachromasy** *or* **metachromatism 1** the property exhibited by certain cationic (i.e. basic) dyes of shifting their absorption spectra towards shorter wavelengths in the presence of substances with appropriately arranged anionic groups. **2** the property exhibited by some biological structures or materials, when treated with certain cationic (basic) dyes, of staining with a different colour from that of the dye used. Biological materials showing metachromasia include heparin, chondroitin sulfate, sulfatides, gangliosides, and polyphosphates. —**metachromatic** *adj.*; **metachromatically** *adv.*

metachromatic granule any cytoplasmic granule that exhibits **metachromasia** (def. 2). **Volutin granules** are an example.

metachromatic leukodystrophy any of a group of hereditary diseases of humans characterized by disintegration of myelin and the accumulation of **metachromatic lipid** – galactosyl sulfatide (cerebroside sulfate) – in the white matter of the central nervous system, in peripheral nerves, and in some visceral organs. It is mainly caused by mutations in the gene for the lysosomal enzyme arylsulfatase A (locus 22q13.31-qter) and the disorder is characterized clinically by progressive paralysis and dementia, usually in the first few years of life, and is transmitted as an autosomal recessive trait. *See also* **saposin**.

metachromism *see* **metachromasia**.

metadrenaline *an alternative name for* **metanephrine**.

metagon a stable, virus-like RNA particle postulated to occur in certain mate-killer strains of *Paramecium aurelia*, in which it acts like messenger RNA.

metahalone any of a class of metaphase inhibitors having the general formula 5-halopyridine-2-one. [From meta(phase) halone.]

metakentrin *an obsolete name for* **luteinizing hormone**.

metal 1 any of a large class of chemical elements that, with certain exceptions, are ductile, malleable solids with a characteristic lustre, are good conductors of electricity and heat, and generally form simple cations. **2** any substance or material consisting entirely or predominantly of one or more **metals** (def. 1). **3** consisting or made of metal. —**metallic** *adj.*

metal-chelate affinity chromatography a form of **affinity chromatography** in which immobilized weak bases such as imidazole coordinated with ions such as Ni^{2+} are used to concentrate recombinant proteins bearing **His-tags**.

metal-ion buffer a **buffer** (def. 1, 2) for metal ions. It comprises a solution of a metal salt, with a suitable complexing agent. The concentration of free metal ions is maintained by the equilibrium with an excess of the complexed form.

metallo+ *comb. form* denoting metal, or characterized by the presence of metal.

metalloantigen any metal-labelled antigen for use in a **metalloimmunoassay**.

metallocarboxypeptidase *recommended generic name for* any **carboxypeptidase** of the sub-subclass EC 3.4.17, for which a bivalent cation is an essential component of the catalytic mechanism.

metallocene *the generic name for* π-bonded bis(η-cyclopentadienyl) coordination compounds of certain metals, e.g. ferrocene, bis(η-cyclopentadienyl)iron, [Fe(C$_5$H$_5$)$_2$].

metallochaperone any of a family of homologous proteins that bind and transport specific metals in the cytoplasm to their sites of utilization. They contain a metal-binding Cys–X–X–Cys motif within a four-stranded antiparallel beta sheet below two alpha helices. Six copies of this motif are present in the N-terminal region of Menkes and Wilson ATPases of the *trans* Golgi network. In yeast and human cells three metallochaperones are required to deliver copper to the Golgi apparatus, mitochondria, and cytosol respectively.

metallochromic describing a dye, indicator, or stain that exhibits a distinctive colour change when complexed with metal ions; the colour change is often specific for a particular metal.

metalloendopeptidase *recommended generic name for* any **endopeptidase** of the sub-subclass EC 3.4.24, for which a bivalent cation is an essential part of the catalytic mechanism.

metalloenzyme any enzyme that is a **metalloprotein**.

metallohapten any metal-labelled hapten for use in a **metalloimmunoassay**.

metalloimmunoassay a form of **immunoassay** in which ions of an appropriate transition metal, tightly complexed to the antigen or hapten in question, are used as the distinguishable label. The amount of metal present in the free and antibody-bound forms is determined by a suitable analytical method such as emission, absorption, or fluorescence spectrometry.

metallopeptidase *see* **metallocarboxypeptidase, metalloendopeptidase**.

metalloporphyrin any compound formed by chelation of a porphyrin with a metal ion, e.g. hemoglobins (with iron), cytochromes (with iron), and chlorophylls (with magnesium).

metalloprotease *a former name for* metalloendoproteinase (*see also* **metallocarboxypeptidase, matrix metalloproteinase**).

metalloprotein any protein containing one or more molecular proportions of a metal bound to it, either directly or as part of a prosthetic group.

metalloproteinase *see* **metallocarboxypeptidase, metalloendopeptidase**.

metalloselenonein a synthetic analogue of the copper metallothionein peptide of *Neurospora crassa* that contains selenocysteine residues in place of cysteine.

metallothionein any of a group of small proteins with extraordinarily high contents both of one or more metals (cadmium, copper, zinc) and of cysteine. They occur widely, from microorganisms to humans, and have highly conserved structures. Mammalian metallothioneins are of 6.8 kDa, consist of 61 amino-acid residues, including 20 cysteine residues, and contain about 7 g atoms of metal per molecule. The apoprotein may be termed **thionein**.

metaprotein *an old term for* the products formed by acid or alkaline hydrolysis of proteins that are soluble in weak acids or alkalis, but insoluble in neutral solutions.

metal proteinase *a former name for* metalloproteinase (*see* **metallocarboxypeptidase, metalloendopeptidase**).

metal response element *abbr.*: MRE; a sequence in the promoter region of metallothionein genes that respond to the presence of metals. *See also* **response element**.

metal response element binding protein *abbr.*: MREBP; a mammalian protein that activates basal and heavy metal-induced transcription of metallothionein genes by binding to the metal response element in their promoter regions. MREBP1 contains a DNA-binding region of six C2H2 zinc fingers and three functional activation domains. Knockout of the mouse gene results in death of embryos with variable degrees of liver damage.

metamer *(in chemistry) a former term for* any one of two or more **constitutional isomers** that are of the same chemical class, e.g. ethylamine, CH$_3$–CH$_2$–NH$_2$, and dimethylamine, CH$_3$–NH–CH$_3$.

metamere *(in morphology)* any of the linearly arranged similar body segments in animals that exhibit **metamerism** (def. 2); a somite.

metameric 1 *(in chemistry)* of or pertaining to a **metamer**; pertaining to the phenomenon of, or displaying **metamerism** (def. 1). **2** *(in morphology)* of, or pertaining to, a **metamere**; consisting of metameres; segmented; of, or pertaining to, **metamerism** (def. 2).

metamerism 1 *(in chemistry)* the occurrence as **metamers** of two or more chemical substances of the same molecular formula. **2** *(in morphology)* the condition of an animal body characterized by the presence of a series of similar segments or **metameres**; segmentation.

metameter a quantity derived from an observation, and independent of all parameters, that conveys the magnitude of the phenomenon observed. A metameter is frequently used to simplify the expression or analysis of dose–response relationships.

metamorphosis 1 a complete change in physical form; a structural transformation. **2** *(in zoology)*: the rapid, postembryonic, structural transformation from a larva into an adult that occurs in certain animals, e.g. the change from a tadpole to a frog. —**metamorphose** *vb.*; **metamorphic** *adj.*

metanephrine *or* **metadrenaline** 3-*O*-methylepinephrine; an inactive metabolite of epinephrine, produced by the action of **catechol** *O*-**methyltransferase**. Small quantities are normally found in the urine, and larger amounts occur in the urine of patients with catecholamine-secreting tumours.

metaphase the stage in **mitosis**, following **prometaphase**, during which the chromosomes become arranged so that their centromeres all lie in one plane and their longitudinal axes are at right angles to the spindle axis; the corresponding phases in **meiosis**.

metaphosphatase *an alternative name for* **1** endopolyphosphatase (EC 3.6.1.10). **2** exopolyphosphatase (EC 3.6.1.11).

metaphosphate *the traditional name for* any anion or any (partial or full) salt or ester derived from the condensed acid metaphosphoric acid, polytrioxophosphoric(v) acid, $(HPO_3)_n$, where n is a small integer, commonly 3, 4, or 6. Metaphosphoric acid and its derivatives are thought not to exist in the monomeric state, although they are often represented as such for convenience. *See also* **polyphosphate**.

metaprotein *an old term for* the products formed by acid or alkaline hydrolysis of proteins that are soluble in weak acids or alkalis but insoluble in neutral solutions.

metarhodopsin either of two intermediates, metarhodopsin I (absorption maximum 480 nm) and metarhodopsin II (absorption maximum 380 nm), formed during the photolysis of **rhodopsin**.

metastable 1 a describing any body or system existing in a state of apparent equilibrium when undisturbed, or subject to very small perturbation, but capable of passing to a more stable state when subjected to greater perturbation. **b** describing such a state or such an equilibrium. **2** describing an excited state of an atom, ion, nucleus, or other quantum-mechanical system that is relatively long-lived because the transitions to a lower energy state are forbidden transitions. **3** *symbol*: m; describing any comparatively stable excited state of a radioisotope that decays to a more stable state by isomeric transition, with the emission of gamma radiation, e.g. the metastable nuclide $^{99m}_{43}\text{Tc}$, which is formed by decay of $^{99}_{42}\text{Mo}$ and which in turn decays by transition to its nuclear isomer $^{99}_{43}\text{Tc}$ with a half-life of 6 h and emission of one quantum of gamma radiation.

metastasis *(pl.* **metastases***)* the transfer of disease, especially tumour cells, from one part of the body to another by way of the natural passages (blood vessels, lymphatics) or by direct continuity. —**metastasize** *or* **metastasise** *vb.*; **metastatic** *adj.*

metazoa the plural of **metazoan** (def. 1) *or* **metazoon**.

Metazoa a subkingdom of animals consisting of multicellular animals whose cells are organized into tissues and their activities coordinated by a nervous system. The group includes all animals except the Protozoa and Parazoa.

metazoal *a variant form of* **metazoan** (def. 2).

metazoan 1 *or* **metazoon** *(pl.* **metazoa***)* any animal of the subkingdom Metazoa. **2** *or* **metazoal** *or* **metazoic** of, characteristic of, or belonging to the Metazoa.

metazoic *a variant form of* **metazoan** (def. 2).

metazoon *(pl.* **metazoa***) an alternative term for* **metazoan** (def. 1).

[Met⁵]enkephalin *or* **[Met]enkephalin** *abbr. for* [5-methionine]enkephalin.

meter 1 an instrument or device for measuring (and recording) the quantity of something (either instantaneously or cumulatively). **2** to measure (something) by means of a **meter** (def. 1). **3** to supply (something) in a measured or regulated amount. **4** *the US spelling of* **metre**.

+meter *comb. form* denoting an instrument or device for measuring (something indicated or specified). —**+metric** *adj.*

metestrus *see* **estrous cycle**.

meth+ *a variant form of* **metho+** *(before a vowel)*.

methadone (hydrochloride) 6-dimethylamino-4,4-diphenyl-3-heptanone hydrochloride; a narcotic analgesic used in cases of opioid dependence.

R-(–)-enantiomer

methaemoglobin *a variant spelling (esp. Brit.) of* **methemoglobin**.

methaemoglobinaemia *a variant spelling (esp. Brit.) of* **methemoglobinemia**.

methane *or* **marsh gas** a colourless, odourless, flammable gas, CH_4; it is the simplest alkane and is the principal constituent of natural gas found in association with mineral oil. It is formed by the anaerobic decay of vegetable matter and in large amounts by ruminants and other herbivores.

methano+ *comb. form.* **1** indicating methane. **2** *(in chemical nomenclature)* denoting the presence of a $-CH_2-$ bridge in a polycyclic hydrocarbon. **3** indicating organisms or compounds involved in metabolism of methane, e.g. **methanotroph**, methanopterin.

methanoate *see* **formate**.

methanogen any microorganism that produces methane. —**methanogenic** *adj.*

methanogenesis the microbial production of methane by anaerobic breakdown of carbon-containing compounds.

methanogenic the ability to form methane from CO_2, a property of some anaerobic bacteria. The pathway involves the coenzyme **tetrahydromethanopterin** *(abbr.:* H₄MPT*)*, and involves formation of 5-formyl-H₄MPT, from which successively methenyl-H₄MPT, methylene-H₄MPT, and methyl-H₄MPT are formed in reactions analogous to those for **folate coenzymes**. From the last of these, methyl is transferred to **coenzyme M** to form methyl-coenzyme M, from which methane is formed in a reductive step involving **factor** $\mathbf{F_{430}}$. *See also* **coenzyme F₄₂₀**.

methanol methyl alcohol; CH_3-OH; the lowest member of the alkanols. It is a colourless, flammable, mobile, poisonous liquid, widely used as a solvent. Numerous esters and ethers of methanol occur naturally, especially in plants. *Other names*: wood alcohol; wood spirit; carbinol. —**methanolic** *adj.*

methanol oxidase *see* **alcohol oxidase**.

methanolysis alcoholysis using methanol.

MetHb *abbr. for* methemoglobin (i.e. **methemoglobin**).

methemalbumin *or (esp. Brit.)* **methaemalbumin** a complex of metheme with albumin, which imparts a brownish colouring to plasma. It is formed as the result of *in vitro* hemolysis.

methemoglobin *or (esp. Brit.)* **methaemoglobin** *abbr.:* MetHb; *an older name for* **ferrihemoglobin**.

methemoglobinemia *or (esp. Brit.)* **methaemoglobinaemia** any clinical condition characterized by an excess of **ferrihemoglobin** (i.e. methemoglobin) in the erythrocytes. It accompanies a number of **hemoglobinopathies**, and also results from administration of large amounts of sulfonamides. A hereditary enzymopenic methemoglobinemia is caused by mutations in the gene for cytochrome b_5 reductase. Ferrihemoglobin imparts a brown colour to urine.

methene *an old name, still commonly used in biochemistry, for* the trivalent diatomic group, $=CH-$, attached to two other groups in an organic molecule. The name now recommended is **methyne**.

methene bridge the **methene** group that links consecutive pyrrole rings in the molecular structure of tetrapyrroles and related compounds.

methenyl+ *prefix (to a trivial biochemical name)* denoting substitution of a **methene** group into the indicated parent compound (at the positions specified by appropriate locants); e.g. 5,10-methenyltetrahydrofolate. *Compare* **methylene+**.

methenyltetrahydrofolate cyclohydrolase EC 3.5.4.9; *systematic name*: 5,10-methenyltetrahydrofolate 5-hydrolase (decyclizing); one of the enzymes involved in transformations between **folate coenzymes**. It catalyses the hydrolysis of 5,10-methenyltetrahydrofolate to form 10-formyltetrahydrofolate. One of three enzymes (*see also* **methylenetetrahydrofolate dehydrogenase (NADP⁺)** and **formate–tetrahydrofolate ligase**) necessary for the biosynthesis of purines, thymidylate, methionine, histidine, pantothenate, and formyl tRNA-Met.

5,10-methenyltetrahydrofolate synthetase *see* **5-formyltetrahydrofolate cyclo-ligase**.

MeTHFA *abbr. for* 5-methyltetrahydrofolate.

methicillin *see* **meticillin (sodium)**.

methine *a former name for* **methyne**.

methioninate 1 methionine anion; the anion, $CH_3–S–[CH_2]_2–CH(NH_2)–COO^-$, derived from **methionine**. 2 any salt containing methionine anion. 3 any ester of methionine.

methionine the trivial name for α-amino-γ-methylmercaptobutyric acid; 2-amino-4-(methylthio)butanoic acid; $CH_3–S–[CH_2]_2–CH(NH_2)–COOH$; a chiral α-amino acid. L-Methionine (*symbol*: M or Met), (*S*)-2-amino-4-(methylthio)butanoic acid, is a coded amino acid found in peptide linkage in proteins; codon: AUG (or also AUA in various species of mitochondria). D-Methionine (*symbol*: D-Met *or* DMet) is not known to occur naturally. In mammals, methionine is glucogenic, and is an essential dietary amino acid, although either enantiomer will serve (the D-form being oxidatively deaminated to the α-oxo acid, which is then reaminated with inversion).

L-methionine

methionine adenosyltransferase *abbr.* MAT; EC 2.5.1.6; *other names*: *S*-adenosylmethionine synthetase; AdoMet synthetase; an enzyme that catalyses the reactions:

L-methionine + ATP = *S*-adenosyl-L-methionine + PP$_i$

$PPP_i = PP_i + P_i$

Three isozymes (MATI–III) are formed by the two genes whose products are α1 (395 amino acids) and α2 (394 amino acids), which share 84% sequence identity. MATI is (α1)$_4$, while MATIII is (α1)$_2$, both being present in adult liver. MATII is present in many tissues and in fetal (but not adult) liver. Mutations in the α1 gene lead to isolated hypermethioninemia.

methionine cycle *see* **methionine salvage pathway**.

methionine dioxide *symbol*: MetO$_2$; methionine *S,S*-dioxide; $CH_3–SO_2–[CH_2]_2–CH(NH_2)–COOH$; an oxidation product of methionine, *formerly known (incorrectly) as* **methionine sulfone**.

[5-methionine]enkephalin *abbr.*: [Met]enkephalin *or* [Met⁵]enkephalin; the recommended designation for the trivial name of the pentapeptide Tyr-Gly-Gly-Phe-Met. *See* **enkephalin**.

methionine oxide *symbol*: MetO; methionine *S*-oxide; $CH_3–SO–[CH_2]_2–CH(NH_2)–COOH$; an oxidation product of methionine and an intermediate in the formation of **methionine dioxide**. *It was formerly known (incorrectly) as* **methionine sulfoxide**.

methionine salvage pathway *or* **methionine cycle** *or* **Yang cycle** a metabolic cycle occurring in plant cells in which methionine is synthesized from carbons originating in the ribose of ATP, the sulfur atom being conserved throughout the cycle. In outline, the cycle consists of reactions in which, starting with L-methionine, *S*-adenosylmethionine (I) is formed in a reaction with ATP. I is then converted by 1-aminocyclopropane-1-carboxylate synthase, EC 4.4.1.14, to 5′-methylthioadenosine (II) and 1-aminocyclopropane-1-carboxylate, a precursor of ethylene (*see* **ethylene-forming enzyme**). 5′-Methylthioadenosine is also produced during synthesis of polyamines. Adenine is then cleaved from methylthioadenosine to form 5-methylthioribose from which 5-methylthioribose-1-phosphate (III) is formed by reaction with ATP. III is then converted through a series of intermediates to 2-keto-4-methylthiobutyrate from which methionine is formed by a transamination reaction. [After S.F. Yang.]

methionine sulfone an old (incorrect) name for **methionine dioxide**.

methionine sulfoxide an old (incorrect) name for **methionine oxide**.

methionine sulfoximine [3-amino-3-carboxypropyl]methyl sulfoximine; an inhibitor of glutamine synthetase (**glutamate–ammonia ligase**) in plants, and identified as the toxic agent in flour treated with the bleaching agent NCl₃ (agene, hence the term **agenized flour**). It interferes with the action of γ-amino-*n*-butyric acid in the brain. This is associated with the ability of methionine sulfoximine to cause fits in a wide variety of animals – ferrets are particularly susceptible. Although methionine sulfoximine has not been clearly proven to be toxic to humans, agene is no longer used for bleaching flour.

methionine synthase *see* **homocysteine methyltransferase**.

methioninium methionine cation; the cation, $CH_3–S–[CH_2]_2–CH(NH_3^+)–COOH$, derived from **methionine**.

methionino the alkylamino group, $CH_3–S–[CH_2]_2–CH(COOH)–NH–$, derived from **methionine**.

methionyl the acyl group, $CH_3–S–[CH_2]_2–CH(NH_2)–CO–$, derived from **methionine**.

methionyllysylbradykinin *or* Met-Lys-bradykinin *see* **bradykinin**.

methionyl-tRNA formyltransferase EC 2.1.2.9; *systematic name*: 10-formyltetrahydrofolate:L-methionyl-tRNA *N*-formyltransferase. An enzyme that catalyses a reaction between 10-formyltetrahydrofolate, L-methionyl-tRNA, and H₂O to form tetrahydrofolate and *N*-formylmethionyl-tRNA. *N*-formylmethionyl-tRNA is involved in the initiation of prokaryote **translation**.

metho+ *or (before a vowel)* **meth+** *(comb. form) a variant of* **methyl+** in the names of certain chemical compounds.

methotrexate 4-amino-10-methylfolic acid; *other name*: amethopterin; a folic acid analogue that is a potent competitive inhibitor ($K_i < 10^{-9}$ M) of **dihydrofolate reductase**. It is useful in the treatment of acute leukemia and choriocarcinoma. It acts by blocking nucleotide synthesis and thus also DNA synthesis and has similar activity to **aminopterin**. The crystal structure of a complex of methotrexate and dihydrofolate reductase from *Lactobacillus casei* is known.

methoxatin *or* **pyrroloquinoline quinone** 4,5-dihydro-4,5-dioxo-1*H*-pyrrolo-[2,3-*f*]quinoline-2,7,9-tricarboxylic acid; a heat-stable fluorescent cofactor implicated in the action of primary alcohol dehydrogenases in methylotrophic bacteria and possibly more generally associated with the oxidation of single-carbon compounds. *See* **quinoprotein**.

methoxy *symbol*: OMe *or* MeO; the alkoxy group, $CH_3–O–$, derived from methanol by loss of a hydrogen atom.

3-methoxy-4-hydroxymandelate acid *see* **hydroxymethoxymandelate**.

5-methoxypsoralen an intercalating reagent for DNA; it forms covalent cross-links on irradiation. *See also* **psoralen**.

methoxyvinylglycine *abbr.*: MVG; L-2-amino-4-methoxy-*trans*-3-butenoic acid; an inhibitor of ethylene biosynthesis in plants.

methuselah a gene of *Drosophila* for a long N-terminal type B G-protein-coupled membrane receptor that regulates the ageing process. Its reduced expression increases resistance to various types of stress (e.g. starvation, high temperature, dietary toxins) and increases lifespan by 35%, but null mutations are lethal to embryos.

methyl *symbol*: Me; the alkyl group, $CH_3–$, derived from methane.

methyl- *prefix (in chemical nomenclature)* used as a locant to indicate in or at a methyl group.

methyl-accepting chemotaxis protein *abbr.*: MCP; any of several similar integral plasma membrane proteins that may be present in the cell membranes of particular strains of certain motile bacteria, e.g. *Escherichia coli* and *Salmonella typhimurium*, and appear to mediate their responses to corresponding classes of chemotactic agent by undergoing enzymic methylation/demethylation of the γ-carboxylate group(s) of one or more glutamic residues (for the enzymes responsible *see* **che**). The type of ligand determines the reaction, which in turn determines the action of the receptor on the bacterial flagellar motor, so that the effect can be either attractant or repellent. The following examples are from *E. coli*. MCP1 is a receptor for attractants such as L-serine and other amino acids and is also responsible for avoidance of a wide range of repellents, including leucine, indole, and weak acids. MCP2 is a receptor for L-aspartate, dicarboxylic acids, and maltose (via interaction with a periplasmic maltose-binding protein), and is also responsible for avoidance of cobalt and nickel. MCP3 interacts with periplasmic ribose- or galactose-binding proteins and mediates movement towards these sugars. MCP4 interacts with periplasmic dipeptide-binding proteins and mediates movement towards such peptides.

N^6-methyladenine a methylated base of DNA; this and **5-methylcytosine** are the only modified bases in the DNA of cellular organisms.

methyl alcohol *a common name for* **methanol**.

methylase *a common name for* **methyltransferase**.

N-methyl-D-aspartate *abbr.*: NMDA; an excitatory amino acid and a neurotransmitter. It acts as agonist for the *N*-methyl-D-aspartate receptor, a subclass of glutamate ionotropic receptors (*see* **glutamate receptor**).

methylaspartate ammonia-lyase *recommended name for* methylaspartase; *see* **methylaspartate mutase**.

methylaspartate mutase EC 5.4.99.1; *systematic name*: L-*threo*-3-methylaspartate carboxy-aminomethylmutase; *other name*: glutamate mutase. An enzyme that catalyses the reaction:

L-*threo*-3-methylaspartate = L-glutamate.

This is the first step of the glutamate-fermentation pathway of the anaerobic bacteria *Clostridium cochlearium*, *C. tetanomorphum*, and others. The product, (2*S*,3*S*)-3-methylaspartate, is converted to mesaconate by the enzyme methylaspartate ammonia-lyase (EC 4.3.1.2). Methylaspartate mutase requires a cobamide coenzyme and has been studied as an example of such a reaction. The enzyme has two subunits (large and small, sometimes called E and S).

methylate 1 to introduce (one or more) methyl groups into (a substance). **2** *(in commerce)* to mix or adulterate with methanol. *See also* **methylation of DNA**. —**methylation** *n*.

methylated-DNA–protein-cysteine *S*-methyltransferase EC 2.1.1.63; *other names*: 6-*O*-methylguanine-DNA methyltransferase; O^6-methylguanine-DNA-alkyltransferase. It catalyses the reaction between DNA (containing 6-*O*-methylguanine) and protein L-cysteine to form DNA (without 6-*O*-methylguanine) and protein *S*-methyl-L-cysteine. As the acceptor is in the enzyme itself this is a 'suicidal' reaction. Bacterial and mammalian enzymes have common motifs.

methylation of DNA the occurrence in vertebrate DNA of varying amounts of 5-methylcytosine which arises from methylation of cytosine bases where they occur in the sequence CpG. The methylation status of DNA correlates with its functional activity: inactive genes are more heavily methylated. *See also* **CpG islands**.

3-methylcholanthrene a carcinogenic hydrocarbon, found in coal tar. It induces synthesis of cytochrome P450 mRNA.

methyl-coenzyme M reductase EC 2.8.4.1; coenzyme-B sulfoethylthiotransferase; a nickel-containing enzyme that catalyses the methane-releasing reaction in methanogenic bacteria. Severe enzyme deficiency from many mutations leads to developmental delay, seizures, and psychiatric manifestations, hyperhomocystinemia and homocystinuria, premature atherosclerosis, and demyelination. *See* **factor F_{430}**.

methyl CoM reductase coenzyme-B sulfoethylthiotransferase, EC 2.8.4.1, an enzyme in methanogenic archaea that catalyses the final reaction of methanogenesis: the reaction of methyl coenzyme M and coenzyme B, with the formation of the mixed disulfide and release of methane. Contains the nickel-containing **factor F_{430}**.

methylCpG-binding protein 2 *abbr.*: MECP2; a ubiquitously expressed, very abundant nuclear protein (486 amino acids) of human tissues that is associated with 5-methylcytosine-rich heterochromatin. It helps recruit histone deacetylases and other proteins. Mutations result in an X-chromosome-linked neurodegenerative disorder of girls that starts 6–18 months after birth and is called Rett syndrome.

methylcrotonyl-CoA carboxylase EC 6.4.1.4; an enzyme of the pathway for the degradation of leucine. It catalyses the formation of 3-methylglutaconyl-CoA from ATP, 3-methylcrotonyl-CoA, and HCO_3^- with release of ADP and orthophosphate. Biotin is a cofactor. The 3-methylglutaconyl-CoA is then converted to 3-hydroxy-3-methylglutaryl-CoA.

5-methylcytosine a methylated base of DNA; this and **N^6-methyladenine** are the only modified bases in the DNA of cellular organisms. In T-even phages all the C residues of DNA are 5-hydroxymethylcytosine.

methyldopa L-2-amino-2-methyl-3-(3,4-dihydroxyphenyl)propionic acid; a compound used as an antihypertensive. It is a competitive inhibitor of **aromatic L-amino acid decarboxylase** (dopa decarboxylase), but its effects on blood pressure may result from its metabolism to α-methylnorepinephrine, which stimulates presynaptic α_2 receptors, thus decreasing norepinephrine release.

methylene 1 *the recommended name for* the bivalent triatomic group CH_2 in a molecule, whether attached to two other atoms by single bonds or to one other atom by a double bond. **2** *a former name for* **carbene** (def. 1).

methylene+ *prefix (to a trivial chemical name)* denoting substitution of a **methylene** group into the indicated parent compound (at the positions specified by appropriate locants); e.g. 5,10-methylenetetrahydrofolate (which may be shortened to 5,10-CH_2-H_4folate). *Compare* **methenyl+**.

***N*,*N*'-methylenebisacrylamide** $(CH_2:CHCONH_2)_2CH_2$; a crosslinking agent used in a ratio (w/w) of between 1:49 and 1:19 with acrylamide in the preparation of polyacrylamide gels.

Methylene Blue *or* **Swiss Blue** *or* **Basic Blue 9** CI 52015; tetramethylthionine chloride; 3,7-bis(dimethylamino)phenothiazinium chloride; a water- and ethanol-soluble blue dye; A_{max} (in water) at 663–667 nm. It is reducible to a leuko (i.e. colourless) compound, which is reoxidizable by dioxygen, and hence is useful as a redox indicator; $E^{o'}$ (pH 7.0, 30°C) = +0.011 V. It is useful also as a stain in bacteriology and histology, e.g. being preferentially taken up by melanin-containing cells, and can be used to target therapeutic agents to such cells.

3,4-methylenedioxymethamphetamine *abbr.*: MDMA; a phenethylamine derivative and analogue of **mescaline** and **amphetamine** that is a potent CNS stimulant and hallucinogen; *popularly known as* **ecstasy**. A controlled substance in the US Code of Federal Regulations.

S-(+)-enantiomer

2-methyleneglutarate mutase EC 5.4.99.4; *systematic name*: 2-methyleneglutarate carboxy-methylenemethylmutase. An enzyme that catalyses the reaction:

2-methyleneglutarate = 2-methylene-3-methylsuccinate.

It is a key enzyme in the fermentation of nicotinate to ammonia, propionate, acetate, and CO_2 by the strict anaerobe *Clostridium barkeri*, but has been much investigated as an example of an enzyme mechanism based on the adenosylcobalamin coenzyme. 2-Methylene-3-methylsuccinate is converted to 2,3-dimethylmaleate by methylitaconate Δ-isomerase (EC 5.3.3.6). The enzyme is a homotetramer of 70 kDa subunits.

methylene-interrupted double bonds double bonds separated by one methylene group. The term is used especially to describe the double bonds in naturally occurring long-chain fatty acids; in many cases these are all of the *Z* configuration, with each pair separated by one methylene group.

methylenetetrahydrofolate dehydrogenase (NADP⁺) EC 1.5.1.5; *systematic name*: 5,10-methylenetetrahydrofolate: NADP⁺ oxidoreductase. *other name*: C1-tetrahydrofolate synthase, cytosolic; an enzyme involved in the interconversion of folate coenzymes that catalyses the reaction:

5,10-methylenetetrahydrofolate + NADP⁺ =
5,10-methenyltetrahydrofolate + NADPH.

One of three enzymes (*see also* **methenyltetrahydrofolate cyclohydrolase**, and **formate–tetrahydrofolate ligase**) necessary for the biosynthesis of purines, thymidylate, methionine, histidine, pantothenate, and formyl tRNA-Met. The human enzyme is a cytosolic trifunctional homodimer of 934 amino acids; the N-terminal domain (amino acids 1–304) has methylenetetrahydrofolate dehydrogenase and cyclohydrolase activity; the C-terminal section (305–934) forms the formyltetrahydrofolate synthetase domain (*other names*: mitochondrial (precursor) NAD-dependent methylenetetrahydrofolate dehydrogenase (EC 1.5.1.5); methenyltetrahydrofolate cyclohydrolase (EC 3.5.4.9)).

methylesterase any of various enzymes that catalyse the hydrolysis of methyl esters. For example, **juvenile-hormone esterase** (EC 3.1.1.59) of insects, *other name*: JH-esterase, catalyses the hydrolysis of methyl (2*E*,6*E*)-(10*R*,11*S*)-10,11-epoxy-3,7,11-trimethyltrideca-2,6-dienoate to form (2*E*,6*E*)-(10*R*,11*S*)-10,11-epoxy-3,7,11-trimethyltrideca-2,6-dienoate and methanol; i.e. it demethylates the insect juvenile hormones JH_1 and JH_2. It is similar to certain other carboxylesterases and lipases. The **protein-glutamate methylesterase** of *Escherichia coli* is involved in chemotaxis in this organism (*see* **che**).

methylglucosamine *N*-methyl-L-glucosamine; 2-deoxy-2-(methylamino)-L-glucose; a component of **streptomycin**.

3-methylglutaconic aciduria any condition in which there is excess 3-methylglutaconic acid in urine. Type I is due to deficiency of a specific hydratase that converts 3-methylglutaconyl-CoA to 3-hydroxy-3-methylglutaryl-CoA in mitochondria during catabolism of leucine. Type II (*see* **Barth syndrome**) is due to mutations in **tafazzin**, and like types III and IV is associated with normal 3-methylglutaconyl-CoA hydratase activity.

*O*⁶-**methylguanine** a modified and, in principle, highly mutagenic base in DNA (it base pairs with T). It has a special repair enzyme, **methylated-DNA–protein-cysteine *S*-methyltransferase**.

7-methylguanine a modified base found in transfer RNA. *See* **guanine** (for numbering), **7-methylguanosine**.

7-methylguanosine a modified nucleoside that forms a **cap** at the 5'-terminus of eukaryotic mRNA. *See also* **7-methylguanine**.

methylidyne *the recommended name for* the trivalent diatomic group CH≡ attached to a single other group in a molecule of an organic compound. *Compare* **methine**.

methylmalonate $CH_3CH(COO^-)_2$; isosuccinate 2-methylpropanedioate; the dianion of methylmalonic acid; as **methylmalonyl-CoA** it is an intermediate in the conversion of propionyl-CoA to succinyl-CoA. Note: methylmalonate should be distinguished from methyl malonate (two words), which is propanedioic acid dimethyl ester.

methylmalonic acidemia *or* **methylmalonic aciduria** any autosomal recessive disease in which abnormally large amounts of methylmalonate occur in the blood and urine. The condition may be caused by abnormalities in either **methylmalonyl-CoA epimerase** or **methylmalonyl-CoA mutase** apoenzymes, or by three different abnormalities in adenosylcobalamin synthesis.

methylmalonyl-CoA 2-carboxypropanoyl-CoA; it can exist as the enantiomers (*R*)- and (*S*)-methylmalonyl-CoA; (*S*)-methylmalonyl-CoA is an intermediate in the **beta oxidation** of odd-numbered fatty acids in animals. Propionyl-CoA is the major end product; for further metabolism, this is converted to succinyl-CoA by, first, carboxylation to (*S*)-methylmalonyl-CoA, which is converted by **methylmalonyl-CoA epimerase** to (*R*)-methylmalonyl-CoA, which is then converted to succinyl-CoA by **methylmalonyl-CoA mutase**.

methylmalonyl-CoA epimerase EC 5.1.99.1; an enzyme that interconverts (*R*)-methylmalonyl-CoA to (*S*)-methylmalonyl-CoA. It is sometimes incorrectly referred to as methylmalonyl-CoA racemase (chiral centres in the CoA moiety are not affected).

methylmalonyl-CoA mutase EC 5.4.99.2; *systematic name*: (*R*)-2-methyl-3-oxopropanoyl-CoA CoA-carbonylmutase; an enzyme that converts (*R*)-methylmalonyl-CoA to succinyl-CoA; adenosylcobalamin is a cofactor. This mitochondrial matrix enzyme is required for degradation of several amino acids, odd-chain fatty acids, and cholesterol via propionyl-CoA to the tricarboxylic-acid cycle; deficiency results in **methylmalonic acidemia**. In some bacteria the enzyme is a heterodimer; both subunits have homology with the human protein. Its role is in the synthesis of propionate from tricarboxylic-acid-cycle intermediates.

methylmercuric hydroxide a powerful reagent used for the denaturation of the secondary structure of RNA molecules, but limited in its use owing to its toxicity.

2-methyl-1,4-naphthalenedione 2-methyl-1,4-naphthoquinone; *see* **menadione**.

Methyl Orange a dye used as an indicator, changing from red to orange-yellow over the pH range 3.1–4.4.

ethyl *N*-succinimidyl carbonate; a reagent used for the introduction of the **MSOC** amino-protecting group.

methylotroph any organism that can use certain single-carbon compounds, in a lower oxidation state than CO_2, as sole carbon source. —**methylotrophic** *adj.*; **methylotrophy** *n.*

methylotrophic yeast any yeast belonging to either of the genera *Hansenula* or *Pichia*. They are potentially valuable recombinant-DNA expression hosts as certain oxidase genes (*MOX* and *AOX*) are vastly overexpressed and their promoters are suitable for the expression of foreign proteins. *See also* **alcohol oxidase, methylotroph**.

3-methyl-2-oxobutanoate dehydrogenase (lipoamide) EC 1.2.4.4; *other names*: 2-oxoisovalerate dehydrogenase; branched-chain α-keto-acid dehydrogenase; branched-chain α-keto-acid decarboxylase; an enzyme present in a complex that catalyses the oxidative decarboxylation of oxo acids derived from branched-chain amino acids. It is similar in mechanism to the **pyruvate dehydrogenase complex**. The E_1 component of the complex, it catalyses the release of CO_2 from the oxo acid and transfer of the resulting acyl group to lipoamide. Thus, for the oxo acid from valine, it catalyses the reaction of 3-methyl-2-oxobutanoate with lipoamide to form *S*-(2-methylpropanoyl)dihydrolipoamide with release of CO_2. Thiamine diphosphate is a coenzyme. *See also* **[3-methyl-2-oxobutanoate dehydrogenase (lipoamide)] kinase**.

[3-methyl-2-oxobutanoate dehydrogenase (lipoamide)] kinase EC 2.7.1.115; *other name*: branched-chain α-keto-acid dehydrogenase kinase; an enzyme that catalyses the phosphorylation by ATP of [3-methyl-2-oxobutanoate dehydrogenase (lipoamide)] with release of ADP, thereby inactivating the enzyme complex.

[3-methyl-2-oxobutanoate dehydrogenase (lipoamide)]-phosphatase EC 3.1.3.52; *other name*: branched-chain oxo-acid dehydrogenase phosphatase; an enzyme that catalyses the hydrolysis of phosphate from [3-methyl-2-oxobutanoate dehydrogenase (lipoamide)] phosphate.

Methyl Red a dye used as an indicator, changing from red to yellow over the pH range 4.2–6.3.

methylsulfonyl nitrophenyl carbonate 2-(methylsulfonyl)ethyl 4-nitrophenyl carbonate; a reagent used for the introduction of the **MSOC** amino-protecting group.

methylsulfonyl succinimidyl carbonate 2-(methylsulfonyl)-

5-methyltetrahydrofolate *abbr.*: MeTHFA *or* 5-CH_3-H_4folate; a member of the tetrahydrofolate group of coenzymes. It is an intermediate in the methylation of vitamin B_{12}, which can then react with homocysteine to form methionine. The methyl group of MeTHFA is derived from C-3 of serine via 5,10-methylene H_4folate.

methyltetrahydrofolate trap hypothesis a hypothesis to explain the accumulation of 5-methyltetrahydrofolate (5-CH_3-H_4folate) in the sera of patients with vitamin B_{12} deficiency. It suggests that, because the only route by which 5-CH_3-H_4folate can be converted to H_4folate (required for single-carbon transfer reactions) is by the vitamin B_{12}-dependent 5-methyltetrahydrofolate–homocysteine *S*-methyltransferase (EC 2.1.1.13) reaction, a large proportion of the body's available folate becomes trapped as 5-CH_3-H_4folate, leading to folate deprivation.

methylthioribose 5-methylthio-D-ribose; an intermediate in the **methionine salvage pathway**.

methyltransferase any enzyme of sub-subclass EC 2.1.1, members of which catalyse the transfer of a methyl group. *See also* **che**, O^6-**methylguanine**.

4-methylumbelliferyl *abbr.*: MU; a group that is attached in glycosidic linkage to various sugars to form substrates useful in the assay of glycosidases. The highly fluorescent 4-methylumbelliferone is released on enzyme action, and can be measured in alkaline solution by its fluorescence (max. 448 nm, excitation at 364 nm), which is quenched in the glycosidic substrate. *See also* **MUG**.

4-methylumbelliferone

methyne *or (formerly)* **methine** the name now recommended for the trivalent diatomic group, =CH–, attached to two other groups in an organic molecule. Such a group is commonly still known in biochemistry as **methene**. *Compare* **methylidyne**.

meticillin (sodium) *or* **methicillin** a semisynthetic antibiotic of the **penicillin** type that is lactamase resistant.

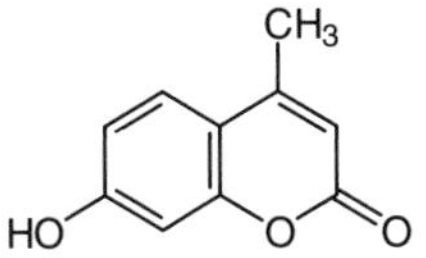

metJ the gene for the repressor (co-repressor is *S*-adenosylmethionine) of the methionine biosynthetic operon, and also *metJ* itself, in *Escherichia coli*. The repressor protein undergoes dimeric binding to DNA by a motif other than a helix-turn-helix.

Met-Lys-bradykinin *a short name for* methionyllysylbradykinin; *see* **bradykinin**.

MetMb *abbr. for* metmyoglobin (i.e. **ferrimyoglobin**).

metmyoglobin *abbr.*: MetMb; *an old name for* **ferrimyoglobin**.

'me too drug' a drug that uses the same therapeutic mechanism as an existing drug, offering no significant additional benefit. Notwithstanding the negative connotations of the term 'me too', such products compete directly with those currently on the market, and hence may help to push prices down.

metre *or (US)* **meter** *symbol*: m; the SI base unit of length, equal to the length of the path travelled by light in a vacuum during a time interval of 1/299 792 458 of a second.

metre–kilogram–second system (of electrical units) *abbr.*: mks system *or* MKS system; a system of electrical (physical) units based on the metre, kilogram, and second as fundamental units of measurement. It has now been superseded by the **SI** system of units.

metric 1 of, or relating to, the metre or the metric system. **2** quantitative.

+metric *see* +meter, +metry.

metric system a system of weights and measures, devised in France (1793–95), based on the principle that each of the physical quantities length, volume, and mass should be defined in terms of one unit whose multiples and submultiples are related by powers of ten and that all three units should be simply interrelated. The units were, respectively, the metre, the litre (one cubic decimetre), and the gram (the mass of one cubic centimetre of water at its maximum density). It has now been supplanted by the International System of Units (*see* **SI**).

metridin 1 a 36-amino acid, 3.97 kDa toxin from *Metridium senile* (brown sea anemone). **2** *or* **metridium proteinase A** EC 3.4.21.3; *other name*: sea anemone protease A; an enzyme that catalyses the preferential cleavage: Tyr-|-Xaa > Phe-|-Xaa > Leu-|-Xaa; it has little action on Trp-|-Xaa.

metridium proteinase A *see* **metridin** (def. 2).

metrizamide *the trivial name for* 2-[3-acetamido-5-(*N*-methylacetamido)-2,4,6-triiodobenzamido]-2-deoxy-D-glucose; a dense, synthetic, nonionic substance, highly soluble in water, that is capable of forming (nonautoclavable) aqueous solutions of densities up to 1.3 g cm^{-3}, useful as media for **isopycnic centrifugation** of biological particles.

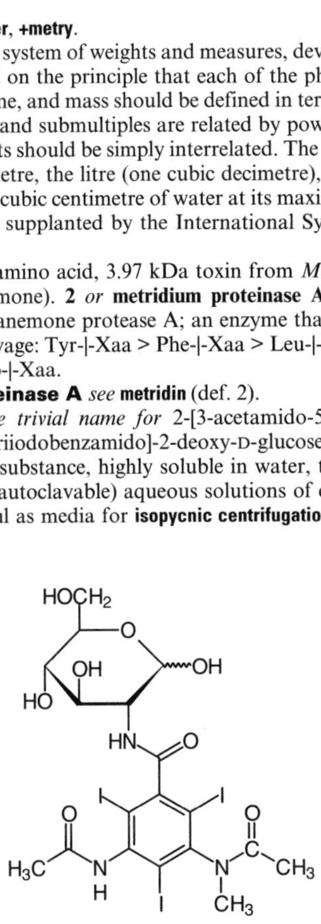

metrizoate *the trivial name for* the 3-acetamido-5-(*N*-methylacetamido)-2,4,6-triiodobenzoate anion; a dense, synthetic substance

used primarily as its sodium salt, which is highly soluble in water and capable of forming autoclavable aqueous solutions of densities up to 1.4 g cm^{-3}, or more, such solutions being useful as media for **isopycnic centrifugation** of biological particles.

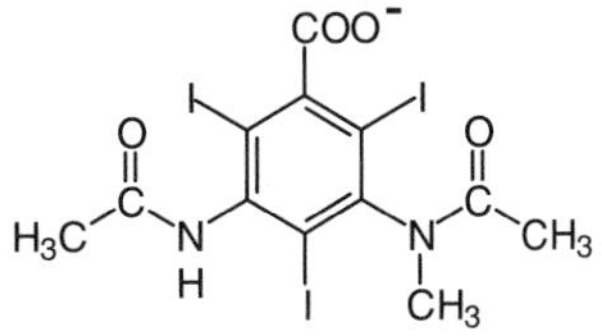

metronidazole 2-methyl-5-nitroimidazole-1-ethanol; an antibiotic used to treat a variety of bacterial and parasitic infections, such as amebiasis, trichomoniasis, and giardiasis. It is a prodrug, which is converted by reduction to a cytotoxic radical.

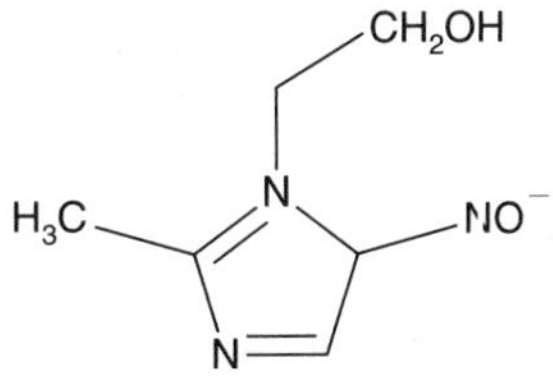

metronidazole radical

+metry *comb. form* denoting the process or science of measuring (something indicated or specified). —**+metric** *adj.*

metyrapone 2-methyl-1,2-di(3-pyridyl)propan-1-one; an inhibitor of the enzyme 11β-hydroxylase (steroid 11β-monooxygenase; EC 1.14.15.4); this enzyme is involved in the synthesis of cortisone, cortisol and aldosterone, its inhibition causing a rise in **corticotropin** secretion as a result of release from negative feedback on the pituitary by cortisol. Metyrapone is used in **challenge** tests of hypothalamic-pituitary function and sometimes in the treatment of **Cushing's syndrome**.

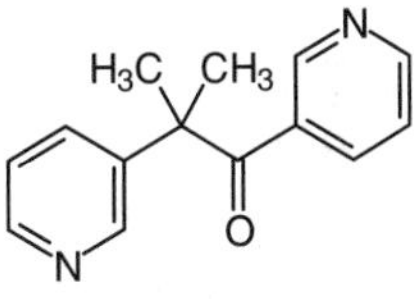

MeV *symbol for* megaelectronvolt (i.e. 10^6 electronvolts).

Mevacor *a proprietary name for* **lovastatin**.

mevaldate 1 the anion OHC–CH$_2$–C(CH$_3$)(OH)–CH$_2$–COO$^-$ derived from mevaldic acid; its (*R*)-enantiomer occurs as an enzyme-bound intermediate during the conversion of hydroxymethylglutaryl-coenzyme A to mevalonate in the biosynthesis of polyprenyl compounds. **2** any salt or ester of mevaldic acid.

mevalonate 1 the anion, HO–[CH$_2$]$_2$–C(CH$_3$)(OH)–CH$_2$–COO$^-$, derived from mevalonic acid; its (*R*)-enantiomer is a strategic intermediate derived from hydroxymethylglutaryl-CoA in the biosynthesis of polyprenyl compounds. **2** any salt or ester of mevalonic acid.

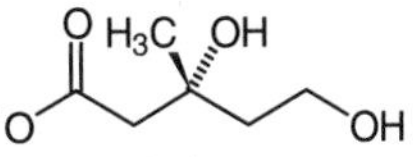

(*R*)-mevalonate

mevalonate kinase a hepatic peroxisomal enzyme (396 and 395 amino acids in human and rat, respectively) that phosphorylates mevalonate to form mevalonic acid-5-phosphate in the metabolic pathway that leads to cholesterol and other isoprenoid compounds. Severe deficiency of the enzyme leads to **mevalonic aciduria**. Less severe deficiency (5–15% of normal activity) leads to **hyper-IgD syndrome**.

mevalonic aciduria a rare, severe, autosomal recessive disease with increased plasma levels of mevalonate and pronounced mevalonic aciduria, and characterized by dysmorphic features, hepatosplenomegaly, and severe developmental delay. It is caused by several mutations in the gene at 12q24 for **mevalonate kinase** that lead to severe deficiency of this enzyme. *Compare* **hyper-IgD syndrome**.

mevalonolactone mevalonic acid lactone; it is hydrolysed in cells to mevalonate, and is often the form in which mevalonate is administered to animals or cells, either radioactively or non-radioactively as a metabolic precursor.

mevalonoyl the acyl group, $HO–[CH_2]_2–C(CH_3)(OH)–CH_2–CO–$, derived from mevalonic acid.

mevastatin *or* **compactin** 6-demethylmevinolin; a substance isolated from *Penicillium citrinum*; it is a potent inhibitor of **HMG-CoA reductase** with actions similar to **lovastatin**.

Mevinacor *a proprietary name for* **lovastatin**.
mevinolin *an alternative name for* **lovastatin**.
Meyerhof pathway *see* **Embden–Meyerhof pathway**.
Meyerhof quotient an index of the effect of the presence or absence of dioxygen on glycolysis or fermentation. It is: [(rate of anaerobic fermentation) – (rate of aerobic fermentation)]/(rate of dioxygen uptake). Assuming 6 mol ATP are formed per mol dioxygen used, a Meyerhof quotient of 6 would mean that aerobic and anaerobic energy supply are the same; a lower value indicates an increase in ATP-supplying and ATP-requiring processes on transition from anaerobic to aerobic conditions. [After Otto Fritz Meyerhof (1884–1951), German physiologist and biochemist.]

mfd *abbr. for* microfarad.
MflI a type 2 **restriction endonuclease**; recognition sequence: [AG]↑GATC[TC]. *Xho*II is an **isoschizomer**.
mg *symbol for* milligram (i.e. 10^{-3} gram).
Mg *symbol for* magnesium.
Mgm1 a yeast dynamin-like protein that forms a ring on the outer surface of the inner mitochondrial membrane to bring about fission of the organelle. Homologues occur in animals and plants.
MHC *abbr. for* major histocompatibility complex.
MHC class II deficiency the deficiency or absence of MHC class II antigens on lymphocytes and macrophages, which leads to abnormal susceptibility of humans to infections. It results from mutations in any component of the heterotrimeric **RFX** complex, which binds a part of the response element in the promoter region of MHC class II genes. *See* **major histocompatibility complex**.

mho a practical unit of electric conductance equal to the reciprocal of the impedance measured in ohms. For circuits containing only noninductive resistance the **siemens** is the preferred unit.
MHS *abbr. for* major histocompatibility system (i.e. **major histocompatibility complex**).
MIC *abbr. for* minimum inhibitory concentration.
mica substrate technique a technique used in preparing specimens of macromolecules for electron microscopic examination. Freshly cleaved mica is used to support the specimen during shadowing with a metal; a backing film of carbon is then applied by evaporation normal to the shadowed surface, and both films are transferred to the grid.
micelle 1 an aggregate, of colloidal dimensions, of oriented molecules of amphipathic substances existing in equilibrium in solution with the chemical species from which it is formed; micelles are generally electrically charged. In aqueous solution the individual molecules of the micellar aggregate are oriented with their polar groups pointing towards the aqueous medium and their hydrophobic moiety directed into the centre of the micelle. **2** a hypothetical ordered region in a natural fibre such as cellulose. —**micellar** *adj.*
micellization *or* **micellisation** transformation into micelles.
micF a regulatory gene in the osmoregulatory system in *Escherichia coli*; *micF* RNA and *ompF* mRNA (omp = outer membrane protein) are induced by interaction of *ompR* with a common divergent, central promoter region. *micF* RNA is partly complementary to the Shine–Dalgarno region of *ompC* and thus represses its expression by translational control.
Michaelian of, relating to, or described by **Michaelis kinetics**.
Michaelis kinetics *or* **Michaelis–Menten kinetics** *or* **Henri–Michaelis–Menten kinetics** a model to explain the kinetics of an enzyme-catalysed reaction,

$$E + S \underset{k_{-1}}{\overset{k_{+1}}{\rightleftharpoons}} ES \overset{k_{cat}}{\rightarrow} E + P$$

where E is enzyme, S is substrate, ES is an intermediate complex (the **Michaelis complex**), and P is product. The initial velocity of the reaction, v, when [P] = 0, is given by the **Michaelis equation** (or **Michaelis–Menten equation** or **Henri–Michaelis–Menton equation**):

$$v = V[S]/(K_m + [S]) = d[P]/dt$$

in which V is the limiting or maximal reaction rate (i.e. that occurring when the enzyme is saturated with substrate), and the **Michaelis constant**, K_m, is given by:

$$K_m = (k_{-1} + k_{cat})/k_{+1}.$$

If $k_{cat} \ll k_{+1}$, then K_m approximates to K_s, the enzyme–substrate dissociation constant (usually known as the **substrate dissociation constant**). K_m and V are known as the **Michaelis parameters**. *Compare* **Dalziel coefficient**. [After V. Henri (1872–1940), French physical chemist; Leonor Michaelis (1875–1949), German physician and biochemist; and Maude Menten (1879–1960), Canadian physician and biochemist.]
Michaelis pH function any one of a set of functions relating hydron concentration, [H⁺], to the acid dissociation constants for different stages of dissociation of a given polybasic acid.

Consider first the simplest example, that of a symmetrical dibasic acid, AH_2 (e.g. succinic acid). This can exist in three stages of dissociation, represented by the equilibria

$$AH_2 \overset{K_1}{\rightleftharpoons} AH^- \overset{K_2}{\rightleftharpoons} A^{2-}$$

where K_1 and K_2 are the acid dissociation constants for the respective stages of dissociation. For such a dissociation there are three Michaelis pH functions, f_0, f_1, and f_2, which are defined thus:

$$f_0 = 1 + K_1/[H^+] + K_1K_2/[H^+]^2$$

$$f_1 = 1 + [H^+]/K_1 + K_2/[H^+]$$

$$f_2 = 1 + [H^+]/K_2 + [H^+]^2/K_1K_2$$

The total concentration of acid of all forms, $[A]_t$, is given by

$$[A]_t = [AH_2] + [AH^-] + [A^{2-}];$$

by substitution for K_1 and K_2 it can be shown that

$$[A]_t = f_0[AH_2] = f_1[AH^-] = f_2[A^{2-}]$$

Hence it will be apparent that the reciprocals of the respective pH functions give the fractional concentrations of the various ionic forms of the acid at a given pH value. The theory of pH functions is independent of the chemical nature of the ionizing groups: the functions apply equally to molecules with identical ionizing groups (e.g. dicarboxylic acids, as above), to ampholytes (e.g. amino acids), and to molecules or parts of molecules in which the nature of the ionizing groups may not be known (e.g. the active centres of enzymes). For an unsymmetrical dibasic acid there are potentially four ionic forms, four dissociation constants, and thus four pH functions, but these are important in practice only if the pK values of the two dissociating groups are fairly close, as could be the situation at an active centre of an enzyme. If, on the other hand, the pK values are widely separated, as is the case with a typical amino acid, then there is effectively only one intermediate form and a set of three pH functions as above may be considered to apply.

micr+ *see* **micro+** (def. 2).

micra a plural of **micron**.

micRNA *abbr. for* messenger-RNA-interfering complementary RNA (i.e. **antisense RNA**).

micro 1 (concerning something) very small (e.g. micro method). **2** short for a microcomputer.

micro+ 1 *symbol*: μ; the SI prefix denoting 10^{-6} times. **2** *or (sometimes before a vowel)* **micr+** denoting very small or concerned with very small things (e.g. microbiology, microsome); abnormally small, underdeveloped (e.g. microcyte); on a small scale or concerned with small quantities (e.g. microanalysis, microbalance); of a small area or volume, or small part of a larger one (e.g. microenvironment). *Compare* **macro+**. **3** indicating amplification or magnification of something (e.g. microphone, microscope). **4** involving the use of a microscope (e.g. micrograph).

microaerophilic 1 describing an aerobic environment with a lower partial pressure of dioxygen than that under normal atmospheric conditions. **2** describing an organism whose maximal rate of growth occurs in such an environment.

microalbuminuria the excretion of plasma albumin in the urine at a rate above the normal range of values (5–20 μg min^{-1}) but below the threshold level (300 μg min^{-1}) for detection by routine dipstick methods and measurable only by radioimmunoassay (or other techniques of comparable sensitivity). The condition is associated with a general increase in vascular permeability and forms a sensitive indicator of diabetic nephropathy.

microanalysis any technique whereby very small amounts of substance may be analysed qualitatively or quantitatively. The term was originally used to indicate analysis of substances in the range 1–10 mg. —**microanalytic** *or* **microanalytical** *adj.*

microarray *or* DNA chip *or* DNA microarray *or* gene expression array a glass slide or bead whose surface bears numerous microscopic samples of DNA or other reagent in an orderly pattern. Microarrays are most commonly used for the simultaneous measurement of the expression levels of large numbers of genes. For example, entire cloned cDNA sequences or oligonucleotides can be bound to the slide, either by means of a robotic spotter or by direct synthesis on the array. A single slide may contain tens of thousands of spots, or addresses, each containing a copy of a specific DNA sequence. cDNA samples that have been reverse transcribed from a cellular RNA extract are then fluorescently labelled and hybridized to the array. The array is scanned with a fluorescent scanner, and the data collected and automatically analysed to quantify the relative amount of fluorescence and hence the relative levels of expression of each gene. *See also* **GeneChip array**.

microassay any technique whereby very small amounts of substance may be assayed. *Compare* **microanalysis**.

microbalance any balance designed to weigh very small quantities, typically between 1 μg and 1 mg.

microbe *a common term for* a microorganism, especially a pathogenic bacterium. —**microbial** *adj.*

microbial collagenase EC 3.4.24.3; *other names*: Clostridium histolyticum collagenase; clostridiopeptidase A; collagenase A; colla-

genase I; an enzyme that catalyses the digestion of native collagen in the triple-helical region at Xaa-|-Gly bonds. Some preference is shown for Gly at P3 and P1′, Pro and Ala at P2 and P2′, and hydroxyproline, Ala, or Arg at P3′ (*see* **peptidase P-sites**).

microbiological of, pertaining to, or concerned with microbiology; effected by microorganisms.

microbiological assay any assay for a substance, especially an amino acid or vitamin, that is based on the extent of growth of a microorganism for which the analyte in question is an essential growth factor.

microbiology the science concerned with microorganisms, and with their interactions with other organisms and the environment. —**microbiologist** *n.*

microbody a cytoplasmic organelle, spherical or oval in shape, and 0.1–1.5 nm across, that is bounded by a single membrane and contains oxidative enzymes, especially those utilizing H_2O_2. Microbodies include **glyoxysomes** and **peroxisomes**.

microbore high-pressure liquid chromatography *or* microbore **HPLC** *a variant of* **high-pressure liquid chromatography** in which a column of only about 1 mm diameter is used. The advantages of the technique include greater sensitivity and a much reduced solvent consumption.

microcalorimeter a **calorimeter** designed to measure continuous power outputs down to 1 μW, or heat pulses down to 200 μJ. —**microcalorimetry** *n.*; **microcalorimetric** *adj.*

microcapsule any very small thin-walled capsule of plastic, wax, etc., containing a liquid, powder, etc., from which the contents may be released (by heat, impact, radiation, dissolution, or other means) in order to fulfil some specific function. *See also* **microencapsulation**.

microcarrier any of a number of preparations of finely particulate, nontoxic, nonrigid, usually transparent material of density close to unity, e.g. dextran-based beads, that enable the growth of animal cells in suspension culture, especially on a large scale. Microcarriers permit the growth of anchorage-dependent cells as a monolayer on the surface of the particles. Generally, the cells may be easily harvested, e.g. by allowing the particles to settle and then treating with a proteinase such as trypsin.

microcell 1 a small, experimentally produced eukaryotic cell-like structure in which a limited amount of genetic material is contained in a micronucleus surrounded by a rim of cytoplasm and an intact cell membrane. It is useful as a vector for the transfer of a small number of chromosomes into a normal cell by virus-induced cell fusion. **2** any small **cell** (def. 2), commonly with a capacity less than 1 mL.

microchemistry chemistry on a very small scale, commonly handling less than 1 mg or 1 mL of material and often requiring special small apparatus. —**microchemical** *adj.*

microcin any of a number of microbial peptide toxins that have antibiotic action. They are of diverse structure and action. Examples include microcin E492, a 6 kDa polypeptide antibiotic. It forms cation-selective channels in planar phospholipid bilayers, leading to a loss of membrane potential. Another example is microcin B17 from *E. coli*. This is a glycine-rich peptide that inhibits DNA gyrase (*see* **(DNA) topoisomerase** type II). Microcin is classified in the TC system under number 1.C.58.

micrococcal of, pertaining to, or produced by micrococci; being a member of the genus *Micrococcus*.

Micrococcus a genus of Gram-positive, strictly aerobic, usually nonmotile coccoid bacteria of the family Micrococcaceae. The cells are commonly 1–2 μm in diameter.

microconcentrator a commercial device for the concentration and desalting of 'small' volumes (i.e. ≤2 mL) of biological samples.

microcrystal any very small crystal, especially one that is visible only under a microscope. —**microcrystalline** *adj.*

microcurie *symbol*: μCi *or (formerly)* μc; a unit of (radio)activity or of radioactive material equal to 10^{-6} **curie** or 3.7×10^4 becquerel.

microdeletion a DNA/chromosomal deletion that is not detectable by conventional techniques such as **Southern blotting** or microscopy.

microdensitometry the measurement of **absorbance** (i.e. optical density) over a very small finite area, such as that viewed by an optical microscope. —**microdensitometric** *adj.*; **microdensitometer** *n.*

microdialyser *or (sometimes, esp. US)* **microdialyzer** a device for

microdialysis, especially one designed to minimize loss and dilution of samples.

microdialysis dialysis of a sample of small volume (commonly of less than 1 mL).

microdiffusion analysis *see* **Conway microdiffusion method**.

microdissection the dissection of small organisms or parts of organisms by means of mechanically controlled instruments viewed through an optical microscope.

microelectrode any electrode in which the sensitive element is very small and also often needle-like. It is useful for investigation of electrical phenomena in individual cells or in very small volumes.

microencapsulation the act or process of enclosing substances, solutions, etc. in **microcapsules**.

microenvironment a very small domain of the environment, such as that surrounding a single molecule or a functional group of a molecule, a single cell or organism, or a small group of cells or organisms. The term is used especially to distinguish zones of local difference in the overall environment.

microfarad *abbr.*: mfd; *symbol*: μF; a unit of capacitance equal to 10^{-6} **farad**.

microfibre *or (US)* **microfiber** any fibre having a diameter in the range 0.1–1 μm, such as used in making glass microfibre filters.

microfibril any small fibril occurring in biological material that is distinguishable only by electron microscopy, especially one of a set that make up a fibril, e.g. in cellulose. —**microfibrillar** *adj.*

microfibrillar polysaccharide one of three groups of cell-wall polysaccharides characterized by solubility differences: the main component in the plant kingdom is **cellulose** but **chitin** is found in some fungi, β-1,4-mannans in some green algae (e.g. Codiaceae), and β-1,3-xylans in other green algae (e.g. Bryopsidaceae).

microfilament a type of **actin filament** that contributes to the cytoskeleton of eukaryotic cells. Microfilaments are elongated structures consisting of two strands of globular **actin** monomers twisted into a helix with 13.5 molecules per turn. Microfilaments form the major component of the cell's contractile machinery and are implicated in cell division, cell movement, muscle contraction, nerve outgrowth, tubular gland formation, gastrulation, and neurulation. *Compare* **intermediate filament**, **microtubule**.

microfilter 1 any small filter, especially as used in microchemistry. **2** *see* **microporous filter**.

microfiltration filtration with a **microporous filter**.

microflora 1 *a collective name for* very small plants. **2** the flora of a **microhabitat**. **3** *a collective name for* all the microorganisms of a tissue or organ of an animal.

microfluorimeter an instrument for **microfluorimetry**.

microfluorimetry the measurement of intensity of fluorescence over a very small finite area, such as a single cell viewed under an optical microscope. —**microfluorimetric** *adj.*

microfold cell *see* **M cell**.

microforge a device by means of which micropipettes, very small glass tools, etc. can be made while being viewed under a low-power microscope.

microfungus (*pl.* **microfungi**) any fungus whose vegetative thallus or fruiting body is too small to be seen easily with the unaided eye. *Compare* **mould**.

microglobulin 1 any plasma **globulin** molecule or globulin fragment of less than ≈40 kDa, e.g. **Bence-Jones protein**, β$_2$-**microglobulin**. **2** *a term sometimes used to refer to* 7S immunoglobulins such as IgG. *Compare* **macroglobulin**.

β$_2$-microglobulin *or* **betamicroglobulin** a human plasma protein, of 11.8 kDa and containing 99 amino-acid residues, that in size, primary structure, and tertiary structure strongly resembles a single domain of the constant region of immunoglobulin molecules: it is a constituent of class I **major histocompatibility complexes**. Plasma levels may be elevated in hepatobiliary disease, chronic active hepatitis, and alcohol-induced liver cirrhosis.

microgram 1 *symbol*: μg *or (esp. in pharmacy)* mcg; a unit of mass equal to 10^{-9} kilogram or 10^{-6} gram; *formerly called* gamma (*symbol*: γ). **2** *an alternative name (sometimes) for* **micrograph**.

micrograph *or (sometimes)* **microgram** a graphic reproduction of the image of an object as seen through a microscope.

microhabitat any habitat that is small or limited in extent and differs in character from a larger, surrounding habitat.

microheterogeneous describing a preparation of a biopolymer that superficially appears to be homogeneous but that, on refined analysis, shows slight differences in size, sequence, charge, state of aggregation, or other properties of its molecules. Such differences are believed, at least in some instances, to be attributable either to artefactual changes or to genetic differences. —**microheterogeneity** *n.*

microincineration a technique for examining the distribution of minerals in tissue sections or microorganisms. The specimen is placed on a slide, the organic material burned away in a furnace, and the nature and position of the mineral ash determined by microscopy.

microinject to inject liquid into a single cell or nerve fibre using a micropipette. —**microinjection** *n.*

microiontophoresis **iontophoresis** (def. 2) applied to very small areas of tissue.

micro-Kjeldahl method an adaptation of the **Kjeldahl method** for determining small amounts of nitrogen, in the range 0.1–1.0 mg. A correspondingly small flask, commonly known as a **micro-Kjeldahl flask**, is used for the digestion, and a **Markham still** may be used for the steam distillation.

microlitre *or (esp. US)* **microliter** *symbol*: μL *or* μl; a unit of volume equal to 10^{-6} litre; *formerly called* **lambda** (*symbol*: λ).

micromanipulation the techniques and practice of performing very delicate operations, such as microdissection, microinjection, or the isolation of single cells, under an optical microscope, usually with the aid of a **micromanipulator**.

micromanipulator an instrument used in conjunction with an optical microscope for holding and manipulating very small instruments and specimens in the field of view.

micrometer 1 any instrument or device for measuring very small distances or angles. **2** *or* **micrometer gauge** a device for measuring small distances, thicknesses, diameters, etc. with great accuracy. **3** *the US spelling of* **micrometre**.

micrometer burette a **micrometer syringe** adapted for use as a burette.

micrometer pipette a **micrometer syringe** adapted for use as a pipette.

micrometer screw a screw with a fine thread of accurate and consistent pitch.

micrometer syringe a syringe for the accurate delivery of small volumes of liquid in which the plunger of the syringe is driven by a calibrated micrometer screw.

micrometre *or (US)* **micrometer** *symbol*: μm; a unit of length equal to 10^{-6} metre; *formerly called* **micron** (*symbol*: μ).

micromicro+ *symbol*: μμ *a former name for* **pico+**.

micromodification alteration of the covalent structure of a macromolecule, especially a histone, by the addition of various groups, e.g. acetyl, methyl, phosphate, after the initial biosynthetic process is terminated.

micromolar *symbol*: μM *or* μM; describing a solution containing one micromole of solute per litre of solution, or a specified multiple or fraction thereof; used also of an amount-of-substance concentration similarly expressed.

micromole *symbol*: μmol; a unit of amount of substance equal to 10^{-6} mole.

micromolecule any (type of) molecule composed of a relatively small number of atoms. In practice the term is usually taken to mean any (type of) molecule that is capable of passing through the pores of dialysis tubing (as generally used) and that thus has a mass less than about 10 kDa. —**micromolecular** *adj.*

micron (*pl.* **microns** *or* **micra**) *symbol*: μ; *a former term for* **micrometre**.

micronucleus a eukaryotic cell nucleus with less than the normal complement of chromosomes, produced experimentally by incubation of a cell with colchicine. —**micronuclear** *adj.*

micronutrient any chemical element that is required by living organisms in tiny quantities only, also known as **trace element**. The term is often extended to include organic compounds such as vitamins.

microorganism any noncellular or unicellular (including colonial) organism, most of which are too small to be seen with the unaided eye. Microorganisms comprise bacteria (including cyanobacteria), lichens, microfungi, protozoa, rickettsiae, virinos, viroids, and viruses, and also some algae; all prokaryotes are included. *See also* **microbe**.

microperoxisome (*sometimes*) a type of small **peroxisome**, 150–250 nm in diameter, found in all mammalian cells.

microphthalmia transcription factor *see* **MITF**.

micropinocytosis (*sometimes*) a form of **pinocytosis** occurring in higher animals, especially in the transfer of dissolved macromolecules from the blood to the cytoplasm of cells, in which the pinosomes formed are very small (**micropinosomes**) and can only be observed in the electron microscope.

micropipette *or* (*US*) **micropipet** any pipette calibrated to contain or deliver a volume in the range 5–500 µL.

micropore any pore (in a gel or porous solid) whose width does not exceed about 2.0 nm. *Compare* **macropore**, **mesopore**. —**microporous** *adj*.

microprocessor a single integrated electronic circuit designed to carry out a specific set of operations, e.g. as the central processor of a computer or a control system.

Microprocessor the smaller of two nuclear complexes containing **Drosha** and **Pasha** and involved in processing **pri-miRNA** to **pre-miRNA**.

microRNA *see* **miRNA**.

microsatellite a repeating DNA segment that is similar to a **minisatellite** except that the overall range of size is smaller and the base core repeat unit involves a two or four nucleotide pair repeat motif. They are so named because of their resemblance to **satellite DNA** in having tandem repeats. The number of copies of the repeats varies not only between individuals but also between the two chromosomes of a pair. Their highly polymorphic nature has made them invaluable for analysis of pedigrees and in linkage analysis.

microscope an instrument consisting of a lens or a combination of lenses that uses light or other electromagnetic radiation to make enlarged images of objects that are invisible to, or not easily seen by, the unaided eye. *See also* **electron microscope**, **phase-contrast microscopy**.

microscopic 1 of a size that is not visible (in detail) to the unaided eye. **2** concerned with small units. **3** relating to the behaviour of individual molecular entities. **4** *or* **microscopical** of, pertaining to, or using a microscope. *Compare* **macroscopic**. —**microscopically** *adv*.

microscopic equilibrium constant any constant describing the equilibrium of the interconversion of two particular molecular entities. *Compare* **intrinsic constant**, **macroscopic equilibrium constant**.

microscopic reversibility a thermodynamic principle requiring that at equilibrium the reaction pathway for the reverse of any particular chemical reaction is the exact opposite of the pathway for the forward reaction, i.e. both forward and backward reactions must pass through the same transition states or intermediates and occur with equal frequency.

microscopy the art and practice of using a microscope; investigation with a microscope.

microsecond *symbol*: µs; a unit of time equal to 10^{-6} second.

microsequenator an apparatus for microsequencing (*see* **microsequence**).

microsequence to **sequence** (def. 3) a biopolymer using a very small sample, e.g. 5–100 pmol of a protein.

microsolute *an operational term used to describe* any diffusible solute in a solution that also contains nondiffusible macromolecular solute(s).

microsomal of, pertaining to, or containing **microsomes**; being in the **microsomal fraction** of a homogenate of a cell or tissue.

microsomal fraction *an operational term for* the subcellular fraction of a homogenate of a eukaryotic cell or tissue that, on differential centrifugation, is sedimented at $10^5 g$; the fraction from some tissues, such as liver, although consisting largely of microsomes, commonly also contains free ribosomes and fragments of the plasma membrane, of the Golgi apparatus, of mitochondria, and of other subcellular structures. The morphological constituents differ according to the tissue from which the fraction was derived.

microsomal triglyceride transfer protein a protein of the endoplasmic reticulum involved in the assembly and secretion of apolipoprotein B-containing plasma lipoproteins in intestinal mucosa and liver. It is a heterodimer of an 88 kDa subunit and a pro-

tein disulfide isomerase. Mutations in the gene for the first subunit, at 4q22-q24 (which encodes an 894-residue protein), are associated with **abetalipoproteinemia**.

microsome any of the small, heterogeneous, artefactual, vesicular particles, 50–150 nm in diameter, that are formed when some eukaryotic cells are homogenized and that sediment on centrifugation at $10^5 g$. Microsomes are formed mostly from disrupted endoplasmic reticulum membranes but some arise from the plasma membrane. Those formed from rough endoplasmic reticulum are studded with ribosomes on the outside (**rough microsomes**); those formed from smooth endoplasmic reticulum and from plasma membrane have no adhering ribosomes (**smooth microsomes**).

microspectrophotometer an instrument, consisting of a microscope and a spectrophotometer, used for measuring light of selected frequencies passing through, or reflected from, a very small specimen, e.g. a cell or a subcellular structure.

microspectrophotometry spectrophotometry applied to a very small specimen, using a microspectrophotometer. —**microspectrophotometric** *adj*.

microsphere one of a preparation of spherical particles of known and uniform size, usually in the range 10–40 µm in diameter, and prepared from, e.g., serum albumin or an ion-exchange resin, that may be labelled with an appropriate radionuclide and used in studies of the circulation of the blood in health or disease.

microspore the smaller of the two types of spores produced by the heterosporous plants. In seed plants the microspore is the cell from which the pollen grain develops. —**microsporous** *adj*.

microstructure *an alternative term for* **fine structure**.

microsyringe a syringe for delivering volumes of gases or liquids in the range 0.1–500 µL.

microtitre plate *or* (*esp. US*) **microtiter plate** any rectangular plate usually made of polycarbonate and containing wells in a gridlike pattern. Most usually the plate has 96 wells arranged in the form of eight rows of 12 columns, each well having a capacity of a few hundred microlitres. They are commonly used for **enzyme-linked immunosorbent assay** (ELISA) or other immunological reactions, or for the growth of small cultures of microorganisms or eukaryotic cells.

microtome an instrument for cutting thin sections of objects, e.g. tissues or cells, preparatory to microscopic observation.

microtrabecular lattice an irregular, three-dimensional lattice of very fine strands, up to 15 nm in diameter, sometimes termed **microtrabeculae**, that can be visualized in the cytoplasmic ground substance of eukaryotic cells by high-voltage electron microscopy. It forms part of the cytoskeleton.

microtubule any of the long, generally straight, hollow tubes of internal diameter 12–15 nm and external diameter 24 nm found in a wide variety of eukaryotic cells. Each consists (usually) of 13 protofilaments of polymeric **tubulin**, staggered in such a manner that the tubulin monomers are arranged in a characteristic helical pattern on the surface, and with the α/β axes of the tubulin parallel to the long axis of the tubule. Microtubules exist in equilibrium with a pool of tubulin monomers in the cytoplasm and can be rapidly assembled and disassembled in response to various physiological stimuli; there are also a number of **microtubule-associated proteins**. Most microtubules in a cell appear to be initiated at **microtubule-organizing centres** and some may also become attached to the **kinetochores**. Microtubules are involved in force generation in cilia and flagella, where they exist in a characteristic 9 + 2 array (a ring of nine doublet microtubules with two single microtubules in the centre), and the integrity of the microtubule network appears to be necessary for certain aspects of mitotic spindle function and for the saltatory movement of cell organelles. Taxol prevents depolymerization of microtubules, and is thus antimitotic. *Compare* **intermediate filament**, **microfilament**.

microtubule-associated protein *abbr*.: MAP; any protein that binds to **microtubules**, and modifies their properties, including proteins that induce polymerization of purified tubulin and become associated with the newly formed microtubules. Many different kinds have been found, including structural proteins and motor proteins. MAP 1 is a structural protein involved in the filamentous crossbridging between microtubules and other skeletal elements. The role of MAP2 is unclear; in neuronal cells, it is confined to the cell body and dendrites. It changes structurally during development by alternative splicing mechanisms. **Dynein** is an example of a motor protein MAP. *Compare* **tau protein.** *See also* **MAP kinase.**

microtubule-organizing centre a region near the centre of a eukaryotic cell consisting typically of two **centrioles** at right angles to each other and surrounded by a complex of associated proteins. The centrioles and their satellite proteins serve as initiating sites for the assembly of microtubules.

microvillin a protein of the microvilli of rat mammary cells, M_r 200 000.

microvillus (*pl.* **microvilli**) any of the very small, finger-like projections that occur on the exposed surfaces of epithelial cells, especially where cellular function requires a maximal surface area for absorption or secretion, as in the proximal renal tubule and the small intestine, where they constitute the brush border. The projections are about 1–2 μm long and 100 nm in diameter. The core of a microvillus contains a bundle of about 40 actin filaments of uniform polarity with their 'barbed' (+) ends anchored in a cap of ill-defined amorphous material in the microvillar tip. At the base the actin filaments extend into the perpendicular network, also made up largely of actin, known as the terminal web. In the core the actin filaments are bound together by the actin-binding proteins fimbrin and villin. The renal and intestinal microvilli contain a wide range of hydrolases that are integral membrane proteins; their hydrophilic domains, containing the catalytic sites, face outwards from the cell. —**microvillar** *adj.*

microvolume a very small volume, commonly in the range 1–100 μL.

microwave radiation electromagnetic radiation of wavelengths in the range 1–300 mm (frequencies 1–300 GHz), lying between infrared radiation and most radiofrequency radiation.

Midas *abbr. for* metal-ion-dependent adhesion site; a region that binds Ca^{2+} in the N-terminal region of the β chain of some **integrins.**

middle molecule *a term sometimes used for* any compound in the range 350–2000 Da that accumulates in the body fluids of persons with uremia.

middle T antigen *see* **T antigen** (def. 1).

midkine *other names:* neurite outgrowth-promoting protein; midgestation and kidney protein; amphiregulin-associated protein (*abbr.:* ARAP); a heparin-binding growth/differentiation factor with neurite-promoting activity. It is a cell-surface or basement-membrane protein, induced by retinoic acid and similar to **pleiotrophin.** The midkine gene product is primarily detected in the embryonic period, whereas pleiotrophin is expressed most strongly in the early neonatal period.

midnight zone (*in sequence analysis*) a region of identity where sequence comparisons fail completely to detect structural similarity. Falling in the lower reaches of the **twilight zone**, it represents the theoretical limit of sequence-based comparison methods.

midpoint 1 the point on a line, or a graph, that is equidistant from either end of the line or from two inflections in the graph. **2** a point in time that is half way between the beginning and the end of an event, or a series of events.

midpoint potential the **electrode potential**, at a given pH and temperature and 1 atm H_2, at the midpoint of a redox titration when the activities of the reductant and the oxidant are equal.

MIF *abbr. for* **1** (macrophage) migration inhibition factor. **2** melanotropin release-inhibiting factor (i.e. **melanostatin**). **3** Müllerian inhibitory factor.

mifepristone *see* **RU-486.**

migratase *an old name for* **mutase**; e.g. lysolecithin migratase is an old name for lysolecithin acylmutase (EC 5.4.1.1).

migrate (*in chemistry*) (of atoms, ions, or molecules) to move from one region of a solution to another, or from a solution in one compartment to a solution in another compartment, under the influence of an electric or a centrifugal field or by diffusion. —**migration** *n.*

migration inhibition factor *abbr.:* MIF; *see* (**macrophage**) **migration inhibition factor.**

MIH *abbr. for* melanotropin release-inhibiting hormone (i.e. **melanostatin**).

MIL the avian homologue of *RAF*; v-*mil* is the oncogene of the avian

retrovirus Mill-Hill-2, which also carries v-*myc*. It codes for a serine/threonine kinase of the *RAF/MOS* subfamily of unknown function but exhibiting homology with Src and related proteins (*see* **src**).

milieu (*pl.* **milieus** *or* **milieux**) environment or setting.

milieu intérieur (*French*) the internal environment of a multicellular organism that surrounds the tissues and cells.

milk 1 the white liquid produced and secreted by the mammary glands of mature female mammals for the nourishment of their young. The milk of certain mammals, especially cows, is used as a food by humans, either as liquid milk or in the production of butter, cheese, etc. **2** any of various natural or manufactured liquids that have the appearance or consistency of milk, e.g. coconut milk, milk of magnesia. **3** to squeeze or draw milk from; to yield milk. — **milky** *adj.*

milk factor *an alternative name for* **Bittner factor.**

milk sugar *an old name for* D-**lactose.**

Miller index (*in crystallography*) any of a set of three integers that determine the position of a face, or internal plane, of a crystal in relation to three crystallographic axes. They are determined on the basis of the reciprocals of the intercepts of the face or plane on the crystallographic axes. [After W. H. Miller (1801–80).]

milli+ 1 *symbol:* m; an SI prefix denoting 10^{-3} times. **2** *prefix* denoting one-thousandth.

millicurie *symbol:* mCi *or* (*formerly*) mc; a unit of (radio)activity or of radioactive material equal to 10^{-3} **curie** or 3×10^7 becquerel.

milliequivalent *symbol:* mequiv *or* meq *or* mEq; a unit of amount of substance equal to 10^{-3} **equivalent** (def. 3); (mmol recommended).

milligram *symbol:* mg; a unit of mass equal to 10^{-6} **kilogram** or 10^{-3} gram.

millilitre *or* (*US*) **milliliter** *symbol:* mL *or* ml; a unit of volume equal to 10^{-3} **litre**. It is now identical to the cubic centimetre; i.e. 1 mL = 1 $cm^3 = 10^{-6}$ m^3.

millimetre *or* (*US*) **millimeter** *symbol:* mm; a unit of length equal to 10^{-3} **metre.**

millimetre of mercury *symbol:* mmHg; a non-SI unit of pressure defined as the pressure exerted by a column exactly 1 mm high of a fluid of density that is exactly 13.5951 g cm^{-3} in a place where the acceleration of free fall is exactly 980.665 cm s^{-2}. Its use is deprecated. 1 mmHg = 1 torr = 133.32 pascal.

millimicro+ *a former name for* **nano+.**

millimicron (*pl.* **millimicrons** *or* **millimicra**) *symbol:* mμ; *a former name for* **nanometre.**

millimolar *symbol:* mM *or* mM; describing a solution containing one millimole per litre of solution, or a specified multiple or fraction thereof; the term is also used of an amount-of-substance concentration similarly expressed.

millimole *symbol:* mmol; a unit of amount of substance equal to 10^{-3} **mole.**

Millipore 1 *the proprietary name for* a range of microporous and other types of membrane filters manufactured by the Millipore Corporation. **2** a water purification system, manufactured by the same company and based on reverse osmosis.

millisecond *symbol:* ms; a unit of time equal to 10^{-3} **second.**

Millon's reaction the formation of red complexes when compounds containing the hydroxybenzene group are heated with **Millon's reagent**. The reaction is given by proteins that contain one or more tyrosine residues. [After Auguste Millon (1812–67), French chemist.]

Millon's reagent a 15% solution of mercuric sulfate in 15% sulfuric acid. This is added to the test sample solution, which is then heated in a boiling water bath for 10 min. After cooling, a 1% aqueous solution of sodium nitrite is added to the cooled reaction mixture. It gives **Millon's reaction** with compounds containing the hydroxybenzene group.

Mills perspective representation *or* **Mills representation** a diagrammatic way of unambiguously representing the structural formulae of carbohydrates and their polycyclic derivatives to show on a plane surface the relative configurational arrangements of substituents and bridges. It is particularly useful in instances where the **Haworth representation** is inconvenient, e.g. for depicting saccharides with substituents that bridge two positions on the ring. The carbohydrate in question, or its derivative, is drawn with single rings and fused rings in the plane of the surface, and with bonds to substituents or bridges above or below the plane of the rings drawn with dark lines (often wedge-shaped) or broken lines, respectively (as is customary in representations of steroids and terpenes). Hydrogen atoms in the angular positions at ring junctions are always shown but those at points of substitution may be omitted. Like **Fischer projections**, but unlike Haworth representations, Mills representations may be rotated in the plane of the paper without alteration of their significance. It was proposed primarily for carbohydrate structures, but resembles systems used for other types of molecule, e.g. steroids.

β-D-glucose

milrinone   1,2-dihydro-6-methyl-2-oxo-5-(4-pyridinyl)nicotinonitrile; a phosphodiesterase inhibitor with vasodilating and cardiostimulant activity.

mimetic imitating closely.

+mimetic *suffix (esp. in pharmacology)* denoting having a similar effect to the substance or action indicated.

min *symbol for* minute.

Min *abbr. for* Minute; a system of bacterial proteins responsible for correct localization of the **Z ring** to the mid-cell region of the plasma membrane inner surface in preparation for cell division. MinC and MinD form a complex on the inner cell-membrane surface and inhibit Z ring assembly. MinE relieves this inhibition in the mid-cell region.

mince 1 to disrupt tissue with a **tissue grinder** to the level at which there is extensive, but incomplete, rupture of cells. **2** the preparation resulting from the action of a coarse tissue grinder.

mineral 1 any naturally occurring inorganic substance of a type often obtained by mining, e.g. coal, stone, petroleum. **2** any of a class of naturally occurring inorganic substances, of definite chemical composition and physical properties and often crystalline in character. **3** of, pertaining to, or being a **mineral** (def. 1, 2).

mineralize *or* **mineralise** (of an organic material) to convert into, or impregnate with, a mineral or minerals. —**mineralization** *or* **mineralisation** *n*.

mineralocorticoid *or* **mineralocorticosteroid** any naturally occurring or synthetic hormonal **corticosteroid** that acts primarily on water and electrolyte balance by promoting the renal retention of sodium ions and excretion of potassium ions. Mineralocorticoids are produced in the outer layer (zona glomerulosa) of the adrenal cortex. **Aldosterone** is the most potent of the naturally occurring mineralocorticoids. *Compare* **glucocorticoid**.

mineralocorticoid receptor a nuclear receptor (984 amino acids) of the steroid hormone type that binds and mediates the biological effects of aldosterone. Loss-of-function mutations result in pseudohypoaldosteronism type 1. Gain-of-function mutations lead to an early-onset type of hypertension that is exacerbated by pregnancy.

mini+ *comb. form* denoting miniature; of smaller than usual dimensions.

miniature inverted-repeat transposable element *abbr.*: MITE; a transposable element containing almost identical sequences of ≈400 bp flanked by characteristic inverted repeats of ≈15 bp. MITEs are too small to encode protein; they are ubiquitous in genome sequences, accounting for 5% or more of some genomes (e.g. rice).

minicell 1 a small cell-like body arising from a polar region of a rod-shaped bacterium by aberrant cell division. It is spherical or near-spherical in shape and devoid of chromosomal DNA; hence it does not grow or divide and is normally unable to synthesize nucleic acid or protein (but may do so if a plasmid is introduced). **2** an experimentally produced eukaryotic cell in which an intact cell nucleus is surrounded by scant cytoplasm and a cell membrane. It is prepared by enucleation of a normal mononucleate cell.

miniexon a small exon, as found in the troponin T gene.

minigastrin a form of the peptide hormone **gastrin** that contains 14 amino-acid residues. It has been isolated from tumour tissue and corresponds to residues 5–17 of human gastrin 1, having the sequence LEEEEEAYGWLDFH-amide.

minima a plural of **minimum**.

minimal of, being, or constituting the minimum; the least possible.

minimal medium a (synthetic) culture medium containing only those compounds essential for the growth of a wild-type organism and that is not able to support growth of an auxotroph.

minimum (*pl.* **minimums** *or* **minima**) **1** the least quantity, degree, or value possible or permissible. **2** (*in mathematics*) a value of a function that is less than that of any neighbouring value. **3** of, or pertaining to, a minimum or minimums.

minimum evolution a principle of phylogenetic reconstruction based on minimization of the sum of the lengths of tree branches. From all possible tree topologies over a defined set of taxa, the minimum evolution topology is the one that requires the shortest overall branch lengths.

minimum inhibitory concentration *abbr.*: MIC; the lowest concentration of a drug, or other substance, that just inhibits the growth of a given test organism (bacterium, etc.) in defined conditions.

minimum lethal dose *abbr.*: MLD; the minimum dose of a toxic substance, or organism, that will kill, within a specified time, all the animals in a test group to which it is administered.

minimum (relative) molecular mass *or (formerly)* **minimum molecular weight** the (relative) molecular mass of a compound as determined by the assay of one of its structural elements, e.g. a metal atom, a ligand, a terminal residue, etc., assuming that there must be at least one such structural element in each molecule of the compound.

minisatellite a repeat DNA segment comprising short head-to-tail tandem repeats giving **variable number tandem repeat**-type polymorphisms, with approximate size between 1 and 30 kb. The name derives from the fact that repeat DNA segments were originally found in **satellite DNA**, but the overall size of the minisatellite segments and the length of individual repeats are smaller, and satellite peaks are not seen on centrifugation.

minivial a very small vial, especially a low-capacity container that can be inserted into a normal-sized sample vial used in liquid scintillation counting, to enable the (radio)activity of small-volume samples to be determined more efficiently.

minor base *see* **minor nucleoside**.

minor groove the smaller and shallower of the two grooves on the surface of a double-helical DNA molecule. *Compare* **major groove**.

minor nucleoside a nucleoside containing any of a group of purine or pyrimidine bases that occur in small amounts in most nucleic acids but in relatively large amounts in **transfer RNA** molecules from all organisms; they are sometimes known as '**odd bases**'. The modifications range from the simple methylation of bases or ribose sugar residues to very complex substitutions, take place after transcription of the transfer RNA genes, and are thought to be related to the stability of the tertiary structure. A typical modified nucleoside is **pseudouridine**. *See also* **wyosine**.

minoxidil 6-(1-piperidinyl)-2,4-pyrimidinediamine-3-oxide; an antihypertensive agent and hair growth promoter. It is activated by sulfate conjugation to the active metabolite minoxidil sulfate.

minoxidil sulfate

minuend *(in mathematics)* the number (or quantity) from which another number (or quantity), the subtrahend, is to be subtracted.

minus end the end of a microtubule or actin filament at which the addition of monomers occurs least readily; the 'slow-growing' end of a microtubule or actin filament. The minus end of an actin filament is also known as the pointed end. *Compare* **plus end.**

minus 10 region and **minus 35 region** regions in the **promoter** defined by the number of nucleotides 5′ from the first nucleotide transcribed, i.e. on the **coding strand.** The first nucleotide (the start site) of a transcribed DNA sequence is denoted as +1, the second +2; the nucleotide preceding the start site is denoted as –1. The most conserved sequence upstream (towards the 5′ terminus) of DNA in *Escherichia coli* is a hexamer centred at the –10 position (*see* **Pribnow box**). It has a consensus sequence of TATAAT in which the leading TA and final T are highly conserved. –35 is the centre point of a consensus sequence involved in RNA polymerase recognition, TTGACA.

minus strand 1 the complementary strand to the parental polynucleotide strand – the **plus strand** (def. 1) – in bacteriophages whose genomes consist of duplex RNA. **2** any polynucleotide strand (RNA or DNA) in a viral genome that is complementary to the virus-specified mRNA – the **plus strand** (def. 2). **3** *(sometimes)* an alternative term for **coding strand** of genetic duplex DNA.

minute 1 *symbol*: min; a non-SI unit of time equal to 60 seconds. **2** *symbol*: ′ ; a unit of plane angle equal to (π/10 800) radians, or 0.291 milliradians. There are 60 seconds (of arc) in one minute (of arc), and 60 minutes in one degree (of arc). **3** *(in genetics)* an empirical unit of distance between genes in a bacterial chromosome, being a measure of the time taken, in minutes, for the transfer of a particular gene during bacterial conjugation relative to an arbitrarily standardized selected origin.

7-minute phosphate *or* **10-minute phosphate** *see* **labile phosphate.**

MIP *abbr. for* macrophage inflammatory protein.

mi protein a basic helix-loop-helix leucine-zipper protein that, as a homodimer, binds promoter regions of melanocyte-specific genes (e.g. tyrosinase). Mouse and human proteins (both ≈415 amino acids) share 95% sequence homology. At least a dozen mutations cause a type of Waardenburg syndrome that includes white forelock, heterochromia iridis, and congenital hearing loss. The protein is named after the microphthalmia-associated transcription factor of mouse.

miRNA *abbr. for* microRNA; *other name*: tiny RNA; any of various RNA species that are 21–25 nt long and may be single- or double-stranded. They have been found in animals, including *Drosophila* and *Caenorhabditis elegans*, and in plants. The term encompasses small interfering RNA (**siRNA**) and small temporal RNA (**stRNA**), as well as miRNA proper. miRNAs are transcribed as parts of longer RNA molecules and processed in the nucleus by the **dsRNA** ribonuclease **Drosha** to hairpin structures 70–100 nucleotides long. These are transported to the cytoplasm where they are digested to 21–23-mers by the dsRNA ribonuclease **Dicer**. Single-stranded miRNAs bind to complementary sequences in mRNA thereby inhibiting translation. Over 100 miRNA species have been found in *Arabidopsis thaliana*, of which some 16 are similar to species characterized from animals. They exhibit diverse developmental and tissue-specific expression profiles and are involved in a wide range of gene regulatory processes.

mirror *(in informatics)* an exact copy of a server made at a different location, to spread the load of traffic and hence facilitate access to that server. Busy servers may have many mirrors (e.g. ExPASy).

misacylate *or* **mischarge** to acylate (a transfer RNA molecule) with an aminoacyl group other than the specific one. —**misacylation** *n.*

mischarge *an alternative term for* **misacylate.**

miscible (of two or more particular liquids) capable of being mixed together to give a homogeneous liquid. Particular liquids may be completely miscible or miscible only in certain proportions. *See also* **proofreading.** —**miscibility** *n.*

mismatch repair a system for the correction of errors introduced during DNA replication when an incorrect base, which cannot form hydrogen bonds with the corresponding base in the parent strand, is incorporated into the daughter strand. **Excinucleases** recognize a pair of non-hydrogen-bonded bases and cause a segment of polynucleotide chain to be excised from the (less methylated) daughter strand, thereby removing the mismatched base. The resulting gap is then filled by the actions of DNA polymerase I and DNA ligase.

mispairing the occurrence of a base in one polynucleotide strand of duplex DNA that is not complementary to the corresponding base in the other strand.

missense codon a particular codon that has been altered so that it codes for an amino-acid residue different from the one for which it normally codes.

missense mutation any mutation causing a base substitution in a gene that results in the incorporation of an incorrect amino-acid residue at a specific position in the polypeptide gene product. *Compare* **nonsense mutation, samesense mutation.**

mistranslation the insertion at a specific position in a polypeptide chain of an amino-acid residue that is not indicated by the corresponding codon in the mRNA molecule.

MIT *abbr. for* (3-)monoiodotyrosine.

Mitchell hypothesis the hypothesis, formulated principally by the British biochemist Peter Dennis Mitchell (1920–92), that mitochondrial systems can be driven by proton and electrical gradients that build up across membranes, a mechanism now referred to as **chemiosmotic coupling.**

MITE *abbr. for* miniature inverted-repeat transposable element.

MITF *abbr. for* microphthalmia transcription factor; a protein (419 amino acids) that shows 91% sequence homology in mouse and human. It is a basic helix-loop-helix leucine zipper transcription factor expressed in the neural crest, eye, and ear during development. Mutations lead to microphthalmia in mouse, and to Waardenburg syndrome type 2 in human.

mitochondria the plural of **mitochondrion.**

mitochondriac a biochemist with a chronic and unusually intense interest in mitochondria.

mitochondrial carrier any of a family of at least 13 metabolite transporters of the inner mitochondrial membrane. They consist of ≈300 amino acids organized into three homologous repeats, each of ≈100 residues and containing two transmembrane segments. They probably function as homodimers and serve as antiporters (e.g. for oxaloacetate/citrate, glutamate/aspartate, malate/2-oxoglutarate, citrulline/ornithine, and ATP/ADP pairs). *See also* **mitochondrial carrier protein.**

mitochondrial carrier protein any of various proteins involved in the translocation of small molecules across the mitochondrial inner membrane. Many are translocases using an **antiport** mechanism, and include carriers for pyruvate (antiport molecule is OH⁻), phosphate (malate), citrate (malate), phosphate (OH⁻), ADP (ATP), aspartate (glutamate), and malate (2-oxoglutarate); e.g., ADP/ATP translocase 1 of yeast; this is an integral inner membrane protein, with a tripartite structure, each repeated segment comprising ~100 residues and containing two transmembrane domains, one if which is more hydrophobic than the other. *See also* **OGCP.**

mitochondrial disease any disease that results from a mutation in mitochondrial DNA. Such diseases are transmitted maternally, are usually of early onset and rapidly progressive, and may involve organs that are seemingly unrelated embryologically or physiologically. Common clinical features include blindness, deafness, myopathy, cardiomyopathy, encephalopathy, diabetes mellitus, renal dysfunction, or liver disease. Among the best characterized are:

MELAS (mitochondrial encephalomyopathy, lactic acidosis, and strokelike episodes); MERFF (myoclonus epilepsy with ragged-red fibres); **Leber's hereditary optic neuropathy**; **Kearns–Sayre syndrome**; **Leigh syndrome**; aminoglycoside-induced deafness; and diabetes mellitus and deafness. The genetic causes of some of these are heterogeneous.

mitochondrial DNA *abbr.*: mtDNA; DNA that is contained, replicated, and expressed in mitochondria. mtDNA is double-stranded in almost all organisms studied; that from mammals has a contour length of $\approx$5 μm, about 15 kilobases, while that from protozoa and fungi is about five times larger. Human mtDNA (16 569 bp) codes for two types of rRNA molecules, 22 types of tRNA molecules, and 13 proteins. mtDNA is transmitted maternally, i.e. with the mitochondria of oocytes. The code in mtDNA differs in some respects from that in nuclear DNA (*see* **genetic code**). *See* **mitochondrial disease**.

mitochondrial fission protein any protein involved in fission of mitochondria in eukaryotes. In yeast, three proteins (Fis1, Dnm1, and Mdv1) form the constriction ring around the outer membrane, while one protein (Mgm1) forms a ring on the outer surface of the inner membrane.

mitochondrial fraction *an operational term* for the more rapidly sedimenting components (for example 10 000–15 000 *g* for 30 min) of the **cytoplasmic fraction** of a homogenate of eukaryotic cells or tissue. It generally consists predominantly of mitochondria but may also contain varying numbers of secretion granules, lysosomes, microbodies, or other intracellular organelles.

mitochondrial plasmid DNA a distinct class of genetic elements, found in certain wild-type strains of *Neurospora*, that show virtually no sequence homology with standard mitochondrial DNA (mtDNA) in DNA–DNA hybridization experiments and achieve high copy number without suppressive behaviour towards wild-type mtDNA.

mitochondrial matrix the gel-like material, with considerable fine structure, that lies in the matrix space, or lumen, of a **mitochondrion**. It contains many enzymes, including ones involved in fatty-acid oxidation and enzymes of the tricarboxylic-acid cycle, amino-acid metabolism, urea cycle, and porphyrin synthesis.

mitochondrial processing peptidase EC 3.4.99.41; an enzyme that catalyses the cleavage of the leader peptide from precursor proteins imported into the mitochondrion, typically ones with Arg in position P2 (*see* **peptidase P-sites**). These enzymes resemble in sequence **premurein-leader peptidase**. In *Neurospora crassa* the enzyme consists of a catalytic subunit and an enhancer subunit. Both are transmembrane proteins. The enzyme requires the presence of divalent cations for activity.

mitochondrial RNA *abbr.*: mtRNA; any RNA molecules that are complementary to genes in mitochondrial DNA.

mitochondrial state *see* **respiratory state (of mitochondria)**.

mitochondrial trifunctional protein *see* **acyl-CoA dehydrogenase**.

mitochondrion (*pl.* **mitochondria**) a semiautonomous, self-replicating organelle that occurs in varying numbers, shapes, and sizes in the cytoplasm of virtually all eukaryotic cells. It is the site of tissue respiration (*see* **respiratory chain, oxidative phosphorylation**). Conventionally, mitochondria are represented as elongated cylinders with a diameter of 0.5–1.0 μm; however, in living cells they show great mobility and plasticity, in some cells forming long, moving chains while in others being fixed in position near sites of high ATP consumption; e.g. in cardiac muscle they are packed between the myofibrils while in a sperm they are tightly wrapped around the flagellum. Mammalian hepatocytes each contain 1000–2000 mitochondria, roughly 20% of the cell volume. A mitochondrion has two functionally distinct membrane systems, the outer one completely surrounding the whole organelle and the inner one being infolded into **cristae**. These membrane systems define two compartments: the mitochondrial matrix and the intermembrane space. The matrix contains enzymes for the oxidation of pyruvate and fatty acids and for the tricarboxylic-acid cycle; it also contains the mitochondrial DNA and the enzymes and structures necessary for expression of the mitochondrial genes. The inner membrane contains the enzymes of the respiratory chain, ATP synthetase, and specific transport proteins. The intermembrane space contains a

number of kinases. The outer membrane, which is permeable to molecules of up to 10 kDa, contains monoamine oxidase, cytochrome b_5, a number of transferases, and a fatty-acid elongation system. The mitochondrion was formerly known as a **chondriosome** and by numerous other names. *See also* **cercidosome**. —**mitochondrial** *adj.*

mitochrome a chromoprotein, or mixture of chromoproteins, released from mitochondria *in vitro* after ageing or other degradative processes. They are hemoproteins of ~105 kDa, with 1 atom of Fe per molecule, and are thought to inhibit *in vitro* oxidative phosphorylation.

mitogen any substance or agent that induces or stimulates **mitosis**. —**mitogenic** or **mitogenetic** *adj.*

mitogenic protein *abbr.*: MP; *an alternative name for* **interleukin 1**.

mitogillin EC 3.1.27.–; *other names*: allergen I/A; IgE-binding ribotoxin; a purine-specific ribonuclease that attacks 28S rRNA. The protein is a powerful allergen, an inhibitor of eukaryotic protein synthesis, and has anti-tumour activity. Its action is similar to **restrictocin** and ribonuclease U2.

mitomycin C a 334 Da antibiotic and the principal member of the aziridine antibiotics (mitomycins) produced by *Streptomyces caespitosus*. It is toxic to bacteria and mammalian cells and possesses antineoplastic activity. When administered, mitomycin C undergoes intramolecular rearrangement and reduction within cells to form a bifunctional alkylating agent that irreversibly cross-links the two strands of duplex DNA molecules, thereby inhibiting DNA synthesis.

mitoplast a structure consisting of the inner membrane and the matrix of a **mitochondrion**. It is formed by removing the outer membrane and the intermembrane material by treatment of a mitochondrion with nonionic detergents under defined conditions.

mitoribosome any of the **ribosomes** found in mitochondria. They resemble those of bacteria and are sensitive to chloramphenicol.

mitosis *or* **karyokinesis** the division of a eukaryotic cell nucleus to produce two daughter nuclei that contain identical numbers of chromosomes and that are identical genetically to the parent nucleus except where crossing over or mutation has occurred. It is divided into five main stages: **prophase**, **prometaphase**, **metaphase**, **anaphase**, and **telophase**. Mitosis is normally accompanied by **cytokinesis** leading to division of the cell. *Compare* **meiosis**. —**mitotic** *adj.*

mitotic coefficient *or* **mitotic index** the proportion of cells in a population that are in **mitosis** at a given time.

mitotic recombination *or* **somatic crossing over** crossing over between homologous chromosomes during mitotic division of a diploid somatic cell. The result of such a recombination will be imperceptible if the cell is homozygous at the loci exchanged but may produce a marked effect if the dividing cell is heterozygous at these loci. *See also* **Holliday junction recombination**.

mitotic spindle *see* **spindle**.

mixed acid fermentation a particular mode of bacterial fermentation of glucose that is characteristic of many members of the Enterobacteriaceae, e.g. *Escherichia*, *Salmonella*, and *Shigella*. It yields, as principal products, acetic, formic, lactic, and succinic acids, in varying proportions according to the organism and growth conditions.

mixed anhydride any **acid anhydride** formed between two different acids.

mixed-bed deionizer an apparatus for removing electrolytes from impure water (or an aqueous solution) and containing an intimate mixture of anion- and cation-exchange resins. It is frequently fitted with a device for measuring the electrolytic conductivity of the effluent.

mixed-bed resin *a common but imprecise term for* an intimate mixture of anion- and cation-exchange resins as used in a mixed-bed deionizer.

mixed-function oxidase *or* **mixed-function oxygenase** *an older term for* **monooxygenase**.

mixed indicator 1 a mixture, in definite proportions, of two acid–base indicators, or of an indicator and a dye, chosen so as to produce a sharp colour change over a narrow range of pH values. **2** a mixture, in definite proportions, of two or more acid–base indicators chosen so as to produce a series of colour changes over an extended range of pH values.

mixed inhibition *or* **mixed-type inhibition** inhibition of the activity of an enzyme that yields complex kinetic data that cannot be clearly categorized as either competitive or non-competitive. This can result e.g. when the enzyme–substrate–inhibitor complex undergoes the catalytic process slowly, or where binding of the inhibitor changes both k_{cat} and K_m.

mixed inhibitor *or* **mixed-type inhibitor** any inhibitor that produces **mixed inhibition** of the activity of an enzyme.

mixed labelled describing a specifically isotopically labelled compound that has more than one type of modified atom, e.g. $[^{13}C]H_3-CH_2[^{18}O]H$.

mixed-lymphocyte reaction *or* **mixed-leukocyte reaction** the reaction that occurs when leukocytes from two genetically differing individuals are cultured together. Small lymphocytes are transformed into immature blast cells, with synthesis of DNA, to an extent that is related to the degree of disparity between the histocompatibility antigens of the cell donors. This reaction forms the basis of a test for the acceptability of an allograft.

mixed triacylglycerol *or (formerly)* **mixed triglyceride** any **triacylglycerol** that contains three types of acyl residue.

mixed-type inhibition *an alternative term for* **mixed inhibition**.

mixed-type inhibitor *an alternative term for* **mixed inhibitor**.

mixed valency describes coordination compounds and clusters in which a metal is present in more than one level of oxidation. In metal clusters the valence electrons are often completely delocalized, allowing efficient electron-transfer processes.

mixotropic series a series of solvents arranged on the basis of their ability to form hydrogen bonds. It is based on the extent to which two given solvents in this series will mix; near neighbours in the series are **isopartitive**.

mixture *(in chemistry)* homogeneous or heterogeneous material in any physical state that consists of two or more chemical components that have no firm chemical bonding between them and that are physically separable.

MK *abbr. for* menaquinone.

mks system *or* **MKS system** *abbr. for* metre–kilogram–second system (of electrical units).

mL *or* **ml** *symbol for* millilitre.

MLD *abbr. for* minimum lethal dose.

M line *see* sarcomere.

MluI a type 2 **restriction endonuclease**; recognition sequence: A↑CGCGT.

mM *or* **mM** *symbol for* millimolar.

M macroglobulin a paraprotein of IgM type that is produced by individuals with Waldenström's macroglobulinemia.

MMDB *abbr. for* molecular modelling database; *see* **structure database**.

mmHg *symbol for* the conventional **millimetre of mercury**.

MMLV *abbr. for* Moloney murine leukemia virus; a virus that encodes a reverse transcriptase with high processivity used in the generation of cDNA.

mmol *symbol for* millimole.

MMP *abbr. for* matrix metalloproteinase.

MMPI *abbr. for* matrix metalloproteinase inhibitor.

MMTV *abbr. for* mouse mammary tumour virus.

Mn *symbol for* manganese.

mnemonic enzyme mechanism *or* **mnemonical enzyme mechanism** an enzyme mechanism proposed to explain cooperativity to-

wards substrate shown by monomeric enzymes with only one substrate-binding site. In this mechanism free enzyme is considered to exist in two conformations, I and II, that are only relatively slowly interconvertible and that differ in their substrate-binding affinities; the binding of substrate induces a change of the enzyme to a third conformation, III, which reverts to conformation I when product is released. The behaviour of the system depends on the rate of interconversion of I and II and on the relative rates at which these forms can bind substrate. If the binding by I is more rapid than that by II, the system will show positive cooperativity, i.e. the enzyme has 'remembered' the catalytically more effective conformation; if the converse, the system will show negative cooperativity.

MNSs system one of the blood-group systems of humans, in which the antigens are the **glycophorins** of the erythrocytes. It was the second human-blood-group system to be discovered, the first being the ABH system (*see* **ABH antigens**). The M and N antigenic sites are poorly immunogenic and antibodies to M and N antigens do not normally occur in the blood. The determinants are located at the N terminus of glycophorins A and B (GPA and GPB); polymorphism between M and N antigens on GPA is attributed to amino-acid substitutions at residues 1 (Ser/Leu) and 5 (Gly/Glu), but the N-terminal Ser^1 and Leu^1 are the primary determinants for the M and N antigens respectively. GPB carries 'N' (N-like activity), S/s, and U blood-group antigens; 'N' activity resides at the N terminus of GPB, the sequences of GPA and GPB between residues 1 and 26 being identical; the S/s antigenic determinants arise from Met/Thr polymorphism at residue 29 of GPB; and the U antigen, present in all Caucasians and most Blacks, resides between residues 33 and 39.

Mo *symbol for* molybdenum.

mob the genetic notation for genes involved in the mobilization of a **plasmid** from cell to cell.

mobile element *an alternative term for* **transposon**.

mobile phase the phase that moves in a chromatographic system; it can include the portion of the starting sample present in this phase at any given time.

mobility 1 the property of being mobile. **2** *see* **electrophoretic mobility**.

mobilization *or* **mobilisation 1** the act or process of mobilizing. **2** the process by which a conjugative plasmid brings about the transfer from one cell to another of DNA to which it is not stably and covalently linked.

mobilize *or* **mobilise** to bring (resources or reserves) into use for a particular purpose, e.g. to mobilize fat by lipolysis or glycogen by glycogenolysis.

mobilizing plasmid *or* **mobilising plasmid** *an alternative name for* **conjugative plasmid**.

Möbius strip *or* **Moebius strip** any one-sided continuous surface with a single bounding curve made by twisting a ribbon through 180° and joining the ends together. When divided lengthwise it forms another Möbius strip of double the length, and is a model for the replicative behaviour of a circular chromosome that contains a single twist. [After its inventor, August Ferdinand Möbius (1790–1868), German mathematician.]

Mo cell a relatively small cell, containing round, osmophilic granules, that is believed to produce **motilin**. Mo cells occur in the upper small intestine of humans, mainly in the crypts; they are sparse in the lower small intestine.

MoCo *abbr. for* **molybdenum cofactor**.

modal class *(in a statistical distribution)* the class that contains the **mode** (def. 2).

mode 1 any particular way or manner of existing or of doing something. **2** *(in a statistical distribution)* the value of the variable occurring with greatest frequency. **3** any of the several wave frequencies that an oscillator can generate, or to which a resonator can be tuned to respond. —**modal** *adj.*; **modality** *n.*

modeccin a 57 kDa lectin present in the turnip-like root of the southern African plant, *Modecca digitata*. It is highly toxic to animals, its prime action being on the liver, and to animal cells in culture. It consists of two polypeptide subunits in disulfide linkage, the A (smaller) subunit being biologically indistinguishable from the A subunits of ricin and abrin. *See also* **ribosome-inactivating protein**.

model 1 a three-dimensional representation of an object, structure, etc., sized to be conveniently seen, handled, and studied. **2** a conceptual representation of a particular phenomenon, system, or set of experimental observations as an aid to understanding and as an object for test or for further experimentation. **3** an animal that, often as a result of mutation, mimics a pathological condition and can be used to study the pathogenesis of the condition. *See also* **transgenic**. **4** a mathematical representation of a particular phenomenon, system, or set of experimental observations as an aid to calculation and prediction. **5** to construct or create a **model** (def. 1, 2, 3). —**modelled** *or (sometimes US)* **modeled** *adj.*; —**modelling** *or (sometimes US)* **modeling** *n.*

model organism an organism used in research to exemplify its type and to represent more complex organisms in which similar phenomena are thought to or do occur. Examples include the bacterium *Escherichia coli*, the yeast *Saccharomyces cerevisiae*, the worm *Caenorhabditis elegans*, the fruit fly *Drosophila melanogaster*, the flowering plant *Arabidopsis thaliana*, and the rodent *Mus musculus*.

model system any biological or biochemical system that is used for study because it is considered to be representative of one or more other (often more complex) systems in which similar phenomena occur or are believed to occur.

moderator *or* **modulator** any substance that changes the rate of an enzymic or other reaction, either positively or negatively.

modification 1 *(in biochemistry) see* **covalent modification**. **2** *(in genetics)* any nonheritable change in the phenotype of an organism in response to variation of its environment; it includes **host-controlled modification**. *Compare* **mutation**.

modification enzyme any enzyme that effects **covalent modification**, and specifically the DNA methylase component of the **restriction-modification system**. Such methylases are conventionally named in the same way as **restriction endonuclease**s with a prefix 'M' and in this dictionary they are listed, where appropriate, under the corresponding nuclease (e.g. M.*Alu*I appears under *Alu*I).

modification gene *see* **restriction–modification system**.

modification methylase *a common name for* any of the enzymes that methylate specific sites on a DNA molecule.

modification–restriction system *an alternative name for* **restriction–modification system**.

modified LDL a low-density lipoprotein (LDL) particle that has been modified *in vitro* by acetylation, treatment with malondialdehyde, cigarette smoke, or oxidation by copper. Similar changes *in vivo* have been implicated in atherogenesis. Such particles bind to a modified LDL (or scavenger) receptor on macrophages and are internalized, thereby transforming the macrophages into foam cells.

modifier *an alternative term for* **effector**.

modifier gene *or* **modifying gene** any gene that affects the phenotypic expression of a gene or genes at other loci.

modify 1 to alter the structure, character, activity, etc. of (something). **2** to convert a chemical compound into an analogue, especially by substitution. *See also* **modification**.

modular primer an unligated tandem arrangement of **primers** (usually two or three) used for the **polymerase chain reaction** and related experiments.

modulating codon any codon that controls the frequency with which a **cistron** or cistrons are transcribed.

modulation 1 adjustment or regulation of the degree or activity of something. **2** regulation of the frequency with which a **cistron** or cistrons is transcribed. **3** control of the activity of a regulatory enzyme by an effector. **4** *(in physics)* the process of changing the amplitude, frequency, or phase of a wave by combination with another wave.

modulator 1 something that modulates or brings about modulation. **2** any molecule that acts as an **effector**. **3** *(in pharmacology)* a ligand that increases or decreases the action of an agonist by combining with a distinct (allosteric) site on the receptor macromolecule. **4** *an alternative name for* **moderator**.

modulator protein *an alternative name for* **calmodulin**.

module a standardized or self-contained part or unit of construction, especially one that can be used in alternative combinations with others; e.g. one of a set of such assemblies of electronic components that comprise a larger piece of equipment or one of a number of autonomous protein folding units (or building blocks) that

are often used to confer a variety of complex functions on a parent protein, either by multiple combinations of the same unit or by combinations of different units to form mosaics; the spread of protein modules is believed to involve genetic shuffling mechanisms (e.g. a **kringle** or SH3 domain). —**modular** *adj.*

MODY *abbr. for* maturity-onset diabetes of the young.

moesin *abbr. for* membrane-organizing extension spike protein; a protein, isolated from bovine uteri, containing a domain similar to one in **ezrin**, band 4.1 and related proteins. Moesin occurs in many cell types and tissues, and is involved in connections of major cytoskeletal structures to the plasma membrane.

Moffitt–Yang equation *or* **Moffitt equation** a phenomenological equation, related to the **Drude equation**, that describes the anomalous variation of optical rotatory dispersion with wavelength shown by polypeptides containing helical sections. It is used to provide an estimate of the alpha-helical content of polypeptides and proteins. If λ is the wavelength of observation and $[m']_\lambda$ is the reduced **mean residue rotation** at that wavelength,

$$[m']_\lambda = \{a_0\lambda_0^2/(\lambda^2 - \lambda_0^2)\} + \{b_0\lambda_0^4/(\lambda^2 - \lambda_0^2)^2\},$$

where a_0, b_0, and λ_0 are empirical constants; a_0 is a function of contributions to the optical rotation from both the helix and the residue side-chains, as well as from the solvent, and b_0 and λ_0 are related principally to the amount of helical structure in the molecule. A value of λ_0 is chosen to give the best straight line when $[m'](\lambda^2 - \lambda_0^2)$ is plotted against $1/(\lambda^2 - \lambda_0^2)$, a good fit usually being obtained if $\lambda_0 = 212$ nm when $\lambda = 350$–600 nm. The slope of this line, $b_0\lambda_0^4$, gives a value of b_0, whence, by reference to the values found for non-helical and fully helical synthetic polypeptides an estimate may be made of the alpha-helical content of a polypeptide or protein under investigation. [After W. Moffitt and Jen Tsi Yang (1922–).]

Mohr pipette a cylindrical pipette calibrated for delivery from the zero mark to a particular chosen graduation mark. [After Carl Friedrich Mohr (1806–1879), German chemist.]

Mohr's salt ammonium ferrous sulfate; di-ammonium iron(II) sulfate hexahydrate; $(NH_4)_2SO_4 \cdot FeSO_4 \cdot 6H_2O$.

m.o.i. *abbr. for* multiplicity of infection; the ratio of viral particles to organisms or cells exposed to infection.

moiety either of two, usually distinctive, component parts of a complex molecule; e.g. the steroid and saccharide moieties of a cardiac glycoside.

mol *symbol for* mole.

mol. *abbr. for* **1** molecular. **2** molecule.

μmol *symbol for* micromole.

molal 1 denoting the concentration of a solute (or a solution as a whole) expressed as the amount of substance of the solute divided by the mass of solvent. It is measured in moles per kilogram. **2** denoting an amount-of-substance **concentration** measured in such terms. *Compare* **molar**.

molality *symbol: m or b*; a measure of concentration expressed as the number of moles (of a specified solute) per kilogram of solvent. Molality is independent of temperature. *See also* **molal**.

molar 1 denoting an extensive physical quantity that is measured per mole (of a specified substance). It is usually denoted by the subscript m in the symbol; e.g. **molar volume** is the volume of one mole of a substance, and the symbol is V_m. *Compare* **specific**. **2** denoting an intensive physical quantity that is measured per mole (of a specified substance); i.e. divided by concentration. Use of the term molar in this sense is no longer recommended, except in the terms **molar absorption coefficient** and **molar conductivity**. **3** denoting the concentration of a solute (or a solution as a whole), expressed as the amount of substance of the solute divided by the mass of the solution. The recommended units are moles per litre but the convenient symbols M and M are frequently used. *Compare* **molal**.

molar absorbancy index *a former name for* **molar absorption coefficient**.

molar absorption coefficient either the **molar decadic absorption coefficient** (*symbol: ε*) or the **molar Napierian absorption coefficient** (*symbol: κ*); these are, respectively, the decadic **absorbance** or the Napierian absorbance of a substance, or of a solute in solution, (at a specified wavelength (or frequency) and temperature) divided by the optical path length (in cm) of the medium and by the amount-of-substance concentration of the substance; i.e. the decadic or Napierian absorption coefficients of a medium whose concentration is expressed in moles per litre. *Former names*: molar absorbancy index; molar absorptivity; molar extinction coefficient; molecular extinction coefficient. *See also* **absorption coefficient**.

molar absorptivity *a former name for* **molar absorption coefficient**.

molar activity an **activity** (def. 1) of a substance or material divided by its amount of substance (in moles). *Compare* **specific activity**.

molar catalytic activity the **catalytic activity** of an enzyme, under specified conditions, divided by the amount of substance (in moles), either of enzyme catalytic centres or of multi-centre molecules; in the latter case it is important to specify whether the measurements refer to catalytic centres or to molecules. It may be expressed in katals per mole of enzyme. If the conditions specified are such that the enzyme is saturated with substrate, the expression (which then replaces the obsolete term turnover number) will be numerically equal to the **catalytic constant**, in s^{-1}, for decomposition of the enzyme–substrate complex into enzyme and products, although its interpretation may be complex. *Compare* **molar enzymic activity**, **specific catalytic activity**.

molar concentration *an old but now incorrect term for* **amount-of-substance concentration** measured in moles per litre. *See* **molar** (def. 2).

molar conductivity *symbol*: Λ ;the **electrolytic conductivity** of a (specified) electrolyte divided by its amount-of-substance concentration. The SI derived unit is the $S\ m^2\ mol^{-1}$.

molar decadic absorption coefficient *symbol*: ε; *see* **molar absorption coefficient**.

molar ellipticity *or (formerly)* **molecular ellipticity** *symbol*: $[\Theta]$; an expression of circular dichroism, $\Delta\varepsilon$, at a specified wavelength, λ. It is given by the relation: $[\Theta]_\lambda = 3299\ \Delta\varepsilon_\lambda$.

molar enzymic activity *(obsolete)* the phenomenological coefficient relating the **enzymic activity** (under specified conditions) to the molar amount of enzyme substance; it is expressed in katals per mole of enzyme. It has now been superseded by **molar catalytic activity**.

molar extinction coefficient *a former name for* **molar absorption coefficient**.

molarity *an older (still widely used) term for* **amount-of-substance concentration** expressed in $mol\ dm^{-3}$ or $mol\ L^{-1}$. *See* **concentration** (def. 1b), **molar** (def. 3).

molar mass the mass of a substance divided by its **amount of substance**; it is commonly expressed in grams (sometimes kilograms) per mole. *Compare* **molecular mass**, **relative molecular mass**.

molar Napierian absorption coeffient *symbol*: κ; *see* **molar absorption coefficient**.

molar optical rotation *symbol*: $[m]$ *or* $[M]$; the observed optical rotation, α_{obs}, of a substance, or of a solute in solution, (at a specified wavelength (or frequency) and temperature) divided by the optical path length (in dm), l, of the medium and by its amount-of-substance concentration (in $mol\ dm^{-3}$), c. It is given by:

$$[m] = \alpha_{obs}/cl = [\alpha]_\lambda^\theta M_r/100$$

in $deg\ dm^2\ g^{-1}$, where $[\alpha]_\lambda^\theta$ is the **specific optical rotation**.

molar optical rotatory power *symbol*: α_m; it is given by the relation $\alpha_m = \alpha/cl$ in $rad\ m^2\ mol^{-1}$, where α is the observed rotation in radians, c is the amount-of-substance concentration in $mol\ dm^{-3}$, and l is the optical path length of the sample in m. *See also* **specific optical rotatory power**.

molar volume *symbol*: V_m; the volume occupied by a substance, under specified conditions of temperature and pressure, divided by its **amount of substance**; it is commonly expressed in litres per mole.

mold *the US spelling of* **mould**.

mole *symbol*: mol; the SI base unit of **amount of substance**, defined as the amount of substance of a system that contains as many elementary entities as there are atoms in 0.012 kilogram of carbon-12; the elementary entities must be specified and may be atoms, molecules, ions, electrons, other particles, or specified groups of such particles. One mole of a compound has a mass equal to its **relative molecular mass** in grams.

molecular 1 of or pertaining to a molecule or molecules. **2** consisting of or existing as molecules; (especially of certain gaseous ele-

ments) present or reacting as intact molecules rather than as atoms, ions, etc., e.g. molecular oxygen (i.e. dioxygen). **3** of dimensions comparable to those of molecules. **4** *(in physical chemistry)* an *obsolete term meaning* divided by molecular weight or expressed per gram-molecule; **molar** now used. **5** considered in terms of or at the level of molecules, or in terms of chemical substances or structures and their interconversions; e.g. molecular biology, molecular medicine.

+molecular *comb. form* indicating the **molecularity** of a chemical reaction (when qualified by a multiplicative prefix), e.g. unimolecular, bimolecular, termolecular.

molecular beacon *or* **molecular platform** a protein that contains a variety of domains for specific proteins and serves as a platform for the assembly of specific proteins that mediate a particular function. *See* **REST**.

molecular biology the science that studies the chemical and physical basis of biological specificity and variation, especially with regard to the structure, replication, and expression of genes, and to the structure, interaction, and physiological function of gene products. —**molecular biologist** *n*.

molecular chaperone *see* **chaperone**.

molecular clock the principle that the relationship between evolutionary distance as determined from macromolecular sequences and time is approximately linear, a concept consistent with the neutral theory of evolution in the presence of more or less constant functional constraints.

molecular cloning *or* **DNA cloning** the isolation in a bacterial cell of a fragment of any heterologous DNA in covalent linkage with a replicon (plasmid, phage, etc.) in order to obtain a homogeneous population (i.e. clone) of DNA molecules from the progeny of such a cell. The DNA to be cloned may be a fragment of genomic DNA obtained by restriction endonuclease digestion, or a complementary DNA (cDNA). The foreign DNA is inserted into the **cloning vector**, which has been cleaved with a restriction enzyme; the gaps between vector and insert DNA are sealed with DNA ligase. In cloning using plasmid vectors, plasmids are then reintroduced into host cells by **transformation**; markers on the cloning vector allow selection of plasmid-containing cells and identification of those cells that contain recombinant plasmids (rather than 'empty' vectors). To identify clones containing particular sequences, DNA from individual colonies may be screened with a labelled DNA probe by **colony hybridization**. In cloning using bacteriophage vectors, the ligated insert and vector are packaged into phage capsids *in vitro,* and then these are used to infect host bacteria.

molecular disease any disease caused by an abnormality in a single protein, usually an enzyme. This may have an abnormal structure that makes it functionally less efficient or harmful to the organism, or it may have a normal structure but be present in reduced amount. Examples are the many inborn errors of metabolism and **sickle-cell anemia**. Overproduction of proteins, e.g. of oncogene products, may also lead to a pathological situation.

molecular distillation distillation at very low pressures (at or below $1.3\ N\ m^{-2}$) when the mean free path of the molecules is of the same order as the distance between the surface of the liquid being distilled and the distillate; the molecules can therefore travel rapidly from the liquid to the distillate with relatively few collisions, and with a reduced risk of thermal denaturation or decomposition. The technique is useful in the separation and purification of fat-soluble vitamins and other natural products.

molecular drive *or* **concerted evolution** a theory to explain the observed greater homogeneity of members of a multigene family within a species compared with their homologous counterparts in a sister species. It holds that unequal crossing over and other mechanisms of molecular turnover between homologous chromosomes spread a particular variant of a gene in a multigene family from one chromosome to the next. The variant gene is spread through the population by sexual reproduction until it reaches fixation.

molecular dynamics the study of the relative intramolecular differences in the motions of the components of (macro)molecules.

molecular ellipticity *a former term for* **molar ellipticity**.

molecular endocrinology endocrinology at the molecular level.

molecular entity any chemically or isotopically distinct atom, molecule, ion, ion pair, radical, radical ion, complex, conformer, etc., capable of existence as a separately distinguishable entity.

molecular enzymology *(sometimes)* the aspects of enzymology concerned with kinetics and mechanisms.

molecular evolution the process whereby changes in DNA or RNA accumulate over time via a range of different mechanisms, giving rise to changes in, for example, genes, gene products, gene expression, or the frequency of genes within populations.

molecular exclusion chromatography *a less common name for* gel-permeation chromatography.

molecular extinction coefficient *a former name for* **molar absorption coefficient**.

molecular formula a chemical formula giving the kinds and number of atoms in one molecule of a specified substance.

molecular fragmentography *an alternative (less satisfactory) name for* **selected-ion monitoring**.

molecular genetics the branch of genetics that attempts to describe genetic events in molecular terms.

molecular graphics *functioning as sing.* the use of computer programs for visualizing (macro)molecules, refining structures and molecular design, protein engineering, etc. Suites of programs that are commonly used include Insight, Ribbons, Midas, and O.

molecular hybridization *an alternative name for* **hybridization analysis**.

molecular ion any ion produced, especially in a mass spectrometer, when a molecule loses or gains an electron, forming a molecular cation or molecular anion, respectively.

molecularity the number of molecular entities involved in a simple, one-step, chemical reaction (it is not used to describe an overall reaction with more than one step). In the two-step reaction: $A + B \rightarrow C \rightarrow D$, the molecularity of the first step is two, i.e. it is bimolecular, and the molecularity of the second step is one, i.e. it is unimolecular.

molecular mass the mass of one molecule of a specified substance expressed in (unified) atomic mass units or daltons. *Compare* **relative molecular mass, molecular weight**.

molecular matchmaker any of a class of proteins, also called matchmaker proteins, that cause a conformational change in one or both components of a DNA–protein binding pair to promote formation of a metastable DNA–protein complex. ATP hydrolysis occurs in the process. Such matchmaking is used extensively in repair, replication, and transcription of DNA.

molecular medicine *(sometimes)* the branch of medicine concerned with the biochemical mechanisms of disease processes and therapy.

molecular mimicry 1 in the context of ligand receptor binding (*see* **ligand (def. 6)**) a molecule, unrelated to the ligand that by accident binds to the receptor. **2** in the etiology of autoimmune diseases, the hypothesis that molecules from a microorganism mimic those in an animal host, leading to activation of self-reactive lymphocytes.

molecular motor a protein or protein complex that generates force upon hydrolysis of ATP. Such motors include **myosins, kinesins, dyneins**, and **DNA helicases**.

molecular orbital a wave equation for one electron describing its movement in the effective field provided by the nuclei and all other electrons of a molecular entity of more than one atom.

molecular pathology the use of techniques of molecular biology to enhance the understanding, diagnosis, and therapy of disease (mainly with respect to humans).

molecular pharmacology the use of techniques of molecular biology to enhance the understanding of the mechanism of action of existing drugs, and with the help of molecular graphics to predict the structure of novel drugs, especially compounds that might bind to proteins of known structure.

molecular pK any one of the pK values of an acid with two or more dissociations. *Compare* **group pK, titration pK**.

molecular recognition the specific **recognition** of one molecule by another; it often includes **chiral recognition**.

molecular ruler 1 any rigid oligoproline crosslinking reagent of defined length. Such reagents are useful in probing the structures of biological macromolecules in solution. **2** any molecule that regulates length, e.g. **nebulin**.

molecular sieve any of a class of substances or materials, whether naturally occurring or artificial, that contain pores of molecular dimensions and of a closely defined range of diameters. They are useful for the separation of substances within a particular range of molecular dimensions on the basis of selective exclusion.

molecular-sieve chromatography any form of chromatography in which a separation is based mainly upon differences in molecular size and/or shape of the substances being separated. It includes **gel-permeation chromatography** and separations using **zeolites**.

molecular-sieve electrophoresis *(sometimes)* an alternative term for polyacrylamide gel electrophoresis.

molecular switch a protein that can change from an active to an inactive form by altering its conformation as a result of binding to a specific ligand or of chemical modification.

molecular tinkering the evolutionary process of transforming pre-existing genes or parts of genes to produce new roles or functions, or of combining molecular systems to give new, often more complex, systems. For example, in mammals, lysozyme, an ancient hydrolytic enzyme that destroys bacterial cell walls, has been recruited as the milk protein α-lactalbumin. Similarly, in birds and reptiles, the ancestral ornithine-urea cycle enzyme argininosuccinate lyase has been recruited as a water-soluble eye-lens structural protein, δ-crystallin.

molecular weight *abbr.*: mol. wt. *or* M.W.; *a former name for* **relative molecular mass**.

molecular weight cut-off *or* **cut-off** *a jargon term for* the nominal maximal value of the relative molecular mass of a macrosolute that can pass through an ultrafiltration membrane of a particular grade.

molecular weight marker *or* **molecular weight standard** a *jargon term for* any soluble polymer of accurately known relative molecular mass that serves as a reference in estimations of M_r values of polymers by techniques such as gel permeation chromatography or gel electrophoresis. Generally a set of such substances, covering a range of M_r values, is used.

molecule a structural unit of matter consisting of one or more atoms; the smallest discrete part of a specified element or compound that retains its chemical identity and exhibits all its chemical properties.

mole fraction *an alternative name for* amount-of-substance fraction. *See* **concentration**.

mole percent the quantity mole fraction multiplied by 10^{-2}; usage no longer recommended.

Molisch test a qualitative test for free or combined aldohexoses. It depends on the formation of a purple colour when the furfural derivatives, formed when aldohexoses are heated with strong acids (e.g. H_2SO_4), react with sulfonated 1-naphthol. [After Hans Molisch (1856–1937), Austrian chemist.]

mollicute *see* **mycoplasma**.

Molscript a computer program, written by Per Kraulis, for displaying molecular 3D structures such as proteins.

molt *the US spelling of* **moult**.

molting hormone *the US spelling of* moulting hormone (i.e. **ecdysterone**).

mol. wt. *abbr. for* molecular weight (alternative to M.W.).

molybdate 1 the bivalent anion, MoO_4^{2-}, derived from molybdic acid. **2** any salt of molybdic acid.

molybdenum *symbol*: Mo; a metallic transition element of group 6 of the (IUPAC) **periodic table**; atomic number 42; relative atomic mass 95.94. It consists of a mixture of eight stable nuclides of which the commonest is that of mass number 98. It forms molybdates (molybdenum(VI)) and can also exist in various lower oxidation states. Molybdenum is of low abundance in the Earth's crust and is an essential trace component of living material, where it participates in numerous redox reactions. It is a constituent of component I (or molybdoferredoxin) of the **nitrogenase** (EC 1.18.6.1) complex from legume nodule bacteria (bacteroids), which converts nitrogen into ammonia. There are many artificial radionuclides of molybdenum, of which the most widely used is molybdenum-99.

molybdenum cofactor a cofactor of **molybdoenzymes** of human cells. It consists of a molybdenum atom complexed by vicinal sulfhydryl groups on a unique pterin species called **molybdopterin**. In some bacteria it is present as a dinucleotide. Synthesis of the cofactor starts with GTP and proceeds by a pathway independent from that for folate or biopterin synthesis. It has been well studied in *E. coli* but homologous genes have been found in the human genome.

molybdenum cofactor deficiency a severe autosomal disease of humans in which activity of the three **molybdoenzymes** is lacking and leads to accumulation and increased excretion of sulfite, thiosulfate, *S*-sulfocysteine, taurine, xanthine, and hypoxanthine. Clinically the presentation is very similar to that of sulfite oxidase deficiency. It is caused by a defect in the synthesis of **molybdenum cofactor**.

molybdo+ *comb. form* denoting containing molybdenum, e.g. molybdoprotein.

molybdoenzyme any enzyme that contains tightly bound **molybdenum cofactor**. Examples from human cells are aldehyde oxidase (EC 1.2.3.1), sulfite oxidase (EC 1.8.2.1), and xanthine dehydrogenase (EC 1.17.1.4). A very rare combined deficiency of these three enzymes presents as xanthinuria.

molybdoferredoxin *an alternative name for* component I of **nitrogenase**.

molybdophosphoric acid *or* **phosphomolybdic acid** a heteropolyacid of approximate formula $24MoO_3 \cdot P_2O_5 \cdot xH_2O$, where *x* is approximately 50. It is useful in reagents for alkaloids, xanthine, uric acid, creatinine, etc.

molybdopterin *or* **pyranopterin** a pterin derivative that binds molybdenum or tungsten in enzymes. *See* **molybdenum cofactor**.

mon+ *a variant form of* **mono+** (sometimes before a vowel).

Mon *symbol for* the montanoyl group.

monactin *see* **nonactin**.

monatomic *or* **monoatomic** (of a molecular entity) composed of only one atom.

monellin an ~11 kDa protein consisting of two polypeptide chains, linked non-covalently, isolated from the fresh fruits of the African shrub, *Dioscoreophyllum cumminsii*, known as serendipity berry. It is intensely sweet-tasting, being 3000 times sweeter than sucrose on a mass basis, and has sequence similarities with **thaumatin**.

monensin a 671 Da, open-chain macrotetrolide-like antibiotic, with many chiral centres, isolated from *Streptomyces cinnamonensis*. It forms stable complexes with monovalent cations, especially with Na$^+$, and exhibits high Na$^+$/K$^+$ selectivity. These complexes are soluble in organic solvents and thus monensin can act as an **ionophore** for monovalent cations. It is used as a digestion-modifying feed additive in ruminants.

Monera the kingdom containing all prokaryotic organisms (i.e. bacteria and cyanobacteria) in the five-kingdom system of classification. —**moneran** *adj., n.*

mongolism *an older name for* **Down's syndrome**.

monitor 1 any instrument or device for checking, controlling, measuring, or keeping a record of some varying biological activity or chemical or physical property, e.g. absorbance, (radio)activity, concentration, pH, respiration. **2** to serve as a monitor of (some activity or property); to use a monitor.

mono+ *or* (*sometimes before a vowel*) **mon+** *comb. form* **1** denoting one, single, onefold, once, alone; **uni+**. **2** (*in adjectives describing physical attributes*) homogeneous; e.g. monochromatic, monodis-

perse, monoenergetic. **3** (*in generic names of chemical compounds*) denoting that member of a series (a) having only one specified substituent or other chemical modification, e.g. monoacylglycerol, monoamine; or (b) comprised of only one constitutional unit, e.g. monomer, monosaccharide. **4** *informal term* (*in specific names of chemical compounds*) indicating or emphasizing the presence in a molecule of a single specified atom, group, or residue; e.g. monoiodotyrosine, adenosine monophosphate, monosodium dihydrogen phosphate. **5** (*in adjectives describing compounds or substances*) having a single function of the kind specified; e.g. monoacidic, monobasic, monodentate, monohydric. **6** (*in names of processes or enzymes*) concerned with or acting on, one or only one chemical group of the kind specified; e.g. monoiodination, monoesterase.

monoacidic (of a **base** (def. 1)) able to accept only one proton per molecule in solution.

monoacylglycerol *or* (*formerly*) **monoglyceride** any ester of glycerol in which any one of its hydroxyl groups has been acylated with a fatty acid, the others being non-esterified.

monoacylglycerol pathway a pathway in which diacylglycerols and triacylglycerols are formed by sequential acylation of monoacylglycerol by acyl-CoA. It is important during fat absorption since monoacylglycerol is extensively absorbed from the gut; however, it is also present in a large number of other mammalian tissues. There is little evidence for this pathway in plants.

monoamine oxidase *abbr.*: MAO; *a common name for* amine oxidase (flavin-containing), EC 1.4.3.4; an enzyme, widely distributed in animal tissues, that catalyses the oxidative deamination of primary amines by a reaction between dioxygen and R–CH$_2$–NH$_2$ to form R–CHO, NH$_3$, and H$_2$O$_2$. It also acts on secondary and tertiary amines. It is important in the catabolism of neuroactive amines, e.g. epinephrine, norepinephrine, serotonin (5-hydroxytryptamine), and tyramine. Mammals have two genes for MAO, both on the X chromosome and near the Norrie disease gene. MAO-A (human, 520 amino acids) contains FAD and is a transmembrane protein of the outer mitochondrial membrane in most tissues. It shares 70% sequence identity with MAO-B (527 amino acids). Deficiency of either or both can be associated with deletion of the Norrie disease gene. *See* **Norrin**.

monoamine oxidase inhibitor any compound that inhibits **monoamine oxidase** (*abbr.*: MAO) and is useful as a thymoleptic (antidepressant) drug. MAO inhibitors act by increasing the concentrations of **norepinephrine** and **serotonin**, which, because of the block on their metabolism, accumulate in the synaptic gap and exert an increased postsynaptic effect. They were discovered as a result of the euphoric side effect of the antitubercular drug **isoniazid**. There are two main groups: (1) the **cyclopropylamines**, including tranylcypromine (*trans*-phenyl-cyclopropylamine); and (2) the **propargylamines**, including pargyline (*N*-benzyl-*N*-methyl-2-propynylamine. **Harmine** and related carboline alkaloids are reversible MAO inhibitors but are not used therapeutically.

monoatomic *a variant spelling of* **monatomic**.

monobactam any of a family of *N*-sulfonated monocyclic β-lactam antibiotics produced by a range of bacterial species.

monobasic 1 (of an acid) able to yield only one hydrogen ion per molecule in solution. **2** (of a salt or ester of a polybasic acid) derived by the replacement of only one of its dissociable hydrogen atoms per molecule.

monochromatic 1 *or* **homochromatic** involving, possessing, or relating to a single colour; consisting of, relating to, or emitting light (or other electromagnetic radiation) of a single frequency or wavelength. **2** *an alternative term for* **monoenergetic**. *Compare* **heterochromatic** (def. 1), **polychromatic**.

monochromator a device used for producing light (or other electromagnetic radiation) of essentially one frequency or wavelength, which may be fixed or selectable.

monocistronic containing as much genetic information as is contained in a single **cistron**.

monocistronic messenger RNA any messenger RNA molecule that carries one initiation site and encodes the synthesis of one polypeptide chain. It is characteristic of eukaryotes, which do not possess **polycistronic messenger RNA**.

monoclonal 1 of, pertaining to, being, formed by, or derived from a single **clone** (def. 1). **2** *(sometimes)* an informal name for **monoclonal antibody**.

monoclonal antibody *or (informal)* **monoclonal** an immunoglobulin secreted by a single clone of antibody-producing cells, either *in vivo* or in culture. Such antibodies are **monospecific**; they can thus be useful as immunochemical reagents for the detection or assay of particular antigens, or used in therapy. *See also* **humanize, hybridoma**.

monocomponent insulin a commercial preparation of natural insulin that has been highly purified by chromatography and is thus made essentially free of proinsulin, intermediates in the conversion of proinsulin to insulin, and breakdown products of insulin, as well as any unrelated contaminants.

monocotyledon *or* **monocot** any flowering plant having a single cotyledon (seed leaf) in the embryo. Such plants comprise the class Monocotyledonae.

monocyclic having a single ring of atoms in the molecule.

monocyte a large leukocyte of the **mononuclear phagocyte system**, 16–22 μm in diameter (in humans), with pale-staining cytoplasm and a reniform or horseshoe-shaped nucleus. Found in bone marrow and the bloodstream, monocytes are derived from pluripotent stem cells and become macrophages when they enter the tissues.

monodisperse describing a colloidal system in which the dispersed phase consists of particles all of (nearly) the same size. —**monodispersity** *n.*

Monod–Wyman–Changeux model (for allosteric interactions) *abbr.*: MWC model; a model describing the nature of **allosteric** interactions in proteins. It requires an allosteric protein to be an oligomer, the protomers of which are associated in such a way that they all occupy equivalent positions. Each protomer has only one stereospecific binding site for each ligand. The protein can exist in either of two conformational states, the T form (**tense form**), the predominant form when unligated, and the R form (**relaxed form**); these states are in equilibrium. The affinity of the R form for ligand is higher than that of the T form. All binding sites in each state are deemed to be equivalent and to have identical dissociation constants, K_R and K_T for the R and T forms, respectively. The sigmoidal binding curve for any allosteric protein and a given ligand can be calculated from the allosteric constant, L, equal to the ratio [T form]/[R form] for the unligated states and the two dissociation constants. [After Jacques Lucien Monod (1910–76), French microbiologist and molecular biologist, Jeffries Wyman (1901–95), US biochemist, and Jean-Pierre Changeux (1936–), French biochemist.] *See also* **mnemonic enzyme mechanism**.

monoenergetic *or* **monochromatic** (of moving particles and sometimes also photons) all having the same kinetic energy.

monoenoic denoting any alkenyl carboxylic acid containing one carbon–carbon double bond per molecule.

monoester any simple ester; i.e. any **ester** formed by condensation of one molecular proportion of an alcohol or phenol with one of an oxoacid.

monoesterase *see* **phosphomonoesterase**.

monofunctional having only one function or one reactive chemical group.

monoglyceride *a former name for* **monoacylglycerol**; its use is discouraged as it does not convey the intended meaning.

monognotobiotic *an alternative term for* **axenic**.

monohydric *or* **monohydroxy** describing any chemical compound containing one hydroxyl group per molecule. It is used especially of alcohols.

monoiodinated reacted with or containing only one atom of iodine per molecule.

monokine *an alternative name for a* **cytokine** produced by monocytes.

monolayer 1 *an alternative name for* **monomolecular layer**. **2** a single layer of cells grown or growing in culture.

monomer 1 any substance that can provide one or more (in number or species) of the monomeric units of an **oligomer** (def. 1) or of a **polymer** (def. 1); a molecule of such a substance. **2** *a loose term for* any of the component molecules (identical or non-identical) formed by the complete dissociation of a macromolecule with quaternary structure. **3** *(in molecular biology)* **a** any protein that is made up of

non-identical structural units. **b** any of the structural units formed by dissociation of an oligomeric protein, and corresponding to a **protomer** in the undissociated protein. —**monomeric** *adj.*

monomeric unit a group of atoms, derived from a molecule of a given **monomer** (def. 1), that comprises any one species of constitutional unit of a polymer.

monomolecular 1 relating to, consisting of, or involving a singular **molecular entity**. **2** (of a reaction) having a **molecularity** of one; unimolecular.

monomolecular layer *or* **monolayer** a layer of a substance or substances that is one molecule thick.

mononuclear 1 (of a cell) having one nucleus. **2** (of a metal-ion–ligand complex) containing a single central metal atom.

mononuclear phagocyte system a cell system in higher animals that comprises all the highly phagocytic mononuclear cells and their precursors. It includes the free macrophages, precursor cells and promonocytes of the bone marrow, monocytes of the bone marrow and the blood, and the tissue macrophages. The term is a replacement for **reticuloendothelial system**, now held to lack precision.

mononucleosis *see* **Epstein–Barr virus**.

mononucleotide any **nucleotide** (def. 1); the term is used especially as the generic name of the constitutional repeating unit of all oligonucleotides, polynucleotides, and nucleic acids. A mononucleotide consists of the (3′- or 5′-)phosphate of a single ribonucleoside or deoxyribonucleoside. The term is extended generally to include: (1) nucleoside oligophosphates, e.g. adenosine 5′-triphosphate (ATP); (2) nucleoside diphosphate sugars, e.g. uridine 5′-diphosphate glucose (UDPG); (3) nucleoside 2′,3′- and 3′,5′-(cyclic)phosphates, e.g. adenosine 3′,5′-phosphate (cyclic AMP); (4) nucleoside phosphates derived from artificial heterocyclic bases, e.g. 5-iodo-2′-deoxycytidine 5′-triphosphate, or from ones that do not occur naturally in nucleic acids, e.g. inosine 5′-phosphate (i.e. inosinic acid; hypoxanthine riboside 5′-phosphate; IMP); and (5) certain analogous compounds such as those containing, in place of a residue of ribose or deoxyribose, a residue of ribitol, e.g. flavin mononucleotide (i.e. riboflavin 5′-phosphate; FMN), of dideoxyribose, e.g. 2′,3′-dideoxyadenosine 5′-triphosphate (i.e. ddATP), or of another pentose, e.g. cytosine arabinoside 5′-triphosphate (i.e. *ara*-CTP). *Compare* **oligonucleotide, polynucleotide**.

monooxygenase any oxidoreductase enzyme that brings about the incorporation of one atom of oxygen from dioxygen into the donor, and catalyses reactions of the type:

$$\text{substrate} + O_2 =$$
$$\text{oxidized substrate} + H_2O \; (+ \; CO_2 \text{ in many cases}).$$

The following are examples. (1) Lysine 2-monooxygenase; EC 1.13.12.2; it catalyses the oxidation by dioxygen of L-lysine to 5-aminopentanamide, CO_2, and H_2O; FAD is a coenzyme. (2) Lysine 6-monooxygenase; EC 1.13.12.10; it catalyses the oxidation by dioxygen of L-lysine to N^6-hydroxy-L-lysine and H_2O. (3) Tryptophan 2-monooxygenase; EC 1.13.12.3; it catalyses the oxidation by dioxygen of L-tryptophan to indole-3-acetamide, CO_2, and H_2O. (4) *myo*-Inositol oxygenase; EC 1.13.99.1; it catalyses the oxidation by dioxygen of *myo*-inositol to D-glucuronate and H_2O; iron is a cofactor. (5) Phenol 2-monooxygenase; EC 1.14.13.7; *other name*: phenol hydroxylase; it catalyses a reaction between phenol, NADPH, and O_2 to form catechol, $NADP^+$, and H_2O; FAD is a coenzyme. It is an enzyme of bacterial aromatic substrate utilization. (6) *trans*-Cinnamate 4-monooxygenase; EC 1.14.13.11; *other names*: cinnamic acid 4-hydroxylase; cinnamate 4-hydroxylase; CA4H; it catalyses a reaction between *trans*-cinnamate, NADPH, and O_2 to form 4-hydroxycinnamate, $NADP^+$, and H_2O. It is a cytochrome P450–thiolate enzyme. *See also* **monophenol monooxygenase**.

monophenol monooxygenase EC 1.14.18.1; *other names*: tyrosinase; phenolase; monophenol oxidase; cresolase. An enzyme involved in the formation of melanins and other polyphenolic pigments, etc. It catalyses a reaction between L-tyrosine, L-dopa, and O_2 to form L-dopa, dopaquinone, and H_2O; copper is a cofactor. It has four motifs, three for the Cu site.

monophosphatidyl glycerol any [1,2-diacyl-*sn*-glycero-3-(phospho-1-*sn*-glycerol)] lipid present in chloroplasts of higher plants.

monophyletic group a group of species (or nodes in a phylogenetic tree) that share a common ancestor that is exclusive to those species (or nodes). For example, a clade is a monophyletic group.

monoploid (of cells or individuals) having a single set of chromosomes; i.e. in a polyploid series, having the fundamental **haploid** chromosome number; true haploid.

monosaccharide *the generic name of* the simplest carbohydrates. Monosaccharides cannot be hydrolysed to give smaller carbohydrates. They are polyhydric alcohols containing either (in **aldoses**) an aldehyde group or (in **ketoses**) a keto group and with from three to ten or more carbon atoms. Monosaccharides form the constitutional repeating units of oligo- and polysaccharides. The names and structures of the common aldoses (from triose to hexose) form the basis for prefixes used to describe other compounds containing a set of >CHOH groups. For structures, see illustration (note that if X = CHO and Y = CH₂OH these structures also depict the common D aldoses). An example of prefix use is the ketoheptonic acid phosphate, DAHP, **3-deoxy-D-*arabino*-(2)-heptulosonate 7-phosphate**. By extension the stereochemical relationships are applied to substituents other than OH, e.g. in **threonine**. *See also* **glycose** (def. 1), **heptose, hexose, pentose, tetrose, triose**.

monose *an obsolete term for* **monosaccharide**.

monosodium glutamate *abbr.*: MSG; the monosodium salt of L-glutamic acid, widely used as a flavouring agent for its meaty taste, and the most active component of soy sauce. Consumption by adults of more than about 3.0 g can cause **Chinese restaurant syndrome**, characterized by a feeling of tightness, facial pressure, burning sensation, and headache. The D-enantiomer does not have the characteristic taste and up to 7.0 g can be consumed without ill effect.

monosome 1 any chromosome in a functionally diploid chromosome set that lacks a homologue. **2** a structure consisting of a single ribosome bound to a molecule of messenger RNA.

monosomic (of cells, tissues, or individuals) having one or more **monosomes** (def. 1).

monospecific (of an antibody or an antiserum) only able to react with a single specified antigen or antigenic determinant.

monoterpene any **terpene** containing or derived from two isoprene units and thus containing ten carbon atoms. Monoterpenes may be acyclic or mono- or dicyclic.

monotonic (of a mathematical function) varying in such a manner that it continuously either increases or decreases as the independent variable increases. —**monotonically** *adv.*

monotopic an integral membrane protein that crosses the membrane once. *Compare* **polytopic**.

monovalent *an alternative term for* **univalent**.

monoxenic *see* **synxenic**.

monozygotic 1 (of twins, triplets, etc.) derived from a single fertilized ovum. **2** any one of a set of monozygotic individuals. *Compare* **dizygotic**.

montanate **1** *the trivial name for* octacosanoate, CH₃–[CH₂]₂₆–COO⁻, the anion derived from montanic acid (i.e. octacosanoic acid), a saturated straight-chain higher fatty acid. **2** any mixture of free montanic acid and its anion. **3** any salt or ester of montanic acid. Montanate esters of long-chain alcohols occur in montan wax, beeswax, and some other natural waxes.

montanoyl *symbol*: Mon; *the trivial name for* octacosanoyl,

D-glycero　　D-erythro　　D-threo　　D-arabino　　D-lyxo

D-ribo　　D-xylo　　D-allo　　D-altro　　D-galacto

D-gluco　　D-gulo　　D-ido　　D-manno　　D-talo

CH_3–$[CH_2]_{26}$–CO–, the acyl group derived from montanic acid (i.e. octacosanoic acid).

Monte Carlo method a way of solving a mathematical problem by sampling methods. The procedure involves the construction of a stochastic model of the mathematical process and the performance of sampling experiments upon it. Such methods have application in molecular graphics aimed at extracting the fundamental principles of macromolecular conformation. [After the resort in Monaco famous for its casino.]

Mops *or* **MOPS** *abbr. for* 3-(*N*-morpholino)propanesulfonic acid; a **Good buffer substance**, pK_a (20°C) = 7.2.

morpholine

morgan a unit of genetic length along a chromosome, defined as the distance between two loci that show, on average, one crossover per meiotic event; it is equal to 100 **map units** (or Morgan's units). [After Thomas Hunt Morgan (1866–1945), US geneticist and zoologist.]

Morgan's unit *an alternative name for* map unit. *Compare* **morgan**.

Mörner's test a qualitative test for tyrosine in which a green coloration is produced when a tyrosine-containing sample is boiled in a solution of formaldehyde in strong sulfuric acid. [After K.A.H. Mörner (1855–1917), Swedish chemist.]

morph+ *a variant form of* **morpho+** (before a vowel).

+morph *comb. form* indicating of a specified form, shape, or structure. —**+morphic** *or* **+morphous** *adj.*; **+morphy** *n.*

morphiceptin a tetrapeptide amide fragment of the milk protein β-casein, structure Tyr-Pro-Phe-Pro–NH_2, and the N-terminal tetrapeptide of casomorphin. It is a potent and specific agonist for μ-opiate receptors, its effects on the receptor being blocked by naloxone.

morphine 7,8-didehydro-4,5-epoxy-17-methylmorphinan-3,6-diol; an opiate and the most important opium alkaloid; it is a powerful, habit-forming narcotic analgesic. In the opium poppy, *Papaver somniferum*, it is derived from two molecules of tyrosine via (*S*)-reticuline as intermediate. Many enzymes, equivalent to those involved in its biosynthesis, have been found in mammalian liver. The diacetate ester, diacetylmorphine, is known as **diamorphine** or **heroin**. This is strongly habit forming and its manufacture, importation, and use are controlled in a number of countries. *See* **opioid receptor**.

morpho+ *or (before a vowel)* **morph+** *comb. form* indicating form or structure.

morphogen a diffusible substance responsible for some aspect of generation of form or structure during development of an organism. Its effects are concentration-dependent. A good example is retinoic acid.

morphogenesis the ensemble of the processes of development of a part, organ, or organism. —**morphogenetic** *or* **morphogenic** *adj.*

morpholine tetrahydro-1,4-oxazine; a heterocyclic compound with a six-membered oxygen- and nitrogen-containing ring. A number of *N*-substituted morpholines are useful as biological buffers, e.g. **Mes**, **Mops**.

morpholino antisense oligonucleotide a nonionic DNA analogue with an altered backbone that is resistant to nucleases. Such nucleotides can base pair with mRNA and can block gene expression and produce phenotypic effects in early development when injected into zebrafish, sea urchin, or *Xenopus* embryos.

morphology the branch of biology concerned with the outer form and inner structure of living organisms.

Morquio syndrome *another name for* mucopolysaccharidosis IV. *See* **mucopolysaccharidosis**.

MORT1 *see* FADD.

mortalin *other name*: 75 kDa glucose-regulated protein (*abbr.*: GRP 75); a protein that resembles **heat-shock proteins** in sequence, and may be a **chaperone**. It is a member of the heat-shock protein 70 family, which functions as a mortality marker in fibroblasts. Mortal fibroblasts contain a protein, MOT-1, uniformly distributed in the cytosol, while, in immortal fibroblasts, another form of the protein, MOT-2, is concentrated in perinuclear or juxtanuclear regions. In MOT-2, there are differences at residues 618 (Val to Met) and 624 (Arg to Gly). The cytosolic form induces cellular senescence; the perinuclear form does not.

MOS a gene family encoding serine/threonine kinases, e.g. the human and mouse protein pp39^{mos}; v-*mos* is the oncogene of the acutely transforming murine Moloney sarcoma virus (hence 'mos' from mo + s). *MOS* mRNA is detectable only in germ cells in testes and ovaries, and is implicated in meiosis; directly or indirectly it stabilizes **M-phase promoting factor** and activates **MAP kinase**. The protein has homology with Src (*see* **src**).

mosaic 1 a design or pattern on a surface, made up of numerous separate pieces of more or less the same area but differing in some characteristic, e.g. colour, shape, texture. **2** *an alternative name for* **chimera** (used especially of animals). **3** a protein comprising a number of different **modules**. —**mosaicism** *n.*

mosaic evolution within a population of organisms, the phenomenon of varying rates of evolution in different functional systems. For example, in hominid evolution, the dental, locomotor, and neurological systems have evolved at markedly different rates.

mosaic model *see* **fluid-mosaic model**.

Mössbauer spectroscopy a spectroscopic technique specific to particular atomic nuclei. The basis of the Mössbauer effect is the recoilless resonance absorption of γ-rays. For biochemical studies the relevant isotope for study is the stable isotope ^{57}Fe. Gamma rays are produced by a radioactive source of ^{57}Co, and the energy of the γ-rays emitted by the radioactive source is shifted by the Doppler shift obtained by moving the source with respect to the absorber. A more recent variation of the method uses synchrotron radiation. Parameters derived from the Mössbauer spectrum, which provide information on the chemical state of the iron are the isomer shift, the nuclear quadrupole shift and the nuclear hyperfine splitting due to interaction with unpaired electrons. [After Rudolf Mössbauer (1929–), co-winner of the Nobel Prize for Physics 1961 for his researches concerning the resonance absorption of gamma radiation.]

most parsimonious reconstruction (*in phylogeny*) for a given character in a data set, the branching order that requires the fewest mutational events to explain the contemporary sequences. *Compare* **most parsimonious tree**.

most parsimonious tree (*in phylogeny*) for all sequences in a data set, the branching order that requires the fewest evolutionary events to explain the contemporary sequences. *Compare* **most parsimonious reconstruction**.

MOT-1, MOT-2 *see* mortalin.

mother liquor the liquid remaining after a substance has been crystallized or precipitated from a solution.

motif 1 (*in relation to protein structure*) a locally ordered region within the core of a protein molecule, formed by 3-D interaction between two or three segments of the secondary structure (α-helix and/or β-strand) that are near one another along the polypeptide chain. The most important types are: (αα), (αβ), (βββ), and (βαβ). *See also* **domain**. **2** (*in relation to protein sequences or sequence alignments*) a contiguous, conserved region, typically 10-20 amino acids in length, usually denoting a key structural or functional unit within a protein, or group of proteins, and often used to diagnose related sequences using pattern-recognition methods (such as fingerprinting). *Alternative name* **block**.

motile (of organisms or cells) capable of independent locomotion; (of cilia, flagella, etc.) capable of spontaneous movement. —**motility** *n*.

motilin a 22-residue peptide produced by **Mo cells** of the mucosa of the upper small intestine, and having putative hormonal activity. Its plasma level rises when food or alkali are placed in the duodenum; it stimulates contractions of the stomach and secretion of pepsin, but its effects show marked species variation.

motor (*in biology*) pertaining to movement of an organ or part; pertaining to nerves or neurons involved in muscular activity.

motor protein a protein, e.g. **dynein**, **kinesin**, **myosin**, that propels itself along a filament or polymeric molecule. Nucleoside triphosphate hydrolysis is usually necessarily involved. A similar process is thought to be involved in the action of elongation factor G (*see* **EF-G**). These processes are referred to as **chemical–mechanical coupling** (or mechanochemical coupling).

Motrin *a proprietary name for* **ibuprofen**.

mould *or* (*US*) **mold 1** any **microfungus** having a distinctive mycelium; *the common name for* any superficial growth of fungal mycelium. **2** a matrix for shaping a cast.

moult *or* (*US*) **molt 1** to cast off periodically an outer covering of cuticle, feathers, hair, horns, skin, etc. **2** the act, process, or condition of moulting; in invertebrates it is known as ecdysis.

moulting hormone *or* (*US*) **molting hormone** *a common name for* **ecdysterone**.

mouse-convulsion method a sensitive method for the detection and biological assay of insulin, used primarily to control the strength of pharmaceutical preparations of insulin. It is based on the finding that small doses of insulin injected into mice (0.1–1.0 μg per animal) cause convulsions. When the method is used for assay, large equal numbers of animals are injected with either of two dilutions of the test solution or of a standard and the results are analysed statistically.

mouse mammary tumour virus *abbr*.: MMTV; a DNA virus that causes breast cancer in mice. It contains a powerful enhancer DNA sequence in the terminal repeats of the genome. This activates an oncogene called *wnt-1* (formerly called *int-1*), which is closely homologous to the *Drosophila* gene **wingless**. MMTV is often used as an **expression vector** especially in transgenic mice.

moving-boundary electrophoresis an analytical method of electrophoresis carried out using a **Tiselius apparatus**, formerly used for measuring electrophoretic mobility directly.

MOX *abbr. for* methanol oxidase (*see* **alcohol oxidase**).

m.p. *abbr. for* melting point.

MP *abbr. for* mitogenic protein (i.e. **interleukin 1**).

6-MP *abbr. for* 6-mercaptopurine.

MPF *abbr. for* M-phase-promoting factor.

M phase the period of mitosis and cytokinesis in the eukaryotic **cell-division cycle**. During this phase the nucleus and cytoplasm divide.

M-phase-promoting factor *or* **maturation-promoting factor** *or* **p34^{cdc}** *abbr*.: MPF; a protein complex, containing **cyclin** and a protein kinase, that triggers a cell to enter M phase of the **cell-division cycle**.

MPP *abbr. for* mitochondrial processing peptidase.

M6P receptor *abbr. for* **mannose 6-phosphate receptor**.

M protein 1 a structural protein that occurs in the M lines of myofibrils of striated muscle. **2** *an alternative name for* galactoside permease of *Escherichia coli*. **3** a cell-surface protein of Gram-positive

cocci, associated with virulence; it renders the organism resistant to phagocytosis. It is a cell-wall component, consisting of a coiled coil with a membrane anchor.

MPTP *abbr. for* 1-methyl-4-phenyl-1,2,3,6-tetrahydropyridine; a toxic by-product of the synthesis of heroin; it produces severe and permanent parkinsonism in primates, resulting from its oxidation by monoamine oxidase to the quaternary 1-methyl-4-phenyl-pyridinium (MPP$^+$) ion. The latter is a potent inhibitor of complex I of the mitochondrial electron-transport chain and a selective neurotoxin for dopaminergic neurons, by which it is taken up in the substantia nigra via the normal neuronal dopamine re-uptake system.

MRC *abbr. for* Medical Research Council (of the UK).

MRCK *abbr. for* myotonic dystrophy kinase-related Cdc42-binding kinase; a protein-serine/threonine kinase that has sequence homology to **myotonic dystrophy protein kinase** and is activated by **Cdc42**.

MRE *abbr. for* metal response element; *see also* **response element**.

MREBP *abbr. for* metal response element binding protein.

MRF *abbr. for* melanotropin-releasing factor (*see* **MSHRH**).

MRF4 *abbr. for* muscle specific regulatory factor 4; *other name*: myogenic factor 6; a DNA-binding, probably **helix-turn-helix**, protein of the *MYC* family. It induces fibroblasts to differentiate into myoblasts.

MRH *abbr. for* melanotropin-releasing hormone (*see* **MSHRH**).

MRI *abbr. for* magnetic resonance imaging.

MRIF *abbr. for* melanotropin release-inhibiting factor (i.e. **melanostatin**).

MRIH *abbr. for* melanotropin-release inhibiting hormone (i.e. **melanostatin**).

mRNA *abbr. for* messenger RNA.

mRNA binding site (ribosome) a ribosomal binding region for mRNA. In bacteria this involves **ribosomal proteins** S1, S3, S4, S5, S12, S18, S21, and the 3′ end of 16S rRNA.

mRNA splicing the process in which excision of **introns** from the primary transcript of **messenger RNA** (mRNA) is followed by ligation of the two **exon** termini exposed by removal of each intron, so that mRNA consisting only of the joined exons is produced. *See* **splicing** (def. 2).

mRNA surveillance the process occurring in the cytosol of eukaryotic cells whereby mRNA that contains a premature stop codon is detected and submitted to selective degradation by nonsense-mediated mRNA decay. It involves assembly of a surveillance complex on the mRNA about 24 nt upstream of exon–exon junctions. If translation terminates because of an upstream stop codon, translation release factors bind to the surveillance complex and trigger mRNA degradation.

mRNP *abbr. for* messenger ribonucleoprotein.

MRSA *abbr. for* **methicillin**-resistant *Staphylococcus aureus*, an organism resistant to antibiotics owing to the inability of penicillins to enter the cell (and not directly to β-**lactamase**). *Compare* **VRE**.

ms- *abbr. for* meso-.

MS *abbr. for* **1** mass spectrometer or mass spectrometry. **2** magic spot (compound). **3** multiple sclerosis.

MSA *abbr. for* multiplication-stimulating activity.

MSD *abbr. for* macromolecular structure database; *see* **structure database**.

msDNA *abbr. for* multicopy single-strand DNA.

MSEL-neurophysin *see* neurophysin II.

MSG *abbr. for* monosodium glutamate.

MSH *abbr. for* melanocyte-stimulating hormone.

MSHIF *abbr. for* melanocyte-stimulating-hormone release-inhibiting factor (i.e. **melanostatin**).

MSHIH *abbr. for* melanocyte-stimulating-hormone release-inhibiting hormone (i.e. **melanostatin**).

MSHRF *abbr. for* melanocyte-stimulating-hormone releasing factor (i.e. **MSHRH**).

MSHRH *abbr. for* melanocyte-stimulating-hormone releasing hormone; a pentapeptide, possibly identical to the N-terminal moiety of **oxytocin**, that is produced in the hypothalamus. In animals that show an unambiguous response of their melanophores to **melanotropin** this regulating factor is believed to promote melanotropin release. *Recommended name*: melanoliberin; *other names*: melanotropin-releasing factor (*abbr.*: MRF); melanotropin-releasing hormone (*abbr.*: MRH); melanocyte-stimulating-hormone releasing factor (*abbr.*: MSHRF).

MSHRIF *abbr. for* melanocyte-stimulating-hormone release-inhibiting factor (i.e. **melanostatin**).

MSHRIH *abbr. for* melanocyte-stimulating-hormone release-inhibiting hormone (i.e. **melanostatin**).

MS/MS *abbr. for* tandem mass spectrometry.

MSOC *abbr. for* the *N*-(2-methanesulfonyl)ethoxycarbonyl group; it is used as an amino-protecting group, its stability to acid, lability to alkali, and resistance to hydrogenation contributing to its usefulness. It is introduced using such reagents as **methylsulfonyl nitrophenyl carbonate** and **methylsulfonyl succinimidyl carbonate**.

Msp-300 a protein (8204 amino acids) of *Drosophila* that is probably involved in tethering nuclei to the cytoskeleton. Homologous with ANC-1 of *Caenorhabditis elegans* and mammalian Syne proteins, it contains two actin-binding domains, large central domains composed of spectrin repeats, and a C-terminal nuclear envelope transmembrane domain.

MSP *abbr. for* maximum-scoring segment pair.

MspI a type 2 **restriction endonuclease**; recognition sequence: C↑CGG. *Hpa*II and *Hap*II are **isoschizomers**. *Msp*I is insensitive to methylation of the 5′C. *Compare* **HpaII**.

Mst1 *abbr. for* mammalian sterile twenty; a protein serine/threonine kinase that phosphorylates Ser14 of histone H2B. This phosphorylation correlates with the onset of apoptosis. Activity of the kinase is regulated by caspase-3.

MSX-1 *or* **Hox-7** a **homeobox** protein.

mtDNA *abbr. for* mitochondrial DNA.

MTOC *abbr. for* microtubule organizing centre.

mTor *abbr. for* mammalian target of rapamycin; *other name*: FKBP12-rapamycin-associated protein (*abbr.*: FRAP); a protein serine/threonine kinase that is inhibited by FKBP12-rapamycin. mTor phosphorylates the initiation factor eIF-4E and regulates S6 kinase and translation of mRNAs for ribosomal proteins and other components of the translation machinery. Its function is influenced by the intracellular concentration of ATP.

mtRNA *abbr. for* mitochondrial RNA.

MT-SP1 membrane type-serine protease 1; *other name*: matriptase; a human type II transmembrane serine protease (855 amino acids) that is present in the gastrointestinal tract, liver, spleen, thymus, and placenta.

mu *symbol*: μ (lower case) *or* M (upper case); the twelfth letter of the Greek alphabet. For uses *see* **Appendix A**.

MU *abbr. for* 4-methylumbelliferyl.

mu chain *or* **μ chain** the heavy chain characterizing **immunoglobulins** belonging to the class IgM.

mucic acid *an old name for* galactaric acid, the *meso*-aldaric acid derived from both D- and L-galactose.

mucilage any of numerous proteoheteroglycans, frequently containing uronic acids, that occur widely in plants, especially in seed coats. Mucilages are hard when dry; in water, they swell, but do not dissolve, to form slimy masses with adhesive properties. *Compare* **gum**. —**mucilaginous** *adj.*

mucin any of a group of widely distributed glycoproteins of high M_r that are secreted by various animal mucous cells and glands; they occur, e.g., in saliva, gastric juice, and intestinal juice. Mucins form viscous solutions that act as lubricants and protectants of the linings of body cavities and of the skin.

mucinase *an alternative name for* **hyaluronidase**.

mucin domain *other name*: PTS region; a proline and threonine/serine-rich peptide sequence present in an apomucin. It contains 8–169 amino acids, is poorly conserved, and frequently tandemly repeated. It provides the sites for *O*-glycosylation. *See* **mucin**.

muco+ *comb. form* of mucus or mucin; mucoid.

mucoid 1 any **mucin**-like substance or mucoprotein. **2** of the nature of, or resembling, **mucus**. **3** describing a bacterial colony that has a gummy consistency, resulting from the abundant synthesis of capsular material.

mucolipid *an alternative name for* **ganglioside**.

mucolipidosis (*pl.* **mucolipidoses**) any of a group of inherited autosomal recessive diseases of humans characterized by deficiency of a variety of lysosomal hydrolases in connective tissue cells of patients, and in fibroblasts cultured from them. Mucolipidosis I, commonly known as **sialidosis**, is due to a deficiency of neuraminidase that leads to accumulation of sialyl oligosaccharides, and excretion of these in the urine. In mucolipidosis II, the cytoplasm of cultured cells from affected patients contains numerous highly refractile inclusions, visible by phase-contrast microscopy. These are very large lysosomes containing heterogeneous material, mucopolysaccharides, glycolipids, and whorls of membranes, giving this the name **I-cell** (inclusion cell) **disease**. Mucolipidosis III is due to a deficiency of N-acetylglucosamine 1-phosphotransferase.

mucolytic having the ability to disrupt mucus.

muconate *a trivial name for* any one of the three stereoisomers (*cis,cis*, *cis,trans*, or *trans,trans*) of 2,4-hexadienedioate; 1,3-butadiene-1,4-dicarboxylate; ⁻OOC–CH=CH–CH=CH–COO⁻, the dianion derived from muconic acid. The *cis,cis* isomer is an intermediate in the degradation of catechol by microorganisms.

muconolactone Δ-isomerase EC 5.3.3.4; an enzyme that catalyses the reaction:

2,5-dihydro-5-oxofuran-2-acetate =
3,4-dihydro-5-oxofuran-2-acetate

in the pathway for the degradation of catechol to acetyl-CoA.

mucopeptide *(sometimes)* an alternative name for *peptidoglycan*.

mucopolysaccharide any of a group of acid heteroglycans built up of characteristic repeating disaccharide units each consisting of an *N*-acylated hexosamine residue and a uronic-acid residue; many mucopolysaccharides also contain sulfate. The group includes hyaluronic acid (*see* **hyaluronate**), **chondroitin sulfate**, **dermatan sulfate**, **keratan sulfate**, **heparin**, and some **blood-group substances**.

mucopolysaccharide storage disease any of a group of clinically progressive, hereditary diseases of humans characterized by the accumulation of mucopolysaccharides in various tissues. They include the **mucopolysaccharidoses** and the **mucolipidoses**.

mucopolysaccharidosis (*pl.* **mucopolysaccharidoses**) any of a group of clinically progressive, hereditary diseases of humans each of which is characterized by the deficiency of a single lysosomal hydrolase concerned in the degradation of mucopolysaccharides. They have been classified into a number of types. *See* individual entries below.

mucopolysaccharidosis IH *or* **Hurler syndrome** a **mucopolysaccharidosis** caused by a deficiency in lysosomal L-iduronidase (EC 3.2.1.76) that affects the degradation of both dermatan sulfate and heparan sulfate. It is inherited as an autosomal recessive disorder. Subjects excrete excessive amounts of both dermatan sulfate and heparan sulfate in the urine in the ratio of approximately 7:3. The biochemical lesion is the same as in mucopolysaccharidoses IH/S and IS, from which differentiation can only be made by clinical criteria.

mucopolysaccharidosis IH/S *or* **Hurler–Scheie compound syndrome** a **mucopolysaccharidosis** caused by the same deficiency in lysosomal L-iduronidase (EC 3.2.1.76) as is found in mucopolysaccharidoses IH and IS. It is inherited as an autosomal recessive disorder and it is believed to be a phenotype intermediate between these two syndromes, one mutant allele being inherited from the father and a different mutant allele from the mother. The condition can only be differentiated from mucopolysaccharidoses IH and IS by clinical criteria.

mucopolysaccharidosis IS *or* **Scheie syndrome** *or* (*formerly*) **mucopolysaccharidosis V** a **mucopolysaccharidosis** caused by the same deficiency in lysosomal L-iduronidase (EC 3.2.1.76) as is found in mucopolysaccharidoses IH and IH/S. It is inherited as an autosomal recessive disorder and it has a milder clinical course than mucopolysaccharidoses IH and IH/S, from which it can be differentiated only by clinical criteria.

mucopolysaccharidosis II *or* **Hunter syndrome** a **mucopolysaccharidosis** caused by a deficiency in iduronate sulfatase that affects the degradation of both dermatan sulfate and heparan sulfate. It is inherited as an X-linked recessive disorder, and several different mutations of the same X-chromosome locus probably account for the different severities of the disease seen in various families. Subjects excrete excessive and roughly equal amounts of both dermatan sulfate and heparan sulfate in the urine.

mucopolysaccharidosis IIIA *or* **Sanfilippo syndrome A** a **mucopolysaccharidosis** caused by a deficiency in heparan *N*-sulfatase that affects the degradation of heparan sulfate. It is inherited as an autosomal recessive disorder and is indistinguishable clinically from mucopolysaccharidosis IIIB. Subjects excrete excessive amounts of heparan sulfate in the urine.

mucopolysaccharidosis IIIB *or* **Sanfilippo syndrome B** a **mucopolysaccharidosis** caused by deficiency in α-*N*-acetyl-D-glucosaminidase (EC 3.2.1.50) that affects the degradation of heparan sulfate. It is inherited as an autosomal recessive disorder and is indistinguishable clinically from mucopolysaccharidosis IIIA. Subjects excrete excessive amounts of heparan sulfate in the urine.

mucopolysaccharidosis IIIC (*abbr.*: MPS IIIC) *or* **Sanfilippo syndrome C** a mucopolysaccharidosis (MPS) that is clinically similar to MPS IIIA, IIIB, and IIID but caused by deficiency of lysosomal acetyl-CoA:α-glucosaminidine acetyl transferase, which affects the degradation of heparan sulfate. Subjects excrete excessive amounts of heparan sulfate in urine.

mucopolysaccharidosis IIID (*abbr.*: MPS IIID) *or* **Sanfilippo syndrome D** a mucopolysaccharidosis (MPS) that is clinically similar to MPS IIIA, IIIB, and IIIC, but caused by deficiency of *N*-acetyl glucosamine 6-sulfatase, which affects the degradation of heparan sulfate. Subjects excrete excessive amounts of heparan sulfate in urine.

mucopolysaccharidosis IVA (*or* IVa) *or* **Morquio A disease** a mucopolysaccharidosis caused by a deficiency of lysosomal *N*-acetyl-galactosamine-6-sulfatase, leading to an accumulation of keratan sulfate and chondroitin 6-sulfate. It is inherited as an autosomal recessive disorder.

mucopolysaccharidosis IVB (*or* IVb) *or* **Morquio B disease** a mucopolysaccharidosis caused by a deficiency of β-galactosidase leading to accumulation of keratan sulfate. It is inherited as an autosomal recessive disorder.

mucopolysaccharidosis V *a former name for* **mucopolysaccharidosis I S**.

mucopolysaccharidosis VI *or* **Maroteaux–Lamy syndrome** a **mucopolysaccharidosis** caused by a deficiency of lysosomal *N*-acetyl-galactosamine 4-sulfatase that affects the degradation of dermatan sulfate. Three clinically distinguishable forms are known: severe, intermediate, and mild; all are inherited as autosomal recessive disorders, with probably two or more mutant alleles at the mucopolysaccharidosis VI locus, one of which produces the severe form when homozygous and the other the mild form. Subjects excrete excessive amounts of dermatan sulfate in the urine.

mucopolysaccharidosis VII a **mucopolysaccharidosis** caused by a deficiency of lysosomal β-D-glucuronidase (EC 3.2.1.31) that affects the degradation of both dermatan sulfate and heparan sulfate. It is inherited as an autosomal recessive disorder.

mucopolysaccharidosis IX (*abbr.*: MPS IX) *or* **hyaluronidase deficiency** a rare mucopolysaccharidosis caused by deficiency of lysosomal endohexosaminidase (hyaluronoglucosaminidase; EC 3.2.1.35), which is responsible for degradation of hyaluronates primarily to a tetrasaccharide. Subjects excrete excessive amounts of hyaluronate in urine. The gene for the enzyme is at 3p21.2-p21.3.

mucoprotein any glycoprotein containing relatively large amounts of acidic polysaccharide.

mucosa (*pl.* **mucosae**) *or* **mucous membrane** any of the moist membranes that line the alimentary canal, the glandular ducts, and the respiratory, urinary, and genital passages; it may be smooth, corrugated, or covered with villi. Mucous membranes consist of a surface layer of epithelium containing mucus-secreting glands, and an underlying layer of connective and muscular tissue. *Compare* **serosa**. —**mucosal** *adj.*

mucotetraose *symbol*: McOse₄ *or* Mc₄ (if space is limited); the tetrasaccharide Gal(β1-3)Gal(β1-4)Gal(β1-4)Glc.

mucotriaose *symbol*: McOse₃ *or* Mc₃ (if space is limited); the trisaccharide Gal(β1-4)Gal(β1-4)Glc.

mucous relating to, resembling, or secreting **mucus**.

mucous cell any cell that secretes **mucus**.

mucous gland any exocrine gland consisting predominantly of mucus-secreting cells.

mucous membrane *an alternative name for* **mucosa**. *Compare* **serous membrane**. —**mucomembranous** *adj.*

mucus the slimy, sticky, viscous secretion, consisting predominantly of **mucins**, that is secreted by mucous cells of the mucosae and of the external body surface of some animals.

MUG *abbr. for* **1** either 4-methylumbelliferyl-glucuronide or 4-methylumbelliferyl-β-D-galactopyranoside. Both compounds are fluorogenic enzyme substrates that yield 4-methylumbelliferone. *See also* **4-methylumbelliferyl**. **2** murinoglobulin.

MUGal *abbr. for* 4-methylumbelliferyl-β-D-galactopyranoside (alternative to MUG).

mugineic acid a phytosiderophore secreted by the roots of barley plants. It is derived from the α-aminobutyrate moieties of three molecules of *S*-adenosylmethionine by the action of nicotinamine synthase.

mull a suspension of very finely ground solid in an inert liquid, especially paraffin or **Nujol**, that is used in infrared spectroscopy of the solid.

Müllerian inhibiting factor *an alternative name for* **anti-Müllerian hormone**.

multi+ *comb. form* denoting more than one, several. *See also* **poly+**.

multicatalytic endopeptidase complex *see* **proteosome**.

multicellular having, or consisting of, several or many cells.

multichain (*of a protein*) consisting of two or more polypeptide chains connected by non-peptide links.

multichannel analyser any instrument or device for sorting a heterogeneous input into a number of ranges, e.g. of frequency or energy, and determining the relative proportions in each range.

multichannel pipette a plunger-operated pipette used for dispensing fluids in the microlitre range that is designed to deliver 4–8 samples of the same volume in parallel. The tips of the pipette have the same spatial separation as the wells on a **microtitre plate**.

multicopy single-strand DNA *abbr.*: msDNA; DNA generated by bacterial reverse transcriptase and found in certain Gram-negative bacteria. The molecules are unique in being the only known naturally occurring examples of nucleic acids containing an RNA–DNA covalent link.

multidentate (*of a ligand*) having two or more groups through which it can be attached to a central atom. Such a ligand is able to form a bridge or a chelate.

multidimensional chromatography a technique in which a single component separated from a mixture by chromatography on one type of column may be automatically transferred, as a sample, for chromatography on another type of column in order to effect further resolution.

multidrug resistance *see* **MDR**.

multidrug transporter a superfamily of integral membrane proteins found in nearly all cells. They bind to structurally and chemically diverse compounds and remove them from cells using energy derived from ATP or protonmotive force. First discovered in mammalian cells (*see* **MDR**) they also occur in bacteria and yeast. There are four classes: (1) ATP-binding cassette (ABC) transporters; (2) major facilitator (MRF) transporters; (3) small multidrug-resistance (SMR) transporters; and (4) resistance-nodulation (RND) transporters. The first class is ATP-dependent, the other three classes are proton-dependent.

multienzyme *the recommended class name for* any protein that possesses two or more autonomous catalytic functions due to separate catalytic domains (*see* **domain**, def. 3), whether contributed by distinct subunits (*see* **multienzyme complex**), by distinct parts of an individual polypeptide chain (*see* **multienzyme polypeptide**), or by both.

Excluded is any enzyme that can catalyse different reactions at the same catalytic centre. *Compare* **enzyme complex**, **multienzyme system**.

multienzyme cluster *see* **enzyme cluster**.

multienzyme complex *the recommended subclass name for* any multienzyme in which catalytic domains occur on more than one type of polypeptide chain. A multienzyme complex may itself contain a **multienzyme polypeptide**.

multienzyme polypeptide *the recommended subclass name for* any multienzyme in which at least two demonstrably distinct catalytic domains of different types occur on a single polypeptide chain and are encoded by a single gene. A multienzyme polypeptide may itself be a component of a multienzyme complex. *Compare* **multifunctional protein**.

multienzyme system *an operational term for* any system of two or more enzymes functioning sequentially to catalyse the reactions of a metabolic pathway, so that the product of the first enzyme becomes the substrate of the second and so on. Three levels of organization of multienzyme systems may be discerned. First, the individual enzymes are independently in solution in a compartment within a cell and are not directly associated with each other when acting. Second, the individual enzymes are physically associated and function together as an **enzyme complex**, or they occur as a **multienzyme**. Third and most highly organized, the individual enzymes are associated with large, supramolecular structures such as membranes or ribosomes. *See also* **enzyme cluster**, **metabolon**.

multifactorial 1 involving or dependent on more than one factor or causes. **2** *(in genetics)* inheritance depending on more than one gene.

multifunctional protein any protein that combines two or more autonomous functions on one polypeptide chain, the functions being both independently measurable and assignable to separate domains of the protein. Such functions may or may not be catalytic (*compare* **multifunctional protein ade2**). Excluded are allosteric enzymes, proenzymes, and enzymes that can catalyse different reactions at a single reaction centre.

multifunctional protein ade2 a protein that contains **phosphoribosylaminoimidazolesuccinocarboxamide synthase** and **phosphoribosyl-aminoimidazole carboxylase** and is involved in the sixth and seventh steps in *de novo* purine biosynthesis. The human enzyme consists of 425 amino acids; the synthase domain resides in residues 1–260, the carboxylase domain in residues 261–425.

multigenic describing a phenotype that results either from the inheritance of more than one mutant allele, or the interaction of a number of genes and environmental factors.

multiiodinated describing any compound that has been reacted so as to contain two or more atoms of iodine per molecule.

multilamellar occurring as, or made up of, several discrete layers.

multilayered film analyser a device for the rapid determination of any of a range of components present in small volumes of biological fluids. It depends on a single-use reagent slide on which the various chemicals required are present in dry form as a series of thin films. These consist successively of an indicator layer, a semipermeable layer, a reagent layer, and a spreading layer, all mounted between two transparent supports. Typically, a 10 μL sample is applied to the spreading layer, a chemical or enzymatic reaction occurs in the reagent layer, and the semipermeable layer withholds macromolecular substances from the indicator layer, where usually a dye combines with a product of the reaction to form a coloured compound. The intensity of the colour, which is proportional to the quantity of the analyte in the specimen, is measured by reflectance spectrophotometry. The concept has been extended to the use of potentiometry for the measurement of sodium and potassium.

multilocus enzyme any of a set of enzymes, with similar catalytic properties and existing in the same organism, whose amino-acid sequences are encoded in nonallelic structural genes at two or more separate loci. There may be marked variations in the expression of the individual loci in different tissues. A class of **isoenzyme**.

multimer any specific aggregate of (identical or nonidentical) molecular entities held together by noncovalent bonds. —**multimeric** *adj.*; **multimerize** *or* **multimerise** *vb.*; **multimerization** *or* **multimerisation** *n.*

multinucleate *or* **multinuclear** (of a cell) containing several or many nuclei.

multipass (*in protein biochemistry*) describing an integral membrane protein that contains more than one transmembrane segment. *Compare* **single-pass**.

multiphase having several phases; polyphase.

multiple alignment *see* **alignment**.

multiple alignment program *(in sequence analysis)* software for aligning more than two sequences of biopolymers. Evolutionarily rigorous multiple alignment is computationally intractable, and is therefore most often effected by means of heuristic techniques that progressively align pairs of aligned sequences: e.g. Clustal, T-Coffee. *See* **alignment**.

multiple cloning site a short DNA sequence in a cloning vector that incorporates recognition sites for a number of restriction endonucleases, which cut the vector only at this point. Their presence affords a choice of individual sites for cloning DNA or, by using two different sites, the possibility of cloning a DNA fragment in a desired orientation with respect to the vector. Also known as a **polylinker**.

multiple codon recognition the recognition by a single tRNA molecule of more than one type of codon in mRNA.

multiple development technique a chromatographic technique in which the chromatogram is developed more than once with either the same solvent system or a different one with the aim of improving the separation.

multiple drug resistance *see* **MDR**.

multiple forms (of an enzyme) *the recommended collective term to describe* all proteins that possess the same (specified) enzymic activity and occur naturally in a single species. It includes, but is not limited to, **isoenzymes**; the term isoenzyme should be applied only to multiple forms that arise from genetically determined differences in primary structure, and some limit this to those arising from a single gene locus.

multiple myeloma *or* **multiple myelomatosis** *see* **myeloma**.

multiple osteochondromata *an alternative name for* **hereditary multiple exostoses**.

multiple sclerosis *abbr.*: MS; a chronic and usually progressive disease of the nervous system, also known as disseminated sclerosis, in which the myelin sheaths surrounding the axons in the central nervous system are lost. The cause is unknown. It is generally considered to be a T-lymphocyte-mediated autoimmune disorder targeted at myelin-specific components. A major antigen during the inflammatory phase is αB-crystallin. It is accompanied by formation of plaques, microscopic to macroscopic in size and containing activated lymphocytes and monocytes. The disease is characterized by a course of relapses and remissions over a number of years.

multiplet any group of two or more associated spectral lines. In electronic spectra the lines of a multiplet are very closely spaced and appear at low resolution as a single line, whereas in nuclear magnetic resonance spectra the separation may be considerable.

multiplicand *(in mathematics)* a number to be multiplied by another number – the multiplier – in an arithmetic multiplication.

multiplication-stimulating activity *abbr.*: MSA; an activity found in mammalian serum and produced by some animal cells that is required for the multiplication of animal cells in culture.

multiplicity *(in physics)* the number of components in a **multiplet**.

multiplicity of infection the ratio of the number of virus particles to the number of susceptible cells in a given system.

multiplicity reactivation the reconstruction of viable genomes in a cell that is multiply infected with otherwise nonviable viruses.

multiplier 1 *(in mathematics)* any number by which another number – the multiplicand – is multiplied. **2** any instrument or device for multiplying an effect, e.g. photomultiplier.

multiplying prefix a prefix to denote the presence of several identical substituents in a molecule, at different positions. The prefix 'bis' indicates two substituents, for example fructose-1,6-bisphosphate; by contrast the prefix 'di' indicates that both substituents are linked at the same position, e.g. adenosine diphosphate. 'tris' corresponds to three substituents, and subsequently the names are formed by adding the ending 'kis' to the basic multiplying prefix ending in 'a', e.g. 'tetrakis', 'pentakis', 'hexakis'.

multiply labelled describing an isotopically substituted compound that has more than one modified atom of the same element at the same position or at different positions.

multipotent (of a cell or tissue) capable of giving rise to several kinds of cells, tissues, or structures. *Compare* **pluripotent** (def. 2).

multisubstrate enzyme any enzyme that requires two or more substrates in order to catalyse a particular reaction.

multiubiquitin *see* **polyubiquitin**.

multivalent *an alternative term for* **polyvalent**.

multivalent phosphorylation the phosphorylation of any hormonally regulated enzyme at a plurality of sites by two or more distinct kinase enzymes, with a resultant increase in the regulatory potential of the enzyme.

multivesicular body an intracellular structure that is lined by a single membrane and contains a number of inner vesicles that are approximately 50 nm in diameter. It is a form of secondary **lysosome**.

multi-wavelength anomalous dispersion *abbr.*: MAD; a method for solving macromolecular structures containing different elements by X-ray crystallography, in which the phases are resolved by diffraction at X-ray wavelengths near to an absorption edge of the element.

***Mun*I** a type 2 **restriction endonuclease**; recognition sequence: C↑AATTG. *Mfe*I is an **isoschizomer**.

mupirocin a substance produced by *Pseudomonas fluorescens* and active against Gram-negative bacteria. It is used to treat impetigo and skin infections. It inhibits protein synthesis by reversibly and specifically binding to isoleucine transfer-RNA ligase (*see* **transfer RNA**).

Mur *symbol for* a residue (or sometimes a molecule) of **muramic acid**.

muramic acid *symbol*: Mur; 3-*O*-α-carboxyethyl-D-glucosamine; the 3-*O*-D-lactyl ether derivative of D-glucosamine. It occurs naturally in peptidoglycan of bacterial cell walls.

muramidase *an alternative name for* **lysozyme**.

muramyl the acyl group derived from **muramic acid**.

muramyl-dipeptide *abbr.*: MDP; *N*-acetyl-muramyl-D-alanyl-D-isoglutamine; a synthetic, water-soluble, **peptidoglycan** derivative that is capable of replacing whole mycobacteria in complete **Freund's adjuvant**. It is highly active in stimulating antibody production against an antigen injected simultaneously.

murein *an alternative name for* **peptidoglycan**.

murexide ammonium purpurate; 5,5′-nitrilodibarbituric acid monoammonium salt; a compound that forms a purple-red solution in water. It is useful as an indicator in complexometric titration of calcium, nickel, cobalt, copper, etc. with EDTA.

murexide test a colour test for uric acid and some other purines. The (solid) sample is first treated with conc. nitric acid, which is slowly evaporated away; subsequent addition of ammonia solution gives a purple colour if uric acid was present, due to formation of murexide, or a yellow colour that turns to red on heating if xanthine was present.

murine of, belonging to, characteristic of, affecting, transmitted by, or being a member of the Muridae, a family of small rodents that includes mice and rats; of or relating to the mouse genus, *Mus*.

murine toxin *or* *Yersinia* **murine toxin** the product of the *Ymt* gene on a plasmid of *Yersinia pestis* (the causative agent of bubonic plague). It belongs to the phospholipase D family of enzymes and causes hypotension and vascular failure as a result of release from lysing bacteria during the terminal stages of septicemic plague. It is required for survival of the *Yersinia* bacteria in the midgut of the rat flea *Xenopsylla cheopis*.

murinoglobulin *abbr.*: MUG; a monomeric protease inhibitor found in blood plasma of vertebrates and invertebrates, and in egg white of birds and reptiles. It has a characteristic **bait region**. *Compare* α₂ **macroglobulin**.

muropeptide *an alternative name for* **peptidoglycan**.

muscarine a toxic alkaloid (quaternary ammonium base) found in the red variety of *Amanita muscaria* (fly agaric) and some other gill fungi. It is an agonist for one type of **cholinoceptor**.

(+)-enantiomer

muscarinic receptor *see* **cholinoceptor**.

muscimol 5-aminomethyl-3-hydroxyisoxazole; an isoxazole isolated from the fungus *Amanita muscaria*. A hallucinogen and potent CNS depressant, it is a selective agonist of GABA$_A$ receptors.

muscle 1 a tissue made up of various elongated cells that are specialized to contract and thus to produce movement and mechanical work. **2** an organ consisting principally of muscle tissue. *See* **cardiac muscle, smooth muscle, striated muscle**. —**muscular** *adj*.

muscle contraction *see* **sliding filament model**.

muscle hemoglobin *a former name for* **myoglobin**.

muscle sugar *a former name for* **inositol** (def. 1).

muscular dystrophy any genetically determined disease of humans in which there is progressive muscular weakness and evidence of muscle degeneration. All are characterized by creatinuria, and serum creatine phosphokinase activity is elevated in several of them. *See also* **dystrophin**.

musculin *see* **capsulin**.

mushroom sugar *a former name for* **trehalose**.

Musk *abbr. for* muscle-specific kinase; a skeletal muscle-specific receptor-tyrosine kinase found at neuromuscular junctions. It may be involved in clustering of acetylcholine receptors and dystroglycans.

mustard gas sulfur mustard; bis(2-chloroethyl)sulfide; a colourless oily liquid with a very low vapour pressure. It is a powerful vesicant and also a mutagen and carcinogen. *See also* **nitrogen mustard**.

mustard oil any alkyl isothiocyanate, of general formula R–N=C=S. Such compounds have a pungent mustard-like smell and are present in the tissues of a number of plants, either free or as thioglycosides.

mut any of a family of genes whose products are involved in **mismatch repair**. All the following examples are from *Escherichia coli*. The product of *mutH* is a repair endonuclease that is specific for unmethylated GATC sequences. The protein encoded by *mutL* is also involved in repair of mismatches in DNA and is required for activity of the *dam* product. The protein encoded by *mutS*, again involved in mismatch repair (possibly the recognition step), has weak ATPase activity.

mutable able or tending to undergo change or **mutation**. —**mutability** *n*.

mutable gene any gene that spontaneously mutates at a high rate.

mutable site any site on a chromosome at which mutations can occur.

mutagen any physical or chemical agent that is capable of increasing the frequency of **mutation** above the spontaneous, background level. —**mutagenic** *adj*.; **mutagenicity** *n*.

mutagenesis the production of mutations.

mutagenize *or* **mutagenise** to treat (cells or organisms) with a mutagen. —**mutagenized** *or* **mutagenised** *adj*.

mutant 1 a any organism that has arisen by or has undergone **mutation**, or one that carries a mutant gene that is expressed in the phenotype of that organism. **b** a mutant gene. **2** produced by or following a mutation; having the attributes of a **mutant** (def. 1).

mutarotation the spontaneous change in the optical rotation of a freshly prepared solution of a pure stereoisomer, especially a carbohydrate, that is caused by epimerization, or some other structural change.

mutase any enzyme that catalyses (or apparently catalyses) an intramolecular transfer of a chemical group. Such enzymes belong to one of the following EC sub-subclasses: EC 5.4.1, transferring acyl groups; EC 5.4.2, transferring phosphoryl groups; EC 5.4.3, transferring amino groups; EC 5.4.99, transferring other groups.

mutate to undergo or cause to undergo mutation.

mutation 1 the process by which genetic material undergoes a detectable and heritable structural change, or the result of such a change. Three categories are recognized: **genome mutations**, in which addition or subtraction of one or more whole chromosomes occurs; **chromosome mutations**, in which the structure of one or more chromosomes is affected; and **gene mutations**, in which the structure of a gene is altered at the molecular level. **2** any modified gene arising from a **mutation** (def. 1). **3** a **mutant** (def. 1). *See also* **deletion**, **inversion** (def. 4), **point mutation**, **translocation** (def. 2).

mutational heteroduplex *see* **heteroduplex**.

mutation bias bias in the mutation frequencies of different codons, affecting the synonymous to nonsynonymous rate ratio. Mutation bias results in an accelerated rate of amino acid replacement in functionally less constrained regions.

mutation frequency the frequency at which a mutation occurs in a population. It may or may not reflect the **mutation rate**, with which it is sometimes used interchangeably.

mutation frequency decline the significant and irreversible reduction in the frequency of some UV-induced mutations in bacteria during the temporary post-irradiation inhibition of protein synthesis.

mutation matrix *an alternative name for* **substitution matrix**.

mutation pressure the continued production of an allele by mutation that tends to increase the allele's frequency in the gene pool.

mutation rate 1 the probability that a particular mutation will occur per virus, cell, or organism during a reproductive cycle. **2** the frequency at which any mutation or mutational class occurs in a given population; **mutation frequency**.

mutator gene any gene that produces an increase in spontaneous mutation rates. Their action may be limited to certain alleles or to certain mutational pathways.

mutein a mutant protein produced by site-specific mutagenesis or other recombinant DNA technique.

MutH a protein of *E. coli* that is involved, with **MutS** and **MutL**, in DNA mismatch repair. No homologue appears to have been found in eukaryotic cells. It functions as a site-specific endonuclease.

MutL a protein of *E. coli* that is involved, with **MutS** and **MutH**, in DNA mismatch repair. MutL homologues (*abbr*.: MLH) are present in all eukaryotic cells. Mutations in the human homologue (hMLH) are common in hereditary nonpolyposis colon cancer.

muton the smallest unit of genetic material whose alteration may give rise to a mutation; commonly supposed to be a single nucleotide (pair).

MutS a homodimeric mismatch-binding protein of *E. coli* that is involved, with **MutH** and **MutL**, in DNA mismatch repair. MutS homologues (*abbr*.: MSH) exist in all eukaryotic cells. In yeast they are called MSH1 and MSH2. Mutations in the human homologue (hMSH2) are common in hereditary nonpolyposis colon cancer.

MVA *abbr. for* mevalonic acid (use discouraged).

MvaI a type 2 **restriction endonuclease**; recognition sequence: CC↑[AT]GG. *Eco*RII is an **isoschizomer**.

M13 vector a DNA cloning vector based on the genome of the filamentous bacteriophage M13. In some **phagemids** only the M13 origin of replication is used.

MVG *abbr. for* methoxyvinylglycine.

MVP *abbr. for* major vault protein. *See* **vault**.

M.W. *abbr. for* molecular weight (alternative to mol. wt.).

MWC model *abbr. for* Monod–Wyman–Changeux model.

my+ *a variant form of* **myo+** (before a vowel).

MYB a gene family that encodes DNA-binding proteins; v-*myb* is the oncogene of the acutely transforming avian myeloblastosis virus. It is highly conserved between mammalian species. The product, Myb, contains DNA-binding, transcriptional activation, and negative regulatory domains; it recognizes YAAC(G/T)G. It is notably associated with cells of the hemopoietic lineage and is essential for normal hemopoiesis. It is expressed in some human malignant tumours. [From 'myeloblastosis'.]

MYC a family of genes encoding short-lived nuclear regulatory proteins that act as transcription factors, e.g. pp64/pp67 (based on expressed mass). The oncogene, v-*myc*, was originally identified in avian myelocytomatosis virus, MC29, but is also carried by the avian retroviruses CM11 and OK-10. The protooncogene, c-*myc*, is a member of a family of at least seven related genes: *myc*, B-*myc*, N-*myc*, L-*myc*, P-*myc*, R-*myc*, and S-*myc*; B-Myc lacks the basic helix-loop-helix leucine zipper (bHLHLZ) domains of other Myc proteins. The product of c-*myc* binds nonselectively to DNA and (with higher affinity) to CAC[GA]TG. In normal cells *myc* expression is strictly dependent on mitogenic stimuli and is required for cell proliferation as well as to prevent cell differentiation. Induction of c-*myc* in the absence of growth factors is sufficient to drive rodent fibroblasts into the cell cycle but concomitantly induces apoptosis unless specific survival cytokines are present. Structurally, the product Myc contains an N-terminal transactivation domain that activates transcription of the gene with which it interacts, and a C-terminal bHLHLZ motif, which is known to mediate Myc dimerization and sequence-specific DNA binding. Although Myc–Myc dimers can form in solution, Myc homodimers are not detected *in vivo*, dimers forming with other partners, e.g. Max, which has a bHLHLZ motif but no transactivation domain (*see* **MAX**). *In vitro*, Myc–Max (and Myc–Myc) dimers have been found to bind DNA with a core sequence of CACGTG. Unlike Myc, Max is a stable protein expressed constitutively in growing, resting, and differentiating cells, and is capable of binding further proteins, e.g. Mad and Mxii, which compete with Myc for binding. The *myc* gene may be activated to become a transforming gene by proviral insertion, chromosomal translocation, or gene amplification, but generally protein structure is not affected; **Philadelphia chromosome** is one example. The *myc* gene has been implicated in various neoplasms in birds, mice, and humans, oncogenic activation occurring mainly through constitutive and elevated expression of the Myc protein. [From 'myelocytomatosis'.]

myc+ *a variant form of* **myco+** (before a vowel).

mycarose 2,6-dideoxy-3-*C*-methyl-L-*ribo*-hexose; a component of several **macrolide** antibiotics.

mycelia the plural of **mycelium**.

mycelium (*pl.* **mycelia**) the network or mass of discrete hyphae that forms the body (thallus) of a fungus. —**mycelial** *adj*.

myceto+ *or (before a vowel)* **mycet+** *comb. form* denoting fungus.

myco+ *or (before a vowel)* **myc+** *comb. form* denoting fungus.

mycobacterium (*pl.* **mycobacteria**) any bacterium belonging to the genus *Mycobacterium*.

Mycobacterium a genus of Gram-positive, aerobic, nonmotile, non-spore-forming, acid-fast bacteria of the family Mycobacteriaceae. The cells are rodlike and can form filaments or branched structures; their cell walls characteristically contain **mycolic acid**.

Some species are pathogenic to humans and/or other animals, e.g. *M. tuberculosis* (the tubercle bacillus) and *M. leprae* (the leprosy bacillus).

mycobactin any of a family of **siderochromes** of the hydroxamic acid-derivative category, produced by various species of *Mycobacterium*. At least nine have been characterized.

mycolic acid any of a family of α-substituted β-hydroxylated very-long-chain fatty acids of general formula R_2–CH(OH)–CHR$_1$–COOH, where R_1 is a C_{20}–C_{24} linear alkyl group and R_2 is a complex C_{30}–C_{60} group of very variable structure. They occur naturally as esters of arabinose in the cell walls of *Mycobacterium* spp., *Nocardia* spp., and *Corynebacterium* spp. and as esters of trehalose in **cord factor**. *See also* **cyclic fatty acids**.

mycology the scientific study of fungi. —**mycological** *or* **mycologic** *adj.*; **mycologist** *n.*

mycolyl *the generic name for* acyl groups derived from **mycolic acids**.

mycophenolic acid 6-(1,3-dihydro-4-hydroxy-6-methoxy-7-methyl-3-oxo-5-isobenzofuranyl)-4-methyl-4-hexenoic acid; an antibiotic and antitumour material from *Penicillium brevi-compactum*. It was first isolated in 1896, and said by the Australian physiologist Howard Florey (1898–1968) to be 'the first antibiotic produced by a mould to be crystallized'. As the morpholinoethyl ester (mycophenolate mofetil) it is an approved immunosuppressant in renal transplantation. Transformed mammalian cells expressing *E. coli* guanine phosphoribosyltransferase (GPT) are resistant to mycophenolic acid if grown in a medium supplemented with xanthine.

mycoplasma *or (formerly)* **pleuropneumonia-like organism** (*abbr.*: PPLO) any member of the Mycoplasmatales, an order of minute, Gram-negative, prokaryotic microorganisms that are the smallest known free-living organisms. They have no cell walls (except for some saprophytic species), are variable in form, and can pass through bacteria-retaining filters. Some are saprophytic while others are pathogenic to humans, other animals, and plants. Mycoplasmas frequently infect tissue cultures. They were once thought to be the same as **L forms**. A newer classification replaces mycoplasma as a trivial name for members of the entire class by the term **mollicute**, and the class is called Mollicutes. The trivial name 'mycoplasma' is then restricted to members of the genus *Mycoplasma*.

Mycoplasma the main genus of **mycoplasmas**.

mycoprotein a commercially produced protein-containing foodstuff derived from moulds grown on carbohydrate substrates.

mycosamine 3-amino-3-deoxy-D-rhamnose; 3,6-dideoxy-3-amino-D-mannose; an amino sugar that contributes the nitrogen-containing moiety of the polyene antibiotics **amphotericin B** and **nystatin**.

mycose *a common name for* **trehalose**.

mycostatic describing any agent that inhibits the growth of fungi; fungistatic.

mycothiol 1-*O*-(2-[*N*-acetyl-L-cysteinyl]amido-2-deoxy-α-D-glucopyranosyl)-D-*myo*-inositol, the major low-molecular-mass thiol in mycobacteria; it is associated with the protection of *Mycobacterium tuberculosis* from oxidants and antibiotics.

mycotoxin any poisonous substance produced by a fungus that is not a mushroom. Examples include aflatoxin, ochratoxin, and patulin.

myel+ *a variant form of* **myelo+** (before a vowel).

myelin the material that constitutes the **myelin sheaths** surrounding vertebrate nerve axons. It is rich in lipid, particularly in glycolipid. *See also* **myelin protein**.

myelination the process by which a nerve axon acquires a **myelin sheath**.

myelin protein any of the proteins that form part of the **myelin sheath**. In the following descriptions, all examples are of human precursor sequences. **Myelin-associated glycoprotein**, a cell adhesion molecule for postnatal neural development, is a member of the immunoglobulin superfamily; it is an integral membrane protein. **Myelin basic protein**, a protein of the cytoplasmic face of the myelin membrane, is a major protein component of normal myelin of the central nervous system (CNS). Very basic in nature (pI > 10.5), it is a single-chain protein with little structure. Administration of myelin basic protein in complete **Freund's adjuvant** in various experimental animals elicits experimental allergic encephalitis, owing to induction of delayed hypersensitivity to the protein, which is thought to have an important role in causing demyelinating diseases in humans. Three forms are produced by alternative splicing. **P2** is probably a lipid transport protein; it is found in Schwann cells, associated with myelin basic protein. **P0** is an immunoglobulin-like membrane protein of Schwann cells of the peripheral nervous system; it is associated with a hereditary polyneuropathy (Charcot–Marie–Tooth disease type 1b). **Myelin proteolipid protein** (PLP *or* lipophilin) is the major myelin protein from the CNS, playing an important role in the formation, or maintenance, of the multilamellar structure of myelin; it is an integral membrane protein. Alternative splicing leads to a shorter variant. Mutations are associated with Pelizaeus–Merzbacher disease, and other mutations result in dysmyelination, e.g. *jimpy* and *rumpshaker* (mouse), and MD (rat). *See also* **Folch-Lees protein**, **Wolfgram proteolipid protein**.

myelin-proteolipid *O*-palmitoyltransferase EC 2.3.1.100; *other names*: myelin PLP acyltransferase; acyl-protein synthase; an enzyme that catalyses a reaction between palmitoyl-CoA and [myelin proteolipid] to form [myelin proteolipid] *O*-palmitoylprotein and CoA.

myelin sheath the sheath that surrounds the axons of vertebrate nerves. It is formed by Schwann cells in peripheral nerves and by oligodendrocytes in the central nervous system; these cells wrap up to 100 concentric layers of their plasma membrane in a tight spiral around the axons. Myelin sheaths prevent almost all electric current leakage across the covered portion of the axon membrane, thereby contributing to the rapid transmission of nerve impulses along the axons. *See also* **myelin**, **myelin protein**.

myelo+ *or (before a vowel)* **myel+** *comb. form* **1** denoting bone marrow. **2** denoting the spinal cord. **3** denoting **myelin**.

myeloblast a cell of the bone marrow that is the precursor of the **myelocyte**.

myelocyte a nucleated cell of the bone marrow that is a precursor of the **granulocytes** of the blood.

myeloid 1 of or pertaining to bone marrow. **2** of or pertaining to the spinal cord. **3** of or resembling myelocytes.

myeloma *or* **myelomatosis** a tumour caused by the uncontrolled proliferation of a clone of immunoglobulin-producing plasma cells. It leads to multiple deposits of tumour cells in the bone marrow and an excessive production of a single type of (i.e. monoclonal) immunoglobulin molecule, which may belong to any of the major classes of immunoglobulins. In some cases either the light or the heavy chains may be produced in excess or exclusively. *See also* **Bence-Jones protein**, **Waldenström's macroglobulinemia**.

myeloma protein an immunoglobulin that is overproduced by a **myeloma**. They are single immunoglobulin types and form the basis of monoclonal antibody technology. Human examples include **Bence-Jones protein**. *See also* **hybridoma**.

myeloperoxidase *a special name for* **peroxidase** (def. 2), so named following its isolation from the blood of patients with myeloid leukemia, and originally named **verdoperoxidase** on account of its green colour. It is present in phagosomes of neutrophils and monocytes, where it catalyses the reaction:

$$H_2O_2 + 2Cl^- = 2HOCl.$$

The reaction product is strongly bactericidal. The enzyme is a heterotetramer of two α subunits (59 kDa) linked by disulfide bonds, and two β subunits (13 kDa). Each α subunit contains a heme group and a Ca^{2+}-binding region. Both subunits are derived from the same precursor. Several isozymes are produced by alternative splicing. Myeloperoxidase deficiency, resulting from mutation in the gene at 17q22, is not rare. It is frequently asymptomatic but may be accompanied by repeated infection.

Mylar *the proprietary name for* a type of polyethylene terephthalate polyester that forms very strong films of good transparency and high electrical resistance.

myo+ *or (before a vowel)* **my+** *comb. form* denoting muscle.

myo- *prefix* used to distinguish common **inositol**, 1,2,3,5/4,6-cyclohexanehexol (*formerly* meso-inositol), from the other eight possible inositol stereoisomers.

myoadenylate deaminase *see* adenylate deaminase.

myoblast any of the mononucleate precursor cells that fuse to form a multinucleate skeletal muscle cell.

myocardin a transcription factor expressed exclusively in myocardium after birth.

myocardium the middle, muscular, layer of the heart wall. —**myocardial** *adj.*

myocilin *or* TIGR a protein (504 amino acids) expressed in many tissues but having no known function. It contains an N-terminal hydrophobic region, a leucine zipper region, and a region of homology with myosin heavy chain. Mutations in the gene at 1q23 are responsible for autosomal dominant open-angle glaucoma.

myocyte 1 a muscle cell of any type. **2** any contractile cell.

myocyte enhancement factor *abbr.:* MEF; a transcription factor that binds an AT-rich DNA sequence in the promoter region of many muscle-specific genes. There are at least four MEF genes in vertebrates, which are nearly identical in coding N-terminal regions that serve for dimerization and DNA binding.

myoD the gene for myoblast determination protein, a basic helix-loop-helix transcription factor related to ***MYC***; the human equivalent is *MYF3*. It is expressed after **myogenin**.

myofibre *or (esp. US)* **myofiber** a skeletal muscle fibre; it consists of a single, long, multinucleate cell, composed predominantly of a large number of myofibrils (typically 1000 or more).

myofibril any of the long, cylindrical, contractile elements, 1–2 μm in diameter, that comprise the major component of the cytoplasm of the **myofibre** and extend for its entire length. Each myofibril is composed in turn of numerous **myofilaments**. —**microfibrillar** *adj.*

myofilament either of two kinds of thin, contractile, threadlike, interdigitating structures within the myofibrils of **striated muscle**. The so-called thick filaments are about 15 nm in diameter and 1.6 μm in length, and the so-called thin filaments are about 8 nm in diameter and 1.0 μm in length.

myogen *a name sometimes used for* a number of different protein mixtures extractable with water from skeletal muscle of various species of animal. **Myogen A** is a crystallizable complex of **aldolase** and **glycerol-3-phosphate dehydrogenase** that can be prepared from rabbit muscle.

myogenic factor any of various proteins involved in muscle differentiation. They are normally produced only in muscle cells, but have been shown to activate muscle differentiation mechanisms when introduced into other cells. They include MyoD (*see* **myoD**), Myf5, and **myogenin**.

myogenin a protein involved in muscle differentiation (i.e., a **myo-**

genic factor) that induces fibroblasts to differentiate into myoblasts. It is probably a sequence-specific DNA-binding protein, similar to other members of the Myc family of **helix-turn-helix** transcription factors (*see* **MYC**).

myoglobin *or (formerly)* **muscle hemoglobin** *or* **myohemoglobin** *abbr.:* Mb; a 17.5 kDa oxygen-carrying hemoprotein that occurs in muscle, particularly in red fibres, and consists of a single polypeptide chain of 153 amino-acid residues, to which a single ferroheme prosthetic group is non-covalently bound. Myoglobin binds dioxygen non-cooperatively, with an affinity between that of hemoglobin and of cytochrome oxidase. It can thus accept dioxygen from hemoglobin, and store it in the muscle cell for release to cytochrome oxidase when the supply becomes limiting. It also assists the direct transfer of dioxygen from the surface of the myocyte to the mitochondria. Sperm-whale myoglobin was the first protein whose 3-D stucture was determined by X-ray crystallography.

myoglobinuria the excretion of myoglobin in the urine.

myohaematin *a historical term*, used by MacMunn for the 'histohaematin' (i.e. the cytochromes) of striated muscle.

***myo*-inositol** *an alternative name for* **inositol** (def. 1) (*formerly known as* **muscle sugar**). *See also* **myo-**.

1D-*myo*-inositol-trisphosphate 3-kinase EC 2.7.1.127; an enzyme that catalyses a reaction between ATP and 1D-*myo*-inositol 1,4,5-trisphosphate to form 1D-*myo*-inositol 1,3,4,5-tetrakisphosphate with release of ADP. The enzyme requires calcium. *See also* **inositol phosphates**.

myokinase *an alternative name for* **adenylate kinase**.

MyoR *see* **capsulin**.

myosin any of a group of related proteins with ATPase activity that occur in the contractile apparatus of both muscle and non-muscle cells; the myosin of striated muscle is the major component of the thick filaments. Individual myosin molecules consist of two identical **heavy chains** of ~200 kDa and two pairs of light chains (*see* **light chain** (def. 2)) with distinct masses that vary within the range 15–27 kDa according to the source. The N-terminal region of each heavy chain consists of a globular head with which the light chains are associated; the remaining three-quarters of each heavy chain consists predominantly of an alpha helix, which intertwines with its partner to form a long tail. The head region contains binding sites for actin and for ATP, while aggregation with other myosin molecules to form filaments occurs at the tail region. For example, human muscle protein is a hexamer of two heavy chains and four light chains. Different isoforms exist for different muscle types. Light chain 1, the skeletal muscle isoform, contains an **EF hand** motif. Some early preparations of myosin were denoted by letters. Thus, myosin A and myosin B were actin-poor and actin-rich preparations respectively, while myosin T and L-myosin were among the first actin-free preparations. The name myosin was originally given by F. W. Kühne, in 1859, to the substance in muscle press juice that, on standing at room temperature, set to a gel.

myosin-binding C protein a protein that is arranged transversely in seven to nine strips in the A bands of muscle **sarcomeres**. It binds the coiled-coil portion of myosin heavy chains to titin. A gene at 11p11.2 encodes the cardiac isoform, a protein of ≈1200 amino acids that contains immunoglobulin-like and fibronectin repeats. Various mutations cause hypertrophic cardiomyopathy.

myosin-heavy-chain kinase EC 2.7.1.129; an enzyme that catalyses the phosphorylation by ATP of specific threonines of myosin heavy chain with release of ADP, during chemotaxis of *Dictyostelium discoideum* (slime mould). It inhibits myosin thick filament formation.

myosin-light-chain kinase EC 2.7.1.117; *other names*: myosin kinase; smooth-muscle-myosin-light-chain kinase; an enzyme that catalyses the phosphorylation by ATP of myosin light chain with release of ADP. It was one of the first Ca^{2+}/calmodulin-dependent protein kinases to be discovered. Its specificity is restricted to myosin light chain. In skeletal muscle, light-chain phosphorylation modulates tension during contraction. In smooth muscle, it initiates contraction.

myosin-light-chain-phosphatase EC 3.1.3.53; *other names*: myosin light chain kinase phosphatase; myosin phosphatase; an enzyme that catalyses the hydrolysis of phosphate from myosin light-chain phosphate.

myostatin a protein of the transfoming growth factor β superfamily that is expressed in cells of skeletal muscle lineage. Gene knockout in mice, and deficiency in cattle, cause dramatic and widespread increase in skeletal muscle mass. Overexpression or prolonged injection in mice lead to severe loss of muscle and adipose tissue. Myostatins from many vertebrate species have an identical, biologically active C-terminal region. Mutations in cattle produce double-muscling.

myotonia any genetically distinct hereditary disease characterized by muscle hyperexcitability, hypertrophy, and stiffness. Myotonia congenita results from numerous mutations in the gene for the sarcolemmal voltage-gated chloride channel, which normally forms multimeric subunits each of which contains ≈13 transmembrane segments.

myotonia congenita *see* **myotonia**.

myotonic dystrophy a common hereditary muscular dystrophy in which muscle wasting and weakness (i.e. dystrophy) are associated with stiffness (i.e. myotonia). Type 1 is due to expansion of a CTG trinucleotide repeat in the 3′ untranslated region of the gene (at 19q13.2) for muscular dystrophy protein kinase. Type 2 is due to expansion of a CCTG tetranucleotide repeat in intron 1 of the gene (at 3q21) for **ZNF9**.

myotonic dystrophy protein kinase a cytosolic protein-serine/threonine kinase found predominantly in heart, skeletal muscle, brain, and eye. The gene at 19q13.2-q13.3 encodes a protein of 585 amino acids. In persons affected by myotonic dystrophy type 1, the 3′ untranslated region contains 50 to several thousand repeats of the trinucleotide CTG. Normal subjects have less than 37 such repeats, whereas healthy genetic carriers of the disorder have 50 to 100 repeats.

myotrophin a 12 kDa protein of hypertrophied myocardium of spontaneously hypertensive rats that stimulates protein synthesis in cultured adult rat myocytes *in vitro*. It contains **ankyrin** repeats, and binds to **NF-λB**.

myotube the multinucleate structure formed by the fusion of proliferating myoblasts in culture under certain conditions. It is characterized by the presence of certain muscle-specific proteins that are not found in the myoblast and may represent a stage in the development of skeletal muscle cells.

myotubularin a protein encoded by a gene at Xq28, mutations of which cause X-chromosome-linked myotubular myopathy. The protein contains ≈621 amino acids, which include a sequence that corresponds exactly with the consensus sequence for tyrosine phosphatases. It is expressed in skeletal muscle and testes, but not in fetal tissues (e.g. brain, liver, kidney).

Myr *symbol for* the **myristoyl group**.

myricetin 3,3′,4′,5,5′,7-hexahydroxyflavone; a **flavone** that is widely distributed in plants. The name originates from *Myrica*, a genus of shrubs. Myricetin inhibits a number of enzymes, including α-glucosidase, xanthine oxidase, and glyoxalase.

myricyl *or* **melissyl** *the common name for* triacontyl,

CH₃–[CH₂]₂₈–CH₂–, the alkyl group derived from the C₃₀ straight-chain alkane, triacontane. Myricyl palmitate is the major component of beeswax.

myristate 1 *the trivial name for* tetradecanoate, CH₃–[CH₂]₁₂–COO⁻, the anion derived from myristic acid (i.e. tetradecanoic acid), a saturated straight-chain higher fatty acid and a major component of many plant and animal fats. **2** any mixture of free myristic acid and its anion. **3** any salt or ester of myristic acid.

myristin *an old name for* any of the glyceryl esters of myristic acid, especially the triester, **trimyristin**, or trimyristoylglycerol.

myristoleic acid *see* **tetradecenoic acid**.

myristoyl *or (formerly)* **myristyl** *symbol*: Myr; *the trivial name for* tetradecanoyl, CH₃–[CH₂]₁₂–CO–, the acyl group derived from myristic acid (i.e. tetradecanoic acid). It occurs in most animal and plant fats and oils as acylglycerols and in phospholipids. The oils of the seeds of nutmeg (*Myristica fragrans*) and other members of the Myristicaceae are particularly rich sources. *Compare* **myristyl**.

myristoylation the introduction of one or more **myristoyl** groups into an organic compound. Protein *N*-myristoylation refers to the covalent attachment of a **myristoyl** group via an amide bond to the N-terminal Gly residue of a nascent polypeptide. This **post-translational modification** has only been observed in eukaryotes and is irreversible. The *N*-myristoyl proteins are numerous with diverse functions. In some cases the modification is essential for the biological role of the protein, while in other cases it is not essential. The process is catalysed by glycylpeptide *N*-tetradecanoyltransferase (EC 2.3.1.97).

myristyl 1 *the common name for* tetradecyl, CH₃–[CH₂]₁₂–CH₂–, the alkyl group derived from tetradecane; use deprecated. **2** *a former trivial name for* tetradecanoyl (**myristoyl** now used).

myrosinase *an alternative name for* **thioglucosidase**.

MySQL *see* **relational database management system**.

myt5 the gene for bacteriophage T5 5′-exodeoxynuclease (EC 3.1.11.3). It performs exonucleolytic cleavage in the 5′ to 3′ direction to yield 5′-phosphomononucleotides. *See also* **deoxyribonuclease**.

myxamoeba *or (esp. US)* **myxameba** a free-living haploid amoeboid cell occurring in the life cycles of the cellular slime moulds and the myxomycetes. In the former they aggregate to form a slug and in the latter a plasmodium.

myxedema *or (esp. Brit.)* **myxoedema** the syndrome caused by hypothyroidism in adults. It is characterized by mucoid infiltration and coarsening of the skin, weight gain, intolerance to cold, and mental sluggishness.

myxo+ *or (usually before a vowel)* **myx+** *comb. form* indicating mucoid, mucus, or slime.

Myxobacterales the myxobacteria; an order of Gram-negative bacteria that bear extracellular slime and have a characteristic swarming habit that leads to the formation of stalked fruiting bodies. A more recent classification has suggested replacement of Myxobacterales by Myxococcales.

myxobacteria bacteria belonging to the order **Myxobacterales**.

myxoedema *a variant spelling (esp. Brit.) of* **myxedema**.

myxomatosis a highly contagious disease of rabbits, caused by a myxoma virus, part of a subgenus of poxvirus. It is characterized by proliferative changes in the connective tissues.

Myxomycetes a large class of very primitive organisms, also known as the true or acellular slime moulds, found living saprophytically in damp conditions. Their major vegetative stage is a multinucleate migratory mass of protoplasm, from which (usually macroscopic) fruiting bodies arise.

myxothiazol an inhibitor of complex III of the mitochondrial electron transport chain, used as an antifungal agent.

myxovirus any of a group of negative-strand RNA animal viruses, now divided into **orthomyxoviruses** and **paramyxoviruses**.

Nn

n *symbol for* **1** nano+ (SI prefix denoting 10^{-9} times). **2** neutron; may be written as n^0 to indicate its charge.

n *symbol for* **1** any (unknown or unspecified) number, especially an integer. **2** amount of substance; the substance in question may be indicated as a suffix in parentheses or by a subscript. **3** refractive index (of a nonabsorbing material); n_D *symbol for* refractive index measured with light of the sodium D-lines (589.0 and 589.5 nm). **4** order of reaction. **5** order of reflection. **6** haploid number of chromosomes. **7** number density.

n- *prefix indicating* normal (isomer); no longer used.

n_H *symbol for* Hill coefficient (alternative to *h*).

N *symbol for* **1** nitrogen. **2** newton. **3** a residue of the α-amino acid L-asparagine (alternative to Asn). **4** a residue of an incompletely specified base in a nucleic-acid sequence that may be adenine, cytosine, guanine, or either thymine (in DNA) or uracil (in RNA). **5** an undetermined, unidentified, or unspecified nucleoside residue (alternative to Nuc). **6** (when suffixed to the symbol for a sugar molecule or residue) aminodeoxy+; e.g. GlcN is the symbol for 2-amino-2-deoxy-D-glucose (i.e. D-glucosamine). **7** normal (def. 1).

N *symbol for* **1** the SI derived physical quantity named number of entities (e.g. molecules, atoms, ions, formula units). **2** neutron number. **3** Avogadro constant (N_A or *L* now recommended). **4** the total number of receptors per unit area of membrane, per cell, or per unit mass of protein.

N- *prefix (in chemical nomenclature)* indicating substitution on nitrogen; a particular nitrogen atom may be specified by a right superscript.

N_A *symbol for* Avogadro constant (alternative to *L*).

N_L *symbol for* Loschmidt constant.

Na *symbol for* sodium.

NA *abbr. for* **1** nicotinic acid. **2** noradrenaline (i.e. **norepinephrine**). **3** nucleic acid. **4** numerical aperture.

NAA *abbr. for* **N**-acetylaspartate.

NAADP *abbr. for* nicotinic acid adenine dinucleotide phosphate; the form of NADP in which the nicotinamide moiety has been exchanged for nicotinic acid in a reaction catalysed by ADP-ribosylcyclase. In sea urchin eggs, plant cells, and some mammalian tissues – e.g. brain (especially medulla, thalamus, and midbrain), pancreatic acini, and heart – it mobilizes intracellular Ca^{2+} stores via receptors that are distinct from, but which interact with, those for inositol trisphosphate and for ryanodine/cyclic ADP-ribose. It is rapidly degraded within the aforementioned cells.

NAc *symbol for* (when suffixed to the symbol for a sugar molecule or residue) an acetylated **aminodeoxy sugar**; e.g. GlcNAc is the symbol for *N*-acetylglucosamine or 2-acetamido-2-deoxy-D-glucose.

NAC *abbr. for* nascent chain-associated complex; a dimer of α (33 kDa) and β (22 kDa) subunits that associates with nascent polypeptide chains and their ribosomes, and dissociates from both after chain release from the ribosome. It is present in eukaryotic cytosol and lacks peptidylprolyl isomerase activity. *See* **trigger factor**.

nactin any of a group of **macrotetrolide** antibiotics produced by various strains of *Actinomyces*. The group includes **nonactin**, monactin, dinactin, and trinactin.

NAD *abbr. for* nicotinamide-adenine dinucleotide (when its oxidation state is unknown or unspecified). Besides being a coenzyme, NAD is a precursor of NADP, of mono- and poly-ADP-ribosylation of some proteins, and of cADP-ribose. *See also* **NAD⁺, NADH, NADP, NAD(P)**.

NAD⁺ *abbr. for* oxidized nicotinamide-adenine dinucleotide. The positive charge refers to the nicotinamide ring; the overall charge is negative, owing to the two phosphate groups, which are negatively charged at neutral pH.

NAD⁺ ADP-ribosyltransferase *see* **poly(ADP)-ribosyl transferase**.

NADH *abbr. for* nicotinamide-adenine dinucleotide (reduced). The corresponding oxidized state is NAD⁺.

NADH₂ *alternative name for* nicotinamide-adenine dinucleotide (reduced), the oxidized form being NAD. NADH is preferred except in cases where the use of the charge on NAD⁺ would be misleading.

NADH dehydrogenase EC 1.6.99.3; *systematic name*: NADH:(acceptor) oxidoreductase; *other names*: cytochrome *c* reductase; type I dehydrogenase. An enzyme, associated with a respiratory chain, that catalyses a reaction between NADH and an acceptor to form NAD⁺ and a reduced acceptor; FMN or, in plant mitochondrin, FAD, is a prosthetic group.

NADH dehydrogenase (ubiquinone) EC 1.6.5.3; *other names*: ubiquinone reductase; type I dehydrogenase; complex I dehydrogenase; a flavoprotein enzyme that catalyses the reaction:

$$NADH + ubiquinone = NAD^+ + ubiquinol.$$

The flavin is FMN, and the complex also contains at least nine iron–sulfur proteins. This is the catalytic activity of mitochondrial **complex I**.

NADH kinase EC 2.7.1.86; an enzyme that catalyses the formation of NADPH from ATP and NADH with release of ADP.

Nadi reagent a mixture of 1-naphthol and *N,N*-dimethyl-*p*-phenylene diamine that can be oxidized by cytochrome oxidase to form indophenol blue.

NAD⁺ kinase EC 2.7.1.23; an enzyme that catalyses the formation of NADP⁺ from ATP and NAD⁺ with release of ADP.

NADP *abbr. for* nicotinamide-adenine dinucleotide phosphate (when its oxidation state is unknown or unspecified). Besides being a coenzyme, NADP is a precursor of NADP. *See also* **NAD, NADP⁺, NAD(P), NADPH**.

NADP⁺ *abbr. for* oxidized nicotinamide-adenine dinucleotide. The positive charge refers to the nicotinamide ring; the overall charge is negtive, owing to the two phosphate groups, which are negatively charged at neutral pH.

NAD(P) *abbr. for* a coenzyme that may be NAD or NADP.

NAD(P)⁺ *abbr. for* a coenzyme that may be NAD⁺ or NADP⁺.

NAD(P)⁺–arginine ADP-ribosyltransferase EC 2.4.2.31; *other names*: ADP-ribosyltransferase; mono(ADP-ribosyl) transferase; a sarcoplasmic reticulum enzyme that catalyses the formation of N^2-(ADP-D-ribosyl)-L-arginine from NAD⁺ and L-arginine with release of nicotinamide. It couples poly(ADP-ribose) to arginine residues in proteins.

NADPH *abbr. for* nicotinamide-adenine dinucleotide phosphate (reduced). The corresponding oxidized state is NADP⁺.

NADPH₂ *alternative name for* nicotinamide-adenine dinucleotide phosphate (reduced), the oxidized form being NADP. NADPH is preferred except in cases where the use of the charge on NADP⁺ would be misleading.

NAD(P)H *abbr. for* a coenzyme that may be either NADH or NADPH. Use of NAD(P)H₂ to represent NAD(P)H is not recommended.

NADPH:adrenodoxin oxidoreductase EC 1.18.1.2; *other names*: adrenodoxin reductase; *recommended name*: ferredoxin-NADP⁺ reductase; a flavoprotein enzyme that catalyses the reaction:

$$reduced\ ferredoxin + NADP^+ =$$
$$oxidized\ ferredoxin + NADPH.$$

The flavin is FAD.

NAD(P)⁺ stereospecificity in oxidation–reduction reactions involving the nicotinamide coenzymes, NAD⁺ and NADP⁺ reversibly accept hydrogen at the 4 position of the nicotinamide ring. In NADH and NADPH this position is **prochiral**, and the faces of the nicotinamide ring and the c-4 hydrogens of the dihydronicotinamide ring may be labelled according to the *Re/Si* and pro-*R*/pro-*S* conventions, respectively. Then, if the ring is drawn with the –CONH₂ group pointing to the right, the hydrogen projecting out from the plane of the paper is the *pro-R* hydrogen (or H_R) and that projecting back the *pro-S* hydrogen (or H_S). Alcohol dehydrogenases (EC 1.1.1.1) transfer to the *Re* face (i.e. that from which H_R

points out) and use H$_R$. In biochemistry the *pro-R* hydrogen region is also labelled as H$_A$, and the *pro-S* hydrogen as H$_B$. Enzymes that use the *pro-R* hydrogen (H$_A$) are described as A-specific, while those that use the *pro-S* hydrogen (H$_B$) are called B-specific. *See also* **pro-R/pro-S convention**, **Re/Si convention**.

NAD(P)⁺ transhydrogenase (AB-specific) EC 1.6.1.2; *other names*: pyridine nucleotide transhydrogenase; transhydrogenase. An enzyme that catalyses the same reaction as **NAD(P)⁺ transhydrogenase (B-specific)**; however, preparations from heart mitochondria are A-specific with respect to NAD⁺ and B-specific with respect to NADP⁺. *See also* **NAD(P)⁺ stereospecificity**.

NAD(P)⁺ transhydrogenase (B-specific) EC 1.6.1.1; *other names*: pyridine nucleotide transhydrogenase; nicotinamide nucleotide transhydrogenase; a flavoprotein enzyme that catalyses the reaction

$$NADPH + NAD^+ = NADP^+ + NADH$$

(with specificity for H$_B$ of NADPH and NADH). The flavin is FAD. The enzyme is coupled to respiration and acts as a **hydrogen-ion pump**. *See also* **NAD(P)⁺ stereospecificity**.

NAD⁺ pyrophosphorylase *see* **nicotinamide-nucleotide adenylyl-transferase**.

NAD⁺ synthase EC 6.3.1.5; *other name*: NAD⁺ synthetase; an enzyme that catalyses the formation of NAD⁺ from deamido-NAD⁺, ATP, and NH₃ with release of AMP and pyrophosphate. *See also* **NAD⁺ synthase (glutamine-hydrolysing)**; **nicotinamide–nucleotide adenylyltransferase**.

NAD⁺ synthase (glutamine-hydrolysing) EC 6.3.5.1; *other name*: NAD⁺ synthetase (glutamine-hydrolysing); an enzyme that catalyses the formation of NAD⁺ from deamido-NAD⁺, L-glutamine, ATP, and H₂O with release of AMP, pyrophosphate, and L-glutamate. *See also* **nicotinamide– nucleotide adenylyltransferase**.

***Nae*I** a type 2 **restriction endonuclease** recognition sequence: GCC↑GGC.

NAF *abbr. for* Neu-activating factor (*see* **neu**).

NAG *abbr. for* **1** N-acetylglucosamine. **2** (*esp. in clinical chemstry*) N-acetylglucosaminidase; *see* β-N-**acetylhexosaminidase**.

NAGA *abbr. for* N-acetylgalactosamine.

Nagarse (proteinase) *another name for* a species variant of **subtilisin**, EC 3.4.21.62, from *Bacillus subtilis*; an endopeptidase with a broad spectrum of proteolytic activity.

Nagase a *pseudoacronym* for β-N-acetylglucosaminidase (formerly EC 3.2.1.30). This enzyme is now included with EC 3.2.1.52, β-N-acetyl-hexosaminidase.

NAIBS *abbr. for* nonadrenergic imidazoline binding site; *see* **imidazoline receptors**.

Na⁺/I⁻ symporter (*abbr.*: NIS) *an alternative name for* **iodide pump**.

Naka *an alternative name for* **CD36**.

Na⁺,K⁺-ATPase *see* **sodium/potassium ATPase**.

nalidixic acid 1-ethyl-7-methyl-1,8-naphthyridin-4-one-3-carboxylic acid; a synthetic **4-quinolone antibiotic** that is active against many Gram-negative bacteria. It specifically inhibits bacterial DNA gyrase thereby interfering with DNA synthesis. It has no apparent effect on eukaryotic DNA synthesis.

naloxone 1-N-allyl-7,8-dihydro-14-hydroxynormorphinone; one of the most effective opiate antagonists. It binds tightly to opiate receptors in the nervous system, its action depending on the 3-carbon side-chain substitution on the nitrogen. It is used as an antidote to overdosage of narcotics.

naloxone

naltrexone 1-N-cyclopropylmethyl-7,8-dihydro-14-hydroxynor-morphinone; the cyclopropylmethyl analogue of **naloxone**; it has similar properties but is longer acting.

NANA *abbr. for* N-acetylneuraminic acid (i.e. **sialic acid**).

nano+ *symbol*: n; SI prefix denoting 10⁻⁹ times. It was formerly known as millimicro (*symbol*: mµ).

nanogram *symbol*: ng; a unit of mass equal to 10⁻⁹ gram, or 10⁻¹² kilogram.

nanometre *or* (*US*) **nanometer** *symbol*: nm; a unit of length equal to 10⁻⁹ metre. It was formerly known as millimicron (*symbol*: mµ).

nanomolar *symbol*: nm *or* nM; describing a solution containing one **nanomole** per litre of solution, or a specified multiple or fraction thereof; describing an amount-of-substance concentration similarly expressed.

nanomole *symbol*: nmol; a unit of amount of substance equal to 10⁻⁹ mole.

Nanos an RNA-binding protein of *Drosophila* that, with **Pumilio**, keeps mRNA for Hunchback and cyclin B translationally repressed in the posterior thoracic region of the early syncytial embryo.

nanosecond *symbol*: ns; a unit of time equal to 10⁻⁹ second.

nanotechnology the creation and control of objects on nanometre scales. –**nanotechnological** *adj.* –**nanotechnologist** *n.*

NAP1 *abbr. for* nuclear assembly protein 1; a protein (≈12.5 kDa), widely distributed among eukaryotes, that is responsible for chaperoning histones H2A and H2B to DNA during DNA replication. *See also* **CAF1**.

NAP-2 *abbr. for* neutrophil activating peptide 2; one of several peptides derived from **platelet basic protein precursor**. It consists of **LA-PF4** without the first 15 N-terminal amino acids. It is a potent chemoattractant and activator for neutrophils.

NAP cell adhesion a cell adhesion system comprising nectin, afadin, ponsin, and possibly other proteins. It occurs in **adherens junctions** and **tight junctions**.

Naphthalene Black *or* **Naphthol Blue Black** *or* **Amido Black** a dye used for staining protein on cellulose acetate or on starch or agarose gels.

Naphthol Blue Black *an alternative name for* **Naphthalene Black**.

Napierian logarithm *symbol*: $\log_e$ *or* ln; *an alternative name for* natural **logarithm**. [After John Napier (1550–1617), Scots mathematician who invented logarithms.]

narcosis a state of partial or complete loss of consciousness induced by a drug or drugs that have a depressant action on the central nervous system.

narcotic 1 any of a heterogeneous group of drugs that is capable of inducing narcosis and noted for their analgesic action. **2** (*esp. US*) *the trivial name for* any illegal drug or abused drug. **3** of, or relating to, narcosis; of, relating to, or denoting a narcotic drug.

naringenin 4′,5,7-trihydroxyflavanone; the aglycon of **naringin**, and an inducer of *nod* genes in some *Rhizobium* species.

S-(–)-enantiomer

naringin naringenin-7-rhamnoglucoside; a compound that occurs in the flowers, fruit, and rind of the grapefruit, and is the main bitter component of grapefruit juice. *See also* **naringenin**.

NARP *abbr. for* neuropathy, ataxia, and retinitis pigmentosa; a mitochondrial disease that includes developmental delay, retinitis pigmentosa, ataxia, and other neurological symptoms. It is associated with two point mutations in the mitochondrial gene for a subunit of F_0F_1 ATPase.

NAS *abbr. for* **nonsense-mediated alternative splicing**.

nascent just formed or in the process of being formed, especially in a chemical or biochemical reaction.

nascent chain-associated complex *see* **NAC**.

native (*of a biopolymer*) undenatured, i.e. having the structure and biological activity possessed in living material.

native state of a protein or nucleic acid, the form in which it occurs in the intact cell, its three-dimensional structure depending on formation of the appropriate hydrogen bonds.

natriuresis the excretion of sodium ions in the urine.

natriuretic 1 describing any agent that promotes the excretion of sodium ions in the urine. **2** such an agent.

natriuretic hormone any peptide present in plasma or urine that is capable of stimulating sodium excretion by the kidney and of inhibiting membrane Na^+,K^+-ATPase in cells from kidney and some other tissues. *See* **natriuretic peptide**.

natriuretic peptide any of several peptides, secreted by various tissues, that stimulate natriuresis. The first of the group to be identified was **atrial natriuretic peptide** (*or* atrial natriuretic factor) (*abbr.*: ANP; *other name*: type A natriuretic peptide; *former names*: atri-opeptin; vasoactive peptide). Its discovery arose from the observation that extracts of atrial granules could bring about **natriuresis**. Subsequently, **brain natriuretic peptide** (*abbr.*: BNP; *other name*: type B natriuretic peptide) and **type C natriuretic peptide** (CNP) were identified. Members of this family share many common features; they are produced primarily in the heart and specialized areas of the central nervous system, and induce diuresis, natriuresis, and vasorelaxation. The natriuretic effect results from direct inhibition of sodium absorption in the collecting duct, increased glomerular filtration, and inhibition of aldosterone production and secretion. The major constituent of atrial granules, ANP, is synthesized from a precursor, prepronatriodilatin (152 amino acids), which is cleaved to proANP (126 amino acids). The latter is strikingly similar between many species; proANP is cleaved into an N-terminal fragment, amino-acid residues 1–98, and the biologically active hormone, residues 99–126; the amino-acid sequence of ANP (proANP 99–126) is identical in all mammalian species, except at residue 12 (110 of proANP), which is methionine in humans, dogs, and cows, but isoleucine in rats, mice, and rabbits. There are two genes for the precursor (prepronatriodilatin); cleavage of the latter leads to two peptides, ANP:

SLRRSSCFGGRMDRIGAQSGLGCNSFRY

and cardiodilatin-related peptide (CDP):

NPMYNAVSNADLMDFKNLLDHLEEKMPLED.

The synthesis of BNP follows a similar course. In humans and pig, a 32-amino-acid hormone is the circulating form of BNP, sequence (human)

SPKMVQGSGCFGRKMDRISSSSGLGCKVLRRH;

in rodents BNP contains 45 amino acids and is known as iso-ANP. A 17-member ring, formed by a disulfide bridge between the two cysteine residues, is essential for biological activity of ANP and BNP. *See also* **natriuretic-peptide receptor**.

natriuretic-peptide receptor any type 1 membrane glycoprotein that binds **natriuretic peptides** and mediates their intracellular effects. Three receptors have been identified: ANP_A (previously known as GC-A, ANP-R_1, or ANP_B); ANP_B (previously known as GC-B); and ANP_C. Human ANP_A and ANP_B each have one transmembrane domain and intrinsic guanylate cyclase activity in an intracellular domain, which is highly conserved (88% identity). At the ANP_A receptor, atrial natriuretic peptide is about ten times more potent than brain natriuretic peptide, whereas type C natriuretic peptide is 50–500 times more potent than other agonists at the ANP_B receptor. ANP_C is a disulfide-linked homodimer with one transmembrane domain and no guanylate cyclase domain; it acts through other second messenger systems.

natural abundance the number of atoms of a given nuclide in any naturally occurring mixture of the nuclides of a particular element, expressed as a percentage or fraction of the total number.

natural killer cell *see* **NK cell**.

natural language processing *abbr.*: NLP; the computational analysis and interpretation of human language. NLP is used in software that provides automatic translations of text from one language to another, in robotic systems that use human-language-type commands, and in text-mining tools (e.g. to provide summaries or abstracts of large volumes of text).

natural logarithm *or* **Napierian logarithm** *symbol*: $\log_e$ *or* ln; *see* **logarithm**.

natural product any organic substance of biological origin, as opposed to one obtainable only by chemical synthesis. In practice the term is often restricted to mean **secondary metabolite**.

natural selection the principle that the best competitors in any given population of organisms have the best chance of breeding success and thus of transmitting their characteristics to subsequent generations. The members of any population show individual differences – anatomical, physiological, or metabolic – that affect their functional efficiency in a given environment. The less efficient members tend to die out or produce fewer offspring than the more efficient members, which are better adapted to compete for food or other resources, and so produce relatively more offspring. The principle is fundamental to modern concepts of evolution, and was first articulated, independently, by the British naturalists Alfred Russel Wallace (in 1858) and Charles Darwin (in 1859).

natural-selection theory of immunity *or* **Jerne's theory of immunity** a theory of antibody production that postulates the spontaneous presence, in the blood of an animal, of small numbers of antibody molecules against all antigens to which the animal can respond, and delegates to the antigen the sole role of carrying such specific globulin molecules from the bloodstream into cells in which these molecules can induce proliferation of a particular antibody.

Nb *symbol for* niobium.

NBPC *abbr. for* 1-[(*m*-nitrobenzyloxy)methyl]pyridinium chloride; a reagent useful for the immobilization of single-stranded nucleic acids on cellulose.

NBS *abbr. for* **1** National Bureau of Standards (of the USA) (now National Institute for Standards and Technology). **2** *N*-bromosuccinimide.

NCAM *abbr. for* neural cell adhesion molecule.

NCBI *abbr. for* National Center for Biotechnology Information; *see* **bioinformatics institute**.

NCE *abbr. for* novel chemical entity.

N cell a type of cell, found mainly in the upper small intestine in humans, that contains round, homogeneous granules in which **neurotensin** is stored. Such cells are sparse in the lower small intestine.

NC–IUB *abbr. for* Nomenclature Committee of the International Union of Biochemistry; redesignated (in 1992) **NC–IUBMB**.

NC–IUBMB *abbr. for* Nomenclature Committee of the International Union of Biochemistry and Molecular Biology. *Compare* **JCBN**.

***Nco*I** a type 2 **restriction endonuclease**; recognition sequence: C↑CATGG.

NCoR *abbr. for* nuclear hormone corepressor; a protein that, with **SMRT**, causes transcriptional repression by recruiting the histone deacetylase complex Sin 3-HDAC. The NCoR–SMRT dimer binds unliganded retinoic acid and thyroid hormone receptors.

ncRNA *abbr. for* noncoding RNA; strictly, any RNA that is not mRNA, but the term also commonly excludes tRNA and rRNA. The term is used predominantly in relation to eukaryotes, with sRNA (small RNA) being used in relation to prokaryotes. It encompasses many varieties of RNA that have specific, but noncoding, functions. Examples are: miRNA (micro RNA; 21–25 nt), which modulates development in many eukaryotes; sRNA (≈100–200 nt), which are translation regulators in bacteria; 7S RNA (300 nt) of the **signal recognition particle**; U-snRNA (100–215 nt), which are involved in **spliceosomes** in eukaryotes; and RNAs (>10 000 nt) that are involved in gene silencing in higher eukaryotes (*see* **Xist**).

NCS *the proprietary name for* a 0.6 M solution in toluene of a mixture of quaternary ammonium compounds of general formula $[(CH_3)_2R'R''N]^+OY^-$, where R' and R'' are unbranched alkyl groups containing 6 to 20 carbon atoms and Y is a hydrogen atom or a lower alkyl group. It is useful for solubilizing a wide range of specimens of biological origin in organic solvents for liquid scintillation counting.

Nd *symbol for* neodymium.

***Nde*I** a type 2 **restriction endonuclease**; recognition sequence: CA↑TATG.

nDNA *abbr. for* nuclear DNA.

NDP *abbr. for* any (unknown or unspecified) nucleoside diphosphate in which the diphosphate moiety is esterified at the 5' position of the glycose.

Ne *symbol for* neon.

NE *abbr. for* norepinephrine.

nearest-neighbour sequence analysis a technique that employs a $5'$-^{32}P-labelled deoxyribonucleotide substrate in the enzymic synthesis of DNA followed by enzymic degradation of the DNA specifically to 3'-deoxyribonucleotides. From knowledge of the labelled substrate used and of the labelled degradation products formed, the **nearest neighbour frequency**, i.e. the frequency with which a particular nucleotide is found next to a specific neighbour nucleotide in the polynucleotide chain, can be calculated.

near ultraviolet radiation ultraviolet radiation in the wavelength range 200–400 nm.

nebulette a 107 kDa filamentous protein specific for cardiac muscle. The C-terminal 30 kDa region is integrated in the Z disc, and is highly homologous with that of nebulin. Defects in nebulette can cause cardiomyopathy.

nebulin a giant protein (800 kDa) that is specific for the skeletal muscle of vertebrates; it acts possibly as a 'molecular ruler' for length regulation of the thin filament. The protein appears to have small repeats of approximately 35 amino acids.

nebulize *or* **nebulise** to convert a liquid into a mist or a spray of fine droplets. —**nebulization** *or* **nebulisation** *n*.

nebulizer *or* **nebuliser** a device for nebulizing a liquid (*see* **nebulize**); an atomizer.

necrobiosis the death of a cell or a group of cells within a tissue, whether a normal occurrence or due to a pathological process.

necropsy *another name for* **autopsy**.

necrosis the death of a portion of a tissue as a result of disease or injury. The cells swell, their plasma membranes become disrupted, and the cell contents are released into the extracellular space, where they often trigger an inflammatory response. The process is unregulated and should be distinguished from **apoptosis**.

necrotizing encephalopathy *an alternative name for* **Leigh syndrome**.

nectary a gland that secretes the sugary fluid called nectar. Nectaries occur in many animal-pollinated flowers, to attract insects, birds, and other pollinating agents.

nectin-1 a transmembrane cell-cell adhesion molecule forming part of the **NAP cell adhesion** system in adherens and tight junctions. The nectin-1 gene is mutated in cases of **cleft lip/palate**-ectodermal dysplasia syndrome. It is the main cell surface receptor for alpha-herpes viruses.

Nedd *abbr. for* NPC-expressed, developmentally down-regulated genes (NPC is an abbreviation of neural precursor cells). These genes are down-regulated during mouse brain development. The products are proteins presumed to be involved in embryonic development and differentiation of the nervous system. Nedd-2 protein is a cysteine endopeptidase that may be involved in controlling cell death. Nedd-4 is ubiquitously expressed in mammals. It contains multiple WW domains, which bind with great affinity to the β and γ subunits of epithelial Na channel (ENaC), and a ubiquitin-ligase domain. Nedd-8 (called Rub 1 in yeast) is a ubiquitin-like molecule in mammals that becomes conjugated with cullin 1 and promotes its degradation in proteasomes.

neddylation the process whereby Nedd-8 (or Rub 1 of yeast), a ubiquitin-like protein conjugate to some proteins to target them for degradation by proteasomes or for denedddylation by COP signalosomes. It is important in controlling the auxin response pathway in plants.

Needleman–Wunsch alignment algorithm a dynamic programming algorithm used in sequence alignment that employs a 2D search matrix, such that its speed and storage requirements are of the order $N \times M$ when sequences of length N and M are aligned. The algorithm has large memory requirements owing to its traceback step, where the matches of the optimal alignment are reconstructed, making it too slow for routine database searches. *See* **alignment algorithm**. *See also* **Smith–Waterman alignment algorithm**.

NEFA *abbr. for* nonesterified fatty acid. *Compare* **FFA**.

negative 1 (of an electric charge) having the same polarity as an electron. **2** (of a chemical entity) having a negative electric charge; having an excess of electrons. **3** (of a point in an electric circuit) having a lower potential than some other point to which a potential of zero is assigned. **4** *(in photography)* an exposed and developed photographic film or plate showing an image in reversed tones, or complementary colours, to those of the object photographed. **5** a quantity less than zero. —**negativeness** *or* **negativity** *n*.

negative chromatography *a term sometimes used to denote* methods of purification of substances by **affinity chromatography** in which a specific contaminant in a sample interacts with and is selectively retained by the adsorbent, especially as opposed to **positive chromatography**.

negative-contrast technique a technique used in preparing specimens for electron microscopic examination in which the specimen is mixed with an electron-dense material that penetrates the interstices of the specimen but not the material of the specimen itself. The specimen thus appears transparent against an opaque background. It is commonly called **negative staining**.

negative control the prevention of some biological activity caused by the presence of a specific chemical entity.

negative cooperativity a form of **cooperative ligand binding** in which binding of one ligand to one site on a (macro)molecule decreases the affinity at other sites for the subsequent binding of other ligands.

negative effector *see* **effector**.

negative elongation factor *see* **NELF**.

negative feedback *see* **feedback** (def. 1).

negative-ion mass spectrometry a version of **mass spectrometry** in which negative rather than positive ions, generated from the sample by chemical ionization, are separated and measured in an appropriate instrument.

negative staining *see* **negative-contrast technique**.

negative strand *another name for* **minus strand** (def. 2); in some RNA viruses, RNA is a minus strand which does not directly code for protein. Such viruses have a replicase packaged into the capsid to enable them to synthesize the plus strands necessary for replication.

negative-strand virus *or* **negative-stranded virus** any RNA virus in which the genome consists of single-stranded RNA (i.e. minus strand) of base sequence complementary to that of the virus-specified mRNA (which is the positive, or plus, strand). Such viruses, which comprise class V in the **Baltimore classification**, must make a positive RNA strand before making viral proteins. *See also* **minus strand** (def. 2), **plus strand** (def. 2).

neighbour-joining a method for inferring phylogenetic trees based on the principle of **minimum evolution**. The method resolves the phylogeny in a stepwise fashion by selectively joining pairs of taxa that minimize the sum of branch lengths. Neighbour joining is computationally efficient and is an appropriate reconstruction approach where evolution has not proceeded in a strictly clocklike manner.

NELF *abbr. for* negative elongation factor; a protein complex that inhibits transcriptional elongation by binding to RNA polymerase. It consists of five subunits (NELF-A to NELF-E). NELF-A (66 kDa, 528 amino acids) and shares 21% sequence identity with hepatitis delta antigen (HDAg-S). NELF-E (46 kDa) is identical with a putative RNA-binding protein.

NEM *abbr. for* 1 *N*-ethylmaleimide. 2 nemaline myopathy.

nemaline myopathy *abbr.*: NEM; a slowly progressing myopathy characterized by rodlike structures, composed largely of α-actinin and actin, in skeletal muscle fibres. Nem1 (gene locus at 1q22-q23) is an autosomal dominant disease and is associated with a missense mutation in α-tropomyosin. NEM2 (gene locus at 2q22) is an autosomal recessive disease whose gene locus is that for nebulin.

nematic (of a substance) being in or having a mesophilic state in which the molecules are arranged as in a series of parallel threads but not in layers. The molecules can rotate about their axes and can move in the plane orthogonal to the line of the thread. *Compare* **smectic**. *See also* **liquid crystal**.

Nembutal *a proprietary name for* pentobarbitone sodium; pentobarbital sodium.

NEMO *abbr. for* nuclear factor κB essential modulator; a 48 kDa zinc finger protein that is the regulatory subunit of the IκB kinase complex, which is essential for NF-κB signalling.

N-end rule a rule stating that the *in vivo* half-life of a protein is a function of its N-terminal residues.

neo *abbr. for* neomycin.

neo+ *comb. form* 1 denoting new, recent. 2 *(in chemistry)* denoting an isomer of new (and therefore initially unknown) structure.

neo[R] *abbr. for* neomycin resistance.

neo-Darwinism the reworking of evolutionary theory in the light of discoveries about genetics as a result of the work of Fisher, Haldane, Sewall Wright, and others. The behaviour of genes in natural populations is understood in terms of the four fundamental evolutionary forces: **natural selection**, **genetic drift**, gene flow, and **mutations**.

neoendorphin any of a small group of [Leu[5]]enkephalin-related peptides. α Neoendorphin is the decapeptide [Leu[5]]enkephalinyl-Arg-Lys-Tyr-Pro-Lys, and β neoendorphin is the nonapeptide [Leu[5]]enkephalinyl-Arg-Lys-Tyr-Pro; both are hypothalamic opioid peptides. *See also* **proenkephalin**.

neoflavone any of a group of 4-phenyl-1,2-benzopyrones found in certain plants. Their occurrence is probably restricted to the Guttiferae and Leguminosae. —**neoflavonoid** *n.*

neoglycoprotein *a generic name for* any originally carbohydrate-free protein that has been chemically derivatized by covalent attachment of carbohydrate.

neokestose *O*-β-D-fructofuranosyl-(2→6)-α-D-glucopyranosyl-β-D-fructofuranoside; an isomer of **kestose** isolated from the sap of the sugar maple and produced artificially, together with kestose, by the action of yeast or mould invertase preparations on sucrose.

neolignan any of a class of phenylpropanoid dimers in which the phenylpropane units are linked head-to-tail instead of tail-to-tail as in **lignans**; e.g. eusiderin from the heartwood of members of the Magnoliaceae.

neomycin an aminocyclitol antibiotic complex produced by *Streptomyces fradiae*. It consists of: neomycin A (*or* **neamine**; a degradation product of neomycin B and neomycin C), neomycin B, and neomycin C. Neomycin causes misreading during protein synthesis in bacteria but its target site on the ribosome is apparently different from that of streptomycin. **Neomycin resistance** is conferred by the enzyme aminoglycoside 3′-phosphotransferase. *See also* **neosamine C**.

neonatal of, relating to, or affecting a newborn animal. For human infants the neonatal period is considered to be the first month after birth.

neonatal jaundice a physiological predominantly unconjugated hyperbilirubinemia with clinical jaundice, affecting about half of all human neonates during the first five days of life. It results from increased bilirubin production and delayed maturation of liver **UDP-glucuronosyltransferase** activity. Maternal–fetal Rh blood group incompatibility and hereditary hyperbilirubinemia syndromes exaggerate the condition, which, if untreated, can lead to kernicterus and brain damage.

neonate any newborn animal.

neoplasia 1 the formation of a **neoplasm** or neoplasms. 2 the formation of new tissue.

neoplasm any new and morbid formation of tissue; a tumour.

neoplastic of, or relating to, neoplasia or to a neoplasm.

neoprene poly(2-chloro-1,3-butadiene); a synthetic rubber that is resistant to oils and organic solvents and is useful for the manufacture of flexible tubing, washers, etc. for laboratory and many other uses.

neopterin the D-*erythro* enantiomer of **biopterin**; its biosynthetic precursor.

neosamine C 2,6-diamino-2,6-dideoxy-D-glucose; a constituent of **neomycin** C and some other antibiotics including **kanamycin**.

neosome the suggested secondary particulate biogenic precursor of the ribosome, intermediate between the **eosome** and the completed ribosome. Neosomes in *Escherichia coli* are particles with sedimentation coefficients of 43S and 30S.

neotenin an alternative name for **juvenile hormone**.

neoxanthin a carotenoid **xanthophyll** present in the leaves of all higher plants.

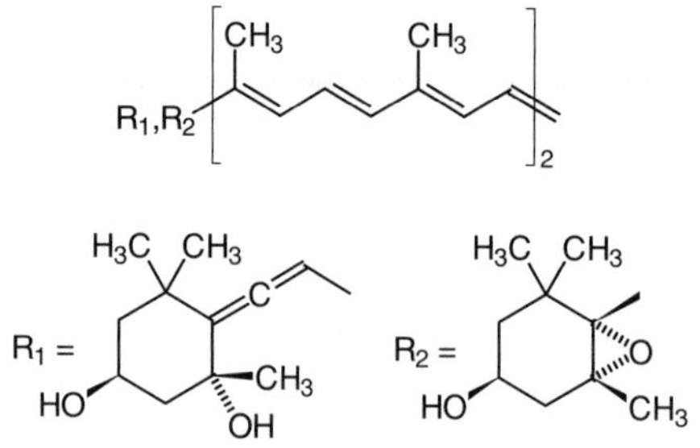

neoxanthine

nephelometry a technique in which the light dispersed by a suspension is measured orthogonally to the direction of the incident light; the amount of scattered light is dependent on the number and size of the particles in the light path. *Compare* **turbidimetry**. —**nephelometric** *adj.*; **nephelometrically** *adv.*

nephr+ *a variant form of* **nephro+** *(before a vowel).*

nephrectomy the surgical removal of one or both kidneys. —**nephrectomize** *vb.*

nephro+ *or (before a vowel)* **nephr+** *comb. form* denoting kidney or kidneys.

nephrocalcin *or* **osteocalcin-related protein precursor** a calcium-binding protein found in bone. It is expressed only in kidney; in humans and other mammals, it is also found as a urinary acidic glycoprotein that strongly inhibits formation of calcium oxalate crystals in renal tubules. It contains γ-carboxyglutamic acid residues; a modified protein lacking these is found in patients with calcium oxalate renal calculi.

nephrogenic diabetes insipidus a rare X-linked recessive genetic disorder involving renal insensitivity to vasopressin resulting in polyuria and polydipsia; it can also be acquired as a result of toxic damage to the tubules. The genetic form is associated with mutations in the V_2 receptor (*see* **vasopressin receptor**), two examples being substitution of Asp for Ala at position 132 and a frameshift mutation at position 246 leading to a premature stop codon and a truncated receptor.

nephron any of the structural and functional urine-secreting units that occur in the kidney. Typically each nephron is made up of a glomerulus, a proximal convoluted tubule, a loop of Henle, and a distal convoluted tubule.

nephropontin *see* **osteopontin**.

neplanocin A a cyclopentenyl analogue of adenosine; a natural antibiotic that inhibits *S*-adenosylhomocysteine hydrolase and vaccinia virus multiplication.

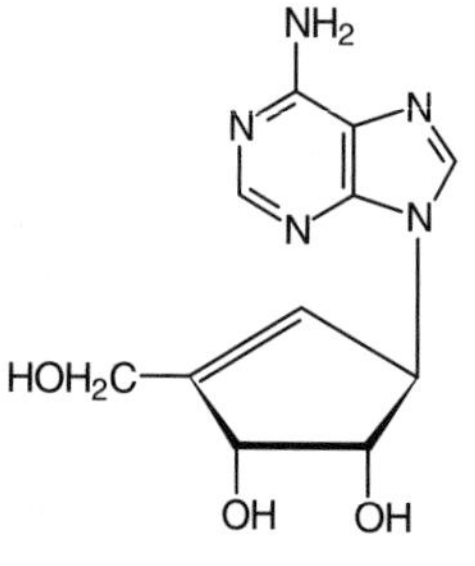

neprilysin EC 3.4.24.11; *other names*: neutral endopeptidase; endopeptidase 24.11; enkephalinase; a member of the zinc metalloendopeptidase family. It is an integral plasma membrane enzyme and is involved in the metabolism of a number of regulatory peptides including enkephalins, tachykinins, and natriuretic and chemotactic peptides.

Ner *symbol for* the nervonoyl (i.e. (*Z*)-tetracos-15-enoyl) group.

NER *abbr. for* nucleotide excision repair.

Nernst distribution law when, at a constant temperature, a solute distributes itself between two immiscible phases, then the ratio of its concentrations in the two phases is constant, and is described by the relation $C_1/C_2 = K$, where C_1 and C_2 are, respectively, the amount-of-substance concentrations of the solute in phases 1 and 2 at equilibrium, and K is the distribution constant. The law only applies in dilute solution. [After Walther Hermann Nernst (1864–1941), German physical chemist.]

Nernst equation a the potential, E, of an electrode reversible to one ion in a solution containing that ion is given by:

$$E = E^{\ominus} + (RT/zF)\ln a,$$

where $E^{\ominus}$ is the standard potential of the electrode (relative to the standard hydrogen electrode), a is the activity of the ion to which the electrode is reversible, z is the charge number of the ion, R is the (molar) gas constant, F is the Faraday constant, and T is the thermodynamic temperature. This equation can be extended to the e.m.f. of an electrolytic cell. **b** For a redox reaction, the potential of an electrode (usually platinum or gold) immersed in a redox solution, E_h, is given by:

$$E_h = E^{\ominus} + (RT/nF)\ln(a_{ox}/a_{red}),$$

where $E^{\ominus}$ is the standard redox potential of the system at a given pH (usually pH 7.0), n is the number of electrons transferred in the reaction, and a_{ox} and a_{red} are respectively the activities of the oxidized and reduced species in the solution. **c** If two solutions are separated by a membrane permeable to only one of two (or more) ions and the activities of the ions are different on either side of the membrane then a membrane potential, E_A, will be generated where

$$E_A = (RT/zF)\ln(a_L/a_R),$$

where a_L and a_R are the activities respectively of the permeable ion on the left and right sides of the membrane.

Nernst heat theorem *another name for* a particular formulation of the third law of thermodynamics (*see* **thermodynamics**).

nerolidol peruviol; 3,7,11-trimethyl-1,6,10-dodecatrien-3-ol; a widespread acyclic sesquiterpene.

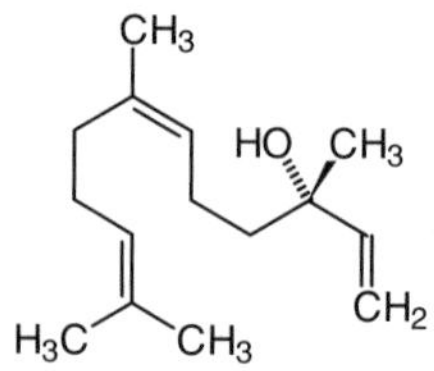

(*S*)-(+)-(*Z*)-enantiomer

nerve any of the cordlike bundles, consisting of **nerve fibres** and **glia** encased in a connective-tissue sheath, that connect the central nervous system to other parts of the body. They may be motor, sensory, or of mixed function.

nerve cell *an alternative name for* **neuron**.

nerve fibre the axonal process of a **neuron** together with its covering sheath. Bundles of nerve fibres running together make up a **nerve**.

nerve gas any of various poisonous gases or volatile liquids that act by inhibiting the passage of impulses through the nervous system and neuromuscular junctions. They are commonly irreversible inhibitors of **acetylcholinesterase**.

nerve growth factor 1 any of a number of polypeptides that exert a trophic effect on neurons. They play a part in the development and maintenance of sensory neurons in dorsal root ganglia and sympathetic neurons in peripheral sympathetic ganglia. **2** *abbr.*: NGF; the first known member of a family of polypeptides that act as growth factors for neurons. In addition to the general properties of such nerve growth factors, it also stimulates growth and differentiation of B lymphocytes, and stimulates histamine release from mast cells. The biologically active form of NGF contains two identical polypeptide chains of 120 amino-acid residues and known sequence and exerts its action through specific receptors in the neuronal plasma membrane. It is produced by neurons, astrocytes, and Schwann cells, but also by fibroblasts, epithelial cells, activated macrophages, and smooth muscle cells. It exists as an inactive complex of two α subunits, two β subunits, and two γ subunits. The active form consists of the two β subunits, and is known as βNGF. Related proteins subsequently discovered include **brain-derived neurotrophic factor** and **neurotrophins**.

nerve growth factor receptor any of a family of plasma membrane integral proteins that bind nerve growth factors. Low-affinity and high-affinity binding is observed; high affinity binding, and action on effector systems, requires the presence of both the *trk* protooncogene product (p140^{prototrk}) and a 75 kDa low-affinity receptor glycoprotein (p75^{NGFR}). The latter can bind **nerve growth factor** (def. 2), **brain-derived neurotrophic factor**, and **neurotrophins** 3 and 4, but it is apparent that considerable specificity exists in transduction of the signal depending on which member of the nerve growth factor family and which cell type is involved.

nerve impulse *see* **impulse** (def. 3).

nervon *or* **nervone** the **galactocerebroside** in which the acyl group is **nervonoyl**. It was originally isolated from nervous tissue.

nervonate 1 *numerical symbol*: 24:1 (15); *the trivial name for* (*Z*)-tetracos-15-enoate, $CH_3–[CH_2]_7–CH=CH–[CH_2]_{13}–COO^-$ (*cis* isomer), the anion derived from nervonic acid, (*Z*)-tetracos-15-enoic acid a monounsaturated straight-chain higher fatty acid. **2** any mixture of free nervonic acid and its anion. **3** any salt or ester of nervonic acid. *See also* **nervonoyl**.

nervonoyl *symbol*: Ner; *the trivial name for* (*Z*)-tetracos-15-enoyl, $CH_3–[CH_2]_7–CH=CH–[CH_2]_{13}–CO–$ (*cis* isomer), the acyl group derived from nervonic acid, (*Z*)-tetracos-15-enoic acid. It occurs as the acyl moiety of the cerebroside **nervon** in brain, as a major constituent of milk phosphatidylcholines and sphingolipids in, e.g., pig and human, and in sphingolipids of other tissues.

nervous of, pertaining to, mediated by, or containing nerves; neural.

nervous system the extensive network of cells specialized to carry information, in the form of nerve impulses, to and from all parts of the body. Nervous systems occur in all orders of multicellular animals other than sponges.

nervous tissue tissue from any part of the nervous system.

nesidioblast a precursor cell of a pancreatic islet cell.

nesidioblastoma a tumour of nesidioblasts.

nesidioblastosis a pathological condition in which individual blastic pancreatic duct cells differentiate abnormally into islet cells scattered among the exocrine tissue. The condition is sometimes associated with severe infantile hyperinsulinism and hypoglycemia.

nesslerize *or* **Nesslerize** to treat (a sample) with **Nessler's reagent**; to test for or estimate (ammonia) by such means. *Also*: **nesslerise**, **Nesslerise**. —**nesslerization** *or* **nesslerisation** *n*.

Nessler's reagent an alkaline aqueous solution of potassium tetraiodomercurate(ii). It gives a yellow or brown colour or precipitate when added to solutions containing ammonia or ammonium ions and is useful in the detection and quantitative estimation of microgram quantities of ammonia in aqueous solution. [After Julius Nessler (1827–1905), German agricultural chemist.]

nested deletion the systematic truncation of a DNA molecule from one end to produce a series of fragments that have an identical sequence at one end and different sequences at the other.

nested PCR primers a set of oligonucleotide primers used for the amplification of DNA by the polymerase chain reaction (PCR) in which the outermost 5′ and 3′ pair are used in the first phase of amplification and a second pair is designed to prime within that PCR product to produce a shorter amplified sequence. Greater specificity of amplification is expected from this use of two pairs of primers.

nestin an intermediate-filament protein whose name derives from neuroepithelial stem cells, in which it is specifically expressed. It shares significant similarity with other intermediate-filament proteins, including a set of heptad repeats.

netrin any of a family of proteins first isolated from chick brain as diffusible chemoattractants that guide commissural axons to their targets during development. Netrin 1 is expressed at high levels in the floor plate during the period when commissural axons are growing towards the floor plate, while netrin 2 is similarly expressed at

lower levels in the ventral two-thirds of the spinal chord. D nectrin is a *Drosophila* homologue, expressed in midline cells of the central nervous system when commissures are forming. The proteins are related to Unc proteins (e.g., *see* **unc-86**) of *Caenorhabditis elegans*.

netropsin a *Streptomyces*-derived basic peptide antibiotic that selectively binds to A–T base pairs in the small groove of right-handed B conformations of DNA. It induces A to B conformational changes in DNA.

Netscape a popular **browser** available for a variety of computers and operating systems.

net sensitivity *see* **flux control coefficient**.

net synthesis increase in the total mass (of a reaction product), as opposed to any apparent increase estimated by incorporation of labelled precursor (which may arise from exchange).

neu an oncogene first described in association with a neuroblastoma after treatment of rats with ethylnitrosourea. The *neu* protooncogene (also known as c-*erbB2*, *NGL*, and *HER2*) encodes a 185 kDa transmembrane glycoprotein with tyrosine kinase activity, related to but distinct from the **epidermal growth factor receptor** (EGFR). The *neu* gene product, Neu, possesses a cysteine-rich extracellular domain, a transmembrane domain, and an intracellular tyrosine kinase domain. EGF does not bind to Neu, but a factor known as **Neu-activating factor** (NAF) binds specifically to Neu and will bind to EGFR. Neu was originally identified in brain, and may play a critical role in neurogenesis. The rat *neu* oncogene is activated constitutively by a point mutation in the transmembrane domain. The association of *neu* with human cancer through gene amplification, or overexpression of the protein product, has been supported by work with breast and ovarian cancer.

Neu 1 *symbol for* a residue (or sometimes a molecule) of **neuraminic acid**. **2** the product of the *neu* gene.

NeuAc *or* (*formerly*) **AcNeu** *symbol for* a residue (or sometimes a molecule) of any isomer of acetylneuraminic acid, i.e. **sialic acid**. *See also* **NeuNAc**, **NeuOAc**, **neuraminic acid**.

Neu-activating factor *see* **neu**.

Neuberg ester *an old name for* D-fructose 6-phosphate. *See* **fructose 6-phosphate**.

Neuberg's first form of fermentation the normal anaerobic **alcoholic fermentation** of carbohydrate by yeast to yield two moles each of ethanol and carbon dioxide per mole of glucose fermented. [After Carl Alexander Neuberg (1877–1956); German biochemist.]

Neuberg's second form of fermentation the anaerobic fermentation of carbohydrate by yeast in the presence of sodium hydrogen sulfite. the acetaldehyde formed normally, which would act as a hydrogen acceptor for the reoxidation of NADH, is trapped as its bisulfite addition product, and its place is taken by dihydroxyacetone phosphate (i.e. glycerone phosphate), which is reduced to *sn*-glycerol 3-phosphate; this is then dephosphorylated to glycerol. Thus one mole each of glycerol, acetaldehyde–bisulfite complex, and carbon dioxide is formed per mole of glucose fermented.

Neuberg's third form of fermentation the anaerobic fermentation of carbohydrate by yeast in alkaline conditions. The acetaldehyde formed normally, which would act as a hydrogen acceptor for the reoxidation of NADH, undergoes a **Cannizzaro reaction** to form acetic acid and ethanol, and its place is taken by dihydroxyacetone phosphate (e.g. glycerone phosphate), which is reduced to *sn*-glycerol 3-phosphate; this is then dephosphorylated to glycerol. Thus one mole each of glycerol and carbon dioxide, and one half mole each of ethanol and acetic acid, are formed per mole of glucose fermented.

NeuGc *symbol for* a residue (or sometimes a molecule) of *N*-glycoloylneuraminic acid. *See also* **sialic acid**.

NeuNAc *or* **Neu5Ac** *symbol for* a residue (or molecule) of *N*-acetylneuraminic acid (i.e. **sialic acid**). It accumulates in lysosomes in Salla disease and infantile sialic acid storage disease, and in cell cytosol in sialuria.

NeuOAc *symbol for* a residue (or molecule) of *O*-acetylneuraminic acid. *See* **sialic acid**.

neur+ *a variant form of* **neuro+** (*before a vowel*).

neural of, or pertaining to, a nerve or nerves, or to the nervous system.

neural cell adhesion molecule *abbr*.: NCAM; any type I membrane glycoprotein of the immunoglobulin superfamily that is involved in neuron–neuron adhesion, neurite fasciculation, outgrowth of neurites, etc. NCAMs have poly-α2,8-sialic acid as a post-translational modification. The human protein NCAM-L1 shows strong homology with a *Drosophila* protein, **neuroglian**.

neural crest a group of embryonic cells, derived from the roof of the neural tube, that migrate to different locations and give rise to various types of adult cells.

neural network *or* **artificial neural network** a general class of machine-learning algorithms, inspired by the neural structure of the brain, in which multiple simple processing units are connected by adaptive weights.

neuraminate 1 the anion derived from **neuraminic acid. 2** any salt or ester of neuraminic acid.

neuraminic acid *symbol*: Neu; the monosaccharide 5-amino-3,5-dideoxy-D-*glycero*-D-*galacto*-nonulosonic acid; it may be regarded as the aldol-condensation product of pyruvic acid and *N*-acetyl-D-mannosamine. It is not found free in nature, but is the parent structure of a family of aminodeoxysugars containing nine or more carbon atoms that occur widely, especially in glycoproteins and gangliosides; they include the **sialic acids**.

α-anomer

neuraminidase *a former name for* **sialidase.**

neuraminosyl *a trivial name for* the glycosyl group derived from **neuraminic acid** by removal of the hydroxyl group from the anomeric carbon atom of the cyclic structure.

neuraminoyl *a trivial name for* the acyl group derived from **neuraminic acid** by removal of the hydroxyl group from the carboxyl group.

neuregulin any member of a family of closely related proteins implicated as regulators of neural and muscle development, and differentiation and oncogenic transformation of mammalian epithelia. They include the Neu differentiation factor (*see* **neu**) and the **heregulins**. They exert their effects through binding to, and activating, receptors of the ErbB2/Neu receptor family. Neuregulins stimulate proliferation of Schwann cells, increase the rate of synthesis of acetylcholine receptors in cultured muscle cells, and are concentrated at nerve–muscle synapses.

neurexin any of a family of cell-surface receptors expressed from three genes in the brain and concentrated in neurons. They are implicated in axon guidance and synaptogenesis. Some are probably signalling molecules. They are highly polymorphic from alternative splicing, and include secreted variants. They contain single transmembrane regions and extracellular domains with repeat sequences similar to those in the globular C-terminal region of laminin A (*see* **laminin**) and **agrin**. Each gene contains two independent promoters so that transcription produces two classes of mRNA, which encode proteins with different N termini, the larger being referred to as α-neurexins and the shorter as β-neurexins. The former contain three epidermal growth factor domains, an O-linked sugar domain, a transmembrane domain, and a short cytoplasmic tail. The α- and β-neurexins differ in their N-terminal sequences but have identical C termini, including the sugar-rich, transmembrane, and cytoplasmic domains. The crystal structure of the laminin A-like sequence ($\approx$200 residues) of neurexin I is similar to that of the pentraxins and some lectins, being a beta sandwich consisting of 14 beta strands.

neurilemma *or* **neurolemma** the thin outer sheath of cells surrounding the axon (and myelin sheath) of larger peripheral nerves or surrounding the axoplasm of unmyelinated nerves.

neurite *a general term for* a dendrite or an axon.

neuritic plaque an extracellular mass of β-amyloid filaments that is associated with dystrophic axons and dendrites, microglia, and astrocytes. In **Alzheimer's disease** such plaques occur in areas of the brain that serve cognition and memory.

neuro+ *or (before a vowel)* **neur+** *comb. form* denoting a nerve or the nervous system.

neuroactive (of an agonist or antagonist) having an effect on the nervous system.

neurocalcin an EF-hand Ca^{2+}-binding protein found in retina and brain neurons, involved in the calcium-dependent regulation of rhodopsin phosphorylation.

neurocan a chondroitin sulfate proteoglycan of brain (rat protein: 1235 amino acids). It is developmentally regulated with regard to mass, concentration, carbohydrate composition, sulfation, and localization.

neurochemistry the study of the chemical composition of, and the chemical reactions occurring in, nervous tissue. —**neurochemical** *adj.*

neurocrine 1 describing or relating to a neuron that secretes an agonist into a synapse with another neuron; describing or relating to such an agonist. **2** such an agonist, e.g. a **neurotransmitter**.

NeuroD1 *abbr. for* neurogenic differentiation factor 1; *other name*: BETA 2; a transcription factor, required for normal development of pancreatic islets, that regulates the insulin gene. Mutations are associated with type 6 maturity-onset diabetes of the young.

neuroendocrine 1 describing or relating to a neuron that secretes an agonist into the bloodstream; describing or relating to such an agonist. **2** such an agonist; a **neurohormone**.

neuroendocrine convertase EC 3.4.21.93; *abbr.*: NEC; *other name* prohormone convertase. Any of the enzymes that act at paired dibasic amino acids in precursors of hormones and other **polyproteins**, to cleave the precursor into individual mature proteins. The catalytic activity depends on the Asp-His-Ser **charge-relay system.** Substrates for NEC1 include **proopiomelanocortin, renin, enkephalin, dynorphin, somatostatin,** and **insulin.** NEC2 and NEC3 have also been described.

neuroendocrinology the study of the interrelations between the nervous system and the endocrine system. —**neuroendocrinological** *adj.*

neurofascin a chick neurite-associated glycoprotein implicated in axon extension; it is an axonal surface recognition molecule of the immunoglobulin superfamily. At the N terminus, it has six immunoglobulin-like motifs, followed by four fibronectin type III motifs, a proline-alanine-threonine-rich region, a transmembrane domain, and a 113-residue cytoplasmic domain.

neurofibril any of the very fine fibres, visible by light microscopy, that occur in the cell bodies, axons, and dendrites of neurons.

neurofibrillary tangle an intraneuronal inclusion composed of bundles of paired helical filaments consisting mainly of **tau protein.** Neurofibrillary tangles are present in the cerebral cortex of virtually all patients with Alzheimer's disease.

neurofibromatosis any of a number of disorders associated with the presence of multiple **neurofibromas**, benign tumours consisting of a mixture of Schwann cells and fibroblasts. Type 1 is a common autosomal dominant disorder that results from mutations in **neurofibromin.** Type 2 is also autosomal dominant but rare and results from mutations in merlin (also called schwannomin).

neurofibromin *other name*: neurofibromatosis-related protein (NF-I); a GTPase activating protein (*see* **GAP** (def. 2)), the product of the *Nf1* tumour-suppressor gene, that stimulates the GAPase activity of the **RAS** oncogene product, thus inactivating it. It appears to be implicated in the pathogenesis of **neurofibromatosis.**

neurofilament a type of **intermediate filament** found in neurons.

neurogenin a transcription factor that promotes neuronal differentiation by inducing neuron-specific genes and by repressing glial cell-specific genes.

neuroglia *an alternative name for* **glia.**

neuroglian a probable cell-adhesion molecule, with structural similarities to **neural cell-adhesion molecule** L1. It is found in neurons and glia of the developing nervous system. It has a single transmembrane domain, with extracellular immunoglobulin-like and fibronectin-type domains.

neuroglobin an oxygen-binding protein of vertebrate brain that may be neuroprotective against ischemia. In mouse retina (which has a very high oxygen demand) its concentration is a hundred times higher than in the brain.

neurohormone any organic compound produced by neurons and released into the circulation to act as a chemical messenger; the term is sometimes extended to all **neurotransmitters** whether released or not. —**neurohormonal** *adj.*

neurohumour *or (US)* **neurohumor** a neurohormone or a neurotransmitter. —**neurohumoral** *adj.*

neurohypophysis (*pl.* **neurohypophyses**) the posterior lobe of the pituitary gland, *also called* **neural lobe** *or* **pars nervosa**. It secretes the peptide hormones oxytocin and vasopressin, and also possibly some others. *Compare* **adenohypophysis**. —**neurohypophyseal** *or* **neurohypophysial** *adj.*

neuroinformatics *see* **informatics**.

neurokinin A *or* **neuromedin L** *or* **substance K** the peptide hormone

His-Lys-Thr-Asp-Ser-Phe-Val-Gly-Leu-Met

derived from **protachykinin β precursor**. It is a member of the **tachykinin** family and is a potent bronchoconstrictor. The related peptide [βAla^8]neurokinin A 4–10,

H-Asp-Ser-Phe-Val-βAla-Leu-Met-NH$_2$,

is a potent and selective **substance K receptor** agonist.

neurokinin B *or* **neuromedin K** a member of the **tachykinin** family that has neuroendocrine and neuroexcitatory functions, especially in the olfactory, gustatory, and visceral systems. It is a potent bronchioconstrictor. It has the sequence

Asp-Met-His-Asp-Phe-Phe-Val-Gly-Leu-Met-Nh$_2$

Neurokinin B acts through a G-protein-coupled receptor, known as NK-3, linked to the phosphatidylinositol cycle.

neurolemma *a variant spelling of* **neurilemma**.

neuroleptic any drug used to treat major psychosis, e.g. schizophrenia or mania. Such drugs include phenothiazines, e.g. **chlorpromazine**, and butyrophenones, e.g. **haloperidol**.

neuroleukin a protein produced by activated T lymphocytes that induces immunoglobulin secretion and is a neurotrophic factor. It is identical with monomeric **glucose-6-phosphate isomerase**.

neurolysin EC 3.4.24.16; *other name*: neurotensin endopeptidase; an enzyme that catalyses the preferential cleavage in neurotensin of Pro^{10}-|-Tyr^{11}.

neuromedin B a decapeptide first isolated from porcine spinal cord. Its structure,

Gly-Asn-Leu-Trp-Ala-Thr-Gly-His-Phe-Met,

has close sequence homology to **bombesin**, with which it shares mitogenic and growth-promoting properties. In rat brain its concentration is highest in the olfactory bulb, but is also high in the hypothalamus, hippocampus, and spinal cord. The B in its name indicates its similarity to bombesin.

neuromedin C a neuropeptide similar in its structure,

Gly-Asn-His-Trp-Ala-Val-Gly-His-Leu-Met,

and in its actions to **neuromedin B** and **bombesin**.

neuromedin L *an alternative name for* **neurokinin A**.

neuromedin N a hexapeptide whose C-terminal tetrapeptide is identical to that of **neurotensin**.

neuromodulator any endogenous agent that influences the function of a **neurotransmitter** but has no direct excitatory or inhibitory postsynaptic actions. The term usually implies long-latency and long-lasting effects (i.e. seconds to days).

neuromuscular of, pertaining to, or affecting both nervous and muscular tissue.

neuromuscular junction the place of contact between the motor end-plate of the fibre of a motor neuron and the membrane of a muscle fibre supplied by the neuron. Impulses are transmitted across the gap by diffusion of a neurotransmitter. *Compare* **synapse**.

neuron *or* **neurone** *or* **nerve cell** a cell that is specialized for the transmission of nerve impulses. Neurons are the structural and functional units of the nervous system. Each consists of an enlarged portion, the cell body or perikaryon, containing the nucleus and from which project a variable number of threadlike processes. Of these

the short branched dendrites convey impulses towards the cell body, while a longer single nerve fibre, or axon, conveys impulses away from the cell body. The axon is unbranched except at the nerve ending. —**neuronal** *adj.*

neuronal PAS domain protein 2 *see* **NPAS2**.

neuronatin a protein that is highly expressed in neonate mammalian brain, and may be involved in terminal brain differentiation.

neuropeptide (*sometimes*) any peptide with neuroendocrine activity.

neuropeptide K a polypeptide hormone, DADSSIEKQVALLKA-LYGHGQISHKRHKTDSFVGLM–NH$_2$, derived from **protachykinin β precursor**. It exhibits a wide spectrum of activities overlapping those of other members of the **tachykinin** family, with which it shares the C-terminal sequence FXLM-NH$_2$.

neuropeptide Y *or* (*formerly*) **melanostatin** *abbr.*: NPY; a 36-residue peptide neurotransmitter, structure YPSKPDNPGE DA-PAEDMARYYSALRHYINLITRQRY–amide (human, porcine has Leu^{17}), found in the autonomic nervous system and in relatively high concentrations in adrenal glands, brain, and heart. It is a potent stimulator of feeding; leptin inhibits NPY gene expression and release. NPY also regulates secretion of **gonadotropin releasing hormone**. It shows considerable sequence homology with **peptide YY** and **pancreatic polypeptide**. [Leu^{31}, Pro^{34}]NPY is a selective agonist for a Y$_1$ receptor subclass. *See also* **neuropeptide-Y receptor**.

neuropeptide-Y receptor *abbr.*: NPY receptor *or* Y receptor; a G-protein-associated membrane receptor that binds **neuropeptide Y** (NPY) and inhibits adenylate cyclase and Ca^{2+} channels. Four types have been recognized. A fifth type, Y5, is suggested to mediate the effects of NPY on feeding.

neurophysin any cystine-rich, single-chain polypeptide present in the neurohypophysis. Neurophysin I binds **oxytocin, with which it** shares a common precursor. For example, in rat the precursor protein consists of 125 amino acids, of which residues 1–19 are the signal peptide, 20–28 are oxytocin, and 32–125 are neurophysin I. Neurophysin II binds **vasopressin**, and again shares a common precursor. *See also* **vasopressin-neurophysin 2-copeptin precursor**.

neurophysin II *another name for* **MSEL-neurophysin**.

neuropilin *abbr.*: NP; an integral membrane protein that, like **plexin**, is a coreceptor for some semaphorins.

neuroscience any of the sciences dealing with the nervous system; such sciences collectively.

neurosecretion 1 the process of secretion of a substance by a (specially adapted) nerve cell. **2** a product of such a secretion; a **neurocrine** (def. 2) or **neuroendocrine** (def. 2). —**neurosecretory** *adj.*

neuroserpin a serpin that is closely homologous to α$_1$-antitrypsin and is secreted by neurons, particularly during their growth phase.

neurosome (*sometimes*) a mitochondrion of a nerve cell.

Neurospora a genus of ascomycete microfungi. The species *N. crassa* has been widely used in biochemical and genetic studies. It is commonly found as a mould on bread that has been kept for too long.

neurotensin a tridecapeptide, Glp-Leu-Tyr-Glu-Asn-Lys-Pro- Arg-Arg-Pro-Tyr-Ile-Leu-OH, found in mammalian brain and gut, especially in **N cells**, packaged within secretory vesicles. It is released from the hypothalamus into the circulation. Although initially shown to induce hypotension, it is now known to have a wide variety of pharmacological actions, including muscle contraction. The precursor gives rise to neurotensin and two derivatives, neuromedin and neuromedin N.

neurotensin endopeptidase *see* **neurolysin**.

neurotoxin any toxin that acts specifically or primarily on nervous tissue. —**neurotoxic** *adj.*

neurotransmission the transmission of nerve impulses from a neuron to another excitable cell via a neurotransmitter.

neurotransmitter any chemical substance released at the distal end of the axon of a neuron in response to the arrival of a nerve impulse that, by diffusing across a **synapse** or other junction, is capable of transmitting the impulse to another neuron, to a muscle cell, or to another excitable cell or, in the case of inhibitory neurotransmitters, of inhibiting the transmission. The neurotransmitter is stored in synaptic vesicles in the axon terminal; it is released into the synaptic cleft on arrival of the action potential and diffuses to and stimulates receptors in the membrane of the post-synaptic cell. An essential component of the mechanism is a means to inactivate the neurotransmitter, e.g. by acetylcholinesterase action on acetylcholine. Both excitatory (*see* **cholinoceptor**, **purinoceptor**, and **glutamate receptor**) and inhibitory (*see* **GABA receptor** and **glycine receptor**) neurotransmitters are known.

neurotrophic factor *see* **brain-derived neurotrophic factor**.

neurotrophin *abbr.*: NT; any of a group of growth factors, similar to **nerve growth factor** but with different tissue specificities. The first to be discovered was named neurotrophin 3, as two proteins with neurotrophic properties had already been discovered (nerve growth factor and **brain-derived neurotrophic factor**). Neurotrophins have properties similar to these latter. Neutrophins 3 to 6 have been described.

neurotropic having a particular affinity for, or affecting nervous tissue.

neurotrypsin a serine protease (875 amino acids) present in presynaptic nerve endings and expressed mainly in cerebral cortex and motor nuclei of the brainstem. It appears at 7–15 weeks of human fetal life. A truncating mutation is associated with a nonsyndromic form of mental retardation.

neurturin a secreted protein that is structurally related to glial cell-derived neurotrophic factor and persephin, and, like them, binds a specific GPI-anchored membrane protein that recruits and signals through RET.

neutral 1 being neither acidic nor basic (or alkaline). **2** having no net charge or electric potential.

neutral amino acid any amino acid containing equal numbers of potentially anionic and cationic groups, as distinct from acidic and basic amino acids.

neutral density filter *an alternative term for* **neutral filter**.

neutral fat *see* **neutral lipid**.

neutral filter *or* **neutral density filter** (*in photography and photometry*) a filter that attenuates all frequencies of light uniformly so that the relative spectral energy distribution of the transmitted light is the same as that of the incident light.

neutrality the condition, quality, or state of being **neutral** (def. 1, 2) or intermediate.

neutralize *or* **neutralise** to render **neutral** (def. 1, 2) or ineffective; to counteract. —**neutralization** *or* **neutralisation** *n*.

neutral lipid *an operational term for* any lipid that is soluble only in solvents of very low polarity. Neutral lipids are divided into two main groups: (1) acylglycerols (glycerides), i.e. fatty-acid esters of glycerol; and (2) waxes, i.e. fatty-acid esters of long-chain monohydroxy alcohols.

Neutral Red an indicator dye that changes colour from red to orange-brown over the pH range 6.8–8.0.

neutral solution any aqueous solution in which the activities of the hydrogen and hydroxide ions are equal. Such a solution in which water is the sole solvent has a pH of 7.00 at 25° C.

neutron *symbol*: n; an elementary particle of zero charge; nucleon number 1; rest mass 1.6749×10^{-27} kg (1.0086 Da); spin 1/2. Neutrons are present in the nuclei of all atoms (except the hydrogen atom), where they are stable; when free, a neutron has a mean life of 932 s, decaying to a proton and an electron.

neutron activation analysis *an alternative name for* **radioactivation analysis**.

neutron diffraction a technique that is similar in principle to **X-ray diffraction** by a crystal, and is applicable particularly to biopolymers and particulate structures. Because hydrogen atoms make a much more substantial contribution to the scattering of neutrons than to X-ray scattering the technique is useful for refining partially deter-

mined structures obtained by the use of X-rays. In rare cases it has been used for determination of absolute configuration.

neutron flux (density) the number of free neutrons passing through unit volume multiplied by their mean velocity.

neutron number *symbol*: N; a dimensionless physical quantity, having an integral value for any given nuclide, equal to the number of neutrons contained in the nucleus of an atom of the nuclide. Hence, it is equal to the arithmetical difference between the **nucleon number**, A, of the nuclide and the **proton number**, Z, of the relevant chemical element; i.e. $N = A - Z$.

neutron scattering *see* **small-angle scattering**.

neutrophil *or* neutrophil leukocyte *or* neutrophilic leukocyte the most numerous **polymorphonuclear leukocyte** found in the blood; a phagocytic cell of the myeloid series that is distinguished by the presence of cytoplasmic azurophil granules and other granules that take up neither acidic nor basic dyes. It plays a major role in the inflammatory response, undergoing chemotaxis towards sites of infection or wounding.

Newman projection a two-dimensional representation of a molecule in which the atoms are arranged as they would be seen if the molecule were viewed from one end along the carbon–carbon bond nearest to the observer. *See also* **conformation**.

newton *symbol*: N; the SI derived unit of force, defined as the force required to give a mass of one kilogram an acceleration of one metre per second per second; i.e. 1 N = 1 kg m s^{-2} = 1 J m^{-1} (= 10^5 dynes).

Newtonian flow the type of flow displayed by a **Newtonian fluid**.

Newtonian fluid any fluid whose (coefficient of) viscosity is independent of flow rate, conforming to **Newton's law of viscosity**. [After (Sir) Isaac Newton (1642–1727), English astronomer, mathematician, philosopher, and physicist.]

Newton's law of viscosity a law, derived by Isaac Newton, that describes the flow of almost all fluids of low relative molecular mass and also that of some solutions of macromolecules. A fluid moving between two parallel plates, in the x direction with a velocity v, is thought of as a number of infinitesimal layers, each of which slides along the adjacent one, the frictional resistance between adjacent layers generating an orthogonal velocity gradient, in the y direction. The frictional force, f, between the fluid layers is then proportional to the area, A, of the layers and to the velocity gradient between them, such that:

$$f = \eta \, A(\mathrm{d}v/\mathrm{d}y),$$

where the constant η is known as the **coefficient of viscosity** (or **viscosity** for short) at the temperature of measurement; its reciprocal, $1/\eta$, is the **fluidity**.

new yellow enzyme *an obsolete name for* the enzyme D-amino-acid oxidase (EC 1.4.3.3), a yellow flavoprotein (FAD) with wide specificity for D-amino acids and acting also on glycine. It was so-named to distinguish it from **old yellow enzyme**. *See* **amino-acid oxidase**.

New Zealand obese mouse *abbr.*: NZO mouse; one of a number of strains of genetically obese mice that are used in metabolic and endocrine research. The obesity is moderate in character and there is an associated hyperinsulinemia and hyperglycemia.

nexin a protein found in the **axoneme** of eukaryotic cilia and flagella. It forms interconnections between the microtubule outer doublets that surround the inner central pair of microtubules. *See also* **glial-derived nexin**.

nexin 1 *other names*: protease nexin I (*abbr.*: PN-I); a glycoprotein that promotes neurite extension, and is an extracellular **serpin**-like serine protease inhibitor.

nexin 2 the secreted isoform of the Alzheimer's amyloid β-protein precursor (*see* **β-amyloid peptide**). *See also* **Alzheimer's disease**.

nexus (*in cytology*) an alternative name for **gap junction**.

NF-1 *see* **CTF**, **neurofibromin**.

NF2 *abbr. for* neurofibromatosis type-2 tumour suppressor; *see* **merlin**.

NF-AT *abbr. for* nuclear factor of activated T cells; a transcription factor complex that activates promoters of the genes for interleukins 2 and 4. It consists of cytoplasmic and nuclear components. The cytoplasmic component, made up of four subunits (NF-ATc1 to NF-ATc4), binds to and is dephosphorylated by calcineurin, then enters the nucleus where it meets the nuclear component (NF-ATn, or NF-AT interacting protein).

NF-κ B a transcription factor for eukaryotic RNA polymerase II promoters. In humans the active protein is a heterodimer of two DNA-binding subunits (p50) which belong to the same family as the **dorsal protein**. It is found in cytoplasm as an inactive complex with an inhibitory subunit, IκB; activation of NF-κB involves its release from its inhibitory subunit. A 105 kDa protein is the precursor of the p50 subunits. These bind to the **kappa-B** consensus sequence, GGRN-NYYCC, e.g. GGGGACTTTCC in mouse and human.

NF-E1 see **GATA-1**.

ng symbol for nanogram.

NGF abbr. for nerve growth factor.

L-NHA abbr. for N^G-hydroxy-L-arginine.

NheI a type 2 **restriction endonuclease**; recognition sequence: G↑CTAGC.

NHS abbr. for N-hydroxysuccinimide.

Ni symbol for nickel.

niacin an alternative name for **nicotinic acid**.

niacinamide an alternative name for **nicotinamide**.

NIAP abbr. for neuronal inhibitor of apoptosis protein. See **IAP**.

nibrin a protein that is missing in cell lysates from persons suffering from **Nijmegen breakage syndrome**. It is phosphorylated by ATM, a protein kinase whose deficiency due to mutation is the cause of **ataxia telangiectasia**.

nicastrin a type 1 transmembrane glycoprotein (709 amino acids) that associates with presenilins 1 and 2 and modulates the release of β-amyloid peptide from β-amyloid precursor protein.

NICE element abbr. for nodule infected cell expression element, a name given to an AT-rich element in *Sesbania rostrata* leghemoglobin glb3.

NICER elements a name suggested for a family of genes that are nerve growth factor-inducible, cyclic AMP-extinguishable, and retrovirus-like, hence the acronym.

nick 1 an interruption in the covalent continuity of one strand of a double-stranded nucleic-acid molecule. **2** to produce a nick in a nucleic-acid molecule. *See also* **nick translation**.

nickase an endodeoxyribonuclease present in phage-infected *Escherichia coli* cells that gives rise to single-strand breaks (nicks) in duplex DNA.

nickel symbol: Ni; it is a ferromagnetic, metallic transition element of group 10 of the IUPAC **periodic table**; atomic number 28; relative atomic mass 58.69. It consists of a mixture of five stable nuclides of which the two most abundant are those of mass numbers 58 and 60; its commonest valence is ii. Nickel's abundance in the earth's crust is 0.018% and it is an essential trace element for living organisms. Several artificial radionuclides of nickel are known, including nickel-56 (γ-emitter; half-life 6.1 days) and nickel-63 (emits β⁻ radiation, 0.066 MeV; half-life 100 years).

nickel–iron hydrogenase any of a family of hydrogenases (EC 1.12..) present in some bacteria, which have an active site comprising a nickel and an iron atom, the iron being bound by two cyanide and one carbonyl ligands. They also contain other iron–sulfur clusters, and in some cases flavin.

nicking–closing enzyme or **nicking-and-closing enzyme** see type I DNA topoisomerase.

nick translation a procedure for preparing radioactively labelled DNA, especially for use as probes in hybridization experiments. Single-stranded nicks are introduced into unlabelled DNA, usually by limited treatment with deoxyribonuclease I, exposing 3′-hydroxyl termini. *Escherichia coli* DNA polymerase I (at low temperature, ≈15°C) is then used successively to incorporate radioactive nucleotides (present as dNTPs in the reaction mixture) to form a new, radioactively labelled polynucleotide strand while the 5′→3′ exonuclease activity of the polymerase concomitantly hydrolyses the 5′ termini ahead of the growing strand, liberating 5′-mononucleotides. The reaction thus progressively replaces nucleotide residues by their labelled counterparts in a DNA duplex, which is otherwise unchanged except for the translation of the nick, i.e. its progressive displacement towards the 3′ end of the strand. (The

word 'translation' in the name may be misleading as the function involved is replication.)

Nicol prism an optical device constructed of two prisms of calcite cemented together with Canada balsam, used for producing plane polarized light and/or determining its orientation. It has now largely been superseded by Polaroid filters, etc. [After William Nicol (1768–1851), British physicist.]

nicotianamine the metabolic precursor of the phytosiderophores avenic acid (of oats) and mugineic acid (of barley). It is derived from the α-aminobutyrate moieties of S-adenosylmethionine.

nicotinamide the recommended trivial name for pyridine-3-carboxamide, the amide of **nicotinic acid**; also called **niacinamide**. It is a member of the B complex of vitamins, and is nutritionally equivalent to nicotinic acid. It occurs widely in living organisms, usually in combined form.

nicotinamide-adenine dinucleotide abbr.: NAD⁺ or NAD; if the oxidation state is not to be specified; nicotinamide(1)-β-D-riboside(5′)diphospho(5′)adenosine; an atypical dinucleotide containing a phosphoric anhydride linkage in place of the usual phosphodiester linkage between the component mononucleotide units, i.e. between nicotinamide mononucleotide and adenylic acid. It is the specific coenzyme in numerous oxidoreductase enzyme reactions, although in a number it functions interchangeably with **nicotinamide-adenine dinucleotide phosphate**. In these reactions it stereospecifically accepts a hydrogen atom at position 4 of the pyridine ring, plus an electron, to form **nicotinamide-adenine dinucleotide (reduced)**. It acts also as a donor in the ADP-ribosylation of some eukaryotic and prokaryotic proteins. The α anomer is inactive as a coenzyme. *Former names*: codehydrogenase I; coenzyme I (Co I); cozymase; diphosphopyridine nucleotide (DPN). *See also* **nicotinamide-adenine dinucleotide (reduced)**.

nicotinamide-adeninedinucleotidephosphate abbr.: NADP⁺; nicotinamide(1)-β-D-riboside(5′)diphospho(5′)adenosine 2′-phosphate; a coenzyme with a structure and coenzymic mechanism related to those of **nicotinamide-adenine dinucleotide**, although it functions as the sole coenzyme for fewer oxidoreductases and tends to be involved more in synthetic than in energy-yielding reactions. *Former names*: codehydrogenase II; coenzyme II (Co II); phosphocozymase; triphosphopyridine nucleotide (TPN). *See also* **nicoti-namide-adenine dinucleotide phosphate (reduced)**.

nicotinamide-adenine dinucleotide phosphate (reduced) *abbr.*: NADPH; 3-aminocarbonyl-1,4-dihydropyridine(1)-β-D-ribo-side (5′)diphospho(5′)adenosine 2′-phosphate; the reduced form of the coenzyme **nictoinamide-adenine dinucleotide phosphate**. As in **nicotinamide-adenine dinucleotide (reduced)** the carbon atom at position 4 of the 1,4-dihydropyridine ring is prochiral. *See also* **NAD(P)⁺ stereospecificity**. *Former names*: reduced codehydrogenase II; reduced coenzyme II (Co IIH₂); reduced phosphocozymase; reduced triphosphopyridine nucleotide (TPNH).

nicotinamide-adenine dinucleotide (reduced) *abbr.*: NADH; 3-aminocarbonyl-1,4-dihydropyridine(1)-β-D-riboside(5′)diphospho(5′)adenosine; the reduced form of the coenzyme **nicotinamide-adenine dinucleotide**. The carbon atom at position 4 of the dihydropyridine ring is prochiral; removal of one of the two hydrogen atoms from position 4 (or introduction of a hydrogen atom at this position into the oxidized form) through the agency of an NAD-linked oxidoreductase is stereospecific, the two hydrogen atoms being designated *pro-R* (or H_A) or *pro-S* (or H_B), and the two corresponding aspects of the (dihydro)pyridine ring frequently being known as the *A* face and the *B* face. The α anomer is inactive as a coenzyme. *Former names*: reduced codehydrogenase I; reduced coenzyme I (Co IH₂); reduced cozymase; reduced diphosphopyridine nucleotide (DPNH). *See also* **NAD(P)⁺ stereospecificity**.

nicotinamide mononucleotide *symbol*: NMN; nicotinamide(1)-β-D-riboside 5′-phosphate; the mononucleotide produced when **nicotinamide-adenine dinucleotide** acts as a donor of an adenylyl group in the activation of DNA ligase.

nicotinamide nucleotide any nucleotide that contains combined nicotinamide, e.g. nicotinamide-adenine dinucleotide (reduced), nicotinamide-adenine dinucleotide phosphate (reduced).

nicotinamide–nucleotide adenylyltransferase EC 2.7.7.1; *other name*: NAD⁺ pyrophosphorylase; an enzyme that catalyses the formation of NAD⁺ from ATP and nicotinamide ribonu-cleotide with release of pyrophosphate and NAD⁺. *See also* **NAD⁺ synthase, NAD⁺ synthase (glutamine-hydrolysing)**.

nicotinate pyridine 3-carboxylate; the anion derived from **nicotinic acid**.

nicotinate mononucleotide 3-carboxylatopyridine(1)-β-D-ribo-side 5′-phosphate; an intermediate in the biosynthesis of nicoti-namide-adenine dinucleotide. It reacts with ATP to form de-samido-NAD, which is subsequently converted to NAD by a glutamine-dependent amidation of the carboxyl group.

nicotine (*S*)(−)-3-(1-methyl-2-pyrrolidinyl)pyridine; the principal example of a number of related very poisonous and addictive alka-loids extractable from the dried leaves of various species of tobacco, especially *Nicotiana tabacum*. It is an agonist of the nicotinic class of **cholinoceptors**. —**nicotinic** *adj*.

nicotinic acetylcholine receptor *see* **cholinoceptor**.

nicotinic acid *abbr.*: NA; *the recommended trivial name for* pyridine 3-carboxylic acid, *also known as* niacin. It is a member of the B complex of vitamins, nutritionally equivalent to **nicotinamide**. At physiological pH values it exists virtually exclusively as the nicoti-nate anion.

nidogen *or* **entactin** a sulfated glycoprotein that is widely distributed in basement membranes, and tightly associated with **laminin**.

Niemann–Pick disease *an alternative name for* **sphingomyelin lipido-sis**. [After Albert Niemann (1880–1920), German physician.]

Niemann–Pick protein C a membrane protein of late endosomes and the *trans* Golgi network that when mutated causes Nie-mann–Pick disease types C and D (*see* **sphingomyelin lipidosis**). The gene at 18q11 encodes a protein of 1278 amino acids, which con-tains 13 transmembrane segments and a leucine-zipper domain. Five of the transmembrane segments have significant homology with the sterol-sensing domain of **SCAP**, HMG-CoA reductase, and Patched.

nif the symbol for any gene controlling bacterial nitrogen fixation; individual genes are distinguished by suffixed capital letters, e.g. *nifB*, *nifQ*. In *Klebsiella pneumoniae* the *nif* cluster has at least 18 genes, arranged in seven distinct operons with no other interspersed genes; the total length of the *nif* region is ≈24 kb. The two polypep-tide subunits of component I of **nitrogenase** and the polypeptide subunit of component II are each controlled by a single *nif* gene; the production and insertion of iron–molybdenum cofactor (FeMo-co) into component I appears to require three to six genes; the produc-tion of mature component II is controlled by two to four genes; two electron-transporting proteins are coded for by one gene each; and three or more genes are concerned with regulating expression of others. Six genes are required for the production and insertion of iron–molybdenum cofactor (Fe-Mo-co) into component I. Two genes, *nifS* and *nifU*, are required for the assembly of iron–sulfur clusters; *nifS* is a cystein desulfurase, EC 2.8.1.7, which provides the sulfide, and *nifU* provides a scaffold for assembly of the clusters.

nifedipine 1,4-dihydro-2,6-dimethyl-4-(2-nitrophenyl)-3,5-pyridine-dicarboxylic acid dimethyl ester; one of the dihydropyridine cal-cium antagonists. It is a calcium channel blocking agent inhibiting the cellular influx of calcium associated with the plateau phase of the action potential. It acts as a vasodilator, reducing blood pressure and increasing coronary blood flow.

nigericin a 725 Da, open-chain macrotetrolide-like antibiotic, with many chiral centres, isolated from *Streptomyces hygroscopicus*. It forms stable complexes with monovalent cations, especially with K+; these complexes are soluble in organic solvents, and thus nigericin can act as an ionophore for monovalent cations. It has some structural resemblance to **monensin**.

nigerose the disaccharide 3-*O*-α-D-glucopyranosyl-D-glucose.

night blindness *or* **nyctalopia** the inability to see in dim light or at night. It is due to deficiency of vitamin A or to degeneration of the **rod cells** in the retina of the eye, which primarily mediate vision in dim light.

NIH *abbr. for* National Institutes of Health (of the USA).

NIH shift *or* **hydroxylation-induced migration** an intramolecular migration that often occurs in the hydroxylation reactions of alkylaromatic compounds catalysed by certain mono- or dioxygenases, whereby a hydrogen atom or a substituent atom or group at the position of hydroxylation moves to an adjacent carbon atom in the aromatic ring. For example, in the hydroxylation of phenylalanine to tyrosine, catalysed by phenylalanine 4-monooxygenase (EC 1.14.16.1), the hydrogen atom at C-4 in the ring can be shown by tritium labelling to migrate to C-3. [After NIH, i.e. National Institutes of Health, where the discoverers were based.]

NIH-3T3 a continuous cell line established from NIH Swiss mouse embryo cultures. These cells are adherent, have fibroblast morphology, and are useful for DNA transfection and transformation studies.

Nijmegen breakage syndrome a human syndrome of mental retardation, increased susceptibility to cancer, immunodeficiency, and chromosomal instability. It is due to mutation in the gene (at 8q11) for **nibrin**, which is missing in cell lysates from patients. It is sometimes considered to be a variant of **ataxia telangiectasia**. [Named after Nijmegen, a city in the Netherlands where it was first indentified.]

NIMR *abbr. for* National Institute for Medical Research (of the UK).

ninein *see* **centrosome**.

ninhydrin *the trivial name of* triketohydrindene hydrate; 2,2-dihydroxy-1,3-indanedione; a pale-yellow crystalline substance, useful as a sensitive chromogenic reagent for the detection and quantification of compounds containing free amino or imino groups. The common amino acids (and the free amino groups of peptides and proteins, and various other amino compounds) give a red, violet, or blue coloration, the shade depending somewhat on conditions, whereas the imino acids proline and hydroxyproline give a yellow colour. The partially reduced ninhydrin derivative, hydrindantin, is often added to the reagent solution, as it reacts with free ammonia that is formed as a side product, and in combination with ninhydrin forms the coloured product, thus enhancing the sensitivity of the reaction.

NIPK *abbr. for* neuronal cell death-inducible protein kinase; *other name*: tribbles homologue protein 3 (*abbr.*: TRB3). One of three mammalian homologues of the *Drosophila* protein **tribbles**. It contains 354 amino acids, inhibits Akt/PKB – a key component of insulin receptor signalling – and its expression in liver is increased by fasting. TRB1, TRB2, and TRB3 share 45% amino acid sequence homology and strongly resemble tribbles.

Nir *symbol for* the ribonucleoside ribosylnicotinamide.

NIS *abbr. for* Na+/I− symporter. *See* **iodide pump**.

nisin a pore-forming toxin produced by the bacterium *Lactococcus lactis*. It has a molar mass of 5.8–6.3 kDa and creates opening in the affected membrane that permit passage of small solutes. It is used commercially for protection of foods against bacterial infection. In the TC system it has number 1.C.20.

Nissl substance *a collective term for* the dense regions of granular endoplasmic reticulum and free polysomes observable by microscopy within the cytoplasm of the cell body of vertebrate neurons. [After Franz Nissl (1860–1919), German neuropathologist, who first observed it.]

NIT domain *abbr. for* nitrate- and nitrite-sensing domain; a protein domain (≈250 amino acids, with several conserved charged residues) that is present in various receptor components of signal-transducing pathways in a variety of bacterial species. It may be intracellular or extracellular and lies between two transmembrane segments. It is predicted to form 10 alpha helices.

nitrate 1 the NO_3^- ion, derived from nitric acid. 2 any salt or ester of nitric acid.

nitrate assimilation *an alternative term for* assimilatory nitrate reduction; *see* **nitrate reduction**.

nitrate dissimilation *an alternative term for* dissimilatory nitrate reduction; *see* **nitrate reduction**.

nitrate reductase 1 any of the enzymes **nitrate reductase (NADH)** (EC 1.6.6.1), **nitrate reductase (NAD(P)H)** (EC 1.6.6.2), or **nitrate reductase (NADPH)** (EC 1.6.6.3) that reduce nitrate to nitrite coupled to the oxidation of the coenzymes indicated. They are widespread among plants and microorganisms, where they participate in assimilatory nitrate reduction. All are flavoproteins, usually containing one atom of molybdenum per molecule. 2 the microbial enzyme, nitrate reductase (EC 1.7.99.4), that reduces nitrate to nitrite coupled to the oxidation of various electron donors in dissimilatory nitrate reduction. 3 the microbial enzyme, **nitrate reductase (cytochrome)** (EC 1.9.6.1), that reduces nitrate to nitrite coupled to the oxidation of a ferrocytochrome to the corresponding ferricytochrome. *See also* **nitrate reduction**.

nitrate reduction any reaction or sequence of reactions in which nitrate ion is converted to other less highly oxidized inorganic nitrogenous substances, e.g. to nitrite ion, nitrous oxide, dinitrogen, ammonium ion, or amino acids. In enzymology the term is restricted to the enzymic reduction of nitrate ion to nitrite ion, catalysed by a **nitrate reductase** enzyme. In biology, nitrate reduction is of two types, distinguishable by the purposes served and by the enzymes involved. **assimilatory nitrate reduction** is the process occurring in higher plants, algae, and fungi, in which nitrogen from nitrate is converted into organic nitrogen-containing compounds by the probable reaction sequence: nitrate → nitrite → [hyponitrite] → hydroxylamine → ammonium → amide group of glutamine → amino acids, etc. **dissimilatory nitrate reduction** or **nitrate respiration** is the process occurring in some microorganisms under anaerobic conditions in which nitrate functions as a terminal electron acceptor in respiratory metabolism. The nitrogen may either be converted to ammonium (which is not assimilated), a process designated **ammonification of nitrate**, or be converted to various gaseous compounds such as nitric oxide, nitrous oxide, and/or dinitrogen, a process designated **denitrification**. *Compare* **nitrification**.

nitrene the chemical entity NH or any of its substitution derivatives. It contains an electrically neutral mono-coordinate nitrogen atom with four nonbonding electrons; two of these electrons are paired while the other two have spins that may be either antiparallel (singlet state) or parallel (triplet state). *Other names*: azacarbene; azene; imene; imine radical.

nitric oxide nitrogen monoxide; NO; a colourless gas, only slightly soluble in water, that reacts readily with O_2 to form nitrogen dioxide, NO_2; it can form the stable cation NO^+. It accounts for the action of endothelium-derived relaxing factor (EDRF). Nitric oxide is a biogenic messenger with potent actions including vasodilation, inhibition of platelet aggregation, and reduction of endothelium adhesion; it also has anticoagulant and fibrinolytic actions. It is formed biosynthetically from L-arginine by **nitric-oxide synthase** in most mammalian tissues in varying amounts, including vascular endothelial cells, the central and peripheral nervous system, immune cells, lung, or liver. It activates guanylate cyclase, producing cyclic GMP. It is oxidized to nitrite, then to nitrate, which is present in the blood.

nitric-oxide synthase EC 1.14.13.39; *abbr.*: NOS; *systematic name*: L-arginine,NADPH:oxygen oxidoreductase (nitric-oxide-forming). An enzyme that catalyses a reaction between L-arginine, NADPH, and O_2 in which citrulline, $NADP^+$, and nitric oxide are formed; heme, FAD, FMN, and tetrahydrobiopterin are required as cofactors; the heme centre has the spectral properties of a cytochrome P450; N^ω-hydroxy-L-arginine is an intermediate, and acts as a substrate for the enzyme. Three isoforms are known: **nNOS**, of neuronal tissues, and **eNOS**, of vascular endothelial cells, are both expressed constitutively and are Ca^{2+}-dependent; **iNOS**, inducible in macrophages and liver cells, is not Ca^{2+}-dependent, but binds calmodulin with extremely high efficiency. All have consensus sequences for the binding of NADPH, FAD, FMN, and calmodulin, and a phosphorylation site for cyclic-AMP-dependent protein kinase A; eNOS has an N-terminal consensus sequence for myristoylation. Inhibitors include L-N^5-(1-iminoethyl)ornithine, N^G-monomethyl-L-arginine, N^G-nitro-L-arginine.

nitrification the process by which certain soil bacteria (nitrifying bacteria) oxidize ammonium ions in decomposing organic material into nitrate ions. *Nitrosomonas* spp. first convert ammonium to nitrite, and *Nitrobacter* spp. then oxidize nitrite to nitrate. Most higher plants can assimilate nitrate much more readily than they can ammonium. *Compare* **denitrification**.

nitrify 1 (of nitrifying bacteria) to oxidize ammonium to nitrate. **2** to treat (soil) with nitrates. —**nitrified** or **nitrifying** *adj*.

nitrifying bacteria obligately aerobic bacteria that are capable of oxidizing ammonia to nitrite, and nitrite to nitrate. *See* **nitrification**. *Compare* **denitrifying bacteria**.

nitrite 1 the NO_2^- ion, derived from nitrous acid. **2** any salt or ester of nitrous acid.

nitrite reductase 1 the enzyme **nitrite reductase (NAD(P)H)** (EC 1.6.6.4) that reduces nitrite to ammonium coupled to the oxidation of either NADH or NADPH. It is a flavin-dependent enzyme that also contains **siroheme**; it participates in assimilatory **nitrate reduction** in higher plants, algae, fungi and bacteria. **2** the enzyme **nitrite reductase (cytochrome)** (EC 1.7.2.1) that reduces nitrite to nitric oxide coupled to the oxidation of ferrocytochrome *c* to ferricytochrome *c*. It is a cuproprotein and participates in denitrification processes in some microorganisms. **3** the enzyme **ferredoxin–nitrite reductase** (EC 1.7.7.1) that reduces nitrite to ammonium coupled to the oxidation of reduced ferredoxin; the enzyme in spinach leaves contains one molecule of siroheme and one iron–sulfur [2Fe-2S] cluster per molecule. It participates in assimilatory nitrate reduction, the reduced ferredoxin being derived from the light reactions of photosynthesis. **4** the microbial enzyme, nitrite reductase (EC 1.7.99.3), that reduces nitrite to nitric oxide coupled to the oxidation of various electron donors in dissimilatory nitrate reduction; the enzymes from *Achromobacter cycloclastes* and *Pseudomonas denitrificans* contain copper.

nitro the chemical group –NO_2 in a molecule of an organic compound.

nitro+ *prefix (in chemical nomenclature)* **1** indicating the presence of a **nitro** group in an organic compound. **2** *(in chemical technology)* indicating that a chemical compound is a nitrate ester.

Nitroblue Tetrazolium *see* **Nitrotetrazolium Blue**.

nitrogen 1 *symbol*: N; a nonmetallic element of group 15 of the IUPAC **periodic table**; atomic number 7; relative atomic mass 14.0067. There are two natural stable nuclides of relative masses 14 (99.635 atom percent) and 15 (*see* **nitrogen-15**); its commonest valences are iii and iv. Nitrogen occurs naturally in the form of diatomic molecules of the gas dinitrogen (see below) and in combined form in numberless organic molecules in all forms of life. *See also* **nitrogen-13**. **2** **dinitrogen** *or* **molecular nitrogen** *symbol*: N_2; a diatomic nonflammable gas. It constitutes 75.5% by mass, or 75.06% by volume, of the Earth's atmosphere, and undergoes exchange with organically combined nitrogen via the processes of the **nitrogen cycle**. Liquid nitrogen, b.p. 77.36 K, is commonly used as a coolant or freezing agent.

nitrogen-13 the artificial radioactive nitrogen nuclide $^{13}_7N$. It emits a positron (β^+-particle) (1.19 MeV max.) and no gamma radiation; it has a half-life of 9.96 min.

nitrogen-15 heavy nitrogen; the stable, naturally occurring nitrogen nuclide$^{15}_7N$; relative abundance in nitrogen 0.365 atom percent. It is extensively used as a tracer in studies of nitrogen metabolism and in **nuclear magnetic resonance**.

nitrogenase an enzyme complex that catalyses the ATP-dependent fixation of atmospheric nitrogen, summarized by the equation:

$$N_2 + 10H^+ + 8e^- + 16ATP \rightarrow 2NH_4^+ + H_2 + 16ADP.$$

Depending on the bacterial species, the electrons may be donated either by reduced ferredoxin (this enzyme is EC 1.18.6.1) or flavodoxin (EC 1.19.6.1). For each molecule of nitrogen reduced, at least one molecule of hydrogen is released. Either type of nitrogenase will use ferredoxin or flavodoxin as donor *in vitro*. Nitrogenase consists of two oxygen-sensitive components, known as MoFe protein (*other names*: component I; dinitrogenase; molybdoferredoxin) and Fe protein; (*other names*: component II, dinitrogenase reductase; azoferredoxin). Component I is a tetramer made up of two polypeptide chains of about 50 and 60 kDa, each present twice; the tetramer contains two P-clusters (Fe_8S_7) and two Fe-Mo cofactors (*abbr.*: FeMo-co). The Fe-Mo cofactor has the formula Fe_7MoS_9. The Mo is coordinated by 3 sulfide ions, a histidine N and 2 oxygen atoms of homocitrate. The extremely oxygen-sensitive component II is a dimer of identical, ~30 kDa, polypeptide chains; each binding a molecule of ATP. The two subunits of the dimer are bridged by a single [4Fe-4S] cluster. Electrons from a donor, such as NADPH, are transferred, via reduced ferredoxin or flavodoxin, to component II. Then, in the presence of Mg^{2+} they are passed to component I, with concomitant hydrolysis of 4–5 molecules of ATP per two electrons transferred. Eventually they are used to reduce dinitrogen to ammonium. Nitrogenase also catalyses ATP-dependent partial reductions of some other small-molecules that are isoelectronic with dinitrogen, e.g. ethyne, azide, cyanide, and dinitrogen monoxide. The reduction of ethyne to ethene, which can be measured by gas chromatography is used as a convenient assay for nitrogenase activity. Alternative nitrogenases are known that contain vanadium instead of molybdenum, or just iron as the only transition metal; these appear to be structurally and functionally analogous to the molybdenum-containing nitrogenase although they produce more H_2 per N_2 reduced.

nitrogenase reductase *another name for* dinitrogenase reductase; i.e. component II of **nitrogenase**.

nitrogen balance 1 *(in physiology)* the **balance** (def. 1) of the nitrogen content of the nutrients and the excretory products of an organism. **2** *(in agriculture)* the net loss or gain of the nitrogen content of soil that results from removal of nitrogen by cropping, leaching, etc. and addition of nitrogen by nitrogen fixation or addition of nitrogenous fertilizers.

nitrogen base *an alternative term for* **nitrogenous base**.

nitrogen cavitation *an alternative name for* **pressure homogenization**.

nitrogen cycle the series of chemical processes by which various inorganic and organic nitrogenous compounds are interconverted and maintained in equilbrium in the biosphere. It consists essentially of **nitrogen fixation**, **nitrification**, and **denitrification**, together with assimilatory and dissimilatory **nitrate reduction** and the interconversion of nitrogenous organic matter and ammonium. Individual reactions may be effected by microorganisms, plants, or animals, with contributions from nonbiological processes such as manufacture of nitrogenous fertilizers, the action of lightning, and emission of combustion products.

nitrogen equilibrium the state of **nitrogen balance** (def. 1) when nitrogen intake equals nitrogen excretion.

nitrogen fixation any biological or nonbiological process that converts atmospheric dinitrogen into inorganic nitrogenous compounds; the meaning is often extended to include the biological assimilation of ammonium so produced. Biological nitrogen fixation occurs by the action of the enzyme **nitrogenase**, found in certain species of bacteria – the nitrogen-fixing bacteria.

nitrogen mustard any of a group of bis(2-chloroalkyl)amines, as typified by the volatile base bis(2-chloroethyl)ethylamine, $CH_3–CH_2–N(CH_2–CH_2Cl)_2$, a nitrogen-containing analogue of **mustard gas**. They are potent alkylating agents, vesicants, and mutagens. They cross-link DNA chains and have been used as immunosuppressive and antineoplastic agents.

nitrogenous 1 of or relating to **nitrogen**. **2** of or relating to any substance containing nitrogen in combined form.

nitrogenous base *or* **nitrogen base** *a generic term for* ammonia or any trivalent compound of nitrogen that can be regarded as a derivative of ammonia; all except quaternary ammonium compounds are weakly basic in character. In biochemistry, the term usually denotes any purine or pyrimidine; now generally shortened to base.

nitrogen regulatory protein *abbr.*: NR protein; any of several proteins of a regulatory system for transcriptional control of the glutamine-ammonia ligase, and for its regulation at the structural level. Transcriptional regulation is mediated by the proteins NR_I and NR_{II}: NR_I, when phosphorylated, acts as a transcriptional activator of the gene *glnA*; NR_{II} is a kinase that converts NR_I to NR_I-P, the phosphorylated, active form. Another protein, P_{II}, prevents this phosphorylation, thereby preventing transcriptional activation. P_{II} can be uridylylated at Tyr^{51} to P_{II}-UMP by a **uridylyl transferase**, and then does not inhibit NR_I phosphorylation, so activating transcription. This occurs when the level of nitrogen is low. Regulation of glutamate–ammonia ligase at the structural level involves adenylylation by adenylyl transferase, which inactivates it. Adenylyl transferase is activated by P_{II}. In the presence of P_{II}-UMP, adenylyl transferase is inactivated, and glutamate–ammonia ligase reverts to a non-adenylylated, active form. Uridylylation of P_{II} thus results in activation of transcription of glutamate–ammonia ligase, and its activation at the structural level.

nitroglycerin 1,2,3-propanetriol trinitrate; a compound that is used industrially in the manufacture of dynamite, and pharmacologically as an antianginal and coronary vasodilator. Its action in the latter case results from its conversion to **nitric oxide**.

nitrophenyl a group that is linked to a variety of compounds to form artificial substrates for enzymes. The nitrophenol released by the enzyme action can then be measured spectrophotometrically at 400 nm (4-nitrophenol) or 420 nm (2-nitrophenol) to follow the enzyme reaction, the esterified form having a maximal absorption at a much lower wavelength. Typical substrates are 4-nitrophenylacetate (for carboxylesterase, lipase), 4-nitrophenyl-*N*-acetylgalactosaminide (galactosaminidase), 4-nitrophenyl-*N*-acetylglucosaminide (glucosaminidase), and 2-nitrophenylbutyrate (cholinesterase).

nitrophorin *abbr.*: NP; a monomeric hemoprotein ($\approx$20 kDa) secreted in saliva of the blood-sucking reduviid bug *Rhodnius prolixus*. There are four nitrophorins (NP 1 to 4), which are homologous and contain a conserved eight-stranded beta barrel. They carry nitric oxide, which causes smooth muscle relaxation and vasodilation in the host. They also bind histamine and reduce the inflammatory and immune responses in the host. **NP2** (also called prolixin-S) inhibits activation of clotting factor X.

nitroprusside test a test for free thiol groups based on the red coloration given on reaction of thiol-containing compounds with ammoniacal sodium nitroprusside, sodium nitrosopentacyanoferrate(iii), $Na_2[Fe(CN)_5NO]$. Nitroprusside also reacts with ketones and is used in tests for ketone bodies in urine. The proprietary Acetest tablets contain a mixture of glycine, sodium nitroprusside, disodium phosphate, and lactose. In the presence of glycine, acetoacetate or acetone forms a blue– purple colour. A positive reaction indicates the presence of $\geq$5 mg dL^{-1} and 10 mg dL^{-1} for urine and blood respectively.

nitros+ *a variant form of* **nitroso+** (before a vowel).

***N*-nitrosamide** any organic compound in which a **nitroso** group is attached to an amide nitrogen, e.g. *N*-methyl-*N*-nitrosourea. *See also **N*-nitrosamine**.

***N*-nitrosamidine** any organic compound in which a **nitroso** group is attached to an amidine nitrogen, e.g. *N*-methyl-*N'*-nitro-*N*-nitrosoguanidine. *See also **N*-nitrosamine**.

***N*-nitrosamine** any organic compound of general formula R_1R_2NNO. They are stable compounds formed by the action of nitrous acid on secondary amines with concurrent monodealkylation, and also on some tertiary amines. Most nitrosamines are strongly carcinogenic and their metabolites are strong mutagens. ***N*-nitrosamides** and ***N*-nitrosamidines** are unstable nitroso compounds that decompose in neutral or alkaline conditions to produce powerful alkylating agents. They are widely used as experimental mutagens. They are carcinogenic in any tissue with which they come into contact.

nitroso the chemical group –NO in a chemical compound.

nitroso+ *or (before a vowel)* **nitros+** *prefix (in chemical nomenclature)* indicating the presence of a **nitroso** group in a chemical compound; in *C*-nitroso compounds the –NO is attached to carbon (e.g. nitrosobenzene) and in *N*-nitroso compounds it is attached to nitrogen.

***S*-nitroso-*N*-acetylpenicillamine** *abbr.*: SNAP; an *S*-nitrosothiol reagent that stimulates guanylate cyclase and mimics the action of nitric oxide.

S-nitroso-*N*-acetyl-d-penicillamine

***S*-nitrosoglutathione** a nitrosating reagent that breaks down to nitric oxide and a glutathione radical. It induces the cellular effects of nitric oxide, including smooth muscle relaxation and inhibition of platelet aggregation.

Nitrotetrazolium Blue *or* **Nitroblue Tetrazolium** a dye very similar in structure to Tetrazolium Blue (*see* **tetrazolium salt**), differing only in that each of the two tetrazolium groups has a nitrophenyl group in place of a phenyl group.

nitrous acid the very unstable monobasic acid HNO_2. It reacts with amines, including purine and pyrimidine amines, to form the corresponding hydroxy compounds, and is thus mutagenic.

nitroxide the divalent group =N–O• in a molecule of an organic compound. It has an unpaired electron, and is useful in spin-labelling, its derivatives being stable free radicals.

NK cell *abbr. for* natural killer cell; a large granular lymphocyte that recognizes structures on the surface of virally infected cells, which it kills, probably through the mediation of a **perforin**.

NKSF *abbr. for* NK stimulating factor. *See* **interleukin 12**.

Nle *symbol for* the α-amino acid norleucine (now known as 2-aminohexanoic acid, *symbol*: Ahx).

NLM *abbr. for* National Library of Medicine.

NLP *abbr. for* natural language processing.

NLS *abbr. for* nuclear localization signal.

nm *symbol for* nanometre.

nM *symbol for* nanomolar.

NMD *abbr. for* nonsense-mediated decay.

NMDA *abbr. for* *N*-methyl-D-aspartate.

NMDA receptor *see* **glutamate receptor**.

NMN *abbr. for* nicotinamide mononucleotide.

nmol *symbol for* nanomole.

NMP *abbr. for* any (unknown or unspecified) nucleoside monophosphate in which the phosphate group is esterified at the 5′ position of the glycose moiety.

NMR *or* **nmr** *abbr. for* nuclear magnetic resonance.

Nn *symbol for* the nonyl group.

Nno *symbol for* the nonanoyl group.

nNOS *see* **nitric-oxide synthase.**

NNRTI *abbr. for* non-nucleoside reverse transcriptase inhibitor.

No *symbol for* nobelium.

nociceptin *an alternative name for* **orphanin FQ.**

nociceptive causing pain or injury; reacting to a painful stimulus.

nociceptor any receptor that reacts to painful (i.e. nociceptive) stimuli; a pain-sense organ.

Nod a family of human intracellular proteins involved in detection of invasive bacteria and activation of the NF-κB transcription factor pathway. In epithelial cells, Nod1 is essential for sensing intracellular Gram-negative bacteria through a tripeptide motif in the bacterial peptidoglycan.

NodD a transcription factor of *Rhizobium* bacteria that, on binding an inducer secreted by infected root cells in a host plant, induces *nod* genes in the bacterium.

node 1 a knob, lump, or swelling; a constriction, especially when serially repeated (along the length of something). 2 *(in botany)* the point on a stem to which a leaf or a lateral branch is attached. 3 *(in physics)* a point of zero or minimum displacement in a standing wave or a system of such waves. *See also* **node of Ranvier.** —**nodal** *adj.*

node of Ranvier any of the constrictions in the myelin sheath that occur at regular intervals along the length of a myelinated nerve fibre. [After Louis Antoine Ranvier (1835–1922), French pathologist.]

Nod factor any bacterial product, made by enzymes encoded in *nod* genes of *Rhizobium* species, that directs nodulation in the host plant. Nod factors are lipopolysaccharides consisting of a short linear backbone of β1-4-linked *N*-acetylglucosamine residues, the first of which is usually *N*-acylated by a C_{16}, C_{18}, or C_{20} fatty acid of varying levels of unsaturation. The factors cause rapid cell division in infected root cortical cells of legumes to form the nodules involved in nitrogen fixation.

***nod* gene** any gene in *Rhizobium* bacteria that is responsible for early nodule formation in nitrogen-fixing plants. The expression of *nod* genes depends on an inducer (e.g. genistein or daidzein) from the host and an endogenous transcription factor (NodD) in the bacterium that is a receptor for the inducer. Many *nod* genes encode enzymes that direct the synthesis of bacterial **Nod factors** responsible for inducing nodulation in the host.

NOD mice *abbr. for* nonobese diabetic mice.

nodoc *a name sometimes used for* **anticodon** (formed by reversing the letters of the word codon).

nodulation factor a lipooligosaccharide secreted by *Rhizobium* species as an extracellular signal that induces nodule formation in the roots of a host legume plant. It consists of a linear chain of 3–6 (usually 5) *N*-acetylglucosamine residues joined β1-4 and with a (usually C_{18}) fatty acid attached to the N atom of C2 of the nonreducing end.

nodule bacterium or **nodule-forming bacterium** *a common name for* any bacterium of the genus *Rhizobium*, found as endosymbionts in root nodules. The bacterium specifically infects root cells of Leguminosae, and causes the development of **root nodules**, in which it loses its outer membrane and becomes dependent on plant metabolism. It is then morphologically differentiated from the free-living form and known as a bacteroid. In return it fixes nitrogen for the plant. *See* **nitrogenase.**

nodulin any protein that is produced within root nodules of nitrogen-fixing plants in response to **Nod factors** secreted by *Rhizobium* bacteria in infected cells. Their roles include acting as membrane channels for transport across the symbiosome membrane; as enzymes involved in metabolism of carbon and nitrogen; or in establishing a low-oxygen environment (e.g. leghemoglobin). Nodulins are described as early or late depending on when they are expressed during nodule formation.

NOE *abbr. for* nuclear Overhauser effect.

NOESY *abbr. for* nuclear Overhauser effect spectroscopy. *See* **NMR.**

nogalamycin an antitumour antibiotic material obtained from *Streptomyces nogalater*. It contains the carbohydrate component **nogalose.**

nogalose 6-deoxy-3-*C*-methyl-L-mannose 2,3,4-trimethyl ether; a carbohydrate component of **nogalamycin.**

noggin a dorsalizing factor localized to **Spemann's organizer** in *Xenopus laevis* embryos.

Nogo an abundant **myelin protein** that is produced by oligodendrocytes in the central nervous system. Like myelin-associated glycoprotein it inhibits neuronal regeneration. It has two transmembrane segments and an extracellular loop of 66 residues (sometimes called Nogo-66) that binds to the **Nogo receptor.**

Nogo receptor a phosphatidylinositol (PI)-anchored peripheral cell surface protein of neurons in the central nervous system. It binds the 66-residue extracellular loop of **Nogo**, and also myelin-associated glycoprotein and oligodendrocyte myelin glycoprotein. *See* **myelin protein.**

noise any undesired disturbance in the signal from an instrument, or other system, that degrades the amount of useful information extractable. *Compare* **signal-to-noise ratio.**

noise analysis the technique whereby the amplitude and frequency of random fluctuations in the membrane potential of an excitable cell are used to investigate changes in conductance of single ion channels induced, e.g., by a drug or neurotransmitter.

nojirimycin any of a group of carbohydrate-like alkaloids or imino sugars originally found in a *Streptomyces* filtrate but also found in plants such as *Morus alba* and *Angylocalyx pynaertii*. Nojirimycins are inhibitors of glycosidases and glycosyltransferases. Derivatives such as *N*-butyldeoxynojirimycin inhibit: (1) α-glucosidase, involved in the processing of N-linked glycoprotein chains; and (2) glucosylceramide transferase, which catalyses the first step in one of the biosynthetic pathways leading to glycosphingolipids and gangliosides.

nominal describing an approximate value assigned to some quantity as a guide to its real value (which may vary according to circumstances).

nomogram or **nomograph** any chart or diagram consisting of scaled lines and used for facilitating calculations, especially one consisting of three lines scaled for different related variables such that a straight line joining chosen values of two variables will intersect the third scale line at the related value of the third variable. —**nomography** *n.*; **nomographic** *adj.*

non+ 1 *prefix* denoting negation; absence of; opposite or reverse of; e.g. nonabsorbable, nonacylated, nonconductor, noncyclic, nonmetal, nonspecific. 2 *a variant form of* **nona+** *(sometimes before a vowel).*

nona+ or *(sometimes before a vowel)* **non+** *comb. form* denoting nine, ninefold, nine times.

nonacosa+ *comb. form* denoting twenty-nine, twenty-nine fold, twenty-nine times.

nonacosane a linear C_{29} hydrocarbon, $CH_3-[CH_2]_{27}-CH_3$.

nonactin a neutral macrotetrolide antibiotic produced by several *Streptomyces* spp. It transports alkali cations through membranes with high selectivity for K^+, but also for NH_4^+. It is one of a related series including monactin, dinactin, and trinactin.

nonadeca+ *comb. form* denoting nineteen, nineteen fold, nineteen times.

nonaethyleneglycol mono-*n*-dodecyl ether a nonionic $C_{12}E_9$ detergent; *see* $C_x E_y$.

nonanoate *or* **pelargonoate 1** the anion $CH_3–[CH_2]_7–COO^-$, derived from nonanoic acid (i.e. pelargonic acid). **2** any salt or ester of nonanoic acid.

nonanoyl *or* **pelargonoyl** *symbol*: Nno; the univalent acyl group $CH_3–[CH_2]_7–CO–$, derived from nonanoic acid (i.e. pelargonic acid).

nonaqueous lacking water; applied especially to solvents whether polar or nonpolar.

nonbonding electrons *an alternative name for* **lone pair**.

noncoded amino acid *or* **uncoded amino acid** any α-amino-acid residue occurring in a polypeptide or protein for which no codon exists: e.g., γ-carboxyglutamic acid, hydroxylysine, hydroxyproline, oxyproline. Residues of these amino acids are formed from residues of the corresponding coded amino acids by post-translational modification of the polypeptide chain.

noncoding region *or* **noncoding sequence** any segment of a nucleic-acid molecule for which there is no recognized gene product; it may sometimes play a structural role in the organization of the molecule. *See also* **intron**.

noncoding strand *or* **template strand** a term that is variously, and confusingly, applied to either of the two strands of genomic duplex DNA. It is now the term recommended for denoting the strand that is complementary in its sequence of bases to messenger RNA (mRNA) transcribed from the DNA; it thus functions as the template for synthesis of mRNA, which then contains the sequence present in the **coding strand** of duplex DNA (except for uracil in RNA serving in place of thymine in DNA). In some bacteria and viruses, alternate segments of both strands of duplex DNA may be noncoding. *Other names*: antisense DNA; antisense strand; complementary strand; minus strand; transcribing strand. *Compare* **complementary DNA**.

noncompetitive antagonism *(in pharmacology)* the phenomenon occurring when agonist and antagonist can be bound simultaneously, in which antagonist binding reduces or prevents the action of the agonist; e.g. channel block of the nicotinic receptor, or inhibition by adrenoceptor antagonists of the response to tyramine.

noncompetitive inhibition inhibition of an enzyme in which inhibition results from binding of the inhibitor at a different site on the enzyme surface from that at which the substrate binds, without interference with substrate binding.

nonconjugative plasmid any plasmid that cannot bring about transfer of DNA by **conjugation**.

nonconservative replication of DNA *or* **dispersive replication of DNA** *see* **replication of DNA**.

noncoupled solute translocation the translocation of a solute across a membrane that is not coupled to the translocation of another solute, in either the same or the opposite direction. *See* **facilitated diffusion, uniport**.

noncovalent describing an interaction between atoms and/or molecules that does not involve the formation of a covalent bond between them.

noncyclic electron flow *(in photosynthesis) see* **photophosphorylation**.

noncyclic photophosphorylation *see* **photophosphorylation**.

nondiffusible material material that is not diffusible in the restricted sense of its being incapable of passing through a diffusion barrier, especially through a semipermeable membrane used in dialysis. *Alternative terms recommended by IUPAC*: retentate, dialysis residue, or residue.

nondisjunction *(in genetics)* failure of a chromosome pair to separate at the first meiotic division or for two chromatids to separate at the second meiotic division, so that both daughter chromatids pass into the same daughter cell. Autosomal trisomies are thought to originate from fertilization of eggs with two daughter chromatids.

nondissymmetric *a term formerly used to describe* an achiral molecule that possesses an alternating axis of symmetry but that (unlike a symmetric molecule) lacks a simple axis of symmetry; the converse of **dissymmetric**. An important example is **citric acid**, which has a one-fold alternating axis but no simple axis; thus its pair of carboxymethyl groups on carbon-3 are not geometrically equivalent. *See also* **prochiral**.

nonelectrolyte any substance that is neither an electrolyte nor a conductor of electrons.

nonessential amino acid any amino acid that can be synthesized in the body in an amount corresponding to need, and hence for which there is not an absolute dietary requirement. The extent to which a particular amino acid is non-essential in any one species depends on the stage of development and physiological state of the subject. The following L-amino acids are non-essential for the maintenance of nitrogen equilibrium in an adult human: alanine, arginine, asparagine, aspartic acid, cysteine, glutamic acid, glutamine, glycine, histidine, hydroxyproline, proline, serine, and tyrosine. Arginine and histidine are, however, essential in growing children.

nonesterified fatty acid *abbr.*: NEFA; **1** *an alternative name for* **free fatty acid**. **2** *a collective term for* the total of free fatty acids in a sample, especially of blood.

nonheme iron *or (esp. Brit.)* **nonhaem iron** any iron in a biological sample or structure that is not chelated in heme groups. In non-heme-iron proteins, the iron is often present in iron–sulfur clusters.

nonhistone protein *or* **nonhistone chromosome protein** any member of a highly heterogeneous group of 10–150 kDa tissue-specific proteins associated with DNA in cell nuclei. They are thought to play a part in the control of gene expression. *See also* **high mobility group proteins**.

nonhomologous (of chromosomes or chromosome segments) not **homologous** (def. 4).

nonideal describing any gas or solution that does not behave like an ideal gas or ideal solution, respectively. —**nonideality** *n*.

Nonidet P-40 *the proprietary name for* a nonionic detergent, (octylphenoxy)-polyethoxyethanol, used for solubilizing membrane proteins. It is no longer commercially available but can be replaced with IGEPAL CA-630, which is chemically identical. *Compare* **Triton**.

non-insulin-dependent diabetes *an alternative name for* type 2 **diabetes mellitus**.

noninvasive 1 not **invasive**. **2** *(of a clinical investigative technique or procedure)* carried out without penetrating the skin.

nonionic having or yielding no ions; the term is used especially of **nonionic detergents**.

nonionic detergent any detergent whose molecules cannot ionize.

non-lactose-fermenter *a jargon term for* any genus, species, or strain of microorganism that does not ferment lactose.

nonmediated transport *an alternative name for* **unmediated transport**.

non-Newtonian fluid any fluid that does not conform to **Newton's law of viscosity**, i.e. one whose (coefficient of) viscosity is not independent of flow rate.

non-nucleoside reverse transcriptase inhibitor *abbr.*: NNRTI; an antiretroviral drug, used in AIDS therapy, that binds to the catalytic site of the HIV-1 reverse transcriptase, preventing RNA conversion to DNA. Owing to the rapid development of resistance, it is often used in combination with other drugs. NNRTIs have no activity against HIV-2. *Compare* **nucleoside reverse transcriptase inhibitor**.

nonoate A compound of structure $R_1R_2NN(O)=NOR_3$, which releases nitric oxide in water. Different derivatives have different rates of NO release e.g. DETA NONOate, (*Z*)-1-[2-(2-aminoethyl)-*N*-(2-ammonioethyl)amino]diazel-1-ium-1,2-diolate.

nonorthologous gene displacement during evolution, the displacement of a particular gene by an unrelated (i.e. nonorthologous) but functionally analogous gene.

nonose any **aldose** having a chain of nine carbon atoms in the molecule.

nonoverlapping code *see* **overlapping code**.

nonpenetrance *(in genetics)* the lack of the expected phenotype in individuals who are heterozygous for an autosomal dominant gene.

nonpermissive restrictive; not **permissive**.

nonpermissive cell any cell that does not support infection by a particular lytic virus; in a small percentage of instances the virus may transform a nonpermissive cell.

nonpolar 1 lacking polarity. 2 or **apolar** *(of a chemical group or substance)* lacking a significant permanent dipole moment.

nonpolar bond 1 any covalent bond that is not appreciably polarized. 2 *(sometimes)* an alternative term for **hydrophobic interaction**.

nonpolar solvent *a loose term denoting* any solvent with low ionizing power and hence low solvating power.

nonproductive binding *or* **dystopic binding** the binding of a substrate in an unreactive mode at the active site of an enzyme, leading to the formation of a nonproductive, or **abortive complex**.

nonproductive complex *an alternative name for* **abortive complex**.

nonprotein amino acid any amino acid that does not naturally occur as a constituent residue of proteins. Over 200 such amino acids are known to be produced in higher plants; although they often contribute to the nitrogen pool of the plants, their specific function is unknown. Non-protein amino acids also occur in animals, but to much less extent, the most prominent examples being **argininosuccinate**, **citrulline**, and **ornithine**.

nonprotein nitrogen *abbr.*: NPN; the part of the nitrogen content of a substance (or fluid) that is not contained in the proteins present. The term is used especially of blood or serum, wherein urea accounts for about half the total NPN.

nonradioactive having no radioactivity; in practice the term is often used to refer to substances having a normal or natural nuclidic composition, ignoring any contribution from minimal proportions of naturally occurring radionuclides such as the carbon-14 content of natural organic materials.

nonradioactive labelling any method for coupling a distinctive and detectable molecular entity to a DNA, RNA, protein, or other molecule, that permits detection of this molecule by nonradioactive means. Such labelling might involve, for example, the introduction of haptens that can subsequently be detected by anti-hapten antibodies coupled to an enzyme that can be detected by colorimetric, fluorimetric, or chemiluminescent methods.

nonradioactive tracer *(sometimes)* a labelled substance that can be detected and estimated by virtue of some property other than radioactivity (such as colour, fluorescence, or a difference in stable-isotope ratio).

nonreducing end the end of a linear oligosaccharide or polysaccharide that does not carry a potential hemiacetal or hemiketal (i.e. reducing) group.

nonribosomal peptide synthesis *abbr.*: NRPS; the type of peptide synthesis employed by actinomycetes, bacilli, and filamentous fungi in manufacturing complex peptide antibiotics, such as cyclosporin, gramicidin, erythromycin, and vancomycin. No mRNA, tRNA, or ribosomes are required. Instead, a multimodular nonribosomal peptide synthetase determines the composition and sequence of the peptide to be synthesized.

nonsecretor *see* **secretor gene**.

nonselectively labelled describing a labelled compound in which the position(s) and the number of the labelling nuclide(s) are both undefined.

nonsense codon *or* **nonsense triplet** any one of the nucleotide triplets UAA (ochre), UAG (amber), and UGA (opal) that signal termination of translation and release of the polypeptide chain from the site of synthesis. A codon that is nonsense in one strain of an organism may nevertheless code for an amino acid in another strain containing an appropriate suppressor gene (*see* **nonsense suppressor**). UGA sometimes codes for selenocysteine and, in mitochondria, for Trp. In mitochondria AGA is a nonsense codon rather than UGA. *Compare* **missense codon**.

nonsense-mediated alternative splicing *abbr.*: NAS; a form of alternative splicing of a pre-mRNA containing a nonsense(i.e. stop) mutation in an internal exon, in which alternative splicing occurs so as to produce a stable mRNA that lacks the mutation. *Compare* **nonsense-mediated decay**.

nonsense-mediated decay the selective degradation in cytoplasm of eukaryotic mRNA that contains a premature stop codon. This reduces the production of abnormally truncated and potentially harmful proteins. The decay is triggered by a surveillance complex. *See* **mRNA surveillance**.

nonsense mutation any mutation in a DNA molecule that results in an alteration of a sense codon to a **nonsense codon** in the mRNA transcript, the translation of which leads to the formation of a truncated polypeptide gene product. *Compare* **missense mutation**.

nonsense suppressor any tRNA molecule that is able to insert an amino-acid residue into a growing polypeptide chain at a nonsense codon in an mRNA molecule. It has an altered anticodon, produced as a result of a **suppressor mutation**.

nonshivering thermogenesis *abbr.*: NST; a mechanism of rapid heat production in response to cold that does not involve muscular contractions. It is important during the neonatal period, during cold adaptation of a number of small mammals, and during the arousal period of animals that hibernate. Heat is produced in the mitochondria of brown adipose tissue in response to the presence in the inner mitochondrial membrane of a specific 32 kDa protein. This protein uncouples the respiratory chain and is inducible by norepinephrine. It is known as **brown fat uncoupling protein**.

nonspecific serine/threonine protein kinase EC 2.7.11.1 a serine/threonine kinase that catalyses phosphorylation of several proteins, using ATP; other names glycogen synthase *a* kinase; cAMP-dependent protein kinase; calcium-dependent protein kinase C. It is activated by phosphorylation by cyclic AMP-dependent protein kinase.

nonsuppressible insulin-like activity *abbr.*: NSILA; **insulin-like activity** that is not suppressible by specific antisera to insulin. The activity is now attributed to **insulin-like growth factors**.

nonsynonymous mutation a nucleotide substitution in a protein-coding gene that results in an amino acid substitution in the translation product. The ratio of nonsynonymous to synonymous substitutions may be used to detect positive Darwinian selection.

nontranscribing strand (of duplex DNA) *see* **coding strand**.

nonulose any **ketose** having a chain of nine carbon atoms in the molecule.

nonyl *symbol*: Nn; the alkyl group $CH_3[CH_2]_7CH_2-$, derived from nonane.

nonyl glucoside *n*-nonyl-β-D-glucopyranoside; a mild, nonionic detergent, CMC 6.5 mm. *Compare* **heptyl glucoside**, **hexyl glucoside**, **octyl glucoside**.

Noonan syndrome *or* **cardiofaciocutaneous syndrome** an autosomal dominant disorder in which cardiac anomalies are associated with short stature, facial abnormalitiy, and webbed neck. It is associated with mutations in PTNP11 (protein tyrosine phosphatase 11). [After Jacqueline Noonan (1921–), US cardiologist.]

nopaline *N*-(1-carboxy-4-guanidinobutyl)glutamic acid; N^2-(1,3-dicarboxypropyl)arginine; a rare amino-acid derivative that is produced by a certain type of crown-gall tissue (*see* **crown-gall disease**); one of a group of **opines**. The genes responsible for the synthesis of nopaline are part of the **T-DNA** from a **Ti-plasmid**.

nopalinic acid *or* **ornaline** *N*-(1-carboxy-4-aminobutyl)glutamic acid; N^2-(1,3-dicarboxypropyl)ornithine; one of a group of **opines** found in plant tumours occurring in **crown-gall disease**.

nopalinic acid

nor+ *prefix (in chemical nomenclature)* denoting **1** replacement of one side-chain methyl group by a hydrogen atom, e.g. norepinephrine, 19-nortestosterone. Replacement of two or three methyl groups is denoted by the prefix dinor+ or trinor+, respectively (in terpene nomenclature the prefix nor+ formerly denoted removal of all replaceable methyl groups; hence, e.g., 8,9,10-trinorbornane was formerly known as norbornane). **2** elimination of one methylene group from a chain or ring of carbon atoms; it is used especially for steroids, e.g. 23-nor-5β-cholanoic acid, A-nor-5β-androstane, norophthalmic acid. Elimination of two or three methylene groups is denoted by the prefix dinor+ or trinor+, respectively. **3** an isomer with an unbranched hydrocarbon chain, e.g. norleucine, norvaline; use discouraged.

noradrenaline *an alternative name for* **norepinephrine**.

noradrenergic *(sometimes)* **1** an (adrenergic) nerve cell that itself liberates noradrenaline (i.e. **norepinephrine**). **2** a cell or receptor that is stimulated preferentially by norepinephrine. —**noradrenergically** *adv.*

norepinephrine *or* **noradrenaline** *or* **levarterenol** (–)-(3,4-dihydroxyphenyl)-2-aminoethanol; a hormone secreted by the adrenal medulla, and a neurotransmitter in the sympathetic peripheral nervous system and in some tracts in the central nervous system. It is also the demethylated biosynthetic precursor of epinephrine in the adrenal medulla. It has a powerful lipolytic action, mediated by cyclic AMP, and is an agonist for **adrenoceptors**, inducing arteriolar constriction and raised systolic and diastolic arterial blood pressure.

R-(–)-enantiomer

norepinephrine transporter *or* **noradrenaline transporter** an integral membrane protein of neuronal synapses that contains 12 transmembrane segments and is responsible for reuptake of most of the norepinephrine released at synapses. It is inhibited by cocaine, amphetamines, and tricyclic antidepressants, all of which cause features of **orthostatic intolerance**.

Norit *the proprietary name for* a commercially produced activated carbon, useful as an adsorbent.

norleucine *symbol*: Ahx *or (formerly)* Nle; 2-aminohexanoic acid; a non-protein α-amino acid. *See also* **nor+** (def. 3).

norm+ *a variant form of* **normo+** *(before a vowel)*.

normal 1 not deviating from an established pattern; approximating to an average, or within the usual range (of values). **2** *(in organic chemistry)* symbol: n; (of a solution) containing one **equivalent** (def. 4) of a specified substance (of specified equivalence factor) in one litre of solution. Use now not recommended. **3** *(in organic chemistry)* a chemical compound that has an unbranched chain of carbon atoms; the names of such compounds may be prefixed by '*n*', as in *n*-butane. However, this is not the currently recommended terminology. **4** *(in biology)* not diseased or not having been exposed to

an experimental procedure. **5** *(in mathematics)* (of a straight line or plane) at a right angle (to another straight line or plane).

normal distribution *or* **Gaussian distribution** any continuous frequency distribution of events, observations, etc. characterized by a symmetrical bell-shaped curve of the frequency of *x*, f(*x*), against the value of *x*, that is represented by the equation:

$$f(x) = [1/\sigma(2\pi)^{0.5}]e^{-(x-\mu)^2/2\sigma^2}$$

where μ is the mean and σ is the standard deviation.

normality *(in chemistry)* the chemical concentration of a solution expressed as the number of **equivalents** (def. 4) of a specified substance (of specified equivalence factor) per litre. Use now not recommended.

normalize *or (esp Brit.)* **normalise 1** *(for a relational database)* to organize data into tables in such a way as to minimize duplication and hence increase efficiency. **2** *(for a probability distribution)* to divide all probabilities by a constant factor so that their sum over all possible alternatives is 1. **3** *(for microarray data)* to adjust the measured intensities to correct for experimental bias. **4** *(for a cDNA library)* to generate a library such that all the genes in the library are represented at the same frequency. –**normalized** *adj.*; –**normalization** *n.*

normalized library *see* **normalize**.

normal range *(in clinical chemistry)* the range of values of a particular characteristic within which 95 percent of observations made on presumed normal healthy individuals fall. It is useful as a reference in the investigation of disease.

normal saline *an alternative and less suitable name for* **physiological saline**.

normal solution a solution in which the amount-of-substance concentration of the **equivalent** (def. 3) of the reagent is one mole per litre; the equivalence factor of the reagent should be specified. Not now recommended. *See also* **normal** (def. 2).

normal temperature and pressure *abbr.*: NTP *or* N.T.P. *or* ntp; *a former term for* **standard temperature and pressure**.

normetadrenaline *an alternative name for* **normetanephrine**.

normetanephrine *or* **normetadrenaline** 3-*O*-methylnorepinephrine; an inactive metabolite of norepinephrine produced by the action of **catechol *O*-methyltransferase**.

normo+ *or (before a vowel)* **norm+** *comb. form* denoting normal; used especially in biology and medicine to indicate within the normal range. *See also* **eu+**.

normoblast a normal precursor cell to an erythrocyte, characterized by condensation of the nucleus into a homogeneous densely staining body. Normoblasts are normally found only in hemopoietic tissue. —**normoblastic** *adj.*

normocyte an erythrocyte of normal colour, shape, and size. —**normocytic** *adj.*

normoglycemia *or (esp. Brit.)* **normoglycaemia** an alternative term for **euglycemia**. —**normoglycemic** *or (esp. Brit.)* **normoglycaemic** *adj.*

normotensive having an arterial blood pressure within the normal range for the particular group to which the subject belongs.

Norrie disease *see* **Norrin**.

Norrin the protein (133 amino acids) encoded by the gene associated with the X-chromosome-linked Norrie disease. The human protein is highly expressed in retina, cerebellum, and olfactory epithelia, and shares 94% sequence identity with the mouse homologue. The secreted portion (residues 25–133) contains a cysteine-rich C-terminal domain. Numerous mutations result in congenital blindness, frequently associated with mental retardation and deafness, and occasionally with hypogonadism and microcephaly.

norsteroid any steroid that, relative to a parent compound, lacks either one axial methyl group or one methylene group in a side chain, or that has one ring contracted by one carbon atom. *See also* **nor+**.

Northern blotting *or* **Northern transfer** any **blotting** in which the primary electrophoresis is performed with RNA, which is subsequently hybridized to radioactive DNA. The term was coined to distinguish the process from **Southern blotting**, which is performed with DNA.

norvaline *symbol*: Avl (from aminovaleric) *or* Ape (from aminopentanoic) *or (formerly)* Nva; 2-aminovaleric acid (now the preferred trivial name); 2-aminopentanoic acid; a non-protein α-amino acid. *See also* **nor+** (def. 3).

NOS *abbr. for* **nitric-oxide synthase** .

noso+ *or (before a vowel)* **nos+** *comb. form* indicating disease.

nosocomial 1 associated with a hospital. **2** a disorder (e.g. an infection) acquired during a hospital stay that is unrelated to the primary condition of the patient. A nosocomial infection can refer to hospital workers as well.

nosogenesis *an alternative term for* **pathogenesis.** —**nosogenic** *or* **nosogenetic** *adj.*

nosography a written systematic classification and description of diseases.

nosology the branch of medicine dealing with the classification of diseases.

notatin *a name sometimes still used for* glucose oxidase. Notatin was so called because it is produced in considerable quantity by some microfungi, e.g. *Penicillium notatum.*

notch a gene, found in *Drosophila*, that encodes a protein essential for the proper differentiation of ectoderm. Expression of the *notch* gene is under the control of neurogenic genes. The gene product is a transmembrane protein (2703 amino acids) with 36 EGF-like repeats.

notexin a **phospholipase** A₂ (EC 3.1.1.4) from snake venom that acts as a presynaptic neurotoxin. The reaction is Ca²⁺-dependent.

***Not*I** a type 2 **restriction endonuclease**; recognition sequence: GC↑GGCCGC.

novel chemical entity *abbr.*: NCE; a molecule with a novel structure that is newly identified as a potential drug (and hence has not been described before in the literature).

noviose 6-deoxy-5-*C*-methyl-L-*lyxo*-hexose 4-methyl ether; a carbohydrate component of **novobiocin**.

novobiocin an antibiotic produced by *Streptomyces spheroides, S. niveus*, and certain other species; it is active against Gram-positive and some Gram-negative bacteria. In susceptible organisms it is a specific inhibitor of **(DNA) gyrase**, interacting with the B subunit, and thereby interfering with DNA synthesis.

noxa (*pl.* **noxae**) anything that is harmful to living organisms. —**noxious** *adj.*

Np *symbol for* neptunium.

NP *abbr. for* **1** neuropilin. **2** nitrophorin.

NP-40 polyethyleneglycol-*p*-isooctylphenyl ether; a polyoxyethylene nonionic detergent, CMC 0.05–0.3 mm. *Proprietary name*: Nonidet P40.

NPAS2 *abbr. for* neuronal PAS domain protein 2; a basic helix-loop-helix transcription factor that has two **PAS domains**, each of which contains a heme group. It forms a dimer with **BMAL1** through which it binds to a specific DNA motif and directs transcription of over a hundred genes, including those for **period** and cryptochrome proteins. NPAS2 is homologous with **clock** and like it controls circadian activity and is regulated by the redox state of NAD(P). It has

the same role in the forebrain as clock has in the hypothalamus. CO binds the heme of NPAS2 and prevents it from dimerizing with BMAL1.

NPC *abbr. for* neural precursor cell (*see* **Nedd**).

NPN *abbr. for* nonprotein nitrogen.

NPY *abbr. for* neuropeptide Y.

Nramp1 *abbr. for* natural resistance-associated protein 1; a transmembrane protein characteristic of macrophages and associated with natural resistance to intracellular parasites (e.g. *Leishmania*, some mycobacteria) in mice. The human homologue is **NRAMP1**, in which structural variation is associated with susceptibility to pulmonary tuberculosis and possibly leprosy. NRAMP1 is homologous with **NRAMP2**, a proton-coupled divalent-metal transporter for Fe²⁺ in the brush border of the proximal small intestine. Mutations in Nramp2 cause microcytic anemia in mice and the Belgrade rat (which is anemic). Mutation in human NRAMP2 causes hemochromatosis in a subset of such patients.

nrdb *abbr. for* nonredundant database; *see* **composite database**.

N repeat *an alternative name for* **asparagine repeat**.

nRNA *abbr. for* nuclear RNA.

NR protein *abbr. for* nitrogen regulatory protein.

NRPS *abbr. for* nonribosomal peptide synthesis.

NRSF *abbr. for* neuronal restricted silencing factor. *See* **REST**.

NRTI *abbr. for* nucleoside reverse transcriptase inhibitor.

***Nru*I** a type 2 **restriction endonuclease**; recognition sequence: TCG↑CGA.

ns *symbol for* nanosecond (i.e. 10⁻⁹ s).

NSF *abbr. for* National Science Foundation (of the USA).

NSF protein *abbr. for* *N*-ethylmaleimide-sensitive fusion protein.

NSILA *abbr. for* nonsuppressible insulin-like activity.

NS-1 myeloma cells a cell line used for the production of monoclonal antibodies by fusion with splenocytes. NS-1 cells are deficient in the gene for **HGPRT** (HPRT) and are killed in the presence of aminopterin. In hybridomas the lack of HGPRT in NS-1 cells is genetically complemented by the gene from mouse splenocytes, and as a consequence they survive selection in culture medium containing **HAT**.

NSP *abbr. for* *N*-succinimidyl propionate, a propionylating agent.

NST *abbr. for* nonshivering thermogenesis.

NT *abbr. for* neurotrophin.

N terminus *or sometimes* **N terminal** the end of any peptide chain in any (poly)peptide or protein at which the 2-amino function of a constituent α-amino acid (or the 2-imino function of an α-imino carboxylic acid) is not attached in peptide linkage to another amino-acid residue. This function may bear an acyl group or other function in non-peptide linkage; when it is a free amino group (and also protonated and positively charged at neutral or acidic pH values), the chain end may then be alternatively termed the **amino terminus** (*or* NH₂ **terminus**) *or* **amino terminal** (*or* NH₂ **terminal**). However, 'amino terminus' is often used loosely for any N terminus. *Compare* **C terminus**. —**N-terminal** *adj.*

NTP 1 *abbr. for* any (unknown or unspecified) nucleoside (5′-) triphosphate. **2** *(in clinical biochemistry)* *abbr. for* 5′-nucleotidase (EC 3.1.3.5). **3** *or* N.T.P. *or* ntp; *abbr. for* normal temperature and pressure; stp (i.e. standard temperature and pressure) now used.

NTPase *abbr. for* nucleoside triphosphatase.

NTP polymerase *see* **polynucleotide adenylyltransferase**.

nu *symbol*: ν (lower case) *or* N (upper case); the thirteenth letter of the Greek alphabet. For uses *see* **Appendix A**.

nu body *or* **ν body** *or* **deoxyribosome** a type of spheroid particle seen on electron-microscopic examination of chromatin; now thought to be the same as a **nucleosome**.

Nuc *symbol for* an undetermined, unidentified, or unspecified nucleoside residue; alternative to N.

nucle+ *a variant for of* **nucleo+** *(before a vowel)*.

nuclear 1 of, pertaining to, or contained within the nucleus of a cell. **2** of, pertaining to, or concerning an atomic nucleus. **3** of, pertaining to, or operated by energy derived from fission or fusion of atomic nuclei.

nuclear area *an alternative name for* **nucleoid** (def. 2).

nuclear assembly protein 1 *see* **NAP1**.

nuclear atom *or* **central atom** (of a coordination entity) the atom to which all other atoms are directly attached.

nuclear body (*abbr.*: NB) *or* **PML nuclear body** *or* **PML oncogenetic domain** a spatial domain or subcompartment in the mammalian nucleus that contains **PML** bound to several other proteins. It becomes disrupted in acute promyelocytic leukemia, associated with formation of the PML-RARα fusion protein, and by some viruses, neoplasia, and inflammation. There is some evidence that nuclear bodies are sites for protein degradation.

nuclear division *or* **nuclear fission** the division of the nucleus of a cell, especially by **mitosis** or **meiosis**.

nuclear emulsion a fine-grained photographic emulsion designed for recording the tracks of subatomic particles within it.

nuclear envelope *or* **nuclear membrane** the membrane that surrounds the nucleus of a eukaryotic cell. It is a double membrane composed of two lipid bilayers separated by a gap of width 20–40 nm (the perinuclear space). The outer nuclear membrane is continuous with the endoplasmic reticulum of the cell and is sometimes studded with ribosomes. At specialized regions (**nuclear pores**) the outer nuclear membrane is connected to the inner nuclear membrane.

nuclear filament-related protein *abbr.*: NUF1; an essential component of the nucleoskeleton, with a role in cross-linking filaments or anchoring other molecules.

nuclear fission 1 *(in biology)* an alternative term for **nuclear division**. **2** *(in physics) see* **fission** (def. 1).

nuclear fraction *an operational term for* the material deposited by low-speed centrifugation of a homogenate of eukaryotic cells or tissue. It consists predominantly of nuclei together with unbroken cells.

nuclear fusion 1 *(in biology)* the **fusion** (def. 3) of the nuclei of a heterokaryon leading to the formation of a hybrid cell. **2** *(in physics) see* **fusion** (def. 4).

nuclear *g*-factor *symbol* g_N or nuclear g-value, a parameter that defines the energy of interaction of a nuclear spin with an applied magnetic field.

nuclear *g*-value alternative term for nuclear g-factor. Compare **G-value**.

nuclear isomer *(in physics) see* **isomer** (def. 2).

nuclear lamina the electron-dense layer lying on the nucleoplasmic side of the inner membrane of a cell nucleus. The polypeptides of the lamina are thought to be concerned in the dissolution of the nuclear envelope and its re-formation during mitosis.

nuclear localization signal *abbr.*: NLS; a sequence of amino acids seven to nine residues long within a protein, normally at sites away from the N or C terminus, that acts as a signal for the localization of the protein within the nucleus. Characteristically, NLSs are rich in basic residues, but include also one or more prolines.

nuclear magnetic resonance *abbr.*: NMR *or* nmr; the phenomenon that occurs when atomic nuclei that possess a magnetic moment are placed in a constant magnetic field of high intensity and are simultaneously exposed to electromagnetic radiation of an appropriate frequency in the radiofrequency region, whereby the nuclei are caused to change from a low-energy to a high-energy state with the absorption of energy. Nuclei that display this phenomenon include those of the nuclides protium, tritium, carbon-13, nitrogen-15, fluorine-19, and phosphorus-31; all these nuclei have nonzero spins and, therefore, magnetic moments. *See also* **nuclear magnetic resonance spectroscopy**.

nuclear magnetic resonance spectroscopy *abbr.*: NMR spectroscopy *or* nmr spectroscopy; absorption spectroscopy in the radiofrequency region based on the phenomenon of **nuclear magnetic resonance**. Unless otherwise qualified the term usually denotes nuclear magnetic resonance spectroscopy of protium atoms, also known as proton magnetic resonance spectroscopy (*abbr.*: pmr

spectroscopy). Nuclei of the same nuclide in different chemical environments give rise to distinguishable spectral lines, the positions, intensities, widths, and multiplicities of which give considerable information about the structure of complicated molecules containing such nuclei. The technique is also useful for determination of the concentrations of particular substances in complex environments, including that of an intact mammal. *Compare* **electron spin resonance spectroscopy**, **magnetic resonance imaging**.

nuclear magneton *symbol*: μ_N; a quantity of magnetic moment, given by $\mu_N = eh/4\pi m_p = (m_e/m_p)\mu_B$, where e is the elementary charge, h is the Planck constant, m_e and m_p are the rest masses of an electron and a proton respectively, and μ_B is the **Bohr magneton**. It is a fundamental physical constant, of value $5.050\ 786\ 6(17) \times 10^{-27}$ J T^{-1}.

nuclear matrin *see* **matrin**.

nuclear matrix *or* **nuclear scaffold** the insoluble protein material left in the nucleus after a series of extraction procedures.

nuclear medicine the branch of medicine concerned with the use of radioactive nuclides for diagnosis or treatment of disease, or for medical research.

nuclear membrane *an alternative name for* **nuclear envelope**.

nuclear mitotic apparatus protein *abbr.*: NuMA; one of a family of abundant coiled-coil nuclear proteins located in the nucleus during interphase, but rapidly redistributes to the developing spindle poles when the nuclear envelope disassembles in prometaphase.

nuclear Overhauser effect *or* **nuclear Overhauser enhancement** *abbr.*: NOE; in nuclear magnetic resonance spectroscopy, a phenomenon in which saturation of the resonance of a nucleus or electron by the simultaneous application of a radiofrequency field causes, under appropriate conditions, a considerable increase in the polarization of the interacting nuclei and hence in the NMR signal produced.

nuclear Overhauser effect spectroscopy *abbr.*: NOESY; in **nuclear magnetic resonance spectroscopy** advantage can be taken of the fact that the energy states of the spins of two nuclei subject to the **nuclear Overhauser effect** (NOE) are slightly perturbed by the mutual dipolar coupling (the effects of J-coupling are ignored). A perturbation of the populations of the states of one of the spins will lead to a transfer of the populations between the upper and lower energy states of the second spin, thereby modifying the resonance intensity. The NOE can be either positive or negative depending on the nuclei involved and the rapidity of the relative motions of the two spins. The strength of the dipolar coupling is distance dependent. Thus measurements of the NOE between two nuclei can lead to the estimation of the distance between them. Conversely, if the distance between the two nuclei is fixed (e.g. in a bond), then the NOE measurement can yield an indication of the dynamic motions in the molecule. Two-dimensional spectroscopy, known as 2D NOESY, is used to elucidate dipolar coupling networks present in complex biomolecules, using the NOEs between individual ^{1}H nuclei in the molecule. The network of NOEs obtained provides a basis for the determination of the three-dimensional structure in solution. Applications include the structures of small proteins and nucleic acids as well as protein–drug and protein–nucleic acid complexes.

nuclear pore any opening in the **nuclear envelope** of a eukaryotic cell, where the inner and outer nuclear membranes are joined. The nuclear pores are the central feature of a complex in higher eukaryotes and are estimated to contain about 100 different polypeptides. The complex constitutes a diffusion channel about 9 nm in diameter through which proteins enter the nucleus and ribonucleoproteins leave. Small proteins enter freely but the entry of larger proteins is controlled by their containing one or more clusters of basic amino acids which comprise the nuclear localization sequence (*abbr.*: NLS). This involves binding to the nuclear envelope, followed by energy-dependent transit through the complex. Four soluble proteins have been shown to assist in the process: karyopherin α, *other name*: importin α (the NLS receptor), karyopherin β, *other name*: importin β (which mediates binding to the nuclear pore complex), a small guanosine triphosphatase (*abbr.*: GTPase), Ran, and p10/NTF2 (a protein that may coordinate the Ran-dependent association and disassociation reaction). Ran cycles between a GDP- and GTP-bound form, which might account for the consumption of energy for the protein import. The exchange factor for Ran (RCC1) is located inside the nucleus whereas its GTPase-activating protein (Rab-GAP1 or RAN1) is cytoplasmic. Much less is known about the mechanism whereby the export of ribonucleoprotein is controlled. RNA itself does not seem to leave the nucleus.

nuclear quadrupole moment a parameter, given the symbol eQ, describing the effective shape of the ellipsoid of electric charge distribution of an atomic nucleus with spin greater than $\frac{1}{2}$. The nuclear electric quadrupole interacts with the electric field gradient eq at the nucleus, and its influence is observed in NMR, ENDOR, and Mössbauer spectroscopy.

nuclear receptor any intracellular protein that is specific for a ligand and elicits a response by binding to nuclear DNA without mediation of a second messenger. Some 48 are encoded in the human genome. They all contain a poorly conserved ligand-independent activation domain, a highly conserved DNA-binding domain that contains two zinc finger motifs, and a ligand-specific activation domain. They fall into three classes: (1) classic endocrine receptors (with high affinity for glucocorticoid, mineralocorticoid, steroid sex hormones, or thyroid hormones, and for 1,25-dihydroxy vitamin D); (2) adopted orphan receptors (e.g. those for retinoic acid and bile acids); and (3) orphan receptors that have no known physiological ligand.

nuclear remodelling factor *see* **NURF**.

nuclear rest *see* **nucleoid** (def. 4).

nuclear run-on *or* **nuclear run-off** a method in which nuclei isolated from cells are allowed to complete transcription of genes in the presence of radioactively labelled ribonucleotide triphosphates. Specific transcripts can be detected, and quantified, by hybridization with immobilized cDNA probes by dot blotting.

nuclear transformation transformation of a nuclide; the change of one nuclide into another.

nuclease any enzyme within subclass EC3.1, that catalyses the hydrolysis of ester linkages in nucleic acids. Nucleases with sugar specificity are referred to as ribonucleases or deoxyribonucleases; **endonucleases** create internal breaks, while 5'- and 3'-**exonucleases** remove nucleotide residues from the indicated end. Some nucleases are specific for double- or single-stranded nucleic acids.

nucleate 1 *or* **nucleated** possessing a nucleus. **2** to form, to act as, or to provide with, a nucleus or nuclei. **3** any salt of a nucleic acid.

nucleation 1 the act or process of nucleating or forming a nucleus. **2** *(in chemistry)* **a** the aggregation of molecules to form a polymer or a new phase within a medium. **b** (*also known as* **seeding**) the addition of a small quantity of finely divided crystalline material to a supersaturated solution in order to promote crystallization.

nuclei the plural of **nucleus**.

nucleic acid *abbr.:* NA; any single- or double-stranded polynucleotide of molar mass in the range 20 kDa to 40 GDa or more. Nucleic acids are either **deoxyribonucleic acids** (DNA) or **ribonucleic acids** (RNA). The phosphoric residue linking any constituent mononucleotide residue to the next bears one free hydroxyl group, which is weakly acidic. They are universal constituents of living matter and are concerned with the storage, transmission, and transfer of genetic information.

nucleics *an informal collective term sometimes used for* nucleotides, nucleosides, and nucleobases, whether constituents of nucleic acids or compounds of low M_r, and including their metabolic relatives and synthetic analogues.

nuclein the name given by J. F. Miescher (1844–95) to an unusual phosphorus-containing material that it isolated from the nuclei of pus cells, now known to have been **nucleoprotein**.

nucleinase *a name once proposed to designate* the enzymes now known as **nucleases**.

nucleo+ *or* (*before a vowel*) **nucle+** *comb. form* denoting **1** nucleus or nuclear. **2** nucleic acid.

nucleobase any nitrogenous base that is a constituent of a nucleoside, nucleotide, or nucleic acid.

nucleobindin a DNA-binding protein containing a signal peptide, leucine zipper, and basic amino-acid-rich regions. It was derived from a cell line from the lupus erythematosus-prone mouse MRL/T.

nucleocapsid the structure within a virus that comprises the proteinaceous capsid and the genomic nucleic acid. In enveloped viruses the nucleocapsid is surrounded by an envelope.

nucleocidin 4'-C-fluoroadenosine 5'-sulfamate; an analogue of adenylate produced by *Streptomyces calvus*, and the first naturally occurring derivative of a fluorosugar to be described. It has antibiotic activity against trypanosomes and inhibits translation in both eukaryotic and prokaryotic systems.

nucleocytoplasmic (of a structure or process) occurring between the nucleus and the cytoplasm; (of a property) relating the nucleus to the cytoplasm.

nucleodisome a structure consisting of two **nucleosomes** and isolable under certain conditions from some forms of chromatin.

nucleohistone any nucleoprotein consisting of DNA and histone(s).

nucleoid 1 resembling a nucleus. **2** the part or parts of a prokaryotic cell containing the genome and functionally equivalent to the nucleus of a eukaryotic cell. The meaning is sometimes extended to include the electron-dense inner zone observable in certain viruses. *Other names*: bacterial nucleus; chromatinic body; karyoid; nuclear area; nuclear body; nucleoid body; prokaryon. **3** the nucleus-like structure liberated from a eukaryotic cell by mild lysis. It contains histone-free supercoiled DNA within a flexible cage of RNA and protein. **4** *or* **nuclear rest** the nuclear residue sometimes observable in an erythrocyte. **5** the very dense core of a **microbody**.

nucleolar DNA *or* **ribosomal DNA** the DNA comprising the **nucleolar organizer(s)**.

nucleolar organizer *or* **nucleolus organizer** the portion of the genome of a eukaryotic cell that is associated with the **nucleolus** and contains the genes from which 45S ribosomal-precursor RNA is transcribed.

nucleolar phosphoprotein B23 *or* **nucleophosmin** *or* **numatrin** a monomeric, or, under some conditions, disulfide-linked dimeric, nucleolar protein that binds to single-stranded nucleic acids, and may be bound to rRNA in the nucleolus.

nucleolin *or* **protein C23** the major nucleolar protein of growing eukaryotic cells believed to function in pre-rRNA processing and ribosome assembly. It is a phosphoprotein found associated with intranucleolar chromatin and peri-ribosomal particles.

nucleolus (*pl.* **nucleoli**) a small, dense body one or more of which are present in the nucleus of eukaryotic cells. It is rich in RNA and protein, is not bounded by a limiting membrane, and is not seen during mitosis. Its prime function is the transcription of the nucleolar DNA into 45S ribosomal-precursor RNA, the processing of this RNA into 5.8S, 18S, and 28S components of ribosomal RNA, and the association of these components with 5S RNA and proteins synthesized outside the nucleolus. This association results in the formation of ribonucleoprotein precursors; these pass into the cytoplasm and mature into the 40S and 60S subunits of the ribosome. —**nucleolar** *adj.*

nucleolytic capable of causing hydrolysis of nucleic acid.

nucleon a neutron or a proton.

nucleonics the branch of physics concerned with the practical applications of nuclear phenomena.

nucleon number *or* **mass number** *symbol*: A; a dimensionless physical quantity, having an integral value for any given nuclide, equal to the total number of nucleons (i.e. protons and neutrons) contained in the nucleus of an atom of the nuclide; it also is the integer nearest to the numerical value of the relative atomic mass of that nuclide and is equal to the arithmetical sum of the **neutron number**, N, of the nuclide and the **proton number**, Z, of the relevant chemical element: i.e. $A = N + Z$. The nucleon number may be attached to the symbol for the chemical element as a left superscript, or to the name of the element as a suffix; e.g. ^{14}C, or carbon-14.

nucleophile *or* **nucleophilic reagent** any reagent that is preferentially attracted to a region of low electron density in a chemical reaction. *Compare* **electrophile**.

nucleophilic 1 of, pertaining to, or being a **nucleophile**; having or involving an affinity for regions of low electron density in a chemical reactant. **2** describing a chemical reaction in which a nucleophilic reagent participates.

nucleophilic catalysis catalysis by a **Lewis base**, which donates an electron pair to the reactant.

nucleophilic displacement *an alternative term for* **nucleophilic substitution reaction.**

nucleophilicity the relative reactivity of a nucleophile, measured in terms of the relative rate constants of different nucleophiles towards a common reactant.

nucleophilic reagent *an alternative term for* **nucleophile.**

nucleophilic substitution reaction *or* **nucleophilic displacement** a chemical reaction in which a **nucleophile** effects heterolytic substitution in another reactant, both bonding electrons being supplied by the nucleophile.

nucleophosmin *see* **nucleolar phosphoprotein B23.**

nucleoplasm all the protoplasm contained within the nuclear envelope of a living cell. —**nucleoplasmic** *adj.*

nucleoplasmin a heat-stable acidic protein present in the nucleoplasm of many cell types. It forms complexes with histones, and functions in the assembly of nucleosomes in the formation of chromatin.

nucleoporin any protein that is a component of the nuclear pore complex. *See* **nuclear pore.**

nucleoprotamine any nucleoprotein consisting of DNA and protamine(s).

nucleoprotein any complex of a protein with a deoxyribonucleic acid (**deoxyribonucleoprotein**) or a ribonucleic acid (**ribonucleoprotein**).

nucleosidase any enzyme that hydrolyses the *N*-glycosidic bond in a **nucleoside** (or in a nucleotide or some other derivative of a nucleoside). Such enzymes are classified within the sub-subclass EC 3.2.2, i.e. enzymes that hydrolyse *N*-glycosyl compounds.

nucleoside **1** *symbol* (for a residue of an undetermined, unknown, or unspecified nucleoside): N *or* Nuc; any **glycosylamine** that is a component of a nucleic acid and that consists of a nitrogenous base linked either to β-D-ribofuranose (forming a ribonucleoside, as in RNA) or to 2-deoxy-β-D-ribofuranose (forming a deoxyribonucleoside, as in DNA). The base is either a purine (linked at N-9) or a pyrimidine (linked at N-1, or in the case of pseudouridine at C-5). **2** any compound that is a glycosylated heterocyclic nitrogenous base; the glycose moiety commonly is ribose, but is sometimes a different sugar or a modified sugar; e.g. cytarabine (1-β-D-arabinofuranosylcytosine).

nucleoside antibiotic any of a group of **nucleosides** (def. 2) that are not components of nucleic acids but are produced by various microorganisms and fungi (especially *Streptomyces* spp.), and have antibiotic activity. They contain unusual nitrogenous bases and/or unusual sugars (or sugar derivatives), and may also contain one or more additional components. Examples include **cordycepin**, **nucleocidin**, **puromycin**, and **showdomycin**.

nucleoside cyclic phosphate *or* **cyclic nucleotide** any nucleotide that consists of a nucleoside doubly esterified with one [ortho]phosphate molecule on its glycose moiety (usually at the 3' and 5'positions, but sometimes at the 2'- and 3'-positions) e.g. cyclic AMP.

nucleoside diphosphatase EC 3.6.1.6; *systematic name*: nucleoside diphosphatase phosphohydrolase; an unspecific enzyme that hydrolyses a number of nucleoside diphosphates to a nucleoside monophosphate and orthophosphate. Substrates include IDP, GDP, UDP, and D-ribose 5-phosphate. There is also the more specific **guanosine-diphosphatase**, EC 3.6.1.42; *systematic name*: GDP phosphohydrolase, which hydrolyses guanosine diphosphate to guanylic acid and orthophosphate.

nucleoside diphosphate any nucleotide that consists of a nucleoside esterified with diphosphate on its glycose moiety (usually at the 5' position, when the symbol is NDP).

nucleoside diphosphate reductase EC 1.17.4.1; *recommended name*: ribonucleoside-diphosphate reductase; *systematic name*: 2'-deoxyribonucleoside-diphosphate:oxidized thioredoxin 2'-oxidoreductase; *other name*: ribonucleotide reductase. A nucleotide reductase enzyme, present in most organisms, that catalyses the formation of a 2'-deoxyribonucleoside diphosphate, oxidized thioredoxin, and H_2O from a ribonucleoside diphosphate and reduced thioredoxin. It requires iron and ATP. The enzyme from *Escherichia coli* is formed from B2 protein (dimer of β chains; Fe-binding subunit, 3-D structure known) and B1 (dimer of α chains), which contains a disulfide/sulfhydryl redox system; it is the enzyme responsible for the biosynthesis of deoxyribonucleotides from ri-

bonucleotides, e.g. in *de novo* biosynthesis. The B2 dimer contains two Fe(iii) ions (one on each monomer) that form a $Fe^{3+}-O-Fe^{3+}$ group. This forms a free radical centre (involving an enzyme tyrosyl residue) that abstracts a hydrogen from the ribose (ultimately from the 2' position, but the mechanism is complex). This leads to loss of the 2'–OH. Hydride ions are lost from enzyme sulfhydryl groups to react with these OH• groups to form water, resulting in formation of disulfide groups. The hydrogen abstracted by the free radical centre is returned to what now becomes 2'-deoxyribose. The enzyme's disulfide groups are reduced in turn by thioredoxin. There is an allosteric site on the B1 subunit dimer; ATP acts at this site as an activator and dATP as an inhibitor, in a complex interaction with a second site that determines the specificity of the enzyme towards different nucleotides. A **nucleoside triphosphate reductase** is produced when the organisms are growing anaerobically.

nucleosidediphosphosugar *or* **nucleoside diphosphate sugar** *or* **nucleotide sugar** any **nucleotide** (def. 2) in which the distal phosphoric residue of a nucleoside 5'-diphosphate is in glycosidic linkage with a monosaccharide (or a derivative of a monosaccharide).

nucleoside monophosphate *see* **nucleoside phosphate.**

nucleoside phosphate *or* **nucleoside monophosphate** any nucleotide that consists of a nucleoside esterified with [ortho]phosphate on its glycose moiety (usually at the 5'-position, when the symbol is NMP).

nucleoside reverse transcriptase inhibitor *abbr.*: NRTI; a modified natural nucleoside used as an antiretroviral drug in AIDS therapy. It suppresses retrovirus replication by interfering with the viral reverse transcriptase, causing premature chain termination during DNA synthesis. It is often used in combination with other drugs. *Compare* **non-nucleoside reverse transcriptase inhibitor.**

nucleoside-triphosphatase *abbr.*: NTPase; EC 3.6.1.15; an unspecific diphosphate phosphohydrolase enzyme that catalyses the hydrolysis of the distal diphosphate bond in various nucleoside triphosphates, with formation of [ortho]phosphate and the corresponding nucleoside diphosphate. It acts also on certain other diphospho-compounds.

nucleoside triphosphate any nucleotide that consists of a nucleoside esterified with triphosphate on its glycose moiety (usually at the 5'-position, when the symbol is NTP).

nucleoside triphosphate reductase EC 1.17.4.2; *recommended name*: ribonucleoside-triphosphate reductase; *systematic name*: 2'-deoxyribonucleoside-triphosphate:oxidized thioredoxin 2'-oxidoreductase. A nucleotide reductase enzyme found in *Lactobacillus leichmannii* (and presumably related organisms). It catalyses the formation of a 2'-deoxyribonucleoside triphosphate, oxidized thioredoxin, and H_2O from a ribonucleoside triphosphate and reduced thioredoxin. It requires cobalt and ATP.

nucleosomal remodelling and histone deacetylase *see* **NURD.**

nucleosome a particle that forms the primary packing unit of DNA in chromatin. Nucleosomes give electron micrographs of decondensed chromatin a 'beads-on-a-string' appearance, and are released on mild digestion of eukaryotic nuclei with micrococcal endonuclease. Each consists of a segment of duplex DNA, 160–240 base pairs long, with associated **histone**; about 146 base pairs of the DNA comprise a **core particle** and the remainder form the linker DNA. The core particle is a disklike structure, 11 nm in diameter and 5.7 nm thick, consisting of two molecules each of histones H2A, H2B, H3, and H4 with approximately two superhelical turns of double-stranded DNA wound round it. In the intact nucleosome, an average of one molecule of histone H1 is on the exterior of each core particle and appears to mediate the packing together of adjacent nucleosomes in condensed chromatin. *See also* **nu body.**

nucleosome code the hypothesis that the local concentration and combination of differentially modified core histones of nucleosomes determines the euchromatin or heterochromatin state of DNA. *See* **histone code.**

nucleosome remodelling factor *see* **NURF.**

nucleotidase any phosphoric monoester hydrolase enzyme that catalyses the hydrolysis of a **nucleotide** (def. 1, 2) to yield a nucleoside and orthophosphate. The principal ones are classified under three EC numbers (1) EC 3.1.3.31; *recommended name*: nucleotidase; enzymes having a wide specificity for 2'-, 3'-, and 5'-nucleotides, and that also hydrolyse glycerol phosphate and 4-nitrophenyl phosphate; (2) EC 3.1.3.6; *recommended name*: 3'-nucleotidase; an enzyme having a wide specificity for 3'-ribonucleotides; these enzymes hydrolyse a 3'-ribonucleotide to a ribonucleoside and orthophosphate. Some such enzymes are not sugar specific. (3) EC 3.1.3.5; *recommended name*: 5'-nucleotidase; an enzyme having a wide specificity for 5'-ribonucleotides. These enzymes hydrolyse a 5'-ribonucleotide to a ribonucleoside and orthophosphate.

nucleotide 1 *or* **mononucleotide** *or* **nucleoside [mono]phosphate** any compound that consists of a **nucleoside** (def. 1) esterified with [ortho]phosphate at either the 3'- or the 5'-hydroxyl group of its glycose moiety (ribonucleosides giving **ribonucleotides**, and deoxyribonucleosides giving **deoxyribonucleotides**). Nucleotides are the constitutional units into which nucleic acids are broken down by partial hydrolysis and from which they are considered to be built up. **2** any compound consisting of a **nucleoside** (def. 2) that is esterified with [ortho]phosphate or an oligophosphate at any hydroxyl group on its glycose moiety. In this sense the term includes **nucleoside cyclic phosphate**, **nucleoside diphosphate**, **nucleoside diphosphosugar**, **nucleoside triphosphate**, and pyridine nucleotide. **3** any compound containing a moiety of a nucleotide (def. 1, 2). Included are: any carrying an additional phosphoric group at another position on the glycose moiety; any formed (actually or conceptually) from two or more such nucleotides, whether the same or not, joined together in phosphoric-ester linkage (to form an **oligonucleotide** or a **polynucleotide**); and, in certain instances, any formed from such a nucleotide joined through a phosphoric anhydride link either to a second nucleotide or to some other phosphate ester. Exceptionally, the term includes also certain analogous compounds (especially flavin mononucleotide, and hence also flavin-adenine dinucleotide) that, being derivatives of ribitol, are not true nucleoside phosphates or else are not formed exclusively from them.

nucleotide-binding fold *or* **nucleotide-binding pocket** a common structural feature of numerous dehydrogenase enzymes utilizing either of the nucleotide coenzymes NAD or NADP as the acceptor. The coenzyme fits in a pocket made by the folding of the polypeptide chain. Many such enzymes combine specifically with **blue agarose** owing to putative structural similarity of the bound dye (Reactive Blue 2) to the coenzyme, a property that can be exploited in the separation and purification of these enzymes by affinity chromatography. A similar property is shown by some other nucleotide-binding enzymes (and also by certain other proteins that do not bind nucleotides).

nucleotide coenzyme any **nucleotide** (def. 2, 3) that is either (1) a coenzyme (e.g. nicotinamide-adenine dinucleotide), or (2) a substrate-carrier in an anabolic reaction (e.g. uridine diphosphate glucose).

nucleotide excision repair *abbr.*: NER; the form of DNA repair in which ultraviolet light-induced photoproducts (e.g. pyrimidine dimers), chemical adducts (e.g. induced by carcinogens such as benzpyrene), or interstrand crosslinks are removed and the defect is repaired. This involves excision of an oligonucleotide that contains the lesion and filling of the resulting single-strand gap. In *E. coli*, the repair is effected by UvrABC excinuclease, the excised oligonucleotide being ≈12 nt long. In eukaryotes, some 16 gene products are required, and the excised oligonucleotide is 24–32 nt long. Defective NER is the cause of xeroderma pigmentosum and of Cockayne syndrome. *See also* **excision repair**.

nucleotide-gated channel *see* **ion channel**.

nucleotide reductase *see* **nucleoside diphosphate reductase**, **nucleoside triphosphate reductase**.

nucleotide sugar *an alternative name for* **nucleoside diphosphate sugar**.

nucleotide unit the repeating unit of a polynucleotide chain, defined typically by the sequence of atoms from the phosphorus atom at the 5' end of any one of its component glycose residues to the oxygen atom at the 3' end of the same residue and including all the atoms of the glycose residue and its attached phosphoric residue and nucleobase; atypically, the nucleotide unit at the 5' end of a chain may lack a phosphate group. Nucleotide units in a given chain are numbered sequentially from the chain's 5' end irrespective

of the presence or absence of a phosphate group on the 5'-terminal unit.

nucleotidyl the acyl group formed from a nucleoside [mono]phosphate by removal of a hydroxyl group from its phosphoric group.

nucleotidyltransferase an enzyme of subclass EC 2.7.7; such enzymes transfer a nucleotidyl group to a reactant. Possession of such activity is a common mechanism of drug resistance (especially to **aminoglycoside antibiotics**) in bacteria. An example is gentamicin 2″-nucleotidyltransferase (EC 2.7.7.46); *systematic name*: NTP:gentamicin 2″-nucleotidyltransferase; *other name*: 2″-aminoglycoside nucleotidyltransferase. It catalyses a reaction between NTP and gentamicin to form pyrophosphate and 2″-nucleotidylgentamicin. This enzyme can be isolated from *Escherichia coli* and *Klebsiella pneumoniae* plasmids, and confers resistance to kanamycin, gentamicin, dibekacin, sisomicin, neomycin, and tobramycin by adenylating the 2″-OH group of these antibiotics.

nucleus (*pl.* **nuclei**) **1** *(in biology)* the most conspicuous organelle of a eukaryotic cell; it contains the chromosomes and (except for mitochondria and chloroplasts) is the sole site of DNA replication and of RNA synthesis in the cell. Usually a spheroidal body, it is separated from the cytoplasm by the nuclear envelope. **2** *(in physics)* **atomic nucleus** the positively charged central part of an atom, with which is associated almost the whole of the mass of the atom but only a very small part of its volume. It is composed of nucleons, i.e. protons and neutrons. **3** *(in organic chemistry)* a characteristic arrangement of atoms, especially a ring structure, that occurs in a series of related organic compounds; e.g. the benzene nucleus. **4** a particle on which a crystal, droplet, or bubble forms in a fluid.

nuclide any individual species of atom, whether stable, metastable, or unstable, that is characterized by the constitution of its nucleus and by its **nucleon number**, **proton number**, and nuclear energy state; its mean life must be long enough for the species to be observable. Nuclides having the same proton number but different nucleon numbers or different nuclear energy states are termed isotopic nuclides, or **isotopes**. Nuclides having the same nucleon number but different proton numbers are termed isobaric nuclides, or **isobars**. Nuclides having the same neutron number but different proton numbers are termed isotonic nuclides, or **isotones**. A given nuclide may be specified by attaching its nucleon number either to the name of the relevant chemical element as a suffix, or to the symbol for the chemical element as a left superscript; the proton number may be attached to the symbol as a left subscript; e.g. carbon-14, ^{14}C, or $^{14}_{6}$C. A metastable nuclide (i.e. one with an elevated nuclear energy state) may be differentiated from its counterpart with the same nucleon number by attaching the suffix m to the nucleon number; e.g. technetium-99m or $^{99m}_{43}$Tc. (Absence of a nucleon number attached to its name or symbol implies the presence of all isotopes of the element in natural abundance.) —**nuclidic** *adj.*

nude a mutant strain of mouse or rat that lacks a **thymus gland** and hence is devoid of **T lymphocytes** and is hairless. It is caused by a mutation in the *nude* gene, which encodes a transcription factor of the winged helix type. *See* **forkhead**.

NUF1 *abbr. for* nuclear filament-related protein.

Nujol *a proprietary name for* a high-boiling-point fraction of liquid paraffin widely used for making mulls for infrared spectroscopy. It is also used medicinally as a laxative, and was referred to by the poet e. e. cummings in *Poem, or Beauty hurts Mr. Vinal*: '...little liver pill-/hearted-Nujol-needing-There's-A-Reason...'.

null a minimum or zero value, especially of an electric current.

null cell *an alternative term for* **null lymphocyte**.

nullisomic 1 describing a cell, organism, strain, etc. that lacks one or more pairs of homologous chromosomes. **2** such a cell, organism, strain, etc.

null lymphocyte *or* **null cell** a **lymphocyte** that is not recognizable as either a B lymphocyte or a T lymphocyte.

null method *or* **zero method** a method of measurement in which a zero reading in a sensitive, but not necessarily calibrated instrument, e.g. a galvanometer, enables an unknown value of some property to be deduced by comparison with reference values.

NuMA *abbr. for* nuclear mitotic apparatus protein.

numatrin *another name for* **nucleolar phosphoprotein B23**.

number 1 a concept or expression of quantity reckoned in ones or units; a particular value of such a quantity (which may be integral or zero). **2** the symbol or group of symbols used to represent a number (def. 1); numeral; figure. **3** a numeral or string of numerals used for identification (of something, especially a member of a sequence). **4** to assign numbers to (members of a series); to label with a number or numbers.

number-average molar mass *see* **average molar mass.**

number-average relative molecular mass *or (formerly)* **number-average molecular weight** *see* **average relative molecular mass.**

number concentration *symbol: C; see* **concentration** (def. 1d).

number fraction *symbol: δ*; the number of defined particles or elementary entities of a specified component of a system divided by the total number of defined particles in the system.

numerical aperture *(in optics) abbr.*: NA; the arithmetic product of the refractive index of the medium in which the objective (lens) of a microscope is situated and the sine of half the angle of view of the objective. The resolving power of the lens is proportional to its numerical aperture.

NURD *abbr. for* nucleosomal remodelling and histone deacetylase; a protein complex found in *Xenopus* and mammals and involved in nucleosome remodelling by some as yet unknown mechanism. It contains histone deacetylases (HDAC1 and HDAC2), retinoblastoma-associated proteins (RbAp48 and RbAp46), metastasis-associated proteins (MTA1 and MTA2), and a subunit (MBD3) that contains a methyl-CpG-binding domain. The largest component (Mi2) contains two chromodomains, a zinc finger, and a distinctive helicase/ATPase domain.

NURF *abbr. for* nucleosome remodelling factor; a tetrameric protein complex of *Drosophila* that acts with **GAGA factor** to disrupt nucleosome spacing near the promoter of the Hsp70 gene. One component of NURF is an ATPase protein called initiation switch (**ISWI**).

Nurr 1 an orphan nuclear receptor expressed almost exclusively in the central nervous system, where it is confined to dopaminergic neurons. In mice, it seems to be absolutely required for generation of these neurons.

nurse cell a cell that is connected by cytoplasmic bridges to an oocyte and thereby conveys macromolecules to the growing oocyte.

nus any of three *Escherichia coli* genes that are involved in transcriptional termination. The product NusA binds to RNA polymerase, RNA, **rho**, and the antitermination protein pN, the product of the *N* gene. It may also bind to NusB. It participates in the termination and antitermination of transcription. NusB is involved in the transcription termination process at certain sites during normal bacterial growth and is essential for the function of pN. NusG is involved in transcription antitermination.

NutraSweet *a proprietary name for* **aspartame.**

nutrient 1 any chemical substance that can be used by an organism to sustain its metabolic activities. **2** serving as or conveying nourishment. **3** *(in microbiology)* describing a solid or liquid medium that is able to support the growth of a range of nutritionally undemanding chemoorganotrophs.

nutrition 1 the supply of substances (i.e. nutrients) required by an organism, directly or indirectly, for its metabolic activities, i.e. for the provision of energy, for growth, and/or for the renewal of degraded components. **2** the act or process of supplying or receiving nutrients. **3** the scientific study of the nutrient requirements of particular organisms and of the supply of those nutrients. —**nutritional** *adj.*

nutritional mutant a mutant strain of a microorganism (alga, bacterium, or fungus) that requires a supply of one or more growth factors in addition to those required by the wild type, i.e. a mutation in a **prototroph** to form an **autotroph.**

nutritive 1 containing nutrient(s). **2** of, concerned in, or promoting nutrition.

nux vomica the dried seeds of the Indian tree, *Strychnos nux-vomica*. They are a source of **strychnine**, **brucine**, and other poisonous alkaloids.

Nva *symbol for* the α-amino acid **norvaline** (now known as 2- aminovaleric acid, *symbol*: Avl).

Nycodenz *the proprietary name for N,N'*-bis(2,3-dihydroxypropyl)-

5-[*N*-(2,3-dihydroxypropyl)acetamido]-2,4,6-triiodoisophthalamide; a dense, synthetic, nonionic substance, highly soluble in water, that is capable of forming (autoclavable) aqueous solutions of densities up to 2.1 g mL^{-1}. It is one of several synthetic water-soluble triiodobenzene derivatives designed for use as intensifiers in radiology that have found use in isopycnic gradient fractionation of biological particles.

nyctalopia *the medical term for* **night blindness.**

nylon any of a class of synthetic thermoplastic substances, characterized chemically as long-chain polyamides and formed by condensation polymerization of dicarboxylic acids with diamines. They have high melting points, are insoluble, tough, resistant to fracture, and exhibit low friction. Hence they are useful in the manufacture of laboratory apparatus for which these properties are advantageous. Nylon can be formed into fibres or sheets and used for coatings, fabrics, and many other products. It is also used as an alternative to nitrocellulose in hybridization membranes.

nystatin a **mycosamine**-containing polyene antifungal antibiotic complex (three components) produced by *Streptomyces noursei* and other *Streptomyces* spp. It is very similar to **amphotericin B** in its properties and mechanism. Two proprietary names are Fungicidin and Mycostatin.

nystatin A$_1$

N-Z amines a proprietary enzymic hydrolysate of casein, used in bacterial growth media.

NZCYM a culture medium for the growth of *E. coli* and suitable for the propagation of bacteriophage lambda. It contains **N-Z amines**, casamino acids, yeast extract, MgSO$_4$, and NaCl.

NZM a culture medium for the growth of *E. coli* that contains **N-Z amines**, MgSO$_4$, and NaCl.

NZO mouse *abbr. for* New Zealand obese mouse.

NZY a culture medium for the growth of *E. coli* that contains **N-Z amines**, yeast extract, and NaCl.

Oo

o+ *prefix (in peptide hormone nomenclature)* denoting ovine.

o- *prefix (in chemical nomenclature)* denoting **ortho-**. Compare **m-, p-**.

O *symbol for* **1** oxygen. **2** a residue of the ribonucleoside orotidine (alternative to Ord).

O- *prefix (in chemical nomenclature)* indicating substitution on oxygen; a particular oxygen atom in a molecule may be specified by suffixing an appropriate superscript.

OA *abbr. for* ovalbumin.

OAA *abbr. for* oxaloacetic acid, or oxaloacetate.

OAADPr *abbr. for* 1-*O*-acetyl-ADP-ribose.

OAG *abbr. for* 1-oleoyl-2-acetyl-glycerol.

Oakley–Fulthorpe technique a simple double immunodiffusion method that can be used for quantitative determination of antigen concentrations; it is essentially a development of the **Oudin technique**. The antiserum in 1 agar is placed at the bottom of a cylindrical tube above which is placed a layer of 1 agar gel in 1 saline, and 0.5 *o*-cresol (as a preservative). Subsequently a bacterial filtrate (containing antigen) is layered onto the agar–saline gel and the tube then incubated at 37 °C. Disks or lines of flocculation of antigen–antibody complexes are produced in the agar–saline gel, from whose positions the concentration of antigen(s) may be determined. [After C. L. Oakley and A. J. Fulthorpe, who described the technique in 1953.]

O antigen 1 any of the lipopolysaccharide–protein somatic antigens of the cell walls of Gram-negative bacteria, especially those of the Enterobacteriaceae. They are useful in the serological classification of *Salmonella*, *Shigella*, and other bacteria. **2** *(sometimes)* an alternative name for H antigen (present on erythrocytes of blood group O); *see* **blood-group substance**.

OAT *abbr. for* ornithine-δ-aminotransferase; EC 2.6.1.13; an enzyme that uses pyridoxal phosphate as coenzyme to catalyse the reaction:

ornithine + 2-oxoglutarate = pyrroline-5-carboxylate + glutamate.

It is widespread in mammalian tissues and especially active in liver, kidney, and small intestinal mucosa. It is a homohexamer of the mitochondrial matrix. The crystal structure of the human enzyme (each subunit contains 439 amino acids) comprises a total of 12 alpha helices and 14 beta strands, with a central domain that contains 8 alpha helices and 7 beta strands. Numerous mutations lead to enzyme deficiency and gyrate atrophy of the choroid and retina with hyperornithinemia and hyperornithinuria.

+oate *suffix (in chemical nomenclature)* denoting the carboxylate group, –COO⁻.

ob *see* **ob/ob**.

obelin a luminescent protein obtained from the jellyfish, *Obelia geniculata*. It is similar to **aequorin** in properties and use.

obese having an excessive amount of body fat. Obesity is the second largest cause of preventable death in the developed world and a major risk factor for vascular disease, diabetes mellitus type 2, and some cancers. —**obesity** *n*.

obestatin a peptide encoded by the **ghrelin** gene that opposes ghrelin's effects on food intake. The 23-residue peptide, has the structure FNAPFDVGIKLSGAQYQQHGRAL-amide in the rat. Treatment of rats with the peptide depresses food intake, inhibits jejunal contraction and results in decreased body weight gain. Ghrelin and obestatin are both derived from the ghrelin gene propeptide by posttranslational cleavage and modification. Obestatin binds to the orphan **G-protein-coupled receptor** GPR39.

OB fold *abbr. for* oligonucleotide/oligosaccharide binding fold; a protein sequence (≈110 residues) present in most prokaryotic and eukaryotic ssDNA-binding proteins. It does not have a definite sequence motif but forms a highly curved five-stranded beta sheet that closes in on itself to form a beta barrel. The BRCA2 protein contains three OB folds.

objective *(in optics)* the lens or system of lenses in an optical instrument (especially a compound microscope) that is positioned nearest to the object to be viewed.

object-oriented database *abbr.*: OODB; a database in which data are stored as abstract objects, with abstract relationships between them. The data representations are flexible (e.g. including character strings, digitized images, tables) and objects may be grouped together.

oblate 1 having an equatorial diameter longer than the polar diameter. **2** describing an **ellipsoid of rotation** (including a spheroid) generated by rotation of an ellipse about its minor axis. **3** describing any protein of **oblate** (def. 1) dimensions, as observed in hydrodynamic measurements. *Compare* **prolate**.

obligate by necessity, or without option; used especially of the nature of the environment or of the mode of life of an organism; e.g. obligate aerobe (*see* **aerobe**), obligate anaerobe (*see* **anaerobe**). *Compare* **facultative**.

obligate heterozygote an individual in a family who is proven to carry one copy of a recessive allele by having had affected progeny who inherited two copies of the mutant allele, one from each parent.

O blood-group substance *see* **ABO system**.

ob/ob the genotype symbol *for* any strain of mouse with an inherited form of severe obesity, due to a homozygous autosomal recessive mutation on chromosome 6. Such animals display marked hyperglycemia and insulin resistance. A protein named **leptin**, or OB protein, has been implicated in the control of fat mass and is a product of the *ob* mutant gene, being deficient in *ob/ob* mice.

obscurin a protein of the Z band of skeletal muscle that interacts with titin.

obtuse angle any plane angle greater than 0.5π rad or 90° and less than π rad or 180°.

Oc *symbol for* the octyl group.

ocatin a storage protein that constitutes over 60% of the soluble protein of oca tubers (*Oxalis tuberosa*). It is homologous with a group of pathogenesis-related proteins and has antibacterial and antifungal activity.

Occam's razor *a variant spelling of* **Ockham's razor**.

occlude 1 to block or obstruct (a passage, tube, opening, etc.). **2** *(in chemistry)* to take up and retain something in the interstices of a solid; e.g. to absorb a gas into a metal or to entrap liquid within crystals during their formation. —**occlusion** *n*.

occludin an integral membrane protein that localizes in tight junctions in chick liver; four putative transmembrane domains reside in the N-terminal half of the protein.

occult concealed or hidden; not apparent to the unaided eye.

occult blood blood that is present in a sample in such small quantities, e.g. in feces, that it can only be detected by chemical testing or by microscopy.

occult virus a virus whose presence in a host is inferred but whose isolation or detection cannot be achieved.

occupancy 1 the state or condition of being filled or occupied. **2** the number of similar sites occupied, usually expressed as a fraction or percentage of the total number available. It can refer, e.g., to the number of receptors on a cell surface occupied by a particular agonist or to the number of atomic orbitals of a particular subshell filled by electrons.

occupation theory of agonist action a theory stating that the magnitude of the response to an agonist is directly proportional to the fraction of specific receptors occupied by molecules of the agonist; it assumes that the binding of agonist to receptor is reversible. According to this theory the response will increase to a maximal value when all the receptors are occupied. *Compare* **rate theory of agonist action**.

ochratoxin any one of several **mycotoxins** produced by the fungus *Aspergillus ochraceus*, other *Aspergillus* spp., and *Penicillium* spp.; the major component, ochratoxin A, is a complex, chlorine-containing derivative of L-phenylalanine that inhibits phosphoenolpyruvate carboxykinase (EC 4.1.1.49) and may cause fatty liver. These toxins may occur on contaminated foodstuffs such as corn (maize), peanuts, storage grains, etc.

ochratoxin A

ochre codon or **ochre triplet** *symbol*: UAA; one of the three terminator codons or **nonsense codons** found in messenger RNA.

ochre mutant any cell or virus carrying an **ochre mutation**.

ochre mutation a mutation that gives rise to an **ochre codon** at an abnormal position in a messenger RNA molecule.

ochre suppressor any of a number of mutations in *E. coli* resulting in changed anticodons of tRNAs that suppress an **ochre codon** (UAA) in mRNA. This allows insertion of one of several alternative amino acids into a polypeptide at that site. Examples are *supC* (Tyr) and *supG* (Lys). Some ochre suppressors (*supC* and *supG*) also suppress **amber codons**.

ochronosis the presence of brown-black pigment in the skin, cartilage, and other tissues due to **alcaptonuria**.

OCIF *abbr. for* osteogenic inhibitory factor. *See* **osteoprotegerin**.

Ockham's razor or **Occam's razor** a maxim stating that when explaining something there should be a minimal number of assumptions. *Entia non sunt multiplicanda praeter necessitatem* (entities should not be multiplied except from necessity). [After William of Ockham (1285–1349), English philosopher.]

Oco *symbol for* the octanoyl group.

OCS *the proprietary name for* a commercial xylene–fluor mixture suited to the liquid-scintillation counting of a wide range of toluene- and xylene-soluble organic samples, especially specimens solubilized in **NCS**.

oct+ *a variant form of* **octa+** (sometimes before a vowel).

OCT (*in clinical chemistry*) *abbr. for* ornithine carbamoyltransferase.

OCT a family of genes encoding transcription factors for eukaryotic RNA polymerase II promoters. The proteins contain a **POU domain** and contribute to the naming of that domain. They are leucine zipper proteins and bind to **octamer** sequences.

octa+ or **octo+** or (*sometimes before a vowel*) **oct+** *comb. form* denoting eight, eightfold, eight times.

octacosa+ *comb. form* denoting twenty-eight, twenty-eightfold, twenty-eight times.

octacosanoate *see* **montanate**.

octacosanoyl *see* **montanoyl**.

octadeca+ *comb. form* denoting eighteen, eighteenfold, eighteen times.

octadecadienoic acid any straight-chain fatty acid having eighteen carbon atoms and two double bonds per molecule. Linoleic acid (*see* **linoleate**) is the all-*Z*-(9,12)-isomer and is a constituent of most vegetable oils and animal fats. The all-*E*-isomer of this acid is linolelaidic acid, a constituent of the seeds of *Chilopsis linearis*, which also contain the all-*E*-(10,12)-isomer. A number of other isomers have been synthesized chemically.

octadecanoic acid *the systematic name for* stearic acid; *see* **stearate**.

octadecapentaenoic acid any straight-chain fatty acid having eighteen carbon atoms and five double bonds per molecule. The all-*Z*-(3,6,9,12,15)-isomer occurs naturally in dinoflagellates.

octadecatetraenoic acid any straight-chain fatty acid having eighteen carbon atoms and four double bonds per molecule. Several isomers occur naturally, including the 3*E*,9*Z*,12*Z*,15*Z*-isomer, found in seed oil, and the all-*Z*-(6,9,12,15)-isomer, found in fish oil. The 9*Z*,11*E*,13*E*,15*Z*-isomer and the all-*E*-(9,11,13,15)-isomer are α- and β-parinaric acid respectively, from *Parinarium laurinum*; both isomers are used as fluorescent probes.

octadecatrienoic acid any straight-chain fatty acid having eigh-

teen carbon atoms and three double bonds per molecule. A number of isomers occur naturally, the most prominent being the all-*Z*-(9,12,15)-isomer, (9,12,15)-linolenic acid; the all-*Z*-(6,9,12)-isomer is (6,9,12)-linolenic acid (*see* **linolenate**). *See also* **columbinic acid**, **eleostearate**.

octadecenoic acid any straight-chain fatty acid having eighteen carbon atoms and one double bond per molecule. (*Z*)-Octadec-9-enoic acid (i.e. oleic acid; *see* **oleate**) is the isomer found widely, often in high proportion, in plant and animal lipids, especially in olive oil. The 5*E*- and 6*Z*- (petroselenic acid; *see* **petroselenate**) isomers are found in seed oils; the 6*E*- isomer is named **petroselaidic acid**, and the 9*E*-isomer is elaidic acid, a common constituent of fats and oils (*see* **elaidate**). The 11*E*- and 11*Z*-isomers are, respectively, *trans*- and *cis*-vaccenic acid (*see* **vaccenate**).

octaethyleneglycol mono-*n*-decyl ether a nonionic $C_{10}E_8$ detergent (*see* $C_x E_y$).

octaethyleneglycol mono-*n*-dodecyl ether a nonionic $C_{12}E_8$ detergent (*see* $C_x E_y$).

octahedral coordination coordination in which six ligands are disposed about a central atom as if at the vertices of an octahedron enclosing that atom.

octahedron (*pl.* **octahedra**) any solid geometrical figure having eight plane triangular faces, 12 edges, and six tetrameric vertices. In a regular octahedron the faces are congruent equilateral triangles and may be considered to consist of two equal pyramids opposing each other on the same square base. —**octahedral** *adj.*

octamer 1 an eight-base sequence element that is common in eukaryotic **promoters**. The consensus sequence is ATTTGCAT. It binds a number of transcription factors (*see* **OCT**). 2 any polymer of a protein having eight subunits, e.g. β-lactoglobulin at pH 4.7 below 4°C. 3 an assembly of eight **histones** containing two each of H2A, H2B, H3, and H4, that forms the histone core of the nucleosome.

octanoate or (*formerly*) **caprylate** 1 *systematic and preferred name for* the anion, CH_3–$[CH_2]_6$–COO^-, derived from octanoic acid (formerly known as caprylic acid). 2 any mixture of octanoic acid and its anion. 3 any salt or ester of octanoic acid.

octanoyl or (*formerly*) **capryloyl** or **caprylyl** *symbol*: Oco; the univalent acyl group, CH_3–$[CH_2]_6$–CO–, derived from octanoic acid (formerly known as caprylic acid).

octet 1 any group of eight. 2 (*in chemistry*) (esp. valency) a stable group of eight electrons constituting an electron shell of an atom.

octo+ *a variant form of* **octa+**.

octonic acid any monocarboxylic **aldonic** acid formally derived from an octose by oxidation at C-1.

octopamine β-hydroxytyramine; 1-(*p*-hydroxyphenyl)-2-aminoethanol; the D(–)-enantiomer is a biogenic amine, about one-tenth as active as norepinephrine, formed by the β-hydroxylation of tyramine by dopamine β-hydroxylase. It is found in the salivary glands of *Octopus* spp. and of *Eledone moschata*.

R-(–)-enantiomer

octopine 1 *N*-(1-carboxy-4-guanidinobutyl)-L-alanine; N^2-(1-carboxyethyl)-L-arginine; D-octopine (i.e., N^2-(D-1-carboxyethyl)-L-arginine) is an **opine** found in the tumours of crown-gall disease in plants and in the muscles of certain invertebrates. The genes responsible for the synthesis of octopine are part of the **T-DNA** from a **Ti-plasmid**. 2 any opine that, like octopine itself, is an *N*-substituted derivative of alanine. Octopine was first isolated from the muscles of *Octopus*, but is found in other cephalopod species and lamellibranchs. The guanidine group can undergo phosphorylation to phosphooctopine, which acts as a **phosphagen**. *Compare* **octopinic acid**.

D-octopine

octopinic acid N-(1-carboxy-4-aminobutyl)-L-alanine; N^2-(1-carboxyethyl)-L-ornithine; D-octopinic acid (i.e., N^2-(D-1-carboxyethyl)- L-ornithine) is an **opine** found in the tumours of crown-gall disease of plants; it is a member of the **octopine** (def. 2) family.

octose any **aldose** having a chain of eight carbon atoms in the molecule.

octreotide a synthetic octapeptide analogue of somatostatin, with which it shares four residues essential for activity. It has a much longer half-life *in vivo* (1–2 h) than somatostatin (1–2 min) and has been used in the therapy of breast, ovarian, prostatic, gut, endocrine, and pituitary tumours.

octulose any **ketose** having a chain of eight carbon atoms in the molecule.

octyl *symbol*: Oc; the alkyl group, CH_3–$[CH_2]_6$–CH_2–, derived from octane.

octyl glucoside *n*-octyl β-D-glucopyranoside; a mild nonionic detergent, aggregation number 84, CMC 20–25 mm. The α-glucoside isomer has also been used, e.g. for crystallization of membrane proteins. The lack of absorbance at 228 nm is an advantage in the use of these compounds. *Compare* **heptyl glucoside, hexyl glucoside, nonyl glucoside, octyl maltoside.**

octyl maltoside *n*-octyl β-D-maltopyranoside; a mild nonionic detergent, CMC 23 mm. *Compare* **octyl glucoside.**

octyl thioglucoside *n*-octyl 1-thio-β-D-glucopyranoside; a mild nonionic detergent, CMC 9 mm, with properties similar to other alkyl glucosides.

ocular 1 of or concerned with the eye or vision. **2** *or* **eyepiece** the lens or system of lenses in an optical instrument (especially a compound microscope) that is positioned nearest to the eye.

ocytocin *(sometimes)* a variant spelling of **oxytocin.**

od *or* **o.d.** *abbr. for* outside diameter.

OD *or* **O.D.** *abbr. for* optical density (now called **absorbance**).

odorant-binding protein any of a family of some 10 proteins that bind and transport odorant molecules across the mucus layer that covers odour receptors in the nasal mucosa. They are homodimers that contain an odorant-binding domain and a dimerization region.

odorant receptor any G-protein-coupled membrane receptor that binds and elicits the biological response of an odorant molecule. Several hundred genes for such receptors are present in the human genome. The receptors function through G_{aolf}, which activates adenylate cyclase, or G_{aP}, which activates phospholipase C. They are present in cells of the olfactory system and also outside it (there are over 50 different receptors in spermatogenic cells).

Oe *symbol for* oersted.

oedema *an alternative spelling (esp. Brit.) of* **edema.**

oenanthate *an alternative spelling (esp. Brit.) of* enanthate (*see* **heptanoate**).

oenology *an alternative spelling (esp. Brit.) of* **enology.**

oersted *symbol*: Oe; the cgs unit of magnetizing force or magnetic field strength;

1 oersted = $10^3/4\pi$ amperes per metre (= 79.58 A m^{-1}).

[After Hans Christian Oersted (1777–1851), Danish physicist and chemist who discovered the phenomenon of electromagnetism.]

17β-oestradiol *an alternative spelling (esp. Brit.) of* **17β-estradiol.**

OEt *symbol for* the ethoxy group, CH_3–CH_2–O– (alternative to EtO).

OFAGE *abbr. for* orthogonal field alternating gel electrophoresis; a type of **pulsed field gel electrophoresis** used for separating megabase-sized DNA molecules.

off-rate the constant, k_{-1}, for the dissociation step in any binding equilibrium, e.g. LB → L + B in the equilibrium L + B = LB, where L represents ligand and B the molecule it binds; the rate of dissociation of a ligand from the bound complex. The dimensions of k_{-1} are time^{-1}. *Compare* **dissociation constant, on-rate.**

OGCP *abbr. for* oxoglutarate/malate carrier protein; a mitochondrial inner membrane integral protein that plays an important role in several processes, including the malate– aspartate shuttle, gluconeogenesis from lactate, and nitrogen metabolism. *See also* **mitochondrial carrier proteins.**

+ogen *a variant form of* **+gen** (after a consonant).

Ogston concept *or* **three-point attachment hypothesis** a concept formulated to explain the inherent differing reactivities, often expressed in an enzymic reaction, of identical chemical groups in a **prochiral** molecule. It states that there must be at least three points of attachment of a substrate molecule to the active site of an enzyme. It is similar to the **Easson–Stedman model.** [After Alexander George Ogston (1911–96), British biophysicist who formulated it in 1948.]

Oguchi disease *see* arrestin.

25-OHD₃ *abbr. for* 25-hydroxycholecalciferol; calcidiol; use not recommended.

1,25-(OH)₂D₃ *abbr. for* 1,25-dihydroxycholecalciferol; **calcitriol;** use not recommended.

24,25-(OH)₂D₃ *abbr. for* 24(R),25-dihydroxycholecalciferol; (24R)-hydroxycalcidiol; use not recommended.

25,26-(OH)₂D₃ *abbr. for* 25,26-dihydroxycholecalciferol; 26-hydroxycalcidiol; use not recommended.

1,24,25-(OH)₃D₃ *abbr. for* 1,24(R),25-trihydroxycholecalciferol; **calcitetrol;** use not recommended.

OHHL *abbr. for* N-(3-oxohexanoyl)-L-homoserine lactone.

ohm *symbol*: Ω; the SI derived unit of electric resistance, defined as the resistance between two points on a conductor through which a current of one ampere flows as the result of a potential difference of one volt applied between the points, the conductor not being the source of any electromotive force; i.e. 1 Ω = 1 VA^{-1}. [After Georg Simon Ohm (1787–1854), German physicist.]

Ohm's law a law stating that under constant conditions the current, I, flowing through a given conductor is proportional to the potential difference, U, applied across it. The law is often expressed in the form $U = IR$, where the proportionality constant, R, is the resistance of the conductor.

+oic *suffix (in chemical nomenclature)* denoting the carboxyl group, –COOH. It is added to the systematic name of an unbranched alkane (or substituted alkane) to denote its modification to the corresponding alkanoic (i.e. carboxylic) acid by formal conversion of a methyl group to a carboxy group; e.g. hexanoic acid, CH_3–$[CH_2]_4$–

COOH, formally derived from hexane, CH_3–$[CH_2]_4$–CH_3. It is used also in the trivial names of certain carbocyclic and heterocyclic carboxylic acids; e.g. benzoic acid, furoic acid.

+oid *suffix forming adjectives and (associated) nouns* **1** denoting likeness to, or having the form of (something specified); e.g. colloid (i.e. gluelike substance). **2** generated by rotation of (the geometrical figure specified); e.g. ellipsoid (i.e. solid figure derived from an ellipse). **3** belonging to the class or group represented by (the type member specified); e.g. steroid (i.e. compound belonging to the same chemical family as sterols). —**+oidal** *adj.*

oil any neutral, flammable substance that is liquid at room temperature and is characteristically soluble in relatively nonpolar solvents but only sparingly soluble in aqueous solvents. There are three main groups: (1) animal and vegetable oils, which usually consist predominantly of triacylglycerols but may contain varying amounts of fatty-acid esters of other alcohols; (2) mineral oils, derived from petroleum, coal, shale, etc., which consist predominantly of hydrocarbons; and (3) **essential oils**.

oil body *an alternative name for* **spherosome**.

oil-immersion objective an objective lens in a light microscope designed to be used with a layer of a special oil between the object (or its coverslip) and the lens. The oil is of the same refractive index as the glass of the lens. This arrangement maximizes the numerical aperture and hence the resolving power of the lens.

Oil Red O a stain for lipid it is widely used for staining lipoproteins on cellulose acetate electrophoretic strips.

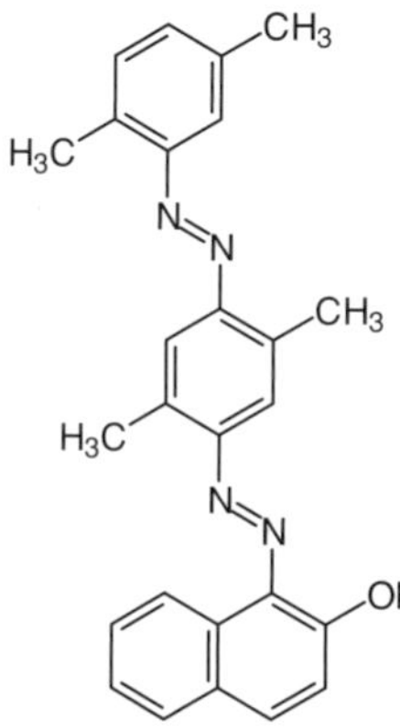

oilseed rape *or* **colza** *or* **canola** *the common name for Brassica campestris, a species of rape grown as a source of rapeseed oil.*

okadaic acid a polyether fatty acid, M_r = 804.9, that is a potent inhibitor of **protein phosphatases**, especially the PP-1 and PP-2A types. It is a tumour promoter, and is implicated as the causative agent of diarrhetic shellfish poisoning. Its name derives from *Hilichondria okadaii*, a marine sponge that feeds on *Prorocentrum lima*, a dinoflagellate from which it is isolated. See structure below.

Okazaki fragment *or* **Okazaki piece** *or* **Okazaki segment** any of the relatively short polydeoxyribonucleotides that are formed, concomitantly with the continuous replication of one of the two strands of a duplex DNA molecule, as intermediates during the discontinuous replication of the other strand. Such fragments are synthesized from mononucleotides in a direction opposite to that of the movement of the replication fork; then, as replication proceeds, they are covalently joined through the action of a polynucleotide ligase to form a long daughter polynucleotide chain, the lagging strand. These fragments appear to be 1000–2000 residues long in *Escherichia coli* and 100–200 residues long in mammalian cells; a proportion of them may have RNA primer attached. [After Reiji Okazaki (1930–1975), Japanese molecular biologist.]

OK cell *abbr. for* opossum kidney cell; any cell of a culture derived from this organ and used to study α_{2C} adrenoceptors.

+ol *suffix (in chemical nomenclature)* indicating the presence of a hydroxyl group attached to a carbon atom.

OLA *abbr. for* oligonucleotide ligation assay.

+olate *suffix* used as an alternative to **+oxide** (def. 1).

old yellow enzyme *an obsolete name for* the enzyme NADPH dehydrogenase (EC 1.6.99.1), a yellow flavoprotein (FMN in yeast, FAD in plants). *Compare* **new yellow enzyme**.

Ole *symbol for* the oleoyl (i.e. (Z)-octaden-9-enoyl) group.

oleandrose 2,6-dideoxy-3-*O*-methyl-*arabino*-hexose; the L enantiomer is a component of some **cardiac glycosides**.

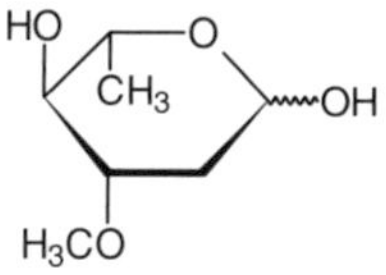

L-oleandrose

oleate 1 *numerical symbol*: 18:1(9); *the trivial name for* Z-octadec-9-enoate, CH_3–$[CH_2]_7$–CH=CH–$[CH_2]_7$–COO⁻ (*cis* isomer), the anion derived from oleic acid (i.e. Z-octadec-9-enoic acid), a monounsaturated straight-chain higher fatty acid. *See also* **octadecenoic acid**, **oleic family**. **2** any mixture of free oleic acid and its anion. **3** any salt or ester of oleic acid. *Compare* **elaidate**, **vaccenate**.

oleate desaturase *see* **phosphatidylcholine desaturase**.

olefin *an older name for* an alkene. —**olefinic** *adj.*

oleic family (of polyunsaturated fatty acids) a series of polyenoic acids in which the hydrocarbon chain terminates with the alkenyl grouping CH_3–$[CH_2]_7$–CH=CH– (as in oleic acid); there are three other series, i.e. the **linoleic family**, **linolenic family**, and **palmitoleic family**. The oleic family can be synthesized from acetyl-CoA: palmitoyl-CoA is first synthesized by the fatty-acid synthase (*see* **fatty-acid synthase complex**); next, by chain elongation and action of **desaturases**, stearoyl-CoA, oleoyl-CoA, and longer chain fatty acyl-CoA compounds are synthesized; then from their respective CoA derivatives, the cognate fatty acids are formed. The series includes (*all-Z*)-eicosa-5,8,11-trienoic acid. In **essential fatty acid deficiency** in mammals, longer chain polyunsaturated fatty acids of the oleic family are synthesized, and partially substitute for long-chain polyunsaturated fatty acids of the linoleic family. *See also* **eicosenoic acid**, **eicosatrienoic acid**, **Mead acid**.

olein *an old name for* any of the glyceryl esters of oleic acid (*see* **oleate**), especially the triester, **triolein** (i.e. trioleoylglycerol).

oleo (*US*) *abbr. for* oleomargarine.

oleomargarine *an alternative name (esp. US) for* **margarine**.

oleosin any of a family of integral membrane proteins (15–25 kDa) of the endoplasmic reticulum in cells of oilseeds that define the accumulation of triacylglycerols and formation of oil bodies (or spherosomes). Oleosins each consist of a conserved central hydrophobic region (70–80 residues) flanked by hydrophobic termini.

oleosome *an alternative name for* **spherosome**.

oleoyl *or (formerly)* **oleyl** *symbol*: Ole; *the trivial name for* (Z)-octadec-9-enoyl, CH_3–$[CH_2]_7$–CH=CH–$[CH_2]_7$–CO– (*cis* isomer), the acyl group derived from oleic acid (i.e. (Z)-octadec-9-enoic acid). It occurs very widely in natural lipids. *Compare* **elaidoyl**, **vaccenoyl**. *See also* **oleic family**.

1-oleoyl-2-acetylglycerol *abbr.*: OAG; an activator of protein kinase C.

oleum fuming sulfuric acid; a solution of SO_3 in concentrated $H_2SO_4 \rightarrow H_2S_2O_7$.

oleyl 1 *a common name for* Z-octadec-9-enyl, CH_3–$[CH_2]_7$–CH=CH–$[CH_2]_8$– (*cis* isomer), the monoalkenyl group derived from Z-octadec-9-ene; use deprecated. **2** *an old and misleading trivial name for* Z-octadec-9-enoyl; oleoyl now used.

olfactomedin a major glycoprotein component of the extracellular matrix of olfactory neuroepithelium.

olfactory receptor any integral membrane protein thought to be involved in effecting the sense of smell. They are G-protein-associated receptors, and the human genome contains several hundred genes for such receptors.

+olide *suffix (to the chemical name of an aliphatic hydrocarbon)* denoting conversion to a lactone having the same number of carbon atoms as the parent hydrocarbon.

oligo abbr. (colloquial) for **oligonucleotide**.

oligo+ or (before a vowel) **olig+** comb form denoting **1** few, little, not much, or not many. **2** (in chemistry and biochemistry) a small number of (the component specified); e.g. **oligopeptide**. **3** (in pathology) deficiency or insufficiency of (some thing or some attribute); e.g. **oliguria**. Compare **multi+**, **poly+**.

oligo-2′,5′-adenylate synthetase an enzyme of the sub-subclass EC 2.7.7 that binds double-stranded RNA and polymerizes ATP into pppA(2′p5′A)$_n$ oligomers; the latter activate the latent RNase L causing it to cleave single-stranded RNAs. The enzyme is induced by **interferons**. It is a homotetramer, associated with many different subcellular fractions. In humans, two forms, E16 and E18, are produced by alternative splicing of the same gene.

oligodendrocyte a type of glial cell that forms and supports the myelin sheath around axons in the central nervous system of vertebrates. In contrast to the **Schwann cell**, which performs this role in peripheral neurons for a single axon, one oligodendrocyte is involved in the myelination of several axons.

oligo dT or **oligo(dT)** a short single-stranded sequence of deoxythymine (dT), 12–18 nucleotides or longer, used for priming reactions catalysed by reverse transcriptase. The transcript is primed in the poly(A) tail of mRNA molecules.

oligo dT cellulose or **oligo(dT)-cellulose** short single-stranded sequences of deoxythymine (oligo dT) immobilized on cellulose particles and used as an affinity chromatography matrix for the purification of poly(A)$^+$ RNA.

oligo-1,6-glucosidase (EC 3.2.1.10), other names: sucrase-isomaltase; limit dextrinase; isomaltase; an enzyme that catalyses the hydrolysis of 1,6-α-D-glucoside linkages in isomaltose and dextrins produced from starch and glycogen by α-amylase. A similar enzyme (EC 3.2.1.48) is a type 2 transmembrane protein of the human intestinal brush border. This contains two catalytic domains that share 40 homology. Congenital sucrase-isomaltase deficiency is rare and is genetically heterogeneous.

oligohistidine the homopolypeptide of histidine; the term is used in referring to a **histidine tail**.

oligomer 1 (in chemistry and biochemistry) any substance or type of substance that is composed of molecules containing a small number – typically two to about ten – of constitutional units in repetitive covalent linkage; the units may be of one or of more than one species. Depending on the branch of chemistry concerned, an oligomer may or may not be considered to be a variety of **polymer**. **2** (in molecular biology and enzymology) any multimeric protein that contains a finite, relatively small, number of identical subunits (i.e. protomers) that are not in covalent linkage but in a state of reversible association with each other; the protomers may themselves each consist either of single polypeptide chains (i.e. monomers) or of two or more reversibly associated (identical or nonidentical) monomers. This definition is based exclusively on the identity of the subunits and does not restrict the number of polypeptide chains each identical subunit may contain. —**oligomeric** adj.; **oligomerize** or **oligomerise** vb.; **oligomerization** or **oligomerisation** n.

oligomycin any **macrolide** antibiotic produced by an actinomycete similar to Streptomyces diastatochromogenes. Oligomycin B is an inhibitor of oxidative phosphorylation in mitochondria, interfering with the passage of hydrons through the F$_0$ component of **H$^+$-transporting ATP synthase**, and of photophosphorylation in chloroplasts. Oligomycin D (**rutamycin**) is useful as an antifungal agent.

oligomycin sensitivity-conferring protein abbr.: OSCP; a protein component of mitochondrial ATP synthase that is required for the binding of the F$_1$ component to the F$_0$ component of the enzyme system. See also **H$^+$-transporting ATP synthase**.

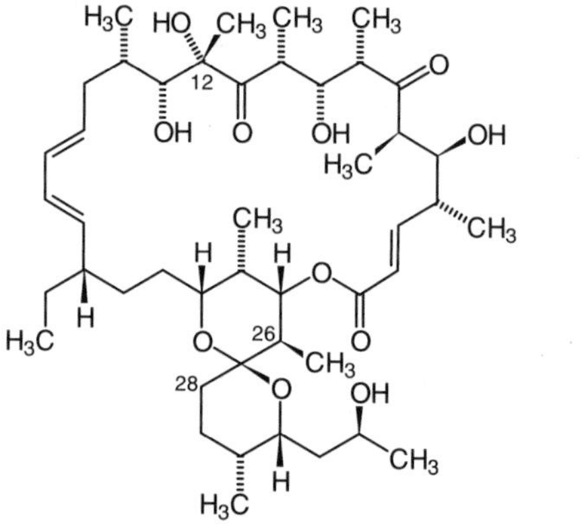

oligomycin A
oligomycin B: 28-oxo
oligomycin C: 12-deoxy
oligomycin D: 26-demethyl

oligonucleotide abbr. (colloquial): oligo; any molecule that contains a small number of nucleotide units connected by phosphodiester linkages between (usually) the 3′ position on the glycose moiety of one nucleotide and the 5′ position of the glycose moiety of the adjacent one. The number of nucleotide units in these small single-stranded nucleic acids (usually DNA) is variable but often in the range of 6 to 24 (hexamer to 24mer), although for some purposes it might be as long as 50mer. They are widely used as primers for reactions catalysed by DNA polymerases.

oligonucleotide-directed mutagenesis a method for introducing a mutation into a cloned gene in which the desired sequence alteration is incorporated into an oligonucleotide. The mutant oligonucleotide is annealed to its complementary sequence in the cloned gene, where it acts as a primer for DNA polymerase-catalysed transcription of a complementary strand of the gene of interest, and the plasmid in which it is cloned. A number of methods have been developed to ensure that the nonmutagenized strands of the original duplex are not propagated when used to transform E. coli cells.

oligonucleotide ligation assay abbr.: OLA; a method for detecting genetic variants that differ at only one base pair. A pair of oligonucleotides are designed so that they are contiguous when annealed to the template DNA strand and with the innermost 3′ end of one oligonucleotide in register with the point of sequence variation. Oligonucleotides designed to detect the normal sequence are then ligated by the action of a thermostable DNA ligase. Another pair of oligonucleotides are designed to be ligated when annealed to the mutant or polymorphic sequence. There will be a mismatch be-

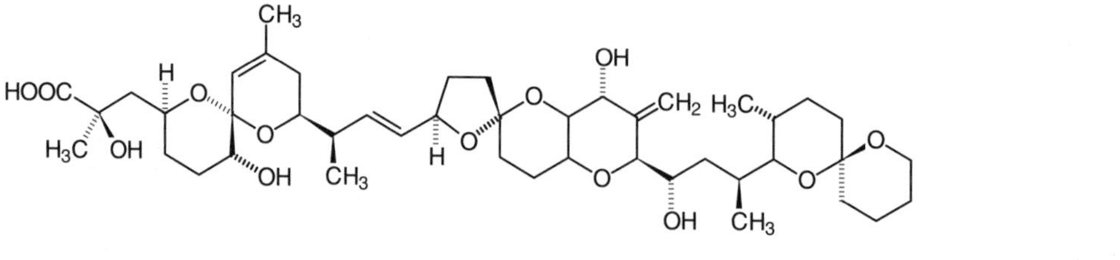

okadaic acid

tween the innermost 3' end of the 'normal' oligonucleotide and the mutant sequence, and between the 'mutant' oligonucleotide and the normal sequence so that in either case they will not be ligated to the contiguous oligonucleotide of the pair. Biotin at the outer 3' end of the oligonucleotide pair and a hapten at the outermost 5' end of the oligonucleotide pair are covalently linked if ligation takes place. Reaction products are immobilized by transferring them to streptavidin coated microtitre plates, washed and probed with anti-hapten antibody coupled to an enzyme such as alkaline phosphatase. A positive enzyme-catalysed reaction gives evidence of ligation of a particular oligonucleotide pair and hence for the cognate DNA sequence.

oligopeptide any molecule that contains a small number (2 to about 20) of amino-acid residues connected by peptide linkages.

oligophosphate 1 any ion or salt of an **oligophosphoric acid** — an acid containing commonly two or three, but sometimes four or more, phosphoric residues connected by anhydride linkages. **2** any mixture of a free oligophosphoric acid and its various anions. **3** *or* **oligophosphoric ester** any ester of an oligophosphoric acid.

oligophrenin-1 the protein product of a human gene associated with a form of X-chromosome-linked mental retardation.

oligosaccharide any molecule that contains 2 to about 20 monosaccharide residues connected by glycosidic linkages.

oligosaccharin any of a group of regulatory oligosaccharides, including a fungal heptaglucoside and a plant oligogalacturonide, that are released from cell walls and control functions such as growth, development, reproduction, and defence against disease. They also include xyloglucans.

N-oligosaccharyltransferase an enzyme that transfers an oligosaccharide with high mannose content from a lipid-linked oligosaccharide donor to an asparagine acceptor site on newly synthesized polypeptides to form glycoproteins. It has three subunits: **ribophorins** I and II, and a 48 kDa subunit.

oligotrophic describing a habitat, especially a lake or other mass of water, that is poor in nutrients capable of supporting the growth of aerobic plants and microorganisms, and (in the case of a lake, etc.) that hence contains abundant dissolved dioxygen at all depths.

olivomycin one of several related components of a complex of antibiotics produced by *Streptomyces olivoreticuli*. It has been investigated for antineoplastic activity. It is used as a dye specific for G–C base pairs, and stains the R bands first demonstrated in Giemsa staining.

+oma (*pl.* **+omas** *or* **+omata**) *suffix* denoting a tumour.

+ome *suffix forming nouns* **1** denoting an inclusive class of cellular constituents (e.g. metabolome, proteome, transcriptome) by analogy with genome. **2** denoting an entire set or a whole sphere of activity; universal. *See also* **+omics**. –**+omic** *adj.*; –**+omically** *adv.*

OMe *symbol for* the methoxy group, $CH_3–O–$ (alternative to MeO).

omega *symbol*: ω (lower case) *or* Ω (upper case); the twenty-fourth and last letter of the Greek alphabet. For uses *see* **Appendix A**.

Omega *an alternative name for* **Mega-8**.

omega-oxidation *or* ω-oxidation a minor oxidative metabolic pathway of fatty acids, particularly those of medium chain-length, in which the terminal methyl group of such an acid is oxidized first to a hydroxymethyl group and ultimately to another carboxyl group.

omega protein *or* ω-protein *an alternative name for* type I DNA topoisomerase.

omeprazole *the nonproprietary pharmaceutical name for* 5-methoxy-2- [(4-methoxy-3,5-dimethyl-2-pyridinyl)methyl]sulfinyl-1*H*-benzimidazole, a gastric proton-pump inhibitor. It blocks both basal and stimulated secretion of acid in the stomach by reversibly inhibiting H^+/K^+-transporting ATPase (EC 3.6.1.36) in the plasma membranes of parietal cells within the oxyntic glands of the gastric mucosa. It is useful clinically in the treatment of erosive reflux esophagitis, benign peptic ulcer, and Zollinger–Ellison syndrome. *Proprietary name*: Losec.

omepraxole

omicron *symbol*: o (lower case) *or* O (upper case); the fifteenth letter of the Greek alphabet.

+omics *suffix forming nouns* used to denote rigorous, systematic analyses of the omes: e.g., proteomics, analysis of the proteome using 2D gel electrophoresis and mass spectrometry to separate, identify and characterize all the constituent proteins; metabolomics, quantitative analysis of all the small molecular weight metabolites present inside a cell (or other sample); pharmacogenomics, analysis of drug responses with respect to genomic information (e.g., polymorphisms), aiming to customize therapies and reduce adverse drug reactions; toxicogenomics, analysis of the toxic effects of potential new drugs on gene expression patterns in target cells or tissues, aiming both to prioritize compound pipelines (to eliminate expensive failures in drug development), and to reveal genetic signatures that may predict toxicity in these compounds. *See* ome.

OMIM *abbr. for* Online Mendelian Inheritance in Man; *see* **specialist database**.

ommatidium any of many identical structural units of the compound eye found in many invertebrates. In *Drosophila*, each consists of eight photoreceptor cells and 12 accessory cells. In sevenless and bride-of-sevenless mutants, one of the photoreceptor cells fails to develop.

ommatin a type of **ommochrome**.

ommatochrome *an alternative name for* **ommochrome**.

ommin a type of **ommochrome**.

ommochrome *or* ommatochrome any yellow, brown, red, or violet natural polycyclic pigment especially common in the Arthropoda, particularly in the ommatidia of the compound eye. Ommochromes contain a phenoxazine nucleus and are formed biologically from tryptophan via kynurenine. They are divided into two groups: **ommatins**, which are alkali-labile, of low relative molecular mass, and rather weakly coloured; and **ommins**, which are alkali-stable, of higher relative molecular mass, and more strongly coloured than the ommatins.

OMP *abbr. for* orotidine monophosphate; *see* **orotidine phosphate**.

OMP decarboxylase *abbr. for* orotidine-5'-phosphate decarboxylase, a catalytic activity present in a domain of the multifunctional enzyme **uridine 5'-monophosphate synthase**.

+onate *noun suffix* denoting the anion, a salt, or an ester of an aldonic acid. *See also* **uronate**.

onco+ *comb. form* denoting **1** tumour. **2** volume.

oncofetal parvalbumin *see* **oncomodulin**.

oncogen any agent that causes **oncogenesis**; the term embraces **carcinogen**.

oncogene any gene associated with the causation of cancer. The first oncogenes were isolated from acutely transforming **retroviruses**; the retroviral oncogenes (v-*onc*) were then found to be related to, and probably derived from, normal cellular genes (c-*onc*), which came to be termed **protooncogenes**; these in most cases are involved with intrauterine growth control. Protooncogenes become oncogenes as a result of (1) deletions of portions of the protooncogene; (2) mutation of the protooncogene resulting in a protein that is constitutively active; (3) translocation of the protooncogene or a part of it to produce a hybrid gene and a hybrid protein with aberrant function; or (4) amplification of the gene resulting in inappropriate expression of the product. Protooncogenes function in normal cells in growth and differentiation pathways. Some have products that act as growth factors (e.g. *sis*, *int-2*) or growth-factor receptors (e.g. *fms*, *neu*, *kit*). Others encode signal-transduction molecules normally activated by growth-factor receptors (e.g. *ras*, *raf*, *mos*, and *src* family members). Others encode proteins that function in the nucleus (e.g. *myc*, *myb*, *fos*, *jun*) where they influence the expression of other genes. For further details see under individual oncogenes. Strictly speaking, those genes carried by DNA tumour viruses, such as **SV40** and **adenoviruses**, should also be construed as oncogenes. Unlike the retroviral genes, however, they appear to have no cellular counterparts but instead operate by subverting the actions of cellular genes (such as protooncogenes) that normally function to control cell growth. *See also* **FMS, FOS, int-2, JUN, KIT, MOS, MYB, MYC, neu, RAF, RAS, sis, src**.

oncogenesis the production or formation of a neoplastic tumour, whether benign or malignant; the term embraces **carcinogenesis**.

oncogenic (of an agent) causing the formation of a neoplastic tumour or tumours; of or pertaining to **oncogenesis**; the term embraces carcinogenic. —**oncogenicity** n.

oncology the branch of medicine concerned with the study and treatment of neoplastic tumours. —**oncological** adj.; **oncologist** n.

oncolysis the destruction of neoplastic tumours or tumour cells, whether spontaneous or in response to treatment. —**oncolytic** adj.

oncomodulin other names: parvalbumin beta; oncofetal parvalbumin; an **EF-hand** Ca^{2+}-binding protein that may function in a similar manner to **calmodulin**. Expression is limited to early embryonic stages, the placental cytotrophoblasts and neoplastic tissues.

oncoprotein any protein that is the product of an **oncogene**. There are some 100 potential oncogenes in the human genome, and their products have various functions. These include growth-factor-related functions; as protein tyrosine kinases (receptor, membrane-associated, or cytosolic); as nonprotein tyrosine kinase receptors; as membrane-associated G proteins; as cytosolic protein serine/threonine kinases; as cell-cycle regulators; as transcription factors; in DNA repair; or mitochondrial-membrane related.

oncornavirus an old name for a group of RNA tumour viruses now classified in the subfamily Oncovirinae. *See* retrovirus.

oncostatin M abbr.: OSM; a polypeptide cytokine, produced by activated T cells and PMA-treated monocytes, that inhibits the growth of a variety of cancer cells, but stimulates growth of AIDS-Kaposi's sarcoma-derived cells and normal fibroblasts. It is a member of the **interleukin 6** subfamily. It exists as a monomer.

oncotic pressure the **colloid osmotic pressure** of plasma proteins; it is important as an index of the flow of water between the blood and the fluid in the tissues.

+one noun suffix denoting ketone.

one-carbon compound an alternative name for **C₁ compound**.

one cistron–one polypeptide hypothesis an alternative name for the **one gene–one polypeptide hypothesis**.

one gene–one antigen hypothesis a hypothesis stating that each of the genes in a tissue transplant governing the host's reaction to it is responsible for the manufacture of a particular transplantation antigen and for that antigen alone. It was first formulated by the US geneticist George Wells Beadle (1903–89) and US microbiologist Edward Lawrie Tatum (1909–75).

one gene–one enzyme hypothesis a hypothesis stating that every enzyme is synthesized under the control of a specific gene. It was subsequently refined into the **one gene–one polypeptide hypothesis**.

one gene–one polypeptide hypothesis or **one cistron–one polypeptide hypothesis** a hypothesis stating that each naturally occurring polypeptide is synthesized biologically under the control of a specific gene, which is transcribed into a unique species of messenger RNA.

one gene–one reaction hypothesis a hypothesis stating that every specific biochemical reaction is under the ultimate control of a different gene. It was later redefined as the **one gene–one enzyme hypothesis**.

+onic suffix denoting the formal conversion of an aldose to the corresponding aldonic acid; e.g. glucose gives rise to gluconic acid.

+onium noun suffix denoting a cationic molecular entity in which a heteroatom is bonded to one more univalent ligand than is normal for a neutral molecule containing that heteroatom and therefore bears a formal positive charge; e.g. ammonium, H_4N^+; dimethylammonium, $(CH_3)_2H_2N^+$; diphenyliodonium, $(C_6H_5)_2I^+$; trialkylsulfonium, R_3S^+ (where R = alkyl). *See also* **carbonium ion** (def. 2).

ONPG abbr. for ortho-nitrophenyl-β-D-galactopyranoside; a chromogenic substrate used for the detection and assay of β-galactosidase activity.

on-rate the constant, k_{+1}, for the association step $L + B \rightarrow LB$ in the equilibrium $L + B = LB$, where L represents ligand and B the molecule it binds; the rate of association of ligand to form a complex. The dimensions of k_{+1} are $time^{-1}$. *Compare* **association constant, off-rate**.

Onsager theory an extension of the **Debye–Hückel theory** of dielectrics. *See also* **Kirkwood theory**. [After Lars Onsager (1903–76), Norwegian-born US physical chemist.]

ontogeny or **ontogenesis** the sequence of events in the development of an individual organism during its lifetime. *Compare* **phylogeny**. —**ontogenic** or **ontogenetic** adj.

ontology (in logic) the formal specification of the terms and concepts used in a field or domain of knowledge providing definitions of those terms and of the relationships between them. –**ontological** adj.; –**ontologically** adv.

oo+ comb. form denoting ovum or egg.

oocyte (in zoology) a cell of an animal ovary that undergoes meiosis to form an ovum. A **primary oocyte** is formed by differentiation of an oogonium and then undergoes the first division of meiosis to form a polar body and a secondary oocyte. Following fertilization of the egg, the **secondary oocyte** undergoes the second meiotic division to form the mature ovum and a second polar body.

OODB abbr. for object-oriented database.

oogenesis the complete process of formation and maturation of an ovum from a primordial female germ cell. —**oogenetic** adj.

oogonium (pl. oogonia) **1** (in zoology) a cell of an animal ovary that undergoes repeated mitosis, eventually forming **oocytes**. **2** (in botany) the female reproductive organ in certain algae and fungi. —**oogonial** adj.

OP abbr. for organophosphorus. *See* **organophosphate**.

opal codon or **opal triplet** symbol: UGA; one of the three terminator codons or **nonsense codons** in a molecule of messenger RNA.

opalescence a milky iridescence, similar to that of opal. In some solutions it is due to the reflection of light by very fine suspended particles. —**opalescent** adj.; **opalesce** vb.

opal mutant any cell or virus carrying an **opal mutation**.

opal mutation a mutation that gives rise to an **opal codon** in a molecule of messenger RNA.

opal suppressor a mutation, supU (Trp), in E. coli that codes for an altered tRNA so that its anticodon can recognize the **opal codon** (UGA) in mRNA thereby allowing insertion of tyrosine into a polypeptide chain at that site.

open chain (in chemistry) **1** a chain of atoms in a molecule that is not formed into a ring, i.e. one that has two (or more) free ends. **2** **open-chain** an alternative term for **acyclic**.

open circle any circular duplex DNA structure containing one or more single-stranded nicks. Such a structure will dissociate at high pH into two (or more) single-stranded structures.

open circuit (in physics) an incomplete electrical circuit, i.e. one with infinite impedance.

open-loop system a type of system, e.g. a metabolic pathway or an electrical circuit, in which control is not exerted by feedback, each operation or activity being affected only by those earlier in the sequence. *Compare* **closed-loop system**.

open reading frame *abbr.*: ORF; a series of codon triplets, deduced from a DNA sequence, that include a 5′ **initiation codon** running through to a **termination codon**, and representing a putative or known gene. For a stretch of DNA to represent a gene it must, apart from including appropriately placed initiation and termination codons, be of a suitable length and should fulfil certain criteria indicating that the correct **reading frame** has been selected. Methods for checking this include the **RYN method**.

open system *(in thermodynamics)* any geometrically defined volume that can exchange both energy (e.g. heat or work) and matter with its surroundings. *Compare* **adiabatic system**, **closed system**.

open tetrapyrrole *see* **phycobilin**.

operand a quantity or function on which a mathematical operation is to be performed.

operator 1 *(in genetics)* a DNA sequence within an **operon** (under negative promoter control) to which a specific repressor protein can bind, thereby regulating the functioning of a **structural gene** or group of genes. **2** *(in mathematics)* a symbol (i.e. an alphabetic character, term, or special symbol) representing a particular operation to be performed on an operand or operands; e.g. Σ (= sum of), d/dx (= differentiate with respect to the variable x), + (= add to).

operon a unit of coordinated and regulated gene activity found in prokaryotes, by means of which the control of the synthesis of a protein or a group of (usually functionally associated) proteins is determined. It consists of a segment of genomic DNA containing a **structural gene** or a linear sequence of structural genes (which are transcribed as a single unit), together with one or more regulatory regions such as an **operator** (def. 1). *See also* **Jacob–Monod model**, **lac operon**.

ophidin *a trivial name for* 2-methylcarnosine; N^α-(β-alanyl)-2-methylhistidine; a constituent of snake muscle.

ophiobolin a class of sesterterpene produced only by fungi; there are six types based on the ophiobolane structure shown.

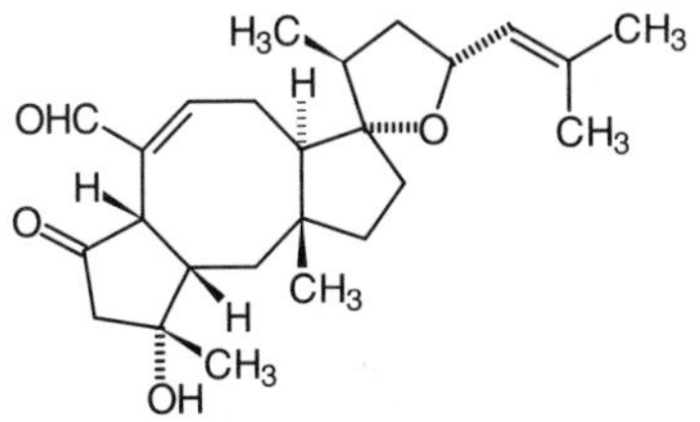

ophthalmic of, or pertaining to, the eye.

ophthalmic acid N-[N-(γ-glutamyl)-α-aminobutyryl]glycine; an analogue of glutathione in which the thiol group is replaced by a methyl group. It was originally isolated from the lens of the eye.

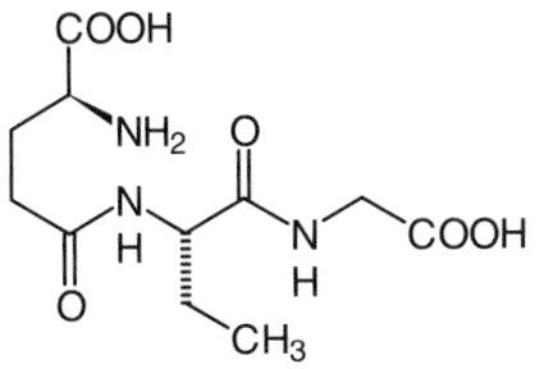

opiate 1 any of a group of narcotic drugs structurally related to and including **morphine** all of which are derived from opium. The group includes **codeine**, **papaverine**, and **thebaine**. The term is sometimes extended to include the **opioids** (def. 1) and any other naturally occurring, semisynthetic, or synthetic narcotic substance with similar properties and whose effects may be reversed by recognized mor-

phine antagonists. **2** consisting of or containing **opium**; having narcotic properties similar to those of opium.

opiate receptor a membrane receptor for opiates (*see* **opioid receptor**.)

opine any of a class of unusual derivatives of basic amino acids not found in normal plant tissue, but found in plant tissue that has been transformed by oncogenic plasmids from agrobacteria. The class includes the **nopaline** and **octopine** families.

opiocortin *an alternative name for* **opiomelanocortin**.

opioid 1 any nonalkaloid having **opiate**-like pharmacological effects that can be reversed by **naloxone** or other recognized morphine antagonists. **2** describing any substance having such properties.

opioid peptide any naturally occurring peptide that is an **opioid** (def. 1). Five main groups are recognized: (1) [Leu]enkephalin and [Met]enkephalin; (2) dynorphin, α and β neoendorphin, and several other peptides that arise (or are presumed to arise) from **enkephalin** precursors; (3) α, β, γ, and δ **endorphins**, all of which are formed from β lipotropin; (4) various pronase-resistant peptides present in body fluids, e.g. β **casomorphin** (in bovine milk); (5) sundry other peptides whose opiate-like action appears to be indirect. [*Note*: in the past, the terms 'endorphin' and 'opioid peptide' were often used synonymously; it is now recommended that the former should be restricted to those opioid peptides that are derived from β lipotropin.]

opioid receptor any G-protein-associated membrane receptor that interacts in a highly selective manner with an opiate or an **opioid** (def. 1) and that mediates its prime pharmacological action; types μ, δ, and κ, and possibly σ have been distinguished on the basis of their response to β endorphin (End), dynorphin A (Dyn), [Met]enkephalin (Met), and [Leu]enkephalin (Leu); the potency being, for μ: End > Dyn > Met > Leu; for δ: End = Leu = Met > Dyn; and for κ: Dyn >> End >> Leu = Met. Morphine and its analogues are primarily μ agonists, but also have activity on δ and κ receptors. μ and δ depress cyclic AMP levels and open a G-protein-modulated K^+ channel, κ inhibits a G-protein-modulated Ca^{2+} channel. Some opioids also bind to other receptors known as σ, but morphine has a very low affinity for these receptors and naloxone does not block its action.

opiomelanocortin *or* **opiocortin** a polypeptide, present in the intermediate lobe of the pituitary gland, that includes within it the entire sequences of the peptide hormones **corticotropin**, α and β melanocyte stimulating hormone, β and γ **lipotropin**, and β **endorphin**. It is formed from **proopiomelanocortin**.

opium the milky, air-dried exudate from incised, unripe seed capsules of the opium poppy, *Papaver somniferum*, and some other species. Opium contains approximately 20 alkaloids, the most significant of which are **codeine**, **morphine**, **papaverine**, and **thebaine**.

OPLC *abbr. for* overpressure(d) layer chromatography.

OPRTase *see* orotate phosphoribosyltransferase.

opsin 1 any of a group of hydrophobic glycoproteins that occur in the visual pigments of vertebrates in equimolecular combination (as Schiff's bases) either with 11-*cis*-retinal – in **iodopsin** and **rhodopsin** – or with 3,4-didehydro-11-*cis*-retinal – in **cyanopsin** and **porphyropsin**. These opsins are of ~35–40 kDa, and consist of single-chain polypeptides glycosylated with two branched oligosaccharides. *See also* **photopsin**, **scotopsin**. **2** any of a group of proteins of ~25 kDa that occur in combination with retinal in the purple-membrane pigments present in some strains of *Halobacterium halobium*; e.g., bacterioopsin (in bacteriorhodopsin), haloopsin (in halorhodopsin), and slow-opsin (in slow-cycling rhodopsin).

opsonify *a less common word for* **opsonize**.

opsonin any blood-serum protein that, when combined with microorganisms or other particulate material (e.g., foreign erythrocytes), increases the latter's susceptibility to phagocytosis. The two main classes are immunoglobulin G antibodies, and the C3b and IC3b fragments formed from the C3 component of **complement**; other classes include **fibronectin** and the soluble mannose-binding protein of serum. An opsonin molecule acts by linking the surface of the particulate matter to a specific receptor on the surface of a phagocytic cell; commonly, this receptor is either an Fc receptor (recognizing the Fc domain of an IgG immunoglobulin molecule) or a CR3 receptor (recognizing the IC3b fragment of complement). —**opsonic** *adj*.

opsonize *or* **opsonise** *or (less commonly)* **opsonify** to render (microorganisms or other particulate material) more susceptible to phagocytosis by coating with **opsonin**. —**opsonization** *or* **opsonisation** *or* **opsonification** *n.*

optic 1 *or* **optical** of or pertaining to the eye or the sense of sight; aiding vision. **2** *a less common word for* **optical** (def. 1, 2).

optical 1 *or* **optic** of, pertaining to, or involving light. **2** *or* **optic** of or pertaining to **optics**. **3** *a less common word for* **optic** (def. 1).

optical activity the ability (of chiral substances or their solutions, and chiral crystals) to rotate the plane of polarization of a transmitted beam of plane-polarized light. There is usually a correspondence between **chirality** of molecules and optical activity. *See also* **chiroptical**.

optical antipode an older term for **enantiomer**.

optical density *a term (no longer recommended) for* **absorbance**.

optical isomer *an older term for* **enantiomer**.

optical isomerism *an older term for* **enantiomerism**.

optically active having or displaying optical activity; the term is often used interchangeably with **chiral**.

optically inactive lacking optical activity (through being either **achiral** or **racemic**).

optical purity the ratio (usually expressed as a percentage) of the **specific optical rotation** of a sample of an enantiomeric substance to the specific optical rotation of a pure single enantiomer of the substance. *Compare* **enantiomeric purity**.

optical rotation *or* **rotation 1** the phenomenon of rotation of the plane of polarization of plane-polarized light on its transmission by (solutions of) chiral substances, and by chiral crystals. By convention, optical rotation is designated as rightwards, or positive – and termed **dextrorotation** – when it is clockwise as seen by an observer looking towards the light source, and designated as leftwards, or negative – and termed **levorotation** – when it is anticlockwise. **2** *or* **optical rotatory power** the ability of (a solution of) a chiral substance to effect such rotation, an effect that is normally associated with **optical activity**. **3** *symbol*: α; any measure, expressed in angular degrees, of such ability under specified conditions; *see* **molar optical rotation, specific optical rotation**.

optical rotatory dispersion *abbr.*: ORD; the variation of **optical rotation** with the wavelength or frequency of the transmitted light. The shape of an optical rotatory dispersion spectrum of a solution of a macromolecule often gives useful information about the conformation of the solute. *See also* **circular dichroism, Cotton effect**.

optical rotatory power *an alternative term for* **optical rotation** (def. 2).

optical system any arrangement of diffraction gratings, lenses, mirrors, prisms, etc., frequently within a specific instrument and used for a specific purpose.

optical tweezers a technique for trapping small particles or macromolecules in a strongly focused laser beam. It is used for examination of individual molecules or cellular components, such as contractile proteins, and the forces they produce.

optics 1 *(functioning as sing.)* the branch of physics concerned with the nature, properties, and behaviour of (visible) light and with the characteristics of optical devices, instruments, and systems. The term is extended to other kinds of electromagnetic or particulate radiation in so far as they have similar properties to light. **2** *a collective term for* an arrangement of optical components (in an instrument); the optical properties of such an arrangement.

optineurin a 66 kDa protein of unknown function that is expressed in retina, brain, and various other tissues, and is present in the aqueous humour of eyes of many vertebrate species. It interacts with huntingtin, Rab8 (a Ran-associated protein), and transcription factor IIIA (*see* **TFIII**). It is thought to be neuroprotective in the eye and optic nerve. Several mutations are associated with an autosomal dominant adult-onset glaucoma. [From optic neuropathy-inducing protein.]

OPTLC *abbr. for* overpressured thin-layer chromatography. *See* **overpressure(d) layer chromatography**.

Oracle *see* **relational database management system**.

oral of or pertaining to the mouth; administered by mouth.

Orb the *Drosophila* homologue of CPEB of *Xenopus laevis*. It is re-

quired for translation of the mRNA for oskar in oocytes. *See* **cytoplasmic polyadenylation element**.

orbit 1 a closed circular or near-circular course taken by a moving body (often around something). **2** the depression in the skull that contains the eye. **3** to move in or as if in an **orbit** (def. 1). *See also* **orbital**.

orbital 1 any **wave function**, Ψ, of an electron in a molecular entity. It is characterized by a particular energy value and a unique set of three quantum numbers; the counterpart in wave mechanical theory of an electron orbit in Niels Bohr's concept of atomic structure. Any particular orbital may be occupied either by one electron or by two electrons of opposed spins, or it may be unoccupied. The square of the value of Ψ at any point can be interpreted as the probability of the electron being located at that point, or as the average fractional charge density at that point. *See also* **atomic orbital, molecular orbital, pi orbital, sigma orbital**. **2** of, relating to, occupying, or moving in or as if in an **orbit** (def. 1). **3** belonging to the **orbit** (def. 2).

orbital shaker a device designed to impart an orbital motion to flasks, tubes, etc. with the purpose of mixing their contents and facilitating the transfer of gases from the gaseous to the liquid phase. It is useful especially for growing cultures of cells.

ORC *abbr. for* origin of replication complex *or* origin recognition complex; *other name*: prereplication complex. A protein complex in eukaryotes, comprising a hexamer of subunits ORI1 to ORI6, that recognizes and binds to the **origin of replication** region of chromosomal DNA to initiate replication. A replisome is assembled on it. It is regulated by several proteins involved in control of the cell cycle.

orcinol test *or* **Bial's test** a colorimetric test specific for aldopentoses. The sample is heated with strong acid to convert any aldopentose present to furfural (i.e. 2-furaldehyde), which is then reacted with orcinol (i.e. 5-methyl-1,3-benzenediol) in the presence of ferric chloride to give a blue-green compound extractable into amyl alcohol. The test is used for the determination of RNA and the detection of **pentosuria**. [After Manfred Bial (1870–1908), German physician.]

ORD *abbr. for* optical rotatory dispersion.

Ord *symbol for* a residue of the ribonucleoside orotidine (alternative to O).

order 1 *(in chemistry)* sequence (of residues in a polymer). **2** **order of reaction**. **3** *(in taxonomy)* a category consisting of a number of similar families (sometimes only one family). One or more orders make up a class. **4** *(in mathematics)* the number of times a function must be differentiated to obtain a particular derivative. **5** *(in mathematics)* the number of columns or rows in a determinant or a square matrix. **6** **order of magnitude**.

ordered mechanism *or* **ordered pathway** (for an enzymic reaction involving two substrates, $A + B \rightleftharpoons C$) a mechanism in which the enzyme, E, must necessarily first bind to one substrate, e.g. $E + A \rightleftharpoons EA$, and then to the second substrate, thus $EA + B \rightleftharpoons EAB$, this ternary complex then reacting to form the product, $EAB \rightleftharpoons E + C$. *Compare* **random mechanism**.

order of magnitude 1 a tenfold range of magnitude extending from a particular value either upwards or downwards. **2** a magnitude expressed to the nearest power of ten.

order of reaction *symbol*: n; the sum of the exponents of the concentration terms in the rate equation for an elementary chemical reaction considered in one direction only. It may also be applied to certain composite reactions.

order of reflection *symbol*: n; the integer, n, in the equation of **Bragg's law**.

ordinate the vertical, or y, coordinate in a plane rectangular (Cartesian) coordinate system. *Compare* **abscissa**.

ORESTES an alternative name for **ORF ESTs**.

orexin an alternative name for **hypocretin**.

ORF *abbr. for* open reading frame.

ORF ESTs *or* **ORESTES** *abbr. for* open reading frame expressed sequence tags; expressed sequence tags (ESTs) that provide sequence information along the whole length of a transcript, rather than just its ends. The method involves low-stringency polymerase chain reaction (PCR) to produce cDNA libraries, samples of which are then sequenced.

organ any part of the body of a multicellular organism that is adapted and/or specialized for the performance of one or more vital functions. The term is sometimes extended to include organelles of unicellular organisms.

organ culture a category of **tissue culture**, in which an organ, part of an organ, or an organ primordium, after removal from an animal or plant, is maintained *in vitro* in a nutrient medium with retention of its structure and/or function.

organelle any discrete structure in a unicellular organism, or in an individual cell of a multicellular organism, that is adapted and/or specialized for the performance of one or more vital functions.

organic 1 of, pertaining to, or derived from an organism or organisms. **2** of, pertaining to, or affecting an organ or organs. **3** of, relating to, or being a compound of carbon, whether or not of natural origin. **4** of any element contained in an organic compound. **5** any organic compound.

organic chemistry the branch of chemistry concerned with the compounds of carbon in covalent linkage, whether natural or artefactual, including most compounds of biological origin but excluding some simple carbon compounds, e.g. oxides, sulfides, and cyanides. *See also* **organic compound**.

organic compound any compound containing carbon in covalent linkage; various simple carbon compounds, especially carbon dioxide, carbonic acid, and salts and ions of carbonic acid, are usually excluded.

organification defect a defect in the iodination of tyrosine residues in thyroglobulin and in their coupling to produce thyroid hormones. It causes a congenital form of hypothyroidism. The reactions are catalysed by thyroperoxidase.

organiser *a variant spelling of* **organizer**.

organism any unicellular or multicellular prokaryote or eukaryote, usually extended to include various noncellular nucleic-acid-containing infective agents such as viruses.

organizer *or* **organiser** *(in embryology)* any part of an embryo that, through substances produced by it, stimulates the morphological development and differentiation of other parts.

organo+ *comb. form denoting* **organic** (def. 1, 2, 3).

organogenesis the formation and development of organs in an organism. —**organogenetic** *adj*.

organoleptic affecting, involving, or making use of the sense organs, especially those of smell and taste.

organomercurial 1 *or* **organomercury** of, pertaining to, or being an organic compound containing mercury, especially one in which a mercury atom is linked directly to one or more carbon atoms. **2** *or* **organomercury compound** any such compound.

organometallic 1 *or* **organometal** of, pertaining to, or being a metal-containing organic compound, especially one in which a metal atom is linked directly to one or more carbon atoms. **2** *or* **organometal compound** any such compound.

organophosphate any organic compound containing a phosphoric (including fluorophosphoric, thiophosphoric, etc.) residue in amide, anhydride, ester, or thioester linkage. The best known is diisopropylphosphofluoridate, which inhibits serine proteases (including acetylcholinesterase, crucial for nerve transmission) by blocking an essential serine residue. Organophosphorus compounds were originally developed as nerve gases, and later used as agricultural insecticides and as anthelmintics. *See also* **aminoethylbenzenesulfonylfluoride, diisopropyl fluorophosphonate, phenylmethylsulfonyl fluoride**.

organotroph *an alternative term for* **heterotroph(e)**.

organotrophic 1 heterotrophic; *see* **heterotroph(e)**. **2** of, pertaining to, or influencing the formation or nourishment of bodily organs.

organyl any organic substituent group, regardless of functional type, that has a single free valence at a carbon atom; e.g. ethyl, CH_3CH_2-; glycyl, $H_2NCH_2C(=0)-$; ribosyl, $C_5H_9O_4-$.

ORI *abbr. for* origin of replication.

oriC an AT-rich region of 245 bp on the *E. coli* chromosome where DNA replication is initiated. Elements of the sequence are highly conserved among Gram-negative bacteria. A complex of DNA proteins (up to 30 subunits) binds to *oriC* causing the duplex DNA to melt using energy released by ATP hydrolysis.

origin *(in mathematics)* the point of intersection of a set of (usually two or three) coordinate axes; the point whose coordinates are all zero. —**original** *adj*.

origin of replication *abbr.*: ORI; a site on DNA at which DNA replication begins. Eukaryotic chromosomes contain many such sites, whereas prokaryotic chromosomes, mitochondrial DNA, and chloroplast DNA contain only one. ORIs are AT-rich. Most contain repetitive nucleotide sequence motifs that vary in sequence, length, and distribution in different organisms; most also contain inverted repeats of various sizes. *See* **replication of DNA**.

origin of replication complex *see* ORC.

origin recognition complex *see* ORC.

O-ring a toroidal ring, usually of circular cross-section, made of neoprene or similar material and useful as a fluid- or gas-tight seal between particular parts of a device while permitting relative movement between them.

Orlistat *another name for* **Xenical**.

Orn *symbol for* a residue of the α-amino acid L-ornithine.

ornaline *an alternative name for* **nopalinic acid**.

ornithine *symbol*: Orn; α,δ-diaminovaleric acid; 2,5-diaminopentanoic acid; an amino acid only rarely found in proteins. L-Ornithine is an intermediate in the biosynthesis of arginine (*see* **ornithine–urea cycle**) and, in plants, of pyrrolidine and tropane alkaloids; it was first found as its dibenzoyl conjugate, **ornithuric acid**. D-ornithine is a component of **bacitracin** A.

ornithine acetyltransferase *see* **glutamate N-acetyltransferase**.

ornithine-δ-aminotransferase *see* **OAT**.

ornithine carbamoyltransferase *abbr. (in clinical chemistry)*: OCT; EC 2.1.3.3; *systematic name*: carbamoyl-phosphate:L-ornithine carbamoyltransferase; *other names*: ornithine transcarbamylase; citrulline phosphorylase. An enzyme of the **ornithine–urea cycle** that catalyses a reaction between carbamoyl phosphate and L-ornithine to form L-citrulline and orthophosphate. It is a homotrimer of the mitochondrial matrix. More than a hundred point mutations and gene deletions cause an X chromosome-linked enzyme deficiency in male humans and congenital **hyperammonemia**. Crystal structures of *E. coli* and human enzymes are nearly isomorphous and contain a total of 14 alpha helices and 9 beta strands organized into two domains.

ornithine decarboxylase EC 4.1.1.17; *systematic name*: L-ornithine carboxy-lyase; an enzyme that catalyses the decarboxylation of L-ornithine to putrescine and CO_2; pyridoxal-phosphate is a coenzyme. It is the first enzyme of polyamine biosynthesis; its activity increases in response to many agonists, especially those stimulating cell growth. *See* **antizyme**.

ornithine transacetylase *see* **glutamate N-acetyltransferase**.

ornithine transporter 1 a member of the mitochondrial carrier family of transport proteins that reside in the inner mitochondrial membrane. The human protein (301 amino acids) consists of three homologous repeats, each of ≈100 residues and containing two transmembrane segments. Several mutations of ornithine transporter 1 are associated with the HHH syndrome (hyperammonemia, hyperornithinemia, and homocitrullinuria).

ornithine–urea cycle *or* **ornithine cycle** *or* **arginine–urea cycle** *or* **urea cycle** a cyclic metabolic pathway, present in mammals and other ureotelic animals, that converts waste nitrogen, in the form of highly toxic ammonium, to essentially nontoxic urea, which may then be excreted. Urea is formed by hydrolysis of L-arginine to L-ornithine, the cycle being completed by conversion of ornithine to L-citrulline by carbamoylation with carbamoyl phosphate (previously formed by enzymic combination of carbon dioxide and ammonium with phosphate from ATP), reaction of the citrulline with L-aspartate to form L-argininosuccinate, and cleavage of this to produce fumarate and regenerate arginine. Two enzymes of the cycle (carbamoylphosphate synthetase and ornithine carbamoyltransferase) reside in the mitochondrial matrix, the rest being cytosolic. Deficiency of any enzyme of the cycle leads to hyperammonemia. The cycle is also sometimes called the **Krebs urea cycle** or the **Krebs–Henseleit cycle**, after its discoverers. The role of argininosuccinate was revealed by the US biochemist, Sarah Ratner (1903–99), in 1954.

ornithuric acid N,N'-dibenzoylornithine; a detoxication product of benzoic acid found in the excreta of birds.

Oro *symbol for* a residue of the pyrimidine base orotate.

orosomucoid *or* **α₁-acid glycoprotein** a ≈40 kDa, water-soluble plasma protein; it consists of a single polypeptide chain, with two intrachain disulfide bonds and carrying five branched, highly sialylated oligosaccharide units attached to the *β*-carboxyl groups of asparagine residues. Certain alleles of the gene encoding this protein have been found to be significantly increased in patients with different types of carcinoma.

orotate **1** *symbol*: Oro; uracil-6-carboxylate; 1,2,3,6-tetrahydro-2,6-dioxo-4-pyrimidine carboxylate; the anion of orotic acid (i.e. uracil-6-carboxylic acid). It is an intermediate in the biosynthesis of pyrimidine nucleotides from aspartate. **2** any mixture of orotic acid and its anion. **3** any salt or ester of orotic acid.

orotate phosphoribosyltransferase *abbr.*: OPRTase; EC 2.4.2.10; *other names*: orotidylic acid phosphorylase; orotidine-5′-phosphate pyrophosphorylase; an enzyme of the pathway for *de novo* biosynthesis of pyrimidines. It catalyses the formation of orotidine 5′-phosphate from orotate and 5-phospho-*α*-D-ribose 1-diphosphate, releasing pyrophosphate. Orotidine 5′-phosphate is decarboxylated to uridine 5′-phosphate by **orotidine-5′-phosphate decarboxylase**. The human protein consists of 480 amino acids and contains domains for OPRTase (residues 1–214) and orotidine 5′-phosphate decarboxylase (residues 221–480). Deficiency causes the recessive disease **orotic aciduria** type 1.

orotic aciduria a rare hereditary disorder of pyrimidine synthesis in humans that is characterized by retarded growth, hyperchromic anemia unresponsive to the usual hematinics, and an excessive excretion of orotate in the urine. It is due to deficiency of activity in the bifunctional **orotate phosphoribosyltransferase** (EC 2.4.2.10) and **orotidine-5′-phosphate decarboxylase** (EC 4.1.1.23), now called uridine 5′-monophosphate synthase. This protein is bicatalytic in mammals, but the two activities occur on separate proteins in prokaryotes and some eukaryotes. The condition results in the subject being a 'pyrimidine auxotroph' and can be effectively treated by administration of uracil, which both rectifies the pyrimidine deficiency and indirectly leads, through the effect of cytidine triphosphate on aspartate transcarbamoylase, to inhibition of orotate synthesis.

orotidine *symbol*: O *or* Ord; 6-carboxylatouridine; orotate riboside; 1-*β*-D-ribofuranosylorotate; a ribonucleoside occurring primarily as its 5′-phosphate, orotidylic acid but produced by certain mutants of *Neurospora crassa* and excreted in relatively large amounts in the urine of cancer patients treated with 6-azauridine. It is not known to be a component of ribonucleic acid.

orotidine monophosphate *abbr.*: OMP; *an alternative name for* any **orotidine phosphate**, but in particular for orotidine 5′-phosphate.

orotidine phosphate orotidine monophosphate (*abbr.*: OMP); any phosphoric monoester or diester of orotidine. Of the three possible monoesters – the 2′-phosphate, the 3′-phosphate, and the 5′-phosphate – and the two possible diesters – the 2′,3′-phosphate and the

3′,5′-phosphate – only **orotidine 5′-phosphate** is known to occur naturally (the locant therefore being omitted if no ambiguity may arise).

orotidine 5′-phosphate *or* **5′-orotidylic acid** *or* **5′-phosphoorotidine** *or* **5′-*O*-phosphonoorotidine** *symbol*: Ord5′*P*; *alternative recommended names for* orotidine monophosphate (*abbr.*: OMP), orotidine 5′-(dihydrogen phosphate), orotate ribonucleotide. (The locant is commonly omitted if there is no ambiguity as to the position of phosphorylation.) It is the immediate biosynthetic precursor of uridine monophosphate, uridylic acid, UMP, to which it is converted by decarboxylation. Thus it is important also as an intermediate in the biosynthesis of other pyrimidine nucleotides from UMP.

orotidine-5′-phosphate decarboxylase EC 4.1.1.23; *other name*: orotidylic decarboxylase; an enzyme that catalyses the decarboxylation of orotidine 5′-phosphate to UMP in a reaction carried out by the multifunctional enzyme **5′-monophosphate synthase**.

orotidine-5′-phosphate pyrophosphorylase *see* **orotate phosphoribosyltransferase**.

orotidylate **1** either the monoanion or the dianion of orotidylic acid. **2** any mixture of orotidylic acid and its anions. **3** any salt or ester of orotidylic acid.

orotidylic acid *the trivial name for* any phosphoric monoester of orotidine. The position of the phosphoric residue on the ribose moiety of a given ester may be specified by a prefixed locant – *see* **orotidine phosphate**. However, 5′-orotidylic acid is the ester commonly denoted, its locant usually being omitted if no ambiguity may arise. 5′-Orotidylic acid is also an alternative recommended name for **orotidine 5′-phosphate**.

orotidylic acid phosphorylase *see* **orotate phosphoribosyltransferase**.

orotidylic decarboxylase *see* **orotidine-5′-phosphate decarboxylase**.

orphanin FQ *or* **nociceptin** an endogenous heptadecapeptide with an N-terminal phenylalanine and a C-terminal glutamine residue. It is derived from a precursor protein (186 amino acids) homologous to preprodynorphin and to preproenkephalin. It binds to a G-protein-coupled membrane receptor related to the opioid receptors. When injected into the cerebral ventricles of mice it produces hyperalgesia and inhibits locomotor activity.

orphan receptor a receptor whose physiological ligand is unknown.

ortho (*followed by* to) at a position in a benzene ring adjacent to a particular (specified) substituent.

ortho- *abbr.*: *o*-; *prefix* (*in chemical nomenclature*) denoting an isomer of a disubstituted monocyclic aromatic compound in which the substituents are attached at positions 1 and 2 of the ring. *Compare* **meta-**, **para-**.

ortho+ *or* (*before a vowel*) **orth+** *comb. form* **1** straight or upright; perpendicular (to) or at a right angle (to); correct. **2 a** (*in inorganic chemistry*) denoting the most hydrated acid of a series of oxoacids formed from any one of certain anhydrides, or the anion or a salt formed from such an acid. The term is now approved only for orthoboric acid, H_3BO_3; orthosilicic acid, H_4SiO_4; orthophosphoric acid, H_3PO_4; orthoperiodic acid, H_5IO_6; and orthotelluric acid, H_6TeO_6. *Compare* **meta+** (def. 2), **pyro+**. **b** (*in organic chemistry*) similarly denoting hypothetical acids of general formula $RC(OH)_3$ or $C(OH)_4$ or real esters of such acids, e.g. trimethyl orthoacetate, $CH_3-C(OCH_3)_3$; tetramethyl orthocarbonate, $C(OCH_3)_4$.

ortho- and **peri-fused** describing a polycyclic compound in which one ring contains two, and only two, atoms in common with each of two or more rings of a contiguous series of *ortho*-**fused** rings. Such compounds have *n* common faces and <2*n* common atoms.

orthochromatic **1** (of a dye or stain) imparting, or capable of imparting, its own colour to a structure. **2** (of a structure) assuming, or capable of assuming, the same colour as that of a dye or stain with which it is treated. *Compare* **metachromatic**. —**orthochromatically** *adv.*

orthodox conformation the conformation adopted by a mitochondrion in the resting state when the [ATP]/[ADP] ratio is high. This is the usual conformation of mitochondria seen in electron micrographs of thin sections of intact tissue.

***ortho*-fused** describing a polycyclic compound in which any two adjacent rings contain two, and only two, atoms in common. Such compounds have *n* common faces and 2*n* common atoms.

orthogonal **1** (of axes, lines, planes, surfaces, etc.) mutually at right angles; perpendicular to one another. **2** (*in statistics*) (of a set of variates) being, or regarded as being, statistically independent.

orthograde *see* **anterograde**.

orthologue *or* (*esp. US*) **ortholog** a gene, protein, or biopolymeric sequence that is evolutionarily related to another by descent from a common ancestor, having diverged as a result of a speciation event. Orthologues usually perform the same or highly similar functions in their different host species; e.g., human and bovine rhodopsins. –**orthologous** *adj.*

orthomyxovirus any of a group of RNA animal viruses, in class V of the **Baltimore classification**, consisting of enveloped, pleomorphic particles 80–120 nm in diameter. The nucleocapsids are helical, 6–9 nm in diameter, and of variable length, and contain segmented RNA (minus strand); this is transcribed by a virion polymerase into complementary RNA molecules, which act as messengers. The group includes the influenza A, B, and C viruses.

orthophosphate *an alternative name for* **phosphate** (used especially to distinguish from **metaphosphate** or **pyrophosphate**). *See also* **ortho+** (def. 2a).

orthostatic intolerance the combination of lightheadedness, fatigue, altered mentation, and fainting with postural tachycardia and excessive plasma norepinephrine concentration. The symptoms may be produced by cocaine, amphetamines, or tricyclic antidepressants, or by an autosomal dominant loss-of-function mutation in the **norepinephrine transporter**.

orthotopic transplant any tissue graft transplanted into the normal position of that tissue in the body of the recipient. *Compare* **heterotopic transplant**.

oryzain a group of proteases expressed during seed germination in rice. Three types (α, β, and γ) have been identified. γ-Oryzain (362 amino acids) is homologous with aleurain of barley, and with SAG (senescence-associated gene)-encoded cysteine proteases of *Arabidopsis*, maize, and *Brassica* species.

Os *symbol for* osmium.

osamine *an alternative name for* **aminodeoxysugar**.

osazone any 1,2-bis(arylhydrazone); i.e., any condensation product of the type RNH–N=CR–CR′=N–NHR formed between an α-dicarbonyl compound (e.g., a 2-ketoaldose) and hydrazine, or a substituted hydrazine. α-Hydroxy- and α-amino-carbonyl compounds will also form osazones through oxidation of the corresponding [mono]hydrazones initially formed. *See also* **phenylosazone**.

oscillator an instrument or device for producing oscillations, especially an electronic device that produces an alternating output of known frequency.

oscillin a 33 kDa protein identified in sperm that induces oscillations in intracellular Ca^{2+} concentrations in mammalian eggs. It is considered to be involved in Ca^{2+} oscillations that serve as the essential trigger for egg activation and early development of the embryo.

oscillograph an instrument for producing a graphical record of an oscillating value.

oscilloscope an instrument for producing a temporary visual record of an oscillating value. A **cathode-ray oscilloscope**, based on the **cathode-ray tube**, provides a visual image of electrical signals, in the form of a graph of the signal over time.

OSCP *abbr. for* oligomycin sensitivity-conferring protein.

+ose *noun suffix* **1** originally indicating related to glucose, as either an isomer (e.g. galactose) or a polymer (e.g. cellulose, maltose), now also extended to other sugars (e.g. arabinose, fructose, sucrose) or classes of sugars (e.g. aldose, furanose, ketose, hexose, pentose, pyranose). **2** indicating the primary products of hydrolysis of a biopolymer; e.g. proteose, peptose.

Ose (*in glycolipid nomenclature*) *symbol for* a monosaccharide residue. A numerical subscript denotes the number of connected monosaccharide units in the oligosaccharide component, and a prefix of two or three characters represents the trivial name of the oligosaccharide; e.g. GbOse$_3$ (i.e. globatriaose), nLcOse$_4$ (i.e. neolactotetraose), McOse$_3$Cer (i.e. mucotriosylceramide).

oside *an alternative term for* **glycoside**.

+oside *noun suffix* denoting **glycoside**.

+osis *noun suffix* denoting **1** a process or state (e.g. metamorphosis, narcosis, necrosis, osmosis). **2** a diseased condition (e.g. nephrosis, tuberculosis). **3** an increase or excess (of something) (e.g. glycogenosis, leukocytosis). *See also* **+otic**.

oskar one of the **maternal effect genes** in *Drosophila melanogaster*. It organizes the germplasm and directs localization of the germ-cell determinants. Oskar protein is required to keep *oskar* RNA and *staufen* (another maternal effect gene) protein at the posterior pole. Three other genes (*capu*, *spir*, and *stau*) are required for the initial localization of the *oskar* product to the posterior pole of the oocyte.

osM *or* **osM** *symbol for* osmolar (osmol L^{-1} now recommended).

OSM *abbr. for* oncostatin M.

osmiophilic (of a structure or substance) having an affinity for, or staining readily with, osmium tetroxide; e.g. osmiophilic globules (plastoglobuli) in plastids.

osmium tetroxide osmium(VIII) oxide; OsO_4; a volatile, evil-smelling, moderately water-soluble, electron-dense substance used for fixing specimens for microscopy, especially electron microscopy, and used chemically in the hydroxylation of alkenes. The vapour is highly poisonous.

osmo+ *comb. form* denoting **1** osmosis. **2** smell, odour.

osmoceptor *see* **osmoreceptor**.

osmol *symbol for* osmole.

osmolal **1** describing a solution that contains one **osmole** (of a specified solute), or an indicated number of osmoles, per kilogram of solvent; used also of a specified solute in such a solution. **2** describing an amount-of-substance concentration expressed in osmoles (of a specified solute) per kilogram of solvent. *Compare* **osmolar**. *See also* **osmolality**.

osmolality the concentration of osmotically active particles in an aqueous solution expressed in **osmoles** per kilogram of solvent. For a nondissociating substance the osmolality is equal to the **molality**, whereas for a substance that dissociates in aqueous solution the osmolality is greater than the molality to an extent dependent on the number of particles into which the substance dissociates and on the position of the equilibrium. *Compare* **osmolarity**.

osmolar **1** describing a solution that contains one **osmole** (of a specified solute), or an indicated number of osmoles, per litre of solution; used also of a specified solute in such a solution. **2** *abbr.*: osM or osM; describing an amount-of-substance concentration expressed in osmoles (of a specified solute) per litre of solution. *Compare* **osmolal**. *See also* **osmolarity**.

osmolarity the concentration of osmotically active particles in an aqueous solution expressed in **osmoles** per litre of solution. For a nondissociating substance the osmolarity is equal to the molarity, whereas for a substance that dissociates in aqueous solution the osmolarity is greater than the molarity to an extent dependent both on the number of particles into which the substance dissociates and on the position of the equilibrium. *Compare* **osmolality**.

osmole *symbol*: osmol; the amount of substance that contains, or gives rise to, one mole of osmotically active particles when in aqueous solution.

osmometer **1** *or* **membrane osmometer** any device for demonstrating the phenomena of osmosis and osmotic pressure; any instrument for measuring osmotic pressure. **2** an apparatus for measuring the acuteness of the sense of smell. —**osmometric** *adj.*; **osmometry** *n.*

osmophile any organism that grows preferentially (or only) in media of relatively high osmotic pressure, such as ones with a high salt or sugar content. —**osmophilic** *adj.*

osmoreceptor *or* **osmoceptor** a sensory cell, or group of such cells, that is specialized to react to changes in environmental osmotic pressure, e.g. of blood plasma or of a tissue fluid. Osmoreceptors are located in the hypothalamus, in the main arteries, and at other sites.

osmoregulation the sum of the processes by which organisms regulate their internal contents of water and solutes. —**osmoregulatory** *adj.*

osmosis the spontaneous net flow of solvent by diffusion through a semipermeable membrane from a phase where the solvent has a higher chemical potential to one where the solvent has a lower chemical potential. The flow continues until the chemical potential of the solvent becomes the same on both sides of the membrane or until it is countered by a difference in hydrostatic pressure either generated or applied. —**osmotic** *adj.*; **osmotically** *adv.*; **osmose** *vb.*

osmotic coefficient *symbol*: ϕ_m (on a molality basis) *or* ϕ_x (on a mole fraction basis); the ratio of the actual to the ideal osmotic pressure of a given solution at a specified temperature; i.e. a representation of the departure of the solution from ideality.

osmotic pressure *symbol*: Π; the excess hydrostatic pressure that must be applied to a solution in order to reduce to zero the net flow of pure solvent into it by osmosis. The osmotic pressure of a dilute solution is equal to the pressure the solute would exert if it could exist as a gas at the same temperature, T, and the same amount-of-substance concentration, c; thus it may be expressed by a relation analogous to the gas equation, i.e. $\Pi = \nu cRT$ where ν is the (average) number of particles into which one solute molecule is dissociated under the given conditions, and R is the gas constant. Osmotic pressure is a **colligative property** of a solution. *See also* **colloid osmotic pressure, oncotic pressure**.

osmotic shock any disruption or other disturbance caused to suspended cells or cell organelles by a sudden change in the osmotic pressure of their suspending medium.

osmotin a basic protein abundant in tobacco cells grown in a salt-rich medium. The 26 kDa mature protein accumulates in the cell vacuole in response to biotic stimuli (i.e. phytohormones, viral or fungal infection) and abiotic stimuli (i.e. water deprivation, cold, UV irradiation, wounding).

osmotolerant describing a cell or organism that is tolerant to a wide range of osmotic pressures in its environment.

osone any 2-ketoaldose; use not recommended.

ossein the organic material, consisting mainly of collagen, that remains after treatment of bone with dilute acid to dissolve out the inorganic components.

ossify to make into bone or to become bone. —**ossification** *n.*

osteo+ *or* (*before a vowel*) **oste+** *or* (*sometimes*) **ost+** *comb. form* denoting bone.

osteoblast a mesenchymal cell that secretes the organic matrix of bone (i.e. osteoid, consisting chiefly of collagen); as the matrix becomes calcified the osteoblast becomes trapped and is known as an osteocyte.

osteocalcin *other names*: γ-carboxyglutamic acid-containing protein; bone GLA-protein (*abbr.*: BGP); a protein that binds Ca^{2+} and apatite; the binding depends on γ-carboxyglutamic acid residues (three per molecule) formed by vitamin K-dependent carboxylation.

osteochondrodysplasia a very heterogeneous disorder of skeletal development that is frequently associated with dwarfism, spinal deformity, and joint abnormalities. There are numerous forms, of which many arise from mutations in collagen types II, IX, X, and XI; in receptors for somatotropin and fibroblast growth factors; or in sulfate transport protein.

osteochondromatosis *an alternative name for* **hereditary multiple exostoses**.

osteoclast a large, multinucleate **macrophage** that erodes bone matrix and is able to tunnel deep into the substance of compact bone. Osteoclasts are formed by the fusion of blood monocytes and play an important role in the remodelling of bone and the removal of dead bone.

osteocyte an osteoblast when it has become trapped in a lacuna of

the bone matrix and is no longer able to divide or secrete appreciable amounts of new matrix. Osteocytes are connected by fine cellular processes that pass through the bone canaliculi.

osteogenesis the processes of bone formation; ossification.

osteogenesis imperfecta *or* **'blue eyes-broken bones disease'** *or* **brittle bone disease** a genetically heterogeneous group of human diseases distinguished by the severity of bone fragility, scleral coloration, hearing loss, abnormal dentition, and soft-tissue dysplasia involved. Four autosomal dominant phenotypes are recognized: type 1 (mild) has blue sclerae and bone fragility; type 2 (prenatal lethal) has dark sclerae, severe bone fragility, and bone deformity; type 3 (deforming) has light sclerae, bone fragility, progressive deformity, imperfect dentition, and short stature; type 4 (mild deforming) has light sclerae, imperfect dentition, and mild short stature. Over 200 mutations in type I collagen genes are responsible. Those that abolish synthesis of pro-α-1 (I) or pro-α-2 (I) can produce most phenotypes, while splice junctions in either gene produce different phenotypes depending on their position. Partial deletions in either gene usually result in type 4.

osteoid 1 the unmineralized bone matrix secreted by osteoblasts; it consists chiefly of collagen. **2** resembling bone.

osteolysis the breakdown and dissolution of bone, especially through disease. —**osteolytic** *adj.*

osteomalacia a clinical condition of adults characterized by softening and deformation of the bones due to a failure to maintain their mineralization. It has a wide variety of possible causes, of which dietary deficiencies of vitamin D, calcium, or phosphate are the commonest.

osteonectin *see* **SPARC**.

osteopetrosis *or* **osteosclerosis** *or* **marble bone disease** a rare, heterogeneous disorder in which there is decreased osteoclastic resorption of bone and widespread expansion of bone mass. *Compare* **osteoporosis**.

osteopontin *other names*: bone sialoprotein 1; urinary stone protein; secreted phosphoprotein 1 (*abbr.*: SPP-1); nephropontin; uropontin; a bone-specific sialoprotein produced by cells in osteoid matrix, but independently discovered from several different functions. It can form a bridge between cells and the mineral in the matrix, and is found at high concentrations on the cement lines, where bone formation follows resorption. It is also found in the kidney, and plays a major role in urinary stone formation. In arterial smooth muscle cells, its level increases in atherosclerotic plaques. It is secreted, binds calcium, and binds covalently to fibronectin. It has an **RGD** integrin-binding sequence, and is one of the major ligands for the vitronectin receptor. It is phosphorylated on serine, and is an *O*- and *N*-glycosylated protein. There are four forms from alternative splicing.

osteoporosis a condition characterized by a general or local reduction in bone mass with little alteration in gross composition. This causes the affected bones to become porous, brittle, and liable to fracture. Recognized forms include: (1) postmenopausal; (2) senile; (3) secondary – to other conditions or to drugs (e.g. corticosteroids, barbiturates, anticonvulsants), chronic alcoholism, and cigarette smoking; (4) idiopathic juvenile. *Compare* **osteopetrosis**.

osteoprotegerin *or* **osteogenic inhibitory factor** (*abbr.*: OCIF) a protein belonging to the tumour necrosis factor (TNF) superfamily that is present on, or secreted by, osteoblasts. It binds **RANKL** and prevents this from binding to **RANK** on osteoclasts, thereby preventing bone resorption. Overexpression of osteoprotegerin in mice leads to **osteopetrosis**, with excessive calcification of bone.

osteostatin a polypeptide that consists of residues 107–139 of **parathyroid hormone-related protein**. It has effects on bone resorption and on brain, although its physiological role is unclear.

Ostwald dilution law a relationship, deduced in 1888 by the Russian-born German chemist Friedrich Wilhelm Ostwald (1853—1932), between the amount-of-substance concentration, c, and the degree of dissociation, α, of a weak electrolyte. It is given by:

$$\alpha^2 cl(1 - \alpha) = K,$$

where K is the apparent dissociation constant of the electrolyte.

Ostwald pipette a pipette for measuring small volumes of liquids, especially viscous liquids. It features a capillary stem and delivery jet and a shoulderless bulb near the jet. Various modifications have been described.

Ostwald viscometer a simple **capillary viscometer**.

+osyl *suffix* indicating a **glycosyl** group.

+otic *suffix forming adjectives and nouns* **1** affected by (e.g. necrotic) or relating to (e.g. osmotic). **2** causing (e.g. mitotic, narcotic). — **+otically** *adv.*

otoferlin a protein that is homologous with spermatogenesis factor (fer-1) of *Caenorhabditis elegans*, and is related to dysferlin and myoferlin of humans. It consists of 1977 amino acids, with a transmembrane segment close to the C-terminus, the rest being intracellular. Several mutations cause an isolated profound deafness.

otogelin the product of a candidate gene for an autosomal recessive form of deafness. It is predicted to be a secreted glycoprotein (2910 amino acids) that, like α-tectorin, is specific to acellular membranes of the inner ear. It consists of a central region that is rich in threonine, serine, and proline residues, flanked on the N-terminal side by four vWF domains, and on the C-terminal side by several vWF-like domains.

ouabain *or* G-strophanthin a **cardiac glycoside** whose carbohydrate moiety is L-rhamnose, obtained from the seeds of *Strophanthus gratus*. It is a specific inhibitor of the membrane-bound sodium pump Na^+,K^+-ATPase (*see* **sodium/potassium ATPase**) of animal cells.

Ouchterlony technique a development of **double immunodiffusion**, effected in two dimensions so that comparisons can be made between two or more antigen-containing samples in their reaction with one antiserum. The particular pattern of precipitation lines or arcs formed enables conclusions to be drawn regarding the identity, partial identity, or nonidentity of antigens in the various samples examined. [After O. Ouchterlony, French immunologist.]

Oudin technique a simple, single immunodiffusion technique, effected in one dimension, in which a solution of an antigen-containing sample is placed above a column of antiserum incorporated in a gel in a tube; antigen diffuses into the gel, where it reacts with antibody to form one or more precipitation bands. [After Jacques Oudin (1923–).]

ounce *symbol*: oz; a unit of mass, equal in the avoirdupois system to one-sixteenth of a pound (= 28.3495 grams). This must be distinguished from the **troy ounce** (equal to 480 grains, 31.1035 grams).

outer membrane 1 (*in eukaryotes*) the more external of the two membranes that surround a chloroplast, a mitochondrion, or a nucleus. **2** (*in prokaryotes*) the layer of lipopolysaccharide, lipoprotein, and protein that lies external to the peptidoglycan in the cell wall of a Gram-negative bacterium.

outer membrane protein P68 *see* **pertactin**.

outlier (*in statistics*) a data point that falls far outside the values of others in a data set. The nature of outliers should be understood prior to further analysis: those arising from measurement errors will distort the statistics and hence interpretation of the data; those that are genuine, however, may point to unexpected and hence highly interesting results.

ov+ *a variant form of* **ovi+** (before a vowel).

ova the plural of **ovum**.

ovalbumin *abbr.*: OA; a ≈44.5 kDa glycoprotein, also called egg al-

bumin, that is the major protein of egg white from hens' eggs. Each molecule consists of a single polypeptide chain, and carries a single oligosaccharide chain containing only mannose and N-acetylglucosamine residues. It is included in the **serpin** family. It is synthesized by the oviduct under hormonal control. Although a secreted protein, it has no N-terminal signal peptide.

ovalocytosis an abnormality of erythrocytes that makes them oval or elliptical rather than circular. Southeast-Asian ovalocytosis, which is common in Melanesia, Indonesia, Papua New Guinea, and Philippines, is caused by mutations in band 3 protein (*see* **band protein**).

ovarian hormone any of the numerous steroid hormones secreted by the ovary.

ovary 1 (*in zoology*) the gonad of a female animal; vertebrates have a pair of ovaries. The ovary contains oocytes from which ova are produced, and it also secretes various steroid hormones. Its activity is cyclical and is controlled by gonadotropin(s). **2** (*in botany*) the enlarged basal part of a carpel or a syncarpous gynecium of a flower that contains one or more ovules. —**ovarian** *adj.*

overgrowth 1 excessive growth; hyperplasia. **2** a later but more vigorous growth by another strain or species of organism that supplants one already existing.

Overhauser effect *or* Overhauser enhancement *see* **nuclear Overhauser effect**.

overlap peptide a fragment, produced by partial hydrolysis of a polypeptide or protein, that contains both the C-terminal part of the sequence of another peptide fragment and the N-terminal part of a further fragment, allowing those two to be placed in the correct position in the overall sequence of the parent compound during protein sequencing procedures.

overlapping code a code in which one element of a code-word forms part of an adjacent code-word in an encoded piece of information. A triplet code (such as the **genetic code**, where the code-words are codons) may be either: a **fully overlapping code**, in which one code-word consists of the elements at positions n, $(n+1)$, $(n+2)$, and the next code-word consists of the elements at positions $(n+1)$, $(n+2)$, $(n+3)$; or a **partially overlapping code**, in which adjacent code-words comprise elements at n, $(n+1)$, $(n+2)$, and those at $(n+2)$, $(n+3)$, $(n+4)$. In a **nonoverlapping code**, such as the genetic code proved to be, no element of one code-word forms part of an adjacent code-word.

overlapping gene any gene whose polynucleotide sequence overlaps, wholly or in part, that of another gene in the same stretch of a genome. The codons may or may not be in the same **reading frame**. In mitochondria and some viruses, some mRNAs are synthesized by the transcription of the same DNA using different reading frames.

overpressure(d) layer chromatography *abbr.*: OPLC; a chromatographic technique that gives some of the advantages of **high-pressure liquid chromatography** to **thin-layer chromatography**, with resultant increases in both speed and efficiency of separation. The sorbent layer is completely covered by a flexible membrane under external pressure and the solvent is pumped continuously through the sorbent layer.

overshoot the extent to which a control system causes the variable being controlled to go initially beyond its final equilibrium position.

ovi+ *or* **ovo+** *or* (before a vowel) **ov+** *comb. form* denoting ovum or egg.

oviduct (*in zoology*) the tube through which ova are conveyed from the ovary to the exterior or to some intermediate organ such as the uterus. In mammals it is known more commonly as the **Fallopian tube** or **uterine tube**.

oviductin *other name*: gp43 processing protease; a high molecular weight glycoprotein that is secreted by the non-ciliated secretory cells of the oviduct, and later transferred to the zona pellucida of the oocyte during oviductal transit. It processes a 43 kDa glycoprotein of the oocyte envelope during its transit through the duct.

ovine *abbr.*: o *or* (*formerly*) O; of, belonging to, characteristic of, obtained from, or being sheep or a sheep; resembling a sheep, sheeplike.

ovo+ *a variant form of* **ovi+**.

ovoflavin *or* oviflavine *a former name for* **riboflavin**.

ovoinhibitor a contaminant of crude **ovomucoid** that, unlike pure ovomucoid, inhibits chymotrypsin as well as trypsin.

ovomucoid a glycoprotein found in egg white from hens' eggs; it is an inhibitor of trypsin.

ovothiol a novel thiohistidine compound from sea-urchin eggs that confers $NAD(P)H-O_2$ oxidoreductase activity on ovoperoxidase.

ovotransferrin *an alternative name for* **conalbumin**.

ovulate to discharge an egg or eggs from an ovary. —**ovulation** *n*.

ovule 1 (*in botany*) a structure in the ovary of a seed plant that develops into a seed after fertilization of the egg cell within it. **2** (*in zoology*) a small, especially immature, ovum or egg.

ovum (*pl.* **ova**) (*in zoology*) strictly, an unfertilized egg cell, i.e. female gamete, which contains a haploid nucleus and which on fertilization has the unique ability to develop into a new individual. In many animals, mammals included, **oogenesis** is arrested and meiosis is not completed until after fertilization of the secondary **oocyte**, hence in these animals an ovum in the strict sense cannot exist. However, the term is often applied instead either to the secondary oocyte or to the product resulting from the second meiotic division after fertilization.

oxa+ *or* (*sometimes before a vowel*) **ox+** *prefix* (*in chemical nomenclature*) denoting replacement of a methylene group in a specified acyclic or monocyclic hydrocarbon by an oxygen atom. For certain natural products, (e.g. tetrapyrroles), the meaning is extended to denote replacement of a noncarbon atom (with any attached hydrogen atoms) by an oxygen atom.

oxalacetate *an alternative name for* **oxaloacetate**.

oxalate 1 *the trivial name for* ethanedioate, $^-OOC-COO^-$, the dianion of oxalic acid, i.e. ethanedioic acid. It occurs in many plants, especially sorrels of the genera *Oxalis* and *Rumex*, and is highly toxic to animals. Normally, small amounts are formed in humans and excreted in the urine; larger amounts occur in patients with **hyperoxaluria**. The calcium salt is virtually insoluble in water, a property that is valuable in the determination of calcium and in the use of oxalate *in vitro* as an anticoagulant. **2** any mixture of free oxalic acid and its mono-ions and dianions. **3** any salt or ester of oxalic acid.

oxalo the univalent acyl group, $HOOC-CO-$, derived from oxalic acid by loss of one hydroxyl group. *See also* **oxalyl** (def. 1).

oxalo+ *prefix* indicating the presence of an **oxalo** group.

oxaloacetate *or* (*rarely*) **oxalacetate** *abbr*.: OAA; **1** *the trivial name for* oxosuccinate (*formerly* ketosuccinate); oxobutanedioate; $^-OOC-cO-CH_2-COO^-$ (keto form) or $^-OOC-c(OH)= CH-COO^-$ (enol form); the dianion of oxaloacetic acid. It is an important intermediate in metabolism, especially as a component of the **tricarboxylic-acid cycle**. The stereochemistry of the **citrate (*Si*)-synthase** reaction is shown below. **2** any mixture of free oxaloacetic acid and its mono- and dianions. **3** any salt or ester of oxaloacetic acid.

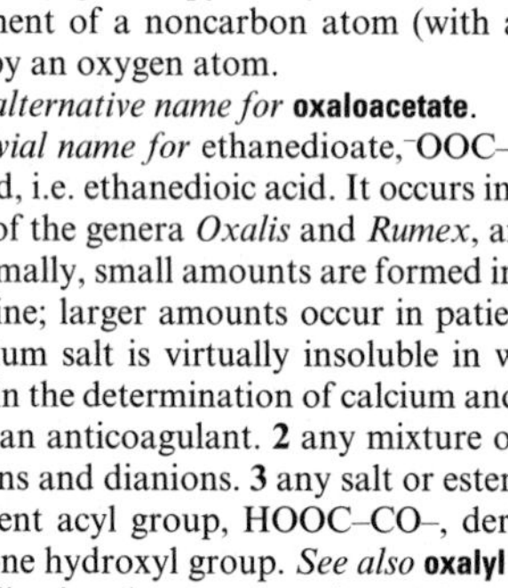

$$\overset{*}{C}H_3\overset{*}{C}OSCoA$$

HOOC
HOOC—C=O
CH₂

↓

HOOC, CH₂COOH
HOOC—C
CH₂ OH

oxalosis the renal and extrarenal deposition of calcium oxalate (*see* **oxalate**) seen in patients with **hyperoxaluria**.

oxalosuccinate 1 *the trivial name for* 2-oxotricarballylate (*formerly* α-ketotricarballylate); 1-oxopropane-1,2,3-tricarboxylate; $^-OOC-CO-CH(COO^-)-CH_2-COO^-$; the trianion of oxalosuccinic acid. It is an enzyme-bound intermediate in the conversion of isocitrate to 2-oxoglutarate by **isocitrate dehydrogenase (NAD⁺)** in the **tricarboxylic-acid cycle**. **2** any mixture of free oxalosuccinic acid and its mono- and dianions. **3** any salt or ester of oxalosuccinic acid.

oxaluria *an alternative name for* **hyperoxaluria**.

oxalyl 1 (*in biochemistry*) (*sometimes*) *an alternative name for* **oxalo**. **2** (*in chemistry*) the divalent diacyl group, $-CO-CO-$, derived from oxalic acid by loss of both hydroxyl groups.

OXBOX a 13 bp sequence (present in the region –123 to –674 bp) of the heart and skeletal muscle gene for the adenine nucleotide translocator and also in the promoter region of the gene for human ATP synthase β subunit. It appears to be a positive translational element.

oxene a reactive species of oxygen involved in the hydroxylation reaction catalysed by cytochrome P450; it is iron-bound oxygen which is electron deficient but electrically neutral $-P450-Fe^{3+}-O$. It reacts with the bound substrate to release the hydroxylated substrate with the regeneration of $P450-Fe^{3+}$.

oxidant the chemical species that accepts one or more electrons in a redox reaction; i.e. the species that undergoes reduction in such a reaction. *Compare* **reductant**.

oxidase any **oxidoreductase** that catalyses a reaction in which dioxygen is the electron acceptor (i.e. **oxidant**).

oxidase test a test for the presence of cytochrome *c* oxidase in bacteria that depends on the oxidation of tetramethyl-*p*-phenylenediamine to a violet-coloured product. It is useful for primary characterization, especially of Gram-negative organisms.

oxidation 1 the action or process of reacting with oxygen, especially the addition of oxygen to a substance. **2** the loss or removal of hydrogen from a substance. **3** the loss or removal of one or more electrons from a molecular entity, with or without concomitant loss or removal of a proton or protons. In this sense, oxidation is the opposite of, and is always coupled to, **reduction**. **4** the increase to a more positive value of the **oxidation number** of an atom, whether the atom is uncharged or charged, and whether free or covalently bound. —**oxidative** *adj*.

α-oxidation *see* **alpha-oxidation**.

β-oxidation *see* **beta-oxidation**.

ω-oxidation *see* **omega-oxidation**.

oxidation level *or* **oxidation state** the status of an organic compound in terms of whether it can be considered to undergo oxidation or reduction (or neither) upon conversion to a related compound. If conversion involves oxidation, the parent compound is said to have a higher oxidation level than the product; if conversion involves reduction, the parent compound is said to have a lower oxidation level. The term is thus used in a relative rather than an absolute sense. *See also* **oxidation number**.

oxidation number *or* **oxidation state** a number assigned to a particular atom in a molecular entity that represents, actually or notionally, the charge on that atom. Thus, atoms of pure elements have an oxidation number of zero; in molecules of simple electrovalent compounds the oxidation number of each component atom is equal to the charge on the respective ion; and in molecules of covalent compounds the notional charge on a particular atom is derived by assigning electrons to the more electronegative partner in each bond involving the atom in question. By convention, hydrogen is considered to have an oxidation number of +1 when in combination with nonmetals. The oxidation number of a carbon atom can range from –4 (e.g. in CH_4) through zero (e.g. in CH_2O or solid C) to +4 (e.g. in CO_2). The oxidation number of an atom may be indicated by a Roman numeral; e.g. iron(III) oxide, ferric oxide, Fe_2O_3. The algebraic sum of the oxidation numbers of all the atoms in any molecular entity must equal the charge on that entity.

oxidation potential a measure of the tendency of a substance to oxidize spontaneously through loss of electrons. It is the standard potential of the half-reaction: $A \rightarrow A^{n+} + ne$ and is equal to the **reduction potential** but of opposite sign.

oxidation–reduction *an alternative term for* **redox**.

oxidation–reduction indicator *an alternative term for* **redox indicator**.

oxidation–reduction potential *an alternative term for* **redox potential**.

oxidation state 1 *an alternative term for* **oxidation number**. **2** *an alternative term for* **oxidation level** (less preferred).

oxidative deamination deamination with concomitant oxidation, as in the enzymic conversion, regardless of the nature of the electron acceptor, of an α-amino acid to an α-oxo acid. *Compare* **reductive amination**.

oxidative decarboxylation decarboxylation with concomitant oxidation, as in the enzymic conversion, regardless of the nature of the electron acceptor, of an α-oxo acid (actually or formally) to a carboxylic acid containing one less carbon atom.

oxidative metabolism *an alternative term for* **respiration** (def. 1).

oxidative pentose phosphate pathway *or* **hexose monophosphate shunt** a route additional to glycolysis in plants that generates: (1) NADPH when it is not being formed by photosynthesis; (2) ribose 5-phosphate required in the biosynthesis of nucleosides; and (3) erythrose 4-phosphate required for biosynthesis of shikimic acid, a key intermediate in the formation of aromatic rings. *See also* **pentose phosphate pathway**.

oxidative phosphorylation *or* **respiratory-chain phosphorylation** the phosphorylation of ADP to ATP that accompanies the oxidation of a metabolite through the operation of the **respiratory chain**. The overall reaction may be summarized by the equation:

$$AH_2 + 0.5O_2 + xP_i + xADP \rightarrow A + (x+1)H_2O + xATP$$

where AH_2 is an oxidizable metabolite, A is the product of the oxidation, P_i is inorganic orthophosphate, and x has a value between 1 and 3 depending on the nature of AH_2 and the functional state of the respiratory chain. The proton gradient built up across the inner mitochondrial membrane by the proton pumps of the respiratory chain drives ATP synthesis by the **H+-transporting ATP synthase**. *See also* **chemiosmotic coupling hypothesis**.

oxidative stress a state of metabolic imbalance within cells that favours pro-oxidant substances (e.g. superoxide, hydrogen peroxide, hypochlorous acid, nitric oxide, peroxynitrite) rather than antioxidants (e.g. glutathione, ascorbic acid) and antioxidant enzyme systems (e.g. superoxide dismutases, catalase). This leads to oxidative damage to all classes of the major biomolecules. Depending on the degree of imbalance, a cell may die or it might survive in a changed state. Such stress can be important in causing a wide variety of degenerative states, including atherosclerosis, ischemia/reperfusion injury in heart and brain, mutagenesis, and chronic inflammatory disease.

oxide 1 *(in inorganic chemistry)* any binary compound of oxygen and some other element. **2** *an alternative generic name for* any cyclic ether that is an **oxirane**; such a compound is named as the oxide of the alkene from which it may be formed through addition of an atom of oxygen across the double bond; e.g. ethylene oxide, C_2H_4O. **3** *a term used in some languages for* **ether**. **4** any compound in which an oxygen atom has been attached to a heteroatom of a precursor compound. This definition includes compounds of general formulae: RC≡NO (nitrile oxides; e.g. benzonitrile oxide), >C=N(O)R or R,R′,R″NO (*N*-oxides, amine oxides; e.g. trimethylamine oxide); and R,R′SO (*S*-oxides, sulfoxides; e.g. methionine *S*-oxide; dimethyl sulfoxide).

+oxide *or* **+olate** *suffix* denoting **1** an anion formed by loss of a proton from the hydroxyl group of an alcohol or phenol; e.g. ethoxide ion, $C_2H_5O^-$. **2** a compound of such an anion with a cation; e.g. sodium phenoxide (i.e. sodium phenolate), C_6H_5ONa. *See also* **epoxide, hydroperoxide, hydroxide, nitroxide, peroxide, sulfoxide, superoxide**.

oxidize *or* **oxidise 1** to undergo or effect **oxidation**. **2** to form or cause to form a superficial layer of oxide on a metal. *Compare* **reduce**. —**oxidization** *or* **oxidisation** *n*.

oxidizer *or* **oxidiser 1** *a less common name for* **oxidizing agent**. **2** a device for effecting a particular oxidation.

oxidizing agent *or* **oxidising agent** any substance or mixture used for effecting **oxidation**.

oxidoreductase *the systematic name for* any enzyme of class EC 1. These enzymes catalyse oxidation–reduction (i.e. **redox**) reactions and the class comprises **dehydrogenases, oxidases**, and **reductases**.

oxidoreduction *an alternative term for* **redox**.

oximeter any photometric device for (continuously) measuring the oxyhemoglobin/hemoglobin ratio in blood, either *in vivo* or *in vitro*.

oxine *the trivial name for* 8-hydroxyquinoline; 8-quinolinol. With many species of metal ion this compound forms water-insoluble

chelate complexes and thereby is useful in analysis. It is also a disinfectant, and, as the copper chelate, is an effective fungicide.

oxirane 1 *the class name for* any compound containing a three-membered saturated ring comprising two carbon atoms and one oxygen atom. **2** *the systematic name for* the type member of this class, known commonly as ethylene oxide, C_2H_4O.

oxo 1 the chemical group consisting of an oxygen atom, =O, doubly bonded to another atom (often of carbon) in a molecular structure. When bonded to a carbon atom it forms the carbonyl group, >C=O. *Compare* **oxy**. **2** *or (formerly)* **oxo+** modifying the classname of a chemical compound to indicate the presence of an oxo group as a substituent in place of two hydrogen atoms; e.g. **oxocarboxylic acid**. The position of substitution may be specified by a predfixed locant; e.g. 17-oxo-steroid. *See also* **keto** (def. 3). *Compare* **oxo** (def. 1).

oxo+ *or (sometimes before a vowel)* **ox+ 1** *prefix (in chemical nomenclature)* indicating substitution of a doubly bonded oxygen atom into a specified compound or group by the (formal or actual) oxidation of a methylene group, >CH_2, to a keto group, >C=O, or of a methyl group, –CH_3, to a formyl group, –CH=O, with the production of a ketone or an aldehyde respectively; e.g. oxoglutaric acid (*see* **oxoglutarate**), 3,6-dioxohexanoic acid (*or* 5-formyl-3-oxopentanoic acid). *Compare* **oxo** (def. 2). **2** *comb. form* denoting presence of one or more (singly or doubly bonded) oxygen atoms in a molecular structure; e.g. **oxoacid, oxonium**.

oxoacid 1 *the recommended term for* any compound whose molecular structure contains oxygen, at least one other element, and at least one atom of hydrogen bound to oxygen, and also that yields a conjugate base (*see* **oxoanion**) by loss of a hydron. Examples include: hypochlorous acid, ClOH; sulfurous acid, O=S(OH)$_2$; nitric acid, (O=)NOH; [ortho]phosphoric acid, O=P(OH)$_3$; and also any phosphonic acid, RP(=O)(OH)$_2$, and any carboxylic acid, RC(=O)OH (where R = H or hydrocarbyl). *See also* oxocarboxylic acid. **2** *or* **oxo acid** *or* **oxyacid** *or* **oxy-acid** the traditional name, now obsolete, for any acid having oxygen in its acidic function, as opposed to a 'hydracid' where oxygen is lacking (e.g. hydrochloric acid, HCl).

oxo acid 1 *see* **oxocarboxylic acid**. **2** *a variant of* **oxoacid** (def. 2).

3-oxoacid CoA-transferase EC 2.8.3.5; *other name*: succinyl-CoA:3-ketoacid-CoA transferase; an enzyme that catalyses the reaction:

succinyl-CoA + 3-oxo acid = succinate + 3-oxoacyl-CoA.

In mammals it converts plasma acetoacetate that has been taken up by tissues into acetoacetyl-CoA for further metabolism.

3-oxoacyl-[acyl-carrier protein] reductase (NADH) EC 1.1.1.212; an enzyme that catalyses the reduction by NADH of 3-oxoacyl-[acyl-carrier protein] to (3*R*)-3-hydroxyacyl-[acyl-carrier protein] with formation of NAD$^+$. This is one of the steps of fatty acid synthesis in plants. *Compare* **fatty acid synthase complex**.

oxoanion *or (formerly)* **oxo anion** *or* **oxyanion** *or* **oxy-anion** any of the possible anions formable from a molecule of an **oxoacid** (def. 1) by loss of one or more hydrons.

oxobutanedioate *see* **oxaloacetate**.

oxocarboxylic acid *the recommended name for* any **carboxylic acid** that contains an aldehydic group attached to, or/and one or more ketonic groups contained in, its principal chain or parent ring system, i.e. any into which one or more oxo groups have been substituted. The term is commonly shortened to **oxo acid**, but the full name should be used wherever confusion with **oxoacid** is possible. The older term keto acid is often still used (*see* **keto** (def. 3)).

17-oxogenic steroid *or (formerly)* **17-ketogenic steroid** any urinary corticosteroid or its metabolites that can be estimated by the **Zimmermann reaction** after oxidation to the corresponding 17-oxo steroid.

oxoglutarate *or (formerly)* **ketoglutarate; 1** 2-oxoglutarate *or* α-oxoglutarate 2-oxopentanedioate; $^-$OOC–[CH$_2$]$_2$–CO–COO$^-$; the dianion of 2-oxoglutaric acid (i.e. 2-oxopentanedioic acid), an α-oxo (carboxylic) acid. It is a key constituent of the tricarboxylic-acid cycle and a key intermediate in amino-acid metabolism. In the cycle it is formed by oxidative decarboxylation of isocitrate and converted into a further oxidative decarboxylation reaction to succinyl-coenzyme A. It may also be formed from or converted into glutamate (and thus several other amino acids indirectly) by transamination, or formed by degradation of lysine via glutarate and 2-hydroxyglutarate. The older name **α-ketoglutarate** is often still used. **2** 3-oxoglutarate or β-oxoglutarate acetonedicarboxylate; 3-oxopentanedioate; $^-OOC-CH_2-CO-CH_2-COO^-$; the dianion of 3-oxoglutaric acid (i.e. 3-oxopentanedioic acid), a β-oxo (carboxylic) acid. It is a product of the action of the mould *Aspergillus niger* on citrate.

oxoglutarate decarboxylase *see* **oxoglutarate dehydrogenase (lipoamide)**.

oxoglutarate dehydrogenase complex an enzyme complex, closely analogous in mechanism to the **pyruvate dehydrogenase complex**. It accomplishes the overall reaction:

$$2\text{-oxoglutarate} + \text{CoA} + \text{NAD}^+ =$$
$$\text{succinyl-CoA} + CO_2 + \text{NADH}.$$

It consists of three activities: **oxoglutarate dehydrogenase (lipoamide)**, the E_1 component; **dihydrolipoamide *S*-succinyltransferase**, the E_2 component; and **dihydrolipoamide dehydrogenase**, the E_3 component (this being identical to the E_3 component of the pyruvate dehydrogenase complex).

oxoglutarate dehydrogenase (lipoamide) EC 1.2.4.2; *other names*: oxoglutarate decarboxylase; α-ketoglutaric dehydrogenase; an enzyme that catalyses the reaction between 2-oxoglutarate and lipoamide to form *S*-succinyldihydrolipoamide with release of CO_2. This is the E_1 component of the **oxoglutarate dehydrogenase complex**. Thiamine diphosphate is a coenzyme.

oxoglutarate/malate carrier protein *see* **OGCP**.

2-oxoisovalerate dehydrogenase *see* **3-methyl-2-oxobutanoate dehydrogenase (lipoamide)**.

oxolinic acid 5-ethyl-5,8-dihydro-8-oxo-1,3-dioxolo[4,5-*g*]quinoline-7-carboxylic acid; a **4-quinolone antibiotic** with antibacterial activity through inhibition of bacterial DNA gyrase.

oxonium 1 *or (formerly)* **hydronium** *or* **hydroxonium** the ion, H_3O^+, formed by covalent linkage of a hydron to a single water molecule. *Compare* **H⁺(aq)**. **2** any organic entity in which an oxygen atom is linked to three groups (one or two of which may be –H) and bears a formal single positive charge.

5-oxoprolinase an enzyme of microorganisms, plants, and most mammalian tissues that catalyses the reaction:

$$5\text{-oxoproline} + \text{ATP} = \text{glutamate} + \text{ADP} + P_i.$$

A rare enzyme deficiency in humans produces excessive 5-oxoprolinuria but no other significant symptoms or signs.

(5-)oxoproline *symbol*: Glp *or* <Glu; *the recommended trivial name for* pyroglutamic acid; pyrrolidone carboxylic acid; 5-oxopyrrolidine-2-carboxylic acid; a **noncoded amino acid**. It occurs at the N terminus of certain peptide hormones, and is an intermediate in the γ-glutamyl cycle. *Compare* **oxyproline**.

oxo steroid *or* **keto steroid** any steroid having an oxo substituent in its molecular structure, whether in the ring system (as part of a ketonic group) or in a side chain (as part of either a ketonic or an aldehydic group). The position of substitution may be specified by a prefixed locant. The term **17-oxo steroid** (or 17-keto steroid) usually refers specifically to any of a group of androgens and their metabolites that are present in urine and that can be extracted and estimated by the **Zimmermann reaction**. The amount of steroids excreted in the urine has been used as an index of androgen production in the body.

oxosuccinate *see* **oxaloacetate**.

oxphos *a colloquial term for* **oxidative phosphorylation**.

OXT *abbr. for* oxytocin.

oxy the chemical group consisting of an oxygen atom, –O–, singly bonded to two carbon atoms in a molecular structure, e.g. of an epoxide, an ester, or an ether. *Compare* **oxo**.

oxy+ *prefix* denoting **1** *(in organic chemical nomenclature)* the presence of an oxygen atom directly attached (by single covalent bonds) to two moieties of the same indicated parent compound, with formation of a symmetrical ether, where there is a functional group or substituent in the parent compound that has priority over 'ether' for citation as the principal characteristic group; e.g. 3,3′-oxydipropanoic acid (*not* bis(3-carboxypropyl) ether). *See also* **epoxy+**. **2** *(obsolete in organic chemical nomenclature but retained in pharma-*

ceutical nomenclature) substitution of a hydroxyl group for a hydrogen atom; e.g. oxyproline (*now* hydroxyproline), oxytetracycline. **3** *(in inorganic chemical nomenclature)* formation of a 'basic' (i.e. double) salt containing the oxide anion, O^{2-}, and a specified second anion; e.g. bismuth oxychloride (i.e. bismuth chloride oxide). **4** *or* oxy- presence of oxygen in the acidic function of an acid or in the corresponding moiety of a derived anion (obsolete; *see* **oxoacid** (def. 2), **oxoanion**). *comb. form* indicating. **5** dioxygen; e.g. oxybiotic, oxyhemoglobin, oxyhydrogen. **6** oxidation; e.g. oxycellulose. **7** acid; e.g. **oxyntic**.

oxyacid *or* **oxy-acid** *a variant of* **oxoacid** (def. 2).

oxyanion *or* **oxy-anion** *(formerly)* *a variant of* **oxoanion**.

oxybiontic *or* **oxybiotic** living in the presence of oxygen; aerobic.

oxycellulose any of the various materials obtained by the oxidation of cellulose. Such materials are useful as cation-exchangers or for fabrication into lint and gauzes.

oxygen 1 *symbol*: O; a nonmetallic element of group 16 of the (IUPAC) **periodic table**; atomic number 8; relative atomic mass 15.9994. Naturally occurring oxygen consists of a mixture of three stable nuclides of relative masses 16 (99.759 atom percent), 17 (0.037 atom percent), and 18 (0.204 atom percent). The latter, commonly known as **heavy oxygen**, is useful as a tracer in studies of chemical and biochemical reaction mechanisms. The commonest valence is two. Oxygen is the most abundant element in the biosphere: in the atmosphere it occurs elementally in the form of molecules of the diatomic gas dioxygen, which comprises 20.95 by volume of dry air, and as traces of the gas **ozone** (i.e. trioxygen). The atmosphere also contains small amounts of the oxygen compounds carbon dioxide and water (vapour); almost half by weight of the surface of the Earth consists of oxygen in combined form, principally as water and silicates. **2** *or* **molecular oxygen** *symbol*: O_2; *the common name for* the gas correctly known as **dioxygen**.

oxygen-15 the radioactive nuclide of oxygen, $^{15}_{8}O$. It is prepared in a cyclotron by bombarding $^{17}_{7}N$ with deuterons, the products being $^{15}_{8}O$ and neutrons. It emits a positron, β^+ (0.511 MeV), and has a half-life of 2 min; it is used in positron emission **tomography**.

oxygenase an enzyme that catalyses a reaction in which molecular oxygen is incorporated into an organic compound (the donor). Monooxygenases (EC class 1.13) act upon single donors, and dioxygenases (EC class 1.14) act upon paired donors.

oxygenate 1 to mix, supply, or treat with dioxygen; to saturate a fluid, especially blood, culture medium, etc., with dioxygen. **2** to cause the combination of an oxygen-carrying or oxygen-storing protein with dioxygen; to undergo such a reaction. —**oxygenation** *n.*

oxygenator an apparatus for oxygenation, especially of blood.

oxygen cycle the sum total of the processes via which atmospheric dioxygen is exchanged with the oxygen combined in organic compounds and carbon dioxide. Photosynthetic organisms, using solar energy, generate dioxygen from water and liberate it into the atmosphere, while heterotrophic cells take up atmospheric dioxygen to oxidize organic compounds and thereby provide energy. *Compare* **carbon cycle**.

oxygen debt the oxygen deficit that develops when a normally aerobic organism or tissue cannot increase its oxygen uptake sufficiently to match a temporarily increased energy requirement, as in muscular exercise. After the increased energy requirement ceases, this deficit has to be made good by a continuing increased rate of oxygen uptake until normal aerobic metabolism is restored.

oxygen demand *see* **biochemical oxygen demand**, **chemical oxygen demand**.

oxygen electrode a device for measuring dioxygen activity in a liquid or gas. It consists essentially of a silver–silver chloride anode in saturated potassium chloride solution and a platinum or gold cathode. When an electric potential is applied between these electrodes dioxygen is electrolytically reduced at the cathode, the current flowing being proportional to the dioxygen activity when the applied potential is 0.5–0.8 V. The electrodes are either directly exposed to the sample or, in the **Clark electrode**, separated from it by a thin dioxygen-permeable membrane in order to prevent deposition of interfering substances on the cathode.

oxygen-evolving complex *abbr.*: OEC; a protein-bound complex of manganese and calcium in photosystem II that is responsible for splitting water to oxygen and reducing equivalents. *See* **photosystem**.

oxygen quotient *(sometimes)* the **metabolic quotient** with respect to dioxygen, *symbol*: Q_{O_2}.

oxygen toxicity toxicity resulting from excessive exposure to gaseous dioxygen. An atmosphere containing 100% oxygen is toxic to animals and cultured animal cells, bacteria, fungi, and plants. In humans, breathing 80–100% oxygen for 8 hours or more causes irritation of the respiratory passages, and after 24–48 hours it causes lung damage. In infants it causes bronchopulmonary dysplasia and retrolental fibroplasia (replacement of the retina by fibrous tissue). Hyperbaric oxygen is toxic to the central nervous system.

oxyhemoglobin *symbol*: HbO_2; the fully oxygenated form of **hemoglobin**. It is bright red in colour, and has a characteristic absorption spectrum.

oxyhydrogen reaction *an alternative name for* **knallgas reaction**.

oxymyoglobin *symbol*: MbO_2; the fully oxygenated form of **myoglobin**. It is bright red in colour, and has a characteristic absorption spectrum.

oxyntic acid-secreting; applied especially to the hydrochloric acid-secreting parietal cells of the gastric mucosa.

oxyntomodulin *an alternative name for* glucagon-37; *see* **enteroglucagon**.

oxyproline *an older name for* **hydroxyproline**. *Compare* **5-oxoproline**.

oxysome *an alternative name for* **elementary particle** (def. 2).

oxysterol receptor *or* **liver X receptor** (*abbr*.: LXR) either of two orphan receptors, LXRα and LXRβ, whose endogenous ligands are oxysterols. Both occur in the enterohepatic system, with LXRα also present in adipose tissues and LXRβ being more widely distributed. Both form obligate heterodimers with **retinoid X receptor** (which binds 9-*cis*-retinoic acid), and the dimer binds to specific hormone response elements on DNA.

oxytetracycline 5-hydroxytetracycline; an antibiotic that is similar in action to **tetracycline**. Proprietary names include Terramycin.

oxytocin *or (sometimes)* **ocytocin** *abbr*.: OXT; a heterodetic cyclic nonapeptide amide,

$$\text{Cys}^1\text{-Tyr-Ile-Gln-Asn-Cys}^6\text{-Pro-Leu-Gly-NH}_2$$

(Cys^1 and Cys^6 form a disulfide bond). It is one of the two (or possibly more) hormones secreted by the neurohypophysis of the pituitary gland. It facilitates the ejection of milk by stimulating the myoepithelial cells in the mammary gland, and it may also aid parturition by stimulating contraction of the uterus. It is synthesized as a polypeptide precursor that is cleaved to oxytocin and neurophysin 1 (*see* **neurophysin**). Oxytocin was the first peptide hormone to be chemically synthesized. [*Note*: the spelling ocytocin is preferred on etymological grounds, the name being derived from the Greek *okytokos*, fast birth, rather than from the Greek *oxys*, acid, sharp; however, the spelling oxytocin is in wide use, esp. in the English language.] *See also* **vasopressin**.

oxytocin receptor any of a number of membrane proteins that bind oxytocin, and mediate its intracellular effects. Structurally, they are of the **7TM** type, characteristic of G protein-coupled receptors, and activate the **phosphatidylinositol cycle**.

+oyl *suffix (in chemical nomenclature)* denoting conversion (of a mono- or dicarboxylic acid) to the corresponding mono- or divalent acyl group by removal of hydroxyl from the carboxyl group; e.g. the univalent acyl group, hexanoyl, $CH_3\text{--}[CH_2]_4\text{--}CO\text{--}$, is derived from hexanoic acid, $CH_3\text{--}[CH_2]_4\text{--}COOH$. It is used also when the carbonyl oxygen atom of the acyl group has been replaced by $=NH$, $=N\text{--}NH_2$, $=N\text{--}OH$, or $=S$. *See also* **+yl**.

oz *symbol for* ounce.

ozone *symbol*: O_3; trioxygen; a colourless, highly reactive gas present in the stratosphere, where it is formed under the influence of short-wavelength ultraviolet radiation. It is also normally present in trace amounts in air, and is formed by the action of sunlight on hydrocarbons and nitrogen oxides – present in vehicle emissions. The latter process can lead to harmful accumulations of ozone at or near ground level. In contrast, stratospheric ozone protects the biosphere from highly damaging ultraviolet radiation, especially UV-C. Ozone is also produced by antigen-specific or monoclonal antibodies from excited dioxygen released by phagocytes or generated during inflammation. No enzyme is known that catabolizes ozone.

ozonolysis a method used for locating a double bond in an alkene. The alkene is treated with ozone, and the resulting ozonide is then cleaved to form two carbonyl compounds, one from each moiety of the parent compound.

Pp

p *symbol for* **1** pico+ (SI prefix denoting 10^{-12} times). **2** the negative logarithm .to the base ten, as in pH, pK. **3** proton; it may be written as p$^+$ to indicate its charge. **4** a phosphoric residue (alternative to *P*). **5** the protein product of a gene; it is prefixed to the product's molecular mass to designate a particular entity, e.g. **p53**. **6** (*in atomic spectroscopy*) electron state $l = 1$.

p21 a protein (21 kDa) inhibitor of cyclin-dependent kinases of the G_1 phase of the cell-division cycle. It is induced by **p53** and is believed to contribute to cell-cycle arrest in the presence of damaged DNA, and to facilitate withdrawal from the cell cycle of cells undergoing terminal differentiation.

p53 a 53 kDa nuclear phosphoprotein originally discovered attached to large T-antigen in cells transformed with **simian virus 40**. It is a probable human cell cycle regulator and *trans*-activator (*see* **trans-acting**) that acts negatively to regulate cellular division by controlling a set of genes required for this process; one of the genes activated is an inhibitor of **cyclin-dependent protein kinases**. p53 acts as a tumour suppressor in some (not all) tissues; mutations in the p53 gene are the most common genetic alterations in human cancers. Wild-type p53 can suppress or inhibit the transformation of cells in culture either by viral or cellular oncogenes, and introduction of wild-type cDNA into a transformed cell in culture stops growth. Hence the p53 gene is classed as a **tumour-suppressor gene**. Three functional domains of p53 have been defined: an N-terminal transcriptional activation domain (residues 1–42), a central DNA-binding domain (residues 120–290), and a C-terminal regulatory domain (residues 311–393). The DNA-binding domain recognizes a Dna motif in genes that are activated by p53, and mutations in p53 that are associated with human cancers generally cluster in its DNA-binding domain.

p56 *see* **Lck**.

p *symbol for* **1** pressure (alternative to *P*). **2** partial pressure of a gas. The gas may be specified by a subscript or in parentheses; e.g. p_{O2} or $p(O_2)$. **3** pyranose form (in monosaccharide symbolism); e.g. Glc*p*, glucopyranose; Gal*p*A, galactopyranuronic acid. **4** pitch height (of a helix). **5** (bold italic) (electric) dipole moment (of a molecule), alternative to μ (bold italic). **6** (bold italic) momentum. **7** probability (alternative to *P*).

p- *prefix (in chemical nomenclature)* denoting **para-**. *Compare* **m-**, **o-**.

p_{50} *symbol for* the partial pressure of a gas at which a binding protein is 50% saturated with it.

p_{LR} *symbol for* the proportion of receptors or binding sites occupied by ligand L.

p_{LR^*} *symbol for* the proportion of receptors in which ligand L occupies its binding site(s), and which are in an active state.

p_R *symbol for* the proportion of receptors or binding sites free of ligand.

P *symbol for* **1** peta+ (SI prefix denoting 10^{15} times). **2** poise. **3** phosphorus. **4** a residue of the α-amino (strictly, cyclic α-alkylamino) acid proline (alternative to Pro). *abbr. for* **5** phosphate or a phosphoric group, as in ATP. **6** (*in genetics*) parental generation. **7** a site near a peptidase cleavage site (*see* **peptidase P-site**).

P_i *symbol for* inorganic phosphate.

P0 a myelin protein.

P1 a temperate bacteriophage whose natural hosts are *Shigella* spp. but which was used formerly for generalized transduction in *Escherichia coli* and is currently used as the basis for the **P1 cloning vector**, which can carry up to 100 kbp insert DNA. P1 prophage is a plasmid.

P2 a myelin protein.

P65 *see* **synaptotagmin**.

P81 *see* **ezrin**.

P450 *or* P-450 any of a group of widely occurring heme-containing proteins that embraces various separately classified enzymes. They are named from the location of the Soret band (*see* **cytochrome absorption bands**) of the reduced CO complex at 450 nm. The group includes unspecific monooxygenase (EC 1.14.14.1), camphor 5-monooxygenase (EC 1.14.15.1), alkane 1-monooxygenase (EC 1.14.15.3), steroid 11β-monooxygenase (EC 1.14.15.4), **cholesterol monooxygenase (side-chain-cleaving)** (EC 1.14.15.6), prostacyclin synthase (EC 5.3.99.4), **thromboxane-A synthase** (EC 5.3.99.5), and probably chloride peroxidase (EC 1.11.1.10). This type of enzyme is also known as **cytochrome P450** (see this for more detail), and considered as a *b*-type **cytochrome**. The name heme-thiolate protein has been proposed on the grounds of a thiolate ligand at the heme being responsible for the group's unusual spectral and catalytic properties.

P680 *or* **pigment 680** a type of chlorophyll *a*, having an absorption maximum at 680 nm, that constitutes the reaction centre of **photosystem** II in most species. It has a redox potential of >+0.81 V; the oxidized form, P680$^+$, is a powerful oxidant that can abstract electrons from water with the evolution of dioxygen. *See also* **chlorophyll**.

P700 *or* **pigment 700** a type of chlorophyll *a*, having an absorption maximum at 700 nm, that constitutes the reaction centre of **photosystem** I. It has a redox potential of approximately +0.5 V, its electron being used ultimately to reduce NADP via ferredoxin. *See also* **chlorophyll**.

P870 the special pair of bacteriochlorophyll *a* in the **photosynthetic reaction centre** of bacteria, so named because they have an absorption maximum of 870 nm.

P960 the special pair of bacteriochlorophyll *b* in the **photosynthetic reaction centre** of bacteria, so named because they have an absorption maximum of 960 nm.

P *symbol for* **1** pressure (alternative to *p*). **2** radiant power (alternative to Φ). **3** probability (alternative to *p*). **4** parity (of a wave function).

Pa *symbol for* pascal.

pA_2 the negative logarithm to the base 10 of the molar concentration of antagonist that makes it necessary to double the concentration of agonist needed to elicit the original submaximal response.

$p[A]_{50}$ the negative logarithm to the base 10 of the $[A]_{50}$ of an agonist. pEC_{50} is the corresponding term for the **EC_{50}** of an agonist (often used interchangeably with $[A]_{50}$).

PAB *or* **PABA** *abbr. for para*-aminobenzoate; *para*-aminobenzoic acid. *See* **p-aminobenzoic acid**.

PABA test an indirect test of pancreatic function in which the synthetic peptide *N*-benzoyl-L-tyrosyl-*p*-aminobenzoic acid (BT-PABA) is administered orally, usually with a small amount of [^{14}C]-*p*-aminobenzoic acid (PABA). Normally the peptide is hydrolysed by chymotrypsin to PABA, which is absorbed and excreted in the urine, but in pancreatic insufficiency hydrolysis is much reduced.

PABP-2 *abbr. for* polyadenylation binding protein 2; a nuclear protein present in all tissues, but especially in skeletal muscle, that is needed for elongation of the poly(A) tail at the 3′ untranslated region of eukaryotic mRNA. The human gene contains a (GCG) repeat for a polyalanine stretch at the N-terminus of the protein. The repeat number is increased in oculopharyngeal muscular dystrophy.

PACAP *abbr. for* pituitary adenylate cyclase-activating polypeptide; a preganglionic neuropeptide that stimulates expression of genes for chromogranins A and B (all of which have a cAMP response element in their promoter) in neuroendocrine cells. When released by melanopsin-containing retinal axons it transmits photic information.

pacemaker reaction any metabolic reaction whose rate depends on factors other than the amounts of enzyme or substrate. An enzyme catalysing such a reaction may be termed a **pacemaker enzyme**.

pachyman *see* **callose**.

pachytene the third phase of prophase I in **meiosis**.

package 1 (*in cell biology*) to surround (one or a number of macromolecules) with a membrane in the formation of a mature secretory granule. **2** (*in virology*) to enclose the nucleic-acid core of a virus particle with a protein shell (the capsid) to form a mature virion; to encapsidate.

packaging cell line any mammalian cell line modified for the production of recombinant retroviruses. They express essential viral genes that are lacking in the recombinant retroviral vector.

packaging extract a lysate of *E. coli* in which all but one of the capsid and tail proteins of bacteriophage lambda have been expressed. A mixture of two such lysates lacking different proteins is capable of packaging recombinant lambda genomes with high efficiency *in vitro*.

packing *(in chromatography)* the material (whether active solid, liquid held on solid support, or swollen gel) that is introduced into a column and that consists of or contains what is to become the stationary phase during a chromatographic separation.

packing density the ratio of the minimum (or actual) volume of an object, such as a molecule in a crystal or in solution, to the total volume it occupies. For a macromolecule or region of a macromolecule, it is the ratio of the volume enclosed by the van der Waals envelope of all atoms in the molecule or region to the total volume of the molecule or region. Theoretical values are 0.74 for close-packed spheres, 0.91 for infinite cylinders, and 1.0 for a continuous solid.

packing fraction the ratio of the relative atomic mass of a particular nuclide minus its nucleon number divided by its nucleon number. It is positive for the lightest and heaviest nuclides and negative for nuclides of intermediate mass.

Paclitaxel *a proprietary name for* **taxol**.

p21-activated protein kinase *see* **PAK**.

PAF *abbr. for* platelet-activating factor.

PAF-AH *abbr. for* platelet-activating factor acetylhydrolase. *See* **lipoprotein-associated phospholipase A2**.

PAGE *abbr. for* polyacrylamide gel electrophoresis.

PAGIF *abbr. for* polyacrylamide gel isoelectric focusing. *See* **polyacrylamide gel electrophoresis**.

PAH *abbr. for* para-aminohippurate; *para*-aminohippuric acid.

pair *see* base pair.

paired the gene for a sequence-specific DNA-binding homeobox segmentation protein in *Drosophila melanogaster*. It is one of a number of **pair-rule genes**, each of which specifies a simple alternation with a repeat distance of two segments. Their protein products are characterized by a paired-box DNA-binding domain of ≈128 amino acids, which consists of two subdomains each resembling a homeobox domain.

paired box *see* **paired**, **Pax** genes.

pair-rule gene any of the eight genes in *Drosophila* that regulate the segment polarity genes. Examples are *hairy*, *runt*, and *paired*.

pairwise alignment *see* **alignment**.

pairwise alignment algorithm *see* **alignment algorithm**.

PaJaMa experiment *the nickname for* a bacterial mating experiment carried out at l'Institut Pasteur, Paris, that showed that induction and repression of the enzyme β-galactosidase in *Escherichia coli* are regulated by two closely linked genes, one of which produces a cytoplasmic repressing substance that blocks the expression of the other. [From the names of the experimenters: Pardee, and French molecular biologists François Jacob (1920–) and Jacques Monod (1910–76).]

PAK *abbr. for* p21-activated protein kinase; any of a family of protein serine/threonine kinases activated by p21 and Cdc42/Rac (a Rho GTPase involved in regulation of the actin cytoskeleton), that are differentially expressed in mammalian cells. PAKs are upstream modulators of protein kinase cascades involved in regulation of apoptosis and in morphogenesis. They have an N-terminal p21-Cdc42/Rac-binding domain and a C-terminal catalytic domain. PAK1 localizes with actin filaments of the cytoskeleton. PAK2 is cleaved into its two domains by certain caspases. A mutation in PAK3 results in an X-chromosome-linked form of mental retardation.

palindrome 1 any linear arrangement of symbols, such as letters or digits, that has the same sequence from either end; e.g. noon, radar; 75311357; 'Madam, in Eden I'm Adam'. **2** *(in molecular biology)* **palindromic sequence** a DNA sequence with a twofold rotational axis of symmetry (**dyad symmetry**). Confusingly the term is used in at least two senses: (1) to describe a region of local twofold rotational symmetry in duplex DNA, e.g.

5′-GAATTC-3′ 3′-CTTAAG-5′

and (2) to describe a sequence in a single polynucleotide strand that contains an inverted repeat, e.g. 5′-TGA-AGT-3′. It is also applied to a region of a single- or a double-stranded polynucleotide that is capable of forming one or more **hairpins** containing palindromes of the first type; an example is shown at **cruciform**. Examples of palindromes are **operators** and **restriction sites** but, in general, any stem–loop in RNA must have a corresponding palindrome in the DNA from which the RNA was transcribed. —**palindromic** *adj.*

pallidin any of a group of endogenous carbohydrate-binding proteins (**lectins**) produced by cells of the slime mould *Polysphondylium pallidum* during differentiation. Similar to **discoidin**, they may be involved in cell adhesion.

palmitate 1 *numerical symbol*: 16:0; *the trivial name for* hexadecanoate, CH_3–$[CH_2]_{14}$–COO^-, the anion derived from palmitic acid (i.e. hexadecanoic acid), a saturated straight-chain higher fatty acid. **2** any mixture of free palmitic acid and its anion. **3** any salt or ester of hexadecanoic acid. *See also* **palmitoyl**.

palmitate synthase *see* **fatty-acid synthase complex**.

palmitic acid hexadecanoic acid; a saturated straight-chain fatty acid having sixteen carbon atoms per molecule, and a major component of plant and animal fats. *See* **palmitate**.

palmitoleate 1 *numerical symbol*: 16:1(9); *the trivial name for* (Z)-hexadec-9-enoate, CH_3–$[CH_2]_5$–$CH{=}CH$–$[CH_2]_7$-COO^-, the anion derived from palmitoleic acid, (Z)-hexadec-9-enoic acid, a mono-unsaturated straight-chain higher fatty acid. **2** any mixture of free palmitoleic acid and its anion. **3** any salt or ester of 9-hexadecenoic acid. *See also* **palmitoleoyl**.

palmitoleic family (of polyunsaturated fatty acids) a series of polyenoic acids in which the hydrocarbon chain terminates with the alkenyl grouping CH_3–$[CH_2]_2$–$CH{=}CH$– (as in palmitoleic acid (*see* **palmitoleate**)); there are three other series of such acids, i.e. the **linoleic family**, **linolenic family**, and **oleic family**. Members of the palmitoleic family can be synthesized from palmitic acid via palmitoleic acid by chain elongation and/or desaturation, but in mammals not from linoleic, (9,12,15)-linolenic, or oleic acids; the series includes *cis*-vaccenic acid.

palmitoleoyl *symbol*: ΔPam; *the trivial name for* (Z)-hexadec-9-enoyl, CH_3–$[CH_2]_5$–$CH{=}CH$–$[CH_2]_7$–CO– (*cis* isomer), the acyl group derived from palmitoleic acid (i.e. (Z)-hexadec-9-enoic acid). It is a relatively minor component of plant and animal lipids. The chain is synthesized metabolically by the action of a Δ⁹-desaturase on palmitoyl-CoA.

palmitoyl *symbol*: Pam; *the trivial name for* hexadecanoyl, CH_3–$[CH_2]_{14}$–CO–, the acyl group derived from palmitic acid (i.e. hexadecanoic acid). This is one of the major fatty-acid components of plant and animal lipids; together with stearic acid it represents a high proportion (often around 30%) of the fatty-acid content of dietary and other lipids. It acts as a precursor in mammals for a family of unsaturated fatty acids in which all double bonds are nine or more carbons from the methyl terminus. The chain is synthesized metabolically from acetyl-CoA as palmitate (in mammals) or palmitoyl-CoA (in yeast), the product of fatty-acid synthase. See **fatty-acid synthase complex**.

palmitoyl-protein thioesterase EC 3.1.2.22. a lysosomal enzyme that cleaves the thioester linkage formed by a palmitoyl moiety with a cysteine side chain near a transmembrane segment of an integral membrane protein. Mutations in this gene on 1p32 cause infantile neuronal lipofuscinosis and progressive myoclonic epilepsy.

palmityl 1 *or* **cetyl** *a common name for* hexadecyl, CH_3–$[CH_2]_{14}$–CH_2–, the alkyl group derived from hexadecane. **2** *a former name (now incorrect) for* **palmitoyl**.

palytoxin any of a group of structurally very similar potent neurotoxins isolated from the marine zoanthid *Palythoa toxica* or other *Palythoa* spp. Each consists of a single carbon chain of 115 carbon atoms containing 64 chiral centres and seven double bonds, and terminating in nitrogen-containing groups; there are 128 carbons in all and a further double bond in a methylene group. The chain bears numerous hydroxyl groups and a number of methyl groups, and in several places it is folded and formed into pyran rings. Palytoxin is the most deadly nonproteinaceous material isolated, having 50 times the toxicity of **tetrodotoxin** in mice. The number of stereoisomers is $2^{64} = 18446744073709551620$ (not including possible E/Z variations at the double bonds!).

Pam *symbol for* **palmitoyl** (i.e. hexadecanoyl).

PAM *abbr. for* **1** point accepted mutation. *See* **PAM matrix**. **2** peptidylglycine α-amidating monooxygenase. *See* **peptide amidation**.

PAM matrix *abbr. for* point accepted mutation matrix; *see* **point accepted mutation**, **substitution matrix**.

pancreas a compound gland, occurring in the abdominal cavity of most vertebrates, that has both exocrine and endocrine functions. The major (exocrine) part consists of acinar tissue, which secretes **pancreatic juice** into the upper part of the gut (duodenum in mammals). Within this acinar tissue are scattered numerous (endocrine) **islets of Langerhans**, containing notably A cells, secreting **glucagon**, B cells, secreting **insulin**, and D cells, secreting **somatostatin**. —**pancreatic** *adj.*

pancreastatin a 49-residue peptide hormone that inhibits glucose-induced insulin release and exocrine pancreatic secretion. It was originally isolated from porcine pancreas, where it is colocalized with insulin, glucagon and somatostatin, but it is part of the sequence of **chromogranin** A and is thus found in tissues where the chromogranin gene is expressed, including most endocrine cells.

pancreatic cholera *an alternative name for* **Verner–Morrison syndrome**.

pancreatic diabetes *an older term for* insulin-dependent **diabetes mellitus**.

pancreatic DNase I *an alternative name for* **deoxyribonuclease I**.

pancreatic hormone *or* **pancreatic polypeptide** *abbr.*: PP; a polypeptide present in the **PP cells** of the pancreas. Avian pancreatic polypeptide, known as aPP (from chicken and turkey pancreas), and bovine, human, ovine, and porcine homologues, known as bPP, hPP, oPP, and pPP, respectively, have been identified. All contain 36 amino-acid residues with C-terminal tyrosinamide and show considerable sequence homology. The plasma PP level rises rapidly after feeding, especially with protein foods, and with fasting, exercise, and hypoglycemia. The hormone decreases food absorption and pancreatic exocrine secretion. The sequence of human PP is: APLEPVYPGDNATPEQMAQYAADLRRYINMLTR-PRY(NH₂).

pancreatic islet *an alternative name for* **islet of Langerhans**.

pancreatic juice a slightly alkaline digestive juice secreted by the exocrine pancreas into the upper part of the small intestine. It contains numerous enzymes and inactive enzyme precursors including α-amylase, chymotrypsinogen, **lipase**, procarboxypeptidase, proelastase, prophospholipase A₂, ribonuclease, and trypsinogen. Its high concentration of hydrogencarbonate ions helps to neutralize the acid digesta from the stomach.

pancreatic lipase a Ca^{2+}-requiring triacylglycerol lipase (EC 3.1.1.3) that is secreted into the intestine by the exocrine pancreas when this is stimulated by cholecystokinin in response to ingestion of food. It degrades triacylglycerols, partially or completely, to fatty acids and glycerol in the intestine; it acts at an ester–water interface. It is a glycoprotein with sequence similarity with other lipases.

pancreatic polypeptide *abbr.*: PP; *another name for* **pancreatic hormone**.

pancreatic ribonuclease see **ribonuclease**.

pancreatic thread protein *or* **pancreatic stone protein** (*abbr.*: PSP); a C-type lectin that may inhibit spontaneous calcium carbonate precipitation, and one of the major secretory proteins of the human exocrine pancreas. It is found in acinar cells of pancreas and (in smaller amounts) in the brain. It is a major soluble protein of human pancreatic calculi, and a Ca^{2+}- binding phosphoprotein present in zymogen granules of pancreatic acinar cells, and secreted in pancreatic juice of normal subjects and calculus formers. It inhibits calcium carbonate precipitation from the juice. The protein is rich in aromatic amino acids. Between pH 5.4 and 9.2, it undergoes reversible fibril formation. It is also typically characteristic of infant brains. There is increased expression of PSP-like proteins in Down's syndrome and Alzheimer's disease.

pancreatic trypsin inhibitor *abbr.*: PTI; a serine proteinase inhibitor, *correctly known as* **pancreatic secretory trypsin inhibitor**. A protein isolated from bovine pancreas, it forms a crystalline complex with trypsin and is relatively heat-stable in trichloroacetic acid solution. It controls trypsin activation of zymogens. The bovine protein has the sequence: NILGREAKCTNEVNGCPRIYN-PVCGTDGVTYSNECLLCMENKERQTPVLIQKSGPC. The 3-D structure is also known, and it is a favoured model for testing protein-folding algorithms. It was also formerly known as **Kunitz and Northrop inhibitor**.

pancreozymin *abbr.*: PZ; a hormone having secretagogic activity on the exocrine pancreas and extractable from duodenal mucosa. It is identical with **cholecystokinin**.

PANK *abbr. for* pantothenate kinase.

panning *(in cell biology)* a method for enriching a population of cells, or phages, displaying peptides or fragments of antibodies by allowing them to bind to ligands immobilized on a solid surface such as that of a **microtitre plate**. Washing the wells of the plate removes nonadherent or weakly adherent cells or phages, leaving those adhering as a result of stronger interactions, to be collected for further study. Several cycles of enrichment are used.

panose an oligosaccharide, Glc(α1-6)Glc(α1-4)Glc; it is found in the fungus *Aspergillus niger*.

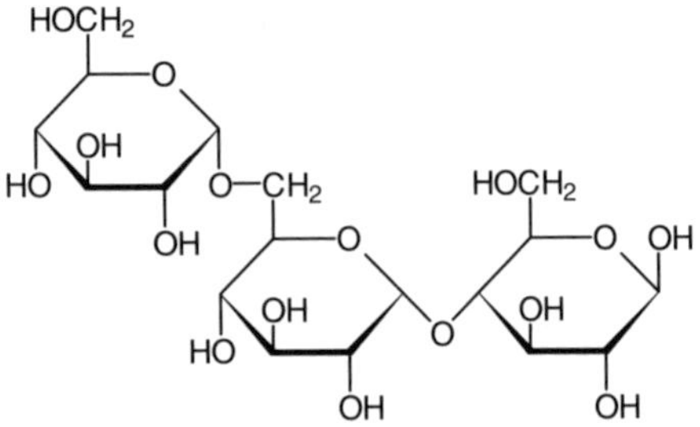

panose

pantetheine *N*-pantothenylcysteamine; the D-enantiomer is a growth factor for *Lactobacillus bulgaricus*, and is an intermediate in the pathway for the biosynthesis of coenzyme A in mammalian liver and some microorganisms.

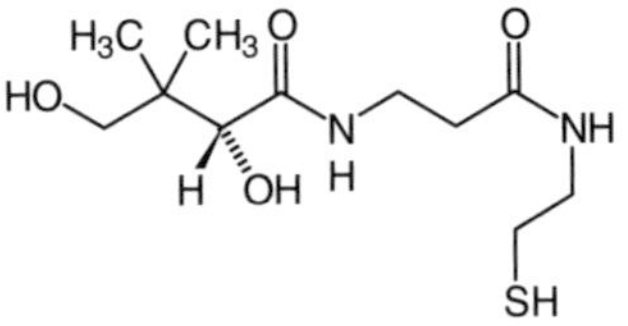

pantoate **1** the anion, $HO–CH_2–C(CH_3)_2–CH(OH)–COO^-$, derived from pantoic acid (i.e. (R)-2,4-dihydroxy-3,3-dimethylbutyric acid). **2** any salt or ester of pantoic acid.

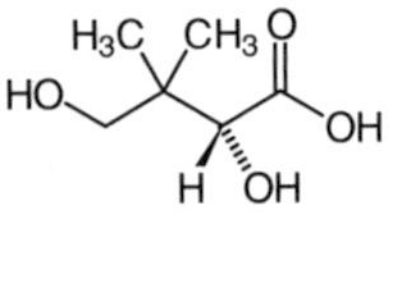

pantoic acid

pantophysin a member of the **physin** family of proteins. It is ubiquitously expressed, is homologous with synaptophysin, and is present in constitutive transport vehicles.

pantothenate **1** the anion, HO–CH$_2$–C(CH$_3$)$_2$–CH(OH)–CO–NH–CH$_2$–CH$_2$–COO$^-$ (i.e. *N*-pantoyl-β-alanine anion), derived from pantothenic acid, a B vitamin complex; only the D-(*R*)-enantiomer is biologically active. **2** any salt of pantothenic acid.

pantothenic acid

pantothenate kinase *abbr.*: PANK. EC 2.7.1.33; a kinase enzyme that is specific for phosphorylation of pantothenate to 4'-phosphopantothenate, being the first reaction in the biosynthesis of coenzyme A. Many mutations in the gene for PANK2 (locus at 20p13) produce enzyme deficiency and Hallervorden–Spatz syndrome, a neurodegenerative disease involving accumulation of iron in the brain.

pantothenyl the acyl group, HO–CH$_2$–C(CH$_3$)$_2$–CH(OH)–CO–NH–CH$_2$–CH$_2$–CO–, derived from pantothenic acid; coenzyme A is a D-(*R*)-pantothenyl derivative.

pantoyl **1** the acyl group, HO–CH$_2$–C(CH$_3$)$_2$–CH(OH)–CO–, derived from pantoic acid, 2,4-dihydroxy-3,3-dimethylbutyric acid.

PaO *abbr. for* pheophorbide *a* oxygenase.

PAP *abbr. for* peroxidase–antiperoxidase.

papain *or* **papaya peptidase I** EC 3.4.22.2; a cysteine endopeptidase obtained from the latex, leaves, and unripe fruit of the papaya (or pawpaw) tree, *Carica papaya*. It will preferentially hydrolyse peptide bonds at the carbonyl end of Arg, Lys, and Phe residues (but never Val), with preference for large hydrophobic residues at the P2 position. It also has esterase, thiolesterase, transamidase, and transesterase activity. Papain is unusually stable to elevated temperatures and to denaturing agents, and consists of a single polypeptide chain of 212 amino-acid residues.

papaverine 1-[(3,4-dimethoxyphenyl)methyl]-6,7-dimethoxyisoquinoline; a constituent of **opium** that acts as a smooth muscle relaxant; this action is thought to be due to phosphodiesterase inhibitory activity and blockade of membrane calcium channels. Like codeine and morphine it is a metabolic derivative of (*S*)-reticuline.

paper chromatogram the visible result of a chromatographic separation effected on paper.

paper chromatography *abbr.*: PC; a technique of **chromatography**, applicable to microgram quantities of soluble substances, in which specially prepared filter paper, **chromatography paper**, forms the support for the stationary phase. The latter is commonly a film of water held by adsorption on the cellulose fibres of the paper and in equilibrium with a water-immiscible liquid or liquid mixture, which forms the mobile phase. A solution of a sample is applied near one end of a strip of paper, allowed to dry, and the mobile phase allowed to flow over the strip from that end. If the sample consists of only one component, development using an appropriate method of detection will reveal a single spot at a characteristic distance from the origin relative to that travelled by the mobile phase; in general, a mixture will give rise to a number of spots each occupying a different and characteristic position. Separations occur essentially by partition between the mobile and stationary phases, with some contribution, under certain conditions, attributable to adsorption or ion-exchange onto the paper support. An elaboration of the technique uses a square of paper on which two successive separations are effected at right-angles using mobile phases of differing separatory powers to produce a two-dimensional pattern of spots on the paper. With suitable treatment of the paper and an appropriate choice of the composition of the two phases, reversed-phase partition chromatography on paper may be effected. Since cellulose is chiral, some enantiomer separations are possible.

paper electrophoresis a type of **zone electrophoresis** on (filter) paper.

Papovaviridae a family of DNA viruses, most or all of which are, under suitable conditions, oncogenic in vertebrate hosts. The virion, 45–55 nm in diameter, consists of a nonenveloped icosahedral capsid with 72 capsomeres, and contains a circular DNA genome.

papovavirus any virus belonging to the family **Papovaviridae**. The name derives from papilloma, polyoma, and vacuolating agent, an early name for **simian virus 40**.

PAPP-A *abbr. for* pregnancy-associated plasma protein A; a plasma protein first found in the serum of pregnant women but also present in men. It is a zinc metalloprotease secreted by fibroblasts and cleaves **insulin-like growth factor binding protein** (IGFBP)-4 to release insulin-like growth factor (IGF).

PAPS *abbr. for* 3'-phosphoadenosine-5'-phosphosulfate; i.e. adenosine 3'-phosphate 5'-phosphosulfate.

par+ *a variant form of* **para+** (before a vowel).

PAR *abbr. for* protease-activated G-protein-coupled receptor; any G-protein-coupled membrane receptor of platelets and endothelial cells that becomes activated on binding thrombin and cleavage of the extracellular segment of the receptor. Most mouse embryos that lack PAR1 die in mid-gestation.

para 1 characterized by or relating to (substitution at) two ring-carbon atoms separated by two others in a monocyclic aromatic compound. **2** (*followed by* to) at a position in a benzene ring next but two to a particular (specified) substituent. *See also* **para-**.

para+ *or* (*before a vowel*) **par+** *prefix* indicating beside or near; beyond; resembling; defective or abnormal.

para- *abbr.*: *p*-; *prefix* (*in chemical nomenclature*) denoting an isomer of a disubstituted monocyclic aromatic compound in which the substituents are attached at positions 1 and 4 of the ring. *Compare* **meta-**, **ortho-**.

parabiont either individual of a pair united in **parabiosis**.

parabiosis the state or condition of two individual animals being functionally united, either naturally, as in Siamese twins, or as a result of an experimental procedure. —**parabiotic** *adj.*

paracasein an alternative name (*esp. US*) for **casein** (def. 2).

paracentric (*in cytogenetics*) describing an inversion of part of a chromosome that does not span the centromere, and is thus confined to one arm of the chromosome. *Compare* **pericentric**.

paracetamol *see* acetaminophen.

paracrine 1 describing or relating to a regulatory cell that secretes an agonist into intercellular spaces in which it diffuses to a target cell other than that which produces it; describing or relating to such an agonist. **2** such an agonist; a **paramone**. *Compare* endocrine, neurocrine, neuroendocrine.

paracrystal any assemblage of molecules or particles having a certain degree of order but one that is insufficient for it to be considered a true crystal. —**paracrystalline** *adj.*; **paracrystallinity** *n.*

paradaxin *or* **pardaxin** any of several ichthyotoxic, neurotoxic peptides, isolated from the defence secretion of the Red Sea Moses sole fish, *Paradachirus marmoratus*, or the Pacific sole, *Paradachirus pavoninus*, that act as a shark repellent. Each consists of 33 amino acids, with hydrophilic C-terminal and hydrophobic N-terminal regions, and with physical and pharmacological similarities to **melittin**, but no sequence similarity.

paraffin 1 *an obsolete term for* **alkane**. **2** *or* **paraffin oil** *or* **kerosene** *or* **kerosine** any liquid mixture, consisting mainly of alkanes, with a boiling point range of 150–300 °C and relative density range of 0.78–0.82. **3** *an alternative name for* **paraffin wax**. *Compare* **liquid paraffin**.

paraffin wax any waxy mixture of higher alkanes, obtained as a residue from the distillation of petroleum, with a relative density of ≈0.9 and a melting point range of 45–65 °C.

Parafilm *the proprietary name for* a flexible, moisture-proof, thermoplastic tissue useful for temporarily sealing small containers.

parafollicular cell *an alternative name for* **C cell** (when found in the thyroid gland).

parafusin an evolutionarily conserved phosphoglycoprotein involved in exocytosis. It is rapidly dephosphorylated via a Ca^{2+}-dependent process when secretagogues induce exocytosis in competent cells.

parahematin *or* *(esp. Brit.)* **parahaematin** *a former name for* **ferrihemochrome**.

parahemophilia a bleeding disorder due to deficiency of factor V. *See* **blood coagulation**.

parallel 1 *(in biochemistry)* describing a pair of linear structures, such as two polynucleotide or polypeptide chains, that are polarized or asymmetric in the same direction. **2** *(in physical chemistry)* denoting the spins of a pair of electrons, occupying the same atomic or molecular orbital, that are described by the same spin quantum number. *Compare* **antiparallel**. **3** *(in electricity)* describing an electrical component that is connected to the same two points in a circuit as another component; describing two or more components that are so connected.

parallel dichroism *see* **dichroic ratio**.

parallelism the quality or state of being parallel; similarity in corresponding details or in evolutionary pattern.

paralogue *or* *(esp. US)* **paralog** a gene, protein, or biopolymeric sequence that is evolutionarily related to another by descent from a common ancestor, having diverged as a result of a gene duplication event within an organism. Paralogues usually perform different but related functions within that organism as, for instance, human red and green opsins. –**paralogous** *adj*.

paramagnetism the property displayed by substances that have a positive but small magnetic susceptibility, due to the presence in them of atoms with permanent magnetic dipoles caused by unpaired electron spins (with a contribution from the orbital motion of the electrons). These dipoles tend to align themselves in the direction of an applied magnetic field but no permanent magnetism is conferred on such substances. *Compare* **diamagnetism, ferromagnetism**. —**paramagnetic** *adj*.

paramecium *(pl. paramecia)* any freshwater protozoan belonging to the genus *Paramecium*. Typically 50–300 μm in length, they are ovoid and uniformly ciliated. *See also* **kappa particle**.

parameter 1 *(in mathematics)* an unknown quantity that is a constant in a particular context but that may have different values in other (similar) contexts; e.g. the coefficient b in the general equation, $y = bx + c$, representing a family of straight lines. **2** *(in statistics)* a numerical characteristic of a population (as opposed to that of a sample of such a population). **3** any distinguishing characteristic of something, especially one to which a measured value is or can be ascribed. —**parametric** *adj*.

paromomycin an aminoglycoside antibiotic produced by *Streptomyces paromomycinus*. Its actions are similar to those of **neomycin**.

paramone *a proposed name for* **paracrine** agonist, derived as a contraction of 'paracrine hormone'.

paramylon a storage polysaccharide occurring in *Euglena gracilis*. It comprises a backbone of $\beta1{\rightarrow}3$-glucopyranose units with side chains joined by $\alpha1{\rightarrow}6$ linkages.

paramyosin a two-chain, coiled-coil, α-helical ~200 kDa protein that forms the core protein of the thick filaments of a number of invertebrate muscles.

paramyxovirus any RNA animal virus of the family Paramyxoviridae (distinguish from the genus Paramyxovirus), in class V of the **Baltimore classification**. **Paramyxoviruses** consist of enveloped, pleomorphic particles ≈150 nm in diameter with helical nucleocapsids, 1000 nm long and 18 nm in diameter. They contain single-stranded RNA (minus strand) of 7 MDa plus virion polymerase. The group includes viruses causing mumps, measles, Newcastle disease of chickens, and distemper of dogs, and the morbilliviruses, pneumoviruses, and parainfluenza viruses.

paranemic describing the nature of the coiling of a double-stranded helix in which the two strands are not intertwined, i.e. they may be separated without uncoiling. *Compare* **plectonemic**.

paraoxonase aryldialkylphosphatase, EC 3.1.8.1; an enzyme that hydrolyses organophosphates. The name derives from its ability to hydrolyse paraoxon, diethyl-p-nitrophenyl phosphate, a cholinesterase inhibitor that is used as a pesticide. The enzyme has antioxidant properties and its physiological substrates are probably the products of lipid peroxidation.

parapepsin I *an alternative name for* **pepsin B**.

parapepsin II *an alternative name for* porcine **gastricsin** (i.e. pepsin C).

paraplegin a mitochondrial **AAA protease** (88 kDa) that is highly homologous with yeast metalloproteases and has proteolytic and chaperone functions at the inner mitochondrial membrane. It is encoded on nuclear DNA. Mutated paraplegin results in hereditary spastic paraplegia and respiratory chain defects.

paraprotein *name originally given to* any plasma protein that gave an abnormal band on electrophoresis, derived from **para+** + protein. Usually, such a protein is a monoclonal immunoglobulin derived from neoplastic plasma cells, and present at abnormally high concentration in the blood plasma. Such proteins are seen as a discrete band in the gamma-globulin region, but they may appear elsewhere if the paraprotein is IgA or IgM. Examples are proteins characteristic of a **myeloma**, e.g. **Bence-Jones protein**, amyloid proteins, **Waldenstrom's macroglobulinemia**, or cryoglobulins. Some paraproteins are not abnormal, e.g. the immunoglobulin that arises as a result of a severe bacterial infection.

paraproteinemia *or* *(esp. Brit.)* **paraproteinaemia** the presence in blood plasma of **paraprotein**.

Paraquat methyl viologen; 1,1′-dimethyl-4,4′-bipyridynium dichloride; a nonselective herbicide whose action requires direct contact with the plant. It is highly toxic to humans and other animals if ingested in undiluted form.

parasite any organism that spends all or part of its life cycle in (**endoparasite**) or on (**ectoparasite**) another living organism of a different species (its **host**), from which it obtains nourishment and/or protection, and to which it is usually detrimental. *Compare* **symbiont**.

parastatin a 73-amino-acid peptide hormone, M_r 11000, that inhibits **parathyrin** secretion. Porcine parastatin has the amino-acid sequence:

LSFRAPAYGFRGPGLQLRRGWRPSSREDSVEAGLPL-
QVRXYLEEKKEEEGSANRPEDQELESLSAIEAELEK.

Activity resides in a 19-amino-acid N-terminal fragment; that from rat has the sequence: LSFRARAYGFRDPGPQLRR. It is synthesized as part of the sequence of **chromogranin** A.

paratartaric acid *an old name for* racemic DL-tartaric acid. The resolution of this acid into the D- and L-enantiomers by Pasteur was a landmark experiment in stereochemistry. *See* **tartaric acid**.

parathion O,O-diethyl O-p-nitrophenyl phosphorothioate; a compound that acts as an inhibitor of cholinesterase through covalent bonding to the esteratic site of the enzyme. It is used as a chemical warfare agent and insecticide.

Parathormone *a proprietary name for* **parathyrin**.

parathyrin *or* **parathyroid hormone** *(abbr.:* PTH); a 9.5 kDa polypeptide hormone of 83 amino-acid residues (bovine). It is derived from a 12.5 kDa precursor. Parathyrin acts in conjunction with **calcitonin** and other hormones to control calcium and phosphate metabolism; it elevates the blood calcium level by dissolving the salts in bone and preventing their renal excretion. Its effects are mediated by a G-protein-associated membrane receptor. Several mutations in the receptor cause a rare form of short-limbed dwarfism (or metaphyseal chondrodysplasia). Mutations that lead to absence or dysfunction of the $G_{S\alpha}$ subunit cause pseudohypoparathyroidism type 1a, in which there is also resistance to a wide variety of other hormones whose effects are mediated by this subunit. *Proprietary name*: Parathormone.

parathyroid 1 situated near the thyroid gland. **2** of, pertaining to, or produced by the parathyroid gland. **3** the parathyroid gland.

parathyroid gland an endocrine gland of higher vertebrates; there are typically two pairs, located near or within the thyroid. They develop from the gill pouches, and secrete **parathyrin**.

parathyroid hormone *abbr.*: PTH; *an alternative name for* **parathyrin**.

parathyroid hormone-related protein *abbr.*: PHRP; a protein (141 amino acids) that is homologous in the N-terminal region with parathyroid hormone, has effects similar to this, and binds the same receptor. It is present in breast and milk, kidney, brain, and other tissues, especially during intrauterine life, and seems to function largely in paracrine fashion. It is secreted by many tumours and is the major cause of malignancy-associated hypercalcemia. It may be a polyhormone because residues 1–36 are parathyroid hormone-like, residues 38–94 effect calcium transport in the placenta and bicarbonate transport in the kidney, and residues 107–139 constitute **osteostatin**. Homozygous gene knockout in mice is often lethal during embryogenesis or produces abnormalities throughout the skeleton.

paratope the part of an antibody, formed by the hypervariable loops of the **variable regions**, that binds to an **epitope**.

paratose 3,6-dideoxy-D-glucose; 3,6-dideoxy-D-*ribo*-hexose; a monosaccharide that occurs in type A O-antigen chains of the lipopolysaccharide on the outer membrane of certain species of *Salmonella*.

Parazoa a subkingdom of multicellular invertebrate animals comprising the sponges (phylum Porifera).

parenchyma 1 *(in botany)* in higher plants, any soft tissue consisting of thin-walled, relatively undifferentiated living cells. **2** *(in zoology)* the tissue constituting the essential or specialized part of an organ, as distinct from supportive tissue, blood vessels, etc. —**parenchymal** *adj.*; **parenchymatous** *adj.*

parenchymal liver cell *see* **hepatocyte**.

parent 1 any organism that has given rise to another such organism, whether sexually or asexually. **2** a precursor of some derived entity; applied, e.g., to a cell, a virus, a chromosome or molecule of DNA, or a radionuclide. **3 parent compound** *(in chemical nomenclature)* the member of a group of related compounds that has the simplest chemical structure and that forms the basis for naming the others. —**parental** *adj.*

parental imprinting *see* **imprinting**.

parenteral 1 located or occurring outside the intestine. **2** any route other than via the gastrointestinal tract, especially by injection. —**parenterally** *adv.*

PARG *abbr. for* poly(ADP-ribose) glycohydrolase.

parietal 1 of, pertaining to, or forming part of the wall of a body cavity or similar structure; e.g. the hydrochloric-acid-secreting **parietal cells** of the gastric mucosa, also known as oxyntic cells. **2** of or relating to the parietal bones of the skull.

parietal layer *see* **serous membrane**.

parinaric acid any octadeca-9,11,13,15-tetraenoic acid; two isomers are found in *Parinarium laurinum*, namely the (9Z,11E,13E,15Z)-isomer (α- or *cis*-parinaric acid) and the (9E,11E,13E,15Z)-isomer (β- or *trans*-parinaric acid). Both are useful as probes in membrane studies, *trans*-parinaric acid especially, as it partitions preferentially into the less fluid regions of membrane lipids with increased quantum yield. They inhibit neutrophil elastase.

α-parinaric acid

β-parinaric acid

parkin a **ubiquitin–protein ligase** whose substrate is a glycosylated form of α-**synuclein** (α-Sp22). The substrate accumulates in parkin-deficient brain. The parkin gene at 6q25.2-q27 encodes a protein (465 residues) whose N-terminal 76 residues are homologous with ubiquitin and whose C-terminal region contains two **RING fingers**. Mutations in the gene are associated with an autosomal recessive form of **Parkinson's disease**.

Parkinson's disease *or* **parkinsonism** a progressive disorder of the central nervous system characterized by tremor and impaired muscular coordination, thought to be due to defective dopaminergic transmission in some parts of the brain. There is a loss of dopaminergic neurons connecting the substantia nigra with the striatum, resulting in loss of dopaminergic inhibition in the striatum, with resulting cholinergic hyperactivity. Therapy is aimed at replacement of dopamine, achieved by administration of L-**dopa**, which is converted to dopamine within the brain (dopamine itself does not cross the blood–brain barrier). *See also* **MPTP**. [After James Parkinson (1755–1824), British palaeontologist and surgeon, who first described the disease in 1817.]

Park nucleotide UDP-*N*-acetylmuramyl pentapeptide; an intermediate in the biosynthesis of peptidoglycan in the cell walls of most bacteria. [After James Theodore Park (1922–).]

P450 arom *an alternative name for* **aromatase**.

paromomycin an antibiotic that, like streptomycin, increases the error rate of translation by binding to the 30S ribosomal subunit of prokaryotes and distorting their A site.

parotid gland either of a pair of **salivary glands** situated near each ear in mammals. The duct runs forwards and empties into the oral cavity.

paroxysmal nocturnal hemoglobinuria an acquired hemolytic anemia in which a clone of erythrocytes, derived from a hemopoietic stem cell that has acquired a somatic mutation, undergoes complement-mediated chronic intravascular hemolysis with consequent hemoglobinuria. It arises following any of numerous mutations in the gene at Xp22.1 for the enzyme that attaches the glucosamine moiety to phosphatidylinositol in the production of GPI-anchored proteins. This causes a lack on erythrocytes of CD55 and CD59, which normally protect the erythrocytes against the action of complement. Deficiency of CD55 alone is a rare cause of the disorder.

PARP *abbr. for* **1** procyclin. **2** proline/arginine-rich protein.

PARP-1 *abbr. for* poly(ADP-ribose) polymerase 1; in mammals, a highly conserved protein (113 kDa) that is strongly activated by DNA strand breaks, and participates in modulating DNA base excision repair, apoptosis, and necrosis. It includes a DNA-binding domain with 23 zinc finger motifs, and a poly(ADP-ribose) polymerase catalytic domain. Mouse gene knockouts for this protein are remarkably resistant to myocardial infarction, stroke, shock, diabetes mellitus, and neurodegeneration. *See also* **PARP-2**.

PARP-2 *abbr. for* poly(ADP-ribose) polymerase 2; in mammals, a 62 kDa protein that catalyses poly(ADP-ribose) formation, being activated by DNA strand breaks. Its activity supplements that of **PARP-1** in mouse gene knockouts for this protein.

parse to analyse amino acid sequences or nucleotide base sequences during the construction of multiple sequence alignments when determining homology.

parsimony 1 *(in computing)* an algorithm used for determining sequence relationships in macromolecules. **2** *(in genetics)* the construction of evolutionary trees from protein and nucleic-acid sequences. It maximizes the genetic likeness associated with common ancestry, whilst minimizing the incidence of convergent mutation, and provides a way to find the best genealogical hypothesis from sequence data.

parthenocarpy the formation of fruit without the setting of seeds. It occurs naturally, but can be induced by spraying flowers with gibberellic acid (*see* **gibberellin**) to produce crops of seedless fruit.

parthenogenesis the development of a new individual from an unfertilized female gamete. —**parthenogenetic** *adj.*; **parthenogenetically** *adv.*

partial agonist an agonist that is unable to evoke the maximal response of a biological system, even at a concentrations sufficient to saturate the specific receptors.

partial derivative *(in mathematics)* the derivative of an expression containing two or more independent variables that is obtained by differentiating the expression with respect to only one of them, the remaining variable or variables being considered constant.

partial hydrolysate any mixture of monomers and oligomers resulting from a limited degree of chemical hydrolysis of a biopolymer or from its exposure to a specific hydrolytic enzyme.

partial inhibition a type of nonlinear enzyme inhibition in which saturation of the enzyme with inhibitor does not decrease the rate of the enzyme-catalysed reaction to zero.

partial inverse agonist an **inverse agonist** that evokes a submaximal response in a biological system, even at a concentration sufficient to saturate the specific receptors. At high concentrations it will diminish the actions of a full inverse agonist. *Compare* **antagonist**, **partial agonist**.

partially overlapping code *see* **overlapping code**.

partially overlapping gene *see* **overlapping gene**.

partial molar quantity a function of any particular extensive property (e.g. enthalpy, entropy, Gibbs energy, heat capacity, internal energy, volume) of a specified component of a single-phase multicomponent system that describes how that property varies, at constant pressure and temperature, with variation of the concentration of the component when the change in the property is expressed per mole of the component. Partial molar Gibbs energy is more usually known as **chemical potential**. *Compare* **partial specific quantity**.

partial pressure *symbol*: p_B or $p(B)$ (for a gas B); the pressure exerted by one particular gas in a mixture of nonreacting gases. It is defined as the pressure the gas would exert if it alone were present and occupied the whole volume of the mixture at the same temperature, and is equal to the product of the amount-of-substance fraction of the gaseous component in question and the total pressure of the gaseous system. *Compare* **tension**. *See also* p_{50}.

partial pressures, law of *see* **Dalton's law of partial pressures**.

partial specific quantity a function of any particular extensive property (e.g. enthalpy, entropy, Gibbs energy, heat capacity, internal energy, volume) of a specified component of a single-phase multicomponent system that describes how that property varies, at constant pressure and temperature, with variation of the concentration of the component when the change in the property is expressed per unit mass of the component. *Compare* **partial molar quantity**.

partial specific volume (of a solute) *symbol*: v_B; an instance of a **partial specific quantity**, defined as the change in volume of a solution on addition of a small amount of a solute B, expressed per unit mass of added solute and extrapolated to infinite dilution. The cgs unit (in which values are often listed) is $cm^3 g^{-1}$ and the SI unit is $m^3 kg^{-1}$.

particle 1 an extremely small piece or portion of (esp. solid or colloidal) matter. **2 elementary particle**. —**particulate** *adj.*

particulate enzyme any enzyme that is bound to an insoluble cellular or subcellular structure or that has been artefactually or artificially attached to particulate material.

particulate radiation any form of radiation consisting of a stream of particles, especially alpha radiation, beta radiation, or neutron radiation emanating from a radioactive source.

partition 1 *(in chemistry)* or **distribution** the equilibration of a solute between two phases (usually of differing composition); the result of such a process. **2** *(in genetics)* the events that lead to the separation of chromosomes or to the segregation of plasmids into daughter cells in prokaryotic organisms.

partition chromatography any form of chromatography in which separation of the components of a mixture is based mainly on differences between the solubilities of the components in the stationary phase (in gas chromatography) or on differences between the solubilities of the components in the mobile and stationary phases (in liquid chromatography).

partition coefficient or **distribution coefficient** *symbol*: α or K_d; the ratio of the equilibrium concentrations of a pure substance dissolved in two phases that are in contact, at a specified temperature. The partition coefficient is independent of the concentrations provided that no chemical interaction occurs with either (or both) of the phases and that the presence of the solute does not affect their immiscibility.

partition function *symbol*: Z; a dimensionless mathematical function, derived by the application of statistical mechanical theory, that can be used to derive the average distribution of the internal energy of a system in terms of its partition between all its molecular inhabitants.

partition law or **distribution law** the principle that a solute, when added to a system consisting of two phases, will, at a given temperature, distribute itself between the two phases in proportion to its solubility in each of the two phases separately. **Henry's law** is a particular case of the partition law. *See also* **Nernst distribution law**.

partonomy *(in an ontology)* a 'parts' hierarchy into which the components of discrete objects are organized, often expressed as 'has a' relationships: e.g. apolipoprotein A has 38 kringle domains.

parturition the act or process of giving birth to offspring.

parvalbumin any of a group of closely related ~12 kDa muscle proteins that bind calcium ions. Parvalbumin α is a muscle **EF-hand** Ca^{2+}-binding protein that may be involved in muscle relaxation. Parvalbumin β is known as **oncomodulin**.

parvovirus any of a group of small, animal viruses, in class II of the **Baltimore classification**, consisting of a naked capsid, 20–25 nm in diameter, and containing linear, single-stranded DNA of 1.2–1.8 MDa. Parvoviruses are highly resistant to chemical and physical agents.

parvulin a **peptidylprolyl isomerase** isolated from *Escherichia coli*. It does not have any function as an **immunophilin**.

PAS *abbr. for* **1** *p*-aminosalicylic acid. **2** periodic acid–Schiff; *see* **periodic acid–Schiff reaction**.

pascal *symbol*: Pa; the SI derived unit of pressure or stress. It is equal to a force of one newton per square metre; i.e.

$$1 \text{ Pa} = 1 \text{ N m}^{-2} = 1 \text{ J m}^{-3}.$$

[After Blaise Pascal (1623–62), French mathematician and philosopher, who discovered the relationship between barometric pressure and altitude.]

PASCAL 1 *see* **bibliographic database**. **2** *see* **programming language**.

PAS domain a structural domain of ≈130 residues that occurs in proteins that respond to stimuli such as light, redox potential, O_2, or small aromatic molecules. It is found in archaea, bacteria and eukaryotes, and named after the proteins in which they occur (period circadian protein, Ah receptor nuclear translocator protein and single-minded protein), PAS domains act as signal sensors in many signalling proteins. Several PAS domain-containing proteins detect their signal by means of an associated cofactor (e.g., haem, flavin, a 4-hydroxycinnamyl chromophore). PAS domains are often associated with PAC domains, with which they are directly linked, together forming the PAS fold. The linking region between these domains is highly divergent; in human PAS kinase, this region is flexible, adopting different conformations depending on the bound ligand.

Pasha (Partner of Drosha) a dsRNA-binding protein which interacts with **Drosha** in the **Microprocessor** complex to initiate miRNA production in the nucleus. Also named DGCR8 (**Di George syndrome** chromosomal region 8).

passage *(in biology)* the action or process of (serially) subculturing cells or microorganisms, especially with a view to adapting them to a changed environment.

passage number *(in cell culture)* the number of subcultures performed after the original isolation of the cells from a primary source.

passenger *or* **passenger DNA** any specific DNA fragment introduced into a **cloning vector** or other cloning vehicle.

passive diffusion *an alternative name for* **simple diffusion** (used especially to emphasize its distinction from **active transport**). *Compare* **facilitated diffusion**.

passive immunity immunity in an individual due to the presence of antibody or primed lymphocytes derived from another, already immune, individual.

Pasteur effect *or (formerly)* **Pasteur–Meyerhof effect** either the phenomenon that occurs in facultatively anaerobic cells whereby oxygen inhibits glycolysis or fermentation, or its converse whereby the rate of glycolysis or fermentation increases when oxygen is excluded. *Compare* **Crabtree effect**. [After Louis Pasteur (1822–95), French chemist, microbiologist, and immunologist.]

pasteurization *or* **pasteurisation** a method of heat-treating milk or similar fluids to improve storage qualities and destroy pathogenic bacteria, without markedly altering taste and nutritional characteristics. Pasteurization of milk involves heating to 65°C for 30 minutes or to 72°C for 15 minutes followed by rapid cooling to below 10 °C. —**pasteurize** *or* **pasteurise** *vb.*

Pasteur pipette *or* **pasteur pipette** a simple, disposable, ungraduated, dropping or transfer pipette constructed by drawing out a short length of narrow glass tubing to form a long fine tip and operated commonly by means of an attached rubber or PVC bulb or teat. Analogous devices are available moulded in low-density polyethylene, with integral bulb and sometimes calibrated. *Also (esp. US):* **Pasteur pipet, pasteur pipet**.

Pasteur reaction the putative chemical reaction once considered to link fermentation and respiration and to account for the **Pasteur effect**.

patatin any of a family of glycoproteins that form 40% of the soluble protein of potato tubers. They serve as potato somatic storage proteins, and also have enzyme activity involved in host resistance. They have lipid acylhydrolase activity.

patch an intermediate stage in the formation of a **cap** (def. 2) on the surface of a cell, e.g. of a lymphocyte.

patch-clamp technique a technique whereby the conductance change of a single ion channel of an excitable membrane is measured directly. A very small electrode tip is sealed onto a patch of cell membrane, thereby making it possible to record the flow of current through individual ion channels in the patch. *See also* **Neher, Sakmann**.

Patched a human protein encoded by a homologue of the segment polarity gene *patched* of *Drosophila*. It is a multipass transmembrane receptor for **sonic hedgehog** (or Shh), with a covalently linked C-terminal cholesterol molecule and a palmitoylated N-terminal cysteine residue. Patched functions through a neighboring G-protein receptor-like protein called smoothened. In the absence of Shh, Patched inhibits smoothened and thus inhibits transcription of genes for Patched, transforming growth factor β, and Wnt proteins. The Patched gene at 9q22.3 encodes a protein of 1296 amino acids. Mutations, mostly of the nonsense variety, cause an autosomal dominant naevoid basal cell carcinoma syndrome, and are also associated with some sporadic medulloblastomas.

paternal imprinting the epigenetic silencing of an allele inherited from the father. *See* **imprinting**.

path *see* **light path**.

+pathic *see* **+pathy**.

patho+ *or (before a vowel)* **path+** *comb. form* denoting disease or clinical disorder.

pathobiochemistry 1 the study of the biochemical processes associated with disease. **2** the processes themselves.

pathogen any agent, especially any living organism, that can cause disease.

pathogenesis the generation and production, or the origin and development, of a disease. —**pathogenetic** *adj.*

pathogenesis-related proteins proteins synthesized in plants in response to attack by fungal or bacterial pathogens, presumably involved in the plant defence mechanism. They belong to an evolutionarily related family of small, eukaryotic, extracellular proteins, dubbed CRISP (cysteine-rich secretory proteins).

pathogenic (capable of) causing or producing disease. —**pathogenicity** *n.*

pathogenicity island a region of a bacterial genome containing clusters of genes linked to its pathogenic behaviour. For example, the *Escherichia coli* LEE pathogenicity island contains genes for gut wall attachment. Such islands are not usually present in all strains of a given bacterial species and are thought to have been acquired by horizontal transfer.

pathological *or* **pathologic 1** of, or pertaining to, pathology. **2** causing, relating to, involving, or caused by disease.

pathology 1 the branch of medical science concerned with the causes and nature of disease and with the effects of disease on the structure and functioning of the organism. **2** the sum of the changes that occur in an organism as the result of a specific disease. —**pathologist** *n.*

pathophysiology 1 the study of the physiological processes associated with disease. **2** the processes themselves.

pathway 1 the chain of reactions undergone by a given molecular entity or class of such entities in a particular ecosystem. **2 metabolic pathway**.

pathway database a database that links gene products and compounds in the cells of an organism to components on particular pathways. Such databases are used to facilitate gene function assignment and metabolic reconstruction. Examples include: BIND, which stores information on biomolecular interactions, molecular complexes, and pathways; EcoCyc, which houses metabolic, regulation, and signal transduction data on *Escherichia coli*; KEGG (Kyoto Encyclopedia of Genes and Genomes), which links human genes with networks of interacting molecules in the cell; WIT (what is there?), which houses metabolic, gene function, and phylogenetic data from multiple genomes; and HPRD (human protein reference database), an integrated pathway database of human reference proteins that contains domain architectures, PTMs, interaction networks, and disease associations for the human proteome.

pathway engineering the use of **genetic engineering** to change either the kinetics or the products of a metabolic pathway.

+pathy *comb. form* denoting **1** feeling, sensitivity. **2** disease of a (specified) part or kind. **3** a method of treating disease. —**pathic** *adj.*

pattern *or* **signature** an abstract representation of conserved features of a multiple sequence alignment that can be used to provide a diagnostic signature for related sequences that share the same or similar features. Examples of patterns are a regular expression, profile, hidden Markov model (HMM), fingerprint, and block. The term is also sometimes used confusingly as a synonym for regular expression.

pattern database *or* **secondary database** *or* **signature database** *(in sequence analysis)* a database that houses diagnostic patterns, or signatures, usually derived from conserved regions of multiple sequence alignments of protein families or domains. Areas of conservation are encoded, for example, in the form of regular expressions, fingerprints, blocks, or hidden Markov models (HMMs). Examples of such databases include PROSITE, PRINTS, Blocks, and Pfam. They tend to include varying levels of annotation describing the protein families they encode and technical details concerning how the pattern was derived.

pattern hit initiated-BLAST *abbr.*: PHI-BLAST; *see* **database-search program**.

pattern recognition 1 *(in sequence analysis)* the recognition of characteristic patterns of conservation in sequences using some form of predefined signature or discriminator – e.g. a regular expression, fingerprint, or hidden Markov model (HMM). **2** *(in data mining)* the discovery of unknown patterns in large data sets, for example by using principal component analysis, partial least squares, or neural network approaches.

patulin 4-hydroxy-4*H*-furo[3,2-*c*]pyran-2(6*H*)-one; a carcinogenic metabolite with antibiotic activity isolated from several fungi, particularly *Aspergillus* and *Penicillium* spp.

pauci+ *comb. form* indicating few.

paucidisperse describing a colloidal system whose dispersed phase consists of particles of only a few sizes. **—paucidispersity** *n*.

Pauli exclusion principle *or* **exclusion principle** *or* **Pauli principle** the principle that no two electrons in an atom can be described by the same set of four quantum numbers. [After Wolfgang Pauli (1900–58), Austrian-born US physicist.]

Paulus filtration method a method for studying the binding of a ligand microsolute to a macrosolute. It is applicable to small amounts of material and is particularly useful for a system having a relatively low binding constant. It involves equilibration of a series of mixtures of macrosolute and ligand in a suitable solvent, then separation of unbound ligand from macrosolute-plus-bound-ligand by ultrafiltration through anisotropic membrane filters of an appropriate retentivity. The amount of microsolute retained with the macrosolute on each filter is then estimated by, e.g., radioactivity measurement or another sensitive technique. [After H. Paulus.]

Pauly's reagent a chromogenic reagent for imidazoles and phenols, consisting of a freshly prepared solution of diazotized sulfanilic acid (or other aromatic amine). In the presence of alkali, this reacts with histidine to give a red colour and with tyrosine to give an orange colour. It is useful for the identification of these amino acids on paper chromatograms.

PAUP *abbr. for* phylogenetic analysis using parsimony; *see* **phylogeny package**.

Pavlovian conditioning the modification of behaviour, whereby a previously neutral stimulus becomes associated with an already existing reflex, so that it will, by itself, evoke a response. This new reflex is said to be conditional. After Ivan P. Pavlov (1849–1936), Russian physiologist and experimental psychologist, who described the effect experiments on dogs. He showed that reflex responses, like salivation, became conditional upon specific previous experiences of the animal, such as light or a tone (but probably not a bell).

Pax *abbr. for* paired box; one of a family of transcriptional regulators that contain a paired-box domain and are specifically expressed during the development of a wide range of structures and organs. There are nine *Pax* genes in mouse and in human. They are expressed primarily in the neural tube and brain in the early embryo. *Pax1* and *Pax9* are closed related, as are the groups *Pax3* and *Pax7*, and *Pax2*, *Pax5*, and *Pax8*. PAX2 protein (392 amino acids) is also expressed in the eye and kidney; mutations cause the renal-coloboma syndrome. PAX3 (836 amino acids) controls MyoD expression in limb buds; numerous mutations cause Waardenburg syndrome. PAX6 (422 amino acids) is identical in sequence in mouse and in human, and contains a homeobox domain and a C-terminal region rich in proline, serine, and threonine. It is also expressed in pancreatic islet cells. Numerous mutations cause aniridia. PAX8 is implicated in embryonic development of the thyroid gland; mutations cause thyroid dysgenesis and congenital hypothyroidism.

paxillin an adhesion protein, associated with **vinculin**, that becomes phosphorylated in response to a number of stimuli, phosphorylation being associated with focal adhesion. It is also associated with dysfunction of **cadherin**-dependent cell–cell contacts, and cell spreading.

Pb *symbol for* lead.

PBD *see* **biphenylylphenyloxadiazole**.

PBG *abbr. for* porphobilinogen.

PBI *abbr. for* protein-bound iodine.

pBluescript *a proprietary name for* a phagemid cloning vector incorporating the bacteriophage f1 origin of replication. It has 21

unique restriction sites in a multiple cloning sequence (MCS) flanked by promoters for T3 and T7 RNA polymerases.

P body *abbr. for* processing body; any of a few discrete cytoplasmic foci in yeast cells that contain proteins that activate or catalyse decapping of mRNA, mRNA degradation products, and a $5'{\to}3'$ mRNA exonuclease. P bodies are essential for metabolism of mRNA and are also present in mammalian cells.

PBP *abbr. for* pencillin-binding protein.

pBR322 a **cloning vector**.

PBS *abbr. for* phosphate-buffered saline.

PC *abbr. for* **1** paper chromatography. **2** phosphocreatine. **3** phosphatidylcholine. **4** personal computer.

pCa *symbol for* the negative logarithm of the concentration (or strictly the activity) of calcium ions in a specified solution.

PCA *abbr. for* perchloric acid.

P-cadherin placental **cadherin**.

PCAF *abbr. for* p300/CBP-associated factor; a **histone acetyltransferase** that associates with p300/CBP and acetylates lysine residues in the N-terminal region of histone H3 and H4.

PCB *abbr. for* **1** pyruvate carboxylase. **2** polychlorinated biphenyl.

p300/CBP-associated factor *see* **PCAF**.

PCCase *abbr. for* propionyl-CoA carboxylase.

pcDNA 3.1 a proprietary vector designed for the expression of genes in mammalian cells using the cytomegalovirus promoter. It has the SV40 origin of replication and carries the gene for neomycin resistance (neo^R).

P cluster an iron-sulfur cluster, comprising a [4Fe-4S] cluster linked to a [4Fe-3S] cluster, that is present in each αβ dimer of component 1 of nitrogenase.

PCM1 *or* **pericentriolar material** a protein of the **centrosome** that recruits centrin, pericentrin, ninein, and dynactin just before centrosomal duplication.

PCMB *abbr. for* *p*-chloromercuribenzoate, or *p*-chloromercuribenzoic acid.

PCMBS *abbr. for* *p*-chloromercuribenzenesulfonate, or *p*-chloromercuribenzenesulfonic acid.

pCMV-Script *a proprietary name for* a phagemid vector designed for the expression of proteins in mammalian cells under the control of the cytomegalovirus immediate early promoter. It has both Col E1 and SV40 origins of replication and confers resistance to neomycin and kanamycin.

pCMV-SPORT any of a series of proprietary phagemid vectors used for expression of proteins in mammalian cells under the control of the cytomegalovirus promoter. The multiple cloning sequence is flanked by T7 and SP6 promoters.

PCNA *abbr. for* proliferating cell nuclear antigen; *other names*: processivity clamp; sliding clamp. A protein subunit (261 amino acids) of prokaryotic **DNA polymerase-δ**. It forms a homotrimer that surrounds DNA and functions as a sliding clamp in a **replisome**. Each subunit contains two similar structural domains – each consisting of a beta sandwich and two alpha helices at the inner edge – that interact with DNA. The structure is very similar to that of the β subunit of the *E.coli* DNA polymerase III holoenzyme. PCNA is also required in the gap-filling step of DNA synthesis in nucleotide excision repair. It binds and is inhibited by p21.

PCP *abbr. for* 1-(1-phenylcyclohexyl)piperidine (*see* **phencyclidine**).

PCR *abbr. for* polymerase chain reaction.

p.d. *abbr. for* (electric) potential difference.

Pd *symbol for* palladium.

PDB *abbr. for* protein data bank; *see* **structure database**.

PDBj *abbr. for* protein data bank Japan; *see* **structure database**.

PDBsum *abbr. for* protein data bank summary; *see* **specialist database**.

PDBu *abbr. for* phorbol 12,13-dibutyrate, a **phorbol ester**.

PDD *abbr. for* phorbol 12,13-didecanoate, a **phorbol ester**.

PD-ECGF *abbr. for* platelet-derived endothelial cell growth factor.

PDGF *abbr. for* platelet-derived growth factor.

PDI *abbr. for* protein disulfide isomerase.

pDNA *abbr. for* plasmid DNA.

P domain *or* **trefoil domain** a domain of highly conserved cysteine residues, disulfide-bonded in the pattern 1–5, 2–4, 3–6, associated with highly conserved alanine, glycine, and tryptophan residues. Such domains are found in secretory polypeptides from animals.

PDS *abbr. for* postsynaptic density.

Pdx1 *abbr. for* pancreatic duodenal homeobox 1; a mammalian homologue of a homeobox gene first described in *Xenopus*. It is expressed in all pancreatic cell types during early development, but only in pancreatic islet B cells and duodenal mucosa in adults. The protein activates promoters of insulin and somatostatin genes. Homozygous gene knockout in mice leads to neonatal death and lack of a pancreas.

PDZ domain a common homologous sequence of ~90 amino acid residues that was first detected in mammalian postsynaptic protein (PSD-95), Drosophila discs-large (Dlg), and mammalian tight junction protein (Z0-1), from which the acronym is derived. It forms a beta-sandwich containing 5–6 beta strands and two alpha helices and binds C-terminal protein sequences. From 1 to 13 copies may be present in a protein.

Pe *symbol for* the pentyl group.

PE *abbr. for* **1** phosphatidylethanolamine. **2** potential energy.

peak 1 a sharp increase followed by a sharp decrease in a physical quantity that varies with distance, frequency, time, etc. **2** *(in chromatography)* the portion of a differential chromatogram recording the changes in detector response or eluate concentration while a given component of the applied mixture emerges from a chromatographic column. **3** the maximum value of a **peak** (def. 1, 2).

pEC$_{50}$ *see* p[A]$_{50}$.

PEC-60 a 60-amino-acid regulatory polypeptide abundant in intestinal tissue and also found in the central nervous system. It inhibits the formation of cyclic AMP and activates Na^+,K^+-ATPase. It is structurally related to pancreatic trypsin inhibitor and inhibits insulin secretion.

PECAM *abbr. for* platelet endothelial cell adhesion molecule; a class of **adhesion molecules** represented by PECAM-1 (*or* CD31), a member of the immunoglobulin superfamily. It is constitutively expressed on endothelial cells, platelets, granulocytes, monocytes, and T cells, and participates in homophilic binding (i.e. binding to itself). It is required for transmigration of leukocytes across the endothelium during extravasation.

pectate any salt or ester of **pectic acid**.

pectic acid the form of **pectin** (def. 1) that is essentially free of methyl ester groups.

pectic substance *an alternative name for* **pectin** (def. 1).

pectin 1 *or* **pectic substance** any mixture of complex, colloidal, macromolecular plant **galacturonans** containing a large proportion of D-galactopyranosyluronic acid residues in ($\alpha1\rightarrow4$) linkage, the carboxyl groups of which may be esterified to varying degrees by methyl groups or be partially or completely converted to salts; some D-arabinose and D-galactose residues are present, and other sugars are sometimes present. M_r is usually in the range 1×10^4–4×10^5. Pectins are important matrix polysaccharides of plant cell walls, especially those of higher plants, sometimes constituting as much as one-third of the dry matter; they also occur in some plant juices. Certain fruits are particularly rich in pectin. **2** *(in commerce)* a whitish, water-soluble powder prepared from the rind of citrus fruits or apple **pomace** by extraction with dilute acid and consisting essentially of **pectin** (def. 1). It is used *inter alia* in the preparation of jams because of its ability to form gels in the presence of sucrose at high concentrations.

pectinase *see* **polygalacturonase**.

pectinesterase EC 3.1.1.11; *other names*: pectin methylesterase, pectin methoxylase; an enzyme that hydrolyses pectin to methanol and **pectic acid**. Found in plant cell walls, it may be concerned with cell-wall metabolism during fruit ripening.

pectinic acid the form of **pectin** (def. 1) in which a high proportion of the carboxyl groups are esterified with methanol.

pectin lyase EC 4.2.2.10; an enzyme that catalyses a *trans* elimination to convert pectin to oligosaccharides with unsaturated (C-4, C-5) terminal residues. Although often described as 4-deoxy-6-methyl-*α*-D-galact-4-enuronosyl groups, they are more accurately termed 4-deoxy-6-methyl-*β*-L-*threo*-hex-4-enopyranouronosyl groups.

pectin sugar *an old name for* L-arabinose.

pedigree the biological relationships of the members of a family often extending over several generations. In medical genetics it consists of a diagram of a family history showing the status of individuals with respect to a particular hereditary condition.

PEG *abbr. for* polyethylene glycol.

PEGylation *or* **pegylation** the process of modifying a protein by treatment with polyethylene glycol (PEG). *Compare* **PIGylation**.

PEI *abbr. for* the poly(ethyleneimine) group,

pelargonate *another name for* **nonanoate**.

pelargonidin *see* **anthocyanidin**.

pelargonoyl *another name for* **nonanoyl**.

pellagra a disease of animals, including humans, arising from a dietary deficiency of **nicotinic acid**, especially in the absence of sufficient dietary tryptophan to generate adequate amounts of nicotinamide coenzymes metabolically. This situation can occur when the diet consists largely of maize (corn). Pellagra is characterized by scaly dermatitis on exposed skin surfaces, diarrhoea, and depression.

pellagra-preventative factor *abbr.*: PP factor *or* PPF; *a former name for* **nicotinic acid** or nicotinamide.

pellet a small compressed mass of material, especially such a mass deposited at the bottom of a tube by centrifugation of a suspension.

pellicle a thin skin, membrane, or film.

pellicular 1 of, relating to, or being in the form of a pellicle. **2** describing a type of column packing, used in high-pressure liquid chromatography, in which the stationary phase is present as a thin layer on the surface of tiny glass beads.

Peltier effect the phenomenon whereby heat is given out, or absorbed, when an electric current passes across a junction between two dissimilar materials. *See* **thermocouple**. *Compare* **Seebeck effect**. [After Jean Peltier (1785–1845), French physicist.]

penatin *an obsolete name for* **glucose oxidase**.

pendrin *an alternative name for* **iodide-chloride transporter**.

penetrance *(in genetics)* the degree to which a particular genotype is expressed as the phenotype. If every individual carrying a dominant mutant gene shows the mutant phenotype, the gene is said to show complete penetrance. Penetrance can be related to age so that the mutant allele might show low penetrance in the young but increasing penetrance in older individuals.

penetratin *or* **Trojan peptide** a natural or synthetic polypeptide that is rich in basic amino acids and mediates uptake of nucleic acids, bioactive peptides, phage particles, and liposomes into mammalian cells. Examples include the HIV Tat protein and *Drosophila* Antennapedia protein.

penicillamine 3-mercapto-D-valine; *β,β*-dimethylcysteine; 2-amino-3-methyl-3-mercaptobutyric acid; the D-enantiomer is a characteristic degradation product of penicillin-type antibiotics. A potent chelator of heavy metal ions, it is used clinically in the treatment of **Wilson's disease** and of poisoning by metals such as copper, lead, and mercury.

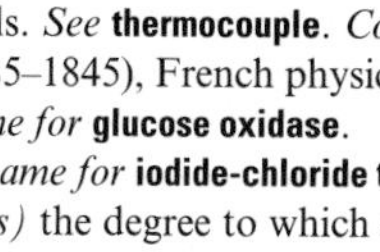

penicillicacid 3-methoxy-5-methyl-4-oxo-2,5-hexadienoic acid (exists in equilibrium with lactone form); a toxic antibiotic metabolite obtained from several fungi. It bears no relationship to penicillin or its degradation products. (Illustrated opposite.)

penicillin 1 *the generic name for* a range of related antibiotics, whether produced naturally during the growth of various microfungi of the genera *Penicillium* and *Aspergillus* or of semisynthetic origin, that contain the condensed *β*-lactam– thiazolidine ring system:

penicillin G acylase *see* **penicillin acylase**.

penicillin V acylase *see* **penicillin acylase**.

penicilloic acid any compound of structure:

and produced by cleavage of the β-lactam ring of **penicillins** (def. 1) by enzymic or alkaline hydrolysis.

penta+ or *(before a vowel)* **pent+** *comb. form* denoting five, fivefold, five times.

pentacosa+ *comb. form* denoting 25, 25-fold, 25 times.

pentadeca+ *comb. form* denoting 15, 15-fold, 15 times.

pentafunctional arom polypeptide a polypeptide that catalyses five consecutive enzymatic reactions in prechorismate polyaromatic amino-acid biosynthesis, i.e the second to sixth steps in the biosynthesis from chorismate of the aromatic amino acids (the **shikimate pathway**). It occurs in some yeasts and other fungi and in *Euglena gracilis*. It contains: **3-dehydroquinate synthase**, **3-dehydroquinate dehydratase**, **shikimate 5-dehydrogenase**, **shikimate kinase**, and **EPSP synthase**. In yeast the complete protein consists of 1588 amino acids; residues 1–392 = 3-dehydroquinate synthase; residues 404–866 = EPSP synthase; residues 887–1060 = shikimate kinase; residues 1061–1293 = 3-dehydroquinate dehydrogenase; and residues 1306–1588 = shikimate dehydrogenase.

pentagastrin *the trivial name for* (*N-tert*-butyloxycarbonyl-β-alanyl)-L-tryptophanyl-L-methionyl-L-aspartyl-L-phenylalaninamide; a synthetic analogue of **gastrin** that consists of the latter's C-terminal tetrapeptide amide sequence with an added protected N-terminal β-alanyl residue. It is used in the **pentagastrin test** as a gastrin analogue to stimulate gastric-acid secretion, which is then measured. The test results are too variable to be of great use in diagnosis. *See also* **calcitonin**.

pentakisphosphate *see* **multiplying prefix**.

pentamer 1 any oligomer consisting of or derived from five monomers. **2** the group of five capsomeres in an icosahedral virus capsid.

pentanoate *another name for* **valerate**.

pentanoyl *another name for* **valeryl**.

pentaose a pentasaccharide in which all the residues are of one type and uniform linkage. *Compare* **pentose**.

pentaric acid any **aldaric acid** obtained by oxidation of a **pentose** at C-1 and C-5.

pentasaccharide any oligosaccharide composed of five monosaccharide units. Pentasaccharides occur in small amounts in plants, e.g. **verbascose** in soybeans.

penta-snRNP the preassembled **spliceosome** complex that contains the five small nuclear RNAs (snRNA).

pentatricopeptide repeat a peptide of 35 amino acids found in multiple copies in some proteins. Their function is unknown but might be involved with RNA stabilization. The repeat seems to be greatly expanded in plants.

pentaxin *or* **pentraxin** any of a family of proteins with a discoid appearance under electron microscopy, having five non-covalently bound subunits. The family includes **C-reactive protein**, serum amyloid P-component, and neuronal pentaxins.

pentitol any **alditol** having a chain of five carbon atoms in the molecule.

pentobarbital 5-ethyl-5-(1-methylbutyl)barbiturate; a major barbiturate drug. *One proprietary name is* Nembutal.

pentolinium test a test in which **pentolinium**, a sympathetic ganglion blocking drug, is used to reduce catecholamine secretion; failure to achieve reduction may indicate a **pheochromocytoma**.

penton a capsomere that is surrounded by five other capsomeres in an icosahedral virus capsid.

where R is any of a variety of acyl groups such as: phenylacetyl, C_6H_5–CH_2–CO– (in **penicillin G**, **penicillin II**, **benzylpenicillin**); phenoxyacetyl, C_6H_5–O–CH_2–CO– (in **penicillin V**); and D-adipyl, HOOC–CH(NH_2)–[CH_2]$_3$–CO– (in **penicillin N**, *also called* **cephalosporin N**). Only those enantiomers having the D-configuration at C-2 of the thiazolidine ring show antibiotic activity. The parent substance, **6-aminopenicillanic acid**, in which R is H, has little intrinsic antibiotic activity but is the naturally occurring substrate for the semisynthesis of artificial penicillins, e.g. **ampicillin**, in which R is aminophenylacetyl, C_6H_5–CH(NH_2)–CO– . All penicillins are readily inactivated by chemical or enzymic hydrolysis of the strained four-membered β-lactam ring, cold dilute acid yielding **penicillamine**, and alkali or penicillinases and other β-lactamases forming the corresponding substituted **penicilloic acids**. The penicillins generally have extremely low toxicity to animals and high antibacterial activity, especially against Gram-positive organisms, the spectrum of activity varying somewhat with the nature of the group R. They inhibit cross-linking of peptidoglycan in the cell wall of growing bacterial cells, and render the cells sensitive to osmotic lysis. **2** the (hypothetical) substance 6-formamidopenicillanic acid, used only in combination in trivial names of certain **penicillins** (def. 1), e.g. benzylpenicillin.

penicillin acylase EC 3.5.1.11; *recommended name*: penicillin amidase; an enzyme that catalyses the hydrolysis of penicillin to a fatty-acid anion and 6-aminopenicillinate. Several enzymes differ in specificity for the fatty acid. For example, **penicillin V acylase** (*or* penicillin V amidase) is a homotetramer of 338 amino acids from *Bacillus sphaericus*. **Penicillin G acylase** (*or* penicillin G amidohydrolase) from *Escherichia coli* consists of 846 amino acids. This forms a periplasmic protein that consists of α and β subunits, representing residues 27–235 and 290–846, respectively. *See also* **6-aminopenicillanic acid**.

penicillin amidase *the recommended name for* **penicillin acylase**.

penicillinase a trivial name for any β-**lactamase** that acts preferentially on penicillins.

penicillin-binding protein *abbr.*: PBP; any of various bacterial proteins that are capable of binding penicillin(s) covalently. Such proteins appear to be located predominantly, if not solely, in the cytoplasmic membrane of all species of bacteria in which peptidoglycan occurs in the cell wall. These proteins are believed to catalyse reactions in the later stages of cell-wall synthesis, and there is a strong presumption that at least one of them catalyses the insertion of nascent peptidoglycan into the cell wall. In Gram-negative bacteria, some penicillin-binding proteins are known to be involved in the maintenance of cell shape or in cell elongation or septation. The

pentonic acid any monocarboxylic acid (aldonic acid) formally derived from a pentose by oxidation at C-1.

pentosan a **glycan** that yields only pentoses on hydrolysis.

pentose any **aldose** having a chain of five carbon atoms in the molecule. *Compare* **pentaose**.

pentose cycle *an alternative name for* **pentose phosphate pathway**.

pentose nucleic acid *abbr.*: PNA; *a former term for* **ribonucleic acid**.

pentose phosphate pathway *or* **hexose monophosphate pathway** a complex series of cytosolic metabolic reactions by which glucose 6-phosphate is oxidized with formation of carbon dioxide, ribulose 5-phosphate, and reduced NADP, the ribulose 5-phosphate then entering a series of reactions in which a number of sugar phosphates containing between three and seven (or eight) carbon atoms are successively interconverted. One consequence of these interconversions is the nonoxidative regeneration of glucose 6-phosphate, which then can enter the pathway for another cycle of oxidative decarboxylation and skeletal rearrangement, the eventual result being the complete oxidation of one molecule of glucose with the formation of six molecules of carbon dioxide and 12 molecules of NADPH. The pathway thus provides (in some organisms or tissues) an alternative route to (or shunt of) part of the glycolytic pathway; it also provides ribose 5-phosphate for nucleotide synthesis and (in some organisms) other intermediates for biosynthesis. Of special importance is D-erythrose 4-phosphate, required for the biosynthesis of aromatic amino acids and many other materials by the **shikimate** pathway. In higher plants many of the pathway enzymes function in the reverse direction as part of the **reductive pentose phosphate cycle**. *Other names*: hexose monophosphate shunt; pentose cycle; phosphogluconate (oxidative) pathway; Warburg–Dickens pathway; Warburg–Dickens–Horecker pathway. *See also* **glucose-6-phosphate dehydrogenase, phosphogluconate, phosphogluconate dehydrogenase (decarboxylating), transaldolase, transketolase**.

pentosuria the presence of excessive amounts of pentose in the urine. In **essential pentosuria**, daily excretion of 1–4 g L-xylulose (normal <60 mg) occurs because of deficiency of L-xylulose reductase (EC 1.1.1.10), an enzyme of the glucuronate pathway that converts L-xylulose to xylitol. In **alimentary pentosuria** excretion of up to 100 mg of L-arabinose or L-xylose occurs following ingestion of unusually large amounts of such fruits as cherries, grapes, and plums. Both forms are clinically benign.

pentraxin *see* **pentaxin**.

pentulose any **ketose** having a chain of five carbon atoms in the molecule. There are two enantiomeric pairs of pentuloses having the carbonyl group at C-2, these being the D- and L-isomers of ribulose, *erythro*-2-pentulose, and of xylulose, *threo*-2-pentulose.

pentyl *symbol*: Pe; the alkyl group, $CH_3–[CH_2]_3–CH_2–$, derived from pentane.

Pep-1 a synthetic 21-amino-acid peptide used experimentally for protein delivery into cells. It contains a tryptophan-rich region for hydrophobic interaction and a lysine-rich region that improves delivery and solubility of the vector. The Pep-1–protein complex dissociates on entry into a cell.

PEP *abbr. for* phospho*enol*pyruvate.

PEPCK *abbr. for* phospho*enol*pyruvate carboxykinase.

pepinase a pepstatin-insensitive lysosomal protease (547 amino acids) that also has carboxypeptidase activity. Deficiency of this enzyme results in a late-onset form of neuronal lipofuscinosis.

peplomer a virus-coded proteinaceous knob or spike, numbers of which project from the peplos (i.e., **envelope** (def. 3)) of an enveloped virus particle. Peplomers may have cell-receptor, hemagglutinating, or neuramidinase activity.

peplos *an alternative name for* **envelope** (def. 3).

pepsin 1 any of the closely related aspartic proteinase enzymes of 31–36 kDa that constitute the principal proteinase activity of gastric juice of vertebrates. The family includes **pepsin A, pepsin B**, pepsin C (i.e. **gastricsin**), and **pepsin D**. Pepsins have maximal activity at pH 2–3; they preferentially catalyse hydrolysis of peptide bonds formed from hydrophobic amino-acid residues, the detailed specificity varying somewhat between the different pepsins. **2** *an alternative name for* **pepsin A**.

pepsin A *or* **pepsin** *the recommended name for* the predominant enzyme (EC 3.4.23.1) of the **pepsin** family. It is a phosphoprotein (in pig, 327 amino acids) formed by limited proteolysis of pepsinogen A. It preferentially cleaves peptide bonds flanked on either side by hydrophobic amino acids; for example, in the B chain of insulin it cleaves Phe[1]-|-Val[2], Gln[4]-|-His[5], Glu[13]-|-Ala[14], Ala[14]-|-Leu[15], Leu[15]-|-Tyr[16], Tyr[16]-|-Leu[17], Gly[23]-|-Phe[24], Phe[24]-|-Phe[25], and Phe[25]-|-Tyr[26] bonds (*compare* **pepsin B**). The unphosphorylated form of porcine pepsin A is called pepsin D.

pepsin B *or* **parapepsin I** a pig enzyme (EC 3.4.23.2) of the **pepsin** family. It is a phosphoprotein (332 amino acids) formed by limited proteolysis of pepsinogen B. It has little activity towards hemoglobin as substrate (*compare* **gastricsin**), but degrades gelatin. Its specificity on the B chain of insulin is more restricted than **pepsin A**; it does not cleave Phe[1]-|-Val[2], Gln[4]-|-His[5], or Gly[23]-|-Phe[24].

pepsin C *an alternative name for* **gastricsin**.

pepsin D unphosphorylated pig **pepsin A**.

pepsinogen a precursor of a **pepsin** enzyme; pepsinogens A, B, and C give rise to **pepsin A, pepsin B**, and pepsin C (i.e. **gastricsin**) respectively. Pepsinogens are secreted by the chief cells of the gastric mucosa of vertebrates. Their active site is masked at neutral pH, and exposed as the pH falls after secretion into the stomach. The first bond breakage causes the release of residues 1–16, leaving **pseudopepsin**. A further 28 residues are removed in stages to form the active enzyme. Pepsinogen C is also secreted by the mucosa of the proximal duodenum. Human pepsinogen A consists of 388 amino acids; residues 63–388 are pepsin.

pepstatin A *N*-isovaleryl-L-valyl-L-valyl-3-hydroxy-6-methyl-4-aminoheptanoyl-L-alanyl-3-hydroxy-6-methyl-4-aminoheptanoic acid; a peptide, isolated from broth cultures of *Streptomyces* spp., that inhibits pepsin and other aspartic proteases, notably cathepsin D, renin, and fungal acid proteases.

peptaibophol *or* **peptaibol** any peptide amide antibiotic, 15–24 residues long and containing up to 40% α-aminoisobutyric acid residues. The group includes emerimicin, zervamicin IIA, antiamoebin I, sizukacillin A, and **alamethicin**, all of which alter the permeability of phospholipid bilayers. The N termini are acylated and the C-terminal carboxyl groups are in amide linkage with phenylalaninol.

peptic 1 of, pertaining to, caused by, or containing **pepsin**. **2** of, pertaining to, or promoting digestion.

peptidase *or* **peptide hydrolase** *or* **protease** any enzyme that hydrolyses peptide bonds. The group includes the **exopeptidases**, such as the aminopeptidases and carboxypeptidases, (sub-subclasses EC 3.4.11–19) and the **endopeptidases** (sub-subclasses EC 3.4.21–99). The 1992 revision of *Enzyme Nomenclature* (published for IUBMB by Academic Press) should be consulted for more detail.

peptidase P-sites sites adjacent to or near the cleaved site in peptidase substrates; in describing the specificity of peptidases, it is useful to have a convention for identifying the residues on either side of the cleaved bond; the following model is often followed (the cleaved, or scissile bond is indicated as -|-, and the N terminus is as usual to the left):

$$-P3-P2-P1-|-P1'-P2'-P3'-.$$

Compare **peptidyl site**.

peptide any compound containing two or more amino-acid residues joined by amide bond(s) (*see* **peptide bond**) formed from the carboxyl group of one amino acid (residue) and the amino group of the next. The term peptide usually applies to compounds in which the amide bond(s) are formed between C-1 of one α-amino acid (residue) and N-2 of another, but it includes compounds in which the residues are linked by other amide bonds. *See also* **oligopeptide, polypeptide**. —**peptidic** *adj*.

peptide amidation a post-translational modification in which the amido group of a glycine residue is cleaved so as to produce the amide group on the carboxyl end of the preceding residue. The reaction requires two enzymes: (1) peptidylglycine α-amidating monooxygenase (*abbr.*: PAM; EC 1.14.17.3), which is a copper-containing membrane protein that changes the glycine to a hydroxyglycine residue, in a reaction requiring O_2 and ascorbate; and (2) peptidylglycine α-amidating lyase, EC 4.3.2.5, which produces the peptidylamine and glyoxylate. Many bioactive peptides and peptide hormones exhibit full biological activity only when their C-terminal group is amidated. *See* **thyrotropin-releasing hormone.**

peptide antibiotic any antibiotic containing amino-acid residues, some of which are often of the D-form, linked to one another by peptide bonds, to hydroxy acids by depsipeptide bonds, and frequently also to additional components; the structure is often cyclic. The group includes the **bacitracins**, **gramicidins**, **polymyxins**, and **tyrocidins.**

peptide bond *or* **peptide linkage** any amide bond formed between two amino acids (or amino-acid residues). The term usually denotes the amide bond (sometimes called the **eupeptide bond**) formed between an α-amino group of one amino acid and an α-carboxyl group of another amino acid, but it also includes any amide bond (sometimes called an **isopeptide bond**) formed from an amino group and a carboxyl group, either or both of which are in other positions in the contributing amino acids. The peptide bond is planar. *See also* **Ramachandran plot.**

peptide chain *or* **polypeptide chain** any discrete linear sequence of amino-acid residues linked by eupeptide bonds, especially in a larger molecular structure. *See* **peptide bond.**

peptide HI *abbr.*: PHI; *other name*: PHI-27; a 27-residue peptide amide found in intestinal tissue, brain, respiratory tract, and pancreas. It stimulates insulin and pancreatic exocrine secretion, causes vasodilation, increases intestinal fluid transport, relaxes smooth muscle, and is a potent prolactin-releasing factor. It shows sequence homology with gastric inhibitory polypeptide, glucagon, secretin, and vasoactive intestinal polypeptide; human PHI has the sequence

HADGVFTSDFSKLLGQLSAKKYLESLM-NH$_2$

(porcine PHI has Arg^{12} and Ile^{27}; bovine has Tyr^{10}, Arg^{12}, and Ile^{27}). It is named for the histidine and isoleucine residues (single-letter codes H and I) at the N and C termini, respectively, of the porcine and bovine peptides.

peptide histidine valine 42 *abbr.*: PHV 42; a peptide with N-terminal histidine and C-terminal valine, derived from residues 81–122 of prepro-vasoactive intestinal peptide.

peptide hormone *or* **polypeptide hormone** any peptide with hormonal activity in animals, whether endocrine, neuroendocrine, or paracrine. Such substances form a very diverse group physiologically, and the boundary between peptide hormones and protein hormones is somewhat indistinct.

peptide hydrolase *another name for* **peptidase.**

peptide library an assembly of peptides with different sequences, that have many applications in testing for interactions with proteins such as antibodies, peptidases, and protein kinases, nucleic acids, or other ligands. The peptides may be synthesised by combinatorial chemistry, or by phage display. They are typically arranged in an array for testing with fluorescent markers. Applications include epitope mapping, bioactivity screening, and other biochemical studies.

peptide linkage 1 *an alternative name for* **peptide bond. 2** chemical linkage between structural components of a molecule by means of peptide bonds.

peptide map a pattern, characteristic of a particular polypeptide or protein, produced by partial hydrolysis, followed by two-dimensional fractionation of the resultant peptides (and amino acids) by chromatography and/or electrophoresis.

peptide nucleic acid a synthetic DNA analogue with an *N*-(2-aminoethyl)glycine backbone that mimics DNA in forming a **heteroduplex.**

peptidergic 1 describing any nerve that is activated by a peptide agonist. **2** describing any nerve that acts by releasing a peptide agonist from its nerve endings, whether into a synapse with another neuron, into intercellular fluid, or directly into the bloodstream.

peptide synthesis 1 *(in biology) see* **protein and peptide biosynthesis. 2** *(in chemistry)* the synthesis, by chemical means, of peptides. The first step is the protection, in the amino-acid reactants, of reactive groups not involved in peptide bond formation (*see* **protecting groups**); peptide-bond formation is then achieved by activation of the carboxyl group, which is reacted with the free α-amino group of another amino acid. A method frequently used for activation of the

carboxyl group is to react it with *N*-hydroxysuccinimide or a derivative to form an acyl succinimido ester, which reacts under mild conditions with free amino groups. Normally, and always in automated methods, the carboxyl group of the first amino acid is attached to a resin, by a bond that is easily detached later, so that the desired product, the growing peptide chain, remains attached to the resin which facilitates its purification; this technique, called **solid-phase peptide synthesis**, was pioneered by **Merrifield.**

peptide unit *(in crystallography)* any sequence of atoms in a peptide chain that may be represented by the general formula –CHR–CO–NH–, where R denotes the side chain of an amino-acid residue. *Compare* **amino-acid residue.**

peptide YY *abbr.*: PYY; a 36-residue peptide amide originally found in porcine upper small intestinal tissue. It is a gut hormone that inhibits exocrine pancreatic secretion. Members of this family regulate numerous physiological processes, including appetite, gastrointestinal transit, anxiety, and blood pressure. It shares sequence similarity with **neurotensin** and **pancreatic hormone**, having the amino-acid sequence:

YPIKPEAPGEDASPEELNRYYASLRHYLNLVTRQRYNH$_2$

It is named after the single-letter code, Y, for the tyrosine residue at its N terminus and the tyrosine amide at its C terminus.

peptido+ *comb. form* indicating the presence of peptide linked to some other (specified) component(s).

peptidoglycan *or* **murein** any glycoconjugate found only in bacterial cell walls and consisting of strands of glycosaminoglycan cross-linked by oligopeptides to form a huge and rigid network. There are characteristic differences between peptidoglycans of eubacteria and those of archaebacteria. The glycosaminoglycan strands are formed of alternating residues, usually *N*-acetylated or *N*-glycoloylated, of D-glucosamine and either muramic acid (in Bacteria) or L-talosaminouronic acid (in archaea); the carboxyl groups of the muramic-acid residues are commonly joined in amide linkage to oligopeptides containing residues of both L- and D-amino acids whereas those of the L-talosaminouronic-acid residues are linked to peptides consisting of L-amino acids only. The peptide side-chains of adjacent polysaccharide strands are sometimes crosslinked between a carboxyl group of one peptide moiety and an amino group of another, either directly by a peptide bond or indirectly by an interpeptide bridge consisting of one to several amino-acid residues.

peptidomimetic describing a substance that imitates a peptide. It is used of drugs (e.g. ritonavir and saquinavir) that inhibit HIV protease because they resemble its natural substrate.

peptidyl the acyl group formed from a peptide by removal of the hydroxyl group from its C-terminal carboxyl group.

peptidylglycine α-amidating monooxygenase *abbr.*: PAM. *See* **peptide amidation.**

peptidyl α-hydroxyglycine α-amidating lyase an enzyme of the Golgi complex that catalyses the conversion in a polypeptide of an α-hydroxyglycine residue into a peptidyl amine and glyoxylate. This is the second reaction of C-terminal **peptide amidation.**

peptidylprolyl isomerase *abbr.*: PPIase; EC 5.2.1.8; *systematic name*: peptidylproline *cis*–*trans*-isomerase; an enzyme that irreversibly converts prolyl residues within a polypeptide chain from a *trans* to a *cis* configuration. It accelerates protein folding. At least two families are known. One is identical to **cyclophilin.** Another class of immunophilins that exhibit isomerase activity are the FK506-binding proteins (*see* **FKBP**). The physiological role of these proteins is unresolved. The peptide Ala-Pro inhibits the reaction.

peptidyl site *or* **P-site** *or (more correctly)* **peptidyl-tRNA site** the ribosomal site to which a newly elongated peptidyl-tRNA molecule is transferred from the aminoacyl site during the sequence of polypeptide biosynthesis reactions, and at which it displaces the tRNA molecule remaining from the previous elongation step. *Compare* **peptidase P-site.** *See also* **ribosome.**

peptidyltransferase EC 2.3.2.12; the ribosomal enzyme that catalyses the elongation step in polypeptide synthesis:

peptidyl-tRNA$_1$ + aminoacyl-tRNA$_2$ =
tRNA$_1$ + peptidyl-aminoacyl-tRNA$_2$.

It is an activity of the large ribosomal subunit involving, in bacteria,

ribosomal proteins L2, L3, L4, L15, and L16, and especially 23S rRNA, which is therefore a ribozyme. The activity of the 23S rRNA is removed by ricin. *See also* **peptidyl site**.

peptidyl-tRNA the form in which a growing polypeptide chain is held to a ribosome during polypeptide biosynthesis. The polypeptide is esterified through its C-terminal carboxyl group to the 3′-hydroxyl of the adenosine residue at the 3′ end of a tRNA. *See* **protein synthesis**.

peptize *or* **peptise** to disperse (an aggregate) to form a colloidally stable suspension or emulsion; especially to convert (a gel) into a sol. —**peptization** *or* **peptisation** *n*.

peptolide (*sometimes*) any naturally occurring cyclodepsipeptide; *see* **depsipeptide**.

peptone any soluble mixture of products, 0.6–3 kDa, of the partial enzymic (or acid) hydrolysis of proteinaceous material. It is used as a constituent of many microbiological culture media.

peptonize *or* **peptonise** 1 to convert (proteinaceous material) into **peptone**. 2 to combine with peptone; e.g., peptonized iron. —**peptonization** *or* **peptonisation** *n*.

per+ *comb. form* denoting 1 through, throughout, beyond; e.g. perennial. 2 *(in chemistry)* **a** (of an indicated element) being in a higher than usual state of oxidation; e.g. periodate. *See also* **peracid** (def. 1). **b** containing a higher than normal proportion of (an indicated element); e.g. peroxide. *See also* **peroxoacid**, **peroxy acid**. **c** (in the trivial names performic, peracetic, and perbenzoic acids) **peroxy+**. **d** exhaustive replacement of hydrogen (except any in a characteristic group) in an organic compound by (an indicated halogen); e.g. perfluoropentane, C_5F_{12}. *See also* **perhydro**.

PER *abbr. for* period.

peracid *or* **per-acid** 1 *a common name for* any inorganic acid in which the central element of the acid is in its highest state of oxidation. 2 *a common name for* **peroxoacid**. 3 *a common name for* **peroxy acid**.

per cent *or* **percent** *symbol*: %; containing or consisting of a specified numerical proportion (of some component in relation to a whole) expressed as a number fraction multiplied by 100; bearing a specified numerical relationship (to some quantity) similarly expressed. When the term per cent or the symbol % is applied to a solution, one of the designations w/w, w/v, or v/v may be appended according as the units used are mass per mass, mass per volume, or volume per volume, respectively.

percentage proportionality expressed per cent; it is used especially to qualify a substantive, e.g. percentage error.

perchloric acid *abbr.*: PCA; chloric(vii) acid, $HClO_4$; a compound used in chemical analysis as an oxidizing agent.

Percoll *the proprietary name for* a polydisperse preparation of colloidal silica particles rendered nontoxic to cells by a coating of polyvinylpyrollidone. Aqueous dispersions of the material can be used to generate density gradients for the separation of cells, viruses, or cell organelles by sedimentation.

percutaneous through the skin; used especially of the route of administration of drugs applied to the skin.

perforin a lymphocyte pore-forming protein; it is produced by cytotoxic T cells when these are in contact with target cells, being stored in secretory vesicles, and released by local exocytosis. With Ca^{2+}, it polymerizes into transmembrane tubules that form pores in the target-cell membrane, and is thus capable of lysing non-specifically a variety of target cells.

performic acid *the trivial name for* peroxyformic acid, H–C(=O)–O–OH; a strong oxidizing agent, useful in protein chemistry for cleavage of any disulfide bond, R–S–S–R′, to the corresponding pair of sulfonic acid groups, $R–SO_3H + R′–SO_3H$, e.g. a cystine residue to two cysteic-acid residues. Tryptophan residues are also oxidized, and methionine residues are converted to methionine *S,S*-dioxide residues.

perfusate any fluid used for a perfusion.

perfuse to pass fluid through an organ or tissue, especially artificially. —**perfusion** *n*.

perhydro+ *prefix (in chemical nomenclature)* signifying conversion of the indicated unsaturated organic compound to the corresponding fully hydrogenated compound; used especially where the parent

compound is markedly unsaturated; e.g. perhydroanthracene, $C_{14}H_{24}$ (formed from anthracene, $C_{14}H_{10}$).

peri+ *comb. form* enclosing, surrounding; near.

periaxin a PDZ domain-containing protein that when mutated causes a severe demyelinating peripheral neuropathy called Dejerine–Sottas syndrome.

pericellular surrounding a cell or cells.

pericentric describing an inversion of part of a chromosome that spans the centromere, with breakpoints in both the arms. *Compare* **paracentric**.

pericentrin *see* **centrosome**.

pericentriolar material *see* **PCM1**.

perifusate any fluid used for a perifusion (*see* **perifuse**).

perifuse to pass a fluid around isolated cells or cell clusters in suspension. —**perifusion** *n*.

peri-fused *see* **ortho-** and **peri-fused**.

perikaryon *or* **cell body** the cytoplasm immediately surrounding the nucleus of a neuron.

perilipin a protein found in adipocytes, exclusively at the surface of the lipid storage droplets that serves as the major A-kinase substrate. It is a major hormonally regulated adipocyte-specific phosphoprotein.

perinuclear space the zone, 20–40 nm wide, between the two lipid bilayers of the **nuclear envelope** of a eukaryotic cell.

period 1 a specified or defined length of time, especially the time occupied by one complete cycle of some regularly recurring phenomenon or event. 2 *(in chemistry)* any of the horizontal rows of chemical elements in the **periodic table**. 3 *abbr.*: PER; either of two **PAS** domain-containing proteins (PER1 and PER2) involved in regulating circadian rhythms. They are induced by the clock-BMAL1 dimer in the suprachiasmatic nucleus in hypothalamus, and by the NPAS2-BMAL1 dimer in forebrain nuclei. Their activity is regulated by a protein-serine/threonine kinase, and they act as negative regulators of the circadian control system, which was discovered in *Drosophila* mutants with abnormal circadian cycles. Homologues occur in other organisms including humans. *See also* **BMAL1**, **clock**, **NPAS2**.

periodate 1 any anion derived from a **periodic acid**, especially the anion, IO_4, derived from metaperiodic acid. 2 any salt of a periodic acid, especially of metaperiodic acid or orthoperiodic acid.

periodate oxidation a method for the investigation of the structure of organic molecules, based upon the ability of aqueous periodate (or periodic acid) to cleave specifically and quantitatively any carbon–carbon single bond where one carbon atom carries a hydroxyl group and the other carries an amino, carbonyl, carboxyl, or hydroxyl group, these groups being oxidized and the periodate reduced to iodate. The method is particularly useful in the structural analysis of carbohydrates. *See also* **periodic acid–Schiff reaction**.

periodic 1 of, relating to, or resembling a **period** (def. 1 or 2); happening or occurring in periods; intermittent; cyclical. 2 *(in chemistry)* denoting any of several oxoacids of iodine containing a greater proportion of oxygen than iodic acid. *See* **per+** (def. 2a), **periodic acid**. — **periodically** *adv*.

periodic acid any of a number of iodic(vii) acids formed by the combination of iodine heptoxide, I_2O_7, with water in different molecular proportions, especially **orthoperiodic acid** (formerly known as paraperiodic acid), H_5IO_6; **metaperiodic acid**, HIO_4, exists mainly as its salts. These acids and their salts are strong oxidizing agents.

periodic acid–Schiff reaction *abbr.*: PAS reaction; a chromogenic reaction used as a sensitive test for carbohydrate, either histochemically or in chromatography. It is based on the periodate oxidation of *vicinal* –CHOH– groups to –CHO groups and the subsequent reaction of these with **Schiff reagent** to give a characteristic magenta colour.

periodic classification *see* **periodic system**.

periodicity the recurrence (of some phenomenon or event) at constant intervals in space or time; a measure of the frequency of such recurrence.

periodic law *or* **Mendeleev's law** the principle that the properties of the chemical elements are a periodic function of their atomic numbers (originally, of their atomic weights).

periodic system *or* **periodic classification** any classification of the chemical elements on the basis of the **periodic law**.

periodic table a mode of representing the **periodic system**; it is a tabular arrangement of the chemical elements in horizontal rows and vertical columns in order of increasing proton number row by row, each row, or **period**, consisting of those elements from an alkali metal to the rare gas of next highest proton number (except for the last, incomplete, period) and each column, or **group**, consisting of a family or set of those elements with similar chemical properties. Over the years, numerous versions of such a tabular arrangement have been proposed and used, the three commonest formats having 8, 18, or 32 columns of elements. Of these, the ones most widely adopted latterly have been variations of the so-called long periodic table, in which there are 18 columns. The particular version of the long table illustrated on the endpapers is the one currently recommended by IUPAC and adopted for use elsewhere in this dictionary. In this IUPAC periodic table, as similarly in other versions of the long table, the 15 elements of proton numbers 57 to 71, the **lanthanoids**, occupy a single position in period 6 of group 3, and those of numbers 89 to 103, the **actinoids**, a single position in period 7 of group 3, practical considerations requiring these elements to be listed separately at the bottom of the table.

In previous versions of the long table it was usual for 14 of the columns to be formed into seven groups, numbered I to VII and each divided into A and B subgroups. However, these subgroup designations were not always applied the same way round, with the consequence that two alternative conventions for the long table have concurrently had widespread support – one commonly in Europe (and hitherto followed in principle by IUPAC), the other in North America and elsewhere. Furthermore, the three columns containing the elements iron, cobalt, and nickel, originally designated together as group VIII, and the eighteenth column, containing the noble gases and originally designated as group O, have sometimes been labelled group VIIIA and group VIIIB respectively in the European version of the table, whereas in the North-American version the noble gases have in recent years been assigned usually to group VIIIA. The resultant confusion led IUPAC to recommend in 1990 that the columns of the long periodic table now be serially numbered as groups 1 to 18. The correspondence of the group designations in the new IUPAC periodic table with those of the European and North-American conventions in recent use is also shown in the endpapers.

peripheral (cell-membrane) protein *or* **extrinsic protein** *an operational term for* any protein that can be dissociated unchanged from a cell membrane by a mild treatment, such as increase in ionic strength of the medium, or addition of a chelating agent to it. Such proteins are presumed to be loosely attached to one or other surface of the cell membrane. *Compare* **integral (cell-membrane) protein**.

peripheral nervous system the part of an animal's nervous system other than the **central nervous system**, and generally located in the peripheral, or outer, parts of the body. It comprises the cranial nerves and spinal nerves and their branches, and includes the **autonomic nervous system**. It serves to link effector organs and sensory receptors with the central nervous system.

peripherin a glycoprotein, named from 'peripheral vision', that is essential for eye-disk morphogenesis; it is probably an adhesion molecule. Dysfunctional forms of peripherin precipitate retinopathies such as **retinitis pigmentosa**.

peripherin/RDS *see* **peripherin**.

periplasm *or* **periplasmic space** the region in a bacterial cell between the cell membrane and the cell wall.

peristalsis a wavelike sequence of involuntary muscular contraction and relaxation that passes along a tubelike structure, such as the intestine, impelling the contents onwards. —**peristaltic** *adj*.

peristaltic pump a pump that functions, by analogy with the action of **peristalsis**, by the repeated compression of a length of elastic tubing in such a way as to cause the contents of the tubing to move in one direction in a pulsating fashion but with an overall flow rate that is constant and regulated by the rate of pulsation and the tubing diameter.

peritoneum the thin translucent membrane, consisting of a single layer of flattened mesothelial cells and fibrous connective tissue, that lines the abdominal cavity and covers most of the viscera. —**peritoneal** *adj*.; **peritoneally** *adv*.

PERK *abbr. for* perturbed folding-activated kinase; *other names*: pancreatic ER-stress kinase; eIF-2 kinase. A protein serine/threonine kinase of the endoplasmic reticulum (ER) membrane, found especially in pancreatic cells. Its luminal regulatory domain normally binds BiP (*see* **binding protein**), which inhibits the cytosolic catalytic domain. When unfolded proteins accumulate in the lumen (i.e. under ER stress), BiP is released and the catalytic domain is activated. One substrate is the initiation factor eIF-2, which on phosphorylation suppresses protein synthesis. PERK-deficient mice develop marked hyperglycemia soon after birth. Inactivating mutations in humans result in a form of diabetes mellitus type I, which is associated with multiple epiphyseal abnormalities, called Wolcott–Rallison syndrome.

Perl *see* **programming language**.

perlecan *another name for* **heparan sulfate proteoglycan core protein**. *See also* **syndecan**.

permeability 1 the quality or degree of being permeable or porous. **2** *or* **magnetic permeability** (of a substance) *symbol*: μ; the ratio of the **magnetic induction**, B, to the magnetic field strength, H, at any point within the substance, i.e. $\mu = B/H$. The SI unit is the henry per metre or newton per square metre.

permeability coefficient a measure of the ability of a specified permeant to diffuse across a permeability barrier. It is expressed as the amount of substance of the permeant passing through a unit area of the barrier in a unit time when a unit difference in amount-of-substance concentration is applied across it.

permeability of vacuum *symbol*: μ_0; a fundamental physical constant, of value $4\pi \times 10^{-7}$ H m^{-1} exactly.

permeability transition pore a pore believed to form at sites of contact between the inner and outer mitochondrial membranes. It probably consists of the voltage-dependent anion channel (VDAC) of the outer membrane, the adenine nucleotide translocator of the inner membrane, and cyclophilin D of the matrix. When open the channel is permeable to particles of <1500 Da. Its opening probably dissipates the H$^+$ gradient and uncouples oxidative phosphorylation.

permeabilize *or* **permeabilise** to render permeable. In cell biology, agents such as **saponin**, streptolysin O (*see* **streptolysin**), or an applied electric field are used to open holes in the plasma membrane of cells. —**permeabilization** *or* **permeabilisation** *n*.

permeable capable of being permeated; pervious.

permeant 1 able to permeate or to diffuse into or through. **2** a substance having such properties.

permease a historical designation of some specific transport proteins; **transporter** is preferable.

permeate to pass through, or cause to pass through, by diffusion; to pervade. —**permeation** *n*.

permeation chromatography any form of chromatography in which a permeable stationary phase is used and in which separation of the components of a mixture is based mainly upon selective exclusion effects, such as differences in molecular size and/or shape (in **molecular-sieve chromatography**) or differences in charge (in **ion-exclusion chromatography**). *See also* **gel-permeation chromatography**.

permissible allowable. —**permissibility** *n*.; **permissibly** *adv*.

permissive 1 allowing; tolerant. **2** (*in virology*) describing a line or strain of cells that allows replication of a specified infecting virus with consequent cell lysis. **3** (*in genetics*) describing a condition, e.g. temperature, or set of conditions that permits the growth of a conditional lethal mutant (*see* **conditional lethal mutation**). —**permissiveness** *n*.; **permissively** *adv*.

permissive temperature *see* **temperature-sensitive mutation**.

permittivity *see* **relative permittivity**.

permselective selectively permeable to certain (classes of) molecular entities.

permutation 1 the process or act of permuting. **2** any of the possible arrangements of the members of a specified set of conditions, numbers, residues, etc.

permute to change the sequence or arrangement of.

Permutit *a proprietary name for* ion-exchange materials and equipment utilizing **permutites**.

permutite any of a class of artificial zeolite ion-exchangers.

pernicious anemia or (*esp. Brit.*) **pernicious anaemia** a form of macrocytic anemia characterized by defective production of erythrocytes and the presence of megaloblasts in the bone marrow, and sometimes accompanied by neurological changes. It is caused by a deficiency of vitamin B_{12} due to either a dietary deficiency of the vitamin or a failure of production of **intrinsic factor**.

per os *Latin* by mouth; used especially to denote the administration of drugs, etc. by the oral route.

peroxidase 1 any oxidoreductase enzyme of the sub-subclass EC 1.11.1. These act on hydrogen peroxide as (electron) acceptor with the formation of water. **2** *the recommended name of* the enzyme EC 1.11.1.7, a hemoprotein that catalyses the reaction of hydrogen peroxide with a wide variety of donors, i.e.

$$\text{donor} + H_2O_2 = \text{oxidized donor} + 2\ H_2O.$$

Apart from removing H_2O_2, it participates in the oxidation of toxic reductants and in lignin metabolism. In plants there are isoforms in different tissues. **—peroxidatic** *adj.*

peroxidase–antiperoxidase *abbr.*: PAP; a soluble complex of horseradish **peroxidase** (def. 2) and an antibody to this peroxidase, used in immunochemical procedures in conjunction with specific primary and bridging second antibodies to give a sensitive method of detecting complementary antigens enzymatically.

peroxide 1 (*in inorganic chemistry*) **a** the O_2^{2-} ion. **b** any member of a class of metallic oxides, such as sodium peroxide, Na_2O_2, that contain the peroxide ion and generate hydrogen peroxide on acidification. The name is sometimes used erroneously for certain higher oxides that do not contain the O_2^{2-} ion; e.g. lead peroxide (lead(iv) oxide, PbO_2), manganese peroxide (manganese(iv) oxide, MnO_2), nitrogen peroxide (nitrogen(iv) oxide, NO_2 or N_2O_4). **c** *short for* hydrogen peroxide, H_2O_2; an oxidizing agent. **2** (*in organic chemistry*) **a** any compound of general formula R'–O–O–R'', where the groups R' and R'' are two chains, two rings, or a ring and a chain, and including cases when R' = R''; e.g. ethyl phenyl peroxide, benzoyl (strictly, dibenzoyl) peroxide. In hydrogen peroxide, R' = R'' = H. **b** *see* **hydroperoxide**.

peroxin *or* **PEX protein** any protein in yeast or mammalian cells that mediates the import of peroxisomal integral membrane and matrix proteins and is necessary for peroxisome biogenesis. Fifteen PEX genes are known in humans. They encode receptors for the two peroxisome target signals, plus cytosolic and peroxisomal proteins. Mutations in several of the genes lead to **Zellweger's syndrome** and infantile **Refsum's disease**. *See* **peroxisome targeting sequence**.

peroxiredoxin (*abbr.*: Prx) *or* **thioredoxin peroxidase** *or* **alkyl peroxide reductase** any member of an abundant and widespread family of antioxidant enzymes, capable of removing hydrogen peroxide and alkyl hydroperoxides from cells. The enzymes contain an active-site cysteine, which is oxidized by peroxide to a sulfenate, which is subsequently reduced by a thiol-containing protein such as thioredoxin.

peroxisomal bifunctional enzyme a monomeric peroxisomal enzyme (mainly in kidney and liver) with the activities of three steps in the **beta-oxidation system**, i.e. **enoyl-CoA hydratase, 3,2- trans-enoyl-CoA isomerase**, and **3-hydroxyacyl-CoA dehydrogenase** activities.

peroxisomal disorder any disorder in which peroxisomes are absent or severely reduced in number, are incapable of importing a group of resident proteins, or lack a particular resident enzyme activity. Examples include **Zellweger's syndrome**, and adult and infantile forms of **Refsum's disease**.

peroxisome a type of eukaryotic cell organelle that is rich in enzymes that act on or generate hydrogen peroxide, especially the enzymes catalase, (*S*)-2-hydroxy-acid oxidase, and D-amino-acid oxidase. Peroxisomes occur in mammalian kidney and liver, and in yeasts and certain protozoans. They contain over 50 matrix enzymes involved in degradation of very-long-chain, branched, or polyunsaturated fatty acids, in the conversion of cholesterol to bile acids, and in biosynthesis of other phospholipids and isoprene compounds. They are particularly numerous in leaf photosynthetic cells of plants, photosynthesizing via the Calvin (C_3) pathway; in plants they are closely associated with chloroplasts and mitochondria,

which is consistent with their function in **photorespiration**. **—peroxisomal** *adj.*

peroxisome proliferator-activated receptor *see* **PPAR**.

peroxisome targeting sequence *abbr.*: PTS; either of two highly conserved peptide sequences that target proteins to the peroxisomal matrix. PTS1 is the C-terminal sequence –Ser–Lys–Leu. It is present in the majority of such proteins and is recognized by PEX5, which contains six or seven tetratricosapeptide repeats in its C-terminal half. PTS2 is the N-terminal consensus sequence –Arg–Leu–Xaa$_5$–His–Leu–. It is present in a small minority and is recognized by PEX7, which contains six WD40 repeats. *See* **peroxin**.

peroxo+ *prefix (in inorganic chemical nomenclature)* indicating substitution (in the corresponding acid) of an –OH group by the –O–OH group; e.g. peroxomonosulfuric acid, HO–SO_2–O–OH.

peroxoacid (*in inorganic chemistry*) any acid that contains the –O–OH group. *Compare* **peroxy acid**.

peroxonitrite *the recommended term for* the anion of peroxynitrous acid, $O=N$–OO^-; a highly reactive oxidizing species formed by reaction of nitrogen monoxide (NO) with superoxide, or nitroxyl ion with O_2. It has been implicated in cellular damage resulting from the **respiratory burst**.

peroxy+ *prefix or infix (in carboxylic-acid nomenclature)* indicating substitution of the –OH group in the carboxyl group by the –O–OH group; e.g. peroxypropionic acid, CH_3–CH_2– CO–O–OH. *See also* **per+** (def. 2c).

peroxy acid (*in organic chemistry*) any acid that contains the –CO–O–OH group. *Compare* **peroxoacid**.

peroxyformic acid *see* **performic acid**.

peroxynitrite *see* **peroxonitrite**.

perpendicular dichroism *see* **dichroic ratio**.

persalt any salt of a **peroxoacid** or a **peroxy acid**.

perseitol D-*glycero*-D-*galacto*-heptitol; a substance found in avocado and the alligator pear.

persephin a secreted protein that is structurally related to **glial cell-derived neurotrophic factor** and neurturin and like these binds a specific GPI-anchored membrane protein that recruits and signals through **RET**.

perspective formula a type of **projection formula** in which the bonds of (a part of) a molecule tend to be depicted with increasing boldness according to their nearness to the viewpoint, and vice versa, in order to give an impression of perspective. *See* **Haworth representation, Mills perspective representation**.

Perspex *a proprietary name for* poly(methyl methacrylate).

pertactin *other name*: outer membrane protein P68; an agglutinogen of *Bordetella* spp. that binds to eukaryotic cells, and is involved in the pathogenicity in whooping cough. It contains the **RGD** sequence motif.

perturb to cause a selective and measurable change in the state of equilibrium of a system, or in its associated properties, through the action or influence of a specific external agent; to undergo perturbation. **—perturbable** *adj.*

perturbant an agent (capable of) causing perturbation.

perturbation 1 the action or process of perturbing; an instance of this. **2** the condition of being perturbed. **3** a cause or factor of disturbance; perturbant.

pertussis toxin a toxic protein complex produced by *Bordetella pertussis*, the bacterium responsible for whooping cough in young children. The toxin causes inactivation of the iα subunit of the **G-protein** through catalysing ADP-ribosylation of a cysteine residue that is the NAD-dependent ADP ribosyltransferase (EC 2.4.2.–). The toxin consists of five proteins (S1–S5) organized in two functional subunits, A and B. Subunit A comprises S1, which has the enzymic activity, and irreversibly uncouples the GTP-binding proteins $G_{i\alpha}$ and **transducin** from their membrane receptors. Subunit B binds to the membrane receptors; it has the composition S2S3S5S4$_2$, dimers S2S4 and S3S4 being held together by S5.

pervaporate to vaporize a liquid from one side of a semipermeable membrane with which it is in contact on the other side, as a means of concentrating a solution or suspension in that liquid. **—pervaporation** *n.*

pervious permeable.

PEST hypothesis the hypothesis that the presence of one or more regions rich in proline (P), glutamic acid (E), serine (S), and threonine (T) in the primary structure of a protein confers susceptibility to rapid intracellular proteolysis.

pesticide any substance, natural or artificial, used to kill pests.

PET *abbr. for* positron emission tomography; *see* **tomography**.

peta+ *symbol*: P; SI prefix denoting 10^{15} times.

PETG *abbr. for* polyethyleneterephthalate glycol; a plastic developed as an alternative to PVC for use in packaging; in contrast to PVC it does not produce chloride fumes when burned, giving off only CO_2 and H_2O. Flexible, clear, robust, and light, it is also suitable for contact with food, since no plasticizers or stabilizers are used in its manufacture.

petite 1 describing a strain of yeast (or other) cells derived from a spontaneous mutant in which certain respiratory enzymes are defective or deficient, with consequent very slow growth and formation of unusually small colonies. **2** describing such a colony or mutant.

Petri dish *or* **Petri plate** a shallow, flat-bottomed, circular dish, usually of glass or transparent plastic and 5–15 cm in diameter, with a loosely fitting, overlapping cover. It is used especially for plate cultures of bacteria and moulds on solid (gelled) media, and for tissue culture. [After Richard Julius Petri (1852–1921), German bacteriologist.]

petroselaidic acid *see* **octadecenoic acids**.

petroselenate *numerical symbol*: 18:1(6); *the trivial name for* Z-octadec-6-enoate; a component of the fatty-acid hydrolysate of umbelliferous seed oils.

petroselenic acid *see* **octadecenoic acids**.

pET vector a plasmid vector designed for the overexpression of cloned genes in *E. coli*. Expressed proteins are fused to a **His-tag** or other fusion partner incorporated to facilitate purification. Expression is under the control of the bacteriophage T7 promoter and is caused by inducing the T7 RNA polymerase in the host cell, often **BL21** (DE3), with **isopropyl β-thiogalactoside** (IPTG).

PEX5 *see* **peroxin; peroxisome targeting sequence**.

PEX7 *see* **peroxin; peroxisome targeting sequence**.

PEX protein *an alternative name for* **peroxin**.

pEX vector a plasmid vector designed for overexpression of cloned genes in *E. coli*. Sequences are cloned at the 3′ end of *lacZ* and the product is a fusion protein with β-galactosidase at the N-terminus. Some proteins expressed in this way are less susceptible to intracellular proteolysis.

Peyer's patch any of numerous subepithelial aggregates of lymphoid tissue found in the intestinal mucosa. The M cells that overlie them transmit bacterial and viral protein antigens, which induce proliferation of the lymphoid cells and the secretory immune response. [After Johann Peyer (1653–1712), Swiss anatomist.]

Pfam *see* **pattern database**.

pFc′ fragment a protein fragment obtained (together with **F(ab′)₂ fragment**) by pepsin hydrolysis of an immunoglobulin molecule. That from human IgG is a 27 kDa protein, which is a non-covalently bonded dimer consisting of the C-terminal moieties of the heavy chains, and resembles the **Fc fragment** without the residues containing the linking disulfide bond.

PFGE *abbr. for* **pulsed field gel electrophoresis**.

PFP *abbr. for* pore-forming peptide *or* pore-forming protein; the term embraces cytotoxic peptides such as staphylococcal δ-toxin, **alamethicin**, **magainins**, and **cecropins**; pore-forming proteins include staphylococcal α-toxin, **streptolysin-O**, cnidarian toxins, **perforin**, and **aerolysin**.

PFK *abbr. for* phosphofructokinase.

pfu *abbr. for* plaque-forming unit; a measure of the abundance of viruses or bacteriophages based on their ability to form zones of infection (plaques) on a monolayer of host cells.

Pfu polymerase a thermostable DNA polymerase from *Pyrococcus furiosus* valued for its 3′→5′ exonuclease activity that imparts a **proofreading** function. Transcripts produced with *Pfu* polymerase have a lower error rate than can be obtained with enzymes such as *Taq* polymerase, which lack 3′→5′ exonuclease activity.

pg *symbol for* picogram (i.e. 10^{-12} gram).

pG *symbol for* titration pK.

PG *abbr. for* **1** phosphatidylglycerol. **2** prostaglandin. **3** protegrin.

PGA *abbr. for* **1** phosphoglyceric acid. **2** pteroylglutamic acid. **3** prostaglandin A; *see* **prostaglandin**.

PGAH₄ *abbr. for* tetrahydropteroylglutamic acid.

PGAM *abbr. for* phosphoglycerate mutase.

PGB *abbr. for* prostaglandin B; *see* **prostaglandin**.

PGD *abbr. for* **1** phosphogluconate dehydrogenase (EC 1.1.1.43). **2** prostaglandin D (*see* **prostaglandin**).

PGE *abbr. for* prostaglandin E; *see* **prostaglandin**.

pGEM vector a proprietary plasmid vector used for cloning polymerase chain reaction (PCR) products. Some DNA polymerases (such as *Taq* polymerase) add a single deoxyadenosine 5′ phosphate to the 3′ ends of amplified fragments and these can be ligated to the single 3′ thymidine overhangs of the linear pGEM vector.

pGEX vector any of various proprietary plasmid vectors used for the expression of foreign proteins in *E. coli*. Genes of interest are cloned downstream, and in the same reading frame, from the gene for glutathione-*S*-transferase (GST), and the product is a fusion protein with GST at the N-terminus. The presence of GST facilitates purification of the fusion protein by affinity chromatography on columns of immobilized glutathione, and its immunodetection using anti-GST antisera. Immediately downstream from GST and upstream from the protein of interest is a short stretch of sequence containing a motif recognized by thrombin (factor Xa). Proteolytic digestion of the fusion protein cleaves the protein of interest from GST. Expression of GST fusion proteins is under the control of the *tac* promoter and is induced with **isopropyl β-thiogalactoside** (IPTG).

PGF *abbr. for* prostaglandin F; *see* **prostaglandin**.

PGF synthase *abbr. for* prostaglandin-F synthase.

PGG *abbr. for* prostaglandin G (*see* **prostaglandin**).

PGH *abbr. for* prostaglandin H (*see* **prostaglandin**).

PGH synthase *see* **prostaglandin-endoperoxide synthase**.

PGI *abbr. for* prostaglandin I (i.e. **prostacyclin**).

PGK *abbr. for* phosphoglycerate kinase.

pGL vector any of a number of proprietary plasmid vectors used for the analysis of promoter or enhancer function. The putative promoter is cloned upstream from the firefly luciferase gene, whose expression can be detected with great sensitivity if the promoter is active in mammalian cells transfected with the pGL construct.

P glycoprotein *or* **MDR protein 1** a glycoprotein associated with drug resistance; it is a transmembrane ATP-dependent transporter that pumps hydrophobic substances out of cells. Drug-resistant cells demonstrate increased expression of the P-glycoprotein gene as one mechanism for their resistance to, e.g., chemotherapeutic anticancer drugs. *See also* **MDR**.

PGT (*in clinical chemistry*) *abbr. for* phosphoglucomutase.

pH a dimensionless quantity notionally defined as the negative decadic logarithm of the molal activity of hydrogen ions in a solution. Since the activity of a single species of ion is not measurable, pH is defined operationally according to the equation:

$$pH(X) = pH(S) + (E_s - E_x)F/(RT \ln 10)$$

where pH(S) and pH(X) are the pH values of a standard solution, S, and of the test solution, X; F is the Faraday constant, R the gas constant, and T the thermodynamic temperature; and E_s and E_x are, respectively, the electromotive forces of the two galvanic cells:

reference electrode | KCl (conc.) ‖ soln. S | H_2(g) | Pt,

and

reference electrode | KCl (conc.) ‖ soln. X | H_2(g) | Pt,

both cells being at the same temperature, with identical reference electrodes and bridge solutions, KCl (aq, $m \geq 3.5$ mol kg⁻¹). A **pH standard** solution, S, is chosen from a set of five covering the range pH 3.5 to pH 9, an assigned value of pH(S) at the required temperature being obtained from published tables. To a good approximation, the hydrogen electrodes in the two cells may be replaced by other hydrogen-ion-responsive electrodes, such as glass or quinhydrone electrodes. Originally pH was considered as the negative decadic logarithm of the molar concentration of hydrogen ions, a practice retained to the present day as representing a convenient approximation.

pH* *a notation commonly used to denote* the apparent pH of a solution containing ^2H_2O (i.e. D_2O), as measured electrometrically using a glass electrode without correction for the isotope effect on the electrode.

Ph *symbol for* the phenyl group.

PH-30 *or* **PH30** a sperm surface protein, with **disintegrin** and metalloprotease domains; that from *Cavia cobaya* consists of two subunits, α and β.

PHA *abbr. for* phytohemagglutinin (def. 1).

pH-activity profile *or* **pH profile** a plot of enzyme activity against pH, over a range in which the activity is reversibly affected by pH.

phaeo+ *an alternative spelling (esp. Brit.) of* **pheo+**.

phaeochromocytoma *an alternative spelling (esp. Brit.) of* **pheochromocytoma**.

phaeofarnesin *an alternative spelling (esp. Brit.) of* **pheofarnesin**.

phaeomelanin *an alternative spelling (esp. Brit.) of* **pheomelanin**.

phaeophorbide *an alternative spelling (esp. Brit.) of* **pheophorbide**.

phaeophytin *an alternative spelling (esp. Brit.) of* **pheophytin**.

phage *abbr. for* bacteriophage.

phage conversion *an alternative name for* **bacteriophage conversion**.

phage display library a special example of a gene library encoding fusion proteins of a foreign polypeptide sequence and a coat protein of phage M13 or (more usually) the closely related phage fd. When cloned, the phage is then said to display the foreign protein (in the coat proteins). There are two common versions of the technique. In one, the library consists of random oligonucleotides (e.g. 7-mers). Clones expressing oligopeptides that react with a specific antibody can then be selected using the antibody. In the other version, the library consists of oligonucleotides encoding a range of single-chain variable-domain antibodies. This approach is used where an antibody is required for an oligopeptide, which can thus be used to detect appropriate clones.

phage immunity the immunity conferred on a **lysogenic bacterium** by the presence of a **prophage** and a consequence of the expression of a gene such as *cI* in **lambda phage**.

phage induction *or* **prophage induction** the stimulation of prophage in lysogenic bacteria to enter the vegetative state. It is accomplished by the action of UV light, X-rays, nitrogen mustards, etc.

phage lambda *see* **lambda phage**.

phagemid a chimeric phage–plasmid **cloning vector** containing origins of both double-stranded and single-stranded replication. *See also* **plasmid**.

phage T4 an icosahedral DNA-containing **bacteriophage** that infects *Escherichia coli*. Its single molecule of double-stranded DNA is 174 kb in length. Various enzymes prepared from this phage are commonly used in genetic engineering; these include ligase, polymerase, and polynucleotide kinase.

phage typing a procedure for classifying bacteria according to their susceptibility to various bacteriophages.

phagocyte any cell that characteristically engulfs particles from its surroundings into its cytoplasm by **phagocytosis**. —**phagocytic** *adj.*

phagocyte NADPH oxidase *abbr.*: phox; a phagosomal and plasma membrane complex that transfers electrons from NADPH, via FAD and heme, to O_2 to produce superoxide (O_2^-) and other microbicidal reactive species. The process constitutes the **respiratory burst**. The complex assembles and becomes active when phagocytes are activated in response to inflammatory stimuli. It consists of membrane components (gp91-phox and p22-phox, together called **cytochrome b_{245}**) and cytosolic components (p47-phox, p67-phox, and a Ras-related GTPase). Deficiency or dysfunction of any of the four phox proteins causes chronic granulomatous disease. Examples: human p47-phox, 390 amino acids; p67-phox, 526 amino acids.

phagocytosis a type of endocytosis whereby certain cells – phagocytes – can engulf external particulate material by extension and fusion of pseudopods around each particle. The particles are initially contained within phagocytic vacuoles (**phagosomes**), which then fuse with primary lysosomes to effect digestion of the particles. Phagocytosis is employed by certain cells of the immune system, notably macrophages and neutrophils, to engulf and destroy target cells, tissue debris, and other particles, and by amoeboid protozoans. *Compare* **pinocytosis**.

phagolysosome an organelle formed by the fusion of a **phagosome** and a **lysosome**. Immediately after fusion there is a brief rise in the pH of the phagolysosome and neutral proteases and cationic proteins are active. Subsequently the pH falls and acid proteases become active. Indigestible substances remain in the phagolysosomes forming residual bodies.

phagosome an inclusion in a eukaryotic cell that arises from the ingestion of particulate material by **phagocytosis**. The phagosomes subsequently fuse with **lysosomes** to form a **phagolysosome**.

phalloidin *see* **phallotoxin**.

phallotoxin any bicyclic heptapeptide present in the highly poisonous agaric fungus, *Amanita phalloides*, commonly called the death-cap fungus or deadly agaric, and in some related species; **phalloidin** is the best-known member. All phallotoxins but one are toxic, though not as intensely so as the congeneric and much more slowly acting **amatoxins**; high doses of phalloidin in mice or rats cause death in a few hours. The common elements include residues of *cis*-4-hydroxy-L-proline, a D-amino acid (which is either threonine or β-hydroxyaspartic acid), and L-tryptathionine (which forms a thioether bridge across the molecule). The majority of phallotoxins bind specifically to F actin and so prevent its depolymerization to G actin, thereby blocking the turnover of microfilaments and inhibiting cell movement.

phane nomenclature a method for naming organic structures by assembling names that describe component parts of a complex structure. A relatively simple skeleton for a parent hydride structure is modified by 'amplification', a process that replaces one or more special atoms (superatoms) of the simplified skeleton by multiatomic structure(s).

phantom indel a spurious insertion or deletion that arises when physical irregularities in a sequencing gel cause the reading software either to call a base too soon, or to miss a base altogether.

pharmaceutic *an alternative name for* **pharmaceutical** (def. 2).

pharmaceutical 1 of, or pertaining to, drugs or pharmacy. **2** *or* **pharmaceutic** any medicinal substance, mixture, or formulation. —**pharmaceutically** *adv.*

pharmaceutics *an alternative name for* **pharmacy** (def. 1).

pharmaco+ *comb. form* indicating a drug or drugs.

pharmacodynamics *functioning as sing.* the branch of pharmacology dealing with the effects of drugs on the body, i.e. with the physiological, therapeutic, and toxicological responses to drugs with particular regard to the extent and time course of such effects. —**pharmacodynamic** *adj.*; **pharmacodynamically** *adv.*

pharmacogenetics *functioning as sing.* the branch of pharmacology dealing with the genetic mechanisms underlying individual differences in responses to drugs. —**pharmacogenetic** *adj.*; **pharmacogenetically** *adv.*

pharmacogenomics the analysis of drug responses with respect to genomic information (e.g. polymorphisms), aiming to customize therapies and reduce adverse drug reactions.

pharmacognosy the study of crude drugs of natural origins. —**pharmacognostic** *adj.*

pharmacokinetics *functioning as sing.* the branch of pharmacology dealing quantitatively with the movement of drugs within the body, i.e. with the absorption, distribution, metabolism, and elimination of drugs. —**pharmacokinetic** *adj.*; **pharmacokinetically** *adv.*

pharmacological *or (esp. US)* **pharmacologic 1** of, pertaining to, or used in pharmacology. **2** (of an effect of a naturally occurring agonist) caused by an unphysiological concentration; (of a concentration or dose of a naturally occurring agonist) having an unphysiological effect. —**pharmacologically** *adv.*

pharmacology the science or study of drugs – their origin, characteristics, identification, biological effects, and modes of action. *Distinguish from* **pharmacy**.

pharmacophore the cluster of atoms in a drug that endows it with its pharmacological properties.

pharmacopoeia *or (US, sometimes)* **pharmacopeia 1** an authoritative book, often an official one, containing descriptions of medicinal drugs with their uses, preparation, methods of assay, formulae, dosages, etc. **2** the whole range of medicinal drugs in use or available. —**pharmacopoeial** *or* **pharmacopoeic** *adj.*

pharmacy 1 *or* **pharmaceutics** the practice of preparing and dispensing drugs. **2** a dispensary (for drugs). *Distinguish from* **pharmacology**.

pharming the production of medicinally active substances by genetic manipulation of farm animals or plants. [From a combination of 'pharmaceutical' and 'farming'.]

phase 1 any distinct or characteristic period or stage in a sequence of events, occurrences, or processes, whether cyclical or unidirectional; e.g. the phases in the life cycle of an organism or the phases of the **cell-division cycle** (*see also* **anaphase**, **gap phase**, **metaphase**, **prophase**, **telophase**). **2** (*in physical science*) **a** the totality of those parts of a heterogeneous material system that are identical in chemical composition and physical state, and that are separated by an interface from the rest of the system; e.g. solid, liquid, or vapour phases, or the discrete phases into which immiscible liquids separate after mixing, etc. *See also* **epiphase**, **hypophase**, **liquid-crystalline phase**, **phase transition**. **b** a particular stage in the cycle of a periodic phenomenon, process, or quantity; the fraction of such a cycle that has been completed at a specific reference time, expressed in circular measure (or its equivalent in angular degrees). Two or more waveforms (or two or more similar sequences of events, etc.) that are in corresponding phases at the same time are described as 'in phase', while ones not exhibiting such correspondence are described as 'out of phase'. —**phasic** *adj*.

phase-contrast microscopy *or* **phase microscopy** a form of light microscopy in which a colourless object having a different refractive index from that of the medium is viewed by converting differences in refractive index into visible differences in light intensity. The technique takes advantage of the quarter-wave retardation of the illuminating light that occurs on its diffraction by the object, imposes a similar change of phase on the undiffracted light, and forms an image by interference between the two beams of light.

phaseolin 1 a crystallizable globulin found in seeds of the kidney bean, or French bean, *Phaseolus vulgaris*, and the lima bean, *Phaseolus lunatus*. It is a major seed-storage protein. α and β types are known. **2** *a former name for* **phaseollin**. **3** *an alternative name for* **legumain**.

phase rule a generalization, valuable in the study of equilibria between **phases** (def. 2), expressed by the equation:

$$F = C - P + 2$$

where C is the number of chemically distinguishable components present, P is the number of phases in equilibrium, and F is the number of degrees of freedom of the system, i.e. the number of independently variable intensive properties that must be specified in order to describe completely the state of the system.

phaseshift mutation *an alternative term for* **frameshift mutation**.

phase transition (of lipids) the transition of a lipid between any of the various phases it can adopt. The transition is dependent on factors such as temperature, purity of the lipid, state of hydration, and, if not pure, the composition of the mixture. In the nonhydrated state, a crystalline polar lipid can on heating, prior to a complete melting, adopt a **liquid-crystalline phase**, in which the polar groups are still associated in sheets, so that the overall structure consists of lipid bilayers, but the fatty-acyl chains undergo increased thermal vibrations leading to considerable disorder; the temperature at which this transition occurs is known as the **phase-transition temperature**. In the presence of water, below the phase-transition temperature, phospholipids adopt a gel phase, in which

water penetrates between the sheets of polar headgroups, but the fatty-acyl chains are highly ordered. On increasing the temperature through the phase-transition temperature, the lipid adopts a hydrated liquid-crystalline phase with disordered fatty-acyl chains. The term **fluidity** has been used to provide a qualitative description of these changes, the bilayer being said to become more fluid as the temperature increases. *See also* **hexagonal phase**.

phase variation (*of flagellin*) *see* **flagellin**.

phasmid 1 a gene-cloning vector consisting of an artificial combination of a plasmid with a phage such that its genome contains functional origins of replication of both; it may thus be propagated either as a plasmid or as a phage in appropriate host strains. A phasmid is formed by insertion of one or more plasmid molecules into a phage genome. *Compare* **cosmid**, **phagemid**. **2** either of a pair of lateral caudal organs characteristic of parasitic nematodes of the class Phasmidia; they are thought to function as chemoreceptors. **3** any plant-eating insect of the order Phasmida, which includes leaf insects and stick insects.

PhD enzyme a phospholipase D enzyme belonging to a family characterized by the HKD catalytic motif (i.e. $HXKX_4DX_6GG/S$, using the single-letter code for amino acids). PhD enzymes are present in certain viruses and in bacteria, fungi, plants, and animals.

PHD finger *abbr. for* plant homeodomain zinc finger; a particular type of zinc finger, widely distributed in eukaryotes and found in many chromatin regulatory factors. It is characterized by a C4HC3-type motif, and adopts a dimetal(zinc)-bound α+β fold. This fold is highly similar to that found in another class of zinc finger domains, which function in membrane recruitment of cytosolic proteins by binding to phosphatidylinositol 3-phosphate.

PH domain *abbr. for* pleckstrin homology domain.

Phe *symbol for* a residue of the α-amino acid L-phenylalanine (alternative to F).

phenazine methosulfate *abbr.*: PMS; N-methylphenazonium methosulfate; 5-methylphenazinium methyl sulfate; $[C_{13}H_{11}N_2]^+$ $[CH_3SO_4]^-$; an artificial electron carrier, $E^{o'} = +0.080$ V, useful in the study of redox enzyme systems.

phencyclidine 1-(1-phenylcyclohexyl)piperidine (*abbr.*: PCP); a potent psychotomimetic agent, known to drug users as angel dust, with analgesic and anesthetic actions. It acts as a ligand for the **sigma receptor** and is a blocker of the NMDA-receptor ionophore (*see* **glutamate receptor**). A controlled substance in the US Code of Federal Regulations.

phene any phenotypic character under genetic control.

phenelzine β-phenylethylhydrazine, (2-phenethyl)hydrazine; a nonselective monoamine oxidase inhibitor used as an antidepressant.

phenobarbital 5-ethyl-5-phenylbarbiturate; a major **barbiturate** drug.

phenocopy a nonhereditary modification of the phenotype of an organism, induced by nutritional or other environmental factors, that mimics a phenotype controlled by a specific gene.

phenol 1 *the common and trivial name for* hydroxybenzene, C_6H_5OH. **2** any compound containing one or more hydroxyl groups directly attached to an aromatic carbocycle. Unlike hydroxyalkyl compounds, phenols are weakly acidic.

phenolase 1 *an alternative name for* **catechol oxidase** (EC 1.10.3.1). **2** *an alternative name for* **monophenol monooxidase** (EC 1.14.18.1). **3** more loosely, any enzyme of the sub-subclass 1.10.3. Such enzymes cause browning of damaged plant tissues when they come into contact with flavonoids, as **catechins** are converted into polymerized brown products.

phenol-chloroform extraction a method for the efficient extraction of nucleic acids free from contaminating proteins. Phenol saturated with Tris-EDTA buffer is mixed with an equal volume of chloroform (or sometimes a 24:24:1 v/v/v mixture with isoamyl alcohol) and is shaken to form an emulsion with the solution containing nucleic acids. On phase separation the nucleic acids occupy the upper aqueous phase and denatured protein occurs at the interface.

phenolic 1 of, containing, derived from, or having the character of phenol or a phenol. **2** describing a hydroxyl group linked to an aromatic carbocycle.

phenol 2-monooxygenase *see* **monooxygenase**.

phenolphthalein an indicator dye that changes from colourless to red over the pH range 8.3–10.0. The β-D-glucuronide (λ_{max} = 420 nm) is used as a substrate for the assay of β-glucuronidase, hydrolysis leading to release of phenolphthalein, which can be measured in alkaline solution by its red colour (λ_{max} = 550 nm).

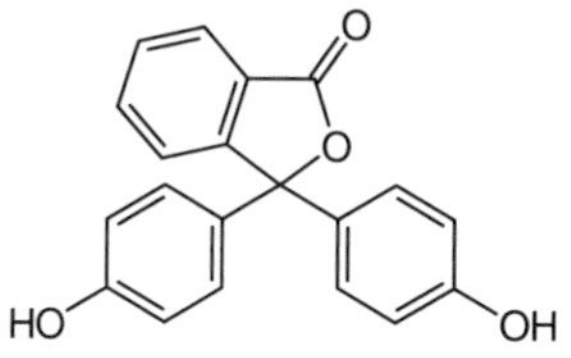

Phenol Red an indicator dye that changes colour from yellow to red over the pH range 6.8–8.4.

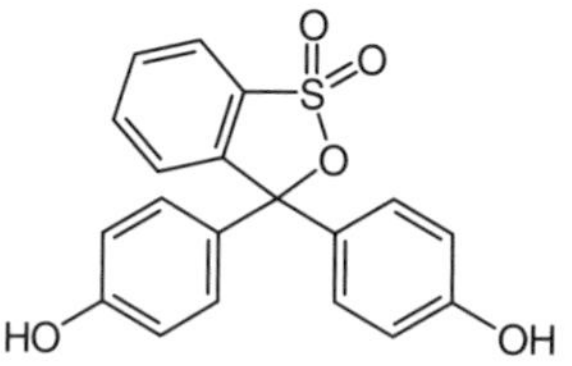

phenome the whole of the phenotypic characteristics of an organism. *Compare* **genome**. —**phenomic** *adj*.

phenomenology a branch of a science dealing with the description and classification of natural phenomena rather than with their explanation or cause. —**phenomenological** *adj*.

phenon any set of organisms grouped together by the methods of numerical taxonomy.

phenothiazine 1 a chemical structure that forms the basis, by *N*-substitution, of a group of **neuroleptics**, such as **chlorpromazine**. **2** any of the group of drugs based on the phenothiazine structure. The neuroleptic effects are those now most commonly exploited, but antiemetic, antihistamine and anthelmintic effects have also been of use.

phenotype 1 the totality of the observable functional and structural characteristics of an organism as determined by interaction of the genotype of the organism with the environment in which it exists. **2** any particular characteristic or set of characteristics of an organism so determined. **3** a group of organisms exhibiting the same set of such characteristics. *Compare* **genotype**. —**phenotypic** *adj*.

phenoxide 1 the anion, $C_6H_5O^-$, derived from phenol by the loss of a proton. **2** any salt of phenol.

phenoxy *symbol*: OPh *or* PhO; the C_6H_5O- group, derived from phenol by the loss of a hydrogen atom.

phentolamine an antagonist for both α_1 and α_2 adrenoceptors. It is an antihypertensive agent and vasodilator.

phenyl *symbol*: Ph; the aromatic carbocyclic group C_6H_5-, derived from benzene.

phenylacetaldehyde dehydrogenase EC 1.2.1.39; an enzyme of the pathway of phenylalanine degradation. It catalyses the oxidation by NAD^+ of phenylacetaldehyde to phenylacetate with formation of NADH.

phenylalaninate 1 phenylalanine anion; the anion, $C_6H_5-CH_2-CH(NH_2)-COO^-$, derived from **phenylalanine**. **2** any salt containing phenylalanine anion. **3** any ester of phenylalanine.

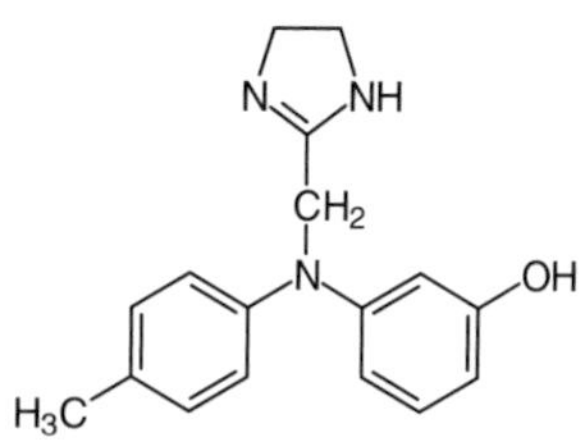

phentolamine

phenylalanine *the trivial name for* β-phenylalanine; α-aminohydrocinnamic acid; α-amino-β-phenylpropionic acid; 2-amino-3-phenylpropanoic acid; $C_6H_5-CH_2-CH(NH_2)-COOH$; a chiral α-amino acid. L-phenylalanine (*symbol*: F *or* Phe), (*S*)-2-amino-3-phenylpropanoic acid, is a coded amino acid found in peptide linkage in proteins; codon: UUC or UUU. In mammals, it is an essential dietary amino acid, and is both glucogenic and ketogenic. Residues of D-phenylalanine (*symbol*: D-Phe *or* DPhe), (*R*)-2-amino-3-phenylpropanoic acid, occur in a number of peptide antibiotics, e.g. bacitracin A, gramicidin S, polymyxins B_1 and B_2, and tyrocidins A, B, and C.

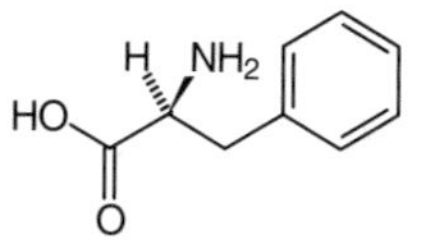

L-phenylalanine

L-phenylalanine ammonia-lyase EC 4.3.1.5; *systematic name*: L-phenylalanine ammonia-lyase; an enzyme, widely distributed in plants, that deaminates L-phenylalanine to *trans*-cinnamic acid, which is rapidly converted to **lignin**, a major polymer of the plant cell wall.

phenylalanine hydroxylase *see* **phenylalanine 4-monooxygenase**.

phenylalanine 4-monooxygenase EC 1.14.16.1; *systematic name*: L-phenylalanine,tetrahydrobiopterin:oxygen oxidoreductase (4-hydroxylating); *other names*: phenylalanine 4-hydroxylase; phenylalaninase. An iron-containing enzyme that catalyses a reaction between L-phenylalanine, tetrahydrobiopterin, and dioxygen to form L-tyrosine, dihydrobiopterin, and H_2O. It is thus of major importance for the provision of L-tyrosine. Over 400 mutations result in enzyme deficiency in humans and consequent **phenylketonuria**.

phenylalaninium phenylalanine cation; the cation, $C_6H_5-CH_2-CH(NH_3^+)-COOH$, derived from phenylalanine.

phenylalanino the alkylamino group, $C_6H_5-CH_2-CH(COOH)-NH-$, derived from phenylalanine.

phenylalanyl the acyl group, $C_6H_5-CH_2-CH(NH_2)-CO-$, derived from phenylalanine.

phenylalanyl chain (*sometimes*) the **B chain** of any of those species of insulin whose amino-acid sequences were the first to be determined. It was so called by virtue of the N-terminal phenylalanine residue occurring in those instances.

phenylephrine (*R*)-(−)-1-(3-hydroxyphenyl)-2-methylaminoethanol; a highly selective α_1 adrenergic agonist. It is a vasoconstrictor, used in treating hypotension; it also acts as a decongestant.

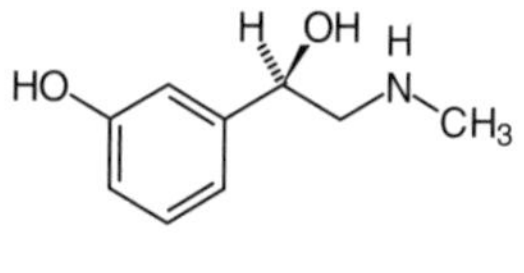

R-(−)-enantiomer
R-(−)-enantiomer

phenylethanolamine *N*-methyl transferase *abbr.*: PNMT; EC 2.1.1.28; an enzyme catalysing *N*-methylation of norepinephrine in the synthesis of epinephrine. *s*-adenosyl methionine is the methyl donor.

phenylhydrazone any condensation product of the type R R''C=N-NHC$_6$H$_5$, formed between carbonyl compounds of the general formula R R''C=O and phenylhydrazine. Phenylhydrazones, in particular 2,4-dinitrophenylhydrazones, are useful in the characterization of aldehydes and ketones. *See also* **phenylosazone**.

phenylketonuria (*abbr.*: PKU) *or* **phenylpyruvic oligophrenia** an autosomal recessive inherited disorder of phenylalanine metabolism characterized by the presence of phenylpyruvate in the urine and excessively high levels of phenylalanine and its metabolites in the blood and tissues. It can be treated by restricting the amount of phenylalanine in the diet. If untreated in young children, severe mental retardation occurs. This form, **classical phenylketonuria**, is due to the absence or deficiency of **phenylalanine 4-monooxygenase**, an enzyme that uses tetrahydrobiopterin as second donor. Hyperphenylalaninemia also results from deficiency in dihydropteridine reductase (EC 1.6.99.7) or phenylalanine(histidine) aminotransferase (EC 2.6.1.58).

phenylmethylsulfonyl fluoride *or* **phenylmethanesulfonyl fluoride** *abbr.*: PMSF; an inhibitor of serine proteases; it is highly toxic, but less so than diisopropylphosphofluoridate. PMSF is insoluble and unstable in aqueous solutions, so is usually dissolved in dioxan. It inhibits trypsin and chymotrypsin irreversibly, sulfonylating the active-site histidine, but does not inhibit cholinesterase. *See also* **aminoethylbenzenesulfonyl fluoride**.

phenylosazone any 1,2-bis(phenylhydrazone); i.e., any condensation product of the type C$_6$H$_5$NH–N=CR'–CR'=N–NH–C$_6$H$_5$ formed between an α-dicarbonyl compound of the general formula R'–CO–CO–R' and phenylhydrazine; α-hydroxy- and α-aminocarbonyl compounds will also form phenylosazones through oxidation of the corresponding [mono]phenylhydrazones initially formed. Phenylosazones, in particular 2,4-dinitrophenylosazones, are useful in the characterization of aldehydes and ketones, especially monosaccharides. *See also* **phenylhydrazone**.

phenylpropanoid any substance in plants derived from phenylpropane units variously linked and interlinked to form polymers of high molecular mass, especially **lignins**.

phenylpyruvate decarboxylase EC 4.1.1.43; an enzyme of the pathway of phenylalanine degradation. It catalyses the formation of phenylacetaldehyde from phenylpyruvate with release of CO$_2$.

phenylpyruvic oligophrenia *an alternative name for* **phenylketonuria**.

pH 5 enzyme a protein fraction from rat liver, soluble at pH 7 and isolated by precipitation at pH 5, that was used in early studies of protein biosynthesis *in vitro*. It is now known to be a mixture of amino acid–tRNA ligases.

pheo+ *or (esp. Brit.)* **phaeo+** *comb. form* denoting dark-coloured.

pheochromocytoma *or (esp. Brit.)* **phaeochromocytoma** *or* **chromaffinoma** a tumour of tissues derived from embryological neuroectoderm, often of the adrenal medulla. It is characterized by its ability to synthesize and secrete catecholamines, often in excessive amounts. The tumour cells commonly show a strong **chromaffin** (*or* pheochrome) reaction.

pheofarnesin *or (esp. Brit.)* **phaeofarnesin** any pigment derived by demetallation of a chlorophyll in which the 17-propionic-acid group is esterified with farnesol.

pheomelanin *or (esp. Brit.)* **phaeomelanin** a red and/or yellow pigment produced in mammalian melanocytes from tyrosine and cysteine. It is formed as an alternative to eumelanin by switching of the melanogenesis pathway at the dopaquinone stage. How the switch operates is unclear.

pheophorbide *or (esp. Brit.)* **phaeophorbide** any pigment derived

by removal of metal from a chlorophyll in which the 17-acrylic-acid or 17-propionic-acid group is unesterified or has been de-esterified.

pheophytin *or (esp. Brit.)* **phaeophytin** any pigment derived by removal of metal from a chlorophyll in which the 17-propionic-acid group is esterified with phytol.

pheromone a substance, or characteristic mixture of substances, that is secreted and released by an organism and detected by a second organism of the same or a closely related species, in which it causes a specific reaction, such as a definite behavioural reaction or a developmental process; a type of **ectocrine**. The mating pheromones of fungi and insects are examples. *Compare* **allomone**, **kairomone**.

pheromone-binding protein any of a group of insect proteins that bind to, and thus solubilize, **pheromones**, which are typically hydrophobic.

pheromone receptor a fungal transmembrane G protein-coupled receptor.

PHEX *abbr. for* phosphate-regulating gene with homologies to endopeptidases on the X chromosome; a putative endopeptidase that is mutated in X-linked hypophosphatemia. *See* **fibroblast growth factor**-23.

PHHI *see* **sulfonylurea receptor**.

phi *symbol*: φ *or* ϕ (lower case) *or* Φ (upper case); the twenty-first letter of the Greek alphabet. For uses *see* **Appendix A**.

PHI *or* **PHI-27** *abbr. for* peptide HI.

PHI-BLAST *abbr. for* pattern hit initiated-BLAST; *see* **database-search program**.

Philadelphia chromosome (Ph^1) a specific chromosomal abnormality in the blood cells of patients with chronic myeloid leukemia. It usually results from a fusion of *bcr* on chromosome 22 with c-*myc* (*see* **MYC**) on chromosome 9 leading to expression of the latter. The Philadelphia chromosome is a very small chromosome consisting of the short arm and a small piece of the long arm of chromosome 22 fused to a short piece from the end of the long arm of chromosome 9 (the modified chromosome 9 that results being called 9q$^+$). It is named after the city in which it was first recorded. *See also* **abl**.

philosin a pore-forming toxin with approximately 100 amino acid residues, produced by the ant *Myrmecia pilosula*. It is classified in the TC system under number 1.C.51.

pH indicator *see* **indicator** (def. 3), **mixed indicator**.

phi X174 *or* φX174 a single-stranded DNA bacteriophage that infects *Escherichia coli*. It contains 5386 nt and was the first DNA to be sequenced, by British biochemist Frederick Sanger (1918–) in 1976.

phloem the main food-conducting tissue of plants, via which nutrients and photosynthate are conveyed between different sites. Phloem comprises two types of cells: sieve-tube elements, which are joined end-to-end to form sieve tubes; and adjacent companion cells. *Compare* **xylem**.

phloretin 3-(4-hydroxyphenyl)-1-(2,4,6-trihydroxyphenyl)-1-propanone; the aglycon of **phlorizin**, with which it shares many biochemical and pharmacological properties.

phlorizin *or* **phlorhizin** *or* **phloridzin** *or* **phlorrhizin** phloretin-2'-β-glucoside; a dihydrochalcone glucoside occurring in all parts of the apple tree except the mature fruit. It inhibits intestinal and renal glucose–Na$^+$ cotransporters, the mediated transport of glucose into erythrocytes and hepatocytes, the glucose 6-phosphate transporter in hepatocytes, and photophosphorylation. It is used experimentally to produce glucosuria.

phloroglucinol 1,3,5-trihydroxybenzene; 1,3,5-benzenetriol; a compound that is useful as a reagent for the qualitative detection of pentoses and pentosans; when these are boiled with a solution of

phloroglucinol in hydrochloric acid, a cherry-red colour is produced. Galactose also reacts in this way.

phlorrhizin *a variant spelling of* **phlorizin**.

pH meter any instrument for determining pH potentiometrically, usually with a glass electrode plus a calomel reference electrode, or with a combination electrode.

phoA the gene for *E. coli* alkaline phosphatase. The signal sequence of PhoA may be incorporated in expression vectors used for secretion of foreign proteins into the periplasmic space of *E. coli*.

phorbol a complex tetracyclic compound, $C_{20}H_{28}O_6$, based on a cyclopropabenzazulene skeleton. Various **phorbol esters** occur in croton oil (from *Croton tiglium*, of the family Euphorbiaceae, hence the name) and act as potent **cocarcinogens**.

phorbol ester any esterified form of **phorbol**. Several phorbol esters possess activity as tumour promoters and activate the mechanisms associated with cell growth. These actions have been widely attributed to the activation of protein kinase C, probably by substituting for diacylglycerol. A much-studied ester is phorbol 12-myristate 13-acetate (PMA; *or* 12-*O*-tetradecanoyl-phorbol-13-acetate, TPA); other active esters include phorbol 12,13-dibutyrate (PDBu) and phorbol 12,13-didecanoate (PDD). Activity depends on the β-configuration of the hydroxyl at C-4. Several 4α-phorbol esters have been utilized in control studies.

+phore *noun suffix* denoting something, especially a chemical substance or group, that carries or produces; e.g. ionophore, chromophore.

+phoresis *noun suffix* denoting carriage of molecular entities by some agency; e.g. electrophoresis, ionophoresis.

phosducin a protein that binds the β and γ subunits of **transducin** in outer and inner segments of retinal rod cells. Light-induced changes in cyclic nucleotide levels modulate the phosphorylation of phosducin by cyclic AMP kinase.

+phosph+ *a variant form of* **+phospho+** (sometimes before a vowel).

phosphacan a chondroitin sulfate proteoglycan of brain that interacts with neurons and neural cell-adhesion molecules. It is an extracellular variant of a receptor-type protein tyrosine phosphatase.

phosphagen any of a group of guanidine phosphates that occur in muscle, and can be used to regenerate ATP from ADP during muscular contraction by a reaction catalysed by **creatine kinase**, or by an analogous enzyme of sub-subclass EC 2.7.3. Included in the group are phosphoagmatine (*see* **agmatine**), phosphoarginine, phosphocreatine, phosphoglycocyamine (*see* **glycocyamine**), phospholombricine (*see* **lombricine**), phosphooctopine (*see* **octopine**), and phosphotaurocyamine (*see* **taurocyamine**).

phosphataemia *a variant spelling (esp. Brit.) of* **phosphatemia**.

phosphatase *or* **phosphomonoesterase** any phosphoric monoester hydrolase of sub-subclass EC 3.1.3. Some phosphatases are very specific while others are relatively unspecific. *See also* **acid phosphatase, acylphosphatase, alkaline phosphatase, fructose-bisphosphatase, glucose-1-phosphatase, glucose-6-phosphatase, inorganic pyrophosphatase, protein phosphatase**.

phosphatased *a jargon term* indicating dephosphorylated by treatment with a **phosphatase**.

phosphatase inhibitor protein an inhibitor of protein phosphatase 1 (*see* **protein phosphatase**) involved in the regulation of glycolysis; it is itself a phosphoprotein.

phosphate 1 the trivalent anion, PO_4^{3-}, derived from phosphoric

acid; orthophosphoric acid; tetraoxophosphoric(v) acid; $PO(OH)_3$. **2** any mixture of [ortho]phosphoric acid and its anions. Once widely used in the laboratory as a buffer in the physiological range of pH values; pK_{a2} (37°C, 0.05 mol L^{-1}) = 6.84. **3** any (partial or full) salt or ester of [ortho]phosphoric acid. Phosphate salts and esters play key roles in the structure and function of all forms of life. **4** any anion or salt of a mixed anhydride of [ortho]phosphoric acid with another acid; e.g. acetyl phosphate (i.e. acetic phosphoric anhydride). **5** the univalent group $-O-P(O)(OH)_2$ (irrespective of its state of ionization). *See also* **diphosphate, inorganic phosphate, metaphosphate, pyrophosphate, triphosphate**.

phosphate bond 1 an ester linkage in a phospho[mono]ester, phosphodiester, or phosphotriester. **2** an anhydride linkage in a phosphoric anhydride. **3** the connecting phosphoric residue in a phosphodiester.

phosphate bond energy *a term formerly and incorrectly used for* the **bond energy** (def. 2) of a **phosphate bond** (def. 1, 2).

phosphatemia *or (esp. Brit.)* **phosphataemia** the level of **phosphate** (def. 2) in the blood; in human plasma its normal range is 0.8–1.4 mmol L^{-1}. The level is regulated by **calcitriol**, which stimulates absorption of dietary phosphate as well as calcium, and by **parathyrin**, which decreases phosphate resorption by the kidney. Calcitriol also stimulates bone resorption, which tends to lower the plasma phosphate level, but in states of hypervitaminosis D the effect on the gut predominates and the plasma level is increased. *See also* **hyperphosphatemia, hypophosphatemia**.

phosphate potential the concentration of ATP in any system divided by the product of the concentrations of ADP and inorganic phosphate in that system. *Compare* **adenylate energy charge**.

phosphatidal *a former term for* **plasmenyl**.

phosphatidate any anion, salt, or ester of a **phosphatidic acid**.

phosphatidate cytidylyltransferase EC 2.7.7.41; *systematic name*: CTP:phosphatidate cytidylyl transferase; *other names*: CDP-diglyceride synthetase; CDPdiglyceride pyrophosphorylase. An enzyme that catalyses a reaction between CTP and phosphatidate to form pyrophosphate and CDPdiacylglycerol. It is an integral membrane protein of the endoplasmic reticulum; it forms the intermediate involved in the synthesis of, e.g., phosphatidylinositol, by reaction with inositol, or phosphatidylserine, by reaction with serine.

phosphatidate phosphatase EC 3.1.3.4; *systematic name*: 3-*sn*-phosphatidate phosphohydrolase; an enzyme that removes phosphate from phosphatidic acid to yield an *sn*-1,2-diacylglycerol.

phosphatide *an old term, no longer recommended for* **glycerophospholipid**.

phosphatidic acid any derivative of glycerol phosphate in which both the remaining hydroxyl groups of the glycerol moiety are esterified with fatty acids; the commonest are derivatives of *sn*-glycerol 3-phosphate.

phosphatidyl *symbol*: Ptd; an acyl group derived from a **phosphatidic acid**.

phosphatidylcholine *abbr.*: PC; any glycerophospholipid in which a phosphatidyl group is esterified to the hydroxyl group of **choline**. Phosphatidylcholines are important constituents of cell membranes and lipoproteins.

phosphatidylcholine desaturase EC 1.3.1.35; *other names*: oleate desaturase; linoleate synthase; an enzyme in plants that catalyses the reaction:

1-acyl-2-oleoyl-*sn*-glycero-3-phosphocholine + NAD$^+$ =
1-acyl-2-linoleoyl-*sn*-glycero-3-phosphocholine + NADH.

phosphatidylcholine–dolichol *O*-acyltransferase *see* **lecithin–cholesterol acyl transferase**.

phosphatidylcholine–sterol *O*-acyltransferase EC 2.3.1.43; *other names*: lecithin–cholesterol acyltransferase (*abbr.*: LCAT); phospholipid–cholesterol acyltransferase; a plasma enzyme that catalyses the transfer of a fatty-acyl group from phosphatidylcholine to a sterol to form a sterol ester and 1-acylglycerophosphocholine. In mammals it is important in plasma lipoprotein metabolism, particularly in esterifying cholesterol, which is then transferred to very-low-density lipoprotein as cholesteryl ester. *Compare* **sterol *O*-acyltransferase**.

phosphatidylethanolamine *abbr.*: PE; any glycerophospholipid in which a phosphatidyl group is esterified to the hydroxyl group of **ethanolamine**. Phosphatidylethanolamines are important constituents of cell membranes.

phosphatidylinositol *abbr.*: PI; *symbol*: PtdIns; 1-(3-*sn*-phosphatidyl)-1d-*myo*-inositol; *the recommended name for* any glycerophospholipid in which its *sn*-glycerol 3-phosphate residue is esterified to the 1-hydroxyl group of 1d-*myo*-inositol. The older term **phosphoinositide** is now used as a generic term for any inositol-containing glycerophospholipid. The corresponding names for the monophosphorylated and diphosphorylated derivatives of phosphatidylinositol are **phosphatidylinositol 4-phosphate** and **phosphatidylinositol 4,5-bisphosphate**, respectively. The older names diphosphoinositide and triphosphoinositide for these derivatives are no longer used.

phosphatidylinositol 4,5-bisphosphate *or (formerly)* **triphosphoinositide** *abbr.*: PIP$_2$ or PI-4,5-P$_2$; *symbol*: PtdIns(4,5)P_2 or PtdIns-4,5-P_2; 1-(3-*sn*-phosphatidyl)-1d-*myo*-inositol 4,5-bisphosphate; any glycerophospholipid in which a phosphatidyl group is esterified to the 1-hydroxyl group of 1d-*myo*-inositol 4,5-bisphosphate.

1-phosphatidylinositol-4,5-bisphosphate phosphodiesterase *see* **phospholipase**.

phosphatidylinositol cycle *or (formerly)* **phosphoinositide cycle** a metabolic cycle that mediates the actions of various hormones and other effectors at the membranes of mammalian cells. Binding of an effector to its specific receptor at the membrane causes a G-protein-coupled stimulation of a phospholipase C, called 1-phosphatidylinositol-4,5-bisphosphate phosphodiesterase. This causes hydrolysis of phosphatidylinositol 4,5-bisphosphate, I, to produce the two intracellular messengers inositol 1,4,5-trisphosphate, II, which causes release of calcium ions from intracellular stores, and *sn*-1,2-diacylglycerol, III, which activates protein kinase C. III is then converted via phosphatidate (by the action of **diacylglycerolkinase** and **phosphatidate cytidylyltransferase**) to **CDPdiacylglycerol**, IV, while II is converted via inositol bisphosphate and inositol monophosphate to inositol. Inositol and IV then react together (enzyme is **CDPdiacylglycerol-inositol 3-phosphatidyltransferase**) to form phosphatidylinositol, from which I is regenerated by stepwise phosphorylation by specific kinases to complete the cycle. Alternatively, II may be converted to another messenger, inositol 1,3,4,5-tetrakisphosphate, V, which may be implicated in the influx of extracellular calcium ions; V is then degraded to inositol via inositol 1,3,4-trisphosphate, which is inactive.

phosphatidylinositolglycan the compound in which phosphatidylinositol is linked (6-1α) to the core glycan GlcN4-1αMan6-1αMan2-1αMan6-P-ethanolamine. Other groups may be attached to the core glycan. It forms the membrane anchor for **PIG-tailed proteins**, in which the protein's C-terminus forms an amide linkage with the phosphoethanolamine moiety. *See* **paroxysmal nocturnal hemoglobinuria**.

phosphatidylinositol glycan protein *abbr.*: PIG protein; *N*-acetylglucosaminyl-phosphatidylinositol biosynthetic protein; a protein required for synthesis of *N*-acetylglucosaminylphosphatidylinositol, the very early intermediate in the biosynthesis of the **GPI anchor**. Another protein required for this synthesis is PIG-F.

1-phosphatidylinositol 3-kinase EC 2.7.1.137; *systematic name*: ATP:1-phosphatidyl-1d-*myo*-inositol 3-phosphotransferase. An enzyme that catalyses the phosphorylation by ATp of 1-phosphatidyl-1d-*myo*-inositol to form 1-phosphatidyl-1d-*myo*-inositol 3-phosphate with release of ADP. It forms part of the complex that associates on activation of certain growth factor (e.g. PDGF) receptors, binding via SH2 domains (*see* **SH domain**). The molecule consists of a catalytic subunit (p110) and a regulatory subunit (p85: two forms, α and β, in mammals). p85 mediates association of p110 to the plasma membrane.

1-phosphatidylinositol 4-kinase EC 2.7.1.67; an enzyme involved in the biosynthesis of membrane **phosphoinositides**. It catalyses the phosphorylation by ATP of 1-phosphatidyl-1d-*myo*-inositol to form 1-phosphatidyl-1d-*myo*-inositol 4-phosphate with release of ADP.

phosphatidylinositol-3-phosphatase EC 3.1.3.64; an enzyme that catalyses the hydrolysis of the 3-phosphate from phos-

phatidylinositol-3-phosphate, forming phosphatidylinositol and orthophosphate.

phosphatidylinositol 4-phosphate *or (formerly)* **diphosphoinositide** *abbr.*: PIP or PI 4-P; *symbol*: PtdIns4*P* or PtdIns-4-*P*; 1-(3-*sn*-phosphatidyl)-1d-*myo*-inositol 4-phosphate; any glycerophospholipid in which a phosphatidyl group is esterified to the 1-hydroxyl group of 1d-*myo*-inositol 4-phosphate.

1-phosphatidylinositol-4-phosphate kinase EC 2.7.1.68; *other name*: diphosphoinositide kinase; an enzyme of the pathway for the biosynthesis of membrane **phosphoinositides**. It catalyses the phosphorylation by ATP of 1-phosphatidyl-1d-*myo*-inositol 4-phosphate to form 1-phosphatidyl-1d-*myo*-inositol 4,5-bisphosphate with release of ADP.

1-phosphatidylinositol phosphodiesterase *see* **phospholipase**.

phosphatidylinositol synthase *see* **CDP-diacylglycerol–inositol 3-phosphatidyltransferase**.

phosphatidylserine any glycerophospholipid in which a phosphatidyl group is esterified to the hydroxyl group of L-serine. Phosphatidylserines are important constituents of cell membranes.

phosphatidylserine decarboxylase EC 4.1.1.65; an enzyme that converts phosphatidylserine to phosphatidylethanolamine, of major importance in the synthesis of that phospholipid. It consists of an α and β chain, formed by cleavage of a proenzyme.

phosphatidylserine synthase *an alternative name for* **CDPdiacylglycerol-serine O-phosphatidyltransferase**.

phosphatonin a heat-labile, trypsin-sensitive factor that is presumed to be secreted by some tumours and to increase phosphaturia and produce phosphate wasting.

phosphinothricin *abbr.*: PPT; CH$_3$PO$_2$CH$_2$CH$_2$CHNH$_2$COOH; a glutamate analogue herbicide that is relatively nontoxic for animals.

phosphite 1 any anionic form of phosphorous acid; trioxophosphoric(III) acid; P(OH)$_3$. **2** any mixture of phosphorous acid and its ionized forms. **3** any (partial or full) salt or ester of phosphorous acid.

phosphite-triester method a method for the chemical synthesis of oligonucleotides in which a suitably protected phosphite diester of general formula Nuc-3′–O–P(OR)X, where R is a methyl or another alkyl group, X is a reactive group such as chloro, dimethylamino, or morpholino, and Nuc is a residue of an appropriate nucleoside, is first condensed with the 5′-hydroxyl of a protected nucleoside (usually anchored through its 3′-hydroxyl to an insoluble support) to form a phosphite triester. Sequential addition of further nucleoside units is then carried out in the same way, the phosphite triester formed at each stage being oxidized to the corresponding phosphate triester before the next cycle of reactions. Protecting groups are removed by hydrolysis at the completion of the synthesis.

+phospho+ *or (sometimes before a vowel)* **+phosph+** *(in chemical nomenclature)* prefix or infix denoting a phosphoric residue linked to one, two, or three heteroatoms in an organic compound. *See also* **phosphono+**.

phosphoacetylglucosamine mutase EC 5.4.2.3; *systematic name*: *N*-acetyl-D-glucosamine 1,6-phosphomutase; *other name*: acetylglucosamine phosphomutase. An enzyme that catalyses the reaction:

$$N\text{-acetyl-D-glucosamine 1-phosphate} =$$
$$N\text{-acetyl-D-glucosamine 6-phosphate}.$$

It participates in the formation of UDP-*N*-acetylglucosamine: glucosamine 6-phosphate → *N*-acetylglucosamine-6-phosphate →*N*-acetylglucosamine-1-phosphate, thence by reaction with UTP to UDP-*N*-acetylglucosamine + pyrophosphate, thence also to sialic acid.

3′-phosphoadenosine 5′-phosphosulfate *an alternative recommended name for* **adenosine 3′-phosphate 5′-phosphosulfate**.

3′-phosphoadenylyl sulfate *symbol*: *P*Ado*PS*; *abbr.*: PAPS; *alternative recommended names for* active sulfate; adenosine 3′-phosphate 5′-phosphosulfate; adenosine 3′-phosphate 5′-*P*-phosphatosulfate; 3′-phospho-5′-adenylic sulfuric monoanhydride; a naturally occuring mixed anhydride. It is synthesized from **adenosine 5′-phosphosulfate** by phosphorylation with ATP through the action of adenylylsulfate kinase (EC 2.7.1.25). It is an intermediate in the formation of a variety of sulfo compounds in biological systems. For example, in animals it is involved in sulfate transfer in the formation of sulfatides and in the synthesis of chondroitin sulfate and other sulfated polysaccharides, while in bacteria, by a process analogous to that involving adenosine 5'-phosphosulfate. In plants, it is reduced by phosphoadenylyl-sulfate reductase (thioredoxin), EC 1.8.4.8 to yield adenosine 3',5' bis(phosphate) and sulfite. The latter can then undergo further reduction by sulfite reductase to sulfide, from which cysteine may be synthesized.

phosphoamino acid any phosphorylated amino acid. The term often refers to *O*-phosphate esters of residues in proteins, e.g. those of serine, threonine, or tyrosine, that are released as phosphoamino acids on hydrolysis. Phosphohistidine can be formed as part of an enzyme mechanism (e.g., **phosphoglycerate mutase**), and phosphohomoserine is an intermediate in threonine biosynthesis. The guanidine group can also be phosphorylated to form a phosphono derivative (*see* **phosphagen**).

phosphoarginine *the recommended trivial name for* N^ω-**phosphonoarginine**; the commonest **phosphagen** in invertebrates. The older name, arginine phosphate, is not recommended.

phosphocreatine *abbr.*: PC; *the recommended trivial name for* N^ω-**phosphonocreatine**; the commonest **phosphagen** in vertebrates. The older name, creatine phosphate, is not recommended.

phosphodiester any diester of phosphoric acid; i.e. any phosphoric ester containing the bridging group –PO(OH)– between two oxygen atoms in other groups (which may be the same or different) or at different positions in the same group. Examples include cyclic nucleotides, glycerophospholipids, and nucleic acids.

phosphodiesterase any esterase that catalyses the hydrolysis of either (but not both) of the ester linkages in a phosphodiester, each linkage being hydrolysed only by an enzyme specific for that linkage. Such enzymes are classified in various sub-subclasses within the subclass EC 3.1, hydrolases, especially in the sub-subclass EC 3.1.4, phosphoric diester hydrolases. The term is often applied to **nucleases** without sugar specificity. *See also* **cyclic AMP phosphodiesterase, phosphodiesterase I, phospholipase C.** —**phosphodiesteratic** *adj.*

phosphodiesterase I *the recommended name for* an enzyme, EC 3.1.4.1, that catalyses the hydrolytic removal of 5'-nucleotides successively from the (nonphosphorylated) 3' end of an oligonucleotide; *other name:* 5'-exonuclease.

phosphodiesterase II an enzyme, *now named* **spleen exonuclease** (EC 3.1.16.1; previously EC 3.1.4.18), that catalyses the hydrolytic removal of 3'-nucleotides successively from the (nonphosphorylated) 5' end of a (preferably single-stranded) ribonucleic or deoxyribonucleic acid; *other names:* 3'-exonuclease; spleen phosphodiesterase.

phosphodiester method a method for the chemical synthesis of oligonucleotides in which, essentially, a 5'-phosphate of an appropriate [mono]nucleotide, after suitable protection of other functional groups, is condensed with the 3'-hydroxyl of a protected nucleoside, [mono]nucleotide, or previously formed oligonucleotide, using dicyclohexylcarbodiimide or an arylsulfonyl chloride as a coupling agent. The reaction sequence is then repeated with another

nucleotide 5'-phosphate. This method is being superseded by the **phosphite-triester method** and **phosphotriester method**.

phospho*enol*pyruvate *abbr.*: PEP *or* P-*enol*pyruvate; 2-hydroxy-2-propenoate phosphate (ester); 2-(phosphonooxy)-2-propenoate; the phosphate ester formed from the hydroxy group of the enol tautomer of pyruvate. Its hydrolysis to pyruvate gives a particularly large negative change in standard free energy ($-\Delta G^o = 62$ kJ mol^{-1} at pH 7). It is a key intermediate in glycolysis and gluconeogenesis, and in the biosynthesis of aromatic amino acids.

phospho*enol*pyruvate carboxykinase *abbr.*: PEPCK; any enzyme that catalyses the reaction:

$$X–P + oxaloacetate = X + phosphoenolpyruvate + CO_2,$$

where X–P, depending on source, can be GTP or ITP (in EC 4.1.1.32), ATP (in EC 4.1.1.49), or pyrophosphate (in EC 4.1.1.38). (1) **Phospho*enol*pyruvate carboxykinase (GTP)**, EC 4.1.1.32; *systematic name*: GTP:oxaloacetate carboxy-lyase (transphosphorylating); *other names*: phospho*enol*pyruvate carboxylase; phosphopyruvate carboxylase; the rate-limiting gluconeogenic enzyme in eukaryotes. It has different forms in cytosol and mitochondria. It catalyses a reaction between GTP and oxaloacetate to form GDP, phospho*enol*pyruvate, and CO_2. (2) **Phospho*enol*pyruvate carboxykinase (ATP)**, EC 4.1.1.49; *systematic name*: ATP:oxaloacetate carboxy-lyase (transphosphorylating); *other names*: phospho*enol*pyruvate carboxylase; phosphopyruvate carboxylase (ATP); it catalyses a similar reaction to (1) but with ATP/ADP not GTP/GDP. In many bacteria it is a Ca^{2+}-regulated rate-limiting gluconeogenic enzyme. *See also* **gluconeogenesis.**

phospho*enol*pyruvate carboxylase EC 4.1.1.31; *systematic name*: orthophosphate:oxaloacetate carboxy-lyase (phosphorylating); an enzyme that catalyses a reaction between orthophosphate and oxaloacetate to form phospho*enol*pyruvate, H_2O, and CO_2. It is an allosteric homotetramer associated with the **tricarboxylic-acid cycle** in plants; effectors include fatty acids, acetyl-CoA, and fructose 1,6-bisphosphate. It forms oxaloacetate for metabolism by this cycle in plants. system system

phospho*enol*pyruvate-dependent sugar phosphotransferase system *abbr.*: PTS; a system for bacterial sugar transport; it comprises two main components: **protein-N^{p}phosphohistidine–sugar phosphotransferase** and **phospho*enol*pyruvate–protein phosphotransferase**. It also functions also as a chemoreceptor monitoring the environment for changes in sugar concentration.

phospho*enol*pyruvate mutase EC 5.4.2.9; *systematic name*: phospho*enol*pyruvate 2,3-phosphonomutase; *other name*: phospho*enol*pyruvate phosphomutase. An enzyme that catalyses the conversion of phospho*enol*pyruvate to 3-phosphonopyruvate.

phospho*enol*pyruvate–protein phosphotransferase EC 2.7.3.9; *systematic name*: phospho*enol*pyruvate:protein-L-histidine N^π-phosphotransferase; *other name* enzyme I of the **phospho*enol*pyruvate-dependent sugar phosphotransferase system**; a soluble bacterial enzyme, part of a system for the transport of hexoses across the cell membrane, that catalyses the phosphorylation by phospho*enol*pyruvate of a low-M_r, heat-stable protein, **HPr**.

phosphoester any [mono]ester of phosphoric acid; i.e. any phosphoric ester containing the group –PO(OH)$_2$ linked to an oxygen atom in a single other group.

phosphofructokinase 1 *abbr.*: PFK; EC 2.7.1.11; *recommended name*: 6-phosphofructokinase; *other names*: phosphofructokinase 1; phosphohexokinase. An enzyme that catalyses the phosphorylation of D-fructose 6-phosphate to D-fructose 1,6-bisphosphate at the expense of ATP, CTP, ITP, or UTP. Phosphofructokinase 1 is an allosteric homotetrameric enzyme of glycolysis. There are three genes for human subunits: M (muscle – 799 amino acids), L (liver – 799 amino acids), and P (platelet). Erythrocytes contain five isoenzymes. Mutations in the M or L gene

cause the rare glycogen disease type VII (Tarui disease). *E. coli* has two entirely unrelated isoenzymes, I and II, presumably of independent evolution. **2** or **phosphofructokinase 2** EC 2.7.1.105; *recommended name*: 6-phosphofructo-2-kinase; an enzyme that catalyses the phosphorylation of D-fructose 6-phosphate to D-fructose 2,6-bisphosphate at the expense of ATP. Phosphofructokinase 2 is part of the same bifunctional protein as **fructose-2,6-bisphosphatase; phosphorylation** inhibits the kinase activity. **3** or **fructose-1-phosphate kinase** EC 2.7.1.56; *recommended name*: 1-phosphofructokinase; an enzyme that catalyses the phosphorylation of D-fructose 1-phosphate to D-fructose 1,6-bisphosphate at the expense of ATP, GTP, ITP, or UTP.

phosphoglucoisomerase *an alternative name for* **glucose-6-phosphate isomerase**.

phosphoglucomutase *abbr. (in clinical chemistry)*: PGT; EC 5.4.2.2; *systematic name*: α-D-glucose 1,6-phosphomutase; *other name*: glucose phosphomutase. A **phosphotransferase** enzyme that catalyses the interconversion of α-D-glucose 1-phosphate and α-D-glucose 6-phosphate; α-D-glucose 1,6-bisphosphate is an intermediate, formed by phosphoryl transfer from the enzyme. It links the metabolism of polysaccharides to glycolysis. There are three isoenzymes in humans.

phosphogluconate 6-phospho-D-gluconate; an intermediate in the **Entner–Doudoroff pathway** and in the **pentose phosphate pathway**. It is a product of the combined action of **glucose-6-phosphate dehydrogenase** and **6-phosphogluconolactonase**, and is a substrate for **phosphogluconate dehydrogenase (decarboxylating)**.

COO⁻
|
H—C—OH
|
HO—C—H
|
H—C—OH
|
H—C—OH   O
| ‖
H₂C—O—P—O⁻
              |
              O⁻

phosphogluconate dehydrogenase (decarboxylating) *abbr.*: 6PGD; EC 1.1.1.44; *systematic name*: 6-phospho-D-gluconate:NADP⁺ 2-oxidoreductase (decarboxylating); *other names*: phosphogluconic acid dehydrogenase; 6-phosphogluconic dehydrogenase; 6-phosphogluconic carboxylase. An enzyme that catalyses a reaction between 6-phospho-D-gluconate and NADP⁺ to form D-ribulose 5-phosphate, CO_2, and NADPH. It participates in the **pentose phosphate pathway**, and is significant in maintaining NADP⁺ in the reduced state.

phosphogluconate (oxidative) pathway *an alternative name for* pentose phosphate pathway.

6-phosphogluconolactonase EC 3.1.1.31; *systematic name*: 6-phospho-D-glucono-1,5-lactone lactonohydrolase; an enzyme of the **pentose phosphate pathway** that catalyses the hydrolysis of 6-phosphogluconolactone to 6-phosphogluconic acid.

phosphoglycerate 1 2-phospho-D-glycerate; D-glycerate 2-phosphate; (*R*)-2,3-dihydroxypropanoate 2-(dihydrogen phosphate). **2** 3-phospho-D-glycerate; D-glycerate 3-phosphate; (*R*)-2,3-dihydroxypropanoate 3-(dihydrogen phosphate). Both are important intermediates in glycolysis. 3-Phosphoglycerate is a precursor in serine biosynthesis. *Compare* **glycerophosphate**, **phosphoglyceride**. *See also* **phosphoglycerate mutase**.

phosphoglycerate dehydrogenase EC 1.1.1.95; *systematic name*: 3-phosphoglycerate:NAD⁺ 2-oxidoreductase; an enzyme that converts 3-phospho-D-glycerate to 3-phosphohydroxypyruvate in the presence of NAD⁺. It participates in the synthesis of serine from 3-phosphoglycerate.

phosphoglycerate kinase *abbr.*: PGK; EC 2.7.2.3; *systematic name*: ATP:3-phospho-D-glycerate 1-phosphotransferase; an enzyme that catalyses reaction:

1,3-bisphosphoglycerate + ADP = 3-phosphoglycerate + ATP.

The forward reaction is the first one to generate ATP during glycolysis; the reverse reaction participates in CO_2 assimilation during photosynthesis in chloroplasts. It exists in tissue-specific or (in protozoa, etc.) organelle-specific isoforms. Enzyme deficiency in humans results in chronic hemolytic anemia, neurological dysfunction, and myopathy.

phosphoglycerate mutase *abbr.*: PGAM; EC 5.4.2.1; *systematic name*: D-phosphoglycerate 2,3-phosphomutase; *other names*: phosphoglycerate phosphomutase; phosphoglyceromutase. An enzyme of the glycolytic and gluconeogenic pathways that catalyses the reaction:

2-phospho-D-glycerate = 3-phospho-D-glycerate

(the forward direction, as written, being the direction in which the enzyme functions in gluconeogenesis). A phosphotransferase enzyme for which 2,3-bisphospho-D-glycerate acts as a reaction primer. The active enzyme has a phosphoryl group active site (histidine is phosphorylated as part of the mechanism) and transfers phosphoryl to the substrate. *See also* **phosphohistidine**.

phosphoglycerate phosphomutase *another name for* **phosphoglycerate mutase**.

3-phospho-D-glyceric phosphoric monoanhydride *see* **3-phospho-D-glyceroyl phosphate**.

phosphoglyceride *another (less correct) name for* **glycerophospholipid**.

3-phospho-D-glyceroyl phosphate *abbr.*: D-glyceric-1,3-P_2; *one of the recommended names for* (*R*)-[2,3-dihydroxypropanoyl dihydrogen phosphate 3-(dihydrogen phosphate)], a key intermediate in the glycolytic pathway, commonly but incorrectly known as 1,3-diphosphoglycerate or 1,3-bisphosphoglycerate. *Alternative recommended names*: 3-phospho-D-glyceric phosphoric monoanhydride; (D-glyceroyl phosphate) 3-phosphate. This compound should be distinguished from **2,3-bisphospho-D-glycerate**.

phosphohexose isomerase *an alternative name for* **glucose-6-phosphate isomerase**.

phosphohistidine N^π-phosphohistidine; it is formed as a residue in certain proteins, being an intermediate in the **phosphoglycerate mutase** reaction and the **phospho*enol*pyruvate**dependent sugar phosphotransferase system.

phosphoinositide *another old term, no longer recommended, for* **phosphatidylinositol** but now much used to denote all phosphatidylinositol derivatives as a class.

phosphoinositide cycle *another term for* **phosphatidylinositol cycle**.

phosphoketolase EC 4.1.2.9; a thiamine diphosphate-protein of the phosphoketolase pathway that catalyses phosphorolysis of D-xylulose 5-phosphate and orthophosphate to acetyl phosphate and D-glyceraldehyde 3-phosphate.

phosphoketolase pathway any metabolic pathway, usually in a microorganism, that involves the enzyme **phosphoketolase**, e.g. in **heterolactic fermentation** or in the fermentation of glucose by *Bifidobacterium* spp.

phosphokinase *a former term for* **kinase**.

phospholamban a homopentameric sarcoplasmic reticulum membrane protein that regulates the calcium pump of cardiac, slow-twitch skeletal, and smooth muscle. It is phosphorylated by protein kinase A in response to β-adrenergic stimulation. The phosphorylated protein increases the rate of the sarcoplasmic reticulum Ca^{2+}-ATPase.

phospholemman a 72-residue cell-membrane protein that forms a chloride-selective ion channel. It is a substrate for both protein kinase C and cyclic AMP-dependent protein kinase. It is rapidly phosphorylated after β-adrenergic and α-adrenergic stimulation of the heart, and in response to insulin in muscle. It has a single transmembrane domain.

phospholipase any enzyme that is a lipase and that catalyses the hydrolysis of a glycerophospholipid. Phospholipases have been subdivided as follows.

Phospholipase A1 EC 3.1.1.32; *the recommended name for* any carboxylic ester hydrolase that removes the acyl residue at C-1 of the glycerol moiety.

Phospholipase A2 *the recommended name for* any carboxylic ester hydrolase that removes the acyl residue at C-2 of the glycerol moiety; e.g. EC 3.1.1.4; *systematic name*: phosphatidylcholine 2-acylhydrolase; *other names*: lecithinase A; phosphatidase; phosphatidolipase. The enzyme requires calcium. Snake-venom enzymes have been much studied and have broad specificity for phosphoglyceride classes. In mammals there are two such enzymes: (1) a pancreatic enzyme, which is released from the exocrine pancreas on stimulation by cholecystokinin, has broad specificity for phosphoglycerides, and hydrolyses phospholipids of digested food in the intestine in conjunction with the action of bile salts to form micelles; and (2) a membrane-bound enzyme, which releases arachidonate for prostanoid synthesis, and is in many cases stimulated by activated hormone and growth-factor receptors.

Phospholipase B EC 3.1.1.5; *recommended name*: lysophospholipase; *other names*: lecithinase B; lysolecithinase. These are enzymes that remove the acyl group from a lysoglycerophospholipid.

Phospholipase C *abbr.*: PLC; *the recommended name for* any phosphoric diester hydrolase that splits the bond between the phosphorus atom and the oxygen atom at C-1 of the glycerol moiety; *other names*: lipophosphodiesterase I; lecithinase C; *Clostridium welchii* α-toxin. Two major types have been studied. both occur in bacteria and animals: (1) one type acts on phosphatidylcholine, releasing diacylglycerol (EC 3.1.4.3); (2) the other is **1-phosphatidylinositol phosphodiesterase** (EC 3.1.4.10), which acts on phosphoinositides; it catalyses the conversion of 1-phosphatidyl-D-*myo*-inositol to D-*myo*-inositol 1,2-cyclic phosphate (which is hydrolysed in animals but not bacteria to inositol 1-phosphate) and diacylglycerol. **1-Phosphatidylinositol-4,5-bisphosphate phosphodiesterase** (EC 3.1.4.11; *other names*: triphosphoinositide phosphodiesterase; phosphoinositide-specific phospholipase C) catalyses the hydrolysis of 1-phosphatidyl-D-*myo*-inositol 4,5-bisphosphate to D-*myo*-inositol 1,4,5-trisphosphate and diacylglycerol. These enzymes play a role in signal transduction. There are at least six forms of the latter, which is required for production of the second messenger molecules **diacylglycerol** and **inositol 1,4,5-trisphosphate**. The activation involves a G_α **G-protein** subunit.

Phospholipase D EC 3.1.4.4; *the recommended name for* any phosphoric diester hydrolase that splits the bond between the phosphorus atom of a glycerophospholipid and the oxygen atom of the nitrogenous base; *other names*: lipophosphodiesterase II; lecithinase D; choline phosphatase.

phospholipid *or (formerly)* **phospholipide** any lipid containing phosphoric acid as mono- or diester.

phospholipid–cholesterol acyltransferase *see* **phosphatidylcholine–sterol O-acyltransferase**.

phospholipid exchange protein any of a number of cytosolic proteins that bind phospholipids, and transport them within the cell. Each type of protein exhibits a degree of specificity for the phospholipid bound, and each protein molecule carries one or two phospholipid molecules. These proteins function to exchange phospholipid classes between different intracellular membrane sites.

phospholipid-hydroperoxide glutathione peroxidase an enzyme, EC 1.11.1.12, containing selenocysteine, which catalyses the reduction of lipid hydroperoxides, such as those produced by lipoxygenase.

phospholipid transfer protein *abbr.*: PLTP; a plasma protein (human: 476 amino acids) that mediates transfer of phospholipid and cholesterol from triacylglycerol-rich lipoproteins to high-density lipoproteins (HDL). It is expressed in many tissues (including liver) and is homologous with cholesterol ester transfer protein (CETP) and lipopolysaccharide-binding protein.

phospholipid translocase a protein that accelerates the transfer of phospholipid (*see* **flip-flop**) between the inner and outer lipid leaflets of the plasma membrane bilayer.

phosphomannomutase EC 5.4.2.8; *systematic name*: D-mannose 1,6-phosphomutase; an enzyme that catalyses the reaction:

D-mannose 1-phosphate = D-mannose 6-phosphate.

One of its functions is to convert mannose 6-phosphate to mannose 1-phosphate, from which GDPmannose is synthesized; this is a precursor of **fucose**, and is an intermediate in the synthesis of the **mannose** core of oligosaccharides.

phosphomonoester any monoester of phosphoric acid.

phosphomonoesterase 1 *an alternative name for* **phosphatase**. **2** *an alternative name for* either **acid phosphatase** or **alkaline phosphatase**.

phosphonic acid *see* **phosphono+**.

phosphono+ *comb. form* denoting a phosphonic residue, –PO(OH)$_2$, derived from **phosphonic acid**, HPO(OH)$_2$. In biochemical usage this name is usually reserved for the –PO(OH)$_2$ group attached to a carbon atom, as in **phosphonoacetate**, (HO)$_2$OP–CH$_2$COO⁻; and **2-phosphonoethylamine**, or **ciliatine**, (HO)$_2$OP–CH$_2$CH$_2$NH$_2$; the group is commonly termed phospho+ when linked to a heteroatom, as in phosphocreatine and phosphogluconate.

N$^\omega$-phosphonoarginine *see* **phosphoarginine**.

N$^\omega$-phosphonocreatine *see* **phosphocreatine**.

phosphonomycin *or* **fosfomycin** (1*R*,2*S*)-(1,2-epoxypropyl)phosphonate; an antibiotic produced by *Streptomyces* and *Pseudomonas* strains. It binds irreversibly to UDP-*N*-acetylglucosamine 1-carboxyvinyl transferase (EC 2.5.1.7), an enzyme catalysing the first stage in the synthesis of bacterial cell-wall peptidoglycan from UDP-*N*-acetylglucosamine, which it thereby inhibits. It penetrates the cell by utilizing the L-α-glycerophosphate transport system. Loss of this system is a primary reason that resistant strains emerge.

phosphonopeptide any of a series of oligopeptide derivatives, designed as inhibitors of bacterial cell-wall biosynthesis, that have a C-terminal residue of L-1-aminoethylphosphonic acid (a phosphorus-containing analogue of L-alanine).

phosphopentomutase EC 5.4.2.7; *systematic name*: D-ribose 1,5-phosphomutase; *other name*: phosphodeoxyribomutase. An enzyme that catalyses the reaction:

D-ribose 1-phosphate = D-ribose 5-phosphate.

phosphoprotein any protein containing one or more phosphoric residues directly attached, usually by ester linkage, to amino-acid residues.

phosphoprotein phosphatase 1 EC 3.1.3.16; *systematic name*: phosphoprotein phosphohydrolase; any enzyme that removes a phosphate group from a protein by hydrolysis. Many different enzymes have been identified with varying specificity with respect to the primary structure of the substrate protein. *See* **protein phosphatase**. **2** *another name for* **serine/threonine-specific protein phosphatase**.

phosphor any substance (capable of) exhibiting **phosphorescence** (def. 1).

phosphor+ *a variant form of* **phosphoro+** (before a vowel).

phosphoramidate *or* **amidophosphate** *or* **phosphoric amide 1** any ion, salt, or ester of phosphoramidic acid; amidophosphoric acid; (OH)$_2$PO–NH$_2$. **2** any compound containing the group (OH)$_2$PO–NH– or (OH)$_2$PO–N<.

phosphoramidite any compound having the general structure R$_3$R$_4$NP(R$_2$)–O–R$_1$. Phosphoramidites are useful in the synthesis of oligonucleotides or for the addition of groups to the 5′-phosphate of nucleotides or oligonucleotides. R$_1$ is often cyanoethyl-; R$_3$ and R$_4$ are often isopropyl with R$_2$ in 3′-nucleotide linkage.

phosphoramidon *N*-(α-L-rhamnopyranosyl-oxyphospho)-L-leucyl-L-tryptophan; a specific inhibitor of **thermolysin**. It is an inhibitor of the metalloendopeptidase, EC 3.4.24.11, that is responsible for cleaving and inactivating **natriuretic peptide**, with corresponding pharmacological effects.

phosphoresce to exhibit **phosphorescence** (def. 1).

phosphorescence 1 a type of luminescence consisting of the emission by a substance of electromagnetic radiation, especially visible light, as a result of the absorption of energy derived from exciting radiation of shorter wavelength or from incident subatomic particles (especially electrons or alpha particles), and that persists for more than approximately 10 ns after excitation ceases, the persistence being a consequence of the excited electrons decaying via a **metastable** (def. 2) state; the property of emitting such radiation. **2** the radiation so emitted. **3** *a common but erroneous term for* **bioluminescence** (which occurs in the absence of exciting radiation). *Compare* **fluorescence**. —**phosphorescent** *adj*.

phosphorescence spectrophotometry measurement of the wavelengths and the intensities of the light emitted from a phosphorescent sample that is excited by monochromatic (or nearly so) exciting radiation.

phosphori+ *a variant form of* **phosphoro+** (def. 3).

phosphoriboisomerase *see* **ribose-5-phosphate epimerase**.

phosphoribosylamine–glycine ligase EC 6.3.4.13; *other names*: phosphoribosylglycinamide synthetase; glycinamide ribonucleotide synthetase (*abbr.*: GARS); an enzyme of *de novo* purine biosynthesis that catalyses the reaction between ATP, 5′-phosphoribosylamine, and glycine to form 5′-phosphoribosylglycinamide (glycinamide ribotide, *abbr.*: GAR) with release of ADP and orthophosphate. The human protein (1010 amino acids) is multifunctional, with domains for three catalytic activities required for purine biosynthesis (amino-acid numbers in brackets): GARS (1–433), **phosphoribosylformylglycinamidine cyclo-ligase** (434–809), and **phosphoribosylglycinamideformyl-transferase** (810–1010), although variations with only one or two of the domains occur through alternative splicing.

phosphoribosylaminoimidazolecarboxamide formyltransferase EC 2.1.2.3; *other name*: AICAR transformylase; inosinicase; IMP synthetase; an enzyme of *de novo* **purine biosynthesis** that catalyses the formation of 5′-phosphoribosyl-5-formamido-4-imidazolecarboxamide from 10-formyltetrahydrofolate and 5′-phosphoribosyl-5-amino-4-imidazolecarboxamide (5-aminoimidazole-4-carboxamide ribotide; *abbr.*: AICAR) forming also tetrahydrofolate. In chicken the protein (593 amino acids) is multifunctional, having activities for AICAR transformylase and **IMP cyclohydrolase**; the latter resides in the C-terminal region.

phosphoribosylaminoimidazole carboxylase EC 4.1.1.21; *other name*: AIR carboxylase; an enzyme of the pathway for *de novo* purine biosynthesis that catalyses the formation of 1-(5′-phosphoribosyl)-5-amino-4-imidazolecarboxylate from 1-(5′-phosphoribosyl)-5-aminoimidazole (5′-aminoimidazole ribotide, *abbr.*: AIR) and CO_2. The reaction requires neither biotin nor ATP. For an example, *see* **multifunctional protein ade2**.

phosphoribosylaminoimidazolesuccinocarboxamide synthase EC 6.3.2.6; *other name*: SAICAR synthetase; an enzyme that catalyses the formation of 1-(5′-phosphoribosyl)-4-(*N*-succinocarboxamide)-5-aminoimidazole(*N*-succinylo-5-aminoimidazole-4-carboxamide ribonucleotide; *abbr.*: SAICAR) from ATP, 1-(5′-phosphoribosyl)-4-carboxy-5-aminoimidazole and L-aspartate with release of ADP and orthophosphate. The reaction forms part of the pathway for the *de novo* synthesis of purines. For an example, *see* **multifunctional protein ade2**.

(5-)phospho(-α-D-)ribosyl diphosphate *symbol*: PPRib*P* or P*Rib*PP; 5-phospho-α-D-ribofuranosyl diphosphate; α-D-ribose 1-diphosphate 5-phosphate; a compound that is essential in the biosynthesis of various nitrogen-containing ring compounds or their derivatives, especially of histidine, and of purines and pyrimidines as their nucleotides. It is commonly but incorrectly known as phosphoribosyl pyrophosphate (*abbr. (not recommended)*: PRPP).

phosphoribosylformylglycinamidine cyclo-ligase EC 6.3.3.1; *other name*: phosphoribosyliminoamidazole synthetase; an enzyme of *de novo* synthesis of purines that catalyses the reaction between ATP and 2-formamido-N^1-(5-phosphoribosyl)acetamidine to form 1-(5-phosphoribosyl)-5-aminoimidazole with release of ADP and orthophosphate.

phosphoribosylformylglycinamidine synthase EC 6.3.5.3; *other names*: phosphoribosylformylglycinamidine synthetase; FGAM synthase; an enzyme of the pathway for *de novo* purine biosynthesis that catalyses the formation of 5′-phosphoribosylformylglycinamidine from ATP, 5′-phosphoribosylformylglycinamide, L-glutamine, and H_2O with release of ADP, orthophosphate, and L-glutamate.

phosphoribosylglycinamide formyltransferase EC 2.1.2.2; *other names*: 5′-phosphoribosylglycinamide transformylase; GAR transformylase (*abbr.*: GART); an enzyme of *de novo* purine biosynthesis that catalyses the formation of 5′-phosphoribosyl-*N*-formylglycinamide (formylglycinamide ribotide; *abbr.*: FGAR) from 10-formyltetrahydrofolate, 5′-phosphoribosylglycinamide (glycinamide ribotide, *abbr.*: GAR), and tetrahydrofolate. In chicken there is a multifunctional protein with activities for **phosphoribosylamine–glycine ligase** (GARS)/**phosphoribosylformylglycinamidine cyclo-ligase** (AIRS)/GART.

phosphoribosyl pyrophosphate *abbr.*: PRPP; *a common name for* **(5-)phospho(-α-D-)ribosyl diphosphate**.

phosphoribosyl pyrophosphate synthetase (*abbr.* PRS) *see* **ribose-phosphate pyrophosphokinase**.

phosphoribosyltransferase EC 2.4.2.7–2.4.2.12, 2.4.2.17; any enzyme that catalyses the biosynthesis of a mononucleotide from 5-phospho-α-D-ribosyl diphosphate (PRPP) in a **purine salvage pathway** for nucleoside biosynthesis. Phosphoribosyltransferases catalyse a reaction between a base and 5-phospho-α-D-ribosyl diphosphate to form a D-ribonucleotide and pyrophosphate; the base may be (number indicates no. in sub-subclass EC 2.4.2) adenine (7), hypoxanthine (8), uracil (9), orotate (10), nicotinate (11), nicotinamide (12), or ATP (17). Examples include: hypoxanthine/guanine phosphoribosyltransferase (**HGPRT**) from human; and ATP phosphoribosyltransferase, involved in histidine biosynthesis, from *Escherichia coli*. Other phosphoribosyltransferases of sub-subclass EC 2.4.2 involve the following (no. in sub-subclass EC 2.4.2): thranilate (18), dioxotetrahydropyrimidine (20), and xanthine (22).

phosphoric *symbol*: *P* or (in the one-letter convention for nucleotides) p; *the general name for* a residue of [ortho]phosphoric acid – i.e. tetraoxophosphoric(v) acid, $PO(OH)_3$ – whether singly, doubly, or triply attached to other groups and whether or not any residual hydroxyl groups are deprotonated.

phosphorimaging a technique for recording a radioactive or chemiluminescent image such as is produced in Southern, Northern, or Western blotting. The image is captured electronically and can be viewed on a computer screen. It is ten times more sensitive than using X-ray film for autoradiography.

phosphorimetry the measurement of **phosphorescence**.

phosphoro+ *or (before a vowel)* **phosphor+** *comb. form* **1** phosphorus. **2** [ortho]phosphoric acid. **3** *or* **phosphori+** phosphorescence.

phosphoroclastic describing a chemical reaction in which a carbon–carbon bond is split apparently by the action of inorganic phosphate with attachment of a phosphoric group to one of the resulting fragments.

phosphorolysis any reversible lytic process or reaction undergone by an acyl compound (I) or glycosyl compound (II) in which nucleophilic displacement by phosphate (anion) occurs, with formation of the corresponding acyl or sugar phosphate and uptake of a hydron by the other moiety of the original compound, as in the generalized equations:

$$(I) \ R\text{–}CO\text{–}OR' + PO_4H_2^- + H^+ = R\text{–}CO\text{–}O\text{–}PO_3H_2 + HOR'$$

or

$$(II) \ R\text{–}O\text{–}R' + PO_4H_2^- + H^+ = R\text{–}O\text{–}PO_3H_2 + HOR'.$$

phosphorolytic of, or relating to, phosphorolysis; capable of undergoing phosphorolysis.

phosphorothiolate linkage an analogue of a phosphodiester linkage between nucleotides in which one of the nonbridging oxygens is replaced with a sulphur atom. DNA strands containing phosphorothiolate linkages can be more resistant to the action of nucleases.

phosphorous 1 of, containing, or resembling phosphorus. **2** a residue of phosphorous acid – i.e. trioxophosphoric(iii) acid, $P(OH)_3$ – whether singly or doubly attached to other groups and whether or not any residual hydroxyl groups are deprotonated.

phosphorus *symbol*: P; a nonmetallic element of group 15 of the (IUPAC) **periodic table**; atomic number 15; relative atomic mass 30.974; it exists naturally as a single stable nuclide, **phosphorus-31**. Phosphorus occurs only in the combined state, almost exclusively as phosphates. Its commonest covalencies are iii and v. Phosphorus compounds are vital components of all living materials. *See also* **phosphorus-32, phosphorus-33**.

phosphorus-31 the stable nuclide $^{31}_{15}P$; its relative abundance in natural phosphorus is 100 atom percent.

phosphorus-32 the artificial radioactive nuclide $^{32}_{15}P$. It emits a relatively high-energy electron or β^- particle (1.709 MeV max.) and no gamma radiation. It has a half-life of 14.3 days.

phosphorus-33 the artificial radioactive nuclide $^{33}_{15}P$. It emits a moderate-energy electron or β^- particle (0.249 MeV max.) and no gamma radiation. It has a half-life of 25.4 days.

phosphorus:oxygen ratio *abbr.*: P:O ratio; the number of atoms of inorganic phosphorus (as phosphate) incorporated into organic compounds (usually ATP) per atom of dioxygen consumed during **oxidative phosphorylation**. The value observed depends on the nature of the substrate being oxidized and the state of preservation of mitochondrial function. If the latter is high, a P:O ratio near 3 is observed during the oxidation of NADH, near 2 for succinate, fatty acyl CoA, 3-phosphoglycerate. *See also* **chemiosmotic coupling hypothesis, tight coupling**.

phosphoryl 1 the trivalent group –P(O)= derived from phosphoric acid. **2** *a name used frequently for* the univalent group –P(O)(OH)$_2$; this usage is not recommended; a phosphoric ester may be named, e.g., choline *O*-(dihydrogenphosphate) or *O*-phosphonocholine. However, phosphoryl is accepted except in derived terms such as the names of enzymes (e.g. phosphorylase) or processes (e.g. phosphorylation). *See also* **phosphono+**.

phosphorylase 1 *the name recommended for* any enzyme, classified as EC 2.4.1.1, that catalyses the phosphorolytic removal of the nonreducing terminal glucose residue from a glucan according to the general reaction:

Glcα1-[4Glcα1-]$_n$ 4Glc + orthophosphate = Glcα1-[4Glcα1-]$_{n-1}$-4Glc + Glcα1-phosphate.

These enzymes vary somewhat in regard to their preferred substrate; the name should be qualified in each instance by prefixing to it the name of the natural substrate (e.g. starch phosphorylase). In humans, glycogen phosphorylase exists as three isoenzymes – of liver, muscle, and brain – all of which contain pyridoxal phosphate. There are two forms: **phosphorylase** *a*, the more active, phosphorylated form; and **phosphorylase** *b* (or dephosphophosphorylase), the less active, unphosphorylated form. The conversion of *b* to *a* is effected with ATP by the enzyme **phosphorylase kinase**, and the conversion of *a* to *b* is brought about by the enzyme **phosphorylase phosphatase** (EC 3.1.3.17). Muscle phosphorylase deficiency results in glycogen disease type V (McArdle disease), and liver phosphorylase deficiency results in glycogen disease type VI (Hers disease). **2** *a generic name recommended for* those glycosyltransferase enzymes within subclass EC 2.4 that utilize orthophosphate as acceptor. Some appear to be specific, e.g. maltose phosphorylase (EC 2.4.1.8) or uridine phosphorylase (EC 2.4.2.3), others less so, e.g. 1,3-β-oligoglucan phosphorylase (EC 2.4.1.30). *See also* **α-glucan phosphorylase**.

phosphorylase kinase *see* glycogen phosphorylase kinase.

phosphorylase-limit dextrin the polysaccharide produced by the action of starch phosphorylase on the branched polysaccharide **amylopectin**.

phosphorylase phosphatase EC 3.1.3.17; an enzyme that catalyses the hydrolysis of phosphorylase *a* to two molecules of the less active phosphorylase *b* and four of orthophosphate. *See also* **protein phosphatase**.

phosphorylate to effect **phosphorylation**.

phosphorylation the act or process of introducing a **phosphoric** group into a molecule, usually with the formation of a phosphoric ester, a phosphoric anhydride, or a phosphoric amide. Biochemical

phosphorylation reactions are important in the trapping of energy, the formation of biosynthetic intermediates during metabolic processes, and in the control of numerous enzymes and other proteins. *See also* **oxidative phosphorylation, photophosphorylation, substrate-level phosphorylation**.

phosphorylation state ratio *symbol*: R_p; a measure of the phosphorylating power of the adenylate system within cells. It is given by:

$$R_p = [ATP]/[ADP][P_i].$$

Compare **(adenylate) energy charge**.

phosphorylcholine–glyceride transferase *see* **diacylglycerol cholinephosphotransferase**.

phosphorylcholine transferase *see* **cholinephosphate cytidylyltransferase**.

phosphorylethanolamine transferase *see* **ethanolamine-phosphate cytidylyltransferase**.

phosphoryn the major phosphate-rich protein of **dentine**. It contains about 40 residue % of phosphoserine and 35 residue % of aspartic acid. It is isolated from incisors, and is involved in bone mineralization. Soluble forms, and a form complexed with collagen in dentine, are known.

phosphoserine *O*-phospho-L-serine; a phosphoamino acid found in peptide linkage in proteins. Phosphorylation of serine and threonine residues is catalysed by protein kinases specific for serine or threonine residues, and phosphorylation and dephosphorylation mechanisms have a regulatory role in the protein function. *See also* **protein kinase, protein phosphatase**.

3-phosphoshikimate 1-carboxyvinyltransferase EC 2.5.1.19; *systematic name*: phosphoenolpyruvate:3-phosphoshikimate 5-*O*-(1-carboxyvinyl)-transferase; *other name*: 5-enolpyruvyl shikimate-3-phosphate (EPSP) synthase; an enzyme of the **shikimate pathway** that catalyses the formation of O^5-(1-carboxyvinyl)-3-phosphoshikimate (5-enolpyruvyl shikimate 5-phosphate) from phosphoenolpyruvate and 3-phosphoshikimate with release of orthophosphate. *See* **pentafunctional arom polypeptide**.

phosphosphingolipid *see* **sphingolipid**.

phosphothreonine *O*-phospho-L-threonine; a phosphoamino acid found in peptide linkage in proteins. Phosphorylation of serine and threonine residues is catalysed by protein kinases specific for serine or threonine residues, and phosphorylation and dephosphorylation mechanisms have a regulatory role in the protein function. *See also* **protein kinase, protein phosphatase**.

phosphotransferase any enzyme of sub-subclasses EC 2.7.1 to 2.7.4, and EC 2.7.9; phosphotransferases catalyse transfer of a phosphate group from a donor to an acceptor, and are classified according to whether the transfer is to an alcohol group (EC 2.7.1), a carboxyl group (EC 2.7.2), a nitrogenous group (EC 2.7.3), a phosphate group (EC 2.7.4) or to paired acceptors (EC 2.7.9 – only two of these are currently classified, transferring phosphate from ATP to both pyruvate and orthophosphate in one case and to pyruvate and water in the other, yielding AMP in both instances). Enzymes of sub-subclass EC 2.7.6 that transfer diphosphate are classified as **diphosphotransferases**.

phosphotransferase system a system for transferring substrates into bacteria (*see* **phosphoenolpyruvate-dependent sugar phosphotransferase system**).

phosphotriester any triester of phosphoric acid.

phosphotriester method (of oligonucleotide synthesis) a method for the chemical synthesis of oligonucleotides in which a suitably protected phosphodiester of general formula Nuc–3′–O–P(O)(OR)OH, where R is an aryl group and Nuc is a residue of an appropriate nucleoside, is first condensed with the 5′-hydroxyl group of another nucleoside (usually anchored through its 3′-hydroxyl group to an insoluble support) to form a phosphotriester. Sequential addition of further nucleoside units is then carried out in the same way, the 5′-protecting group being removed at each stage before the next cycle of reactions. All remaining protecting groups are removed at the completion of the synthesis.

phosphotyrosine *O*-phospho-L-tyrosine; a phosphoamino acid found in peptide linkage in proteins, especially in proteins involved in effector mechanisms in growth stimulation, including growth-factor receptors. *See also* **protein kinase**, **protein phosphatase**.

phosphotyrosine antibodies *or* **antiphosphotyrosine antibodies** antibodies used in cell-signalling research to detect proteins in which tyrosines are phosphorylated by the action of tyrosine kinases.

phosphovitin *or* **phosvitin** a phosphoprotein (subunit size 25 kDa) that, together with **lipovitellin**, constitutes the major part of the mass of the egg-yolk proteins in oviparous vertebrates. It has the highest phosphorus content of all known phosphoproteins (10% P), and approximately half of its amino-acid residues are serine, all of which are *O*-phosphorylated. Phosphovitin and lipovitellin are derived from a common precursor protein, **vitellogenin**.

phot the cgs unit of illuminance, equal to one lumen per square centimetre; 1 phot = 10^4 lux.

phot+ *a variant form of* **photo+** (sometimes before a vowel).

photagogikon (*pl.* **photagogika**) the light-emitting compound of any luminescent system regardless of binding of the compound or of its state of ionization. *See also* **photogen**.

photic 1 of, or relating to, light. **2** of, or relating to, the reaction to light of, or the production of light by, living organisms. **3** describing the uppermost zone in a body of water that receives sufficient light to permit photosynthesis. —**photically** *adv.*

photo+ *or* (*sometimes before a vowel*) **phot+** *comb. form* denoting **1** of, pertaining to, produced by, or responding to light. **2** involving a photographic process.

photoactivate to render active or reactive by means of light. —**photoactivation** *n.*; **photoactivatable** *or* **photoactivable** *adj.*

photoactive (*of a substance*) capable of giving a physical or chemical response to light or other electromagnetic radiation.

photoactive yellow protein a blue light photoreceptor, containing FAD as chromophore, implicated in a negative phototactic response in photosynthetic bacteria.

photoaddition any chemical addition reaction promoted by light or other electromagnetic radiation.

photoaffinity label *or* **photolabel** an **affinity label** carrying a separate photoactivatable group that on illumination reacts with the protein being labelled to form a covalent link.

photoautotroph *or* **photolithotroph** any organism that is both an **autotroph** and a **phototroph**. —**photoautotrophic** *adj.*; **photoautotrophy** *n.*

photobiochemistry the photochemistry of biochemical processes. —**photobiochemical** *adj.*; **photobiochemically** *adv.*; **photobiochemist** *n.*

photobiology the study of the effects of radiation, mainly of visible and ultraviolet light, on living organisms. —**photobiological** *adj.*; **photobiologist** *n.*

photobleaching the loss of colour or fluorescence through the action of incident visible or near-ultraviolet radiation.

photocatalysis 1 the acceleration of a chemical reaction by light. **2** the catalysis of a photochemical reaction. —**photocatalytic** *adj.*; **photocatalytically** *adv.*

photocathode (*in an electronic device*) a cathode from which electrons are emitted when exposed to electromagnetic radiation of suitable frequency.

photocell *or* **photoelectric cell** any type of transducer capable of converting incident electromagnetic radiation to an electric signal of corresponding magnitude. It may function by generating a potential difference or a current, or by changing its conductance.

photochemical action spectrum the relationship between the efficiency of a **photochemical reaction** and the frequency of the illuminating light.

photochemical reaction any chemical reaction that is initiated or accelerated by the absorption of visible or near-ultraviolet radiation.

photochemistry the branch of chemistry concerned with the effect of electromagnetic, especially visible and ultraviolet, radiation on chemical processes and with the uptake or production of photons during chemical reactions. —**photochemical** *adj.*; **photochemically** *adv.*

photochromism *or* **phototropy** the property possessed by certain substances of changing colour reversibly when exposed to electro-magnetic radiation of appropriate frequency. *Compare* **electrochromism**, **thermochromism**. —**photochromic** *adj.*

photoconductive describing an electrical device that is capable of changing its conductance on exposure to light or other electromagnetic radiation. —**photoconductivity** *n.*

photodamage deleterious chemical change caused by the action of light, especially ultraviolet light.

photodensitometer *see* **densitometer** (def. 1).

photodetector a device, e.g. a **photocell**, that responds to incident light.

photodimerization *or* **photodimerisation** the formation of dimers by a photochemical reaction. It is of particular importance in the formation of **pyrimidine dimers** in DNA. —**photodimerize** *or* **photodimerise** *vb.*

photodiode any semiconductor **diode** (def. 1) that generates a potential difference or changes its conductivity when exposed to light.

photodissociation the splitting of a molecule into smaller molecules, radicals, or atoms on absorption of a quantum of electromagnetic radiation.

photodynamic involving, causing, or promoting a deleterious reaction to light, particularly ultraviolet light. Such reactions usually involve oxidation. Damage is caused by **singlet oxygen** produced, e.g., by illumination of chlorophyll in chloroplasts. Normally the singlet oxygen is quenched by the carotenoid pigments coexisting with the chlorophylls in chloroplasts. Hence, any mutation that deletes carotenoids from chloroplasts is lethal, because the chlorophylls are destroyed by singlet oxygen.

photoelectric of, pertaining to, or concerned with electric or electronic effects caused by electromagnetic radiation, especially light.

photoelectric cell *an alternative name for* **photocell**.

photoelectron an electron that is ejected from an atom on interaction of the atom with a photon of sufficient energy.

photoelectron micrograph any photograph of an object taken by **photoelectron microscopy**.

photoelectron microscopy an electron-microscopic technique in which **photoelectrons** emitted from a specimen exposed to ultraviolet radiation are accelerated and focused to produce an image. It is the counterpart in the electron microscope of fluorescence microscopy in the light microscope, the resolution being correspondingly enhanced.

photoelectron spectroscopy a technique in which a specimen (a gas sample or the surface of a solid or a liquid) is exposed to monochromatic electromagnetic radiation in the ultraviolet or X-ray regions and the energy spectrum of the emitted **photoelectrons** recorded. This provides information about the atomic composition of the specimen and also about the types of orbitals to which the constituent electrons of its atoms and molecules are assigned.

photoemission the emission of electrons from a surface as a result of bombardment by photons.

photoexcitation the act or process of raising an atom or molecule to a higher energy state by interaction with a photon.

photogen 1 *an older term for* **photagogikon**. **2** *a term proposed for* the specific biochemical precursor of a photagogikon.

photographic emulsion *or* **emulsion** any medium that is sensitive to light or other electromagnetic radiation, used for preparing photographic film, paper, etc., and consisting usually of a suspension of a silver salt (especially the bromide or chloride) in gelatin; a coating of such a suspension applied to a substrate.

photoheterotroph *or* **photoorganotroph** any organism that is both a **heterotroph** and a **phototroph**. —**photoheterotrophic** *adj.*; **photoheterotrophy** *n.*

photoinactivation the irreversible abolition of specific biological activity by the action of light.

photoinhibition the reversible abolition of specific biological activity by the action of light.

photoionization *or* **photoionisation** the ionization of a molecular entity by the loss of an electron on interaction of the entity with a photon of sufficient energy.

photokinesis any change in the undirected movement of an organism in response to variation in light intensity. —**photokinetic** *adj.*

photolabel *an alternative term for* **photoaffinity label**.

photolabile (of a substance) unstable in the presence of light.

photolithotroph *an alternative term for* **photoautotroph**. —**photolithotrophic** *adj.*; **photolithotrophy** *n.*

photoluminescence luminescence as a result of absorption of electromagnetic radiation. The emitted light is always of lower frequency than that of the incident radiation. —**photoluminescent** *adj.*; **photoluminesce** *vb.*

photolyase *see* **photoreactivation**.

photolyse *or (sometimes, US)* **photolyze** to subject to or to undergo **photolysis**.

photolysis the cleavage of one or more covalent bonds in a molecular entity resulting from absorption of energy from light or other electromagnetic radiation; any photochemical process in which such a cleavage is an essential part, e.g. laser photolysis; flash photolysis. —**photolytic** *adj.*; **photolytically** *adv.*

photolyze *a US spelling of* **photolyse**.

photometer any instrument used to measure the intensity of light or other electromagnetic radiation; includes **spectrophotometer**.

photometry the measurement of the intensity of light or other electromagnetic radiation; it includes spectrophotometry (*see* **spectrophotometer**). —**photometric** *adj.*; **photometrically** *adv.*

photomicrograph a photograph taken through a light microscope; it is sometimes, incorrectly, termed a microphotograph. —**photomicrographic** *adj.*; **photomicrographically** *adv.*; **photomicrography** *n.*

photomultiplier a photocell consisting of a **photocathode** linked to an **electron multiplier**. Photons striking the photocathode cause the emission of electrons, which are then amplified by the electron multiplier to give an enhanced signal.

photon *symbol (in nuclear reactions, etc.)*: γ; a quantum of light (or other electromagnetic radiation) considered as a particle with zero rest mass, zero charge, and energy $h\nu$ joule, where h is the Planck constant and ν the frequency of the radiation in hertz.

photon counter a device, consisting of a photomultiplier and an associated scaler, that is capable of responding to and recording individual photons.

photonicking photolysis of a nucleic-acid molecule with production of a **nick**.

photoorganotroph *an alternative term for* **photoheterotroph**. —**photoorganotrophic** *adj.*; **photoorganotrophy** *n.*

photooxidize *or* **photooxidise** to oxidize by the use of, or as a result of, radiation in the visible and ultraviolet wavelengths. —**photooxidizable** *or* **photooxidisable** *adj.*

photoperiod 1 the length of daylight in a 24-hour period. 2 the duration of daily exposure of an organism to illumination; the length of this period that favours optimum functioning. —**photoperiodic** *adj.*; **photoperiodically** *adv.*

photoperiodism *or* **photoperiodicity** the response of an organism to the relative durations of light and dark in a 24-hour period.

photophore any light-emitting organ of an animal.

photophosphorylation *or* **photosynthetic photophosphorylation** the metabolic processes by which photosynthetic organisms use light energy to convert ADP to ATP without the concomitant reduction of dioxygen to water that occurs in oxidative phosphorylation (*see* **photosynthesis**). There are two distinct electron transport mechanisms: in **noncyclic photophosphorylation**, which involves both photosystems I and II, ATP synthesis is linked to the transport of electrons from water to NADP$^+$, with production of NADPH and dioxygen; in **cyclic photophosphorylation**, which involves only photosystem I, ATP synthesis is driven by a proton gradient generated across the thylakoid membrane.

photopia adjustment of the eye to vision in light of moderate or high intensity, permitting perception of colour; it is considered to involve mainly the **cone cells** of the retina. *Compare* **scotopia**. —**photopic** *adj.*

photopigment any pigment whose chemical state depends on the intensity of the light falling on it.

photopolymerize *or* **photopolymerise** to effect polymerization by the action of light. —**photopolymerization** *or* **photopolymerisation** *n.*

(6–4)photoproduct a type of **pyrimidine dimer**.

photoprotein any protein actively involved in the emission of light by a living organism.

photopsin an **opsin** that is the apoprotein of an **iodopsin**.

photoreactivating enzyme *see* **DNA photolyase**, **photoreactivation**.

photoreactivation the process by which genetic damage caused to organisms by ultraviolet radiation can be reversed by subsequent illumination with visible or near-ultraviolet light. The pyrimidine dimers formed in DNA by the action of ultraviolet light bind **DNA photolyase**, which absorbs light, maximally at 380 nm, and catalyses the restoration of two pyrimidine residues in the DNA.

photoreactive (of a substance or reagent) reactive in a photochemical reaction. —**photoreactivity** *n.*

photoreceptor any biological structure that responds to incident electromagnetic radiation, particularly visible light. It especially refers to a cell that, on absorbing light, generates a nerve impulse or impulses.

photoreduction 1 chemical reduction brought about by electromagnetic radiation, especially visible light. 2 a process occurring at low light intensities in certain algae whereby dihydrogen functions in place of water as an electron donor for the reduction of carbon dioxide and oxygen is not evolved; at higher light intensities normal photosynthesis takes place. 3 reduction in size photographically.

photorespiration a light-dependent catabolic process occurring concomitantly with photosynthesis in plants (especially **C$_3$ plants**) whereby dioxygen is consumed and carbon dioxide is evolved. The substrate is glycolate formed in large quantities in chloroplasts from 2-phosphoglycolate generated from ribulose 1,5-bisphosphate by the action of **ribulose-bisphosphate carboxylase**; the glycolate enters the peroxisomes where it is converted by **glycolate oxidase** to glyoxylate which undergoes transamination to glycine. This then passes into the mitochondria where it is decarboxylated. C$_3$ plants may lose 20–40% of photosynthetically fixed carbon by photorespiration, whereas **C$_4$ plants** lose relatively little. C$_4$ plants defeat photorespiration in two ways. First, they recapture released CO$_2$ in the cytoplasm using **phospho*enol*pyruvate carboxylase** which has a very high affinity for CO$_2$. Although this does not directly affect photorespiration, it prevents loss of CO$_2$. Second, they concentrate CO$_2$ near ribulose-biphosphate carboxylase, and this increases the CO$_2$:O$_2$ ratio, which has the effect of inhibiting the oxygenase activity. —**photorespiratory** *adj.*

photosensitivity 1 the property of being reactive or unusually reactive to incident photons, especially to those of visible light. 2 *(in medicine)* an abnormal, local or generalized reaction to sunlight, due to the accumulation in the skin of **photodynamic** substances, either endogenous or exogenous. —**photosensitive** *adj.*

photosensitize *or* **photosensitise** 1 to render photosensitive; *see* **photosensitivity**. 2 to render a target molecule more reactive (e.g. more easily oxidized) by the addition to it of a second type of molecule, often a dye, that absorbs light energy and transfers it to the target molecule. —**photosensitization** *or* **photosensitisation** *n.*

photosensitizer *or* **photosensitiser** any substance that photosensitizes or is photodynamic.

photosuicide inhibitor a type of **suicide inhibitor** that, after reaction with an enzyme or other appropriate protein, is capable of being selectively photodecomposed to a highly reactive species that can irreversibly modify the protein in the neighbourhood of its active site.

photosynthate any product of the dark reactions of photosynthesis.

photosynthesis 1 *(in biology)* the synthesis by organisms of organic chemical compounds, esp. carbohydrates, from carbon dioxide using energy obtained from light rather than from the oxidation of chemical compounds (*compare* **chemosynthesis**). In higher plants, green algae, and cyanobacteria, water is the electron donor and dioxygen is evolved; in other photosynthetic bacteria, however, various other simple substances (e.g. H$_2$S) may act as electron donors, dioxygen is not formed, and the photosynthetic pathway is simpler. Photosynthesis comprises two separate processes: the light reactions and the dark reactions of photosynthesis. The system in higher plants, green algae, and cyanobacteria brings about an overall reaction in which electrons are released from water molecules and are transferred through complexes of chlorophylls, cytochromes, proteins, and small molecules to NADP$^+$, reducing it to NADPH. Protons are released from water and form a chemiosmotic gradient that results in the formation of ATP by a process similar to that operating in mitochondria (*see* **chemiosmotic coupling hypothesis**). These are the **light reactions**. The NADPH and ATP formed by these reactions then drive the **dark reactions** i.e. the **reductive pentose phosphate cycle**. The electrons and protons are released from water by the oxygen-evolving complex (*abbr.*: OEC), and dioxygen is formed in the process. The sequence of reactions involves the OEC, from which electrons pass to photosystem II (*see* **photosystem**), thence to **plastoquinone** reducing it to a quinol which then reduces the cytochrome b_6–f complex. This in turn reduces **plastocyanin** which passes electrons to photosystem I (*see* **photosystem**) which then reduces NADP$^+$ to NADPH. In photosynthetic bacteria, a simpler system exists, in which water is not split, dioxygen is not evolved, and NADP$^+$ is not reduced directly. In this system, photons excite molecules of bacteriochlorophyll b in a **photosynthetic reaction centre** to transfer electrons through a cyclic electron-transfer system. Cyclic electron flow can also occur under some conditions in plants, and involves only photosystem I, electrons being released by photon activation from photosystem I and returned to it through cytochrome b_6–f and plastocyanin, with pumping of protons to generate ATP. Under these conditions there is no reduction of NADP$^+$. **2** (*in chemistry*) any synthesis by a photochemical reaction. —**photosynthetic** *adj.*; **photosynthetically** *adv.*

photosynthetic phosphorylation *an alternative term for* **photophosphorylation**.

photosynthetic quotient a ratio of the rate of dioxygen evolution by a cell or organism capable of photosynthesis to that of carbon dioxide uptake. It is an index of photosynthetic activity relative to respiration (including photorespiration).

photosynthetic reaction centre the photosynthetic complex in bacteria, consisting essentially of a transmembrane protein of three subunits, denoted L, M, and H. Collectively the three subunits generate 11 transmembrane helices. Although the L and M helices (which have binding sites for the many prosthetic groups) are different, the groups themselves are arranged with near twofold symmetry because of similarities in the folds of the subunits. The L and M subunits bear bacteriochlorophyll, bacteriopheophytin, ubiquinone, and menaquinone molecules. The bacteriochlorophylls will be either a or b types, as will be the bacteriopheophytins. Here we describe a system with b types. Of the four bacteriochlorophyll b molecules, two constitute a 'special pair' with an Mg^{2+}–Mg^{2+} distance of 7 Å and in hydrophobic environments; other prosthetic groups are **bacteriopheophytin** b (two molecules), one nonheme Fe(ii), one **ubiquinone**, and one **menaquinone**. There is cyclic electron flow through a cytochrome bc_1 complex, cytochrome c, and the bacteriochlorophyll–bacteriopheophytin complex, protons being pumped from the bacterial cytosol to the periplasmic space, and then returning to the cytosol through an ATP synthase with synthesis of ATP. The system differs from that in plants in that there is only one type of reaction centre, no splitting of water, and no direct reduction of NADP$^+$. Its essential features are that photons excite electrons in the 'special pair' and these excited electrons then flow through the bacteriochlorophyll–bacteriopheophytin complex, ubiquinone, and menaquinone, dissipating energy in bringing about protein pumping to form the chemiosmotic gradient that results in ATP formation. They then return to the 'special pair' via cytochromes bc_1 and c. This protein (and the complex from *Rhodopseudomonas viridis*) represents the first transmembrane protein to have its 3-D structure determined in atomic detail by X-ray crystallography – by J. Deisenhofer, German biophysicist Hartmut **Michel** (1948–), and German biochemist Robert **Huber** (1937–). *See also* **P870**, **P960**, **photosystem** (def. 2).

photosystem 1 either of the two functionally distinct but cooperating systems, designated photosystem I (*abbr.*: PSI) and photosystem II (*abbr.*: PSII), between which are distributed the various molecules that effect the light reactions of **photosynthesis**; photosystems are located in the **thylakoid** membranes of the chloroplasts of higher plants and green algae or in the free thylakoid membranes of cyanobacteria. Photosystem I consists of a transmembrane multisubunit complex, of about 10 subunits in all, consisting of a heterodimer complex of PsaA and PsaB which binds P700, together with the chlorophyll acceptor A_0, the phylloquinone acceptor A_1, and Fx, a 4Fe–4S centre. PsaC, which is bound to this complex, contains two additional 4Fe–4S centres required for ferredoxin reduction. PsaC of pea is similar to bacterial-type 4Fe–4S **ferredoxins**.

Other subunits are involved in binding soluble ferredoxin, ferredoxin-NADP oxidoreductase and regulation of electron flow. Approximately 100 chlorophyll a molecules are bound to the minimal heterodimer unit. Photosystem II is formed from a heterodimer of the D_1 and D_2 proteins which bind P680, together with the pheophytin acceptor, two plastoquinone acceptors Q_a and Q_b, and a non-heme iron atom. The minimal purified unit contains about six chlorophyll a molecules. **Cytochrome b_{559}** is always associated with PSII. A large number of other subunits including light-harvesting chlorophyll-protein complexes, proteins involved in stabilizing the Mn complex, and regulatory subunits are required to form a functional photosystem II. **2** any functionally analogous but simpler single photosystem present in photosynthetic bacteria. That from *Rhodopseudomonas viridis* contains three hydrophobic subunits, a special pair of bacteriochlorophyll b (the reaction-centre bacteriochlorophyll P960), two molecules of bacteriopheophytin b (*see* **bacteriopheophytin**), a bound iron atom and one molecule each of **ubiquinone** and a **menaquinone**. There are four heme c-type cytochromes in this case (but not in all). The complete structure has been determined by X-ray crystallography. *See also* **photosynthetic reaction centre**.

phototaxis the movement of a motile cell or organism towards (**positive phototaxis**) or away from (**negative phototaxis**) a source of light. —**phototactic** *adj.*

phototoxic able to cause physiological damage by making the skin abnormally sensitive to light. —**phototoxicity** *n*.

phototransistor a type of junction transistor that responds to incident light or other electromagnetic radiation by the generation of an electric current, which is then amplified.

phototroph *or* **phototrophe** any organism whose principal or only source of primary energy is sunlight. The great majority of such organisms are **photoautotrophs** (*or* **photolithotrophs**), which use light energy to synthesize their organic requirements solely from inorganic precursors; these include all green plants and algae, cyanobacteria, and green and purple sulfur bacteria. Relatively few phototrophs are **photoheterotrophs** (*or* **photoorganotrophs**), which require a supply of organic precursors to manufacture their own organic components; examples include certain purple nonsulfur bacteria. *Compare* **chemotroph**. —**phototrophic** *adj.*; **phototrophy** *n*.

phototropic 1 involving or showing **phototropism**. **2** involving or showing phototropy (i.e. **photochromism**).

phototropin a blue-light receptor protein, containing FMN as chromophore, that regulates different aspects of growth and development in plants.

phototropism the orientation or growth of a cell or (a part of) an organism in relation to a source of light. The term is used especially of plants or plant organs.

phototropy *an alternative term for* **photochromism**.

phototube a type of photocell consisting of a vacuum tube or gas-filled tube in which a photocathode responds to incident light or other electromagnetic radiation by emission of electrons.

phox *abbr. for* phagocyte NADPH oxidase.

php *see* programming language.

pH paper absorbent paper, usually in strip form, that is impregnated with a colour-changing pH indicator dye. It is useful for estimating the approximate pH of a solution.

pH profile *see* pH-activity profile.

phragmoplast a structure formed in the developing plant cell wall from the endoplasmic reticulum and spindle fibres. It is observed during **cytokinesis** in plant cells as a cylindrical structure, composed of microtubules, between the two sets of daughter chromosomes. The microtubules are those of the two sets of polar microtubules of the **spindle**. These microtubules transport small Golgi-derived vesicles to the equatorial (or central) region, which fuse there to form the early **cell plate**, which then grows out to join up with the cell wall.

PHRAP *see* sequence assembly software.

PHRED *see* sequence assembly software.

phrenosin a **cerebroside** in which the acyl constituent is cerebronate. *See* **lignocerate**, of which cerebronate is the 2-hydroxy derivative.

PHRP *abbr. for* parathyroid hormone-related protein.

pH standard any of a set of six standard solutions which may be used as primary pH standards; they are: potassium hydrogen tartrate, saturated at 25°C, pH 3.557 at 25°C; 0.1 mol/kg potassium dihydrogen citrate, pH 3.776 at 25°C; 0.025 mol/kg KH_2PO_4 + 0.025 mol/kg Na_2HPO_4, pH 6.865 at 25°C; 0.008695 mol/kg KH_2PO_4 + 0.03043 mol/kg Na_2HPO_4, pH 7.413 at 25°C; 0.01 mol/kg disodium tetraborate, pH 9.180 at 25°C; 0.025 mol/kg $NaHCO_3$ + 0.025 mol/kg Na_2CO_3, pH 10.012 at 24°C.

pH-stat an instrument for maintaining a constant pH in an aqueous liquid mixture in which hydrogen ions are being liberated or taken up. It is frequently arranged to measure the amount of reagent, as a function of time, that has to be added to maintain a predetermined pH.

Pht< or **–Pht–** *symbol for* the (divalent) phthaloyl group.

Pht– *symbol for* the (univalent) phthalyl group.

phthalate 1 *the trivial name for* 1,2-benzene dicarboxylate, the dianion of phthalic acid. **2** any mixture of phthalic acid and its mono- and dianions. **3** any salt or ester of phthalic acid.

phthaloyl *symbol*: Pht< *or* –Pht–; the divalent acyl group, $-OC-C_6H_4-CO-$, derived from phthalic acid.

phthalyl *symbol*: Pht–; the univalent acyl group, $HOOC-C_6H_4-CO-$, derived from phthalic acid.

pH unit the difference between any two successive integral values of pH.

phyco+ *comb. form* denoting seaweed or other algae.

phycobilin any **bilin** found in **phycobiliproteins**. The two most common are **phycocyanobilin** and **phycoerythrobilin**. They are linear tetrapyrroles or **open tetrapyrroles**, i.e. chains of four pyrroles, the terminal pyrroles not being linked to each other to form the cyclic tetrapyrrole structure. The structure of phycocyanobilin is shown; phycoerythrobilin is similar, except that the ethyl group on ring A is $-CH=CH_2$, and the methyne group between rings A and B is a methylene group.

phycobiliprotein *or* **biliprotein** any phycobilin–protein conjugate found in algae of the phyla Cryptophyta and Rhodophyta, and in cyanobacteria. Phycobiliproteins are intensely coloured, fluorescent, water-soluble globular proteins. The two most common types, **phycocyanin** and **phycoerythrin**, are composed of two kinds of subunit, α (10–20 kDa) and β (14–30 kDa), each of which carries about three molecules of covalently bound **phycobilin** (phycocyanobilin in phycocyanin, phycoerythrobilin in phycoerythrin), and is present *in vivo* most commonly as aggregates of formula $(\alpha,\beta)_n$, where n is 3 or 6. Phycocyanin (blue: A_{max} at 618 nm) predominates in Cyanophyta, and phycoerythrin (red: A_{max} at 545 nm) predominates in Rhodophyta. A third type, **allophycocyanin** (pale blue: A_{max} at 650 nm), present in small amounts in both Cyanophyta and Rhodophyta, has only one type of protein subunit (~15.5 kDa), carries one molecule of covalently bound phycocyanobilin, and is present *in vivo* as aggregates of six subunits. It is a heterodimer of α and β chains. See also **phycobilisome**.

phycobilisome any of the granules, approximately 32 nm × 48 nm, consisting of highly aggregated **phycobiliproteins**, that are attached in arrays to the external face of a **thylakoid** membrane in algae of the phyla Cyanophyta and Rhodophyta, where they function as light-harvesting devices in photosynthesis. Excitation energy in the phycobilisome flows in the sequence:

phycoerythrin → phycocyanin → allophycocyanin,

whence it passes to the antenna chlorophyll of photosystem II. In Cyanophyta it is also known as a **cyanosome**.

phycocyanin a blue phycobilin; *see* **phycobiliprotein**.

phycocyanobilin a type of **phycobilin**; *see also* **phycobiliprotein**.

phycoerythrin a red phycobilin; *see* **phycobiliprotein**.

phycoerythrobilin a type of **phycobilin**; *see also* **phycobiliprotein**.

phycofluor any artificial conjugate of a **phycobiliprotein** with a molecule conferring biological specificity (e.g., avidin, immunoglobulin, protein A). Such conjugates are useful in fluorescence immunoassay and fluorescence-activated cell sorting, by virtue of the distinctive and intense fluorescence of the phycobiliprotein moiety.

phycotoxin any poisonous substance produced by algae.

phyl+ *a variant form of* **phylo+** (sometimes before a vowel).

phyla the plural of **phylum**.

PHYLIP *abbr. for* phylogeny inference package; *see* **phylogeny package**.

phyllo+ *or (before a vowel)* **phyll+** *comb. form* denoting leaf.

phyllocerulein *or (esp. Brit.)* **phyllocaerulein** *see* **cerulein**.

phylloquinone *abbr.*: K; *the recommended name for* vitamin K_1; 2-methyl-3-phytyl-1,4-naphthoquinone. See **vitamin K**.

phylo+ *or (sometimes before a vowel)* **phyl+** *comb. form* denoting race or tribe.

phylogenetic analysis study of the evolutionary relationships between genes, populations, species, etc., usually by constructing **phylogenetic trees**. Phylogenetic analysis may also be applied to individual characters, in order to chart the evolution of particular traits or conditions.

phylogenetic tree a graphical representation of the (putative) evolutionary relationships of a group of organisms.

phylogenomic analysis the application of phylogenetic analysis methods at the level of whole genomes. It is used, for example, to chart the evolutionary history of individual genes (and their products), of genome segments, or of complete genomes.

phylogeny the evolutionary history of an organism or group of related organisms. In biochemistry, such evolutionary relationships are based on similarities and differences in nucleic acid or protein sequences. —**phylogenetic** *or* **phylogenic** *adj*.

phylogeny package a software suite for phylogenetic analysis, which typically includes parsimony, maximum-likelihood, and distance methods, together with various tree-drawing tools. Examples include PAUP (phylogenetic analysis using parsimony) and PHYLIP (phylogeny inference package).

phylogram a phylogenetic tree that indicates the relationships between taxa and also conveys a sense of time or rate of evolution.

phylum (*pl.* **phyla**) a primary taxon of animals or plants (in plants also termed division). It consists of a number of classes (or sometimes a single class).

physalaemin Glp-Ala-Asp-Pro-Asn-Lys-Phe-Tyr-Gly-Leu-Met-NH_2; an 11-residue peptide amide **tachykinin** from the skin of the amphibian *Physalaemus fuscumaculatus*. It is more potent than **eledoisin** and **substance P** in some assay systems, and is useful in the study of receptors for substance P.

physeteric acid *see* tetradecenoic acid.

physical 1 of, or pertaining to, matter and energy; of material things. **2** of, or relating to, natural science, especially of nonliving systems. *Compare* **biological**. **3** having reference to or used in physics. —**physically** *adv*.

physical adsorption *an alternative term for* **physisorption**.

physical biochemistry the application of physical chemistry to biochemical substances, structures, and systems. Often used synonymously with **biophysical chemistry**.

physical chemistry the branch of chemistry concerned with the relationship between the physical properties of substances and their chemical properties, reactions, and structures.

physical map a map that identifies the physical location of genes on a chromosome, in contrast to a **linkage map**, which is generated by measuring recombination frequencies. Physical length can be measured in base pairs, kilobases, or megabases. Such maps are obtained using a variety of methods, e.g. pulsed field gel electrophoresis, fluorescence *in situ* hybridization, contig mapping using polymorphic markers, and by sequencing entire genomes.

physico+ *comb. form* denoting physical or of physics.

physicochemical of, pertaining to, or using the methods of physical chemistry. —**physicochemically** *adv*.

physics *functioning as sing.* the branch of science concerned with the properties of matter and of energy, and with the interactions and interconversions between matter and energy.

physin any of various membrane proteins that have similar architecture to, and are homologous with, the **gyrin** family. They occur in several types of transport vesicle. Examples include pantophysin, synaptophysin, and synaptobrevin.

physiological 1 of, pertaining to, or used in physiology. **2 normal** (def. 4); not **pathological** (def. 2) or **pharmacological** (def. 2). —**physiologically** *adv*.

physiological antagonist a drug that produces the opposite effect to a specified agonist by an action that is unrelated to that of the agonist.

physiological chemistry *an older term for* **biochemistry** (especially that of humans and higher animals).

physiological conditions environmental conditions simulating those under which the (normal) functions of a cell, organ, or tissue can be expressed.

physiological saline *or (sometimes)* **normal saline** a (sterile) solution of 0.9% w/v (0.154 mol L^{-1}) sodium chloride in water. It is approximately isotonic with mammalian blood and lymph, and is useful in the temporary maintenance of living cells and tissues, and as a vehicle for injections.

physiology the science dealing with the functioning of cells, tissues, organs, and organisms, and with the chemical and physical phenomena concerned.

physisorption *or* **physical adsorption** *or* **van der Waals adsorption** the type of **adsorption** (def. 1) in which only van der Waals forces are involved and hence in which no significant changes occur in the electronic orbital patterns of the interacting substances. *Compare* **chemisorption**.

physostigmine *or* **eserine** a toxic alkaloid extracted from Calabar beans (*Physostigma venenosum*). It inhibits **cholinesterase** by binding to the anionic site of the enzyme, forming a carbamoyl ester of the serine hydroxyl, which is only slowly hydrolysed. It has been used in West Africa in witch ordeals.

phyt+ *a variant form of* **phyto+** (before a vowel).

phytanic acid 3,7,11,15-tetramethylhexadecoanoic acid; a branched-chain fatty acid that is formed metabolically from **phytol**, a constituent of chlorophyll, by bacteria in the ruminant stomach. In humans it is derived from dairy products, meat, ruminant fat, fish, and free phytol. Its oxidation in animals depends on the **alpha oxidation** pathway, since the methyl group on C-3 (the beta carbon) blocks beta oxidation. After alpha oxidation this methyl group is on C-2 of the product, so that beta oxidation can occur, releasing propionyl-CoA as a product. It accumulates secondarily in various peroxisome biogenesis disorders (e.g. **Zellweger syndrome**) and primarily in **Refsum disease** (classical or adult).

phytanoyl-CoA hydroxylase EC 1.14.11.18; an enzyme of liver peroxisomes that catalyses the reaction:

$$\text{phytanoyl-CoA} + O_2 + \text{2-oxoglutarate} =$$
$$\text{2-hydroxyphytanoyl-CoA} + CO_2 + \text{succinate}.$$

The reaction requires Fe^{2+} and ascorbate, and is the first step in the catabolism of **phytanic acid**. Enzyme deficiency results in classical or adult **Refsum's disease**.

phytase either of the two enzymes, designated 3-phytase (EC 3.1.3.8) and 6-phytase (EC 3.1.3.26), that catalyse the hydrolytic removal of a phosphoric residue from **phytate** at the 3-position or the 6-position, respectively; most commonly it signifies 6-phytase.

phytate any anion of **phytic acid**, any mixture of free phytic acid and its ionized forms, or any salt of phytic acid. It is the major phosphorus compound in plants and is important for phosphate storage in cereal grains, especially as its calcium–magnesium salt, **phytin**. In higher animals phytin binds dietary Ca^{2+} and Mg^{2+}, making them less available.

+phyte *noun suffix* denoting **1** plant (of the habitat or type specified). **2** pathological outgrowth. —**+phytic** *adj.*

phytic acid *the trivial name for myo*-inositol hexakis(dihydrogen phosphate). *See* **phytate**.

phytin *see* **phytate**.

phyto+ *or (before a vowel)* **phyt+** *comb. form* denoting plant.

phytoagglutinin *or (rarely)* **phytohemagglutinin** any plant **lectin**. The majority are glycoproteins, although some, e.g. **concanavalin A**,

contain no carbohydrate. They accumulate mainly in seeds, in cytoplasm of cotyledons and embryonic cells, although they are synthesized in leaves.

phytoalexin any substance with antibiotic activity that is produced by plant tissues in response to infection, especially with a fungus. A variety of substance are regarded as phytoalexins. They include compounds as divergent as **phaseollin**, some sesquiterpenes and long-chain acetylenic alcohols.

phytochelatin any of a group of peptides that bind heavy metals (Cd, Zn, Cu, Pb, Hg) in thiolate coordination complexes, and can be isolated from plant cell suspension cultures. The structure is of the type (γ-glutamyl-cysteinyl)$_n$-glycine, where n is 2 to 11.

phytochelatin synthase EC 2.3.2.15; *recommended name*: γ-Glu-Cys-dipeptidyltransferase; an enzyme that catalyses the reaction:

$$n\text{GSH} = (\gamma\text{-Glu–Cys})_n\text{Gly} + \text{glycine}_{(n-1)}$$

where n = 2–11. It is induced by, and active only in the presence of, heavy metals (e.g. Cd, Cu, Hg, As), and is present in plants, some microorganisms, and the nematode *Caenorhabditis elegans*. Examples: *Arabidopsis thaliana* (485 amino acids); *Schizosaccharomyces pombe* (414 amino acids); *C. elegans* (371 amino acids); and *Dictyostelium discoideum* (626 amino acids). All are rich in cysteine and show 40–50% sequence identity in their N-terminal regions.

phytochemistry the branch of chemistry concerned with plants, especially with **secondary metabolites**.

phytochrome a chromoprotein present in trace amounts in all higher plants, involved in regulating light-dependent processes, such as leaf development and flowering. It consists of a **phycobilin**-related chromophore covalently linked to a polypeptide of ~124 kDa. Phytochrome exists in two forms, P_R (A_{max} at 666 nm) and P_{FR} (A_{max} at 730 nm), which are indefinitely interconvertible on absorption of light in the red and far-red regions of the spectrum, respectively. The spectral changes are the result of *cis–trans* isomerization in the phycobilin. The active form, P_{FR}, plays a central role in light-promoted modulation of gene expression, which is of importance in plant growth and development in every phase of the life cycle; conversion of P_{FR} to P_R cancels these responses. The phytochromes are homodimeric proteins, with a linear tetrapyrrole chromophore covalently linked to a polypeptide.

phytoecdysone *(sometimes)* any **ecdysteroid** found in plants.

phytoene a C_{40} carotenoid, derived by head-to-head combination of two molecules of geranylgeranyldiphosphate catalysed by phytoene synthase within the plastids of leaves of many plant species. It is the precursor of ζ-carotene, lycopene, β-carotene, and various xanthophylls.

phytoferritin a protein, similar to **ferritin** of animals, that occurs almost exclusively in plastids, especially in those of plant storage organs.

phytoglycogen plant glycogen; a highly branched polymer of D-glucose, with a structure similar to that of animal **glycogen**, found in fungi, corn (maize), and rice. A similar substance is found in bacteria and cyanobacteria.

phytohemagglutinin *or (esp. Brit.)* **phytohaemagglutinin 1** *abbr.*:

PHA; *or* **phytomitogen** the **lectin** from the red kidney bean, *Phaseolus vulgaris*; a potent mitogen. **2** *a rare alternative name for* **phytoagglutinin**.

phytohormone *or* **plant hormone** any plant growth factor or similar substance that regulates plant development. The group includes **abscisic acid**, **auxins**, **ethylene**, **cytokinins**, and **gibberellins**.

phytokinin *an alternative term for* **cytokinin**.

phytol *a* trivial name for *trans-* (or *E-*) (7*R*,11*R*)-3,7,11,15-tetramethyl-2-hexadecen-1-ol; the pure substance is a colourless oily liquid. Phytol occurs in combined form in chlorophyll and vitamin K_1. In animals it is oxidized to **phytanic acid**, which then undergoes alpha oxidation. *See also* **phytyl**.

phytolysosome a **lysosome** of plant cells.

phytomitogen *an alternative name for* **phytohemagglutinin** (def. 1).

phytoplankton plant plankton; consisting of diatoms, dinoflagellates, and other microscopic algae.

phytoporphyrin any plant **porphyrin**.

phytosiderophore any of various organic iron-chelating molecules secreted by the roots of different species of the grass family (including oat, barley, wheat, and rice). The iron–phytosiderophore complex enters the roots through an iron transporter in the plasma membrane. The two that have been characterized are the nonprotein amino acids mugineic acid (of barley) and avenic acid (of oats). Both are derived from the *α*-aminobutyrate moieties of three *S*-adenosylmethionine molecules by the action of nicotianamine synthase.

phytosphingosine 2*S*,3*S*,4*R*-2-amino-1,3,4-octadecanetriol; a constituent of many plant sphingolipids.

phytosterol any sterol occurring in higher plants; they differ from animal sterols in having C_1 or C_2 residues at C-24 and/or a double bond at C-22.

phytotoxic toxic to a plant.

phytotoxin 1 any toxin produced by a plant. **2** any substance that is toxic to a plant or plants.

phytotron a laboratory where plants can be grown and studied under a variety of controlled conditions.

phytyl the alkyl group, $(CH_3)_2$–CH–$[CH_2]_3$–CH(CH_3)–$[CH_2]_3$– CH(CH_3)–$[CH_2]_3$–C(CH_3)=CH–CH_2–, (*E*)-(7*R*,11*R*)-3,7,11,15-tetramethyl-2-hexadecenyl, derived from **phytol**.

pi 1 *symbol*: π (lower case) *or* Π (upper case); the sixteenth letter of the Greek alphabet. For uses *see* **Appendix A**. **2** *(in mathematics) symbol*: π; a transcendental number that is the ratio of the circumference of a circle to its diameter. It has the approximate value 3.141 592 653 59.

pI *symbol for* isoelectric point.

PI *abbr. for* phosphatidylinositol.

pi adduct *or* **π adduct** any adduct formed by coordination and involving donation of an electron pair from a **pi orbital** into a **sigma orbital**, from a sigma orbital into a pi orbital, or from a pi orbital into a pi orbital. It is also known, less correctly, as a **pi complex**.

PIAS *abbr. for* protein inhibitor of activation by STAT; a human nuclear protein that inhibits transcriptional activation by STAT3 (*see* **STAT**). It contains a SAP (DNA-binding) domain and a zinc finger domain. Orthologues are present in other vertebrates.

pi bond *or* **π bond** the type of covalent bond formed between atoms by electrons occupying a **pi orbital**.

picein (wax) an unreactive thermoplastic material, composed of a mixture of hydrocarbons of very low vapour pressure, that adheres well to glass and metal. It is useful as a sealing substance in the assembly of vacuum systems.

Pichia *see* **methylotrophic yeasts**.

picket-fence (porphyrin) complex any synthetic iron–porphyrin complex in which the porphyrin has been derivatized with four large groups attached to one side of the porphyrin ring, to each of the inter-pyrrole carbon atoms, so as to form a protective enclosure for the binding of a dioxygen molecule as in **myoglobin** or **hemoglobin**.

pico+ *symbol*: p; SI prefix denoting 10^{-12} times.

picogram *symbol*: pg; a unit of mass equal to 10^{-15} kilogram or 10^{-12} gram.

picometre *or* *(US)* **picometer** *symbol*: pm; a unit of length equal to 10^{-12} metre.

picomolar *symbol*: pm *or* pM; describing a solution containing one picomole of solute per litre of solution, or a specified multiple or fraction thereof; used also of a substance concentration similarly expressed.

picomole *symbol*: pmol; a unit of amount of substance equal to 10^{-12} mole.

pi complex *or* **π complex** *a less suitable name for* **pi adduct**.

picoplankton plankton having cell dimensions of between 0.2 and 2 μm and including bacteria, cyanobacteria, some algae, and certain microscopic protozoans. Picoplankton appears to be ubiquitous in open waters of lakes and seas and probably exceeds in total biomass any other group of pelagic organisms.

picornavirus any RNA animal virus consisting of naked, icosahedral 27 nm capsids with single-stranded infectious RNA (plus strand) of 2.7 MDa. The group includes human poliovirus, some common-cold viruses, and foot-and-mouth disease virus (of cattle and other cloven-footed mammals).

picosecond *symbol*: ps; a unit of time equal to 10^{-12} second.

4-pi counter any type of radiation counter that detects radiation emitted through a solid angle of 4π steradians, i.e. in all directions from a source.

picrotoxin *or* **cocculin** a preparation containing an equimolar mixture of **picrotoxinin** ($C_{15}H_{16}O_6$) and **picrotin** ($C_{15}H_{18}O_7$) obtained from the seeds of *Anamirta cocculus* (fishberries). It contains a potent ligand for the chloride ionophore of the $GABA_A$ receptor, binding to the channel protein and inhibiting Cl^- transport without displacing GABA; this endows it with potent convulsant activity. It can thus also act as an antidote to barbiturates, which enhance GABA binding. It is displaced from its binding site by anticonvulsants.

piebaldism *see* **KIT**.

pi electron *or* **π electron** an electron in a **pi orbital**.

piericidin A an antibiotic that is a structural analogue of **ubiquinone** and a classical inhibitor of mitochondrial electron transport similar in action to **rotenone**. It is a potent inhibitor of NADH:ubiquinone oxidoreductase (*see* **NADH dehydrogenase**), acting at or near the ubiquinone catalytic site.

pig 1 any nonruminant, omnivorous ungulate mammal of the family Suidae, especially the domestic pig, *Sus scrofa*. **2** *informal* short for **guinea pig**.

PIG *abbr. for* phosphatidylinositolglycan.

PIG-A *abbr. for* phosphatidylinositol glycan protein class A.

pigeon any stout-bodied bird of the family Columbidae, especially the domesticated or feral pigeon, *Columba livia*. Pigeon-breast muscle has been extensively used as a convenient tissue for the study of oxidative metabolism.

pigment 1 any general or particular colouring matter in living organisms. **2** any natural or artificial substance used to impart colour. **3** to colour with pigment.

pigment 680 *see* **P680**.

pigment 700 *see* **P700**.

pigmentation 1 natural coloration of organisms. **2** an increase in such coloration.

pigment system *an alternative name for* **photosystem**.

pigmenturia the presence of a component that imparts an abnormal colour to urine. The pigment may be endogenous (e.g. bilirubin diglucuronide, porphyrin, myoglobin, hemoglobin, blood, homogentisic acid) or exogenous (e.g. beet, cascare, senna, levodopa, riboflavin, methylene blue).

PI-G-tailed proteins proteins that are anchored to the membrane through linkage to phosphatidylinositol-glycan.

PIGylation the biosynthetic process whereby **PIG-tailed** proteins are produced.

pili the plural of **pilus**.

pilin any protein subunit of a bacterial **pilus**.

pilocarpine an alkaloid extracted from *Pilocarpus jaborandi*; it has muscarinic cholinomimetic activity.

pilosulin any of a family of peptidic channel-forming toxins, with about 100 amino acid residues, produced by various animals, particularly ants. Pilosulin is classified in the TC system under number 1.C.51.

pilot plant a prototype apparatus set up to develop and scale up a new manufacturing process.

pilus (*pl. pili*) a hairlike appendage, about 25 nm in diameter and 10–12 μm long, present sometimes on cells of certain Gram-negative bacteria, especially members of the Enterobacteriaceae and *Pseudomonas* spp., in addition to any flagella. The two main types, distinguishable morphologically and antigenically, are **common pilus** and **sex pilus**, both of which may be present on the same cell. Up to several hundred common pili per cell may be present, either distributed over the entire surface or concentrated at a pole; their function is not known. Only one or two sex pili commonly are present; they appear to be required for bacterial **conjugation**. Both types of pili are thought to be tubes or tubelike structures formed from one or two helically wound strands, each of which consists of a linear arrangement of molecules of the protein pilin. The term pilus is often used interchangeably with **fimbria**, but may be used specifically to mean sex pilus (in which case fimbria then denotes common pilus).

PIM a gene family encoding serine/threonine kinases of unknown function named from proviral integration site MuLV-induced murine T-cell lymphomas. They are highly conserved between human and mouse, and are expressed at high levels in hemopoietic tissues, testes and ovaries, and embryonic stem cells.

pI marker an isoelectric point marker; *see* isoelectric point.

PIN *abbr. for* protein inhibitor of neuronal nitric oxide synthase; a highly conserved 10 kDa protein from rat hippocampus that inhibits neuronal nitric oxide synthase.

pineal body *or* **pineal gland** a small, pine-cone-shaped structure situated deep between the cerebral hemispheres of the mammalian brain. It elaborates and secretes melatonin.

pineal hormone an older name for melatonin.

ping-pong mechanism an alternative name for **double-displacement (enzyme) mechanism**.

pinocytic an alternative term for pinocytotic (see **pinocytosis**).

pinocytic vesicle an alternative term for **pinocytotic vesicle**.

pinocytosis the process whereby cells take in liquid material from their external environment; literally 'cell drinking'. Liquid is enclosed in vesicles, formed by invagination of the plasma membrane. These vesicles then move into the cell and pass their contents to endosomes. Pinocytosis is the type of **endocytosis** occurring in most types of cells; hence the term is often applied interchangeably with endocytosis to such cells (*compare* **phagocytosis**). —**pinocytotic** or **pinocytic** *adj*.

pinocytotic vesicle *or* **pinocytic vesicle** a cellular vesicle serving **pinocytosis**; an alternative term to **endocytotic vesicle** when containing nonparticulate material.

pinosome *an alternative name for* **endosome** (def. 1) (when containing nonparticulate material).

pi orbital *or* **π orbital** a bonding molecular orbital formed by the in-phase, sideways-on overlap of parallel p atomic orbitals, one from each of two atoms, to produce a charge-density distribution that is mainly between the atomic nuclei but maximal in two zones, one on either side of and parallel to the internuclear axis. When occupied by two electrons of opposed spins it becomes the second bond in a double covalent bond (two such occupied orbitals, mutually orthogonal, providing the second and third bonds of a triple bond).

PIP *or* **PI 4-P** *abbr. for* phosphatidylinositol 4-phosphate.

PIP$_2$ *or* **PI 4,5-P$_2$** *abbr. for* phosphatidylinositol 4,5-bisphosphate.

Pipes *or* **PIPES** *abbr. for* 1,4-piperazinediethanesulfonic acid; a **Good buffer substance**; pK_a (20°C) = 6.8.

pipette *or (US)* **pipet 1** any device for transferring or delivering (usually measured) volumes of liquid and commonly consisting of a slender, calibrated glass or plastic tube with a finely drawn-out delivery tip and (for larger volumes) a centrally placed bulb. It may be filled by applied suction or by capillarity (according to its bore), and emptied by drainage or by applied pressure. **2** to transfer or deliver a (usually measured) volume of liquid with a **pipette** (def. 1). — **pipettable** *adj*.

pipettor any device for the rapid, repetitive dispensing of a predetermined volume of a liquid.

pipsyl *symbol*: Ips; *p*-iodophenylsulfonyl; I-C$_6$H$_4$-SO$_2$; a chemical group with which free amino groups may be derivatized by reaction with pipsyl chloride, *p*-iodophenylsulfonyl chloride. *Compare* **tosyl**.

pipsylate to convert (an amino compound) to a **pipsyl** derivative. — **pipsylation** *n*.

Pirani gauge an electric gauge for low gas pressures that functions over the range 10^{-2} to 10^{-5} mm Hg. It depends on the principle that an electrically heated wire loses heat by conductivity, and thus varies in electric resistance, to an extent empirically dependent on the pressure of the surrounding gas.

PIR-PSD *abbr. for* protein identification resource-protein sequence database; *see* **sequence database**.

piscine of, relating to, or being a fish.

pitch height (of a helix) *symbol*: *p*; the linear displacement along the axis of a helix corresponding to one complete turn of the helix.

PIT any of three closely related homeobox genes for positive transcription factors (Pit). Pit-1 activates the genes for somatotropin, prolactin, and the β subunit of thyrotropin, binding specifically to the consensus sequence 5'-TAAAT-3'. It contains a **POU domain** and contributes to the naming of the domain. Deficiency of Pit-1 leads to a syndrome caused by combined lack of somatotropin, prolactin, and thyrotropin. PITX2 (271 amino acids) and PITX3 (302 amino acids) are homologous with Pit-1. Mutations in either result in ocular defects.

Pitocin *a proprietary name for* (a solution of) **oxytocin**.

Pitressin *a proprietary name for* (a solution of) **vasopressin**.

pituitary 1 pertaining to or produced by the **pituitary gland**. **2** the pituitary gland.

pituitary adenylate cyclase-activating polypeptide *see* **PACAP**.

pituitary function *see* **combined test of pituitary function**.

pituitary gland *or* **hypophysis** (**cerebri**) the most important endocrine gland of vertebrates. It is situated below the brain, to which it is attached by a stalk, and measures about 8 mm by 12 mm in an adult human. It consists of two parts: the anterior lobe, or **adenohypophysis**, and the posterior lobe, or **neurohypophysis**; both lobes secrete a range of hormones.

Pituitrin *a proprietary name for* an aqueous extract of bovine neurohypophysis; it contains both oxytocin and vasopressin.

PIVKA *abbr. for* protein induced by vitamin K absence; examples are the abnormal proteins produced in vitamin K-deficient animals corresponding to coagulation factors II, VII, IX, and X (*see* **blood**

coagulation) in which some glutamate residues have not been carboxylated to γ-carboxyglutamate residues.

pixel one of the very small units that make up the image on a cathode-ray tube. [From 'pix' (i.e. pictures) + el(ement).]

pK *symbol for* the decadic logarithm of the reciprocal of a (thermodynamic) dissociation constant, K; i.e.

$$pK = \log 1/K = -\log_{10}K.$$

The type of constant may be indicated by a subscript: e.g. **pK**$_a$, a dissociation constant of an acid; **pK**$_b$, a dissociation constant of a base; **pK**$_2$, a second dissociation constant. For the dissociation of an acid, HA, to its ions, H^+ and A^-,

$$K = [H^+][A^-]/[HA].$$

Thus when $[A^-] = [HA]$, i.e. when the acid is half-dissociated, $K = [H^+]$, and hence pK is then equal to pH. Formerly symbolized as p_k. *See also* **group pK**, **Henderson–Hasselbalch equation**, **molecular pK**, **titration pK**.

pK′ *symbol for* the decadic logarithm of the reciprocal of an apparent (concentration) dissociation constant, K'.

pK$_2$ *see* **pK**.

pK$_a$ *see* **pK**.

pK$_b$ *see* **pK**.

pK$_w$ *symbol for* the decadic logarithm of the reciprocal of the ion product of water, K_w.

PK (*in clinical chemistry*) *abbr. for* pyruvate kinase.

PKB *abbr. for* **protein kinase B**.

PKC *abbr. for* protein kinase C.

PKD 1 repeat a protein sequence (80–90 amino acids) that occurs 15 times in the N-terminal extracellular region of polycystin 1. It forms seven beta strands organized in two sheets with a well-defined hydrophobic core.

PKR *abbr. for* RNA-activated protein kinase; a protein kinase that contains two dsRNA-binding motifs and is activated by binding dsRNA and by homodimerization. It is induced by interferon, and it phosphorylates the α subunit of the initiation factor eIF-2, the regulatory IκB subunit of the NF-κB transcription factor, and the tumour suppressor p53.

PKU *abbr. for* phenylketonuria.

PL *abbr. for* **1** placental lactogen (i.e. **choriomammotropin**). **2** pyridoxal.

placebo (*pl.* **placebos** *or* **placeboes**) a tablet, capsule, or liquid medicine devoid of any specific pharmacological activity, but appearing identical to a preparation containing a biologically active substance. Placebos are used in human medicine for their psychological effects, and as controls in experimental therapeutics for elucidating the clinical actions of a drug.

placenta (*pl.* **placentas** *or* **placentae**) a highly vascularized organ within the uterus of a pregnant animal by which the embryo is attached to the uterine wall and through which the embryo can exchange solutes with the maternal circulation. The placenta also functions as an endocrine gland, secreting *inter alia* **choriogonadotropin**, **choriomammotropin**, and certain steroid hormones. —**placental** *adj.*

placental lactogen *abbr.*: PL; *an alternative name for* **choriomammotropin**.

placental ribonuclease inhibitor *or* **placental ribonuclease/angiogenin inhibitor** a 15 kDa protein with 15 leucine-rich repeats found in mammalian tissues and conveniently extracted from placenta. It binds with a very high affinity to many RNases thereby inhibiting them. It is used in experiments with purified RNA to minimize degradation by contaminating RNases. It also inhibits angiogenin.

plakoglobin *or* **desmoplakin III** a protein of **cell junctions**. It is thought to play a central role in the function of desmosomes and intermediate junctions.

planar **1** of, or pertaining to, a **plane**. **2** lying in one plane; flat.

planchet *or* **planchette** a small, shallow dish used to contain a specimen for determination of its radioactivity with an end-window or windowless counter.

Planck constant *symbol*: h; a fundamental constant of proportionality between the frequency, ν, of electromagnetic radiation and the energy, E, of a quantum of the radiation; i.e. $h = E/\nu$. It has a value

of $6.626\ 075\ 5(40) \times 10^{-34}$ J s. [After Max Karl Ernst Ludwig Planck (1858–1947), German physicist.]

plane **1** a surface in which a straight line joining any two contained points lies entirely within the surface; a flat surface. **2** an imaginary surface of such a kind within a real structure or lying conceptually in space. **3** lying wholly in one plane; flat or level; planar.

plane of polarization *or* **plane of polarisation** (of linearly **polarized light** or other electromagnetic radiation) the plane containing the electric vectors of the vibrations.

plane of symmetry a plane through a three-dimensional structure, e.g. a crystal or molecule, that divides the object into two parts that are mirror images of each other.

plane-polarized *or* **plane-polarised** *an alternative term for* **linearly polarized**. *See* **polarized light**.

planimeter a mechanical integrating instrument for measuring the area of an irregular plane figure. It consists of a moveable tracing arm, which is made to follow the boundary of the figure, and a dial on which the arm's motion is recorded. —**planimetry** *n.*

plankton *the collective term for* the small or microscopic plants (**phytoplankton**) and animals (**zooplankton**), that drift freely in the surface waters of lakes, seas, and oceans.

plant any organism of the kingdom Plantae. Plants are characterized by their ability to effect photosynthesis and the possession of rigid cell walls that contain cellulose.

plant cell-wall protein any of four major classes of proteins that form networks in plant cell walls. They are: **hydroxyproline-rich glycoproteins** (HPRG); **proline-rich proteins** (PRP); **glycine-rich proteins** (GRP); and arabinogalactan proteins (AGP).

planteobiose the disaccharide, 6-*O*-α-D-galactopyranosyl-D-fructose; a unit of the trisaccharide **planteose**.

α-anomer

planteose *O*-α-D-galactopyranosyl-(1→6)-*O*-β-D-fructofuranosyl-(2→1)-α-D-glucopyranoside; a trisaccharide isolated from defatted seeds of *Plantago ovata*.

plant glucosyltransferase *an alternative name for* **D-enzyme**.

plant hormone *an alternative term for* **phytohormone**.

plantibody a human antibody produced by genetically engineered tobacco plants.

plaque **1** any small disk-like object, patch, or zone. **2** a macroscopic or microscopic rounded clear zone in a layer of cells or bacterial lawn that results from the killing or lysis of adjacent cells by the action of a virus or other agent. **3 a** (a patch of) fibrous or lipid material on the inner surface of an artery. **b** an area of tissue degeneration with distinctive histological features characteristic of certain (especially neurological) diseases, e.g. **kuru**, **multiple sclerosis**. **c** a deposit of β-amyloid, numbers of which are typically found in the brains of patients with Alzheimer's disease. **4** (a patch of) deposit closely adherent to the surface of a tooth that contains a mixed microbial flora and is composed largely of extracellular bacterial polysaccharide.

plaque-forming unit *see* **pfu**.

plaque-lift a technique used in recombinant DNA technology for screening DNA libraries constructed in bacteriophages such as lambda. Plaques of recombinant phage are allowed to form on a lawn of bacteria spread on a nutrient agar plate. Then a nitrocellulose (or nylon) membrane is pressed against the surface thereby adsorbing phage from the plaques in a replica pattern. The membrane can then be screened with a DNA probe, or antibody for an expressed protein if appropriate, and the desired clones identified.

+plasia or **+plasy** *comb. form* indicating development or growth. **—+plastic** *adj.*

plasm+ *a variant form of* **plasmo+** (before a vowel).

+plasm *comb. form* denoting the colloidal material of which living cells are composed. **—+plasmic** *adj.*

plasma 1 the proteinaceous fluid in which the cells of blood or lymph are suspended; the meaning is sometimes extended to include also the analogous fluid in which the fat droplets of milk are suspended. *See also* **blood plasma**. **2** *a less common word for* **protoplasm**. **3** *(in physics)* any region of highly ionized gas (as in a gas-discharge tube, a hot flame, or a thermonuclear reaction) containing approximately equal numbers of electrons and positively charged ions. It differs from ordinary gas in being a good conductor of electricity and being affected by a magnetic field. **—plasmatic** or **plasmic** *adj.*

plasma albumin *an alternative name for* **serum albumin**.

plasma cell or **plasmacyte** or **plasmocyte** a fully mature antibody-secreting B lymphocyte, found in lymphoid tissue and in connective tissue liable to encounter foreign material. It has an eccentrically placed nucleus, basophilic cytoplasm, and an unusually large amount of rough endoplasmic reticulum.

plasma cholesterol *(in medical biochemistry)* the concentration of cholesterol in blood plasma. In human blood, cholesterol occurs as a component of all plasma lipoproteins, but especially **very-low-density lipoprotein**, **low-density lipoprotein**, and **high-density lipoprotein**. A correlation exists between the level of plasma cholesterol and the occurrence of **atheroma** and heart attacks; this is associated with the role of low-density lipoprotein, which has the highest concentration of cholesterol and distributes cholesterol to the tissues. However, the ratio of high-density lipoprotein cholesterol to low-density lipoprotein cholesterol must be taken into account, as the former is thought to have a protective function in cycling cholesterol from peripheral tissues back to the liver. The circulating level of cholesterol may be altered by diabetes, thyroid hormone imbalance, diet, and drugs. It is thought that the level of circulating total cholesterol should be maintained below about 5 mmol L^{-1}.

plasma clearance *see* **clearance** (def. 2).

plasmacyte *an alternative name for* **plasma cell**.

plasmacytoma a tumour of plasma cells; most such tumours secrete myeloma proteins belonging to the same category as monoclonal antibodies.

plasma kallikrein *see* **kallikrein**.

plasmalemma *an alternative term for* **cell membrane**.

plasma lipid transfer protein a plasma protein that mediates transfer and exchange of lipids (i.e phospholipid, cholesterol ester, triacylglycerol) between plasma lipoproteins. **Phospholipid transfer protein** and **cholesterol ester transfer protein** belong to this group; they are homologous with each other and with **lipopolysaccharide-binding protein**.

plasmalogen any glycerophospholipid in which the glycerol moiety bears an O-(1-alkenyl) group at position 1. The term embraces **plasmenic acid**.

plasmalogen synthase EC 2.3.1.25; an enzyme that catalyses the formation of plasmenylcholine from acyl-CoA and 1-O-alk-1-enyl-sn-glycero-3-phosphocholine with release of CoA.

plasma membrane *an alternative term for* **cell membrane**.

plasmanic acid a derivative of sn-glycerol 3-phosphate in which the glycerol moiety is etherified with an alkyl (or alkenyl other than 1-alkenyl) group at position 1 and esterified with an acyl group at position 2; i.e. any 2-acyl-1-O-R-sn-glycerol 3-phosphate where R = alkyl or n-alkenyl when n > 1. *Compare* **plasmenic acid**.

plasmanyl the trivial name for any phosphoric acyl group derived from a **plasmanic acid**. The term is useful for naming phosphoric diesters of plasmanic acid, e.g. plasmanylethanolamine, 2-acyl-1-O-alkyl-sn-glycerol 3-phosphoethanolamine. *Compare* **plasmenyl**.

plasmapheresis or **plasmaphoresis** or **plasmapharesis** or **plasmaphaeresis** a technique for decreasing the protein content of blood plasma circulating in the body by withdrawing blood, separating the cells by centrifugation, and then returning them to the circulation as a suspension in a suitable protein-free medium.

plasma protein 1 any of the numerous proteins present in or derived from blood plasma. **2** such proteins collectively. In mammalian plasma the concentration of plasma protein is 60–80 g L^{-1}; of this, ≈60% is serum albumin, ≈18% is immunoglobulin G, ≈9% is β_1-lipoprotein, and ≈4% is fibrinogen.

plasma thromboplastin antecedent *an alternative name for* factor XI; *see* **blood coagulation**.

plasmenic acid a derivative of sn-glycerol 3-phosphate in which the glycerol moiety is etherified with a 1-alkenyl group at position 1 and esterified with an acyl group at position 2; i.e. any 2-acyl-1-O-(1-alkenyl)-sn-glycerol 3-phosphate. *Compare* **plasmanic acid**, **plasmalogen**. [Contraction of plasmalogenic acid.]

plasmenyl or *(formerly)* **phosphatidal** the trivial name for any phosphoric acyl group derived from a **plasmenic acid**. The term is useful for naming phosphoric diesters of plasmenic acid, e.g. plasmenylethanolamine, 2-acyl-1-O-(1-alkenyl)-sn-glycerol 3-phosphoethanolamine. *Compare* **plasmanyl**.

plasmid strictly any extrachromosomal genetic element; however, mitochondrial and chloroplast DNA, yeast killer, and other cases are commonly excluded, hence the term plasmid generally refers to an extrachromosomal covalently continuous double-stranded DNA molecule that occurs in bacteria and (rarely) other microorganisms. Natural plasmids vary from <5 kb to over 100 kb in size. They are classified in a number of different ways: the **copy number** describes the ratio of plasmid/chromosome molecules, and is greatly influenced by the nature of the origin of replication. Plasmids under stringent control have a copy number of one and those under relaxed control may have copy numbers of hundreds per cell. The **host range** describes the variety of species that can act as host (*see also* **incompatibility**). Plasmids capable of transfer are said to be conjugal; wide host-range, conjugal plasmids are described as promiscuous. Plasmid genes include drug resistance (**resistance plasmids**), genes for catabolic enzymes (catabolic plasmids), genes for antibiotic synthesis, and genes for toxin production. Plasmids are popular choices for **cloning vectors**. and is greatly influenced by the nature of the origin of replication. Plasmids under stringent control have a copy number of one and those under relaxed control may have copy numbers of hundreds per cell. The **host range** describes the variety of species that can act as host (*see also* **incompatibility**). Plasmids capable of transfer are said to be conjugal; wide-host-range conjugal plasmids are described as promiscuous. Plasmid genes include drug resistance (**resistance plasmids**), genes for catabolic enzymes (**catabolic plasmids**), genes for antibiotic synthesis, and genes for toxin production. Plasmids are popular choices for **cloning vectors**.

plasmid chimera a hybrid DNA molecule that has been constructed *in vitro* by joining fragments of separate plasmids and that forms a new, biologically functional replicon when inserted into a cell.

plasmid curing the elimination of a plasmid from a cell culture by treatment with **Acridine Orange** at a concentration insufficient to inhibit chromosome replication but sufficient to inhibit plasmid replication.

plasmid engineering genetic engineering when a plasmid is used as a cloning vector.

plasmid incompatibility *see* **incompatibility**.

plasmid segregation *see* **segregation** (def. 3).

plasmid transconjugant *see* **transconjugant**.

plasmid vector *see* **cloning vector**.

plasmin or **fibrinolysin** or **fibrinase** EC 3.4.21.7; a serine proteinase that converts the insoluble fibrin of a blood clot into soluble products. It is formed from **plasminogen** by proteolysis, the heavier A chain being derived from the N-terminal part of plasminogen and linked by a disulfide bond with the lighter B chain, which contains the enzymically active site. It exhibits preferential cleavage at Lys-|-Xaa > Arg-|-Xaa bonds, with a higher selectivity than trypsin. *See also* **blood coagulation**.

plasminogen or **profibrinolysin** a ≈90 kDa, single-polypeptide glycoprotein found in blood plasma. It is the inactive precursor of **plasmin**, into which it may be converted by hydrolysis of one peptide bond (Arg^{560}-|-Val^{561}) brought about, e.g., by the action of **plasminogen activator** or **streptokinase**. *See also* **blood coagulation**.

plasminogen activator any serine proteinase that converts plas-

minogen into plasmin. Numerous variants occur in different tissues, and can be classified immunologically into **urokinase-type** and **tissue-type** plasminogen activators (*abbr.*: UPA and TPA, respectively). Example 1, urokinase-type (EC 3.4.21.73); this cleaves the Arg^{560}-|-Val^{561} bond in plasminogen to form plasmin. It is found in urine, and is used clinically for therapy of thrombotic disorders. Example 2, tissue-type (EC 3.4.21.68); this cleaves the Arg^{560}-|-Val^{561} bond in plasminogen to form plasmin. Cleavage after Arg^{310} by plasmin or trypsin results in a two-chain form of the molecule; the two halves are held together by one or more disulfide bonds. TPA is active in tissue remodelling and destruction, in fibrinolysis, and in cell migration. Example 3, found in certain pathogens, is the activator EC 3.4.21.–, *also known as* coagulase/fibrinolysin precursor; it is a homopentamer occurring in the outer membrane. *See also* **urokinase**.

plasmo+ *or (before a vowel)* **plasm+** *comb. form* denoting of, pertaining to, related to, or derived from plasma.

plasmocyte *an alternative name for* **plasma cell**.

plasmodesma (*pl.* **plasmodesmata** *or* **plasmodesmas**) a fine cytoplasmic channel, found in all higher plants, that connects the cytoplasm of one cell to that of an adjacent cell. It is a roughly cylindrical, membrane-lined channel of diameter 20–40 nm, through the centre of which runs a narrower cylindrical **desmotubule**; the latter appears to be continuous with elements of the endoplasmic reticulum membrane of each of the connected cells. The function of plasmodesmata appears to be similar to that of **gap junctions** between animal cells. They permit passage of molecules of M_r less than about 800.

plasmodium (*pl.* **plasmodia**) 1 any member of the protozoan genus *Plasmodium*. Certain species parasitize humans causing malaria, while others parasitize other mammals. Their life cycle is complex, requiring, at different stages, an insect host and a mammalian host. 2 the motile multinucleate mass of protoplasm formed by the true **slime moulds** (Myxomycetes).

plasmolipin a protein of the myelin and lymphocyte (MAL) family of **tetraspanins**. It may be involved in vesicle trafficking and membrane linking.

plasmolysis temporary shrinkage of the protoplasm of a plant or bacterial cell away from the cell wall, caused by the withdrawal of water from the cell. —**plasmolyse** *vb.*

plasmon 1 the sum total of the extrachromosomal hereditary determinants in a cell. 2 in physics, a quasi-particle resulting from the quantization of plasma oscillations in a conducting material such as a metal. A surface plasmon occurs at the interface of a metal with material such as a protein, and interacts strongly with light. Plasmons are exploited as a way of observing binding of macromolecules to metal surfaces in **surface plasmon resonance**.

plasmoptysis the localized extrusion of cytoplasm through the cell wall of a bacterium.

plasmosome *an alternative name for* **nucleolus**.

+plast *comb. form* denoting an organized subcellular particle from a living organism.

plastein a gel that may be formed on treating a partial hydrolysate of a protein with an endopeptidase. It was once thought to be a demonstration of the synthetic capability of proteases.

plastic 1 any of a large and diverse group of polymeric materials, either synthetic or semisynthetic, that may be obtained following moulding, extrusion, etc. at a stage in their manufacture when soft or liquid. 2 made of or containing plastic; of the nature of plastic. 3 (of a material) capable of being formed or moulded. 4 (*in biology*) of or pertaining to an ability to adapt to the environment; able to change, develop, or grow. —**plasticity** *n.*

+plastic *see* +plasia.

plasticity-related gene-1 *see* **PRG-1**.

plasticizer *or* **plasticiser** any substance added to another to render the latter less brittle and/or more mouldable and flexible.

plastid any of the organelles found in the cytoplasm of eukaryotic plants that contain DNA, are bounded by a double membrane, and develop from a common type, the **proplastid**. Plastids also contain pigments and/or storage materials. They include **aleuroplasts**, **amyloplasts**, **chloroplasts**, **chromoplasts**, **elaioplasts**, and etioplasts.

plastidome the total complement of **plastids** of a cell.

plastocyanin a soluble protein of eukaryotic plants that contains one atom of copper per molecule, and is blue in its oxidized form. It occurs in the chloroplast at a molar concentration equal to that of P700, where it functions as a one-electron carrier (E'_o = +0.43 V) between the cytochrome b_6-f and P700 of photosystem I. It is loosely bound to the inner thylakoid membrane surface.

plastogene any of the hereditary determinants of plastids.

plastoglobuli droplets, 50–220 nm in diameter, that occur close to thylakoid membranes in chloroplasts of senescing leaves. Rich in lipids, particularly **plastoquinone**, they are particularly numerous when thylakoids are disrupting, and appear to be liberated therefrom.

plastome *or* **plastom** the total complement of the **plastogenes** of a cell.

plastoquinol *abbr.*: PQH$_2$; any hydroquinone derived from the corresponding plastoquinone by a two-electron reduction.

plastoquinone *abbr.*: PQ; any 2,3-dimethyl-5-multiprenylbenzoquinone; plastoquinones occur mainly in plants; intact prenyl units in the side chain (commonly nine in the plastoquinone of higher plants) may be designated by a suffixed numeral, e.g. plastoquinone-9 (*abbr.*: PQ-9). The plastoquinone–plastoquinol redox system (E'_o ≈ +0.1 V) functions in the electron-transport system, transferring electrons from photosystem II to the cytochrome b_6-f complex.

plate 1 (*in biology*) a Petri dish or other shallow vessel containing so-called solid or semisolid culture medium; a culture grown on this. 2 (*in chemistry*) any one of the horizontal perforated plates that comprise the fractionating component of a **plate column** for fractional distillation, or the theoretical equivalent of such a plate. *See also* **theoretical plate**. 3 (*in biology*) *or* **plate out** to inoculate a **plate** (def. 1) with microbial or other cells, or with material (suspected of) containing such cells. 4 to coat (a metal or other substrate) with a thin layer of (other) metal.

plateau (*pl.* **plateaux** *or* **plateaus**) 1 a part of a graph in which the value of the ordinate shows little or no change with increasing value of the abscissa. 2 a period of (relative) stability in the course of a progression. 3 to reach or pass through a plateau.

plate column *see* **plate** (def. 2).

plate height *see* **height equivalent to a theoretical plate**.

platelet *or* **thrombocyte** the smallest of the blood cells. It is an anucleate biconvex disk, 2–3 μm in diameter, formed by division of the cytoplasm of a megakaryocyte. Platelets have an exterior coat rich in glycoprotein, a normal cell membrane, and a well-developed system of microtubules some of which are arranged as an equatorial ring. Their main functions are in hemostasis, by (1) adhering to damaged blood vessels and aggregating to form a plug; (2) releasing various vasoconstrictive agonists, e.g. **epinephrine**, **serotonin**, and **thromboxane A$_2$**; (3) activating or binding certain coagulation factors; and (4) releasing factor XIII, **platelet-derived growth factor**, and **platelet factor IV**; blood clotting then occurs around the aggregated platelets. *See* **blood coagulation**.

platelet-activating factor *abbr.*: PAF *or* PAF-acether; 1-*O*-alkyl-2-acetyl-*sn*-glycerol 3-phosphocholine, where 'alkyl' may be hexadecyl or octadecyl; a potent general inflammatory mediator released from a variety of cells in response to a number of stimuli, and causing aggregation of platelets, activation of polymorphonuclear leukocytes, monocytes, and macrophages, mediation of increased vascular permeation, hypotension, decreased cardiac output, and stimulation of uterine contraction. It is synthesized in the cell in either of two ways: (1) from 1-*O*-alkyl-*sn*-glycerol-3-phosphate by addition of an acetyl group and release of phosphate, forming 1-alkyl-2-acetyl-*sn*-glycerol, which then reacts with CDPcholine to form PAF and CMP; or (2) by a remodelling pathway, in which the 2-fatty acyl group is removed from a phosphatidylcholine by phospholipase A$_2$ and replaced with acetyl. A number of analogues act as PAF agonists, e.g. the stable analogues in which the 2-acetyl group is replaced by an ethyl, methyl, or benzyl ether group; replacement of the choline moiety by (*N*-methylpyrrolidino)-ethanolamine greatly increases its effect. PAF antagonists include gossypol, hexanolamino-PAF, *bis*(methylthio)gliotoxin, and 2,5-*bis*(3,4,5-trimethoxyphenyl)-1,3-dioxolane.

platelet-activating factor acetylhydrolase (*abbr.*: PAF-AH) *an alternative name for* **lipoprotein-associated phospholipase A2**.

platelet basic protein precursor *abbr.*: PBP; the precursor of several peptides generated by alternative processing. For example, the human protein contains a signal, platelet basic protein, **LA-PF4**, β-**thromboglobulin** and **NAP-2**.

platelet-derived endothelial cell growth factor *abbr.*: PD-ECGF; a **cytokine** produced by platelets, fibroblasts, and smooth muscle cells. It is a monomer of 471 amino acids. A related protein, **gliostatin**, is produced by astrocytes. Both proteins stimulate growth of endothelial cells *in vitro* and angiogenesis *in vivo*, and promote survival and differentiation of CNS neurons; they inhibit the growth of astrocytes and glial cells. PD-ECGF proteins initiate thymidine phosphorylase activity.

platelet-derived growth factor *abbr.*: PDGF; a **cytokine** of ~30 kDa, composed of two disulfide-bonded polypeptide chains, denoted A and B, which are non-identical but homologous. It is believed that human PDGF is an AB heterodimer, but pig PDGF is a homodimer of B chains. PDGF is a potent mitogen for cells of mesenchymal origin (endothelial cells, epithelial cells, some fibroblasts, smooth muscle cells) and for glial cells; it has chemotactic properties. It is released by platelets on wounding, and stimulates nearby cells to grow and repair the wound; it is also produced by endothelial cells, fibroblasts, monocytes, macrophages, smooth muscle cells, and neurons. PDGF-like proteins are produced by numerous human tumour cell lines, and the B chain is almost identical to part of the transforming protein produced by simian sarcoma virus (*see* **sis**). Homodimers of A and B chains (both can bind to the receptor) are implicated in transformation processes. *See also* **platelet-derived growth factor receptor**.

platelet-derived growth factor receptor *or* PDGF receptor a tyrosine kinase dimer, of αα, ββ, or αβ subunits, belonging to the immunoglobulin superfamily. It is a transmembrane cell surface receptor, the intracellular kinase sequence being split by an inserted sequence. Binding of platelet-derived growth factor (PDGF) stimulates tyrosine-kinase activity, which leads to phosphorylation of tyrosine residues.

platelet endothelial cell adhesion molecule *see* **PECAM**.

platelet factor IV a polypeptide secreted by platelets when they adhere at a site of injury. It is a **chemokine** that binds heparin strongly, and is a powerful attractant for neutrophils and monocytes.

platform protein *an alternative name for* **molecular beacon**.

plating *see* **plate** (def. 3, 4).

platysome (*sometimes*) the core particle of a **nucleosome**.

PLC *abbr. for* **1** preparative-layer chromatography. **2** phospholipase C; *see* **phospholipase**.

pleated sheet *see* **beta-pleated sheet**.

pleckstrin *other name*: platelet p47 protein; a platelet protein that is the major substrate for Ser/Thr phosphorylation by protein kinase C. Its function is not known. *See also* **pleckstrin homology domain**.

pleckstrin homology domain *abbr.*: PH domain; a domain, first noted in the platelet protein **pleckstrin**, containing about 100 residues. Its function is unknown, but it has since been found in many (more than 60) proteins, especially those associated with intracellular signal transduction downstream of cell-surface receptors. It binds inositol-1,4,5-triphosphate.

plectin a coiled-coil protein that links intermediate filaments to hemidesmosomes. It is expressed in epidermis and skeletal muscle and is very similar in sequence to bullous pemphigoid antigen 1e. Null mutations in the gene are associated with epidermolysis bullosa with muscular dystrophy.

plectonemic coiling a type of coiling of a double-stranded helix in which the two strands are intertwined, i.e. they may not be separated without uncoiling.

pleiotropin *or* **pleiotrophin** *other name*: heparin-binding growth factor 8; a heparin-binding mitogenic protein with neurite extension activity for osteoblasts and brain cells.

pleiotropism *or* **pleiotropy** the phenomenon in which a single gene is responsible for producing multiple, distinct, apparently unrelated phenotypic effects; an instance of such a phenomenon. The meaning is sometimes extended to include multiple, distinct, unrelated actions or effects of a single polypeptide agonist. —**pleiotropic** *adj.*

pleiotypic relating to the process by which a single stimulus to a living cell elicits multiple, unrelated responses.

pleomorphism *or* **pleomorphy** the quality of occurring in, or exhibiting more than one distinct form. *Compare* **polymorphism**. — **pleomorphic** *adj.*

pleuropneumonia-like organism *abbr.*: PPLO; *an obsolete name for* **mycoplasma** or mollicute.

Plexiglas *or* **Plexiglass** *a proprietary name for* **poly(methyl methacrylate)**.

plexin any of a family of transmembrane proteins that form part of **semaphorin** receptor complexes. Their intracellular region binds directly to **RhoGTPases** and to Rho guanine nucleotide-exchange factors. Plexins are widespread in embryonal and adult tissues. They regulate axonal pathfinding in the developing nervous system. In *Drosophila*, plexin A is a receptor for semaphorin 1A. In vertebrates, plexin A is a receptor for secreted semaphorin 3, whereas plexins B, C, and D bind other semaphorins. Their extracellular N-terminal ≈500 residues are homologous with the same region of semaphorin and hepatocyte growth factor receptors, but their intracellular ≈600 residues show no known homology and might bind a cytosolic protein-tyrosine kinase.

PLI *abbr. for* pulsed laser interferometry.

+ploid *adj. and n. comb. form* indicating the specific multiple of homologous sets of chromosomes in cell. —**+ploidy** *n.*

ploidy the number of homologous sets of chromosomes in a cell, or in each cell of an organism.

P-loop *see* **Walker motif**.

plot 1 a diagram showing the relationship between two or more variables. **2** to construct such a plot.

plotter an instrument for automatically making a **plot** (def. 1).

PLP *abbr. for* **1** pyridoxal phosphate. **2** proteolipid protein; *see* **myelin proteins**.

PLTP *abbr. for* phospholipid transfer protein.

plug flow a type of flow of fluids in which there is no mixing in the direction of flow.

plumper a disposable, nonscratching plastic device for adding a small volume (5–50 μL) of a reagent to, and mixing it with, a liquid in a small container such as a spectrophotometer cuvette.

pluripoietin granulocyte colony stimulating factor (G-CSF) precursor. Pluripoietin α is a second human hemopoietic colony stimulating factor produced by the human bladder carcinoma cell line 5637. *See* **colony stimulating factor**.

pluripotent 1 (of a biomolecule) capable of being converted (biologically) into more than one active product. **2** (of a cell) capable of differentiating into more than one alternative type of mature cell.

plus and minus screening a strategy used previously for isolating genes for gene products that are present in one type of cell and absent in another. It involves comparing the results of plaque or **colony hybridization** of replicate plates of a cDNA clone library from the first cell type with cDNA probes derived from the mRNAs of both cell types used independently. Any plaque or colony that hybridizes to one probe and not the other is a candidate of interest. The objective can now be achieved more conveniently using subtractive PCR methods.

plus end the end of a microtubule or actin filament at which addition of monomers occurs most readily; the 'fast-growing' end of a microtubule or actin filament. The plus end of an actin filament is also known as the **barbed end**.

plus strand 1 the parental polynucleotide strand in bacteriophages whose genomes consist of duplex RNA. It is the template used for the production of the complementary minus strand. **2** virus-specified messenger RNA, or any given strand of nucleic acid (RNA or DNA) whose base sequence is equivalent to that of virus-specified mRNA. Any strand having a base sequence complementary to virus-specified mRNA is described as a minus strand. **3** (*sometimes*) *an alternative term for* **coding strand** (of genomic duplex DNA). *Compare* **minus strand**.

PLZF *abbr. for* promyelocytic leukemia zinc finger protein; a transcription factor that is expressed in many tissues during development and may be involved in hemopoietic cell differentiation. It contains an N-terminal domain for protein–protein interaction and nine C-terminal zinc fingers, Krüppel-type (i.e. C_2H_2). The N-terminal domain binds the nuclear receptors RAR and RXR and inhibits their binding to DNA. Translocation of the PLZF gene at

11q23.1 with that for RARα results in a PLZF-RARα fusion protein in some acute promyelocytic leukemia patients.

pM *or* **pm** *symbol for* picomolar.

Pm *symbol for* promethium.

PMA *abbr. for* phorbol 12-myristate 13-acetate; *see* **phorbol ester**.

PmaCI a type 2 **restriction endonuclease**; recognition sequence: CAC↑GTG.

pMAL vector any of a range of plasmid vectors used for the expression of foreign proteins in *E. coli*. Genes are cloned downstream from the gene for the *E. coli* maltose binding protein (MBP) *malE* and in the same reading frame. The resulting fusion protein can be purified on columns of cross-linked amylose. The leader sequence of MalE facilitates export to the periplasm. A short sequence motif at the point of fusion is recognized by factor Xa, which can be used to cleave the protein of interest from MBP.

pMB1 a plasmid replicon widely used in its original or modified form in the construction of synthetic plasmids used in molecular biology. Plasmids with pMB1 have copy numbers of 15–30.

pMC9 a plasmid encoding the *lac* repressor *I^q* and the gene for resistance to ampicillin. It is carried by *E. coli* strains such as Y1090 to control the expression of foreign genes cloned in bacteriophage lambda vectors and transcribed from *lac* or *tac* promoters.

PMF *abbr. for* protonmotive force, *or sometimes (facetiously), after its discoverer,* Peter **Mitchell** force. [After Peter Dennis Mitchell (1920–92), British biochemist.]

PML *abbr. for* promyelocytic leukemia protein; a protein that occurs free in nucleoplasm and in association with other proteins in **nuclear bodies**. The gene at 15q21 encodes a protein that contains a RING finger and a coiled-coil domain, and resembles **COP1** in structure, cell distribution, and dynamics. A translocation between the PML gene and the RARα gene produces the fusion protein PML-RARα and disrupts the nuclear bodies. Cytoplasmic aggregates of PML occur in virally infected, neoplastic, or inflamed tissue.

PML nuclear body *see* **nuclear body**.

PMN *abbr. for* **1** pyridine mononucleotide. **2** polymorphonuclear (leukocyte).

pmol *symbol for* picomole.

PMR *or* **pmr** *abbr. for* proton magnetic resonance; *see* **nuclear magnetic resonance spectroscopy**.

PMS *abbr. for* **1** phenazine methosulfate. **2** pregnant mares' serum.

PMSF *abbr. for* phenylmethylsulfonyl fluoride.

PMSG *abbr. for* pregnant mares' serum gonadotropin.

PN-1 *see* **nexin 1**.

PNA *abbr. for* pentose nucleic acid (i.e. ribonucleic acid).

PNAD *abbr. for* protein NH$_2$-terminal arginine deamidase; an enzyme, discovered in the pig, that catalyses deamidation of N-terminal arginine residues in peptides.

pneumo+ *or* **pneumono+** *or (before a vowel)* **pneum+** *or* **pneumon+** *comb. form* of or pertaining to a lung or lungs; respiratory.

pneumococcus (*pl.* **pneumococci**) any strain of bacterium belonging to the species *Streptococcus pneumoniae*. —**pneumococcal** *adj*.

pneumococcus capsule swelling reaction *an alternative name for* **Quellung reaction**.

p[NH]ppA *symbol for* adenosine 5'-[β,γ-imido]triphosphate (alternative to Ado*PP*[NH]*P*).

PNK *abbr. for* polynucleotide kinase.

PNMT *abbr. for* phenylethanolamine *N*-methyltransferase.

PNP *abbr. for* **1** *para*-nitrophenol. **2** purine-nucleoside phosphorylase.

Po *symbol for* polonium.

pocket *(in molecular biology)* a cavity or hollow in a three-dimensional structure, e.g. a substrate pocket of an enzyme.

podocalyxin a sialated glycoprotein that is prominently expressed on renal glomerular epithelial cells and found in patients with glomerulonephritis.

podophyllotoxin a **lignan** found in the May apple, or American mandrake (*Podophyllum peltatum*). It is known in folk medicine as a cathartic and poison, and is useful in topical treatment of warts, including venereal warts; it may have antiviral action. In mammalian cells it inhibits assembly of tubulin into microtubules, and also nucleoside transport. It is used in the synthesis of **Etoposide** and **Teniposide**.

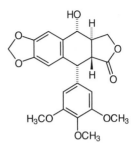

podophyllotoxin

+poiesis *noun comb. form* indicating an act or process of making (something specified), e.g. erythropoiesis, hemopoiesis. —**+poietic** *adj*.

poikilo+ *comb. form* indicating various, variable.

poikilocyte an erythrocyte that is distorted in shape and usually larger than normal, found in the blood of persons with pernicious anemia and some other anemias.

poikilosmotic describing an organism whose internal osmotic pressure varies with that of its environment. *Compare* **homoiosmotic**.

poikilothermic *or* **poikilothermal** describing an organism whose body temperature varies with that of the environment. *Compare* **homoiothermic**. —**poikilotherm** *n*.

point 1 *(in mathematics)* **a** any dimensionless geometric element whose position in space is defined by its coordinates. **b** a location or position, especially on a diagram, graph, etc. **2** a decimal point. **3** a **point of inflection**.

point accepted mutation *symbol*: PAM; a unit that quantifies the amount of evolutionary change in a protein sequence. 1PAM marks the evolutionary distance in which 1% of amino acids in a protein sequence have changed; 250PAMs represents 80% change.

point group 1 a notation used in describing the geometrical arrangement of the symmetry of globular oligomeric proteins. Such molecules may belong to any of the three point groups having **cyclic symmetry**, **dihedral symmetry**, or **cubic symmetry**. **2** *(in crystallography)* a group of symmetry operations that leave unmoved at least one point within the object to which they apply. It can describe single molecules or actual crystals.

point ligand any of two or more individual atoms or groups linked covalently to a specified carbon atom in an organic compound. Point ligands may be identical or nonidentical, but they may not occupy the same position in space.

point mutation any **mutation** (def. 1) involving a single base change. *See also* **transition**, **transversion**.

point of inflection any point on a graph where the direction of curvature changes.

poise *symbol*: P; the cgs unit of dynamic viscosity; it has the dimensions of g s^{-1} cm^{-1}; 1 P = 0.1 N s m^{-2}. [After Jean-Louis Marie Poiseuille (1799–1869), French physician and physicist.]

Poiseuille's equation an equation describing the behaviour of a liquid flowing under pressure through a capillary tube of length *l* and radius *r*. The volume flowing per second, d*V*/d*t*, is given by:

$$\mathrm{d}V/\mathrm{d}t = \pi P r^4/8\eta\, l,$$

where *η* is the viscosity of the liquid and *P* is the pressure difference of the liquid along the tube over a distance *l*. [After Jean-Louis Marie Poiseuille (*see* **poise**).]

poison 1 *(in toxicology)* any substance that, in relatively small amounts, can impair the functioning of, or cause structural damage to a cell or organism. **2** *(in chemistry)* any substance that, at relatively low concentrations, inhibits a chemical reaction or inactivates a catalyst. **3** to add or administer poison to (a cell, organism, catalyst, etc.).

pokeweed mitogen *abbr.*: PWM; a plant lectin (phytoagglutinin) extractable from the roots of pokeweed, *Phytolacca americana*, a North American plant. It is used in immunology and cell biology, has an affinity for *N*-acetyl-*β*-D-glucosamine oligomers, and agglutinates leukocytes and erythrocytes (of all ABO blood groups). It also has mitogenic properties, which appear to reside in five (or more) separable components; one of these activates both T and B lymphocytes, and the others are specific for T lymphocytes.

poky mutant a mutant strain of *Neurospora* with a lower rate of growth than that of wild-type strains, due to absence or deficiency of certain components of the respiratory chain. The *poky* gene resides in mitochondrial DNA and is transmitted to sexually derived progeny only when the female parent exhibits the poky phenotype.

pol 1 a retroviral gene that encodes a polyprotein containing **retropepsin**, **reverse transcriptase**, and ribonuclease H (*see* **ribonuclease**). **2** as *polA*, *polB*, *polC*, genes that encode DNA polymerases I to III.

Pol the gene product of **polA**.

polA the structural gene in *Escherichia coli* for DNA polymerase I (Pol or Pol I; *see* **DNA polymerase**), which serves as a repair polymerase and as an exonuclease in the generation of **Okazaki fragments**. The enzyme belongs to family A of DNA polymerases and has three characteristic domains: 5′-exonuclease (amino acids 1–323), 3′-exonuclease (amino acids 324–517), and polymerase (amino acids 521–928). Klenow polymerase (amino acids 324–928) contains the latter two domains. *See* **Klenow enzyme**.

polar 1 pertaining to, or possessing, a **pole** or poles. **2** (*in chemistry*) (of a group or a molecular entity) having an uneven distribution of electrons and hence a permanent dipole moment. *See also* **polarity**.

polar body (*in biology*) either of the two small cells that form as a result of meiosis of a primary oocyte during its development to a mature ovum. One is formed in the first division, and the other in the second division. At each division, the cytoplasm divides unequally, the polar body being much smaller than the developing oocyte. At the second division the polar body and the developing oocyte each contains a haploid set of chromosomes. The polar bodies eventually degenerate.

polar bond 1 a covalent bond in which the electron pair(s) of the bond are held with greater force by one of the two bonded atoms leading to the existence of a permanent dipole moment in the group or molecular entity in question. **2** an ionizable bond between two oppositely charged ions of a salt.

polar coordinate (*in mathematics*) either of two numbers that locate a point in a plane, one being the distance along a line to the point from the origin and the other being the angle this line makes with a fixed line passing through the origin.

polar group *see* **polar** (def. 2).

polarimeter 1 an instrument for measuring the degree of polarization of light. **2** an instrument for measuring the rotation of the plane of vibration of linearly polarized light as a result of passing through a given length of optically active material. —**polarimetric** *adj.*; **polarimetry** *n.*

polarity 1 the state, or condition, of having **poles**. **2** having a differential distribution of a property or properties along an axis; *see* **polar**. **3** having opposing physical properties at different points, e.g. north and south magnetic poles, positive and negative electric charges, hydrophobic and hydrophilic regions. **4** the state of being electrically either positive or negative. **5** (of a solvent) **ionizing power**. **6** (of an organism) the tendency to develop differentially along an axis; e.g. plants exhibit polarity in developing roots downwards and stems upwards.

polarity gradient (*in genetics*) the quantitative effect of a **polar mutation** in one gene on the expression of later genes in the operon.

polarity mutant a mutant gene formed by a **polar mutation**.

polarizability or **polarisability** the quality of being able to be polarized.

polarization or **polarisation 1** inducing polarity. **2** the quality of being or becoming polarized.

polarization microscope or **polarisation microscope** an alternative name for **polarizing microscope**.

polarize or **polarise 1** to acquire or cause to acquire **polarity**. **2** to cause the polarization of light or other electromagnetic radiation; *see* **polarized light**. **3** to cause the electrical resistance of an electric cell to increase through the accumulation of electrolysis products at an electrode.

polarized light or **polarised light** light (or other electromagnetic radiation) in which the angular orientations of the electric vectors of its transverse vibrations are nonrandom. In **linearly** (or **plane-**) **polarized light** the electric vectors all lie in a given plane containing the light beam; to an observer looking along the light beam the vectors appear as a straight line through the axis of propagation. In **circularly polarized light** the angular orientations of the electric vectors are such that a given point on the vectors generates a (left-handed or right-handed) helix whose axis is the direction of the beam; to an observer looking along the light beam this point appears to describe the circumference of a circle, the light being defined as right or left circularly polarized depending on whether the point appears to the observer looking towards the source to be rotating clockwise or anticlockwise, respectively. Circularly polarized light may be considered to consist of two superimposed linearly polarized waves of equal amplitude but differing in phase by an odd number of quarter-wavelengths and having their electric vectors mutually at right angles; conversely, if right and left circularly polarized waves of equal amplitude are superimposed the resultant is a linearly polarized wave. In **elliptically polarized light** the angular orientations of the electric vectors are such that a given point on the vectors generates a (left-handed or right-handed) flattened helix whose axis is the direction of the beam; to an observer looking along the light beam this point appears to describe the perimeter of an ellipse, the degree of such polarization being characterized by the **ellipticity** and by the angular orientation of the major axis of the ellipse. Elliptically polarized light may be considered to result from the superimposition of two linearly polarized waves of equal amplitude and having their electric vectors mutually at right angles but where the phase difference is neither zero nor exactly an even or odd number of quarter-wavelengths; alternatively, it may be considered to result from the superimposition of right and left circularly polarized waves of differing amplitude.

polarizer or **polariser** an optical device for producing linearly polarized light, e.g. a **Nicol prism** or a sheet of **Polaroid** (def. 1).

polarizing microscope or **polarization microscope** a microscope that is useful for examining the birefringence properties of small objects, or for visualizing objects by virtue of their birefringence properties. *Also:* polarising microscope, polarisation microscope.

polar microtubule *see* **mitotic spindle**, **prometaphase**.

polar molecule a molecule in which there is a polarized distribution of positive and negative charges due to an uneven distribution of electrons. Polar molecules are likely to be soluble in water.

polar mutation a mutation in one gene that reduces the expression of genes further from the promoter in the same operon.

polarogram the current–voltage diagram obtained by **polarography**.

polarograph 1 an instrument used for **polarography**. **2** to subject (a sample) to analysis by polarography.

polarography a method for quantitative or qualitative determination of small concentrations of reducible or oxidizable substances based on the nature of the current–voltage curve obtained when a test solution is subjected to electrolysis with a steadily increasing voltage between, e.g., a dropping mercury cathode and a pool of mercury serving as anode. The current–voltage curve obtained (called a **polarographic wave**) is sigmoidal; the potential at the midpoint (called the **half-wave potential**) is characteristic of the substance being oxidized or reduced and the limiting current is related to its concentration. —**polarographic** *adj.*

Polaroid a proprietary name for **1** a thin, transparent material that linearly polarizes light transmitted by it (*see* **polarized light**). It consists of oriented, doubly refracting microcrystals of herapathite (quinine iodosulfate) in a matrix of cellulose acetate. **2** a photographic process or camera producing a nearly instantaneous, permanent photograph.

polar solvent any solvent of relatively high **ionizing power**.

polar zipper a motif in proteins that is considered to promote binding between protein subunits, or between proteins, as a result of interactions between polar residues, which are aligned in the form of a 'zip fastener'. The polar groups may consist of repeats of glutamines, Asp–Arg or Glu–Arg repeats, or repeats of glutamates and either lysines or histidines. The concept derives from the previously coined term **leucine zipper**.

pole 1 either end of an axis of rotation of a sphere or spheroid body, especially of the Earth. **2** *(in biology)* **a** either of the differentiated zones at opposite ends of an axis in a cell, organ, or organism. **b** *(in mitosis or meiosis)* either of the opposite ends of the **spindle**. **3** *(in physics)* **a** any of the zones in a magnetized body, commonly two in number and near its extremities, where the lines of magnetic flux appear to originate (**north pole**) or end (**south pole**). **b** either of the (positively or negatively charged) terminals of an electric battery, or of the corresponding terminals of a d.c. electric generator; either of two points, positions, or regions carrying opposite electric charges in a piece of material, system, or (polar or polarized) molecular entity. *Compare* **dipole**. *See also* **polar**.

policeman a device used for detaching adherent particles of precipitate, cells, etc. from the wall of a vessel and consisting commonly of a glass stirring-rod with a short piece of rubber tubing fitting tightly over one end, but sometimes of a small snipe feather cemented into the end of a capillary tube.

polio virus a member of the *Picornaviridae* genus *Enterovirus*. which has 60 copies each of four subunits as coat proteins, VP1,2,3,4,produced synthesized as a polyprotein. They are arranged with the symmetry of an icosahedron. The genome is a positive sense RNA. The polymerase used in the replication of the virus is a multisubunit enzyme of four subunits only one of which is virus specific, the other subunits being provided by the host. Two kinds of vaccine have been used, killed virus, Salk, and attentuated live, Sabin. The latter sometimes can revert to the wild strain in the gut.

poly+ *comb. form* **1** denoting more than one; many or much; more than the usual number or amount; often used interchangeably with **multi+**. **2** indicating heterogeneity of a physical attribute e.g. polychromatic, polydisperse. *Compare* **mono+** (def. 2). **3** *(in chemistry and biochemistry)* **a** denoting more than one of a specified component or substituent, e.g. polyamine, polyene, polyester; in names of large molecules, usually denoting more than 10–20 of the component specified, especially in contrast to **oligo+** (def. 2), e.g. polynucleotide, polypeptide, polysaccharide. *Compare* **mono+** (def. 3). **b** denoting polymer of a specified substance, e.g. polyglycine, polystyrene. **c** indicating more than one function of the kind specified, e.g. polyacidic, polybasic, polydentate, polyhydric. *Compare* **mono+** (def. 5).

poly(A) *abbr. for* polyadenylate.

polyacidic (of a **base** (def. 1)) able to accept more than one proton per molecule in solution.

polyacrylamide (gel) *a common name for* any hydrogel containing a 3-D matrix of randomly cross-linked polymers of acrylamide (2-propenamide; $CH_2=CH–CO–NH_2$). The cross-linking agent commonly used is N,N'-methylene bisacrylamide, $(CH_2=CH–CO–NH)_2CH_2$, polymerization being initiated by various alternative methods for generating free radicals. The pore size and rigidity of the gel depend on the concentration of acrylamide, and the proportion of cross-linking agent used.

polyacrylamide gel electrophoresis *abbr.*: PAGE; a technique of **zone electrophoresis** that uses polyacrylamide gel as supporting medium. It is especially useful for the separation of biopolymers and fragments thereof. A two-dimensional electrophoresis has been used to resolve as many as 1000 proteins: in the first step the polypeptides are separated by **isoelectric focusing electrophoresis** and in the second step by **SDS (polyacrylamide) gel electrophoresis**. The 2-D process was devised by P. H. O'Farrell, after whom it is often named.

polyacrylamide gradient gel electrophoresis *an alternative name for* **gradient gel electrophoresis**.

polyadenylate 1 *abbr.*: poly(A); any homopolymer of adenylate (i.e. adenosine 5′-monophosphate). Most eukaryotic mRNAs are terminated at the 3′ end by a tract of poly(A) 80–250 residues long (*see* **poly(A) tail**). **2** to add a stretch of polyadenylate to a molecule. —polyadenylation *n*.

polyadenylation binding protein *see* **PABP-2**.

polyadenylation element a nucleotide sequence in the 3′ untranslated region of eukaryotic mRNA that contains a highly conserved AAUAAA followed by 10–30 nucleotides from where cleavage occurs and poly(A) formation starts. It is recognized by a complex

that contains **cleavage and polyadenylation specificity factor** (CPSF), **poly(A) polymerase**, and **PABP-2**.

poly(ADP-ribose) glycohydrolase *abbr.*: PARG; an enzyme that hydrolyses poly(ADP-ribose) to ADP-ribose units. Highly conserved in mammals, it contains an N-terminal putative regulatory domain, a central nuclear localization signal, and a C-terminal catalytic domain. There are multiple isoforms derived from a single gene.

poly(ADP-ribose) polymerase *see* **PARP-1**; **PARP-2**.

poly(ADP-ribosyl) a structure found attached to histones H1 and H2B, to other eukaryotic chromosomal proteins, and to some mitochondrial proteins. It consists of a polymer of 20–50 ADP-D-ribosyl units linked by glycosidic bonds; linkage is formed by transfer of the ADP-ribosyl unit of NAD^+, linking its ribosyl C-1 position to the C-2′ position of the ribosyl of the terminal ADP-ribose of the growing chain. The attachment to protein appears, at least in some instances, to be through an ester linkage between the terminal ribose group and certain free carboxyl groups in the protein. *See also* **ADP-ribosylation**; **NAD(P)$^+$-arginine ADP-ribosyltransferase**.

poly(ADP-ribosyl) transferase EC 2.4.2.30; *recommended name*: NAD$^+$ADP-ribosyltransferase; *systematic name*: NAD$^+$:poly(adenine-diphosphate-D-ribosyl)-acceptor ADP-D-ribosyl-transferase; *other names*: poly(ADP-ribose) polymerase; poly(adenosine diphosphate ribose) polymerase; ADP-ribosyltransferase (polymerizing); poly(ADP-ribose) synthetase. An enzyme responsible for ADP-ribosylation of acceptors; it catalyses a reaction between NAD^+ and (ADP-D-ribosyl)$_n$-acceptor to form nicotinamide and (ADP-D-ribosyl)$_{n+1}$-acceptor. It often has zinc as cofactor (as in example); the acceptor is a histone, a transcription factor, a DNA repair protein, or the enzyme itself. The modification is dependent on DNA and is involved in the regulation of cellular processes such as differentiation, proliferation, and tumour transformation and also in the regulation of the molecular events involved in the recovery of a cell from DNA damage. Mammals have four homologues (PARP-1, PARP-2, V-PARP, and tankyrase). In *Drosophila*, application of a heat-shock stimulus causes PARP to accumulate and become active at heat-shock genes. Ecdysone also causes accumulation of PARP at chromosome puffs. *See also* **NAD(P)$^+$-arginine ADP-ribosyltransferase**. *Compare* **ADP-ribosylation**.

polyamide any polymer in which the constitutional units are joined by amide or thioamide links. *Compare* **nylon**, **polypeptide**.

polyamine any organic compound containing two or more amino groups. **Putrescine**, **spermine**, and **spermidine** are important animal polyamines. Polyamines are synthesized by various routes. In animals, **putrescine** is the decarboxylation product of **ornithine**. Decarboxylated S-adenosylmethionine (*see* **S-adenosylmethionine decarboxylase**) is a substrate in a reaction in which propylamine is transferred to putrescine, to form spermidine, and again to spermidine to form spermine. *See also* **ornithine decarboxylase**.

poly(amino acid) any homopolymer, or mixture of homopolymers, of a particular (specified) amino acid.

polyampholyte any amphoteric **polyelectrolyte**.

polyanion any macromolecular ion that carries multiple negative charges.

poly(A) polymerase *abbr.*: PAP; *the recommended name for* **polynucleotide adenylyltransferase**.

poly(A) tail a sequence of adenylyl residues at the 3′ end of eukaryotic mRNA. Almost all mature eukaryotic mRNAs have 3′-poly(A) tails of 40–250 nucleotides, those of histones being a notable exception. The primary transcript is first cleaved 10–30 nucleotides past a highly conserved AAUAAA sequence, then the poly(A) tail is generated from ATP through the action of **polynucleotide adenylyltransferase and several other proteins**. In practical terms the poly(A) tail on mRNA facilitates its ready isolation from total cellular RNA by affinity chromatography on oligo(dT)-cellulose.

polyatomic (of a molecular entity) composed of two or more atoms.

polybasic (of an **acid** (def. 1)) able to give up more than one proton per molecule in solution.

Polybrene *a proprietary name for* 1,5-dimethyl-1,5-diazaundecamethylene polymethobromide; *other name*: hexadimethrine bromide; a positively charged polymer with coagulant properties. It causes nonspecific agglutination of red blood cells and has been used in the transfection of eukaryotic cells.

poly(C) *abbr. for* polycytidylate.

polycarbonate any polymer in which the structural units are linked by carbonate ester groups, usually formed by the polymerization of diphenylols with phosgene (i.e. carbonyl chloride). Such polymers are tough thermoplastic resins.

polycation any macromolecular ion that carries multiple positive charges.

polychlorinated biphenyl *abbr.*: PCB; any of various compounds that are based on the biphenyl ring structure and chlorinated at various positions. PCBs were formerly used in industrial and commercial applications such as transformers and hydraulic equipment, and as plasticizers in paints, plastics, and rubber products. They persist in the environment. Manufacture in the USA ceased in 1977.

polychromatic 1 relating to, or possessing, various or varying colours; consisting of, relating to, or emitting light (or other electromagnetic radiation) of two or more frequencies or wavelengths. **2** of, relating to, or showing **polychromatophilia**.

polychromatophilia the quality of being stainable by more than one type of dye, especially by both acid and basic dyes.

polycistronic messenger RNA *or* **multicistronic messenger RNA** *or* **polygenic messenger RNA** any messenger RNA molecule that carries more than one initiation site and encodes the synthesis of two or more (usually related) gene products. It is formed by transcription of a single operon and is characteristic of prokaryotes.

polyclonal of, pertaining to, being formed by, or derived from two or more **clones** (def. 1).

polyclonal antibody an antibody that arises from the immune response to an immunogen. Each **epitope** on the immunogen has the potential for causing the formation of a clone of **B lymphocytes** each capable of synthesizing an antibody specific to the particular epitope.

Polycomb one of a group of proteins in *Drosophila* that regulate the expression patterns of many developmental genes. Polycomb group proteins are involved in chromatin-based gene silencing and enable established patterns of gene expression to be transmitted over multiple rounds of cell division. *Compare* **trithorax**.

Polycomb a family of genes in *Drosophila* that select which homeotic selector genes should not be expressed. They encode homologous regulatory proteins that bind to the chromatin of the genes they control.

polycyclic having more than one ring of atoms in the molecule; used especially of molecules with condensed ring systems.

polycystin either of two proteins produced by different genes for autosomal dominant polycystic kidney disease (PKD). **Polycystin 1** is a membrane-associated glycoprotein (4303 amino acids) involved in cell–extracellular matrix interactions, and expressed in fetal, adult, and polycystic kidney. It consists of an extracellular N-terminal region of ≈3000 residues (containing 16 PKD1 repeats), 11 transmembrane helices, and a C-terminal coiled-coil region and sites for phosphorylation of serine/threonine and tyrosine. It shows ≈30% homology with polycystin 2. More than 60 mutations result in PKD type 1, which affects infants and is severe. **Polycystin 2** is an integral membrane protein (968 amino acids) that contains six transmembrane helices and intracellular N- and C-termini. The N-terminal region contains a coiled-coil region and an EF hand for Ca^{2+} binding. Polycystin 2 is more closely related to voltage-gated calcium channels than to polycystin 1. It is expressed ubiquitously but especially in fetal kidney and lung. More than 50 mutations cause PKD type 2, which affects adults.

polydentate *an alternative term for* **multidentate**.

polydeoxynucleotide *or* **polydeoxyribonucleotide** any **polynucleotide** composed solely of **deoxynucleotide** residues.

polydisperse describing a colloidal system whose disperse phase consists of particles of many different sizes. —**polydispersity** *n.*

polyelectrolyte any macromolecular substance that, on dissolution in water or another ionizing solvent, dissociates to give polyanions or polycations together with an equivalent of oppositely charged ions, which may be simple ions or themselves polyions.

polyene any unsaturated organic compound containing two or more carbon–carbon double bonds.

polyene antibiotic any member of a group of structurally related antibiotics produced by *Streptomyces* spp. They possess a large lactone ring containing 4–7 conjugated double bonds (generally all-*trans*) and substituted by aliphatic groups, by a residue of **mycosamine** or another amino sugar, and by numerous hydroxyl groups. The polyene antibiotics include **amphotericin B** and **nystatin**; they interact with sterols in cell membranes and hence have antifungal but not antibacterial activity and tend to be toxic to mammals. *Compare* **macrolide**.

polyenoic acid any alkenyl carboxylic acid containing two or more double bonds per molecule.

polyester any synthetic polymer in which the constitutional units are joined by ester bonds; proprietary materials include Dacron and Mylar. Polyesters are extensively used in textiles, ropes, etc.

polyethylene *an alternative name (esp. US) for* **polythene**. *See also* **polyethylene glycol**.

polyethylene glycol *abbr.*: PEG; any polymer of general formula $H(OCH_2CH_2)_n OH$, where $n \geq 4$; a suffixed number may be used as a guide to the average relative molecular mass, e.g. polyethylene glycol 4000. Polyethylene glycols are clear viscous liquids or white solids, useful as embedding media for microscopy, as bases for ointments, and in radioimmunoassays. Treatment of proteins with PEG increases their functional stability and decreases their clearance and immunogenicity when injected in therapy.

poly(ethyleneimine) a water-soluble synthetic linear polymer of ethyleneimine (i.e. aziridine) of 30–40 kDa. It is useful for the selective precipitation of proteins and nucleic acids. It is also used as a substituent in hydroxylated chromatographic support matrices, as in PEI-cellulose, to form anion exchangers. *Proprietary name*: Polymin P.

polyfructosan *former term for* **fructan**.

poly(G) *abbr. for* polyguanylate.

polygalactan *former term for* **galactan**.

polygalacturonase EC 3.2.1.15; *other name*: pectinase; an enzyme that hydrolyses 1,4-α-D-galactosiduronate linkages in pectin and other galacturonans in a random fashion. It participates in cell-wall degradation in plants.

polygene any one of a group of genes, each of which individually exerts a small effect on the phenotype but that together control a quantitative character such as stature.

polygenic describing a character or trait that is controlled by polygenes. **Polygenic diseases** result from the interaction of multiple genes and include common diseases such as diabetes mellitus, coronary heart disease, and hypertension. Some genes can have major effects whereas others might have relatively minor effects. *Compare* **multigenic**.

polyglucosan *a former term for* **glucan**.

polyglutamine *or* (*sometimes*) **polyglutamine tract** a stretch of amino acids in a protein that is entirely composed of glutamine. Gene sequences giving rise to this structure may have $(CAG)_n$ repeats and can be subject to replication slippage. A consequence is expansion of polyglutamine tracts in succeeding generations eventually causing symptoms of disease. Huntingtin is a well-known example. *See* **Huntington's disease**.

polyglutamine disease any disease of humans caused by the expansion of a CAG trinucleotide repeat in a gene. On expression, the region that contains the elongated stretch of glutamine repeats misfolds and forms antiparallel beta strands, with consequent production of nuclear and cytoplasmic inclusions. Examples include Huntington's disease, spinobulbar muscular atrophy, and various forms of spinocerebellar ataxia.

polygon any closed plane figure bounded by three or more (nonintersecting) straight lines. —**polygonal** *adj.*

polyhedron (*pl.* **polyhedrons** *or* **polyhedra**) any solid figure bounded by four or more plane faces, each of which is a polygon. —**polyhedral** *adj.*

polyhormone a polypeptide that is a precursor to more than one biologically active peptide; an example is **proopiomelanocortin**.

polyhydric describing any chemical compound containing two or more hydroxyl groups per molecule; used especially of alcohols.

poly(hydroxybutyrate) a linear polymer of approximately 1500 (*R*)-3-hydroxybutyrate residues held together in ester linkage. It is found in the cells of a number of prokaryotic genera where it occurs in membrane-bound granules and acts as a storage material.

poly(I) *abbr. for* polyinosinate.

polyion any macromolecular ion that carries multiple charges.

polykaryocyte a multinucleate cell. —**polykaryotic** *adj.*

polyketide *or (sometimes)* **acetogenin** any natural product synthesized via linear poly-β-ketones, which are themselves formed by repetitive head-to-tail addition of acetyl (or substituted acetyl) units indirectly derived from acetate (or a substituted acetate) by a mechanism similar to that for fatty-acid biosynthesis but without the intermediate reductive steps. In many cases, acetyl-CoA functions as the starter unit and malonyl-CoA as the extending unit. Various molecules other than acetyl-CoA may be used as starter, often with methylmalonyl-CoA as the extending unit. The poly-β-ketones so formed may undergo modification by alkylation, cyclization, glycosylation, oxidation, or reduction. Polyketides include: coniine (of hemlock) and orsellinate (of lichens) – acetyl-CoA; flavanoids and stilbenes – cinnamoyl-CoA; tetracyclines – amide of malonyl-CoA; urushiols (of poison-ivy) – palmitoleoyl-CoA; erythronolides – propionyl-CoA and methylmalonyl-CoA as extender. Polyketide synthases are large multidomain proteins that contain phosphopantetheine.

polylabelling labelling where more than one atom of a molecule has been isotopically substituted.

polylinker any **linker** containing multiple restriction sites. *See* **multiple cloning site**.

poly-L-lysine any polypeptide composed entirely of L-lysine. Solutions of poly-L-lysine are used to coat microscope slides to improve the adhesion of tissue sections, or to improve adhesion of oligonucleotides in the preparation of microarrays for gene expression profiling. The attachment and spreading of some mammalian cell types in culture is promoted by coating surfaces with poly-L-lysine.

polymannan *a former term for* **mannan**.

polymer 1 *(in chemistry and biochemistry)* any substance that is composed of molecules containing a large number of constitutional units (or 'mers') that are in repetitive covalent linkage and that may be of one or more than one species. Polymers are generally considered to comprise at least ten mers, although sometimes the term is taken to imply simply 'more than one' mer. Hence 'polymer' may or may not embrace **oligomer**, depending on the branch of chemistry or biochemistry concerned. **2** *(in molecular biology and enzymology)* *(sometimes)* an alternative term for **multimer**. *See also* **heteropolymer**, **homopolymer**. —**polymeric** *adj.*; **polymerize** *or* **polymerise** *vb.*; **polymerization** *or* **polymerisation** *n.*

polymerase a general name for any transferase enzyme that catalyses the formation of biological **polymers** (def. 1), especially of polynucleotides. *See* **DNA polymerase**, **RNA polymerase**.

polymerase accessory protein any of a group of proteins that are associated with **DNA polymerases**. Examples include DNA polymerase α binding factor from yeast, and phage T4 accessory proteins.

polymerase chain reaction *abbr.*: PCR; a method whereby a specific sequence of nucleotides within a double-stranded DNA is amplified. The sequence is identified by the use of short synthetic oligonucleotide primers that are complementary to the two terminal regions of the DNA sequence to be amplified; these oligonucleotides are extended by the thermostable *Taq* DNA polymerase on the DNA template. The effect is that the new chains span the region delimited by the two chosen termini. There are three phases in a single cycle of a PCR reaction: the nascent chains are heat denatured at 94°C; the temperature is then decreased to allow annealing of the primers to the template strands; and then the temperature is increased to 72°C for extension. The theoretical yield of amplified product is 2^n where n is the number of cycles. More than a 10^9-fold **amplification** of the DNA stretch required may be obtained. The method was first described by Kary B. Mullis in 1984. The use of the thermostable *Taq* (or other) polymerase from a thermophilic bacterium avoids the need to add more polymerase at each cycle because the enzyme is not inactivated at the temperature of denaturation of the DNA. PCR amplification has become an indispensable tool in a great variety of applications. Clinically it may be used for the rapid diagnosis of infectious diseases and the detection of rare mutations associated with cancer. Forensically, the DNA from even a single hair or sperm can be used to identify the donor. RNA may be amplified by first converting it to cDNA through the use of reverse transcriptase. *See also* **hot start PCR, inverse PCR, touchdown PCR, whole genome PCR**.

polymer drug a polymer that contains either chemically bonded drug molecules or pharmacologically active moieties. Polymer drugs are typically used to provide targeted drug delivery to an organ and controlled release of an active drug at the target organ.

polymer functionalization the introduction of chemical groups into a polymer to exert specific chemical, physical, biological, pharmacological, or other functions. *See* **functional polymer**.

polymetaphosphate *(sometimes)* an alternative name for **polyphosphate**.

poly(methyl methacrylate) a clear, easily worked, relatively nontoxic thermoplastic, useful as a substitute for glass and in the construction of many pieces of scientific apparatus. *Proprietary names*: Lucite, Perspex, Plexiglass, etc.

Polymin P *a proprietary name for* **poly(ethyleneimine)**.

polymorph *see* **polymorphonuclear leukocyte**.

polymorphic *or* **polymorphous** having, assuming, or passing through various different forms or styles; exhibiting or undergoing **polymorphism**.

polymorphism 1 *(in biology)* **a** the occurrence of two or more distinct kinds of individual (**morphs**) belonging to the same species within a population that is interbreeding freely. **b genetic polymorphism**. **2** *(in chemistry)* the existence of the same substance in two or more different crystalline forms. *See also* **enantiomorph**.

polymorphonuclear 1 describing a cell whose nucleus is multipartite or **polymorphic**. **2 a polymorphonuclear leukocyte**.

polymorphonuclear leukocyte *or* polymorphonuclear *or* polymorph *abbr.*: PMN; any white blood cell of the myeloid series that in its mature form has a lobed and variably shaped nucleus and granular cytoplasm; included are basophil leukocyte, eosinophil leukocyte, and neutrophil leukocyte. The term is often used erroneously for neutrophil leukocyte solely.

polymorphous *a variant form of* **polymorphic**.

polymyxin any of a family of antibiotics isolated from cultures of *Bacillus polymyxa* and active against most Gram-negative bacteria. Individual polymyxins are designated by suffixed letters, e.g. polymyxin B. All are decapeptides containing five or six residues of L-2,4-diaminobutyric acid. Their biosynthesis does not involve ribosomes (*see* **protein and peptide biosynthesis**). The sequence of seven residues at the C-terminal end is formed into a heterodetic ring through an isopeptide link to one of the diaminobutyric acid residues, while the N-terminal residue is acylated with a 6-methylheptanoyl or 6-methyloctanoyl group. Polymyxins bind to the plasma membrane of sensitive organisms, causing leakage of cytoplasmic components; however, they also attack the cell membranes of eukaryotic cells and therefore have limited clinical usefulness; they are used in antibiotic ointments. Polymyxin B inhibits protein kinase C. *See also* **colistin**.

CH₃   O
| ||
CH(CH₂)₄C—L-DAB-L-Thr-L-DAB
| | |
CH₂CH₃   NH₂   NH₂

└─L-DAB-L-DAB-D-Phe-L-Leu
       |
       NH₂

L-Thr——L-DAB——L-DAB
| | |
NH₂   NH₂

polymyxin B1
DAB = diaminobutyric acid

polynucleate *or* **polynuclear** *(of a cell)* containing several or many nuclei; **multinucleate**.

polynucleotidase *an alternative name for* **polynucleotide phosphatase.**

polynucleotide any one-stranded homo- or heteropolymer of at least about ten **nucleotide units** connected by phosphodiester linkages between (usually) the 3′ position on the glycose moiety of one nucleotide unit and the 5′ position on the glycose moiety of the adjacent one, or any two-stranded molecule comprised of two such one-stranded molecules held together by hydrogen bonds. Comprehends the term **nucleic acid.**

polynucleotide adenylyltransferase EC 2.7.7.19; *other names*: NTP polymerase; RNA adenylating enzyme; poly(A) polymerase (*abbr.*: PAP); an enzyme that adds the **poly(A) tail** to mRNA. It catalyses a reaction of the type:

$$n \text{ ATP} + (\text{nucleotide})_m =$$
$$n \text{ pyrophosphate} + (\text{nucleotide})_{m+n}.$$

PAP I and II contain an RNA-binding domain, a catalytic region, two nuclear localization signals, and a C-terminal serine/threonine-rich region. Clones of truncated forms have been obtained, some of which are expressed in tissues. The enzyme from *Escherichia coli* is also known as **plasmid copy number protein.** *See also* **polyadenylate.**

polynucleotide kinase *abbr.*: PNK; any enzyme belonging to subclass EC.2.7 that phosphorylates the 5′ end of nucleic acids. The reaction is between ATP and the 5′-OH of RNA or DNA to form ADP and RNA or DNA-5′-phosphate.

polynucleotide ligase *an alternative name for* DNA ligase (ATP); *see* **DNA ligase.**

polynucleotide ligase (NAD⁺) *an alternative name for* DNA ligase (NAD⁺); *see* **DNA ligase.**

polynucleotide phosphatase *or* **polynucleotidase** either of the enzymes polynucleotide 3′-phosphatase (EC 3.1.3.32) or polynucleotide 5′-phosphatase (EC 3.1.3.33).

polynucleotide phosphorylase *an alternative name for* polyribonucleotide nucleotidyltransferase (EC 2.7.7.8).

polyol any polyhydric alcohol.

polyol pathway (*sometimes*) a metabolic pathway by which non-phosphorylated sugars and sugar alcohols are interconverted.

polyomavirus any member of a genus (*Polyomavirus*) of mouse viruses of the Papovaviridae family. They are widely distributed and may be oncogenic under appropriate conditions.

polyose *an obsolete term for* any polymer consisting of an unknown or unspecified number of covalently linked monose residues.

polypeptide 1 any peptide comprising more than about 10–20 amino-acid residues connected by peptide linkages. Unlike a **protein**, it usually lacks appreciable tertiary structure in solution and is not liable to irreversible denaturation. **2** (*loosely*) an alternative term for **protein**. **3** describing another entity of a polypeptide nature, e.g. polypeptide chain, polypeptide hormone.

polypeptide hormone *an alternative term for* **peptide hormone.**

polyphase *or* **polyphasic 1** (of matter) consisting of two or more phases. **2** *or* **multiphase** (of an electrical system or device) having, generating, or using two or more alternating electric voltages of the same frequency but differing from each other in phase. **3** (of a process) consisting of or occurring in a number of separate stages.

polyphenol oxidase EC 1.10.3.1; *recommended name* catechol oxidase; any copper-containing oxidoreductase that catalyses the oxidation of mono and *o*-diphenols to *o*-diquinones. These enzymes are found in plants, localized in membranes of plastids, but their function is unknown. However, they are responsible for brown discoloration ('browning') in fruit and vegetables on storage, as their products give rise, probably nonenzymatically, to polyphenolic melanin-like compounds. Their gene sequences in different plants are very similar, but in relation to bacterial, fungal, or animal enzymes (*see* **monophenol monooxygenase**) only the putative copper-binding sites are conserved.

polyphosphate *or* (*sometimes*) **polymetaphosphate 1** any ionic form of polyphosphoric acid, $H_{n+2}P_n O_{3n+1}$ (where *n* is an unknown or unspecified integer greater than 1), or any mixture of polyphosphoric acid and its ionized forms. Polyphosphate occurs in the **volutin granules** observed in many microorganisms, where it is synthesized by the action of the enzyme **polyphosphate kinase** and is

believed to serve as a phosphate store. **2** any (partial or full) salt or ester of polyphosphoric acid. *See also* **metaphosphate.**

polyphosphate kinase EC 2.7.4.1; a microbial enzyme that catalyses the synthesis of **polyphosphate** by transfer of the terminal phosphoric residue of ATP to diphosphate (i.e. pyrophosphate) or to a polyphosphate one residue smaller than the product.

polyphosphoinositide *or* **polyphosphatidylinositide** *a former inappropriate general name for* any [mono]phosphate, bisphosphate, trisphosphate, or tetrakisphosphate of phosphatidylinositol; use not recommended.

polyploid describing a cell (or organism) that has three, four, five, or more times the haploid number of chromosomes in its nucleus. **—polyploidy** *n.*

polyprenol *a recommended term for* any **prenol** in which there are more than four isoprene residues; the isoprene units may be all-*trans* or a mixture of *cis* and *trans*. The number of such units in a particular polyprenol may be indicated by a multiplicative prefix, e.g. undecaprenol. Polyprenols and their derivatives are widely distributed in living organisms. *See also* **dolichol.**

polyprenyl the univalent alkyl group derived from a **polyprenol** by loss of its hydroxyl group. The number of isoprene residues in a particular polyprenyl derivative may be indicated by a multiplicative prefix, e.g. undecaprenyl diphosphate.

polyprotein any polypeptide that, on cleavage, gives rise to two or more biologically active proteins, e.g. **proopiomelanocortin.**

polyQ *an alternative term for* **glutamine repeat.**

polyribonucleotide any polynucleotide composed solely of ribonucleotide residues.

polyribonucleotide nucleotidyltransferase EC 2.7.7.8; *other name*: polynucleotide phosphorylase. An enzyme that degrades mRNA, phosphorylating single nucleotides successively from the 3′ end to yield a nucleoside diphosphate and a shortened mRNA.

polyribonucleotide synthase (ATP) *see* **RNA ligase (ATP).**

polyribosome *an alternative name for* **polysome.**

polysaccharide *an alternative name for* **glycan**; i.e. any linear or branched polymer consisting of monosaccharide residues. Important polysaccharides include glycogen, starch, hyaluronic acid, and cellulose.

poly-α2,8-sialic acid a linear homopolymer ($n \approx 100$) of α2,8-linked sialic acid molecules present mainly on **neural cell adhesion molecules** (NCAMs). Its biosynthesis is mediated by polysialyl transferase, which uses CMP-sialic acid as substrate. It is abundant in fetal brain during development, and in adult brain in areas of synaptic plasticity, and is a marker for several tumours.

polysome *or* **ergosome** *or* **polyribosome** any structure consisting of two or more ribosomes attached to different points on the same strand of messenger RNA. Hence, a polysome is a complex involved in the cellular synthesis of the polypeptide specified in the message. Polysomes effect the synthesis of all proteins, but in the case of secretory proteins they become attached to the **rough endoplasmic reticulum.**

polysomic 1 describing a cell or organism that has one or more normal chromosomes in excess of the usual diploid or polyploid complement; being such a chromosome. **2** a polysomic organism. **—polysomy** *n.*

polystyrene phenylethene homopolymer; a clear, brittle, rigid thermoplastic with high electrical resistivity. A proportion of divinylbenzene may be added as a cross-linking agent during polymerization in order to produce polystyrene resins of various porosities for use in the preparation of polystyrene ion-exchange resins.

polysubstituted having more than one chemical substituent. **—polysubstitution** *n.*

poly(T) *abbr. for* polythymidylate.

polytene chromosome a type of chromosome in a **polyploid** cell, formed when multiple copies of homologous chromosomes are aligned side by side to give a giant chromosome in which distinct chromosome bands are readily visible.

polyteny *an alternative name for* **endomitosis**. **—polytene** *adj.*

polytetrafluoroethylene *abbr.*: PTFE; tetrafluoroethene homopolymer; a polymer consisting of chains of ($\approx$2000) linked $-CF_2-$ units prepared by polymerization of tetrafluoroethylene, $F_2C=CF_2$. It is a nonflammable thermoplastic, extremely chemically resistant, of high electrical resistivity, and with an extremely low coefficient of friction. It is useful as tubing and sheeting in chemical laboratory work, as a lining to reaction vessels, as gasket material, and as packing for pumps and bearings. *Proprietary names*: Fluon, Teflon, etc.

polythene *or (esp. US)* **polyethylene** ethene homopolymer; a versatile thermoplastic polymer of ≈1.5–100 kDa, prepared by polymerization of ethylene. **High-density polyethylene**, a linear polymer, is fairly rigid and crystalline, while **low-density polyethylene**, a branched polymer, is softer and more flexible; both forms are chemically resistant and have high electrical resistivity.

polytopic describing an integral membrane protein that completely traverses the membrane two or more times.

polytrioxophosphate(v) *the recommended name for* **metaphosphate**.

poly(U) *abbr. for* polyuridylate.

polyUb *abbr. for* polyubiquitin. *See* **multiubiquitin**.

polyubiquitin 1 the form in which **ubiquitin** (def. 1) is synthesized, a precursor with a number of head-to-tail repeats. **2** *an alternative name for* **multiubiquitin**.

polyUb recognition factor a protein subunit of the 19S proteasome that binds multiubiquitin (polyubiquitin) chains tightly. When tested individually only one subunit (called S5a) qualified for the role.

polyunsaturated describing a substance containing more than one carbon–carbon multiple bond per molecular entity or residue; applied especially to a fatty acid containing more than one multiple bond in the hydrocarbon chain of its molecular structure, and to a fat or oil derived from one or more such fatty acids. *See also* **polyunsaturated fatty acid**. —**polyunsaturation** *n*.

polyunsaturated fatty acid *abbr.*: PUFA; any fatty acid that is **polyunsaturated**; the multiple bonds are most frequently all ethylenic and usually methylene-interrupted, the configuration of their substituents generally being Z (i.e. *cis*); more rarely is an acetylenic bond present. In animal metabolism, polyunsaturated fatty acids belong to families denoted by the position of the double bond nearest to the methyl group and often also by the name of the fatty acid from which the remainder can be derived; hence **linoleic family**, **linolenic family**, **oleic family**, and **palmitoleic family**. In recent years, emphasis has been placed on a role of some polyunsaturated fatty acids in lowering plasma cholesterol, and potentially reducing the risk of heart disease. Foremost among these have been those of the *n*-6 family such as linoleic acid. A recommendation has been made that the diet should contain these polyunsaturated fats in a ratio of 2:1 to saturated fats. *See also* **essential fatty acid, fatty-acid nomenclature, fish oil**.

polyuria the passing of large volumes of urine.

polyuronide *an old term for* **glycuronan**.

polyvalent *or* **multivalent 1** *(in chemistry)* **a** having a numerical valency greater than unity. **b** capable of exhibiting more than one valency value. **2** *(in immunology)* **a** (of an antibody molecule or fragment) having at least two combining sites. **b** (of an antiserum or other antibody preparation) capable of combining with two or more different antigens. **c** (of a vaccine) capable of eliciting immunity to more than one (strain of) microorganism, toxin, or other antigenic agent.

polyvinyl acetate *abbr.* PVA; ethenoic acid homopolymer; a synthetic polymer obtained from ethyne and acetic acid, used in adhesives and paints and in the manufacture of polyvinyl alcohol.

polyvinyl alcohol *abbr.*: PVA; ethenol homopolymer; a polymer prepared from polyvinyl acetates by replacement of the acetate groups by hydroxyl groups; the commercial products have differing contents of residual acetyl groups and hence differing physical characteristics. Polyvinyl alcohols are used in adhesives and can also be compounded into elastomers, which are useful in the manufacture of artificial sponges, hydrocarbon-resistant tubing, etc.

polyvinyl chloride *abbr.*: PVC; chloroethene homopolymer; a plastic solid, soluble in many organic solvents, that discolours on exposure to light or heat unless containing stabilizers. It is useful for the manufacture of rubber substitutes and electric wire coverings.

polyvinylidine fluoride *abbr.*: PVDF; a synthetic polymer, $(CH_2–CF_2)_n$; membranes made of PVDF are used to bind proteins

in Western blotting, protein sequencing, peptide mapping, and in analysis of amino acid composition.

polyvinyl pyrollidone *abbr.*: PVP; 1-ethenyl-2-pyrridinone homopolymer; commercial preparations are ≈10 kDa to ≈700 kDa; it is soluble in water to give colloidal solutions. PVP is used in the preparation of enzymes from plants; it removes phenolic compounds that can destroy enzyme activity.

polyxenic *see* synxenic.

pomace the pulpy residue remaining after extraction of juice from fruits such as apples, or after extraction of oil from seeds, nuts, or similar materials.

POMC *abbr. for* proopiomelanocortin.

Pompe's disease *an alternative name for* type II glycogen (storage) disease; *see* **glycogen disease**.

Ponceau S *or* **Fast Ponceau 2B** a dye used for staining protein, especially on cellulose acetate strips.

pontal (of an atom or group) bridging.

ponticulin an F-actin-binding transmembrane glycoprotein of plasma membranes of the cellular slime mould *Dictyostelium discoideum*. It is a major membrane protein of this organism, and a protein of the major high-affinity actin–membrane link. It is directly involved in actin polymerization.

pool *see* metabolic pool.

POPOP 1,4-bis(5-phenyloxazol-2-yl)benzene; a secondary fluor used in liquid scintillation counting, originally as a wavelength shifter and more recently to decrease sensitivity to the action of quenching agents.

population 1 *(in biology)* any group of individual organisms of the same species inhabiting a given area; the number of such individual organisms. **2** *(in statistics)* or **universe** any finite or infinite collection of individuals or items from which a sample has been drawn.

population genetics the study of the distribution of genes (alleles) in populations and the factors that change the frequency of genes and genotypes from generation to generation. It is based on the concept of the **Hardy–Weinberg law**.

population standard deviation *see* standard deviation.

P:O ratio *abbr. for* phosphorus:oxygen ratio.

porcine *abbr. (in compound abbrs.)* p *or (formerly)* P; of, characteristic of, or relating to a pig or pigs; resembling a pig.

porcupine a segment-polarity gene of *Drosophila* that encodes a member of the membrane-bound *O*-acyltransferase superfamily and is involved in Wingless signalling. The protein appears to be localized in endoplasmic reticulum.

pore an opening in a membrane that allows nonselective diffusional passage of small solutes across the membrane. In addition to pores formed spontaneously by thermal and electrical movement of lipid molecules there are the effects of extraneous (frequently toxic) substances that form pores.

pore-forming toxins subclass of transport proteins of the TC system, designated 1.C. It contains nearly 60 families, including **colicins**, δ-**endotoxins**, α-**hemolysins**, **aerolysins**, diphtheria, tetanus and botulinum toxins, **cytohemolysins**, **magainins**, **cecropin**, **melittin**, **defensins**, **nisins**, **mastoparans**, **pilosulins**, and **dermaseptins**. They are produced by bacterial, animal, and plant cells and make irreversible openings or lesions in the membranes of other cells.

porin 1 any member of a class of similar proteins of ~15 kDa (monomer) that can be isolated from the outer membranes of certain Gram-negative bacteria. Trimers of these proteins are thought to make these membranes porous to hydrophilic substances of ≤600 Da by forming in them water-filled channels that allow transmembrane diffusion of uncharged small solutes. The proteins are usually homotrimers, typically with β-barrel folds and including short α-helical domains. **2** any of the transmembrane proteins typically of the mitochondrial outer membrane in eukaryotes that form aqueous channels through the lipid bilayer, allowing passage of small molecules. These constitute a structural family, with high β structure content. Human mitochondrial membrane porin (voltage-dependent anion-selective channel protein) allows diffusion of small hydrophilic molecules; the channel adopts an open conformation at low or zero membrane potential, and a closed conformation at potentials above 30–40 mV.

porosity 1 the state or quality of being **porous**. **2** the degree to which a material or substance is porous.

porosome nuclear pore complex.

porous having pores; allowing the passage or uptake of gases and/or liquids, especially by diffusion rather than by bulk flow.

porphin *or* **porphine** *an alternative name for* **porphyrin** (def. 1).

porphobilinogen *abbr.*: PBG; 5-(aminomethyl)-4-(carboxymethyl)-1*H*-pyrrole-3-propanoic acid; the biosynthetic precursor of all porphyrins. It is excreted in relatively large amounts in the urine in some porphyrias.

porphobilinogen deaminase *see* **hydroxymethylbilane synthase**.

porphobilinogen synthase EC 4.2.1.24; *systematic name*: 5-aminolevulinate hydro-lyase (adding 5-aminolevulinate and cyclizing); *other names*: δ-aminolevulinic acid dehydratase (*abbr.*: ALAD); aminolevulinate dehydratase. An enzyme that catalyses the conversion of two moles of 5-aminolevulinate to one mole of porphobilinogen + 2 H_2O in the pathway for the synthesis of **porphyrins**. It is an abundant cytosolic homooctamer in mammalian cells. Each subunit contains a sulfhydryl group and a Zn atom. The crystal structure of the yeast enzyme consists of a TIM barrel. ALAD is identical with the 24 kDa proteasome inhibitor. Enzyme deficiency leads to a rare form of neurovisceral porphyria.

porphyr+ *a variant form of* **porphyro+** (before a vowel).

porphyria any metabolic disorder characterized by the excretion of abnormally large amounts of various porphyrins and/or their biosynthetic precursors. Porphyrias are currently classified as primary (if inherited) or secondary (if acquired). The primary disorders fall into two broad categories: (a) the neurological/psychiatric forms and (b) the forms associated with cutaneous photosensitivity, though these two types of symptoms may in some cases overlap. The acquired disorders are much more common. There is no clear-cut association of enzyme defects with specific porphyrias. Elevated blood levels of intermediates in porphyrin biosynthesis are normally present during attacks. Elevated intermediates that are found, together with (in parentheses) associated enzymes and conditions are (1) aminolevulinic acid (or ALA) (porphobilinogen (PBG) synthase: PBG synthase deficiency porphyria; *also called*: ALA dehydratase: ALA dehydratase deficiency); (2) ALA and PBG (PBG deaminase; acute intermittent porphyria, variegate porphyria; (3) uroporphyrin (uroporphyrinogen decarboxylase; porphyria cutanea tarda, acute intermittent porphyria) (4) coproporphyrin (coproporphyrinogen oxidase; variegate porphyria, hereditary coproporphyria). Acquired porphyria may be due to lead exposure, some malignancies, and chronic ethanol exposure. It has been suggested that the British king, George III, suffered from a form of porphyria leading to psychiatric symptoms. *See* **erythropoietic porphyria**.

porphyrin 1 *the systematic name for* the fundamental skeleton, $C_{20}H_{14}N_4$, of the macrocyclic **tetrapyrroles**; *often known also as* **porphin** *or* **porphine**. It consists of a ring of four pyrrole nuclei linked each to the next at their α positions (i.e. those adjacent to their nitrogen atoms) through a methine group, –CH= (*compare* **corrin**). The structure is aromatic and tautomeric, two hydrogen atoms being associated with any two of its four nitrogen atoms (though for the purposes of nomenclature the name 'porphyrin' implies that the saturated nitrogen atoms are at positions 21 and 23 in the molecule unless there is a specific indication otherwise). Many di- or trivalent metal ions are chelated by porphyrin through these four central nitrogen atoms. [*Note*: It was recommended in 1986 that the widely used Fischer system of (incomplete) numeration for the porphyrin ring be discontinued in favour of one in which all the atoms are numbered, the C atoms 1 to 20 and the N atoms 21 to 24.] The systems for numbering tetrapyrrole carbons are shown at **bacteriochlorophyll**. Porphyrin itself is a synthetic substance, deep purple in colour. **2** any member of a large group of naturally occurring or synthetic derivatives or analogues of porphyrin (def. 1), of which **protoporphyrin IX** and its derivatives, especially the chlorophylls and hemes, are the most important.

porphyrinogen 1 5,10,15,20,22,24-hexahydroporphyrin; *see* **porphyrin**. **2** *a general name for* the corresponding hexahydro derivative of any **porphyrin** (def. 2).

porphyrinuria the excretion of abnormally large amounts of porphyrins in the urine. *See also* **porphyria**.

porphyro+ *or (before a vowel)* **porphyr+** *comb. form* denoting purple.

porphyropsin a purple, light-sensitive, visual pigment in the rod cells of the retinas of freshwater and migratory fishes, and some amphibians, consisting of a rod-type **opsin** combined (as a Schiff's base) with 3,4-didehydro-11-*cis*-retinal. It has an absorption maximum at ~522 nm, and is analogous in function to **rhodopsin** in other vertebrates.

portable (*in molecular biology*) describing DNA sequences that can be inserted into recombinant molecules for the purposes of gene expression; e.g. portable promoter, portable Shine–Dalgarno sequence, etc.

portal 1 relating to a blood vessel, or system of blood vessels, that carries blood from one capillary bed to another, especially the hepatic portal vein. **2** relating to a porta, such as the hepatic porta (porta hepatis). **3** the route of entry into the body of a drug, microorganism, beam of radiation, etc. (or sometimes of its exit therefrom).

porter any transporter of solutes between aqueous phases on either side of a lipid membrane. The term includes **antiporter, symporter,** and uniporter (*see* **uniport**).

positional cloning a strategy for the identification of a gene of unknown function responsible for a genetic disease. The phenotype is correlated with a chromosomal site, the DNA of that region is cloned and compared with that in normal individuals. If a single gene is involved, it is sequenced, and the protein structure deduced. An example is the discovery of the genetic basis of cystic fibrosis and the identification of the **cystic fibrosis transmembrane conductance regulator** (CFTR).

positional information information supplied to or possessed by cells according to their position in a multicellular organism.

positional isomer any of two or more constitutional isomers that differ in the position of a particular substituent or group.

position-specific iterated-BLAST *abbr.:* PSI-BLAST; *see* **database-search program.**

position-specific scoring matrix *abbr.:* PSSM; *(in sequence analysis)* a position-specific scoring table that encapsulates the residue information within sets of aligned sequences. The scores may be simply the observed residue frequencies, or they may be the log-odds scores for finding a particular amino acid in a target sequence at a particular position.

positive 1 having a value greater than zero. **2** *(in mathematics)* of the same value but of opposite sense to that regarded as negative. **3** (of the results of a test) indicating the presence of the material or condition tested for. **4** (of an electric charge) having the opposite polarity to the charge of an electron. **5** (of a molecular entity) having a positive electric charge; having a deficiency of electrons. **6** (of a point in an electric circuit) having a higher potential than some other point to which a potential of zero is assigned. **7** *(in biology)* signifying growth or movement towards a (specified) stimulus. **8** *(in photography)* an exposed and developed photographic film, plate, or print showing an image in colours or tones corresponding to those of the object photographed. *Compare* **negative.** —**positiveness** *n.*; **positivity** *n.*

positive chromatography *(sometimes)* any method of purification of substances by affinity chromatography in which a substance of interest in a sample interacts with and is selectively retained by the adsorbent, especially as opposed to **negative chromatography.**

positive control the initiation or enhancement of some biological activity by the presence of a specific molecular entity.

positive cooperativity cooperative ligand binding in which binding of one ligand to one site on a macromolecule increases the affinity at other sites for the subsequent binding of the same or ligands.

positive correlation *see* **correlation** (def. 2).

positive Darwinian selection the relatively rare phenomenon whereby selective pressure favours change. A commonly used statistical test for positive Darwinian selection is comparison of the rate of synonymous with nonsynonymous substitutions: where nonsynonymous substitutions occur more frequently than synonymous substitutions, this is evidence that natural selection has favoured amino acid change.

positive effector *see* **effector.**

positive electron *an alternative name for* **positron.**

positive feedback *see* **feedback** (def. 1).

positive-negative selection a strategy used in making gene knockouts designed to enrich for homologous recombinants and select against random integration of the targeting vector. Cells transfected with the targeting vector incorporating two **selectable markers** are first selected for their resistance to an antibiotic such as G418 (positive selection) and then for the loss of a second marker, such as HSV thymidine kinase (HSV*tk*), which confers sensitivity to ganciclovir (negative selection). HSV*tk* placed at one end of the linear targeting vector is lost on homologous recombination but is retained on random integration.

positive staining a technique used for preparing specimens for electron microscopic examination in which charged biological macromolecules or structures are rendered electron-dense by allowing heavy-metal ions of opposite charge to bind to them. *Compare* **negative-contrast technique.**

positron *or* **positive electron** *symbol:* β⁺; an elementary particle that

is the antiparticle of the electron. It has the same mass as an electron and a numerically equal but positive electric charge.

positron emission tomography *see* **tomography.**

post+ *or (sometimes)* **post-** *prefix* denoting after in time or sequence; behind, posterior to.

postabsorptive describing the period immediately after absorption of food.

posterior after in position, sequence, or time; behind; following.

posterior lobe the posterior lobe of the pituitary gland; i.e. the **neurohypophysis.**

postgreSQL *see* **relational database management system.**

postlysosome a degenerate **telolysosome** that has lost its enzymes.

postmitochondrial fraction *or* **postmitochondrial supernatant** *or* **supernatant fraction** *an operational term for* the part of a homogenate of eukaryotic cells or tissue remaining after sedimentation of the mitochondrial fraction by centrifugation. It consists essentially of the **microsomal fraction** and the cytosol.

post-mortem 1 occurring after death. **2** a post-mortem examination; necropsy, or **autopsy.**

postnatal describing the period after birth or giving birth.

postpartum describing the period immediately after birth or giving birth.

postprandial describing the period immediately after intake of food.

postreplication repair *see* **recombinational repair, mismatch repair.**

postreplicative describing the period following replication of a cell or of a nucleic-acid molecule.

postribosomal fraction *or* **postribosomal supernatant** *or* **supernatant fraction** *an operational term for* the part of a homogenate of eukaryotic cells or tissue remaining after sedimentation of the ribosomal (microsomal) fraction by centrifugation. It consists essentially of the cytosol.

postsynaptic 1 describing a neuron or a neuronal membrane that is on the efferent side of a synapse. **2** describing the membrane of a muscle cell at a motor end-plate.

postsynaptic density *abbr.:* PDS; a structure, visible on microscopy, that is associated with a postsynaptic membrane. It consists of a variety of scaffolding proteins and signalling enzymes, such as neuronal nitric oxide synthase, Src, Ca^{2+}-calmodulin-dependent protein kinase, and calcineurin. The protein PDS-95 also couples NMDA receptors to some of these enzymes.

postsynaptic membrane *see* **synapse.**

post-transcriptional describing any phenomenon or process occurring after completion of transcription of genomic nucleic acid. —**post-transcriptionally** *adv.*

post-transcriptional gene silencing *an alternative term for* **RNA silencing.**

post-translational describing any phenomenon or process occurring after completion of translation of messenger RNA into polypeptide; postsynthetic. —**post-translationally** *adv.*

post-translational modification *or* **post-translational processing** *abbr.:* PTM; any enzyme-catalysed change to a protein made after it is synthesized by the translation of messenger RNA. Examples are proteolytic cleavage and C-terminal peptide amidation. **glycosylation, methylation, phosphorylation,** and **prenylation.**

Pot-1 *abbr. for* protection of telomeres-1; a protein, widely distributed in eukaryotes, that binds to the G-rich DNA of telomeres.

potash 1 *the common name for* potassium hydroxide or potassium carbonate. **2** *the name for* potassium in the common names of certain of its salts.

potassium *or (in scientific and medical Latin)* **kalium** *symbol:* K; a soft, silvery white, highly reactive, metallic element of group 1 of the (IUPAC) **periodic table**; atomic number 19; relative atomic mass 39.0983. Natural potassium is a mixture of stable nuclides of mass numbers 39 (93.2 atom percent) and 41 (6.7 atom percent) together with a minute proportion of the radionuclide of mass number 40 (0.012 atom percent). Potassium is widely distributed in all living organisms, besides occurring in various minerals; it is one of the most abundant elements of the Earth's crust. It forms a monovalent cation, K^+. Several artificial radioactive isotopes are known. In the body, potassium is a major intracellular cation, with little in the extracellular compartment; the range in normal human plasma is

3.6–5.0 mmol L^{-1}, with serum 0.2–0.3 mmol L^{-1} higher due to release from platelets on clotting. The cellular uptake of potassium is stimulated by insulin, and its excretion by the kidney is indirectly controlled by aldosterone, which stimulates sodium reabsorption in the distal tubules, potassium being excreted by passive transfer due to the electrochemical gradient thereby created. It is actively pumped into cells by **sodium/potassium-ATPase**. *See also* **hyperkalemia, hypokalemia**.

potassium-40 the naturally occurring radioactive nuclide $^{40}_{19}$ K. It has a half-life of 1.28×10^9 years, emitting beta particles (i.e. electrons) of 1.34 MeV max. and gamma radiation of 1.46 MeV.

potassium-42 the artificial radioactive nuclide $^{42}_{19}$ K. It has a half-life of 12.36 h, emitting high-energy beta particles (i.e. electrons) of 3.52 MeV max. (82%) and 1.97 MeV max. (18%), together with gamma radiation of 1.52 MeV (18%) and 0.31 MeV (0.15%).

potassium channel *or* **K$^+$ channel** any type of structure that permits controlled (gated) passage of potassium ions, predominantly efflux, through membranes. In excitable cells the role of potassium channels is to set the resting potential, participate in restoring it after depolarization, and attenuate and regulate potentials generated by other excitatory inputs. They are classified into: (1) those primarily responding to membrane polarization, termed voltage-dependent, voltage-gated, or delayed rectifier (*see* **rectification**) K$^+$ channels; transmembrane (typically six transmembrane domains, designated S1 to S6) proteins; (2) Ca^{2+}-activated K$^+$ channels, which are activated by increase of intracellular Ca^{2+} concentration; (3) receptor-coupled channels, which may be activated or inhibited by receptor occupancy; these may have an intrinsic ion channel as part of the receptor, or act through a G-protein that activates an independent channel; and (4) ATP-sensitive K$^+$ channels that close in response to increased intracellular ATP/ADP ratios generated by increased metabolic flux. Types 3 and 4 are known as inward rectifier channels, in that they favour inward movement of K$^+$. Such channels have two transmembrane domains rather than the six of voltage-gated channels. However, all K$^+$ channels have a loop known as the H5 or P region, which is thought to line the conducting pore. In voltage-gated channels, this lies between segments S5 and S6, and in inward rectifier channels between the two transmembrane domains.

potency *(in pharmacology)* an expression of the activity of a drug or other agent, in terms of either the concentration or the amount needed to produce a defined effect. It is an imprecise term that should always be further defined. *Compare* **EC$_{50}$, IC$_{50}$**.

potential *(in physics)* **1** the work done in bringing unit positive charge, unit positive pole, or unit mass from infinity to a specified point in an electric, magnetic, or gravitational field, respectively. **2** **electric potential**. **3 electric potential difference**.

potential difference *abbr.*: p.d.; **1** generally, the difference in **potential** (def. 1) between any two points. **2** *see* **electric potential difference**.

potential divider a chain of resistors (or inductors or capacitors) connected in series, across which an electric potential may be applied and from which a desired fraction of the potential may then be obtained from a contact suitably placed in the chain. *See also* **potentiometer** (def. 3).

potential energy *abbr.*: PE; *symbol*: E_p or V or Φ; the energy of a body or system consequent upon its physical state or its position in an electric, magnetic, or gravitational field. It is expressed as the work necessary to bring it to its present state or position from a reference state or position, to which a potential energy value of zero is assigned.

potential energy barrier *or* **energy barrier** **1** the difference between the bond energy at which a molecule dissociates and the energy of that molecule in the ground state. **2** the difference between the energy of the ground state of an enzymically activated reaction complex and the sum of the energies of the ground states of the reactants forming the activated complex.

potential gradient **1** the potential difference per unit distance between a point of higher and one of lower (electric) potential; the algebraic differential of this at a particular point. **2** (in cell biology) the difference in chemical and/or electrostatic potential across a membrane.

potentiate to cause to be (more) potent, especially to increase the potency of a drug or other agent. —**potentiation** *n*.

potentiometer **1** a precision instrument for measuring an electric potential difference of constant polarity without drawing current from the circuit being examined. In its simplest form, it consists of a length of wire of uniform resistance and across which a constant potential is applied. By means of a sliding contact the proportion of this applied potential is found that will balance the potential being measured as determined by a null reading on a galvanometer. **2** a direct-reading electronic device for measuring electric potential difference or electromotive force without drawing appreciable current. **3** *or (informal)* **pot** a development of a potential divider consisting of a resistor with a movable contact and used, especially in electronics, either as a potential divider or as a variable resistor. —**potentiometry** *n*.; **potentiometric** *adj*.; **potentiometrically** *adv*.

potentiometric titration *or (sometimes)* **electrometric titration** titration in which an electrode reversible to one of the ionic components of the analyte or titrant is immersed in the analyte solution and the electric potential of the electrode (relative to that of an inert reference electrode) is measured during addition of titrant. Because the potential of an electrode is a logarithmic function of the activity of the ion to which the electrode is reversible, an abrupt change in its potential is observable as an equivalence point is approached.

potocytosis a mechanism for small molecule transport across eukaryotic cell membranes that uses **caveolae** rather than clathrin-coated pits.

Potter–Elvehjem homogenizer a device for preparing small quantities of cell or tissue homogenates. It consists of a cylindrical pestle (of glass, Perspex, or Teflon) that fits closely into a hard-glass test tube. Cells or small pieces of tissue in a suspending medium are placed in the tube and the pestle is rotated by means of an electric motor while the tube is slowly moved up and down by hand, a shear thereby being imparted to the contents of the tube. [After van Rensselaer Potter (1911–2001), US biochemist, and Conrad Arnold Elvehjem (1901–62).]

POU domain a bipartite DNA-binding domain composed of a POU-specific domain (POU$_s$, ≈75 amino acids) and a POU-homeodomain (POU$_h$, 60 amino acids) that are highly conserved, are joined by a hypervariable linker sequence, and form helix-turn-helix structural motifs. It occurs in many **transcription factors, and** was first recognized in three families of transcriptional regulators, Pit (*see* **PIT**), Oct, and Unc (*see*, e.g., **unc-86**). POU domain genes include the Brn family and others. The recognition sequences are A/T-rich, including the highly conserved **octamer** sequence 5'-ATG-CAAAT-3'. The domain is also involved in protein–protein binding. Mutations in two POU-domain proteins (POU3F4 and POU4F3) result in isolated deafness.

pound *symbol*: lb; the avoirdupois unit of mass, now defined as 0.453592338 kg exactly.

powder pattern *or* **powder diagram** a diffraction pattern obtained from a specimen of a crystalline material in finely divided form. The specimen is rotated in a parallel beam of monochromatic X-rays, electrons, neutrons, etc. so that all planes of the crystals assume all possible orientations with respect to the incident beam.

power **1** *symbol*: P; a measure of the rate of doing work, expressed as energy divided by time. The derived SI unit of power is the **watt**. **2** *(in optics)* a measure of the ability of a lens or mirror to magnify, expressed as the reciprocal of the focal length of the lens or mirror in metres. **3** *(in mathematics)* **a** an alternative name for **exponent** or index; e.g. 2 to the power 3 = 2^3). **b** one more than the cardinal number of times that a number or expression has to be multiplied by itself to give a required number; e.g. 2 to the power 3 = $2 \times 2 \times 2$. **c** one more than the ordinal number of times that a particular number has to be multiplied by itself to give a required number; e.g. the third power of 2 is 8 = $2 \times 2 \times 2$.

power calculation a statistical test used to assess how many samples have to be analysed to maximize success in answering a hypothesis, while minimizing the use of resources. Power calculations are used in genetic association studies.

poxvirus any member of the Poxviridae, a family of large, double-stranded DNA viruses, in class I of the **Baltimore classification**. They consist of a single molecule of DNA of 130–240 MDa together with

30 or more structural proteins, a number of enzymes, and lipid. Their hosts range from insects to mammals; examples include the viruses of variola (smallpox) and vaccinia (cowpox) – but not of varicella (chickenpox) – in humans, and the virus of myxomatosis in rabbits.

Pp *symbol for* the propionyl group.

PP *abbr. for* pancreatic polypeptide.

PP$_i$ *symbol for* inorganic pyrophosphate. *See* **diphosphate**.

PPAR *abbr. for* peroxisome proliferator-activated receptor; any of a subfamily of nuclear receptors for hormones. Mice produce three isoforms. **PPARα** predominates in liver, kidney, and heart. It binds leukotriene B$_4$ and heterodimerizes with retinoid receptor (RXR) to activate numerous genes. PPARα and PPARγ are activated by the hypolipidemic agent tetradecylthioacetic acid. **PPARγ** is the dominant transcription factor of adipocyte precursors. It binds thiazolidinediones (synthetic antidiabetic agents). When phosphorylated by MAP kinase, PPARγ inhibits adipogenesis. The ligand binding site of PPAR is homologous to that of the human estrogen receptor. In rats, but not in humans, clofibrate causes peroxisomal proliferation.

P particle a liberated, mature **kappa particle**.

ppb *or* **p.p.b.** *abbr. for* parts per billion, i.e. parts per 10^9.

PPCA *abbr. for* **protective protein/cathepsin C**.

PP cell *or (formerly)* **F cell** a type of cell containing granules that store pancreatic polypeptide (**pancreatic hormone**). In adult humans such cells are located exclusively in the pancreas (79% in the islets, 19% in the acini, and 2% in the ducts).

pp[CH$_2$]pA *symbol for* adenosine 5′-[α,β-methylene]triphosphate; an analogue of ATP in which the α,β-bridging oxygen atom has been replaced by a methylene group.

PPF *or* **PP factor** *abbr. for* pellagra-preventative factor; an early name for **nicotinic acid** or **nicotinamide**.

ppGpp *symbol for* guanosine 3′-diphosphate 5′-diphosphate (*also known as* **magic spot I**). *See* **relA, spoT**.

p22-phox *see* phagocyte **NADPH oxidase**.

p47-phox *see* phagocyte **NADPH oxidase**.

p67-phox *see* phagocyte **NADPH oxidase**.

PPIase *abbr. for* peptidylprolyl isomerase.

PPLO *abbr. for* pleuropneumonia-like organism (i.e. **mycoplasma**).

ppm *or* **p.p.m.** *abbr. for* parts per million, i.e. parts per 10^6.

PPO *abbr. for* 2,5-diphenyloxazole; a substance useful as a primary fluor in liquid scintillation counting.

pppGpp *symbol for* guanosine 3′-diphosphate 5′-triphosphate (*also known as* **magic spot II**). *See* **relA**.

PPRibP *symbol for* (5-)phospho(-α-D-)ribosyl diphosphate.

P protein a membrane protein of **melanosomes** that functions in ion transport into the melanosome and in maintaining its acidic pH. The protein consists of 838 amino acids and contains 12 putative transmembrane sequences. Its gene is at 15q11.12, and numerous mutations are associated with a form of oculocutaneous **albinism**.

P-protein *abbr. for* phloem protein; any structural protein of phloem, the principal food-conducting tissue of vascular plants. P-proteins are present usually at low concentration in phloem exudates, when they are called **sieve tube exudate proteins** (STEPs).

P-protein body any crystalline or amorphous accumulation of protein that may be present during sieve element differentiation in vascular plants. The bodies may be in equilibrium with soluble monomeric proteins.

PPS (*in clinical chemistry*) *abbr. for* pepsin A.

ppt 1 *or* **ppt.**; *abbr. for* precipitate. 2 *or* **p.p.t.** *abbr. for* parts per trillion, i.e. parts per 10^{12}.

PPT *abbr. for* 1 preprotachykinin (*see* **protachykinin β precursor, tachykinin**). 2 phosphinothricin.

PQ *abbr. for* plastoquinone.

PQH$_2$ *abbr. for* dihydroplastoquinone (i.e. **plastoquinol**).

PQQ *abbr. for* pyrroloquinoline quinone (i.e. **methoxatin**). *See* **quinoprotein**.

Pr *symbol for* 1 praseodymium. 2 the propyl group.

Pr^i *symbol for* the isopropyl group, (CH$_3$)$_2$CH–.

prazosin *or* **furazosin** a quinazoline antagonist of α$_1$ adrenoceptors; it has anti-hypertensive properties.

pre+ *or* **pre-** *prefix* 1 before in time, position, sequence, rank, etc.;

compare **pro+** (def. 1). 2 denoting a substance that moves faster (or further) than a specified substance in a separative procedure (e.g. prealbumin). Usage in this sense for naming proteins is now deprecated in order to avoid confusion with def. 3. 3 denoting a (usually metabolic) precursor, whether endogenous (e.g. prehormone, prekallikrein, pre-messenger RNA, preprotein, pretyrosine) or exogenous (e.g. precarcinogen), of the (class of) substance specified; *Compare* **pro+** (def. 3); *see also* **prepro+**.

PRE *abbr. for* proton relaxation enhancement.

prealbumin *an operational (and confusing) term for* either of two protein-containing zones observed to migrate more rapidly towards the anode than does serum albumin on zone electrophoresis of serum at pH 8.6 by some techniques. The faster-moving of these zones is now known to be **transthyretin**, and the other to be acidic α$_1$-glycoprotein (i.e., **orosomucoid**).

prebiotic of or relating to the period before the appearance of living organisms on Earth.

precancerous describing a tissue or condition that is not yet malignant but is expected to form a cancer.

precarcinogen (*sometimes*) a chemical carcinogen that can be converted by metabolism into another of no less potency. *Compare* **proximate carcinogen**.

precarcinomatous describing a tissue that is expected to form a carcinoma.

P receptor *see* **purinoceptor**.

precerebellin a cerebellum-specific protein that is the precursor of cerebellin (a 16-residue peptide flanked by Val-Arg and Glu-Pro), which is a protein of the postsynaptic structures of Purkinje cells. Concentrations of cerebellin are low at birth, and (in humans) peak at days 5–15. Its sequence is conserved in mammals and birds.

precess to undergo, or cause to undergo, **precession**.

precession (slow) gyration of the axis of rotation of a spinning body caused by application of a torque such that the axis of rotation generates a cone.

precipitant an agent or a reagent that induces the formation of a precipitate.

precipitate 1 to (cause to) come out of solution or suspension. 2 substance or material that is precipitated. —**precipitation** *n*.

precipitating antibody *an alternative name for* **precipitin**.

precipitin *or* **precipitating antibody** an antibody that is capable of reacting with a soluble antigen with the formation of an insoluble antigen–antibody complex.

precipitin reaction the reaction together of antigen and antibody near their equivalence point to form a cross-linked precipitate. Quantitative analysis of this interaction gives both the antibody content of an immune serum and an indication of the valency of the antigen. If the reaction occurs in a gel then the reactants form precipitin arcs.

precision 1 a measure of the sharpness (or certainty) of a measurement – a precise measurement may nevertheless be inaccurate (e.g. if there are unrecognized errors in the measurement). 2 *or* **selectivity** (*in informatics*) the proportion of the predictions made by an algorithm that are correct (i.e. discrimination of true matches from false): true positives/(true positives + false positives).

precision-bore (of tubing or a tubular object) having a lumen whose diameter varies only within extremely narrow limits.

prednisolone 1,4-pregnadiene-11β,17α,21-triol-3,20-dione; a semisynthetic steroid with glucocorticoid action. *Compare* **prednisone**.

prednisolone

prednisone 1,4-pregnadiene-17α,21-diol-3,11,20-trione; a synthetic steroid with glucocorticoid action. *Compare* **prednisolone**.

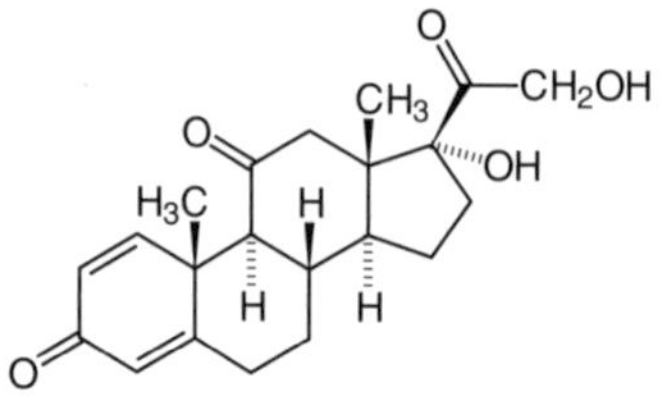

preelectrophoresis a preliminary treatment of an electrophoresis gel in which a potential difference is applied in order to remove undesirable substances such as excess polymerization reagents.

preflashing the hypersensitization of photographic emulsion by a preliminary exposure to a brief (≈1 ms) flash of light in order to increase its sensitivity to low levels of radiation and to render its response more nearly linear. It is useful especially in fluorography and (auto)radiography.

prefoldin *or* **Gim complex** (*abbr. for* genes involved in microtubule synthesis complex) a chaperone that cooperates in the proper folding of actin and tubulin by binding to their nascent polypeptide chains. In eukaryotic cytosol it is a ≈90 kDa complex of two α and four β subunits. Its structure is that of six alpha-helical coiled-coil tentacles arising from a beta-barrel body. It does not associate with the ribosome synthesizing the nascent chain and operates in an ATP-independent manner.

pregnancy-associated α2-glycoprotein *see* **pregnancy zone protein**.
pregnancy-associated plasma protein A *see* **PAPP-A**.
pregnancy zone protein *abbr.*: PZP; *other names*: pregnancy-associated α2-glycoprotein; α2-pregnoglobulin; a homotetrameric disulfide-linked proteinase inhibitor of the α2-macroglobulin family that inhibits all four classes of proteinase by a 'bait-and-trapping' mechanism (*see* **bait region**). One of the major human pregnancy-associated plasma proteins, it may reach levels of 1.0–1.4 g L^{-1} just before term. It has extensive homology with human α2-macroglobulin. In the sequence (example below) amino acids 1–24 form the signal, 26–1482 are the pregnancy zone protein, and 685–735 the bait region.

pregnane a C$_{21}$ steroid to which **pregnanediol** is related.

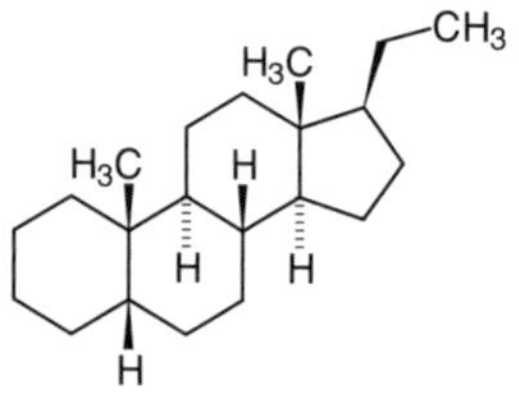

5β-pregnane

pregnanediol *a common name for* the steroid 5β-pregnane-3α,20α-diol; (20S)-5β-pregnane-3α,20-diol; the principal urinary metabolite of progesterone, occurring mainly as its 3-glucuronide. Its level in the urine increases during the luteal phase of the menstrual cycle and especially in the latter part of pregnancy, the excretion rate being a convenient (if approximate) index of the functional state of the corpus luteum and placenta.

pregnant mares' serum gonadotropin *abbr.*: PMSG; a clinically useful impure gonadotropin preparation extracted from the serum of pregnant mares – such serum has high activities of both follicle-stimulating hormone and luteinizing hormone. *Compare* **urogonadotropin**.

pregnenolone *a common name for* the steroid Δ^5-pregnen-3β-ol-20-one; 3β-hydroxypregn-5-en-20-one; a key intermediate on the biosynthetic route from cholesterol to progesterone. It has gluco-

pregnanediol

corticoid activity and has been used in the treatment of rheumatoid arthritis. It is the product of **cholesterol monooxygenase (side-chain-cleaving)**.

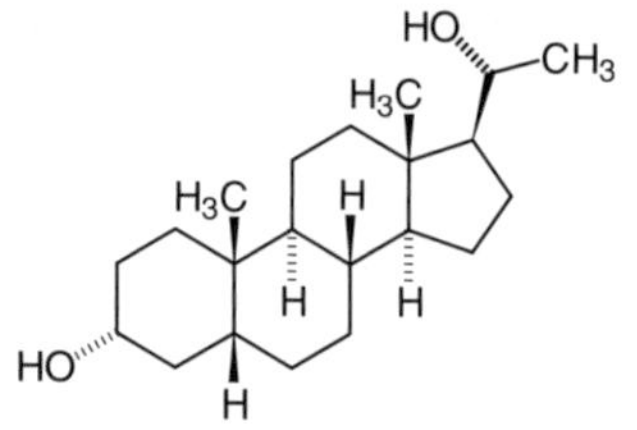

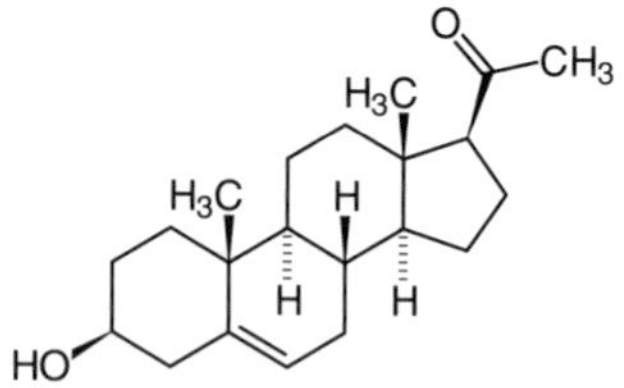

α2-pregnoglobulin *see* **pregnancy zone protein**.

prehybridization a procedure used for the treatment of nitrocellulose or nylon membranes following Northern or Southern transfer and before the use of labelled nucleic acid probes to detect specific sequences on the blot. The intention is to block the surface of the membrane to decrease non-specific binding of the probe. A variety of blocking agents can be used.

preincubate to bring a reaction mixture or a cell or tissue preparation to (chemical, metabolic, thermal, etc.) equilibrium before starting the reaction to be investigated. —**preincubation** *n.*

pre-initiation complex the complex formed on a gene promoter by RNA polymerase II and the general transcription factors, especially TFII, before initiation of transcription of the gene.

prekallikrein a plasma glycoprotein of 88 kDa (bovine), and a **blood coagulation** factor. It is the precursor of the protease **kallikrein**, to which it is converted by the action of factor XIIa. The protein contains four **apple domains**. The active site contains His, Asp, and Ser residues characteristic of a **charge-relay system**.

prelysosomal vacuole *or* **prelysosome** *a former name for* **endosome** (def. 1).

pre-messenger RNA *abbr.*: pre-mRNA; the class of primary gene transcripts from which the corresponding messenger RNA molecules are subsequently formed by post-transcriptional processing; any member of this class. It comprises a substantial part of **heterogeneous nuclear RNA**.

pre-miRNA the product of **Drosha** activity on **pri-miRNA** and a substrate for **Dicer**.

pre-mRNA *abbr. for* pre-messenger RNA.

pre-mRNA processing protein (*abbr.*: PRP protein) *or* **pre-mRNA splicing factor** any of the proteins involved in processing the primary transcript RNA. Many of them function as components of a **spliceosome**. Most are components of snRNP particles, which are numbered U1, U2, etc. The following examples are mostly from yeast (*Saccharomyces cerevisiae*): PRP2 is an ATP-binding RNA helicase required for the first cleavage ligation reaction; PRP5 is an ATP-binding RNA helicase required for unwinding of pre-mRNA, and is part of the spliceosome, a so-called **DEAD-box** helicase; PRP8 is a U5 snRNP protein; PRP9 is required for U2 snRNP to bind to pre-mRNA; PRP15 is an RNA-dependent ATPase required for branch-point recognition; PRP16 is an RNA helicase (another DEAD-box helicase); PRP18 is a component of the U4/U5/U6 snRNP, and is a homodimer; PRP19 is a U2-binding protein of unknown function; PRP21 is probably involved in retention of pre-mRNA in the nucleus; PRP22 is a DEAD-box helicase involved in release of spliced mRNA from the spliceosome; PRP28 is a DEAD-box helicase involved in U4/U5/U6 SnRNA; PRP38 is involved in maintenance of U6 snRNA; PRP39 facilitates interaction between U1 snRNP and the 5'-splice site.

premurein-leader peptidase EC 3.4.99.35; *other names*: prolipoprotein-signal peptidase; signal peptidase II; SPase II. An enzyme that cleaves a single bond between Gly^{20} and alkylated Cys^{21} in a bacterial lipoprotein, detaching the leader peptide.

premutation an expanded triplet repeat region of intermediate size in a gene, that does not of itself cause a disease. However, it is likely to increase in length during meiosis and to cause the disease in subsequent generations. In **fragile X syndrome** the $(CGG)_n$ repeat may increase in length during meiosis in the mother, more so than in the father. The risk that there will be expansion is proportional to the number of repeats and is 99% for repeat lengths of greater than 90. *See* **anticipation**.

prenol *the recommended name for* any **isoprenoid** of general formula $H–[CH_2–C(CH_3)=CH–CH_2–]_n –OH$; i.e. any primary monohydroxy alcohol whose carbon skeleton consists of two or more isoprene residues linked head to tail. *See also* **polyprenol**.

prenyl the univalent alkyl group derived from a **prenol** by loss of its hydroxyl group. *See also* **polyprenyl**. *Compare* **isopentenyl**.

prenylation the enzymic addition of **prenyl** moieties to proteins as a post-translational modification; **geranyl, farnesyl,** or **geranylgeranyl** groups may be added. The process involves reaction of a prenyl diphosphate with a cysteinyl sulfhydryl group near the C terminus of the protein, to give a prenyl–S–Cys moiety. Characteristically a Cys-ali-ali-Xaa sequence is recognized by the transferase that mediates the reaction, where Cys is cysteine, ali is an aliphatic amino acid, and Xaa is any amino acid. When Xaa is serine, alanine, or methionine the protein is farnesylated; when Xaa is leucine it is geranylgeranylated, but this is variable; for example, in the case of Rab 3A (*see* **RAB**), the Cys-Xaa-Cys motif leads to geranylgeranylation on both cysteine residues and methyl esterification of the terminal carboxyl, while in other members of the Rab family Cys-Cys directs geranylgeranylation without esterification. The function of the process appears often to be regulation of protein–protein and protein–membrane interaction.

prenyltransferase *see* **dimethylallyl*trans*transferase**.

PR-enzyme *an old name for* **1** **phosphorylase phosphatase** (EC 3.1.3.17). [PR stood originally for prosthetic-group removing, later for phosphorylase rupturing, and later still for phosphate removing.] **2** deoxyribodipyrimidine photo-lyase (EC 4.1.99.3). [PR stood for photoreactivating.]

preparative ultracentrifuge any ultracentrifuge designed for preparation or purification of tangible quantities of macromolecular substances, subcellular fractions, etc.

pre-part *or* pre-peptide *an alternative name for* **signal peptide**.

prephenate a trivial name for 2-oxo-3(1-carboxylato-4-hydroxy-2,5-cyclohexadiene-1)-propanoate; a nonaromatic intermediate in the branched biosynthetic pathway (the **prephenate pathway**) from chorismate to phenylalanine or tyrosine in most autotrophic organisms. *See also* **arogenate**.

prephenate dehydratase EC 4.2.1.51; *systematic name*: prephenate hydrolyase (decarboxylating); an enzyme that catalyses the formation of phenylpyruvate from prephenate with release of H_2O and CO_2 in a pathway for the biosynthesis of phenylalanine. In some cases (e.g. *Escherichia coli*) the activity is in the N-terminal domain of a bifunctional enzyme that also contains **chorismate mutase** (another enzyme of the phenylalanine biosynthetic pathway). *See also* **arogenate**.

prephenate dehydrogenase any enzyme that converts prephenate to 4-hydroxyphenylpyruvate with release of CO_2 and reduction of either NAD^+ or $NADP^+$. They function in the pathway for the biosynthesis of aromatic amino acids, to initiate the branch leading to tyrosine biosynthesis. They include prephenate dehydrogenase, EC 1.3.1.12; *systematic name*: prephenate:NAD^+ oxidoreductase (decarboxylating); *other name*: hydroxyphenylpyruvate synthase; an enzyme that also has **chorismate mutase** activity, so can convert chorismate to hydroxyphenylpyruvate. There is also prephenate dehydrogenase ($NADP^+$), EC 1.3.1.13; *systematic name*: prephenate:$NADP^+$ oxidoreductase (decarboxylating).

prephenate pathway a pathway for the synthesis of phenylalanine and tyrosine from chorismate, utilizing prephenate as an intermediate. Prephenate is formed from chorismate by **chorismate mutase**. It is converted by **prephenate dehydrogenase** to 4-hydroxyphenylpyruvate, from which tyrosine is formed by transamination with glutamate as the amino-group donor. For the synthesis of phenylalanine, phenylpyruvate is formed by **prephenate dehydratase**, and is then converted to phenylalanine by transamination, glutamate again acting as amino-group donor. *See also* **arogenate**.

pre-piece *an alternative name for* **signal peptide**.

prepolypeptide *or* **presecretory polypeptide** the transient, primary translation product of a eukaryotic messenger RNA for any polypeptide destined to be secreted by a cell, irrespective of whether an intermediary **propolypeptide** is formed; the polypeptide counterpart of **preprotein**. *See also* **prepropolypeptide**.

prepro+ *or* prepro- *compound prefix* denoting the presence in a secretory polypeptide or protein of the N-terminal signal sequence and any other sequence than the polypeptide or protein of interest.

preprocessing enzymic conversion of a preproprotein (or prepropolypeptide) to a proprotein (or propolypeptide).

preprohormone the presecretory precursor of a polypeptide or protein hormone, which contains the N-terminal signal sequence, the sequence of the hormone of interest, and any other sequence. Examples are: preproglucagon, preproinsulin, preproparathyrin, prepro-VIP.

preprophase band a dense band of microtubules, 1–3 μm wide, that appears just beneath the cell membrane before prophase of cell division in higher plants. It disappears as mitosis begins, yet it somehow determines the plane of orientation of the new cell plate forming in late telophase and marks the zone of the parental cell wall where fusion with the growing cell plate ultimately occurs.

prepropolypeptide the prepolypeptide of any propolypeptide: e.g., preproenkephalin and prepromelittin.

preproprotein the preprotein of any proprotein: e.g., preproalbumin (serum) and prepro-von Willebrand factor. Note that preprolactin is not a preproprotein, as prolactin is not a protein.

preprotachykinin *abbr.*: PPT; *see* **protachykinin β precursor**.

preprotein the transient, primary translation product of a eukaryotic messenger RNA for any protein destined to be secreted by a cell, irrespective of whether an intermediary **proprotein** is formed. It is detectable only in a cell-free protein-synthesis system, as in an intact cell its N-terminal presequence (i.e., **signal peptide**) is cleaved off before translation has been completed. *See also* **prepolypeptide**, **preproprotein**.

prereplication complex *another name for* origin of replication complex. *See* **ORC**.

presecretory polypeptide *an alternative name for* **prepolypeptide**.

presenilin either of two multipass transmembrane proteins, mutation of which results in familial **Alzheimer's disease**.

presequence *an alternative name for* **1** signal peptide. **2** transport sequence (of a protein imported into mitochondria), especially in those instances where the sequence is not removed by proteolytic cleavage prior to translocation through the membrane, but is removed at its final destination.

pressor *(in physiology)* **1** of, relating to, or bringing about an increase in arterial blood pressure. **2** any drug or other agent that has a pressor effect.

pressure *symbol*: p *or* P; the force per unit area acting on a surface; i.e. $p = F/A$. The pressure at a point in a fluid is equal to the force per unit area on an infinitesimal plane situated at that point, and in a fluid at rest it is the same in all directions. The SI derived unit of pressure is the **pascal**; the **bar** is also acceptable (*see* **standard atmosphere**). *See also* **millimetre of mercury, torr**.

pressure dialysis dialysis in which a hydrostatic pressure is applied to the material being dialysed in order to minimize its initial dilution by solvent entering by osmosis.

pressure homogenization *or* **nitrogen cavitation** a method of disrupting cells that depends on equilibrating a cell suspension with a gas, e.g. nitrogen, at high pressure (up to 7 MPa) and then suddenly reducing the pressure, whereupon the gas dissolved in the cytoplasm is explosively released.

pressure-jump method a relaxation technique in which change of pressure is the disturbing factor. *See* **relaxation kinetics**.

pre-steady-state kinetics *or* **transient-phase kinetics** the kinetics of a catalysed reaction during the phase in which the concentration of the catalyst–reactant complex rises from zero to its steady-state value. For an enzyme-catalysed reaction this phase usually occupies a very brief period of time (from a second to a microsecond or less) and special techniques are required for its study. *See also* **rapid reaction kinetics**.

prestin an integral membrane motor protein responsible for the electromotility of outer hair cells of the mammalian cochlea. It contains 12 transmembrane segments and is located in the basolateral membrane, where it undergoes structural rearrangement in response to transmembrane voltage changes. Intracellular chloride and bicarbonate ions bind to prestin and trigger the structural change.

pre-swollen describing a commercial gel preparation that has already been equilibrated with water or another liquid phase.

presynaptic 1 *(in anatomy)* **a** designating a neuron that transmits an impulse directly to a synapse. **b** of or pertaining to that part of such a neuron close to the synapse. **2** *(in cytology)* prior to meiotic synapsis.

presynaptic membrane *see* **synapse**.

presynaptic nerve ending *see* **synapse**.

pre-transfer RNA *abbr.*: pre-tRNA; the class of primary gene transcripts from which the corresponding transfer RNA molecules are subsequently formed by post-transcriptional processing; any member of this class.

pretransition temperature the specific temperature at which a lipid bilayer being heated shows the first signs of **melting** (def. 2), before it reaches its transition temperature. Special techniques (e.g. employing spin-labelled or fluorescent probes) are required for its detection and study.

pretyrosine an *obsolete name for* **arogenate**.

previtamin *a less common word for* **provitamin**.

PRF *abbr. for* Protein Research Foundation; *see* **sequence database**.

PRG-1 *abbr. for* plasticity-related gene-1; a gene expressed in rat hippocampal neurons during active axonal outgrowth. It encodes a membrane-associated phospholipid phosphatase that modulates the extracellular phospholipid concentration and regulates axonal outgrowth.

Pribnow box a promoter site in *Escherichia coli* and phage genes that specifically binds **RNA polymerase**. The Pribnow box has the consensus sequence TATAAT and is centred at –10 (ten nucleotides on the 5′ side of the first nucleotide transcribed, which is designated +1).

PRibPP *symbol for* (5-)phospho(-α-D-)ribosyl diphosphate.

primaeval soup *a variant spelling of* primeval soup; *see* **primordial soup**.

primapterin *or* **7-biopterin** the 7-isomer of **biopterin** that is formed nonenzymatically from tetrahydrobiopterin. The reaction is prevented by a bifunctional protein (103 amino acids) that has **pterin-4α-carbinolamine dehydratase** (PCD) activity and is also the dimerization cofactor for hepatocyte nuclear factor 1α (DCoH). Several mutations in PCD result in hyperphenylalaninemia and primapterinuria.

primaquine one of several 8-aminoquinoline antimalarial drugs. It is active against the exo-erythrocytic form of the parasite, and can effect radical cures as well as serving as a preventive medicine. Primaquine sensitivity is a hemolytic anemia that occurs in persons deficient in glucose 6-phosphate dehydrogenase.

primary *(in chemistry)* **1 a** describing an alkyl compound (e.g. an alkanol) in which the functional group (e.g. a hydroxyl group) is attached to a carbon atom linked to only one other. **b** describing the carbon atom bearing the functional group in such a compound. **2** describing an amide or an amine in which a single appropriate group is attached to a nitrogen atom. **3** describing a salt formed by replacing only one of the ionizable hydrogen atoms of a tribasic acid by another cation. *Compare* **secondary, tertiary**.

primary cell wall the outer layer of the wall of a plant cell. It is produced by deposition of cellulose microfibrils against the cell plate separating two daughter cells newly formed by division. When cell extension is complete, it is supplemented by deposition of the secondary cell wall (*see* **secondary cell wall deposition**).

primary culture a culture of cells (or tissue) that have been obtained directly from their place of origin.

primary database an *alternative name for* **sequence database**.

primary electron an electron incident on a surface that causes **secondary electrons** to be emitted.

primary F′-containing strain *see* **F′ plasmid**.

primary fluor *or* **primary scintillator** *or* **primary solute** *(in liquid scintillation counting)* the fluorescent substance to which radiation energy is directly tranferred from excited molecules of the solvent. *Compare* **secondary fluor**.

primary granule an *alternative name for* **azurophil granule**.

primary hormone *a term proposed to embrace* any hormone that acts rapidly and uses a cyclic nucleotide as second messenger.

primary immune response the response produced in an animal when it is first exposed to an antigen. It consists characteristically of a slow production of antibody molecules and the priming of lymphoid tissue in readiness for the production of a **secondary immune response** on subsequent challenge with the same antigen.

primary kinetic isotope effect an **isotope effect** on a reaction rate that is attributable to isotopic substitution of an atom to which a bond is made or broken in the rate-limiting step or in a pre-equilibrium step of a specified reaction.

primary lysosome *or* **protolysosome** a **lysosome** that is a precursor of a **secondary lysosome**.

primary metabolism the ensemble of metabolic activities that are common to most if not all living cells and are necessary for growth, maintenance, and survival. *Compare* **secondary metabolism**.

primary metabolite any intermediate in, or product of, **primary metabolism**.

primary response *see* **primary immune response**.

primary scintillator an *alternative name for* **primary fluor**.

primary solute an *alternative name for* **primary fluor**.

primary solvent *(in liquid scintillation counting using mixed solvents)* the solvent to which energy of the radiation emitted by the specimen is first transferred as excitation energy.

primary standard 1 any standard of measurement of nationally or internationally agreed and specified value that is used as the basis of a unit of measurement. **2** a specimen of a particular substance that is of extremely high purity and stability and that serves as an ultimate reference to which may be compared the purity of any secondary standard used in analysis.

primary structure the first order of complexity of structural organization exhibited by polypeptide and protein molecules, and by polynucleotide and nucleic-acid molecules. When applied to a segment of a polypeptide chain, or to a polypeptide or protein molecule, it refers to the linear sequence of the amino-acid residues of the polypeptide chain(s), without regard to spatial arrangement, apart from configuration at the α-carbon atoms (and excluding the positions of any disulfide bonds). When applied to a segment of a polynucleotide chain, or to a polynucleotide or nucleic-acid molecule, it refers to the linear sequence of the nucleotide residues of the polynucleotide chain(s), without regard to spatial arrangement. *Compare* **secondary structure, tertiary structure, quaternary structure**.

primary transcript the RNA as synthesized by RNA polymerase directly from a gene, unaltered by any further modification, such as **splicing**, that it may undergo.

primase 1 any RNA polymerase that synthesizes an RNA primer needed for the initiation of DNA synthesis. Such enzymes vary greatly in structure, specificity, and mode of regulation. **2** *or* **DNA primase** the DNA-directed **RNA polymerase** (EC 2.7.7.–) of the bacterial **primosome**. In *Escherichia coli*, primase is the product of the *dnaG* gene while in T4 phage the primase is the product of gene 61. In either case, the primase is only active in the presence of other proteins which create a complex called a **primosome**. The complex participates in the unwinding of parental DNA strands ahead of the replication fork, and synthesizes short RNA primers for the **Okazaki fragments**.

primate any member of the Primates, an order of placental mammals. Primates are characterized by flexible pentadactyl limbs, opposable first digits, good stereoscopic vision, and, in the higher apes, highly developed brains. Included are lemurs, tarsiers, monkeys, apes, and humans.

prime 1 the printer's character, ′, used (singly, doubly, etc.) as a mark to discriminate between characters or symbols representing things that are or would otherwise be indistinguishable. In chemical nomenclature it is used to distinguish between (1) identical locants referring to different moieties of the same molecule; or (2) two or more unspecified but differing univalent groups represented by the same symbol R. In physical chemistry it is used to differentiate the symbol for a quantity measured under arbitrary conditions from that for the corresponding quantity referred to the (thermodynamic) standard state; e.g. K', apparent equilibrium constant, from K, thermodynamic equilibrium constant; E_o', standard redox potential at a specified pH, from E_o, standard redox potential at a hydron activity of unity. **2** *(in biochemistry)* to provide a metabolic reaction with a primer; to bring about activation of a process or substrate. **3** *(in immunology)* to induce a **primary immune response** in an animal; to activate a cell specifically with respect to a given antigen. **4** *(in mathematics)* **a** having no other integral factor than itself and one. **b prime number**.

prime number any integer that possesses no integral factors and hence is divisible only by itself and 1.

primer 1 something used to **prime** (def. 2, 3). **2** an oligonucleotide required as the starting point for the stepwise synthesis of a polynucleotide from mononucleoside triphosphates by the action of a nucleotidyltransferase. **3** a short-chain fatty acyl-CoA, e.g. acetyl-CoA or butyryl-CoA, required for the initiation of the stepwise enzymic synthesis, from two-carbon units, of long-chain fatty acids as their CoA derivatives. *See also* **glycogenin**.

primer extension a technique used to map the 5′ ends of mRNA transcripts and thereby identify the exact position of the transcription start site in the sequence of a gene. A labelled antisense oligonucleotide primer is hybridized to mRNA close to the 5′ end and extended by the action of reverse transcriptase. The length of the resulting cDNA reflects the distance between the labelled oligonucleotide and the 5′ end of the mRNA template.

primeval soup or *(esp. Brit.)* **primaeval soup** *an alternative name for* **primordial soup**.

primeverose 6-*O*-β-D-xylosyl-D-glucose; a disaccharide occurring in **madder**.

pri-miRNA *abbr. for* primary transcript **miRNA**, a substrate for **Drosha**.

primordia *the plural of* **primordium**.

primordial existing at or from the beginning, primeval; earliest formed, primitive.

primordial soup or **primordial slime** or **primeval soup** or *(esp. Brit.)* **primaeval soup** the aqueous solution of organic compounds that is considered to have been the environment in which the synthesis of macromolecules first occurred and cellular life originated. *See also* **coacervate droplet**.

primordium *(pl. primordia)* an organ or part at the earliest stage of its development; a rudiment.

primosome a protein complex that is essential for the activity of the

primase responsible for the synthesis of RNA primers in the replication of DNA in *Escherichia coli*. *See* **replication complex, replication of DNA**.

PRINTS *see* **pattern database**.

prion a protein originally identified as the transmissible pathogenic agent responsible for prion diseases, such as **scrapie, kuru, bovine spongiform encephalopathy, Creutzfeld–Jakob disease**, and **Gerstmann–Sträussler syndrome**. The agent is now widely accepted to be a variant of the normal **prion protein**; it is believed that the variant folds abnormally, and induces the normal protein also to fold abnormally, typically into beta conformations that form aggregates. *See also* **fatal familial insomnia**.

prion protein *abbr.*: PrP^C; a membrane protein of 33–35 kDa that is encoded in the mammalian genome; it contains, in humans, 253 amino-acid residues (near the same number in other species), and has a glycosylinositol phospholipid anchor. A variant, termed PrP^{Sc}, with the same amino-acid sequence, is found in the brains of sheep and other mammals affected by **spongiform encephalopathy**. PrP^C is rich in α-helix, while PrP^{Sc} contains much β-sheet. PrP^{Sc} associates to form aggregates ('rods') in the central nervous system; such aggregates are seen in many manifestations of spongiform encephalopathy. Polymorphism, e.g. at position 171, occurs in several species and in subgroups in one species, and may be important with respect to disease susceptibility.

prism 1 any polyhedron of which two faces are polygons in parallel planes and the other faces are parallelograms. **2** a crystal form of which three or more faces are parallel to one axis. **3** a solid, bounded in part by two nonparallel plane faces and made of material transparent in the relevant region of the spectrum, that is used in optics to invert an image or to disperse or deviate a beam of light.

PRKAG3 *abbr. for* protein kinase AMP-dependent gamma chain 3. *See* **AMPK.**.

PRL *abbr. for* prolactin.

pro+ or **pro-** *prefix* **1** before in time or position; anterior; projecting; *compare* **pre+** (def. 1). **2** for, on behalf of, substituting for, favouring, encouraging, etc. **3** *(in biochemistry)* denoting (inactive or less active) biosynthetic precursor of the (class of) proteinaceous agent specified, to which it is convertible by proteolytic cleavage in a specific manner. Examples are: proalbumin, prohormone, proinsulin, proprotein, and prothrombin (but not prolactin!). *Compare* **pre+** (def. 3). *See also* **+gen**.

Pro *symbol for* a residue of the α-amino (strictly, cyclic α-alkylamino) acid L-proline (alternative to P).

pro- *prefix* in symbols designating prochirality; *see* **pro-R/pro-S convention**.

proaccelerin *an alternative name for* factor V; *see* **blood coagulation**.

probability *(in statistics)* *symbol*: p or P; a measure of the likelihood of the occurrence of a given event, expressed as the ratio of the number of times it occurs in a series of observations to the total number of observations, or as its decimal equivalent.

probable error an older measure of sampling variability such that one-half of the events, observations, etc. in a **normal distribution** lie between the mean value minus the probable error and the mean value plus the probable error. The probable error is equal to the standard error multiplied by 0.6745.

proband *an alternative term for* **propositus**.

probe 1 something that searches into, examines, or tests; e.g. an instrument or tube used for exploring a cavity or sampling a fluid, or an electric conductor introduced into an electric circuit or a cavity resonator to provide electric coupling to an external circuit. **2** *(in biochemistry)* a chemical species, or a chemical group attached to a carrier molecule, used to investigate the chemical nature of some other particular chemical species or the physical nature of a particular environment, especially within an intact cell; a **reporter group**. Typically some measurable physical property of the probe (e.g. fluorescence, electron spin resonance) alters in a way that can be correlated with changes in the environment or in the macromolecule. Examples are a group having paramagnetic properties, for detection using electron spin resonance (*see*, e.g., **TEMPO**), and a fluorescent lipid molecule used to report membrane fluidity (*see*, e.g., **parinaric acid**). **3** *(in molecular biology)* or **gene probe** or **hybridization probe** an oligo- or polynucleotide that is complementary to an oligonucleotide or nucleic-acid sequence under investigation. The probe is normally labelled in a way that permits its ready detection, e.g. with a radioactive isotope or a fluorescent constituent. Hybridization of the probe to the oligonucleotide sequence under investigation allows that sequence to be detected. The technique finds much use in screening gene libraries, detecting oligonucleotides on blots of electrophoresis gels, and in other associated technologies. **4** to examine with or as with a probe.

procainamide a local anaesthetic similar in action to **procaine** but having a less easily hydrolysed amide linkage, which gives it longer duration of action.

procainamide azide a local anaesthetic; it binds the 43 kDa postsynaptic protein (μ chain *or* rapsyn) of the nicotinic **cholinoceptor** and is useful as a photoaffinity label.

procaine *or* **novocaine** the prototype injectable local anaesthetic. It brings about changes in the physical properties of the lipid bilayer towards a more fluid state, decreasing ordering of the fatty-acyl side chains.

procapsid a precursor of the virion of certain viruses, consisting of a viral capsid without its specific nucleic acid.

procarboxypeptidase component III a protein, resembling **elastase** but without enzyme activity, found in the pancreatic juice of ruminants in noncovalent association with zymogens, carboxypeptidse A, etc. It probably prevents denaturation of the latter in the acid of the duodenum.

procaryon *a variant spelling of* **prokaryon**.

procaryote *a variant spelling of* **prokaryote**.

process *(in anatomy)* a thin prominence or protuberance of a structure, or an extension or outgrowth of a cell.

processed gene a **pseudogene** that lacks the intervening sequence or sequences of the homologous functional gene. It usually contains a short segment consisting entirely or almost entirely of adenine nucleotides, attached near the 3′ end, and is thought to originate by reverse transcription of an mRNA template with incorporation of the cDNA so formed into genomic DNA. A processed gene is sometimes called a **retrogene**.

processing *(in molecular biology)* **covalent modification** either of the primary transcription product (i.e. RNA) of a DNA or of the primary translation product (i.e. polypeptide) of a messenger RNA. *See also* **post-translational modification**.

processing body *see* **P body**.

processive *(in molecular biology)* or **nondistributive** describing (the action of) any enzyme or catalytic complex that progressively synthesizes or degrades a biopolymer by effecting several or many cycles of the same type of reaction or reaction sequence without dissociating from the template or intermediate product (as appropriate) between catalytic events. *Compare* **distributive**.

processivity (of a nucleic acid polymerase) the average number of nucleotides inserted before the enzyme dissociates from its template. In the case of DNA polymerases, repair enzymes (such as Pol

I in *Escherichia coli*) have low processivity replication enzymes (such as Pol II and its associated proteins in *E. coli*) have high processivity.

processivity clamp *an alternative name for* proliferating cell nuclear antigen. *See* **PCNA**.

processosome a hypothetical ribonucleoprotein associated with processing of pre-RNA in the nucleolus.

prochiral describing a molecule, or a centre or axis in a molecule, that in itself is achiral but is capable of becoming **chiral** by a replacement reaction. In general, a prochiral molecule has a structure *Caabc*, where the '*a*' groups (or atoms) are identical. Replacement of one '*a*' by another group or atom, '*d*', leads to a chiral molecule. Thus, in ethanol, replacement of one ^{1}H of the methylene group by ^{2}H leads to a chiral structure which can be shown to be optically active. In a prochiral molecule *Caabc*, the chemically like '*a*' groups may be differentiated one from another in a reaction with a chiral molecule, especially an enzyme. An achiral molecule containing a double bond or a planar cyclic ring system is also prochiral if separate addition of an achiral reagent to each face leads to two enantiomeric products. *See also* **meso-carbon atom**, **pro-R/pro-S convention**. —**prochirality** *n*.

procholecystokinin precursor a member of the gastrin/ cholecystokinin family, and a precursor for five different **cholecystokinins**.

prochromacin the sequence that comprises residues 79–439 of human chromogranin A. It is bacteriostatic and antifungal by forming membrane ion channels, and is the precursor of four active peptides: chromacin (residues 176–197), pancreastatin (250–301), catestatin (352–372), and parastatin (357–428). *See* **chromogranin**.

Procion *the proprietary name for* a range of reactive dyes that includes Procion Blue H-B, Procion Red H-E3B, as well as a number of other dyes useful in biochemistry, especially in **dye-ligand chromatography**.

procoagulant *(sometimes)* any of factor V, factor VII, or factor VIII, each of which acts to accelerate the conversion of prothrombin to thrombin during **blood coagulation**.

procollagen an ~150 kDa triple-helical protein that is an intermediate in the biosynthesis of **tropocollagen**, into which it is converted proteolytically by removal of the additional sequences at both the N and the C termini of each of the three peptide chains. *Compare* **protocollagen**.

procollagen C-endopeptidase *an alternative name for* bone morphogenetic protein 1.

procollagen N-endopeptidase EC 3.4.24.14; *other name*: procollagen N-proteinase; an enzyme that catalyses the cleavage of the N-propeptide of collagen chain α-1(I) at Pro-|-Gln and of chain α-2(II) at Ala-|-Gln. Mutations at these sites lead to Ehlers–Danlos syndrome type VII. Inactivating mutations in the enzyme lead to dermatosparaxis of cattle.

procollagen-lysine 5-dioxygenase EC 1.14.11.4; *other names*: procollagen-lysine,2-oxoglutarate 5-dioxygenase (*abbr.*: PLOD); lysine hydroxylase; lysine,2-oxoglutarate 5-dioxygenase; an enzyme that catalyses the formation of hydroxylysine residues in collagen. The reaction is between dioxygen, procollagen L-lysine, and 2-oxoglutarate to form procollagen 5-hydroxy-L-lysine, succinate, and CO_2. Iron and **ascorbic acid** are cofactors. There are three genes for PLOD in humans. Several inactivating mutations in PLOD1 cause Ehlers–Danlos syndrome type VI. *See also* **procollagen-proline dioxygenase**.

procollagen-proline 4-dioxygenase EC 1.14.11.2; *systematic name*: procollagen-L-proline,2-oxoglutarate:oxygen oxidoreductase (4-hydroxylating); *other names*: proline,2-oxoglutarate-4-dioxygenase; protocollagen hydroxylase; proline hydroxylase; prolyl 4-hydroxylase. An enzyme that catalyses a reaction between dioxygen, procollagen L-proline, and 2-oxoglutarate to form procollagen *trans*-4-hydroxy-L-proline, succinate and CO_2. Hydroxylation only occurs during translation. **Ascorbic acid** and Fe^{3+} are also required as cofactors, and the reaction is one of the few well-defined roles for ascorbic acid; it seems likely that scurvy, as a result of vitamin C deficiency, is due to defective activity of this enzyme, which plays a key role in collagen synthesis. The enzyme contains two α chains and two β chains; the β chain is the multifunctional **protein disulfide isomerase** and hence has microsomal triacylglycerol transfer protein activity. It resides in the endoplasmic reticulum lumen. Two forms of the α subunit are produced by alternative splicing from the same gene. See also **procollagen-lysine 5-dioxygenase**.

proconvertin *an alternative name for* factor VII; *see* **blood coagulation**.

procorticotropin *or* **procortin** *an alternative name for* **proopiomelanocortin**.

procyclin *or* **PARP** a major surface antigen of procyclic forms of trypanosomes. It is GPI-anchored.

ProDom *see* **cluster database**.

prodrug a drug that is inert but has pharmacological effects after **bioactivation**. An example is **ganciclovir**.

produce *(in mathematics)* to extend a line or a plane.

product 1 *(in biochemistry and chemistry)* something formed in a reaction. **2** *(in mathematics)* the result of multiplying together two or more numbers, quantities, or expressions.

product inhibition the inhibition of an enzymic reaction caused by increased concentration of one or more products of that reaction.

productive (capable of) yielding a product, result, or benefit.

productive binding *or* **eutopic binding** the binding of a substrate in a reactive mode at the active site of an enzyme. *Compare* **nonproductive binding**.

productive complex any enzyme–substrate complex in which the substrate is bound to the enzyme in a manner that renders catalysis possible so that products can be formed. *Compare* **abortive complex**.

productivity *(in biotechnology) symbol*: *r*; the mass of product formed per unit reactor volume per unit time; often per unit of enzyme or biomass. It is measured typically in $kg\ m^{-3}\ h^{-1}$.

prodynorphin *see* **dynorphin**.

proenkephalin a protein precursor for several neuropeptides, including various **enkephalins** formed in brain and adrenal medulla. For example, in the human, proenkephalin A precursor yields four copies of [Met]enkephalin (100–104; 107–111; 136–140, 210–214) and single copies of each of [Leu]enkephalin (230–234), [Met]enkephalin–Arg–Gly–Leu (octapeptide) (186–193), and [Met]enkephalin–Arg–Phe (heptapeptide) (261–267), which arise by enzymatic cleavage of the gene product by cleavage at paired basic amino acids. Another human protein, proenkephalin B precursor (*other name*: β-neoendorphin–dynorphin precursor) yields single copies of α-neoendorphin, dynorphin, [Leu]enkephalin, rimorphin, and leumorphin, also by cleavage at paired basic amino acids.

proenzyme *an alternative name for* **zymogen**.

pro-E/pro-Z convention if the two chemically-like hydrogen atoms of *ab*C=CH$_2$ are separately replaced by the achiral, *d*, an achirotopic stereogenic element is generated and two (diastereoisomeric) alkenes, *ab*C=CH*d* are fomed that are *E* and *Z* stereoisomers (*see* **E/Z convention**). The replacement has converted a prostereogenic element into which one which is stereogenic; the situation is comparable to the change, prochiral → chiral, although a chiral element is not involved. The two chemically-like hydrogen atoms are designated as *pro-E* and *pro-Z*, and are abbreviated as H$_e$ and H$_z$. Replacement of a *pro-E* hydrogen with ^{2}H yields an *E* diastereoisomer and the similar replacement of a *pro-Z* hydrogen yields the *Z* form. Example: the two hydrogen atoms of phospho*enol*pyruvate; that on the same side of the double bond as the COOH group is H$_E$, that on the same side as the OPO$_3$H$_2$ group is H$_Z$. Replacement of ^{1}H$_E$ by ^{2}H yields (*E*)-[3-^{2}H$_1$]phospho*enol*pyruvate; the *Z* diastereoisomer results if ^{1}H$_Z$ is replaced by ^{2}H. A similar situation is found in the side chain of **chorismic acid**.

proestrus *see* **estrous cycle**.

profibrinolysin *an alternative name for* **plasminogen**.

profile *(in sequence analysis)* a scoring table that encodes the residue conservation within an aligned domain, used to provide a diagnostic signature for that domain. Profiles provide differential scores/penalties for matches, deletions, or insertions, and for transitions between these states. They are included as a supplement to the regular expressions in the PROSITE database. *Compare* **hidden Markov model**.

profilin any of a family of globular proteins (123–139 amino acids) that prevent the polymerization of actin. Profilins are ubiquitous in eukaryotes and consist of a five-stranded beta sheet and three small

alpha helices. They bind dynamin and clathrin and localize with them in axons and dendrites, and bind diaphanous in hair cells of the inner ear. They compete with cofilin for ADP-actin, and promote exchange of the ADP for ATP. In mice profilin is essential for life. Profilin binds PtdInsP_2 and inhibits the γ isoform of phospholipase C (*see* **phospholipase**), an inhibition that is overcome by receptor kinase-dependent tyrosine phosphorylation.

proflavin 3,6-diaminoacridine; an **acridine** dye that intercalates between adjacent base pairs in duplex DNA. It is useful as an antibacterial agent, by inhibiting the biosynthesis of both DNA and RNA, possibly through intercalation into preexisting DNA, and as a frameshift mutagen for bacteriophages.

progeria any of two rare hereditary syndromes characterized by premature ageing in children (*see* **Hutchinson–Gilford syndrome**) or in adults (*see* **Werner syndrome**).

progesterone *the common name for* pregn-4-ene-3,20-dione; the principal steroid hormone of the corpus luteum, secreted during the latter half of the estrous cycle and acting upon the endometrium to prepare it for embryo implantation. During pregnancy, secretion occurs also from the placenta, increasing markedly as pregnancy proceeds, when it acts to maintain the uterus, inhibit further release of ova, and promote proliferation of acini within the mammary gland. Progesterone occurs also in the adrenal cortex and testis, and in some plants, e.g. *Hollatrhena floribunda*. It is synthesized from **pregnenolone** by 3β-hydroxy-Δ^5-steroid dehydrogenase, EC 1.1.1.145, and is an intermediate in the formation of androgens, estrogens, glucocorticoids, and mineralocorticoids.

progestin 1 *an alternative name for* **progestogen**. **2** natural **progesterone**, especially in unpurified form.

progestin receptor 1 a nuclear receptor of the steroid hormone type that binds and mediates the biological effects of progesterone. **2** a G-protein-coupled membrane protein that inhibits adenylate cyclase and mediates the effects of progesterone that are too rapid to be produced via transcription. The α isoform of animals occurs in reproductive tissues, while the β isoform occurs in brain.

progestogen *or* **progestagen** *or* **progestin** any natural or synthetic compound with the hormonal activity of progesterone. In oral contraception regimes, a progestogen is normally combined with an estrogen. The effect is to inhibit ovulation. A preferred method is to use a gestagen, which acts by an effect on the endometrium; these have fewer side effects than progestogens.

proglucagon the prohormone of **glucagon** and **enteroglucagon**. In mammals it consists of ≈170 residues, which contain the entire 29-residue sequence of glucagon in the N-terminal region, and two glucagon-like peptides, designated GLP-1 and GLP-2, of 37 and 35 residues respectively, within the C-terminal region. The 69-residue N-terminal portion, containing the glucagon sequence, is identical with **glicentin**. Angler-fish proglucagon is similar but smaller, and exists in two forms: a 14 kDa form (lacking GLP-2 and a 13-residue connecting peptide) and a 12 kDa form (of unknown sequence).

program 1 a sequence of coded instructions for controlling a computer or other electrical or electromechanical device, especially for enabling a computer to manipulate data in a predetermined way; an equivalent of this in an electronic or magnetic storage medium. **2** to constrain to function or operate in a particular way, especially by providing with a program or other (set of) instructions. **3** to write (and test) a computer program for a specific application. —**programmable** or sometimes (esp. US) **programable** adj.

programmable messenger a term proposed for a cytosolic protein messenger molecule in a hypothesis advanced to explain the multiple actions on their target cells of hormones whose cellular receptors are linked to oligomeric GTP-regulatory proteins in cell membranes. According to this concept, the activation of GTP-regulatory proteins (by GTP and the respective hormones) causes release of their α subunits from the membranes as cytosolic proteins, which become the primary messengers of hormone action and which thereupon are transformed by various protein-modifying enzymes (e.g. kinases and proteases) to yield new structures with selective affinities for different effectors, each responsive to its specific activating system. A variety of responses of the target cells thus may be elicited: accordingly, these messengers are considered programmable.

programmed ribosomal frameshifting a mechanism whereby different proteins may result from a single mRNA molecule, due to the presence of alternative sites within the mRNA that are recognized for initiation and termination of translation.

programming language a technical language in which computer programs are written. Different languages have evolved for different applications. Low-level languages tell the computer directly what operations to execute (e.g. machine and assembly languages); higher-level languages are used for scientific computation, business data processing, computer science education, symbol processing, and systems processing (e.g. Fortran, Cobol, Pascal, Lisp, and C, respectively); high-level scripting languages include awk, javascript, perl, php, python, sh, and tcl; object-oriented programming languages facilitate modular software design, and include Visual Basic, Java, and C++.

progress curve a graph showing the change with time in the concentration or amount of an analyte as a chemical reaction proceeds.

progressive alignment see **multiple alignment program**.

prohibitin any member of a family of highly conserved integral membrane proteins of the inner mitochondrial membrane that modulate the action of m-**AAA proteases**. Prohibitins contain a domain that is also present in stomatin and flotillin.

prohormone a natural (poly)peptide or protein precursor of a hormone.

prohormone convertase or **prohormone converting enzyme** any member of a family of endopeptidases that cleave prohormones within **dibasic cleavage sites** to form active polypeptides. They reside in acidic secretory vesicles. The C-terminal basic residue that remains is then removed by a carboxypeptidase. As EC 3.4.23.17 its recommended name is proopiomelanocortin converting enzyme. See also **neuroendocrine convertase**.

prohormone converting enzyme another name for **prohormone convertase**.

proinsulin the prohormone of **insulin**; it is a single-chain polypeptide of ≈9 kDa with three intrachain disulfide bridges, the number of amino-acid residues varying somewhat with species (78 in dog, 81 in cattle and sheep, 84 in cod and pig, 86 in horse, human, and rat). It is cleaved into insulin and C (or connecting)-peptide. Synthesized in the pancreatic B cells, it is normally present at about 5% of the concentration of insulin, and is secreted in small amounts into the blood.

projection formula any two-dimensional diagrammatic representation of the (three-dimensional) structural formula of a molecule (or part thereof) as if seen by projection onto a plane surface from a particular viewpoint in space. Bonds with different spatial orientations relative to the observer may be depicted distinctively. The term includes **conformational formula**, **Fischer projection**, **Newman projection**, and **perspective formula**.

prokaryon or **procaryon** (pl. **prokarya** or **procarya**) a primitive nucleus; i.e. genomic DNA in a defined structure, but lacking a nu-

clear membrane or other nuclear inclusions. Such an arrangement is found in bacteria (including actinomycetes) and cyanobacteria. See also **nucleoid** (def. 2). —**prokaryous** or **procaryous** adj.; **prokaryosis** or **procaryosis** n.

prokaryosis the condition of possessing a prokaryon or prokarya.

prokaryote or **procaryote** any organism in which the genomic DNA is not enclosed by a nuclear membrane within the cells; i.e. any organism whose cells possess a **prokaryon**. The prokaryotes comprise the bacteria, cyanobacteria, and archaea and are classified in their own kingdom, variously named Monera or Procaryotae. Compare **eukaryote**. —**prokaryotic** or **procaryotic** adj.

prolactin abbr.: PRL; a hormone secreted by the anterior pituitary that stimulates lactation. Human prolactin (hPRL) is a 199-residue protein, 23 kDa in its monomeric form (**little prolactin**), although dimeric (**big prolactin**, 48–56 kDa) and polymeric (>100 kDa) forms circulate in blood. The prolactin gene belongs to a family of genes, along with those encoding placental lactogen and human growth hormone (see **somatotropin**). Secretion is episodic; it is inhibited by dopamine, which binds to lactotrope D_2 receptor (see **dopamine receptor**) to inhibit adenylate cyclase, and is probably stimulated by **thyrotropin-releasing hormone** and **vasoactive intestinal peptide**. Prolactin synthesis is positively regulated by estrogen acting at the level of the prolactin gene. Receptors for prolactin are located in breast, liver, ovary, testis, and prostate, but its main site of action is the mammary gland, which is stimulated to develop by estrogen, progesterone, prolactin, and placental mammotropic hormones. Lactation is stimulated by prolactin but is prevented during pregnancy by the inhibitory action of estrogen and progesterone, the levels of which fall after parturition, allowing prolactin to initiate lactation. See also **prolactin receptor**, **proliferin**.

prolactin receptor a membrane protein of the cytokine receptor family. The prolactin receptor is associated with a small protein **calcyclin**.

prolamin or (formerly) **prolamine** any of a group of simple proline-rich proteins, found especially in the seeds of cereal plants, that are insoluble in water, neutral salt solutions, and absolute ethanol, but soluble in dilute acid or alkali, and in 70–80% aqueous ethanol. The group includes the **gliadins** of wheat and rye, **hordein** of barley, and **zein** of maize. Compare **glutelin**.

prolate 1 having a polar diameter longer than the equatorial diameter. **2** describing an **ellipsoid of rotation** generated by the rotation of an ellipse about its major axis. Compare **oblate**.

prolidase another name for **X-Pro dipeptidase**. Distinguish from **prolinase**.

proliferating cell nuclear antigen see **PCNA**.

proliferin or **mitogen-regulated protein** a prolactin-related protein, the mRNA of which appears in several murine cell lines during growth. One of several proteins of the somatotropin/prolactin family that provide a growth stimulus to target cells in maternal and fetal tissues during the development of the embryo. The mRNA occurs in placenta, which secretes proliferin as a glycoprotein distinct from placental lactogen.

prolinase the recommended name for **Pro-X dipeptidase**. Distinguish from **prolidase**.

prolinate 1 proline anion; the anion, $(C_4H_7NH)–COO^-$, derived from **proline**. **2** any salt containing proline anion. **3** any ester of proline.

proline the trivial name for pyrrolidine-2-carboxylic acid, $(C_4H_7NH)–COOH$; a chiral cyclic N-alkylated α-amino acid. L-proline (symbol: P or Pro), (S)-pyrrolidine-2-carboxylic acid, is a coded amino acid found in peptide linkage in proteins; codon: CCA, CCC, CCG, or CCU; by virtue of its more constrained molecular structure relative to most other α-amino acids, it usually does not participate in α-helix formation in globular proteins, but does occur in transmembrane helices, in which it introduces a pronounced kink. In mammals, it is a non-essential dietary amino acid, and is glucogenic. It is formed from **glutamate 5-semialdehyde** by reduction by NADPH. D-Proline (symbol: D-Pro or DPro), (R)-pyrrolidine-2-carboxylic acid, is not known to occur naturally. See also **hydroxyproline**.

L-proline

proline/arginine-rich protein *abbr.*: PARP; a 24 kDa proline- and arginine-rich protein from bovine cartilage, closely related to the N-terminal domain of collagen α1 (XI).

proline dehydrogenase EC 1.5.99.8; *other name*: proline oxidase; a flavoprotein enzyme that catalyses the reaction:

L-proline + acceptor + H_2O =
(S)-1-pyrroline-5-carboxylate + reduced acceptor.

The flavin is FAD. In *Escherichia coli* the enzyme is bifunctional, having the activities of both proline dehydrogenase and **1-pyrroline-5-carboxylate dehydrogenase**. In humans, the enzyme (516 amino acids) is located with the inner mitochondrial membrane and is present in liver, kidney, and brain. Enzyme deficiency results in hyperprolinemia type 1.

proline dipeptidase *see* X-Pro dipeptidase.

proline hydroxylase *see* procollagen-proline dioxygenase.

proline-rich proteins *abbr.*: PRP; one of the four major classes of **plant cell-wall proteins**. They contain proline-rich repeat sequences (Pro–Pro–Val–Tyr–Lys in soybean), are poorly glycosylated, and are thought to form rodlike molecules.

prolinium proline cation; the cation, $(C_4H_7NH_2^+)$–COOH, derived from proline.

prolino the pyrrolidino group, HOOC–(C_4H_7N)–, derived from proline.

prolixin-S *an alternative name for* nitrophorin 2.

prolyl the acyl group, (C_4H_7NH)–CO–, derived from proline.

prolyl dipeptidase *see* Pro-X dipeptidase.

L-prolylglycine dipeptidase *see* Pro-X dipeptidase.

prolyl hydroxylase *see* procollagen-proline dioxygenase.

prolyl oligopeptidase EC 3.4.21.26; *other names*: prolyl endopeptidase; post-proline cleaving enzyme; post-proline endopeptidase. An enzyme that catalyses the hydrolysis of Pro-|-Xaa >> Ala-|-Xaa in oligopeptides. It is a cytoplasmic (in eukaryotes) serine endopeptidase, also found in some bacteria.

prometaphase the stage in **mitosis** or **meiosis**, following **prophase**, during which the nuclear envelope is disrupted and breaks into membrane vesicles, and the spindle microtubules enter the nuclear region. **Kinetochores** mature on each **centromere** and attach to some of the spindle microtubules; these are then called **kinetochore microtubules** while those to which kinetochores do not attach are called **polar microtubules**; microtubules outside the spindle are called **astral microtubules**. Kinetochore microtubules begin the process of aligning chromosomes in one plane halfway between the poles.

promiscuous DNA any DNA sequence that occurs in more than one of the genetic systems (i.e. nuclear, mitochondrial, or plastid) of eukaryotic cells. Many such DNA segments have been found in yeast, fungi, plants, and animals.

promoter *symbol*: P; a DNA sequence located 5' to a gene that indicates the site for the initiation of transcription. It may influence the amount of mRNA produced and the tissue specificity of expression. The symbol may be subscripted to indicate particular promoters, e.g. P_{RM}, a promoter for the *cI* gene, RM standing for repressor maintenance. In prokaryotes consensus sequences centred at –10 (i.e. **Pribnow box**) and –35 nucleotides upstream from the transcription start site are recognized by **RNA polymerase**. Eukaryotic promoters have sequence elements that are recognized by general transcription factors such as TFIIA, TFIIB, TFIIC, TFIID (i.e. **TATA-binding protein**), TFIIE, TFIIH, and TFIIJ that form a transcription initiation complex with RNA polymerase II. **TATA boxes** (–25), **CAAT boxes** (–100), and **GC boxes** are often found in the promoters of housekeeping genes. In addition, eukaryotic promoters may have binding sites (**response elements**) for transcription factors

found in only some cell types. Promoters for genes transcribed by RNA polymerase III may be downstream from the start site.

promoter escape *or* **promoter clearance** the melting of dsDNA downstream of the promoter region effected by a subunit of transcription factor TFIIH that is an ATPase and acts as a molecular wrench that twists the two strands apart.

promoter ligand an additional ligand that is included in the loading buffer in **affinity chromatography**, to promote specific binding to the immobilized ligand.

promoter melting the denaturation or separation of the two strands of DNA of the promoter region on binding a 3'→5' DNA helicase subunit of transcription factor TFIIH.

promutagen any substance that, when acted upon by the cell's metabolism, becomes a **mutagen**.

promyelocytic leukemia protein *see* PML.

promyelocytic leukemia zinc finger protein *see* PLZF.

Pronase *a proprietary name for* a mixture of various exo- and endo-peptidases, obtained from *Streptomyces griseus*, that is able to hydrolyse virtually any protein almost completely to free amino acids. It is used for hydrolysis of many proteins, for digestion of mucins, in isolation of intact DNA or RNA from cells or viruses, and in dispersal of certain types of mammalian tissue cells, e.g. chondrocytes.

pronucleus (*pl.* **pronuclei**) the nucleus of either the ovum or the spermatozoon following fertilization. Thus, in the fertilized ovum, there are two pronuclei, one originating from the ovum, the other from the spermatozoon that brought about fertilization; they approach each other, but do not fuse until just before the first cleavage, when each pronucleus loses its membrane to release its contents. —**pronuclear** *adj*.

proof a standard by which the strength of distilled alcoholic solutions (or vinegars) is judged. In the UK it is the strength of a mixture of alcohol and water having a specific gravity of 0.91984, and containing 0.495 of its weight or 0.5727 of its volume of absolute alcohol. Degrees of proof express the percentage of proof spirit present; hence 70° proof spirit contains $0.7 \times 57.27\%$ alcohol. In the US, proof is a number that is twice the percentage (by volume) of alcohol present; hence 70° proof spirit contains $0.5 \times 70\%$ alcohol.

proofreading (*in genetics*) verification of the exact nature of a sequence of bases in DNA or amino-acid residues in a protein. Proofreading of DNA sequence occurs in transcription, during the action of DNA polymerase, when incorporation of an incorrect nucleotide causes mismatch in pairing with the complementary base on the template strand; this leads to immediate excision of the incorrect nucleotide, retreat of the enzyme to the previous position, and another attempt to insert the correct nucleotide. During translation, charging of the tRNA with its aminoacyl group is verified by its fit in the binding site; if incorrect, the aminoacyl-tRNA is hydrolysed to tRNA and amino acid and the whole process is repeated until successful. Examples of such proofreading during translation include rejection of threonine by valine–tRNA ligase, and of valine by isoleucine–tRNA ligase. *See also* mismatch repair.

Pro(3-OH) *symbol for* a residue of the α-amino acid (*trans*-)3-hydroxy-L-proline (alternative to 3Hyp).

Pro(4-OH) *symbol for* a residue of the α-amino acid (*trans*-)4-hydroxy-L-proline (alternative to 4Hyp).

proopimelanocortin *abbr.*: POMC; a 241-residue prohormone protein, *also called* **corticotropin–lipotropin precursor**, synthesized in the intermediate and anterior lobes of the pituitary. POMC is cleaved in the anterior pituitary by a **prohormone convertase**, PC1 to yield an N-terminal peptide (residues 1–80), **corticotropin** (residues 110–148) and β lipotropin (β-LPH, residues 148–241). The segment 83–107 is known as **joining peptide**. Prohormone convertase, PC2, then cleaves corticotropin to yield α-**melanocyte-stimulating hormone** (α-MSH; residues 110–122), and cleaves β-LPH to yield γ-LPH (residues 151–208) and β **endorphin** (residues 211–241). Derived hormones (cleavage points and some sizes differ from porcine above): corticotropin, 39 amino acids; **corticotropin-like intermediate lobe peptide**, 21 amino acids; β endorphin, 31 amino acids; lipotropins β-LPH (89 amino acids) and γ-LPH (56 amino acids); melanocyte-stimulating hormones α-MSH (13 amino acids), β-MSH (18 amino acids), and γ-MSH (11 amino acids); [Met]enkephalin (*see* enkephalin), 5 amino acids. In the intermediate lobe, POMC synthesis is regulated mainly by dopamine and serotonin; in the anterior lobe by glucocorticoids and corticotropin-releasing hormone. The prohormone convertases are related to **furin** and **kexin**.

proopiomelanocortin converting enzyme EC 3.4.23.17; *the recommended name for* **prohormone convertase**.

propargylamine *see* **monoamine oxidase inhibitors**.

propellant *or* **propellent** any substance that through pressure or combustion or other means causes propulsion, as, e.g., the use of compressed gases in spray cans.

propeptide that part of a peptide chain that follows the signal sequence, and is cleaved off in formation of the mature protein.

properdin *see* **complement**.

prophage a form of the genome of a bacteriophage in which the nucleic acid is integrated into the genome of the phage's bacterial host, a state known as **lysogeny**. **Induction** to produce mature phage can occur, with accompanying lysis of the bacterium. For example, in **lambda phage**, transcription of the prophage is repressed by the *cI* gene product, the repressor, which binds to the O_L and O_R operators thereby preventing use of the P_L and P_R promoters. The product of the *cII* gene, CII, is crucial in determining the choice between lysogeny and lysis; if CII is active lysogeny is favoured, but if CII is not active (due to breakdown by host proteases) lysis is favoured.

prophage induction *see* **phage induction**.

prophage-mediated conversion *another name for* **bacteriophage conversion**.

prophage reactivation the phenomenon in which the survival of UV-irradiated **lambda phage** is higher in a host carrying a homologous heteroimmune prophage than in a host that is nonlysogenic or is lysogenic with a nonhomologous prophage.

prophase the initial stage of **mitosis** or **meiosis** in which the chromosomes are condensed, becoming apparent within the nucleus, but are not yet attached to a mitotic spindle. By early prophase, the single centrosome contains two centriole pairs; at late prophase, the centrosome divides and the two asters move apart. In meiosis, the chromosomes come together in homologous pairs, and undergo crossing-over. This occurs during the first of the two meiotic cell divisions, and is divided into several stages. In the first, zygotene, a complex structure develops between the two sister chromatids of each pair, which may play a part in **recombination**. They remain in the complex during a stage known as pachytene, then separate at the diplotene stage.

propheromone convertase Y a protease present in yeast with specificity towards paired basic residues (-Lys-Arg-, -Arg-Arg-, or -Arg-Lys-) and no activity towards single basic residues.

propidium iodide a phenanthridine derivative (λ_{max} excitation 536 nm, emission 617 nm) whose fluorescence intensity increases greatly when intercalated in the DNA double helix. It is widely used in flow cytometry in the analysis of ploidy.

propionate the anion, $CH_3–CH_2–COO^-$, derived from propionic acid, in any salt; designation for the **propionyl** radical in naming an ester.

propionibacteria bacteria belonging to the genus *Propionibacterium*. They are characterized by their activity in fermenting glucose or lactic acid to propionic acid; acetic acid is also a major product. *See also* **propionic-acid fermentation**.

propionic acid propanoic acid; $CH_3–CH_2–COOH$; a carboxylic acid, important in the energy metabolism of ruminants, in which it is formed by rumen bacteria. It is the basis of a series of nonsteroidal anti-inflammatory agents, known as substituted propionic acids; *see* **ibuprofen**.

propionic-acid fermentation fermentation of glucose or lactate by *Propionibacterium* spp. (*see* **propionibacteria**) to either propionate or acetate; pyruvate is a common intermediate for both transformations. In propionate formation, pyruvate first reacts with methylmalonyl-CoA to form propionyl-CoA and oxaloacetate. The latter is converted by reactions of the tricarboxylic-acid cycle to succinate, which then reacts with the propionyl-CoA to form propionate and succinyl-CoA. The succinyl-CoA is converted to methylmalonyl-CoA, which is then available to react with a further molecule of pyruvate. Conversion of pyruvate to acetic acid involves oxidative decarboxylation to acetyl-CoA, reaction of this with orthophosphate to form acetyl phosphate, and reaction of the latter

with ADP to form ATP and acetic acid. When lactate is the initial substrate, this is first converted to pyruvate by lactate dehydrogenase. This enzyme is a flavoprotein, that is oxidized by a cytochrome *b*.

propionyl *symbol*: Pp; *the trivial name for* propanoyl, $CH_3–CH_2–CO–$, the univalent acyl group derived from propionic (i.e. propanoic) acid. *Compare* **propyl**.

propionyl-CoA carboxylase *abbr.*: PCCase; EC 6.4.1.3; an enzyme that catalyses a reaction between ATP, propanoyl-CoA, and HCO_3^- to form (*S*)-methylmalonyl-CoA, ADP, and orthophosphate; biotin is a coenzyme. It is an enzyme in the catabolic pathways of odd-chain fatty acids and isoleucine, occurring in the mitochondrial matrix. Defects cause propionic acidemia, which is inherited as a recessive trait. The human enzyme consists of six α chains (biotin binding) and six β chains.

proplastid the precursor of a **plastid**.

propolypeptide the inactive form in which some polypeptide hormones, or other agonists, are produced; it is activated by removal of part of the sequence by a protease.

proportional counter any device for measuring radioactive disintegrations by means of a gas-filled ionization chamber, e.g. Geiger counter, operating in the **proportional region**.

proportional region a range of voltages applied to a gas-filled ionization chamber (e.g. Geiger counter) over which the gas amplification is greater than unity. In radioactive counters of this type, measurement depends on ionization of the gas in the ionization chamber by emitted radiation. Up to a certain applied voltage, ion production and thus the current through the chamber depends solely on collisions between emitted radiation and gas molecules; at higher voltages, the current is enhanced by additional collisions, so that sensitivity is increased, but the current remains proportional (hence the name proportional region) to the rate of emission of radiation. *Compare* **Geiger region**.

propositus *or* (*fem.*) **proposita** (*pl.* **propositi**) *or* **proband** the first individual to present as a patient, in a series of cases related (usually) to a heritable disorder, and from whom the line of descent is traced.

propranolol an important β-adrenoceptor antagonist; the (*S*) form is the active structure. It is used therapeutically as an antihypertensive, anti-anginal, and anti-arrhythmic. *One proprietary name*: Inderal (as hydrochloride).

(*S*)-form

propressophysin the polypeptide precursor of **vasopressin** and **neurophysin** (*see* **vasopressin–neurophysin 2–copeptin precursor**). Mutation of the gene is associated with familial neurogenic diabetes insipidus.

proprotein an inactive form of a protein, having an additional sequence that is removed by a protease to yield the active protein; a classic example is **proinsulin**. A **zymogen** is a form of proprotein, sometimes referred to as a proenzyme.

propyl *symbol*: Pr; *n*-propyl; the alkyl group, $CH_3–CH_2–CH_2–$, derived from propane. *Compare* **isopropyl**, **propionyl**.

pro-R/pro-S convention stereochemical descriptors for paired, chemically-like ligands at a prochiral centre; they derive from an extension of the **sequence rule** used to assign *R,S* descriptors to a chiral centre. In the prochiral (but achiral) structure, *Caabc*, the chemically-like ligands are the two atoms (or groups of atoms), *a*. For the sake of example assume that the sequence rule gives the priority sequence, $a > b > c$. In the molecular model (or projection formula) arbitrarily label the *a* groups as *a'* and *a''* and arbitrarily assign higher priority to one of them, e.g., $a' > a''$. The modified priority sequence, $a' > a'' > b > c$, is then considered with the model viewed from the side remote from *c*. If this gives a right-handed ordering for $a' \to a'' \to b$, thus leading to R chirality for the arbitrarily created chiral element, the promoted a' is designated the *pro-R* group; conversely, if the arbitrarily created chiral element reveals an S ordering, the promoted group is *pro-S*. Example: glycerol, $HOCH_2$–$CHOH$–CH_2OH, contains two chemically-like CH_2OH groups, and the normal priority sequence is $OH > CH_2OH > H$. In the **Fischer projection** formula with the central carbon atom having the OH group to the right and H to the left, arbitrarily assign the top $C'H_2OH$ group priority over that at the bottom, $C''H_2OH$, giving the modified sequence, $OH > C'H_2OH$ (top) $> C''H_2OH$ (bottom) $> H$. When the model is viewed from the side remote from the lowest priority atom (H) the ordering, $OH \to C'H_2OH$ (top) $\to CH_2OH$ (bottom) is right-handed so that the arbitrarily created chiral element has R chirality; hence the top $C'H_2OH$ is the *pro-R* group. Subscript letters R and S can be used to indicate whether an atom or group of atoms is *pro-R* or *pro-S*. This is most commonly done for the two hydrogens of a methylene group, with H_R indicating the *pro-R* hydrogen and H_S the *pro-S* hydrogen. For a methylene group, replacement of ^1H_R by ^2H or ^3H gives a centre with R chirality; similar replacement of ^1H_S gives a centre with S chirality. *See also* **enantiotopic**.

pros a designator of position of a nitrogen in the histidine imidazole ring, being that nitrogen nearest to the side chain; often abbreviated to π, as in N^π. *Compare* **tele**.

prosaposin the polyprotein precursor (524 amino acids) of the **saposins**. A mutation in the translation initiation codon leads to a very rare prosaposin deficiency (and thus deficiency of all saposins), which is rapidly fatal in infancy.

PROSITE *see* **pattern database**.

prosolin *see* **stathmin**.

prosome *another name for* **proteasome**.

Prospero a transcription factor localized in the basal cortical crescent of *Drosophila* neuroblasts that enters the nucleus after cell division to activate ganglion mother cell genes, repress neuroblast-specific genes, and cause differentiation into neurons.

prostacyclin *or* **prostaglandin I** (*abbr.*: PGI) any of a group of prostanoids synthesized by **prostaglandin-I synthase** (EC 5.3.99.4) from prostaglandin H – which itself is the product of **prostaglandin-endoperoxide synthase** – as prostaglandin I_2, I_1, or I_3, depending on whether the substrate for the enzyme is arachidonate, dihomo-(6,9,12)-linolenate, or eicosa-5,8,11,14,17-pentaenoate respectively. PGI_2 is the major product *in vivo*; it is produced by endothelial cells of blood-vessel walls, has potent platelet anti-aggregatory properties resulting from stimulation of adenylate cyclase, and induces vasodilation. It has a normal half-life of 2 min, being degraded by hydrolysis to 6-oxo-prostaglandin $F_{1\alpha}$; this is then converted to the 6,15-dioxo-13,14-dihydro-$PGF_{1\alpha}$, and follows a similar degradative path to other prostaglandins. More stable but still active analogues have been synthesized, e.g. carbacyclin. (*S*)-15-hydroperoxyeicosatetraenoate inhibits prostaglandin-I synthase. *See also* **prostaglandin**.

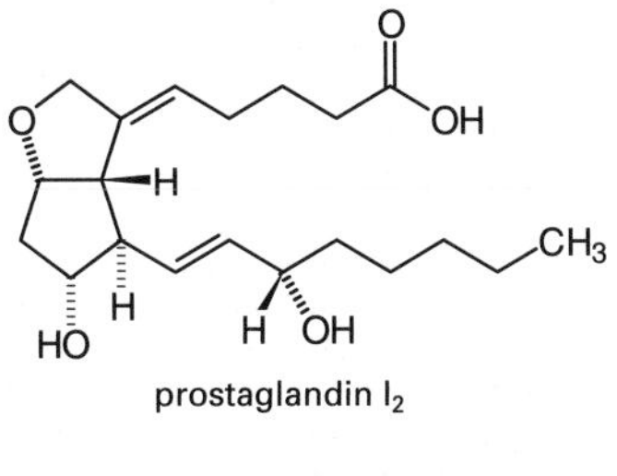

prostaglandin I_2

prostacyclin synthase *see* **prostaglandin-I synthase**.

prostaglandin *abbr.*: PG; any of a group of biologically active metabolites of arachidonate (PG_2 series), dihomo-γ-linolenate (PG_1 series), or eicosa-5,8,11,14,17-pentaenoate (PG_3 series). They characteristically contain a cyclopentane ring due to formation of a bond between C-18 and C-12 of the fatty acid; hydroxy or oxo substitution at positions 9 and 11 is a major distinguishing feature between prostaglandin classes. They possess a protean array of highly potent biological activities that defy simplification. Their pharmacological effects include: vasodilation (PGE_2 and **prostacyclin**) which is short-acting, and vasoconstriction ($PGF_{2\alpha}$ and **thromboxane** A_2); oxytocic activity ($PGF_{2\alpha}$ can be used to terminate pregnancy in the second trimester and PGE_2 induces labour at term); bronchodilation (PGE series) and constriction (PGF series). They are involved in inflammatory reactions, and the enzymes that synthesize them are targets for antiinflammatory agents such as aspirin and non-steroidal antiinflammatory drugs such as ibuprofen. They are also involved in the regulation of cell proliferation in normal and neoplastically transformed cells. Prostaglandins are formed from the precursors prostaglandin G_1, G_2, or G_3, the intermediate products of **prostaglandin-endoperoxide synthase**, depending on the substrate fatty acid; the same enzyme converts these to prostaglandins H, from which other enzymes form prostaglandins D, E, and F, and also **prostacyclins** (or prostaglandins I). These are the bioactive products of the system, but they have a half-life of minutes, being rapidly converted to metabolites with much weaker, or often inhibitory properties. The initial metabolites are the 15-oxo-13,14-dihydroprostaglandins, produced by the sequential actions of 15-hydroxyprostaglandin dehydrogenases (e.g. EC 1.1.1.196 and 1.1.1.197) and 15-oxoprostaglandin 13-reductase (EC 1.3.1.48); these are further degraded by beta oxidation and omega oxidation to urinary metabolites such as 7α-hydroxy-5,11-dioxotetranorprostane-1,16-dioate (from PGE_2) and 5,7-dihydroxy-11-oxotetranorprostane-1,16-dioate. The structures of prostanoic acid, with its carbon numbering, and of PGG_2, PGH_2, PGD_2, PGE_2 and $PGF_{2\alpha}$ are shown.

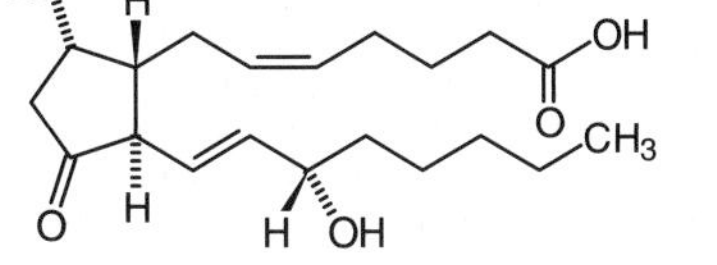

prostaglandin D_2

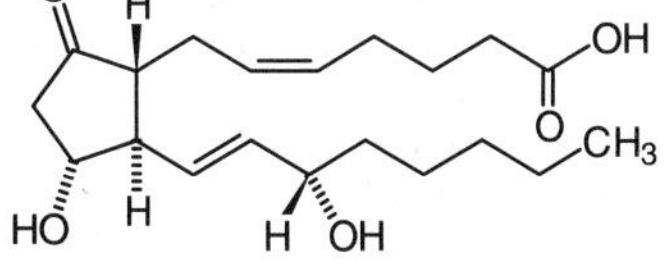

prostaglandin E_2

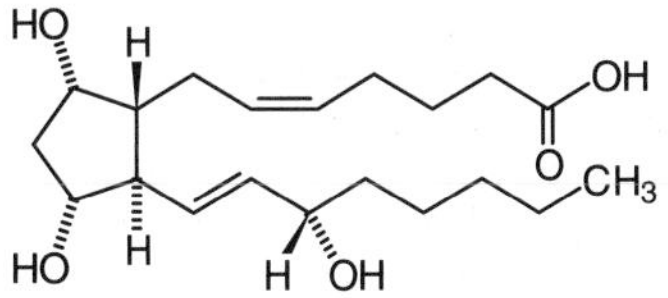

prostaglandin $F_{2\alpha}$

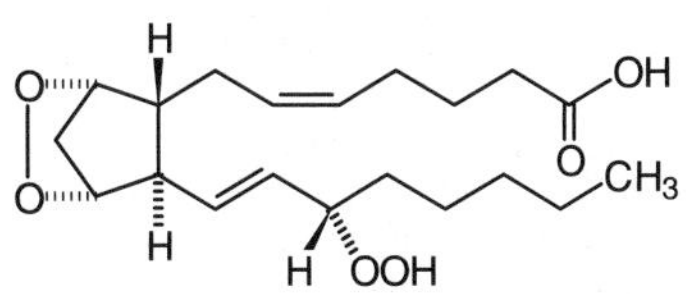

prostaglandin G_2

prostaglandin H₂

prostanoic acid

prostaglandin-D synthase *abbr.*: PGD synthase; EC 5.3.99.2; *other name*: prostaglandin-H_2 D-isomerase; an enzyme that catalyses the conversion of (5*Z*,13*E*)-(15*S*)-9α,11α-epidioxy-15-hydroxyprosta-5,13-dienoate to (5*Z*,13*E*)-(15*S*)-9α,15-dihydroxy-11-oxoprosta-5,13-dienoate (prostaglandin D). Glutathione is a cofactor in some cases.

prostaglandin-endoperoxide synthase EC 1.14.99.1; *other names*: prostaglandin synthase; prostaglandin G/H synthase. An enzyme whose reaction embraces the activity formerly called **cyclooxygenase**; it catalyses a reaction between dioxygen, arachidonate and two molecules of reduced glutathione (GSH) to form prostaglandin H_2, oxidized glutathione (GSSG), and H_2O. The enzyme acts as both a dioxygenase and as a peroxidase. It catalyses the formation of the precursor, prostaglandin G_2 (PGG₂), for the biosynthesis of all prostanoid compounds, by forming an endoperoxide bridge between C-9 and C-11, simultaneously bonding C-8 and C-12 to form the cyclopentane ring, and also forming the hydroperoxide at C-15. The latter is reduced by a reduced glutathione (GSH)-dependent peroxidase activity of the enzyme to a hydroxyl group to form prostaglandin H_2; these, and their further metabolites, prostaglandins D, E and F, are highly potent bioactive compounds with a great diversity of actions (*see* **prostaglandin**). The enzyme can also act on dihomo-(6,9,12)-linolenic acid, to form a series of prostaglandins (denoted by a subscript 1, e.g. PGE₁) lacking the C-5 double bond, and on eicosa-5,8,11,14,17-pentaenoic acid, to form the 3 series of prostaglandins (e.g. PGE₃), with the additional double bond at C-17. Two isoenzymes, COX-1, constitutively expressed, and COX-2, inducible in inflammation by cytokines, are highly homologous membrane proteins, but selective inhibition of COX-2 is possible due to its larger inhibitor-binding site. Aspirin acetylates an active-site serine residue in both, the IC_{50} being 10 to 100 times lower for COX-1 than for COX-2. Its inhibiton of COX-1 is the basis of its antithrombotic effects. *See also* **prostacyclin, thromboxane**.

prostaglandin-E₂ 9-reductase EC 1.1.1.189; an enzyme that catalyses the reduction by NADPH of (5*Z*,13*E*)-(15*S*)-11α,15-dihydroxy-9-oxoprosta-5,13-dienoate (PGE₂) to (5*Z*,13*E*)-(15*S*)-9α,11α,15-trihydroxyprosta-5,13-dienoate (PGF₂α) with formation of NADP⁺. The monomeric cytoplasmic human enzyme catalyses reduction of many carbonyl compounds. In addition to the activity of prostaglandin-E₂ 9-reductase (also termed prostaglandin 9-ketoreductase), it also contains two other activities: carbonyl reductase (NADPH) EC 1.1.1.184, and 15-hydroxyprostaglandin dehydrogenase (NADP⁺), EC 1.1.1.197.

prostaglandin-E synthase EC 5.3.99.3; *other names*: prostaglandin-H_2 E-isomerase; endoperoxide isomerase; an enzyme that catalyses the conversion of (5*Z*,13*E*)-(15*S*)-9α,11α-epidioxy-15-hydroxyprosta-5,13-dienoate to (5*z*,13*e*)-(15*s*)-11α,15-dihydroxy-9-oxoprosta-5,13-dienoate (PGE₂). Glutathione is a cofactor.

prostaglandin-F synthase EC 1.1.1.188; *other name*: prostaglandin-D_2 11-reductase; an enzyme that catalyses the formation of (5*Z*,13*E*)-(15*S*)-9α,11α,15-trihydroxyprosta-5,13-dienoate and NADP⁺ from (5*Z*,13*E*)-(15*S*)-9α,15-dihydroxy-11-oxoprosta-5,13-dienoate and NADPH.

prostaglandin-I synthase EC 5.3.99.4; *other name*: prostacyclin synthase; an enzyme that catalyses the conversion of (5*Z*,13*E*)-(15*S*)-9α,11α-epidioxy-15-hydroxyprosta-5,13-dienoate to (5*Z*,13*E*)-(15*S*)-6,9α-epoxy-11α,15-dihydroxyprosta-5,13-dienoate (PGI₂, prostacyclin). This is a heme-thiolate protein.

prostanoate the anion of prostanoic acid, 2-octylcyclopentaneheptanoic acid; the 8*S*,12*S* form is the parent compound for prostanoid nomenclature. For structure, *see* **prostaglandin**.

prostanoid any compound based on or derived from the **prostanoate** structure, including the **prostaglandins**, **prostacyclins**, and **thromboxanes**.

prostasin a serine protease with trypsin-like activity, found in human seminal fluid.

prostate-specific antigen *abbr.*: PSA; a serine endopeptidase belonging to the family of glandular **kallikreins** with a His-Asp-Ser **charge-relay system**. It is virtually prostate specific. Levels may be elevated in prostate hypertrophy but the release from prostate cancer is higher than from normal tissue. It is a much better marker than **acid phosphatase**, being elevated at an earlier stage of the tumour. Its physiological substrate is the predominant protein of the seminal vesicle coagulum.

prostatic of, pertaining to, or related to the prostate gland.

prosthetic group a non-protein group that is combined specifically with a protein, in stoichiometric proportion.

prostin a prostatic serine endopeptidase that is not related to the kallikreins. *Compare* **prostate-specific antigen**.

protachykinin β precursor a preproprotein from which certain polypeptide hormones are cleaved, not all by the same processing pathway. For example, from the human precursor, the derived peptides are **substance P**, **neurokinin A**, and **neuropeptide K**. *See also* **tachykinin**.

protamine any of a group of small peptides, containing arginine, alanine, and serine, that are isolated from the sperm of fish (e.g., salmon, herring, sturgeon, trout, and mackerel), but are present in the sperm of all animals. They have a histone-like function in the chromatin of sperm, compacting sperm DNA into a condensed complex.

protamine sulfate a sulfated derivative of a **protamine**. Protamine sulfates combine with, and inactivate, **heparin**, and are used to prevent bleeding in heparin overdose.

protease any enzyme that catalyses the hydrolysis of peptide bonds in a protein or polypeptide. The specific names **endopeptidase**, **aminopeptidase**, and **carboxypeptidase** are recommended. *See also* **proteinase**.

protease-activated G-protein-coupled receptor *see* **PAR**.

protease inhibitor 4 *see* **kallistatin**.

protease nexin I *see* **nexin 1**.

proteasome or **proteosome** EC 3.4.99.46 *recommended name*: multicatalytic endopeptidase complex; *other names*: multicatalytic proteinase (complex); ingensin; macropain; prosome; lens neutral proteinase. Proteasomes are large multi-subunit protease complexes that selectively degrade intracellular proteins that are tagged for destruction by ubiquitination. They control cellular processes, such as metabolism and the cell cycle, through signal-mediated proteolysis of key enzymes and regulatory proteins. They also operate in the stress response, by removing abnormal proteins, and in the immune response, by generating antigenic peptides. The proteolytic activity is due to a 26S protease (a 2 MDa complex), whose ability to degrade proteins generally depends on both the ubiquitination of the substrate and the presence of ATP. At the core is a 20S particle that carries the catalytic activity. When first isolated, it was called 'multicatalytic proteinase' because of its ability to cleave peptide bonds carboxy-terminal to basic, hydrophobic, and acidic residues. The 20S particle does not require ATP for activity. The crystal structure of this core from the archaean *Thermoplasma acidophilum* has been

droxy-9-oxoprosta-5,13-dienoate (PGE₂). Glutathione is a cofactor.

determined. The core alone is inactive because it requires two specific subunits of a 19S particle to mediate recognition of ubiquitin–protein conjugates. The core subunits are arranged in four heptameric rings, stacked to form a hollow cylinder with the protease activity on the inside. The active site nucleophile is threonine. The protein substrate first must be unfolded and the disulfide bonds reduced before entry into the cylinder. In *Thermoplasma*, the two outer rings are composed of seven α subunits, and each inner ring contains seven β subunits. The enzyme catalyses the cleavage at Xaa-|- bonds in which Xaa carries a hydrophobic, basic, or acidic side chain. Components of the human proteasome include: c2, c3, c5, c7, c8, c9, β chain, c13, δ chain, ε chain, mecl-1, τ chain, and ζ chain.

protectin *see* **CD59**.

protecting group *or* **protective group** *or* **blocking group** any chemical group added to a reactive centre of a molecule to prevent it from reacting during chemical treatments in which it is not intended to participate. Such groups are commonly used during chemical synthesis of macromolecules or shorter polymers of amino acids or nucleotides. Reagents contributing protecting groups should react readily under mild conditions, with high specificity, and lead to addition of a group that can be removed highly specifically under mild conditions. For protein or amino-acid protecting groups *see* **aminoethyl, Boc, citraconylation, DABITC, dabsyl, dansyl, Fmoc, maleylation, pipsyl, silylation, succinylation, tosyl, triphenylmethyl group**.

protection of telomeres-1 *see* **Pot-1**.

protective protein/cathepsin C *abbr.*: PPCA; a lysosomal multifunctional protein that has separate protective properties (for lysosomal β-galactosidase and neuraminidase) and catalytic properties (being a serine protease). The gene at 20q13.1 encodes a glycoprotein of 480 residues, which is processed to form a 32 kDa polypeptide joined by a disulfide bridge to another of 20 kDa. A serine on the heavy chain, and a histidine and an aspartate on the light chain constitute the catalytic triad. It is active as cathepsin A at acid pH, and as an amidase (e.g. for substance P, endothelin I, and neurokinin) and an esterase at neutral pH. At least a dozen point mutations in the gene are known to cause **galactosialidosis**, in which there is a triple enzyme deficiency.

protegrin *abbr.*: PG; any of a group of leukocyte antimicrobial peptides, denoted PG-1 to PG-5, that are active against *Escherichia coli*, *Listeria monocytogenes*, and *Candida albicans in vitro*. Like **tachyplesins**, they contain two intramolecular cystine disulfide bonds, and share similarity with **defensins**.

protein any of a large group of organic compounds found as major macromolecular constituents of living organisms. All enzymes are proteins, although catalytic activity is shown by some nucleic-acid molecules (*see* **ribozyme**). A protein is a linear polymer of **amino acids** linked by peptide bonds in a specific sequence. In the biosynthesis of the polypeptide chain, any of 20 different amino acids may be incorporated, according to the genetic instructions of the cell (*see* **protein and peptide biosynthesis**). The amino-acid residues may subsequently be modified so that the chains may contain a much wider variety of residues, amounting to nearly 200. The modifications may involve the covalent attachment of various groups, such as carbohydrates and phosphate; these are the simple proteins. Other substances may be more loosely associated with the polypeptide chains, such as heme or lipid, giving rise to **conjugated proteins**. Four hierarchies of structure have been defined for polypeptide chains, termed **primary structure, secondary structure, tertiary structure,** and **quaternary structure**. Four classes of protein have been recognized, which differ according to their content of **alpha helices** and **beta sheets**: all-alpha; all-beta; alpha/beta; and alpha + beta. Some important protein types are listed separately in this dictionary, and many others are listed individually: e.g., contractile proteins (**actin, myosin, tropomyosin, tubulin**), heat-shock proteins, hemoproteins (**cytochromes, hemoglobin, myoglobin**), lipoproteins, membrane proteins (**ATPases, ion channels, rhodopsin**), non-heme proteins (**blood coagulation factors, serum albumin, serum globulin**), nuclear proteins (**histones**), plasma proteins, and structural proteins (**collagen, crystallin, keratin**). Nutritional aspects of protein metabolism include the fact that higher animals cannot synthesize certain of the amino acids required for protein synthesis; these are referred to as **essential amino acids**. Plants and most microorganisms can synthesize all of the amino acids required for protein synthesis, and many others. The nutritional value of different proteins depends on their composition. Some proteins are of poor nutritional value, owing to lack of essential amino acids, e.g. gliadin, which lacks lysine, and zein, which lacks lysine and tryptophan. —**proteinaceous** *adj.*

14-3-3 protein any of a highly conserved family of proteins, widespread in higher eukaryotes, that seem to act as regulators in signal transduction or phosphorylation. The name derives from the migration position on DEAE–cellulose columns, and on starch-gel electrophoresis, of certain brain proteins that were the first members of the family to be studied. Members of this family have been implicated in regulation of **protein kinase C**, transcriptional regulation in plants, exocytosis, cell-cycle regulation, and interaction with **Raf** in the **MAP kinase** cascade. For example, 14-3-3 protein β (*other name*: protein kinase C inhibitor protein-1) from sheep activates tyrosine and tryptophan hydroxylases in the presence of Ca^{2+}/calmodulin-dependent protein kinase II. The bovine equivalent has an identical sequence: it is a homodimer.

protein A a bacterial cell-wall protein, found in many strains of *Staphylococcus aureus*, that can bind, without masking the antigen-binding site, to the Fc region of IgG. It binds weakly to human IgA, IgE, and IgM, but fails to bind to human IgG_3 and IgD; it does not bind to cell Fc receptors. The purified protein, or a preparation of cells of *S. aureus* that carry protein A, is widely used as an immunochemical reagent, often in an immobilized form on affinity columns, for isolation and purification of immunoglobulins and, by forming ternary complexes, of antigens and immune complexes. It is also used as a binding reagent in enzyme-linked immunoassay, in histochemistry, and in blotting techniques.

protein and peptide biosynthesis the biochemical synthesis of proteins. The synthesis of most peptides in living cells depends on the presence of **ribosomes**. This includes the small peptide hormones, such as **oxytocin** and **vasopressin**, which are synthesized as precursor proteins. Other peptides are not synthesized by means of ribosomes. Thus **glutathione** is synthesized by soluble enzymes in two steps. Many biologically active peptides, especially cyclic structures, are products of microorganisms. These include such antibiotic peptides as **gramicidin S**, the **tyrocidines**, and the **polymyxins, which are synthesized by** interacting multienzymes in which activated aminoacyl groups are transferred onto –SH groups to form intermediate thioesters. The multienzymes are unusual in size, ranging from 120 to 1700 kDa. They consist of repeating modules, each responsible for the incorporation of one amino-acid residue. The genes are arranged in clusters associated with genes encoding auxiliary proteins for the synthesis of precursors, modifying enzymes, exporting proteins, and regulatory systems. In addition to the amino acids and ATP, *S*-adenosylmethionine is required as methyl group donor if *N*-methylated bonds are involved. Due to their complexity (cyclosporin synthase for example, integrates 40 reactions on one polypeptide chain), the multienzyme systems have low turnover numbers. Cell-free systems for the production of various peptides and their analogues have been developed.

For synthesis involving ribosomes the first step is the activation of the amino acids to form amino-acid **adenylates**. The amino acids then become attached by an ester linkage to **transfer RNA** (tRNA). The template is **messenger RNA** (mRNA), which associates with the ribosomes themselves. The **translation** of the triplet codons in mRNA to provide a polypeptide involves three steps: chain initiation, chain elongation (*see* **elongation factor**), and chain termination (*see* **release factor**). Chain initiation normally starts with the codon for methionine, **AUG, and involves** a unique initiator tRNA. Many protein factors play a part in the three processes and five molecules of ATP and GTP participate. The ribosomes participate in a cycle during which they subdivide into their two subunits only to reassemble. Hence the synthesis of a peptide bond is a surprisingly expensive process in terms of energy. The targeting of the newly synthesized proteins to their final destination is termed **protein kinesis**. The mechanism of ribosomal protein synthesis is very similar in all types of cell but there are subtle differences in the number and properties of the soluble factors, particularly concerning chain initiation: in prokaryotes this is by way of formylmethionine while in eukaryotes, apart from mitochondria, this involves methionine.

proteinase an enzyme that catalyses the hydrolysis of peptide bonds more readily in an intact protein than in small peptides. **Endopeptidase** is now recommended.

proteinase K EC 3.4.21.64; *recommended name*: endopeptidase K; *other name*: *Tritirachium* alkaline proteinase. A serine proteinase enzyme that catalyses the hydrolysis of keratin, and of other proteins with subtilisin-like specificity.

protein A-Sepharose an affinity chromatography medium for the purification or precipitation of immunoglobulins made by covalent coupling of **protein A** to agarose beads.

proteinase V8 *see* **glutamyl endopeptidase**.

protein B a cell-surface protein of group B *Streptococcus* species that binds specifically to the Fc region of human IgA, without affecting its antigen-binding capacity, but not to other immunoglobulins or serum proteins. It is used in enzyme- linked immunoassay and blotting techniques.

protein biosynthesis *see* **protein and peptide biosynthesis**.

protein body a membrane-delimited aggregate of prolamins that accumulates within regions of the endoplasmic reticulum in the endosperm of cereals (barley, maize, oats, wheat, etc.). The membrane surface often contains trapped polysomes as a result of the nascent prolamin molecules aggregating with those already stored. *Compare* **protein storage vacuole**.

protein-bound iodine *abbr.*: PBI; that fraction of iodinated molecules bound to protein in serum. It is due largely to thyroxine (T_4), 99.97% of which in human serum is present bound to the binding proteins T_4-binding globulin, T_4-binding prealbumin and albumin, but also to T_3, 99.7% of which is bound to the same proteins. Determination of PBI (now replaced by immunoassay) was used for estimation of blood thyroid hormones as a basis for assessment of thyroid function.

protein C a **vitamin K**-dependent glycoprotein that is the **zymogen** of a serine endopeptidase in normal blood plasma; M_r 62 000 (human), 54 300 (bovine). It is formed from a precursor, autoprothrombin IIA, which is cleaved into a light chain and a heavy chain, held together by a disulfide bond to form protein C. Thrombin cleaves a tetradecapeptide from the amino end of the heavy chain; this reaction, which occurs at the surface of endothelial cells, is strongly promoted by thrombomodulin; protein C then becomes the enzyme protein C (activated), EC 3.4.21.69. This activated form specifically hydrolyses factors Va and VIIIa to inactive forms (in a reaction requiring phospholipid, calcium, and **protein S**), thereby acting as a potent anticoagulation factor; it contains 10–12 γ-carboxyglutamyl residues in the N-terminal 40 residues. A congenital deficiency of protein C (usually autosomal dominant) is often not compatible with survival, or leads to hemorrhagic disease of the newborn. Protein C is so named because it eluted from columns as the third of four peaks labelled A, B, C and D.

protein C23 *see* **nucleolin**.

protein C-binding protein *see* **apolipoprotein H**.

protein degradation the hydrolytic catabolism of proteins that occurs in the intestinal tract during digestion or intracellularly as part of regular turnover or apoptosis. Eukaryotic cells contain several proteolytic systems. The **vacuolar pathway** involves **endosomes** and **lysosomes** for the breakdown of extracellular proteins that enter cells by endocytosis. Lysosomes also degrade membrane proteins and proteins with long half-lives. In well-nourished organisms the lysosomes are nonselective but in some organs (e.g. liver and kidney) after prolonged fasting they are selective for cytosolic proteins that contain the pentapeptide sequence –KFERQ– or a related one. The **cytosolic system** is largely ubiquitin- and ATP-dependent and is mediated by **proteasomes**. Cytosolic **procaspases** are activated early during apoptosis. **Calpains** are also cytosolic. In the lumen of the endoplasmic reticulum, misfolded or unfolded proteins are degraded by proteases that function at high Ca^{2+} concentrations. Bacteria, mitochondria, and chloroplasts contain **AAA proteases** that degrade nonintegrated membrane proteins. *See* **PETS hypothesis**; **N-end rule**.

protein disulfide-isomerase *abbr.*: PDI; EC 5.3.4.1; *other name*: disulfide rearrangase; an enzyme that enhances the rate of inter-

change in the groups attached to two or more intrachain or interchain disulfide bonds in a protein. First described by Anfinsen in 1966 in connection with the reversible denaturation of ribonuclease and called 'rearrangase', it is a homodimer of the endoplasmic reticulum lumen and contains two domains homologous to **thioredoxin** of *Escherichia coli*. It is also the triacylglycerol transfer protein (MTP) that facilitates the incorporation of lipids into newly synthesized core **lipoproteins** within the endoplasmic reticulum. It serves as the β subunit of **procollagen-proline dioxygenase**.

protein engineering the process whereby a protein of any desired primary structure can (in principle) be created by the use of recombinant DNA technology. The ability to express DNA in a bacterium such as *Escherichia coli* creates the potential to synthesize a protein with the desired primary structure. Bacteria cannot effect post-translational modifications so if these also are required a eukaryotic host cell such as yeast must be employed. In principle there are no problems with the proteins of bacterial cells but **introns** present in the genome of eukaryotic cells present a potential problem. This can be overcome by preparing complementary DNA, **cDNA**, which is synthesized by the use of the **mRNA** as template and the enzyme **reverse transcriptase**. There are several ways whereby the DNA for the required protein may be prepared by modification of the base sequence of the natural DNA. DNA may be used that contains inserts of synthetic oligonucleotides, or stretches replaced by these. Oligonucleotide-directed mutagenesis may be used to modify a single base to produce a protein with a single amino-acid replacement. This is done by preparing an oligonucleotide primer with the modified sequence which will base pair with the natural DNA around the site to be altered; a single base substitution still permits satisfactory pairing. The primer is elongated by DNA polymerase and the double-stranded circle closed by **DNA ligase**. A specific deletion can be obtained by cleaving a plasmid at two sites with a restriction enzyme and religating to form a smaller circle. A smaller deletion can be obtained by cutting a plasmid at a single site. The ends of the linear DNA are then digested with an exonuclease that removes nucleotides from both strands and the shortened piece of DNA is religated.

protein expression map a map of protein expression in particular tissues or species, under particular conditions, generated from 2D electrophoresis data. Such maps are often used to compare samples exposed to different conditions, and hence to identify groups of proteins that have related functions or whose expression is interdependent. As 2D gels can reveal PTMs, the resulting maps can provide more relevant information about cellular dynamics than corresponding expression maps at the mRNA level.

protein family a set of proteins that share a common evolutionary history, often arising via duplication of an ancestral gene. Families, which tend to be defined via their sequence, functional, and/or structural relationships, can be arranged hierarchically, whereby closely related proteins with a recent common ancestor comprise subfamilies, and larger groups of distant homologues comprise superfamilies.

protein folding the process by which a polypeptide chain assumes its 3D conformation.

protein folding disease any disease that results from the abnormal folding of a protein. **Prion** diseases are examples of such diseases that can be inherited or acquired; **Alzheimer's disease** is a different type of disease, caused by alternative folding of proteins derived from the amyloid precursor.

protein fusion *an alternative name for* **gene fusion**.

protein G a bacterial cell-wall protein isolated from group G streptococci. Its properties and uses are similar to those of **protein A**, but it does not bind to IgM, IgD, or IgA; it does bind to all subclasses of human IgG, and animal immunoglobulins, with which protein A does not react well.

protein gene product any one of the proteins, or protein subunits, synthesized biochemically on the basis of information encoded in a genome, whether generated at **translation** or by **post-translational modification**; the pattern of protein gene products formed, or capable of being formed, by an organism is unique to any one genotype. Individual proteins are often referred to using the symbol p, followed by the relative molecular mass, e.g. p53. The protein may be further identified by adding an indication of its origin as a superscript, e.g. p21^{ras}. A protein gene product may also be referred to by the name of the gene that encodes it, but normal font is used instead of italics, with a capital initial followed by lower case characters, e.g. Src protein encoded by *SRC* gene.

protein-glutamate methylesterase a chemotaxis protein encoded by *cheB* (*see* **che**).

protein–glutamine γ-glutamyltransferase the product of one gene in invertebrates, but any of nine in vertebrates. The best known is **factor XIIIa** (EC 2.3.2.13) (*see* **blood coagulation**). It forms inter- and intramolecular links between a γ-carboxyl group of a glutamine residue of a protein and the ε-amino group of a lysine residue, releasing ammonium ion. Its natural substrate is the **fibrin** monomer, and its function is to link these into polymers.

 Transglutaminase K *other names*: transglutaminase 1; keratinocyte transglutaminase; an enzyme that catalyses formation of cross-linkages in the cell envelope (*see* **involucrin**) that replaces the plasma membrane of fully differentiated keratinocytes where it is anchored to the inner surface of the membrane. Several mutations at 14q11.2 are responsible for an autosomal recessive form of ichthyosis.

 Transglutaminase 2 *other name*: tissue transglutaminase; a constitutive enzyme in endothelial and smooth muscle cells, and inducible (e.g. by retinoic acid) in other cell types. It is localized to plasma membrane, cytoskeleton, and nucleus. The human enzyme (686 amino acids) is a G-protein (called Gh/TG2) that mediates transmembrane signalling by stimulating phospholipase C, and cell-surface adhesion following externalization by cells. It is involved in generating stimulatory gluten peptides in celiac disease.

protein intron *see* **intein**; **splicing** (def. 3).

protein kinase any of a number of enzymes that phosphorylate one or more hydroxyl or phenolic groups in proteins, ATP being the phosphoryl-group donor. Two classes are recognized: those that phosphorylate seryl or threonyl hydroxyls, and those that phosphorylate tyrosyl phenolic groups. The first class contains enzymes that were classically identified as regulating pathways of intermediary metabolism (e.g. glycogen **phosphorylase kinase**); subsequently, protein kinase A (**cyclic-AMP-dependent protein kinase**) and **protein kinase C** were recognized in this class. Enzyme activity of the second class is found in some cytokine receptors such as **platelet-derived growth factor receptor** and **epidermal growth factor receptor** and in oncogene products such as that of *SRC*. Over 110 unique gene products are known to make up the protein kinase superfamily; important motifs include GXGXXG...AHK and APE...DXWSXG, which are common to both serine/threonine- and tyrosine-specific activities. The tyrosine-specific kinase activity associated with the cytoplasmic domains of certain cytokine receptors forms part of an important signal-transduction mechanism implicated in cell growth (*see* **tyrosine kinase**).

protein kinase A *an alternative name for* **cyclic-AMP-dependent protein kinase**.

protein kinase B (*abbr.*: PKB) *or* Akt a cytosolic protein-serine/threonine kinase that plays a significant role in insulin and growth factor signalling, cellular survival, and transformation. It contains a pleckstrin homology domain for binding phosphatidylinositides, and is a major target for phosphatidylinositol trisphosphate kinase and for phosphatidylinositide-dependent protein kinase. The activities of these enzymes are provoked by various tyrosine kinase receptors for growth factors (e.g. those for platelet-derived growth factor, epidermal growth factor, fibroblast growth factor, and insulin). There are three isozymes (Akt 1, 2, and 3) in humans and rodents. Akt 2 (PKBβ) is enriched in insulin-responsive tissues. Mice deficient in Akt 2 are insulin-resistant and develop a diabetes mellitus-like syndrome. Substrates for Akt 2 include: **GSK-3**, ribosomal S6 kinase, several transcription factors (including those of the forkhead family), and APC protein (*see* **APC**).

protein kinase C *abbr.*: PKC; any **protein kinase** enzyme that requires anionic phospholipid for activity, and is regulated by diacylglycerol and Ca²⁺. Such enzymes phosphorylate serine and threonine residues. Several isoforms have now been characterized: α, β₁, β₂, γ, δ, ε, ζ, θ, η, ι, and μ. Four conserved regions are recognized,

C₁, C₂, C₃, and C₄; C₃ and C₄ comprise the C-terminal half of the molecule and bear the active site, including the ATP-binding sequence GXGXXG...K. PKC-ε is Ca²⁺-independent, and δ, ε, and ζ lack the homologous C₂ region. Activation of PKC occurs when plasma membrane receptors coupled to phospholipase C are themselves activated, releasing diacylglycerol. Phorbol ester tumour promoters can substitute for diacylglycerol in enzyme activation, which may indicate their mechanism of action, and much evidence points to PKC activation being involved in cell-growth stimulation. Inhibitors of PKC attract considerable interest, and include **bisindolyl-maleimide**, 1-O-hexadecyl-2-O-methylglycerol, **melittin**, **phloretin**, **polymyxin B**, and **staurosporine**.

protein kinesis the directed movement of proteins from their site of synthesis to specific targets within cells and the modulation by proteins of interactions among intracellular organelles. It is a factor in generating subcellular compartments and determining the way in which reactions are compartmentalized to control metabolic processes. It embraces the mechanisms for ensuring that proteins reach their site of action and that the amounts and locations of particular proteins are controlled during development, during the cell-division cycle, in combating pathological states of the cell, and during apoptosis.

protein mapping the resolution of a mixture of proteins into a two-dimensional pattern (i.e. map) by successive use of two differing separative procedures, as in isoelectric focusing followed by gradient gel electrophoresis, the pattern being rendered visible by an appropriate detection method.

protein phosphatase any enzyme that hydrolyses phosphate groups from proteins. With protein kinases, these enzymes provide an important mechanism for regulating cellular activity. Protein phosphatases fall into three groups. **Class I enzymes** are mainly specific for seryl- or threonyl-phosphate. Phosphatase 1 is ATP,Mg²⁺-dependent (AMD phosphatase); it consists of a catalytic subunit of 38 kDa and a regulatory (or glycogen-binding) subunit of 126 kDa, which on phosphorylation activates the phosphatase. Phosphatase 2A is polycation-stimulated (PCS), being directly stimulated by protamine, polylysine, or histone H1; it constitutes a subclass of several enzymes activated by different histones and polylysine. These have a very high affinity for the tumour promoter **okadaic acid**. Human phosphatase 2A has a catalytic subunit of 309 amino acids. This enzyme can modulate the activities of phosphorylase kinase *b*, casein kinase 2, mitogen-stimulated S6 kinase, and MAP-2 kinase. It is a cytoplasmic enzyme that exists in several oligomeric forms, all of which contain a catalytic (C) subunit (36–38 kDa) associated with one or both of two putatively regulatory subunits, the A subunit (61–65 kDa) and the B subunit (51–55 kDa). Phosphatase 2B (or **calcineurin**) consists of a subunit of 61 kDa, which contains the catalytic site and a high-affinity calmodulin-binding site, and a Ca²⁺-binding subunit of 19 kDa. Phosphatase 2C is a Mg²⁺-dependent monomeric 43 kDa protein; it has been purified as myosin light chain phosphatase, but is active against the phosphorylated forms of glycogen synthase, 6-phosphofructo-1-kinase, pyruvate kinase, fructose 1,6-bisphosphatase, and 6-phosphofructo-2-kinase. **Class II enzymes** hydrolyse tyrosylphosphates, and are generally designated 'phosphatase 3'. Some are cytoplasmic, and others are membrane-bound and may have receptor properties. **Class III enzymes** have dual specificity.

protein-N^π-phosphohistidine–sugar phosphotransferase EC 2.7.1.69; *systematic name*: protein-N^π-phosphohistidine:sugar N^π-phosphotransferase; *other name* enzyme II; any membrane-bound bacterial enzyme that forms part of the **phospho*enol*pyruvate-dependent sugar phosphotransferase system** for the transport of hexoses across the cell membrane. It involves phosphorylation of a protein substrate (9.5 kDa), a phosphocarrier known as HPr. Individual enzymes are specific for a range of sugars. They catalyse the reaction:

$$\text{protein } N^\pi\text{-phosphohistidine (HPr) + sugar =}$$
$$\text{protein histidine + sugar phosphate.}$$

protein S a single-chain glycoprotein occurring in blood, M_r 69 000. It is a vitamin K-dependent protein (not a protease) that promotes the binding of **protein C** to platelets, and functions as a cofactor for the anticoagulant activity of activated protein C. It is named after Seattle, where it was discovered.

protein sorting the secretory and endocytic processes by which proteins are directed to specific locations inside or outside the cell.

protein sorting signal a motif in the sequence of a protein which determines the compartment to which it is directed in the cell.

protein splicing *see* **splicing**.

protein storage vacuole *or* **vacuolar protein body** a specialized vacuole in the reserve tissue of seeds that contains 7S (vicilin-type) or 11S (legumin-type) globulins. The proteins are synthesized on the endoplasmic reticulum and transported to storage vacuoles via the Golgi apparatus, where the 7S type are glycosylated but the 11S rarely so. Protein deposition occurs during seed maturation, whereas breakdown occurs during germination to provide amino acids to the new seedling. The vacuoles also contain other proteins such as lectins and enzyme inhibitors. *Compare* **protein body**.

protein synthesis 1 (*in cell biology*) *see* **protein and peptide biosynthesis**. **2** (*in chemistry*) *see* **peptide synthesis** (def. 2).

protein targeting to glycogen *see* **PTG**.

protein thiotemplate mechanism *see* **thiotemplate mechanism**.

protein truncation test *abbr*.: PTT; a method for the detection of mutations that lead to the premature termination of translation and hence a truncated protein product. Preferably mRNA is the template for amplification by reverse transcription-polymerase chain reaction (RT-PCR) using a forward primer that incorporates the promoter for T7 RNA polymerase and a **Kozak** translation initiation sequence. The product is used in a transcription and translation system *in vitro* to produce a radiolabelled polypeptide that can be detected by autoradiography following SDS-PAGE. Alternatively, a c-*myc* epitope tag can be encoded in the forward primer enabling the detection of the protein product by anti-MYC antibodies following Western blotting. The sizes of the wild-type and mutant gene products can be compared alongside each other on the gel.

proteinuria the presence of protein in urine. Normal excretion of protein in human urine is less than 0.1 mg every 24h, about half of which is **Tamm–Horsfall glycoprotein**, and about 20% is albumin. Protein may appear in urine due to extrarenal factors such as fever, strenuous exercise, or burns; direct involvement of the kidney occurs in the nephrotic syndrome, which may arise from several causes, such as diabetic nephropathy, cancer of the kidney, or various forms of glomerulonephritis of unknown etiology.

protein Z a major protein of barley endosperm albumin. It is structurally similar to **serpin**, but its main role is as a storage protein, being especially rich in lysine.

proteoclastic *an obsolete term for* **proteolytic**.

proteoglycan a molecule consisting of one or more **glycosaminoglycan** chains attached to a core protein.

proteolipid *a term variously applied to describe* certain hydrophobic membrane proteins, or any other protein, that has a lipid moiety covalently bound to one or more of its constituent amino acids. Many such proteins are soluble in chloroform–methanol, a property that was significant in their first identification.

proteoliposome a **liposome** into which a specific protein, or group of proteins, has been incorporated.

proteolysis degradation of a protein, usually by hydrolysis at one or more of its peptide bonds. —**proteolytic** *adj*.

proteolytic enzyme *see* **peptidase**.

proteome *see* **'ome**.

proteomics *see* **'omics**.

proteosome *a variant spelling of* **proteasome**.

prothrombin *an alternative name for* factor II; *see* **blood coagulation**.

prothymosin *see* **thymosin**.

protide 1 *symbol*: ^{1}H ; a **hydride** (def. 1) ion derived from an atom of **protium**. **2** any **hydride** (def. 2) formed from protium. **3** *or* **protid** *a generic name embracing* protein, peptide, or amino acid, suggested by analogy with glucide and lipid(e). Its use in English is largely restricted to translations or abstracts from French. —**protidic** *adj*.

protio+ *comb. form* indicating that one or more atoms of hydrogen in a hydrogen-containing chemical compound consists exclusively of **protium**.

Protista a taxonomic grouping of eukaryotic organisms that are unicellular, coenocytic, or multicellular but that are not classified as fungi, animals, or embryophytes.

protium *symbol*: ^{1}H; hydrogen-1, the nuclide of hydrogen, $^1_1$H, of relative atomic mass 1.008. It is a stable isotope, having a relative abundance in natural hydrogen of 99.98 atom percent. The name is used especially when distinction from **deuterium** or **tritium** is required.

protoalkaloid an alkaloid that does not contain a heterocyclic ring, e.g. **damascenine** in *Nigella* spp.

protobiont one of the postulated first precursors of cells, supposed to have arisen when a boundary or membrane formed around one or more macromolecules possessing catalytic properties. *Compare* **coacervate droplet**.

protochlorophyllide Mg^{2+}-vinylpheoporphyrin A_5 methyl ether; an intermediate in the synthesis of chlorophyll, it is formed from Mg^{2+}-protoporphyrin, and converted to **chlorophyllide** by NADPH:protochlorophyllide oxidoreductase (EC 1.3.1.33). Protochlorophyllides may be suffixed with the letter of the related chlorophyll, e.g. protochlorophyllide *a*.

protochlorophyllide holochrome a pigment protein complex that carries out a light-catalysed reduction of **protochlorophyllide** to **chlorophyllide** with an external hydrogen donor, which is probably strongly bound NADPH. The enzyme involved in regenerating NADPH is **NADPH-photochlorophyllide oxidoreductase**; it is a single polypeptide, M_r 36000, and is only one component of the complex protein entity of the holochrome.

protocol a set of rules governing the way data are exchanged between computers. Computers with very different hardware and operating systems can communicate provided the same protocol is implemented on each. Whereas **API**s ensure portability, protocols ensure interoperability.

protocollagen **collagen** chains in which proline and lysine residues have not been hydroxylated. Protocollagens may be produced by experimental inhibition of the relevant hydroxylases, either by use of a chelating agent or by imposed anaerobiosis. *Compare* **procollagen**.

protoheme *or* (*esp. Brit.*) **protohaem** the iron–porphyrin prosthetic group of the **cytochromes *b***.

protohemin *see* **hemin**.

protokaryon *an obsolete term for* **prokaryon**.

protokaryote *an obsolete term for* **prokaryote**.

protolysosome *see* **primary lysosome**.

protomer any of the subunits of an oligomeric protein that are identical. *See also* **monomer**, **oligomer**. —**protomeric** *adj*.

proton 1 *symbol*: ^{1}H$^+$; the cation derived from an atom of **protium**; a hydrogen ion of nucleon number unity; used in this sense especially when distinction between ^{1}H$^+$, ^{2}H$^+$ (i.e. deuteron), and ^{3}H$^+$ (i.e. triton) is needed; any solvated species is included. **2** *symbol*: H$^+$; *commonly used term for* the cation derived from any atom of natural **hydrogen** (of which a small proportion is **deuterium**) by loss of its orbital electron; any solvated species is included. *See also* **hydrogen ion**, **hydron**. **3** *symbol*: p; a nuclear particle of nucleon number unity, having a charge equal and opposite to that of an **electron** and having a mass of 1.007 276 470 (12) Da. —**protonic** *adj.*

proton magnetic resonance *abbr.*: PMR *or* pmr; *see* **nuclear magnetic resonance spectroscopy.**

proton magnetogyric ratio *symbol*: γ_p; a fundamental physical constant, of value $2.675\ 221\ 28(81) \times 10^8$ s^{-1} T^{-1}. *See also* **gyromagnetic ratio.**

protonmotive indicating, or related to an ability to transport protons (i.e. hydrons).

protonmotive force the driving force that leads to a gradient of protons (hydrons) across a membrane, for example of the mitochondrial inner membrane or the chloroplast thylakoid membrane. The protonmotive force was named by P.D. Mitchell, by analogy with electromotive force, and is conventionally given the symbol, $\Delta m_H +$ and expressed in volts. It comprises two components, the electrostatic potential $\Delta\psi$ and the pH gradient, ΔpH. It may be produced by coupling to electron transfer reactions, or to the hydrolysis of ATP.

protonmotive Q cycle a hypothetical mechanism put forward to explain the **protonmotive** and **redox** properties of certain components of the **respiratory chain** in the **coupling membrane** of bacteria, mitochondria, and chloroplasts. The **Q cycle** applies to the part of the respiratory chain involving passage of electrons through cytochrome *b*. Essentially, the cycle accepts two hydride ions from a reduced flavoprotein (e.g. of complex I or complex II of the respiratory chain; *see* **complex I, II, III, IV**) passes electrons singly to cytochrome c_1 and releases the two hydrons to the intermembrane space. **Ubiquinone** is the molecule that accepts the hydride ions, and releases the electrons and hydrons. Cytochromes b_{566} and b_{562} are involved in shuttling electrons between different oxidation states of ubiquinone during the cycle. The property of ubiquinone that it can exist as a **semiquinone** is important in the switch from two-electron to one-electron transfer.

proton number *or* **atomic number** *symbol*: Z; a dimensionless physical quantity, with a unique integral value for each chemical element, that is equal to the number of protons contained in the nucleus of each atom of the element irrespective of the mass of any particular atom. It also is equal to the number of unit positive charges on the nucleus of each atom of a given chemical element, to the number of orbital electrons surrounding the nucleus of an unionized atom of that element, and, for a specific nuclide of that element, to the arithmetical difference between the nuclide's **nucleon number**, A, and its **neutron number**, N; i.e. $Z = N - A$. The proton number of an element determines the position of the element in the periodic table. It may be attached to the symbol for the element as a left subscript, e.g. $_6$C.

proton pump *another term (esp. in pharmacology) for* **hydrogen-ion pump.**

proton relaxation enhancement *abbr.*: PRE; a technique that yields information concerning (macro)molecular motion. It depends on the enhancement of the solvent water-proton relaxation rates on addition of a macromolecule.

protooncogene a normal gene, usually concerned with the regulation of cell proliferation, that can be converted into a cancer-promoting **oncogene** by mutation. For example, in the case of the Rous sarcoma virus the viral oncogene is v-*src* (*see* **src**). This is closely related to a normal gene, c-*src*. The appearance of v-*src* is believed to have followed the acquisition of a copy of c-*src* by an ancestor of the Rous sarcoma virus. Through many generations this ancestral gene was mutated and became dominant. Protooncogenes encode proteins involved in all aspects of controlling cell proliferation, including growth factors, growth-factor receptors, nuclear factors regulating gene expression proteins that generate second messengers, and protein kinases.

protoplasm the living contents of a cell; i.e. the matter contained within (and including) the plasma membrane, usually taken to exclude large vacuoles and masses of secretory or ingested material. In eukaryotes it includes the **nucleus** and **cytoplasm.**

protoplast a spherical, osmotically sensitive plant cell without its cell wall but retaining an intact cell membrane. The cell wall is degraded by digestion with different hydrolytic enzymes, as a mixture or sequentially, e.g. cellulase, hemicellulase. Protoplasts are used to create hybrid cells via protoplast fusion. In some cases they are totipotent, and are thus capable of yielding plants. Such protoplasts can be used in plant transformation.

protoporphyrin IX the specific substrate for the enzyme **ferrochelatase**, which catalyses the insertion of iron to form **protoheme**. It is also probably the substrate for the insertion of magnesium in the formation of chlorophylls in plants.

prototroph any strain of a microorganism (alga, bacterium, or fungus) that does not require any substances in its nutrition additional to those required by the wild type. *Compare* **auxotroph.** —**prototrophic** *adj.*

Protozoa a phylum or subkingdom comprising single-celled eukaryotic microscopic organisms, usually classified as animals. *Compare* **Metazoa, Parazoa.**

provirus a virus, e.g. a **retrovirus**, that is integrated into the chromosome of its host cell and can be transmitted through the generations without causing lysis of the host cell.

provitamin a compound which is converted to an active vitamin in the body. In practice the term is only applied to **carotenes.**

Pro-X dipeptidase EC 3.4.13.8; *recommended name*: prolinase; *other names*: prolyl dipeptidase; iminodipeptidase; prolinase; L-prolylglycine dipeptidase; an enzyme that catalyses the hydrolysis of Pro-|-Xaa dipeptides. It also acts on hydroxyprolyl analogues.

proximal protein any protein that binds early in the assembly of complex structures, such as ribosomes. Their presence in the partially assembled structure is required for the binding of **distal proteins.**

proximate carcinogen *or* **proximate carcinogenic metabolite** any metabolite of a **chemical carcinogen** that is itself carcinogenic; the term may embrace **ultimate carcinogen.**

Prozac *see* **fluoxetine.**

PrP *abbr. for* prion protein.

PRP protein *abbr. for* **pre-mRNA processing protein.**

PRPP *abbr. (not recommended) for* phosphoribosyl pyrophosphate (*see* **(5-)phospho(-α-D-)ribosyl diphosphate**).

PRPP synthetase *another name for* **ribose-phosphate pyrophosphokinase.**

PR protein *abbr. for* pathogenesis-related protein.

pRSETA one of a family of phagemid vectors designed for high levels of expression of proteins in *E. coli*. The gene of interest is cloned downstream from the promoter for T7 RNA polymerase. When propagated in BL21 (DE3) cells the expression of T7 RNA polymerase can be induced with isopropyl β-thiogalactoside (IPTG), which then transcribes the cloned gene.

prunasin (–)-(*R*)-mandelonitrile β-D-glucoside; a **cyanogenic** glycoside.

ps *symbol for* picosecond.

P-selectin *see* **selectin.**

pseudoachondroplasia *see* **achondroplasia.**

pseudoalkaloid any **alkaloid** in which the skeleton is partly derived from a terpene.

pseudoallele any of two (or more) mutations that are allelic functionally but not structurally. *See* **allele.**

pseudoaxial (*in chemistry*) *abbr.*: a′; *see* **conformation.**

pseudocholinesterase EC 3.1.1.8; *recommended name*: cholinesterase; *other name*: choline esterase II. An enzyme, found in blood and other tissues, that catalyses the hydrolysis of a variety of choline esters and a few other compounds. *Compare* **acetylcholinesterase.**

pseudoequatorial *(in chemistry)* abbr.: e′; *see* **conformation**.

pseudoexon a portion of an **exon** that is duplicated in an adjacent **intron**.

pseudogene *symbol*: ψ; a sequence in DNA that is related to a functional gene but cannot be transcribed. This may be due to mutational changes that preclude it from being expressed in the form of functional products or the lack of some flanking control region in the promoter. Pseudogenes may or may not contain the introns of the functional gene. Examples include pseudogenes for α, β, and ζ globulins in several species, $V_κ$ chains of immunoglobulins, actin, and tubulin. **Processed pseudogenes**, or LINES (*see* **LINE**), occur in a genome somewhere other than the normal position of the corresponding functional gene and so cannot be expressed because of alterations in regulatory DNA sequences, or coding sequences, or both.

pseudoglobulin a globulin that is soluble in solutions of low ionic strength, or in pure water. *Compare* **euglobulin**.

pseudohypoaldosteronism type II *see* **WNK**.

pseudointron an **intron** that contains 5′ and 3′ splice sites that are appropriate but not utilized in the excision of the intron.

pseudoisoenzyme any one of the epigenetically derived multiple forms of an enzyme.

pseudo-operator a functional **operator** element within a structural protein-coding region of DNA.

pseudopepsin *see* **pepsinogen**.

pseudoplasmodium *see* ***Dictyostelium discoideum***.

pseudopodium (*pl.* **pseudopodia**) a temporary protrusion or retractile process of a cell, associated with flowing movements of the protoplasm and serving for locomotion and feeding. —**pseudopodial** *adj.*

pseudouridine *symbol*: Ψ; 5-β-D-ribofuranosyluracil; a component of **transfer RNA** (tRNA) containing a C–C bond between the ribose (C-1) and uracil (C-5). Typical of the **minor nucleosides** present in tRNA, it is formed by modification of uridine in the original transcript; the chemistry of this process is unclear. When pseudouridine occurs in the anticodon of transfer RNA it base pairs with adenine. Mutation in the gene for dyskerin, a putative pseudouridine synthase, results in dyskeratosis congenita. *Compare* **uridine**.

pseudouridine synthase E.C.4.2.1.70; *other name*: pseudouridylate synthase; any of the enzymes (e.g. ψS1, PUS1) that catalyse the conversion of uridine residues in RNA to pseudouridines. They include the dyskeratosis congenita gene product, **dyskerin**. Members of the family differ in their specificity for RNA substrates.

pseudoxanthoma elasticum a hereditary disorder of connective tissue that affects skin, retina, and blood vessels. There is accumulation of calcium in these and in other tissues, with the formation of yellowish lumps in the skin. Several mutations in an ABC transporter (ABCC6) cause the disorder.

***Psh*AI** a type 2 **restriction endonuclease**; recognition sequence: GACNN↑NNGTC.

***Psh*BI** a type 2 **restriction endonuclease**; recognition sequence: AT↑TAAT. *Vsp*I is an **isoschizomer**.

psi 1 *symbol*: ψ (lower case) *or* Ψ (upper case); the twenty-third letter of the Greek alphabet. For uses *see* **Appendix A**. **2** *or* **p.s.i.** *abbr. for* pounds per square inch (a unit of pressure in the Imperial and US systems of units).

PSI *abbr. for* photosystem I (*see* **photosystem**).

PSI-BLAST *abbr. for* position-specific iterated-BLAST; *see* **database-search program**.

psicofuranine 6-amino-9-D-psicofuranosylpurine; an antibiotic produced by *Streptomyces hygroscopicus*, found to have antitumour activity. *See also* **D-psicose**.

D-psicose *a nonsystematic name for* the keto-hexose D-*ribo*-2-hexulose. The hydroxyl at C-3 has the opposite configuration to that of D-fructose. It is a component of **psicofuranine**.

α-D-pyranose form

PSII *abbr. for* photosystem II (*see* **photosystem**).

psilocin 3-[2-(dimethylamino)ethyl]-1*H*-indol-4-ol; a hallucinogenic alkaloid extracted from the mushroom *Psilocybe mexicana*.

P-site *abbr. for* a peptidyl-tRNA binding site of **ribosome**; in bacteria this involves the **ribosomal proteins** L2, L14, L18, L24, L27, and L33, and a region close to the 3′ end of 16S rRNA.

psoralen 6-hydroxy-5-benzofuranacrylic acid δ-lactone; a furocoumarin present as a **phytoalexin** in many plants. It has photosensitizing and phototoxic effects in humans and animals, and is used in photochemistry, intercalating into DNA and forming cross-links on irradiation. *See also* **methoxypsoralen**.

PSP *abbr. for* pancreatic stone protein (*see* **pancreatic thread protein**).

***Psp*1406I** a type 2 **restriction endonuclease**; recognition sequence: AA↑CGTT. *Acl*I is an **isoschizomer**.

pSPORT 1 one of a proprietary series of vectors for cloning cDNA in *E. coli*.

PSSM (*usually pronounced 'possum'*) *abbr. for* position-specific scoring matrix.

***Pst*I** a type 2 **restriction endonuclease**; recognition sequence: CTGCA↑G.

pSV vector a plasmid vector designed for the expression of cloned sequences in mammalian cells driven by the SV40 early promoter and enchancer.

psychosine 1-*O*-β-D-galactosylsphingosine; it is formed from UDP-galactose and sphingosine by sphingosine β-galactosyltransferase, EC 2.4.1.23. It is acylated to a cerebroside by reaction with a long-chain acyl-CoA.

psychrometer *or* **wet-and-dry bulb hygrometer** a type of hygrometer consisting of two thermometers, one with a wet bulb and one with a dry bulb. The cooling effect of evaporation from the wet bulb creates a temperature difference between the two thermometers, from which the relative humidity can be calculated.

psychrophile an organism that grows best at low temperatures (below about 20 °C). —**psychrophilic** *adj.*

P system a system of blood groups in which the antigenic determinants are oligosaccharides. The antigens are detected on erythrocytes, erythroblasts, platelets, megakaryocytes, fibroblasts, and endothelial cells; they are designated P^k, P, and P_1. The common phenotypes are P_1 (antigens P_1 and P positive) and P_2 (P antigen positive); P^k phenotypes are rare. Individuals with the p phenotype lack all of the above determinants, having only the core structure lactosylceramide, and have anti-P and anti-P^k antibodies. The determinants are:

Pk (Gb$_3$Cer): Gal(α1-4)Gal(β1-4)GlcCer
Pk (Gb$_4$Cer): GalNAc(β1-3)Gal(α1-4)Gal(β1-4)GlcCer
P1 : Gal(α1-4)Gal(β1-4)GlcNAc(β1-3) Gal(β1-4)GlcCer.

Pt *symbol for* platinum.

PTA *abbr. for* plasma thromboplastin antecedent (i.e. factor XI); *see* **blood coagulation**.

PTB domain *abbr. for* phosphotyrosine-binding domain. *See* **SH2 domain**.

PTC *abbr. for* **1** plasma thromboplastin component (i.e. factor IX); *see* **blood coagulation**. **2** phenylthiocarbamoyl.

Ptd *symbol for* the phosphatidyl group.

PtdIns *symbol for* phosphatidylinositol.

PtdIns4*P* or **PtdIns-4-*P*** *symbol for* phosphatidylinositol 4-phosphate.

PtdIns(4,5)*P*₂ or **PtdIns-4,5-*P*₂** *symbol for* phosphatidylinositol 4,5-bisphosphate.

pteridine pyrazino[2,3-*d*]pyrimidine; the parent structure of pterins and the **pteroyl** group. *See also* **xanthopterin**.

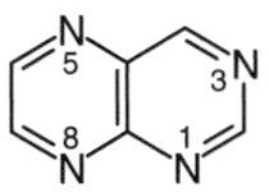

pterin-4α-carbinolamine dehydratase *abbr.*: PCD. EC 4.2.1.96; an enzyme activity of liver and other tissues that is part of a bifunctional protein (*abbr.*: PCD/DCoH; 103 amino acids) that also acts as the dimerization cofactor for hepatocyte nuclear factor 1α (*abbr.*: DCoH). The enzyme converts 4α-carbinolamine tetrahydrobiopterin (formed during the conversion of phenylalanine to tyrosine) to quinonoid dihydrobiopterin by elimination of water. It also prevents nonenzymatic conversion of tetrahydrobiopterin to **primapterin**. Several mutations in the protein lead to deficiency of PCD activity, with consequent hyperphenylalaninemia and primapterinuria. *See* **biopterin**.

pteroyl 4-[[(2-amino-1,4-dihydro-4-oxo-6-pteridinyl)methyl]amino]-benzoyl; the acyl group of pteroic acid, and the structural basis of **folate** and its derivatives.

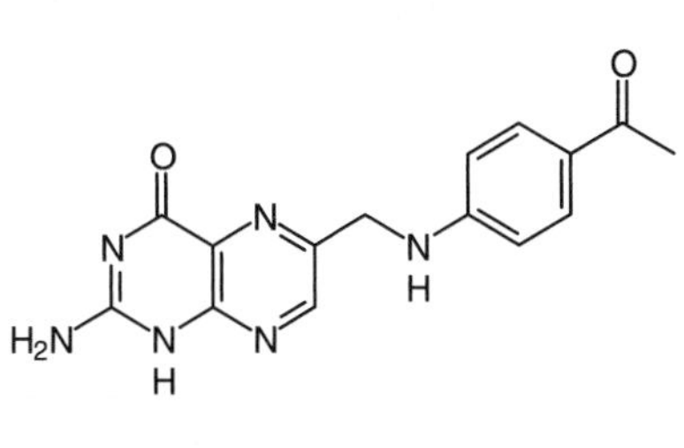

PTFE *abbr. for* polytetrafluoroethylene.

PTG *abbr. for* protein targeting to glycogen; a protein (294 amino acids), expressed predominantly in insulin-sensitive tissues, that binds to the catalytic subunit of **protein phosphatase** 1 and localizes it to glycogen. It also forms complexes with phosphorylase kinase, phosphorylase *a* (*see* **phosphorylase**), and **glycogen synthase**.

PTGS *abbr. for* post-transcriptional gene silencing. *See* **RNA silencing**.

PTH *abbr. for* **1** parathyroid hormone (i.e. **parathyrin**). **2** phenylthiohydantoin; as in PTH-amino acid (i.e. the phenylhydantoin derivative of an amino acid).

PTI *abbr. for* pancreatic trypsin inhibitor.

PTM *abbr. for* post-translational modification.

PTNP11 *abbr. for* protein tyrosine phosphatase noinreceptor type 11. *See* **Noonan syndrome**. .

Ptprz *abbr. for* protein tyrosine phosphatase receptor z. *See* **Helicobacter pylori**.

pTriEX a proprietary plasmid vector that enables the expression of genes in bacterial, insect, and mammalian cells from a single plasmid.

PTS *abbr. for* **1** phospho*enol*pyruvate-dependent sugar phosphotransferase system. **2** peroxisome targeting sequence.

PTS receptor either of two proteins that recognize a **peroxisome targeting sequence** (PTS) in targeting proteins to the peroxisomal matrix. Both are PEX proteins (*see* **peroxin**); PEX5 binds PTS1, while PEX7 binds PTS2.

PTT *abbr. for* protein truncation test.

ptyalin a substance in saliva, identified as α-**amylase**.

P-type transporter a member of a family of membrane transport proteins, found in bacteria, archaea and eukaryotes, that catalyse cation uptake and/or efflux driven by ATP hydrolysis. Most are multisubunit complexes with a large subunit serving the primary ATPase and ion translocation functions. Distinct bacterial enzymes specific for K$^+$ or Mg^{2+} (uptake), Ca^{2+}, Ag$^+$, Zn^{2+}, Co^{2+}, Pb^{2+}, Ni^{2+}, and/or Cd^{2+} (efflux) and Cu^{2+} or Cu$^+$ (uptake or efflux, depending on the system) have been characterized.

Pu *symbol for* plutonium.

PubMed *see* **information retrieval software**.

pUC vector any of a series of plasmid cloning vectors having an ampicillin resistance gene and part of the *lacZ* gene encoding the α-peptide containing a multiple cloning site (MCS). Different members of the series (e.g. pUC18, pUC19) differ in the orientation of the MCS. They are high copy number plasmids and so can be purified by rapid, small-scale methods of isolation. A related series of bacteriophage M13 vectors, the M13mp vectors, contain the same MCS and are useful for DNA sequencing. *See also* **Messing vector**.

PUFA *abbr. for* polyunsaturated fatty acid.

puff any discrete, localized expansion occurring at specific sites in polytene chromosomes, e.g. the giant salivary gland chromosomes in Diptera. Typically a few micrometres long, they are believed to be a gene or group of genes that are actively synthesizing RNA, which accounts for a considerable part of the mass of the puff. *See also* **Balbiani ring**.

puffer fish *see* **Fugu**.

pulldown technique a method for investigating protein–protein interactions in which a protein of interest is expressed as a fusion protein with glutathione-*S*-transferase (GST) or an epitope tag such as 6 × His or c-Myc. This is used to explore previously unknown interactions with other proteins in a mixture such as a cell extract. Protein complexes formed with the fusion protein are precipitated either with specific antibodies or with a suitable affinity adsorbent, and the nature of the bound proteins is identified by gel electrophoresis and mass spectrometry.

'pull' model *a nickname for* the **Monod–Wyman–Changeux model** (for ligand-induced changes in protein conformation) because, in this model, ligands influence the equilibrium between enzyme states by sequestering the enzyme in ligand-bound forms. *Compare* **'push' model**.

pullulan an extracellular **glucan** elaborated by *Aureobasidium pullulans* (yeast form) and formed by other fungi. It is a linear structure with both (1→4)-α-D- and (1→6)-α-D-linkages. Pullulan is used as a non-caloric material, a water-retention agent, and as a film and fibre former.

pullulanase 1 isopullulanase EC 3.2.1.57; an enzyme that catalyses the hydrolysis of **pullulan** to isopanose (i.e. 6-α-maltosylglucose). **2 neopullulanase** EC 3.2.1.135; an enzyme that catalyses the hydrolysis of pullulan to panose (i.e. 6-α-D-glucosylmaltose). **3** *see* α-**dextrin endo-1,6-α-glucosidase**.

pulmonary surfactant a secretion, produced by type II cells of alveoli and bronchial mucosa, that forms a monolayer at the alveolar air–water interface to lower the surface tension. It consists primarily of dipalmitoyllecithin, unsaturated lecithin and phosphatidylglycerol, and surfactant proteins B and C. The surfactant's components are essential at birth to convert a fluid-filled lung to one filled with air. They are cleared by the type II cells and macrophages for recycling or degradation. Secretion and clearance are increased by exercise.

pulse-chase experiment a technique whereby cells growing in culture are exposed for a short period (pulse) to a radiolabelled molecule such as an amino acid and then transferred to a medium containing an excess of the nonlabelled substance for a longer period (chase). By removing samples at intervals the fate of intracellular components labelled during the pulse can be ascertained.

pulsed field gel electrophoresis *abbr.*: PFGE; a type of **gel electrophoresis** in which fragments of DNA in the megabase size range can be separated by continuously altering the angle at which the electric current is applied.

pulse radiolysis a technique whereby reactive chemical species are generated by pulses of high-energy radiation and the subsequent reactions of these species are followed spectrophotometrically as a function of time.

Pumilio a protein of *Drosophila* that binds the Nanos response elements in the 3′ untranslated region of the *hunchback* mRNA and recruits the proteins Nanos and Brain tumor to suppress its translation. It contains eight repeats of a ≈36 residue trihelical sequence that assemble into an arched structure. Through its concave surface this arch binds the RNA motifs, which have a core UGU trinucleotide sequence; while through the outer surface of repeats 7 and 8 it binds with Nanos and Brain tumor, respectively. A homologue in *Xenopus*, called XPum, is present throughout oogenesis. There is also a human homologue.

pump *(in cell biology)* any transmembrane protein that drives the active transport of ions and small molecules across the lipid bilayer. *See also* **hydrogen-ion pump**, **sodium pump**.

PUMP-1 *see* **matrilysin**.

pump-1 protease *see* **matrilysin**.

Punnett square a matrix showing all possible combinations of alleles in a genetic cross. [After Reginald Punnett (1875–1967), British geneticist.]

purine 1 1*H*-purine; 7*H*-imidazo[4,5-*d*]pyrimidine; an organic nitrogenous base, sparingly soluble in water. **2** any of a class of derivatives of 1*H*-purine; these form one of the two classes of nitrogen-containing ring compounds found in DNA and RNA, which include **adenine** and **guanine**. *See also* **purine biosynthesis**.

purine biosynthesis the biosynthesis of purine ribonucleotides. The pathway begins with the formation of 5-phospho-α-D-ribosyl diphosphate (5-phosphoribosyl-α-D-pyrophosphate, *common abbr.*: PRPP) by ribose-phosphate pyrophosphokinase, EC 2.7.6.1. This is converted to β-5-phosphoribosylamine (i) by reaction with glutamine, catalysed by amidophosphoribosyltransferase, EC 2.4.2.14; reaction of I with glycine, with hydrolysis of ATP to ADP and Pᵢ, catalysed by **phosphoribosylamine–glycine ligase**, EC 6.3.4.13 (glycinamide ribonucleotide synthetase, *common abbr.*: GAR synthetase), brings about the formation of N^1-(5-phospho-D-ribosyl)glycinamide (II, glycinamide ribotide, *common abbr.*: GAR); II reacts with 10-formyltetrahydrofolate to form 5′-phosphoribosyl-*N*-formylglycinamide (III, *common abbr.*: FGAR), the enzyme being **phosphoribosylglycinamide formyltransferase**, EC 2.1.2.2 (GAR transformylase); III is converted to 5′-phosphoribosyl-formylglycinamidine (IV, formylglycinamidine ribotide, *common abbr.*: FGAM) by **phosphoribosylformylglycinamidine synthase**, EC 6.3.5.3 (*common abbr.*: FGAM synthetase) by reaction with glutamine with ATP hydrolysis to ADP and Pᵢ. IV is converted to 1-(5-phosphoribosyl)-5-aminoimidazole (V, 5-aminoimidazole ribotide, *common abbr.*: AIR), with hydrolysis of ATP to ADP and Pᵢ, by **phosphoribosylformylglycinamidine cyclo-ligase**, EC 6.3.3.1 (*common abbr.*: AIR synthetase); V is carboxylated to 1-(5-phosphoribosyl)-5-amino-4-imidazolecarboxylate (VI, carboxyaminoimidazole carboxamide, *common abbr.*: CAIR), by **phosphoribosylaminoimidazole**

carboxylase, EC 4.1.1.21 (*common abbr.*: AIR carboxylase); VI reacts with L-aspartate, with hydrolysis of ATP to ADP and Pᵢ, to form 1-(5-phosphoribosyl)-5-amino-4-(*N*-succinocarboxamide)-imidazole (VII, 4-(*N*-succinylocarboxamide)-aminoimidazole ribotide, *common abbr.*: SACAIR) catalysed by **phosphoribosylaminoimidazolesuccinocarboxamide synthase**, EC 6.3.2.6 (*common abbr.*: SACAIR synthetase); VII is converted, with loss of fumarate, to 1-(5-phosphoribosyl)-4-carboxamide-5-aminoimidazole (VIII, aminoimidazolecarboxamide ribotide, *common abbr.*: AICAR) by adenylosuccinate lyase, EC 4.3.2.2 (VII is an alternative substrate for this enzyme); VIII reacts with 10-formyltetrahydrofolate to form 5′-phosphoribosyl-5-formamido-4-imidazolecarboxamide (IX, formaminoimidazolecarboxamide ribotide, *common abbr.*: FAICAR), catalysed by **phosphoribosylaminoimidazolecarboxamide formyltransferase**, EC 2.1.2.3 (*common abbr.*: AICAR transformylase); IX undergoes ring closure catalysed by IMP cyclohydrolase, EC 3.5.4.10, to form inosine monophosphate (IMP), from which other purine nucleoside monophosphates are formed. Thus it is converted to AMP via adenylosuccinate, in steps catalysed by adenylosuccinate synthase, EC 6.3.4.4, and adenylosuccinate lyase, EC 4.3.2.2. It is converted to GMP via xanthosine monophosphate, in steps catalysed by **IMP dehydrogenase**, EC 1.1.1.205, and **GMP synthase**, EC 6.3.4.1. A **salvage pathway** converts purine bases released during nucleic-acid degradation to nucleoside monophosphates. The enzymes adenine phosphoribosyltransferase and hypoxanthine–guanine phosphoribosyltransferase are responsible for these reactions.

purine-nucleoside phosphorylase *abbr.*: PNP; EC 2.4.2.1; *systematic name*: purine-nucleoside:orthophosphate ribosyltransferase; *other name*: inosine phosphorylase. An enzyme that catalyses the phosphorolysis of a purine nucleoside to a purine and α-D-ribose 1-phosphate. It functions in the degradation of purines; decreased activity leads to accumulation of dGTP, which affects DNA replication. Enzyme deficiency is associated with moderate immunodeficiency.

purinergic describing a type of nerve in which synaptic transmission is brought about by purine-like transmitter substances. *See* **purinoceptor**. *Compare* **adrenergic**, **cholinergic**.

purine salvage pathway a pathway by which the purine bases arising from degradation of RNA are used for the biosynthesis of nucleotides. Two **phosphoribosyltransferases** are involved, one for adenine (adenine phosphoribosyltransferase, APRT) and one for guanine or hypoxanthine (hypoxanthine/guanine phosphoribosyltransferase, HGPRT). *See also* **hypoxanthine phosphoribosyltransferase**.

purinoceptor *or* **purinergic receptor** a family of receptors (P₁, P₂ₓ, P₂ᵧ, P₂z, P₂ₜ, and P₂ᵤ) classified according to the relative potency of purine nucleotides – ATP, ADP, AMP, and UTP – in stimulating them and by their response to either adenosine (P₁) or ATP (P₂). P₂ᵧ and P₂ᵤ are G-protein-associated membrane receptors that regulate turnover of phosphoinositides. P₂ₓ, P₂z, and P₂ₜ modulate the opening of an intrinsic cation channel but P₂ₜ also stimulates release of intracellular Ca²⁺ and reduces synthesis of cyclic AMP.

puromycin a nucleoside antibiotic produced by the soil actinomycete *Streptomyces alboniger*. It interrupts protein synthesis by virtue of its structural similarity to the 3′ end of aminoacyl-tRNA: it enters the A-site (*see* **aminoacyl site**) on the ribosome, and is incorporated into the growing polypeptide chain, thereby causing premature release of the puromycinyl polypeptide from the ribosome.

purothionin any of a group of basic peptides, isolated from grain and of M_r ~5000, that contain a large number of cysteine residues, and are toxic to bacteria and fungi by inhibiting DNA transcription and RNA translation.

purple membrane part of the cell membrane of *Halobacterium halobium* containing **bacteriorhodopsin**.

'push' model *a nickname for* the **Koshland model** (for ligand-induced changes in protein conformation) because, in this model, binding of the ligands results in a conformational change, and does not simply select a pre-existing state. *Compare* **'pull model'**.

putidaredoxin a ferredoxin, isolated from *Pseudomonas putida*, that is an electron carrier in a hydroxylase system acting on camphor and closely related monoterpenes. *Compare* **adrenodoxin**.

puromycin

putrescine 1,4-diaminobutane; the polyamine formed by decarboxylation of **ornithine** and the metabolic precursor of **spermidine** and **spermine**.

PVA *abbr. for* **1** polyvinyl alcohol. **2** polyvinyl acetate.

p-value (*in sequence analysis*) a statistical measure of the significance of a match obtained by searching a database with a query sequence. The p-value denotes the probability of a match occurring with a score better than or equal to the score of the retrieved match, relative to the expected distribution of scores that result when random sequences of the same length and composition as the query are compared with the database. The closer the p-value to 0, the more significant the match.

PVC *abbr. for* polyvinyl chloride.

PVDF *abbr. for* polyvinylidine fluoride.

PVP *abbr. for* polyvinyl pyrrolidone.

PvuI a type 2 **restriction endonuclease**; recognition sequence: CGAT↑CG.

PvuII a type 2 **restriction endonuclease**; recognition sequence: CAG↑CTG.

PWM *abbr. for* pokeweed mitogen.

pYAC4 a yeast artificial chromosome vector for cloning *Eco*RI fragments of 50–100 kb.

pycnodysostosis a disorder of bone matrix remodelling caused by **cathepsin K** deficiency in osteoclast lysosomes, which accumulate partially degraded collagen type I fibrils. The syndrome is characterized by short stature, dysmorphism, osteosclerosis of long and of flat bones, hypoplasia of craniofacial bones and phalanges, pathological fractures, and dental abnormalities. Several point mutations in the cathepsin K gene are responsible.

pycnometer an apparatus for measuring the density of liquids, and also sometimes of solids. It consists of a container whose volume can be accurately measured, and that can be filled with great precision, so that when filled with the liquid in question it can be weighed to determine the density. —**pycnometric** *adj*.

pYD1 a proprietary yeast vector designed to target and display recombinant proteins at the surface of *Saccharomyces cerevisiae* as a fusion protein with the α-agglutinin subunit AGA2.

Pyd *symbol for* a residue of an unspecified pyrimidine nucleoside (alternative to Y).

pyr-1 the symbol for the pyrimidine 1 gene, which encodes **CAD protein** in several eukaryotes.

pyran *the trivial name for* either of the hypothetical isomers α-pyran, 2*H*-pyran, oxacyclohexa-3,5-diene, or γ-pyran, 4*H*-pyran, oxacyclohexa-2,5-diene, c_5H_6o. These pyrans are not known as such, but are the parent structures of various known pyranoid compounds such as dihydropyran, α- and γ-pyrones, and pyranoses.

pyranoid pyran-like, i.e. describing a molecular structure that consists of or contains a ring of five carbon atoms and one oxygen atom (as in **pyran**).

pyranopterin *preferred name for* the pterin component of the molybdenum and tungsten cofactors.

pyranose *symbol*: *p*; any monosaccharide or monosaccharide derivative whose molecule contains a **pyranoid** ring. *Compare* **furanose**, **septanose**.

pyridine-nucleotide coenzyme any coenzyme containing a nicotinamide nucleotide residue, e.g. nicotinamide-adenine dinucleotide (NAD+); now known as **nicotinamide-nucleotide coenzymes**.

pyridoxal *abbr.*: PL; 3-hydroxy-5-(hydroxymethyl)-2-methyl-4-pyridine carboxaldehyde; a compound with **vitamin B₆** activity.

pyridoxal kinase EC 2.7.1.35; an enzyme that catalyses the formation of pyridoxal 5′-phosphate from ATP and pyridoxal with release of ADP.

pyrimidine **1** 1,3-diazine; an organic nitrogenous base, sparingly soluble in water. **2** any of a class of derivatives of pyrimidine (def. 1); these form oPZP ne of the two categories of nitrogen-containing ring compounds found in DNA and RNA, which include **cytosine**, **thymine**, and **uracil**. *See also* **pyrimidine biosynthesis**.

pyrimidine biosynthesis *or* **pyrimidine ribonucleotide biosynthesis** pyrimidines are formed from carbamoyl phosphate (see **carbamoylphosphate synthase**) after conversion to carbamoyl aspartate in a reaction with L-aspartate catalysed by **aspartate transcarbamylase**, EC 2.1.3.2. Carbamoyl aspartate is converted to orotate via dihydroorotate, in steps catalysed by **dihydroorotase**, EC 3.5.2.3, and dihydroorotate dehydrogenase, EC 1.3.99.11. Orotate is phosphorylated to orotidine monophosphate (OMP) in a reaction with 5-phospho-α-D-ribosyl diphosphate catalysed by **orotate phosphoribosyltransferase**, EC 2.4.2.10. OMP is converted to uridine monophosphate by **orotidine 5′-phosphate decarboxylase**, EC 4.1.1.23.

pyrimidine dimer any structure formed within a strand of DNA from two neighbouring pyrimidine residues through covalent cross-linking induced by ultraviolet irradiation. Two types are known. (1) Addition of one pyrimidine ring to the other at each end of their respective 5,6-double bonds generates a **cyclobutadipyrimidine**, a fully substituted cyclobutane. Such a dimer is formed most readily from two thymine residues (*see* **thymine dimer**), but may be formed also from two cytosine residues, or from one cytosine and one thymine residue. For each of these varieties there are four possible configurational isomers: if the pyrimidine rings have become linked in a parallel manner, i.e. by 5–5′ and 6–6′ bonds, the configuration is designated by the prefix *syn*-, but if they are linked in an antiparallel manner, i.e. by 5–6′ and 6–5′ bonds, it is designated by *anti*-; further, if the pyrimidine rings are both on the same side of the plane of the cyclobutane ring, the dimer is described as being in the 'dog-house configuration', designated by the prefix *cis-*, but if the rings are on opposite sides of the cyclobutane plane, the dimer is said to be in the 'chair configuration', designated by *trans-*. The presence of a cyclobutadipyrimidine in a DNA strand causes a local distortion of the DNA duplex and blocks the action of DNA polymerase, thus preventing replication; cell death may ensue unless repair is effected by **photoreactivation**. (2) Alternatively, addition may take place between position 6 of one pyrimidine residue and position 4 of an adjacent one, with creation of a single covalent link and generation of a **(6–4)photoproduct**. Such a dimer is usually formed between position 6 of a cytosine or thymine residue on the 5′ side of a dinucleotide segment and position 4 of a cytosine residue on its 3′ side. The (6–4)photoproduct is thought to be the main type of lesion in DNA responsible for mutations; it is not susceptible to photoreactivation, but can be eliminated by other DNA-repair mechanisms, e.g. **excision repair**.

pyrimidine-nucleoside phosphorylase 1 uridine phosphorylase EC 2.4.2.3; an enzyme that catalyses the phosphorolysis of uridine to uracil and α-D-ribose 1-phosphate. **2 thymidine phosphorylase** EC 2.4.2.4; an enzyme that catalyses the phosphorolysis of thymidine to thymine and 2-deoxy-D-ribose 1-phosphate.

pyrin *see* **familial Mediterranean fever.**

pyro+ *or (before a vowel)* **pyr+** *comb. form* **1** of, relating to, caused, or produced by fire. **2** *(in chemistry)* indicating material derived by the action of heat; it often indicates an acid formed by loss of a molecule of water in a heating process, and hence also some anhydrides, e.g. pyrophosphate.

pyrocatechol *or* **pyrocatechin** *or (formerly)* **catechol** 1,2-benzenediol; *o*-dihydroxybenzene. *Compare* **catecholamine, *o*-benzoquinone, hydroquinone.**

pyrolysis the decomposition or dissociation of a chemical compound by the application of heat. *See also* **thermolysis.** —**pyrolytic** *adj.*; **pyrolyse** *or (sometimes, US)* **pyrolyze** *vb.*

pyrophosphatase 1 *an imprecise name for* **inorganic pyrophosphatase.** **2** *the generic name for* any enzyme within sub-subclass EC 3.6.1 (enzymes acting on diphosphate bonds); e.g. ATP pyrophosphatase (EC 3.6.1.8).

pyrophosphate *an older name for* **diphosphate** (def. 1–5) (use no longer recommended although it is still used as an alternative in such terms as inorganic pyrophosphate, pyrophosphatase, etc.). *See also* **inorganic diphosphate.**

pyrophosphokinase *the general name for* any enzyme belonging to the sub-subclass EC 2.7.6, diphosphotransferases, which transfer a diphosphate group from ATP to a named substrate; e.g. GTP pyrophosphokinase (EC 2.7.6.5), which converts guanosine 5′-triphosphate, GTP to guanosine 3′-phosphate 5′-triphosphate, pppGpp.

pyrrolidonecarboxylic acid *symbol*: <Glp; pyroglutamic acid; 5-oxoproline; pyrrolid-2-one-5-carboxylic acid. *See* **thyrotropin-releasing hormone.**

L-form

1-pyrroline-5-carboxylate dehydrogenase EC 1.5.1.12; an enzyme that catalyses a reaction between 1-pyrroline-5-carboxylate (*abbr.*: P5C), NAD$^+$, and H$_2$O to form L-glutamate and NADH. P5C is an intermediate in the conversion of glutamate, proline, and ornithine. In *Escherichia coli* the protein also has **proline dehydrogenase** activity.

pyrroloquinoline–quinone *abbr.*: PQQ; *an alternative name for* **methoxatin.**

pyruvate *trivial name for* 2-oxopropanoate, CH$_3$COCOO$^-$; it is found in virtually every type of cell and species, being an important intermediate in many pathways including the catabolism of glucose, and the biosynthesis of amino acids and carbohydrates.

pyruvate carboxylase *abbr.*: PCB; EC 6.4.1.1; *systematic name*: pyruvate:carbon dioxide ligase (ADP-forming); *other name*: pyruvic carboxylase; an enzyme of mitochondrial matrix that catalyses a reaction between ATP, pyruvate, and HCO$_3^-$ to form oxaloacetate, ADP, and orthophosphate; biotin and manganese or zinc are cofactors. It requires acetyl-CoA as an allosteric activator. Enzyme deficiency results in lactic acidosis, which may be severe and lethal in infancy, or less severe and lead to delayed development.

pyruvate decarboxylase EC 4.1.1.1; *systematic name*: 2-oxo-acid carboxy-lyase; *other names*: α-carboxylase, pyruvic decaboxylase; α-ketoacid carboxylase. An enzyme that catalyses the decarboxylation of a 2-oxo acid to an aldehyde and CO$_2$; thiamine diphosphate is a coenzyme.

pyruvate decarboxylase kinase *or* **pyruvate dehydrogenase kinase** a subunit of the mammalian **pyruvate dehydrogenase complex**. It phosphorylates three serine residues on pyruvate dehydrogenase (lipoamide) resulting in loss of activity, and is stimulated by elevated levels of acetyl-CoA and NADH.

pyruvate dehydrogenase complex an enzyme complex that catalyses the overall reaction between pyruvate, NAD$^+$, and coenzyme A to form acetyl-CoA, NADH, and CO$_2$. The complex consists of four activities: (1) **pyruvate dehydrogenase (lipoamide)**, designated E$_1$, catalysing a reaction between pyruvate and lipoamide to form *S*-acetyldihydrolipoamide and CO$_2$; (2) **dihydrolipoamide *S*-acetyltransferase**, designated E$_2$, catalysing a reaction between *S*-acetyldihydrolipoamide and coenzyme A to form acetyl-CoA and dihydrolipoamide; (3) **dihydrolipoamide dehydrogenase**, designated E$_3$, which oxidizes dihydrolipoamide to lipoamide; and (4) a lipoyl-containing catalytic **protein X**, which can also effect acyl transfer and which binds E$_2$ and E$_3$. The complex is regulated by the phosphorylation and dephosphorylation of the E$_1$ component, phosphorylation being catalysed by **pyruvate decarboxylase kinase** and dephosphorylation being catalysed by **[pyruvate dehydrogenase (lipoamide)] phosphatase** (or pyruvate decarboxylase phosphatase), EC 3.1.3.43. Elevated levels of NADH and acetyl-CoA inhibit the complex by stimulating the kinase. Defects in E$_1$, E$_2$, E$_3$, and protein X or the phosphatase result in lactic acidosis, which may be fatal early after birth, produce profound psychomotor retardation, or carbohydrate-induced episodic ataxia, depending on the locus of the defect.

pyruvate dehydrogenase (lipoamide) EC 1.2.4.1; *systematic name*: pyruvate:lipoamide 2-oxidoreductase (decarboxylating and acceptor-acetylating); *other names*: pyruvate decarboxylase; pyruvate dehydrogenase; pyruvic dehydrogenase. A component enzyme of the **pyruvate dehydrogenase complex**. It catalyses a reaction between pyruvate and lipoamide to form *S*-acetyldihydrolipoamide and CO$_2$; thiamine diphosphate is a coenzyme. The enzyme in humans consists of two α (41 kDa each) and two β (36 kDa each) subunits. The α subunit is catalytic and binds the pyrophosphate end of the thiamine pyrophosphate, while the β subunit binds the thiazolidine ring of the coenzyme. Numerous mutations (mostly derived *de novo* in the gene on the X chromosome) in the α subunit result in enzyme deficiency and severe lactic acidosis.

[pyruvate dehydrogenase (lipoamide)]-phosphatase EC 3.1.3.43; an enzyme that catalyses the hydrolysis of phosphate from [pyruvate dehydrogenase (lipoamide)] phosphate. This hydrolysis activates the enzyme activity. Enzyme deficiency leads to inactivation of the α subunit of the appropriate dehydrogenase and consequent lactic acidosis.

pyruvate kinase EC 2.7.1.40; *systematic name*: ATP:pyruvate 2-*O*-phosphotransferase; *other names*: phospho*enol*pyruvate kinase; phospho*enol* transphosphorylase. An enzyme that catalyses the formation of pyruvate and ATP from phospho*enol*pyruvate and ADP; it requires Mg^{2+} and K$^+$. In the glycolytic pathway, it effects the second phosphorylation of ADP to ATP. In many cases the enzyme is a homotetramer. In humans there are two genes for the enzyme: PKLR, which is expressed in liver and erythrocytes, and PKM, which is expressed in muscle and other tissues. Numerous mutations in PKLR result in an autosomal recessive form of hemolytic anemia.

[pyruvate kinase]-phosphatase EC 3.1.3.49; an enzyme that catalyses the hydrolysis of phosphate from [pyruvate kinase] phosphate in order to activate pyruvate kinase.

pyruvate–malate cycle a metabolic cycle involving malate dehydrogenase (oxaloacetate-decarboxylating) (the **malic enzyme**) that occurs during fatty-acid biosynthesis. In this situation, acetyl-CoA is transported into the cytoplasm as citrate, formed by mitochondrial citrate synthase; the citrate is cleaved in the cytoplasmic compartment to acetyl-CoA, which participates in fatty-acid synthesis, and oxaloacetate. No transport system exists for taking this oxaloacetate back into the mitochondrion; however, it can be reduced to malate by cytoplasmic NAD^+-linked malate dehydrogenase, and the malate so formed can be converted to pyruvate by malic enzyme, EC 1.1.1.38; the pyruvate can then be transported into the mitochondrion, to be carboxylated to oxaloacetate by **pyruvate carboxylase**.

pyruvic–malic carboxylase *an alternative name for* **malic enzyme**.

python *see* **programming language**.

PYY *abbr. for* peptide YY.

PZ *abbr. for* pancreozymin.

PZP *abbr. for* pregnancy zone protein.

Qq

q *symbol for* quartet (in nuclear magnetic resonance spectroscopy).

q *symbol for* **1** heat (alternative to *Q*). **2** partition function (individual entity). **3** specific humidity.

Q *symbol for* **1** a residue of the *α*-amino acid L-glutamine (alternative to Gln). **2** a residue of the ribonucleoside **pseudouridine** (replacing Ψ for computer work). **3** coenzyme Q (i.e. **ubiquinone**); not recommended. The number of prenyl residues may be indicated by a suffixed arabic number, e.g. Q-10. **4** queuosine.

Q *symbol for* **1** heat (alternative to *q*). **2** electric charge. **3** radiant energy (alternative to *W*). **4** *or* *Q*$_v$ quantity of light. **5** partition function (whole system).

Q$_{10}$ *symbol for* temperature coefficient of a reaction.

Q$_{CO2}$ *symbol for* metabolic quotient in respect of discharge of carbon dioxide; it is expressed in terms of microlitres CO_2 evolved per milligram dry mass of material per hour.

Q$_{O2}$ *symbol for* metabolic quotient in respect of uptake of dioxygen; it is expressed in terms of microlitres O_2 absorbed per milligram dry mass of material per hour.

Q$_S$ *symbol for* a measure of the activity of an enzyme, especially a respiratory enzyme, defined as the number of microlitres (at stp) of gaseous substrate S used up per hour per milligram enzyme. It could be extended to nongaseous substrates by taking 1 µmol as equivalent to 22.4 µL.

QAE *abbr. for* the quaternized aminoethyl; denoting the group diethyl-(2-hydroxypropyl)aminoethyl; $(C_2H_5)_2(CH_3-CHOH-CH_2)N^+-CH_2-CH_2-$; a strongly basic substituent sometimes used in hydroxylated chromatographic support matrices, as in QAE-Sephadex, to form anion exchangers. *Compare* **DEAE**.

Q cycle *an alternative name for* **protonmotive Q cycle**.

Q enzyme *an alternative name for* amylopectin branching enzyme; *see* **branching enzyme**.

Q gas a mixture of 98.7% helium and 1.3% butane, used as filling for gas-flow Geiger counters.

QH• *symbol for* ubisemiquinone.

QH$_2$ *symbol for* ubiquinol.

Q repeat *an alternative name for* **glutamine repeat**.

QTL *abbr. for* quantitative trait locus.

quad denoting four kinetically important substrates or products in an enzyme mechanism. *See* **reactancy of enzymes**.

quadri+ *or* (*sometimes*) **quadru+** *or* (*before a vowel*) **quadr+** *comb. form* four; fourth or one-fourth; square.

quadruplex DNA *another name for* **tetraplex DNA**.

quadrupole mass spectrometer A type of **mass spectrometer** in which the molecular ions are separated by mass filter comprising four metal rods with an alternating high frequency voltage (quadrupole field).

quadrupole, nuclear electric *see* **nuclear quadrupole moment**.

quality factor *or* (*formerly*) **relative biological effectiveness** an index of the ability of a given type of ionizing radiation to cause biological damage, depending on the density of the ionization produced relative to that produced by gamma radiation. For all X- (and gamma) radiation and beta particles (+ or –) likely to be encountered in the use of radioisotopes it is assigned a value of unity, for thermal neutrons usually a value of 3, and for all alpha particles a value of 10. *See also* **dose equivalent**.

quanta the plural of **quantum**.

quantal 1 of, pertaining to, or being a **quantum** or something that is quantized; of or pertaining to the quantum theory. **2** of a biological effect or response in which there are only two categories, conditions, or states (e.g. all or none, alive or dead). —**quantally** *adv.*

quantitative electrophoresis (*sometimes*) an alternative name for **crossed immunoelectrophoresis**.

quantitative trait locus *abbr.*: QTL; a region of a chromosome containing genes that are believed to make a significant contribution to the expression of a complex phenotypic trait. Quantitative traits (QTs) are typically affected by more than one gene and by the environment, and include several medically significant traits (e.g.

blood pressure, obesity). Once mapped to small chromosomal regions (e.g. using linkage maps), specific genes can be isolated by other molecular methods.

quantity *or* **physical quantity** (*in systems of measurement*) any property whose value can be expressed as the product of a numerical value and a unit. In the International System of Units (*see* **SI**) there are seven **base quantities**, physical quantities regarded as being dimensionally independent; these are: length, mass, time, electric current, thermodynamic temperature, amount of substance, and luminous intensity; all are arbitrarily defined. **Derived units** are expressed algebraically in terms of the base units. Symbols for physical quantities are printed in italic type; they are generally single Roman or Greek letters, and may be lower case or capital.

quantity of electricity an alternative name for **electric charge**.

quantize *or* **quantise** to restrict (a physical quantity) to a set of discrete values, characterized by quantum numbers. A quantized physical quantity cannot vary continuously but must change in steps. —**quantization** *or* **quantisation** *n.*

quantum (*pl.* **quanta**) **1** a (specific) quantity or amount of something. **2** a unit amount of a physical quantity, especially of electromagnetic energy, such that all other amounts are integral multiples of it. According to the **quantum theory**, a quantum of energy released during radiation emission or taken up during radiation absorption is equal to $h\nu$, where h is the Planck constant and ν the frequency of the radiation in hertz. A quantum of light (or other electromagnetic energy) is termed a **photon**.

quantum efficiency 1 (*in a luminescent system*) the fraction of the molecules in a particular excited state that emit luminescence. **2** (*in biochemistry*) an alternative term for **quantum yield** (def. 1).

quantum mechanics a mathematical theory developed from the quantum theory (*see* **quantum**) as a replacement for classical mechanics in order to explain satisfactorily the behaviour of atoms, molecules, and elementary particles in terms of observable quantities such as the intensities and frequencies of spectral lines.

quantum number any of a set of small integral or half-integral numbers that give the various possible values of a quantized property of a system. Quantum numbers are used especially in describing the properties of elementary particles, such as their charge (values of 0, +1, or –1) or spin (values of +1/2, or –1/2), and in specifying the energy states of electrons in atomic orbitals (each of which is characterized by a unique set of four quantum numbers).

quantum requirement (*in photosynthesis*) the inverse of **quantum yield** (def. 2); i.e. the number of quanta of light energy required to bring about the release of one molecule of dioxygen. The theoretical value is eight, but the value observed in an experimental system is generally somewhat higher, depending on conditions.

quantum theory *see* **quantum**.

quantum yield 1 *abbr.*: *Q*; (*in a luminescent system*) the probability of luminescence occurring in given conditions, expressed by the ratio of the number of photons, i.e. quanta of light, emitted by the luminescing species to the number absorbed. *Compare* **quantum efficiency**. The measurement of *Q* requires the counting of photons, because

$$Q = \text{photons emitted/photons absorbed.}$$

It requires highly specialized instrumentation, but a less rigorous method may be employed in which the fluorescence of a fluor is compared with that of a fluor of known *Q*. Then

$$Q_x = [I_x Q_s A_s]/[I_s A_x]$$

where Q_s is the quantum yield of the standard, I_s and I_x are the fluorescence intensities of the sample and standard and A_s and A_x are the percentage of absorption of each solution at the exciting wavelength. **2** (*in photosynthesis*) the inverse of **quantum requirement**; i.e. the fractional number of molecules of carbon dioxide reduced per photon absorbed.

quartz a colourless, often transparent, glassy form of silica, SiO_2, used in optical devices and cuvettes because of its transparency to near-ultraviolet radiation. It is enantiomorphous, the right- and left-handed forms rotating the plane of polarized light in opposite directions.

quasi-species (*in virology*) the population of closely related but different viral genomes that evolves over time through spontaneous mutations within an individual infected with a single genotype. Diversification through the emergence of quasi-species is one mechanism by which viruses evade the host immune response.

quaternary 1 consisting of four (items, parts, etc.); by or in fours. **2** (*in chemistry*) consisting of or containing an atom bound to four atoms or groups. **3** of the fourth order; fourth in a series or hierarchy.

quaternary ammonium compound any compound that can be regarded as derived from ammonium hydroxide or an ammonium salt by replacement of all four hydrogen atoms of the NH_4^+ ion by organic groups. Certain compounds of this class – ones in which one of the organic groups is a long-chain (C_8–C_{18}) alkyl group and the other three are shorter-chain alkyl or other groups – have the properties of cationic detergents and are powerful antimicrobial agents. They are bacteriostatic at low concentrations and bactericidal at higher concentrations, being generally more active against Gram-positive than Gram-negative organisms. They display relatively low toxicity to higher animals and humans, and hence are widely used as antiseptics and disinfectants.

quaternary structure the fourth order of complexity of structural organization exhibited by protein, nucleic acid, and nucleoprotein molecules. It refers to the arrangement in space of the subunits of a multimeric macromolecule, and the ensemble of its intersubunit contacts and interactions, without regard to the internal geometry of the subunits. Hence, it is possessed only when the molecule in question is made up of at least two (identical or non-identical) subunits that are (at least potentially) separable, i.e. are not linked by covalent bonds.

Quellung reaction or **Neufeld Quellung reaction** or **pneumococcus capsule swelling reaction** the swelling of the capsule observed when specific antibody is mixed with suitable bacterial cells (e.g. pneumococci). It is probably due to deposition of antibody on the outside of the capsule, making the latter clearly visible. [From the German, *Quellung*, swelling; described by Ferdinand Neufeld (1869–1945), German bacteriologist, in 1902.]

quench correction (*in liquid-scintillation counting*) a correction to be applied to an observed count rate to allow for losses resulting from **quenching** (def. 2).

quenched flow *see* **rapid-reaction kinetics**.

quencher a substance capable of reducing or destroying luminescence through deactivation of an excited chemical species. In biological systems this usually occurs by chemical reaction (e.g. complex formation), by exchange of energy on collision, or by resonance energy transfer.

quenching 1 the reduction or destruction of luminescence, especially fluorescence, of a sample by addition or inclusion of a **quencher**. Fluorescence quenching is the basis of a useful technique for studying the binding of small molecules to proteins or other macromolecules. **2** the reduction by whatever means of the efficiency of energy transfer from beta particles to the photomultiplier(s) in a liquid-scintillation counter. Such means include: (1) **dilution quenching** – reduction of the probability of scintillation occurring through dilution of the scintillator by the sample; (2) **chemical** (or **impurity**) **quenching** – absorption by a component of the sample of some of the energy of the beta particles without emitting photons, or (without fluorescing and in competition with the primary fluor) of some of the photons emitted by excited solvent molecules; and (3) **colour quenching** – absorption by a coloured component of the sample of some of the photons emitted by the secondary fluor. **3** the termination of a chemical or enzymic reaction by addition of another reagent or sudden change of the conditions.

quercetin 3,3′,4′,5,7-pentahydroxyflavone; a member of the **flavonoids**, or **flavones**, widely distributed in plants, often as **glycosides**; it is the aglycon of **quercitrin**, **rutin**, and other materials. Known in folk medicine for its anti-inflammatory properties, it inhibits the **lipoxygenase** pathway, has some antiplatelet and anti-

thrombotic activity *in vivo*, inhibits mast-cell degranulation, and possesses anti-asthmatic activity. It inhibits many enzymes, including protein kinases, and also inhibits DNA, RNA, and protein synthesis. A number of its actions may reflect its property as a nucleoside antagonist.

quercitrin quercetin-3-L-rhamnoside. *See also* **quercetin**.

query sequence (*in bioinformatics*) a sequence used as a probe to search a database.

queuosine *symbol*: Q; any one of a series of nucleosides found in tRNA and having an additional pentenyl ring added via an NH group to the methyl group of 7-methylguanosine. The pentenyl ring may carry other substituents.

Qui *symbol for* a residue (or sometimes a molecule) of the deoxyaldohexose **quinovose**.

Quin-2 *abbr. for* 2-[(2-amino-5-methylphenoxy)methyl]-6-methoxy-8-aminoquinoline-*N*,*N*,*N*′,*N*′-tetraacetic acid; a molecule functionally related to the Ca^{2+}-chelating agent **EGTA**, but containing fluorescent groups. The quantum yield of fluorescence is greatly enhanced on binding Ca^{2+}, leading to its use as a quantitative indicator of Ca^{2+} concentration. It is normally used as its tetraacetoxy ester, which readily crosses the plasma membrane and is then hydrolysed within the cell by esterases to Quin-2, the polar nature of which causes it to be retained as an intracellular indicator. *See also* **Fura-2, INDO-1**.

quinacrine or **mepacrine** or **atebrine** 6-chloro-9-(4-diethylamino-1-methylbutylamino)-2-methoxyacridine; a derivative of **acridine**, formerly used as an antimalarial and anthelmintic. It fluoresces powerfully in UV light and can be used as a fluorochrome to label DNA in chromosomes.

quinacrine mustard a compound derived by chloro-substitution of the terminal methyls of **quinacrine** ethylamino groups, and possessing properties of an alkylating agent; this enhances its value as a DNA-labelling agent.

quinhydrone an addition compound of one mole each of hydroquinone and *p*-benzoquinone. It is a reddish-brown crystalline substance with a dark-green lustre, useful as an antioxidant, in photography, and as a redox system in the **quinhydrone electrode**.

quinhydrone electrode a half-cell consisting of a bright platinum (or sometimes gold) electrode immersed in a test solution to which has been added a little **quinhydrone**. It has an electric potential related to the pH value of the solution and thus is useful as a secondary hydrogen electrode for the electrometric measurement of pH.

quinic acid the D enantiomer ($1\alpha,3\alpha,4\alpha,5\beta$)-1,3,4,5-tetrahydroxycyclohexanecarboxylic acid occurs commonly in plants, either free or as esters (e.g. **chlorogenic acid**), and may make up as much as 2–10% of the dry weight of leaves. It may be formed by the **shikimate pathway**. Its catabolism to protocatechuic acid has been extensively studied in *Aspergillus nidulans* and *Neurospora crassa*.

quinidine β-quinine; 6'-methoxycinchonan-9-ol; a stereoisomer of **quinine** noted for its cardiac anti-arrhythmic effects.

quinine 6'-methoxycinchonan-9-ol; a bitter-tasting alkaloid obtained especially from the bark of cinchona, any of a number of trees or shrubs of the tropical genus *Cinchona* (especially the Javan species *C. lidgeriana* and various South American species). It was formerly much used (and sometimes is still needed) for the prevention and treatment of malaria. It binds to the DNA of the malarial parasite *Plasmodium* spp., and is thought thereby to inhibit biosynthesis of the parasite's nucleic acids. Quinine and drinks containing it (e.g. tonic water) help to prevent cramp. *Compare* **quinidine**.

quinoid *or* **quinonoid** of, derived from, or resembling a **quinone** (def. 2).

quinol *an alternative name for* **hydroquinone**.

quinolinic acid *or* **pyridine dicarboxylic acid** a degradation product of tryptophan that, as the substrate for quinolate phosphoribosyltransferase, is the precursor of nicotinate mononucleotide in the biosynthesis of the coenzymes NAD and NADP.

4-quinolone antibiotic *a generic term for* any of a wide range of synthetic antibacterial compounds whose molecular structures contain a 4-oxo-1,4-dihydroquinoline or a 4-oxo-1,4-dihydrocinnoline nucleus; e.g. **cinoxacin**, **ciprofloxacin**, **nalidixic acid**, **oxolinic acid**. Their antibiotic activity derives from interaction with the A subunit of bacterial DNA gyrase (*see* type II **DNA topoisomerase**); this results in trapping of bacterial DNA in the complex it forms with the A sububit. Rejoining of the DNA strands broken by the enzyme is thereby arrested.

quinone **1** *a common name for* **p-benzoquinone**. **2** any member of a class of diketones derivable from aromatic compounds by conversion of two CH groups into CO groups with any necessary rearrangement of double bonds. Simple quinones are usually *p*-quinones, i.e. derivatives of *p*-benzoquinone (*compare* **hydroquinone** (def. 2)), but they may also be *o*-quinones, i.e. derivatives of **o-benzoquinone** (*compare* **catechol** (def. 2)). Naturally occurring quinones form a large, varied, and widespread group of compounds; they include numerous pigments, various electron carriers (e.g. **plastoquinone**, **ubiquinone**), and vitamins K. *See also* **semiquinone**.

+quinone *comb. form* denoting a **quinone** (def. 2); e.g. benzoquinone, anthroquinone, naphthoquinone.

quinone cofactor *see* **quinoprotein**.

quinonoid *an alternative term for* **quinoid**.

quinoprotein *other name*: quinoenzyme; any enzyme protein that contains a **quinone cofactor**, i.e. the freely dissociable pyrroloquinoline quinone (PQQ) or methoxantin, or the covalent species topaquinone (TPQ), tryptophan tryptophylquinone (TTQ), or lysine tyrosylquinone (LTQ) that are derived by post-translational modification of certain proteins. PQQ is characteristic of oxidoreductase enzymes within the sub-subclass EC 1.1.99, which includes alcohol dehydrogenase (acceptor) (EC 1.1.99.8) of methylotrophic bacteria, glucose dehydrogenase (pyrroloquinoline-quinone) (EC 1.1.99.17), and a number of dehydrogenases acting on other specified (types of) alcohols or polyols. TPQ occurs in bovine serum copper amine oxidase, TTQ in bacterial amine dehydrogenase, and LTQ in bovine aorta lysyl oxidase (EC 1.4.3.13).

quinovosamine 2-amino-2,6-dideoxy-D-glucose; an amino derivative of **quinovose** present in bacterial polysaccharides.

quinovose symbol: Qui; *a trivial name for* 6-deoxy-D-glucose. α-Quinovose occurs, e.g., as its 6-sulfo derivative (6-deoxy-α-D-glucopyranosyl 6-*C*-sulfate) in the sulfoquinovosyl diacylglycerols, which form one of the two major groups of the glycerolipids of chloroplasts. The sulfoquinovosyl derivative also occurs in some **cardiac glycosides**.

quinternary structure the arrangement of separate molecules in macromolecular complexes, such as in protein–protein or protein–nucleic acid interactions.

quisqualate receptor *see* α-amino-3-hydroxy-5-methyl-isoxazolepropionic acid receptor.

quorum sensing the process by which bacteria detect their population cell density and use the information to regulate gene expression accordingly.

Rr

r 1 *symbol for* **ribo+** (def. 1); it is used as a prefix in (one- or three-letter) symbols for ribonucleosides for emphasis or clarity, especially to distinguish from 2′-deoxyribonucleosides. **2** *symbol for* reaction. **3** *abbr. for* rat (as a prefix to an abbreviation for the name of a peptide hormone); denoting derived (or as if derived) from a rat or rats; e.g. rGH.

r *symbol for* **1** concentration ratio. **2** rate of concentration change. **3** radius. **4** interatomic distance.

R *symbol for* **1** a residue of the α-amino acid L-arginine (alternative to Arg). **2** a residue of an incompletely specified base in a nucleic-acid sequence that may be either adenine or guanine. **3** a residue of an unspecified purine nucleoside (alternative to Puo). **4** an unspecified univalent group in a formula of an organic compound. The group must be attached by means of carbon and derived from an aliphatic, carbocyclic, or heterocyclic compound. Up to three such groups, when different, may be designated R, R′, and R″, or R^1, R^2, and R^3; for more than three such groups, superscript numerals only should be used. **5** röntgen (a non-SI unit of radiation exposure). **6** resonance effect (in electron displacement).

R *symbol for* **1** electric resistance. **2** thermal resistance. **3** gas constant. **4** molar refraction.

R *stereochemical descriptor* denoting *rectus*; i.e. the absolute configuration of a chiral compound in which the priority (obtained using the Cahn–Ingold–Prelog sequence rules) of the substituent groups on the chiral centre decreases in a clockwise (right-handed) direction. For use as a prefix in chemical nomenclature, the descriptor is placed in parentheses and connected with a hyphen, e.g.(*R*)-alanine. For chiral compounds in which the relative but not the absolute configuration is known, *rectus* configuration is denoted by the prefix (*R**)-. *See* **sequence rule**. *Compare* **S**.

R_B *symbol for* R_B value.

R_f or **R_F** *symbol for* R_f value.

R_p *symbol for* phosphorylation state ratio.

R_s *symbol for* saturation ratio.

RAB a gene encoding a Rab protein – any of a family of GTP-binding proteins similar to Ras proteins (*see* **RAS**) and probably involved in membrane traffic, etc. Rab proteins contain dicysteine motifs near their C termini to receive a geranylgeranylyl moiety.

rabbit aorta-contracting substance *abbr.*: RCS; a name given (in 1969) to an unstable factor released from guinea-pig lung that caused contraction of isolated strips of rabbit aorta. The activity was probably due to the presence of thromboxane A_2 (*see* **thromboxane**).

rabbit reticulocyte system a cell-free system, from lysed rabbit reticulocytes, that translates added eukaryote mRNAs from a wide variety of heterologous sources. Endogenous mRNA is destroyed using a Ca^{2+}-dependent ribonuclease, micrococcal endonuclease; this enzyme is inactivated by adding EGTA. Heterologous mRNA is then added and is translated by the system.

rac- *abbr. for* racemo-.

Rac either of two **RhoGTPases** that are cytosolic components of the **phagocyte NADPH oxidase** system, Rac 1 and Rac 2, which show 92% homology. They are involved in activation of MAP kinase and JNK kinases by integrins.

RacA *abbr. for* remodelling and anchoring of chromosome A; a protein of *Bacillus subtilis* and related spore-forming species that anchors the origin of replication regions of sister chromosomes to the cell poles during sporulation.

RACE *abbr. for* rapid amplification of cDNA ends; a variation on the **polymerase chain reaction** designed for amplification of sequence corresponding with the 5′ or 3′ ends of specific mRNAs. The former is important for the location of the transcription start site as well as providing full-length cDNA sequences.

racemase any enzyme that interconverts the two enantiomers of a chiral compound; such enzymes are categorized in subclass EC 5.1 (which also includes **epimerases**). If more than one chiral centre is present, a racemase must invert configuration of all the chiral cen-

tres. Thus, the name methylmalonyl-CoA racemase, sometimes used for methylmalonyl-CoA epimerase (EC 5.1.99.1), is incorrect – the enzyme does not change configurations of the chiral centres of the CoA moiety.

racemate any homogeneous phase containing equimolar amounts of enantiomeric molecules.

racemic *symbol*: (±)- *or (formerly) dl*-; denoting the presence of equimolar amounts of the dextrorotatory and levorotatory enantiomers of a compound, regardless of whether the (optically inactive) product formed is crystalline, liquid, or gaseous. A homogeneous solid phase composed of equimolar amounts of enantiomeric molecules is termed a **racemic compound**, and a mixture of equimolar amounts of enantiomeric molecules present as separate solid phases is termed a **racemic mixture**; *see also* **racemate**. A (±)-product may be resolved into its (+)- and (–)-components. The term is derived from **racemic acid** [from Latin *racemus°*, bunch of grapes], the optically inactive mixture of (+)- and (–)-tartaric acids sometimes found during the manufacture of wine. By extension, the term is applied to DL- and (*RS*)-mixtures. *See* **D/L convention**, **optical isomerism**, **sequence rule**. *See also* **racemo-**.

racemize *or* **racemise** to induce the formation of a mixture of the two possible enantiomers from one of the pure isomers of a chiral compound. —**racemization** *or* **racemisation** *n*.

racemo- *abbr.*: rac-; *prefix (in chemical nomenclature)* designating a racemate. It may be used as an alternative to the symbol (±)- or (for certain classes of racemic compounds) the symbol DL-. *Compare* **meso-**.

rachitic of, pertaining to, or afflicted with **rickets**.

rachitis *an alternative name for* **rickets**.

rachitogenic tending to cause **rickets**.

rad 1 *symbol for* radian. **2** *symbol*: rad *or* (to avoid possible confusion with **rad** (def. 1)) rd; the cgs unit of absorbed dose of ionizing radiation, equivalent to an absorption of 0.01 joule of energy per kilogram (= 100 ergs per gram) of irradiated material. Whereas the rad represents, in practice, about the same amount of energy as the **röntgen** (the exact equivalence depending on the material irradiated), it applies to any type of ionizing radiation in any medium. 1 rad = 0.01 Gy. **3** *abbr. for* radiation.

radial 1 of, resembling, or relating to a radius or ray; in rays. **2** emanating from the centre of a circle or a common central point. **3** spreading out uniformly in all directions on a surface. —**radially** *adv*.

radian *symbol*: rad; the SI supplementary unit of plane angle. One radian is equal to the angle subtended at the centre of a circle by an arc whose length is equal to the radius. It is dimensionless. There are 2π radians in a circle; 1 rad ≈ 57.295 78°.

radiant energy *symbol*: *Q or W*; the energy of any form of electromagnetic radiation. The SI derived unit is the **joule**.

radiant intensity *symbol*: *I*; the **radiant power** of radiation from a point source per unit of solid angle; i.e. $I = d\Phi/d\Omega$, where Φ is the radiant power and Ω is the solid angle measured in steradians. The SI unit is the watt per steradian (Wsr^{-1}).

radiant power *or* **radiant flux** *symbol*: *P or Φ*; the rate of flow of energy of radiation; i.e. $\Phi = dQ/dt$, where *Q* is the radiant energy. The SI derived unit is the watt.

radiation 1 the emission of energy from a source and its transmission as particles, rays, or waves, especially as electromagnetic rays or waves, sound waves, or streams of subatomic particles. **2** energy so transmitted, especially the corpuscular and electromagnetic rays emitted in the decay of radionuclides. **3 ionizing radiation**. **4** divergence from a common point, especially radially from a central point or source.

radiation absorbed dose *see* **absorbed dose**, **rad** (def. 2).

radiation biology *an alternative name for* **radiobiology**.

radiation-chemical reaction any chemical reaction that is initiated by the absorption of radiation but is distinguishable from a **photochemical reaction** by its lack of specificity (i.e. by giving rise to a variety of reaction products) and by occurring at higher radiation energies (i.e. by absorption either of electromagnetic radiation from the mid-ultraviolet region onwards or of high-energy particulate radiation) and thus by resulting always in the formation of detectable ions.

radiation chemistry the branch of chemistry concerned with **radiation-chemical reactions**. *Compare* **radiochemistry**.

radiation damage the totality of the harmful changes induced in a cell or tissue by radiation. The harmful effects may include formation of **thymine dimers** in DNA or of radicals from polyunsaturated acyl groups in lipids.

radiation dose equivalent *see* **dose equivalent**.

radiation hybrid map *abbr*.: RH map; a map of irradiated chromosomal sites derived from hybrid cells containing fragments of irradiated chromosomes. RH maps provide important markers, allowing the construction of precise sequence-tagged site (STS) maps (with the STSs positioned according to their frequency of separation by radiation-induced breaks). These are useful in ordering genetic loci along chromosomes and in studying multifactorial diseases.

radiation inactivation method a technique, based on **target theory**, that enables the size of the functional unit of a biologically active macromolecule (or of its complexes with other molecules) to be estimated; it can be used on unpurified (e.g. membrane-bound) material. Determinations are made of the biological activity remaining in a sample after exposure to various large doses of ionizing radiation (electrons or X- or gamma radiation of at least 1 MeV), and the value, D_{37}, of the radiation dose, expressed in Mrad, required to reduce the activity to 37% of (i.e. e^{-1} times) its original value is obtained; the relative molecular mass, M_r, of such a macromolecule may then be found from the empirical relationship $M_r = 6.4 \times 10^5/D_{37}$, where D_{37} is determined at 30°C.

radiation source a quantity of radioactive material used as a source of ionizing radiation.

radical (*in chemistry*) **1** any molecular entity, charged or uncharged, that possesses an unpaired electron (but normally excluding any paramagnetic metal ion); often formed by homolysis of a covalent bond. Radical character is indicated in a formula by a centred dot symbolizing the unpaired electron and placed (if possible) beside the symbol for the atom of highest spin density; e.g. HOO•, •CH$_3$. The term **free radical** is now preferably restricted to any radical that does not form part of a **radical pair**. **2** *a former name for* **group** (def. 2).

radical anion *see* **radical ion**.

radical cation *see* **radical ion**.

radical centre the atom in any polyatomic **radical** (def. 1) on which the unpaired electron is largely localized.

radical ion any **radical** (def. 1) that carries an electric charge. One carrying a negative charge is termed a **radical anion** (e.g. the benzene radical anion, C_6H_6 •$^-$); one carrying a positive charge is termed a **radical cation** (e.g. the benzene radical cation, C_6H_6 •$^+$).

radical pair 1 any two **radicals** (def. 1) in close proximity in liquid solution, with a solvent cage. They may have been formed simultaneously, e.g. by homolysis, or have come together by diffusion. Whilst together, correlation of their unpaired electron spins occurs and manifests itself as a distortion of nuclear magnetic resonance spectra. **2** a cation and anion radical produced transiently in the primary photochemical process of photosynthesis. *See* **photosystem**.

radical SAM refers to enzymes that bind S-adenosyl-L-methionine (SAM) next to a [4Fe-4S] cluster to generate a reactive free-radical species.

radical scavenger any substance that can react readily with, and thereby eliminate, radicals. In biological systems or materials, radical scavengers may function as **antioxidants** or may protect from damage by ionizing radiation. Among the important radical scavengers that function in animal tissues at physiological concentrations are the water-soluble substances ascorbate, glutathione, and the purine bases, and the fat-soluble substances tocopherols, the retinols, the carotenes, and the ubiquinones.

radioactivate to convert a stable nucleus into one undergoing radioactive decay by bombardment with high speed particles, including protons, alpha particles, neutrons, and deuterons. These are formed in a cyclotron or synchrotron from a radioactive source,

and accelerated through a potential difference of a few thousand volts by a magnetic field before being focused on the substance to be converted.

radioactivation analysis *or* **activation analysis** a method for the qualitative and/or quantitative estimation of the chemical elements in a sample. It depends on the identification and determination of the radionuclides formed when the sample is bombarded with neutrons or other particles.

radioactive having the properties associated with a **radioactive isotope**.

radioactive concentration the **activity** (def. 2) per unit quantity of any material in which a **radionuclide** occurs. *See also* **specific activity**.

radioactive decay *see* **decay** (def. 1b).

radioactive isotope *or* **radioisotope** any isotope of a chemical element whose nucleus is unstable and emits alpha, beta, or gamma rays. The product is another element, which may be stable or unstable; if the latter, it decays further. Thus, the principal isotope of uranium, ^{238}U, emits an alpha particle and decays to ^{234}Th, which emits a beta particle forming ^{234}Pa; this chain continues to the formation of ^{206}Pb, which is stable.

radioactive label *or* **radiolabel** a **radioactive isotope** incorporated into a molecule to confer on it the property of radioactivity.

radioactive tracer *or* **radiotracer** a radioactive substance that is added to a metabolic system in quantities (mass) too small to perturb the system, in order to follow, by isolating the radioactive intermediates formed, the sequence of transformations undergone normally by the same (nonradioactive) molecule under similar conditions.

radioactivity 1 the emission of radiation from certain **nuclides** by the spontaneous transformation of their **nuclei**. **2** (*loosely*) the radioactive material itself or its emitted radiation. *See* **activity** (def. 2).

radioallergosorbent test *abbr*.: RAST; a test used primarily for quantifying the levels of antigen-specific immunoglobulin E (IgE) in serum. A specific allergen, which has been covalently coupled to an insoluble **Sephadex** carrier, is incubated with serum containing the unknown amount of IgE. The carriers are then washed and further incubated with radiolabelled antibodies to IgE, the amount of radioactivity bound being a measure of the amount of IgE. The advantage of this method over the **radioimmunosorbent test** is that IgE specific to a particular antigen can be selectively quantified.

radioassay any type of **assay** in which a radioactive isotope is used in the measurement. The term embraces, e.g., **immunoradiometric assay**, **radioimmunoassay**, and competitive protein-binding assay.

radioautogram *or* **radioautograph** *or* **autoradiogram** the pattern formed after development of an X-ray film that has been exposed to a paper, gel, or thin-layer chromatogram containing radioactive compounds.

radioautograph 1 *an alternative term for* **radioautogram**. **2** to prepare a radioautogram. —**radioautography** *n*.

radiobiology *or* **radiation biology** the study of the interaction between radiation and biological material.

radiocarbon any radioactive isotope of carbon, usually carbon-11 or carbon-14.

radiochemical purity *abbr*.: RCP; the proportion of the total **activity** (def. 2) that is present in the stated chemical form in a radioactive material.

radiochemistry the branch of chemistry concerned with the compounds of radioactive nuclides. *Compare* **radiation chemistry**.

radiochromatogram a chromatogram of substances some or all of which contain a radioactive isotope.

radiofrequency radiation electromagnetic radiation encompassing wavelengths in the region 1 mm (300 GHz) to 30 km (10 kHz) (frequencies in parentheses). This range includes microwave radiation and lies above that of infrared radiation.

radiography the use of X-rays to examine the internal structures of a body. An X-ray film is placed behind the body, through which X-rays are passed. After exposure, there are areas of the film in which development has not occurred due to absorption of the X-rays by structures of the body, thus yielding a pattern of these structures. —**radiographer** *n*.; **radiographic** *adj*.; **radiographically** *adv*.

radioimmunoassay *abbr*.: RIA; a highly sensitive method of assay of nonradioactive material. A known amount of antibody (Ab) directed against the substance (antigen, Ag) to be assayed is saturated with a mixture of Ag and radioactive Ag (Ag*), so that total Ag (Ag + Ag*) is in excess. When nonradioactive Ag is added to Ab along with Ag*, Ag and Ag* compete for binding to Ab, so that less Ag* will be found in the antibody–antigen complex as the ratio Ag:Ag* increases. If the antibody–antigen complex is then separated from free antigen, the total amount of antigen (the substance being assayed) can be calculated from the ratio Ab-Ag*:Ag* and the known **titre** of antibody. It is often convenient to attach the antigen to plates so that the amount of ligand bound to the plate is proportional to the amount of test antibody.

radioimmunosorbent test *abbr.*: RIST; a test used primarily for quantifying total serum immunoglobulin E (IgE) levels. Antibodies specific for IgE, covalently linked to an insoluble dextran, are incubated first with a known amount of radiolabelled IgE and then with the serum containing the unknown amount of IgE. From the decrease in the amount of radiolabel bound to the immobilized antibodies, due to competition by the unlabelled IgE in the serum, the total amount of IgE in the serum can be estimated.

radioiodinate to incorporate one or more atoms of **radioiodine** (usually iodine-125 or iodine-131) into a molecule.

radioiodine any radioactive isotope of **iodine**, notably iodine-125 (half-life 60 days) and iodine-131 (half-life 8 days).

radioisotope a **radioactive isotope**.

radiolabel 1 a **radioactive label**. **2** to incorporate a radiolabel.

radioligand any **ligand** containing a radioactive isotope; *see also* **Scatchard plot**.

radiology the study and use of radiation and radioactive substances, especially in the diagnosis and treatment of disease. It includes the use of diagnostic X-rays in **radiography**, the therapeutic use of ionizing radiation to treat disease (**radiotherapy**), and the experimental and diagnostic use of radioisotopes introduced into the body. —**radiological** *adj.*; **radiologically** *adv.*; **radiologist** *n.*

radiolysis the cleavage of chemical bonds by high-energy radiation. —**radiolytic** *adj.*; **radiolyse** *vb.*

radiometer 1 any instrument used to measure radiant energy. **2 Crookes radiometer** a specific type of radiometer in which the different behaviour of a polished surface, which reflects energy, and a blackened surface, which absorbs it, is utilized to cause a rotor to turn at a rate related to the intensity of radiation. [After William Crookes (1832–1919), British chemist and physicist.] —**radiometric** *adj.*; **radiometry** *n.*

radiomimetic 1 describing any drug or other chemical substance, typically an alkylating agent, whose effects in living systems (e.g. carcinogenic, immunosuppressive, mutagenic) resemble those produced by **ionizing radiation**; more loosely, describing any such agent able to cause gene and chromosome mutations. **2** a radiomimetic (def. 1) substance or agent.

radionuclide a radioactive **nuclide**. —**radionuclidic** *adj.*

radionuclidic impurity any radionuclide, other than the nuclide that should be present, in a radioactive material.

radionuclidic purity *or* **radiopurity** (of a radioactive material) the proportion of the total **activity** (def. 2) that is in the form of the stated **radionuclide**.

radiotherapy *see* **radiology**.

radiotracer *an alternative name for* **radioactive tracer**.

radixin a protein of the erythrocyte band 4.1 family present in a variety of eukaryotic cells; it shares about 75% similarity with human **ezrin**. It is an actin barbed-end capping protein, highly concentrated in the undercoat of the cell-to-cell **adherens junction** and the cleavage furrow in the interphase and mitotic phase.

radyl (*in chemical nomenclature*) a term used to indicate any acyl, alkyl, or alkenyl group where this cannot be specified in detail.

RAF any gene encoding a Raf protein, of which there are at least three: Raf-1, A-Raf, and B-Raf. These are protooncogene cytoplasmic serine/threonine protein kinases (EC 2.7.1.–) of the Ras/Mos family; they are homologous with members of the protein kinase C family in both their conserved C-terminal catalytic and regulatory N-terminal domains. Human Raf-1 has binding domains for ATP, phorbol ester, and diacylglycerol. This protein is involved in transduction of mitogenic signals from the cell membrane to the nucleus. In the mouse, *raf* is the transforming gene of the mouse sarcoma

virus, 3611, v-Raf having a truncated regulatory domain that causes constitutive activation of kinase activity. The c-Raf-1 protein kinase, a 74 kDa protein, has been found in many mammalian cell types. Raf-1 is an important mediator of signals involving cell growth, transformation, and differentiation and is activated by a wide variety of extracellular stimuli. It has **MAP kinase** activity. In many cell types, Raf functions immediately downstream of the G-protein Ras (*see* **RAS**) in the activation pathway; Ras mediates translocation of Raf-1 to plasma membrane, where it is activated. In some cells Raf may be phosphorylated and activated by a protein kinase C. The physiological substrate for Raf kinase is a protein kinase (MEK) of yeast that is required for meiotic recombination and which on phosphorylation can activate MAP kinase.

raffinose *or* **gossypose** *or* **melitose** *or* **melitriose** the **trisaccharide** O-α-D-galactopyranosyl-(1→6)-O-α-D-glucopyranosyl-(1→2)-β-D-fructofuranoside; it occurs in plants almost as commonly as sucrose, being present, e.g., in cereal grains, cotton seeds, and many legumes. It is the first member of a series in which galactosyl residues are attached to sucrose, others being the tetrasaccharide **stachyose** and the pentasaccharide **verbascose**; these are all synthesized from sucrose by transfer reactions in which an α-D-galactopyranoside of *myo*-inositol (i.e. galactinol) is the donor of the galactosyl residues.

raft a sphingolipid- and cholesterol-rich microdomain in a plasma membrane. Rafts are similar in composition to caveolae, but lack caveolin, and are present in all cell types. They segregate and concentrate GPI-anchored, doubly acylated peripheral membrane proteins and transmembrane proteins. They may be precursors of caveolae.

RAG *abbr. for* recombination-activating gene; either of two genes (locus at 11p13) that encode the lymphocyte-specific RAG1 and RAG2 proteins. RAG1 is a zinc finger protein that forms heterotetramers with RAG2. In cooperation, they bring together recombination signal sequences on the genes for V(D)J segments of the immunoglobulins. Rare inactivating mutations in either gene lead to T and B lymphocyte deficiency and severe combined immunodeficiency (SCID).

ragged-red fibers aerobic skeletal muscle fibers that accumulate enlarged and abnormal mitochondria containing highly-ordered inclusions when they are ischemic or anoxic. The mitochondrial aggregates appear red on staining with Gomori modified Trichrome. They are usually associated with mitochondrial disease. *See* **MERRF**.

Raji a lymphoblastoid cell line derived from an African patient with Burkitt's lymphoma.

r.a.m. *abbr. for* relative atomic mass.

ram *abbr. for* ribosomal ambiguity.

Ramachandran plot *or* **conformational map** a plot that shows the ranges of bond angles that are permissible and the main types of structure adopted by a polypeptide chain (e.g., α-helix, β-sheet). As the peptide bond, –C′O–NH–, is planar, a polypeptide chain has only two degrees of freedom per residue, the twist about the $C^α$–N bond axis, ϕ, and that about the $C^α$–C′ axis, ψ. Some conformations will be easily achieved; others will be impossible, because they would bring neighbouring unbonded atoms or groups within van der Waals contact distances. The Ramachandran plot shows the permitted range within a plot of ϕ against ψ. [After G. N. Ramachandran (1922–).]

Raman scattering the inelastic scattering of a photon by a molecule, producing in the scattered light weak radiation of frequencies not present in the incident light. The sample does not have an absorption band at the wavelength of the incident light (*compare* **fluorescence**) and the frequency differences between the weak Raman lines in the spectrum and the exciting line are characteristic of the scattering substance and independent of the frequency of the exciting line. [After Chandrasekhara Venkata Raman (1888–1970), Indian physicist.]

Raman spectrum a plot of **Raman scattering** over a range of frequencies.

Ramos a lymphoblast cell line derived from a Caucasian patient with Burkitt's lymphoma.

RAMP *abbr. for* receptor activity modifying protein.

Ran a small nuclear G-protein involved with **karyopherins** that exists in two conformational states depending on whether it binds GTP (RanGTP) or GDP (RanGDP). RanGTP causes disassembly of complexes of cargo plus **importin-β** but is required for assembly of cargo-**exportin** complexes. RanGTP concentrations are higher in the nucleus than the cytoplasm thus favouring disassembly of import complexes and formation of export complexes. *See also* **RAN**.

RAN a gene encoding a small nuclear G-protein, first identified as a well-conserved gene distantly related to *H-RAS*. Its product, Ran, interacts with **RCC1**, and the complex is involved in regulating cell-cycle progression and messenger RNA transport. In the absence of RCC1, GDP–Ran predominates and this results in activation of **M-phase-promoting factor** and chromatin condensation. The presence of RCC1 activates nucleotide exchange on Ran, increasing the amount of GTP–Ran and preventing premature chromatin condensation. *See also* **nuclear pore, Ran**.

random coil the condition in solution of a linear polymer that exhibits little side group interaction, or resistance to rotation about the bonds in its chain. There are no preferred conformations. Quantitative expressions have been developed for the average radius of gyration. The term has formerly been incorrectly applied to various states of protein chains.

random copolymer a polymer of more than one species of monomer, in which the sequence of species within the polymer is not predetermined by controlled steps during synthesis, but depends only on random collisions between the reactant molecules. An example is the sequence of bases in RNA synthesized from nucleoside diphosphates by **polynucleotidephosphorylase**.

randomization or **randomisation** *(in experimental design)* the generation of data sets with different arrangements of the data, within the constraints of the experimental design. Randomization obviates the possibility of a systematic relationship between the data and any measurable characteristic of the procedure by which the data were sampled.

random mechanism or **random pathway** (for an enzymic reaction involving two substrates, A + B ⇌ C) a mechanism in which the enzyme, E, may first react with either substrate and the resulting enzyme–substrate complex then reacts with the second substrate, e.g. A + E ⇌ EA then EA + B ⇌ EAB, or, B + E ⇌ EB then EB + A ⇌ EAB, subsequently EAB ⇌ E + C. *Compare* **ordered mechanism**.

random priming a technique in which a mixture of short (typically 6 bp) oligonucleotides primes reactions catalysed by DNA polymerases. It is used in the synthesis of labelled DNA probes and is one of the options when priming cDNA synthesis catalysed by reverse transcriptase. Because all combinations of six nucleotides are present in the mixture of hexamers, priming is ensured at some point in the template sequence without prior need for specific sequence information.

random sample a selection of part of a population made in such a way that no individual is favoured over any other individual. *See also* **sample, statistical sample**.

ranitidine a competitive antagonist at histamine H_2 receptors, being the major and pharmacologically more selective analogue of the imidazole derivative, **cimetidine**. It inhibits histamine- and pentagastrin-evoked gastric acid secretion and cardiac stimulation and also inhibits gastric secretion evoked by physiological stimuli, agonists at muscarinic cholinoceptors, and gastrin. It is used in the treatment of gastric and duodenal ulcers. *One proprietary name*: Zantac (hydrochloride).

RANK *abbr. for* receptor activator of NF-κB; a member of the tumour necrosis factor receptor family that is present on osteoclasts and is the receptor for **RANKL**, which is present on osteoblasts and their precursor cells. The interaction of RANK with RANKL causes osteoclasts to differentiate and become active unless osteo-

protegerin (which also binds RANKL) is present. Mutations in RANK cause familial expansile osteolysis.

RANKL *abbr. for* RANK ligand; *other names*: osteoprotegerin ligand; TNF-related activation-induced cytokine (*abbr.*: TRANCE). RANKL is released from its transmembrane precursor present on osteoblasts and their precursor cells, and the transmembrane precursor of RANKL, which is released by the action of an ADAM metalloprotease. It is the ligand for RANK on osteoclasts.

RANTES *abbr. for* regulated upon activation, normal T-cell expressed and secreted chemokine; *other names*: small inducible cytokine A5; a chemokine secreted by T cells and macrophages on stimulation by mitogens, that acts as a chemoattractant, and stimulates basophils and eosinophils.

ranunculeic acid *another name for* **columbinic acid**.

rapamycin an immunosuppressant (and antifungal agent) from *Streptomyces hygroscopicus*; it is used to reduce rejection of organ transplants. Unlike cyclosporin A it is antiangiogenic and thus anticancer. It is a lipid-soluble macrocyclic lactone, with the unusual feature of a cyclohexane ring. It competes with **FK506** for a common binding site on **FKBP**. Moreover, it inhibits the signalling pathway of interleukin 2 and certain other cytokines, and also **mTor**. *See also* **cyclosporin**.

RAPD PCR *abbr. for* random polymorphic DNA PCR; a variation of the **polymerase chain reaction** in which many short (≈10 nucleotide) primers are used. The products of the reaction from different sources can lead to recognition of polymorphisms and physical maps.

rape or **colza** a herbaceous plant, *Brassica napus*, of the mustard family, Cruciferae; it resembles a turnip, and the seeds yield valuable **rapeseed oil**.

rapeseed oil oil derived from the seed of **rape**; it is widely used as a lubricant and cooking fuel, and for the manufacture of animal foodstuffs, soap, and synthetic rubber. It contains **erucic acid**, which has cardiotoxic effects varying with species; rape varieties of low erucic acid content have recently been developed.

rapid-reaction kinetics the kinetic analysis of an enzyme reaction before the steady state is reached. The **stopped-flow method** is commonly used, in which two independent sources of substrate and enzyme are mixed immediately before flow through a spectrophotometer cell. An opposing syringe stops the flow immediately before activating the detection and recording system, such that reaction times of a few milliseconds can be achieved. Characteristically, reaction rates may be appreciably faster than during the steady state. In the **quenched flow** modification of the method, a further reagent is added to stop the reaction immediately before detection and recording. *See also* **relaxation kinetics**.

Rapoport–Lübering shunt a unique feature of erythrocyte glycolysis in which the glycolytic intermediate 3-phosphoglyceryl phosphate is converted by bisphosphoglycerate mutase to 2,3-bisphosphoglycerate, which is in turn converted to 3-phosphoglycerate and inorganic phosphate by bisphosphoglycerate phosphatase, both reactions being catalysed by a single enzyme molecule. The shunt bypasses the near equilibrium phosphoglycerate kinase reaction, one of the two ATP-generating reactions of glycolysis. Since 2,3-bisphosphoglycerate is an important intracellular modulator of hemoglobin function, regulation of the Rapoport–Lübering shunt balances the energy needs and oxygen-transporting function of the red cell. [After Samuel Mitja Rapoport (1912–2004) and J. Lübering.]

RAR *abbr. for* retinoic acid receptor.

rare cutter any of a class of restriction endonucleases that cleave infrequently within human DNA and then mostly at sites that contain a CpG dinucleotide within the recognition sequence.

rare earth *(strictly)* an oxide of a **rare earth metal**, or *(more loosely)*, a rare earth metal itself.

rare earth metal *or* **rare earth element** scandium, yttrium, or a **lanthanoid**.

RAS an oncogene (v-*ras*) discovered in the retroviruses Harvey and Kirsten rat sarcoma viruses. There are three closely related (85% homology) oncogenes, Ha-*ras*, Ki-*ras*, and N-*ras*. Their cellular counterparts encode 21 kDa guanine nucleotide-binding proteins that regulate growth and differentiation in nearly every eukaryotic cell studied. There are two subfamilies, the Rho and Rac proteins (involved in relaying signals from the cell-surface receptors to the actin cytoskeleton) and the Rab family (involved in regulating the traffic of intracellular transport vesicles). Ras proteins undergo modification of the N and C termini that localize them to the plasma membrane. Ras is considered active in the GTP-bound form and inactive in the GDP-bound form. The intrinsic rate of nucleotide exchange and GTPase activity is very slow. Ras•GDP predominates in resting cells, but when Ras is activated by occupied receptors, specific guanine nucleotide exchange factors (GEFs), e.g. **Sos**, enhance nucleotide exchange leading to Ras•GTP, GTP exceeding GDP in cytoplasm. Ras•GTP then activates downstream effector proteins such as Raf-1 (*see* **raf**). Activated GTPase activating protein (GAP) causes the formation of inactive Ras•GDP. The retroviral oncogene, v-*ras*, encodes a protein that differs from the c-*ras* product in a single residue. If this affects amino acids 12, 13, 59, 61, or 63 the protein fails to hydrolyse its bound GTP and so persists abnormally in its active state, constitutively transmitting an intracellular signal for cell proliferation. Alternatively, some transforming mutations increase the rate at which Ras can exchange guanine nucleotides, e.g. when amino acids 116, 117, 119, or 146 are affected, again increasing the amount of Ras•GTP. Activated *ras* oncogenes occur in up to 30% of all human tumours, with the most common mutations being at codons for amino acids 12 and 13. In some types of malignancy, the figure is nearly 100%. Two genes corresponding to H-*ras* occur in yeast; they encode **G-proteins** that inhibit **adenylate cyclase**.

rasiRNA *abbr. for* repeat-associated siRNA. Transcription from opposing promoters found in repetitive DNA elements leads to the formation of long dsRNAs. These are cleaved by **Dicer** into siRNAs which are unwound and used in the RNA-induced transcriptional silencing complex (RITS) which directs silencing of chromatin in the region from which the dsRNA was transcribed. Silencing is thought to involve methylation of lysine 9 of histone H3.

RasMol *see* **visualization software**.

RAST *abbr. for* radioallergosorbent test.

rate coefficient *see* **rate constant**.

rate constant *or* **rate coefficient** *symbol*: k; a constant relating the concentration of a reactive species to the velocity of a reaction; thus for a first order reaction (*see* **order of reaction**) such as: EA → E + A, the rate of reaction, v, is proportional to the concentration of the reactant EA, so $v = k[EA]$, where k is the first-order rate constant. For a second order reaction, e.g. E + A → EA, the rate is proportional to the product of the reactant concentrations, i.e. $v = k[E][A]$, where k is the **second-order rate constant**.

rate-determining step *see* **rate-limiting step**.

rate-limiting step the slowest step in a reaction sequence; i.e. the step with the smallest rate constant.

rate of consumption *see* **rate of reaction**.

rate of conversion *symbol*: $\dot{\xi}$; for a specified chemical reaction, the rate of increase of the extent of reaction at any moment in time, regardless of the conditions under which the reaction is carried out: $\dot{\xi} = d\xi/dt$, where ξ is the extent of reaction and t is time. In the case of a reaction with one-to-one stoichiometry, such as a typical simple enzyme-catalysed reaction, the extent of reaction is equal to the amount of a reactant consumed or the amount of a product formed. Rate of conversion is an extensive quantity (*see* **extensive property**), its dimensions are (amount of substance) divided by time, and it is expressed in mol s^{-1} (*compare* **rate of reaction**). It is therefore independent of the volume of the system. Where this volume remains constant, as during a typical enzyme-catalysed reaction, the rate of conversion is equal to the rate of reaction multiplied by the volume: $\dot{\xi} = vV$. The term was formerly known in physical chemistry as rate of reaction. *See also* **catalytic activity**.

rate of formation *see* **rate of reaction**.

rate of reaction *or* **velocity of reaction** *or* (*sometimes*) **reaction rate** **1** *symbol*: v; for a specified enzyme-catalysed reaction of a typical kind: one involving a single substrate, S, a single product, P, one-to-one stoichiometry between them, and constant-volume conditions, the **rate of consumption** of substrate or the **rate of formation** of product at any moment in time is given by:

$$v = -d[S]/dt = d[P]/dt,$$

where [S] and [P] are respectively the concentrations of substrate and product, and t is time. A similar set of relationships applies where there are several reactants in one-to-one stoichiometry, but if the overall stoichiometry is not one-to-one – a situation rarely arising in enzyme kinetics – the relationships are more complicated. Where intermediates are formed in amounts comparable to those of the reactants, there is time-dependent stoichiometry and a rate of reaction for the complete reaction cannot be specified. Rate of reaction is an intensive quantity (*see* **intensive property**), its dimensions are (amount-of-substance concentration) divided by time, and it is expressed in mol L^{-1} s^{-1} (*compare* **rate of conversion**). *See also* **initial velocity**, **limiting rate**. **2** (*formerly, in physical chemistry*) the extensive quantity now termed **rate of conversion**.

rate theory of agonist action a theory of agonist action stating that the response to an agonist is proportional to the number of receptor–agonist associations per unit time, i.e. to the rate of receptor occupation, the binding of agonist to receptor being reversible. The response is therefore maximal at time zero and falls exponentially to an equilibrium value.

rational drug design the design of drug molecules based on the 3D structure of the target molecule. Owing to its complexity, the approach has had only limited success, but it has worked for some targets where the molecular mechanism of activation or inactivation is well understood (e.g. some enzymes), or where structural modifications are readily engineered (e.g. antibodies).

raw sequence (*in bioinformatics*) in the early stages of the sequencing process, sequence reads that are unassembled and carry no annotation.

Rayleigh ratio *symbol*: R; the ratio of the intensity of light scattered by particles (molecules) in solution at an angle θ from the incident beam, I_θ, multiplied by the square of the distance, r, from the particle to the detector, divided by the intensity of the incident light, I_o, i.e. $R = I_\theta r^2/I_o$. The **reduced Rayleigh ratio**, $R_\theta = I_\theta r^2/I_o(1 + \cos^2\theta)$. Many literature references do not distinguished between these two. *See also* **Rayleigh scattering**. [After John William Strutt, Lord Rayleigh (1842–1919), British mathematician and physicist.]

Rayleigh scattering the elastic scattering of photons of light by molecules or atoms of the substance through which the light is passing. When light passes through a liquid or gas, atoms or molecules are polarized by the electric vector of the light, thus inducing a rapidly fluctuating dipole in the atoms or molecules in the light path. The fluctuating dipole leads to the emission of electromagnetic waves in various directions of the same frequency as the incident radiation, and this emission is seen as scattered light. *See also* **Rayleigh ratio**.

Rb *symbol for* rubidium.

RB a human **tumour suppressor gene** that encodes a nuclear phosphoprotein, **retinoblastoma protein**. Inactivation or loss of *RB* on both copies of the chromosome (13q14) leads to **retinoblastoma**.

R band *see* **Giemsa's stain**.

RbAp *abbr. for* retinoblastoma-associated protein. *See* **Sin 3-HDAC**.

rBAT *abbr. for* related to basic amino acid transporter. *See* **cystinuria**.

RBC *or* **rbc** *or* **r.b.c.** *abbr. for* red blood cell. *See* **erythrocyte**.

Rbe *abbr. for* relative biological effectiveness (i.e. **quality factor**).

R-binder *an alternative name for* **haptocorrin**.

RBS *abbr. for* Réné Borghgraef solution; *the proprietary name for* any of a range of general-purpose surfactants and decontaminating agents, especially for laboratory use, that have combined detergent and chelating properties. They are identified with a number or other designation, e.g. RBS 25, RBS 35, RBS solid.

RCC1 a chromatin-bound guanine nucleotide release (or exchange) factor. Its function is required for the normal coupling of the completion of DNA synthesis and the initiation of mitosis. It is also necessary for RNA export from the nucleus, spindle formation, and formation of the nuclear envelope during cell division. It binds to and acts on the nuclear G-protein, **Ran**. [From regulator of chromosome condensation.]

RCF *abbr. for* relative centrifugal force.

R chromophore a **chromophore** that gives rise to low-intensity absorption bands at long wavelengths. *Compare* **K chromophore**. [From German *Radikal*.]

RCP *abbr. for* radiochemical purity.

RCS *abbr. for* rabbit aorta-contracting substance.

RCSB *abbr. for* Research Collaboratory for Structural Bioinformatics; *see* **structure database**.

rd *symbol for* rad (def. 2) (alternative to rad).

Rd *abbr. for* rubredoxin.

RDBMS *abbr. for* relational database management system.

rDNA 1 *abbr. for* ribosomal DNA. **2** *(sometimes)* *abbr. for* recombinant DNA; use not recommended because of possible confusion with ribosomal DNA.

re- *prefix (in chemical nomenclature)* denoting *rectus* (i.e. right, or clockwise) configuration of a face of a trigonal atom. *See* **re/si convention**.

Re *symbol for* rhenium.

Re *symbol for* Reynolds number.

RE *abbr. for* (eukaryotic) response element.

reactancy of enzymes the number of reactants involved in an enzyme-catalysed reaction in a given direction. The numbers of kinetically important substrates or products in an enzyme mechanism are designated by Uni (1), Bi (2), Ter (3), or Quad (4). Thus, e.g., UniUni indicates one substrate and one product, and Ter Bi indicates three substrates and two products.

reactant any material participating in a chemical reaction in a given direction.

reaction *(in chemistry)* any process in which molecules interact with one another, leading to chemical or physical change.

reaction centre *see* **photosystem**. *Compare* **photosynthetic reaction centre**.

reaction intermediate a transient chemical species, with a lifetime appreciably longer than a molecular vibration time, formed directly or indirectly from the reactants of a chemical (often enzymic) reaction, and that further reacts (directly or indirectly) to give the reaction products.

reaction rate *(sometimes)* an alternative term for **rate of reaction**.

reaction time an alternative name for **latent period** (def. 1).

reactivate to restore activity to any enzymic or cellular system.

reactive describing any molecular entity or centre that is capable of reacting readily. Thus, a **reactive centre** is the point in the molecule at which electron configurations favour a reaction.

reactive oxygen species *abbr.:* ROS; a highly active oxygen species, such as **superoxide** ion, hydroxyl radical, hydrogen peroxide (H_2O_2), or peroxynitrite. Most ROS are generated as toxic by-products of oxidative phosphorylation in mitochondria. H_2O_2 is also produced by various reactions in peroxisomes, and by phagocyte NADPH oxidase, thyroperoxidase, and xanthine oxidase in specific cell types. Cells protect themselves through the superoxide dismutase, glutathione peroxidase, and catalase activities, and the antioxidants (e.g. glutathione, carotene) they contain. Chronic exposure to ROS causes oxidative damage to mitochondrial and cellular proteins, lipids, and nucleic acids. *See* **oxidative stress**.

reactivity the degree to which a molecule or centre is reactive.

reading *(in molecular biology)* the process of one-way linear decoding of a nucleotide sequence during, e.g., **transcription** or **translation**.

reading frame any sequence of bases in DNA that codes for the synthesis of either a protein or a component polypeptide. The point of initiation of reading determines the frame, i.e. the way in which the bases will be grouped in triplets as required by the genetic code. For example, the sequence

$$-ATGGTCA-$$

could be grouped

$$-A,TGG,TCA-$$

or

$$-AT,GGT,CA-$$

or

$$-ATG,GTC,A-$$

depending on the point at which the reading frame begins. In double-stranded DNA there six potential reading frames. *See also* **open reading frame**.

readthrough the continuation of DNA transcription beyond a normal termination sequence, due to a failure of RNA polymerase to recognize the termination signal. Readthrough may also occur in translation, where a mutation has caused a termination codon to be converted to a codon for an amino acid (a sense codon); this is the underlying etiology of several **hemoglobinopathies**. *See also* **readthrough protein**.

readthrough protein a product formed by translation of a messenger RNA when one of its termination codons is misread, or is recognized by a **suppressor tRNA**. Polypeptide-chain synthesis then continues until the next termination codon is reached, resulting in the production of a variant protein with a C-terminal extension.

reagent 1 any chemical substance that reacts or participates in or is necessary for a reaction. **2** a substance used for the detection or determination of another substance by chemical or microscopical means, usually in solution; e.g. the **Folin–Ciocalteu reagent**.

reagent grade a level of purity of a chemical; chemicals of this quality conform to the specifications established by the Committee on Analytical Reagents of the American Chemical Society and may often be sufficiently pure to use as primary standards in an analytical laboratory, with information concerning the actual percentage of impurities or at least the maximum limit of impurities being given on the label.

reagin 1 a heat-labile antibody-like substance in the serum of many people suffering from atopic allergies, and identical with immunoglobulin E (IgE). The **immediate hypersensitivity** (type I) generated by the binding of IgE to its specific antigen on the surface of mast cells involves the release of cell mediators via the reaginic or anaphylactic mechanism. **2** a special kind of antibody response produced by syphilitics that has specificity for **cardiolipin** antigens.

real-time *(in computing)* indicating that a computer monitors a process continuously as it occurs, in distinction to the analysis of recorded data.

real-time PCR a variant of the **polymerase chain reaction** (PCR) in which the quantity of amplified DNA is recorded in each cycle by monitoring the fluorescence intensity of a DNA-binding dye such as **SYBR Green**. An advantage is that the kinetics of the reaction can be monitored and, with reference to the behaviour of standards, the amount of PCR product quantified. The abbreviation RT-PCR for real-time PCR is to be deprecated. Real-time PCR is frequently used for quantitation of reverse transcriptase PCR in which case real-time RT-PCR is the appropriate terminology. *See* **RT-PCR**.

rearrangase *or* **S–S rearrangase** *an alternative name for* **protein disulfide isomerase**.

rearrangement reaction a reaction in which molecular bonds are rearranged without the loss of any atom from the molecule, e.g. as in the conversion of glucose 6-phosphate to fructose 6-phosphate.

recA a gene in *Escherichia coli* encoding a protein, RecA, involved in **genetic recombination** and in the SOS response for DNA repair (*see*

SOS repair). The protein has two activities: (1) a DNA-dependent ATPase in the presence of single-stranded DNA that catalyses the ATP-dependent uptake of single-stranded DNA to form the heteroduplex required in recombination. (2) In the presence of DNA fragments it becomes a protease (RecA*; EC 3.4.99.37) that cleaves a specific Ala-|-Gly bond in certain repressor molecules, including its own repressor (*see* **lexA**). This allows all the SOS genes to be activated. *See also* **exodeoxyribonuclease V**, **recombinational repair**, **recombinase**.

recalcitrance *or* **persistence** the ability of a substance to remain in a particular environment in an unchanged form.

recall *or* **sensitivity** (*in informatics*) the proportion of the true members of a set that are correctly identified by a prediction algorithm: true positives/(true positives + false negatives).

recBC a gene in *Escherichia coli* for **exodeoxynuclease V**, an ATP-dependent nuclease specific for double-stranded DNA; it mediates **genetic recombination**.

recDNA (*sometimes*) *abbr. for* recombinant DNA.

receiver operating characteristic *or* **receiver operator characteristic** *abbr.*: ROC; (*in signal detection theory*) a graphical representation of the relationship between true and false hits for a binary classifier system as its discrimination threshold is varied. ROC curves are used, for example, in bioinformatics to evaluate prediction algorithms: results in the top left of the plot denote 100% sensitivity (all true positives found) and 100% specificity (no false positives found); completely random results lie on a straight line at 45° from the origin, indicating that, as the threshold is raised, equal numbers of true and false positives are matched.

receptor any cellular macromolecule that binds a hormone, neurotransmitter, drug, other agonist, or intracellular messenger to initiate a change in cell function. Receptors are concerned directly and specifically in chemical signalling between and within cells. Cell-surface receptors, e.g. the **acetylcholine receptor** and the **insulin receptor**, are located in the plasma membrane, with their ligand-binding site exposed to the external medium. Receptors may also be present in membranes of the sarcoplasmic reticulum (e.g. ryanodine receptor), T tubule (e.g. dihydropyridine receptor), or endoplasmic reticulum (e.g. inositol trisphosphate, or IP_3 receptor). Intracellular receptors (e.g. steroid-hormone receptors) bind ligands that enter the cell across the plasma membrane. These can be located in the cytosol or nucleus. Receptors can be catalytic, enzyme-associated, or G-protein-associated.

receptor activator of NF-κB *see* **RANK**.

receptor activity modifying protein *abbr.*: RAMP; a single-transmembrane-domain protein required for expression of calcitonin gene-related peptide or adrenomedullin receptor phenotypes from the calcitonin receptor-like receptor gene (CRLR). For full activity, CRLR forms heteromers with RAMP and receptor component protein.

receptor assay 1 the determination of receptor numbers and the dissociation constant of receptor–ligand binding. The binding at various concentrations of ligand is measured, often by use of a radioactive ligand, and is analysed by means of a **Scatchard plot**. **2** the determination of the concentration of a ligand by utilizing a purified receptor in a manner similar to that used in **radioimmunoassay**; in this case the receptor, rather than an antibody, performs the role of specific binding agent.

receptor cross-talk interaction between two receptors such that one receptor, when activated, modulates the action of the other. This type of interaction often involves the production by the first receptor of a **second messenger**, which then modulates the action of the second receptor.

receptor-destroying enzyme an enzyme that brings about chemical change in a (usually membrane-bound) receptor, leading to inactivation of the receptor. Knowledge of the specificity of the enzyme provides information about the nature of the groups or linkages exposed to enzyme attack, and thus to receptor structure or orientation in the membrane. Examples are **sialidase** and **trypsin**.

receptor–effector complex the complex that results when an occupied receptor binds to another protein (the **effector**), the activation of which by the occupied receptor leads to signal transduction.

receptor-mediated describing a response that requires the intermediary action of a receptor, especially one located in a membrane.

receptor-mediated endocytosis a form of **endocytosis** involving the internalization (i.e. passage into the cell interior) of stretches of membrane into which receptors have clustered after binding of ligand. The cytoplasmic surfaces of these stretches of membrane are coated with **clathrin** and are termed **coated pits**. *See also* **endosome**.

receptor transformation *see* **transformation** (def. 3).

recessive describing an **allele** that lacks effect when accompanied in the same diploid by a **dominant** form of the same gene. Most (not all) defective mutant genes are recessive to their normal functional counterparts. —**recessivity** *n*.

RecET a **recombinase** from the Rac prophage of *E. coli* consisting of *recE*-encoded exonuclease VIII and the RecT protein. Together they catalyse recombination between linear duplex DNA and homologous circular single-stranded DNA. RecET is used for **recombineering**.

RecG an *E. coli* helicase involved in DNA recombination, acting with RuvC to resolve **Holliday junctions**.

recipient bacterium *see* **bacterial conjugation**.

reciprocal 1 expressed in the reverse direction, in response to an initial application of any force or stimulus. **2** (*in mathematics*) the inverse value, obtained when the values of numerator and denominator are reversed; the reciprocal of x is obtained by evaluating $1/x$.

reciprocal plot a plot in which the units of both the ordinate and the abscissa are reciprocals of the values that have been determined. *Compare* **Lineweaver–Burk plot**.

reciprocal translocation a chromosomal rearrangement in which the ends of, usually, two nonhomologous chromosomes are exchanged. If the reciprocal translocation is balanced there is no loss or gain of genetic material after the exchange.

recircularization *or* **recircularisation** the restoration of circularity to a plasmid vector following the insertion of recombinant DNA. The plasmid used for gene cloning is a small circular molecule. This is cut with a **restriction endonuclease** to create linear DNA molecules. The foreign DNA fragment with the same cohesive ends is then annealed to the cut plasmid to form a recombinant DNA circle. The DNA molecule containing the foreign DNA insert is then covalently sealed with **DNA ligase**. Recircularization also occurs during the replication of some DNA phages by a **rolling circle** replication mechanism.

recognition any specific binding interaction between molecules. Such interaction between two molecules is based on the existence of **recognition sites**, one (or more) on each of the interacting molecules, that are tailored to fit in such a way that highly specific binding occurs; i.e. the two molecules uniquely 'recognize' each other. *See also* **chiral recognition**, **molecular recognition**.

recognition sequence the part of the sequence of a macromolecule that is able to recognize and bind to a particular sequence in another macromolecule.

recognition site 1 (*in proteins*) a site for **recognition** of another protein, involving a specific sequence of amino acids; the latter may be in a single chain (e.g., a target site for protein phosphorylation), or may involve amino acids in separate chains brought into juxtaposition by quaternary structure. **2** (*in nucleic acids*) a site for recognition of a protein, involving a specific sequence of bases; *see* **restriction endonuclease**. **3** (*in pharmacology*) a region of a receptor macromolecule to which an endogenous agonist binds.

recombinant clone a population of bacteria or other host derived from a single cell into which a vector containing **recombinant DNA** has been introduced.

recombinant DNA a fragment of deoxyribonucleic acid (DNA) that has been inserted into a cloning vector, thereby leading to its use in the isolation of a clone of cells characterized by the presence of the fragment. The term is derived from the concept that insertion into the vector is a form of **genetic recombination**. [*Note*: The abbreviation rDNA has sometimes been applied to recombinant DNA, but this use is discouraged (rDNA having been pre-empted for ribosomal DNA); while alternatives such as recDNA or rtDNA have been suggested, IUBMB considers a standard abbreviation unnecessary.] *See also* **recombinant DNA technology**.

recombinant DNA technology an area of biotechnology concerned with the manipulation of **recombinant DNA**. It has many important applications, including (1) DNA sequencing, which may lead to the ability to predict the primary structure of a protein where this is the product of a cloned gene, (2) the synthesis of recombinant protein by a suitable expression system, and (3) the production of DNA **probes** (def. 3) for use in hybridization techniques. The technology has led, e.g., to the production of human insulin for the treatment of diabetics, human growth hormone for administration to children with growth defects, and a vaccine for hepatitis B. DNA probes are essential for the detection of aberrant genes which are the cause of numerous diseases and the technique has the potential for gene therapy. *See also* **genetic engineering**.

recombinant protein a protein encoded by a gene – **recombinant DNA** – that has been cloned in a system that supports expression of the gene and translation of messenger RNA (*see* **expression system**). Modification of the gene by recombinant DNA technology can lead to expression of a mutant protein. Proteins coexpressed in bacteria will not possess post-translational modifications, e.g. phosphorylation or glycosylation; eukaryotic expression systems are needed for this (see, e.g., **baculovirus**).

recombinase 1 an enzyme that recognizes specific sites involved in the joining of V-region gene segments in L- and H-chain gene segment pools in the generation of antibody. **2** RecA protein (*see* **recA**) is also a recombinase, the activity of which is increased in *Escherichia coli* following damage to DNA. RecA binds to a region of single-stranded DNA and catalyses the formation of duplex DNA; hydrolysis of ATP is involved. A recombinase recognizes and binds to each of two recombinant sites on different DNA molecules or within the same DNA. One DNA strand in each site is cleaved at a specific point within the site, and the recombinase becomes covalently linked to the DNA at the cleavage site through a phosphotyrosine (sometimes a phosphoserine) bond. The transient protein–DNA linkage preserves the phosphodiester bond lost in cleaving the DNA; cofactors such as ATP are unnecessary in subsequent steps. The cleaved DNA strands are rejoined to new partners, with new phosphodiester bonds created at the expense of the protein–DNA linkage. The result of the initial breakage and rejoining process is the **Holliday junction**. To complete the reaction, the process must be repeated at a second point within each of the two recombination sites. In summary the recombinase can be viewed as a site-specific endonuclease and ligase in one package.

recombination *see* **genetic recombination**, **recombinant DNA**, **immunoglobulin synthesis**.

recombination activating gene *see* **RAG**.

recombinational heteroduplex *see* **heteroduplex**.

recombinational repair a mechanism to restore the integrity of DNA when replication of DNA stops at a **thymine dimer** but resumes after the dimer is passed, creating a gap. The undamaged parental strand recombines into the gap, a process promoted by RecA protein (*see* **recA**). The new gap in the parental strand is filled by the action of **DNA polymerase** and **DNA ligase**. The thymine dimer itself is not repaired in this process but time is allowed for the excision system to repair this damage later.

recombineering *or* **recombination-mediated genetic engineering** a method for manipulating large DNA molecules such as **BAC**s by promoting homologous recombination with linear DNA (such as products of the polymerase chain reaction). *E. coli* strains carrying recombinases such as **RecET** or bacteriophage lambda Red functions are used (*see* **Red recombination system**).

recoverin an **EF-hand** Ca^{2+}-binding protein, apparent molecular mass 26 kDa; *also known as* **cancer-associated retinopathy protein**. It selectively localizes to the retina and pineal gland, and is implicated in the activation of retinal rod guanylate cyclase by rhodopsin, possibly being involved in the blocking of the phosphorylation of rhodopsin, by forming an inhibitory complex with rhodopsin kinase. It has been identified as an autoantigen in cancer-associated retinopathy, a putative autoimmune process.

recovery the portion of material or activity that remains after any extraction or purification, usually expressed as a percentage of the amount that was present as starting material.

RecQ-like protein 2 *an alternative name for* **BLM protein**.

recruitment the action by any cell, cell constituent, or molecule that

brings about the cooperation of any other component in performance of any function.

recrystallize *or* **recrystallise** to dissolve a crystalline substance and subsequently crystallize it from the solution, usually to increase the purity of the substance. —**recrystallization** *or* **recrystallisation** *n*.

rectangular hyperbola a **hyperbola** whose asymptotes are at right angles. It is sometimes said that, for enzymes that follow so-called **Michaelis kinetics**, the plot of initial rate against substrate concentration has the shape of the curve of a rectangular hyperbola. It is true that the Michaelis–Menten equation does yield a curve that is part of a rectangular hyperbola, but the axes of the hyperbola must be rotated through 45°, and displaced, if they are to coincide with the axes of the Michaelis–Menten plot.

rectification 1 (*in physics*) the process of obtaining a direct current from an alternating electrical supply. **2** (*in physiology*) the regulation of membrane conductance by the voltage across the membrane; strong rectification implies that ionic channels carrying current are open at some membrane potentials and shut at others. Conductance is low at very negative membrane potentials, but increases as the potential becomes less negative. **3** (*in chemistry*) the process of purifying a liquid by distillation.

rectus *Latin* right; used to describe the right, or clockwise, configuration of one of the two faces of a compound containing a trigonal carbon atom. It is designated by the prefix *re-*. Compare **sinister**. See *also* **re/si convention**.

recycle to pass again through a system one or more times after return to a starting condition or state.

RED *abbr. for* RNA-dependent deaminase. See **RNA editing enzyme**.

red drop a phenomenon in photosynthesis, in which the **quantum yield** is relatively constant in light of wavelengths between 500 and 600 nm, but drops dramatically at longer wavelengths.

redistil *or* (*US*) **redistill** to distil again that which has already been distilled, usually as an aid to purification of liquids.

red marrow *see* **bone marrow**.

red muscle dark, skeletal, voluntary muscle that is relatively rich in myoglobin and cytochromes, and derives energy for contraction from a high oxidative capacity, in contrast to **white muscle**, which has a high glycolytic capacity. The distinction between muscle types should more strictly be based on muscle-fibre types. Type I fibres have a high oxidative capacity based on a high activity of tricarboxylic-acid cycle and respiratory chain enzymes. Type IIA fibres have high oxidative and high glycolytic capacity. Type IIB fibres have low oxidative and high glycolytic capacity. The proportions of these fibres vary in different muscle types, and can be changed by training in athletes.

redox *abbr. for* reduction–oxidation; see **oxidation–reduction**.

redox balance the ratio of the number of reductions to the number of oxidations in a given metabolic pathway. Compare **redox state**.

redox couple the oxidized (electron acceptor) and the reduced (electron donor) forms of the species of a given **redox reaction**; e.g. NAD^+ and $(NADH + H^+)$. The potential of any redox couple in solution can be measured against a reference electrode, and a table of relative potentials constructed. See **redox potential**.

redox hypothesis a now discounted model for the **Pasteur effect** in glycolysing muscle extracts. Addition of redox indicators that changed the redox state in the direction of oxidation decreased the rate of lactate formation; it was therefore suggested that the effect of oxygen might be connected with a change of the redox state in the same direction.

redox indicator *see* **indicator** (def. 3).

redox potential *or* oxidation–reduction potential *symbol*: E_o; for a particular **redox couple** at a given pH and temperature, the potential of an electrode (usually bright platinum) immersed in a solution containing 1 M oxidant and 1 M reductant, relative to a standard **hydrogen electrode**. The more 'reducing' the couple, the more negative is E_o; the convention in biochemistry is that the more 'oxidizing', the more positive is E_o. Some standard potentials, measured against the hydrogen electrode at pH 7, are: $NAD^+/NADH + H^+$, –0.32 V; fumarate/succinate, +0.032 V; cytochrome $c(Fe^{3+})$/cytochrome $c(Fe^{2+})$, +0.25 V; $O_2 + 2H^+/H_2$, +0.82 V.

redox reaction a chemical reaction in which the oxidation of one reactant is coupled to the reduction of a second reactant, electrons being transferred from the reactant being oxidized to that being reduced.

redox state the ratio of the concentration of the oxidized member of a **redox couple** to the concentration of the reduced member; e.g. $[NAD^+]/[NADH]$ or $[Fe^{3+}]/[Fe^{2+}]$. The term should only be applied to a particular redox system and not to a heterogeneous system.

Red recombination system a bacteriophage lambda-encoded recombination system. Full recombination potential requires the products of three genes: *exo*, *bet*, and *gam*. See **recombineering**.

red shift any shift of the peaks of a spectrum to longer wavelengths (i.e. towards the red end of the spectrum), including a shift from the red to infrared. *Compare* **blue shift**.

red tide a bloom of the dinoflagellate *Ptychodicus brevis* (*Gymnopodium breve*), an organism that produces **brevetoxins** and causes epidemics of food poisoning.

reduce to cause or undergo **reduction**.

reduced folate carrier *abbr*.: RFC; any of several homologous membrane-associated glycoproteins that bind reduced folates and methotrexate with low affinity. The gene at 21q22.2 has multiple promoters and produces several isoforms, including the putative intestinal folate transporter.

reduced Rayleigh ratio *see* **Rayleigh ratio**.

reducing agent any substance acting as a **reductant**. *Compare* **oxidizing agent**.

reducing end the end of an oligo- or polysaccharide molecule that carries a potential hemiacetal or hemiketal (reducing) group.

reducing sugar any **sugar** that will reduce heated alkaline solutions of cupric salts, such as **Fehling's** or **Benedict's solutions**, due to the presence of a hemiacetal or hemiketal group.

reductant the chemical species that donates an electron in an **oxidation–reduction reaction**; i.e. the species that undergoes **oxidation** in such a reaction. *Compare* **oxidant**.

reductase a donor:acceptor oxidoreductase enzyme (EC class 1) when named after the more oxidized alternative substrate and when O_2 is not a substrate. See **oxidoreductase**.

reduction the chemical process by which oxygen is withdrawn from, hydrogen is added to, or (more generally) an electron is added to (with or without addition of a proton) a molecular entity. *Compare* **oxidation**. —**reductive** *adj.*; **reduce** *vb.*

reduction division the first division of **meiosis**.

reduction potential the electrode potential that measures the tendency of one half of a **redox reaction** to proceed so as to gain electrons. See **oxidation potential**, **redox potential**.

reductive alkylation the reaction of amino groups with an alkyl carbonyl group, followed by reduction of the resulting **Schiff base** with a reducing agent such as sodium borohydride. See **reductive amination**.

reductive amination the process whereby ammonia or a primary or secondary amine reacts with ketones and aldehydes in the presence of reducing agents such as H_2 and a metal catalyst, or more conveniently, $NaBH_4$ or its derivative $NaBH_3CN$, to produce primary, secondary, or tertiary amines. In a formal sense, the amine is alkylated in this reaction while the carbonyl group is reduced, hence the term reductive amination, or alternatively **reductive alkylation**. Amino acids have been made from oxo acids by this process. A variation of reductive amination, termed the **Eschweiler–Clarke reaction**, involves the methylation of an amine to the tertiary amine stage using formic acid as the reducing agent and formaldehyde as the methylating reagent.

reductive carboxylic-acid cycle *or* **reductive tricarboxylic-acid cycle** a metabolic cycle used by members of the Chlorobiaceae for CO_2 fixation. The usual operation of the tricarboxylic-acid cycle is reversed, with three enzyme replacements, and other supplementary reactions are used. Initially one CO_2 reacts with succinyl-CoA to form oxoglutarate; oxoglutarate synthase (EC 1.2.7.3) replaces the usual **oxoglutarate dehydrogenase complex**. A second CO_2 converts oxoglutarate to isocitrate with use of **isocitrate dehydrogenase**, and isocitrate proceeds to citrate. Citrate is then cleaved by ATP citrate *(pro-S)* lyase to acetyl-CoA and oxaloacetate; the latter recycles to succinyl-CoA with fumarate reductase (NADH) replacing succinate dehydrogenase. These operations convert two CO_2 to acetyl-CoA. A net gain of oxaloacetate is possible by conversion of acetyl CoA and one CO_2 to pyruvate (pyruvate synthase), pyruvate to phospho*enol*pyruvate (phospho*enol*pyruvate synthase) and **phospho*enol*pyruvate** and one CO_2 to oxaloacetate (**phospho*enol*pyruvate carboxylase**). The overall result is that four CO_2 yield a net gain of the four-carbon oxaloacetate.

reductive methylation a specific form of **reductive alkylation**, in which the alkyl group is methyl. Formaldehyde is often used as a reagent in this case.

reductive pentose-phosphate cycle *or* **Calvin cycle** *other names*: carbon reduction cycle; C_3 cycle/pathway; more appropriately, Calvin–Bassham cycle; the cyclic set of reactions, occurring in chloroplasts in the majority of higher plants (**C_3 plants**), that results in the fixation of CO_2 as glucose using the ATP and the NADPH formed in the light reactions of photosynthesis. Briefly, CO_2 reacts with ribulose 1,5-bisphosphate under the action of **ribulose-bisphosphate carboxylase** (Rubisco), EC 4.1.1.39, to form two molecules of 3-phosphoglycerate; these are phosphorylated (by ATP) to 1,3-bisphosphoglycerate, which in turn is reduced (by NADPH) to glyceraldehyde 3-phosphate. This is then converted by aldolase, transketolase, and other enzymes to fructose 6-phosphate and ribulose 5-phosphate; the latter being phosphorylated (ATP) to ribulose 1,5-biphosphate, completing the cycle. *See also* **Hatch– Slack pathway**.

reductone any reducing substance with a chemical structure containing the stabilized enediol –C(OH)=C(OH)–. Reductones are generated when monosaccharides are heated for prolonged periods with strong alkali, whereby the enols originally produced are broken down to strongly reducing fragments. Another notable example is ascorbic acid, and in some contexts reductones are described as '**apparent ascorbate**', i.e. the reductive acids that are present in certain processed foods and react similarly to ascorbate, interfering with its determination.

reelin a large extracellular matrix protein encoded by the gene that, when defective in mouse, produces the neurological mutant called reeler. It is a serine protease that degrades fibronectin, laminin, and collagen, and is secreted by the earliest neurons during development. It binds receptors (e.g. for ApoE and for very low-density lipoproteins) on adjacent neurons to govern the overall patterning of cortical neurons into layers and to control neuronal migration.

re-entrant loop a segment of a membrane protein that loops between two neighbouring transmembrane helices and is partly outside and partly inside the membrane. It may be extracellular or intracellular.

Re-face (of an atom or molecule) *see* **Re/Si convention**.

reference something that acts as a standard in relation to quality, quantity, or type against which other values or characteristics can be compared; e.g. reference spectrum.

reference electrode any electrode, selected by convention, against which the electronegativity of compounds is measured. The standard biochemical reference is the **hydrogen electrode**, normally a platinum electrode in hydrochloric acid through which hydrogen gas is passed. A cell is constructed such that another electrode is immersed in a test solution and the potential difference between these two half-cells is measured. For convenience, however, the reference electrode is often a **calomel electrode**, and after measurements have been made against this, the electronegativity against the hydrogen electrode can be calculated from the known difference between this and the hydrogen electrode.

reflect to change the path of electromagnetic radiation, when it meets a surface, such that it returns partially or completely in a direction on the same side of the surface as the source.

reflectometer *see* **densitometer** (def. 1).

reflex *(in physiology)* an automatic or involuntary response of the nervous system to a stimulus. A reflex may be a monosynaptic spinal reflex (e.g. a simple motor response to a sensory stimulus), involving transmission through the spinal cord only, or polysynaptic and involving synaptic transmission in the brain. A **conditioned reflex** is one that is learned through repeated association of a stimulus with either rewarding (e.g. food) or aversive (e.g. punishment) events.

refolding *(for proteins)* the restoration of 3-D structure in a manner that may or may not achieve a full return to the native state following **denaturation**.

refraction the change in direction undergone by a wave of electromagnetic radiation on passing through a boundary between two media in which its velocity differs. **Snell's law** states that $n_1\sin\theta_1 = n_2\sin\theta_2$, where n_1 and n_2 are the **refractive indices** of the two media, and θ_1 and θ_2 are the angles between the radiation and the normal to the boundary in the corresponding media. [After Willebrord Snell (1591–1696), Dutch astronomer and mathematician.]

refractive index *abbr.*: RI; *symbol*: *n*; for a given medium, a measure of the degree of **refraction** induced by the medium when light or other electromagnetic radiation passes through it. The refractive index of a medium is the ratio of the speed of light in a vacuum, c_0, to the speed of light in the medium, c; i.e. $n = c_0/c$. For many practical purposes the speed of light in air may be used instead of that in a vacuum.

refractivity a measure of the extent to which a medium will cause the deviation of a ray of light or other electromagnetic radiation entering its surface. If the **refractive index** of a medium is *n*, its refractivity is $(n - 1)$. The **specific refractivity** is given by $(n - 1)/\rho$, where ρ is the density of the medium. The **molecular refractivity** is the specific refractivity multiplied by the relative molecular mass.

refractometer an instrument for measuring **refractive index**.

refractory 1 (of a clinical condition) resistant to treatment. **2** (of a cell or tissue) resistant to stimulation, as by a drug, hormone, or nerve impulse, especially during a period following previous stimulation. *See also* **tachyphylaxis**. **3** (of a substance or material) resistant to heat, corrosion, mechanical deformation, or other aggressive agents causing alteration. **4** a substance or material having such resistance.

Refsum's disease (classical) *or* **Refsum's disease (adult)** *other name*: phytanic acid storage disease; a rare autosomal recessive inherited disease starting in adulthood with neurological symptoms, including retinitis pigmentosa, peripheral neuropathy, cerebellar ataxia, and nerve deafness. It is caused by several mutations that lead to deficiency of peroxisomal phytanoyl-CoA hydroxylase, with consequent accumulation of phytanic acid in tissues and serum of patients. The symptoms are reduced by restriction of dietary dairy products and meat from ruminants, which contain phytanic acid. Infantile Refsum's disease is a form of **Zellweger syndrome** and results from mutations in the PEX1, 2, 5, or 12 genes. *See* **alpha oxidation**.

Regan isozyme *or* **Regan isoenzyme** *other name*: carcino-placental alkaline phosphatase; an isoenzyme of alkaline phosphatase that is occasionally present in the plasma of patients with malignant tumours.

regeneration 1 (*in chemistry*) the process of restoring to a state of preparedness for use, as in the case of an ion-exchange resin washed with acid or alkali to restore it to the H^+ or OH^- form. **2** (*in biology*) the growth of new tissues or organs to replace ones lost or damaged, e.g. through injury or surgery. In some tissues, such as liver, epithelial tissue, or pancreas, regeneration occurs by simple duplication of cells, with the **regeneration time** varying from one week to a year. In other systems, such as epidermis, renewal is from specialized **stem cells**. Regenerating liver can be used as a model in which systems involved in proliferation are more highly activated than in normal liver. It is a unique property of liver that it is possible to remove two thirds of the organ yet it regenerates to form an organ of similar total weight to the original, albeit with somewhat different organization of the lobes. Many plants can regenerate a complete plant from a shoot segment or a single leaf or from tissue culture cells.

regeneration time the time taken for a tissue or organ to regain its former size or weight following damage or removal. For example, if two-thirds of the liver of a rat is removed surgically the liver regains its former weight in about two weeks. The proliferation of cells can be measured by the use of [³H]thymidine to label the cells synthesizing DNA.

regex *abbr. for* regular expression.

regio+ *comb. form* denoting, or relating to, an area or site.

regioselective pertaining to a chemical change that occurs with greater frequency at one site than at a number of other possible sites of similar type, usually involving a structural or positional isomer.

regiospecific describing a chemical reaction that occurs specifically

in only one of two possible ways, but not including stereospecificity (*see* **stereospecific**).

registration peptide one of the postulated extensions of the α chains of **tropocollagen**, when synthesis has just been completed; two α1 extensions and one α2 extension interact to form a stable assembly with the correct chains brought together into register.

regression coefficient *see* **regression line**.

regression line a line fitted to a series of two variable quantities, *x* and *y*, so as to minimize the sum of the squares of the distances parallel to the *y* axis of the observations from the line. The regression line of *y* on *x* has the equation (**regression equation**): $y = a + bx$ where $a = \sum y - b\sum x/n$ and *b*, the **regression coefficient**, is the slope of the line and is given by:

$$b = (n\sum xy - \sum x\sum y)/(n\sum x^2 - (\sum x)^2)$$

where *n* is the number of pairs of observations. In an analogous manner a regression equation, line, and coefficient may be obtained for *x* on *y*.

regucalcin a hepatic Ca^{2+}-binding protein, and also a GTP-binding protein, that modulates the hormonal regulation of plasma membrane Ca^{2+},Mg^{2+}-ATPase.

regular expression *abbr.*: regex; a shorthand representation of the conservation in an aligned sequence motif. For proteins, the single-letter code is used to denote allowed or forbidden residues (or residue groups) at each position; wildcards (x) denote positions that are not conserved. For example, a regex that detects possible sites of *N*-glycosylation is N-**P**-[S,T]-{P}, which means that Asn is completely conserved at the first position, Pro is forbidden at the second and fourth, and Ser or Thr are allowed at the third. Regexes can provide characteristic signatures for protein families; they form the basis of the PROSITE and eMOTIF databases. *See also* **rule**.

regulation (*in metabolism*) control of the rate of a metabolic system, especially in intact organs and cells, as a result of interaction from components of related systems. A distinction is sometimes made between **control** and regulation, with the latter described as the maintenance of a variable (e.g. rate or concentration) at a constant value over time, in spite of fluctuations in external conditions. It is therefore linked to homeostasis. —**regulatory** *adj.*

regulator gene *or* **regulatory gene** any gene that does not contribute structural information to the proteins whose synthesis it controls, but that determines the production of a specific cytoplasmic substance, a **repressor**, that inhibits information transfer from a **structural gene** (or genes) to protein. It may control the synthesis of several different proteins, which are determined by a coordinated group of structural genes (*see* **operon**). Although, when it was put forward, the concept of the regulatory gene was helpful, it is now realized that the genome codes for many different proteins that control the transcription of the structural genes (*see*, e.g., **attenuators**, **transcription factors**), and that can be classed as gene-activator proteins; these will possess two domains, one to recognize a specific regulatory DNA sequence, the other to contact the transcription machinery.

regulator of G-protein signalling protein *see* **RGS protein**.

regulator-site hypothesis a hypothesis to explain the control of insulin release from pancreatic islet B cells by glucose. It postulates that the glucoreceptor is an enzyme that is activated (or inhibited) by glucose and catalyses the production from a precursor of an activator, which controls the release process. *Compare* **substrate-site hypothesis**.

regulatory enzyme an enzyme that regulates the rate of a metabolic pathway; under regulating conditions, it is the enzyme of lowest activity in the pathway. Commonly, the rate is altered either by variation in the amount of regulatory enzyme (as a result of increased or decreased synthesis or degradation), or by variation in its activity by covalent modification (often by phosphorylation or dephosphorylation by kinases or phosphatases) or by allosteric modifiers. Classic examples are **phosphofructokinase** and glycogen phosphorylase (*see* **phosphorylase**).

regulatory factor for X box *see* **RFX**.

regulatory gene *see* **regulator gene**.

regulatory sequence a sequence in a DNA molecule concerned with regulating the expression of a gene; e.g. a **promoter** or **operator**.

regulatory site a specific site on an enzyme that is distinct from the substrate-binding site and that binds a feedback inhibitor or other regulatory molecule. *Compare* **allosteric site**.

regulatory subunit any of the subunits of a heteromultimeric enzyme that has a regulatory rather than a catalytic function.

regulon a group of genes, whether linked in clusters or unlinked, that respond to a common regulatory signal.

rehydrate to restore water to material that has been lyophilized (*see* **freeze-dry**) or otherwise dehydrated.

reiterated describing nucleotide sequences that occur many times within a genome. These occur especially in higher organisms. *See also* **repetitive sequences**.

REL a gene family encoding transcription factors of the κB family; v-*rel* is the oncogene of avian reticuloendothelial virus strain T (REV-T), encoding p69^{v-rel}. Rel and v-Rel form homo- or heterodimers that bind to **NF-κB** motifs. v-Rel has lost two N-terminal amino acids and 118 C-terminal residues compared with Rel.

relA a gene in *Escherichia coli*, mutations in which can lead to a 'relaxed' phenotype in which amino-acid starvation fails to switch off ribosomal RNA biosynthesis. The gene product is an enzyme active in the presence of ribosomes and uncharged tRNA, and called GTP pyrophosphokinase, EC 2.7.6.5; *systematic name*: ATP:GTP 3'-pyrophosphotransferase; *other names*: pppGpp synthetase I; **stringent factor**; guanosine 3',5'-polyphosphate synthase; it catalyses a reaction between ATP and GTP to form guanosine 3'-diphosphate 5'-triphosphate and AMP. The product is referred to as pppGpp and is converted to ppGpp (guanosine 3'-diphosphate 5'-diphosphate), which is the active nucleotide in evoking the **stringent response**.

relational database a database in which data are stored in tables. Different aspects or properties of the data are embodied in different tables, but tables may contain overlapping information.

relational database management system *abbr*.: RDBMS; software that manages a **relational database**, and is optimized for rapid and flexible retrieval of data (e.g. MySQL, postgreSQL, Oracle).

relative atomic mass *or (formerly)* **atomic weight** *abbr*.: r.a.m.; *symbol*: A_r. **1** (of an atom) the ratio of the mass, m_a, of the atom in question, to the atomic mass constant, m_u, which is one-twelfth of the mass of the nuclide ^{12}C; i.e. $A_r = m_a/m_u$. A_r is a mass ratio, a pure number with no units. **2** (of an element) the ratio of the average mass, $\bar{m}_a$, of the atoms in a specified sample of the element in question, E, to the atomic mass constant, m_u; i.e. $A_r(E) = \bar{m}_a(E)/m_u$. Since, in general, elements are mixtures of isotopes of differing nucleon numbers (mass numbers), an average mass of the atoms is required in this definition. Furthermore, the isotopic composition of many elements can vary with the origin of the sample and its treatments. Hence there are limits to the precision with which the values of the relative atomic masses of many elements can be given.

relative band rate *or* **relative band speed** *see* **R_B value**.

relative biological effectiveness *abbr*.: RBE; *an alternative term for* **quality factor**.

relative centrifugal force *symbol*: G; the force acting on a particle in a centrifugal field expressed in multiples of the acceleration due to gravity, g. $G = \omega^2 r$, where ω is the angular velocity in rad s^{-1} and r the radial distance from the axis of rotation in m. ($G = 1440 \times \pi^2 \omega^2 r$ where ω is in revolutions per minute and r in cm.)

relative density *symbol*: d; the ratio of the mass density of a substance, ρ, at a given temperature, to the mass density of a reference substance (usually water), $\rho°$ at a reference temperature (usually 4°C); i.e. $d = \rho/\rho°$.

relative formula mass the relative molecular mass as calculated from the formula of a compound.

relative molecular mass *or* **relative molar mass** *or* **molecular weight** *abbr*.: r.m.m.; *symbol*: M_r; the ratio of the mass of a molecule, m_f, to the atomic mass constant, m_u, which is one-twelfth of the mass of the nuclide ^{12}C; i.e. $M_r = m_f/m_u$. M_r is a mass ratio, a pure number with no units. The convention for indicating the size of a protein is thus 'M_r' followed by a number without units, e.g. M_r 10000; the equivalent **molecular mass**, i.e. its mass in daltons, would be 10 kDa. *See also* **dalton**, **relative atomic mass**, **unified atomic mass unit**.

relative permittivity *symbol*: ε_r; *formerly called* **dielectric constant**, the ratio of the permittivity of a substance, ε, to the permittivity of vacuum, ε_0, i.e. $\varepsilon_r = \varepsilon/\varepsilon_0$.

relaxation *(in chemistry)* the process of self-adjustment of a molecular system to a new state of minimum free energy following a disturbance.

relaxation kinetics a technique used to study rapid reactions in which a mixture at equilibrium is rapidly adjusted to a new set of conditions. The system is then observed as it approaches its new equilibrium, a process known as **relaxation**. The temperature-jump method is often used, whereby an electric discharge is used to increase the temperature by several degrees in a period of about 1 μs. *See also* **rapid-reaction kinetics**.

relaxation spectrometry the use of spectrometry to follow the kinetics of a reaction undergoing relaxation from a perturbation. *See* **relaxation kinetics**.

relaxed control a mechanism of control of plasmid replication in which replication of plasmids greatly exceeds that of the chromosome. *Compare* **stringent control**.

relaxed DNA circular DNA that is not supercoiled (*see* **supercoil**), and instead has an unwound loop. The supercoiled form is energetically favoured over the relaxed form. Since the supercoiled DNA is more compact it moves faster when centrifuged or upon electrophoresis, and so the two forms of DNA can be separated.

relaxed form *abbr*.: R form; one of two alternative quaternary structures of a protein, the other being the **tense form** (or taut form, *abbr*.: T form). These forms derive from the concept of two conformational states of a protein, which differ in their affinities for a ligand. Factors affecting the conformational equilibrium also influence the affinity of ligand-binding, and vice versa. Such factors may include the ligand itself, which could enhance or reduce the affinity of the protein, giving rise to **positive cooperativity** or **negative cooperativity** of binding respectively. The T form has a lower affinity for the substrate. For example, oxyhemoglobin is R, and deoxyhemoglobin is T; the binding of O_2 to the heme group causes the iron atom to move with respect to the heme plane, and this favours the move from T to R. *See also* **allosteric constant**, **Monod–Wyman–Changeux model (for allosteric compounds)**.

relaxed mutant a bacterial mutant that lacks the *rel* (*relaxed*) gene, which exerts a critical control over the **stringent response**. The *rel*⁻ mutant is unresponsive to amino-acid starvation, so that rates of RNA synthesis, for example, continue as before. *See also* **relA**.

relaxin a protein hormone (M_r ~5600) that is responsible in most mammals (including humans) for dilation of the symphysis pubis, softening of the cervix, and various other physiological changes prior to parturition. It is synthesized and stored in the corpus luteum. Like **insulin**, the hormone consists of two polypeptide chains linked by two disulfide bonds, with an intrachain disulfide bridge also present. Although otherwise dissimilar in primary structure, the two chains have very similar tertiary structures. As with insulin, synthesis is thought to occur via an inactive, single-chain precursor, in this case **prorelaxin**, and storage is as granules in vesicles. *See also* **insulin-related growth factors**.

relaxin-like factor a protein, homologous with **relaxin**, that is synthesized by Leydig cells in the testis and by theca cells of ovarian follicles. It binds one of the G-protein-coupled membrane receptors for relaxin. Its physiological role is unclear.

release factor *abbr*.: RF *or (formerly)* R factor; a protein that participates in the release of a nascent polypeptide chain from a ribosome. There are two classes of release factor, one codon-specific and the other non-codon-specific. In bacteria, two codon-specific factors have been identified: RF-1, specific for UAG/UAA; and RF-2, specific for UGA/UAA. In eukaryotes, one such factor has been found, eRF-1. These factors act at the ribosomal A-site, and require polypeptidyl-tRNA at the P-site. In bacteria, the actions of RF-1 and RF-2 are aided by the non-codon-specific RF-3, which binds to guanine nucleotides. In eukaryotes, a similar factor, eRF-3, has been identified. The genes encoding RF-1 (*prfA*) and RF-2 (*prfB*) have been identified. Mutations in these genes often cause misreading of termination signals, increased frameshifting, and temperature-sensitive growth of the cells. *See also* **ribosome recycling factor**.

release-inhibiting factor *or* **release-inhibiting hormone** any polypeptide that inhibits the release of a hormone or hormones. An example is **somatostatin**.

releasing hormone *or* **releasing factor** any neuropeptide synthesized in the hypothalamus that is carried to the anterior pituitary where it regulates the production and secretion of the tropic hormones, e.g. thyrotropin, somatotropin, or the gonadotropins. In the anterior pituitary releasing hormones act via the **adenylate cyclase** system and are then degraded by proteolysis.

Relish an NF-κB type of transcription factor of *Drosophila* that activates genes for antibacterial peptides. REL-110 is cleaved into an N-terminal REL-68, which enters the nucleus, and a C-terminal REL-49, which remains in the cytoplasm.

rem *symbol:* rem; a non-SI unit of (radiation) **dose equivalent**, equal to 0.01 j kg^{-1} of irradiated human (or other mammalian) tissue. It is being superseded by the **sievert**; 1 rem = 10^{-2} sievert. [Acronym for röntgen equivalent man (or mammal).]

REMI *abbr. for* restriction enzyme-mediated integration (*see* **insertional mutagenesis**).

remnant a fragment of a molecule, or other entity, remaining after partial degradation, especially **chylomicron** remnant.

remnant hyperlipidemia *or* **type III hyperlipidemia** *or* **familial betadysproteinemia** a combined hypercholesterolemia and hypertriglyceridemia due to excessive chylomicron and very-low-density lipoprotein (VLDL) remnants in plasma. It is often associated with cutaneous xanthomas and premature vascular disease. Hypothyroidism, obesity, and diabetes mellitus are frequently associated with the condition. Homozygosity for apoE-2 is a predisposing factor, and rare point mutations in apoE are the cause of an autosomal dominant form. *See* **hyperlipidemia**.

renal pertaining to, or of, the kidney.

renal rickets a condition occurring in older children and characterized by the skeletal changes of **rickets**, due to impairment of renal function allowing excessive loss of calcium and phosphate into the urine.

renaturation the return of denatured macromolecules (protein, DNA, etc.) to the conformation they maintained before denaturation. In the case of proteins, reversible denaturation may be brought about by disulfide reducing agents and urea; for nucleic acids, by heat or salts. *See also* **native state**.

renewal (of tissue) *see* **regeneration** (def. 2).

Renilla reniformis a species of soft coral, or sea pansy, that contains a **luciferase** enzyme (also found in the related species *R. mullerei*) responsible for producing a characteristic green fluorescence. The luciferase catalyses the oxidative decarboxylation of coelenterazine to produce coelenteramide and light emission with a spectral peak at 480 nm. The luciferase is used as a reporter gene in vectors for promoter assays.

renin *or* **angiotensin-forming enzyme** *or* **angiotensinogenase** EC 3.4.23.15; a glycoprotein aspartyl proteinase formed from prorenin in plasma and kidney. It is also produced by the submaxillary gland of certain strains of mice. It is highly specific for the Leu-|-Leu bond in **angiotensinogen**, which it cleaves to generate angiotensin I. In humans, renin is derived from a 406-amino-acid **preproprotein**; residues 1–23 are the signal peptide, and 24–66 are the propeptide. *Compare* **rennin**.

renin–angiotensin system a system involved in the control of blood pressure. **Renin**, secreted into the blood by the kidney in response to osmotic changes, anoxia, or kidney trauma, acts on angiotensinogen to release angiotensin I (AI) (*see* **angiotensin**). **Angiotensin-converting enzyme** forms angiotensin II (AII) from AI by release of His–Leu from the C-terminal end. AII stimulates release of the mineralocorticoid **aldosterone** from the glomerulosa cells of the adrenal cortex. Various nonapeptide analogues of AI inhibit converting enzyme and are used in the treatment of renin-dependent hypertension. *See also* **captopril**.

rennet an extract obtained from the stomachs of young mammals living on milk. It contains rennin (i.e. **chymosin**), an enzyme that clots milk and is used in cheese-making.

rennin *an alternative name for* **chymosin** (which is preferred because of possible confusion with **renin**). *See also* **rennet**.

R-enzyme *or* **starch debranching enzyme** an enzyme in plants that catalyses hydrolysis of α1-6-glucan branch points in starch, in the reaction: branched α1-4(α1-6)-glucan + H_2O = linear α1-4-glucans. The α1-6 linkage acted on carries an α1-4-glucan chain about four

residues long that cannot be cleaved by starch phosphorylase. These short chains become substrates for **D-enzyme**.

Reoviridae a family of double-stranded RNA viruses (class III in the **Baltimore classification**) with a segmented genome, infecting vertebrates, arthropods, and plants. The virion consists of a nonenveloped icosahedral particle, 60–80 nm across, and containing 10–12 segments of RNA together with **RNA-directed RNA polymerase**; the latter serves to transcribe the negative strand of each RNA molecule into mRNA. This family comprises six groups: the orthoreoviruses, orbiviruses, rotaviruses, cytoviruses, phytoreoviruses, and Fijiviruses. No pathology is associated with the orthoreoviruses infecting humans, but rotaviruses cause gastroenteritis, and orbiviruses cause the disease blue tongue in sheep.

reovirus *see* **Reoviridae**.

rep an obsolete unit of **absorbed dose** of **ionizing radiation** of any kind, equal to the dose of radiation that, upon absorption by a given mass of living tissue, causes the same energy deposition as 1 röntgen of X- or gamma radiation, i.e. about 96 ergs per gram (0.0096 J kg^{-1}) of tissue. *See also* **rad**, **rem**. [Acronym for röntgen equivalent physical.]

Rep *abbr. for* Rab escort protein; either of two proteins (Rep-1 and Rep-2) that are ubiquitously expressed, mediate binding of Rab proteins (*see* **RAB**), and are involved in membrane trafficking. Rep-1 (95 kDa) is a subunit of the heterotrimeric geranylgeranyl transferase that prenylates Rab proteins (*see* **prenylation**). Deficiency arises from many mutations and causes choroideremia. Rep-2 (658 amino acids) is produced by an intronless gene and is homologous with Rep-1.

REP *abbr. for* repetitive extragenic palindromic sequence.

RepA the protein product of the *Escherichia coli* gene *repA*, which has a role in the replication of DNA in *E. coli*. Although the principal **DNA helicase** in *E. coli* is DnaB helicase, there is also a different 68 kDa monomeric protein, which has some helicase activity during certain life-cycle stages of the replication of single-stranded phages. The contribution of RepA protein to the host cell is less clear, but the rate at which *E. coli* replication forks propagate is reduced in *rep$^-$* mutants.

repairase an enzyme concerned in the repair of an interrupted polynucleotide chain.

repair enzyme any enzyme responsible for the excision of **pyrimidine dimers**, which are formed on exposure of DNA to ultraviolet light, when adjacent pyrimidine residues on a DNA strand become covalently linked. Three enzymic activities are essential for the repair: (1) the complex of three proteins encoded by the *uvrABC* genes (*see* **uvr**) that detects the distortion produced by the pyrimidine dimer: (2) the UvrABC enzyme complex also has another activity, described as a specific exinuclease of the UvrABC complex, which cuts the damaged DNA strand eight nucleotides away from the dimer on the 5′ side and four nucleotides away on the 3′ side; (3) DNA polymerase I enters the gap to effect repair synthesis. Gaps are finally sealed by **DNA ligase**. *See also* **DNA repair enzyme**, **excision repair**, **recombinational repair**, **mismatch repair**.

repairosome the nucleotide excision repair (NER) complex of yeast that contains the majority of NER proteins.

repeated genes *see* **repetitive sequences**.

repeat-induced point (mutation) *see* **RIP**.

repeat sequence *or* **repeat 1** *(in DNA) see* **repetitive sequences**. **2** *(in proteins)* any sequence of the same amino-acid residue, or group of amino-acid residues, that is repeated in a polypeptide: e.g., the protein huntingtin, involved in **Huntington's disease**, contains a poly- glutamine tract, which can vary in size from 11–86 repeat units; and the primary structure of **collagen** is characterized by a G-x-P repeat unit. **3** the repeats of sugar residues often found in complex polysaccharides.

repeat unit a major periodicity in the structure of a molecule as deduced from X-ray diffraction patterns.

repetitive extragenic palindromic sequences *or* **palindromic units** palindromic sequences in repetitive DNA in *Escherichia coli* and *Salmonella typhimurium*. These are never found within a coding sequence. They are topoisomerase substrates and potential transcription terminators.

repetitive sequence a repeated sequence of nucleotides found in the DNA of eukaryotes. There are various types. (1) **Satellite DNA** typically consists of tandem (head-to-tail) repeats of a few hundred nucleotides. It is so called because those first discovered had an unusual ratio of bases that made it possible to separate the satellite DNA from the bulk of the cell DNA in a centrifuge. (2) **Minisatellite DNA** consists of shorter tandem repeats. The **restriction fragment length polymorphisms** linked to many genes have short sequences of bases in common. rDNA (*see* **ribosomal DNA**) probes for these sequences are useful in the preparation of a profile of the genome of an individual (fingerprinting). (3) **Transposons** (transposable elements) also contain repeated sequences, often numbering several hundred copies. Primate DNA has large numbers of two types of transposable elements, L1 transposable elements and **Alu sequences**, which comprise about 300 bases. (4) **Repeated genes** for rRNA and histones, so that cells can synthesize large amounts of these vital products. In many cases repetitive sequences seem to be maintained solely by their ability to replicate within the genome (the '**selfish DNA hypothesis**'). Rather than conferring benefits their behaviour can sometimes result in loss of fitness to the host, as in some human genetic diseases; mutations can arise due to insertion of transposable elements, to chromosomal rearrangements induced by recombination between repeated sequences, or to amplification of microsatellite sequences.

replacement vector a **cloning vector** in which part of the normal genome is replaced by foreign DNA. For example, the lambda phage EMBL4 contains only two *Bam*HI cleavage sites and so the middle segment, or stuffer fragment, of the DNA can be removed leaving 72% of the normal genome. A suitably long DNA replacement between the two arms of lambda DNA enables such a recombinant DNA molecule to be packaged into a lambda particle. It contrasts with an **insertion vector** where the foreign DNA is incorporated at a single restriction site.

replica (*in electron microscopy*) a reproduction of the surface of an object. There are various ways by which the surface of an object, e.g. membrane of an organelle, may be examined by transmission electron microscopy by means of a replica. Such a replica may be obtained by evaporating a thin film of a heavy metal such as platinum on to the dried specimen. The metal is sprayed from an oblique angle in order to deposit a coating of uneven thickness because a shadow effect is created that gives the image a three-dimensional appearance. The organic material may be dissolved away to provide a replica of the surface of the specimen. Two other methods, **freeze-etching** and **freeze-fracturing**, employ frozen sections.

replica plating a technique for the production of identical patterns of bacterial colonies on a series of agar plates. The method uses velveteen or other fabric stretched on a pad, which is then pressed onto a plate of bacterial cultures so that some bacteria from each colony are transferred to the pad, preserving the spatial pattern of the colonies (in mirror image). If this pad is then pressed lightly onto a fresh culture plate, it will transfer colonies in the same formation to the new plate. This new plate may be prepared using media that will be selective for certain mutants, fermentation reactions, or antibiotic sensitivity. A similar technique can be used in the screening of **hybridomas** for antibody production.

replicase *a loose term for* certain polymerase enzymes; e.g. certain viral **RNA-directed RNA polymerases** can be referred to as RNA replicases.

replicate 1 to make an exact copy of something, as in the **replication of DNA**. **2** (*in electron microscopy*) to make a **replica**.

replication bubble *or* **replication eye** *an alternative term for* **theta structure**.

replication complex the complex of replication proteins that are linked together as a single large unit, of total mass 10^6 daltons, that moves rapidly along the DNA during **replication of DNA**. It allows DNA to be synthesized on both sides of the replication fork in a coordinated and efficient manner. That part of the complex that contains **DNA primase** and **DNA helicase** is known as the **primosome**.

replication-error mutation a mutation that arises during DNA replication as a result of the incorrect copying of the sequence of nucleotides in the parent DNA. Probably fewer than one mistake occurs in 10^9 nucleotides added.

replication fork *or* **replicating fork** the Y-shaped region of a repli-

cating DNA molecule or chromosome, resulting from separation of the DNA strands and in which synthesis of new strands takes place. The process requires the participation of **helicases** that open the duplex.

replication of DNA the means whereby new strands of deoxypolynucleotides are synthesized. The structure of DNA, i.e. two strands of complementary base pairs, led Watson and Crick to propose that the two strands could unwind and each act as the template for a new strand, thus forming two helices, each containing one parental strand and one new one. The elegant **Meselson–Stahl experiment** demonstrated that DNA replication was indeed semiconservative. Unwinding of parental DNA requires more than 20 proteins in a complex known as a **primosome**. All DNA-dependent DNA polymerases synthesize the new strands in the direction 5′ to 3′ using as substrates deoxynucleoside triphosphates with the production of pyrophosphate. The new strands are antiparallel to the parental template strand. One strand, the **leading strand** (*or* continuous strand), is synthesized in a continuous fashion whereas the other strand, the **lagging strand** (*or* discontinuous strand), is synthesized discontinuously to give **Okazaki fragments** (1000–2000 nucleotides long in prokaryotes and 100–200 nucleotides long in eukaryotes) which are eventually joined by **DNA ligase**. The DNA polymerases cannot start a chain and must rely on a priming device; this is provided by RNA **primase**, which again synthesizes an RNA primer in the direction 5′ to 3′ using nucleoside triphosphates as substrate. The continuous strand needs to be primed only once whereas the discontinuous strand must be primed repeatedly. Three principal **DNA polymerases** for *Escherichia coli* were discovered in the laboratory of US biochemist Arthur **Kornberg (1918–)**. **DNA polymerase I**, in addition to possessing synthetic activity *in vitro*, is also a 3′–5′ **exonuclease** which gives it an editing function. It also has 5′–3′ nuclease activity. **DNA polymerase II** participates in DNA repair but is not required for replication. **DNA polymerase III** is the predominant polymerase for replication of *E. coli* DNA *in vivo*. Eukaryotic cells have at least four DNA polymerases.

replicative able to be replicated. *See also* **replicative form**.

replicative form *abbr.*: RF; the intracellular form of viral nucleic acid that is active in replication. A single-stranded viral nucleic acid, be it RNA or DNA, is first converted into a double-stranded form – the **replicative intermediate** – to enable the viral strand to be replicated preferentially. In the case of a **retrovirus** a hybrid of RNA and DNA is formed.

replicative intermediate *see* **replicative form**.

replicon 1 a structural gene that controls the initiation and replication of DNA. **2** by association, the whole DNA sequence (such as the bacterial chromosome) that is replicated under the influence of a single replicon system.

replisome a multiprotein complex that includes two DNA polymerase molecules for synthesis of the leading and lagging strands at the site of origin of DNA replication. *See* **replication of DNA**.

replitase a particulate fraction that can be isolated from the nuclei of **S phase** Chinese hamster embryo fibroblasts and containing some of the enzyme activities required for DNA synthesis.

reporter gene a gene used to disclose the function of potential regulatory sequences. The sequences are placed in plasmids, upstream of the reporter gene; these chimeric plasmids are then introduced into cells and the expression of the reporter gene is measured as an index of the function of the regulatory sequences under the conditions being investigated. The *lacZ* (*β*-**galactosidase**) gene, the *CAT* gene (*see* **chloramphenicol *O*-acetyltransferase**), and the gene encoding firefly **luciferase** have been used in this way.

reporter group a chemical group that acts as a **probe**.

repress to check or restrain; the effect may be partial or total. *See* **repression**, **repressor**.

repressible describing a metabolic reaction that is capable of undergoing **repression**.

repression (*in genetics*) the act of inhibiting gene expression. An example is **catabolite repression** of the *lac* operon by glucose, mediated through binding of cyclic AMP to **catabolite (gene) activator protein** (CAP). By extension, the gene product itself (enzyme or other protein) and thus any associated metabolic reaction is said to be repressed. *See also* **repressor**.

repressor *(in genetics)* a molecule that specifically prevents the transcription of regulatory genes under control of an **operator**. When the repressor binds the operator, transcription cannot occur. An inducer induces transcription by binding to the repressor to form a complex that has a low affinity for the operator. The classic example is the lactose repressor protein, which controls the expression of the **lac operon** in *Escherichia coli*. The protein binds to a specific sequence of 21 nucleotide pairs (called the operator) that overlaps an adjacent RNA polymerase binding site (the promoter) at which RNA synthesis begins. In the presence of lactose, allolactose is formed, which binds to the repressor. This induces a conformational change that causes the repressor to loosen its hold on the DNA so that the gene is derepressed. In other examples, the binding of a signalling molecule may increase the affinity with which the repressor binds to the DNA; e.g. the *trp* operon, encoding enzymes that produce tryptophan in *E. coli* (*see* **trp gene**). *See also* **repression**.

reproducibility the ability to carry out a series of identical experiments to yield results that fall within a small range of error derived by statistical analysis (*see* **standard deviation**, **standard error of the mean**). The limits of error that are acceptable depend on the limitations of experimental technique and the inherent variability of the system under investigation.

reproduction *(in biology)* the production by an organism of new individuals that are more or less similar to itself. Reproductive strategies can be divided broadly into two categories: sexual reproduction, which requires the participation of male and female **gametes** and gives rise to genetic variation among the offspring; and asexual reproduction, which does not involve gametes and leads to the formation of offspring that are genetically identical to the parent. The latter takes various forms; examples include **budding**, **fragmentation**, **parthenogenesis**, **spore** formation, and **vegetative reproduction**.

reptation a term invoked to rationalize the movement and thermodynamics of linear polymers, dividing the chain conceptually into sections (**reptons**), which form the basis for statistical mechanical calculations of the relaxation time of the polymer and viscosity of polymer solutions. Large DNA molecules move through the pores of agarose gels by reptation.

repton *see* **reptation**.

RER *abbr. for* rough endoplasmic reticulum.

reseal to join again the ends of ruptured components, as in, e.g., the resealing of plasma membranes after whole cells have been rendered permeable by various techniques, or of fractionated membrane fragments, which may reseal to form closed vesicles. The term can also refer to ligation of the cut ends of DNA after insertion of foreign DNA into a vector.

reserpine (3β,16β,17α,18β,20α)-11,17-dimethoxy-18[(3,4,5-trimethoxybenzoyl)oxy]yohimban-16-carboxylic acid methyl ester; an alkaloid isolated from the roots of certain species of *Rauwolfia* and noted for its antihypertensive and CNS depressant actions. These arise from its inhibition of the intraneuronal Na^+, Mg^{2+}-ATPase pump, which maintains the neuronal vesicular stores of biogenic monoamines (**dopamine**, **norepinephrine**, and **serotonin**).

reserve polysaccharide any polysaccharide that is stored to provide an energy source at times of need. Examples are starch granules in plants, and glycogen granules in animals.

Re/Si convention a nomenclatural system for specifying the **stereoheterotopic** faces of a trigonal atom using the **sequence rule** in two dimensions. This is used to establish a priority sequence for the three ligands, e.g. of C*abc*. The plane containing C and the first atoms of the ligands is then inspected from an arbitrarily chosen side. Starting with the highest priority ligand, it is determined whether the priority sequence is clockwise, hence *re* (from Latin *rectus*, right), or counterclockwise, hence *si* (from Latin *sinister*, left); e.g. for acetaldehyde, CH_3CHO, if the face inspected shows a clockwise ordering, $O \rightarrow CH_3 \rightarrow H$, that face is designated *Re*. There are a number of extensions of this general principle, and the specialist literature should be consulted for details.

residence time the averaged time that a site is occupied by any designated molecule.

resident protein *see* **luminal protein**.

residual body a membrane-lined cytoplasmic inclusion characterized by undigested residues (e.g. membrane fragments or whorls, ferritin-like particles). The term embraces **telolysosome** and (hypothetical) **post-lysosome**.

residue 1 any of the incorporated amino-acid moieties in a peptide or protein. 2 material that remains after any procedure to remove something.

residue mass the relative molecular mass of a residue; i.e. the sum of the **relative atomic masses** of all the atoms composing the residue.

residue weight the molecular weight of a residue; i.e. the sum of the **atomic weights** of all the atoms composing the residue. **Residue mass** is the preferred term.

resin 1 *or* **rosin** any of a group of naturally occurring amorphous solids or semisolids, typically light yellow to dark brown, brittle, and insoluble in water. They are found as plant exudates, often on trees such as pines and firs. 2 the polymer base of certain ion-exchange materials used in chromatography. *See* **ion-exchange polymer**.

resistance 1 the ability of a living organism, particularly a bacterium, to resist the effect of a disadvantageous environment or substance, especially an antibiotic. A bacterium may possess one or more resistance plasmids. 2 *see* **electric resistance**.

resistance donor a strain of an organism, usually a bacterium, that is capable of transferring resistance, usually by conjugation, to a suitable recipient.

resistance factor *abbr.*: R factor; *an alternative name for* **resistance plasmid**.

resistance plasmid *abbr.*: R plasmid; a **plasmid** that carries genetic information for resistance to antibiotics or other antibacterial drugs. *Other name*: resistance factor (*abbr.*: R factor).

resistin *or* **FIZZ 3** a protein, related to FIZZ 1, that is secreted by white adipose tissue into the bloodstream. The mouse protein contains two identical subunits (each of 114 amino acids) joined by a disulfide bond. There are three murine and two human homologues, which show ≈53% sequence identity. In obese mice, serum levels of FIZZ 3 are high and are decreased by thiazolidinedione drugs, which increase sensitivity to insulin in various tissues. Antibodies to resistin lower blood glucose levels and improve insulin sensitivity in these mice. *See* **FIZZ**.

resolution 1 the degree of separation between two compounds being subjected to a transport method such as chromatography or sedimentation equilibrium (i.e. being resolved); complete resolution yields two compounds each uncontaminated by the other. 2 *(in physics and biophysics)* the extent to which closely juxtaposed objects can be distinguished as separate objects using an optical instrument, e.g. a microscope. The degree of resolution is dependent on the **resolving power** of the system. The fineness of detail with which objects can be seen is limited by the wavelength of light used. Only those objects similar in dimensions to the wavelength of the light (or larger) can be resolved. The wavelength of X-rays is from about 10^{-8} m to 10^{-11} m and they can thus be used to resolve structures at the atomic level. Thus a structure can be said to have been determined at 3 Å resolution, 7 Å resolution, etc.

resolvase an enzyme that, together with **transposase**, is involved in the replicative transposition of genetic elements in *Escherichia coli*. The 3′ ends of the target chromosome, after the first cutting and splicing, serve as replicative primers for copying both the gaps and the two strands of the transposable element (which contains transposase) itself. Ligase action generates a cointegrate, a large circular structure containing both donor and target chromosomes with two freshly replicated copies of the transposable element. Resolvase now catalyses the site-specific recombination between the two elements, resulting in one copy of the transposable element inserted in each of the two chromosomes.

resolving power 1 *(in chromatography)* the power to resolve individual compounds (*see* **resolution**(def. 1)). It depends on (1) the efficiency of the system employed; for columns this is given by the **theoretical plate** number; and (2) the relative affinities of different compounds for the stationary and mobile phases. **2** *(in physics)* **a** the ability of an optical instrument, e.g. a microscope or telescope, to produce separate images of objects placed close together. **b** the ability of a spectrometer to separate adjacent peaks in a spectrum.

resonance 1 *(in physics)* the induction or amplification of one vibration by another vibration of a periodicity equal to the natural oscillation frequency of the first, which is then said to resound. In an electronic circuit, resonance occurs when an applied emf equals the natural frequency of oscillation of the circuit (used, e.g., in tuning a radio receiver). **2** *(in chemistry)* a phenomenon shown by any molecular entity to which more than one **contributing structure** can be assigned, when these structures are of similar energy and differ only in the distribution of the valence electrons; the entity then oscillates between the two structures and adopts, in effect, an intermediate state. Resonance always results in a different distribution of electrons than would be the case if there were no resonance. *See also* **resonance energy, resonance hybrid. 3** *see* **electron spin resonance spectroscopy, nuclear magnetic resonance.** —**resonant** *adj.*; **resonate** *vb.*

resonance energy the difference in energy of an actual molecule and the **contributing structure** of lowest energy.

resonance hybrid the structure that results when a chemical entity exhibits **resonance** (def. 2), such as occurs by the sharing of one or more electrons over several bonds. It has a lower energy than any of the possible formal nonresonance **contributing structures**, giving additional stability of which the so-called **resonance energy** is a measure. A classic example is carbon dioxide, which may be represented by the **Lewis formula** for each as $\ddot{O}::OC::\ddot{O}$, $:\ddot{O}:C:::O^+$, and $+:O:::C:$ $\ddot{O}:-$. *Compare* **mesomerism.**

resonance Raman spectroscopy a spectroscopic technique for characterization and assignment of vibrations directly connected with a **chromophore**, as well for the assignment of the chromophore. The excitation frequency is applied close to the absorption maximum of the chromophore.

resonance stabilization *or* **resonance stabilisation** stabilization of molecular structure by resonance; *see also* **resonance hybrid.**

resonate *see* **resonance.**

resorb to absorb again. —**resorbent** *adj.*; **resorption** *n.*; **resorptive** *adj.*

resorcinol *m*-dihydroxybenzene; a compound used in the **Seliwanoff test** for ketoses.

resorption *see* **resorb.**

RESOURCERER a database for annotating commonly available microarray resources.

respiration 1 the process of gaseous exchange between an organism and its environment. In plants, microorganisms, and many small animals, air or water makes direct contact with the organism's cells or tissue fluids, and the processes of diffusion supply the organism with dioxygen and remove carbon dioxide. In larger animals the efficiency of gaseous exchange is improved by specialized respiratory organs, such as lungs and gills, which are ventilated by breathing mechanisms. These organs possess respiratory surfaces, across which gases are exchanged with the blood or other transport medium. This then carries dioxygen to the body tissues, and removes carbon dioxide. **2** *or* **oxidative metabolism** the various energy-yielding reactions of cells or organisms that require oxygen as the final electron acceptor. *See* **respiratory chain.** —**respiratory** *adj.*

respiratory burst a phase of elevated oxygen consumption that occurs in neutrophils, monocytes, and macrophages shortly after phagocytosing material. This is associated with increased activity of the hexose monophosphate shunt and production of **NADPH** (*see* **pentose phosphate pathway**). An NADPH oxidase complex passes electrons from the NADPH to O_2 creating the reactive superoxide (O_2^-) ion and H_2O_2 in the vacuole. Patients with chronic granulomatous disease never produce a respiratory burst. The patients suffer an accumulation of mononuclear cells at sites of chronic inflammation, forming granulomas. *See* **phagocyte NADPH oxidase.**

respiratory chain *or* **electron-transport chain** the sequence of enzymes and other proteins within the mitochondrion and prokaryotic cell membranes by which substrates, including NADH and succinate, are oxidized by dioxygen. It is fundamental to tissue **respiration** (def. 2) which has the function of maintaining tissue metabolism through **oxidative phosphorylation**. The chain can be separated into a number of particles (*see* **respiratory complex**) each of which contains individual elements of the chain. The mechanism involves, in the case of NADH, removal by NADH dehydrogenase of a proton plus two electrons. In the case of succinate, there is a transfer of two hydride ions (i.e. two protons plus two electrons) by succinate dehydrogenase to ubiquinone; other flavoproteins can similarly oxidize fatty acyl-CoA or β-hydroxybutyrate or α-glycerophosphate. The protons are initially released to the intermembrane space, and the electrons are transferred by the appropriate flavoprotein to ubiquinone, from which they are transferred singly in sequence to cytochrome b, cytochrome c_1, cytochrome c, and cytochrome a/a_3 at which point they reduce molecular oxygen to promote formation of water with the protons that were formed earlier. Many other components are also involved. In tightly coupled mitochondria, the protons released to the intermembrane space reach the mitochondrial matrix (the space in which they react with molecular oxygen) by transfer through the inner mitochondrial membrane by the F_0F_1 complex of the mitochondrial ATPase (*see* **H$^+$-transporting ATP synthase**), with concomitant phosphorylation of ADP to ATP. *See also* **chemiosmotic coupling hypothesis.**

respiratory complex any of several groups of physically associated components of the **respiratory chain**. The concept arose during experiments to purify individual components, when it was found that membrane particles could be isolated each with a consistently characteristic composition. Four main complexes have been identified: complex I, which contains primarily NADH dehydrogenase; complex II, which contains succinate dehydrogenase; complex III, which contains cytochromes b and c_1; and complex IV, which contains cytochrome a/a_3 together with copper-containing proteins. Complexes I, II, and III all contain additionally a number of proteins rich in iron and sulfur. The F_o and F_1 subunits of the ATPase associated with the respiratory chain have been termed complex V (*see* **H$^+$-transporting ATP synthase**). *See also* **complex I, II, III, IV.**

respiratory control the regulation, in coupled mitochondria, of the rate of oxidative phosphorylation by the level of ADP. As the ADP level increases, and the ATP level falls, electron flow through the **respiratory chain** and coupled ATP synthesis are enhanced. The **respiratory control ratio**, or **respiratory control index**, represents the ratio of the rate of oxygen consumption of mitochondria in the presence of both substrate and ADP to the rate in the presence of substrate but the absence of ADP; i.e. the ratio of the rate in respiratory state 3 to respiratory state 4 (*see* **respiratory state**).

respiratory pigment any of a number of pigmented proteins, associated with respiratory processes, that derive their colour from a prosthetic group, usually heme in red proteins such as cytochrome c, or flavin nucleotides in flavoprotein dehydrogenases, which are yellow.

respiratory quotient *or* **respiratory ratio** *abbr.*: RQ; a ratio of the rate of carbon dioxide evolution by any cell, tissue, or organism to its rate of dioxygen uptake.

respiratory state *or* **state of the respiratory chain** any one of the possible steady states of the phosphorylating **respiratory chain** in mitochondria. In state 1 both ADP and respiratory substrates are lacking. State 2 is a 'standard state' in which ADP has been added to exhaust the endogenous respiratory substrate; in state 2′ substrate has been added to exhaust the endogenous ADP. In state 3, the 'active' state of rapid respiration, all required components are present and the respiratory chain itself is the rate-limiting factor. State 4, an anaerobic 'resting' state, is characterized by a low respiration rate. In state 5, only oxygen is lacking, the respiration rate then being at zero.

State	[O_2]	[ADP]	Substrate level	Respiration rate	Rate-limiting factor
1	>0	low	low	slow	ADP
2	>0	high	0	slow	substrate
2′	>0	low	high	slow	ADP
3	>0	high	high	fast	respiratory chain
4	>0	low	high	slow	ADP
5	0	high	high	0	O_2

respirometer an instrument for measuring the uptake of oxygen during incubation experiments. The **Barcroft apparatus** and the **Warburg apparatus** were examples of this type of instrument.

response the events brought about by a **stimulus**, especially in a cell, tissue, or organism. See **stimulus–response coupling**, **sensitization**, **desensitization**.

response coefficient symbol: R; an overall measure of the control exercised by a specific enzyme inhibitor, at its concentration, I, on the overall flux, F, of a multienzyme system. Hence,

$$R_I^F = (\partial F/F)/(\partial I/I)$$

or

$$R = \text{fractional change in flux/fractional change in inhibitor.}$$

A response coefficient can also be determined for the effect of I on the concentration of a metabolite S. Thus

$$R_I^S = (\partial S/S)/(\partial I/I)$$

It is the product of the **elasticity coefficient** and the **flux control coefficient**.

response element abbr.: RE; a DNA sequence (sometimes called a 'module') that is common to promoters/enhancers of genes whose expression is coordinately regulated (see **coordinate induction/repression**). Examples: the **estrogen response element** (ERE) has the consensus sequence GGTCANNNTG[A/T]CC; the glucocorticoid response element (GRE) has the consensus sequence TGGTACAAATGTTCT; the **heat-shock response element** (HSE) has the consensus sequence GNNGAANNTCCNNG; the **metal response element** (MRE) has the consensus sequence CGNCCCG-GNCNC; the **serum response element** (SRE) has the consensus sequence CATATTAGG; and the **thyroid response element** (TRE), which is common to several enhancers, has the consensus sequence TGACTCA. RE-1 (or NRSE) is a conserved 21–23 bp response element present in neuron-specific genes. See **REST**.

response regulator part of a **two-component signal transduction system**. Response regulators of bacterial sensory transduction systems generally comprise a receiver module domain covalently linked to an effector domains. In the presence of an effector molecule, a **histidine sensor kinase** phosphorylates an aspartate residue in the receiver module. Effector domains include DNA binding and/or catalytic units that are regulated by phosphorylation of the receiver module.

response time the interval between initiation of a process and the commencement of the action that it induces.

responsiveness (in physiology) the ability of a cell, tissue, or organ to undergo a response. Maximal response normally corresponds to the healthy in vivo state, so that the degree of response is an indication of the integrity of the preparation. However, negative regulators may be removed during tissue preparation, thus giving enhanced activity in vitro.

REST abbr. for response element RE-1 silencing transcription factor; other name: neuronal restricted silencing factor (abbr.: NRSF). A transcription factor (116 kDa) that contains a central DNA-binding domain comprising eight zinc fingers and flanked by two repressor domains. It binds a 21–23 bp conserved response element called response element-1 (RE-1), or neuronal restricted silencing element (NRSE). This element is present in neuron-specific genes, and binding of REST represses the expression of such genes. The N-terminal domain of REST associates with a histone deacetylase

and other proteins (depending on the genetic context). The C-terminal domain associates with a corepressor (called CoREST), which serves as a **molecular beacon** for the assembly of other components of the repressor machinery.

restin Reed–Sternberg cell intermediate filament-associated protein; other name: cytoplasmic linker protein-170 α-2 (abbr.: CLIP-170); a protein that is highly expressed in such cells in Hodgkin's lymphoma, and having a large coiled-coil α-helical domain similar to those in myosins.

resting potential the potential difference across the membrane of a cell in the unstimulated state. See also **membrane potential**.

resting state an inactive or unstimulated state, especially of a cell. See also **cell-division cycle**.

restricted DNA DNA that has been cut by a **restriction endonuclease** within a bacterium; bacteria are protected from the propagation within them of foreign DNA, usually viral DNA, by the possession of restriction endonucleases which degrade the invading DNA. Each nuclease recognizes a specific sequence of 4–8 nucleotides in the DNA. The cleaved sequences, where they occur in the genome of the bacterium itself, are protected from cleavage by methylation at the amino group of an adenine, or either the 5 position of a cytosine or the amino group of a cytosine. The newly replicated strand of the bacterial host DNA, which is protected from degradation by the methylated parent strand with which it forms a duplex, is modified before the next cycle of replication (see **restriction–modification system**). Such sequences in the viral DNA are usually not so methylated and hence are degraded first by the restriction enzymes and then by bacterial exonucleases. Invading DNAs are only rarely modified and enabled to reproduce in their new host. Its progeny are, however, no longer modified in the way that permits them to propagate. See also **host-controlled modification**.

restricted phage or restricted bacteriophage see restricted virus.

restricted virus a virus whose host range is limited by **host-controlled modification** so that when it infects a related but nonidentical host cell its DNA is degraded by the new host cell's **restriction endonucleases**.

restriction (in molecular biology) the action or use of **restriction endonucleases**. See also **restrictive**.

restriction analysis another term for **restriction mapping**.

restriction endonuclease or restriction enzyme or restriction nuclease any of a group of enzymes, produced by bacteria, that cleave molecules of DNA internally at specific base sequences. They therefore act to 'restrict' the replication of foreign, usually viral, DNA entering the bacterial cell (see **restricted DNA**). Some of them also methylate the host bacterial DNA to protect it from degradation. There are three known types of restriction enzymes. Types I and III each carry activity for both the endonuclease and methylase function whereas type II enzymes do not carry the methylase activity. Type I enzymes cleave the DNA at a random site located at least 1000 bp from the recognition sequence, type III do so 24–26 bp distant from the recognition site. In contrast, type II enzymes cleave DNA at specific sites within the recognition sequence. This makes type II enzymes indispensable for DNA manipulation. Around 2500 species of type II restriction enzymes with some 200 differing sequence specificities have been characterized. A restriction endonuclease is named by a three-letter abbreviation of the species from which it was originally isolated, followed by letters and/or numerals indicating the strain or characteristic, e.g. EcoRI is derived from a strain of Escherichia coli with an R (drug resistance) marker and is enzyme I thus described.

Type I (EC 3.1.21.3) restriction and modification enzymes are complex multifunctional systems that require ATP, S-adenosylmethionine, and Mg^{2+} as cofactors; in addition to their endonucleolytic and methylase activities, they are potent DNA-dependent ATPases. The subunit composition, comprising three subunits designated R, M, and S, is similar to that of ATPases. They catalyse the endonucleolytic cleavage of DNA to give random double-stranded fragments with terminal 5′-phosphates; ATP is simultaneously hydrolysed.

Type II enzymes have activity as site-specific deoxyribonucleases (EC 3.1.21.4), which catalyse the endonucleolytic cleavage of DNA to give specific double-stranded fragments with terminal 5′-phosphates; Mg^{2+} is a cofactor. Such enzymes are the basis of **restriction mapping**.

Type III enzymes have activity as site-specific deoxyribonucleases (EC 3.1.21.5). The protein has two subunits, products of the *res* (from 'restriction') and *mod* (from 'modification') genes; these are, respectively, responsible for binding the system-specific DNA recognition site and the methylation of one of the adenosyl residues in the recognition site. The more widely used restriction enzymes appear in the body of the dictionary and their activity is indicated by the following convention. As an example *Aat*II has the recognition sequence GACGT↑C meaning that the sequence on one strand is 5′-GACGTC-3′and that the cleavage site is on the 3′ side of the T preceding the ↑ symbol. This is thus an example of an enzyme that will result in restriction fragments with a 3′ 'sticky end'. Some enzymes have an ambiguous recognition sequence, e.g. *Acc*I, and for these this dictionary uses a convention of square brackets. Thus the sequence in this example is GT↑[AC][GT]AC meaning that the third nucleotide is either A or C and the fourth is either G or T. A different type of ambiguity is illustrated by Bgl.

restriction enzyme *an alternative name for* **restriction endonuclease**.

restriction fragment a fragment of DNA produced by the action of a **restriction endonuclease**. It usually refers to the product of a type II enzyme that recognizes specific base sequences in double-helical DNA and cleaves both strands of the duplex at specific sites.

restriction fragment length polymorphism *abbr.*: RFLP; **polymorphism** within a population of organisms of the size of the fragments of DNA produced by the action of a type II **restriction endonuclease**. It results from a corresponding polymorphism of **restriction sites** in the genome, due to slight differences in base sequence between individual members of the population. RFLP is exploited in constructing physical maps of the genome, and widely used for localizing specific genes and detecting genetic differences between closely related individuals. Such polymorphisms can be detected by Southern blotting using a DNA probe for a particular locus to reveal differences in the sizes of fragments produced by a particular endonuclease. The analysis of restriction fragments can reveal the presence of a mutation that may itself cause disease or be closely linked to one that does; for example, genetic diseases such as sickle-cell anemia, cystic fibrosis, and Huntington disease can be detected by RFLP analyses. *See* **restriction mapping**.

restriction map a pattern obtained as the result of **restriction mapping**.

restriction mapping the use of type II **restriction endonucleases** to cleave genomic or other DNA to produce a characteristic pattern of fragments. The fragments are resolved by gel electrophoresis and visualized. Every DNA molecule produces a unique pattern, which accords with the specificity of the particular restriction enzyme. By comparing the patterns obtained from two or more different DNAs, differences between closely similar DNA molecules, such as arise from mutations, may be detected. The use of DNA probes, such as cDNA, to identify particular fragments may often be of value. By using restriction enzymes singly and in pairs it is possible to deduce from the sizes of resulting fragments the order of the restriction sites on the DNA and the distances between them.

restriction–modification system any system whereby host DNA is methylated to render it resistant to the action of the cell's **restriction endonucleases** (*see* **restricted DNA**); methylation involves the adenine or cytosine residues at the restriction sites. Several enzymes are implicated: (1) EC 2.1.1.72; *recommended name*: site-specific DNA-methyltransferase (adenine-specific); *other names*: N-6 adenine-specific DNA methylase; modification methylase; it catalyses a reaction between *S*-adenosyl-L-methionine and DNA adenine to form *S*-adenosyl-L-homocysteine and DNA 6-methylaminopurine. (2) EC 2.1.1.73; *recommended name*: site-specific DNA-methyltransferase (cytosine-specific); *other names*: C-5 cytosine-specific DNA methylase; modification methylase; it catalyses a reaction between *S*-adenosyl-L-methionine and DNA cytosine to form S-adenosyl-L-homocysteine and DNA 5-methylcytosine.

restriction negative describing an organism that does not contain any **restriction endonuclease**.

restriction site a sequence of bases in a DNA molecule that is recognized by a **restriction endonuclease**.

restrictive 1 describing a strain or line of cells that does not permit the replication of a particular viral DNA, and is thus resistant to infection. **2** describing a set of conditions, such as a temperature range, in which a mutant phenotype is expressed, thus limiting the growth of the organism or causing its death. *Compare* **permissive**. *See also* **conditional lethal mutation**, **temperature-sensitive mutation**.

restrictive conditions *see* **conditional lethal mutation**.

restrictive temperature *see* **temperature-sensitive mutation**.

restrictocin an antitumour polypeptide (16 kDa) from *Aspergillus* spp. It cleaves the large RNA of the eukaryotic 60S ribosomal subunit and inhibits protein synthesis in picornavirus-infected cells.

resveratrol a stilbene **phytoalexin** produced by the grapevine and present in the skin of red grapes, red wine, and other plant sources. It is a potent antioxidant and claims have been made for its antitumour activity.

RET *abbr. for* rearranged during transfection; a human transforming oncogene derived through fibroblast transfection with T-cell lymphoma DNA. It encodes a receptor-tyrosine kinase present in the central and peripheral nervous systems. Several isoforms are produced by alternative splicing. RET is recruited by GPI-anchored membrane proteins specific for the structurally related secreted proteins glial cell-derived neurotrophic factor, neurturin, and persephin. It signals through the adaptor protein GRB-19. Activating mutations in RET are associated with multiple endocrine dysplasia, and inactivating mutations with Hirschsprung disease.

retentate *or* **dialysis residue** *or* **residue** *(in* **dialysis***)* the moiety of the dialysed material that has not traversed the membrane; the liquid containing this moiety (which may or may not still be entirely in solution). Some authorities prefer the term **nondiffusible material**.

retention coefficient *(in column chromatography)* the ratio of the **void volume** to the **elution volume** for a specific solute. It is a measure of the partition of the solute between the column material and the solvent, and is independent of column size.

retention time the time taken by a compound to traverse the length of a **gas–liquid chromatography** column. The retention time is a characteristic of each compound under defined conditions. It can be used to aid in identification within a series of homologues, e.g. fatty acids, which are characterized by their **carbon number** derived from the retention time.

retention volume a parameter of liquid chromatography, related to the volume at which each compound elutes under defined conditions.

retGC-1 *abbr. for* retinal guanylate cyclase 1. *See* **guanylate cyclase**.

RETGC protein *abbr. for* retinal guanylyl cyclase protein; *other name*: photoreceptor-specific membrane protein; either of two type I integral membrane proteins (RETGC1 and RETGC 2) of the retina that contain an N-terminal ligand-binding region, a transmembrane helix, an intracellular protein kinase homology domain, and a guanylyl cyclase domain. They form heterodimers, do not bind natriuretic peptides, but are activated by retinal guanylyl cyclase activator proteins (GCAP1 and GCAP2) that belong to the calmodulin family. Mutations in the RETGC1 gene cause Leber's congenital amaurosis.

reticulate 1 *or* **reticular** resembling a net; in the form of a network. **2** to create in the form of a net or network. —**reticulation** *n.*

(S)-reticuline a benzoisoquinoline that is the metabolic precursor of several tetrahydrobenzylisoquinoline alkaloids including (from *Papaver somniferum*): codeine, noscapine, papaverine, and morphine. It is derived from two molecules of tyrosine and three molecules of *S*-adenosylmethionine, which contribute their methyl groups.

reticulocalbin a Ca^{2+}-binding luminal protein of the endoplasmic reticulum. It has six **EF-hands**, a 20-residue N-terminal **signal sequence**, and the C-terminal sequence **HDEL**.

reticulocyte an immature **erythrocyte** that can synthesize protein, ≈90% being hemoglobin. It retains intracellular organelles, including mitochondria, and histologically stains for the presence of ribosomes. The name derives from the reticulate appearance of the stained cells.

reticulocyte lysate the preparation obtained by suspending **reticulocytes** in distilled water so that the cells are lysed by osmotic shock.

The reticulocytes are usually obtained from rabbits in which **reticulocytosis** has been induced by the administration of phenylhydrazine. The residual RNA is usually removed by the addition of a ribonuclease that is calcium-dependent and is thereby inactivated by the addition of EDTA. Such a lysate is often used for the *in vitro* translation of added mRNA.

reticulocytosis enhanced levels of reticulocytes in the blood, often due to experimentally-induced increased production. *See also* **reticulocyte lysate.**

reticuloendothelial system *a former name for* **mononuclear phagocyte system.**

reticulum 1 the endoplasmic reticulum, a netlike structure of membranes found in the cytoplasm. **2** the most anterior chamber of the ruminant stomach.

retina the light-sensitive layer lining the inside of the posterior wall of the eye in vertebrates and some classes of molluscs. The inner nervous layer of the vertebrate retina is transparent and contains photoreceptive **cone cells** and/or **rod cells** together with their associated nerve cells and fibres, glia, and blood vessels; the outer layer, next to the choroid, is darkly pigmented and prevents back-reflection of light.

retinal 1 pertaining to the retina. **2** *the recommended trivial name for* (2*E*,4*E*,6*E*,8*E*)-3,7-dimethyl-9-(2,6,6-trimethylcyclohex-1-en-1-yl)nona-2,4,6,8-tetraen-1-al; *other names*: retinal$_1$; retinaldehyde; vitamin A aldehyde; retinene. A compound that plays an important role in the visual process in most vertebrates (*compare* **3,4-didehydroretinal**). In the retina this (all-*E*)-isomer, all-*trans*-retinal, is converted by **retinal isomerase** to its (2*E*,4*Z*,6*E*,8*E*)-stereoisomer, 11-*cis*-retinal, which then combines (as a Schiff's base) with an opsin to form (in cone cells) **iodopsin, porphyropsin,** or **cyanopsin,** or (in rod cells) **rhodopsin.** Interaction with a photon brings about dissociation of the retinal from the opsin to produce all-*trans*-retinal, thereby completing the visual cycle. Retinal is formed in the retina by enzymic oxidation of **retinol**; in addition, most vertebrates can convert dietary *α*-, *β*-, and *γ*-carotenes to retinal by oxidative cleavage of the polyprenyl chain at its midpoint. [*Note*: (1) The numbering system of the systematic name is different from that of the trivial name, the latter following the rules used for carotenoids. (2) The fuller name **retinaldehyde** is recommended if confusion with def. 1 may occur.]

all-*trans*-retinal

retinal$_1$ *(sometimes) an alternative name for* **retinal** (def. 2) (especially to distinguish it from **retinal$_2$**; not recommended).

retinal$_2$ *(sometimes) an alternative name for* **3,4-didehydroretinal** (especially to distinguish it from **retinal$_1$**; not recommended).

retinal cone *see* **cone cell.**

retinaldehyde *see* **retinal.**

retinal isomerase EC 5.2.1.3; *systematic name*: all-*trans*-retinal 11-*cis*-*trans*-isomerase; *other name*: retinene isomerase; an enzyme that catalyses the conversion of all-*trans*-retinal to 11-*cis*-retinal. *See also* **retinal** (def. 2).

retinal rod *see* **rod cell.**

retinene or **retinene$_1$** *an alternative name for* **retinal** (def. 2) (not recommended).

retinene$_2$ *an alternative name for* **3,4-didehydroretinal** (not recommended).

retinene isomerase *an alternative name for* **retinal isomerase.**

retinitis pigmentosa a relatively common genetic disorder which in humans affects 1 in 4000 people and is caused by the premature death of photoreceptors. The result is progressive night blindness, contraction of the visual field and finally complete loss of sight.

One form of autosomal dominant disease results from an abnormality on chromosome 3. In this, there is a single amino-acid substitution in the **rhodopsin** molecule caused by a base transversion of C to A, which results in the change of a proline to a histidine residue at position 23 in the **opsin** molecule, causing the characteristic retinal degeneration. Another severe form of the disease, affecting up to around 20% of some populations, is X-linked; it usually becomes manifest within the first two decades of life and progresses to blindness within 10–20 years. The gene, named *RPGR*, for this form has now been found on the short arm of the X-chromosome. The predicted 90 kDa protein product of the *RPGR* gene bears significant homology to the **RCC1 family.**

retinoate 1 the anion of **retinoic acid. 2** any salt or ester of retinoic acid.

retinoblastoma a childhood cancer of the developing retina. The offspring of surviving victims have a high incidence of this disease as well as several other types of malignancies. Retinoblastoma is associated with the inheritance of a copy of chromosome-13 from which a particular segment has been deleted. It develops through a mutation in a retinoblast (a retinal precursor cell). The affected chromosomal segment contains the *RB* gene which specifies a factor that restrains uninhibited cell proliferation. Thus the *RB* gene is a **tumour suppressor gene.** *See also* **retinoblastoma protein.**

retinoblastoma protein *abbr.*: pRb; the product of the retinoblastoma gene, **RB.** pRb is a 105 kDa DNA-binding protein that is localized in the nucleus of normal retinal cells, but absent in **retinoblastoma** cells. The protein is phosphorylated in the cell cycle by protein kinase **CDC2.** It appears that the interaction of pRb with various **transcription factors** suppresses cellular proliferation.

retinoic acid *the recommended trivial name for* (2*E*,4*E*,6*E*,8*E*)-3,7-dimethyl-9-(2,6,6-trimethylcyclohex-1-en-1-yl)nona-2,4,6,8-tetraenoic acid; *other names*: **tretinoin**; vitamin A acid; it is formed from **retinal** by enzymic oxidation in kidney and other tissues, and is active in preventing some of the symptoms of vitamin A deficiency. It is a morphogen and regulator of differentiation during embryogenesis. *See* **retinoic acid receptor, vitamin A.**

all-*trans* form

retinoic acid orphan receptor *an alternative name for* **retinoid receptor.**

retinoic acid receptor a nuclear receptor for **retinoic acid** that mediates its effects on cells. These receptors bind retinoic acid and directly regulate gene expression. The proteins belong to the steroid/thyroid/retinoic acid family of nuclear receptors for hormones. The ligand-binding domain is in the C-terminal region, the zinc-finger DNA-binding domain in a central region, and a function-modulating domain in the N-terminal region. Binding of hormone induces changes in receptor conformation that control transcriptional activation and repression, and also regulates homo- or heterodimerization. In the absence of ligand, these receptors repress basal gene expression, probably through co-repressor proteins. *α, β, γ* and *δ* types of receptor are known.

retinoid any isoprenoid, whether naturally occurring or not, that contains or is derived from four prenyl groups linked in a head-to-tail manner. All retinoids may formally be derived from a monocyclic parent compound with 20 carbon atoms containing five carbon–carbon double bonds and a functional group at the terminus of its acyclic moiety.

retinoid receptor (*abbr.*: RXR) or **retinoic acid orphan receptor** a nuclear receptor that binds more tightly to other retinoids than to retinoic acid. Three isoforms (α, β, and γ) are known. Phytanic acid binds to all three. Retinoid receptors form heterodimers with **RAR** or with **PPARs,** and are widely distributed in brain and spinal cord.

retinol *the recommended trivial name for* (2*E*,4*E*,6*E*,8*E*)-3,7-dimethyl-9-(2,6,6-trimethylcyclohex-1-en-1-yl)nona-2,4,6,8-tetraen-1-ol; *other names*: vitamin A; vitamin A alcohol; axerophthol. The predominant form of **vitamin A** in higher vertebrates and marine fishes (*compare* **3,4-didehydroretinol**). It is stored primarily in the liver, mostly as the ester, retinyl palmitate.

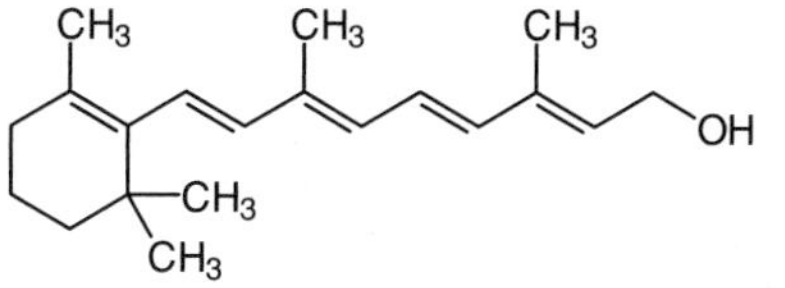

all-*trans* form

retinol-binding protein a protein, secreted by liver, that binds retinol and as a homodimer forms a complex with transthyretin in plasma. The crystal structure shows eight antiparallel beta strands, which form a retinol-binding pocket.

retinoyl the acyl group formally derived from **retinoic acid**.

retinyl the alkyl group formally derived from **retinol** by loss of its hydroxyl group.

retro+ *prefix* signifying backwards, or in the reverse direction.

retroendocytosis the process by which some particles, e.g. low-density lipoproteins, are released, after receptor-binding and endocytosis, back into the medium rather than being degraded within lysosomes.

retrogene a segment of DNA that has arisen by reverse transcription of the genetic material in an RNA virus (**retrovirus**) and incorporation of the DNA so formed into the host cell's genomic DNA.

retrogradation the spontaneous irreversible change of a liquid colloidal solution to an insoluble or gelled state. The term is used especially of aqueous amylose solutions in which microcrystalline precipitates of aligned, hydrogen-bonded amylose molecules occur on standing.

retrograde 1 moving backwards; reverse or inverse; tending towards a former state or condition; deteriorating. **2** to move backwards or in reverse, as in retrograde transport.

retrogress to move backwards or in a less favourable direction; to deteriorate or degenerate. —**retrogression** *n.*; **retrogressive** *adj.*

retroinhibition *an alternative name for* **feedback inhibition**.

retropepsin EC 3.4.23.16; a **pepsin**-type endopeptidase from **HIV** that catalyses the hydrolysis of Xaa-|-Xbb bonds, where Xaa is a hydrophobic residue, and Xbb is variable, but often is Pro. It is part of the polyprotein product of the *pol* gene.

retroposon *or* **retrotransposon** a transposable element that uses a mechanism indistinguishable from part of a retroviral life cycle. Retroposons are present in many organisms, such as yeast, *Drosophila*, and mammals. A good example is the Ty1 element of yeast: after the formation of a double-stranded DNA by **reverse transcriptase** it is integrated into a random site on the chromosome.

retroregulation the regulation of translation of messenger RNA (mRNA) by a sequence that is downstream of the region undergoing translation. For example, an mRNA of phage lambda carrying the sequence of gene *int* is not normally translated; mutation in a site called *sib*, located downstream of the *int* coding sequence, allows translation of the *int* mRNA.

retrotransposon *an alternative name for* **retroposon**.

Retrovir *see* **zidovudine**.

retroviral oncogene *see* **oncogene**, **retrovirus**.

retrovirus *or* **ribodeoxyvirus** *or* **RNA–DNA virus** any virus belonging to the family Retroviridae. This comprises single-stranded RNA animal viruses whose genome is diploid, consisting of two copies of the RNA, which is transcribed by the virion-associated enzyme **reverse transcriptase** into DNA. This double-stranded DNA, or **provirus**, can then integrate into the host genome and be passed from parent cell to progeny cell as a component of the host genome. The virion is an enveloped, roundish particle, ≈100 nm in diameter, with a helical nucleocapsid containing 6 MDa of RNA (in two iden-

tical molecules) joined to their 5′ ends and hydrogen-bonded into loops (70S RNA), plus small (45S) RNA from the host. The genome is flanked by long terminal repeat sequences and comprises at least three genes: *gag*, which encodes the inner capsid protein; *pol*, which codes for a polyprotein containing **retropepsin**, ribonuclease H (*see* **ribonuclease**), and **reverse transcriptase**; and *env*, which codes for the spikes on the envelope. A subfamily of retroviruses, known as the Oncovirinae (or the RNA tumour viruses), contain an additional **oncogene**, which confers the ability to form tumours in the host. These oncoviruses include the Rous sarcoma virus and other leukosis viruses of birds, and viruses causing leukemia in various mammals (not in humans). They have also been called **oncornaviruses**, **rousviruses**, and **leukoviruses**. Other subfamilies include the Spumavirinae (spumaviruses) and the Lentivirinae (lentiviruses); **HIV** is a member of the latter. The retroviruses are in class VI of the **Baltimore classification**.

Rett syndrome *see* **methylCpG-binding protein 2**.

reversal an evolutionary change that results in a derived character state reverting to its ancestral state.

reverse cholesterol transport the efflux of cholesterol (usually paralleled by efflux of phospholipid) from cells. It occurs, via a specific ABC transporter (which is defective in Tangier disease), onto high-density lipoprotein (HDL) particles bound to the cell surface. The cholesterol is eventually delivered to the liver. The process accounts for the antiatherogenic properties of HDL.

reverse dialysis a technique for the concentration of macromolecules in solution. The (dilute) solution is placed in a bag made of a semipermeable material, through which the macromolecules cannot pass but water and other small molecules can. The bag is placed in a bed of dry water-soluble polymer (which cannot enter the membrane) such as polyethylene glycol. Water, together with other small molecules, is then drawn out of the bag into the external phase until equilibrium is reached, thereby concentrating the solution of macromolecules in the bag. *Compare* **dialysis**.

reversed phase *or* **reverse phase** a system used in **adsorption chromatography** (often thin-layer or high-pressure liquid chromatography), in which the **stationary phase** consists of a material with a hydrophobic surface. In such a system, molecules migrate in order of polarity, the most polar migrating fastest. In column chromatography, mobile phases are used in a sequence commencing with more polar solvents and proceeding in decreasing order of polarity, the reverse of the sequence used in normal phase chromatography. The stationary phase often consists of silicic acid in which all hydrophilic groups are masked by silanization.

reversed rocket immunoelectrophoresis a variant of **rocket immunoelectrophoresis** in which antibodies are forced by electrophoresis through antigen-containing agarose gel, the heights of the 'rockets' so formed being related to the antibody concentrations. It is useful for titre determinations in large series of antibody samples.

reverse genetics *another name for* **positional cloning**. A term coined to describe the study of the biological consequences to an organism of artificial alteration of its DNA. The changes may be either gross (e.g. deletion, insertion, or transposition of segments of DNA effected by recombinant DNA technology) or limited to specific loci (e.g. base substitution through site-directed mutagenesis).

reverse gyrase an ATPase enzyme with a **DNA helicase**-like domain and a type-I (**DNA**) **topoisomerase** in the same protein. It is able to induce positive supercoiling into DNA, a property of DNA that is found only in bacteria that express this enzyme.

reverse mutation *or* **back mutation** a heritable change in a mutant gene that restores a nucleotide sequence to that which obtained prior to mutation. *Compare* **reversion**.

reverse osmosis the pressure-induced flow of solvent through a semipermeable membrane in a direction that enhances any existing difference in the activities (or concentrations) of the nonpermeant solutes on the two sides of the membrane (*compare* **osmosis**). The principle can be used in the preparation of purified water from tap water or sea water.

reverse phase *see* **reversed phase**.

reverse pinocytosis *see* **exocytosis**.

reverse transcriptase EC 2.7.7.49; *recommended name*: RNA-directed DNA polymerase; *systematic name*: deoxynucleoside-triphosphate:DNA deoxynucleotidyltransferase (RNA-directed); *other names*: DNA nucleotidyltransferase (RNA-directed); revertase. A DNA polymerase enzyme, found particularly in retroviruses and possibly in normal animal cells, that uses either DNA or RNA as a template. It catalyses a reaction between a deoxynucleoside triphosphate and DNA_n to form pyrophosphate and DNA_{n+1}; this protein also has ribonuclease H (RNase H) activity. When the single-stranded RNA of the retrovirus enters a cell, the enzyme first makes a DNA copy of one of the **long terminal repeats** of the viral genomic RNA strand using a tRNA as a primer. The stretch of RNA used as a template is then digested by the RNase H activity of the enzyme, releasing the newly synthesized stretch of DNA, which then pairs with a complementary stretch of viral RNA to form a DNA–RNA hybrid. This heteroduplex is then used by the same enzyme to make a double-stranded DNA, which circularizes and integrates into a host cell chromosome. Reverse transcriptase lacks a proofreading exonuclease function and is highly error prone. It can thus evolve even within a single patient, and this interferes with long-term therapy. Reverse transcriptase is used *in vitro* to make cDNA from mRNA in a reaction requiring deoxynucleoside triphosphates (dNTPs) and a primer such as **oligo dT**(12–18). Enzymes from avian myeloblastosis virus (AMV) and Moloney murine leukemia virus (MMLV) are suitable for this purpose. **Telomerase** is a template-bearing reverse transcriptase.

reverse transcription the process carried out by **reverse transcriptase**. *See also* **transcription**.

reverse turn (*in protein structure*) a structural feature that allows the polypeptide chain to turn back through about 180°. *See also* **beta bend**.

reversible competitive antagonist (*in pharmacology*) an antagonist for which the rate of dissociation from its receptor is relatively high, such that by increasing the concentration of an agonist the receptor **occupancy** (def. 2) of the antagonist falls. At sufficiently high concentrations of agonist a maximal tissue response can be obtained, even in the presence of the antagonist, such that the extent of antagonism is determined by the relative concentrations of agonist and antagonist.

reversible inhibition (of an enzyme or other system) a type of inhibition that occurs when the equilibrium between free inhibitor and the inhibitor–enzyme (or system) complex lies sufficiently far towards the free inhibitor state for the inhibitor to be readily removed, leaving an active enzyme (or system).

reversible reaction a reaction in which the **equilibrium constant** is such that the reaction can be made to proceed at a detectable rate in either direction under appropriate conditions.

reversion the restoration, by a second mutation, of the genomic base sequence of a mutant virus or organism to that of the parental strain. Less strictly, any return to the original **phenotype**. *See also* **back mutation**, **reverse mutation**.

reversion frequency the proportion of the total number of cells in a population that undergo **reversion**.

reversion spectroscope a **spectroscope** in which two spectra alongside each other, one of which is reversed by optical means, are formed by the entering light. The two spectra may be moved longitudinally relative to each other by moving a micrometer screw. If two absorption bands are aligned when a particular substance, e.g. oxyhemoglobin, is viewed, the amount by which the screw has to be turned to align the same band when a mixture of the same substance and another with a similar absorption spectrum, e.g. carbonmonoxyhemoglobin, is viewed is related to the proportion of the second substance present.

revertant 1 a mutant that has partially or completely regained the wild-type phenotype. **2** a gene that has undergone **reversion**.

revertase *an alternative name for* **reverse transcriptase**.

rexinoid a synthetic agonist of **retinoid X receptors**.

Reye syndrome an acute noninflammatory encephalopathy in children that is associated with fatty degeneration of the liver and hyperammonemia. Many cases are caused by inborn errors of fatty acid oxidation. There is strong epidemiological evidence for an association with salicylate ingestion and influenza B virus infection. [After R.D.K. Reye (1912–77), Australian pathologist.]

Reynolds number *symbol*: *Re*; a nondimensional parameter of fluid motion through a cylindrical tube or past an obstruction. It is given by the expression vl/ν, where *l* has the dimensions of length, *v* is the fluid's velocity, and ν is the kinematic viscosity. In general, low Reynolds numbers show that viscous forces predominate in controlling the flow, while high Reynolds numbers indicate that inertial forces are more important. [After Osborne Reynolds (1842–1912), Irish physicist and engineer.]

RF *abbr. for* **1** release factor. **2** replicative form.

R factor 1 *abbr. for* resistance factor (*see* **resistance plasmid**). **2** (*formerly*) *abbr. for* release factor.

RFC *abbr. for* reduced folate carrier.

rfDNA *see* **DNA**.

RFLP *abbr. for* restriction fragment length polymorphism.

R form 1 *abbr. for* relaxed form (of a protein). **2** R-form (hyphenated) *abbr. for* rough form; describing bacterial colonies with a jagged perimeter.

RFX *abbr. for* regulatory factor for X box; a DNA-binding heterotrimeric protein complex that binds the X box (a nucleotide sequence within the 150 bp response element) of the promoters of MHC class II genes. The three subunits are RFX5 (75 kDa), RFXAP (36 kDa), and **RFXANK**, and the protein forms higher-order complexes with proteins that bind other regions within the response element. Mutations in any component of the RFX complex cause MHC class II deficiency in humans.

RFXANK a 33 kDa protein component of the **RFX** complex. It contains three ankyrin repeats and is required for efficient DNA-binding and transcriptional activation of MHC class II gene promoters by the RFX complex.

RGD the tripeptide Arg–Gly–Asp, which is present in **fibronectin** and some other extracellular matrix proteins and is recognized by some **integrins**.

R gene *abbr. for* resistance gene; any gene in plants that imparts resistance against viruses, bacteria, fungi, aphids, or nematodes. Examples are known in potato, tobacco, flax, tomato, maize, barley, and sugar beet, among others. Many of the proteins (**R proteins**) they encode contain leucine-rich repeats, which mediate protein–protein interactions. Some also contain leucine zipper, coiled-coil, or other sequences. Many R proteins are cytoplasmic, and some are transmembrane.

RGG box a peptide sequence (20–25 amino acids) present in some proteins, often in association with the RNA-binding domains found in FMR1. It contains repeats of Arg–Gly–Gly, often interspersed with aromatic residues.

RGK-2 *see* **tonin**.

RGS protein *abbr. for* regulator of G-protein signalling protein; any protein containing a conserved sequence (≈120 amino acids) that binds the α subunit of a trimeric **G-protein** and enhances its GTPase activity; i.e. they are GTPase-activating proteins (GAPs). There are over 25 in mammals. They frequently contain other homology domains that increase the range of proteins that they bind. RGS-PX1 regulates $G_{S\alpha}$, RGS7 putatively binds the C-terminal region of polycystin-1, RGS9 regulates $G_{T\alpha}$ (**transducin**), and RGS12 regulates $G_{O\alpha}$.

rH the negative logarithm of the concentration (expressed as pressure in atmospheres) of hydrogen gas in equilibrium with a redox system; i.e.

$$rH = -\log[H_2].$$

Also,

$$rH = 2(E_h F)/(RT\ln 10) + pH,$$

where *F* is the Faraday constant, *R* the gas constant, *T* the thermodynamic temperature, and E_h the **redox potential** at the pH of the system under study.

Rh 1 *abbr. for* Rhesus factor. **2** *symbol for* rhodium.

Rha *symbol for* rhamnose.

Rhabdoviridae a family of single-stranded RNA viruses infecting vertebrates, arthropods, amoebae, and plants. The virions are 70–80 nm elongated particles, with a lipid envelope. They contain negative-strand RNA together with RNA-dependent RNA polymerase, which transcribes the genome directly to mRNA – they are class IV viruses in the **Baltimore classification**. Rhabdoviruses are usually transmitted by arthropods; they include the plant rhabdoviruses, the vesiculoviruses (e.g. the vesicular stomatitis virus of cattle), the lyssaviruses, which include the rabies virus, and sigma virus of *Drosophila*. [From Greek *rhabdos*, rod.]

rhabdovirus see **Rhabdoviridae**.

L-rhamnosamine 2-amino-2,6-dideoxy-L-mannose; an amino sugar that occurs in some bacterial lipopolysaccharides.

rhamnose *symbol*: Rha; 6-deoxy-L-mannose; a monosaccharide (hexose) that occurs commonly in plant glycosides, in polysaccharides of gums and mucilages, and in bacterial polysaccharides; it also occurs in some plant cell-wall polysaccharides, and in flavonoids. Free rhamnose is found in some plants. The D-enantiomer occurs in some bacterial capsular polysaccharides. *See also* **L-rhamnose isomerase**.

L-rhamnose isomerase EC 5.3.1.14; *systematic name*: L-rhamnose ketol-isomerase; an enzyme that catalyses the interconversion of L-rhamnose and L-rhamnulose. It initiates the catabolism of rhamnose.

rhamnulose 6-deoxyfructose; it is formed by L-**rhamnose isomerase**.

rheology the science concerning the deformation and flow of matter. —**rheological** *adj*.

rheoseme *or* **flow signal** a chemical messenger or transmitter substance (other than a neurotransmitter) carried from a site of generation to site of action within a cell (e.g. a neuron) by cytoplasmic flow; a translocated **bioseme** or **second messenger**.

rheostat a device in which a controllable resistance is introduced into an electrical circuit to vary the output to permit an apparatus to be regulated within desired limits.

Rhesus factor *abbr.*: Rh; a protein associated with the rhesus (Rh) system of blood groups in humans. The system was so designated as a result of the finding that rabbit and guinea-pig antibodies raised against erythrocytes from *Macacus rhesus* monkeys agglutinated erythrocytes from 85% of the random human population. Initially, an antigen, designated D, was isolated from the erythrocytes of Rh-positive individuals; those lacking D were classified Rh-negative. Subsequently, another antigen, LW (Landsteiner-Weiner), was identified, which, although phenotypically similar to D, is encoded by a gene that segregates independently from *Rh*. C and E antigens have also been found. The antigenic determinant is associated with the protein structure. In hemolytic disease of the newborn resulting from rhesus incompatibility, antibodies from a Rh-negative mother, often formed during the birth of a previous Rh-positive child whose blood has entered the mother's bloodstream, pass across the placenta and cause destruction of fetal red cells.

rheumatoid factors autoantibodies to the IgG Fc region, also known as antiglobulins, that are the hallmark of patients with rheumatoid **arthritis**.

rhinovirus any small RNA virus belonging to the Picornaviridae family, members of which are responsible for the common cold in humans, and foot-and-mouth disease in cattle.

Rhizobium a genus of soil bacteria whose members reside symbiotically in specialized root nodules of leguminous plants. The bacteria are active in **nitrogen fixation** providing the plant with organic nitrogen and obtaining carbon compounds in return. Many *Rhizobium* species contain plasmids that enable them to colonize their host plant species.

rhizomorph a stout, rootlike mass of hyphae, produced by some fungi, that serves to generate vegetative spread.

RH map *abbr. for* radiation hybrid map.

rho *symbol*: ρ (lower case) *or* P (upper case); the seventeenth letter of the Greek alphabet. For uses see **Appendix A**. *See also* **rho protein, Rho protein**.

Rhodamine any of a series of dyes based on 9-(2-carboxyphenyl)xanthine. Rhodamine 6G, *N*-[9-(2-carboxyethylphenyl)-6-(diethylamino)-3*H*-xanthen-3-ylidene]ethylethanamine, is used as a spray to reveal lipids as pink fluorescent spots after thin-layer chromatography, and as the isothiocyanate derivative as a label for proteins and nucleotides. Rhodamine 6G has λ_{max} excitation 525 nm and emission 555 nm.

Rhodamine 6G

rhodanese EC 2.8.1.1; *recommended name*: thiosulfate sulfur-transferase; *systematic name*: thiosulfate:cyanide sulfur-transferase; *other names*: thiosulfate cyanide transsulfurase; thiosulfate thiotransferase. An enzyme that catalyses the displacement of sulfite from thiosulfate by cyanide ion, forming sulfite and thiocyanate and thus detoxifying cyanide. It has two domains of closely similar conformation yet very different sequence. [From the use of rhodanate as a name for thiocyanate.]

rhodanid any **thiocyanate** produced by the metabolism of mustard-oil glycosides (glycosinolates) in damaged leaf tissues.

rhodium *symbol*: Rh; a silvery-white metallic transition element of group VIII of the (IUPAC) **periodic table**; atomic number 45; relative atomic mass 102.9. It occurs mainly in oxidation state III. A rare member of the platinum metals, it is resistant to nitric acid and aqua regia.

rhodniin a thrombin inhibitor found in the saliva of the blood-sucking reduviid bug *Rhodnius prolixus*.

rhodo+ *or* (before a vowel) **rhod+** *comb. form* indicating a red colour, or the odour of a rose. [From Greek *rhodos*, rose.]

rhodopsin *or* **visual purple** a brilliant purplish-red, light-sensitive visual pigment found in the rod cells of the retinas of most vertebrates (exceptions being freshwater and migratory fish, and some amphibians). It is a **7TM protein** similar to other members of the G protein-coupled receptor family. It consists of a rod-type **opsin** combined (as a Schiff's base) with 11-*cis*-retinal (see **retinal**). It has an absorption maximum at ~500 nm. *Compare* **porphyropsin**. A single amino-acid substitution in the rhodopsin molecule causes one form of **retinitis pigmentosa** in humans. *See also* **bacteriorhodopsin, halorhodopsin, slow-cycling rhodopsin**.

rho factor *an alternative name for* **rho protein**.

RhoGEF *abbr. for* Rho guanine nucleotide exchange factor; a protein that enhances the exchange of GDP for GTP in a **Rho protein** and contains a GEF domain (≈200 amino acids) followed by a pleckstrin homology domain. **FGD1 protein** is a RhoGEF for Cdc42.

RhoGTPase any of a subfamily of Ras-related GTPases that are involved in the assembly of actin microfilaments into stress fibres and in formation of the contractile ring during cell division. They have crucial roles in metazoan development because they regulate cell shape, adhesion, and migration. There are at least ten mammalian RhoGTPases (RhoA to RhoE; Rac1, Rac2, and RacE; Cdc42 and TC10), which share ≈50% homology and become prenylated at the C-terminal CAAX motif. They cycle between (active) RhoGTP and (inactive) RhoGDP forms, the relative proportions being regulated by **Rho guanine nucleotide-exchange factors** (RhoGEFs).

Rho guanine nucleotide exchange factor see **RhoGEF**.

rho protein *or* (*sometimes*) **rho factor** a bacterial ATP-dependent helicase that facilitates transcription termination. Binding of rho to the nascent RNA is followed by activation of rho's RNA-dependent ATPase activity, and release of the mRNA from the DNA template; it recognizes and binds to C-rich regions of the RNA transcript. In *Escherichia coli*, this protein is the site of action of the antibiotic bicyclomycin, which binds to rho protein and inhibits its ATPase activity. *Compare* **Rho protein**.

Rho protein a small Ras-related GTPase (see **RAS**), involved in controlling the polymerization of actin into filaments, and subsequent organization. *RHOB* is an early response gene transiently induced in fibroblasts by platelet-derived growth factor, or epidermal growth factor; its over-expression can be tumorigenic. *Compare* **rho protein**.

RI *abbr. for* **1** refractive index. **2** ribonuclease inhibitor.

RIA *abbr. for* radioimmunoassay.

Rib *symbol for* ribose.

ribaric acid one of the **pentaric acids**. *See also* **aldaric acid**.

ribavarin A guanosine nucleotide analogue, 1-β-D-ribofuranosyl-1,2,4-triazole-3-carboxamide. Mainly used against syncytial virus in young children but also effective with **α-interferon** against **hepatitis C**. It is probable that the antiviral action is due to the monophosphate derivative competitively inhibiting inosine-5′-monophosphate dehydrogenase.

ribitol *or* **adonitol** a **pentitol** derived formally by reduction of the –CHO group of D- or L-ribose. It is a *meso* compound and does not require a D or L configurational descriptor. It occurs free in some plants and is a component of **riboflavin**.

ribitol teichoic acid a complex polymer of **ribitol** residues linked by phosphate groups, substituted by glucosyl or acetylglucosamine residues and ester-linked to D-alanine. It is found in the walls of Gram-positive bacteria.

ribo+ *comb. form* **1** of, relating to, or containing **ribose**, especially as opposed to deoxyribose. **2** containing a ribose residue linked through its anomeric carbon. It is used in **riboflavin**, inappropriately, for a ribitol residue.

ribo- *prefix (in chemical nomenclature)* indicating a particular configuration of three contiguous >CHOH groups (as found in the acyclic form of D- or L-ribose). *See also* **monosaccharide**.

ribocharin an acidic protein specifically associated with the granular component of the nucleolus and nucleoplasmic 65S particles. These particles contain the nuclear 28S rRNA, and may represent the precursor to the large ribosomal subunit.

ribodeoxyvirus any RNA virus that replicates via DNA; such viruses are usually known as **retroviruses**. *Compare* **deoxyribovirus**.

riboflavin *an alternative name for* **vitamin B₂**.

ribofuranosyl *an alternative term for* **ribosyl**.

ribonuclease *abbr.*: RNase *or (sometimes)* RNAase; any of a group of nuclease enzymes that cleave phosphodiester bonds in RNA. With the exception of RNase P, which is a **ribozyme**, they are all proteins. RNAses are classified into **exoribonucleases**, catalysing sequential cleavage of mononucleotides from either a free 5′ or 3′ terminus, or both, or **endonucleases**, cleaving bonds within the polyribonucleotide chain. Endonucleases, which can also cleave circular molecules of RNA, fall into further categories according to their specificity: EC 3.1.13, exoribonucleases producing 5′-phosphomonoesters; EC 3.1.14, exoribonucleases producing other than 5′-phosphomonoesters; EC 3.1.15, exonucleases active with either ribo- or deoxyribonucleic acids and producing 5′-phosphomonoesters; EC 3.1.16, exonucleases active with either ribo- or deoxyribonucleic acids and producing other than 5′-phosphomonoesters; EC 3.1.26, endoribonucleases producing 5′-phosphomonoesters; EC 3.1.27, endoribonucleases producing other than 5′-phosphomonoesters; EC 3.1.30, endonucleases active with either ribo- or deoxyribonucleic acids and producing 5′-phosphomonoesters; and EC 3.1.31, endonucleases active with either ribo- or deoxyribonucleic acids and producing other than 5′-phosphomonoesters. Ribonucleases include enzymes having mechanisms of (A) phosphotransferases, (B) phosphodiesterases, or (C) phosphorylases. In group C are polynucleotide phosphorylase and polynucleotide pyrophosphorylase. The feature of group A is the involvement of the 2′-OH group in an intramolecular attack at the adjacent phosphodiester bond, in addition to conferring specificity for one or more of the four bases. The nucleoside 2′:3′-cyclic phosphates are obligate intermediate products. In group B, a direct attack of water on the 3′:5′-phosphodiester bond is catalysed, so that this group includes the enzymes that hydrolyse both RNA and DNA and those that cleave diester bridges to form only 5′-nucleotide products. Thus group A may be classed as cyclizing and group B as non-cyclizing enzymes. The group A enzymes do not depend for their activity on divalent cations whereas the group B enzymes require Mg^{2+} or Ca^{2+}. Within each group there are many types, as described below.

Exoribonuclease II EC 3.1.13.1; *other name*: ribonuclease II; reaction: exonucleolytic cleavage in the 3′- to 5′-direction to yield 5′-phosphomononucleotides. **Exoribonuclease H** EC 3.1.13.2; reaction: exonucleolytic cleavage to 5′-phosphomonoester oligonucleotides in both 5′- to 3′- and 3′- to 5′-directions. **Poly(A)-specific ribonuclease** EC 3.1.13.4; reaction: exonucleolytic cleavage of poly(A) to 5′-AMP. **Yeast ribonuclease** EC 3.1.14.1; reaction: exonucleolytic cleavage to 3′-phosphomononucleotides.

Endoribonucleases cleave the polynucleotide internally. This first group (sub-subclass EC 3.1.26) produce exclusively 5′-phosphomonoesters. *Physarum polycephalum* ribonuclease EC 3.1.26.1; reaction: endonucleolytic cleavage to 5′-phosphomonoester. **Ribonuclease alpha** EC 3.1.26.2; reaction: endonucleolytic cleavage to 5′-phosphomonoester. **ribonuclease III** EC 3.1.26.3; *other names*: RNase O; RNase D; reaction: endonucleolytic cleavage to 5′-phosphomonoester. **Ribonuclease H** EC 3.1.26.4; *other names*: endoribonuclease H (calf thymus); calf thymus ribonuclease H; reaction: endonucleolytic cleavage to 5′-phosphomonoester; it acts on RNA–DNA hybrids (hence the H in the name) and is required for removing RNA from **Okazaki fragments**; *see also* **RNA-directed DNA polymerase**. **Ribonuclease P** EC 3.1.26.5; reaction: endonucleolytic cleavage of RNA, removing 5′-sequence of extra nucleotides from tRNA precursor; it is a **ribozyme**; *see* **RNase P. Ribonuclease IV** EC 3.1.26.6; *other names*: endoribonuclease IV; poly(A)-specific ribonuclease; reaction: endonucleolytic cleavage of poly(A) to fragments terminated by 3′-hydroxyl and 5′-phosphate groups. **Ribonuclease P4** EC 3.1.26.7; reaction: endonucleolytic cleavage of RNA, removing 3′-sequence of extra nucleotides from tRNA precursor. **Ribonuclease M5** EC 3.1.26.8; *other name*: 5S rRNA maturation nuclease; reaction: endonucleolytic cleavage of RNA, removing 21 and 42 nucleotides, respectively, from the 5′- and 3′-termini of a 5S rRNA precursor. **Ribonuclease M16** cleaves the 5′ and 3′ termini of a 16S rRNA precursor. **Ribonuclease M23** cleaves the 5′ and 3′ termini of a 23S rRNA precursor. **Ribonuclease (poly(U)-specific)** EC 3.1.26.9; reaction: endonucleolytic cleavage of poly(U) to fragments terminated by 3′-hydroxyl and 5′-phosphate groups. **Ribonuclease IX** EC 3.1.26.10; reaction: endonucleolytic cleavage of poly(U) or poly(C) to fragments terminated by 3′-hydroxyl and 5′-phosphate groups (*see* **Rrp**).

The second group of endoribonucleases (sub-subclass EC 3.1.27) yield products other than 5′-phosphomonoesters. **Ribonuclease T2** EC 3.1.27.1; *other name*: ribonuclease II; reaction: two-stage endonucleolytic cleavage to 3′-phosphomononucleotides and 3′-phosphooligonucleotides with 2′,3′-cyclic phosphate intermediates. *Bacillus subtilis* ribonuclease EC 3.1.27.2; reaction: endonucleolytic cleavage to 2′,3′-cyclic nucleotides. **Ribonuclease T1** EC 3.1.27.3; *other names*: guanyloribonuclease; *Aspergillus oryzae* ribonuclease; ribonuclease N1; ribonuclease N2; reaction: two-stage endonucleolytic cleavage to 3′-phosphomononucleotides and 3′-phosphooligonucleotides ending in G-p with 2′,3′-cyclic phosphate intermediates. **Ribonuclease U2** EC 3.1.27.4; reaction: two-stage endonucleolytic cleavage to 3′-phosphomononucleotides and 3′-phosphooligonucleotides ending in Ap or Gp with 2′,3′-cyclic phosphate intermediates. **Pancreatic ribonuclease** EC 3.1.27.5; *other names*: ribonuclease; ribonuclease I; ribonuclease A; a notably thermostable enzyme that catalyses the reaction: endonucleolytic cleavage to 3′-phosphomononucleotides and 3′-phosphooligonucleotides ending in Cp or Up with 2′,3′-cyclic phosphate intermediates. A glycosylated form, **ribonuclease B**, has the same specificity. *Enterobacter* ribonuclease EC 3.1.27.6; reaction: endonucleolytic cleavage to 3′-phosphomononucleotides and 3′-phosphooligonucleotides with 2′,3′-cyclic phosphate intermediates. **Ribonuclease F** EC 3.1.27.7; reaction: endonucleolytic cleavage of RNA precursor into two, leaving 5′-hydroxyl and 3′-phosphate groups. **Ribonuclease V** EC 3.1.27.8; *other name*: endoribonuclease V; reaction: hydrolysis of poly(A), forming oligoribonucleotides and ultimately 3′-AMP. *Compare* **deoxyribonuclease**.

ribonuclease inhibitor *abbr.*: RI; a porcine protein that inhibits pancreatic nuclease. It contains 17 leucine-rich repeats (each ≈24 residues) that form hydrophobic beta strands. The crystal structure is available.

ribonuclease protection assay a method for measuring the abundance of a specific mRNA in which a short (~100 nucleotide) radioactively labelled antisense RNA probe is hybridized with a population of RNA molecules. Treatment with ribonucleases A and T1 digests all single-stranded RNA leaving double-stranded RNA intact. Autoradiography or phosphorimaging of acrylamide electrophoresis gels, on which the digestion products are run, can provide quantitative information about mRNA abundance.

ribonucleate 1 *abbr.*: RNA; any anionic form of ribonucleic acid. **2** any salt of ribonucleic acid.

ribonucleic acid *abbr.*: RNA; pentose nucleic acid (*abbr.*: PNA); ribose nucleic acid; yeast nucleic acid. *See* **RNA**.

ribonucleoprotein *abbr.*: RNP; any conjugated protein that contains RNA as the non-protein moiety. Various specific kinds of RNP are listed with their abbreviations at **RNP**.

ribonucleoside any **nucleoside** (def. 1) in which the glycose moiety is β-D-ribofuranose. The link is from C-1 of ribose to N-9 of a purine or to N-1 of a pyrimidine. To distinguish between the numbering systems of the base and sugar, the numbers of the sugar atoms are characterized by the addition of 'prime', i.e. 1′ to 5′. The various products have trivial names, e.g. adenosine, guanosine, cytidine, and uridine. Strictly, the term ribonucleoside should be confined to structures occurring in nucleic acids, but it can be used for any base–ribose compound. By extension, it can be applied to **pseudouridine** (which occurs in tRNA) with a C–C link between ribose and uracil.

ribonucleoside-diphosphate reductase *the recommended name for* **nucleoside-diphosphate reductase**.

ribonucleoside-triphosphate reductase *the recommended name for* **nucleoside-triphosphate reductase**.

ribonucleotide any ribonucleoside in ester linkage to phosphate, commonly at the 5′ position of the ribose moiety, although some 2′- and 3′-esters and some (cyclic) 2′,3′-, or 3′,5′-diesters are known. The 5′-ribonucleotides are metabolically important; they may be esters of [mono]phosphate, diphosphate, or triphosphate, e.g. guanosine 5′-[mono]phosphate, cytidine 5′-diphosphate, adenosine 5′-triphosphate. *See* **nucleotide**.

ribonucleotide reductase any enzyme that catalyses the reduction of ribonucleotides to deoxyribonucleotides during biosynthesis. *See* **nucleoside-diphosphate reductase**, **nucleoside-triphosphate reductase**.

ribophorin either of two type I single-pass transmembrane glycoproteins, ribophorins I and II, that form two subunits of **N-oligosaccharyltransferase**. They were first isolated bound to ribosomes on the rough endoplasmic reticulum. They form part of the ribosome receptor.

riboprobe any labelled RNA molecule used as a hybridization probe. Riboprobes are conveniently produced by *in vitro* transcription and are particularly useful for detecting specific mRNAs by hybridization *in situ*, and for ribonuclease protection assays.

ribose *symbol*: Rib; *ribo*-pentose. The D enantiomer, as β-D-ribofuranose, forms the glycose moiety of all ribonucleosides, ribonucleotides, and ribonucleic acids, and of ribose phosphates, various glycosides, some coenzymes, and some forms of vitamin B_{12}.

β-D-ribofuranose

ribose nucleic acid *a former name for* **ribonucleic acid**.

ribose-5-phosphate epimerase EC 5.3.1.6; *systematic name*: D-ribose-5-phosphate ketol-isomerase; *other names*: phosphopentoisomerase; phosphoriboisomerase. An enzyme that catalyses the reaction:

$$D\text{-ribose 5-phosphate} = D\text{-ribulose 5-phosphate}.$$

It functions in pentose metabolism, especially the **pentose phosphate pathway**.

ribose-phosphate pyrophosphokinase EC 2.7.6.1; *other name*: phosphoribosyl pyrophosphate synthetase (*abbr.*: PRS); a conserved enzyme that catalyses the formation of 5-phospho-α-D-ribose-1-diphosphate (*abbr.*: PRPP; *other name*: phosphoribosyl pyrophosphate) from ATP and D-ribose 5-phosphate with release of AMP; it is Mg^{2+}-dependent. PRPP is an intermediate in the biosynthesis of histidine and tryptophan, and in the *de novo* biosynthesis of purines and pyrimidines.

riboside the product of the formation of a glycosidic bond between ribose and another group, either through the nitrogen of an amine or a hydroxyl group, i.e. via *N*- or *O*-glycosidic linkages. *See also* **pseudouridine**. *Compare* **ribonucleoside**.

ribosomal ambiguity *abbr.*: ram; the effect of any mutation (e.g. in ribosomal protein S4 or S5) in bacterial genomes that affects selection of aminoacyl-tRNA species during protein biosynthesis.

ribosomal DNA *abbr.*: rDNA; the genetic DNA coding for ribosomal RNA (*see* **ribosome**).

ribosomal protein *abbr.*: r-protein; any of the protein components of the ribosomal subunits. They are typically small, and some are basic. Ribosomal proteins are referred to as S (small subunit) proteins and L (large subunit) proteins; in *Escherichia coli*, there are 21 S proteins, designated S1–S21, and 34 L proteins, designated L1–L34. In eukaryotes (e.g., in rat liver), there are 33 S proteins and 49 L proteins. *See also* **ribosome**.

ribosomal protein S6 kinase *see* **RSK**.

ribosomal RNA *abbr.*: rRNA; *see* **ribosome**.

ribosomal RNA processing *see* **Rrp**.

ribosome an intracellular organelle, about 200 Å in diameter, consisting of RNA and protein. It is the site of protein biosynthesis resulting from translation of messenger RNA (mRNA). It consists of one large and one small subunit, each containing only protein and RNA. Ribosomes and their subunits are characterized by their sedimentation coefficients, expressed in **Svedberg units** (*symbol*: S). The prokaryotic ribosome (70S) comprises a large (50S) and a small (30S) subunit, while the eukaryotic ribosome (80S) comprises a large (60S) subunit and a small (40S) subunit. Several ribosomes may bind to one mRNA to form a **polysome**. Two sites on the large subunit are involved in translation, namely the **aminoacyl site** (A site) and **peptidyl site** (P site). Ribosomes from prokaryotes, eukaryotes, mitochondria, and chloroplasts have distinct **ribosomal proteins**. Prokaryotic ribosomes contain 35–45% protein, while eukaryotic ribosomes contain 50% protein. **Ribosomal RNA** (rRNA) is the RNA part of a ribosome. In prokaryotes, a 30S subunit contains one 16S RNA molecule with associated proteins, while a 50S subunit contains one 5S and one 23S RNA molecule. In eukaryotes, the 60S subunit contains one 28S, one 5.8S, and one 5S RNA molecule, while the 40S subunit contains one 18S RNA molecule. The structures are known. In *Escherichia coli*, the rRNAs form part of a large transcript containing transfer RNA (tRNA) molecules, which is then cleaved during processing. In eukaryotes, the genes for the 5.8S, 18S, and 28S molecules reside in the **nucleolus**, and code for a 40S transcript, which is then cleaved during processing; this includes splicing that occurs in the absence of protein. The gene for the 5S rRNA is a separate transcription unit outside the nucleolus. *See also* **mRNA binding site**, **protein biosynthesis**.

ribosome-inactivating protein *abbr.*: RIP; any of a group of proteins, obtained from plants, that inactivate eukaryotic ribosomes. They are of two types. Type I proteins include agrostin, dianthrin, **dodecandrin**, **gelonin**, **luffin**, nomordin, **saporin-2**, and **tritin**. They are enzymes, EC 3.2.2.22, that catalyse endohydrolysis of the *N*-glycosidic bond at one specific adenosine on 28S rRNA. They are plant protein toxins, and possibly have antiviral activity. Type II ribosome-inactivating proteins are enzymes with a similar action to the type I proteins, but have an additional covalently bound chain with lectin properties. They include **abrin**, **modeccin**, **ricin**, viscumin, and **volkensin**.

ribosome receptor the binding site on the endoplasmic reticulum (ER) membrane for ribosomes bearing nascent polypeptides destined for secretion or incorporation into ER, Golgi apparatus, or plasma membrane. It consists of several proteins (including the ribophorins) oriented with their C-terminal segments on the cytoplasmic ER surface. The proteins bind directly to the large ribosomal subunit, and may also function in translocation of the nascent polypeptide into the ER lumen.

ribosome recycling factor *abbr.*: RRF; *other name*: ribosome releasing factor; a factor required by bacteria for the recycling of ribosomes. It dissociates the termination complex after the polypeptide has been released from the peptidyl-tRNA by release factor.

riboswitch any part of an mRNA molecule (but usually in the 5′ untranslated region) that can change its tertiary structure by interacting with some ligand and so inhibit translation.

ribosyl *or* **ribofuranosyl** any **glycosyl** group formally derivable from the α- or the β-furanose forms of D- or L-ribose. (The pyranose forms of ribose are exhibited only by the uncombined sugar.)

ribosylate to add a ribosyl residue. *Compare* **ADP-ribosylation**.

ribosylthymine *an alternative name for* **ribothymidine**.

ribosylthymine monophosphate *an alternative name for* **ribothymidine monophosphate**.

ribosylthymine phosphate *an alternative name for* **ribothymidine phosphate**.

ribothymidine *or* **ribosylthymine** *symbol*: T *or* Thd; 5-methyluridine; thymine riboside; 1-β-D-ribofuranosyl-5-methyluracil; a rare ribonucleoside, found as a minor component of many species of transfer RNA, where it is present as a residue of its 5′-phosphate. *See also* **ribothymidine 5′-phosphate**. *Compare* **thymidine**.

ribothymidine monophosphate *or* **ribosylthymine monophosphate** *abbr.*: TMP; *the common name for* any ribothymidine phosphate, particularly for ribothymidine 5′-phosphate.

ribothymidine phosphate *or* **ribosylthymine phosphate** ribothymidine monophosphate *or* ribosylthymine monophosphate (*abbr.*: TMP); any phosphoric monoester or diester of ribothymidine. Of the three possible monoesters – the 2′-phosphate, the 3′-phosphate, and the 5′-phosphate – and the two possible diesters – the 2′,3′-phosphate and the 3′,5′-phosphate – only **ribothymidine 5′-phosphate** is of any significance (the locant therefore being omitted if no ambiguity may arise).

ribothymidine 5′-phosphate *or* **ribosylthymine 5′-phosphate** *or* **5′-ribothymidylic acid** *symbol*: Thd5′P *or* pT; *the alternative recommended names for* ribothymidine monophosphate or ribosylthymine monophosphate (*abbr.*: TMP); 5′-ribothymidylyl phosphate; thymine riboside 5′-(dihydrogen diphosphate); thymine (mono)ribonucleotide. (The locant is commonly omitted if there is no ambiguity as to the position of phosphorylation.) A residue of ribothymidylic acid occurs in many species of transfer RNA, where it is formed *in situ* by methylation of the uracil moiety of an existing uridylic acid residue. This modification is effected by tRNA (uracil-5-)-methyltransferase (EC 2.1.1.35), the methyl group being derived from *S*-adenosyl-L-methionine.

ribothymidylate 1 either the monoanion or the dianion of **ribothymidylic acid**. **2** any mixture of ribothymidylic acid and its anions. **3** any salt or ester of ribothymidylic acid.

ribothymidylic acid *the trivial name for* any phosphoric monoester of ribothymidine. The position of the phosphoric residue on the ribose moiety of a given ester may be specified by a prefixed locant – *see* **ribothymidine phosphate**. However, 5′-ribothymidylic acid is the ester commonly denoted, its locant usually being omitted if no ambiguity may arise. 5′-ribothymidylic acid is also an alternative recommended name for ribothymidine 5′-phosphate.

ribothymidylyl the ribothymidine[mono]phospho group; the acyl group derived from **ribothymidylic acid**.

ribovirus *an alternative name for* **viroid**.

ribozyme an RNA molecule with catalytic activity. Such activity was first demonstrated in 1981 by Canadian-born US biochemist Sidney **Altman** (1939–) for **RNase P**. In 1982 the US biochemist Thomas R. **Cech** (1947–) discovered the catalysis by RNA of the reactions involved in the **splicing** of ribosomal RNA from *Tetrahymena*. In this process a 414-nucleotide **intron** is removed from a 6.4 kb precursor to yield the mature 26S rRNA. A guanine nucleotide is essential in the reaction, binding to the RNA and then attacking the 5′ splice site to form a phosphodiester bond with the 5′ end of the intron. The 414-nucleotide intron is then released and undergoes further splicing reactions to produce a linear RNA of 395 nucleotides which has lost 19 nucleotides and is named L19. L19 is catalytically active and acts both as a nuclease and a polymerase. The rate of conversion of a pentacytidylate into longer and shorter oligomers is about 10^{10} times the uncatalysed rate. Mg^{2+} plays an

essential role. Much smaller ribozymes have been demonstrated, e.g. the **viroids** which infect plants and undergo self-splicing after replication. Comparisons of nucleotide sequences in the vicinity of specific cleavage sites suggest that the active site has a **hammerhead** secondary structure consisting of three helical regions radiating from a central core of unpaired nucleotides. mRNA precursors in the mitochondria of yeast and fungi also undergo self-splicing as do some RNA precursors in chloroplasts. Such reactions can be classified according to the nature of the unit that attacks the upstream splice site. Group I are mediated by a guanosine cofactor. In group II the attacking moiety is the 2′-OH of a specific adenylate of the intron. Such reactions resemble those that occur in the **spliceosomes**. Ribozymes are especially interesting in respect of theories on the origin of life because of the popular idea that the first macromolecules were composed of RNA. *See also* **ribonuclease**.

(Rib5)ppA *symbol for* adenosinediphosphoribose (alternative to Ado*PP*Rib *or* A5′pp5Rib).

D-ribulose *a nonsystematic name for* the ketopentose D-*erythro*-2-pentulose.

α-D-furanose form

D-ribulose 1,5-bisphosphate *abbr.*: RBP or RuBP; *former name*: ribulose diphosphate (*abbr.*: RDP or RuDP); a key metabolite of the **reductive pentose phosphate cycle**. It is formed in a reaction in which ribulose 5-phosphate is phosphorylated by ATP, catalysed by phosphoribulokinase, EC 2.7.1.19, and is a substrate of **ribulose-bisphosphate carboxylase**.

ribulose-bisphosphate carboxylase EC 4.1.1.39; *systematic name*: 3-phospho-D-glycerate carboxy-lyase (dimerizing); *other names*: ribulose-bisphosphate carboxylase/oxygenase (*abbr.*: Rubisco); the key enzyme in the fixation of CO_2 in the **reductive pentose phosphate cycle**; a copper protein. One molecule of D-ribulose 1,5-bisphosphate reacts with one molecule of CO_2 to yield two molecules of 3-phospho-D-glycerate, which can be converted into a hexose product (fructose 6-phosphate). If, as is possible, D-ribulose 1,5-biphosphate reacts with O_2 instead of CO_2, the products are one molecule each of 3-phospho-D-glycerate and 2-phosphoglycolate. This enzyme comprises up to 50% of leaf proteins and is the most abundant protein in the biosphere.

D-ribulose 5-phosphate the 5-phosphate ester of D-ribulose; it is a component of the **pentose phosphate pathway**, the product of 6-phosphogluconate dehydrogenase. It is converted to ribulose 1,5-bisphosphate in the **reductive pentose phosphate cycle** of photosynthesis.

L-ribulose-phosphate 4-epimerase EC 5.1.3.4; an enzyme that catalyses the reaction:

L-ribulose 5-phosphate = D-xylulose 5-phosphate.

Richner–Ranhart syndrome *see* **tyrosine transaminase**.

ricin a highly toxic lectin obtained from the seeds of the castor oil plant, *Ricinus communis*. It is a 493 amino-acid glycoprotein dimer comprising an A chain (M_r 66000) and a B chain (M_r 34000) linked by a disulfide bond. Ricin binds to the cell via interaction between the B chain and galactose groups on cell-surface receptors; then the disulfide bond is cleaved allowing the A chain to enter the cell. The A chain then binds to the 60S ribosomal subunit and, acting as an *N*-glycosidase, causes depurination of a specific adenosine residue in the 28S rRNA that is essential for peptidyltransferase activity, thereby halting protein translation. *See also* **abrin**, **ribosome-inactivating protein**.

rickets *or* **rachitis** a disturbance of calcium (and phosphate) metabolism in young growing animals (including humans), resulting from a deficiency of **vitamin D** or of its activation. It is characterized by softening of the bones and skeletal deformities, particularly bowing of the legs. — **rachitic** *adj.*

RID *abbr. for* radial immunodiffusion.

RIEP *abbr. for* rocket immunoelectrophoresis.

Rieske protein EC 1.10.99.1; *recommended name*: plastoquinol-plastocyanin reductase; *systematic name*: plastoquinol: oxidized-plastocyanin oxidoreductase; one of the four subunits of the cytochrome b_6–f complex. An enzyme that catalyses a reaction between plastoquinol-1 and two molecules of oxidized plastocyanin to form plastoquinone and two molecules of reduced plastocyanin. It is a high-potential 2Fe–2S protein. Apart from the Rieske protein, the subunits of the b_6–f complex consist of **cytochromes b_6** and ***f***, and a 17 kDa peptide; it is similar to other iron–sulfur proteins (e.g., in mitochondria).

rifampicin *or* **rifampin** an antibiotic, M_r 823, obtained as a semisynthetic derivative of rifamycin B, differing only in the substituents on the ring marked X in the structure for that compound. It inhibits bacterial transcription in sensitive bacteria (*see* **rifamycin**) at extremely low concentrations (0.01 µg ml^{-1}) while having no effect on eukaryotic transcription at 10^4 times that dose. Mitochondrial and chloroplast transcription may be affected at higher doses. It has application in molecular biology for removing plasmids from bacteria in a process known as **plasmid curing**.

rifamycin any anti-transcription antibiotic produced by *Streptomyces mediterranei* that is active against Gram-positive bacteria and *Mycobacterium tuberculosis* but much less effective against Gram-negative organisms. With the **streptovaricins** they are referred to as **ansamycins** – which contain a chromophoric naphthohydroquinone nucleus spanned by a long aliphatic bridge. They inhibit bacterial but not eukaryotic RNA synthesis via binding to the β subunit of bacterial DNA-dependent RNA polymerase and prevention of chain initiation; chain elongation is not affected once the initiation of RNA chain synthesis progresses beyond the third diester bond.

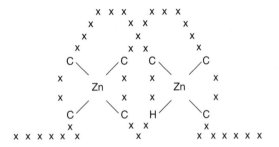

rifamycin B

rigid *(in lipid biochemistry)* a term sometimes used to describe the ordered phase of a lipid bilayer below the phase transition temperature. *See* **phase transition (of lipids)**.

rimorphin a 13-residue peptide that is a major Leu-enkephalin-containing peptide in tissues that contain **dynorphin** and α-neo-dynorphin. It yields three **enkephalin** sequences.

rim protein *see* **ABCR**.

Rim protein *an alternative name for* ABC transporter retina. *See* **Stargardt macular dystrophy**.

ring chromosome an intrachromosomal rearrangement in which either the chromosome is circularized by joining of the telomere ends, or the ends of the p and q arms break off with subsequent formation of a ring and the consequent loss of genetic material. Ring chromosomes can behave in one of three ways at cell division, the outcome depending on the numbers of sister chromatid exchanges (SCE). An even number of SCEs in the same direction, or none, results in symmetrical segregation of chromatids. An even number of SCEs in different directions will lead to interlocked rings. An odd number of SCEs will lead to the formation of a continuous ring

from two parallel chromatids that is twice the size of the original rings.

ring closure the formation of a covalent bond that converts a linear molecule into a ring structure.

Ringer–Locke solution *a less common name for* **Locke's solution**.

Ringer's solution *or* **Ringer solution** any of a number of aqueous solutions of mixed inorganic salts, of composition determined empirically by the British physician and physiologist Sydney Ringer (1835–1910), or adapted therefrom, that have been useful as fluid replacements *in vivo* or in place of serum as perfusing or suspending media for organs, tissues, or cells *in vitro*. Ringer's own mixtures (reported over the period 1882–87) were devised mostly for maintaining normal contractility of perfused isolated frog heart. One of the most effective solutions contained (in g L^{-1}, approx.) NaCl (7.0), KCl (0.04), CaCl$_2$ (anhydrous, 0.09), and NaHCO$_3$ (0.5). *See also* **Krebs' mammalian Ringer, Locke's solution, Tyrode's solution**.

Ringer–Tyrode solution *a less common name for* **Tyrode's solution**.

RING finger a **zinc finger** motif defined as

$$CX_2CX_{9-27}CXHX_2CX_2CX_{6-17}CX_2C$$

and present in a family of nuclear proteins. RING is an acronym for 'really interesting new gene'.

ring opening the breaking of a bond that converts a ring structure to an open-chain form.

RIP *abbr. for* **1** repeat-induced point (mutation); a phenomenon found in *Neurospora*; when more than one copy of a DNA sequence is present in a haploid genome that goes through a cross, both ectopic and endogenous copies are subject to point mutation at a high rate. RIP occurs in the dikaryotic ascogenous hyphae. Because transformation of *Neurospora* typically leads to gene amplification, RIP provides a highly efficient and selective method of introducing mutations *in vivo* in any gene cloned *in vitro*. **2** ribosome-inactivating protein. **3** Fas receptor interacting protein; a cytosolic protein kinase that contains a **death domain** and participates in the activation of NF-κB and in promoting apoptosis.

RISBASES proteins that share sequence similarity with members of the ribonuclease family, but are thought to act as ribonucleases. Examples are bovine seminal RNase and **angiogenin**. [From ribonuclease with special biological action.]

RISC *abbr. for* RNA-induced silencing complex; a ribonucleoprotein complex that targets its perfectly (or partially) complementary mRNA for either cleavage or translational repression. Short interfering RISC (siRISC) is programmed by siRNA, and micro RISC (miRISC) by miRNA. Both siRISC and miRISC are of two types, cleaving and noncleaving. It is likely that this specificity is determined by one of the **argonaute** proteins.

RIST *abbr. for* radioimmunosorbent test.

ristocetin a glycopeptide antibiotic complex from *Nocardia lurida*. Ristocetin A, M_r 2066, acts primarily on Gram-positive bacteria by inhibiting the biosynthesis of peptidoglycans of cell walls. It is particularly useful against infections of the gut when administered by injection, but not when taken orally, and has toxic side effects including kidney damage.

RITS *abbr. for* RNA-induced transcriptional gene silencing *or* RNA-induced initiation of gene silencing.

Rittenberg tube an evacuable two-legged tube provided with a stopcock and used for the conversion of ^{15}N-labelled ammonium salts into dinitrogen prior to mass-spectrometric analysis. It has been jocularly known, especially in Israel, as **Rittenberg's trousers**. [After David Rittenberg (1906–70), US biochemist.]

Rituximab a monoclonal antibody active against non-Hodgkin's lymphona (NHL). In some cases of NHL there is present a chemokine CD20, which arises due to the translocation of a gene from chromosome 14 to 18. The oncogene is *BCL2*, which is responsible for the overexpression of CD20. Thus rituximab is an anti-CD20 agent. The higher the grade of NHL the less likely it is to have CD20 receptors. The proprietary name is Mabthera.

R loop the structure formed when an RNA of complementary sequence stably base pairs with one of the DNA strands of double-stranded DNA, as happens at the start of **transcription**.

r.m.m. *abbr. for* relative molecular mass.

RMS *abbr. for* root mean square.

RMSD *abbr. for* root mean square deviation.

Rn *symbol for* radon.

RNA *abbr. and common name for* **ribonucleate** (def. 1) or **ribonucleic acid**; one of the two main types of **nucleic acid**, consisting of a long, unbranched macromolecule formed from ribonucleotides, the 3′-phosphate group of each constituent ribonucleotide (except the last) being joined in 3′,5′-phosphodiester linkage to the 5′-hydroxyl group on each ribose moiety of the next one. The presence of a free 2′-hydroxyl group on each ribose moiety renders these phosphodiester bonds susceptible to hydrolytic attack by alkali, in contrast to those of **DNA**. The RNA chain has polarity, with one **5′ end** and one **3′ end**. Two purines, adenine and guanine, and two pyrimidines, cytosine and uracil, are the major bases usually present. In addition, **minor bases** may occur; **transfer RNA**, however, contains unusual bases in relatively large amounts. The sequence of bases carries information, whereas the sugar and phosphate groups play a structural role. RNA is fundamental to **protein biosynthesis** in all living cells. **Messenger RNA** (mRNA) is responsible for carrying the coded genetic 'message', transcribed from DNA, to sites of protein assembly at the **ribosomes**. The latter are composed of **ribosomal RNA** (rRNA) plus proteins. A third species, transfer RNA (tRNA), is instrumental in importing amino acids to the assembly site, according to instructions carried by mRNA. In some viruses, RNA is the genetic material instead of DNA. Specific forms, functions, molecules, or preparations of RNA may be designated by prefixes or suffixes, thus: A-RNA *or* RNA-A, **A-form** of RNA; cRNA, **complementary RNA**; dsRNA, double-strand(ed) RNA; hnRNA *or* H-RNA *or* HnRNA, **heterogeneous nuclear RNA**; mRNA, **messenger RNA**; micRNA, **messenger-RNA-interfering complementary RNA**; mtRNA, **mitochondrial RNA**; nRNA, nuclear RNA; rRNA, **ribosomal RNA**; sRNA *or* S-RNA, **soluble RNA**; scRNA, small cytoplasmic RNA; snRNA, **small nuclear RNA**; snoRNA, **small nucleolar RNA**; ssRNA, single-strand(ed) RNA; tRNA, **transfer RNA**; tcRNA, **translational-control RNA**; Z-RNA *or* RNA-Z, **Z-form** of RNA.

RNA I a short 108 nucleotide plasmid-encoded transcript that is antisense to the 5′ end of RNA II, with which it forms a duplex. It acts as a negative regulator of plasmid replication as it inhibits the action of RNase H on **RNA II**. Pairing between RNA I and RNA II is facilitated by the Rop protein.

RNA II a plasmid-encoded transcript that initiates 555 bp upstream from the Col E1 origin of replication. Its cleavage by RNase H leaves a primer with a 3′-hydroxy end that is used for initiating DNA synthesis during plasmid replication.

RNA-activated protein kinase *see* PKR.

RNA adenylating enzyme *see* polynucleotide adenylyltransferase.

RNAase *(sometimes) abbr. for* ribonuclease; RNase is more usual.

RNA-dependent DNA polymerase *another name for* RNA-directed DNA polymerase (*see* **reverse transcriptase**).

RNA-dependent RNA replicase *see* RNA-directed RNA polymerase.

RNA-directed RNA polymerase EC 2.7.7.48; *other name*: RNA nucleotidyltransferase (RNA-directed); an enzyme that catalyses a reaction between a nucleoside triphosphate and RNA_n to form RNA_{n+1} and pyrophosphates; it is required for the replication of RNA viruses; in the case of eukaryotic viruses, the enzyme normally comprises several subunits.

RNA–DNA virus *a former name for* **retrovirus**.

RNA editing *see* **editing** (def. 2).

RNA editing enzyme *or* **RNA-dependent deaminase** (*abbr.*: RED);

any enzyme that catalyses deamination of a nucleotide base (e.g. cytosine to uracil; adenine to inosine) at a specific site in RNA. *See* **APOBEC1**.

RNA helicase an ATP-dependent enzyme that is similar to **DNA helicase** except that its substrate is RNA. An example is the *dbpA* gene product from *Escherichia coli*, which is specific for 23S rRNA.

RNAi *abbr. for* RNA interference.

RNA-induced silencing complex *see* RISC.

RNA-interacting module any peptide sequence in a polypeptide that binds RNA, e.g. G-patch and some zinc fingers (of C2H2 and C4 types).

RNA interference (*abbr.*: RNAi) *or* **post-transcriptional gene silencing** a general mechanism for silencing the transcript of an active gene in many organisms. The interference is initiated by **siRNA** (as part of an RNA-induced silencing complex; *see* **RISC**) that is specific for the nucleotide sequence of the target transcript. There is no evidence that this type of silencing occurs in normally growing animals, but it has been demonstrated in *Caenorhabditis elegans* and is used by plants to combat virus infection.

RNA ligase (ATP) EC 6.5.1.3; *other name*: polyribonucleotide synthase (ATP); an enzyme from bacteriophage T4 that catalyses ATP-dependent ligation of single-stranded RNA or DNA onto the 5′-phosphate terminus of single-stranded RNA or DNA molecules:

$$ATP + (ribonucleotide)_n + (ribonucleotide)_m = ADP + pyrophosphate + (ribonucleotide)_{m+n}.$$

RNA polymerase any enzyme that utilizes ATP, GTP, CTP, and UTP to synthesize RNA from a DNA or RNA template. DNA-dependent RNA polymerases include (i) the enzymes responsible for transcription of DNA and (ii) **primases**, which synthesize the RNA primers required for replication of DNA. An **RNA-directed RNA polymerase** (replicase or RNA synthetase) is involved in the replication of RNA viruses.

DNA-dependent (or DNA-directed) RNA polymerase (EC 2.7.7.6); *other name*: RNA nucleotidyltransferase (DNA-directed); it catalyses a reaction between a nucleoside triphosphate and RNA_n to form RNA_{n+1} and pyrophosphate. This EC class number includes the following examples.

In prokaryotes one DNA-dependent RNA polymerase transcribes all classes of DNA. The core *Escherichia coli* enzyme consists of three types of subunit, α, β, and β′, and has the composition $α_2ββ′$; the holoenzyme contains an additional σ subunit or sigma factor. **Sigma factors** are responsible for specific promoter recognition and transcription initiation. Following transcription initiation, the sigma factor dissociates from the core enzyme. The β subunit is probably involved in binding ribonucleotide substrates; it is the target for **rifamycin** and **streptolydigin** antibiotics. The β′ subunit may bind template DNA.

Phage RNA polymerases are much smaller and simpler than bacterial ones: the polymerases from phage T3 and T7 RNA, e.g., are single polypeptide chains of <100 kDa.

Eukaryotes have three types of DNA-dependent RNA polymerase in the nucleus, each producing a different type of RNA. RNA polymerase I, located in the nucleolus, synthesizes precursors of most ribosomal RNAs; RNA polymerase II, located in the nucleoplasm, synthesizes messenger RNA precursors; and RNA polymerase III, also located in the nucleoplasm, synthesizes the precursors of 5S rRNA, the transfer RNAs, and other small RNAs. These enzymes can be distinguished by their sensitivity to α-amanitin (*see* **amanitin**); class I enzymes are not inhibited by this octapeptide, whereas class II enzymes are rapidly inhibited by low concentrations; the sensitivity of class III enzymes varies depending on species. **RNA polymerase I** has 13 subunits totalling over 600 kDa; it recognizes a single type of promoter and requires at least two ancillary factors, UBF1 and SL1; it accounts for 50–70% of the RNA polymerase activity in eukaryotic cells. **RNA polymerase II** from all sources contains two large subunits of 215 and 139 kDa and several smaller polypeptides. The 215 kDa subunit has homology with the prokaryotic β′ subunit and contains a C-terminal domain (CTD) with multiple repeats of the consensus sequence YSPTSPS; the CTD is involved in initiation. The 139 kDa subunit has homology with the bacterial β subunit; the third largest subunit is related to the bacterial α subunit. Three of the remaining subunits are common to RNA polymerases I, II, and III. Most promoters used by RNA polymerase II contain the **TATA box**, which is recognized by the TATA-binding protein (TBP), a component of the **transcription factor** TFIID. Several other accessory factors are required for transcription by RNA polymerase II. **RNA polymerase III** has 14 subunits totalling 700 kDa. The promoters it uses may be upstream of the transcribed gene (for snRNAs) or within the gene (for tRNAs and 5S RNA). Three accessory factors, TFIIIA (a **zinc-finger** protein), TFIIIB, and TFIIIC, are involved in transcription from internal promoters.

Mitochondria and chloroplasts also contain RNA polymerase activity that synthesizes the RNAs required for protein synthesis in these organelles. These polymerases are smaller than the nuclear enzymes and more like prokaryotic RNA polymerases.

[RNA-polymerase]-subunit kinase EC 2.7.1.141; *other name*: CTD kinase (*abbr.* for C-terminal domain kinase); an enzyme that catalyses a reaction between ATP and [DNA-directed RNA polymerase] to form phospho-[DNA-directed RNA polymerase] and ADP.

RNA primer a segment of RNA (60 nt long in prokaryotes, depending on species) that is complementary to a segment of the template DNA and is formed by DNA **primase** during the initiation of DNA replication. *See* **replication of DNA**.

RNA processing the splicing, 5′-capping, and 3′-polyadenylation by which eukaryotic pre-mRNA is converted into mature mRNA. All three processes are associated with the C-terminal domain of the largest subunit of the **RNA polymerase** II complex. *See* **cap**; **poly(A) tail**; **splicing** (def. 2).

RNase *usual abbr. for* ribonuclease; *compare* **RNAase**.

RNase III a dsRNase such as **Dicer** and **Drosha** involved in the production of siRNA and miRNA respectively. *See also* **ribonuclease**.

RNase A *abbr. for* ribonuclease A (pancreatic ribonuclease). *See* **ribonuclease**.

RNase B *abbr. for* ribonuclease B; a glycosylated form of ribonuclease A with the same specificity. *See* **ribonuclease**.

RNase H *abbr. for* ribonuclease H. *See* **ribonuclease**.

RNase MRP a site-specific endoribonuclease for mitochondrial RNA processing (hence 'MRP'). It contains an essential RNA component and cleaves mitochondrial RNA (mtRNA) transcripts that are complementary to the D-loop region of vertebrate mtDNA.

RNase ONE *a proprietary name for* RNase I from *E. coli* that catalyses hydrolysis of phosphodiester bonds after all four ribonucleotides.

RNase OUT *a proprietary name for* a formulation for removing RNase and DNase from laboratory surfaces.

RNase P *abbr. for* ribonuclease P. *See* **ribonuclease**.

RNase T1 *abbr. for* ribonuclease T1; a single-stranded RNA (or DNA) endonuclease from *Aspergillus* that cleaves after guanine residues. *See* **ribonuclease**.

RNA silencing *or* cosuppression *or* post-transcriptional gene silencing (*abbr.*: PTGS); **RNA interference** in plants that is induced by viruses as a general antiviral defence mechanism. Many plant viruses encode proteins that are suppressors of silencing. When triggered locally, it can spread via a mobile silencing signal.

RNasin a proprietary recombinant ribonuclease inhibitor, similar to human placental RNase inhibitor, that acts by forming stable noncovalent complexes with ribonuclease.

RNA synthetase an RNA-directed **RNA polymerase**.

RNA tumour virus a common designation for **retrovirus**, as this is the only group of RNA viruses known to cause cancer.

RNP *abbr. for* ribonucleoprotein. Specific kinds of RNP may be designated by prefixes, thus: hnRNP, **heterogeneous nuclear ribonucleoprotein**; scRNP, **small cytoplasmic ribonucleoprotein**; snRNP, **small nuclear ribonucleoprotein**.

RNPase *or* **RNP disruptase** a putative enzyme that disrupts RNA–protein interactions.

RNP disruptase *an alternative name for* **RNPase**.

Ro an RNA-binding protein and autoantigen in human systemic lupus erythematosus (SLE). Mice that lack Ro develop antibodies against ribosomes and chromatin and have glomerulonephritis and skin photosensitivity as in SLE. Ro also enables the nematode

Caenorhabditis elegans to withstand environmental stress, and some bacteria to withstand UV irradiation.

Robertsonian translocation any translocation in which the long arms of two acrocentric chromosomes fuse at the centromere. In humans, chromosomes 13, 14, 15, 2, and 22 are most commonly involved, especially 13 and 14. Robertsonian fusion of the long arms of chromosomes 21 and 14 is responsible for about 4% of Down syndrome cases. [After W.R.B. Robertson (1881–1941), US geneticist.]

Robison ester *an old name for* D-glucose 6-phosphate. *Compare* **Neuberg ester**. [After R. Robison, British chemist (1883–1941).]

Robo *or* **roundabout** any of a conserved family of receptors present on the growth cones of *Drosophila* axons that do not cross the midline. They consist, in sequence, of five immunoglobulin-like domains, three fibronectin repeats, one transmembrane segment, and four cytoplasmic motifs. On binding to a slit protein they regulate intracellular GTPase-activating proteins (GAPs). In zebrafish Robo2 is called astray.

robot any device for performing computational or physical tasks automatically. For example, in computing, a robot is a program that trawls the Web for information, indexing it, say, for a search engine. In biology, robot machines execute the numerous mundane, repetitive tasks associated with high-throughput technologies (e.g. genome sequencing, drug screening). –**robotic** *adj.*; –**robotism** *or* **robotry** *n.*

robotics the automation of laboratory procedures in which there are a number of repetitive operations. In genomics and proteomics such operations may be performed by a robotics workstation.

ROC *abbr. for* receiver operating characteristic.

Rochelle salt potassium sodium tartrate tetrahydrate; it is used as a constituent of **Fehling's solution**. [After La Rochelle, French port.]

rocket immunoelectrophoresis *abbr.*: RIEP; a one-dimensional **electroimmunodiffusion**, *also known as* **electroimmunoassay**, used for quantitative analysis of proteins. Exact volumes of protein (antigen) solution are applied in an agarose gel of uniform thickness containing the corresponding antibody. An electric field causes the migration of both antigen and antibody. They react to form elongated curved precipitation lines, known as rockets because of shape, which are proportional in size to the amount of antigen. *See also* **crossed immunoelectrophoresis**.

rod cell *or* **rod** a rod-shaped light-sensitive cell of the vertebrate retina. Rod cells contain **rhodopsin** or (in freshwater and migratory fish and some amphibians) **porphyropsin**, and are responsible for vision in dim light. They occupy all of the margin of the retina and are absent from the fovea. Each is a highly specialized cell with an outer and inner segment, a cell body, and a synaptic region where the rod passes a chemical signal to retinal nerve cells. Phototransduction occurs in the outer segment, which contains a stack of rhodopsin-rich membrane disks. The plasma membrane of the outer segment contains cyclic-GMP-gated Na^+ channels. These are kept open in the dark by cyclic GMP molecules bound to the channels. Metarhodopsin II, i.e. rhodopsin activated by a photon, activates **transducin**, a **G-protein**, which then activates a cyclic GMP phosphodiesterase, reducing the cellular level of cyclic GMP to initiate a nerve impulse in the optic nerve. *Compare* **cone cell**.

rod gel a gel used in a form of polyacrylamide gel electrophoresis, in which the gel is cast in cylindrical glass tubes. Now largely superseded by slab-gel electrophoresis.

roentgen *a variant spelling of* **röntgen**.

Rohypnol *proprietary name for* flunitrazepam, a **benzodiazepine** central nervous system depressant administered to treat insomnia. It is used in drug-facilitated rape ('date rape').

roller bottle a cylindrical container used for the culture of monolayers of adherent cells. *See* **roller culture**.

roller culture a method of cell culture using roller bottles in which adherent cells are bathed with culture medium as the bottle is rotated. This method of culture is used when large numbers of cells are required.

rolling circle a mechanism for the replication of circular double-stranded DNA. One strand is nicked, then, using the unbroken strand as a template, the DNA is replicated around the circle. Replication does not cease when one circumference has been replicated, but continues around the circumference several more times (hence the name 'rolling'). Consequently, multimers of the replicon are produced in a long single strand with the appearance of a tail; this may then be cut into unit replicons, or into multimers containing tandemly repeated copies of the original unit; these may remain as single strands or be transcribed to yield a complementary strand.

röntgen *or* **roentgen** *symbol* R; the unit of **exposure** (def. 3) to X- or gamma radiation, expressed in terms of the amount of ionization caused in air. Defined originally (1928, 1937) as that quantity of radiation having a corpuscular emission per 0.001293 gram of air (equivalent of 1 cm^3 of dry air at stp) producing, in air, ions carrying one electrostatic unit quantity of electricity of either sign. Expressed in SI units, 1 R causes the production in air of ions (of one sign) carrying a charge of 2.58×10^4 C kg^{-1}; this corresponds to the formation of 1.61×10^{15} ion-pairs per kg of air and to an absorption of energy by air equal to 0.00869 J kg^{-1}. The energy absorption by water or tissue from 1 R of x- or gamma-radiation is about 0.0096 J kg^{-1}. Prior to 1956, the röntgen was used in clinical work to express both exposure and **absorbed dose**. *See also* **rad** (def. 2), **rem**, **rep**. [After Wilhelm Konrad von Röntgen (1845–1923), German physicist, discoverer of X-rays.]

röntgen equivalent man *or* **röntgen equivalent man** *see* **rem**.

röntgen equivalent physical *see* **rep**.

röntgen rays *a former name for* **X-rays**.

room temperature ambient temperature, often used for reactions that can be carried out on the bench without devices to maintain constant temperature. Normally within several degrees of 20 °C.

root effect the lowering of ligand affinity and of cooperativity of ligand binding (in hemoglobin) brought about by a low pH. [After R. W. Root who reported it in 1931.]

root mean square *abbr.* RMS; the positive square root of the mean of the squares of a series of values of x_i:

$$\text{RMS value} = \sqrt{\left(\frac{x_1^2 + x_2^2 + \dots x_n^2}{n}\right)}$$

Compare **standard deviation**.

root mean square deviation *abbr.*: RMSD; the square root of the sum of squared deviations of the observed values from the mean, divided by the number of observations. The RMSD gives a measure of the dispersion of points in a data set, or, in molecular structure determination, a measure of the similarity of two molecular conformations.

Rop *abbr. for* Rho-like protein of plants.

ROS *abbr. for* **1** rod outer segment; *see* **rod cell**. **2** reactive oxygen species.

ROS a gene family encoding receptor-tyrosine kinases of unknown function; v-*ros* is the oncogene of the acutely transforming avian sarcoma virus UR2 (named after the University of Rochester). The vertebrate gene product, Ros, is similar to the *Drosophila* **sevenless protein**; it appears to be associated with a phosphoinositide response (*see* **phosphatidylinositol cycle**); Ros-1 is a member of the Src family.

rosetting a method of isolating cells by allowing them to associate with red blood cells. For example, lymphocytes become surrounded with red cells, forming a 'rosette', and may then be isolated by sedimentation through **Ficoll** gradients.

Rossmann fold a super-secondary motif in protein structure comprising two adjacent β-α-β units, creating a β-α-β-α-β domain, and often containing a hydrophobic core between the closely packed sheet and helices. The Rossmann fold occurs in a variety of proteins, but was first described as the NAD(P)-binding domain in a family of dehydrogenases.

rota-evaporate to evaporate a solution in a rotating round-bottomed flask in a temperature-controlled water bath, using an apparatus that can at the same time apply a vacuum to the flask interior. —**rota-evaporation** *n*.

rotamer *see* **rotational isomer**.

rotary evaporator an apparatus used for **flash evaporation**; *see also* **rota-evaporate**.

rotational correlation time *see* **correlation time**.

rotational diffusion rotation of a solute molecule around its centre of mass as a result of **Brownian movement**.

rotational frictional coefficient *see* **frictional coefficient**.

rotational isomer molecules of identical **configuration** may be distinguished as having different **conformations** after rotation about a bond. If an assembly of atoms X–A–B–Y (X and Y not being collinear with A–B) is viewed along the axis A–B, the X–A and B–Y bonds will make an angle with each other which may be 0° (superposition), 180° or some intermediate angle. This angle is termed the torsion angle, designated θ or ψ. Conformations are then described as synperiplanar (*sp*), synclinal (*sc*), anticlinal (*ac*) or antiperiplanar (*ap*) according as the torsion angle is within ± 30° of 0°, ± 60°, ± 120°, or ± 180°, respectively. Formerly, the terms *cis*, *gauche* and *trans* were in use for synperiplanar, synclinal and antiperiplanar respectively.

rotational strength *or* **rotatory strength** *symbol*: R; an index of the degree to which an **absorption band** exhibited by an optically active solute displays **circular dichroism**. It may be evaluated for the absorption band in question by means of the expression:

$$R = 2.296 \times 10^{-44} \int \left(\frac{\Delta \varepsilon_\lambda}{\lambda}\right) d\lambda$$

where λ is a wavelength of observation and Δe_k is the difference between the molar (decadic) absorption coefficients of the left and right circularly polarized light beams at that wavelength. Rotational strength is thus essentially the absorption coefficient difference integrated over the entire absorption band.

rotation axis a crystal or an oligomeric protein molecule is said to possess an *n*-fold rotation axis if it presents exactly the same appearance after a rotation about that axis of 360°/*n*. *Compare* **screw axis**.

rotation–reflection axis *see* **symmetry**.

rotenone an insecticide and the active principle in derris. It is present in plants of the genera *Derris*, *Lonchocarpus*, *Tephrosia*, and *Mundulea*, used by indigenous peoples as fish poisons (*Derris* in Asia, *Lonchocarpus* in South America, and *Tephrosia* more extensively). Rotenone inhibits mitochondrial respiration on the oxygen side of the nonheme iron of NADH dehydrogenase. Extracts of rotenone-containing plants are used as insecticidal sprays and dusts (derris dust). *Compare* **piericidin A**.

rotor **1** the part of a centrifuge that holds the material to be subjected to centrifugal force. It generally has compartments that house tubes or bottles, and these may either be set at a fixed angle with respect to the axis of rotation, or designed to allow the containers to swing out into a horizontal position during centrifugation. Centrifugal force is generated by rotating the rotor about its axis. **2** the part of the **H⁺-transporting ATP synthase** that rotates during flow of hydrons through the F₀ part of the complex. *Compare* **stator**.

Rotor syndrome a rare autosomal recessive benign predominantly conjugated hyperbilirubinemia. Unlike **Dubin–Johnson syndrome** it shows no liver pigmentation. A defect (of unknown nature) in canalicular multiorganic anion transporter (cMOAT) is thought to be responsible. [After Arturo Belleza Rotor (1907–88), Philippine physician.]

Rot value *or* **R₀t value** a measure of the kinetic complexity of RNA based on its hybridization ability. The notation is a modification of R_0t, where R_0 is the initial RNA concentration and t is time. It is analogous to the Cot value (*see* **C₀t**) for DNA. $R_0t_{1/2}$ is the value corresponding to half complete association between complementary polynucleotides.

roughanic acid all-*cis*-7,10,13-hexadecatrienoic acid; a plant fatty acid present in storage lipids.

rough endoplasmic reticulum *abbr.*: RER; a region of the **endoplasmic reticulum** associated with ribosomes and involved in the synthesis of secreted and membrane-bound proteins. The nascent protein destined for secretion with its **signal peptide** associates with a receptor for the **signal recognition particle** and then translocates through the RER lipid bilayer by means of proteins, including **TRAM,** which form a **translocon**. The signal peptide is removed by the **signal peptidase**. The process is aided by **heat-shock proteins**, which serve as **chaperones**.

roundabout *an alternative name for* **Robo**.

Round-up *see* glyphosate.

rousvirus *see* retrovirus.

rP the negative decadic logarithm of the molar concentration of inorganic phosphate; i.e. $rP = -\log[P_i]$. If, for the equilibrium

$$X + P_i = XP,$$

$$K = [X][P_i]/[XP]$$

then

$$rP = -\log K - \log([XP]/[X]).$$

RP-HPLC *abbr. for* reversed phase high-pressure liquid chromatography; *see* **reversed phase**.

R plasmid *abbr. for* resistance plasmid.

RPM *abbr. for* Redhat Package Manager; a utility designed by the authors of the Redhat distribution of **Linux** but also implemented by other distributors of Linux. RPM files are compressed and can be found in certain **software archives**.

RPMI *abbr. for* Roswell Park Memorial Institute (Buffalo, USA).

r-protein *abbr. for* ribosomal protein.

R protein *see* R gene.

RP-TLC *abbr. for* reversed phase thin-layer chromatography; *see* **reversed phase**.

RQ *abbr. for* respiratory quotient.

RRM *abbr. for* RNA recognition motif; a conserved 80-amino-acid sequence present in a family of proteins that bind RNA. It has a β-α-β-β-α-β structure.

rRNA *abbr. for* ribosomal RNA; *see* **ribosome**.

Rrp *abbr. for* ribosomal RNA processing; any of a group of exonucleases that are involved in processing of RNA precursors in yeast, where they form part of the exosome core in the cytoplasm. Some Rrps catalyse a phosphorylase reaction (Rrp41p, Rrp42p, Rrp43p, Rrp45p, and Rrp46p), while others catalyse a hydrolase reaction (Rrp4p, Rrp40p, and, in the nucleus, Rrp6p).

(RS)- *prefix (in chemical nomenclature)* denoting a racemate containing a single chiral centre. *See* **sequence rule**.

R/S convention the system of configuration specification of chiral compounds according to the **sequence rule**.

R segment *abbr. for* repeat segment; part of retroviral RNA consisting of direct repeats, i.e. repeats having the same orientation; the size of the repeat varies in different strains of virus, from 80 to 100 nucleotides.

RSK *abbr. for* ribosomal protein S6 kinase; any of a family of protein serine/threonine kinases that, on binding of certain growth factors to their tyrosine-kinase receptors, translocate from cytoplasm (where they may phosphorylate ribosomal protein S6) to the nucleus (where they phosphorylate several transcription factors). RSK2 is the same as CREB (protein) kinase.

RSKG-5 *see* tonin.

R state *see* R form.

rtDNA *(sometimes) abbr. for* **recombinant DNA**; use not recommended.

RTP *symbol for* any unspecified purine nucleoside 5′-triphosphate.

RT-PCR 1 *abbr. for* reverse transcriptase-polymerase chain reaction *or* reverse transcription-polymerase chain reaction; a quantitative technique for amplifying DNA transcripts of cellular mRNA. **2** *sometimes abbr. for* **real-time PCR**; use not recommended.

rtTA *abbr. for* reverse tetracycline-controlled *trans*-activator; a chimeric transcriptional activator based on a modified DNA-binding domain of the tetracycline controlled *trans*-activator fused to the *trans*-activator domain of VP16 and used in the **Tet-On** system.

rTth a proprietary recombinant thermostable DNA polymerase originally from *Thermus thermophilus*. It has proofreading activity and is used in the generation of long (>4 kb) polymerase chain reaction (PCR) products. It also has efficient reverse transcriptase activity in the presence of manganese.

Ru *symbol for* ruthenium.

RU-486 *or* **RU-38486** mifepristone; 11β-(4-dimethylaminophenyl)-17β-hydroxy-17α-prop-1-ynylestra-4, 9-dien-3-one; a steroidal progesterone-receptor antagonist that prevents implantation of a fertilized ovum in the uterus. It is an effective contraceptive even when taken after coitus.

rubber a *cis*-1,4-polyisoprene with M_r varying from about 1×10^5 to 1×10^6, representing polymers of about 1500 to 60000 **isoprene** residues. It is found in the latex of about 300 different flowering plants, but only *Hevea brasiliensis* is a commercial source. A nonelastic rubber from *Manilkara bidentata* latex is known as **balata**. It is a *trans*-1,4-polyisoprene.

ruberythric acid the glycoside; 2-alizarin β-D-primeveroside, a major component of the plant dyestuff madder.

primeverose—O

OH

Rub 1 homologue of Nedd-8 *see* Nedd.

rubidium *symbol*: Rb; a silvery-white metallic element of group 1 of the IUPAC **periodic table**; atomic number 37; relative atomic mass 85.48. Highly reactive, it must be stored in an inert atmosphere. Rb^+ may be used experimentally as a K^+ analogue (sometimes as substitute, other times as competitor).

Rubinstein–Taybi syndrome *see* CBP/p300.

Rubisco *abbr. for* ribulose-bisphosphate carboxylase/oxygenase. It has **ribulose bisphosphate carboxylase** activity, and in a side-reaction with dioxygen it produces 2-phosphoglycolate (*see* **photorespiration**).

RuBP *abbr. for* ribulose 1,5-bisphosphate.

rubredoxin *abbr.*: Rd; any of a group of bacterial **iron–sulfur proteins** that are small, highly conserved, contain non-heme iron, and are non-enzymic. They lack acid-labile sulfur, and are characterized by having either one or two iron atoms each entirely in mercaptide coordination, i.e. surrounded by four cysteine residues or sulfur-containing ligands. They are electron acceptors for, e.g., hydrogenase-linked redox processes and hydroxylase systems. Rubredoxins of M_r 6 000–19 000 have been found in a number of bacteria. The ferric form is intense red; the ferrous form is colourless. *Compare* **ferredoxin**.

rudimentary a gene that in *Drosophila* is the equivalent of *pyr-1* in a number of other eukaryotes. It encodes a **CAD protein**.

ruffled edge *or* **ruffling** the appearance of the leading edge of a moving cell in culture, caused by the extension of microspikes and lamellipodia, which when not attached to the substratum have the appearance of ruffles in flimsy dress fabric.

rule *(in sequence analysis)* a short regular expression (typically 4–6 residues in length) used to identify generic (non-family-specific) patterns in protein sequences. Rules tend to be used to encode particular functional sites (e.g. sugar attachment sites, phosphorylation sites, glycosylation sites). Their small size means that they do not provide good discrimination, and can only indicate that a certain functional site *might* exist in a sequence.

Runt a protein of *Drosophila* that is involved in sex determination, segmentation, and neurogenesis. Its DNA-binding domain is homologous with that of Lozenge (of *Drosophila*) and AML1 (of human). Several Runt-related proteins occur in mammals.

RUNT domain *or* **runt homology domain** (*abbr.*: RHD) a 128-amino-acid DNA-binding domain occurring in **Runt** and other proteins required for the formation of hematopoietic, skeletal, gastrointestinal, and nervous systems in a variety of organisms. RHD proteins form a complex with core binding factor β (CBFβ).

runt gene a *Drosophila* **pair-rule gene** encoding a transcription factor involved in segmentation.

Russell's viper venom the venom of Russell's viper, *Vipera russellii*, a snake of SE Asia. It is a source of enzymes especially phospholipases, proteases, and ribonucleases.

rusticyanin a blue, copper-containing, single-chain polypeptide obtained from a thiobacillus. It undergoes reduction during oxidation of Fe^{2+} with dioxygen. It is stable in acid, labile in alkali, and contains no arginine.

rutamycin *or* **oligomycin D** 26-demethyloligomycin A; an agent that is strongly inhibitory to a limited number of yeast and other fungal species. It inhibits mitochondrial ATPase F_1, preventing phosphoryl group transfer, but has no effect on isolated F_1 units. *See also* **oligomycin**.

rutamycin

ruthenium red ruthenium oxychloride ammoniated; $[(NH_3)_5-Ru-O-(NH_3)_4-O-Ru-(NH_3)_5]Cl_6 \cdot 4H_2O$. A brownish-red, water-soluble inorganic compound, used in microscopy and electron microscopy as a stain for acid mucopolysaccharides. It inhibits the influx of Ca^{2+} ions into mitochondria at extremely low concentrations.

rutinose the disaccharide, 6-*O*-β-L-rhamnosyl-D-glucose; it occurs in rutin, the 3′-rutinoside of **quercetin**.

Ruv any of a family of three proteins involved in branch migration of a **Holliday junction**. In *E. coli*, Ruv A (209 amino acids) forms a homotetramer that binds the junction. Its crystal structure resembles a four-petalled flower and has four negatively charged projections at the centre. The junction fits in the concave face and is pierced at its centre by the projections. Ruv B (37 kDa) is a DNA-dependent ATPase that forms a homohexameric ring that binds dsDNA on

each side of the Ruv A tetramer. Ruv C (19 kDa) is a nuclease that cleaves the crossover strands of the junction.

R_B value *or* **relative band rate** *or* **relative band speed** *(in chromatography)* *symbol*: R_B; the ratio of the distance travelled by a zone of the substance of interest to the distance simultaneously travelled by a reference substance B. *See also* R_f **value**.

R_f value *or* **R_F value** *or* **band rate** *or* **band speed** *symbol*: R_f *or* R_F; *(in chromatography)* the ratio of the distance travelled by the centre of a zone of the substance of interest to the distance simultaneously travelled by the mobile phase. In paper and thin-layer chromatography, it is generally determined from the distance travelled by the substance from the origin and the distance travelled by the eluent front from the same point. [From 'rate relative to front'.]

RXR *abbr. for* **retinoid receptor**.

ryanodine a toxic alkaloid with insecticidal activity, isolated from *Ryania speciosa*. *See also* **ryanodine receptor**.

ryanodine receptor *abbr.*: RYR; any of a group of receptor proteins that bind **ryanodine**. Although first discovered in muscle, they occur in other cell types, and function as intracellular ligand-gated Ca^{2+}-release channels. In muscle, they provide communication between transverse tubules (**T-tubules**) and sarcoplasmic reticulum (SR); contraction of skeletal muscle is triggered by release of Ca^{2+} from SR following depolarization of T-tubules. The Ca^{2+} channel is modulated by Ca^{2+}, Mg^{2+}, and calmodulin. Three types are known: RYR1, the skeletal muscle type; RYR2, the cardiac muscle type; and RYR3, a highly truncated protein found in some nonmuscle cells. Cyclic ADP-ribose activates the RYR2 but not the RYR1 receptor. In muscle the receptors are tetrameric integral transmembrane glycoproteins. The C-terminal region is on the SR lumen side and contains four transmembrane helices per subunit; the N-terminal region (the major part of the molecule for RYR1 and 2) protrudes into the cytosol, where it forms a foot region in close proximity to the voltage-gated dihydropyridine receptors in T tubules. Deficiency is associated with malignant hyperthermia.

Ryk *abbr. for* related-to-tyrosine-kinase; a subfamily of the receptor tyrosine kinase superfamily. Members of Ryk have a short extracellular region that is related to the WIF domain. Homologues occur in *Drosophila*, *Caenorhabditis elegans*, and in most mammalian tissues.

RYN method a method for determining the plausibility of an **open reading frame**. A computational analysis of the sequence determines the frequency of the favoured codons R (purine), Y (pyrimidine), and N (anything). If these are found to be in the proportions expected from that normally found, it lends support to the fact that the correct **reading frame** has been selected, and that the open reading frame is correct.

RYR *abbr. for* ryanodine receptor.

Ss

s *symbol for* **1** second (def. 1). **2** solid state (in a chemical equation). **3** atomic orbital corresponding to an orbital angular momentum quantum number (l) = 0. **4** singlet (in nuclear magnetic resonance spectroscopy). **5** strong absorption (in infrared spectroscopy).

s *symbol for* **1** sedimentation coefficient. **2** solubility. **3** *(in monosaccharide nomenclature)* the **septanose** form; it is suffixed to the symbol for the corresponding monosaccharide; e.g. glucoseptanose is symbolized as Glcs. **4** specific entropy (i.e. **entropy**/mass). **5** spin quantum number.

s- *abbr.: for* ***sec-***.

S *symbol for* **1** sulfur. **2** siemens. **3** Svedberg unit ($S_{w,20}$ is the symbol under standard conditions). **4** a residue of the α-amino acid L-serine (alternative to Ser). **5** a residue of an incompletely specified base in a nucleic-acid sequence that may be either cytosine or guanine. **6** a residue of the ribonucleoside **thiouridine** when the position of the thiol group in its thiouracil moiety is unknown or unspecified (alternative to sU *or* Srd). **7** substitution (reaction mechanism).

^{2}S *symbol for* a residue of the ribonucleoside 2-thiouridine (alternative to s^2U); *see* **thiouridine**.

^{4}S *symbol for* a residue of the ribonucleoside 4-thiouridine (alternative to s^4U); *see* **thiouridine**.

S_N1 *abbr. for* unimolecular nucleophilic substitution.

S_N2 *abbr. for* bimolecular nucleophilic substitution.

$S_{w,20}$ *see* **S** (def. 3).

S *symbol for* **1** entropy. **2** spin quantum number (of a system). **3** strangeness quantum number. **4** area (alternative to A).

S *stereochemical descriptor* denoting *sinister*; i.e. the absolute configuration of a chiral compound in which the priority (obtained using the Cahn–Ingold–Prelog sequence rules) of the substituent groups on the chiral centre decreases in an anticlockwise (left-handed) direction. For use as a prefix in chemical nomenclature, the descriptor is placed in parentheses and connected with a hyphen; e.g. (*S*)-alanine. For chiral compounds in which the relative but not the absolute configuration is known, *sinister* configuration is denoted by the prefix (*S**)-. *See* **sequence rule**. *Compare* ***R***.

S- *prefix (in chemical nomenclature)* denoting substitution on a sulfur atom in a compound; a particular sulfur atom may be specified by a right superscript.

-S *conformational descriptor* designating the **skew conformation** of a six-membered ring form of a monosaccharide or monosaccharide derivative. Locants of ring atoms that lie on the side of the structure's reference plane from which the numbering appears clockwise are indicated by superscripts preceding the letter, and those that lie on the other side of the reference plane by subscripts following the letter; e.g. 1,2-*O*-ethylidene-α-D-glucopyranose-$^1S_3$. *See also* **conformation**.

s.a. *abbr. for* specific activity.

SAA *abbr. for* serum amyloid A; *see* **amyloid A protein**.

SABP *abbr. for* secretory actin-binding protein.

saccharase any hydrolase that hydrolyses the fructose (**fructosaccharases**) or glucose (**glucosaccharases**) end of appropriate oligosaccharides, liberating fructose or glucose respectively. The term is also used for **invertase** (which is both a fructosaccharase and a glucosaccharase).

saccharate *see* **glucarate**.

saccharic acid 1 *an old name for* **aldaric acid**. **2** *an old name for* glucaric acid; *see* **glucarate**.

saccharide any carbohydrate, especially a simple sugar, i.e. a **monosaccharide**, **disaccharide**, or short-chain **oligosaccharide**. *See also* **polysaccharide**.

saccharification the hydrolysis of sugar derivatives or complex carbohydrates to soluble, fermentable sugars.

saccharimeter an instrument for measuring the amount of sugar in a solution, especially a **polarimeter** used for such a purpose. *Compare* **saccharometer**.

saccharin 2,3-dihydro-3-oxobenzisosulfonazole; *O*-sulfobenzimide; an artificial, nonnutritive sweetener that is several hundred times sweeter than sucrose. Rats receiving high doses produce tumours of

the urinary tract but there is no clear-cut evidence of such an effect in humans.

saccharogenesis the process of producing soluble, fermentable sugars from a complex carbohydrate. —**saccharogenic** *adj.*

saccharogenic method a method for assaying amylase and similar enzymes that hydrolyse polysaccharides, based on determination of the amount of simple sugar product formed by the enzyme.

saccharometer an instrument for measuring the amount of sugar in a solution, especially a **hydrometer** specially calibrated for the purpose. *Compare* **saccharimeter**.

Saccharomyces a genus of budding yeasts that reproduce asexually by budding or sexually by conjugation. They are used especially as simple model organisms in the study of eukaryotic cell biology, and in genetic engineering. *Saccharomyces cerevisiae* is used in bread-making and in the production of alcoholic beverages and industrial alcohol. It divides vegetatively in either the haploid or diploid phase. This permits the isolation of recessive mutations in the haploids and complementation testing in the diploids. Transformation procedures that permit DNA uptake by either **protoplasts** or whole yeast cells have been developed and cloning vectors constructed. In cases where a recombinant **glycoprotein** is desired, yeast has proved very useful since it can, unlike bacteria, effect post-translational **glycosylation**. It has been used for the production of a vaccine for hepatitis B.

saccharopine N^6-[1,3-dicarboxypropyl]-L-lysine; an intermediate in the **aminoadipic pathway** for lysine biosynthesis (e.g. in *Saccharomyces* spp., hence the name) and for lysine catabolism (e.g. in mammalian liver). The reactions involved are:

$$\text{lysine} + \text{2-oxoglutarate} \rightleftharpoons \text{saccharopine} \rightleftharpoons \text{glutamate} +$$
$$\text{2-aminoadipate 6-semialdehyde;}$$

they are catalysed by **saccharopine dehydrogenases**, which are characterized as lysine-forming or glutamate-forming, and require either NAD^+/NADH or $NADP^+$/NADPH as coenzyme; *see* **saccharopine dehydrogenase (NAD$^+$, L-glutamate-forming)** and **saccharopine dehydrogenase (NAD$^+$, L-lysine forming)**. The latter is also called lysine-2-oxoglutarate reductase. In mammals both are part of the bifunctional **aminoadipic semialdehyde synthase**. *See* **familial hyperlysinemia**.

saccharopine dehydrogenase (NAD$^+$, L-glutamate-forming) EC 1.5.1.9; an enzyme involved in the synthesis and degradation of lysine. It catalyses the oxidation by NAD^+ of N^6-(L-1,3-dicarboxypropyl)-L-lysine to L-glutamate and 2-aminoadipate 6-semialdehyde with formation of NADH. In bovine and human liver, this activity, with that of lysine-2-oxoglutarate reductase, forms part of the bifunctional aminoadipic semialdehyde synthase. Another saccharopine dehydrogenase uses the same reaction mechanism, in which the coenzyme is $NADP^+$: EC 1.5.1.10, saccharopine dehydrogenase ($NADP^+$, L-glutamate-forming).

saccharopine dehydrogenase (NAD$^+$, L-lysine-forming) *or* lysine–2-oxoglutarate reductase EC 1.5.1.7; an enzyme involved in the synthesis and degradation of lysine. It catalyses the reaction between L-lysine, 2-oxoglutarate, and NADH to form N^6-(L-1,3-dicarboxypropyl)-L-lysine (**saccharopine**), NAD^+, and H_2O. In mammals this activity, with that of saccharopine dehydrogenase (NAD^+, L-glutamate forming), is part of the bifunctional aminoadipic semialdehyde synthase safety cabinet.

saccharopinuria a rare, autosomal recessive, generally benign condition in which **saccharopine** excretion in urine is prominent. It is a variant of **familial hyperlysinemia**.

saccharose *an old name for* **sucrose**.

Sac1p a single-pass transmembrane protein (623 amino acids) of yeast that is a polyphosphoinositide phosphatase.

SACS *abbr. for* **spastic ataxia of Charlevoix-Saguenay**.

sacsin a protein encoded by the single-exon gene (12 794 bp) for **spastic ataxia of Charlevoix-Saguenay**. A homologue is present in mouse. It contains two copies of a ubiquitous ATPase domain, a histidine kinase A domain, and Hsp70 DNAJ domain, and at residues 3701–3817 a domain (*see* **HEPN domain**) for which sacsin is the prototype.

Saethre–Chotzen syndrome the most common autosomal dominant craniosynostosis disorder in humans. Many mutations in **TWIST** protein cause the craniofacial and limb abnormalities.

SAF *abbr. for* **scaffold attachment factor**.

safety cabinet a cabinet providing an area at bench level in which pathogenic agents may be handled or stored while preventing their entry into the atmosphere outside the cabinet; at the same time, sterility within the cabinet is normally ensured. Usually the cabinet is ventilated with a stream of air, which is extracted through a **HEPA filter** system that traps any pathogens. Cabinets are classified according to the level of protection they offer. **Class I** cabinets have an inward flow of air through the front aperture providing protection for personnel and the environment. **Class II** cabinets are ventilated by a downward flow of filtered air; this is extracted so as to ensure also an inward flow through the front aperture, the extracted air being recirculated through the filter and providing protection for personnel, sample, and environment. **Class III** cabinets are for more potentially hazardous materials; there is no front aperture, the operator manipulating the interior components using gloves mechanically sealed into the front of the cabinet. They provide absolute personal, environmental, and sample protection.

SAG *abbr. for* senescence-associated gene.

SAGA complex a histone acetyltransferase complex ($\approx$2 MDa) that can be recruited by transcription activators in yeast and in mammals. It preferentially acetylates histone H3 and contains at least 14 subunits.

SAGE *abbr. for* **serial analysis of gene expression**.

SAH *abbr. for* S-adenosyl-L-homocysteine; *see* **S-adenosylhomocysteine**.

SAICAR *or* **SACAIR** *abbr. for* 1-(5-phosphoribosyl)-5-amino-4-(*N*-succinocarboxamide)-imidazole.

SAICAR synthetase *see* **phosphoribosylaminoimidazole-succinocarboxamide synthase**.

Sakaguchi test a test for guanidines, i.e arginine and peptides that contain it; in alkaline solution they give a red colour with the **Sakaguchi reagent**, which contains α-naphthol and sodium hypochlorite.

salbutamol *other name*: albuterol; 4-hydroxy-3-hydroxymethyl-α-[*tert*-butylamino)methyl]-benzyl alcohol, α^1-[[1,1-dimethylethyl)-amino]methyl]-4-hydroxy-1,3-benzenedimethanol; a β-adrenergic agonist whose β_2-**adrenoceptor** activity is about 60 times that of **isoproterenol**.

salicylic acid 2-hydroxybenzoic acid; a water-soluble derivative of phenylalanine in plants, where it is formed by hydroxylation of benzoic acid. It initiates thermogenesis in the voodoo lily (in which process it has been called calorigen) and is associated with resistance to infection by fungi, bacteria, and viruses by inducing pathogenesis-related proteins. It is the active ingredient (as methylsalicylate, saligenin, or their glycosides) responsible for the analgesic action of plants such as willow, poplar, and meadowsweet, which have been used for this purpose for over two millennia. It is the active ingredient of the very versatile drug acetylsalicylic acid (*see* **aspirin**).

saline 1 of, concerned with, or containing common salt (sodium chloride). **2** of, or containing, salts of alkali metals or magnesium. **3** an aqueous solution of sodium chloride and (sometimes) other salts, of defined osmolarity, prepared for intravenous injection or perfusion; usually 0.9% w/v sodium chloride. *See also* **physiological saline**.

saliva the mixed secretions of the salivary glands and of the mucous membrane of the mouth. It contains an α-**amylase** (ptyalin), mucin, various inorganic ions, urea, buffers, lysozyme, etc., and has a pH near 7. It is secreted in response to food, which it moistens and lubricates; the amylase may also hydrolyse ingested starch. *See also* **sialic**. —**salivary** *adj.*

salivary gland any of the various glands that discharge **saliva** into the mouth. In humans there are three pairs: the parotid, sublingual, and submaxillary glands. In snakes they include the venom glands.

salivary juice the collective secretions of the salivary glands; i.e. **saliva**. Its secretion was studied by the Russian physiologist Ivan Pavlov (1849–1936), and led him to the discovery of the conditioned reflex.

Salkowski test a test for cholesterol. When concentrated sulfuric acid is added to a chloroform solution of cholesterol, the chloroform layer shows a red to blue colour and the acid layer shows a green fluorescence. [After Ernst Leopold Salkowski (1844–1923), German physiological chemist.]

Salla disease an autosomal recessive disorder, rather frequent in Finland, in which a variable degree of psychomotor retardation accompanies accumulation of free sialic acid (and glucuronic acid) in lysosomes and excessive excretion of sialic acid in urine. Patients have a near normal lifespan. It is caused by mutations in the gene for **sialin**, and is a milder form of infantile sialic acid storage disease.

salmine a 6–7 kDa **protamine**, found in salmon sperm.

salmon any of various fish of the family Salmonidae, which live in coastal waters of the North Atlantic and North Pacific and breed in rivers. Their flesh is comparatively rich in oils with a high content of long-chain *n*-3 **fatty acids**; typical composition (major fatty acids, expressed as % of total fatty acids): 16:0, 10; 16:1(*n*-7), 5; 18:0, 4; 18:1 (*n*-9), 24; 18:2(*n*-6), 5; 18:3(*n*-3), 5; 20:1(*n*-9), 1; 20:4(*n*-6), 5; 20:5(*n*-3), 5; 22:6(*n*-3), 17.

Salmonella a genus of rod-shaped, motile, aerobic bacteria including the agents responsible for typhoid fever (*S. typhi*), some types of food poisoning, and other enteric diseases.

***Salmonella* mutagenesis test** *or* **Ames test** a method of screening chemicals for carcinogenicity by their mutagenic effect on selected strains of *Salmonella typhimurium* – there being a high correlation between mutagenicity and carcinogenicity. The strains normally used are those that require histidine. They are plated out on histidine-deficient medium after being mixed with the potential mutagen. Mutations induced by the test substance that restore the ability to synthesize histidine are revealed by colonies growing on the plate.

salmon sperm DNA a preparation of DNA used for prehybridization, and hybridization, of membranes used for Northern or Southern blotting of mammalian RNA and DNA. It is chosen because its evolutionary distance from mammals makes it unlikely to bind DNA probes under the conditions of hybridization.

salt 1 any member of a class of compounds formed, together with water, by reaction of an acid with a base (most commonly an inorganic acid and metallic base). **2** *or* **common salt** sodium chloride.

saltation 1 the action of jumping or leaping. **2** proceeding in a discontinuous manner by leaps interspersed by periods of rest. **3** an irreversible and inherited change in a cell or organism due to mutation. —**saltatory** *adj.*

saltatory conduction the passage of a nerve impulse across a gap, as between the nodes of Ranvier in a myelinated nerve fibre.

saltatory evolution evolution by sudden changes or by periods of active change with intervening inactive periods.

saltatory replication a hypothetical, sudden lateral replication of DNA thought to be responsible for the production of families of similar nucleotide sequences.

salt bridge 1 a tube, usually closed with porous plugs, that is filled with a solution of a salt (often potassium chloride, commonly saturated) and used to obtain electrical contact between two electrolytic half-cells without intermingling of their respective electrolytes. **2** any electrostatic bond, between positively and negatively charged groups on amino-acid residues of a protein, that contributes to the stability of the protein structure.

salt fractionation the differential precipitation of proteins, or nucleic acids, from solution by the addition of neutral salts, often ammonium sulfate. *Compare* **salting out**.

salting in the increase in the solubility of certain proteins (e.g., globulins) and dipolar ions (e.g., cystine), produced in solutions of low ionic strength by an increase in the concentration of neutral salts.

salting out the decrease in the solubility of proteins, gases, uncharged molecules, and, in sufficiently concentrated solutions, electrolytes, produced in solutions of high ionic strength by an increase in the concentration of neutral salts.

salting-out chromatography a method of chromatography whereby water-soluble organic compounds are separated by ion-exchange chromatography by use of an aqueous salt solution as eluent.

salting-out constant *symbol*: K'_S; the slope of the straight line relating the logarithm of the solubility of a substance in neutral salt solutions and the ionic strength. It is given by

$$K'_S = -[\log(s/s^*)]/I,$$

where *s* is the solubility in mol dm^{-3} at ionic strength *I*, and *s** is the hypothetical solubility at zero ionic strength.

salt link *or* **salt linkage** an **ionic bond**.

salvage pathway any metabolic pathway that utilizes for biosynthetic purposes compounds formed in catabolism. Thus, free purine and pyrimidine bases may be converted to the corresponding ribonucleotides. In the case of purines, the ribose phosphate moiety of 5-phospho-α-D-ribosyl diphosphate (I) is transferred to the purine; thus

adenine + I = adenylate + PP$_i$.

See also **purine biosynthesis**.

SAM *abbr. for* *S*-adenosylmethionine.

SAMase *abbr. for* *S*-adenosylmethionine cleaving enzyme (*S*-adenosylmethionine hydrolase; EC 3.3.1.2).

same-sense mutation *an alternative name for* **silent mutation**.

sample a small, separated portion of the whole. In experimentation it is often essential that the sample be fully representative of the characteristics of the whole (and always so for a **statistical sample**).

sample applicator a device to aid application of samples to chromatography plates, etc.

sample well a hole cut in an electrophoresis gel in which the sample to be analysed is placed.

sampling error *(in statistical analysis)* the chance difference between a measurement obtained from a **sample** and the corresponding measurement obtained from the whole population.

Sandhoff's disease *see* **gangliosidosis**.

sandwich assay *or* **sandwich technique** a type of immunoassay in which antibody against the antigen to be assayed is bound to a solid surface (e.g. plastic). After addition of solution containing antigen the fixed antigen is washed. A second antibody, which is radioactive or fluorescent, is added, sandwiching the antigen; after removal of excess, the amount of bound label is determined. The second antibody may be specific for a different epitope on the antigen, thus enhancing overall specificity, or for the first antibody bound to an antigen.

sandwich complex a 2:1 macrocycle:cation complex in which the metal ion is held between two parallel macrocyclic **complexones**.

sandwich technique *an alternative name for* **sandwich assay**.

sandwich test an immunohistochemical procedure employing a **sandwich assay** to detect an antibody within a cell.

Sanfilippo syndrome A *an alternative name for* **mucopolysaccharidosis type IIIA**.

Sanfilippo syndrome B *an alternative name for* **mucopolysaccharidosis type IIIB**.

Sanfilippo syndrome C *the alternative name for* **mucopolysaccharidosis IIIC**.

Sanfilippo syndrome D *the alternative name for* **mucopolysaccharidosis IIID**.

Sanger method 1 (for identifying and estimating N-terminal amino-acid residues of polypeptides) a method in which free unprotonated amino groups react with 1-fluoro-2,4-dinitrobenzene (FDNB). The dinitrophenylamino groups formed are stable to the acid used to hydrolyse the peptide bonds; the yellow arylated (DNP-) amino acids so released can then be identified by chromatography and estimated spectrophotometrically. **2** (for polypeptide sequencing) a general method for deriving the primary sequence of a polypeptide chain, based on selective hydrolytic degradation of the chain into smaller peptides. The sequence of each smaller peptide may then be found by use of FDNB to label its N-terminal amino acid, together with total hydrolysis to give the amino-acid composition and further stepwise degradation by use of carboxypeptidase or by **Edman degradation**. By selection of degradation procedures to ensure overlap of sequences between the various smaller peptides, the overall sequence of the starting material may be determined. **3** (for DNA sequencing) *an alternative name for* **chain-termination method**. [After Frederick Sanger (1918–), British biochemist.]

Sanger's reagent 2,4-dinitrofluorobenzene (*abbr.*: DNFB); 1-fluoro-2,4-dinitrobenzene (*abbr.*: FDNB); a substance used in structural protein chemistry to arylate free amino groups. *See* **Sanger's method** (def. 1, 2).

sanitizer a substance that significantly reduces the bacterial population in the inanimate environment but does not destroy or eliminate all microorganisms. *Compare* **disinfectant, sterilant**.

SAP *abbr. for* **1** Sin 3-associated protein; *see* **Sin 3-HDAC**. **2** serum amyloid P-component. **3** sphingolipid activator protein. **4** *see* **SAP domain**.

SAP domain *abbr. for* SAF-A/B, Acinus, and PIAS domain; a peptide sequence (35 amino acids) predicted to form two DNA-binding amphipathic alpha helices and present in proteins involved in chromosomal organization (e.g. SAF-A/B, **Acinus**, and **PIAS**).

sapecin an insect **defensin** that is active against (mainly Gram-positive) bacteria. It has a high affinity for **cardiolipin**.

SAPK *abbr. for* stress-activated (*or* stress-associated) protein kinase.

sapogenin any of the **aglycon** moieties of **saponins**; they may be steroids or triterpenes.

saponifiable fraction the portion of total lipid that, after treatment with hot alkali, is soluble in water and insoluble in ether.

saponifiable lipid any **lipid** that can be hydrolysed with alkali to give soap (fatty-acid salts) as one product; they include any **glycolipid, cholesteryl ester**, or **sphingolipid**.

saponification the hydrolysis of an **ester** by an alkali to give the alcohol and salt of the acid. Originally the term meant the hydrolysis of a neutral fat by alkali (e.g. potassium hydroxide) into glycerol and a soap. —**saponify** *vb.*

saponification equivalent the amount in grams of an ester that consumes one gram-equivalent of alkali in **saponification**. *Compare* **saponification number**.

saponification number the number of milligrams of potassium hydroxide consumed in the complete **saponification** of one gram of any particular fat, oil, or wax. *Compare* **saponification equivalent**.

saponin any of the glycosides, widely distributed in plants, that are powerful **surfactants**. Each saponin consists of an aglycon moiety (i.e. **sapogenin**), which may be a steroid or a triterpene, and a sugar moiety, which may be glucose, galactose, a pentose or methylpentose, or an oligosaccharide. All saponins foam strongly when shaken with water. They are membrane-active, powerful hemolytic agents and are used at low concentrations to **permeabilize** cells. They are highly toxic on injection, but not on ingestion as they are not absorbed through the gut. *See also* **digitonin**.

saponin detoxifying enzyme an enzyme that prevents activity of antifungal **saponins**.

sapor that which is sensed by taste; any quality relating to taste or savour. —**saporific** *or* **saporous** *adj.*

saporin a type I **ribosome-inactivating protein**, EC 3.2.2.22, from the soapwort, *Saponaria officinalis*.

saposin *or* **sphingolipid activator protein** any of a group of peptide cofactors of enzymes for the lysosomal degradation of sphingolipids. They stimulate various enzymes, including glucosylceramidase, galactosylceramidase, cerebroside-sulfatase, α-galactosidase, β-galactosidase, and sphingomyelin phosphodiesterase. Saposins A to D are known. Deficiency of any may cause a mild form of a **sphingolipidosis** that normally results from deficiency of an enzyme the saposin activates. They contain characteristic domains, known as saposin A-type or B-type domains. As examples, from the human precursor, saposin A and saposin C together stimulate hydrolysis of glucosylceramide by β-glucosylceramidase (EC 3.2.1.45) and of galactosylceramide by β-galactosylceramidase (EC 3.2.1.46). Saposin C deficiency causes a variant of **Gaucher's disease**. Saposin B is a homodimer, and forms a solubilizing complex with the substrates of the sphingolipid hydrolases; deficiency causes a variant of **metachromatic leukodystrophy**. Saposin D is a specific activator of sphingomyelin phosphodiesterase (EC 3.1.4.12), and deficiency causes a variant of **Tay–Sachs disease** (G_{M2} gangliosidosis). *See also* **swaposin**.

saprophyte any plant, fungus, or microorganism that lives on dead or decaying organic matter. —**saprophytic** *adj.*

saquinavir a peptidomimetic antiviral drug that contains phenyl and other bulky groups and inhibits HIV protease by binding to its active site.

SAR *abbr. for* scaffold-associated region; *other name*: matrix-associated region (*abbr.*: MAR). A DNA sequence of interphase chromosomes that binds them to the nuclear matrix.

SARA *abbr. for* Smad anchor for receptor activation; a protein that recruits Smad2 to transforming growth factor β receptors for phosphorylation by their protein serine/threonine kinase. *See* **Smad**.

sarafotoxin any of the toxic peptides from the burrowing asp, *Atractaspis engaddensis*. They cause cardiac arrest when injected into mice, and are strong vasoconstrictors. All have 21 amino acids, with four cysteine residues in identical positions and tryptophan at the C terminus. They are homologous with **endothelins** and have endothelin receptor agonist action. Sarafotoxin S6D has the sequence

CTCKDMTDKECLYFCHQDIIW.

See also **bibrotoxin**.

Saran Wrap *proprietary name for* a vinylidene chloride copolymer cling-film used for gel drying, food storage, etc.

α-sarcin a cytotoxin produced by *Aspergillus giganteus*. It has ribonuclease activity, specific for purines in both single-stranded and double-stranded RNA. It inhibits protein synthesis, cleaving a single phosphodiester linkage between guanosine and adenosine residues in 28S rRNA in the 60S' subunit of the eukaryotic ribosome, resulting in loss of 393 nucleotides from the 3′ end of 28S rRNA. This single site of action in the intact ribosome contrasts with its action on purified 28S rRNA, in which it causes extensive degradation, cutting on the 3′ side of every adenine or guanine.

sarcoglycan a complex of proteins found in skeletal and cardiac muscle, associated with the dystrophin glycoprotein complex but of

unknown function. It consists of four type I transmembrane proteins: α, (50 kDa), also known as adhalin; β (43 kDa); γ (35 kDa); and δ (35 kDa). The α, γ, and δ proteins are expressed exclusively in muscle, whereas the β protein occurs predominantly in muscle. Mutation in any of them causes limb-girdle muscular dystrophy.

sarcolemma the outer membrane of a muscle fibre. It consists of the plasma membrane, a covering basement membrane (about 100 nm thick and sometimes common to more than one fibre), and the associated loose network of collagen fibres. The term was first used to designate the line visibly marking the outer fibre edge.

sarcoma a malignant tumour of connective tissue or of its derivatives. —**sarcomatous** *adj.*

sarcomere the functional unit of a myofibril of vertebrate muscle, about 2.3 μm long. In **striated muscle** the sarcomeres of many parallel myofibrils are positioned such that myosin **thick filaments** are aligned in register across the myofibril (having a dense appearance in microscopic preparations), as also are the actin **thin filaments** (which have a lighter appearance), resulting in the alternation of light, isotropic bands known as **I bands**, and dark, anisotropic bands known as **A bands**; the I band is bisected by a very dense, narrow **Z line**, while the central, less dense region of the A band is known as the **H zone**, which in turn is bisected by the dark **M line**, or midline, the location of specific proteins that link adjacent thick filaments to each other. The Z lines are due to attachment sites for thin filaments.

sarcoplasm all of the contents of a muscle fibre except nuclei. The sarcoplasm is pervaded by the **sarcoplasmic reticulum**. *Compare* **cytoplasm**. —**sarcoplasmic** *adj.*

sarcoplasmic reticulum *abbr.*: SR; a fine reticular membranous network that pervades the **sarcoplasm** of a muscle fibre. Like the endoplasmic reticulum the SR is continuous over large portions of the cell, it has a large surface-to-volume ratio, and is continuous with and/or homologous to the nuclear envelope. At the level of the I band (*see* **sarcomere**), the SR forms dilated sacs, or cisternae. Opposite the I bands and Z lines of every sarcomere two SR terminal sacs face each other across a narrow space, through which run **T tubules**. At the level of the A–I junction longitudinally orientated SR sacs run along the fibril, forming a pallisade around the A band. The SR contains a Ca^{2+}-translocating ATPase that moves Ca^{2+} into the SR during ATP hydrolysis; ATP resynthesis is associated with movement of Ca^{2+} out of the SR vesicles.

sarcosine *N*-methylglycine; an intermediate in the conversion of choline, via betaine and dimethylglycine, to glycine, and in the synthesis of choline from glycine. *See also* **sarcosinemia**.

sarcosine dehydrogenase EC 1.5.99.1; a mitochondrial flavoprotein enzyme that catalyses the reaction:

$$\text{sarcosine} + \text{acceptor} + H_2O =$$
$$\text{glycine} + \text{formaldehyde} + \text{reduced acceptor}.$$

The flavin is FAD. The acceptor is electron-transfer flavoprotein.

sarcosinemia *or* (*esp. Brit.*) **sarcosinaemia** a rare inherited metabolic disease of humans, also found in ethylnitrosourea-induced mutants of mice, giving elevated plasma and urine levels of **sarcosine**. Primary sarcosinemia results from deficiency of **sarcosine dehydrogenase** (EC 1.5.99.1). Sarcosinemia may also be caused by folate deficiency (*see* **folic acid**) in humans, and occur in glutaric aciduria type II, but in primary sarcosinemia elevated levels of other organic acids are not found. Inheritance is probably autosomal recessive.

sarcosome a subcellular organelle found in muscle tissue with similar biochemical properties as a **mitochondrion** in other tissues. In muscles that are largely anaerobic, such as skeletal muscles of vertebrates and invertebrates, sarcosomes are found near the I bands (*see* **sarcomere**). In predominantly aerobic muscles, such as bird and insect flight muscles and vertebrate heart muscle, sarcosomes are found near the A bands.

sarcotubule *another name for* **T tubule**.

sarin methylphosphonofluoridic acid 1-methylethyl ester; an **organophosphate** (def. 2) that inhibits acetylcholinesterase and other serine esterases. It is a nerve gas but is useful in the treatment of leprosy and has some effect on AIDS.

Sarkosyl sodium lauroyl sarcosinate; a mild anionic surfactant used for solubilizing membrane proteins.

D-sarmentose 2,6-dideoxy-3-*O*-methyl-D-*xylo*-hexose; a substance found as a component of some **cardiac glycosides**.

SARS *abbr. for* severe acute respiratory syndrome. *See* **coronavirus.**.

satellite colony any of numerous small bacterial colonies that can surround a large antibiotic-resistant colony when grown on nutrient agar plates. When the antibiotic is ampicillin the secretion of β-lactamase by the resistant colony depletes the antibiotic in the surrounding medium thereby permitting nonresistant bacteria to grow in that zone.

satellite DNA a fraction of DNA, amounting to 10% or more of total DNA in most eukaryotic cells, with a highly repetitive nucleotide sequence. Its base composition gives rise to a 'satellite' band having a different buoyant density when DNA preparations, following **shear** force degradation, are centrifuged in cesium chloride density gradients. Mouse satellite DNA contains about a million copies of a sequence of 300 bp with a lower (G+C) content than that of the moderately repetitive and unique DNA sequences. *See also* **minisatellite**.

satellite RNA small RNAs (≈350 bases) also called **virusoids**, found in plants; they are **encapsidated** by some plant viruses. *See also* **ribozyme**.

satellite virus a small virus that occurs in association with another virus, upon which it is dependent. **Satellite tobacco mosaic virus** (STMV) is a 172 Å diameter particle containing a 1059 nt RNA that encodes only two proteins, the 159-residue viral coat protein and a 58-residue protein of unknown function. STMV can only multiply in cells coinfected with tobacco mosaic virus.

saturable capable of undergoing **saturation** (def. 5). —**saturability** *n.*

saturated fatty acid any **fatty acid** in which the carbon–carbon bonds of the alkyl chain are exclusively single bonds. The predominant natural ones are **palmitate** and **stearate**; others include **laurate**, **myristate**, arachidate (*see* **arachidic acid**), behenate (*see* **behenoyl**), **lignocerate**, and cerotate (*see* **ceretoyl**). They can all be derived metabolically from acetyl-CoA, in animals and plants. *See also* **fatty-acid nomenclature**.

saturates *an informal term for* **saturated fatty acids**.

saturation 1 (of a chemical compound) the point at which all the valence bonds of the component atoms are satisfied, especially the state of an organic compound that contains only single carbon–carbon bonds. **2** (of a solution) the state in which it has the greatest concentration of the solute that can remain in stable equilibrium with undissolved solute at a given temperature and pressure. **3** (of a gas) the state in which there is the greatest concentration of the vapour associated with a liquid or solid that can remain in stable equilibrium with unvaporized liquid or solid, respectively, at a given temperature and pressure. **4** (of an active site of an enzyme or transporter) the point at which the site is fully occupied with ligand. Saturable enzyme or receptor systems exhibit hyperboliform or sigmoid curves of activity versus concentration (*see* **saturation kinetics**). **5** the act or process of bringing to or towards saturation; especially the addition of hydrogen to triple and double bonds.

saturation analysis *or* **binding assay** a development of **isotope dilution analysis** used for the measurement of very low concentrations of biologically active substances in small samples of complex mixtures such as biological fluids. The essential steps are: (1) the substance, S, is allowed to react with a fixed amount of a specific binding agent under conditions such that the latter is always fully saturated with S; (2) the partitioning of S between the free and bound states is determined by the use of an indicator and by physical separation of the free from the bound form; and (3) the distribution of the indicator is compared with distributions obtained using a series of known amounts of S. The indicator is a form of pure S labelled with a radioisotope of high specific activity; alternatively, fluorescent, enzymic, or other labels may be used if they give the required sensitivity. The specific binding agent is generally an antibody, as in **enzyme immunoassay**, **fluoroimmunoassay**, or **radioimmunoassay**, or a naturally occurring macromolecular carrier or cellular receptor, as in **competitive protein-binding assay**.

saturation fraction *symbol*: y; the ratio of the concentration of bound ligand to the total concentration of the binding agent. Hence, in a binding reaction where a compound P binds X to form PX, then $y = [PX]/[P_{total}]$. A plot of y against $[X]$ is known as the **adsorption isotherm** for this system.

saturation kinetics the kinetics of a reaction (or of **mediated transport**) when the velocity of the reaction (or of the transport) increases to a maximal value (plateau) as the concentration of the reactant (or of a component being transported) is increased.

saturation ratio *or* **cooperativity index** *symbol*: R_s; (of an allosteric protein) the ratio of the concentration of the substrate (or other small molecule) necessary to produce 90% to that required to produce 10% saturation of the specific sites on a protein. It is a convenient measure of the extent of cooperativity. *See also* **cooperative ligand binding**.

saxiphilin a protein, derived from the bullfrog, that is related to transferrin and binds saxitoxin with great affinity.

saxitoxin *abbr.*: STX; a paralytic poison, produced by the marine dinoflagellates *Gonyaulax catenalla* and *G. excavata*. It can be accumulated by shellfish, making them toxic when eaten. Its action is similar to that of **tetrodotoxin**, with which it competes for sodium channels.

Sb *symbol for* antimony.

SBH *abbr. for* sequencing by hybridization; a hypothetical method yet to be implemented for the very rapid sequencing of DNA. It is based on the idea of having all possible octamers (there are 63536) on a small hybridization screen called a DNA chip. As the octamers form nested arrays, a DNA fragment (of a few hundred bases) would only hybridize to some of these and the sequence could be deduced computationally.

SBP2 *abbr. for* SECIS-binding protein 2; a mammalian protein (≈94 kDa) that binds the selonocysteine insertion sequence (**SECIS**) of mRNA for selenoproteins and is required for selenocysteine insertion into the latter. It binds through a 32-amino acid sequence that it shares with several ribosomal proteins. SBP2 also binds the mRNA at a region 5′ to the SecUDA codon. *See also* **SelB**.

SBS model *or* **SBS** *abbr. for* side-by-side model.

sc *abbr. for* synclinal (def. 2).

s.c. *abbr. for* subcutaneous.

Sc 1 *symbol for* scandium. **2** *abbr. for* secretory component (*see* **secretory piece**).

SCAD *abbr. for* short-chain acyl-CoA dehydrogenase *or* butyryl-CoA dehydrogenase; *see* **beta-oxidation system**.

SCAF *abbr. for* SR-like CTD-associated factor; *other name*: CTD-associated SR-like protein (*abbr.*: CASP); any protein that is homologous with splicing regulatory (SR) proteins and binds to the C-terminal domain (CTD) of the largest subunit of RNA polymerase II. *See* **SR protein**.

scaffold (*in gene sequencing*) an ordered set of **contigs** that are in the right order but do not necessarily form a contiguous stretch of sequence.

scaffold attachment factor *abbr.*: SAF; either of two nuclear proteins (SAF-A and SAF-B) of humans that bind AT-rich chromosomal regions termed scaffold attachment regions. They bind DNA through an N-terminal **SAP domain**. SAF-A also binds ATP and is a target for cleavge by caspase during apoptosis. SAF-B (or SAF-145) also contains a conserved domain seen in proteins involved in splicing of pre-mRNA.

scaffolding protein any of a group of proteins that have specific binding sites for, and are instrumental in determining the structure and assembly of, large 3-D structures, such as viruses. Such proteins are integral to the **self-assembly doctrine** elaborated by Caspar and Klug.

scalar 1 a physical quantity, such as mass or time, that is characterized only by its magnitude and does not contain any concept of direction. **2** describable by a number representing a point on a scale undirected in space. *Compare* **tensor**, **vector** (def. 1).

scaler an electronic circuit or device that counts electrical pulses by aggregating the pulses and giving one output pulse for a predetermined number of input pulses.

scalloped the DNA-binding component of the selector complex for wing formation in *Drosophila*.

SCAM *abbr. for* substitute cysteine accessibility method; a method used in studying integral membrane proteins, in which specific amino acids are mutated to cysteine and their accessibility is then tested by thiol-reacting chemicals.

SCAMP *abbr. for* secretory carrier membrane protein; a Ca^{2+}-binding protein with leucine zipper and zinc finger. SCAMPs are components of post-Golgi membranes, and function in recycling proteins between the Golgi and the cell surface.

scanner an instrument or device used to measure or sample the distribution of some quantity or condition in a particular system, area, or region, e.g. to measure the distribution of colour or radioactivity on a chromatogram or electrophoretogram, or to measure the radial distribution of light absorbance in a cell in an ultracentrifuge.

scanning electron microscope *abbr.*: SEM; *see* **electron microscope**.

scanning proton microprobe an instrument in which a sample is bombarded by a high-energy (1–4 MeV) proton beam, causing emission of X-rays that can be characteristically attributed to certain elements. The technique permits the detection of elements down to about 1 ppm.

SCAP *abbr. for* SREBP-cleavage activating protein; a membrane protein of the endoplasmic reticulum and Golgi apparatus that regulates cleavage and release of the transcriptional domain of **SREBP** by **S1P** and **S2P** acting in sequence. It consists of N-terminal transmembrane segments (of which five constitute a sterol-sensing domain), and a cytoplasmic C-terminal region that contains five **WD40** repeats and interacts with the regulatory domain of SREBP. Cholesterol and polyunsaturated fatty acids, acting via the sterol-sensing domain, inhibit cleavage of SREBP and release of its transcriptional domain.

Scatchard plot a method of analysis of the binding of ligand to a macromolecule (or receptor), based on the **Scatchard equation**, $\bar{v}/[L] = (n - \bar{v})K$, where $\bar{v}$ is the average number of ligand molecules bound per binding macromolecule, [L] is the free ligand concentration, n is the number of binding sites on the macromolecule and K is the intrinsic association constant. A plot is made of $\bar{v}/[L]$ on the y-axis and $\bar{v}$ on the x-axis. If the binding sites are identical and independent, a straight line is obtained of slope $-K$ with intercepts on the x-axis of n and on the y-axis of Kn. If there is interaction between the binding sites, or there are several classes of independent sites, the plot is not linear. [After George Scatchard (1892–1973), US physical chemist.]

scatter to disperse in random direction(s) by impact or other interaction, especially light, electrons, neutrons, or X-rays. The measurement of scattered rays or particles may lead to useful analysis, as in **light scattering**.

scatter diagram a graph upon which two measured characteristics of each individual of a population are represented by one dot.

scatter factor *an alternative name for* **hepatocyte growth factor**.

scavenger *(in chemistry)* a substance that reacts with or binds a trace component, such as a trace metal ion, or that traps a reactive **reaction intermediate**.

Scc *abbr. for* separase cysteine cleaved; any of a small family of proteins that are cleaved by **separases**. Scc1 and Scc3 are components of the **cohesin** complex.

SCF *abbr. for* stem cell factor.

Schardinger dextrin *an alternative name for* **cyclodextrin**. [After Franz Schardinger (1853–1920), who first described such compounds in 1908.]

Schardinger enzyme *an alternative name for* **xanthine oxidase**.

Schardinger reaction the decolorization of methylene blue by milk in the presence of formaldehyde; the reaction is catalysed by **xanthine oxidase** (or Schardinger enzyme) in the milk.

Scheie syndrome *an alternative name for* **mucopolysaccharidosis I S**. [After Harold G. Scheie (1909– 1990).]

schema *pl.* **schemata** *(in informatics)* **1** a specification or description of the types of data represented in a database, the attributes they possess, and the relationships between them. **2** a diagram representing this specification.

Schiff base *or* Schiff's base any **imine** having the more limited general structure $R_2C=NR'$ where R may be any organyl group or H and R′ is any organyl group. It may be formed by condensation of the carbonyl group of an aldehyde or ketone with the amino group of a primary amine according to the equation:

$$R_2C=O + H_2NR' = R_2C=NR' + H_2O$$

Often considered to be synonymous with **azomethine**. [After Hugo Schiff (1834–1915), German chemist.]

Schiff reagent a solution of leucofuchsin – fuchsin bleached with sulfurous acid – that produces a red colour when reacted with an aldehyde. The reagent is used in the **Feulgen reaction** and the **periodic acid–Schiff reaction**.

Schiff test **1** (for aldehydes) a test based on the red colour produced when aldehydes react with **Schiff reagent**. **2** (for uric acid) a test based on the reduction of silver ions (silver nitrate solution) to metallic silver by uric acid in sodium carbonate solution. **3** (for urea) a test based on the formation of a purple colour when urea is treated with a concentrated solution of furfural in concentrated hydrochloric acid solution. Allantoin, but not uric acid, also gives a positive test.

Schild equation the expected relationship between the **concentration ratio**, r, and the concentration of a reversible competitive antagonist, B. It is given by $r = [B]/K_B + 1$, where K_B is the dissociation equilibrium constant for the combination of B with the receptor. *Compare* **Gaddum equation**. [After Heinz Otto Schild (1906–84), Austrian-born British pharmacologist.]

Schild plot a plot of log (**concentration ratio** – 1) versus log concentration of antagonist. For a reversible competitive antagonist the plot has a gradient of 1. The intercept on the abscissa then defines the concentration of antagonist required to reduce the response of the target tissue (or organism) by half (the dissociation equilibrium constant: $-\log K_B = pA_2$ when the slope is exactly unity and there are no complicating factors; *see* **pA₂**).

Schilling test a test for cobalamin (vitamin B_{12}) malabsorption that depends on the oral administration of ^{57}Co- or ^{58}Co-labelled **cyanocobalamin** and measurement of the urinary excretion of the radioactive label during the ensuing 24 hours. [After Robert Frederick Schilling (1919–), US hematologist and nutritionist.]

Schindler disease *see* α-N-acetylgalactosaminidase deficiency.

schizencephaly a rare disorder of human brain development that produces seizures, mental retardation, and blindness or speech defects. One or both cerebral hemispheres contain a full-thickness cleft and missing portions. It results from mutation in the homeodomain protein **EMX2**. –schizencephalic *adj*.

schizogeny (*in botany*) the separation of cells that occurs when new structures (e.g. aerenchyma or oil glands) are differentiating.

schizolysigeny (*in botany*) a combination of **schizogeny** and **lysigeny**.

schizophrenia any mental disorder characterized by a special type of personality disintegration, random thought processes, and impaired relation to reality. Antipsychotic (neuroleptic) drugs such as **chlorpromazine**, widely used in the symptomatic treatment of all forms of schizophrenia, are considered to act by blocking post-synaptic dopamine D_2 receptors, though most are not totally selective in this action. —**schizophrenic** *adj., n*.

Schizosaccharomyces pombe the fission yeast that is used to brew the African beer called *pombe*. The rod-shaped cells grow by elongation, and the spores germinate and divide as haploids. The membranous organelles, e.g. the Golgi apparatus, are closely similar to those of higher eukaryotes, in marked contrast to those of budding yeast. It is suitable for analysing the relationship between cell size and the cell cycle because: (1) the cells divide symmetrically, unlike budding yeast; and (2) the yeast grows in length but not in diameter.

Schlesinger test a qualitative test for **urobilin** in urine. The urobilinogen in the urine is first oxidized by iodine to urobilin, then a zinc–urobilin complex is formed by addition of zinc acetate in ethanol; the complex has a yellow-green fluorescence and an absorption band at 507 nm.

Schlieren method an optical method of exhibiting inhomogeneities in refractive index in transparent media, dependent on the deflection of a ray of light from its undisturbed path when it passes through a medium in which there is a solute inducing a gradient of refractive index normal to the ray. The curvature of the ray is proportional to the refractive-index gradient in the direction normal to the ray. It is used in biochemistry particularly in measurements of diffusion, sedimentation, and electrophoretic mobility of proteins and other substances. [From German *Schliere*, streak.]

Schmidt–Thannhauser procedure a method for the extraction and determination of DNA, RNA, and phosphoproteins. Material insoluble in dilute, cold trichloroacetic acid is dried, digested with alkali, and subsequently precipitated with a strong acid. The total phosphorus in the precipitate represents DNA, the soluble inorganic phosphorus represents the phosphoproteins, and the soluble organic phosphorus represents the RNA. [After Gerhard Schmidt (1901–81) and Siegfried Joseph Thannhauser (1885–1962), German-born US biochemist.]

Schultz–Dale reaction an *in vitro* anaphylactic reaction in which the uterus or ileum of a sensitized animal (guinea pig) contracts specifically when a small amount of the sensitizing antigen is added to the bathing medium. The reaction is caused by histamine and similar substances that are released when the antigen combines with cellular antibody. [After H. H. **Dale**.]

Schütz–Borrisow rule *or* **Schütz law** an empirical rule according to which the velocity of an enzyme reaction is proportional to the square root of the enzyme concentration. It was first developed for pepsin but applies only to the crude enzyme under certain limited conditions; similar results have been obtained with some other proteinases. The effect is due to the presence of a reversibly dissociating pepsin inhibitor in crude pepsin preparations.

Schwann cell a uninucleate cell of the **neurilemma** surrounding a myelinated neuronal axon that is responsible for the formation and maintenance of the **myelin sheath** in peripheral myelinated neurons. [After Theodor Schwann (1810–82), German anatomist and physiologist.]

Schwannomin *see* merlin.

SCID *abbr. for* severe combined immunodeficiency; a heterogeneous group of primary immunodeficiency disorders that result from profound functional impairment of B and T lymphocytes. It presents with severe or repeated infections (especially by opportunistic microorganisms). About 50–60% of patients are boys, the disorder being X-linked and caused by numerous mutations in the gene for interleukin 2 receptor γ chain (IL-2Rγ) at Xp13.1 (*see* interleukin 2). Autosomal recessive forms are due to deficiency of adenosine deaminase, purine nucleoside phosphorylase, JAK3, IL-7Rα, RAG1, RAG2, ZAP-70, or Lck. *See* **adenosine deaminase deficiency**.

SCIF *abbr. for* secondary cytotoxic T-cell-inducing factor (i.e. **interleukin 2**).

scillabiose the disaccharide 4-*O*-β-D-glucopyranosyl-L-rhamnose, found in a **glycoside** of *Scilla maritima*.

scinderin *or* **adseverin** a Ca^{2+}-dependent cytosolic actin filament-severing protein present in chromaffin cells, platelets, and a variety of secretory cells.

scintillant 1 (of a phosphor) exhibiting scintillation. **2** *a jargon term for* **scintillator**.

scintillate 1 to give off sparks, to sparkle. **2** (of a phosphor) to emit quanta of light discontinuously, by fluorescence, when struck by a charged particle or a high-energy photon.

scintillation 1 the act of scintillating. **2** the flash of light produced in a phosphor when it is struck by a charged particle or a high-energy photon.

scintillation autoradiography *an alternative term for* **fluorography** (def. 2).

scintillation cocktail *or* **scintillation fluid** any mixture, containing a **scintillator**, to which **beta emitters** can be added for the purpose of **scintillation counting**; the radiation excites the scintillator to emit light. Many scintillation cocktails are based on hydrophobic solvents, since for the most part scintillators sparingly soluble in aqueous solutions have been used. The first scintillator to be widely used was 2,5-diphenyloxazole (PPO), often with 1,4-bis(5-phenyl-2-oxazolyl)benzene (PoPoP), which is excited by the light emitted by PPO to emit light of a higher wavelength more suitable for measurement. Many other scintillators, especially biodegradable compounds and others more compatible with water, have been developed. *See also* **biphenylylphenyloxadiazole, bismethylstyrylbenzene**.

scintillation counter an instrument for the detection and measurement of the intensity of high-energy radiation. Instruments for detecting gamma radiation contain a detector, normally a crystal of sodium iodide containing a small amount of thallium, and this is encased in a metal 'top hat' lined with a white reflector, the open end being sealed with Perspex. When the sodium iodide absorbs γ rays it emits a flash of light. For the counting of β⁻ emissions, the radioactive substance is dissolved or suspended in a **scintillation cocktail** contained in a glass or plastic vial, which is inserted into a chamber in the counter. Sometimes a solid scintillant is used for this also, e.g, for the effluent from chromatography columns. In the case of both types of instrument, the flashes of light produced are detected by a photomultiplier and converted into pulses of electric current which may be counted by an electronic counter.

scintillation fluid *an alternative term for* **scintillation cocktail**.

scintillation spectrometer *another name for* **scintillation counter**.

scintillator *or* **scintillant** any phosphor that fluoresces when struck by a charged particle or a high-energy photon. *See* **scintillation cocktail**.

scintillon a subcellular, crystal-like structure, found in bioluminescent organisms, that emits light on acidification in the presence of oxygen.

scissile capable of being split easily or cut smoothly, especially of an easily cleaved chemical bond.

scission a division or splitting; an act of dividing a nonisometric object orthogonally to its longest axis, especially a chemical bond or a fibrous molecule; **cleavage** (def. 2).

scleral pertaining to the sclera, the fibrous outer layer of the eyeball.

scleroderma *or* **systemic sclerosis** a chronic autoimmune disease in which skin, subcutaneous tissues, and various internal organs accumulate several types of collagen in the extracellular matrix. Autoantibodies to nuclear antigens are present in the serum.

scleroprotein any of a group of simple, generally fibrous, proteins that are insoluble in aqueous solutions; examples include **keratin** and silk **fibroin**.

sclerosis hardening of a tissue as a sequel to inflammation; especially an increase of connective tissue in the nervous system. *See also* **arteriosclerosis**. —**sclerotic** *adj*.

SCOP *abbr. for* structural classification of proteins; *see* **structure classification database**.

scopolamine *see* hyoscine.

scorbutic of, pertaining to, or affected with **scurvy**.

scoring matrix *or* **weight matrix** *(in sequence analysis)* a table of all possible pairwise relationships between nucleotide or amino acid symbols, used by database search and sequence alignment programs to quantify the similarity between sequences. The simplest scoring matrix is an identity or unitary matrix, in which only pairwise identities receive a positive score (nonidentical characters receive zero score). *Compare* **substitution matrix**.

scotophobin a 15-residue peptide that accumulates in the brains of rats trained to avoid the dark; administration is said to elicit a similar response in untrained animals. It is thought to be the first memory-directing substance isolated and identified.

scotopia adaptation to darkness; the adjustment of the eye to vision in dim light or the dark. It is considered to involve mainly retinal **rod cells**. *Compare* **photopia**. —**scotopic** *adj*.

scotopsin an **opsin** that is the apoprotein of a **rhodopsin**.

SCP *abbr. for* **1** single-cell protein. **2** sterol carrier protein.

SCP2 *abbr. for* sterol carrier protein 2.

SCPx *or* **SCP2/3-ketothiolase** a protein (547 amino acids) of the peroxisomal matrix that is processed therein into 3-ketothiolase (residues 1–404) and to **sterol carrier protein 2** (SCP2; the remaining residues). It is essential for beta oxidation of pristanic acid by catalysing the final step.

scrambalase a membrane protein that acts as a procoagulant by prompting Ca²⁺-dependent migration of phosphatidylserine across the membrane bilayer of activated platelets. A defect in the protein may be associated with a bleeding disorder.

scrapie a spongiform encephalopathy of sheep and goats, and the first to be identified of a group once thought to be 'slow virus' infections occurring naturally and caused by an unconventional agent. Scrapie has similarities to two other animal diseases, transmissible mink encephalopathy and **bovine spongiform encephalopathy**, and three transmissible human dementias, **kuru**, **Creutzfeldt–Jakob disease**, and **Gerstmann–Sträussler syndrome**. All these diseases are now accepted as being caused by alterations in the **prion proteins** in the nervous system. The disease has been known since its first reported occurrence in merino sheep in Spain in 1732, which was rapidly followed by reports of other cases in Britain and Germany. The incubation period for the scrapie agent is usually 1–3 years. Different breeds of sheep exhibit markedly different susceptibilities to scrapie. As many as 20 different strains of scrapie agent, each with distinct properties in terms of incubation period in a given host and location of damage within the brain have been identified and transferred to mice. It would appear that the strains are passed to the progeny, so any explanation of the nature of the infective agent and process must take this into account.

screen 1 to protect from; esp. from harmful agents or radioactivity using sheets of lead, glass, or plastic; or from radiofrequency interference using a copper sheath or wire net. 2 to sift data, often using a limited or general assay, for selection of a particular phenomenon or result; samples selected by preliminary screening can be isolated for more detailed study. 3 *(in clinical chemistry)* to carry out any diagnostic procedure on a large cohort of patients with a view to detecting hitherto asymptomatic disease; *see also* **genetic screening**. 4 any device or material that acts to screen.

screw axis a crystal or an oligomeric protein molecule is said to possess an *n*-fold screw axis if it presents exactly the same appearance after rotation about that axis of 360°/*n*, coupled with a translation parallel to the axis. *Compare* **rotation axis**.

scRNA *abbr. for* small cytoplasmic RNA.

scRNP *abbr. for* small cytoplasmic ribonucleoprotein.

SCU *abbr. for* synonymous codon usage.

scurfin *an alternative name for* **FOXP3**

scurfy an X-linked disease of mice caused by mutations in **FOXP3** and resulting in a multiorgan autoimmune and inflammatory syndrome.

scurvy avitaminosis C; a deficiency disease caused by lack of **ascorbic acid** (vitamin C) in the diet. It is characterized by anemia, sponginess and ulceration of the gums, hemorrhages into the skin and subcutaneous tissues, and delayed wound healing.

scute a gene in *Drosophila* for **AS-C protein**.

scutelarin EC 3.4.21.60; *other name*: taipan activator (being from the venom of the taipan (snake), *Oxyuranus scutellatus*). An enzyme that catalyses (like factor Xa of blood coagulation) the cleavage of Arg-|-Thr and Arg-|-Ile in prothrombin to form **thrombin** and two inactive fragments. *Compare* **scutellarein**.

scutellarein 4′,5,6,7-tetrahydroxyflavone; the aglycon of **scutellarin**. *Compare* **scutelarin**.

scutellarin the glucuronide of **scutellarein**, from *Scutellaria* spp. *Compare* **scutelarin**.

scyrp *colloquialism for* scRNP (i.e. **small cytoplasmic ribonucleoprotein**).

SD *abbr. for* standard deviation.

SDS *abbr. for* sodium dodecyl sulfate.

SD sequence *abbr. for* Shine–Dalgarno sequence.

SDS–PAGE *abbr. for* SDS–polyacrylamide gel electrophoresis.

SDS–polyacrylamide gel electrophoresis *or* SDS gel electrophoresis *or* detergent gel electrophoresis *abbr.*: SDS–PAGE; a rapid method for resolving a protein into its subunits and determining their relative molecular masses, based on the ability of **sodium dodecyl sulfate** to cause dissociation of oligomeric proteins and to bind to the subunits to form complexes of constant charge-to-mass ratio. On **zone electrophoresis** in polyacrylamide gel, such complexes often have mobilities directly related to their M_r values, which may thus be obtained by comparison with standards.

Se *symbol for* selenium.

SE *abbr. for* standard error; *see* **standard error of estimate of the mean**.

SEA a family of genes for membrane receptor tyrosine kinases homologous with the **insulin receptor** family. The acutely transforming virus, avian erythroblastosis virus (*abbr.*: AEV) strain 13, contains the oncogene v-*sea*; this transforms fibroblasts and erythroblasts, but not avian myeloid cells. Sea, the product of c-*sea*, the cellular homologue of v-*sea*, phosphorylates Shc (*see* **SHC**) on tyrosine residues.

sealase *a former name for* DNA ligase (ATP); *see* **DNA ligase**.

search engines Internet sites that have facilities for text searching. Depending on the way they work they are variously referred to as Web crawlers, spiders, robots, etc. For general purposes the search engine http://www.google.com or one of its mirrors such as http://www.google.co.uk is an excellent starting point.

sea-urchin-hatching proteinase *see* envelysin.

sebaceous gland any of the single or branched glands in the skin that secrete **sebum** into the hair follicles.

sebum the oily secretion of the sebaceous glands in the skin. It contains lipids such as waxes, squalene, and triacylglycerols; the triacylglycerols are broken down by anaerobic bacteria during secretion to form fatty acids. Sebum helps to form an effective barrier against water loss.

sec *symbol for* secant (def. 1).

sec. *abbr. and obsolete symbol for* **second** (def. 1).

sec- *abbr.*: *s*-; *prefix (in chemical nomenclature)* signifying **secondary** (def. 1a).

Sec *symbol for* a residue of the α-amino acid L-selenocysteine.

secant 1 *abbr. and symbol*: sec; the reciprocal of the **cosine** of an angle. 2 a straight line cutting a curve at two or more points.

SECIS *abbr. for* selenocysteine insertion sequence; a putative stem-loop nucleotide sequence required for decoding the SecUGA codon of eukaryotic mRNA for selenoproteins. It is located in the 3′ untranslated region of the mRNA that binds **SBP2** and **SelB**

seco- *prefix (in chemical nomenclature)* denoting cleavage of a ring with addition of one or more hydrogen atoms at each terminal group thus created. The numbering of the skeletal atoms of the parent molecule is retained.

seco-carotenoid a **carotenoid** in which a ring has been opened by breakage of a carbon–carbon bond; e.g. semi-*β*-carotenone (5,6-seco-*β*-*β*-carotene-5,6-dione).

second 1 *symbol*: s; the SI base unit of time, equal to the duration of 9 192 631 770 periods of the radiation corresponding to the transition between the two hyperfine levels of the ground state of the caesium-133 atom. 2 *symbol*: ″ a unit of plane angle equal to 1/60 of a minute; i.e. 4.84814 microradian (π/648000 rad). See **angle** (def. 1).

secondary *(in chemistry)* 1 a *prefix*: sec- (*abbr.*: *s*-); describing an alkyl compound (e.g. an alkanol) in which the functional group (e.g. a hydroxyl group) is attached to a carbon atom linked to two others. b describing the carbon atom bearing the functional group in such a compound. 2 describing an amide or amine in which two appropriate groups are attached to a nitrogen atom. 3 describing a salt formed by replacing two of the ionizable hydrogen atoms of a tribasic acid by one or two other cations of appropriate valency. *Compare* **primary**, **tertiary**.

secondary bond any of the weak attractive forces between atoms and/or molecules, especially when these stabilize a three-dimensional structure. It includes **hydrogen bonds**, **hydrophobic interactions**, and **van der Waals forces**.

secondary cell wall (*in plants*) the microfibrils and noncellulosic polysaccharides deposited on the primary cell wall. It provides additional conducting and supporting tissue and often makes up the greater part of the mature cell wall.

secondary charge effect an effect seen in the hydrodynamic study of charged macromolecules when the positive and negative ions of the supporting electrolyte have different sedimentation (or diffusion) coefficients, leading to the establishment of local electric fields due to charge separation. These local fields act as an additional force on the macromolecules, either accelerating or retarding their movement.

secondary cytotoxic T cell-inducing factor *abbr.*: SCIF; *an alternative name for* **interleukin 2**.

secondary database *an alternative name for* **pattern database**.

secondary deficiency a nutritional deficiency occurring when the dietary content of some essential nutrient is known to be adequate in normal conditions. It may be caused by malabsorption from the gut due to disease, diminished synthesis of a vitamin by the intestinal flora during treatment with an antibacterial agent, or increased requirement, e.g. in pregnancy.

secondary electron an electron emitted from a solid as a result of the impact of another electron. If the energy of the incident primary electron is sufficient, more than one secondary electron may be emitted, an effect exploited in an **electron multiplier**. *See also* **photomultiplier**.

secondary emission the emission of **secondary electrons**.

secondary F′-containing strain *see* **F′ plasmid**.

secondary fluor a second fluorescent agent that absorbs the fluorescent light emitted by a **primary fluor** and is thus excited to emit light at a higher wavelength, which may be more convenient for detection.

secondary hyperlipidemia *see* **hyperlipidemia**.

secondary immune response a response to an **immunogen** to which an animal has previously been exposed. It is characterized by the rapid production of large amounts of **antibody** over a few days, followed by an exponential decline. *Compare* **primary immune response**.

secondary lysosome *or* **phagolysosome** a **lysosome** that results from fusion of a **primary lysosome** with a **phagosome** or a vesicle containing endocytosed matter. *Compare* **telolysosome**.

secondary messenger *a variant form of* **second messenger**.

secondary metabolism the formation of end products of metabolism – **secondary metabolites** – that often have no readily apparent use in the producing organism.

secondary metabolite any of a group of very diverse natural products that have a restricted taxonomic distribution, usually possess no obvious function in cell growth, and are often synthesized by cells that have stopped dividing; they usually belong to closely related chemical families, usually of $M_r < 1500$, but some bacterial toxins are considerably larger. An example of a secondary metabolite is penicillin.

secondary response *see* **secondary immune response**.

secondary structure 1 (*of a protein*) the arrangement of the polypeptide primary structure into locally-organised, hydrogen-bonded structures, in particular α-helices and β-sheets; *see* **alpha helix**, **beta sheet**. **2** (*of a nucleic acid*) the arrangement of one or two polynucleotide chains in bihelical structures, stabilized by hydrogen bonds between complementary bases (*see* **complementary base sequence**); *compare* **clover-leaf structure**. *See also* **primary structure**, **quaternary structure**, **tertiary structure**.

second law of thermodynamics *see* **thermodynamics**.

second messenger *or* **secondary messenger** a mediator that is caused to accumulate in an effector (target) cell by the action of a **hormone**, **growth factor**, or other agonist, and that brings about the action of that agonist on the cell. The major second messenger systems include: synthesis of cyclic AMP by **adenylate cyclase** (the first of such systems to be discovered); synthesis of cyclic GMP by **guanylate cyclase**; opening of **ion channels**; the phosphoinositide sys-

tem, (*see* **phosphatidylinositol cycle**) involving activation of phosphoinositide-specific phospholipase C (*see* **phospholipase**), or **1-phosphatidylinositol 3-kinase**; the release of ceramide from sphingolipids by sphingomyelinase; and protein phosphorylation by serine/threonine-specific or tyrosine-specific **protein kinases**. An important component in several of these involves coupling of activated receptor to the effector system through a **G-protein**.

second-order rate constant *see* **rate constant**.

second-order reaction *see* **order of reaction**.

seco-steroid a **steroid** in which a ring has been opened by breakage of a carbon–carbon bond.

secretagogue *or* **secretogogue** an agent that stimulates secretion (usually of a hormone, juice, or other glandular product). *See also* **+agogue**.

secretase an enzyme that cleaves the extracellular domain of type I or II transmembrane proteins, releasing that part of the protein into the circulation. Cleavage normally occurs near the extracellular face of the membrane. Substrates for secretases include Alzheimer's amyloid precursor protein (APP), angiotensin-converting enzyme, transforming growth factor-α, and the tumour necrosis factor ligand and receptor superfamilies. α-Secretase is a zinc-dependent protease; β-secretase (*also called* memapsin 2; BACE2) is a type I transmembrane protein with an aspartyl protease domain; γ-secretase requires senilins for activity and cleaves the transmembrane segment of APP. All three cleave APP but only β- and γ-secretases are required for release of β-amyloid peptide.

α-secretase *see* **ADAM-10**.

secrete to elaborate and emit a **secretion** from a cell or, particularly, from a gland.

secretin an intestinal hormone, discovered in 1902 by William Maddock Bayliss (1860–1924, British biologist) and Ernest Henry Starling (1866–1927, British physiologist) (the first hormone to be discovered). It has the sequence (human):

$$\text{HSDGTFTSELSRLREGARLQRLLQGLV-NH}_2$$

(3.04 kDa), and is formed from a precursor synthesized in the S cells of the small intestinal mucosa. It stimulates the release of bicarbonate, enzymes, and potassium ions from the pancreas and stimulates the secretion of enzymes and electrolytes in the gut; it also inhibits HCl production by the stomach. Secretin lends its name to a superfamily that includes **gastric inhibitory peptide**, **GHRH**, **glucagon**, **peptide HI**, and **vasoactive intestinal polypeptide**.

secretin receptor any G-protein-associated membrane receptor that binds **secretin** and mediates its intracellular effects by activating adenylate cyclase. Secretin receptors are not structurally related to other G-protein-coupled receptors.

secretion 1 the process of elaborating, segregating, and emitting a substance or juice from a cell, organ, or organism; e.g. insulin by the B cells of the islets of Langerhans, or urine by the kidneys. There are three basic types of secretion: **apocrine**, **holocrine**, and **merocrine**. **2** a product of **secretion** (def. 1) from a cell, organ, or organism.

secretogogue *a variant spelling of* **secretagogue**.

secretogranin I *another name for* chromogranin B (*see* **chromogranin**).

secretor any person who secretes water-soluble glycoproteins with A, B, or H blood-group specificities into their mucous secretions, e.g. saliva and gastric juices. Over 80% of humans are secretors, having a **secretor gene** in their genomes.

secretor gene a gene determining **secretor** phenotype in humans. Secretor status is determined by a pair of allelic genes, *Se* and *se*. Secretors are either *Se/Se* or *Se/se* while nonsecretors are *se/se*. These alleles occur at a different locus from the ABO locus and do not affect the A, B, and H specificities of the red blood cells.

secretory describing a molecule or molecular fragment destined for secretion from a cell, such as secretory protein, polypeptide, etc.

secretory actin-binding protein *abbr.*: SABP; *also called* prolactin-inducible protein; gross cystic disease fluid protein 15; a glycoprotein that has been isolated in several different studies (as the alternative names indicate), but whose function remains unknown. As the prolactin-inducible protein, it was identified as a highly specific and sensitive marker of primary and metastatic breast cancer.

secretory component *another name for* **secretory piece**.

secretory granule a membrane-bound subcellular vesicle that is formed from the Golgi apparatus and contains a protein destined for secretion. The membrane of the granule fuses with the plasma membrane, and its protein load is exteriorized. Processing of the contained protein may take place in secretory granules. *Compare* **zymogen granule**.

secretory piece (*abbr.*: SP) *or* **secretory component** (*abbr.*: Sc) the 58 kDa polypeptide that joins the two identical monomers comprising the secretory form of **immunoglobulin A**. It has strong affinity for mucus, thus prolonging retention of IgA on mucous surfaces, and may protect the IgA against proteolytic destruction in the digestive tract. *Other names*: T chain; transport piece.

secretory protein I *abbr.*: SP-I; a glycoprotein of the parathyroid gland that is co-secreted *in vitro* with **parathyrin**. It consists of at least two homologous glycoproteins of similar size (M_r 72 000 and 70 000), related to chromogranin A (*see* **chromogranin**).

sector any portion of a circle bounded by two radii and the included arc. —**sectoral** *adj*.

sector cell *or* **sector-shaped cell** a cell used in an analytical ultracentrifuge to contain the sample. It is sector-shaped when viewed parallel to the rotation axis. Since the accelerating force is radial, sedimenting molecules move along radii and this cell shape minimizes the collision of sedimenting molecules with the cell walls.

securin a protein that binds to the N- and C-terminal regions of **separase** and inhibits its protease activity. Mutations in the securin gene are associated with defects in the separation of sister chromatids during anaphase of cell division.

sediment 1 to undergo **sedimentation**. **2** the accumulation of material resulting from sedimentation.

sedimentation the setting of solid particles (or solute) through a liquid (or solution) under the influence of a gravitational or centrifugal field.

sedimentation coefficient symbol: *s*; the rate of sedimentation of a particle in an ultracentrifuge or other system. For a particle or solute sedimenting in a system at constant applied field, temperature, and pressure, $s = v/a$, where *a* is the acceleration of free fall or centrifugation and *v* is the velocity of sedimentation. In centrifugation $s = v/\omega^2 r$, where ω is the angular velocity in rad s^{-1} and *r* is the distance of the particle or solute from the axis of rotation. It has the dimensions of time. The sedimentation coefficients of macromolecules and cellular particles are often expressed in **svedbergs**.

sedimentation equilibrium a technique in which compounds of interest are centrifuged for a period sufficiently long for the system to come to equilibrium. The technique has a number of applications. In one, used for molecules that have a density greater than the solvent, centrifugation is carried out at relatively slow speeds so that sedimentation of the molecule is slow enough to be counterbalanced by diffusion. At equilibrium, the concentration is lower at the meniscus and higher at the bottom of the cell, but is unchanging. The relative molecular mass, M_r, of the compound in question, often a protein, can be calculated from the expression:

$$m_r = 2RT \ln(c_2/c_1)/\omega^2(1 - v\rho)(r_2^2 - r_1^2),$$

where *R* is the gas constant, *T* is the thermodynamic temperature, *v* is the specific volume of the component, ρ is the density of the solution, ω is the angular velocity of the rotor, and c_1 and c_2 are the concentrations of the component at distances r_1 and r_2 from the centre of rotation. Ideally, determinations should be done at several concentrations and extrapolated to zero concentration.

sedoheptulose *a nonsystematic name for* D-*altro*-2-heptulose.

α-D-pyranose form

sedoheptulose-bisphosphatase EC 3.1.3.37; a chloroplast enzyme of the **reductive pentose phosphate cycle** that catalyses the reaction:

$$\text{sedoheptulose 1,7-bisphosphate} + H_2O =$$
$$\text{sedoheptulose 7-phosphate} + \text{orthophosphate}.$$

sedoheptulose 7-phosphate the 7-phosphate ester of **sedoheptulose**; it is a component of the **pentose phosphate pathway**.

SEE1 a cysteine protease (360 amino acids) of senescing leaves of maize. It is homologous to aleurain (of barley) and γ-oryzain (of rice).

Seebeck effect the phenomenon whereby an electromotive force is generated in a circuit containing junctions between dissimilar metals if these junctions are not all at the same temperature. *See* **thermocouple**. *Compare* **Peltier effect**. [After Thomas Johann Seebeck (1770–1831), Estonian-born German physicist.]

SEG *see* **sequence filter**.

Segawa syndrome *an alternative name for* **dopamine-responsive dystonia**.

segmentation 1 (of an ovum) *see* **cleavage** (def. 3). **2** (of an animal's body) *see* **metamerism**.

segmentation genes any of the 15 genes that are active during early development of *Drosophila* whose products are involved in dividing the embryo into a series of segments. There are three categories: segment polarity genes; pair-rule genes; and gap genes.

segment polarity gene any of the 16 or so genes in *Drosophila* that define the anterior and posterior compartments of segments. Examples are *engrailed*, *wingless*, and *hedgehog*.

segregation (*in genetics*) **1** the separation of homologous genetic elements (i.e. allele pairs) during meiosis in diploid cells. **2** the separation of sister chromatids during mitosis; by extension, the separation of any two independent genetic elements during cell division in prokaryotes. **3** (of plasmids) the partitioning of plasmids into daughter cells at cell division. Naturally occurring plasmids contain a partitioning function, *par*, which ensures correct partitioning into the two daughter cells.

segresome a membrane-bound, cytoplasmic aggregate of a dye, formed either by passage of the dye into a pre-existing **phagosome** or **lysosome**, or by the *de novo* formation of a membrane around the dye aggregate.

Seitz filter a filtration apparatus that removes microorganisms from a liquid. It formerly consisted of a flat pad of asbestos, or asbestos and cellulose, suitably mounted between the sample container and the receiving flask, filtration being assisted by reducing the pressure in the receiving flask. It has now been superseded by a sintered-glass filter. [After Seitz-Filter-Werke, T. and G. Seitz Gmbh & Co., Bad Kreuznach, Germany.]

selachyl alcohol 3-(octadec-9-enyloxy)-1,2-propanediol; a 1-octadec-9-enyl ether of glycerol, and a hydrolysis product of **ether lipids**.

SelB *or* **eEFsec** a specific eukaryotic elongation factor that conveys selenocysteine tRNA (Sec-tRNAsec) to the ribosome and binds the selenocysteine insertion sequence (**SECIS**) of the mRNA for a selenoprotein. Mouse SelB (583 amino acids) contains four GTP-binding sites in the N-terminal portion, which also binds the tRNA. It functions with **SBP2**. In eubacteria SelB and SBP2 form part of a bifunctional protein.

SELDI *abbr. for* surface-enhanced laser desorption and ionization; a commercial mass spectrometry technique for capturing peptides and proteins on an affinity surface (either chemical or biological) and subsequently quantifying and identifying the proteins. *See also* **tandem mass spectrometry**.

selectable marker a gene whose expression confers selective advantage on a cell or organism when exposed to an antibiotic. The marker is often an enzyme whose action modifies the antibiotic thereby making it inactive. Markers for use in eukaryotes are usually derived from bacteria and have no eukaryotic counterparts.

selected-ion monitoring *or* **mass fragmentography** *or* **molecular fragmentography** a type of **gas chromatography/mass spectrometry** in which the intensities of one or more molecular or fragment ions selected from the mass spectrum of a particular compound are monitored continuously as a means of quantifying that compound.

selectin any of a group of leukocyte surface molecules that act as leukocyte adhesion molecules; they are classed as CD62 in the **CD marker** system. All have **lectin** family carbohydrate-binding domains and EGF-repeats, with four motifs. They are expressed on the surface of platelets and endothelial cells as well as leukocytes. **E selectin** (CD62E), M_r 107–115 000 in various glycosylated forms, is expressed on endothelial cells, being induced by cytokines such as interleukin 1β and tumour necrosis factor α in 3–6 hours. It mediates endothelial cell binding to neutrophils, monocytes, and some memory T cells. **L selectin** (*or* lymph node homing receptor; CD62L), M_r 74 000 (on lymphocytes), is constitutively expressed on monocytes, lymphocytes (thymocytes, NK cells), neutrophils, eosinophils, and basophils. It mediates the adherence of lymphocytes, neutrophils, and lymphocytes to endothelium, including that of lymphocytes to endothelial cells of high endothelial venules in peripheral lymph nodes. It binds to **MadCAM** and CD34. **P-selectin** (*or* PADGEM; *abbr. for* platelet activation dependent granule external membrane protein), M_r 140 000, is expressed at the surface of activated platelets and endothelial cells, and is inducible within minutes by thrombin, histamine, or peroxides from α granules of platelets or Weibel–Palade bodies of neutrophils. It mediates the adhesion of neutrophils and monocytes to platelets and endothelial cells.

selection *(in microbiology)* a laboratory method in which a mixture of microorganisms is cultured under particular growth conditions that permit only cells with certain characteristics to survive, thereby enabling their isolation.

selection pressure the intensity with which natural selection operates; it is often measured by the change in **gene frequency** per generation due to the effects of natural selection.

selectively labelled indicating an isotopically labelled molecule in which the position(s) of the labelling is known, but the number of labelling nuclides at that (those) position(s) is not known; e.g. selectively labelled ethanol is designated as $[1,2\text{-}^2H]CH_3\text{–}CH_2\text{–}OH$, which indicates that some, or all, of the hydrogen atoms on carbon atoms 1 and 2 are ^2H atoms.

selective medium a culture medium designed either to encourage or to inhibit the growth of certain types of organisms, especially microorganisms.

selective theory of immunity *or* **elective theory of immunity** any theory that ascribes to antigens a selective function in immunity by which antigens select cells genetically prepared to respond specifically and cause them to multiply and differentiate. They include the **clonal-selection theory** and the **natural-selection theory of immunity**. *Compare* **instructive theory of immunity**.

selectivity *(in informatics)* an alternative name for **precision** (def. 2).

selector a gene whose product controls the formation and identity of a discrete set of cells in an embryo that forms a particular structure (e.g. eye, leg, or wing). One type includes the *Hox* genes, whose products differentiate between homologous morphological fields; another type consists of field-specific genes (e.g. *eyeless* or *Distalless* in *Drosophila*), which direct the formation of entire complex structures.

selector gene any of a class of genes encoding transcription factors that specify cell type and tissue, organ, and regional identity in animals such as *Drosophila,* in which they were first described.

selenium *symbol*: Se; a metalloid element of group 16 of the (IUPAC) **periodic table**; atomic number 34; relative atomic mass 78.96. Its main oxidation states are –2, +2, +4, and +6, with properties similar to those of sulfur, but more metallic. It occurs in various allotropic forms (grey, red, black); the grey form shows some electrical conductivity, which is enhanced by light, and it is used in 'selenium cells' to measure light intensity. The most abundant isotope is selenium-80 (mass 79.916; relative abundance 49.8%); others are selenium-74 (0.9%), selenium-76 (9.0%), selenium-77 (7.6%), selenium-78 (23.5%), and selenium-82 (9.4%). Radioactive isotopes are **selenium-72**, **selenium-73**, **selenium-75**, and **selenium-79**. Selenium is an essential trace element, required for the formation of **selenoproteins**, notably **glutathione peroxidase**. Gut absorption is poor, but selenium deficiency is rare (seen, e.g., in some parts of China with low soil selenium), and manifests as myopathy (especially cardiomyopathy). Measurement of red-cell glutathione peroxidase activity gives an index of selenium status.

selenium-72 a radioactive nuclide of selenium, $^{72}_{34}Se$, mass 71.927; it emits gamma radiation (0.046 Mev) and has a half-life of 8.4 days.

selenium-73 a radioactive nuclide of selenium, $^{73}_{34}Se$, mass 72.927; it emits a β$^+$ particle (1.32 Mev) and gamma radiation (0.361 Mev) and has a half-life of 7.1 h.

selenium-75 a radionuclide of selenium, $^{75}_{34}Se$, mass 74.923; it emits gamma radiation (0.481 Mev) and has a half-life of 118.5 days.

selenium-79 a radionuclide of selenium, $^{79}_{34}Se$, mass 78.918; it emits an electron (β$^-$ particle, 0.16 Mev), no gamma radiation, and has a half-life of 6.5×10^4 years.

selenoamino acid an amino acid in which selenium replaces a sulfur atom; *see* **selenocysteine**, **selenomethionine**.

selenocysteine *symbol*: U or Sec; $H\text{–}Se\text{–}CH_2\text{–}CH(NH_2)\text{–}COOH$; an essential component of glutathione peroxidase and some other proteins (*see* **selenoprotein**). For its synthesis, a special tRNA, charged with serine, responds to a UGA codon. The serine is converted to selenocysteine just before its use on the ribosome. It is assumed that the neighbouring base sequence indicates a special sense for this codon, which is normally a 'stop' codon.

selenocysteine insertion complex *abbr.*: SIC; the complex of requirements for recoding an in-frame UGA codon as selenocysteine (Sec) in the mRNA for a selenoprotein. It comprises: a Sec codon UGA; a selenocysteine insertion sequence (**SECIS**) in the 3′ untranslated region of the mRNA; the selenocysteine tRNA (Sec-tRNAsec); the elongation factor **SelB**; and the SECIS-binding protein **SBP2**.

selenomethionine $CH_3\text{–}Se\text{–}[CH_2]_2\text{–}CH(NH_2)\text{–}COOH$; (*symbol*: Sem) a selenoamino acid used as an antimetabolite, competing with methionine; it is also found in proteins; *see* **selenoprotein**. For the solution of the phases in x-ray crystallography by single-wavelength anomalous dispersion (SAD), it is introduced into proteins by growth of bacteria on medium containing selenomethionine instead of methionine.

selenoprotein a protein containing selenium, almost invariably as **selenocysteine**, although **selenomethionine** can occur, apparently as a random substitute for methionine. Well-known selenoproteins include the mammalian **glutathione peroxidases** and tetraiodothyronine 5′-deiodinase, and bacterial formate dehydrogenases and glycine reductase.

selenoprotein P a plasma glycoprotein (42 kDa) of mammals that is rich in Lys and His residues and contains 10 selenocysteine residues per monomer. Its function is unknown.

selenoprotein W a protein (10 kDa) of skeletal and cardiac muscle that contains one selenocysteine residue per monomer. Its function is unknown.

SELEX *abbr. for* systematic evolution of ligands by exponential enrichment.

self 1 the individuality of a person or thing; the totality of components intrinsic to an individual or thing. **2** *(in immunology)* the components of an individual's own tissues that can be distinguished from foreign substances by the individual's immune system, and towards which **immunological tolerance** is shown.

self-absorption the absorption of any emission by the emitting molecules.

self antigen any (potentially) antigenic molecule originating in an individual that is recognized as nonforeign by the individual's immune system, and towards which **immunological tolerance** is normally shown. *See also* **autoimmunity**, **self-tolerance**.

self-assembly the formation of a complex entity from more simple, identical units without intervention from any external agency.

self-assembly doctrine a doctrine elaborated by Aaron Klug (1926–), Lithuanian-born British molecular biologist, and Caspar to explain the assembly of small viruses and subcellular organelles. It is postulated that the structure assembles as a result of specific binding sites on proteins that interact to form a scaffold on which other components form.

self-catabolite repression repression of gene expression by a **catabolite** formed within the same organism.

self-cloning the cloning of a gene within the species from which it was derived, a technique that in some countries attracts less stringent government restrictions than when a gene is cloned in another species, being considered the less hazardous of the two approaches with respect to the genetic environment.

self determinant a **determinant** originating from the tissue of an organism and thus treated as **self** (def. 2) by the immune system of that organism (except in the case of autoimmune pathology; *see* **autoimmunity**).

self-fertilization the union of male and female gametes of the same organism.

self-hybridization hybridization between nucleic acids that exhibit complementarity.

selfing *(in microbiology)* the production of wild-type 'transductants' by infection of an auxotrophic strain with a transducing phage grown on the same auxotroph as the donor.

selfish DNA hypothesis a hypothesis that accounts for the presence in eukaryotic genomes of large amounts of repetitive DNA(*see* **repetitive sequences**). Because there is no evidence that such DNA serves any useful purpose for its host, it has been termed **selfish DNA** or **junk DNA**, and is suggested to be a molecular parasite that has disseminated itself throughout the genome. It is a mystery why natural selection has not eliminated 'selfish DNA'.

self tolerance the state in which the immune system of an individual lacks reactivity against potentially antigenic molecules originating within that individual. Self tolerance develops during perinatal life. *See also* **acquired tolerance**, **autoimmunity**, **immunological tolerance**.

Seliwanoff test a test for ketohexoses based on the production of a red colour when the sample is boiled with a solution of 0.05% w/v resorcinol in 3 m hydrochloric acid. The test relies on the fact that ketoses are dehydrated more rapidly than aldoses to yield furfural products. [After Feodor Feodorowitsch Seliwanoff (1859–1938), Russian chemist.]

Sem *symbol for* a residue of L-selenomethionine.

SEM *or* **S.E.M. 1** *abbr. for* standard error of estimate of mean. **2** *abbr. for* scanning electron microscope; *see* **electron microscope**.

sem-5 the gene for Sem-5, sex muscle abnormal protein 5, in *Caenorhabditis elegans*. Sem-5 acts in vulval induction and sex myoblast migration, contains SH2 and SH3 domains (*see* **SH domain**), and is phosphorylated by some tyrosine kinase receptors. Mutations of the *sem-5* gene block its function in the worm.

Sema *abbr. for* semaphorin.

semantide *or* **semantophoretic molecule** *or* **informational macromolecule** any of the macromolecules that carry the information of the genes or a transcript (or a translation) thereof. The genes themselves are **primary semantides**, messenger RNA molecules are **secondary semantides**, and most polypeptides are **tertiary semantides**.

semantophoretic molecule *an alternative name for* **semantide**.

semaphorin *(abbr.: Sema)* or **collapsin** any of a family of secreted or membrane-bound proteins (≈750 residues) that are expressed in various embryonal and adult tissues. In the developing nervous system they are released by cells that surround the paths of axon migration and can repel (e.g. Sema 3A and Sema 3F) or attract growing axons. They may also have non-neuronal functions. Their extracellular N-terminal regions (≈500 residues) are homologous to those of hepatocyte growth factor receptors and of **plexins**. Semaphorins are overexpressed during invasive or metastatic progression of tumours. Plexins are components of semaphorin receptor complexes.

semen the fecundating fluid of male higher animals, consisting of spermatozoa suspended in seminal plasma – secretions of the accessory glands.

semenogelin one of two major gel-forming proteins of semen. These proteins have 80% of their amino-acid residues in common, and their genes are localized to the long arm of chromosome 20. They are hydrolysed by a **kallikrein**-like protease.

semi+ *prefix* **1** half. **2** partial or partially; intermediate.

SEMI *abbr. for* specific endogenous mitotic inhibitor; *see* **chalone**.

semialdehyde any organic compound formed by the reduction of one carboxyl group of a dicarboxylic acid to an aldehyde group, as in glutamic semialdehyde.

semiconductor any material, e.g. germanium or silicon, whose electrical conductivity lies between that of conductors and insulators and increases with increasing temperature. Its crystal structure

has atomic bonds that allow the conduction of electric current by either positive or negative carriers when the appropriate additives (dopants) are present.

semiconservative replication of DNA *see* **replication of DNA**.

semiconstitutive mutant a mutant in which negative regulation is partially lost, so that a gene is expressed in the absence of inducer, but at a lower level than in the fully constitutive mutation, yet in the presence of inducer the expression reaches the fully constitutive level. *See also* **constitutive mutation**.

semi-micro 1 describing or concerning something moderately small; applied to a scale of operation, an item of equipment, a quantity of material, etc. **2** *(in analytical chemistry)* denoting the analysis of amounts of substance in the range 10–100 mg. *Compare* **micro** (def. 1). *See also* **microanalysis**.

seminal 1 a pertaining to seed. **b** of, or relating to, semen. **2** germinating new concepts; fundamentally original.

seminal plasma the suspending fluid in which semen is ejaculated.

seminalplasmin *another name for* **caltrin**.

seminolipid 1-*O*-alkyl-2-*O*-acyl-3(β-3'-sulfogalactosyl) glycerol; the major sulfogalactoglycerolipid of mammalian testes and sperm.

semiochemical any chemical substance that delivers a message or signal between two organisms. There are two classes: (1) **pheromones**, in which information transfer is intraspecific; and (2) allelochemics, in which information transfer is interspecific. *See also* **allomone**, **kairomone**.

semiotics *functioning as sing.* **1** the practice of symptomatology; a term that became obsolescent, but has been reintroduced in connection with CT, NMR, and other imaging techniques. **2** the philosophical theory of signs and symbols, having three branches: syntactics, semantics, and pragmatics.

semipermeable membrane a membrane that allows the passage of only certain solutes, usually small molecules, but is freely permeable to solvent.

semipolar double bond *an alternative name for* **dipolar bond**.

semiquinone any free radical derived from a quinone or quinoid compound by removal of a single hydrogen atom. *Compare* **quinone**.

semisolid medium *(in cell culture)* a medium of consistency between that of a solution and a full gel.

semisynthesis the partial synthesis of a chemical compound by chemical or/and enzymic modification of one isolated from natural sources or produced biotechnologically (e.g. by fermentation). The procedure has been used successfully in the manufacture of artificial penicillins and cephalosporins. —**semisynthetic** *adj.*

semisystematic name *or* **semitrivial name** *(in organic chemistry)* a name of a molecule in which only a part is used in a systematic sense; e.g. (meth)ane, (but)ene, (calci)diol (in each of which only the part of the name outside the parentheses is systematic). Most names in organic chemistry are of this type. *Compare* **systematic name**, **trivial name**.

senescence the process of ageing. —**senescent** *adj.*

senescence-associated gene *abbr.:* SAG; any gene expressed during leaf senescence, fruit ripening, death of floral parts or unpollinated fruit tissues, and in seeds of germinating plants. Their products include cysteine proteases, aspartic proteases, other enzymes, polyubiquitin, metallothioneins, and pathogenesis-related proteins. They may be expressed early or late in senescence, in an organ-specific manner, or show enhanced expression due to stress (e.g. drought). Some respond to ethylene, others respond to abscisic acid.

senile aged, to a point of severely decreased capability. —**senility** *n.*

senna a preparation of the dried leaves of *Cassia angustifolia* or *C. senna* used as a laxative. It contains glycosides of hydroxyanthracene (sennoside B), which are metabolized to purgative anthraquinones by colonic bacteria.

sense DNA *see* **coding strand**.

sense strand (of duplex DNA) *see* **coding strand**.

sensitivity 1 an index of the ability of any analytical method or other detection procedure to make quantitative determinations at very low levels. **2** *(in metabolic control theory)* the susceptibility of flux through a pathway to change in response to change in the level of one of the component enzymes, expressed quantitatively in terms of the **flux control coefficient**. An alternative meaning arises when an allosteric ligand acts at several points in a pathway, where a plot of the fractional modification in flux versus ligand concentration yields a sigmoidal curve for a positive effector; the sensitivity in this case can be derived from the slope of the sigmoidal curve. **3** *(in informatics)* an alternative name for **recall**.

sensitivity coefficient *former name for* **flux control coefficient**.

sensitization *or* **sensitisation 1** enhanced reactivity to any treatment (such as exposure to an agonist) of any tissue, cell, or molecular system resulting from prior experience or treatment (often by the same agonist). **2** the conditioning of an animal by antigen in such a way that **hypersensitivity** reactions occur on subsequent exposure to the antigen. **3** the attachment of antibody to sheep red blood cells to make them susceptible to lysis by complement in the **complement-fixation test**. Red blood cells treated in such a way are commonly referred to as **sensitized cells**. —**sensitize** *or* **sensitise** *vb*.

sensitizer *or* **sensitiser** *see* **allergen**.

sensor any device that records the level of a substance of interest or under investigation, using physical parameters (heat, conductivity) or chemical methods. Sensors are often applied to continuous monitoring in flow systems; **biosensors** utilize the reactions of enzymes immobilized on a probe and often coupled with a second substrate that yields a coloured or fluorescent product.

sensor kinase *abbr*.: SK; a histidine kinase that forms one element of the **two-component regulatory systems** that perform signal transduction in microorganisms and plants. Sensor kinases are often membrane proteins that respond to environmental changes.

sensory rhodopsin either of two phototaxis proteins (SRI or SRII) of plasma membranes of halobacteria, similar in structure to the light-driven proton pump bacteriorhodopsin, that influence bacterial migration via signals to the flagellar motor. Their action is mediated by integral membrane proteins (HtrI and HtrII respectively) that extend far into the cytoplasm.

sensory transducer any component of bacterial systems involved in signal transduction. An example is the *ntrB/ntrC* system, which occurs in several bacteria. In *Escherichia coli* under nitrogen limitation the product of the *ntrB* gene, NtrB (EC 2.7.1.–), activates NtrC – the *ntrC* gene product – by phosphorylating it; in nitrogen excess it reverses this. NtrC contains an ATP-binding domain that interacts with **sigma factor** σ^{54}. Another example is sensory rhodopsin I transducer from archaebacteria. This is a typical seven-transmembrane-domain **opsin**-type protein used for sensing light. *See also* **che**.

sentrin *an alternative name for* **SUMO**.

sepacin *see* **separase**.

separase a cysteine peptidase, EC 3.4.22.49, that cleaves Sec1 of the **cohesin** complex to cause sister chromatid separation at the anaphase stage of the cell cycle. Most separases are large molecules (180–250 kDa) that have a catalytic domain homologous to that of the **caspases**, and a highly conserved C-terminal region. Separases are inhibited by binding to **securin**.

separation *(in genetics)* the break-up of a cointegrate into two or more **replicons** of similar sizes. *Compare* **excision**.

Sephacel *proprietary name for* a gel-filtration medium based on **cellulose** cross-linked using **epichlorohydrin** so that a mesh results, the size of the apertures in the mesh being determined by the extent of cross-linking.

Sephacryl *proprietary name for* a gel-filtration medium prepared from allyldextran covalently cross-linked using *N,N*-methylene bisacrylamide, to yield a mesh; the size of the apertures in the mesh depend on the degree of cross-linking.

Sephadex *proprietary name for* a gel-filtration medium based on **dextran** cross-linked using **epichlorohydrin** so that a mesh results, the size of the apertures in the mesh being determined by the extent of cross-linking.

Sepharose *proprietary name for a* gel-filtration medium based on **agarose** cross-linked using **epichlorohydrin** to form a mesh, the size of the interspaces being determined by the extent of cross-linking.

septanose the form of any **monosaccharide**, or monosaccharide derivative, that contains a ring consisting of six carbon atoms and one oxygen atom. In monosaccharide symbolism the septanose form

may be indicated by the suffix *s*; e.g. Glc*s* symbolizes glucoseptanose. *Compare* **furanose**, **heptose**, **pyranose**.

septate divided by, or involving, a **septum** or septa.

septation the process of forming the **septum** that cuts a cell into two at the end of the **cell cycle** in bacteria.

septic shock the clinical state of hypotension and multiorgan dysfunction that results from systemic infection by Gram-negative bacteria. Bacterial lipopolysaccharide endotoxins stimulate the host's immune system, causing the secretion of **tumour necrosis factor-α** (cachectin) and other factors.

septum 1 *(in anatomy)* a wall of tissue, sheet of cells, or other membrane-like structure that partitions two compartments. **2** a disk of silicone rubber, PTFE, or similar material used in any apparatus to seal an injection port. A syringe needle can pierce the septum for the purpose of injection, and on withdrawal the elasticity of the septum ensures that the hole reseals.

Sequenase *proprietary name for* a chemically modified T7 DNA polymerase with high **processivity**. It is the usual choice for DNA sequencing by the **chain-termination method**. *See also* **ThermoSequenase**.

sequenator an apparatus for determining the sequence of monomeric residues (e.g. amino acids) in an ordered linear polymer (e.g. an oligopeptide) by repeating a chemical process.

sequence 1 the ordinal arrangement of the constituent parts of a biopolymer, e.g. the order of amino-acid residues in a polypeptide chain, or of the nucleotide residues in a polynucleotide chain; the known arrangement of such units in any biopolymer or fragment. **2** any particular segment, occurring in, or derived from, a biopolymer, having a known order of its constituent parts; a synthetic polymer composed of parts equivalent to those occurring naturally, and present in a particular known order. **3** to determine the order of residues in a biopolymer or fragment. —**sequential** *adj*.

sequence analysis package a suite of programs for the computational analysis of protein and/or nucleic acid sequences, including, for example, tools for fragment assembly, mapping, translation, automatic and manual sequence alignment, phylogenetic analysis, database searching, pattern recognition, property visualization, composition analysis, and RNA secondary structure plotting. One of the first academic packages was GCG, which then became a commercial product; a freely available alternative is EMBOSS, the European Molecular Biology Open Software Suite.

sequence assembly *or* **assembly** the process by which the overlapping DNA sequence readings from a single sequencing experiment (usually fragments ~500–800bp in length) are correctly ordered along the genome. *See also* **sequence assembly software**.

sequence assembly software software for computational **sequence assembly**, including programs for base calling from sequencing gels (e.g. PHRED) and for ordering the resulting overlapping DNA sequence fragments (e.g. PHRAP).

sequence compression *see* **band compression**.

sequence database *or (sometimes)* **primary database** a database that houses biomolecular sequences (nucleotide or protein) and associated annotation (relating to organism, species, function, mutations linked to particular diseases, functional or structural features, bibliographic references, etc.): e.g. EMBL, GenBank, and DDBJ are the primary nucleotide sequence databases; PIR-PSD, Swiss-Prot, TrEMBL, JIPID, and PRF are primary protein sequence databases.

sequence filter *(in bioinformatics)* a program that masks regions of local compositional bias in biological sequences, usually as a pre-processing step for database-search programs to reduce the number of spurious matches and thereby increase the likelihood that the highest scoring matches are the result of common descent. Examples include SEG, DUST, and XNU.

sequence hypothesis the hypothesis, formulated by F. H. C. **Crick** in 1958, that 'the specificity of a piece of nucleic acid is expressed solely by the sequence of its bases, and that this sequence is a (simple) code for the amino-acid sequence of a particular protein.'

sequence ladder the ladder-like bands of separated components occurring in the electrophoresis gel that result from DNA sequencing experiments.

sequencer an apparatus for automatically determining the sequence of monomeric residues in an ordered linear polymer.

sequence rule *or* **Cahn–Ingold–Prelog rule** *or* ***R S* convention** (of configuration specification) a system for specifying the absolute molecular chirality of a chemical compound. This system considers the sequence of the proton numbers of the atoms directly attached to a chiral centre(*see* **chirality**); if two or more of these atoms have the same proton number, their substituents are then taken into account, the group with substituents of highest proton number taking precedence. When a multiple bond to a substituent is present, the substituent is arbitrarily considered to occur twice, or thrice; an isotope of higher nucleon number takes precedence over one of lower number. For a tetrahedral centre (e.g. a carbon atom) the substituent with the lowest priority is placed away from the observer (behind the chiral atom) with the other three projecting towards the observer; if these three substituents lie in a clockwise (right-handed) array of decreasing priority, then the chiral centre is designated with the stereochemical descriptor *R*, denoting Latin *rectus*, right; if the substituents lie in an anticlockwise (left-handed) array of decreasing priority, then the chiral centre is designated with the stereochemical descriptor *S*, denoting Latin *sinister*, left. For the name of a compound consisting of molecules with a single chiral centre the descriptor *R* or *S*, with a preceding locant if needed, is formed into a prefix by placing it in parentheses and adding a linking hyphen; a racemate with a single chiral centre in its molecules is labelled with the prefix (*RS*)-. When a molecule contains more than one chiral centre the prefix to the name is composed of multiple descriptors, each preceded by an appropriate locant; e.g. (2*R*,3*R*)-2,3-dihydroxybutanedioic acid (= L$_g$-tartaric acid). If only the relative, but not the absolute, configurations of two or more chiral centres in a molecule are known, these centres may be designated with the descriptors *R** and *S** (spoken as *R* star, *S* star), preceded when necessary by appropriate locants, on the arbitrary assumption that the chiral centre first cited (which usually is the one with the lowest locant) has an *R* configuration; for examples, compare the systematic names of **allothreonine** and **threonine**. *See also* **D/L** convention. Extensions of the sequence rule are the ***E/Z* convention**, the **pro-*E*/pro-*Z* convention**, the **pro-*R*/pro-*S* convention** and the ***Re/Si* convention**. [After Robert Sidney Cahn (1899–1981), British natural-products chemist, editor, and scientific semanticist, (Sir) Christopher Kelk Ingold (1893–1970), British physical organic chemist, and Vladimir Prelog (1906–98), Bosnian-born Swiss organic chemist, who jointly developed the system and described it in a series of three papers published in 1951, 1956, and 1966.]

sequence-tagged site *abbr.*: STS; a unique short, single-copy DNA sequence that characterizes a mapping landmark on the genome. Such sequences provide unambiguous identification of DNA markers generated by the Human Genome Project (*see* **HUGO**). STSs bind to **YACs**. If two YACs contain the same STS they must overlap. The YACs can be built up into an overlapping series of clones (a **contig**) covering an entire gene.

sequencing the act, or process, of determining the **sequence** (def. 1) of proteins or nucleic acids. *See* **chain-termination method**, **chemical cleavage method**, **solid-phase technique**.

sequencing by hybridization *see* **SBH**.

sequencing centre a research institute that provides a major focus for genome mapping and sequencing, often also providing genome analysis software for free use by the community. Notable examples include the Sanger Institute in the UK, and The Institute for Genome Research (TIGR) and the Whitehead Institute for Biomedical Research (WIBR) in the USA.

sequencing gel a polyacrylamide gel run to resolve oligonucleotides produced in a DNA sequencing procedure. *See* **chain-termination method**, **chemical cleavage method**.

sequential analysis the analysis of a protein or nucleic-acid **sequence** (def. 1).

sequential mechanism an enzyme reaction involving two or more substrates in such a way that all the substrates must be bound to the enzyme to form the central complex before any products can be released. A sequential mechanism may be an **ordered mechanism** or a **random mechanism**.

sequestering agent *see* **chelating agent**.

sequestration 1 the process of removal or separation. **2** (*in chemistry*) the effective removal of ions from solution by **coordination** with another kind of ion or molecule to form a firm complex molecule with different properties from the original ion. *Compare* **chelation**. **3** (*in cell biology*) the intracellular enclosing of material in an organelle. —**sequester** *vb*.

Ser *symbol for* a residue of the α-amino acid L-serine (alternative to S).

SER *abbr. for* smooth endoplasmic reticulum (*see* **endoplasmic reticulum**).

sera a plural of **serum**.

SERCA *abbr. for* sarcoplasmic/endoplasmic reticulum **calcium-transporting ATPase**, EC 3.6.1.38; a family of Ca^{2+} pump integral membrane protein isoforms of sarcoplasmic or endoplasmic reticulum. They catalyse the hydrolysis of ATP coupled with Ca^{2+} transport from the cytosol.

serglycin a proteoglycan of secretory vesicles of leukocytes. In rat, the core protein (104 amino acids) contains a central domain of 49 alternating Ser and Gly residues. Some 15 chondroitin sulfate/dermatan sulfate chains are linked to the Ser residues in this domain.

serial analysis of gene expression *abbr.*: SAGE; a method for detecting and quantifying gene expression. Short sequence tags (10–14bp), each serving as a unique marker for its transcript, are extracted from defined locations at the 3′ end of cDNAs, concatenated, inserted into vectors, and cloned in bacteria. The concatemers are sequenced and the tags counted. The relative numbers of each tag provide an estimate of the expression level of the corresponding transcript. Also, tags are matched with their corresponding genes where the gene sequences are known. SAGE projects usually generate an order of magnitude more tags than expressed sequence tag (EST) sequencing projects, allowing statistically more accurate and more sensitive expression estimates to be made.

serial dilution a method of achieving dilution of a liquid by sequentially transferring a sample of predetermined volume into another, empty, vessel (usually a tube), which is then made up with diluent to the same total volume as the first. The overall dilution achieved is $(v/s)^x$ -fold, where v is the chosen total volume in each tube, s is the size of the sample, and x is the number of times the procedure is repeated.

serinate 1 serine anion; the anion, HO–CH$_2$–CH(NH$_2$)–COO$^-$, derived from **serine**. **2** any salt containing serine anion. **3** any ester of serine.

serine the trivial name for α-amino-β-hydroxypropionic acid; 2-amino-3-hydroxypropanoic acid; HO–CH$_2$–CH(NH$_2$)–COOH; a chiral α-amino acid. L-Serine (*symbol*: S *or* Ser), (*S*)- 2-amino-3-hydroxypropanoic acid, is a coded amino acid found in peptide linkage in proteins; codon: AGC or AGU; UCA, UCC, UCG, or UCU. In mammals, it is a non-essential dietary amino acid, and is glucogenic. D-Serine (*symbol*: D-Ser *or* DSer), (*R*)-2-amino-3-hydroxypropanoic acid, in both free and combined forms occurs in silkworms and earthworms; residues of D-serine occur in certain peptide antibiotics, e.g. polymyxins D$_1$ and D$_2$. [*Note*: the absolute configuration of D-serine (and hence that at the α-carbon atom of other α-amino acids of the D series) is identical to that of D-glyceraldehyde (and hence to that at the reference carbon atom of other monosaccharides of the D series).]

L-serine

L-serine dehydratase EC 4.2.1.13; *systematic name*: L-serine hydro-lyase (deaminating); *other names*: serine deaminase; L-hydroxyaminoacid dehydratase. An enzyme that catalyses a reaction between L-serine and water to form pyruvate and ammonium ion. It is a pyridoxal-phosphate enzyme. The enzyme for gluconeogenesis from serine in *Escherichia coli* is a 4Fe–4S protein.

serine hydroxymethyl transferase *see* **glycine hydroxymethyltransferase**.

serine-like carboxypeptidase *see* **serine endopeptidase**.

serine *C*-palmitoyltransferase EC 2.3.1.50; an enzyme involved in the biosynthesis of **sphingoid** bases. It catalyses the formation of 3-dehydro-D-sphinganine from palmitoyl-CoA and L-serine, releasing CoA and CO_2. Pyridoxal-phosphate is a coenzyme.

serine peptidase any enzyme that has a residue of serine (and one each of aspartate and histidine) at the active centre and is irreversibly inhibited by organic fluorophosphates (which phosphorylate the serine's hydroxyl group). Serine peptidases are of two types: (1) **serine-type carboxypeptidases**, and (2) **serine endopeptidases**, subsubclass EC 3.4.21. During catalysis by serine peptidases an acylenzyme intermediate is formed by esterification of the hydroxyl group of the reactive serine with the carbonyl group of a sensitive peptide bond of the substrate. The serine endopeptidase group includes **chymotrypsin**, most **elastases**, **kallikrein**, **thrombin**, **trypsin**, and a number of microbial proteinases. *See also* **serine-type carboxypeptidase**.

serine/threonine kinase a **protein kinase** that phosphorylates serines or threonines on its target protein.

serine/threonine-specific protein phosphatase EC 3.1.3.16; *recommended name*: phosphoprotein phosphatase; *other names*: protein phosphatase-1; protein phosphatase-2A; protein phosphatase-2B; protein phosphatase-2C. This name includes enzymes that hydrolyse the serine- or threonine-bound phosphate group from phosphoproteins.

serine transhydroxymethylase *see* **glycine hydroxymethyltransferase**.

serine-type carboxypeptidase EC 3.4.16.1; any enzyme of broad specificity that hydrolyses protein with release of a C-terminal amino acid.

serinium serine cation; the cation, $HO–CH_2–CH(NH_3^+)–COOH$, derived from **serine**.

serino+ *prefix* denoting the group, $HO–CH_2–CH(NH–)–COOH$, derived from **serine** by loss of hydrogen from the amino group.

sero+ *comb. form* denoting a connection within or origin in **serum**.

serological pertaining to, or originating in, serum; connected with the study of serum.

serology the study of **serum**; in particular, the immunological phenomena involving circulating antibody that may be observed *in vitro*.

serosa a **serous membrane**, especially the peritoneal covering. —**serosal** *adj*.

serotonergic 1 describing a nerve that is activated by **serotonin**. **2** describing a nerve that acts by the release of serotonin from the nerve endings. *Compare* **adrenergic**, **aminergic**, **cholinergic**, **peptidergic**, **purinergic**.

serotonin 5-hydroxytryptamine (*abbr*.: 5-HT); a monoamine neurotransmitter occurring in the peripheral and central nervous systems and having hormonal properties; it is also found in mast cells, platelets, brain, and the enterochromaffin cells of the gastrointestinal tract, and causes vasoconstriction, increased vascular permeability, and contraction of smooth muscle. Its reuptake via a 5-HT transporter is sensitive to antidepressants and to cocaine. Serotonin is involved in mood control, fear, and anxiety. It is formed from tryptophan, after hydroxylation to 5-hydroxytryptophan, by tryptophan 5-monooxygenase (EC 1.14.16.4; requires tetrahydrobiopterin), followed by decarboxylation by aromatic-L-amino-acid decarboxylase (EC 4.1.1.28). It is a substrate for **monoamine oxidase**.

serotonin receptor *see* **5-hydroxytryptamine receptor**.

serotransferrin *see* **transferrin**.

serous 1 relating to or resembling **serum** (or similar watery fluid). **2** producing a fluid resembling serum.

serous membrane *or* **serosa** a thin connective-tissue membrane lining certain closed body cavities and reflected over the viscera. The peritoneum of the abdomen, pleura of the chest, and pericardium of the heart are all serous membranes. Each consists of a **parietal layer**, lining the walls of the cavity, and a **visceral layer** covering the organs. The two are continuous, forming a closed sac. *Compare* **mucous membrane**.

serpentine domain any series of transmembrane segments of a single polypeptide chain of a membrane protein that traverses the lipid bilayer several times, as, e.g., in **7TM proteins**.

serpin any of a superfamily of proteins, many of which inhibit serine proteases, and are highly similar to classical serine protease inhibitors, such as **$α_1$-antitrypsin** or antithrombin. The superfamily is now considered to extend beyond compounds that are known to be protease inhibitors and, based on similarity, includes proteins such as **ovalbumin** and **angiotensinogen**, which have no inhibitory function. [From serine protease inhibitor.]

serpocidins members of the serine protease (with chymotryptic activity) superfamily that are present in neutrophil **azurophil granules**. They include: **cathepsin G**, which is active against proteins of Grampositive and Gram-negative bacteria; and **azurocidin**, **elastase**, and proteinase 3. The latter three are encoded by a gene cluster at 19pter, and are active against proteins of Gram-negative bacteria.

Sertoli cell *or* **sustentacular cell** a supporting cell of the mammalian testis that surrounds and nourishes developing sperm cells. [After Enrico Sertoli (1842–1910), Italian histologist.]

serum (*pl*. **sera** *or* (*esp. US*) **serums**) the liquid extruded after blood has clotted and the clot has retracted (**syneresis** (def. 2)). *Compare* **plasma**.

serum albumin *or* **plasma albumin** the smallest and most abundant of the plasma proteins; M_r 65 000–68 000 for the bovine form, 69 000 for the human. It is involved in the regulation of osmotic pressure, and in the transport of sparingly soluble metabolites. The serum albumins belong to a multigene family of proteins that includes **α-fetoprotein** and **vitamin D binding protein**. Human serum albumin has three homologous domains, which assemble to form a heart-shaped molecule. It is an all-α protein, containing 28 helices, ranging from five to 31 amino-acid residues, formed into six subdomains; two each of these subdomains form one of the major subdomains. Bovine (serum) albumin (*abbr*.: BSA) and human (serum) albumin (*abbr*.: HSA) are widely used as standards in protein estimations; these include fatty acids, amino acids (such as tryptophan), steroids, metals (e.g., calcium, copper, zinc), and hydrophobic drugs. *See also* **albumin**.

serum amyloid A *abbr*.: SAA; any plasma protein whose synthesis in liver is induced by interleukins 1 and 6 and other cytokines. SAA1 and SAA2 (each of 104 amino acids) are acute-phase reactants. Their cleavage releases the N-terminal polypeptide (76 residues), which forms fibrils and accumulates in reactive amyloidosis. SAA4 is not an acute-phase reactant, does not form fibrils, and is a normal constituent of high-density lipoproteins (HDL). Their function is unknown.

serum amyloid P-component *abbr*.: SAP; a precursor of amyloid component P. Amyloid component P is found in basement membranes, and is associated with **amyloid** deposits.

serum converting enzyme a *former name for* **angiotensin converting enzyme**.

serum globulin any **serum protein** that is also a **globulin**. An early classification of globulins was based on their mobility on electrophoresis. **Serum albumin** moves the fastest; behind albumin on paper electrophoresis or cellulose acetate strips are two bands known as the $α_1$- and $α_2$-globulin bands, followed by a β-globulin band, with the γ-globulins forming a diffuse band near the origin. The $α_1$-globulins include thyroxine- binding protein, transcortin, $α_1$-glycoprotein, α-lipoprotein, and antitrypsin; the $α_2$-globulins include haptoglobin, $α_2$-glycoprotein, macroglobulin, and ceruloplasmin; the β-globulins include transferrin, β-lipoprotein, and β-glycoprotein; the γ-globulins comprise the immunoglobulins.

serum protein any protein contained in serum; the term is usually applied to those proteins present in high quantities. The most prominent of these is **albumin**, which is present in the normal serum of human adults in the range 34–48 g L^{-1}. *See also* **serum globulin**.

serum response element *abbr*.: SRE; a stretch of DNA, usually of the sequence CCATATTAGG, that regulates promoters for genes that are activated by addition of serum to cell cultures.

serum spreading factor *see* **vitronectin**.

seryl the acyl group, $HO–CH_2–CH(NH_2)–CO–$, derived from **serine**.

sesqui+ *comb. form* **1** denoting one-and-a-half times; in a ratio of three parts to two. **2** (*in chemistry*) indicating a compound having one-and-a-half molecular proportions of the specified element group, or unit, either in relation to a second component, as in sesquioxide (e.g. chromium sesquioxide, Co_2O_3), or in relation to a reference compound, as in **sesquiterpene**; it also sometimes denotes a compound intermediate in constitution between two others, as in sesquicarbonate (e.g. sodium sesquicarbonate, $Na_2CO_3 \cdot NaHCO_3 \cdot 2H_2O$).

sesquiterpene any **terpene** formed from three isoprene units, i.e. having 15 carbon atoms per molecule (a monoterpene has 10 carbon atoms and is formed from two isoprene units).

sesquiterpene lactone any sesquiterpene (15 carbon atoms) containing a lactone structure. Several are derived from *Artemisia* spp., including **artemesinin** (qinghaosu) from *A. annua* (qing hao), which has been used for almost 2000 years as an antimalarial.

artemesinin

sester+ *comb. form* **1** denoting two-and-a-half times; in a ratio of five parts to two or 10 parts to 4. **2** *(in chemistry)* indicating a compound having two-and-a-half molecular proportions of the specified group or unit in relation to a reference compound, as in **sesterterpene**.

sesterterpene any **terpene** formed from five isoprene units, i.e. having 25 carbon atoms per molecule (a monoterpene has 10 carbon atoms and is formed from two isoprene units). An **ophiobolin** is an example.

sev *symbol for* seventess gene; *see* **Seventess protein**.

Sevag method a method for removing proteins from nucleic acids. The proteins are precipitated after repeated denaturation by shaking with a solution of octanol in chloroform.

Seventess protein *abbr.*: Sev; a **tyrosine kinase** receptor, encoded by the *seventess (sev)* gene, that participates in differentiation of R7 photoreceptors in *Drosophila*. Sev has a ligand, **Boss**, the product of the *bride of seventess (boss)* gene; it is expressed by R8 photoreceptor cells and internalized in a Sev-dependent manner.

seven-minute phosphate *see* **labile phosphate**.

seven-transmembrane-domain protein *or* **seven-transmembrane-segment protein** *or* **seven-transmembrane-helix protein** *or* **7TM protein** a protein that contains, in a single polypeptide chain, seven hydrophobic domains that traverse the lipid bilayer. This structural characteristic is typically used to describe the structures of receptors that couple to effector systems through **G proteins**; these are also sometimes referred to as **serpentine receptors**. The structures of bacteriorhodopsin (a 7TM protein but not a G protein-coupled receptor (GPCR)) and of bovine rhodopsin (which is a GPCR) have been determined to high resolution; similar structural characteristics have been attributed to several other proteins, largely on the basis of sequence comparison. Thus, for many proteins, the name '7TM protein' is a tentative one, and the implication that the transmembrane domains are helical is equally so.

seven-up a receptor required for photoreceptor cells during eye development. It is a **zinc-finger** DNA-binding protein with two forms (1 and 2) generated by alternative splicing.

severe combined immunodeficiency *see* **SCID**.

sex chromatin a condensed mass of **chromatin** that represents an inactivated X chromosome. In mammalian nuclei, each X chromosome apart from the activated one forms a sex chromatin body or **Barr body**.

sex chromosome a chromosome involved in sex determination. In many species, these are the **X chromosome** and **Y chromosome**. They are distinguished from the **autosomes** i.e. all the other chromosomes of that species. In some species, however, e.g. fish of the genus

Xiphophorus, sex determinants are spread over a number of chromosomes.

sex determination the specification of the sex of an individual organism by genes. In many animal species, these genes are located on the **X chromosome** and the **Y chromosome**. Their effects are manifested as a result of secretion of **androgens** and **estrogens**.

sex-determining region of Y chromosome *see* **SRY**.

sex factor *or* **fertility factor** *or* **F factor** *or* **F agent** an episomal genetic element (*see* **episome**) in bacteria that enables the cell to act as a genetic donor. Cells of such donor strains are able to form stable unions with cells devoid of such factors and to transfer to them extrachromosomal material, including the sex factor, and sometimes segments of bacterial chromosomes (*see* **conjugation** (def. 4)). The **F plasmid** is the prototypical sex factor, responsible for conjugation in certain strains of *Escherichia coli*.

sex hormone any steroid hormone that is responsible for controlling sexual characteristics and reproductive function. **Estrogens** and **progesterone** are the female sex hormones, while **androgens** are the male sex hormones. They are produced mainly by the ovaries or testes, with contributions from the adrenals and placenta.

sex hormone-binding globulin (*abbr.*: SHBG) *or* **testosterone-binding globulin** a plasma glycoprotein of the globulin group, secreted by liver, that transports the sex hormones testosterone, estradiol, or their derivatives. Its level increases rapidly early in pregnancy and following estrogen therapy.

Sex lethal symbol: *Sxl*; a *Drosophila* gene whose product regulates sexual differentiation. It is so called because it kills flies if it is inappropriately active in males or inactive in females. The functional gene product is expressed in embryos that have the chromosomal constitution 2X:2A where A represents a single set of autosomes. Flies with an X:A ratio of 1.0 express SXL and develop as females whereas those with a ratio of 0.5 do not produce SXL and develop as males. The presence of SXL in females triggers sex-specific alternative mRNA splicing that leads to the production of sex-specific *doublesex* (*dsx*) and *transformer* (*tra*) products.

sex linkage the linkage of a gene producing a certain phenotype (often unrelated to primary or secondary sexual characters) with a **sex chromosome** (usually an X chromosome). Such a gene, or the character it determines, is said to be **sex linked**. Mutation of genes on the X chromosome is often apparent only in male offspring, which have only one X chromosome. The mutation may be masked in the female unless the mutation also occurs on the other X chromosome.

sex pilus (*pl.* **sex pili**) a long (2–3 μm but can be up to 20 μm) hollow proteinaceous projection from the surface of a bacterium that permits **conjugation** and the transfer of an episome from a donor 'male' cell to a recipient 'female' cell. The **F plasmid** of *E. coli* is an episome that carries genes for the expression of the **F-pilus** and is transferred as single-stranded DNA during conjugation.

sexual conjugation *see* **conjugation** (def. 4).

S-form *abbr. for* smooth form; describing bacterial colonies with a **smooth** appearance. *Compare* **R form** (def. 2).

SGD *abbr. for* *Saccharomyces* genome database; *see* **genome database**.

SGF *abbr. for* skeletal growth factor.

SGOT *abbr. for* serum glutamic–oxaloacetic transaminase (not recommended; *see* **aspartate transaminase**).

SGPT *abbr. for* serum glutamic–pyruvic transaminase (not recommended; *see* **alanine transaminase**).

sh *see* **programming language**.

shadow casting a procedure used in the preparation of specimens for electron microscopic examination. The specimen is exposed in a vacuum to a heavy metal evaporated with an electrically heated tungsten filament. Because the metal atoms travel in straight lines, if evaporation is from an acute angle the metal will condense on only one side of the features of the specimen leading to variations in electron density over the specimen, creating a 'shadow' effect.

SHAGGY a protein serine/threonine kinase of *Drosophila* that is highly homologous with mammalian **GSK-3**.

shaker a superfamily of voltage-gated channels for K^+, Na^+, and Ca^{2+} found in animal cells, named for the *Drosophila* mutant from which the first K^+ channel was cloned. Shaker proteins are homotetramers, each subunit containing six transmembrane segments. Segment 4 has a regular pattern of lysine or arginine residues in the third position. This forms the voltage sensor. Between segments 5 and 6 is a loop (the P domain) that is forced back into the membrane by two proline residues at each end (which act as helix breakers), and also contains a conserved TXGYGD motif. The four P domains form the ion pore . *See* **AKT1**.

sham operated describing a control procedure in which all the operations employed in a surgical technique are imitated with the exception of the experimental lesion that is the subject of study.

SHAR *abbr. for* shell archive; a file format that is similar in some respects to a **TAR** file but does not require a special program to extract the archive because of a facility in the UNIX Bourne shell. The command (UNIX or Linux) to extract files from an archive programs.shar is: sh programs.shar.

Shardinger enzyme *see* **xanthine oxidase**.

shark's tooth comb a plastic device for creating wells in an electrophoretic gel. It is shaped like a comb with the teeth coming to a point.

Sharples super centrifuge a, now superseded, efficient centrifuge with a flow-through rotor in which the sedimentation path was short. It is used for harvesting bacteria, blood cells, etc. The version powered with an electric motor turned at 21 000–23 000 rpm while the air-turbine-driven version turned at 50 000 rpm. The instruments of this type have been superseded by others that can operate continuous flow rotors.

SHBG *abbr. for* sex hormone-binding globulin.

SHC a gene family first identified by screening cDNA libraries for genes encoding SH2-containing proteins (*see* **SH domain**). The mammalian gene encodes two overlapping, widely expressed proteins with a C-terminal SH2 domain that binds phosphotyrosine residues on activated growth factor receptors. Phosphorylated Shc binds **Grb2**, and may form a link between certain tyrosine kinases and the Grb2/Ras signalling pathway.

SH domain *abbr. for* Src homology domain; the **SH2 domain** is a protein module, consisting of ~100 residues, found in many proteins involved in signal transduction. Apart from Src, which contains the prototype SH2 domain (*see* **src**), these include, among many others, phospholipase C, products of proto- oncogenes such as **ABL**, **FYN**, and **FGR**, GTPase-activating protein, and protein tyrosine phosphatases. The SH2 domain has also been found in structural proteins such as **fodrin**. One function of the domain is to bind to the autophosphorylated site of growth factor receptors, e.g. PDGF and EGF receptors. Proteins with SH2 domains are frequently found also to have SH3 domains. The **SH3 domain** is composed of ~50 amino acids, and was originally identified in Src, but has since been found within a number of other proteins. SH3 domains mediate protein–protein interactions through their ability to recognize and bind proteins containing specific amino-acid sequences rich in proline and hydrophobic residues. To date, there are known to be at least two general classes of ligand to which SH3 domains can bind: class I, which comprises proteins containing the consensus sequence PxLPPZP; and class II, with the sequence XPPLPxR (where x is any amino acid, and Z appears to be a residue specific to the type of SH3 domain involved). Given the nature of the proteins possessing SH3 domains and SH3-binding domains, it is clear that the protein interactions mediated by these motifs are important to the operation and maintenance of cytoskeletal architecture and intracellular signalling pathways.

SH2 domain *see* **SH domain**.

SH3 domain *abbr. for* Src homology 3 domain. *See* **SH domain**. *See also* **src**.

shear 1 to disrupt or fracture (an entity) using forces acting parallel to a plane, rather than perpendicularly, as with tensile or compressive forces. For example, shear forces can be used to disrupt DNA or other macromolecules by enclosing the molecules between two closely applied plane surfaces moving rapidly in opposite directions. Homogenization techniques depend on shear forces. **2** the deformation or fracture produced by such forces.

shear gradient *see* **viscosity**.

shear rate *see* **viscosity**.

shear stress *see* **viscosity**.

Sherman paradox the seemingly paradoxical situation in fragile-X

syndrome whereby the risk of mental retardation appears to increase in successive generations and the daughter of an unaffected male carrier is more likely to have affected offspring than the mother of the unaffected male carrier. This phenomenon is now called **anticipation**, and is explained by the expansion of CGG repeats during maternal meiosis. [After Stephanie L. Sherman, US geneticist.]

shibire the protein homologue of **dynamin** in *Drosophila*.

shikalkin the racemic mixture of **alkannin** and its (+)(*R*) enantiomer, shikonin.

shikimate the anion of shikimic acid; [–]3α,4α,5β-trihydroxy-1-cyclohexene-1-carboxylic acid; (3*R*,4*S*,5*R*)-trihydroxy-1-cyclohexenecarboxylic acid; an important intermediate, in microorganisms and plants, in the **shikimate pathway** for the synthesis of phenylalanine, tyrosine, tryptophan, and many other substances (e.g. tocopherol, vitamin K, folic acid, ubiquinone, enterobactin, and some antibiotics). The enzyme **3-phosphoshikimate 1-carboxyvinyltransferase** (EPSP synthase) in higher plants is specifically inhibited by the compound **glyphosate**, which inhibits growth and is an effective and readily degraded broad-spectrum herbicide. Shikimic acid was first isolated from the plant *Illicium religiosum* (star anise) in 1885. The name is derived from the Japanese name for this plant, *shikimi-no-ki*. *See also* **shikimate kinase**.

shikimate 5-dehydrogenase EC 1.1.1.25; *systematic name* shikimate: NADP+ 5-oxidoreductase; an enzyme of the **shikimate pathway** that catalyses the reduction by NADPH of 3-dehydroshikimate to shikimate with formation of NADP$^+$. With the now-accepted chemical numbering for shikimate, it is in fact a 3-dehydrogenase. *See also* **pentafunctional arom polypeptide**.

shikimate kinase EC 2.7.1.71; *systematic name*: ATP:shikimate 3-phosphotransferase; an enzyme of the **shikimate pathway** that catalyses the phosphorylation by ATP of shikimate to shikimate 3-phosphate with formation of ADP. The product is the precursor of 5-*enol*pyruvylshikimate 3-phosphate (5-*O*-(1-carboxyvinyl)-3-phosphoshikimate) from which **chorismate** is formed. *See* **pentafunctional arom polypetide** for another example. *See also* **shikimate**.

shikimate pathway a multi-branched metabolic pathway that leads from carbohydrate precursors via shikimate to: (1) phenylalanine and tyrosine, via chorismate and prephenate(*see* **arogenate**); (2) tryptophan, via chorismate and anthranilate; (3) ubiquinone, plastoquinones, and tocopherols, via chorismate; (4) many naphthoquinones (including the vitamins K), 3-carboxy aromatic amino acids, salicylate, and dihydroxybenzoate (for enterobactin formation) via chorismate and isochorismate; and (5) certain non-aromatic cyclohexane compounds, via shikimate. The carbohydrate precursors are phospho*enol*pyruvate and D-erythrose 4-phosphate; they yield 2-dehydro-3-deoxy-D-*arabino*-heptonate 7-phosphate (I, more commonly termed 3-deoxy-D-*arabino*-heptulosonate 7-phosphate, *abbr.*: DAHP) by action of EC 4.1.2.15, 2-dehydro-3-deoxyphosphoheptonate aldolase (DAHP synthase). I is converted to 3-dehydroquinate (II) by 3-dehydroquinate synthase, EC 4.6.1.3, and II is then converted to 3-dehydroshikimate (III) by 3-dehydroshikimate dehydratase, EC 4.2.1.10. Shikimate is then formed from III by **shikimate 5-dehydrogenase**, EC 1.1.1.25, and converted to shikimate 3-phosphate (IV) by **shikimate kinase**, EC 2.7.1.71. Phospho*enol*pyruvate is used again to convert IV to 5-*enol*pyruvylshikimate 3-phosphate (V, *abbr.*: EPSP) by 3-phosphoshikimate 1-carboxyvinyltransferase, EC 2.5.1.19, and V yields chorismate (VI) by action of **chorismate synthase**, EC 4.6.1.4. Chorismate leads to prephenate by action of **chorismate mutase**, EC 5.4.99.5, and undergoes isomerization by isochorismate synthase, EC 5.4.99.6, to form isochorismate. Organization of the various enzymatic activities varies considerably in plants, bacteria, yeasts, and other fungi. In yeasts and other fungi, and also in *Euglena gracilis*, the conversion of DAHP to EPSP requires a complex pentafunctional polypeptide specified by the *arom* gene cluster (*see* **pentafuctional arom polypeptide**).

shikonin the (+)-(*R*) enantiomer of **alkannin**.

Shine–Dalgarno sequence a ribonucleotide sequence for the formation of the preinitiation complex between a 30S ribosomal subunit and an mRNA molecule. It is a seven-nucleotide, pyrimidine-rich sequence near the 3′ terminus of 16S (prokaryotic) or 18S eukaryote ribosomal RNA, found in many organisms, that forms base pairs with a complementary three- to seven-nucleotide, purine-rich sequence (the Shine–Dalgarno sequence) preceding the initiation codon in many mRNA molecules, thereby allowing the initiating ribosomal subunit to bind the mRNA so as to discriminate between AUG or GUG (i.e. initiation) triplets and to select the one at the beginning of a cistron. [After John Shine (1946–) and Lynn Dalgarno (1935–).]

SHIRPA a protocol for phenotype assessment designed to analyse behavioural changes in mice caused by **ENU mutagenesis**. The first four letters of the acronym are from the names of participating institutions in the UK.

shockate the preparation that is obtained when (bacterial) cells are suspended in an osmotically stabilizing medium and lysed by the sudden addition of water, followed by immediate centrifugation.

short-chain describing an aliphatic compound with a chain length of less than 10 carbons.

short-chain acyl-CoA dehydrogenase an enzyme that catalyses the following reaction in fatty-acid oxidation:

$$RCH_2CH_2COSCoA + FAD + RCH = CHCOSCoA + FADH_2$$

in which the acyl moiety (R) contains 4–6 carbon atoms. It is a homotetrameric flavoprotein of the mitochondrial matrix, with each subunit containing 388 amino acids. Several mutations lead to enzyme deficiency, which manifests in infancy as hypoglycemia, acidosis, hyperammonemia, and short-chain organic aciduria.

short interfering RNA *abbr.*: siRNA; a double-stranded RNA, 21–23 bp long, that can specifically silence gene expression. The most efficient gene silencing effect is given by having a 19 nucleotide region complementary to the mRNA and projections of two nucleotides at the 3′ ends.

short-patch repair excision repair of DNA in which the excised DNA is about 20 nt long. It contrasts with **very-short-patch repair**, for mismatches between specific bases, and **long-patch repair**, in which stretches of ≈1500 nt or longer are involved. *See also* **DNA repair enzyme**, *uvr*.

shotgun cloning a method, first used for cloning an entire genome, in which the DNA is randomly broken into fragments of approximately the average size of genes. The fragments are then cloned into suitable vectors, to give a **genome library**.

shotgun collection *see* genomic library.

shotgun sequencing a strategy for sequencing whole genomes. Fragmented genomic DNA is cloned into a plasmid vector and randomly picked clones are sequenced using primers annealing to the plasmid. Sequences are compared and aligned using contig building software. Enormous computational power is required for sequencing large genomes in this way.

showdomycin 2-(β-D-ribofuranosyl)-maleimide; an antibiotic produced by *Streptomyces showdoensis* and active against several Gram-positive bacteria. It is used as a thiol-reactive agent. *Compare* **N-ethylmaleimide**.

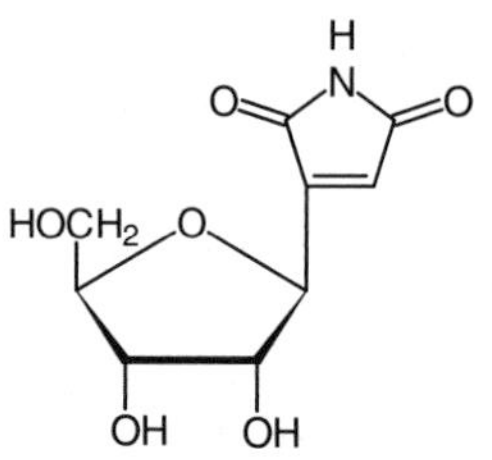

shrimp alkaline phosphatase an enzyme that catalyses hydrolysis of 5′-phosphate from DNA (or RNA) regardless of whether the DNA has the 5′ end protruding or recessed, or is blunt-ended.

shRNA *abbr. for* short-hairpin RNA.

shuttle an intracellular metabolic cycle concerned in the transfer of, e.g., reducing equivalents or carbon chains across membranes. Different carriers operate each way across the membrane. Examples include the **3-hydroxybutyrate cycle**, the **glycerol-3-phosphate cycle**, and the **malate–aspartate cycle**.

shuttle vector a DNA molecule, e.g. a plasmid, that can replicate in two different host organisms and can therefore be used to 'shuttle' or convey genes from one to the other.

shuttle vesicle an endocytotic vesicle that transfers proteins across an epithelial cell from one face to another e.g. from the sinusoidal to the bile-canalicular face of a hepatocyte.

sI *symbol for* a residue of the ribonucleoside thioinosine; mercaptopurine ribonucleoside (alternative to Sno).

Si *symbol for* silicon.

si- *stereochemical descriptor* denoting *sinister* (i.e. counterclockwise, or 'left-handed') configuration of face of a trigonal atom. *See* **Re/Si convention**.

SI *abbr. for* Système International (d'Unités); International System (of Units); a rationalized coherent system of metric units, in which the magnitude of any physical quantity may be expressed, that is recommended for use by scientists throughout the world. It is based on seven **SI base units**: ampere, candela, kelvin, kilogram, metre, mole, and second. There are also 22 named **SI derived units**, which are expressed in terms of the base units, and two **SI supplementary units**, radian and steradian, which are regarded as dimensionless derived units. A series of prefixes signify decimal multiples and submultiples of any SI unit.

Sia *symbol for* the sialoyl group.

sial+ *a variant of* **sialo+** (before a vowel).

sialagogue *or* sialogogue any agent that stimulates the flow of saliva. —**sialagogic** *or* sialogogic *adj.*

sialate the anion of **sialic acid**.

sialic of, or relating to, saliva or a salivary gland.

sialic acid any of the *N*- or *O*-substituted derivatives of **neuraminic acid** found as components of complex carbohydrates, e.g. *N*-acetylneuraminic acid (5-(acetylamino)-3,5-dideoxy-D-*glycero*-D-*galacto*-non-2-ulosonic acid, (*abbr.*: Neu5Ac; *see also* **NeuNAc**). Linkage of sialic acids is through C-2 to monosaccharides at the ends of glycosylation chains of glycolipids and glycoproteins. Neu5Ac can also be linked at its C-8 and C-9 hydroxyl groups to another Neu5Ac residue, often forming multiple repeats as in mammalian **NCAM**. Hydroxyl groups at C-4, -7, -8, and -9 can be *O*-acylated (*see also* **NeuOAc**). Other common derivatives are *N*-glycoloylneuraminic acid (in mammals, but not normally in humans; *abbr.*: NeuGc), and N-**ketodeoxyoctanooylneuraminic acid** (e.g. in bacteria). Sialic acid accumulates in lysosomes in Salla disease and infantile sialic acid storage disease, and in cell cytosol in sialuria.

sialidase EC 3.2.1.18; *recommended name*: exo-α-sialidase; *former name*: neuraminidase; a glycosidase enzyme that hydrolyses 2,3-, 2,6-, and 2,8-glucosidic linkages joining terminal nonreducing *N*- or *O*-acylneuraminyl residues to galactose, *N*-acetylhexosamine, or *N*- or *O*-acylated neuraminyl residues in oligosaccharides, glycoproteins, or glycolipids.

sialidosis *another name for* **mucolipidosis** I.

sialin *or* anion and sugar transporter (*abbr.*: AST); an integral protein (495 amino acids) of lysosomal membranes, that contains 12 transmembrane segments and permits hydron-driven exit of sialic acid (and glucuronic acid) into the cytosol. It is expressed mostly in cardiac and skeletal muscle, liver, kidney, brain, and placenta. Mutations result in the allelic **Salla disease** and infantile sialic acid storage disease.

sialo- *or* (before a vowel) sial+ *comb. form* **1** of, or relating to, saliva or a salivary gland; sialic. **2** denoting the presence in a compound of one or more residues of sialic acid. *Compare* **asialo+**.

sialoadhesin a macrophage-restricted sialic-acid-dependent glycoprotein receptor of several forms. It contains 17 immunoglobulin-like domains. It is an adhesion molecule that recognizes oligosaccharides terminating in NeuAc(α2-3)Gal in N-linked and O-linked glycans.

sialogogue *a variant spelling of* **sialagogue**.

sialosyl the glycosyl group derived by removal of the exomeric hydroxyl group from sialic acid or sialate.

sialoyl the acyl group derived from sialic acid by removing the hydroxyl group from its carboxyl group.

sialuria a very rare, autosomal dominant disease in which coarse features, hepatosplenomegaly, and developmental delay accompany massive excretion of free sialic acid in urine. Sialic acid accumulates in cell cytosol as a result of impaired regulation of sialic acid synthesis by **UDP-*N*-acetylglucosamine 2-epimerase** by the allosteric inhibitor CMP-sialic acid.

SIC *abbr. for* selenocysteine insertion complex.

sickle cell an abnormal red blood cell that contains mostly HbS or HbS and another hemoglobin variant. Such cells occur chiefly in **sickle-cell anemia** and in the sickle-cell trait. At low oxygen tensions such cells adopt a crescent shape reminiscent of the blade of a sickle. In HbS there is replacement of Glu by Val at position 6 in the hemoglobin β chain. In its deoxygenated state the HbS forms long rodlike helical fibres, which deform the red cells because the β6 Val can pack into a pocket in the EF corner (between helices E and F) of another β chain, a pocket which is lost in conformational changes resulting from oxygenation.

sickle-cell anemia an inherited disease of humans caused by homogyzosity for the gene for HbS (*see* **sickle cell**). It is particularly common in populations indigenous to or originating from the Mediterranean region, parts of Saudi Arabia and India, and in those of African origin. The resultant abnormal red blood cells, called **sickle cells**, are removed from the circulation at a faster rate than normal, leading to anemia. There is no satisfactory treatment, and early death often results. Heterozygotes have **sickle-cell trait**; these survive longer and benefit from greater resistance against the malaria parasite, which is adversely affected by sickle-cell trait during the red-cell phase of its life cycle. When the heterozygous sickle cells pass through capilliaries, they sickle, causing them to lose K^+ and killing the malaria parasite, which needs high levels of K^+. *See also* **glucose-6-phosphate dehydrogenase**.

side-by-side model *abbr.*: SBS model *or* SBS; a model of duplex DNA consisting of an alternating arrangement of right-handed and left-handed segments of duplex DNA, or other double-stranded polynucleotide, with five base pairs per segment. This produces a structure similar to the **Watson–Crick model of DNA** except that the two antiparallel strands are side by side rather than intertwined in a double helix.

side chain 1 any part of an organic molecule based on an aliphatic carbon chain that can be regarded as branching from a main chain. **2** the part of an amino acid that extends from the α-carbon atom, and is unique to that amino acid.

sideramine one of the categories of **siderochrome**. *See also* **sideromycin**.

siderochrome *see* **siderophore**.

sideromycin any iron-containing antibiotic compound elaborated by certain actinomycetes. Sideromycins are structurally related to sideramines (*see* **siderochrome**), which can antagonize their antibiotic action.

siderophage *an alternative name for* **siderophore D**.

siderophilin *an alternative name for* serotransferrin, serum **transferrin**.

siderophore *or* **siderochrome** any of various low-molecular-mass Fe(III)-chelating substances elaborated by aerobic or facultatively anaerobic microorganisms, especially when growing in iron-deficient media. Fe^{3+}–siderochrome complexes have very high stability constants (10^{30}–10^{50} dm^3mol^{-1}). They are taken up by specific transport proteins in the microorganisms; subsequent intracellular release of iron requires enzymic action. Siderophores are of two main categories, one or both of which may be formed by any particular organism: (1) catechol derivatives, e.g. **enterobactin**; and (2) hydroxamic acid derivatives, e.g. **mycobactin** (*see also* **ferrichrome**). The transport systems for ferrichrome or ferric enterobactin in certain strains of bacteria appear to be identical with the membrane receptors for colicin M or B respectively (*see* **colicin**), and in the former case for the antibiotic albomycin and various phages as well.

siderosome a membrane-bound organelle found in spleen macrophages. Siderosomes contain aggregates of **hemosiderin** and may be derived from **lysosomes**.

siemens *symbol*: S; the SI derived unit of **electric conductance**, defined as the electric conductance between two points on an electric conductor when an electric potential difference of one volt causes an electric current of one ampere to flow; i.e. $1 S = 1 \Omega^{-1}$. It was formerly called the **mho** or the reciprocal ohm. [After (Ernst) Werner von Siemens (1816–92), German electrical engineer, industrialist, inventor, and metrologist.]

sievert *symbol*: Sv; **1** the SI derived unit of (radiation) **dose equivalent**, equal to one joule per kilogram of living tissue; i.e. $1 Sv = 1 J kg^{-1}$. Also, $1 Sv = 100$ rem. *Compare* **gray**. **2** *or* **intensity millicurie** an obsolete non-SI unit of intensity of gamma radiation, numerically equal to 8.4 röntgens per hour. [After Rolf Maximillian Sievert (1896–1966), Swedish radiation physicist.]

sieve tube exudate protein *abbr.*: STEP; any of numerous types of proteins present usually at low concentration in phloem exudates. They include enzymes of glycolysis, gluconeogenesis, and other aspects of carbohydrate metabolism, structural proteins (**P-proteins**), transcription factors, and lectins.

sievorptive chromatography a form of **chromatography** in which absorption and molecular-sieve properties are combined.

***Si*-face** (of an atom or molecule) *see* **Re/Si convention**.

SIF cell *abbr. for* small intensely fluorescent cell.

sigma *symbol*: σ (lower case) *or* Σ (upper case); the eighteenth letter of the Greek alphabet. For uses *see* **Appendix A**.

sigma bond *see* **orbital**.

sigma factor an **RNA polymerase** initiation factor occurring in bacteria; a protein that promotes attachment of the RNA polymerase, of which it is a subunit, to specific initiation sites, and is then released. The main sigma factor of *Escherichia coli* is σ^{70}. An unrelated family is σ^{54}, with associated regulatory proteins; members of this are specific to certain operons. Other sigma factors are involved in response to heat shock and respiratory stress; examples from *E.coli* include σ^E and σ^{32}. Major switches in transcriptional control in bacteria, e.g. sporulation in bacilli, and transcriptional control in certain phages, involve modification to, or replacement of, sigma factors.

sigma orbital *see* **orbital**.

sigma receptor any receptor that is activated by powerful psychotomimetic drugs, e.g. **phencyclidine**. Sigma receptors appear to mediate disorientated and depersonalized feelings.

sigma subunit a component of bacterial RNA polymerase; *see* **sigma factor**.

sigmoid *or* **sigmoidal** S-shaped; especially as in a sigmoid flexure (e.g. of the large intestine); or a sigmoid plot, i.e. one similar to a skewed letter S, that is with slope increasing initially slowly, progressively increasing more rapidly, and then decreasing towards a plateau. —**sigmoidicity** *n*.

sigmoid kinetics *(in enzymology)* a type of kinetics of enzyme-catalysed reactions characterized by a sigmoid plot of rate versus substrate concentration. Such kinetics are characteristic of **allosteric** enzymes, in that progressive activation occurs with increasing substrate concentration due to **cooperativity** in binding of substrate. The presence of allosteric modifiers will alter the inflection of the curve, activators shifting it towards that found in Michaelis–Menten kinetics (*see* **Michaelis kinetics**).

signal 1 any variable property or parameter that serves to convey information. **2** the message potentially to be transmitted by any agonist interacting with a cell membrane receptor. **3** a **signal peptide**.

signalase *see* **signal peptidase**.

signal averaging a method used to improve **signal-to-noise ratios**, particularly in spectroscopy, by summing repeated scans of the information. The signals are thereby reinforced while the background noise tends to average out.

signal hypothesis a model, now largely substantiated, for the synthesis of secretory proteins on ribosomes bound to the rough **endoplasmic reticulum**. It holds that secretory proteins have mostly hydrophobic residues at the N terminus, which act as a signal to mediate binding to the ER membrane. It was proposed in 1971 by the US cytologist Gunter Blobel (1936–) and the US cell biologist David Domingo Sabatini (1931–), and by the Argentinian-born British biochemist César Milstein (1927–) and coworkers in 1972. *See* **signal peptidase**, **signal peptide**, **signal recognition particle**.

signalling molecule any molecule, e.g. **hormone**, **cytokine**, or **second messenger**, that cues the response of a cell to the behaviour of other cells or objects in its environment.

signal peptidase any endopeptidase that removes the **signal peptide** after passage of a protein through an endoplasmic reticulum (ER), mitochondrial, or chloroplast membrane. In prokaryotes there are two types: **1** signal peptidase I, EC 3.4.99.36; *recommended name*:

leader peptidase; *other name*: prelipoprotein signal peptidase; an endopeptidase that cleaves N-terminal signal peptides from secreted proteins and periplasmic membrane proteins. It is a single-pass protein of the inner membrane. **2 signal peptidase II**, EC 3.4.99.35; *recommended name*: premurein-leader peptidase; an endopeptidase that selectively cleaves the signal peptide from bacterial membrane lipoproteins, hydrolysing the protein on the N-terminal side of a specific Cys residue.

In mammals, signal peptidases of the ER (EC 3.4.–.–) consist of: SPC 18 and SPC 21 (single-pass type I transmembrane proteins, each with a luminal serine endopeptidase domain); SPC 22/23 (a single-pass glycoprotein); and SPC 25 and SPC 12 (which span the membrane twice and have termini in the cytosol). Signal peptidases for processing proteins targeted to mitochondrial and chloroplast membranes are poorly understood.

signal peptidase II *see* **premurein-leader peptidase**.

signal peptide *or* **leader peptide** *or* **pre-part** *or* **prepeptide** *or* **prepiece** *or* **presequence** (def.1) *or* **signal** any sequence of amino-acid residues that, when linked to a newly synthesized protein, directs the protein to a location among the organelles of a eukaryotic cell, or from the cytoplasm to the periplasmic space of prokaryotic cells. The term originally applied to the N-terminal linked peptide of secretory proteins that is bound by the **signal recognition particle**, which transfers the ribosome to the endoplasmic reticulum. Such peptides usually comprise 20–30 amino-acid residues, and are characterized by a central region of hydrophobic amino acids preceded by one or more basic amino-acid residues. An interesting feature is that, at the junction of the signal peptide and the mature protein, there is no consensus sequence of amino-acid residues, so the basis for the specificity of the protease at this position remains unclear (*see* **signal peptidase**). The term signal peptide is now more generally applied to any peptide that is responsible for the location of a newly synthesized protein within the organelles of eukaryotic cells (*see* **protein kinesis**). Thus it applies to the mitochondrial proteins synthesized in the cytoplasm, nuclear proteins, and luminal proteins that are retained within the cisternae of the endoplasmic reticulum. In addition to the signal peptides described, which may be located either at the N terminus or C terminus of the nascent protein, and which are subsequently cleaved, single-spanning membrane proteins also contain a signal sequence within the protein known as a **stop-transfer sequence**. This serves to anchor the protein within the membrane. The orientation of the protein within the membrane may be either type I or type II, depending on whether the N terminus of the resulting protein is extracytoplasmic (type I) or cytoplasmic (type II). *See also* **signal recognition particle**.

signal peptide protease *abbr.*: SPP; a presenilin-type aspartic protease that catalyses intramembrane cleavage of some signal peptides after they have been removed from a preprotein. It is an integral membrane protein of the endoplasmic reticulum and contains seven transmembrane segments. In humans it generates signal sequence-derived lymphocyte antigen-E epitopes, which are recognized by the immune system.

signal recognition particle *abbr.*: SRP; a ribonucleoprotein particle of 325 kDa, composed of a 7S (300 nucleotide) RNA molecule and a complex of six different polypeptides. For proteins destined for the endoplasmic reticulum, as they emerge from the large ribosomal subunit, the SRP binds both to the N-terminal signal peptide and to the ribosome. This binding arrests further translation, preventing the proteins from being released into the cytosol. The SRP–ribosome complex then diffuses to the endoplasmic reticulum, where it is bound to the **signal recognition particle receptor**, which allows resumption of protein synthesis, and facilitates the passage of the growing polypeptide chain through the **translocon**. Through a process involving GTP hydrolysis, the SRP–SRP receptor complex dissociates, and SRP returns to the cytosol. Of the six polypeptides of SRP, the 54 kDa subunit (SRP54) is the central player. It contains an N-terminal GTPase domain and a C-terminal domain that binds directly to the signal peptide and the SRP RNA.

signal recognition particle receptor *abbr.*: SRP receptor; a transmembrane heterodimeric protein (α subunit, 69 kDa; β subunit, 30 kDa), also known as **docking protein**, located in the rough endoplasmic reticulum. Both subunits contain GTPase domains

with which **signal recognition particle** interacts. In the presence of GTP and SRP receptor, SRP is released from the ribosome–nascent chain complex. The α subunits have a negatively charged N-terminus and a positively charged but phosphorylated cytoplasmic tail (C terminus). *See also* **signal peptide**.

signal-to-noise ratio the proportional difference between the average values for a specific property under investigation (signal) and the average of nonspecific values (noise) in any record. *See also* **signal averaging**.

signal transduction the process by which an extracellular signal (chemical, electrical, or mechanical) is converted into a cellular response. Typically, interaction of a hormone, growth factor, or other agonist with a specific membrane receptor leads to signal amplification by synthesis within the cell of one or more **second messengers**, or to activation of other downstream cascades, e.g. by phosphorylation of proteins. Chemical agonists that cross the cell membrane (e.g. steroid hormones) produce a cellular response without such amplification of the signal. Electrical signals flowing down axonal membranes lead to release of neurotransmitters at synapses. Their plasma membrane receptors are similar to those for hormones and growth factors, or are ion channels. For phototransduction *see* **photosynthesis**, **visual cascade**.

signature *see* **pattern**.

signature database *an alternative name for* **pattern database**.

significant *(in statistics)* reaching a degree of **probability** that experimental values lie on a distribution different from that of control values, such that the investigator will conclude that experimental and control samples differ, using tests such as Student's *t*-test or the chi-square test. The degree of probability that will be accepted as indicating a significant difference depends on the type of investigation being conducted. If the tests indicate that the probability that the results are due to chance is $\geq 5\%$ (for biological or medical studies) or $\leq 1\%$ or, in some cases, $\leq 0.1\%$ (for biochemical studies), differences are considered significant. —**significance** *n*.

silaffin a peptide that is implicated in the biogenesis of diatom biosilica. Silaffins are rich in serine and lysine residues that are, respectively, phosphorylated and linked to long-chain polyamines. Silaffin-1A from *Cylindrotheca fusiformis* has the sequence (before modification): SSKKSGSYSGSKGSK.

silane 1 monosilane; silicon tetrahydride; SiH_4; the silicon-containing analogue of methane. **2** any of a class of saturated silicon hydrides. Such compounds are silicon-containing analogues of alkanes or cycloalkanes; linear or branched silanes are of general formula $Si_n H_{2n+2}$, and cyclic silanes, termed cyclosilanes, are of general formula $[SiH_2]_n$ ($n > 2$). The value of n, when >1, may be indicated by a specific or general numerical prefix; e.g. disilane, cyclohexasilane, oligosilane, polysilane. Silane and disilane are gases with a repulsive odour; higher silanes are liquids or solids. Silanes are spontaneously flammable in oxygen or air. **3** *a term often loosely applied to* any hydrocarbyl or other derivative of a silane.

silanization treatment of a hydrophilic surface with a reactive **silane** (def. 3) such as dimethyldichlorosilane, trimethylchlorosilane, octyltrichlorosilane, or an octadecyltrialkoxysilane in order to render it more hydrophobic. Such a treatment is useful in particular for converting hydrophilic **silanol** (def. 4) groups on the surfaces of glassware or dried silica-gel particles into lipophilic alkylsiloxane moieties. When applied to glass, it reduces adsorption of polar compounds and minimizes the activation of lymphocytes, platelets, or other cells that can occur when untreated glass vessels are used; other benefits include promotion of drainage of aqueous liquids and minimization of electrical leakage (*see also* **siliconize**). Silanization has been applied to particles of dried silica gel for use in reversed-phase adsorption chromatography (*see also* **end capping**) and to diatomaceous earth used as the support for the stationary phase in reversed-phase liquid–liquid partition chromatography or in some variants of gas–liquid (partition) chromatography. *Compare* **silylation**. —**silanize** *vb.*; **silanized** *adj.*

silanol 1 silyl hydroxide; H_3SiOH; the silicon-containing analogue of methanol. **2** any monohydroxy derivative of a **silane** (def. 2); such compounds are silicon-containing analogues of alkanols and are of general formula $Si_n H_{2n+1}OH$. **3** *a commonly used term for* any Si-hydrocarbyl derivative of silanol, silanediol, or silanetriol; such derivatives are respectively of general formula R_3SiOH, $R_2Si(OH)_2$, or $RSi(OH)_3$. **4** the trivalent group $->Si–OH$ occurring in hydrated silica (so-called silicic acid). Such groups occur also on the surfaces of dried silica-gel particles and of silica-based glasses under normal conditions (i.e. in equilibrium with undried air).

sildenafil see **Viagra**.

silent mutation or **synonymous mutation** or **same-sense mutation** a nucleotide substitution that does not result in an amino acid substitution in the translation product owing to the redundancy of the genetic code. The term 'silent' is also applied to a protein gene product arising from such a mutation, e.g. **silent hemoglobin**. *Other names*: silent site mutation; silent nucleotide change; silent polymorphism.

silent polymorphism an alternative name for **silent mutation**.

silica silicon(IV) oxide; silicon dioxide; SiO_2; a colourless or white solid used in many chromatographic procedures; silicic acid contains **silanol** (def. 4) groups to some of which water is bound. For chromatographic use, it is heated to $100\,°C$ to drive out this water. Heating above $170\,°C$ causes formation of oxide linkages $(Si–O–Si)$, and at very high temperatures $(800–1000\,°C)$ it forms glass. Silicic acid is available in a variety of particle sizes. For chromatography, these range from 10 to 500 μm, the larger sizes being used for column chromatography at normal pressures, the smaller sizes for thin layer chromatography and high-pressure liquid chromatography. **Silica gel** is silicic acid precipitated as granules, about 3 mm diameter; it is hygroscopic and used as a drying agent, which can be re-used after heating. A powder of very fine silica particles, 0.007–0.014 μm, known as **fumed silica**, has thixotropic properties and has been used to create a gel in organic solvents, e.g. in scintillation cocktails to maintain hydrophilic materials in suspension during counting (now superseded by cocktails that themselves form a gel with water). Silica is found in certain algae, notably in the hard shell, or test, of diatoms (*see* **silaffin**); it also serves to reinforce the stems of the primitive vascular plants known as horsetails.

siliceous or **silicious** of, pertaining to, or containing **silica**.

silicone any oligomeric or polymeric **siloxane** containing the repeating structural unit $–O–SiR_2–$, where R is a hydrocarbyl (e.g. methyl or phenyl) group. The molecule is usually linear, but can be branched, crosslinked, or cyclic. Silicones are odourless and generally colourless, and are characterized by high hydrophobicity and by remarkably high chemical and thermal stability. Silicones may be oils, greases, gums, elastomers, or resins, and have a wide range of clinical, electrical, engineering, industrial, and laboratory applications. In commercial usage, the term often includes organosilicon compounds used for **silanization** and **silylation**.

siliconize to render (filter-paper, glassware, textiles, etc.) water-repellent by application of a coating of an appropriate silicone. *See also* **silanization**. —**siliconized** *adj.*; **siliconizing** *n*.

siloxane any saturated silicon-oxygen hydride with unbranched or branched chains of alternating silicon and oxygen atoms in the molecule. Unbranched siloxanes are of general formula $H_3Si–[O–SiH_2]_n$ $–SiH_3$; each is named after its parent **silane** (def. 2), e.g. disiloxane ($n = 0$), tetrasiloxane ($n = 2$). Hydrocarbyl derivatives are commonly included. *See also* **silicone**.

silver staining a sensitive method for detecting protein in polyacrylamide gels. Silver ions are reduced to metallic silver in the presence of the developing reagent, formaldehyde, in a reaction promoted by the protein. Amounts of less than 1 μg of protein can be detected. A modification of the protocol can be used for the sensitive detection of DNA in polyacrylamide gels.

silyl 1 $H_3Si–$, the group derived from **silane** (def. 1) by loss of a hydrogen atom. **2** any hydrocarbyl derivative, $R_3Si–$, of the **silyl** (def. 1) group.

silylation the process of converting an organic compound into its **silyl** (def. 2) derivative by attachment of trimethylsilyl (or similar organosilicon) groups. Reagents frequently used are trimethylchlorosilane and 1-(trimethylsilyl)imidazole, although others have particular applications. Reactive compounds include those possessing amino, carboxyl, hydroxyl, or thiol groups, for which silylation affords selective protection during synthetic work (*see* **protecting group**). Silylation is widely used for conversion of mixtures of related compounds into derivatives capable of separation

and analysis by gas chromatography/mass spectrometry. *Compare* **silanization**. —**silylate** *vb.*; **silylator** *n.*; **silylated**, **silylating**, *adj*.

simian characteristic of, resembling, or belonging to the apes.

simian immunodeficiency virus *abbr*.: SIV; a lentivirus, common in the sooty mangabey (a type of monkey) and many African green monkeys. SIV_{sM} is closely related to – and probably gave rise to – HIV-2 (*see* **HIV**). HIV-1 is closely related to SIV from chimpanzees. Infection of sooty mangabeys and African green monkeys does not cause disease, but SIV infection occurs in macaques in captivity and is pathogenic.

simian virus 40 *abbr*.: SV40; an icosahedral virus containing circular duplex DNA of about 5.2 kb associated with four histones (H4, H2A, H2B, H3) in a minichromosome. It infects monkeys and apes and induces tumours in other species, its DNA being integrated into the host genome. It has two types of life cycle: in permissive cells (usually monkey permanent cell lines) replication occurs in normal infection; in nonpermissive cells there is no lytic infection, but growth transformation can occur. The virus is useful as an **expression vector** in recombinant DNA technology, having good promoter properties: it can be used either as a virion or as a construct maintained transiently in host cells at high copy number in unintegrated plasmid-like DNA molecules. *See also* **small T-antigen**.

similarity (*in bioinformatics*) the extent to which biopolymeric sequences share identical or related characters (nucleotides or amino acids) at equivalent positions, usually expressed as a percentage. For protein sequences, the degree of relatedness of amino acids is quantified in substitution or mutation matrices – e.g. PAM (point accepted mutation), BLOSUM – allowing similarity scores to be assigned in pairwise comparisons.

similarity matrix *see* **scoring matrix**.

simple diffusion or **passive diffusion** a type of **diffusion** (def. 2) across a membrane in the absence of any specific transporting agent or carrier. *Compare* **facilitated diffusion**, **active transport**.

simple radial immunodiffusion *see* **single radial immunodiffusion**.

simple sequence repeat *abbr*.: SSR; a DNA sequence comprising repeats of identical or highly similar short motifs. SSRs are polymorphic and include minisatellites, microsatellites, and nontandem motif arrangements (the latter are called **cryptically simple sequences**). SSRs are ubiquitous in the human genome and account for at least 0.5% of genomic DNA.

SIMS *abbr*. for secondary ion mass spectrometer.

sin *abbr*. for sine.

Sin 3 a mammalian nuclear protein that contains four paired amphipathic helix domains and associates with two histone deacetylases, retinoblastoma protein-binding proteins, p53, BRCA1, c-Ski, and possibly other proteins.

sinalbin a glucoside found in white or yellow mustard. *See also* **sinapine**.

sinapine 2[[3-(4-hydroxy-3,5-dimethoxyphenyl)-1-oxo-2-propenyl]-oxy]-*N,N,N*-trimethylethanaminium; a component of **sinalbin**; it is also found in black mustard and other seeds.

sinaptobrevin *a variant spelling of* **synaptobrevin**.

sine *abbr.*: sin; a trigonometric function, being the ratio of the side of a right-angled triangle opposite a given angle to the hypotenuse; the sine of an obtuse angle is numerically equal to that of its supplement.

SINE *abbr. for* short interspersed nucleotide element; a highly repeated sequence of DNA occurring interspersed in a mammalian genome; it is similar to a **LINE** but is shorter – typically less than 500 bp long. The major SINE in the human genome is the **Alu sequence**. SINEs appear to be processed **pseudogenes** derived from genes that encode small cytoplasmic RNAs including tRNAs. Thus the sequences of SINES are often homologous to the sequences of such RNAs.

Singer–Nicholson model *an alternative name for* **fluid-mosaic model**.

single-cell protein *abbr.*: SCP; a protein-rich material produced by the culture of bacteria, yeasts, other fungi or algae, and extracted for use as a protein supplement in animal or human foods.

single-copy plasmid a plasmid that replicates at the same pace as the bacterial chromosome, and thus maintains equality in numbers with the chromosome.

single nucleotide polymorphism *abbr.*: SNP; a single nucleotide position in a genome sequence at which alternative alleles are present at an appreciable frequency within a population. SNPs are important in biomedical research because they can be used as markers for disease-associated mutations, and may account for the variations in individual responses to bacteria, viruses, toxins, and drugs.

single-pass *(in protein biochemistry)* describing an integral membrane protein that contains only one transmembrane segment. *Compare* **multipass**.

single radial immunodiffusion *or* **Mancini method** an **immunodiffusion** technique in which holes (wells) are cut in a sheet of antibody-containing agar in a glass dish or on a slide. The wells are filled with antigen solution, which is then left to diffuse radially outwards; in doing so, it reacts with antibody to form a ring of precipitate. The diameter of the circle of precipitate is proportional to the concentration of antigen.

single-spanning describing an integral membrane protein having a single hydrophobic sequence embedded in the membrane lipid bilayer. Such a sequence anchors the protein in the membrane.

single-strand binding protein *abbr.*: SSB protein; any of a group of proteins found in bacteria and mitochondria that have a greater affinity for **single-stranded DNA** than for double-stranded species. They are essential for replication, recombination, and repair. They bind tightly and cooperatively, and do not catalyse associated activities, such as the DNA-dependent ATPase activities of **helicases** and **topoisomerases**. They have been called 'unwinding', 'melting' and 'helix-destabilizing' proteins, but each of these names is rather misleading, as these proteins have multiple functions.

single-strand conformation polymorphism analysis *abbr.*: SSCP analysis; a method for detecting mutations by the rate at which single-stranded DNA migrates on electrophoresis gels, under conditions in which the conformation adopted by the DNA is critical. The technique involves amplification of a DNA sequence of interest using the **polymerase chain reaction** (PCR); the DNA is denatured to form single strands and then applied to the gel under nondenaturing conditions. The three-dimensional conformation adopted by ssDNA is sequence dependent and sensitive to single base changes, and this in turn affects the rate of migration in the gel. Wild-type DNA produces two bands of characteristic mobilities that act as references with which to compare the rate of migration of DNA containing a mutation or polymorphism. Heterozygotes characteristically produce four bands on the gel, two from the wild-type allele and two from the mutant or polymorphic allele.

single-stranded DNA *abbr.*: ssDNA; a form of **DNA** consisting of a single chain of deoxyribonucleotides. ssDNA occurs naturally in certain circumstances: (1) transiently, during replication (*see* **replicate**), when the duplex helical molecule of double-stranded (ds) DNA is unwound in the replication fork, initiated by **helicase**; (2) during replication of circular DNA, when in some circumstances only one strand is replicated, yielding a single-stranded product; as replication proceeds, the synthesized DNA unwinds as a single-stranded 'tail'. Replication may proceed around the unbroken circular DNA strand for more than one cycle, giving rise to the term **rolling circle**; (3) in certain filamentous **coliphages**, such as M13, f1, and fd, which contain a single-stranded circular DNA molecule.

singleton *(in gene sequencing)* a nucleotide sequence that has no overlaps with other fragments at a given stringency and is not therefore contained in any fragment assembly for a given gene.

singlet oxygen a dioxygen molecule in which two 2p electrons have similar spin. It is more highly reactive than the form in which these electrons are of opposite spin. Singlet oxygen is produced in chloroplasts that, due to mutation, lack carotenoids; excited triplet chlorophyll reacts with oxygen to yield singlet oxygen, which is normally quenched by carotenoids. In the absence of carotenoids, lethal photodynamic damage occurs.

single-turnover kinetics the kinetics of an enzyme-catalysed reaction in conditions under which formation of the enzyme–substrate complex can be measured. Such conditions are required for the detection and analysis of transient enzyme–substrate and enzyme–intermediate forms in the pre-steady state (*see* **pre-steady state kinetics**), and contrast with measurement of the multiple recycling of the enzyme in the steady state. Although the reaction is proceeding normally, only the first part is studied, leading to the expression 'single-turnover conditions'. In this kind of study, a substrate is chosen for which the enzyme–substrate complex can be distinguished from substrate alone, e.g. by having a different absorption spectrum. The rate of formation of the complex can then be determined, often by stopped-flow methods.

singly labelled describing a **specifically labelled** compound containing only one isotopically modified atom in a uniquely specified position, e.g CH_3–$CH[^2H]$–OH.

Sin 3–HDAC *abbr. for* Sin 3–histone deacetylase; a protein complex that in mammals consists of **Sin 3**, two histone deacetylases (HDAC 1 and 2), two retinoblastoma-associated proteins (RbAp46 and RbAp48), and two Sin 3-associated proteins (SAP18 and SAP30)

sinigrin 1-thio-β-D-glucopyranose 1-[*N*-(sulfo-oxy)-3-butenimidate] monopotassium salt; a β-D-thioglucoside occurring in mustard seeds and horseradish roots; it is used as a substrate for thioglucosidase.

sinister *Latin* left; used to describe the left, or counterclockwise, configuration of one of the two faces of a compound containing a trigonal carbon atom. It is designated by the prefix *si*-. *Compare* **rectus**. *See also* **Re/Si convention**.

sintered glass *or* **frit** small particles of glass that have been heated so that they just fuse but without extensive melting. The resultant material is porous, and is often used in glass filtration apparatus or as a support to retain packing materials in chromatographic columns.

Sir *abbr. for* silent information regulator; any of a family of highly conserved NAD-dependent histone/protein deacetylases, present in nearly all living cells, that catalyse the reaction:

$$\text{acetyl-protein} + \text{NAD} =$$
$$O\text{-acetyl-ADP ribose} + \text{protein} + \text{nicotinamide}.$$

In bacteria, Sir-2 controls acetyl-CoA synthetase by deacetylating a residue in the active site. In yeast it is involved in gene silencing, chromosomal stability, and ageing. Sir-2-like proteins are called *sirtuins*. Seven homologues (SIRT1 to SIRT7) occur in humans.

sirna a bioinformatics computer program used for the design of siRNA duplexes.

siRNA *abbr. for* small interfering RNA; a form of microRNA (miRNA) that associates with RNA helicase and RNase molecules to form a complex called **RISC**. The siRNA then unwinds and directs precise, sequence-specific degradation of mRNA. siRNAs are 21–23 bp long, generated by **Dicer** from a dsRNA precursor, and are highly specific for the nucleotide sequence of their target mRNA. They require **Argonaute** to function.

siroheme *or (esp. Brit.)* **sirohaem** a heme molecule discovered in bacterial sulfite reductase, EC 1.8.99.1. It is a tetrahydroporphyrin, having adjacent, reduced pyrrole rings. It is also a component of nitrite reductase (EC 1.6.6.4) from *Nitrobacter*.

siroheme synthase *or (esp. Brit.)* **sirohaem synthase** a multifunctional bacterial protein that contains **uroporphyrin-III C-methyltransferase, ferrochelatase,** and precorrin-2 oxidase activities. The latter activity converts precorrin-2 into **siroheme** by oxidation and Fe^{2+} chelation.

sirtuin *see* Sir.

sirup *a variant spelling (esp. US) of* syrup.

sis an oncogene (v-*sis*) from simian osteosarcoma virus that is also expressed in human connective tissue tumours. The protooncogene, c-*sis*, encodes the B chain of **platelet-derived growth factor** (PDGF), and v-*sis* encodes an almost identical B chain, differing only in the amino acids at its N and C termini. Cells transformed by v-*sis* produce the B chain, which stimulates the PDGF receptor in an autocrine fashion.

site-directed mutagenesis *or* **site-specific mutagenesis** the use of cDNA technology to generate a **point mutation** at a predetermined position in a gene. It may be achieved in a number of ways but is commonly performed by **oligonucleotide-directed mutagenesis**.

site-1 protease *see* SP1.

site-2 protease *see* SP2.

site-specific DNA methyltransferase *see* DNA methylase.

sitosterol 22,23-dihydrostigmasterol; a widely distributed plant sterol related to **stigmasterol**.

β-sitosterol

SI units *see* SI.

SIV *abbr. for* simian immunodeficiency virus.

six-frame translation (*in sequence analysis*) translation of a DNA sequence taking into account the three possible reading frames in each direction of the strand, giving rise to three forward and three reverse translations.

six-loop-six structure the topology of an integral membrane protein in which six N-terminal transmembrane segments are separated from six segments in the C-terminal region by a hydrophilic loop. It is characteristic of many transporters in eukaryotes and prokaryotes.

sizing column a **gel-filtration** column that is used to provide information about the M_r (size) of proteins.

sizing gel a **gel-filtration** medium, especially one used in a sizing column.

sizukacillin *see* peptaibophol.

Sjl protein *abbr. for* synaptojanin-like protein; any of three proteins of yeast that have phosphatidylinositol-5-phosphatase activity: Sjl1p (946 amino acids); Sjl2p (1183 amino acids); and Sjl3p (1107 amino acids). The first two seem to function in endocytosis.

Sjögren–Larsson syndrome an autosomal recessive disorder in which congenital ichthyosis is associated with mental retardation, retinopathy, spastic diplegia or tetraplegia, and accumulation of fatty alcohols in tissues. It results from mutations that cause deficiency of the enzyme fatty aldehyde:NAD^+ oxidoreductase. [After Torsten Sjögren (1859–1939), Swedish physician, and Tage Konrad Leopold Larsson (1905–), Swedish scientist.]

Sjögren syndrome an autoimmune disease, usually of middle-aged women, in which lymphocytes infiltrate and destroy salivary and lacrimal glands, and autoantibodies for several ribonucleoproteins appear in serum. It is often associated with other autoimmune diseases. [After Henrik C. Sjögren (1899–1986), Swedish ophthalmologist.]

S2 kallikrein *see* tonin.

SKALP *abbr. for* skin-derived antileukoproteinase; *other name* elafin; a serine protease inhibitor; it probably inhibits elastase-mediated tissue proteolysis.

skatole 3-methylindole; a bacterial metabolic product of indoleacetic acid that contributes to the odour of feces.

skeletal growth factor a protein that stimulates bone growth. Skeletal growth factor from human bone (hSGF) stimulates growth in human bone cells and embryonic chick bones in culture. The activity involves **insulin-like growth factors** (IGFs) in association with IGF-binding proteins (IGFBPs).

skeletin (*formerly*) *an alternative name for* desmin.

skeleton 1 (*in zoology*) any structure that provides support and protection for the soft tissues and organs of the body. An **exoskeleton**, such as the cuticle in arthropods, lies external to the body tissues, while an **endoskeleton**, such as the bony or cartilaginous skeleton of vertebrates, lies within the body. **2** (*in chemistry*) the carbon chain of a molecule, including any noncarbon elements that necessarily link carbon moieties of the chain. —**skeletal** *adj*.

skew 1 asymmetric, biased, distorted. **2** oblique, slanting. **3** (of a statistical distribution) not symmetrical about the mean. **4** to become or make skew; to distort. *See also* **skew conformation**. —**skewness** *n*.

skew conformation 1 any conformation of a six-membered ring form of a monosaccharide or monosaccharide derivative when three adjacent ring atoms and the remaining nonadjacent ring atom are coplanar. Such compounds may be designated by the conformational descriptor -*S* added to the name. The reference plane is so chosen that the lowest-numbered carbon atom in the ring, or failing that the atom numbered next above it, is exoplanar. Skew conformation is the median conformation through which one **boat conformation** passes during conversion to another boat conformation in the cycle of flexible forms; it is similar to **twist conformation** (def. 1). *See also* **conformation**. **2** *a less common alternative term for* synclinal conformation; *see* **conformation**.

Ski *abbr. for* super killer; any of a group of yeast cytoplasmic proteins that normally repress translation of an endogenous RNA virus. When Ski protein is mutated, a super killer phenotype results from overexpression of a toxin by the virus. Ski2p, Ski3p, Ski7p, and Ski8p are associated with the exosome core in the cytoplasm.

SKI a gene related to v-*ski*, the oncogene of Sloan-Kettering viruses; it has no marked homology with other oncogenes. It does not bind DNA but interacts with a histone deacetylase complex, Sin3, and Smads 2 and 3.

slab gel a gel, usually polyacrylamide or starch, that is formed as a sheet rather than a rod.

slaty a coat colour mutation in mice allelic for the *tyrp-2* gene that causes dark-grey (slate-coloured) melanin to be formed. The enzyme affected is **dopachrome Δ-isomerase**.

SLC *abbr. for* solute carrier; any member of the solute carrier gene series, which currently comprises 43 families and 298 transporter genes. SLC7A7, for example, is a gene for the cationic amino acid carrier light chain belonging to family SLC7.

sleep peptide any peptide that brings about sleep when infused into the cerebrospinal fluid (CSF) of animals. Delta-sleep- inducing peptide, isolated from CSF, has the structure Trp-Ala-Gly-Gly-Asp-Ala-Ser-Gly-Glu. A large number of endogenous substances isolated from the brain can likewise induce sleep, but it is not known if this is their physiological function.

sliding clamp *an alternative name for* proliferating cell nuclear antigen. *See* **PCNA**.

sliding filament model a model of muscle contraction in which contraction results from **thin filaments** sliding past **thick filaments**. The activation involves enhanced binding of Ca^{2+} by troponin C (*see* **troponin**) associated with the thin filament. In the resting state, troponins I and T inhibit the actin–tropomyosin complex of the thin filament from binding to myosin; increase in Ca^{2+} concentration causes troponin C (each molecule of which binds up to four Ca^{2+} ions) to release this inhibition, allowing the myosin head-groups to bind actin. Then, in the presence of ATP, the thin filaments slide over the thick filament myosin molecules to cause contraction. The ATPase of myosin heads hydrolyses the ATP; the latter must be continuously produced for contraction to continue.

sliding 1 the mode of action of *lac* **repressor** in *Escherichia coli*. It first binds at a non-operator site, then slides along the DNA until it finds the target site, i.e. the **operator**. **2** the displacement of a nucleosome along a track of DNA. *See* **sliding clamp, sliding filament model**.

slime mould *or (US)* **slime mold** any of a group of organisms that exhibit features of both fungi and protozoa. Characteristically they exist as slimy masses on decaying wood or in damp soil, and have the ability to switch from an amoeboid single-celled phase, called a **myxamoeba**, to an aggregated or multicellular phase from which fruiting bodies may arise. In the so-called **true slime moulds**, the myxamoebae form an acellular mass called a plasmodium, whereas in the **cellular slime moulds** the myxamoebae aggregate to form a multicellular pseudoplasmodium, or 'slug'. A well-known example of a cellular slime mould is *Dictyostelium discoideum*; when the food supply is exhausted, its myxamoebae aggregate to form a small (1–2 mm) slug, containing up to 10^5 cells, which migrates to seek an area of food abundance. When nutritional conditions are favourable the slug undergoes differentiation to form fruiting bodies, from which spores are released. Germination of a spore leads to the formation of a myxamoeba. *D. discoideum* has a relatively small genome (about 10 times larger than that of a bacterium) and is popular with geneticists, providing insights into the nature of the inducible behaviour leading to aggregation, and the rapid cell division and differentiation that precede the formation of a fruiting body.

slip a term used for situations in which the mitochondrial enzyme, **cytochrome c oxidase**, may carry out its reaction without pumping protons into the intermembrane space.

slit one of a group of proteins, secreted by cells at the midline in the *Drosophila* embryo, that binds to a **Robo** receptor on growth cones of axons and prevents them from crossing the midline. Slit proteins are also present in non-neuronal tissues.

slope $\delta y/\delta x$, where results are expressed graphically as a continuous line, the ordinate (vertical axis) representing y, the abscissa (horizontal axis) representing x.

slow-cycling rhodopsin a **retinal**-containing, energy-transducing pigment present in small amounts in some of the strains of *Halobacterium halobium*, in which purple membranes occur. It is distinct from **bacteriorhodopsin** and **halorhodopsin**, and is believed to play a role in the characteristic variation in the phototactic response of such organisms at different wavelengths. Illumination with yellow light, or with blue or near-ultraviolet light, results respectively in accumulation of forms with absorption maxima at 375 nm and 590 nm.

slow reacting substance (of anaphylaxis) *or* **slow reacting substance A** *abbr.*: SRS *or* SRS-A; a substance that mediates anaphylaxis; it was identified as a polar, lipid-like compound synthesized in mammalian tissues as a consequence of the combination of antibody with antigen. It increases vascular permeability and induces contraction of some types of smooth muscle, especially bronchial muscle. It is slow in action relative to histamine – hence its name – and is not antagonized by antihistamines. **Leukotrienes C_4, D_4,** and **E_4** are the active components.

SLP-76 *abbr. for* SH2-containing leukocyte protein 76 kDa; an adaptor protein involved in platelet degranulation, secretion of interleukin 2, and formation of lymphatic vessels. It is phosphorylated on tyrosine residues by **Syk**.

slurry a fluid suspension of particles having the consistency of cream.

Sly syndrome *see* **mucopolysaccharidosis VII**.

Sm *symbol for* samarium.

***Sma* I** a type 2 **restriction endonuclease**; recognition sequence CCC↑CGG.

Smac *abbr. for* second mitochondria-derived activator of caspases. *See* **Diablo**.

Smad any of several proteins that act as transcription factors. The name is derived from a fusion of the gene names *sma* (in *Caenorhabditis elegans*) and *Mad* (in *Drosophila*). Smads fall into one of three classes: (1) receptor-regulated (R-Smads: 1, 2, 3, 5, and 8); (2) common-partner (Co-Smads: 4 and 4β); and (3) inhibitory (I-Smads: 6 and 7). R-Smads and Co-Smads share two highly homologous sequences in the N-terminal and C-terminal regions: one is important for complex formation, regulation of transcription, and interaction with receptor serine/threonine protein kinases (e.g. bone morphogenetic protein receptors); the other binds DNA and associates with other transcription factors. R-Smad-Co-Smad dimers enter the nucleus following activation by ligand binding to the receptors.

small cytoplasmic ribonucleoprotein *abbr.*: scRNP; *other name*: 'scyrp'; any particulate complex of proteins with molecules of **small cytoplasmic RNA**, found in the cytoplasm. The function of scRNPs is unknown.

small cytoplasmic RNA *abbr.*: scRNA; any of various low-molecular mass RNA molecules (100–300 nt) found in the cytoplasm as components of **small cytoplasmic ribonucleoprotein**.

small Hsp one of a family of small heat-shock proteins. Like the alpha-crystallin family they consist of 12- to 43-kDa proteins. They tend to form large multimeric complexes. Their chaperone activity is confined to the binding of unfolded intermediates, thereby protecting them from irreversible aggregation. Unlike Hsp70 and Hsp90, the function of small Hsp is not dependent on ATP.

small intensely fluorescent cell *abbr.*: SIF cell; a cell found in autonomic ganglia and containing dopamine. The name derives from the appearance of such cells after formaldehyde condensation in fluorescence microscopy.

small leucine rich proteoglycans a group of small interstitial proteoglycans which have a high degree of homology in their protein core sequence. Each has between 10 and 12 highly conserved leucine rich tandem repeats which makes up the central portion of the core protein. The current members of the group are **fibromodulin** and **lumican**, which are both keratan sulphate substituted, and **decorin** and **biglycan** which are both chondroitin sulphate substituted.

small modulating RNA *abbr.*: smRNA; small (~20 bp) noncoding double-stranded RNAs with possible functions in regulating cell-type-specific gene expression.

small nuclear ribonucleoprotein *abbr.*: snRNP (hence often colloquially termed 'snurp'); any of various particulate complexes of proteins with molecules of **small nuclear RNA** (snRNA), found in the eukaryotic nucleus; the particles are ~250 kDa. They are named according to the snRNAs they contain: the U4/U6 snRNP, e.g., contains U4 and U6 snRNA. They are involved in RNA splicing (*see* **spliceosome**). Patients with systemic lupus erythematosus make antibodies directed against one or more snRNPs.

small nuclear RNA *abbr.*: snRNA; any of the RNA molecules (100–300 nt) that contain a uridine-rich region, are designated U1, U2, ... U12 RNA, and are found in the eukaryotic nucleus as components of **small nuclear ribonucleoprotein**. *See* **Sm protein, spliceosome**.

small nucleolar ribonucleoprotein *abbr.*: snoRNP (hence often colloquially termed 'snorp'); any of various particulate complexes of proteins with molecules of **small nucleolar RNA**, found in the eukaryotic nucleolus. They are possibly the same as **processosomes**.

small nucleolar RNA *abbr.*: snoRNA; any small RNA that is associated with the eukaryotic nucleus as a component of **small nucleolar ribonucleoproteins** and participates in the processing of ribosomal RNA in cell nuclei. snoRNAs have a long stretch of sequence complementarity to conserved sequences in mature mRNA. Six have been identified in metazoan cells (87–280 bases) and 12 in yeast (125–605 bases). They seem to assemble with a large multi-RNA RNP complex.

small serine proteinase inhibitor *see* **smapin**.

small T-antigen *see* **T-antigen** (def. 1).

smapin *abbr. for* small serine proteinase inhibitor; any protein (up to 100 amino acids) of nematodes that contains five disulfide bonds and inhibits serine proteinases.

S_1 mapping *or* **S_1 nuclease mapping** a technique based on the ability of **S_1 nuclease** to remove single-stranded (ss) DNA or ssRNA from DNA/RNA hybrids. It has been used to detect coding regions in DNA that has been hybridized – with cDNA for a particular mRNA or the mRNA itself – since any ssDNA formed by introns is removed by the S_1 nuclease leaving only the coding DNA. The enzyme can be used similarly to identify the 5' end of a mRNA.

SMART *abbr. for* simple modular architecture research tool; *see* **derived database**.

Smc *abbr. for* structural maintenance of chromosomes; any of a family of proteins involved in binding to and separation of sister chromatids during the cell cycle. Smc1 and Smc3 are components of the **cohesin** complex. Smc2 and Smc4 are components of the **condensin** complex.

SMCC *abbr. for* SRB- and MED-containing cofactor complex; a multiprotein complex (1.5 MDa, ≈25 polypeptides) that contains several homologues of components of the yeast transcriptional coactivator mediator. Its subunit composition is almost identical to that of **TRAP**.

SmcX a protein encoded on the X chromosome that escapes X-chromosome inactivation in mouse and human. It belongs to the ARID family of DNA-binding regulatory proteins (*see* **ARID protein**). A homologue (called **SmcY**) is encoded on the Y chromosome.

SmcY a protein encoded on the Y chromosome that is homologous to **SmcX**. It belongs to the ARID family of DNA-binding regulatory proteins (*see* **ARID protein**). In mouse, it is essential for expression of male-specific minor histocompatibility antigens.

smectic (of a substance) being or having a **mesomorphic** state in which the molecules are arranged in parallel layers. The molecules in any layer can move in the plane of that layer but not into adjacent layers. *Compare* **nematic**. *See also* **liquid crystal**.

smectic mesophase *see* **liposome, liquid crystal**.

Smith–Waterman alignment algorithm (*in sequence analysis*) a global dynamic programming algorithm optimized for local alignment. A modification of the **Needleman–Wunsch alignment algorithm**, it aligns similar segments of sequence pairs and exploits a substitution matrix with negative values (otherwise, if non-negative values are used, it behaves as a global alignment algorithm). *See* **alignment algorithm**.

SMN protein *abbr. for* survival motor neuron protein; a protein (294 amino acids) that is widely expressed in cell nuclei, where it interacts with components of the spliceosome. Mutations, including gene deletions, cause autosomal recessive spinal muscular atrophy.

smooth describing a bacterial strain having shiny (smooth) colonies. *See* **smooth–rough variation**.

smooth endoplasmic reticulum *abbr.*: SER; *see* **endoplasmic reticulum**.

smooth muscle *or* **involuntary muscle** muscle that lacks striations, hence giving a 'smooth' appearance under the microscope; it produces slower, longer lasting contractions than **striated muscle** and is present in visceral organs such as stomach, bladder, intestine, and uterus, and the walls of arteries. The cells are mononucleate, and contain both **thick filaments** and **thin filaments** but not in the pattern

that leads to striations. Contraction is regulated by the sarcoplasmic (cytosolic) free Ca^{2+} concentration, which increases from resting levels of 120–270 nm to 500–700 nm in response to stimuli such as membrane depolarization, or α-adrenergic or muscarinic agonists. Ca^{2+} bound to calmodulin activates **myosin-light-chain kinase**, which phosphorylates myosin at Ser^{19} of each of the two 20 kDa light chains. This causes cycling of myosin cross-bridges along actin filaments and the development of force (*see* **sliding filament model**). Calponin, caldesmon, and **protein kinase C** may also play a role in fine-tuning the contractile state.

smooth–rough variation *or* **S→R variation** any type of cell-surface variation that occurs in bacteria, usually changing the appearance of colonies from that of the original. In the Enterobacteriaceae the colonies are typically shiny (smooth) and change to matt (rough), an S→R variation that may be accompanied by loss of surface antigens, reduction or loss or virulence, and changes in susceptibility to certain bacteriophages.

Sm protein any of a family of proteins that participate in splicing of pre-mRNA. Seven members assemble into a ring, their inner surface binding the uridine-rich region of small nuclear RNAs that are shared among small ribonucleoproteins.

smRNA *abbr. for* small modulating RNA.

SMRT *abbr. for* silencing modifier of retinoid and thyroid hormone; a protein that binds **NCoR** and unliganded receptors for retinoic acid and thyroid hormone. This complex then binds the histone deacetylase complex **Sin 3-HDAC**, which represses transcription.

sn- *prefix* signifying stereospecifically numbered; it is used in designating the configuration of derivatives of prochiral compounds (especially glycerol and its derivatives) to ensure differentiation from conventional numbering. Thus sn-1 denotes C-1 of the glycerol moiety of a phospholipid under the **stereospecific numbering** system. C-2 and C-3 are similarly denoted *sn-2* and *sn-3*.

Sn *symbol for* tin.

SNAP *abbr. for* **1** S-nitroso-N-acetylpenicillamine. **2** synaptosome-associated protein; a number of these exist. The nomenclature has become somewhat confusing, since SNAPs exist with different but closely related roles. Thus SNAP-25 functions as a SNAP receptor, or **SNARE**, while other SNAP proteins such as α-SNAP (*see* def. 3) function to bind NSF proteins (*see* **N-ethylmaleimide**) to SNAREs such as SNAP-25. **3** soluble NSF attachment proteins, such as α-SNAP, β-SNAP, etc. These are cytoplasmic proteins that mediate binding of **N-ethylmaleimide-sensitive fusion protein** to membranes. *See also* **SNARE, SNARE hypothesis**.

SNARE *abbr. for* SNAP receptor; **syntaxin, synaptobrevin**, and SNAP-25 (*see* **SNAP** (def. 2)) act as SNAREs; they bind in an ATP-dependent manner to NSF–SNAP complexes (SNAP in this case being the NSF attachment protein (*see* **N-ethylmaleimide**)). Syntaxin, synaptobrevin, and SNAP-25 form a complex that is much more effective than the single proteins acting alone, suggesting that a core complex bridges the synaptic vesicle and plasma membranes, and may function in the docking and fusion process. *See also* **SNARE hypothesis**.

SNARE hypothesis a model of vesicular fusion that involves two types of **SNARE**; v-SNAREs on vesicles and t-SNAREs on target membranes. Each transport vesicle has its own v-SNARE, which recognizes a cognate t-SNARE on a target membrane and mediates docking; fusion is then mediated by N-ethylmaleimide-sensitive fusion protein and an attachment protein (**SNAP** (def. 3)).

Snell's law *see* **refraction**.

SNF *abbr. for* sucrose nonfermenting; a defect in *Saccharomyces cerevisiae* that prevents it from metabolizing sucrose because of failure to activate transcription of the sucrase gene. *See* **SWI/SNF**.

Sno *symbol for* a residue of the ribonucleoside thioinosine; mercaptopurine ribonucleoside; alternative to sI.

snoRNA *abbr. for* small nucleolar RNA.

snoRNP *abbr. for* small nucleolar ribonucleoprotein.

snorp *colloquialism for* snoRNP, i.e. **small nucleolar ribonucleoprotein**.

SNP *abbr. for* single nucleotide polymorphism.

snRNA *abbr. for* small nuclear RNA.

snRNP *abbr. for* small nuclear ribonucleoprotein.

S_1 nuclease EC 3.1.30.1; *recommended name*: *Aspergillus* nuclease S_1; an endonuclease from *Aspergillus oryzae* that degrades single-stranded (ss) RNA or ssDNA to 5' mononucleotides; it has no apparent base specificity. It is useful for the removal of unpaired regions in hybridization technology (*see also* **S₁ mapping**). The enzyme is a monomer, and contains three zinc atoms.

SNuPE *a proprietary name for* single nucleotide primer extension; a method for detecting **single nucleotide polymorphisms** in which a primer is annealed to the target sequence with its 3′ end immediately before the polymorphic site. The identity of the polymorphic nucleotide is revealed by which of the four different fluorescently labelled dideoxynucleoside triphosphates, incorporated by DNA polymerase, it forms a base pair with .

snurp *colloquialism for* snRNP, i.e. **small nuclear ribonucleoprotein**.

snurposome any morphologically and biochemically distinct granule in the amphibian oocyte nucleus that contains one or more types of small nuclear ribonucleoproteins (snRNP).

soap any salt formed between a hydroxide of a metal (usually an alkali or alkaline earth) and a higher fatty acid; such compounds are useful as **detergents** (def. 2). It can also mean the salt of any ionized lipid **amphipath**, e.g. of a long-chain amine or sulfate or of a sterol carboxylate. The common soaps were traditionally the mixed sodium salts (giving **hard soaps**) or the mixed potassium salts (giving **soft soaps**) of oleic, palmitic, and stearic acids, prepared by **saponification** of animal or vegetable fats or oils with sodium or potassium hydroxide respectively. Modern soap compositions are more complex, with solid or liquid soaps containing various additional detergents.

soapstone *see* talc.

SOB medium a nutritionally rich growth medium for the culture of recombinant strains of *E. coli*.

SOC medium a nutritionally rich growth medium for the culture of recombinant strains of *E. coli*. It is identical to **SOB medium** except that it contains 20 mM glucose.

SOD *abbr. for* superoxide dismutase.

soda any of various compounds of sodium, especially sodium hydroxide (NaOH; **caustic soda**), sodium hydrogencarbonate (sodium bicarbonate, NaHCO₃; **baking soda**), sodium carbonate decahydrate (Na₂CO₃·10H₂O; **washing soda**), or sodium monoxide (Na₂O).

soda glass common glass, made from a mixture of silica, sodium oxide (soda), and lime, in approximate proportions SiO₂ 70%, Na₂O 15%, CaO 10%. Its low melting point led to the production of more heat-resisting glasses, e.g. Pyrex. *See also* **borosilicate (glass)**.

sodium *symbol*: Na; a soft, silvery-white, highly reactive alkali metal of group 1 of the (IUPAC) **periodic table**; atomic number 11; relative atomic mass 22.997. Sodium is widely distributed and is one of the most abundant elements of the Earth's crust, notably as sodium chloride (salt), especially in sea water. Apart from the nonradioactive, and most abundant, isotope, sodium-23, there are two radioactive isotopes, **sodium-22** and **sodium-24**. In biochemistry, sodium (Na⁺) is the principal extracellular cation, and the one that largely determines the extracellular fluid volume. In humans the normal range for plasma sodium is 135–145 mmolL⁻¹. Sodium ions are actively pumped out of cells by **sodium/potassium ATPase**. Their concentration is regulated by **aldosterone**, which increases sodium reabsorption in the kidney distal tubule. Sodium was formerly known as 'natrium', hence the symbol Na and the term **natriuretic**.

sodium-22 a radioactive nuclide of sodium, $^{22}_{11}$Na. It emits a β⁺ particle (1.83 MeV) and gamma radiation (1.275 MeV). Its half-life is 2.605 years.

sodium-24 a radioactive nuclide of sodium, $^{24}_{11}$Na. It emits an electron (β⁻ particle, 1.39 MeV) and gamma radiation (2.75 MeV). Its half-life is 15.02h.

sodium–alanine symporter *see* **sodium/neutral amino acid cotransporter**.

sodium–calcium ion exchanger a plasma membrane protein **antiporter** of muscle and nerve cells that promotes the exchange diffusion of sodium and calcium ions. Influx of Na⁺ to the cytosol drives the efflux of Ca²⁺ from the cell. The affinity for Ca²⁺ is low, so that it only functions at high intracellular Ca²⁺ concentrations, transporting Ca²⁺, e.g., during excitation–contraction coupling.

sodium channel any structure that permits controlled (gated) passage of sodium ions through membranes. Voltage-gated channels are involved in conduction and transmission of nerve impulses,

whereas transmitter-gated channels respond only to neurotransmitters (*see* **cholinoceptor**). Voltage-gated channels are integral proteins in the plasma membranes of excitable cells. They sense the membrane electric field, opening channels to permit an inward flux of sodium ions, thereby causing the membrane depolarization that is essential for the propagation of the **action potential**. The *Electrophorus* electroplax protein is a single chain of about 2000 amino-acid residues and contains four homologous repeats, each of which has six putative transmembrane segments with extensive nonmembrane cytosolic and extracellular loops. It shares significant homology with other ion channels, particularly the α1 subunit of the Ca²⁺ channel. A cytoplasmic loop has been implicated in the rate of inactivation of the channel. The protein bears about 500 sugar residues (mostly *N*-acetylglucosamine and sialic acid) and 50 fatty-acyl chains (palmitoyl and stearoyl). The mammalian channels are heterooligomeric, consisting of a large α subunit similar to the *Electrophorus* protein, and several other smaller peptides; most of the functional properties reside in the α subunit. The α subunit is the target for polypeptide neurotoxins, e.g. from scorpions; other sodium-channel blockers include **saxitoxin**, **tetrodotoxin**, and **veratridine**.

sodium dodecyl sulfate *or* (*esp. Brit.*) **sodium dodecyl sulphate** *abbr.*: SDS; sodium lauryl sulfate; CH₃(CH₂)₁₀CH₂OSO₃·Na; the sodium salt of the 1-dodecanol half-ester of sulfuric acid (1-dodecanol was formerly known as lauryl alcohol). It is an anionic detergent and wetting agent; aggregation number 62 (0–0.1 m Na⁺), 101 (>0.1 m Na⁺); CMC 7–10 mm (0–0.05 m Na⁺), 1–2 mm (0.1–0.2 m Na⁺). The amphipathic dodecylsulfate anion is a powerful hydrophobic-bond breaker, hence it is much used in protein chemistry as, e.g., a denaturant, a dissociant of oligomers, and a solubilizer of membrane components. *See also* **SDS–polyacrylamide gel electrophoresis**.

sodium–hydrogen ion exchanger a plasma membrane protein that functions as a Na⁺/H⁺ **antiporter**. Influx of Na⁺ drives the efflux of H⁺. It functions to maintain intracellular pH, to which it is sensitive, its activity increasing as pH falls.

sodium-ion channel *see* **sodium channel**.

sodium/neurotransmitter transporter any of a family of integral membrane glycoproteins, having 12 putative transmembrane helices, that are involved in sodium-dependent transport of neurotransmitters: e.g., the sodium- and chloride-dependent GABA transporter 1 from human, which is the target for cocaine and other psychomotor stimulants.

sodium/neutral amino-acid cotransporter an integral membrane protein responsible for the Na⁺-dependent uptake of the neutral amino acids alanine, serine, glycine, cysteine, and proline. Transporters with specificities for different types of amino acid are **system A**, **system L**, and **system N**.

sodium/potassium ATPase EC 3.6.1.37; *recommended name*: Na⁺/K⁺-exchanging ATPase; *systematic name*: ATP phosphohydrolase (Na⁺/K⁺-transporting); *other names*: sodium pump; Na⁺,K⁺-ATPase. An integral membrane protein that catalyses the hydrolysis of ATP to ADP and orthophosphate. Na⁺,K⁺-ATPases pump out approximately 3 Na⁺ ions for every 2 K⁺ ions that enter, with hydrolysis of 1 ATP. They consist of two large α subunits and two smaller β subunits. The α subunits bear the active site and penetrate the membrane, while the β subunits carry oligosaccharide groups and face the cell exterior only. **Ouabain** is a specific inhibitor.

sodium pump any **transporter** of sodium ions across membranes, especially enzymes of the Na⁺,K⁺-ATPase type (*see* **sodium/potassium ATPase**). The Na⁺/K⁺ ratio across the plasma membrane of eukaryotic cells (outside: Na⁺ 140 mm, K⁺ 10 mm; inside: Na⁺ 10 mm, K⁺ 100 mm) requires that sodium ions be continually pumped out of the cell in exchange for potassium ions if the gradient is to be maintained; this is an energy-requiring process. *See also* **active transport**.

software archive a site, usually accessed by **anonymous FTP**, from which public-domain software can be downloaded. Some software is in binary, i.e. is designed to be loaded directly into the local computer; some code is written in a high-level language, such as C, C++, or PERL, and needs to be compiled locally. Typically software for Windows PCs and Apple Macintosh computers is of the former type, and more generic code and software for UNIX and

Linux systems is of the latter and requires compiling, typically with a supplied **Makefile** or (in the case of Linux) as an **RPM** file.

sol any fluid colloidal system composed of two or more components. The dispersed phase may be either particles of solid or droplets of liquid, and the continous phase either a liquid or a gas (*see* **aerosol**).

solanesol a 45-carbon polyprenol present in tobacco leaves.

solation the (usually reversible) act or process of forming a **sol** from a **gel**. *Compare* **gelation**. —**solate** *vb*.

solenoid 1 a cylindrical coil of wire that acts as a magnet when carrying an electric current. **2** a tightly coiled higher-order structure of **chromatin** consisting of stacks containing six nucleosomes per turn. **3** the coiled structure of some proteins.

solid a physical state of matter in which the constituent atoms, ions, or molecules have no translatory motion though they vibrate about fixed positions. In **crystalline solids** the constituents are arranged in a definite array, whereas in **amorphous solids** they are not. Solids, especially crystalline solids, commonly have characteristic melting temperatures at which they change to **liquids**.

solid-phase peptide synthesis *see* **peptide synthesis**.

solid-phase technique any technique in which reagents are immobilized on a support or in some other way insolubilized. In **radioimmunoassay** the insolubility or binding properties of an antigen–antibody complex are commonly exploited to remove labelled antigen. Enzymes immobilized on columns (**immobilized enzymes**) have many applications. In **peptide synthesis**, the growing peptide chain may be immobilized on a column. So-called solid-phase extraction systems are in effect extractions using batch column chromatography techniques, in which the material to be extracted is bound to a column, then removed by **batch elution**. In peptide **sequencing** a peptide immobilized on a column may be degraded residue by residue, facilitating recovery of the undegraded remaining peptide at each cycle.

solid scintillation fluorography a technique for visualizing radioactive materials in a specimen (e.g. a chromatogram) by covering the specimen with a solid **fluor** and allowing the ejected photons to darken an applied photographic emulsion. *Compare* **fluorography** (def. 2).

solid-state *(in electronics)* describing any system in which current flow takes place entirely through solid materials, such as semiconductors and transistors, rather than through vacuum devices such as thermionic valves.

solubility 1 the attribute of being soluble, of being able to dissolve. **2** the amount of a substance (**solute**) that can dissolve in a given amount of another substance (**solvent**) at a given temperature and pressure.

solubility product *symbol*: K_{sp}; for a solute $A_x B_y$ that ionizes in a stated solvent according to the equilibrium $A_x B_y \rightleftharpoons xA + yB$, at a given temperature and pressure, $K_{sp} = [a_A]^x [a_B]^y$, where $[a_A]$ and $[a_B]$ are the activities of the two ions when the solution is saturated by the solute. In unsaturated solutions the product $[a_A]^x [a_B]^y$ can have any value less than K_{sp}, but if further amounts of either ion are added from whatever source, such that K_{sp} is exceeded, precipitation from solution will occur to the extent that the relationship $[a_A]^x [a_B]^y = K_{sp}$ is restored. In dilute solutions the concentrations C_A and C_B of A and B may be used as a good approximation.

solubilize *or* **solubilise** to bring into solution any material, especially complexes that normally exist in the cell as part of membrane structures. Such materials usually have hydrophobic domains, and solubilization depends on masking these with detergents or similar substances, to bring the material into micellar suspension.

soluble 1 capable of being dissolved in or as if in a fluid. **2** (of, e.g., a problem or mathematical equation) capable of being solved.

soluble collagen *see* **tropocollagen**.

soluble RNA *abbr*.: S-RNA *or* sRNA; *an old term for* **transfer RNA** (a 'soluble' RNA fraction from disrupted liver cells that did not sedi-

ment after centrifuging at 100 000 *g* for several hours led to the discovery of transfer RNA).

solute a dissolved substance, particularly a component of a **solution** that is present in a smaller amount than the **solvent**.

solution 1 a homogeneous molecular mixture of two or more substances, usually of dissimilar molecular structures, i.e. of a solid in a liquid. **2** the action or process of solving a problem. **3** the result of solving a problem, equation, etc.

solvation the (loose) combination of solvent with molecules or ions of the solute.

solvation effect the effect of solvent molecules on the behaviour of (especially) macromolecules in solution, especially during centrifugation.

solvent 1 a substance or mixture, usually a liquid, that is able to dissolve other substances to form a solution. Typically it has the same physical state as the solution itself, and is the component of greatest extent in the solution. **2** *(in chromatography)* the liquid used to pass through a chromatography column, paper, or other support to 'develop' the chromatogram.

solvent-perturbation method a technique in which some physical property of a macromolecule is measured in a polar and in a nonpolar solvent. For example, it can be determined whether an amino-acid residue in a protein is internal or external by measuring the absorption spectra of the protein in the two types of solvents. Difference spectroscopy is conveniently applied in such instances.

soma (*pl.* **somata** *or* **somas**) **1** the whole of any plant or animal organism but excluding the germ cells. **2** the body, as opposed to the mind. *See also* **somatic cell**. —**somatic** *adj*.

somat+ *a variant form of* **somato+** (before a vowel).

somatic cell any cell of an organism other than a germ cell.

somatic cell genetic disorder a genetic disorder in which defects are restricted to specific somatic cells, e.g. cancer cells. This contrasts with germ-line genetic disorders – single gene, multifactorial, and chromosome – in which the abnormality is present in all cells, including the germ cells.

somatic cell hybrid a hybrid cell formed from the fusion of different cells, usually from different species. Somatic cell hybrids obtained by the fusion of human and rodent cells are frequently used for gene mapping. *See* **cell fusion**.

somatic cell hypermutation The process in the lifetime of a **B lymphocyte** whereby **mutations** arise in the **immunoglobulin** polypeptides. The rate of mutation is especially high in the region around the **IgG** genes. The mechanism becomes activated as B cells differentiate and is associated with **class switching** and so IgG molecules usually vary more from **germline** sequences than **IgM**.

somatic crossing over *an alternative name for* **mitotic recombination**.

somato+ *or* (before a vowel) **somat+** *comb. for* denoting the body, or soma.

somatocrinin *another name for* growth hormone releasing hormone (**GHRH**).

somatogenic 1 originating in the **soma** (def. 1), or body tissues. **2** of organic, rather than mental, origin.

somatoliberin *the recommended name for* growth hormone releasing hormone (**GHRH**).

somatomammotropin human chorionic somatomammotropin; *see* **choriomammotropin**.

somatomedin *generic name for* **insulin-like growth factors** (*abbr*.: IGF); in the human they comprise IGF-IA, IGF-IB, and IGF-II. Somatomedins were characterized by: (a) activity on cartilage (incorporation of sulfate and thymidine), (b) insulin-like activity in adipose and muscle, and (c) mitogenic activity in cell culture. Somatomedin A is IGF-II, and somatomedin C is IGF-I (A and B). The term 'somatomedin' should be used only as a generic term, IGF-I (A or B) and IGF-II being used for the specific peptides. Somatomedin B, released from **vitronectin**, is a serum factor of unknown function, is increased in response to **somatotropin**, and is a proteinase inhibitor.

somatostatin *or* **somatotropin-release inhibiting factor** (*abbr*.: SRIF) *or* **growth hormone-release inhibiting hormone** (*abbr*.: GH-RIH); a hypothalamic cyclic 14-residue peptide that inhibits the release of the pituitary growth hormone **somatotropin** and the release of glucagon and insulin from the pancreas of fasted mammals; it also inhibits the release of gastrointestinal hormones and other secretory proteins. **Somatostatin 14** has the sequence Ala-Gly-Cys-Lys-Asn-Phe-Phe-Trp-Lys-Thr-Phe-Thr-Ser-Cys, with a disulfide link between cysteines 3 and 14. In **somatostatin 28** the 14-residue peptide is extended by the N-terminal residues SANSNPALAPRE-RKAG–. Both forms are derived from the 92 amino-acid precursor, **prosomatostatin**, by tissue-specific proteolytic processing. Somatostatin is widely distributed in the central nervous system and in peripheral tissues such as stomach, intestine, and pancreas. It has effects that are tissue specific, and can function as a neurotransmitter. *See also* **somatostatin receptor**.

somatostatin receptor any G-protein-associated membrane protein that binds different forms of **somatostatin** and mediates their effects. Some receptors act through G_i to decrease intracellular Ca^{2+} concentrations and inhibit the voltage-dependent current, I_{Ca}. Somatostatin 14 enhances and somatostatin 28 reduces potassium current, I_K. In *Rattus norvegicus* the type 1 receptor has higher affinity for somatostatin 14 than 28 and is coupled to phosphotyrosine phosphatase and sodium–hydrogen-ion exchanger. Types 2 and 3 are receptors for somatostatins 14 and 28 and inhibit adenylate cyclase. Type 4 is a receptor for somatostatin 14 and inhibits adenylate cyclase. Type 5 is a receptor for somatostatin 28 and inhibits adenylate cyclase. Hybridization studies indicate these are members of a larger family of somatostatin receptors; they are G-protein-coupled adenylate cyclase activators.

somatotrope *or* **somatotroph** a type of cell found in the lateral wings of the anterior pituitary and comprising about 50% of the hormone-producing cells of that part of the gland. Somatotropes are responsible for the production of **somatotropin, and their** adenomas cause **acromegaly**.

somatotroph *see* **somatotrope**.

somatotrophic 1 stimulating and maintaining the growth and development of any cell or tissue. **2** *(sometimes)* of or pertaining to **somatotropin**.

somatotropic 1 of, or relating to, **somatotropin**. **2** stimulating the mechanisms connected with growth or development of a cell or tissue. *Compare* **somatotrophic**.

somatotropic hormone *see* **somatotropin**.

somatotropin *or* **growth hormone** (*abbr*.: GH); *also called* **somatotropic hormone** (*abbr*.: STH); a hormone secreted by the anterior pituitary whose main effect is to stimulate growth. The human hormone (hGH) is mostly synthesized as the precursor of a single-chain, 191-residue protein, M_r 22000, but 5–10% is a form of M_r 20000 resulting from alternative splicing that deletes codons for amino acids 32–46 from the mRNA. Genes encoding hGH, **prolactin**, and **choriomammotropin** form a family with high nucleotide homology with five coding exons. Somatotropin secretion is pulsatile, especially during fasting, when cycles of about four hours' duration occur; it is stimulated by fasting or hypoglycemia, and hormonally by glucagon, GHRH, estrogens, and arginine vasopressin, and during the first phase of sleep; secretion is inhibited by hyperglycemia, somatostatin, the insulin-like growth factor IGF-I, and glucocorticoids. It is secreted from storage granules. *In vivo*, somatotropin tends to promote synthesis of protein, and lipolysis or reduced fat deposition, and growth and mitosis of cartilage. Deficiency leads to pituitary dwarfism types I and IV. Its actions *in vitro* include stimulation of amino-acid uptake in muscle, and lipolysis in adipose tissue, but many of its actions *in vivo* may depend on its action in stimulating synthesis and secretion of **somatomedins**. hGH has 85% homology with human choriomammotropin and 35% homology with human prolactin. *See also* **GHRH, GHRH receptor, hypopituitarism**.

somatotropin-release inhibiting factor *abbr*.: SRIF; *an alternative name for* **somatostatin**.

somatrem methionyl human **somatotropin** (growth hormone); somatotropin with an additional N-terminal methionine. It results from the production of somatotropin using cDNA technology in bacterial clones; methionine is involved in the initiation of bacterial protein synthesis.

+some *comb. form* denoting body, esp. intracellular particle.

somite any of a series of paired blocks of mesoderm that form during early development in vertebrate embryos and lie on either side of the notochord. —**somital** *or* **somitic** *adj*.

sonicate to expose any material or sample to ultrasonic pressure waves (at or around 20 kHz). The immersion of vessels in a water bath and their exposure to ultrasound is used as a means of cleaning (*see* **sonication bath**). The insertion of a **sonicator** probe into a vessel containing a suspension of membranes or other large assemblies causes their dispersion to smaller aggregates or assemblies.

sonication bath a water bath that is equipped with an oscillator, sited beneath the bowl, to produce ultrasound waves that permeate the water. It may be used for cleaning glass or other vessels or for degassing solvents.

sonicator a device that produces **ultrasonic** pressure waves. It usually consists of a stainless-steel rod, of diameter from 1–2 mm to about 2 cm, that is activated by an oscillator to produce ultrasound waves and is inserted into a vessel to **sonicate** a sample. *See also* **sonication bath**.

sonic hedgehog proteins analogues of **hedgehog protein** found in **zebrafish** and chickens, and involved (in the latter case) in limb-bud pattern formation. The hedgehog proteins thus appear to be vital to pattern formation in several metazoan groups. The name sonic hedgehog is an allusion to a children's animated cartoon character.

sophorose the disaccharide 2-*O*-β-D-glucopyranosyl-D-glucose; it occurs in a glycoside of *Sophora japonica*.

sorbin a heat-stable peptide isolated from porcine intestine that increases water and sodium absorption in the intestine and gall-bladder.

sorbitol *a nonsystematic name for* D-glucitol, one of the ten stereoisomeric hexitols. It can be derived from glucose chemically or metabolically (by NADPH-requiring aldose reductase; *other name*: aldose reductase, EC 1.1.1.21) by reduction of the aldehyde group, and is converted to D-fructose by L-iditol 2-dehydrogenase (EC 1.1.1.14). It is widely distributed in algae and higher plants, and was first discovered (1872) in the juice of berries from mountain ash (*Sorbus aucuparia*). It about half as sweet as sucrose, and is often used in sweeteners in conjunction with saccharin. Industrially it is of great importance in the food, pharmaceutical, paper, and textile industries. Its formation from excessive glucose concentrations in diabetics may have pathological implications in lens tissue, which cannot convert it to fructose.

sorbose *a trivial name for* the ketohexose *xylo*-2-hexulose; the enantiomer, L-sorbose, is present in fermented juice of mountain-ash berries, being formed by the bacterial oxidation of **sorbitol**. It is produced by fermentation and used in the manufacture of **ascorbic acid**.

sorcin a protein, encoded in multidrug-resistant cells (*see* **MDR**), that binds calcium with high affinity. *See also* **grancalcin**.

Soret band an absorption band at about 400 nm, characteristic of porphyrins. *See also* **cytochrome absorption bands**. [After Louis Soret (1827–90), Swiss physicist.]

sorocarp *see* **Dictyostelium discoideum**.

sos *symbol for* son of sevenless, a gene for a guanine-nucleotide releasing protein involved in neuronal development in *Drosophila*. When bound to the activated Sev receptor (*see* **Sevenless protein**) the *sos* product, Sos, activates *ras* by stimulating release of GDP, thereby promoting uptake of GTP.

SOS repair *or* **error-prone repair** a metabolic alarm system, occurring in bacteria, that helps the cell to save itself in the presence of potentially lethal stresses such as UV irradiation, thymine starvation, or the inactivation of genes essential for replication. Responses include mutagenesis, filamentation (cell elongation in the absence of division), activated excision repair, and activation of latent phage genomes. Mutagenesis occurs because under SOS conditions the gaps formed opposite **thymine dimers** can be filled by replication rather than by daughter-strand transfer. Replication of the dimer template is extremely inaccurate. The proteins involved include RecA protein (*see* **recA**) and the products of two other RecA-inducible genes; expression is regulated by the **lexA** gene product.

SOURCE *abbr. for* Stanford Online Universal Resource for Clones and Ests; a unified genomic resource for functional annotations, ontologies, and gene expression data.

Southern blotting a procedure for transferring denatured DNA from an agarose gel to a solid support membrane such as nitrocellulose or nylon. It depends on tenacious binding by nitrocellulose of single-stranded (ss) DNA but not of double-stranded (ds) DNA. The gel is soaked in 0.5 m NaOH, which converts the dsDNA to ssDNA. The gel is then overlaid with a sheet of nitrocellulose paper, which in turn is covered by a thick layer of paper towels and the entire assembly is compressed by a heavy weight. The liquid in the gel is forced (blotted) through the nitrocellulose so that the single-stranded DNA binds to it at the same position it had in the gel. The transfer to nitrocellulose can also be accomplished by **electrophoretic transfer**. Nylon is used in preference to nitrocellulose. DNA bands may be located using a radioactive or otherwise labelled probe. The name led to a nomenclature for other types of **blotting** (*see also* **Northern blotting**, **South-western blotting**, **Western blotting**). [After E. M. Southern (1938–), British molecular biologist.]

South-western blotting a technique in recombinant DNA technology for detecting plaques expressing **fusion proteins** where part of the sequence is for a protein that binds to a particular DNA sequence. After treatment of the plaques to ensure release of the protein, **plaque-lift** is carried out onto a nitrocellulose (or other) **transfer membrane**, which is then incubated with a radiolabelled duplex DNA oligonucleotide that binds to the protein, which can then be detected by **autoradiography**. The term was derived by extension through the cardinal points from **Southern blotting**.

Soviet gramicidin an alternative name for **gramicidin S**.

SOX *abbr. for* SRYHMG-box gene; any of a family of genes encoding transcription factors that are related to SRY and contain HMG-type DNA-binding domains. SOX1 and SOX2 proteins are crucial for activating crystallin genes in the lens but are expressed in many tissues. SOX9 protein (509 amino acids) is widely expressed in fetal and adult tissues and is essential for formation of cartilage and testes. Many mutations result in campomelic dysplasia. *SOX10* is expressed in melanocytes and neural crest derivatives during development, and mutations result in Waardenburg syndrome type 4.

Soxhlet extractor an apparatus for exhaustive extraction of soluble components from a sample of solid material by recycling a solvent through it, whereby solvent vapour from a boiler is continually condensed and the condensate allowed to percolate through the sample in a solvent-permeable thimble, from which it is returned, usually periodically via a siphon, to the boiler, where the eluted substances accumulate. [After Franz Soxhlet (1848–1926), Belgian chemist, who devised it.]

soybean or (*esp. Brit.*) **soya bean** a leguminous plant, *Glycine max*, originating in China (known from at least 2500 BC) but cultivated widely in Asia, Europe, and America for forage and for its seeds. The bean has an oil content of 18–22%, and is also rich in protein — soybean protein is a human food and is the basis of tofu. *See also* **soybean oil**, **soybean trypsin inhibitor**.

soybean oil an edible oil obtained from the seeds of the soybean plant and valued for its high content of linoleic acid. The composition is typically as follows: 18:2 ω6, 44–62%; 18:3 ω3, 4–11%; 18:1 ω9, 19–30%; 16:0, 7–14%.

soybean trypsin inhibitor *abbr.*: STI; either of two types of trypsin inhibitor, the Bowman–Birk and the Kunitz inhibitors, which have given their names to classes of proteinase inhibitor. The two soybean inhibitors differ from each other in size, structure, and properties. The **Bowman–Birk inhibitor** contains 71 amino acids and seven disulfide bonds. There are two active sites, one specific for trypsin and the other for chymotrypsin. Bowman–Birk inhibitors are found in many cereals and legumes, and may represent storage proteins. **Kunitz inhibitors** comprise three allelic proteins, A, B, and C, and two Kunitz-type inhibitors, named KTI1 and KTI2. The A, B, and C proteins have 181 amino acids and two disulfide bonds. They are heat-stable, resistant to proteases, and form a 1:1 stoichiometric complex with trypsin that is highly stable (K_a $5×10^9$). They weakly inhibit chymotrypsin and are inactive against other endopeptidases, but modification of Arg^{63} to Trp converts them to strong chymotrypsin inhibitors with little action on trypsin. The Kunitz family embraces trypsin, chymotrypsin, cathepsin D, and subtilisin inhibitors (*see* **Kunitz inhibitor**).

sp *abbr. for* synperiplanar; *see* **conformation**.

sp. (*pl.* **spp.**) *abbr. for* species (singular).

sp-300 a protein (8204 amino acids) of *Drosophila* that has the same general structure as **ANC-1** of *Caenorhabditis elegans* and the Syne proteins of mammals. It contains multiple spectrin repeats and a nuclear envelope localization domain, and structurally consists mostly of coiled coil.

Sp *abbr. for* isomer of a thionucleotide.

Sp1 *abbr. for* specificity protein 1; a transcriptional activator protein of humans that targets genes by binding to GC-rich sequences in certain promoters. It contains glutamine-rich activation domains that bind components of transcription factor TFIID, including the glutamine-rich component TAFII 130. It interacts with human mutant huntingtin, which carries an expanded glutamine repeat.

SP *abbr. for* **1** secretory piece. **2** substance P. **3** spasmolytic polypeptide.

SP-1 *abbr. for* secretory protein 1.

S1P *abbr. for* site-1 protease; a transmembrane serine protease of the subtilisin family. It occurs in the Golgi apparatus, with its catalytic domain in contact with the lumenal loop of **SREBP**. S1P is inhibited by binding of sterol to **SCAP**, and cleaves SREBP into its transcriptional and regulatory domains. Release of the transcriptional domain for entry into the nucleus needs a cleavage by **S2P**.

S2P *abbr. for* site-2 protease; a highly hydrophobic zinc metalloprotease of the Golgi apparatus that releases the transcriptional domain of **SREBP** after **S1P** has separated it from the regulatory domain.

[S]pA *symbol for* adenosine 5′-thiophosphate (alternative to AdoP[S].

space or **volume of distribution** (*in tracer kinetics*) an apparent volume obtained as the amount of radioactivity retained divided by the concentration of the tracer in the plasma. *See* **apparent exchangeable mass**, **compartment**.

spacer 1 (*in molecular biology*) a sequence of bases of unknown function occurring in DNA and lying between sequences of transcribed DNA; spacer DNA is not usually transcribed. The term also refers to some stretches of **primary transcript**, e.g. a region in the mammalian primary transcript for rRNA, lying between 18S RNA and 28S RNA, that is not found in mature rRNA. **2** (*in chromatography*) an alternative term for **spacer arm**.

spacer arm or **spacer** the (usually) hydrocarbon chain interposed between the specific ligand and the supporting matrix in **affinity chromatography**.

spacer gel (*in electrophoresis*) a small section of gel, polymerized above a polyacrylamide (or other) resolving gel, that lacks resolving power and serves to concentrate the material being electrophoresed as a tight band at the top of the resolving gel.

Spalt a transcription factor protein (1355 amino acids) that contains nine zinc fingers. In mouse it is expressed in the central nervous system, limb buds, palate, teeth, kidneys, and genitals. In *Drosophila* it is involved in wing imaginal disc formation.

SPARC *abbr. for* secreted protein acidic and rich in cysteine; *also called* **osteonectin**; an abundant non-collagenous protein of bone and cartilage secreted by osteoblasts but widely expressed in cells of tissues undergoing morphogenesis or wound repair and secreted into the basement membrane, where it interacts with other proteins, e.g. collagen and thrombospondin, and appears to have a role in regulating cell growth. It contains an **EF-hand** that binds one calcium ion and an acidic region that binds other calcium ions more loosely.

spare receptor any of the receptors in excess of those required to evoke a given response to a drug, hormone, or other agonist. The size of the population of spare receptors (**receptor reserve**) can vary with the tissue and with the magnitude of the response being measured. For different drugs, the receptor reserve varies with drug efficacy.

sparging the bubbling of gas into a liquid medium. This technique may be applied to fermenters to introduce air directly into a culture, or it may be used to replace one dissolved gas by another, e.g. in displacing oxygen with an inert gas in liquid chromatography solvents.

sparse matrix (*in sequence analysis*) a **scoring matrix** in which most of the elements or cells have zero score. Sparse matrices perform with high selectivity.

sparsomycin an antibiotic inhibitor of peptidyltransferase activity of prokaryotic ribosomes.

spasmolytic polypeptide *abbr*.: SP. *See* **trefoil factor**.

spastic ataxia of Charlevoix-Saguenay *abbr*.: SACS; an autosomal recessive early-onset neurodegenerative disease with a high prevalence in the Charlevoix-Saguenay region of Quebec, Canada. It is caused by mutations in the gene (locus at 13q12) for **sacsin**, which contains a single exon encoding 4265 amino acids.

spatula a small, flattened or curved utensil made of metal (often stainless steel), bone, or wood and used to dispense crystalline or powdered material.

Spd *abbr. for* sphingoid.

specialist database *or* **boutique database** a database that maintains a specialist data set, dedicated, for example, to a specific gene family, molecular or macromolecular property, disease or diseases. For example, GCRDB (G Protein-Coupled Receptor Database) and MEROPS house information relating specifically to the superfamily of G protein-coupled receptors and to peptidases, respectively; BRENDA stores information pertaining to enzymes; DIP houses information on experimentally determined protein–protein interactions; PDBsum (Protein Data Bank summary) summarizes information relating to protein structures in the PDB (Protein Data Bank); OMIM (Online Mendelian Inheritance in Man) contains information on human genes and genetic disorders.

speciation 1 the evolutionary process by which divergent species of animals or plants are formed from a common ancestral stock. **2** the chemical form or compound in which an element occurs in both non-living and living systems. It may also refer to the quantitative distribution of an element.

species (*pl.* **species**) *abbr*.: sp. (*sing.*) *or* spp. (*pl.*); a fundamental taxonomic category ranking below a genus and consisting of a group of closely related individuals that can interbreed freely to produce fertile offspring.

species-specific confined to the one or several species referred to.

specific 1 relating to a particular or specified thing. **2** (as a modifier of certain quantitative terms) indicating a relationship of the quantity to mass and dimension in terms of standard (now normally SI) units; e.g. **specific catalytic activity**, **specific enzyme activity**, **specific optical rotation**. **3** of, or relating to, a **species**. *See also* **specificity**.

specific acid catalysis a **homogeneous catalysis** in which the catalysts are free hydrons, H^+ (or hydronium ions, H_3O^+). The catalysis is not affected by other acidic species present. *Compare* **general acid-catalysis**.

specific activity the **activity** (def. 1, 2) of, e.g., a radionuclide or an enzyme, expressed as a function of mass.

specifically labelled describing an isotopically labelled compound in which both the position of labelling and the number of labelling nuclides at that position are known. Such a molecule is designated by a convention exemplified by $CH[^2H_2]-CH_2-O[^2H]$.

specific base catalysis a **homogeneous catalysis** in which the catalysts are hydroxide ions, OH^-. The catalysis is not affected by other basic species present. *Compare* **general base-catalysis**.

specific catalytic activity the **catalytic activity** of an enzyme under specified conditions divided by the mass of an enzyme or enzyme preparation. It is a quantity useful for denoting the extent of purification of an enzyme; it may be expressed in katals per kilogram.

specific dynamic action the increase in heat production that occurs in animals after the ingestion of food, particularly of protein.

specific enzymic activity the phenomenological coefficient that relates **enzymic activity** under specified conditions to the mass of an enzyme or enzyme preparation. It may be expressed in katals per kilogram of protein. Now obsolete, it has been superseded by **specific catalytic activity**. *Compare* **molar enzymic activity**.

specific gravity *a former name for* **relative density**.

specific heat capacity *symbol*: c_p (at constant pressure) *or* c_v (at constant volume); the quantity of **heat** required to raise the **temperature** of unit mass of a substance by one degree. It is expressed (usually) in $J K^{-1} kg^{-1}$.

specificity the degree to which an association between two molecular units may be considered unique; in many applications it is judged by the magnitude of an effect (enzyme–substrate, ligand–receptor interaction, or drug action). It is measured under defined conditions, in relation to the magnitude of effect of other substances with similar action, and when the substance is in direct competition with other such substances. The competitive success of other substances is normally quantified by kinetic analysis. *See* **competitive inhibition**, **specificity constant**.

specificity constant the ratio of the catalytic constant, k_{cat}, to the Michaelis constant, K_m. It can aid in comparing the activity of an enzyme in its action on different substrates. From **Michaelis kinetics**, using the Michaelis–Menten equation,

$$v = (k_{cat}/K_m)[E]_o[S]$$

when [S] is very small (since then, little enzyme has bound substrate, so $[E] \approx [E]_o$, and [S] can be neglected in relation to K_m). Thus for substrate A,

$$v_a = (k_{cat}/K_m)_A[E]_o[A],$$

and for substrate B,

$$v_b = (k_{cat}/K_m)_B[E]_o[B].$$

So

$$v_a/v_b = (k_{cat}/K_m)_A \times [A] / (k_{cat}/K_m)_B \times [B],$$

and, at the same substrate concentrations, the activity of the enzyme against the different substrates reduces to a comparison of the respective specificity constants (k_{cat}/K_m); K_m is not of itself a sufficient index of specificity.

specific optical rotation *symbol* $[\alpha]_\lambda^\theta$; the optical rotation of light of a specified wavelength (or frequency) achieved by a solution at a specified temperature. It is given by the relation $[\alpha]_\lambda^\theta = \alpha/l\,\rho$ where α is the observed rotation in degrees, l is the length of the sample tube (in dm), ρ is the density of the liquid (in gml^{-1}), λ is the wavelength of the light (or other electromagnetic radiation) used, and θ is the Celsius temperature. For a solution of concentration c (in g solute per 100 ml solution), $[\alpha]_\lambda^\theta = \alpha/lc$ in degrees $dm^2\ g^{-1}$. *Compare* **molecular optical rotation**.

specific refractivity *see* **refractivity**.

specific viscosity *symbol*: η_{sp}; the fractional change in **viscosity** produced by adding a solute to a solvent. It is given by the equation:

$$\eta_{sp} = (\eta / \eta_0) - 1 = Avc + Bv^2c^2 + Cv^3c^3 + \ldots\ldots.$$

where η_0 is the viscosity of the solvent, η is the intrinsic viscosity of the solution, A, B, and C are shape-dependent constants, c is the concentration of the solute, and v is the specific volume of one solute molecule.

specific volume *symbol*: v; for a pure substance, the volume, V, of a sample, divided by the mass, m, of the sample; $v = V/m = 1/\rho$, where ρ is the **mass density** of the substance. The SI unit is $m^3\ kg^{-1}$. *See also* **partial specific volume**.

'specifier' protein a protein that, when it combines with an enzyme or enzyme–substrate, modifies the catalytic specificity of the system. An example is α-lactalbumin, which modifies the substrate (acceptor) specificity of *N*-acetyllactosamine synthase (EC 2.4.1.90); *other name*: UDPgalactose:*N*-acetylglucosamine β-D-galactosyltransferase) from *N*-acetylglucosamine to glucose.

spectinomycin an aminocyclitol antibiotic secreted by *Streptomyces spectabilis*. It is widely active against Gram-negative bacteria, including *Neisseria gonorrhoeae*, but less so against Gram-positive bacteria. It inhibits protein synthesis by binding to the 30S ribosomal subunit.

spectra the plural of **spectrum**.

spectrin a protein that is the major constituent of the erythrocyte cytoskeletal network. It associates with band 4.1 (*see* **band protein**) and **actin** to form the cytoskeletal superstructure of the erythrocyte plasma membrane. It is composed of non-homologous chains, α and β, which aggregate side-by-side in an antiparallel fashion to form dimers, tetramers, and higher polymers. Anchorage to the cytoplasmic face of the plasma membrane is mediated by another protein, **ankyrin**, which links β spectrin to transmembrane protein band 3, and by the binding of band 4.1 to **glycophorin**. Interaction between spectrin and other proteins is thought to be responsible for maintenance of the biconcave shape of human erythrocytes, for regulation of plasma membrane components, and for maintenance of the lipid asymmetry of the plasma membrane.

spectrin repeat *see* **spectrin**.

spectrofluorometer an instrument used for measuring the **fluorescence** emitted by compounds excited by incident radiation produced within the instrument. The excitation radiation is produced by a lamp, and filters or diffraction grating monochromators are employed to isolate both the incident and emitted radiation to narrow wavebands. The emitted radiation is detected by photomultipliers.

spectrogram a photographic record of a **spectrum** produced by a **spectrograph**.

spectrograph a spectrometer or spectroscope with a facility for producing a photographic record of the spectrum under consideration. *Compare* **spectroscope**. —**spectrographic** *adj.*; **spectrographically** *adv.*; **spectrography** *n.*

spectrometer any instrument used to produce and examine a **spectrum**, such as a **spectrophotometer** or a **spectroscope**. —**spectrometric** *adj.*; **spectrometry** *n.*

spectrophotometer an apparatus to measure the proportion of the incident light (or other electromagnetic radiation) absorbed (or transmitted) by a substance or solution at each wavelength of the spectrum. —**spectrophotometric** *adj.*; **spectrophotometry** *n.*

spectroscope an instrument for studying spectra, either qualitatively or quantitatively. —**spectroscopic** *or* **spectroscopical** *adj.*; **spectroscopically** *adv.*

spectroscopy the investigation of the chemical composition, molecular structure, or atomic structure of a substance or solution by observation of its interaction with electromagnetic radiation using a spectroscope.

spectrum (*pl.* **spectra**) (*in physics*) an arrangement of the components of a complex electromagnetic radiation (light, etc.) or sound in order of frequency or energy, thereby showing a distribution of energy among the components. —**spectral** *adj.*

Spemann's organizer an embryonic signalling centre located in the dorsal lip of the blastopore. It plays a crucial role in the organization of the formation of the main body axis of a developing embryo. [After Hans Spemann (1869–1941), German zoologist and embryologist.]

spermaceti *or* **cetaceum** a white, translucent, waxy solid, m.p. 42–50°C, consisting mainly of the ester cetyl palmitate and obtained from oil in the head of the sperm whale (*Physeter catodon*) and related cetaceans. It is used in the manufacture of cosmetics, ointments, soaps, etc.

spermadhesin *or* **AQN1** a secretory protein of male accessory glands that is found on the sperm surface, and mediates sperm binding to the zona pellucida.

spermatogenesis the formation of spermatozoa in the male reproductive system. The sperms are formed from male germ cells, **sper-**

matogonia, which line the inner wall of the seminiferous tubules in the testis. A single spermatogonium divides by mitosis to form **primary spermatocytes**, each of which undergoes the initial division of meiosis to form two **secondary spermatocytes**. Each of these then undergoes the second meiotic division to form two **spermatids**, which mature into spermatozoa without further cell division by a process called **spermiogenesis**. All these cell types are nourished and supported by neighbouring **Sertoli cells**. Spermatogenesis also requires the presence of testosterone-secreting **Leydig cells**.

spermatogenesis factor *see* **dysferlin**.

spermidine *N*-(3-aminopropyl)-1,4-diaminobutane; $NH_2(CH_2)_3$-$NH(CH_2)_4NH_2$; one of the polyamines.

spermidine synthase EC 2.5.1.16; *other names*: putrescine aminopropyltransferase; aminopropyl-transferase; an enzyme that catalyses the formation of **spermidine** and 5′-methylthioadenosine from *S*-adenosylmethioninamine and putrescine.

spermine *N*,*N*-bis(3-aminopropyl)-1,4-diaminobutane; $NH_2(CH_2)_3$-$NH(CH_2)_4NH(CH_2)_3NH_2$; one of the polyamines.

spermine synthase EC 2.5.1.22; *other name*: spermidine aminopropyltransferase; an enzyme that catalyses the formation of **spermine** and 5′-methylthioadenosine from *S*-adenosylmethioninamine and spermidine.

sperm tail *see* **cilium**.

SPF *abbr. for* specific-pathogen-free.

Sph *symbol for* sphingosine.

S phase the phase of the eukaryotic cell cycle in which DNA is synthesized. *See* **cell-division cycle**.

spherocyte a red blood cell that has lost its normal biconcave shape owing to damage to the surface membrane or to some disorder of metabolism. Spherocytes appear smaller and darker than normal red cells.

spherocytosis the condition of having **spherocytes** in the blood. Hereditary spherocytosis in humans is a chronic hemolytic anemia that results from defects in ankyrin, band 3 protein, α spectrin, β spectrin, or 4.2 protein.

spheroid 1 a solid figure that is similar to but not identical with a sphere. **2** an **ellipsoid of rotation**, especially one in which the major axis of its parent ellipse is only slightly longer than its minor axis. —**spheroidal** *adj.*

spheroplast the globular form of a bacterial cell in which the cell-wall structure has been modified (e.g. by growth in the presence of penicillin) rather than totally removed. The term also denotes a form of yeast cell obtained by partial hydrolysis of the cell wall. A prefix may be used to indicate the method of induction, e.g. penicillin-spheroplast. *Compare* **protoplast**.

spherosome an organelle, generally spherical or oblate and 0.4–3 μm in diameter, that occurs in lipid-storage tissues in plants, e.g. endosperm of castor-bean seeds. Spherosomes are enclosed within a phospholipid monolayer (a half-unit membrane) with embedded oleosins and other proteins, and contain triacylglycerols formed by enzymes in the endoplasmic reticulum (ER) and partitioned into the interior of the ER membrane at sites defined by the presence of oleosins.

sphinganine D-*erythro*-2-amino-1,3-octadecanediol; D-*erythro*-dihydrosphingosine; the metabolic precursor of **sphingosine**. It is formed from palmitoyl-CoA and serine by the action of **serine C-palmitoyltransferase**, which forms 3-dehydro-D-sphinganine, which is then reduced to sphinganine by 3-dehydrosphinganine reductase, EC 1.1.1.102. *See also* **sphingolipids**, **sphingoid**.

sphingoid *or* **sphingoid base** *symbol*: Spd; any of a class of compounds comprising **sphinganine**, and its homologues and stereoisomers, and derivatives of these compounds.

sphingolipid any lipid that contains the long-chain amino diol, **sphingosine**, or a closely related base (i.e. a **sphingoid**). A fatty acid is bound in an amide linkage to the amino group and the terminal hydroxyl may be linked to a phosphate ester or a carbohydrate. The predominant base in animals is sphingosine while in plants it is **phytosphingosine**. The main classes are: (1) phosphosphingolipids (also known as sphingophospholipids), of which the main representative is **sphingomyelin**; and (2) **glycosphingolipids**, which contain at least one monosaccharide and a sphingoid, and include the **cerebrosides**

and **gangliosides**. Sphingolipids have a structural role in cell membranes, and may be involved in the regulation of protein kinase C.

sphingolipid activator proteins *see* saposins.

sphingolipidosis (*pl.* **sphingolipidoses**) any inherited pathological condition in humans in which one or more sphingolipids accumulate in the tissues because of a deficiency of the corresponding degradative enzyme. They include **ceramidase** deficiency, **Fabry's disease**, **galactosylceramide lipidosis**, G_{M1} and G_{M2} **gangliosidoses, Gaucher's disease, metachromatic leukodystrophy**, and **sphingomyelin lipidosis**.

sphingomyelin *N*-acyl-4-sphingenyl-1-*O*-phosphorylcholine; any of a class of phospholipids in which the amino group of **sphingosine** is in amide linkage with one of several fatty acids, mostly C_{20} or higher and either saturated or monounsaturated, while the terminal hydroxyl group of sphingosine is esterified to phosphorylcholine. Sphingomyelins are found in large amounts in brain and other nervous tissue. They do not occur in plants.

sphingomyelinase *see* sphingomyelin phosphodiesterase.

sphingomyelin lipidosis *or* **Niemann–Pick disease** a genetic disease of humans in which **sphingomyelin** accumulates in the tissues due to a deficiency of the enzyme **sphingomyelin phosphodiesterase**. It is associated with mental retardation.

sphingomyelin phosphodiesterase EC 3.1.4.12; *other names*: acid sphingomyelinase; neutral sphingomyelinase; a lysosomal enzyme that catalyses the hydrolysis of sphingomyelin to *N*-acylsphingosine and choline phosphate. Deficiency of the enzyme is the cause of Niemann–Pick disease (**sphingomyelin lipidosis**), in which there is an accumulation of sphingomyelin.

sphingophospholipid *see* sphingolipid.

sphingosine *symbol*: Sph; *trans*-D-*erythro*-2-amino-4-octadecene-1,3-diol; $CH_3(CH_2)_{12}CH=CH–CHOH–CHNH_2–CH_2OH$; a long-chain amino diol sphingoid base that occurs in most **sphingolipids** of animal tissues. *See also* **phytosphingosine**.

sphingosine *N*-acyltransferase EC 2.3.1.24; an enzyme of the pathway for synthesis of sphingolipids that catalyses the acylation of sphingosine by acyl-CoA to form *N*-acylsphingosine (ceramide) and CoA.

spicule a roughly cone-shaped structure or tissue element, such as that projecting from the membrane of a red blood cell. Calcareous or siliceous spicules are found in the skeletons of sponges and corals.

spin (*in physics*) **1** the quantized angular momentum of an elementary particle, in the absence of orbital motion. **2** the quantized angular momentum of an atomic nucleus, including contributions from the orbital motions of nucleons. **3** the quantized angular momentum of an electron that adds to the orbital angular momentum, thus producing fine structure in line spectra.

spinal and bulbar muscular atrophy (*abbr.*: SBMA) *or* **Kennedy's disease** an X-linked progressive neuromuscular disorder that results in proximal muscle weakness, muscle atrophy, and involuntary contractions or fasciculations. It is caused by expansion of a CAG trinucleotide repeat in the androgen receptor gene. [After William Kennedy, US neurologist.]

spinal cord the part of the central nervous system that is enclosed in the vertebral column. It connects the innervation of most parts of the body via the spinal nerves to the brain.

spinal muscular atrophy an autosomal recessive disorder in which degeneration of lower motor neurons leads to progressive atrophy of limb and trunk musculature. It is caused by mutations in a neuronal apoptosis inhibitory protein. *See* **IAP**.

spinal nerves the pairs of nerves that arise from the spinal cord and are distributed to various peripheral parts of the body. In humans there are 31 pairs.

spin column chromatography *or* **spun column chromatography** a small-scale chromatography procedure in which fluid is moved through the column by centrifugal force. It has the advantage that several columns can be run simultaneously in the centrifuge rotor.

spin coupling *or* **spin–spin coupling** (*in magnetic resonance spectroscopy*) the influence of a nuclear or electron spin on another. In NMR the nuclear spin of one nucleus is coupled via the bonding electrons to the coupled nucleus, which experiences two different microenvironments and thus resonates at two slightly different frequencies. This results in the splitting (**spin–spin splitting**) of the

resonance line into two lines (a doublet). In EPR spectroscopy the electron-nuclear coupling is known as **hyperfine coupling**. Electron–electron couplings can also occur, and lead to splitting or broadening of the EPR spectrum.

spin density the fraction of unpaired electron spin that is in the vicinity of a given atomic nucleus in a molecule.

spindle the array of microtubules and associated molecules that forms between the opposite poles of a eukaryotic cell during mitosis or meiosis and serves to move the duplicated chromosomes apart. During mitosis the spindle starts to form outside the nucleus while the chromosomes are condensing during prophase. At the start of M phase, the two **centrosomes** that have been formed move to opposite sides of the nucleus, and form the two poles of the mitotic spindle. Microtubules grow away from the centrosomes towards the chromosomes, the minus ends being attached to the centrosome, and the plus ends growing towards the **kinetochore**. These microtubules eventually bind to kinetochores and are thus known as kinetochore microtubules. Other microtubules, known as polar microtubules, grow towards the chromosomal region, but eventually bind directly to other polar microtubules growing from the opposite centrosome. these push the poles of the spindle apart. Each centrosome eventually becomes attached through the kinetochore microtubules to one of the sister chromatids of each chromosome, and pulls these away to effect the chromatid separation that occurs at anaphase. A similar series of events occurs at the corresponding phases of meiosis.

spin hamiltonian a formulation in **quantum mechanics** used in the determination of the energy levels of the spin of an electron or a nucleus in **electron spin resonance spectroscopy** or, less commonly, in **nuclear magnetic resonance spectroscopy**. *See also* **Hamiltonian operator**.

spin immunoassay *or* **free-radical assay technique** (*of immunoassay*) (*abbr.*: FRAT) a technique in which a **spin label** is attached to a specimen of the substance to be assayed and an antibody is raised against the substance. On mixing the antibody with the labelled substance there is a change in the electron spin resonance (ESR) spectrum because of the inhibition of tumbling caused by the antibody 'immobilizing' the labelled substance. On adding some of the unlabelled substance there is competition for the antibody, and the changes in the ESR spectrum are partially reversed to an extent dependent on the concentration of the added, unlabelled, substance.

spin label 1 a synthetic paramagnetic organic free radical, usually having a molecular structure and/or chemical reactivity that facilitates its attachment or incorporation at some particular target site in a macromolecule, or assemblage of macromolecules. **2 spin-label** to effect labelling of a substance or structure with a spin label (def. 1).

spin-lattice relaxation time *or* **longitudinal relaxation time** (*in magnetic resonance spectroscopy*) *symbol*: T_1; the time taken for the upper nuclear or electron spin state to dissipate its excess energy. Its reciprocal is the spin-lattice relaxation rate (*see* **correlation time**).

spinner culture a method of culturing cells that uses the rotation of an impeller to maintain cells in suspension.

spinocerebellar ataxia *abbr.*: SCA; any of a heterogeneous group (SCA1 to SCA10) of neurological disorders caused by variable degrees of degeneration of the cerebellum, brainstem neurons, and spinocerebellar tracts. Most subtypes result from expansion of the number of translated or untranslated CAG trinucleotide repeats in a gene for an ataxin, atrophin-1, or the α1A subunit of a neuronal Ca^{2+} channel. *See* **hemiplegic migraine**.

spin polarization *or* **spin polarisation** a phenomenon in which the unpaired electron spin on one atom or part of a molecule is transferred to another atom, brought about by interaction between the unpaired electron on the first atom and the paired electrons on the second atom (which lacks unpaired electrons) causing one of the paired electrons to have a lower energy than the other. *See also* **chemically induced dynamic nuclear polarization**.

spin probe *see* electron paramagnetic resonance spectroscopy.

spin–spin broadening an increase in line width in **electron spin resonance spectroscopy**, caused by interactions between neighbouring dipoles.

spin–spin coupling *see* spin coupling.

spin–spin coupling constant (*in nuclear magnetic resonance spec-*

troscopy) a measure of the specific interaction of one nucleus with another that gives rise to a splitting of the resonance bands produced by each nucleus.

spin–spin splitting splitting in the lines of a magnetic resonance spectrum brought about by the interaction of the magnetic moment of the nucleus or unpaired electron in a molecule with those of neighbouring nuclei or electrons. It is used in the assignment of particular nuclear resonances and in the determination of molecular conformations. *See also* **hyperfine splitting**.

spin state the orientation, in a strong magnetic field, of an unpaired electron or nuclear spin, e.g. parallel or antiparallel to the field direction.

spin trapping a technique in which a reactive free radical is trapped by an addition reaction to produce a more stable radical, detectable by **electron spin resonance spectroscopy**, whose hyperfine coupling parameters permit identification of the initial radical trapped.

spiro compound any compound containing two rings joined by one common atom, designated the **spiro atom**, as in the case of the *γ*-lactone ring of **spironolactone**.

spironolactone 7-(acetylthio)-17-hydroxy-3-oxo-pregn-4-ene-21-carboxylic acid *γ*-lactone; a synthetic steroid lactone that is used as a diuretic. It is structurally similar to **aldosterone**, and competitively inhibits the renal tubular action of this hormone. It is metabolized to canrenone, the active compound, by removal of the acetylthio group and insertion of a 6,7 double bond.

spleen exonuclease *see* **phosphodiesterase II**.

spleen tyrosine kinase *see* **Syk**.

splenin *see* **thymopoietin**.

splenopentin a synthetic peptide that corresponds to residues 32–36 of splenin (*see* **thymopoietin**).

spliceosome a ribonucleoprotein complex, containing RNA and **small nuclear ribonucleoproteins** (snRNPs) that is assembled during the **splicing** of the messenger RNA **primary transcript** to excise an **intron**. snRNPs designated U1, U2, U4, U5, and U6 are variously involved; U1 may bind the GU consensus sequence of the **5′ splice site** and U5 with U4/U6 probably binds to the 3′ splice site consensus sequence, AG. U2 may be part of a component binding the UACUA[A]C consensus sequence ([A] is invariant) at a site slightly upstream from the 3′ splice site, known as the branch site. The mechanism appears to involve cutting at the 5′ side of the GU consensus sequence, which curls back to form a G5′–2′A bond with the invariant A of UACUA[A]C. This linkage generates the loop of the so-called **lariat**. Cutting out the 3′ splice site follows on the 3′ side of AG to release the complete intron (in the form of the lariat), and the cut ends of the exons are then ligated.

splice site the sequence of bases at each end of an **intron** that determines the point of **splicing**. The 5′ splice site (site at the 5′ end of the intron) is termed the donor site, consensus sequence GU, while the 3′ splice site is termed the acceptor, consensus sequence AG.

splice variant *or* alternatively spliced form any of various proteins of different length that arise through translation of mRNAs that have not included all available exons in the template DNA.

splicing 1 (*of DNA*) the covalent linkage, by the action of a **DNA ligase** (EC 6.5.1.1, requires ATP, or EC 6.5.1.2, requires NAD⁺), of two fragments of duplex DNA at complementary single-stranded terminations. **2** (*of RNA*) the enzymic process in eukaryotic cells by which **introns** are excised from heterogeneous nuclear RNA (hnRNA) following its transcription from DNA, and the cut ends are rejoined

to form messenger RNA (mRNA), in which the message is continuous, before its translation into polypeptide. In yeast, the splicing of transfer RNA involves cutting at the splicing junctions by an endonuclease, followed by a process involving ATP, whereby the two exon sequences are joined. Group I mitochondrial RNA is spliced by an autocatalytic mechanism, the RNA itself having the ability to catalyse its own splicing; nuclear mRNA precursor – the primary transcript – is spliced by a complex assembly incorporating snRNP and called a **spliceosome**; it involves the formation of a **lariat**. Splicing of group II mitochondrial RNA also involves the formation of a lariat, but without formation of a spliceosome. Splicing in which the cut ends are joined directly within the same RNA molecule is called **cis-splicing**; **trans-splicing** refers to situations in which the cut end from the splicing site rejoins to a splicing site on another RNA molecule. **3** (*of protein*) a post-translational modification of protein that includes two concerted proteolytic cleavages and one ligation, resulting in excision of an inner sequence of the original polypeptide chain to form one protein, and ligation of the two outer sequences to form a second protein. It is thought to be an autocatalytic process. The excised inner sequence is known as an **intein**, the ligated outer sequences as an **extein**. *See also* **gene splicing**.

splicing factor 2 *abbr.*: SF2; a protein, present in nuclear extracts, that functions in an early step in splicing pre-mRNA. It is resistant to mild heat treatment, but inactivated by *N*-ethylmaleimide. It consists of two peptides, P32 and P33, and is involved in ensuring accuracy of splicing. *See also* **pre-mRNA processing protein**.

splicing regulatory protein *see* **SR protein**.

split gene a gene that consists of coding (exon) and noncoding (intron) sequences.

spoke protein a protein associated with the filamentous structure of kinetochore microtubules in the mitotic spindle. Soluble, and cytoplasmically localized during interphase, it becomes primarily associated with the mitotic spindle in mitosis.

spongiform encephalopathy any disease of humans and other animals characterized by the presence of many vacuoles in brain tissue, giving it a spongy appearance. The group includes **bovine spongiform encephalopathy**, **scrapie** in sheep, and **Creutzfeldt–Jakob disease**, **Gerstmann–Straussler syndrome**, and **kuru** in humans. Some of these are classified as **transmissible spongiform encephalopathies**, being caused by infectious agents and carried from one individual to another and in some cases from one species to another. *See also* **prion**, **prion protein**.

spontaneous generation *see* **abiogenesis**.

spontaneous hypoglycemia *see* **glycemia**.

spore a small, often microscopic, reproductive unit consisting of one or several cells. Spores are produced by fungi, bacteria, certain plants, and protozoa. They may variously serve as a means of rapid vegetative propagation, or as a dormant stage in the organism's life cycle. Spores have very low metabolic activity and give rise to a vegetative cell upon germination; they are usually more heat resistant than vegetative cells and are adapted for dispersal. Spores may be sexual or asexual.

sporulation the formation of **spores**.

spot 1 (*in chemistry*) *or* spot test to test for a substance by applying a small amount of detection reagent to a surface. **2** (*in molecular biology*) to detect, using radioactive oligonucleotide probes, oligonucleotides applied in spots onto hybridization filters (**spot blots**).

spoT a gene in *Escherichia coli* for the enzyme guanosine-3′,5′-bis(diphosphate) 3′-pyrophosphohydrolase; EC 3.1.7.2; *other name*: (ppGpp)ase; it catalyses the hydrolysis of guanosine 3′,5′-bis(diphosphate) to guanosine 5′-diphosphate and pyrophosphate as a part of the **stringent response**. The name is jocular, ppGpp and pppGpp being called 'magic spot nucleotides'. *See also* **relA**, **stringent factor**.

spp. *abbr. for* species (plural).

SPP-1 *see* **osteopontin**.

SPP *abbr. for* signal peptide protease.

[S]ppA *symbol for* adenosine 5′-*β*-thiodiphosphate (alternative to AdoP*P*[S]).

SP6 phage promoter a 20 bp sequence from bacteriophage SP6 that is sometimes incorporated into the design of plasmid vectors to permit *in vitro* **transcription** of cloned inserts using SP6 RNA polymerase.

[S]pppA *symbol for* adenosine 5'-γ-thiotriphosphate (alternative to Ado*PPP*[S]).

spreading factor *see* **hyaluronidase**.

SP6 RNA polymerase a single subunit enzyme from bacteriophage SP6 that transcribes genes downstream from the **SP6 phage promoter**.

SQDG *abbr. for* sulfoquinovosyldiacylglycerol (*see* **sulfonolipid**).

SQL *abbr. for* symbolic query language; a **query language** used with relational databases, e.g. **SYBASE**. It is often pronounced 'sequel'.

squalamine an aminosterol from the dogfish shark, *Squalus acanthias*. It has antibiotic properties against Gram-negative and Gram-positive bacteria and protozoa. It is an adduct of **spermidine** and a sulfate bile salt.

$$R = H_3\overset{+}{N}-(CH_2)_4-\overset{+}{N}H_2-(CH_2)_3-$$

squalene 2,6,10,15,19,23-hexamethyl-2,6,10,14,18,22-tetracosahexaene; the linear triterpene precursor of all cyclic triterpenoids and the sterols. Although first isolated from shark-liver oils, it is widely distributed in plant and animal cells as a metabolic intermediate. It is formed from acetyl-CoA, via **mevalonate**, **isopentenyl diphosphate**, **geranyl diphosphate**, and farnesyl diphosphate. The latter is converted to presqualene diphosphate, which is then reduced to squalene by NADPH with loss of pyrophosphate; both steps are catalysed by **farnesyltransferase**. [From Latin *squalus*, marine fish, later specifically used to mean shark.]

squalestatin a fermentation product of *Coelomycete* spp. and a selective inhibitor of squalene synthase. It lowers serum cholesterol.

sr *symbol for* steradian.

Sr *symbol for* strontium.

SR *abbr. for* sarcoplasmic reticulum.

src an oncogene (v-*src*), originally discovered in the Rous sarcoma virus, whose product is pp60$^{v\text{-}src}$, a 60 kDa protein that undergoes phosphorylation/dephosphorylation. The **protooncogene**, c-*src*, codes for a related protein; both proteins are **tyrosine kinases**. The C-terminal 19 residues of pp60$^{c\text{-}src}$ are replaced by 12 other residues in pp60$^{v\text{-}src}$. The Src family is the largest of the nonreceptor tyrosine kinase families. It includes Src, Yes, Fyn (*see* **FYN**), Lck (*see* **LCK**),

Lyn (*see* **lyn**), Blk (*see* **blk**), Hck, Fgr (*see* **FGR**), and Yrk, which range in size from 53 to 64 kDa. They all have an N-terminal myristoylation site, a unique domain, an SH2 domain and an SH3 domain (*see* **SH domain**), a tyrosine kinase domain, and a C-terminal regulatory domain. In vertebrates, the kinase activity is usually inhibited by the phosphorylation of a tyrosine residue within the kinase domain. In v-*src* the region encoding this tyrosine is deleted leading to v-Src having a higher constitutive tyrosine kinase activity than c-Src. Other mutations associated with oncogenic activation map to the SH2 and SH3 domains, suggesting that Src may be inactivated by the binding of its SH2 domain to the phosphotyrosine residue in its negative regulatory domain. The SH3 domain may also participate in this repression.

Src homology domain *see* **SH domain**.

Srd *symbol for* a residue of the ribonucleoside thiouridine (when the position of the thiol group in its thiouracil moiety is unknown or unspecified; alternative to S or sU).

SRE *abbr. for* serum response element.

SREBP *abbr. for* sterol-response-element binding protein; any of three transcription factors that are membrane proteins of the endoplasmic reticulum and Golgi apparatus. Each is complexed with **SCAP** (SREBP-cleavage activating protein), which serves as a sterol sensor. SREBP is cleaved by two Golgi-specific proteases, **S1P** and **S2P**, which release the transcription domain for entry into the nucleus and transcription of genes that contain a 10 bp sterol-response element in their promoter region. These include genes for LDL receptor and enzymes of cholesterol, fatty acid, and triacylglycerol synthesis. SREBP is a polypeptide of ≈1100 amino acids. Its N-terminal half contains a sequence of acidic residues (for binding of accessory transcriptional activators), a basic helix-loop-helix-leucine zipper sequence (for DNA binding and for homodimer or heterodimer formation), and a transmembrane segment. The C-terminal half contains one transmembrane segment and a regulatory domain that interacts with the cytoplasmic domain of SCAP. Cholesterol and polyunsaturated fatty acids, acting via the sterol-sensing domain of SCAP, block the release of the transcriptional domain of SREBP by S1P and S2P. In animals SREBP-1a and -1b are encoded by the same gene using alternative splicing; SREBP-2 is encoded by a different gene.

SREBP-cleavage activating protein *see* **SCAP**.

SRF *abbr. for* serum response factor; a nuclear, dimeric transcription factor. It binds **Elk-1**, which is constitutively bound to the **serum response element**.

SRIF *abbr. for* somatotropin release-inhibiting factor (i.e. **somatostatin**).

sRNA *or* **S-RNA** *abbr. for* soluble RNA (an early name for **transfer RNA**).

SRP *abbr. for* signal recognition particle.

SR protein *abbr. for* splicing regulatory protein; any protein that contains one or more N-terminal RNA-recognition domains that bind RNA with sequence specificity, and a C-terminal Arg-Ser-rich (or RS) domain for protein binding. They function in pre-mRNA splicing and assemble on exonic splicing enhancer sequences. When they bind the C-terminal domain (CTD) of the largest subunit of RNA polymerase II they are called **SCAF** or CASP.

SR-related protein any protein that is related to, but different structurally and functionally from, proteins of the SR family. They function as splicing factors but are not part of small nuclear ribonucleoproteins. They bind to exonic splicing-enhancer sequences on pre-mRNA. *Compare* **SR protein**.

SRS *or* **SRS-A 1** *abbr. for* slow reacting substance (of anaphylaxis). **2** *abbr. for* Sequence Retrieval System; *see* **information retrieval software**.

S→ R variation *see* **smooth–rough variation**.

SRY *abbr. for* sex-determining region of Y chromosome; *other name*: testis-determining gene; a gene on the Y chromosome in humans that encodes a DNA-binding protein (240 amino acids) that induces differentiation of Sertoli cells in the developing gonad so that they produce and secrete anti-Müllerian hormone, which causes regression of female internal genitalia. It also induces Leydig cells to secrete the androgen necessary for development of male genitalia. The DNA-binding domain (≈80 amino acids) is homologous with that of HMG proteins. Many mutations in the gene cause familial XY gonadal dysgenesis (i.e. a female phenotype in the presence of X and Y chromosomes).

ss *abbr. for* single-stranded.

SSB protein *abbr. for* single-strand binding protein.

SSC a solution used in DNA hybridization procedures. A solution of 20× SSC contains 3.0 M NaCl and 0.3 M sodium citrate.

SSCP analysis *abbr. for* single-strand conformation polymorphism analysis.

ssDNA *see* DNA.

SSPE a solution used in DNA hybridization procedures. A solution of 20× SSPE contains 3.0 M NaCl, 0.2 M NaH_2PO_4, and 0.02 M EDTA.

SSR *abbr. for* simple sequence repeat.

St *symbol for* stokes (a cgs unit of kinematic viscosity).

Sta *symbol for* statine.

stability constant *symbol*: K_s; an equilibrium constant for the reversible formation of a complex chemical compound from two or more simpler entities; the reciprocal of the **dissociation constant**. In some cases, an **apparent stability constant** (*symbol*: K'_s), constrained with respect to certain variables (e.g. pH) is determined (for the distinction see **equilibrium constant**). The stability constant is also known as the **association constant** (*symbol*: K_{ass}) in cases where ions associate into a substance, e.g. for the reaction $A^+ + B^- \rightleftharpoons AB$, the (concentration) stability or association constant is given by:

$$K_s = [AB]/[A^+][B^-] = K_{ass}.$$

stable expression *(in genetics)* the production of a recombinant protein sustained over a period of weeks or months. *Compare* **transient expression**.

stable factor *an alternative name for* factor VII; *see* **blood coagulation**.

stable isotope an **isotope** that is not radioactive.

stachyose *or* **lupeose** the tetrasaccharide, *O*-α-D-galactopyranosyl-(1→6)-*O*-α-D-galactopyranosyl-(1→6)-*O*-α-D-glucopyranosyl-β-D-fructofuranoside, isolated from *Stachys tuberifera* rhizomes. It is widely distributed, usually associated with **raffinose** (*see* that entry for synthesis) and **sucrose**.

stacking gel *an alternative name for* **spacer gel**.

staggered conformation *see* **conformation**.

stainless steel steel that is resistant to corrosion due to its high content of chromium and, often, nickel. It is highly resistant to all organic acids and weak mineral acids but is slowly attacked by concentrated mineral acids. It does not rust.

Stains-all *a proprietary name for* 1-ethyl-2-[3-(3-ethylnaphtho[1,2-*d*]thiazolin-2-ylidene)-2-methylpropenyl]naphtho[1,2-*d*]thiazolium bromide; a cationic carbocyanine metachromatic dye used as a semiselective stain for locating nucleic acids, proteins, and polysaccharides. RNA stains bluish-purple, DNA blue, protein red, and phosphoprotein blue.

stamp collecting according to Lord Rutherford (1871–1937), one of the two branches of science (the other being physics).

standard *(in analytical chemistry)* **1 primary standard** a chemical that can be obtained in an unvarying state (e.g. of purity and/or stability) and used as an approved example against which other examples of the same chemical can be compared. **2 secondary standard** a chemical that can be used for comparative purposes following calibration against a primary standard. Although its properties are less stringent than a primary standard, its ready availability makes for convenience of use. *See also* **standard atmosphere, standard state, standard temperature and pressure**.

standard atmosphere an internationally established reference for

pressure, defined as 101 325 pascals. The **atmosphere** (def. 3) (*symbol*: atm), equal to 101 325 pascals, was formerly used as a unit of pressure; its use is now discouraged. *See also* **standard temperature and pressure**.

standard curve a plot of any appropriate parameter against known amounts (or concentrations) of detectable substance. Such a curve can be used to determine the amount (or concentration) of the same substance in an unknown solution given a reading for the parameter in the unknown solution.

standard deviation *abbr.*: SD; *symbol*: s (for a sample) *or* σ (for a distribution); the common measure of dispersion of a series of observations of x that is normally distributed. It is the root-mean-square average of the deviations of the observations from their mean and is in the same units as those of the observations. The standard deviation, s_x, is given by:

$$s_x = \sqrt{\left[\frac{1}{n} \sum_{i=0}^{n} (x - \bar{x})^2 \right]}$$

Where $\bar{x}$ is the arithmetic average of deviations of x and n is the number of observations; s_x^2 is called the variance of the sample of observations of x. Some authorities prefer the definition:

$$s_x = \sqrt{\left[\frac{1}{n-1} \sum_{i=0}^{n} (x - \bar{x})^2 \right]}$$

so that s_x^2 is then the best estimate of the variance, σ^2, of the population from which the sample of observations of $\bar{x}$ was made. *See also* **degrees of freedom, standard error of estimate of the mean**.

standard error of estimate of the mean *or* **standard error of the mean** *abbr.*: SEM; a measure of the reliability of an estimate of the **mean** of a population. It is equal to the **standard deviation**, s_x, of the observations divided by the square root of the number of observations, n; i.e. SEM = $s_x/\sqrt{n}$.

standard free energy *see* **Gibbs energy**.

standardize *or* **standardise** to relate any instrument, vessel, or reaction to an absolute standard. —**standardization** *or* **standardisation** *n*.

standard pressure *symbol*: p; an established reference for pressure. Since 1982 the recommended standard pressure for reporting thermodynamic data has been set at 10^5 pascals (1 bar). Before 1982 it was taken to be 101 325 pascals (1 atmosphere).

standard state a set of conditions, established by convention, under which a standard reference is determined or calculated. *See* **chemical potential, enthalpy, entropy, Gibbs energy, hydrogen electrode, standard pressure, standard temperature and pressure**.

standard temperature and pressure *abbr.*: stp *or* s.t.p. *or* STP; the standard conditions for the comparison of properties of gases. Standard temperature is now 298.15 K (formerly 273.15 K, i.e. 0°C) and **standard pressure** is 10^5 pascals. *Former name*: normal temperature and pressure (*abbr.*: ntp *or* n.t.p. *or* NTP).

stanniocalcin a peptide of bony fishes that counteracts hypercalcemia and stimulates phosphate resorption in kidney.

Staphylococcus a genus of Gram-positive bacteria, commonly associated with wounds and generally causing localized infections. *Staphylococcus aureus* forms yellow or orange pigments and yields plasmids that are useful in recombinant DNA experimentation: *S. aureus* plasmids coding for resistance to various antibiotics can be transfected into *Escherichia coli* or into *Bacillus subtilis* where they are stable, replicate, and express antibiotic resistance. They have provided the basis from which improved vectors have been developed.

staphylokinase a noncatalytic protein (136 amino acids) produced by *Staphylococcus* spp. that binds to and activates plasminogen. It has no sequence similarity to streptokinase, and is used to dissolve blood clots.

StAR *abbr. for* steroidogenic acute regulatory; a protein of the mitochondrial matrix that is phosphorylated in response to cAMP released by corticotropin in the adrenal cortex and gonads. It plays a role in the entry of cholesterol into the matrix. Point mutations in the gene for StAR at 8p11.2 (which encodes 286 amino acids) result in congenital lipoid adrenal hyperplasia.

star activity the relaxation of specificity under nonoptimal conditions exhibited by some restriction endonucleases. They then cleave DNA sequences that are similar to but not identical with their defined recognition sequence.

starch the most important reserve polysaccharide found in plants. It has two components: **amylose**, which is a homopolymer of glucose units linked only by $\alpha(1 \rightarrow 4)$ bonds; and **amylopectin**, also a glucose homopolymer (branched) that contains $\alpha(1 \rightarrow 6)$ bonds as well as $\alpha(1 \rightarrow 4)$ bonds. Starch is synthesized in chloroplasts during active periods of photosynthesis, accumulating as starch grains. It is then converted to sucrose and translocated to storage organs, such as seeds and tubers, where it is re-formed. *Compare* **glycogen**.

starch (bacterial glycogen) synthase EC 2.4.1.21; *recommended name*: starch synthase; *systematic name*: ADPglucose:1,4-α-D-glucan 4-α-D-glucosyltransferase; *other name*: ADPglucose–starch glucosyltransferase. An enzyme that catalyses a reaction between ADPglucose and $(1,4$-α-D-glucosyl$)_n$ to form $(1,4$-α-D-glucosyl$)_{n+1}$ and ADP. It uses ADPglucose instead of the UDPglucose of **glycogen synthase**.

starch debranching enzyme *an alternative name for* **R-enzyme**.

starch phosphorylase *see* α-glucan phosphorylase.

starch synthase EC 2.4.1.11; *recommended name (example)*: glycogen (starch) synthase (the nature of the synthetic product should be included in the name); *systematic name*: UDPglucose:glycogen 4-α-D-glucosyltransferase; *other name*: UDPglucose–glycogen glucosyltransferase. An enzyme that catalyses a reaction between UDPglucose and $(1,4$-α-D-glucosyl$)_n$ to form $(1,4$-α-D-glucosyl$)_{n+1}$ and UDP. *See also* **glycogen synthase**. *Compare* **starch (bacterial glycogen) synthase**.

starch syrup *an alternative name for* **corn syrup**.

Stargardt macular dystrophy *or* **juvenile macular degeneration** an autosomal recessive condition that is the most common form of macular degeneration. It manifests in the pre-teen years and is caused by mutations in an ABC transporter protein, called ABCR (ABC transporter retina, or Rim protein; 2273 amino acids), that is found in rod photoreceptor outer segments. [After Karl Stargardt (1875–1927), German ophthalmologist.]

stargazin a protein (38 kDa) with four transmembrane segments that is essential for bringing AMPA receptor (an ionophoric glutamate receptor) to the postsynaptic membrane. Its intracellular C-terminal region binds to the postsynaptic protein PSD-95. It is homologous with the γ subunit of voltage-gated Ca^{2+} channels of muscle. Defects in the protein are the cause of stargazer mutant mouse.

Start an important checkpoint in the G_1 phase of the eukaryotic **cell-division cycle**. Passage through Start commits the cell to enter S phase.

start codon the trinucleotide **codon** AUG that codes for *N*-formylmethionine, the first amino-acid residue in the synthesis of all prokaryotic and mitochondrial proteins.

start kinase a complex of cdc2 (*see cdc gene*) and a G_1 **cyclin** that is a serine/threonine protein kinase; its activation is necessary for the cell to pass **Start**. The identity of the G_1 cyclin modifies the specificity of the complex.

start point the base pair in DNA at which the first nucleotide is incorporated into an RNA transcript. It is usually a purine, and often is the central base of the sequence CAT. It is sited at the downstream end of the **promoter**. *Compare* **start codon**.

start protein any protein that is involved uniquely at the start of the cell cycle. For example, protein Cdc10 from *Schizosaccharomyces pombe* (which has a close homologue in *Saccharomyces cerevisiae*) is probably a transcriptional regulator for *cdc10* and also for *cdc2*, which encodes a component of **start kinase**.

starve to deprive of nutrients. In animal experiments a minimum of 24 h deprivation is generally understood. Cultures may be starved of specified single nutrients.

stasis arrest or stagnation of flow.

+stasis *comb. form* denoting lack of movement; cessation of flow. —**+static** *adj*.

STAT *abbr. for* signal transducer and activator of transcription; a family of eukaryotic transcription factors that mediate the response to many cytokines and growth factors. Upon receptor activation, STAT proteins dimerize, translocate to the nucleus, and bind to specific promoter sequences on target genes. Seven different genes have been so far identified in mammals. Example: STA1_HUMAN, 750 amino acids (87.33 kDa).

+stat *comb. form* indicating a device for maintaining a particular property at a constant value, e.g. **cryostat**, **thermostat**. —**+static** *adj*.

stathmin *other names*: phosphoprotein p19; oncoprotein p18; prosolin; a ubiquitous, phylogenetically conserved, cytoplasmic protein that serves as a phosphorylated intermediate in diverse second-messenger pathways. Stathmin undergoes phosphorylation; both phosphorylation and expression of the protein are regulated throughout development, in response to extracellular signals that control cell proliferation and differentiation. It is a major phosphorylation substrate in neurons.

+static *see* **+stasis**, **+stat**.

statics *functioning as sing.* the branch of applied mathematics dealing with the interaction of forces in nonmoving systems.

statin 1 a phosphoprotein present specifically in the nuclei of nonproliferating cells. It forms a complex with a 45 kDa serine/threonine protein kinase, and has structural similarity to the elongation factor EF-la. **2** any of a class of drugs used to lower blood cholesterol levels through inhibition of hydroxymethylglutaryl-CoA reductase. Examples are Lipitor (atorvastatin) and Zocor (simvastatin). The effect of these drugs can be enhanced by the simultaneous administration of an agent, such as **cholestyramine**, that prevents the reabsortion of bile acids, thus increasing the conversion of cholesterol to bile acids. **3** a variant spelling of **statine**.

+statin *word ending (in hormone nomenclature)* denoting a hypothalamic release-inhibiting factor (or hormone). *Compare* **+liberin**.

statine *or* **statin** *symbol*: Sta; [$3S,4S$]-4-amino-3-hydroxy-6-methylheptanoic acid; a novel amino acid contained in positions 3 and 5 of **pepstatin A**.

stationary phase 1 *(in chromatography)* the immobile phase to which the analyte adsorbs or partitions from the mobile phase. **2** *(in fermentation)* the phase of an *in vitro* culture of microorganisms or eukaryotic cells that follows the exponential growth phase and in which there is little or no growth.

statistically significant *see* **significant**.

statistical mechanics the theoretical analysis of molecular properties based on quantum theory.

statistical sample a finite selection of items from a population of which the measurement of a variable will give a close approximation to the distribution of that variable in the whole population. *See also* **random sample**, **sample**.

statistics 1 *functioning as sing.* the science of collecting, analysing, interpreting, and presenting quantities of numerical data. **2** *functioning as pl.* a collection of quantitative data. —**statistical** *adj*.

stator the part of the **H⁺-transporting ATP synthase** that is held fixed relative to the F_0 part during flow of hydrons through the latter. *Compare* **rotor**.

statyl the acyl group derived from **statine**.

Staufen a dsRNA-binding protein of *Drosophila* that is involved in localizing maternal mRNA in oocytes and the embryo. It contains five dsRNA-binding motifs (**dsRBMs**). Mouse and human homologues lack the first dsRBM, bind tubulin, and interact with the rough endoplasmic reticulum and with polysomes.

staurosporine an antibiotic from *Streptomyces* spp., M_r 466.2, best known as a potent inhibitor of protein kinase C.

Ste *symbol for* the stearoyl group, $CH_3–[CH_2]_{16}–CO–$.

steady state a condition in which the properties of any part of a system are constant during a process or reaction; i.e. where the rate of formation or increase of a particular quantity is balanced by its rate of removal or decrease.

steady-state fluorescence anisotropy a method in which a sample containing a fluorescent compound is irradiated with plane-polarized light, the emitted fluorescence being analysed to determine its degree of polarization both in the plane of the excitation radiation and in a plane at right angles to this. Only molecules of the fluorescent compound that are aligned in the plane of the incident radiation can absorb radiation and be excited to emit fluorescence. If, in the interval between absorption of the radiation and the emission of fluorescence, the molecule moves out of the plane of polarization of the incident radiation, the emitted radiation will lose polarization (be depolarized) to a degree that depends on the extent to which the molecule had moved. The technique is used to measure the degree to which fluorescent probes are tumbling in a sample. If the probes are held in a relatively rigid position, the extent of depolarization will be small, if tumbling freely it will be high.

steady-state kinetics the analysis of an enzyme reaction in the steady state, i.e. after the pre-steady state and when the intermediates have reached a steady concentration. The duration of the steady state is relatively short, and this state is in any case an approximation since the accumulation of products and the depletion of substrate occur continuously from the start of the reaction. *See* **Michaelis kinetics**.

steapsin *an old term for* pancreatic lipase.

stearate 1 *a trivial name for* octadecanoate, $CH_3–[CH_2]_{16}–COO^-$, the anion derived from stearic acid, or octadecanoic acid, a saturated straight-chain higher fatty acid. **2** any mixture of free stearic acid and its anion. **3** any salt or ester of stearic acid. *See also* **stearoyl**.

stearin 1 *an old name for* any of the glyceryl esters of stearic acid, especially the triester, **tristearin** (tristearoylglycerol). **2** *a commercial name for* a grade of stearic acid containing other fatty acids, especially palmitic acid.

stearoyl *symbol*: Ste; *numerical symbol* 18:0; *a trivial name for* octadecanoyl, $CH_3–[CH_2]_{16}–CO–$, the acyl group derived from stearic acid (*see* **stearate**). A major fatty-acid component of plant and animal lipids; tallow and other hard animal fats contain up to 35%. Together with palmitate it represents a high proportion of the saturated fatty-acid content of dietary and other lipids. It is synthesized from palmitoyl-CoA by a microsomal chain-elongation enzyme. It is incorrectly referred to as **stearyl**.

stearoyl-CoA desaturase EC 1.14.99.5; *systematic name*: stearoyl-CoA,hydrogen donor:oxygen oxidoreductase; *other names*: acyl-CoA desaturase; fatty-acid desaturase; Δ^9-desaturase. An enzyme that catalyses a reaction between dioxygen, an electron donor and stearoyl-CoA to form oleoyl-CoA. It is an iron protein with a short half-life, and is the only inducible component of the endoplasmic reticulum fatty acyl-CoA desaturase system. It utilizes electrons from reduced cytochrome b_5.

stearyl 1 *a common name for* octadecyl, $CH_3–[CH_2]_{16}–CH_2–$, the alkyl group derived from octadecane; octadecyl is recommended. **2** *an old trivial name for* octadecanoyl; **stearoyl** is now recommended.

steatite *see* **talc**.

steato+ *comb. form* indicating presence or relation to fatty substances.

steatorrhoea the presence of abnormally large amounts of fat in the feces.

steatosis fatty degeneration of tissue.

steel factor *see* **stem cell factor**.

steel factor receptor *see* **kit**.

stefin *see* **cystatin**.

Stelazine *see* **trifluoperazine**.

stellacyanin a blue, copper-containing mucoprotein of low molecular weight obtained from the Japanese lacquer tree, *Rhus vernicifera*. It functions as an electron-transfer protein, accepting electrons from the cytochrome b_6–f complex, and donating them to **photosystem** I, and has been much studied for the structure of the copper-binding site.

STEM *abbr. for* scanning transmission electron microscope.

stem cell any member of the various groups of reserve cells that replace cells destroyed during the normal life of the animal, e.g. blood cells, epithelial cells, spermatogonia, and skin cells. Stem cells can divide without limit; after division, the stem cell may remain as a stem cell or proceed to terminal differentiation.

stem cell factor (*abbr.*: SCF) or **mast cell growth factor** or **steel factor** a **cytokine** that stimulates the proliferation of mast cells. It is produced by endothelial cells, fibroblasts, bone marrow cells, and Sertoli cells. A homodimer glycoprotein with type I membrane and soluble forms, it enhances the proliferation of both myeloid and lymphoid hemopoietic progenitors. Its receptor is the membrane receptor-tyrosine kinase encoded by the protooncogene c-*kit* (*see* **KIT**).

stem cell marker any molecule, especially one expressed at the cell surface, that is used to distinguish stem cells from other cells; e.g. CD34.

stem name any generic name for a category of similar compounds, enzymes, organisms, etc. that is used also as an affix in forming the specific names of individual members of such a category.

steno+ or (*before a vowel*) **sten+** *comb. form* indicating narrowness.

stenohaline describing an organism that is unable to tolerate wide variations in environmental osmotic pressure.

stenothermous describing an organism that is unable to tolerate wide variations in environmental temperature.

STEP *abbr. for* sieve tube exudate protein.

step gradient a discontinuous gradient, formed in a tube by layering solutions of different densities carefully one on the other. In column chromatography the same effect is achieved by pumping buffers of different composition sequentially through the column. *See* **gradient**.

steradian *symbol*: sr; the SI supplementary unit of solid angle. It is equal to the angle subtended at the centre of a sphere by an area of surface equal to the square of the sphere's radius. The surface of a sphere subtends an angle of 4π sr at its centre.

stercobilin or **stercobilin IXα** a linear red-brown tetrapyrrole formed in heme catabolism. Structurally it is an oxidation product of **stercobilinogen**. Metabolically, it is formed from urobilinogen by intestinal bacteria and is the major pigment of feces.

stercobilinogen *an alternative name for* 10,23-dihydrostercobilin; *former name*, stercobilinogen IXα. A product of heme catabolism; it is a **urobilinogen** molecule in which the double bonds between the methyl and ethyl substituents on the two end rings have been reduced.

sterculic acid *see* **cyclic fatty acid**.

stereo+ *comb. form* denoting solid or three-dimensional.

stereobase unit the smallest set of one, two, or more successive **configurational base units** that prescribes repetition in a polymer molecule.

stereochemistry the branch of chemistry dealing with the three-dimensional properties of molecules; the study of the properties of **stereoisomers**. *See* **cis–trans** isomerism, **Re/Si** convention, sequence rule, stereospecific numbering.

stereoheterotopic a term describing chemically-like ligands (or faces of double bonds) whose separate replacement with an achiral ligand (or for faces, addition of an achiral reagent) gives rise to two stereoisomers. If the stereoisomeric products are **enantiomers**, the

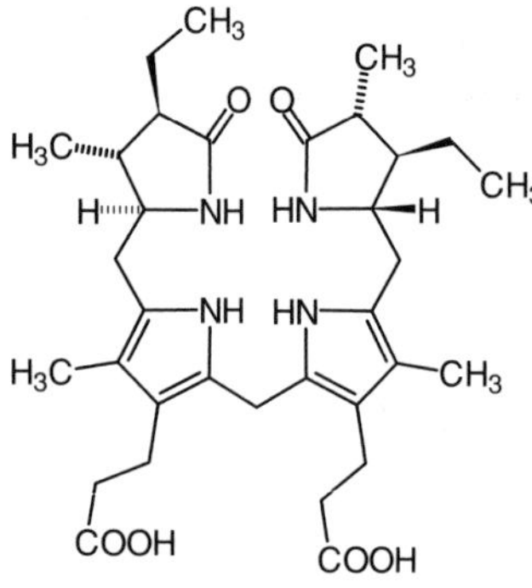

stercobilinogen

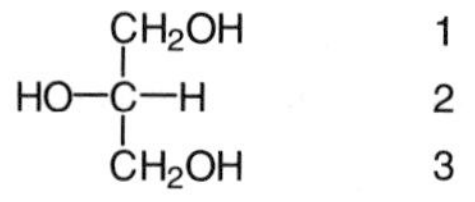

(1) glycerol

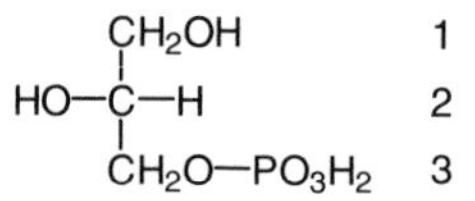

(2) sn-glycerol-3-phosphate

(3) citric acid

ligands (faces) are enantiotopic; if they are diastereoisomers, the ligands (faces) are diastereotopic. for more extensive descriptions, see **diastereotopic**, **enantiotopic**.

stereoisomer any of two or more **isomers** that have the same molecular constitution and differ only in the three-dimensional arrangement of their atomic groupings in space. Stereoisomers may be **diastereoisomers** or **enantiomers**. *Compare* **constitutional isomer**. —**stereoisomeric** *adj.*; **stereoisomerism** *n.*

stereology the body of mathematical methods dealing with the interpretation of the structure and dimensions of three-dimensional objects from information contained in two-dimensional representations, such as photomicrographs of tissue sections.

stereoregular polymer a regular polymer composed of molecules having a regular repetition of identical **stereobase units** at all stereoisomeric sites of the main chain.

stereoselective polymerization the formation of a polymer from a mixture of stereoisomeric monomer molecules by preferential incorporation of one species into a growing polymer chain.

stereospecific (of an enzymatic or other reaction system) acting on or producing a particular **stereoisomer**; stereochemically specific. —**stereospecificity** *n.*

stereospecific numbering a numbering system applied mostly to prochiral compounds, *Caabc*, with two similar atoms, or groups of atoms, *a*. It numbers the prochiral *a* groups and, by extension, their chiral derivatives. A stereospecifically numbered compound is designated by the prefix *sn-*. The Hirschmann system was devised particularly so that the ^{14}C-labelled glycerol giving rise to [3,4-^{14}C$_2$]glucose (in rats) could be termed [1-^{14}C] glycerol and thus be related, biosynthetically and by nomenclature, to [1-^{14}C]lactate. It uses the basic *R/S* system (*see* **sequence rule**) to set up a model, which is viewed away from the group with the lowest sequence priority. If this is *c* in *Caabc*, a counterclockwise ordering is examined for the remaining three groups, *baa*, starting with *b*. The first *a* group to be encountered receives the lower number. The numbering for glycerol is shown in structure 1. By extension, the glycerol phosphate formed by **glycerol kinase** becomes *sn*-glycerol 3-phosphate (structure 2). This structure has also been described as D-glycerol 3-phosphate, L-glycerol 3-phosphate, and L-α-glycerophosphate. The Hirschmann nomenclature has found wide acceptance in the lipid field. The Hanson and Hirschmann system expands the *R/S* system in a logical fashion (rather than using the arbitrary counterclockwise ordering described for glycerol). The two similar groups, *a*, of *Caabc* are first given *pro-R* and *pro-S* assignments. Then, following the established '*R* precedes *S* rule', *pro-R* precedes *pro-S*, and the *pro-R* group is assigned the lower number. This system has been applied to citric acid as shown in structure 3. Application of this system to glycerol unfortunately leads to the product of glycerol kinase being *sn*-glycerol 1-phosphate. Although this system has a logical basis in terms of established nomenclature, it seems unlikely that the *sn* system for glycerol and derivatives will be brought into line.

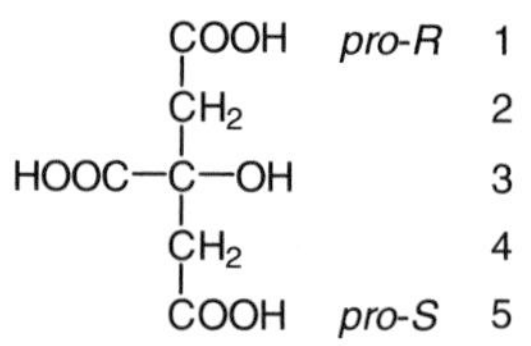

stereospecific polymerization a polymerization in which a **tactic polymer** is formed.

steric relating to effects involving arrangements of atoms in space.

steric hindrance 1 prevention of the free rotation of an atom or group in a molecular entity with respect to a connected one due to the relative sizes and/or spatial disposition of the atoms or groups in question. 2 restriction or prevention of the reaction of a molecular entity with another due to the relative sizes and/or spatial disposition of atoms or groups in either or both of the reactants.

sterilant a substance that destroys or eliminates all forms of microbial life in the inanimate environment, including bacteria, bacterial spores, fungi, fungal spores, and viruses. *Compare* **disinfectant**, **sanitizer**.

sterile 1 (of inanimate objects) free from microbiological or other life forms. 2 (of a living organism) incapable of sexual reproduction.

sterilize *or* **sterilise** 1 to treat in such a way as to kill undesired microorganisms without damaging the material under treatment. *Compare* **disinfection**. 2 to render incapable of sexual reproduction. —**sterilization** *or* **sterilisation** *n.*

Stern–Volmer equation an expression relating **fluorescence quenching** to the concentration of the quenching substance, *c*, and the fluorescence lifetime, τ. It is: $F_o/F = 1 + kc\,\tau$, where F_o is the fluorescence intensity when no quencher is present, *F* is the fluorescence intensity in the presence of quencher, and *k* is a constant that depends on the system; the concentration of the absorbing molecule must be the same in both quenched and unquenched systems. The **Stern–Volmer plot** is a plot of F_o/F against concentration of quenching substance, from the slope of which, if the fluorescence lifetime is known, the quenching constant can be obtained. [After Otto Stern (1888–1969) and M. Volmer].

steroid any of a large group of substances that have a ring system based on 1,2-cyclopentanoperhydrophenanthrene. The group includes such natural products as **bile acids**, **corticosteroids**, **sex hormones**, **sterols**, and various plant steroids. *See also* **vitamin D**.

steroid diabetes the combination of decreased glucose tolerance, fasting hyperglycemia, and glucosuria, found in patients exposed to excessive quantities of **glucocorticoids**. It may thus accompany **Cushing's syndrome**.

steroid-hormone receptor any intracellular receptor that binds steroid hormones and mediates their effects; the family also includes receptors for thyroid hormones, retinoids, and vitamin D. After the steroid enters the cell, the steroid–receptor complex is translocated to the nucleus and binds to a specific DNA sequence, a **response element**. The receptors contain a modulating N-terminal domain, a zinc finger DNA-binding domain, and a C-terminal steroid-binding domain.

steroid Δ-isomerase EC 5.3.3.1; *systematic name*: 3-oxosteroid $\Delta^5–\Delta^4$-isomerase; *other name*: $\Delta^5–\Delta^4$-ketosteroid isomerase. An enzyme involved in microbial transformation of steroids; it catalyses the reaction:

$$3\text{-oxo-}\Delta^5\text{-steroid} = 3\text{-oxo-}\Delta^4\text{-steroid.}$$

steroid 21-monooxygenase EC 1.14.99.10; *other names*: steroid 21-hydroxylase; cytochrome P450 21A1; an enzyme that catalyses a 21-hydroxylation of progesterone to 11-deoxycorticosterone and of 17-hydroxyprogesterone to 11-deoxycortisol in the fasciculata/reticularis region of the adrenal cortex. It is a heme–thiolate protein. Mutations that decrease its activity result in congenital adrenal hyperplasia: there is a failure to synthesize cortisol, and this results in overproduction of **corticotropin**, 17-hydroxyprogesterone, and androstenedione.

steroid 11β-monooxygenase EC 1.14.15.4; *other names*: steroid 11β-hydroxylase; cytochrome P45011B1 (*or* CYP11B1); cytochrome P45011B2 (or CYP11B2). Either of the enzymes found in the mitochondrial matrix of adrenal cortical cells that are involved in production of cortisol and aldosterone. CYP11B1 converts 11-deoxycorticosterone to corticosterone and 11-deoxycortisol to cortisol, in the zona fasciculata and zona reticularis. CYP11B2 also has **aldosterone synthase** activity, converting corticosterone to aldosterone in the zona glomerulosa. Mutations in CYP11B1 lead to congenital adrenal hyperplasia, whereas mutations in CYP11B2 lead to aldosterone deficiency.

steroid 17β-monooxygenase EC 1.14.99.9; *other names*: steroid 17β-hydroxylase; cytochrome P45017. An enzyme of the endoplasmic reticulum occurring in the zona fasciculata and zona reticularis of adrenal cortex and in the gonads. It catalyses the following conversions:

(1) progesterone → 17β-hydroxypregnenolone →
    17β-dehydroepiandrosterone;
(2) progesterone → 17-hydroxyprogesterone →
    androstenedione.

Both reactions involve sequential 17β-hydroxylation and 17,20-lyase reactions. Mutations that decrease both activities results in congenital adrenal hyperplasia.

steroidogenesis the biosynthesis of steroids, but often referring to the genesis of other steroids from cholesterol.

steroidogenic factor 1 *abbr.*: SF-1; a protein of the nuclear receptor family of zinc finger transcription factors that regulates tissue-specific expression of P450 steroid hydroxylases, 3β-hydroxysteroid dehydrogenase, ACTH receptor, StAR, and anti-Müllerian hormone. It is expressed in gonads and adrenal glands from the earliest stages of development, and in the hypothalamus and anterior pituitary gland of mouse and human. Gene knockout in mice results in lack of adrenal glands and gonads.

steroid 5α-reductase *or* 3-oxo-5α-Δ⁴-dehydrogenase either of two enzymes that catalyse the reaction: testosterone + NADPH + H⁺ → 5α-dihydrotestosterone + NADP⁺. One of the enzymes, 5α-reductase 1, has an alkaline pH optimum and is found in liver and skin. The other, 5α-reductase 2, has an acid pH optimum and is found in the male urogenital tract. Many mutations resulting in 5α-reductase 2 deficiency cause an autosomal recessive form of testicular feminization syndrome.

sterol any **steroid** that contains one or more hydroxyl groups and a hydrocarbon side-chain in the molecule.

sterol O-acyltransferase EC 2.3.1.26; *other names*: cholesterol acyltransferase; sterol-ester synthase; an intracellular enzyme that catalyses the formation of cholesterol ester from acyl-CoA and cholesterol. It is involved in the storage of cholesterol esters in phagocytic cells, a process that plays a part in producing lipid-laden cells in arterial walls. There are two isozymes in yeast. The enzyme belongs to the superfamily of membrane-bound O-acyltransferases.

sterol carrier protein 2 *abbr.*: SCP2; a protein (123 amino acids) required for intracellular sterol transport. It may be a cofactor for the endoplasmic reticulum enzyme 3β-steroid-Δ⁷-reductase (EC 1.3.1.21), which converts 7-dehydrodesmosterol to desmosterol in the final stages of cholesterol biosynthesis. SCP2 is produced from a short transcript that forms part of a long transcript encoding SCPx, and also by cleavage of SCPx in the peroxisomal matrix – the C-terminal region of SCPx contains SCP2.

sterol 27-monooxygenase *or* sterol 27-hydroxylase *or* **cytochrome P45027** *or* **CYP27**. A mitochondrial cytochrome P450 enzyme (498 amino acids) that catalyses hydroxylation of the C27 in the side chain of cholesterol or of dihydroxycholestane. This is the first step in the degradation of this side chain in the biosynthesis of bile acids. Many mutations lead to enzyme deficiency and cerebrotendinous xanthomatosis.

sterol-response-element binding protein *see* **SREBP**.

sterol-sensing domain a hydrophobic domain of proteins that binds cholesterol. *See* **SREBP**.

steryl-sulfatase EC 3.1.6.2; *other names*: steroid sulfatase (sulfatase); steryl-sulfate sulfohydrolase; arylsulfatase C; an enzyme that catalyses the hydrolysis of 3-β-hydroxyandrost-5-en-17-one 3-sulfate to 3-β-hydroxyandrost-5-en-17-one and sulfate.

stevioside 13-[(2-*O*-β-ᴅ-glucopyranosyl-α-ᴅ-glucopyranosyl)oxy]kaur-16-en-18-oic acid β-ᴅ-glucopyranosyl ester; a glycoside extracted from the leaves of *Stevia rebaudiana* (yerba dulce). It is used as a sweetening agent, being about 300 times as sweet as sucrose.

STH *abbr. for* somatotropic hormone; *see* **somatotropin**.

STI *abbr. for* soybean trypsin inhibitor.

Stickler syndrome *or* **arthro-ophthalmodystrophy** an autosomal dominant chondrodysplasia in which early joint degenerative disease is accompanied by disorders of the eye and deafness. Mutations in genes for collagens XIα-1, XIα-2, and IIα-1 are responsible. [After Gunnar Stickler (1925–), US physician.]

sticky end *see* **cohesive end**.

stiff-man syndrome *or* **stiff-person syndrome** a rare progressive rigidity of the body musculature. In ≈60% of patients autoantibodies to glutamate decarboxylase are present in plasma and cerebrospinal fluid, and in ≈58% of patients diabetes mellitus type 1 is also present. The muscular hyperactivity is suppressed by sleep, anesthesia, GABA agonists, or curare.

stigmasterol 3β-hydroxy-24-ethyl-Δ⁵,²²-cholestadiene; a **phytosterol** structurally related to β-sitosterol (*see* **sitosterol**) and found in plant oils.

stilbene *or* **1,2-diphenylethene** an unsaturated hydrocarbon, $C_6H_5CH=CHC_6H_5$, several derivatives of which occur in plants, where they protect against bacterial and fungal pathogens, discourage herbivory, inhibit seed germination, and function as growth inhibitors and in dormancy. Combrestatin and resveratrol are substituted stilbenes.

stilbestrol *or (esp. Brit.)* **stilboestrol** 4,4′-dihydroxystilbene; a synthetic nonsteroid compound, first prepared in 1938, that became a source of a range of compounds with estrogenic activity, of which *α,β*-diethylstilbene (*see* **diethylstilbestrol**) is the most active. Stilbestrol has been used as a growth promoter in farm animals but is now banned in most countries. It is still used to treat hormonal disorders in small animals. It was formerly used to treat prostatic cancer but there is some evidence for its carcinogenicity.

trans-stilbestrol

still an apparatus for the distillation of liquids, including the preparation of distilled water. It consists of a heated vessel, a condenser, and a receiver. A continuous apparatus incorporates an automatic feed for the liquid in question, a heating element, and a thermoregulator.

stimulate *(in physiology)* to provide a **stimulus** to any tissue or cell leading to its excitation or to initiation of a sequence of events with some defined endpoint, involving an electrical event, activation of a metabolic pathway, or initiation of cell division or growth. —**stimulation** *n.*; **stimulative** *adj.*

stimulus *(in physiology)* any event or phenomenon, such as radiation, electrical potential, or addition of molecules that leads to excitation of a tissue or cell.

stimulus–response coupling a molecular mechanism, often involving **signal transduction**, that transmits a stimulus from the exterior of a cell's plasma membrane, in many cases through **second messenger** pathways, to effector systems that bring about a response. For example, in **stimulus–secretion coupling** the effector system brings about a secretion event such as degranulation of mast cells or release of hormone (e.g. insulin) from storage granules to the cell exterior.

stimulus–secretion coupling *see* **stimulus–response coupling**.

stochastic arrived at by skilful conjecture; e.g. a stochastic model, a stochastic process.

stock solution a solution of a reagent, at a stable or convenient concentration, from which appropriate dilutions can be made at the time of use.

Stock system *(in inorganic nomenclature)* a system of designating the **oxidation number** of an element by placing a Roman numeral in parentheses immediately following the name of the element or, where symbols are used, in superscript to the right of the symbol; e.g. manganese(IV) oxide, $Mn^{IV}O_2$.

stoichiometry *or (rarely)* **stoicheiometry** the quantitative relationship of the reactants and products of a chemical reaction in the proportions that they appear in the chemical equation describing the reaction. —**stoichiometric** *or* **stoicheiometric** *adj.*

stokes *symbol*: St; the cgs unit of **kinematic viscosity**; its use is now discouraged. In SI units, kinematic viscosity is measured in square metres per second; $1\ St = 10^{-4}\ m^2\ s^{-1}$. [After (Sir) George Gabriel Stokes (1819–1903), British mathematician.]

Stokes' law of fluorescence a law stating that the wavelength of the emitted fluorescent light, λ_{em}, is greater than that of the light exciting the fluorescence, λ_{ex}. This relationship is true for most fluorescent materials; those for which it does not hold are termed **anti-Stokes**.

Stokes' law of viscosity a law stating that the frictional coefficient, f, of a spherical particle of radius r, moving through a liquid of viscosity η, is given by $f = 6\pi\eta r$.

Stokes' loss the loss of excitation energy available for fluorescence due to collision of molecules in the first excited state, S_1, with their neighbours, which results in a lower vibration level of S_1.

Stokes' shift the difference in wavelength between the excitation and emission maxima for a particular fluorescent substance. In quantitative form, Stokes' shift is $10^7(1/\lambda_{ex} - 1/\lambda_{em})$, where λ_{ex} and λ_{em} are the corrected maximum wavelengths for excitation and emission expressed in nanometres.

stoma (*pl.* **stomata**) any of the pores in the epidermis of plants, particularly in leaves, through which gaseous exchange takes place.

stomatin *or* **band 7.2b protein** a major membrane protein of human erythrocytes that contains a central domain also found in prohibitins and flotillin. It is lacking in an autosomal dominant form of hemolytic anemia (hereditary stomatocytosis).

stop codon any of the trinucleotide **codons**, UGA, UAG, and UAA, that signal the termination of translation of a messenger RNA molecule and the release of the nascent polypeptide chain. *See also* **termination codon**.

stopped-flow technique a method in which two solutions are caused to flow into a mixing chamber and then into an observation chamber, after which the flow is stopped abruptly and the chemical reaction is followed, usually spectroscopically, as the mixture in the observation chamber ages. Critical features include the quality of mixing, the speed with which the mixed solution fills the observation chamber, and the geometry of the observation chamber in relation to the sensitivity of observation. *See also* **rapid-reaction kinetics**.

STOP protein *abbr. for* stable tubule only polypeptide; a protein that blocks the endwise dissociation of microtubules.

stop-transfer protein *or* stop-transfer sequence *see* signal peptide.

storage granule any small **organelle**, bounded by a lipid-bilayer membrane, that contains stored material. For example, mast-cell storage granules contain histamine, while those of pancreatic B cells contain insulin; in both cases the stored material is destined for secretion. This involves a stimulus that causes fusion of the granule membrane with the plasma membrane, followed by release of the granule contents by **exocytosis**.

storage polysaccharide any polysaccharide that serves as a form of stored energy in living organisms. Storage polysaccharides include starch, phytoglycogen (e.g. in maize), and fructosans (e.g. inulin) in plants, and glycogen in animals.

stp *or* **s.t.p.** *or* **STP** *abbr. for* standard temperature and pressure.

straight-chain describing any chain of carbon atoms in which none of the carbon atoms is directly bonded to more than two other carbon atoms.

strain 1 a group of related individuals having certain characters that distinguish the members from other such groups within the same species or variety; a race. **2** a line of organisms descended or derived from a particular ancestral individual. **3** *(in physics)* the temporary or permanent deformation of a body resulting from an applied stress.

streptavidin a tetrameric biotin-binding protein (subunit M_r 14 500), produced by *Streptomyces avidinii*, that binds up to four molecules of **biotin** per molecule. It is used in enzyme-linked immunosorbent assay (ELISA), radioimmunoassay, immunocytochemistry, and protein blotting. It is also used to detect DNA probes that contain bound biotin. The streptavidin may be labelled with **FITC**, gold, peroxidase, or another agent. Alternatively, a biotin-labelled detection system (such as alkaline phosphatase) can be bound to streptavidin through another of its biotin-binding sites. Streptavidin has the advantage over **avidin** of having a near-neutral isoelectric point (7.25–7.45), with consequently less nonspecific binding. *See also* **biotinylation**.

streptidine 1,3-diguanido-2,4,5,6-cyclohexanetetrol; the non-carbohydrate component of **streptomycin**.

streptobiosamine 5-deoxy-2-*O*-[2-deoxy-2-(methylamino)-α-L-glucopyranosyl]-3-*C*-formyl-L-lyxose; a disaccharide component of **streptomycin**, containing **streptose** and *N*-methylglucosamine.

R = NHCH₃

Streptococcus a genus of Gram-positive, facultatively or obligately anaerobic cocci or coccoid bacteria that produce **streptokinase**, **streptolysin**, and **streptodornase**.

streptodornase any of at least four deoxyribonucleases produced by streptococci. They have the characteristics of **deoxyribonuclease I** (EC 3.1.21.1), cleaving DNA to 5′-phosphodinucleotide and 5′-phosphooligonucleotide end-products. [From streptococcal deoxyribonuclease.]

streptokinase a noncatalytic protein secreted by certain streptococci that binds to and activates **plasminogen, thus converting it** to plasmin, which in turn lyses fibrin clots. It has been used therapeutically after heart attacks and strokes. It shows no sequence similarity to staphylokinase.

streptolydigin an antibiotic, derived from *Streptomyces lydicus*, that is particularly potent against anaerobes and some Gram-positive aerobes. It inhibits **RNA polymerase, thereby** reducing the rate of transcription, and slowing the rate of phosphodiester bond formation without affecting the fidelity of transcription. This contrasts with **rifampicin**, which inhibits initiation. *Compare* **rifamycin**.

streptolysin either of two **hemolysins** produced by certain streptococci (particularly of group A). **Streptolysin-O** (oxygen-sensitive) is produced in serum-free medium. Its hemolytic activity is destroyed by oxygen (but restored by certain mild reducing agents, e.g. 2-mercaptoethanol) and inhibited by free cholesterol. It is immunogenic, the level of circulating antibody (**antistreptolysin-O**, ASO) being useful in the diagnosis of streptococcal infections. Streptolysin-O can also lyse lysosomes. It resembles O-toxin of *Clostridium perfringens*, and can be used as an agent to permeabilize cells. **Streptolysin-S** (serum-dependent) is oxygen-insensitive and its production is promoted by serum proteins. Its hemolytic activity is inhibited by phosphatidyl ethanolamine at low concentration. It is nonimmunogenic, and is also toxic to leukocytes.

Streptomyces a genus of filamentous, spore-forming, Gram-positive bacteria that are largely responsible for the 'earthy' smell of soil but are chiefly remarkable for the diversity of secondary metabolites that they produce, including clinically useful antibiotics, e.g. **streptomycin**. They are amenable to genetic manipulation with many plasmid and phage vectors; this provides a unique opportunity for overexpression and for the design and synthesis of novel and 'hybrid' antibiotics. *See also* **streptomycete**.

streptomycete strictly, any member of the bacterial genus *Streptomyces*. However, the term may include certain related types, including *Streptoverticillum* spp.

streptomycin an **aminoglycoside antibiotic** produced by the soil bacterium *Streptomyces griseus*, having a relatively broad spectrum of activity but particularly active against mycobacteria (especially *Mycobacterium tuberculosis*), enterobacteria, and staphylococci. It binds the 30S ribosomal subunit and distorts the A site (**aminoacyl site**), thereby causing miscoding or prevention of binding of aminoacyl-tRNA. The components of streptomycin are **streptidine**, **streptose**, and *N*-methyl-L-glucosamine.

streptose L-streptose; 5-deoxy-3-*C*-formyl-L-lyxose; a component of **streptomycin** and **streptobiosamine**.

streptovaricin any **ansamycin** antibiotic obtained from *Streptomyces spectabilis*. *Compare* **rifamycin**.

Streptovaricin	W	X	Y	Z	R
A	OH	OH	Ac	OH	COOCH$_3$
B	H	OH	Ac	OH	COOCH$_3$
C	H	OH	H	OH	COOCH$_3$
D	H	OH	H	H	COOCH$_3$
E	H	O=	H	OH	COOCH$_3$
G	OH	OH	H	OH	COOCH$_3$
J	H	OAc	H	OH	COOCH$_3$
K	OH	OAc	H	OH	COOCH$_3$

streptozocin   2-deoxy-2-(3-methyl-3-nitrosoureido)-D-glucopyranose; a broad-spectrum antibiotic, antineoplastic, and diabetogenic substance isolated from *Streptomyces achromogenes*. It has a rapid and specific cytotoxic effect on the B cells in pancreatic islets, and is used to induce experimental diabetes in animals and for chemotherapy of insulinoma or other islet-cell neoplasms in humans. It acts as a methylating agent for DNA.

α-D-streptozocin

stress-activated protein kinase (*abbr.*: SAPK) *an alternative name for* **JNK**.

stress-associated protein kinase *abbr.*: SAPK; *an alternative name for* **JNK**.

striated muscle muscle in which the **thick filaments** and **thin filaments** are arranged in microscopically visible bands producing a striped appearance. Such an arrangement is typical of skeletal muscle and cardiac muscle. *Compare* **smooth muscle**. *See also* **sarcomere**.

strict aerobe *see* **aerobe**.

strict anaerobe *see* **anaerobe**.

stringency (*in molecular biology*) the rigour with which the ability of complementary DNA sequences to recognize each other (and thus to hybridize) is tested. It is affected by temperature, ionic strength, and the presence of destabilizing agents such as formamide. As temperature is increased it becomes less likely that more distantly related sequences will hybridize; hence, high stringency is associated with high temperature. Low ionic strength also gives high stringency.

stringent very rigorous. *See also* **stringency**, **stringent control**, **stringent plasmid**, **stringent response**.

stringent control a mechanism of control of plasmid replication whereby the plasmid is replicated only once for each time the chromosome is replicated, thus regulating the plasmid copy number.

stringent factor a protein associated with ribosomes, and encoded by the gene ***relA***, that is associated with the **stringent response**. It is an enzyme that donates a pyrophosphate group to 3′ of either 5′-GTP or GDP to produce pppGpp and ppGpp, which probably act as allosteric modulators mediating the stringent response. *See also* **relaxed mutant**, **spoT**.

stringent plasmid a plasmid that has a limited number of copies per cell.

stringent response a response by bacteria to poor growth conditions (e.g. lack of amino acids) in which the cell shuts down protein synthesis and other metabolic activity. *See also* **relaxed mutant**, **stringent factor**.

strip 1 to remove adherent, admixed, or combined substances or material from a fundamental, underlying, or required substance or structure. **2** (*in molecular biology*) **a** to remove polysomes from rough endoplasmic reticulum *in vitro* to produce **stripped membranes** (*see* **degranulation** (def. 2)). **b** to remove ribosomal proteins from a ribosome to produce a **core particle**. **c** to remove the amino acid from an aminoacyl-transfer RNA molecule to produce **stripped transfer RNA**. **d** to remove hybridized **probe** (def. 3) from a **transfer membrane** prior to reprobing. **3** (*in biochemistry*) to remove endogenous 2,3-bisphosphoglycerate from a hemoglobin solution to produce **stripped hemoglobin**. **4** to remove liquid components from a gaseous mixture.

stripped hemoglobin a hemoglobin solution from which endogenous 2,3-bisphosphoglycerate has been removed, e.g. by gel filtration, or dialysis at high NaCl concentrations.

stripped membrane rough endoplasmic reticulum from which polysomes have been removed *in vitro* by treatment with EDTA.

stripped transfer RNA a molecule of aminoacyl-transfer RNA from which the amino acid has been hydrolytically removed.

stRNA *abbr. for* small temporal RNA; a form of microRNA (*see* **miRNA**) that is ≈22 nt long and derived from dsRNA by the endonuclease activity of **Dicer** and **Argonaute**. It is highly conserved in certain animals and controls developmental events.

stroma (*pl.* **stromata**) **1** fibrous connective tissue or other intercellular material that forms the structural framework of an organ, as opposed to the functional tissue (parenchyma). **2** the spongy framework of protein strands within a red blood cell in which the hemoglobin is packed. **3** (*in botany*) the space enclosed by the double membrane of a **chloroplast** but excluding the thylakoid space. It contains DNA, ribosomes, and some temporary products of photosynthesis. **4** (*in mycology*) a mass of interwoven hyphae in which fruiting bodies (perithecia) develop. —**stromal** *adj.*

stromelysin EC 3.4.24.17; *other names*: matrix metalloproteinase-3 (*abbr.*: MMP-3); transin-1; an enzyme that catalyses the degradation of fibronectin, laminin, and some gelatins and collagens. Zn^{2+} and Ca^{2+} are cofactors. It is a glycoprotein of the extracellular matrix. Stromelysin 1 is secreted by fibroblasts; stromelysin 2 is secreted by keratinocytes.

strong acid an **acid** that remains ionized except at the lowest pH values, i.e. one that has a large **acid dissociation constant**.

strong base a **base** that remains ionized except at the highest pH values, i.e. one that has a large **basic dissociation constant**.

strophanthin a highly toxic glycoside mixture produced by *Strophanthus kombe*, with actions similar to those of **digoxin** and **ouabain**.

structural gene any gene that carries the genetic information for the structure of a protein. One or more structural genes may be contained in an **operon**. *Compare* **regulator gene**.

structural genomics the systematic effort to determine the 3D structures of all proteins encoded in complete genomes.

structural polysaccharide any polysaccharide of plant cell walls that performs a structural function. There are three solubility classes: **pectins**, extractable with dilute acid; **hemicelluloses**, extractable with aqueous alkali; and **microfibrillar polysaccharides**.

structure-based alignment the alignment of protein sequences made on the basis of structure comparison. Structure-based alignments do not always respect patterns of sequence conservation and may consequently differ from sequence alignments. They should not be considered the 'right' answer (or 'gold standard') – they merely reflect a different model of protein comparison, usually used for different purposes from those of sequence-based alignments. *See* **structure comparison**.

structure classification database a database in which protein structures are classified according to their geometrical and evolutionary similarities. Structures tend to be classified at the level of their individual domains, but multidomain structures can also be classified into evolutionary families and superfamilies. Examples of such databases include SCOP (structural classification of proteins) and CATH (class architecture topology homology).

structure comparison the comparison of protein folds by superposition of equivalent alpha or beta carbon atom coordinates. Equivalence of positions is largely determined by comparing the similarities of the properties or relationships of the residues or secondary structures. The quality of the structural superposition is measured in terms of the root mean square deviation (RMSD) (typically 4Å or less for similar folds). Structure comparison is important, e.g., for detecting relationships between proteins that are not evident at the sequence level owing to their extreme divergence.

structure database a database that houses the 3D structural coordinates of biological macromolecules, as determined by X-ray crystallography, nuclear magnetic resonance (NMR), electron microscopy and/or molecular modelling. The PDB (Protein Data Bank) is the global repository for macromolecular structure information of the RCSB (Research Collaboratory for Structural Bioinformatics). The MSD (Macromolecular Structure Database), derived in part from the PDB, is the structure database of the European Bioinformatics Institute (EBI). The MMDB (Molecular Modelling Database) is the NCBI's (National Center for Biotechnology Information) subset of 3D structures obtained from the PDB, excluding theoretical models. The PDBj (Protein Data Bank Japan) is the subset of PDB from the Institute for Protein Research in Osaka University.

structure factor a parameter describing the scattering of electromagnetic radiation (e.g. X-rays) by an electron. It is the ratio of the radiation scattered by any real sample to that scattered by a single electron at the origin.

strychnine a bitter alkaloid obtained from seeds of the Indian tree *Strychnos nux-vomica*. It is highly toxic, and is sometimes used to poison moles, rodents, and other vermin. It causes convulsions by antagonizing the inhibitory transmitter, glycine, in the spinal cord (*see* **glycine receptor**). *See also* **brucine**.

STS *abbr. for* sequence-tagged site.

Stuart factor *or* **Stuart–Prower factor** *an alternative name for* factor X; *see* **blood coagulation**. [After a Mr Stuart, in whom the deficiency was first described.]

Student's *t* test a statistical test, based on a distribution due to William Sealy Gosset (1876–1937) writing in 1908 under the pseudonym 'Student', that can be applied to small samples of data, *x*. In this distribution *t* depends only on the sample size, *N*, and not on the variance of the population; as *N* increases, the distribution curve of *t* approaches a normal distribution curve more closely.

STX *abbr. for* saxitoxin.

sU *symbol for* a residue of the ribonucleoside thiouridine (when the position of the thiol group in its thiouracil moiety is unspecified or unknown; alternative to S or Srd).

s²U *symbol for* a residue of the ribonucleoside 2-thiouridine (alternative to ²S).

s⁴U *symbol for* a residue of the ribonucleoside 4-thiouridine (alternative to ⁴S).

sub *jargon term for* **subculture** (def. 1).

sub+ *prefix* **1** beneath, under, at the bottom (of). **2** immediately below in order, rank, or status; secondary. **3** less than, partial(ly), imperfect(ly). **4** in place of. **5** denoting a subordinate part or subdivision of a whole. **6** (*in chemical nomenclature*) designating a compound that contains less than the usual proportion of a specified element; e.g. carbon suboxide, C_2O_3. *Compare* **infra+**, **super+**.

subcellular describing organelles, functional units, or other components that are found within cells or derived from cells.

subcellular organelle a pleonasm that is commonly used to describe **organelles** (which by definition are subcellular).

subclone to clone into a new vector a subfragment of part of a larger cloned DNA, after excising the required fragment using restriction endonucleases.

subculture 1 to prepare a fresh culture from a small sample of an existing culture. **2** a culture prepared in such a way.

subcutaneous *abbr.*: s.c.; below the skin.

suberin the polymeric substance that covers the epidermal cell layer in the underground parts of plants (e.g. roots and tubers), and in stem epidermis, cork cells of the periderm, and surfaces of wounded cells. It is a polyester containing very-long-chain ω-hydroxy fatty acids and dicarboxylic acids, and phenolics such as *p*-coumaric and ferulic acids. The core is lignin-like and has attached long-chain hydrocarbons.

sublime to vaporize from the solid without passing through a liquid phase. —**sublimation** *n.*

submaxillaris protease an endopeptidase isolated from the submaxillary gland of mice. It cleaves Arg-|-Xaa bonds.

submaxillary relating to the region beneath the upper jaw (maxilla).

submaximal below the maximal.

submitochondrial particle any particle derived by disruption of mitochondria. *See* **H⁺-transporting ATPase, complex I, II, III, IV**.

submolecular describing any chemical or chemical component that exists or functions at a lower level of complexity than that of intact individual molecules. The term may be used in reference to electrons, ions, free radicals, or other molecular fragments. *Compare* **supramolecular**.

subnatant 1 lying under. **2** a liquid lying under a **supernatant** or under solid material. *Compare* **infranatant**.

suboptimal below or failing to reach the optimal condition.

subsite a site, e.g. on a protein, that has a specific function within the overall function of a larger site.

substance K *see* **neurokinin A**.

substance K receptor *or* **neurokinin A receptor** *or* **NK-2 receptor** a G-protein-associated membrane receptor that binds substance K and mediates its intracellular effects by activating the **phosphatidylinositol cycle**.

substance P *abbr.*: SP; an 11-residue amide,

H-Arg-Pro-Lys-Pro-Gln-Gln-Phe-Phe-Gly-Leu-Met-NH₂,

found in most tissues of the mammalian body but especially in nervous tissue (both central and peripheral) and in gut tissues. It also occurs in the central nervous system of many other vertebrtates. Derived from **protachykinin β precursor**, it is a **tachykinin**; it causes hypotension and vasodilation, and stimulates contraction of the intestine and secretion of saliva. *See also* **tachykinin receptor**.

substituent group *or* **substituent** (*in organic-chemical nomenclature*) any atom or **group** (def. 1) that has replaced or can be considered to have replaced (i.e. been substituted for) a hydrogen atom in a parent molecule; for bivalent or tervalent groups there are corresponding replacements of two or three hydrogen atoms.

substitute cysteine accessibility method *see* **SCAM**.

substituted mechanism a mechanism of an enzyme reaction in which a substrate reacts with the enzyme to produce a modified enzyme and a product. The modified enzyme then reacts with a second substrate to produce a second product and regenerate the enzyme, e.g.

$$E + S_1 \rightleftharpoons E^* + P_1,$$

then

$$E^* + S_2 \rightleftharpoons E + P_2,$$

where E is the enzyme, E* the modified enzyme, S_1 and S_2 the substrates, and P_1 and P_2 the products.

substitution 1 *(in genetics)* a type of point mutation in which one base pair is replaced by another. In a **transition**, a base is replaced by another of the same chemical class (e.g. a purine by a purine, or pyrimidine by another pyrimidine); in a **transversion**, a base is replaced by one of a different type (e.g. purine by a pyrimidine, or vice versa). **2** *(in sequence analysis)* the replacement of a nucleotide or amino acid residue by another at a given position in an alignment. For protein sequences, if the aligned residues have similar physicochemical properties, the substitution is said to be conservative (e.g. Ile and Leu).

substitution matrix *or* **amino acid exchange matrix** *or* **mutation matrix** a symmetrical 20×20 matrix containing values proportional to the probability that amino acid *i* mutates to amino acid *j* for all pairs of amino acids; the diagonal denotes the odds for self-conservation. The most widely used mutation or substitution matrices are the PAM (point accepted mutation) and BLOSUM (Blocks substitution matrix) series. PAM matrices were calculated from aligned sets of closely related sequences, and larger evolutionary distances were extrapolated from the original 1-PAM matrix. BLOSUM matrices were derived from alignments of larger, more diverse sets of sequences. Where the alignments are large enough to be statistically significant, the resulting matrices should reflect the true probabilities of mutations occurring during a particular evolutionary period.

substitution reaction the replacement of a substituent (normally hydrogen), usually at a carbon atom, by any other group.

substrate 1 a substance that is acted upon, especially by an enzyme; a molecule or structure whose transformation is catalysed by an enzyme. **2** *or* **substratum a** the base upon which an organism grows or lives, e.g. soil or rock. **b** the material upon which a microorganism grows or is placed to grow.

substrate activation the increase of an enzyme's activity by its substrate by an **allosteric** mechanism. *Compare* **substrate inhibition**.

substrate cycle interconversion of reactant and product by two enzymes, one of which (A) catalyses the reaction in one direction while the second enzyme (B) catalyses the reaction in the opposite direction, e.g.

$$\text{fructose 6-phosphate} \ldots \overset{A}{\underset{B}{\rightleftharpoons}} \text{f fructose 1,6-bisphosphate.}$$

As **flux** (def. 3) (i.e. the rate of reaction catalysed by A minus rate of reaction catalysed by B) through a system varies, the rate of each enzyme reaction may vary independently with concomitant changes in cycling rate. The existence of a substrate cycle increases sensitivity of control, since even if rates of forward and reverse reaction are high, changes in one or both may make relatively larger changes to flux; e.g. if rate of A = 9 and rate of B = 8, then flux = 1; if rate of A doubles to 18 and rate of B remains 8, then flux = 10 (i.e. increases tenfold).

substrate dissociation constant *or* **substrate constant** *symbol* K_s; the parameter of an enzymatic reaction that is the equilibrium (dissociation) constant for the reaction ES = E + S; i.e. $K_s = [E][S]/[ES]$.

substrate inhibition the inhibition of an enzyme's activity by its substrate by an **allosteric** mechanism. *Compare* **substrate activation**.

substrate-level phosphorylation the formation of ATP (or GTP) in a coupled reaction that is not linked to an electron transport system, e.g. coupled to the deacylation of succinyl coenzyme A or the conversion of 3-phosphoglyceroyl phosphate to 3-phosphoglycerate.

substrate-site hypothesis a hypothesis for the control of insulin release from pancreatic islet B cells by glucose. It postulates that the glucoreceptor is either a carrier transporting glucose into the cell or an enzyme phosphorylating glucose, and the signal for insulin release is a metabolite in the pathway of metabolism of glucose in the B cell. *Compare* **regulator-site hypothesis**.

substratum (*pl.* **substrata**) **1** any layer (stratum) or structure that lies beneath another, or that forms a base or underlying support for something else. **2** *an alternative term for* **substrate** (def. 2).

subtligase any engineered derivative of **subtilisin** that efficiently ligates peptides. The two mutations converting the peptidase, subtilisin, to a ligase involve Ser[221] to Cys and Pro[225] to Ala.

subtilin *or* **subtilin C** an antibiotic produced by *Bacillus subtilis* and active against Gram-positive bacteria, *Neisseria* spp. (*N. catarrhalis*, *N. gonorrhoeae*), and some pathogenic fungi.

subtilisin EC 3.4.21.62; a bacterial **serine endopeptidase** enzyme from *Bacillus subtilis* and other *Bacillus* spp. (e.g. commercially from *B. licheniformis*). It has broad specificity for peptide bonds, with preference for a large uncharged residue at P1 (*see* **peptidase P-sites**) and low activity for glutamyl residues; it hydrolyses peptide amides. The enzyme has similar active-site geometry to the trypsin family but otherwise no homology. *See also* **Nagarse proteinase**.

subtractive cloning a technique for isolating genes expressed in one cell population but not in another. Nucleic acid (typically first-strand cDNA) from which one seeks to isolate differentially expressed genes is hybridized to a large excess of nucleic acid (typically poly(A)$^+$ mRNA) from a source that does not express those genes. Only sequences common to the two sources will hybridize and these are removed (typically by chromatography on hydroxyapatite) leaving a pool of nucleic acid enriched in the sequences of interest. This pool can be used to make a cDNA library from which clones of interest may be isolated.

subtractive hybridization a method to identify DNA or RNA that is present in one sample but not in others, and forming the basis of **subtractive cloning**.

subtrahend *(in mathematics)* the number to be subtracted from another number (the minuend).

subunit any polypeptide component within a protein, or other distinct biochemical entity, whether identical to or different from any other components, that is separable from such other components without rupture of covalent bonds. The term is sometimes used in a more restricted sense as referring to any chemically or physically identifiable submolecular entity within a protein, whether identical to, or different from, other components.

subunit-exchange chromatography a form of chromatography in which protein subunits are immobilized on a solid matrix and allowed to interact with protein subunits in solution. Quantification of subunit exchange between the matrix and solution may be used in analysing the association–dissociation properties of the protein system. The method also provides a powerful, specific purification method.

Suc– *symbol for* **succinyl** (def. 1).

–Suc– *or* **Suc<** *symbol for* **succinyl** (def. 2).

succinamic acid *a trivial name for* 3-carbamoylpropanoic acid, the monoamide of succinic acid.

succinamide *a trivial name for* butanediamide, the diamide of succinic acid.

succinate 1 *a trivial name for* butanedioate; ethanedicarboxylate; the dianion of succinic acid. It is an important intermediate in metabolism and a component of the **tricarboxylic-acid cycle**. **2** any mixture of free succinic acid and its mono- and dianions. **3** any salt or ester of succinic acid.

succinate–CoA ligase (GDP-forming) EC 6.2.1.4; *other name*: succinyl-CoA synthetase (GDP-forming); an enzyme of the tricarboxylic-acid cycle that catalyses the formation of succinyl-CoA from GTP, succinate, and CoA with release of GDP and orthophosphate.

succinate dehydrogenase EC 1.3.99.1; *systematic name*: succinate:(acceptor) oxidoreductase; *other names*: fumarate reductase/dehydrogenase. A bacterial enzyme, or a degraded entity from succinate dehydrogenase (ubiquinone), a major component of mitochondrial complex II (*see* **respiratory complex**). It catalyses the oxidation of succinate by an acceptor to form fumarate and a reduced acceptor; it does not react with ubiquinone or free FAD.

succinate dehydrogenase (ubiquinone) EC 1.3.5.1; *systematic name*: succinate:ubiquinone oxidoreductase; *other name*: succinic dehydrogenase. An enzyme that catalyses the oxidation of succinate by ubiquinone to form fumarate and ubiquinol. It is an iron–sulfur flavoprotein (FAD), containing at least four different subunits: a flavoprotein, an iron–sulfur protein, and two hydrophobic anchor proteins, one of which is cytochrome b_{556}. It is located in the mitochondrial inner membrane, where it is a major component of complex II of the respiratory chain.

succinate-semialdehyde dehydrogenase EC 1.2.1.24; a homotetrameric enzyme of mitochondrial matrix that catalyses the reaction:

$$\text{succinate semialdehyde} + NAD^+ + H_2O =$$
$$\text{succinate} + NADH.$$

Succinate semialdehyde can undergo transamination to form γ-aminobutyric acid. However, the transaminase primarily functions to convert γ-aminobutyrate to succinate-semialdehyde, which is then converted to succinate by succinate-semialdehyde dehydrogenase in the catabolism of γ-aminobutyrate. Mutation results in 4-hydroxybutyric aciduria, which is associated with retarded mental, motor, and language development. *See also* **4-aminobutyrate transaminase**.

succinic dehydrogenase *see* **succinate dehydrogenase (ubiquinone)**.

succinimide *a trivial name for* butanimide; 2,5-pyrrolidinedione; the imide of succinic acid. For structure *see* **+imide**.

succinoglycan *or* **EPS I** an acidic exopolysaccharide produced by *Rhizobium* spp. that is important for invasion of the nodules that the bacteria elicit on their host, *Medicago sativa*. Succinoglycan is a high-molecular-weight polymer composed of repeating octasaccharide subunits. These subunits are synthesized on membrane-bound isoprenoid lipid carriers, beginning with a galactose residue followed by seven glucose residues, and modified by the addition of acetate, succinate, and pyruvate.

succinyl 1 *(in biochemistry) symbol*: Suc–; the univalent acyl group 3-carboxypropanoyl, $HOOC–[CH_2]_2–CO–$, derived from succinic acid by loss of one hydroxyl group. **2** *(in chemistry) symbol*: –Suc– *or* Suc<; the bivalent acyl group butanedioyl, $–CO–[CH_2]_2–CO–$, derived from succinic acid by loss of both hydroxyl groups.

succinylation the introduction of one or more **succinyl** (def. 1) groups into a substance by acylation, either chemically, e.g. with succinic anhydride, or enzymically, e.g. with succinyl-coenzyme A. It is used to acylate a protein's free lysine (and other) amino groups to change their charge at neutral pH from positive to negative and to render adjacent peptide bonds resistant to hydrolysis by trypsin; it can also cause disaggregation of a multimeric protein. The reaction occurs readily at pH 7–10 (at 0–25 °C), but it is irreversible, unlike the otherwise similar processes of **citraconylation** and **maleylation**, and so is of limited usefulness in protein sequence work.

succinylcholine chloride *see* **suxamethonium**.

succinyl-CoA synthetase (GDP-forming) *see* **succinate–CoA ligase (GDP-forming)**.

***O*-succinylhomoserine** an intermediate in methionine biosynthesis in plants and bacteria. It is formed by succinylation (utilizing succinyl-CoA) of homoserine. It reacts with cysteine with loss of succinate to form cystathionine.

sucralose 4,1′,6′-trichloro-4,1′,6′-trideoxy-*galacto*-sucrose; a chlorinated sucrose, tasting sweeter than sucrose itself.

sucrase 1 *recommended name*: sucrose α-glucosidase; EC 3.2.1.48; *systematic name*: sucrose α-D-glucohydrolase; an α-D-glucosidase-type enzyme from intestinal mucosa that hydrolyses sucrose and maltose. It is also known as **sucrase–isomaltase** by virtue of being an oligo-1,6-glucosidase, EC 3.2.1.10, displayed towards isomaltose. **2** any bacterial or plant enzyme of sub-subclass EC 2.4.1 (hexosyl-

transferases) that synthesizes homopolysaccharides from sucrose by transfer of either its fructose moiety or its glucose moiety respectively to a fructan or to a glucan, with release of the opposing moiety as free D-glucose or D-fructose. The recommended names are formed by adding 'sucrase' to a prefix indicating the polysaccharide product. These enzymes include: (1) **amylosucrase**; EC 2.4.1.4; *other name*: sucrose–glucan glucosyltransferase; *systematic name*: sucrose:1,4-α-D-glucan 4-α-D-glucosyltransferase; it synthesizes amylose; (2) **dextransucrase**; (3) **inulosucrase**; EC 2.4.1.9; *other name*: sucrose 1-fructosyltransferase; *systematic name*: sucrose:2,1-β-D-fructan 1-β-D-fructosyltransferase; it synthesizes inulin; and (4) **levansucrase**; EC 2.4.1.10; *other name*: sucrose 6-fructosyltransferase; *systematic name*: sucrose:2,6-β-D-fructan 6-β-D-fructosyltransferase; it synthesizes levan.

sucrase–isomaltase *see* **oligo-1,6-glucosidase**.

sucrase–isomaltase deficiency *see* **oligo-1,6-glucosidase**.

sucrose *or* **saccharose** the trivial name for O-β-D-fructofuranosyl-$(2\rightarrow1)$-α-D-glucopyranoside; a sweet-tasting, nonreducing disaccharide. It is usually isolated industrially in crystalline form from sugar cane, *Saccharum officinarum*, or from sugar beet, *Beta vulgaris*.

sucrose gradient a density gradient that consists of different concentrations of sucrose layered in a centrifuge tube. It is used to separate molecules or particles on the basis of their sedimentation coefficients. *See also* **isopycnic centrifugation**.

sucrose-phosphate synthase EC 2.4.1.14; *systematic name*: UDPglucose:D-fructose-6-phosphate 2α-D-glucosyltransferase; *other names*: UDPglucose–fructose-phosphate glucosyltransferase; sucrosephosphate–UDP glucosyltransferase. An enzyme that catalyses a reaction between UDPglucose and D-fructose 6-phosphate to form sucrose 6-phosphate and UDP.

sucrose phosphorylase EC 2.4.1.7; *systematic name*: sucrose:orthophosphate α-D-glucosyltransferase; *other name*: sucrose glucosyltransferase. An enzyme that catalyses the phosphorylysis of sucrose to D-fructose and α-D-glucose 1-phosphate.

sucrose synthase EC 2.4.1.13; *systematic name*: UDP glucose:D-fructose 2α-D-glucosyltransferase; *other names*: UDPglucose–fructose glucosyltransferase; sucrose–UDP glucosyltransferase. An enzyme that catalyses a reaction between UDPglucose and D-fructose to form sucrose and UDP.

Sudan a family of dyes that are soluble in organic solvents and are used for staining lipids. **Sudan Red** is the 2,2′-demethyl analogue of **Oil Red O**. **Sudan Black B** is used for staining chromosomes, Golgi apparatus, and leukocyte granules.

suffix tree a data structure that allows rapid solution of problems involving character strings (sequences) and facilitates substring searches. The tree is built by removing the first character of a string, storing the resulting suffix, removing the first character of the result, storing this, and so on, until the result is a single character.

sugar any, usually sweet, soluble disaccharide or small oligosaccharide carbohydrate. More specifically it is applied to **sucrose**, the sugar of commerce. The name is sometimes used as a synonym for **carbohydrate**. *See also* **beet sugar**, **blood sugar**, **cane sugar**, **corn sugar**, **fruit sugar**, **grape sugar**, **malt sugar**, **milk sugar**, **mushroom sugar**, **reducing sugar**.

sugar acid any **aldaric acid**, **aldonic acid**, or **uronic acid**.

sugar alcohol *an old name for* **alditol**.

sugar beet the plant *Beta vulgaris*, which has a large root with a high sugar content, for which it is extensively cultivated.

sugar transporter any of various families of transmembrane proteins responsible for sugar transport into cells. **GTR1** causes facilitated diffusion of glucose, whereas XylE protein in *Escherichia coli* acts as an H⁺–xylose symporter (*see* **xylE**). Both belong to the 12-transmembrane domain superfamily. *See* **glucose transporter, PTS**.

suicide enzyme *see* **suicide repair (of DNA)**.

suicide inhibitor a type of enzyme inhibitor that is relatively unreactive and becomes reactive only after interaction with the enzyme's active site. *See also* **photosuicide inhibitor**.

suicide repair (of DNA) a response in bacterial and mammalian cells to exposure to a nontoxic alkylation dose. Such exposure causes the rapid induction of enzymes that remove the O^6-methylguanine formed by the alkylating agent. In this process the methyl group is transferred to the enzyme, which is thereby inactivated (i.e. it is a **suicide enzyme**).

suicide substrate 1 a substrate that causes an enzyme to be modified in a way that permanently inactivates it. **2** a compound that is converted by a metabolic pathway into an inhibitor of the pathway, as with the conversion of fluoroacetate to fluorocitrate. If the pathway is of sufficient importance, the cell thereby 'commits suicide'. *See also* **lethal synthesis**.

sulfa drug any **sulfonamide** compound that is clinically useful in the treatment of bacterial infections.

sulfamethoxazole 4-amino-*N*-(5-methyl-3-isoxalolyl)benzenesulfonamide; a **sulfonamide** antibacterial agent. As a mixture with **trimethoprim** it has the name **co-trimoxazole** (*proprietary names*: Bactrim; Septrim).

sulfane a chain of divalent sulfur atoms, as in hydrogen per-sulfide (HSSH) or polysulfanes (H_2S_x ; e.g. H_2S_5 is pentasulfane).

sulfanilamide *p*-aminobenzenesulfonamide; *see* **sulfonamide**.

sulfatase *see* **arylsulfatase**.

sulfate *or (esp. Brit.)* **sulphate** the anion SO_4^{2-}.

sulfate adenylyltransferase EC 2.7.7.4; *other names*: sulfate adenylate transferase; ATP-sulfurylase; sulfurylase; an enzyme that catalyses the formation of **adenosine 5′-phosphosulfate** (adenylylsulfate) from ATP and sulfate with release of pyrophosphate.

sulfatide a **galactosyl cerebroside** with a sulfate radical esterified at the 3′ position of the galactose moiety. Sulfatides are formed by **galactosylceramide sulfotransferase**.

sulfatide lipidosis *an alternative name for* **metachromatic leukodystrophy**.

sulfation factor *an alternative name for* **somatomedin**.

sulfenic acid a compound having the structure RSOH (e.g. cysteine sulfenic acid), in which the sulfur is oxidized by addition of one oxygen atom. *Compare* **sulfinic acid**.

sulfhemoglobin any of a group of poorly characterized derivatives of **hemoglobin**, often formed *in vivo* together with ferrihemoglobin. They are incapable of carrying oxygen and cannot be converted back to hemoglobin.

sulfhydryl *or* **thiol** the chemical group –SH; it is found in the amino acid **cysteine** and in other molecules. Two sulfhydryls can be oxidized to a disulfide bond, –S–S–.

sulfinic acid a compound having the structure RS(=O)OH (e.g. cysteine sulfinic acid), in which the sulfur is oxidized by addition of two oxygen atoms. *Compare* **sulfenic acid**.

sulfite *or (esp. Brit.)* **sulphite** the anion SO_3^{2-}.

sulfite oxidase EC 1.8.2.1; the terminal enzyme in the degradation of sulfur-containing amino acids in animals. It catalyses the oxidation of sulfite to sulfate and is present in the mitochondrial intermembrane space in many tissues but not in blood or skeletal muscle. The crystal structure of the chicken-liver enzyme contains a domain for binding heme, another for molybdenum cofactor, and a third for dimerization. *See* **sulfite oxidase deficiency**.

sulfite oxidase deficiency a severe, frequently fatal, autosomal recessive human disease that presents with convulsions early after birth, encephalopathy, and accumulation and increased excretion of sulfite, thiosulfate, and *S*-sulfocysteine. It produces neuronal cell loss and demyelination, and results from various mutations that inactivate **sulfite oxidase**.

sulfite reductase *abbr.* SiR. **1 sulfite reductase (NADPH)** EC 1.8.1.2; *systematic name*: hydrogen sulfide:NADP⁺ oxidoreductase; the enzyme responsible for interconversion of hydrogen sulfide and sulfite in plants and most bacteria. It catalyses the reduction of sulfite, e.g. released from **3′-phosphoadenosine 5′-phosphosulfate** by thioredoxin. The reaction is:

$$H_2S + 3\ NADP^+ + 3\ H_2O = \text{sulfite} + 3\ NADPH;$$

cofactors are FAD, FMN, and siroheme. In plants and most microorganisms the H_2S reacts with *O*-acetylserine to form cysteine and acetate. The enzyme also oxidizes H_2S to sulfite. **2 sulfite reductase** An enzyme, EC 1.8.7.1; *systematic name*: hydrogen-sulfide:ferredoxin oxidoreductase. In plants it is a siroheme containing iron–sulfur protein that catalyses the reaction:

$$\text{sulfite} + 6\ \text{reduced ferredoxin} = H_2S + 6\ \text{ferredoxin} + 3H_2O.$$

sulfmethemoglobin the complex formed by combination of sulfide with the ferric ion of **ferrihemoglobin** in poisoning by sulfide.

sulf(o)+ *or (esp. Brit.)* **sulph(o)+** of, or pertaining to, or containing a sulfur atom or group.

sulfoindocyanine *or (esp. Brit.)* **sulphoindocyanine** a fluorescent dye, such as **Cy3** and **Cy5**, used in labelling proteins and DNA.

sulfokinase *or* **aryl sulfotransferase** *or* **phenol sulfotransferase** EC 2.8.2.1; an enzyme that catalyses transfer of a sulfate radical from 3′-phosphoadenylylsulfate to a phenol to form an aryl sulfate.

sulfolipid any lipid containing sulfur, especially a **sulfatide**. *See also* **sulfonolipid**.

sulfonamide 1 the amide of a sulfonic acid; the sulfonamide group is –SO₂–NH₂. **2** (*in biology and medicine*) *or* **sulfa drug** any of a group of antibacterial substances derived from **sulfanilamide** (*p*-aminobenzenesulfonamide), many of which have a substituent on the amide nitrogen.

sulfanilamide

sulfonate an anion of a sulfonic acid, having the structure RS(=O)(O⁻)2, in which the sulfur is oxidized by addition of one oxygen atom.

sulfone the oxidation product that results when one oxygen atom is added to sulfur bonded to two alkyl chains.

sulfonolipid *or* **plant sulfolipid** 1,2-diacyl-3-(6-sulfo-α-D-quinovosyl)-*sn*-glycerol; sulfoquinovosyldiacylglycerol (*abbr.*: SQDG); the only sulfolipid present in higher plants and the only known lipid with a sulfonic-acid linkage; the predominant fatty acid is α-linolenic acid. Sulfonolipid is localized in the chloroplasts, mainly in the membrane of the thylakoids.

sulfonylurea any oral hypoglycemic drug that contains a sulfonic acid–urea nucleus modified by chemical substitution. The group includes tolbutamide, chlorpropamide, gliburide, and glipizide. A sulfonylurea receptor is the regulatory subunit for the ATP-dependent K⁺ channel (**KATP**) of the pancreatic B-cell plasma membrane. The drugs also bind intracellular proteins on membranes of secretory granules of these cells, which may mediate part of their effect.

sulfonylurea receptor 1 *abbr.*: SUR-1; an integral membrane protein of pancreatic B cells that binds sulfonylurea drugs with high affinity and regulates the function of ATP-sensitive K channel (*also called* K_{ATP}). It binds ATP and ADP, belongs to the ABC transporter family, and in hamster contains 1582 amino acids. Several mutations in SUR-1 have been associated with persistent hyperinsulinemic hypoglycemia of infancy (PHHI).

sulfoxide the oxidation product that results when two oxygen atoms are added to sulfur bonded to two alkyl chains.

sulfur *or (esp. Brit.)* **sulphur** *symbol*: S; a yellow nonmetallic element of group 16 of the IUPAC **periodic table**; atomic number 16; relative atomic mass 32.066. Its main oxidation states are –2, +4, and +6.

Sulfur exists in a number of forms including crystalline (rhombic, monoclinic, rhombohedral) and amorphous forms. The relative abundance of stable isotopes is: sulfur-32 (mass 31.972) 95.02%; sulfur-33 (mass 32.971) 0.75%; sulfur-34 (mass 33.968) 4.21%. There are two radioactive isotopes, **sulfur-35** and **sulfur-38**. Sulfur is important as a constituent of the amino acids methionine and cysteine, and as a constituent of **glutathione**. It also occurs in **sulfolipids**. Sulfate is the terminal electron acceptor for anaerobic respiratory metabolism in certain bacteria e.g. *Desulfovibrio* spp.

sulfur-35 the radioactive nuclide, $^{35}_{15}$S, with a mass of 34.969. It emits an electron (β^- particle, 0.167 MeV), no gamma radiation, and has a half-life of 87.4 days.

sulfur-38 the radioactive nuclide, $^{38}_{16}$S, with a mass of 37.971. It emits an electron (β^- particle, 3.0 MeV) and gamma radiation (1.94 MeV), and has a half-life of 2.87 h.

sulfurylase *see* **sulfate adenylyltransferase**.

sulph(o)+ *a variant spelling (esp. Brit.) of* **sulf(o)+**.

sum 1 the total value of a set of individual values. **2** to determine a sum.

sumatriptan 3-[(2-dimethylamino)ethyl]-*N*-methyl-1*H*-indole-5-methanesulfonamide; an agonist for 5-HT$_{1D}$ receptors (*see* **5-hydroxytryptamine receptor**). It constricts cerebral blood vessels, and is used in the treatment of migraine and cluster headaches. *Proprietary name*: Imigran (succinate).

summation the determination of the **sum** (def. 1) of a series.

SUMO *abbr. for* small ubiquitin-related modifier; *other name*: sentrin; a ubiquitin-like protein that becomes conjugated by an isopeptide bond with several proteins (e.g. RanGAP1, PML, IκB in mammals). Human SUMO (101 amino acids) shares 18% sequence identity with ubiquitin, and a globular fold consisting of beta sheets wrapped around an alpha helix. Conjugation (**sumoylation**) is between Gly-97 (uncovered by removal of the C-terminal tetrapeptide by a SUMO-specific protease) and the side chain of a lysine in the target protein. It occurs predominantly in the nucleus and regulates nuclear transport, signal transduction, the stress response, and cell-cycle progression; it does not involve proteasomes. Deconjugation (**desumoylation**) is catalysed by SUMO-specific proteases.

sumoylation *see* **SUMO**.

SUMT *abbr. for* *S*-adenosyl-L-methionine:uroporphyrin-III *C*-methyltransferase (*see* **uroporphyrin-III *C*-methyltransferase**).

Sunday driver a mutation in *Drosophila* that mixes up intracellular traffic.

super+ *prefix* **1** above, over, at the top (of). **2** of greater size or extent, or of higher quality. **3** beyond the usual or normal standard; exceeding. **4** *(in chemical nomenclature)* designating a compound that contains more than the usual proportion of a specified element. *Compare* **sub+, supra+**.

superantigen an antigen with the ability to activate a large proportion (up to 20%) of T lymphocytes in a given individual through interaction with the variable domain of the β chain of the lymphocyte's antigen receptor. This contrasts with conventional foreign antigens the peptides of which, attached to the major histocompatibility complex (MHC), have to be specifically recognized by the T-cell receptor. With superantigens the specificity resides in the β chain of the receptor not the MHC peptides. This accounts for the fact that many more than usual T lymphocytes are activated. Superantigens are present in a number of bacteria and viruses; the best example is the *Staphylococcus* enterotoxin. Superantigens have been implicated in **autoimmunity**.

supercoil *or* **superhelix** *or* **coiled coil** the form of DNA that can be envisaged as resulting if a **duplex** were turned through a hairpin bend and the two duplex rails twisted one around the other. DNA *in vivo* generally is supercoiled, and as it has a closed structure, a supercoil may be envisaged as a rubber band twisted to a helical shape. In **negative supercoiling** the DNA twists around its axis in the opposite direction from the clockwise turns of the right-handed double helix; this allows relief of torsional pressure, and the DNA is said to be **underwound**. Twisting in the other direction (**positive supercoiling**) increases torsional pressure, winding the double helix more highly; the DNA is said to be **overwound**. **Superhelix density**, σ, is defined as the number of turns of the superhelix, τ, per turn of the duplex, β; i.e. $\sigma = \tau/\beta$ (β represents about 10 base pairs). **Linking**

number is the number of times the duplex strands cross over each other (in projection onto a plane surface). The number of superhelical turns is called the **writhing number** (W).

supercritical fluid chromatography a technique that can be regarded as an extension of **gas–liquid (partition) chromatography** in which the mobile phase is a relatively dense gas such as CO_2 (or SF_6, NH_3, or Xe). The gas must be pumped (at pressures up to 6000 psi) through a capillary column that is coated with stationary phases similar to those used in gas chromatography. The system depends on the partition of the compounds being separated between the stationary and mobile phases and has the advantage that compounds with high boiling points partition appreciably into the gas phase at temperatures well below their boiling points so that such compounds can be separated at lower temperatures than with less dense gases.

superfamily any group of genes and their cognate proteins that are related usually by sequence homology.

superfold 1 a common fold found in nonhomologous proteins (e.g. TIM barrels). **2** *(sometimes)* an alternative term for **super-secondary structure**.

superfolding *(of proteins)* the formation of small domains, beyond the level of simple secondary structures, that are energetically stable and contribute as units to the full tertiary structure of a protein. They may consist of **alpha helices** or **beta sheets** or both. They may be formed during the early stages of protein folding, before the protein has fully folded. For example, **beta barrels** are **super-folds**.

superfusion 1 the pouring of a liquid onto or over something. **2** *(in physiology and medicine)* a technique in which blood, plasma, or other fluid is allowed to drip onto, or flow over, the surface of an organ or piece of tissue.

supergene a group of linked genes (*see* **linkage** (def. 1)) that are usually inherited as a unit.

superhelix an alternative name for **supercoil**.

superhelix density *see* **supercoil**.

superinfection the proliferation of one or more pathogenic microorganisms arising from the disturbance of a mixed population of microorganisms normally sustained at non-pathogenic levels by secretion of endogenous antibiotics. Such disturbance can be induced by administration of exogenous antibiotic agents.

superinfection inhibition interference with the entry or establishment of an entering plasmid by a resident plasmid.

supernatant 1 floating on the surface of a liquid or lying over something. **2** *(or* **supernate***)* **a** a liquid lying above another liquid, a sediment, or a settled precipitate. **b** the fluid phase that remains after removal of solid material by centrifugation. A **cell supernatant** is the fluid phase that remains after centrifugation of a cell homogenate at high speed (characteristically at 100 000 *g* for 30 min). *Compare* **infranatant**.

supernate an alternative name for **supernatant** (def. 2).

superoxide 1 the anion, O_2^-, formed by addition of one electron to **dioxygen**. **2** any compound containing the superoxide anion.

superoxide dismutase *abbr.*: SOD; EC 1.15.1.1; *systematic name*: superoxide:superoxide oxidoreductase; *other names*: cytocuprein; erythrocuprein; hemocuprein. Any of a group of metal-containing enzymes that bring about the **dismutation** of superoxide radicals to form dioxygen and hydrogen peroxide as follows:

$$O_2^- \bullet + O_2^- \bullet + 2H^+ = O_2 + H_2O_2.$$

It is important in removing the highly toxic superoxide radical. Some (e.g. those from *Neurospora*, yeast, erythrocytes, heart, peas) contain copper and zinc; others (e.g. those from *Escherichia coli*, *Streptomyces mutans*) contain manganese or iron. Eukaryotic cells have copper–zinc types in cytoplasm and manganese types in mitochondria. The copper–zinc types comprise a well-conserved family with four motifs.

Superscript *proprietary name for* a reverse transcriptase derived from Moloney murine leukemia virus (MMLV) and modified by protein engineering to have reduced RNase H activity.

super-secondary structure *or (sometimes)* **superfold** the arrangement of protein secondary structures into discrete folded units – e.g. beta barrels, β-α-β units, and Greek keys. *Compare* **superfold** (def. 1).

supported reagent any reagent linked to a **support medium**.

support medium any (usually solid or insoluble) medium to which a cell, reagent, or functional group can be attached.

suppression 1 *(in molecular biology)* the cancellation, or reversal of the effects, of one mutation by another mutation, except where the second mutation causes a single base change at the same point as an earlier **point mutation** (in which case the term **reversion** is used). In **intergenic suppression**, the effects of a mutation in one gene are reversed by a mutation in another gene, usually because the second gene codes for a mutant tRNA (*see* **frameshift suppressor**). In **intragenic suppression** a mutation is cancelled by a second mutation close to the first in the same gene, normally within the same triplet, giving rise to a codon that is compatible with translation to a functional protein product. As an example, if the first mutation changed AGT (leading to the RNA codon for Ser) to ATT (giving a nonsense codon), a suppressor mutation might change the triplet to ATA (giving the RNA codon for Tyr); if the change Ser to Tyr led to a functional protein, suppression would be achieved. *See also* **amber suppressor, ochre suppressor, opal suppressor. 2** *(in immunology)* any attenuation of an immune response. The existence of a subpopulation of T cells, **suppressor cells**, is held to be responsible for attenuation of the effect of **helper cells**. A different mechanism is probably responsible for the reduction in antibody production arising from injection of antibody prior to administration of antigen. Allotype suppression refers to suppression of the production of one allotype (e.g. expression of the allotype inherited from the mother, and suppression of the allotype of the father).

suppressor cell *or* **suppressor T cell** *or* **T-suppressor cell** any **T lymphocyte** that inhibits (*see* **suppression** (def. 2)) the stimulation of the immune response by **helper cells**. Classically, suppressor cells belong to the **CD8+** subset in the human. Induction of the suppressor function may derive from helper cells.

suppressor gene a gene that can reverse the effect of a specific type of mutation in another gene. *See* **suppression** (def. 1).

suppressor mutation a secondary mutation that totally or partially restores a function lost due to a primary mutation at another locus. *See* **suppression** (def. 1).

suppressor tRNA any species of **transfer RNA** (tRNA) that recognizes a **termination codon** in messenger RNA. It adds an amino-acid residue to what is normally the C terminus of a nascent polypeptide chain; hence termination of synthesis of the peptide is suppressed and a **readthrough** protein is consequently formed. Suppressor tRNA may occur naturally or be produced by mutagenesis.

supra+ *prefix* above, over, greater than. *Compare* **infra+, super+**.

supramolecular describing any chemical system or component thereof that exists and functions at a higher level of complexity than intact individual molecules. In biochemistry the term may refer to multienzymes, microtubules, organelles, biological membranes, viral envelopes, cell walls, ribosomes, or other cellular elements; hence, e.g., supramolecular assembly, supramolecular complex, supramolecular structure. *Compare* **submolecular**.

suprarenal gland *an obsolete term for* **adrenal gland**.

suprarenin *a former name for* adrenaline (i.e. **epinephrine**).

Sur *symbol for* a residue of the thiopyrimidine base **thiouracil**.

SURE *abbr. for* Stop Unwanted Rearrangement Events; a proprietary strain of *E. coli* developed to allow cloning of DNA sequences that cannot be cloned in other *E. coli* strains.

SURF1 *abbr. for* surfeit gene 1 product; one of the proteins encoded by a cluster of six housekeeping genes (locus at 9q34). It is targeted to the inner mitochondrial membrane, contains 300 amino acids and two transmembrane segments, and is required for normal activity of complex IV of the respiratory chain in humans. Mutations affect mostly the brain and produce **Leigh syndrome**.

surface-active agent *an alternative term for* **surfactant**.

surface concentration *see* **concentration** (def. 1).

surface plasmon resonance a phenomenon that occurs when light is reflected off thin metal films, which may be used to measure interaction of biomolecules on the surface. An electron charge density wave arises at the surface of the film when light is reflected at the film under specific conditions. A fraction of the light energy incident at a defined angle can interact with the delocalized electrons in the metal film (plasmon) thus reducing the reflected light inten-

sity. The angle of incidence at which this occurs is influenced by the refractive index close to the *backside* of the metal film, to which target molecules are immobilized. If ligands in a mobile phase running along a flow cell bind to the surface molecules, the local refractive index changes, leading to a change in SPR angle which is proportional to the mass being immobilised. This can be monitored in real time by detecting changes in the intensity of the reflected light, producing a *sensorgram*.

surface-stress theory a theory that relates mechanical stress to enlargement of the bacterial cell wall during growth. The cell wall withstands stress from hydrostatic pressure by a stress-bearing, two-dimensionally linked peptidoglycan sacculus that surrounds the cell. During enlargement, enzymes that enlarge the wall must at the same time maintain the hydrostatic integrity of the cell by withstanding stress.

surface tension *symbol*: Y *or* σ *or* γ; the property that makes a liquid behave as if it had an elastic skin on its surface – at an interface with a gas or an immiscible liquid – as a result of intermolecular forces tending to pull the molecules in the surface into the interior of the liquid. It is defined as the force acting tangential to the surface on one side of a line of unit length in the surface, and is expressed in newtons per metre. It can also be defined as the work required to produce unit increase in surface area, expressed in joules per square metre.

surfactant *or* **surface-active agent** any substance, such as a **detergent** or an **emulsifier**, that can bring about a reduction of the **surface tension** of a liquid, allowing it to foam, to penetrate porous solids more easily, and to wet surfaces of nonporous solids or of immiscible liquids.

surfactant protein *abbr.*: SP; any of four proteins secreted by type II cells of alveoli and bronchial mucosa as constituents of **pulmonary surfactant**. SP-A and SP-D are **collectins**, having similar structures and serving mostly in defence against microorganisms. SP-B and SP-C are structurally unrelated and help stabilize the phospholipids of the secretion. SP-B is an amphipathic 79-residue polypeptide. SP-C is a hydrophobic 33–34-residue polypeptide. All are derived from larger precursors. Many mutations in the gene at 2p12-p11.2 for SP-B, and more rarely in the genes for the other proteins, are associated with respiratory distress syndrome of infants (especially in those born prematurely), and also in adults.

surfactin a surface-active antibiotic and nonimmunogenic hemolysin secreted by *Bacillus subtilis*. It consists of a nonapeptide (containing two residues each of L- and D-leucine) in cyclic depsipeptide linkage with β-hydroxy-ω-methylmyristic acid. It is classified in the TC system under the number 1.C.12.

surfeit gene *see* SURF1.

surrogate genetics a branch of molecular genetics in which DNA is modified *in vitro* and then reintroduced into a genetic system – the 'surrogate' system – e.g. by injection into *Xenopus* oocytes, by transfection of cells, or by the creation of transgenic animals. The effect of the modification on the expression of the gene provides insights into the role in the genetic system of particular base sequences.

surveyor substrate one of a system of fluorescent substrates designed to measure distances within the active centre of an enzyme and also to survey other aspects of the topography of the active centre during the catalytic event.

survivin a member of the Inhibitor of Apoptosis gene family implicated in regulation of cell division and suppression of apoptosis. It is over-expressed in human cancers, being concentrated in mitotic spindle. In normal tissue, survivin is expressed only during the development of the embryo and fetus, where it is believed to play a role in controlling apoptosis.

Sushi domain *or* **complement control protein module** *or* **short consensus repeat** a polypeptide domain of ~60 amino acids that contains disulfide bonds between the first and third, and second and fourth cysteines, as found in β_2 glycoprotein-1 of plasma, in complement proteins, and in **blood coagulation** factor XIII.

suspension culture the culture of cells in medium without adherence to a stationary substrate. Relatively few animal cell types are amenable to suspension culture (classically **HeLa cells** can be so grown). An intermediate situation utilizes beads to which the cells adhere; the beads can be maintained in suspension and collected by sedimentation

Su(var) *abbr. for* suppress variegation; any of a group of genes in *Drosophila* and yeast that suppress euchromatin/heterochromatin variegation in the higher structure of chromatin. Among them are genes for many enzymes that modify residues in the N-terminal regions of core histones of nucleosomes. Their action leads to gene silencing and assembly of heterochromatin. Su(var) genes are opposed by the products of **E(var)** genes. Some human homologues encode histone methyltransferases that selectively methylate histone H3 at lysine residue 9.

suxamethonium succinylcholine chloride, 2,2′-[(1,4-dioxo-1,4-butanediyl)bis(oxy)]bis[*N,N,N*-trimethylethanaminium] dichloride; a nicotinic receptor ligand that causes sustained activation of the receptors leading to blockade of synaptic transmission ('depolarizing block'). It is used as a neuromuscular blocking agent in surgery. It acts by mimicking acetylcholine at the neuromuscular junction, but disengagement from the receptor site and subsequent breakdown are slower than for acetylcholine. It is hydrolysed by serum cholinesterase and recovery is spontaneous. **Decamethonium** was the first drug of this type to be developed. Both of these compounds depend for their action on the N^+–N^+ distance of 1.1 nm. *Compare* **curare, tubocurarine**.

Sv *symbol for* **1** sievert. **2** svedberg (alternative to S).

SV40 *abbr. for* simian virus 40.

svedberg *or* **Svedberg unit** *symbol*: S *or* Sv; a coherent non-SI unit used as a measure of **sedimentation coefficient**; 1 S = 10^{-13} s.

Svedberg centrifuge a centrifuge, designed and built by Swedish colloid chemist Theodor Svedberg (1884–1971) in 1923, that was the forerunner of the **ultracentrifuge**.

Svedberg equation an equation relating the relative molecular mass, M_r, of a solute to its sedimentation velocity in an applied centrifugal field. If s is the sedimentation coefficient, D is the diffusion coefficient, $\bar{v}$ is the partial specific volume of the solute, ρ is the density of the solution, R is the gas constant, and T is the thermodynamic temperature, then

$$M_r = sRT/D(1 - \bar{v}\rho).$$

swainsonine $8\alpha,\beta$-indolizidine-$1\alpha,2\alpha,8\beta$-triol; an inhibitor of a number of α-mannosidases, particularly Golgi mannosidase II (*see* **Golgi mannosidase**), an important enzyme in the synthesis of the core oligosaccharide of a number of membrane glycoproteins. It inhibits the growth of tumours and prevents metastases in mice. [After its source, *Swainsona canescens*] .

swaposin a feature of some proteins in which domains, characteristically found in one protein, are found arranged in a different order in other proteins. The phenomenon was first commented on in relation to certain plant aspartic peptidases, which contain saposin-like domains swapped into a different order from that found in **saposins** (hence the name). In another example, some **pleckstrin homology domains** are found to be interrupted by inserts of **SH domains**.

sweetener any compound intended as a substitute for sucrose in sweetening food or drink, and aimed at avoiding the possible harmful effects of sucrose. Such compounds therefore should be nonnutritive and nonharmful but with a taste and aftertaste similar to those of sucrose. They include **aspartame, lactitol, mannitol, monellin, saccharin, stevioside, sucralose**, and **volemitol**. Sodium cyclamate was used as a sweetener until it was banned in the USA in 1970 because of its carcinogenicity.

sweetness one modality of the sense of taste, characterized by the effect of certain sugars such as sucrose and fructose.

SWI *abbr. for* mating-type switching gene; a gene of yeast that is required for transcriptional regulation of an endonuclease involved in mating-type switching. *See* **SWI/SNF**.

swinging-bucket rotor a type of centrifuge head in which the centrifuge tubes are placed in buckets that swivel on trunnions so that at speed the tube is oriented with its axis parallel to the centrifugal force (i.e. horizontal). Such a head is essential for zonal centrifugation through density gradients as angled tubes induce mixing. *Compare* **angle head**.

SWI/SNF *abbr. for* switching/sucrose nonfermenting; a protein complex ($\approx$2.0 MDa) that contains $\approx$11 polypeptides and is present in all eukaryotic cells. It causes ATP-dependent nucleosome remodelling and stimulates binding of transcription factors to DNA. It was first characterized in yeast as essential for expression of genes for mating type switching and for sucrase. Inositol tetrakisphosphate stimulates its action. *See* **NURF**.

Swiss-Prot *see* **sequence database**.

switch 1 *(of a gene)* to activate (switch on) or inactivate (switch off) transcription. **2** *(of a protein)* to change from an active to an inactive form by changing conformation, e.g. as a result of binding to a specific ligand or of chemical modification. **3** *(in development)* to transfer from synthesis of a protein appropriate for a fetus to that appropriate for postnatal life. **4** *(in immunity)* to change synthesis of one class of immunoglobulins to that of another. *See* **molecular switch**.

switch peptide the flexible covalent connection between the **variable region** and **constant region** of an immunoglobulin molecule.

switch recombinase *see* **homothallic switching endonuclease**.

swivelase *see* **DNA swivelase**.

SWRP *abbr. for* systemic wound response protein; a group of defence proteins in plants whose synthesis is induced by the peptide hormone **systemin** in response to wounds inflicted mechanically or by herbivorous insects. At the target tissue systemin activates a signalling cascade that produces jasmonic acid, which is the inducer of the SWRP genes.

Sxl *symbol for* **sex lethal** gene of *Drosophila*.

SYBR Gold a proprietary cyanine dye used for fluorescent staining of nucleic acids with great sensitivity.

SYBR Green *a proprietary name for* one of two asymmetrical cyanine dyes whose fluorescence is greatly increased by binding to DNA. Sybr Green I is used for the measurement of low concentrations of DNA and in monitoring DNA amplification in real-time polymerase chain reaction.

Syk *abbr. for* spleen tyrosine kinase; a protein tyrosine kinase that phosphorylates the adaptor protein SLP-76. It and SLP-76 are involved in platelet degranulation, secretion of interleukin 2, and generation of lymphatic vessels. Mice with mutations in Syk or SLP-76 develop arteriovenous malformations.

sym+ *a variant form of* **syn+** (before b, m, or p).

sym- *prefix (in chemical nomenclature)* denoting symmetric; e.g. *sym*-trinitrobenzene.

symbiont an organism that lives as a partner in a **symbiosis**.

symbiosis (*pl.* **symbioses**) a long-term association between individuals belonging to two different species. The term is often used in a restricted sense to denote associations that are beneficial to one or both partners, although strictly it refers equally to neutral or harmful associations. —**symbiotic** *adj.*

symbiosome a vacuole within root cells of nitrogen-fixing plants that forms on infection by *Rhizobium* species and engulfs one or more bacteria, which become transformed to bacteroids. The delimiting membrane (peribacteroid membrane) acquires newly synthesized proteins (nodulins), which form membrane channels, transport proteins, enzymes of carbon and nitrogen metabolism, or leghemoglobin.

symmetry precise correspondence between two halves with respect to both size and shape. A molecule has an *n*-fold **axis of symmetry** if rotation around this axis through an angle of 360°/*n* produces an arrangement indistinguishable from the original (*n* being a small integer). For example, water has a twofold (*n*=2) axis of symmetry. A plane of symmetry is a plane drawn through a molecule such that a mirror placed in this plane forms an image of one half of the molecule that is superimposable on the other half. For example, a plane of symmetry is present in *meso*-**tartaric acid**. An alternating axis of symmetry or rotation–reflection axis is an axis such as is present in a molecule when rotation around the axis through an angle of $360°/n$ followed by reflection in a plane perpendicular to the axis produces a molecule superimposable on the original molecule. For example, a compound with such a twofold (n=2) axis is fumaric acid. A centre of symmetry is a position (point) such that if a straight line is drawn from every atom to the point, and is continued beyond the point, an equivalent atom is found at a distance equal to that between the point and the original atom. A molecule with such a centre has a twofold alternating axis of symmetry. An example of a molecule with a centre of symmetry is fumaric acid.

sympathectomy the surgical section of sympathetic nerve fibres.

sympathetic nervous system one of the two divisions of the **autonomic nervous system** the other being the parasympathetic nervous system. Most, but not all, sympathetic nerves release **norepinephrine** as a neurotransmitter.

sympatheticomimetic *an alternative term for* **sympathomimetic**.

sympathomimetic *or* **sympatheticomimetic 1** describing a substance or action that stimulates the **sympathetic nervous system**; **adrenergic**. **2** a sympathomimetic agent, especially a sympathomimetic amine.

symport *or sometimes* **cotransport** a transport process in which two substances (usually a cation and an uncharged molecule) move across the membrane on a specific transporter, the process representing a downhill movement of one of them (the cation) and an uphill movement of the other (the uncharged molecule). In some cases, a cation and an anion are transported together when no external source of energy is required. The proteins involved may be designated as **symporters**.

syn- *abbr. for* synclinal (def. 1); *see also* **conformation**.

syn+ *comb. form* signifying similarity, association, or concerted action.

Synacthen *see* **tetracosactide**.

synaeresis *a variant spelling of* **syneresis**.

synapse *or (rare)* **synapsis** the junction between a nerve fibre of one neuron and another neuron; the site of interneuronal communication. As the nerve fibre approaches the synapse it enlarges into a specialized structure, the **presynaptic nerve ending**, which contains mitochondria and **synaptic vesicles**. At the tip of the nerve ending is the **presynaptic membrane**; facing it, and separated from it by a minute cleft – the **synaptic cleft** – is the **postsynaptic membrane** of the receiving neuron. On arrival of nerve impulses, the presynaptic nerve ending secretes **neurotransmitters** into the synaptic cleft. These diffuse across the cleft and transmit the signal to the postsynaptic membrane. *See also* **neuromuscular junction**. —**synaptic** *adj.*

synapsin I a major neuron-specific protein present in virtually all synaptic vesicles. Phosphorylated at multiple sites by Ca^{2+}- calmodulin-dependent and cyclic AMP-dependent protein kinases, it coats synaptic vesicles, and binds to the cytoskeleton. Synapsin IB is derived by alternative splicing; synapsins II are related.

synapsis 1 the pairing of chromosomes during **meiosis**. **2** *a less common word for* **synapse**.

synaptic cell adhesion molecule *see* **SynCAM**.

synaptic cleft the extracellular space, typically some 20 nm across, that separates the presynaptic nerve ending from the postsynaptic membrane of the receiving neuron (or other target cell) in a **synapse**. Chemical mediation of nerve transmission occurs across the synaptic cleft.

synaptic vesicle a vesicle, 20–65 nm in diameter, occurring within the cytoplasm of the presynaptic nerve ending and often closely related to the presynaptic membrane (*see* **synapse**). Synaptic vesicles store **neurotransmitters** and release them by fusing with the presynaptic membrane.

synaptic vesicle-associated membrane protein *another name for* **synaptobrevin**.

synaptobrevin *or* **sinaptobrevin** *or* **synaptic vesicle-associated membrane protein** (*abbr.*: VAMP) a SNAP receptor and member of the physin family involved in the docking and fusion complex (*see* **SNARE**) of synaptic vesicles and playing a central role in neural exocytosis. Synaptobrevin isoforms 1 and 2 (VAMP-1 and VAMP-2) are known. Both are expressed in all the main rat tissues, but differential expression is found, e.g. VAMP-1 is restricted to exocrine

pancreas and kidney tubular cells, VAMP-2 predominates in islets of Langerhans and kidney glomerular cells.

synaptogyrin any of a family of membrane proteins (29 kDa) of synaptic vesicles of neurons and neuroendocrine cells. It contains four transmembrane segments, localizes (and shares some sequence homology) with synaptophysin, and like it is phosphorylated on tyrosine residues. *See also* **gyrin**.

synaptojanin a protein, found in nerve terminals in brain, that has inositol 5-phosphatase activity, and binds to the SH3 domain of **amphiphysin**, colocalizing with that protein in double immunofluorescence micrographs. It appears to participate with **dynamin** in synaptic vesicle recycling. It shares similarity with the yeast protein Sac1, which is genetically implicated in phospholipid metabolism and the function of the actin cytoskeleton.

synaptonemal complex an elaborate structure that holds paired chromosomes together during synapsis in prophase I of meiosis and promotes genetic recombination. It consists of a ladder-like protein core (comprising central elements joined to lateral elements by transverse filaments) with corresponding regions of the two sister chromatids aligned on opposite sides. FKBP6 is one of the protein components.

synaptophysin an integral membrane glycoprotein of the physin family that occurs in presynaptic vesicles of neurons and in similar vesicles of the adrenal medulla. It is a marker protein for neuroendocrine cells and neoplasms.

synaptosome any of the discrete particles (nerve-ending particles) formed from the clublike presynaptic nerve endings that resist disruption and are snapped or torn off their attachments when brain tissue is homogenized in media isosmotic to plasma. In synaptosomes the main structural features of the nerve endings are preserved.

synaptosome-associated protein *see* SNAP (def. 2).

synaptotagmin an abundant integral membrane protein of synaptic vesicles and chromaffin granules, with a proposed role in membrane interactions during trafficking of synaptic vesicles at the active zone of the synapse. The primary structure contains two copies of a repeat that is homologous to the regulatory domain of protein kinase C.

synarchic regulation *see* **synarchy**.

synarchy cooperation between two intracellular messengers (e.g. cyclic AMP and Ca^{2+}) that act in concert to activate intracellular pathways.

SynCAM *abbr. for* synaptic cell adhesion molecule; a brain-specific type I integral membrane protein that mediates homophilic cell adhesion at synapses. The longest transcript encodes 456 amino acids, organized into three extracellular immunoglobulin-like domains, a transmembrane segment, and an intracellular **PDZ domain**.

syncatalytic synchronous with the catalytic process.

synchrotron an apparatus used in nuclear physics that accelerates particles of atomic size through an electric field. The particles are initially charged by an ion source, and their path may be linear (**linear accelerator**) or of spiral form (**cyclotron**). The acceleration imparts a high kinetic energy to the particles, which can lead to structural changes in the particles. One use of a synchrotron, therefore, is the production of radioactive or other nuclides for use in biology or medicine. A synchrotron radiation source is a circular storage ring in which electrons are accelerated in a magnetic field.

synclinal 1 *abbr.*: *syn*-; describing one of the two possible conformations of the base and sugar ring in a nucleotide, the other being the **anticlinal** conformation. *See* **conformation**. **2** *or (sometimes)* **gauche** *or* **skew** *abbr.*: *sc*-; describing a particular staggered conformation of a molecule in which the **torsion angle** is within ±60° of 0°. *See* **conformation**.

syncytium (*pl.* **syncytia**) a mass of cytoplasm containing several nuclei enclosed within a single plasma membrane. Syncytia are normally derived from single cells that fuse or fail to complete cell division. Examples are muscle fibres, formed by fusion of **myoblasts**, and residual bodies formed during **spermatogenesis**.

SYND2 *abbr. for* syndecan-2. *See* **syndecan**.

syndapin 1 a cytosolic protein that contains an SH3 domain, binds dynamin of the cytoskeleton, and is involved in vesicle budding at the plasma membrane. It also binds the brain-specific **synaptojanin** and **synapsin I**.

syndecan a type I integral membrane proteoglycan that links the cytoskeleton to the interstitial matrix. It bears heparin sulfate and chondroitin sulfate, and binds to bFGF (*see* **fibroblast growth factor**) with high affinity, and also to components of the extracellular matrix. Its intracellular domain binds to the actin cytoskeleton.

syndein *an alternative name for* **ankyrin**.

syndesine a di(α-amino acid) isolated from hydrolysates of certain **collagens**, in which it forms covalent cross-links between adjacent polypeptide strands. It is formed by **aldol condensation** between the side chains of a residue of **allysine** and one of **hydroxyallysine**; two isomers are possible.

syndiotactic having or relating to a regular alternation of differences in stereochemical structure in the repeating units of a polymer. *Compare* **isotactic**.

syndrome a set of concurrent symptoms characteristic of a particular pathological state.

syndrome X *see* **hyperinsulinemia syndrome**.

Syne *abbr. for* synaptic nuclei expressed; either of two proteins found enriched at the nuclear envelope of myonuclei clustered at neuromuscular junctions. They are probably involved in anchoring nuclei to the cytoskeleton. Syne-1 (human: 8739 amino acids) and Syne-2 (6855 amino acids) each contain actin-binding domains, large central domains of spectrin repeats, and a C-terminal nuclear envelope transmembrane domain. Mouse homologues have been partially characterized. Syne proteins are homologous with ANC-1 (of *Caenorhabditis elegans*) and Msp-300 (of *Drosophila*).

synemin a protein, M_r 230 000, found in the Z disk of skeletal and cardiac muscle cells, and in **intermediate filaments** of some non-muscle cells. It binds tightly to **desmin**.

synenkephalin *or* **proenkephalin 1–70** a derivative of **proenkephalin** secreted as an intact molecule or as a part of precursors in the adult brain and adrenal medulla. It has the immunoreactivity of proenkephalin, but does not contain the enkephalin sequence.

syneresis *or* **synaeresis 1** (*in chemistry*) the separation of liquid from a gel caused by contraction of the gel structure. **2** (*in pathology*) the gradual contraction of a blood clot that occurs after its formation, leading to a more solid mass that seals the ruptured blood vessel.

synergism 1 the phenomenon whereby the effect of two or more agonists together is greater than the sum of their effects when used individually. **2** the concerted action of two muscles to enhance their combined effects. *Compare* **antagonism**. —**synergetic** *or* **synergistic** *adj.*

synergist 1 an agent that increases the effectiveness of another agent. **2** a muscle that acts in concert with another muscle to increase its effect. *Compare* **antagonist**.

synexin *or* **annexin VII** a cytosolic Ca^{2+}-dependent, lipid-binding protein belonging to the **annexin** family. Originally isolated from adrenal medullary cells, isoforms have since been found in skeletal muscle, brain, and liver, and also in *Dictyostelium discoideum*. Adrenal medulla and liver have isoforms of 47 kDa and 51 kDa, while skeletal muscle has only the 51 kDa isoform. Synexin promotes membrane fusion, aggregates isolated chromaffin granules, and forms voltage-dependent calcium channels in artificial and natural membranes.

syngamy the union of the nuclei of two gametes to form the single nucleus of the zygote during fertilization; sexual reproduction. —**syngamic** *or* **syngamous** *adj.*

syngeneic of or pertaining to cell types in an experimental mammalian chimera that are antigenetically similar. *Compare* **allogeneic**.

synologue *or* (*esp. US*) **synolog** a gene transferred to a species from another during or after the fusion of the two species. Examples are genes transferred to the eukaryotic cell from the endosymbionts that have evolved into the mitochondria and chloroplasts. *Compare* **xenologue**.

synology the transfer of genes between organelles and the nucleus. –**synological** *or* **synologic** *adj.*; –**synologically** *adv.*

synonym codon a **synonym triplet** in messenger RNA.

synonymous codon usage *abbr.*: SCU; the frequency with which a codon is used relative to its synonymous alternative(s) in a partic-

ular species, gene, or group of genes. Synonymous alternatives are not used with equal frequency; *see* **codon usage bias**.

synonymous mutation *an alternative name for* **silent mutation**.

synonym triplet any of two or more trinucleotides of a nucleic acid that can specify the same amino acid. In messenger RNA the term is synonymous with **synonym codon**.

synovia *or* **synovial fluid** the fluid surrounding a joint or filling a tendon sheath.

synovial membrane *or* **synovium** the membrane of mesothelial cells and connective tissue that surrounds a joint; it secretes the **synovia**.

synovium *an alternative name for* **synovial membrane**.

synperiplanar *see* **conformation**.

syntaxin a SNAP receptor (**SNARE**) involved in docking of synaptic vesicles at the presynaptic zone of a **synapse**.

synteny the state in which genes or other loci reside together in a linkage group, i.e. on the same chromosome. Conserved synteny denotes that homologous genes occur in the same linkage groups across the genomes of different species. –**syntenic** *adj.* –**syntenically** *adv.*

synthase an enzyme that catalyses a condensation reaction in which no nucleoside triphosphate is required as an energy source. *Compare* **synthetase**.

synthase kinase glycogen synthase α kinase. *See* **non-specific serine/threonine protein kinase**.

synthesis (*in chemistry*) the production of a more complex molecule from simpler reagents by one or a series of reactions. —**synthesize** *or* **synthesise** *vb.*

synthesizer *or* **synthesiser** an instrument that automates a chemical synthesis, especially of peptides or oligonucleotides.

synthetase an enzyme that catalyses a condensation reaction in which a nucleoside triphosphate is required as an energy source. *Compare* **synthase**.

synthetic 1 of, or pertaining to, **synthesis**. **2** (of a substance or material) chemically synthesized as opposed to prepared from recognizable natural materials. Such a product is identical chemically to that isolated from the natural source.

synuclein a brain presynaptic protein that is expressed also in other tissues, but at very low levels. It is highly conserved, rodent and zebrafish α synucleins being 95% and 86% identical to human. Some mutations are linked to familial **Parkinson's disease**.

synxenic pertaining to the growth of organisms of a single species in the presence of one or more other species. Cultures in which one, two, three, or more other species are involved are described, respectively, as **monoxenic**, **dixenic**, **trixenic**, or **polyxenic**. *Compare* **axenic**. —**synxenity** *n.*; **synxenize** *or* **synxenise** *vb.*

synxenite the principal organism being studied in a **synxenic** culture.

syringomycin a cyclic lipodepsipeptide-containing nonprotein amino acids and forming cation-permeable pores. It is produced by the Gram-negative bacterium *Pseudomonas syringae* and is classified in the TC system under number 1.D.2.

syringopeptin a channel-forming cyclic lipodepsipeptide similar to **syringomycin**. It is classified in the TC system under number 1.D.3.

syrup *or* **sirup 1** a thick, sticky, viscous liquid consisting of a concentrated aqueous solution of sucrose. **2** the concentrated juice of a fruit or plant, especially of sugar cane. **3** any solution that is thick, sticky, and viscous.

system A a transport system for the uptake of neutral amino acids by animal cells. It prefers amino acids having short, polar or linear side chains, including the nonmetabolizable 2-aminoisobutyric acid. Transport is sodium ion-dependent and is reduced by lowered extracellular pH. **System ASC**, prominent in erythrocytes, is distinguished from system A on the basis of pH insensitivity and higher stereospecificity. *See also* **sodium/neutral amino acid cotransporter**, **system L**, **system N**.

system ASC *see* **system A**.

systematic evolution of ligands by exponential enrichment *abbr.*: SELEX; an iterative method for selecting high-affinity **aptamers** using directed molecular evolution to enrich oligonucleotide mixtures with respect to their ability to bind a target.

systematic name 1 (*in organic-chemical nomenclature*) a name of a molecule that is composed wholly of specially coined or selected syllables, with or without numerical affixes and locants; e.g. oxazole, pentane, cyclohex-2-en-1-ol. Such a name unequivocally specifies the complete chemical structure of a particular molecular entity. *Compare* **semisystematic name**, **trivial name**. **2** *(in enzyme nomenclature)* a name of an enzyme that describes its catalytic action as exactly as possible, thus identifying the enzyme precisely and permitting a code number to be assigned to it. Thus all proteins having the same catalytic property share the same systematic name and code number. The systematic name consists of the name of the substrate or, in the case of a bimolecular reaction, of the two substrates separated by a colon (with small and equal spaces before and after the colon); and the nature of the catalysed reaction, ending in '+ase'.

systematics *functioning as sing.* **1** the study of systems, especially the principles of classification. **2** *(in biology)* the discipline that deals with classification and nomenclature.

Système International *abbr.*: SI; Système International d'Unités; a system of units based on the metric system. *See* **SI units**.

systemic concerning the whole body of an animal or a whole plant rather than an individual part.

systemic circulation the blood vessels supplying all parts of the body except the lungs.

systemin an 18-residue polypeptide, isolated from tomato leaves, that acts as a potent activator of plant protease inhibitor genes. It induces synthesis of two protease inhibitors in tomato and potato leaves. It is the only plant polypeptide hormone-like signalling molecule presently known. It has the sequence

AVQSKPPSKRDPPKMQTD,

and is proteolytically processed from a 200-residue precursor, **prosystemin**. Plant protease inhibitors are induced by wounding, e.g. by insects or pathogens, and may be part of the plant defence mechanisms.

system L a transport system responsible for the uptake of neutral amino acids by animal cells. It prefers branched-chain and aromatic amino acids such as leucine, isoleucine, valine, phenylalanine, and the nonmetabolizable analogue 2-aminobicyclo-[2,2,1]-heptane-2-carboxylic acid (BCH). The transport mechanism is sodium-ion-independent and may be stimulated by lowered extracellular pH. It is present in a wide variety of animal cell types, including avian and mammalian. *See also* **sodium/neutral amino-acid cotransporter**, **system A**, **system N**.

system N a sodium-ion-dependent transport system responsible for the uptake of asparagine, glutamine, and histidine amino acids by animal cells. *See also* **sodium/neutral amino acid cotransporter**, **system A**, **system L**.

systems biology the study of biology as an interconnected system of dynamic, interdependent genetic, metabolic, cellular, and pathway events (rather than as a set of individual molecular components), in an attempt to understand how emergent properties appear at increasing levels of complexity and hence how biological systems function as an integral whole.

systole a rhythmical repeated contraction, as of the chambers of the heart or of a pulsating vacuole. *Compare* **diastole**. —**systolic** *adj.*

Tt

t 1 *symbol for* **a** tert+ (i.e. **tertiary**) in a chemical name (alternative to *t*). **b** triton (def. 2). **2** *abbr. for* terminator sequence.

t *symbol for* **1** time. **2** Celsius temperature (alternative to θ). **3** transference number.

t- *abbr. for* **1** tert-. **2** trans-.

$t_{1/2}$ *symbol for* half-life. *See also* **$T_{1/2}$**.

T *symbol for* **1** tera+ (SI prefix denoting 10^{12} times). **2** tritium. **3** tesla. **4** a residue of the α-amino acid L-threonine (alternative to Thr). **5** a residue of the base thymine in a nucleic-acid sequence. **6** a residue of the ribonucleoside ribosylthymine (*not* thymidine, i.e. deoxyribosylthymine; alternative to Thd). **7** twist number.

T *symbol for* **1** thermodynamic temperature. **2** transmittance. **3** period. **4** kinetic energy (alternative to E_K or K).

-*T* conformational descriptor designating the **twist conformation** (def. 2) of a five-membered ring form of a monosaccharide or monosaccharide derivative. Locants of ring atoms that lie on the side of the structure's reference plane from which the numbering appears clockwise are indicated by left superscripts and those that lie on the other side of the reference plane are indicated by right subscripts; e.g. 1,2-*O*-isopropylidene-β-L-idofuranose-$^3T_2$. *See also* **conformation**.

$T_{1/2}$ *symbol for* half-life of a radionuclide (alternative to $t_{1/2}$).

T_1 *symbol for* spin-lattice relaxation time, longitudinal relaxation time.

T_2 *symbol for* spin-lattice relaxation time, transverse relaxation time.

T_m or **t_m** or **Θ_m** *symbol for* melting temperature.

T_t *symbol for* transition temperature.

T1 *abbr. for* bacteriophage T1, a **T-odd phage**.

T2 1 or **T_2** *symbol for* L-diiodothyronine. **2** *abbr. for* bacteriophage T2, a **T-even phage**.

T3 1 or **T_3** *symbol for* L-triiodothyronine. **2** *abbr. for* bacteriophage T3, a **T-odd phage**.

T4 1 or **T_4** *symbol for* L-thyroxine. **2** *abbr. for* bacteriophage T4, a **T-even phage**.

T6 *abbr. for* bacteriophage T6, a **T-even phage**.

T7 *abbr. for* bacteriophage T7, a **T-odd phage**.

tA *symbol for* N^6-(threonylcarbamoyl)adenosine; *see* **hypermodified base**.

Ta *symbol for* tantalum.

TACE *see* **ADAM-17**.

tachometer an instrument that measures the speed at which any object is rotating. —**tachometric** *adj.*; **tachometry** *n.*

tachy+ *comb. form* denoting fast or rapid.

tachykinin any of a group of closely related vasoactive peptides characterized by a rapid stimulant effect on vascular (and extravascular) smooth muscle, producing hypotension (*compare* **bradykinin**). They also have a direct effect on nervous tissue, and stimulate salivary and lachrymal secretion. Naturally occurring tachykinins are oligopeptide amides containing 10–12 amino-acid residues, and terminating in the common sequence -Phe-Xaa-Gly-Leu-Met-NH$_2$ (where Xaa = Phe, Tyr, Ile, or Val). Mammalian tachykinins include neuromedin K, **neurokinin A** (formerly substance K) and **substance P**; others (e.g., **eledoisin**, **physalaemin**) occur in amphibians and octopods. Substance P and neurokinin A are generated from a preprotachykinin protein precursor (*see* **protachykinin β precursor**). The primary mRNA transcript is differentially processed into three mature mRNAs: α, β, and γ; the β and γ preprotachykinin mRNAs are predominant in the rat, each encoding substance P and neurokinin A.

tachykinin receptor any G-protein-associated membrane protein that binds **tachykinins** and mediates their transmembrane action. There are three classes: (1) NK$_1$ (407 amino acids, human), agonist potency: substance P (SP) > neurokinin A (NKA) > neurokinin B (NKB); (2) NK$_2$ (398 amino acids, human), agonist potency: NKA > NKB >> SP; and (3) NK$_3$ (452 amino acids, rat), agonist potency: NKB > NKA > SP. In all, ligand binding activates phospholipase C.

tachyphylaxis a decline in the response to repeated application of an agonist. *See also* **desensitization, fade**.

tachyplesin any peptide antibiotic of arthropods, 17 or 18 amino acids long and with a C-terminal arginine α-amide. Tachyplesins are abundant in hemocyte debris of horseshoe crabs (*Tachypleus* spp.), and are active against Gram-positive and Gram-negative bacteria, and fungi such as *Candida albicans*. It is classified in the TC system under number 1.C.34.

***tac* promoter** a constructed hybrid **promoter** that contains the –35 sequence of the *trp* promoter and the –10 sequence of the *lacZ* promoter. *See* **minus 10 region**.

Tacrolimus *see* **FK506**.

-tactic refers to the ability of a motile cell or organism to move along a gradient, e.g. of a chemical (chemotactic) or light (phototactic).

tactic polymer a regular polymer in whose molecules there is an ordered structure with respect to the configurations around at least one main-chain site of stereoisomerism per **constitutional base unit**.

tactile capable of exciting the sense of touch; relating to the sense of touch (as a sense organ).

tactoid a *paracrystal*-like aggregate; the term may be applied to the rodlike filaments of hemoglobin that form in **sickle cells**, or to assemblies of virus particles (e.g. of tobacco mosaic virus).

TAE a buffer for agarose gel electrophoresis comprising 40 mM Tris-acetate and 1 mM EDTA.

tafazzin one of a family of acyl-CoA synthetases that is important in cardiolipin synthesis. Mutations in the gene locus (at Xq28) produce **Barth syndrome**.

tag 1 to identify by a marker; for example, a cell may be tagged by attachment of a (fluorescent or radiolabelled) antibody. **2** an identifier used in tagging.

Tagamet *see* **cimetidine**.

D-tagatose a *nonsystematic name for* D-*lyxo*-2-hexulose, a hydrolysis product of some plant gums.

α-D-pyranose form

Tagit a *proprietary name for* N-succinimidyl 3-(4-hydroxyphenyl) propionate; it is used in the preparation of **Bolton and Hunter reagent** for tagging peptides and proteins with radioiodine.

tail 1 an extension growing or appending from a point of origin or attachment. For example, the fatty-acyl groups of a phospholipid and the glycerol to which they are bonded are referred to as the tail. *Compare* **headgroup**. **2** *(in chromatography)* a region of more slowly moving or decomposed material extending back from a band towards the **origin**. *See also* **poly(A) tail, tailing**.

tailing 1 the addition of a stretch of identical nucleotides to the ends of a restriction fragment using **terminal deoxynucleotidyl transferase**. **2** the spreading out of a band or spot in a chromatographic process.

TAIL PCR *abbr. for* thermal asymmetric interlaced polymerase chain reaction; a method for finding previously unknown sequences adjacent to known sequences in genomes. It is useful for recovering sequences flanking sites of insertional mutagenesis caused by integration of a plasmid, T-DNA, or transposon.

Takahara disease *an alternative name for* **acatalasemia**.

Takifugu rubripes *see* **fugu**.

Tal *symbol for* a residue, or sometimes a molecule, of **talose**.

talaric acid one of the **hexaric acids**; *see also* **aldaric acid**.

talc a soft mineral composed of hydrated magnesium silicate, $Mg_3Si_4O_{10}(OH)_2$, with small quantities of metal impurities. In its powdered form it is known as talcum powder, or French chalk, and in granular form as soapstone (steatite). Powdered talc is used as a lubricant, as a filler in paper and other products, and in cosmetics, etc.

talcum powder powdered **talc**.

talin a protein involved in connections between the cytoskeleton and the plasma membrane. A phosphoprotein, it binds strongly to **vinculin**, and less strongly to **integrins**. It is a founder member of the **TERM** family.

talo+ *or (before a vowel)* **talos+** *comb. form of* **talose**.

talo- *prefix (in chemical nomenclature)* indicating a particular configuration, a set of (usually four) contiguous CHOH groups as in the acyclic form of D- or L-**talose**. *See also* **monosaccharide**.

talopeptin *N*-(6-deoxy-α-L-talopyranosyl-oxyphospho)-L-leucyl-L-tryptophan; an inhibitor of several metallopeptidases, including **thermolysin**. *Compare* **phosphoramidon**.

talosaminouronic acid a component of the **peptidoglycan** of archaea.

talose *symbol*: Tal; *the trivial name for* the aldohexose *talo*-hexose, which differs from glucose in the configuration of the groups at C-2 and C-4; there are two enantiomers.

α-D-talose

Tamm–Horsfall glycoprotein *or* **uromodulin** the principal glycoprotein of human urine. It occurs in other species, e.g. the cow and rat. The molecule is filamentous, with an M_r of several million, and is composed of subunits of M_r ~80 000; it contains about 25% carbohydrate and a small amount of lipid. An abnormally high M_r is observed in cystic fibrosis. *See also* **proteinuria**. [After Igor Tamm (1922–71), US virologist, and Frank Lappin Horsfall (1906–71).]

tamoxifen [*Z*]-2-[4-(1,2-diphenyl-1-butenyl)-phenoxy]-*N*,*N*-dimethylethanolamine; an estrogen antagonist used in the treatment of breast cancer and suggested for prophylaxis in women with a family history of breast cancer. It can behave as a pure estrogen agonist, a partial agonist, or an antagonist, depending on the species of animal, the target organ examined, and the endpoint measured. It is a **genotoxic** carcinogen in the rat.

tan *abbr. for* tangent (def. 1).

tandem affinity purification *abbr.*: TAP; a separation technique designed to isolate protein complexes, whereby an IgG-binding domain sequence tag is added to the gene for a target protein, and the protein is synthesized with the section encoded by the tag. Proteins with the domain are separated from a mixture using an affinity column containing IgG beads, to which they stick. If the target protein is part of a complex, proteins associated with it are also retained on the column.

tandem-crossed immunoelectrophoresis *see* **crossed immunoelectrophoresis**.

tandem mass spectrometry *abbr.*: MS/MS; a separation technique in which samples are first ionized and then subjected to two or more sequential stages of **mass spectrometry** analysis using either laser-based methods – e.g. MALDI (matrix-assisted laser desorption/ionization) and SELDI (surface-enhanced laser desorption and ionization) – or electrospray-based (e.g. ESI) methods.

tandem repeat 1 (of DNA sequences) an arrangement in which two or more copies of a particular base sequence are situated immediately one after the other. Tandem repeats are commonly found in **telomeres**, the repeated sequence being characteristic of species and source of DNA, and of relatively short length (up to about 10 bases). They also occur in **highly repetitive DNA**. **2** (of genes) a cluster of adjacent multiple copies of the same gene. The genes for ribosomal RNA (rRNA) are characteristically contained in tandem clusters, sometimes referred to as **ribosomal DNA** (rDNA). *See also* **concatemer**.

Tanford–Kirkwood theory a theory of protein titration curves based on a model in which the charges are taken to be discrete unit charges located at fixed positions on the protein molecule. [After Charles Tanford (1921–) and John Gamble Kirkwood (1907–59).]

tangent 1 *abbr.*: tan; a function of an angle, being, in a right-angled triangle, the ratio of the side opposite the given angle (if acute) to that of the side opposite the other acute angle. The tangent of an obtuse angle is numerically equal to that of its supplement but of opposite sign. **2** a straight line that touches, but does not cut, a curve or a curved surface.

tangential-flow filtration a convenient and rapid method of concentrating cells, cell organelles, or other particulate material, or for eliminating such material from suspensions. In a special apparatus, the suspension flows over ('tangential flow') a semipermeable membrane; only material smaller than the membrane pores will pass through the membrane, forming the filtrate, leaving larger matter to be collected (retentate).

Tangier disease a rare autosomal recessive condition in humans in which there is a deficiency or complete absence of **high-density lipoprotein**. Homozygotes have very low levels of **apolipoprotein A-I**. There is accumulation of cholesterol esters in cells of the mononuclear phagocyte system. The condition is relatively benign as organ function is not affected. [After Tangier Island, Chesapeake Bay; the disease was first found in inhabitants of this island.]

tankyrase a 142 kDa protein associated with telomeres in human cells, and also present in the Golgi apparatus, the nuclear pore complex, and mitotic centrosomes. It has a catalytic domain homologous to that of **PARP-1**, and 24 tandem repeats of the ankyrin motif.

It does not require DNA for activity and has **TEP-1** as substrate, causing release of the latter from telomeric DNA.

tannic acid *or* **tannin** *or* **gallotannin** *or* **gallotannic acid** *an imprecise term used for* a mixture of hydrolysable **tannins** (def. 1).

tannin 1 any polyphenolic plant product that can be used to tan animal skins to produce leather. Tannins are divided into **hydrolysable tannins** (the larger group) and **condensed tannins**. The former consist of a polyhydric alcohol, usually glucose, esterified with either gallic acid or hexahydroxydiphenic acid. Condensed tannins are polymers of flavans, never contain sugar residues, and are often termed **flavolans**. Tannins form insoluble tannates with proteins, which forms the basis of **tanning** (def. 1). The release of tannins during the extraction of plant enzymes can lead to formation of insoluble enzyme tannates. The addition of polyvinylpyrrolidone may help to avoid this problem. Good tanning agents should have an M_r of between 500 and 3000 and contain sufficient phenolic hydroxyl groups (1–2% of M_r) to form effective cross-links with protein. **2** *a common name for* **tannic acid**.

tanning 1 a technique, known by the Egyptians around 3000 BC, for producing leather from animal skins by the use of **tannin** (def. 1). Modern methods using chrome salts are faster, taking a few hours rather than weeks. **2** the treating of red blood cells with tannic acid. Such cells are used in a test for anti-thyroglobulin antibodies, after they have been treated with thyroglobulin. The tanning enhances sensitivity.

T antigen 1 any early gene product of a genetic locus of certain transforming DNA viruses, including adenovirus, simian virus 40 (SV40), and polyoma virus. The substance originally detected in nuclei of transformed cells by immunofluorescence (hence 'T' for tumour or transformation) led to the identification of several early proteins that are needed for replication. Two products are formed by SV40, known as **large T antigen** (*or* T antigen) and **small t antigen** (*or* t antigen). Polyoma virus produces large T antigen, middle T antigen, and small t antigen. Large T antigen of SV40 alone is sufficient to immortalize primary rodent cells and transform established rodent cell lines. It is a multifunctional protein that is a potent transcriptional activator which activates promoters containing a TATA box or initiator element and at least one upstream transcription-factor-binding site. Middle T antigen of polyoma virus is essential for polyoma-virus-mediated tumorigenesis and cell transformation. It activates cellular tyrosine kinases, including c-Src, c-Yes, and Fyn, and binds phosphatidylinositol 3-kinase and protein phosphatase 2A. Small t antigen of SV40 has a helper effect in SV40 transformation. It binds protein phosphatase 2A and regulates the phosphorylation and transcriptional transactivation of **CREB protein**. It stabilizes p53 in a manner that correlates with enhanced transformation efficiency of SV40. **2** an antigen found on normal human erythrocytes that is unreactive unless the cells are first treated with neuraminidase. **3** one of the principal protein antigens in the cell wall of *Streptococcus* spp. **4** *abbr. for* transplantation antigen (i.e. a class I MHC antigen); see **major histocompatibility complex**.

TAP *abbr. for* tandem affinity purification.

tapeworm any parasitic flatworm (phylum Platyhelminthes) belonging to the class Cestoda.

Taps *or* **TAPS** *abbr. for* N-tris(hydroxymethyl)methyl-3-aminopropanesulfonic acid; 3-[2-hydroxy-1,1-bis(hydroxymethyl)ethyl]-amino propanesulfonic acid; a **Good buffer substance**, pK_a (20°C) = 8.4.

***Taq* DNA polymerase** a heat-stable DNA polymerase from the thermophilic bacterium *Thermus aquaticus*. It is widely used in the **polymerase chain reaction** (PCR). The enzyme lacks $3' \rightarrow 5'$ exonuclease activity and therefore has no proofreading function. It has a template-independent terminal transferase activity with a prefer-

ence for the addition of deoxyadenosine to the 3′ ends of PCR products.

Tar a protein that serves as a multifunctional receptor of *E. coli*, used for detection of aspartate and temperature. This capability involves methylation at up to eight sites in the protein's cytoplasmic domain.

TAR *abbr. for* tape archive; a file format (not necessarily on a tape) used on UNIX (or Linux) computers. TAR files usually have the extension .tar; they are not **compressed** but might be compressed in which case the extension would become .tar.Z or .tar.gz. The program tar is needed to extract the files from the archive.

tare 1 to compensate for the weight of a container, etc. when weighing, by adjusting the balance to zero with the empty container in place before adding the material to be weighed. **2** the weight of the tared vessel or material.

target the object at which an action or process is directed. For example, a **target cell** is the cell under attack by a cytolytic or killer cell or other cytolytic agent, and a **target tissue** is the tissue that is acted on by any hormone, growth factor, or other agonist. *See also* **drug target**.

target discovery *or* **target identification** the process of identifying a biological entity (usually a protein or gene) that interacts with, and whose activity is modulated by, a particular compound. Once a suitable target has been identified, it may then go on to **target validation**.

targeting subunits protein subunits that specify the location and catalytic and regulatory properties of protein phosphatases and kinases.

target repeat *see* **transposon**.

target site the site where a **restriction endonuclease** cleaves a DNA molecule; usually the **recognition site** (def. 2).

target theory *(in radiation chemistry)* a theory stating that all single hits on biologically active macromolecules by particles or quanta of ionizing radiation result in damaged and inactive molecules, that no partially active molecules are formed, and that all unhit molecules retain full activity.

target validation the process by which the role of a target (usually a specific gene or protein) is confirmed within a pathogenic process.

tartaric acid 2,3-dihydroxybutanedioic acid; 2,3-dihydroxysuccinic acid; an **aldaric acid** formed by oxidation of a **tetrose** at C-1 and C-4. The L$_g$(+) (*see* **d/l convention**), or 2*R*,3*R*, enantiomer occurs widely in plants, especially grape juice, and in fungi and bacteria; the D$_g$(–) (or 2*S*,3*S*) enantiomer and **meso** form have limited distribution in plants. The racemic mixture, DL-tartaric acid, sometimes occurs during wine manufacture; it was originally known as **paratartaric acid**. The first determination of an absolute configuration by anomalous dispersion of X-rays was made on sodium rubidium L(+)-tartrate in 1951. The property of **optical activity** of some substances was first described by **Pasteur** in experiments with crystals of sodium ammonium tartrate.

L-form

meso-form

Tarui disease *an alternative name for* **glycogen disease** type VII.

tasiRNA *abbr for* trans-acting **siRNA**, siRNAs that direct cleavage of endogenous mRNAs having a region of complementarity to the siRNAs but little overall resemblance to the genes from which the siRNAs were transcribed.

tat a membrane protein complex, found in bacteria and plants, that carries folded proteins (with their cofactors) in their intact state through energy-transducing membranes. The system recognizes the **twin-arginine** motif in the protein to be transported.

TATA-binding protein *abbr.*: TBP; a component of transcription factor IID (*see* **TFII**). *See also* **RNA polymerase**.

TATA box *or* **Hogness box** a base-sequence element common in eukaryotic promoters, characterized by the consensus TATAAAA. It binds a general transcription factor and specifies the position where **transcription** is initiated.

tau 1 *symbol*: τ (lower case) *or* T (upper case); the nineteenth letter of the Greek alphabet. For uses *see* **Appendix A**. **2** *an alternative name for* transcription factor TFIIIC; *see* **TFIII**.

tau protein a microtubule-associated protein that promotes microtubule assembly, and stabilizes microtubules. It constitutes at least a part of the paired helical filament (PHF) core found in **Alzheimer's disease**.

tau-protein kinase EC 2.7.11.26; a protein serine/threonine kinase that phosphorylates **tau protein**.

taurine 2-aminoethanesulfonic acid; it is derived metabolically from cysteine by oxidation of the sulfhydryl group to sulfoxide followed by decarboxylation of the resulting cysteic acid. *N*-conjugated with cholic acid, it forms the bile salts **taurochenodeoxycholate, taurocholate**, and **taurodeoxycholate**.

taurochenodeoxycholate the anionic form of the amido conjugate of **taurine** and **chenodeoxycholate**. It is a major **bile acid**.

taurocholate the anionic form of taurocholic acid, *N*-cholyltaurine, a major **bile salt** of humans and other mammals. It is useful as an anionic bile-salt detergent (aggregation number 4; CMC 3–11 mM). Taurine conjugates of other bile acids are known, e.g. **taurochenodeoxycholate** and **taurodeoxycholate**.

taurocyamine *N*-(aminoiminomethyl) taurine, found in marine polychaete worms. The guanidine group can be phosphorylated to form phosphotaurocyamine, which acts as a **phosphagen**.

taurodeoxycholate the anionic form of taurodeoxycholic acid, *N*-(7-deoxy)cholyltaurine. It is useful as an anionic bile-salt detergent (aggregation number 6; CMC 1–4 mM).

taurolipid any lipid consisting of a non-hydroxy fatty acid esterified with a hydroxy group of tri-, tetra-, or pentahydroxystearic acid in amide linkage with taurine. Taurolipids are characteristic of the protozoan *Tetrahymena*.

tauryl the acyl group, $H_2N–CH_2–CH_2–SO_2–$, derived from taurine.

taut form *abbr.*: T form; *see* **tense form, Monod–Wyman–Changeux model for allosteric interaction**.

tautomer any of the alternative structures that may exist as a result of **tautomerism**.

tautomerism a form of constitutional isomerism in which a structure may exist in two or more constitutional arrangements, particularly with respect to the position of hydrogens bonded to oxygen. The isomers, or tautomers, are readily interconvertible and exist in equilibrium. The best-known example is **keto–enol tautomerism**, in which the structures X–C(=O)–CH$_2$–X (keto) and X–C(OH)=CH–X (enol) exist, in proportions determined by the equilibrium position for the conditions pertaining (normally strongly favouring the keto form). Other examples include quinone–quinol and **amino–imino tautomerism**. —**tautomeric** *adj.*

-taxis refers to the movement of a motile cell or organism along a gradient, e.g. of a chemical (chemotaxis) or light (phototaxis).

taxol an antitumour and antileukemic agent isolated from the bark of the Pacific yew, *Taxus brevifolia*. It is used clinically to treat refractory ovarian cancer. It inhibits mitosis and blocks the cell cycle at G or M phases through enhancement of the polymerization of **tubulin** and the consequent stabilization of the **microtubules**. Taxol is manufactured semisynthetically from a material present in *T. baccata*. A total chemical synthesis has also been achieved. A compound with similar properties, isolated from the myxobacterium *Sporangium cellulosum*, is **epothilone**; this is more water-soluble than taxol and can be produced in quantity by fermentation. Taxol and its analogues are known as **taxanes**. One proprietary name is Paclitaxel. A proprietary name of a semisynthetic analogue is Docetaxel.

taxon (*pl.* **taxa**) a named taxonomic group of any rank; a particular taxon comprises organisms having certain genetic characters in common. Strictly, the term should be used to denote only named groups, e.g. Chordata, Hominidae, etc., not as a generic term for the ranks themselves, e.g. phylum, family, etc.

Tay–Sachs disease *or* **G$_{M2}$ gangliosidosis** an inherited neurological metabolic disorder resulting from deficiency of the lysosomal enzyme, **hexosaminidase** (form A, deficiency in α chain). This leads to the accumulation of its substrate, ganglioside G$_{M2}$ (*see* **ganglioside**), in the brain. The disease is relatively common (1 in 2500 live births) among Ashkenazy Jews. There is motor and mental deterioration with an exaggerated startle response to sound; by 18 months of age, most patients are blind, deaf, and spastic; they die by the age of 3 years. *See also* **saposin**. [After Warren Tay (1843–1927), British physician, and Bernard Sachs (1858–1944), US neurologist.]

Tb *symbol for* terbium.

TBARS *abbr* for thiobarbituric acid-reactive substances, compounds of low-molecular-mass, particularly malondialdehyde, that are formed during the decomposition of lipid peroxidation products. They are measured in order to estimate the degree of cellular damage caused by oxygen radicals.

TBE a buffer for agarose gel electrophoresis that comprises 90 mM Tris-borate and 1 mM EDTA.

TBP *abbr. for* TATA-binding protein (*see* **TFII**, **RNA polymerase**).

tBu *abbr. for* tertiary butyl (alternative to Bu^t).

TBW *abbr. for* total body water; *see* **body water**.

Tc *symbol for* technetium.

TCA *abbr. for* **1** tricarboxylic acid. **2** trichloroacetic acid.

T-cap a protein of the Z disc of skeletal and cardiac muscle that is associated with the N-terminal portion of **titin**. Defects in T-cap or in titin lead to dilated cardiomyopathy.

TψC arm the base-paired segment of the clover-leaf model of **transfer RNA** to which the loop containing the base sequence TψC is attached (ψ = pseudouridine).

T cell *an alternative name for* **T lymphocyte**.

T_H cell *abbr. for* **helper cell**.

T-cell growth factor *abbr.*: TCGF; *an alternative name for* **interleukin 2**.

T-cell receptor a glycoprotein of the **immunoglobulin** superfamily. It is present on T-cell (**T-lymphocyte**) membranes and resembles class II MHC antigens (*see* **major histocompatibility complex**) but its α and β chains are disulfide bonded. T-cell receptors recognize target antigen in combination with MHC, transduce the signal into the T cell, probably through mediation by associated CD3 molecules (see **CD marker**), and bring about T-cell activation. A small number of adult T-cell receptors have γ and δ chains. This class of receptor appears first in ontogeny.

T-cell-replacing factor *abbr.*: TRF; *a former term for* some interleukins. *See* **interleukin 1**, **interleukin 5**.

TCGF *abbr. for* T-cell growth factor (i.e. **interleukin 2**).

T chain *an alternative name for* **secretory piece**.

tcl *see* programming language.

T-coffee *see* multiple alignment program.

TCP-1 *abbr. for* T-complex protein 1; a **molecular chaperone** that forms two stacked rings, 12–16 nm in diameter and is associated with other proteins in a 850–900 kDa complex. It is similar in function to Hsp60 (a **heat-shock protein**) in mitochondria and the **groEL** product in bacteria (*see also* **chaperonin**).

TC system a classification system of membrane transport proteins, using a five-digit designation for each of the defined and sequenced transport proteins. It comprises several classes (1 Pores and channels; 2 Electrochemical-potential-driven transporters; 3 Primary active transporters; 4 Group translocators; 5 Transmembrane electron carriers; 8 Accessory factors involved in transport; 9 Incompletely characterized transport systems), each class containing several subclasses; the subclasses are divided into families, the families into subfamilies and these then contain all the relevant individual proteins. Thus, for example, bacterial Ag^+-exporting ATPase is TC 3.A.3.5.4.

T-DNA *abbr. for* transfer DNA; that part (23 kb) of the *Agrobacterium* **Ti plasmid** that is incorporated into the genome of infected plant cells. It codes for an **opine**, usually **nopaline** or **octopine**, which the transformed cells produce and the bacteria then use. Each end of the T-DNA is flanked by a nearly identical 25 bp sequence; the righthand one is essential for the transfer process.

TDP *abbr. for* **1** ribosylthymine diphosphate *or* ribothymidine diphosphate. **2** thymidine diphosphate (dTDP is recommended).

Tdr *or* **TdR** *or* **TDR** *abbr. for* thymine deoxyriboside (i.e. thymidine). Use deprecated; the symbol dThd is recommended.

TdT *abbr. for* **terminal deoxynucleotidyltransferase**.

Te *symbol for* tellurium.

TE a buffer commonly used for dissolving DNA and comprising 10 mM Tris-chloride and 1 mM EDTA at pH 7.4, 7.6, or 8.0.

TEA *abbr. for* **1** triethanolamine; N(CH_2CH_2OH)_3, a base used in the preparation of buffers, pK_a (25°C) 7.90. **2** tetraethylammonium; used (as the chloride) to block sympathetic and parasympathetic ganglionic transmission.

TEAE *abbr. for* the triethylaminoethyl group, (C_2H_5)_3NCH_2CH_2–. It is used in hydroxylated chromatographic support matrices, e.g. cellulose, Sephadex, to form anion-exchange columns.

α-tectorin a protein (2155 amino acids) that, like otogelin, is specific to the cochlear tectorial membrane. It contains a C-terminal transmembrane segment, an N-terminal domain similar to one in **entactin**, and four vWF domains. Several mutations are associated with an autosomal recessive form of isolated deafness.

teichoate the anionic form of **teichoic acid**.

teichoic acid any polymer occurring in the cell wall, membrane, or capsule of Gram-positive bacteria and containing chains of glycerol phosphate or ribitol phosphate residues. Additionally, carbohydrates are linked to glycerol or ribitol, and some –OH groups are esterified with D-alanine.

tektins a family of similar proteins that are adapted to the assembly of a variety of eukaryotic cell structures. Tektins have common features that adapt them to sensitive and specific associations with molecules of their own kind; they have similar amino-acid compositions: they include actin, actin-like proteins, some microtubule proteins, mitochondrial structural protein, and erythrocyte membrane proteins.

tele a designator of the position of one of the nitrogens of the histidine imidazole ring, indicating the nitrogen furthest from the side chain. It is often symbolized by τ, as in $N^τ$. Compare **pros**.

teleology the doctrine or study of ends or final causes, especially as related to the evidences of design or purpose in nature; the use of design, purpose, or utility as an explanation of any natural phenomenon. An example of a teleological concept in biology is that organs are made to react for the good of the organism as a whole.

teleonomy the idea that the existence of a structure or a function in an organism must have conferred an evolutionary advantage upon that organism.

teleost any member of the infraclass Teleosti, comprising fishes with rayed fins and a swim bladder; it includes most of the extant bony fishes.

telestability the transmission of three-dimensional stability from a stable part of a macromolecule to an inherently less stable part.

telethonin a protein that colocalizes with actin in the Z disc of skeletal muscle, where it interacts with **FATZ** and, through its N-terminal 140 residues, with the same region of titin. Two nonsense mutations in the gene at 17q12 result in autosomal recessive limb-girdle muscular dystrophy type 2G.

telokin the C-terminal sequence of smooth muscle **myosin-light-chain kinase** expressed as an independent protein. It binds calmodulin.

telolysosome a **lysosome** that has accumulated so much indigestible residue that it is unable to accumulate further material to be digested. The structure still contains enzymes but these are no longer renewed. It is a type of **residual body**.

telomerase EC 2.7.7.–; *other name*: telomerase reverse transcriptase (*abbr.*: TERT); a DNA polymerase enzyme involved in **telomere** formation, and in telomere elongation to maintain the telomere sequences during replication. The catalytic subunit contains several sequence motifs characteristic of reverse transcriptase and at least one telomerase-specific motif. Activity is undetectable in most human somatic cells but is present in germ-line cells, in ≈90% of tumours, and in immortalized cell lines. The enzyme recognizes the G-rich strand of an existing telomere repeat sequence, elongating it in the 5' to 3' direction. The yeast *Candida albicans* contains two highly homologous catalytic subunits each of 867 amino acids. In fission yeast, the core enzyme contains Est2p before the S phase, and becomes activated late in S phase when Est1p and Est3p are recruited to the core (*see* **EST1**). The RNA component includes a template for synthesizing the G-rich repeating sequence, and in humans is bound by dyskerin.

telomerase-associated protein a human protein of 2629 amino acids that interacts with the RNA component of **telomerase**. It contains two sequence repeats in the N-terminal region, a central NTP-binding domain, and WD40 repeats in the C-terminal region.

telomere the DNA–protein structure that seals either end of a chromosome. Telomeric DNA consists of simple tandemly repeated sequences specific for each species. In human germline cells the 3' end of each chromosome contains 1000–1700 repeats of the hexanucleotide TTAGGG, and the 5' end contains the complementary repeats. In somatic cells the repeat number decreases with each round of replication. Telomeric sequences are added to the 3' terminus of chromosomal DNAs by **telomerase**, thereby maintaining telomere length. **TEP-1** binds to telomeric repeats. —**telomeric** *adj*.

telomere-specific protein-1 *see* **TEP-1**.

telomeric repeat-binding factor (*abbr.*: TRF) *an alternative name for* **TEP-1**.

telopeptide any of the peptides that protrude from the triple-helical body of **tropocollagen**, and have a composition different from the triple-helical portions. They are removed by proteases, with concomitant breakage of intramolecular interchain bonds. They are thought to occupy near-terminal positions. The antigenic response to injected heterologous tropocollagen is directed against these telopeptide appendages external to the triple helix.

telophase the final stage of **mitosis** or **meiosis** in which the nuclei form in the daughter cells.

TEM or **T.E.M.** abbr. for **1** transmission electron microscope; see **electron microscope**. **2** Tris–EDTA–mercaptoethanol (buffer).

TEMED abbr. for N,N,N',N'-tetramethylethylenediamine; N,N,N',N'-tetramethyl-1,2-diaminoethane; a polymerizing agent used in the preparation of polyacrylamide gels. It stabilizes free radicals produced by ammonium persulfate.

temperate phage a **bacteriophage** that, following infection of its host bacterium, may either enter the lytic cycle, like a **virulent phage**, or establish a symbiotic relationship with the host cell that results in its perpetuation in the descendants of the bacterium. Bacteria carrying temperate phage are said to be **lysogenic**, while the carried phage is called a **prophage**.

temperature abbr.: temp.; a fundamental physical quantity of a body, region, or system that determines the direction of net heat flow between it and its surroundings. If two systems have the same temperature they are in thermal equilibrium and there is no net heat flow between them. If they have different temperatures they are not in thermal equilibrium and heat flows from the system of higher temperature to that of lower temperature. Essentially there are two different methods of measuring temperature. One is to use a scale of values set up between arbitrary fixed points based on reproducible temperature-dependent events. For example, the **Celsius temperature** scale uses a freezing point and boiling point of water as fixed points, to which are assigned the values 0 and 100, respectively; the scale between them is divided into 100 degrees. Although widely used for many practical purposes, it lacks a theoretical basis and is unsuitable in many scientific contexts. Instead, the concept of **thermodynamic temperature** is used. This is defined in terms of the second law of thermodynamics (see **thermodynamics**) and is independent of any particular substance. However, in practice thermodynamic temperatures cannot be measured directly. Consequently, a practical realization of the thermodynamic temperature scale is used, called the **International Practical Temperature Scale** (IPTS). This is based on a number of fixed points, and is measured in **kelvins**; it defines temperature down to 13.81 kelvins, the triple point of equilibrium hydrogen.

temperature coefficient of a reaction symbol: Q_{10}; the factor by which the velocity of a chemical reaction is increased by raising the temperature, t, by 10°C. Usually it is the ratio of the rate of reaction at $(t + 10°)$ to that at t, denoted by $Q_{10} = k_{t+10}/k_t$, where k_{t+10} and k_t are the rate constants at the respective temperatures. The Q_{10} value of homogeneous chemical reactions is generally in the range 2–3; many enzyme-catalysed reactions and physiological processes exhibit, over a limited range of temperatures, a Q_{10} value of ≈2. See also **Arrhenius equation**.

temperature jump a technique used in **relaxation kinetics** for the study of enzyme mechanisms in which the temperature is raised rapidly (e.g. over about 1 μs), the reaction being monitored during this and the subsequent (very short) period during which the system relaxes to a new equilibrium.

temperature-sensitive mutation a mutation that is manifest in only a limited range of temperature. The mutant carrying such a mutation behaves normally at the lower, **permissive temperature**, but when grown at the lower or higher, **restrictive temperature**, it shows the mutant phenotype.

template a macromolecule whose structure serves as a pattern for the synthesis of another macromolecule. The term applies particularly to a nucleic acid on which another nucleic acid of complementary base sequence is synthesized, as in genetic **transcription**, or in the case of messenger RNA, on which a polypeptide is synthesized (genetic **translation**).

template RNA an alternative name for **messenger RNA**.

template strand see **noncoding strand**.

TEMPO abbr. for 2,2,6,6-tetramethylpiperidinooxy; 2,2,6,6-tetramethylpiperidine 1-oxyl; a stable free radical used as a spin label in electron spin-resonance spectroscopy.

temporal gene any gene that 'programs' a system in time, determining the activation of genes controlling different developmental stages in various cell types.

tenascin a glycoprotein of the extracellular matrix that is widely distributed and prominent during embryonic development, being produced by mesenchymal and glial cells. In adults it is present in small amounts in the central nervous system, cartilage, soft-tissue stroma, and close to basement membranes, but is prominent in a variety of tumours and in areas of wound healing. It consists of six identical (or similar) polypeptide chains, which radiate from a central knob, like the spokes of a wheel, with a knob at the free end of each. Several isoforms (1573 to 2203 amino acids) result from alternative splicing from one gene, and differ by tissue and time of appearance during development. The central knob contains the hydrophobic N-termini and their eight cysteine residues, which play a part in hexamer formation. Each chain contains ≈15 epidermal growth factor repeats and up to 15 ≈90-residue fibronectin repeats, and a C-terminal 210-residue segment rich in basic amino acids. Tenascin interacts with the cell-surface transmembrane proteoglycan **syndecan** and with fibronectin. It promotes or inhibits cell adhesion depending on cell type. Other names: hexabrachion; cytotactin; neuronectin; gmem; ji; myotendinous antigen; glioma-associated extracellular matrix antigen; gp150–225.

teniposide a semisynthetic lignan derivative prepared from **podophyllotoxin**. It acts like **etoposide**, but is a stronger cytotoxin.

ten-minute phosphate see **labile phosphate**.

tense form or **taut form** abbr.: T form; the form of an allosteric protein that has a lower affinity for a ligand than the R form (**relaxed form**). See **Monod–Wyman–Changeux model** (for allosteric interaction).

tenside any of the group of surface-active compounds widely known as **detergents**. The name derives from their effects on surface tension.

tension 1 the concentration of a gas expressed as the partial pressure of the gas. **2** the force exerted by stretching any material. **3** voltage, electromagnetic force, or potential difference, as in high tension, low tension.

tensor 1 *(in mathematics)* a quantity expressing the ratio in which the length of a **vector** is increased. *Compare* **scalar. 2** *(in anatomy)* a muscle that stretches or tightens some part while not altering the direction of the part.

tentoxin cyclo[L-leucyl-*N*-methyl-(*Z*)-dehydrophenylalanyl-glycyl-*N*-methyl-L-alanyl]; a cyclic tetrapeptide isolated from *Alternaria alternata*. It induces chlorosis in some plant species when applied to the germinating seedling.

TEP-1 *abbr. for* telomere-specific protein-1; *other name*: telomeric repeat binding factor (*abbr.*: TRF); a protein (439 amino acids) associated with telomeres, where it is a substrate for tankyrase, which causes its release from telomeric DNA. It is also a component of the large ribonucleoprotein complexes known as **vaults**, where it is a substrate for V-PARP. It contains an N-terminal aspartate and glutamate-rich region, two nuclear localization signals, and a sequence strongly homologous to the DNA-binding repeats of MYB.

Ter *see* reactancy of enzymes.

tera+ *symbol*: T; SI prefix denoting 10^{12} times.

terato+ *or (before a vowel)* **terat+** *comb. form* denoting congenital abnormality.

teratocarcinoma a **teratoma** that includes carcinomatous cells.

teratogen a substance that produces malformations in embryos. —**teratogenic** *adj.*

teratoma a tumour composed of disorganized tissues that are foreign embryologically to the tissue normally found at the site. Teratomas most frequently occur in testis or ovary, and may be derived from remnants of multipotential embryological cells that can differentiate into many cell types.

TERM a family of F-actin binding proteins, named from **talin** and the related proteins **ezrin**, **radixin**, and **moesin**.

terminal amino acid the amino-acid residue at the end of a peptide or polypeptide chain. The **C-terminal amino acid** is the one that has a free α-carboxyl group (*see* **C terminus**); the **N-terminal amino acid** is the one that has a free α-amino group (*see* **N terminus**).

terminal deletion *see* deletion.

terminal deoxynucleotidyltransferase *abbr.*: TdT; EC 2.7.7.31; *recommended name*: DNA nucleotidylexotransferase; *systematic name*: nucleoside-triphosphate:DNA deoxynucleotidylexotransferase; *other names*: terminal transferase; terminal deoxyribonucleotidyltransferase. An enzyme, commonly isolated from calf thymus, that catalyses the successive transfer of deoxynucleotide residues from the corresponding nucleoside triphosphates to the 3′ end of an oligo- or polydeoxynucleotide. It is a template-independent, Mg^{2+}-dependent **DNA polymerase**.

terminal oxidase an enzyme in a respiratory chain that transfers electrons from other carriers such as quinol or cytochrome *c* to oxygen, forming water..

terminal tip *an alternative name for* **leaf** (def. 3).

terminal transferase *see* terminal deoxynucleotidyltransferase.

terminal web the cortical network of **cytoskeleton** filaments under the plasma membrane. Actin filament bundles may be attached, and it contains a dense network of **spectrin** molecules.

termination codon *or* **terminator codon** any **codon** that signals the termination of a growing polypeptide chain. *See also* **amber codon, ochre codon, stop codon.**

termination factor any protein that participates in the termination of either transcription (*see* **rho protein**) or translation (*see* **release factor**).

terminator (of transcription) a DNA sequence lying just downstream of the coding segment of a gene that is recognized by RNA polymerase as a signal to stop synthesizing messenger RNA during **transcription**.

termolecular reaction *see* molecularity.

ternary consisting of three parts.

ternary complex any complex formed between three chemical entities, e.g. a complex formed between an enzyme and two substrate molecules.

terpene any lipid that consists of multiples of **isoprene** (C_5H_8) units. Terpenes may be acyclic, cyclic, or multicyclic, saturated or unsaturated, and can be substituted by various functional groups, e.g. OH. Terpenes containing 10, 15, 20, 30, 40, or more carbon atoms per molecule are known as **mono-, sesqui-, di-, tri-, tetra-,** or **polyterpenes** respectively. The class includes alcohol, aldehyde, and carboxylic acid derivatives of terpene hydrocarbons. Isoprene is the only naturally occurring **hemiterpene**, i.e. having five carbons, but the phosphorylated compounds Δ^2- and Δ^3-isopentenyl diphosphates are key intermediates in terpene biosynthesis. *See also* **polyprenyl.**

terpenoid 1 resembling a **terpene** in chemical structure. **2** any member of a class of compounds characterized by an isoprenoid chemical structure (as in a **terpene**) and often including their oxygenated and hydrogenated derivatives.

Terramycin a proprietary name for **oxytetracycline**.

tert- *abbr. for* t-; *prefix (in chemical nomenclature)* signifying **tertiary** (def. 1).

tertiary *(in chemistry)* **1 a** *prefix*: *tert-* (*abbr.*: *t-*); describing an alkyl compound (e.g. an alkanol) in which the functional group (e.g. a hydroxyl group) is attached to a carbon atom linked to three others. **b** describing the carbon atom bearing the functional group in such a compound. **2** describing an amide or an amine in which three appropriate groups are attached to a nitrogen atom. **3** describing a salt formed by replacing all three of the ionizable hydrogen atoms of a tribasic acid by one, two, or three other cations of appropriate valency. *Compare* **primary, secondary.**

tertiary structure the level of protein structure at which an entire polypeptide chain has folded into a 3-D structure. The tertiary structure results from interactions between amino-acid residues that may be widely separated in the primary structure, but may be brought into proximity by the folding of the polypeptide chain. The forces of interaction between residues include hydrogen bonds and electrostatic attraction. The structure is stabilized in some proteins by disulfide bonds. In multichain proteins, the term tertiary structure applies to individual chains. *Compare* **primary structure, secondary structure, quaternary structure.**

Tes *or* **TES** *abbr. for* N-tris(hydroxymethyl)methyl-2-aminoethanesulfonic acid; 2-{[2-hydroxy-1,1-bis(hydroxymethyl)ethyl]amino}-ethanesulfonic acid; a **Good buffer substance**; pK_a (20°C) = 7.5.

tesla *symbol*: T; the SI derived unit of **magnetic flux density**. It is equal to one weber of magnetic flux per square metre; i.e. $1 T = 1 Wb m^{-2}$. [After Nikola Tesla (1856–1943), Croatian-born US physicist.]

testes-specific poly(A) polymerase *abbr.*: TPAP; a cytoplasmic poly(A) polymerase (70 kDa) that is specific to testes, and is involved in poly(A) tail extension of mRNAs expressed by haploid-specific genes. Male mice that lack the enzyme are infertile because spermiogenesis is arrested at the rounded spermatid stage.

testican a multidomain glycoprotein of testes.

testicular of, or relating to, the **testis**.

testicular feminization *or* **testicular feminisation** an inherited condition in which genetic males have developmental abnormalities ranging from a complete external female phenotype to partially masculinized ambiguous genitalia. The commonest form is X-linked and results from any of several hundred mutations in the androgen receptor that impair its sensitivity to the hormone. An autosomal recessive form results from many mutations that produce steroid 5α-reductase 2 deficiency and lack of 5α-dihydrotestosterone (*see* **steroid 5α-reductase**).

testis (*pl.* **testes**) the male gonad; in vertebrates there is typically a pair of testes, producing spermatozoa and synthesizing **androgens**.

testosterone 4-androsten-17β-ol-3-one; the principal and most potent androgen. It is formed from androstenedione by 3α(17β)-hydroxysteroid dehydrogenase (NAD$^+$) (EC 1.1.1.239), in Leydig cells of the testis. Testosterone is responsible during embryonic development for virilization, and promotes and maintains male secondary sex characteristics throughout life. It is also synthesized in the ovary from which some is secreted into the blood, but most is converted to estradiol. Its actions are in large part mediated by **dihydrotestosterone**, to which it is converted in target cells. *See also* **androgen receptor.**

testosterone

TetA a membrane-associated protein of Gram-negative bacteria that exports tetracycline from the bacterial cell thereby conferring antibiotic resistance. It is under the control of the **tetracycline repressor** (TetR). *See* **tetracycline resistant**.

tetanus 1 a disease resulting from infection with the bacterium *Clostridium tetani*. *See also* **tetanus toxin**. **2** *or* **tetany** a sustained contraction of a muscle. —**tetanal** *or* **tetanoid** *adj*.

tetanus toxin the protein toxin of *Clostridium tetani*, the causative agent of tetanus. It is a heterodimer of light and heavy chains. It blocks neural exocytosis and thus norepinephrine release, possibly as a result of a Zn^{2+}-dependent proteolysis of **synaptobrevin** by the light chain, but also by inhibition of a neuronal transglutaminase (protein–glutamine γ-glutamyltransferase, EC 2.3.2.13). The protein is synthesized as a precursor of 1314 amino acids. This is cleaved to yield the light chain (residues 1–456) and heavy chain (residues 457–1314).

tetanus toxoid the **toxoid** prepared from *Clostridium tetani* toxin and included in anti-tetanus immunogens.

tetany *an alternative term for* **tetanus** (def. 2).

tetO 1 a 42 bp operator sequence from the tetracycline resistance operon of *E. coli* (Tn10). **2** a gene in Gram-negative bacteria encoding a tetracycline resistance protein that abolishes the effect of the antibiotic on protein synthesis.

Tet-Off *a proprietary name for* a system for regulating the expression of transgenes *in vivo* by manipulating the concentration of **tetracycline**. The gene of interest is placed downstream from **tetO** tetracycline response elements. The tetracycline *trans*-activator tTA (a fusion protein incorporating the N-terminal 1–203 amino acids of the **tetracycline repressor** protein TetR and the C-terminal 127 amino acids of the HSV VP16 activation domain) binds to tetO response elements and activates transcription. Addition of tetracycline or doxycycline causes dissociation of tTA from the operator thereby turning off transcription of the gene of interest.

Tet-On *a proprietary name for* an adaptation of the **Tet-Off** system in which a modified *trans*-activator, **rtTA**, binds to **tetO** sequences in the presence of doxycycline thereby activating transcription of the gene of interest.

TetR *abbr. for* tetracycline repressor.

tetra+ *or (before a vowel)* **tetr+** *comb. form* **1** denoting four, fourfold, every fourth; *see also* **tetrakis+** (def. 1). **2** *(in chemical nomenclature) (distinguish from* **tetrakis+** (def. 2)) **a** indicating the presence in a molecule of four identical specified unsubstituted groups; e.g. tetrachloroethylene, tetramethylammonium chloride. **b** indicating the presence in a molecule of four identical inorganic oxoacid residues in linear anhydride linkage; e.g. sodium tetraborate.

tetracosa+ *or (before a vowel)* **tetracos+** *comb. form* indicating twenty-four.

tetracosactide *or* **tetracosactrin** *or* **cosyntropin** *generic names for* $β^{1–24}$-corticotropin, a synthetic polypeptide identical to the N-terminal 24 residues of **corticotropin**, and having all of its hormonal activity. It is sometimes known as 1–24 ACTH. Proprietary names include Synacthen.

tetracosapeptide a 24-residue peptide.

tetracycline 1 any of a group of broad-spectrum antibiotics that are also effective against rickettsial organisms, mycoplasmas, and certain protozoa. Their bacteriostatic activity depends on the direct inhibition of protein synthesis. They are active against protein synthesis on bacterial 70S or eukaryotic 80S ribosomes although 70S ribosomes are rather more sensitive. The tetracyclines are much more effective in intact bacteria than in eukaryotic cells because of selective absorption by bacterial cells. They inhibit the binding of aminoacyl-tRNA to the acceptor site on the ribosome. **2** any antibiotic based on the tetracycline nucleus (*see* **polyketide**). They include aureomycin (7-chlortetracycline); terramycin (5-oxytetracycline); declomycin (6-demethyl-7-chlorotetracycline); doxycycline (6-deoxy-5-oxytetracycline). *See also* **tetracycline resistant**.

tetracycline repressor *abbr.*: TetR; a homodimer that binds to tandem DNA operators through N-terminal alpha helix-turn-alpha helix domains thereby blocking expression of the associated *tetA* gene and its own gene *tetR*. *See* **TetA**, **tetracycline resistant**.

tetracycline resistant *abbr.*: TC^r; resistant to the effects of the antibiotic **tetracycline**. Three different mechanisms of tetracycline resistance are known – tetracycline efflux, ribosome protection, and tetracycline modification – and over 60 tetracycline resistance genes have been sequenced. In enteric bacteria two **transposons**, Tn1721 and Tn10, commonly carry the genes for tetracycline resistance. Tn10 encodes two proteins: a membrane-located protein (TetA, M_r 36 000) that appears to mediate resistance and a repressor protein, TetR (*see* **tetracycline repressor**), associated with the inducibility of tetracycline resistance. The membrane protein pumps the tetracycline out of the cells against a concentration gradient. Foreign DNA may be fused to the tetracycline-resistance gene in a cloning vector thereby enabling bacterial clones containing the foreign DNA to be selected from those lacking resistance to the antibiotic. The plasmid pBR322 confers resistance to both tetracycline and ampicillin. Insertion of a foreign DNA at certain sites inactivates either or both genes.

tetracycline sensitive *abbr.*: TC^s; sensitive to the effects of **tetracycline**.

tetradeca+ *or (before a vowel)* **tetradec+** *comb. form* indicating fourteen.

tetradecadienoic acid any 14-carbon straight-chain fatty acid having two double bonds. Tetradeca-3*E*,5*Z*-dienoic acid (**megatomic acid**) and the (3*Z*,5*Z*)-isomer are sex attractants for *Attagenus megatoma* and for *A. elongatulus* respectively.

tetradecapeptide a 14-residue peptide.

tetradecenoic acid any 14-carbon monounsaturated fatty acid. Tetradec-9*Z*-enoic acid (**myristoleic acid**) is the most common naturally occurring form, but the 4*Z*-isomer occurs in tohaku oil and the 5*Z*-isomer (**physeteric acid**) in sperm-whale oil.

tetraenoic describing an acid having four double bonds. In long-chain fatty acids of the *n*-6 or *n*-3 families (*see* **fatty acid**), these will be **methylene-interrupted double bonds**.

tetragonal symmetry an arrangement, e.g. of a crystal or of ligands around a metal ion in solution, having a single fourfold axis of **symmetry**.

tetrahydro+ *comb. form* indicating the presence of four hydrogen atoms at positions of specific interest (in relation to the parent compound, or as specified by a locant).

tetrahydrobiopterin *see* **biopterin**.

tetrahydrofolate *abbr.*: H$_4$folate *or* (not recommended) THF; *N*-[4-{[(2-amino-3,4,5,6,7,8-hexahydro-4-oxo-pteridin-6-yl)methyl]amino}benzoyl]-L-glutamate; folate bearing additional hydrogens at positions 5, 6, 7, and 8 of the pterin moiety. It is formed by the action of **dihydrofolate reductase** on dihydrofolate with NADPH as co-substrate.

tetrahydrofolate

tetrahydrofuran *abbr.*: THF; tetramethene oxide; an organic solvent that is useful for bringing into aqueous solution compounds that are otherwise sparingly soluble in water. It is a cyclic ether that is moderately soluble in water.

tetrahydromethanopterin *abbr.*: H$_4$MPT; a derivative of 7-methylpterin (2-amino-4-hydroxy-7-methylpteridine) that is involved in methane formation in **methanogenic** bacteria. It undergoes a series of interconversions involving N^5 and N^{10}, analogous to those of the **folate coenzymes**, with formation of formyl-H$_4$MPT, methenyl-H$_4$MPT, methylene-H$_4$MPT, and methyl-H$_4$MPT.

Tetrahymena a genus of ciliated protozoans, members of which are used in studies of ciliary axonemes, self-splicing RNA, and telomere reproduction.

tetrakis+ *prefix* **1** denoting four times, quadrupled; *see also* **tetra+** (def. 1). **2** *(in chemical nomenclature)* *(distinguish from* **tetra+** *(def. 2))* **a** indicating the presence in a molecule of four identical specified groups each substituted in the same way; e.g. tetrakis(2-hydroxyethyl)ammonium chloride. **b** indicating the presence at separate positions in a molecule of four identical inorganic oxoacid residues; e.g. inositol 1,3,4,5-tetrakisphosphate.

tetrakisphosphate *see* **multiplying prefix**.

tetraloop a tetranucleotide **hairpin loop**.

tetramer a molecular complex having four components or subunits. If the subunits are identical, the term **homotetramer** is used; if any one is different from any other, the complex is called a **heterotetramer**.

tetraose tetrasaccharide; a stem name for any oligosaccharide consisting of four of the same residues per molecule in one type of linkage, as in cellotetraose. *Distinguish from* **+tetraose** and **tetrose**.

+tetraose *suffix* used to indicate a glycolipid containing an oligosaccharide moiety composed of four (identical or different) sugar residues, as in gangliotetraose. *Distinguish from* **tetraose**.

tetrapeptide a four-residue peptide.

tetraplex DNA or **quadruplex DNA** a four-stranded DNA structure

adopted by sequences rich in guanine bases. Two major classes are known. The first involves the folding back of a repetitive sequence of guanines resulting in antiparallel strands. The other is characterized by the association of four independent parallel strands. *See also* **triplex DNA**.

tetrapyrrole a structure found in **heme** and other **porphyrins**, **chlorophylls**, **bacteriochlorophylls**, **phytochromes**, **phycobilins**, and **corrinoids** (e.g. vitamin B$_{12}$). It consists of four pyrroles variously substituted and linked to each other through carbons at the α position; in the case of hemes, these links are formed by one-carbon units (methylidene group). Characteristically, a metal is found coordinated to the pyrrole nitrogens, including iron (in hemes), magnesium (in chlorophylls), copper (in **turacin**, the pigment of turaco bird feathers), and nickel (in **factor F$_{430}$**, a prosthetic group in methanogenic bacteria). Synthesis is via aminolevulinic acid, which is synthesized from glycine and succinyl-CoA in animals and nonphotosynthetic organisms by 5-aminolevulinate synthase. For further steps in the pathway, *see* **porphobilinogen synthase**, **hydroxymethylbilane synthase**, **uroporphyrinogen-III synthase**, and **uroporphyrinogen decarboxylase**.

tetraric acid an **aldaric acid** derived by oxidation of a **tetrose** at C-1 and C-4. There are three tetraric acids, D-, L-, and *meso-* **tartaric acids**.

tetrasaccharide an oligosaccharide with four **monosaccharide** units.

tetrasomic describing any polysomic cell, tissue, or individual in which one chromosome is represented four times in an otherwise **diploid** chromosome complement. *Compare* **monosomic**, **nullisomic**, **trisomic**. —**tetrasomy** *n.*

tetraspanin any of a family of integral membrane proteins that contain four transmembrane helices and have intracellular N- and C-termini. They contain <350 amino acid residues and occur mostly in vertebrates.

tetratricosapeptide repeat *abbr.*: TPR; a degenerate consensus sequence of 34 amino acids found in multiple copies in several fungal and other proteins. These sequences participate in RNA synthesis regulation, protein import, and *Drosophila* development. Each sequence has two regions that are stereochemically complementary in α-helical conformation, and form a hole and a knob surrounded by hydrophobic residues.

tetrazolium salt any compound that is readily reduced to highly coloured, sparingly soluble formazan compounds. Tetrazolium salts are employed in the assay of some dehydrogenases, and include Blue Tetrazolium and Nitroblue Tetrazolium. **1H-tetrazole** is used in the chemical synthesis of oligonucleotides.

tetritol any alditol having a chain of four carbon atoms in the molecule.

tetrodotoxin *abbr.*: TTX; *also known as* **fugu toxin**; a toxin found in the liver and ovaries of the **fugu**, or puffer fish (order Tetraodontiformes, from which the toxin derives its name). It is a specific, reversible blocker of voltage-gated sodium channels, and can cause death by increasing paralysis. The LD$_{50}$ in mice is 8 µg kg^{-1}. It does not block potassium or chloride channels.

tetrose any aldose having a chain of four carbon atoms in the molecule. *Distinguish from* **tetraose**.

tetrulose any ketose having a chain of four carbon atoms in the molecule.

T-even phage any one of the **bacteriophages** T2, T4, or T6 of *Escherichia coli*. These DNA phages share a number of morphological, biochemical, and other features that distinguish them as a group from **T-odd phages** and are important historically in research on phages.

Texas Red a red-emitting fluorophore and a sulfonyl chloride derivative of sulforhodamine 101. It is used with **FITC** in dual-label flow microfluorimetric and fluorescent microscopic studies because its excitation and emission spectra do not overlap with those of fluorescein.

text mining the automated process by which large volumes of unstructured, natural-language text are analysed in order to pinpoint and extract user-specified information.

TF *abbr. for* **1** tissue factor. **2** transcription factor.

TFII any of a group of **transcription factors** for eukaryotic RNA polymerase II promoters. **TFIIA** is needed for TFIID binding. **TFIIB** binds to RNA polymerase. **TFIID** has two components, a TATA-binding protein that binds to the **TATA box**, and one of a number of accessory factors; it shows weak homology with bacterial **sigma factors**. **TFIIE** binds to RNApolymerase; it also shows weak homology with bacterial sigma factors. It is a tetramer of two α chains and two β chains.

TFIII any of a group of **transcription factors** for eukaryotic RNA polymerase III promoters. **TFIIIA** is a zinc-finger protein (*see* **zinc finger**) that binds both DNA and the 5S RNA transcript. **TFIIIB** is an RNA polymerase-binding protein. **TFIIIC** (*or* **tau**) is a heterotrimer.

TFA *abbr. for* trifluoroacetic acid.

T form *abbr. for* tense form (*or* taut form).

TG1 a derivative of the *E. coli* strain JM101 used for the propagation of bacteriophage M13 vectors.

TGF *abbr. for* transforming growth factor.

Th *symbol for* thorium.

thalassemia *or (esp. Brit)* **thalassaemia** any disease resulting from an inherited defect in the rate of synthesis of one of the types of globin subunits of **hemoglobin**. Such defects lead to ineffective erythropoiesis, hemolysis, and a variable degree of anemia. In **alpha thalassemia** the defect is in the production of α globin and there is concomitant excessive β- and γ-globin production leading to the formation of various abnormal hemoglobins. In **beta thalassaemia** the formation of β globin is defective with an excessive synthesis of α globin and the continued production of fetal hemoglobin ($\alpha_2\gamma_2$).

thalidomide *N*-(2,6-dioxo-3-piperidyl)phthalimide; α-phthalimidoglutarimide; a synthetic tranquillizer and sedative. It was first used (in the 1950s) as a racemate, which causes severe fetal abnormalities such as the absence or shortening of one or more limbs. The L(*S*) enantiomer apparently has the teratogenic activity (*see* **teratogen**), while the D(*R*) enantiomer has the desired sedative property. The situation is unclear since thalidomide racemizes rapidly *in vivo*.

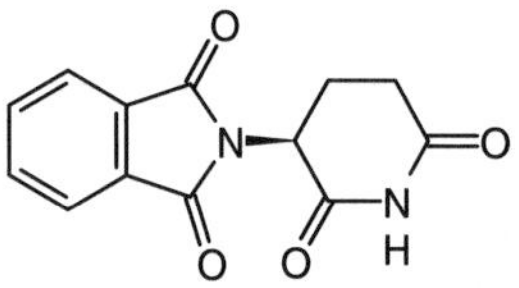

(*S*)-form

THAM *abbr. for* tris(hydroxymethyl)aminomethane; *see* **Tris**.

thanatogene a hypothetical gene whose expression is supposed to lead to cell death in terminally differentiated cells. [From Greek *thanatos*, death.]

thanatophoric dysplasia an autosomal dominant form of severe dwarfism that is usually lethal at or soon after infancy. It results from mutations in fibroblast growth factor receptor 3.

thapsigargin a tumour-promoting sesquiterpene lactone. It inhibits the calcium pump **SERCA** and increases cytosolic Ca^{2+}.

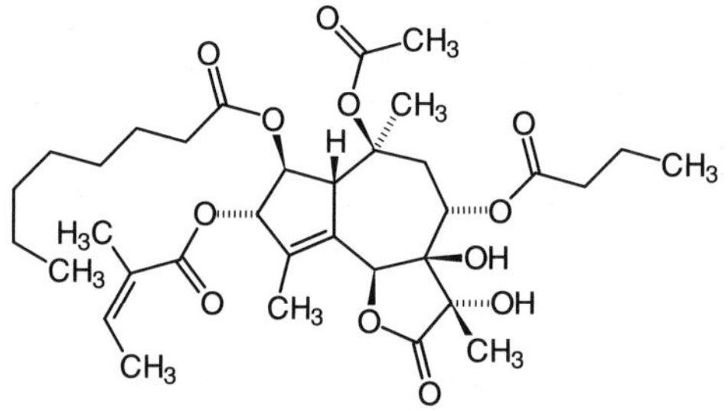

thapsigargin

thaumatin a sweet-tasting, basic, 20 kDa protein isolated from the berries of the bushy African plant, *Thaumatococcus danielli*; there are five different forms. Its sweetness in aqueous solution is 2500 times greater than that of sucrose; it has sequence similarities with **monellin**.

Thd *symbol for* a residue of the ribonucleoside ribothymidine; ribosylthymine; thymine riboside (alternative to T). *Compare* **dThd**.

Thd5′P *symbol for* ribothymidine 5′-phosphate.

Thd5′PP *symbol for* ribothymidine 5′-diphosphate (alternative to ppT).

Thd5′PPP *symbol for* ribothymidine 5′-triphosphate (alternative to pppT).

thebaine one of the alkaloids present in opium.

T-helper cell *see* **helper cell**.

theobromine 3,7-dimethylxanthine; the principal alkaloid of the cacao bean. Its actions are similar to those of **theophylline**. *See also* **caffeine**.

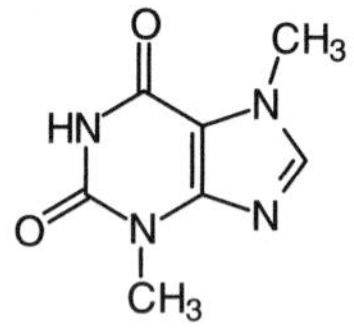

theophylline 1,3-dimethylxanthine; a competitive inhibitor of cyclic AMP phosphodiesterase. It has diuretic and cardiac stimulatory action, and relaxes smooth muscle; it is a bronchodilator. *See also* **caffeine**, **theobromine**.

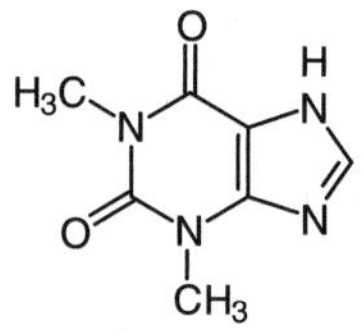

Theorell–Chance mechanism (*in enzyme kinetics*) an enzyme mechanism in which it is obligatory for substrates to bind in a certain order and in which no kinetically significant ternary complex is formed. [After Axel Hugo Theodor Theorell (1903–82), Swedish biochemist, and Britton Chance (1913–), US biochemist.]

theoretical plate a term used in the analysis of chromatography column performance; the higher the number of theoretical plates achieved, the greater the efficiency of the column. The number of theoretical plates is derived by dividing the peak width by the retention time. The term derives from the concept that the column can notionally be divided into a large number of 'slices' or 'plates' at each of which the separation process will progress. *See also* **effective theoretical plate number**.

therapeutic index the ratio of **LD50** (median lethal dose) to **ED50** (median effective dose), used in quantitative comparison of drugs.

therm+ *a variant form of* **thermo+** (before a vowel).

thermal analysis *see* **differential thermal analysis, differential scanning calorimetry.**

thermal conductance *symbol*: G; the ability of a material to conduct heat, measured as the ratio of the heat flow rate, Φ, to the temperature difference, i.e. $G = \Phi/\Delta T$. SI units are watts per kelvin.

thermal conductivity *symbol*: λ *or* k; a measure of the ability of a material to transmit heat. It is the heat flux in $W\ m^{-2}$ divided by the temperature gradient; i.e. $\lambda = J_q\ /(dT/dl)$, where J_q is the heat flux and dT is the thermodynamic temperature difference over the length, dl. It is measured in watts per metre per kelvin.

thermal denaturation profile a curve relating some physical property of a biopolymer in solution (e.g. viscosity, optical absorbance, or optical rotation) to temperature. The shape of the curve indicates the occurrence of physical changes in the biopolymer, e.g. separation of the two strands of double-stranded nucleic acid or alteration in the tertiary structure of a protein. *See also* **denaturation, melting temperature.**

thermal equilibrium the condition of a system when there is no tendency for heat to flow within it, or to or from its surroundings; all parts of the system and its surroundings are at the same temperature.

thermistor a semiconductor whose resistance is sensitive to temperature. [From thermal resistor.]

thermitase EC 3.4.21.66; *other name*: thermophilic *Streptomyces* serine protease; a Ca^{2+}-binding serine proteinase from *Thermoactinomyces vulgaris*, and a homologue of **subtilisin**. It is inhibited by *p*-mercuribenzoate, which reacts with the cysteine residue near the active-site histidine.

thermo+ *or (before a vowel)* **therm+** *comb. form* indicating a relationship to heat or heat transference.

thermobarometer a control manometer and flask used in a **Warburg apparatus**, or other constant-volume-type respirometer, to compensate for changes in temperature and barometric pressure during the course of an experiment.

thermochromism any reversible change of colour of a substance with change of temperature. *Compare* **electrochromism, photochromism. —thermochromic** *adj.*

thermocouple a device consisting of two pieces of dissimilar metals joined together at a point, their free ends being connected to an instrument to measure electromotive force (emf). The emf produced at the metal–metal junction is a function of the junction temperature, hence a thermocouple may be used to measure temperature. The phenomenon is known as the **Seebeck effect.**

thermodynamicase *(sometimes)* an alternative name for **protein disulfide isomerase.**

thermodynamic equilibrium constant *see* **equilibrium constant.**

thermodynamics the study of the laws that govern the conversion of energy from one form to another, the direction in which heat will flow from one system to another, and the availability of energy to do work. The **zeroth law of thermodynamics** states that if two systems are each in thermal equilibrium with a third system, then they must be in thermal equilibrium with each other. The **first law of thermodynamics** or the **law of conservation of energy** specifies that in an isolated, adiabatic system the sum of the internal energies of all kinds, U, remains constant, whatever may be the changes within the system itself. If a system of constant mass gains heat, q, and has work, w, done on it then a change in its internal energy, ΔU, occurs such that $\Delta U = q + w$. (Where the system loses heat, $q < 0$, and where work is done by it, $w < 0$.) The **second law of thermodynamics** is concerned with the restrictions in the direction of energy flow in natural systems and takes account of the irreversibility of natural processes. It specifies that the entropy of any system, S, unlike energy, is not conserved but increases with time in any spontaneous process in an isolated system. In the limiting case of a reversible process, the increase in entropy of a system, dS, is defined as the heat absorbed by the system, dq_{rev}, divided by the thermodynamic temperature, T; i.e. $dS = dq_{rev}/T$. The **third law of thermodynamics** is concerned with the fact that the difference between the change in enthalpy, ΔH, and the change in the free energy, ΔG, of a system diminishes as the temperature of the system diminishes, and the value of $(\Delta H - \Delta G)$ tends to zero as the thermodynamic temperature tends to zero. This requires that the change in the heat capacity of the system (at constant pressure), ΔC_p, and ΔS must fall to zero at 0 K.

thermodynamic temperature or *(formerly)* **absolute temperature** *symbol*: T; a fundamental physical quantity – one of the seven SI base quantities – defined in terms of the laws of thermodynamics, independently of any particular substance. The SI unit is the **kelvin**. *See also* **temperature.**

thermogenesis the generation of heat by specialized tissues in mammals and in flowers or inflorescences of plants, e.g. arum lilies (*Arum maculatum*). In mammals, this function is performed by **brown adipose tissue. —thermogenic** *adj.*

thermogenin *another name for* **brown fat uncoupling protein.**

thermolabile unstable at elevated temperatures. *Compare* **thermostable. —thermolability** *n.*

thermolysin EC 3.4.24.27; *other name*: *Bacillus thermoproteolyticus* neutral proteinase; a heat-stable metalloendoproteinase produced by *B. thermoproteolyticus*; it contains four zinc and calcium ions. It catalyses the preferential cleavage: Xaa-|-Leu > Xaa-|-Phe.

thermolysis the cleavage of one or more covalent bonds caused by exposing a chemical compound to a raised temperature. *See also* **pyrolysis. —thermolytic** *adj.*

thermometer any instrument for measuring temperature. All thermometers depend on the change with temperature of some easily measured physical property such as the expansion of mercury (or other liquids) or the change in electrical resistance. *See also* **thermocouple.**

thermometry the branch of physics concerned with thermometers and the measurement of temperature.

thermoneutron activation *an alternative name for* **radioactivation.**

thermoosmosis the flow of solvent, especially water, across a membrane as a result of a temperature gradient across the membrane.

thermophile an organism, especially a microorganism, that can tolerate high temperatures and that grows optimally at temperatures above 45°C. *Compare* **mesophile, psychrophile. —thermophilic** *adj.*

thermoregulator any device or mechanism (including physiological) that controls the temperature of an environment. **—thermoregulatory** *adj.*

ThermoSequenase *proprietary name for* a DNA polymerase modified by protein engineering to have the thermal stability required for **cycle sequencing.**

thermospray *or* **electron spray mass spectrometer** (*abbr.*: ESMS); an instrument that directly applies a proportion of the eluent of a liquid chromatograph to a mass spectrometer for analysis of resolved components.

thermostable maintaining characteristics or function under a change (normally increased) of temperature. *Compare* **thermolabile. —thermostability** *adj.*

thermostat **1** an automatic device for regulating temperature, especially maintaining a constant temperature. Such devices often work by cutting off the supply of heat when the required temperature is exceeded and restoring it when the temperature falls below the required value. **2** to provide with or to control using a thermostat. **—thermostatic** *adj.*

thermotitration the determination of the endpoint of a reaction, or the point at which the reaction reaches equilibrium, by measurement of heat of reaction.

THESIT an alternative name for $C_{12}E_9$ (*see* $C_x E_y$), or Lubrol PX (*see* **Lubrol).**

theta *symbol*: θ (lower case) *or* Θ (upper case); the eighth letter of the Greek alphabet. For uses *see* **Appendix A.**

theta structure *or* **replication eye** *or* **replication bubble** the structure that forms during replication of circular duplex DNA by the opening up of the two strands. Under electron microscopy it looks like a Greek theta.

L-thevetose 6-deoxy-4-*O*-methyl-L-glucose; a component of some cardiac glycosides.

THF *abbr. for* **1** tetrahydrofuran. **2** tetrahydrofolate (H_4folate preferred).

thia+ *or (before a vowel)* **thi+** *comb. form* containing sulfur, especially in place of carbon. *See also* **thio+**.

thiamine *or (sometimes)* **thiamin** *or (formerly)* **aneurin(e)** *an alternative name for* **vitamin B₁**.

thiamine diphosphate *or (sometimes)* **thiamin diphosphate** *or (formerly)* **thiamin pyrophosphate** *abbr.*: TPP; *see* **vitamin B₁**.

thick filament a longitudinal assembly of **myosin** molecules occurring in muscle. The thick filaments are the major constituents of the A band of the **sarcomere** in skeletal muscle; they interact with the **thin filaments** to bring about muscle contraction (*see* **sliding filament model**).

thimet oligopeptidase a thiol-activated metallopeptidase (78.5 kDa) that prefers substrates 6–17 residues long, and cleaves them 3–6 residues from the C-terminus. It is present in cytosol or vesicles, has a neutral pH optimum, and is inactivated by oxidation.

thin filament a molecular complex of **actin**, **troponin** (composed of troponins I, T, and C), and **tropomyosin**, found in muscle tissue. In striated muscle the thin filaments span the I band and interdigitate between the ends of the **thick filaments** in the A band. Interaction between the thin and thick filaments brings about muscle contraction (*see* **sliding filament model**).

thin-film dialysis a type of **dialysis** that maintains the **diffusate** and the **retentate** in thin layers in intimate contact with the dialysis membrane in order to lessen the contribution to the overall rate of the process made by diffusion of solutes in free solution. Very rapid removal or exchange of small molecules can occur, especially if the volume ratio of diffusate to retentate is high. With appropriate membranes the technique may be used for analytical studies of the size and conformation of peptides and proteins in solution.

thin-layer chromatography *abbr.*: TLC; chromatography on a thin layer of any of various powdered materials, including silicic acid, aluminium oxide, or cellulose. The layer, characteristically between 0.05 and 1 mm thick, is made to adhere to a sheet of glass, aluminium, or plastic. Thicker layers accept heavier loading, while thinner layers resolve more efficiently and rapidly. Resolving power is high, runs are rapid, and, because support materials such as silicic acid or aluminium oxide are inert to them, corrosive sprays, such as sulfuric acid, can be used to convert organic materials to carbon (a process known as charring), revealing them as black spots.

thio+ *or (sometimes before a vowel)* **thi+** *comb. form* **1** relating to or containing sulfur. **2** *(in chemical nomenclature)* denoting replacement of oxygen in a compound by (bivalent) sulfur. In trivial names of compounds it is used as a prefix (as in thiouracil; *compare* **uracil**), whereas in the systematic names of individual compounds it is used as an affix and is usually placed before the affix denoting an oxygen atom or an oxygen-containing group (as in ethanethiol; *compare* **ethanol**), although for organic sulfides, if named analogously to ethers by substitutive nomenclature, it replaces oxy+ (as in 3-(methylthio)propionate. *Compare* **thia+**. *See also* **mercapto+**.

thioacetal *see* **acetal**.

thiobarbituric acid 4,6-dihydroxy-2-mercaptopyrimidine. It reacts with aldehydes, particularly malondialdehyde, produced by oxidation of unsaturated fatty acids, to form a fluorescent red adduct.

thioctic acid *a less common name for* **lipoic acid**.

thiocyanate 1 the anion, N≡C–S⁻, derived from thiocyanic acid. **2** any salt of thiocyanic acid. **3** any organic compound containing the monovalent group –S–C≡N. *Compare* **isothiocyanate**.

thioester any organic compound formed, actually or formally, by the elimination of the elements of water either (1) between the carboxyl group of a carboxylic acid and a thiol, or (2) between the thiocarboxyl group of a thiocarboxylic acid and an alcohol or phenol. Thioesters formed by mechanism (1) are of general formula R¹–CO–SR², whereas those formed by mechanism (2) may be of general formula either R¹–CO–SR² or R¹–CS–OR².

thioether *an old name for* any organic sulfur-containing compound of general formula R¹–S–R². Sulfide is now the recommended generic name.

thiogalactoside transacetylase *see* **galactoside acetyltransferase**.

thioglucosidase EC 3.2.3.1; *systematic name*: thioglucoside glucohydrolase; *other name*: sinigrase. An enzyme that hydrolyses thioglucosides to the constituent thiol and sugar, e.g. sinigrin to glucose and *N*-(sulfoxy)-3-mercapto-3-butenimidate.

thioglycoside *or* **S-glycoside** any glycoside formed from a sugar and a thiol compound, e.g. **sinigrin**.

thiohemiacetal *see* **acetal**.

thiohemiketal *see* **acetal**.

thiohypoxanthine *an alternative name for* **mercaptopurine**.

thioinosine *symbol*: sI *or* Sno; 6-thioinosine; thiohypoxanthine riboside; mercaptopurine ribonucleoside; 9-β-D-ribofuranosyl-6-mercaptopurine; a synthetic nucleoside. *See also* **mercaptopurine**.

thioketal *see* **acetal**.

thiokinase any ATP-requiring enzyme of sub-subclass EC 6.2.1 (acid thiol ligases) that synthesizes CoA thioesters of various fatty acids. These enzymes (recommended names given first) are: (1) **acetate–CoA ligase**, which acts on acetate, propanoate, and propenoate; (2) **butyrate–CoA ligase**; EC 6.2.1.2; *other names*: fatty acid thiokinase (medium chain); butyryl-CoA synthetase; butanoate–CoA ligase (AMP forming); this acts on C₄ to C₁₁ acids and on corresponding 3-hydroxy- and 2,3- or 3,4-unsaturated acids; (3) **long-chain-fatty-acid–CoA ligase**; EC 6.2.1.3; *other names*: fatty acid thiokinase (long chain); acyl-CoA synthetase; acid:CoA ligase (AMP forming); this acts on a wide range of long-chain saturated fatty acids, depending on the tissue of origin of the enzyme; and (4) **acetate–CoA ligase (ADP-forming)**; also acts on propionate and (very slowly) on butanoate.

thiol 1 *or (formerly)* **mercaptan** any organic compound containing –SH as the principal group directly attached to a carbon atom. **2** *or* **sulfhydryl** the (covalently linked) –SH group in any chemical compound.

+thiol *suffix (in organic chemical nomenclature)* indicating the presence of a **thiol** group attached to a carbon atom. *Compare* **mercapto+**.

thiolactomycin an antibiotic isolated from *Nocardia* spp. It selectively inhibits straight-chain fatty-acid synthases of the multienzyme complex type (as in *E. coli*) but not the multicatalytic protein type (as in *Saccharomyces cerevisiae*).

thiolase a misleading generic name for certain enzymes of differing chain-length specificity within sub-subgroup EC 2.3.1 (acyltransferases) that catalyse the **thiolysis** by coenzyme A of CoA thioesters of β-oxo fatty acids according to the general equation:

RCO–CH₂CO–CoA + CoA = RCO–CoA + CH₃CO–CoA.

Thiolases are thiol enzymes, their *S*-acyl derivatives being formed as intermediates in the reactions catalysed. Although the overall reactions are reversible, the equilibrium positions greatly favour cleavage. These enzymes include: **acetyl-CoA C-acetyltransferase** (EC 2.3.1.9) and **acetyl-CoA C-acyltransferase** (EC 2.3.1.16).

thiolate 1 the anion, R–S⁻, derived from a **thiol** (def. 1). **2** any salt of a thiol. **3** to introduce a thiol group into an organic compound. —**thiolation** *n.*

thiol proteinase *an old name for* **cysteine endopeptidase**.

thioltransacetylase A *see* **dihydrolipoamide S-acetyltransferase**.

thiolysis lysis of a covalent organic compound by a thiol; a reaction analogous to **alcoholysis** in which bond cleavage is mediated by nucleophilic attack by the –SH group of an alkanethiol. The term is applied to the **thiolase**-catalysed cleavage of a β-oxoacyl-CoA by coenzyme A in each turn of chain-length reduction during fatty-acid catabolism. —**thiolytic** *adj.*

thionein the apoprotein of a **metallothionein**.

thionin any of a group of small proteins, characterized by two conserved motifs, that occur in plants and are toxic to animals. The toxic principle of leaf thionin in barley is formed from residues 29–34 of a 137-amino-acid precursor. Thionins act by forming pores in the membranes of the affected organism. They are classified in the TC system under number 1.A.44.

thionucleoside any *N*-glycoside of any **thiopurine** or **thiopyrimidine**.

thionucleotide any phosphoric ester of any **thionucleoside**.

thiopurine any **purine** in which a thiol group has replaced a hydroxyl group. *Compare* **mercaptopurine**.

thiopyrimidine any **pyrimidine** in which a thiol group has replaced a hydroxyl group.

thioredoxin any of the widely occurring heat-stable hydrogen-carrier proteins of $\approx$12 kDa that serve as hydrogen donors in various reduction reactions, notably of ribonucleoside diphosphates to the deoxy analogues, and in the light-dependent reductive activation of various chloroplast enzymes (including several in the pentose-phosphate cycle). Chloroplast thioredoxin occurs in *f* and *m* forms, which show specificity in their enzyme-activating properties; a third form (designated *c*), of unknown function, occurs outside chloroplasts. Thioredoxins can exist in disulfide (i.e. oxidized) and dithiol (i.e. reduced) states, interconversion being catalysed by ferredoxin–thioredoxin reductases. *See also* **thioredoxin reductase (NADPH)**.

thioredoxin reductase (NADPH) EC 1.6.4.5; a flavoprotein (FAD) enzyme that catalyses the reaction:

$$\text{NADPH} + \text{oxidized thioredoxin} =$$
$$\text{NADP}^+ + \text{reduced thioredoxin.}$$

thiotemplate mechanism *or* **protein thiotemplate mechanism** the mechanism of biosynthesis of *Bacillus brevis* peptide antibiotics, e.g. **gramicidin S** and **tyrocidin** A, in which the precursor amino acids, after activation, are bound as thioesters to the synthetase enzymes.

thiouracil *symbol*: Sur; 2-thiouracil; 4-hydroxy-2-mercaptopyrimidine; 4-oxy-2-thiopyrimidine; a thiopyrimidine found in seeds of members of the cabbage family (e.g. *Brassica* spp.) and as a minor base in certain types of ribonucleic acid, especially transfer RNA (*see also* **thiouridine**). Thiouracil, 6-methylthiouracil, and 6-propylthiouracil are used to treat hyperthyroidism and thyrotoxicosis. They inhibit biosynthesis of the thyroid hormones at the peroxidase-catalysed iodination and coupling steps.

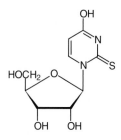

thiouridine *symbol*: S *or* sU *or* Srd; either of two constitutional isomers occurring as minor nucleosides in certain types of ribonucleic acid, especially transfer RNA (tRNA): **2-thiouridine** (2-thiouracil riboside; 1-β-D-ribofuranosyl-4-oxy-2-thiopyrimidine; *symbol*: ^{2}S *or* s^{2}U), or its 5-methyl, 5-(methylaminomethyl), or 5-(methoxycarbonylmethyl) derivatives, which are found in some tRNA species in the first position of the anticodon; and **4-thiouridine** (4-thiouracil riboside; 1-β-D-ribofuranosyl-2-oxy-4-thiopyrimidine; *symbol*: ^{4}S *or* s^{4}U), which is found in some tRNA species elsewhere than in the anticodon.

2-thiouridine

third law of thermodynamics *see* **thermodynamics**.

thixotrope a gel that exhibits **thixotropy**.

thixotropy the property shown by certain fluids and gels of becoming less viscous when subjected to stress, e.g. by shaking or stirring, and of becoming more viscous again when the stress is relieved. —**thixotropic** *adj.*

Thr *symbol for* a residue of the α-amino acid L-threonine (alternative to T).

THR a gene family encoding thyroid receptors; part of a superfamily including receptors for steroids, retinoic acid, and vitamin D$_3$; *see* **erbA**.

threading a fold-recognition method that identifies folds compatible with the sequences of proteins of unknown fold by threading the target sequence through template structures held in a reference library. In this process, the side chains are replaced and the main chain is kept constant, and the compatibility of the target with the template is estimated by means of pair potentials and solvation potentials. The approach is used when the target shows little or no similarity to any other sequence.

threitol active erythritol, *threo*-1,2,3,4-tetrahydroxybutane, *threo*-1,2,3,4-butanetetrol; an enantiomeric pair of polyols, configurationally isomeric with **erythritol**. D-Threitol occurs in certain fungi, e.g. the honey fungus, *Armillaria mellea*.

threo- a configurational prefix to the systematic name of a polyhydric alcohol, especially a monosaccharide, used to indicate the particular stereochemical configuration of a set of two contiguous >CHOH groups that occurs in D-threose (see **threose** for structure) or L-threose, e.g. D-*threo*-2-pentulose for D-xylulose. *Compare* **erythro-**.

threoninate **1** threonine anion; the anion, CH$_3$–CH(OH)–CH(NH$_2$)–COO$^-$, derived from **threonine**. **2** any salt containing threonine anion. **3** any ester of threonine.

threonine *the trivial name for* α-amino-β-hydroxybutyric acid; (2*R**,3*S**)-2-amino-3-hydroxybutanoic acid; CH$_3$–CH(OH)–CH(NH$_2$)–COOH; an α-amino acid with two chiral centres. Because molecules of threonine possess a second chiral centre, at C-3, in addition to the chiral centre at C-2 common to all α-amino acids other than glycine, the enantiomers L-threonine, (2*S*,3*R*)-2-amino-3-hydroxybutanoic acid (*symbol*: T *or* Thr), and D-threonine, (2*R*,3*S*)-2-amino-3-hydroxybutanoic acid (*symbol*: D-Thr *or* DThr), are diastereoisomeric with those of **allothreonine**, (2*R**,3*R**)-2-amino-3-hydroxybutanoic acid. [*Note*: the enantiomers of threonine may also be named semi-systematically as derivatives of threose: L$_s$-threonine in amino-acid nomenclature is synonymous with 2-amino-2,4-dideoxy-D$_g$-threonic acid in carbohydrate nomenclature, and D$_s$-threonine with 2-amino-2,4-dideoxy-L$_g$-threonic acid (the subscript letters s or g being added to the configurational prefixes where there might be uncertainty regarding the reference centre of chirality; *see* **D/L convention**).] L-Threonine is a coded amino acid found in peptide linkage in proteins; codon: ACA, ACC, ACG, or ACU. In mammals, it is an essential dietary amino acid, is glucogenic, and its deficiency causes fatty liver. The fungal toxins phalloidin and phalloin contain one residue of D-threonine per molecule.

L-threonine

threonine aldolase *see* **glycine hydroxymethyltransferase**.

threonine dehydratase EC 4.2.1.16; *systematic name*: L-threonine hydro-lyase (deaminating); *other names*: threonine deaminase; L-serine dehydratase; serine deaminase. A pyridoxal-phosphate enzyme that catalyses a reaction between L-threonine and water to form 2-oxobutanoate, NH_3, and water; it acts also on L-serine. In *Escherichia coli* type 1 is responsible for the first step of isoleucine biosynthesis (serine is an alternative but poorer substrate); type 2 is a catabolic enzyme that acts on both serine and threonine, and is properly considered as a hydroxy-amino-acid deaminase.

threonine synthase EC 4.2.99.2; *systematic name*: O-phospho-L-homoserine phospho-lyase (adding water). A pyridoxal-phosphate enzyme that catalyses the formation of L-threonine from O-phospho-L-homoserine and water with release of orthophosphate. It shows sequence homology with the serine dehydratase/threonine dehydratase group.

threoninium threonine cation; the cation, CH_3–CH(OH)–$CH(NH_3^+)$–COOH, derived from **threonine**.

threonino the alkylamino group, CH_3–CH(OH)–CH(COOH)–NH–, derived from **threonine**.

threonin-O^3-yl the alkyloxy group, HOOC–CH(NH$_2$)–CH(CH$_3$)–O–, derived from **threonine**.

threonyl the acyl group, CH_3–CH(OH)–CH(NH$_2$)–CO–, derived from **threonine**.

threose *the trivial name for* the aldotetrose *threo*-tetrose; it has D and L enantiomers, which are respectively diastereoisomeric with those of erythrose. *See also* **threo-**.

α-D-threose

threshold the value that must be achieved or exceeded for an effect to become discernible. For example, the **kidney threshold of excretion**, for any substance, is the plasma concentration at which not all of the substance that appears in the filtrate can be reabsorbed, so that it appears in the urine.

THRF *abbr. for* thyrotropic hormone releasing factor; i.e. **thyrotropin-releasing hormone**.

THRH *abbr. for* thyrotropic hormone releasing hormone; i.e. **thyrotropin-releasing hormone**.

thrombin EC 3.4.21.5; *other name*: fibrinogenase; a serine proteinase, M_r 33 700–33 900, the natural coagulant of **fibrinogen**. Thrombin appears in the blood following activation of the coagulation system, as a result of proteolysis of **prothrombin**. It catalyses the preferential cleavage of Arg-|-Gly, activating fibrinogen to **fibrin** and releasing fibrinopeptides A and B. Its effects are mediated by the **thrombin receptor**. *See also* **blood coagulation**.

thrombin receptor a protein of platelets and vascular endothelial cells that binds **thrombin** and mediates its intracellular effects. It is a G-protein-coupled receptor. The precursor of human thrombin receptor is a glycoprotein of 425 amino acids. The signal peptide comprises amino acids 1–26, and amino acids 27–41 are removed by thrombin, after which the new N terminus functions as a tethered ligand and the receptor is thereby activated.

thrombocyte *an alternative name for* **platelet**.

thrombogenic 1 tending to promote the formation of a blood clot

(i.e. a **thrombus**). **2** having a precursor function for, or tending to promote the formation of **thrombin**.

β-thromboglobulin a protein (79 amino acids) derived from **platelet basic protein precursor** by removal of the first four N-terminal residues from **LA-PF4**. Released as a homotetramer into plasma by activated platelets, it mediates several actions in the inflammatory response, including chemotaxis of leukocytes. It is the precursor of NAP2 (68 amino acids).

thrombokinase *an alternative name for* factor Xa; *see* **blood coagulation**.

thrombolysis the degradation of a blood clot (i.e. **thrombus**) as a result of physiological or pharmacological action.

thrombolytic 1 tending to promote **thrombolysis**. **2** a drug or other substance promoting thrombolysis.

thrombomodulin *abbr.*: TM; *other name*: fetomodulin; a specific endothelial cell receptor that binds thrombin, the complex converting protein C to activated protein C. It is a type I membrane protein.

thrombophilia any tendency to intravascular thrombosis. The commonest form is resistance to activated **protein C** due to a mutation in blood coagulation factor V.

thromboplastin *an alternative name for* factor III; *see* **blood coagulation**.

thrombospondin *or* **glycoprotein G** a thrombin-sensitive major glycoprotein located in the alpha granules of human platelets. It is secreted on thrombin activation, and binds to the platelet membrane in the presence of Ca^{2+}. It is an adhesion molecule that mediates cell–cell and cell–matrix interactions, and can bind to fibrinogen, fibronectin, laminin, and type V collagen. It is a trimeric, disulfide-linked protein and contains three types of repeating amino-acid sequence: the first is 57 amino acids long and shares similarity with circumsporozoite protein from a malarial parasite; the second is 50–60 amino acids long and shares similarity with **epidermal growth factor**; the third occurs as a continuous eight-fold repeat of a 38-residue sequence; structural similarity with **calmodulin** indicates that these repeats constitute multiple calcium-binding sites.

thromboxane any of a class of **prostanoids** synthesized by **thromboxane-A synthase** (EC 5.3.99.5) from prostaglandin H (the product of prostaglandin-endoperoxide synthase) as thromboxane A (TXA). Depending on whether the substrate of the enzyme is dihomo-(6,9,12)-linolenate, arachidonate, or eicosa-5,8,11,14,17-pentaenoate, the product is TXA_1, TXA_2, or TXA_3 respectively. TXA_2 is the major thromboxane *in vivo*. It is synthesized in platelets when these are activated by other aggregatory compounds and strongly reinforces the aggregatory reaction. TXA_2 is converted in minutes to the inactive thromboxane B_2 (TXB_2) by addition of water across the endoperoxide. Oxidation of the 11-hydroxy group of TXB_2 to an oxo group yields 11-dehydro-thromboxane A_2, a major long-lived metabolite in human plasma, useful for measurement as an index of thromboxane production. Beta-oxidation of TXB_2 yields 2,3-dinor-thromboxane B_2, the major human urinary metabolite. **Dideoxy-epoxymethanoprostaglandin F$_{2α}$** is a stable analogue that mimics TXA_2.

TXA$_2$

thromboxane-A synthase EC 5.3.99.5; *other names*: thromboxane synthase; thromboxane synthetase (*abbr.*: TXS); an enzyme that catalyses the formation from (5Z,13E)-(15S)-9α,11α-epidioxy-15-hydroxyprosta-5,13-dienoate of (5Z,13E)-(15S)-9α,11α-epoxy-15-hydroxythromba-5,13-dienoate (**thromboxane A$_2$**). It is a heme-thiolate protein. 1-(7-carboxyheptyl)imidazole and benzylimidazole inhibit thromboxane-A synthase, as does the drug U-63557A, sodium furegrelate.

thrombus (*pl.* **thrombi**) a blood clot forming within a blood vessel. Thrombus formation may occur as a normal response during physiological blood coagulation. It may also occur pathologically, triggered by **atheroma**, and is the immediate cause of ischemia in a heart attack. Thrombi may also form, especially in veins, at other times, e.g. after surgery, and if they reach the lungs cause pulmonary embolism.

Thunberg method an early method for detecting and estimating **dehydrogenases** from the rate and extent of decolorization of Methylene Blue in solution in the presence of enzyme and substrate. The apparatus, known as a **Thunberg tube**, is a stoppered glass tube (sometimes with a side-arm for addition of reagents) that may be evacuated through the stopper, and then sealed by a twist of the stopper. [After Torsten Ludvig Thunberg (1873–1952), Swedish physiological chemist.]

Thx *symbol for* a residue of the α-amino acid and thyroid hormone L-thyroxine (alternative to T$_4$).

Thy *symbol for* a residue of the pyrimidine base thymine (alternative to T).

Thy-1 an allelic variable molecule, encoded on mouse chromosome 9, that is present on all T lymphocytes and also occurs in brain. It is a GPI-anchored membrane glycoprotein of the immunoglobulin superfamily, with a simple structure homologous to the **variable region** of an immunoglobulin molecule. *See also* **thymopoietin**.

thylakoid a saclike vesicle that bears the photosynthetic pigments in photosynthetic organisms. In prokaryotes the thylakoids are of various shapes and are attached to the plasma membrane. In eukaryotes they are flattened, membrane-bound dislike structures located in the **chloroplasts**; in higher plants they form dense stacks called **grana**. Isolated thylakoid preparations can carry out photosynthetic electron transport and the associated phosphorylation.

thymic hormones *see* **thymopoietin, thymosin**.

thymidine *or* **deoxyribosylthymine** *symbol*: dT *or* dThd; thymine 2′-deoxyriboside; 1-(2-deoxy-β-D-ribofuranosyl)-5-methyluracil; a deoxynucleoside very widely distributed almost entirely as phosphoric esters in deoxynucleotides and deoxyribonucleic acid, DNA. *Compare* **ribothymidine**. Because thymidine is absent from ribonucleic acid, the incorporation of radioisotope-labelled thymidine into cells is the basis of a procedure for measuring the rates of DNA synthesis and cell growth.

thymidine 5′-diphosphate *symbol*: dThd5′*PP* or ppdT; *the recommended name for* thymidine diphosphate (*abbr.*: dTDP), 5′-diphosphothymidine, 5′-thymidylyl phosphate, thymidine 5′-(trihydrogen diphosphate). It is an intermediate in the formation of thymidine 5′-triphosphate from thymidine 5′-phosphate. Derived compounds are the rare **nucleosidediphosphosugars** thymidinediphosphorhamnose and its biosynthetic precursor thymidinediphosphoglucose.

thymidine kinase EC 2.7.1.21; *systematic name*: ATP:thymidine 5′-phosphotransferase; an enzyme that catalyses the phosphorylation of thymidine to thymidine 5′-phosphate. It acts on thymidine as a first step in its further metabolism. Mitochondrial and cytosolic forms are known, the latter being more active in proliferating than in resting cells.

thymidine monophosphate *abbr.*: dTMP; *an alternative name for* any **thymidine phosphate**, but in particular for thymidine 5′-phos-

phate, especially when its distinction from thymidine (5′-)diphosphate and thymidine (5′-)triphosphate requires emphasis.

thymidine phosphate *symbol*: dThd*P*; thymidine monophosphate (*abbr.*: dTMP); any phosphoric monoester or diester of thymidine. There are two monoesters – thymidine 3′-phosphate and **thymidine 5′-phosphate** – and one diester – thymidine 3′,5′-phosphate – although thymidine 5′-phosphate is the ester commonly denoted (the locant being omitted if no ambiguity may arise). Thymidine 3′-phosphate (*symbol*: dThd3′*P*) (*or* thymidine 3′-monophosphate (*abbr.*: 3′dTMP) *or* 3′-thymidylic acid) is released from deoxyribonucleic acid by certain endo- or exonucleases, e.g. deoxyribonuclease II (EC 3.1.22.1) or spleen exonuclease (EC 3.1.16.1). The monophosphoric diester thymidine 3′,5′-phosphate (*symbol*: dThd-3′,5′-*P*) (*or* thymidine 3′,5′-(cyclic)phosphate *or* cyclic thymidine 3′,5′-monophosphate (*abbr.*: 3′,5′-cyclic dTMP or cyclic dTMP or cdTMP)) is an intermediate during the acid hydrolysis of deoxyribonucleic acid.

thymidine 3′-phosphate *see* **thymidine phosphate**.

thymidine 3′,5′-phosphate *see* **thymidine phosphate**.

thymidine 5′-phosphate *or* 5′-thymidylic acid *or* 5′-phosphothymidine *or* 5′-*O*-phosphonothymidine *symbol*: dThd5′P; *recommended names for* thymidine monophosphate (*abbr.*: dTMP), thymidine 5′-(dihydrogen phosphate), thymine (mono)deoxynucleotide. (The locant is commonly omitted if there is no ambiguity as to the position of phosphorylation.) It is synthesized (very indirectly) from **uridine 5′-phosphate**, its immediate precursor being 2′-deoxyuridine 5′-phosphate – *see* **thymidylate synthase**. Resynthesis of dTMP from free thymidine formed by degradation of DNA can also be effected via a **salvage pathway**.

thymidine phosphorylase *see* **pyrimidine-nucleoside phosphorylase**.

thymidine 5′-triphosphate *symbol*: dThd5′*PPP* or pppdT; *the recommended name for* thymidine triphosphate (*abbr.*: dTTP), 5′-triphosphothymidine, 5′-thymidylyl diphosphate, thymidine 5′-(tetrahydrogen triphosphate); a substrate for DNA synthesis. It is formed from thymidine 5′-phosphate by nucleoside-phosphate kinase (EC 2.7.4.4) and nucleoside-diphosphate kinase (EC 2.7.4.6), which transfer in turn the terminal phosphoric residues from two molecules of ATP.

thymidyl either of the chemical groups formed by the loss of a 3′- or a 5′-hydroxyl group from the deoxyribose moiety of thymidine.

thymidylate 1 either the monoanion or the dianion of thymidylic acid. 2 any mixture of thymidylic acid and its anions. 3 any salt or ester of thymidylic acid.

thymidylate synthase EC 2.1.1.45; *systematic name*: 5,10-methylenetetrahydrofolate:dUMP *C*-methyltransferase; an enzyme responsible for the formation of thymidylate, dTMP, by the *de novo* pathway. It catalyses reductive methylation of 2′-deoxy-5′-uridylate, dUMP, by reaction with 5,10-methylenetetrahydrofolate to form dihydrofolate and dTMP. Thymidylate synthase is a target for the antineoplastic agents **fluorouracil** and fluorodeoxyuridine. These drugs are metabolized to fluorodeoxyuridylate, which is a potent irreversible inhibitor of the enzyme. *See also* **thymineless death**.

thymidylic acid any phosphoric monoester of thymidine. The position of the phosphoric residue on the deoxyribose moiety of a given ester may be specified by a prefixed locant – *see* **thymidine phosphate**. However, 5′-thymidylic acid is the ester commonly denoted, its locant usually being omitted if no ambiguity may arise. 5′-Thymidylic acid is a recommended name for **thymidine 5′-phosphate**.

thymidylyl the thymidine[mono]phospho group; the acyl group derived from **thymidylic acid**.

thymin *a former name for* **thymopoietin**.

thymine *symbol*: T *or* Thy; 5-methyluracil; 2,4-dihydroxy-5-methylpyrimidine; one of the two major pyrimidine bases present (as thymidine) in DNA but not in RNA other than (as ribothymidine) in transfer RNA, where it is a minor base.

thymine dimer a **pyrimidine dimer** formed within a DNA strand from two adjacent thymine residues by **photodimerization**. It is most commonly of the cyclobutadipyrimidine type.

thymineless death the death of an animal or bacterial cell following thymine (or thymidylate) deprivation. Such a deprived cell may continue to produce protein and RNA, but can no longer synthesize DNA in the absence of thymidylate. It eventually loses its vitality and dies, perhaps because it is also unable to repair altered or damaged DNA. The phenomenon may be demonstrated with a thymine-requiring auxotrophic bacterial mutant by omitting thymine from the medium, or with any cell by adding **fluorouracil** or its metabolite fluorodeoxyuridine, which inhibit **thymidylate synthase**. A rapidly metabolizing animal cell, such as a tumour cell, is particularly susceptible to thymineless death.

thymine deoxyriboside *an alternative name for* **thymidine**.

thymine riboside *an alternative name for* **ribothymidine**.

thyminose *a former name for* 2-deoxy-D-ribose (*see* **deoxyribose**).

thymo+ *comb. form* of, or pertaining to, the thymus.

thymocyte any lymphocyte found in the thymus.

thymocyte mitogenic factor *abbr.*: TMF; *an alternative name for* **interleukin 2**.

thymocyte-stimulating factor *abbr.*: TSF; *an alternative name for* **interleukin 2**.

thymoma a tumour of thymic tissue.

thymonuclease *another name for* **deoxyribonuclease I**.

thymonucleic acid *an alternative term for* **thymus nucleic acid**.

thymopentin a synthetic pentapeptide that corresponds to residues 32–36 of **thymopoietin**.

thymopoietin *abbr.*: TP; *former name*: thymin; any one of three polypeptide hormones that result by alternative splicing of the same gene, and originally isolated from bovine thymus. They have pleiotropic actions on prothymocytes, inducing expression of the differentiation antigens such as **Thy-1**, and are important in T-cell development and function. They impair neuromuscular transmission after injection in mice, due to an effect on the nicotinic acetylcholine receptor. A synthetic pentapeptide corresponding to residues 32–36, named **thymopentin** and with the sequence Arg-Lys-Asp^{34}-Val-Tyr, has full activity. **Splenin**, a corresponding peptide in spleen, contains a similar pentapeptide, called **splenopentin**; bovine splenin has Glu in place of Asp^{34}, and human splenin has Ala in the corresponding position. These replacements affect the biological activity.

thymosin any of the polypeptide hormones of thymic origin that are involved in differentiation of T lymphocytes in the thymus. They include α_1 and β_4 thymosins, which are of known structure. Prothymosin-α is an oncoprotein that inhibits apoptosis by inhibiting formation of the complex (of seven molecules of Apaf-1 with seven of cytochrome *c*) that activates caspase 9. Thymosin β_4 (human, bovine, rat) has the sequence SDKPDMAEIEKFDKSKLKK-TETQEKNPLPSKETIEQEKQAGES.

thymulin a zinc-requiring immunomodulatory thymic nonapeptide that induces T-cell differentiation. Its level is decreased in immunodeficiency and in autoimmune disease.

thymus a bilobed glandular organ located in the mediastinum in the anterior thorax. In mammals the thymus is encapsulated and divided into lobules, each consisting of cortex and medulla; the cortex consists mainly of lymphocytes (thymocytes). It is responsible for populating the blood, lymph, and thymus-dependent areas with **T lymphocytes** during the neonatal period. Although large at birth it shrinks thereafter and may be difficult to identify in adults. It is important in **cell-mediated immunity**. —**thymic** *adj.*

thymus nucleic acid *a former name for* **deoxyribonucleic acid**, so called because the thymus was the first particularly rich source of a nucleic acid thought to be characteristic of animal tissues and distinct from that believed to be characteristic of plant tissues (called **yeast nucleic acid**).

thyro+ *comb. form* of, or pertaining to, the **thyroid gland**.

thyrocalcitonin *see* **calcitonin**.

thyroglobulin a protein in the **thyroid gland**, tyrosyl residues of which are modified to form the principal thyroid hormone, **thyroxine**, which is released by proteolysis. Thyroglobulin, a dimer of molecular mass ~660 kDa, is the main constituent of the so-called colloid of the thyroid follicles. Iodination of certain tyrosyl residues results in the formation of 3,5-iodotyrosyl residues; the iodinated phenolic moiety of one of these tyrosyls (in the free form, this would be a 2,6-iodophenyl group) is then transferred to the phenolic oxygen of another iodinated tyrosyl to form a tetraiodo compound; this is released by proteolysis as thyroxine (T$_4$) (some triiodo forms are also produced). The proteolysis occurs as colloid migrates from the basal to the apical surfaces of the thyroid cell. The colloid containing the modified thyroglobulin is taken into vesicles by phagocytosis to form periodic acid–Schiff positive (PAS) vesicles. Dense granules containing proteolytic enzymes then fuse with the vesicles to digest the thyroglobulin with the loss of PAS-positive material. The vesicles containing triiodothyronine and thyroxine are then extruded.

thyroidectomy the surgical removal of the thyroid gland.

thyroid gland a ductless gland of vertebrates, located near the larynx in most species, that secretes the hormones **thyroxine** and **triiodothyronine**. It contains closely packed follicles, or acini, each consisting of a sac that has a layer of cuboidal cells resting on an outer basement membrane. Each follicle is full of thyroglobulin-containing colloid. This is synthesized and secreted into the colloid space by the follicular cells; microvilli extend from the apical surface of the follicular cells into the colloid. In between the follicles are nests of parafollicular cells, or C cells, that synthesize **calcitonin**. Enlargement of the thyroid is called **goitre**. The thyroid concentrates iodide, a property that can be used for thyroid imaging, and for the treatment of cancer of the thyroid.

thyroid hormone either of the compounds secreted by the **thyroid gland**, namely **thyroxine** and **triiodothyronine**.

thyroid hormone receptor a receptor for **triiodothyronine** that belongs to the superfamily of eukaryotic transcription factors and the steroid/thyroid/retinoic acid family of nuclear receptors for hormones. There is a ligand-binding domain in the C-terminal region, a zinc-finger DNA-binding domain in the central region, and a function-modulating domain in the N-terminal region. Binding of hormone induces changes in conformation that control transcriptional activation and repression, and also regulates homo- or heterodimerization. In the absence of ligand, these receptors repress basal gene expression, probably through co-repressor proteins. Numerous mutations in the gene for the receptor result in resistance to thyroid hormone.

thyroid-hormone receptor-associated protein *see* **TRAP**.

thyroid response element *abbr.*: TRE; *see* **response element**.

thyroid-stimulating hormone *abbr.*: TSH; *recommended name*: thyrotropin; *other name*: thyrotropic hormone; a hormone, secreted by the adenohypophysis, that stimulates the activity of the thyroid gland. The bovine hormone is a glycoprotein (M_r 28 300) of α and β chains. The α chain has an amino-acid sequence identical with the α chain of **luteinizing hormone** and follicle-stimulating hormone. TSH increases uptake of iodide by the thyroid, the rate of conversion of diiodotyrosine to thyroxine, and the release of thyroid hormones. The release of TSH is influenced by **thyrotropin-releasing hormone**.

thyroliberin *recommended name for* **thyrotropin-releasing hormone**.

thyronine 4-(4-hydroxyphenoxy)phenylalanine; the notional parent structure of thyroid hormones (*see* **thyroglobulin**).

thyronine

sues. Although in humans the plasma levels of T_4 are 20-to 50-fold higher than those of T_3, T_4 is converted to T_3 in tissues and the effects are mediated by T_3 binding to receptors in the cell nucleus, leading to mRNA and protein synthesis. The normal range in human blood is 60–150 nmol L^{-1} (total of bound and free), 9–26 pmol L^{-1} (free). The chemical constitution of thyroxine was deduced by the British biochemist Sir Charles R. Harington (1897–1972), and confirmed by synthesis (1927).

thyroperoxidase a glycosylated integral membrane protein and enzyme of the apical surface of thyroid follicle cells, involved in the formation of thyroxine from thyroglobulin. It contains 933 amino acids and a transmembrane segment near the C-terminus. The N-terminal region contains the catalytic site and a heme moiety, and shares 42% sequence homology with myeloperoxidase. It catalyses the peroxidative iodination of specific tyrosine residues of thyroglobulin to monoiodotyrosine and diiodotyrosine, a process called **organification**. It also catalyses coupling of two of these diiodotyrosine residues to form thyroxine, a process called **coupling**. Both processes require the presence of hydrogen peroxide. Several mutations lead to congenital goitre and hypothyroidism due to the failure in organification.

thyrotoxicosis *see* **hyperthyroidism**.

thyrotroph a thyrotropic cell.

thyrotrophic having the effect of nourishing the thyroid gland, stimulating its growth and maintaining its size.

thyrotropic or *(esp. Brit.)* **thyrotrophic** having a modulating effect on the thyroid gland, especially in stimulating the synthesis of thyroid hormones. *See* **thyrotropin**.

thyrotropic hormone *an alternative name for* **thyroid-stimulating hormone**.

thyrotropic hormone releasing factor *(abbr.:* THRF*) or* **thyrotropic hormone releasing hormone** *(abbr.:* THRH*) an alternative name for* **thyrotropin-releasing hormone**.

thyrotropin *the recommended name for* **thyroid-stimulating hormone**.

thyrotropin receptor a G-protein-associated membrane protein that binds thyrotropin (i.e. **thyroid-stimulating hormone**) and mediates its intracellular effects. Different forms are produced by alternative splicing. Mutations are associated with hyperfunctioning thyroid adenomas. The hyperthyroidism of **Graves' disease** results from autoantibodies that stimulate the thyroid through the thyrotropin receptor.

thyrotropin-releasing factor *abbr.:* TRF; *an alternative name for* **thyrotropin-releasing hormone**.

thyrotropin-releasing hormone *abbr.:* TRH; *recommended name:* thyroliberin; *other name:* thyrotropin-releasing factor (*abbr.:* TRF); a hormone, released by the mammalian hypothalamus, that regulates the secretion of **thyroid-stimulating hormone** (TSH) by the adenohypophysis. It is present not only in primitive vertebrates but also in *Amphioxus* (a provertebrate) and in nerve ganglia of snails. It is the tripeptide pyroGlu–His–Pro–NH2, the cyclized glutamate and prolylamide being essential for activity. The pyroGlu is derived from a glutamine residue in the precursor, with release of ammonia, and the C-terminal amine from a glycine residue in the precursor (*see* **C-terminal peptide amidation**). It is released from a polyprotein precursor that contains six copies of the tetrapeptide sequence required to generate the hormone. TRH occurs in other parts of the brain, in pancreatic islets, and in the gastrointestinal tract; it probably serves as a neurotransmitter or neuromodulator, being known in mammals to alter sleep patterns, increase motor activity and blood pressure, and affect norepinephrine turnover. *Other names:* thyrotropic hormone releasing factor (*abbr.:* THRF); thyrotropic hormone releasing hormone (*abbr.:* THRH).

thyroxine *abbr.:* Thx *or* T_4; 4-(4-hydroxy-3,5-diiodophenoxy)-3,5-diiodophenylalanine; 3,3′,5,5′-tetraiodothyronine; the main hormone secreted by the thyroid gland, although its activity is lower than that of **triiodothyronine** (T_3). It is synthesized from L-tyrosyl residues of **thyroglobulin**. Thyroxine is essential for normal metabolism and physical development. It stimulates the metabolic rate, causing increased oxygen consumption and heat production in tis-

thyroxine-binding globulin *abbr.:* TBG; a major thyroid hormone transport protein found in mammalian serum. It is synthesized in liver, and variations in its concentration in serum result in similar variations in the level of circulating thyroid hormones. Alterations in its affinity for thyroid hormone also affect circulating thyroid hormone levels, decreased affinity being accompanied by decreased serum concentration. In the human, this variation may result from non-thyroidal illness, the presence of drugs (salicylates and phenytoin block thyroid- binding sites), or genetic variation, many variants being known.

Ti *symbol for* titanium.

TIC complex *abbr. for* peptide-translocating channel on inner chloroplast membrane; a protein complex of the inner chloroplast membrane that forms a peptide-translocating channel. The major component is TIC 110. *Compare* **TOC complex**.

tight coupling a term applied to mitochondrial function implying a high degree of efficiency in the yield of ATP (or other functional response) in relation to electron flow. The nearer the yield approaches the optimal value, the tighter the coupling is said to be. Tight coupling is obtained in isolated mitochondria only if gentle procedures, preserving membrane integrity, are employed during isolation. *See also* **chemiosmotic coupling hypothesis**, **phosphorus:oxygen ratio**.

tight junction or **zona occludens** (in vertebrate tissues) a beltlike region of very close contact between the plasma membranes of adjacent cells, such that the intercellular space is completely occluded. Tight junctions occur in epithelia and brain endothelia, and are effective barriers to the passage of water and solutes. *Compare* **gap junction**.

tight turn *see* **beta turn**.

TIGR *abbr. for* The Institute for Genome Research; *see* **sequencing centre**.

TIM *abbr. for* **1** triose-phosphate isomerase. **2** translocase of inner mitochondrial membrane; a protein complex that forms a channel in the inner mitochondrial membrane for proteins targeted for import into the organelle. It comprises four transmembrane proteins (Tim33, Tim23, Tim17, and Tim14) and a peripheral protein (Tim44). Tim23 and Tim17 contain 3–4 transmembrane segments each. The portion of Tim23 in the intermembrane space contains heptad repeats for coiled-coil formation and is sensitive to the electrochemical gradient across the inner membrane. TIM and **TOM** are believed to be apposed across the intermembrane space. *See also* **DPP**.

TIM barrel a protein structure in which 8 α/β segments form a barrel, as first described for **triose-phosphate isomerase**.

time *symbol:* t; one of the seven SI base physical quantities, usually indicating duration or a precise moment. The SI base unit of time is the **second**.

time constant a characteristic time taken by a system or a process to respond to a perturbation.

time course a graph plot of a parameter (that changes with time) scaled on the ordinate and time scaled on the abscissa.

time-of-flight mass spectrometer a type of **mass spectrometer** in which molecular ions from an ion source are ejected by a laser flash or electrical pulse, and are separated by their speed of travel, which is inversely proportional to their **mass-to-charge ratio**.

time-resolved fluorescence a method in which a sample is irradiated using a laser delivering a short pulse of light, after which the exponential decay of fluorescence is measured using a photomultiplier. Lanthanides such as europium emit fluorescence over a period of microseconds and measurements can be made after the relatively short-lived autofluorescence of proteins has decayed. It is used as the basis of DELFIA. *See also* **steady-state fluorescence anisotropy**.

time-resolved fluorescence spectrometry a method in which a sample is irradiated using a laser delivering a short (sub-nanosecond) pulse of light, after which the exponential decay of fluorescence is measured using a photomultiplier. The sensitivity of detection must be at the single photon level ('single photon counting'), and fluorescence lifetimes as short as a few nanoseconds can be determined. However, a disadvantage of the technique is that many measurements over possibly several hours may be needed to obtain reliable and sufficient data. *See also* **steady-state fluorescence anisotropy**.

TIMP *abbr. for* tissue inhibitor of metalloproteases; any protein that inhibits metalloproteinases. TIMP-1, *or* EPA (erythroid potentiating activity), mediates erythropoiesis. It forms complexes with enzymes such as collagenases, bringing about irreversible inactivation. *See also* **matrix metalloproteinase**.

tintometer an apparatus formerly used for determining the colour of a solution by comparison with a graded colour scale.

tiny RNA *see* **miRNA**.

Ti plasmid *abbr. for* tumour-inducing plasmid; any large conjugative plasmid found in the soil bacterium *Agrobacterium tumefaciens* and responsible for **crown-gall disease** of broad-leaved plants – a segment of the Ti plasmid, the **T-DNA**, is found in the genome of the tumour tissue of affected plants. With appropriate modifications the Ti plasmid can carry foreign DNA sequences into the genome of a susceptible plant.

Tiselius apparatus an apparatus for performing **moving boundary electrophoresis** in which there is a U-tube with sliding joints. This enables the protein (or other) solution and the buffer to be placed in separate parts of the U-tube and sharp boundaries made between them when the parts are slid into position to complete the U-tube. [After Arne Wilhelm Kaurin Tiselius (1902–71), Swedish protein chemist.]

tissue any collection of cells that is organized to perform one or more specific function. *Compare* **organ**.

tissue culture 1 the technique or process of growing or maintaining tissue cells (**cell culture**), whole organs (**organ culture**), or parts of an organ, from an animal or plant, in artificial conditions. **2** any living material grown or maintained by such a technique.

tissue extract *an alternative name for* factor III; *see* **blood coagulation**.

tissue factor *abbr.*: TF; *or* factor III; a transmembrane glycoprotein, mainly of subendothelial cells, that initiates **blood coagulation** by forming a complex with factor VII or VIIA. The complex activates factors IX or X. In normal hemostasis tissue factor initiates cell-surface assembly and propagation of the coagulation protease cascade.

tissue grinder any device for disrupting tissue by a crushing or shearing action. The term may embrace devices ranging from a meat grinder (mincer), mechanical pestle and mortar, or mill, to the more precise **Potter–Elvehjem homogenizer** or the **Dounce homogenizer**.

tissue kallikrein *see* **kallikrein**.

titer *the US spelling of* **titre**.

titin *or* **connectin** a giant 521 kDa protein that forms a single molecule elastic filament, extending from the M line to the Z line in the striated muscle sarcomere, and is one of the largest polypeptides yet described. The sequence consists mainly of repeats of two types of ~100-amino-acid motifs (termed class I and class II) that share similarity with the fibronectin type-III domain and **immunoglobulin superfamily** C2 domain, respectively. There is also a domain characteristic of protein kinases near the C terminus. It is thought to play an important role in sarcomere alignment during muscle contraction. *See also* **twitchin**.

titratable acidity 1 a measure of the acidity of a urine sample, expressed as the volume of 0.1-M NaOH required to neutralize, usually to a phenolphthalein endpoint, a 24-hour volume of urine. **2** a measure of the acidity of any solution, etc.

titrate 1 to add acid of known concentration to base of unknown concentration (or vice versa) until the point of equivalence is reached, from which can be determined the concentration of the unknown. **2** more generally, to add any standard solution to an unknown until some detectable equivalence point is reached, enabling the amount of the unknown to be determined. **3** (*esp. in microbiology or immunology*) to determine the maximum dilution at which activity can be detected. *See* **titre** (def. 2). —**titratable** *adj.*; **titration** *n*.

titration curve originally, a curve relating pH to the equivalents of strong base added per equivalent of acid in the solution. Similar curves are obtained when the logarithm of ligand concentration is plotted against the degree of association (or dissociation) of a ligand to its acceptor, or when redox potential is plotted against the degree of oxidation (or reduction) of a redox system.

titration pK *symbol*: pG; a kind of **pK** that characterizes the titration curve of a multibasic acid. If this acid has n hydrons that titrate over the pH range of interest, then its titration curve is identical to that of an equimolar mixture of n (hypothetical) monobasic acids. The pK values of these acids are the titration pK values of the multibasic acid. They differ little from the **molecular pK** values if the differences between sequential molecular values exceed unity. *See also* **group pK**.

titre *or* (*esp. US*) **titer 1** a value found by titration; *see* **titrate** (def. 1, 2). **2** (*in microbiology and immunology*) a measure of the concentration or activity of an active substance, e.g. an antibody, in a solution, usually expressed as the highest dilution of the solution in which the activity can be detected. By convention, if the highest dilution giving activity is 100-fold, the titre is said to be 100.

titrimetry the technique of measurement by titration, particularly using instrumentation and automation.

Tl *symbol for* thallium.

TLC *abbr. for* thin-layer chromatography.

tld a gene in *Drosophila* for **Tolloid** protein, which is required for normal dorsal development. Mutations in *tld* lead to a partial transformation of dorsal ectoderm into ventral ectoderm.

TLR *abbr. for* Toll-like receptor.

T lymphocyte *or* **T cell** any lymphocyte that undergoes maturation and differentiation in thymus. T lymphocytes are responsible for immune reactions involving cell–cell interaction, i.e. **cell-mediated immunity**. Subsets include: (1) cytotoxic T-lymphocytes, directly responsible *in vivo* for the death of cells identified for elimination; they include nonspecific killer cells (**NK cells**) and specific cytotoxic T-lymphocytes primed by antigen activation; (2) **helper cells**, which collaborate with **antigen-presenting cells** in the initiation of an immune response; and (3) **suppressor cells**, which down-regulate the response of helper cells.

Tm *symbol for* thulium. *Compare* **T_m**.

TMA *abbr. for* transcription-mediated amplification.

TMB-8 *abbr. for* (8-diethylamino)octyl-3,4,5-trimethoxybenzoate; an agonist that blocks intracellular Ca^{2+} channels.

TM domain *abbr. for* transmembrane domain.

TMF *abbr. for* thymocyte mitogenic factor; i.e. **interleukin 2**.

TMM *abbr. for* too many mouths; a protein (496 amino acids) of *Arabidopsis* that is expressed in proliferative postprotodermal cells and is thought to be a receptor. It contains a transmembrane segment at the C-terminus and ten uninterrupted leucine-rich repeats in the central region. Mutations in the gene produce abnormal clusters of stomata on the leaf surface.

TMP *abbr. for* **1** ribosylthymine monophosphate *or* ribothymidine monophosphate, *the common names for* ribosylthymine 5′-phosphate, ribothymidine 5′-phosphate, 5′-ribothymidylyl phosphate, thymine riboside 5′-phosphate. **2** *(sometimes)* thymidine monophosphate. dTMP is recommended.

7TM protein *an alternative name for* **seven-transmembrane-domain protein**.

TM7 receptor *or* **TM7 protein** any plasma membrane receptor or other protein that contains seven transmembrane segments, each of 20–25 amino acids. Included are the family of G-protein associated membrane receptors and the **TRP channels**.

tmRNA *abbr. for* transfer-messenger RNA; *other names*: 10SaRNA; SsrA; a bacterial RNA (260–430 nt long, depending on species) that contains a tRNA domain that can be charged with an alanine residue at the 3′ CCA end, and a short open reading frame (ORF) of nine codons. It binds at the A site of a ribosome that has stalled during translation because of some defect in the mRNA, adds its Ala to the nascent chain, then moves to the P site of the ribosome, the chain being transferred to the tmRNA (a process called **trans-translation**). Translation ends at the stop codon of the ORF, the defective protein being targeted for degradation.

TMR spectroscopy *abbr. for* topical magnetic resonance spectroscopy.

TMS *abbr. for* **1** tetramethylsilane; a water-soluble reference substance used in nuclear magnetic resonance spectroscopy. **2** the trimethylsilyl group; the symbol $Me_3Si–$ is preferred.

TMV *abbr. for* tobacco mosaic virus, an RNA plant virus.

Tn *abbr. for* **1** troponin. **2** transposon.

TNBS *abbr. for* 2,4,6-trinitrobenzenesulfonate; a reagent that combines readily with amino groups. It may be used either to protect these groups or to introduce a chromophore. It is removable using hydrazine.

Tn-C *abbr. for* troponin C.

TNF *abbr. for* tumour necrosis factor.

TNF-α converting enzyme *abbr.* TACE. *other names* TNF-α convertase, ADAM-17; a member of the ADAM family of metalloprokinases that cleaves the TNF-α precursor present in plasma membranes of activated macrophages and T cells to release soluble TNF-α, which is a cachectin.

TNF receptor-1-associated death-domain protein *see* **TRADD**.

TNF receptor-associated protein 2 *abbr.*: TRAF 2; an adaptor protein that is recruited by **TRADD** as part of the signalling complex that forms on the intracellular domain of tumour necrosis factor (TNF) receptor 1 following its binding by TNF. It recruits two inhibitor of apoptosis proteins (IAP1 and IAP2) that have ubiquitin–protein ligase activity. It may activate a protein kinase that activates the transcriptional activator JUN.

Tn-I *abbr. for* troponin I.

TNMR *or* **tnmr** *abbr. for* tritium nuclear magnetic resonance (spectroscopy).

TNS *abbr. for* 2-*p*-toluidylnaphthalene-6-sulfonate; a **fluor** used in **extrinsic fluorescence** studies of proteins.

Tn-T *abbr. for* troponin T.

tobacco acid pyrophosphatase an enzyme used for removing the 5′ cap from mature mRNA by hydrolysis of the diphosphate bond.

TOC complex *abbr. for* peptide-translocating channel on outer chloroplast membrane; a protein complex of the outer chloroplast membrane that forms a peptide-translocating channel. The core component is TOC 75. *Compare* **TIC complex**.

tocopherol *a generic term for* di- and trimethyltocols; α-tocopherol is 5,7,8-trimethyltocol, and is active as **vitamin E**; β-tocopherol is 5,8-dimethyltocol; γ-tocopherol is 7,8-dimethyltocol; and δ-tocopherol is 8-methyltocol.

α-tocopherol

T-odd phage any one of the **bacteriophages** T1, T3, T5, or T7 of *Escherichia coli* that share morphological, biochemical, and other features that distinguish them from **T-even phages**.

togavirus *or (formerly)* **arbovirus** any RNA animal virus consisting of an enveloped, icosahedral, 50–70 nm capsid with single-stranded infectious RNA (plus strand) of 4 MDa. Togaviruses include mosquito- and tick-borne viruses causing encephalitis and yellow fever.

tolaasin a channel-forming cyclic lipodepsipeptide similar to **syringomycin**. It is produced by *Pseudomonas tolaasi* and causes the brown blotch disease of the cultivated mushroom *Agaricus bisporus*. It is classified in the TC system under number 1.D.4.

tolbutamide 1-butyl-3-(*p*-tolylsulfonyl)urea, *N*-[(butylamino) carbonyl]-4-methylbenzenesulfonamide; the prototype of sulfonylurea drugs, which exert a hypoglycemic action by stimulating release of insulin from the pancreas and are used to treat type II **diabetes mellitus**.

TolC a homotumeric protein that forms a 12-stranded beta-barrel that spans the **outer membrane** of *E. coli* and has a long 12-stranded, mostly alpha-helical, barrel that extends far into the periplasm. It permits exit of large secreted proteins and of smaller toxic compounds. Each channel consists of a homotrimer, with each subunit containing 471 amino acids.

TolC channel funnel a structure that functions to export substances out of the Gram negative bacteria by the Type I secretory pathway. It forms a contiguous channel with an inner membrane transporter, providing a direct pathway through the periplasm and outer membrane. TolC channels eliminate a wide variety of toxic compounds and drugs.

tolerance **1** the progressive attenuation of the response to an agent (usually a drug) whereby increasing concentrations of the agent are required to maintain the response. Underlying mechanisms confer either functional (or pharmacodynamic) tolerance, where the loss of response is due to desensitization of effector mechanisms, or metabolic (or pharmacokinetic) tolerance, whereby elimination of the agent is accelerated usually by induction of catabolic enzymes or other inactivating mechanism. **2** the ability of an organism to grow and thrive in an unfavourable environment. **3** an allowable deviation from a standard. **4 immunological tolerance**.

tolerogen an antigen that, in given circumstances, induces **immunological tolerance** when administered to an animal. Tolerogens may, when presented in different circumstances, induce an immune response. —**tolerogenic** *adj.*

Toll a developmental protein in *Drosophila melanogaster*, the product of *toll* gene, that is required for embryonic dorsal–ventral polarity. It is a type I transmembrane signalling receptor protein with a **leucine-rich repeat** segment.

Toll-like receptor *abbr.*: TLR; any of a family of receptors that are homologous to Toll of *Drosophila* and occur on leukocytes and epithelial cell surfaces. They recognize components of pathogenic microorganisms and trigger secretion of cytokines such as tumour necrosis factor and interleukins 6 and 8. About ten have been identified in humans. TLR2 binds bacterial peptidoglycan and lipoprotein; TLR4 binds Gram-negative lipopolysaccharides; TLR5 binds flagellin; and TLR9 binds the CpG motif of bacterial DNA.

tolloid a developmental protein in *Drosophila melanogaster* encoded by the *tld* gene; it is a member of the **astatin** subfamily of metalloproteases.

tol plasmid a plasmid that specifies the degradation of aromatic hydrocarbons. *Pseudomonas putida* containing the pWWO plasmid grows on toluene, *m*- and *p*-xylene, *m*-ethyltoluene, and 1,2,4 trimethylbenzene. Enzymes required for the metabolism of these compounds are encoded by the *xyl* operons.

TOM *abbr. for* translocase of outer mitochondrial membrane; a complex of transmembrane proteins in the outer mitochondrial membrane that consists of receptors for proteins targeted for import into the organelle and a general insertion pore (GIP). The receptors are the heterodimers Tom70–Tom37 and Tom22–Tom20. The GIP consists of multiple copies of Tom40, Tom7, Tom6, and Tom5. Tom40 is predicted to form a beta barrel, whereas all the others have one transmembrane segment each. The cytosolic parts of Tom70 and Tom20 contain tetratricosapeptide repeats, which bind and bring their respective heterodimers together. This information is based on studies performed on the yeast complex. It is believed that TOM and **TIM** are apposed across the intermembrane space.

tomography a radiographic technique that can image a thin plane or section of a body. An X-ray beam is focused on an area of a body, the beam is then rotated, and successive images are stored and subsequently analysed by computer (**computer-assisted tomography**, or CAT scanning). **Positron emission tomography** (or PET scanning) involves the detection of X-rays emitted from radionuclides that decay by positron emission and are located within the patient's body, e.g. from ^{15}O (**oxygen-15**) after administration of $H_2^{15}O$.

ton 1 *or (esp. US)* **long ton** a unit of mass equal to 2240 **pounds** avoirdupois (1016.05 kg). **2** *or (esp. US)* **short ton** *or* **net ton** a unit of mass equal to 2000 pounds avoirdupois (907.18 kg). **3** metric ton; *see* **tonne**.

TonB-dependent transporter any 22-stranded beta-barrel transporter protein of the outer membrane of Gram-negative bacteria that is dependent for its function on contact with the TonB complex of the inner membrane. Examples are the ferric enterobactin receptor/transporter (FepA) and the ferrichrome receptor/transporter (FhuD), which permit entry of iron chelates and of vitamin B_{12}.

tone 1 *(in physiology)* *or* **tonus** the state of sustained partial tension adopted by muscles in order to maintain posture, etc. It is brought about by the contraction of only a certain proportion of the muscle fibres at any given time. **2** *(in clinical medicine)* the state or degree of firmness of a muscle on palpation.

tonicity 1 the **osmotic pressure** of a solution; often used in the sense of the ratio of the osmotic pressure of a solution to that of a given reference solution. **2** (of a muscle) the condition of possessing **tone** or tension. *See also* **hypertonic, hypotonic, isotonic**.

tonin EC 3.4.21.35; *other names*: esterase 1; S2 kallikrein; RGK-2; RSKG-5; a serine protease, present in several rat tissues, that can cleave angiotensinogen and angiotensin-tetradecapeptide renin substrate to produce angiotensin II directly. It also acts on the Phe-|-His bond of angiotensin I to form angiotensin II.

tonne *or* **metric ton** *symbol*: t; a unit of mass equal to 10^3 kg. 1 tonne = 0.984 207 UK ton or 1.102 31 short (US) tons.

tonometer 1 an apparatus, usually a closed glass vessel, used to equilibrate a liquid, especially blood, with a particular gas of known partial pressure. When equilibrium is established the phases are separated and the gas in question determined in both phases. **2** a device for measuring vapour pressure.

tonus *an alternative term for* **tone** (def. 1).

too many mouths *see* **TMM**.

tooth-lid factor *a name originally given to* **epidermal growth factor**.

top+ *a variant form of* **topo+** (before a vowel).

TOPA *abbr. for* 3,4,6-trioxyphenylalanine; 6-hydroxydopa. *See also* **TOPA-quinone**.

TOPA-quinone 3,4,6-trihydroxyphenylalanine quinone; 6-hydroxydopa quinone; the covalently bound redox prosthetic group of bovine plasma amine oxidase and some other copper-containing oxidases. It is formed by a post-translational modification involving hydroxylation of tyrosine contained in an active site with the consensus sequence, Asn-Tyr-Asp/Glu.

TOPA-quinone

topical relating to a particular place or part of the body; local; (of a medicament, nutrient, etc.) applied locally rather than systemically. —**topically** *adv.*

topical magnetic resonance spectroscopy *abbr.*: TMR spectroscopy; a method for obtaining high-resolution **nuclear magnetic resonance** spectra from a selected place in a larger specimen (e.g. a specific organ of a living body). The sensitive volume is generated by superimposing high-order magnetic field gradients onto the main magnetic field of the spectrometer in such a manner as to define a central region of uniform field and adjustable size, surrounded by rapidly changing fields. Signals may be detected from individual nuclei, and the signals can be attributed to specific metabolites, enabling measurement of the level of these metabolites *in situ*.

topo+ *or (before a vowel)* **top+** *comb. form* denoting place or region.

topochemical 1 of, or pertaining to, **topochemistry**. **2** of, pertaining to, or constituting a locally confined chemical reaction. **3** describing a combined tactile and chemical sense in some animals that is important in the perception of odours in relation to track or position.

topochemistry the characteristic chemical activity of a **topomer**.

topogenesis *an alternative term for* **morphogenesis**, especially in regard to biomolecular structures. —**topogenic** *adj.*

topogenic sequence any part of the sequence of a nascent polypeptide that is involved in getting the mature protein to its proper position in the cell. *See* **signal peptide**.

topogenic signal any feature of the structure of a macromolecule that directs it to a particular organelle in a cell. The term includes **topogenic sequence**. *See* **mannose 6-phosphate**.

topography 1 the science and practice of describing the arrangement in space of shapes and surfaces, and of components in relation to one another. **2** the detailed description, or representation on a map, diagram, etc., of the surface features of a three-dimensional object, or part of one, with reference to the underlying structure. *Compare* **topology**. —**topographic** *or* **topographical** *adj.*; **topographically** *adv.*

topoinhibition the inhibition of a cellular process by an action at the cell membrane, especially through contact with, or proximity to, other cells.

topoisomer *or* **topological isomer** any of two or more **isomers** of a macrocyclic molecule, or system of two or more macrocyclic molecules or molecular subunits, that differ in the degree of knotting of a loop or the degree of interlocking of rings. Topoisomers of a given molecule, or system of molecules, display the same connectivity, bond orders, and configurations but can be interconverted only by breaking and reforming a covalent bond; they have different **linking numbers**. *Compare* **topomer**. —**topoisomeric** *adj.*; **topoisomerism** *n.*

topoisomerase *see* **DNA topoisomerase**.

topological bond the force equivalent to a chemical bond that may be conceived as holding together any two interlocked rings of a catenated molecular structure or two loops of a knot in a cyclic molecule. Such a bond is the property not of a pair of atoms but of the complete molecule, the strength of the union being that of the weakest chemical bond in the entire structure.

topological isomer *an older name for* **topoisomer**.

topological isomerism *an alternative term for* topoisomerism; *see* **topoisomer**.

topological winding number *see* **supercoil**.

topology 1 *(in molecular biology)* the study of deformability and deformation of structures, including macromolecules. The basic concept is founded on whether a particular figure or surface can be continuously deformed without rupture and remaking of, e.g., a chemical bond. **2** *(in botany)* the study of localities where plants are found. **3** *(in mathematics)* the branch of geometry that deals with general types of shape rather than with particular shapes and sizes; the study of relationships that remain invariant under one-to-one transformation. *Compare* **topography**. —**topological** *or* **topologic** *adj.*; **topologically** *adv.*

topomer any of two or more **conformers** of a macromolecule that have differing **topography** (def. 2). Any topomer (or its exact mirror image) is characterized by a unique set of interactions between any two residues of the complete sequence of the macromolecule. *Compare* **topoisomer**.

topotactic describing a chemical reaction in which the product has an ordered geometric relationship to the reactant.

top yeast *see* **bottom yeast**.

torin *or* **calpromotin** a protein of human erythrocytes that forms a ten-subunit torus that interacts with the plasma membrane, with stomatin and with cyclophilin. Each subunit contains 198 amino acids.

toroid 1 an alternative term for **torus**. **2** the solid enclosed by a torus. —**toroidal** *adj.*

torr *symbol*: torr; a non-SI unit of pressure, used especially in connection with low pressures. It is defined as equal to 1/760 of an atmosphere; i.e. 1 torr = 101 325/760 Pa = 133.322 Pa. Its use is now discouraged, in favour of the **pascal**. [After Evangelista Torricelli (1608–47), Italian physicist.]

torsin an **AAA protease** of the endoplasmic reticulum lumen. Mutation leads to autosomal dominant torsion dystonia.

torsion 1 the twisting of something about an axis by the application of equal and opposite rotational forces acting in parallel planes. **2** the twist so produced. —**torsional** *adj.*

torsion angle *symbol*: θ *or* ω; consider an assembly of attached atoms represented by the generalized partial structure X–A–B–Y, where neither X nor Y is collinear with A and B, then the torsion angle is the smaller angle subtended by the bonds X–A and Y–B in a plane projection obtained by viewing the assembly along the axis A–B. Depending on the direction of rotation, right or left, of the bond to the furthermost atom, Y, relative to the bond to the atom nearest the viewer, X, the angle is considered positive or negative, respectively. The multiplicity of the bonding of the various atoms is not relevant. A torsion angle may similarly be defined for a partial structure X–A–Z–B–Y where Z represents one or more additional atoms and A, Z, and B are collinear. The torsion angles for the backbone of a polypeptide chain are symbolized as follows: that for rotation about the N–C^α bond is denoted by ϕ; that for rotation about the C^α–C′ bond is denoted by ψ; that for rotation about the C′–N bond is denoted by ω. The torsion angles for the backbone of a polynucleotide chain are as shown below. *See also* **conformation**, **Ramachandran plot**, **beta turn**.

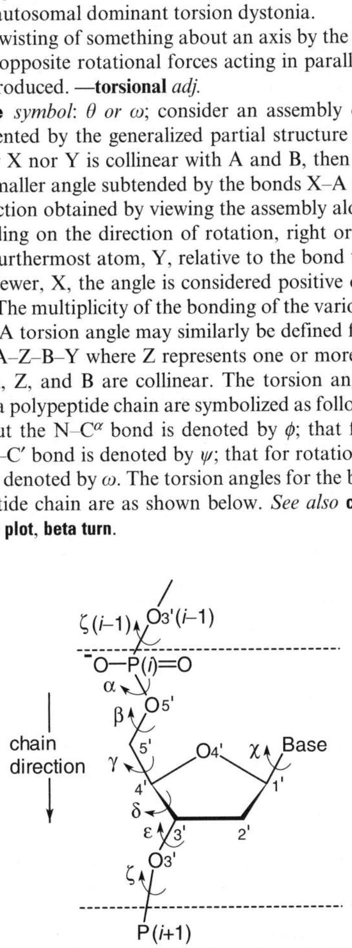

designation of torsion angles for polynucleotides

torsion balance an instrument, used for the rapid weighing of small amounts of substances, in which the sample is balanced by the torsion of a wire and the amount of torsion that has to be applied to restore balance is indicated as a scale calibrated in units of mass.

torso a gene for a tyrosine kinase (EC 2.7.1.112) receptor in *Drosophila melanogaster*. The receptor is a type I membrane glycoprotein and is involved in the determination of anterior and posterior structures in the embryo.

torulin *an obsolete name for* extracts of yeast that cure polyneuritis in pigeons fed on a diet of polished rice, no doubt due to the vitamin B_1 content of yeast extract.

torus (*pl.* **tori**) a shape generated by rotating a circle or disk about an axis that is displaced outside the diameter of the circle or disk, but in the same plane; i.e. shaped like a ring doughnut (donut). *See also* **toroid**. —**toric** *adj.*

Tos *symbol for* the tosyl group.

tosyl *symbol*: Tos; *a trivial name for* the toluene-4-sulfonyl group, CH_3–C_6H_4–SO_2–, with which amino groups may be modified by allowing them to react with *p*-toluenesulfonyl chloride. *Compare* **pipsyl**.

total body water *abbr.*: TBW; *see* **body water**.

totipotent (of a cell) having the potentiality to develop in any way possible, given its particular genetic constitution, and thus to form a new organism or to regenerate any part of an organism.

touchdown PCR a variant of **polymerase chain reaction** (PCR) in which the annealing temperature is decreased by, say, 1°C every second cycle until a touchdown annealing temperature is reached for completion of the amplification. The advantage is that it enriches for products containing correct matches between primers and template, and minimizes spurious priming during amplification.

Tourette syndrome a relatively common autosomal dominant condition, characterized by chronic motor and vocal tics, that usually begins in childhood. Dopamine drugs enhance the symptoms, whereas certain dopamine receptor antagonists decrease them. The gene product responsible is unknown. [After Georges Gilles de la Tourette (1857–1904), French physician.]

toxic 1 poisonous. **2** of, relating to, or caused by a **toxin** or poison.

toxicity 1 the quality or state of being poisonous or toxic. **2** the relative degree of being poisonous or toxic; the potency of a **toxin**.

toxicogenic *an alternative term for* **toxigenic**.

toxicogenomics the analysis of the toxic effects of potential new drugs on gene expression patterns in target cells or tissues. It aims both to prioritize compound pipelines (to eliminate expensive failures in drug development), and to reveal genetic signatures that may predict toxicity in these compounds.

toxicology the science dealing with the effects of poisons on living organisms. The term is sometimes restricted to the effects of synthetic poisons and the term **toxinology** used when dealing with naturally occurring poisons.

toxic shock syndrome a syndrome of acute high fever, hypotension, diarrhoea, and a skin rash, that occurs in women usually during menstruation and with use of certain highly absorbent types of tampons. It is occasionally fatal, and is caused by infection from contamination by certain strains of *Staphylococcus aureus* that produce **toxic-shock syndrome toxin-1** (TSST-1).

toxic shock syndrome toxin-1 *abbr.*: TSST-1; a protein (193 amino acids) secreted by certain strains of *Staphylococcus aureus*, that is responsible for the **toxic shock syndrome**. It is a potent inducer of interleukins 1 and 2 and tumour necrosis factor.

toxify to poison; to render poisonous. —**toxification** *n.*

toxigenic *or* **toxicogenic** *or* **toxinogenic** (especially of bacteria and fungi) producing toxin.

toxin any of various specific poisonous substances that are formed biologically. Not all such poisons are so termed, their classification as toxins being somewhat arbitrary and tending to vary with the discipline concerned; furthermore, the term is sometimes extended to include synthetic poisonous substances. Various types of toxin may be designated according to the source of the toxin, e.g. **endotoxin**, **exotoxin**, **mycotoxin**, **phycotoxin**, **phytotoxin**, and **zootoxin**, or according to the specific or prime site of action of the toxin, e.g. **hepatotoxin** and **neurotoxin**.

toxinogenic *an alternative term for* **toxigenic**.

toxinology *see* **toxicology**.

toxoid an **exotoxin** that has been modified (e.g. by formalin treatment) so that its toxicity has been lost while its antigenicity (both immunogenicity and reactivity) is retained.

toxophore a term introduced by the German biochemist Paul Ehrlich (1854–1915) to denote the specific chemical group in a toxin molecule that is responsible for its toxic effect. Unlike the **haptophore**, it is destroyed on conversion of the toxin to a **toxoid**.

TP *abbr. for* thymopoietin.

TPA *abbr. for* **1** tissue-type plasminogen activator; *see* **plasminogen activator**. **2** 12-*O*-tetradecanoyl-phorbol-13-acetate (*or* phorbol 12-myristate 13-acetate; *see* **PMA**).

TPAP *abbr. for* testes-specific poly(A) polymerase.

TPCK *abbr. for* L-(1-tosylamido-2-phenyl)ethylchloromethyl ketone; a modifying reagent for histidine residues in peptidases such as trypsin.

T3 phage promoter a short DNA sequence recognized by the RNA polymerase from bacteriophage T3, e.g. –17 AATTAACC-CTCACTAAAGGAAGA +6.

T7 phage promoter a short DNA sequence recognized by the RNA polymerase from bacteriophage T7, e.g. –16 TATACGACT-CACTATAGGGAGA +6.

TPI *(formerly) abbr. for* **1** triphosphoinositide; use PtdInsP_2 (or PIP$_2$) instead. **2** phosphatidylinositol 4,5-bisphosphate; use PtdIns-4,5-P_2 instead. *See* **phosphoinositide**.

TPN *abbr. for* triphosphopyridine nucleotide; obsolete, use **NADP$^+$**.

TPNH *abbr. for* reduced triphosphopyridine nucleotide; obsolete, use **NADPH**.

TPP *abbr. for* thiamine diphosphate (*formerly* thiamin pyrophosphate); *see* **vitamin B$_1$** .

TPQ *abbr. for* topaquinone trioxyphenylalanine quinone; a quinone cofactor produced by post-translational modification of a tyrosyl residue in some quinoproteins.

TPR *abbr. for* tetratricosapeptide repeat.

tpr-met an oncogene formed by the translocation of the *tpr* (translocated promoter) locus of chromosome 1 and *met* (the gene for **hepatocyte growth factor receptor**).

T protein *see* **glycine cleavage enzyme**.

tra symbol for **transformer**, a gene in *Drosophila*.

trace (*in sequencing*) a DNA sequence chromatogram that constitutes the primary data source for large-scale sequencing projects. Within the trace, coloured peaks represent each of the bases; the relative heights of the peaks determine which particular bases are read off to generate the raw nucleotide sequence.

tracee (*in tracer kinetics*) the non-radioactive compound of which the tracer is the radioactive equivalent; coined to represent unambiguously what has been variously called **stable material**, **mother substance**, etc. *Compare* **compartment**.

trace element *or* **microelement** any element that is present in the body of an organism at relatively minute concentrations (often arbitrarily defined as <0.005%), and consequently is required in the organism's diet in extremely small amounts. In higher animals the most important trace elements include: iron, zinc, copper, manganese, iodine, cobalt, molybdenum, selenium, and chromium. The extended list includes lead, arsenic, antimony, nickel, bromine, fluorine, boron, silicon, aluminium, silver, cadmium, bismuth, mercury, vanadium, titanium, indium, barium, strontium, and lithium. The term may be extended to cover elements of primarily toxicological interest.

trace metal any **trace element** that is a metal.

tracer an element or compound containing atoms that can be distinguished from their normal counterparts by physical means, e.g. by virtue of their radioactivity, mass spectrum, colour, etc. Tracers are used in biochemistry for following the metabolic pathways of substances in an organism, cell, or cell-free system.

track an individual lane resolving the components of a single sample on a slab electrophoresis gel.

tracking dye a dye that is added to a sample on an electrophoresis gel to be the fastest component and thus to indicate visually how far the samples have run at any instant in time.

TRADD *abbr. for* TNF receptor-1-associated death-domain protein; a cytosolic protein that interacts with tumour necrosis factor (TNF) receptor 1 after TNF binding and leads to activation of **NF-κB** and apoptosis. *See* **tumour necrosis factor receptor**.

TRAF 2 *abbr. for* TNF receptor-associated protein 2.

tragacanth 1 any thorny leguminous shrub of the genus *Astragalus*. **2** a gum obtained especially from *A. gummifer* and used extensively in commerce. It is a complex mixture of polysaccharides, including D-galacturonic acid, D-xylose, L-fucose, and D-galactose.

trailer sequence a sequence following the termination signal at the 3′ end of **messenger RNA**.

training set a data set that is characteristic of, or encapsulates information on, the problem to be solved, and is used as input for learning algorithms, such as neural nets and hidden Markov models (HMMs).

trait any observable, phenotypic feature of an individual.

TRAM *or* **TRAM protein** *abbr. for* translocating chain associating membrane protein; a transmembrane protein that spans the **rough endoplasmic reticulum** of mammalian cells, and plays a role in the translocation of nascent proteins destined for export from the endoplasmic reticulum.

TRAMP *abbr. for* tyrosine-rich acidic matrix protein; *other names*: 22 kDa extracellular matrix protein; dermatopontin; a protein of the extracellular matrix that may have various binding and adhesion functions.

TRANCE *an alternative term for* **RANKL**.

tranexamic acid *trans*-4-(aminomethyl)cyclohexanecarboxylic acid; it inhibits binding of plasminogen and plasmin to fibrin thereby inhibiting fibrinolysis.

trans+ *or* (*sometimes before s-*) **tran+** *comb. form* denoting across, beyond, on the other side, transverse; transfer, interchange. *See also* **trans-acting**.

trans- *abbr.*: *t-*; *prefix* (*in chemical nomenclature*) denoting a *trans*-isomer; *see* **cis–trans isomerism**.

trans-acting describing a regulatory genetic element whose effects are insensitive to its position. Examples include the genes for bacterial repressors. *Compare* **cis-acting**.

trans-activation gene activation by means of a **trans-acting** mechanism.

transacylase *an older name for* **acyltransferase**.

transaldolase EC 2.2.1.2; *other names*: dihydroxyacetone transferase, glyceronetransferase; an enzyme of the pentose phosphate pathway that catalyses the transfer of a 3-carbon glycerone moiety from sedoheptulose 7-phosphate to glyceraldehyde 3-phosphate forming erythrose 4-phosphate and fructose 6-phosphate. It has a broad substrate specificity.

transamidation a chemical reaction in which the amino group of an amide is exchanged for another amino group, i.e.:

$$R-CO-NHA + BNH_2 = R-CO-NHB + ANH_2.$$

Transamidation is catalysed by such proteolytic enzymes as **trypsin**, **chymotrypsin**, **papain**, and **subtilisin**.

transamidination a chemical reaction, catalysed by a transamidinase (aminotransferase) enzyme of sub-subclass EC 2.1.4, in which the amidino moiety of arginine is reversibly transferred to an amino group to form a different guanidino compound. An example is **glycine amidinotransferase**.

transaminase *an alternative name for* **aminotransferase**. *See also* alanine transaminase, aspartate transaminase, transamination.

transamination a chemical reaction, catalysed by any **aminotransferase** (transaminase) enzyme, in which the α-amino group of one amino acid is transferred to the α-carbon atom of an α-oxo acid. As a result, the α-oxo acid is transformed into an amino acid, and the amino acid that donated the amino group is converted to the corresponding α-oxo acid:

$$R'-CH(NH_2)-CO_2H + R''-CO-CO_2H =$$
$$R'-CO-CO_2H + R''-CH(NH_2)-CO_2H$$

Transamination is important in the catabolism of amino acids, in which the amino group is eventually or directly transferred to oxaloacetate (to form aspartate) or to α-oxoglutarate (to form glutamate). If transferred to aspartate, the amino group enters the **urea cycle** at the argininosuccinate step; if transferred to glutamate, the amino group is removed by glutamate dehydrogenase, yielding ammonia, which then enters the urea cycle. Transamination is also involved in amino-acid synthesis. Pyridoxal phosphate is the coenzyme of aminotransferases. Transamination was discovered by the Russian biochemists Alexander Evseyevich Braunstein (1902–86) and Maria Grigorievna Kritzmann (1904–71).

transbilayer signifying across or through the membrane **lipid bilayer**, as in **transbilayer distribution**, transbilayer movement.

transbilayer distribution the distribution of components between the two leaflets of a membrane lipid bilayer, but often referring specifically to lipids. In the plasma membrane this distribution is asymmetric: phosphatidylcholine and sphingomyelin are relatively enriched in the outer leaflet, and phosphatidylethanolamine and phosphatidylserine are more predominant in the inner leaflet (i.e. the cytoplasmic face).

transcarboxylase 1 methylmalonyl-CoA carboxyltransferase; EC 2.1.3.1; *systematic name*: (*S*)-2-methyl-3-oxopropanoyl-CoA:pyruvate carboxyltransferase; a biotinyl-protein, containing cobalt and zinc, that is present in prokaryotes and catalyses the transfer of a carboxyl group from methylmalonyl-CoA [(*S*)-2-methyl-3-oxopropanoyl-CoA] to pyruvate forming oxaloacetate and propanoyl-CoA. **2** a component of the prokaryote form of **acetyl-CoA carboxylase** that transfers the carboxyl group from *N*-carboxybiotin to acetyl-CoA.

transcarboxylation the transfer of a carboxyl group from one compound to another, e.g. as catalysed by a **transcarboxylase**.

transcobalamin *abbr.*: TC; any of three plasma glycoproteins (TC I, TC II, and TCIII) that are secreted by the liver and bind cobalamin. TC II binds newly absorbed cobalamin and transports it to the tissues, where the complex binds a specific receptor, is endocytosed, and the cobalamin released in lysosomes. In the postabsorptive state most cobalamin is bound to TC I. Deficiency of TC II results in congenital macrocytic anemia, which progresses to include immunodeficiency and neurological complications.

transconjugant a bacterial cell that has received genetic material from another bacterium by **conjugation**. A transconjugant should be referred to as a **recombinant** only if the transferred genetic material has been inserted into a pre-existing **replicon** in the recipient. If the transferred material is perpetuated *per se* as a plasmid, then the cell is called a **plasmid transconjugant**. If such material is not a plasmid and does not become a resident replicon, it will fail to replicate; the bacterium will be called an **abortive transconjugant**.

transcortin a specific plasma α_1-globulin, M_r 52 000, that binds **cortisol** at one binding site, and serves to transport the hormone. It also binds corticosterone and certain other steroid hormones; aldosterone is bound only weakly. Transcortin is synthesized in the liver. The normal serum concentration of transcortin is 3–4 mg dL^{-1}; saturation occurs at a cortisol serum concentration of about 28 µg dL^{-1}, other cortisol being less tightly bound to albumin or (about 5%) unbound. *Other names*: corticosteroid-binding globulin; corticosteroid-binding protein; cortisol-binding globulin (*abbr.*: CBG); cortisol-binding protein.

transcribe 1 to copy. **2** to use DNA as a template for the synthesis

of mRNA, or, in some cases (e.g. as in retrovirus infection), to synthesize DNA from an RNA template. *See* **transcription**.

transcribing strand *an alternative name for* **noncoding strand** (of duplex DNA).

transcript the product of a **transcription** process; the **primary transcript** is the immediate product of an RNA polymerase.

transcriptase *an alternative name for* **DNA-dependent RNA polymerase**.

transcription the synthesis of either RNA on a template of DNA or DNA on a template of RNA. The latter is important in the replication of **retroviruses** by the enzyme **reverse transcriptase**. In the transcription of double-stranded (ds) DNA the RNA is formed on a template comprising only one strand, the transcribed strand. This is also called the antisense strand since the other strand is termed the sense strand. Transcription involves many steps. (1) The region of DNA to be transcribed must be unwound approximately one turn to form an open loop to allow the transcribed strand to form a DNA–RNA hybrid with the 3′ end of the newly synthesized RNA. (2) The mRNA is transcribed by **RNA polymerase** from nucleoside triphosphates. The enzyme binds to the **promoter**, which is a region of DNA on the transcribed strand that has a characteristic base sequence (e.g. **TATA box**), usually located upstream of the start of transcription. However, in the case of eukaryotic RNA polymerase III, the promoter may sometimes be located downstream of the start point. In prokaryotes, for transcription to proceed, a complex must be formed on the dsDNA with the promoter and an accessory factor, **sigma factor**. In eukaryotes, an extensive and complex system of **transcription factors** regulates the process. While prokaryotes appear to utilize only a single RNA polymerase, eukaryotes possess three RNA polymerases; RNA polymerase I is located in the nucleolus and is responsible for the synthesis of ribosomal RNA. This polymerase accounts for the greatest quantity of RNA synthesized. RNA polymerase II is situated in the **nucleoplasm** and is responsible for the synthesis of heterogeneous nuclear RNA, for **primary transcript** from which messenger RNA is formed. This polymerase is thus instrumental in determining the nature of the protein profile of the cell. RNA polymerase III is also in the nucleoplasm and synthesizes transfer RNA. (3) In eukaryotes RNA polymerase II advances along the DNA template. The DNA unwinds ahead of the growing RNA and rewinds behind it thereby stripping the newly synthesized RNA from the transcribed strand of the DNA. The origin of the energy requirements for this movement is unclear. The movement is terminated by specific sequences in the transcribed DNA strand. In some cases **rho protein** plays a part in termination. (4) **Transcription factors**, which may be *cis*-acting or *trans*-acting, control transcription. In prokaryotes changes are very rapid since transcription and translation are tightly coupled. In eukaryotes such changes are slower. The activity of eukaryotic transcription factors is enhanced by proteins called **enhancers**, which may exert their influence over distances of several thousand base pairs. They may be **upstream**, **downstream**, or even in the midst of a transcribed gene. They may also be tissue specific. In the case of eukaryotes where the gene is split by **introns** the entire gene is transcribed and the introns are subsequently excised from the primary transcript by splicing.

transcriptional repressor any **transcription factor** (usually bacterial) that down-regulates gene expression by binding to an **operator**.

transcriptional silencing the repression of gene expression in a chromosomal region, which can span several or many genes. It can spread along a chromosome until insulator sequences are reached. Silencing can limit the expression of transgenes inserted into genomes randomly.

transcriptional territory any large group of adjacent genes that are expressed together but are not related by structure or function of their products.

transcriptional unit the segment of DNA within which the **transcription** occurs that eventually leads to RNA formation. *Compare* **cistron, gene**.

transcription elongation complex the complex of proteins required for elongation of the transcript during transcription in eukaryotes. It includes several transcription factors, RNA polymerase, elongin and other elongation factors, chromatin modelling complexes, and RNA processing proteins.

transcription factor *abbr.*: TF; any protein required to initiate or regulate **transcription**; the term includes **enhancers** as well as the general transcription factors. Transcription factors bind to DNA sequence elements upstream of the **start point** and thus participate in the formation of the complex that includes RNA polymerase and initiates transcription. *See, e.g.* **NF-KB, Oct, TFII, TFIII.**

transcription initiation complex the complex of over 100 proteins (3.5 MDa) required for initiation of transcription in eukaryotes. It includes several transcription factors, RNA polymerase, Mediator, chromatin remodelling complexes, histone acetylases, and transcriptional activator proteins.

transcription *in vitro an alternative name for* **in vitro** **transcription**.

transcription-mediated amplification *abbr.*: TMA; a method for RNA amplification that uses RNA transcription by an RNA polymerase (such as T7 RNA polymerase) and DNA synthesis by reverse transcriptase to produce RNA transcripts from either DNA or RNA templates. It takes place under isothermal conditions and amplification of the order of 10^9-fold is possible in less than 30 min. Oligonucleotide primers are specific for the target sequence at the 3′ end and have a promoter for the RNA polymerase at the 5′ end.

transcription start site *abbr.*: TSS; the location at the 5′ end of a gene, adjacent to the **promoter**, at which the RNA polymerase complex binds to the DNA and initiates the process of transcription of that gene into mRNA. The precise context of the TSS depends on the gene, its host organism, the type of polymerase involved, and other factors. *Compare* **start point**.

transcriptome the entire set of mRNAs transcribed from a cell's genome. *Compare* **proteome**. –**transcriptomic** *adj.*

transcriptomics the study of transcriptomes, particularly how transcript patterns are affected by development, disease, or environmental factors such as hormones, drugs, etc. *Compare* **proteomics**.

transcytosis the transport of molecules across a cell, especially across a **polarized cell**, such as an intestinal epithelial cell, which has a basolateral and an apical membrane of different composition, which thus provides a spatially orientated transport system. The molecules undergoing transcytosis are usually contained within vesicles. One example is the absorption of maternal antibodies from the gut in newborn animals. The antibodies are taken into epithelial cells by endocytosis at the apical surface, and are transported through the cell before being released at the basal surface into the blood. Transcytosis also occurs in thyroid cells: following endocytosis from the thyroid follicle, thyroglobulin is transported across the cell in vesicles with concomitant proteolysis, with release of thyroid hormone into a blood capillary at the opposite face of the cell (*see* **thyroid gland**). —**transcytotic** *adj.*; **transcytose** *vb.*

transdeamination *a former term for* the coupling of a transamidation reaction and the dehydrogenation of L-glutamate whereby an amino group of an amino acid is transferred to α-oxoglutarate by a transaminase enzyme and the resulting L-glutamate is dehydrogenated to yield α-oxoglutarate, ammonia, and a reduced coenzyme molecule.

trans-dominant mutation a mutation in a regulatory gene that can control the expression of a structural gene on another chromosome.

transducer any device that converts a quantity of energy from one form into another.

transducin *abbr.*: G_T or G_t; a heterotrimeric **G protein** of the retina, and a component in the visual transduction pathway. In the vertebrate light response, inactive heterotrimeric transducin with bound GDP ($G_{t\alpha\beta\gamma}$·GDP) binds the photoactivated rhodopsin, metarhodopsin II, and releases GDP. GTP then binds to the α subunit, which then separates from both the receptor and the $\beta\gamma$ heterodimer. $G_{t\alpha}$ then activates cyclic GMP phosphodiesterase, and the resulting fall in cyclic GMP concentration closes cyclic GMP-gated cation channels and hyperpolarizes the retinal rod cell, generating the nerve impulse.

transducon any large protein complex that contains components of a signalling cascade anchored to one or more scaffold proteins.

transduction 1 (*in microbiology*) the transfer of genetic information to a bacterium from a bacteriophage or between bacterial or yeast cells mediated by a phage vector. In **generalized transduction** any of the donor genes may be transduced, whereas in **specialized transduction**, i.e. by a **lysogenic** bacteriophage, only those genes at one of the ends of the prophage can be transduced. **Abortive transduction** occurs when phage DNA is not incorporated into the bacterial genome, but can nevertheless be transmitted in the phage to one of the daughter cells. **2** (*in biochemistry*) the activation of an intracellular effector pathway as a result of receptor activation by an extracellular signal. *See also* **second messenger, signal transduction**.

transesterification a chemical reaction in which the alcohol moiety of an ester is exchanged for another alcohol:

$$R{-}CO{-}OA + BOH = R{-}CO{-}OB + AOH$$

It is catalysed by such proteolytic enzymes as trypsin, chymotrypsin, papain, and subtilisin, and by esterases.

trans-fatty acid any fatty acid that contains ethylenic bonds in the *E* configuration (i.e. *trans*). They can occur as the result of biohydrogenation in ruminant animals (cows, sheep, and goats) or during partial hydrogenation/isomerization of unsaturated fats (both vegetable and animal) to harden fats. Because of the association with risk of heart disease, the levels of *trans* fatty acids in margarines are kept to a minimum (at least in the UK and Denmark). The reasons why trans fatty acids are associated with increased risk of heart disease is uncertain – they lower HDL cholesterol compared with the corresponding *cis*-isomers, but not compared with carbohydrate.

transfection originally the process of infection of competent bacterial cells by free bacteriophage or plasmid nucleic acid that results in the subsequent production of normal bacteriophage or plasmids in the infected bacterial cell. It has come to mean the process of bringing about genetic alteration of any cell or organism using recombinant DNA technology.

transferase an enzyme that catalyses the transfer of a group – e.g. a methyl group, glycosyl group, acyl group, or phosphorus-containing group – from one compound (the donor) to another compound (the acceptor). Transferases, of which there are many, are EC class 2 enzymes.

transference number *or* **transport number** *symbol*: *t*; the fraction of the current being carried by a given type of ion when an electric current is passing through an electrolytic solution.

transfer factor a lymphokine extractable from leukocytes and thought to transfer the ability to elicit a **delayed hypersensitivity** reaction from one person or animal to another.

transfer gene any gene carried by a **conjugative plasmid** that is responsible for the donor phenotype.

transfer membrane a membrane used with **blotting** techniques in recombinant DNA technology or protein studies. Materials of interest are transferred from an electrophoretic gel to the transfer membrane either by capillary attraction or by using an electric potential (**electroelution**). Transfer membranes are of three types: (1) nitrocellulose – mixed esters of cellulose acetate and cellulose nitrate; high cellulose nitrate improves binding of biomolecules while a low percentage of cellulose acetate improves handling properties; nucleic acids can be fixed to nitrocellulose membranes by baking; (2) nylon, which is physically more robust and reusable, and to which nucleic acids can be fixed by exposure to ultraviolet light; or (3) polyvinylidene-based membranes.

transfer-messenger RNA *see* **tmRNA**.

transfer potential *another name for* **group transfer potential**.

transfer reaction *another name for* **group transfer reaction**.

transferrin any of a class of monomeric, two-domain (76–81 kDa), metal-binding glycoproteins, widely distributed in physiological fluids and cells of vertebrates, with the characteristic property of a stringent association of their metal-combining properties with an anion-binding requirement; iron-binding transport proteins that bind two Fe^{3+} in association with bicarbonate (or other anion). They are an essential component in cell culture media, transporting iron into cells through the **transferrin receptor**. The transferrin family includes proteins such as **lactoferrin**, **melanotransferrin** (with electrophoretic polymorphisms), ovotransferrin (*see* **conalbumin**) and **serotransferrin** (*other names*: siderophilin; β-1-metal binding globulin).

transferrinjection a method of gene transfer in which DNA complexed to transferrin–polycation conjugates is introduced into cells by receptor-mediated endocytosis.

transferrin receptor a type I transmembrane glycoprotein of the plasma membrane that, on binding transferrin, undergoes endocytosis for transfer of Fe^{3+} ions into the cell. It consists of two identical single-pass subunits joined extracellularly by a disulfide bond. The intracellular domain is phosphorylated and palmitoylated, i.e. acylated with a palmitoyl group by a process analogous to **myristoylation**. Synthesis of the receptor is regulated by binding of aconitase to mRNA when intracellular iron concentrations are low, stimulating receptor synthesis, possibly by stabilizing the mRNA.

transfer RNA *abbr.*: tRNA; *other names*: acceptor RNA; amino-acid-accepting RNA; *(formerly)* soluble RNA (*abbr.*: S-RNA *or* sRNA). Any relatively small **RNA** molecule (73–93 nt) that mediates the insertion of an amino acid at the correct point in the sequence of a nascent polypeptide during protein synthesis. The amino-acid residue is attached by ester link to the adenosine of a CCA sequence at the 3′ terminal of tRNA at either the 2′- or 3′-OH on the ribose. Specific tRNA molecules are present for the various amino acids. The amino acid is attached by a specific amino acid–tRNA ligase, which first forms an aminoacyl-adenylate, and then transfers the aminoacyl group to the tRNA. Each tRNA possesses an **anticodon** of three nucleotides that base pairs in an antiparallel fashion with the codons of mRNA. There are 61 sense codons but only about 41 different tRNAs in the cytoplasm and even fewer in mitochondria due to flexibility in the base pairing between codon and anticodon (*see* **wobble hypothesis**). The sequence of tRNA for phenylalanine – determined by US biochemist Robert Holley (1922–93) – indicated a clover-leaf secondary structure and the anticodon to be on a loop lacking base pairing. The tertiary structures that have been determined for tRNAs have supported the clover-leaf structure. tRNA is characterized by the presence of many unusual **minor bases** whose function has not been completely established. *See also* **initiator transfer RNA, suppressor tRNA**.

transform 1 to bring about **transformation**. **2** *(in mathematics)* **a** a process for deriving one mathematical entity from another. **b** the transformed entity itself.

transformant a bacterial cell that has undergone **transformation** (def. 1), i.e. one that contains integrated donor genes that can be detected by plating on media selective for some or all of the donor genes.

transformation 1 *or* **genetic transformation** the intraspecific and interspecific transfer of genetic information by means of 'naked' extracellular DNA in bacteria. In transformation, in contrast to **transduction**, the DNA not integrated into the recipient's genome cannot function and manifest its genetic information phenotypically. **2** characteristic and inherited changes produced in cells in culture when they have been treated with certain viruses (which may be DNA- or RNA-containing), or with chemical carcinogens, or with X-rays, or when they have been subjected to genetic modification by DNA technology. In animal cells, these changes include certain specific alterations in morphology (especially of the nucleus) and stainability, and the cells usually show a loss of **contact inhibition** and can give rise to neoplastic growth on injection into animals. The term also refers to corresponding changes occurring in cells *in vivo*. **3** *or* **blast transformation** *or* **lymphocyte transformation** the morphological and other changes that occur in both B and T lymphocytes when they are cultured in the presence of an antigen to which they are primed, or that occur nonspecifically in B lymphocytes when they are exposed to a mitogen (e.g. certain lectins). The lymphocytes increase in size and develop a basophilic ribosome-rich cytoplasm, a prominent nucleolus, and a paler-staining nucleus. After about three days these cells resemble **blast cells**. **4** the alteration produced in certain specific hormone-receptor proteins, present in the extranuclear region of hormone-responsive cells, when the hormone binds to them. This alteration is accompanied by the migration of the hormone–receptor complexes to the nucleus. **5** *or* **nuclear transformation** (of an atomic nucleus) the change of one **nuclide** to another.

transformer *symbol*: tra; a gene in *Drosophila* that is required for all aspects of female somatic sexual differentiation. Binding of the female SXL protein to the transformer pre-mRNA transcript results in splicing to remove exon 2, which contains a stop codon present in

the male transcript. In this way females express a functional transformer protein, and males a nonfunctional one.

transforming growth factor *abbr.*: TGF; either of two types of mitogenic **cytokine**, TGFα and TGFβ. **TGFα** was isolated and purified on the basis of its ability to stimulate cell growth in soft agar. It is produced by macrophages, brain cells, and keratinocytes, and is mitogenic for fibroblasts, is angiogenic *in vivo*, and induces epithelial development. It is a glycoprotein and a member of the **epidermal growth factor** (EGF) family, interacting with the EGF receptor. TGFα comprises 50 amino acids with the following sequence (residues invariant in the EGF family are underlined): VVSFHND<u>C</u>PDSHTQF<u>C</u>F<u>H</u>G<u>T</u><u>C</u>RFLVQEDKPA<u>C</u>V<u>C</u>HSG<u>Y</u>V-GAR<u>C</u>EHADLLA. **TGFβ** is a multifunctional peptide that controls proliferation, differentiation, and other functions in many cell types. It is produced by fibroblasts, platelets, monocytes, chondrocytes, and osteoblasts. A glycoprotein with EGF repeats, at least five types (TGFβ1 to TGFβ5) are known; all have 112 amino acids except TGFβ5, which has 114. They have a great many biological activities and can be mitogenic or antiproliferative depending on cell type. Platelets contain large amounts, but these proteins are also found in a variety of cell types. Alternative splicing results in different forms of TGFβ1 in platelets and fibroblasts.

transforming growth factor receptor any of the membrane proteins that bind transforming growth factor β (TGFβ) and mediate its effects. Three types of receptor for TGFβ have been distinguished: types I and II have high affinity for TGFβ1, with lower affinity for TGFβ2; type III has high affinity for TGFβ1, 2, and 3. Types I and II have a single transmembrane domain with a cytoplasmic protein serine/threonine domain. The human type III receptor has a single transmembrane domain and little cytoplasmic structure. It is derived from an 849-amino-acid precursor, of which residues 1–16 are the signal, 17–781 form the extracellular domain (heavily glycosylated), and 808–849 are intracellular. It may present TGFβ to the signalling receptors.

transforming principle *a historical name for* the purified DNA that, when incorporated into a bacterial cell, brings about the **transformation** (def. 1) of that cell. The term was used, most famously, by Oswald Theodore Avery (1877–1955, US bacteriologist and immunologist) and colleagues in their work establishing that DNA, not protein, was responsible for genetic transformation. They used a wild strain of *Pneumococcus*, known as S (for smooth) strain because of the kind of colony that it forms, which is pathogenic; and a mutant strain, known as R (for rough) strain, which is non-pathogenic. When mice were injected with live R strain or heat-killed S strain, they remained healthy, but when injected with a combination of live R strain and heat-killed S strain they died. These workers concluded that the live R strain could be transformed into a pathogenic strain by material from the heat-killed S strain, and were able to demonstrate that the material was DNA.

transfuse to transfer blood from one animal into the circulation of another animal, or to inject liquid into the circulation of an animal to maintain the circulating fluid volume. —**transfusion** *n*.

transgene a gene that is inserted, using cDNA technology, into the germ line in a manner that ensures its function, replication, and transmission as a normal gene. *See also* **transgenesis, transgenic**.

transgenesis the creation in plants or animals of a stably incorporated gene or genes derived from another cell or organism which can be passed on to successive generations.

transgenic describing an organism harbouring in its germ line a gene that has been introduced using cDNA technology (*see* **complementary DNA**). In animals, the transgene is introduced during transient culturing of an egg or early embryo, which is then implanted into a foster mother. Several techniques are available: cells may be transferred from one embryo to another (to produce a **chimera**); **pluripotent** cells can be introduced into an embryo; cells infected with a retrovirus can be introduced; or cDNA can be introduced by microinjection into the pronucleus of a fertilized egg. In plants, plasmids from *Agrobacterium tumefaciens* have been widely used to produce transgenic varieties. The **Ti plasmid** is a natural vector for genetically engineering plant cells, because it can transfer its **T-DNA** to the plant genome. Expression of the introduced genes can be controlled by suitable promoters, such as the cauliflower mosaic virus promoter.

transgenomes large complexes of DNA found inside recipient cells following transfection of the latter with calcium phosphate-precipitated DNA.

transgenosis the transfer of genetic information between unrelated organisms and its expression (including gene maintenance, transcription, translation, and function), especially when donor and recipient cells are widely separated by evolution (e.g. bacterial and eukaryotic cells respectively) and when the mechanisms of gene transfer and maintenance are obscure. *Compare* **transgenesis**.

transglutaminase *an alternative name for* **protein–glutamine** γ-glutamyltransferase.

transglycosylation a chemical reaction in which one glycosyl residue is transferred from a glycoside onto a receptor molecule so as to form a new glycosidic linkage, e.g. in the biosynthesis of polysaccharides. The process involves breaking simple glycosidic linkages and reforming them into different types of glycosidic linkages. **Glycosyltransferases** are in subclass EC 2.4; also included in the term are glycosyl transfers to water and to inorganic phosphate.

transient 1 temporary, transitory, short-lived. **2** a sudden, short-lived disturbance in a system. **3** a short-lived chemical species such as an excited atom or molecule or a free radical.

transient expression the phenomenon whereby following transfection of mammalian cells with recombinant DNA a high level of expression of the protein of interest may be possible for only a few days. *Compare* **stable expression**.

transient-phase kinetics *an alternative name for* **pre-steady-state kinetics**.

transient receptor potential channel *see* **TRP channel**.

trans isomer *see* **cis–trans isomer**.

transition *(in molecular biology)* a type of mutation of DNA that occurs by replacement of one purine by another purine, or one pyrimidine by another pyrimidine whether by chemical change or by substitution. *Compare* **transversion**.

transition element *or* **transition metal** any element whose atom has an incomplete d-subshell of extranuclear electrons, or which gives rise to a cation or cations with an incomplete d-subshell. Such elements fall into groups 3 to 11 inclusive of the IUPAC **periodic table**. (Elements of group 12, which formerly were also considered as transition elements, have complete d-subshells and are now excluded.) The first series of transition elements is in period 4 of the periodic table and comprises those of proton numbers 21 to 29 inclusive, i.e. the elements scandium, titanium, vanadium, chromium, manganese, iron, cobalt, nickel, and copper; the second series is in period 5 and comprises those of proton numbers 39 to 47 inclusive, i.e. the elements yttrium, zirconium, niobium, molybdenum, technetium, ruthenium, rhodium, palladium, and silver; the third group is in period 6 and comprises those of proton numbers 57 to 79 inclusive, i.e. the **lanthanoids** together with the elements hafnium through to gold; and the fourth series is in period 7 and contains those of proton numbers 89 to 103 inclusive, i.e. the **actinoids**, and may well be found also to contain a number of elements of proton number 104 or more that yet remain to be characterized or discovered. The lanthanoids and actinoids are designated **inner transition elements** of periods 6 and 7 respectively; their atoms also have incomplete inner f-subshells of electrons. All transition elements are metals, they often have more than one valency state, and they form many coordination complexes, which frequently are coloured.

transition state *or* **activated complex** *(in enzymology)* the state of most positive molar **Gibbs energy** through which an assembly of atoms must pass on going from the reactants of a single-step chemical reaction to the products. This is the state that occurs immediately after binding of the reactants to the enzyme's active site; it is during this time that transient bonds between reactants and enzyme are first formed, and initial electronic interactions occur. It is followed by a state in which more stable intermediates form, prior to the eventual formation of products.

transition-state theory the theory that relates the rate of an enzymic reaction to the difference in **Gibbs energy** between the **transition**

state and **ground state** ($\Delta G^{\ddagger}$). If x is the ground state and $x^{\ddagger}$ the transition state, then the rate of decomposition of x is given by:

$$-d[x]/dt = \nu[x^{\ddagger}] = [x](kT/h)\exp(-\Delta G^{\ddagger}/RT),$$

where ν is the vibrational frequency of the bond that is breaking, h is the Planck constant, k the Boltzmann constant, R the gas constant, and T the thermodynamic temperature.

transition temperature *symbol*: T_t; (of a membrane lipid) the surface temperature at which a heated lipid bilayer undergoes transition from one physical state (e.g. liquid crystalline) to another (liquid). *See also* **pretransition temperature, phase transition (of lipids)**.

transit peptide any ~4 kDa peptide sequence at the N or C terminus (or both) of cytoplasmically synthesized precursor proteins of certain chloroplast proteins that contains the information to ensure both the specific post-translational transport of the protein into the chloroplast and its localization within the organelle. The transit peptide is removed within the chloroplast.

transit time the time taken by a coenzyme or metabolite in travelling between two enzymes. If the two enzyme reactions were infinitely fast, the overall velocity would be determined by the reciprocal of the transit time.

transketolase EC 2.2.1.1; *systematic name*: sedoheptulose-7-phosphate:D-glyceraldehyde-3-phosphate glycolaldehydetransferase; *other name*: glycolaldehyde transferase. An enzyme that catalyses the transfer of a glycolaldehyde ($HOH_2C–CHO–$) moiety from xylulose 5-phosphate to ribose 5-phosphate to form glyceraldehyde 3-phosphate and sedoheptulose 7-phosphate, in the pentose phosphate pathway. It requires thiamine diphosphate (TPP) as coenzyme; a glycolaldehyde–TPP intermediate is formed transiently during the reaction. It has broad substrate specificity. Measurement of transketolase activity in a red cell hemolysate, in the presence and absence of thiamin, gives an indication of the thiamin status of the individual (by revealing the degree of activation of the enzyme in the absence of thiamin).

translation 1 *(in molecular biology)* the process by which a particular sequence of bases in messenger RNA (mRNA) determines a sequence of amino acids in a polypeptide chain during **protein and peptide biosynthesis**. One or more specific base triplets (codons) code for each of the 20 amino acids (*see* **genetic code**). Each codon of the mRNA molecule is recognized by the corresponding anticodon on the **transfer RNA** (tRNA) to which the amino acid is attached. The tRNA binds to a site on the large subunit of the **ribosome**, which is associated with the small subunit to which the mRNA is bound. The first codon (the initiation codon) in each mRNA molecule is AUG, which codes for formylmethionine (in prokaryotes and mitochondria) or methionine (in eukaryotes). The methionine or formylmethionine, or just the formyl group, is removed before the chain is completed. Completion of translation is signalled by one of three termination codons. **Release factors** effect the release of the complete polypeptide chain and the dissociation of the ribosomal subunits. **2** movement or displacement laterally in space without rotation or change of orientation.

translational frictional coefficient *see* **frictional coefficient**.

translational fusion *an alternative name for* **gene fusion**.

translation in vitro *an alternative name for* **in vitro translation**.

translation start site the mRNA codon at which the ribosome initiates the process of translation of the sequence into an amino acid sequence. The start codon is usually AUG (coding for methionine); in prokaryotes it may also be GUG (leucine). In prokaryotes, ribosome binding to the AUG is mediated by the **Shine–Dalgarno sequence**. In eukaryotes, recognition of the AUG is mediated by the **Kozak sequence**, which the 40S ribosomal subunit locates by scanning the message starting from the 5′ end.

translocase a system catalysing translocation of a substrate across an osmotic barrier. *See* **EF-G, translocator**.

translocation a vectorial transfer process, e.g. of a solute across a membrane. *See also* **chromosomal translocation**. —**translocational** *adj*.

translocator *or* **porter 1** a system catalysing a secondary translocation reaction, i.e. one not involving primary bond exchanges between different pairs of chemical groups or the donation and acceptance of electrons. **2** any molecular group that is attached to a substrate during transport of the latter across a membrane.

translocon a translocation complex that is the specific site of protein translocation across the endoplasmic reticulum. It involves the **signal recognition particle receptor**. The component proteins, called **translocon-associated proteins** (*abbr.*: TRAP), may be involved in ensuring retention in the endoplasmic reticulum of resident proteins. TRAP proteins form a heterotetramer of α, β, γ, and δ subunits.

translucent transmitting light (or other radiation) with some scattering or diffusion. Consequently, objects cannot be clearly distinguished through translucent materials. *Compare* **transparent**. —**translucence** *or* **translucency** *n*.

transmembrane domain *abbr.*: TM domain; a region of a protein sequence that traverses a membrane. For alpha-helical structures this is normally a span of 20–25 residues; for beta-structures the span is generally shorter, typically ≈12 residues. Alpha-helical TM domains are often predicted by the presence of characteristic peaks and troughs in **hydropathy profiles**.

transmembrane protein any protein that traverses the membrane, especially the plasma membrane. For the different types, *see* **membrane protein**.

transmethylation the transfer of a methyl or methylene group, as frequently occurs in **one-carbon metabolism**. Methyltransferases are in sub-subclass EC 2.1.1, and enzymes concerned with one-carbon metabolism are in sub-subclass EC 2.1.2. *See also* **C_1 compound**, folate coenzymes, *S*-adenosylmethionine.

transmissible dementia *see* **prion**.

transmissible spongiform encephalopathy *abbr.*: TSE; *see* spongiform encephalopathy.

transmission 1 the act, process, or an instance of transmitting. **2** the proportion of radiant energy that is transmitted perpendicularly through a substance or solution; *compare* **transmittance**. **3** the delivery of a nerve stimulus across a **synapse**, e.g. by a neurotransmitter.

transmission coefficient *(in nuclear physics)* the probability that a particle striking a nucleus penetrates it.

transmission electron microscope *abbr.*: TEM; *see* **electron microscope**.

transmission factor *an alternative name for* **transmittance**.

transmittance *or* **transmission factor 1** *(in chemistry)* symbol: *T*; the ratio I/I_0, where I_0 is the intensity of electromagnetic radiation falling on a body or substance and *I* the intensity after transmission through it. It is a dimensionless physical quantity. **2** *(in physics)* symbol: τ; the fraction of radiant energy that, having entered an absorbing substance, reaches its far boundary. It is given by

$$\tau = \Phi_{tr}/\Phi_0 = \exp(-\kappa c l),$$

where Φ_{tr} and Φ_0 are the radiant powers of the transmitted and incident radiation, κ is the molar napierian absorption coefficient, *c* the speed of light in the medium, and *l* the path length.

transmitter-gated channel *see* **ion channel**.

transparent transmitting light with little scattering or diffusion so that objects beyond the transparent material can be clearly distinguished. The term can be used in appropriate circumstances to indicate materials allowing the passage of other electromagnetic, particulate, or sonic radiation without disturbance. *Compare* **translucent**. —**transparency** *n*.

transpeptidase 1 *an alternative term for* **peptidyltransferase**. **2** the enzymic activity of a number of proteinases that catalyse, apart from hydrolysis of peptide links, the replacement of one terminal amino-acid residue in a peptide by another amino acid or other suitable molecule. **3** certain aminoacyltransferase enzymes, e.g. D-glutamyl transpeptidase, D-glutamyltransferase (EC 2.3.2.1), and glutamyl transpeptidase (i.e. γ-**glutamyltransferase**, EC 2.3.2.2).

transpeptidation a reaction catalysed by a **transpeptidase**.

transphosphoribosidase *see* **adenine phosphoribosyltransferase**.

transplantation antigen *an alternative term for* **histocompatibility antigen**.

transport the movement of a material from one place to another, especially the movement of substances around the body (e.g. in blood) or across a biological membrane, or of electrons along a series of carriers. This term implies a positive agency, compared with **passive diffusion**. *See also* **active transport**, **antiport**, **facilitated diffusion**, **symport**, **transport protein**.

transporter a membrane protein catalysing the passage of molecules across a membrane.

transport method *see* **transport process**.

transport number *an alternative term for* **transference number**.

transport piece *see* **secretory piece**.

transport process any process, such as sedimentation, diffusion, or electrophoresis, that is irreversible in the thermodynamic sense, the system being removed from a state of equilibrium. Such processes are of particular value in providing information about the dimensions, shapes, and masses of macromolecules.

transport protein a protein that is instrumental in transporting material, often in a specific manner, across a biological membrane, or within a biological fluid (e.g., blood). *See* **TC system**.

transposable element *another name for* **transposon**.

transposase an enzyme that is responsible for the transposition of **transposable elements** or **transposons**. Transposases are involved in site-specific DNA recombination required for transposition in bacteria and other organisms.

transpose 1 to alter the position of; to interchange or place in a different order. **2** *(in mathematics)* **a** to interchange the rows and columns of a **matrix** (def. 8). **b** the matrix resulting from interchanging the rows and columns of a given such matrix.

transposition 1 the process of transposing or the state of being transposed. **2** something that has been transposed; especially a segment of a chromosome or a piece of a DNA molecule.

transposition sequence *an alternative name for* **translocation sequence**.

transposon *or* **transposable element** a specific DNA sequence that is transferred from one **replicon** to another. Transposons were first identified as a result of work on **transposable elements** and also as spontaneous insertions in bacterial operons. The simplest are referred to as **insertion sequences**, being designated IS followed by a number related to the sequence in which they were identified, preceded by an indication of the site of insertion followed by a double colon (thus λ::IS1 indicates an IS1 element inserted into phage lambda). Each transposon is flanked by repeated sequences, due to short sequences copied from the original gene called **target repeats** or **direct repeats**, and also due to short inverted repeats that are a property of the insert. More complex transposons, called **composite transposons**, may have a central portion having a variety of markers, and flanked by IS elements that are responsible for identifying the transposon for transposition. These are designated Tn followed by a number. In eukaryotes, transposons constitute much of the **repetitive sequences** in the genome, with up to a few hundred repeat copies. They are flanked by terminal inverted sequences; the terminal sequences in each strand characteristically represent the complementary bases of the inverted sequence, thus TCAG......CGTA. Some transposons move physically from one site to another, while others are replicated, with one copy remaining at the original site and a duplicate inserting elsewhere. Both mechanisms involve a **transposase** enzyme; the former also requires a **resolvase** enzyme.

transthyretin a thyroid-binding protein occurring in the bloodstream, and having an unusual structure. The molecule is a homotetramer, each monomer comprising two four-stranded β-sheets, and having the shape of a prolate ellipsoid. Two of the monomers dimerize, and the dimers themselves dimerize to form a structure with an internal channel. Defects in transthyretin cause various forms of amyloidosis and hyperthyroxinemia.

trans-translation the transfer of a nascent polypeptide from a stalled bacterial ribosome onto a **tmRNA** followed by translation of the open reading frame (ORF) of the tmRNA.

transudate any fluid, with its solutes, that has passed through a membrane or through the interstices of a system.

transude to pass or be passed through a membrane or the interstices of a system as a **transudate**. —**transudation** *n*.

transverse relaxation a process in magnetic resonance spectroscopy, such as EPR and NMR, that describes the decay of transverse (x,y) magnetization. The characteristic spin-spin relaxation time is normally given the symbol T_2.

transversion a type of mutation of DNA involving the insertion of a purine in place of a pyrimidine, or vice versa. *Compare* **transition**.

***trans*-zeatin-producing protein** an isopentenyl transferase (dimethylallyl transferase) of sub-subclass EC 2.5.1. It introduces the isopentenyl moiety (2-methylbut-2-en-1-ol group) into zeatin.

TRAP *abbr. for* **1** translocon-associated protein; *see* **translocon**. **2** thyroid-hormone receptor-associated protein; a multiprotein complex of human cells that is associated with ligand-bound thyroid hormone receptor. It has an almost identical subunit composition to that of **SMCC**. It is the equivalent of the yeast transcriptional coactivator mediator.

trap hypothesis 1 a mechanism for the binding of proteinases by α_2-**macroglobulin** (α_2M). It proposes that, when α_2M interacts with a proteinase, hydrolysis of one or more peptide bonds in its **bait region** occurs, whereupon α_2M undergoes a conformational change that entraps the proteinase molecule, thus hindering reaction between the proteinase and other large substrate molecules. **2** *see* **methyltetrahydrofolate trap hypothesis**.

TRASH domain *abbr. for* trafficking, resistance, and sensing heavy metals domain; a well-preserved protein-sequence motif ($\approx$38 residues) that contains three cysteine residues and is predicted to form a treble-clef finger (three or four beta strands with a C-terminal alpha helix) for metal coordination. It is present in many prokaryotes and several vertebrate proteins, alone or with other motifs.

Trasylol *a proprietary name for* **aprotinin**.

TRE *abbr. for* **1** TPA response element, a response element for phorbol esters. **2 thyroid response element**. *See* **response element**.

Treacher Collins syndrome an autosomal dominant disorder of craniofacial development caused by mutation of the 'treacle' gene *TCOF1* (locus 5q32-q33.1). The treacle gene product is a nucleolar phospoprotein that participates in the 2′-O-methylation of pre-rRNA. [After Edward Treacher Collins (1862–1919), British ophthalmologist.]

tree length the total number of steps required to map a data set onto a phylogenetic tree, or the sum of the estimated or actual branch lengths in a tree.

trefoil domain *an alternative name for* **P domain**.

trefoil factor (*abbr.*: TFF) *or* **trefoil peptide** a family of abundant, highly stable and conserved polypeptides ($\approx$60 amino acids long) secreted mainly by the mammalian gastrointestinal mucosa. Each member contains six conserved cysteine residues, predicted to form three intrachain loops (*see* **P domain**). There are three species in human: **p52** (52 amino acids), which is produced by gastric fundus and is highly expressed in cancers and around ulcerative disease of the upper intestinal tract; **SP** (spasmolytic polypeptide); and **ITF** (intestinal trefoil factor), which is secreted by goblet cells of musosa of intestine and colon especially at sites of injury, inflammation, or dysplasia. They are thought to be involved in repair of mucosal damage.

trefoil motif *an alternative name for* **P domain**.

trehalose α-D-glucopyranosyl-α-D-glucopyranoside; α,α-trehalose; a nonreducing disaccharide, *also known as* **mushroom sugar** *or* **mycose**, that acts as a reserve carbohydrate in certain fungi (especially yeasts), algae, and lichens. It is cleaved by trehalase but not by most other α-glucosidases. It is a component of **cord factors**, which are a mixture of trehalose 6,6′-mycolates (*see* **mycolic acid**). α-D-Glucopyranosyl-β-D-glucopyranoside (α,β-trehalose) and β-D-glucopyranosyl-β-D-glucopyranoside (β,β-trehalose) occur naturally, but only very rarely. The carbohydrate is a **compatible solute** that acts as a cryoprotectant in many microorganisms, and is used industrially as a preservative in foods and pharmaceuticals.

TrEMBL *abbr. for* translated EMBL; *see* **sequence database**.

tremerogen any of a group of oligopeptides that contain non-amino-acid components and are the sex hormones of the jelly fungi. They induce formation of a conjugation tube in compatible cell types. They may contain a farnesyl group bound to cysteine.

Tresyl *proprietary name* indicating the 2,2,2-trifluoroethane sulfonyl group; Tresyl-activated Sepharose can be used in the preparation of affinity gels, the sulfonyl groups being readily displaceable by amines, thiols, phenolic groups, or imidazole, which remain bound to the gel as the stationary component; e.g.

$$\text{support–O–SO}_2\text{–CH}_2\text{–CF}_3 + \text{RSH} =$$
$$\text{support–S–R + HOSO}_2\text{–CH}_2\text{–CF}_3.$$

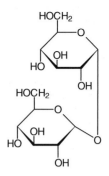

trehalose

tretinoin *a WHO-approved nonproprietary name for* **retinoic acid**, especially for pharmacological use.

TRF *abbr. for* **1** thyrotropin-releasing factor; i.e. **thyrotropin-releasing hormone**. **2** T-cell-replacing factor; *see* **interleukin 5**. **3** telomeric repeat-binding factor. *See* **TEP-1**.

TRF-III *or* **TRF-m** *see* **interleukin 1**.

TRH *abbr. for* thyrotropin-releasing hormone.

tri+ *comb. form* **1** denoting three, threefold, thrice, every third; *see also* **tris+**. **2** (*in chemical nomenclature*) (*distinguish from* **tris+** (def. 2)) **a** indicating the presence in a molecular entity of three identical specified unsubstituted groups, e.g. trichloroethylene, trimethylamine. **b** indicating the presence in a molecular entity of three identical oxoacid residues in linear anhydride linkage; e.g. adenosine 5′-triphosphate.

triac (*informal*) *abbr. for* 3,3′,5-triiodothyroacetic acid, a catabolite of **thyroxine**.

triaconta+ *or* (*before a vowel*) **triacont+** *comb. form* denoting thirty or thirty times.

triacontanoic acid melissic acid; a 30-carbon straight-chain aliphatic acid; m.p. 93.6°C (methyl ester 71.5°C).

triacontapeptide a 30-residue peptide.

triacylglycerol *or* (*formerly*) **triglyceride** any triester of glycerol; the three fatty acids may all be the same, or differ in any permutation. Triacylglycerols are important components of plant oils, animal fats, and animal plasma lipoproteins. The degree of unsaturation of the component fatty acids determines the melting point: plant oils, which typically contain a high percentage of (poly)unsaturated fatty acids, are liquid, while animal fats, being more highly saturated, are solid or semisolid. *See also* **lipoprotein** and individual plant oils and animal fats.

triacylglycerol lipase EC 3.1.1.3; *systematic name*: triacylglycerol acylhydrolase; *other names*: lipase; triglyceride lipase; tributyrase. An enzyme that catalyses the reaction:

triacylglycerol + H_2O = diacylglycerol + fatty-acid anion.

There are various types ranging from those found in microorganisms to those such as **pancreatic lipase**, and the hormone-sensitive lipase of mammalian adipose tissue. That from the yeast *Geotrichum candidum* is a 544-amino-acid globulin with mainly alpha helix and turns; it contains one modified glutamine residue (pyrrolidone carboxylic acid).

triad a group of three; e.g. three amino acid residues can constitute a catalytic triad.

triadin a major membrane glycoprotein of the sarcoplasmic reticulum triad junction of skeletal and cardiac muscle. The luminal domain has a highly conserved region, and can bind both the ryanodine receptor and calsequestrin.

+triaose *suffix* used in nomenclature to indicate a molecule containing an oligosaccharide moiety composed of three sugar residues (which may be the same or different), as in gangliotriaose. *Distinguish from* **triose**.

triazine-dye affinity chromatography a type of **dye-ligand chromatography** in which the coupled dye is a triazinyl compound. It is not **affinity chromatography** in the original sense.

Triazol reagent a proprietary reagent used for the extraction of RNA from tissues.

tribbles a protein of *Drosophila* that inhibits mitosis in early development by binding to String (a Cdc25 homologue) and promoting its ubiquitination and degradation by proteasomes. It contains a truncated kinase motif, but lacks an ATP-binding site and *in vitro* kinase activity. The mammalian homologues TRB1, 2, and 3 strongly resemble tribbles. *See* **NIPK**.

tributyltin chlorotributylstannane; tri-N-butyltin chloride, SnBu₃; an organometallic compound, used primarily as a biocide in antifouling paints. It is extremely toxic to aquatic life and causes severe reproductive effects in aquatic organisms.

TRiC *see* **TCP-I**.

tricaprin a trivial name for **tridecanoin**, the triacylglycerol containing three caproyl (decanoyl) residues. *See also* **tricaprin**, **tricaprylin**.

tricaproin a trivial name for **trihexanoin**, the triacylglycerol containing three capric (hexanoic) ester residues. *See also* **tricaprin**, **tricaprylin**.

tricaprylin a trival name for **trioctanoin**, the triacylglycerol containing three caprylic (octanoic) ester residues. *See also* **tricaprin**, **tricaproin**.

tricarboxylic acid *abbr.*: TCA; any organic compound having three carboxylic-acid groups. The name often has the connotation of relationship with **citric acid**, as in the **tricarboxylic-acid cycle**.

tricarboxylic-acid cycle (*abbr.*: TCA cycle) *or* **citric-acid cycle** *or* **Krebs cycle** a nearly universal metabolic pathway in which the acetyl group of acetyl coenzyme A is effectively oxidized to two CO_2 and four pairs of electrons are transferred to coenzymes. The acetyl group combines with oxaloacetate to form citrate, which undergoes successive transformations to isocitrate, 2-oxoglutarate, succinyl-CoA, succinate, fumarate, malate, and oxaloacetate again, thus completing the cycle. In eukaryotes the tricarboxylic acid is confined to the mitochondria. *See also* **glyoxylate cycle**.

trichloroacetic acid *abbr.*: TCA; Cl₃CCOOH; an organic acid widely used for the precipitation of protein from tissue and other cell extracts. A final concentration of 3–9% (w/w) is normally used. TCA can be extracted into diethyl ether.

trichloroethyl alcohol 2,2,2-trichloroethanol; an active metabolite of chloral hydrate with CNS depressant actions.

trichohyalin a structural protein produced and retained in cells of the inner root sheath and medulla of the hair follicle. In the sheep, 75% of its amino acids are glutamate or glutamine, arginine or lysine; the gene is also expressed in epithelia, hoof, and rumen. Trichohyalin associates in regular arrays with keratin intermediate filaments.

trichostatin A an antifungal antibiotic from *Streptomyces platensis*. It is a reversible inhibitor of histone deacetylase.

trichothiodystrophy a syndrome of neurological dysfunction, premature ageing, photosensitivity, and brittle hair and nails. It is caused by a defect in **nucleotide excision repair** as a result of mutations in the xeroderma pigmentosum complementation proteins XPB (782 amino acids) and XPD (760 amino acids). These are DNA helicases, the largest subunits of transcription factor TFIIH, and essential for transcription initiation.

tricine *or* TRICINE *trivial name for* N-tris[(hydroxymethyl)methyl]-glycine; N-[2-hydroxy-1,1-bis(hydroxymethyl)ethyl]glycine; a **Good buffer substance**; pK (20°C) = 8.15.

tricorn protease a protease (1071 amino acids) with trypsin-like specificity that is abundant, with proteasomes, in archaebacteria such as *Thermoplasma acidophilum*. It forms a homohexameric toriid that resembles a tricorn hat, and cleaves oligopeptides produced by proteasomes into di-, tri-, and tetrapeptides. Aminopeptidases with different specificities are usually associated with it.

tricosa+ *or (before a vowel)* **tricos+** *comb. form* denoting twenty-three or twenty-three times.

tricosapeptide a 23-residue peptide.

trideca+ *or (before a vowel)* **tridec+** denoting thirteen or thirteen times.

tridentate 1 having three teeth or toothlike projections. **2** (of a **ligand**) chelating a metal ion by means of three donor atoms.

triene any alkatriene or substituted alkatriene.

+triene(+) *or (before a vowel)* **+trien+** *(in chemical nomenclature)* infix or suffix in systematic names denoting the presence in a molecular structure of an unsaturated aliphatic hydrocarbon chain containing three (conjugated or unconjugated) double bonds. The position of each double bond may be indicated by a locant for its lowest numbered carbon atom; e.g. cycloheptatriene, icosa-8,11,14-trienoic acid.

trienoic indicating a compound having a carboxylic-acid group and an aliphatic chain with three double bonds.

trienoyl indicating the acyl group of a trienoic acid.

triflavin a **disintegrin** from the snake *Trimeresurus flavoviridis*, containing an Arg-Gly-Asp (**RGD**) motif, that inhibits fibrinogen interaction with platelet receptors, and thereby inhibits platelet aggregation.

trifluoperazine 10-[3-(4-methyl-1-piperazinyl)propyl]-2-(trifluoromethyl)-10H-phenothiazine; an antagonist with high affinity for D₂ **dopamine receptors**. It is used in the treatment of manic and schizophrenic psychoses and as an inhibitor of **calmodulin** action. *One proprietary name*: Stelazine (dihydrochloride).

trifluoroacetic acid *abbr.*: THA; CF₃COOH; a strong organic acid used: (1) as a catalyst in organic synthesis; (2) to cleave Cbz– and tBoc– groups from protected amino acids; and (3) as the anhydride, to prepare trifluoroacetyl derivatives for the protection of alcohols and amines, and as volatile derivatives for gas–liquid chromatography.

trifunctional protein an enzyme protein complex, present on the matrix side of the inner mitochondrial membrane, that has long-chain 2-enoyl-CoA hydratase, long-chain 3-hydroxyacyl-CoA dehydrogenase, and long-chain 3-ketoacyl-CoA thiolase activities. That from rat liver consists of four α subunits (79 kDa), which bear the first two aforementioned activities, and four β subunits (51 kDa), which bear the third activity. Deficiency is associated with massive excretion of long-chain 3-hydroxycarboxylic acids. In **type 1 deficiency**, probably a defect of the α subunits, only the dehydrogenase activity is lacking. In **type 2 deficiency**, probably a defect of β subunits, all three activities are lacking.

trigger 1 to induce (often rapidly) into activity. **2** a mechanism or agent that brings about such an action. The term is commonly used in connection with membrane receptor activation and signal transduction, as in trigger mechanism.

trigger factor a 48 kDa protein of eubacteria that binds ribosomes in 1:1 stoichiometry, and also binds nascent ≈57-residue polypeptides by hydrophobic interaction. It dissociates from the nascent chain after this is released from the ribosome. It has **peptidylprolyl isomerase** activity but binds nascent chains independently of their proline content. *See* **NAC**.

triglyceride *a former name for* **triacylglycerol**; this usage is discouraged because 'triglyceride' is properly a generic term for compounds having three glyceryl residues (e.g. **cardiolipin**).

trigonelline nicotinic acid *N*-methylbetaine; 3-carboxy-1-methylpyridium inner salt; a metabolite of nicotinic acid. It is excreted in the urine after the consumption of relatively large amounts of nicotinic acid. It is also produced in alfalfa root exudates, where it induces *nod* genes in some *Rhizobium* species.

trihydric (of a chemical compound) containing three hydroxyl groups per molecule; trihydroxy; used especially of alcohols.

triiodothyronine *abbr.*: T$_3$; 3,3,5-triiodo-L-thyronine; one of the hormones secreted by the thyroid gland. Plasma levels of T$_3$ are much lower than those of **thyroxine** (T$_4$) (20- to 50-fold), but the physiological activity of T$_3$ is higher. Much of T$_4$ is converted to T$_3$ within cells. T$_3$ binds to a **thyroid-hormone receptor**, which binds to **thyroid response element**, a transcription-regulatory element. *See also* **thyroglobulin**.

trilaurin a triacylglycerol in which all the fatty-acyl moieties are **lauroyl** (dodecanoyl).

trimer a molecular complex having three components or subunits. If these are identical, it is referred to as a **homotrimer**; if any one is different from any other, it is called a **heterotrimer**. —**trimeric** *adj.*

trimethoprim 2,4-diamino-5-(3,4,5-trimethoxybenzyl)pyrimidine; a specific inhibitor of formylation and a competitive inhibitor of **dihydrofolate reductase**. The trimethoprim resistance gene can be used as a selectable marker for cloning vectors in animals or plants. *See also* **sulfamethoxazole**.

trimethylaminuria *see* **fish-odour syndrome**.

trimyristin a triacylglycerol in which all the fatty-acyl moieties are **myristoyl** (tetradecanoyl).

trinactin a neutral macrotetralide antibiotic with an action similar to that of **nonactin**.

trinitroglycerin glyceryl trinitrate; trinitroglycerol; the nitrate moiety is metabolized to nitrite and, subsequently, to **nitric oxide** and nitrosothiol, which activate guanylate cyclase. The resulting increase in cyclic GMP synthesis may explain the potent vasodilator effects of this and similar organic nitrates.

trinor+ *prefix (in a trivial name of a compound)* indicating replacement of three methyl groups by hydrogen. *See also* **nor+**.

trinucleotide repeat an increase in copy number of a normally polymorphic triplet. Some genetic disorders, e.g. **fragile X syndrome**, result from this.

triol any aliphatic chain having three hydroxyl groups.

triolein a triacylglycerol in which all the fatty-acyl moieties are **oleoyl** (*cis*-9-octadecenoyl). *See also* **triolein breath-test**.

triolein breath-test a test for intestinal malabsorption of fat. When ^{14}C-labelled triacylglycerol (10 μCi) is administered by mouth, a proportion of the ^{14}C will be breathed out as $^{14}CO_2$, which is trapped in a solution of hyamine carbonate for scintillation counting.

triose 1 any monosaccharide having a chain of three carbon atoms in the molecule, i.e. the chiral enantiomers D- and L-**glyceraldehyde**, and their achiral constitutional isomer **glycerone**. **2** trisaccharide; used as a stem name for any oligosaccharide consisting of three residues of the same sugar in a single type of linkage, as in maltotriose. *Distinguish from* **+triaose**.

triose phosphate D-glyceraldehyde 3-phosphate, glycerone phosphate, or a mixture thereof. These two compounds are interconvertible by the enzyme **triose-phosphate isomerase**.

triosephosphate dehydrogenase (NADP⁺) *see* **glyceraldehyde-3-phosphate dehydrogenase (NADP⁺) (phosphorylating)**.

triose-phosphate isomerase *abbr.*: TIM; EC 5.3.1.1; *systematic name*: D-glyceraldehyde-3-phosphate ketol-isomerase; *other names*: triosephosphate mutase; phosphotriose isomerase. The sugar **isomerase** enzyme that is important in the glycolytic pathway, and catalyses the aldose–ketose isomerization of the two **triose phosphates**, i.e. the reaction:

$$\text{D-glyceraldehyde 3-phosphate} = \text{glycerone phosphate.}$$

trioxsalen 2,5,9-trimethyl-7*H*-furo[3,2-*g*][*I*]-benzopyran-7-one; 4,5′,8-trimethylpsoralen (*abbr.*: TMP); a pigmentation agent (photosensitizer). *See also* **psoralen**.

tripalmitin a triacylglycerol in which all the fatty-acyl moieties are **palmitoyl** (hexadecanoyl).

tripeptide a three-residue peptide.

triphenylmethyl group *abbr.*: trityl; a group used in chemical syntheses to protect nitrogen, sulfur, and, particularly, oxygen functions as their trityl derivatives. It is stable under many conditions, and readily removed by mild acid. It can be introduced by reagents such as triphenylmethyl chloride.

triphosphate 1 the pentavalent ion, $P_3O_{10}^{5-}$, derived from triphosphoric acid, decaoxotriphosphoric(v) acid, $(HO)_2PO-O-PO(OH)-O-PO(OH)_2$ (an acid not known in the free state). **2** any mixture of the anions of triphosphoric acid. **3** any (partial or full) salt or ester of triphosphoric acid. **4** any organic compound containing three phosphoric residues linked linearly by oxygen atoms. **5** name of the univalent group $-O-PO(OH)-O-PO(OH)-O-PO(OH)_2$ (irrespective of its state of ionization). **6** *(formerly)* any organic compound containing three independent phosphoric residues attached to different parts of the molecule; now incorrect, use **trisphosphate**.

triphospho+ 1 *prefix* to a chemical name indicating the presence in ester linkage of a triphosphoric residue. **2** *(formerly) prefix* to a chemical name indicating the presence of three independent phosphoric residues in ester linkages; now incorrect, use **trisphospho+**.

triphosphoinositide *abbr.*: TPI; *a former name for* **phosphatidylinositol 4,5-bisphosphate** (*symbol*: PtdIns-4,5-P_2). Current nomenclature requires bisphosphate, trisphosphate, etc. for the substituent phosphates on the inositol moiety. Other phosphatidylinositol bisphosphates are known, e.g. PtdIns-phosphatidylinositol 3,4-biphosphate.

triphosphopyridine nucleotide (oxidized) *abbr.*: TPN⁺; *an obsolete name for* **nicotinamide-adenine dinucleotide phosphate**.

triphosphopyridine nucleotide (reduced) *abbr.*: TPNH; *an obsolete name for* **nicotinamide-adenine dinucleotide phosphate (reduced)**.

triphosphoric *symbol*: *PPP* or (in the one-letter convention for nucleotides) ppp; *general name for* a residue of triphosphoric acid, $(HO)_2PO-O-PO(OH)-O-PO(OH)_2$, whether singly or multiply attached through oxygen to other groups and whether or not any residual hydroxyl groups are dissociated.

triple bolus test *see* **combined test of pituitary function**.

triple helix 1 (*in proteins*) a superhelically coiled arrangement of three polypeptide chains, e.g. as in **tropocollagen**. **2** *or (sometimes)* **triplex** *see* **triplex DNA**.

triple membrane *see* **unit membrane**.

triple point the point on a phase diagram at which three phases may coexist.

triplet 1 *(in molecular biology)* any sequence of three purine and/or pyrimidine bases in nucleic acids that specifies a particular amino acid. It is synonymous with **codon** in mRNA or **anticodon** in DNA or tRNA. *See also* **synonym triplet**. **2** a collection of three of a kind. **3** any one of three offspring born from a single pregnancy.

triplet code the **genetic code**, which is based on **triplets** (def. 1) of purine and pyrimidine bases.

triplet repeat any stretch of DNA that has the same three nucleotides (triplet) repeated in tandem fashion.

triplet repeat disorder a genetic disease resulting from the expansion of triplet repeats, such as Huntington's disease (CAG), fragile X syndrome (CCG), and myotonic dystrophy (CTG).

triplex DNA a three-stranded structure formed by duplex DNA in association with an oligonucleotide, when purine or pyrimidine bases occupy the major groove of the DNA double helix adopting **Hoogsteen base pairing** with the Watson–Crick base pairs. Triplex DNA may be involved as an intermediate in **crossing over**. *See also* **tetraplex DNA**.

triploid 1 describing a cell, tissue, or organism having three times the haploid (monoploid) number of chromosomes. **2** such a cell, tissue, or organism. —**triploidy** *n*.

TRI reagent a proprietary reagent formulated to facilitate the extraction of RNA, DNA, and protein from tissues.

tris+ *prefix* **1** denoting three times, thrice, tripled; *see also* **tri+** (def. 1). **2** *(in chemical nomenclature)* *(distinguish from* **tri+** (def. 2)) **a** indicating the presence in a molecular entity of three identical specified groups each substituted in the same way; e.g. tris(2-hydroxyethyl)amine. **b** indicating the presence at separate positions in a molecular entity of three identical inorganic oxoacid residues; e.g. inositol 1,4,5-trisphosphate.

Tris *or* **TRIS** *or* **THAM** tris(hydroxymethyl)aminomethane; 2-amino-2-hydroxymethyl-1,3-propanediol; a compound widely used as a biological buffer substance in the pH range 7–9; pK_a (20°C) = 8.3; pK_a (37°C) = 7.82.

trisaccharide an **oligosaccharide** with three **monosaccharide** units.

Trisacryl 1 *a proprietary name for* N-acryloyl-2-amino-2-hydroxymethyl-1,3-propanediol; CH_2=CH–CO–NH– $C(CH_2OH)_3$; a compound used in the manufacture, by copolymerization with a cross-linking agent such as N,N'-diallyl L-tartardiamide (*abbr.*: DATD) or N,N'-methylenebisacrylamide, of a range of strongly hydrophilic artificial support materials of controlled porosity that are useful for gel-permeation chromatography. Copolymerization with a cationic or anionic derivative as a third component yields support materials of utility in ion-exchange chromatography. Such materials are thermally, chemically, and biologically stable, and resistant to high pressures. **2** *as modifier* describing a range, or any member of a range, of such chromatographic support materials; e.g. Trisacryl polymers, Trisacryl ion-exchangers, Trisacryl gels, Trisacryl GF05, DEAE–Trisacryl M.

triskelion a hexamer of **clathrin** molecules, composed of three heavy chains and three light chains. In the **electron microscope**, triskelions appear to have three bent arms each linked into a central point. [From triskele, a symbolic figure consisting of three legs radiating from a common centre.]

trisomic 1 describing a cell, tissue, or organism having three copies of one or more chromosome(s) in an otherwise diploid chromosome set. In humans, trisomy 21 (i.e. presence of three copies of chromosome 21) is asssociated with **Down's syndrome**. **2** such a cell, tissue, or organism. —**trisomy** *n*.

trisphosphate any compound containing three independent phosphoric residues in ester linkages, at positions indicated by locants; e.g. inositol 1,4,5-trisphosphate. *Compare* **triphosphate**.

trisphospho+ *prefix* to a chemical name indicating the presence of three independent phosphoric residues in ester linkages, at positions indicated by locants. *Compare* **triphospho+**.

tristearin a triacylglycerol in which all the fatty-acyl moieties are **stearoyl** (octadecanoyl).

trithorax one of a group of proteins in *Drosophila* that regulate the expression patterns of many developmental genes. Trithorax group proteins are involved in countering chromatin-based gene silencing to maintain gene activity, and enable established patterns of gene expression to be transmitted over multiple rounds of cell division. *Compare* **polycomb**.

tritiate to combine or label with **tritium**; to replace one or more atoms of hydrogen with tritium. Tritium may be introduced into specific sites by chemical synthesis but ^{3}H-labelled steroids are prepared by the reduction of unsaturated precursors with $^3H_2$, the number of ^{3}H atoms introduced depending upon the degree of unsaturation of the precursor and the catalyst used. Biosynthesis is rarely used for ^{3}H-labelling since intramolecular migration of ^{3}H atoms is often caused by enzymes.

tritin a type I **ribosome-inactivating protein** from wheat.

tritium the radioactive isotope of hydrogen, $^3_1$H (i.e. the nucleus contains two neutrons and one proton). It decays to produce β⁻ particles of low energy (0.018 MeV); half-life 12.26 years. *Compare* **deuterium**, **protium**.

tritium nuclear magnetic resonance spectroscopy *abbr.*: tritium NMR spectroscopy *or* TNMR spectroscopy; the application of **nuclear magnetic resonance spectroscopy** to determine the positions and extents of tritium substitution in substances (generally of known structure) that have been labelled with tritium (^{3}H). The technique depends upon an equality of the displacements of the resonance frequencies (chemical shifts) of tritium and protium (^{1}H) nuclei in corresponding environments relative to those occurring with the tritiated and nontritiated forms, respectively, of a reference substance. The method is quick, nondestructive, and applicable to very small amounts of material. It is thus particularly useful for determining the pattern of labelling in substances that are costly, complex in structure, and/or unstable. The technique is also used to follow the chemical and steric course of biochemical reaction sequences.

triton 1 *(in chemistry) symbol*: ^{3}H⁺; the cation derived from an atom of **tritium**; a **hydron** of nucleon number 3. **2** *(in nuclear physics) symbol*: t; a particle having a charge equal and opposite to that of an electron and having a mass of 3.0155 Da.

Triton *the proprietary name for* any of a series of polyoxyethylene ethers of certain alkylphenols that are surfactants and classed as nonionic detergents. Those used most commonly are based on the formula *tert*-octyl-C_6H_4-$[OCH_2CH_2]_x$OH; e.g. in Triton X-45, $x \approx$ 5; in Triton X-100, $x = 9$ or 10 (aggregation number 100–155, CMC 0.2–0.9 mM); in Triton X-405, $x \approx 40$; Triton CF is a benzyl-polyoxyethylene *tert*-octylphenyl ether; Triton N-101 is a polyoxyethylene nonylphenyl ether (C_9H_{19}-C_6H_4-$[OCH_2CH_2]_x$ OH), with $x = 9$ or 10.

tritosome a **lysosome** that is loaded with Triton. Tritosomes have an isopycnic density in sucrose solution of about 1.14, and can thus be separated readily from other cell constituents.

tritriaconta+ *or (before a vowel)* **tritriacont+** *comb. form* denoting thirty-three or thirty-three times.

triturate 1 to rub or grind (solid material) to a very fine powder either dry or in a liquid. **2** a finely ground powder or an even mixture of finely ground powders, especially powdered drugs.

trituration 1 the act of triturating or the condition of being triturated; *see* **triturate** (def. 1). **2** the composing of a dental amalgam.

trityl *trivial name for the* **triphenylmethyl group**.

trivial name *(in chemical nomenclature)* a name of a chemical species that has no part of it used in a systematic sense; e.g. xanthophyll, urea. It generally has historical or colloquial origins, and refers to attributes such as appearance or source. It makes no pretension to describe molecular structure. The adjective 'trivial' has no perjorative meaning, numerous compounds of biochemical interest having approved or recommended trivial names. *Compare* **descriptive name, systematic name**.

trixenic *see* **synxenic**.

trk *abbr. for* tropomyosin receptor kinase; a protooncogene encoding a transmembrane tyrosine kinase receptor that is selectively expressed in certain neurons of neural crest origin, or, as its counterpart *trkB*, in cells of the neuroepithelium. It is required for high-affinity binding of nerve growth factor, and neurotrophins 3 and 4. It was discovered as the transforming factor of a colon carcinoma and was originally known as *onc*D. The oncogenic activity results from a genomic rearrangement that produces a hybrid protein consisting of part of a nonmuscle tropomyosin sequence and the transmembrane and cytoplasmic domains of Trk.

tRNA *abbr. for* transfer RNA.

T3 RNA polymerase an enzyme from bacteriophage T3 that will transcribe sequences downstream from the **T3 phage promoter** and is used for *in vitro* **transcription**.

T7 RNA polymerase an enzyme that will transcribe sequences downstream from the **T7 phage promoter** and is used for *in vitro* **transcription**.

Trojan horse inhibitor *see* **affinity label**.

Trojan peptide *an alternative name for* **penetratin**.

tropane 8-methyl-8-azabicyclo[3.2.1]octane; the basic ring system in tropane alkaloids.

+trope *suffix* denoting turning.

+troph *suffix* denoting an organism with a specified mode of nutrition, e.g. **autotroph, chemotroph**.

trophic of, or concerning, nourishment, nutrition. *Compare* **tropic** (def. 1).

+trophic *suffix* denoting nourishment, nutrition; e.g. organotrophic, saprotrophic. *See also* **+tropic**.

+trophin *noun suffix* denoting a substance having a **trophic** function. *See also* **+tropin**.

tropho+ *or (before a vowel)* **troph+** *comb. form* denoting food or nourishment.

trophoblast the outer part of the blastocyst stage of a mammalian conceptus. Its outermost layer consists of syncytial cells concerned with penetration and implantation into the uterine endometrium; these cells secrete extracellular hydrolases that facilitate movement of the blastocyst into the previously organized endometrial structure.

+trophy *noun suffix* **1** denoting food or nourishment. **2** denoting growth or size.

tropic 1 *or (sometimes, esp. Brit.)* **trophic** having a stimulating effect (especially of a hormone, etc.). **2** of or relating to a **tropism**.

+tropic *or (sometimes, esp. Brit.)* **+trophic** *suffix* denoting a stimulatory function; e.g. gonadotropic. *See also* **+trophic, +tropin**.

tropic acid α-phenyl-β-hydroxypropionic acid; (*S*)3-hydroxy-2-phenylpropanoic acid; a compound that occurs in ester form in **atropine** and other tropane alkaloids.

+tropin *or (sometimes, esp. Brit.)* **+trophin** *noun suffix* denoting a

substance having a stimulatory function, especially a hormone produced by the adenohypophysis. *See also* **+trophin**.

tropine tropan-3α-ol; 2-hydroxytropane; *N*-methyl-8-azabicyclo[3.2.1]octan-3-ol; **atropine** is the ester with (±)-tropic acid. Tropine is a *meso* structure. The –OH group is *trans* to the CH₃N– group. In the geometrical isomer, **pseudotropine**, this relationship is *cis*.

tropinone *N*-methyl-8-azabicyclo[3.2.1]octan-3-one; the oxidation product of **tropine**.

tropism movement directed towards some positive stimulus, or away from some negative stimulus.

+tropism *noun suffix* denoting a specific type of **tropism** as indicated by the prefixed word.

tropocollagen *or* **soluble collagen** a rod-shaped molecule, 3000 Å long and 15 Å in diameter, consisting of two α1 and one α2 polypeptide chain, each of about 1000 amino-acid residues, of which about a third are glycine, and 21% are proline plus hydroxyproline; they also contain hydroxylysine. The three polypeptide chains wind around each other to form a superhelical cable, the chains of which are hydrogen-bonded to each other. Tropocollagen fibres spontaneously associate, with their N-terminal sequences overlapping C-terminal sequences of adjacent chains in a staggered array, to form **collagen** molecules. *Compare* **procollagen**. *See also* **registration peptide**.

tropolone 2-hydroxy-2,4,6-cycloheptatrien-1-one; a structure found in some mould metabolites and in **colchicine**.

tropomodulin a cytoskeletal tropomyosin-regulatory protein that binds to the end of erythrocyte tropomyosin and blocks its head-to-tail association along actin filaments; it does not bind to F actin. Its modulation of the spectrin–actin complex affects the viscoelastic properties of erythrocytes.

tropomyosin a protein found in smooth, skeletal and cardiac muscle. It comprises α and β subunits, and is bound to **actin**, forming a component of the **troponin** complex that regulates calcium-dependent binding of actin and myosin and, consequently, muscle contraction. Tropomyosin forms a characteristic coiled coil, with a prominent seven-residue periodicity. Different isoforms are produced by differential splicing: e.g., isoforms of α-tropomyosin differ in striated and smooth muscle.

tropomyosin kinase EC 2.7.1.132; an enzyme that catalyses the phosphorylation of tropomyosin to form *O*-phosphotropomyosin.

tropone 2,4,6-cycloheptatrien-1-one.

tropone

troponin *abbr.*: Tn; a central regulatory protein of striated muscle contraction, comprising three subunits: **troponin T** (Tn-T), which binds to **tropomyosin**; **troponin I** (Tn-I), which regulates actino-myosin ATPase; and **troponin C** (Tn-C), which binds Ca^{2+} and, with Ca^{2+} bound, abolishes the inhibition of the binding of myosin to actin filaments produced by the other two troponins.

+tropy *noun suffix* denoting turning.

TROSY *abbr. for transverse relaxation optimized spectroscopy*: a method in NMR spectroscopy that gives sharper peaks for proteins of high molecular mass.

trp a family of tryptophan biosynthetic genes in bacteria. The following examples of *trp* gene products are from *Escherichia coli*. (1) TrpA and TrpB are the α and β subunits of tryptophan synthase (EC 4.2.1.20); it catalyses the formation of L-tryptophan from L-serine and 1-(indol-3-yl) glycerol 3-phosphate; glyceraldehyde 3-phosphate is also formed. The enzyme is a tetramer (α_2,β_2); the subunits have half-activities (α for aldol cleavage of indoleglycerol phosphate and β for tryptophan synthesis). (2) TrpC is a monomeric bifunctional enzyme: the 'TrpF domain' is *N*-(5'-phosphoribosyl)anthranilate isomerase (EC 5.3.1.24); it catalyses the interconversion of *N*-5'-phosphoribosyl-anthranilate and 1-(2-carboxyphenylamino)-1-deoxy-D-ribulose 5-phosphate. The 'TrpC domain' is **indole-3-glycerol-phosphate synthase** (EC 4.1.1.48; *abbr.*: IGPS); it catalyses the decarboxylation of 1-(2-carboxyphenylamino)-1-deoxy-D-ribulose 5-phosphate to form 1-(indol-3-yl) glycerol 3-phosphate, CO_2, and H_2O. (3) TrpE/G, **anthranilate synthase** (EC 4.1.3.27); it catalyses a reaction between chorismate and L-glutamine to form anthranilate, pyruvate, and L-glutamate. This enzyme is a tetramer with two molecules of component I (TrpE) and two of component II (TrpG). (4) TrpR, aporepressor (*trp* is co-repressor) for the operators of the *trp* operon, *trpR* itself, and *aroH*; a homodimer of known 3-D structure.

Trp *symbol for* a residue of the α-amino acid L-tryptophan (alternative to W).

Trp-2 tyrosinase-related protein 2; this is a product of the *tyrp-2* gene. *See also* **slaty**.

TRP channel *abbr. for* transient receptor potential channel; any of a family of membrane ion channels, first characterized as *Drosophila* TRP, which is responsible for the depolarization of photoreceptor cells in response to light. Mammalian homologues belong in three classes: short TRP (*or* STRP), long TRP (*or* LTRP), and osm-9-like (*or* OTRP). **TRP channel 2** (*or* LTRPC2) is expressed in many human tissues including immunocytes, where it is heptahelical and has a long cytoplasmic tail. It is activated by adenosine diphosphoribose (ADPR) and by NAD (the precursor of ADPR), and is inhibited by ATP. It is a nonspecific cation channel that permits Ca^{2+} influx into cells and has ADPR pyrophosphatase activity. **LTRPC7** is widely expressed in mouse tissues and is an ion channel and protein kinase. The cold and menthol (CMR1) and the heat and capsaicin (TRPM8) receptors are also members of the TRP family.

true cholinesterase *an alternative name for* **acetylcholinesterase**.

true negative an entity that is not a true member of a data set and is not recognized by a discriminator for that data set. *Compare* **false negative**.

true positive a match that is recognized by a discriminator for a given data set and is a true member of that data set. *Compare* **false positive**.

truncate 1 to shorten by cutting. **2** to terminate prematurely. —**truncation** *n.*

truncated hemoglobin any of a family of small oxygen-binding heme proteins that are widely distributed in eubacteria and protozoa, and nearly ubiquitous in the plant kingdom. They are 20–40 residues shorter than nonvertebrate hemoglobins and are distinct from bacterial flavohemoglobins, plant symbiotic and nonsymbiotic hemoglobins, and animal hemoglobins. They consist of four alpha helices rather than the eight of animal hemoglobins.

Trypan Blue a blue dye that is water-soluble and possesses sulfate and amino groups, whose charge prevents it from penetrating the plasma membrane of healthy cells. Dead cells or cells with damaged membranes take it up, making it a useful stain for determining cell

viability (*see* **dye-exclusion test**). It does, however, stain healthy macrophages.

trypanosomatid any protozoan of the order **Kinetoplastida**, which includes **trypanosomes** and other genera.

trypanosome any member of the genus *Trypanosoma* (order **Kinetoplastida**), comprising parasitic flagellate protozoans. Infection causes trypanosomiasis in a range of animals. In humans there are two main forms: African trypanosomiasis (or sleeping sickness), which is caused by *T. brucei* and transmitted by tsetse flies, and South American trypanosomiasis (or Chagas' disease), which is caused by *T. cruzi* and transmitted by insects such as the assassin bug. Invasion of the nervous system is more common with the African form than with the South American form, which mainly affects the heart.

trypanothione the oxidized form of N^1,N^6-bis(glutathionyl)spermidine, derived from the trypanosomatid *Crithidia fasciculata*, a parasite of insects. Members of the **Kinetoplastida** lack **glutathione reductase**, and **glutathione** is converted to the conjugate, trypanothione. *See also* **trypanothione reductase**.

trypanothione reductase EC 1.6.4.8; *systematic name*: NADPH:trypanothione oxidoreductase; an enzyme unique to **trypanosomatids**, and hence a potential chemotherapeutic target. It catalyses the reaction:

$$NADPH + trypanothione =$$
$$NADP^+ + reduced\ trypanothione;$$

FAD is a cofactor. *See also* **trypanothione**.

trypsin EC 3.4.21.4; a serine endopeptidase occurring in two forms, known as α and β trypsin. These result from proteolytic cleavage, β being formed directly from the precursor, **trypsinogen**, and α and other forms by further cleavage. Trypsin is of major importance in protein digestion in the small intestine, trypsinogen being a constituent of **pancreatic juice**. Trypsin is also widely used as a proteolytic agent experimentally, e.g. in protein sequencing. It preferentially cleaves at Arg-|-Xaa and Lys-|-Xaa. *See also* **trypsin inhibitor**.

trypsin inhibitor any naturally occurring substance that inhibits the activity of **trypsin**. *See* **pancreatic trypsin inhibitor**, **soybean trypsin inhibitor**.

trypsinize *or* **trypsinise** to treat something, especially intact cells or membrane preparations, with trypsin. Trypsinization is used to release cells from adhesion to tissue-culture plates, and to indicate domains of integral membrane proteins (unattacked bonds might indicate that they are shielded by inclusion within the membrane). —**trypsinization** *or* **trypsinisation** *n.*

trypsinogen the pancreatic **zymogen** precursor of **trypsin**. It is secreted in **pancreatic juice** and in the duodenum is cleaved by **enteropeptidase**, which cleaves the Lys^8-|-Ile^9 bond in trypsinogen, to yield β trypsin.

tryptamine β-3-indolylethylamine; 3-(2-aminoethyl)indole; 2-(1*H*-indol-3-yl)ethanamine; $(C_8H_6N)CH_2–CH_2–NH_2$; a biogenic amine, formed by enzymic decarboxylation of tryptophan. Of widespread occurrence in mammalian tissues, it is found also in plants, and is a product of microbial degradation of proteinaceous material.

tryptase EC 3.4.21.59; *other names*: mast cell tryptase; mast cell protease II; skin tryptase; lung tryptase; pituitary tryptase. An enzyme that cleaves Arg-|-Xaa, Lys-|-Xaa, but with more restricted specificity than **trypsin**. It is a trypsin-like serine endopeptidase released from the secretory granules upon mast-cell activation. Cytotoxic T-lymphocyte tryptase, or CTL tryptase, is **granzyme A**.

tryptic of, or relating to, trypsin; effected or produced by the action of trypsin; especially **tryptic digestion**, the partial hydrolysis of a protein by trypsin.

tryptophan or *(formerly)* **tryptophane** the trivial name for β-3-indolylalanine; α-amino-β-3-indolepropionic acid; 2-amino-3-(1*H*-indol-3-yl)propanoic acid; $(C_8H_6N)CH_2$–$CH(NH_2)$–COOH; a chiral α-amino acid. L-tryptophan (*symbol*: W or Trp), (*S*)-2-amino-3-(1*H*-indol-3-yl)propanoic acid, is a coded amino acid found in peptide linkage in proteins; codon: UGA (only in mitochondria) or UGG. In mammals, it is an essential dietary amino acid, and is glycogenic. The peptide antibiotic, **tyrocidin** C, contains one residue per molecule of D-tryptophan (*symbol*: D-Trp or DTrp), (*R*)-2-amino-3-(1*H*-indol-3-yl)propanoic acid.

L-tryptophan

tryptophanal α-amino-β-3-indolepropionaldehyde; 2-amino-3-(1*H*-indol-3-yl)propanal; $(C_8H_6N)CH_2$–$CH(NH_2)$–CHO; the α-amino-aldehyde analogue of **tryptophan**.

tryptophanase 1 EC 4.1.99.1; *recommended name*: tryptophanase; *systematic name*: L-tryptophan indole-lyase (deaminating). An enzyme that catalyses a reaction between water and L-tryptophan to form indole, pyruvate, and NH_3. It is responsible for indole formation in *Escherichia coli* and other enteric bacteria and bacilli. The reaction is comparable to that of β-tyrosinase (*see* **tyrosinase** (def. 2)). **2** *an alternative name for* **tryptophan 2,3-dioxygenase**.

tryptophanate 1 tryptophan anion; the anion, $(C_8H_6N)CH_2$–$CH(NH_2)$–COO⁻, derived from **tryptophan**. **2** any salt in which tryptophan is anionic. **3** any ester of tryptophan.

tryptophan 2,3-dioxygenase EC 1.13.11.11; *systematic name*: L-tryptophan:oxygen 2,3-oxidoreductase (decyclizing); *other names*: tryptophan pyrrolase; tryptophanase; tryptophan oxygenase; tryptamine 2,3-dioxygenase. An enzyme that catalyses a reaction between dioxygen and L-tryptophan to form L-formylkynurenine. It is a heme protein in mammalian liver, and it shows broad specificity towards tryptamine and other derivatives.

tryptophan 5-hydroxylase EC 1.14.16.4; *recommended name*: tryptophan 5-monooxygenase; *systematic name*: L-tryptophan, tetrahydrobiopterin:oxygen oxidoreductase (5-hydroxylating); *other name*: tryptophan oxygenase. An enzyme that catalyses a reaction between dioxygen, L-tryptophan, and tetrahydrobiopterin to form 5-hydroxy-L-tryptophan, dihydrobiopterin, and water. This is the rate-limiting enzyme in the biosynthesis of the neurotransmitter **serotonin (5-hydroxytryptamine)**.

tryptophanium tryptophan cation; the cation, $(C_8H_6N)CH_2$–$CH(NH_3^+)$–COOH, derived from **tryptophan**.

tryptophan 2-monooxygenase EC 1.13.12.3; *systematic name*: L-tryptophan:oxygen 2-oxidoreductase (decarboxylating); *other name*: tryptophan oxygenase. An enzyme that catalyses a reaction between dioxygen and L-tryptophan to form indole-3-acetamide, CO_2, and water. This is the first step in **auxin** biosynthesis.

tryptophan 5-monooxygenase the recommended name for **tryptophan 5-hydroxylase**.

tryptophano the alkylamino group, $(C_8H_6N)CH_2$–$CH(COOH)$–NH–, derived from **tryptophan**.

tryptophanol 3-(2-amino-3-hydroxypropyl)indole; 2-amino-3-(1*H*-indol-3-yl)propanol; $(C_8H_6N)CH_2$–$CH(NH_2)$–CH_2OH; the α-amino-alcohol analogue of **tryptophan**.

tryptophan oxygenase *an alternative name for* **1 tryptophan 2,3-dioxygenase**. **2 tryptophan 2-monooxygenase**. **3 tryptophan 5-hydroxylase**.

tryptophan pyrrolase *an alternative name for* **tryptophan 2,3-dioxygenase**.

tryptophan repressor *see* **trp**.

tryptophan synthase *see* **trp**.

tryptophan tryptophylquinone *abbr.*: TTQ; a quinoprotein cofactor of methylamine dehydrogenase, EC 1.4.99.3, an enzyme of methylotropic bacteria. TTQ consists of a tryptophanyl side-chain cross-linked to the tryptophyl 6,7-dione side-chain of a modified tryptophan residue.

tryptophyl the acyl group, $(C_8H_6N)CH_2$–$CH(NH_2)$–CO–, derived from **tryptophan**.

TSE *abbr. for* transmissible spongiform encephalopathy.

TSF *abbr. for* thymocyte-stimulating factor, i.e. **interleukin 2**.

TSH *abbr. for* thyroid-stimulating hormone.

TSS *abbr. for* transcription start site.

TSST-1 *abbr. for* toxic shock syndrome toxin-1.

T state *see* **T form**.

tTA *abbr. for* tetracycline regulated *trans*-activator. *See* **Tet-Off**.

TTP *abbr. for* **1** ribosylthymine triphosphate *or* ribothymidine triphosphate, *common names for* ribosylthymine 5′-triphosphate, ribothymidine 5′-triphosphate, 5′-ribothymidylyl triphosphate, thymine riboside 5′-triphosphate. **2** *(sometimes)* thymidine triphosphate. dTTP is recommended.

TTQ *abbr. for* tryptophan tryptophylquinone; a quinone cofactor produced by post-translational modification of two tryptophan residues in some quinoproteins.

TTSP *abbr. for* type II transmembrane serine protease; any of a class of single-pass transmembrane proteins that have an extracellular C-terminal region containing a serine protease domain. Other domains can also be present. Human examples include **corin** and **enteropeptidase**.

T tubule *abbr. for* transverse tubule; a type of tubule in a muscle fibre that arises as an invagination of the plasma membrane and penetrates the fibre, running mostly transversely among the fibrils and surrounding each of them with a ring. Each ring is shared by adjacent fibrils and the whole forms an intercommunicating system of tubules across the fibre. The function of these tubules is to pass the excitation signal from the muscle cell plasma membrane to the sarcomeres, ensuring rapid and synchronous activation. *See also* **sarcoplasmic reticulum**.

TTX *abbr. for* tetrodotoxin.

tuberin a protein that is widely expressed and shares a region of homology with GTPase-activating proteins. It forms a largely cytosolic complex with **hamartin**. It may be a tumour suppressor. Many mutations in the gene at 16p13.3 produce tuberous sclerosis type 2 (TSC2). The equivalent gene is inactive in the Eker rat and in the *gigas* mutant of *Drosophila*.

tuberous sclerosis an autosomal dominant multisystem disorder that frequently manifests as seizures, behavioural abnormalities, various skin lesions, and renal tumours and cysts. It is caused by mutations in **tuberin** and in **hamartin**.

tubocurarine (+)-tubocurarine is the active principle of **curare**; it is a nicotinic cholinergic receptor competitive antagonist, and is used as a muscle relaxant. It contains groups that are arranged in space in a rigid framework like the atoms of acetylcholine and are important in its action. It competes with acetylcholine at the neuromuscular junction and causes the endplate potential to fall dramatically.

tubulin a widely distributed globular protein, profilaments of which form the structural elements of a **microtubule**. It is a heterodimer of two chains, α and β, each ~50 kDa. The stable form of isolated tubulin is a dimer in which α–β pairing is believed to predominate; such a heterodimer is also the basic component of the microtubule. The dimer binds two molecules of GTP, one at an exchangeable site on the β chain and one at a non-exchangeable site on the α chain. The GTP is hydrolysed to GDP at the plus end more slowly than assembly of heterodimers, to give a GTP cap at that end. GDP–tubulin is disassembled more readily than GTP–tubulin, so the rate of GTP hydrolysis could regulate tubulin assembly. At least 6 isoforms are now known, designated α, β, γ, δ, ε, and ζ.

***α*-tubulin acetylase** *see* **tubulin *N*-acetyltransferase**.

(+)-tubocurarine

tubulin *N*-acetyltransferase EC 2.3.1.108; *other name*: α-tubulin acetylase; an enzyme that catalyses the reaction:

acetyl-CoA + α-tubulin L-lysine =

CoA + α-tubulin N^6-acetyl-L-lysine.

tubulinyl-Tyr carboxypeptidase EC 3.4.17.17; *other names*: carboxypeptidase-tubulin; soluble carboxypeptidase; an enzyme that catalyses the cleavage of the Glu-|-Tyr bond to release the C-terminal tyrosine residue from the native tubulin. It is inactive on carbobenzoxy-Glu-Tyr.

tuftelin *or* **enamelin** an acidic enamel protein of teeth, with a composition similar to tuft proteins. It is found in association with the crystal component, and secreted at a very early stage of enamel formation in vertebrate teeth.

tuftsin the tetrapeptide Thr-Lys-Pro-Arg, from **leukokinin** (leukophilic human IgG). It is present in the Fc fragment of IgG, binds to and stimulates the bactericidal, phagocytic, and tumoricidal activities of macrophages and granulocytes, and may also stimulate the migration of polymorphonuclear leukocytes. [After Tufts University, where it was discovered.]

tumble 1 to rotate in a plane vertical to the axis of rotation; 'head over heels'. Proteins and other macromolecules tumble in solution whereas integral membrane proteins rotate on an axis normal to the plane of the membrane but cannot tumble through the membrane surface. **2** in bacterial chemotaxis, to change direction by a rapid reversal of the direction of flagellar rotation, the bacterium then 'tumbling' in response to Brownian motion.

tumor *a variant spelling (esp. US) of* **tumour**.

tumorigenic describing a substance or treatment that has the tendency to cause the formation of a **tumour**.

tumour *or (esp. US)* **tumor** a mass of proliferating cells lacking, to varying degrees, normal growth control. Tumours may be benign, i.e. growing slowly, or malignant, i.e. growing aggressively, invading neighbouring tissues, and spreading to distant parts of the body by releasing cells into the bloodstream or lymph. These cells form secondary tumours (*see* **metastasis**). *See also* **adenoma**, **carcinoma**, **lymphoma**, **sarcoma**. —**tumorous** *or* **tumoral** *adj*.

tumour initiator any compound that causes the formation of a tumour.

tumour necrosis factor *abbr.*: TNF; either of two related proteins: TNF-α (*or* cachectin) is produced mainly by monocytes and macrophages, whereas TNF-β (*or* lymphotoxin) is produced by lymphoid cells. They show ≈30% sequence homology, bind to the same cell-surface receptors, and exist as homotrimers. They have a multiplicity of actions. In binding to their receptors, present on virtually all cells examined, they activate a large array of cellular genes and also multiple signal-transduction pathways, kinases, and transcription factors. Their genes are single-copy genes, closely linked within the MHC cluster. *See* **tumour necrosis factor** α; **tumour necrosis factor** β, **tumour necrosis factor receptor**.

tumour necrosis factor α (*abbr.*: TNF-α) *or (formerly)* **cachectin** a **cytokine** that is produced by macrophages, monocytes, endothelial cells, neutrophils, smooth muscle cells, activated lymphocytes, and astrocytes. It is a transmembrane glycoprotein and cytotoxin with a variety of functions, including the ability to mediate the expression of genes for growth factors, cytokines, transcription factors, and receptors. It can cause cytolysis of certain tumour cell lines, it has been implicated in the induction of cachexia (but *see* **cachectic factor**), it is a potent pyrogen, causing fever by direct action or by stimulation of interleukin 1 secretion, and it can stimulate cell proliferation and induce cell differentiation under certain conditions. The molecule is a homotrimer. *See also* **tumour necrosis factor**.

tumour necrosis factor β (*abbr.*: TNF-β) *or (formerly)* **lymphotoxin** a **cytokine** produced by T lymphocytes that is cytotoxic for a wide range of tumour cells. It binds to the same ligands as **tumour necrosis factor** α. *See also* **tumour necrosis factor**.

tumour necrosis factor receptor *abbr.*: TNF receptor; either of two distinct receptors for **tumour necrosis factor** (TNF), one of M_r 55 000 and the other of 75 000. The two types have related extracellular domains, based on cDNA sequencing, and these are also related to the extracellular domains of nerve growth factor receptor and some other cell-surface molecules. These form a receptor family characterized by four domains with regularly spaced cysteine residues; each has a single transmembrane domain; the intracellular domains show no significant homologies. Either TNF receptor type alone appears sufficient for high-affinity binding and full biological activity, and binds either **tumour necrosis factor** α or **tumour necrosis factor** β with high affinity. The signal transduction mechanism remains speculative, but receptor activation leads to a number of events, including activation of protein kinase C, protein kinase A, phospholipase A_2, and phosphatidylcholine-specific phospholipase C, and accumulation of cyclic AMP and diacylglycerol.

tumour promoter any compound that sensitizes tissues to the action of a **tumour initiator**, or predisposes tissue to become tumorous in any other way.

tumour-suppressor gene a normal gene that suppresses tumorigenesis. Certain cancers are associated with mutant tumour-suppressor genes; *see* (e.g.) **p53**, **RB**.

TUNEL *abbr* for terminal deoxyribonucleotidyl transferase (TdT)-mediated dUTP nick end labelling. A method of labelling cleaved DNA molecules by enzymically attaching a 2′-deoxyuridyl tag to the 3′ end and detecting it with a fluorescent antibody. Used for recognizing apoptosis *in situ*.

tunnel effect the phenomenon in which an electron or a proton, because of its small mass, does not behave kinetically according to the laws of classical mechanics. These particles can sometimes penetrate an energy barrier that is greater than their kinetic energy.

turacin *see* **tetrapyrrole**.

turanose 3-*O*-α-D-glucopyranosyl-D-fructose; an early product of the acid hydrolysis of **melezitose**.

turbid (of a solution) opaque due to the presence of suspended particles. —**turbidity** *n*.

turbidimeter an instrument for measuring the degree to which turbidity hinders the passage of light through a solution.

turbidimetry the quantitative determination of the concentration of substances in suspension by measurement of the decrease in light transmission caused by the suspended particles. *Compare* **nephelometry**.

Turcot syndrome the clinical combination of a primary tumour of the central nervous system and multiple colorectal adenomas or familial predisposition to colorectal cancer. It is associated with germline mutations in *APC*, in *mutL* homologue 1 (h*MLH1*), or in postmeiotic segregation homologue 2 (h*PMS2*). [After Jacques Turcot (1914–), Canadian surgeon.]

turnaround sequence (in DNA) a sequence of nucleotides between two **inverted repeats**. A turnaround sequence will cause a 'bubble' of noncomplementary nucleotides at one end of a **hairpin**.

Turner syndrome a chromosomal abnormality in females characterized by inheritance of only one X chromosome, or one normal X chromosome and one with a structural abnormality. The frequency in liveborn girls is about 1/5000. The syndrome is associated with short stature, webbing of the neck, a failure to menstruate, and infertility. [After Henry H. Turner (1892–1970), US endocrinologist.]

turnover number an obsolete expression of enzymic activity, variously defined on different occasions as the number of substrate molecules converted to product per catalytic site per minute, or as the number of moles of substrate converted per mole of enzyme per minute or second; it has the dimension of reciprocal time. *Compare* **catalytic constant, katal**.

turret a multiple cuvette holder that rotates in a spectrophotometer.

T-vector cloning a method for cloning polymerase chain reaction (PCR) products that depends on the addition of deoxyadenosine to the 3′ ends of the products by the terminal transferase activity of *Taq* polymerase. Vectors with single dideoxythymidine residues at the 3′ ends of the linearized vector form AT base pairs with the PCR products and are ligated efficiently before bacterial transformation and cloning.

Tween *the proprietary name for* a series of surfactants based on polyoxyethylene sorbitan esters. For example, Tween 20 is polyoxyethylene sorbitanmonolaurate (CMC 0.059 mM). Similarly the sorbitan moiety in other Tweens is: Tween 40, sorbitanmonopalmitate; Tween 60, sorbitanmonostearate; Tween 65, sorbitantristearate; Tween 80, sorbitanmonooleate (aggregation number 58, CMC 0.012 mM); Tween 85, sorbitantrioleate.

twilight zone a region of identity (~0–20%) within which sequence alignments appear plausible to the eye but are not statistically significant (i.e. the same alignment could have arisen by chance). *Compare* **midnight zone**.

twin-arginine a motif, consensus sequence SRRXFLK, found in the signal peptides for certain proteins, that is required for transport of the protein in the folded state, complete with cofactors, by the **tat** complex. Transport of the protein is driven by the membrane potential.

Twinkle a mammalian mitochondrial protein, similar to the primase/helicase of bacteriophage T7, that is required for integrity of the mitochondrial genome. Mutations in humans cause autosomal dominant progressive external ophthalmoplegia (*abbr.*: PEO).

twist a *Drosophila* gene for a **myc** family homodimeric **helix-turn-helix** transcription factor involved in the establishment of germ layers.

TWIST the human orthologue (202 amino acids) of *twist* protein (490 amino acids) of *Drosophila*. It is a basic helix-loop-helix transcription factor. *twist* protein induces myogenesis during development. Mutations in TWIST cause **Saethre–Chotzen syndrome**.

twist conformation 1 a median conformation of a nonplanar, six-membered, saturated ring compound through which one **boat conformation** passes during conversion into another boat conformation, or through which a boat conformation passes during conversion into a **chair conformation** or vice versa. It is similar to **skew conformation** (def. 1). **2** any conformation of the five-membered ring form of a monosaccharide or monosaccharide derivative when three adjacent ring atoms are coplanar and the other two lie on opposite sides of the plane. The conformational descriptor -*T* may be added to the name of such a compound. *See also* **conformation**.

twist number *or* **twisting number** *symbol*: T; the total number of base pairs in a DNA molecule divided by the number of bases per turn of the double helix. It is one of the properties that identifies **topological isomers** of DNA. *See also* **linking number, writhing number**.

twitchin the protein product of *unc-22*, one of about 40 genes important for muscle assembly and function in the nematode, *Caenorhabditis elegans*. Twitchin is a giant myosin-associated protein (*compare* **titin**) that is autophosphorylated on threonine; it contains repeats that are similar to **fibronectin** type III. The protein was named for a mutation that results in twitching. It has a counterpart in *Aplysia californica*, and presumably other invertebrates.

two-carbon fragment 1 any group of atoms containing two carbon atoms. **2** a metabolically active acetyl radical, as in acetyl coenzyme A.

two-component regulatory system one of the signal transduction systems in microorganisms and plants that consists of a membrane **sensor kinase** (SK) and a cognate response regulator (RR). Such systems respond to different types of environmental signal by autophosphorylation of a conserved histidine residue in SK and subsequent transfer to aspartate in RR, which then activates transcription of target genes.

two-component signal transduction a system used by bacteria and lower eukaryotes to adapt to the prevailing environmental and nutrition conditions. A typical system comprises a sensor **histidine kinase** and a cognate **response regulator**. In response to a particular stimulus, the activated sensor kinase phosphorylates its cognate response regulator. This activated response regulator then effects changes in gene expression.

two-dimensional chromatography a chromatographic method in which separation is first effected in one direction and, after rotation of the chromatogram through 90°, then in a second direction, often with a different solvent (or a different pH). **Two-dimensional electrophoresis** works on the same principle.

two-dimensional polymer a macromolecular structure in which approximately parallel chains extending in one direction are cross-linked by another series of approximately parallel chains extending in a second direction. Bacterial cell-wall peptidoglycans are two-dimensional polymers, with glycan chains extending in one direction, cross-linked to peptide chains.

TXA *abbr. for* thromboxane A (*see* **thromboxane**).

TXB *abbr. for* thromboxane B (*see* **thromboxane**).

Tylenol *see* **acetaminophen**.

Tylosin a **macrolide** class antibiotic, similar to **erythromycin**, used in animal feeds and in studies on the structure of **ribosomes**.

tyndallization *or* **tyndallisation** a method of sterilization in which the sample is heated at 80–100°C on successive days so as to kill cells on each occasion, but to allow spores (which are not killed by heating) to produce susceptible cells during the intervening periods of incubation. [After John Tyndall (1820–93), British natural philosopher, alpinist, and popularizer of science.]

type 1,2,3 copper different classes of copper-binding sites in proteins. In type 1, or blue copper centres, the copper is coordinated to at least two imidazole nitrogens from histidine and one sulfur from cysteine. They are characterized by small copper hyperfine couplings and a strong visible absorption in the Cu(II) state. In type 2, or non-blue copper sites, the copper is mainly bound to imidazole nitrogens from histidine. Type 3 copper centres comprise two spin-coupled copper ions, bound to imidazole nitrogens.

type III system a protein secretory system found in Gram-negative bacteria that is responsible for the secretion of **Yop** proteins. The system is known as type III because it was the third type of secretory system found in the membrane of Gram-negative bacteria.

type II transmembrane serine protease *see* **TTSP**.

typing (*in immunology*) the identification of the classes of histocompatibility antigens, blood groups, or other surface markers present in a sample of serum, or on white blood cells or red blood cells.

Tyr *symbol for* a residue of the α-amino acid L-tyrosine (alternative to Y).

tyramine tyrosamine; *p*-(β-aminoethyl)phenol; 2-(4-hydroxyphenyl)ethylamine; a biogenic amine with adrenergic action, formed by enzymic decarboxylation of tyrosine. It is widespread in mammalian tissues, found in several plant species and in **ergot**, and is a product of microbial action in the intestine, in the putrefaction of animal tissue, and in the ripening of cheese.

Tyr(I₂) *or* (if the context does not imply the locants) **Tyr(3,5-I₂)** *symbol for* a residue of the α-amino acid 3,5-diiodo-L-tyrosine (*see* **diiodotyrosine**).

tyrocidine *or* **tyrocidin** any peptide antibiotic produced by *Bacillus brevis* that is a cyclic decapeptide. The structure of tyrocidine A is (amino acids are L unless otherwise stated) cyclic –Leu–D–Phe–Pro–Phe–D–Phe–Asn–Gln–Tyr–Val–Orn–. Tyrocidine B has L-Tyr in place of L-Phe, and in tyrocidine C D-Trp replaces the D-Phe attached to the Asn. They are similar in structure and action to **gramicidin S**, increasing the permeability of bacterial membranes, promoting leakage of cytoplasmic constituents, and permitting entry of ions larger than those normally transported. *Bacillus brevis* is immune from this action. They have little application in medicine. Biosynthesis is independent of ribosomes (*see* **protein and peptide biosynthesis**).

Tyrode's solution *or* **Tyrode solution** *or* **Ringer–Tyrode solution** a balanced salt solution modified from **Locke's solution**, initially for experiments on mammalian intestine. The solution usually contains (in g L^{-1}, approx.): NaCl (8.0); KCl (0.2); $MgCl_2 \cdot 6H_2O$ (0.1); $CaCl_2$ (anhydrous, 0.2); $NaHCO_3$ (0.1); $NaH_2PO_4 \cdot H_2O$ (0.05); and glucose (1.0). [After Maurice Vejux Tyrode (1878–1930), US pharmacologist, who devised this solution in 1910.]

tyrosinase 1 *an alternative name for* **monophenol monooxygenase**. **2** β-**tyrosinase** EC 4.1.99.2; *recommended name*: tyrosine phenol-lyase; an enzyme that catalyses a reaction between L-tyrosine and water to form phenol, pyruvate, and NH_3. Pyridoxal-phosphate is a coenzyme. The protein contains copper and is a homotetramer, similar to **tryptophanase** (def. 1).

tyrosinate 1 tyrosinate(1–); tyrosine monoanion. In theory, the term denotes any ion, or mixture of ions, formed from **tyrosine**, and having a net charge of –1, although the species $HO–C_6H_4–CH_2–CH(NH_2)–COO^-$ predominates in practice. **2** *the systematic name for* tyrosinate(2–); tyrosine dianion; $^-O–C_6H_4–CH_2–CH(NH_2)–COO^-$. **3** any salt containing an anion of tyrosine. **4** any ester of tyrosine.

tyrosine *the trivial name for* β-(*p*-hydroxyphenyl)alanine; α-amino-*p*-hydroxyhydrocinnamic acid; α-amino-β-(*p*-hydroxyphenyl) propionic acid; 3-(4-hydroxyphenyl)alanine; 2-amino- 3-(4-hydroxyphenyl)propanoic acid; $HO–C_6H_4–CH_2– CH(NH_2)–COOH$; a chiral α-amino acid. L-Tyrosine (*symbol*: Y *or* Tyr), (*S*)-2-amino-3-(4-hydroxyphenyl)propanoic acid, is a coded amino acid found in peptide linkage in proteins; codon: UAC or UAU. In mammals, it is an essential dietary amino acid, and is both glucogenic and ketogenic. One residue per molecule of D-tyrosine (*symbol*: D-Tyr *or* DTyr), (*R*)-2- amino-3-(4-hydroxyphenyl)propanoic acid, occurs as its *O*-methyl ether in the antibiotic cycloheptamycin.

L-tyrosine

tyrosine decarboxylase EC 4.1.1.25; an enzyme that catalyses the decarboxylation of L-tyrosine to **tyramine**. Pyridoxal-phosphate is a coenzyme.

tyrosine hydroxylase EC 1.14.16.2; *recommended name*: tyrosine 3-monooxygenase; *systematic name*: L-tyrosine, tetrahydrobiopterin:oxygen oxidoreductase. An enzyme that catalyses a reaction between dioxygen, L-tyrosine, and tetrahydrobiopterin to form 3,4-dihydroxy-L-phenylalanine, dihydrobiopterin, and water; it requires ferric ion. It is the rate-limiting enzyme in the synthesis of **catecholamine** neurotransmitters. In humans four types are produced by alternative splicing. *See* **DOPA-responsive dystonia**.

tyrosine kinase EC 2.7.1.112; *recommended name*: protein-tyrosine kinase; *other names*: tyrosylprotein kinase; protein kinase (tyro-sine); hydroxyaryl-protein kinase. An enzyme that transfers the terminal phosphate of ATP to a specific tyrosine residue on its target protein; i.e. it catalyses the phosphorylation by ATP of protein-tyrosine to form protein-tyrosine phosphate with release of ADP. The enzyme molecule has five motifs. The tyrosine kinase family is of profound importance due to its involvement in the signalling processes associated with mitogenic stimulation, cell growth, and oncogenesis. The tyrosine kinases fall into two major classes, those that are cell membrane receptors and integral membrane proteins, and those that are intracellular and function downstream of the receptors. The membrane receptors have been grouped into subclasses according to their structure (all individual polypeptides have a single transmembrane domain, with the kinase activity in an intracellular domain); examples include subclass I, proteins having two cysteine-rich repeat sequences in the extracellular domain of a monomeric receptor, a group that includes the **epidermal growth factor receptor**; subclass II, disulfide-linked heterotetrameric $\alpha_2\beta_2$ structures with extracellular cysteine-rich repeat sequences similar to subclass I, an example being the **insulin receptor**; subclass III, proteins with five immunoglobulin-like repeats in the extracellular domain, and with a non-kinase insert in the kinase region, typified by the **platelet-derived growth factor receptor**; and subclass IV, proteins with three immunoglobulin-like repeats in the extracellular domain, with a non-kinase insert in the kinase region. This classification continues to be extended. The intracellular non-receptor tyrosine kinases include, among many, the products of **FGR**, **FYN**, and *src*. *See also* **protein kinase**.

tyrosine 3-monooxygenase *the recommended name for* **tyrosine hydroxylase**.

tyrosine transaminase *or* **tyrosine aminotransferase** (*abbr.*: TAT) EC 2.6.1.5; *systematic name*: L-tyrosine:2-oxoglutarate aminotransferase; an enzyme that catalyses the reaction:

$$\text{L-tyrosine} + \text{2-oxoglutarate} =$$
$$\text{4-hydroxyphenylpyruvate} + \text{L-glutamate}.$$

It is a pyridoxal-phosphate enzyme and it catalyses the first step in tyrosine catabolism; defects lead to tyrosinemia type II (**Richner–Ranhart syndrome**).

tyrosinium tyrosine cation; the cation, $HO–C_6H_4–CH_2–CH(NH_3^+)–COOH$, derived from **tyrosine**.

tyrosino the alkylamino group, $HO–C_6H_4–CH_2–CH(COOH)–NH–$, derived from **tyrosine**.

tyrosin-O^4-yl the aryloxy group, $HOOC–CH(NH_2)–CH_2–C_6H_4–O–$, derived from **tyrosine**.

tyrosyl the acyl group, $HO–C_6H_4–CH_2–CH(NH_2)–CO–$, derived from **tyrosine**.

tyrphostin any inhibitor of protein-tyrosine kinases derived from the substituted benzylidenemalononitrile, shown as the core structure; A, B, and C can be $–NO_2$, $(CH_3)_3C–$, –OH, or –H. In tyrphostin 25, [(3,4,5-trihydroxyphenyl)-methylene]-propanedinitrile, D is –CN. Some tyrphostins uncouple oxidative phosphorylation and inhibit lipoxygenases.

tyrphostin 25

tyvelose the aldohexose 3,6-dideoxy-D-mannose, 3,6-dideoxy-D-*arabino*-hexose, a monosaccharide of various lipopolysaccharides produced by some Gram-negative bacteria, e.g. certain serotypes of *Salmonella typhi*, where it contributes to the antigenic specificity of the organism's O antigens. The L-enantiomer is called **ascarylose**.

Uu

u **1** *symbol for* unified atomic mass unit. **2** *see* **IU**.

u *symbol for* **1** speed (alternative to *v*, *w*, or *c*). **2** (bold italic) velocity (alternative to *v*, *w*, or *c*).

U *symbol for* **1** a residue of the base uracil in a nucleic-acid sequence. **2** a residue of the ribonucleoside uridine (alternative to Urd). **3** uniformly labelled. **4** enzyme unit (obsolete; superseded by **katal**). **5** uranium.

U *symbol for* **1** electric potential difference (alternative to ΔV or $\Delta\phi$). **2** internal energy.

U46619 a stable thromboxane A_2 analogue and mimetic; *another name for* **9,11-dideoxy-11α,9α-epoxymethanoprostaglandin F$_{2\alpha}$**.

U937 a cell line derived from a patient with histiocytic lymphoma.

UAA a codon in mRNA for chain termination; the **ochre codon**.

UAC a codon in mRNA for L-tyrosine.

UAG a codon in mRNA for chain termination; the **amber codon**.

UAS *abbr. for* upstream activator sequence; the counterpart in yeast of an **enhancer**.

UAU a codon in mRNA for L-tyrosine.

Ubbelohde viscometer *see* **capillary viscometer**.

UBE3A *symbol for* a gene encoding ubiquitin protein ligase E3A, which functions by attaching the small protein ubiquitin to proteins targeted for degradation within proteasomes. Mutations in the gene are responsible for some cases of Angelman syndrome. The paternal allele is imprinted.

UBIP *abbr. for* ubiquitous immunopoietic polypeptide.

ubiquilin a protein that binds to **presenilins**, promotes their accumulation, and appears to increase their synthesis. It is related to ubiquitin in structure, and like ubiquitin is present in the neuropathological lesions characteristic of Alzheimer's disease (neurofibrillary tangles) and Parkinson's disease (Lewy bodies). The responsible gene maps to a region thought to contain a gene for late-onset Alzheimer's disease.

ubiquinol *abbr.*: QH$_2$; reduced **ubiquinone**.

ubiquinol–cytochrome *c* reductase EC 1.10.2.2; *systematic name*: ubiquinol:ferricytochrome-*c* oxidoreductase; *other name*: complex III. An enzyme complex that catalyses the oxidation of ubiquinol by two molecules of ferricytochrome *c* to form ubiquinone and two molecules of ferrocytochrome *c*. It contains cytochromes b_{562}, b_{566}, and c_1, and a Rieske iron–sulfur protein.

ubiquinone *or (formerly)* **coenzyme Q** *abbr.*: Q; a lipid-soluble electron-transporting coenzyme that is an essential component in the **respiratory chain**. The structure is shown below; *n* has values from 6 to about 12, as indicated by the abbreviation Q$_n$. Ubiquinone is synthesized in mammals from tyrosine, and in bacteria and other organisms by the **shikimate pathway**. *See also* **protonmotive Q cycle**.

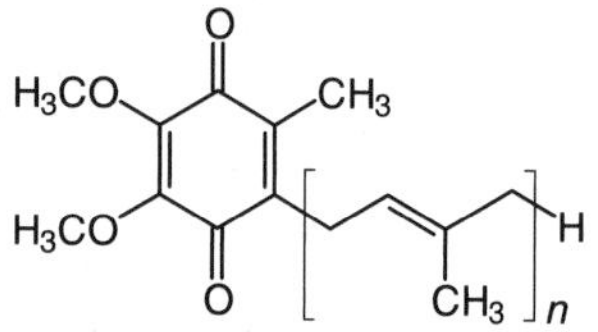

ubiquinone reductase *see* **NADH dehydrogenase (ubiquinone)**.

ubiquitin 1 a small, highly conserved, heat-stable protein of 76 amino-acid residues, first isolated from bovine thymus, but subsequently found in the cells of all tissues studied, in animals, yeast, bacteria, and higher plants. Located in both nuclei and cytoplasm, it derives its name from its widespread distribution. It undergoes an ATP-dependent reaction with proteins that results in condensation of its terminus with lysine amino groups; this process is mediated by a large multiprotein complex, the 26S **proteasome**. For example, cyclin degradation in the control of the cell cycle is triggered by ubiq-

uitination. Thus, the cell has two means of protein degradation: ubiquitination and proteolysis in lysosomes. Such modified proteins are degraded rapidly. Ubiquitin is synthesized as a polyubiquitin precursor, with exact head-to-tail repeats, the number of repeats differing between species. In some species, there is a final amino acid after the last repeat (e.g., Val in human). Some ubiquitin genes encode a single copy of ubiquitin fused to a ribosomal protein. The following sequence is conserved in numerous species:

MQIFVKTLTGKTITLEVEPSDTIENVKAKIQDKEGIP-
PDQQRLIFAGKQLEDGRTLSDYNIQKESTLHLVLRLRGG.

2 *a former name for* **ubiquitous immunopoietic polypeptide**.

ubiquitin-activating enzyme E1 an enzyme that activates ubiquitin by first adenylylating, with ATP as co-substrate, its C-terminal glycine residue and thereafter linking this to itself via the side chain of a cysteine residue, releasing AMP. The protein is a monomer and can accommodate two ubiquitin molecules.

ubiquitination *see* **ubiquitin** (def. 1).

ubiquitin-conjugating enzyme E2 *or* **ubiquitin-carrier protein** an enzyme that accepts the ubiquitin attached to **ubiquitin-activating enzyme E1** and, with **ubiquitin–protein ligase E3**, transfers it to the acceptor protein. Numerous isoforms exist within a species, with specificity for an E3 isozyme and an acceptor protein. Some E2 isoforms transfer the ubiquitin directly to the acceptor protein without the need for E3.

ubiquitin–protein ligase E3 EC 6.3.2.19; *other names*: ubiquitin-conjugating enzyme; ubiquitin-activating enzyme; an enzyme that couples ubiquitin to a protein lysine to form protein N-ubiquityllysine with hydrolysis of ATP to ADP and pyrophosphate. Several isozymes with specificity for an E2 and the acceptor protein exist with a species.

ubiquitous immunopoietic polypeptide *abbr.*: UBIP; a polypeptide that has lymphocyte-differentiating properties, and can stimulate adenylate cyclase. It is involved in maintenance of chromatin structure, regulation of gene expression, stress response, and ribosome biogenesis. It was formerly called ubiquitin. *See also* **proteasome**.

Ubl protein *abbr. for* ubiquitin-like protein. *See* **SUMO**

UCA a codon in mRNA for L-serine.

UCC a codon in mRNA for L-serine.

UCG a codon in mRNA for L-serine.

UCP *abbr. for* brown fat uncoupling protein.

UCU a codon in mRNA for L-serine.

Udo *symbol for* the undecanoyl group, CH$_3$–[CH$_2$]$_9$–CO–.

UDP *abbr. for* uridine (5′-)diphosphate.

UDP-*N*-acetylglucosamine 6-dehydrogenase EC 1.1.1.136; an enzyme that catalyses the reaction:

UDP-*N*-acetyl-D-glucosamine + 2 NAD$^+$ + H$_2$O =
UDP-*N*-acetyl-2-amino-2-deoxy-D-glucuronate + 2 NADH.

UDP-*N*-acetylglucosamine–dolichyl-phosphate-*N*-acetylglucosamine phosphotransferase EC 2.7.8.15; an enzyme that catalyses a reaction between UDP-*N*-acetyl-D-glucosamine and dolichyl phosphate to form *N*-acetyl-D-glucosaminyl-diphospho-dolichol and UMP. This is the first reaction in the synthesis of oligosaccharide lipids.

UDP-*N*-acetylglucosamine 2-epimerase EC 5.1.3.14; an enzyme that catalyses the reaction:

UDP-*N*-acetyl-D-glucosamine =
UDP-*N*-acetyl-D-mannosamine.

This is the initial reaction in the synthesis of sialic acid, and the enzyme is allosterically inhibited by CMP-sialic acid. The activity resides in the N-terminal 416 residues of the bifunctional protein (727 amino acids), which has *N*-acetylmannosamine kinase activity in the C-terminal portion. Mutations that impair binding of the inhibitor to the epimerase portion cause the rare disease sialuria.

UDP-*N*-acetylglucosamine 4-epimerase EC 5.1.3.7; an enzyme that catalyses the reaction:

$$\text{UDP-}N\text{-acetyl-D-glucosamine} = \text{UDP-}N\text{-acetyl-D-galactosamine.}$$

UDPG *abbr. for* uridinediphosphoglucose; UDPGlc is preferred (to avoid possible ambiguity in context with galactose).

UDPGal *or* **UDPgalactose** *abbr. for* uridinediphosphogalactose.

UDPgalactose 4-epimerase *abbr.*: GALE; *an alternative name for* UDPglucose 4-epimerase.

UDPGlc *(preferred) abbr. for* uridinediphosphoglucose; *see also* UDPG.

UDPglucose 6-dehydrogenase EC 1.1.1.22; an enzyme that catalyses the reaction:

$$\text{UDPglucose} + 2\ \text{NAD}^+ + \text{H}_2\text{O} = \text{UDPglucuronate} + 2\ \text{NADH.}$$

UDPglucose 4-epimerase *the recommended name and systematic name for* the isomerase EC 5.1.3.2; *other names*: UDPgalactose 4-epimerase; *(abbr.*: GALE); *(formerly)* galactowaldenase; an enzyme that interconverts uridinediphosphoglucose and uridinediphosphogalactose. The mechanism involves oxidation of the C-4 hydroxyl group of the hexose moiety of either uridinediphospho-sugar to an oxo group by a tightly bound molecule of NAD⁺; the resultant NADH is then reoxidized by the intermediate oxo derivative with inversion of configuration of the resulting hydrogen and hydroxyl groups at C-4. Mutations that produce enzyme deficiency result in a benign anomaly when it affects only blood cells, or in a severe but rare form of galactosemia when it affects all tissues. The difference may reflect the degree of molecular instability produced by the mutation. The human gene encodes 348 amino acids. *Compare* UDPglucose–hexose-1-phosphate uridylyltransferase.

UDPglucose–hexose-1-phosphate uridylyltransferase EC 2.7.7.12; *systematic name*: UDPglucose:α-D-galactose-1-phosphate uridylyltransferase; *other names*: uridylyltransferase; galactose-1-phosphate uridylyltransferase. An enzyme that interconverts uridinediphosphoglucose and uridinediphosphogalactose. It is important in the metabolism of galactose by catalysing the exchange reaction:

$$\text{UDPglucose} + \alpha\text{-D-galactose 1-phosphate} = \alpha\text{-D-glucose 1-phosphate} + \text{UDPgalactose.}$$

More than 150 mutations cause deficiency of the enzyme, which is associated with a form of **galactosemia**. In *Escherichia coli* the enzyme comprises 890 amino acids (102.31 kDa) and modifies the **nitrogen regulatory protein P$_{\text{II}}$**. *Compare* UDPglucose 4-epimerase.

UDPglucose pyrophosphorylase *an alternative name for* UTP–glucose-1-phosphate uridylyltransferase.

UDPglucuronate 4-epimerase EC 5.1.3.6; an enzyme that catalyses the reaction:

$$\text{UDP-D-glucuronate} = \text{UDP-D-galacturonate.}$$

UDPglucuronate 5′-epimerase EC 5.1.3.12; an enzyme that catalyses the reaction:

$$\text{UDP-D-glucuronate} = \text{UDP-L-iduronate.}$$

NAD⁺ is a cofactor.

UDPglucuronosyltransferase EC 2.4.1.17; *abbr.*: UDPGT *or* UGT; *recommended name*: glucuronosyltransferase; an enzyme involved in the formation of **glucuronosides**. It catalyses a reaction between UDPglucuronate and an acceptor to form UDP and acceptor β-D-glucuronoside. The numbering of variants uses several systems; the one followed here has UDPGT followed by a number, letter, and sometimes a number, e.g. UDPGT1F, UDPGT2B1O. Human examples include: UDPGT1a (bilirubin-specific isozyme 1); UDPGT1b; and UDPGT1c (*or* UGT13).

UDPGP *abbr. for* UDPglucose pyrophosphorylase (i.e. **UTP– glucose-1-phosphate uridylyltransferase**).

UGA 1 a codon in nonmitochondrial mRNA for chain termination; the **opal codon. 2** a codon in mitochondrial mRNA for L-tryptophan.

UGC a codon in mRNA for L-cysteine.

UGG a codon in mitochondrial mRNA, and the only codon in non-mitochondrial mRNA, for L-tryptophan.

UGU a codon in mRNA for L-cysteine.

+ulose *noun suffix* denoting a **ketose**; it may be expanded to **+ulofuranose** or **+ulopyranose** according to whether the cyclic form of the sugar is five-membered or six-membered respectively.

+ulosonic acid *noun suffix* denoting a **ketoaldonic acid**; it may be expanded to **+ulofuranosonic acid** or **+ulopyranosonic acid** according to whether the cyclic form of the sugar acid is five-membered or six-membered respectively.

ultimate carcinogen *or* **ultimate carcinogenic metabolite** any metabolite of a **chemical carcinogen** that reacts with cellular constituents so as to initiate carcinogenesis. *See also* **proximate carcinogen**.

ultimobranchial body one of a pair of glandlike structures derived from the last branchial (gill) pouches. They occur in all vertebrates except cyclostomes. In most mammals they become embedded in the respective lobes of the thyroid gland, forming **C cells**, which secrete the hormone **calcitonin**.

ultra+ *prefix* **1** beyond in space; on the other side (of). *Compare* **infra+, trans+. 2** beyond a specified (esp. in scientific usage, lower) limit, range, value, etc. *Compare* **super+. 3** beyond the common, ordinary, natural, etc.; extreme(ly). *Compare* **hyper+.**

ultracentrifuge any centrifuge that operates at high rotational velocities generating high gravitational fields. Ultracentrifuges can be operated at known speeds with small variations and without temperature fluctuations. They may be either analytical or preparative.

ultradian (of a biological activity) occurring less often than once per 24-hour period. *Compare* **circadian, infradian.**

ultrafilter a filter of small pore size suitable for **ultrafiltration**.

ultrafiltration a separation process whereby a solute of molecular size significantly greater than that of the solvent molecule is removed from the solvent by the application of a hydraulic pressure, which forces only the solvent to flow through a suitable membrane (usually one having a pore size in the range 0.001–0.1 μm), thereby concentrating the solute.

ultramicrotome a device for preparing very thin sections of cells or tissues, after fixing and embedding or freezing, for inspection in the transmission electron microscope.

ultrasonic of, pertaining to, or producing **ultrasound** waves.

ultrasonication the process of exposing material to **ultrasound**. It is used for cell disruption, for denaturation of proteins, and for cleaning solid surfaces. *See also* **sonicate.**

ultrasonic disintegrator an apparatus for disintegrating cells and other material, or for cleaning solid surfaces, by exposing them to **ultrasound**. *See also* **sonicator.**

ultrasound propagated waves, of the same nature as sound but inaudible, with frequencies of >20 kHz. Frequencies of ≈25 kHz can be used for disintegrating cells and subcellular components and also in cleaning baths for the removal of deposits on a wide variety of objects. In physiotherapy the range of 0.5–3 MHz is used for the relief of painful muscles by producing warmth and by changing the permeability of cell membranes, which allows for the removal of fluid. For diagnostic purposes the range of 3–12 MHz is commonly used and is a popular method of non-invasive imaging, e.g. in pregnant women. There is also ultrasonic microscopy, which mimics light microscopy and employs sound in the GHz range. Ultrasound is also being developed for the disintegration of tumours.

Ultraspiracle *abbr.*: USP; a nuclear hormone receptor of *Drosophila* that forms heterodimers with the ecdysteroid receptor and with *Drosophila* hormone receptor 38 (DHR38).

ultrastructure the fine structure of an organ, tissue, cell, or subcellular particle, especially that demonstrable with an electron microscope.

ultraviolet (radiation) *abbr.*: UV *or* uv; electromagnetic radiation of wavelengths in the range 13.6–400 nm (frequencies 750 THz–22.1 PHz). This range overlaps the wavelengths of soft X-radiation and ends just across the short-wavelength limit of visible light. Ultraviolet radiation is often divided into **near ultraviolet** (200–400 nm) and **far ultraviolet** (or **vacuum ultraviolet**) (<200 nm);

radiation in the far ultraviolet is absorbed by O_2 and requires the use of evacuated apparatus or flushing with N_2.

ultraviolet-endonuclease an enzyme involved in the repair of ultraviolet-irradiated DNA in microorganisms. *See also* **deoxyribonuclease (pyrimidine dimer)**, *uvr*.

ultraviolet reactivation *see* **W reactivation**.

Ultrogel *the proprietary name for* a range of bead-form rigid support materials used in **gel filtration**, **affinity chromatography**, and related procedures. It consists of a lattice of polyacrylamide in a matrix of agarose gel.

umbelliferone 7-hydroxycoumarin; a common component of plant extracts, used as a fluorescent pH indicator and as a fluorescent label.

UML *abbr. for* universal markup language; *see* **markup language**.

UMP *abbr. for* uridine monophosphate; i.e. uridine phosphate.

UMP synthase the combination of enzymes that form uridine 5′-phosphate from orotate: these are **orotate phosphoribosyltransferase** (forming orotidylic acid) and **orotidine 5′-phosphate decarboxylase**. *See also* **uridine-5′-monophosphate synthase**.

umu bacterial genes encoding proteins that are involved in the error-prone **SOS repair** process, probably by allowing bypass synthesis across a damaged template.

un+ *prefix* **1** (used with adjectives, adverbs, and nouns) not; opposite of; contrary to. **2** (used with verbs) reversal of an action or state; release; removal from.

unary operator an arithmetic or logical operator (e.g. NOT) that operates on only one variable. *Compare* **binary operator**.

unc-86 a gene of *Caenorhabditis elegans* that encodes a transcription factor containing the **POU domain** and contributes to the naming of the domain.

uncharacterized protein family *abbr.*: UPF; a related group of protein sequences whose function has not been experimentally determined and that shows no similarity to other, functionally characterized, sequences.

uncoating enzyme an enzyme that catalyses the removal of the protein coat from a virus.

uncoded amino acid *an alternative term for* **noncoded amino acid**.

uncompensated acidosis an **acidosis** in which there is a fall of the blood pH, due to the inadequacy of the physiological compensating mechanisms.

uncompensated alkalosis an **alkalosis** in which there is a rise of the blood pH, due to the inadequacy of the physiological compensating mechanisms.

uncompetitive inhibition the inhibition of an enzyme system when the inhibitor does not combine with the free enzyme, but only with one of the enzyme–substrate intermediates.

unconventional myosin any of about 14 isoforms of muscle **myosin** (myosin II) that function as motor proteins and are also found in many nonmuscle cell types. Each consists of a globular head (an actin-activated ATPase) and a short tail (which binds membranes including those of organelles and vesicles). Movement of the head over actin microfilaments is responsible for cytokinesis, cytoplasmic streaming, and axonal transport. **Myosin I** is present in most nonmuscle cells. **Myosin Va** is responsible for transport of melanosomes in pigment cells and of vesicles in neurons. Mutations in its gene at 15q21 are associated with Griscelli syndrome (hypopigmentation of skin and hair and immunological deficiency). **Myosin VIIa** is a protein of 2215 amino acids. Some mutations in its gene at 11q13.5 are associated with a form of familial nonsyndromic deafness, whereas others result in Usher syndrome type I (congenital deafness with vestibular dysfunction and retinitis pigmentosa). **Myosin XV** is 3530 amino acids long and present in the sensory cells of the inner ear. Mutations in its gene at 17p11.2 result in a form of familial nonsyndromic deafness.

uncoupled 1 disconnected; unfastened. **2** (in biochemistry) describing an organism, tissue, cell, or mitochondrion in which phosphorylation is dissociated from oxidation in the **respiratory chain**. *See also* **uncoupling**.

uncoupler *an alternative name for* uncoupling agent (*see* **uncoupling**).

uncoupling (in biochemistry) the dissociation of phosphorylation from oxidation in the **respiratory chain**, so that electron transport proceeds without the esterification of phosphate and without the need for inorganic phosphate. An **uncoupling agent** (or uncoupler) is a substance that causes uncoupling of phosphorylation from electron transport at one or more points in the electron-transport chain. A classical example is **dinitrophenol**. Uncoupling agents probably act by causing leakage of protons across the mitochondrial inner membrane thereby destroying the proton gradient that drives oxidative phosphorylation. *See also* **chemiosmotic coupling hypothesis**.

uncoupling protein *see* **brown fat uncoupling protein**.

undec+ *or* **undeca+** *prefix* denoting eleven.

undecaprenol *or* **bactoprenol** a polyunsaturated long-chain (C_{55}) primary alcohol containing 11 **prenyl** units.

undecaprenyl phosphate *or* **undecaprenol phosphate** a very hydrophobic phosphate ester of **undecaprenol**. It acts as a membrane-soluble carrier to which intermediates are combined in the course of the biosynthesis of bacterial cell-envelope components, e.g. peptidoglycan and lipopolysaccharide. *Other names*: bactoprenyl phosphate; carrier lipid; C_{55} carrier lipid; C_{55} lipid carrier.

underivatized *or* **underivatised** (of a substance) in the naturally occurring state.

undulin a non-collagenous glycoprotein of the interstitial extracellular matrix, associated with collagen. Isolated from skin or placenta, it is found mainly between densely packed mature collagen fibrils as bundles of uniform wavy fibres. It is a member of the fibronectin–tenascin family, and is related to collagen type XIV. It is associated with dense collagen matrices in soft tissues, and is likely to be involved in the supramolecular organization of interstitial collagens. Human undulin 2 shares sequence similarity with fibronectin type III.

ung *see* **uracil-DNA N-glycosylase**.

Uni *see* **reactancy of enzymes**.

uni+ *comb. form* consisting of, having, or relating to one.

unicode a set of characters that, unlike **ASCII**, occupy 16 bits and can therefore encode many different alphabets.

unidentified reading frame *abbr.*: URF; an open reading frame that encodes a protein of unknown function.

unified atomic mass unit *or* (formerly) **atomic mass unit** (*abbr.*: amu *or* a.m.u.) *symbol*: u; a non-SI unit of mass equal to one-twelfth of the mass of one atom of carbon-12; i.e.

$$1\ \text{u} = m_a(^{12}\text{C})/12 \approx 1.660\ 540 \times 10^{-27}\ \text{kg}.$$

It is identical with **dalton**.

uniformly labelled *symbol*: U; describing a **labelled compound** in which the label is distributed in a statistically uniform, or nearly uniform, manner between all the possible positions in the molecule.

UniGene *see* **cluster database**.

unimolecular reaction *see* **molecularity**.

uniparental disomy the inheritance of two copies of a chromosome from one parent.

uniport a membrane transport process in which a single kind of substrate (ion or uncharged molecule) is transported across the membrane, either by the mechanism of mediated diffusion where the substrate equilibrates at the two membrane sides (e.g., monosaccharides in mammalian erythrocytes; TC 2.A.1.1.12), or by active transport where an external source of energy is used to achieve movement against a chemical potential or an electrochemical potential gradient (e.g., plasma membrane H$^+$-transporting ATPase TC 3.A.3.3.1 or EC 3.6.3.6). Such transport systems may be designated as **uniporters**.

UniProt *abbr. for* universal protein sequence database; *see* **composite database**.

unique (sequence) DNA *or* **single-copy DNA** any unique and specifically ordered nucleotide sequence in DNA that occurs only once in a genome. *Compare* **repetitive sequences**.

unitary matrix *an alternative name for* identity matrix. *See* **scoring matrix**.

unit cell the smallest portion of a crystal that, by translation in three dimensions, is able to account for the structure of the entire crystal. The array of unit cells, arranged along parallel straight lines, makes up the **crystal lattice**.

unit membrane *or* **double membrane** *or* **triple membrane** any membrane, irrespective of cellular location, that on staining with osmium or other heavy metals appears in section in the electron microscope as a pair of parallel dark lines separated by a less dense layer, the whole being 6–10 nm thick. Unit membranes include the plasma membrane and the membranes of organelles such as mitochondria, nuclei, and endoplasmic reticulum.

unit of replication *see* replicon.

unit of transcription *see* transcriptional unit.

univalent 1 *or* **monovalent** (of a chemical species) having a **valency** of one. **2** an unpaired chromosome seen during meiosis when bivalents are also present. Such a chromosome either lacks a homologue or results from **asynapsis**. —**univalency** *n*.

UniVec *see* **vector database**.

universal buffer mixture any solution containing several buffering systems chosen so as to provide a relatively high buffer capacity over a wide pH range.

universal donor an individual with O-type blood; such blood may be transfused into any recipient. *See* **ABO system**.

universal primers oligonucleotides used as primers in DNA sequencing reactions that are specific for plasmid sequences flanking the cloned DNA insert. Because sequences such as the T3, T7, or SP6 promoters are commonly used in this position, primers annealing to them can be used to obtain sequences of inserts cloned in a variety of vectors.

universal recipient an individual with AB-type blood; such individuals may be transfused with blood from any donor. *See* **ABO system**.

unmediated transport *or* **nonmediated transport** the movement of a solute across a barrier by **diffusion** (def. 1).

unpaired electron an electron in an atom or molecule whose spin is not paired with the oppositely directed spin of another electron in the same atom or molecule. Only systems containing unpaired electrons will give an electron spin resonance signal. *See* **electron spin resonance spectroscopy**.

unrooted tree a **phylogenetic tree** in which the evolutionary ancestor is unknown.

unsaponifiable lipid *or* **unsaponifiable material** part of a lipid sample that is not solubilized by **saponification**. It consists principally of steroids and terpenes.

unsaturated 1 (of a solution) able to dissolve more of the solute in question. **2** (of an organic compound) containing double bonds; *see also* **polyunsaturated**. **3** (of any chemical system, e.g. an enzyme or antibody) not fully saturated with ligand (e.g. substrate or antigen).

unsaturated-fatty-acid oxidation the degradation of unsaturated fatty acids to CO_2 and H_2O; this is achieved largely by enzymes of the **beta-oxidation system**, but enzymes specific for unsaturated fatty acids are also required. Thus, when beta-oxidation reactions have removed carbons to the point at which the resulting acyl-CoA molecule has a Δ^3-*cis* double bond (which is not a substrate for **enoyl-CoA hydratase)**, an enoyl-CoA isomerase (*see* **dodecenoyl-CoA Δ-isomerase**) converts this bond to a Δ^2-*trans* double bond. Similarly, when an acyl-CoA with a Δ^4-*cis* double bond is produced, which again is not a substrate for enoyl-CoA hydratase, **2,4-dienoyl-CoA reductase** followed by enoyl-CoA isomerase yield the substrate with a Δ^2-*trans* double bond.

untranscribed spacer a region of genomic DNA, lying between genes or groups of genes, that is not transcribed. *See* spacer (def. 1).

untranslated region *abbr*.: UTR; any part of an mRNA molecule not coding for a protein; e.g. in eukaryotes the **poly(A) tail**.

unweighted pair-group method with arithmetic averaging *abbr*.: UPGMA; a distance-based, hierarchical clustering method used to create phylogenetic trees under the molecular clock assumption (i.e. assuming constant rates of evolution over different lineages).

unwinding protein *or* **unwindase** any of the proteins concerned in DNA replication that have a higher affinity for single-stranded DNA than for double-stranded DNA and hold the strands apart during replication.

UPA *abbr. for* urokinase-type plasminogen activator; *see* **plasminogen activator**.

UPF *abbr. for* uncharacterized protein family.

UPGMA *abbr. for* unweighted pair-group method with arithmetic averaging.

U5′pp1Gal *symbol for* uridinediphosphogalactose (alternative to Urd*PP*Gal *or* Urd-5′*PP*-Gal).

U5′pp1Glc *symbol for* uridinediphosphoglucose (alternative to Urd*PP*Glc *or* Urd-5′*PP*-Glc).

up-promoter a mutation that increases **promoter** strength in the expression of a gene.

up-regulation 1 the process of reducing or suppressing the response of a cell to a stimulus. **2** a process in which the expression of a gene is increased.

upsilon *symbol*: υ (lower case) *or* Y (upper case); the twentieth letter of the Greek alphabet.

upstream in or towards positions in a DNA molecule lying in the 5′ direction relative to the start of transcription of a gene.

upstream activator sequence a sequence in yeast DNA that is located 5′, and a few hundred base pairs from, the transcriptional initiator site of a protein-coding gene. When bound by one or more specific proteins it stimulates transcription of the gene. In higher eukaryotes the same role is played by **enhancers**. *See* **USA**.

Ur *or* **UR** *abbr. for* uridine (the symbol Urd is recommended).

Ura *symbol for* a residue of the pyrimidine base **uracil**.

URA3 *symbol for* the gene in *Saccharomyces cerevisiae* for orotidine-5′-phosphate decarboxylase. Null mutants are uracil auxotrophs. Vectors encoding *URA3* can complement yeast strains carrying the *ura3* mutation.

uracil *symbol*: U *or* Ura ; 2,4-dioxopyrimidine; 2,4-pyrimidinediol; a pyrimidine base occurring in RNA (but not in DNA).

uracil-DNA N-glycosylase an enzyme (sub-subclass EC 3.2.2) that removes from DNA any uracil resulting from misincorporation of dUTP that has escaped the action of dUTPase (EC 3.6.1.23, dUTP pyrophosphatase). The enzyme hydrolyses the bond at C-1 of the deoxyribose to yield free uracil and DNA containing an apyrimidinic site; another enzyme, apyrimidine endonuclease, recognizes this site and cleaves the phosphodiester bond on the 5′ side of the deoxyribose moiety. Uracil-DNA N-glycosylase may represent the evolutionary basis for the presence of T (not U) in DNA. Cytosine deamination leads to U residues, which would generate transition mutations on replication. The enzyme is a monomer, and is highly conserved in all species. In *Escherichia coli* it is encoded by the *ung* gene, and mutations in this can be used in certain forced site-directed mutagenesis protocols.

urate oxidase EC 1.7.3.3; *systematic name*: urate:oxygen oxidoreductase; *other name*: uricase. An enzyme that catalyses the reaction: urate + O_2 → intermediate(s) that lead(s) to **allantoin**. It is a conserved peroxisomal copper-containing enzyme. Some mammals (e.g. humans, gorillas) that lack the enzyme are prone to **gout**.

Urd *symbol for* a residue of the ribonucleoside **uridine** (alternative to U).

Urd*P* *symbol for* any uridine phosphate.

Urd2′*P* *symbol for* uridine 2′-phosphate.

Urd-2′3′-*P* *symbol for* uridine 2′,3′-phosphate.

Urd3′*P* *symbol for* uridine 3′-phosphate.

Urd-3′5′-*P* *symbol for* uridine 3′,5′-phosphate.

Urd5′*P* *symbol for* uridine 5′-phosphate.

Urd5′*PP* *symbol for* uridine 5′-diphosphate (alternative to ppU).

UrdPPGal or **Urd-5′PP-Gal** *symbol for* uridinediphosphogalactose (alternatives to U5′pp1Gal).

UrdPPGlc or **Urd-5′PP-Glc** *symbol for* uridinediphosphoglucose (alternatives to U5′pp1Glc).

Urd5′PPP *symbol for* uridine 5′-triphosphate (alternative to pppU).

urea the water-soluble compound $H_2N–CO–NH_2$, produced in the liver via the **ornithine–urea cycle**. It is the main nitrogen-containing (urinary) excretion product in **ureotelic** animals. It is used as a source of nonprotein nitrogen in ruminant livestock feeds.

urea cycle *see* **ornithine–urea cycle**.

urease EC 3.5.1.5; *systematic name*: urea amidohydrolase; a nickel-protein enzyme that catalyses the hydrolysis of urea to carbon dioxide and ammonia. A carbamoylated lysine provides an oxygen ligand to each nickel, thus explaining a requirement for CO_2 as an activator of urease apoenzyme. In *Proteus vulgaris* urease is a decamer of three subunits, probably in two $\alpha\beta_2\gamma_2$ complexes.

urea transporter *abbr.*: UT; any facilitated urea transporter in kidney. In rat kidney, **UT1** (929 amino acids) consists of homologous halves, each containing ten transmembrane segments, and intracellular N- and C-termini. It is present in the collecting duct and is involved in vasopressin-regulated urea resorption. **UT2** is present in the proximal renal tubules, and **UT3** is present in the descending vasa recta.

ureido the $NH_2–CO–NH–$ group, derived from urea by loss of a hydrogen atom.

ureidosuccinase EC 3.5.1.7; an enzyme that hydrolyses *N*-carbamoyl-L-aspartate (ureidosuccinate) to L-aspartate, CO_2, and NH_3.

ureogenic describing an organism (or species) that possesses the full complement of **ornithine–urea cycle** enzymes, and that potentially, but not necessarily, can biosynthesize significant amounts of urea.

ureosmotic describing an organism (or species) that adjusts urea production or retention so as to maintain osmotic equilibrium, rather than as a means of disposing of surplus nitrogen.

ureotelic describing an organism (or species) in which **urea** is the principal end product of the degradation of nitrogen-containing compounds, e.g. mammals and marine fishes.

URF *abbr. for* unidentified reading frame.

+uria *comb. form* signifying (in relation to a substance specified) presence unusually, or at a supranormal concentration, in urine. Hence also used to denote a morbid condition so marked, e.g. **phenylketonuria**.

uric acid 2,6,8-trioxypurine; purine-2,6,8-triol; the end product of purine metabolism in certain mammals (including humans) and the main nitrogenous excretory product in **uricotelic** animals. It is formed from xanthine by **xanthine oxidase**. Uric acid is very sparingly soluble in water, and deposits as monosodium urate crystals in the tissues in gout.

uridine

uridine 2′,3′-(cyclic)phosphate *see* **uridine phosphate**.

uridine 3′,5′-(cyclic)phosphate or **uridine 3′,5′-cyclophosphate** *see* **uridine 3′,5′-phosphate**.

uridine 5′-diphosphate *symbol*: Urd5′PP or ppU; *the recommended name for* uridine diphosphate (*abbr.*: UDP); 5′-diphosphouridine; 5′-diphosphouridine; 5′-uridylyl phosphate; uridine 5′-(trihydrogen diphosphate); a universally distributed nucleotide that occurs both in the free state and as a component of a number of important **uridinediphosphosugars**. It is the intermediate in the formation of uridine 5′-triphosphate, UTP, from uridine 5′-phosphate, UMP.

uridine diphosphate galactose *symbol*: Urd-5′-PP-Gal or U5′pp1Gal or UDPGal; uridine 5′-(α-D-galactopyranosyl diphosphate); *recommended name*: uridine(5′)diphospho(1)-α-D-galactose; an important intermediate in the metabolism of D-galactose and in the formation of numerous derivatives. It is formed by the reaction of galactose 1-phosphate with UDP-glucose (catalysed by **UDPglucose–hexose-1-phosphate uridylyltransferase**), or with UTP (catalysed by **UTP–hexose-1-phosphate uridylyltransferase**).

uric acid (lower left structure)

uricase *an alternative name for* **urate oxidase**.

uricosuric (of a drug) tending to promote urinary excretion of uric acid.

uricotelic describing an organism (or species) in which uric acid is the principal end product of the degradation of nitrogen-containing compounds. Uricotelic animals include birds, terrestrial reptiles, and insects. The low solubility of uric acid in water means that it is present as a solid in the excreta of these animals, enabling excretion with little water loss.

uridine *symbol*: U *or* Urd; uracil riboside; 1-β-D-ribofuranosyluracil; a ribonucleoside very widely distributed but occurring almost entirely as phosphoric esters in ribonucleotides and ribonucleic acids. *Compare* **pseudouridine**.

uridine diphosphate glucose *symbol*: Urd-5′-PP-Glc *or* U5′pp1Glc *or* UDPGlc *or* UDPG; uridine 5′-(α-D-glucopyranosyl diphosphate); *recommended name*: uridine(5′)diphospho(1)-α-D-glucose; an important intermediate in the formation of other sugars from glucose. It is formed by the reaction of glucose 1-phosphate with UTP, catalysed by **UTP–glucose-1-phosphate uridylyltransferase**.

uridinediphosphosugar *abbr.*: UDPsugar; *the recommended name for* **uridine diphosphate sugar**; any **nucleosidediphosphosugar** in which the distal phosphoric residue of uridine 5'-diphosphate is in glycosidic linkage with a monosaccharide (or a derivative of a monosaccharide). UDPsugars are important as coenzymes for reactions involving the transfer of monosaccharide residues in the biosynthesis of oligosaccharides and polysaccharides (especially in plants) and in the formation of the oligosaccharide moieties of **glycoconjugates**. The main UDPsugars are: UDP-L-arabinose, UDP-D-galactose (*see* **uridine diphosphate galactose**), UDP-D-galacturonate, UDP-D-glucose (*see* **uridine diphosphate glucose**), UDP-D-glucuronate, and UDP-D-xylose, all of which are (directly or indirectly) interconvertible where they occur together; UDP-L-iduronate, which is formed by epimerization of UDP-D-glucuronate; UDP-N-acetyl-D-galactosamine and UDP-N-acetyl-D-glucosamine, which are interconvertible; and UDP-N-acetylmuramate, which is synthesized from UDP-N-acetyl-D-glucosamine. Certain UDPsugars, particularly UDP-D-glucose and UDP-D-glucuronate, are coenzymes for the formation in animals or plants of glycosides with aglycons such as bilirubin, phenols, plant pigments and sterols, and steroid hormones.

uridine kinase EC 2.7.1.48; *other name*: uridine monophosphokinase; an enzyme that catalyses the phosphorylation by ATP of uridine to uridine 5'-phosphate with formation of ADP.

uridine monophosphate *abbr.*: UMP; *an alternative name for* any uridine phosphate, particularly for uridine 5'-phosphate when its distinction from uridine (5'-)diphosphate and uridine (5'-)triphosphate requires emphasis.

uridine 5'-monophosphate synthase *abbr.*: UMP synthase; a multifunctional enzyme that in higher eukaryotes carries out the following two reactions: (1) orotate phosphoribosyltransferase, EC 2.4.2.10; *other names*: orotidylic acid phosphorylase, orotidine-5'-phosphate pyrophosphorylase; *systematic name*: orotidine-5'-phosphate:pyrophosphate phospho-α-D-ribosyltransferase; a reaction of the pathway for *de novo* biosynthesis of pyrimidine nucleotides. It catalyses a reaction between orotate and 5-phospho-α-D-ribose 1-diphosphate to form orotidine 5'-phosphate and pyrophosphate. (2) Orotidine-5'-phosphate decarboxylase, EC 4.1.1.23; *systematic name*: orotidine-5'-phosphate carboxy-lyase; a decarboxylation reaction that forms uridine 5'-monophosphate and CO_2 from orotidine 5'-phosphate.

uridine phosphate *symbol*: Urd*P*; uridine monophosphate (*abbr.*: UMP); any phosphoric monoester or diester of uridine. There are three monoesters – uridine 2'-phosphate, uridine 3'-phosphate, and **uridine 5'-phosphate** – and two diesters – uridine 2',3'-phosphate and **uridine 3',5'-phosphate** – although uridine 5'-phosphate is the ester commonly denoted (the locant being omitted if no ambiguity may arise). Uridine 2',3'-phosphate (*symbol*: Urd-2',3'-*P*), *also named* uridine 2',3'-(cyclic)phosphate *or* cyclic uridine 2',3'-monophosphate (*abbr.*: 2',3'-cyclic UMP), is formed as an intermediate during the alkaline hydrolysis of ribonucleic acid. This diester then readily hydrolyses to a mixture of the two monoesters uridine 2'-phosphate (*symbol*: Urd2'*P*), *also named* uridine 2'-monophosphate (*abbr.*: 2'UMP) *or* 2'-uridylic acid *or* uridylic acid a, and uridine 3'-phosphate (*symbol*: Urd3'*P*), *also named* uridine 3'-monophosphate (*abbr.*: 3'UMP) *or* 3'-uridylic acid *or* uridylic acid b.

uridine 2'-phosphate *see* **uridine phosphate**.

uridine 2',3'-phosphate *see* **uridine phosphate**.

uridine 3'-phosphate *see* **uridine phosphate**.

uridine 3',5'-phosphate *symbol*: Urd-3',5'-*P*; *the recommended name for* cyclic uridine 3',5'-monophosphate (*abbr.*: 3',5'-cyclic UMP *or* cyclic UMP *or* cUMP); uridine 3',5'-cyclophosphate; uridine 3',5'-(cyclic)phosphate; a monophosphoric diester of uridine found in most mammalian tissues. It is a cyclic nucleotide similar to **adenosine 3',5'-phosphate**, cyclic AMP. It has been reported to inhibit the growth of certain transplantable tumours.

uridine 5'-phosphate *or* **5'-uridylic acid** *or* **5'-phosphouridine** *or* **5'-O-phosphonouridine** *symbol*: Urd5'*P*; *alternative recommended names for* uridine 5'-monophosphate (*abbr.*: 5'UMP); uridine 5'-(dihydrogen phosphate); uridine (mono)ribonucleotide. (The locant is commonly omitted if there is no ambiguity as to the position of phosphorylation.) It is synthesized from aspartate via the decar-

boxylation of orotidine 5'-phosphate (i.e. orotidylic acid). Resynthesis of UMP from free uridine formed by degradation of ribonucleic acid can be effected via a **salvage pathway**. UMP is a key intermediate in the formation of uridine 5'-triphosphate, UTP, and hence in the biosynthesis of other pyrimidine nucleotides.

uridine phosphorylase *see* **pyrimidine-nucleoside phosphorylase**.

uridine 5'-triphosphate *symbol*: Urd5'*PPP or* pppU; *the recommended name for* uridine triphosphate (*abbr.*: UTP); 5'-triphosphouridine; 5'-uridylyl diphosphate; uridine 5'-(tetrahydrogen triphosphate); a nucleotide present in all cells. It is formed from uridine 5'-phosphate via uridine 5'-diphosphate by the successive action of nucleoside-phosphate kinase (EC 2.7.4.4) and nucleoside-diphosphate kinase (EC 2.7.4.6), which transfer in turn the terminal phosphoric residues from two molecules of ATP. UTP is the starting point for the biosynthesis, first of cytidine triphosphate, next of cytidine diphosphate, and then of various deoxyribonucleotides of cytosine, uracil, and thymine. It is important also for the formation of uridinediphosphogalactose and uridinediphosphoglucose through the action of certain **uridylyltransferases**, and thence of a number of other uridinediphosphosugars.

uridyl 1 any chemical group formed by the loss of a 2'-, a 3'-, or a 5'-hydroxyl group from the ribose moiety of uridine. **2** *a misnomer for* **uridylyl**.

uridylate 1 either the monoanion or the dianion of uridylic acid. **2** any mixture of uridylic acid and its anions. **3** any salt or ester of uridylic acid.

uridylic acid *the trivial name for* any phosphoric monoester of uridine. The position of the phosphoric residue on the ribose moiety of a given ester may be specified by a prefixed locant; *see* **uridine phosphate**. However, 5'-uridylic acid is the ester commonly denoted, its locant usually being omitted if no ambiguity may arise. 5'-Uridylic acid is also an alternative recommended name for **uridine 5'-phosphate**.

uridyl transferase *another name for* **UDPglucose–hexose-1-phosphate uridylyltransferase**. This is a misnomer; *compare* **uridyl** *with* **uridylyl**.

uridylyl the uridine[mono]phospho group; the acyl group derived from uridylic acid. *Distinguish from* **uridyl**.

uridylylate to introduce uridylyl groups into a compound (e.g. a protein, a ribonucleic acid, or a sugar phosphate), generally through the action of a uridylyltransferase. —**uridylylated** *adj.*; **uridylylation** *n.*

uridylyl-removing enzyme a putative enzyme present in *Escherichia coli* that detaches uridylyl groups from uridylylated nitrogen regulatory protein P_{II} by hydrolysis, with formation of uridylate. *See also* **nitrogen regulatory protein**, UDPglucose–hexose-1-phosphate **uridylyltransferase**.

uridylyltransferase *an alternative name for* UDPglucose–hexose-1-phosphate uridylyltransferase.

urinalysis the chemical and physical analysis of urine.

urinary of, or pertaining to, urine.

urinary stone protein *see* **osteopontin**.

urine the aqueous excretion produced by the kidneys in higher animals. It serves as a vehicle for the elimination of certain metabolic waste products, e.g. **urea** and **uric acid**, and for the removal of excess ions and water. In mammals and fishes it is fluid, while in birds and reptiles it is semi-solid.

urinometer a **hydrometer** specifically designed and calibrated for measuring the specific gravity of (mammalian) urine.

URL *abbr. for* uniform resource locator; a method for identifying WWW sites that is 'uniform', i.e. independent of any specific protocol. URLs have a convention of the type <type>// of which <type> would typically be http or ftp. Here is a fairly typical URL:

> http://www.oup.co.uk/;

in this case

> http://www.oup.co.uk/index.html

is the **home page**.

urobilin *or* **urobilin IXα** a yellow linear tetrapyrrole. It is formed in kidney and is an oxidation product of **urobilinogen**; the central methylene group linking pyrrole units has been converted to –CH=. Urobilin gives urine its yellow colour. *See also* **stercobilin**.

urobilinogen *or* **mesobilirubinogen** *or* **urobilinogen IXα** a colourless linear tetrapyrrole formed in heme catabolism by reduction of free **bilirubin** released from its glucuronide by intestinal bacterial hydrolases. The four pyrrole rings are linked by three methylene groups. *See also* **stercobilinogen**.

urocanase EC 4.2.1.49; a cytosolic enzyme of liver that converts *trans*-urocanic acid to 4-imidazole-5-propionic acid. Enzyme deficiency results in the rare autosomal recessive anomaly **urocanic aciduria**, which may be associated with mental retardation.

urocanic aciduria *see* **urocanase**.

urocanoate 4-imidazoleacrylate; 3-(1*H*-imidazol-4-yl)propenoate: an intermediate in the degradation of L-histidine by **histidine ammonia-lyase** with the elimination of ammonium.

urocanic acid

urocortin a neuropeptide that has 45% sequence identity with corticotropin-releasing hormone (**CRH**). It binds CRH receptors and is more potent than CRH in suppressing appetite.

urodilatin a 32-amino-acid peptide, isolated from human urine, that is identical in sequence to human **natriuretic peptide** (ANP), but with an N-terminal Thr-Ala-Pro-Arg tetrapeptide extension. It has actions similar to ANP.

uroflavin 1 *or* **uroflavine** *a former name for* **riboflavin**. 2 *an alternative name for* **aquaflavin**.

urogastrone a protein present in human urine that potently inhibits gastric secretion. It is **epidermal growth factor**.

urokinase a proteolytic enzyme found in urine, blood, and kidney cells, that converts **plasminogen** to **plasmin**. *See* **plasminogen activator**.

uromodulin *another name for* **Tamm–Horsfall glycoprotein**.

uronate 1 the anion of uronic acid. 2 any mixture of a uronic acid and its anion. 3 any salt or ester of a uronic acid.

uronic acid any monocarboxylic acid formally derived by oxidizing to a carboxyl group the terminal hydroxymethylene group either of an **aldose** having four or more carbon atoms in the molecule, or of any glycoside derived from such an aldose. The carbon atom of the (potential) aldehydic carbonyl group is numbered C-1.

uropontin *see* **osteopontin**.

uroporphyrin-III *C*-methyltransferase EC 2.1.1.107; *systematic name*: *S*-adenosyl-L-methionine:uroporphyrin-III *C*-methyltransferase (*abbr.*: SUMT); *other names*: urogen-III methylase; uroporphyrin-III methylase. An enzyme that catalyses a reaction between two molecules of *S*-adenosyl-L-methionine and uroporphyrin III to form two molecules of *S*-adenosyl-L-homocysteine and sirohydrochlorin. It forms part of **siroheme synthase**, which is involved in the biosynthesis of siroheme and cobalamin.

uroporphyrinogen decarboxylase EC 4.1.1.37; *systematic name*: uroporphyrinogen-III carboxy-lyase. A cytosolic enzyme that catalyses the decarboxylation of uroporphyrinogen-III to copro-

porphyrinogen with loss of four molecules of CO_2; this converts the four carboxymethyl groups of uroporphyrinogen to methyl groups, and is a step in heme synthesis. The crystal structure of the human enzyme (367 amino acids) is available. Mutations in the gene at 1p34 result in enzyme deficiency and the autosomal dominant diseases porphyria cutanea tarda and hepatoerythropoietic porphyria. *See* **porphyria**.

uroporphyrinogen-III synthase EC 4.2.1.75; *systematic name*: hydroxymethylbilane hydro-lyase (cyclizing); *other names*: uroporphyrinogen-III cosynthetase; uroporphyrinogen-III cosynthase. A cytosolic enzyme that catalyses the conversion of hydroxymethylbilane to uroporphyrinogen-III and H_2O. There are two isoforms of the human enzyme (both contain 265 amino acids), derived from the same gene. Many mutations lead to enzyme deficiency and erythropoietic porphyria.

urothione a thienopterin metabolic degradation product of molybdopterin. It is present in human urine, but is undetectable in molybdenum cofactor deficiency.

URR *abbr. for* **untranslated region**.

ursodeoxycholate *or* **ursodiol** 3α,7β-dihydroxy-5β-cholan-24-oate; an epimer of **chenodeoxycholate** found in the bile of bears (hence the name, from Ursidae), as a conjugate with taurine. It prevents the synthesis and absorption of cholesterol and can lead to the dissolution of gallstones.

ursodiol *an alternative name for* **ursodeoxycholate**.

USA *abbr. for* upstream stimulatory activity; the functional characteristic of a protein that binds an **upstream activator sequence** in a yeast gene and stimulates transcription of the gene by RNA polymerase II.

Usher syndrome an autosomal recessive syndrome of retinitis pigmentosa associated with various hearing disorders, depending on the type of syndrome: **type I** involves vestibular dysfunction and congenital deafness; **type II**, the commonest, involves partial hearing loss; and **type III** is associated with progressive hearing loss. Six genes are associated with subtypes of type I (that for myosin VIIA in type IB, and that for harmonin in type IC); one gene is responsible for type II and encodes an extracellular matrix protein (≈1500 amino acids) of retina and other tissues; and one gene is responsible for type III. [After Charles Howard Usher (1865–1942), British ophthalmologist.]

U-snRNA *abbr. for* uridylate-rich small nuclear RNA; a family of snRNA molecules, often designated U1 to U6 snRNA, that contain 100–215 nucleotides and a 'capped' 5′ end. They usually occur as small nuclear ribonucleoprotein particles. Their function may be related to splicing heterogeneous nuclear RNA (hnRNA) to mature mRNA.

USP *abbr. for* United States Pharmacopeia.

USPHS *abbr. for* United States Public Health Service.

UT *abbr. for* urea transporter.

uterine metalloendopeptidase *see* **matrilysin**.

uteroglobin *or* **blastokinin** a steroid-inducible secreted protein regulator of progesterone concentrations reaching the blastocyst and a potent inhibitor of phospholipase A$_2$. It is a disulfide-linked homodimer.

uterus (*pl.* **uteri**) a muscular part of the reproductive tract of female mammals (except Monotremata) that contains the developing embryo and fetus during pregnancy. —**uterine** *adj*.

UTP *abbr. for* uridine (5′-)triphosphate.

UTP–glucose-1-phosphate uridylyltransferase EC 2.7.7.9; *sys-*

tematic name: UTP:α-D-glucose-1-phosphate uridylyltransferase; *other names*: UDPglucose pyrophosphorylase; glucose-1-phosphate uridylyltransferase. An enzyme that forms uridinediphosphoglucose; it catalyses a reaction between UTP and α-D-glucose 1-phosphate to form UDPglucose and pyrophosphate.

UTP–hexose-1-phosphate uridylyltransferase EC 2.7.7.10; *systematic name*: UTP:α-D-hexose-1-phosphate uridylyltransferase; an enzyme that can form uridinediphosphogalactose. It catalyses a reaction between UTP and α-D-galactose 1-phosphate to form UDPgalactose and pyrophosphate. The sugar substrate α-D-glucose 1-phosphate can also act (but more slowly) as an acceptor, forming uridinediphosphoglucose. Deficiency of the enzyme in infancy causes a mild form of **galactosemia**.

UTP–xylose-1-phosphate uridylyltransferase EC 2.7.7.11; *other name*: xylose-1-phosphate uridylyltransferase; an enzyme that catalyses a reaction between UTP and α-D-xylose 1-phosphate to form UDPxylose and pyrophosphate.

UTR *abbr. for* untranslated region.

utrophin *or* **dystrophin-related protein** (*abbr.*: DRP) a protein that is highly similar to **dystrophin**, and found in normal adult human muscle, primarily at the neuromuscular junction. In the absence of dystrophin, in Duchenne muscular dystrophy, it is also present in the sarcolemma. It is an actin-binding protein related to α-**actinin**. Both gene and protein sequences resemble those for dystrophin. Over-expression of utrophin in a dystrophin-deficient (*mdx*) mouse demonstrates that it can functionally replace dystrophin; it has been suggested that this could offer a therapeutic strategy for Duchenne muscular dystrophy.

UUA a codon in mRNA for L-leucine.

UUC a codon in mRNA for L-phenylalanine.

UUG a codon in mRNA for L-leucine.

UUU a codon in mRNA for L-phenylalanine.

UV *or* **uv** *abbr. for* ultraviolet (radiation).

UV crosslinking a method whereby short exposure (1 min) to ultraviolet (UV) light (254 nm) causes RNA or DNA immobilized on nitrocellulose or nylon membranes to become covalently attached to the membrane.

uvomorulin *or* **epithelial cadherin** *or* **E-cadherin** a type I membrane protein and **cadherin** of non-neural epithelial tissues.

uvr any of four genes in *Escherichia coli* that are induced by the **SOS repair** system and whose products are involved in the repair of damage to DNA caused by UV light and DNA-reactive chemicals. Three of these genes, *uvrA*, *uvrB*, and *uvrC*, encode the three subunits of the ABC excision nuclease; this recognizes the structural distortion created by a **thymine dimer** and cleaves the damaged strand at two sites, about 12 nt apart, leaving a gap that is subsequently filled by polymerase and ligase action. Because the enzyme cuts at two sites it is not a classical endonuclease, but is referred to as an **excinuclease**. The fourth gene, *uvrD*, encodes a DNA helicase with ATPase activity, similar to the *rep* (*see* **RepA**) gene product; it promotes the ATP-dependent unwinding of the DNA duplex during repair.

UWGCG *abbr. for* University of Wisconsin Genetics Computer Group. *See* **GCG**.

Vv

v *symbol for* **1** specific volume. **2** volume (alternative to V). **3** velocity of reaction (i.e. **rate of reaction**); a subscript (in small capital letters) may be added (e.g. v_A) to denote rate of consumption, or rate of formation, of a specified reactant (e.g. A). **4** initial velocity of an enzymic reaction; *see also* $\mathbf{v_0}$. **5** (bold italic) velocity (alternative to u, w, or c).

$\mathbf{v_0}$ *symbol for* initial velocity (i.e. initial rate) of an enzymic reaction. The subscript may be omitted when rates at other times are not in question.

$\mathbf{v_i}$ *symbol for* chemical flux through the (enzymic) step with rate constant k_i. Similarly, flux through steps with rate constants k_{-i} or k_{ij} are denoted by v_{-i} and v_{ij} respectively.

$\mathbf{v_{max}}$ *symbol for* the true maximum value of the **rate of reaction**. *Compare* $\mathbf{V_{max}}$.

V *symbol for* **1** volt. **2** a residue of the α-amino acid L-valine (alternative to Val). **3** vanadium. **4** *the recommended symbol for* **limiting rate**.

V *symbol for* **1** electric potential. **2** potential energy. **3** volume (alternative to v). **4** *the recommended symbol for* **limiting rate**.

V_m *symbol for* molar volume.

V_M *symbol for* hold-up volume.

V_{max} *symbol for* maximum velocity (i.e. **limiting rate**) of an enzyme reaction (alternative to V).

Vac *symbol for* vaccenic acid.

Vac A a toxin secreted by *Helicobacter pylori* that is the major virulence factor in human gastric ulcer formation.

vaccenate 1 *numerical symbol*: 18:1(11); *the trivial name for* either of the isomers (*cis*-vaccenate and *trans*-vaccenate) of octadec-11-enoate, CH_3–$[CH_2]_5$–CH=CH–$[CH_2]_9$–coo^-, the anions derived from *cis*-vaccenic acid, (Z)-octadec-11-enoic acid, and *trans*-vaccenic acid, (E)-octadec-11-enoic acid, a stereoisomeric pair of monounsaturated straight-chain higher fatty acids. *See also* **octadecenoic acid**. **2** any mixture of either *cis*- or *trans*-vaccenic acid and its corresponding anion. **3** any salt or ester of either isomer of vaccenic acid.

vaccenoyl *symbol*: Vac; *the trivial name for* either of the isomers (*cis*-vaccenoyl and *trans*-vaccenoyl) of octadec-11-enoyl, CH_3–$[CH_2]_5$–CH=CH–$[CH_2]_9$–CO–, the acyl groups derived from *cis*-vaccenic acid, (Z)-octadec-11-enoic acid, and *trans*-vaccenic acid, (E)-octadec-11-enoic acid. The *cis*-vaccenoyl group is a major fatty-acyl component of many bacterial lipids and a constituent of the oils of marine organisms. The *trans*-vaccenoyl group is present in a minor proportion of the acylglycerols in the body and milk fats of ruminants.

vaccination the process of administering a **vaccine** to produce immunity.

vaccine any preparation of immunogenic material suitable for the stimulation of active immunity in animals without inducing disease. Vaccines may be based on dead or attenuated microorganisms; altered toxins (toxoids); or viruses.

vaccinia virus a DNA virus and the type species of the genus *Orthopoxvirus* (or 'vaccinia subgroup'), which also includes the viruses responsible for cowpox, buffalopox, catpox, mousepox, and (formerly) smallpox (variola virus). Vaccinia infection usually causes only mild symptoms. Because of its serological similarity to the smallpox virus, vaccinia virus was formerly used as the basis of a live vaccine to protect humans against smallpox. Engineered vaccinias with surface antigens of hepatitis B, influenza virus, and vesicular stomatitis virus (which kills cattle, horses, and pigs) have produced useful vaccines in animal tests. However, adverse reactions with vaccinia-virus vaccines are well documented.

vacuolar apparatus the cellular digestive system, consisting of **lysosomes** and closely related, hydrolase-free, vacuolar structures. It is concerned with the digestion of both endogenous and exogenous materials.

vacuolar ATPase an enzyme of the synaptic vesicle membrane that pumps protons into the vesicle as it loses acetylcholine during nerve transmission; subsequently, the hydrons are released as the vesicle is recharged with acetylcholine. The enzyme (at least 10 subunits) resembles **H+-transporting ATP synthase**, but functions as a proton pump.

vacuolar sorting protein or **vacuolar protein sorting-associated protein** (*abbr.*: vps); any of numerous proteins that are involved in the vesicle-mediated transfer of enzymes to the yeast vacuole. Many share similarity with proteins from higher eukaryotes.

vacuole a closed structure, found only in eukaryotic cells, that is completely surrounded by unit membrane and contains liquid material. Plant cells contain a conspicuous vacuole, which may occupy up to 95% of the cell volume. In these cells it has a role similar to that of lysosomes of animal cells, being rich in hydrolytic enzymes (especially various types of protease). In seed reserve tissues, specialized protein storage vacuoles contain abundant storage protein deposited during seed maturation. Upon germination, a specific protease is imported to supply amino acids to the new seedling. *Compare* **vesicle**.

vacuum (*pl.* **vacuums** or **vacua**) a space containing a gas at low pressure; i.e. a space in which there are relatively few atoms or molecules. A **perfect vacuum** would contain no atoms or molecules, but this is unobtainable owing to the vapour pressure of materials containing any vaccum. Vacuums may be classified into **low** (or **soft**) **vacuum**, $> \approx 10^{-2}$ Pa; **high** (or **hard**) **vacuum**, 10^{-2}–10^{-7} Pa; and **ultrahigh vacuum**, $< 10^{-7}$ Pa.

vacuum evaporation a technique for depositing a thin film of some solid material on a surface by evaporating the material at a high temperature in a vacuum. Atoms leaving the hot material travel directly to the cool surface to form a coating without colliding with other molecules in the gas phase; they condense on the cool surface and build up a layer of solid. *See also* **shadow casting**.

vacuum ultraviolet *see* ultraviolet radiation.

vagusstoff *the German name for* 'active factor'; a factor first demonstrated in 1903 (by German-born US physiologist Otto **Loewi** (1873–1961)) to be secreted into plasma after stimulation of the vago-sympathetic trunk and to inhibit cardiac function. It was later identified as **acetylcholine** by British physiologist (Sir) Henry Hallett Dale (1875–1968).

Val *symbol for* **1** the α-amino acid L-valine (alternative to V). **2** the (*cis*- or *trans*-) vaccenoyl group.

valency or (*esp. US*) **valence 1** the numerical combining power of an atom (and thus of a chemical element), ion, or chemical group, equal to the number of hydrogen atoms that the atom, ion, or group could combine with or replace. *See also* **covalence**, **electrovalence**. **2** the number of antigenic determinants with which one antibody molecule, or fragment thereof, will combine.

valerate 1 *the trivial and preferred name for* pentanoate; the anion, CH_3–$[CH_2]_3$–COO^-, derived from valeric acid (i.e. pentanoic acid). **2** any mixture of free valeric acid and its anion. **3** any salt or ester of valeric acid.

valeryl *symbol*: Vl; *the trivial name for* pentanoyl; the acyl group, CH_3–$[CH_2]_3$–CO–, derived from valeric acid (i.e. pentanoic acid).

valinate 1 valine anion; the anion, $(CH_3)_2CH$–$CH(NH_2)$–COO^-, derived from **valine**. **2** any salt containing valine anion. **3** any ester of valine.

valine *the trivial name for* α-aminoisovaleric acid; α-amino-β-methylbutyric acid; 2-amino-3-methylbutanoic acid; $(CH_3)_2CH$–$CH(NH_2)$–COOH; a chiral α-amino acid. L-valine (*symbol*: V or Val), (S)-2-amino-3-methylbutanoic acid, is a coded amino acid found in peptide linkage in proteins; codon: GUA, GUC, GUG, or GUU. In mammals, it is an essential dietary amino acid, and is glucogenic. Residues of D-valine (*symbol*: D-Val or DVal), (R)-2-amino-3-methylbutanoic acid, occur in a number of peptide antibiotics: e.g., in several members of the actinomycin family; gramicidins A, B, C, and D; and valinomycin (the latter also contains residues of L-valine).

L-valine

valine–isoleucine aminotransferase *see* **valine–3-methyl-2-oxo-valerate aminotransferase**.

valine–3-methyl-2-oxovalerate aminotransferase EC 2.6.1.32; *recommended name*: valine–3-methyl-2-oxovalerate transaminase; *other names*: valine–isoleucine transaminase; valine– isoleucine aminotransferase; an enzyme that catalyses the reaction:

L-valine + (*S*)-3-methyl-2-oxopentanoate =
3-methyl-2-oxobutanoate + L-isoleucine.

See also **transamination**.

valinium valine cation; the cation $(CH_3)_2CH–CH(NH_3^+)–$ COOH, derived from **valine**.

valino the alkylamino group, $(CH_3)_2CH–CH(COOH)–NH–$, derived from **valine**.

valinomycin a cyclic 12-residue depsipeptide antibiotic that consists of three moieties, each having one molecule each of L-valine, D-α-hydroxyvaleric acid, D-valine, and L-lactic acid. The D-valine carboxyl group is directly linked to the α-carbon of L-lactic acid. Valinomycin is produced by *Streptomyces fulvissimus* and is active especially against *Mycobacterium tuberculosis*. It forms an ionophore by folding to produce a hydrophobic surface with an interior that binds Rb^+, Cs^+, NH_4^+, or, especially, K^+. It interferes with oxidative phosphorylation by rendering the mitochondrial membrane permeable to K^+ ions.

valium *see* **diazepam**.

valproic acid 2-propylpentanoic acid; $(CH_3–CH_2–CH_2)_2CH–$ COOH; an antiepileptic, also used as sodium valproate (*one proprietary name*: Epilim).

valyl the acyl group, $(CH_3)_2CH–CH(NH_2)–CO–$, derived from **valine**.

VAMP *abbr. for* vesicle-associated membrane protein (i.e. **synaptobrevin**).

vanadyl ribonucleoside complex an inhibitor of ribonucleases sometimes used in the isolation of RNA from cells and tissues.

vancomycin a complex glycopeptide, $C_{66}H_{75}Cl_2N_9O_{24}$, from *Streptomyces orientalis*. It inhibits a specific step in the synthesis of the peptidoglycan layer in the Gram-positive bacterium *Staphylococcus aureus*. It is synthesized by a multimodular nonribosomal peptide synthetase. *See also* **vancosamine**.

vancosamine the aminodeoxysugar 3-amino-2,3,6-trideoxy-3-*C*-methyl-L-*lyxo*-hexose; the carbohydrate component of **vancomycin**.

van den Bergh reaction a reaction for the determination of **bilirubin** that depends on the formation of a red azo dye by bilirubin and diazotized sulfanilic acid. A **direct reaction** in aqueous solution is given by bilirubin glucuronide, and an **indirect reaction** is given in alcoholic solution by free bilirubin. [After A. A. Hijmans van den Bergh (1869–1943), Dutch physician.]

van der Waals force a long-range force between molecules or submolecular groups, effective from >50 nm to interatomic spacings. Such forces are always present, but weak, and often compete with electrostatic forces. In general they do not follow a simple power law; e.g. the force may be attractive at large separations but repulsive at small ones, or vice versa. The van der Waals free energy between two molecules or submolecular groups usually depends on their mutual orientation, i.e. these forces tend to align two molecules so as to minimize their free energy of interaction. [After Johannes Diderik van der Waals (1837–1923), Dutch physicist.]

van der Waals radius one half of the internuclear distance between atoms at equilibrium when the long-range attractive forces balance the short-range repulsive forces. It is larger than the atom's covalent radius and is about equal to the ionic radius of monovalent ions.

Van Dyke protein *an old name for* **neurophysin**.

vanilloid receptor *see* **capsaicin**.

vanillylmandelate *see* **hydroxy-methoxymandelate**.

Van Slyke apparatus 1 a volumetric apparatus formerly widely used for the determination of oxygen and carbon dioxide contents of blood samples. 2 an apparatus for the gasometric determination of aliphatic amino-nitrogen, e.g. in proteins, by measurement of the nitrogen gas evolved on reaction with nitrous acid. [After Donald Dexter Van Slyke (1883–1971), US biochemist.]

van't Hoff equation *or* **van't Hoff isochore** a relation describing the temperature variation of an equilibrium constant, K, of a reaction in terms of the change in heat content, $\Delta H'$, i.e. in terms of the heat of reaction at constant pressure. It is: $d\ln K/dT = \Delta H/RT^2$, where R is the gas constant and T the thermodynamic temperature. [After Jacobus Henricus van't Hoff (1852–1911), Dutch physical chemist.]

van't Hoff law the principle that the osmotic pressure exerted by a solute is equal to the pressure the solute would exert if it were an ideal gas at the same temperature as the solution and of the same volume. Hence $\Pi V = RT$, where Π is the osmotic pressure, V is the volume of solution containing 1 mole of solute, R the gas constant, and T the thermodynamic temperature.

van't Hoff limiting law a relation giving the limiting value for the osmotic pressure exerted by an ideal solute:

$$\lim (c \to 0) \, \Pi = RT/M_r,$$

where Π is the osmotic pressure, c the solute concentration, M_r the relative molecular mass of the solute, R the gas constant, and T the thermodynamic temperature.

varanic acid *or* **3α,7α,12α,24β-tetrahydroxycholestanic acid** an intermediate (as the CoA ester) in the beta oxidation of the side chain of trihydroxycholestanoyl-CoA to choloyl-CoA in the synthesis of cholic acid. It accumulates in some peroxisomal defects, e.g. in D-bifunctional enzyme deficiency. *See* **peroxisomal bifunctional enzyme**.

variable 1 liable to change; having a range of possible values; liable to deviate from an established type. 2 something that is subject to variation; a quantity or function that can assume any of a set of specified values; a symbol for any unspecified number or quantity.

variable arm *an alternative name for* **extra arm** of tRNA.

variable number tandem repeats *abbr.*: VNTR; a multiallelic DNA polymorphism that results from insertions or deletions of DNA between two **restriction sites**. The existence of these repeats is exploited in **genetic profiling**.

variable region *symbol*: V; a region within a chain of an **immunoglobulin** molecule having an amino-acid sequence that characteristically varies between molecules of the same immunoglobulin class. There are typically four variable regions: the N-terminal halves of the two **light chains** (V_L) and the N-terminal halves of the **Fd fragments** (V_H) of the two heavy chains. The amino-acid sequences of these variable regions determine the structure of the antibody-combining site. *Compare* **constant region**.

variance a measure of the dispersion of a population, the mean square average of the deviations from the **mean**. It is the square of the **standard deviation**, σ, of that population.

variant 1 something that deviates from the norm; used especially of abnormal proteins (e.g., hemoglobins). 2 different; showing variation; used especially of different amino-acid residues in the corresponding positions of polypeptide chains from different sources.

variant Creutzfeldt–Jakob disease *abbr.*: vCJD; *see* **Creutzfeldt–Jakob disease**.

Vasa a protein of *Drosophila* that is required to activate the mRNA for oskar in oocytes.

vascular of, or relating to vessels that conduct fluid from one part of an organism to another, e.g. the blood vessels in animals, or the xylem and phloem in plants.

vascular cell adhesion molecule *see* **VCAM**.

vascular endothelial growth factor *abbr.*: VEGF; *also known as* **vascular permeability growth factor** *or* **vasculotropin**; a protein secreted by epithelial cells, macrophages, and smooth muscle cells, that is active in angiogenesis and endothelial cell growth; it induces endothelial proliferation and vascular permeability. Related to **platelet-derived growth factor**, it is a disulfide-linked homodimer.

vascular endothelial growth factor receptor *abbr.*: VEGF receptor; a type I membrane protein that contains seven immunoglobulin-fold domains in the extracellular segment. It is related to the platelet-derived growth factor receptor, with an intracellular bipartite tyrosine kinase domain.

vascular permeability growth factor *see* **vascular endothelial growth factor**.

vasculature the **vascular** system in animals.

vasculotropin *see* **vascular endothelial growth factor**.

vaso+ *comb. form* denoting a blood vessel.

vasoactive describing any agent that affects the diameter of blood vessels. *See also* **vasoconstrictor**, **vasodilator**.

vasoactive intestinal contractor *see* **endothelin**.

vasoactive intestinal polypeptide *abbr.*: VIP; a basic 28-residue peptide amide, similar to **gastric inhibitory polypeptide**, **glucagon**, and **secretin**, that is widely distributed in the body, mainly in nervous tissue (especially of the gut) and appears to be a neurotransmitter in **peptidergic** autonomic nerves. It has potent hypotensive and vasodilatory effects. It has the sequence

HSDAVFTDNYTRLRKQMAVKKYLNSILN–NH$_2$.

vasoactive intestinal polypeptide receptor *or* **vasoactive intestinal peptide receptor** any G-protein-associated membrane protein that binds **vasoactive intestinal polypeptide** (VIP) and mediates its intracellular effects. That of humans contains 362 residues. It activates adenylate cyclase; other endogenous ligands include peptide HI (PHI), GHRH, and secretin, the order of potency being VIP > PHI > GHRH > secretin.

vasoconstrictor a vasoactive agonist that causes **vasoconstriction**, i.e. a decrease in the diameter of blood vessels.

vasodilator a vasoactive agonist that causes **vasodilation**, i.e. an increase in the diameter of blood vessels.

vasopressin *or* **antidiuretic hormone** (*abbr.*: ADH); either of two nonapeptide hormones secreted by the neurohypophysis and similar to **vasotocin** and **oxytocin**. [8-Arginine]vasopressin (*abbr.*: [Arg^8]vasopressin *or* AVP) has the sequence

Cys-Tyr-Phe-Gln-Asn-Cys-Pro-Arg-Gly-amide,

with a disulfide bridge joining residues at positions 1 and 6. It raises blood pressure by constricting the peripheral arterioles and capillaries. It is also an antidiuretic, accelerating water reabsorption from the proximal region of the distal convoluted tubules of the kidney. AVP is released from a large precursor molecule, **vasopressin–neurophysin 2–copeptin precursor**, which includes its associated carrier protein (**neurophysin**). It is packaged into granules where the hormone is cleaved off and secreted into the blood in free form as required. Deficiency of AVP is the cause of autosomal dominant neurohypophyseal diabetes insipidus (ADNDI). The analogue, [8-lysine]vasopressin (*abbr.*: [Lys^8]-vasopressin), in which lysine is substituted for arginine at position 8, is formed, either singly or together with arginine vasopressin, in the pig family.

vasopressin–neurophysin 2–copeptin precursor the precursor of **vasopressin**, neurophysin 2 (*see* **neurophysin**), and copeptin. Copeptin is a glycopeptide (38 amino acids) that is released during processing of the precursor.

vasopressin receptor any G-protein-associated membrane receptor that binds [8-arginine]vasopressin (AVP; *see* **vasopressin**) and mediates its intracellular effects; peptides related to AVP, including **oxytocin**, are also bound. There are two types: V$_1$ – subdivided into V$_{1a}$ (vascular/hepatic) and V$_{1b}$ (anterior pituitary) – and V$_2$ (kidney); these can be distinguished pharmacologically on the basis of selective agonists. V$_2$ receptors are restricted to the kidney and are encoded at the locus for **nephrogenic diabetes insipidus**. V$_1$ receptors are coupled to phosphoinositide-specific phospholipase C, and V$_2$ receptors to adenylate cyclase. AVP is a more potent agonist for both V$_1$ and V$_2$ receptors than oxytocin. *See also* **oxytocin receptor**.

vasostatin either of two segments of **chromogranin A**. Vasostatin I comprises residues 1–76, lowers vascular tension, reduces secretion of parathyroid hormone, and is bacteriostatic and antifungal. Vasostatin II comprises residues 1–115 and has the first two properties of vasostatin I.

vasotocin a vasoactive peptide, closely related structurally to argi-nine **vasopressin** and **oxytocin**, that is secreted by the posterior pituitary gland in birds, reptiles, and some amphibians. It has the sequence

Cys-Tyr-Ile-Gln-Asn-Cys-Pro-Arg-Gly-amide,

with a disulfide bond between the two cysteines.

VASP *abbr. for* vasodilator-stimulated phosphoprotein; a substrate for cyclic AMP- and cyclic GMP-dependent protein kinases in platelets. It is associated with actin filaments and focal adhesions, and contains an EVH1 homology sequence that binds a proline-rich motif.

VAST *abbr. for* vector alignment search tool; *see* **database-search program**.

vault a large ribonucleoprotein complex of unknown function localized primarily in the cytoplasm of eukaryotic cells. Vaults contain **V-PARP**, MVP (major vault protein, which is also a substrate for V-PARP), and **TEP-1** (a protein of the telomerase complex). They increase in amount in multidrug-resistant cancers.

VCAM *abbr. for* vascular cell adhesion molecule; a class of **adhesion molecules** belonging to the immunoglobulin superfamily. The class is exemplified by VCAM-1 (CD106; molecular mass 90–110 kDa). This is expressed on endothelial cells, macrophages, dendritic cells, fibroblasts, and myoblasts, and is inducible within 6–10 h by interleukin 1β, interleukin 4, tumour necrosis factor α, or interferon γ. It binds to VLA-4 (*see* **VLA**) and integrin α$_4$β$_7$ (*see* **integrin**), and mediates adhesion of lymphocytes, monocytes, and eosinophils to activated endothelium.

vCJD *abbr. for* variant **Creutzfeldt–Jakob disease**.

VCP *abbr. for* valosin-containing protein; the human homologue of yeast Cdc48. *See* **cdc gene**.

V(D)J recombinase *see RAG*.

VecScreen *see* **vector-screening program**.

vector 1 *or* **vector quantity** a quantity specified by its direction and sense as well as by its magnitude. **2** *(in pathology)* an organism that carries an organism of a different species from one place to another, especially one that transfers a **pathogen** between hosts, either mechanically (**mechanical vector**) or by playing a specific role in the pathogen's life cycle (**biological vector**). *Compare* **vehicle** (def. 2). **3** *(in genetics) see* **cloning vector**.

vector database a repository of known vector sequences used by vector-screening programs to identify sequence contaminants; an example is UniVec.

vectorette PCR a method based on the **polymerase chain reaction** (PCR) used for amplifying an unknown DNA sequence adjacent to a known DNA sequence. It was previously used in **chromosome walking** but still finds application in recovering sequences adjacent to mutations caused by insertional mutagenesis. The double-stranded vectorette is ligated to the ends of restriction fragments thereby introducing a known sequence that can be used to prime one side of a PCR reaction, the other being primed on the known genomic sequence.

vectorial enzyme any enzyme, such as one fixed in a membrane, that is directional in its action, e.g. **sodium/potassium ATPase**.

vectorial processing processing of a fully formed precursor polypeptide into a mature protein when used to drive the movement of the protein from one aqueous compartment (e.g. the cytoplasm) through a membrane into a second aqueous compartment (e.g. an intramitochondrial or intrachloroplastal space). *Compare* **vectorial translation**.

vectorial translation the stage of **translation** in the biosynthesis of any protein destined for export from a cell, where a nascent polypeptide being produced on a **polysome** bound to the membrane of the rough endoplasmic reticulum in a eukaryote is conducted through the membrane by its N-terminal **signal peptide**. *Compare* **vectorial processing**.

vector-screening program a program that uses a database of known vector sequences to screen sequences for vector contaminants; an example is VecScreen.

vegetal hemisphere glycoprotein 1 *abbr.*: Vg-1; a glycoprotein of the TGF-β family (*see* **transforming growth factor**) that facilitates the differentiation of either mesoderm or endoderm.

vegetal pole the end at which most of the yolk is located e.g. in an amphibian egg; the end opposite the **animal pole**.

vegetative describing cells or tissues that are engaged in nutrition and growth rather than sexual reproduction, and excluding dormant forms.

vegetative bacteriophage the intrabacterial genome of a **bacteriophage** when it is noninfective but controlling the synthesis by the host of components needed to make new infective bacteriophage particles.

vegetative reproduction 1 (in plants) a form of **asexual reproduction** in which new individuals develop from specialized multicellular organs (e.g. bulbs, corms, tubers) that break off from the parent plant. **2** (sometimes) a term used to describe **budding** (def. 2) and similar forms of asexual reproduction in animals such as coelenterates and sponges.

VEGF abbr. for vascular endothelial growth factor.

vehicle 1 any inert medium with which a biologically active substance (e.g. a drug) is mixed or in which it is dissolved, suspended, or emulsified in order to increase its bulk and/or to facilitate its absorption after administration. **2** any agent (animate or inanimate) of transmission, especially of infection; see also **vector** (def. 2).

vein 1 (in animals) the blood vessels conducting blood from the tissues. **2** (in plants) a vascular bundle in a leaf.

velocity of reaction an alternative name for **rate of reaction**.

venom any toxic animal secretion that is produced by discrete glands and is delivered by a specific mechanism for defensive or offensive purposes.

venombin either of two serine proteinases found in snake venom. **Venombin AB** EC 3.4.21.55; other name: gabonase (from the venom of the gaboon viper, Bitis gabonica); it cleaves Arg-|-Xaa bonds in fibrinogen to form fibrin and release fibrinopeptides A and B; it is not inhibited by antithrombin III or hirudin. **Venombin A** EC 3.4.21.74; it cleaves Arg-|-Xaa bonds in fibrinogen to form fibrin and release fibrinopeptide A, but further reaction depends on species; it is a trypsin-like enzyme.

venous of, or relating to, a **vein**, or to the blood in the veins.

ventilation the passage of air into and out of the respiratory tract, or other confined space.

Ventolin see **sambutamol**.

Vent polymerase or **Deep Vent polymerase** the proprietary name for thermostable DNA polymerases used as alternatives to *Taq* **DNA polymerase** in the **polymerase chain reaction**. They have a higher fidelity than the *Taq* DNA polymerase, derived in part from their exonuclease activity. They derive from the archaean *Thermococcus litoralis*, isolated from a submarine thermal vent, hence the enzyme's name.

ventral situated towards the belly surface of an animal.

verapamil α-3-[2-(3,4-dimethoxyphenyl)ethyl]methylamino propyl-3,4-dimethoxy-α-(1-methylethyl)benzeneacetonitrile; iproveratril; D365; a Ca²⁺-blocking agent used in the treatment of cardiac disorders and to inhibit the movement of calcium ions across biological membranes. It acts like **nifedipine, which** binds to the dihydropyridine binding site of vascular voltage-operated **calcium channels**; verapamil probably acts at another related site. See also **D600**.

veratridine a toxic alkaloid, $C_{36}H_{51}NO_{11}$, found in the seeds of *Schoenocaulon officinalis* (of the lily family) and in the rhizome of *Veratrum album*. It blocks the development of the action potential involved in axonal conduction, binding to the Na⁺ channels and blocking them in the open configuration.

veratidine

veratryl alcohol 3,4-dimethoxyphenol, a compound derived from the breakdown of lignin. Is a substrate for lignin peroxidase, becoming oxidized to a cation radical that attacks the lignin structure.

v-erbA related protein see **COUP**.

verbascose the tetrasaccharide, O-α-D-galactopyranosyl-(1→6)-O-α-D-galactopyranosyl-(1→6)-O-α-D-glucopyranosyl-(1→2)-β-D-fructofuranoside. It occurs in roots of the mullein, *Verbascum thapsus*.

verdoperoxidase see **myeloperoxidase**.

verification 1 establishment of the correctness of a theory, fact, etc. **2** a term used to describe the deacylation of misacylated tRNA molecules by the specific aminoacyl-tRNA synthase.

Verner–Morrison syndrome or **pancreatic cholera** a syndrome of humans characterized by severe watery diarrhoea, achlorhydria or hypochlorhydria, weight loss, and dehydration. It is caused by tumours (**VIPomas**), mostly of pancreatic origin (some are neurogenic), that produce **vasoactive intestinal polypeptide** (VIP), the circulating levels of which are raised. [After John Victor Verner (1927–), US physician and Ashton Byrom Morrison (1922–), US pathologist.]

Veronal a proprietary name for **barbital**.

verotoxin a pentameric pore-forming toxin (5 × 89 amino acid residues) belonging to the shiga toxin B family (TC 1.C.54).

Versene a proprietary name for the tetrasodium salt of **edetate**.

versican or **large fibroblast proteoglycan** or **chondroitin sulfate proteoglycan core protein 2** a protein involved in intercellular signalling. The N-terminal part is similar to a glial hyaluronate-binding protein; the middle contains glucosaminoglucan attachment sites; the C-terminal part contains EGF-like repeats (see **epidermal growth factor**) and a lectin-like domain.

vertical rotor a rotor for use in an **ultracentrifuge**. The wells holding the tubes are drilled parallel to the axis of rotation and at right angles to the lines of the centrifugal force.

very-long-chain acyl-CoA dehydrogenase an acyl-CoA dehydrogenase that is specific for acyl moieties 14–22 carbons long. The human enzyme contains 615 residues and FAD, and occurs as a homodimer associated with the matrix side of the inner mitochondrial membrane. Many mutations result in deficiency and a clinical presentation that resembles **Reye syndrome**.

very-low-density lipoprotein abbr.: VLDL; a class of lipoproteins found in blood plasma in many animals (data normally relate to humans). VLDL is also known as pre-beta lipoprotein because its electrophoretic mobility is slightly greater than that of low-density, or beta-lipoprotein. VLDL particles have diameter 25–70 nm, solvent density for isolation (g mL^{-1}) <1.006, hydrated density (g mL^{-1}) 0.97. Their approximate composition (% by weight) is 5–8% unesterified cholesterol, 11–14% esterified cholesterol, 20–23% phospholipid, 44–60% triacylglycerol, 4–11% protein. Their apolipoprotein composition (% by weight total apolipoprotein) is 36.9% B-100, 49.9% C-I + C-II + C-III, 13% E-II + E-III + E-IV. They are synthesized in liver and are the precursor of low-density lipoprotein.

very-low-density-lipoprotein receptor *abbr*.: VLDL receptor; a membrane protein that binds VLDL, accumulates in clathrin-coated pits, and internalizes VLDL. It has a large extracellular domain, a single transmembrane domain, and a small intracellular domain. In human the precursor protein consists of 873 amino acids; residues 1–27 are the signal, 28–797 constitute the extracellular domain, and 820–873 are the cytoplasmic domain.

vesicle a closed structure, found only in eukaryotic cells, that is completely surrounded by a unit membrane but, unlike a **vacuole**, contains material that is not (or is not known to be) in the liquid state.

Vg-1 *abbr. for* vegetal hemisphere glycoprotein 1.

VHDL *abbr. for* very-high-density lipoprotein.

VHL *abbr. for* von Hippel–Lindau protein; a 213-residue protein encoded at 3p25 by the gene for the **von Hippel–Lindau syndrome** of hereditary renal and other cancers. VHL binds **cullin**-2 and then **elongins** B and C, and the complex translocates to the nucleus. By sequestering elongins B and C it prevents formation of the heterotrimeric transcription factor elongin (which consists of elongins A, B, and C). But its proper function may be to target other nuclear proteins to ubiquitin-mediated degradation. Numerous mutations in the VHL gene are associated with sporadic and hereditary renal carcinoma, indicating that VHL is a tumour suppressor. [After Eugen von Hippel (1867–1939), German ophthalmologist, and Arvid Lindau (1892–1958), Swedish pathologist.]

viability a measure of the capacity of a cell for metabolism and division. It is often estimated experimentally by determining the number of cells that exclude normally membrane-impermeant dyes, e.g. **Trypan Blue**.

Viagra *a proprietary name for* sildenafil; 5-[2-ethoxy-5-(4-methylpiperazin-1-yl)phenyl]-1-methyl-3-propyl-1,6-dihydro-7*H*-pyrazolo[4,3-d]pyrimidin-7-one; used for the oral treatment of male erectile dysfunction. It selectively inhibits a specific phosphodiesterase that hydrolyses cyclic GMP; the latter mediates the action of nitric oxide in normal erectile function, an action which is thus potentiated.

VIC *abbr. for* vasoactive intestinal contractor; *see* **endothelin**.

vic- *prefix (in chemical nomenclature)* denoting the presence of **vicinal** (def. 2) substituents.

vicianose the disaccharide 6-*O*-α-L-arabinosyl-D-glucose, obtained from a glycoside in seeds of *Vicia angustifolia*.

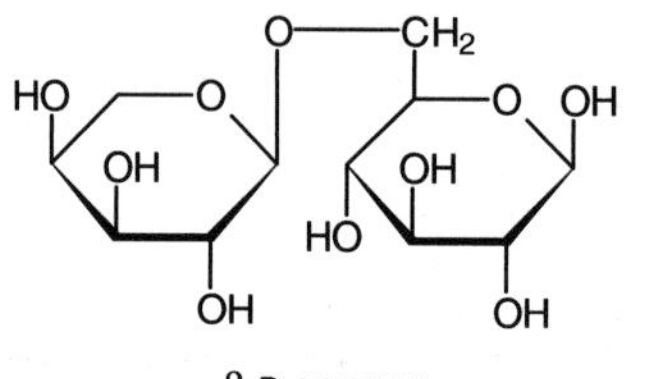

β-D-anomer

vicillinpeptidohydrolase *see* **legumain**.

vicinal 1 adjacent, neighbouring. **2** *(in organic chemistry)* describing two (usually identical) atoms or groups attached one to each of two linked carbon atoms in a molecule. The presence of vicinal substituents is denoted by the prefix *vic-* attached to the name of the compound. *Compare* **geminal**.

vicilin a major storage protein in seeds of *Pisum sativum* (pea); it comprises 124 amino acids, M_r 170 000. Legumin-like and vicilin-like globulins are the main storage proteins of most angiosperms and gymnosperms.

vidarabine adenine arabinoside; 9-*β*-D-arabinofuranosyladenine; a

purine nucleoside, a metabolite of hypoxanthine arabinoside, produced by *Streptomyces antibioticus*. Its triphosphate derivative inhibits viral DNA polymerase and may be incorporated into viral DNA; it also inhibits DNA synthesis.

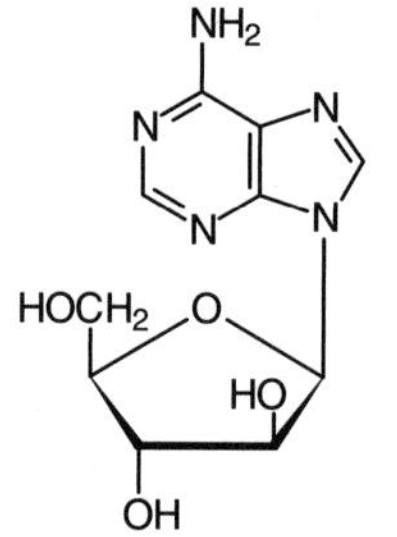

Vif *abbr. for* viral infectivity factor; a protein of HIV-1 that is essential for production of infectious viruses by T lymphocytes and macrophages.

vigabatrin γ-vinyl-γ-aminobutyric acid; γ-vinyl GABA; 4-amino-5-hexenoic acid; an irreversible inhibitor of the enzyme GABA transaminase (**4-aminobutyrate transaminase**), which metabolizes the neurotransmitter γ-amino-*n*-butyric acid.

vigilin a protein derived from chickens that contains 14 **KH domains**. There is a human homologue.

villin a Ca^{2+}-regulated actin-binding protein occurring in the microvilli of intestinal epithelial cells and kidney proximal tubule cells. Villin consists of a large core fragment, the N-terminal portion, and a small head-piece, the C-terminal portion; the head-piece binds F-actin (*see* **actin**) strongly in both the presence and absence of Ca^{2+}; it is similar to **gelsolin**.

vimentin a protein found in class III **intermediate filaments** in mesenchymal and other non-epithelial cells, and in the Z disk of skeletal and cardiac muscle cells. It is a phosphoprotein, phosphorylation being enhanced during cell division.

vinblastine an alkaloid, C$_{46}$H$_{58}$N$_4$O$_9$, obtained from *Vinca rosea*, that binds to spindle microtubules and arrests metaphase of mitosis. It is used as an antineoplastic agent.

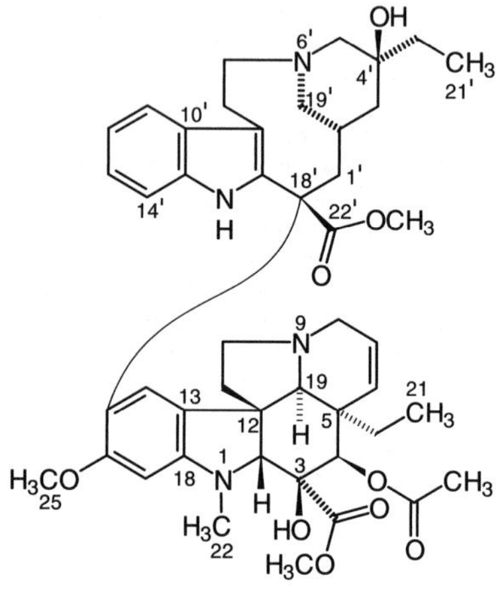

vinblastine

vincristine an alkaloid, C$_{46}$H$_{56}$N$_4$O$_{10}$, obtained from *Vinca rosea* and structurally related to **vinblastine**, that binds to spindle microtubules and arrests metaphase of mitosis. It is used as an antineoplastic agent.

vincristine

vinculin a cytoskeletal protein associated with the cytoplasmic face of the focal adhesion plaques that anchor actin microfilaments to the plasma membrane and that attach a cell to the substratum. It associates with **talin** in binding **integrins** to the cytoskeleton. Vinculin is phosphorylated (on serines, threonines, and tyrosines) and acylated by myristic acid and/or palmitic acid. Vinculin and metavinculin are produced by alternative splicing of the same gene. Metavinculin differs from vinculin in having a 68-residue additional domain near the C terminus.

vindoline the major alkaloid from *Vinca rosea*; $C_{25}H_{32}N_2O_6$. It is without physiological activity but constitutes a pentacyclic moiety of **vinblastine** and **vincristine**.

vinyl ether alk-1-enyl ether; a class of compounds characterized by a bond formed between a long-chain aldehyde and one (in the case of monoalk-1-enyl ethers) or two (for dialk-1-enyl ethers) of the hydroxy groups of the glycerol moiety of a glycerolipid; only the *sn*-1 isomers occur naturally, and chains of 16 or 18 carbons are usual. Such glycerophospholipids are known as **plasmalogens**; ethanolamine glycerophospholipids characteristically contain larger amounts of plasmalogens than other classes, and heart tissue is rich in plasmalogens.

γ-vinyl GABA see **vigabatrin**.

viosamine 4-amino-4,6-dideoxy-D-glucose; a component of the lipopolysaccharide of some strains of *Escherichia coli*.

VIP *abbr. for* vasoactive intestinal polypeptide.

VIP36 *abbr. for* vesicular integral membrane protein 36 kDa; a mannose-specific lectin protein that resides in the Golgi apparatus.

VIPoma a tumour producing **vasoactive intestinal polypeptide** (VIP). It is often associated with a watery diarrhoea and other pathological signs known as **Verner–Morrison syndrome**.

viral infectivity factor see **Vif**.

viral particle see **virion**.

viral RNA RNA that constitutes the genome of certain viruses – the RNA viruses. Examples are poliovirus, which has single-stranded RNA, and cytoplasmic polyhedrosis virus of the tussock moth, which has 10 double-stranded RNA molecules. All **retroviruses** contain RNA as the genome.

virino a former name for a putative type of particulate, noncellular, submicroscopic, transmissible, pathogenic agent, typified by the agent responsible for **scrapie**. The infective particle was believed to consist of a small nucleic-acid molecule covered with host-specified protein. See **prion**.

virion *or* **viral particle** a complete virus particle, found extracellularly and capable of surviving in crystalline form and of infecting a living cell. It comprises the nucleic-acid **core** and the protein **capsid**; the latter may be enclosed by an **envelope** in some viral families.

virogene see **oncogene hypothesis**.

viroid any of about 30 autonomously replicating plant pathogens that consist of single-stranded (ss) circular RNA (about 350 nt) containing much self-complementarity. They have no genes, so do not express protein. It is assumed that the viroid RNA acts as its own template, subverting host cell enzymes to carry out the replication. The product consists of concatamers of daughter strands that undergo self-cleavage to form new viroid RNA genomes. It is likely that hammerhead structures (see **ribozyme**) are involved. An example is potato spindle tuber viroid (PSTV). This is similar to human hepatitis delta, which is part viroid and part virus in that it replicates like a viroid and part of its genome is homologous with PSTV.

virology the scientific study of viruses and the diseases caused by them. —**virological** *adj.*; **virologist** *n.*

viropexis the process by which virus particles are adsorbed onto the surface of a cell and then taken into the cytoplasm.

virtual organization see **VO**.

virtual screening *an alternative name for* **combinatorial chemistry**.

virulence the capacity of a microorganism or virus to cause disease in a host under certain conditions. —**virulent** *adj.*

virulent phage a bacteriophage that invariably lyses the bacterium it infects. Virulent phages have no alternative but to follow the **lytic cyle**.

virus a noncellular infective agent that reproduces only in an appropriate host cell. Viruses are typically smaller than bacteria, and can infect animal, plant, or bacterial cells (see **bacteriophage**); many are important agents of disease. The infective particle (**virion**) consists of a core of nucleic acid (either DNA or RNA) surrounded by a protein **capsid** and, in some cases, an outer envelope. A virus can interact with its host cell in several different ways. In a lytic interaction the cell's replicative and protein biosynthetic apparatus is directed by the viral genes to produce numerous progeny virions, which are released upon disintegration of the host cell (see **lytic cycle**). However, in a persistent infection, low levels of virus production may be sustained without destruction of the host cell. In a **latent infection**, the virus resides in a host cell without replicating, although it may be provoked to enter the lytic phase by some stimulus. Oncogenic viruses, which include both DNA viruses and RNA viruses (**retroviruses**) carry **oncogenes** and are potentially capable of transforming their host cells into tumour cells. Animal viruses are classified according to the nature of their genome and the mechanism of synthesis of messenger RNA (see **Baltimore classification**).

virusoids see **satellite RNA**.

visceral layer see **serous membrane**.

viscometer *or* **viscosimeter** an apparatus for measuring the viscosity or resistance to flow of a liquid.

viscosity *symbol*: η; the resistance to flow of a fluid due to the sum of the effects of adhesion and cohesion. Consider a liquid between two parallel plates, one of which is moving in the *x* direction with a velocity *v*. The liquid is thought of as a number of layers, each of which slides along the adjacent layer; the frictional resistance between adjacent layers generates a velocity gradient in the *y* direction. The deformation of the liquid produced by the velocity gradient is known as **shear** (def. 1). The frictional force, *f*, between the liquid layers is proportional to the area, *A*, of the layers and to the velocity gradient between them, such that $f = \eta\, A(\mathrm{d}v/\mathrm{d}y)$, where η is the **coefficient of viscosity** or simply the viscosity; $f/A = F$, the **shear stress**; and $\mathrm{d}v/\mathrm{d}y = G$, the **shear gradient** or **shear rate**. If η is independent of *F* or *G*, the fluid is a **Newtonian fluid**; if η is a function of *F* or *G*, the fluid is a **non-Newtonian fluid**.

viscotoxin a pore-forming toxin, permitting passage of small molecules across biomembranes. It is produced by various plants. It contains 45–47 amino acid residues organized in two antiparallel α-helices and two antiparallel β-strands. It is classified in the TC system under number 1.C.44.

visinin a calcium-binding protein specific to cone photoreceptors; a chicken isoform homologous to bovine **recoverin** has three Ca^{2+}-binding sites.

Visual Basic see **programming language**.

visual cascade the sequence of reactions occurring after the absorption of a photon by visual pigment (e.g. **rhodopsin**). It eventually leads to a rapid fall in cyclic GMP levels, which closes Na$^+$ channels in the membrane of the visual receptor and activates the nerve impulse in the optic nerve. The activated pigment stimulates the G-protein **transducin,** which initiates the signal-transduction cascade.

visualization software 1 any software that facilitates analysis and interpretation of data using simple visual metaphors (e.g. graphs, trees, 3D images). **2** *(in structure analysis)* software that allows visualization of 3D representations of molecular structures from multiple perspectives, using multiple modes (e.g. backbone trace, space-filled, surface) and multiple colour schemes; examples include MAGE, RasMol, and Cn3D. **3** *(in sequence analysis)* software that allows visualization of sequence properties (such as hydropathy profiles) or visualization and editing of sequence alignments: e.g. CINEMA (colour interactive editor for multiple alignments) and JalView.

visual pigment any of the homologous photosensitive retinal-containing transmembrane proteins of the retina that are responsible for vision. The rods contain **rhodopsin,** while the cones contain **cyanopsin, chloropsin,** or **porphyropsin.**

visual purple *see* **rhodopsin.**

vitalism the doctrine that the phenomena of biology are due to a vital principle distinct from physicochemical forces, and cannot be explained by the laws of physics and chemistry alone. The accidental discovery, in 1890, by the brothers Büchner that a cell-free extract of yeast could catalyse alcoholic fermentation did much to convince adherents that such a doctrine was untenable. Also important was Wöhler's synthesis of urea from ammonium cyanate in 1828.

vital staining a technique in which living cells or tissues are exposed to a stain that does not kill them. Dyes such as Trypan Blue are excluded from live cells but stain dead cells thereby allowing the proportion of each in a population to be counted using a hemocytometer. Alternatively, the tetrazolium dye MTT stains live cells blue without killing them.

vitamer any of two or more chemically and metabolically related compounds that display qualitatively the same biological activity as a vitamin. For example, retinol, retinal, retinoic acid, 3,4-didehydroretinol, and various carotenes are all vitamers of vitamin A.

vitamin an organic substance that is present in foodstuffs (or in intestinal bacteria) in relatively minute amounts, is distinct from the main organic components of food (i.e. protein, carbohydrate, and fat), and is needed for the normal nutrition of the organism (or species) in question. Deficiency or absence of any vitamin causes a specific deficiency disease. **Vitamins A, D, E,** and **K** are the fat-soluble vitamins; **vitamins B$_1$, B$_2$, B$_6$, B$_{12}$,** and **C,** plus **biotin, choline, folic acid, nicotinic acid,** and panthothenic acid (*see* **pantothenate**), are the water-soluble vitamins.

vitamin A 1 *a common name for* **retinol. 2** *a generic name for* any retinoid exhibiting qualitatively the biological activity of retinol. Included, besides retinol (relative activity 100) are: **retinal** (91); **retinoic acid** (≈65); **3,4-didehydroretinol** (40); **3,4-didehydroretinal**; and **3,4-didehydroretinoic acid** (the activities relative to retinol being measured by the standardized growth test on vitamin A-deficient rats). Vitamin A is fat-soluble, and is present in various lipid-containing animal preparations, especially liver and dairy products. Vitamin A activity is displayed also by various carotenes; these are provitamins, being converted to retinal in the small intestine in most vertebrates. The vitamin A of commerce was originally an extract from fish-liver oils, which contain large amounts of retinol, principally as its palmitate; it is now largely derived chemically. Proper vitamin A nutrition is necessary for: (1) somatic function, including growth and differentiation, e.g. of epithelial structures and bone; (2) reproduction, including spermatogenesis, oogenesis, placental development, and embryonic growth; and (3) visual processes (*see* **retinal, 3,4-didehydroretinal**). Vitamin A deficiency causes stunted growth, thickening of various membranes by keratinization, and night blindness, xerophthalmia, keratomalacia, and eventually complete blindness. In moderate excess, vitamin A is highly toxic.

vitamin A$_1$ *an alternative name (not recommended) for* **retinol.**

vitamin A$_2$ *an alternative name (not recommended) for* **3,4-didehydroretinol.**

vitamin A acid *or* **vitamin A$_1$ acid** *an alternative name (not recommended) for* **retinoic acid.**

vitamin A$_2$ acid *an alternative name (not recommended) for* **3,4-didehydroretinoic acid.**

vitamin A alcohol *or* **vitamin A$_1$ alcohol** *an alternative name (not recommended) for* **retinol.**

vitamin A$_2$ alcohol *an alternative name (not recommended) for* **3,4-didehydroretinol.**

vitamin A aldehyde *or* **vitamin A$_1$ aldehyde** *an alternative name for (not recommended)* **retinal.**

vitamin A$_2$ aldehyde *an alternative name (not recommended) for* **3,4-didehydroretinal.**

vitamin B$_1$ *or* **thiamine** *or (sometimes)* **thiamin** *or (formerly)* **aneurin(e)** 3-(4-amino-2-methylpyrimidin-5-ylmethyl)-5-(2-hydroxyethyl)-4-methylthiazolium chloride; a water-soluble vitamin, present in fresh vegetables and meats, especially liver. Deficiency in humans causes beriberi, manifest as cardiopathy and neuropathy; in pigeons, a characteristic head-drop is an overt sign of deficiency. **Thiamine diphosphate** (*abbr.*: TPP), formerly known as **cocarboxylase,** is a coenzyme for oxidative decarboxylation and other enzymic reactions, including ones catalysed by pyruvate and oxo-glutarate dehydrogenases, pyruvate decarboxylase, and transketolase.

vitamin B$_2$ *or* **riboflavin** 6,7-dimethyl-9-ribitylisoalloxazine; *formerly also called* lactoflavin, hepatoflavin, ovoflavin, uroflavin, vitamin G, vitamin H. A water-soluble vitamin, present in most foods but especially in milk and meat products. It is a component of the **flavin nucleotide** coenzymes (e.g. FAD, FMN). Isolated from milk and eggs by Kuhn and co-workers in 1933 and chemically synthesized by them and others in 1935. Significant deficiency seldom occurs, but causes changes of the tongue and lips and sores in the corners of the mouth.

vitamin B₅ *(rarely) an alternative name for* pantothenic acid; *see* **pantothenate**.

vitamin B₆ *or* **pyridoxine** 2-methyl-3-hydroxy-4,5-bis(hydroxymethyl)pyridine; a water-soluble vitamin; **pyridoxal** and pyridoxamine are vitamers. It is present in many foods, including meats (largely as pyridoxal or pyridoxamine), and plant sources (largely as pyridoxine) such as cereals, lentils, nuts, and vegetables. Pyridoxal phosphate is a coenzyme for enzyme reactions including those for transamination, amino-acid decarboxylation, racemization, and modifications to amino-acid side chains. Deficiency is a minor problem and signs are difficult to produce.

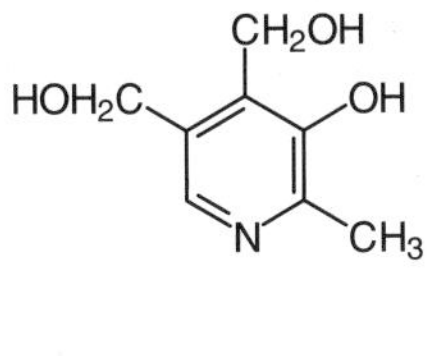

vitamin B₆

vitamin B₁₂ *or* **cobalamin** *or* **cobamide** a water-soluble vitamin with several related forms characterized by possession of a **corrin** nucleus containing a cobalt atom. Four of the cobalt coordinate bonds are liganded by nitrogens of the four pyrrole rings found in the corrin nucleus; the fifth position is occupied by the ribonucleotide of 5,6-dimethylbenzimidazole, which is also linked via a 1-amino-2-propanol moiety to one of the pyrrole rings; the sixth position is variously occupied by a cyanide group (cyanocobalamin, the form in which the molecule was first isolated, the cyanide group having been introduced during extraction), or a methyl group (**methylcobalamin**, which is a coenzyme), or a 5′-deoxyadenosyl group (**deoxyadenosylcobalamin**, the other coenzyme form). Vitamin B₁₂ is present only in foods of animal origin, especially liver, fish, and

cyanocobalamin

eggs. The coenzyme forms are needed for: (1) reactions that require the deoxyadenosyl form, exemplified by **methylmalonyl-CoA mutase**; and (2) reactions that require the methyl form, exemplified by 5-methyltetrahydrofolate-homocysteine *S*-methyltransferase (EC 2.1.1.13) (*see* **5-methyltetrahydrofolate**). In normal individuals, deficiency rarely occurs, but is precipitated by lack of **intrinsic factor**, a protein secreted by gastric mucosa and required for efficient absorption of the vitamin. Lack of intrinsic factor leads to **pernicious anemia**, investigation of which first led to the discovery of cobalamin.

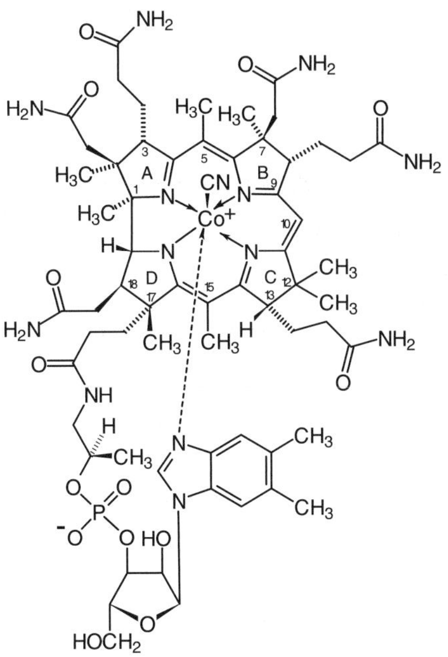

vitamin B₁₂ᵣ reductase *see* **cob(II)alamin reductase**.

vitamin C ascorbic acid; a water-soluble vitamin, and the first vitamin to be recognized. Awareness of the ability of fresh vegetables and fruit to prevent **scurvy** is attributed to the British physician James Lind (1716–94), who published *A Treatise on Scurvy* in 1757. The active compound was first isolated in 1928 by Albert Szent-Györgyi (1893–1986; Hungarian-born US biochemist) as 'hexuronic acid', a strong reducing substance from rat adrenals and citrus fruits. This was shown four years later to be identical with the anti-scorbutic substance isolated by Zilva and associates. Vitamin C is widely distributed in vegetables and fruit, but easily destroyed by cooking. *See* **ascorbic acid** for structure. *See also* **l-gulonolactone oxidase**.

vitamin D any steroid that exhibits the biological activity of calciol (= vitamin D₃ – *see below*); the antirachitic vitamin. Such compounds are fat-soluble and structurally related to either the parent hydrocarbon, **cholestane**, or its 24-methyl derivative, ergostane, and all are 9,10-seco-sterols, i.e. 3-hydroxysteroids in which bond breakage has occurred between C-9 and C-10 in ring B of the steroid nucleus. Variants of vitamin D, with their recommended trivial names in parentheses, are: **vitamin D₂** *or* calciferol (**ercalciol** *or* **ergocalciferol**), formed in plants by the action of UV light on the endogenous provitamin **ergosterol** and produced industrially by UV-irradiation of ergosterol isolated from yeast; **vitamin D₃** (**calciol** *or* **cholecaliferol**), occurring naturally in particularly high concentrations in the liver oils of certain fish, especially cod, halibut, or tunny, and produced nonenzymically in the skin of animals by the action of a UV component (230–313 nm) of sunlight on the provitamin 7-dehydrocholesterol; **vitamin D₄** (22,23-dihydroercalciol *or* (24*S*)-methylcalciol or (*formerly*) 22,23-dihydroergocalciferol), produced by irradiation of 22,23-dihydroergosterol; **vitamin D₅** ((24*S*)-ethylcalciol); and **vitamin D₆** ((22*E*)-(24*R*)-ethyl-22,23-didehydrocalciol). Of these variants, calciol (D₃) and ercalciol (D₂) are the most significant. Vitamin D activity is also displayed by several synthetic derivatives or isomers of calciol (or of ercalciol) and by the 25-hydroxylated and 1α,25-dihydroxylated metabolites **calcidiol** and **calcitriol** (or **ercalcidiol** and **ercalcitriol**). Calcidiol (or ercalcidiol) is formed in liver from calciol (or ercalciol) and converted in kidney to calcitriol (or ercalcitriol), the physiologically active form of vitamin D₃ (or D₂). Vitamin D mediates the absorption of calcium and phosphate ions from the intestine and promotes the mineralization of bone. The deficiency conditions of **rickets** (in children) and **osteomalacia** (in adults) may be treated by administration of the vitamin, commonly as D₂. Overdosage results in disturbances of calcium metabolism and ultimately leads to decalcification of bone. *See also* **calcitetrol**. For structures see entries for individual compounds.

vitamin D binding protein *abbr.*: DBP; a multifunctional protein found in plasma and other body fluids, and on the surface of many cell types. In plasma it carries the vitamin D sterols. DBP binds membrane-bound immunoglobulin on the surface of B lymphocytes, and IgG Fc receptor on the membranes of T lymphocytes.

vitamin E α-tocopherol; a fat-soluble vitamin considered to act primarily as an antioxidant. Other tocopherols (*see* **tocopherol** for structure) have vitamin E activity. Vitamin E is thought to prevent the formation of lipid autoxidation products by trapping free radicals and thus preventing the chain reaction that leads to further peroxide formation. It is present in vegetable-seed oils, and also in meat, milk, and eggs. Deficiency occurs only in pathological states in which absorption is affected, and can result in premature red cell death. A relative deficiency may occur when the intake of polyunsaturated fatty acids is high unless the diet is supplemented with α-tocopherol.

vitamin F *(formerly)* a collective term for **essential fatty acids**.

vitamin G *an old name for* riboflavin (i.e. **vitamin B₂**).

vitamin H 1 *(sometimes)* an alternative name for **biotin**. **2** *an old name for* riboflavin (i.e. **vitamin B₂**).

vitamin H₁ *(sometimes)* an alternative name for **p-aminobenzoic acid**.

vitamin K 2-methyl-3-phytyl-1,4-naphthoquinone; phylloquinone; 3-phytylmenadione; vitamin K₁; a fat-soluble vitamin required for the synthesis of prothrombin and certain other **blood coagulation** factors. The **menaquinones** (vitamins K₂) are also active. Vitamin K is found in green leafy vegetables, meat, and milk. Deficiency is uncommon because it is produced by gut bacteria, but gut sterilization may trigger symptoms (decrease in plasma prothrombin) if intake is defective. Vitamin K is a cofactor in the post-translational formation of 4-carboxyglutamyl residues in certain proteins, e.g. **blood coagulation** factors VII, IX, X, and especially prothrombin, during which an epoxidase reaction converts vitamin K₁ to its epoxide form, which is converted back by a reductase. This reductase can be inhibited by **warfarin**, which thus reduces vitamin K₁ levels, thereby reducing prothrombin formation and so acting as an anticoagulant.

vitamin-K-epoxide reductase (warfarin-sensitive) EC 1.1.4.1; an enzyme that catalyses the formation of 2-methyl-3-phytyl-1,4-naphthoquinone, oxidized dithiothreitol, and H_2O from 3-epoxy-2,3-dihydro-2-methyl-3-phytyl-1,4-naphthoquinone and 1,4-dithiothreitol.

vitamin-like dietary factor any biomolecule that resembles the vitamins in its biological properties and is considered to be produced in inadequate amounts within an organism so that dietary supplementation may be recommended. The term includes: inositol, carnitine, lipoic acid, p-aminobenzoic acid, and ubiquinone.

vitellogenin the precursor protein of **lipovitellin** and **phosphitin**, a major egg-yolk protein synthesized in the liver of chickens and *Xenopus*. It has an M_r of 135 000, and is split in the ovary into its constituent proteins. In *Xenopus*, two genes, for vitellogenin I and II, are spread over more than 21 kbp of DNA, their primary transcripts consisting of 6 kbp of **pre-mRNA** with 33 **introns**. A related protein is found in *Drosophila melanogaster*.

vitronectin *or* **serum spreading factor** an adhesion protein, present in plasma and on the external surface of membranes of tissues, that promotes the adhesion of cells in tissue culture. The $\alpha_v\beta_3$ **integrin** binds vitronectin, fibrinogen, laminin, von Willebrand factor, fibronectin, and thrombospondin. The human precursor encodes an N-terminal signal, vitronectin, and somatomedin B (*see* **somatomedin**).

Vl *symbol for* the valeryl group.

VLA a family of **integrins** all having the β₁ subunit. They comprise VLA-1, $\alpha_1\beta_1$ (β subunit also called VLAβ, glycoprotein IIa or gpIIa, CD29; α subunit also called CD49a), which binds laminin (and collagen); VLA-2, $\alpha_2\beta_1$ (α subunit also called gpIa, CD49b), which binds collagen (and laminin); VLA-3, $\alpha_3\beta_1$ (α subunit also called CD49c), which binds fibronectin, laminin, and collagen; VLA-4, $\alpha_4\beta_1$ (α subunit also called CD49d), which binds VCAM-1 and fibronectin; VLA-5 (*or* fibronectin receptor), $\alpha_5\beta_1$ (CD49e), which binds fibronectin; VLA-6 (*or* laminin receptor), $\alpha_6\beta_1$ (CD49f), which binds laminin; $\alpha_7\beta_1$, which binds laminin; $\alpha_8\beta_1$, ligand doubtful; and $\alpha_v\beta_1$ (α subunit also called vitronectin receptor subunit, CD51), which binds fibronectin.

VLDL *abbr. for* very-low-density lipoprotein.

VLGR-1 *abbr. for* very large G-protein-coupled receptor 1; a large cell-surface protein (≈5500 amino acids) that is highly expressed in the developing central nervous system of mouse. The C-terminal region contains seven transmembrane segments and is G-protein-coupled. The extracellular region contains a tandem array of 20 repeats similar to those present in Calx of *Drosophila* (*see* **Calx**) and in their middle, seven repeats like those present in the C-terminal region of **epitempin**. The human gene is located in a region linked to a form of familial febrile convulsions.

VMA *abbr. for* vanillylmandelic acid; *see* **hydroxymethoxymandelate**.

VO *abbr. for* virtual organization; a concept derived from computer science and comprising people from a variety of actual organizations. One example of a VO would be those involved in different aspects of a large engineering design project. Another would be those involved in, for example, a large genome annotation project.

void volume 1 *(in column chromatography)* the volume of the mobile phase, i.e. total **bed volume** of the column minus the volume occupied by the support particles. **2** *(in gel chromatography) or* **exclusion volume** the volume of the mobile phase passing through the gel required to elute a molecule that never entered the stationary phase.

volemitol D-*glycero*-D-*talo*-heptitol; a polyol widely distributed in plants, fungi, and lichens, and used as a sweetening agent.

volkensin a ricin-like toxic glycoprotein from the roots of *Adenia volkensii*, a Kenyan shrub. It consists of two subunits, A (M_r 29 000) and B (M_r 36 000), linked by disulfide bonds.

volt *symbol*: V; the SI derived unit of electric potential or electromotive force. It is equal to the difference in electric potential between two points on a conductor carrying a constant current of one ampere when the power dissipated between the points is one watt. 1 V = 1 W A⁻¹ = 1 J C⁻¹. [After Alessandro Volta (1745–1827), Italian physicist.]

voltage-gated ion channel a cell-membrane channel whose opening is governed by the membrane potential.

voltage-gated K⁺ channel *or* **KCNQ** any of a family of membrane proteins that form K⁺ seletive channels that contain six transmembrane segments and intracellular N- and C-termini. The channels open when a change in voltage occurs across the membrane. Mutations in KCNQ1 are common in hereditary **long QT syndrome**. Mutations in KCNQ2 and KCNQ3 cause autosomal dominant familial neonatal convulsions. KCNQ4 mutations cause a form of hereditary deafness.

voltage sensor the fourth transmembrane segment (S4) of shaker-type ion channels. It contains lysine or arginine in every third position and is involved in opening the channel at the permissive voltage. *See* **shaker**.

voltameter *an old term for* **coulometer**.

voltametry *an old term for* **coulometry**.

voltammetry an electroanalytical procedure for identifying and determining the concentrations of various ions in solution by studying the relation between a varied electric potential applied between two electrodes immersed in a test solution and the resultant faradaic current. It includes **anodic stripping voltammetry** and **polarography**. —**voltammetric** *adj.*; **voltammetrically** *adv.*

volume of distribution *symbol*: V_d; the theoretical volume of fluid in which a specified dose of drug is distributed, based on measurement of the concentration of free drug (i.e. that not bound to plasma proteins) in plasma. $V_d = Q/C_p$, where Q is the amount of drug administered and C_p is the concentration in plasma.

volutin granule *or* **metachromatic granule** an electron-dense cytoplasmic inclusion found in various bacteria, yeasts, and other microorganisms. It exhibits **metachromasia** on staining with certain basic dyes, contains polymetaphosphate, and is believed to serve as a phosphate store. *See also* **Babes–Ernst body**.

von Gierke's disease *see* **glycogen disease**.

von Willebrand factor *abbr.*: vWFa; a blood protein involved in clotting. It performs two important hemostatic functions: it links specific platelet membrane receptors to collagen, and possibly other components of subendothelial connective tissue, so mediating platelet adhesion to sites of vascular damage; and it binds to, and stabilizes, factor VIII (*see* **blood coagulation**). Deficiency results in bleeding from skin, owing to failure of platelet adhesion, with a prolonged bleeding time, and also results in spontaneous bleeding into joints and soft tissues as a result of the secondary factor VIII deficiency, caused by the instability of factor VIII. The primary translation product, synthesized by vascular endothelial cells and megakaryocytes, is a protein of 2813 amino acids (309.29 kDa). The human precursor protein encodes a signal peptide, a peptide that is cleaved off and is known as **von Willebrand antigen II**, and the mature protein itself; cysteine is the most abundant amino acid. Before the mature factor is secreted from cells of origin, many intermolecular disulfide bridges between vWF molecules are formed that lead to a complex that may vary in mass from dimers of ~500 kDa to species of more than 20 megadaltons. Inherited defects of vWF affect nearly 1% of the population, but most are not clinically significant. The majority of patients have a quantitative deficiency of vWF, with autosomal dominant inheritance, and are known as type I. Type II variants have an abnormal vWF. In vWF deficiency (*abbr.*: vWD), type IIA intermediate and large complexes are absent. Mutations cluster near Tyr^{842}- |-Met^{843}, a site that is sensitive to proteolysis. vWD type III is rare, and is a severe autosomal recessive disease characterized by virtual absence of vWF, in some cases known to be the result of total or partial deletion of the vWF gene. Unactivated platelets bind vWF through gpIb–IX complex, deficiency of which leads to a bleeding disorder, **Bernard-Soulier syndrome**. After activation, platelets bind vWF through the **integrin** gpIIb–IIIa complex, which is defective in **Glanzmann thrombasthenia**, another bleeding disorder. [After Erik Adolph von Willebrand (1870–1949), Finnish physician.]

V-PARP *abbr. for* vault-poly(ADP-ribose) polymerase; in mammals, a 193 kDa protein that contains a domain homologous to the catalytic domain of **PARP-1**. It binds to, and uses as substrate, MVP of **vaults**, and is also localized to the mitotic spindle. It does not appear to require DNA for activity.

V8 proteinase *see* **glutamyl endopeptidase**.

vps *abbr. for* **vacuolar sorting protein**.

vps34 *or* **vacuolar sorting protein 34** a protein-trafficking protein involved in the sorting and segregation of vacuoles. It is a homologue of **1-phosphatidylinositol 3-kinase**, having that enzymic activity, which may be the mechanism by which it acts. It is, in turn, activated via phosphorylation by a serine/threonine protein kinase, vps15.

VRE *abbr. for* **vancomycin**-resistant *Enterococcus*. It is a drug-resistant pathogen (*compare* **MRSA**) found in hospital populations of bacteria. The resistance arises from modification of the peptidoglycan precursor that is the target for vancomycin.

Vrille a basic leucine zipper transcription factor of *Drosophila* that represses Clock.

v-yes *see* **YES**.

Ww

w *symbol for* **1** mass fraction. **2** work (alternative to *W*). **3** speed (alternative to *u*, *v*, or *c*). **4** (bold italic) velocity (alternative to *v*, *u*, or *c*).

W *symbol for* **1** watt. **2** a residue of the α-amino acid L-tryptophan (alternative to Trp). **3** a residue of an incompletely specified base in a nucleic-acid sequence that may be adenine or may be thymine (in DNA) or uracil (in RNA). **4** a residue of the minor nucleoside **wyosine** (alternative to Y). **5** tungsten.

W *symbol for* **1** weight (alternative to *G* or *P*). **2** work (alternative to *w*). **3** radiant energy (alternative to *Q*). **4** writhing number.

W3 *abbr. for* World Wide Web. *See* **WWW**.

Waardenburg syndrome an autosomal dominant syndrome of developmental defects in and around the eyes, pigmentary defects in eyes and hair, and congenital deafness. Types 1 and 3 are caused by mutations in *PAX-1* and *PAX-3*. Type 3 also shows limb muscle hypoplasia. Type 2 is caused by mutations in *MITF*; type 4 is associated with **Hirschsprung disease**, and is caused by mutations in *SOX10*. [After Petrus Waardenburg (1886–1979), Dutch ophthalmologist.]

WAGR *abbr. for* Wilm's tumour, aniridia, ambiguous genitalia, and mental retardation; a contiguous gene syndrome resulting from a deletion in chromosome 11p13.

WAK1 *abbr. for* wall-associated kinase 1; a single-pass plasma membrane protein of plants, whose N-terminal region interacts with the cell wall while the intracellular region is a protein kinase.

Walden inversion the inversion of the configuration at a chiral centre in a bimolecular nucleophilic substitution reaction. [After its discoverer, Paul Walden (1863–1957), Latvian chemist.]

Waldenström's macroglobulinemia a lymphoid tumour that arises from the most mature B lymphocytes and invariably produces IgM, which appear as a **paraprotein**. [After Jan Goesta Waldenström (1906–1996), Swedish physician.]

Walker motif either of two sequence motifs in proteins that bind and hydrolyse nucleoside triphosphates. Walker-A (or P-loop) consists of the sequence A/GX4GKT/S (in one-letter code), is flanked by a beta strand and an alpha helix, and loops around the triphosphate moiety. Walker-B consists of X4D (where X is almost exclusively a hydrophobic residue), occurs at the end of a beta strand, and interacts with the Mg^{2+} ion of the triphosphate moiety. [After John Walker (1941–), British biochemist who discovered them in H^+-transporting ATPase of mitochondria.]

wall effect 1 *(in centrifugation)* the collision of sedimenting molecules with the wall of the cell. This occurs because the centrifugal field is radial, so can be avoided by using sector-shaped cells. **2** *(in chromatography)* the curving and spreading of a zone as it migrates down a chromatographic column because of inhomogeneities in solvent flow near the wall of the column.

Warburg apparatus a sensitive, constant-volume respirometer for measuring gas exchange of cells, homogenates, or tissue slices. It consists of a small flask, with one or more side arms for the addition of reagents, and a centre well, connected to a U-tube manometer of about 1 mm^2 internal cross-section, fitted with a reservoir for manometer fluid. The vessel is immersed in a constant temperature bath and shaken continually to equilibrate the gas in the liquid phase, where the reaction is taking place, with the gas phase, where the reaction is being measured. It was used extensively for metabolic studies before the wide availabilty of radioactive tracers, but is now virtually obsolete. [After Otto Heinrich Warburg (1883–1970), German physiologist and biochemist.]

Warburg–Dickens–Horecker pathway *or* **Warburg–Dickens pathway** *an alternative name for* **hexose monophosphate pathway**.

Warburg effect 1 the relative overproduction of lactate that occurs in many tumours. **2** the inhibitory effect of high oxygen concentrations on photosynthesis.

Warburg's enzyme *an alternative name for* old yellow enzyme; *see* **yellow enzyme**.

warfarin 3-(α-phenyl-β-acetylethyl)-4-hydroxycoumarin; 1-(4′-hydroxy-3′-coumarin-3′-yl)-1-phenylbutan-3-one; an anticoagulant and rodenticide. It inhibits the reductase that converts the epoxide form of **vitamin K** to its reduced form, which is a cofactor for the production of the 4-carboxyglutamyl residues of, among other proteins, the **blood coagulation** factors II (i.e. prothrombin), VII, IX, and X. [From the name of the patentees, Wisconsin Alumni Research Foundation, + coumarin.] *See also* **vitamin-K-epoxide reductase (warfarin-sensitive)**.

Waring blendor *the proprietary name for* a blender used in the preparation of tissue homogenates. It consists of a hydrodynamically designed vessel (of tetrafoil cross-section) in the bottom of which a specially shaped blade rotates at high speed. [After the manufacturers, Waring Products Corporation.]

washing soda *see* **soda**.

WASP *abbr. for* Wiskott–Aldrich syndrome protein; any of a small family of proteins including WASP1 (whose expression is limited to lymphocytes and platelets) and N-WASP (which is more widely expressed but especially in neurons). They bind small GTPases and phospholipids through their N-terminal regions and are homologous in their C-terminal regions. They activate the **Arp2/3 complex** and help in actin filament nucleation. Over 20 mutations of a locus at Xp11.23-p11.22 for WASP (501 amino acids) are associated with the Wiskott–Aldrich syndrome of severe thrombocytopenia and defective immunity.

Wasserman reaction a **complement-fixation test** for human blood and cerebrospinal fluid, widely used to detect syphilitic infection, although false positive reactions are often given by yaws, leprosy, paroxysmal hemoglobinuria, and malarial infection. The antigen on which the test is based is **cardiolipin**. [After August von Wassermann (1866–1925), German bacteriologist.]

Watanabe rabbit a breed of rabbits suffering from high blood cholesterol levels as a result of mutation of the low-density lipoprotein (LDL) receptor gene. It is used as an animal model for familial hypercholesterolemia.

water balance (of an organism) the **water intake** (as a liquid, the water in food, and water of oxidation) less the **water output** (water in urine, feces, or other excreta, water expended in saturating respired air, water lost through the integument).

waterfall sequence *an alternative name for* **cascade sequence**.

water regain (of an ion-exchange resin) for cationic resins, the weight of water taken up by 1 g dry resin in the hydrogen form; for anionic resins, the weight of water taken up by 1 g dry resin in the chloride form.

water-soluble B *an old name for* a fraction rich in water-soluble vitamins prepared from egg yolk.

Watson–Crick model of DNA a model for one form of **DNA** consisting essentially of two antiparallel helical polynucleotide strands coiled around the same axis to form a double helix. The deoxyribose–phosphate backbones lie on the outside of the helix and the purine and pyrimidine bases lie approximately at right angles to the axis on the inside of the helix. The diameter of the helix is 2.0 nm and there is a residue on each chain every 0.34 nm in the *z* direction. The angle between each residue on the same strand is 36°, so that the structure repeats after 10 residues (3.4 nm) on each strand. The two strands are held together by hydrogen bonds between pairs of bases that are complementary and opposing (i.e. on different strands). Adenine pairs with thymine and guanine with cytosine. The two chains are therefore complementary. Watson and Crick's paper, published in *Nature* in 1953, also contained a prescient statement: 'It has not escaped our notice that the specific pairing we have postulated immediately suggests a possible copying mechanism for the genetic material.' This structure has stood the test of time. It is difficult to exaggerate the importance of the short paper. To quote from L. Stryer's *Biochemistry* (fourth edition), 'This brilliant accomplishment ranks as one of the most significant in the history of biology because it led the way to an understanding of gene function in molecular terms.' *See also* **double helix**. [After James Dewey Watson (1928–), US biologist, and Francis Harry Crick (1916–2004), British molecular biologist.]

Watson strand the upper strand in representations of double-stranded DNA sequences. Open reading frames (ORFs) may occur on either strand of a genomic DNA duplex and it is convenient to distinguish between the strands in order to specify specific ORFs. ORFs in the Watson strand of the yeast genome, for example, are given the suffix W, as in the systematic name YLR451W for the *LEU3* gene on yeast chromosome XII. *Compare* **Crick strand**. [After J.D. **Watson**.]

watt *symbol*: W; the SI derived unit of power or radiant flux; it is the power dissipated when one joule is expended in one second; i.e. 1 W = 1 J s^{-1}. [After James Watt (1736–1819), British engineer.]

wave function *symbol*: ψ or w or Φ or ϕ; the amplitude of the wave associated with a particle, e.g. an electron, as obtained from the Schrödinger wave equation. The square of the value of ψ at any point is proportional to the probability of finding the particle at that point.

wavelength *symbol*: λ; the distance between successive points of equal phase of a wave. Wavelength is equal to the velocity of the wave motion divided by its frequency. For electromagnetic radiation, $\lambda = c_0/\nu$, where c_0 is the velocity of light *in vacuo* and ν the frequency of the radiation.

wavenumber *or* **repetency** for an electromagnetic radiation (e.g. light), the reciprocal of the **wavelength**; i.e. the number of waves in unit distance. *Compare* **kayser**.

wax 1 any lipid fraction from living organisms or from crude petroleum that is a plastic substance, hard when cold, easily moulded when warm, and insoluble in water. **2** any fatty-acid ester of a long-chain monohydroxy alcohol.

wax synthase *see* **long-chain-alcohol *O*-fatty-acyltransferase**.

Wb *symbol for* weber.

WBC *or* **wbc** *or* **w.b.c.** *abbr. for* white blood cell(s); *see* **leukocyte**.

WC *abbr. for* white collar.

WD40 *or* **WD repeat** *or* **WD40 repeat** a conserved protein sequence of ≈40 amino acids that typically ends in tryptophan–aspartate (WD in the one-letter code for amino acids). The WD40 protein family comprises many regulatory or adaptor proteins involved in signal transduction, and its members may have one or several WD40 repeats. WD40 mediates protein–protein interaction. Examples are: the β subunits of trimeric G-proteins, the β subunit of protein phosphatase 2A, Apaf-1, and SCAP.

WDHA syndrome *abbr. for* watery diarrhoea, hypokalemia, plus achlorhydria syndrome; i.e. **Verner–Morrison syndrome**.

WD repeat *see* **WD40**.

weak interactions noncovalent interactions between molecules and parts of molecules, including charge–charge interactions, those involving permanent dipoles, van der Waals forces, and hydrogen bonds. They are particularly important in protein and nucleic-acid structure.

weber *symbol*: Wb; the SI derived unit of magnetic flux. It is equal to the flux that, linking a circuit of one turn, produces in it an emf of one volt as it is reduced to zero at a uniform rate in one second; i.e. 1 Wb = 1 V s. [After Wilhelm Eduard Weber (1804–91), German physicist.]

wee1 a gene in *Schizosaccharomyces pombe* encoding a protein kinase (EC 2.7.1.112) that can phosphorylate tyrosine and serine/threonine residues. In yeast it phosphorylates a tyrosine of cdc2; in the human it phosphorylates a tyrosine and a threonine of the equivalent protein. It has conserved catalytic domains of the serine/threonine type of protein kinase, and is a negative regulator for entry to mitosis; it closely resembles the corresponding human homologue. *See also* ***cdc* gene**.

weight average molecular mass *old name for* mass average molar mass; *see* **average molar mass**.

weight matrix *an alternative name for* **scoring matrix**.

Weigle reactivation *see* **W reactivation**.

Weiss unit a measure of T4 DNA ligase activity based on ATP–PP$_i$ exchange. One unit is the amount of enzyme required to catalyse the exchange of one nanomole of ^{32}P from ^{32}PP$_i$ into ATP as **Norit-adsorbable** material in 20 min at 37°C. ATP and ^{32}P labelled ATP are adsorbed by Norit and recovered by filtration in this bound state for counting of radioactivity.

Werner syndrome a rare autosomal recessive disorder characterized by premature formation of cataracts, scleroderma-like skin, short stature, greying hair and hair loss, a generalized appearance of premature ageing, and a high risk of malignancy. It is caused by mutations in the Werner syndrome gene product (*see* **WRNp**). [After Otto Werner (1879–1936), German physician.]

Wernicke–Korsakoff syndrome a combination of **Wernicke's disease** (gliosis and capillary proliferation leading to ocular manifestations and polyneuropathy) and **Korsakoff's psychosis** (difficulty in recording and retaining new impressions). Wernicke's disease is due to vitamin B$_1$ deficiency, and manifests particularly in patients with a genetic abnormality in transketolase resulting in its reduced affinity for thiamine diphosphate. In such patients, even mild B$_1$ deficiency, as often occurs in alcoholics, may produce symptoms. [After Karl Wernicke (1848–1905), German alienist, and Sergei Sergeyevich Korsakoff (1854–1900), Russian neurologist.]

Western blotting a technique for **blotting** proteins onto nitrocellulose, nylon, or other **transfer membrane** after they have been resolved by gel electrophoresis. The proteins can then be detected by autoradiography (if labelled), or through binding to fluorescently labelled, ^{125}I-labelled, or enzyme-linked (*see* **ELISA**) antibody, lectin, or other specific binding agent. The name derives by extension through the cardinal points from **Southern blotting**, the first such blotting technique, with a capital initial by similar extension.

Whitaker plot a plot that, ideally, gives a linear correlation between the logarithm of the relative molecular mass of a macromolecule and the ratio of the **elution volume**, V, to the **void volume**, V_o, when the macromolecules in solution are passed through a suitable **gel filtration** column; i.e. $\log M_r = V/V_o$. [After John Robert Whitaker (1929–), US biochemist.]

white adipose tissue *or* **white fat** a tissue developed particularly in mammals and birds to store fat as triacylglycerols to supply energy to the whole animal. The cells contain a single large droplet of fat that, in the nourished state, fills most of the cell. The nucleus lies in the narrow band of cytoplasm around the periphery. Adipose tissue is found in many parts of the body but is concentrated under the skin (subcutaneous) and around internal organs (heart and kidney). *Compare* **brown adipose tissue**.

white blood cell *see* **leukocyte**.

white coat effect the effect of the presence of the investigator on the accurate measurement of the blood pressure of a subject.

white collar *abbr.*: WC; a circadian blue-light photoreceptor that binds to the promotor of the clock component called frequency in the fungus *Neurospora crassa*. It is a light sensor and a DNA-binding transcription factor. WC-1 and WC-2 are zinc-finger proteins that function as a complex.

white fat *see* **white adipose tissue**.

white muscle pale, skeletal, usually involuntary muscle that is relatively deficient in myoglobin and cytochromes. *Compare* **red muscle**.

whole genome PCR a method whereby genomic DNA is converted to a form that can be amplified by the **polymerase chain reaction** (PCR). DNA is reduced to small fragments by restriction enzyme digestion or sonication, and adaptors are ligated to their ends. Oligonucleotides specific for the adaptors can be used for PCR amplification of the fragments. The technique can be used to identify individual fragments capable of binding transcription factors or other DNA-binding proteins using antibodies specific for the protein concerned.

WIBR *abbr. for* Whitehead Institute for Biomedical Research; *see* **sequencing centre**.

widow *(in a sequence alignment)* an amino acid residue or nucleotide base isolated from neighbouring residues or bases by spurious gaps at each side, usually the result of overzealous gap insertion by automatic alignment programs.

WIF-1 *abbr. for* Wnt-inhibitory factor-1. *See* **Wnt**.

WIF domain *see* Wnt inhibitory factor 1..

wild-type the **phenotype** that is characteristic of most of the members of a species occurring naturally and contrasting with the phenotype of a **mutant**. *Compare* **auxotroph**.

Williams–Beuren syndrome an autosomal dominant contiguous gene deletion disorder involving over 17 genes on chromosome 7 and characterized by a broad spectrum of abnormalities. The gene at 7q11.23 for FKBP6 (a component of the synaptonemal complex) is frequently part of the deletion. [After J.C.P. Williams (1900–), New Zealand cardiologist and Alois J. Beuren (1919–1984), German cardiologist.]

Wilm's tumour *or* **nephroblastoma** a malignancy in human infants (incidence 1 in 10000) associated with protein **WT33**. [After Marx Wilms (1867–1918), German pathologist.]

Wilson ATPase *see* **Wilson disease**.

Wilson's disease *or* **hepatolenticular degeneration** a rare autosomal recessively inherited disease characterized by degenerative changes in the brain, particularly in the basal ganglia, and cirrhosis of the liver. There are excessive deposits of copper in the liver, brain, cornea, and kidney; the serum copper is usually low and there is a **ceruloplasmin** deficiency. Patients may be effectively treated with British antilewisite (**dimercaprol**) or other chelating agents. [After (Samuel Alexander) Kinnier Wilson (1878–1937), US-born British neurologist.]

Wilzbach method a method for the random labelling of a compound with tritium by exposing it to isotopically pure tritium gas in a sealed container for several weeks. The exchange is promoted by the radiation.

winding angle (of double-stranded DNA) the angle of rotation between successive base pairs.

winding number the number of turns in a closed-circular double-stranded DNA measured relative to a surface on which the DNA lies. *See also* **supercoil**.

windowless counter a counter for radioactive (or X-ray) ionization in which the sample (or source) is not separated from the detector by a membrane or 'window'.

wingless a gene in *Drosophila melanogaster* that encodes a protein in the **WNT family**, probably a growth factor that acts on neighbouring cells to regulate the **engrailed** gene. A homologue is present in the brachiopod *Terebratulina retusa*.

Wiskott–Aldrich syndrome a syndrome of thrombocytopenia, eczema, and immunodeficiency, with increased risk of malignancy. It is caused by mutations in a protein (*see* **WASP**) encoded on the X chromosome. [After Alfred Wiskott (1898–1978), German paediatrician, and Robert Anderson Aldrich (1917–), US paediatrician.]

Wiskott–Aldrich syndrome protein *see* **WASP**.

WIT *abbr. for* what is there; *see* **pathway database**.

WNK any of a family of protein serine/threonine kinases, first described in the rat, that have cysteine in place of lysine at a key position in the active site and contain two coiled-coil domains. The human WNK1 (gene on chromosome 12p) shows 86% sequence identity with the rat protein. Large deletions in the first intron increase expression of the protein and cause pseudohypoaldosteronism type II (an autosomal dominant hypertension, hyperkalemia, and metabolic acidosis). hWNK4 (gene on chromosome 17) is highly homologous with hWNK1. Several missense mutations cause pseudohypoaldosteronism type II. Both proteins localize to the distal nephron, hWNK1 being cytoplasmic, while hWNK4 is associated with tight junctions.

Wnt *abbr. for* (*Drosophila*) Wingless/Int-1; any of a family of secreted signalling glycoproteins found throughout the animal kingdom. They are important for patterning and development in the embryo and for regulation of cell proliferation thereafter. Wnt-1 is expressed in the central nervous system. On binding to G-protein-coupled receptors of the **Frizzled** family in the plasma membrane, they cause inactivation of **GSK-3**, which is complexed with APC protein, axin, and β-catenin, and targeted for ubiquitination and proteosomal degradation. In the extracellular space they are inhibited by binding Wnt-inhibitory factor-1 (WIF-1).

WNT family a family of secreted glycoproteins (e.g. WNT-2 from human and mouse) that are signalling molecules in tissue development. They act through G-protein-associated membrane receptors of the frizzled family to inhibit glycogen synthase kinase 3. The *WNT-1* gene is closely homologous to **wingless**.

Wnt inhibitory factor 1 *abbr.*: WIF-1; a secreted protein that contains an N-terminal WIF domain (≈150 amino acids), which binds and inhibits **Wnt** proteins; five EGF-like repeats; and a short hydrophilic domain in the C-terminal region. Homologues are present in many animals. The WIF domain is related to the extracellular domain of **Ryk** receptor tyrosine kinases. .

wobble base the third base in the anticodon of **transfer RNA** (5′ end). It can bind to one of several possible bases at the 3′ end of the codon in mRNA. *See also* **wobble hypothesis**.

wobble hypothesis the hypothesis that in **translation** (during protein synthesis) a less strict specificity in the base pairing of the 5′ base of the anticodon of **transfer RNA** (tRNA) allows it to make alternative hydrogen-bonding with the third base of the codon of messenger RNA (mRNA) beyond the usual G–C and A–U pairings. Thus, anticodon U may recognize A or G and anticodon G may recognize C or U. Although the hypothesis has been fully supported by experiment, UGG and AUG are the only examples of the first type of wobble, and no examples of the second type are known. The presence of bases other than U,A,G, or C at the anticodon 5′ position also contributes to more general pairing at this position. Thus, bases such as **wyosine**, **queuosine**, and **inosine** may be present (I always substitutes for A) with less strict pairing constraints. The degeneracy at the third codon position is consistent with the fact that about 40 different tRNA molecules interact with the 61 possible sense codons.

Wolcott–Rallison syndrome *see* **PERK**.

Wolfgram proteolipid protein one of the three principal classes of central nervous system **myelin protein**, extractable by chloroform: methanol (2:1, v/v), and consisting of three main fractions of 235, 54, and 62 kDa. *Compare* **Folch–Lees protein**. [After F. Wolfgram, who first discovered it.]

Wolf–Hirschhorn syndrome a syndrome of multiple malformations with mental and developmental defects. It is caused by deletion of the gene for **NELF**-A. [After Ulrich Wolf (1933–), German physician and Kurt Kirschhorn (1926–), US physician.]

Wolman disease a rapidly fatal autosomal recessive disease of infancy in which there is massive accumulation of cholesterol esters and triacylglycerols in most tissues. It is associated with mutations that lead to complete elimination in the activity of lysosomal **acid lipase**. [After Moshe Wolman (1914–), Israeli pathologist.]

work *symbol*: W *or* w; the transfer of energy expressed as the product of a force, F, and the distance, s, through which its point of application moves in the direction of the force; i.e. $W = \int F\,ds$. The SI unit of work is the joule. The manifestation of work in productive form usually involves transduction by a mechanical device of the action of a force through a distance. A relationship between the work done on (or by) a system of constant mass, the heat it gains (or loses), and the change in the internal energy of the system is provided by the first law of thermodynamics (*see* **thermodynamics**).

World Wide Web *abbr.*: WWW *or* W3; the world-wide network of information (including biochemical and molecular biological data) stored on computers linked to the Internet, and made accessible for visualization and interrogation by browser software.

wortmannin *or* **MS-54** *or* **KY 12420** a potent and specific inhibitor of phosphatidylinositol 3-kinase, derived from the fungus *Penicillium funiculosum*.

W reactivation *or* **Weigle reactivation** *or* (*formerly*) **UV reactiva-**

tion the phenomenon whereby the survival of UV-irradiated bacteriophage is higher in a UV-irradiated host than in an unirradiated host. [After Jean Jacques Weigle (1901–68), who discovered it and described it in 1953.]

writhing number *symbol*: W; an index of the superhelical coiling of DNA, termed **writhe**. The writhing number and the **twist number** between them determine the **linking number** (L). The writhing number does not have a precise quantitative definition, but represents the degree of supercoiling. A decrease in L involves some decrease in supercoiling, and an increase in L an increase in supercoiling. *See also* **supercoil**.

WRNp *abbr. for* Werner syndrome gene product; *other name*: RECQL3. A RecQ helicase (1432 amino acids) of humans that has $3' \to 5'$ exonuclease activity and binds to the ssDNA-binding protein replication protein A. Over a score of mutations cause **Werner syndrome**.

wt *or* **wt.** *abbr. for* weight.

WT33 a protein that is related to **early growth response proteins**, and is associated with **Wilm's tumour**; it is probably a tumour suppressor. Certain mutations are also associated with Denys–Drash syndrome. Its sequence contains alternative splice sites and a zinc finger.

w/v *or* **wt/vol.** *abbr. for* weight per unit volume (of a solution or mixture).

w/w *or* **wt/wt** *abbr. for* weight per unit weight (of a solution or mixture).

WW domain *or* **WWP domain** a small motif characterized by two conserved tryptophan residues (spaced ~20-23 amino acid residues apart) and a conserved proline, hence its alternative name. It is found in various disparate proteins, including vertebrate YAP protein, mouse NEDD-4, **utrophin** and **dystrophin**, the protein encoded by the gene responsible for **Duchenne muscular dystrophy**. The domain contains ~35-40 residues, which fold as a triple-stranded β-sheet, and may be repeated several (up to 4) times. It binds proteins that contain specific proline motifs ([AP]-P-P-[AP]-Y), and bears some resemblance to **SH3 domains**. It is often associated with other domains characteristic of proteins involved in signal transduction.

WWE motif *see* **deltex**.

wwPDB *abbr. for* Worldwide Protein Data Bank; *see* **composite database**.

WWP domain *another name for* **WW domain**.

WWW *abbr. for* **World Wide Web**.

wyosine *symbol*: W *or* Y; any modified form of guanosine present in the anticodon loop of some species of **transfer RNA**. They are based on the structure N^2-(2-methyletheno)guanosine, and carry varying side chains on the five-carbon imidazo ring. The structure of one such compound is shown. *See also* **minor nucleoside**.

Xx

x *symbol for* **1** amount-of-substance fraction (i.e. mole fraction) or number fraction for condensed phases. *Compare* **y. 2** a cartesian space coordinate, complementary to *y* and *z*.

X *symbol for* **a** a residue of an unknown, unspecified, or (specified) unusual α-amino acid (alternative to Xaa). **b** a residue of the ribonucleoside **xanthosine** (alternative to Xao).

X *symbol for* (electric) reactance.

X-537A *an alternative name for* **lasaloacid**.

Xaa *symbol for* a residue of an unknown, unspecified, or (specified) unusual α-amino acid (alternative to X).

Xan *symbol for* a residue of the purine base **xanthine**.

xanthan gum a capsular complex heteropolysaccharide formed by strains of the pseudomonad bacterium *Xanthomonas campestris* and believed to promote adhesion of the organism to its plant host. It is widely used in the cosmetic, food, and other industries as a crystallization inhibitor, emulsifier, gelling agent, etc.

xanthine *symbol*: Xan; 2,6-dihydroxypurine; 2,6-dioxopurine; 3,7-dihydro-1*H*-purine-2,6-dione; a purine formed in the metabolic breakdown of guanine. *Compare* **hypoxenthine**. *See also* **allopurinol**.

xanthine oxidase EC 1.1.3.22; *the recommended name for* hypoxanthine oxidase; *systematic name*: xanthine:oxygen oxidoreductase; *former name*: Schardinger enzyme. A flavoprotein (FAD) enzyme containing two 2Fe–2S clusters and molybdenum cofactor, abundant in milk and liver, that catalyses the oxidation of **xanthine** to urate and superoxide (or hydrogen peroxide) and also the oxidation of hypoxanthine to xanthine. The enzyme has two forms: (1) xanthine dehydrogenase, which has **NAD** as coenzyme, and urate and NADH as products; and (2) xanthine oxidase, which reacts with dioxygen and has urate and H_2O_2 as products. The dehydrogenase is converted reversibly to the oxidase by oxidation of sulfhydryl groups or irreversibly by proteolysis.

xanthinuria the excretion of excessive amounts of xanthine in the urine. It is sometimes due to a hereditary disease in which there is a gross deficiency of **xanthine oxidase** activity in the tissues. *See* **molybdoenzymes**.

xantho+ or *(before a vowel)* **xanth+** *comb. form* denoting yellow.

xanthoma a yellow fatty deposit below the skin or associated with a tendon. Usually multiple, xanthomas are due to extravascular phagocytosis of chylomicrons, LDL-derived cholesterol, or other sterols, by macrophages of skin or subcutaneous tissue. They usually occur in a variety of hyperlipidemias and in cerebrotendinous xanthomatosis but may also occur with a normal plasma lipid profile. –**xanthomatous** *adj*.

xanthophyll any **carotenoid** in which one or more oxygenated functions are present in the molecule.

xanthoproteic test a test for protein in which a yellow colour (or precipitate from solution) forms on addition of conc. nitric acid; the colour turns orange when made alkaline.

xanthopterin a yellow pterin pigment of insect wings, e.g. of the common sulfur butterfly. *Compare* **leukopterin**.

xanthopterin

xanthosine *symbol*: X *or* Xao; xanthine riboside; 9-β-D-ribofuranosylxanthine; a nucleoside formed by the deamination of guanosine.

xanthosine monophosphate *abbr.*: XMP; any xanthosine phosphate, particularly xanthosine 5′-phosphate when its distinction from xanthosine (5′-)diphosphate and xanthosine (5′-)triphosphate requires emphasis.

xanthosine phosphate xanthosine monophosphate (*abbr.*: XMP); any phosphoric monoester or diester of xanthosine. Of the three possible monoesters and the two possible diesters only **xanthosine 5′-phosphate** occurs naturally (the locant being omitted if no ambiguity may arise).

xanthosine 5′-phosphate *or* **5′-xanthylic acid** *or* **5′-phosphoxanthosine** *or* **5′-O-phosphonoxanthosine** *symbol*: Xao5′P; *alternative recommended names for* xanthosine monophosphate (*abbr.*: XMP); xanthosine 5′-(dihydrogen phosphate); xanthine (mono)ribonucleotide. (The locant is commonly omitted if there is no ambiguity as to the position of phosphorylation.) An intermediate in the biosynthesis of guanosine 5′-phosphate (i.e. 5′-guanylic acid, or GMP), it is formed from inosine 5′-phosphate (i.e. 5′-inosinic acid) by an NAD$^+$-dependent dehydrogenase; a glutamine-requiring amidotransferase reaction (catalysed by **GMP synthase**) then converts XMP to GMP.

xanthylic acid *the trivial name for* any phosphoric monoester of xanthosine. The position of the phosphoric residue on the ribose moiety of a given ester may be specified by a prefixed locant (*see*

xanthosine phosphate). However, 5′-xanthylic acid is the ester commonly denoted, its locant usually being omitted if no ambiguity may arise.

Xao *symbol for* a residue of the ribonucleoside **xanthosine** (alternative to X).

Xao5′P *symbol for* xanthosine 5′-phosphate.

Xao5′PP *symbol for* xanthosine 5′-diphosphate (alternative to ppX).

Xao5′PPP *symbol for* xanthosine 5′-triphosphate (alternative to pppX).

Xba I a type 2 **restriction endonuclease**; recognition sequence: T↑CTAGA.

X box a small nucleotide sequence within the 150 bp response element of the promoter of all MHC class II genes. It binds the **RFX** complex.

X cell *see* **X-Y-Z cell theory**.

X chromosome a **sex chromosome**. It is often found paired in the **homogametic** sex, which in many species is the female, and single in the **heterogametic** sex, in many species the male. The X chromosome carries a large number of genes that control many aspects of development and function. A mutant gene so carried is called **X linked**. Since males possess only one X chromosome an abnormal gene that it carries cannot be paired with a normal allele; the male is then said to be hemizygous for that gene. The female with two identical members of a pair of X-linked genes is said to be homozygous for that gene; the female with one mutant gene is called a heterozygote or carrier for the abnormal allele. One X chromosome is inactivated during early embryonic development. Since the progeny of each cell have the same inactivated X chromosome, a proportion of cells have the paternal X chromosome in an active state while in the other cells the maternal X chromosome is active. Hence the female who is heterozygous for an abnormal gene will have cells with the normal gene and cells with the mutant gene. She is said to be a mosaic. *Compare* **Y chromosome**. *See also* **sex determination**.

X-chromosome inactivation *or* **X inactivation** the inactivation in female mammals of one of the two **X chromosomes** during early embryonic development. This phenomenon occurs randomly in cells of the female embryo, which is thus a mosaic in respect to whether it the maternal or paternal X chromosome is functional. X inactivation means that the dosage of genes in female cells is comparable with that of males, who inherit only one X chromosome. *See also* **Lyon hypothesis**.

XDP *abbr. for* xanthosine (5′-)diphosphate.

Xe *symbol for* xenon.

Xeloda *another name for* **Capecitabine**.

Xenical (Orlistat) A slimming drug that acts by inhibiting lipase in the gut so that fat is not absorbed. As a result about 30% of the fat in a meal is indigested.

xenin a 25-residue peptide isolated from human gastric mucosa. The six C-terminal amino acids are identical to those of the octapeptide, **xenopsin**, present in the skin of amphibians. It stimulates exocrine pancreatic secretion, its concentration rising in plasma after a meal.

xeno+ *or (before a vowel)* **xen+** *comb. form* indicating something strange or foreign.

xenobiochemistry the biochemistry of organic compounds foreign to the organism.

xenobiotic 1 any substance that does not occur naturally but interacts with the metabolism of any organism. **2** of, or relating to, substances that are foreign to living systems.

xenobiotic-response element *abbr.*: XRE; a DNA-regulatory sequence that binds transcription factors that activate gene expression for enzymes that detoxify xenobiotics, including halogenated and other aryl hydrocarbons. *See also* **Ah**.

xenogeneic *or* **xenogenic** originating in a different species, especially of a tissue transplant. *See also* **xenograft**.

xenograft a tissue graft between individuals of different species.

xenologue *or (esp. US)* **xenolog** a gene acquired by horizontal transfer from another species (but not including gene transfer to eukaryotic cells from their endosymbionts, or hence from organelles to the nucleus). *Compare* **synologue**.

Xenopus laevis the South African clawed toad, the oocytes and embryos of which have been extensively exploited in molecular genetics and vertebrate embryology.

xenotope an antigenic **epitope** that is foreign to an animal.

xenotransplantation the transfer of living cells, tissues, and/or organs from one species to another, especially from nonhuman animals into humans for medical purposes. –**xenotransplant** *n., vb.*

xenotype a class or group of related **xenotopes**; i.e. an antigenic **epitype** that is foreign to a particular animal.

xeroderma pigmentosum an autosomal recessive hereditary skin disease of humans in which homozygotes show a marked tendency to develop skin cancers following exposure to sunlight. It is due to a reduced ability to excise UV-induced cyclobutane **pyrimidine dimers** from DNA.

xerophthalmia progressive drying and wrinkling of the conjunctiva of the eye. One cause is **vitamin A** deficiency. ·

X-gal *abbr. for* 5-bromo-4-chloro-3-indolyl-β-D-galactoside, a colourless substrate that is hydrolysed by β-galactosidase to galactose and a blue product. It is used with **cloning vectors** containing part of the *lacZ* gene to select recombinants in molecular cloning: the peptide produced by the *lacZ* fragment complements the defective β-galactosidase enzyme in host cells. Insertion of foreign DNA into the vector inactivates the *lacZ* gene, so host cells containing recombinant plasmids will grow as white colonies on medium containing X-gal, whereas host cells containing 'empty' vectors will have a functional β-galactosidase and thus produce blue colonies.

Xho I a type 2 **restriction endonuclease**; recognition sequence: C↑TCGAG.

Xho II a type 2 **restriction endonuclease**; recognition sequence: R↑GACTY.

xi *symbol*: ξ (lower case) *or* Ξ (upper case); the fourteenth letter of the Greek alphabet. For uses *see* **Appendix A**.

XIAP *abbr. for* X-linked inhibitor of apoptosis.

XIC *abbr. for* X inactivation centre. *See* **Xist**.

X-inactive-specific transcript *see* **Xist**.

Xiphophorus a genus of fish, that includes the South American minnow, some species of which have been much used for the study of inherited characteristics involved in carcinogenesis. Crossing of certain species, e.g. backcrossing *X. maculatus* (platyfish) with *X. helleri* (swordtail: recurrent parent) leads to spontaneous development of melanoma.

Xis protein *or* **excisionase** the product of the *xis* gene in lambda phage. It is essential in the excision of phage DNA from **prophage**. *See also* **integrase**.

XIST *or (in mouse)* **Xist** *abbr. for* X-inactive-specific transcript; the noncoding RNA(16 500 nt) transcript of the X inactivation centre (XIC) on the X chromosomes of female mammals. It is confined to the nucleus and associated with the heterochromatin (or Barr body) in female interphase nuclei. Xist accumulates before inactivation of the X chromosome.

XL1-Blue a proprietary strain of *E. coli* allowing blue-white colour screening for recombinants and the single-stranded rescue of phagemid DNA.

XL1-Blue MRF′ a proprietary strain of *E. coli* lacking K12 restriction systems and suitable for cloning methylated DNA such as that from eukaryotes.

Xma I a type 2 **restriction endonuclease**; recognition sequence: C↑CCGG.

XML *abbr. for* extensible markup language. *See* **markup language**.

XMP *abbr. for* xanthosine monophosphate; i.e. xanthosine (5′-)phosphate.

XNU *see* **sequence filter**.

xotch a transmembrane protein in *Xenopus* that is similar in sequence, and presumably in role, to the product of **notch**. It is uniformly expressed in early embryos. It contains 36 EGF-like domains and three notch-type repeats.

X-Pro dipeptidase EC 3.4.13.9; *other names*: proline dipeptidase; imidodipeptidase; prolidase; peptidase D; γ-peptidase. A cytosolic homodimeric enzyme that catalyses the hydrolysis of Xaa-|-Pro dipeptides; it also acts on aminoacyl-hydroxyproline analogues, but has no action on Pro-|-Pro. It requires manganese. Enzyme deficiency leads to imidodipeptiduria.

XPS *abbr. for* X-ray photoelectron spectroscopy; *see* **photoelectron spectroscopy**.

XPum a homologue of Pumilio, in *Xenopus laevis*.

X-ray crystallography the study of the geometric forms of crystals by **X-ray diffraction**. Myoglobin was the first protein for which the structure was solved by X-ray crystallography.

X-ray diffraction a method that uses the diffraction pattern obtained by passing X-rays through crystals, or other regular molecular arrays, to measure interatomic distances and to determine the three-dimensional arrangement of atoms (or molecules) in the structure studied. For the pattern to be sharp, the wavelength of the radiation used must be shorter than the regular spacing between the elements of the structure. X-rays typically have wavelengths of only a few tenths of a nanometre, making them suitable for biological samples. X-ray diffraction is used extensively in the determination of structures of proteins, nucleic acids, and other natural products (an early landmark was the determination of the structure of penicillin). In order to resolve the phase differences, much use has been made of **isomorphous replacement** with heavy metal atoms.

X-ray photoelectron spectroscopy *abbr.*: XPS; *see* **photoelectron spectroscopy**.

X-rays *or (sometimes, esp. US)* **x-rays** *or (formerly)* **röntgen rays** electromagnetic radiation emitted by atoms during extranuclear loss of energy of incident radiation (e.g. high-energy electrons, gamma radiation) or by atoms of certain radionuclides during transformation by **electron capture**. X-rays have wavelengths in the range 1 pm–10 nm (frequencies 3 PHz–3 EHz), which lies above that of gamma radiation and overlaps the wavelengths of the far ultraviolet. Characteristic X-rays have specific wavelengths of a particular nuclide or target element. *See also* **bremsstrahlung**, **synchrotron**.

XRE *abbr. for* xenobiotic-response element.

XTP *abbr. for* xanthosine (5′-)triphosphate.

Xul *symbol for* a residue (or a molecule) of the ketopentose xylulose.

xyl+ *a variant (sometimes before a vowel) of* **xylo+**.

Xyl *symbol for* a residue (or a molecule) of the aldopentose xylose.

xylan any homopolysaccharide of **xylose**, consisting mainly of β-1,4-linked xylopyranose units with various substituent groups, that occurs in the cell walls of higher plants, and in some marine algae. Xylans form the major glycans of the **hemicellulose** in the fibrous parts of plants.

xylaric acid the **aldaric acid** derived from **xylose**.

xylE a gene in *Escherichia coli* encoding a xylose transport protein that contains 12 transmembrane segments and is an H^+–xylose symporter. The gene is induced by xylose. *See also* **sugar transporter**.

xylem the vascular tissue that conducts water and mineral salts from the roots to the aerial parts of a plant and provides it with mechanical support.

xylene cyanol a dye used as a mobility marker in agarose gel electrophoresis.

L-xylitol reductase EC 1.1.1.10; an enzyme of the glucuronate pathway that catalyses the reaction:

$$xylitol + NADP^+ = L\text{-xylulose} + NADPH + H^+.$$

Deficiency of the enzyme results in essential pentosuria.

xylo+ *or (sometimes before a vowel)* **xyl+** *comb. form indicating the* trivial name for the aldopentose xylose.

xylo- *prefix (in chemical nomenclature)* indicating a particular configuration of a set of three contiguous >CHOH groups, as in the acyclic form of D- or L-xylose. *See* **monosaccharide**.

xylose *symbol*: Xyl; *xylo*-pentose; a constituent of plant polysaccharides (*see* **xylan**); the naturally occurring enantiomer is always D-xylose. In humans most ingested xylose is excreted unchanged in the urine. This is the basis for the **xylose absorption test**, the most widely used test of carbohydrate absorption.

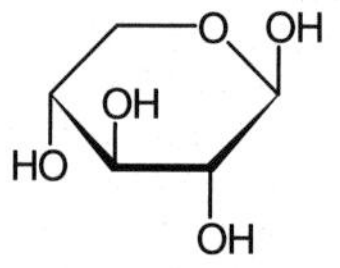

β-D-pyranose form

xylose isomerase EC 5.3.1.5; *systematic name*: D-xylose ketol-isomerase; an enzyme that catalyses the reaction:

$$D\text{-xylose} = D\text{-xylulose};$$

magnesium ions are a cofactor. The 3-D structure is known, and it is a **beta-barrel** protein.

xylose-1-phosphate uridylyltransferase *see* **UTP–xylose-1-phosphate uridylyltransferase**.

xylulose *symbol*: Xul; *the trivial name for* the ketopentose *threo*-2-pentulose; D-xylulose is the naturally occurring enantiomer. Excessive excretion of xylulose in the urine is called essential pentosuria.

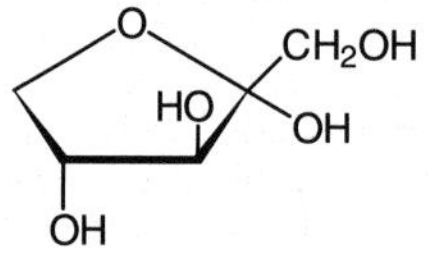

α-D-xylulose

D-xylulose 5-phosphate the 5-phosphate ester of D-xylulose and a component of the **pentose phosphate pathway**. It is formed from D-ribulose 5-phosphate by ribulose-phosphate 3-epimerase (EC 5.1.3.1); D-xylulose 5-phosphate then reacts with D-ribose 5-phosphate, catalysed by **transketolase**, to give glyceraldehyde 3-phosphate and sedoheptulose 7-phosphate.

X-Y-Z-cell theory a theory of immunocyte maturation that proposes that the first committed immunocyte, the **X cell**, is transformed by antigen into a memory cell, the **Y cell**, in the primary response, and is further stimulated in the secondary response to become an antibody producer, the **Z cell**.

Yy

y *symbol for* yocto+ (SI prefix denoting 10^{-24} times).

y *symbol for* **1** amount-of-substance fraction (i.e. mole fraction) or number fraction for gases. *Compare* **x**. **2** a Cartesian space coordinate, complementary to x and z.

Y *symbol for* **1** yotta+ (SI prefix denoting 10^{24} times). **2** a residue of the α-amino acid L-tyrosine (alternative to Tyr). **3** a residue of an incompletely specified base in a nucleic-acid sequence that may be either cytosine or thymine. **4** a residue of an unspecified pyrimidine nucleoside (alternative to Pyd). **5** a residue of the minor nucleoside **wyosine** (alternative to W). **6** yield. **7** yttrium.

Y *symbol for* admittance (reciprocal of impedance).

YAC *abbr. for* yeast artificial chromosome.

Yang cycle *see* **methionine salvage pathway**.

Yb *symbol for* ytterbium.

Y cell *see* **X-Y-Z-cell**.

Y chromosome a **sex chromosome** found only in the **heterogametic** sex and usually different in size from the **X chromosome**. In many animals it carries the testis-determining factor that triggers male embryonic development.

yeast any of a group of unicellular fungi that reproduce asexually – by budding or fission – and sexually – by the production of ascospores. Yeast cells may occur singly or in short chains, and some species produce a mycelium. The term 'yeast' is often used to mean members of the genus *Saccharomyces*, e.g. *S. cerevisiae*), which is an example of a budding yeast, or *Schizosaccharomyces pombe*, which is an example of a fission yeast.

yeast artificial chromosome *abbr.*: YAC; a specialized cloning vector that contains a **centromere**, an autonomously replicating sequence (ARS), a pair of **telomeres**, selectable marker genes, and the fragment of DNA to be cloned. Usually, genomic DNA is digested to produce fragments containing the genes, which are separated by **pulsed field gel electrophoresis**; the large fragments are then ligated into YACs. They are capable of propagation in yeasts, where they function as 'artificial chromosomes', being efficiently replicated. YACs are essential in large mapping projects, such as the Human Genome Project (*see* **HUGO**), as they are stable and can carry vast DNA inserts of 100 to 1000 kbp.

yeast cloning any method of cloning using yeast as the host. Expression of eukaryote genes by yeasts and other microbial eukaryotes is subject to eukaryotic regulatory systems, and the products undergo post-translational modification in contrast to when they are expressed in prokaryotes. Thus, yeast clones are often used for the expression of glycoproteins.

yeast eluate factor *a former term for* **vitamin B₆** .

yeast filtrate factor *a former term for* pantothenic acid. *See* **pantothenate**.

yeast nucleic acid *a former term for* ribonucleic acid.

yeast two-hybrid system a system that exploits two hybrid proteins, X, fused to the GAL4 DNA-binding domain, and Y, fused to the GAL4 activation domain. Plasmids encoding each of the hybrid proteins are introduced together into yeast, which leads to the expression of both hybrids. The system is used to test whether protein Y binds to protein X, for example in investigating the role of newly discovered proteins. If the proteins do bind, the resulting complex will be bound to the upstream activation sequence for the yeast *GAL* genes by the GAL4 DNA-binding domain, and the presence of the GAL4 activation domain will lead to transcriptional activation of the **reporter gene**, i.e. β-galactosidase. The system has the advantage over immunoprecipitation that if Y binds to X, the gene encoding it is already available as a clone. The hybrid containing X is referred to as the **bait**.

yellow enzyme 1 old yellow enzyme, EC 1.6.99.1; NADPH dehydrogenase; a flavoprotein (FMN in yeast, FAD in plants). **2** new **yellow enzyme** *an alternative name for* D-amino-acid oxidase; *see* **amino-acid oxidase**.

yellow fever virus There are two kinds of fever; **jungle** which is a disease of monkeys which is spread by infected mosquitoes; rare in humans; and **urban**, a disease of humans caused by infected mosquitoes *Aedes aegypti*. The virus is a member of the *Flaviviridae* with a sense plus RNA genome translated to a polyprotein of envelope, membrane and capsid proteins. Vaccination is by live-attenuated 17D strain.

yellow marrow *see* **bone marrow**.

Yersinia murine toxin *see* **murine toxin**.

YES v-*yes* is the oncogene of the Yamaguchi sarcoma virus, and has a cellular counterpart, c-*yes*, the product of which is a member of the *src* family of non-receptor tyrosine kinases, having a myristoylation site, and SH2, SH3 and tyrosine kinase domains.

yield *symbol*: Y; a ratio expressing the efficiency of a mass conversion process. The yield coefficient is defined as the amount of cell mass or product formed related to the consumed substrate or to the intracellular ATP production.

yin-yang hypothesis a model of biological regulation based on the opposing actions of certain cyclic nucleotides. It arises from the observation that hormones or other biologically active substances that promote the cellular accumulation of cyclic GMP produce cellular responses that are antagonistic to ones occurring when the concentration of cyclic AMP is increased in the same tissues or cells. Hence cyclic GMP and cyclic AMP may have opposing or antagonistic regulatory influences in certain biological systems, by analogy with the oriental concept of yin and yang symbolizing a dualism between opposing natural forces.

+yl *suffix (in chemical nomenclature)* denoting a free valence derived from the loss of a hydrogen atom from a parent hydride, e.g. methyl, pentanyl. It also applies to particular acyl groups, that are exceptions to the rule that acyl groups are named by changing the '-ic acid' or '-oic acid' ending to '-oyl'. These exceptions are acetyl, malonyl, succinyl, propionyl, butyryl, oxalyl, and glutaryl.

yocto+ *symbol*: y; SI prefix denoting 10^{-24} times.

yohimbine an indole alkaloid, $C_{21}H_{26}N_2O_3$, with α_2-adrenoceptor antagonist activity. It is produced by *Corynanthe johimbe* and *Rauwolfia serpentina*.

Yop *abbr. for* Yersinia outer protein; any of a number of proteins, known as virulence proteins, originally found on the surface of Gram-negative bacteria of the genus *Yersinia*; they are now known to be characteristic of plant-pathogenic proteins that are secreted by bacteria of genera as diverse as *Pseudomonas*, *Xanthomonas*, and *Erwinia*. They are also produced by animal pathogens, such as *Shigella flexneri*, *Salmonella typhimurium*, and *Escherichia coli*. They are secreted by the **type III system**. Unlike other secreted proteins, they lack a signal sequence.

yotta+ *symbol*: Y; SI prefix denoting 10^{24} times.

Yphantis method *an alternative name for* **meniscus depletion method (sedimentation equilibrium)**.

Zz

z *symbol for* zepto+ (SI prefix denoting 10^{-21} times).

z *symbol for* **1** charge number of an ion. **2** a Cartesian space coordinate, complementary to x and y.

Z *symbol for* **1** zetta+ (SI prefix denoting 10^{21} times). **2** a residue of either of the α-amino acids L-glutamic acid or L-glutamine when the state of amidation is uncertain (alternative to Glx). It may also be used to represent a residue of any other substance, such as 4-carboxy-L-glutamic acid (as alternative to Gla) or 5-oxo-L-proline (as alternative to Glp), that yields glutamic acid on acid hydrolysis of peptides. **3** the **benzyloxycarbonyl** group (alternative to Cbz).

Z *symbol for* **1** proton number. **2** impedance.

(Z)- *prefix (in chemical nomenclature)* denoting a geometric isomer in which the highest priority substituent groups, determined according to the Cahn–Ingold–Prelog rules (*see* **sequence rule**), are located on the same side of a double bond. *See also* **E/Z convention**.

zalcitabine *or* 2′,3′-dideoxycytidine (*abbr.*: ddC); an inhibitor of reverse transcriptase that is used in the treatment of HIV-1 infection.

Zantac *see* **ranitidine**.

ZAP *abbr. for* zinc finger antiviral protein; a protein (776 amino acids) of rat that contains four zinc fingers (of C_4H type). When expressed in fibroblasts in cell culture it confers specific antiviral activity.

ZAP-70 *or* zeta chain-associated tyrosine kinase a protein-tyrosine kinase (70 kDa) that associates with the ζ subunit that functions as part of the **T-cell receptor** and **CD3** complex in lymphocytes. It contains two SH2 domains, which interact with immunoglobulin tyrosine-based motifs on ζ or other subunits of these complexes. Rare inactivating mutations in the ζ chain gene cause **severe combined immunodeficiency** (SCID).

zaragozic acid any of a family of fungal metabolites that are potent inhibitors of squalene synthase.

zaragozic acid A

Z-average relative molecular mass *see* **average relative molecular mass**.

Z cell *see* **X-Y-Z cell theory**.

Z disk *or* **Z line** a platelike region of a muscle **sarcomere** to which the plus ends of actin filaments are attached. It is seen as a dark transverse line in micrographs.

Z-DNA *abbr. for* Z form of DNA.

zeatin *trans*-zeatin; (*E*)-2-methyl-4-(1*H*-purin-6-ylamino)-2-buten-1-ol; a plant hormone of the **cytokinin** class. Its derivative, *trans*-zeatin riboside (N^6-(*trans*-4-hydroxy-3-methyl-*cis*-2-buten-1-yl)adenosine) is also a cytokinin. The *cis* form is a rare component in some RNA structures.

zebrafish *Brachydanio rerio*; a small, highly fertile species of characinoid teleost fish with a shiny blue body and four longitudinal yellowish stripes along its sides. It is much used in studies on the molecular biology of differentiation and development in metazoan animals.

Zeeman effect the splitting of a single line in a spectrum, indicative of the degeneration of the excited state of a particular chromophore, into two or more components of slightly different wavelengths brought about by the application of an external magnetic field. [After Pieter Zeeman (1865–1943), Dutch physicist.]

zein the principal storage protein of maize (corn) seeds. It is relatively deficient in the essential amino acids lysine and tryptophan; tryptophan deficiency can occur in human populations dependent on maize as the major protein source. The zein genes are clustered.

Zellweger syndrome *or* **cerebrohepatorenal syndrome** a heterogeneous disorder of peroxisome biogenesis that results from mutations in at least 11 loci for PEX proteins (*see* **peroxin**), most commonly for PEX1. Typically patients have multiple congenital anomalies of the face and head, ongoing metabolic disturbances with accumulation of very-long-chain fatty acids, and usually die in infancy. Infantile **Refsum's disease** is one form of the syndrome. [After Hans Ulrich Zellweger (1909–1990), US paediatrician.]

zeolite any porous alkali metal- or alkaline earth-aluminium silicate that shows ion-exchange properties. Zeolites may be used for water softening or as molecular sieves. —**zeolitic** *adj.*

zepto+ *symbol*: z; SI prefix denoting 10^{-21} times.

zero-order kinetics kinetics shown by a **zero-order reaction**.

zero-order reaction a chemical reaction whose velocity is independent of the concentration of one or more of the reactants. *See* **order of reaction**.

zeroth law of thermodynamics *see* **thermodynamics, laws of**.

zervamicin *see* **peptaibophol**.

zeta *symbol*: ζ (lower case) *or* Z (upper case); the sixth letter of the Greek alphabet. For uses *see* **Appendix A**.

zeta chain 1 a transmembrane protein (16 kDa) of most T cells that contains three intracellular immunoreceptor tyrosine-based activation motifs (ITAMs) to which protein tyrosine kinases (e.g. **ZAP-70**) bind. It forms a dimer of disulfide-linked chains that are tightly associated with components of CD3 and T-cell receptors. **2** the α-globin-like globin that with γ-globin forms Hb Portland I, and with ε-globin forms Hb Gower I, in the first few weeks of human embryonic development. *See* **hemoglobin**.

zeta chain-associated tyrosine kinase *an alternative name for* **ZAP-70**.

zeta potential *symbol*: ζ; the potential at the surface of shear of a charged particle; for macromolecular ions it is the potential at the surface of the **hydrodynamic particles** formed by such ions.

zeta protein *an alternative name for* cellular **fibronectin**.

zetta+ *symbol*: z; SI prefix denoting 10^{21} times.

zeugmatography a technique in which the addition of carefully controlled inhomogeneous magnetic fields enables measurements of **nuclear magnetic resonance** (NMR) to be made at selected regions of heterogeneous samples and hence the formation of two-dimensional maps of the distribution of almost any property measurable by NMR.

Zfn *abbr. for* zinc finger nuclease; a synthetic chimeric protein that contains three C_2H_2 zinc fingers and a nonspecific DNA cleavage domain.

Z form of DNA *abbr.* Z-DNA or DNA-Z; a form of DNA consisting of a left-handed double helix in which the pyrimidine residues are always in the *anti* orientation and the purine residues in the *syn*. So-called because of the zig-zag appearance of its conformation. *See also* **A form** (def. 1), **B form**, **C form**.

Zfp *abbr. for* zinc finger protein.

zidovudine 3′-azido-3′-deoxythymidine (*abbr.*: AZT); an analogue of thymidine, the phosphorylated form of which is an inhibitor of **reverse transcriptase** in **retroviruses**; it also terminates DNA synthesis. Zidovudine undergoes phosphorylation in human T cells to a nucleoside 5′-triphosphate, which competes with thymidine triphosphate (TTP) and serves as a chain-terminating inhibitor of **HIV** reverse transcriptase. It is used clinically to treat patients with HIV infection and AIDS. *Proprietary name*: Retrovir.

Zimm–Crothers viscometer a modified **Couette viscometer** in which the inner cylinder is a self-centring float containing a steel pellet. The inner cylinder is rotated by an external rotating magnetic field, the speed of rotation being dependent, *inter alia*, on the viscosity of the fluid. The instrument only measures the viscosity of a sample relative to a standard.[After Bruno H. Zimm (1920–) US chemist and Donald M. Crothers (1937–) US biophysical chemist.]

Zimmermann reaction the reaction of *m*-nitrobenzene in strongly alkaline solution with the methylene group at position 16 in 17-ketosteroids to give a purple colour with an absorption maximum at 520 nm. It may be used to estimate such steroids.

Zimm plot a graphical double extrapolation used to analyse most macromolecular light-scattering data. Measurements are made of Kc/R_θ at various concentrations as a function of the scattering angle, 2θ, where K is an optical constant, c is the concentration, and R_θ is the Rayleigh ratio. A series of plots of Kc/R_θ against $(\sin^2\theta + K'c)$ are made, where K' is an arbitrary constant chosen to provide a convenient spread. The data are then extrapolated separately to $c = 0$ and $\theta = 0$. The intersection of the two extrapolated curves yields the relative molecular mass.

zinc *symbol* Zn; a metal of group 12 of the IUPAC **periodic table**; relative atomic mass 65.38; atomic number 30. The principal ore is zincblende, ZnS. Its oxidation state is Zn(II). Zinc oxide is used as an antiseptic. Naturally occurring nuclides include (relative abundance follows each) $^{64}_{30}$Zn (48%), $^{66}_{30}$Zn (28%), $^{68}_{30}$Zn (18%). Radionuclides include **zinc-62**, **zinc-65**, and **zinc-72**. Dietary deficiency in the rat is manifested by retarded growth, alopecia, and lesions in the skin, esophagus, and cornea. Deficiency is rare in normal individuals, but can manifest itself in alcoholics and in patients with renal disease or malabsorption. Zn^{2+} is an essential cofactor for many enzymes. Recent work suggests zinc is bactericidal.

zinc-62 the artificial radioactive nuclide $^{62}_{30}$Zn; it emits a positron (β^+ particle) (0.66 Mev), and γ-radiation of four energies (0.041, 0.25, 0.51, and 0.60 Mev). It has a half-life of 9.26 h.

zinc-65 the artificial radioactive nuclide $^{65}_{30}$Zn; it emits a positron (β^+ particle) (0.325 Mev) and γ-radiation (1.12 Mev). It has a half-life of 244.1 days.

zinc-72 the artificial radioactive nuclide $^{72}_{30}$Zn; it emits an electron (β^- particle) (0.3 Mev) and γ-radiation of three energies (0.016, 0.145, and 0.192 Mev). It has a half-life of 1.938 days.

zinc finger a polypeptide motif present in many DNA-binding proteins, and having the consensus sequence

$$-CX_{2-4}CX_3FX_5LX_2HX_3H-.$$

The motif takes its name from its zinc-binding site, which forms a finger-like loop in the peptide, also described as the **Cys₂/His₂ finger**, the Cys and His residues coordinating a zinc ion. Zinc fingers are most notably present in **transcription factors** (e.g., TFIIIA and SP1), which have several repeats of the motif; they are required for binding of the transcription factor to DNA.

ZIP 1 *abbr. for* the leucine zipper domain. *See also* **bZIP**. **2** a family of membrane proteins involved in import of zinc into the cytoplasm in

eukaryotes. They are predicted to contain eight transmembrane segments, of which S4 and S8 are highly conserved and contain conserved histidine residues, which may be involved in metal binding.

Z line *another name for* **Z disk**.

Zn *symbol for* zinc.

ZNF9 a protein that contains seven zinc finger motifs, binds RNA, and is expressed in skeletal and heart muscle. Expansion of a CCTG tetranucleotide repeat in intron 1 of the gene causes myotonic dystrophy type 2.

Zollinger–Ellison syndrome a disease syndrome caused by excessive secretion of **gastrin**, either from (type 1) **G-cell** hyperplasia in the antrum of the stomach, or from (type 2) a benign or malignant pancreatic islet tumour (**gastrinoma**). The manifestations include multiple peptic ulcers, or peptic ulcers in unusual sites, with a marked tendency to bleeding, often associated with hyperparathyroidism. [After Robert M. Zollinger (1903–) and Edwin H. Ellison (1918–70).]

zona adherens *see* **desmosome**.

zonadhesin a protein of spermatozoa, probably present in the acrosome, that binds to ZP3 of the zona pellucida of mammalian ova.

zona fasciculata *see* **adrenal gland**.

zona glomerulosa *see* **adrenal gland**.

zonal centrifugation *an alternative name for* **density-gradient centrifugation**.

zonal centrifuge a centrifuge designed to allow large-scale and continuous fractionation by **density-gradient centrifugation**.

zonal rotor any centrifuge rotor of high capacity used preparatively in a **zonal centrifuge**.

zona occludens *an alternative name for* **tight junction**.

zona pellucida a layer of glycoproteins that surrounds a mammalian ovum and disappears before implantation. It is secreted by cells of the Graafian follicle. *See also* **ZP**.

zona pellucida sperm-binding protein *abbr.*: ZP; any of the sulfated transmembrane glycoproteins (ZP1, ZP2, ZP3) that together form the zona pellucida of the oocyte. ZP3 is a sperm receptor.

zona reticularis *see* **adrenal gland**.

zone electrophoresis a method of **electrophoresis** in which the protein (or other) solution is placed at the starting position as a thin band or spot in an inert supporting medium (paper, starch gel, polyacrylamide gel, etc.), containing buffer solution. An electric potential is then applied to the supporting medium, causing the proteins (or other substances) to migrate to give distinct bands or zones. These may be located *in situ* by staining, light absorption, etc., or by analysis after elution of discrete pieces of the supporting medium.

zone precipitation a technique in which protein is precipitated as a zone in a gel-filtration column by eluting with a gradient of a protein-precipitating agent. *See* **gel-permeation chromatography**.

zone spreading the broadening of a zone in electrophoresis or chromatography due to eddy currents or other disturbances in the supporting medium.

zoo blot a method for detecting conservation of DNA sequence during evolution. A segment of DNA being investigated is used as a probe to hybridize against a series of DNA samples from various species, and hence establish whether the DNA sequence has been conserved during evolution.

zoonosis (*pl.* **zoonoses**) any disease that can be transmitted to humans from other animals in natural circumstances.

zooplankton *see* **plankton**.

zootoxin any poisonous substance formed by animal cells.

Zovirax *a proproetary name for* **acyclovir**.

ZP *abbr. for* zona pellucida protein; any of three glycoproteins (ZP1, ZP2, and ZP3) present in the zona pellucida of mammalian ova. ZP3 binds the sperm proteins p47, Sp56, and zonadhesin, which may be present in the acrosome.

Zr *symbol for* zirconium.

Z ring a contractile ring of proteins that forms in the mid-region of the inner side of the plasma membrane of bacteria as part of cell division, normally towards the end of replication. Formation is initiated by **FtsZ** and depends on the presence of the **Min** system for correct localization.

Z scheme a term applied to two-system photosynthetic light reactions, utilizing both **photosystems** I and II. It describes the mode of photosynthesis used by higher plants and algae, and by cyanobacteria. As it is normally plotted, with redox potential (on the ordinate) against time, it shows there is a sudden jump on excitation by a photon of a reaction centre, followed by a slow decline through electron transport components, and gives a (horizontal) Z-like appearance, hence the name.

zuotin a nuclear Z-DNA-binding protein of *Saccharomyces cerevisiae*.

zwischenferment *an obsolete term for* the enzyme–coenzyme complex of **glucose-6-phosphate dehydrogenase**.

Zwittergent *the proprietary name for* any of a group of zwitterionic detergents containing a **betaine** and a sulfonate group; such molecules retain their amphoteric (zwitterionic) properties over a very wide range of pH. The formula is *n*-alkyl *N*,*N*-dimethyl-3-ammonio-1-propanesulfonate. In Zwittergent 3-08, alkyl is octyl (CMC 330 mm); 3-10, decyl (aggregation number 41, CMC 25–40 mm); 3-12, dodecyl (aggregation number 55, CMC 2–4 mm); 3-14, tetradecyl (aggregation number 83, CMC 0.1–0.4 mm); 3-16, hexadecyl (aggregation number 155, CMC 0.01–0.06 mm).

zwitterion *an alternative name for* **dipolar ion**.

zwitterionic buffer a **buffer solution** whose acid or base component is a zwitterion (i.e. **dipolar ion**). The term is used especially of solutions of **Good buffer substances** and newer equivalents.

zygote any fertilized ovum before it has undergone cleavage. —**zygotic** *adj*.

zygotene the second phase of prophase I in **meiosis**.

zymase the name originally given to the heat-labile, nondiffusible fraction of a crude extract of brewers' yeast that, with the addition of the heat-stable, diffusible fraction (i.e. **cozymase**), would enable the alcoholic fermentation of glucose to occur in a cell-free system. It was later shown to consist of a mixture of enzymes, including those of the **glycolytic pathway**.

zymogen *or* **proenzyme** *or (formerly)* **proferment** the inactive precursor of an enzyme, often convertible to the enzyme by partial proteolysis. The term is applied especially to catalytically inactive forms of pancreatic enzymes such as trypsinogen, chymotrypsinogen, proelastase, and procarboxypeptidase; these are cleaved to release the active enzyme after their secretion from granules (**zymogen granules**) in the acinar cells of the pancreas.

zymogen granule a membrane-bound, cytoplasmic **secretory granule** visible by light microscopy. Zymogen granules are formed in the Golgi apparatus of enzyme-secreting cells and serve to store a **zymogen**. The term is used especially of a secretory granule containing the proenzyme of a digestive enzyme.

zymogram 1 any record of a **zone electrophoresis** separation in which enzymes in a sample have been separated and their positions, and relative amounts, revealed by an **activity stain**. **2** any table showing the results of carbohydrate-fermentation tests carried out in the process of identifying a microorganism.

zymology the science dealing with fermentations. —**zymological** *adj*.; **zymologist** *n*.

zymosan a crude preparation of yeast cell walls, consisting chiefly of polysaccharide, that activates the alternative pathway of the complement system in the presence of properdin.

zymosterol cholesta-8,24-dien-3β-ol; a sterol present in yeasts e.g. *Candida albicans*, *Saccharomyces cerevisiae*.

zymurgy the branch of chemistry dealing with the applications of fermentation processes, especially to brewing and wine-making.

Zyvox *another name for* Linezolid.

zyxin a protein component of cell substratum and cell–cell adherens junctions. It is a component of adhesion plaques and the termini of stress fibres near the point of association with the cytoplasmic face of the adhesive membrane; it interacts with α-actinin.

Appendices

Appendix A

The Greek alphabet

The Greek alphabet

Name	Capital letter	Lower-case letter	English transliteration
alpha	A	α	a
beta	B	β	b
gamma	Γ	γ	g
delta	Δ	δ	d
epsilon	E	ε	e
zeta	Z	ζ	z
eta	H	η	ē or e
theta	Θ	θ	th
iota	I	ι	i
kappa	K	κ	k
lambda	Λ	λ	l
mu	M	μ	m
nu	N	ν	n
xi	Ξ	ξ	x
omicron	O	o	o
pi	Π	π	p
rho	P	ρ	r
sigma	Σ	σ or ς (at end of word)	s
tau	T	τ	t
upsilon	Y	υ	u
phi	Φ	φ	ph
chi	X	χ	kh
psi	Ψ	ψ	ps
omega	Ω	ω	ō or o

Single-character symbols

α (*or* α) **1** a heavy chain of immunoglobulin A. **2** maximal agonist effect. **3** alpha particle.

α (*or* α) **1** angle of optical rotation (α_{obs} is used for observed rotation and $[\alpha]$ for specific optical rotation *or* specific optical rotatory power; α_m is used for molar optical rotatory power). **2** absorptance. **3** (linear) napierian absorption coefficient. **4** Bunsen coefficient. **5** partition coefficient. **6** plane angle. **7** *(in organic chemical nomenclature)* a locant for a substituent on the first atom from the atom carrying the principal function in an acyclic structure or in a side chain attached to a cyclic structure. **8** an anomeric configuration (of a sugar molecule or residue; *see* **anomer**). **9** winding number (of a helical duplex DNA molecule).

β (*or* β) beta particle (β^- is used for an electron and β^+ for a positron).

β (*or* β) **1** buffer value. **2** pressure coefficient. **3** plane angle. **4** *(in organic chemical nomenclature)* a locant for a substituent on the second atom from the atom carrying the principal function in an acyclic structure or in a side chain attached to a cyclic structure. **5** an anomeric configuration (of a sugar molecule or residue; *see* **anomer**). **6** number of secondary turns (in an unconstrained helical duplex DNA molecule).

γ (*or* γ) a heavy chain of immunoglobulin G. **2** a photon of gamma radiation. **3** *(obsolete)* microgram (μg now used).

γ (*or* γ) **1** activity coefficient (γ_c is used for activity coefficient when expressed on a concentration basis, γ_m when expressed on a molality basis, γ_x when expressed on a mole fraction basis, and $\gamma_{\pm}$ is used for mean ionic activity coefficient). **2** electrical conductivity (alternative to κ or σ). **3** mass concentration *or* mass density (alternative to ρ). **4** plane angle. **5** surface tension *or* interfacial tension (alternative to σ). **6** gyromagnetic ratio. **7** *(in organic chemical nomenclature)* a locant for a substituent on the third atom from the atom carrying the principal function in an acyclic structure or in a side chain attached to a cyclic structure.

δ (*or* δ) **1** a heavy chain of immunoglobulin D. **2** (*in mathematics*) an infinitesimal change or difference.

δ (*or* δ) **1** chemical shift. **2** number fraction. **3** *(in organic chemical nomenclature)* a locant for a substituent on the fourth atom from the atom carrying the principal function in an acyclic structure or in a side chain attached to a cyclic structure.

Δ (*or* Δ) (*in mathematics*) a finite change or difference.

Δ (*or* Δ) (*in older organic chemical nomenclature*) a locant for a double bond (a right superscript number designates the lower-numbered carbon atom; see **fatty-acid nomenclature**); no longer recommended for denoting unsaturation in names of individual compounds although it may still be used in generic terms (e.g. Δ^5-steroid) and in symbols (e.g. Δ_2Ach).

ε (*or* ε) a heavy chain of immunoglobulin E.

ε (*or* ε) **1** molar (decadic) absorption coefficient. **2** permittivity (ε_r is used for relative permittivity and ε_0 for permittivity of vacuum). **3** *(in organic chemical nomenclature)* a locant for a substituent on the fifth atom from the atom carrying the principal function in an acyclic structure or in a side chain attached to a cyclic structure. **4** *(in chemical nomenclature)* hapticity of a ligand (less common alternative to η; *see* **hapto**).

ε_a degree of activation (of an enzymic reaction).

ε_i degree of inhibition (of an enzymic reaction).

ε^v_s elasticity coefficient (of a specific enzymic step in a metabolic system).

ζ (*or* ζ) electrokinetic potential *or* zeta potential.

η (*or* η) **1** viscosity (alternative to μ). **2** *(in chemical nomenclature)* hapticity (of a ligand; more common alternative to ε; *see* **hapto**).

θ (*or* θ) **1** Bragg angle (*see* **Bragg's law**). **2** *(in surface chemistry)* contact angle. **3** plane angle. **4** torsion angle (in a linear sequence of four attached atoms; alternative to ω). **5** Celsius temperature (alternative to t).

Θ_m melting temperature (alternative to T_m or t_m).

$[\Theta]$ molar ellipticity.

κ (*or* κ) one type of light chain of any class of immunoglobulin (the other being λ).

κ (*or* κ) **1** electrical conductivity (alternative to γ or σ). **2** electrolytic conductivity *or* (*formerly*) specific conductance. **3** molar napierian absorption coefficient. **4** magnetic susceptibility (alternative to χ). **5** *(in inorganic chemical nomenclature)* an affix in the name of a polydentate chelate complex indicating single attachments of a polydentate ligand to a coordination centre. A right superscript numerical index indicates the number of such attachments.

λ (*or* λ) **1** one type of light chain of any class of immunoglobulin (the other being κ). **2** *(obsolete)* microlitre (μl now used).

λ (*or* λ) **1** *(in thermodynamics)* absolute activity (*see* **activity**, def. 4). **2** disintegration constant *or* decay constant. **3** wavelength. **4** thermal conductivity (alternative to k).

Λ (*or* Λ) molar conductivity (of an electrolyte).

μ (*or* μ) a heavy chain of immunoglobulin M. **2** micro+ (SI prefix denoting 10^{-6} times). **3** *(obsolete)* micron (now called micrometre; symbol: μm).

μ (*or* μ) **1** *(in chemical nomenclature)* a bridging ligand. **2** viscosity (alternative to η). **3** chemical potential (μ^o or $\mu^{\ominus}$ is used for standard chemical potential and $\tilde{\mu}$ is used for electrochemical potential). **4** electric mobility (alternative to u). **5** *(in magnetism)* permeability (μ_r is used for relative permeability). **6** magnetic dipole moment (bold italic) (alternative to $\boldsymbol{m}$). **7** electric dipole moment (bold italic) (alternative to $\boldsymbol{p}$).

μ_0 permeability of vacuum (a fundamental physical constant).

μ_B Bohr magneton (a fundamental physical constant).

μ_e electron magnetic moment (a fundamental physical constant).

μ_N nuclear magneton (a fundamental physical constant).

μ_p proton magnetic moment (a fundamental physical constant).

ν (*or* ν) **1** stoichiometric number (of a chemical reactant). **2** frequency (alternative to f for electromagnetic radiation). **3** kinematic viscosity. **4** endocyclic torsion angle (in a sugar residue).

ξ **1** *(in organic chemical nomenclature)* unknown configuration at a chiral centre. **2** extent of reaction.

$\dot{\xi}$ rate of conversion.

π (*or* π) a transcendental number that is the ratio of the circumference of a circle to its diameter (3.141 592 653 59...).

π (*or* π) **1 pros**. **2** (*or* π^s) surface pressure.

Π (*or* Π) **1** osmotic pressure (Π_D may be used to represent colloid osmotic pressure). **2** binding potential (of a ligand).

ρ (*or* ρ) **1** density *or* mass density (ρ_A or ρ_S is used for surface density). **2** mass concentration *or* mass density (alternative to γ). **3** radiant energy density (alternative to w). **4** charge density. **5** resistivity.

σ (*or* σ) **1** surface tension *or* interfacial tension (alternative to γ). **2** electrical conductivity (alternative to γ or κ). **3** wavenumber in a medium. **4** surface charge density. **5** *(in surface chemistry)* area per molecule. **6** *(in nuclear magnetic resonance spectroscopy)* shielding constant. **7** superhelix density (of a helical duplex DNA molecule).

Σ (*or* Σ) *(in mathematics)* a summation sign.

τ (*or* τ) **1 tele**. **2** bond angle. **3** relaxation time *or* time constant (alternative to T). **4** transmittance *or* transmission factor (alternative to T). **5** mean life. **6** number of superhelical turns (in a helical duplex DNA molecule).

φ (*or* φ) **1** osmotic coefficient (ϕ_m is used for osmotic coefficient when expressed on a molality basis and ϕ_x when expressed on a mole fraction basis). **2** electric potential (alternative to V). **3** fluidity. **4** plane angle. **5** phase difference. **6** volume fraction. **7** Dalziel coefficient. **8** torsion angle (from the anomeric carbon atom to the oxygen atom of a glycosidic bond). **9** quantum yield (alternative to Φ). **10** winding angle (of intercalated helical DNA).

Φ (*or* Φ) **1** heat flow rate. **2** radiant power *or* radiant (energy) flux (alternative to P). **3** magnetic flux. **4** potential energy (alternative to E_p or V). **5** quantum yield (alternative to ϕ).

χ (*or* χ) **1** electronegativity. **2** magnetic susceptibility (alternative to κ; χ_m is sometimes used, but it should be reserved for molar magnetic susceptibility). **3** surface electric potential. **4** exocyclic torsion angle (in a sugar residue).

ψ (*or* ψ) **1** replacement of the –CO–NH– group in a peptide by another (indicated) grouping. **2** torsion angle (from the oxygen atom of a glycosidic bond to the glycosylated molecule). **3** wavefunction (alternative to Ψ).

Ψ (*or* Ψ) **1** a residue of the ribonucleoside pseudouridine (alternative to Ψrd). **2** electric flux. **3** wave function (alternative to ψ).

ω (*or* ω) **1** angular frequency. **2** angular velocity. **3** solid angle. **4** torsion angle (in a linear assembly of four attached atoms; alternative to θ). **5** *(in organic chemical nomenclature)* a locant for the terminal atom or group in an acyclic organic compound. **6** *(in notation for polyunsaturated fatty acids)* ω-x signifies that a double bond is located on the xth carbon; unlike standard chemical nomenclature it is counted from the distal (methyl carbon) atom.

Ω (*or* Ω) ohm.

Ω (*or* Ω) solid angle.

Multi-character biochemical symbols commencing with a Greek character

αLnn   the linolenoyl *or* (9,12,15)-linolenoyl *or* *(formerly)* α-linolenoyl (i.e. (*all-Z*)-octadeca-9,12,15-trienoyl) group (alternative to Lnn).

βAad   a residue of the β-amino acid L-β-aminoadipic acid (i.e. L-3-aminohexanedioic acid).

βAla   a residue of the β-amino acid β-alanine (i.e. 3-aminopropanoic acid).

γLnn   the (6,9,12)-linolenoyl *or* *(formerly)* γ-linolenoyl (i.e. (*all-Z*)-octadeca-6,9,12-trienoyl) group.

Δ$_2$Ach   the (*all-Z*)-eicosa-8,11-dienoyl group.

Δ$_3$Ach   the (*all-Z*)-eicosa-5,8,11-trienoyl group.

Δ$_4$Ach   the arachidonoyl (i.e. (*all-Z*)-eicosa-5,8,11,14-tetraenoyl) group.

ΔPam   the palmitoleoyl (i.e. (*Z*)-hexadec-9-enoyl) group.

εAcp   *see* **εAhx**.

εAhx   *or* *(formerly)* **εAcp** a residue of the ε-amino acid 6-aminohexanoic acid (formerly known as ε-aminocaproic acid).

Ψrd   a residue of the ribonucleoside pseudouridine (alternative to Ψ).

Appendix B
Sequence-rule priorities

The names of some common ligands are listed alphabetically below, the numbers preceding them indicating the relative priorities assigned to the ligands under the **sequence rule**. A higher number denotes greater preference; for example, a hydroxy group (numbered 57) takes precedence over an amino group (numbered 43). It should be noted that any change to the structure of a listed ligand may alter the order of preference. Where a ligand consists of or contains a heavy or radioactive isotope of hydrogen, the relative priorities are tritium > deuterium > protium. Racemates with a single chiral centre are labelled (*RS*); extensions of the sequence rule are used in the *E/Z*, *pro-R/pro-S*, *pro-E/pro-Z*, and *Re/Si* nomenclature systems.

64 acetoxy
36 acetyl
48 acetylamino
21 acetylenyl
10 allyl
43 amino
44 ammonio ($^+H_3N-$)
37 benzoyl
49 benzoylamino
65 benzoyloxy
13 benzyl
60 benzyloxy
41 benzyloxycarbonyl
50 benzyloxycarbonylamino
75 bromo
42 *tert*-butoxycarbonyl
15 butyl
16 *sec*-butyl
19 *tert*-butyl
38 carboxy
74 chloro
17 cyclohexyl
52 diethylamino
51 dimethylamino
34 2,4-dinitrophenyl
28 3,5-dinitrophenyl

59 ethoxy
40 ethoxycarbonyl
13 ethyl
46 ethylamino
68 fluoro
35 formyl
63 formyloxy
62 glycosyloxy
17 hexyl
11 hydrogen
57 hydroxy
76 iodo
19 isobutyl
18 isopentyl
20 isopropenyl
14 isopropyl
69 mercapto (HS–)
58 methoxy
39 methoxycarbonyl
12 methyl
45 methylamino
71 methylsulfinyl
66 methylsulfinyloxy
72 methylsulfonyl
67 methylsulfonyloxy

70 methylthio (CH$_3$S–)
11 neopentyl
56 nitro
27 *m*-nitrophenyl
33 *o*-nitrophenyl
24 *p*-nitrophenyl
55 nitroso
16 pentyl
61 phenoxy
22 phenyl
47 phenylamino
54 phenylazo
18 propenyl
14 propyl
29 1-propynyl
12 2-propynyl
73 sulfo (HO$_3$S–)
25 *m*-tolyl
30 *o*-tolyl
23 *p*-tolyl
53 trimethylammonio
32 trityl
15 vinyl
31 2,6-xylyl
26 3,5-xylyl

The periodic table of the elements

Group

Period	1	2	3	4	5	6	7	8	9	10	11	12	13	14	15	16	17	18
1	1 H																	2 He
2	3 Li	4 Be											5 B	6 C	7 N	8 O	9 F	10 Ne
3	11 Na	12 Mg											13 Al	14 Si	15 P	16 S	17 Cl	18 Ar
4	19 K	20 Ca	21 Sc	22 Ti	23 V	24 Cr	25 Mn	26 Fe	27 Co	28 Ni	29 Cu	30 Zn	31 Ga	32 Ge	33 As	34 Se	35 Br	36 Kr
5	37 Rb	38 Sr	39 Y	40 Zr	41 Nb	42 Mo	43 Tc	44 Ru	45 Rh	46 Pd	47 Ag	48 Cd	49 In	50 Sn	51 Sb	52 Te	53 I	54 Xe
6	55 Cs	56 Ba	57–71 La–Lu	72 Hf	73 Ta	74 W	75 Re	76 Os	77 Ir	78 Pt	79 Au	80 Hg	81 Tl	82 Pb	83 Bi	84 Po	85 At	86 Rn
7	87 Fr	88 Ra	89–103 Ac–Lr	104 Rf	105 Db	106 Sg	107 Bh	108 Hs	109 Mt	110 Ds	111 Rg	112 Uub	113 Uut	114 Uuq	115 Uup	116 Uuh	117 Uus	118 Uuo

6 Lanthanoids	57 La	58 Ce	59 Pr	60 Nd	61 Pm	62 Sm	63 Eu	64 Gd	65 Tb	66 Dy	67 Ho	68 Er	69 Tm	70 Yb	71 Lu
7 Actinoids	89 Ac	90 Th	91 Pa	92 U	93 Np	94 Pu	95 Am	96 Cm	97 Bk	98 Cf	99 Es	100 Fm	101 Md	102 No	103 Lr

Correspondence of recommended group designations to other designations in recent use

IUPAC 1990 recommendations	1	2	3	4	5	6	7	8	9	10	11	12	13	14	15	16	17	18
Usual European convention	IA	IIA	IIIA	IVA	VA	VIA	VIIA	VIII (or VIIIA)			IB	IIB	IIIB	IVB	VB	VIB	VIIB	0 (or VIIIB)
Usual N. American convention	IA	IIA	IIIB	IVB	VB	VIB	VIIB	VIII (or VIIIB)			IB	IIB	IIIA	IVA	VA	VIA	VIIA	VIIIA (or 0)